现代汉英词典
A MODERN CHINESE-ENGLISH DICTIONARY

海峰出版社有限公司
Hai Feng Publishing Co. Ltd

牛津大学出版社
Oxford University Press

现代汉英词典

主　编　段世镇

出　版　牛津大学出版社
香港鲗鱼涌糖厂街和域大厦东翼18楼
海峰出版社有限公司
香港中环德辅道中71号永安集团大厦1502室
印　刷　阳光印刷制本厂有限公司
香港柴湾安业街三号六字楼G及H座
版　次　1989年4月第一版
1992年11月第二版
1994年1月第三版
1994年7月第四版
1996年1月第五版

国际书号　ISBN 0 19 585189 7

主编简介

段世镇教授1960年毕业于北京外国语学院英语系后，留校从事了多年英语教学工作。1979-1983年，曾在联合国粮食与农业组织出版处翻译科任定稿人，审定联合国各种会议文件、年鉴及专业书刊，例如《联合国粮农组织商品回顾与展望》、《食品和营养政策手册》、《防止易腐农产品的损失》等。回国后，担任外语教学与研究出版社副社长、社长，并任大学出版社协会北京地区分会副理事长。

About the Chief Editor

Duan Shizhen was graduated in English at Beijing Foreign Studies University in 1960. He stayed on to teach in the English Department for many years. In 1979 he went to Rome to join the staff of the Food and Agriculture Organization of the United Nations, working as a revisor in the Translation Service. His duties included, among other things, finalizing the Chinese version of FAO publications, with as wide a range as evidenced by such titles as *FAO Commodity Review and Outlook*, *Manual on Food and Nutrition Policy*, *Soil Erosion by Wind*, *Aerial Photo Interpretation in Soil Survey*, etc. Back in China, he was appointed Deputy Director, and subsequently Director, of Foreign Language Teaching and Research Press. At present he is concurrently Vice Chairman of the Council of the Beijing Area Branch of the National Association of University Presses.

编者的话

最近十年来，世界在政治、经济、法律、科学、技术、文化、艺术、生活等各方面，都发生了巨大的变化，中国和世界各国的交流也日趋频繁，广大读者迫切需要一本与时俱新的汉英词典，能充分收入在此期间出现的大量新词、新语和新义。满足读者的这一需要，就是我们编纂这部词典的目的。

本书收有常用汉字单字条目四千八百余个，多字条目五万余个。所有条目都有英语释义和对译，必要时并举例说明用法。许多多字条目之下附有合成词和固定搭配，所收词汇单位共约八万余个。总计书中提供的英译汉语单词、短语、句子十多万个。

本书在选材、释义、例证诸方面，力求丰富充实，准确实用。生僻罕见的字、词、语，一概不收，尽量省下篇幅多收当前常用的词语；特别着重选收在其它汉英词典中难以找到的新词、新义、新用法，例如，“经济体制改革”、“企业自主权”、“合资企业”、“外向型经济”、“商检证明书”、“破产法”、“侵权行为”、“公证处”、“艾滋病”、“试管婴儿”、“生物工程”、“微电脑”、“耐用消费品”、“游乐园”、“碰碰车”、“霹雳舞”、“汽车拉力赛”等等。

本书在编排方面力求合理，方便读者使用。其特点是：

一、全部单字条目、多字条目、附列词组，均按汉语拼音音序排列。

二、多于一个义项的单字条目和多字条目，每个义项开头都注有简明扼要的汉语释义，可以便利读者迅速准确地找到所需的英语译义。

三、有些多字条目，末尾还列出了相关的参考词汇。例如在“花草”词条下，收有四十种常见花草的汉英对照名称，以供参考。全书共有一百三十组相关参考词汇。

由于各种汉语字典、词典的汉字部首设置各异，本书只收汉语拼音音节索引和汉字笔画索引。

在本书编纂过程中，我们得到了各方面的热心帮助和支持。我们参阅了海内外上百种词典等工具书，在“主要参考书目”中，不可能一一列举，我们在此谨向所有的有关编著者一总致谢。

还要感谢我社参加本词典编纂和校对等工作的诸位同仁，他们在长时间的工作中付出大量心血。

我们高兴地看到，本书问世数月之后，就准备出版香港版，并作一些增补和订正。为了方便香港和海外的读者，香港版中的汉字笔书索引所收的汉字包括简体字和繁体字。另外还增加了几种附录。

在这里要特别感谢北京外国语学院语言研究所所长许国璋教授为香港版作序，感谢香港八龙出版文化服务公司总编辑萧滋先生为出版香港版所给予的鼓励和协助。

一部词典要想具有强大的生命力，惟有不断增补修订。我们希望，本书每次印刷都能有所进步。我们在此诚恳希望广大读者不吝指教。

Editors' Remarks

The past decade has witnessed stupendous and far-reaching changes in every field of life in the world — politics, economy, law, science, technology, culture, art, and everyday life. Along with domestic developments, China's international contact and exchange have been growing at an ever-increasing rate. So it is not surprising that a need has been keenly felt both at home and abroad for Chinese-English dictionaries that will keep abreast of the changing times by incorporating the many thousand new words and phrases, or new meanings of old terms, which have found their way into the Chinese vocabulary. It was to fill this need, if only partially, that *A Modern Chinese-English Dictionary* was conceived and brought to completion.

The Dictionary contains (1) 4,800 single-character entries and (2) 50,000 multiple-character entries. The first group are words of one character as well as forms which are used as word-building elements while the second are words of more than one character and set phrases which function as lexical units.

All entries, single-character or multiple-character, are appropriately defined, provided with their English equivalent(s), and, where necessary, further illustrated with examples to show usage. Many multiple-character entries are followed by derivatives, compounds, and fixed collocations, bringing the total number of lexical items so treated up to 80,000. Altogether well over 100,000 Chinese words, phrases and sentences are given English translation between the covers of this volume.

Informativeness, accuracy and utility have always been

aimed at in the choice of lexical items and their treatment. Obsolete words, terms and expressions of rare occurrence have been excluded to save space for those in current use. Special pains have been taken to include newly introduced terms not as a rule found in other Chinese-English dictionaries. Terms such as 经济体制改革 (the restructuring of economic systems), 企业自主权 (decision-making power of enterprises), 合资企业 (joint venture), 外向型经济 (export-oriented economy), 商检证明书 (commodity inspection certificate), 破产法 (insolvent law; law of bankruptcy), 侵权行为 (act of tort; infringement act), 公证处 (notary public office), 艾滋病 (Aids), 试管婴儿 (test-tube baby), 生物工程 (genetic engineering), 微电脑 (personal computer), 耐用消费品 (durable consumer goods), 游乐园 (amusement park), 碰碰车 (bumper car), 霹雳舞 (break dance), 汽车拉力赛 (rally), and many more, each has a place in the Dictionary.

The following features are calculated to help convenience of use:

(1) All entries and subsidiary items within each entry are consistently arranged in alphabetical order of the *Pinyin* system.

(2) Each sense division of a polysemous word, either under a single-character entry or under a multiple-character entry, begins with a brief explanation in plain Chinese, which hopefully will make it easier for the user to locate a wanted meaning with its English rendering.

(3) Where appropriate, semantically related terms are given as a group. Thus, under the entry 花草 (flowers and plants) is found a grouping of 40 common ornamental flowers and plants. The Dictionary has 130 such groupings.

Existing schemes of indexing Chinese characters by radicals are notoriously complicated and confusingly diversified and hardly any of them is likely to be of great help to our users. So, in addition to a Syllabic Index of *Pinyin*, the much simpler

Stroke Index (by type and number of strokes) is given instead.

In the course of their work the editorial team has been fortunate enough to receive generous support and assistance from many quarters. Scores of dictionaries and reference works, published in China and elsewhere, have been drawn on with thanks and it would be impossible to mention all of them individually in the Acknowledgements. The editors wish to express here generally their indebtedness to all whose works have been of assistance to them.

Warm appreciation is also extended to those members of the Press who gave unfailing help in compiling, proofreading and other tasks during the long months of work on the project.

It is gratifying to see that, a few months after the appearance of the Dictionary in Beijing in November, 1988, a new Hong Kong edition is soon to be available with some additions and corrections. For the convenience of readers in Hong Kong and overseas areas, both the simplified and the unsimplified traditional forms of Chinese characters are given in the Stroke Index in this edition. Several new appendices have also been added.

Special thanks are here expressed to Professor Xu Guozhang, Director of the Research Centre of Linguistics, Foreign Studies University, Beijing, who has written a preface to the new edition.

Hearty thanks are also due to Mr Shaw Tze, Editor in Chief, 8 Dragons Publishing & Cultural Services, Hong Kong, for his good offices in arranging for the Dictionary to come out in Hong Kong.

A lexicographical work, by its very nature, derives its viability only from unremitting efforts at revision and renewal. It is hoped that each new printing of the present work will bring some improvements. We sincerely invite criticisms and suggestions from all who have occasion to use the Dictionary. We promise prompt response.

Preface

This Modern Chinese-English Dictionary, now offered in Hong Kong in modern print and with a scientific apparatus, is a work of several distinct features. Being a student of language and, in consequence, of bilingual lexicography, I might perhaps venture a view about what reasonable standards to set for a good Chinese-English dictionary and, when that has been made clear, see whether the present work comes close to it.

Now I think there are two important things we must ask of such a dictionary. One, it must have a well-researched Chinese lexicon to start from, so that a reader looking up its pages from Chinese to English, will not find a wanted Chinese item missing. Two, the English near-equivalents there provided, must be, as far as possible, sociolinguistically acceptable, so that a successful finder can operate readily and happily.

As if in response to the call of standards, the MCED has a Chinese lexicon planned in three layers:

(1)a first layer of 4,800 one-character words (法 etc.);
(2)a second layer of 50,000 two-character words (including some longer terms) (法网，法定人数 etc.);
(3)a third layer of 55,000 set phrases and expressions (e.g. 难逃法网；法律面前，人人平等).

of the three layers, (1) forms the base of the Chinese lexis, while (2) and (3) are its rich offsprings, their phrasal formations giving the language a vitality and power not perhaps possible with single-character words. It is these, too, which bear the ever-expanding information load so characteristic of our times. Finally, (1) and (2) and (3) make a total of about 100,000 lexical items, a figure which a dedicated student of Chinese might well aim at —and why not, if he is to learn the language real well.

Additionally, there is

(4) the thesaurus groupings, a handy word-finder for the user who can think of a group his wanted word (or expression) is related to but may not have the word at his processor touch. Thus under 法庭 (court) one may find:

开庭审理	hold a hearing
公诉	public prosecution
起诉	prosecute; bring a case to court

and 36 other entries.

Such groupings by semantic fields are of course a standard feature of thesaurus books. but its adoption by the MCED shows the editors' intention to meet the word-desperate reader halfway.

So the whole lexicon is one built on four supports, each of a different function, presenting a stock of 100,000 items whose numerical strength and adequacy of coverage ensure a fair chance of success for the anxious searcher.

It now remains to cite evidence for the quality of the Dictionary's English work. And the first thing one notices is the fine sense distinctions one meets on first approaching a word. Witness the 4 senses given of the word 好歹:

sense 1 good and bad; what is good and what is bad: 不知好歹 unable to tell what's good or bad

sense 2 mishap; disaster: 万一她有个好歹 if something

should happen to her

sense 3 in any case; anyhow: 好歹我先开个头。Let me make a start, anyhow.

sense 4 no matter in what way: 别再做什么了，好歹吃点儿就得了。Don't cook us anything more. We'll have whatever there is.

Now 好歹 is an everyday word whose many meanings tend to shade one into another and a distinct definition of each is by no means easy—and the difficulty of providing the English for each illustrative example matching in tone and brevity is obvious. How well our editors have done can be safely left to the reader to judge.

With regard to the translation of what may be called "story words", the editors seem to have adopted a 2-tier method—one literal, one metaphorical. Thus:

	literal	metaphorical
一网打尽	catch the whole lot in a dragnet	round up the whole gang at one fell swoop
一意孤行	cling obstinately to one's course	be bent on having one's own way
一反常态	depart from one's normal behaviour	act out of character
一箭双雕	shoot two hawks with one arrow	kill two birds with one stone

Here the problem seems to be two-fold: one to tell the story accurately enough to satisfy the inquisitive learner (though he should know that the native speaker seldom bothers about etymological meanings), the other, pragmatically far more important, to transcend the story and put across the message as straightforwardly as possible. And our editors seem to have done well enough on both accounts.

Translating from one culturally disparate language to another could be at once a trying experience and a joy—trying because the implicatures tend to diverge so much that a competent version generally turns out to be dull and ordinary, and a joy when a brilliant version is arrived at by good luck. The MCED editors seem to have excelled by being uniformly competent and very often brilliant. A few more examples:

(1) 话不投机半句多。When the conversation gets disagreeable, to say on more word is a waste of breath.
(2) 话里有话 The words mean more than they say; there's more to it than what is said.
(3) 听他的话碴儿，这事好办。From what he says, that'll be easily done.
(4) 听他的话音，他准是反对的。Judging from his tone, there's no doubt he's against it.

Now of the four, (1) and (2) are clearly brilliant: no shade of meaning is lost in the English rendering, nor worked up to become a clumsy mouthful. It is very, very nearly what it is in Chinese—short, brisk, stimulating, but because it now appears in a new idiom, it gets a vigor which startles even those who feel at home in both languages. (3) and (4) are admittedly less than brilliant, but even here one ought to be grateful for the way 话碴儿 (message) and 话音 (tone) are so neatly distinguished.

Finally, a look at MCED's Acknowledgements will tell us how wide the editors have cast their net for a good word haul. Altogether 56 works are mentioned as among the more important consulted, of which the published vocabularies of law, economics, foreign exchange, banking, agriculture, being key issues in China's development, naturally occupied a large part of the editors' attention. Many of these terms collected were actually once translations from English, but as they have been accepted into the national vocabulary in the course of a decade, few people using them today seem aware of their origin. The editors saw the importance of having these included — and they form a sizeable portion of the lexicon. Not surprisingly, many of such terms have found their ways into the fashionable talk of China's growing crop of entrepreneurs. Bilingually involved executives will be delighted.

Xu Guozhang

Professor of English and Linguistics
Research Centre of Linguistics
Foreign Studies University
Beijing, PRC

目 录

体例说明

1. 本词典所收词条一律按汉语拼音字母顺序排列,同音词按声调顺序排列;声调相同时按笔画排列。多字条目的第二个字相同时,按第三个字排列,并依此类推。对于“八、不、七、一”等有两种或两种以上声调的字,该字的单字条目和以该字为第一个字的多字条目均按第一声(阴平)排列。
2. 为节省篇幅,只注汉语拼音音节,单字词条和多字词条一律不注汉语拼音。单字词条按音节单列。多字词条第二个字同形异调时,一般不另分条,只分义项。
3. 冷僻的词语、现代汉语中不单独使用的字和词、一部分语气助词和不甚通用的方言一般不收。
4. 多义项的词条用①②③…分列,并在①②③…等圈码之后用方括号“[ ]”注明所属学科或在圆括号“( )”内用汉语注明释义。
5. 词条中的词组、例语、例句等按先词组后动宾搭配、先例语后例句分类,再依汉语拼音顺序排列。
6. 词组、例语、例句等一律用斜线“/”隔开,词组、例语、例句中与本词条相同的部分一律用波浪号“~”代替。
7. 在词组、例语、例句英语释义或译文中,圆括号“( )”里的词一种是可以省略的词,一种是该词的缩略语。词组、例语、例句译文中尖括号“〈 〉”里的词可以替代前面紧相邻的词。
8. 词条中菱形号“◇”后面的是不单列多字条目的合成词。
9. 词条中圆圈号“○”后面的是本词条的参考词汇,参考词汇的排列有三种:一是根据词条中的释义顺序,如“三废”条;二是分类,如“汽车”条;三是按汉语拼音顺序,如:“蔬菜”条。
10. 因各种汉语字典、词典的部首设置各异,本书只收汉语拼音音节索引和汉字笔画索引。

学科术语略语表

(按汉语拼音字母顺序排列)

[测]	测绘
[地]	地质学；地理学
[电]	电学；电工
[动]	动物；动物学
[法]	法律
[方]	方言
[纺]	纺织印染
[副]	副词
[工美]	工艺美术
[化]	化学；化工
[机]	机械
[简]	简称
[建]	建筑
[交]	交通运输
[教]	教育
[介]	介词
[经]	经济
[旧]	旧时用语
[剧]	戏剧
[军]	军事
[口]	口语
[矿]	矿物；矿业
[连]	连词
[量]	量词
[林]	林业
[逻]	逻辑学
[骂]	骂人的话
[牧]	畜牧业
[农]	农业
[气]	气象学
[商]	商业
[摄]	摄影
[生]	生物
[生化]	生物化学
[生理]	生理学；人体解剖
[史]	历史
[书]	文言文或书面语
[数]	数学
[水]	水利
[套]	套语
[体]	体育
[天]	天文学
[外]	外交
[微]	微生物
[无]	无线电
[物]	物理学
[象]	象声词
[讯]	电信
[谚]	谚语
[药]	药物；药物学
[冶]	冶金
[医]	医学
[印]	印刷
[邮]	邮政
[渔]	渔业
[语]	语言学
[乐]	音乐
[哲]	哲学
[植]	植物；植物学
[纸]	造纸
[助]	助词
[自]	自动化
[宗]	宗教

说明：凡属可以望文知意的多字学科术语，如：[棒球]、[佛教]、[计算机]、[铁道]等，本表不再列入。

主要参考书目

(按汉语拼音字母顺序排列)

辞海(合订本)
当代汉英词典
电影、电视、录像词典
国际汉英成语大辞典
汉英词典
汉英综合科学技术词汇
简明不列颠百科全书(中文版)
拉汉脊椎动物名称
拉汉兽类名称
拉汉英种子植物名称
朗曼现代英语词典
民族乐器制作概述
牛津杜登图解词典
牛津高级英语词典
人体解剖图谱
人体解剖学图谱
实用汉英词典
实用汉语图解词典
苏联百科词典
图书专业词汇
外汇业务常用词汇
现代汉语词典
现代汉语通用词典
现代科学技术词典
新英汉词典
兄弟民族服饰资料
印刷书业常用词汇
英汉道路工程词汇
英汉电工词汇
英汉法律词典
英汉工程技术辞汇
英汉海洋学词汇
英汉航空工程词典
英汉化学化工词汇(及各补编)
英汉计算机词典
英汉舰船科技词汇
英汉经济综合词典
英汉军语词汇
英汉农业词典
英汉气象学词汇
英汉汽车技术词典
英汉数学词汇
英汉水文学词汇
英汉体育常用词汇
英汉天文学词汇
英汉铁路词典
英汉土木建筑工程词汇
英汉物理学词汇
英汉心理学名词
英汉医学词汇
英汉银行业务词典
英汉油矿词典
英汉造纸词汇
英和大词典
英华大词典
远东英汉大词典
中国高等植物图鉴
最新实用汉英词典

汉语拼音音节索引

dī 滴嘀镝堤提低 dí 涤嘀镝嫡笛敌的 dǐ 底诋砥抵骶邸 dì 帝谛蒂碲缔弟递地棣第的
diān 颠癫巅掂 diǎn 踮点典碘 diàn 淀靛奠垫店惦玷电佃殿癜
diāo 貂凋碉雕鲷刁叼 diào 调掉吊钓
diē 跌爹 dié 谍碟喋蝶牒迭叠
dīng 丁疔玎耵酊叮盯钉 dǐng 顶鼎 dìng 定锭订钉
diū 丢铥
dōng 东冬氡 dǒng 董懂 dòng 动冻栋洞恫
dōu 都兜 dǒu 斗抖陡 dòu 斗窦豆痘逗
dū 都嘟督 dú 毒渎读黩犊牍独 dǔ 堵睹赌笃肚 dù 度渡镀杜肚蠹妒
duān 端 duǎn 短 duàn 断段煅椴锻缎
duī 堆 duì 兑对队
dūn 敦墩吨蹲 dǔn 趸盹 dùn 炖顿囤钝盾遁
duō 咄多哆掇 duó 度踱夺 duǒ 朵垛躲 duò 惰驮舵垛剁跺堕

E(220–224 页)

ē 阿婀 é 额蛾锇俄鹅娥讹 ě 恶 è 恶垩噩厄苊扼轭呃遏愕萼颚鹗腭鳄饿
ēn 恩蒽 èn 摁
ér 而鸸儿 ěr 耳铒饵尔迩 èr 二贰

F(225–268 页)

fā 发 fá 罚乏伐阀筏 fǎ 法砝 fà 发珐
fān 帆番幡藩翻 fán 烦樊蕃繁凡矾钒 fǎn 反返 fàn 泛范梵贩饭犯
fāng 方芳坊妨钫 fáng 房防坊妨 fǎng 访仿舫纺 fàng 放
fēi 非扉霏菲蜚鲱绯飞妃 féi 腓肥 fěi 斐悱诽菲匪榧蜚翡 fèi 吠痱废肺沸费镄狒
fēn 分芬酚吩氛纷 fén 坟焚 fěn 粉 fèn 粪愤奋分份
fēng 丰封烽蜂峰锋风疯枫砜 féng 逢缝 fěng 讽 fèng 奉俸凤缝
fó 佛
fǒu 否
fū 夫麸呋肤敷孵跗fú 芙扶福辐幅蝠孚浮蜉俘伏茯符凫服弗拂氟 fǔ 府腐俯拊辅脯抚斧釜 fù 傅缚富副讣赴赋父复覆蝮馥腹鲋付附驸阜服负妇

G(269–333 页)

gā 夹嘎旮伽咖 gá 钆
gāi 该赅 gǎi 改 gài 盖丐芥钙概
gān 干杆酐矸竿肝甘泔坩苷柑尴 gǎn 赶杆秆擀感敢橄 gàn 干绀
gāng 缸肛冈刚钢纲 gǎng 港岗 gàng 杠钢
gāo 高膏篙羔糕睾 gǎo 槁搞镐稿缟 gào 膏告诰锆
gē 割哥歌戈搁疙鸽咯胳 gé 革葛嗝膈镉隔蛤阁格搁 gě 舸葛 gè 个各硌铬
gěi 给
gēn 根跟 yén 哏 gěn 艮 gèn 亘
géng 庚羹耕更 gěng 耿埂梗哽鲠 gèng 更
gōng 宫工攻功供恭公肱觥弓躬 gǒng 巩汞珙拱 gòng 贡共供
gōu 篝佝勾沟钩 gǒu 苟枸狗 gòu 媾彀诟垢够勾构购
gū 沽辜轱咕估姑菇骨箍呱孤菰 gú 骨 gǔ 汩鼓臌古钴牯贾蛊骨谷股毂 gù 雇顾故固痼锢梏
guā 栝刮瓜呱胍 guǎ 寡剐 guà 褂挂卦
guāi 掴乖 guǎi 拐 guài 怪
guān 官棺倌关冠鳏观 guǎn 管馆 guàn 冠灌罐鹳观盥贯惯掼
guāng 光桄 guǎng 广犷 guàng 逛
guī 规圭闺硅鲑归皈瑰龟 guǐ 庋晷鬼诡轨癸 guì 桂柜贵跪桧刽鳜
gǔn 衮滚磙绲 gùn 棍
guō 郭聒过蝈锅 guó 国 guǒ 果裹 guò 过

H(334–398 页)

hā 哈铪 há 蛤 hǎ 哈 hà 哈
hāi 嗨咳 hái 骸孩还 hǎi 海胲 hài 害亥氦骇
hān 犴顸鼾憨酣蚶 hán 寒含焓函涵 hǎn 罕喊 hàn 汉汗旱悍焊捍翰瀚憾撼颔
hāng 夯 háng 吭航行绗 hàng 沆巷
hāo 蒿薅 háo 豪壕嚎号貉嗥毫蚝 hǎo 好 hào 耗号浩皓好
hē 诃呵嗬喝 hé 涸阂核劾阖翮河何合貉颌盒禾和 hè 鹤赫荷壑吓褐喝和贺
hēi 黑嘿
hén 痕 hěn 很 hèn 狠恨
hēng 亨哼 héng 恒横珩桁衡 hèng 横
hng 哼
hōng 烘哄轰 hóng 鸿虹红洪宏弘泓 hǒng 哄 hòng 哄
hóu 骺侯瘊喉猴 hǒu 吼 hòu 厚候后
hū 糊乎烀呼忽 hú 壶胡湖糊煳葫蝴猢核囫鹄狐弧槲 hǔ 浒虎琥唬 hù 糊户戽护扈怙瓠互笏
huā 化花哗 huá 划滑猾华哗铧 huà 话画划化桦
huái 怀槐踝 huài 坏
huān 獾欢 huán 还环寰 huǎn 缓 huàn 浣宦豢患涣焕换唤幻
huāng 荒慌 huáng 黄磺簧皇惶煌蝗篁鳇隍 huǎng 谎恍晃幌 huàng 晃
huī 麾挥辉灰恢诙徽 huí 回茴蛔 huǐ 悔毁 huì 汇溃讳彗慧卉惠蕙喙秽贿会烩荟绘诲晦
hūn 荤昏阍婚 hún 浑魂馄 hùn 混诨
huō 豁劐攉锪 huó 活和 huǒ 火钬伙夥 huò 豁

祸霍藿获或惑和货

J(399–494 页)

jī 激齑跻迹绩积击基奇畸唧犄唧羁稽几讥机叽肌饥汲芨鸡姬缉 jí 疾蒺嫉脊瘠吉籍棘辑集及极岌级急即亟 jǐ 脊济挤虮麂己给 jì 寂济荠剂鲚计髻蓟技伎妓寄冀觊季悸祭鲫系既记忌纪际继

jiā 家镓夹佳加痂袈嘉茄 jiá 夹荚颊蛱戛 jiǎ 甲岬钾胛假 jià 稼嫁价假架驾

jiān 煎兼间肩笺菅歼缄监坚鲣尖艰犍奸 jiǎn 简剪茧柬拣减碱检捡睑俭 jiàn 渐间涧践贱溅饯荐鉴监见舰剑箭僭件建键毽健腱

jiāng 将浆姜江豇僵疆 jiǎng 桨奖讲 jiàng 酱将匠降绛强糨犟

jiāo 浇交茭胶郊教椒娇骄焦蕉礁鵁 jiáo 嚼 jiǎo 铰佼皎狡饺绞搅矫侥缴脚角剿 jiào 窖觉校较教酵叫轿

jiē 秸结接揭皆阶嗟街节疖 jié 洁诘拮桔结截劫节捷睫竭羯碣桀杰孑 jiě 解姐 jiè 戒诫介疥芥界借解届

jīn 津禁襟巾今矜金筋斤 jǐn 堇谨紧锦仅尽 jìn 进晋觐禁噤近劲浸尽烬荩

jīng 京惊鲸旌粳精睛腈荆兢晶泾茎经 jǐng 井肼阱警儆景颈 jìng 竟境镜竞净静靖敬痉径胫劲经

jiǒng 窘炯迥

jiū 阄揪究鸠赳纠 jiǔ 酒韭九久灸 jiù 就鹫厩救旧臼柏舅疚柩咎

jū 车疽狙鞠掬拘驹居锔 jú 菊橘局锔 jǔ 举沮龃咀枸 jù 聚巨炬拒距遽具惧俱飓句据踞剧锯

juān 圈涓捐娟镌 juǎn 卷 juàn 眷卷圈倦绢隽

juē 撅 jué 觉厥蕨橛蹶矍攫爵角谲决诀抉掘崛倔绝 juě 蹶 juè 倔

jūn 军皲均菌龟君 jùn 菌浚竣峻俊骏

K(495–525 页)

kā 喀咖 kǎ 卡咔胩咯

kāi 开锎揩 kǎi 慨楷铠凯

kān 刊堪戡勘龛看 kǎn 槛侃坎莰砍 kàn 瞰看

kāng 康慷糠 káng 扛 kàng 亢炕抗钪伉

kǎo 考烤栲拷 kào 铐靠犒

kē 颏磕瞌珂苛窠颗科蝌 ké 咳壳 kě 渴可 kè 刻克氪可课骒客恪

kěn 肯啃恳垦

kēng 坑吭铿

kōng 空 kǒng 恐倥孔 kòng 空控

kōu 抠眍 kǒu 口 kòu 寇扣筘叩

kū 窟枯骷哭 kú 苦 kù 库裤酷

kuā 夸 kuǎ 垮 kuà 挎跨胯

kuài 会侩脍块快筷

kuān 宽髋 kuǎn 款

kuāng 匡诓框哐筐 kuáng 狂诳 kuàng 矿旷框眶况

kuī 窥亏盔岿 kuí 奎喹蝰魁葵睽 kuǐ 傀 kuì 愧溃愦聩匮馈

kūn 坤昆醌 kǔn 捆 kùn 困

kuò 廓扩阔括蛞

L(526–587 页)

lā 垃拉邋 lá 拉 lǎ 喇 là 落蜡腊辣瘌

lái 来莱铼 lài 赖癞籁

lán 阑谰镧兰栏拦褴蓝篮岚 lǎn 懒览揽缆罱 làn 滥烂

láng 郎廊榔琅锒狼 lǎng 朗 làng 浪莨

láo 捞牢劳痨铹醪 lǎo 老铑佬姥 lào 涝烙落酪

lè 乐勒鳓

lēi 勒擂 léi 雷镭累 lěi 蕾磊累垒 lèi 泪类擂累肋

léng 棱 lěng 冷 lèng 愣

lī 哩 lí 离篱厘狸罹梨犁黎藜 lǐ 礼李里理锂俚鲤 lì 立莅丽俪栗历励利痢例栎砾隶力历沥呖荔

liǎ 俩

lián 帘廉镰臁怜联连涟莲鲢 liǎn 敛脸 liàn 恋楝炼练潋殓链

liáng 梁凉椋良粮莨量 liǎng 两 liàng 凉谅晾亮踉辆量

liāo 撩 liáo 聊寮燎撩嘹僚鹩獠缭寥疗辽 liǎo 蓼潦了钌 liào 料撂镣了钌

liē 咧 liě 咧 liè 列烈裂趔劣猎鬣

lín 麟遴磷辚嶙鳞林淋霖琳临邻 lǐn 凛檩 lìn 淋吝赁躏

līng 拎 líng 凌菱鲮陵绫羚零玲聆龄囹铃伶翎灵 lǐng 令领岭 lìng 另令

liū 溜熘 liú 刘浏流琉硫留瘤榴镏馏 liǔ 柳绺 liù 六碌溜遛馏鹨陆

lóng 龙茏聋砻笼隆 lǒng 垄拢笼 lòng 弄

lōu 搂 lóu 娄耧楼喽蝼 lǒu 搂篓 lòu 漏瘘镂露陋

lú 庐炉芦卢颅鸬鲈 lǔ 卤虏掳鲁橹镥 lù 鹿漉麓辘路露鹭戮录禄碌绿陆

lǘ 驴 lǚ 旅膂偻屡缕捋铝侣履 lǜ 率虑滤律氯绿

luán 栾峦銮挛鸾孪 luǎn 卵 luàn 乱

lüè 掠略

lūn 抡 lún 沦轮伦纶 lùn 论

luō 捋罗 luó 螺骡罗逻萝箩锣 luǒ 瘰裸 luò 荦摞洛落骆络

M(588–629 页)

mā 摩抹妈 má 麻 mǎ 马玛码吗蚂 mà 蚂骂

mái 霾埋 mǎi 买 mài 麦卖迈脉

mān 颟 mán 蛮埋瞒蔓鳗馒 mǎn 满螨 màn 曼漫慢谩蔓幔镘

máng 忙芒茫盲 mǎng 莽蟒

māo 猫 máo 毛牦锚猫矛茅蝥 mǎo 卯铆 mào

茂冒帽耄貌贸
méi 没糜煤媒玫枚霉莓梅酶眉楣镅 měi 美镁每 mèi 寐昧魅妹媚
mēn 闷 mén 门扪钔 mèn 闷焖
mēng 蒙 méng 虻蒙朦萌盟 měng 懵蒙蠓锰猛 mèng 梦孟
mī 咪眯 mí 靡糜麋迷谜醚弥猕 mǐ 靡米眯脒 mì 泌秘蜜密嘧幂觅
mián 眠棉绵 miǎn 湎腼缅免冕勉 miàn 面
miāo 喵 miáo 苗描瞄 miǎo 秒渺藐淼 miào 庙妙
miē 咩乜 miè 灭蔑篾
mín 民 mǐn 悯敏泯抿
míng 冥暝螟明盟鸣名茗铭 mǐng 酩 mìng 命
miù 谬
mō 摸 mó 磨蘑摩魔模摹膜 mǒ 抹 mò 磨末沫茉抹秣莫漠寞貘蓦陌墨默脉没
mōu 哞 móu 谋牟眸 mǒu 某
mú 模 mǔ 亩牡母拇姆 mù 墓暮幕募慕木沐目苜钼睦牧穆

N(630-649 页)

ná 拿镎 nǎ 哪 nà 捺衲呐钠纳那 na 哪
nǎi 乃氖奶 nài 奈萘耐
nán 南喃男难 nǎn 赧蝻 nàn 难
nāng 囔 náng 囊 nǎng 攮
nāo 孬 náo 硇挠蛲 nǎo 恼脑瑙 nào 闹
né 讷
něi 馁 nèi 内
nèn 嫩
néng 能
ní 霓鲵尼泥呢铌 nǐ 拟你 nì 溺逆匿腻泥昵
niān 拈蔫 nián 黏粘鲇年 niǎn 撵捻碾 niàn 念埝
niáng 娘 niàng 酿
niǎo 袅鸟茑 niào 尿脲
niē 捏 niè 颞蹑镊蘖孽涅啮镍
nín 您
níng 宁柠拧狞凝 nǐng 拧 nìng 宁佞
niú 牛 niǔ 忸扭狃纽 niù 拗
nóng 农浓脓 nòng 弄
nú 奴驽 nǔ 努 nù 怒
nǚ 女钕
nuǎn 暖
nüè 疟虐
nuó 挪 nuò 懦糯诺锘

O(650 页)

ōu 讴鸥欧殴 ǒu 耦藕偶呕 òu 沤怄

P(651-680 页)

pā 趴 pá 扒耙琶筢爬 pà 怕帕
pāi 拍 pái 排徘牌 pǎi 排迫 pài 派哌
pān 攀 pán 蹒盘磐 pàn 判畔叛襻盼
pāng 滂膀乓 páng 旁磅螃膀彷庞 pǎng 耪 pàng 胖
pāo 泡抛 páo 炮袍咆刨狍 pǎo 跑 pào 泡疱炮
pēi 呸胚 péi 培赔锫陪 péi 沛配辔佩
pēn 喷 pén 盆 pèn 喷
pēng 烹抨澎 péng 澎膨蓬篷朋棚硼鹏 pěng 捧 pèng 碰
pī 坯砒批纰披霹劈 pí 琵枇毗蚍啤蜱脾皮疲铍 pǐ 否痞匹癖劈 pì 屁媲辟譬僻
piān 篇偏翩片 pián 胼蹁便 piàn 片骗
piāo 漂剽飘缥 piáo 瓢嫖 piǎo 漂瞟 piào 票漂嘌
piē 撇瞥氕 piě 苤撇
pīn 拼姘 pín 频贫 pǐn 品 pìn 聘牝
pīng 娉乒 píng 瓶屏平评坪苹萍鲆凭
pō 泊钋坡颇泼 pó 婆 pǒ 叵钷笸 pò 朴破迫魄
pōu 剖
pū 扑仆噗铺 pú 菩蒲葡莆璞镤仆 pǔ 普谱镨朴圃蹼 pù 铺瀑曝

Q(681-730 页)

qī 期欺栖桤漆戚槭嘁七沏妻凄萋蹊 qí 齐蛴脐鳍其麒旗棋祈歧奇崎骑畦 qǐ 启企乞起杞岂绮 qì 泣弃契葺碛砌器气汽憩讫迄
qiā 掐 qiǎ 卡 qià 髂洽恰
qiān 悭谦牵签铅千迁扦钎阡愆 qián 潜前乾荨掮虔黔钱钳 qiǎn 浅遣谴 qiàn 歉茜堑欠芡嵌纤
qiāng 锵羌蜣枪戗呛镪腔 qiáng 墙蔷樯强 qiǎng 羟抢强襁 qiàng 呛戗
qiāo 敲悄橇跷缲锹劁 qiáo 翘乔荞桥侨憔瞧 qiǎo 悄巧 qiào 窍壳撬翘鞘峭俏
qiē 切 qié 茄 qiě 且 qiè 妾锲挈怯惬切窃
qīn 亲钦侵 qín 秦勤芹禽擒噙琴 qǐn 寝 qìn 沁
qīng 青清蜻鲭倾轻氢 qíng 情晴氰擎 qǐng 请苘顷 qìng 亲庆磬罄
qióng 穹穷琼茕
qiū 秋楸丘蚯 qiú 酋求球裘囚泅犰
qū 趋祛区躯驱蛆曲黢屈 qú 渠蠼 qǔ 取娶龋曲 qù 阒趣去觑
quān 悛圈 quán 拳鬈颧蜷权全痊醛诠泉 quǎn 犬 quàn 券劝
quē 阙炔缺 qué 瘸 què 却鹊榷雀确
qún 麇群裙

R(731-753 页)

rán 髯然燃 rǎn 染冉
rāng 嚷 ráng 瀼瓤 rǎng 壤攘嚷 ràng 让
ráo 荛桡饶 rǎo 扰 rào 绕
rě 惹 rè 热
rén 人壬仁 rěn 荏稔忍 rèn 认任妊刃韧纫

rēng 扔 réng 仍
rì 日
róng 容溶熔榕戎绒荣蝾茸融 rǒng 冗
róu 柔糅鞣揉蹂 ròu 肉
rú 濡蠕儒孺如茹铷 rǔ 汝辱乳 rù 溽褥蓐缛入
ruǎn 朊阮软
ruǐ 蕊瑞蚋锐
rùn 闰润
ruò 若偌箬弱

S(754–856 页)

sā 撒仨 sǎ 洒撒 sà 飒萨脎
sāi 塞噻腮鳃 sài 塞赛
sān 三叁 sǎn 散伞 sàn 散
sāng 丧桑 sǎng 搡嗓 sàng 丧
sāo 臊搔骚缫 sǎo 扫嫂 sào 扫臊
sé 涩塞瑟啬色铯
sēn 森
sēng 僧
shā 杉沙痧鲨砂纱杀刹煞 shǎ 傻 shà 霎厦歃煞
shāi 筛 shǎi 色 shài 晒
shān 潸扇芟苫山舢衫杉钐膻珊栅删姗 shǎn 闪
shàn 擅嬗善膳鳝缮禅扇骟苫赡疝讪
shāng 商熵墒伤觞 shǎng 赏晌 shàng 尚绱上
shāo 烧鞘梢捎稍艄 sháo 勺芍杓韶 shǎo 少
shào 捎哨潲少劭
shē 奢赊猞 shé 折蛇舌 shě 舍 shè 涉社设赦慑
摄舍射麝
shēn 深莘申砷呻伸绅身参 shén 神什 shěn 审
婶 shèn 慎甚肾胂蜃渗
shēng 声生甥笙牲升 shéng 绳 shěng 省
shèng 盛乘剩胜圣
shī 湿诗蓍师狮嘘失施尸虱鲺 shí 实识十什石
拾时鲥食蚀 shǐ 史使驶矢屎始 shì 室市柿
铈视式试拭弑示士仕恃侍螫世莳事誓逝势
是嗜释噬适似饰氏
shōu 收 shǒu 守首手 shòu 瘦兽寿受授绶售狩
shū 梳疏蔬枢叔淑菽输殊倏抒舒书 shú 孰熟塾
秫赎 shǔ 数暑署薯曙蜀黍鼠属 shù 树漱
庶数术述束戌竖恕
shuā 刷 shuǎ 耍 shuà 刷
shuāi 衰摔 shuǎi 甩 shuài 率帅
shuān 闩栓拴 shuàn 涮
shuāng 霜孀双 shuǎng 爽
shuí 谁 shuǐ 水 shuì 说税睡
shǔn 吮 shùn 瞬顺
shuō 说 shuò 数朔蒴硕烁铄
sī 斯澌厮撕嘶思锶私司丝 sǐ 死 sì 肆寺四驷似
嗣伺饲巳祀
sōng 松嵩 sǒng 悚怂耸 sòng 宋送诵讼颂
sōu 溲搜嗖馊 sǒu 薮嗾 sòu 嗽
sū 苏酥 sú 俗 sù 宿溯塑诉素嗉速簌夙肃
suān 酸 suàn 蒜算
suī 虽 suí 随绥 suǐ 髓 suì 碎遂邃燧隧岁穗祟
sūn 孙 sǔn 损笋隼榫
suō 莎桫娑蓑羧梭唆缩 suǒ 索琐唢锁所

T(857–905 页)

tā 它铊溻塌踏趿他她 tǎ 塔鳎獭 tà 挞拓榻沓踏
tāi 胎 tái 薹台苔抬鲐 tài 泰太态酞钛肽
tān 坍贪滩摊瘫 tán 痰谈坛昙檀潭弹 tǎn 毯忐
袒坦钽 tàn 探叹炭碳
tāng 汤铴羰趟耥 táng 唐溏糖塘搪堂棠樘螳镗
膛 tǎng 淌倘躺 tàng 烫趟
tāo 涛掏饕叨滔韬绦 táo 逃桃淘陶 tǎo 讨 tào
套
tè 铽特
téng 疼誊藤腾
tī 梯锑剔踢体 tí 啼蹄鹈题提 tǐ 体 tì 涕剃绨替
嚏屉
tiān 添天 tián 恬填田钿甜 tiǎn 舔腆
tiāo 挑 tiáo 条调迢苕 tiǎo 挑 tiào 眺跳
tiē 帖萜贴 tiě 帖铁 tiè 帖
tīng 烃厅听 tíng 亭停庭 tǐng 挺铤艇
tōng 通 tóng 童瞳同茼桐酮铜 tǒng 筒桶捅统
tòng 痛通
tōu 偷 tóu 头投 tòu 透
tū 突凸秃 tú 屠图涂途荼徒 tǔ 土吐钍 tù 吐兔
tuān 湍 tuán 团
tuī 推 tuí 颓 tuǐ 腿 tuì 蜕退褪
tūn 吞 tún 屯囤豚臀 tùn 褪
tuō 脱拖托 tuó 沱坨柁鸵陀驼鼍驮 tuǒ 椭妥
tuò 拓唾

W(906–945 页)

wā 挖洼哇蛙 wá 娃 wǎ 瓦 wà 袜瓦
wāi 歪 wǎi 崴 wài 外
wān 豌剜蜿弯湾 wán 完烷玩顽丸纨 wǎn 莞宛
惋碗婉挽晚绾 wàn 蔓万腕
wāng 汪 wáng 亡王 wǎng 枉罔惘魍网往
wàng 忘妄望旺往
wēi 威煨偎微逶巍危 wéi 为韦违围圩惟唯帷维
桅 wěi 伪苇伟纬鲔唯委萎猥尾娓 wèi 为
未味畏喂胃谓位尉慰蔚卫
wēn 温瘟 wén 文蚊纹闻 wěn 紊稳刎吻 wèn
问
wēng 翁嗡 wèng 蕹瓮
wō 涡窝莴蜗喔 wǒ 我 wò 沃斡卧硪肟握龌
wū 污巫诬乌呜钨屋 wú 无芜梧鼯蜈毋 wǔ 武
妩五伍捂午忤舞侮 wù 误恶痦悟焐晤兀
务雾勿芴物坞鹜骛

X(946–1012 页)

xī 曦熹嘻嬉昔惜熙析晰蜥西硒牺锡吸奚溪蹊翕
希烯稀悉蟋息熄夕汐矽膝犀 xí 席檄袭媳

习 xǐ 喜洗铣徙玺 xì 阋隙系戏细
xiā 瞎虾呷 xiá 峡侠狭辖黠匣狎遐霞瑕暇 xià 下吓夏
xiān 籼仙氙先酰鲜掀锨纤 xián 涎舷弦闲娴咸贤衔嫌 xiǎn 显险铣鲜藓 xiàn 宪羡霰献县现苋腺馅陷限线
xiāng 襄镶相厢箱香乡 xiáng 详翔降 xiǎng 享鲞想响飨 xiàng 项巷相向象橡像
xiāo 哓骁消宵逍霄硝削销哮嚣潇萧箫 xiáo 淆 xiǎo 晓小 xiào 校效孝肖啸笑
xiē 楔些蝎歇 xié 鞋挟携谐偕邪斜协胁 xiě 写血 xiè 亵泻械泄卸谢榭蟹懈邂屑
xīn 辛锌新薪心馨欣 xín 寻 xìn 芯信衅
xīng 兴星惺腥猩 xíng 形刑型行 xǐng 醒擤省 xìng 兴幸悻杏性姓
xiōng 兄凶汹匈胸 xióng 雄熊
xiū 羞馐修休鸺 xiǔ 宿朽 xiù 袖秀锈绣臭溴嗅
xū 需吁须虚墟嘘 xú 徐 xǔ 许诩栩醑 xù 畜蓄叙酗煦恤旭序絮婿绪续
xuān 宣萱喧轩 xuán 旋玄悬 xuǎn 癣选 xuàn 渲楦旋炫眩绚
xuē 削靴 xué 学穴噱 xuě 雪鳕 xuè 谑血
xūn 勋熏醺 xún 驯循旬询巡寻鲟 xùn 训蕈汛讯迅殉徇逊

Y(1013—1099 页)

yā 丫椏压呀鸦押鸭 yá 涯崖牙芽蚜 yǎ 哑雅 yà 亚氩压揠轧砑
yān 湮嫣淹阉腌烟咽胭殷 yán 颜言阎炎研盐严芫檐岩延筵沿 yǎn 演偃奄掩眼俨衍鼹 yàn 宴谚艳堰燕酽厌雁赝唁咽砚焰验
yāng 央泱殃秧 yáng 羊洋佯阳杨扬 yǎng 痒氧养仰 yàng 漾恙样怏
yāo 要腰夭妖邀幺吆约 yáo 尧肴窑谣遥摇鳐 yǎo 窈杳咬舀 yào 疟药要鹞钥耀
yē 耶椰掖噎 yé 爷 yě 冶野也 yè 夜液掖腋页叶业曳
yī 衣铱依一壹医伊咿 yí 宜颐咦胰姨遗仪移疑怡贻饴 yǐ 旖椅倚蚁乙钇已尾酏迤以 yì 意癔薏镱臆亦奕益溢谊抑易邑义议刈轶屹逸肄毅疫役忆亿艺呓译驿翼异
yīn 音因洇铟姻殷阴荫 yín 淫寅吟龈银 yǐn 饮引蚓隐瘾 yìn 窨荫印茚饮
yīng 应鹰莺英罂婴樱嘤鹦缨 yíng 赢荧萤营萦潆蝇盈楹迎 yǐng 瘿影颖 yìng 应硬映
yōng 庸鳙雍壅臃痈拥佣 yǒng 永泳咏甬涌蛹踊俑勇 yòng 用佣
yōu 忧优幽悠 yóu 游莜尤疣鱿犹由油柚蚰铀邮 yǒu 莠酉有铕友黝 yòu 又右囿柚釉鼬幼蚴诱
yū 淤迂 yú 于盂舆虞娱愚隅渝愉逾榆余腴鱼渔
yǔ 宇语雨与屿羽予 yù 育玉芋阈域郁与誉寓遇浴裕欲愈喻御狱预鹬驭
yuān 渊冤鸢鸳 yuán 元芫园鼋辕猿原源员圆援缘 yuǎn 远 yuàn 院愿怨
yuē 约 yuè 悦阅越跃岳月钥乐
yūn 晕 yún 云耘芸纭匀 yǔn 殒陨允 yùn 晕运酝愠蕴韵孕熨

Z(1100—1195 页)

zā 匝咂扎 zá 砸杂 zǎ 咋
zāi 灾栽哉甾 zǎi 宰载崽仔 zài 载再在
zān 糌簪 zán 咱 zǎn 趱攒 zàn 暂錾赞
zāng 赃脏臧 zàng 葬藏脏
zāo 糟遭 záo 凿 zǎo 枣早澡藻蚤 zào 灶燥噪躁造皂
zé 责啧则泽择 zè 仄
zéi 贼
zěn 怎 zèn 谮
zēng 曾憎增 zèng 甑赠锃综
zhā 渣楂扎咋 zhá 闸铡炸扎札轧 zhǎ 眨砟 zhà 栅乍痄炸诈蚱榨
zhāi 斋摘 zhái 宅择 zhǎi 窄 zhài 寨债
zhān 占沾粘毡谵瞻 zhǎn 盏斩崭展搌辗 zhàn 湛蘸栈占站战颤绽
zhāng 章璋樟蟑彰獐张 zhǎng 掌长涨 zhàng 瘴嶂幛障丈杖仗涨帐胀
zhāo 着朝招昭 zháo 着 zhǎo 沼找爪 zhào 肇罩兆笊召诏照
zhē 遮折蜇 zhé 谪折哲蛰辙辄 zhě 褶者锗赭 zhè 这蔗鹧柘 zhe 着
zhēn 榛斟甄真砧贞桢侦箴珍胗针 zhěn 枕缜疹诊轸畛 zhèn 鸩震振赈镇朕阵
zhēng 正症怔征丁争挣睁峥筝铮狰蒸 zhěng 整拯 zhèng 郑正症证政帧诤挣
zhī 汁之芝支枝肢只织知蜘栀指脂 zhí 职直植殖值执指踯侄 zhǐ 止趾只枳咫纸旨酯指 zhì 滞治志痣痔栉至窒桎蛭致掷挚置制秩智雉稚质炙
zhōng 中衷忠钟螽终 zhǒng 种肿踵 zhòng 中种仲众重
zhōu 州洲舟周粥 zhóu 轴妯 zhǒu 肘 zhòu 咒皱绉昼骤
zhū 朱诛珠株蛛铢侏诸猪潴 zhú 烛逐竹 zhǔ 主煮嘱瞩 zhù 苎贮注柱蛀住驻祝著助筑铸
zhuā 抓 zhuǎ 爪
zhuān 专砖 zhuǎn 转 zhuàn 转传赚篆撰
zhuāng 妆装庄桩 zhuàng 撞壮状
zhuī 椎锥追 zhuì 赘坠
zhūn 谆 zhǔn 准
zhuō 捉拙卓桌 zhuó 着琢啄浊灼酌
zī 咨资姿滋孳吱龇孜辎锱鲻 zǐ 梓紫姊子籽仔 zì 字恣自

zōng 宗棕鬃踪综 zǒng 总 zòng 粽纵
zǒu 走 zòu 奏 zū 租 zú 卒族足 zǔ 诅祖阻组
zuān 钻 zuǎn 纂 zuàn 钻
zuǐ 嘴 zuì 醉最罪
zūn 尊遵鳟 zǔn 撙
zuō 作 zuó 琢昨作 zuǒ 左佐 zuò 凿坐座柞作做

汉字笔画索引

六画

七画

十二画

十三画

十七画

二十一画

二十二画

二十三画

二十四画

二十五画

二十六画

二十七画

二十八画以上

A

ā

阿

【阿的平】[药] atabrine
【阿斗】a weak-minded person; fool
【阿尔巴尼亚】Albania ◇ ～人 Albanian / ～语 Albanian (language)
【阿尔卑斯山】the Alps
【阿尔法】alpha ◇ ～粒子 [物] alpha particle / ～射线 [物] alpha ray
【阿尔及利亚】Algeria ◇ ～人 Algerian
【阿尔泰山】Altay Mountains
【阿飞】hooligan; young rowdy
【阿富汗】Afghanistan ◇ ～人 Afghan
【阿根廷】Argentina ◇ ～人 Argentine
【阿訇】ahung; imam
【阿拉伯】Arabian; Arabic; Arab ◇ ～半岛 the Arabian Peninsula; Arabia / ～国家 Arab countries; Arab states / ～胶 gum arabic; gum acacia / ～人 Arab / ～数字 Arabic numerals / ～语 Arabic
【阿拉木图】Alma-Ata
【阿拉斯加】Alaska
【阿曼】Oman ◇ ～人 Omani
【阿芒拿】[化] ammonal
【阿门】[宗] amen
【阿米巴】[动] amoeba, ameba ◇ ～痢疾 amoebic dysentery
【阿米妥】[药] amytal
【阿片】[药] opium ◇ ～制剂 opiate
【阿司匹林】[药] aspirin
【阿托品】[药] atropine
【阿姨】[方] ① (姨母) auntie ② (对母辈妇女的称呼) auntie ③ (对保育员或保姆的称呼) nurse (in a family)
【阿谀奉承】flatter and toady
【阿月浑子】[植] pistachio
【阿扎尼亚】Azania ◇ ～人 Azanian

啊

[叹] eh; oh; o; ah

锕

[化] actinium (Ac)

【锕系元素】[化] actinides; actinide elements

āi

哀

①(悲伤; 悲痛) grief; sorrow: ～哭 wail / 喜怒～乐 joy, anger, grief and happiness ②(悼念) mourning: 志～ express one's mourning for the deceased ③(怜悯) pity: 乞～告怜 piteously beg for help

【哀兵必胜】an army burning with righteous indignation is bound to win
【哀愁】sad; sorrowful
【哀辞】lament
【哀悼】grieve over sb.'s death; mourn over sb.'s death; lament sb.'s death: 向死者家属表示深切的～ express one's heartfelt condolences to the family of the deceased
【哀而不伤】felt but not sentimental
【哀告】beg piteously; supplicate
【哀歌】a mournful song; dirge; elegy
【哀号】cry piteously; wail
【哀鸿遍野】a land swarming with famished refugees; disaster victims everywhere
【哀怜】feel compassion for; pity
【哀鸣】a plaintive whine; wail
【哀求】entreat; implore: 苦苦～ piteously entreat
【哀伤】distressed; grieved; sad
【哀思】grief; sad memories (of the deceased): 寄托～ give expression to one's grief
【哀叹】lament; bemoan; bewail
【哀痛】deep sorrow; grief
【哀怨】sad; plaintive
【哀乐】funeral music; dirge

锿

[化] einsteinium (Es)

哎

[叹] hey; ah

【哎哟】[叹] ouch; ow

埃

① (尘埃) dust ② [物] angstrom (A)

【埃及】Egypt ◇ ～人 Egyptian
【埃塞俄比亚】Ethiopia ◇ ～人 Ethiopian

挨

① (靠近; 紧接着) be next to; get close to: ～着窗口坐 sit by the window ② (顺着次序; 逐一) by turns; in sequence

【挨次】in turn; in order; one after another; one by one: ～入场 file in / ～上车 get on the bus one after another
【挨个儿】[口] in turn; one by one: ～检查身体 have medical check-ups in turn
【挨近】be near to; get close to
【挨门挨户】go from door to door: ～访问 make a house- to- house call

唉

① (答应的声音) right; yes ② (失望或无可奈何的感叹声) alas

【唉声叹气】heave deep sighs; moan and groan; sigh in despair

ái

癌 [医]cancer; carcinoma: 肠～ cancer of the bowels／肺～ cancer of the lung／肝～ cancer of the liver; liver cancer／皮肤～cutaneum carcinoma／绒膜～ chorionic carcinoma／乳腺～ mammary cancer／食道～ cancer of the esophagus／胃～ cancer of the stomach／转移性～ metastatic carcinoma／子宫～ cancer of the womb／子宫颈～ carvical carcinoma ◇致～物质 carcinogen; carcinogenic substance
【癌变】[医] canceration
【癌病】[医] carninomatosis
【癌扩散】[医] metastasis of cancer; proliferation of cancer
【癌前期病变】[医] precancerous lesion
【癌切除术】[医] carcinectomy; carcinomectomy
【癌细胞】[医] cancer cell
【癌症】[医] cancer
【癌肿】[医] a cancerous swelling

呆
【呆板】stiff; inflexible; rigid; stereotyped: ～的公式 a rigid formula／动作～ stiff and awkward movements

皑 pure white; snow white
【皑皑】pure white: ～白雪 an expanse of white snow

挨 ①(遭受; 忍受) suffer; endure: ～饿 go hungry; suffer from hunger／～浇 be caught in the rain／～骂 get a scolding; get a dressing down／～批评 be criticized ②(困难地度过) drag out ③(拖延) delay; play for time; stall
【挨打】take a beating; get a thrashing; come under attack: ～受饿 be beaten and starved
【挨饥受冻】suffer cold and hunger

ǎi

霭 haze; mist: ～暮 evening haze

蔼 friendly; amiable
【蔼然可亲】kindly; amiable; affable

嗳 [叹](表示不同意或否定) ～, 别客气了。Come on. Don't be so polite.
【嗳气】[医] belch; eructation

矮 ①(身材短) short ②(高度小的; 级别、地位低的) low: ～墙 a low wall／～一级 a grade lower
【矮凳】low stool; taboret
【矮墩墩】dumpy; pudgy; stumpy
【矮秆品种】[农]short-stalked variety; short-straw variety
【矮个儿】a person of short stature; a short person
【矮林】brushwood; coppice
【矮胖】short and stout; dumpy; roly-poly
【矮生果树】dwarf fruit trees
【矮松】[植]pinon pine
【矮小】short and small; low and small: ～的房屋 a small, low house／身材～ short and slight in figure
【矮星】[天] dwarf (star)
【矮种马】pony
【矮壮素】[农] cycocel
【矮子】a short person; dwarf
【矮子里拔将军】pick a general from among the dwarfs; choose the best person available

ài

隘 ①(狭窄) narrow: ～巷 alley; a narrow lane ②(险要的地方) pass: 要～ a strategic pass
【隘路】defile; narrow passage
【隘口】(mountain) pass

艾 ①[植] argy wormwood ②(停止) end; stop: 方兴未～ be just unfolding
【艾蒿】[植] argy wormwood
【艾绒】moxa
【艾滋病】Aids (Acquired Immune Deficiency (Syndrome)
【艾炷】moxa cone ◇～灸 moxibustion

砹 [化] astatine (At)

碍 be in the way of; hinder; obstruct: ～于情面 for fear of hurting sb.'s feelings; just to spare sb.'s feelings／有～于团结 be harmful to unity
【碍口】be too embarrassing to mention
【碍事】①(不方便; 有妨碍) be in the way; be a hindrance ②(严重; 大有关系) be of consequence; matter: 不～ nothing serious; It doesn't matter.
【碍视弯路】blind curved road
【碍手碍脚】be in the way; be a hindrance
【碍眼】be unpleasant to look at; be an eyesore; offend the eye

爱 ①(有很深的感情) love; affection: ～祖国 love one's country／母～ maternal love／相

～ fall in love ② (喜欢) like; be fond of; be keen on: ～游泳 be fond of swimming ③ (爱惜; 爱护) cherish; hold dear; take good care of; treasure ④ (常常发生某种行为,容易发生某种变化) be apt to; be in the habit of: ～发脾气 be apt to lose one's temper; be short-tempered / 铁～生锈 iron rusts easily

【爱不释手】fondle admiringly

【爱财如命】love money like one's very life

【爱称】diminutive; term of endearment; pet name

【爱戴】love and esteem

【爱丁堡】Edinburgh

【爱尔兰】Ireland ◇ ～人 the Irish; Irishman / ～语 Irish

【爱抚】show tender care for

【爱国】love one's country; be patriotic: ～一家, ～不分先后 All patriots belong to one big family, whether they rally to the common cause early or late. ◇ ～华侨 patriotic overseas Chinese / ～人士 patriotic personage / ～同胞 patriotic fellow-countrymen / ～卫生运动 patriotic health campaign / ～心 patriotic feeling; patriotism / ～者 patriot / ～志士 ardent patriot / ～主义 patriotism

【爱好】① (具有浓厚的兴趣并积极参加) like; love; be fond of; be keen on: ～和平 love peace; be peace-loving / ～京剧 be keen on Beijing opera ② (喜爱) interest; hobby

【爱好者】amateur; enthusiast; fan; lover (of art, sports, etc.): 体育～ sports enthusiast; sports fan / 无线电～ radio amateur / 音乐～ music-lover

【爱护】cherish; take good care of; treasure: ～儿童 take good care of children; bring up children with loving care / ～公物 take good care of public property

【爱克斯光】X ray; Roentgen ray ◇ ～机 X-ray apparatus / ～透视 X-ray examination; fluoroscopy / ～照片 roentgenogram / ～诊断 X-ray diagnosis; roentgen diagnosis

【爱理不理】look cold and indifferent; be standoffish

【爱怜】show tender affection for

【爱恋】be in love with; feel deeply attached to

【爱面子】be concerned about face-saving; be sensitive about one's reputation

【爱莫能助】willing to help but unable to do so

【爱慕】admire; adore: ～虚荣 be vain / 相互～ adore each other

【爱琴海】Egean Sea

【爱情】love (between man and woman)

【爱人】① (丈夫或妻子) husband; wife ② (恋爱中男女的一方) sweetheart

【爱沙尼亚】Esthonia

【爱神星】Eros

【爱斯基摩人】(因纽特人) Eskimo

【爱屋及乌】love for a person extends even to the crows on his roof; love me, love my dog

【爱惜】cherish; treasure; use sparingly: ～人力物力 use manpower and material sparingly / ～荣誉 cherish the good name of / ～时间 be economical of one's time; make the best use of one's time

【爱憎】love and hate: ～分明 be clear about what to love and what to hate

嗳 (表示悔恨,懊恼) oh

暧 [书] (of daylight) dim

【暧昧】① (态度、用意含糊; 不明白) ambiguous; equivocal: 态度～ assume an ambiguous attitude ② (行为不光明, 不可告人) dubious; shady: ～关系 dubious relationship

ān

安 ① (安定) peaceful; calm; quiet; tranquil: ～睡 sleep peacefully / 心神不～ feel uneasy or perturbed ② (使安定) set (sb.'s mind) at ease; calm ③ (心满意足) be satisfied; rest content: ～于现状 be content with things as they are; be satisfied with the existing state of affairs ④ (平安; 安全; 安好) safe; secure; in good health: ～抵 arrive safely; arrive safe and sound / 欠～ be slightly indisposed; be unwell ⑤ (使有适合的位置) place in a suitable position; find a place for ⑥ (安装; 设立) install; fit; fix: ～窗玻璃 put in a windowpane / ～电灯 install electric lights / 门上～锁 fit a lock on the door ⑦ (加上) bring (a charge against sb.); give (sb. a nickname): ～罪名 bring charges against ⑧ (存着; 怀有) harbour (an intention): ～坏心 harbour evil intentions ⑨ [电] ampere

【安邦定国】bring peace and stability to country

【安不忘危】mindful of possible danger in time of peace; In times of security do not be unprepared for the possibility of danger.

【安步当车】walk over leisurely instead of riding in a carriage; walk rather than ride

【安瓿】[药]ampoule; ampul; ampule

【安插】place in a certain position; assign to a job; plant: ～亲信 put one's trusted followers in

key positions
【安道尔】Andorra ◇ ～人 Andorran
【安定】① (生活、形势等平静正常) stable; quiet; settled: ～的生活 a stable life; a settled life / ～团结 stability and unity / 职业～ security of employment; job security ② (使安定) stabilize; maintain: ～人心 reassure the public; set people's minds at rest / ～社会秩序 maintain social order
【安定药】[药] tranquillizer
【安顿】① (使有着落; 安排妥当) arrange for; find a place for; help settle down ② (安稳) peaceful; undisturbed
【安放】lay; place; put in a certain place
【安分】be law-abiding; not go beyond one's bounds; know one's place
【安分守己】abide by the law and behave oneself; know one's place
【安抚】appease; pacify; placate: ～政策 pacification policy
【安哥拉】Angola ◇ ～人 Angolan
【安好】safe and sound; well
【安家】set up a home; settle down ◇ ～费 settling-in allowance
【安静】quiet; peaceful: 保持～。 Keep quiet!
【安居乐业】live and work in peace and contentment
【安康】good health
【安乐】peace and happiness ◇ ～窝 cosy nest / ～椅 easy chair
【安乐死】euthanasia
【安眠】sleep peacefully ◇ ～药 sleeping pill; sleeping tablet; hypnotic; soporific
【安眠酮】[药] hyminal; methaqualone
【安民告示】notice to reassure the public; an advance notice (of the agenda)
【安乃近】[药] analgin
【安宁】① (秩序正常) peaceful; tranquil ② (心情安定) calm; composed; free from worry
【安宁片】[药] meprobamate
【安排】arrange; fix up; plan: ～本年度的生产 plan this year's production / ～参观游览 arrange visits and sightseeing trips
【安培】[电] ampere ◇ ～定律 Ampere's law / ～计 amperemeter / ～小时 ampere-hour
【安全】safe; secure: ～到达 arrive safely / ～第一 safety first / ～行车 safe driving / 交通～ traffic safety / 保证～生产 ensure safety in production ◇ ～玻璃 safety glass / ～操作 safe operation / ～措施 safety measures; safety precautions / ～带 seat belt; safety belt; safety strap / ～岛 safety island; pedestrian island / ～阀 safety valve / ～感 sense of security / ～高度 safe altitude; safe height / ～工程 safety engineering / ～规程 safety regulations; safety code; safety rule / ～胶片 safety film / ～角 safety angle / ～界 safety limit / ～净空 safe clearance / ～帽 safety helmet / ～门 exit / ～设施 safety devices; safety equipment; safety installations / ～梯 emergency staircase; fore escape / ～剃刀 safety razor / ～停车距离 safe stopping distance / ～停止装置 safety stop / ～通行证 safe-conduct / ～网 safety netting / ～系数 safety coefficient; safety factor / ～系统 fail-safe / ～着陆 safe landing
【安全灯】①[矿] safety lamp ②[摄] safelight
【安全期】[生理] safe period ◇ ～避孕法 rhythm method
【安然】①(平安) safely: ～脱险 be out of danger ②(没有顾虑; 很放心) peacefully; at rest: ～入睡 go to sleep peacefully
【安然长逝】expire calmly
【安然逃脱】get-off; go scot-free
【安然无恙】safe and sound; (escape) unscathed
【安如磐石】as solid as a rock
【安如泰山】as secure as Mount Taishan; as solid as a rock
【安设】install; set up
【安身】make one's home; take shelter
【安身立命】settle down and get on with one's pursuit
【安神】① (使心神安定) calm the nerves; soothe the nerves ② [中医] relieve uneasiness of body and mind
【安神药】sedative; tranquillizer
【安生】① (安定) peaceful; restful ② (安静; 不生事) quiet; still
【安适】quiet and comfortable
【安替比林】[药] antipyrine
【安土重迁】be attached to one's native land and unwilling to leave it; hate to leave a place where one has lived long
【安妥】[药] antu
【安危】safety; safety and danger: 不顾个人～ heedless of one's personal safety
【安慰】comfort; console ◇ ～赛 consolation event; consolation match
【安稳】smooth and steady
【安息】①(安静地休息; 入睡) rest; go to sleep ②(悼念用语) rest in peace

【安息日】[宗] Sabbath (day)
【安息香】① [植] storax; snowbell ② [药] benzoin
【安闲】leisurely; peaceful and carefree: ～自在 leisurely and carefree
【安详】composed; serene; unruffled: 举止～ behave with composure
【安歇】go to bed; retire for the night
【安心】①(放心) feel at ease; be relieved; set one's mind at rest ②(心情安定) keep one's mind on sth.: ～工作 keep one's mind on one's work; work contentedly
【安逸】easy; easy and comfortable: 贪图～ seek an easy life; love comfort
【安营】camp; pitch a camp
【安营扎寨】camp; pitch a camp
【安于现状】satisfaction with things as they are
【安葬】bury (the dead) ◇ ～费 funeral expenses
【安枕无忧】sleep without anxiety
【安之若素】bear with equanimity; hardship with equanimity; regard with equanimity; wrongdoing with equanimity
【安置】arrange for; find a place for; help settle down
【安装】install; erect; fix; mount: ～电话 install a telephone / ～机器 install machinery / ～扩音器话筒 set up a microphone / ～蒸馏塔 erect a distillation column

鞍 saddle
【鞍部】saddle (of a hill or mountain)
【鞍马】①(体育器材或项目) pommelled horse; side horse ② (骑马或战斗生活) saddle and horse: ～劳顿 travel-worn / ～生活 life on horseback
【鞍子】saddle

桉 eucalyptus
【桉树】eucalyptus
【桉树脑】[化] cineole

氨 [化] ammonia: 合成～ synthetic ammonia
【氨苯磺胺】[药] sulphanilamide
【氨茶碱】[药] aminophylline
【氨化】ammoniation ◇ ～过磷酸钙 [农] ammoniated superphosphate
【氨基】[化] amino; amino-group ◇ ～塑料 amino-plastic / ～酸 amino acid
【氨基比林】[药] aminopyrine
【氨碱法】[化] ammonia soda process
【氨硫脲】[药] thiacetazone
【氨水】[化] ammonia water; aqua ammoniae

谙 know well: 素～水性 be a skillful swimmer
【谙练】conversant; proficient; skilled

庵 ①(小草屋) hut: 草～ a thatched hut ②(尼姑住的佛寺) nunnery; Buddhist convent

鹌
【鹌鹑】quail

ǎn

埯 ① (点种时挖的小坑) a hole to sow seeds in ② (挖小坑点种) dibble: ～豆 dibble beans ③ [量] (用于点种的作物) 一～花生 a cluster of peanut seedlings

揞 apply (medicinal powder to a wound)

铵 [化] ammonium: ～矾 ammonium alum

àn

案 ①(长桌) table; desk: 条～ a long narrow table ② (案件) case; law case: 办～ handle a case / 破～ clear up a criminal case; solve a criminal case ③(案卷; 记录) file; record: 有～可查 be on record; be on file ④(提出计划、办法或建议的文件) a plan submitted for consideration; proposal: 决议草～ a draft resolution / 提～ proposal; motion
【案板】kneading board; chopping board
【案秤】counter scale
【案件】law case; case; legal case: 公诉～ case of public charge 〈prosecution〉/ 民事～ civil case / 刑事～ criminal case / 自诉～ case of private charge 〈prosecution〉
【案卷】records; archives; dossier; files
【案例】case; example of case
【案情】case; details of a case: 了解～ investigate the details of a case◇ ～陈述 state of the case
【案情大白】The riddle of a puzzling case has been completely unraveled.
【案头】on one's desk ◇ ～日历 desk calendar
【案文】text: 协商～ negotiating text
【案验】investigate the evidence of a case
【案由】cause of action; main points of a case
【案子】①(长桌)long table; counter: 乒乓球～ ping-pong table / 肉～ meat counter② (案件) case; law case

按 ① (用手或指头压) press; push down: ～电钮 press a button; push a button / ～门铃 ring a doorbell / ～手印 put one's thumbprint (on a document, etc.) ②(压住；搁下) leave aside; shelve: ～下此事不提 leave this aside for the moment ③(抑制) control; restrain ④(用手压住不动) keep a tight grip on; keep one's hand on: ～住操纵杆 keep a tight grip on the control lever ⑤(依照) according to; in accordance with; in the light of; on the basis of: ～成本 at cost / ～股分配 issues to shareholders / ～年代顺序 in chronological order / ～我的意思 in my opinion / ～姓氏笔画为序 in the order of the number of strokes in the surnames / ～质定价 fix the price according to the quality ⑥(考查；核对) check; refer to: 有原文可～。There's the original to refer to. ⑦(加按语) note: 编者～ editor's note

【按比例】pro rata; in proportion: ～发展 develop in proportion; proportional development / ～分配 pro rata allotment / ～回扣 pro rata rebate / ～均摊 pro rata average

【按兵不动】not throw the troops into battle; take no action; bide one's time

【按部就班】follow the prescribed order; keep to conventional ways of doing things

【按程序】[法] amenable to process

【按键】button: 波段选择～ push-button for selecting wavebands / 重放～ playback button / 倒带～ re-wind button / 录音～ record button / 抹音～ erase cut-out button / 暂停～ pause button

【按劳分配】distribution according to work

【按理】according to reason; in the ordinary course of events; normally

【按脉】feel the pulse; take the pulse

【按摩】massage ◇～疗法 massotherapy

【按捺】control; restrain: ～不住激动的心情 be unable to hold back one's excitement

【按年摊付】yearly installment

【按钮】push button ◇～控制 push-button control; dash control

【按期】on time; on schedule: ～交货 deliver goods on schedule

【按其是非曲直】on its merits

【按人口平均】per capita; per head

【按时】on time; on schedule: ～到达 arrive on time

【按说】in the ordinary course of events; normally; ordinarily

【按图索骥】look for a steed with the aid of its picture; try to locate sth. by following up a clue

【按蚊】anopheles; malarial mosquito

【按下葫芦浮起瓢】hardly has one gourd been pushed under water when another bobs up; solve one problem only to find another cropping up

【按需分配】distribution according to need

【按语】note; comment

【按月付款销售】sales on monthly installment

【按照】according to; in accordance with; in the light of; on the basis of: ～惯例 according to the custom; according to the practice; ad usum / ～规定 according to the rules / ～实际情况 in the light of actual conditions / ～习惯 according to the usage; ad usum / ～原则办事 act according to principles / ～自愿原则 on a voluntary basis

胺 [化] amine

【胺化】amination

【胺盐】amine salt

黯 dim; gloomy

【黯然】[书] ①(阴暗的样子) dim; faint: ～失色 be overshadowed; be eclipsed; pale into insignificance ②(情绪低落的样子) dejected; downcast; low-spirited: ～神伤 feel dejected; feel depressed

暗 ①(黑暗) dark; dim; dull: ～绿 dark green / ～紫色 dull purple ②(隐藏不露的) hidden; secret ③(糊涂) unclear; hazy: 对情况若明若～ have only a vague idea of the situation

【暗暗】secretly; inwardly; to oneself: ～跟踪 secretly follow sb.

【暗堡】bunker

【暗藏】hide; conceal: ～的敌人 convert enemy / ～枪枝 conceal firearms

【暗娼】unlicensed prostitute; unregistered prostitute

【暗潮】undercurrent

【暗袋】[摄] camera bag (for changing film)

【暗淡】dim; dismal; faint; gloomy: ～的景象 a dismal picture / ～的颜色 a dull color

【暗地里】secretly; inwardly; on the sly

【暗地埋伏】ambush

【暗访民情】make secret inquires into the condition of the people

【暗害】①(暗杀) kill secretly ②(陷害) stab in the back

【暗含着】imply

【暗号】secret signal; secret sign; countersign;

cipher
【暗合】agree without prior consultation; (happen to) coincide
【暗河】underground river
【暗盒】[摄] magazine; cassette
【暗疾】unmentionable disease; a disease one is ashamed of
【暗记儿】secret mark
【暗间儿】inner room
【暗箭】an arrow shot from hiding; attack by a hidden enemy; a stab in the back: ～伤人 stab sb. in the back; injure sb. by underhand means
【暗箭难防】Hidden arrows are difficult to guard.
【暗礁】submerged reef; submerged rock
【暗流】undercurrent
【暗码】private mark
【暗盘出卖】resale through illegal channels
【暗杀】assassinate: ～未遂 attempted assassination ◇～者 assassin
【暗伤】①(内伤) internal injury; invisible injury ②(不显露的伤害) internal damage; invisible damage
【暗射地图】blank map
【暗示】①(示意;启示) drop a hint; hint; suggest ②[心] suggestion: ～疗法 suggestion therapy
【暗室】[摄] darkroom
【暗适应】[心] dark adaptation
【暗送秋波】make eyes at sb.; make secret overtures to sb.; stealthily give sb. the glad eye
【暗算】plot against: 遭人～ fall a prey to a plot
【暗锁】built-in lock
【暗滩】hidden shoal
【暗探】detective; secret agent
【暗无天日】complete darkness; total absence of justice
【暗喜】feel happy or delighted secretly
【暗线光谱】[物] dark-line spectrum
【暗箱】[摄] camera bellows; camera obscura
【暗想】muse; ponder; turn over in one's mind
【暗星云】dark nebula
【暗笑】laugh in one's sleeve; snigger; snicker
【暗影】①(阴影) shadow ②[天] umbra
【暗语】code word
【暗中】①(黑暗之中) in the dark: ～摸索 grope in the dark ②(背地里) in secret; on the sly; surreptitiously: ～操纵 pull strings from behind the scenes／～串通 collude with; conspire／～反对 veiled resistance／～破坏 disrupt furtively／～支持 give secret support to
【暗转】[剧] blackout
【暗自】inwardly; secretly; to oneself: ～庆幸 congratulate oneself; consider oneself lucky

岸 ①(水边的陆地) bank; coast; shore: 海～ coast; seashore／江～ the bank of a river; a river bank／上～ go ashore ②(高傲)[书]lofty: 傲～ haughty
【岸标】shore beacon
【岸然】in a solemn manner: 道貌～ be sanctimonious
【岸上交货】landed terms
【岸线】water front

āng

肮
【肮脏】dirty; filthy: ～的阴沟 a filthy sewer／～的勾当 dirty work; a foul deed

áng

昂 ①(仰着头,抬起头) hold (one's head) high ②(高涨) high; soaring
【昂昂】high-spirited; brave-looking
【昂贵】expensive; costly
【昂然】upright and unafraid
【昂首阔步】stride forward with one's chin up; stride proudly ahead
【昂扬】high-spirited: 斗志～ have high morale; be full of fight; be militant

àng

盎
【盎然】abundant; exuberant; full; overflowing: 趣味～ full of interest／生机～ overflowing with vigor; exuberant／春意～。Spring is in the air.
【盎司】ounce

āo

熬 cook in water; boil: ～白菜 stewed cabbage

凹 concave; hollow; sunken; dented: ～凸不平 full of bumps and holes; uneven
【凹岸】[地] concave bank
【凹版】[印] intaglio; gravure: 照相～ photogravure ◇～印刷 intaglio printing; gravure printing／～印刷机 intaglio press; gravure press／～制版 gravure plate-making
【凹槽】beard; [木] fillister
【凹度】concavity
【凹面镜】concave mirror

【凹透镜】concave lens
【凹凸镜】concave and convex lenses
【凹凸轧花】embossing
【凹凸印刷】embossing; die stamping ◇ ～机 embossing press; die stamping press
【凹陷】hollow; sunken; depressed: 双颊～ sunken cheeks; hollow cheeks / 地面～。The ground caved in.
【凹字】characters cut in bas-relief

áo

鏖 [书] engage in fierce battle
【鏖战】[书] fight hard; engage in fierce battle

熬 ① (煮) boil; decoct; stew: ～药 decoct medicinal herbs / ～粥 cook gruel ② (忍受) endure; hold out
【熬红眼睛】stay up till one's eyes become bloodshot
【熬煎】suffering; torture: 受尽～ be subjected to all kinds of suffering
【熬夜】stay up late; stay up all night

遨 stroll; saunter
【遨游】roam; travel: ～太空 roam in the heavens; travel in space

嗷
【嗷嗷待哺】cry piteously for food

螯 chela; pincers
【螯虾】crayfish; crawdad; crawfish

翱 [书] take wing
【翱翔】hover; soar ◇ ～飞行 soaring flight / ～机 sailplane; soaring glider

ǎo

袄 coat; jacket: 棉～ a cotton-padded jacket / 皮～ a fur coat

ào

傲 ①(自高自大) pround; haughty ②(不屈的性格) refuse to yield to; brave; defy
【傲岸】[书] proud; haughty: ～不群 proud and aloof / ～的青松 a proud and towering pine
【傲骨】lofty and unyielding character
【傲慢】arrogant; haughty: 态度～ adopt an arrogant attitude; put on airs
【傲气】air of arrogance; haughtiness: ～十足 extremely haughty; full of arrogance
【傲然】loftily; proudly; unyieldingly
【傲视】despise; look down upon

奥 ①(含义深; 不容易理解) profound; abstruse; difficult to understand ②[物] oersted
【奥长石】oligoclase
【奥地利】Austria ◇ ～人 Austrian
【奥林匹克运动会】the Olympic Games
【奥纶】orlon
【奥秘】profound mystery
【奥妙】profound; secret; subtle
【奥斯特】[物]oersted (Oe)
【奥陶纪】[地]the Ordovician period
【奥托发动机】[机]Otto engine

澳
【澳大利亚】Australia ◇ ～人 Australian
【澳大利亚抗原】[微] Australia antigen; hepatitis-associated antigen; Au antigen
【澳抗】[微] Au antigen

懊 ①(悔恨) regretful; remorseful ②(烦恼) annoyed; vexed
【懊悔】feel remorse; regret; repent
【懊恼】annoyed; upset; vexed
【懊丧】dejected; depressed; despondent

坳 col; a depression in a mountain range: 珠峰北～ the North Col of Mount Qomolangma

拗
【拗口】hard to pronounce; awkward-sounding
【拗口令】tongue twister
【拗陷】[地] depression

B

bā

八 eight

【八宝】eight treasures ◇ ～菜 eight-treasures pickles; assorted soy-sauce pickles / ～饭 eight-treasure rice pudding

【八倍体】[生] octoploid

【八边形】[数] octagon

【八成】①(十分之八) eighty per cent: ～新 eighty per cent new; practically new ②(大概; 看样子) most probably; most likely: ～他不来了。Most probably he isn't coming.

【八度】[乐] octave

【八方】the eight points of the compass; all directions

【八分音符】[乐] quaver; eighth note

【八哥儿】[动] myna

【八股】eight-part essay; stereotyped writing

【八卦】the Eight Diagrams

【八级风】[气] force 8 wind; fresh gale

【八级工】eighth-grade worker; top-grade worker

【八级工资制】eight-grade wage scale; eight-grade wage system

【八角】①[植] anise; star anise ②(调料) aniseed ③(八角形的) octagonal ◇ ～帽 octagonal cap / ～亭 octagonal pavilion / ～形 octagon

【八角枫】[植] alangium

【八九不离十】pretty close; very near; about right: 猜个～ make a very close guess

【八开】[印] octavo; 8vo ◇ ～本 octavo

【八路军】the Eighth Route Army

【八面玲珑】be smooth and slick (in establishing social relations)

【八面体结构】octahedral structure

【八气缸发动机】[机] eight cylinder engine

【八仙】the Eight Immortals (in the legend): ～过海, 各显神通 like the Eight Immortals crossing the sea, each one showing his or her special prowess

【八仙桌】square table for eight people

【八项注意】eight points for attention

【八小时工作制】eight-hour day

【八月】①(阳历) August ②(阴历) the eighth month of the lunar year; the eighth moon ◇ ～节 the Mid-Autumn Festival

【八折】twenty per cent discount

【八字】character "八": ～步 a measured gait with the toes pointing outwards / ～脚 splayfoot / ～还没见一撇儿 not even the first stroke of the character "八" is in sight; there's not the slightest sign of anything happening yet

巴 ①(盼望) hope earnestly; wait anxiously: ～望 look forward to ②(紧贴; 粘住) cling to; stick to: 爬山虎～在墙上 the ivy clings to the wall / 粥～锅了 the porridge has stuck to the pot ③[方](挨着) be close to; be next to: 前不～村, 后不～店 with no village ahead and no inn behind; stranded in an uninhabited area ④[物](压强单位) bar: 毫～ millibar / 微～ microbar

【巴巴多斯】Barbados ◇ ～人 Barbadian

【巴比合金】babbitt (metal)

【巴比伦】Babylon

【巴比妥】[药] barbital; barbitone

【巴布亚新几内亚】Papua New Guinea ◇ ～人 Papua New Guinean

【巴不得】be only too anxious (to do sth.); eagerly look forward to; earnestly wish: 我～这样。I wish it were so.

【巴旦杏】[植] almond; Prunus amygdalus

【巴豆】[植] (purging) croton ◇ ～霜 [中药] defatted croton seed powder

【巴尔干】Balkan ◇ ～半岛 the Balkan Peninsula / ～国家 Balkan states; the Balkans

【巴哈马】the Bahamas ◇ ～人 Bahamian

【巴基斯坦】Pakistan ◇ ～人 Pakistani

【巴结】curry favor with; fawn on; make up to; underlings fawn on their bosses

【巴克夏猪】Berkshire (swine)

【巴拉圭】Paraguay ◇ ～人 Paraguayan

【巴勒斯坦】Palestine ◇ ～人 Palestinian

【巴黎公社】the Paris Commune

【巴黎绿】[化] Paris green

【巴林】Bahrain ◇ ～人 Bahraini

【巴龙霉素】[药] paromomycin

【巴拿马】Panama ◇ ～帽 Panama hat / ～人 Panamanian / ～运河 the Panama Canal

【巴儿狗】①(一种供玩赏的狗) pekingese ②(谄媚者) sycophant; toady

【巴松管】[乐] bassoon

【巴望】look forward to: 我们都～你来。We are all looking forward to your coming.

【巴西】Brazil ◇ ～人 Brazilian

【巴掌】palm; hand: 打他一～ give him a slap / 拍～ clap hands

扒 ①(攀住) cling to; hold to: ～在墙上 catch hold of the top of the wall ②(刨; 挖; 拆) dig up; pull down; rake: ～土 rake earth / 把旧房～了

pull down the old house ③(拨动) push aside: ～开芦苇 push aside the reeds ④(脱掉; 剥下) strip off; take off: ～兔皮 skin a rabbit / ～衣裳 strip off clothes

【扒钉】cramp

【扒拉】push lightly: 把压在苗上的土～开 flick the earth off the seedlings

疤 scar

【疤痕】scar

芭

【芭蕉】[植]banana; *bajiao*

【芭蕉扇】palm-leaf fan

【芭蕾舞】ballet ◇ ～设计 choreography / ～演员 ballet dancer; (女) ballerina

吧

[象]～的一声, 弦断了。The string broke with a snap. / ～～两声枪响。Crack! Crack! Two shots rang out.

【吧嗒】①[象]～一声, 门关上了。The door clicked shut. ②(嘴唇开合作声) smack one's lips (in surprise, alarm, etc.): 他～了两下嘴, 一声也不言语。He smacked his lips but did not utter a word. ③(抽旱烟) pull at (a pipe, etc.)

笆 basketry

【笆斗】round-bottomed basket

bá

拔

①(拉出; 抽出) pull out; pull up: ～草 pull up weeds; weed / ～剑 draw one's sword / ～牙 pull out a tooth; extract a tooth ②(吸出) draw; suck out: 把火～一～。Put a chimney on the stove to make the fire draw. ③(挑选) choose; pick; select: 选～ select (from candidates) ④(向高提) lift; raise: ～起嗓子直嚷 shout at the top of one's voice ⑤(超出) stand out among; surpass: 出类～萃 stand out among one's fellows; be out of the common run ⑥(夺取) capture; seize: 连～敌人三个据点 capture three enemy strongholds in succession

【拔除】pull out; remove: ～敌军哨所 wipe out an enemy sentry post / ～杂草 weed out the rank grass

【拔钉锤】claw hammer

【拔钉钳】nail drawer; nail-extractor

【拔毒】[中医] draw out pus by applying a plaster to the affected part

【拔罐子】[中医] cupping

【拔河】[体] tug-of-war

【拔火罐儿】detachable stove chimney

【拔尖儿】①(出众) tiptop; top-notch: 学习成绩～ be a top-notch student ②(突出个人) push oneself to the front

【拔节】[农] jointing ◇ ～期 jointing stage; elongation stage

【拔锚】weigh anchor

【拔苗助长】try to help the shoots grow by pulling them upward; spoil things by excessive enthusiasm

【拔染】[纺] discharge ◇ ～剂 discharging agent; discharge / ～印花 discharge printing

【拔丝】①[机] wire drawing: ～机 wire drawing bench; wire drawing machine ②(熬糖拔丝) candied floss: ～山药 hot candied yam

【拔腿就跑】immediately take to one's heels; start running away at once

【拔牙钳】dental forceps; extracting forceps

【拔秧】[农] pull up seedlings (for transplanting): ～机 seedling collector; seedling lifter

【拔营】[军]strike camp

【拔桩机】pile-drawing machine; pile extractor; pile puller

跋

①(翻山越岭) cross mountains: ～山涉水 scale mountains and ford streams; travel across mountains and rivers ②(写在书籍或文章后的短文) postscript

【跋扈】domineering; bossy

【跋涉】trek; trudge: 长途～ make a long and difficult journey; trek a long way; trudge over a long distance

【跋文】postscript

bǎ

把

①(用手握住) grasp; hold: ～犁 handle a plough / ～住栏杆 hold on to railing / ～着手教 take sb. by the hand and teach him how to do sth. ②(把持) control; dominate; monopolize ③(看守; 把守) guard; watch: ～门 guard a gate ④(车把) handle: 自行车～ the handlebar of a bicycle ⑤[量] 一～茶壶 a teapot / 一～刀 a knife / 一～花 a bunch of flowers / 一～米 a handful of rice / 一～椅子 a chair / 加～劲 make an extra effort; put on a spurt / 拉他一～ give him a tug; give him a hand; lend him a hand / 有一～年纪 be getting on in years ⑥(大约) about; or so: 百～人 some hundred people / 个～月 about a month; a month or so ⑦[介] ～方便让给别人, ～困难留给自己 take difficulties on oneself and leave what is easy to others / ～房间收拾一下 tidy up the room /

～衣服洗洗 wash the clothes
【把柄】handle: 给人抓住～ give sb. a handle (against oneself)
【把持】control; dominate; monopolize: ～一部分权力 seize a certain amount of power / ～一切 monopolize everything
【把舵】hold the rudder; hold the helm; steer; take the helm
【把杆】[芭蕾舞] barre
【把关】check on; guard a pass: 层层～ make checks at all levels / 把好质量关 guarantee the quality (of products)
【把酒】fill a wine cup for sb.; raise one's wine cup
【把式】①[武术] *wushu*: 练～ practise *wushu* ②(有技术的人) person skilled in a trade: 车～ carter ③(技术) skill: 学会全套～ learn all the skills
【把守】guard: ～城门 guard a city gate / 分兵～ divide up one's forces for defense
【把手】grip; handle; knob: 装上～ fit a handle on
【把水搞混】muddy the water
【把头】labor contractor; gangmaster
【把握】①(抓住) grasp; hold: ～时机 seize the opportunity; seize the right time / 透过现象, ～本质 see through the phenomena to grasp the essence ②(可靠性) assurance; certainty: 没有成功的～ have no certainty of success / 我们有必胜的～。 We have full confidence in our victory; We are assured of our victory. / 做这项工作, 他很有～。 He's quite sure he can do this job.
【把戏】①(杂耍) acrobatics; jugglery ②(花招) cheap trick; game: 耍鬼～ play dirty tricks
【把兄弟】sworn brothers

钯 [化] palladium (Pd)

靶 target: 打～ shooting practice; target practice ◇碟～ clay pigeon; clay target bird / 环～ round target / 人像～ man target; silhouette target; silhouette / 十环～ decimal target / 移动～ moving target
【靶场】shooting range; range
【靶船】[军] target ship
【靶垛】target butt; target mound
【靶壕】pit; marking pit
【靶机】[军] drone; target drone
【靶心】bull's-eye
【靶纸】target sheet
【靶子】target

bà

霸 ①(古代诸侯联盟的首领) chief of feudal princes; overlord ②(暴君; 恶霸) despot; tyrant; bully: 恶～ local despot; local tyrant ③(霸权; 霸权主义) hegemonist power; hegemonism; hegemony: 反～斗争 the struggle against hegemonism / 永不称～ never seek hegemony ④(霸占) dominate; lord it over; tyrannize over: 军阀割据, 各～一方。 The country was torn by warlordism, with each warlord dominating a region.
【霸道】①(蛮横) high-handed; overbearing: 横行～ play the bully ②(药、酒等利害; 猛烈) potent; strong
【霸权】hegemony; supremacy
【霸权主义】hegemonism
【霸王】(封建领主) overlord
【霸王鞭】①(彩色短棍) a rattle stick used in folk dancing ②(民间舞蹈) rattle stick dance
【霸占】forcibly occupy; seize: ～别国领土 forcibly occupy the territory of another country
【霸主】hegemon; overlord: ～地位 hegemony; overlordship / 海上～ maritime overlord

坝 ①(拦截水的坝) dam ②(防洪水的堤坝) dike; embankment ③(沙滩) sandbar ④(平地、平原, 多用于地名) flatland; plain

罢 ①(停止) cease; stop: 欲～不能 try to stop but cannot; cannot refrain from carrying on ②(免去; 解除) dismiss: ～职 dismiss; remove from office ③(完了) finish: 说～, 他就走了。 With these words he left.
【罢黜】① (免职) dismiss from office ② (排斥) ban; reject
【罢工】strike; go on strike: 静坐～ sit-down strike ◇～权 the right to strike
【罢官】dismiss form office
【罢教】teachers' strike
【罢课】students' strike
【罢了】①[助]你就是不想去～。 You just don't want to go, that's all. ②(表示容忍;算了) 他不肯也～, 连个回信也不给。 I wouldn't have minded his refusing, but he didn't even answer my letter.
【罢论】abandoned idea: 此事已作～ the idea has already been dropped
【罢免】recall ◇ ～权 right of recall
【罢市】shopkeepers' strike
【罢手】give up

【罢休】give up; let the matter drop: 不达目的，决不～。We'll not stop until we reach our goal.

鲅

【鲅鱼】Spanish mackerel

爸 pa; dad; father

【爸爸】papa; dad; father

耙 ①(农具) harrow ②(耙地) draw a harrow over (a field); harrow a field

【耙齿】teeth of a harrow

把 ①(器具上便于握持的部分) grip; handle: 茶壶～儿 the handle of a teapot / 枪～儿 rifle butt ②(花、叶、果实等的柄) stem (of leaf, flower or fruit)

【把子】handle: 刀～ the handle of a knife

bāi

掰 break off with the fingers and thumb: ～玉米 break off corncobs / ～着手指算 count on one's fingers / 把饼～成两半 break the cake in two

【掰腕子】hand wrestling

bái

白 ①(白色) white: 几根～发 a few white (grey) hairs / 皮肤～ have a fair complexion ②(弄清楚) clear: 真相大～。Everything is clear now. (或: The whole truth has come out.) ③(没有加其他东西) pure; plain: ～开水 plain boiled water ④(徒然) in vain; for nothing: ～忙了半天 go to a lot of trouble for nothing / ～跑了一趟 make a fruitless trip ⑤(无代价) free of charge; gratis: ～送 give away free (of charge) ⑥(象征反动) White: ～军 the White army ⑦(戏剧中的道白) spoken part in opera, etc.: 独～ soliloquy; monologue / 对～ dialogue ⑧(空白) blank: ～纸 a blank sheet of paper / 一穷二～ poor and blank

【白矮星】[天] white dwarf (star)

【白班儿】[口] day shift

【白报纸】newsprint

【白璧微瑕】a flaw in white jade; a slight blemish

【白璧无瑕】flawless white jade; impeccable moral integrity

【白醭】mould (on the surface of vinegar, soy sauce, etc.)

【白布】plain white cloth; calico

【白菜】Chinese cabbage

【白痴】①(指白痴病) idiocy ②(患白痴病人) idiot

【白炽】white heat; incandescence ◇ ～灯 incandescent lamp

【白搭】[口] no use; no good: 和他辩也是～。It's no use arguing with him.

【白带】[医] leucorrhea; whites; leukorrhea

【白蛋白】[生化] albumin

【白道】[天] moon's path

【白癜风】[医] vitiligo

【白垩】chalk

【白垩纪】[地] the Cretaceous period

【白俄罗斯】Belorussia; Byelorussia ◇ ～人 Belorussian

【白矾】alum

【白匪】White bandits

【白费】waste: ～口舌 waste one's breath / ～力气 waste one's energy / ～时间 waste time / ～心思 bother one's head for nothing

【白粉病】[农] powdery mildew

【白宫】the White House

【白姑鱼】[动] white Chinese croaker

【白果】[植] ginkgo; gingko

【白鹤】[动] white crane

【白喉】[医] diphtheria

【白狐】[动] arctic fox

【白化病】albinism ◇ ～人 albino

【白话】vernacular ◇ ～诗 free verse written in the vernacular / ～文 writings in the vernacular

【白桦】[植] white birch

【白芨】[中药] the tuber of hyacinth bletilla

【白僵蚕】[中药] the larva of a silkworm with batrytis

【白金】platinum

【白鲸】[动] beluga; white whale

【白净】(of skin) fair and clear

【白酒】spirit usu. distilled from sorghum or maize; white spirit

【白卷】blank examination paper: 交～ hand in an examination paper unanswered

【白开水】plain boiled water

【白口铁】white iron

【白蜡】white wax; insect wax ◇ ～虫 wax insect / ～树 Chinese ash

【白兰地】brandy

【白兰瓜】[植] honey dew melon

【白痢】①[中医] dysentery characterized by white mucous stool ②[兽医] white diarrhoea

【白鲢】[动] silver carp

【白脸】white face; face painting in Beijing opera, etc, traditionally for the villain: 唱～ play the villain; pretend to be harsh and severe
【白磷】[化] white phosphorus
【白蛉】sand fly ◇ ～热 [医] sand-fly fever
【白榴石】[矿] leucite
【白鹭】[动] egret
【白茫茫】a vast expanse of whiteness: 一片～的大雪 a vast expanse of white snow
【白茅】[植] cogongrass ◇ ～根 [中药] cogongrass rhizome
【白煤】①[方](无烟煤; 硬煤) anthracite; hard coal ②(水力) white coal; waterpower
【白米】(polished) rice ◇ ～饭 (cooked) rice
【白棉纸】stencil tissue paper
【白面】wheat flour; flour
【白面书生】pale-faced young scholar
【白描】①[美术] line drawing in traditional ink and brush style ②(文字简练的写作风格) simple, straightforward style of writing
【白木耳】tremella
【白内障】[医] cataract ◇ ～摘除术 cataract extraction
【白砒】[化] white arsenic; arsenic trioxide
【白皮书】white paper; white book
【白皮松】[植] lacebark pine
【白区】White area (the Kuomintang-controlled area)
【白屈菜】[植] greater celandine
【白热】white heat; incandescence
【白热化】turn white-hot: 争论达到了～的程度。The debate became white-hot.
【白人】white man or woman
【白刃】naked sword ◇ ～战 bayonet charge; hand-to-hand combat
【白日做梦】daydream; indulge in wishful thinking
【白肉】plain boiled pork
【白色】white ◇ ～据点 White stronghold; stronghold of reaction / ～恐怖 white terror / ～政权 white regime
【白色人种】the white race
【白色体】[植] leucoplast
【白芍】[中药] (peeled) root of herbaceous peony
【白手起家】build up from nothing; start from scratch
【白薯】[植] sweet potato
【白苏】[植] common perilla
【白糖】(refined) white sugar
【白陶】[考古] white pottery (of the Shang Dynasty, c. 16th-11th century B.C.)
【白体】[印] lean type
【白天】daytime; day: ～黑夜 day and night
【白铁】galvanized iron: ～皮 galvanized iron sheet
【白铜】copper-nickel alloy
【白头偕老】live to ripe old age in conjugal bliss; remain a devoted couple to the end of their lives
【白头翁】①[中药] the root of Chinese pulsatilla ②[动] Chinese bulbul
【白钨矿】scheelite
【白细胞】[生理] white blood cell; leukocyte; leucocyte ◇ ～减少 [医] leukocytopenia; leukopenia / ～增多 [医] leukocytosis
【白鲜】[植] shaggy-fruited dittany ◇ ～皮 [中药] the root bark of shaggy- fruited dittany
【白鹇】[动] silver pheasant
【白熊】[动] polar bear; white bear; ice bear
【白血病】[医] leukemia
【白血病性肉瘤】[医] leukosarcoma
【白血球】[生理] white blood cell; leukocyte; leucocyte
【白鲟】[动] Chinese paddlefish
【白眼】supercilious look: ～看人 treat people superciliously; look upon people with disdain / 遭人～ be treated with disdain
【白羊座】[天] Aries
【白杨】[植] white poplar
【白药】[中药] *baiyao*, a white medicinal powder for treating hemorrhage, wounds, bruises, etc.
【白夜】[天] white night
【白衣战士】warrior in white; medical worker
【白蚁】termite; white ant
【白银】silver
【白鱼】[动] whitefish
【白云母】[矿] muscovite; white mica
【白云石】[矿] dolomite
【白纸黑字】(written) in black and white
【白种】the white race
【白昼】daytime
【白字】wrongly written or mispronounced character: ～连篇 pages and pages of wrongly written characters

bǎi

百 ①(一百; 十个十) hundred ②(比喻很多) numerous; all kinds of: ～花盛开 a hundred flowers in bloom / ～忙之中 in the midst of pressing affairs; despite many claims on one's time / 数以～计 by hundreds

【百般】in a hundred and one ways; in every possible way; by every means: ～抵赖 try by every means to deny / ～刁难 create all sorts of obstacles; put up innumerable obstacles / ～奉承 flatter sedulously / ～劝说 try to persuade in every possible way
【百倍】a hundredfold; a hundred times: 展望未来,信心～ look to the future with full confidence
【百步穿杨】shoot an arrow through a willow leaf a hundred paces away; shoot with great precision
【百尺竿头,更进一步】make still further progress
【百川归海】all rivers flow to the sea; all things tend in one direction
【百读不厌】be worth reading a hundred times: 这部书～。 You never get tired of reading this book.
【百端待举】a thousand things remain to be done; numerous tasks remain to be undertaken
【百发百中】a hundred shots, a hundred bull's- eyes; every shot hits the target; shoot with unfailing accuracy
【百废俱兴】all neglected tasks are being undertaken; full-scale reconstruction is under way
【百分】pec cent; percent: ～之六 6 per cent / 获～之九的利息 get 9 per cent interest
【百分比】percentage: 按～计算 in terms of percentage
【百分率】percentage; per cent
【百分之百】a hundred per cent; out and out; absolutely: 有～的把握 be a hundred per cent sure; be absolutely certain / 这是～的谎话! That's an out- and- out lie!
【百分制】hundred-mark system
【百感交集】all sorts of feelings well up in one's heart; a multitude of feelings surges up
【百合】[植] lily
【百花齐放,百家争鸣】let a hundred flowers blossom and a hundred schools of thought contend
【百花齐放,推陈出新】let a hundred flowers blossom, weed through the old to bring forth the new
【百花争艳】all flowers vying for beauty in full bloom
【百货】general merchandise: ◇ ～商店 department store; general store / 日用～ articles of daily use
【百科全书】encyclopedia
【百孔千疮】riddled with gaping wounds; afflicted with all ills
【百口难辩】difficult to get at the truth
【百里挑一】one in a hundred; cream of the crop
【百炼成钢】be tempered into steel: 在革命斗争中～ be tempered in revolutionary struggle
【百灵】[动] lark
【百忙中】while fully engaged; in the thick of things
【百米赛跑】100-meter dash
【百慕大】Bermuda ◇ ～人 Bermudan
【百年】①(指很多年) a hundred years; a century: ～不遇的大水灾 the biggest flood in a century ②(指死亡) lifetime: ～之后[婉] when sb. has passed away; after sb.'s death ◇ ～纪念 centenary; centennial
【百年大计】a project of vital and lasting importance: 基本建设是～,要求质量第一。 Capital construction projects, which are to last for generations, call for good quality above everything else.
【百日咳】[医] whooping cough; pertussis ◇ ～疫苗 pertussis vaccine
【百十】a hundred or so: ～户人家 about a hundred households
【百事通】knowledgeable person; know-all
【百思不解】remain puzzled after pondering over sth. a hundred times; remain perplexed despite much thought
【百听不厌】worth hearing a hundred times: 这个故事～。 You never get tired of hearing this story.
【百万】million: ～雄师 a million bold warriors ◇ ～吨级 megaton / ～富翁 millionaire
【百闻不如一见】it is better to see once than hear a hundred times; seeing for oneself is a hundred times better than hearing from others
【百无禁忌】no prohibitions of any kind; Nothing is under taboo.
【百无聊赖】bored to death; bored stiff; overcome with boredom
【百无一失】no danger of anything going wrong; no risk at all
【百姓】common people
【百页】[建] louver ◇ ～窗 louver window
【百叶窗】shutter; blind; jalousie
【百叶箱】[气] thermometer screen
【百业萧条】all business languishes
【百依百顺】docile and obedient; all obedience
【百战百胜】fight a hundred battles, win a hun-

dred victories; emerge victorious in every battle; be ever-victorious

【百折不挠】keep on fighting in spite of all setbacks; be undaunted by repeated setbacks; be indomitable

【百周年纪念】centenary; centenary celebration; centennial

【百足之虫,死而不僵】a centipede does not topple over even when dead; a centipede dies but never falls down; old institutions die hard

柏 cypress

【柏树】[植] cypress cypress

【柏油】pitch; tar; asphalt ◇ ～混凝土 tar concrete / ～路 asphalt road / ～喷洒机 tar sprayer

【柏子仁】[中药] the seed of Oriental arborvitae

摆 ①(安放;陈列) put; place; arrange: ～在我们面前的任务 the task confronting us / 把工具～整齐 put ⟨set⟩ the tools in order / 把碗筷～好 set the table; lay the table / 把问题～到桌面上来 place the problem on the table; bring the issue out into the open / 把药瓶～在架子上 put the medicine bottles on the shelf ②(陈述) lay bare; state clearly: ～矛盾 lay bare the contradictions / 把有利和不利的条件～一～ set forth the advantages and disadvantages ③(显示;炫耀) put on; assume: ～老资格 flaunt one's seniority; put on the airs of a veteran / ～威风 give oneself airs; put on airs ④(摆动) ～来～去 swing ⟨sway⟩ to and fro / sway; wave: 他～手叫我走开。 He waved me away. ⑤[机] pendulum: 单～ simple pendulum / 复～ compound pendulum

【摆布】order about; manipulate: 任人～ allow oneself to be ordered about; be at the mercy of others

【摆荡吊环】[体] swinging rings

【摆动】swing; sway: 柳条迎风～。 The willows swayed in the breeze. / 指示针来回～。The pointer flickered. ◇ ～轴 oscillating axle / ～装置 pendulous device

【摆渡】ferry: 把旅客～过河 ferry the passengers across the river ◇ ～船 ferry boat

【摆架子】put on airs; assume great airs

【摆阔气】parade one's wealth; be ostentatious and extravagant

【摆擂台】① (比武) make open challenges to fights ②(挑战) make challenges to a contest

【摆龙门阵】[方] chat; gossip; spin a yarn

【摆轮】balance (of a watch or clock); balance wheel

【摆门面】keep up appearances

【摆弄】①(反复移动或拨动) move back and forth; fiddle with: 你别来回～那儿盆花了。 Don't move those flower pots back and forth. ②(摆布;利用) order about; manipulate

【摆平】treat... fairly: 对双方要～ treat both sides equally; be square to both sides

【摆设】furnish and decorate (a room): 屋里～得很雅致。 The room is tastefully furnished.

【摆设儿】ornaments; furnishings: 小～ knick-knacks

【摆事实,讲道理】present the facts and reason things out

【摆摊子】①(在路边或市场陈列货物出售) set up a stall ②(铺张) maintain a large staff or organization

【摆脱】cast off; shake off; break away from; free oneself from; extricate oneself from: ～困境 extricate oneself from a predicament / ～贫穷落后的状态 be lifted out of poverty and backwardness / ～殖民主义的桎梏 cast off the yoke of colonialism / ～追踪者 shake off one's pursuers

【摆样子】do sth. for show: 他的自我批评不是～的。 His self- criticism was not made for show.

【摆针】the balancing point; the axis

【摆钟】pendulum clock

【摆轴】balance staff; pendulum shaft

【摆子】[方] malaria: 打～ have malaria

bài

败 ①(失败,跟'胜'相对) be defeated; lose: ～下阵来 lose a battle / 敌军必～ the enemy is bound to fail / 主队以二比三～于客队。 The home team lost to the visitors 2 to 3. ②(打败) defeat; beat: 大～侵略军 inflict a severe defeat on the invading troops ③(失败, 跟'成'相对) fail: 成～ success or failure ④(搞坏) spoil: 事情可能就～在他手里。 He may spoil the whole show. ⑤(解除;消除) counteract: ～毒 counteract a toxin ⑥(破旧;腐烂;凋谢) decay; wither: 枯枝～叶 dead twigs and withered leaves

【败北】[书] suffer defeat; lose a battle

【败笔】①(用于书画) a faulty stroke in calligraphy or painting ②(用于诗文) a faulty expression in writing

【败坏】ruin; corrupt; undermine: ～风纪 demoralization / ～风俗 corrupt morals; exert a

bad moral influence / ～名誉 discredit; defame / ～社会风气 corrupt social values / 道德～ morally degenerate
【败火】[中医] relieve inflammation or internal heat
【败绩】[书] be utterly defeated; be routed
【败家荡产】squander a patrimony and ruin the family
【败家子】spendthrift; wastrel; prodigal
【败局】lost game; losing battle
【败类】scum of a community; degenerate: 民族～ scum of a nation
【败露】(of a plot, etc.) fall through and stand exposed: 阴谋终于～。In the end the conspiracy was brought to light.
【败落】decline (in wealth and position): 一个封建家庭的～ the decline of a feudal family / 逐渐～ on the decline; on the wane
【败诉】lose a lawsuit ◇ ～案件 defeat suit / ～方 party losing the lawsuit / ～人 defeated suitor; loser of a lawsuit
【败退】retreat in defeat: 敌军节节～。Again and again the enemy retreated in defeat.
【败胃】spoil one's appetite
【败兴】disappointed: ～而归 come back disappointed
【败血症】[医] septicemia
【败仗】lost battle; defeat: 打～ be defeated in battle; suffer a defeat
【败阵】be defeated on the battlefield; be beaten in a contest: ～而逃 lose the field and take to flight
【败子回头】return of the prodigal son

拜 ①(崇拜) do obeisance: ～佛 prostrate oneself before the image of Buddha; worship Buddha ②(表示敬意的礼节) make a courtesy call: 回～ pay a return visit ③(恭敬地与对方确立某种关系) acknowledge sb. as one's master, godfather, etc.: ～他为老师 take him as one's teacher ④[敬] ～读大作 have the pleasure of perusing your work / ～谢 express one's thanks
【拜把子】become sworn brothers
【拜别】[敬] take leave of
【拜倒】prostrate oneself; fall on one's knees; grovel: ～在某人的脚下 grovel oneself at the feet of sb.; lie prostrate oneself at the feet of sb.
【拜读】have the honor to read
【拜访】pay a visit; call on: ～亲友 visit 〈call on〉 relatives and friends / 专诚～ make a special trip to call on sb. / 正式～ formal visit
【拜会】pay an official call; call on: 告别～ farewell call / 礼节性～ courtesy call / 私人～ personal visit; personal call
【拜火教】Zoroastrianism; Mazdaism
【拜见】pay a formal visit; call to pay respects
【拜金主义】money worship
【拜年】pay a New Year call; wish sb. a Happy New Year: 给军属～ pay New Year calls to servicemen's families
【拜师】formally acknowledge sb. as one's master; take sb. as one's teacher
【拜寿】congratulate an elderly person on his birthday; offer birthday felicitations
【拜托】[敬] request sb. to do sth.: ～您捎个信儿给他。Would you be kind enough to take a message to him?
【拜望】[敬] call to pay one's respects; call on
【拜物教】fetishism: 商品～ commodity fetishism / 信奉～ practice fetishism
【拜谒】①(拜见) pay a formal visit; call to pay respects ②(瞻仰) pay homage (at a monument, mausoleum, etc.)

稗 ①(稗草) barnyard grass ②[书](逸闻琐事) insignificant; unofficial
【稗官野史】books containing anecdotes
【稗子】[植] barnyard grass; barnyard millet

bān

斑 ①(斑点) spot; speck; speckle; stripe: 霉～ speck of mould / 锈～ rusty spot / 油～ oil stains; grease spots ②(有斑点的;有斑纹的) spotted; striped
【斑白】grizzled; greying: 两鬓～ greying at the temples / 头发～ with streaks of white hair
【斑斑】full of stains or spots: 血迹～ blood-stained
【斑驳陆离】variegated
【斑翅山鹑】partridge
【斑点】spot; stain; speckle
【斑海豹】harbor seal; common seal; hair seal
【斑鸠】turtledove; turtle dove
【斑斓】bright-colored; gorgeous; multicolored: 五彩～ a riot of color
【斑羚】goral
【斑马】zebra
【斑蝥】[动] Chinese blister beetle; cantharides
【斑铜矿】bornite
【斑纹】stripe; streak: 老虎的～ tiger's stripes
【斑岩】[地] porphyry

【斑疹】[医] macula ◇～伤寒 typhus
【斑竹】mottled bamboo
班 ①(班组) class; team: 上～ go to work / 作业～ work team ②(工作的班次) shift; duty: 上夜～ be on night shift / 三～倒 work in three shifts / 轮～护理病人 take turns tending the sick ③[军] squad ④[量](交通运输工具的班次) 末～车 last bus, train, etc.(of the day) / 搭下一～火车 take the next train
【班车】regular bus (service)
【班次】①(学校的) classes ②(交通运输的) number of runs or flight: 增加货车～ increase the number of runs of freight trains
【班底】ordinary members of a theatrical troupe
【班房】[口] jail: 坐～ be (put) in jail
【班机】airliner; regular air service: 京沪～ scheduled flights between Beijing and Shanghai
【班级】classes and grades in school
【班轮】regular passenger or cargo ship; regular steamship service
【班门弄斧】display one's slight skill before an expert
【班师】[书] withdraw troops from the front; return after victory
【班图人】Bantu
【班图语】Bantu (language)
【班委会】class committee
【班长】①(学生班) class monitor ②[军] squad leader
【班主任】a teacher in charge of a class
【班卓琴】banjo
【班子】①(剧团的旧称) theatrical troupe ②(组织)organized group: 领导～ a leading body; leader group / 生产～ a team in charge of production / 专业～ a special, full-time group
【班组】teams and groups (in factories, etc.)
扳 pull; turn: ～成平局 equalize the score / ～倒 pull down / ～道岔 pull railway switches / ～枪栓 pull back the bolt of a rifle
【扳不倒儿】[口] tumbler; roly-poly
【扳道员】[交] pointsman; switchman
【扳机】trigger
【扳手】①(工具) spanner; wrench ②(器具上用手扳的部分) lever (on a machine)
【扳子】spanner; wrench
颁 promulgate; issue
【颁布】promulgate; issue; publish: ～法令 promulgate a decree; issue a decree
【颁发】①(发布) issue; promulgate: ～嘉奖令 issue an order of commendation ②(授与) award: ～奖金 bestow a prize / ～奖章 award a medal / ～证书 award a certificate
【颁行】issue for enforcement
般 sort; kind; way: 百～ in every possible way / 这～ such; this kind of / 暴风雨～的掌声 stormy applause; thunderous applause / 兄弟～的情谊 fraternal feelings
搬 ①(移动位置) take away; move; remove: ～掉绊脚石 remove a stumbling block / 把桌子～走 take the table away ②(迁移) move (house): 他早就～走了。He moved out long ago. ③(搬用；照抄) apply indiscriminately; copy mechanically: 生～硬套 copy mechanically and apply indiscriminately / 照～照抄 copy slavishly
【搬家】move (house): 我们下星期～。We're moving next week.
【搬弄】①(翻动;搬动) move sth. about; fiddle with: 别～枪栓。Don't fiddle with the rifle bolt. ②(卖弄) show off; display: ～学问 show off one's erudition
【搬弄是非】sow discord; tell tales; make mischief
【搬运】carry; transport: ～货物 transport goods / ～行李 carry luggage ◇～费 transportation charges / ～工人(车站等) porter; (码头) docker
瘢 scar
【瘢痕】[医] scar
【瘢痕瘤】[医] cheloid; keloid; keloma
【瘢痕性毛囊炎】[医] folliculitis keloidalis

bǎn

板 ①(片状物体) board; plank; plate: 玻璃～ plate glass; glass top (of a desk) / 钢～ steel plate / 混凝土～ concrete slab / 切菜～ chopping block ②(店铺的门板) shutter: 上～儿 put up the shutters ③(乒乓球) bat; (板羽球) battledore ④[乐](打击乐器) *ban*; clappers ⑤[乐](节拍) an accented beat; time; measure ⑥(硬结) hard: 地～了,锄不动。The ground is too hard to hoe. ⑦(呆板) stiff; unnatural: 我这张像片照得太～了。I look too stiff in this picture. ⑧(表情严肃) stop smiling; look serious: ～起面孔 put on a stern expression / ～着脸 keep a straight face

【板壁】wooden partition
【板擦儿】blackboard eraser
【板材】[林] board; plates
【板锉】flat file
【板凳】wooden bench; wooden stool
【板斧】broad axe
【板鼓】[乐] a small drum for marking time
【板规】[机] plate gauge
【板胡】[乐] *banhu* fiddle; a bowed stringed instrument with a thin wooden soundboard
【板结】harden: ～的土壤 hardened and impervious soil
【板锯】bladed saw; hand plate saw
【板块学说】plate tectonics
【板栗】Chinese chestnut
【板梁桥】[建] plate girder bridge
【板皮】[林] slab
【板球】criket
【板书】writing on the blackboard
【板刷】scrubbing brush
【板条】[建] lath
【板条箱】crate
【板瓦】[建] pan tile; plate tile
【板鸭】pressed salted duck; dried salted duck
【板牙】[机] screw die; threading die ◇ ～扳手 stock and die
【板烟】plug (of tobacco)
【板岩】[地] slate
【板眼】①(民族音乐的节拍) measure in traditional Chinese music ②(条理;层次) orderliness: 他说话有板有眼。 Whatever he says is well presented. / 她做事很有～。 She is very methodical in her work.
【板油】leaf fat; leaf lard
【板羽球】①(球类) battledore and shuttlecock ②(球) shuttlecock
【板纸】paperboard; board: 草～ strawboard / 牛皮～ kraft board
【板桩】[建] sheet pile
【板子】board; plank

版 ①(印刷用) printing plate; printing block: 铜～ copperplate / 制～ plate making ②(图书的版本) edition: 初～ first edition / 绝～ out of print / 新～ new edition / 修订～ revised edition / 原～ original edition / 再～ second edition ③(报纸的页面) page (of a newspaper): 头～新闻 front-page news
【版本】edition
【版次】order (number) of the edition
【版画】woodcut; woodblock; engraving
【版刻】carving; engraving
【版面】①(每一页的整面) space of a whole page ②(版面编排) layout of a printed sheet; make-up of a printed sheet ◇ ～设计 layout
【版权】copyright: ◇ ～法 copyright law / ～诉讼 copyright action / ～所有 all rights reserved / ～页 copyright page; colophon / 国际～公约 Universal Copyright Convention
【版式】format
【版税】royalty (on books)
【版图】domain; territory: ～辽阔 vast in territory

bàn

瓣 ①(花瓣) petal ②(种子或果实的瓣) segment or section (of a tangerine, etc.); clove (of garlic) ③(人或动物器官的瓣膜) valve; lamella: 鳃～ gill lamella / 三尖～ tricuspid valve ④(小块;碎块) fragment; piece: 摔成几～儿 be broken into several pieces
【瓣阀】[机] clack valve
【瓣膜】[生理] valve

半 ①(一半) half; semi-: ～年 six months / ～小时 half an hour / 一个～月 a month and a half; one and a half months / 增加一倍～ increase by 150% ②(在…中间) in the middle; halfway: ～山腰 halfway up a hill / ～夜 midnight ③(比喻很少) very little; the least bit: 他连～句话都不说。 He wouldn't breathe a word. ④(不完全;部分) partly; about half: ～开化的 semicivilized / ～开玩笑地说 say sth. half jokingly / ～熟练的 semiskilled / 房门～开着。The door was left half open.
【半百】fifty (years of age): 年过～ over (above) fifty / 年近～ getting on for fifty; approaching fifty
【半…半…】～饥～饱 underfed / ～吞～吐 hesitate to speak one's mind; be mealymouthed / ～文～白 half literary, half vernacular / ～心～意 half-hearted / ～信～疑 half-believing, half-doubting; not quite convinced / ～真～假 half- genuine, half- sham; partly true, partly false
【半半拉拉】[口] incomplete; unfinished: 这点活儿干完算了,别剩下～的。 Let's finish the job off. Don't leave a lot of loose ends hanging over.
【半辈子】half a lifetime: 大～ the greater part of one's life
【半壁江山】half of the country

【半边莲】[中药] Chinese lobelia
【半边天】①(天空的一部分) half the sky: 妇女能顶～。 Women hold up half the sky. ②(新社会之妇女) women of the new society; womenfolk
【半波整流体】half-wave rectifier; single-wave rectifier
【半…不…】半死不活 half-dead; more dead than alive / 半新不旧 no longer new; showing signs of wear / 半生不熟 half cooked
【半场】[体] ①(球赛时间) half of a game or contest ②(场地) half-court: ～紧逼 [篮球] half-court press
【半成品】semi-manufactured goods; semi-finished articles; semi-finished products
【半导体】semiconductor ◇ ～存储器 semiconductor store; simiconductor memory / ～二极管 diode; semiconductor diode; semi-diode / ～集成电路 semiconductor integrated circuit / ～器件 semiconductor devices / ～收音机 transistor radio; transistor receiver / ～元件 semiconductor element / ～闸流管 thyristor
【半岛】peninsula
【半点】the least bit: 没有～慌张 not the least bit flurried
【半吊子】①(知识肤浅; 不通事理) dabbler; smatterer ②(吊儿郎当的人) tactless and impulsive person
【半费】half the fee; fifty per cent reduction in fees
【半封建】semi-feudal
【半复赛】[体] eighth-finals
【半工半读】part work, part study; work-study program
【半公开】semi-overt; more or less open
【半官方】semi-official: 据～人士称 according to semi-official sources
【半饥半饱】underfed; half-starving
【半机械化】semi-mechanization
【半价】half price: ～出售 sell at half price
【半截】half (a section): ～香肠 half of a sausage / 话只说了～儿 finish only half of what one has to say
【半斤八两】six of one and half a dozen of the other; not much to choose between the two; tweedledum and tweedledee
【半径】radius
【半决赛】[体] semifinals
【半空中】in mind air; in the air: 悬在～ hang in midair
【半劳动力】one able to do light manual labor only; semi- ablebodied or part- time (farm) worker
【半流体】semifluid
【半路】halfway; midway; on the way
【半路出家】become a monk or nun late in life; switch to a job one was not trained for
【半履带式货车】half-track truck
【半履带式拖拉机】half-track tractor
【半年度报告】semi-annual reports
【半年决算】half-yearly closing
【半票】half-price ticket; half fare
【半瓶醋】dabbler; smatterer
【半旗】half-master: 下～ fly a flay at half- mast
【半球】hemisphere: 北～ the Northern Hemisphere / 东～ the Eastern Hemisphere / 南～ the Southern Hemisphere / 西～ the Western Hemisphere
【半人马座】[天] Centaur
【半日制学校】half- day school; double- shift school
【半身不遂】[医] hemiplegia
【半身像】①(照片) half-length photo or portrait ②(雕像) bust
【半生】half a lifetime: 前～ first half of one's life
【半失业】semi- employed; partly employed; underemployed
【半时】[体] half: 上～ first half / 下～ second half
【半熟练】semi-skilled
【半数】half the number; half: ～以上 more than half
【半衰期】[物] half-life
【半天】① half of the day: 后～ afternoon / 前～ morning ②(长时间) a long time; quite a while: 他～说不出话来。 He remained tongue-tied for a long time.
【半透明】translucent; semitransparent ◇ ～体 [物] translucent body / ～纸 onionskin
【半途】halfway; midway: ～拆伙 part company halfway
【半途而废】give up halfway; leave sth. unfinished
【半脱产】partly released from productive labor; partly released from one's regular work
【半文不值】not worth a farthing 〈penny〉; entirely worthless
【半文盲】semiliterate
【半无产阶级】semi-proletariat
【半夏】[中药] the tuber of pinellia

【半血亲关系】relationship of the half-blood
【半夜】midnight; in the middle of the night: 会议一直开到～。The meeting went on far into the night.
【半夜三更】in the depth of night; late at night
【半音】[乐] semitone ◇ ～音阶 chromatic scale
【半影】[物] penumbra ◇ ～锥 penumbra cone
【半元音】[语] semivowel
【半圆】semicircle: ～形(的) semicircular
【半月瓣】[生理] semilunar valve
【半月刊】semimonthly; fortnightly
【半载】[交] half load
【半支莲】[植] sun plant
【半殖民地】semi-colony ◇ ～半封建社会 semi-colonial, semi-feudal society
【半制浆】[纸] half stuff; semi-pulp
【半中腰】[口] middle; halfway: 山的～有一座亭子。A pavilion is halfway up the hill./ 他的话说到～就停住了。He broke off in the middle of a sentence.
【半自动】semi- automatic ◇ ～步枪 semi-automatic rifle / ～控制 semi-automatic control / ～装置 semi-automatic plant
【半自动化】semi-automatization
【半自耕农】semi-tenant peasant; semi-owner peasant

拌 mix: ～鸡丝 shredded chicken salad / ～饲料 mix fodder / ～匀 mix thoroughly
【拌和】mix and stir; blend
【拌浆机】grout mixer; machine grout mixer
【拌面】noodles served with soy sauce, sesame butter, etc.
【拌种】[农] seed dressing ◇ ～机 seed dresser
【拌嘴】bicker; squabble; quarrel

伴 ①(同伴) companion; partner: 旅～ a travelling companion / 作～ keep sb. company ②(伴同;伴随) accompany
【伴唱】①(从旁歌唱) vocal accompaniment ②(配合表演) accompany (a singer)
【伴侣】companion; mate; partner
【伴生气】[石油] associated gas
【伴随】accompany; follow: 孩子们～着欢乐的乐曲,跳起舞来。Accompanied by cheerful music, the children began to dance.
【伴舞】be a dancing partner
【伴星】[天] companion (star)
【伴音】[电视] audio ◇ ～电平 audio level / ～监听器 audio monitor / ～信号 audio signal
【伴奏】accompany (with musical instruments): 钢琴～ piano accompaniment ◇ ～者 accompanist

绊 (cause to) stumble; trip: ～手～脚 be in the way / 差点儿～了我一交。I tripped and almost fell. / 他被树根～了一下。He stumbled over the root of a tree.
【绊脚石】stumbling block; obstacle

扮 ①(化装成) be dressed up as; play the part of; disguise oneself as: 他在戏里～一位老农。In the opera he plays the part of an old peasant. / 侦察员～作一个商人。The scout disguised himself as a merchant. ②(表情装成) put on (an expression): ～鬼脸 make grimaces; make faces
【扮相】the appearance of an actor or actress in costume and makeup
【扮演】play the part of; act: 在影片中～奶奶 play (the part of) a granny in the film
【扮装】makeup

办 ①(办理;处理) do; handle; manage; tackle; attend to: 我有点事得～一～。I have something to attend to. / 怎么～? What's to be done? / 这点事她一个人～得了。She can handle this by herself. ②(创办;经营) set up; run: 县～工厂 a county-run factory / 勤俭～一切事业 run all enterprises with diligence and frugality / 村里新～了一所中学。A new middle school has been set up in the village. ③(采购;置备) buy a fair amount of; get sth. ready: ～年货 do New Year shopping; do shopping for the Spring Festival ④(惩治) punish (by law); bring to justice: 严～ punish severely
【办案】handle a case
【办报】run a newspaper
【办法】way; means; measure: 采取有效的～ adopt effective measures / 找出克服困难的～ find a way to overcome a difficulty
【办公】handle official business; work (usu. in an office) ◇ ～费 administrative expenses / ～时间 office hours / ～室 office / ～厅 general office / ～桌 desk; bureau
【办酒席】prepare a feast
【办理】handle; conduct; transact: ～进出口业务 handle imports and exports / ～手续 go through the formalities go through procedure / 我来～这桩事。I'll take up this matter.

【办事】handle affairs; work: ～公正 be fair and just in handling affairs / ～认真 be conscientious in one's work / 按原则～ act according to principles ◇ ～处 office; agency / ～机构 administrative body; working body / ～员 office worker
【办喜事】manage a wedding; prepare for a happy occasion
【办学】run a school ◇ ～方式 way of running schools / ～方针 guiding principle for running a school
【办罪】punish

bāng

邦 nation; state; country: 邻～ a neighboring country / 友～ friendly country; fraternal country
【邦交】relations between two countries; diplomatic relations: 断绝～ sever diplomatic relations / 恢复～ resume diplomatic relations / 建立～ establish diplomatic relations
【邦联】confederation

梆
【梆子】①(打更用的) watchman's clapper ②(乐器) wooden clappers
【梆子腔】①(戏曲) Chinese local operas performed to the accompaniment of wooden clappers ②(梆子曲调) the music of such operas

帮 ①(帮助) help; assist: ～他搬行李 help him with his luggage / 互～互学 help each other and learn from each other ②(船帮;车帮) side (of a boat, truck, etc.) ③(菜帮) outer leaf (of cabbage, etc.) ④(集团;群伙) gang; band; clique: 匪～ bandit gang ⑤[量]来了一～孩子。Here comes a group of children.
【帮办】①(协助办理) assist in managing: ～军务 assist in handling military affairs ②(官职) deputy: 副国务卿～ Deputy Under Secretary (of the U.S. Department of States) / 助理国务卿～ Deputy Assistant Secretary (of the U.S. Department of State)
【帮厨】help in the mess kitchen
【帮倒忙】be more of a hindrance than a help
【帮工】①(帮助干活) help with farm work ②(帮工的人) helper
【帮会】secret society; underworld gang
【帮忙】help; give a hand; lend a hand; do a favor; do a good turn: 帮大忙 be a big help; give a lot of help / 来找人～ come for help / 要我～么? Can I help you?
【帮派】faction ◇ ～体系 factionalist setup
【帮腔】①[乐] vocal accompaniment in some traditional Chinese operas ②(帮别人说话) speak in support of sb.; echo sb.; chime in with sb.: 给帝国主义～ chime in with imperialism; take up the refrain of imperialism
【帮手】helper; assistant
【帮闲】hang on to and serve the rich and powerful by literary hack work, etc. ◇ ～文人 literary hack
【帮凶】accomplice; accessory
【帮助】help; assist: 没有多大～ be of little help; not be much help
【帮子】①(菜帮) outer leaf ②(鞋帮) upper

bǎng

榜 ①(张贴的名单) a list of names posted up: 光荣～ honor roll / 发～ publish the list of successful candidates ②(文告) announcement; notice
【榜样】example; model: 给我们树立了～ set an example to 〈for〉 us / ～的力量是无穷的。A fine example has boundless power / 学习雷锋好～。Follow the good example of Lei Feng.

膀 ①(胳膊) upper arm; arm ②(肩膀) shoulder: ～阔腰圆 broad-shouldered and solidly-built; hefty; husky ③(翅膀) wing (of a bird)
【膀臂】①[方](胳膊) upper arm; arm ②(得力助手) reliable helper; right-hand man
【膀子】①(胳膊) upper arm; arm: 光着～ stripped to the waist ②(翅膀) wing: 鸭～ duck wings

绑 bind, tie: ～个三脚架 tie three sticks together to make a tripod
【绑匪】kidnapper
【绑架】①(用强力把人劫走) kidnap ②[农] staking: 给黄瓜～。stake the cucumbers
【绑票】kidnap (for ransom)
【绑腿】leg wrappings; puttee

bàng

谤 [书](诽谤) slander; defame; vilify

磅 ①(重量单位) pound ②(磅秤) scales ③(称重量) weigh: ～体重 weigh oneself or sb. on the scales / 把棉花过～ weigh the cotton on the scale ④[印] point (type): 5～字太小。5-point type is too small.
【磅秤】platform scale; platform balance

镑 pound (a currency)

傍 draw near; be close to: 依山～水 be situated at the foot of a hill and beside a stream／船～了岸。 The boat drew alongside the bank.

【傍晚】toward evening; at nightfall; at dusk

棒 ①(木棍) stick; club; cudgel: 垒球～ softball bat ②(好;强)[口] good; fine; excellent; strong: ～小伙子 a strong young fellow／字写得～ write a good hand

【棒槌】wooden club (used to beat clothes in washing)

【棒磨机】[矿] rod mill

【棒球】baseball ◇ ～棒 bat／～场 baseball field／～击球手 batter／～投手 pitcher ○安全上垒 safe／本垒 home base／本垒板 home plate／本垒打 home run; homer／边线 foul line／不合法击球 illegally batted ball／不合法投球 illegal pitch／侧身投球法 set up position／出局 out; down; away／传球 throwing／触击 swing／触杀 touch out／得分 run; score／防守 fielding／封杀 force out／滑垒 sliding／攻队 offensive team／后攻队 home team／挥击 bunting／击球 batting; hitting／击球员 batter; hitter／击球员区 batter's box／击球员准备区 on deck circle／夹杀 run down／接球 catching／接杀 put-out／接手 catcher／接手区 catcher's box／界内球 fair ball／界外球 foul ball／垒垫 bag／内场 infield; diamond／内场手 infielder／跑垒 base running／跑垒员 base-runner／三击出局 struck-out／三杀 triple play; triple kill／守场员 fielder／守队 defensive team／守垒员 baseman／双杀 double play; double kill／四球安全上垒 base on balls／偷垒 stealing／投手 pitcher／投手投球 pitching／外场 outfield／外场手 outfielder／先攻队 visiting team／游击手 shortstop／正面投球法 wind-up position

【棒糖】sucker; lollipop

【棒子】①(棍子) stick; club; cudgel ②(玉米) maize; corn ◇ ～面 cornmeal; corn flour／～面粥 cornmeal porridge (mush)

蚌 freshwater mussel; clam

bāo

包 ①(裹起来) wrap: 把东西～起来 wrap things up／～饺子 make dumplings／纸～不住火。 You can't wrap fire in paper. ②(包好的东西) bundle; package; pack; packet; parcel: 邮～ postal parcel; postal packet ③(袋子) bag; sack: 书～ satchel; school bag ④[量] 一大～衣服 a big bundle of clothes／一～香烟 a packed of cigarettes／两～大米 two sacks of rice ⑤(肿起的疙瘩) protuberance; swelling; lump: 脑门上碰了个～ have〈get〉a bump on one's forehead／腿上起了个～ have a swelling in the leg ⑥(包围) surround; encircle; envelop ⑦(包括) include; contain: 无所不～ all-inclusive; all-embracing ⑧(全部承担) undertake the whole thing: 这事由我～了吧。 Just leave it all to me. ⑨(担保) assure; guarantee: ～你满意。 You'll like it, I assure you; Satisfaction guaranteed. ⑩(约定专用) hire; charter: ～机 a chartered plane／～一只船 hire a boat; charter a boat／～一桌酒席 order a feast

【包办】take care of everything concerning a job; run the whole show; monopolize everything: ～代替 take on what ought to be done by others; run things all by oneself without consulting others ◇ ～婚姻 arranged marriage

【包庇】shield; harbor; cover up: ～坏人坏事 harbor evildoers and cover up their evil deeds／～罪犯 harbor criminals; hide up criminals／互相～ shield each other

【包藏】contain; harbor; conceal: ～祸心 harbor evil intentions; harbor malicious intent

【包产】make a production contract; take full responsibility for output quotas ◇ ～合同 contract for fixed output／～指标 targets stated in a contract for fixed output

【保产到户】fixing of farm output quotas for each household

【包场】book a whole theatre or cinema; make a block booking

【包抄】outflank; envelop: 从两翼～逃敌 outflank the fleeing enemy on both wings

【包乘制】[交] responsible crew system

【包乘组】[交] (responsible) crew

【包虫病】echinococcosis; hydatid disease

【包饭】get or supply meals at a fixed rate; board: 在附近的饭馆里～ board at a nearby restaurant／在学校～ board at school

【包缝机】cup seaming machine

【包袱】①(包东西的布) cloth-wrappers ②(布包起来的包) a bundle wrapped in cloth ③(负担) millstone round one's neck; load; weight; burden: 思想～ a load on one's neck; a weight on one's mind

【包干】be responsible for a task until it is com-

leted: 财政～ be responsible for one's finance／大～ all-round contract system／分片～ divide up the work and assign a part to each individual or group／经费～ take responsibility for one's surplus or deficits
【包干制】a system of payment partly in kind and partly in cash
【包工】contract for a job ◇～头 labor contractor／～制 piece work system; contract system
【包管】assure; guarantee: ～他会来。 Surely he'll come.／～退换。 Merchandise will be exchanged if found unsatisfactory.
【包裹】①(包;包扎) wrap up; bind up ②(包扎成的包) bundle; package; parcel: 邮政～ postal parcel; postal packet／包好～ wrap a parcel／解开～ undo a parcel ◇～单 parcel form／～收据 parcel receipt
【包含】contain; embody; include: 他的建议～不少合理的因素。 His proposal contains much that is reasonable
【包涵】[套] excuse; forgive; bear with: 我唱得不好,请多多～。 Excuse (me for) my poor singing.
【包金】cover with gold leaf; gild
【包茎】[医] phimosis
【包括】include; consist of; comprise; incorporate: 房租每月四元,水电费～在内。 The rent is 4 *yuan* a month, including water and electricity.／他的发言～了他所有的论点。 His statement embraces all his arguments.
【包揽】undertake the whole thing; take on everything: ～词讼 engage in pettifoggery
【包罗】include; cover; embrace: 民间艺术～甚广。 Folk art covers a wide range.
【包罗万象】all-embracing; all-inclusive
【包赔】guarantee to pay compensations
【包皮】①(包装的材料) wrapping; wrapper ②[生理] prepuce; foreskin ◇～环切术 circumcision／～炎 posthitis
【包身工】indentured laborer
【包围】surround; encircle: 以农村～城市 encircle the cities from the rural areas ◇～圈 ring of encirclement
【包厢】box
【包销】have exclusive selling rights; be the sole agent for a production unit or a firm
【包圆儿】[口] ①(全部买下) buy the whole lot ②(全部承担) finish up; finish off: 剩下的活儿我一个人～了。 I'll finish off what's left of the work.
【包扎】wrap up; bind up; pack: ～伤口 bind up a wound
【包扎机】bundling machine
【包治百病】guarantee to cure all diseases: ～的药方 remedy for all ills; panacea; cure-all
【包装】pack; package ◇～车间 packing department／～费 packing charges／～机 packer／～清单 packing list／～设计 packing design／～箱 packing box; packing case／～纸 wrapping paper; packing paper
【包子】steamed stuffed bun: 肉～ steamed bun with meat stuffing
【包租】fixed rent for farmland (to be paid no matter how bad the harvest might be)

炮 quick-fry; sauté: ～羊肉 quick-fried mutton

苞 ①(花苞) bud: 含～未放 still in bud ②[书] (丛生而繁密) luxuriant; profuse; thick: 竹～松茂 bamboos and pines growing in profusion

胞 ①(胞衣) afterbirth ②(同胞) born of the same parents: ～兄弟 full brothers; blood brothers
【胞衣】[中医] afterbirth

孢

【孢子】[植] spore ◇～虫 sporozoite／～生殖 sporogony／～体 sporophyte／～形成 sporogenesis／～叶 sporophyll／～植物 cryptogam

剥 shell; peel; skin: ～花生 shell peanuts／～去他的画皮 rip off his mask／～兔皮 skin a rabbit／～香蕉 peel a banana
【剥得精光】stripped off all belongings
【剥皮】skin; peel off the skin

龅

【龅牙】bucktooth

褒 praise; honor; commend
【褒贬】①(评论好坏) pass judgment on; appraise: ～人物 pass judgment on people／不加～ make no comment, complimentary or otherwise; neither praise nor censure ②(批评;指责) speak ill of; cry down: 别在背地里～人。 Don't speak ill of anybody behind his back.
【褒奖】praise and honor; commend and award
【褒扬】praise; commend
【褒义】commendatory ◇～词 commendatory term

báo

雹 hail

【雹暴】hailstorm

【雹灾】disaster caused by hail

【雹子】hail; hailstone

薄 ①(与"厚"相对) thin; flimsy: ～纸 thin paper / 复写用的～纸 flimsy ②(淡) weak; light: 酒味很～。This is a light wine. ③(冷淡) lacking in warmth; cold: 待他不～ treat him quite well ④(不肥沃) infertile; poor: ～地 poor land

【薄板】[冶] sheet metal; sheet: 不锈钢～ stainless sheet steel ◇ ～轧机 sheet rolling mill

【薄饼】thin pancake

【薄片】slice; flake

【薄壳结构】[建] shell structure

【薄页纸】tissue paper

bǎo

宝 ①(珍贵的东西) treasure: 无价之～ Priceless treasure ② (珍贵的) precious; treasured: ～刀 a treasured sword ③ [敬] ～眷 your wife and children; your family

【宝宝】darling; baby

【宝贝】①(珍贵的东西) treasured object; treasure ②(对小孩的爱称) darling; baby ③[讽] good-for-nothing or queer character: 这人真是个～! What a fellow! ④[动] cowrie

【宝贵】valuable; precious: / ～经验 valuable experience / ～文物 precious cultural relics / ～意见 valuable suggestion

【宝剑】a double-edged sword

【宝库】treasure-house: 知识～ treasure-house of ideas

【宝蓝】sapphire blue

【宝石】precious stone; gem ◇ ～玻璃 cameo glass

【宝塔】pagoda ◇ ～菜[植] Chinese artichoke / ～筒子[纺] cone

【宝物】treasure

【宝藏】precious (mineral) deposits: 发掘地下～ unearth buried treasure; tap mineral resources

【宝座】throne

保 ①(保护;保卫) protect; defend: ～家卫国 protect our homes and defend our country ②(保持) keep; maintain; preserve: ～暖 keep warm / 丢卒～车 sacrifice a pawn to preserve the chariot ③(保证) guarantee; ensure: ～质～量 guarantee both quality and quantity / 旱涝～收 ensure stable yields despite drought or waterlogging ④(担保) stand guarantor for sb.; stand surety for sb.: ～外就医[法] be released on bail for medical treatment / ～外执行[法] serve a sentence on bail / 取～候审 [法] release upon bail pending trial ⑤(保人) guarantor

【保安】ensure public security; ensure safety ◇ ～措施 security measures / ～规程 safety regulations / ～人员 security personnel / ～装置 protective device

【保镖】bodyguard

【保不住】most likely; more likely than not; may well: ～会下雨。Most likely it's going to rain.

【保藏】keep in store; preserve: 食品～ food preservation

【保持】keep; maintain; preserve: ～安静 keep quiet / ～警惕 maintain vigilance; be on the alert / ～冷静的头脑 keep a cool head; keep cool / ～跳高纪录 retain the high jump record / ～优良品质 retain the fine qualities / ～中立 remain neutral; maintain neutrality / 跟群众～密切联系 keep close to the masses

【保存】preserve; conserve; keep: ～精力 conserve one's vitality (strength) / ～实力 preserve one's strength; conserve one's forces

【保单】guarantee slip

【保管】①(保藏; 管理) take care of: 把工具～好 take good care of the tools / 图书～工作 the care of library books ②(担保) certainly; surely: 他～不知道。He certainly doesn't know. ◇ ～费 storage charges; storage fee / ～室 storeroom / ～员 storeman; stockman; storekeeper

【保护】protect; safeguard: ～环境,防止污染 protect the environment against pollution / ～人 guardian / ～人民的利益 safeguard the people's interests / ～视力 preserve eyesight / ～现场 keep intact the scene of a crime or accident ◇ ～措施 safeguarding (protective) / ～地 protectorate; dependent territory / ～关税 protective tariff / ～国 protectorate / ～疗法 protective therapy / ～贸易政策 policy of protection; protectionism / ～伞 umbrella / ～色 [动] protective coloration / ～涂剂 protective coating / ～性拘留 protective custody; protective detention / ～主义 protectionism

【保皇党】royalists

【保加利亚】Bulgaria ◇ ～人 Bulgarian / ～语 Bulgarian (language)

【保甲制度】the *Bao-Jia* system; neighborhood administrative system (an administrative system organized on the basis of households by which the KMT regime enforced its rule at the primary level

【保驾】escort the Emperor: 放心吧,我给你~。Don't worry. I'll escort you.

【保健】health protection; health care: 妇幼~ maternal and child hygiene; mother and child care ◇ ~按摩 keep-fit massage / ~操 setting-up exercises / ~费 health subsidies / ~机构 health institution / ~护士 health nurse / ~事业 public health work / ~网 health care network / ~箱 medical kit / ~员 health worker / ~站 health station; health center / ~组织 health (care) organization / ~专家 sanitarian

【保洁箱】litter-bin

【保留】①(保存不变) continue to have; retain: 他还~着战争年代的革命朝气。He still retains the revolutionary fervor of the war years. ②(暂时留着不处理) hold bach; keep back; reserve: ~以后再答复的权利 reserve the right to reply at a later date / 持~意见 have reservations ◇ ~地 reservation / ~剧目 repertory; repertoire / ~条款 reservation clause / 无~地同意 agree unreservedly; agree without reservation

【保龄球】①[运动] bowling: 草地~ flat green bowls; lawn bowling / 十瓶(柱)~ tenpin bowling ②[球] bowling bowl◇ ~场地 bowling alley / ~木瓶(柱) bowling pin

【保密】maintain secrecy; keep sth. secret: 这事绝对~。This must be kept absolutely secret; This is strictly confidential. ◇ ~级别 security classification / ~审查 inquiry on the protection of secrets / ~条例 security regulations / ~文件 classified document

【保苗】[农] keep a full stand of seedlings

【保姆】①(为人照顾小孩的) (children's) nurse ②(为人做家务的) housekeeper

【保全】①(使不受损失) save from damage; preserve: ~革命力量 preserve revolutionary strength / ~名誉 preserve one's reputation / ~面子 save face ②(保养;维修) maintain; keep in good repair ◇ ~工 maintenance worker

【保墒】[农] preservation of soil moisture

【保释】[法] release on bail; bail: 不准~ refuse bail / 准予~ accept bail ◇ ~金 bail; recognizance / ~人 bail; bailman

【保守】①(保持使不失去) guard; keep: ~党和国家机密 guard Party and state secrets ②(守旧的) conservative: ~观点 conservative point of view / ~思想 conservative ideas; conservative thinking ◇ (英国)~党 the Conservative Party / ~疗法 [医] conservative treatment / ~派 conservatives / ~主义 conservatism

【保送】recommend sb. for admission to school, etc.

【保泰松】[药] phenylbutazone

【保卫】defend; safeguard: ~国家主权和领土完整 safeguard state sovereignty and territorial integrity / ~祖国 defend one's country ◇ ~部门 public security bodies / ~工作 security work / ~科 security section

【保温】heat preservation ◇ ~材料 thermal insulation material / ~层 [建] (thermal) insulating layer / ~车 [铁道] refrigerator wagon; refrigerator car / ~瓶 vacuum flask; vacuum bottle; thermos

【保险】①(保险业务) insurance: 船舶定期~ time hull insurance / 海损~ marine insurance / 海洋运输~ maritime transportation insurance / 货物~ cargo insurance / 强制~ compulsory insurance / 人寿~ life insurance / 再~ reinsurance ②(稳妥可靠) safe: 骑车太快可不~。It's not safe to cycle too fast. ③(保证;担保) be sure; be bound to: 他明天~会来。He is sure to come tomorrow. / ~能行! It's bound to work.◇ ~带 safety belt / ~单 insurance policy / ~刀 safety razor / ~对象 insured object / ~法 insurance law / ~范围 insurance coverage / ~粉 [纺] sodium hydrosulphite / ~杠 [汽车] bumper / ~公司 insurance company / ~柜 safe / ~机 safety (of a firearm) / ~金 insurance money / ~金额 insurance amount, insured amount; sum insured / ~客户 policy-holder / ~类别 branch of insurance / ~赔款 insurance proceeds / ~赔偿 insurance indemnity / ~期 insurance period / ~申请 insurance application / ~箱 safe; strong-box / ~装置 safety device / ~总额 coverage

【保险费】(insurance) premium: 初期~ initial premium / 全部~ in-full premium / 预付~ deposit premium

【保险丝】[电] fuse; fuse-wire: ~烧断了。The fuse has blown.

【保修】guarantee to keep sth. in good repair: 这只表~一年。This watch is guaranteed for a year.

【保养】①(保护调养) take good care of one's health: 好好~你的身体。Take good care of your health. ②(保护修理) maintain; keep in good repair: 机器～ maintenance of machinery; upkeep of machinery ◇ ～费 maintenance cost; upkeep / ～工 maintenance worker / ～间隔期 reserve maintenance period

【保佑】bless and protect: 不靠老天～ don't rely on blessings from heaven

【保育】child care; child welfare ◇ ～员 child-care worker; nurse / ～院 nursery school

【保障】ensure; guarantee; safeguard: ～供给 ensure supply / ～人民言论自由 guarantee freedom of speech for the people / 在新社会职业有～。In the new society jobs are secure.

【保真度】[电子] fidelity: 高～ high fidelity

【保证】pledge; guarantee; assure; ensure: ～不再发生类似事件 guarantee against the occurrence of similar incidents / ～完成任务 pledge 〈guarantee〉 to fulfill a task ◇ ～书 written pledge; guarantee; guaranty; letter of guarantee

【保证金】① earnest money; cash deposit ②[法] bail

【保证人】[法] ①(担保人) guarantor ②(保释人) bail; recognizance

【保质保量】quality and quantity guaranteed

【保重】take care of oneself: 多多～。 Take good care of yourself; Look after yourself.

饱 ①(吃饱) have eaten one's fill; be full: 吃～喝足 eat and drink one's fill / 我～了。 I've had enough. ② (饱满) full; plump: 谷粒很～。 The grains are quite plump. ③(充分地) fully; to the full: ～尝旧社会的辛酸 taste to the full the bitterness of life in the old society / ～览海岛的美丽风光 drink in the beauty of the island scenery ④(满足) satisfy: 一～眼福 have the opportunity to feast one's eyes on sth.

【饱汉不知饿汉饥】the well-fed don't know how the starving suffer

【饱和】saturation ◇ ～差[气] saturation deficit; saturation deficiency / ～点 saturation point / ～轰炸 saturation bombing / ～剂[化] saturant / ～器 saturator / ～压力 saturation pressure / ～溶液[化] saturated solution

【饱经沧桑】have seen much of the changes in human life; have been through the vicissitudes of life

【饱经风霜】weather-beaten; having experienced the hardships of life: ～的海员 seasoned sailor / ～的面容 a weather-worn face

【饱满】full; plump: 颗粒～的小麦 plump-eared wheat / 精神～ full of vigor; energetic ◇ ～度[农] plumpness (of seeds) / ～感[心] satiety

【饱食终日,无所用心】eat three square meals a day and do no work; be sated with food and remain idle

【饱学】learned; erudite; scholarly: ～之士 an erudite person; a learned scholar; a man of learning

鸨 ①[动] bustard ②(鸨母) procuress

葆 [书] ①(草茂盛) luxuriant growth ② (保持; 保护) preserve; nurture: 永～青春 keep alive the fervor of youth

堡 fort; fortress

【堡垒】fort; fortress; stronghold; blockhouse: 封建～ a stronghold of feudalism ◇ ～战 blockhouse warfare

【堡礁】barrier reef

bào

报 ①(报告) report; announce; declare: ～火警 report a fire / ～上级批准 report sth. to the higher authorities for approval ②(回答) reply; respond; reciprocate: ～以热烈的掌声 respond with warm applause ③(报答) recompense; requite: 无以为～ be unable to repay a kindness / 以怨～德 requite kindness with ingratitude; return evil for good ④(报纸) newspaper: 晨～ morning paper / 日～ daily paper / 晚～ evening paper / 星期日～ Sunday newspaper ⑤(刊物) periodical; journal: 画～ pictorial / 学～ college journal / 周～ weekly ⑥(文字报道或墙报) bulletin; report: 喜～ report of success, a happy event, etc.; good news / 战～ war bulletin ⑦(电报) telegram; cable: 发～ transmit a telegram; send a telegram

【报案】report a case to the security authorities

【报表】forms for reporting statistics, etc.; report forms: 月～ monthly report

【报偿】repay; recompense

【报仇】revenge; avenge: 为兄弟～ avenge one's brothers

【报仇雪恨】avenge oneself and cancel out one's hatred

【报酬】reward; remuneration; pay: 不计～ not concerned about pay; irrespective of remuneration / 逐步提高劳动～ gradually increase payment for labor

【报春】be a harbinger of spring; herald spring
【报答】repay; requite: ～某人的好意 repay sb. for his kindness
【报单】taxation form; declaration form
【报到】report for duty; check in; register: 向部里～ report for duty at the ministry / 向大会秘书处～ check in at the secretariat of the congress / 新生已开始～。The new students have started registering.
【报道】①(报导) report (news); cover: ～大选的情况 report on the general election / ～会议情况 cover the conference / 据～ it is reported that ②(新闻稿) news report; story
【报恩】pay a debt of gratitude
【报废】report sth. as worthless; discard as useless; scrap: 使～矿井复生 reopen an abandoned mine / 这架机器太旧,快～了。This machine is so old it will soon have to be scrapped. ◇～零件 faulty part; scrapped parts
【报分】[体] call the score
【报复】make reprisals; retaliate: 图谋～ nurse thoughts of revenge ◇～行为 vindictive act / ～性措施 retaliatory measures / ～性打击 vindictive blow; retaliatory strike
【报告】①(告诉) report; make known: 向上级～ report to the higher authorities / 现在～新闻。Here is the news. ②(讲演;书面申请或总结) report; speech; talk; lecture: 调查～ investigation report / 动员～ mobilization speech / 时事～ report on current events / 总结～ summing-up report / 作～ give a talk or lecture ◇～会 public lecture / ～人 speaker; lecturer / ～文学 reportage
【报关】declare sth. at customs; apply to customs: 进口～ customs entry
【报馆】[旧] newspaper office
【报国】dedicate oneself to the service of one's country
【报户口】apply for a residence permit: 报临时户口 apply for a temporary residence permit / 给新生婴儿～ register the birth of a child
【报话机】handie-talkie
【报夹】newspaper holder
【报架】newspaper rack; newspaper shelf
【报价】[经] quoted price
【报捷】report a success; announce a victory
【报界】the press; journalistic circles; the journalists: 向～发表谈话 make a statement to the press
【报警】report (an incident) to the police; give an alarm: ～装置 warning device / 鸣钟～ sound the alarm bell
【报刊】newspapers and periodicals; the press
【报考】enter oneself for an examination
【报名】enter one's name; sign up: ～参加百米赛跑 enter one's name for the 100-meter dash / ～参军 enter one's name for the army ◇～单 entry form; application form
【报幕】announce the items on a (theatrical) program ◇～员 announcer
【报丧】give obituary notice; announce sb.'s death
【报社】general office of a newspaper; newspaper office
【报失】report the loss of sth. to the authorities concerned
【报时】give the correct time ◇～器 chronopher / ～台 (telephone) time inquiry service
【报数】number off: ～!(口令) Count off!
【报税】declare dutiable goods; make a statement of dutiable goods ◇～单 taxation form; duty declaration form
【报摊】news-stand; news stall
【报头】masthead (of a newspaper, etc.); name plate
【报务员】telegraph operator; radio operator
【报喜】announce good news; report success: ～不报忧 report only the good news and not the bad; hold back unpleasant information
【报销】①(报销帐目) submit an expense account; apply for reimbursement: 向财务科～ submit an expense account to the treasurer's office ②(报废销帐) hand in a list of expended articles ③[口](消灭) write off; wipe out
【报晓】herald the break of day; be a harbinger of dawn
【报效】render service to repay sb.'s kindness
【报信】notify; inform: 通风～ tip off information
【报应】[宗] retribution; judgment
【报章】newspapers ◇～杂志 newspapers and magazines
【报帐】render an account; submit an expense account; apply for reimbursement: 修理费用可以～。Costs of repairs may be reimbursed.
【报纸】① newspaper ②(白报纸) newsprint ◇～夹 newspaper holder

暴 ①(突然而且猛烈) sudden and violent: ～饮～食 eat and drink too much at one meal ②(凶狠) cruel; savage; fierce: 残～ brutal ③(急

躁) short-tempered; hot-tempered: 脾气～ have a hot temper ④(鼓起来) stick out; stand out; bulge: 急得头上的青筋都～出来了 be so agitated that the veins on one's forehead stand out
【暴病】sudden attack of a serious illness: 得～ be suddenly seized with a severe illness
【暴跌】steep fall (in price); slump: 股票价格～ a slump in the prices of stocks
【暴动】insurrection; rebellion
【暴发】①(突然发作) break out: 山洪～。Torrents of water rushed down the mountain. ②(突然发财得势) suddenly become rich or important; get rich quick ◇ ～户 upstart
【暴风】storm wind; storm (force 11 wind)
【暴风雪】snowstorm; blizzard
【暴风雨】rainstorm; storm; tempest: ～般的掌声 thunderous applause / 革命的～ a storm of revolution; a revolutionary tempest
【暴风骤雨】violent storm; hurricane; tempest: 其势如～ with the force of a hurricane
【暴光】[摄] exposure ◇ ～表 exposure meter / ～宽容度 exposure latitude
【暴洪】a sudden, violent flood; flash flood
【暴君】tyrant; despot
【暴力】violence; force: 采用～ resort to violence ◇ ～革命 violent revolution / ～机关 organ of violence / ～行为 act of violence
【暴利】sudden huge profits: 牟取～ reap staggering profits; reap colossal profits
【暴戾】[书] ruthless and tyrannical; cruel and fierce
【暴戾恣睢】[书] extremely cruel and despotic
【暴烈】violent; fierce: 性情～ have a fiery temper
【暴露】expose; reveal; lay bare: ～帝国主义的侵略本性 bare the aggressive nature of imperialism / ～目标 give away one's position / ～无遗 be thoroughly exposed / ～在光天化日之下 be exposed to the light of day ◇ ～文学 literature of exposure
【暴乱】riot; rebellion; revolt: 反革命～ a counterrevolutionary rebellion / 镇压～ put down a riot; suppress a riot
【暴怒】violent rage; fury
【暴虐】brutal; tyrannical
【暴殄天物】a reckless waste of grain, etc.
【暴跳如雷】stamp with fury; fly into a rage
【暴徒】ruffian; thug
【暴行】savage act; outrage; atrocity
【暴雨】torrential rain; rainstorm
【暴躁】irascible; irritable: 性格～的人 person with a violent temper; hot-tempered person
【暴涨】rise suddenly and sharply: 河水～。The river suddenly rose. / 物价～。Prices soared. (或: Prices skyrocketed.)
【暴政】tyranny; despotic rule
【暴卒】die of a sudden illness; die suddenly

爆 ①(猛然破裂) explode; burst: 车胎～了。The tire's burst. ②(用猛火炒菜) quick-fry; quick-boil: 油～肚儿 quick-fried tripe
【爆发】erupt; burst out; break out: 火山～ volcanic eruption / 人群中～出一片欢呼声。The crowd burst into cheers. / 战争～。War broke out. ◇ ～力 [体] explosive force
【爆管】cartridge igniter; squib
【爆裂】burst; crack: 豌豆过熟就会～。Pea pods burst open when overripe.
【爆米花】puffed rice
【爆破】blow up; demolish; dynamite; blast: ～敌人的碉堡 blow up an enemy pillbox / 连续～ successive demolitions ◇ ～弹 blasting cartridge / ～手 dynamiter / ～筒 bangalore (torpedo) / ～音[语] plosive / ～英雄 demolition hero; ace dynamiter / ～组 demolition team
【爆音】[航空] sonic boom; shock-wave noise
【爆玉米花】popcorn
【爆炸】explode; blow up; detonate: ～性的局势 an explosive situation / ～一个核装置 detonate a nuclear device / 炸弹～了。A bomb exploded. ◇ ～成形[冶] explosive forming / ～地震学 explosion seismology / ～极限 explosive limit / ～力 explosive force / ～物 explosive / ～性新闻 startling news
【爆竹】firecracker: 放～ let off firecrackers / ～没响。The firecracker didn't go off.

豹 leopard; panther
【豹猫】leopard cat

抱 ①(用手臂抱) hold or carry in the arms; embrace; hug: 把小孩子～起来 take a child in one's arms ②(初次得到) have one's first child or grandchild: 她快～孙子了。She'll soon be a grandmother. ③(领养) adopt (a child): 他的女儿是～的。His daughter is adopted. ④[方](结合在一起) hang together: ～成一团 gang up; hang together ⑤(心里存着) cherish; harbor: ～必胜的信心 hold a firm belief in victory / ～很大希望 entertain high hopes / ～正确的态度 adopt a correct attitude; take a correct

attitude / 不～幻想 cherish no illusions ⑥[量] 一～草 an armful of hay / 这颗树有一～粗。You can just get your arms around this tree ⑦ (孵) hatch (eggs); brood
【抱病】be ill; be in bad health: ～工作 go on working in spite of ill health
【抱不平】be outraged by an injustice (done to another person): 打～ defend sb. against an injustice
【抱残守缺】cherish the outmoded and preserve the outworn; be conservative
【抱成一团】hold together to form a clique
【抱粗腿】latch on to the rich and powerful
【抱佛脚】clasp Buddha's feet; profess devotion only when in trouble; make a hasty last-minute effort: 平时不烧香,急来～ never burn incense when all is well but clasp Buddha's feet when in distress; do nothing until the last minute
【抱负】aspiration; ambition: 很有～ have high aspirations; cherish high ambitions
【抱撼终身】regret sth. all one's life
【抱恨】have a gnawing regret: ～终天 regret sth. to the end of one's days
【抱愧】feel ashamed
【抱歉】be sorry; feel apologetic; regret: 叫你久等了,很～。 Very sorry to have kept you waiting.
【抱屈】feel wronged
【抱头鼠窜】cover the head and sneak away like a rat; scurry of like a frightened rat; scamper of like a frightened rat
【抱头痛哭】weep in each other's arms; cry on each other's shoulder
【抱窝】sit (on eggs); brood; hatch: 母鸡～了。 The hen is sitting.
【抱薪救火】carry faggots to put out a fire; adopt a wrong method to save a situation and end up by making it worse
【抱养】adopt (a child)
【抱怨】complain; grumble: 不要总是～。 Don't always complain.

刨 ①(刨光;刨平) plane sth. down; plane: ～木板 plane a board ② (刨子或刨床) plane; planer; planing machine
【刨冰】water ice (powdered or in shavings)
【刨程】[机] planing length
【刨齿】[机] gear-shaping ◇ ～机 gear shaper
【刨床】planer; planing machine
【刨刀】[机] planer tool
【刨工】① (工种) planing ② (技术工人) planing machine operator; planer
【刨花】wood shavings ◇ ～板[建] shaving board
【刨刃儿】plane iron
【刨式磨床】planer-type grinder
【刨式铣床】planer-type miller
【刨子】plane

鲍
【鲍鱼】①[动] abalone ②(咸鱼)[书] salted fish: 如入～之肆,久而不闻其臭。 It's like staying in a fish market and getting used to the stink; long exposure to a bad environment accustoms one to evil ways.

bēi

杯 ①(杯子;杯) cup: 玻璃～ glass / 茶～ teacup / 大～ mug / 大玻璃～ tumbler / 啤酒～ beer mug / 葡萄酒～ wine glass / 香槟酒～ champagne glass / 小酒～ liqueur glass / 一～茶 a cup of tea / 一～啤酒 a mug of beer ②(杯状的锦标) (prize) cup; trophy: 银～ silver cup
【杯弓蛇影】mistaking the reflection of a bow in the cup for a snake; extremely suspicious
【杯盘狼藉】wine cups and dishes lying about in disorder after a feast
【杯水车薪】trying to put out a burning cartload of faggots with a cup of water; an utterly inadequate measure
【杯子】cup; glass

悲 ①(悲伤) sad; sorrowful; melancholy: ～不自胜 be overcome with grief ②(怜悯) compassion: 慈～ compassionate; merciful
【悲哀】grieved; sorrowful
【悲惨】miserable; tragic: ～的过去 the bitter past / ～的结局 tragic ending / ～的遭遇 a tragic experience
【悲愁】sad; heavy-hearted; low-spirited
【悲悼】mourn; grieve over sb.'s death
【悲愤】grief and indignation: ～填膺 be filled with grief and indignation
【悲歌】①(悲壮曲调) sad melody; stirring strains ②(悲壮的或哀痛的歌) elegy; dirge; threnody ③(悲壮地歌唱) sing with solemn fervor
【悲观】pessimistic: ～情绪 pessimism / 持～看法 take a pessimistic view; take a gloomy view / 感到～失望 feel disheartened ◇ ～主义 pessimism

【悲欢离合】joys and sorrows, partings and reunions; vicissitudes of life
【悲剧】tragedy
【悲鸣】utter sad calls; lament:发出～ emit a plaintive cry; give a plaintive cry
【悲泣】weep with grief
【悲伤】sad; sorrowful
【悲叹】sign mournfully; lament
【悲天悯人】bemoan the state of the universe and pity the fate of mankind
【悲痛】grieved; sorrowful: 感到深切的～ be deeply grieved; be filled with deep sorrow / 化～为力量 turn grief into strength
【悲喜交集】mixed feelings of grief and joy; grief and joy intermingled; joy tempered with sorrow
【悲喜剧】tragicomedy
【悲壮】solemn and stirring; moving and tragic: ～的歌曲 a solemn and stirring song

背 ①(用脊背驮) carry on the back: ～着孩子 carry a baby on one's back ②(负担) bear; shoulder (responsibility; burden; etc.)
【背包】① knapsack; rucksack; infantry pack; field pack ②[军] blanket roll
【背包袱】have a weight on one's mind; have a load on one's mind; take on a mental burden: 你不要因此～。 Don't let it weigh on your mind.
【背带】①(裤子的;裙子的) braces; suspenders ②(枪的) sling ③(背包的) straps
【背负】bear; carry on the back; have on one's shoulder
【背黑锅】[口] be made a scapegoat; be unjustly blamed
【背篓】a basket carried on the back
【背债】be in debt; be saddled with debts

卑 ①(低下) low: 地势～湿 low-lying and damp ②(低劣) inferior: ～不足道 not worth mentioning ③(谦恭) [书] modest; humble: ～辞厚礼 humble words and handsome gifts
【卑鄙】base; mean; contemptible; despicable: ～的诽谤 mean slander / ～勾当 a dirty deal / ～手段 contemptible means; dirty tricks / ～行为 a base action; a mean action; sordid conduct
【卑鄙无耻】mean and shameless
【卑躬屈节】bow and scrape; cringe; act servilely; obsequiously act
【卑贱】①(地位低下) lowly ②(卑鄙下贱) mean and low
【卑劣】base; mean; despicable: ～手法 a mean trick; a despicable trick / ～行径 base conduct
【卑怯】mean and cowardly; abject: ～行为 abject behavior
【卑下】base; low

碑 an upright stone tablet; stele: 墓～ tombstone / 人民英雄纪念～ the Monument to the People's Heroes / 西安～林 the Forest of Steles in Xi'an
【碑额】the top part of a tablet
【碑记】a record of events inscribed on a tablet
【碑碣】[书] an upright stone tablet; stele
【碑帖】a rubbing from a stone inscription (usu. as a model for calligraphy)
【碑文】an inscription on a tablet

běi

北 ①(北方) north: ～风 a north wind / 城～ north of the city / 华～ north China / ～屋 a room with a southern exposure ②[书](打败仗) be defeated: 敌军连战皆～。 The enemy suffered repeated defeats.
【北半球】the Northern Hemisphere
【北冰洋】the Arctic (Ocean)
【北大荒】the Great Northern Wilderness (in northeast China)
【北大西洋公约组织】the North Atlantic Treaty Organization (NATO)
【北斗星】the Big Dipper; the Plough
【北伐军】the Northern Expeditionary Army
【北伐战争】the Northern Expedition
【北方】①(北) north ②(北部地区) the northern part of the country; the North: ～的风俗习惯 northern habits and customs ◇ ～话 northern dialect / ～人 Northerner
【北国】[书] the northern part of the country; the North: ～风光 northern scenery
【北海】the North Sea
【北寒带】the north frigid zone
【北回归线】[地] the Tropic of Cancer
【北极】①(地轴的北端) the North Pole; the Arctic Pole ②(北磁极) the north magnetic pole ◇ ～地区 Arctic regions; north polar regions / ～光 [天] northern lights; aurora borealis / ～狐 Arctic fox / ～圈 the Arctic Circle / ～星 Polaris; the North Star; the polestar / ～熊 polar bear
【北京】Beijing
【北京人】[考古] Peking Man (*Sinanthropus pekinensis*)

【北京鸭】Beijing duck: 北京烤鸭 roast Beijing duck
【北美洲】North America
【北温带】the north temperate zone
【北洋军阀】the Northern Warlords

bèi

焙 bake over a slow fire: ～干 dry over a fire / ～制 cure sth. by drying it over a fire
【焙烧】roast; bake ◇ ～炉 roaster

倍 ①(倍数) times; -fold: 四～ four times; fourfold / 大一～ twice as big; twice the size / 增长了五～ increase by 500% / 产量成～增长。Output has doubled and redoubled. / 二的五～是十。Five times two is ten. / 十是五的两～。Ten is twice as much as five. ②(加倍) double; twice as much: 勇气～增 with redoubled courage
【倍频器】[电子] frequency multiplier
【倍数】[数] multiple
【倍塔】beta ◇ ～粒子 [物] beta particle / ～射线 [物] beta ray
【倍增器】[电子] multiplier: 光电～ photo-electric multiplier
【倍增系数】multiplication constant; multiplication factor

蓓
【蓓蕾】bud

悖 [书]①(相反) be contrary to; go against: ～理 contrary to reason / 并行不～ parallel and not contrary to each other; not mutually exclusive ②(违背道理) perverse; erroneous
【悖论】[哲] paradox
【悖谬】[书] absurd; preposterous
【悖入悖出】ill-gotten, ill-spent

辈 ①[书] (某类人) people of a certain kind; the like: 无能之～ people without ability ②(行辈;辈分) generation: 老一～的无产阶级革命家 proletarian revolutionaries of the older generation / 上一～ the last generation; the preceding generation / 下一～ the coming generation; the younger generation / 他俩同～。They belong to the same generation. ③(一生) lifetime: 后半～儿 the latter part of one's life
【辈出】come forth in large numbers: 人材～ people of talent coming forth in large numbers / 英雄～的时代 an age of heroes
【辈分】seniority in the family or clan; position in the family hierarchy: 她的～比我小。She ranks as my junior in the clan.
【辈子】all one's life; lifetime

背 ①(脊背) the back of the body: ～痛 backache ②(反面;背面) the back of an object: 手～ the back of the hand / 椅～ the back of a chair ③(背部对着) with the back towards: ～着太阳坐 sit with one's back to the sun ④(转向后面) turn away: 把脸～过去 turn one's face away ⑤(躲避;瞒) hide sth. from; do sth. behind sb.'s back: ～着人说话 talk behind sb.'s back / 没有什么～人的事 have nothing to hide from anyone ⑥(背诵) recite from memory; learn by heart; learn by rote: ～台词 speak one's lines / 死～课文 learn a lesson by rote ⑦(违背;违反) act contrary to; violate; break: ～约 violate an agreement; break one's promise; go back on one's word ⑧(听觉不灵) hard of hearing: 耳朵有点～ be a bit hard of hearing ⑨[口] (倒霉) unlucky
【背包】packsack; knapsack: 打～ make one's pack; make up knapsack
【背城借一】make a last-ditch stand before the city wall; fight to the last ditch; put up a desperate struggle
【背道而驰】run in the opposite direction; run counter to
【背地里】behind sb.'s back; privately; on the sly
【背风】out of the wind; on the lee side; leeward: 向～处航行 steer leeward ◇ ～波 lee wave / ～处 lee side; sheltered side
【背光】be in a poor light; do sth. with one's back to the light ◇ ～性 [生] negative phototopism
【背后】①(后面) behind; at the back; in the rear: 从～袭击敌人 attack the enemy from the rear / 房子～ at the back of the house / 门～ behind the door ②(不当面) behind sb.'s back: ～说坏话 backbite; speak ill of sb. behind his back
【背井离乡】leave one's native place (esp. against one's will)
【背景】background; backdrop: 历史～ historical background; historical setting / 时代～ background characteristic of times / 舞台～ stage background ◇ ～反射 background reflectance / ～亮度 background luminance
【背景知识】cultural background: 英语学习～ cultural background for English study

【背静】quiet and secluded
【背靠背】back to back; without one's presence
【背阔肌】[生理] latissimus dorsi
【背离】deviate from; depart from: ～革命原则 deviate from the revolutionary principle
【背面】the back; the reverse side; the wrong side: 信封的～ the back of an envelope / 请阅～ please turn over (P.T.O.); see overleaf
【背叛】betray; forsake: ～马克思列宁主义 betray Marxism-Leninism / 忘记过去就意味着～。Forgetting the past means betrayal.
【背鳍】[动] dorsal fin
【背弃】abandon; desert; renounce: ～原来的立场 abandon one's original stand / ～自己的诺言 go back on one's word
【背日性】[植]negative heliotropism
【背山面海】with hills behind and the sea in front
【背山面水】facing the water with hills behind
【背时】[方] ①(不合时宜) behind the times ②(倒霉) unlucky
【背书】①(背诵) recite a lesson from memory; repeat a lesson ②(票据背面的签字) [经] endorsement (on a check)
【背水一战】fight with one's back to the river; fight to win or die
【背诵】recite; repeat from memory
【背斜】[地] anticline ◇ ～层 anticlinal strata / ～理论 [地] anticlinal theory
【背心】a sleeveless garment: 汗～ vest; singlet / 毛～ sleeveless woollen sweater / 棉～ cotton-padded waistcoat / 西服～ waistcoat
【背信弃义】break faith with sb.; be perfidious: ～的行为 a breach of faith; perfidy / ～地撕毁协定和合同 perfidiously tear up agreements and contracts
【背阴】in the shade; shady ◇ ～处 shady spot
【背影】a view of sb.'s back; a figure viewed from behind: 凝望着他逐渐消失的～ gazing at his receding figure
【背约】break an agreement; go back on one's word; fail to keep one's promise
【背着手】with one's hands clasped behind one's back

贝 ①(软体动物) shellfish ②(古代用贝壳做的货币) cowrie: 虎斑～ tiger cowrie
【贝雕】[工美] shell carving ◇ ～画 shell carving picture
【贝壳】shell ◇ ～学 conchology
【贝类】shellfish; molluscs
【贝母】[中药] the bulb of fritillary
【贝宁】Benin ◇ ～人 Beninian
【贝丘】[考古] shell mound

钡 [化] barium (Ba)
【钡餐】[医] barium meal

备 ①(具备,具有) be equipped with; have: 德才兼～ have both ability and political integrity / 各种农业机械无一不～ be equipped with all sorts of farm machinery ②(备有) prepare; get ready: 把料～齐 get all the materials ready ③(防备) provide against; prepare against; take precautions against: 以～万一 prepare against all eventualities ④(设备;装备) equipment: 军～ military equipment; armaments ⑤(完全;尽) fully; in every possible way: ～受欢迎 enjoy great popularity; be very popular / ～受虐待 be subjected to every kind of maltreatment / 艰苦～尝 suffer untold hardships
【备案】put on record; put on file; enter (a case) in the records
【备查】for future reference
【备而不用】have sth. ready just in case; keep sth. for possible future use
【备耕】make preparations for ploughing and sowing
【备荒】prepare against natural disasters
【备件】spare parts ◇ ～箱 spare part case
【备考】(an appendix, note, etc.) for reference
【备课】prepare lessons
【备料】①(准备生产所需材料) get the materials ready ②(准备牲畜饲料) prepare feed (for livestock) ◇ ～场 charge make-up area / ～车间 workshop for preparing the feed
【备取】be on the waiting list (for admission to a school)
【备忘录】①[外](外交文书) memorandum; aide-mémoire ②(供记事用的笔记本) memorandum book
【备用】reserve; spare; alternate ◇ ～航空站 alternate airport / ～机器 standby machine / ～款项 reserve funds / ～轮胎 spare tire / ～燃料 guarantee fuel / ～燃油箱 reserve fuel tank / ～系统 backup system / ～线路 spare circuit
【备战】prepare for war: 扩军～ arms expansion and war preparations
【备至】to the utmost; in every possible way: 关怀～ show sb. every consideration / 颂扬～

praise profusely
【备注】remarks ◇ ～栏 remarks column
惫 exhausted; fatigued: 疲～ tired out
被 ①(被子) quilt: 棉～ cotton-wadded quilt ②[介] 他爸爸是～人害死的。 His father was murdered by sb.. ③[助] ～捕 be arrested; be under arrest / ～选为主席 be elected chairman
【被保险人】assured; insured
【被保险物】subject of insurance
【被逼】be forced or compelled
【被剥削阶级】exploited class
【被乘数】[数] multiplicand
【被除数】[数] dividend
【被袋】bedding bag
【被单】(bed) sheet ◇ ～布 sheeting
【被动】passive: 变～为主动 regain the initiative / 陷于～地位 land oneself in a passive position; be thrown into passivity ◇ ～加工贸易 passive improvement trade / ～式 [语] passive form / ～语态 [语] passive voice
【被服】bedding and clothing (esp. for army use) ◇ ～厂 clothing factory
【被俘】be captured; be taken prisoner ◇ ～人员 captured personnel
【被告】[法] defendant; the accused ◇ ～席 defendant's seat; dock
【被告人权利】[法] rights of accused
【被害人】[法] the injured party; the victim
【被加数】[数] summand
【被减数】[数] minuend
【被拒付者】protesting party
【被开方数】[数] radicand
【被里】the underneath side of a quilt
【被面】the facing of a quilt: 绣花～ an embroidered quilt cover
【被难】be killed in a disaster, political incident, etc.
【被迫】be compelled; be forced; be constrained: 敌人～放下武器。 The enemy were compelled to lay down their arms.
【被侵略者】victim of aggression
【被褥】bedding; bedclothes
【被上诉人】[法] appellee
【被套】①(装被褥的袋子) bedding bag ②(为拆洗方便,缝成袋状的被里和被面) quilt cover ③(棉被的胎) cotton wadding for a quilt
【被统治者】the ruled
【被窝儿】a quilt folded to form a sleeping bag
【被选举权】the right to be elected
【被压迫民族】oppressed nation
【被子】quilt
【被子植物】angiosperm

bēn

贲
【贲门】[生理] cardia (of stomach)
奔 ①(急跑) run quickly: ～马 a galloping horse ②(紧赶) hurry; hasten; rush: ～赴前线 hurry to the front / ～向共产主义明天 march on towards the Communist future ③(逃跑) flee: 东～西窜 flee in all directions
【奔波】rush about; be busy running about: 在两地之间～ shuttle back and forth between two places
【奔驰】run quickly; speed: 火车向前～。The train sped on. / 骏马在草原上～。Sturdy steeds gallop on the grasslands.
【奔窜】flee about; scuttle of: 四处～ flee about in all directions
【奔放】bold and unrestrained; untrammelled: ～不羁的风格 a bold and flowing style / 热情～ overflowing with enthusiasm
【奔赴】go to; hurry off to; hurry off for; rush to
【奔流】①(流得很快) flow at great speed; pour: ～入海 flow into the sea / 铁水～ molten iron pouring out in a stream ②(急流的水) racing current
【奔忙】be busy rushing about; bustle about: 为革命日夜～ be busy day and night working for the revolution
【奔命】rush about on errands; be kept on the run: 疲于～ be tired out by too much running around; be kept constantly on the run
【奔跑】run
【奔丧】hasten home for the funeral of a parent or grandparent
【奔逃】flee; run away: 四散～ flee in all directions; flee helter-skelter; stampede
【奔腾】①(跳跃奔跑) gallop: 犹如万马～ like ten thousand horses galloping ahead ②(比喻汹涌) surge forward; roll on in waves: 洪水～而下 the floodwater rolled downwards / 革命的洪流～向前。 The tide of revolution is surging ahead.
【奔袭】[军] long-range raid
【奔泻】(of torrents) rush down; pour down: 怒涛滚滚,～千里。 An angry torrent rolls thunderously on for a thousand *li*.

【奔走】①(急走) run ②(到处活动) rush about; be busy running about: ～呼号 go around campaigning for a cause / ～相告 run around spreading the news; lose no time in telling each other the news

锛 ①(锛子) adze ②(用锛子削平木料) cut with an adze

【锛子】adze

běn

本 ①(草木的茎根) the root or stem of a plant: 水有源,木有～。A stream has its source; a tree has its root. ②(事物的根本) foundation; basis; origin～ ③(本钱) capital; principal: ～小利微 have tiny funds and small earnings / 还～付息 pay back the capital 〈principal〉 plus interest ④(本来;原来) original: ～意 original idea; real intention ⑤(自己方面的) one's own; native: ～厂 this factory ⑥(现今的) this; current; present: ～决议 this resolution / ～月 this month; the current month / ～周 this week; the current week ⑦(按照;根据) according to; based on: ～着政策办事 act according to policy ⑧(本子) book: 日记～ diary / 帐～儿 account book / 照相～ photograph album ⑨(版本) edition; version: 孤～ unique copy / 精装～ de luxe edition / 廉价～ cheap edition; paperback / 马克思《资本论》英译～ the English translation of Marx's Capital / 平装～ paperback / 普及～ popular edition / 善～ rare book / 上演～ stage version / 微型～ miniature edition / 袖珍～ pocket edition / 样～ sample copy / 赠～ gift ⑩[量] 两～书 two books ⑪[电影] reel

【本本主义】book worship; bookishness

【本草】a book on Chinese (herbal) medicine; Chinese materia medica ◇《～纲目》*Compendium of Materia Medica*

【本初子午线】the first meridian; the prime meridian

【本底】[物] background: 放射性～ radioactive background ◇ ～噪声 background noise

【本地】this locality: ～风光 local color / ～风俗 local customs / ～货 local goods; native goods / ～口音 local accent / ～新闻 local items; local news / 我是～人。I'm a native of this place. (或: I was born here.)

【本分】one's duty: 尽～ do one's duty; do one's bit / 为人民服务是我们的～。To serve the people is our duty.

【本固枝荣】when the root is firm, the branches flourish

【本国】one's own country: ◇ ～产品 home products / ～语 native language; mother tongue / ～资源 national resources

【本行】one's line; one's own profession: 搞建筑是他是～。Architecture is his line.

【本家】a member of the same clan; a distant relative with the same family name

【本届】current; this year's: ～毕业生 this year's graduates / ～联合国大会 the current session of the U.N. General Assembly

【本金】capital; principal

【本科】undergraduate course; regular college course ◇ ～学生 undergraduate

【本来】①(原有的) original: ～的意思 original meaning; original intention / 事物～的辩证法 the dialectics inherent in things ②(原先;先前) originally; at first ③(理所当然) it goes without saying; of course: ～就该这样办。Of course it should be handled that way.

【本来面目】true colors; true features: 恢复历史的～ restore historical truth

【本垒】[棒、垒球] home base

【本利】①(资本和利润) investment and profit; capital and profit ②(本金和利息) principal and interest

【本领】skill; ability; capability: 掌握为人民服务的～ master the skills needed for serving the people / 组织生产的～ ability to organize production

【本末】①(事情从头至尾的经过) the whole course of an event from beginning to end; ins and outs: 详述～ tell the whole story from beginning to end ②(底部和顶部) the fundamental and the incidental

【本末倒置】take the branch for the root; put the incidental before the fundamental; put the cart before the horse

【本能】instinct: 好奇～ instinct of curiosity / 阶级～ class instinct / 模仿～ instinct of imitation ◇ ～行为 instinctive behavior

【本年度】this year; the current year: ～国家预算 the national budget for this fiscal year / ～计划 this year's plan / ～收入 current annual revenue / ～盈余 surplus for this year / ～支出 current annual expenditure

【本票】cashier's cheque

【本钱】capital

【本人】I; me; myself; oneself; in person: 必须你～来。You must come in person.

【本色】①(本来面貌) true qualities; inherent qualities; distinctive character: 劳动人民的～ the true qualities of the laboring people ②(原来的颜色) natural color ◇ ～布 grey (cloth) / ～清漆 flat-finish varnish

【本身】itself; in itself: 运动～就是矛盾。 Motion itself is a contradiction.

【本生灯】Bunsen burner

【本事】①(故事情节) source material; original story ② (本领) skill; ability; capability

【本题】the subject under discussion; the point at issue: 请不要离开～。 Please keep to the point.

【本体】[哲] noumenon; thing-in-itself ◇ ～感觉 proprioception / ～论 ontology

【本土】①(乡土) one's native country 〈land〉 ②(指殖民国家本土) metropolitan territory

【本位】①(货币本位) standard: 金～ gold standard ②(本单位) one's own department or unit: ～工作 the work of one's own department; one's own job; one's own work ◇ ～号[乐] natural

【本位主义】selfish departmentalism; departmental selfishness

【本文】①(所指的这篇文章) this text, article, etc. ②(原文) the main body of a book

【本息】principal and interest

【本乡本土】one's native land; one's hometown

【本性】natural instincts; natural character; nature; inherent quality: ～难移。 It is difficult to alter one's character. (或: The leopard can't change his spots.)

【本义】[语] original meaning; literal sense: 这个词不能按～去理解。 This word should not be taken in its literal sense.

【本意】original idea; real intention: 我～不想去。 Originally I didn't want to go.

【本影】[物] umbra

【本源】origin; source

【本着】in line with; in conformity with; in the light of: ～我们一贯的立场 in line with our consistent stand / ～为人民服务的精神 motivated by a desire to serve the people

【本职】one's job; one's duty: 做好～的工作 do one's own job well

【本质】essence; nature; innate character; intrinsic quality: ～差别 an essential distinction / ～方面 an essential aspect / 非～方面 a non-essential aspect / ～和现象 the essence and the outward appearance / 透过现象看～ see through the appearance to the essence

【本子】book; notebook: 笔记～ notebook / 改～ go over students' written exercises; correct papers

【本族语】native language; mother tongue

苯 [化] benzene; benzol: ～中毒 benzene poisoning; benzolism

【苯胺】[化] aniline ◇ ～革 aniline leather / ～燃料 aniline dyes / ～印刷 aniline printing; flexography / ～紫 mauve

【苯巴比妥】[药] phenobarbital; phenobarbitone; luminal

【苯酚】[化] phenol

【苯海拉明】[药] diphenhydramine; benadryl

【苯甲酸】[化] benzoic acid

【苯妥英纳】[药] phenytoin sodium; dilantin

【苯乙烯】[化] styrene: 聚～ polystyrene

畚 [方] scoop up with a dustpan

【畚箕】[方] a bamboo or wicker scoop; dustpan

bèn

奔 ①(走向目的地) go straight towards; head for: 直～实验室 head straight for the laboratory ②(年纪接近) approach; be getting on for: 他是～六十的人了。 He's getting on for sixty.

【奔命】[口] be in a desperate hurry

【奔头儿】sth. to strive for; prospect: 大有～ have a bright prospect

笨 ①(智力差) stupid; dull; foolish: ～人 a stupid person; fool / 脑子～ stupid; slow-witted ②(不灵巧) clumsy; awkward: 他这人～手～脚。 He is clumsy. (或: His fingers are all thumbs.) ③(笨重) cumbersome; awkward; unwieldy: 这把锄头太～。 This is an awkward hoe.

【笨蛋】[骂] fool; idiot

【笨口拙舌】awkward in speech

【笨鸟先飞】clumsy birds have to start flying early; the slow need to start early

【笨重】heavy; cumbersome; unwieldy: ～的家具 heavy furniture; cumbersome furniture / ～的体力劳动 heavy manual labor

【笨拙】clumsy; awkward; stupid: ～的伎俩 stupid tricks / 动作～ clumsy in movement; awkward in movement

bēng

崩 ①(倒塌) collapse: 山～ landslide; landslip ②(破裂) burst: 把气球吹～了 burst a balloon / 他们谈～了。 Their negotiations broke down. ③(崩裂的东西击中) be hit by sth. bursting: 爆竹～了他的手。 The firecracker went off in his hand. ④[口](枪毙) execute by shooting; shoot

【崩溃】collapse; crumble; fall apart: 旧世界正在～。 The old world is crumbling. / 殖民主义的～ the collapse of colonialism

【崩裂】burst apart; break apart; crack: 炸药轰隆一声,山石～。 Boom! The dynamite sent the rocks flying.

【崩漏】[中医] uterine bleeding

【崩落】[地]avalanche

【崩塌】collapse; crumble: 山洞～了。 The cave fell in.

【崩陷】fall in; cave in

嘣 [象] 爆竹～地一响。 The firecracker went bang. / 我心里～～直跳。 My heart is thumping.

绷 ①(拉紧) stretch tight; draw tight: 弓弦～得很紧。 The bowstring is stretched taut. / 在绷子上～块绸子 stretch a piece of silk on an embroidery frame ②(弹起) spring; bounce: 盒子一打开,弹簧就～出来了。 When the box was opened, the spring jumped out. ③(长针疏缝) baste; tack; pin

【绷床】[杂技] trampoline

【绷带】bandage

【绷簧】[方] spring

【绷绣制品】tambor work

【绷子】embroidery frame; tambour; hoop

【绷子床】bed of framed matting

béng

甭 [方] don't; needn't: ～客气。 Don't stand on ceremony. (或: Make yourself at home.) / ～再说了。 Don't say any more. / 他～提有多高兴了。 Needless to say, he was extremely happy.

běng

绷 [口] ①(板着)～着脸 look displeased; pull a long face ②(勉强支撑) strain oneself: 咬住牙～住劲 clench one's teeth and strain one's muscles

bèng

迸 spout; spurt; burst forth: 火星乱～ sparks flying in all directions / 血从伤口～出 blood gushing from the wound

【迸发】burst forth; burst out: 大厅里～出一阵笑声。 There was an outburst of laughter in the hall. / 热烈的掌声,有如春雷～。 Applause broke out like spring thunder.

【迸裂】split; burst (open): 脑浆～ have one's brains dashed out

泵 pump: 高扬程～ high lift pump / 离心～ centrifugal pump / 水～ water pump; pump

【泵抽水量】discharge of pump

【泵房】pump house

【泵机组】pump assembly

【泵排量】pumpage; pump delivery

绷 ①(裂开) split open; crack: 玻璃～了一条缝儿。 The glass has a crack in it. ②[口] very: ～脆 very crisp / ～硬 hard as a rock; stiff as a board

蹦 leap; jump; spring: 他使劲一～就过了沟。 With one powerful leap he crossed the ditch.

【蹦蹦跳跳】bouncing and vivacious

bī

逼 ①(逼迫) force; compel; drive: ～使对方采取守势 force one's opponent onto the defensive ②(强迫索取) press for; extort: ～租 press for payment of rent ③(逼近) press on towards; press up to; close in on: 直～城下 press up to the city wall

【逼宫】(of ministers, etc.) force the king or emperor to abdicate

【逼供】extort a confession

【逼供信】obtain confessions by compulsion and give them credence

【逼近】press on towards; close in on; approach; draw near: ～敌主力 close in on the main force of the enemy / 我军已～运河。 Our troops were pressing on towards the canal.

【逼迫】force; compel; coerce

【逼人】pressing; threatening: ～太甚 drive sb. too hard; put too much pressure on sb. / 寒气～。 There is a cold nip in the air. / 时间～。 Time presses. / 形势～。 The situation spurs us on.

【逼上梁山】be driven to join the Liangshan Mountain rebels; be driven to revolt; be forced to do sth. desperate

【逼视】look at from close-up; watch intently: 光采夺目,不可～ shine with dazzling brilliance

【逼死】hound sb. to death

【逼肖】[书] bear a close resemblance to; be the very image of
【逼债】press for payment of debts; dun
【逼真】①(极像真的) lifelike; true to life: 这幅湘竹画得十分～。This painting of mottled bamboos is really true to life. ②(真切) distinctly; clearly: ～的描述 vivid description / 听得～ hear distinctly ◇ ～度[电子] fidelity
【逼租】press for rents; press sb. to pay rents

bí

荸
【荸荠】water chestnut

鼻 nose
【鼻出血】nosebleed
【鼻窦】[生理] paranasal sinus ◇ ～炎 nasosinusitis
【鼻化元音】nasalized vowel; orinasal
【鼻尖】nose tip
【鼻镜】[医] rhinoscope: 电光～ nasoscope
【鼻孔】nostril
【鼻梁】bridge of the nose
【鼻毛】vibrissa
【鼻粘膜】schneiderian membrance; tunica mucosa nasi
【鼻血】[医] nosebleed; epistaxis
【鼻腔】[生理] nasal cavity
【鼻青脸肿】a bloody nose and a swollen face; badly battered: 打得～ have one's face bashed in; be beaten black and blue
【鼻儿】[方] ①(小孔) a hole in an implement, utensil, etc., for sth. to be inserted into; eye: 门～ bolt staple / 针～ the eye of a needle ②[口](哨声;笛声) whistle: 火车拉～了。The engine is whistling.
【鼻塞】have a stuffy nose
【鼻饲法】[医] nasal feeding
【鼻涕】nasal mucus; snivel: 流～ have a running nose
【鼻息】breath: 听见均匀的～声 hear sb.'s regular and even breathing / 仰人～ be slavishly dependent on others
【鼻咽癌】[医] nasopharyngeal darcinoma
【鼻烟】snuff ◇ ～盒 snuffbox / ～壶 snuff bottle
【鼻炎】[医] rhinitis
【鼻翼】wing of nose
【鼻音】[语] nasal sound: 说话带～ speak with a twang ◇ ～化 nasalization
【鼻渊】[中医] nasosinusitis
【鼻韵母】[语] (of Chinese pronunciation) a vowel followed by a nasal consonant
【鼻针疗法】[中医] nose-acupuncture therapy
【鼻中隔】[生理] nasal septum
【鼻子】nose: 高～ high-bridged nose; high nose / 塌～ snub nose; pug nose / 鹰钩～ aquiline nose; Roman nose / 抠～ pick one's nose / 牵着～走 lead by the nose
【鼻祖】the earliest ancestor; originator (of a tradition, school of thought, etc.)

bǐ

鄙 ①(粗俗;低下) low; mean; vulgar: 卑～ mean; despicable / 粗～ coarse; vulgar ②[谦] my: ～意 my humble opinion; my idea ③[书](轻视;看不起) despise; disdain; scorn: 可～ despicable
【鄙薄】despise; scorn: ～技术工作 despise technical work
【鄙陋】superficial; shallow: ～无知 shallow and ignorant
【鄙弃】disdain; loathe: ～这种庸俗作风 disdain such vulgar practices
【鄙人】[谦] your humble servant; I
【鄙视】despise; disdain; look down upon: ～某人 despise sb.; hold sb. in contempt / ～体力劳动 despise manual labor
【鄙俗】vulgar; philistine

笔 ①(写字、画图的用具) pen: 粉～ chalk / 钢～ fountain pen / 毛～ (Chinese) writing brush / 铅～ pencil / 铱金～ fountain pen with iridium nib / 圆珠～ ball-point pen / 蘸水～ pen / 自来水～ fountain pen / 下～ set pen to pater; put pen to paper ②(笔法) technique of writing, calligraphy or drawing: 文～ style of writing ③(用笔写出) write: ～之于书 put down in black and white / 代～ write sth. for sb. ④(笔画) stroke; touch ⑤[量] 一～钱 a sum of money; a fund / 写得一～好字 write a good hand / 我们有三～帐要算。We have three scores to settle.
【笔底下】ability to write: ～不错 write well / ～来得快 write with ease; write with facility
【笔调】(of writing) tone; style: 讽刺的～ a satirical tone / 轻松的～ a light touch / 通俗的～ a popular style / 优美的～ an elegant style
【笔法】technique of writing, calligraphy or drawing: ～工整 neat and orderly strokes
【笔锋】①(毛笔的尖端) the tip of a writing brush ②(笔势;文章的锋芒) vigor of style in

writing; stroke; touch: ～犀利 write in an incisive style; wield a pointed pen
【笔杆】①(笔的手拿部分) the shaft of a pen or writing brush; pen-holder ②(指笔) pen: 耍～ wield the pen
【笔杆子】an effective writer: 他是我们组的～。He is one of the most effective writers of our group.
【笔画】strokes of a Chinese character: "王"字的～有四个。The character 王 has four strokes. ◯点 dot／横 horizontal／竖 vertical／撇 left-falling stroke／捺 right-falling stroke／折 turning stroke／钩 hook stroke
【笔迹】a person's handwriting; hand: 对～ identify sb.'s handwriting ◇～鉴定 handwriting verification／～鉴定人 handexpert／～证据 proof of handwriting
【笔记】①(用笔做记录) take down (in writing) ②(做的记录) notes: 对～ check notes／记～ take notes ③(著作的体裁) a type of literature consisting mainly of short sketches ◇～本 notebook／～小说 literary sketches; sketchbook
【笔架】pen rack; penholder
【笔尖】nib; pen point; the tip of a writing brush or pencil
【笔力】vigor of strokes in calligraphy or drawing; vigor of style in literary composition: ～雄健 powerful strokes
【笔录】①(用笔记录) put down (in writing); take down ②(记录的文字) notes; record
【笔录供词】deposition
【笔帽】the cap of a pen; the cap of pencil
【笔名】pen name; pseudonym: 用～写作 write under a pen name
【笔墨】pen and ink; words; writing: 我们激动的心情难以用～来形容。Words can hardly describe how excited we were. ◇～官司 written polemics; written controversy; a battle of words／～生涯 literary career; writer's life
【笔石】[古生物] graptolite
【笔试】written examination
【笔顺】order of strokes observed in calligraphy
【笔算】do a sum in writing; written calculation
【笔谈】conversation by writing
【笔挺】①(很直地) (standing) very straight; straight as a ramrod; bolt upright ②(烫得很平) well-ironed; trim: 穿着一身～的制服 be dressed in a trim uniform
【笔筒】pen container; brush pot
【笔头】①(笔尖) nib; pen point ②(写作技巧和能力) ability to write; writing skill ③(书面的) written; in written form: ～练习 written exercises
【笔误】a slip of the pen
【笔下】the wording and purport of what one writes
【笔心】①(铅笔心) pencil lead ②(圆珠笔心) refill (for a ball-point pen)
【笔译】written translation
【笔战】written polemics
【笔者】the author; the writer
【笔直】perfectly straight; straight as a ramrod; bolt upright: ～的马路 straight avenues／～的树 upright tree／～走 go straight on; go straight ahead／身子挺得～ stand straight as a ramrod; draw oneself up to one's full height

匕

【匕首】dagger

比

①(比较) compare; contrast: ～得上 can compare with; compare favorably with ②(较量) emulate; compete; match: ～力气 compare strength／咱俩～～谁先跑到。Let's have a race and see who gets there first. ③(比拟) draw an analogy; liken to; compare to: 把工作～作战斗 liken one's work to a battle ④(比画) gesture; gesticulate: 连说带～ gesticulate as one talks ⑤[介] 人民的生活一年～一年好。The life of the people is getting better and better each year. ⑥(比率) ratio; proportion: 反～ inverse ratio; inverse proportion／约为一与三之～ a ratio of about one to three ⑦(得分对比) to (in a score): 甲队以二～一胜乙队。Team A beat team B (by a score of) two to one.／现在是几～几? What's the score?
【比比皆是】can be found everywhere
【比方】analogy; instance: 打～ draw an analogy／拿盖房子作～ take for instance the building of a house／这不过是个～。This is only by way of analogy.
【比分】[体] score: 拉开～ widen the lead／场上～是三比二。The score is 3 to 2.／双方～十分接近。. It's a close game.
【比高低】compare so as to see who is superior
【比画】gesture; gesticulate: 他～着讲。He made himself understood with the help gestures.
【比价】price relations; parity; rate of exchange: 工农业产品～ the price parities between indus-

trial and agricultural products / 粮棉～ parity between grain and cotton / 英镑和美元的～ the rate of exchange between the pound sterling and the U.S. dollar

【比较】①(对比) compare; contrast: 把译文和原文～一下 check the translation against the original ②[介] ～去年有显著的增长 show a marked increase over last year ③[副] fairly; comparatively; relatively; quite; rather: ～大 relatively big / ～简单 fairly simple / ～全面地处理问题 deal with problems in a relatively comprehensive way / 这里条件～艰苦。Conditions are rather tough here. ◇ ～级[语] comparative degree

【比较法】comparative law: ～的研究 study of comparative law

【比较法学】comparative jurisprudence

【比较利益】[经] comparative advantage

【比较损益表】comparative profit and loss statement

【比利时】Belgium ◇ ～人 Belgian

【比例】① proportion: 反～ inverse proportion / 正～ direct proportion / 按～发展 develop in proportion; proportionate development / 不合～ out of proportion ②(比较倍数关系) scale: 按～绘制 be drawn to scale

【比例尺】①[测] scale: 这张地图的～是四十六万分之一。The scale of the map is 1:460000. ②(绘图用尺) architect's scale; engineer's scale

【比例代表制】proportional representation

【比例式】proportion expression

【比例税】proportional tax

【比例税率】flat rate

【比例图】scale map

【比量】take rough measurements (with the hand, a stick, string, etc.)

【比邻】①(近邻) neighbor; next-door neighbor ②(接近) near; next to: 跟车站～的那个工厂 the factory next to the railway station

【比率】ratio; rate ◇ ～分析 ratio analysis

【比目鱼】flatfish; flounder

【比拟】①(比较) compare; draw a parallel; match: 荒唐的～ an absurd parallel / 无可～ beyond compare; incomparable; matchless ②(修辞手法) analogy; metaphor; comparison: 这种～是不恰当的。It is inappropriate to draw such a parallel.

【比丘】Bhiksu; Buddhist monk

【比丘尼】Bhiksuni; Buddhist nun

【比热】[物] specific heat

【比容】[物] specific volume

【比如】for example; for instance; such as

【比赛】match; competition: 长跑～ long-distance race / 国际～ international / 射击～ shooting contest / 网球～ tennis tournament / 象棋～ chess tournament / 自行车～ bicycle race / 足球～ football match / 参加～ take part in the competition ◇ ～地点 venue / ～规则 rules of the game; rules of a contest / ～项目 event

【比色分析】[化] colorimetric analysis

【比色计】[化] colorimeter

【比上不足,比下有余】fall short of the best but be better than the worst; can pass muster

【比湿】[气] specific humidity

【比试】①(较量高低) have a competition: 不信咱俩～～。If you don't believe me, let's have a competition and see. ②(做出某种动作的姿势) measure with one's hand or arm; make a gesture of measuring

【比武】[军] demonstration of and competition in military stills

【比翼】fly wing to wing: ～双飞 pair off wing to wing; fly side by side ◇ ～鸟 a pair of lovebirds; a devoted couple

【比喻】metaphor; analogy; figure of speech: 这只是一个～说法。This is just a figure of speech.

【比照】①(按照) according to; in the light of: 我们可以～其它工厂的做法拟定计划。We can draw up our plan in the light of the experience of other factories. ②(对比) contrast: 两相～ contrasting the two

【比值】specific value; ratio

【比重】①(比值) proportion: 工业在整个国民经济中的～ the proportion of industry in the national economy as a whole ②[物] specific gravity ◇ ～计 [物] hydrometer / ～选种[农] specific gravity selection (of seeds)

吡

【吡啶】[化] pyridine

【吡拉明】[化] pyrilamine; mepyramine

【吡咯】[化] pyrrole

【吡唑】[化] pyrazole

秕 (of grain) not plump; blighted

【秕糠】chaff; worthless stuff

【秕子】blighted grain

妣 [书] deceased mother: 先～ my deceased mother

彼 ①(那;那个,跟'此'相对) that; those; the other; another: ～时 at that time / 由此及～ proceed from one to the other ②(对方) the other party: 要知己知～。You must know both your opponent and yourself.

【彼岸】[佛教] the other shore; Faramita

【彼此】①(双方) each other; one another: ～呼应 support each other; act in coordination with each other / 消除～间的误会 clear up misunderstandings between each other ②[套] 您辛苦啦! —～～! You must have taken a lot of trouble about it. — So must you!

【彼一时,此一时】 that was one situation, and this is another; times have changed

bì

闭 ①(关;合) shut; close: ～上眼 close one's eyes / ～嘴! Hold your tongue! (或: Shut up!) ②(堵塞不通) stop up; obstruct: ～住气 hold one's breath

【闭关锁国】 close the communications at the borders

【闭关政策】 closed-door policy

【闭关自守】 close the country to international intercourse

【闭合】 close ◇ ～电路 closed circuit / ～度 closure / ～生态 [宇航] closed ecology

【闭会】 close a meeting; end a meeting; adjourn a meeting: 委员会～期间 when the committee is not in session

【闭经】[医] amenorrhoea

【闭口不谈】 refuse to say anything about; avoid mentioning

【闭口无言】 remain silent; be tongue-tied; be left speechless

【闭路电视】 closed-circuit television

【闭门羹】吃～ be denied entrance; find the door slammed in one's face / 飨以～ slam the door in sb.'s face

【闭门思过】 shut oneself up and ponder over one's mistakes

【闭门造车】 make a cart behind closed doors; work behind closed doors; divorce oneself from the masses and from reality and act blindly

【闭目塞听】 shut one's eyes and stop up one's ears; be out of touch with reality

【闭幕】①(演出结束) the curtain falls; lower the curtain: 在观众热烈掌声中～。The curtain fell to the loud applause of the audience. ②(会议结束) close; conclude ◇ ～词 closing address; closing speech / ～式 closing ceremony

【闭塞】①(堵塞) stop up; close up: 鼻孔～ with one's nose stuffed up ②(交通不便) hard to get to; out-of-the-way; inaccessible: 以前这一带交通～。In the past this district was very hard to get to. ③(消息不灵通) unenlightened: 耳目～ uninformed; ignorant / 消息～ ill-informed ④[电] blocking ◇ ～信号[铁道] block signal

【闭塞眼睛捉麻雀】 try to catch sparrows with one's eyes blindfolded; act blindly

【闭音节】[语] closed syllable

【闭元音】[语] close vowel

【闭月羞花】 outshine the moon and put the flowers to shame

敝 ①[书](破烂) shabby; worn-out; ragged: ～衣 ragged clothing; shabby clothes; worn-out clothes ②[谦](与自己有关的事物) my; out; this: ～处 my place / ～校 my school / ～姓陈。My name is Chen.

【敝屣】[书] worn-out shoes; a worthless thing: 弃之如～ cast away like a pair of worn-out shoes

【敝帚自珍】 value one's own old broom; cherish sth. of little value simply because it is one's own

蔽 cover; shelter; hide: ～风雨 shelter from the wind and rain / 浮云～日 with the sun screened off by floating clouds / 衣不～体 be dressed in rags; have nothing but rags on one's back

弊 ①(蒙骗;占便宜) fraud; abuse; malpractice: 舞～ practice fraud; engage in corrupt practices / 作～ cheat; be dishonest ②(害处;毛病) disadvantage; harm: 有利有～。There are both advantages and disadvantages.

【弊病】①(弊端) malady; evil; malpractice: 社会～ social evils ②(毛病) drawback; disadvantage: 这种做法～不少。This method has quite a few drawbacks 〈disadvantages〉.

【弊端】 malpractice; abuse; corrupt practice

【弊多利少】 the disadvantages outweigh the advantages

币 money; currency: 外～ foreign currency / 银～ silver coin

【币值】currency value: ～稳定 be a stable currency.
【币值调整因素】currency adjustment factor
【币制】currency system; monetary system ◇ ～改革 currency reform; monetary reform

必 ①(必定;必然) certainly; necessarily; surely: 骄兵～败。 An army puffed up with pride is bound to lose. / 他～定来。 He will certainly come. ②(必须) must; have to: ～读书目 a list of required reading
【必备条件】essential condition
【必不得已】have no choice but to; be under the necessity of
【必不可少】indispensable; absolutely necessary; essential
【必得】must; have to: 你～去一趟。 You simply must go.
【必定】be bound to; be sure to: 反动派～失败。 Reactionaries are bound to fail. / 那是～会发生的。 It will happen, as sure as a gun.
【必恭必敬】reverent and respectful; extremely deferential
【必然】①(确定不移) inevitable; certain: ～结果 inevitable outcome / ～联系 positive connection / ～趋势 inexorable trend ②[哲] necessity ◇ ～规律 inexorable law / ～王国 realm of necessity / ～性 necessity; inevitability; certainty
【必然因果关系】positive causal relationship
【必修课】a required course; a obligatory course
【必须】must; have to: ～指出 it must be pointed out that / 共产党员～勇于批评和自我批评。 A Communist must be bold in criticism and self-criticism. / 我们～遵守革命纪律。 We must abide by revolutionary discipline.
【必需】essential; indispensable: 发展工业所～的原料 raw materials essential for industrial development / 用在最～的地方 be used where it is most needed
【必需品】necessities; necessaries: 生活～ daily necessities
【必要】necessary; essential; indispensable: ～的损失赔偿 necessary damages / 国家为这个农场提供了～的资金。 The state provided the requisite capital for this farm. ◇ ～产品[经] necessary product / ～劳动[经] necessary labor / ～前提 prerequisite; precondition / ～条件 essential condition; prerequisite / ～性 necessity
【必由之路】the road one must follow or take; the only way

铋 [化] bismuth (Bi)

秘
【秘鲁】Peru ◇ ～人 Peruvian

碧 ①[书](青玉) green jade ②(青绿色) bluish green; blue: ～波 blue waves / ～海 the blue sea / ～空 a clear blue sky; an azure sky / ～草如茵 a carpet of green grass
【碧蓝】dark blue
【碧绿】dark green
【碧瓦】green, glazed tile
【碧血】blood shed in a just cause: ～丹心 righteous blood and loyal heart
【碧玉】jasper

庇 shelter; protect; shield
【庇护】shelter; shield; put under one's protection; take under one's wing ◇ ～权 right of asylum / ～所 sanctuary; asylum
【庇荫】①(遮住阳光) give shade ②(比喻包庇) shield

毕 ①(完结;完成) finish; accomplish; conclude: 阅～请放回原处。 Please replace after reading. ②[书](完全) fully; altogether; completely: 原形～露 show one's true colors
【毕恭毕敬】reverent and respectful; extremely deferential
【毕竟】[副] after all; all in all; when all is said and done; in the final analysis: 她的缺点同她的成绩相比,～是第二位的。 Compared with her achievements, her shortcomings are, after all, only secondary.
【毕其功于一役】accomplish the whole task at one stroke
【毕生】all one's life; lifetime: ～精力 energy throughout one's life / ～事业 lifework; work of a lifetime
【毕肖】[书] resemble closely; be the very image of: 画得神情～ paint a lifelike portrait of sb.
【毕业】graduate; finish school ◇ ～班 graduating class / ～典礼 graduation (ceremony); commencement / ～分配 job assignment on graduation / ～鉴定 graduation appraisal / ～论文 graduation thesis; graduation dissertation / ～设计 graduation project / ～生 graduate / ～实习 graduation field work / ～证书 diploma; graduation certificate

筚 a bamboo or wicker fence: 蓬门～户 a house with a door of wicker and straw; a humble abode

【筚路蓝缕】[书] drive a cart in ragged clothes to blaze a new trail; endure great hardships in pioneer work

毙 ①(死) die; get killed: 倒～ drop dead ②(枪毙) kill or execute by shooting; shoot

【毙命】meet violent death; get killed: 两名匪徒当场～。Two of the bandits were killed on the spot.

陛 [书] a flight of steps leading to a palace hall

【陛下】(直接称呼) Your Majesty; (间接称呼) His ⟨Her⟩ Majesty

裨 [书] benefit; advantage: 无～于事。It won't help matters. (或: It won't do any good.)

【裨益】[书] benefit; advantage; profit: 不无～ not without benefits / 大有～ be of great benefit / 毫无～ of no avail

婢 slave girl; servant-girl

【婢女】slave girl; servant-girl

蓖

【蓖麻】[植] castor-oil plant ◇ ～蚕 castor silkworm / ～油 castor oil / ～子 castor bean

辟 ①[书] (君主) monarch; sovereign: 复～ restore a monarchy; restoration ②[书] (排除) ward off; keep away

【辟邪】exorcise evil spirits

避 ①(躲开;回避) avoid; evade; shun: ～而不答 avoid making a reply / ～而不见 avoid meeting sb.; evade a meeting with sb. / ～而不谈 evade the question; avoid the subject; keep silent about the matter ②(防止) prevent; keep away; repel

【避车台】[铁路] refuge platform

【避弹坑】[军] foxhole

【避风】① take shelter from the wind ②(避开风头) lie low; stay away from trouble ◇ ～港 haven; harbor

【避讳】taboo on using the personal names of emperors, one's elders, etc.

【避讳】①(忌讳) a word or phrase to be avoided as taboo; taboo ②(回避) evade; dodge: ～这个问题 evade the issue

【避开】avoid; evade; keep away from: ～危险地带 avoid a dangerous zone

【避坑落井】dodge a pit only to fall into a well; out of the frying pan into the fire

【避雷器】lightning arrester

【避雷针】lightning rod

【避免】avoid; refrain from; avert: ～不必要的牺牲 avoid unnecessary sacrifice / ～错误 avoid mistakes / ～轻率行动 refrain from any rash action / 设法～了一场事故 succeed in averting an accident

【避难】take refuge; seek asylum ◇ ～港 port of refuge / ～国 country of refuge / ～权 the right of asylum / ～所 refuge; sanctuary; asylum; haven

【避人耳目】avoid being noticed; elude observation

【避其锐气,击其惰归】avoid the enemy when he is fresh and strike him when he is tired and withdraws

【避强打弱】[军] evade the strong and attack the week

【避实就虚】stay clear of the enemy's main force and strike at his weak point

【避世】retire from the world; live a sequestered life

【避暑】①(到凉爽地方休息) be away for the summer holidays; spend a holiday at a summer resort ②(避免中暑) prevent sunstroke ◇ ～胜地 summer resort / ～药 medicine for preventing sunstroke; preventive against sunstroke

【避蚊剂】mosquito repellent

【避嫌】avoid doing anything that may arouse suspicion; avoid arousing suspicion

【避雨】take shelter from rain

【避孕】contraception: 器具～ instrumental contraception / 药物～ medical contraception ◇ ～措施 contraceptive; contraception measures / ～工具 contraceptive device / ～环 intrauterine contraceptive ring / ～栓 contraceptive suppository / ～套 condom / ～丸药 the pill / ～药膏 contraceptive jelly / ～用品 contraceptives

【避重就轻】avoid the important and dwell on the trivial; keep silent about major charges while admitting minor ones

璧 *bi*, a round flat piece of jade with a hole in its center

【璧还】[敬] ①(用于归还原物) return (a borrowed object) with thanks ②(辞谢赠品) decline

(a gift) with thanks
【璧谢】[敬] decline (a gift) with thanks

壁 ①(墙) wall ②(围绕物体的壁) sth. resembling a wall: 细胞～ cell wall ③(峭壁) cliff: 峭～ a precipitous cliff; precipice ④(壁垒) rampart; breastwork: 作～上观 watch the fighting from the ramparts; be an onlooker
【壁报】wall newspaper
【壁橱】a built-in wardrobe or cupboard; closet
【壁灯】wall lamp; bracket light
【壁虎】[动] gecko; house lizard
【壁画】mural (painting); fresco: 敦煌～ the Dunhuang frescoes
【壁垒】rampart; barrier: 关税～ tariff wall / 贸易～ trade barrier / 哲学中的两大～ two rival camps in philosophy; two diametrically opposed philosophical theories
【壁垒分明】be diametrically opposed; be sharply divided
【壁垒森严】closely guarded; strongly fortified
【壁立】(of cliffs, etc.) stand like a wall; rise steeply: ～的山峰 a sheer cliff / ～千尺 a sheer rise of a thousand feet
【壁炉】fireplace ◇～台 mantelpiece
【壁虱】[动] ① tick ②[方] bedbug
【壁毯】tapestry (used as a wall hanging)
【壁纸】[建] wallpaper

臂 ① arm: 右～ the right arm / 助一～之力 give sb. a hand ②(指上臂) upper arm
【臂膀】arm
【臂纱】(black) armband: 戴～ wear a black armband
【臂章】① armband; armlet ②[军] shoulder emblem; shoulder patch

biān

蝙
【蝙蝠】bat

鳊
【鳊鱼】bream

编 ①(编织) weave; plait: ～辫子 plait one's hair / ～草席 braid a mat / ～柳条筐 weave wicker baskets ②(组织;排列) organize; group; arrange: ～班 group into classes / 把他～在我们班吧。 Put him in our class. ③(编辑) edit; compile: ～词典 compile a dictionary / ～教材 compile teaching material / ～杂志 edit a magazine ④(创作) write; compose: ～儿童歌曲 compose songs for children / ～剧本 write a play ⑤(编造;瞎编) fabricate; invent; make up; cook up: 这事儿是他～出来的。 He made the whole thing up. ⑥(书按内容划分的单位) part of a book; book; volume: 上～ Book I; Volume I; Part I / 下～ Book II; Volume II; Part II
【编次】order of arrangement
【编导】①(编写和导演戏剧) write and direct (a play, film, etc.) ②(编剧和导演的人) (戏剧) playwright-director; (舞剧) choreographer-director; (电影) scenarist-director
【编队】① form into columns; organize into teams ②[军] formation (of ships or aircraft) ◇～飞行 formation flight; formation flying / ～轰炸 formation bombing
【编号】①(按顺序编号数) number: 给运动员～ give numbers to the athletes 〈players〉 ②(编定的号数) serial number
【编辑】①(加工整理) edit; compile: ～图书索引 compile an index (of books) ②(做编辑工作的人) editor; compiler: 总～ editor-in-chief ◇～部 editorial department / ～人员 editorial staff / ～委员会 editorial board
【编剧】①(创作剧本) write a play, scenario, etc. ②(编剧的人)(戏剧) playwright; (电影) screenwriter; scenarist
【编码】coding ◇～程序 coded program / ～方案 code scheme / ～规则 code rule
【编目】①(编制目录) make a catalogue; catalogue: 新到的图书正在～。 The new books are being catalogued. ②(编制成的目录) catalogue; list ◇～部 cataloguing department / ～员 cataloguer
【编年史】annals; chronicle
【编年体】annalistic style (in historiography)
【编排】arrange; lay out: 文字和图片的～ the layout of picture and articles / 杂志的～ the layout of a magazine
【编入】enroll; put in; classify: 他被～飞行班。 He was enrolled in the flying class.
【编审】①(编辑和审定) read and edit ② (做编审工作的人) copy editor
【编外人员】non-permanent staff; extra-organizational personnel
【编写】①(整理写成) compile: ～教科书 compile a textbook ②(创作) write; compose: ～歌剧 compose an opera / ～剧本 write a play
【编选】select and edit; compile
【编译】translate and edit; translate and compile: ～人员 translator and compiler

【编印】compile and print; publish
【编造】①(将资料组织排列起来) compile; draw up; work out: ～表册 compile statistical tables / ～预算 draw up a budget ②(捏造) fabricate; invent; concoct; make up; cook up: ～谎言 fabricate lies / ～情节 falsify the details of an event; invent a story; make up a story ③(凭想象创造) create out of the imagination
【编者】editor; compiler ◇ ～按 editor's note; editorial note
【编织】weave; knit; plait; braid: ～草席 weave a straw mat / ～地毯 weave a rug / ～发网 crochet a hair-net / ～毛衣 knit a sweater
【编制】①(指手工艺) weave; plait; braid: ～工艺 basketry / ～竹器 weave bamboo articles ②(根据资料做出) work out; draw up: ～财务报告 draw up financial report; draft financial report / ～教学大纲 draw up a teaching program / ～生产计划 work out a production plan③ (组织设置及人员定额) authorized strength; establishment: 部队～ establishment (for army units) / 缩小～ reduce the staff / 政府机关的～ authorized size of a government body
【编钟】serials bells
【编著】compile; write
【编组】① organize into groups ②[铁道] marshalling ◇ ～场 marshalling yard; classification yard
【编纂】compile: ～词典 compile a dictionary

鞭 ①(鞭子) whip; lash ②(古代兵器) an iron staff used as a weapon in ancient China ③(教鞭) (teacher's) pointer ④(小爆竹) a string of small firecrackers ⑤[书](鞭打) flog; whip; lash: ～马 whip a horse
【鞭策】spur on; urge on: 要经常～自己努力学习。 We should constantly urge ourselves on to study hard.
【鞭长莫及】beyond the reach of one's power; too far away for one to be able to help
【鞭笞】[书] flog; lash
【鞭虫】[动] whipworn
【鞭打】 whip; lash; flog; thrash ◇ ～损伤 whiplash
【鞭痕】welt; whip scar; lash mark
【鞭毛】[动] flagellum ◇ ～虫 flagellate
【鞭炮】① firecrackers ②(成串的小爆竹) a string of small firecrackers
【鞭辟入里】penetrating; trenchant; incisive
【鞭挞】[书] lash; castigate: 影片对旧社会进行了无情的～。 The film mercilessly castigates the old society.
【鞭子】whip

砭 ①[中医] a stone needle used in acu- puncture in ancient China ②(刺;扎) pierce: 冷风～骨。 The cold wind cuts one to the marrow.

边 ①(几何图形的一边) side: 街道两～ both sides of the street / 三角形的一～ one side of a triangle ②(边缘) margin; edge; brim; rim: 衬衣的～ the hem of a shirt / 宽～草帽 a straw hat with a broad brim / 每页～上都有批注。 There are notes in the margin on every page. / 田～地头 edges of fields / 碗～儿 the rim of a bowl ③(边界) border; frontier; boundary: ～城 border town; frontier town / ～寨 borderland village / 戍～ garrison a border region④ (界线) limit; bound: 无～的大海 a boundless sea ⑤(身旁) by the side of; close by: 老师傅身～有两个好徒弟。 The veteran worker has two good apprentices working with him. ⑥(方位词后缀) 东～ in the east / 后～ at the back / 里～ inside / 那～ there / 前～ in front / 外～ outside / 这～ here / 左～ on the left
【边材】[木] sapwood; alburnum
【边陲】[书] border area; frontier
【边地】border district; borderland
【边防】frontier defence border defence ◇ ～部队 frontier guards / ～检查条例 frontier inspection regulations / ～检查站 frontier inspection station / ～军 frontier force / ～哨 border sentry / ～要塞 frontier stronghold / ～战士 frontier guard / ～站 frontier station
【边锋】[足球] wing; wing forward: 右～ right wing / 左～ left wing
【边际】limit; bound; boundary: 不着～ wide of the mark; not to the point; irrelevant / 漫无～ rambling; discursive
【边际生产率说】marginal productivity theory
【边际效用论】[经] the theory of marginal utility
【边疆】border area; borderland; frontier; frontier region: 保卫～ guard the frontier / 支援～建设 support the construction of the border areas
【边角料】leftover bits and pieces (of industrial material)
【边界】boundary; border: 标定～ demarcate boundaries / 划定～ delimit boundaries / 越过～ cross a boundary; cross the border ◇ ～

实际控制线 line of actual control on the border / ～事件 border incident / ～现状 *status quo* on the border; *status quo* of the boundary / ～线 boundary line / ～协定 boundary agreement / ～争端 boundary dispute / ～走向 alignment of the boundary line

【边境】border; frontier: 封锁～ close the frontiers; seal off the borders ◇ ～冲突 border clash; clash conflict / ～检查站 border check point; frontier inspection station / ～贸易 frontier trade / ～市镇 border town / ～事件 border incidents

【边框】frame; rim: 镜子的～ the rim of a mirror

【边料】rim charge

【边门】side door; wicket door; wicket gate

【边民】people living on the frontiers; inhabitants of a border area

【边卡】border checkpoint

【边区】border area

【边塞】frontier fortress

【边线】①[体] sideline ②[棒、垒球] foul line

【边心距】[数] apothem

【边沿】edge; fringe: 森林～ the edge of a forest; fringe of a forest

【边音】[语] lateral (sound)

【边缘】①(沿边的部分) edge; fringe; verge; brink; periphery: 经济破产的～ the verge of economic bankruptcy / 悬崖的～ the edge of a precipice / 战争～政策 brink-of-war policy; brinkmanship ②(靠近界线的) marginal; borderline ◇ ～地区 border district; borders / ～海 marginal sea / ～科学 borderline science / ～学科 borderline subject

【边远】far from the center; remote; outlying: ～地区 an outlying district / ～省份 remote border provinces

biǎn

扁 flat: ～体字 squat-shaped handwriting / 一只～盒子 a flat case; a shallow box / 别把人看～了。Don't underestimate people.

【扁虫】flatworm

【扁蝽】flat bug

【扁锉】flat file

【扁担】carrying pole; shoulder pole:

【扁豆】hyacinth bean: 小～ lentil

【扁钢】flat-rolled steel; flat steel

【扁骨】[生理] flat bone

【扁坯】[冶] slab

【扁平足】flat foot; pes planus

【扁平轧材】flat products

【扁鲨】squatinia

【扁桃】①(扁桃树) almond tree ②(扁桃果实) almond ③[方] flat peach

【扁桃体】[生理] tonsil ◇ ～切除术 tonsillectomy / ～肥大 [医] hypertrophy of tonsils / ～炎 tonsillitis

【扁圆】oblate

匾 ①(木制题字横牌) a horizontal inscribed board ②(绸布做的绣匾) a silk banner embroidered with words of praise: 绣金～ embroidering a silk banner with words of gold ③(竹编器具) a big round shallow basket

【匾额】a horizontal inscribed board

褊 [书] narrow; cramped

【褊急】[书] narrow-minded and short-tempered

【褊狭】[书] narrow; cramped: 居处～ live in cramped quarters / 气量～ small-minded

贬 ①(降低官职) demote; relegate: ～黜 dismiss ②(降价) reduce; devalue: ～价出售 sell a reduced price ③(评价低) censure; depreciate: ～得一钱不值 condemn as worthless

【贬斥】①[书] (降低官职) demote ②(贬低) denounce

【贬词】derogatory term; expression of censure

【贬低】belittle; depreciate; play down: ～其重要性 belittle the importance of sth. / 企图～这一文件的意义 try to play down the significance of the document

【贬义】derogatory sense ◇ ～词 derogatory term

【贬抑】belittle; depreciate

【贬谪】banish from the court; relegate

【贬值】[经] ①(购买力下降) devalue; devaluate ②(兑换率下降) depreciate ◇ ～通货 depreciated currency

biàn

辨 differentiate; distinguish; discriminate: 不～真伪 fail to distinguish between truth and falsehood; be unable to tell the true from the false

【辨别】differentiate; distinguish; discriminate: ～方向 take one's bearings / ～是非 differentiate between right and wrong; discriminate between right and wrong

【辨别力】ability to see things in their true light; discerning power

His handwriting is easy to identify.
【辨析】differentiate and analyse; discriminate: 同义词～ synonym discrimination
【辨正】identify and correct (errors)
【辨症论治】[中医] diagnosis and treatment based on an overall analysis of the illness and the patient's condition
辩 argue; dispute; debate: ～个水落石出 argue a matter out / 真理愈～愈明。The more truth is debated, the clearer it becomes.
【辩白】offer an explanation; plead innocence; try to defend oneself
【辩驳】dispute; refute: 无可～ beyond all dispute; indisputable; irrefutable
【辩才】eloquence: 颇有～ be quite eloquent; have a silver tongue
【辩才无碍】very eloquence
【辩护】①(说明意见或行为正确) speak in defense of; argue in favor of; defend: 不要替他～了。Don't try to defend him. ②[法] plead; defend: 出庭～ (of a lawyer) defend a case in court / 为被告人～ plead for the accused / 被告人有权获得～。The accused has the right to defense. ◇ ～律师 counsel for the defense; defense counsel / ～权 right to defense / ～人 defender; counsel / ～士 apologist / ～制度 system of advocacy
【辩解】provide an explanation; try to defend oneself: 错了就错了,不要～。A mistake is a mistake. Don't try to explain it away.
【辩论】argue; debate: 这件事不必～了. There is no need arguing about this matter. ◇ ～会 a debate / ～终结 closure of debate
【辩证】dialectical: ～的统一 dialectical unity / ～地看问题 look at things dialectically / 事物发展的～规律 the dialectical law of the development of things ◇ ～法 dialectics / ～逻辑 dialectical logic
【辩证唯物主义】dialectical materialism: ～的认识论 the dialectical materialist theory of knowledge / ～观点 a dialectical materialist point of view ◇ ～者 dialectical materialist
辫 plait; braid; pigtail: 发～ braid; plait / 梳小～儿 wear pigtails / 蒜～ a braid of garlic
【辫梢】tip of a plait
【辫子】①(小辫) plait; braid; pigtail: 梳～ wear one's hair in braids; plait the hair ②(把柄) a mistake or shortcoming that may be exploited by an opponent; handle: 揪～ seize on sb.'s mistake or shortcoming; capitalize on sb.'s vulnerable point
苄 [化] benzyl
忭 [书] glad; happy: 不胜欣～ be overjoyed
变 ①(变化;改变) change; become different: 多～的战术 varied tactics / 情况～了。The situation has changed. ②(变成为) change into; become: 旱地～水田。Dry land has been turned into paddy fields. ③(使改变) transform; change; alter: ～废为宝 change waste material into things of value; recycle waste material / ～害为利 turn bane into boon ④(有重大影响的突变) an unexpected turn of events: 兵～ mutiny / 事～ incident / 政～ coup d'état
【变本加厉】become aggravated; be further intensified: ～扩军备战 step up arms expansion and war preparations
【变成】change into; turn into; become: 在一定的条件下,坏事能够～好事。Under given conditions, bad things can be turned into good things.
【变电站】(transformer) substation
【变调】[语] ① modified tone ②(转调) tonal modification
【变动】change; alteration: 人事～ personnel changes / 文字上作一些～ make some changes in the wording; make some alterations in the wording
【变法】[史] political reform ◇ ～维新 Constitutional Reform and Modernization (1898)
【变分法】[数] calculus of variations
【变革】transform; change: ～自然 transform nature / 社会～ social change
【变格】[语] declension
【变更】change; alter; modify: ～程序 alter the procedure / ～作息时间 alter the daily timetable / 所有制方面的～ changes in the system of ownership / 我们的计划有～。We have modified our plan.
【变工】exchange work; exchange labor ◇ ～队 work-exchange team
【变故】an unforeseen event; accident; misfortune: 发生了～。Something quite unforeseen has happened.
【变卦】go back on one's word; break an agreement: 昨天说得好好的,怎么～了? Yesterday you agreed What made you change your mind?
【变化】change; vary: ～无常 constantly chang-

ing; changing all the time / 化学～ chemical change / 气温的～ variations of temperature; fluctuations of temperature / 他发球～多端。He's always changing his way of serving.

【变幻】change irregularly; fluctuate: ～莫测 changeable; unpredictable / 风云～ unexpected gathering of clouds; constant change of events

【变换】①vary; alternate: ～手法 vary one's tactics / ～位置 shift one's position ②[数] transformation

【变价】appraise at the current rate: ～出售 sell at the current price

【变焦距镜头】[摄] zoom lens

【变节】make a political recantation; turn one's coat ◇ ～分子 recanter; turncoat

【变脸】suddenly turn hostile

【变量】[数] variable

【变流器】[电] converter

【变卖】sell off (one's property)

【变频】[电子] frequency conversion ◇ ～管 converter tube / ～器 frequency converter

【变迁】changes; vicissitudes: 时世～ the changes of the times

【变色】①(改变颜色) change color; discolor: 这种墨水不会～。This ink will not change color. ②(改变脸色) change countenance; become angry: 勃然～ suddenly change countenance

【变色龙】chameleon

【变数】[数] variable

【变速】[机] speed change; gearshift ◇ ～比 [汽车] gear ratio / ～器 gearbox; transmission / ～系数 coefficient of variation in speed / ～运动 [物] variable motion

【变态】①[生] metamorphosis ② abnormal; anomalous ◇ ～反应 [医] allergy / ～心理 aberrant personality / ～心理学 abnormal psychology

【变天】①(天气发生变化) change of weather: 太闷热了,看来要～。The weather is bound to change soon, it's so close. ②(反动势力复辟) restoration of reactionary rule ◇ ～账 restoration records

【变通】be flexible; accommodate sth. to circumstances; adapt sth. to circumstances: ～办法 accommodation; adaptation

【变温动物】poikilothermal animal; coldblooded animal

【变戏法】perform conjuring tricks; conjure; juggle

【变相】in disguised form; covert: ～的剥削行为 a covert act of exploitation / ～贪污 embezzlement in disguised form / ～体罚 corporal punishment in disguised form

【变心】cease to be faithful

【变星】[天] variable (star)

【变形】be out of shape; become deformed: 病人的脊椎骨已经～。The patient has a deformed spine. / 这箱子压得～了。The box has been crushed out of shape.

【变形虫】amoeba

【变形动物】amoebula

【变形矫正】straightening

【变形体】[生] plasmodium

【变性】[化] denaturation: ～蛋白质 denatured protein / ～酒精 denatured alcohol

【变压器】[电] transformer: 电源～ mains transformer / 降压～ step-down transformer / 升压～ step-up transformer / 输出～ output transformer / 输入～ input transformer

【变异】[生] variation ◇ ～系数 coefficient of variation / ～性 variability

【变质】①(变坏) go bad; deteriorate: 蜕化～ become morally degenerate / 牛奶～了。The milk has turned sour. / 这肉～了。The meat has gone bad. ②[地] metamorphism ◇ ～岩 metamorphic rock

【变种】① [生] mutation; variety ②(实质相同的思潮或流派) variety; variant: 机会主义的～ a variety of opportunism

【变奏】[乐] variation ◇ ～曲 variations

【变阻器】[电] rheostat

遍 ①(普遍;全部) all over; everywhere: 走～全国 have travelled all over the country / 我们的朋友～天下。We have friends all over the world. ②[量] 请再说一～。Please say it again. / 这本书我看过两～。I've read the book twice.

【遍布】be found everywhere; spread all over: ～全国 can be found all over the country

【遍地】everywhere; all around: ～开花 blossom everywhere; spring up all over the place.

【遍及】extend all over; spread all over

【遍体鳞伤】covered all over with cuts and bruises; beaten black and blue; be a mass of bruises

便 ①(方便;便利) convenient; handy: 日夜服务,顾客称～. Customers find the 24-hour service very convenient. ②(方便的时候) when an opportunity arises; when it is convenient: 悉

听尊～ suit your own convenience／得～请来一趟。 Come whenever it's convenient.③(非正式的;简单的) informal; plain; ordinary: ～宴 an informal dinner／～装 ordinary clothes; everyday clothes ④(排泄屎尿) relieve oneself: 大～ shit; defecate／小～ piss; urinate ⑤(屎尿) piss or shit; urine or excrement: 粪～ excrement; night soil ⑥[副]天一亮她～下地去了。 She went to the fields as soon as it was light.

【便秘】[医] constipation

【便步走】[军] march at ease; route step: ～!(口令) At ease, march! (或: Route step, march!)

【便池】urinal

【便当】convenient; handy; easy

【便道】①(近便小路) shortcut: 抄～走 take a shortcut ②(马路边的人行道) pavement; sidewalk: 行人走～。 Pedestrians walk on the pavement. ③(临时的道) makeshift road.

【便饭】a simple meal; potluck: 跟我们一块儿吃顿～吧。 Come along and take potluck with us.

【便服】①(日常穿的) everyday clothes; informal dress ②(中式服装) civilian clothes

【便函】an informal letter sent by an organization

【便壶】(bed) urinal; chamber pot

【便笺】notepaper; memo; memo pad

【便览】brief guide: 交通～ roadbook／旅游～ guidebook

【便利】①(方便) convenient; easy: 交通～ have convenient communications; have good transport facilities; be conveniently located ②(使方便) facilitate: 为～居民,新建了一个百货商店。 A new general store has been built for the convenience of the residents.

【便帽】cap

【便门】side door; wicket door

【便溺】relieve the bowls

【便盆】bed pan

【便桥】temporary bridge; makeshift bridge

【便人】somebody who happens to be on hand for an errand: 如有～,请把那本书捎来。 Please send the book by anyone who happens to come this way.

【便士】penny: 五～ five pence

【便条】(informal) note

【便桶】chamber pot

【便鞋】cloth shoes; slippers

【便携式】portable: ～穿孔机 port-a-punch／～计算机 portable computer

【便血】[医] having blood in one's stool; passing blood in one's stool

【便宴】informal dinner: 设～招待 give a dinner for sb.

【便衣】①(平常的服装) civilian clothes; plain clothes: ～公安人员 plainclothes public security personnel; public security personnel in plain clothes ②(身着便衣执行任务的军人和警察) plain-clothes man

【便宜行事】act at one's discretion; act as one sees fit

【便于】easy to; convenient for: ～携带 easy to carry

【便中】at one's convenience; when it's convenient: 我替你捎来了一双鞋,望～来取。 I've brought you a pair of shoes. Please come for them whenever it's convenient.

弁 ①(古代男人帽子) a man's cap used in ancient times ②(旧时低级军官) a low-ranking military officer in old China

【弁言】[书] foreword; preface

biāo

标 ①(标志) mark; sign: 界～ landmark／路～ road sign／商～ trade mark／音～ phonetic symbol ②(标明) put a mark; label: ～界 mark a boundary／～上号码 put a number on／～上句号 mark a period ③(奖品) prize; award: 夺～ compete for the first prize ④(投标或招标) tender; bid: 投～ make a tender／招～ invite tenders ⑤(事物的枝节或表面) triviality; superficiality; outward sign: 治～不治本 tackle a problem on the surface, not at the root

【标榜】①(以好听的名义加以宣扬) flaunt; advertise; parade: ～个人自由 flaunting individual freedom／～自己是一个老手 advertise oneself as an old hand ②(吹嘘) boost; praise excessively: 互相～ boost each other

【标本】specimen; sample: 昆虫～ insect specimen

【标本虫】[动] spider beetle

【标兵】①(作为榜样的人或单位) example; model; pacesetter: 被树为～ be cited as a pacesetter ②(标志界线的人) marker

【标尺】①[测] surveyor's rod ②[水] staff gauge ③[军](枪炮尺) rear sight

【标灯】beacon light; beacon

【标点】①(标点符号) punctuation ②(加标点) punctuate ◇～符号 punctuation mark

【标定】demarcate: ～边界线 demarcate a

boundary by setting up boundary markers
【标杆】[测] surveyor's pole
【标高】[测] elevation; level
【标格清高】the appearance is sublime
【标号】grade: 高～水泥 high-grade cement / 水泥～ grade of cement
【标记】sign; mark; symbol: 作为～ serve as a mark
【标价】①(标出货价) mark the price ②(所标明的价钱) marked price ◇～签 price tag
【标量】[物] scalar quantity
【标明】mark; indicate
【标签】label; tag: 价目～ price tag / 贴上～ stick on a label
【标枪】javelin: 掷～ javelin throw
【标售】sell by tender
【标题】title; heading; headline; caption: ～陈旧 the heading is trite / 通栏大字～ banner headline; banner / 小～ subheading; crosshead ◇～音乐 program music
【标新立异】start sth. new in order to be different; do sth. unconventional or unorthodox; create sth. new and original
【标音】mark with phonetic symbols
【标语】slogan; poster
【标志】①(表明特征的记号) sign; mark; symbol; hallmark: 地图上的各种～ various kinds of indications on a map / 交通～ traffic signs ②(表明某种特征) indicate; mark; symbolize ◇～层[地] marker bed
【标致】beautiful; handsome: ～面貌 a handsome face
【标桩】stake
【标准】standard; criterion: 达到～ be up to the standard ◇～部件 normalizer / ～层[地] key bed / ～产品 standardized product / ～大气压 standard atmosphere / ～大气压力 normal atmospheric pressure / ～轨距 standard gauge / ～化石 index fossil / ～件 standard component / ～时 standard time / ～像 official portrait / ～音 standard pronunciation / ～语 standard speech / ～钟 regulator; regulator-clock / ～制 metric system
【标准化】standardize; standardization: 汽车零件是～了的。 The parts of an automobile are standardized. ◇ ～程序 standardization program
【标注】mark

镖 a dartlike weapon
【镖局】[旧] establishment which provides escorts 〈bodyguards〉 for a fee
【镖客】[旧] armed escort

膘 fat: 长～ get fat; put on flesh; flesh out

飙 violent wind; whirlwind

彪
【彪炳】[书] shining; splendid: ～千古 shining through the ages / ～显赫的历史功绩 splendid achievements in history
【彪形大汉】burly chap; husky fellow

biǎo

表 ①(外表;外貌) surface; external; outside ②(表示) show; express; demonstrate: ～决心 express one's determination ③(榜样) model; example ④(表格) table; form; list: 登记～ registration from / 价目～ price list / 填～ fill in a form ⑤(测量某种量的器具) meter; gauge: 电～ electric meter / 水～ water meter / 万用～ volt- ohm meter; multimeter / 温度～ thermometer ⑥(钟表) watch: 怀～ pocket watch / 手～ wristlet watch; wrist watch
【表白】vindicate: ～诚意 assert one's sincerity / ～心迹 bare one's heart; bare true intentions
【表报】statistical tables and reports
【表册】statistical forms; book of tables or forms: 公文报告～ documents, written reports and statistical forms
【表层】surface layer
【表尺】[军] rear sight: 定～ set the rear sight ◇ ～座 rear sight base
【表达】express; show; voice; convey: 非语言所能～ can't be conveyed by words / 无法～ beyond expression / 用英语～思想 express oneself in English ◇～力 expressiveness; power of expression
【表带】watchband; watch strap; bracelet
【表兜】watch pocket; fob
【表高】[航空] indicated altitude
【表格】form; table: 填写～ fill in a form
【表功】boast of one's bit of contribution; claim merit for oneself
【表观】[物] apparent ◇ ～运动 apparent motion / ～质量 apparent mass
【表记】[书] sth. given as a token; souvenir
【表姐妹】(female) cousin
【表决】decide by vote; vote: ～通过 be voted through / 唱名～ vote by roll call; roll-call vote / 交付～ put to the vote; take a vote / 举

手～ vote by a show of hands／口头～ voice vote／起立～ vote by sitting and standing／投票～ vote by ballot ◇ ～程序 voting procedure／～机器 voting machine／～指示牌 vote indicator
【表决权】right to vote; vote: 行使～ exercise the right to vote／无～ have the no right to vote／有～ have the right to vote
【表壳】watchcase
【表里】the outside and the inside; one's outward show and inner thoughts: ～不一 think in one way and behave in another／～如一 think and act in one and the same way
【表链】watch chain
【表露】show; reveal
【表蒙子】watch glass; crystal
【表面】surface; face; outside appearance: ～之词 a superficial statement／地球的～ the surface of the earth; the face of the earth／事物的～ the surface of things ◇ ～活性剂 surface active agent／～硬化 surface hardening／～性 superficiality; superficialness／～张力[物] surface tension
【表面化】become apparent; come to the surface
【表面上】superficial; ostensible; seeming; apparent: ～的动机 ostensible motive／～的相似 superficial resemblance／～的印象 surface impression／从～看来 outwardly; seemingly
【表面文章】①(华而不实的言论) specious writing ②(掩饰)pretense; cover-up
【表面现象】superficial phenomenon: 透过～认识事物的本质 discern the essence of things by looking below the superficial phenomenon
【表明】make known; make clear; state clearly; indicate: ～立场 make known one's position; declare one's stand／～态度 make clear one's attitude／～意图 disclose one's intention; show one's hand／有迹象～… There are indications that...
【表盘】dial plate; dial
【表皮】[生] epidermis; cuticle
【表亲】①(指有表亲关系的人) cousin ②(指表亲关系)cousinship
【表情】expression; countenance; look: 显出高兴的～ wear a cheerful look／愉快的～ cheerful countenance／有～的朗读 reading with expression
【表情达意】communicate views; convey one's ideas or feelings
【表示】show; express; indicate: ～不服罪 plead not guilty／～服罪 plead guilty／～愤慨 express one's indignation／～感谢 express one's thanks／～软弱 show weakness／～态度 define one's attitude／～同情 show sympathy／～同意 give one's adhesion to／～异议 demur／点头～赞同 nod approval／对某人～好感 show sb. favor／～对形势的关切 show concern over the situation
【表饰】watch fob
【表率】model; example: 起～作用 play an exemplary role／做…的～ set a good example to
【表速】[航空] indicated airspeed
【表态】make known one's position; declare where one stands: 明确～ take a clearcut stand／作～性发言 make a statement of one's position
【表土】[农] surface soil; topsoil
【表现】①(表示出来) show; express; display; manifest: 集中～ epitomize; express in a focalized way／～无遗 reveal all without omission ②(行为或作风) show; behavior; conduct: ～好 behave oneself well ③(故意显示自己) show off: 好～ be fond of showing off ◇ ～手法 technique of expression／～形式 form of expression; manifestation
【表现主义】expressionism
【表现型】[生] phenotype
【表象】[心] idea
【表象论】[哲] representationism
【表兄弟】cousin
【表演】①(演出) perform; act; play: ～节目 give a performance; put on a show／精彩～ excellent performance／作充分的～ do enough acting ②(作示范动作) demonstrate ◇ ～唱 singing with actions／～赛 exhibition match
【表扬】praise; commend: ～好人好事 praise good people and good deeds ◇ ～信 commendatory letter
【表意文字】[语] ideograph
【表语】[语] predicative
【表彰】commend; cite; honor: ～先进集体 give commendation to the advanced units

裱

mount; paste up: ～一幅国画 mount a traditional Chinese painting
【裱褙】mount (a picture)
【裱糊】paper; paste paper on: ～房间 paper the room

婊

【婊子】prostitute; whore

biē

憋 ①(抑制或堵住不让出来) keep down; hold back: ～着一肚子气 keep down one's smouldering anger / ～足劲 be store up for a spurt of energy ②(闷) suffocate; feel oppressed: 心里～得慌 feel very much oppressed

【憋气】①(窒息的感觉) feel suffocated ②(有委屈或烦恼而不能发泄) choke with resentment; feel injured and resentful

【憋闷】feel oppressed; be depressed; be dejected

鳖 soft-shelled turtle

【鳖甲】[中药] turtle shell

【鳖裙】calipash

bié

蹩 [方] sprain

【蹩脚】inferior; shoddy; poor; cheap: ～的英语 poor English / ～货 inferior stuff; poor stuff; thirdrate goods / ～演员 poor actor

别 ①(离别) leave; part ～故乡 leave one's native place / 久～重逢 meet after a long separation / 临～赠言 parting advice②(另外) other; another: ～人 other people ③(类别) distinction: 性～ sex (distinction) ④(差别) difference; distinction: 天渊之～ a world of difference ⑤(区别) differentiate; distinguish: ～其真伪 determine whether it's true or false ⑥(用别针别上) fasten with a pin or clip: 用针把文件～起来 pin papers together ⑦(插住) stick in: 皮带上～一支枪 stuck a pistol in one's belt ⑧(不要) don't: ～见怪 don't be offended ⑨(跟"是"字合用,表示揣测) ～是他不来了。May be he isn't coming.

【别称】another name; alternative name

【别出心裁】adopt an original approach; try to be different

【别处】else where; another place

【别动队】①(单独执行任务的小分队) special detachment; commando ②(武装特务组织) an armed secret agent squad ◇ ～员 ranger

【别管】①(不管) no matter ②(不要过问) never mind; leave sb.(sth.) alone: ～闲事。Don't put your finger in the pie. / ～我。 Leave me alone.

【别号】alias

【别具慧眼】have a special insight understanding

【别具一格】have a distinctive style

【别开生面】start something new; break a new path; break fresh ground: 一次～的现场会 an entirely new sort of on-the-spot meeting

【别来无恙】hope that you are well

【别离】take leave of; leave: ～家乡,踏上征途 leave home and start on a long journey

【别名】another name

【别人】other people; others; people: 总是先想到～的利益 always think of other people's interests first

【别生枝节】have new complications

【别树一帜】set up a new banner; found a new school of thought; have a style of one's own

【别墅】villa

【别提】[口] indescribably

【别无他法】having no other way out

【别绪】the sorrow of parting

【别有风味】have a distinctive flavor

【别有天地】a place of unique beauty; scenery of exceptional charm

【别有用心】have ulterior motives; have an axe to grind

【别针】①(妇女装饰用) brooch ②(普通的别针) safety pin; pin

【别致】unique; unconventional; interesting and novel

【别字】wrongly written or mispronounced character: 读～ mispronounce a character / 写～ write a character wrongly / ～连篇。 There are a lot of characters wrongly written.

biě

瘪 shrivelled; shrunken: ～花生 blighted peanuts / 干～ dry and shrivelled; wizened

biè

别

【别扭】①(难对付) difficult to deal with; troublesome ②(意见不相投) can not see eye to eye; not get along well: 闹～ be at odds ③(不流畅) not smooth; awkward

bīn

濒 ①(紧靠) be close to; border on ②(临近) be on the brink of; be on the point of

【濒河之郡】a prefectural city located near the river

【濒临】be close to; border on; be on the verge of: ～太平洋 border on the Pacific Ocean

【濒危】①(接近危险的境地) be in imminent danger ②(病危) be critically ill

【濒于】on the verge of: ～灭亡 be on the brink of ruin; near extinction / ～破产 on the verge of bankruptcy / ～死亡 on the verge of one's grave

宾 guest: 贵～ distinguished guest / 国～ state guest

【宾词】[逻] predicate

【宾格】[语] objective case

【宾馆】guesthouse

【宾客】guests; visitors

【宾朋满座】the house is full of guests

【宾语】[语] object: 间接～ indirect object / 直接～ direct object

【宾至如归】guests fell at home; a home from home

滨 ①(水边) bank; brink; shore: 海～ seashore / 湖～ lakeshore; lakeside ②(靠近水边) by; be close to: ～海 by the sea / 东～大海 bordering the sea on the east

【滨海带】[地] littoral zone

槟

【槟子】[植] *binzi*, a species of apple which is slightly sour and astringent

镔

【镔铁】wrought iron

傧

【傧相】[旧] attendant of the bride or bridegroom at a wedding: 男～ best man / 女～ bridesmaid

缤

【缤纷】[书]in riotous profusion: 落英～ petals falling in riotous profusion / 五彩～ a riot of color

彬

【彬彬君子】a refined gentleman

【彬彬有礼】refined and courteous urbane

bìn

鬓 temples; hair on the temples

【鬓发】hair on the temples: ～皆白 with greying temples; hoary-haired

【鬓角】temples; hair on the temples

殡 ①(停放灵柩) lay a coffin in a memorial hall ②(把灵柩送到埋葬处) carry a coffin to the burial place

【殡车】hearse

【殡殓】encoffin a corpse and carry it to the grave: ～遗骸 dress and prepare the remains a coffin

【殡仪馆】the undertaker's; funeral parlor

【殡葬】funeral and interment

摈 [书] discard; get rid of: ～而不用 reject / ～诸门外 shut sb. out

【摈斥】reject; dismiss: ～异己 dismiss those who hold different opinions

【摈除】discard; get rid of; dispense with: ～繁文缛节 dispense with all unnecessary formalities / ～障碍 remove obstacles

【摈弃】abandon; discard; cast away

髌 kneecap

【髌骨】[生理] kneecap; patella

bīng

冰 ① ice: 一层薄～ a thin layer of ice ②(把东西冰凉) ice; put on the ice: ～一瓶啤酒 ice a bottle of beer / ～镇西瓜 ice watermelon ③(感到寒冷) feel cold: 这水～手。 This water is freezing cold.

【冰雹】hail; hailstone

【冰场】skating rink; ice rink; ice stadium; ice arena

【冰川】glacier ◇ ～湖 glacial lake; tarn / ～舌 glacier tongue / ～学 glaciology / ～作用 glaciation

【冰醋酸】[化] glacial acetic acid

【冰镩】ice chisel

【冰袋】[医] ice bag

【冰蛋】frozen eggs

【冰刀】[体] (ice) skates

【冰岛】Iceland ◇ ～人 Icelander / ～语 Icelandic

【冰点】[物] freezing point ◇ ～测定器 cryoscope

【冰冻】freeze: ～三尺,非一日之寒。 It takes more than one cold day for the river to freeze three feet deep; The trouble has been brewing for quite some time. ◇ ～季节 freezing season / ～区 frost zone / ～食物 frozen food

【冰斗】[地]cirque

【冰帆】[体]iceboating

【冰封】ice-bound

【冰盖】[地] ice sheet

【冰糕】[方]①(冰淇淋) ice cream ②(冰棍儿) ice-lolly; popsicle

【冰镐】ice axe

【冰棍儿】ice-lolly; popsicle; ice-sucker; frozen sucker

【冰河】glacier ◇ ～时代 glacial epoch
【冰肌雪肠】ice as symbol of purity of character
【冰肌玉骨】flesh of ice and bones of jade
【冰窖】icehouse
【冰晶】[气] ice crystal
【冰晶石】cryolite
【冰块】lump of ice; ice cube
【冰冷】ice-cold: ～的态度 icy manner
【冰凉】ice-cold
【冰凝器】[物] cryophorus
【冰排】ice raft; ice floe
【冰片】[中药] borneol
【冰期】[地] glacial epoch; ice age
【冰淇淋】ice cream: 水果～ sundae
【冰碛】[地] moraine ◇ ～物 till / ～岩 tillite
【冰裂缝】crevasse
【冰瀑区】icefall
【冰橇】sled; sledge; sleigh
【冰清玉洁】clear as ice and pure as gem
【冰球】[体] ①(一种冰上运动) ice hockey: ～场 rink / ～队 sextet / ～棍 ice stick / ～运动员 ice hockey player ②(冰球运动用的球) puck
【冰山】iceberg: ～难靠 an ice mountain is hardly reliable / ～已倒 power has collapsed
【冰上表演】ice show
【冰上溜盘】[体] Eisschiessen
【冰上溜石】[体] curling
【冰上曲棍球】[体] bandy
【冰上舞蹈】ice dancing
【冰上运动】ice sport ◇ ～会 ice-sports meet
【冰释】disappear; vanish; be dispelled: 涣然～ be instantly dispelled
【冰霜】①(有节操) moral integrity ②(神情严肃) austerity: 凛若～ severe-looking
【冰塔】[地] serac
【冰炭不相容】as incompatible as ice and hot coals
【冰糖】crystal sugar; rock candy ◇ ～葫芦 candied haws on a stick
【冰天雪地】a world of ice and snow
【冰隙】[地] crevasse
【冰箱】icebox; refrigerator; freezer
【冰消瓦解】melt like ice and break like tiles; disintegrate; dissolve
【冰鞋】skating boots; skates
【冰镇】ice; iced: ～可乐 iced coke / ～啤酒 ice beer
【冰洲石】Iceland spar
【冰柱】icicle
【冰爪】[登山] crampon
【冰砖】ice-cream brick
【冰锥】[登山] ice piton
【冰姿雪魄】of purity of soul

槟

【槟榔】betel (palm); areca
【槟榔因】[有化] arecadeine; arecaine

兵

①(兵器) weapons; arms ②(军人) soldier; fighter ③(部队) troops; army: 防化学～ anti-chemical corps / 工程～ engineer corps / 炮～ artillery / 铁道～ railway corps / 通讯～ signal corps / 装甲～ armored corps ④(关于军事或战争的) military affairs: 用～如神 marvellous in generalship; superb in strategy
【兵变】mutiny
【兵不血刃】the edges of the swords not being stained with blood; win victory without firing a shot
【兵不厌诈】there can never be too much deception in war; in war nothing is too deceitful; all's fair in war
【兵船】man-of-war; naval vessel; warship
【兵多将广】enjoy numerical superiority
【兵法】art of war; military strategy and tactics: ～巧妙 military tactics are skillful
【兵戈】①(兵器) weapons; arms: 不动～ without resorting to force ②(战争) fighting; war: ～扰攘 war-torn
【兵工厂】munitions factory; arsenal
【兵贵神速】speed is precious in war
【兵荒马乱】turmoil and chaos of war
【兵家】military strategist; military commander: ～常事 a commonplace in military operations
【兵舰】warship
【兵力】military strength; armed forces; troops; numerical strength: ～对比 relative military strength / ～分配 distribution of forces / ～优势 numerical superiority / ～转移 transfer of troops / 集中优势～ concentrate a superior force / 五千人的～ an army of five thousand strong
【兵连祸结】ravaged by successive wars; war-torn; war-ridden
【兵临城下】the attacking army has reached the city gates; the city is under siege
【兵马】troops and horses; military forces: ～未动,粮草先行 food and fodder should go before troops and horses; proper preparations should be made in advance ◇ ～俑[考古] wood or clay figures of warriors and horses buried with the dead

【兵痞】army riffraff; army ruffian; soldier of fortune
【兵器】weaponry; weapons; arms
【兵强马壮】strong soldiers and sturdy horses; a well-trained and powerful army
【兵权】military leadership; military power
【兵戎】arms; weapons: ～相见 resort to arms; appeal to arms
【兵士】ordinary soldier
【兵书】a book on the art of war
【兵团】[军] ①(泛指大部队) large unit; corps; formation: 正规～ regular forces / 主力～ main force ②(由几个军组成的单位) army
【兵役】military service: 服～ serve in the army; perform military service ◇ ～法 military service law / ～年龄 military age / ～制 system of military service
【兵营】military camp; barracks
【兵员】soldiers troops
【兵源】manpower resources; sources of troops
【兵站】army service station; military depot
【兵种】arm of the services: 技术～ technical arms

bǐng

禀 ①(禀报) report; petition ②(承受) receive; be endowed with
【禀报】report to a superior
【禀承】in accordance with; obedience to
【禀赋】natural endowment; gift: ～聪明 be gifted with keen intelligence
【禀复】make reply to a superior
【禀告】report
【禀明】explain to a superior 〈elder〉; clarify a matter to a superior 〈elder〉
【禀诉冤屈】file a plaint of grievance
【禀性】natural disposition: ～纯厚 be simple and honest by nature / ～慈祥 be merciful and propitious by disposition / ～难移 human nature is hardly changed / ～善良的人 man of good nature / ～温和的人 man of gentle disposition / ～愚蒙 one's natural disposition is stupid

饼 ①(烤熟或蒸熟的扁圆形面食) a round flat cake: 烙～ unleavened pancake; flapjack / 月～ moon cake ②(形状象饼的东西) sth. shaped like a cake: 豆～ soybean cake / 铁～ [体] discus
【饼铛】baking pan
【饼饵】cakes; pastry
【饼肥】cake
【饼干】biscuit; cracker
【饼子】pancake

屏 ①(抑止) hold ②(除去) reject; get rid of; abandon
【屏除】remove; get rid of; do away with: ～成见 remove prejudices / ～恶习 get rid of bad habits / ～杂念 dismiss distracting thoughts
【屏绝交际】isolate oneself from social activities
【屏绝往来】break off intercourse
【屏气】hold one's breath: ～凝神 deep concentration
【屏弃】discard; abandon; reject: ～不用 casting aside and not using it
【屏息】hold one's breath: ～不动 hardly daring to breathe / ～观之 hold breath and watch / ～静听 listen with bated breath

丙 ①(天干第三位) the third of the ten Heavenly Stems ②(第三) third: ～等 the third grade ③(用作代号) ～种维生素 vitamin C
【丙醇】[化] propyl alcohol
【丙纶】[纺] polypropylene fiber
【丙酮】[化] acetone ◇ ～树脂 acetone resin
【丙烷】[化] propane
【丙烯】[化] propylene
【丙烯画】[美] acrylic painting
【丙烯腈】[化] acrylonitrile; vinylcyanide
【丙烯酸】[化] acrylic acid
【丙种球蛋白】[药] gamma globulin

炳 [书] bright; splendid; remarkable: 彪～ shining; splendid

柄 ①(器物的把儿) handle ②[植] stem ③(权) power; authority
【柄不可授】do not offer the handle to another

秉 ①(拿着) grasp; hold ②(掌握) control; preside over
【秉笔直书】write the truth without fear or favor
【秉承】in accordance with; obedience to: ～某人旨意 subservient to the will of sb.
【秉公】justly; impartially: ～办理 handle a matter impartially; decide case according to law
【秉性】nature: disposition: ～纯朴 be simple and honest by nature
【秉政】hold political power; be in power
【秉烛夜游】take an evening stroll with a lantern

bing

病 ①(疾病) illness; sickness; disease: 常见～ common diseased; common illnesses / 传染～ infectious disease (通过空气); contagious disease (通过接触) / 地方～ endemic disease / 多发～ frequently occurring illnesses / 肺～ lung trouble / 妇女～ gynecologic disease; disease of women / 肝～ liver trouble / 各种儿科～ illnesses of children / 功能性疾～ functional disease / 后天～ acquired disease / 急性～ acute disease / 寄生虫～ parasitic disease / 结核～ tuberculosis; T.B. / 流行～ epidemics; epidemie disease / 慢性～ chronic disease / 皮肤～ skin disease; dermatosis / 胃～ stomach trouble; gastric disease / 先天～ congenital disease / 小～ minor ailment; slight disease / 心脏～ heart trouble / 血液～ blood disease / 职业～ occupational disease / 重～ serious disease ②(生病) be ill; be taken ill; fell unwell: 卧～在床 lie stick in bed ③(有病的) sick; sickly: ～树 sickly tree ④(缺点) fault; defect: 语～ ill-chosen expression

【病包儿】[口] a person who is always falling ill; chronic invalid

【病变】pathological changes

【病程】course of disease

【病虫害】plant diseases and insect pests

【病床】hospital bed

【病从口入】illness finds its way in by the mouth

【病倒】be down with an illness; be laid up

【病毒】[医] virus ◇ ～病 virosis / ～灵 moroxydine / ～性肝炎 virus hepatitis / ～性感冒 virus flu / ～学 virology

【病笃】[书] be critically ill; be terminally ill

【病房】ward (of a hospital); sickroom: 产科～ maternity ward / 隔离～ isolation ward / 内科～ medical ward / 外科～ surgical ward / 查～ make ward rounds; go on rounds in the wards

【病夫】sick man

【病根】①(病源; 没有完全治好的旧病) the lingering effect of a chronic disease; an old complaint ②(比喻能引起失败或灾祸的原因) the root cause of trouble

【病骨】get skinny due to a disease

【病故】die of an illness

【病害】(plant) disease

【病号】sick personnel; person on the sick list; patient: 老～ one who is always ill; chronic invalid ◇ ～饭 patient's diet; special food for patients

【病急乱投医】turn to any doctor one can find when critically ill; try anything when in a desperate situation

【病家】a patient and his family

【病假】sick leave: 给三天～ grant three days' sick leave / 请～ ask for sick leave / 休～ be on sick leave ◇ ～条 certificate for sick leave

【病菌】pathogenic bacteria; germs

【病况】state of an illness; patient's condition

【病理】pathology ◇ ～解剖学 pathological anatomy / ～学 pathology / ～诊断 pathological diagnosis

【病历】medical record; case history ◇ ～室 records room

【病例】case: 非典型～ borderline case / 急诊～ emergency case / 误诊～ missed case ◇ ～报告 case report

【病魔】serious illness: ～缠身 be afflicted with a lingering disease

【病情】state of an illness; patient's condition: ～公报 medical bulletin

【病人】patient; invalid: 急诊～ emergency case / 门诊～ out-patient / 重～ a serious case

【病容】sickly look: 面带～ look ill; look unwell

【病入膏肓】the disease has attacked the vitals; beyond cure

【病史】medical history: 记录～ take sb.'s history

【病室】sickroom; ward

【病势】degree of seriousness of an illness; patient's condition

【病榻】[书] sickbed

【病态】morbid state ◇ ～心理 morbid psychology

【病痛】slight illness; indisposition; ailment

【病危】be critically ill; be terminally ill

【病象】symptom

【病休】sick rest; lie up

【病因】cause of disease; pathogeny

【病友】a friend made in hospital or people who become friends in hospital; wardmate

【病愈】recover

【病员】sick personnel; person on the sick list; patient

【病原】cause of disease; pathogeny ◇ ～体 pathogen / ～学 aetiology

【病院】a specialized hospital: 传染～ infectious diseases hospital; isolation hospital / 精神～ mental hospital

【病灶】focus

【病征】symptom
【病症】disease; illness
【病株】diseased or infected plant
【病状】symptom

并 ①(合在一起) combine; merge; incorporate ②(同时存在或进行) simultaneously; side by side ③[连] and; moreover
【并不】not; not at all; by no means; in no sense
【并存】exist side by side; coexist
【并蒂莲】twin lotus flowers on one stalk; a devoted married couple
【并发】be complicated by; erupt simultaneously
【并发症】[医] complication: 引起～ lead to complications
【并激】[电] shunt excitation ◇ ～电动机 shunt motor / ～绕组 shunt winding
【并驾齐驱】run neck and neck; keep abreast of one another
【并肩】shoulder to shoulder; side by side; abreast: ～前进 advance shoulder to shoulder
【并进】keep abreast of; advance together: 与时代～ keep abreast of the time
【并举】develop simultaneously: 土洋～ employ both simple and sophisticated methods; employ both indigenous and foreign methods
【并卷机】[纺] ribbon lap machine
【并立】exist side by side; exist simultaneously
【并联】[电] parallel connection ◇ ～电路 parallel circuit / ～馈电 parallel feed
【并列】stand side by side; be juxtaposed: ～第二名 be both runners-up; tie for second place ◇ ～分句[语] coordinate clauses / ～句[语] compound sentence
【并排】side by side; abreast: ～坐着 sit in a line
【并且】①(表示同时或先后进行) and; also; and... as well; in addition ②(表示更进一层) besides; moreover; furthermore
【并入】merge into; incorporate in
【并纱】[纺] doubling ◇ ～机 doubling winder
【并条】[纺] drawing ◇ ～机 drawing frame
【并吞】swallow up; annex; merge: ～别国领土 annex the territory of another state / ～小企业 swallow up small enterprises
【并行不悖】both can be implemented without coming into conflict; not be mutually exclusive; run parallel
【并重】lay equal stress on; pay equal attention to: 两者～ lay equal stress on both

摒 get rid of; dismiss; brush aside
【摒挡】[书] arrange; put in order; get ready: ～行李 get one's luggage ready / ～一切 set everything in order
【摒弃】abandon; get rid of: ～前嫌 disregard previous enmity
【摒除】get rid of; renounce

bō

拨 ①(拨动东西) stir; poke; turn: ～电话号码 dial a telephone number / ～灰 stir ashes / ～算盘 work an abacus; tick off the beads on an abacus / ～弦 pluck the strings / ～一下炉火 poke the fire / ～转马头 turn the horse round / 把表～慢一小时 set back the watch one hour ②(调配;分给) allocate; appropriate; set aside ③[量] (用于人的分组等) group; batch: 轮～休息 take rest by turns
【拨付】appropriate (a sum of money); make payment as earmarked
【拨号盘】[电话] dial
【拨火棍】poker
【拨开】push aside
【拨开云雾见青天】dispel the clouds and see the sun; restore justice
【拨款】①(拨给款项) appropriate funds; allocate funds ②(指款项) appropriation: 财政～ financial allocation / 军事～ military appropriations
【拨剌】[象] splash
【拨浪鼓】a drum-shaped rattle; rattle-drum
【拨乱反正】bring order out of chaos; set to rights things which have been thrown into disorder; set wrong things right
【拨弄】①(来回拨动) fiddle with; move to and fro: ～算盘珠 fiddle with the beads of an abacus / 用手不停地～铅笔 fiddle about with a pencil ②(挑拨) stir up; incite; rouse: ～是非 stir up troubles by gossip; stir things up
【拨冗】[套] find time in the midst of pressing affairs
【拨弦乐器】plucked string instrument; plucked instrument
【拨云见日】dispel the clouds and see the sun; restore justice
【拨正】set right; correct: ～航向 correct the course
【拨准】set correctly
【拨子】[乐] plectrum
【拨奏】[乐] *pizzicato*

播 ①(播种) sow; seed: 在田里～小麦 seed the

fields with wheat ②(传播) spread; carry
【播弄】①(摆布) order sb. about ②(挑拨) stir up: ～是非 sow discord; stir things up
【播送】broadcast; transmit: ～国际新闻 broadcast world news / ～音乐 broadcast music / 第一次～ the first transmission / 由无线电～讯息 transmit a message by radio
【播扬】propagate and uphold: ～于外 send forth abroad
【播音】transmit; broadcast: 开始～ go on the air / 停止～ go off the air ◇ ～室 broadcasting studio / ～员 announcer
【播种】①(撒种) sow seeds; sow; seed ②(种植) sowing; seeding ◇ ～机 seeder; planter; grain drill / ～面积 sown area; seeded area / ～期 sowing time; seeding time

钵

①(陶制的器具) earthen bowl ②(钵盂) alms bowl: 沿门托～ begging alms from door to door
【钵盂】Buddhist priest's alms bowl

波

①(波浪) wave: 微～ ripples ② [物] wave: 长～ long wave / 电～ electric wave / 电磁～ electromagnetic wave / 短～ short wave / 光～ light wave / 横～ transverse wave / 声～ sound wave / 中～ medium wave / 纵～ longitudinal wave ③(事情的意外变化) an unexpected turn of events: 风～ storm; disturbance / 一～未平，一～又起 before one upheaval quiets down, there comes another
【波长】wavelength ◇ ～计 wavemeter; cymometer
【波茨坦公告】Potsdam Proclamation (1945)
【波导】[物] wave guide ◇ ～管 wave guide / ～通信 wave guide communication
【波动】①(起伏不定) undulate; fluctuate; rise and fall: 物价～ price fluctuation ②[物] wave (motion) ◇ ～汇率 fluctuating rate / ～力学 wave machanics / ～说[物] wave theory / ～条款 fluctuation clause
【波段】[无] wave band ◇ ～开关 band switch; waver
【波多黎各】Puerto Rico ◇ ～人 Puerto Rican
【波尔多液】[农] Bordeaux mixture
【波峰】[物] wave crest
【波幅】[物] amplitude ◇ ～失真 amplitude distortion
【波谷】[物] trough
【波光】the glistening light of waves
【波及】spread to; involve; affect: ～他人 be involved in
【波江星座】[天] Eridanus; River Po
【波兰】Poland ◇ ～人 Pole / ～语 Polish
【波澜】great waves; billows
【波澜起伏】(of a piece of writing) with one climax following another
【波澜壮阔】surging forward with great momentum; unfolding on a magnificent scale: ～的民族解放运动 the surging national liberation movement / 一首～、气势磅礴的史诗 an epic of magnificent sweep
【波浪】wave: ～式前进 advance wave upon wave / ～式起伏 undulating movement / ～滔天 the billows dashing against the sky
【波利尼西亚】Polynesia ◇ ～人 Polynesian / ～语 Polynesian (language)
【波罗的海】Baltic Sea
【波美比重计】Baume hydrometer
【波美度】[化] Baume degrees
【波平浪静】breakers calm and waves still
【波平如镜】the wave is smooth as a glass
【波谱】[物] spectrum
【波束】[物] beam
【波斯菊】[植] coreopsis
【波斯湾】Persian Gulf
【波涛】great waves; billows: ～滚滚的大海 rolling seas / ～起伏 swellings, heaving and subsiding / ～万顷 a myriad surges / ～汹涌 roaring waves
【波纹】①(小波浪形成的水纹) ripple ②(折皱) corrugation ◇ ～管 [机] bellows; corrugated pipe / ～铁 corrugated iron / ～纸板 corrugated cardboard
【波形】[物] waveform; wave pattern: ～曲线 wavy curve / ～瓦[建] corrugated tile
【波音】[乐] mordent: 逆～ inverted mordent
【波折】twists and turns
【波状热】[医] undulant fever; brucellosis
【波状叶】[植] sinuate leave
【波状云】[气] undulatus

菠

【菠菜】spinach
【菠萝】pineapple

玻

【玻璃】① glass: 板～ plate glass / 窗～ glass sheet; window pane / 雕花～ cut glass / 防弹～ bullet-resisting glass / 钢化～ tempered glass / 光学～ optical glass / 夹丝安全～ wired safety glass / 毛～ ground glass / 冕～ crown glass / 磨沙～ frosted glass / 钠钙硅酸

盐～ soda lime glass／耐火～ fireproof glass／硼硅酸盐～ borosilicate glass／片～ sheet glass／一块～ a piece of glass／压层安全～ laminated glass ②(像玻璃的塑料) nylon; plastic ◇～板 glass plate; plate glass; glass top (of a desk)／～版[印] collotype／～杯 glass; tumbler／～布 glass cloth／～厂 glassworks／～刀 glass cutter; glazier's diamond／～粉 glass dust／～钢 glass fiber reinforced plastic／～棉(绒) glass wool／～片 sheet glass／～器皿 glassware／～纱[纺] organdy／～丝 glass silk／～体[生理] vitreous body／～纤维 glass fiber／～纸 cellophane; glassine／～砖 glass block

【玻璃陶瓷】glassceram: 光晶化～ fotoceram／热晶化～ pyroceram

【玻利维亚】Bolivia ◇～人 Bolivian

【玻色子】[物] boson

【玻意耳定律】[物] Boyle's law

剥

【剥采比】[矿] stripping-to-ore ratio; stripping ratio

【剥夺】deprive; expropriate; strip: ～公民权 deprival of civil right; disfranchisement／～继承权 disherison; disinherit／～权益 outlawry／～人身自由 deprivation of personal liberty／～所有权 expropriate／～政治权利 deprive sb. of his political rights／～政治权利终身 civil death; deprival of political right for life／～自由 deprivation of freedom

【剥离】come off; peel off; be stripped: 表土～ topsoil stripping

【剥落】peel off

【剥皮】skin; peel off the skin

【剥蚀】①(侵蚀) erode; wear away: 风雨的～ eroded by weather ②[地] erode: 冰川～ glacial erosion／海岸～ coast erosion ◇～作用[地] denudation

【剥削】exploit ◇～方式 form of exploitation／～阶级 exploiting class／～收入 income from exploitation／～思想 exploiting outlook／～行为和变相的～行为 overt and covert practices of exploitation／～者 exploiter

【剥啄】[象] tap (on a door or window): ～声 sound of tapping

bó

鹁

【鹁鸽】pigeon

【鹁鸪】wood pigeon

脖 neck

【脖颈儿】back of the neck; nape: 抓住动物的～ seize the animal by the scruff of the neck

勃 suddenly

【勃勃】thriving; vigorous; exuberant: 生气～ full of vitality; alive with activity／兴致～ full of enthusiasm; in high spirits／野心～ driven by wild ambition; overweeningly ambitious

【勃发】①(焕发) thrive; prosper ②(突然发生) break out: 游兴～ be seized with a desire to travel

【勃起】erection

【勃然】agitatedly; excitedly: ～变色 agitatedly change color; be visibly stung／～大怒 fly into a rage

【勃然而起】spring into life; burst into activity

【勃溪】[书] family quarrel; tiff; squabble: ～之声 the noise of quarrel

【勃兴】rise suddenly; grow vigorously: 一个工业城市的～ the vigorous growth of an industrial town

博 ①(多;丰富) rich; abundant; plentiful: ～而不精 have wide but not expert knowledge; know something about everything／地大物～ vast in territory and rich in natural resources ②(取得) win; gain: 聊～一笑 just for your entertainment

【博爱】universal fraternity; universal love: ～主义 the policy of love without distinction; philanthropy

【博茨瓦纳】Botswana ◇～人(单数) Motswana; (复数) Batswana

【博大精深】broad and profound

【博得】win; gain: ～好评 have a favorable reception／～全场喝采 draw loud applause from the audience; bring the house down／～同情 win sympathy／～信任 win the confidence

【博而不精】extensive but shallow in knowledge

【博古通今】conversant with things past and present; erudite and informed

【博览】read extensively: ～群书 well-read

【博览会】fair: 国际～ international fair

【博取】try to gain; court: ～欢心 curry favor／～同情 seek sb.'s sympathy／～微利 seize petty gains／～信任 try to win sb.'s confidence

【博士】①(指学位) doctor: ～学位 doctor's degree; doctorate／授与～学位 confer a doctor's

degree on sb. / 文学～ Doctor of Literature (Litt.D.) / 哲学～ Doctor of Philosophy (Ph.D.) ②(有学问的人) learned scholar
【博闻广识】extensive information and learning
【博闻强记】have wide learning and a retentive memory; have encyclopedic knowledge
【博物】natural science ◇ ～学家 naturalist
【博物馆】museum: 历史～ the Museum of History / 中国革命～ the Museum of the Chinese Revolution ◇ ～学 museology
【博物院】museum: 故宫～ Palace Museum
【博学】learned; erudite: ～多才 be of great learning and great ability / ～多闻 very learned and well-informed / ～鸿儒 a great literate of wide learning / ～之士 learned scholar
【博雅】learned: ～君子 learned and accomplished worthies / ～之士 a scholar of profound knowledge
【博奕】play a game of 'go' chess
【博奕论】game theory

薄

①(轻微) slight; meager; small: ～酬 small reward; meager reward ②(不厚道) unkind; ungenerous; mean: ～待某人 treat sb. ungenerously ③(看不起) despise; belittle: 厚此～彼 favor one and slight the other
【薄饼】thin pancake for wrapping up meat
【薄脆】thin and brittle biscuit
【薄钢板】steel sheet
【薄瘠之地】a thin and lean land
【薄技在身】have a thin skill by oneself
【薄酒】diluted wine
【薄壳】[建] shell: ～屋顶 shell roof
【薄礼】a meager present; my humble gift
【薄利】small profits: ～多销 small profits but quick turnover
【薄命】star-crossed; ill-fated: ～女子负心汉 an unfortunate girl and a heartless man
【薄膜】membrane; film: 塑料～ plastic film ◇ ～电阻 film resistor
【薄暮】[书] dusk; twilight
【薄片】thin slice; thin section ◇ ～分析[地] thin section analysis
【薄情】inconstant in love; fickle: ～郎 a heartless lover
【薄弱】weak; frail: 技术力量～ lack qualified technical personnel / 能力～ lacking in ability / 意志～ weak-willed ◇ ～环节 weak link; vulnerable spot
【薄胎瓷器】eggshell china; eggshell porcelain
【薄雾】mist; haze
【薄油层】[石油] oil sheet

搏

①(搏斗) wrestle; fight; combat; struggle: 肉～ hand-to-hand fight ②(扑上去抓) pounce on ③(跳动) beat; throb: 脉～ pulse
【搏动】beat rhythmically; throb; pulsate
【搏斗】wrestle; fight; struggle: 近距离～ fight at close quarters / 与困难～ wrestle with difficulties
【搏击】strike; fight with hands

膊

arm: 赤～ bare to the waist

钹

[乐] cymbals

泊

be at anchor; moor; berth: 停～ lie at anchor / ～岸 anchor alongside the shore
【泊位】[交] berth: 深水～ deepwater berth

箔

①(苇子等编成的帘子) screen: 苇～ reed screen ②(蚕箔) bamboo tray for rearing silkworms ③(金属薄片) foil; tinsel: 金～ gold foil

帛

[书] silks: 布～ cottons and silks / ～画 painting on silk / ～书 (ancient) book copied on silk

铂

[化] platinum (Pt)

伯

①(伯父) father's elder brother; uncle ②(兄弟排行老大) the eldest of brothers ③(伯爵) earl; count
【伯伯】father's elder brother; uncle
【伯父】father's elder brother; uncle
【伯爵】count; earl: 女～ countess ◇ ～夫人 countess
【伯母】wife of father's elder brother; aunt
【伯仲】[书] ～之间 almost on a par / 相～ be much the same
【伯仲叔季】eldest, second, third and youngest of brothers; order of seniority among brothers

舶

oceangoing ship
【舶来品】imported goods; foreign goods

驳

①(否认别人的意见) refute; argue; contradict: 不值一～ not worth arguing ②(驳船) barge; lighter: 铁～ iron barge / 油～ oil barge ③(驳运) transport by lighter
【驳岸】a low stone wall built along the water's edge to protect an embankment; revetment
【驳斥】refute; denounce: ～谬论 refute a fallacy
【驳船】barge; lighter
【驳倒】demolish sb.'s argument; outargue; re-

fute
【驳回】reject; turn down; overrule: ～某人的请求 overrule sb.'s claim / ～上诉 reject an appeal
【驳壳枪】Mauser pistol
【驳运】transport by lighter; lighter ◇ ～费 lighterage
【驳杂】heterogeneous: 内容～ heterogeneous in content

bǒ

跛 lame: ～了一只脚 lame in one leg / 一颠一～ walk with a limp; limp along
【跛鳖千里】a lame turtle can go a thousand miles by perseverance
【跛行】walk lamely; have a limp
【跛子】lame person; cripple

簸 winnow with a fan; fan: ～谷 winnow away the chaff; fan the chaff
【簸荡】roll; rock
【簸谷机】winnower; winnowing fan
【簸扬】winnow ◇ ～机 winnower

bò

薄
【薄荷】[植] field mint; peppermint ◇ ～醇[化] menthol; peppermint / ～糖 peppermint drops / ～酮 menthone / ～油 peppermint oil

簸
【簸箕】①(簸粮食用) winnowing pan ②(清除垃圾用) dustpan ③(簸箕形的指纹) loop

bǔ

捕 catch; seize; arrest: ～蛇 catch snakes / 被～ be arrested
【捕虫叶】[植] insect-catching leaf
【捕风捉影】chase the wind and clutch at shadows; make groundless accusations; speak or act on hearsay evidence
【捕俘】[军] capture enemy personnel
【捕获】catch; capture; seize: 当场～ catch sb red-handed ◇ ～法[法] law of prize / ～量 catch
【捕鲸船】whaler; whale catcher
【捕鲸小艇】whaleboat
【捕捞】fish for (aquatic animals and plants); catch: ～对虾 catch prawns ◇ ～能力 fishing capacity
【捕拿】arrest; capture; catch: ～犯人 arrest the offender
【捕杀】catch and kill
【捕食】catch and feed on; prey on ◇ ～现象 predation
【捕鼠器】mousetrap
【捕蝇草】Venus's-flytrap
【捕蝇器】flytrap
【捕蝇纸】flypaper
【捕鱼】catch fish; fish: ～过活 catch fish for livelihood ◇ ～法 fishery law / ～权 fishery
【捕捉】catch; seize: ～镜头 seize the right moment to get a good shot / ～昆虫 catch insects / ～战机 seize the opportunity for battle; seize the right moment to strike

哺 ①(喂) feed; nurse ②(咀嚼着的食物) the food in one's mouth
【哺乳】breast-feed; suckle; nurse ◇ ～动物 mammal / ～动物学 mammalogy / ～室 nursing room
【哺养】feed; rear
【哺育】①(喂养) feed ②(培养) nurture; foster

卜 ①(占卜) divination; fortune-telling ②(预料) foretell; predict ③(选择) select; choose: ～居某地 select a place for one's home
【卜辞】oracle inscriptions of the Shang Dynasty
【卜卦】a system of divination
【卜筮】divination

补 ①(修补) mend; patch; repair: ～车胎 mend a puncture / ～锅 tinker pans / 把鞋～一～ have the shoes repaired ②(补充; 填补) fill; supply; make up for: ～空缺 fill up a vacancy / 把缺少的页数～上 supply the missing pages ③(补养) nourish ④(利益; 用处) help; benefit; use: 无～于大局 be of no help to the overall situation
【补白】filler
【补报】①(事后报告) make a report after the event; make a supplementary report ②(报答) repay a kindness
【补偿】compensate; make up: ～损失 indemnify; recover damage ◇ ～保险 insurance by ways of indemnity / ～措施 indemnifying measure / ～电容器 compensation condenser / ～费 compensation / ～贸易 compensation trade / ～赔款 recovery / ～物 indemnity

【补充】①(增加一部分) replenish; supplement; complement; add: ～存货 replenish the stock / ～人力 replenish manpower ②(追加一些) additional; complementary; supplementary ◇ ～成本 supplementary cost / ～读物 supplementary reading material / ～规定 additional regulations / ～判决 supplementary judgment / ～收益 supplementary earning / ～说明 additional remarks / ～条款 subsidiary; supplementary terms
【补丁】patch: 打～ put a patch on; patch up
【补发】supply again; reissue; pay retroactively
【补付】make up a deficiency in payment; make a deferred payment
【补花】[工美] appliqué
【补给】[军] supply; provision: 缺乏～ go short of supply; supplies running short / 医药～品 medical supplies ◇ ～点 supply point / ～基地 supply base / ～品 supplies / ～线 supply line / ～站 depot
【补假】take a deferred holidays
【补角】[数] supplementary angle
【补救】remedy: ～办法 remedial measure; remedy / ～措施 remedial measure / 无可～ be past remedy; irremediable; irreparable
【补考】make-up examination
【补课】make up a missed lesson
【补炉】[冶] fettling
【补苗】[农] fill the gaps with seedlings
【补偏救弊】remedy defects and rectify errors; rectify abuses
【补票】buy one's ticket after the normal time
【补品】tonic
【补其不足】make up a deficiency
【补其缺漏】compensate for the shortage and leakage
【补其所短】make up for one's shortcomings or defects
【补缺】fill a vacancy; supply a deficiency ◇ ～选举 by-election
【补色】complementary color
【补税】①(补交逃税) pay a tax one has evaded ②(交过期的税) pay an overdue tax
【补体】[医] complement ◇ ～结合试验 complement fixation test
【补贴】subsidy; allowance: 出口～ export subsidy / 粮食～ grain subsidy / 旅费～ travelling allowance / 生活～ living allowances / 由国家给予～ be subsidized by the state
【补习】take lessons after school or work ◇ ～班 a class for supplementary schooling / ～教育 supplementary education / ～学校 continuation school
【补修】study for a second time courses one has flunked
【补选】by-election: ～代表 hold a by-election for a deputy
【补血】enrich the blood ◇ ～剂 blood tonic
【补牙】[医] fill a tooth; have a tooth stopped
【补养】take a tonic or nourishing food to build up one's health
【补药】tonic
【补液】[医] fluid infusion
【补遗】addendum
【补益】[书] benefit; help: 有所～ be of some help
【补语】[语] complement
【补种】reseed; resow; replant
【补助】subsidy; allowance: 煤火～ heating allowance / 生活～ extra expenses; cost of living ◇ ～金 grant-in-aid; subsidy
【补缀】①(缝补) mend; patch; darn ②(拼凑) patch up: ～成文 patch up an article
【补足】bring up to full strength; make up a deficiency; fill ◇ ～语[语] a complement; a supplementary clause

卟

【卟吩】[化] porphin

bù

部 ①(部分) part; section: 腹～ abdominal region / 上～ upper part; 下～ lower part ②(单位) unit; department: 编辑～ editorial department ③(军队领导机构) headquarters; department: 总参谋～ Headquarters of the General Staff / 总后勤～ General Logistics Department / 总政治～ General Political Department ④(政府部门) ministry: 国防～ the Ministry of National Defense / 外交～ the Ministry of Foreign Affairs ⑤(部队) troops; forces ⑥[量] 一～电影 a film / 一～好作品 a fine work of literature
【部队】①(军队) army; armed forces ②(军队的一部分) troops; unit; force: 边防～ frontier forces / 步兵～ infantry / 测绘～ topographic troops / 导弹～ missile unit / 地方～ local forces / 地面～ ground forces / 防空～ air defense unit; anti-aircraft unit / 高射炮～ anti-aircraft artillery unit / 工兵～ engineer troops; sappers / 后备～ reserve unit / 后勤～ service troops; logistics unit / 化学兵～ chemi-

cal troops / 火箭～ rocket unit / 机械化～ mechanized force; mechanized unit / 空降～ airborne troops / 雷达～ radar troops / 骑兵～ cavalry / 汽车运输～ motor transport troops / 伞兵～ parachute troops / 水陆两栖～ amphibious troops / 坦克～ tank unit / 特种～ special troops / 通讯兵～ signal troops / 野战～ field forces / 医疗～ medical troops / 作战～ combat unit ◇ ～代号 code designation

【部分】part; section; portion: ～权力 a portion of the power / ～瘫痪 partial paralysis / ～自给 partly self-supporting

【部件】parts; components; assembly ◇ ～分解图 exploded view

【部类】category; division

【部落】tribe ◇ ～社会 tribal society

【部门】department; branch: 交通～ communications department / 政府各～ various government departments / 主管～ the department responsible for the work

【部首】[语] radicals by which characters are arranged in traditional Chinese dictionaries

【部属】①(下级) subordinate ②(部的附属单位) affiliated to a ministry

【部署】①(安排) arrange; map out; lay out: ～计划 map out the plan ②[军] disposition; deployment: ～兵力 dispose troops / 作战～ battle disposition

【部委】ministries and commissions: 国务院各～ ministries and commissions under the State Council

【部位】position; place: 发音时舌的～ the position of the tongue in pronunciation; tongue position / 受伤～ the location of an injury

【部下】①(军队中被统率的人) troops under one's command ②(下级) subordinate

【部长】minister; head of a department: ～级会议 conference at ministerial level ◇ ～会议 Council of Ministers / ～助理 assistant minister

埠 ①(码头) wharf; pier ②(商埠) port: 本～ this port / 商～ a commercial port / 外～ other port

不 [副] ①(表示否定) not no: ～规则 irregular / ～合法 illegal / ～完全 not complete; incomplete / ～严重 not serious / ～愿意 not willing; reluctant / ～正确 incorrect / 临危～惧 be fearless in face of danger ②(做否定回答) no: 他知道吧? ～, 他不知道。 He knows, doesn't he? No, he doesn't. ③(不能) can't; be not able to: ～辨菽麦 unable to tell beans from wheat ④(用在句末表示疑问) 你明儿来～? Are you coming tomorrow? ⑤(在不字的前后, 叠用相同的词表示不在乎或不相干) 什么难学～难学 no matter how hard it is ⑥(跟"就"搭用, 表示选择) 他～是在车间就是在实验室。 He's either in the workshop or in the laboratory.

【不安】①(不安宁) intranquil; unpeaceful; unstable: 动荡～ turbulence and intranquility ②(过意不去) sorry: 这样麻烦您, 真是～。 Sorry to have caused you so much trouble.

【不安定】unsettled; unstable; insecure; precarious: ～的局面 unstable situation / ～的生活 unsettled life

【不安分】discontented with one's lot

【不安心】not settle down to: ～工作 not settle down to one's work

【不谙世故】know the world as little

【不按规定负担抚养费】non-support

【不白之冤】unrighted wrong; unredressed injustice: 蒙受～ suffer an unredressed injustice

【不败】unbeaten; undefeated; invincible: 保持～纪录 hold an unbeaten record / 立于～之地 put oneself in an invincible position

【不备】unprepared; off guard: 乘其～ catch sb. off guard / 伺其～ watch for a chance to take sb. by surprise

【不比】unlike: ～任何人差 second to none; not inferior to anyone

【不必】[副] need not; not have to: ～担心 there is no need to worry / ～说 needless to say; it goes without saying

【不必要】unnecessary; dispensable; uncalled-for: ～的忧虑 uncalled-for worries / 避免～的牺牲 avoid unnecessary sacrifice

【不避艰险】shrink from no difficulty or danger; make light of difficulties and dangers

【不变价格】[经] fixed price; constant price

【不变资本】[经] constant capital

【不便】①(不方便) inconvenient: 交通～ have poor transport facilities; not be conveniently located / 如果对你没有什么～的话 if it is not inconvenient to you ②(不适当) not suitable; inappropriate ③(手边无钱) have no money at hand; be short of cash: 如果你手头～ if you have no cash at hand

【不辨是非】fail to make a distinction between right and wrong

【不辨菽麦】be unable to tell beans from wheat;

have no knowledge of practical matters
【不别而行】take French leave; go away without saying good-bye
【不…不…】①(用于意思相同或相近的词或词素之前,表示否定,稍强调) ～慌～忙 unhurried; calm; leisurely / ～骄～躁 not conceited or rash; free from arrogance and rashness / ～理～睬 ignore; take no notice of / ～清～楚 not clear / ～声～响 quiet; silent ②(用于意思相对的词或词素之前,表示"既不…又不…") ～大～小 neither too big nor too small; just right / ～多～少 not too much and not too little; just right / ～上～下 suspended in mid air; in a fix / ～死～活 neither dead nor alive; lifeless; lethargic / ～盈～亏 break even ③(表示条件) no... without...: ～破～立 no construction without destruction
【不才】without capability
【不测】accident; mishap; contingency: ～风云 unforeseen storm / 如有～ if anything untoward should happen / 险遭～ have a narrow escape / 以防～ be prepared for any contingency
【不曾】never
【不称职】incompetent; unfit for the job
【不称霸】not seek hegemony
【不成】①(不行) won't do ②[助](表示揣度或反问) 难道这就算了～? Is this to be the end, then?
【不成材】good-for-nothing; worthless; ne'er-do-well
【不成大器】he will never amount to much
【不成敬意】just a little token to show my respect to you
【不成事实】it is not a proper matter
【不成体统】behave very badly; wildly
【不成文法】[法] unwritten law
【不诚实】dishonesty
【不承担义务】not commit oneself (to); make no commitment (to)
【不承担责任】irresponsibility
【不承认】derecognition; non-recognition: ～事实 ignore the facts
【不承认主义】policy of nonrecognition
【不逞之徒】desperado
【不齿】[书] despise; hold in contempt: ～于人类的狗屎堆 filthy and contemptible as dog's dung
【不耻下问】not feel ashamed to ask and learn from one's subordinates
【不充分的证据】insufficient evidence
【不出所料】as expected; not beyond expectation: 不出我所料 not beyond my expectation
【不出庭】default of appearance; non-appearance: ～证书 nonappearance certificate
【不揣冒昧】[套] venture to; presume to; take the liberty of
【不存芥蒂】non-existence of the stem of mustard plant
【不辞而别】leave without saying good-bye
【不辞辛苦】make nothing of hardships
【不错】①(正确) correct; right: 假如我理解得～的话 ... if I understand right ②(好) not bad; pretty good: 这片子真～。This film is not bad at all.
【不打不成器】spare the rod and spoil the child
【不打不相识】[谚] from an exchange of blows friendship grows; no discord, no concord
【不打自招】confess without being pressed; make a confession without duress; reveal unintentionally
【不大】①(不经常) not often; seldom; rarely; hardly; scarcely: 他～看电影。 He seldom goes to cinema. ②(程度不深) not very; not quite: ～可能的事 something improbable / ～清楚 not too clear
【不大离儿】[口] ①(差不多) pretty close; just about right ②(不错) not bad
【不待说】needless to say; it goes without saying
【不丹】Bhutan ◇ ～人 Bhutanese / ～语 Bhutanese (language)
【不单】①(不止) not the only ②(不但) not merely; not simply
【不但】[连] not only
【不当】unsuitable; improper; inappropriate: ～裁决 unfair verdict / ～得利 illegal profit; ill-gotten gains / ～逮捕 improper arrest; unjust arrest / ～干涉 unwarranted intervention / ～判决 unjust judgement / ～行为 improper act; undue behavior / ～延误 undue delay / 处理～ not be handled properly / 措词～ wrong choice of words
【不倒翁】tumbler; roly-poly
【不到黄河心不死】[谚] not stop until one reaches the Huanghe River; not stop until one reaches one's goal; refuse to give up until all hope is gone
【不到庭】[法] absence
【不道德】immoral
【不得】①(表示不可能、不能够) cannot; not to: 吃～ not to be eaten / 记～ can't remember / 提～ not to be mentioned ②(表示不应该、不可

以) should not; must not; be not supposed to; not be allowed: 马虎～ mustn't be careless
【不得不】have no choice but to; cannot but; have to
【不得而知】unknown; unable to find out
【不得劲】[口] ①(不顺手) awkward; unhandy ②(不舒服) be indisposed; not feel well
【不得了】①(表示情况严重) terrible; horrible; desperately serious ②(表示程度很深) very; extremely, terribly: 高兴得～ be extremely happy／热得～ be awfully hot
【不得其法】do not know the right way
【不得人心】not enjoy popular support; be unpopular
【不得上诉】not subject to cassation
【不得要领】fail to grasp the main points
【不得已】act against one's will; have no alternative but to; have to: ～而求其次 have to be content with the second best／～而为之 feel constrained to act against one's will
【不登大雅之堂】[书] not appeal to refined taste; be unrefined; be unpresentable
【不等】vary; differ: 大小～ differ in size／数量～ vary in amount
【不等边三角形】[数] scalene triangle
【不等号】[数] sign of inequality
【不等价交换】[经] exchange of unequal values
【不等式】[数] inequality
【不抵抗主义】policy of nonresistance
【不迭】①(来不及;急忙) cannot cope; find it too much: 后悔～ too late for regrets／忙～ hasten ②(不停止) incessantly: 称赞～ praise profusely／叫苦～ complain incessantly
【不定】①(表示不肯定) indefinite; indeterminate ②(不稳定) unsteady; drifting; fitful: 方向～的风 a fitful wind／漂泊～ drifting from place to place／心神～ unsteady in mind ◇ ～变异[生] indeterminate variation／～方程[数] indeterminate equation／～根[植] adventitious root／～冠词[语] indefinite article／～积分[数] indefinite integral／～式[语] infinitive／～芽[植] adventitious bud
【不定期】irregularly scheduled; nonscheduled: ～存款 undated deposit／～航班 nonregular service／～航线 tramp route／～刊物 publication issued at irregular intervals
【不动产】real estate; immovable property; immovables
【不动脑筋】don't use one's brain; don't take the trouble to think
【不动声色】maintain one's composure; stay calm and collected; not turn a hair; not bat an eyelid
【不动摇】firm; stand firm; steadfast
【不冻港】ice-free port; open port
【不独】not only
【不端】improper; dishonorable: 品行～ bad conduct; dishonorable behavior
【不断】unceasing; uninterrupted; continuous; constant: ～调整 continual readjustment／促进生产力的～发展 promote the uninterrupted growth of the productive forces
【不断革命论】the theory of uninterrupted revolution
【不对】①(错误) incorrect; wrong ②(不正常) amiss; abnormal; queer
【不对劲】not in harmony; feeling not at par; listless
【不二法门】the one and only way; the only proper course to take
【不二价】uniform price
【不贰过】not to repeat a previous mistake
【不乏】[书] there is no lack of: ～其人 there is no lack of such people; such people are not rare／～先例 there is no lack of precedents
【不法】lawless; illegal; unlawful: ～活动 illegal activity／～侵害 unlawful infringement／～伤害 unlawful wounding／～收益 illegal interest／～行为 unlawful practice; an illegal act／～行为的责任 delictual liability／～占有 dishonest possession／～之徒 a lawless person
【不凡】out of the ordinary; out of the common run: 自命～ consider oneself a person of no ordinary talent; have an unduly high opinion of oneself
【不妨】there is no harm in; might as well: ～一试 there is no harm in try
【不放心】be anxious for; feel worried about
【不费吹灰之力】as easy as blowing off dust; not needing the slightest effort
【不分彼此】make no distinction between what's one's own and what's another's; share everything; be on very intimate terms
【不分敌我】not to distinguish between the enemy and ourselves
【不分高低】be equally matched
【不分泾渭】make no such distinction between good and evil
【不分巨细】regardless of size
【不分亲疏】irrespective the degree of intimacy
【不分青红皂白】indiscriminately
【不分轻重缓急】without regard to the relative

importance or urgency
【不分胜负】tie; draw; come out even: 一场～的比赛 a drawn game
【不分是非】fail to distinguish right from wrong
【不分昼夜】day and night
【不孚众望】not inspire popular confidence; not popular with the masses
【不符】not agree with; not conform to: 名实～ have an undeserved reputation / 言行～ deeds not matching words / 与事实～ be inconsistent with the facts
【不符合客观实际】at variance with objective reality
【不服】refuse to obey; refuse to accept as final; remain unconvinced by; not give in to: ～裁判 refuse to accept the referee's ruling / ～从 disobey / ～老 refuse to give in to old age / ～气 recalcitrant; unwilling to submit / ～输 refuse to take defeat lying down / ～罪 not admit one's guilt; plead not guilty
【不服上诉】appeal against; lodge
【不服水土】not accustomed to the climate of a new place; not acclimatized
【不附带条件的贷款】untied loan
【不负所托】merit someone's trust
【不负义务的】uncommitted
【不负责任】breach of duty
【不干不净】unclean; filthy: 嘴里～ be foul-mouthed
【不干涉】noninterference; nonintervention: 互～内政 noninterference in each other's internal affairs ◇ ～政策 policy of noninterference
【不干预】non-intervention
【不甘】unreconciled to; not resigned to; unwilling: ～寂寞 unwilling to remain out of the limelight; eager to seek publicity / ～居人后 not willing to be outdone / ～落后 unwilling to lag behind / ～示弱 not to be outdone
【不甘心】not reconciled to; not resigned to: ～于失败 not take one's defeat lying down
【不感兴趣】lose interest in
【不敢】dare not; not dare
【不敢当】[谦] I really don't deserve this; you flatter me.
【不敢领教】too bad to be accepted
【不敢越雷池一步】dare not go one step beyond the prescribed limit
【不敢正视】not dare to face up to; not dare to look things straight in the face
【不公】unjust; unfair: 办事～ be unfair in handling matters
【不公开】not public; not published; private: ～审理 heard in private session / ～审讯 in camera
【不公平】unfair; unjust: ～交易 wrongful dealing
【不公正】injustice: ～行为 injustice
【不攻自破】collapse of itself
【不恭】disrespectful: 却之～ it would be disrespectful to decline / 言词～ use disrespectful language
【不共戴天】will not live under the same sky; absolutely irreconcilable: ～的敌人 sworn enemy / ～之仇 inveterate hatred
【不苟】not lax; not casual; careful; conscientious: ～言笑 serious in speech and manner / 工作一丝～ work most conscientiously
【不够】not enough; insufficient; short of; lack: ～本 at a loss; less than cost price / ～好 not good enough / ～朋友 hardly a friend in need / ～条件 fail to meet the requirements / 准备～ be inadequately prepared
【不辜负】be worthy of; live up to: ～党的信任 be worthy of the trust of the Party / ～希望 justify the hope of; not to disappoint
【不顾】disregard; ignore; in spite of; without giving any thought to: ～大局 ignore the larger issues / ～法律 disregard of law / ～后果 regardless of the consequences / ～情面 without any consideration of personal feelings / ～事实 ignore the facts; have no regard for the truth / ～死活 take no notice of dead or alive / ～一切 up hill and down dale
【不关你事】none of your business
【不关痛痒】irrelevant; insignificant
【不关心】be indifferent to; not concern oneself with
【不管】no matter; whether... or; regardless of: ～结果如何 whatever the consequences / ～哪一个 no matter which; whichever; whichsoever / ～什么地方 wherever; no matter where / ～什么时候 no matter when; whenever / ～谁 whoever; whosoever; no matter who / ～怎样 whatever happens; come what may; in any case
【不管部部长】minister without portfolio
【不管三七二十一】casting all caution to the winds; regardless of the consequences; recklessly
【不光】①(不止) not the only one ②(不但) not only
【不光采】disgraceful; dishonorable; ignominious: ～的行为 a disgraceful act / 充当～的角

色 play a dishonorable role
【不规则】irregular: ～变化[语] irregular inflection / ～动词 irregular verb / ～曲线 irregular curve
【不轨】against the law or discipline: 图谋～ engage in conspiratorial activities
【不过】①(只;仅仅)[副] only; just; merely; nothing but; no more than: ～是个烟幕 no more than a smoke screen ②(转折)[连]but; nevertheless; however; only
【不过尔尔】[书] merely mediocre; just middling
【不过如此】so-so; tolerably passable; barely acceptable
【不过问】keep aloof from; not inquire into; not look into
【不过意】sorry; regretful; apologetic
【不寒而栗】shiver all over though not cold; tremble with fear; shudder: 使人～ make sb.'s flesh creep; give sb. the creeps; make sb.'s hair stand on end
【不含糊】[口] ①(不模棱两可) unambiguous; unequivocal; explicit ②(不平凡) not ordinary; really good; uncommonly
【不好的下场】unhappy end; sad issue
【不好过】①(指生病) unwell; not well; indisposed; out of sorts; under the weather ②(日子不好过) have a very hard time; things are tough
【不好惹】not to be trifled with; not to be pushed around; stand no nonsense
【不好意思】①(害羞) embarrassed; shy; coy; diffident ②(碍于情面不便或不肯) would be rude; find it embarrassing (to do sth.)
【不合】①(不符合) not conform to; be unsuited to; be out of keeping with: ～标准 not up to the standard / ～传统 be out of keeping with traditions / ～规定 not conform to the rules / ～规格 fall short of the specifications / ～乎事实 at variance with the facts ②(不适合) not fit; not suited to: ～他的胃口 be not to his taste
【不合法】wrongful; wrongfulness: ～的民事诉讼 wrongful civil proceeding / ～的刑事诉讼 wrongful criminal proceedings
【不合格】unqualified; substandard: ～的产品 substandard products / ～的医务人员 unqualified medical worker
【不合理】unreasonable; irrational: ～的规章制度 irrational rules and regulations / ～的推论 illogical deduction / 各种～的现象 all sorts of irrational practices / 一个～的要求 an unreasonable demand
【不合逻辑】illogic: ～的推论 paralogism
【不合时宜】untimely; ill-timed; be out of keeping with the times: ～的话 untimely remark
【不合适】improper; inappropriate; out of place: 在～的时间访问了某人 call on sb. at an inopportune hour
【不合作】uncooperative; noncooperative
【不和】discord; dissension; be on bad terms; be at loggerheads (with): 制造～ sow discord
【不怀好意】harbor evil designs; harbor malicious intentions
【不欢而散】part on bad terms; break up in discord
【不慌不忙】unhurriedly; leisurely: ～地走来 come over in an unhurried manner
【不讳】[书] without concealing anything: 供认～ candidly confess; confess everything; make a clean breast of everything / 直言～ speak bluntly; be outspoken; call a spade a spade
【不会】①(不可能) will not; not likely: 人～多 there won't be too many people ②(不能做) can't; be unable to; be incapable of: ～抽烟 do not smoke / ～说谎 be incapable of telling lies
【不惑】without doubt; with full self-confidence
【不羁】[书] unruly; uninhibited
【不吉之兆】an unlucky omen
【不及】①(不如;比不上) not as good as; inferior to ②(来不及) too late: 后悔～ too late for regrets ③(够不上) fall short of; fail to reach: 力所～ beyond one's power
【不及格】fail; flunk: 她英语～。She failed in English.
【不及物动词】[语] intransitive verb
【不即不离】be neither too familiar nor too distant; keep sb. at arm's length
【不急之务】a matter of no great urgency
【不济】[口] not good; of no use: 眼力～ one's eyesight is failing
【不济事】no good; of no use; not of any help
【不计】disregard; not take into account; irrespective of: ～报酬 regardless of pay / ～成败 despite of success or failure / ～个人得失 disregard one's personal gain or loss / ～名利 not mindful of fame and gain
【不计其数】countless; innumerable
【不记名投票】secret ballot
【不加可否】refuse to comment
【不假思索】without thinking; without hesitation; readily; offhand
【不坚定】lack resolution; not to be firm enough: ～分子 waverer; wavering element
【不简单】①(复杂) not simple; fairly complicat-

ed ②(不平凡) unusual; remarkable
【不见】①(未见面) haven't seen; haven't met ②(失踪) disappear; be lost; be missing
【不见得】not likely; not necessarily: ～对 not necessarily correct / ～可能 hardly possible
【不见棺材不掉泪】not shed a tear until one sees the coffin; refuse to be convinced until faced with grim reality
【不见经传】not to be found in the classics; not authoritative; unknown
【不健全】unsound; imperfect; defective: 机构～ organizationally imperfect / 精神～ unsound in mind; mentally defective / 心脏～ have an unsound heart
【不讲道理】be unreasonable; refuse to see reason; be impervious to reason: ～的人 unreasonable person
【不讲情面】without sparing sensibilities; not care to save sb.'s face
【不骄不躁】free from arrogance and impetuosity; neither conceited nor rash
【不结盟】nonalignment ◇ ～国家 nonaligned countries / ～政策 nonalignment policy
【不解】①(不理解) not understand: 迷惑～ be puzzled; be bewildered / ～其意 not understand what he means / ～之谜 an unsolved riddle; enigma; mystery ②(解不开) indissoluble: ～之仇 irreconcilable enmity / ～之缘 an indissoluble bond
【不介入】nonintervention; noninvolvement; nonentanglement: ～政策 policy of nonintervention
【不介意】not mind; not care
【不禁】can't help (doing sth.); can't refrain from: ～大笑 can't help bursting into laughter / ～下泪 can't refrain from falling tears
【不矜不躁】be neither proud nor irascible
【不仅】①(不止) not the only one ②(不但) not only: ～如此 not only that; nor is this all; moreover
【不进则退】not to advance is to go back
【不近人情】not amenable to reason; unreasonable
【不尽然】not exactly so; not necessarily so
【不经一事,不长一智】you can't gain knowledge without practice; wisdom comes from experience
【不经意】carelessly; by accident; thoughtless
【不经之谈】absurd statement; cock-and-bull story
【不景气】①[经](经济不繁荣) depression; recession; slump ②(不兴旺) depressing state
【不胫而走】①(传播快而广) get round fast; spread like wildfire ②(畅销) sell like hot cakes
【不久】soon; before long
【不咎既往】not censure sb. for his past misdeeds; overlook sb.'s past mistakes; let bygones be bygones
【不拘】①(不拘泥;不计较) regardless of; not stick to; not confine oneself to: ～礼节 not stand on ceremony / ～小节 regardless of trivial matters; not bother about small matters / ～形式 not particular about form ②(不论) whatever
【不拘形迹】not stick to formalities; not standing on ceremony
【不拘一格】not stick to one pattern
【不倦】tireless; untiring; indefatigable: 诲人～ be tireless in teaching; teach with tireless zeal
【不绝】without cease
【不绝如缕】①(像细线一样连着) hanging by a thread; very precarious; almost extinct ②(声音微弱) linger on faintly
【不堪】①(承受不了) can't bear; can't stand; be too deplorable to: ～回首 can't bear to look back / ～其苦 can't bear the hardship / ～入目 disgusting / ～设想 worse than anything imaginable / ～一击 can't stand a single blow / ～坐食 can't bear to sit and eat ②(程度深) utterly; extremely: 穿得破烂～ be dressed in rags / 狼狈～ be in a sorry plight / 疲惫～ extremely tired; exhausted / 贫困～ extremely poor
【不亢不卑】neither haughty nor humble; neither overbearing nor servile; neither supercilious nor obsequious
【不可】①(不可以) can not; should not; must not: ～一概而论 must not make sweeping generalizations ②(非…不可) must: 我非去～. I simply must go.
【不可避免】be inevitable: ～的错误 inevitable mistake / ～的事故 inevitable accident
【不可辩驳的事实】an irrefutable fact
【不可剥夺】not to be deprived; inalienable
【不可撤回的诉状】irrevocable indictment
【不可撤销的判决】irrevocable judgment
【不可代替之物】non-fungible thing
【不可多得】rare; exceptional; hard to come by; hard to get: ～的机会 rare opportunity / ～的人才 exceptional talent
【不可分割】indivisible; inseparable: ～的领土 inseparable territory / ～的权利 undivided

right / ～的整体 indivisible whole
【不可告人】not to be divulged; hidden: ～的动机 ulterior motives / ～的勾当 a sinister trick
【不可估量】inestimable; incalculable; beyond measure
【不可接受的证据】inadmissible evidence
【不可救药】①(不能治愈) incurable; beyond cure ②(无可挽救) incorrigible; hopeless
【不可开交】be locked; be tied up: 打得～ be locked in a fierce struggle / 争得～ be engaged in a heated argument
【不可抗拒】irresistible; inexorable: ～的规律 inexorable law / ～的原因 irresistible cause
【不可抗力】[法] force majeure
【不可靠证据】flimsy evidence
【不可理喻】be impervious to reason; won't listen to reason
【不可弥补】irretrievable; irrecoverable; irredeemable: ～的损失 irretrievable loss
【不可名状】indescribable; beyond description
【不可磨灭】indelible: ～的功绩 everlasting merit / ～的贡献 an indelible contribution / ～的印象 indelible impressions
【不可逆转】irreversible; not to be turned back
【不可偏废】not to stress one to the neglect of the other
【不可侵犯权】[外] inviolability
【不可缺少】indispensable; requisite; essential; absolutely necessary
【不可让与的权利】inalienable right
【不可胜数】countless; innumerable
【不可收拾】irremediable; unmanageable; out of hand; hopeless
【不可思议】inconceivable; unimaginable: ～的逻辑 inconceivable logic / ～的神态 a mysterious look
【不可调和】irreconcilable; incompatible
【不可同日而语】cannot be mentioned in the same breath
【不可推卸】bounden; inescapable: ～的责任 inescapable responsibility
【不可挽回】irretrievable; irrecoverable
【不可一世】consider oneself unexcelled in the world; be insufferably arrogant
【不可逾越】impassable; insurmountable; insuperable: ～的鸿沟 an impassable chasm / ～的障碍 an insurmountable barrier
【不可争辩】beyond dispute
【不可知论】[哲] agnosticism ◇ ～者 agnostic
【不可终日】be unable to carry on even for a single day; be in a desperate situation
【不可捉摸】subtle; elusive; intangible; hard to grasp or trace: ～的性格 subtle temperament; elusive character
【不可转让】non-assignable; non-negotiable; non-transferable; unalienable
【不克】be unable to; cannot: ～胜任 be unequal to the job
【不客气】①(使人难堪) be rude to; be hard on: 对某人太～ be too rude to sb. ②(直率) frank; candid; straight forward: ～地告诉你 tell you frankly ③(客套) don't mention it; you're welcome; not at all
【不快】①(不愉快) be unhappy; be displeased ②(不舒服) be indisposed; feel under the weather; be out of sorts ③(慢) slow
【不愧】be worthy of; deserve to be called; prove oneself to be
【不赖】[方] not bad; good; fine
【不稂不莠】useless; worthless; good-for-nothing
【不劳而获】reap without sowing; profit by other people's toil
【不冷不热】neither very warm nor cold; lukewarm: ～的态度 lukewarm attitude
【不理睬】ignore; neglect; take not heed; turn a deaf ear to: ～那些流言蜚语 ignore those gossips
【不力】not do one's best; not exert oneself: 办事～ not do one's best in one's work; be slack in one's work / 领导～ not exercise effective leadership
【不利】①(没好处) unfavorable; disadvantageous; harmful; detrimental ②(不顺利) unsuccessful
【不良】bad; harmful; unhealthy: ～倾向 harmful trends / ～现象 unhealthy tendencies / ～影响 harmful effects / 存心～ harbor evil intentions; have ulterior motives / 消化～ poor digestion ◇ ～导体 non-conductor
【不量力】lack of proper estimation of one's strength; not reckon oneself fairly
【不了】without end
【不了解】do not understand; not familiar with
【不了了之】settle a matter by leaving it unsettled; end up with nothing definite
【不料】unexpectedly; to one's surprise
【不列颠百科全书】Enciclopaedia Britannica
【不吝】[套] not stint; not grudge; without sparing: ～指教 spare no advice; not hesitate to make comments

【不灵】not work; be ineffective: 耳朵～ dull of hearing / 这机器～了。 The machine doesn't work.
【不留情面】be very strict; disregard other's "face" or feelings
【不留意】careless
【不留余地】leave no room; make no allowance (for)
【不露声色】not show one's feelings, intentions, etc.
【不伦不类】neither fish nor fowl; nondescript: ～的比喻 an inappropriate metaphor; a far-fetched analogy
【不论】no matter; irrespective of; regardless of; whether... or...: ～性别年龄 regardless of sex and age; irrespective of sex and age
【不落窠臼】not follow the beaten track; have an original style
【不买帐】not buy it; not go for it
【不满】resentful; discontented; dissatisfied: 对处境～ be discontented with one's circumstances
【不忙】there's no hurry; take one's time
【不毛之地】barren land; desert
【不免】would naturally; unavoidable: ～失望 would naturally be disappointed
【不妙】not going well; anything but reassuring
【不明】not clear; unknown; fail to understand: ～事理 lack common sense / ～是非 confuse right and wrong / ～真相 be unaware of the truth; be ignorant of the facts
【不明飞行物】unidentified flying objects; UFO
【不明确】indeterminate; undefined
【不明智】unwise; ill-advised: ～的行动 unwise action
【不名数】[数] abstract number
【不名一文】without a penny to one's name; penniless; stony-broke
【不名誉】disreputable; disgraceful
【不谋而合】have the same idea by coincidence; happen to hold the same view
【不耐烦】impatient; out of patience
【不男不女】neither a male nor a female
【不能】can not; must not; should not
【不能不】have to; cannot but
【不能赦免】unabsolvable
【不能自拔】be inextricably bogged down; cannot extricate oneself
【不能自圆其说】unable to make out one's case; unable to justify one's own assertion
【不念旧恶】forget old grievances; forgive and forget
【不怕】be not afraid of; not fear: ～艰难困苦 defy difficulties and hardships / ～牺牲 be not afraid of sacrifice
【不配】be unworthy of; be not qualified to
【不偏不倚】even-handed; impartial; unbiased: ～的评价 unbiased appraisal
【不平】①(不平坦) uneven; not level; not smooth ②(不公平) injustice; unfairness ③(因不公平而气愤) indignant; resentful
【不平等待遇】unequal treatment
【不平等条约】unequal treaty
【不平衡】disequilibrium
【不平则鸣】where there is injustice, there will be an outcry; man will cry out against injustice
【不平之鸣】voice of protest
【不破不立】without destruction there can be no construction
【不期而遇】meet by chance; have a chance encounter
【不起诉】nonprosecution
【不巧】unfortunately; as luck would have it
【不切实际】unrealistic; unpractical; impracticable: ～的幻想 unrealistic notions; fanciful ideas / ～的计划 an impracticable plan
【不切题】irrelevant to the subject; off the point; beside the mark
【不求名利】do not care for wealth or fame
【不求上进】have no desire for progress; not seek to make progress
【不求甚解】not seek to understand things thoroughly; be content with superficial understanding
【不屈】unyielding; unbending: 坚强～ iron-willed and unyielding
【不屈不挠】unyielding; indomitable: ～的意志 an inflexible will
【不然】①(不是这样) not so; not the case; no ②(否则) if not so; otherwise; or else
【不人道】inhuman
【不仁】①(不仁慈) not benevolent; heartless ②(失去知觉) numb: 麻木～ insensitive; apathetic
【不忍】cannot bear to: ～坐视 cannot bear to stand idly by
【不认帐】go back on one's word; refuse to acknowledge
【不日】within the next few days
【不容】not tolerate; not allow; not brook: ～辩解 allow of no excuse / ～耽搁 allow of no delay / ～反悔 estoppel / ～讳言 there is no de-

nying that / ～申辩 immune to all pleas / ～置辩的事实 undeniable fact / ～置疑 undoubtedly; beyond doubt
【不如】①(比不上) not equal to; not as good as ②(还是) had better; would rather
【不入虎穴,焉得虎子】how can you catch tiger cubs without entering the tiger's lair; nothing venture, nothing gain
【不三不四】①(不正派) dubious; shady; indecent ②(不象样子) nondescript; neither fish nor fowl
【不善】①(不擅长) not good at; a bad hand at ②(不好) bad; ill; evil
【不设防城市】open city; undefended city; unfortified city
【不声不响】stealthily; furtively
【不生育】infertility
【不胜】①(承担不了) cannot bear; be unequal to ②(非常) tremendously; very deeply: ～感激 be very much obliged; be deeply grateful ③(表示不能做或做不完) too ... to: 防～防 be impossible to prevent / 数～数 too numerous to count
【不胜枚举】too numerous to mention individually
【不胜其烦】be pestered unbearably
【不胜任】incompetent; unfit for the post
【不胜庆幸】have great cause for rejoicing
【不失时机】seize the opportune moment; lose no time
【不失为】can yet be regarded as; may after all be accepted as
【不识】fail to see; be ignorant of; not know; not appreciate: ～大体 fail to see the general interest / ～时务 be ignorant of the times / ～抬举 not know how to appreciate favors
【不时】①(时时) frequently; often ②(随时) at any time
【不时之需】a possible period of want or need: 以备～ for emergency needs; to provide against a rainy day
【不食烟火】stop eating cooked food
【不使用武力】nonuse of force
【不是玩儿的】[口] it's no joke
【不是味儿】①(味道不正) not the right flavor; not quite right; a bit off ②(不对头) fishy; queer; amiss ③(不好受) feel bad; be upset
【不适】unwell; indisposed; out of sorts: 略感～ feel a bit unwell / 全身～ general malaise / 胃部～ have a stomach upset
【不适应】not adapted to; not fit to; not suited to: 对气候～ not accustomed to the climate; not acclimatized
【不守信用】break one's word; not keep one's word; go back on one's word
【不受处罚】get-off
【不受法律保护】outlawry
【不受法律约束的责任】imperfect obligation
【不受欢迎的人】[外] persona non grata
【不受控制】out of control
【不受理】[法] reject a complaint; [外] refuse to entertain
【不受审判】immunity from trial
【不熟】①(未烧熟) not yet done; still raw ②(生疏) unacquainted with; unfamiliar with
【不爽】①(身体或心情不爽快) not well; out of sorts; in a bad mood ②(没有差错) without discrepancy; accurate: 丝毫～ not deviate a hair's breadth; be perfectly accurate; be just right
【不爽快】not frank
【不顺眼】incurring dislike; disagreeable to the eye; look odd: ～的事 something offending the eye
【不死心】unwilling to give up; unresigned; not give up hope
【不送】[套] don't bother to see me out
【不送气】[语] unaspirated ◇ ～音 unaspirated sound
【不俗】original; uncommon; not hackneyed
【不速之客】uninvited guest; unexpected guest; chance comer; casual visitor
【不随意肌】[生理] involuntary muscle
【不碎玻璃】safety glass; shatterproof glass
【不通】①(阻隔) be obstructed; be blocked up; be impassable: 鼻子～ have a stuffed up nose / 此路～。Not a Through Road. / 电话～。The line's dead. ②(转用于抽象事物) 想～ can't figure out why / 行～ won't work ③(文理错误) not make sense; be illogical; be ungrammatical.
【不通情理】unreasonable; impervious to reason
【不同】not alike; different; distinct: 有～癖好的人 people with dissimilar hobbies / 在～的程度上 to varying degrees
【不同步】outer-sync; out of step
【不同凡响】outstanding; out of the ordinary; out of the common run
【不同情节】distinctive circumstances
【不痛不痒】scratching the surface; superficial; perfunctory: ～的批评 superficial criticism /

讲些～的话 make some perfunctory remarks
【不透风】stuffy
【不透明】opaque ◇ ～色 body color／～体 opaque body／～性 opacity
【不透气】①(密封) airtight; air proof ②(通风不良) stuffy; badly ventilated
【不透水】waterproof; watertight; impermeable ◇ ～层[地] impermeable stratum; impervious bed
【不吐气】[语] unaspirated ◇ ～音 unaspirated sound
【不妥】not proper; inappropriate: 觉得有些～ feel that something is amiss
【不外】not beyond the scope of; nothing more than
【不完全】incomplete; imperfect ◇ ～花[植] imperfect flower／～句[语] incomplete sentence／～物权 imperfect real right／～信托 imperfect trust／～叶[植] incomplete leaf／～中立[法] imperfect neutrality
【不完善赠与】imperfect gift
【不为已甚】not be too hard on sb.
【不为所动】remain unmoved
【不畏】defy: ～艰险 defy hardship and danger; take the bull by the horns／～强暴 defy brutal force
【不闻不问】not bother to ask questions or listen to what's said; show no interest in sth.; be indifferent to sth.
【不稳】unstable; unsteady; not firm; insecure: 站～ be not steady on one's legs; can't keep one's balance／局势～。The situation was unstable. ◇ ～平衡[物] unstable equilibrium
【不问】①(不过问) pay no attention to; ignore; disregard; not consider: ～前因后果 not consider the cause and effect／～是非曲直 make no distinction between right and wrong ②(不审讯;不追究) let go unpunished; let off
【不问青红皂白】without forethought
【不无】not without: ～可取之处 not without merit; not altogether unacceptable／～小补 not be without some advantage; be of some help
【不务正业】①(不专心本职工作) be unmindful of one's own work; not attend to one's duties ②(不从事正当行业) not engage in honest work
【不惜】①(不顾惜) not stint; not spare: ～工本 spare neither labor nor money; spare no expense／～一切代价 at all costs; at any cost ②(舍得) not scruple; not hesitate: ～牺牲个人的一切 never balk at any personal sacrifice
【不暇】have no time (for sth.); be too busy (to do sth.): ～顾及 be too busy to attend to sth.
【不下于】①(不少于) as many as; no less than ②(不低于) not inferior to; as good as; on a par with
【不现实】unrealistic; impractical; fanciful: ～的计划 impractical plan／～的态度 unrealistic attitude
【不相称】unbecoming; unsuited; out of proportion: 同某人地位～ be unsuited to sb.'s position
【不相干】be irrelevant; have nothing to do with: ～的话 irrelevant remarks／完全～ altogether irrelevant
【不相容】incompatible: 水火～ incompatible as fire and water; mutually antagonistic
【不相上下】equally matched; about the same: 能力～ of about the same ability; equally able
【不相往来】have no dealings with each other; be not on speaking terms
【不祥】ominous; inauspicious: ～之兆 an ill omen
【不详】①(不清楚) not quite clear: 内容～ contents are not clear ②(不详细) not in detail; unspecified
【不像话】①(言语、行动不合理) unreasonable; nonsensical; absurd ②(坏得无法形容) outrageous; monstrous; scandalous; shocking
【不像样】①(不够标准;外形不好) in poor shape; unpresentable; shapeless ②(用在"得"后作补语) beyond recognition: 破得～ worn to shreds／瘦得～ extremely thin; worn to a mere shadow
【不肖】[书] unworthy: ～之徒 worthless fellow／～子孙 unworthy descendants
【不孝】not in accordance with filial
【不协调】inharmonious; discordant; out of tune
【不屑】disdain to do sth.; think sth. not worth doing; feel it beneath one's dignity to do sth.: ～教诲 disdain to give instruction／～为友 feel shame to be friends with／～一顾 not worth a single glance; cock a snoot at
【不谢】[套] don't mention it; not at all
【不懈】untiring; unremitting; indefatigable: 坚持～ persevere unremittingly／作～的努力 make unremitting efforts; make a sustained effort
【不信任】distrust; not believe in; have no confidence in: ～案 no-confidence motion／～他

人 be distrustful of others / ～投票 vote of no-confidence
【不信邪】not believe in heresy; refuse to be misled by fallacies
【不兴】①(不流行) out of fashion; outmoded ②(不许) impermissible; not allowed: ～这样做。That's not allowed. ③(不能) can't
【不行】①(不可以) will not do; be not allowed ②(不中用) be no good; not work: 这个方法～。This method just doesn't work. ③(不好) not good; poor: 她画画～。She is a poor painter. ④(表示程度很深) awfully; extremely; deeply: 高兴得～ awfully happy / 累得～了 extremely tired; worn-out
【不行了】on the point of death; dying
【不行使应享权利】non-exercise of entitlements
【不省人事】be unconscious; be in a coma
【不幸】①(灾祸) misfortune; adversity: ～中之大幸 a lucky break out of misfortune / 遭受种种～ experienced all sorts of misfortunes ②(不希望发生而发生的) unfortunately: ～而言中。The prediction has unfortunately come true. ④(不幸运) unfortunate; sad: ～的消息 sad news
【不休】endlessly; ceaselessly: 争论～ argue endlessly; keep on arguing
【不修边幅】not care about one's appearance; be slovenly: ～的人 a sloven
【不朽】immortal: ～的功勋 immortal deeds / ～的著作 an immortal masterpiece
【不锈钢】stainless steel
【不虚此行】the trip has not been made in vain; the trip has been well worthwhile; it's been a worthwhile trip
【不许】①(不允许) not allow; not permit; must not: ～说谎。You mustn't tell lies. / ～停车! No Parking! ②(不能) can't
【不宣而战】open hostilities without declaring war; start an undeclared war
【不旋踵】[书] in less time than it takes to turn one's hells; a very brief moment
【不学而能】do a thing easily and naturally
【不学无术】have neither learning nor skill; be ignorant and incompetent: ～之徒 ignoramus; unread and useless fellow
【不寻常】unusual; uncommon; extraordinary: ～的记忆力 phenomenal memory
【不逊】[书] rude; impertinent: 出言～ make impertinent remarks
【不雅观】offensive to the eye; unbecoming
【不亚于】not second to; as good as
【不言不语】keep silent
【不言而喻】it goes without saying; it is self-evident
【不厌】not mind doing sth.; not tire of; not object to: ～其烦 not mind taking all the trouble; take great pains; be very patient / ～其详 go into minute details; dwell at great length
【不扬】bad-looking: 其貌～ have a ugly face
【不要紧】①(不成问题;没有妨碍) unimportant; not serious ②(不碍事) it doesn't matter; never mind ③(表面上似乎不妨碍) it looks all right, but
【不要脸】shamless; brazen; have no sense of shame
【不一】vary; differ: 长短～ differ in length / 观点～ not identical in views / 质量～ vary in quality
【不一定】uncertain; not sure; not necessarily so
【不一而足】by no means an isolated case; numerous: 如此等等,～ and so on and so forth
【不一会儿】in a moment; in a little while; after a while; a few moments later
【不依】①(不听从) not comply; not go along with ②(不放过) not let off easily; not let sb. get away with it
【不宜】not suitable; inadvisable: ～操之过急。It's no good being overhasty.
【不遗余力】spare no pains; spare no efforts; do one's utmost
【不已】endlessly; unceasingly: 赞叹～ praise again and again
【不以为然】object to; take exception to; not approve
【不义之财】ill-gotten wealth
【不亦乐乎】awfully; terribly; exceedingly; extremely: 忙得～ awfully busy
【不易之论】perfectly sound proposition; unalterable truth; irrefutable argument
【不意】[书] ①(不料) unexpectedly ②(没想到) unawareness; unpreparedness: 出其～,攻其无备 catch sb. unprepared; take sb. by surprise
【不翼而飞】①(东西突然不见了) take to itself wings; disappear without trace ②(流传迅速) spread fast; spread like wildfire
【不应由法院解决的纠纷】non-justifiable dispute
【不用】need not: ～客气 don't mention it; don't stand on ceremony / ～说 it goes without saying; needless to say
【不由得】can't help; can not but
【不由分说】allowing no explanation; leaving

no chance to explain
【不由自主】can't help; involuntarily: ～地流下了眼泪 couldn't help shedding tears; couldn't hold back one's tears
【不予】not give; deny; refuse; not grant: ～承认 deny recognition / ～答复 give no reply / ～理睬 ignore; pay no attention to / ～批准 not grant approval / ～受理 off the docket / ～追究 not inquire into the cause 〈responsibility〉
【不育】sterile: ～性 sterility / ～症 sterility; barrenness
【不远千里】make light of travelling a thousand *li*; go to the trouble of travelling a long distance
【不约而同】take the same action or view without prior consultation; happen to coincide
【不悦】unhappy; displeased
【不再】no longer; not any more
【不在】①(指不在家或某处) be out; be not in: 他～。 He is out. ②(死亡的婉辞) died; be dead
【不在此例】not within the rule
【不在此限】not subject to the limits or restrictions; do not apply
【不在犯罪现场】[法] alibi
【不在乎】not mind; not care: 满～ not care a pin
【不在话下】be nothing difficult; be a cinch: 再大的困难也～。 No difficulty amounts to much.
【不在其位】not in the position to
【不在其位, 不谋其政】unwilling to comment on sth. which is not one's own cincern
【不在意】①(不在乎) pay no attention to; take no notice of; not mind ②(疏乎) negligent; careless
【不赞一词】keep silent; make no comment
【不择手段】by fair means or foul; by hook or by crook; unscrupulously: ～的人 a man of no scruples / ～地达到目的 attain one's end by hook or by crook
【不怎么】not very; not particularly
【不怎么样】not up to much; very indifferent; so-so
【不战不和】no war, no peace: ～的局面 a stalemate of "no war, no peace"
【不战而胜】win without fighting a battle; win hands down
【不长进】without improvement or progress
【不折不扣】to the full; one hundred percent: ～地执行指示 carry out the instructions to the full
【不争气】be disappointing; fail to live up to expectations
【不正常】abnormal; irregular: ～的行为 irregular be behavior / ～状态 abnormal state of affairs
【不正当】improper; illegitimate; dishonest; devious: ～的理由 improper reason / ～的手段 dishonest methods; devious means / ～的收益 unlawful profits / ～的职业 illegitimate occupation
【不正之风】unhealthy tendency: 纠正～ overcome unhealthy tendencies
【不知】①(不知道) not know; have no idea of; be ignorant of: ～世事 know nothing about what is going on in the world ②(表示疑问或请求) wonder if: ～你是否可以告诉我… I wonder if you can tell me...
【不知不觉】unconsciously; unwittingly; unknowingly
【不知底细】not know the inside story
【不知分寸】indiscreet; lacking a sense of propriety; not knowing how far one can go
【不知好歹】not know good from bad; not know what's good for oneself
【不知死活】heedless of consequences; daredevil; act recklessly
【不知所措】be at a loss; be at one's wits' end
【不知所云】not know what sb. is driving at; be unintelligible
【不知天高地厚】not know the immensity of heaven and earth; have an exaggerated opinion of one's abilities
【不知羞耻】shameless; unblushing; barefaced; brazenfaced: 不知人间有羞耻事 lost to all sense of shame
【不知自爱】act without self-respect
【不值】not worth: ～识者一笑 beneath the contempt of the discerning / ～一驳 not worth refuting / ～一顾 not worth a single glance / ～一提 unworthy of mentioning / ～一文 not worth a penny; worthless
【不止】①(超出某个范围) more than; not limited to ②(继续不停) incessantly; without end
【不只】not only; not merely
【不至于】①(不会到某种程度) can't go so far as to; not to such an extent as to ②(未必) be unlikely to
【不治之症】incurable disease
【不置可否】decline to comment; not express an opinion; be noncommittal; hedge

【不中听】not worth listening
【不中用】useless; good for nothing; not good for use
【不中意】not to one's liking
【不周延】[逻] undistributed
【不追究责任的保险】no-fault insurance
【不准】not allow; forbid; prohibit: ～入内。No admittance. / ～吸咽! No Smoking!
【不着边际】not to the point; wide of the mark; neither here nor there; irrelevant: ～的长篇大论 a long rambling talk / ～的话 irrelevant remarks
【不着陆飞行】nonstop flight
【不自爱】do not have self-respect
【不自量】not take a proper measure of oneself; overrate one's own abilities
【不自量力】overrate one's own abilities; put a quart into a pint pot
【不自由，毋宁死】Give me liberty or give me death.
【不自在】uneasy; ill at ease; feel uncomfortable
【不足】①(不充足) not enough; insufficient; inadequate: 光线～ insufficient light / 能源～ inadequate energy sources / 信心～ lack confidence / 资源～ inadequate resources / ②(不满某个数目) less than: ～一年 in less than a year ③(不值得) not worth; be beneath: ～道 of no consequence / ～挂齿 not worth mentioning / ～为奇 nothing surprising; no wonder / ～为训 not fit to be regarded as an exampie ④(不能) can't; should not: ～为凭 can't be taken as evidence
【不足法定人数】insufficient quorum
【不作为犯罪】negative crime
【不做声】say nothing; keep silent

钚 [化] plutonium (Pu)

布 ①(棉、麻等织物) cloth; fabric; texteles: 绸～ silk textile / 花～ cotton prints / 棉～ cotton cloth; cotton / 纱～ gauze / 坯～ grey ②(宣布) declare; announce; publish; proclaim ③(散布) spread; disseminate ④(布置) dispose; arrange; deploy
【布帛】cloth and silk; cotton and silk textiles
【布帛菽粟】cloth, silk, beans and grain; food and clothing; daily necessities
【布达拉宫】Potala Palace
【布道】[宗] preach ◇ ～坛 pulpit
【布店】cloth store; draper's; piece-goods store
【布丁】pudding
【布尔什维克】Bolshevik
【布告】notice; bulletin; proclamation: 张贴～ paste up a notice ◇ ～栏 notice board; bulletin board
【布谷鸟】cuckoo
【布基纳法索】Burkina Faso
【布景】①[剧] setting; scenery ②(国画用语) composition ◇ ～设计师 set designer
【布局】①(全面安排) overall arrangement; layout; distribution: 国民经济的～ layout of national economy / 工业的合理～ rational distribution of industry ②(指绘画、作文等) composition ③(指棋子) position
【布局新奇】the plot is novel
【布拉耶盲字】Braille
【布朗运动】[物] Brownian movement
【布雷】lay mines; mine: 在港口～ mine a harbor ◇ ～舰艇 minelayers / ～区 minefield
【布料】cotton material
【布隆迪】Burundi ◇ ～人 Burundian
【布鲁氏菌病】brucellosis; undulant fever
【布满尘埃】covered with dust all over
【布面】cloth cover ◇ ～精装本 clothbound *de luxe*edition
【布匹】cloth; piece goods ◇ ～染色 piece dyeing
【布施】[佛教] alms giving; donation
【布头】leftover of bolt of cloth; odd bits of cloth
【布网船】[军] netlayer
【布纹纸】[摄] wove paper
【布线】[电] wiring ◇ ～图 wiring diagram
【布鞋】cloth shoes
【布衣】commoners ◇ ～之交 a friend one made when one was a commoner or in humble circumstances
【布置】①(陈设) fix up; arrange; decorate: ～展品 arrange exhibits ②(安排) arrange; make arrangements for; give instructions about ～工作 assign work; give instructions about an assignment

怖 fear; be afraid of: 可～ horrible; frightful / 恐～ terror; horror

步 ①(步子) step; pace: ～测 pace off / 没几～路 only a few steps away; within a stone's throw ②(阶段) stage; step ③(地步) situation; condition; state ④(用脚走) walk; go on foot ⑤[棋](移动棋子) move: 一～好棋 an excellent move
【步兵】①(兵种) infantry ②(士兵) infantryman;

rifleman; foot soldier
【步步】step by step; at every step: ～高升 get promotion continuously; attain eminence step by step / ～进逼 press forward steadily / ～留心 be careful of every step
【步步为营】advance gradually and entrench oneself at every step; consolidate at every step
【步测】pacing
【步调】pace; step: ～参差 marching orders are confused / ～一致 keep in step
【步伐】step; pace: ～整齐 (march) in step / 跟上时代的～ keep pace with the times / 加快～ quicken one's steps
【步法】[体] [舞蹈] footwork
【步幅】stride
【步话机】walkie-talkie
【步甲】[动]ground beetle
【步进马达】stepper motor; stepping motor
【步进制】[邮] step-by-step system
【步犁】walking plough
【步履维艰】[书] have difficulty walking; walk with difficulty
【步枪】rifle: 半自动～ semi-automatic rifle / 标准～ standard rifle / 大口径～ big-bore rifle / 气～ air rifle / 速射自动～ quick-firing tommy-gun / 小口径～ small-bore rifle / 运动～ sporting rifle / 自动～ automatic rifle / 自选～ free rifle
【步人后尘】follow in other people's footsteps; trail along behind others
【步入歧途】take the wrong turning
【步哨】sentry; sentinel
【步速】leg speed
【步行】go on foot; walk
【步行虫】[动] ground beetle
【步韵】use the rhyme sequence of a poem
【步骤】step; move; procedure: 解决问题的～ moves towards setting a problem / 采取适当的～ take proper steps
【步子】step; pace: ～轻快 walk with springy steps

簿

簿 book: 登记～ register / 练习～ exercise book / 账～ account book
【簿册】books for taking notes or keeping accounts
【簿籍】account books, registers, records, etc.
【簿记】bookkeeping: 单式～ single-entry bookkeeping / 复式～ double-entry; bookkeeping ◇ ～学 bookkeeping / ～学校 bokkeeping school / ～员 bookkeeper; ledger clerk
【簿子】notebook; book

C

cā

擦 ①(摩擦)rub: ～火柴 strike a match / ～伤膝盖 rub the skin off one's knee; graze one's knee ②(用布、毛巾等擦干净) wipe: ～车 wax a car; wax a bicycle / ～地板 mop the floor; scrub the floor / ～汗 wipe the sweat away / ～净玻璃杯 rub clean a glass / ～皮鞋 polish shoes / ～枪 clean a gun / ～桌子 wipe the table ③(涂抹)spread on; put on: ～粉 powder one's face / ～胭抹粉 paint rouge and powder / 给机器～油 grease a machine; oil a machine / 给伤口～碘酒 apply iodine to a wound ④(一个物体贴近另一个物体很快地过去) brush; shave: ～肩而过 brush past sb. / ～身而过 pass each other so close that they almost rubbed each other ⑤(把瓜果等在礤床儿上擦成细丝儿) scrape (into shreds)

【擦棒球】[棒、垒球] foul tip

【擦边球】[乒乓球] edge ball; touch ball

【擦亮眼睛】remove the scales from one's eyes; sharpen one's vigilance

【擦伤】abrasion; gall; scotch; scratch

【擦伤的】abrasive

【擦拭】clean; cleanse: ～武器 clean weapons

【擦图片】erasing shield

【擦网球】[体] net ball

【擦音】[语] fricative

【擦澡】rub oneself down with a wet towel; take a sponge bath

【擦子】eraser: 橡皮～ rubber

cǎ

礤

【礤床儿】shredder (for vegetables)

cāi

猜 ①(猜想;猜测) guess; conjecture; speculate: ～一～ make a guess; have a guess / ～中 guess right ②(起疑心) suspect

【猜不透】unable to guess; unable to make out

【猜不着】cannot to guess; unable to make out

【猜测】guess; conjecture; surmise

【猜度】surmise; conjecture

【猜度再三】form a judgement again and again

【猜忌】be suspicious and jealous of

【猜谜儿】①(猜谜底) guess a riddle; solve riddles ②(猜测) guess

【猜拳】a finger-guessing game; mora

【猜想】suppose; guess; suspect

【猜疑】harbor suspicions; be suspicious; have misgivings: 互相～ be suspicious of each other; have misgivings about each other

cái

裁 ①(用刀、剪分成若干部分) cut (paper, cloth, etc.) into parts ②(削减;裁员) reduce; cut down; dismiss ③(衡量;判断) judge; decide: ～夺 consider and decide ④(控制;抑止) check; sanction: 经济制～ economic sanction ⑤(安排取舍) mental planning: 别出心～ adopt an original approach; try to be different

【裁并】cut down and merge (organizations)

【裁长补短】cut off the long and compensate the short

【裁撤】dissolve (an organization)

【裁处】make arrangement after due consideration

【裁定】[法] ruling; holding; judge: ～汇率 arbitrated exchange / ～书 award

【裁断】consider and decide

【裁夺】consider and decide

【裁缝】tailor; dressmaker: ～店 the tailor's (shop)

【裁减】reduce; cut down: ～军备 reduction of armaments; reduce military preparations

【裁剪】cut out: ～衣服 cut out garments

【裁决】ruling; adjudication: ～理由 reasons for findings / ～令 adjudication order; judicial order / 依法～ adjudicate according to law

【裁军】disarmament: ～会议 a disarmament conference

【裁判】①(判决和裁定)[法] judgment ②(评判)[体] act as referee; referee ③[体](裁判员) judge (指运动会或竞赛); umpire (排球、乒乓球、羽毛球、网球、棒球); referee (篮球、足球、拳击) ◇ ～权 jurisdiction / ～上的加重[法] juridical aggravation / ～上的减轻[法] juridical extenuation / ～台 referee's platform; referee's stand / ～员 referee; judge; umpire / ～椅 referee's chair / ～长 head judge; chief judge

【裁遣回乡】dismiss and send home

【裁弯取直】curve cut-off; [转] rid of the unnecessary and take the most appropriate

【裁员】cut down the number of persons employed; reduce the staff

【裁纸机】(paper) trimmer; paper cutter

才 ①(才能) ability; talent; gift: 德～兼备 have both ability and political integrity ②(有才

能的人）capable person: 人～ a person of talent; talent ③[副] just; only
【才不出众】one's ability does not exceed the average
【才大心细】have a great talent and an attentive mind
【才德兼备】have both talent and virtue
【才干】ability; competence; talent: 这人很有～。 He is a man of great ability.
【才华】literary of artistic talent: ～出众 of uncommon brilliance; posses exceptional talent
【才略】ability and sagacity (in political and military affairs)
【才貌双全】having both looks and real talent
【才能】ability; talent; [心理] aptitude
【才女】talented woman
【才气】literary talent: ～横溢 brim with talent; have superb talent
【才识】ability and insight: ～过人 be gifted with talent and insight far beyond the average person
【才疏学浅】[谦] have little talent and less learning
【才思】[书] imaginative power; creativeness: ～敏捷 have a facile imagination
【才学】talent and learning; scholarship: ～超群 one's ability and learning surpass the average / ～浅陋 shallow knowledge and small talent
【才智】ability and wisdom
【才子】gifted scholar: ～佳人 gifted scholars and beautiful ladies (in Chinese romances); a handsome scholar and a pretty girl

材 ①(木料) timber: 木～ timber; lumber; wood ②(材料) material: 钢～ steel products / 教～ teaching material / 就地取～ obtain material from local sources; draw on local resources ③(才能) ability; talent; aptitude: 因～施教 teach students in accordance with their aptitude ④(人才) capable person: 培养成～ bring sb. up to be a useful person ⑤(棺材) coffin
【材积】[林] volume (of timber)
【材料】①(原料) material: 建筑～ building material / 原～raw material ②(资料) data; material: 参考～ reference material / 档案～ archival material / 第一手～ firsthand information / 调查～ data; findings / 搜集～ gather material; collect data / 学习～ material for study ③(适于做某种事的人才)makings; stuff: 她不是演戏的～。 She hasn't the makings for an actress. ◇ ～科学 materials science / ～加工 materials processing / ～力学 mechanics of materials

财 wealth; money
【财宝】money and valuables: 发～ get rich
【财帛】wealth; money
【财产】property: ～的抵押权 encumbrance / ～的没收 expropriation of property / ～的转让 alienation of property; cession of property / 公共～ public property / 国家～ state property ◇ ～保险 property insurance / ～不可侵犯 inviolability of property / ～法 law of property / ～估价 assessment / ～监护人 guardian of property / ～明细帐 property ledger / ～清册 inventory of property / ～权 property right / ～税 property tax / ～所有人 owner of property; property owner
【财大气粗】He who has wealth speaks louder than others
【财东】[旧]①(店主) shopowner ②(财主) moneybags
【财阀】financial magnate; plutocrat; tycoon
【财富】wealth; riches: 精神～ spiritual wealth / 社会～ wealth of the society / 自然～ natural wealth
【财经】finance and economics
【财力】financial resources; financial capacity
【财贸】finance and trade; finance and commerce
【财迷】moneygrubber; miser
【财迷心窍】have one's head turned by greed
【财气】luck in making big money
【财权】①(财产所有权)right of property ②(经济大权)economic rights
【财团】financial group: 国际～ consortium ◇ ～法人 juridical person; corporate body
【财务】financial affairs ◇ ～报告 financial statement; financial report / ～代理人 fiscal agent / ～科 finance section / ～年度 fiscal year / ～事项 financial transaction
【财物】property; belongings: 个人～ personal effects
【财源】financial resources; source of revenue: 开辟～ explore more source of revenue
【财源茂盛】The source of wealth is flourishing.
【财运亨通】The luck for wealth is prosperous.
【财政】(public) finance: ～金融危机 financial and monetary crisis / ～收支平衡 balance of revenue and expenditure ◇ ～补贴 financial subsidies / ～部 the Ministry of Finance / ～赤字 financial deficits / ～法规 financial laws

and regulations / ～机关 financial organ; financial administration; fiscal organ; fiscal administration / ～监督 financial supervision / ～经济委员会 Financial and Economic Committee / ～局 Bureau of Finance / ～年度 financial year; fiscal year / ～实力 fiscal solvency / ～税收 fiscal levy / ～收入 revenue / ～厅 Provincial Department of Finance / ～危机 financial crisis / ～政策 financial policy; fiscal policy / ～支出 expenditure

【财主】rich man; moneybags

cǎi

采 ①(摘)pick; pluck; gather: ～茶 pick tea / ～药 gather medicinal herbs / ～珍珠 dive for pearls ②(开采) mine; extract: ～煤 mine coal / ～油 extract oil ③(采取) adopt; select: ～取一系列措施 adopt a series of measures ④(精神) complexion; spirit: 兴高～烈 in high spirits

【采办】but on a considerable scale; purchase

【采场】[矿] stope

【采伐】fell; cut; cut over: ～原始森林 open up a primeval forest ◇ ～迹地 cutover / ～量 cut

【采访】(of a reporter) cover; gather material: ～新闻 gather news / 向某人～ interview sb.

【采风】collect folk songs

【采购】make purchases for an organization or enterprise; purchase: ～建筑材料 purchase building materials ◇ ～员 purchasing agent / ～站 purchasing station

【采光】[建](natural) lighting; daylighting

【采集】gather; collect: ～标本 collect specimens / ～民歌 collect folk songs

【采掘】[矿] excavate ◇ ～设备 equipment for excavation

【采矿】mining: 地下～ underground mining / 露天～ opencut mining; opencast mining ◇ ～工程 mining engineering

【采录】collect and record: ～民歌 collect and record folk songs

【采买】purchase; buy

【采煤】excavate coal; coal mining; coal extraction; coal cutting ◇ ～工人 coal miner / ～工作面 coal face / ～回收率 coal recovery / ～机 coal-winning machine

【采棉机】cotton picker

【采纳】accept; adopt: ～合理的建议 accept rational suggestions

【采暖】[建] heating: 蒸气～ steam heating ◇ ～设备 heating equipment; heating facilities

【采取】adopt; take: ～步骤 take steps / ～攻势 take the offensive / ～坚决的手段 take a firm hand / ～紧急措施 take emergency measures / ～强制手段 resort to compulsion / ～拖延战术 employ stalling tactics / ～主动 take the initiative

【采石场】stone pit; quarry

【采收率】[石油] recovery ratio

【采撷】[书] ①(采摘) pick; pluck ②(采集) gather

【采血针】blood taking needle

【采样】[矿] sampling

【采用】put to use; adopt; use; employ: ～新技术 adopt new techniques

【采油】[石油] oil extraction; oil recovery: 二次～ secondary recovery / 气举～ air-lift recovery ◇ ～队 oil production crew / ～树 Christmas tree

【采摘】pluck; pick: ～苹果 pick apples

【采制】collect and process

【采种】[农] seed collecting

睬 pay attention to; take notice of: 理～ pay attention of; show interest in

踩 step on; trample: ～油门 step on the gas (accelerator) / 把地～平 stamp the ground flat

【踩水】[体]tread water

【踩线】[体]step on the line; footfault

彩 ①(颜色) color: ～云 rosy clouds / 五～ of different colors; multicolored ②(彩色丝绸) colored silk; variegated silk: 张灯结～ decorate with lanterns and colored ribbons ③(称赞夸奖的欢呼声) applause; cheer: 喝～ acclaim; cheer ④(花样; 精彩的成分) variety; splendor: 丰富多～ rich and colorful; rich and varied ⑤(中奖;奖品)prize: 中～ win a prize (in a lottery, etc.) ⑥(负伤流血) blood from a wound: 挂～ be wounded in battle

【彩车】float (in a parade)

【彩绸】colored silk; silk of various colors

【彩带】colored ribbon (streamer)

【彩号】wounded soldier

【彩虹】rainbow

【彩绘】colored drawing or pattern: ～磁器 porcelain decorated with colored drawings

【彩礼】betrothal gifts (from the bridegroom to the bride's family); bride- price

【彩门】decorated gateway

【彩排】dress rehearsal

【彩票】lottery ticket

【彩旗】colored flag; bunting
【彩色】multicolor; color; colour ◇ ～电视 color television / ～电视(接收)机 color receiver / ～电视摄像机 color television camera / ～电视显像管 chromatron; trichromoscope / ～负片 color negative / ～胶片 color film / ～铅笔 color pencil; crayon / ～摄像管 color pick-up tube / ～摄影 color photography / ～相纸 color paper / ～印片法 [电影] technicolor / ～印刷 color printing; multicolor printing / ～影片 color film ○白 white / 橙黄 orange / 翠绿 emerald / 淡蓝 light blue / 淡绿 light green / 靛青 indigo / 粉红 pink / 古铜色 bronze / 黑 black / 红 red / 花青 flower blue / 黄 yellow / 灰 grey / 金色 golden / 蓝 blue / 绿 green / 铅白 lead white / 铅灰色 leaden / 乳金 milk gold / 深红 crimson / 深蓝 dark blue / 深绿 dark green / 石黄 mineral yellow / 石绿 mineral green / 藤黄 rattan yellow; light bright yellow / 天蓝 azure / 象牙白 ivory / 猩红 scarlet / 胭脂 rouge / 银色 silver / 赭黄 sienna / 赭色 umber / 朱红 vermilion / 紫 purple / 紫罗兰色 violet / 棕褐 brown
【彩色缤纷】a riot of color
【彩陶】ancient painted pottery
【彩霞】rosy clouds; pink clouds
【彩釉陶】glazed colored pottery

cài

采

【采邑】fief; benefice

菜

①(蔬菜) vegetable; greens: 咸～ pickles / 种～ grow vegetables ②(泛指副食) food: 上街买～ go to the market to buy food ③(菜肴) dish; course: 荤～ meat dish / 素～ vegetable dish / 做～ prepare the dishes; do the cooking
【菜板】chopping board
【菜帮儿】outer leaves (of a cabbage, etc.)
【菜场】food market
【菜单】menu; bill of fare
【菜刀】kitchen knife
【菜地】vegetable plot
【菜豆】kidney bean
【菜墩子】chopping board
【菜粉蝶】cabbage butterfly
【菜瓜】snake melon
【菜花】①(花椰菜) cauliflower ②(油菜花) rape flower
【菜窖】vegetable cellar; clamp
【菜篮子】①(买菜用的篮子) basket for carrying vegetables and food items ②(指副食品) non-staple food
【菜牛】beef cattle
【菜农】vegetable grower
【菜谱】cookery-book; cookbook
【菜畦】small sections of a vegetable plot; vegetable bed
【菜青】dark greyish green
【菜青虫】cabbage caterpillar
【菜色】famished look; emaciated look: 面有～ look famished
【菜市】food market
【菜蔬】①(蔬菜) vegetables; greens ②(吃饭时各的各种菜) dishes at a meal
【菜苔】bolt (of rape, mustard, etc.)
【菜摊】vegetable stall
【菜心儿】heart (of a cabbage, etc.)
【菜秧】vegetable shoots
【菜肴】cooked food (usu. meat dishes)
【菜油】repeseed oil; rape oil; colza oil
【菜园】vegetable garden; vegetable farm
【菜子】①(蔬菜的种子) vegetable seeds ②(油菜子) rapeseed ◇ ～饼 rapeseed cake / ～油 rape oil; rapeseed oil; colza oil

cān

餐

①(吃饭) eat: 聚～ dine together / 野～ go on a picnic; picnic ②(饭食) food; meal: 便～ light meal / 快～ snack / 冷～ buffet / 晚～ supper / 西～ Western food / 午～ lunch / 早～ breakfast / 正～ dinner / 中～ Chinese food ③(每顿饭) regular meal: 一日三～ three meals a day
【餐车】restaurant car; dining car; diner
【餐风宿露】hardship of travel without shelter
【餐巾】table napkin ◇ ～纸 napkin paper; paper napkin
【餐具】tableware; dinner service; dinner set ◇ ～柜 sideboard / ～锡合金 Minofar
【餐券】meal coupon; meal ticket
【餐厅】①(饭厅,食堂) dining room; dining hall ②(餐馆) restaurant

参

①(参加;加入) join; enter; take part in: ～战 enter a war ②(参考) refer; consult: ～阅 see; consult; compare ③(进见;谒见) call to pay one's respects to: ～谒烈士陵园 pay homage at the mausoleum of the martyred heroes ④(弹劾) impeach an official before the emperor
【参拜】pay respects to; make a formal visit to

【参半】half; half-and-half: 毁誉～ get both praise and blame; be as much praised as blamed / 疑信～ half believing, half doubting
【参观】visit; look around: ～游览 visit places of interest; go sightseeing / 欢迎～。visitors are welcome. ◇～团 visiting group
【参加】①(加入) join; attend; take part in: ～革命 join the revolutionary ranks / ～会谈 take part in talks / ～会议 attend a meeting / ～容易摆脱难 easier to get into than to get out of ②(提出意见) give (advice, suggestion, etc.) ◇～国 acceding State / ～诉讼 intervene; intervention / ～诉讼人 intervenient; intervening party / ～者 participant
【参见】①(参看) see also; cf.: ～第四章。See also Chapter 4. ②(进见;谒见) pay one's respects to (a superior, etc.)
【参军】join the army; join up; enlist
【参看】①(参见) see (also): ～第三十二页 see page 32 / ～下面注释 see note below ②(查阅) consult: 他～了不少有关书刊。He consulted a number of relevant books and periodicals. ③(参考) read sth. for reference
【参考】①(查阅) consult; refer to: ～历史文献 consult historical documents ②(用于帮助了解情况) reference: 仅供～ for reference only ◇～书 reference book / ～书目 a list of reference books; bibliography / ～资料 reference material
【参谋】①[军] staff officer ②(出主意) give advice ◇～长 chief of staff / 总～长 chief of the general staff
【参事】counsellor; adviser
【参数】[数] parameter ◇～变异法 variation of parameters / ～推断 parametric inference
【参天】reaching to the sky; towering; very tall: ～古木 old trees that reach into the skies / ～古树 towering old trees
【参谒】①(进见;谒见) pay one's respects to sb. ②(参拜) pay homage to sb. (before his tomb or image)
【参议员】senator
【参议院】senate
【参与】participate in; have a hand in: ～其事 have a hand in the matter / ～制订规划 participate in the drawing up of a plan / ～者 participator
【参赞】counsellor; councillor: 商务～ commercial counsellor / 文化～ cultural attaché
【参战】enter a war; participate in a war; take part in a war ◇～国 belligerent state
【参照】consult; refer to: ～具体情况 in the light of the specific situation / 互相～ cross-reference ◇～系 reference frame
【参政】participate in government and political affairs
【参酌】consider (a matter) in the light of actual conditions; deliberate

cán

惭 feel ashamed: 大言不～ be shamelessly boastful; brazenly brag
【惭愧】be ashamed: ～万分 be extremely ashamed

蚕 silkworm: 家～ domestic silkworm / 桑～ mulberry-feeding silkworm / 养～ raise silkworms; silkworm breeding; sericulture / 养～业 sericulture / 野～ wild silkworm
【蚕箔】a bamboo tray for raising silkworms
【蚕床】rearing shed
【蚕簇】a small bundle of straw, etc., for silkworms to spin cocoons on
【蚕豆】broad bean
【蚕豆病】[医] favism
【蚕蛾】silk moth
【蚕茧】silkworm cocoon
【蚕农】silkworm raiser; sericulturist
【蚕沙】silkworm excrement
【蚕食】nibble; nibble up ◇～政策 the policy of "nibbling" at another country's territory
【蚕食鲸吞】nibble away like a silkworm or swallow like a whale; seize another country's territory by piecemeal encroachment or wholesale annexation
【蚕丝】natural silk; silk
【蚕蚁】newly-hatched silkworm
【蚕蛹】silkworm chrysalis
【蚕纸】paper with silkworm eggs; silkworm-egg card
【蚕子】silkworm seed (egg)

残 ①(不完整;残缺) incomplete; deficient: ～稿 an incomplete manuscript ②(剩余的) remnant; remaining: ～敌 remnants of the enemy forces / ～冬 the last days of winter ③(伤害;毁坏) injure; damage: 身～志不～ broken in health but not in spirit ④(凶恶) savage; barbarous; ferocious: 凶～ cruel
【残暴】cruel and ferocious; ruthless; brutal; savage
【残杯冷炙】remains of a meal; a scrappy dinner

【残兵败将】remnants of a routed army
【残存】remnant; remaining; surviving
【残而不废】disabled but useful
【残废】①(四肢或双目等失去一部分或丧失机能) maimed; crippled; disabled: ～军人 disabled armyman ②(残废的人) a maimed person; cripple ◇ ～抚恤金 disability pension / ～证明书[军] certificate of disability; disability certificate
【残羹剩饭】remains of a meal; leftovers; crumbs from the table
【残骸】remains; wreckage: 敌机～ the wreckage of an enemy plane
【残害】cruelly injure or kill: ～肢体 cause bodily injury / ～忠良 persecute the faithful and honest
【残货】shopworn goods; damaged goods
【残迹】a remaining trace, sign, etc.; vestiges
【残疾】deformity ◇ ～人 deformed man
【残局】①(快结束的棋局) the final phase of a game of chess ②(失败或动乱后的局面) the situation after the failure of an undertaking or after social unrest: 收拾～ clear up the mess; pick up the pieces
【残酷】cruel; brutal; ruthless: ～暴戾 cruel and fierce / ～行为 brutal act
【残留】remain; be left over
【残年】①(一年将尽的时候) the last days of the year ②(人的晚年) the evening of life; declining years: 风烛～ old and ailing like a candle guttering in the wind / ～晚景 circumstances in one's dealing years
【残篇断简】fragments of ancient texts
【残品】damaged article; defective goods
【残破】broken; dilapidated
【残缺】incomplete; fragmentary
【残气】residual gas
【残忍】cruel; ruthless: ～刻薄 cruel and cold hearted
【残杀】murder; massacre; slaughter: 自相～ mutual slaughter; kill each other
【残生】one's remaining years; surviving span of life
【残碳测定法】conradson method
【残阳】the setting sun
【残余】remnants; remains; survivals; vestiges: ～势力 remaining forces; surviving forces / 封建～ survivals of feudalism
【残月如弓】The waning moon resembles a bow.
【残渣】residue; caput mortuum
【残渣余孽】evil elements from the old society; dregs of the old society
【残照】the setting sun: 西风～ the sun setting in the wild west wind
【残值】scrap value

cǎn

惨 ①(悲惨;凄惨) miserable; pitiful; tragic: ～遭不幸 die a tragic death ②(凶恶;狠毒) cruel; savage: ～无人道 inhuman; brutal ③(程度严重,利害) to a serious degree; disastrously
【惨案】①(政治性的) massacre ②(凶杀) murder case; tragic case; tragedy
【惨白】pale: 脸色～ look deathly pale
【惨败】crushing defeat; disastrous defeat
【惨不忍睹】so tragic that one cannot bear to look at it; too horrible to look at
【惨淡】gloomy; dismal; bleak: 天色～ gloomy sky; gloomy weather / 在～的星光下 in the dim starlight
【惨淡经营】keep (an enterprise, etc.) going by painstaking effort; take great pains to carry on one's work under difficult circumstances
【惨祸】horrible disaster; frightful calamity
【惨境】miserable condition; tragic circumstances; dire straits
【惨剧】tragedy; calamity
【惨绝人寰】tragic beyond compare in this human world; extremely tragic: ～的暴行 atrocity of unparalleled savagery / ～之浩劫 terrible holocaust
【惨然】saddened; grieved
【惨杀】massacre; murder
【惨死】die a tragic death
【惨痛】deeply grieved; painful; bitter: ～的教训 a bitter lesson
【惨无人道】very cruel and inhuman
【惨无天日】too dark, or full of suffering, that it is as if the sun were not in the sky
【惨笑】a wan smile
【惨重】heavy; grievous; disastrous: ～的失败 a disastrous defeat / 伤亡～ suffer heavy casualties / 损失～ suffer heavy losses; suffer grievous losses
【惨状】miserable condition; pitiful sight

càn

灿
【灿烂】magnificent; splendid; resplendent; bright: ～的民族文化 splendid national culture / ～的前景 magnificent prospects;

bright future / ～的阳光 the bright sun; brilliant sunshine

粲 [书] ①(鲜明发光) bright; beaming ②(笑时露出牙齿的样子) smile: 以博一～ just for your amusement

【粲然】①(鲜明发光) bright; beaming ②(笑时露出牙齿) smiling broadly: ～一笑 give a beaming smile; grin with delight

cāng

仓 storehouse; warehouse: 谷～ barn

【仓储】keep grain, goods, etc. in a storehouse

【仓促】hurriedly; hastily; all of a sudden: 走得～ leave in a hurry / ～应战 accept battle in haste / ～做出结论 rush to a conclusion; jump to a conclusion

【仓房】warehouse; storehouse

【仓皇】in a flurry; in panic: ～失措 be scared out of one's wits; be panic-stricken; be disturbed not knowing what to do / ～逃窜 flee in confusion; flee in panic; flee helter-skelter / ～退却 retreat in haste

【仓库】warehouse; storehouse; depository: 清理～ take stock; check warehouse stocks ◇ ～保管员 warehouseman / ～交货 ex warehouse

【仓鼠】hamster

【仓租】warehouse storage charges

沧 (of the sea) dark blue

【沧海】the blue sea; the sea

【沧海横流】the chaos of the world; the changes and disorders of the times

【沧海桑田】seas change into mulberry fields and mulberry fields into seas; time brings great changes to the world; the whirligig of time

【沧海一粟】a drop in the ocean; a drop in the bucket

【沧桑】("沧海桑田")饱经～ have experienced many vicissitudes of life

苍 ①(绿色的)dark green: ～松 green pines ②(蓝色的)blue: ～天 the blue sky ③(灰白色的) grey; ashy: ～髯 a grey beard

【苍白】pale; pallid; wan: ～无力 pale and weak; feeble / 脸色～ look pale

【苍苍】①(灰白) grey: 两鬓～ greying at the temples ②(苍茫)vast and hazy: 天～,野茫茫。Vast is the sky, boundless the wilds.

【苍翠】dark green; verdant: ～的山峦 verdant hills

【苍耳】[植] Siberian cocklebur; clotbur

【苍黄】greenish yellow: 面色～ have a sallow complexion / ～的天空 a somber sky

【苍劲】①(用于树木) old and strong: ～挺拔的青松 hardy, old pines ②(用于书画) vigorous; bold: 笔力～ (write or paint) in bold, vigorous strokes

【苍老】①(老态) old; aged ②(笔力雄健) vigorous; forceful

【苍凉】desolate; bleak

【苍鹭】heron

【苍茫】①(空阔辽远) vast; boundless: ～大地 boundless land ②(没有边际) indistinct: 暮色～ deepening shades of dusk

【苍穹】[书] the vault of heaven; the firmament

【苍天】①(天空) the blue sky ②(天,天神) Heaven

【苍鹰】goshawk

【苍蝇】fly ◇ ～拍子 flyswatter

【苍术】[中药] Chinese atractylodes

【苍松翠柏】pines and cypresses of verdant green

伧 rude; rough

【伧俗】vulgar

舱 ① cabin: 船～ cabin; ship's hold / 货～ hold / 客～ (passenger) cabin / 装～ stow the hold ②[宇航] module: 指令～ command module

【舱口】hatchway; hatch ◇ ～盖 hatch door; hatch cover / ～起重机 hatch crane

【舱门】hatch door; cabin door

【舱面】deck ◇ ～货 deck cargo

【舱内货】underdeck cargo

【舱位】①(铺位或座位) cabin seat or berth ②(船仓) shipping space

cáng

藏 ①(躲藏;隐藏) hide; conceal ②(收藏;储藏) store; lay by

【藏垢纳污】shelter evil people and countenance evil practices: ～之地 a sink of iniquity; a sewer in which all that is evil finds a home

【藏经阁】depositary of Buddhist texts

【藏龙卧虎】"hidden dragons and crouching tigers"; talented men still remained in concealment

【藏匿】conceal; hide; go into hiding

【藏身】hide oneself; go into hiding: ～之处 hiding-place; hideout / 无处～ no place to hide

【藏书】①(收藏书籍) collect books ②(所藏的书

籍) a collection of books; library
【藏头露尾】show the tail but hide the head; hide the head but show the tail; tell part of the truth but not all of it
【藏掖】try to cover up: ～躲闪 dodge and hide
【藏拙】hide one's inadequacy by keeping quiet

cāo

糙 rough; coarse: ～纸 rough paper
【糙米】brown rice; half-polished rice
【糙皮病】pellagra
操 ①(抓在手里,拿)grasp; hold: ～刀 hold a sword, cleaver, etc. in one's hand／稳～胜券 have full assurance of success; be sure to win ②(做;从事) act; do; operate: 重～旧业 resume one's old profession; take up one's old trade again ③(用某种语言或方言说话) speak (a language or dialect) ④(操练) drill; exercise: 体～ gymnastics; exercise ⑤(品行;行为) conduct; behavior: 节～ one's moral principles; personal integrity
【操场】playground; sports ground; drill ground
【操持】manage; handle: ～家务 manage household affairs
【操典】[军]drill regulations; drill manual; drill book
【操舵室】wheelhouse; pilothouse; steering room
【操劳】①(辛苦劳动) work hard: 过度～ overwork oneself; strain oneself; work beyond measure ②(费心料理)take care; look after
【操练】drill; practice
【操切】rash; hasty: ～从事 act with undue haste
【操心】①(费神;担心) worry about; trouble about; take pains ②(费心考虑或料理) rack one's brains
【操行】behavior or conduct of a student
【操演】demonstration; drill
【操之过急】act with undue haste; act too hastily
【操纵】①(控制,开动) operate; control: ～机器 operate machines／无线电～radio control ②(支配,控制) rig; manipulate: ～表决机器 tamper with the voting machine; manipulate the voting／～市场 rig the market／幕后～ manipulate from behind the scenes; pull strings ◇～杆 operating lever; control rod; control stick／～台 control panel; control board
【操纵自如】restraint or lenient as it pleases one
【操作】operate; manipulate ◇～程序 operation sequence／～程序图 flow diagram; flow chart／～方法 method of operation／～规程 operating rules; operating instructions; working order／～人员 attending personnel; operating staff／～手册 operation manual／～说明书 operating manual／～系统程序 operation system management／～性能 [机] serviceability

cáo

槽 ①(器具) trough: 马～ manger／水～ water trough ②(两边高中间凹的部分)[机] groove; slot: 键～ key groove／开～ slotting
【槽车】tank wagon
【槽钢】channel (iron); box iron; channel steel
【槽谷】[地]trough valley
【槽距】[机]slot pitch
【槽口】[机]notch
【槽轮机构】Geneva mechanism
【槽磨机】slot grinder
【槽探】[矿]trenching
【槽头】trough (in a livestock shed): ～兴旺 a manger full of sturdy livestock
【槽牙】molar
嘈 noise; din
【嘈杂】noisy: 人声～ a hubbub of voices／多么～ What a noise!

cǎo

草 ①(植物) grass; straw: ～绳 straw rope ②(草率,不细致) careless; hasty; rough: 字写得很～. The handwriting is very sloppy. ③(草稿) draft: 起～文件 draft a document ④(口语中指雌性家禽、家畜) female: ～鸡 hen／～驴 jenny ass
【草案】draft (of a plan, law, etc.): 决议～ a draft resolution／宪法～ a draft constitution
【草包】①(草袋) straw bag; straw sack ②(无才能的人) idiot; blockhead; good-for-nothing
【草本】herbaceous ◇～植物 herb
【草草】carelessly; hastily: ～过目 read through roughly; give a cursory reading; skim through／～了事 get through with sth. any old way; dispose of a thing carelessly or hastily／～收兵 declare the matter closed before it is thoroughly settled／～收场 hastily wind up the matter
【草测】preliminary survey
【草叉】pitch-fork
【草创】start (an enterprise, etc.): ～时期 initial

stage; pioneering stage
【草丛】a thick growth of grass
【草地】① grassland; meadow: 别踩～ keep off the grass ②(草坪)lawn: ～网球 lawn tennis
【草甸子】grassy marshland; meadow; pelouse; Wiesen
【草垫子】straw mattress; pallet
【草垛】haystack; hayrick
【草房】thatched cottage
【草稿】rough draft; draft
【草菇】straw mushroom
【草荒】farmland running to weeds; more weeds than crops
【草菅人命】treat human life as if it were not worth a straw; act with utter disregard for human life
【草浆】[纸]straw pulp
【草芥】trifle; mere nothing: 视如～ regard as worthless; treat like dirt
【草寇】robbers in the greenwood; bandits
【草兰】cymbidium; orchid
【草料】forage; fodder
【草履虫】[生]paramecium
【草绿】grass green
【草莽】①(草丛)a rank growth of grass ②(草野)uncultivated land; wilderness ◇ ～英雄 a hero of the bush; greenwood hero
【草帽】straw hat
【草帽辫】plaited straw (for making hats, baskets, etc.)
【草莓】strawberry
【草棉】cotton; levant cotton
【草木灰】[农] plant ash
【草木皆兵】every bush and tree looks like an enemy; All grass and trees are mistaken for enemy troops.
【草木犀】[植] sweet clover
【草拟】draw up; draft: ～工作日程 draw up a work program／～一个计划 draft a plan
【草棚】thatched shack; straw shed
【草皮】sod; turf
【草坪】lawn
【草签】initial: ～合同 sign a referendum contract／～文本 initialed text／～协定 initial an agreement
【草石蚕】(宝塔菜) Chinese artichoke
【草食动物】plant-eating animal; herbivore
【草率】careless; perfunctory; rash: ～从事 take any hasty action; do a job carelessly or perfunctorily／～了事 dispose of a matter carelessly
【草酸】[化]oxalic acid
【草头王】king of the bushes (a euphemism for bandit chief)
【草图】sketch (map); draft
【草席】straw mat
【草鞋】straw sandals
【草药】medicinal herbs
【草鱼】grass carp
【草原】grasslands; prairie ◇ ～带[地] steppe belt
【草约】draft treaty; draft agreement; protocol; ad referendum contract
【草泽】grassy marsh; swamp
【草纸】①(用稻草制成的粗纸) rough straw paper ②(卫生纸) toilet paper
【草子】grass seed

cè

测 ①(测量) survey; fathom; measure: ～海深 fathom (the depth of) a sea／～雨量 gauge rainfall; measure rainfall／空～ aerial survey ②(测度) conjecture; infer: 变化莫～ unpredictable; constantly changing
【测标】mark
【测程仪】[交] log
【测电笔】test pencil
【测定】determine: ～船只方位 take a ship's bearings／放射性碳素～年代 radiocarbon dating／示踪～ tracer determination
【测度】①(推测) estimate; infer ②[数] measure
【测风气球】pilot balloon
【测杆】measuring staff; surveying rod
【测高法】altimetry; hypsometry
【测高仪】height indicator
【测候】astronomical and meteorological observation ◇ ～网[气] reseau
【测绘】survey and drawing; mapping ◇ ～板 plotting board／～飞机 air-mapping plane／～员 surveyor; cartographer
【测井】[石油] (well) logging: 电～ electric logging／放射性～ radioactivity logging
【测距】range finding ◇ ～仪 range finder
【测力计】[物]dynamometer
【测量】survey; measure; gauge: ～地形 survey the topography／大地～ geodetic survey／航空～ aerial survey; air survey ◇ ～队 survey party／～学 surveying／～仪表 instrumentation／～仪器 surveying instrument
【测漏】track down a leak; leak hunting
【测深锤】sounding bob; sounding lead
【测深仪】fathometer; depth-sounder

【测湿学】psychrometry
【测试台】testboard; test desk
【测试图】resolution chart; test pattern
【测温学】thermometry
【测向计】goniometer
【测斜仪】inclinometer
【测验】test: ～机械性能 test the performance of a machine / 算术～ arithmetic test; arithmetic quiz
【测云气球】ceiling balloon
【测震学】seismometry
【测字】fortune-telling by analysing the component parts of a Chinese character; divine by means of character; glyphomancy

恻 sorrowful; sad: 凄～ sad; grieved
【恻隐】compassion; pity: ～之心 sense of pity; a heart of sympathy

厕 lavatory; toilet; washroom; W.C.: 公～ public lavatory / 男～ men's; men's room; men's toilet / 女～ women's; women's room; women's toilet
【厕所】lavatory; toilet; W.C.

侧 ①(旁边) side: 右～ the left right side / 左～ the left side ②(向旁边歪斜) incline; lean: ～耳细听 incline the head and listen attentively; prick up one's ears
【侧柏】[植] oriental arborvitae
【侧吹】[冶] side-blown ◇ ～转炉 side-blown converter
【侧根】[植] lateral root
【侧航】[航空] crabbing
【侧击】flank attack; make a flank attack on
【侧记】sidelights
【侧力】side force; lateral force
【侧门】side door; side entrance
【侧面】side; aspect; flank; profile; side face: 从～了解 find out from indirect sources / 问题的一个～ one side of the problem ◇ ～图 side view / ～像 profile
【侧目】sidelong glance: ～而视 look askance at sb. (with fear or indignation)
【侧身】on one's side; sideways: ～而卧 lie on one's side
【侧石】[交] curbstone; curb
【侧视图】[机] end view; lateral view
【侧手翻】[体] cartwheel turn a cartwheel
【侧卫】[军] flank guard
【侧线】①[铁道] siding ②[动] lateral line
【侧卸汽车】side-discharging car; side dump truck; side dumper
【侧向】side direction; crossrange
【侧压力】[物] lateral pressure
【侧芽】[植] lateral bud
【侧翼】[军] flank: ～包围敌人 outflank the enemy; flank the enemy
【侧影】silhouette; profile
【侧泳】sidestroke
【侧枝】side shoot; offshoot
【侧重】lay particular emphasis on

策 ①(计谋;办法) plan; scheme; strategy: 决～ policy making / 失～ take a false step / 献～ submit a scheme ②(赶马前进) whip: ～马前进 whip a horse on
【策动】instigate; engineer; stir up: 阴谋～政变 plot to stage a *coup d'état*
【策动理论】hormic theory
【策反】instigate rebellion within the enemy camp; incite defection
【策划】plan; plot; scheme; engineer: ～阴谋 hatch a plot / 幕后～ plot behind the scenes
【策划者】sponsor; plotter; schemer
【策励】encourage; spur on: 时时刻刻～自己 constantly spur oneself ahead
【策略】①(制定的行动方针和斗争方式) tactics: 研究对敌斗争的～ study the tactics of our struggle against the enemy ②(讲究斗争艺术) tactful: 做事要～一些 do things more tactfully
【策士谋臣】a counsellor and a strategist
【策应】[军] support by coordinated action
【策源地】source; place of origin; hotbed: 战争～ a source of war; a hotbed of war

册 ①(册子) volume; book: 第二～ Volume II / 画～ an album of paintings / 装订成～ bind into book form ②[量] copy
【册封】confer titles of nobility on: ～公爵 confer the title of duke
【册立皇后】be appointed as the empress
【册授勋位】bestow an order of merit
【册页】an album of paintings or calligraphy
【册子】book; volume: 小～ pamphlet; booklet; brochure

cēn

参
【参差】irregular; uneven: ～不齐 uneven; not uniform / ～错落 confused with errors and omission

céng

曾

【曾几何时】before long; not long after; not long since

【曾经】[副]几年前我~见过她一面。I met her once several years ago.

【曾经沧海】have sailed the seven seas; having seen and experienced much; have experienced great things: ~难为水。To a sophisticated person there is nothing new under the sun.

层

①[量] layer; tier; stratum: 一~薄冰 a thin sheet of ice; a thin layer of ice / 一~油漆 a coat of paint ②(楼层) story; floor: 五~楼 a five-story building / 一~ the ground floor (美: first floor) / 二~ the first floor (美: second floor) ③(可分项分步的东西) a component part in a sequence

【层层】layer upon layer; ring upon ring; tier upon tier: ~把关 check at each level / ~包围 surround ring upon ring / ~叠叠 tier upon tier / ~发动 mobilize level by level / ~设防 set up successive lines of defense; erect defensive works in depth

【层出不穷】emerge in an endless stream; layer upon layer without an end

【层次】①(各级机构) administrative levels: ~重叠 overlapping ②(内容的次序) arrangement of ideas (in writing or speech): ~不清 lack unity and coherence ③(分层) gradation: ~分明 gradation is distinct / 颜色的~ gradation of colors

【层见迭出】occur frequently; appear repeatedly

【层理】[地]bedding; stratification

【层流】laminar flow

【层峦迭嶂】peaks rising one higher than another

【层压】[化] lamination: ~玻璃 laminated glass

【层云】[气]stratus

【层子】[物] straton: ~模型 straton model

cèng

蹭

①(摩擦) rub: 把手~破了 have the skin rubbed off one's hand ②(因擦过去而沾上) be smeared with: 小心~油漆。 Mind the fresh paint. ③(慢吞吞地行动) dillydally; loiter: 磨~ dawdle; dillydally

chā

差

①(差别;差异) difference; dissimilarity: 时~ time difference ②(差错) mistake: 偏~ deviation ③[数] difference

【差别】difference; disparity: 年龄~ disparity in ◇ ~待遇 discrimination treatment / ~关税 differential rates of duty; differential duties / ~汇率 discriminatory cross-rates ~选举制 differential voting / ~阈限[心] difference limen; difference threshold

【差错】①(错误) mistake; error; slip ②(意外的变化) mishap; accident

【差动】[机] differential ◇ ~齿轮 differential gear / ~滑轮 differential pulley / ~轴 differential shaft

【差额】difference; balance; margin: 补足~ make up the balance 〈difference〉/ ~表 balance sheet / ~帐 balance account

【差分方程】difference equation

【差价】price difference: 地区~ regional price differences / 季节~ seasonal price differences ◇ ~关税 variable import levy

【差距】①(差别程度) gap; disparity ②[机] difference

【差强人意】just passable; barely satisfactory; fair

【差速器】differential mechanism

【差异】difference; divergence; discrepancy; diversity

【差之毫厘,谬以千里】an error the breadth of a single hair can lead you a thousand *li* astray; a small discrepancy leads to a great error

喳

【喳喳】①(小声说话的声音) whispering sound ②(小声说话) whisper

插

①(放进、挤入、刺进或穿入) stick in; insert ②(中间加进去或加进中间去) interpose; insert

【插班】join a class in the middle of the course

【插翅难飞】unable to escape even if given wings; Even if you were given wings, you couldn't fly away.

【插管】[医] intubate ◇ ~法 intubation

【插花地】land belonging to one production unit but enclosed in that of another

【插话】①(在别人讲话时插话) interpose (a remark, etc.); chip in ②(插曲,大事件中的小故事) digression; episode

【插科打诨】①(指戏曲演员) make impromptu comic gestures and remarks ②(爱开玩笑的;滑稽) jesting; buffoonery

【插口】[电] socket; jack; spigot

【插屏】table screen

【插曲】①[乐] interlude ②(电影插曲) songs in a film or play ③(作品中的一段情节) episode; interlude
【插入】①(插进去) insert ②[电] plug in: ～部件 plug-in unit
【插入语】[语] parenthesis
【插身】①(挤进去) squeeze in; edge in ②(参与) take part in; get involved in
【插手】①(参加) take part; lend a hand ②(参与) have a hand in; poke one's nose into; meddle in; take a hand in
【插条】[植] transplant a cutting; cutting
【插头】[电] plug: ～板 patchboard / 三脚～ three-pin plug
【插图】illustration; plate ◇ ～本 illustrated edition / ～出血[印] bleed
【插销】①(门窗等用) bolt ②[电] plug
【插叙】narration interspersed with flashbacks
【插秧】transplant rice seedlings 〈shoots〉◇ ～机 rice transplanter; seedling planting machine
【插页】inset; insert
【插足】①(放脚) put one's foot in ②(参与) participate (in some activity)
【插嘴】interrupt; chip in
【插座】[电] socket; outlet: 弹簧～ cushion socket

叉 ① fork: 餐～ (table) fork / 干草～ hayfork; pitchfork / 钢～ (steel) fork ②(用叉取东西) work with a fork; fork ③(叉形符号, '×') cross
【叉车】forklift; fork truck; forklift truck
【叉腰】akimbo: 双手～ with arms akimbo
【叉子】fork: 粪～ dung fork

杈 wooden fork; hayfork; pitchfork

chá

察 examine; look into; scrutinize: ～其言,观其行 examine his words and watch his deeds; check what he says against what he does
【察访】make calls and investigate; go about to find out
【察觉】be conscious of; become aware of; perceive: ～到敌人的阴谋 discover the plot of the enemy / ～到危险 sense the danger
【察看】watch; look carefully at; observe: ～地形 survey the terrain / ～某人的脸色 examine sb.'s face
【察言观色】carefully weigh up a person's words and closely watch his expression; watch a person's every mood; examine a person's language and observe his countenance

檫 [植] sassafras

茬 ①(收割后留在地里的根、茎) stubble: 麦～ wheat stubble ②(种植、生长次数) crop; batch: 换～ change crops
【茬口】①(轮作种类和次序) crops for rotation ②(某种作物收割后的土壤) soil on which a crop has been planted and harvested

茶 ①(茶叶) tea: 淡～ weak tea / 红～ black tea / 花～ jasmine tea / 菊花～ chrysanthemum tea / 龙井～ Longjing tea; Dragon Well tea / 绿～ green tea / 茉莉花～ jasmine tea / 浓～ strong tea / 沏～ make tea / 砖～ brick tea ②(某些饮料) certain kinds of drink or liquid food: 杏仁～ almond paste
【茶杯】teacup
【茶场】tea plantation
【茶匙】①(小勺) teaspoon ②[量] teaspoonful
【茶炊】①(茶汤壶) tea-urn ②(俄式) samovar
【茶点】tea and pastries; refreshments
【茶碟儿】saucer
【茶饭不香】have no appetite for food and drinks
【茶房】[旧] waiter; steward
【茶缸子】mug
【茶馆】teahouse
【茶褐色】dark brown
【茶壶】teapot
【茶花】camellia
【茶话会】tea party
【茶会】tea party
【茶几】tea table; teapoy; side table
【茶剂】species; tea
【茶碱】theophylline
【茶晶】citrine; yellow quartz
【茶具】tea set; tea-things; tea service
【茶楼酒肆】tearooms and taverns
【茶末】tea dust
【茶农】tea grower
【茶盘】tea tray; teaboard
【茶钱】①(喝茶用的钱) payment for tea (in a teahouse) ②(小费) tip
【茶色】dark brown
【茶食】cakes and sweetmeats
【茶室】tearoom
【茶树】tea tree
【茶水】tea water; boiled water ◇ ～站 tea-stall
【茶亭】tea-booth; tea-stall; tea-kiosk

【茶托】saucer
【茶碗】teacup
【茶锈】tea stain
【茶叶】tea; tea-leaves ◇ ～罐 tea caddy; canister
【茶油】tea-seed oil; tea oil
【茶余酒后】over a cup of tea or after a few glasses of wine; at one's leisure
【茶园】①(茶场) tea plantation ②(茶馆) a place where tea and soft drinks are served; tea garden
【茶砖】brick tea
【茶座】①(卖茶的地方) teahouse ②(座位) seats in a teahouse or tea garden

搽 put (powder, ointment, etc.) on the skin; apply: ～粉 powder / ～雪花膏 put on vanishing cream / ～药 apply ointment, lotion, etc.

查 ①(检查) check; examine: ～某人的身份 check up on sb. / ～卫生 make a public health and sanitation check 〈inspection〉/ ～血 have a blood test ②(调查) look into; investigate: ～事故的原因 find out the cause of an accident ③(翻检着看) look up; consult: ～档案 look into the archives / ～资料 read up the literature (on a special subject) / ～字典 look up a word in the dictionary; consult a dictionary
【查办】investigate and deal with accordingly: 撤职～ dismiss a person and have him prosecuted
【查抄】make an inventory of a criminal's possessions and confiscate them; confiscate
【查点】check the number or amount of; make an inventory of: ～存货 make an inventory of the goods in stock; take stock / ～人数 check the number of people present; check the attendance
【查对】check; verify: ～材料 check the data / ～数字 verify the figures / ～无误 examined and found correct; verified / ～原文 check against the original (text, manuscript, etc.)
【查房】(查病房) make the rounds of the wards; go the rounds of the wards
【查访】go around and make inquiries; investigate
【查封】seal up; close down: ～黄色书刊 seal up pornographic literature
【查号台】[讯] directory inquiries; information
【查户口】check residence cards; check on household occupants
【查获】hunt down and seize; ferret out; track down: 当场～赃物 seize the spoils on the spot
【查禁】ban; prohibit: ～赌博 ban gambling
【查究】investigate and ascertain (cause, responsibility, etc.): ～责任 find out who should be held responsible
【查勘】survey; prospect: ～地界 survey the boundaries of a piece of land / ～矿产资源 prospect for mineral deposits
【查看】look over; examine: ～水情 look into the water 〈flood〉 situation / ～帐目 examine the accounts
【查考】examine; do research on; try to ascertain
【查明】prove through investigation; find out; ascertain: ～事实真相 find out truth; ascertain the facts / ～有罪 find guilty / 现已～ it has been established that; investigation reveals that
【查票】examine tickets; check tickets
【查铺】[军] go the rounds of the beds at night; bed check
【查讫】checked
【查清】make a thorough investigation of; check up on: ～某人来历 find out sb.'s background; check up on sb.
【查哨】[军] go the rounds of guard posts; inspect the sentries
【查收】①(多用于书信) please find (sth. enclosed) ②(检查后收下) check and accept: ～货物 check and accept the cargo
【查税】tax inspection
【查问】inquire; question; interrogate: ～证人 interrogate a witness
【查无实据】investigation reveals no evidence (against the suspect); investigations show no evidence (for it)
【查询】inquire about: ～地址 inquire sb.'s address / ～行李下落 inquire about (the whereabouts of) the luggage
【查验】check; examine: ～护照 examine a passport / ～行李 inspect the baggage / ～遗嘱 inspection of will
【查夜】①(夜间巡视) go the rounds at night ②(夜间巡逻) night patrol
【查阅】consult; look up: ～技术资料 consult technical data; look up technical literature / ～某案件的卷宗 look up the records of a case
【查帐】check accounts; audit accounts; examine accounts ◇ ～报告 audit report
【查找】[自] seek
【查照】(旧时公文用语) please note: 希～办理。

Please note and take appropriate action.
【查证】investigate and verify; check: ～属实 be checked and found to be true; be verified

楂 short, bristly hair or beard; stubble: 胡子～ a stubbly beard

碴
【碴儿】①(小碎块) broken pieces; fragments: 冰～ small pieces of ice / 玻璃～ fragments of glass ②(器物上的破口) sharp edge of broken glass, china, etc.: 碗～ the sharp edge of a broken bowl ③(嫌隙;争执的事由) the cause of a quarrel; quarrel: 找～打架 pick a quarrel (with sb.) ④(提到的事或人家刚说完的话) sth. just said or mentioned: 答～ make a reply / 接不上～ cannot take the cue

叉 (挡住;卡住) block up; jam

chǎ

镲 [乐] small cymbals

叉 part so as to form a fork; fork: ～着腿站着 stand with one's legs apart

chà

诧 be surprised
【诧异】be surprised; be amazed; be astonished: ～的神色 a surprised look

姹 [书] beautiful
【姹紫嫣红】brilliant purples and reds; beautiful flowers; gaily dressed maidens

差 ①(不相同) differ from; fall short of ②(错误) wrong; mistake ③(缺欠) wanting; short of; missing ④(不好) not up to standard; poor; bad; inferior
【差不多】①(时间、距离上相差有限) almost; nearly; just about ②(程度上相差有限) about the same; similar ③(过得去) just about right; just about enough; not far off; not bad
【差不多的】the average person; ordinary people; common people: 这点知识～都有。Any common people will have that bit of knowledge.
【差点儿】①(质量上稍次) not quite up to the mark; not good enough; slightly inferior to ②(几乎) almost; nearly; on the verge of
【差得多】①(不相同) very different; entirely different ②(相差) way below
【差劲】no good; disappointing; not up to the mark

刹 Buddhist temple; Buddhist monastery
【刹那】instant; split second: 一～ in an instant; in a flash; in the twinkling of an eye

岔 ①(分岔) branch off: ～道 turnout / ～路 branch road / 三～路口 a fork in the road; a junction of three roads ②(拐弯) turn off ③(岔子;事故) accident; trouble
【岔开】①(分岔) branch off; diverge ②(离题) diverge to (another topic); change (the subject of conversation) ③(使分开) stagger; space out
【岔口】fork (in a road)
【岔路】branch road; byroad; side road
【岔气】feel a pain in the chest when breathing
【岔子】①(事故) accident; trouble; something wrong ②(错误) fault ③(岔路) branch road; byroad; side road

汊 branch of a river

衩 vent <slit> in the sides of a garment

杈 branch (of a tree)
【杈子】branch: 树～ a branch of a tree

chāi

差 ①(派遣) send on an errand; dispatch: 因公出～ be away on official business ②(公务;职务) errand; job: 兼～ hold more than one job concurrently
【差遣】send sb. on an errand or mission; dispatch; assign: 听候～ await assignment; be at sb.'s disposal
【差使】①(派遣) send; assign; appoint ②[旧] official post; billet; commission
【差事】①(被去做的事情) errand; assignment ②[旧] official post
【差役】[旧] ①(无偿劳动;劳役) corve'e ②(衙门中当差的人) runner or bailiff in a feudal *yamen*

拆 ①(拆开;打开) tear open; take apart: ～机器 disassemble a machine; take a machine apart; strip a machine / ～信 open a letter / 把这个组～了 break up the group ②(拆卸;拆除) pull down; dismantle: ～房子 pull down a house / ～桥 dismantle a bridge / ～帐篷 strike tents
【拆包机】bale breaker
【拆包钳】bale hoop cutter
【拆除】demolish; dismantle; remove: ～城墙

remove〈demolish〉a city wall／～军事基地 dismantle military bases／～障碍物 remove obstacles
【拆穿】expose; unmask: ～骗局 expose a fraud／～西洋镜 strip off the camouflage; expose sb.'s tricks; nail a lie to the counter
【拆东墙,补西墙】tear down the east wall to repair the west wall; mend the west wall by tearing down the east wall
【拆封】seal off
【拆股】dissolve a partnership
【拆毁】demolish; pull down
【拆伙】dissolve a partnership; part company
【拆开】take apart; open; separate: ～包裹 open a parcel／～发动机 take apart the engine; disassemble the engine
【拆模】[建]form removal; form stripping
【拆墙脚】undermine; pull away a prop
【拆散】①(使成套的物件分散) break (a set): 不要把整套的东西～。Don't break the set. ②(使家庭、集体分散) break up (a marriage, family, etc.): ～鸳鸯 break up a married couple
【拆台】cut the ground from under sb.'s feet; pull the rug from under sb.'s feet; pull away a prop; undercut
【拆洗】①(拆洗衣服) wash (padded coats, quilts, etc.) after removing the padding or lining; unpick and wash ②(拆洗机器) strip and clean: ～打字机 strip and clean a typewriter
【拆线】[医] take out stitches
【拆卸】dismantle; disassemble; dismount
【拆阅】open (a letter, document, etc.) and read

钗 hairpin

chái

柴 firewood
【柴草】firewood; faggot
【柴胡】(北柴胡)[中药] Chinese thorowax
【柴火】firewood; faggot: ～垛 stack of firewood
【柴米油盐】fuel, rice, oil and salt
【柴油】diesel oil ◇ ～电动机车 diesel-electric railcar／～发电机组 diesel-electric set／～机车 diesel locomotive
【柴油机】diesel engine: 船用～ marine diesel engine

豺 jackal
【豺狼】jackals and wolves
【豺狼当道】Wolf stands astride road; bad person in power
【豺狼成性】wolfish; rapacious and ruthless
【豺狼座】[天] Lupus

chān

搀 ①(搀扶) help by the arm; support sb. with one's hand ②(混合) mix; mingle
【搀扶】support sb. with one's hand
【搀和】mix; mingle: 把水和酒精～起来 mingle water with alcohol
【搀假】adulterate: ～的酒 adulterated wine
【搀杂】mix; mingle

觇
【觇标】[测]surveyor's beacon

chán

缠 ①(缠绕) twine; wind: ～线轴 wind thread onto a reel ②(纠缠) tangle; tie up; pester ③(应付)[方] deal with: 这人真难～。This fellow is really hard to deal with.
【缠绵】①(纠缠不已) lingering: ～病榻 be bedridden with a lingering disease／乡思～ be tormented by nostalgia ③(宛转动人) touching; moving
【缠绵悱恻】(of writing) exceedingly sentimental; inextricable and commiserative
【缠绕】① twine; bind; wind ②(纠缠) worry; harass ◇ ～植物 twining plant; twiner
【缠手】troublesome; hard to deal with: 这事有些～。The matter is rather troublesome.
【缠足】foot-binding

禅 [佛教] ①(静坐) prolonged and intense contemplation; deep meditation; dhyana: 坐～ sit in meditation; [道教] umbilicular contemplation ②(佛教事物) Buddhist: 参～ try to reach understanding of Chan
【禅房】meditation abode
【禅机】Buddhist allegorical word or gesture
【禅偈】a gatha〈short verse〉containing a Chan message
【禅林】Buddhist temple
【禅师】Chan master
【禅堂】meditation room
【禅心】meditative mind
【禅悟】realization to truth; awakening to truth
【禅杖】Buddhist monk's staff
【禅宗】the Chan sect; Dhyana; Zen

蝉 cicada
【蝉联】continue to hold a post or title
【蝉蜕】[中药] cicada slough

【蝉翼】cicada's wings ◇ ～纱[纺] organdie

婵

【婵娟】[书] ①(美好) lovely (used in ancient writings to describe women) ②(月亮)the moon

蟾

【蟾蜍】toad

【蟾宫】[书] the moon

【蟾酥】[中药] the dried venom of toads; toad-cake

谗 slander; backbite

【谗害】calumniate or slander sb. in order to have him persecuted; frame sb. up

【谗言】slanderous talk; calumny

馋 greedy; gluttonous: 嘴～ greedy; fond of good food

【馋涎欲滴】mouth drooling with greed: 使他～ make his mouth water

【馋嘴】gluttonous

孱 frail; weak

【孱弱】frail (of physique); delicate (in health)

潺

【潺潺】[象] murmur; babble; purl: ～流水 a murmuring stream

chǎn

产 ①(生育) give birth to; be delivered of; (下崽) breed: 流～ miscarriage (自然的); abortion (人工的) / 难～ a difficult delivery / 顺～ an easy delivery / 死～ still birth / 早～ premature birth / 助～ midwifery / 每胎～仔一二只 have only one or two young at a birth ②(生产, 出产) produce; yield: ～棉区 cotton-producing area / ～油 produce oil; oil-producing ③(物产) product; produce: 土特～ local and special products ④(产业) property; estate: 房地～ real estate; real property / 家～ family possessions

【产床】obstetric table

【产蛋率】laying rate

【产道】[医] birth canal; obstetric canal; parturient canal

【产地】place of production; place of origin; producing area: 甘蔗～ a sugarcane growing area / 金丝猴～ the native haunt of the golden monkey / 原料～ sources of raw materials ◇ ～证明书 certificate of origin

【产犊】[牧]calving

【产儿】newborn baby

【产房】delivery room

【产妇】lying-in woman; puerpera; puerperant: 初～ primipara / 待～ woman expecting confinement

【产羔】[牧] lambing; kidding

【产后】postpartum ◇ ～出血 [医] postpartum hemorrhage

【产假】maternity leave

【产驹】[牧] foaling

【产科】obstetrical department; maternity department ◇ ～病房 obstetrical ward; maternity ward / ～学 obstetrics / ～医生 obstetrician / ～医院 maternity hospital; lying-in hospital

【产量】output; yield: 煤～ output of coal / 亩～ per *mu* yield

【产卵】(鸟、家禽) lay eggs; (鱼、蛙) spawn; (昆虫) oviposit

【产品】product; produce: 工业～ industrial products / 农～ farm produce / 畜～ livestock products ◇ ～成本 cost of goods manufactured / ～成品 finished product / ～范围 range of products / ～结构 product mix / ～交换 exchange of products / ～销售成本 cost of goods sold / ～销售净额 net sales / ～销售税金 tax on sales / ～销售总额 gross sales / ～性能 properties of product

【产婆】midwife

【产前】antenatal: ～检查 antenatal examination

【产钳】[医]obstetric forceps

【产权】property right ◇ ～诉讼 action of real right / ～所有人 owner of title / ～说明书 abstract of title / ～要求 property claim

【产褥期】[医] puerperium

【产褥热】[医] puerperal fever; childbed fever

【产生】①(出现) produce; engender: ～好的结果 produce good results / ～影响 exert an influence; exercise influence over; have bearing on ②(产生于) emerge; come into being

【产物】outcome; result; product: 必然～ inevitable outcome / 集体智慧的～ result of collective wisdom

【产销】production and marketing: ～两旺 Both production and marketing thrive / ～平衡 coordination of production and marketing / ～合一 integration of production and marketing operations / ～直接挂钩 direct contact between the producing and marketing departments

【产业】①(私有财产) estate; property ②(工业生产的) industrial ◇ ～革命 the Industrial Revolution / ～工会 industrial union / ～工人 industrial worker / ～后备军 industrial reserve army; reserve army of labor / ～结构 industrial structure / ～界 industrial circles / ～情报 industrial espionage / ～心理学 industrial psychology
【产值】value of output; output value
【产仔】[牧] farrowing

铲 ①(铲子) shovel: 锅～ slice / 煤～ coal shovel ②(撮取;清除) lift or move with a shovel; shovel: ～煤 shovel coal / 把地～平 scrape the ground even; level the ground with a shovel or spade
【铲草除根】destroy the weeds and dig up the roots
【铲车】[机] forklift (truck)
【铲除】root out; uproot; eradicate
【铲斗】scraper bowl; scraper bucket; scraper pan
【铲路机】road planer
【铲土机】earth scraper, giant shovel
【铲运机】scraper; carry-scraper
【铲子】shovel
【铲装机】shovel loader

阐 explain
【阐发】elucidate
【阐明】expound; clarify: ～案情真相 state the facts of a case / ～观点 clarify one's views
【阐释】explain; expound; interpret
【阐述】expound; elaborate; set forth: 各方～了自己对这一问题的立场。Each side set forth its position on this question.
【阐扬】expound and propagate

谄 flatter; fawn on
【谄媚】flatter; fawn on; toady: ～取宠 flatter for favorite
【谄上欺下】be servile to one's superiors and tyrannical to one's subordinates; fawn on those above and bully those below
【谄笑】ingratiating smile
【谄谀】flatter

chàn

颤 quiver; tremble; vibrate
【颤动】vibrate; quiver: 声带～ vibration of the vocal chords / 树叶在～。The leaves are quivering.
【颤抖】shake; tremble; quiver; shiver: 冻得～ shiver with cold / 全身～ be all of a tremble / 吓得两腿～ shake in one's shoes
【颤巍巍】tottering; faltering
【颤音】①[语] trill ②[乐] trill; shake ◇ ～琴 vibraphone
【颤悠】shake; quiver; flicker

忏 repent
【忏悔】①(认识错误、罪过而感到痛心) repent; be penitent ②[宗] confess (one's sins)

chāng

昌 prosperous; flourishing
【昌明】flourishing; thriving; well-developed: 科学～。Science is flourishing.
【昌盛】prosperous: 一个繁荣～的社会主义国家 a prosperous socialist country

菖
【菖蒲】[植] calamus

猖
【猖獗】be rampant; run wild: ～一时 run wild for a while / 疾病～ infested with various diseases
【猖狂】savage; furious: ～的攻击 a furious attack / ～的挑衅 reckless provocation / ～进攻 savage onslaught / ～忘行 act violently without due consideration

鲳
【鲳鱼】silvery pomfret; butterfish

娼 prostitute
【娼妇】(旧时多用于骂人) bitch; whore
【娼妓】prostitute; streetwalker; woman of the town; street girl

cháng

尝 ①(辨别滋味) taste; try the flavor of: ～～咸淡 have a taste and see if it's salty enough ②(感受) taste; experience; come to know: ～到甜头 become aware of the benefits of; come to know the good of / 艰苦备～ have experienced all the hardships ③(曾经) ever; once: 未～见过此人 have never seen the person
【尝闭门羹】drink the soup of closed door; be denied entrance
【尝试】attempt; try
【尝新】have a taste of what is just in season

偿 ①(归还;抵补) repay; compensate for: ~债 pay a debt; discharge a debt / 补~损失 compensate for the loss ②(满足) meet; fulfill: 得~夙愿 have fulfilled one's long-cherished wish
【偿付】pay back; pay: ~能力 solvency / 延期~[法] moratorium
【偿还】repay; pay back: ~期 maturity / ~请求权 right of recourse / ~债务 pay a debt; meet one's engagements / 如数~ pay back the exact amount
【偿金】indemnity
【偿命】pay with one's life (for a murder); a life for a life
【偿清】clear off: ~债务 clear off one's debts
【偿清夙债】get square with one's creditors
【偿清宿债】satisfy all demands
【偿债能力】debt paying ability; liquidity

常 ①(一般;普通;平常) ordinary; common; normal: 反~ unusual; abnormal / 人情之~ natural and normal / 习以为~ be used to sth.; accustomed to sth. ②(不变的;经常) constant; invariable: 冬夏~青 remain green throughout the year; evergreen ③(时常;常常) frequently; often; usually: ~来~往 exchange frequent visits; pay frequent calls; frequently see each other
【常备不懈】always be on the alert; be ever prepared (against war)
【常备军】standing army
【常常】frequently; often; usually; generally
【常川】frequently; constantly: ~供给 keep sb. constantly supplied / ~往来 keep in constant touch
【常春藤】[植] ivy; Chinese ivy
【常服】suit: 单排扣男式~ single-breasted suit / 男式~ lounge suit; business suit / 女式~ women's suit; twopiece / 双排扣男式~ double-breasted suit
【常规】①(沿袭下来的规矩) convention; rule; common practice; routine: 按照~办事 follow the old routine / 打破~ break with convention ②[医] routine: 尿~ routine urine test; 血~ routine blood test; ◇ ~部队 conventional forces / ~化验 routine test / ~检查 routine examination; routine inspection / ~疗法 routine treatment / ~武器 conventional weapons / ~战争 conventional war
【常轨】normal practice; normal course
【常衡】avoirdupois
【常会】regular meeting; regular session
【常见】common ◇ ~病 common disease; common ailment
【常将有日思无日】Waste not, want not
【常例】common practice
【常量】constant
【常绿林】evergreen forest
【常绿树】evergreen (tree)
【常年】①(终年) throughout the year; perennial: ~坚持体育锻炼 persist in physical training all the year round ②(长期) year in year out ③(平常的年份) average year
【常青】evergreen
【常情】reason; sense
【常染色体】autosome; euchromosome ◇ ~遗传 autosomal inheritance
【常人】ordinary person; the man in the street
【常任】permanent; standing: 安理会~理事国 permanent member of the Security Council ◇ ~代表 permanent delegate; permanent representative
【常山】(黄常山) [中药] antifebrile dichroa; antipyretic dichroa
【常设】standing; permanent ◇ ~机构 standing body; permanent organization / ~秘书处 permanent secretariat / ~委员会 permanent committee; permanent commission
【常识】①(普通知识) general knowledge; elementary knowledge: 卫生~ elementary knowledge of hygiene and sanitation ②(生活经验和见识) common sense ◇ ~课 general knowledge course
【常数】[数] constant: ~项 constant term
【常态】normality; normal behavior or conditions: 恢复~ come to the normal state / 一反~ contrary to one's normal behavior; contrary to the way sb. usually behaves ◇ ~曲线[统计] normal curve
【常委】[简] (常务委员) member of the standing committee
【常温】①(15-20 摄氏度) normal atmospheric temperature ②[动物] homoiothermy ◇ ~动物 homoiothermal animal; warm-blooded animal
【常务】day-to-day business; routine: 主持~ in charge of day-to-day business ◇ ~董事 managing director / ~委员 member of the standing committee / ~委员会 standing committee

【常项】constant
【常言】saying: ～道 as the saying goes / ～说得好 it is well said that
【常用】in common use: ～词语 everyday expressions / ～药材 medicinal herbs most in use
【常住居民】inhabitant
【常驻】resident; permanent ◇ ～大使 resident ambassador / ～代表 permanent delegate; permanent representative; resident representative / ～代表机构 resident representative office / ～联合国代表团 permanent representative to the United Nations / ～记者 resident correspondent / ～使节 permanent envoy / ～外交使团 permanent diplomatic mission

嫦

【嫦娥】the goddess of the moon

徜

【徜徉】[书] wander about unhurriedly

长

长 ①(两点之间距离大) long ②(长度) length ③(时间久) of long duration; lasting: 与世～辞 pass away ④(长处) strong point; forte: 取人之～,补己之短 overcome one's short-comings by learning from others' strong points ⑤(长于) be good at; be strong in: ～于绘画 be good at painting
【长臂猿】gibbon
【长柄锅】skillet
【长波】long wave ◇ ～通信 long-wave communication
【长城】① the Great Wall ②(坚不可摧的堡垒) impregnable bulwark
【长抽短吊】[乒乓球] combine long drives with drop shots
【长处】good qualities; strong points; good points: 每人都有自己的～和短处。 Everyone has his own strong and weak points.
【长蝽】[动] chinch bug: 高粱～ sorghum chinch bug
【长此以往】if things go on like this; if things continue this way
【长存】live forever
【长笛】flute
【长凳】bench
【长度】length: ～系数 length factor
【长短】①(长度) length ②(意外变故) accident; mishap: 免得有个～ lest there should be any mishap ③(是非;好坏) right and wrong; strong and weak points: 她最爱背地里说人～。 She is too fond of gossip.
【长吨】gross ton; long ton
【长方体】cuboid; rectangular parallelepiped
【长方形】rectangle
【长庚星】[天] Venus; Hesperus
【长工】farm laborer hired by the year; long-term hired hand
【长骨】[生理]long bone
【长鼓】long drum
【长号】[乐]trombone
【长河】long river; endless flow: 历史的～ the long process of history
【长话短说】make a long story short
【长活】long-term job (of a farm laborer): 扛～ a long-term hired laborer
【长江】the Changjiang River (Yangtse River)
【长江后浪推前浪】～,世上新人换旧人。 Time make it inevitable that in every profession, young men replace the old. / ～,一代更比一代强, As in the Changjiang River the waves behind drive on those before, so each new generation excels the last one
【长颈鹿】giraffe
【长久】for a long time; permanently: 不是～之计 not a permanent solution; just a makeshift arrangement
【长距离】long distance: ～赛跑 a long-distance race
【长空】vast sky
【长裤】trousers; slacks; pants
【长跨度】long span
【长廊】① a covered corridor or walk; gallery ②(北京颐和园长廊) the Long Corridor
【长毛绒】plush
【长矛】pike
【长眠】[婉] eternal sleep; death: ～地下 dead and buried
【长年】all the year round
【长年累月】year in year out; over the years
【长袍】long gown; robe
【长跑】long-distance race; long-distance running
【长篇大论】a lengthy speech or article
【长篇小说】novel; full-length novel
【长期】over a long period of time; long-term: ～贷款 long-term loan / ～规划 a long-term plan / ～天气预报 a long-range weather forecast / ～投资 long-term investment / ～债券 long-term bonds / 作～打算 take a long view; make long-term plans ◇ ～性 protracted na-

ture
【长期共存，互相监督】long-term co-existence and mutual supervision
【长期以来】for a long time; for quite some time
【长枪】①(长矛) spear ②(长筒的枪) long-barrelled gun
【长驱】(of an army) make a long drive; push deep: ～直入 drive straight in; drive deep in
【长绒棉】long-staple cotton
【长衫】(unlined) long gown
【长舌】fond of gossip
【长蛇阵】single-line battle array: 排成一字～ deploy the troops in a long line; string out in a long line
【长蛇座】[天]Hydra
【长生不老药】elixir
【长石】[矿]feldspar ◇ ～砂岩 arkose
【长逝】pass away; be gone forever
【长寿】long life; longevity: 祝您健康～。I wish you good health and a long life.
【长丝】[纺]filament
【长叹】deep sigh: ～一声 heave a deep sigh
【长条校样】[印] galley proof
【长统袜】stockings
【长统靴】boots
【长途】long-distance: ～跋涉 make a long, arduous journey / ～奔袭 long-distance raid ◇ ～电话 long-distance trunk; long-distance call / ～电话局 long-distance exchange; trunk-line exchange / ～飞行 long-range flight / ～记录台 trunk-record position / ～汽车 long-distance bus; coach / ～运输 long-distance transport
【长网】[纸]fourdrinier wire ◇ ～造纸机 fourdrinier (machine)
【长尾猴】guenon
【长尾叶猴】langur
【长物】anything that may be spared; surplus: 别无～ have nothing other than daily necessities; have no valuable personal possessions
【长效磺胺】[药] sulphamethoxypyridazine (SMP)
【长吁短叹】sighs and groans; moan and groan
【长须鲸】finback
【长夜】long night; eternal night
【长涌】blind rollers; blind seas; roller
【长元音】[语] long vowel
【长远】long-term; long-range: ～的利益 long-term interests / ～规划 a long-term plan; long-range plan / 从～的观点看问题 from a long-term point of view
【长斋】(Buddhists') permanent abstention from meat, fish, etc.
【长征】①(长途旅行；长途出征) expedition; long march ②(中国工农红军二万五千里长征) the Long March
【长轴】[数] major axis
【长足】[书] by leaps and bounds: 取得～的进步 make rapid progress; make great strides forward / 有了～的进展 have made considerable progress

场 ①(打谷场) a level open space; threshing ground: 打～ threshing ②[量] 一～大雨 a downpour / 害了一～病 be ill for a while
【场院】threshing ground

肠 intestines: 大～ large intestine / 小～ small intestine
【肠癌】[医] intestinal cancer
【肠出血】[医] enterorrhagia; intestinal hemorrhage
【肠穿孔】[医] enterobrosis; intestinal perforation
【肠梗阻】[医] ileus; intestinal obstruction
【肠激酶】[生化] enterokinase
【肠绞痛】[医] intestinal colic
【肠结核】[医] enterophthisis; intestinal tuberculosis
【肠痉挛】[医] enterospasm; intestinal spasm
【肠鸣音】gurgling sound
【肠扭转】volvulus
【肠儿】[食品] sausage
【肠绒毛】intestinal villus
【肠疝】[医]enterocele
【肠套叠】[医]intussusception; intestinal intussusception
【肠胃】intestines and stomach; stomach; belly: ～不好 suffer from indigestion / ～气胀 flatulence ◇ ～炎 enterogastritis
【肠系膜】mesenterium; mesentery
【肠炎】[医] enteritis
【肠衣】casing (for sausages)
【肠液】intestinal juice; succus entericus
【肠痈】[中医] appendicitis
【肠粘连】[医] intestinal adhesion
【肠子】intestines

chǎng

场 ①(场地) a place where people gather; ground; field: 会～ meeting-place / 篮球～ basketball court; basketball pitch / 战～

battlefield / 观众进～ spectators enter the stadium, etc. / 运动员人～ athletes enter the arena; athletes enter the sports field ②(农场;养殖场) farm: 国营农～ state farm / 牛奶～ dairy (farm) / 畜牧～ stock farm / 养鸡～ chicken farm / 养鸭～ duck farm / 种马～ stud farm ③(场合) stage; spot; scene: 出～ come on the stage; appear on the scene / 在～ be present at the scene / 不在～ be absent from the scene / 当～抓住 caught on the spot ④(舞台) stage: 上～ come on the stage; (剧本用语) enter / 下～ leave the stage; (剧本用语) exit ⑤(戏剧中的段落) scene: 第二幕第三～ Act II, Scene iii ⑥[量] 一～电影 a film show / 一～球赛 a match; a ball game / 加演一～ give an extra performance or show ⑦[物] field: 磁～ magnetic field / 电～ electric field

【场磁铁】[电]field magnet

【场次】the number of showings of a film, play, etc.

【场地】space; place; site: 本方～ home ground; home court / 比赛～ competition arena; ground; court / 对方～ away ground; away court / 交换～ change sides / 施工～ construction site ◇ ～使用权 right to the use of a site

【场合】occasion; situation: 外交～ a diplomatic occasion

【场记】[剧][电影] ①(记录) log ②(记录员) log keeper; (男) script holder; (女) script girl

【场界灯】[航空]boundary lights

【场理论】[物]field theory

【场面】①(戏剧、电影中的场景) scene (in drama, fiction, etc.); spectacle ②(情景) occasion; scene: 热烈友好的～ a scene of warm friendship / 盛大的～ a grand occasion ③(排场) appearance; front; facade: 撑～ keep up appearances

【场能】field energy

【场频】field frequency

【场强】intensity of field; field intensity

【场所】place; arena: 公共～ a public place / 蚊蝇孳生的～ a breeding ground of flies and mosquitoes / 娱乐～ place of recreation

【场子】a place where people gather for various purposes (e.g. theatre, hall, sports ground, etc.)

厂 ①(工厂)factory; mill; plant; works: 电机～ electrical machinery plant / 电影制片～ studio; film studio / 服装～ clothing factory / 钢铁～ iron and steel works / 化工～ chemical works / 机床～ machine tool plant / 面粉～ flour mill / 酿酒～ winery / 水泥～ cement works / 糖～ sugar refinery / 鞋～ shoe factory / 造船～ ship yard / 针织～ knitwear mill ②(厂子) yard; depot: 煤～ coal yard / 木～ timberyard

【厂标】emblem mark

【厂房】①(房屋) factory building ②(车间) workshop

【厂矿】factories and mines ◇ ～企业 factories, mines and other enterprises; industrial enterprises

【厂商】firm: 承包～ contractor

【厂休日】day of rest for factory workers; workers' day off

【厂长】factory director; factory manager ◇ ～负责制 system of overall responsibility by factory manager

【厂址】the site of a factory; the location of a factory: 选择～ choose a site for building a factory

【厂主】factory owner; millowner

敞 ①(宽绰,没有遮拦) spacious: 宽～ spacious; roomy ②(张开;打开) open; uncovered: ～着门 leave the door open; with the door open

【敞车】[铁道] ①(没有车顶的货车) open wagon; open freight car ②(没有车篷的车) flatcar

【敞怀】with one's coat or shirt unbuttoned ～畅饮 being cheerful and happy, one drinks to capacity

【敞开】①(打开;大开) open wide: ～思想 say what's in 〈on〉 one's mind; get things off one's chest / 把门～ open the door wide ②(尽量;任意) unlimited; unrestricted: ～供应 unlimited supply; open-ended supply

【敞亮】①(宽敞明亮) light and spacious ②(心里豁亮) clear (in one's thinking)

【敞篷车】open car

【敞式车身】open body

【敞胸露怀】bare the breast

氅 cloak: 大～ overcoat

chàng

唱 ①(唱歌) sing: 二重～ duet / 三重～ trio / 四重～ quartet / 五重～ quintet / 表演～ an item combining singing, dancing and acting / 独～ solo / 轮～ round / 齐～ sing in unison ◎女高音 soprano / 花腔女高音

coloratura soprano / 女中音 mezzo soprano / 女低音 alto / 男高音 tenor / 男中音 baritone / 男低音 bass ②(大声叫) call; cry
【唱白脸】wear the white makeup of the stage villain; play the villain; pretend to be harsh and severe
【唱本】the libretto or script of a ballad-singer
【唱词】libretto; words of a ballad
【唱独角戏】play a monodrama; do a thing alone
【唱段】aria: 京剧～ an aria from a Beijing opera
【唱对台戏】put on a rival show; enter into rivalry with; put up a rival show
【唱反调】sing a different tune; deliberately speak or act contrary to; strike up a discordant tune
【唱高调】use high-flown words; affect a high moral tone; chant bombastic words
【唱歌】sing (a song)
【唱工】[剧] art of singing; singing
【唱和】①(一人唱,别人和) one singing a song and the others joining in the chorus: 此唱彼和。 When one starts singing, another joins in. ②(一人做诗词,别人相应作答) one person writing a poem to which one or more other people reply, usu. using the same rhyme sequence
【唱红脸】wear the red makeup of the stage hero; play the hero; pretend to be generous and kind: 一个～,一个唱白脸 one coaxes, the other coerces
【唱机】gramophone; phonograph: 电～ record player
【唱老调】sing the same old song; harp on the same string
【唱名】①(逐一高声念姓名) roll call: ～表决 vote by roll call ②[乐] sol-fa syllables
【唱名法】[乐] sol-fa; solmization: 固定～ fixed-do system / 首调～ movable-do system
【唱片】gramophone record; phonograph record; disc: 密纹～ long-playing record / 放～ play a gramophone record / 灌～ cut a disc
【唱票】call out the names of those voted for while counting ballot-slips ◇ ～人 teller
【唱腔】[剧]music for voices in a Chinese opera
【唱双簧】①(曲艺) give a two-man comic show ②(互相配合) collaborate with each other
【唱头】pickup
【唱戏】act in an opera
【唱针】gramophone needle; stylus
【唱做念打】[剧]singing, acting, recitation and acrobatics

倡 initiate; advocate: 首～ initiate; start
【倡导】initiate;; propose: 在某人的～下 on the initiative of sb.
【倡言】propose; initiate
【倡议】propose; sponsor: ～召开国际会议 propose the calling of an international conference / 在他的～下 at his suggestion ◇ ～权 initiative / ～书 written proposal; proposal / ～者 initiator

怅 disappointed; sorry
【怅怅不乐】disconsolate; feeling gloomy
【怅然】disappointed; upset: ～而返 come away disappointed
【怅惘】distracted; listless

畅 ①(无阻碍;不停滞) smooth; unimpeded: 流～ easy and smooth; fluent ②(痛快;尽情) free; uninhibited: ～饮 drink one's fill
【畅达】fluent; smooth: 交通～ have a good transport and communications network; be easily accessible / 译文～。 The translation reads smoothly.
【畅快】free from inhibitions; carefree: 心情～ have ease of mind
【畅所欲言】speak without any inhibitions; speak one's mind freely; speak out freely
【畅谈】talk frecly and to one's heart's content; speak glowingly of
【畅通】unimpeded; unblocked: 前面道路～无阻 open road ahead
【畅销】be in great demand; sell well; have a ready market: ～国外 sell well on foreign markets / ～全国 sell well all over the country ◇ ～书 best seller
【畅行无阻】pass unimpeded
【畅叙】chat cheerfully (usu. about old times): ～友情 recall old friendship heartily
【畅游】①(畅快地游泳) have a good swim ②(尽情游览) enjoy a sightseeing tour: ～名胜古迹 enjoy a trip to places of historic interest

chāo

焯 scald (as a way of cooking)

绰 grab; take up

抄 ①(誊写;抄写) copy; transcribe: ～写稿件 make a fair copy of the manuscript / 照～原文

make a verbatim transcription of the original ②(抄袭) plagiarize; lift ③(搜查没收) search and confiscate; make a raid upon ④(抓取) go off with; walk off with ⑤(从侧面或近路过去) take a shortcut ⑥(袖手) fold (one's arms): ~着手站在一边 stand by with folded arms ⑦(拿起) grab; take up: ~起一把铁锹 take up a spade
【抄本】hand-copied book; transcript
【抄道】①(走近便的路) take a shortcut ②(近便的路) shortcut
【抄后路】outflank the enemy and attack him in the rear; turn the enemy's rear
【抄获】search and seize; ferret out
【抄家】search sb.'s house and confiscate his property ◇ ~灭门 confiscate the property and exterminate the family
【抄件】duplicate; copy
【抄录】make a copy of; copy
【抄身】search sb.; frisk
【抄送】make a copy for; send a duplicate to
【抄网】[渔] dip net
【抄袭】①(抄别人作品当做自己的) plagiarize; lift: ~行为 (an act of) plagiarism ②(照搬他人经验方法) borrow indiscriminately from other people's experience ③(绕道袭击) launch a surprise attack on the enemy by making a detour
【抄写】copy; transcribe

吵

【吵吵】make a row; kick up a racket

钞

①(钞票) bank note; paper money: 现~ cash ②(誊写下来的文字) collected writings: 诗~ collected poems
【钞票】bank note; paper money; bill

超

①(超过) exceed; surpass; overtake ②(超出寻常的) super-; extra-: ~高温 superhigh temperature ③(超出) transcend; go beyond
【超产】overfulfill a production target 〈quota〉
【超车】overtake other cars on the road: 不准~ No overtaking!
【超尘脱俗】transcend the worldly
【超出】overstep; go beyond; exceed: ~定额 exceed the quota / ~法律权限的 extralegal / ~范围 go beyond the scope 〈bounds〉 / ~票面价值 above par / ~预料 exceed one's expectations
【超导】[物] superconduction ◇ ~电性 superconductivity / ~体 superconductor
【超等】of superior grade; extra fine: ~质量 extra good quality; superfine
【超低空飞行】hedgehopping; minimum altitude flying
【超低频】ultralow frequency
【超低温】ultralow temperature
【超低压轮胎】extra-low pressure tire
【超度】[宗] release souls from purgatory; expiate the sins of the dead ◇ ~众生 save mankind from the sea of misery which is life
【超短波】ultrashort wave
【超短裙】miniskirt
【超额】above quota: ~完成生产指标 overfulfill the production quota; surpass the production target ◇ ~保险 excess insurance / ~利润 superprofit; excess profits / ~赔款 excess of loss / ~剩余价值 excess surplus value / ~损害赔款 excessive damages / ~征税 overtax
【超负荷】excess load; overload
【超高】[水] freeboard
【超高频】[电] ultrahigh frequency (UHF)
【超高压】①(大气压) superhigh pressure ②(电压) extrahigh voltage; extrahigh tension: ~线路带电作业 working on live extrahigh tension power lines
【超过】outstrip; surpass; exceed: ~规定速度 exceed the speed limit / ~历史最高水平 top all previous records / ~世界先进水平 surpass advanced world levels / ~限度 go beyond the limit
【超级】super ◇ ~大国 superpower / ~公路 superhighway / ~客轮 superliner / ~明星 superstar / ~商场 supermarket / ~油轮 supertanker
【超假】overstay one's leave
【超经济剥削】[经] extraeconomic exploitation
【超巨星】[天] supergiant (star)
【超绝】unique; superb; extraordinary: 技艺~ extraordinary skill; superb performance
【超龄】overage: ~团员 overage Youth League member
【超敏反应】hypersensitivity
【超期服役】[军] extended active duty; extended service in the army
【超群】head and shoulders above all others; preeminent: 武艺~ extremely skillful in martial arts
【超然】aloof; detached
【超然物外】hold aloof from the world; be above worldly considerations; stay away from the scene of contention
【超人】①(超过一般人) be out of the common

run: ～的记忆力 exceptionally good memory ②[哲] superman (as defined by Nietzsche)
【超声波】ultrasonic (wave); supersonic (wave) ◇ ～疗法 ultrasonic therapy / ～探伤 supersonic flaw detecting; ultrasonic examination / ～探伤仪 ultrasonic flaw detector / ～诊断仪 supersonic diagnostic set
【超声速飞行】supersonic flight
【超声物理学】ultrasonic physics
【超声学】ultrasonics
【超速】①(超出速度限制) exceed the speed limit ②[力] hypervelocity ◇ ～驾驶 furious driving / ～粒子[物] hypervelocity particle
【超脱】①(不拘泥成规等) unconventional; original ②(超出,脱离) be detached; stand aloof; hold aloof; keep aloof
【超外差】[电子]superheterodyne; superhet ◇ ～收音机 superheterodyne (radio set)
【超现实主义】surrealism
【超小型管】[电子]subminiature tube
【超新星】[天]supernova
【超逸】unconventionally graceful; free and natural
【超音速】supersonic speed ◇ ～喷气机 superjet / ～战斗机 supersonic fighter-plane; supersonic fighter
【超铀元素】transuranium element
【超越】surmount; overstep; transcend; surpass: ～障碍 surmount an obstacle / ～障碍赛 show jumping / ～职权范围 go beyond one's terms of reference; overstep one's authority ◇ ～射击 [军] overhead fire
【超载】[交] overload ◇ ～能力 overload capacity
【超支】overspend: 从不～ never live beyond one's income
【超重】①(超过载重量) overload ②(超过规定的重量) overweight: ～费 heavy lift charge / ～信件 overweight letter / ～行李 excess luggage ③[天] superheavy
【超轴】[铁道] over haulage ◇ ～牵引 trains hauling above-normal tonnage
【超子】[物]hyperon
【超自然】supernatural

cháo

朝 ①(朝廷,跟'野'相对) court; government: 上～ go to court / 在～党 party in power; ruling party ②(朝代) dynasty ③(一个君主的统治时期) an emperor's reign ④(朝见) have an audience with (a king, an emperor, etc.); make a pilgrimage to ⑤(向) facing; towards: ～南走 go southward / ～前走 march ahead / 坐东～西 with a western exposure; facing west
【朝拜】pay respects to (a sovereign); pay religious homage to; worship
【朝代】dynasty
【朝贡】pay tribute (to an imperial court); present tribute
【朝见】have an audience with (a king, an emperor, etc.): 进宫～ be presented at court
【朝觐】①[书] have an audience with ②[宗]go on a pilgrimage (to a shrine or a sacred place)
【朝山】[佛教]make a pilgrimage to a temple on a famous mountain
【朝圣】[宗]pilgrimage; hadj ◇ ～团 a pilgrimage mission
【朝廷】①(君主听政的地方) royal or imperial court ②(以君主为首的中央政府) royal or imperial government
【朝鲜】Korea ◇ ～人 Korean / ～文字 Korean writing / ～语 Korean (language)
【朝向】[建] orientation
【朝阳】①(向着太阳) exposed to the sun; sunny ②(向南) with a sunny, usu. southern, aspect
【朝野】①(朝廷和民间) the court and the commonalty ②(资本主义国家政府与非政府方面) the government and the public
【朝政】affairs of state

潮 ①(潮汐) tide: 大～ spring tide / 低～ low tide; low water / 高～ high tide; high water / 平～ slack water / 小～ neap tide / 早～ morning tide / 涨～了。The tide is flowing. 落～了。The tide is ebbing. ②(社会变动) (social) upsurge; current; tide: 低～ low tide / 工～ workers' strike / 怒～ raging tide / 思～ trend of thought / 革命高～ a revolutionary high tide ③(潮湿) damp; moist
【潮差】tide range
【潮呼呼】damp; dank; clammy
【潮解】[化] deliquescence
【潮流】①(潮汐引起的水流) tide; tidal current ②(社会变动或趋势) trend: 历史～ historical trend / 顺应世界之～ adapt oneself to world trends; go along with world trends
【潮气】moisture in the air; damp; humidity
【潮热】[中医]hectic fever
【潮湿】moist; damp
【潮水】tidewater; tidal water
【潮位曲线】tide curve
【潮汐】morning and evening tides; tide ◇ ～表

tide table / ～测站 tide station / ～摩擦 tidal friction / ～能 tidal energy / ～能发电 tidal power / ～预报 tide prediction
【潮汛】spring tide

嘲 ridicule; deride: 解～ try to explain things away when ridiculed / 冷～热讽 freezing irony and burning satire
【嘲讽】sneer at; taunt
【嘲弄】mock; poke fun at
【嘲笑】ridicule; deride; jeer at; laugh at

巢 nest: 匪～ nest of robbers; den of robbers; bandits' lair / 鸟～ bird's nest
【巢菜】common vetch
【巢础】septum
【巢蛾】ermine moth
【巢框】frame
【巢脾】honeycomb
【巢鼠】harvest mouse
【巢箱】brood chamber with breeding combs
【巢穴】lair; den; nest; hideout

chǎo

炒 stir-fry; fry; sauté: ～黄瓜 sautéed cucumber / ～鸡蛋 scrambled eggs / ～肉丝 stir-fried shredded pork / 蛋～饭 rice fried with eggs / 糖～栗子 chestnuts roasted in sand with brown sugar
【炒菜】①(炒) stir-fry; saute' ②(炒的菜) a fried dish
【炒饭】①(炒) fry rice ②(炒的饭) fried rice
【炒货】roasted seeds and nuts
【炒冷饭】heat leftover rice; rehash; repeat without any new content
【炒米】①(干炒过的米) parched rice ②(蒙古族食品) millet stir-fried in butter ◇ ～花 puffed rice
【炒面】①(油炒面条) *chow mein*; fried noodles ②(炒熟的面粉) parched flour
【炒勺】round-bootomed frying pan

吵 ①(声音杂乱) make a noise: 别～了! Hold your row!; Be quiet! ②(争吵) quarrel; wrangle; squabble: 同某人～架 have a quarrel with sb.
【吵架】quarrel; wrangle; have a row
【吵闹】①(大声争吵) wrangle; kick up a row ②(骚扰声) din; hubbub
【吵嚷】make a racket; shout in confusion; clamor
【吵人】disturb others (by noise)
【吵嘴】quarrel; bicker

chào

耖 ①(农具) a harrow-like implement for pulverizing soil ②(用耖整地) level land with such an implement

chē

车 ①(车辆) vehicle: 火～ train / 军用～ army vehicle / 汽～ motor vehicle; automobile ②(利用轮轴旋转的工具) wheeled machine or instrument: 纺～ spinning wheel / 滑～ pulley / 水～ waterwheel ③(机器) machine: 开～ set the machine going; start the machine / 停～ stop the machine / 试～ trial run; test run ④(用车床切削) lathe; turn: ～机器零件 lathe a machine part / ～光 smooth sth. on a lathe ⑤(用水车取水) lift water by waterwheel
【车把】①(自行车、摩托车的) handlebar ②(手推车等的) shaft
【车把式】cart-driver; carter
【车床】lathe: 多刀～ multicut lathe / 立式～ vertical lathe / 万能～ universal lathe / 卧式～ horizontal lathe ◇ ～卡盘 lathe chuck
【车次】①(列车的编号) train number ②(汽车行车的次第) motorcoach number (indicating order of departure)
【车刀】lathe tool; turning tool
【车到山前必有路】the cart will find its way round the hill when it gets there; things will eventually sort themselves out
【车道】(traffic) lane
【车队】motorcade
【车费】fare
【车工】①(工种) lathe work ②(技工) turner; lather operator
【车钩】[铁道] coupling
【车轱辘话】[方] repetitious talk
【车祸】traffic accident; road accident; automobile accident: ～致人死命 vehicular homicide
【车技】[杂技] trick-cycling
【车架】frame (of a car, bicycle, etc.)
【车间】workshop; shop: 大件装配～ shop for assembling big parts / 锻工～ blacksmith shop; forge shop / 翻砂～ foundry / 工具～ tool shop / 冷锻～ cold hammering shop / 量具～ measuring gauge shop / 木工～ carpentry work-shop / 热处理～ heat treatment shop / 制模～ pattern shop / 装配～ assembly shop; fitting shop ◇ ～主任 workshop director

【车库】garage
【车辆】vehicle; car: 来往～ traffic／ 铁路机车及～ rolling stock ◇ ～门道 prote-coche／ ～事故 car accident／ ～周转率[铁道] average turnround rate of rolling stock
【车轮】wheel (of a vehicle)
【车轮式审讯】grueling trial
【车轮战】the tactic of several persons taking turns in fighting one opponent to tire him out
【车马费】travel allowance
【车皮】railway wagon; railway carriage
【车票】train or bus ticket; ticket
【车前】[植] Asiatic plantain
【车身】automobile body
【车水马龙】incessant stream of horses and carriages; heavy traffic: 门前～。 The countyard is thronged with visitors.
【车速】speed of a motor vehicle
【车胎】tire
【车厢】railway carriage; railroad car
【车削】[机] turning
【车辕】shaft (of a cart, etc.)
【车载斗量】enough to fill carts and be measured by the *dou*; common and numerous
【车闸】brake (of a car, bicycle, etc.)
【车站】station; depot; stop
【车照】licence (of a car, bicycle, etc.)
【车辙】rut
【车轴】axletree; axle
【车子】small vehicle (such as a car, pushcart, etc.)

砗

【砗磲】[动] giant clam; tridacna

chě

扯 ①(拉) pull ②(撕) tear ③(买布等) buy (cloth, thread, etc.) ④(漫无边际地闲谈) chat; gossip
【扯淡】[方] talk nonsense; nonsense
【扯后腿】hold sb. back (from action); be a drag on sb.; be a hindrance to sb.
【扯谎】tell a lie; lie
【扯裂】divulsion
【扯皮】dispute over trifles; argue back and forth; wrangle
【扯碎】discerp

chè

撤 ①(除去) remove; take away: 把盘子、碗～了 clear away the dishes／ 把障碍物～了 remove the barrier ②(撤退) withdraw; evacuate: ～伤员 evacuate the wounded／ 向后～ withdraw; retreat／ 主动～出 withdraw on one's own initiative
【撤兵】withdraw troops
【撤除】remove; dismantle: ～军事基地 dismantle a military base／ ～军事设施 dismantle military installations
【撤防】withdraw a garrison; withdraw from a defended position
【撤换】dismiss and replace; recall; replace
【撤回】①(招回;命令回来) recall; withdraw: ～步哨 withdraw the guard／ ～代表 recall a representative ②(收回) revoke; retract; withdraw: ～声明 retract a statement／ ～诉讼 withdraw a claim; revoke a court action／ ～指控 withdrawal of charge
【撤离】withdraw from; leave; evacuate: ～城市 evacuate a city／ ～阵地 abandon a position
【撤诉】nolle prosequi
【撤退】withdraw; pull out: ～方向 the line of withdrawal／ 安全～ make good one's retreat
【撤席】remove a dinner table
【撤销】cancel; rescind; revoke: ～处分 rescind a penalty; annul a penalty／ ～定货单 recall an order／ ～法令 repeal a decree／ ～合同 abandonment of contract／ ～决议 annul a decision／ ～命令 countermand an order／ ～判决 recall a judgement; recall of judicial decisions／ ～其职务 dismiss a person from his post／ ～起诉 quash an indictment／ ～上诉 abandonment of appeal／ ～宵禁令 lift the curfew／ ～邀请 withdraw an invitation／ ～原计划 rescind the original plan／ ～原判 disaffirmance; disaffirmation
【撤职】dismiss 〈discharge〉 sb. from his post; remove sb. from office: ～查办 discharge sb. from his post and prosecute him
【撤走】withdraw

澈 (of water) clear; limpid

彻 thorough; penetrating: ～夜工作 work all night／ 透～了解 thorough understanding／ 响～云霄 resounding across the skies
【彻底】thorough; thoroughgoing: ～改变 radical change／ ～击败 mop the floor with／ ～抛弃 cast off once for all／ ～破产 complete bankruptcy／ ～胜利 complete victory／ ～失败 thorough defeat／ ～消灭 utterly destroy
【彻骨】to the bone: ～之寒 biting cold／ 寒风～。 The bitter wind chills one to the bone.

【彻头彻尾】out and out; through and through; downright: ～的骗局 a downright fraud; sheer fraud; deception from beginning to end
【彻夜】all night; all through the night: ～不眠 lie awake all night / ～灯火通明 the lights were ablaze all through the night

坼 [书] split open; crack
【坼裂】split open; crack

掣 ①(拽) pull; tug ②(抽) draw: ～签 draw lots
【掣肘】hold sb. back by the elbow; impede; handicap

chēn

嗔 ①(怒;生气) be angry; be displeased: 生～ get angry ②(对人不满) be annoyed (with sb.)
【嗔怪】blame; rebuke
【嗔怒】get angry
【嗔色】angry or sullen look: 微露～ look somewhat displeased

抻 [口]pull out; stretch
【抻面】①(用手抻面条) make noodles by drawing out the dough by hand ②(抻成的面条) hand-pulled noodles

chén

沉 ①(在水里往下落) sink: ～底儿 sink to the bottom ②(多用于抽象事物) keep down; lower: ～下心来 settle down (to one's work, etc.); concentrate (on one's work, study, etc.) / 把脸一～ put on a grave expression; pull a long face ③(程度深) deep; profound: 睡得很～ be in a deep sleep; be fast asleep; sleep like a log ④(分量重;感觉沉重) heavy
【沉沉】①(分量重) heavy: 穗子～地垂下来 The ears hang heavy on the stalks. ②(程度深) deep: ～入睡 sink into a deep sleep / 暮气～ lifeless; lethargic; apathetic / 暮霭～。 Dusk is falling.
【沉甸甸】heavy; heavy and not easy to wield: ～的谷穗 heavy ears of millet / ～的一口袋稻种 a heavy sack of rice seed
【沉淀】sediment; precipitate ◇ ～池[环保] precipitating tank; sedimentation tank / ～剂 [化] precipitating agent / ～物 sediment; precipitate
【沉管】immersed tube
【沉积】[地] deposit: 陆相～ continental deposit / 海相～ marine deposit ◇ ～物 deposit; sediment / ～旋回 cycle of sedimentation / ～岩 sedimentary rock / ～作用 deposition; sedimentation
【沉寂】①(十分寂静) quiet; still: ～的深夜 in the still of (the) night ②(消息全无) no news
【沉降】subside: 地面～ earth subsidence ◇ ～缝[建] settlement joint
【沉浸】immerse; steep
【沉井】[建] open caisson
【沉静】quiet; calm; serene; placid: ～的神色 a serene look / 心情～ be in a placid mood
【沉疴】severe and lingering illness
【沉沦】sink into (vice, degradation, depravity, etc.)
【沉闷】①(天气、气氛等) oppressive; depressing: ～的天气 dull weather; depressing weather ②(心情不舒畅) depressed; in low spirits: 心情～ feel depressed ③(性格不爽朗) not outgoing; withdrawn
【沉迷】indulge; wallow
【沉湎】[书] wallow in; be given to: ～酒色 over-indulge oneself in wine and women / ～于酒 be given to heavy drinking
【沉没】sink
【沉默】①(不爱说笑) reticent; taciturn; uncommunicative: ～寡言的人 a reticent person; a person of few words ②(不说话) silent: 保持～ remain silent / 他～了一会又继续说下去。 After a moment's silence he went on speaking.
【沉溺】indulge; wallow: ～于声色 wallow in sensual pleasures
【沉砂池】[环保] grit chamber
【沉睡】be sunk in sleep; be fast asleep
【沉思】ponder; meditate; be lost in though: ～凝想 think deeply or profoundly
【沉痛】①(深深的悲痛) deep feeling of grief or remorse: 表示～的哀悼 express profound condolences / 怀着～的心情 be deeply grieved ②(深刻,严重) deeply felt; bitter: ～的教训 bitter experience
【沉陷】①sink; cave in ②[建] settlement: 不均匀～ unequal settlement
【沉香】[植] agalloch eaglewood
【沉箱】[建] caisson
【沉吟】mutter to oneself, unable to make up one's mind
【沉鱼落雁之容】a beautiful face which causes fish to sink out of sight and the flying crane to drop down
【沉郁】depressed; gloomy
【沉冤】gross injustice; unrighted wrong

【沉冤莫白】suffer a grievous-wrong; wrongs that cannot be redressed
【沉渣】sediment; dregs
【沉重】①(分量大;程度深) heavy: ~的打击 a heavy blow / ~的脚步 heavy steps / 心情~ with a heavy heart ②(严重) serious; critical: 病情~ critically ill
【沉住气】keep calm; keep cool; be steady
【沉着】cool-headed; composed; steady; calm: ~应战 meet the attack calmly / ~镇静 be steady and calm / 勇敢~ brave and steady
【沉子】[渔] sinker
【沉醉】get drunk; become intoxicated: ~在节日的欢乐里 be intoxicated with the spirit of the festival

忱 [书] sincere feeling; true sentiment: 热~ zeal; warmheartedness / 谢~ thankfulness

辰 ①(日、月、星) celestial bodies: 星~ stars ②(时光;日子) time; day; occasion: 诞~ birthday
【辰砂】cinnabar; vermillion

晨 morning: 清~ early morning; dawn
【晨风】matinal; morning breeze
【晨光】the light of the early morning sun; dawn: ~熹微 first faint rays of dawn
【晨曦】first rays of the morning sun
【晨星】①(清晨的星) stars at dawn: 寥若~ as few as stars at dawn ②[天] morning star

橙
【橙子】orange

臣 official under a feudal ruler; subject: 君~ the monarch and his subjects
【臣服】[书] submit oneself to the rule of; acknowledge allegiance to
【臣民】subjects of a feudal ruler
【臣子】official in feudal times

尘 ①(尘土) dust; dirt: 一~不染 not stained with a particle of dust; spotless ②(尘世) this world: ~俗 this mortal world
【尘埃】dust ◇ ~传染[医] dust infection
【尘暴】[气] dust storm
【尘肺】[医] pneumoconiosis
【尘封】covered with dust; dust-laden
【尘垢】dust and dirt; dirt
【尘世】[宗] this world; this mortal life
【尘俗之见】a commonplace view
【尘土】dust
【尘缘未了】the dusty affinity is not yet finished
【尘嚣】hubbub; uproar

陈 ①(安放;摆设) lay out; put on display ②(叙说) state; explain: 此事当另函详~。The matter will be explained in detail in a separate letter. ③(时间久的;旧的) old; stale
【陈兵】mass troops; deploy troops: ~百万 deploy a million troops / ~边境 mass troops along the border
【陈陈相因】follow a set routine; stay in the same old groove; keep on doing the same thing over and over again
【陈词滥调】hackneyed and stereotyped expressions; clichés
【陈醋】mature vinegar
【陈腐】old and decayed; stale; outworn: ~的词句 stale phrases
【陈规】outmoded conventions: ~陋习 outmoded conventions and bad customs; bad customs and habits
【陈货】old stock; shopworn goods
【陈迹】a thing of the past
【陈酒】old wine; mellow wine
【陈旧】outmoded; obsolete; old-fashioned; out-of-date: ~的词语 obsolete words and expressions / ~的观点 an outmoded notion / ~的设备 obsolete equipment
【陈列】display; set out; exhibit ◇ ~馆 exhibition hall / ~柜 showcase / ~品 exhibit / ~室 exhibition room; showroom
【陈皮】[中药] dried tangerine or orange peel
【陈设】①(摆设) display; set out ②(摆设的东西) furnishings
【陈述】state; declare: ~的理由 alleged cause / ~自己的意见 state one's views ◇ ~句[语] declarative sentence
【陈说】state; explain: ~利害 explain the advantages and disadvantages (of a situation, course of action, etc.)
【陈诉】state; recite: ~委屈 state one's grievances

chèn

衬 ①(在里面托上一层) line; place sth. underneath: ~驼绒的大衣 a fleece-lined overcoat / ~一层纸 put a piece of paper underneath / 里面~一件背心 wear a vest underneath ②(衬在里面的附属品) lining; liner: ~管 liner tube / 钢~ stell liner / 领~ collar lining / 袖~ cuff lining ③(陪衬;衬托) set off

【衬布】lining cloth
【衬裤】underpants; pants
【衬里】lining: 水泥～[机] cement lining
【衬裙】underskirt; petticoat
【衬衫】shirt: 男～ shirt / 女～ blouse; shirt
【衬托】set off; serve as a foil to
【衬衣】underclothes; shirt
【衬纸】slip sheet; interleaving paper

谶 [书]augury
【谶语】a prophecy believed to have been fulfilled

趁 ①(利用机会) take advantage of; avail oneself of ②(利用时间) while: ～你在这儿,咱们谈谈。 Let's have a talk while you are here.
【趁便】when it is convenient; at one's convenience
【趁火打劫】loot a burning house; take advantage of sb.'s misfortune to do him harm
【趁机】take advantage of the occasion; seize the chance: ～捣乱 seize the opportunity to make trouble
【趁空】use one's spare time; avail oneself of leisure time
【趁热打铁】strike while the iron is hot
【趁人之危】take advantage of another's perilous state
【趁势】take advantage of a favorable situation
【趁早】as early as possible; before it is too late; at the first opportunity

称 fit; match; suit: 颜色相～ well matched in color
【称身】fit
【称心】find sth. satisfactory; be gratified: ～如意 have as one wishes; in accord with one's wishes
【称愿】be gratified (esp. at the misfortune of a rival)
【称职】fill a post with credit; be competent

chēng

撑 ①(抵住) prop up; support ②(撑船) push or move with a pole: ～船 pole a boat; punt ③(支持) maintain; keep up ④(张开) open; unfurl: ～伞 open an umbrella / 把麻袋～开 hold open the sack ⑤(充满到容不下的程度) fill to the point of bursting ⑥[机] brace; stay: 角～ corner brace
【撑臂】[机] brace; supporting arm
【撑场面】keep up appearances
【撑持】prop up; shore up; sustain: ～局面 shore up a shaky situation
【撑竿】vaulting pole
【撑竿跳高】pole vault; pole jump ◇ ～运动员 pole-vaulter
【撑条】[机] stay: 横～ cross stay / 斜～ diagonal stay
【撑腰】support; back up; bolster up: ～打气 bolster and pep up / 给某人～ back sb. up

瞠 [书] stare
【瞠乎其后】stare helplessly at the vanishing back of the runner ahead; despair of catching up
【瞠目结舌】stare tongue-tied 〈dumb-founded〉

称 ①(叫;叫做) call ②(名称) name: 俗～ popular name ③[书](说) say; state: 据外交部发言人～ according to the Foreign Ministry spokesman / 连声～好 say "good, good" again and again ④(称赞)[书] commend; praise: 著～ be famous for ⑤(测定重量) weigh
【称霸】seek hegemony; dominate: ～世界 dominate the world
【称便】find sth. a great convenience
【称病】plead illness
【称道】speak approvingly of; commend: 无足～ nothing commendable; nothing praiseworthy / 值得～ be praiseworthy
【称得起】deserve to be called; be worthy of the name of
【称孤道寡】style oneself king; act like an absolute ruler
【称号】title; name; designation
【称呼】①(叫) call; address ②(当面招呼用的名称) form of address
【称快】express one's gratification: ～一时 get satisfaction for a time / 拍手～ clap one's hands with satisfaction
【称量体重】[举重] weighing-in
【称颂】praise; extol; eulogize; pay tribute to
【称王称霸】act like an overlord; lord it over; domineer
【称谓】appellation; title
【称羡】express one's admiration; envy: ～不已 express profuse admiration
【称谢】express one's thanks; thank
【称兄道弟】call each other brothers; be on intimate terms
【称雄】hold sway over a region; rule the roost: 割据～ break away from central authority and

exercise local power; set up separationist rule
【称许】praise; commendation
【称誉】sing the praises of; praise; acclaim
【称赞】praise; acclaim; commend

柽

【柽柳】[植] Chinese tamarisk

蛏 razor clam

【蛏干】dried razor clam
【蛏子】razor clam

chéng

成 ①(完成;成功) accomplish; succeed: 事～之后 after this is achieved ②(成为;变为) become; turn into: 他～了队长。He has become a team leader. ③(成果;成就) achievement; result ④(成年) fully developed; fully grown: ～人 adult ⑤(定形的;现成的) established; ready-made: 既～事实 established fact; *fait accompli* / 现～服装 ready-made clothes ⑥(数量多) in considerable numbers or amounts: ～千上万的人 tens of thousands of people ⑦(答应;许可) all right; O.K. ⑧(有能力) able; capable ⑨(十分之一) one tenth: 增产一～ a 10 % increase in output; output increased by 10 per cent
【成败】success or failure: ～未卜 between cup and lip / ～在此一举 Success or failure hinges on this one action / ～利钝尚难逆料。Whether this will be successful or not is still difficult to predict.
【成本】cost: 固定～ fixed cost / 可变～ variable cost / 间接～ indirect cost / 生产～ production cost / 直接～ direct cost ◇ ～核算 cost accounting / ～价格 cost price / ～会计 cost accounting / ～效益分析 costbenefit analysis / ～帐 cost accounts
【成材】①(可用做木料) grow into useful timber; grow to full size ②(人材) become a useful person
【成材林】standing timber; mature timber
【成虫】[动] imago; adult
【成堆】form a pile; be in heaps
【成方】[中医] set prescription
【成分】①(组成部分) composition; component part; ingredient: 肥料～ the composition of a fertilizer / 化学～ chemical composition / 土壤～ composition of the soil ②(个人成分) one's profession or economic status
【成风】become a common practice; become the order of the day
【成个儿】①(生长到成熟时的大小) be well formed; grow to a good size ②(具有一定的形状) be in the proper form
【成功】succeed; success
【成规】established practice; set rules; groove; rut: 打破～ break down the (established) conventions / 墨守～ stick to conventions; get into a rut
【成果】achievement; fruit; gain; positive result: 科研～ achievements in scientific research / 劳动～ product of one's labor
【成婚】get married
【成活】survive ◇ ～率 survival rate
【成绩】result; achievement; success: ～斐然 the achievement is elegant / ～显著 (make a) remarkable achievement
【成绩册】[体] results
【成绩单】school report card
【成家】(of a man) get married: ～立业 get married and start one's career
【成见】preconceived idea; prejudice: 固执～ prejudiced; biased; opinionated / 消除～ dispel prejudices
【成交】strike a bargain; clinch a deal; conclude a transaction ◇ ～额 volume of business
【成就】①(成绩) achievement; accomplishment; attainment success ②(完成) achieve; accomplish ◇ ～测验 achievement test
【成矿区】metallogenic province
【成矿作用】mineralization
【成立】①(建立),found; establish; set up ②(有根据,站得住) be tenable; hold water: 这个论点不能～。That argument is untenable; The argument does not hold water.
【成例】precedent; existing model
【成殓】encoffin
【成淋巴细胞】lymphoblast
【成龙配套】fill in the gaps to complete a chain (of equipment, construction projects, etc.); link up the parts to form a whole; fitting together of parts
【成眠】[书] fall asleep; go to sleep: 夜不～ lie awake all night
【成名】become famous; make a name for oneself
【成名成家】establish one's reputation as an authority
【成命】order already issued: 收回～ countermand 〈retract〉 an order; revoke a command
【成年】①(成熟;到达成年) grow up; come of

age：未～ be under age ②(成年人) adult；grown-up：～人 an adult；a grown-up ③(整年) year after year：～累月 year in year out；for years on end／～在外 be away all year

【成批】group by group；in batches：～生产 serial production；mass production

【成品】end product；finished product

【成气候】(多用于否定) make good：成不了什么气候 will not get anywhere

【成器】grow up to be a useful person

【成亲】get married

【成全】help (sb. to achieve his aim)

【成群】in groups；in large numbers：～结队 in crowds；in throngs；to band together／三五～ in threes and fours；in small groups

【成人】①(长大) grow up；become full-grown：长大～ be grown to manhood ②(成年人) adult；grown-up ◇ ～教育 adult education

【成人之美】help sb. to fulfill his wish；aid sb. in doing a good deed；help completion of worthy goal

【成仁】[书] die for a righteous cause：～取义 die to preserve one's virtue intact

【成色】①(含纯金银量) the percentage of gold or silver in a coin, etc.；the relative purity of gold or silver ②(质量) quality：看～定价钱 fix the prices according to the quality

【成事】accomplish sth.；succeed：～不足,败事有余 unable to accomplish anything but liable to spoil everything；not good enough to accomplish anything, but more than enough to spoil things

【成熟】ripe；mature：～的经验 ripe experience／～的意见 well-considered opinion／政治上～ politically mature ◇ ～林 mature forest／～期[农] mature period

【成双成对】in pairs, female and male

【成说】accepted theory or formulation

【成诵】able to recite；able to repeat from memory：熟读～ read again and again until one knows by heart；learn by rote

【成套】form a complete set；whole set；complete set：～唱腔 a complete score for voices (in an opera)／～设备 complete sets of equipment／提供～项目和技术援助 supply whole plants as well as technical aid

【成天】[口] all day long；all the time

【成为】become；turn into：～事实 come true

【成文】①(现成的文章) existing writings：抄袭～ copy existing writings；follow a set pattern ②(书面的) written ◇ ～法 written law；statute law；formal law

【成问题】be a problem；be open to question 〈doubt；objection〉

【成像】[物] formation of image；imagery

【成效】effect；result：～甚少 achieve little／～显著 produce a marked effect；achieve remarkable success／初见～ win initial success

【成心】intentionally；on purpose；with deliberate intent you

【成形】take shape

【成型】[机] shaping；forming：爆炸～ explosive forming／冷滚～ cold roll forming

【成性】by nature；become sb.'s second nature

【成药】patent medicine；medicine already prepared by a pharmacy

【成衣】①[旧] tailoring ②(成品服装) ready-made clothes ◇ ～铺 tailor's shop；tailor's

【成因】cause of formation；contributing factor

【成语】[语] set phrase；idiom

【成员】member：～国 member state；member country

【成灾】cause disaster：鼠害～ be infested with rats

【成长】grow up；grow to maturity：在红旗下～ be brought up under the red flag

诚 ①(真诚) sincere；honest：开～相见 treat sb. open-heartedly ②(的确) really；actually；indeed：～非易事 be by no means easy／～然不错 good indeed／～非所料。 It is really unexpected.／～如所言。 It is exactly as you said.／～有此事。 There actually was such a thing.

【诚惶诚恐】with reverence and awe；if fear and trepidation

【诚恳】sincere：～待人 treat others with earnestness

【诚朴】honest；sincere and simple：～的青年 an honest youth

【诚然】true；indeed；to be sure

【诚实】honest：～可靠 honest and dependable；honest and reliable

【诚心】sincere desire；wholeheartedness：～诚意 earnestly and sincerely／一片～ in all sincerity

【诚意】good faith；sincerity：表明～ show one's good faith／毫无～ be not sincere in the least／缺乏～ lack sincerity

【诚挚】sincere；cordial：～的接待 a cordial reception／～的谢意 heartfelt thanks／～友好的气氛 a sincere and friendly atmosphere

城 ①(城墙) city wall; wall: 长～ the Great Wall / ～外 outside the city wall; outside the city ②(城区) city: 东～ the eastern part of the city / 内～ inner city / 外～ outer city ③(城市) town

【城堡】castle

【城池】city wall and moat; city

【城防】the defense of a city: ～巩固。The city is closely guarded. ◇ ～部队 city garrison / ～工事 defense works of a city

【城府】shrewdness; subtlety: ～很深 shrewd and deep; subtle / 胸无～ artless simple and candid

【城根】sections of a city close to the city wall

【城关】the area just outside a city gate

【城隍】[道教] town god ◇ ～庙 town god's temple

【城郊】outskirts of a town: 在北京～ in the suburbs of Beijing

【城里】inside the city; in town: ～人 city dwellers; townspeople

【城楼】a tower over a city gate; gate tower

【城门】city gate

【城门楼】towers over city gates; towers over the wall; battlements of a city wall

【城门失火，殃及池鱼】when the city gate catches fire, the fish in the moat suffer: in a disturbance innocent bystanders get into trouble

【城墙】city wall

【城区】the city proper: ～和郊区 the city proper and the suburbs

【城市】town; city ◇ ～尘埃 urban dust / ～高速铁路 rapid transit / ～公用设施 urban public utilities / ～供水系统 water-supply systems / ～规划 city planning / ～化 urbanization / ～环境 urban environment / ～建设 urban construction / ～居民 city dwellers; urban population / ～垃圾 municipal refuse / ～贫民 urban poor; city poor / ～热岛 urban heat island

【城头】on top of the city wall

【城下之盟】a treaty concluded with the enemy who have reached the city wall; terms accepted under duress; a treaty signed under coercion

【城厢】the city proper and areas just outside its gates

【城乡】town and country; city and countryside; urban and rural: ～并重 laying equal stress on the cities and the countryside / ～差别 difference between town and country

【城镇】cities and towns: ～规划 town planning / ～污水 town sewage

盛 ①(装;舀) fill; ladle: ～饭 fill a bowl with rice / ～汤 ladle out soup wine / 把菜～出来 ladle food from the pot; dish out food ②(容纳) hold; contain: ～一百公斤粮食 hold 100 kilos of grain

【盛酸器】acid receiver

呈 ①(具有;呈现) assume (form, color, etc.) ②(恭敬地送上去) submit; present ③(呈文) petition; memorial

【呈报】submit a report; report a matter: ～备案 report the matter for the record

【呈递】present; submit: ～国书 present credentials; present letter of credence

【呈交诉状】file a petition

【呈请】apply (to the higher authorities for consideration or approval)

【呈文】document submitted to a superior; memorial; petition

【呈现】present (a certain appearance); appear; emerge: ～新的面貌 assume a new aspect

【呈献】respectfully present

程 ①(规矩;法则) rule; regulation: 规～ rules / 章～ rules; constitution ②(程序) order; procedure: 议～ agenda ③(道路;一段路) journey; stage of a journey: 启～ set out on a journey ④(路程) distance: 射～ range (of fire) / 行～ distance of travel

【程度】①(知识、能力的水平) level; degree: 觉悟～ level of political consciousness / 文化～ level of education; degree of literacy ②(事物变化达到的状况) extent; degree: 在不同～上 in varying degrees / 在很大～上 to a great extent / 在一定～上 to a certain extent

【程式】form; pattern; formula: 公文～ forms and formulas of official documents ◇ ～动作 stylized movements / ～化 stylization

【程序】①(进行次序) order; procedure; course; sequence: ～问题 point of order / 法律～ legal procedure / 符合～ be in order / 工作～ working procedure ②[自] program ◇ ～编制 [自] programming / ～法[法] law of procedure; procedural law / ～检查[自] program check / ～教学 programmed instruction or learning / ～控制[自] program control / ～库[自] program library / ～设计[自] programming / ～语言[自] program language

乘 ①(坐;搭乘) ride: ～出租汽车到火车站去 take a taxi to the railway station / ～船旅行 travel by boat / ～飞机旅行 travel by plane / ～公共汽车 ride in a bus; go by bus / ～海轮旅行 travel by ship / ～火车旅行 travel by train ②(利用) take advantage of; avail oneself of: ～敌不备 take the enemy unawares ③[数] multiply: 二～三等于六 two times three is six; 2 multiplied 3 is 6

【乘便】when it is convenient; at one's convenience: 请你～把那本书带给我。 Please bring me the book whenever it's convenient.

【乘法】[数] multiplication ◇ ～表 multiplication table

【乘方】[数]involution; power: *n* 的五次～ the fifth power of *n*: *n* (raised) to the power of 5; n^5

【乘风破浪】ride the wind and cleave the waves; brave the wind and the waves

【乘机】seize the opportunity: ～反攻 seize the opportunity to counterattack

【乘积】[数]product

【乘客】passenger

【乘凉】enjoy the cool; relax in a cool place

【乘龙快婿】a handsome or lucky son-in-law

【乘幂】[数] power

【乘其不备】take advantage of sb.'s unpreparedness; take sb. unawares

【乘人之危】take advantage of sb.'s precarious position

【乘胜】exploit a victory; follow up a victory: ～前进 advance on the crest of a victory; push on in the flush of victory / ～追击 follow up a victory with hot pursuit

【乘数】[数] multiplier

【乘务员】attendant on a train

【乘隙】take advantage of a loophole; turn sb.'s mistake to one's own account: ～而入 take advantage of the crack and enter / 乘敌之隙 exploit the enemy's blunder

【乘兴】while one is in high spirits: ～而来 be present taking advantage of one's good mood / ～而来,败兴而归 set out cheerfully and return disappointed; come with great enthusiasm and return disappointed / ～而来,兴尽而返 arrive in high spirits and depart after enjoying oneself to one's heart's content / ～作了一首诗 improvise a poem while in a joyful mood

【乘虚】take advantage of a weak point in opponent's defense; act when sb. is off guard

【乘虚而入】attack where the enemy is unguarded; get a chance to step in

【乘晕宁】[药]dramamine

惩 punish; penalize

【惩办】punish; chastise; punishment; chastisement: ～和宽大相结合 combine punishment with leniency / ～少数,改造多数的原则 principle of punishing the few and reforming the many / 依法～反革命分子 punish counterrevolutionaries according to law

【惩处】penalize; punish; punishment; administer justice: ～罪犯 inflict a punishment on a criminal / 依法～ punish in accordance with the law

【惩罚】punish; penalize: ～措施 punitive measures / ～性的损害赔偿费 punitive damages / ～性制裁 punitive sanction

【惩戒】punish sb. to teach him a lesson; discipline sb. as a warning; take disciplinary action against: 吊销执照,以示～ revoke sb.'s licence as a punishment

【惩前毖后,治病救人】learn from past mistakes to avoid future ones, and cure the sickness to save the patient

【惩一儆百】punish one to warn a hundred; make an example of sb.

【惩治】punish; mete out punishment to ◇ ～贪污条例 Regulation for Suppression of Corruption

澄 clear; transparent: ～空 a clear, cloudless sky

【澄清】①(清亮) clear; transparent: 湖水碧绿～。 The water of the lake is green and clear. ②(肃清,弄清楚) clear up; clarify: ～事实 clarify some facts / ～误会 clear up a misunderstanding / 要求～ demand clarification ③[化] defecation ◇ ～度 clarity / ～剂 clarifying agent; clarifier

橙 ①(橙子) orange ②(橙色) orange color: ～黄 orange (color)

承 ①(托着;接着) bear; hold; carry ②(承担) undertake; contract (to do a job): ～印 undertake the printing of / ～制棉衣 accept orders for padded clothes ③(客套话:承蒙) be indebted (to sb. for a kindness); be granted a favor ④(继续;接续) continue; carry on: 继～ inherit; carry on

【承办】undertake: ～土木工程 undertake civil engineering projects
【承包】contract (with): ～建桥工程 contract to build a bridge ◇ ～单位 contractor unit / ～工程 undertaking contracted projects / ～合同 contract / ～人 contractor / ～商 contrac- tor / ～制 contracting out system
【承保】accept insurance; under writing ◇ ～单 open cover / ～范围 insurance coverage / ～人 insurer / ～通知书 cover note
【承担】bear; undertake; assume: ～法律责任 bear legal liability / ～风险 acceptance of risk / ～经济责任 bear financial responsibility / ～票据上的义务 liable a bill of exchange / ～一切费用 bear all the costs / ～义务 commit oneself; be committed to; accept the responsibility for / ～由此而产生的一切严重后果 bear responsibility 〈be held responsible〉 for all the serious consequences arising therefrom / ～债务 incurred obligation / ～责任 undertake the responsibility; bear the responsibility / ～重任 take on heavy responsibilities; take a heavy task upon oneself
【承当】take; bear: ～责任 bear the responsibility
【承兑】[商] honor; accept: ～汇票 accept a draft; acceptance bill; acceptance of exchange / 此处～旅行支票。Traveller's checks (are) cashed here. ◇ ～人 accepter
【承继】①(给伯父或叔父做儿子) be adopted as heir to one's uncle ②(把侄子收做儿子) adopt one's brother's son (as one's heir)
【承接】①(接住流下的液体) hold out a vessel to have liquid poured into it ②(接续) continue; carry on: ～上文 continued from the preceding paragraph
【承揽】contract to do a job; hire of work: ～合同 contract of work; contractor's agreement
【承蒙】[套] be indebted (to sb. for a kindness); be granted a favor ◇ ～不弃 meet with your gracious consent / ～错爱 have received undeserved kindness from you
【承诺】promise to undertake; undertake to do sth.; acceptance of offer ◇ ～人 accepter; acceptor
【承情】[套] be much obliged; owe a debt of gratitude
【承认】①(肯定;认可) admit; acknowledge; recognize: ～错误 admit one's mistake; acknowledge one's fault / ～负有责任 admission of liability / ～事实 admission of fact / ～债务 admission of a debt / ～遗嘱 acknowledgement of will / ～有罪 admission of guilt ②(肯定法律地位) give diplomatic recognition; recognize: ～新国家 recognition of a new state / ～新政府 recognition of a new government / 法律～ de jure recognition / 互相～ mutual recognition / 事实上～ de facto recognition
【承上启下】form a connecting link between the preceding and the following (as in a piece of writing, etc.)
【承受】①(承担;经受) bear; support; endure ②(继承) inherit (a legacy, etc.)
【承袭】①(沿袭) adopt; follow (a tradition, etc.) ②(继承) inherit (a peerage, etc.) ◇ ～海 patrimonial sea
【承先启后】inherit the past and usher in the future; serve as a link between past and future
【承运人】carrier
【承载】bear the weight of ◇ ～能力 bearing capacity; load-bearing capacity
【承重】bearing; load-bearing ◇ ～梁 spandrel girder / ～墙 [建] bearing wall; load- bearing wall
【承转】forward (a document to the next level above or below)
【承租人】lessee; tenant

丞 assist assistant officer: 县～ county magistrate's assistant
【丞相】prime minister

chěng

逞 ①(显示;夸耀) show off; flaunt: ～威风 show off one's strength or power; swagger about / ～英雄 pose as a hero ②(得逞;坏主意达到目的) carry out (an evil design); succeed (in a scheme): 得～ succeed in one's schemes ③(纵容;放任) indulge; give free rein to: ～性子 be wayward
【逞能】show off one's skill or ability; parade one's ability: 好～ like to show off
【逞强】flaunt one's superiority: ～好胜 parade one's superiority and strive to outshine others
【逞凶】act violently; act with murderous intent; bluster

骋 [书] ①(跑) gallop: 驰～ gallop about; dash about ②(放开) give free rein to
【骋目】[书] look into the distance: ～远眺 scan distant horizons

chèng

秤 balance; steelyard: 杆～ steelyard; lever scales / 台～ platform balance; platform scale / 弹簧～ spring balance
【秤锤】the sliding weight of a steelyard
【秤杆】the arm of a steelyard; the beam of a steelyard
【秤钩】steelyard hook
【秤毫】the lifting cord of a steelyard
【秤盘】the pan of a steelyard
【秤砣】the sliding weight of a steelyard
【秤星】gradations marked on the beam of a steelyard

chī

痴 ①(愚笨) silly; idiotic: 白～ idiot ②(迷恋某人或某事) crazy about: 书～ bookworm
【痴呆】①(愚笨) dull-witted; stupid ②[医] dementia: 老年性～ senile dementia
【痴迷】infatuated; obsessed; crazy
【痴情】unreasoning passion; infatuation
【痴人说梦】idiotic nonsense; lunatic ravings; talk fantastic nonsense
【痴想】wishful thinking; illusion
【痴心】infatuation: ～女子负心汉 an infatuated girl deserted by a heartless man / 一片～ sheer infatuation
【痴心妄想】wishful thinking; fond dream; hope madly
【痴愚】imbecility; moronity

吃 ①(吃东西) eat; take: ～苹果 eat an apple / ～糖 have some sweets / ～药 take medicine ②(在某个出售食物的地方吃) have one's meals; eat: ～馆子 eat in a restaurant; dine out / ～食堂 have one's meals in the mess ③(以…为生) live on; live off ④(消灭) annihilate; wipe out: ～一个子儿 take a piece (in chess) / 又～掉敌军一个师 annihilate another enemy division ⑤(耗费) exhaust; be a strain: 感到～力 feel the strain (of work, etc.); find a job difficult ⑥(吸收) absorb; soak up: ～墨 absorb ink ⑦(受) suffer; incur: ～批评 be criticized
【吃不服】not be accustomed to eating sth.; not be used to certain food
【吃不开】be unpopular; won't work
【吃不来】not be fond of certain food
【吃不了兜着走】get more than one bargained for; land oneself in serious trouble
【吃不上】①(没有饭吃) be unable to get something to eat ②(赶不上吃饭) miss a meal
【吃不下】not feel like eating; be unable to eat any more
【吃不消】be unable to stand (exertion, fatigue, etc.): 痛得我～。The pain is more than I can stand.
【吃不住】be unable to bear or support
【吃穿】food and clothing: ～不愁 not have to worry about food and clothing
【吃醋】be jealous (usu. of a rival in love)
【吃大户】mass seizure and eating of food in the homes of landlords
【吃刀】[机] penetration of a cutting tool
【吃得开】be popular; be much sought after
【吃得来】be able to eat; not mind eating
【吃得上】①(吃得起) can afford to eat ②(赶上吃饭) be in time for a meal; be able to get meal
【吃得下】be able to eat
【吃得消】be able to stand (exertion, fatigue, etc.)
【吃得住】be able to bear or support
【吃饭】①(进餐) eat; have a meal: 请客～ invite sb. to dinner ②(指生活或生存) keep alive; make a living: 靠工资～ live on one's wages
【吃喝玩乐】eat, drink and be merry
【吃紧】be critical; be hard pressed: 形势～。The situation was critical.
【吃惊】be startled; be shocked; be amazed; be taken aback: 大吃一惊 be flabbergasted
【吃苦】bear hardships: ～耐劳 bear hardships and stand hard work / ～在前，享乐在后 be the first to bear hardships and the last to enjoy comforts
【吃亏】①(受损失) suffer losses; come to grief; get the worst of it ②(在某方面不利) at a disadvantage; in an unfavorable situation
【吃老本】live off one's past gains; rest on one's laurels
【吃里爬外】live off one person while secretly helping another
【吃力】entail strenuous effort; be a strain: ～不讨好的差使 a thankless task
【吃零嘴】take snacks between meals; nibble between meals
【吃奶】suck the breast: ～的孩子 sucking child; sucking / 使尽～的力气 strain every muscle
【吃请】accept an invitation to dinner (extended as a bribe)
【吃软不吃硬】be open to persuasion, but not to coercion

【吃水】①(供食用的) drinking water ②(吸取水分) absorb water ③[航海] draught; draft: 空载～ light draught / 满载～ load draught / 这船～五米。 The ship has a draught of 5 meters. ◇～标志 draft marks / ～线 waterline; line of flotation
【吃素】abstain from eating meat; be a vegetarian
【吃现成饭】lead an idle life; be a loafer or sponger
【吃香】[口] be very popular; be much sought after; be well-liked
【吃一堑,长一智】a fall into the pit, a gain in your wit
【吃斋】practice abstinence from meat (as a religious exercise); be a vegetarian for religious reasons
【吃重】①(责任艰巨;费力) arduous; strenuous ②(载重) carrying capacity; loading capacity

笞 [书] beat with a stick, cane, etc.: 鞭～ flog; whip

鸱
【鸱鸺】[动] owl

魑
【魑魅魍魉】evil spirits; demons and monsters

嗤 sneer
【嗤笑】laugh at; sneer at
【嗤之以鼻】give a snort of contempt; despise

chí

持 ①(拿着;握着) hold; grasp: ～枪 hold a gun / ～保留态度 have reservations / ～相反意见 hold a contrary opinion ②(支持;保持) support; maintain: 维～ keep; maintain / 支～ support; sustain ③(主管;料理) manage; run: 操～ manage; handle / 主～ take charge of; manage ④(对抗) oppose: 相～不下 be locked in stalemate
【持不同政见者】dissident
【持刀抢劫犯】knife robber
【持股公司】holding company
【持骨钳】bone holding forceps
【持家】run one's home; keep house: ～有方 manage the affairs of the family methodically / 勤俭～ be industrious and thrifty in running one's home
【持久】lasting; enduring; protracted: ～和平 lasting peace; enduring peace / ～力 stamina; sustaining power / ～战 protracted war
【持论】present an argument; put a case; express a view: ～公平 state a case fairly / ～有据 put forward a well-grounded argument
【持平】unbiased; fair: ～之论 a fair argument; an unbiased view
【持枪】①hold a gun ②[军] port arms: ～! (口令) Port arms! ◇～歹徒 gunman / ～抢劫 gun robbery / ～抢劫犯 gun robber
【持球】[排球] holding
【持续】continued; sustained: ～时间(期间) duration / ～罪 continuous offense
【持有】hold: ～不同意见 hold differing views / ～护照 hold a passport
【持针钳】needle forceps; needle-holding forceps
【持之以恒】persevere: 刻苦学习,～ study assiduously and perseveringly
【持之有故】have sufficient grounds for one's views
【持重】prudent; cautious; discreet: 老成～ experienced and prudent; dignified and experienced

匙 spoon: 茶～ teaspoon / 汤～ soup spoon
【匙子】spoon

踟
【踟蹰】hesitate; waver: ～不前 hesitate to move forward

迟 ①(慢) slow; tardy ②(晚) late
【迟迟】slow; tardy: ～不表态 not state one's position even after stalling for a long time
【迟到】be late; come late; arrive late: ～五分钟 be five minutes late
【迟钝】slow (in thought or action); obtuse: 反应～ be slow in reacting; react slowly / 脑子～ slow of understanding
【迟缓】slow; tardy; sluggish: 进展～ make slow progress / 行动～ act slowly
【迟脉】[中医]retarded pulse (less than 60 beats per minute)
【迟误】delay; procrastinate: 不得～ admit of no delay
【迟延】delay; retard: ～交付 delay deliverance
【迟疑】hesitate: ～不答 hesitated in giving a reply / ～不决 hesitate to make a decision; be irresolute; be undecided / 毫不～ without a moment's hesitation
【迟早】sooner or later

【迟滞】①(缓慢) slow-moving; sluggish ②[军](行动缓慢) delaying (action)
【迟做比不做好】Better late than never.

池 ①(池塘) pool; pond: 养鱼～ fishpond / 游泳～ swimming pool ②(旁边高中间洼的地方) an enclosed space with raised sides: 花～ flower bed / 舞～ dance floor / 乐～ orchestra pit ③(剧场正厅前部) stalls ④(城池) moat: 城～ city wall and moat; city
【池塘】①(蓄水的坑) pond; pool ②(浴池) a big pool in a bathhouse
【池盐】lake salt
【池沼】pond; pool
【池子】[口]①(蓄水的坑) pond ②(浴池) a big pool in a bathhouse
【池座】the stalls

弛 [书] relax; slacken: 一张一～ tension alternating with relaxation
【弛缓】relax; calm down
【弛张热】[医] remittent fever

驰 ①(跑得很快) speed; gallop ②(传播) spread: ～名 well-known ③[书](向往) turn eagerly towards: 心～神往 let one's thoughts fly to (a place or person); long for
【驰骋】[书] gallop: ～疆场 gallop across the battlefield / ～文坛 play an outstanding role in the literary world; bestride the literary stage
【驰名】known far and wide; well-known; famous; renowned: ～中外 renowned at home and abroad / 世界～ world-famous
【驰驱】①(骑马快跑) gallop ②(奔走效力) do one's utmost in sb.'s service
【驰援】rush to the rescue

chǐ

褫 [书] strip; deprive: ～职 deprive sb. of his post; remove sb. from office
【褫夺】strip; deprive: ～公民权 disfranchise; disfranchisement / ～公权 deprive sb. of civil rights / ～公权终身 deprivation of civil right for life

耻 shame; disgrace; humiliation: 引以为～ regard as a disgrace / 知～ have a sense of shame
【耻骨】[生理] pubic bones; pubis
【耻骨联合】pubic symphysis; symphysis pubis
【耻辱】shame; disgrace; humiliation
【耻笑】hold sb. to ridicule; sneer at; mock
【耻与为伍】feel ashamed to associate with him

齿 ①(牙齿) tooth ②(齿状部分) a tooth-like part of anything: 锯～儿 the teeth of a saw / 梳～儿 the teeth of a comb ③(说到;提到) mention: 不足挂～ not worth mentioning
【齿根】root of tooth
【齿轨铁道】rack railway
【齿冷】[书] laugh sb. to scorn: 令人～ arouse one's scorn
【齿轮】gear wheel; gear: 伞～ bevel gear / 斜～ helical gear / 正～ spur gear ◇ ～插床 gear slotter / ～传动 gear drive / ～间隙 gear clearance / ～磨床 gear grinder / ～箱 gear box / ～组 gear cluster
【齿条】[机] rack
【齿龈】[生理] gums
【齿音】dental

侈 [书] ①(浪费) wasteful; extravagant ②(夸大) exaggerate
【侈谈】talk glibly about; prate about; prattle about

尺 ①(量长度的器具) rule; ruler: 丁字～ T-square / 折～ folding rule ②(像尺的东西) an instrument in the shape of a ruler: 计算～ slide rule / 镇～ bronze paperweight ③(长度单位) *chi*, a unit of length (3 *chi* = 1 meter)
【尺寸】measurement; dimensions; size: ～地方 a foot of land / 加工～ finish size / 量～ take sb.'s measurements / 轮廓～ overall size / 名义～ nominal size / 衣服～ measurements of a garment
【尺动脉】[生理] ulnar artery
【尺牍】①(书信写法示范) a model of epistolary art ②(旧指书信) correspondence (of an eminent writer)
【尺度】yardstick; measure; scale
【尺短寸长】A foot may be too short in one case while an inch may be long enough in another. (或: Every person has his weak points as well as strong points.)
【尺幅千里】a panorama of a thousand *li* on a one *chi* scroll; rich content within a small compass
【尺骨】[生理] ulna
【尺蠖】[动] looper; inchworm; geometer: 桑～ mulberry looper ◇ ～蛾[动] geometrid moth
【尺码】size; measures
【尺子】rule; ruler

chì

炽 flaming; ablaze

【炽烈】burning fiercely; flaming; blazing

【炽热】①(热) red-hot; blazing: ～的阳光 blazing sun ②(极热烈) passionate: ～的情感 passionate feelings

【炽盛】flaming; ablaze; flourishing

赤 ①(红色) red: 面红耳～ get red in the face; be flushed (with excitement, shame or shyness) ②(忠诚) loyal; sincere; single-hearted: ～心 loyalty; sincerity ③(光着;露着) bare: ～背 barebacked / ～身裸体 naked

【赤膊】barebacked: 打～ be stripped to the waist

【赤膊上阵】go into battle stripped to the waist; throw away all disguise; come out into the open

【赤诚】absolute sincerity: ～待人 treat people with absolute sincerity

【赤胆忠心】utter devotion; wholeheartedness; loyalty

【赤道】①[地] the equator ②[天] the celestial equator ◇ ～东风带 deep easterlies; equatorial easterlies / ～面 the equatorial plane / ～无风带 the equatorial calms; doldrums / ～西风带 equatorial westerlies

【赤道几内亚】Equatorial Guinea ◇ ～人 Equatorial Guinean

【赤地千里】thousands of miles of cracked, parched and deserted land; barren lands extending over thousands of miles

【赤豆】red bean

【赤褐色】russet

【赤红】crimson

【赤脚】barefoot

【赤金】pure gold; solid gold

【赤痢】[中医] dysentery characterized by blood in the stool

【赤磷】red phosphorus

【赤露】bare: ～着胸口 with bared chest

【赤裸裸】①(光着身子) without a stitch of clothing; stark-naked ②(毫无掩饰) undisguised; naked; out-and-out

【赤霉素】[生化] gibberellin

【赤贫】in abject poverty; utterly destitute; in extreme poverty

【赤日当空】the red sun is hanging in the sky

【赤芍】[中药] the (unpeeled) root of common peony

【赤手空拳】bare-handed; unarmed

【赤松】[植] Japanese red pine

【赤陶】terra-cotta

【赤条条】have not a stitch on; be stark-naked

【赤铁矿】red iron ore; hematite

【赤铜矿】red copper ore; cuprite

【赤卫队】Red Guards

【赤血盐】[化] potassium ferricyanide; red prussiate of potash

【赤眼蜂】trichogramma

【赤子】a newborn baby: ～之心 the pure heart of a newborn babe; the innocent heart of a child; utter innocence

【赤字】deficit: ～开支 deficit spending / 财政～ financial deficit / 贸易～ trade deficit / 弥补～ make up a deficit; meet a deficit / 预算～ budgetary deficits ◇ ～财政 deficit financing

翅 ①(翅膀) wing ②(鱼翅) shark's fin

【翅膀】wing

【翅果】[植] samara

【翅脉】[动] vein (of the wings of an insect)

叱 [书] loudly rebuke; shout at: 怒～ shout angrily at sb.

【叱喝】shout at; bawl at

【叱骂】scold roundly; curse; abuse

【叱责】scold; upbraid; rebuke

【叱咤风云】commanding the wind and the clouds; shaking heaven and earth; all-powerful

斥 ①(责备) upbraid; scold; denounce; reprimand: 痛～ ②(使离开) repel; exclude; oust: 排～异己 exclude outsiders

【斥力】[物] repulsion

【斥骂】reproach; upbraid; scold: ～某人 give sb. a scolding

【斥退】①(罢免)[旧] dismiss sb. from his post ②(开除) expel from a school ③(喝令退出) shout at sb. to go away

【斥责】reprimand; rebuke; denounce: 厉声～ severely reprimand; excoriate

饬 [书]①(整饬) put in order; readjust: 整～ put in order; strengthen (discipline, etc.) ②(饬令) orderly; well-behaved: 谨～ sober and well-behaved ③(旧时公文用语) order: 严～ issue strict orders

chōng

憧

【憧憧】flickering; moving: 灯影～ the flickering light of a lantern / 人影～ shadows of people moving about / 树影～ flickering shadows of trees

【憧憬】long for; look forward to: ～着幸福的明天 long for the happy days to come / 充满着对未来的～ have a great longing for a bright future

充

①(满;足) sufficient; full: 供应～分 have ample supplies ②(装满;塞住) fill; charge: ～电 charge (a battery) ③(担任;当) serve as; act as ④(冒充) pretend to be; pose as; pass sth. off as: ～好汉 pose as a hero / ～内行 pretend to be an expert

【充斥】flood; congest; be full of

【充当】serve as; act as; play the part of: ～辩护士 play the part of an apologist / ～翻译 act as interpreter

【充电】charge (a battery) ◇ ～器 charger

【充耳不闻】stuff one's ears and refuse to listen; turn a deaf ear to

【充分】full; ample; abundant: ～动员 mobilize to the fullest extent / ～发挥 bring... into full play / ～发挥生产潜力 develop potential productive forces to the full / ～就业 full employment / ～理由 adequate cause; sufficient reason / ～利用 fully utilize; make full use of; turn to full account / ～协商 full consultation / ～证据 ample evidence; satisfactory evidence; sufficient evidence / ～证明 full proof

【充公】confiscate

【充饥】allay one's hunger; appease one's hunger

【充军】[旧] be transported to a distant place for penal servitude; banish

【充满】full of; brimming with; permeated with; imbued with

【充沛】plentiful; abundant; full of: 精力～ full of vim and vigor; vigorous; energetic / 雨水～ abundant rainfall

【充其量】at most; at best; at the maximum

【充气灯泡】gas-filled lamp bulb; nitra-lamp

【充气浮筒】camel

【充气机】aerator

【充气轮胎】pneumatic tire

【充任】fill the post of; hold the position of: 挑选某人～经理 select sb. to fill the post of manager

【充塞】fill (up); cram: ～市场 overstock the market

【充实】①(丰富) substantial; rich: 内容～ substantial in content ②(使充足) substantiate; enrich; replenish: ～库存 replenish the stocks / ～领导班子 strengthen the leading bodies / ～论据 substantiate one's argument

【充数】make up the number; serve as a stopgap

【充血】[医] hyperemia; congestion

【充溢】full to the brim; exuberant; overflowing

【充裕】abundant; ample; plentiful: 经济～ well-off / 时间～ have ample time; have plenty of time

【充足】adequate; sufficient; abundant; ample: 经费～ have sufficient funds; have ample funds / 阳光～ full of sunshine; sunny

舂

pound; pestle: ～米 husk rice with mortar and pestle / ～药 pound medicinal herbs in a mortar

冲

①(通行大道;重要地方) thoroughfare; important place: 要～ hub ②(突破障碍) charge; rush; dash ③(猛烈地撞击) clash; collide: ～突 conflict ④(用开水等浇) pour boiling water on: ～茶 make tea ⑤(冲洗) rinse; flush: 把盘子～一～ rinse the plates / 便后～水。Flush the toilet after use. ⑥[摄] develop: ～胶卷 develop a roll of film ⑦[天] opposition: 大～ favorable opposition

【冲程】[机] stroke: 四～发动机 four-stroke engine

【冲冲】in a state of excitement: 怒气～ in a great rage / 兴～ bursting with enthusiasm; in spirits

【冲刺】[体] spurt; sprint: 向终点线～ make a spurt 〈dash〉 towards the tape / 最后～ a final sprint; a sprint at the finish ◇ ～速度 dash speed

【冲淡】①(稀释) dilute: 把溶液～ dilute the solution ②(使减弱) water down; weaken; play down: ～戏剧效果 weaken the dramatic effect

【冲动】①(神经兴奋) impulse: 出于一时～ act on impulse ②(情感强烈) get excited; be impetuous: 他很容易～。He easily gets excited.

【冲断层】[地] thrust fault

【冲犯】offend; affront

【冲锋】charge; assault: 打退敌人的～ beat back the enemy assault ◇ ～号 bugle call to

charge / ～枪 submachine gun; tommy gun
【冲锋陷阵】 charge and shatter enemy positions; charge the enemy lines; charge forward
【冲服】 take (medicine) after mixing it with water, wine, etc.
【冲毁】 destroy by rush of water: ～堤岸 embarkment was destroyed by rush of waters
【冲昏头脑】 turn sb.'s head; have one's head turned: 胜利～ dizzy with success
【冲击】①(水流撞击) lash; pound ②[军] charge; assault: 向敌人阵地发起～ charge an enemy position ◇ ～波[物] shock wave; blast wave
【冲积】[地] alluviation ◇ ～层 alluvium / ～平原 alluvial plain / ～扇 alluvial fan / ～土 alluvial soil / ～物 alluvial deposits
【冲剂】[中药] medicine to be taken after being mixed with boiling water, wine, etc.
【冲减财政收入】 eat up part of the state's revenue
【冲决】 burst; smash: ～堤防 burst the dikes
【冲口而出】 say sth. unthinkingly; blurt sth. out
【冲垮】 burst; shatter: ～敌军防线 shatter the enemy lines
【冲浪运动】 surfing
【冲力】 impulsive force; momentum
【冲量】[物] impulse
【冲破】 break through; breach: ～重重障碍 break through one barrier after another; surmount all obstacles / ～传统观念的束缚 smash the bonds of tradition
【冲散】 break up; scatter; disperse: ～人群 disperse a crowd
【冲杀】 charge; rush ahead
【冲沙闸】[水] scouring sluice
【冲晒】[摄] develop and print
【冲刷】 erode; scour; wash out; wash away: ～水沟 scour the ditch
【冲塌】 (of floodwater, etc.) cause to collapse; burst: ～堤坝 burst dikes and dams / ～房屋 dash against the houses and wash them away
【冲天】 towering; soaring: 怒气～ in a towering rage ◇ ～炉[冶] furnace cupola; cupola
【冲突】 conflict; clash: ～法 conflict of laws / ～法规 conflict (of) rules / 边境～ a border clash / 防止～ avoid a conflict (with) / 利害～ conflict of interests / 武装～ an armed conflict
【冲洗】①(用水冲) rinse; wash: 用消毒水～伤口 wash a wound with a disinfectant ②[摄] develop
【冲线】[体] breast the tape
【冲要】 strategically important (place)
【冲帐】[会计]①(结帐) strike a balance ②(收支抵销) reverse an entry
【冲撞】①(撞击) collide; bump; ram: 两辆汽车～了。The two cars collided. ②(冲犯) give offense; offend: 我的话～了她。 My words offended her.

忡

【忡忡】 laden with anxiety; careworn: 忧心～ heavyhearted; deeply worried

chóng

虫 insect; worm: 害～ destructive insect; harmful insect / 益～ useful insect; beneficial insect
【虫草】[中药] Chinese caterpillar fungus
【虫害】 insect pest
【虫胶】 shellac ◇ ～清漆 shellac (varnish)
【虫媒花】[植] entomophilous flower
【虫漆】 lacca
【虫情】 insect pest situation ◇ ～测报站 pest forecasting station
【虫牙】 carious tooth; decayed tooth
【虫瘿】[植] gall
【虫灾】 plague of insects
【虫子】 insect; worm: 拉～ pass worms

崇 ①(高) high; lofty; sublime: ～山峻岭 high mountain ridges ②(重视;尊敬) esteem; worship: ～洋迷外 worship and have blind faith in things foreign
【崇拜】 worship; adore: 盲目～ worship blindly / ～偶像 worship of idols; idolatry
【崇奉】 believe in (a religion); worship
【崇高】 lofty; sublime; high: ～的理想 a lofty ideal / ～的威望 high prestige / 顺致最～的敬意。 I avail myself of this opportunity to renew to you the assurances of my highest consideration.
【崇敬】 esteem; respect; revere: 怀着十分～的心情 cherish a feeling of great reverence for
【崇尚】 uphold; advocate: ～勤俭 advocate industry and thrift / ～正义 uphold justice

重 ①(重复) repeat; duplicate ②(再) again; once more: 旧地～游 revisit a once familiar place ③(层) layer: 双～领导 dual leadership
【重版】 republication
【重操旧业】 return to one's old trade
【重瓣胃】 manyplies; omasum; psalterium

【重唱】[乐] an ensemble of two or more singers, each singing one part: 二～ duet
【重重】layer upon layer; ring upon ring: 顾虑～ full of misgivings / 克服～困难 overcome one difficulty after another; surmount numerous difficulties / 受到～剥削 be fleeced right and left / 陷入～包围 be encircled ring upon ring
【重蹈覆辙】follow the same old disastrous road; recommit the same error
【重叠】one on top of another; overlapping: 山峦～ range upon range of mountains ◇ ～影像 superimposed image
【重发球】[体] let service; let
【重返】return: ～家园 return to one's homeland ◇ ～大气层运载工具 reentry vehicle
【重犯】repeat (an error or offense): 吸取教训, 避免～错误 draw a lesson from past errors so as to prevent their recurrence
【重逢】meet again; have a reunion: 久别～ meet again after a long separation / 旧友～ reunion of old friends
【重复】repeat; duplicate
【重合】[数] coincide
【重婚】[法] bigamy; a bigamous marriage: ～罪 bigamy; offense of bigamy
【重见光明】see light again
【重见天日】once more see the light of day; release after imprisonment or great injustice
【重建】rebuild; reconstruct; reestablish; rehabilitate: ～家园 rehabilitate one's homeland; rebuild one's home village or town / 战后的～工作 postwar reconstruction
【重起炉灶】begin all over again; make a fresh start
【重申】reaffirm; reiterate; restate: ～观点 reassert one's views / ～立场 reaffirm one's stand
【重施故技】play the same old trick; repeat a stock trick
【重孙】great-grandson
【重孙女】great-granddaughter
【重弹老调】harp on the same string; sing the same old tune
【重提】bring up again: 旧事～ bring up an old case; recall past events
【重围】tight encirclement: 杀出～ break through a tight encirclement
【重温】review
【重温旧梦】revive an old dream; relive an old experience
【重温旧情】review one's friendship
【重现】reappear
【重新】[副] again; anew; afresh: ～部署 rearrange; redeploy / ～发起进攻 launch a fresh offensive / ～分配 redistribute / ～开放 reopen / ～开始 make a fresh start / ～考虑 reconsider / ～确认(证实) reconfirm / ～委任(指定) reappoint; reappointment / ～上台 return to power / ～审理 rehearing / ～使用 reactivate / ～统一 reunification / ～做人 begin one's life anew; turn over a new leaf
【重修旧好】renew cordial relations; become reconciled; bury the hatchet; pass the sponge over
【重檐】[建] double-eaved roof
【重演】①(重新上演) put on an old play, etc. ②(再次出现) recur; reenact; repeat
【重洋】the seas and oceans: 远隔～ be separated by seas and oceans / 远涉～ travel across the oceans
【重译】retranslate
【重印】reprint ◇ ～本 reprint
【重振军威】restore the prestige of an army; make an army's might felt once again
【重整旗鼓】rally one's forces (after a defeat); regroup for battle
【重奏】[乐] an ensemble of two or more instrumentalists, each playing one part: 四～ quartet

chǒng

宠 dote on; bestow favor on: ～坏孩子 spoil the child / 得～ find favor with sb.; be in sb.'s good graces / 失～ fall out of favor
【宠爱】make a pet of sb.; dote on: 受某人～ win sb.'s favor
【宠儿】pet; favorite
【宠辱不惊】remain indifferent whether granted favors or subjected to humiliation
【宠信】be specially fond of and trust unduly (a subordinate)

chòng

冲 [口] ①(劲头儿足;力量大) with vim and vigor; with plenty of dash; vigorously ②(气味浓烈刺鼻) (of smell) strong ③(朝;向;对) facing; towards ④(凭;根据) on the strength of; on the basis of; because ⑤[机] punching
【冲床】[机] punch (press); punching machine ◇ ～工 puncher
【冲模】[机] die ◇ ～插床 die slotting machine
【冲头】[机] drift; punch pin

【冲压】[机] stamping; punching ◇ ～机 punch
【冲子】punching pin

chōu

抽 ①(把夹在中间的东西取出) take out (from in between) ②(从全部中取出一部分) take (a part from a whole) ③(生长出) (of certain plants) put forth: ～枝 branch out; sprout ④(吸) obtain by drawing, etc.: ～水 pump water / ～血 draw blood (for a test or transfusion) ⑤(收缩) shrink: 这种布一洗就～。 This cloth shrinks in the wash. ⑥(用条状物打) lash; whip; thrash: ～陀螺 whip a top / ～牲口 lash a draught animal
【抽查】selective examination; spot check; spot test
【抽搐】①(痉挛) twitch ②[医] tic
【抽打】lash; whip; thrash
【抽搭】[口] sob: 抽抽搭搭地哭了起来 break into sobs
【抽调】transfer (personnel or material)
【抽动】twitch; spasm; spasmodic jerk
【抽肥补瘦】take from the fat to pad the lean; take from those who have too much and give to those who have too little
【抽风】[医] convulsions
【抽筋】①(抽掉筋) pull out a tendon ②(筋肉痉挛) cramp: 腿～ have a cramp in the leg
【抽空】manage to find time: 抽不出空来 be unable to find time
【抽搦】[医] tic; twitch
【抽气机】air exhauster; air extractor; air pump
【抽泣】sob
【抽签】draw lots; cast lots
【抽球】[体] drive
【抽纱】[工美] drawnwork
【抽身】leave (one's work); get away; absent oneself
【抽水】draw water; pump water: 从河里～ pump water from a river ◇ ～机 water pump; pumper; water raising engine / ～马桶 flush toilet; water closet / ～站 pumping station
【抽税】levy a tax
【抽丝】reel off raw silk from cocoons ◇ ～剥茧 make a painstaking investigation
【抽穗】heading; earing ◇ ～期 heading stage; heading period
【抽缩】shrink; contract
【抽簸】[农] bolting
【抽提】[化] extraction: ～蒸馏 extractive distillation
【抽屉】drawer
【抽头】①(提成) take a percentage of the winnings in gambling ②[电] tap: ～电路 tap circuit ◇ ～聚赌 take a fee from gambling
【抽象】abstract: ～的概念 an abstract concept ◇ ～代数 abstract algebra / ～劳动 abstract labor / ～派 abstractionist school; abstractionism / ～数 abstract number / ～思维 abstract thought / ～艺术 abstract art
【抽薪止沸】take out the firewood to stop the pot boiling; take drastic measures to stop sth.
【抽芯铆钉】self-plugging rivet
【抽芽】put forth buds; bud; sprout
【抽烟】smoke (a cigarette or a pipe)
【抽样】[统计] sample; sampling: 有意～ purposive sampling / 随机～ random sampling ◇ ～单位 sampling unit / ～调查 sampling survey / ～分布 sampling distribution / ～间隔 sampling interval / ～检查 sample survey; curtailed inspection / ～误差 sampling error
【抽噎】sob: 抽抽噎噎 sob intermittently
【抽印】offprint ◇ ～本 offprint
【抽壮丁】press-gang

chóu

畴 [书] ①(田地) farmland: 平～千里 a vast expanse of cultivated land ②(种类) kind; division: 范～ category ②[物] domain
【畴理论】domain theory
【畴辈】people of the same generation 〈position〉

踌
【踌躇】hesitate; shilly-shally: ～不前 hesitate to move forward; hesitate to make a move
【踌躇满志】enormously proud of one's success; smug; complacent

筹 ①(小棍儿;小片儿) chip; counter: 竹～ bamboo ②(筹划;筹措) pare; plan: ～款 raise money; raise funds / 统～ over-all planning
【筹办】make preparations; make arrangements
【筹备】prepare; arrange ◇ ～工作 preparatory work; preparations / ～会议 preparatory meeting; preliminary meeting / ～委员会 preparatory committee
【筹措】raise (money): ～福利基金 raise a welfare fund / ～旅费 raise money for travelling expenses / ～资金 raise funds / ～维艰。 It is not easy to devise means.
【筹划】plan and prepare; project; design: ～不

周 ill-designed; ～周密 well-designed
【筹集】raise (money): ～基金 raise funds
【筹建】prepare to construct or establish sth.: ～研究所 make preparations for the setting up of a research institute
【筹款】raise funds; raise money
【筹码】chip; counter: 政治交易的～ bargaining counters in political deals
【筹募】collect (funds)
【筹商】discuss; consult: ～对策 discuss what countermeasures to take

酬 ①[书](敬酒) propose a toast; toast ②(报答;报酬) reward; payment: 稿～ payment for an article or book written ③(交往) friendly exchange ④(实现) fulfill; realize: 壮志未～ with one's lofty aspirations unrealized
【酬报】requite; reward; repay; recompense
【酬答】①(酬谢) thank sb. with a gift ②(用言语或诗文应答) respond with a poem or speech
【酬金】monetary reward; remuneration
【酬劳】recompense; reward
【酬谢】thank sb. with a gift
【酬应】social intercourse: 不善～ socially inept

愁 worry; be anxious: 不～吃,不～穿 not have to worry about food and clothing
【愁肠】pent-up feelings of sadness: ～百结 weighed down with anxiety; with anxiety gnawing at one's heart / ～寸断。The sorrow is so deep that it seems to have cut the bowels to pieces.
【愁苦】anxiety; distress
【愁眉】knitted brows; worried look: ～不展 with a worried frown; have a gloomy countenance / ～苦脸 have a worried look; pull a long face; a distressed expression
【愁闷】feel gloomy; be in low spirits; be depressed
【愁容】worried look; anxious expression: ～满面 look extremely worried; have a mournful countenance
【愁绪】[书] gloomy mood

仇 ①(仇敌) enemy; foe: 亲痛～快 sadden one's friends and gladden one's enemies ②(仇恨) hatred; enmity: 记～ nurse a grievance / 有～ have a score to settle
【仇敌】foe; enemy
【仇恨】hatred; enmity; hostility: 满腔～ seething with hatred
【仇人】personal enemy: ～相见,分外眼红。When enemies come face to face, their eyes blaze with hate. (或: Meeting enemies open old wounds.)
【仇杀】kill in revenge
【仇视】regard as an enemy; look upon with hatred; be hostile to
【仇隙】[书] bitter quarrel; feud: ～冰消。A grudge has melted away as the ice.

惆
【惆怅】disconsolate; melancholy

稠 ①(浓) thick: 粥很～。The porridge is very thick. ②(稠密) dense: 地窄人～ small in area but densely populated
【稠密】dense: 交通网～ a dense communications network / 人烟～ densely populated; populous
【稠人广众】large crowd; big gathering; a dense crowd

绸 silk fabric; silk: ～伞 silk parasol
【绸缎】silks and satins
【绸缪】sentimentally attached: ～缱绻 bind closely and attach to / 情意～ be head over heels in love
【绸子】silk fabric

chǒu

丑 ①(不好看) ugly; unsightly; hideous: 长得不～ not bad-looking ②(叫人厌恶或瞧不起的) disgraceful; shameful; scandalous: 出～ make a fool of oneself
【丑八怪】[口] a very ugly person
【丑表功】brag shamelessly about one's deeds
【丑恶】ugly; repulsive; hideous: ～表演 a disgusting performance / ～灵魂 an ugly soul / ～面目 ugly features / ～现象 vile practices / ～嘴脸 hideous features
【丑化】smear; uglify; defame; vilify: ～劳动人民的形象 smear the images of the laboring people
【丑剧】farce
【丑角】clown; buffoon
【丑陋】ugly
【丑事】scandal
【丑态】ugly performance; ludicrous performance; buffoonery: ～百出 act like a buffoon; cut a contemptible figure; behave in a revolting manner / ～毕露 show the cloven hoof 〈foot〉
【丑闻】scandal

【丑媳妇总要见公婆】People may hide the worst side of their nature from casual friends, but they cannot hid it from those with whom they live.

chòu

臭 ①(气味难闻) smelly; foul; stinking: ～鸡蛋 a rotten egg / ～味 stink; offensive odor; foul smell / ～不可闻 give off an unbearable stink ②(惹人讨厌的) disgusting; disgraceful: 摆～架子 put on nauseating airs

【臭吃臭喝】drink like a fish and eat like a pig

【臭虫】bedbug

【臭椿】[植] tree of heaven

【臭豆腐】strong-smelling preserved bean curd

【臭烘烘】stinking; foul-smelling

【臭街烂巷】plentiful and cheap of goods

【臭骂】curse roundly; scold angrily and abusively: 挨了一顿～ get a tongue-lashing; get a dressing down

【臭名昭著】of ill repute; notorious

【臭气】bad smell; offensive smell; stink: ～冲天 stinking smell assaulting one's nostrils / ～熏天 stink to high heaven

【臭味相投】People of the same ilk like each other

【臭氧】[化] ozone ◇ ～层 ozonosphere

【臭鼬】[动] skunk

chū

初 ①(开始) at the beginning of; in the early part of: ～夏 early summer / 八月～ early in August; in early August / 年～ at the beginning of the year ②(第一个) first (in order): ～五 the fifth day (of a lunar month) / ～雪 first snow / ～战 first battle ③(刚开始) just; for the first time: ～具规模 begin to take shape / 大病～愈 have just recovered from a serious illness / 感冒～起 with the first symptoms of a cold ④(等级最低的) elementary; rudimentary: ～级词典 elementary dictionary ⑤(原来的) original: ～愿 one's original intention / 和好如～ become reconciled

【初版】first edition

【初步】initial; preliminary; tentative: ～拨款 initial appropriations / ～措施 preliminary measure / ～方案 a tentative program / ～分析 preliminary analysis / ～概算 initial budget estimate / ～估计 preliminary estimates / ～加工 initial processing / ～交换意见 have a preliminary exchange of views / ～设想 a tentative idea / 获得～成果 reap first fruits; get initial results

【初产】primiparity ◇ ～妇 primipara

【初出茅庐】just come out of one's thatched cottage; at the beginning of one's career; young and inexperienced; new-fledged: ～的作家 a fledgling writer

【初创】newly established: ～的企业 a newly established enterprise / ～阶段 initial stage

【初次】the first time: ～登台 appear for the first time on the stage; make one's *dèbut* / ～见面 see sb. for the first time

【初等】elementary; primary ◇ ～教育 primary education / ～数学 elementary mathematics / ～算术 elementary arithmetic

【初犯】①(初犯者) first offender ②(第一次犯罪) first offense; first crime

【初伏】①(头一伏) the first of the three ten-day periods of the hot season ②(头一伏的第一天) the first day of the first period of the hot season

【初稿】first draft

【初婚】first marriage

【初级】elementary; primary; junior ◇ ～班 junior class; elementary course / ～产品 primary products / ～读本 primer / ～农业生产合作社 elementary agricultural producers' cooperative / ～人民法院 basic People's court / ～线圈[电] primary coil / ～小学 lower primary school / ～中学 junior middle school

【初交】new acquaintance

【初亏】[天] first contact (of an eclipse)

【初恋】first love

【初馏塔】[石油] primary tower

【初露锋芒】display one's talent for the first time

【初露头角】make first appearance

【初期】initial stage; early days: 二十世纪～ early in the twentieth century / 解放～ during the initial post-liberation period; just 〈right〉 after liberation / 战争～ in the early days of the war

【初入世途】start in life

【初审】[法] trial of first instance; first trial ◇ ～案件 case of first instance / ～裁决 ruling of first instance / ～法庭 court of first instance / ～法院 court of first instance; court of original jurisdiction; trial court / ～判决 judgement of first instance

【初生之犊】newborn calf: ～不畏虎 Newborn calves are not afraid of tigers; Young people are fearless.

【初试】①(初次试验) first try ②(分两次考试的第一次) preliminary examination
【初速度】[物] initial velocity
【初小】[简](初级小学) lower primary school
【初选】primary election
【初学】begin to learn
【初叶】early years (of a century): 二十世纪～ early in the twentieth century
【初轧机】blooming mill; cogging mill
【初诊】first visit (to a doctor or hospital)
【初值】[数] initial value
【初中】[简](初级中学) junior middle school
【初衷】original intention: 不改～ not change one's original intention

出 ①(从里到外) go or come out: (演员)～场 come upon the stage／～城 go out of town／～国 go abroad ②(超出) exceed; go beyond: ～格 go beyond the limit／～月 after this month; next month／不～三年 within three years ③(拿出) issue; put up; give; offer: ～布告 post an announcement; put up a notice／～考题 set the paper; set the examination questions／～证明 issue a certificate／～主意 offer advice; supply ideas; make suggestions／有钱～钱。Those who have money offer money. ④(生产; 产生) produce; turn out: ～成果 produce result／～煤 produce coal／～人材 turn out talents／～油 yield oil／实践～真知。Genuine knowledge comes from practice. ⑤(发生) arise; happen; take place; occur: ～问题 go wrong; go amiss／防止～事故 prevent accidents／这事～在二十年前。It happened twenty years ago.／这事～在1954年。It took place in the year of 1954. ⑥(发出; 发泄) put forth; vent: ～气 vent one's spleen／～芽 put forth buds; sprout／～疹子 have measles ⑦(显得量多) rise well (with cooking): 这种米～饭。This kind of rice rises well when it's cooked. ⑧(支出) pay out; expend: 量入为～ keep expenditures within the limits of income; cut one's coat according to one's cloth／入不敷～ one's income falling short of one's expenditure; unable to make both ends meet ⑨(剧目) a dramatic piece: 一～戏 an opera; a play
【出版】come off the press; publish; come out ◇～法 law of publication; press law／～合同 contract for publication／～界 the press／～社 publishing house; press／～物 publication／～者 publisher／～自由 freedom of the press; liberty of press
【出榜】①(贴出被录取人的名单) publish a list of successful candidates ②[旧](贴出文告) put up a notice
【出奔】leave; run away
【出殡】carry a coffin to the cemetery; hold a funeral procession
【出兵】dispatch troops
【出操】(go out to) drill; go out to do exercises
【出岔子】go wrong; get into trouble; meet with an accident
【出差】be away on official business; be on a business trip ◇～补贴 travelling allowance／～费 allowances for a business trip
【出产】produce; manufacture
【出厂】(of products) leave the factory ◇～价格 producer price; ex-factory price／～日期 date of production; date of manufacture
【出场】①(登台; 表演) come on the stage; appear on the scene ②(剧本用语) enter ③(运动员进运动场) enter the playing ground; enter the arena ◇～运动员名单 list of players for the match
【出超】favorable balance of trade
【出车】①(开出车辆) dispatch a vehicle ②(开车出去) be out driving a vehicle
【出丑】①(出洋相) make a show of oneself; make a fool of oneself; become a laughing stock ②(丢脸) bring shame on oneself: 当众～ make a fool of oneself before others; be disgraced in public
【出处】source (of a quotation or allusion): 注明～ indicate the source; give references ◇～同上 ibid.; ib. (ibidem)
【出错】make mistakes
【出点子】offer advice; make suggestions: 在背后出坏点子 be directing the show from behind the scenes
【出动】①(队伍外出行动) set out; start off: 待命～ await orders to set out (go into action)／小分队提前～。The detachment set off ahead of schedule②(派出军事力量) call out; send out; dispatch: ～飞机二十架次 fly 20 stories／～军舰 dispatch warships／～伞兵参战 call out paratroops to join the battle ③(行动起来) go into action; turn out
【出尔反尔】go back on one's word; contradict oneself; now yes, now no
【出发】①(离开原地去其他地方) set out; start off; leave: 准备～ get ready to start off ②(从某一方面着眼) start from; proceed from: 从长远的观点～ from a long-term point of view／一

切从人民的利益～ Proceed in all cases from the interests of the people ◇～港 port of departure

【出发点】starting point; point of departure; jumping-off point

【出风头】seek or be in the limelight; cut a smart figure: 喜欢～ like to be in the limelight; seek the limelight

【出伏】ending of the dog days

【出钢】[冶] tapping (of molten steel): 出一炉钢 turn out a heat of steel

【出港】clear a port; leave port ◇～呈报表 bill of clearance / ～证 clearance (papers)

【出阁】(of a woman) get married; marry

【出格】exceed what is proper

【出工】go to work; show up for work

【出恭】go to the lavatory (for a bowel movement)

【出乖露丑】make an exhibition of oneself

【出轨】①(脱离轨道) be derailed; go off the rails ②(言语行动出乎常规之外) overstep the bounds: ～行为 improper behavior

【出国】go abroad; leave one's native land

【出海】go to sea; put out to sea: 捕鱼～ go fishing on the sea

【出汗】perspire; sweat: 出一身汗 break into a sweat; sweat all over

【出航】①(船离开港口) set out on a voyage; set sail ②(飞机离开机场) set out on a flight; take off ◇～航线 outbound course

【出乎能力之外】beyond the capacity of

【出乎意料】exceeding one's expectations; contrary to one's expectations; unexpectedly; to sb.'s surprise

【出乎预料】cap the climax; beyond expectation exceed expectation

【出活】yield results in work; be efficient

【出击】launch an attack; hit out; make a sally: 四面～ hit out in all directions

【出家】become a monk or nun

【出价】offer a price; bid ◇～人 bidder / ～最高的投标人 highest-bidder

【出嫁】(of a woman) get married; marry

【出界】[体] out-of-bounds; outside

【出借】lend; loan

【出境】leave the country: 办理～手续 go through exit formalities / 递解～ send out of the country under escort / 驱逐～ deport ◇～登记 departure registration / ～回执 exit receipt / ～签证 exit visa / ～许可证 exit permit / ～证书 exit certificate

【出局】[棒、垒球] out

【出口】①(说出话来) speak; utter: ～伤人 speak bitingly ②(出口处) exit; way out: 会场～ the exits of a conference hall ③[经] export: ～大米 export rice ◇～补贴 export subsidy / ～货 exports; outbound freight; exportation / ～货物许可证 licence for the export of commodities / ～检疫 export quarantine / ～结关 customs clearance / ～贸易 export trade / ～商品 export commodities / ～税 export duties / ～信贷担保 export credit guarantee / ～许可证 export licence

【出口成章】words flow from the mouth as from the pen of a master

【出口粗野】swear like a trooper

【出来】come out; emerge: 从房间里～ come out of the room / 说～! Out with it!

【出类拔萃】stand out from one's fellows; be out of the common run; be preeminent; prominent: ～的人物 an outstanding figure

【出力】①(尽力;拿出力量) put forth one's strength; exert oneself; exert one's efforts; make great efforts ②(输出功率) output power: ～试验 service test / ～效率 power efficiency

【出列】①[军] leave one's place in the ranks ②(口令) Fall out!

【出猎】go hunting

【出笼】①(从蒸笼中取出) come out of the steamer ②(坏事物出现) come out into the open; come forth; appear ③(大量抛出) put forth in large quantities; dump; inflate (the paper currency)

【出漏子】in trouble

【出路】①(通向外面的路) way out: 找～ find a way out ②(销售货物的去处) outlet

【出乱子】go wrong; get into trouble

【出落】grow (prettier, etc.)

【出马】go into action; take the field; take up a matter: 亲自～ take up the matter oneself; attend to the matter personally; take personal charge of the matter

【出卖】①(卖) offer for sale; sell ②(背叛) sell out; betray; barter away: ～朋友 betray one's friend / ～原则 barter away principles / ～主权 barter away the sovereignty

【出毛病】be 〈go〉 out of order; go wrong; break down: 机器～了。The machine is out of order. / 收音机～了。Something has gone wrong with the radio.

【出门】①(外出) go out ②(离家远行) leave home; be away from home; go on a journey

～不露财 When you go out to buy, don't show your silver.

【出面】act in one's own capacity or one behalf of an organization; appear personally; come forward: ～调解 act as a mediator / 自己不～ keep oneself in the background

【出苗】[农] (of seedlings) emerge; come out; germinate; sprout ◇ ～率 rate of emergence; rate of germination

【出名】①(著名;有名声) famous; well-known ②(出面) lend one's name (to an occasion or enterprise); use the name of

【出没】appear and disappear; come and go; haunt: ～无常 appear and disappear unexpectedly; come and go unpredictably

【出谋划策】give counsel; offer advice; mastermind a scheme: 躲在背后～ mastermind a scheme from behind the scenes

【出纳】①(现金、票据的出纳) receive and pay out money or bills ②(担任出纳的人) cashier; teller ③(图书馆) receive and lend books, etc. ◇ ～台 (图书馆) circulation desk; (银行等) cashier's desk; teller's desk / ～员 cashier; teller

【出偏差】deviate; err on one side or the other

【出品】①(制造) produce; manufacture; make ②(产品) product: 新～ a new product

【出其不备】take one unaware

【出其不意】take sb. by surprise; catch sb. unawares

【出奇】unusually; extraordinarily: 冷得～ be unusually cold

【出奇制胜】defeat one's opponent by a surprise move; make a successful surprise raid

【出气】①(发泄怨愤) give vent to one's anger; vent one's spleen: 在某人身上～ give vent to one's anger on sb. / 出～ blow off steam ②(排气,放气) air out ◇ ～口 gas outlet; air vent

【出勤】①(按时工作) turn out for work: 全体～ full attendance ②(外出办理公务) be out on duty ◇ ～率 rate of attendance; attendance

【出去】go out; get out: ～走走 go out for a walk

【出圈儿】[方] overstep the bounds; go too far: 说话出了圈儿 go too far in what one says

【出缺】(of a high post) fall vacant

【出让】sell (one's own things): 自行车减价～ sell one's bicycle at a reduced price ◇ ～人 transferor

【出人头地】rise head and shoulders above others; stand out among one's fellows; become outstanding

【出人意表】exceeding all expectations; beyond all expectations; come as a surprise: 疗效之佳～。The curative effect far exceeded all expectations.

【出人意外】contrary to one's expectations; come as a surprise

【出任】[书]take up the post of

【出入】①(出去和进来) come in and go out; go out and come in ②(不相符) discrepancy; divergence; inconsistency ◇ ～相抵 income and expenses just balance / ～证 pass (identifying a staff member, etc.)

【出色】outstanding; remarkable; splendid: 干得很～ do a remarkable job; acquit oneself splendidly

【出身】①(家庭出身) class origin; family background: 工人家庭～ come from a worker's family / 穷苦～ be from poor families ②(个人经历) one's previous experience or occupation ◇ ～卑贱 spring from obscurity / ～贫贱 emerge from poverty or obscurity / ～微寒 born of low extraction

【出神】be spellbound; be in a trance; be lost in thought: 听得～ listen with rapt attention

【出神入化】reach the acme of perfection; extremely miraculous; be superb: ～的表演艺术 superb performance

【出生】be born ◇ ～登记 registration of birth / ～地 birthplace / ～率 birthrate / ～日期 date of birth / ～证 birth certificate / ～住所 domicile of origin

【出生入死】go through fire and water; brave untold danger; risk one's life

【出师】①(学徒期满) finish one's apprenticeship ②(出兵) dispatch troops to fight; send out an army

【出使】serve as an envoy abroad; be sent on a diplomatic mission

【出示】show; produce: ～证件 produce one's papers / ～证据 lodge a proof; presentation of evidence; tender evidence / ～证物 offer an exhibit / ～证物人 exhibitor

【出世】①(出生) come into the world; be born ②(产生) renounce the world; stand aloof from worldly affairs

【出事】meet with a mishap; have an accident: ～地点 site of an accident / 出了什么事? What's wrong?

【出手】①(卖出) get (hoarded goods, etc.) off

one's hands; dispose of; sell ②(开始做某件事时表现出来的本领) skill displayed in making opening moves: ～不凡 make skillful 〈masterly〉 opening moves (in *wushu,* chess, etc.) ③(袖子的长短) length of sleeve
【出售】offer for sale; sell: ～一空 be sold out／廉价～ sell sth. at a bargain price
【出数儿】[口] (of rice) rise well with cooking
【出水芙蓉】a lotus flower when fully open
【出台】①(演员上场) appear on the stage; enter ②(出面活动) appear personally
【出逃】run away
【出挑】①(指体格、相貌方面) grow (prettier, etc.) ②(指智能方面) develop (in skill, etc.)
【出粜】sell (grain)
【出铁】[冶] tap a blast furnace; tapping ◇～口 taphole; iron notch
【出庭】appear in court; before the court; enter an appearance ～辩护 defend a case in court／～作证 appear in court as a witness; serve as a witness at court ◇ ～通知 memorandum of appearance; notice of appearance
【出头】①(摆脱困苦) lift one's head; free oneself (from misery, persecution, etc.) ②(出面,带头) appear in public; come forward ③(整数后的零头) a little over; odd: 五米～ five meters and a bit over; a little over five meters／一百～ one hundred odd
【出头露面】appear in public; be in the limelight: ～的人物 a public figure／喜欢～ fond of being in the limelight
【出土】①(挖掘出土) be unearthed; be excavated: ～文物 unearthed relics; unearthed artifacts ②(幼苗出土) come up; come up out of the ground
【出脱】①(货物卖出) manage to sell; dispose of ②(出落) grow (prettier, etc.) ③(开脱罪名) acquit; absolve
【出亡】flee; live in exile
【出席】attend; be present: ～会议 attend a meeting; be present at a meeting／～人数 number of persons present; attendance／～宴会 be present at a banquet
【出息】promise; prospects; future: 有～的姑娘 a high-minded girl／有～的青年 a promising youth／ 这个人真没～。 This chap is a good-for-nothing.
【出险】①(脱险) be or get out of danger ②(发生危险) be in danger; be threatened
【出现】appear; arise; emerge; turn up: ～了新问题。 A new problem has arisen.／～了一种倾向。 A tendency has showed itself.
【出项】item of expenditure; expenses; outlay
【出血】①[医] hemorrhage; bleeding: 大～ massive hemorrhage／ 内～ internal hemorrhage／皮下～ subcutaneous hemorrhage／外～ external hemorrhage／胃～ gastric hemorrhage／ 阴道大量～ bleed copiously from the vagina／ 伤口大量～。 The wound bleeds profusely. ②[印] bleed
【出巡】①(帝王巡行) royal progress ②(出外巡视) tour of inspection
【出芽】sprout; germinate; be budding
【出言不逊】make impertinent remarks; speak insolently; use rude language
【出洋】[旧] go abroad: ～留学 go abroad to pursue one's studies; study abroad
【出洋相】make an exhibition of oneself
【出以公心】keep the public interest in mind; act without any selfish considerations; have the public interest at heart
【出迎】go or come out to meet
【出油井】[石油] producing well
【出游】go on a (sightseeing) tour
【出于】start from; proceed from; stem from; out of: ～本心 from one's heart／～本意 of one's own accord／～不可告人的目的 actuated by ulterior motives／～对同志的关怀 out of concern for one's comrades／～故意 do something out of spite／～同情 out of sympathy／～无奈 as it cannot be helped; there being no alternative／～无意 be not intentional; act unintentionally／～无知 from ignorance／～无知的错误 ignorant error／～责任感 proceed from a sense of duty／～自卫的抗辩 plea of self-defense／～自愿 on a voluntary basis; of one's own accord
【出语犀利】speak daggers
【出狱】be discharged from prison; be released from prison
【出院】leave hospital: 病愈～ be discharged from hospital after recovery ◇ ～处 discharge office／～证明 hospital discharge certificate
【出渣口】[冶] slag notch; cinder notch
【出帐】①(把支出款项登上帐) enter an item of expenditure in the accounts ②(开支,支出) items of expenditure
【出诊】(of a doctor) visit a patient at home; pay a home visit; make a house call
【出征】go on an expedition; go out to battle
【出众】be out of the ordinary; be outstanding: 人才～ a person of exceptional ability／智力

～ with exceptional intelligence
【出走】leave; run away; flee: 仓卒～ leave in a hurry
【出租】hire; let; rent out: ～某物 let out sth. / 房屋～(广告用语) House to Let ◇ ～汽车 taxicab; taxi; cab / ～人 lessor; leasor

chú

厨 kitchen
【厨房】kitchen ◇ ～用具 kitchen utensils; cooking utensils / ～有人好进餐,朝里有人好做官。 The influence of friends in the right place can make a man's lot much easier.
【厨师】cook; *chef*

橱 cabinet; closet: 壁～ built-in cabinet; closet / 书～ bookcase / 碗～ cupboard; pantry cabinet / 五斗～ chest of drawers / 衣～ wardrobe
【橱窗】①(商店玻璃橱窗) show window; display window; showcase; shopwindow ②(宣传橱窗) glass-fronted billboard
【橱柜】①(食具柜) cupboard ②(矮立柜) a cupboard that also serves as a table; sideboard
【橱式写字台】secretary

除 ①(去掉) get rid of; eliminate; remove: 为民～害 rid the people of a scourge ②(不计算在内) except; except (for); but ③(此外) besides; in addition to ④[数] divide: 二～八得四。 Eight divided by two is four; 2 goes into 8 four times. / 十二能被四～尽 12 divides by 4; 4 goes into 12
【除暴安良】weed out the wicked and let the lawabiding citizen live in peace
【除不尽】indivisible
【除草】weeding ◇ ～机 weeder / ～剂 weed killer; herbicide; weedicide
【除尘器】dust remover
【除虫菊】[植] Dalmatian pyrethrum; pyrethrum
【除虫菊酯】pyrethrins
【除臭剂】deodorant
【除此之外】in addition
【除恶务尽】one must be thorough in exterminating an evil; evil must be completely eradicated
【除法】[数] division
【除非】[连]①(表示唯一的条件) only if; only when ②(表示不计算在内) unless
【除根】dig up the roots; cure once and for all; root out; eradicate
【除害安良】remove the evil and quiet the good
【除号】sign of division (÷)
【除旧布新】get rid of the old to make way for the new; do away with the old and set up the new; ring out the old, ring in the new
【除了】①(不计算在内) except (for) ②(此外) besides; in addition to
【除名】remove sb.'s name from the rolls; take sb.'s name off the books; expunge sb.'s name from a list; strike one's name off
【除数】[数] divisor
【除四害】eliminate the four pests (i.e. rats, bedbugs, flies and mosquitoes)
【除外】except; not counting; not including
【除夕】New Year's Eve
【除莠剂】herbicide; weed-killer

锄 ①(农具) hoe ②(松土除草) work with a hoe; hoe: ～草 hoe up weeds; weed with a hoe / ～地 hoe the fields ③(铲除) uproot; eliminate; wipe out
【锄奸】eliminate traitors; ferret out spies: ～工作 elimination of traitors; anti-espionage work
【锄强扶弱】eliminate the bullies and help the down-trodden
【锄式开沟器】hoe coulter
【锄头】hoe

刍 [书]①(饲草) hay; fodder: 反～ ruminate; chew the cud ②(割草) cut grass
【刍秣】fodder
【刍议】[谦] my humble opinion; a rustic opinion

雏 young (bird): ～鸡 chicken: chick / ～鸭 duckling / ～燕 young swallow
【雏鸟】nestling; fledgling
【雏儿】①(幼小的鸟) young (bird) ②(年纪轻、阅历少的人) a young, inexperienced person; fledgling
【雏形】①(未定型的最初形式) embryonic form; embryo ②(缩小的模型) model; miniature

chǔ

楮 [植] paper mulberry

储 store up; keep in reserve; have in reserve
【储备】①(储存备用) store for future use; lay in; lay up: ～过冬饲料 lay up fodder for the winter / ～粮食 store up grain; build up supplies of grain ②(储藏品) reserve; store: 黄金～ gold reserve / 外汇～ foreign exchange reserve

◇ ～基金 reserve fund / ～粮 grain reserves

【储藏】①(保藏) save and preserve; store; keep: 冷冻～ cold storage / 蔬菜～ preservation of vegetables ②(蕴藏) deposit ◇ ～量[矿] reserves; deposit / ～室 storeroom

【储存】lay in; lay up; store up; keep in reserve; stockpile: ～干草以备过冬 lay in hay for the winter / ～战略物资 stockpile strategic materials

【储户】depositor

【储君】crown prince

【储量】[矿] reserves: 可采～ recoverable reserves; workable reserves / 探明～ proved reserves / 远景～ prospective reserves

【储气】gas storage ◇ ～构造 gas-bearing structure / ～罐 gas tank; gas receiver / ～筒 receiver

【储汽筒】steam reservoir

【储蓄】save; deposit: 定期～ fixed deposit / 活期～ current deposit ◇ ～存款 savings deposit / ～存折 savings account book / ～额 total savings deposits / ～所 savings department of a bank; savings bank / ～银行 savings bank

【储油】[石油] oil storage ◇ ～构造 oil-bearing structure

【储油罐】[石油] oil storage tank; oil tank; stock tank: 浮顶～ floating roof tank / 球形～ spherical tank

楚 ①(清楚) clear; neat: 一清二～ perfectly clear ②(痛苦) pang; suffering: 苦～ distress; suffering

【楚材晋用】a great person given important post by another country

【楚楚】clear; tidy; neat: 衣冠～ immaculately dressed

【楚楚动人】lovingly pathetic

【楚楚可观】Being clear and distinct, it is worth seeing.

【楚楚可怜】delicate and touching

【楚河汉界】the border of two opposing powers

杵 ①(捣杵) pestle: ～臼 mortar and pestle ②(棒槌) a stick used to pound clothes in washing ③(戳;捅) poke: 用手指～他一下 give him a poke

础 plinth: ～石 the stone base of a column; plinth

处 ①(相处) get along (with sb.): 容易相～ easy to get along with ②(处于;居) be situated in; be in a certain condition ③(处置;办理) manage; handle; deal with ④(处罚) punish; sentence: ～以两年徒刑 sentence sb. to two years' imprisonment

【处罚】punish; penalize: ～无照驾驶者 penalize unlicensed drivers / 减轻～ mitigate a punishment / 免除～ remit a punishment

【处方】①(开药方) write out a prescription; prescribe ②(药方) prescription; recipe

【处分】take disciplinary action against; punish: 党内～ disciplinary action within the Party / 罚款～ monetary penalty / 免予～ exempt sb. from punishment / 行政～ administrative disciplinary measure / 予以警告～ give sb. disciplinary warning

【处境】unfavorable situation; plight: ～尴尬 be in an awkward situation / ～困难 live in difficult circumstances; be in a sorry plight; be in a predicament; be in a difficult situation / ～狼狈 in poor case / ～危险 be in a dangerous 〈precarious〉 situation; be in peril

【处决】put to death; execute: 依法～ put to death in accordance with the law

【处理】①(安排;解决) handle; deal with; dispose of; manage; settle: ～国家大事 conduct state affairs / ～垃圾 dispose of rubbish / ～日常事务 handle day-to-day work; deal with routine matters / ～污水 dispose of sewage / ～遗留问题 settle the remaining problems ②(变价、减价出售) sell at reduced prices: ～积压商品 sell old stock at reduced prices ③(用特定方法加工) treatment; treat by a special process: 热～ heat treatment / 用硫酸～ treat with sulphuric acid

【处理机】[自] processor

【处理价格】reduced price; bargain price

【处理品】goods sold at reduced prices; shopworn or substandard goods

【处女】①(未出嫁的女子) virgin; maiden: 老～ old maiden; spinster ②(初次) maiden: ～航 maiden voyage; maiden flight / ～作 maiden work; first effort

【处女地】virgin land; virgin soil; uncultivated land

【处女膜】[生理] hymen ◇ ～痕 hymenal caruncles; carunculae hymenales

【处世】conduct oneself in society: ～哲学 philosophy of life / ～之道 ways of life; ways of the world

【处事精明】be clever and smart in attending to business

【处暑】the Limit of Heat (14th solar term)
【处死】put to death; execute; put to execution
【处心积虑】deliberately plan (to achieve evil ends); incessantly scheme; seek by all means
【处刑】[法] condemn; sentence
【处于】be (in): ～窘境 in a tight box / ～困境 be in a hobble / ～逆境 be in adverse circumstances / ～平等地位 be on an equal footing / ～顺境 be in favorable circumstances / ～死地 send a person to his doom / ～优势 have the advantage / ～有利的地位 find oneself in an advantageous position
【处之泰然】take things calmly; remain unruffled; keep one's head
【处治】punish
【处置】①(处理) handle; deal with; manage; dispose of: ～失当 mismanage; mishandle ②(发落;惩治) punish

chù

畜 domestic animal; livestock: ～群 a herd of livestock / 耕～ farm animal / 家～ livestock / 母～ female animals / 役～ draught animals / 幼～ young animals
【畜肥】animal manure
【畜类】domestic animals
【畜力】animal power ◇ ～车 animal-drawn cart / ～割草机 horse mower / ～农具 animal-drawn farm implements / ～牵引 animal traction; animal hauling
【畜生】①(泛指禽兽) domestic animal ②(骂人话) beast; dirty swine
【畜疫】epidemic disease of domestic animals

憷 fear; shrink from
【憷头】[方] shrink from difficulties; be timid

矗 [书] stand tall and upright
【矗立】stand tall and upright; tower over sth.

处 ①(地方) place: 别～ another place; elsewhere / 停车～ parking place car park / 住～ dwelling place; quarters ②(点;部分) point; part: 长～ strong point; forte / 有相同之～ bear a resemblance; have something in common ③[量] 几～人家 several homesteads / 两～印刷错误 two misprints ④(机关;部门) department; office: 登记～ registration office / 联络～ liaison office / 人事～ personnel department; personnel section / 售票～ booking office / 问讯～ inquiry office / 总务～ general affairs department
【处处】everywhere; in all respects: ～掣肘 meet hindrance everywhere / ～碰壁 hit against a wall everywhere / ～严格要求自己 set strict demands on oneself in all respects
【处所】place; location
【处长】the head of a department; section chief

黜 [书] remove sb. from office; dismiss
【黜免】[书] dismiss (a government official)

绌 [书] inadequate; insufficient: 相形见～ prove definitely inferior; pale by comparison

触 ①(接触) touch; contact: 请勿～摸展品。Please don't touch the exhibits. ②(碰) strike; hit: ～雷 strike a mine; touch off a mine; run into a mine ③(触动) touch: ～到痛处 touch a sore spot; touch sb. to the quick ④(感动) move sb.; stir up sb.'s feelings
【触电】get an electric shock: 小心～! Danger! Electricity! Danger! Live wire!
【触动】①(碰;撞) touch sth.; moving it slightly: ～某人的利益 affect sb.'s interests ②(感动;感触) move sb.; stir up sb.'s feeling: 对他～很大 shook him up a lot; gave him quite a shake-up / 有所～ be somewhat moved
【触发】detonate by contact; touch off; spark; trigger: ～热核聚变 trigger thermonuclear fusion / ～乡思 touch off a train of home thoughts; provoke nostalgic longing ◇ ～地雷 trap mine; contact mine / ～电路 trigger circuit / ～器 trigger; flip-flop / ～水雷 contact mine
【触犯】offend; violate; go against: ～法律 violate the law; break the law / ～某人 offend sb. / ～人民的利益 go against the people's interests; encroach on the interests of the people
【触击】[棒、垒球] bunt
【触及】touch: ～灵魂 touch people to their very souls / ～事物本质 get to the essence of a matter
【触礁】run (up) on rocks; strike a reef 〈rock〉
【触角】[动] antenna; feeler; tentacle
【触景生情】the sight strikes a chord in one's heart; the scene brings back memories
【触觉】①[生理] tactile sensation; tactual sensation; sense of touch ②[动] touch reception ◇ ～器官 tactile organ
【触类旁通】grasp a typical example and you will grasp the whole category; comprehend by analogy; draw an analogy

【触媒】[化] catalyst; catalytic agent
【触目】①(接触到视线) meet the eye: ～皆是 can be seen everywhere ②(显眼;引人注目) conspicuous; attracting attention
【触目惊心】startling; shocking; horrifying; a ghastly sight
【触怒】make angry; infuriate; enrage
【触杀剂】[农] contact insecticide
【触身式橄榄球】touch football
【触手】[动] tentacle
【触痛】① touch a tender spot; touch sb. to the quick ②[医] tenderness
【触网】[体] touch net
【触须】[动] cirrus: 无脊椎动物～ palp／鱼类～ barbel
【触靴】[电] contact shoe
【触诊】[医] palpation

chuāi

揣 hide or carry in one's clothes; tuck: ～进口袋 tuck sth. into one's pocket／～在怀里 hide in the bosom; tuck into the bosom
【揣手儿】tuck each hand in the opposite sleeve

搋 rub; knead: ～面 knead dough

chuǎi

揣 [书] estimate; surmise; conjecture
【揣测】guess; conjecture; surmise; surmise; reckon: ～之词 statement out of surmise／纯属～ be mere conjecture
【揣度】estimate; appraise; conjecture
【揣摩】try to fathom; try to figure out; elicit sth. by careful study
【揣情度理】weigh the pro and cons

chuài

踹 ①(向外踢) kick ②(踩) tread; stamp; step in

chuān

穿 ①(破;透) pierce through; penetrate: ～个窟窿 pierce a hole; bore a hole／～云破雾 pierce through fogs and clouds／看～ see through ②(通过) pass through; cross; go through: ～过马路 cross a street／～过隧道 pass through a tunnel／～街过巷 go through streets and alleys／从人群中～过去 thread one's way through the crowd ③(穿着) wear; put on; be dressed in; have...on: ～得很朴素 simply dressed
【穿插】①(交叉) alternate; do in turn: 施肥料和除草～进行 do manuring and weeding in turn ②(交织) interweave; weave in; insert ③[军] thrust deep into the enemy forces: ～到敌后 thrust into the enemy's rear／～分割敌人 penetrate and cut up the enemy forces／～营 deep-thrust battalion／打～ fight a deep-thrust battle
【穿刺】[医] puncture: 肝～ liver puncture／腰椎～ lumbar puncture ◇ ～术 centesis; paracentesis
【穿戴】apparel; dress: ～整齐 be neatly dressed; dress neatly／不讲究～ not be particular about one's dress
【穿甲弹】armor-piercing projectile; armor-piercing shell or bullet; armor piercer
【穿甲炸弹】armor-piercing bomb
【穿卡机】punched-card machine
【穿孔】①bore a hole; punch a hole; perforate ②[医] perforation: 阑尾～ appendicular perforation／胃～ gastric perforation ◇ ～机 punch; perforator／～卡片 punched card; aperture card／～纸带 punched tape; punched paper tape; chadded tape
【穿筘机】[纺] reeding machine
【穿墙套管】wall busing
【穿山甲】①[动] pangolin ②[中药] pangolin scales
【穿梭】shuttle back and forth ◇ ～轰炸 shuttle bombing／～外交 shuttle diplomacy
【穿堂风】draught
【穿堂门】a passageway
【穿小鞋】给某人～ give sb. tight shoes to wear; make things hard for sb. by abusing one's power; deliberately put sb. to trouble
【穿孝】be in mourning; wear mourning
【穿心莲】[中药] creat
【穿衣镜】full-length mirror
【穿越】pass through; cut across
【穿针】thread a needle
【穿针引线】act as a go-between; try to make a match
【穿着】dress; apparel: ～朴素整洁 be plainly but neatly dressed
【穿凿】give a farfetched 〈strained〉 interpretation; read too much into sth.: ～附会 give strained interpretations and draw farfetched analogies

川 ①(河流) river: 高山大～ high mountains and big rivers ②(平;平地) plain: 一马平～ a

vast expanse of flat land; a great stretch of land
【川贝】[中药] tendril-leaved fritillary bulb
【川断续】[植] teasel
【川流不息】flowing past in an endless stream; never-ending; a continuous flow
【川芎】[中药] Ligusticum wallichii
【川资】travelling expenses: ～短缺 short of travelling expenses

氚 [化] tritium (T 或 H^3)
【氚核】[物] triton

chuán

椽 rafter
【椽子】rafter

传 ①(转给; 传递) pass; pass on ～球 pass a ball ②(交给另一代) hand down: 家～秘方 a secret recipe handed down in the family ③(传授) pass on (knowledge, skill, etc.); impart; teach ④(传播) spread: ～为佳话 become a favorite tale / 恶事～千里。 I'll news spreads everywhere. ⑤(传导) transmit; conduct: ～电 conduct electricity / ～热 transmit heat⑥(表达) convey; express: 眉目～情 flash amorous glances ⑦(发出命令叫人来) summon: ～某人出庭 summon sb. to court / ～证人 summon a witness ⑧(传染) infect; be contagious
【传播】①(广泛散布) disseminate; propagate; spread: ～马克思列宁主义 propagate 〈disseminate〉 Marxism- Leninism / ～知识 spread knowledge / 制止病菌～ check the spread of germs ②[物] propagation: ～损耗 propagation loss / 散射～ scatter propagation / 直线～ rectilinear propagation
【传布】disseminate; spread
【传抄】make private copies (of a manuscript, document, etc. which is being circulated)
【传出神经】[生理] efferent nerve
【传达】①(转达) pass on (information, etc.); transmit; relay; communicate: ～命令 transmit an order / ～上级指示 communicate the instructions of the higher level / 听～报告 hear a relayed report ②(机关门口的传达室工作) reception and registration of callers at a public establishment ③(工作人员) janitor ◇ ～室 reception office; janitor's room
【传代】go down to posterity; go down to the future generation
【传单】leaflet; handbill; propaganda sheet
【传导】[物] conduction: 热的～ conduction of heat
【传道】①[旧] propagate doctrines of the ancient sages ②[宗] preach; deliver a sermon
【传递】transmit; deliver; transfer: ～信件 deliver mail / ～信息 transmit messages
【传动】[机] transmission; drive: 变速～ change drive / 齿轮～ gear drive; gear transmission / 液压～ hydraulic drive ◇ ～比 drive ratio; transmission ratio / ～齿轮 transmission gear; drive gear / ～带 transmission belt / ～箱 transmission case / ～轴 transmission shaft / ～装置 gearing
【传粉】[植] pollination ◇ ～媒介 pollination medium
【传感器】[电] sensor; transducer: 激光～ laser sensor
【传呼】notify sb. of a phone call; pass on a message left by phone ◇ ～电话 neighborhood telephone service
【传话】pass on a message: ～给某人 send word to sb.
【传唤】[法] summon to court; subpoena
【传家宝】①(祖传宝物) family heirloom; hereditary treasure: 传家之宝 a treasure that has been handed down in the family ②(作风,传统) cherished tradition; cherished heritage
【传教】[宗] do missionary work ◇ ～士 missionary
【传令】transmit orders; dispatch orders: ～嘉奖 cite sb. in a dispatch
【传票】①[法] (court) summons; subpoena: 发出～ issue a summons ②[会计] voucher
【传奇】①(唐宋短篇小说) tales of marvels ②(传奇故事) legend; romance: ～式的人物 legendary figure; legend
【传球】pass: 低手～ underhand pass / 反手～ reverse pass / 反弹～ bounce pass / 横～ parallel pass; [冰球] cross rink pass; lateral pass / 后～[冰球] back pass / 间接～ indirect pass / 肩上～ shoulder pass / 凌空～ volley pass / 手腕～ snap pass / 头上～ overhead pass / 斜～ diagonal pass / 直接～ direct pass
【传染】infect; be contagious: 接触～ contagion / 空气～ infection through air / 水～ waterborne infection ◇ ～病 infectious disease; contagious disease / ～病院 hospital for infectious diseases / ～期 infective stage / ～途径 routes of infection / ～性飞沫 infectious droplet / ～性肝炎 infectious hepatitis / ～源 source of infection
【传热】conduct heat

【传入神经】[生理] afferent nerve
【传神】vivid; lifelike: ～之笔 a vivid touch (in writing or painting)
【传声器】microphone
【传声清晰度】[无] articulation
【传声筒】①(喊话筒) megaphone; loud hailer ②(照别人的话说,无主见的人) one who parrots another; sb.'s mouthpiece
【传世】be handed down from ancient times: ～珍宝 a treasure handed down from ancient times／～之作 a production that will be handed on from age to age
【传授】pass on (knowledge, skill, etc.); impart; teach: ～技术 pass on one's technical skill; impart one's technical skill
【传输】[电] transmission ◇ ～损耗 transmission loss／～线 transmission line
【传说】①(据别人说) it is said; they say ②(口头流传下来的故事) legend; tradition: 民间～ folklore; popular legend
【传送】convey; deliver ◇ ～带[机] conveyer belt; band carrier; line belt
【传诵】be on everybody's lips; be widely read: 为世人所～ be read with admiration by people all over the world
【传颂】be eulogized everywhere; be on everybody's lips
【传统】tradition: ～观念 traditional ideas／～习惯 traditional customs／～友谊 traditional (ties of) friendship／革命～ revolutionary tradition ◇ ～产业 conventional industries／～剧目 traditional theatrical pieces
【传闻】①(听到流传) it is said; they say ②(流传的事情) hearsay; rumor; talk
【传心术】telepathy
【传讯】[法] summon for interrogation of trial; subpoena; cite
【传言】①(辗转流传的话) hearsay; rumor ②(传话) pass on a message
【传扬】spread (from mouth to mouth): ～四方 spread far and wide
【传阅】pass round for perusal; circulate for perusal: ～有关文件 pass around the relevant documents
【传真】①portraiture ②[讯] facsimile: 无线电～ radio facsimile; radiophotography ◇ ～电报 phototelegraph／～照片 radiophoto
【传种】propagate; reproduce: 选良种马～ select a horse of good strain for propagation (reproduction)
【传宗接代】have a son to carry on his family name

船

boat; ship: 乘～去 go to boat; embark (on a ship) for／翻～ capsize; overturn／上～ board a ship; go on board; embark／下～ disembark／在～上 on board a ship ◇驳～ barge; lighter／捕鲸～ whaler／沉～ wreck／电站～ floating power station／公用事业～ utility ship／火车渡～ train ferry／冷藏～ refrigerator ship／内河轮～ river steamer／商～ merchantman／挖泥～ dredge boat; dredger／引航～ pilot boat／油～ oil tanker／原子破冰～ atomic ice-breaker／远洋轮～ ocean-going vessel／运煤～ coal carrier ○班轮(定期客轮) passenger liner／远洋班轮 ocean-liner／沿海客轮 coastwise passenger ship／汽艇 steam boat; launch／游艇 yacht／拖轮 tug／货轮 freighter; cargo ship／船长 captain／船员(总称) crew／大副 first mate／二副 second mate／海员(水手) sailor; seaman／轮机长 chief engineer／水手长 boatswain／报务员 radio operator／港口 port; harbor／港务局 port office／引水员 pilot／港务监督 Harbor Superintendency Administration／泊地 berth／码头 wharf／中途停靠 call at／驶往 be bound for／方位 breaking／前进 go headway; steer ahead／微速前进 dead slow／慢速前进 slow ahead／常速前进 half ahead／全速前进 full ahead／强速前进 emergency full ahead／后退 go sternway; steer astern／减速 retard; slow down／倒车 reverse the engine／停车 stop the engine／打空车 racing／逆风 against the wind／顺风 before the wind／顺流 downstream／漂流 floating／海图室 chart room／客舱 cabin／头等舱 first-class stateroom／二等舱 second-class stateroom／三等舱 steerage; third-class／货舱 cargo hold／行李舱 luggage hold／淡水舱 fresh-water tank／机舱 engine room／总吨位 gross tonnage／净吨位 net tonnage／排水量 displacement／满载排水量 load displacement／满载吃水线 load line
【船帮】①(船身的侧面) the side of a ship; shipboard ②(船队) merchant fleet
【船边交货】free alongside ship; free from alongside
【船边提货】alongside delivery; shipside delivery
【船舶】shipping; boats and ships ◇ ～登记证书 certificate of registry／～证书 ship's papers
【船埠】wharf; quay

【船舱】①(货舱) ship's hold ②(客舱) cabin
【船厂】shipyard; dockyard
【船到江心补漏迟】it's too late to plug the leak when the boat is in midstream; it will be too late to mend a boat when it has reached the middle of the river
【船到桥头自然直】You will cross the bridge when you get to it.
【船队】fleet; flotilla
【船方】[商] the ship: ～不负担装货费用 free in (F.I.) / ～不负担卸货费用 free out (F.O.) / ～不负担装、卸、理仓费用 free in and out and stowed (F.I.O.S.)
【船夫】[旧]boatman ◇ ～曲 boatmen's song
【船工】boatman; junkman
【船级】ship's classification; ship's class ◇ ～条款 classification clause / ～证书 classification certificate
【船级社】classification society: 劳氏～ Lloyd's Register of Shipping
【船籍港】port of registry; home port
【船家】[旧] boatman
【船壳】hull
【船老大】[方] ①(船工中的领头人) the chief crewman of a wooden boat ②(船工) boatman
【船篷】①(覆盖物) the mat or wooden roofing of a boat ②(帆) sail
【船票】steamer ticket
【船破又遇顶头风】meet unfavorable winds when the boat is broken
【船期】sailing date ◇ ～表 sailing schedule
【船桥】(ship's) bridge
【船蛆】[动] shipworm
【船上交货】free on board
【船身】hull (of a ship): ～险 hull insurance
【船首】stem; bow; prow ◇ ～楼 forecastle
【船台】(building) berth; shipway; slipway; slip: 干式～ dry shipway ◇ ～周期 berth period
【船体】the body of a ship; hull
【船桅】mast
【船尾】stern ◇ ～部 quarter / ～楼 poop / ～轴 stern shaft
【船位】ship's position: 测定～ fix a ship's position (at sea); position finding ◇ ～推算法 dead reckoning
【船坞】dock; shipyard: 浮～ floating dock / 干～ dry dock; graving dock ◇ ～费 dockage
【船舷】side (of a ship or boat)
【船用柴油机】marine diesel
【船用雷达】marine radar
【船用罗盘】mariner's compass
【船员】(ship's) crew; seaman; sailor
【船闸】(ship) lock
【船长】captain; skipper
【船只】shipping; vessels: ～失事 shipwreck / 往来～ shipping traffic
【船钟】ship's bell
【船主】shipowner

chuǎn

喘 ① breathe heavily; gasp for breath; pant ②[医] asthma
【喘气】①(呼吸) breathe (deeply); pant; gasp ②(紧张活动中的短时休息) take a breather
【喘息】①(喘气) pant; gasp or breath: ～未定 before regaining one's breath; before one has a chance to catch one's breath ②(紧张活动中的短时休息) breather; breathing spell; respite
【喘吁吁】puff and blow

chuàn

串 ①(连贯) string together: 把珠子～起来 string the beads together ②[量] string; bunch; cluster: 一～葡萄 a cluster of grapes / 一～钥匙 a bunch of keys / 一～珍珠 a string of pearls ③(错误地连接) get things mixed up: ～行 skip a line; miss a line / (收音机)～台 get two or more (radio) stations at once / 电话～线 get the (telephone) lines crossed ④(勾结) conspire; gang up; ～骗 gang up and swindle sb. ⑤(走动) go here and there; go from place to place; run about; rove: ～亲戚 go visiting one's relatives / 走村～寨 go from village to village⑥(担任戏曲角色) play a part (in a play); act: 客～ be a guest performer
【串并联】[电] series-parallel connection
【串供】act in collusion to make each other's confessions tally
【串话】[讯] cross talk
【串换】exchange; change; swap
【串讲】construe
【串联】① establish ties; contact ②[电] series connection ◇ ～电池组 series battery / ～电阻 series resistance
【串通】gang up; collaborate; collude with; be in collusion with
【串演】play the role of; act the role of
【串珠】a string of beads

chuāng

窗 window: 百叶～ shutters / 花格～ lattice window / 气～ transom / 纱～ screen win-

dow / 上下开关～ sash window
【窗玻璃】windowpane
【窗洞】an opening in a wall (to let in light and air)
【窗格子】window lattice
【窗钩】window catch
【窗户】window; casement
【窗花】paper-cut for window decoration
【窗口】①(窗户跟前) window: 坐在～ sit at the window; sit by the window ②(售票挂号等窗口) wicket; window
【窗框】window frame
【窗帘】(window) curtain: 拉开～ draw the window curtain apart / 挑花～ lace curtain
【窗明几净】with bright windows and clean tables; bright and clean
【窗纱】gauze for screening windows; window screening; window gauze
【窗扇】casement
【窗台】windowsill
【窗台板】window board
【窗子】window

疮 ①sore; skin ulcer: 恶～ malignant sore / 褥～ bedsore / 生～ grow a boil / 头上长～, 脚底流脓 —坏透了 with boils on the head and feet running with pus — rotten from head to foot; rotten to the core ②(外伤) wound: 刀～ a sword wound
【疮疤】①(疤) scar: 脸上的～ a scar on the face ②(痛处) sore: 揭人～ pull the scab right off sb.'s sore; touch sb.'s sore spot; tread on sb.'s corns
【疮痂】[医] scab
【疮口】the open part of a sore
【疮痍满目】everywhere a scene of devastation meets the eye; one sees suffering everywhere

创 wound: ～巨痛深 badly injured and in great pain; in deep distress / 予以重～ inflict heavy casualties (on the enemy) / 重～敌人 cause the enemy heavy losses
【创痕】scar
【创口】wound; cut
【创伤】wound; trauma: 精神上的～ a mental scar; a traumatic experience / 医治战争的～ heal the wounds of war

chuáng

幢
【幢幢】flickering; dancing: 人影～ shadows of people moving about

床 ①bed: 成对～ twin beds / 单人～ single bed / 帆布～ camp bed; cot / 双层～ double-deck bed; bunk bed / 双人～ double bed / 小孩～ child's cot; baby's crib / 行军～ camp bed / 折叠～ folding bed / 铺～ make the bed / 上～ go to bed / 卧病在～ be bed-ridden; take to one's bed; be laid up in bed ②(像床的器具、地面) sth. shaped like a bed: 车～ lathe / 河～ riverbed ③[量] 两～铺盖 two sets of bedding / 一～被 one quilt
【床单】sheet
【床第之言】private talks between husband and wife
【床垫】mattress: 弹簧～ spring mattress
【床架】bedstead
【床铺】bed
【床上用品】bedclothes
【床身】[机] lathe bed
【床头】the head of a bed; bedside ◇ ～灯 bedside lamp / ～柜 bedside cupboard / ～小桌 bedside-table; night table; nightstand
【床头箱】[机] headstock
【床位】berth; bunk; bed
【床罩】bedspread; counterpane ◇ ～布 sheetings

chuǎng

闯 ①(猛冲) rush; dash; charge: ～进来 rush in; break in; force one's way in / 横冲直～ charge about furiously; run amuck ②(闯练;突破) break through; temper oneself (by battling through difficulties and dangers): ～出一条新路来 break a new path / 敢～ dare to break through
【闯祸】get into trouble; bring disaster
【闯江湖】make a living wandering from place to place (as a fortune-teller, acrobat, quack doctor, etc.)
【闯将】daring general; pathbreaker
【闯劲】the spirit of a pathbreaker; pioneering spirit; daring spirit; dash
【闯练】leave home to temper oneself; be tempered in the world
【闯入】burst into; intrude: ～会场 intrude oneself into a meeting

chuàng

怆 [书] sorrowful: ～然泪下 burst into sorrowful tears

创 start (doing sth.); achieve (sth. for the first time); initiate; establish; create: ～高产 achieve higher output / ～纪录 establish 〈make〉 a new record; set a record / ～奇迹 create miracles; work miracles

【创办】establish; set up; found: ～一所学校 establish a school; set up a school / ～一所医院 found a hospital

【创见】original idea; creative idea; brand-new idea; creative viewpoint

【创建】found; establish; set up

【创举】pioneering work; pioneering undertaking: 伟大～ a great beginning

【创刊】start publication ◇ ～号 first number; initial issue; first issue / ～词 inaugural statement on the first issue of a periodical

【创立】found; originate: ～新学派 found a new academic school

【创设】found; create; set up

【创始】originate; initiate ◇ ～人 founder; originator

【创新】bring forth new ideas; blaze new trails; make innovations

【创新霉素】creatmycin

【创业】start an undertaking; do pioneering work; found an undertaking: ～容易守业难。It's easy to open a shop but hard to keep it open.

【创造】create; produce; bring about: ～奇迹 create miracles; work wonders; achieve prodigious feats / ～社会财富 create the wealth of society / ～优异成绩 produce excellent results / ～有利条件 create favorable conditions ◇ ～力 creative power; creative ability / ～性 creativeness; creativity

【创制】formulate; institute; create: ～拼音文字 formulate an alphabetic system of writing

【创作】①(创造作品) create; produce; write: ～美术作品 produce works of art ②(作品) creative work; creation: 划时代的～ epoch-making creative work / 文艺～ literary and artistic creation ◇ ～技巧 artistic technique; craftsmanship / ～经验 creative experience / ～思想 ideas guiding artistic or literary creation

chuī

炊 cook a meal

【炊具】cooking utensils

【炊事】cooking; kitchen work ◇ ～用具 cooking utensils; appliances for food preparation / ～员 a cook or the kitchen staff

【炊事班】cookhouse squad; kitchen squad

【炊烟】smoke from kitchen chimneys: ～袅袅 smoke spiraling from kitchens / ～四起 cooking smoke all round

【炊帚】a brush for cleaning pots and pans; pot-scouring brush

吹 ①(吹气;刮) blow; puff: ～哨子 blow a whistle / ～一口气 give a puff / 把灯～灭 blow out the lamp / 雨打风～ be exposed to the weather ②(吹奏) play (wind instruments): ～长笛 play the flute ③(夸口) boast; brag: ～得天花乱坠 boast in the most fantastic terms / 自～自擂 blow one's own trumpet ④(破裂;不成功) break off; break up; fall through: 他们俩～了。That couple have broken up.

【吹吹打打】beating drums and blowing trumpets; piping and drumming

【吹吹拍拍】boasting and toadying; flattery and touting: ～,拉拉扯扯 resort to boasting, flattery and touting

【吹打乐】[乐] an ensemble of Chinese wind and percussion instruments

【吹风】①(被风吹) be in a draught; catch a chill ②(理发吹风) dry one's hair; dry hair with a blower ③(透露消息) let sb. in on sth. in advance; give a cue ◇ ～会 briefing; background briefing / ～机 blower (for drying hair); hair drier

【吹拂】sway; stir

【吹鼓手】①(吹奏乐器的人) trumpeter; bugler ②(鼓吹者;吹捧者) eulogist

【吹管】[机] blowpipe: 氢氧～ oxyhydrogen blowpipe / 氧乙炔～ oxyacetylene blowpipe

【吹胡子瞪眼】froth at the mouth and glare with rage; foam with rage; snort and stare in anger

【吹号】blow a bugler; blare the call: 吹冲锋号 blare the call to charge

【吹灰之力】the effort needed to blow away a speck of dust; just a small effort: 不费～ as easy as blowing away dust; without the least effort

【吹口】[乐] mouthpiece

【吹冷风】blow a cold wind over; throw cold water on

【吹炼】[冶] blowing

【吹毛求疵】find fault; pick holes; nitpick; cavil at

【吹牛】boast; brag; talk big: ～拍马 boast and flatter / ～的人 boaster; braggart

【吹捧】flatter; laud to the skies; lavish praise on; extol: 把某人～得上了天 extol sb. to the skies / 互相～ flatter each other
【吹塑】blowing; blow molding
【吹嘘】lavish praise on oneself or others; boast of; boast about: 自我～ self-praise
【吹氧】[冶] oxygen blast: ～炼钢 oxygen furnace steel / ～转炉 oxygen-blown converter
【吹制玻璃】blow-molded glass
【吹奏】play (wind instruments) ◇ ～乐 band music; wind music

chuí

槌 mallet; beetle: 鼓～儿 drumstick / 碾～ pestle
【槌球】[体] croquet
垂 ①(下垂) hang down; droop; let fall: ～发 have one's hair hang down / ～首 hang the head ②[敬](长辈或上级对自己的行动) condescend: ～念 show kind concern for (me) / ～询 condescend to inquire ③(流传) go down; hand down; bequeath to posterity: 功～竹帛 be recorded in history in letters of gold / 名～青史 one's name will go down in history; leave a name in history ④(将近) be on the verge of; nearing; approaching: ～老 approaching old age; getting on in years / ～暮 getting dark; drawing near sunset
【垂钓】fish with a hook and line; go angling
【垂泪】shed tears; weep: ～满面 falling tears cover the face
【垂帘听政】hold court from behind a screen; attend to state affairs
【垂柳】[植] weeping willow
【垂暮】[书] dusk; towards sunset; just before sundown: ～之年 in old age
【垂盆草】[中药] stringy stonecrop
【垂青】[书] show appreciation for sb.; look upon sb. with favor
【垂手可得】easy to obtain or get
【垂死】moribund; dying: ～挣扎 be in one's death throes; put up a last-ditch 〈deathbed〉 struggle
【垂体】[生理] hypophysis; pituitary body; pituitary gland ◇ ～后叶 posterior pituitary / ～后叶素[药] pituitrin
【垂头丧气】crestfallen; dejected; downcast; in low spirits
【垂危】critically ill; at one's last gasp
【垂涎】drool; slaver; covet: ～三尺 spittle three feet long; drool with envy; mouth watering three feet down the lips / ～欲滴 make one's mouth watery
【垂线】[数] vertical line; perpendicular line
【垂直】perpendicular; vertical: ～发射 vertical firing; vertical launching / ～俯冲 steep dive; nose dive / ～平面 vertical plane / ～起飞 vertical takeoff ◇ ～贸易 vertical trade / ～起降飞机 vertical takeoff and landing aircraft; VTOL aircraft; vertaplane / ～起降喷气式飞机 vertijet / ～天线 vertical antenna
捶 beat (with a stick or fist); thump; pound: ～背 pound sb.'s back (as in message) / ～鼓 beat a drum / ～门 bang on the door
【捶打】beat; thump: ～衣服 beat clothes (when washing them)
【捶胸顿足】thump one's chest and stamp one's feet; beat one's breast and stamp one's feet (in deep sorrow, etc.)
锤 ①(锤子) hammer: 大～ sledge-hammer / 铁～ iron hammer ②(锤打) hammer into shape; knock with a hammer: ～金箔 hammer gold into foil ③(像锤的东西) weight: 秤～ steelyard weight / 平衡～[机] balance weight / 调节～[机] governor weight
【锤柄】hammer handle
【锤骨】[生理] malleus
【锤炼】①(磨炼) temper; steel and temper ②(刻苦钻研,反复琢磨) refine; polish: ～字句 refine upon the wording and phrasing
【锤式打桩机】monkey driver
【锤子】hammer
陲 frontiers; borders: 边～ frontier; border area

chūn

春 ①(春季) spring: 温暖如～ as warm as spring ②(男女情欲) love; lust ③(生机) life; vitality
【春播】spring sowing ◇ ～作物 spring-sown crops
【春分】the Spring Equinox (4th solar term) ◇ ～点 first point of Aries; spring equinox; March equinox
【春风】spring breeze: 满面～ (a face) beaming with smiles
【春风得意】ride on the crest to success
【春风化雨】life-giving spring breeze and rain — salutary influence of education; the kindly in-

fluence of a good teacher
【春风满面】beaming with satisfaction; radiant with happiness; smile broadly
【春风一度】a sexual intercourse
【春耕】spring ploughing
【春宫】pornography: ～画 pornographic pictures
【春灌】[农] spring irrigation
【春光】sights and sounds of spring; spring scenery: ～明媚 a sunlit and enchanting scene of spring; with spring in all its brightness and charm
【春寒】spring chill
【春花秋实】the progression of seasons
【春华秋实】glorious flowers in spring and solid fruits in autumn
【春化】[农] vernalization
【春回大地】Spring returns to the good earth.
【春季】spring; springtime
【春假】spring vacation; spring holidays
【春节】the Spring Festival
【春困秋乏】One feels dizziness in spring and fatigue in autumn.
【春雷】spring thunder
【春雷霉素】[药] kasugamycin
【春联】Spring Festival couplets
【春令】①(春季) spring ②(春季的气候) spring weather: 冬行～ a springlike winter; a very mild winter
【春梦】spring dream; transient joy
【春眠不觉晓】In spring, one sleeps and wakes up to find it is already day.
【春暖花开】During the warmth of spring all the flowers bloom.
【春情】stirrings of love
【春秋】①(春秋两季;指整个一年) spring and autumn; year: ～多佳日 There are many fine days in spring and autumn ②(指人的年岁) age: ～已高 be advanced in years / ～正富 in the prime of youth ③(编年体的史书) annals; history
【春日霉素】kasugamycin
【春色】spring scenery: ～满园 a garden full of the beauty of spring; a garden overflowing with the beauties of springtime / 水乡～ spring in a waterside village
【春笋】bamboo shoots in spring: 雨后～ springing up like mushrooms
【春天】spring; springtime
【春小麦】spring wheat
【春心】desire for love; longing for love; thoughts of love
【春心荡漾】the surging of lustful desire
【春汛】①[水] spring flood; spring freshet ②[渔] spring (fishing) season
【春药】aphrodisiac
【春意】①(春天的气象) spring in the air; the beginning of spring; the awakening of spring: ～盎然。Spring is very much in the air. ②(春心) thoughts of love
【春游】spring outing
【春雨贵如油】Rain during springtime is as precious as oil.
【春装】spring clothing

椿 [植] ①(香椿树) Chinese toon; mahogany ②(臭椿树) tree of heaven
【椿象】[动] stinkbug; shieldbug

蝽 [动] stinkbug; shieldbug

chún

淳 [书] pure; honest
【淳厚】pure and honest; simple and kind
【淳朴】honest; simple; unsophisticated

醇 ①[书](优质烧酒) mellow wine; good wine ②[书](纯粹) pure; unmixed ③[化] alcohol: ～醛 alcohol aldehyde
【醇和】pure and mild
【醇厚】①(气味;滋味) mellow; rich: 酒味～。The wine is 〈tastes〉 mellow. ②(淳厚) pure and honest; simple and kind
【醇化】①(更纯粹;更美满) refine; purify; perfect ②[化] alcoholization ◇ ～物 alcoholate
【醇解】[化] alcoholysis
【醇酸】[化] alcohol acid; alcoholic acid ◇ ～树脂 alkyd resin
【醇中毒】alcoholism; alcoholic poisoning

鹑 quail

唇 lip: 上～ upper lip / 下～ lower lip
【唇齿相依】be as close as lips and teeth; be closely related and mutually dependent; close interdependence
【唇齿音】[语] labiodental (sound); dentilabial
【唇齿之邦】as states of lips and teeth
【唇膏】lipstick
【唇角】labial angle
【唇裂】[医] harelip; cleft lip
【唇枪舌剑】cross verbal swords; engage in a battle of words; heated verbal exchange

(debate)
【唇舌】words; argument; talking round; persuasion: 费一番～ take a lot of explaining or arguing / 徒费～ a waste of breath
【唇亡齿寒】if the lips are gone, the teeth will be cold; if one (of two interdependent things) falls, the other is in danger; share a common lot
【唇音】[语] labial (sound)

纯 ①(纯净) pure; unmixed: ～白 pure white / ～黑 all black / ～金 pure gold; solid gold / ～毛 pure wool ②(纯粹;单纯) simple; pure and simple: ～属捏造 sheer fabrication ③(纯熟) skillful; practiced; well versed: 功夫不～ not skillful enough
【纯粹】①(成份纯) pure; unadulterated ②(单纯地) solely; purely; only
【纯度】purity; pureness
【纯化】purification
【纯碱】[化] soda ash; sodium carbonate
【纯洁】①(纯粹清白) pure; clean and honest; chaste: ～无私 clean and unselfish ②(使纯洁) purify: ～党的组织 purify the Party organization / ～性 purity
【纯净】pure; clean
【纯利】net profit
【纯朴】honest; simple; unsophisticated: ～敦厚 simple and honest / ～爽朗 honest and frank / 文风～ simplicity of style
【纯色】pure color
【纯收入】net income
【纯熟】skillful; practiced; well versed: 技术～ highly skilled
【纯损】net loss
【纯一】single; simple: 目标～ singleness of purpose
【纯音】[物] pure tone; simple tone
【纯真】pure; sincere: ～无邪 pure and innocent
【纯正】pure; unadulterated: 动机～ have pure motives
【纯种】purebred: ～牛 purebred cattle; pedigree cattle

莼
【莼菜】[植] water shield

chǔn

蠢 ①(愚蠢) stupid; foolish; dull ②(蠢笨;笨拙) clumsy
【蠢笨】clumsy; awkward; stupid
【蠢材】idiot; fool; dumbbell; blockhead
【蠢蠢欲动】ready to start wriggling; ready to make trouble; be restless and about to make trouble
【蠢动】①(虫子爬动) wriggle; squirm ②(骚动;开始敌对行动) create disturbances; carry on disruptive activities; stir up trouble
【蠢货】blockhead; dunce; idiot
【蠢驴】idiot; donkey; ass
【蠢人】fool; blockhead
【蠢事】folly: 干～ commit a folly; play the fool
【蠢猪】idiot; stupid swine; ass

chuō

戳 ①(戳破) jab; poke; stab: 一～就破 break at the slightest touch / 在纸上～了一个洞 poke a hole in the paper ②(因猛戳而损坏或受伤) sprain; blunt: 打排球～了手 sprain one's wrist while playing volleyball / 钢笔尖儿～了 The nib is blunted ③(图章) stamp; seal; chop: 盖～(私章) put one's personal chop on; (公章) affix an official seal
【戳穿】①(刺穿) pierce through; puncture ②(揭穿) lay bare; expose; explode: ～谣言和诡辩 lay bare sb.'s lies and sophistry
【戳记】stamp; seal
【戳子】stamp; seal: 橡皮～ rubber stamp / 在文件上盖个～ put a seal on a document

chuò

辍 [书]stop; cease: ～工 stop work / 时作时～ on and off; by fits and starts
【辍笔】stop in the middle of writing or painting
【辍学】discontinue one's studies; leave off one's study

啜 [书]①(喝) sip; suck: ～茗 sip tea ②(抽噎的样子) sob
【啜泣】sob

绰 [书] ample; spacious: ～有余裕 enough and to spare
【绰绰有余】more than sufficient; enough and to spare; more than enough
【绰号】nickname
【绰约】[书] (of a woman) graceful: ～多姿 graceful and attractive

cī

疵 flaw; defect; blemish: 小～ a trifling defect / 无～ flawless; impeccable
【疵点】flaw; fault; defect; blemish

Cí

茨 [书] ①(用茅草盖屋) thatch (a roof) ②(蒺藜)[植] puncture vine
【茨菰】arrowhead

瓷 porcelain; china: ～碗 china bowl / 裂变～ crackled porcelain / 青～ celadon / 青花～ blue and white porcelain / 细～ fine china
【瓷雕】[工美] porcelain carving
【瓷漆】enamel paint; enamel
【瓷器】porcelain; chinaware: 薄～ eggshell china / 细～ fine china
【瓷实】solid; firm; substantial
【瓷土】porcelain clay; china clay
【瓷釉】porcelain glaze
【瓷砖】ceramic tile; glazed tile

糍
【糍粑】cooked glutinous rice pounded into paste; glutinous rice cake

慈 ①(和善) kind; loving: 心～ tenderhearted; kindhearted ②(母亲)[书] mother: 家～ my mother
【慈爱】love; affection; kindness
【慈悲】mercy; benevolence; pity: ～好善 merciful and benevolent / 发～ have pity; be merciful
【慈姑】[植] arrowhead
【慈航】[佛教] merciful ferry; way of salvation
【慈航普渡】The barge of mercy ferries all the miserable people to the world of bliss; salvation through charity to others
【慈和】kindly and amiable
【慈眉善目】a benignant look
【慈母】loving mother; mother: ～严父 a kind mother and a severe father
【慈善】charitable; benevolent; philanthropic ◇ ～机关 charitable institution; charitable organization / ～家 philanthropist / ～事业 charities; good works; philanthropy / ～为怀 cherish charity
【慈祥】kindly: ～的面容 a kindly face / ～的微笑 kingly smile; loving smile
【慈幼养老】be kind to the young and care for the old

磁 ①[物] magnetism: 地～ terrestrial magnetism / 起～ magnetization ②(瓷) porcelain; china
【磁棒】bar magnet
【磁暴】[物] magnetic storm ◇ ～记录器 magnetic storm monitor
【磁北】the magnetic north
【磁场】[物] magnetic field ◇ ～强度 magnetic field intensity
【磁带】(magnetic) tape ◇ ～盒 cassette / ～录音机 tape recorder / ～录像机 video tape recorder / ～盘 tape reel
【磁带机】[自] magnetic tape station
【磁动势】magnetomotive force
【磁感应】[物] magnetic induction
【磁钢】magnet steel
【磁鼓】magnetic drum; drum
【磁化】[物] magnetization ◇ ～率 magnetic susceptibility / ～器 magnetizer
【磁极】[物] magnetic pole ◇ ～强度 magnetic pole strength ○北极 north pole / 南极 south pole / 同性极 like poles / 异性极 unlike poles
【磁力】[物] magnetic force ◇ ～测定 magnetometry / ～勘探 magnetic prospecting / ～探矿仪 magnetic detector (for ore deposits) / ～探伤器 magnetic flaw detector; magnetic fault finder / ～线 magnetic line of force / ～选矿 magnetic dressing / ～仪 magnetometer
【磁控管】magnetron
【磁流体】[物] magnetic fluid ◇ ～力学 magnetofluid dynamics
【磁盘】disc; disk; magnetic disc ◇ ～存储器 magnetic disc store / ～盒 disk cartridge
【磁偏角】[物] magnetic declination
【磁气】magnetism: 地～ terrestrial magnetism / 动物～ animal magnetism
【磁倾角】magnetic dip; inclination
【磁倾仪】dip circle
【磁石】①[矿] magnetite ②[电] magnet: ～电话机 local battery apparatus / ～发电机 magneto / ～检波器 magneto detector
【磁体】[物] magnetic body; magnet: 反～ diamagnetic body / 顺～ paramagnetic body / 铁～ ferromagnetic body
【磁铁】[物] magnet; ferromagnet: 电～ electromagnet / 马蹄形～ horseshoe-magnet / 人造～ artificial magnet / 天然～ natural magnet / 永久～ permanent magnet ◇ ～矿 magnetite
【磁头】magnetic head
【磁通量】[物] magnetic flux
【磁效应】magnetic effect: 电流～ magnetic effect of electric current
【磁心】magnetic core ◇ ～储存器 magnetic core memory

【磁 性】[物] magnetism; magnetic: 抗～ diamagnetism / 顺～ paramagnetism / 铁～ ferromagnetism ◇ ～水雷 magnetic mine / ～炸弹 magnetic bomb
【磁悬浮】magnetic suspension
【磁 选】[矿] magnetic separation: 湿法～ wet magnetic separation ◇ ～厂 magnetic ore dressing plant / ～法 magnetic method; magnetic process / ～机 magnetic separator; magnetic cobber
【磁针】magnetic needle: ～方位 magnetic bearings / ～偏差 magnetic declination
【磁子】[物] magneton

雌 female
【雌蜂】queen bee
【雌花】[植] female flower; pistillate flower
【雌 黄】①[矿] orpiment ②(胡说八道) 信口～ make irresponsible remarks; wag one's tongue too freely
【雌激素】estrogen; estrogenic hormone
【雌鸟】jenny
【雌禽】hen
【雌蕊】[植] pistil
【雌兽】jenny
【雌酮】estrone
【雌性】female
【雌雄】①(雌性和雄性) male and female ②(胜和负) victory and defeat: 决一～ have a showdown; see who's master
【雌雄同体】[动] her-maphroditism; monoecism
【雌雄同株】[植] monoecism
【雌雄异体】[动] gonochorism; dioecism
【雌雄异株】[植] dioecism

辞 ①(优美的语言) diction; phraseology: 修～ rhetoric ②(古典文体) songs; a type of classical Chinese literature ③(一种古体诗) ballad; a form of classical poetry ④(告别) take leave: 不～而别 leave without saying good-bye / 告～ take one's leave ⑤(辞退) dismiss; discharge ⑥(辞职) resign ⑦(躲避;推托) decline; shirk; evade: 不～艰辛 no shirk any hardships / 不～劳苦 spare no effort; take pains
【辞 别】bid farewell; say good-bye; take one's leave
【辞 呈】(written) resignation: 提出～ submit one's resignation; hand in one's resignation
【辞典】dictionary
【辞工跳槽】leave work and go to another
【辞 令】language appropriate to the occasion: 善于～ gifted with a silver tongue / 外交～ diplomatic language
【辞让】politely decline
【辞书】dictionary; lexicographical work
【辞 岁】bid farewell to the outgoing year; celebrate the lunar New Year's Eve
【辞吐雅致】a refined conversation
【辞退】dismiss; discharge
【辞谢】politely decline; decline with thanks
【辞 行】say good-bye (to one's friends, etc.) before setting out on a journey
【辞 藻】flowery language; rhetoric; ornate diction: 堆砌～ string together ornate phrases
【辞章】①(韵文和诗文总称) poetry and prose; prose and verse ②(修辞) art of writing; rhetoric
【辞职】resign; hand in one's resignation

词 ①[语] word; term: 贬义～ derogatory term / 褒义～ commendatory term / 反义～ antonym / 同义～ synonym ②(讲话) speech; statement: 欢迎～ speech of welcome / 开幕～ opening speech / 台～ lines of an opera or play / 各执一～ each holds to his own statement ③(一种词体) *ci*, a Chinese poetic genre
【词不达意】the words fail to convey the idea; the language fails to express the meaning
【词典】dictionary: 简明～ concise dictionary / 双语～ bilingual dictionary / 袖珍～ pocket dictionary / 查～ consult the dictionary; look up a word in the dictionary ◇ ～学 lexicography
【词法】[语] morphology
【词干】[语] stem
【词根】[语] root
【词汇】[语] vocabulary; words and phrases: ～常用 common words / 科技～ scientific and technical terms ◇ ～表 word list; vocabulary; glossary / ～统计学 lexicostatistics / ～学 lexicology
【词句】words and phrases; expressions: 空洞的～ empty phrases
【词类】[语] parts of speech
【词穷】arguments exhausted; nothing more to say
【词讼】legal cases
【词 素】[语] morpheme: 粘着～ bound morpheme
【词条】entry (in a dictionary)
【词头】[语] prefix
【词尾】[语] suffix

【词形】[语] morphology ◇ ～变化 morphological changes; inflections
【词性】[语] syntactical functions and morphological features that help to determine a part of speech
【词序】[语] word order
【词义】[语] the meaning of a word; sense of a word
【词语】words and expressions; terms
【词源】[语] the origin of a word; etymology
【词缀】[语] affix
【词藻】ornate terms
【词组】[语] word group; phrase

祠 ancestral temple: 宗～ clan hall
【祠堂】ancestral hall; ancestral temple; memorial temple

cǐ

此 ①(这) this: ～处 this place; here / ～等 this kind; such as these ②(此时;此地) now; here: 从～以后 from now on; henceforward / 由～往南 go south from here
【此岸】[佛教] this shore; temporality
【此岸性】[哲] this sideness
【此地】this place; here: ～人 local people
【此地无银三百两】a guilty person gives himself away by conspicuously protesting his innocence; protest one's innocence too much; No 300 taels of silver buried here.
【此后】after this hereafter; henceforth
【此呼彼应】take concerted action
【此间】around here; here: ～已有传闻。It has been so rumored here.
【此刻】this moment; now; at present
【此路不通】dead end; blind alley: ～!(路牌) No thoroughfare. (或: Not a Through Road!)
【此起彼伏】as one falls, another rises; rise one after another; rise here and subside there
【此时】this moment; right now: ～此地 here and now / ～此刻 at this very moment
【此外】besides; in addition; moreover
【此一时,彼一时】this is one situation and that was another; circumstances have changed with the passage of time; things are now different from what they were

cì

次 ①(次序;等第) order; sequence: 车～ train number / 席～ seating arrangement / 依～ in due order; in succession; one by one ②(第二) second; next: ～日 next day / ～子 second son ③(质量较差) second-rate: ～棉 poor quality cotton / 真～ really no good; terrible ③[化] hypo-: ～氯酸 hypochlorous acid ④[量]二十一～列车 No. 21 train / 三～ three times / 首～ first time; first / 进行几～会谈 hold several talks
【次大陆】subcontinent
【次等】second-class; second-rate; inferior
【次第】①(次序) order; sequence ②(一个挨一个地) one after another: ～入座 take seats one after another
【次货】inferior goods; substandard goods
【次级线圈】[电] secondary coil
【次级帐户】secondary account
【次品】substandard products; defective goods
【次轻量级】[举重] featherweight
【次生】secondary: ～海岸 secondary coast / ～矿床 secondary deposit
【次声】[物] infrasonic sound; infrasound
【次数】number of times; frequency
【次序】order; sequence: ～颠倒 not in the right order / 按字母～排列 be arranged in alphabetical order
【次要】less important; secondary; subordinate; minor: ～矛盾 secondary contradiction / ～收益 secondary income / ～问题 secondary questions
【次之】take second place
【次中音号】[乐] tenor horn
【次重量级】[举重] middle heavyweight
【次最轻量级】[举重] flyweight

刺 ①(刺儿) thorn; splinter ②(扎) sting; stab; prick: ～伤 stab and wound ③(暗杀) assassinate: 遇～ be assassinated ④(刺激) irritate; stimulate: ～鼻 irritate the nose ⑤(侦探;打听) detect; spy ⑥(讽刺) criticize: 讽～ satirize
【刺柏】Chinese juniper; Taiwan juniper
【刺鼻】pungent; irritate the nose; assail one's nostrils
【刺刺不休】talk incessantly; chatter on and on
【刺刀】bayonet: 拼～ bayonet-fighting / 上～! (口令) Fix bayonets! / 下～! (口令) Unfix bayonets!
【刺耳】grating on the ear; jarring; ear-piercing; harsh: ～的话 harsh words; sarcastic remarks / ～的叫声 piercing cry
【刺钢丝】(铁丝网) barbed wire
【刺骨】piercing to the bones; piercing; biting: 寒风～。The cold wind chills one to the bone.

【刺槐】[植] locust (tree)
【刺激】①(推动事物起积极变化) stimulate; excite: ～神经 excite a nerve / ～生产 stimulate production / 强～ strong stimulus ②(使激动) provoke; irritate; upset ◇～素 stimulin / ～物 stimulus; stimulant / ～性毒剂 irritant agent
【刺客】[旧] assassin
【刺杀】①(暗杀) assassinate ②[军] bayonet charge: 练～ practice bayonet fighting
【刺伤】stab and wound: 被～ be wounded by stabbing
【刺史】feudal provincial; prefectural governor
【刺探】make roundabout or secret inquiries; pry; spy: ～军情 spy out military secrets; gather military intelligence
【刺网】[渔] gill net: 三层～ trammel net
【刺猬】hedgehog
【刺绣】①(指工艺) embroider ②(指产品) embroidery ◇～品 embroidery
【刺眼】dazzling; offending to the eye: 打扮得～ be loudly dressed / 亮得～ dazzlingly bright

赐

grant; favor; gift: 赏～ grant a reward; bestow a reward
【赐福】blessing
【赐复】kindly favor us with a reply
【赐教】[敬] condescend to teach; grant instruction: 不吝～ please favor me with your instructions; be so kind as to give me a reply
【赐予】grant; bestow

伺

【伺候】wait upon; serve: 难～ hard to please; fastidious

cōng

聪

①(听觉)[书] faculty of hearing: 左耳失～ become deaf in the left ear ②(听觉灵敏) acute hearing: 耳～目明 able to see and hear clearly
【聪慧】bright; intelligent
【聪敏好学】be intelligent and fond of study
【聪明】intelligent; bright; clever: ～才智 intelligence and wisdom / ～活泼 wise and active / ～能干 bright and capable / ～反被～误。Clever people may be victims of their own cleverness; Cleverness may overreach itself. (或: A wise man can be ruined by his own wisdom.) / ～一世,糊涂一时。 clever all one's life but stupid this once; smart as a rule, but this time a fool; A life time of cleverness can be interrupted by moments of stupidity.
【聪颖】intelligent; bright; clever

匆

hastily; hurriedly
【匆匆】hurriedly; in a rush; in haste: 来去～ come and go in haste / 行色～ be in a rush getting ready for a journey; be pressed for time on a journey
【匆促】hastily; in a hurry: ～起程 set out hastily / 时间～ be pressed for time
【匆忙】hastily; in a hurry; in haste

葱

①[植] onion; scallion: 大～ scallion; green Chinese onion / 小～ shallot / 洋～ onion ②(青色) green
【葱白】very light blue
【葱白儿】scallion stalk
【葱翠】fresh green; luxuriantly green: ～的竹林 a green bamboo grove
【葱花】chopped green onion ◇～饼 green onion pancake
【葱茏】verdant; luxuriantly green: ～茂盛 luxuriant growth of vegetation / 草木～ luxuriant vegetation
【葱绿】pale yellowish green; light green; verdant
【葱头】onion
【葱郁】verdant; luxuriantly green: ～的松树林 a verdant pine wood

从

【从容】①(不慌不忙;镇静) calm; unhurried; leisurely: ～不迫 calm and unhurried / ～考虑 consider in an unhurried manner; think over deliberately ②(宽裕) plentiful; sufficient; enough: 时间很～ have enough time to spare; there is plenty of time / 手头不～ be hard for money; be out of cash

枞

[植] fir

cóng

淙

【淙淙】murmuring; gurgling: 流水～ a gurgling stream

从

①(跟随) follow: ～师 be an apprentice / ～俗 follow the general custom / ～征 go on a military expedition ②(顺从;听从) comply with; obey: ～命 comply with sb.'s wish; obey an order ③(从事;参加) join; be engaged in: ～军 join the army; enlist ④(采取某种方针或态度) in a certain manner or according to a certain principle: ～宽处理 be lenient in treatment / ～速 be as quickly as possible / ～严处理 be severe in treatment ⑤(跟随的人) follower; attendant: ～者如云 have a large following / 随～ attendant; retainer; retinue ⑥(从属的;次要的) secondary; accessary: 分别首～ distinguish between the chief culprit and the accessary / 主～ the primary and the secondary ⑦[介](表示起于或经过) from; through: ～长远看 from a long-term point of view; in the long run / ～根本上说 essentially; in essence / ～全局出发 proceed from the situation as a whole / ～实际出发 proceed from the actual situation / ～现在起 from now on / ～这儿往西 go west from here; west of here ⑧[副](用在否定词前面) ever: ～不迟到 be never late / ～未听说过 have never heard of

【从长计议】give the matter further thought and discuss it later; need further consideration

【从此】from this time on; from now on; from then on; henceforth; thereupon

【从…到…】from... to...: 从古到今 from ancient times to the present; from time immemorial / 从上到下 from top to bottom; from the higher levels to the grass roots / 从生到死 from the cradle to the grave / 从头到尾 from beginning to end; from first to last / 从无到有 grow out of nothing / 从小到大 from small to large; expand from a small to a large force / 从早到晚 from dawn to dusk; from morning till night / 从左到右 from left to right

【从动】[机] driven: ～齿轮 driven gear / ～构件 driven member

【从而】[连] thus; thereby

【从犯】accessary criminal; accessary

【从价税】[经] *ad valorem* duties

【从简】conform to the principle of simplicity: 一切～ dispense with all unnecessary formalities

【从井救人】risk one's life or compromise one's own interest without doing others any good; try to do a good deed in the wrong way

【从句】[语] subordinate clause

【从军】join the army; enlist

【从来】always; at all times; all along

【从良成家】reform and settle down

【从量税】[经] specific duties

【从略】be omitted: 此处引文～。 The quotation is omitted here.

【从前】before; formerly; in the past

【从权】as a matter of expediency: ～处理 do what is expedient

【从戎】[书] join the army; enlist: 投笔～ throw aside the writing brush and join the army; renounce the pen for the sword

【从善如流】follow good advice as naturally as a river follows its course; readily accept good advice

【从事】①(投身于) go in for; be engaged in: ～各种活动 be engaged in various activities / ～科研工作 be engaged in scientific research / ～文学创作 take up writing as a profession; be engaged in literary work ②(处理) deal with: 军法～ deal with according to military law; court-martial / 慎重～ act cautiously; steer a cautious course

【从属】subordinate: ～地位 subordinate status

【从速】as soon as possible; without delay: ～处理 deal with the matter as soon as possible; settle the matter quickly

【从头】①(从最初) from the beginning: ～儿做起 start from the very beginning ②(重新) anew; once again: ～儿再来 start afresh; start all over again

【从小】from childhood; as a child

【从刑】[法] accessary punishment

【从中】out of; from among; therefrom: ～调解 mediate between two sides; mediate from among them / ～吸取教训 draw lesson from it / ～渔利 profit from; cash in on

丛 ①(聚集) crowd together: ～生 grow thickly ②(生长在一起的草木) clump; thicket; grove: 草～ a patch of grass / 树～ a clump of trees; grove ③(聚集在一起的人或物) crowd; collection: 论～ a collection of essays; collected essays / 人～ a crowd of people

【丛集】crowd together; pile up: 百感～ all sorts of feelings welling up / 债务～ debts piling up

【丛刊】a series of books; collection

【丛林】jungle; forest ◇ ～热[医] jungle fever / ～战 jungle warfare

【丛生】①(草木聚集在一起生长) (of plants) grow thickly: 荆棘～ be overgrown with brambles ②(疾病等同时发生) (of diseases, evils, etc.) break out: 百病～ all kinds of diseases and ailments breaking out

【丛书】a series of books; collection: 自学～ self-study series

【丛杂】motley

còu

凑 ①(聚集) gather together; pool; collect: ～

钱 pool money／～情况 pool information／～整数 make up a round number; make up an even amount／～足人数 gather together enough people; get a quorum ②(碰；赶；趁) happen by chance; take advantage of: 正～上是雨天。 It happened to be a rainy day. ③(接近) move close to; come close to; press near: ～近点儿。 Come closer, please.
【凑份子】club together (to present a gift to sb.)
【凑合】①(聚集) gather together; collect; assemble ②(拼凑) improvise ③(将就) make do (with) ④(还可以的) passable; not too bad
【凑集】gather together
【凑巧】luckily; fortunately; as luck would have it: 我～在家。 I happened to be at home.
【凑趣儿】①(迎合别人兴趣) join in (a game, etc.) just to please others ②(逗笑取乐) make a joke about; poke fun at
【凑热闹儿】①(和大家一起玩儿) join in the fun ②(添麻烦) add trouble to
【凑手】at hand; within easy reach: 钱不～ have no money at hand
【凑数】make up the number or amount; serve as a stopgap

cū

粗 ①(粗大的) wide (in diameter); thick: ～绳 a thick rope／～树干 thick trunk ②(粗糙的) coarse; crude; rough: ～布 coarse cloth／～黑的手 rough, work-soiled hands／～沙 coarse sand; grit／～盐 crude salt ③(声音大而低) gruff; husky: ～嗓子 a husky voice／～声大气 a deep, gruff voice ④(疏忽；不周密) careless; negligent: ～中有细 usually careless, but quite sharp at times; crude in most matters, but subtle in some ⑤(鲁莽；粗野) rude; unrefined; vulgar: 说话很～ speak rudely; use coarse language ⑥(略微) roughly; slightly: ～具规模 be roughly in shape／～通文字 can read and write a little／～知一二 have a rough idea; know a little
【粗暴】rude; rough; crude; brutal: ～态度 a rude attitude／～行为 crude behavior
【粗笨】①(笨拙) clumsy; awkward: 动作～ clumsy／手脚～ clumsy in doing things ②(笨重) unwieldy; heavy; bulky; cumbersome: ～的家具 unwieldy furniture
【粗鄙】vulgar; coarse: 言语～ vulgar in speech
【粗布】crash; coarse cloth
【粗糙】coarse; rough; crude: ～的译文 rough translation／皮肤～ rough skin／手工～ crudely make; of poor workmanship
【粗茶淡饭】plain tea and simple food; homely fare; simple food
【粗柴油】gas oil
【粗大】①(又粗又大) thick; bulky: ～的手 big strong hands ②(声音) loud: ～的鼾声 thunderous snoring／～的嗓门 a loud voice
【粗帆布】canvas
【粗放】[农] extensive: ～耕作 extensive cultivation
【粗犷】①(粗野；粗鲁) rough; rude; boorish ②(粗豪；豪放) straightforward and uninhibited; bold and unconstrained; rugged: ～的性格 of straightforward, unsophisticated character
【粗豪】forthright; straightforward
【粗花呢】[纺] tweed
【粗话】vulgar language
【粗活】heavy manual labor; unskilled work
【粗加工】[机] rough machining; roughly process; roughing
【粗粮】coarse food grain (e.g. maize, sorghum, millet, etc. as distinct from wheat and rice)
【粗劣】of poor quality; cheap; shoddy
【粗陋】coarse and crude: 这所房子盖得很～。 This is a crudely built house.
【粗鲁】rough; rude; boorish: 说话～ have a rough tongue; say rude things／态度～ rude
【粗略】rough; sketchy: ～估计 a rough estimate／～一看 on cursory examination
【粗麻布】burlap; gunny; sacking
【粗毛羊】coarse-wooled sheep; coarse wool
【粗眉大眼】bushy eyebrows and big eyes
【粗浅】superficial; shallow; simple: ～的道理 a simple truth／～的体会 a superficial understanding／～的知识 elementary knowledge
【粗人】rough fellow; boor; unrefined person
【粗纱】low count yarn
【粗纱机】[纺] fly frame
【粗梳毛纺】[纺] woolen spinning
【粗梳棉纱】[纺] carded yarn
【粗疏】careless; inattentive
【粗率】rough and careless; ill-considered: ～的决定 an ill-considered decision
【粗饲料】coarse fodder; roughage
【粗俗】vulgar; coarse: 举止～ uncouth behavior／说话～ use coarse language; use vulgar language
【粗细】①(粗细的程度)(degree of) thickness ②(粗糙和细致的程度) crudeness or fineness; degree of finish; quality of work
【粗线条】①(粗略) thick lines; rough outline: ～

的描写 a rough sketch ②(作风粗率) rough-and-ready; slapdash
【粗心】careless; thoughtless: ～大意 negligent; careless; inadvertent
【粗野】rough; boorish; uncouth: 比赛中动作～ play rough / 举止～ behave boorishly
【粗轧】[冶] roughing (down) ◇ ～机 roughing mill; breaking-down mill
【粗支纱】[纺] coarse yarn
【粗枝大叶】crude and careless; sloppy; slapdash; done in broad strokes or rough outline
【粗制滥造】manufacture in a rough and slipshod way; turn out rough and slipshod work
【粗制品】semifinished product
【粗重】①(声音粗) rough; harsh; gruff; loud and jarring: ～的嗓音 a gruff voice ②(物体笨重) big and heavy; bulky ③(粗而浓) thick and heavy: 眉毛浓黑～ bushy black eyebrows ④(繁重费力) strenuous; heavy: ～工作 heavy work; heavy manual labor
【粗壮】①sturdy; thickset; brawny; robust: ～的胳臂 brawny arms / ～的小伙子 a sturdy lad / ②(粗大而结实) thick and strong: ～的树干 a thick tree trunk ③(声音大) deep and resonant: 声音～ have a deep, resonant voice

cù

蹴 [书] ①(踢) kick ②(踏) tread: 一～而就 reach the goal in one step; accomplish one's aim in one move
猝 [书] sudden; abrupt; unexpected: ～不及防 be taken by surprise; be caught off guard
【猝倒病】[农] damping off; [医]cataplexy
【猝然】suddenly; abruptly; unexpectedly: ～决定 make a sudden decision
【猝死】sudden death
簇 ①[书](团;堆) form a cluster; pile up: 花团锦～ bouquets of flowers and piles of brocades; rich multicolored decorations ②[量] cluster; bunch: 一～人群 a group of people / 一～鲜花 a bunch of flowers
【簇射】[物] shower: 高能～ energetic shower / 宇宙线～ cosmic-ray shower
【簇新】brand new
【簇拥】cluster round
醋 ①(调味品) vinegar ②(嫉妒) jealousy (as in love affair): ～意 (feeling of) jealousy / 吃～ feel jealous
【醋剂】acetum
【醋精】[食品] vinegar concentrate
【醋栗】[植] gooseberry
【醋酸】[化] acetic acid ◇ ～酐 acetic oxide / ～氢化可的松 hydrocortisone acetane / ～纤维 cellulose acetate / ～盐 acetate / ～乙烯酯 vinyl acetate
【醋酯纤维】acetate fiber
酢 vinegar
【酢浆草】[植] creeping oxalis
蹙 [书]①(紧迫) pressed; cramped: 穷～ in dire straits ②(皱) knit (one's brows): ～额 knit one's brows; frown
促 ①(时间短) short; hurried; urgent: 气～ breathe quickly; be short of breath; pant ②(催;推动) urge; promote: 催～ urge; hurry ③(靠近) close to; near: ～膝 sit knee to knee; sit close together
【促成】help to bring about; facilitate: ～双方取得协议 help to bring about an agreement between the two parties ◇ ～栽培[农] forcing culture
【促进】promote; advance; accelerate: ～团结 promote unity / ～相互了解 further mutual understanding / 互相～ help each other forward ◇ ～剂[化] promoter
【促染剂】[纺] accelerant
【促肾上腺皮质激素】ACTH; corticotropin; adrenocorticotropic hormone; adrenotropic hormone ◇ ～试验 ACHT test; adrenocorticotropic hormone test
【促使】impel; urge; spur: ～我们加倍努力 spur us on to greater effort
【促膝谈心】sit side by side and talk intimately; have a heart-to-heart talk; have intimate chat together

cuān

蹿 leap up
镩 cut or break (ice) with an ice pick
【镩子】ice pick
氽 quick-boil: ～丸子 quick-boiled meat balls with soup

cuán

攒 collect together; assemble
【攒聚】gather closely together
【攒三聚五】(of people) gather in little knots;

gather in threes and fours

cuàn

窜 ①(乱跑;乱逃) run about; flee; scurry: 东逃西~ flee in all directions / 鼠~ scurry like rats; run away like frightened rats ②(改动) change (the wording in a text, manuscript, etc.); alter: 点~ make some alterations (in wording)

【窜犯】raid; make an inroad into

【窜改】alter; tamper with; falsify: ~记录 tamper with the minutes / ~原文 alter the original text / ~原则 adulterate a principle / ~帐目 falsify accounts

【窜扰】harass: ~边境 harass the border area / ~活动 harassment / ~领空 intrude into the air space

【窜逃】flee in disorder; scurry off

篡 usurp; seize: ~权 usurp power

【篡夺】usurp; seize: 妄图~领导权 vainly attempt to usurp the leadership

【篡改】distort; misrepresent; tamper with; falsify: ~历史 distort history

【篡位】usurp the throne

cuī

摧 break; destroy; ruin: 无坚不~ capable of destroying any stronghold; all-conquering

【摧残】wreck; destroy; devastate: ~民主 trample on democracy / ~身体 ruin one's health

【摧毁】destroy; smash; wreck: ~敌人的阵地 destroy the enemy positions

【摧枯拉朽】(as easy as) crushing dry weeds and smashing rotten wood; easily overcome; sweep away all obstacles in the way

【摧折】break; snap

催 ①(催促) urge; hurry; press: ~办 press sb. to do sth. / ~人答复 press for an answer / 扬鞭~马 urge one's horse on with a whip; whip one's horse on / 去~他一下。 Go and hurry him up. ②(使事物的产生或变化加快) hasten; expedite; speed up: 春风~绿。 The spring wind speeds the greening of the plants.

【催逼】press (for payment of debt, etc.)

【催产】expedite child delivery; hasten parturition ◇ ~素 oxytocin; pitocin / ~药 oxytocic

【催促】urge; hasten; press

【催肥】[牧] fatten

【催化】[化] catalysis ◇ ~促进剂 catalytic promoter / ~反应 catalytic reaction / ~剂 catalyst; catalytic agent / ~裂化 [石油] catalytic cracking

【催泪弹】tear bomb; tear-gas grenade

【催泪剂】dacryagogue

【催眠】lull (to sleep); hypnotize; mesmerize ◇ ~暗示 hypnotic suggestion / ~后暗示 posthypnotic suggestion / ~疗法 hypnotherapy / ~曲 lullaby; cradlesong / ~术 hypnotism; mesmerism / ~药 hypnotic; somnificant; soporfic

【催奶】stimulate the secretion of milk; promote lactation ◇ ~剂 galactagogue

【催青】[农] hasten the hatching of silkworms (by adjusting temperature and humidity)

【催生】hasten child delivery

【催熟】[农] accelerate the ripening (of fruit)

【催吐剂】[药] emetic

【催醒剂】[药] analeptic: 中药~ herbal analeptic

【催芽】[农] vernalization; germination

cuǐ

璀

【璀璨】[书] bright; resplendent: ~夺目 dazzling

cuì

淬 temper by dipping in water, oil, etc.; quench

【淬火】quench ◇ ~不足 under hardening; underquenching / ~钢 chilled steel / ~剂 hardening agent; quenching liquid / ~炉 glowing furnace; hardening furnace / ~液 quench bath / ~硬化 quench hardening / ~油 quenching oil

瘁 [书] overworked; tired: 鞠躬尽~ bend oneself to a task and exert oneself to the utmost / 心力交~ be physically and mentally tired

粹 ①(纯粹) pure: ~白 pure white / ~而不杂 pure and unadulterated ②(精华) essence; the best: 精~ essence; quintessence

萃 [书]①(聚集) come together; assemble: 荟~ assemble ②(聚在一起的人或事物) a gathering of people or a collection of things: 出类拔~ outstanding

【萃取】[化] extraction ◇ ~分离设备 extraction and stripping apparatus / ~蒸馏 extractive distillation

啐 spit; expectorate: ～他一口 spit at him / ～在某人脸上 spit in sb.'s face

翠 ①(翠绿色) emerald green; green: ～竹 green bamboos ②(翡翠鸟) kingfisher: 点～ handicraft using kingfisher's feathers for ornament ③(翡翠石) jadeite: 珠～ pearls and jade jewellery

【翠花】flowers worked with kingfisher's feathers

【翠菊】[植] China aster

【翠绿】emerald green; jade green

【翠鸟】kingfisher

【翠绕珠围】be surrounded by beauties

脆 ①(易折;易碎) fragile; brittle: ～金属 brittle metal / 这纸太～。This kind of paper is too fragile. ②(一咬就碎裂) crisp: ～香可口 crisp, fragrant and pleasant to taste / 吃起来又甜又～。 It tastes sweet and crisp. ③(声音清脆) clear; ringing; clear and sharp

【脆骨】gristle (as food)

【脆弱】fragile; frail; weak: ～性 frailty / 感情～ be easily upset

【脆裂】brittle rupture ◇ ～强度 bursting strength

【脆性】[冶] brittleness ◇ ～断裂应力 brittle fracture stress

cūn

村 ①(村庄) village; hamlet ②(粗俗) rustic; boorish: ～野 boorish

【村边杂草】[生态] aletophyte

【村落】village hamlet

【村寨】stockaded village

【村长】village head

【村镇】villages and small towns

【村庄】village; hamlet

【村子】village; hamlet

皴 (of skin) chapped (from the cold); cracked

cún

存 ①(存在;生存) exist; live; survive ②(储存;保存) store; keep: ～粮 store up grain ③(蓄积;聚集) accumulate; collect ④(储蓄) deposit: 把钱～在银行里 deposit money in a bank ⑤(寄存) leave with; check: ～行李 deposit one's luggage / ～自行车 leave one's bicycle in a bicycle park ⑥(保留) reserve; retain: 求同～异 seek common ground while reserving differences ⑦(结存) remain on balance; be in stock ⑧(心里怀着) cherish; harbor: ～着很大的希望 cherish high hopes / 不～幻想 harbor no illusions

【存案】register with the proper authorities; keep on record; register officially for the record: ～备查 file for reference

【存查】file for reference; keep in the files for future reference

【存车处】parking lot (for bicycles); bicycle park; bicycle shed

【存储】[电子] memory; storage ◇ ～二极管 storage diode / ～器 memory; storage / ～容量 memory capacity / ～元件 memory element

【存单】deposit receipt: 定期～ time certificate

【存档】keep in the archives; place on file; file: 把文件～ place a document on file

【存底儿】keep the original draft; keep a file copy

【存而不论】leave the question open

【存放】leave with; leave in sb.'s care

【存根】counterfoil; stub: 支票～ check stub

【存户】depositor

【存货】goods in stock; existing stock: ～倍增 goods in storage are doubled / ～告罄 all our stock has been exhausted ◇ ～表 stock sheet / ～价值 inventory value

【存款】deposit; bank savings: 储蓄～ savings deposits / 定期～ fixed deposit; time deposit / 个人～ personal savings account / 活期～ current deposit; demand deposit ◇ ～利息 interest on deposit / ～收据 deposit receipt / ～人 depositor

【存栏】[牧] livestock on hand: ～总头数 (total) amount of livestock on hand

【存身】take shelter; make one's home: 无处～ find no shelter; have no place to call one's home

【存食】suffer from indigestion

【存亡】live or die; survive or perish: ～绝续的关头 at a most critical moment / ～未卜 to preserve or to ruin cannot be foretold / 危急～之秋 at the critical juncture of life and death / 与阵地共～ defend one's position to the death

【存项】credit balance; balance

【存心】①(念头) cherish certain intentions: ～不良 cherish evil designs 〈intentions〉 ②(故意) intentionally; deliberately; on purpose

【存疑】leave a question open; leave a matter for future consideration

【存衣处】cloakroom

【存在】exist; be: ～决定意识。 Man's social be-

ing determines his consciousness.
【存折】deposit book; bankbook
【存执】counterfoil; stub

cǔn

忖 turn over in one's mind; ponder; speculate
【忖度】speculate; conjecture; surmise: ～人心 conjecture at another's mind
【忖量】①(思量) think over; turn over in one's mind; ponder ②(揣度) conjecture; guess

cùn

寸 ① *cun*, a unit of length (3 *cun* = 1 decimeter) ②(极短;极小) very little; very short; small: ～功 small contribution; meager achievement / ～进 a little progress
【寸步】a tiny step; a single step
【寸步不离】follow sb. closely; keep close to; not to move a step from
【寸步不让】refuse to yield an inch; fight every inch of the way; not to yield a single step
【寸步难行】difficult to move even one step; cannot move a single step
【寸草不留】leave not even a blade of grass; be devastated; be left in complete devastation
【寸草不生】not even a blade of grass grows
【寸地千金】an inch of land is worth a thousand gold
【寸金难买寸光阴】[谚] money can't buy time; time is more precious than gold
【寸进尺退】advance by inch and retreat by foot
【寸丝不挂】not a stitch of clothing on
【寸土必争】fight for every inch of land; contest every inch of land
【寸心】feelings: 聊表～ as a small token of my feelings; just to show my appreciation
【寸有所长】an inch has length

cuō

磋 consult
【磋商】consult; exchange views: 与某人～某事 consult with sb. about sth.
搓 ①(捻) twist: ～麻绳 make cord by twisting hemp fibers between the palms / ～线 twist thread / ～纸捻 roll paper spills ②(摩擦) rub; scrub; rub with the hands: ～手取暖 rub one's hands together to warm them / ～衣服 give the clothes a scrubbing
【搓板】washboard
【搓捻机】bunching machine
【搓绳机】cabling machine
【搓手顿脚】wring one's hands and stamp one's feet; get anxious and impatient
【搓澡】give sb. a rubdown with a damp towel (in a public bathhouse)
蹉
【蹉跎】waste time; idle away: ～岁月 let time slip by without accomplishing anything; idle away one's time / 一再～ let one opportunity after another slip away
撮 ①(聚合;聚拢) gather; bring together ②(用簸箕等把东西聚在一起) gather up; scoop up: ～走一簸箕土 scoop up a dustpan of dirt ③(摘取要点) extract; summarize: ～要 make extracts ④[量] pinch: 一～盐 a pinch of salt / 一小～匪徒 a handful of bandits
【撮合】make a match; act as go-between
【撮弄】①(戏弄) make fun of; play a trick on; tease ②(教唆) abet; instigate; incite
【撮要】①(摘取要点) make an abstract; outline essential points ②(摘取出来的要点) abstract; synopsis; extracts: 论文～ abstract of a thesis

cuó

痤
【痤疮】[医] acne
【痤疮炎】[医] acnitis

cuò

厝
【厝火积薪】put a fire under a pile of fagots — a hidden danger
措 ①(安排;处置) arrange; manage; handle: 不知所～ be at a loss what to do; be at one's wit's end / 惊慌失～ be seized with panic; be seized with panic; be frightened out of one's wits ②(筹划) make plans: 筹～款项 raise funds
【措辞】wording; diction: ～不当 inappropriate wording / ～含糊 ambiguously worded / ～刻薄 use harsh words / ～强硬 strongly worded / ～严厉 couched in harsh terms
【措施】measure; step: 安全～ safety measure / 采取重大～ adopt an important measure / 决定性～ a decisive step / 有效～ effective measure
【措手不及】be caught unprepared; be caught unawares: 使人～ throw sb. off his guard; take sb. by surprise

【措置】handle; manage; arrange: ～得当 be handled properly／～失当 mismanaged; improperly managed

错 ①(参差;错杂) interlocked and jagged; intricate; complex: 犬牙交～ jigsaw-like; interlocking ②(相对摩擦) grind; rub: ～牙 grind one's teeth (in one's sleep) ③(避开;不碰上) alternate; stagger ④(不正确) wrong; mistaken; erroneous: 拿～东西 take sth. by mistake 认～ admit one's mistake ⑤(过错;错处) fault; demerit: 挑～ find fault ⑥(用于否定式;坏,差) bad; poor: 这幅画不～。This picture is not bad.

【错爱】[谦] undeserved kindness

【错案】[法] misjudged case

【错别字】①(写错的字) wrongly written ②(读错的字) mispronounced characters

【错车】one vehicle gives another the right of way

【错处】fault; demerit

【错怪】blame sb. wrongly

【错过】miss; let slip: ～机会 miss an opportunity／～这趟汽车 miss this bus

【错角】[数] alternate angle

【错金】inlaying gold: ～器皿 gold-inlaid ware; metal-inlaid ware

【错觉】illusion; misconception; wrong impression

【错开】stagger: 把休假日～ stagger the holidays

【错乱】in disorder; in confusion; deranged: 颠倒～ topsy-turvy／精神～ mentally deranged; insane

【错落】strewn at random: ～不齐 scattered here and there／苍松翠柏～其间 dotted with green pines and cypresses

【错位】[医] malposition

【错误】①(不正确) wrong; mistaken; erroneous: ～的结论 wrong conclusion／～估计形势 make a wrong estimate of the situation／～路线 an erroneous line／～思想 wrong thinking; a mistaken idea ②(不正确的事物、行为等) mistake; error; blunder; fault: ～百出 riddled with errors; full of mistakes／犯～ make a mistake; commit an error

【错杂】mixed; heterogeneous; jumbled; of mixed content

【错字】①(写错的字) wrongly written character ②(印错的字) misprint ◇～连篇 a link of erroneous characters

【错综复杂】intricate; complex; very complicated: 这个问题～。It is a complicated problem.

挫 ①(挫折) defeat; frustrate: 受～ suffer a setback ②(压下去;降低) subdue; lower: ～敌人的锐气 deflate the enemy's arrogance／～其锋芒 blunt the edge of one's advance

【挫败】frustrate; foil; defeat; thwart

【挫伤】①[医] contusion; bruise ②(损伤积极性等) dampen; deflate; blunt; discourage: ～积极性 dampen 〈deflate〉 one's enthusiasm

【挫折】setback; reverse: 遇到～ meet with setbacks／遭受～ suffer setbacks 〈reverses〉

锉 ①(手工工具) file: 方～ square file／木～ (wood) rasp／圆～ round file ②(用锉加工) make smooth with a file; file: ～光 file sth. smooth

【锉床】rasper

【锉刀】file: 粗～ bastard file／细～ smooth file

【锉屑】filing

D

dā

褡

【褡包】a long, broad girdle
【褡裢】①(长方型口袋) a long, rectangular bag sewn up at both ends with an opening in the middle (usu. worn round the waist or across the shoulder) ②(摔跤运动员的上衣) a jacket, made of several layers of cloth worn by wrestlers

搭 ①(架设) put up; build: ～临时舞台 put up a makeshift stage / ～桥 build a bridge / ～帐篷 pitch a tent ② (挂) hang over; put over: 把洗好的衣服～在绳上 hang the washing on a line ③ (连接) come into contact; join: ～上关系 strike up a relationship with; establish contact with / 前言不～后语 speak incoherently; mumble disconnected phrases ④ (加上; 凑上) throw in more (people, money, etc.); add: 这笔钱～上还不够。 It won't be enough even with this sum thrown in. ⑤ (共同抬起) lift sth. together ⑥ (乘) take (a ship, plane, etc.); travel by; go by: ～长途汽车 travel by coach / ～飞机 go by plane / ～轮船去大连 go to Dalian by boat / ～他们的车走 get a lift in their car / 没有～上公共汽车 fail to catch the bus
【搭伴】join sb. on a trip; travel together: ～旅行 travel in company / 跟他们～ join company with them
【搭乘】travel by (plane, car, ship, etc.)
【搭档】①(协作) cooperate; work together: 咱俩～吧 Let us two team up ②(协作人) partner: 老～ old partner; old work-mate
【搭焊】joint welding
【搭焊机】lap (seam) welder
【搭伙】①(结伴) join as partner: 同他们～去郊游 join them in an outing ②(加入伙食组织) eat regularly in (a mess, etc.) 在食堂～ take meals regularly at the canteen
【搭架子】①(搭起间架) build a framework; get (an undertaking, etc.) roughly into shape ②(摆架子) put on airs
【搭箭】get ready to shoot an arrow
【搭救】rescue; go to the rescue of: 把某人从危险中～出来 rescue sb. from danger; save sb. from danger
【搭配】①(按一定目的安排分配) arrange in pairs or groups ②[语](词语搭配) collocation
【搭棚】put up a shed; a makeshift shelter
【搭腔】①(接话) answer; respond ②(交谈) talk to each other
【搭讪】strike up a conversation with sb.; say something to smooth over an embarrassing situation
【搭线】①(连系) make contact ②(做媒) act as go-between; act as matchmaker

答

【答理】acknowledge (sb.'s greeting, etc.); respond; answer
【答应】①(应声回答) answer; reply; respond ②(许诺) promise; comply with: 他～尽力而为。He promised to do his best. ③(同意) agree; consent: ～她的请求 consent to her request

耷 [书] big-eared

【耷拉】droop; hang down: ～着脑袋 hang one's head

dá

达 ①(通) extend: 四通八～ extend in all directions ②(达到) reach; attain; amount to: 灌溉面积共～二万公顷。 The irrigated area amount to 20000 ha. ③(通达) understand thoroughly: 通情～理 be understanding and reasonable; be sensible ④(表达) express; communicate ⑤(显达) eminent; distinguished: ～官 ranking official
【达草灭】[化]norflurazon
【达成】reach (agreement): ～交易 strike a bargain / ～谅解 come to an understanding / ～协议 reach an agreement
【达旦】until dawn: 通宵～ all through the night
【达到】achieve; attain; reach: ～法定年龄 come of age / ～高潮 reach a high tide; come to a climax / ～目的 achieve the goal; attain the goal / ～世界先进水平 come up to advanced world standards
【达尔文主义】Darvinism
【达观】takc things philosophically ◇ ～者 philosopher
【达官贵人】high officials and noble lords; VIPs
【达姆弹】[军] dumdum (bullet)
【达金氏溶液】[化] Dakin's solution
【达赖喇嘛】Dalai Lama
【达人知命】a wise person understands the will of Heavens
【达意】express one's idea; convey one's ideas: 词不～ the words fail to convey the idea / 抒

情～ express one's thoughts and feelings
【达因】[物] dyne
【达因湿度表】[气象] Dines hygrometer

鞑

【鞑靼】Tartar; Tatar ◇ ～海峡 Tartar Strait / ～人 Tartar / ～语 Tartar language

打 [量] dozen: 半～ half a dozen / 成～包装 pack in dozens / 论～出售 sell by the dozen / 一～铅笔 a dozen pencils

答 ①(回答) answer; reply; respond: ～以微笑 respond with a smile / 笑而不～ smile without replying ②(还报) return (a visit, etc.); reciprocate: ～礼 return a salute
【答案】answer; solution; key: 练习的～ key to an exercise / 找不到问题的～ find no solution to the problem
【答拜】return a courtesy call
【答辩】reply (to a charge, query or an argument): 保留公开～的权利 reserve the right of public reply
【答词】thank-you speech; answering speech; reply: 致～ make a speech in reply
【答对】answer sb.'s question; reply: 被问得没法～ be baffled by the question
【答非所问】give an irrelevant answer
【答复】answer; reply: ～他的询问 reply to his inquiry / 请尽早～。Please reply at your earliest convenience.
【答话】answer; reply: 你为什么不～? Why don't you answer?
【答礼】reciprocate another's courtesy; return a solute
【答数】[数] answer
【答谢】express appreciation (for sb.'s kindness or hospitality); acknowledge ◇ ～宴会 a return banquet

沓 [量] pile; pad: 一～钞票 a wad of bank notes / 一～报纸 a pile of newspapers / 一～信纸 a pad of letter paper

dǎ

打 ①(敲打) strike; hit; knock: ～稻子 thresh rice / ～门 knock at the door / ～钟 ring the bell / 敲锣～鼓 beat gongs and drums ②(破碎) break; smash: 花瓶～得粉碎。The vase was smashed to pieces. / 碗～了。The bowl was broken. ③(攻打) fight; attack: ～硬仗 fight a hard battle ④(建造) construct; build: ～坝 construct a dam / ～田埂 build low ridges between paddy fields ⑤(锻造) make (in a smithy); forge: ～一把刀 forge a knife ⑥(搅拌) mix; stir; beat: ～鸡蛋 beat eggs / ～浆糊 make some paste ⑦(捆) tie up; pack: ～成一捆 tie (things) up in a bundle / ～行李 pack one's luggage; pack up ⑧(编;织) knit; weave: ～草鞋 weave straw sandals / ～毛衣 knit a sweater ⑨(涂沫;画) draw; paint; make a mark on: ～方格儿 draw squares / ～手印 put one's fingerprint on a document / ～一个问号 put a question mark; put a query ⑩(喷撒) spray; spread: ～农药 spray insecticide / 在地板上～蜡 wax the floor ⑪(揭;凿) open; dig: ～井 dig a well; sink a well / ～炮眼 drill a blasting hole ⑫(举着;提起) raise; hoist: ～旗 hold a banner; raise a banner / ～起精神来 raise one's spirits; cheer up / ～伞 hold up an umbrella ⑬(发出) send; dispatch; project: ～电报 send a telegram / ～电话 make a phone call / ～炮 fire a cannon / ～手电 flash a torch / ～信号 signal; give a signal ⑭(付给或领取) issue or receive (a certificate, etc.): ～介绍信 write a letter of introduction (for sb.); get a letter of introduction (from one's organization) ⑮(除去) remove; get rid of: ～蛔虫 take medicine to get rid of roundworms; take worm medicine ⑯(舀取) ladle; draw: ～一盆水 fetch a basin of water / ～粥 ladle gruel / 从井里～水 draw water from a well ⑰(用割、砍等动作来收集) gather in; collect; reap: ～柴 gather firewood ⑱(买) buy: ～酱油 buy soy sauce / ～票 buy a (train, bus, etc.) ticket ⑲(捉) catch; hunt: ～野鸭 go duck-hunting / ～鱼 catch fish ⑳(计算;定出) estimate; calculate; reckon: 成本～一百美元 estimate the cost at 100 dollar ㉑ (拟定) work out: ～草稿 work out a draft ㉒(做) do; engage in: ～短工 work as a day or seasonal laborer; be a temporary worker / ～夜班 go on night shift ㉓(做某种游戏) play: ～篮球 play basketball / ～扑克 play cards / ～秋千 have a swing ㉔(表示身体上的某些动作) ～个跟斗 turn a somersault; do a somersault / ～喷嚏 sneeze / ～手势 make a gesture; gesticulate ㉕ (采取某种方式) adopt; use: ～个比方 draw an analogy ㉖(从) from; since: ～那以后 since then
【打把势】demonstrate pugilistic skills
【打靶】target practice; shooting practice ◇ ～场 target range
【打摆子】[方] suffer from malaria
【打败】①(战胜) defeat; beat; worst ②(失败)

suffer a defeat; be defeated
【打扮】dress up; make up; deck out: ～得整整齐齐 all dressed up / 把自己～成英雄 pose as a hero
【打包】①(包装) bale; pack ②(打开包着的东西) unpack ◇ ～费 packing charges / ～机 baler; baling press
【打包票】vouch for; guarantee
【打包铁皮】baling hoop; steel baling hoop
【打苞】(of wheat, sorghum, etc.) form ears; ear up
【打饱嗝儿】belch after a solid meal
【打抱不平】take up the cudgels for the injured party; defend sb. against an injustice
【打辫子】wear a pigtail; knit a pigtail
【打补丁】put a patch (on a garment, etc.)
【打草稿】prepare a draft
【打草惊蛇】beat the grass and frighten away the snake; act rashly and alert the enemy
【打杈】pruning: 给棉花～ prune cotton plants
【打岔】interrupt; cut in
【打场】thresh grain (on the threshing ground)
【打成一片】become one with; identify oneself with; merge with: 和群众～ be one with the masses; become one with the masses
【打虫】get rid of intestinal parasites by means of drugs
【打错算盘】miscalculate; make a wrong decision
【打打闹闹】fight in jest or for fun; boisterous
【打打谈谈】fight and talk alternately (without reaching a real settlement)
【打蛋机】egg beater
【打倒】overthrow: ～暴君 overthrow a tyrant / ～帝国主义! Down with imperialism!
【打得好】①(该打) deserve the beating; deserve the spanking ②(出色的技术) excellent performance
【打得火热】be on terms of intimacy; be as thick as thieves
【打得落花流水】shatter to pieces
【打底】[纺] bottoming ◇ ～机 padding machine
【打底子】①(起草) sketch (a plan, picture, etc.) ②(奠定基础) lay a foundation
【打底漆】under coating varnish
【打地铺】make a bed on the floor; make a bed on the ground
【打点】①(收拾; 准备) get (luggage, etc.) ready ②(旧时送礼, 请求照顾) bribe
【打点穿孔】spot punch
【打掉】destroy; knock out; wipe out: 把敌人的探照灯～ knock out the enemy's searchlight
【打动】move; touch
【打赌】bet; wager: 同意～ take up a bet; accept a bet
【打短工】work as a casual laborer
【打断】①(打折) break: ～脊梁骨 break the backbone ②(使中断) interrupt; cut short: ～念头 give up an idea / ～思路 interrupt sb.'s train of thought / 别～他的话。 Don't interrupt him.
【打盹儿】doze off; take a nap
【打哆嗦】tremble; shiver
【打耳光】box sb.'s ears; slap sb. in the face
【打发】①(派出去) send; dispatch: ～人请医生 sent for a doctor ②(使离开) dismiss; send away: 把孩子们～出去 sent the children away ③(消磨) while away (one's time)
【打翻】overturn; strike down
【打个照面】meet face to face
【打更】sound the night watches
【打埂】[农] ridging
【打嗝儿】[口] ①(呃逆) hiccup ②(嗳气) belch; burp
【打躬作揖】folded the hands and make deep bows; bow and raise one's clasped hands in salute
【打谷场】threshing ground; threshing floor
【打鼓】①(敲鼓) beat a drum ②(心神不定) feel uncertain; feel nervous: 心里直～ feel extremely diffident
【打官腔】talk like a bureaucrat; stall with official jargon
【打官司】①(进行诉讼) go to court; go to law; engage in a lawsuit ②[口](争吵;口角) squabble: 打不完的官司 endless squabbles
【打光棍】remain a bachelor; stay single
【打滚】roll about: 疼得直～ writhe with pain
【打棍子】come down with the big stick (upon sb.); bludgeon
【打哈哈】make fun; crack a joke: 别拿我们～。 Don't make fun of us.
【打哈欠】yawn
【打鼾】snore
【打寒噤】tremble because of cold; shudder because of cold
【打夯】ramming; tamping ◇ ～机 ramming machine; rammer; tamper
【打火】strike sparks from a flint; strike a light
【打火机】a cigarette lighter
【打火石】a flint

【打击】hit; strike; attack: ～报复 retaliate; take revenge / ～投机倒把活动 crack down on speculation and profiteering / ～歪风 take strong measures against unhealthy tendencies; combat unhealthy tendencies ◇ ～力 hitting power / ～面 scope of attack

【打击乐器】percussion instrument

【打基础】do spade work; prepare oneself for bigger tasks ahead

【打家劫舍】loot; plunder

【打架】come to blows; fight; scuffle

【打尖】①(途中休息吃东西) stop for refreshment when travelling; have a snack (at a rest stop) ②[农](掐去棉花等作物的尖) topping; pinching

【打江山】fight for sovereignty over rivers and mountains; seize political power by force

【打浆】[纸] beating ◇ ～机 beating engine; beater

【打交道】come into contact with; make contact with; have dealings with

【打搅】disturb; trouble: 别～他。 Don't disturb him. / 对不起，～您了。 Sorry to have troubled you.

【打劫】rob; plunder; loot: 趁火～ loot a burning house

【打结】tie a knot ◇ ～器 [化纤] knotter / ～强度 knot strength.

【打进】make one's way into

【打开】①(揭开; 拉开; 解开) open; unfold: ～包袱 untie a bundle / ～盖子 take off the lid / 把门～ open the door ② (开收音机等) turn on; switch on: ～收音机(电灯) turn on the radio (light) ③ (攻开) break through: ～缺口 make a breach / ～铁锁链 break open the iron shackles / ～一条出路 break through an encirclement or fight a way out ④(开展) open up; spread: ～地图 spread out the map / ～局面 open up a new prospect / ～思路 broaden one's scope of mind / ～眼界 widen one's horizon

【打开僵局】break the impasse; find solution for a problem

【打开天窗说亮话】frankly speaking; let's not mince matters

【打瞌睡】doze off; nod

【打垮】strike down; defeat; to beat

【打孔】drill a hole; punch a hole; perforate ◇ ～机 pinhole plotter

【打捆机】bander; tying machine

【打蜡】wax; polish with wax: ～地板 waxed floor / 在地板上～ wax a wooden floor

【打捞】①(寻找并捞出沉在水里的东西) get out of water; salvage: ～沉船 salvage a sunken ship / ～尸体 retrieve a corpse from the water / ～作业 salvage procedure ②(捞取) fishing: ～工具 fishing tool / ～装置 fishing gear / ～浮筒 caisson

【打雷】thunder: 打了个响雷。 There was a loud crash of thunder.

【打擂台】accept the challenge; take up the challenge; pick up the gauntlet

【打冷枪】snipe; fire a sniper's shot; stab in the back: 向某人～ snipe at sb.

【打冷战】①(因寒冷或害怕身体突然抖动) shudder ②(非武力的斗争) fight a cold war

【打量】①(观察) measure with the eye; look sb. up and down; size up ②(以为; 估计) think; suppose; reckon

【打猎】go hunting: 爱好～ be found of hunting; be found of shooting

【打乱】throw into confusion; upset: ～计划 disrupt a plan; upset a scheme

【打落水狗】beat a drowning dog; completely crush a defeated enemy

【打麻将牌】play mah-jong

【打马虎眼】[方] pretend to be ignorant of sth. in order to gloss it over; act dumb

【打骂】beat and scold; maltreatment

【打埋伏】①(伏击) lie in ambush; set an ambush; ambush ②(隐藏; 隐瞒) hold sth. back for one's own use

【打麦】thresh wheat or barley

【打毛线衣】knit a woolen sweater

【打磨】polish; burnish; shine ◇ ～器 [机] sander

【打闹】quarrel and fight noisily

【打内战】be engaged in a civil war; fight a civil war

【打拍子】beat time

【打牌】①(打纸牌) play cards ②(打麻将牌) play mah-jong

【打破】break; smash: ～常规,尽量采用先进技术 break free from conventions and use advanced techniques as much as possible / ～藩篱 break through hedging-in tradition / ～记录 break a record / ～僵局 break a deadlock; find a way out of a stalemate / ～界线 break down barriers / ～平衡 upset a balance / ～洋框框 break with foreign conventions

【打破沙锅问到底】insist on getting to the bottom of the matter

【打铺盖】set up a bed
【打旗语】signal with a flag; use semaphore
【打起精神】cheer up; pluck up courage; with chin up
【打气】①(充气) inflate; pump up: ～筒 bicycle pump; inflator / 给车胎～ inflate a tire; pump up a tire②(鼓动) bolster up the morale; boost the morale; encourage; cheer up
【打钎】drill a blasting hole in rock with a hammer and a drill rod
【打前站】act as an advance party; set out in advance to make arrangements
【打桥牌】play a bridge game
【打情骂俏】tease one's lover by showing false displeasure
【打秋风】seek gratuitous financial help
【打秋千】have a swing; get on a swing; play on the swing
【打趣】banter; tease; make fun of
【打圈子】①(转圈子) circle round; whirl about ②(不直说)speak in a roundabout way; beat about the bush
【打拳】box; practice boxing: 打太极拳 practice shadow boxing
【打群架】gang war
【打如意算盘】expect things to turn out as one wishes
【打入】①(置于) throw into; banish to: ～冷宫 consign to the back shelf; throw into limbo / ～十八层地狱 banish sb. to the lowest depths of hell / 被～地下 be driven underground ②(渗透) infiltrate: ～匪巢 infiltrate the bandits' den
【打散】break up; scatter: 把原来的组～重编 break up the existing groups to form new ones
【打扫】sweep; clean: ～房间 clean a room / ～垃圾 sweep away rubbish / ～战场 clean up the battlefield / 把院子～干净 sweep the courtyard clean
【打闪】lightning; 象～那样一亮 flash like a lightning
【打胜仗】be victorious; win a war
【打食】①(觅食) hunt for food; seek food ②(用药消食) relieve indigestion with a drug
【打手】hired roughneck; hired thug; hatchet man
【打手势】gesticulate
【打手印】make an impression of the hand as a signature
【打算】①(考虑) intend; plan; think; mean: 不～放弃 have no intention to give up / 作最坏的～ be prepared for the worst ②(想法) consideration; calculation: 另有个人～ have an ax to grind; have a purpose of one's own / 没有个人～ with no consideration of personal interests
【打算盘】calculate on an abacus; calculate: 打错算盘 miscalculate / 打小算盘 be calculating; be petty and scheming
【打碎】break into pieces: smash; destroy: ～旧的国家机器 smash the old state machinery; destroy the old state machinery
【打胎】have an (induced) abortion
【打天下】①(夺取政权) struggle to seize state power ②(创立事业) establish an enterprise; start an enterprise; set up an enterprise
【打铁】forge iron; work as a blacksmith: ～的 a blacksmith / ～先得本身硬[谚] one must be ideologically sound and professionally competent to do arduous work
【打听】ask about; inquire about: ～某人的地址 try to find out sb.'s address / ～战友的消息 inquire about one's comrades-in-arms
【打通】get through; open up: ～思想 straighten out sb.'s thinking; talk sb. round / 电话打不通 be unable to get through (on the telephone)
【打头】take the lead: 在赛跑中～ lead in the race
【打头炮】fire the first shot; be the first to speak or act
【打头阵】fight in the van; spearhead the attack; take the lead
【打退】beat back; beat off; repulse: ～敌人的进攻 repulse an enemy attack
【打退堂鼓】beat a retreat; back out
【打围】encircle and hunt down (animals)
【打先锋】fight in the van; be a pioneer
【打响】①(开火) start shooting; begin to exchange fire: 枪没～。The gun failed to shoot off. ②(初步成功) win initial success; make a good start; come off well as a start: ～了春耕第一炮。The spring ploughing got off to a good start.
【打消】give up (an idea, etc.); dispel (a doubt, etc.) ～顾虑 dispel misgivings / ～念头 dismiss the idea; drop the idea ～原计划 drop the original plan; cancel the original plan
【打小报告】inform secretly on a colleague, etc.
【打信号】communicate by signals
【打雪仗】have a snowball fight; throw snowballs
【打鸭子上架】drive a duck onto a perch; make sb. do sth. entirely beyond him

【打掩护】①(军)provide cover for; shield: 为主力部队～ provide cover for the main force ②(庇护) lend shelter to; give refuge to

【打眼】①(钻孔) punch (bore) a hole; drill: ～穿孔 spot punch / ～放炮 drill and blast ②(惹人注意) catch the eye; attract attention

【打样】①(画出设计图样) draw a design ②[印](打校样) make a proof ◇ ～机 proof press

【打印】①(盖印) put a seal on; stamp ②(打字油印) cut a stencil and mimeograph; mimeograph: ～机 printer; dot printer / ～机终端 typewriter terminal / ～文件 mimeograph documents

【打印台】ink or stamp pad

【打油诗】doggerel; ragged verse

【打游击】①[军] fight as a guerrilla: 上山～ join the guerrillas in the mountains; wage guerrilla warfare in the mountains ②[口](没有固定地点的工作或活动) work (eat, sleep, etc.) at no fixed place

【打援】attack enemy reinforcements; ambush enemy reinforcements: 围点～besiege an enemy stronghold in order to strike at the reinforcements

【打圆场】mediate a dispute; smooth things over

【打砸抢】beating, smashing and looting ◇ ～者 smash-and-grabber

【打杂儿】do odds and ends: 出外～ go out charring

【打战】shiver; tremble; shudder: 浑身～ shiver all over

【打仗】fight; go to war; make war: 打恶仗 fight a fierce battle / 打硬仗 fight the enemy head-on; fight a hard-fought battle

【打招呼】①(用语言或动作表示问侯) greet sb.; say hello; tip one's hat ②(事先提醒) give a previous notice; warm

【打折扣】①(降低商品价格) sell at a discount; give a discount: 打八折出售 sell at twenty per cent discount ②(不完全按规定或商定的去做) fall short of a requirement or promise: 说到做到, 不～ carry out one's pledge to the letter

【打针】give or have an injection: 打肌肉针 inject into a muscle / 打静脉针 inject into the veins / 打皮下针 inject under the skin

【打制石器】[考古] chipped stone implement

【打肿脸充胖子】slap one's face until it's swollen in an effort to look imposing; puff oneself up to one's own cost

【打中】hit the mark; hit the target; hit: ～靶心 hit the bull's eye / ～要害 hit on the vital spot; hit where it really hurts / ～一艘敌舰 hit an enemy vessel

【打主意】①(想办法) think of a plan; evolve an idea: 打错主意 miscalculate; make a wrong decision / 打定主意 make up one's mind ②(设法谋取) try to obtain; seek

【打住】come to a halt; (in speech or writing) stop

【打转】spin; rotate; revolve

【打桩】pile driving; piling ◇ ～机 pile driver / 震动～机 vibrating pile driver

【打字】typewrite; type ◇ ～带 typewriter ribbon / ～稿 typescript / ～机 typewriter / ～蜡纸 stencil / ～员 typist / ～纸 typing-paper

【打坐】(of a Buddhist or Taoist monk) sit in meditation

dà

大 ①(不小) big; large; great: ～城市 a big city / ～救星 the great liberator ②(超过一般) heavy (rain, etc.); strong (wind, etc.); loud: ～损失 heavy losses / 声音太～ too loud ③(规模大) general; main; major: ～反攻 general counteroffensive / ～路 main road; highway / ～手术 major operation / ～问题 major issue; big problem ④(大小) size: 象豆粒一样～ the size of a bean; as big as a bean ⑤(年龄) age; old: 你的小孩多～了? How old is your child? ⑥(程度深) greatly; fully: ～吃一惊 be greatly surprised; be quite taken aback / ～～增加 greatly increased; considerably increased / 进步很～ show much progress ⑦(排行第一) eldest: ～房 senior branch of a family / ～哥 eldest brother ⑧[敬] your: ～札 your letter / ～作 your writing ⑨(用在时间前表示强调) ～白天 in broad daylight / ～清早 early in the morning

【大白】①(完全清楚) come out; become known: 真相已～于天下 the truth has become known to all ②[方](白垩) whiting ◇ ～浆 [建] whitewash

【大白菜】Chinese cabbage

【大伯子】husband's elder brother; brother-in-law

【大败】①(打败) defeat utterly; put to rout: ～敌军 inflict a crushing defeat on the enemy ②(被打败) suffer a crushing defeat

【大班】①(幼儿园的)the top class in a kindergarten ②(洋行经理) manager of a foreign firm ③(舞厅领班) captain (of taxi dancers)

【大板车】large handcart
【大办】go in for sth. in a big way
【大半】①(过半数) more than half; greater part; most: ～天 the greater part of a day; most of the day／他们～是青年人。Most of them are young. ②(较大可能性) very likely; most probably: 她～不来了。Most probably she isn't coming.
【大包干】the all-round contract
【大本营】①[军](最高统帅部) supreme headquarters ②(登山营地) base camp: 登山队～ the base camp of a mountaineering expedition
【大便】①(拉屎) defecate; have a bowel movement; shit: ～不通 (suffer from) constipation／～不正常 irregular bowel movement／～干燥 (motions) be hard and dry／～困难 have difficulty in passing one's motions／～失禁 incontinence of feces; be unable to hold one's motions／～四次 have four motions ②(屎) stool; human excrement; shit; feces: 去化验～ have one's stool examined
【大冰期】[地质] Great Ice Age
【大兵】[旧] common soldier
【大兵团】large troop formation ◇ ～作战 large formation warfare; grand tactics; major tactics
【大病】a serious illness: 患～ suffer from a serious illness.
【大饼】a kind of large flatbread; pancake
【大伯】①(伯父) father's elder brother; uncle ②(尊称年长的男人)uncle
【大不了】①(至多) at the worst; if the worst comes to the worst: ～再从头开始。If the worst comes to the worst, we'll start all over again. ②(了不得) [多用于否定式] alarming; serious: 这不是什么～的成就。It's not so remarkable a success.
【大不敬】great disrespect to one's superior; great disrespect to one's seniors
【大步流星】with vigorous strides; at a stride
【大部】greater part: 歼敌～ annihilate the greater part of the enemy
【大材小用】put fine timber to petty use; use talented people for trivial tasks; waste one's talent on a petty job
【大菜】①(主菜) main course ②(西餐) Western meal; meal in Western style
【大肠】[生理] large intestine ◇ ～杆菌 colon bacillus／～型细菌 coliform bacteria
【大氅】overcoat; cloak; cape
【大潮】spring tide
【大车】①(牲口拉的货车) cart; dray: 赶～ drive a cart／胶轮～ rubber tire cart ②(火车司机) engine driver ③(轮船上主要管机器的人) chief engineer (of a ship)
【大车店】an inn for carters
【大臣】minister (of a monarchy)
【大成】great achievement
【大成问题】very questionable; very doubtful; in doubt
【大乘】[佛教] Mahayana; Great Vehicle
【大吃大喝】eat and drink extravagantly; be spendthrift in feasting
【大吃一惊】be startled at; be taken aback; be astounded at
【大出血】[医] massive hemorrhage
【大处落墨】concentrate on the key points
【大处着眼，小处着手】keep the general goal in sight and take the daily tasks in hand
【大吹大擂】make a great fanfare; make a big noise
【大锤】sledgehammer
【大醇小疵】sound on the whole though defective in details
【大慈大悲】infinitely merciful; infinitely compassionate
【大词】[逻] major term
【大葱】green Chinese onion
【大错特错】completely mistaken; absolutely wrong
【大打出手】strike violently; attack brutally
【大大】greatly; enormously: 生产效率～提高 productivity has risen greatly
【大…大…】(表示规模大、程度深) ～吵～闹 kick up a row; make a scene／～红～绿 loud colors／～鱼～肉 plenty of meat and fish; rich food
【大大咧咧】[方] (of a person) careless; casual
【大袋鼠】kangaroo
【大胆】bold; daring; audacious: ～的革新 a bold innovation／～的行动 a daring act
【大刀】broadsword
【大刀阔斧】bold and resolute; drastic
【大道理】major principle; general principle; great truth: 小道理要服从～。Minor principles should be subordinated to major ones.
【大灯】[汽车] headlight
【大敌】formidable enemy; archenemy: ～当前 faced with a formidable foe
【大抵】generally speaking; in the main; on the whole: ～相同 more or less the same
【大地】the earth; mother earth; the world: ～回

春。Spring returns to the earth. (或: Spring is here again.) / 阳光普照～。The sun shines all over the world. ◇ ～测量学 geodesy / ～构造学 geotectology; tectonics / ～水准面 geoid
【大典】①(隆重的典礼) grand ceremony: 开国～ the ceremony to proclaim the founding of a state ②(经典) a body of classical writings; canon:《永乐～》*Yongle Canon*
【大调】[乐] major: C～奏鸣曲 sonata in C major
【大动脉】[生理] main artery; aorta: 南北交通的～ the main artery of communications between north and south
【大豆】soybean; soya bean
【大都市】a large city; a metropolis
【大度】[书] magnanimous: ～包容 magnanimous and tolerant
【大端】[书] main aspects; main features; salient points: 仅举其～ merely point out the main features
【大队】①[军] a military unit corresponding to the battalion or regiment; group ②(人数众多) a large body of: ～人马 a large contingent of troops; a large body of marchers, paraders, etc.
【大多】for the most part; mostly: 他们～是先进工作者。They are mostly advanced workers.
【大多数】great majority; the bulk; vast majority: 绝～ the overwhelming majority / 人口的～ the bulk of the population / 团结～ unite with the great majority / 在～情况下 in most of the cases
【大而无当】large but impractical; unwieldy: ～的计划 a grandiose but impractical plan
【大发雷霆】be furious; fly into a rage; bawl at sb. angrily
【大法】the fundamental law; the constitution of a nation
【大法官】a grand justice
【大凡】generally; in most cases
【大方】[书] expert; scholar: 贻笑～ incur the ridicule of experts
【大方】①(不吝啬) generous; liberal: 用钱～ be openhanded ②(不拘束) natural and poised; easy; unaffected: 举止～ have an easy manner; have poise; carry oneself with ease and confidence ③(不俗气) in good taste; tasteful: 式样～ style in good taste; graceful in style
【大方向】general orientation
【大放厥词】talk a lot of nonsense; spout a stream of empty rhetoric
【大放异彩】yield unusually brilliant results
【大费唇舌】long harangue; ranting speech; a lot of talking
【大分子】[化] macromolecule
【大粪】human excrement; night soil
【大风】①[气] fresh gale ②(强风) gale; strong wind: 外面刮着～。There's a gale blowing. (或: It is blowing hard.) ◇ ～警报 gale warning
【大风大浪】wind and waves; great storms
【大风子】[植] chaulmoogra ◇ ～油 chaulmoogra oil
【大夫】a senior official in feudal China
【大幅度】by a wide margin; by a big margin; substantially: ～持续增产 large continuous increase of crops / ～提高 be increased by a big margin / ～增长 increase by a wide margin; rise by a big margin
【大副】first mate; chief mate; mate; chief officer
【大腹贾】potbellied merchant; rich merchant
【大腹便便】potbellied; big-bellied
【大概】①(大致内容或情况) general idea; broad outline: 只知道个～ have only a general idea ②(大约) general; rough; approximate: ～的估计 a rough estimate / ～的数字 an approximate figure / 作一个～的分析 make a general analysis ③(可能) probably; most likely; presumably
【大干】work energetically; go all out; make an all-out effort: ～快上 get going and go all out
【大纲】outline: 教学～ teaching program / 世界史～ an outline history of the world
【大搞】do sth. vigorously; get sth. into full swing
【大哥】①(最大的哥) eldest brother ②(尊称年纪相仿的男子) elder brother
【大革命】①(大规模革命) great revolution: 法国～ the French Revolution ②(特指中国第一次国内革命战争) the Great Revolution in China (1924–1927)
【大公】grand duke ◇ ～国 grand duchy
【大公无私】①(没有私心) selfless; unselfish ②(公正) perfectly impartial
【大功】great merit; extraordinary service: 立了～ have performed exceptionally meritorious services
【大功告成】(of a project, work, etc.) be accomplished; be crowned with success

【大功率】[电] high-power: ～可控硅 high-power silicon controlled rectifier
【大姑子】husband's elder sister; sister-in-law
【大骨节病】Kaschin-Beck disease
【大鼓】①[乐] bass drum ②[曲艺] *dagu*, versified story sung to the accompaniment of a small drum and other instruments
【大褂】unlined long gown: 蓝布～ a blue cotton gown
【大观】grand sight; magnificent spectacle: 蔚为～ present a magnificent spectacle
【大管】[乐] bassoon
【大规模】large-scale; extensive; massive; mass: ～毁灭性武器 weapon of mass destruction / ～进攻 launch a massive attack / ～生产 large-scale production / ～兴修水利 launch a large-scale water conservancy project / 举行～罢工 stage a massive strike
【大锅饭】food prepared in a large canteen caldron; mess: 吃～ eat in the canteen the same as everyone else; mess together
【大国】power; leading powers; great power
【大国沙文主义】great-nation chauvinism
【大过】serious offense: 记～一次 record a serious mistake
【大海捞针】fish for a needle in the ocean; look for a needle in a haystack
【大寒】Great Cold (24th solar term)
【大喊大叫】①(大声喊叫) shout at the top of one's voice ②(大力宣传) conduct vigorous propaganda
【大汉族主义】Han chauvinism
【大旱望云霓】long for a rain cloud during a drought; look forward to relief from distress
【大好】very good; excellent: ～风光 a superb view / ～河山 beautiful rivers and mountains of a country; one's beloved motherland / ～时光 the golden years; prime of one's life / ～时机 opportune moment; golden opportunity; finest hour / ～形势 excellent situation
【大号】①(大的) large size: ～的鞋 large-size shoes ②[乐] tuba; bass horn ③[敬] your (given) name
【大合唱】cantata; chorus: 《黄河～》 *The Yellow River Cantata*
【大亨】big shot; bigwig; magnate
【大轰大嗡】make a terrific din; raise a hue and cry
【大红】bright red; scarlet
【大后方】rear area
【大后年】three years from now
【大后天】three days from now
【大户】①(有钱人家) rich family ②(人口多) big family: ～人家 a wealthy and influential family; a family of long standing
【大话】big talk; tall talk; boast; bragging: 说～ talk big; brag
【大黄蜂】hornet
【大黄鱼】large yellow croaker
【大回环】[体] giant circle; giant: 单臂～ single-arm circle
【大茴香】[植] anise; star anise
【大会】①(社会团体、政党等召开的会) plenary session; general membership meeting: 全国科学～ National Science Conference ②(群众会) mass meeting; mass rally
【大火】a big fire; a conflagration
【大伙儿】[口] we all; you all; everybody
【大惑不解】be extremely puzzled; be unable to make head or tail of sth.
【大祸】great misfortune; calamity; disaster: ～临头。 A great misfortune has come.
【大计】a major program of lasting importance; a matter of fundamental importance: 百年～ a matter of fundamental importance for generations to come; a major project affecting future generations / 共商～ discuss matters of vital importance
【大蓟】[植] setose thistle
【大家】①(专家) great master; authority: 书法～ a great master of calligraphy; a noted calligrapher ②(大伙) all; everybody: ～的事～管。 Everybody's business should be everybody's responsibility.
【大家庭】big family; community
【大将】①(最高一级将官) senior general ②(高级军官) high-ranking officer
【大将风度】the style of a great general or admiral
【大蕉】plantain
【大教堂】cathedral
【大街】main street; street: ～小巷 streets and lanes / 逛～ go window-shopping; go for a walk in the street
【大节】political integrity
【大捷】great victory
【大姐】①(最大的姐) eldest sister ②(对与自己年龄相仿的女人的尊称) elder sister
【大解】have a bowel movement
【大襟】the front of a Chinese garment which buttons on the right
【大进化】[进化] macroevolution

【大惊失色】get such a big scare that color goes out of one's face; turn pale with fright
【大惊小怪】be surprised at sth. perfectly mormal; be alarmed at sth. perfectly normal; make a fuss
【大净】[伊斯兰教] Ghusl
【大静脉】[生理] vena cava
【大舅子】[口] wife's elder brother; brother-in-law
【大局】overall situation; general situation; whole situation: 顾全～ take the whole situation into account; take the interests of the whole into account / ～已定。The general situation has settled. (或: The result is certain.)
【大举】carry out (a military operation) on a large scale: ～进攻 mount a large-scale offensive; attack in force
【大军】①(主力部队) main forces; army ②(人数众多) large contingent: 百万～ an army a million strong / 筑路～ a large contingent of road builders
【大卡】[物] kilocalorie; large calorie; great calorie
【大开眼界】eye-opening; widen one's horizon; an eye-opener
【大楷】①(手写大楷体汉字) regular script in big characters, as used in Chinese calligraphy exercises ② (大写印刷体) block letters; block-writing
【大考】end-of-term examination; final exam
【大课】a lecture given to a large number of students; enlarged class
【大跨径桥】long-span bridge
【大快人心】affording general satisfaction; most gratifying to the people; to the immense satisfaction of the people
【大块头】a tall and bulky fellow
【大括弧】[印] brace
【大牢】[口] prison; jail
【大老粗】uncouth fellow; uneducated person; rough and ready fellow
【大礼】the most solemn of ceremonies
【大礼拜】alternate Sunday on which one has a day off; fortnightly holiday: 休～ have every other Sunday off
【大礼服】formal dress; ceremonial dress
【大礼堂】hall; auditorium
【大理石】marble
【大力】energetically; vigorously: ～发展教育事业 devote major efforts to developing education / ～支援农业 give energetic support to agriculture
【大力士】a man of unusual strength
【大丽花】[植] dahlia
【大殓】encoffining ceremony
【大梁】[建] girder
【大量】①(数量多) a large number; a great quantity: ～财富 enormous wealth; large fortune / ～库存 huge stocks / ～杀伤敌人 inflict heavy casualties on the enemy / ～生产拖拉机 mass-produce tractors / ～时间 plenty of time / ～事实 a host of facts / 收集～科学资料 collect a vast amount of scientific data / 为国家积累～资金 accumulate large funds for the state ②(气量大; 能容忍) generous; magnanimous: 宽宏～ magnanimous; large-minded
【大料】[方] aniseed
【大楼】multi-storied building: 办公～ office building / 教学～ classroom building / 居民～ apartment house; block of flats
【大陆】continent; mainland ◇ ～边缘 continental margin / ～架 continental shelf / ～漂移说 the theory of continental drift / ～坡 continental slope / ～性气候 continental climate
【大路】the highroad
【大路货】popular goods of dependable quality
【大略】①(大致内容或情况) general idea; broad outline ②(大约) generally; roughly; approximately: ～相同 roughly the same
【大妈】①(伯母) father's elder brother's wife; aunt ②(尊称年长的妇人) aunt
【大麻】①(线麻) hemp ②(用大麻制成的麻醉药) marijuana
【大麻哈鱼】chum salmon; dog salmon (又作"大马哈鱼")
【大麦】barley
【大忙】very busy: ～季节 rush season; busy season
【大猫熊】giant panda
【大帽子】an exaggerated epithet used to categorize a person; unwarranted charge; political label
【大门】entrance door; front door; gate
【大梦初醒】wake up from a dream; wake up to; dawn on
【大米】(husked) rice
【大面儿】[方] general appearance; surface: ～上还过得去 it is basically all right
【大面积】large tracts of land; large area: ～丰收 reap a bumper harvest over large areas ◇ ～烧伤 large-area burns

【大民族主义】big-nationality chauvinism
【大螟】pink rice borer
【大名】①(正式名字) one's formal personal name ②[敬] your (given) name
【大名鼎鼎】famous; celebrated; well-known
【大谬不然】entirely wrong; grossly mistaken
【大模大样】in an ostentatious manner; with a swagger
【大拇指】thumb: 竖起～叫好 hold up one's thumb in approval
【大拿】[口] person with power; boss
【大难】catastrophe; disaster: ～不死 escape from death in a great catastrophe / ～临头 be faced with imminent disaster
【大脑】[生理] cerebrum ◇ ～半球 cerebral hemisphere / ～脚 cerebral peduncle / ～皮层 cerebral cortex / ～性麻痹 cerebral palsy / ～炎 cerebritis; encephalitis
【大鲵】[动] giant salamander
【大逆不道】treason and heresy; worst offense; greatest outrage
【大年】①(丰收年) good year; bumper year; (of fruit trees) on year ②(农历十二月有三十天的年份) a lunar year in which the last month has 30
【大年初一】[口] first day of the lunar year; lunar New Year's Day
【大年夜】[方] lunar New Year's Eve
【大娘】[口] ①(伯母) wife of father's elder brother; aunt ②(尊称年长的妇人) aunt
【大拍卖】sell at a bargain
【大泡】[医] bulla
【大炮】①(口径大的炮) artillery; big gun; cannon ②[口] (好发表激烈意见的人) one who speaks boastfully or forcefully; one who noisily overstates things
【大批】large quantities of; numbers of; amounts of: ～定货 large order / ～现钞 large sum in ready money
【大谱表】[乐] great stave
【大瀑布】cataract
【大漆】lacquer
【大企业主】the big owners of private enterprises
【大气】①(地球周围的气体) atmosphere; air ②(沉重的呼吸) heavy breathing: 跑得直喘～ breathe heavily from running / 吓得连～也不敢出 catch one's breath fear; hold one's breath in fear ◇ ～层 atmospheric layer; atmosphere / ～电 atmospheric electricity / ～干扰 atmospheric interference / ～光学 atmospheric optics / ～环流 atmospheric circulation; general circulation of atmosphere / ～科学 atmospheric sciences / ～空间 airspace / ～密度 atmospheric density / ～探测 atmospheric sounding / ～污染 air pollution; atmospheric pollution / ～压 atmospheric pressure; atmosphere / ～折射 atmospheric refraction; astronomical refraction
【大气候】[气] macroclimate
【大器晚成】great minds mature slowly
【大千世界】[佛教] the boundless universe
【大前年】three years ago
【大前提】[逻] major premise
【大前天】three days ago
【大钱】an old Chinese coin of low denomination
【大清洗】great purge
【大庆】①(大规模庆祝) grand celebration of an important event; great occasion: 十年～ the festive occasion of the 10th anniversary ②(称老人寿辰的敬词) birthday: 八十～ eightieth birthday
【大秋】①(秋天收获季节) harvest season in autumn ②(秋天的收成) crops harvested in autumn; autumn harvest ◇ ～作物 crops sown in spring and reaped in autumn; autumn-harvested crops
【大曲】①(酿酒的曲) yeast for making hard liquor ②(酒名) a hard liquor made with such yeast
【大权】power over major issues; authority: ～独揽 centralize power in one man's hands to deal with major issues; arrogate all authority to oneself / ～在握 hold power in one's hands / ～旁落。power has fallen into the hands of others.
【大全】complete works of; a complete collection of; a complete volume on
【大犬座】[天] Canis Major
【大人】[敬] 父亲～ Dear Father
【大人】①(成人) adult; grown-up ②(地位高的人) Your Excellency; (间接称呼) His Excellency
【大人物】important person; great personage; big shot; VIP
【大容量】high-capacity
【大儒】a scholar who combines profundity with virtue
【大扫除】general cleaning; thorough cleanup: 节日～ thorough cleanup before a holiday
【大嫂】①(大哥的妻子) eldest brother's wife;

sister-in-law ②(对已婚妇女的称呼) elder sister
【大沙洲】[地理] Grand Banks
【大杀风景】spoil the fun; sink the spirits of; mar the pleasure of
【大厦】large building; mansion
【大少爷】①(主人的大儿子) eldest son (of a rich family)②(挥霍的人) a spoilt son of a rich family; spendthrift: ～作风 behavior typical of the spoilt son of a rich family; extravagant ways
【大舌头】[口] a thick-tongued person; one who lisps; lisper
【大赦】amnesty; general pardon: ～一批犯人 grant amnesty to a group of criminals ◇ ～令 bill of oblivion; order of general amnesty
【大婶儿】aunt
【大声疾呼】raise a cry of warning; loudly appeal to the public
【大声辱骂】insult loudly; shout insult at
【大牲口】draught animal
【大失所望】greatly disappointed; to one's great disappointment
【大师】①(造诣深的人) great master; master: 国画～ a great master of traditional Chinese painting ②[佛教] (对和尚的尊称) Great Master
【大师傅】cook; *chef*
【大使】ambassador: 特命全权～ ambassador extraordinary and plenipotentiary / 巡回～ ambassador at large ◇ ～馆 embassy / ～级会谈 talks at ambassadorial level; ambassadorial talks / ～级外交关系 ambassadorial level diplomatic relation / ～衔 ambassadorial rank
【大事】①(重大的事) great event; major important matter; major issue: 当前国际政治中的一件～ a major event in current international politics / 关心国家～ concern oneself with affairs of state / 头等～ a matter of prime importance 〈paramount〉 importance / 完成了一桩～ have accomplished an important task ②(总的形势) overall situation: ～不好。A disaster is imminent. ③(大搞) in a big way: ～宣传 play up; ballyhoo / ～渲染 enormously exaggerate; play up ◇ ～记 chronicle of events
【大事化小，小事化了】turn big problems into small problems and small problems into no problem at all
【大事做不来，小事又不做】be incapable of doing great things, yet disdain minor tasks; disdain minor assignments while being quite unequal to major tasks
【大势】general trend of events: ～所趋，人心所向 the trend of the times and the desire of the people; the general trend and popular feeling / ～已去。The game is up. (或: Not much can be done to save the situation.)
【大是大非】major issues of principle; cardinal questions of right and wrong: 分清～ draw clear distinctions concerning cardinal issues of right and wrong
【大手笔】the work or writing of a great author or calligrapher
【大手大脚】wasteful; extravagant: ～的人 waster; squanderer
【大叔】①(父亲的弟弟) younger brother of one's father; uncle ②(尊称同父亲同辈的年轻男人) uncle
【大书特书】record in letters of gold
【大暑】Great Heat (12th solar term)
【大肆】without restraint; wantonly: ～攻击 wantonly vilify; launch an unbridled attack against; launch an all-out attack against / ～鼓吹 noisily advocate / ～叫嚣 set up a great clamor about; allege vociferously / ～蹂躏 trample on / ～污蔑 slander violently; slander indiscriminately / ～宣扬 indulge in unbridled propaganda for; give enormous publicity to
【大苏打】[化] sodium thiosulfate; sodium hyposulfite; hypo
【大蒜】garlic: 一头～ a head of garlic
【大踏步】in big strides: ～前进 stride along; make great strides forwards
【大谈特谈】talk at length
【大提琴】violoncello; cello
【大题小做】do little about a major issue; treat lightly; make little of
【大体】①(重要的道理) cardinal principle; general interest: 识～，顾大局 have the cardinal principles in mind and take the overall situation into account ②(大致) roughly; more or less; on the whole; by and large; for the most part: ～上正确 generally correct / ～相同 more or less alike; about the same
【大天白日】[口] broad daylight
【大田】land for growing field crops ◇ ～作物 field crop
【大厅】hall
【大庭广众】(before) a big crowd; (on) a public occasion: ～之中 in public; on a public occasion
【大同】Great Harmony (an ideal or perfect so-

ciety): 世界～ universal harmony in the world
【大同小异】largely identical but with minor differences; alike except for slight differences; very much the same
【大头菜】[植] rutabaga
【大头钉】tack
【大头针】pin
【大团圆】①(全家人聚一起) happy reunion ②(团聚的结局) happy ending
【大腿】thigh
【大尉】senior captain
【大无畏】dauntless; utterly fearless; indomitable: ～的英雄气概 dauntless heroism
【大西洋】the Atlantic (Ocean)
【大喜】[口] great rejoicing: 在这～的日子里 in these days of great rejoicing
【大喜过望】be delighted that things are better than one expected; be overjoyed
【大虾】prawn
【大显身手】display one's skill to the full; give full play to one's abilities; distinguish oneself; give a good account of oneself
【大显神通】give full play to one's remarkable skill; give full play to one's remarkable abilities
【大相径庭】[书] be widely divergent
【大小】①(程度大小) big or small: ～水库十座 ten reservoirs of varying sizes ②(辈份高低) degree of seniority: 说话没个～ speak impolitely to elderly people ③(大人和孩子) adults and children
【大校】senior colonel
【大协作】large-scale cooperation; a major pooling of efforts
【大写】①(汉字数字的一种写法) the capital form of a Chinese numeral: ～金额 amount in words ②(拼音字母的一种写法) capitalization: ～字母 capital letter; majuscule
【大兴】go in for sth. in a big way: ～调查研究之风 energetically encourage the practice of conducting investigations and studies / ～水利 large-scale building of water conservancy projects / ～土木 go in for large-scale construction / ～问罪之师 angrily point an accusing finger at sb.; condemn scathingly
【大猩猩】gorilla
【大行皇帝】the late emperor
【大行星】[天] major planet
【大型】large-scale; large: ～彩色记录片 full-length color documentary film / ～企业 large enterprise / ～运输机 giant transport aircraft; air freighter / ～轧钢厂 heavy steel rolling plant
【大熊猫】giant panda
【大熊座】[天] Ursa Major; the Great Bear
【大修】[机] overhaul; heavy repair
【大选】general election
【大学】(高等学校) university; college: 广播电视～ radio broadcasting and television university / 师范～ normal university; teachers' university / 文科～ university of liberal arts / 业余工业～ part-time engineering college / 业余职工～ spare-time college for staff and workers / 综合性～ comprehensive university ◇ ～入学注册 matriculation / ～预科 preparatory course
【大学生】university student; college student: 二年级～ sophomore / 三年级～ junior / 四年级～ senior / 一年级～ freshman / 低年级～ lower grade students / 高年级～ upper grade students; higher grade students ○男生 man student; boy student; schoolboy / 女生 woman student; girl student; schoolgirl / 旁听生 auditor / 研究生 postgraduate (student); graduate student / 应届毕业生 graduating student; this year's graduates / 住宿生 boarder / 走读生 day student; non-resident student
【大雪】Great Snow (21st solar term)
【大循环】[生理] systemic circulation
【大牙】①(槽牙) molar ②(门牙) front tooth: 叫人笑掉～ make a laughing-stock of oneself
【大雅】[书] elegance; refinement; good taste: 不登～之堂 not appeal to refined taste; not in good taste
【大烟】opium ◇ ～鬼 opium addict
【大言不惭】brag unblushingly; talk big
【大盐】crude salt
【大雁】wild goose
【大洋】①(海洋) ocean ②(银元) silver dollar
【大洋洲】Oceania; Oceanica
【大样】①[印] (报纸整版清样) full-page proof ②[建] (工程的细部图) detail drawing
【大摇大摆】strutting; swaggering
【大要】main points; gist: 文章的～ the gist of an article
【大业】great cause; great undertaking: 创～ become pioneers in a great undertaking / 革命～ the great cause of revolution
【大爷】①(伯父) father's elder brother; uncle ②(尊称年长的男子) uncle
【大衣】overcoat; topcoat: 风雪～ anorak; parka / 夹～ light overcoat / 军～ uniform greatcoat / 棉～ cotton-padded overcoat / 男

～ overcoat; greatcoat / 男皮～ long fur coat / 女～ winter coat; overcoat / 女皮～ fur coat; fur
【大姨】[口] mother's eldest sister; aunt
【大姨子】[口] wife's elder sister; sister-in-law
【大义】cardinal principles of righteousness; righteous cause: 深明～ be deeply conscious of the righteousness of a cause
【大义凛然】inspiring awe by upholding justice
【大义灭亲】place righteousness above family loyalty
【大意】①(主要意思) general idea; main points; gist; tenor: 段落～ the gist of a paragraph ②(疏忽) careless; negligent; inattentive: 千万不可粗心～ must never on any account be negligent
【大意失荆州】suffer a major setback due to carelessness
【大音阶】[乐] major scale
【大印】great seal; the seal of power
【大油】[口] lard
【大有好处】be of great benefit to; be of much good
【大有可为】be well worth doing; have bright prospects
【大有区别】poles apart; entirely different
【大有人在】there are plenty of such people; such people are by no means rare
【大有文章】there's something behind all this; there's more to it than meets the eye
【大有作为】there is plenty of scope for one's talents; be able to develop one's ability to the full
【大鱼吃小鱼】big fish swallowing little fish; with the small interests ruined by the big interests
【大雨】heavy rain: ～倾盆 rain cats and dogs; a pouring rain / ～如注 the rain came down in sheets
【大冤案】gross injustice
【大元帅】generalissimo
【大员】[旧] high-ranking official: 委派～ appoint high-ranking officials
【大圆航向】[航空] great-circle course
【大院】courtyard; compound; 居民～ residential compound
【大约】①(约略) approximately; about ②(很可能) probably
【大月】①(阳历) a solar month of 31 days ②(阴历) a lunar month of 30 days
【大跃进】great leap forward
【大运河】the Grand Canal
【大杂烩】hodgepodge; hotchpotch
【大杂院儿】a compound occupied by many households
【大灶】ordinary mess: 吃～ board in ordinary mess
【大展宏图】realize one's ambition; ride on the crest of success
【大张旗鼓】on a grand scale; in a big way
【大丈夫】true man; real man; man
【大枝】[植] bough
【大指】thumb
【大治】great order: ～之年 a year of great order / 天下～ run the country well
【大志】high aim; lofty aim; exalted ambition; high aspirations: 有～ aim high
【大致】roughly; approximately; more or less: ～相同 roughly the same
【大智若愚】a man of great wisdom often appears slow-witted
【大众】the masses; the people; the public; the broad masses of the people ◇ ～歌曲 popular songs / ～科学 popular science / ～文艺 art and literature for the masses; popular literature
【大众化】popular; in a popular style: ～的饭菜 popular low-priced dishes / ～商品 popular commodities / 语言～ use the language of institutions of ordinary people
【大洲】continent
【大主教】archbishop
【大专院校】universities and colleges; institutions of higher education
【大资本家】big capitalist; monopolist
【大资产阶级】the big bourgeoisie
【大字本】[印] large-character edition; large-type edition
【大字标题】banner headline; banner-line; streamer
【大自然】nature: 征服～ conquer nature
【大宗】①(大批的) a large amount; a large quantity: ～款项 a large amount of money; large sums ②(数量最大的产品) staple: 本地产品以茶叶为～。Tea is the staple crop here.
【大作】your work; your script
【大做文章】make a fuss; make a big issue; turn out much propaganda (about)

dāi

呆 ①(呆笨) slow-witted; dull: ～头～脑 dull-looking ②(发愣) blank; wooden: ～～地望

着 stare at sth. blankly / 吓得发～ be stupefied; be scared stiff; be dumbstruck ③(在) stay: ～在家里 stay at home
【呆若木鸡】dumb as a wooden chicken; dumbstruck; transfixed (with fear or amazement)
【呆小病】[医] cretinism
【呆小病患者】[医] cretin
【呆性物质】[化] inert material
【呆帐】bad debt
【呆滞】①(不灵活) dull; inert; lifeless: 脸上表情～ lifeless in facial expression / 目光～ with a dull look in one's eyes ②(不景气) dull; sluggish; stagnant; slack; idle: 避免资金～ prevent capital from lying idle / 贸易～ stagnant in trade / 商业～ slack in business / 市场～ dull market / 销路～ slow in sale
【呆住了】dumbfounded
【呆子】idiot; simpleton; blockhead

dǎi

歹 bad; evil; vicious: 好～不分 cannot distinguish the good from the bad / 为非作～ do evil
【歹毒】vicious; viciousness; malicious; malice
【歹人】evil person; thieves or burglars
【歹徒】scoundrel; ruffian; hoodlums
【歹心】evil intent
【歹意】malice; malicious intent

逮 capture; catch: 猫～老鼠 cats catch mice / 那小偷被警察～住了。The thief has been seized by the policeman.

dài

戴 ①(戴上) put on; wear: ～上手套 put on one's gloves / ～上眼镜 put on glasses ②(戴着) wear; have on: ～有色眼镜看问题 look through colored spectacles / ～着镣铐 in irons / ～着面罩 masked / ～着头盔 helmeted ③(拥护; 尊敬) respect; honor: 爱～ love and respect
【戴高帽子】receive flattery or compliment; flatter
【戴绿帽子】be a cuckold
【戴帽子】①(戴上或戴着帽子) wear one's hat; put on one's hat ②(加罪名) pin the label on sb.; stigmatize; falsely accuse sb. of a crime
【戴胜】[动] hoopoe
【戴孝】wear mourning for a parent, relative, etc.; be in mourning
【戴圆履方】stand between Heaven and Earth
【戴罪立功】atone for one's crimes by doing good deeds; redeem oneself by good service

带 ①(带子) belt; girdle; ribbon; band; tape: 录音～ recording tape / 皮～ leather belt / 丝～ silk ribbon / 鞋～ shoelaces; shoestrings / 腰～ waist band ②(轮胎) tire: 自行车～ bicycle tire ③(地带) zone; area; belt: 绿化地～ greenbelt / 热～ the torrid zone ④(携) take; bring; carry ⑤(顺带; 捎带) do sth. incidentally: 放牛～割草 cut grass while grazing cattle ⑥(呈现; 含有) bear; have: ～有时代的特点 bear the imprint of the times / 面～笑容 wear a smile / 一项～根本性的措施 a measure of fundamental importance ⑦(连着; 附带) having sth. attached; simultaneous: ～叶的橘子 tangerines with leaves on / 连说～笑地走进来 enter laughing and talking / 玉米地里～着种点黄豆 grow some soybeans in the maize field ⑧(引导; 带领) lead; head: ～兵 lead troops; be in command of troops / ～队 lead a group of people; the leader of a group of people ⑨(照顾) look after; bring up; raise: ～孩子 look after children
【带病】in spite of illness
【带病体】[植] carrier
【带材】[冶] strip
【带电】electrified; live; charged ◇ ～导线 live wire / ～粒子 [物] charged particle / ～体 charged body; electrified body / ～作业 live-wire work
【带动】drive; spur on; bring along: 用拖拉机上的发动机～打谷机 use the tractor motor to power the thresher
【带发修行】submit to Buddhist discipline while still wearing one's hair
【带钢】strip steel
【带坏】lead astray; rub bad habits on an innocent campanion
【带劲】①(有劲头) energetic; forceful: 走得更～ walk in more vigorous steps ②(能引起兴趣) interesting; exciting; wonderful: 听起来非常～ simply wonderful to listen to it
【带锯】[机] band saw
【带菌者】[医] carrier
【带累】implicate; involve
【带领】lead; guide: ～一支部队 lead a army
【带路】show the way; lead the way; act as a guide ◇ ～人 guide
【带球走】traveling; running with a ball
【带伤】get wounded; get injured

【带上温和色彩】moderate
【带头】take the lead; be the first; take the initiative; set an example: ～冲锋 lead the charge／～发言 be the first to speak; break the ice／起～作用 play a leading role
【带头羊】bellwether
【带徒弟】train an apprentice; take on an apprentice
【带小数】a whole number with a decimal
【带孝】wear mourning for a parent, relative, etc.; be in mourning
【带笑】smilingly; carrying a smile
【带信儿】bring a massage; bear a massage
【带羞】look shy; look bashful
【带音】[语] voiced
【带鱼】hairtail; ribbonfish
【带状】banding ◇ ～疱疹 [医] berpes zoster; shingles zoster／～分布 zonal distribution／～煤 banded coal／～体[地质] shoestring
【带子】belt; girdle; ribbon; band; tape

大

【大夫】[口] doctor; physician
【大黄】[植] Chinese rhubarb

代

代 ①(代替) take the place of; be in place of: ～人受过 suffer for the faults of another; bear the blame for somebody else ②(代理) acting: ～部长 acting minister ③(历史分期) historical period: 古～ ancient times／汉～ the Han Dynasty ④[地] (地质年代) era: 古生～ the Paleozoic Era ⑤(辈) generation: ～～相传 pass on from generation to generation; hand down from generation to generation
【代办】①(替人办理) do sth. for sb.; act on sb.'s behalf ②[外] chargé d'affaires: 临时～ chargé d'affaires ad interim ◇ ～处 Office of the Chargé d'Affaires
【代办所】agency: 储蓄～ savings agency／邮政～ postal agency
【代笔】write on sb.'s behalf
【代表】①(委派代表) deputy; delegate; representative: 常驻～ permanent representative; permanent delegate／党代会～ delegate to the Party Congress／副～ alternate; deputy; substitute／全国人大～ deputy to the National People's Congress／全权～ plenipotentiary／首席～ chief delegate／双方～ representatives from both sides ②(体现) represent; stand for: ～时代精神 embody the spirit of the era／～无产阶级利益 represent the interests of the proletariat ③(替人办事或讲话) on behalf of; in the name of ◇ ～权 representation／～人物 representative figure; representative personage; typical representative; leading exponent／～资格 qualifications of a representative／～资格审查委员会 (delegates') credentials committee
【代表大会】congress; representative assembly; representative conference: 中国共产党全国～ the National Congress of the Communist Party of China
【代表团】delegation; mission; deputation: 专家～ expert attached to a delegation
【代表作】representative work
【代偿】compensatory ◇ ～功能[医] compensation／～失调 decompensation／～作用 compensation
【代词】[语] pronoun
【代步】[书] ride instead of walk
【代购】buy on sb.'s behalf; act as a purchasing agent ◇ ～代销点 purchasing and marketing agency
【代管】manage, govern or administer on behalf of another
【代号】code name
【代价】price; cost: 不惜任何～ prepared to pay any price; at any cost; at all costs／付出很高～ pay a high price (for)
【代课】take over a class for an absent teacher
【代劳】do sth. for sb.; take trouble on sb.'s behalf
【代理】①(代人负责) act on behalf of someone in a responsible position ～厂长 acting manager of a factory ②(代表别人工作) act as agent; act as procurator: ～关系 agent relation ◇ ～领事 pro-consul／～权 power of attorney; authority of agency; commission／～人 agent; deputy; proxy; procurator; attorney
【代码】code
【代名词】①[语] pronoun ②(同义词) synonym
【代庖】do what is sb. else's job; act in sb.'s place
【代乳粉】milk powder substitute; soybean milk powder
【代入法】Substitution method
【代售】be commissioned to sell sth.
【代书】write legal document for others
【代数】algebra ◇ ～方程 algebraic equation／～式 algebraic expression／～数论 algebraic theory of numbers

【代替】replace; substitute for; take the place of: 以感情～政策 substitute one's personal feelings for the policy

【代位】subrogate ◇ ～继承 representation / ～权 right of subrogation / ～行使 subrogation

【代销】sell goods (for the state) on a commission basis; be commissioned to sell sth. (usu. as a sideline); act as a commission agent ◇ ～店 shop commissioned to sell certain goods; commission agent

【代谢】①(更替;取代) supersession: 新旧事物的～ the supersession of the old by the new ② [生] metabolize: 分解～ catabolism / 组成～ anabolism ◇ ～病 disease of metabolism / ～期 metabolic stage / ～物 metabolite / ～作用 metabolism

【代行】act on sb.'s behalf: ～职权 function in an acting capacity

【代序】an article used in lieu of a preface

【代言人】spokesman; mouthpiece

【代议制】the representative system (of government) ◇ ～机构 representative institution / ～政府 representative government

【代用】substitute: ～材料 substitute materials; ersatz materials ◇ ～品 substitute; ersatz

【代字号】swung dash (～)

【代罪羔羊】a scapegoat

袋

①(口袋) bag; sack; pocket; pouch: 暗～ concealed pocket / 插～ inset pocket / 工具～ tool kit / 旅行～ travelling bag / 睡～ sleeping bag / 贴～ patch pocket / 衣～ pocket / 邮～ mailbag ②[量] 两～水泥 two sacks of cement / 一～面粉 a sack of flour / 一～烟的功夫 time needed to smoke a pipe

【袋囊】[脊椎] marsupium

【袋式除尘器】bag-type collector

【袋式过滤器】breather bag; deep bed filter

【袋兽】marsupial

【袋鼠】kangaroo

【袋蛙】marsupial frog

【袋装】in bags ◇ ～奶粉 milk powder in bags

【袋子】sack; bag

玳

【玳瑁】[动] hawksbill turtle

黛

a black pigment used by women in ancient times to paint their eyebrows

【黛绿】[书] dark green

【黛眉】①(画眉) blacken the eyebrows ②(画了的眉) blackened eyebrows of women

贷

①(贷款) loan: 农～ agricultural loans ②(借入或借出) borrow or lend: 向银行～款 get a bank loan ③(推卸) shift (responsibility); shirk: 责无旁～ be one's unshirkable responsibility; be duty-bound ④(饶恕) pardon; forgive: 严惩不～ punish without mercy

【贷方】[簿记] credit side; credit

【贷款】provide a loan; extend credit to; make an advance to; loan; credit: ～条件 terms of loan / 工业～ industrial loan / 农业～ agricultural loan / 未偿～ outstanding loans / 无息～ interest-free loans / 信用～ fiduciary loan / 专项～ special-purpose loan ◇ ～公司 finance company / ～计划 borrowing plan / ～凭单 credit memo / ～人 accommodator / ～条件 condition for loans / ～限额 basic credit line / ～帐户 loan account

待

①(对待) treat; deal with: ～人诚恳 treat people sincerely; be sincere with people / 宽～俘虏 treat prisoners of war leniently ②(招待) entertain: ～客 entertain a guest ③(等待; 留待) wait for; await: ～查 yet to be investigated / ～机 await an opportunity; bide one's time / 尚～解决的问题 a problem awaiting solution; an outstanding issue / 尚～证实 yet to be confirmed / 有～改进 have yet to be improved ④(需要) need: 自不～言。This goes without saying. ⑤(要; 打算) going to; about to: ～说不说 swallow what one was about to say

【待毙】await death; be a sitting duck

【待哺】wait for feeding

【待产室】labor room; pre-delivery room

【待发】[军] committed; due out

【待机而动】wait for one's opportunity; bide one's time

【待价而沽】wait for the right price to sell; wait for the highest bid

【待见】[口] like; be fond of

【待考】need checking; remain to be verified

【待理不理】lukewarm reception; give the cold shoulder to sb.

【待领】wait for claimant

【待命】await orders: ～出发 await orders to set off / 原地～ stay where one is, pending orders; stand by ◇ ～中断 [自] armed interruption / ～状态 [自] armed state

【待人接物】the way one gets along with people

【待如已出】treat a child as if he were one's

own
【待用】inactive; stand-by ◇ ～程序块 inactive block／～入口 inactive entry／～文件 inactive file
【待业人员】people waiting for employment
【待遇】①(对待；权利) treatment: 优惠的～ preferential treatment／政治～ political treatment／最惠国～ most-favored-nation treatment ②(物质报酬) remuneration; pay; wages; salary: ～菲薄 treatment shabby／物质～ material benefits／优厚的～ excellent pay and conditions
【待月西厢】wait for one's lover in the night; have a nocturnal rendezvous with one's lover
【待运提单】received for shipment B／L
【待字闺中】not betrothed yet
迨 ①[书](等到)wait till ②(趁着) before sth. happens: ～天之未阴雨 before it rains
怠 idle; remiss; slack
【怠惰】idle; lazy; indolent
【怠工】slow down; go slow ◇ ～者 saboteur; slow-down striker
【怠慢】①(冷淡) cold-shoulder; slight: 不要～了客人 see that none of the guests are neglected ②[套](表示招待不周) ～了! I'm afraid I have been a poor host.
【怠情养性】renounce aggressiveness and practice relaxation
【怠速系统】[机] idling system
殆 ①[书](危险) danger ②(几乎) nearly; almost
逮 [书] reach: 力有未～ beyond one's reach; beyond one's power
【逮捕】arrest; take into custody: ～法办 arrest and deal with according to law; bring to justice ◇ ～令 order for attachment, warrant of arrest／～权 right of arrest／～证 arrest warrant

dān

单 ①(一个) one; single: ～扇门 single-leaf door ②(奇数) odd: ～号 odd number／～日 odd-numbered days ③(单独) singly; alone: ～人独马 single-handed ④(只；光) only; alone: 不～ not only ⑤(不复杂) simple: 简～ simple; plain ⑥(薄弱) thin; weak ⑦(只有一层) unlined ⑧(单子) sheet: 床～ bed sheet ⑨(清单) bill; list: 菜～ menu; bill of fare／工资～ pay sheet／价目～ price list／名～ name list／提货～ bill of lading
【单摆】[物] simple pendulum
【单板】[林] veneer
【单板雪橇】skibobbing
【单帮】[旧] a travelling trader working on his own: 跑～ travel around trading on one's own
【单倍体】[生] monoploid; haploid
【单本位制】[经] monometallic standard; monometallism
【单边】[经] unilateral: ～出口 unilateral export／～进口 unilateral import／～贸易 unilateral trade
【单薄】①(衣服薄而且少) thin: 穿得～ be thinly clad ②(弱) thin and weak; frail: 力量～ weak in strength／身体～ have a poor physique ③(内容不充实) insubstantial; flimsy; thin: 论据～ a feeble argument
【单步】one step; single step
【单产】per unit area yield
【单程】one way ◇ ～车票 one-way ticket; single ticket／～费 one way fare／～清棉机[纺] single process scutcher
【单纯】①(简单纯一) simple; pure: 象孩子一样～ as simple as child ②(单一；仅仅) alone; purely; merely: ～追求利润 go merely after profit／不～追求数量 not concentrate on quantity alone ◇ ～词 [语] single-morpheme word／～技术观点 exclusive concern about technique; putting technique above everything else／单纯军事观点 purely military viewpoint
【单词】①[语] individual word; word ②(一个词素的词) single-morpheme word ◇ ～表 word list
【单打】[体] singles: 男子～ men's singles／女子～ women's singles／少年男子～ boys' singles／少年女子～ girls' singles
【单打一】concentrate on one thing only
【单单】only; alone
【单刀】①(短柄长刀) short-hilted broadsword ②[武术](用一把刀表演) single-broadsword event
【单刀直入】come straight to the point; speak out without beating about the bush
【单调】monotonous; dull; drab: 色彩～ dull coloring／声音～ in a monotonous tone
【单斗挖掘机】power shovel
【单独】alone; by oneself; one one's own; single-handed; independent: 采取～行动 take independent action ◇ ～海损 particular average／～检查 individual inspiration／～营业

private business
【单发】[军] single shot ◇ ~射击 single shot
【单方】folk prescription; home remedy
【单方面】one-sided; unilateral: ~撕毁协定 unilaterally tear up an agreement / ~停火 cease-fire carried out by one side only
【单飞】[航空] solo flight
【单峰驼】[动] one-humped camel; dromedary; Arabian camel
【单幅】[纺] single width
【单干】①(单独干活) work on one's own; go it alone; work by oneself; do sth. single-handed ②(个体农户) individual farming
【单缸】one-cylinder; single-cylinder: ~泵 [机] single-cylinder pump; simplex pump
【单杠】[体] ①(体操器械) horizontal bar ②(体操项目) horizontal bar gymnastics ◇ ~表演 performance on horizontal bar
【单个儿】①(独自一个) individually; alone ②(成套或成对中的一个) an odd one
【单轨】single track: ~铁路 single track railway; monorail
【单号】odd numbers (of tickets, seats, etc.)
【单簧管】[乐] clarinet
【单级火箭】single-stage rocket
【单季稻】single cropping of rice
【单价】①[经] unit price ②[化][生] univalent
【单间儿】separate room
【单键】[化] single bond
【单脚跳】[体] hop
【单晶】single crystal
【单晶硅】[电子] monocrystalline silicon
【单晶体】[物] monocrystal
【单镜头反光照相机】single-lens reflex
【单句】[语] simple sentence
【单据】receipts; bills; vouchers; invoices: ~不符 discrepancy in the documents / 付款~ paying voucher / 货运~ shipping documents / 空白~ billhead
【单孔目】[动] Monotremata ◇ ~动物 monotreme
【单跨】[建] single span
【单利】[经] simple interest
【单轮滑车】gin block
【单轮射箭】[体] single round archery: 女子三十米~ women's 30-meter single round archery event
【单门独户】a single isolated house
【单宁酸】[化] tannic acid
【单片微型处理机】chip microprocessors
【单片眼镜】monocle
【单枪匹马】single-handed; all by oneself; alone
【单枪三束彩色显像管】tritron
【单人床】single bed
【单人房间】a single-bed room
【单人舞】solo dance: 跳~ dance a solo
【单日】odd-numbered days (of the month)
【单色】monochromatic ◇ ~电视 monochrome television / ~光 [物] monochromatic light / ~画 monochrome / ~胶印机 single-color offset press / ~性 monochromaticity
【单身】①(没结婚的) unmarried; single ②(不跟家属住在一起) not be with one's family; live alone: ~在外 live alone away from home ◇ ~汉 bachelor / ~宿舍 quarters for single men or women; bachelor quarters
【单生花】[植] solitary flower
【单式】[簿记] single entry: ~簿记 bookkeeping by single entry
【单数】①(奇数) odd number ②[语] singular number
【单丝】[纺] monofilament
【单丝不成线】one strand of silk is not a thread
【单瘫】[医] monoplegia
【单糖】[化] monose; monosaccharide; simple sugar
【单体】[化] monomer
【单位】①(计量单位) unit: ~面积产量 yield per unit area / 长度~ a unit of length / 货币~ monetary unit / 以秒为~计算时间 measure time by the second ②(部门) unit: 基层~ basic unit; grass-roots unit / 生产~ production unit / 行政~ administrative unit ◇ ~圆 [数] unit circle / ~制 system of unit
【单细胞】unicellular ◇ ~动物 unicellular animal
【单弦儿】*danxianr*, story-telling to musical accompaniment
【单线】①(单一的线) single line ②(单独联系) one and only one line (link) ③[交] (单轨) single track ◇ ~铁路 single-track railway; single-track line
【单相思】unrequited love
【单向】one-way; unidirectional ◇ ~电路 one-way circuit / ~交通 one-way traffic
【单项】[体] individual event ◇ ~比赛 individual competition
【单相】single-phase; monophase ◇ ~电动机 single-phase motor / ~合金 single-phase alloy
【单斜】[地] monocline ◇ ~层 monoclinal stratum
【单行本】separate edition; offprint

【单行法规】special regulations; separate regulations
【单行条例】specific regulations
【单行线】one-way road
【单性花】[植] unisexual flower
【单性生殖】[生] parthenogenesis; parthenogenetic propagation; parthenogenetic reproduction
【单循环制】[体] single robin
【单眼】[动] simple eye
【单眼皮】eyelids that do not have a distinct fold along the edges
【单叶】[植] simple leaf
【单一】single; unitary ◇ ～汇率 unified foreign exchange rate / ～经济 single-product economy / ～种植 monoculture; one-crop farming
【单衣】unlined garment
【单翼机】monoplane
【单音词】[语] monosyllabic word; monosyllable
【单元】unit: 十八号楼四～二号 No. 2, Entrance 4, Building 18 / 运算～ arithmetic unit
【单元体】[植] haplont
【单渣操作】[冶] single-slag practice
【单张汇票】sola bill; sola check
【单张纸印刷机】[印] sheet-fed press
【单子】①(记事的纸片) list; bill form: 菜～ bill of fare; menu / 开个～ make out a list / 填写～ fill in a form ②(盖在床上的布) bed sheet
【单子宫】[生理] simplex uterus
【单子叶植物】monocotyledon
【单字】individual character; separate word
【单座飞机】single-seater; single-seater airplane

箪

[书] a bamboo utensil for holding cooked rice
【箪食壶浆】welcome (an army) with food and drink

耽

①(拖延) delay ②[书] (入迷; 沉溺) abandon oneself to; indulge in: ～乐 indulge in pleasure / ～溺 indulge in evil ways; unable to free oneself from bad habits
【耽搁】①(停留) stop over; stay ②(拖延) delay: 不得～ admit of no delay / 毫不～ without delay
【耽误】delay; hold up: ～功夫 waste time / ～了火车 missed the train / ～了整个工程 hold up the whole project; delay the whole project / 把～的时间夺回来 make up for lost time
【耽于】addict; indulge in: ～幻想 indulge in illusion

眈

【眈眈】glaring covetously; eying greedily: 虎视～～ glare like a tiger eyeing its prey

担

①(挑) carry on a shoulder pole: ～水 carry water (with a shoulder pole and buckets) ②(担负; 承当) take on; undertake: ～重担 bear a heavy burden / 不怕～风险 ready to face any danger; not be afraid of running risks
【担保】assure; guarantee; vouch for: 出口信贷～ export credit guarantees ◇ ～充分的债 full secured liabilities / ～期间 guaranty period / ～品 guarantee / ～人 guarantor; guarantee / ～书 deed of security / ～信用状 stand-by letter of credit / ～债权人 secured creditors / ～责任 obligation under bond / ～证书 certificate of security
【担不是】take the blame
【担待】①(负责) take the responsibility ②(原谅) excuse: 请您～一下。 Please excuse me. (或: Please show tolerance for it.)
【担当】take on; undertake; assume: ～重任 take on heavy responsibilities
【担风险】take risks; face the risk (of doint sth.); assume the risk (of doing sth.)
【担负】bear; shoulder; take on; be charged with: ～费用 bear an expense / ～领导工作 hold a leading post / ～损失 afford losses / ～责任 shoulder responsibility
【担搁】delay
【担荷】shoulder a burden; assume a responsibility
【担架】stretcher; litter ◇ ～队 stretcher-team / ～员 stretcher-bearer
【担惊受怕】feel alarmed; be in a state of anxiety
【担任】assume the office of; hold the post of: ～裁判 serve as referee; act as referee / ～会议主席 take the chair / ～工会主席 be the chairman of a trade union
【担心】worry; feel anxious: ～他的健康 worry about his health
【担忧】worry; be anxious
【担子】[真菌] basidium
【担子体】[真菌] basidiophore

丹

red
【丹顶鹤】red-crowned crane
【丹毒】[医] erysipelas

【丹方】folk prescription; home remedy
【丹桂】[植] orange osmanthus
【丹麦】Denmark ◇ ～人 Dane / ～语 Danish (language)
【丹宁酸】tannic acid; gallotannic acid; digallic acid
【丹青】[书] painting: ～妙笔 superb artistry (in painting); the superb touch of a great painter
【丹田】the pubic region: ～之气 deep breath controlled by the diaphragm
【丹心】a loyal heart; loyalty
【丹竹】the red bamboo

dǎn

掸 brush lightly; whisky: ～～衣服 brush the dust off one's clothes / ～掉身上的雪花 brush 〈whisk〉 the snow off one's coat
【掸子】duster: 鸡毛～ feather duster

胆 ①[解] gall bladder ②(胆量) courage; guts; bravery: 丧～ be overwhelmed with fear; be terror-stricken / 壮～ boost sb.'s courage ③(内胆) a bladder-like inner container: 球～ the rubber bladder of a ball / 热水瓶～ the glass liner of a vacuum flask
【胆大】bold; audacious: ～包天 audacious in the extreme / ～妄为 reckless / ～心细 bold but cautious
【胆道】biliary tract
【胆矾】[化] chalcanthite; blue vitriol
【胆敢】dare; have the audacity to
【胆固醇】[生化] cholesterol: ～不正常 have an abnormal blood cholesterol level
【胆管】[解] bile duct ◇ ～炎 cholangitis / ～造影 cholangiography
【胆寒】be terrified; be struck with terror
【胆红素】[生化] bilirubin
【胆黄素】[生化] biliflavin
【胆碱】[生化] choline
【胆量】courage; guts; pluck; spunk: 很有～ have plenty of guts; have plenty of spunk
【胆略】courage and resourcefulness: ～过人 have unusual courage and resourcefulness
【胆囊】[生理] gall bladder ◇ ～切除术 cholecystectomy / ～炎 cholecystitis / ～造影 cholecystography
【胆怯】timid; cowardly
【胆石】[医] cholelith; gallstone ◇ ～病 cholelithiasis
【胆识】courage and insight
【胆酸】[生化] cholic acid
【胆小】timid; cowardly: ～如鼠 as timid as a mouse; chicken-hearted ◇ ～鬼 coward
【胆战心惊】tremble with fear; be terror-stricken: 使人～ strike terror into sb.; be terrifying
【胆汁】[生理] bile ◇ ～疗法 bilitherapy; choletherapy
【胆子】courage; nerve: 放开～ pluck up courage; stop being afraid

dàn

弹 ①(弹子) ball; pellet ②(枪弹; 炮弹) bullet; bomb: 穿甲～ armor-piercing bullet / 达姆～ dumdum bullet / 导～ guided missile / 教练～ drill shell / 芥子气～ mustard bomb / 炮～ shell / 氢～ hydrogen bomb; H-bomb / 燃烧～ incendiary bomb; incendiary bullet / 杀伤～ anti-personnel bullet / 手榴～ grenade / 烟幕～ smoke shell / 曳光～ tracer bullet / 原子～ atom-bomb; A-bomb / 炸～ bomb / 照明～ flare bomb / 中子～ neutron bomb / 子～ cartridge
【弹道】trajectory ◇ ～导弹 ballistic missile / ～弧线 ballistic curve / ～火箭 ballistic rocket
【弹道学】ballistics: 内～ interior ballistics / 外～ exterior ballistics
【弹弓】catapult; slingshot
【弹痕】bullet or shell hole; shot mark
【弹夹】(cartridge) clip; charger
【弹尽粮绝】run out of ammunition and food supplies
【弹壳】shell case; cartridge case
【弹坑】(shell) crater
【弹幕】barrage
【弹盘】cartridge drum; magazine
【弹片】shell fragment; shell splinter; shrapnel
【弹膛】chamber
【弹头】bullet; projectile nose; warhead: ～鉴定 identification of bullet / 导弹～ missile warhead / 热核～ thermonuclear warhead
【弹丸】①(铁丸或泥丸) pellet; shot; bullet ②[书] (小地方) ～之地 a tiny area; a small bit of land
【弹无虚发】every shot tells
【弹匣】[军] magazine
【弹药】ammunition ◇ ～库 ammunition depot; ammunition storehouse / ～手 ammunition man; ammunition bearer / ～所 ammunition supply point; refilling point / ～箱 ammunition chest; cartridge box
【弹着】[军] impact ◇ ～点 point of impact; hitting point / ～观察 spotting / ～观察兵

spotter / ～角 angle of impact / ～区 impact area; objective area
【弹子】①(弹弓射的弹丸) a pellet shot from a slingshot ②(玩具) marble: 打～ play marbles ③[方] (台球) billiards ◇ ～房 billiard room

淡 ①(稀薄) thin; light: ～酒 light wine / 烟味很～。 The tobacco is very mild. ②(不浓) tasteless; weak: ～而无味 tasteless; insipid / ～茶 weak tea ③(颜色浅) light; pale: ～黄 light yellow / ～紫 pale purple; lilac ④(冷淡) indifferent: ～～地答应了一声 answer dryly / ～然处之 treat with indifference; take things coolly ⑤(营业不旺盛) slack; dull ⑥[方] (无关紧要的) meaningless; trivial: 扯～ talk nonsense
【淡泊】not seek fame and wealth
【淡薄】①(密度小; 味不浓) thin; light ②(不浓厚) become indifferent; flag ③(渐渐消失) faint; dim; hazy
【淡菜】[动] mussel; dried mussel meat
【淡出】[电影] fade out
【淡淡】light; slight
【淡化】desalination: 海水～ desalination of sea water
【淡积云】[气] cumulus humilis
【淡季】slack season; dull season; off season
【淡漠】①(冷淡) indifferent; apathetic; nonchalant ②(记忆不清) faint; dim; hazy
【淡墨】light ink
【淡青】light greenish blue
【淡然处之】treat with indifference; regard coolly
【淡入】[电影] fade in
【淡色】light color; delicate shade
【淡水】fresh water ◇ ～湖 freshwater lake / ～养鱼 freshwater fish-farming / ～鱼 freshwater fish
【淡忘】fade from one's memory
【淡雅】simple and elegant; quietly elegant
【淡月】a slack month
【淡竹】[植] henon bamboo

啖 [书] ①(吃) eat ②(给 ... 吃) feed ③(拿利益引诱人) entice; lure

氮 [化] nitrogen (N)
【氮肥】nitrogenous fertilizer
【氮化处理】nitrogen treatment
【氮化物】[化] nitride
【氮血症】azotemia; nitremia

诞 ①(诞生) birth ②(生日) birthday ③(荒唐的) absurd; fantastic: 荒～ fantastic
【诞辰】birthday
【诞生】be born; come into being; emerge: 新中国的～ the birth of New China / 在斗争的烈火中～ emerge from the flames of struggle

石 a unit of dry measure for grain (=1 hectoliter)

旦 ①[书] (天亮) dawn, daybreak ②(天) day: 元～ New Years Day ③(旦角) the female character type in Beijing opera ④[纺] (但尼尔) denier
【旦夕】[书] this morning or evening; in a short while: ～之间 in a day's time; overnight / 危在～ in imminent danger
【旦夕祸福】unexpected good or bad fortune

担 ①(量词) a unit of weight (=50 kilograms) ②(挑子) a carrying pole and the loads on it; load; burden: 货郎～ loads of goods carried on a shoulder pole by an itinerant pedlar ③[量] 一～菜 a load of vegetables / 一～水 two buckets of water (carried on a shoulder pole)
【担担面】Sichuan noodles with peppery sauce
【担子】a carrying pole and the loads on it; load; burden: ～拣重的挑 ready to shoulder the heaviest loads

但 ①(但是) but; yet; still; nevertheless ②(只) only; merely
【但凡】in every case; without exception; as long as
【但求无过】seek only to avoid blame
【但是】[连] but; yet; still; nevertheless
【但书】[法] proviso
【但愿】if only; I wish: ～如此。 I wish it were true. (或: Let's hope so.) / ～天气赶快放晴。 If only it would clear up soon!

蛋 ①(鸟、龟、蛇等产的卵) egg: 鸡～ hen's egg / 下～ lay eggs / 鲜～ fresh eggs ②(蛋形的) an egg-shapped thing: 泥～儿 mud ball
【蛋白】①(卵中透明的胶状物质) egg white; albumen ②(蛋白质) protein ◇ ～酶 [生化] protease; proteinase / ～尿[医] albuminuria
【蛋白石】[地] opal
【蛋白质】protein ◇ ～塑料 protein plastics
【蛋彩画】tempera painting
【蛋粉】powdered eggs; egg powder
【蛋糕】cake
【蛋黄】yolk

【蛋黄粉】yolk powder
【蛋卷】egg roll
【蛋卷冰淇淋】ice cream cone
【蛋壳】eggshell
【蛋品】egg products: ～商业生产 commercial egg production
【蛋清】[口] egg white ◇ ～画 egg tempera
【蛋用鸡】layer

dāng

当 ①(相称) equal: 实力相～ well-matched in strength ②(应当) ought; should; must ③(面对着; 向着) in sb.'s presence; to sb.'s face: ～着他大胆地讲 speak out in his face ④(正在…) just at (a time or place): ～场 on the spot / ～时 at that time ⑤(担任; 充当) work as; serve as; be: ～翻译 serve as interpreter; act as interpreter / ～官做老爷 act like an overlord / 选他～组长 elect him group leader ⑥(承当; 承受) bear; accept; deserve: 一人做事一人～。One should answer for what he does. ⑦(掌管; 主持) direct; manage; be in charge of: ～家 manage household affairs ⑧[象] ～～的钟声 the tolling of a bell; the ding-dong of bells
【当班】be on duty by turn
【当兵】be a soldier; serve in the army
【当场】on the spot; then and there: ～拒绝他们的要求 turn down his request on the spot / ～抓住 catch red-handed; catch in the act
【当场出彩】①(出丑) make a spectacle of oneself ②(露出马脚) give the whole show away on the spot
【当初】originally; at the outset; in the first place; at that time
【当代】the present age; the contemporary era: ～文学 contemporary literature / ～英语 present-day English
【当道】①(路中间) blocking the way ②[贬](掌握政权) be in power; hold sway
【当地】at the place in question; in the locality; local: ～标准时间 local standard time / ～风俗习惯 local customs / ～人民 local people / ～时间 local time
【当地公司】domestic corporation
【当断不断】be indecisive when decision is need
【当归】[中药] Chinese angelica
【当机立断】decide promptly and opportunely; make a prompt decision
【当即】at once; right away: ～表示同意 give one's consent right away
【当家】manage (household) affairs
【当家作主】be master in one's own house; be the master of one's own affairs 〈destiny〉
【当街】facing the street
【当今】now; at present; nowadays: ～之世 in the world of today; at the present time
【当局】the authorities: 地方～ the local authorities / 学校～ the school authorities / 有关～ the authorities concerned / 政府～ the government authorities
【当局者迷, 旁观者清】the spectators see the chess game better than the players; the onlooker sees most of the game
【当空】high above in the sky: 一轮明月～照。A bright moon is shining in the sky.
【当口儿】[口] this or that very moment: 就在这～ at the very moment; just at that time
【当啷】[象] clank; clang
【当量】[化] equivalent (weight): 电化～ electrochemical equivalent / 克～ gram equivalent ◇ ～比例定律 the law of equivalent proportions / ～浓度 equivalent concentration
【当令】in season
【当面】to sb.'s face; in sb.'s presence: ～弄清楚 straighten thing out face to face / ～撒谎 tell a barefaced lie / ～说好话, 背后下毒手 say nice things to sb.'s face, then stab him in the back
【当面锣对面鼓】right in one's face; in one's presence
【当年】①(过去某一时间) in those years; in those days ②(身强力壮时) the prime of life: 她正～。She is in her prime.
【当前】①(在面前) before one; facing one: 大敌～。with a formidable enemy before us ②(目前; 现阶段) present; current: ～的国际形势 the current international situation / ～的中心任务 the central task at present / ～利益 immediate interests / ～世界的主要倾向 the main trend in the world today
【当权】be in power; hold power ◇ ～者 person in power; people in authority
【当然】①(应当这样) as it should be; only natural ②(合情合理) without doubt; certainly; of course; to be sure ③(理所应当) natural: ～同盟军 natural ally ◇ ～继承 natural succession / ～继承人 heir apparent; natural heir
【当仁不让】not pass on to others what one is called upon to do; not decline to shoulder a responsibility
【当时】then; at that time
【当事人】①[法](参加诉讼的一方) party (to a lawsuit); litigant ②(跟事情有关的人) person

concerned; party concerned; interested parties ◇ ～的行为 act of party / ～的一方 one of the parties / ～能力 capacity of party; capacity to be a party
【当头】①(迎头) right overhead; right on sb.'s head; head on: ～一棒 a head-on blow / 给他～一瓢冷水 pour cold water on him ②(到了眼前;临头) facing one; confronting one; imminent: 国难～ confronted with the national crisis
【当头棒喝】a blow and a shout; a sharp warning; a severe warning
【当务之急】a pressing matter of the moment; a task of top priority; urgent matter
【当下】instantly; immediately; at once
【当先】in the van; in the front ranks; at the head: 奋勇～ fight bravely in the van / 一马～ gallop at the head; take the lead
【当心】take care; be careful; look out: ～碰脑袋! Mind your head. / ～有车! Look out! Mind the car.
【当选】be elected: 他～为班长。He was elected monitor. / 他～为委员。He was elected to the committee.
【当政】be in power; be in office
【当之无愧】fully deserve (a title, an honor, etc.); be worthy of
【当中】①(正中) in the middle; in the center: 坐在主席台～ be seated in the center of the rostrum / 河～水流最急。The current is swiftest in the middle of the river. ②(中间;之内) among
【当众】in the presence of all; in public: ～出丑 make an exhibition of oneself / ～认错 acknowledge one's mistakes in public / ～宣布 announce publicly
【当作】look on as; take as; treat as

档 ①(裤裆) crotch (of trousers) ②[解](腿裆) crotch

dǎng

挡 ①(拦住) keep off; ward off; block: ～风 shelter sth. from the wind; keep out the wind / ～雨 keep off the rain; shelter one from the rain ②(遮蔽) block; get in the way of: ～光 be in the light; get in the light / ～路 be in the way; get in the way ③(挡子) fender; blind: 炉～儿 (fire) fender; fire screen / 窗～子 window blind; window shade ④[汽车](排挡) gear: 倒车～ reverse gear / 低速～ bottom gear / 高速～ top gear / 前进～ forward gear
【挡车工】spinner
【挡风玻璃】a windshield
【挡风器】draught deflector
【挡驾】[婉] turn away a visitor with some excuse; decline to receive a guest
【挡箭牌】①(盾牌) shield ②(借口) excuse; pretext
【挡路】be in the way; obstruct traffic
【挡泥板】[汽车] mudguard; fender
【挡土墙】[建] retaining wall
【挡住】block; hamper: ～风沙 shield off the sandstorm / ～去路 block the way / 用扇子～脸 screen one's face with a fan

党 ①(政党) political party; party ②(中国共产党) the Party (the Communist Party of China): ～的生活 Party life / 入～ join the Party / 整～ Party consolidation ③(集团) clique; faction; gang: 死～ sworn follower ④[书](偏袒) be partial to; take sides with: ～同伐异 defend those who belong to one's own faction and attack those who don't ⑤[书](亲族) kinsfolk; relatives: 父～ father's kinsfolk
【党八股】stereotyped Party writing; Party jargon
【党报】party newspaper; organ of the party
【党代表】Party representative
【党的基层组织】Party organizations at the primary level
【党费】party membership dues
【党风】party conduct
【党纲】party program
【党锢】suppression of the conspiratorial cliques
【党棍】a dirty politician who uses his party membership as a means in promoting self-interest
【党籍】party membership: 开除～ expel from the party
【党纪】party discipline: ～国法 party discipline and the law of the state
【党课】Party class; Party lecture: 听～ attend a Party lecture / 讲～ give a Party lecture
【党魁】[贬] party chieftain; party boss
【党龄】party standing: 一位多年～的老党员 a Communist Party member of many years' standing
【党内】within the party; inside the party; inner-party ◇ ～民主 democracy within the Party; inner-Party democracy
【党派】political parties and groups; party

groupings: ～关系 party affiliation
【党旗】party flag
【党人】partisans
【党同伐异】defend those who belong to one's own faction and attack those who don't; be narrowly partisan
【党徒】①(参加某一集团的人) member of a clique or a reactionary political party ②(亲信) henchman
【党团】①(政党和团体) political parties and other organization ②(中国共产党和青年团) the Chinese Communist Party and the Chinese Communist Youth League; the Party and the League: ～员 Party and League members ③(多用于外国议会) 议会～ parliamentary group of a political party
【党外】outside the party ◇～人士 non-Party personages
【党委】Party committee ◇～书记 secretary of the Party committee / ～委员 member of the Party committee / ～制 the Patty committee system
【党务】party work; party affairs
【党小组】Party group
【党校】Party school
【党性】party spirit; party character: ～不纯的表现 a sign of impurity in party spirit
【党羽】[贬] members of a clique; adherents; henchmen
【党员】party member: 预备～ probationary member of the Party ◇～标准 requirements for a party member / ～大会 general membership meeting of a party organization; meeting of all party members
【党章】party constitution
【党政机关】Party and government offices
【党证】party card
【党支部】Party branch: 发挥～的战斗堡垒作用 bring into play the role of fighting bastion of the Party branch ◇～书记 Party branch secretary
【党中央】the Party Central Committee; the central leading body of the Party
【党总支】general Party branch
【党组】leading Party group

dàng

宕 [书] delay: 延～ procrastinate; put off

当 ①(合适; 合宜) proper; right: 用词不～ inappropriate choice of words ②(抵得上) match; equal to ③(作为; 当作) treat as; regard as; take for: 不要把支流～主流。Don't take minor aspects for major ones. ④(认为) think ⑤(事情发生的时间) that very day etc: ～月 the same month; that very month ⑥(典当) pawn: ～衣服 pawn one's clothes; put one's clothes in pawn ⑦(典当物) sth. pawned; pawn; pledge: 赎～ take sth. out of pledge; redeem sth. pawned
【当耳边风】take no serious heed to
【当年】the same year; that very year
【当票】pawn ticket
【当铺】pawnshop ◇～老板 pawnbroker
【当日】the same day; that very day: ～有效 good for the date of issue only
【当时】right away; at once; immediately
【当天】the same day; that very day: ～事～毕。Don't put off today's work for tomorrow. (或: Today's work must be done today.)
【当真】①(信以为真) take seriously ②(果然) really true; really
【当做】treat as; regard as; look upon

档 ①(有格子的架或橱) shelves (for files); pigeonholes: 把文件归～ file a document; place a document on file ②(档案) files; archives: 查～ consult the files ③(起支承作用的木条) crosspiece (of a table, etc.) ④(等级) grade: 低～商品 low-grade goods / 高～商品 high-grade goods
【档案】files; archives; record; dossier ◇～材料 dossier / ～馆 archives / ～管理员 archivist / ～柜 filing cabinet
【档块】link stopper

荡 ①(摇动) swing; sway; wave: ～桨 pull on the oars / ～秋千 play on a swing ②(游荡) loaf: 游～ loaf about ③(冲洗) rinse: 冲～ rinse out; wash away ④(清除) clear away; sweep off: 扫～ mopping up; mopping-up operation ⑤(放纵) loose in morals: 放～ dissolute; dissipated / 淫～ lustful; lascivious ⑥(洼地) shallow lake; marsh: 芦苇～ a reed marsh
【荡船】swingboat
【荡荡悠悠】swing; moving to and fro
【荡涤】cleanse; clean up; wash away: ～旧社会遗留下来的污泥浊水 clean up the filth and mire left over from the old society
【荡妇】a woman of loose moral; vampire
【荡平】mop up and quell; suppress and wipe out

【荡气回肠】very touching; pathetic
【荡然无存】[书] all gone; nothing left
【荡漾】ripple; undulate: 歌声～ the song rose and fell like waves / 湖水～ there were ripples on the lake

dāo

氘 [化] deuterium (H^2 或 D)
【氘核】deuteron
刀 ①(刀子) knife; sword ②(形状像刀的东西) sth. shaped like a knife ③[量](一百张纸) one hundred sheets of paper
【刀把子】①(刀柄) (sword) hilt ②(权柄) military power; power ③(把柄) sth. that may be used against one; a handle
【刀背】the back of a knife blade
【刀笔】writing of indictments, appeals, etc.; pettifoggery ◇ ～吏 petty official who draws up indictments, etc.; pettifogger
【刀兵】①(武器) weapons; arms ②(战争) fighting; war: 动～ resort to arms; resort to force
【刀柄】hilt; knife handle
【刀叉】knife and fork
【刀豆】sword bean
【刀锋】the point or edge of a knife
【刀斧手】executioners
【刀耕火种】slash-and-burn cultivation
【刀光剑影】the glint and flash of cold steel
【刀痕】[机] tool marks
【刀架】[机] tool carrier; tool carriage
【刀尖】① [机] end land; nose of tool ②(刀的尖儿) point of a knife; point a of sword
【刀具】[机] cutting tool; tool
【刀口】①(刀刃) the edge of a knife ②(最能发挥作用的地方) where a thing can be put to best use; the crucial point; the right spot: 把劲儿使在～上 bring efforts to bear on the right spot ③(刀伤) cut; incision
【刀片】①(刮胡刀片) razor blade ②[机](片状零件) (tool) bit; blade
【刀枪】sword and spear; weapons: ～入库，马放南山 put the weapons back in the arsenal and graze the war horses on the hillside; relax vigilance against war
【刀鞘】sheath; scabbard
【刀刃】①(刀口) the edge of a knife ②(最能发挥作用的地方) where a thing can be put to best use; the crucial point: 好钢用在～上 use the best steel to make the knife's edge; use resources where they are needed most
【刀山火海】a mountain of swords and a sea of flames; most dangerous places; most severe trials
【刀下留人】Hold the execution!
【刀形开关】chopper switch; knife-blade switch
【刀子】[口] small knife; pocketknife
叨
【叨唠】talk on and on; chatter away: 别总～那件小事。 Don't rattle over that trifle. / 他～了半天。 He chattered for a long time.
【叨念】chatter incessantly

dǎo

祷 ①(祷告) pray ②(盼望) ask earnestly; beg
【祷告】pray; say one's prayers
蹈 ①(践踏) tread; step: 赴汤～火 go through fire and water; defy all difficulties and dangers / 循规～矩 not step out of bounds; toe the line; stick to convention ②(跳动) skip; trip: 舞～ dance
【蹈常袭故】go on in the same old way; get into a rut
【蹈袭】follow slavishly: ～前人 slavishly follow one's predecessors
倒 ①(横躺下) fall; topple: 摔～ fall over ②(失败；垮台) collapse; fail: 内阁～了。 The cabinet collapsed. ③(倒闭) close down; go bankrupt; go out of business ④(嗓子变低或变哑) (of voice) become hoarse: 他嗓子～了。 He has lost his voice. ⑤(转移) change; exchange: ～车 change trains or buses / ～肩 shift a burden from one shoulder to the other ⑥(腾挪) move around
【倒把】engage in profiteering; speculate: 投机～ engage in speculation and profiteering
【倒班】change shifts; work in shifts; work by turns: 昼夜～ work in shifts round the clock
【倒闭】close down; go bankrupt; go into liquidation: 企业～ bankruptcy of an enterprise
【倒茬】[农] rotation of crops
【倒伏】(of crops) lodging: 抗～力强的稻种 a strain of rice with strong resistance to lodging
【倒戈】change sides in a war; turn one's coat; transfer one's allegiance
【倒戈投敌】turn renegade, be a turncoat
【倒换】①(轮流替换) rotate; take turns: ～着看护伤员 take turns looking after the wounded ②(换掉) rearrange (sequence, order, etc.); replace

【倒买倒卖】fraudulent buying and selling
【倒霉】have bad luck; be out of luck; be down on one's luck: 真～! What lousy luck!
【倒嗓】(of a singer) lose one's voice
【倒手】(of merchandise, etc.) change hands
【倒塌】collapse; topple down: 房子～了。The house collapsed. (或: The house fell down.)
【倒台】fall from power; downfall
【倒头】touch the pillow; lie down: ～就睡 tumble into bed
【倒胃口】spoil one's appetite: 一见它就～ feel sick at the sight of it
【倒向】swing to; move to
【倒运】①(甲地买运到乙地卖) profiteer by buying cheap and selling dear ②[方] (倒霉) be unlucky; be out of luck

岛 island: 安全～ safety island / 海南～ Hainan Island / 小～ islet
【岛国】country consisting of one or more islands; island country
【岛弧】island arcs
【岛宇宙】[天] island universe
【岛屿】islands and islets; islands
【岛状冰山】[海] ice island; iceberg

捣 ①(捶打) pound with a pestle, etc.; beat; smash: ～米 husk rice with a pestle and mortar / ～药 pound medicine in a mortar / ～衣 beat clothes (in washing) / 直～匪巢 drive straight on to the bandits' den ②(搅乱) harass; disturb
【捣蛋】make trouble: 调皮～ be mischievous
【捣锤】stamp hammer
【捣固】make firm by ramming or tamping ◇ ～机 tamping tool
【捣鬼】play tricks; do mischief
【捣毁】smash up; demolish; destroy: ～敌军据点 destroy enemy strongpoints
【捣臼】stamp box
【捣烂】pound sth. until it becomes pulp
【捣乱】make trouble; create a disturbance: 别跟我～! Don't make trouble with me! ◇ ～分子 trouble-maker
【捣碎】pound to pieces ◇ ～机 stamp mill; gravity mill

导 ①(引导) lead; guide: ②(传导) transmit; conduct: ～电 transmit electric current; conduct electricity ③(开导) instruct; teach; give guidance to: 教～ teach; instruct
【导标】[交] beacon
【导弹】guided missile: 弹道～ ballistic missile / 地对地～ surface-to-surface missile / 地对空～ ground-to-air missile; surface-to-air missile / 地对空寻的制导～ homing surface-to-air guided missile / 反导弹～ antimissile missile; contra-missile / 反轰炸机地对空～ ground-to-air bomber destroyer / 防空～ anti-aircraft missile; interceptor missile / 防御性～ defensive guided missile / 红外线寻的制导～ infrared seeker / 机载～ guided aircraft missile; guided air rocket / 截击～ interceptor missile / 空对地～ air-to-ground missile / 空对空～ air-to-air missile / 空对水下～ air-to-underwater missile / 雷达制导～ radar homing missile; radar guided missile / 潜艇发射的～ submarine-launched missile / 寻的～ target-seeking missile / 远程战术～ long-range tactical missile / 中程弹道～ intermediate range ballistic missile / 洲际弹道～ international ballistic missile ◇ ～弹头 missile warhead / ～发射场 missile (launching) site; launching sits / ～发射井 launching silo / ～发射器 missile launcher / ～发射台 (missile) launching pad / ～核潜艇 nuclear submarine armed with guided missiles / ～基地 missile base / ～驱逐舰 guided missile destroyer / ～系统 a guided missile system / ～巡洋舰 guided missile cruiser
【导电】electric conduction ◇ ～玻璃 conductive glass / ～塑料 conductive plastics / ～体 electric conductor / ～性 electric conductivity
【导风板】[航空] baffle
【导管】①[机] (输液管) conduit; pipe; duct: 金属～ metal conduit / 冷却～ cooling duct ②(生物体内输液管) vessel; duct
【导轨】[机] slideway; guide: 刀架～ tool guide ◇ ～磨床 slideway grinder
【导棍】deflector roll; delivery roll
【导航】navigation: 雷达～ radar navigation / 天文～ celestial navigation / 无线电～ radio navigation ◇ ～灯 range lights / ～雷达站 navigation radar station / ～台 guidance station; nondirection radio beacon (NDB) / ～卫星 navigational satellite / ～仪 avigraph; navigator
【导火索】[军] (blasting) fuse
【导火线】①(引线) (blasting) fuse ②(引起事变爆发的事件) a small incident that touches off a big one: 战争的～ an incident that touches off

a war
【导流】[水] diversion ◇ ～隧洞 diversion tunnel
【导轮】[机] guide pulley; pilot wheel
【导论】an introduction
【导尿】[医] catheterization
【导盘】[化纤] godet
【导热】[物] heat conduction ◇ ～能力 capacity of heat transmission / ～系数 thermal conductivity
【导伞】[航] pilot chute
【导师】①(指导老师) tutor; teacher ②(指导大事业的人) guide of a great cause; teacher
【导数】[数] derivative
【导体】[物] conductor: 半～ semiconductor / 超～ superconductor / 非～ nonconductor
【导线】[电] lead; (conducting) wire: 玻璃纤维～ fiberglass wire ◇ ～管 conduit / ～荷载 conductor load
【导向架】[铁路] leading truck
【导向轮】directive wheel; steerable wheel
【导言】introduction (to a piece of writing); introductory remarks
【导演】①(组织演出) direct (a film, play, etc.) ②(担任导演的人) director
【导游】①(向导) conduct a sightseeing tour ②(旅游指南) guidebook ◇ ～图 tourist map
【导源】①(发源) (of a river) have its source ②(由某物发展而来) originate; derive
【导致】lead to; bring about; result in; cause: ～分裂 cause a split; bring about a split / ～战争 lead to war

dào

盗 ①(偷) steal; rob ②(强盗) thief; robber
【盗伐】fell trees unlawfully
【盗匪】bandits; robbers: 肃清～ exterminate banditry
【盗汗】[医] night sweat
【盗卖】steal and sell: ～公物 steal and sell public property
【盗名】steal glory one does not deserve; seek undeserved publicity: 欺世～ steal fame by deceiving the public
【盗墓】rob a tomb; rob a grave ◇ ～人 grave robber
【盗窃】steal: ～国家机密 steal state secrets ◇ ～犯 thief / ～集团 embezzling group / ～行为 act of theft / ～罪 larceny; banditry
【盗取】steal; embezzle
【盗用】embezzle; usurp: ～公款 embezzle public funds / ～名义 usurp a name ◇ ～公章罪 crime of fraudulent use of public seals
【盗贼】robbers; bandits

悼 mourn; grieve: 哀～死者 mourn for the dead
【悼词】memorial speech: 致～ make a memorial speech
【悼念】mourn; grieve over: ～亡友 mourn for a dead friend; mourn over the death of a friend / 沉痛～ mourn with deep grief

道 ①(道路) road; way; path: 山间小～ a mountain path ②(水道) channel; course: 黄河改～ change of course of the Huanghe River ③(方法) way; method: 养生之～ the way to keep fit / 以其人之～, 还治其人之身 deal with a man as he deals with you; pay sb. back in his own coin ④(学术或宗教体系) doctrine; principle: 传～ preach / 孔孟之～ doctrines of Confucius and Mencius; teachings of Confucius and Mencius ⑤(道教) Taoism; Taoist: ～观 a Taoist temple / 老～ a Taoist priest ⑥(迷信组织) superstitious sect: 会～门 superstitious sects and secret societies ⑦(线条) line: 画一条斜～儿 draw a slanting line ⑧[量] (用于某些长条的东西) 万～金光 myriads of golden rays / 一～缝儿 a crack / 一～河 a river ⑨[量] (用于门、墙等) 两～门 two successive doors / 三～防线 three lines of defense ⑩[量](用于命令、题目等) 出五～题 set five questions (for an examination, etc.) / 一～命令 an order ⑪ [量] (表示"次") 上四～菜 serve four courses / 省一～手续 save one step in the process ⑫(说) say; talk; speak: 常言～ as the saying goes / 能说会～ have a glib tongue; have the gift of the gab ⑬ (以为) think; suppose
【道白】spoken parts in an opera
【道班】railway or highway maintenance squad
【道不拾遗】no one pockets anything found on the road; honesty prevails throughout society
【道岔】[铁道] switch; points
【道场】①(做的法事) Taoist or Buddhist rites (performed to save the souls of the dead) ②(做法事的场所) place where such rites are performed
【道床】[铁道] roadbed: 整体～ monolithic roadbed
【道道儿】[方] way; method: 说出个～来 give a convincing explanation / 找到增产的新～ find new ways of increasing production

【道德】morals; morality; ethics: ～败坏 degenerate / 共产主义～ communist morality; communist ethics / 公共～ public morals / 旧～观念 old moral concepts / 商业～ commercial morality; business ethics / 体育～ sportsmanship ◇ ～规范 code of ethics / ～品质 moral character / ～问题 a question of morality / ～哲学 moral philosophy / ～准则 code of ethics
【道钉】[铁道] (dog) spike
【道高一尺,魔高一丈】as virtue rises one foot, vice rises ten; the more illumination, the more temptation
【道姑】Taoist nun
【道观】Taoist temple; Taoist abbey
【道号】Taoist monastic name (of person)
【道贺】congratulate
【道家】Taoist school
【道教】[宗] Taoism ○三才(天地人) the Three Geniuses (the heaven, the earth and the man) / 天机 God's design; hidden plans of Providence / 三清 Taoist trinity / 天数 predestination / 天条 the Laws of God in heaven
【道经】Taoist scriptures
【道具】[剧] stage property; prop ◇ ～管理员 property man
【道口】road junction; level crossing
【道理】①(事物的规律) principle; truth; hows and whys: 革命～ the truth of revolution / 讲解深耕细作的～ explain the principles of deep ploughing and intensive cultivation ②(情理; 理由) reason; argument; sense: 摆事实,讲～ bring out facts and reason things out / 讲不出一点～ unable to come up with any convincing argument; unable to justify oneself in any way
【道林纸】glazed printing paper
【道路】road; way; path: 为两国首脑会谈铺平～ pave the way for summit talks between the two countries / 走前人没有走过的～ break paths none have explored before
【道貌岸然】pose as a person of high morals; be sanctimonious
【道门】sect: 反动～ reactionary illegal religious sect
【道袍】Taoist robe; Taoist priest's robe
【道破】point out frankly; lay bare; reveal: 一语～其中奥秘 lay bare its secret with one remark
【道歉】apologize; make an apology
【道情】[曲艺] chanting folk tales to the accompaniment of simple percussion instruments
【道士】Taoist priest
【道听途说】hearsay; rumor; gossip
【道喜】congratulate sb. on a happy occasion
【道谢】express one's thanks; thank
【道学】①(理学) a Confucian school of philosophy of the Song Dynasty; Neo-Confucianism ②(古板; 迂腐) affectedly moral: 假～ canting moralist; hypocrite ◇ ～先生 Confucian moralist
【道义】morality and justice: ～上的支持 moral support / ～上说不过去 not justified on moral principles
【道砟】[铁道] ballast

到 ①(到达) arrive; reach ②(往) go to; leave for: ～苏州去 go to Suzhou / ～群众中去 go among the masses; go into the midst of the masses ③(止) up until; up to: ～目前为止 up to the present; until now; so far / 从星期日～星期三 from Sunday to Wednesday ④(用作动词的补语,表示动作有结果) 办得～ can be done / 说～做～ be as good as one's word ⑤(周到; 想得到) thoughtful; considerate
【到岸价格】cost, insurance and freight (C.I.F.)
【到场】be present; show up; turn up
【到处】at all places; everywhere: ～流浪 wander from place to place / ～碰壁 run against the wall everywhere / ～煽动 go around agitating
【到达】arrive; get to; reach ◇ ～港 port of arrival / ～站 destination
【到底】①(到尽头) to the end; to the finish: 打～ fight to the finish ②[副](最后) at last; in the end; finally ③[副](究竟) ever; indeed ④[副](终究) after all; in the final analysis
【到顶】reach the summit; reach the peak; reach the limit; cannot be improved
【到会】be present at a meeting; attend a meeting
【到货通知】arrival notice
【到家】reach a very high level; be perfect; be excellent: 她的武术练～了。She is expert at Chinese boxing.
【到来】arrival; advent
【到目前为止】so far; by now; up to now
【到期】become due; mature; expire: 签证下月～。The visa expires next month. ◇ ～股金和股息 due share capital and dividends / ～票据 due bill / ～日 date due
【到任】assume a post; assume office
【到手】in one's hands; in one's possession
【到头】to the end; at an end; throughout: 一年

～ throughout the year / 这条街走～有一个邮局。There is a post office at the end of the street.
【到头来】[副] in the end; finally
【到职】take office; arrive at one's post; assume office

倒 ①(颠倒)upside down; inverted; inverse; reverse: 次序～了。The order is reversed. ②(向相反方向移动) move backward; turn upside down: 把车～回车库 back a car into the garage ③(倾倒) pour; tip: ～垃圾 tip rubbish; dump rubbish / ～一杯茶 pour a cup of tea ④[副](表示跟意料相反) on the contrary: 病人情况不但没有好转，～恶化了。The patient was no better but rather grew worse. ⑤[副](表示事情不是那样) 你说得～容易，做起来可不容易。That's easier said than done. ⑥[副](表示让步) 东西～不坏，就是太贵了些。The thing is not so bad, only it is a bit too dear. ⑦[副](表示催促或追问) 你～去不去呀? Do you want to go or don't you?
【倒背如流】can even recite sth. backwards fluently; know sth. thoroughly by heart
【倒背手】with one's hands behind one
【倒不如】it's better to; no better than
【倒彩】booing; hooting; catcall: 喝～ make catcalls; boo and hoot
【倒车】back a car: 开历史～ turn back the wheel of history; put the clock back
【倒刺】hangnail; agnail
【倒打一耙】make unfounded counter-charges; put the blame on one's victim; recriminate
【倒挡】reverse gear
【倒飞】[航空] upside down flight; inverted flight
【倒挂金钟】[植] fuchsia
【倒灌】flow backward
【倒果为因】take effect for cause
【倒好儿】booing; hooting; catcall: 叫～ make catcalls; boo and hoot
【倒睫】[医] trichiasis
【倒立】①(顶端朝下) stand upside down ②[体] handstand
【倒流】flow backwards: 防止商品运输上的～ avoid transporting goods back to their place of origin
【倒赔】have to pay after expecting to receive
【倒片】[电影] rewind ◇ ～机 rewinder
【倒三角形】[数] del; nabla
【倒摄遗忘】[心] retroactive amnesia; retrograde amnesia
【倒收付息】[经] negative interest
【倒数】count backwards: ～第三行 the third line from the bottom
【倒数计时】[航天] count down
【倒数】[数] reciprocal
【倒算】seize back confiscated property
【倒贴】pay for the upkeep of a lover
【倒退】go backwards; fall back: 坚持进步，反对～ persist in progress and oppose retrogression
【倒相器】inverted amplifier
【倒像】[物] inverted image
【倒行逆施】①(违背时代方向) go against the historical trend; try to put the clock back; push a reactionary policy ②(违背社会正义) perverse acts
【倒叙】flashback
【倒悬】[书] hang by the feet; be in sore straits
【倒因为果】reverse cause and effect; take cause for effect
【倒影】inverted image; inverted reflection in water
【倒栽葱】fall head over heels; fall headlong
【倒置】place upside down; invert: 本末～ attend to the superficial and neglect the essentials / 轻重～ place the unimportant before the important
【倒转】turn the other way round; reverse
【倒装词序】[语] inverted word order
【倒装句】[语] inverted sentence

稻 rice; paddy: 旱～ upland rice / 双季～ double-cropping paddy / 水～ paddy rice / 晚～ late rice / 早～ early rice / 中～ middle-season rice
【稻白叶枯病】bacterial blight of rice
【稻苞虫】rice plant skipper
【稻草】rice straw ◇ ～人 scarecrow
【稻恶苗病】Bakanae disease of rice
【稻秆锈病】rice bunt
【稻谷】paddy ◇ ～分选机 rice sorter / ～脱壳机 rice sheller
【稻糠】rice chaff
【稻壳】rice husk; rice hull
【稻烂秧】seedling blight of rice
【稻粒】grain of rice
【稻螟虫】rice borer
【稻热病】rice blast
【稻穗】ear of rice; spike of rice
【稻田】(rice) paddy; rice field; paddy field ◇

～除草机 paddy field weeder／～皮炎［医］paddy-field dermatitis／～热 rice-field fever／～中耕机 paddy field cultivator
【稻条纹病】cercosporiosis of rice
【稻瘟病】rice blast
【稻纹枯病】sheath and culm blight of rice
【稻秧】rice seedlings; rice shoots
【稻叶黑穗病】leaf smut of rice
【稻种】seed rice
【稻子】［口］rice; paddy
【稻纵卷叶螟】rice leaf roller

dé

锝 technetium (Tc)

得 ①（得到）get; obtain; gain: ～了结核病 have tuberculosis; contract tuberculosis／取～经验 gain experience ②（演算结果）(of a calculation) result in: 四二～八。Four times two makes eight. ③（适合）fit; proper: ～用 fit for use; handy ④［书］（得意）satisfied; complacent: 自～ pleased with oneself; self-satisfied ⑤［口］（完成）be finished; be ready: 饭～了。Dinner is ready. ⑥［口］（表示同意或禁止）～，就这么办。All right! Just go ahead.
【得便】when it's convenient
【得不偿失】the loss outweighs the gain; the game is not worth the candle
【得逞】［贬］have one's way; prevail; succeed
【得宠】［贬］find favor with sb.; be in sb.'s good graces
【得出】reach (a conclusion); obtain (a result): 从中～教训 draw a lesson (from)
【得寸进尺】reach out for a yard after taking an inch; give him an inch and he'll take an ell; be insatiable
【得当】apt; appropriate; proper; suitable: 安排～ be properly arranged／措词～ aptly worded; appropriate wording
【得到】get; obtain; gain; receive: ～成功 achieve success／～广泛运用 find broad application／～教训 draw lessons (from)／～及时治疗 get timely medical treatment／～群众的支持 enjoy the support of the masses／～荣誉 win honor／～优势 gain the upper hand／～最充分发挥 be brought into fullest play
【得道多助，失道寡助】a just cause enjoys abundant support while an unjust cause finds little support
【得法】do sth. in the proper way; get the knack: 管理～ be properly managed／讲授不甚～ not teach in the right way; not teach in the proper way
【得分】score: 连得四分 win four points in a row
【得过且过】muddle along; drift along
【得计】succeed in one's scheme: 自以为～ think oneself clever
【得奖】win a prize; be awarded a prize ◇～单位 prizewinning unit／～人 prizewinner
【得劲】①（舒服）feel well: 他这几天身体不太～。He is not so well these days. ②（方便；好用）fit for use; handy: 这支笔使起来很～。This pen writes easily.
【得空】have leisure; be free
【得了】stop it; hold it
【得力】①（得益）benefit from; get help from: ～于平时勤学苦练 profit from diligent study and practice ②（有能力）capable; competent: ～干部 competent cadre／～助手 capable assistant; right-hand man／办事～ do things efficiently
【得陇望蜀】covet Sichuan after capturing Gansu; have insatiable desires
【得人心】popular; beloved and supported by the people
【得胜】win a victory; triumph: ～归来 return in triumph; return with flying colors
【得失】①（所得和所失）gain and loss; success and failure: ～相当 gains and losses balance each other; break even／从不计较个人～ never give a thought to personal gain or loss ②（利弊）advantages and disadvantages; merits and demerits: ～参半 have both merits and demerits
【得失荣枯】the vicissitudes of life
【得势】①（掌权）be in power ②（有影响）get the upper hand; be in the ascendant
【得手】go smoothly; come off; do fine; succeed
【得体】befitting one's position or suited to the occasion; appropriate: 讲话～ speak in appropriate terms
【得天独厚】be richly endowed by nature; abound in gifts of nature; enjoy exceptional advantages
【得悉】hear of; learn about
【得闲】have leisure; be at leisure
【得新厌旧】disdain the old when one gets the new
【得心应手】①（运用自如）with facility; with high proficiency ②（好用）serviceable; handy
【得一望二】have an insatiable desire to acquire more
【得宜】proper; appropriate; suitable: 措置～

handle properly
【得以】so that... can ...; so that... may ...
【得益】benefit; profit
【得意】proud of oneself; pleased with oneself; complacent: ～扬扬 be immensely proud; look triumphant
【得意门生】favorite pupil
【得意忘形】get dizzy with success; have one's head turned by success
【得鱼忘荃】forget the trap as soon as the fish is caught; forget the means by which the end is attained; forget the things or conditions which bring one success
【得志】achieve one's ambition; have a successful career: 少年～ enjoy success when young / 小人～ villains holding sway
【得罪】offend; displease: 不～人 offend nobody; give offense to no one / 不怕～人 not be afraid of giving offense

德 ①(道德) virtue; morals; moral character: 品～ moral character ②(心意)heart; mind: 同心同～ be of one heart and one mind ③(恩惠) kindness; favor: 感恩戴～ bear a debt of gratitude (for sb.'s kindness) / 以～报怨 return good for evil; repay evil with good / 以怨报～ return evil for good; repay kindness with ingratitude; bite the hand that feeds you
【德薄能鲜】lacking both in virtues and ability
【德才兼备】have both ability and political integrity
【德高望重】be of noble character and high prestige; enjoy high prestige and command universal respect
【德国】Germany ◇ ～人 German
【德行】①(道德和品行) moral integrity; moral conduct ②[方](讽刺人的话) disgusting; shameful
【德意志联邦共和国】Federal Republic of Germany
【德意志民主共和国】German Democratic Republic
【德语】German (language)
【德育】moral education
【德政】benevolent rule
【德治】rule of virtue
【德智体全面发展】all round development of morality, intelligence and physique

de

地 (用在状语的后面) 实事求是～处理问题 handle problems in a practical and realistic way

的 ①(用在定语的后面) 铁～纪律 iron discipline / 我～父亲 my mother / 无产阶级～政党 a party of the proletariat / 已经站起来～中国人民 the Chinese people who have stood up ②(用来造成没有中心词的"的"字结构) 开汽车～ a driver / 菊花开了，有红～，有黄～。 The chrysanthemums are in bloom; some are red and some yellow. ③(用在谓语动词后面，强调动作的施事者、时间、地点等) 是我打～稿子，他上～色。 I made the sketch; he filled in the colors ④(用在陈述句末尾，表示肯定的语气) 你们这两天真够辛苦～。 You've really been working hard the past few days.
【的话】[助] (用在表示假设的分句后面，引起下文) 如果你有事～，就不要来了。 Don't come if you're busy.

得 ①(用在动词后面，表示可能) 她去～，我为什么去不～? If she can go, why can't I? ②(用在动词和补语中间，表示可能) 她回～来么? Can she get back? / 我拿～动。 I can carry it. ③(用在动词或形容词后面，连接表示程度或结果的补语) 唱～不好 not sing well / 冷～打哆嗦 shiver with cold / 笑～肚子痛 laugh till one's sides split / 写～非常好 very well written

děi

得 ①[口] (需要) need: 这一段～重写。 This paragraph needs rewriting. ②(必须；必要) must; have to: 你～认真考虑一下。 You must think it over seriously. ③(表示揣测的必然) will; ought: 你不快走，就～挨淋。 You will get wet if you don't start right off.
【得负责任】should be responsible
【得用功】must be more studious

dēng

灯 ①(照明用具) lamp; lantern; light: 壁～ wall lamp / 床头～ bedside lamp / 电～ electric light / 吊～ chandelier; pendant lamp / 宫～ palace lantern / 落地～ floor lamp / 煤油～ kerosene lamp / 日光～ fluorescent lamp / 台～ desk lamp ②(电子管) valve; tube: 六～收音机 a six-tube radio set ③(燃烧液体的器具) burner: 酒精～ alcohol burner; spirit lamp
【灯标】[航海] beacon light ◇ ～船 light float
【灯彩】①(做彩灯的工艺) colored-lantern making ②(舞台上用的彩灯) colored lanterns
【灯草】rush (used as lampwick)
【灯船】lightship; light vessel

【灯蛾】moths attracted by lamplight ◇～扑火 self-destruction; suicidal act
【灯管】(fluorescent) tube
【灯光】①(灯的光度) the light of a lamp; lamplight ②(舞台或摄影棚的照明设备) (stage) lighting: ～渐暗 lights slowly dim; lights fade to dark / 舞台～ stage lights; lighting ◇～球场 floodlit court, field etc.; illuminated court, field, etc.
【灯红酒绿】red lanterns and green wine; scene of debauchery
【灯花】snuff (of a candlewick)
【灯火】lights: ～辉煌 brilliantly illuminated; ablaze with lights ◇～管制 blackout
【灯节】the Lantern Festival (15th of the first lunar month)
【灯具】lamps and lanterns
【灯笼】lantern
【灯笼裤】knee-length sports trousers; ankle-length sports trousers; knickerbockers
【灯谜】riddles written on lanterns; lantern riddles: 猜～ guess lantern riddles
【灯泡】[口] (electric) bulb; light bulb: 卡口～ bayonet socket bulb / 螺口～ screw socket bulb / 磨砂～ frosted bulb / 乳白～ opal bulb
【灯绳】lamp cord
【灯丝】filament (in a light bulb or valve)
【灯塔】lighthouse; beacon
【灯台】lampstand
【灯头】①(电灯头) lamp holder; electric light socket: 开关～ switch socket / 螺口～ screw socket ②(油灯头) a holder for the wick and chimney of a kerosene lamp
【灯心】lampwick; wick
【灯心草】[植] rush
【灯心绒】[纺] corduroy
【灯油】lamp-oil; kerosene; paraffin oil
【灯语】lamp signal
【灯罩】lampshade; lamp-chimney
【灯座】lampstand

登 ①(由低处到高处) ascend; mount; scale (a height): ～岸 go ashore / ～上峰顶 reach the summit / ～上讲台 mount the platform ②(刊登) publish; record; enter: ～广告 advertise / ～帐 enter an item in an account book ③(谷物登场) be gathered and taken to the threshing ground: 五谷丰～ reap a bumper grain harvest ④(用力踏) press down with the foot; pedal; treadle: ～三轮车 pedal a pedicab ⑤(踩; 踏) step on; tread: ～在窗台儿上擦玻璃 step onto the sill to clean the window ⑥[方](穿) wear (shoes, etc.)
【登报】publish in the newspaper: ～声明 make a statement in the newspaper
【登场】(谷物运到场上) be gathered and taken to the threshing ground
【登场】come on stage ◇～人物 characters in a play; *dramatis personae*
【登峰造极】reach the peak of perfection; have a very high level (of scholastic attainment or technical skill); reach the limit
【登高】ascend a height: ～远眺 ascend a height to enjoy a distant view / ～自卑 reach a high position, one must start from a low position
【登革热】[医] dengue fever
【登基】ascend the throne; be enthroned
【登记】register; check in; enter one's name: 结婚～ marriage registration / 向有关部门～ register with the proper authorities / 在旅馆～住宿 check in at a hotel ◇～簿 register; registry / ～处 registration office; registry office / ～吨位 registered tonnage / ～费 registration fee / ～人 registrant / ～日期 date of registration / ～税 registration duty / ～证 registration / ～证明书 registration certificate
【登临】①(登山临水) climb a hill, a tall building, etc. which commands a broad view ②(旅游) visit famous mountains, places of interest, etc.
【登陆】land; disembark ◇～部队 landing force / ～场 beachhead / ～地点 debarkation point; landing point / ～舰 landing ship / ～母舰 landing-craft carrier / ～艇 landing boat / ～作战 landing operations
【登门】call at sb.'s house: ～拜访 pay sb. a visit / ～答谢 call on sb. to express gratitude
【登山】[体] mountain-climbing; mountaineering ◇～队 mountaineering party; mountaineering expedition / ～上衣 alpine jacket / ～绳 climbing rope / ～运动 mountaineering / ～运动员 mountaineer; alpinist; climber / ～协会 an alpinist club / ～鞋 climbing boot / ～杖 alpin-stock
【登时】immediately; at once; then and there
【登台】mount a platform; go up on the stage: ～演说 deliver a speech from a platform / 想～表演一番 strive to take the stage and perform
【登堂入室】pass through the hall into the inner chamber; reach a higher level in one's studies or become more proficient in one's profession

【登载】publish (in newspapers or magazines); carry: ～在第一版的消息 front-page news

噔 [象] thump; thud: 听见楼梯上～～～的脚步声 hear heavy footsteps on the stairs

蹬

【蹬技】[杂技] juggling with the feet

děng

戥 weigh with a small steelyard

【戥子】a small steelyard for weighing precious metal, medicine, etc.

等 ①(等级) class; grade; rank: 分为三～ classify into three grades / 三～功 third class merit / 一～品 top quality goods ②(相等) equal: ～距离 equal distance / 大小不～ unequal in size / 长短相～ be equal in length ③(等候) wait; await: ～车 wait for a train, bus, etc. / ～上级批准 await approval by the higher authorities ④(等到) when; till ⑤[助](表示复数) 我～ we⑥[助](表示列举未尽) and so on; and so forth; etc.: 购置书籍、纸张、文具～ buy books, stationery and so on / 赴兰州、银川～地视察 go to Lanzhou, Yinchuan and other places on a tour of inspection ⑦[助](列举后煞尾) 长江、黄河、黑龙江、珠江～四大河流 the four large rivers — the Changjiang, the Huanghe, the Heilongjiang and the Zhujiang

【等比】geometric ratio; ratio of equality ◇ ～级数 [数] geometrical progression

【等边】[数] equilateral ◇～三角形 equilateral triangle

【等差】equal difference

【等差级数】arithmetic progression

【等次】place in a series; grade

【等待】wait; await: ～时机 await a favorable opportunity; wait for a chance; bide one's time

【等到】by the time; when: ～他来，我便告诉他。I'll tell him when he comes.

【等等】①(表示列举未尽) and so on; and so on and so forth; etc. ②(等一下) wait a minute

【等而下之】from that grade down; lower down

【等分】divide into equal parts ◇ ～线 bisector

【等风速线】[气] isotach

【等高线】[地] contour (line) ◇～地图 contour map

【等号】[数] equal-sign; equality sign

【等候】wait; await; expect: ～命令 wait for instructions; await orders

【等级】①(级别) grade; rank: 棉花按～收购 pay for cotton according to its grade ②(社会地位) order and degree; social estate: 外交官的～ the rank of a diplomatic officer ◇ ～观念 sense of hierarchy / ～制度 hierarchy; social estate system

【等加速度】[物] uniform acceleration ◇ ～运动 uniform acceleration motion

【等价】of equal value; equal in value ◇ ～交换 exchange of equal values; exchange at equal value / ～物[经] equivalent

【等角线】isogonal line; rhumb line

【等角形】isogon

【等距离】equidistance ◇ ～外交 equidistant diplomacy

【等离子体】[物] plasma ◇ ～激光器 plasma laser / ～加速器 plasmatron / ～物理学 plasma physics

【等量齐观】equate; put on a par

【等幂】idempotent

【等日照线】[气] isohel

【等深线】[地] isobath; bathymetric contour

【等式】[数] equality

【等同】equate; be equal ◇ ～语 equivalent

【等外】substandard; off-grade ◇ ～品 substandard product

【等温线】[气] isotherm

【等闲】①(平常) ordinary; unimportant: ～视之 regard as unimportant; treat lightly; treat casually ②(随便; 轻易) aimlessly; thoughtlessly: ～虚度 pass days in a useless, common way / 大好时光，不可～度过 don't fritter away your precious time

【等效】[电] equivalent ◇ ～电抗 equivalent reactance / ～天线 equivalent antenna

【等压面】[气] isobaric surface; constant pressure surface

【等压线】[气] isobar; isobaric line

【等腰】[数] isosceles ◇ ～三角形 isosceles triangle

【等音】[乐] enharmonic

【等于】①(相等) equal to; equivalent to: 四加二～六。Four plus two equals (is) six. ②(几乎就是) amount to; be tantamount to: 他的回答～拒绝。His answer amounts to a refusal.

【等雨量线】[气] isohyet

【等震线】[地质] isoseismal line

【等值线】[气] isopleth ◇ ～图 isogram

dèng

澄 (of a liquid) settle

【澄清】(of a liquid) settle; become clear: 明矾可以～浊水。 Alum will settle turbid water.
【澄沙】sweetened bean paste
磴 steps on rock
【磴道】rocky mountain path
瞪 open (one's eyes) wide; stare; glare
【瞪眼】①(睁大眼睛) open one's eyes wide; stare; glare: 干～look on helplessly ②(怒目而视) glower and glare at sb.; get angry with sb.: 我生气地瞪了他一眼。 I gave him on angry stare.
镫 stirrup
【镫骨】[生理]stapes; stirrup bone
凳
【凳子】stool: 长～ bench / 方～ square stool / 竹～ bamboo stool

dī

滴 ①(滴落)drip: ～眼药 put drops in one's eyes ②[量] drop: 一～水 a drop of water
【滴虫】trichomonad ◇ ～病 trichomoniasis / ～性阴道炎 trichomonas vaginitis
【滴答】①[象] tick; ticktack; ticktock: 台钟在～～地响。 The clock is ticking. ②(成滴地落下) drip
【滴滴答答】ticktack; pitapat
【滴滴滴】[化] DDD
【滴滴涕】DDT (dichloro-diphenyl-trichloroethane)
【滴定】[化] titration: 比浊～ heterometric titration / 碘量～ iodometry ◇ ～度 titer / ～管 burette / ～剂 titrant
【滴管】dropper
【滴灌】[农] drip irrigation; trickle irrigation
【滴溜溜】going round and round
【滴滤池】[环保] trickling filter
【滴瓶】[化] dropping bottle
【滴水不漏】watertight; leakproof
【滴水槽檐】[建] drip edge
【滴水成冰】(so cold that) dripping water freezes; freezing cold: ～的天气 freezing weather
【滴水穿石】water constantly dripping wears holes in stone; little strokes fell great oaks
【滴水石】dripstone
【滴水挑檐】[建] drip cap
【滴水瓦】projecting tile of eaves
【滴水线檐】[建] drip mould
【滴水檐】[建] dripping eave
嘀
【嘀嗒】tick; ticktack; ticktock
镝 [化] dysprosium (Dy)
堤 dike; embankment
【堤岸】embankment ◇ ～码头 bulkhead wharf
【堤坝】dikes and dams
【堤防】dike; embankment: 加固～ strengthen the dikes ◇ ～工程 dike building; embankment project
提
【提防】take precautions against; be on guard against; beware of: ～感染 take precautions against infections / ～坏人破坏 guard against sabotage by bad elements
【提溜】hold; take in hand
低 ①(不高) low: ～年级学生 students of the junior years / ～声 in a low voice / ～水位 low water level / 代价很～ at a very low price ②(向下垂) let droop; hang down: ～头 hang one's head
【低倍显微镜】low-powered microscope
【低产】low yield: ～田 low-yield land; low-yielding land / ～作物 low-yielding crop ◇ ～油井[石油] stripper well; stripped well
【低潮】low tide; low ebb: 处于～ be at a low tide; be at a low ebb
【低沉】①(阴暗) overcast; lowering: ～的天空 an overcast sky ②(低郁沉重) (of voice) low and deep ③(情绪等低落) low-spirited; downcast
【低垂】hang low
【低地】lowland
【低调】low-key
【低飞】low flight; fly at a low altitude
【低估】underestimate; underrate: ～价值 undervaluation
【低合金钢】[冶] low-alloy steel
【低级】①(初步) elementary; rudimentary; lower: ～阶段 lower stage / ～生物 low forms of life ②(庸俗) vulgar; low: ～趣味 vulgar interests; bad taste
【低贱】low and degrading; humble
【低空】low altitude; low level ◇ ～飞行 low-altitude flying; low-level flying / ～轰炸 low-level bombing / ～扫射 low-level strafing; ground

strafing / ～侦察 low-altitude reconnaissance
【低栏】[体] low hurdles
【低利货款】low-interest loan
【低利息】low interest
【低廉】cheap; low: 物价～ prices are low
【低劣】inferior; low-grade
【低落】low; downcast: 情绪～ be low-spirited
【低能】mental deficiency; feeble-mindedness ◇ ～儿 imbecile; retarded child
【低频】low frequency ◇ ～变压器 low-frequency transformer / ～放大器 low-frequency amplifier
【低气压】[气] low pressure; depression
【低人一等】inferior to others
【低三下四】①(卑贱) lowly; mean; humble ②(奴态) servile; obsequious; cringing
【低烧】[医] low fever; slight fever
【低声】in a low voice; under one's breath; with bated breath
【低声波】[物] infrasonic wave
【低声下气】soft-spoken and submissive; meek and subservient
【低首下心】obsequiously submissive
【低速】low speed: ～驾驶 drive at low speed
【低碳钢】[冶] low-carbon steel
【低头】①(垂下头) lower one's head; bow one's head; hang one's head: ～默哀 bow one's head in silent mourning / ～认罪 hang one's head and admit one's guilt; plead guilty ②(屈服) yield; submit: 决不向困难～ never bow to difficulties
【低洼】low-lying ◇ ～地 low-lying land
【低微】①(声音细小) low: ～的呻吟 low groans ②(地位低) lowly; humble
【低温】①(低温度) low temperature ②[气](低气温) microtherm: ～气候 microthermal climate ③[医](低体温) hypothermia ◇ ～操作 low-temperature operation / ～恒温器[物] cryostat / ～麻醉 hypothermic anesthesia / ～生物学 cryobiology / ～学 cryogenics
【低下】low; lowly: 经济地位～ be of low economic status
【低消耗】low consumption (of raw materials, fuel, etc.): 保持高产、优质、～ maintain a record of high production, good quality and low consumption
【低血糖】[医] hypoglycemia
【低压】①[物](低的压力) low pressure ②[电](低电压) low tension; low voltage ③[气](低气压) low pressure; depression ④[医](低血压) minimum pressure ◇ ～槽[气] trough / ～气流[气] low-pressure air current / ～水银蒸气灯 low-tension mercury-vapor lamp
【低音】[声] bass
【低音提琴】double bass; contrabass
【低云】[气] low clouds

dí

涤 [书] wash; cleanse
【涤除】wash away; do away with; eliminate: ～旧习 do away with old customs
【涤荡】wash away; cleanse
【涤卡】dacron drill; dacron khaki
【涤纶】[纺] polyester fiber: 毛～ modelon / 棉～ trueran

嘀
【嘀咕】①(小声说) whisper; talk in whispers ②(猜疑) have misgivings about sth.; have sth. on one's mind

镝 [书] arrowhead: 鸣～ whistling arrow

嫡 ①(妻子所生) of or by the wife: ～长子 the wife's eldest son ②(血统近的) of lineal descent; closely related
【嫡传】handed down in a direct line from the master
【嫡派】disciples taught by the master himself
【嫡亲】blood relations; close paternal relations: ～弟兄 blood brothers; whole brothers
【嫡嗣】eldest son born of the legal wife of a man
【嫡堂兄妹】cousin-german
【嫡系】①(一线相传) direct line of descent ②(正支) one's own clique ◇ ～部队 troops under one's direct control

笛 ①(竹笛) bamboo flute ②(响声尖锐的发音器) whistle: 汽～ steam whistle
【笛卡儿主义】Cartesianism
【笛膜】bamboo membrane (for Chinese flute)
【笛子】bamboo flute: 吹～ flute; play the flute

敌 ①(敌人) enemy; foe: ～机 an enemy plane / 劲～ a formidable enemy; a foe worthy of one's steel; a worthy opponent ②(对抗) oppose; resist; fight: 所向无～ carry all before one; carry everything before one; be all-conquering / 以寡～众 fight against heavy odds ③(力量相等) match; equal
【敌百虫】[农] dipterex
【敌稗】[农] Stam F-34; dichloropropionanilide

【敌不住】be no match for
【敌草净】[化] desmetryn; methylmercapto
【敌草隆】[化] diuron
【敌敌畏】DDVP; dichlorvos
【敌对】hostile; antagonistic: ～分子 a hostile element / ～阶级 antagonistic classes; hostile classes / ～情绪 hostility; enmity / ～双方 opposing sides; parties to hostilities / ～态度 hostile attitude / ～行动 hostilities / ～行为 a hostile act
【敌国】enemy state
【敌后】enemy's rear area: 建立～根据地 establish base areas behind the enemy lines / 深入～ penetrate into the enemy's rear area
【敌机】an enemy plane; a hostile plane
【敌舰】an enemy war ship
【敌军】enemy troops; the enemy; hostile forces
【敌忾】hatred towards the enemy
【敌乐胺】[化] dinitramine
【敌螨通】[化] dinobuton
【敌情】the enemy's situation: ～的变化 changes on the enemy's side / 侦察～ make a reconnaissance of the enemy's situation
【敌情观念】alertness to the presence of the enemy: ～强 be keenly aware of the enemy's presence / 要有～ must not relax our vigilance against the enemy
【敌人】enemy; foe
【敌视】be hostile to; be antagonistic to; adopt a hostile attitude towards: 互相～的国家 nations inimical to one another
【敌手】①(对手) match; opponent; adversary ②(敌人的掌握) enemy hands: 落入～ fall into enemy hands
【敌台】enemy broadcasting station
【敌探】enemy spy
【敌特】enemy spy; enemy agent
【敌伪】the enemy and the puppet regime ◇ ～人员 enemy and puppet personnel / ～时期 the period of Japanese occupation; during the Japanese occupation
【敌我不分】fail to differentiate between the enemy and ourselves
【敌我界限】line of demarcation between the enemy and ourselves
【敌我矛盾】contradictions between ourselves and the enemy
【敌意】hostility; enmity; animosity
【敌营】enemy camp
【敌占区】enemy-occupied area; enemy-occupied territory
【敌众我寡】we are outnumbered by the enemy

的

【的确】indeed; really
【的确良】[纺] dacron; terylene

dǐ

底 ①(底部) bottom; base: ～价 base price / 井～ the bottom of a well ②(底细) the heart of a matter; ins and outs: 刨根问～ get to the bottom of sth.; get to the root of things / 心里没～ feel unsure of sth. ③(底稿) rough draft ④(底子) a copy kept as a record: 留个～儿 keep a copy on file; duplicate and file (a letter, etc.) ⑤(末尾) end: 年～ the end of a year ⑥(衬托面) ground; background; foundation: 白～红花 red flowers on a white background
【底版】negative; photographic plate
【底本】①(底稿) a copy for the record or for reproduction; master copy ②(校勘时作为依据的本子) a text against which other texts are checked
【底边】base
【底册】a bound copy of a document kept on file
【底层】①(一楼)(美) first floor; (英) ground floor ②(社会最低一层) bottom; the lowest rung
【底肥】base fertilizer
【底稿】draft; manuscript
【底火】①(炉中原有的火) the fire in a stove before fuel is added ②[军](子弹底部发火装置) primer; ignition cartridge
【底价】minimum price; base price
【底架】[机] chassis
【底孔】[水] bottom outlet
【底牌】cards in one's hand; hand: 亮～ show one's hand
【底盘】[汽车] chassis
【底片】negative; photographic plate ◇ ～夹 film holder / ～架 negative rack
【底栖生物】benthon
【底漆】priming paint; primer
【底色】bottom
【底墒】[农] soil moisture (before sowing or planting)
【底视图】bottom view
【底数】①(事情的原委) the truth or root of a matter; how a matter actually stands: 心中有～ know how the matter stands ②[数] base number

【底图】[地] base map
【底土】subsoil; undersoil
【底细】ins and outs; exact details
【底下】①(下面) under; below; beneath: 笔～不错 write well / 手～工作多 have one's hands full / 树～ under the tree ②(以后) next; later; afterwards
【底薪】base pay
【底子】①(底) bottom; base: 鞋～ the sole of a shoe ②(基础) foundation: ～薄 have a poor foundation to start with ③(草稿) rough draft or sketch ④(留底) a copy kept as a record ⑤(剩下的最后部分) remnant: 货～ remnants of stock
【底座】base; pedestal; foundation

诋 slander; defame
【诋毁】slander; defame; vilify; calumniate ◇ ～人格 degeneration of the dignity of man

砥 whetstone
【砥砺】①(磨炼) temper: ～革命意志 temper one's revolutionary will ②(勉励) encourage: 互相～ encourage each other
【砥石】whet-slate; whetstone
【砥柱】baffle

抵 ①(支撑) support; sustain; prop: 用手～着下巴颏儿 prop one's chin in one's hands ②(抵挡) resist; withstand: ～住来自外面的压力 withstand the pressure from outside ③(抵偿) compensate for; make good: ～命 pay with one's life (for a murder, etc.); a life for a life ④(抵押) mortgage: 用房屋做～ mortgage a house ⑤(抵消) balance; set off: 收支相～ income balances expenditure ⑥(相当) be equal to: 他一个人～两个用。 He can do the work of two. ⑦(抵达) reach; arrive at: 日内～京 arrive in Beijing in a day or two
【抵偿】compensate for; make good; give sth. by way of payment for: ～损失 compensate for losses
【抵触】conflict; contradict: ～情绪 resentment; resistance / 与法律相～ contravene the law; go against the law
【抵达】arrive; reach
【抵挡】keep out; ward off; check; withstand: ～风寒 keep out the wind and the cold / ～洪水 keep the flood in check
【抵换】substitute for; take the place of
【抵近射击】[军] point-blank firing
【抵抗】resist; stand up to: ～侵略 resist aggression; stand up against aggression / 奋起～ rise in resistance / 增强对疾病的～能力 build up one's resistance to disease
【抵赖】deny; disavow: ～罪行 deny one's guilt; refuse to admit one's crime / 不容～ brook no denial
【抵免所得税】tax credit
【抵命】pay for life with life; atone by life
【抵消】offset; cancel out; counteract: ～损失 offset the loss / ～药物的作用 counteract the effect of a medicine / ～影响 offset an influence / 相互～ cancel out each other
【抵押】mortgage: 以某物作～ raise a mortgage on sth.; leave sth. as a pledge ◇ ～财物 hypothec / ～法 law of mortagages / ～放款 mortgage loan; secured loan / ～品 security; pledge / ～权 mortgage / ～人 mortgagor / ～条款 mortgage clause / ～物 hostage / ～债券 mortgage debentures / ～证书 mortgage instrument / ～资产 mortgage assets
【抵御】resist; withstand: ～侵略 resist aggression / ～自然灾害 withstand natural calamities / 建立防风林～风沙侵袭 build a shelter belt against sandstorms
【抵债】pay a debt in kind or by labor
【抵制】resist; boycott: ～关税 compensation duty
【抵罪】be punished for a crime ◇ ～者 expiator

骶
【骶骨】sacrum
【骶椎】sacral vertebrae

邸 the residence of a high official: 官～ official residence

dì

帝 ①(上帝) God ②(皇帝) emperor: 称～ proclaim oneself emperor ③(帝国主义) imperialism
【帝俄】tsarist Russia
【帝国】empire: 罗马～ the Roman Empire / 英～ the British empire
【帝国主义】imperialism ◇ ～分子 imperialist element; imperialist / ～者 imperialist
【帝王】emperor; monarch
【帝制】autocratic monarchy; monarchy

谛 ①(仔细) carefully; attentively: ～听 listen attentively ②(道理) meaning; significance: 真

～ true significance; truth

蒂 the base of a fruit

碲 [化] tellurium (Te)

缔 from a friendship; conclude a treaty

【缔合】[化] association

【缔交】①(缔结邦交) establish diplomatic relations ②(订交) from a friendship; contract a friendship

【缔结】conclude; establish: ～邦交 establish diplomatic relations / ～条约 conclude a treaty

【缔约】conclude a treaty; sign a treaty: ～双方 both contracting parties ◇ ～国 signatory (state) to a treaty; party to a treaty; (high) contracting party

【缔造】found; create ◇ ～者 founder

弟 younger brother

【弟弟】younger brother; brother

【弟妇】younger brother's wife; sister-in-law

【弟妹】①(弟弟和妹妹) younger brother and sister ②(弟媳) younger brother's wife; sister-in-law

【弟兄】brothers: 亲～ blood brothers

【弟子】disciple; pupil; follower

递 ①(传送) hand over; pass; give: ～眼色 tip sb. the wink; wink at sb. / 给他～个口信 take a message to him ②(顺次) successively; in the proper order: ～升 promote to the next rank

【递补】fill vacancies in the proper order

【递给】hand over; pass on

【递回】pull-over

【递加】progressively increase; successively increase; increase by degrees

【递减】decrease progressively; decrease successively; decrease by degrees

【递交】hand over; present; submit: ～国书 (of an ambassador) present one's credentials / ～抗议书 lodge a protest / ～入党申请书 submit an application for Party membership / ～一份声明 send in a statement ◇ ～时间 time of delivery

【递解】escort (a criminal) from one place to another: ～回籍 send (a convict, etc.) to his native place under escort

【递进】go forward one by one

【递送】send; deliver: ～公文 deliver a document / ～情报 send out information; pass on information / ～信件 deliver letters

【递推公式】recurrence formula

【递增】increase progressively; increase by degrees

地 ①(地球) the earth ②(陆地) land; soil: 山～ hilly land / 盐碱～ saline and alkaline land ③(田地) fields: 麦～ wheat field / 下～干活儿 go and work in the fields ④(地面) ground; floor: 水泥～ cement floor ⑤(地区) place; locality: ～处山区 be located in a mountain area / 各～党组织 Party organizations of all localities / 全国各～ all parts of the country; throughout the country ⑥(形势; 状况) position; situation: 立于不败之～ be in an invincible position ⑦(衬托面) background; ground: 白～红花的大碗 a big white bowl with a pattern of red flowers on it / 一块白～黑字的木牌 a board with black characters on a white background ⑧(路程) distance: 二十里～ a distance of 20 *li*.

【地巴唑】[药] dibazol

【地板】①(木楼板) floor board ②(室内地面) floor: 水泥～ cement floor

【地堡】[军] bunker; blockhouse; pillbox

【地表】the earth's surface ◇ ～水 surface water

【地鳖虫】[药] ground beetle

【地步】①(处境) condition; plight ②(程度) extent; degree: 发展到公开对抗的～ develop to the point of an open clash / 兴奋到不能入睡的～ be so excited that one can't get to sleep ③(余地) room for action: 留～ leave room for manoeuvre; have some leeway; give oneself elbowroom

【地蚕】cutworm

【地槽】geosyncline ◇ ～学说 theory of geosyncline

【地层】stratum; layer ◇ ～层序 stratigraphic succession; stratigraphic sequence / ～对比 stratigraphic correlation / ～图 stratigraphic map / ～学 stratigraphy

【地产】landed estate; landed property; real estate ◇ ～税 land tax / ～帐 lot and building account

【地秤】weighbridge

【地磁】terrestrial magnetism; geomagnetism ◇ ～场 terrestrial magnetic field; geomagnetic field / ～极 geomagnetic pole / ～记录仪 magnetograph / ～学 geomagnetics / ～仪 magnetometer / ～异常 magnetic anomaly

【地大物博】vast territory and abundant resources; a big country abounding in natural wealth

【地带】district; region; zone; belt: 沙漠～ a desert region / 森林～ a forest region / 危险～ a danger zone / 无人～ no man's land

【地道】①(地下坑道) tunnel: ～口 entrance to a tunnel; subway entrance / ～战 tunnel warfare ②(真名产) from the place noted for the product; genuine: ～的吉林人参 genuine Jilin ginseng ③(真正的) pure; typical: 讲一口～的英语 speak idiomatic English ④(够标准; 实在) well-done; thorough 他干的活儿真～。 He does excellent work.

【地点】place; site; locale: 故事发生的～ the locale of a story / 开会～ place for a meeting; venue

【地电】terrestrial electricity

【地动】[口] quake; earthquake ◇ ～仪 seismograph

【地洞】a hole in the ground; burrow

【地段】a sector of an area; a section of an area: 划分～ divide into sections

【地盾】shield

【地方】①(各行政区的) locality: 党的～组织 Party organizations in the localities / 民族自治～ national autonomous area / 中央和～的关系 the relationship between the central and the local authorities ②(当地的) local: ～武装 local armed forces ③(某一地区) place; space; room: 你是什么～人? Where are you from? ④(部分) part; respect: 剧中最动人的～ the most touching part of the play ◇ ～病 endemic disease / ～法院 district court / ～分权 decentralization / ～观念 localism / ～工业 local industry / ～检察院 local procuratorates / ～军 local forces; regional troops / ～民族主义 local nationalism; local-nationality chauvinism / ～时间 local time / ～税 local taxes / ～戏 local opera; local drama / ～志 local chronicles; annals of local history / ～自治 local self-government

【地方国营】state-owned but locally-administered: ～农场 state farm under local administration / ～企业 locally-administered state enterprise

【地肤】[植] summer cypress ◇ ～子 the fruit of summer cypress

【地瓜】sweet potato

【地光】flashes of light preceding an earthquake

【地广人稀】a scarcely populated area

【地滚球】①[棒、垒球] ground ball; grounder ②(保龄球) bowling

【地核】[地] the earth's core

【地黄】[植] glutinous rehmannia

【地基】①(土层) ground ②[建](基础) foundation

【地极】[地] terrestrial pole

【地脚】lower margin (of a page)

【地脚螺栓】[机] foundation bolt

【地窖】cellar

【地界】the boundary of a piece of land

【地块】[地] massif

【地蜡】[矿] earth wax; ozocerite: 纯～ ceresin wax

【地牢】dungeon

【地老虎】cutworm

【地雷】(land) mine: 触发～ contact mine / 待发～ actuated mine / 反坦克～ anti-tank mine / 化学～ chemical mine / 埋～ plant mines; lay mines / 自发～ automatic action mine◇ ～场 minefield / ～战 (land) mine warfare

【地垒】[地] horst

【地理】①(一个地区的总情况) geographical features of a place: 熟悉～民情 be familiar with the place and its people ②(地理学) geography: 经济～ economic geography / 自然～ physical geography ◇ ～发现 geographical discovery / ～分布 geographical distribution / ～环境 geographical conditions / ～特点 geographical features / ～位置 geographical position / ～学 geography / ～学家 geographer / ～坐标 geographical coordinates

【地力】soil fertility

【地利】①(地理优势) favorable geographical position; topographical advantages ②(土地有利于种植的条件) land productivity

【地沥青】asphalt; bitumen

【地龙】[中药] earthworm

【地龙墙】[建] sleeper wall

【地漏】[建] floor drain

【地幔】[地] (the earth's) mantle

【地貌】the general configuration of the earth's surface; landforms ◇ ～发育 landform evolution / ～图 geomorphologic map / ～循环 geomorphic cycle / ～学 geomorphology / ～学家 geomorphologist

【地面】①(地的表面) the earth's surface; ground: 高出～两米 two meters above ground level ②[建] ground; floor: 水磨石～ *terrazzo* floor ③(地区) region; area; territory ◇ ～部队

ground forces / ～沉降[地] surface subsidence / ～导航设备[航] ground-based navigation aid / ～辐射[气] terrestrial surface radiation / ～灌溉 surface irrigation / ～目标 ground target / ～能见度[气] control-tower visibility; ground visibility / ～炮兵 ground artillery / ～情报接收站 terrestrial reception points / ～天气形势预报图[气] prebaratic chart / ～卫星站 ground satellite station / ～遥测装置 ground telemetering equipment / ～砖[建] floor tile

【地名】place name ◇ ～辞典 dictionary of place names; gazetteer / ～学 toponomy; toponymy

【地盘】territory under one's control; domain: 争夺～ compete for spheres of influence

【地皮】①(建筑用地面) land for building ②(地表面) ground

【地痞】local ruffian; local riffraff

【地平经度】[天] azimuth

【地平经纬仪】[天] altazimuth

【地平纬度】[天] altitude

【地平线】horizon

【地平坐标】[天] horizontal coordinates

【地铺】shakedown

【地契】title deed for land

【地壳】[地] the earth's crust ◇ ～均衡 isostasy / ～运动 crustal movement

【地勤】[航] ground service ◇ ～人员 ground crew; ground personnel

【地球】 the earth; the globe ◇ ～构造学 geognosy / ～化学 geochemistry / ～科学 geoscience; earth sciences / ～同步卫星 geostationary satellite / ～卫星 earth satellite / ～物理学 geophysics / ～物理探矿队 geophysical prospecting party / ～仪 (terrestrial) globe / ～资源卫星 Earth Resources Technology Satellite

【地区】①(较大的地方) area; district; region: 多山～ a mountainous district / 上海～ the Shanghai area ②(行政划分单位) prefecture ◇ ～差价 regional price differences

【地权】land ownership: 平均～ equalization of land ownership

【地热】[地] the heat of the earth's interior; terrestrial heat ◇ ～电力 geothermal power / ～能源 geothermal energy resources / ～学 geothermics

【地声】earthquake sounds

【地史学】historical geology

【地势】physical features of a place; relief; terrain; topography

【地台】[地] platform

【地毯】carpet; rug

【地毯式轰炸】carpet bombing

【地头】①(地边儿) edge of a field ②(目的地) the destination ③[印] lower margin of a page

【地头蛇】 a snake in its old haunts; local villain; local bully

【地图】map ◇ ～集 atlas / ～投影 map projection / ～学 cartography

【地委】prefectural Party committee

【地位】①(人或团体在社会关系中所处的位置) position; standing; place; status: ～平等 equal in status; on an equal footing / 国际～ international standing / 经济～ economic status / 领导～ leading position / 社会～ social position; social status / 一定的历史～ a proper or definite place in history / 政治～ political position; political standing ②(人或物所占的地方) place

【地温】[气] ground temperature; earth temperature ◇ ～表 ground thermometer; earth thermometer / ～梯度 geothermal gradient

【地文学】physical geography; physiography

【地物】surface features

【地峡】isthmus: 巴拿马～ the Isthmus of Panama

【地下】①(地面之下) underground; subterranean: ～仓库 underground storehouse ②(秘密活动) secret (activity); underground: ～党 underground Party; underground Party organization / 搞～工作 do underground work / 转入～ go underground ③(地面上) on the ground: 从～拣起 pick up from the ground / 掉在～ fall on the ground ◇ ～电缆 buried cable; underground cable / ～电影 underground film / ～宫殿 underground palace / ～河流 subterranean river / ～核试验 underground nuclear test / ～茎[植] subterranean stem / ～渗流 underground percolation / ～室 basement; cellar / ～水 groundwater / ～水位 groundwater level; water table / ～铁道 underground (railway); tube; subway / ～资源 hidden resources

【地线】[电] ground wire; earth wire

【地心】[地] the earth's core ◇ ～引力 terrestrial gravity; gravity

【地形】topography; terrain: ～地物 the terrain and its features / ～优越 enjoy topographical advantages ◇ ～测量 topographic survey / ～图 topographic map; relief map / ～学 topography / ～雨[气] orographic rain / ～云[气] orographic cloud / ～侦察 terrain reconnaissance

【地衣】[植] lichen ◇～红 cudbear

【地应力】[地] crustal stress

【地狱】hell; inferno: 人间～ a hell on earth

【地域】region; district: ～辽阔 vast in territory ◇～观念 regionalism

【地缘政治学】geopolitics

【地震】earthquake; seism ◇～波 seismic wave; earthquake wave / ～波曲线 seismogram / ～带 seismic belt / ～工作者 seismologist / ～观测 seismological observation / ～海啸 seismic sea wave; tsunami / ～活动 seismic activity / ～检波器 geophone / ～烈度 earthquake intensity / ～区 seismic area; seismic region / ～台站 seismograph station; seismic station / ～学 seismology / ～仪 seismograph / ～源 the seismic origin; the seismic focus / ～预报 earthquake prediction; earthquake forecasting / ～震级 (earthquake) magnitude / ～资料 the seismic data

【地址】address: 回信～ return address

【地志学】[地] topology

【地质】geology ◇～博物馆 geological museum / ～储量 oil in place / ～调查 geological survey / ～构造 geological structure / ～勘探 geological prospecting / ～勘探队 geological prospecting party / ～科学 geological sciences / ～力学 geomechanics / ～年代学 geochronology / ～时代 geologic age; geologic period / ～图 geologic map / ～学 geology / ～学家 geologist

【地中海】the Mediterranean (Sea)

【地轴】the earth's axis

【地主】①(土地占有者) landlord: 破落～ bankrupt landlord ②(主人) host: 尽～之谊 perform the duties of the host ◇～阶级 the landlord class

【地租】land rent; ground rent; rent: ～剥削 exploitation through land rent / 货币～ money rent / 劳役～ labor rent / 实物～ rent in kind

棣

[书] younger brother

【棣棠】[植] kerria

第

①(用在数词前表示次序) ～一 the first ②(官僚的大宅子) the residence of a high official

【第二把手】number two man; second in command

【第二版】the second edition

【第二产业】secondary industry

【第二次答辩】rejoin; rejoinder

【第二次扣押】second distress

【第二次申诉】repleader

【第二次世界大战】the Second World War; World War II

【第二审】[法] second instance ◇～案件 case of second instance / ～程序 procedure of second instance / ～法院 court of second instance; court of second hearing

【第二现场】secondary scene

【第二线兵力】second-line troops

【第二信号系统】[生理] the second signal system

【第二性】[哲] secondary

【第二主犯】principal in the second degree

【第六感官】sixth sense

【第三产业】tertiary industry

【第三纪】[地] the Tertiary period

【第三世界】the third world

【第三者】a third party (to a dispute, etc.)

【第四纪】[地] the Quaternary Period

【第五纵队】fifth column

【第一】first; primary; foremost: ～号种子选手 No. 1 seeded player / 党委～书记 the first secretary of the Party committee / 获得～名 win first place; get a first; win a championship / 做出～等的工作 do first-rate work

【第一把手】first in command; number one man; a person holding primary responsibility

【第一产业】primary industry

【第一次世界大战】the First World War; World War I

【第一副本】first authentic copy

【第一流】first-rate; first-class

【第一轮投票】first round of voting

【第一审】[法] first instance ◇～程序 procedure of first instance / ～法院 court of first instance

【第一手】firsthand: ～材料 firsthand material

【第一线】forefront; front line; first line: 战斗在～ fight in the forefront ◇～飞机 first-line aircraft

【第一现场】primary scene

【第一信号系统】[生理] the first signal system

【第一性】[哲] primary

【第一主犯】principal in the first degree

的

target; bull's-eye

diān

颠

①(头顶) crown (of the head) ②(高而直立的东西的顶部) top; summit: 山～ mountain top / 塔～ the top of a pagoda ③(颠簸) jolt; bump ④(跌落) fall; turn over; topple down: ～覆 overturn; subvert ⑤(跑) run; go away: 整天跑跑～～ be on the go all day long

【颠簸】jolt; bump; toss: 在山路上～行驶 bump along the mountain road; jolt along the mountain road

【颠倒】①(倒置) put upside down; turn upside down; transpose; reverse; invert: 主次～ reverse the order of importance ②(错乱) confused; disordered: 神魂～ be in a confused state of mind; be infatuated

【颠倒黑白】confound black and white; confuse right and wrong; stand facts on their heads

【颠倒事实】stand facts on their head; reverse the facts

【颠倒是非】confound right and wrong; confuse truth and falsehood; turn things upside down

【颠覆】overturn; subvert ◇ ～活动 subversive activities / ～政府罪 subversion of the government

【颠来倒去】over and over

【颠连】①(困苦) hardship; trouble; difficulty ②(连绵不断) peak upon peak

【颠沛流离】drift form place to place, homeless and miserable; wander about in a desperate plight; lead a vagrant life

【颠扑不灭】be able to withstand heavy battering; irrefutable; indisputable: ～的真理 irrefutable truth

【颠茄】[药] belladonna

【颠三倒四】incoherent; disorderly; confuse

癫 mentally deranged; insane

【癫狂】①[病] demented; mad; insane ②(轻佻) frivolous

【癫痫】[医] epilepsy ◇ ～发作 epileptic attack / ～患者 epileptic

巅 mountain peak; summit: 喜马拉雅山之～ the summit of the Himalayas

掂 weigh in the hand

【掂斤播两】engage in petty calculations; be calculating in small matters

【掂量】①(掂) weigh in the hand ②(斟酌) think over; weigh up

【掂算】estimate; calculate; weigh

diǎn

踮 stand on tiptoe: 他～起脚尖。 He stood on tiptoe.

点 ①(小滴) drop (of liquid): 雨～ raindrops ②(小痕迹) spot; dot; speck: 墨～ ink spots / 污～ stain ③[数] point: 基准～ datum point; datum mark / 两线的交～ the point of intersection of two lines ④(小数点) decimal point; point: 七～二 seven point two ⑤(少量) a little; a bit; some: 吃～东西 take some food / 读～鲁迅 read some of Lu Xun's works ⑥(地点或起步的标志) place; point: 沸～ boiling point / 居民～ residential area / 突破一～ make a breakthrough at one point / 以～带面 promote work in all areas by drawing upon experience gained at key points ⑦(事物的方面或部分) aspect; feature: 从这～上去看 viewed from this aspect / 特～ characteristic feature ⑧(加上点) put a dot: ～三个点表示省略 put three dots to show that something has been omitted ⑨(触) touch on very briefly; skim: 蜻蜓～水 dragon-flies skimming (over) the water ⑩(滴下) drip: ～眼药 put drops in the eyes ⑪(点播) sow in holes; dibble: ～豆子 dibble beans ⑫(查点; 对数) check one by one: ～货 check over goods; take stock / ～页数 count the pages ⑬(指定) select; choose: ～菜 order dishes (in a restaurant) ⑭(指点) hint; point out: 一～他就明白了。 He quickly took the hint. ⑮(引着火) light; burn; kindle: ～灯 light a lamp ⑯(钟点) o'clock: 上午八～钟 eight o'clock in the morning ⑰(规定时间) appointed time: 误～ be hind time; delayed; late / 正～ on time ⑱(点心) refreshments: 茶～ tea and cake; tea / 早～ breakfast

【点拨】teach; instruct

【点播】[农] dibble seeding; dibbling ◇ ～器 dibbler

【点菜】choose dishes from a menu; order dishes (in a restaurant)

【点穿隐情】point out the secret; expose the hidden facts

【点滴】①(零星; 微小) a bit: ～经验 bits of experience / 点点滴滴积累 accumulate bit by bit ②[医] intravenous drip: 打葡萄糖～ have an intravenous glucose drip

【点估计】point estimation

【点焊】spot welding; point welding ◇ ～机 mash welder

【点火】①(把火点着) light a fire ②[机](起动) ignition③(制造事端) stir up trouble ◇ ～系统 ignition system

【点饥】have a snack to stave off hunger
【点将】name a person for a particular job
【点交】hand over item by item
【点金成铁】damage the beauty of the original work; miscorrect a piece of writing
【点睛】provide striking key points
【点名】①(查点人数) call the roll: 晚～ evening roll call ②(指名) mention sb. by name: ～攻击 attack sb. by name ◇ ～册 roll book; roll
【点明】point out; put one's finger on: ～问题的所在 put one's finger on the cause of the trouble
【点破】bring sth. out into the open; lay bare; point out bluntly
【点燃】light; kindle; ignite: ～导火线 light a fuse; ignite a fuse / ～革命之火 kindle the flames of revolution / ～火把 light a torch
【点染】①[美] add details to a painting ②(修饰文字) touch up a piece of writing
【点射】[军] firing in bursts
【点石成金】touch a stone and turn it into gold; turn a crude essay into a literary gem
【点收】check and accept: 按清单～货物 acknowledge receipt of goods after checking them against a list
【点数】check the number (of pieces, etc.); count
【点题】bring out the theme
【点头】nod one's head; nod: ～打招呼 nod to sb. (as a greeting) / ～示意 signal by nodding / ～同意 nod assent / ～之交 nodding acquaintance; bowing acquaintance
【点头哈腰】[口] bow unctuously; bow and scrape
【点心】light refreshments; pastry
【点穴】attack a vital point
【点验】examine item by item
【点阵】[物] lattice ◇ ～理论 lattice theory / ～能 lattice energy
【点阵字符】dot character
【点阵式打印机】dot matrix printer
【点种】①(点播种子) dibble in the seeds ②(点播) dibbling
【点缀】①(衬托; 装饰) embellish; ornament; adorn ②(应景) use sth. merely for show
【点字】braille
【点子】①(小滴) drop (of liquid) ②(小痕迹) spot; dot; speck: 油～ grease spot ③(打击乐器的节拍) beat (of percussion instruments): 鼓～ drumbeat ④(要点) key point: 工作抓到～上 get to grips with the essentials in one's work; put one's finger on the right spot ⑤(主意) idea; pointer: 他的～多。 He is full of ideas.

典 ①(标准) standard; law; canon ②(典范性书籍) standard work of scholarship: 词～ dictionary / 药～ pharmacopoeia ③(典故) allusion; literary quotation: 用～ use allusions ④(典礼) ceremony: 盛～ a grand ceremony; a grand occasion ⑤(主持) be in charge of: ～狱 prison warden ⑥(抵押) mortgage: 被迫～出两亩地 be forced to mortgage two *mu* of land
【典藏】book reservation
【典当】mortgage; pawn ◇ ～业 pawnbroking
【典范】model; example; paragon
【典故】allusion; literary quotation
【典籍】ancient codes and records; ancient books and records
【典礼】ceremony; celebration: 奠基～ cornerstone-lying ceremony / 下水～ launching ceremony
【典型】①(具有代表性的人或事) typical case; typical example; model; type: 理论和实践相结合的～ a model of the integration of theory and practice / 抓～ grasp typical cases ②(个性特征) typical; representative: ～的中国村庄 a representative Chinese village / ～人物 a typical character / ～示范 demonstrate with typical examples / ～事例 a typical instance; a typical case ◇ ～案件 typical case / ～性 typicalness
【典押】mortgage; pawn
【典雅】(of diction, etc.) refined; elegant
【典狱官】custodial officer
【典章】institutions; decrees and regulations

碘 [化] iodine (I)
【碘酊】[药] tincture of iodine
【碘仿】[化] iodoform
【碘甘油】[药] iodine glycerin
【碘化铝】silver iodide
【碘化银】silver iodide
【碘酒】[药] tincture of iodine
【碘缺乏】iodine deficiency
【碘值】iodine value

diàn

淀 ①(沉淀) form sediment; settle; precipitate ②(浅湖) shallow lake
【淀粉】starch; amylum ◇ ～酶 amylase
【淀积作用】[地] illuviation

靛 ①(靛蓝) indigo ②(深蓝色) indigo-blue

【靛蓝】indigo ◇ ～色 indigo-blue
【靛青】indigo-blue
奠 ①(建立) establish; settle ②(祭奠) make offerings to the spirits of the dead
【奠定】establish; settle: ～基础 lay a foundation
【奠都】establish a capital; found a capital
【奠基】lay a foundation ◇ ～礼 foundation stone laying ceremony／～人 founder／～石 foundation stone; cornerstone
【奠仪】a gift of money made on the occasion of a funeral
垫 ①(垫平; 垫高) put sth. under sth. else to raise it or make it level; fill up; pad: ～路 repair a road by filling the holes ②(垫子) pad; cushion; mat: 床～ mattress／鞋～ inner sole; insole／椅～ chair cushion ③(暂时替人付款) pay for sb. and expect to be repaid later
【垫付】pay for sb. and expect to be repaid later
【垫肩】shoulder pad; shoulder padding
【垫脚石】stepping-stone
【垫圈】bed down the livestock; spread earth in a pigsty, cowshed, etc.
【垫款】money advanced for sb. to be paid back later
【垫密片】[机] gasket: 气缸～ cylinder gasket
【垫片】①[机] spacer: 绝缘～ insulation spacer ②(楔形填隙片) shim: 轴承～ bearing shim
【垫平】level up: 把篮球场～ level a basketball court
【垫圈】[机] washer: 开口～ snap washer／锁紧～ locking washer／毡～ felt washer
【垫上运动】mat tumbling; mat work
【垫子】mat; pad; cushion: 茶杯～ teacup mat; coaster／蹭鞋～ doormat／沙发～ sofa cushion／弹簧～ spring mattress／体操～ gym mat
店 ①(店铺) shop; store: 服装～ clothing store／书～ bookshop; bookstore／文具～ stationer's ②(客店) inn: 住～ stop at an inn
【店客】customer
【店面】shop front
【店铺】shop; store
【店钱】inn expense; cost of lodging
【店员】shop assistant; salesclerk; clerk; salesman or saleswoman
【店主】shopkeeper; storekeeper
惦 remember with concern; be concerned about; keep thinking about
【惦记】remember with concern; be concerned about; keep thinking about: ～同志们的安全 be concerned about the safety of one's comrades
【惦念】keep thinking about; be anxious about; worry about
玷 ①(白玉上的斑点) a flaw in a piece of jade ②(使有污点) blemish; disgrace
【玷辱】bring disgrace on; be a disgrace to
【玷污】stain; sully; tarnish: ～某人名誉 smear sb.'s reputation; sully sb.s reputation／～自己的人格 soil one's hands
电 ①(电力) electricity: 导～ conduction／断～ cut off the power supply／放～ discharge／流～ current electricity／静～ static electricity ②(触电) give or get an electric shock ③(电报) telegram; cable: 复～ reply by telegraph／贺～ telegraph one's congratulations to sb.; cable a message of congratulations／急～ urgent telegram
【电棒】[方] (electric) torch; flashlight
【电报】telegram; cable: 打～ send a telegram／传真～ photo telegram; photo telegraph; photogram; facsimile telegraph／公事～ official telegram／国际～ international telegram／国内～ inland telegram／国外～ foreign cable／加急～ urgent telegram／express telegram／密码～ cipher telegram／明码～plain code telegram／无线～ radiogram; radiotelegram／无线电传真～ photoradiogram／新闻～ press cable／用户～ telex／有线～ wire telegram ◇ ～等级 telegram message precedence／～费 telegram charge／～费收据 telegram receipt／～挂号 cable address; telegraphic address／～机 telegraph／～局 telegraph office ○复电费已付 reply paid
【电报纸】telegram form; telegram blank: 填写～ fill in the telegraph form; fill in the telegram form
【电笔】electrography
【电表】①(电气仪表) electric meter; ammeter; voltmeter ②(瓦特小时计) kilowatt-hour meter; watt-hour meter; electric meter
【电冰箱】(electric) refrigerator; fridge; freezer
【电波】electric wave
【电铲】power shovel
【电场】[电] electric field ◇ ～强度 electric field intensity
【电厂】power plant; generating plant

【电唱机】electric gramophone; electric phonograph; record player
【电唱头】pickup
【电唱针】(gramophone) stylus; needle
【电车】①(电动车;有轨电车) tram; tramcar; streetcar ② (无轨电车) trolleybus; trolley ◇ ～钢轨 tram rail / ～售票员 conductor (男); conductress (女)
【电池】(electric) cell; battery: 干～ dry cell / 太阳能～ solar cell ◇ ～组 battery
【电传】telex ◇ ～打字电报机 teletypewriter; teleprinter
【电磁】electromagnetism ◇ ～波 electromagnetic wave / ～感应 electromagnetic induction / ～铁 electromagnet / ～学 electromagnetics
【电导】conductance ◇ ～仪 conductivity gauge
【电灯】electric lamp; electric light ◇ ～泡 electric (light) bulb
【电动】motor-driven; power-driven; power-operated; electric ◇ ～泵 motor-driven pump; electric pump / ～车 electrically operated motor car / ～发电机 motor generator / ～割草机 power-operated mower / ～回转罗盘 electric gyro-compass / ～机 (electric) motor / ～记分牌 electric scoreboard / ～起重葫芦 electric hoist; electric block / ～势 electro-motive force (EMF)
【电动力学】electrodynamics
【电度表】kilowatt-hour meter; watt-hour meter; electric meter
【电镀】electroplate: 无氰～ electroplating without using cyanide
【电费】charges for electricity
【电感】inductance ◇ ～电桥 inductance bridge
【电工】①(电工学) electrical engineering ②(指工人) electrician ◇ ～技术 electrotechnics / ～器材厂 electrical appliances factory / ～学 electrical engineering; electrotechnics
【电感应】induction
【电功率】electric power ◇～计 electrodynamometer ○千瓦小时 kilowatt-hour / 马力 horsepower; h.p.
【电灌站】electric pumping station
【电光】light produced by electricity; lightning ◇ ～工艺[纺] schreinering
【电焊】electric welding ◇ ～工 electric welder / ～机 electric welding machine; electric welder / ～钳 electrode holder / ～条 welding electrode; welding rod
【电荷】electric charge; charge: ～密度 charge density / 负～ negative charge / 正～ positive charge
【电弧】electric arc ◇ ～焊接 arc welding / ～炉 arc furnace
【电化当量】electrochemical equivalent
【电化教育】education with electrical audio-visual aids; audio-visual education program
【电化学】electrochemistry
【电话】①(电话装置) telephone; phone: 长途～ trunk-call; long distance call / 公用～ public telephone; pay telephone / 市内～ local call / 无线～ radio telephone; wireless telephone / 专线～ special line / 自动～ automatic telephone ② (打电话) phone call: 打～ make a phone call; phone sb.; call sb. up; ring sb. up; give sb. a ring ◇ ～簿 telephone directory; telephone book / ～分机 extension (telephone) / ～号码 telephone number / ～会议 telephone conference / ～机 telephone (set) / ～交换台 telephone exchange; telephone switchboard / ～接线员 telephone operator; switchboard operator / ～用户 telephone subscriber / ～增音机 telephone repeater / ～中继线 main line; trunk line / ～中继线号码 general number / ～总机 central exchange ○接通 put a call through / 打不通 cannot get through / 占线 number engaged; The line is busy. / 给人留话 leave a message for someone / 请等一下，别挂上。 Hold on a minute. (或: Hold the line a minute.) / 听不清。 The connection is bad. / 一点声音都听不见。 The line is dead. / 您打错号码了。 Sorry, wrong number. / 电话坏了。The phone is out of order.
【电话间】telephone box; telephone booth; telephone kiosk: 公用～ public telephone booth; public telephone box, public telephone kiosk
【电话局】telephone office; telephone exchange: 长途～ trunk exchange; long distance office
【电汇】telegraphic money order ◇ ～汇率 rate for telegraphic transfer
【电火花】electric spark ◇ ～加工 electric spark machining
【电击】electrical shock
【电机】electrical machinery ◇ ～厂 electrical machinery plant / ～工程 electrical engineering
【电机车】electric locomotive
【电极】electrode: 阳～ anode; positive electrode / 阴～ cathode; negative electrode
【电价键】[化] electrovalent bond
【电剪】electric scissors; electric shears

【电键】telegraph key; key button
【电解】electrolysis ◇ ～池 electrolytic cell / ～分离 electrolytic dissociation / ～炉 electrolytic furnace / ～铜 electrolytic coppers / ～液 electrolyte / ～质 electrolyte
【电介质】dielectric
【电锯】electric saw
【电抗】reactance ◇ ～器 reactor
【电缆】electric cable; cable: 同轴～ coaxial cable
【电烙铁】①(电熨斗) electric iron ②(电焊铁) electric soldering iron
【电离】ionization ◇ ～层[气] ionosphere
【电力】electric power; power ◇ ～工程 electric power project / ～工业 power industry / ～供应 supply of electricity / ～机车 electric locomotive / ～机械 electrical power equipment / ～网 power network / ～系统 power system / ～线 power line; electric line of force / ～消耗 power consumption
【电疗】[医] electrotherapy: 超短波～ ultra-short-wave therapy / 短波～ shortwave therapy ◇ ～法 diathermy
【电料】electrical materials and appliances
【电铃】electric bell
【电流】electric current: 反向～ reverse current / 负载～ load current / 载波～ carrier current ◇ ～计 galvanometer / ～强度 current intensity
【电炉】①(取暖电炉) electric stove; hot plate ② [冶] electric furnace ◇ ～钢 electric steel / ～炼钢法 electric furnace process
【电路】(electric) circuit: 并联～ in parallel; parallel connection / 串联～ in series; series connection / 固体～ solid circuit / 集成～ integrated circuit / 印刷～ printed circuit ◇ ～图 circuit diagram
【电码】(telegraphic) code: 莫尔斯～ Morse code ◇ ～本 code book
【电门】(electric) switch
【电木】[化] bakelite ◇ ～粉 phenolic moulding powder / ～插座 bakelite receptacle; bakelite socket
【电脑】electronic brain; storage system in a computer ◇微～ personal computer
【电钮】push button; button: 按～ press a button; push a button
【电抛光】electropolishing
【电瓶】storage battery; accumulator ◇ ～车 storage battery car; electromobile
【电气】electric ◇ ～工程学 electrical engineering / ～机车 electric locomotive / ～开关 electric switch / ～设备 electrical equipment
【电气化】electrification: 工业～ electrification of industry ◇ ～铁路 electric railway
【电气石】[矿] tourmaline
【电器】electrical equipment; electrical appliance ◇ ～制造厂 electric appliances plant
【电热】electric heat; electrothermal ◇ ～处理 electrothermal treatment / ～丝 heating wire
【电容】electric capacity; capacitance ◇ ～率 permittivity
【电容器】condenser; capacitor: 单联～ single-connection / 可变～ variable capacitor / 双联～ double-connection
【电熔炼】[冶] electric smelting
【电扇】electric fan
【电渗析】[化] electrodialysis
【电石】[化] calcium carbide ◇ ～灯 acetylene lamp / ～气 acetylene
【电视】television; TV: 彩色～ color television / 黑白～ black-and-white television / 看～ watch television ◇ ～播出 television broadcasting; telecasting; videocast / ～播送 television transmission; telecasting / ～电影 telecine / ～发射机 television transmitter / ～剧 TV play / ～讲座 telecourse / ～接收机 television receiver; television set / ～连续剧 TV play series / ～录像 telerecording / ～屏幕 television screen / ～扫描 television scanning / ～摄像转播车 television pickup station / ～摄影机 television camera; telecamera / ～摄影记者 television cameraman / ～实况转播 live television coverage; live telecast / ～实况转播车 outside broadcasting van / ～塔 television tower / ～台 television station / ～天线 television antenna / ～网 television network / ～信道 television channel / ～影片 telefilm / ～转播 television relay / ～转播卫星 television transmission satellite ○显像管 picture tube / 彩色显像管 tricolor tube / 对比度 contrast / 亮度 brightness; brilliance / 色调 hue / 色度 chrome / 行频 horizontal frequency; line frequency / 帧频 vertical frequency; picture frequency / 甚高频 very high frequency (VHF) / 超高频 ultra high frequency (UHF) / 磁带录像机 video tape recorder (VTR)
【电视电话】video telephone; video-phone
【电视雷达导航仪】teleran
【电枢】[电] armature ◇ ～绕组 armature winding

【电刷】[机] brush ◇ ～触点 brush contact
【电台】①(无线电台) transmitter-receiver; transceiver ②(广播电台) broadcasting station; radio station
【电毯】electrical blanket
【电烫】permanent hair styling; permanent wave; perm
【电梯】lift; elevator ◇ ～司机 lift operator; elevator runner
【电筒】(electric) torch; flashlight
【电网】electrified wire netting; live wire entanglement
【电位】(electric) potential ◇ ～差 potential difference
【电文】text (of a telegram)
【电线】(electric) wire ◇ ～杆子 (wire) pole
【电信】telecommunications ◇ ～局 telecommunication bureau / ～业务 telecommunication service
【电刑】electrocution
【电学】electricity (as a science)
【电讯】①(电报发出的消息) (telegraphic) dispatch: 世界各地发来的～ dispatches from all parts of the world ②(无线电讯号) telecommunications ◇ ～稿 news bulletin / ～设备 telecommunication equipment
【电压】voltage ◇ ～表 voltmeter / ～分压器 voltage divider / ～调压器 voltage regulator
【电眼】electric eye; magic eye
【电唁】send a telegram of condolence; send a message of condolence
【电冶金】electrometallurgy
【电椅】electric chair
【电影】film; movie; motion picture: ～在…放映 the film is on at...; the film is showing at... / 放映～ show a film / 拍摄～ shoot a film / 变形镜头式宽银幕～ CinemaScope / 彩色～ color film / 黑白～ black-and-white film / 宽银幕～ wide-screen film / 立体～ stereoscopic film / 全景式宽银幕～ Cinerama / 无声～ silent film / 有声～ sound film / 遮幅式宽银幕～ superscope ◇ ～发行公司 film distribution corporation / ～放映队 film projection unit / ～放映机 (film) projector; cineprojector / ～放映网 film projection network / ～观众 moviegoers; cinema-goer / ～剪辑机 film editing machine / ～胶片 cinefilm; motion-picture film / ～节 film festival / ～剧本 script; scenario / ～剧本作者 scenarist; scenario writer / ～摄影机 cinecamera; film camera / ～摄影师 cinematographer; cameraman / ～审查 censorship of film / ～事业 cinemindustry; the movies / ～说明书 film synopsis / ～演员 film actor or actress / ～译制厂 film dubbing studio / ～院 cinema; movie (house) / ～制片厂 (film) studio / ～周 film week / ～字幕 film caption ○ 导演 director / 摄影师 cameraman / 演员 movie actor (男); movie actress (女) / 制片人 productor / 故事片 feature film / 新闻片 newsreel / 纪录片 documentary film / 大型纪录片 full length documentary film / 科教片 science and educational film / 科学普及片 popular science film / 音乐片 musical film / 体育片 sports film / 动画片 cartoon; animated cartoon / 木偶片 marionette film / 电视纪录片 televised documentary / 翻译片 dubbed film
【电泳】[物] electrophoresis
【电玉粉】[化] urea-formaldehyde molding powder
【电源】power supply; power source; mains: 接上～ connect with the mains / 切断～ cut off the electricity supply ◇ ～变压器 power transformer; mains transformer
【电乐器】electrophone
【电灶】electric cooking stove; electric cooking range
【电渣炉】[冶] electroslag furnace
【电闸】electric brake
【电站】power station
【电针疗法】[中医] acupuncture with electric stimulation; galvano-acupuncture
【电针麻醉】[中医] galvano-acupuncture anesthesia
【电钟】electric clock; electroclock
【电铸版】[印] electortype
【电子】electron: 负～ negatron / 热～ thermal electron / 正～ positron ◇ ～称 electronic-weighing system / ～程序控制 electron program me control / ～电荷 electron charge / ～分色机 electronic color scanner / ～伏特 electron-volt / ～工业 electronics industry / ～管 electron tube; valve / ～管收音机 valve radio set / ～光学 electron optics / ～回旋加速器 betatron / ～加速器 electron accelerator / ～监视 electronic surveillance / ～刻版机 electronic engraving machine / ～流 electron current / ～炉 electronic oven / ～器件 electronic device / ～枪 electron gun / ～壳层 electron shell / ～人工喉 artificial electronic larynx / ～束 electron beam / ～望远镜 electron telescope / ～稳压器 electronic volt-

age regulator / ～物理学 electron physics / ～学 electronics / ～印相机 electronic printer / ～云 electron cloud / ～战 electronic warfare / ～照相术 electro-photography / 无线电～学 radio-electronics
【电子计算机】electronic computer: 微型～ personal computer
【电子显微镜】electron microscope: 八十万倍～ an electron microscope with a magnification of 800000 times
【电阻】resistance ◇ ～表 ohmmeter / ～率 resistivity; specific resistance / ～器 resistor
【电钻】electric drill

佃 rent land (from a landlord)
【佃户】tenant (farmer)
【佃农】tenant-peasant; tenant farmer
【佃租】land rent

殿 ①(高大的房屋) hall; palace; temple: 太和～ the Hall of Supreme Harmony ②(在最后) at the rear
【殿后】bring up the rear
【殿军】①(走在最后的部队) rear guard ②(最后一名优胜者) last winner in a contest; last among the winners; the last of the successful candidates
【殿下】(直接称呼) Your Highness; (间接称呼) His Highness; Her Highness

癜 purplish or white patches on the skin: 白～风 vitiligo / 紫～ purpura

diāo

貂 marten
【貂皮】fur or pelt of marten; marten ◇ ～大衣 marten coat
【貂裘】marten coat
【貂熊】glutton

凋 wither: 常绿不～ remain green all the year round
【凋敝】①(生活困苦) hard; destitute ②(事业衰败) depressed
【凋零】withered, fallen and scattered about
【凋落】wither and fall
【凋谢】①(脱落) wither and fall ②(老死) die of old age: 老成～ the passing away of worthy old people

碉
【碉堡】pillbox; blockhouse
【碉楼】watchtower

雕 ①(刻画) carve; engrave: 瓷～ carved porcelain / 浮～ relief / 石～ stone carving ②(鹫) eagle; vulture
【雕版】[印] wood block for printing
【雕虫小技】insignificant skill; the trifling skill of a scribe; literary skill of no high order
【雕虫篆】a type of calligraphy featuring characters twisting and turning like warms
【雕花】①(在木器上刻花纹) carve patterns or designs on woodwork ②(雕刻成的花纹) carving: ～家具 carved furniture / ～烟盒 engraved cigarette case
【雕刻】carve; engrave: 玉石～ jade carving ◇ ～玻璃 engraved glass / ～刀 carving tool; burin / ～工艺 artistic carving / ～品 carving / ～师 carver; engraver
【雕梁画栋】carved beams and painted rafters; a richly ornamented building
【雕漆】carved lacquerware
【雕塑】sculpture: ～家 sculptor / ～艺术 sculptural arts; statuary
【雕像】statue: 半身～ bust / 大理石～ marble statue / 小～ statuette
【雕鸮】eagle owl
【雕琢】①(雕刻) cut and polish; carve ②(过分修饰) write in an ornate style

鲷 porgy

刁 tricky; artful; sly: 放～ play the fox; try a ruse
【刁悍】cunning and fierce
【刁滑】cunning; crafty; artful
【刁难】create difficulties; make things difficult: 百般～ create obstructions of every description; raise all manner of difficulties; put up innumerable obstacles / 故意～ deliberately make things difficult for others
【刁诈】knavish; crafty
【刁钻】cunning; artful; wily: ～古怪 sly and capricious / 发球～ tricky service

叼 hold in the mouth: 嘴里～着支香烟 with a cigarette dangling from one's lips

diào

调 ①(调动) transfer; shift; move: ～挡 shift gears / ～干部 transfer cadres / ～军队 move troops ②(腔调) accent ③[乐] key: B～ key of B / G 大～ G major / G 小～ G minor ④(腔

调；音调) air; tune; melody: 定～ call the tune / 老～重弹 strike up an old tune ⑤(声调) tone; tune: 降～ falling tone; falling tune / 升～ rising tone; rising tune

【调兵遣将】move troops; deploy forces

【调拨】allocate and transfer (goods or funds); allot: ～款项购置图书 allocate funds for books / ～物资 allocate supplies

【调查】investigate; inquire into; look into; survey: ～原因 investigate the cause / 农村～ rural survey / 作社会～ make a social investigation ◇ ～报告 findings report / ～会 fact-finding meeting / ～机关 investigatory appa-ratus / ～人 investigator; inquirer / ～提纲 outline for investigation; questionnaire / ～团 fact-finding mission / ～委员会 board of inquiry / ～研究 investigation and study / ～证据 investigation of evidence / ～罪行 investigate a crime

【调车场】[铁道] switchyard

【调动】①(更动) transfer; shift: ～工作 transfer sb. to another post ②(调动军队) move (troops); maneuver; muster ③(动员) bring into play; arouse; mobilize: ～一切积极因素 bring every positive factor into play

【调度】①(调遣) dispatch (trains, buses, etc.) ②(调度员) dispatcher ③(安排) manage; control: 生产～ production management ◇ ～室 dispatcher's office; control room / ～员 dispatcher; controller

【调防】[军] relieve a garrison

【调干学员】a college student enrolled from among cadres; cadre student

【调号】①(音调符号) tone mark ②(曲调符号) key signature

【调虎离山】lure the tiger out of the mountains; lure the enemy away from his base

【调换】exchange; change; swap

【调回】recall (troops, etc.)

【调集】assemble; muster: ～兵力 assemble forces / ～二十个师 concentrate twenty divisions

【调令】transfer order

【调门儿】pitch

【调派】send; assign

【调配】allocate; deploy: ～原材料 allocation of raw materials / 合理～劳动力 rational deployment of manpower

【调遣】dispatch; assign: ～军队 dispatch troops / 听从～ (be ready to) accept an assignment

【调任】be transferred to another post

【调式】[乐] mode

【调用】transfer (under a unified plan): ～干部 transfer cadres (to a specific job)

【调运】allocate and transport

【调值】[语] tone pitch

【调职】be transferred to another post

【调子】①(乐曲的调子) tune; melody: 降低～ play down one's tune ②(论调) tone (of speech); note: ～低沉的演说 low-keyed speech

掉 ①(落) fall; drop; shed; come off: ～进圈套 fall into a trap / ～下几滴眼泪 shed a few tears ②(遗失) lose; be missing ③(落在后边) fall behind ④(回；转) turn: 把车头～过来 turn the car round ⑤(降低；减少) drop; reduce: ～分量 reduction in weight / ～价 a drop in price ⑥(用在动词后,表示完成) 擦～ wipe off / 改～坏习气 correct bad habits / 扔～ throw away / 洗～ wash out ⑦(互换) change; exchange: ～座位 change seats; exchange seats; swap places with sb.

【掉包】stealthily substitute one thing for another

【掉队】drop out; drop off; fall behind

【掉换】exchange; change; swap: ～工作 be assigned a new job; be transferred to another

【掉魂】lose one's wits; terrified

【掉泪】come to tears; tears falling

【掉脑袋】get beheaded

【掉色】lose color; fade

【掉书袋】excessive fondness of making literary quotations and historical allusions

【掉头】turn round; turn about: 自行车～ turn bicycle right round

【掉以轻心】lower one's guard; treat sth. lightly

【掉转】turn round: ～枪口 turn one's gun (against one's superiors or old associates) / ～身子 turn round

吊 ①(悬挂) hang; suspend ②(吊起物件向上提或向下放) lift up or let down with a rope, etc. ③(祭奠) condole; mourn: ～丧 pay a condolence call ④(缝) put in a fur lining: ～皮袄 line a coat with fur ⑤(收回) revoke; withdraw: ～销 revoke (a licence, etc.) ⑥(吊车) crane: 塔～ tower crane

【吊钹】[乐] suspension cymbal

【吊车】crane; hoist: 桥式～ overhead crane ◇ ～梁 crane beam

【吊床】hammock

【吊灯】pendent lamp
【吊斗】cableway bucket
【吊儿郎当】careless and casual; slovenly
【吊杆】boom; jib: (船用)起重～ derrick／起重机～ crane boom
【吊杆托架】boom crutch
【吊杠】[体] trapeze
【吊钩】(lift) hook; hanger
【吊古】think of the ancients or ancient events
【吊环】[体] rings: 摆荡～ swinging rings／静止～ still rings
【吊货盘】platform sling; tray sling
【吊货网】cargo net
【吊架】[机] hanger: 平衡～ balance hanger
【吊景】[剧] drop scenery
【吊雷】[军] hanging mine
【吊兰】[植] bracketplant
【吊链】chain sling; sling chain
【吊铺】hanging bed; hammock
【吊桥】①(用钢索悬吊的桥) suspension bridge ②(城门前可吊起、放下的桥) drawbridge
【吊丧】visit the bereaved to offer one's condolences; pay a condolence call
【吊嗓子】train one's voice; exercise one's voice
【吊扇】ceiling fan
【吊死】hang by the neck; hang oneself
【吊索】sling: 钢丝～ wire sling／绳～ rope sling
【吊艇柱】[海] davit
【吊桶】well-bucket; bucket
【吊袜带】garters; suspenders
【吊胃口】tantalize
【吊慰】condole with
【吊文】a funeral oration; a memorial address
【吊线】plumb-line
【吊线板】ceiling plate
【吊线缆】messenger cable
【吊销】revoke; withdraw: ～护照 withdraw a passport／～驾驶执照 revoke a driving licence／～汽车执照 suspend a motor licence
【吊唁】condole; offer one's condolences: ～函电 messages of condolence
【吊钟花】bellflower
【吊装】[建] hoisting

钓 fish with a hook and line; angle
【钓饵】bait
【钓竿】fishing rod
【钓钩】fishhook
【钓具】fishing tackle
【钓丝】fishing line
【钓鱼】angle; go fishing

diē

跌 ①(摔) fall; tumble: ～入小河 tumble into a stream ②(落下) drop; fall: 物价下～. Prices have dropped.
【跌打损伤】injuries from falls, fractures, contusions and strains
【跌宕】①(洒脱) free and easy; bold and unconstrained ②(抑扬顿挫) flowing rhythm
【跌倒】fall; tumble: ～在地 fall to the ground
【跌跌撞撞】dodder along; stagger along
【跌价】go down in price; fall in price; drop in price: 黑白电视机～了。The prices of B&W TV sets have gone down.
【跌交】①(摔跟头) trip and fall; stumble and fall; fall: 跌了一交 have a fall ②(犯错误; 受挫折) make a mistake; meet with a setback
【跌落】fall; drop
【跌伤】get injured by a fall; fall and get hurt

爹 father; dad; daddy; pa: ～娘 father and mother; mum and dad; ma and pa; parents

dié

谍 ①(谍报活动) espionage ②(谍报人员) intelligence agent; spy
【谍报】information obtained through espionage; intelligence report; intelligence ◇ ～员 intelligence agent; spy

碟 small plate; small dish: 一～炒黄豆 a dish of fried soya beans
【碟子】small plate; small dish

喋
【喋喋不休】chatter away; rattle on; talk endlessly
【喋血】bloodshed; bloodbath

蝶 butterfly
【蝶骨】[生理] sphenoid bone
【蝶形花】papilionaceous flower
【蝶泳】butterfly stroke

牒 an official document or note; certificate: 最后通～ ultimatum

迭 ①(轮流) alternate; change ②(屡次) repeatedly; again and again: ～挫强敌 inflict repeated reverses on a formidable enemy
【迭次】repeatedly; again and again: ～磋商 repeatedly consult each other

【迭起】occur repeatedly; happen frequently
【迭用】use alternately
叠 ①(重叠;重复) pile up; repeat: 层峦～嶂 peaks rising one higher than another ②(折叠) fold: ～被子 fold up a quilt / 把信～好 fold the letter
【叠层结构】laminated construction
【叠床架屋】pile one bed upon another or build one house on top of another; needless duplication
【叠句】[语] reiterative sentence
【叠罗汉】[体] pyramid
【叠韵】[语] two or more characters with the same vowel formation
【叠嶂】rows of mountains
【叠字】[语] reiterative locution; reduplication

dīng

丁 ①(成年男子) man: 成～ reach manhood / 壮～ able-bodied man ②(人口) members of a family; population: ～口 population / 添～ have a baby born into the family ③(从事某种职业的人员) a person engaged in a certain occupation: 园～ gardener ④(第四) fourth: ～等 the fourth grade; grade D / ～种维生素 vitamin D ⑤(小块) small cubes(of meat or vegetable); cubes: 黄瓜～ diced cucumber
【丁坝】[水] spur dike; spur
【丁苯橡胶】butadiene styrene rubber
【丁财两旺】be blessed with many male children and great wealth
【丁醇】butanol; butyl alcohol
【丁当】[象] ding-dong; jingle; clatter
【丁点儿】a tiny bit: 这～事不必放在心上。Don't bother about such trifles.
【丁东】[象] tinkle
【丁二烯】butadiene
【丁零】[象] tinkle; jingle: ～～的自行车铃声 the jingling of bicycle bells
【丁零当郎】[象] jinglejangle; cling-clang
【丁宁】urge again and again; warn; exhort: ～周至 give thoughtful advice
【丁是丁,卯是卯】be conscientious and meticulous
【丁烷】[化] butane ◇ ～气 butagas
【丁烯】[化] butene
【丁香】①(花名;树名) lilac ②(香料) clove ◇ ～油 clove oil
【丁字】T-shaped ◇ ～尺 T-square / ～钢 T-steel / ～镐 pickax / ～街 T-shaped road junction / ～梁 T-beam / ～形 T-shaped
疔 malignant boil; malignant furuncle
【疔疽】miliary vesicle under the nose or on either side of the mandible
玎
【玎玲】[象] clink; jingle; tinkle
耵
【耵聍】[药] earwax; cerumen
酊 tincture
【酊剂】tincture
叮 ①(咬) sting; bite: 腿上叫蚊子～了一下 get a mosquito bite on the leg ②(追问) say or ask again to make sure
【叮嘱】urge again and again; warn; exhort
盯 fix one's eyes on; gaze at; stare
【盯梢】shadow sb.; tail sb.
【盯住】keep a close watch(on)
钉 ①(铁钉) nail; tack ②(紧跟不放) follow closely; tail ③(督促) urge; press: ～着孩子做作业 urge the children to do their homework
【钉齿耙】spike-tooth harrow
【钉锤】nail hammer; claw hammer
【钉螺】[动] oncomelania
【钉帽】the head of a nail
【钉耙】(iron-toothed) rake
【钉人】[体] watch an opponent in a game ◇ ～防守 man-for-man defense; man-to-man defense
【钉梢】shadow sb.; tail sb.
【钉鞋】spiked shoes; spikes
【钉子】①(铁钉) nail ②(意外的障碍) snag: 碰～ hit a snag; strike a snag; meet with a rebuff; run against a snag

dǐng

顶 ①(人或物体的最高部分) crown; peak; top: 到～ reach the limit / 山～ mountaintop; hilltop / 秃～ be bald / 屋～ roof ②(用头支撑) carry on the head: 头上～着一罐水 carry a pitcher of water on one's head ③(用头撞击) gore; butt: 这牛爱～人。 This bull gores people. ④(迎着) go against: ～风雪,战严寒 face blizzards and brave severe cold ⑤(支撑) push from below or behind; push up; prop up: 把门～起来 prop up the door ⑥(顶撞) retort; turn down ⑦(担当) cope with; stand up to ⑧(顶替)

take the place of; substitute; replace ⑨(相当) equal; be equivalent to ⑩(最) the most; exceedingly; best: ～小的那个孩子 the youngest child; the smallest child / ～有用 very useful
【顶班】work on regular shifts; work full time
【顶板】[矿] roof: 直接～ immediate roof
【顶吹】[冶] top-blown ◇ ～转炉 top-blown converter
【顶灯】[汽车] dome light
【顶点】①(极点) apex; zenith; acme; pinnacle ②[数] vertex; apex
【顶端】top; peak; apex
【顶多】at (the) most; at best
【顶风】①(迎着风) against the wind: ～冒雨 brave wind and rain / ～骑车 cycle against the wind / 开～船 sail against the wind ②(迎面的风) head wind
【顶峰】peak; summit; pinnacle
【顶骨】[生理] parietal bone
【顶呱呱】tip-top; first-rate; excellent
【顶回去】reject; turn down: 把他们的不合理要求～ reject their unreasonable demand
【顶交】[农] topcross
【顶角】[数] vertex angle
【顶礼】[佛教] prostrate oneself before sb. and press one's head against his feet
【顶礼膜拜】prostrate oneself in worship; make a fetish of; pay homage to
【顶梁柱】pillar; backbone
【顶名】assume sb.'s name
【顶牛儿】lock horns like bulls; clash; be at loggerheads
【顶棚】ceiling
【顶球】[足球] head (a ball)
【顶少】at least
【顶事】be useful; serve the purpose
【顶替】take sb.'s place; replace
【顶天立地】of gigantic stature; of indomitable spirit
【顶头】coming directly towards one: ～风 head wind
【顶头上司】one's immediate superior; one's direct superior
【顶芽】[植] terminal bud
【顶用】be of use; be of help; serve the purpose: 我去也不～。 I can't be of any help even if I go.
【顶针】thimble
【顶踵】from head to heel
【顶住】withstand; hold out against; stand up to: ～风浪 weather a storm / ～逆流 stand up against an adverse current / ～压力 withstand pressure
【顶撞】contradict (one's elder or superior)
【顶嘴】reply defiantly; answer back; talk back: 不许你～! None of your lip!

鼎 an ancient cooking vessel with two loop handles and three or four legs: 三足～ tripod / 四足～ quadripod
【鼎鼎大名】a great reputation
【鼎沸】like a seething caldron; noisy and confused: 人声～ a hubbub of voices
【鼎革】change of dynasty
【鼎力】[套] your kind effort: 多蒙～相助，不胜感谢。 We are extremely grateful to you for the trouble you have taken on our behalf.
【鼎立】stand like the three legs of a tripod; tripartite confrontation; tripartite balance of forces
【鼎盛】in a period of great prosperity; at the height of power and splendors: 春秋～ in the prime of manhood
【鼎足】the three legs of a tripod; three rival powers: ～之势 a situation of tripartite confrontation

dìng

定 ①(稳定) calm; stable: 心神不～ be ill at ease; feel restless ②(决定) decide; fix; set: ～方针 decide on a policy / ～计划 make a plan ③(已确定的) fixed; settled; established: ～评 accepted opinion / ～数 fixed number ④(约定) subscribe to; book(seats, tickets, etc.); order ⑤[书](必定) surely; certainly; definitely: ～可取胜 be sure to win
【定案】①(做最后决定) decide on a verdict; reach a conclusion on a case ②(所做的最后决定) verdict; final decision
【定比定律】[化] the law of definite proportions
【定单】order for goods; order form
【定调子】set the tone; set the keynote; call the tune
【定都】choose a site for the capital; establish a capital: ～北京 make Beijing the capital; decide on Beijing as the capital
【定夺】make a final decision; decide: 这事由你～。 It is up to you to decide.
【定额】quota; norm: 生产～ production quota
【定稿】①(最后审稿) finalize a manuscript, text, etc. ②(审定了的稿件) final version or text

【定冠词】[语] definite article
【定规】established rule or practice; set pattern;
【定婚】be engaged(to be married); be betrothed
【定货】order goods; place an order for goods ◇ ～单 order form
【定计】devise a stratagem; work out a scheme
【定价】①(规定价格) fix a price ②(决定了的价格) fixed price
【定见】definite opinion; set view
【定金】bargain money; earnest money
【定睛】fix one's eyes upon: ～细看 look fixedly and scrutinize
【定居】settle down: 牧民～点 herdsmen's settlement
【定局】①(确定不移的形势) foregone conclusion; inevitable outcome ②(做最后决定) settle finally
【定理】theorem: 基本～ fundamental theorem
【定例】usual practice; set pattern; routine
【定量】①(规定数量) fixed quantity; ration ②(测定物质成分的数量) determine the amounts of the components of a substance ◇ ～分析 [化] quantitative analysis / ～供应 rationing
【定律】law: 能量守衡～ law of the conservation of energy / 万有引力～ the law of universal gravitation / 物质不灭～ law of conservation 〈indestructibility〉 of matter / 物质能量守衡～ matter-energy equivalence
【定论】final conclusion
【定苗】[农] final singling(of seedlings)
【定名】name; denominate
【定期】①(定下日期) fix a date ②(有一定期限的) regular; at regular intervals; periodical: ～付款 payment on terms / ～汇报工作 regularly report back on one's work / ～交货 delivery on term / ～轮换 rotate at regular intervals / ～体格检查 regular physical checkups ◇ ～保险 term insurance / ～存款 fixed deposit; time deposit / ～航线 regular line / ～决算表 periodical statements / ～刊物 periodical publication; periodical / ～徒刑 penal for a fixed time / ～租船 time charter / 不～刊物 nonperiodic publication
【定亲】engagement (arranged by parents); betrothal
【定然】certainly; definitely: ～如此。 It most necessarily be so.
【定神】①(使心神安定) collect oneself; compose oneself; pull oneself together ②(集中注意力) concentrate one's attention
【定时引爆】[军] time fire
【定时炸弹】time bomb
【定时装置】timer equipment
【定位】①(确定的位置) fixed position; location; orientation ②(用仪器测定位置) orientate; position ◇ ～器 [矿] positioner
【定息】fixed interest
【定弦】tune a stringed instrument
【定向】directional ◇ ～爆破 directional blasting / ～变异 directed variation / ～地雷 oriented mine / ～广播 directional broadcasting / ～培育 [农] directive breeding / ～天线 directional antenna / ～系统 orientation system / ～仪 [气] direction finder / ～钻井 directional drilling; directed drilling
【定心】[机] centering ◇ ～装置 centering device
【定心丸】sth. capable of setting sb.'s mind at ease: 吃了～ be reassured
【定形】①[化纤](凝固) setting: 热～ heat setting ②[针织](成型) boarding ◇ ～机 boarding machine
【定型】finalize the design; fall into a pattern ◇ ～反应 stereotyped response
【定性】①(确定问题的性质) determine the nature ②(测定物质的成分及性质) determine the chemical composition of a substance ◇ ～分析 [化] qualitative analysis
【定义】definition: 下～ give a definition; define
【定音鼓】kettledrums; timpani
【定影】[摄] fixing; fixation ◇ ～罐 fixing tank / ～剂 fixer / ～液 fixing bath
【定于】due to; scheduled to
【定语】[语] attribute ◇ ～从句 attributive clause
【定员】fixed number of staff members or passengers
【定则】[物] rule: 右手～ right-hand rule / 左手～ left-hand rule
【定植】[植] field planting net; field fixed net
【定制】have sth. made to order; have sth. custom-made: ～家具 have furniture made to order
【定子】[电] stator ◇ ～绕组 stator winding
【定罪】declare sb. guilty; convict sb. (of a crime): ～量刑 punishment fits the crime
【定座】reservation
【定做】have sth. made to order; have sth. made to measure: ～的衣服 tailor-made clothes; clothes made to measure

锭 ①(块状物) ingot-shaped tablet (of medicine, Chinese ink, etc.): 一～墨 a cake of

Chinese ink / 一～银子 a small ingot of silver ②[纺](锭子) spindle

【锭剂】[药] lozenge; pastille; troche

【锭模】[冶] ingot mold

【锭子】[纺] spindle ◇ ～油 bobbin oil; spindle oil

订 ①(制订) conclude; draw up; agree on: ～合同 enter into a contract; make a contract / ～计划 draw up a plan; work out a plan / ～日期 fix a date / ～生产指标 set a production target / ～条约 conclude a treaty ②(预先约定) subscribe to (a newspaper, etc.); book (seats, tickets, etc.); order (merchandise, etc.) ③(改正错误) make corrections; revise: 修～ revise ④(装订) staple together

【订单】order for goods; order form

【订房间】make room reservation

【订费】subscription (rate)

【订购】order (goods); place an order for sth.

【订户】subscriber

【订婚】be engaged to; be betrothed to

【订货】order goods; place an order for goods ◇ ～单 order form / ～确认书 confirmation of order

【订立】conclude; make ◇ ～合同 conclusion of contract

【订书机】stapler; stapling-machine

【订约】conclude a bargain; enter into an agreement: ～的另一方当事人 other contracting party / ～能力 capacity to contract

【订阅】subscribe to (a newspaper, periodical, etc.)

【订正】make corrections; emend: ～印刷上的错误 correct errors of printing

钉 ①(用钉子固定) nail: ～钉子 drive in a nail / ～马掌 nail on horseshoes / 把窗子～死 nail up a window ②(用针线缝) sew on: ～扣子 sew a button on

diū

丢 ①(遗失) lose; mislay: 我的钱包～了。I've lost my purse. ②(扔) throw; cast; toss: 把菜帮子～给小兔吃 throw the outer leaves to the rabbit ③(搁置;放) put aside; lay aside: ～在脑后 let sth. pass out of one's mind; clean forget; completely ignore

【丢丑】lose face; be disgraced

【丢掉】①(遗失) lose ②(抛弃) throw away; cast away; discard: ～错误观点 discard mistaken views / ～坏习惯 drop a bad habit / ～幻想 cast away illusion

【丢饭碗】lose one's job

【丢开】leave it off; forget for a while

【丢盔卸甲】throw away one's helmet and coat of mail; throw away everything in headlong flight

【丢脸】lose face; be disgraced

【丢面子】lose face

【丢弃】abandon; discard; give up

【丢人】lose face; be disgraced

【丢三落四】forgetful; scatterbrained

【丢失】lose

【丢手】wash one's hands of; give up: ～不管某事 wash one's hands of sth.

【丢下】throw down; lay aside

【丢眼色】wink at sb.; tip sb. the wink

【丢卒保车】give up a pawn to save a chariot; sacrifice minor things to save major ones

铥 [化] thulium (Tm)

dōng

东 ①(东方) east: ～城 the eastern part of the city / ～郊 eastern suburbs / 城～ east of the city / 近～ Near East / 远～ Far east / 中～ Middle East ②(主人) master; owner: 房～ landlord ③(东道) host: 做～ stand treat; stand host; play the host

【东半球】the Eastern Hemisphere

【东北】①(东北方向) northeast ②(中国东北地区) northeast of China; the Northeast: ～平原 Northeast plain

【东北抗日联军】the Anti-Japanese Amalgamated Army of the Northeast

【东北漂流】[海] Northeast Drift Current

【东北信风】[气] northeast trades

【东奔西跑】run around here and there; bustle about; rush about; rush around

【东窗事发】be exposed; come to the light

【东道】one who treats sb. to a meal; host: 做～ play the host; stand treat ◇ ～国 host country / ～主 host

【东倒西歪】①(形容人醉后走路的姿势) walk unsteadily ②(物体不正) dilapidated; rickety

【东渡】take a sea-voyage eastward

【东方】①(东) east ②(指亚洲) the East; the Orient ③(指东欧和苏联等国) the East ◇ ～文化 oriental cultural / ～正教 the Greek Orthodox Church

【东非】East Africa

【东非大裂谷】East African Rift Valley

【东风】①(指春风) east wind ②(革命的力量或气势) driving force of revolution; the East Wind: ～压倒西风。 The East Wind prevails over the West Wind.
【东风吹马耳】like the east wind blowing at the ear of a horse; go in one ear and out the other
【东风带】[气] easterlies
【东扶西倒】difficult to cultivate plants
【东郭先生】Master Dongguo, a naive person who gets into trouble through being softhearted to evil people
【东海】the Donghai Sea; the East China Sea
【东海扬尘】unpredictability of world affairs
【东家】master; boss
【东经】[地] east longitude
【东拉西扯】drag in all sorts of irrelevant matters; talk at random; ramble
【东鳞西爪】odds and ends; bits and pieces; fragments
【东南】①(东南方向) southeast ②(中国东南沿海地区) southeast China; the Southeast
【东南西北】all directions; north, south, east and west
【东南亚】Southeast Asia
【东挪西借】scrape money; borrow all around
【东欧】Eastern Europe
【东拼西凑】scrape together; knock together
【东跑西颠】run about busily
【东山再起】stage a comeback
【东施效颦】blind imitation with ludicrous effect
【东食西宿】go wherever profit is
【东西】①(东边和西边) east and west ②(从东到西) from east to west ③(事物；物体) thing ④(指人或动物) thing; creature
【东…西…】here... there: 东一个,西一个 (of things) be scattered here and there / 东一榔头,西一棒槌 hammer here and batter there; act or speak haphazardly
【东亚】East Asia
【东张西望】gaze around; peer around
【东正教】the Orthodox Eastern Church

冬 ①(冬季) winter ②[象] rub-a-dub; rat-tat; rat-a-tat
【冬菜】preserved, dried cabbage or mustard greens
【冬虫夏草】[中药] Chinese caterpillar fungus
【冬耕】winter ploughing
【冬菇】dried mushrooms (picked in winter)
【冬瓜】wax gourd; white gourd
【冬灌】[农] winter irrigation
【冬烘】shallow but pedantic ◇～先生 pedant
【冬候鸟】winter bird
【冬季】winter ◇～攻势[军] winter offensive / ～施工 winter construction / ～体育运动 winter sports / ～作物 winter crops
【冬眠】[生] winter sleep; hibernation ◇～疗法 hibernation therapy
【冬眠灵】wintermin
【冬青】[植] Chinese ilex
【冬笋】winter bamboo shoots
【冬天】winter
【冬小麦】winter wheat
【冬汛】[渔] winter fishing season
【冬衣】winter clothes
【冬至】the Winter Solstice (22nd solar term)
【冬装】winter dress; winter clothes

氡 [化] radon (Rn)

dǒng

董 ①[书](监督管理) direct; superintend; supervise: ～其成 supervise the project until its completion ②(董事) director; trustee
【董事】director; trustee
【董事会】①(企业的) board of directors ②(学校等的) board of trustees
【董事长】chairman of the board

懂 understand; know: ～礼貌 have good manners / ～英语 know English
【懂得】understand; know; grasp: ～革命道理 understand revolutionary principles
【懂行】know the business; know the ropes
【懂事】sensible; intelligent: ～的孩子 a sensible child

dòng

动 ①(与静相对) move; stir ②(行动) act; get moving: 一举一～ every movement; every action ③(改动) change; alter ④(使用) use: ～脑筋 use one's head ⑤(触动；感动) touch; arouse: ～肝火 flare up / ～感情 be carried away by emotion; get worked up / ～了公愤 have aroused public indignation / 不为甜言蜜语所～ not be swayed by fine words ⑥(吃) eat or drink: 不～荤腥 never touch meat or fish; be a vegetarian
【动笔】take up the pen; start writing
【动宾词组】[语] verb-object word group
【动不动】easily; frequently; at every turn: ～就发脾气 be apt to lose one's temper; often get in-

to a temper / ～就感冒 catch cold easily
【动产】movable property; movables; personal property ◇ ～抵押权 chattels mortgage / ～留置权 chattels lien / ～权 right over movables / ～文据 chattels paper
【动词】[语] verb ◇～不定式 infinitive
【动荡】turbulence; upheaval; unrest: ～的局势 a turbulent situation
【动荡不安】turbulent; in turmoil
【动荡不稳】shaky and unstable
【动电学】electrokinetics
【动工】begin construction; start building
【动滑轮】fall block; movable block
【动画片】animated cartoon; cartoon
【动火】[口] get angry; flare up
【动机】motive; intention: ～不纯 have impure motives / 出于自私的～ be actuated by selfish motives
【动静】①(动作或说话的声音) the sound of sth. astir ②(情况) movement; activity: 发现可疑～ spot something suspicious
【动力】①(各种作用力) motive power; power ②(推动力量) motive force; driving force; impetus: 社会发展的～ the motive force of the development of society ◇ ～灌溉机械 power-driven irrigation machinery / ～来源 power resources / ～设备 power plant / ～学 dynamics; kinetics / ～转向机构 power steering
【动量】[物] momentum: 广义～ generalized momentum ◇ ～矩 moment of momentum / ～守恒定律 the law of conservation of momentum
【动令】[军] command of execution
【动乱】turmoil; disturbance; upheaval; turbulence: ～年代 years of upheaval / ～时期 a time of turmoil; a time of storm and stress / 社会～ social upheaval
【动脉】[生理] artery ◇ ～弓 arch of aorta / ～瘤 aneurysm / ～脉搏 arterial pulse / ～血压 arterial pressure / ～炎 arteritis / ～硬化 arteriosclerosis / ～粥样硬化 atherosclerosis
【动名词】[语] gerund
【动脑筋】consider; think hard
【动能】[物] kinetic energy
【动怒】lose one's temper; flare up
【动气】[口] take offense; get angry
【动情】①(情绪激动) get worked up; become excited ②(产生爱慕的感情) become enamored; have one's (sexual) passions aroused ◇ ～期 estrus; oestrus
【动人】moving; touching: ～的情景 a moving scene / ～的事迹 stirring deeds / ～心弦 be deeply moving; tug at one's heartstrings
【动身】go on a journey; set out on a journey; leave (for a distant place)
【动手】①(开始做) start work; get to work ②(用手接触) touch; handle: ～动脚 get fresh with sb. / 请勿～! No touching! ③(打人) raise a hand
【动手术】①(做手术) perform an operation; operate on sb. ②(被做了手术) have an operation; be operated on
【动态】①(事物发展情况) trends; developments: 国际～ the new developments in international affairs / 科技新～ recent developments in science and technology / 油井～ behavior of an oil well ②(运动变化状态) dynamic state ◇ ～电阻 dynamic resistance / ～平衡 [物] dynamic equilibrium / ～特性 [电] dynamic characteristic
【动弹】move; stir: ～不得 cannot move
【动听】interesting to listen to; pleasant to listen to: ～的故事 attractive story / 娓娓～ be very pleasant to the ear
【动土】break ground; start building
【动窝儿】[方] start moving; make a move
【动武】use force; start a fight; come to blows
【动物】animal: 反刍～ ruminant / 环节～ annelid; annelida / 节肢～ arthropod / 棘皮～ echinoderm / 脊椎～ vertebrate / 脊索～ chordate / 寄生～ parasitic animal / 冷血～ cold-blooded animal / 两栖～ amphibian / 灵长～ primate / 卵生～ oviparous animal / 爬行～ reptile / 啮齿～ rodent / 哺乳～ mammal / 腔肠～ coelenterate / 软体～ mollusk / 蠕形～ verme / 食草～ herbivorous animal / 食肉～ carnivore / 胎生～ viviparous animal / 无脊椎～ invertebrate / 有蹄～ ungulate / 原生～ protozoa / 甲壳～ crustacean ◇ ～传染病 zoonosis / ～胶 [化] animal size; animal glue / ～界 the animal kingdom / ～区系 fauna / ～生态学 animal ecology / ～纤维 animal fiber / ～学 zoology / ～油 animal oil / ～园 zoological garden; zoo / ～志 fauna ○ 狮 lion (pl: lioness) / 虎 tiger (pl: tigress) / 猞猁 lynx / 豹 leopard / 猎豹 cheetah / 美洲狮 cougar; puma / 美洲虎 jaguar / 美洲豹 panther / 黑熊 black bear / 棕熊 brown bear / 北极熊 polar bear / 狼 wolf / 豺 jackal / 狐 fox / 鬣狗 hyena; hyaena / 象 elephant / 斑马 zebra / 长

颈鹿 giraffe / 鹿 deer / 梅花鹿 sika / 驯鹿 reindeer; caribou / 麋鹿 elk / 麝 musk deer / 骆驼 camel / 单峰驼 dromedary camel / 双峰驼 bactrian camel / 美洲驼，大羊驼 llama / 马来貘 Malayan tapir / 大熊猫 giant panda; panda / 小熊猫 lesser panda / 麝香牛 musk ox / 水牛 buffalo / 美洲野牛 bison / 牦牛 yak / 犀牛 rhinoceros / 扭角羚 takin; addax / 羚羊 antelope; chamois / 瞪羚 gazelle / 河马 hippopotamus / 獾 badger / 箭猪 porcupine / 刺猬 hedgehog / 野猪 wild boar / 袋鼠 kangaroo / 蝙蝠 bat / 臭鼬 skunk / 鼹鼠 mole / 松鼠 squirrel / 黄鼠狼 weasel / 貂 ermine; mink; marten / 田鼠 field mouse / 土拨鼠 marmot / 鸭嘴兽 platypus / 猿 ape / 猩猩 orang-outang; orangutan / 大猩猩 gorilla / 黑猩猩 chimpanzee / 猴 monkey / 狒狒 baboon / 山魈 mandrill / 金丝猴 golden monkey / 海马 sea horse / 海狸 beaver / 海豹 seal / 海狮 sea lion / 海象 walrus / 海豚 dolphin / 水獭 otter / 鲸 whale / 蟾蜍 toad / 青蛙 frog / 蜗牛 snail / 龟 tortoise / 鳖 turtle / 鳄鱼 crocodile / 短吻鳄 alligator / 蜥蜴 lizard / 壁虎 house lizard; gecko / 避役 chamaleon / 蝾螈 salamander; newt / 蝎子 scorpion / 蛇 snake / 大蛇 serpent / 蟒蛇 python; boa; baoconstrictor / 眼镜蛇 cobra / 毒蛇 viper; asp / 蝰蛇 adder / 响尾蛇 rattle-snake / 蜈蚣 centipede / 毒蜘蛛 tarantula / 蜘蛛 spider / 蚯蚓 earthworm / 水蛭 leech / 蜻蜓 dragon-fly / 蚱蜢 grasshopper / 蝉 cicada / 螳螂 mantis / 蟋蟀 cricket / 蜜蜂 bee / 蚂蚁 ant / 蟑螂 cockroach / 白蚁 termite [另见鸟类]

【动向】trend; tendency: 新～ new trends

【动心】one's mind is perturbed; one's desire, enthusiasm or interest is aroused

【动刑】subject sb. to torture; torture

【动眼神经】oculomotor nerve

【动摇】shake; vacillate; waver: ～军心 shake the morale / 风吹浪打不～ never waver in the storm and stress of struggle ◇ ～分子 wavering element; vacillating element

【动议】motion: 紧急～ an urgent motion / 提出一项～ put forward a motion

【动用】put to use; employ; draw on: ～大量人力 employ a tremendous amount of manpower / ～公款 draw upon public funds / ～库存 draw on stock

【动员】mobilize; arouse: 总～ general mobilization / 作一番～ give a mobilization talk ◇ ～报告 mobilization speech / ～大会 mobilization meeting / ～令 mobilization order

【动辄】[书] easily; frequently; at every turn: ～得咎 be frequently taken to task; be blamed for whatever one does / ～发怒 fly into a rage on the slightest provocation

【动植物检疫】quarantine of animals and plants

【动作】movement; motion; action: ～缓慢 slow in one's movements / ～敏捷 quick in one's movements / 规定～ compulsory exercise / 优美的舞蹈～ graceful dance movements / 自选～ optional exercise; voluntary exercise

冻 ①(遇冷凝固) freeze: ～肉 frozen meat / 不能让这些白菜～坏。 We mustn't let the cabbages be damaged by frost. ②(汤、汁等凝结的) jelly: 肉～儿 jellied meat ③(受冷或感到冷) feel very cold; freeze; be frostbitten: 我～得慌。 I am freezing.

【冻冰】freeze: 河上～了。 The river is frozen.

【冻疮】chilblain: 生～ have chilblains

【冻豆腐】frozen bean curd

【冻害】[农] freeze injury

【冻僵】frozen stiff; numb with cold

【冻结】①(液体遇冷凝固) freeze; congeal ②(阻止流动或变动) freeze: ～的资产 frozen assets / 工资～ wage freeze ◇ ～帐目 frozen account / ～资本 frozen capital

【冻凝】congeal ◇ ～点 congealing point

【冻伤】frostbite

【冻死】freeze to death; freeze and perish; die of frost

【冻土】frozen earth; frozen ground; frozen soil ◇ ～学 cryopedology

【冻原】[生态] tundra

【冻雨】[气] sleet

栋 ①[书](正梁) ridgepole ②[量] 一～楼房 a building

【栋梁】ridgepole and beam; 国家的～ pillar of the state

洞 ①(穿通或凹入的部分) hole; cavity: 衬衣破了一个～ have a hole in one's shirt / 山～ mountain cave ②(深远;透彻) penetratingly; thoroughly: ～见症结 see clearly the crux of the matter; get to the heart of the problem

【洞察】see clearly; have an insight into: ～是非 see clearly the rights and wrongs of the case / ～下情 fully perceive the feelings of the masses

◇～力 insight; discernment; acumen
【洞彻】understand thoroughly; see clearly
【洞达】understand thoroughly: ～事理 be sensible
【洞房】bridal chamber; nuptial chamber: ◇～花烛 wedding festivities; wedding / ～花烛夜 wedding night
【洞若观火】see sth. as clearly as a blazing fire
【洞悉】know clearly; understand thoroughly
【洞萧】a vertical bamboo flute
【洞晓】have a clear knowledge of: ～其中利弊 have a clear understanding of the advantages and disadvantages
【洞穴】cave; cavern ◇～墓 catacomb
【洞烛其奸】see through sb.'s tricks

恫 fear
【恫吓】threaten; intimidate: 不怕～ defy any intimidation; defy any threat / 虚声～ bluff; bluster

dōu

都 ①[副](总括) all ②(跟"是"字合用, 说明理由) just because of: ～是下雨, 运动会才延期的。 Just because of the rain, the spots meet was put off. ③(甚至) even: 这个我连想～没想过。 I haven't even thought of it. ④(已经) already: 她～九十岁了。 She's already ninety.

兜 ①(口袋一类的东西) pocket; bag: 裤～儿 trouser pocket / 网～ string bag ②(兜住) wrap up in a piece of cloth, etc.: 用毛巾～着几个鸡蛋 carry a few eggs wrapped up in a towel ③(绕) move round: 在场上～圈子跑 run round in the field ④(招揽) canvass; solicit: ～售 peddle ⑤(承担) take upon oneself; take responsibility for sth.: 出了问题我～着。 If anything goes wrong, I'll take responsibility for it.
【兜捕】surround and seize; round up
【兜抄】close in from the rear and both flanks; round up
【兜底】reveal all the details; disclose the whole inside story
【兜风】①(船帆等挡住风) catch the wind ②[方](坐车、船等游逛) go for a drive, ride or sail; go for a spin
【兜揽】①(招揽顾客) canvass; solicit: ～生意 solicit custom; drum up trade ②(把事情往身上拉) take upon oneself(sb. else's work)
【兜圈子】①(绕圈儿) go around in circles; circle ②(说话不直截了当) beat about the bush
【兜售】peddle; hawk
【兜子】pocket; bag

dǒu

斗 ①(容量单位) *dou*, a unit of dry measure for grain (=1 decalitre) ②(量粮食的器具) a *dou* measure ③(形状象斗的东西) an object shaped like a cup or dipper: 漏～ funnel / 烟～ (tobacco) pipe ④[天](北斗星的简称) the Big Dipper: ～柄 the handle of the Dipper
【斗车】trolley (in a mine or at a construction site); tram
【斗胆】[谦] make bold; venture: ～建议 venture to suggest
【斗拱】[建] *dougong*, a system of brackets inserted between the top of a column and a crossbeam
【斗酒百篇】great capacity for drinking and poetry
【斗笠】bamboo hat
【斗篷】cape; cloak
【斗渠】[水] lateral canal
【斗室】a little room
【斗式提升机】[机] bucket elevator
【斗烟丝】pipe tobacco

抖 ①(哆嗦) tremble; shiver; quiver: 浑身直～ tremble all over / 冷得发～ shiver with cold / 气得发～ quiver with anger ②(抖动) shake; jerk: ～开棉被 spread the quilt with a flick / ～一～缰绳 give the reins a jerk / 把衣服上的雪～掉 shake the snow off one's clothes ③(鼓起精神) rouse: stir up: ～起精神 pluck up one's spirits ④[讽](得意) get on in the world
【抖动】shake; tremble; vibrate
【抖搂】①[方](振动) shake off; shake out of sth.: 把包里的东西～出来 shake a bag to empty it ②(揭露) expose; bring to light ③(浪费) waste; squander: 别把钱～光了。 Don't waste all the money.
【抖擞】enliven; rouse: ～精神 brace up; pull oneself together / 精神～ full of energy; full of beans
【抖威风】throw one's weight about

陡 ①(坡度大) steep; precipitous ②(陡然) suddenly; abruptly
【陡槽】[水] chute
【陡度】[物] gradient: 压力～ pressure gradient
【陡竣】high and precipitous
【陡立】rise steeply
【陡坡】steep slope

【陡峭】cliffy; precipitous: ～的山峰 cliffy summit
【陡然】suddenly; unexpectedly: ～下降 fall suddenly

dòu

斗 ①(对打) fight; tussle: 拳～ fist fight; fisticuffs ②(斗争) struggle against; denounce: ～恶霸地主 struggle against despotic landlords; settle scores with despotic landlords ③(比赛争胜) contest with; contend with ④(使动物斗) make animals fight (as a game): ～鸡 cockfighting / ～牛 bullfight / ～蛐蛐 cricketfight ⑤(往一块凑) fit together: ～榫 fit the tenon into the mortise; dovetail / ～一～情况 pool information to size up the situation
【斗法】①(用法术相斗) match magic powers ②(使用计谋) play tricks
【斗鸡眼】cross-eye
【斗气】quarrel just to vent one's spleen
【斗心眼】rival in trickery
【斗争】①(矛盾双方的冲突) struggle; fight; combat: 新与旧的～ conflict between the new and the old / 作不疲倦的～ wage a tireless struggle (against) ②(说理；控诉) accuse and denounce at a meeting ③(努力奋斗) strive for; fight for ◇ ～性 fighting spirit; militancy
【斗志】will to fight; fighting will: ～昂扬 have high morale / 鼓舞群众的～ arouse the fighting will of the masses
【斗智】battle of wits
【斗嘴】squabble; bicker; tiff: 别再～了。Stop bickering!

窦 ①(孔；漏) hole ②[生理](人体某些组织的凹入部分) sinus: 鼻旁～ paranasal sinus

豆 (豆子) legumes; pulses; beans; peas: 扁～ hyacinth beans / 蚕～ broad beans / 豌～ peas
【豆瓣酱】thick broad-bean sauce
【豆包】steamed bun stuffed with sweetened bean paste
【豆饼】[农] soya-bean cake; bean cake
【豆豉】fermented soya beans, salted or otherwise
【豆腐】bean curd; tofu: 冻～ frozen bean curd ◇ ～房 bean-curd plant / ～干 dried bean curd / ～脑儿 jellied bean curd / ～皮 skin of soya-bean milk / ～乳 fermented bean curd
【豆荚】pod
【豆浆】soya-bean milk
【豆角儿】[口] fresh kidney beans
【豆秸】beanstalk
【豆科】[植] the pulse family; bean family; pea family; Fabaceae ◇ ～植物 legume; leguminous plant; fabaceceous plant
【豆蔻】[植] round cardamom
【豆蔻年华】marriageable age
【豆绿】pea green
【豆面】bean flour
【豆娘】[动] damselfly
【豆萁】[方] beanstalk
【豆蓉】fine bean mash, used as stuffing in cakes
【豆沙】sweetened bean paste
【豆象】[动] bean weevil
【豆芽儿】bean sprouts
【豆雁】[动] bean goose
【豆油】soya-bean oil
【豆渣】bean dregs; residue from beans after making soya-bean milk
【豆汁】a fermented drink made from ground beans
【豆制品】bean products
【豆子】beans; peas

痘 ①(天花) smallpox ②(豆状疱疹) pox: 水～ chickenpox
【痘疮】smallpox; variola
【痘痕】pockmark
【痘苗】(bovine) vaccine
【痘疱】[医] pock

逗 ①(引逗) tease; play with: ～孩子玩 play with a child ②(逗笑) provoke (laughter, etc.): amuse ③[方](逗笑) funny: 你真～! How funny you are! ④(停留) stay; stop ⑤(读句中的停顿) a slight pause in reading
【逗号】comma (,)
【逗留】stay; stop: ～一夜 make an overnight stop
【逗弄】tease; kid; make fun of
【逗趣儿】[方] set people laughing (by funny remarks, etc.); amuse
【逗笑儿】[方] amusing
【逗引】tease

dū

都 ①(首都) capital ②(大城市) big city; metropolis
【都城】capital
【都督】(military) governer
【都会】city; metropolis

【都市】city; metropolis
【都市化】urbanization
嘟 ①[象]toot; honk ②[方](撅嘴) pout
【嘟噜】[口]①(量词,连成一簇) bunch; cluster: 一~葡萄 a bunch of grapes ②(向下垂成为一堆) hang down in a bunch ③(发小舌音) trill: 打~儿 pronounce with a trill; trill
【嘟囔】mutter to oneself; mumble
督 superintend and direct: ~战 supervise operations
【督察】superintend; supervise
【督促】supervise and urge: ~大家及时归还图书 urge everybody to return the books on time
【督励】urge and encourage
【督学】[旧] educational inspector

dú

毒 ①(毒物) poison; toxin: 服~ take poison ②(毒品) narcotics: 贩~ traffic in drugs / 吸~ take drugs ③(有毒的) poisonous; noxious; poisoned: ~箭 a poisoned arrow / ~蜘蛛 poisonous spider / 有~气体 noxious gas ④(毒辣;厉害) malicious; cruel; fierce: ~打某人 beat sb. mercilessly; beat sb. up ⑤(用毒物害死) kill with poison
【毒扁豆】[植] calabar bean ◇ ~碱[药] physostigmine; eserine
【毒草】poisonous weeds
【毒虫】poisonous insect; noxious insect
【毒打】beat cruelly; savagely
【毒蛾】tussock moth
【毒饵】poison bait
【毒害】poison (sb.'s mind)
【毒化】poison; spoil: ~会谈的气氛 poison the atmosphere of the talks
【毒计】venomous scheme; deadly trap
【毒剂】toxic; toxicant
【毒箭】poisoned arrow
【毒酒】poisonous wine
【毒辣】sinister; diabolic: 心肠~ harbor a murderous heart
【毒理学】toxicology
【毒瘤】malignant tumor; cancer
【毒品】narcotic drugs; narcotics ◇ ~贩子 drug trafficker / ~管制 drug control / ~买卖 sale of drugs
【毒气】poison gas ◇ ~弹 gas bomb / ~室 gas chamber
【毒区】contaminated area (in chemical warfare); gassed area
【毒杀】kill with poison
【毒杀芬】[农] toxaphene; octachlorocamphene
【毒蛇】poisonous snake; venomous snake; viper
【毒手】violent treachery; murderous scheme: 下~ resort to violent treachery; lay murderous hands on sb.
【毒死】kill with poison; poison
【毒素】[生] toxin
【毒瓦斯】poisonous gas; poison gas
【毒物】poisonous substance; poison
【毒物鉴定】identification of poison
【毒腺】[动] poison gland
【毒刑】cruel corporal punishment; horrible torture
【毒性】toxicity; poisonousness: ~猛烈 with drastic toxicity
【毒蕈】poisonous fungus; toadstool
【毒牙】poison fang; venom fang
【毒药】poison; toxicant
【毒液】venom
【毒瘾】drug addiction
【毒汁】venom
渎 show disrespect; contempt: 亵~ blaspheme; profane
【渎职】malfeasance; dereliction of duty ◇ ~行为 jobbery / ~罪 crime of misconduct in office
读 ①(看着文字发出声音) read; read aloud ②(上学) attend school: ~完大学 finish college
【读本】reader; textbook: 汉语~ a Chinese reader
【读出】read; read-out
【读回】read back
【读入】read in
【读书】①(出声或不出声地看书) read; study: 认真~ study in earnest ②(上学) attend school: 只读过两年书 attend school for only two years ◇ ~班 study class / ~笔记 reading notes / ~人 a scholar; an intellectual
【读熟】learn by heart
【读数】reading: 标度~ scale reading / 温度计~ thermometer reading
【读图】interpret blueprints; interpret drawings
【读物】reading matter; reading material: 儿童~ children's books / 科普~ popular science readings / 课外~ books for outside reading / 通俗~ popular literature

【读音】pronunciation
【读者】reader ◇ ～来信 readers' letters; letters to the editor

黩 act wantonly
【黩武】militaristic; warlike; bellicose: 穷兵～ engage in unjust military ventures ◇ ～主义 militarism / ～主义者 militarist

犊 calf

牍 ①(古代用的书简) wooden tablets or slips for writing (in ancient times) ②(文件；书信) documents; archives; correspondence

独 ①(一个) only; single: ～子 only son ②(独自) alone; by oneself; in solitude: ～居 live a solitary existence / ～坐 sit alone ③(年老没有孩子的人) old people without offspring; the childless
【独霸】dominate exclusively; monopolize: ～一方 lord it over a district; be a local despot
【独白】soliloquy; monologue
【独裁】dictatorship; autocratic rule ◇ ～者 autocrat; dictator / ～政治 autocracy
【独唱】(vocal) solo ◇ ～会 recital (of a vocalist)
【独出心裁】show originality; be original
【独创】original creation: ～一格 create a style all one's own ◇ ～精神 creative spirit / ～性 originality
【独词句】[语] one-member sentence
【独当一面】take charge of a department; take charge of a locality
【独到】original: ～的见解 original view / ～之处 originality
【独断】arbitrary; dictatorial
【独断独行】make arbitrary decisions and take peremptory actions; act arbitrarily
【独夫】a bad ruler forsaken by all; autocrat: ～民贼 autocrat and traitor to the people
【独家代理】sole agency
【独家经营】engage in a line of business without competition
【独家新闻】an exclusive news report; a scoop
【独角戏】monodrama; one-man show: 唱～ put on a one-man show; go it alone
【独具匠心】show ingenuity; have originality
【独具一格】of a unique style; be original in style
【独具只眼】be able to see what others cannot; have exceptional insight
【独揽】arrogate; monopolize: ～大权 arrogate all powers to oneself / ～经济大权 assume arbitrary power in finance
【独力】by one's own efforts; on one's own: ～经营 manage affairs on one's own
【独立】①(单独站立) stand alone ②(自主存亡) independence: 宣布～ proclaim independence ③(不依靠他人) independent; on one's own: ～分析问题和解决问题的能力 ability to analyse and solve problems on one's own ◇～成分[语] independent element / ～公正人 independent surveyor / ～国 independent state / ～核算单位 independent accounting unit / ～结构[语] absolute construction / ～师 independent division / ～团 independent regiment / ～营 independent battalion / ～王国 independent kingdom / ～自治 self-government / 民族～运动 the movement for 〈of〉 national independence
【独立性】independent character; independence: 闹～ assert one's "independence"; refuse to obey the leadership
【独立自主】maintain independence and keep the initiative in one's own hands; act independently and with the initiative in one's own hands
【独轮车】wheelbarrow
【独门独院】a single house which has its own entrance and courtyard
【独木不成林】one tree does not make a forest; one person alone cannot accomplish much
【独木难支】one log cannot prop up a tottering building; one person alone cannot save the situation
【独木桥】①(用一根木头搭成的桥) single-plank bridge; single-log bridge ②(艰难的途径) difficult path
【独木舟】dugout canoe
【独幕剧】one-act play
【独辟蹊径】open a new road for oneself; develop a new style; develop a new method of one's own
【独善其身】pay attention to one's own moral uplift without thought of others
【独身】①(单身) separated from one's family: ～在外 be away from home and family ②(不结婚的) unmarried; single ◇ ～女子 spinster; discovert / ～生活 celibacy / ～主义 celibacy
【独生女】only daughter
【独生子】only son
【独生子女证书】certificate of the only child

【独树一帜】fly one's own colors; develop a school of one's own
【独特】unique; distinctive: ～的风格 a unique style / ～的经历 unique experience / ～的性格 peculiar trait
【独吞】pocket profit without sharing with anyone else
【独往独来】act independently without seeking company
【独舞】solo dance
【独行其事】go one's own way; take one's own way; turn a deaf ear to advice from others
【独眼龙】a person blind in one eye; one-eyed person
【独一无二】unique; unparalleled; unmatched
【独占】have sth. all to oneself; monopolize: ～市场 monopolize the market
【独占鳌头】come out first; head the list of successful candidates; be the champion
【独占花魁】be the lucky man in winning the pretty courtesan's hand
【独占资本】[经] monopoly capital
【独资企业】ventures exclusively with one's own investment
【独自】alone; by oneself
【独奏】(instrumental) solo: 钢琴～ piano solo ◇～会 recital (of an instrumentalist)

dǔ

堵 ①(堵塞) stop up; block up: 把老鼠洞～死 stop up mouseholes ②(闷) stifled; suffocated; oppressed: 心里～得难受 have a load on one's mind / 胸口～得慌 feel suffocated; feel a tightness in the chest ③[量](用于墙) 一～墙 a wall
【堵击】intercept and attack: ～逃敌 intercept the fleeing enemy
【堵截】cut off: ～敌人的供应线 cut off the enemy's line of supply
【堵塞】stop up; block up: ～漏洞 stop up a loophole; plug a hole / 交通～ traffic jam
【堵嘴】gag sb.; silence sb.

睹 see: 目～ see with one's own eyes; be an eyewitness to
【睹物思人】seeing the thing one thinks of the person; the thing reminds one of its owner

赌 ①(赌博) gamble: 禁～ ban gambling ②(争输赢) bet: 打～ make a bet; bet
【赌本】money to gamble with
【赌博】gambling: 一场政治上的～ a political gamble
【赌场】gambling house
【赌棍】hardened gambler; professional gambler
【赌具】gambling paraphernalia; gambling device
【赌客】gambler
【赌窟】gambling-den
【赌气】feel wronged and act rashly
【赌钱】gamble: 喜欢～ fond of gamble
【赌徒】gambler
【赌友】gambling companions
【赌债】a gambling debt
【赌咒】take an oath; swear
【赌注】wager; stake: ～押错了 wager on a wrong horse

笃 ①(忠实) sincere; earnest: 友谊甚～ in devoted friendship ②(沉重) serious; critical: 病～ be dangerously ill; be in a critical condition; be terminally ill
【笃厚】sincere and magnanimous
【笃实】①(忠诚老实) honest and sincere: ～可靠 honest and reliable ②(实在) solid; sound: 学问～ sound scholarship
【笃信】sincerely believe in; be a devout believer in
【笃学】diligent in study; devoted to study; studious

肚 tripe: 拌～丝儿 slices of tripe and cucumber in soy sauce
【肚子】tripe

dù

度 ①(计量长短) linear measure ②(程度) degree of intensity: 湿～ humidity / 硬～ hardness ③(计量单位) degree ④[电](千瓦小时) kilowatt-hour (kwh) ⑤(限度) limit; extent; degree: 劳累过～ be overworked ⑥(宽容的程度) tolerance; magnanimity: 大～包容 regard with kindly tolerance; be magnanimous ⑦(所打算和计较的) consideration: 把生死置之～外 give no thought to personal safety ⑧[量](次) occasion; time: 一年一～ once a year / 再～ a second time; once more ⑨(过) spend; pass 欢～节日 joyously celebrate a festival
【度荒】tide over a lean year
【度假】spend one's holidays; go vacationing
【度量】tolerance; magnanimity: ～大 broad-minded; magnanimous / ～小 narrow-minded
【度量衡】length, capacity and weight; weights

and measures ◇ ～学 metrology
【度密月】honeymoon
【度盘照明】dial light
【度日】subsist (in hardship); eke out an existence: 靠救济～ live on relief
【度日如年】one day seems like a year; days wear on like years
【度数】number of degrees; reading
【度外】outside one's consideration

渡 ①(通过) cross (a river, the sea, etc.): ～河 cross a river / 飞～大西洋 fly (across) the Atlantic ②(渡过) tide over; pull through: ～过难关 tide over a difficulty; pull through ③(运载过河) ferry (people, goods, etc.) across
【渡槽】aqueduct
【渡场】[军] crossing site
【渡船】ferryboat; ferry
【渡过】tide over: ～危机 tide over a crisis
【渡海】sail across a sea
【渡河点】point of crossing
【渡口】ferry; ferry crossing
【渡鸦】[动] raven

镀 plating: ～镍 nickel-plating / 电～ electroplating; galvanizing
【镀层】cladding material; film
【镀铬】chromate treatment
【镀金】①(镀上一层金子) gold-plating; gilding ②(为图虚名而深造或锻炼) get gilded
【镀金属塑料】metalized plastics
【镀铝钢】aluminum-plated steel
【镀铜】copper facing; copperplating ◇ ～钢丝 coppered steel wire
【镀锡】tin-plating; tinning ◇ ～铁皮 tinplate
【镀锌】zinc-plating; galvanizing ◇ ～钢绞线 galvanized stranded wire / ～铁皮 galvanized iron sheet / ～铁丝 galvanized iron wire / ～瓦楞钢板 galvanized corrugated sheet
【镀银】silvering ◇ ～玻璃 silvered glass

杜 ①(杜梨) birch-leaf pear ②(阻塞) shut out; stop; prevent: ～门谢客 close one's door to visitors / 以～流弊 so as to put an end to abuses
【杜蘅】[植] wild ginger
【杜鹃】①[动](布谷鸟) cuckoo ②[植](杜鹃花) azalea ◇ ～座[天] Toucan
【杜绝】stop; put an end to: ～弊端 stop all corrupt practices / ～浪费 put an end to waste / ～危险 wall up the danger
【杜梨】birch-leaf pear
【杜灭芬】[药] domiphen
【杜宇】[动] cuckoo
【杜仲】[中药] the bark of eucommia ◇ ～胶 gutta-percha
【杜撰】fabricate; make up

肚 belly; abdomen; stomach
【肚带】bellyband; girth
【肚皮】[方] belly
【肚脐】navel; belly button
【肚子】belly; abdomen: ～空 with an empty stomach / ～痛 have a stomachache; suffer from abdominal pain / 笑得～痛 laugh till one's sides split / 一～气 absolutely exasperated; full of pent-up anger

蠹 ①(蠹虫) a kind of insect that eats into books, clothing, etc.; moth: 木～ woodworm / 书～ bookworm ②(蛀) moth-eaten; worm-eaten: 户枢不～ a door-hinge never gets worm-eaten
【蠹虫】①(蛀虫) moth ②(坏人) a harmful person; vermin
【蠹鱼】silverfish; fish moth

妒 be jealous of; be envious of; envy
【妒忌】be jealous of; be envious of; envy: 出于～ out of envy; out of jealousy
【妒贤忌能】be jealous of the worthy and able

duān

端 ①(东西的头) end; extremity: 两～ both ends / 岛的西～ the western tip of the island ②(开头) beginning: 开～ beginning ③(项目) point; item: 举其一～ for instance; just to mention one example ④(缘故) reason; cause: 借～ use sth. as a pretext / 无～ without rhyme or reason; unwarranted ⑤(端正) upright; proper: ～坐 sit up straight / 品行不～ improper behavior; misconduct ⑥(平举着拿) hold sth. level with both hands; carry: ～饭上菜 serve a meal / ～盘子 carry a tray
【端丽】neat and beautiful; graceful
【端量】look sb. up and down
【端倪】clue; inkling: ～已见 see an indication / 毫无～ have no inkling at all / 略有～ have an inkling of the matter
【端午节】the Dragon Boat Festival (the 5th day of the 5th lunar month)
【端线】[体] end line
【端详】①(详情) details: 细说～ give a full and detailed account; give full particulars ②(端庄

安详) dignified and serene: 举止～ behave with serene dignity ③(仔细看) look sb. up and down
【端绪】inkling; clue
【端正】①(不歪斜) upright; regular: 五官～ have regular features ②(正派) proper; correct: 品行～ correct in behavior ③(使端正) rectify; correct ～学习态度 take a correct attitude towards study
【端庄】dignified; sedate: 举止～ conduct oneself sedately／容貌～ decorous in appearance
【端子】[电工] terminal

duǎn

短 ①(跟"长"相对) short; brief ②(缺少) lack; owe: ～斤缺两 give short measure／理～ lack sound argument ③(缺点) weak point; fault: 揭人的～儿 pick on sb.'s weakness／说长道～ gossip
【短兵相接】fight at close quarters; engage in hand-to-hand fight
【短波】shortwave
【短不了】①(不能缺少) cannot do without ②(免不了) cannot avoid; have to
【短程】short distance; short range
【短处】shortcoming; failing; fault; weakness
【短传】[体] short pass
【短促】of very short duration; very brief: 呼吸～ be short of breath; gasp; pant／时间～ time is short and pressing
【短大衣】short overcoat
【短刀】a dagger; a short sword
【短笛】[乐] piccolo
【短吨】short ton
【短发】bob; shingle: 一个～姑娘 a girl wearing her hair in a bob; a bobbed hair girl
【短工】casual laborer; seasonal laborer
【短骨】[生理] short bone
【短号】[乐] cornet
【短见】①(见解短浅) shortsighted view ②(自杀) suicide: 寻～ attempt suicide; commit suicide
【短距离】short distance ◇ ～赛跑 short-distance run; dash; sprint
【短距起落飞机】STOL airplane; short takeoff and landing airplane
【短裤】shorts
【短路】[电] short circuit
【短命】die young; be short-lived
【短跑】dash; sprint ◇ ～运动员 dash man; sprinter
【短篇小说】short story
【短片】[电影] short film; short
【短评】short commentary; brief comment
【短期】short-term: 在～内 in a short time; in a brief space of time ◇ ～贷款 short-term loan／～轮训 short-term training in rotation／～票据 short (time) bill／～投资 short term investment
【短浅】narrow and shallow: 目光～ shortsighted／见识～ lacking knowledge and experience; shallow
【短欠】①(欠) owe; be in arrears ②(缺少) be short of
【短枪】short arm; handgun
【短缺】shortage
【短少】deficient; short; missing
【短时记忆】[心] short-term memory
【短视】①(近视) nearsightedness; myopia ②(眼光短浅) lack foresight; be shortsighted
【短统靴】ankle boots
【短途】short distance ◇ ～运输 short-distance transport; short haul
【短袜】socks
【短尾猴】stump-tailed macaque; stump-tailed monkey
【短文】short essay
【短纤维】①(短绒) short- staple: ～棉花 short-staple cotton ②[纺](短化纤) staple (fiber) ◇ ～切断器 staple cutter
【短小】short and small; short; small: ～的序幕 a brief prologue／身材～ of small stature
【短小精悍】①(指人) not of imposing stature but strong and capable ②(指文章) short and pithy; terse and forceful
【短训班】short-term training course
【短音程】[音] minor interval
【短音阶】[音] minor scale
【短语】[语] phrase
【短元音】[语] short vowel
【短暂】of short duration; transient; brief: ～的停留 brief stay; short stay
【短装】be dressed in a Chinese-style jacket and trousers

duàn

断 ①(分成段) break; snap: 喀嚓一声,～成两截 break in two with a snap ②(断绝) break off; cut off; stop: ～敌退路 cut off the enemy's retreat／～水 cut off the water supply ③(戒除) give up; abstain from: ～烟 give up smoking; quit smoking ④(决定) judge; decide: 当机立～

decide promptly and opportunely; make a prompt decision ⑤[书][副](绝对) absolutely; decidedly: ～不可信 absolutely incredible / ～无此理 absolutely untenable; absolutely unreasonable; the height of absurdity

【断案】①(审判案件) settle a lawsuit ②[逻](结论) conclusion (of a syllogism)

【断壁残垣】dilapidated walls; ruins of some buildings

【断编残简】stray fragments of text

【断层】[地] fault: 倾向～ dip fault / 走向～ strike fault ◇ ～带 fault zone / ～地震 fault earthquake / ～湖 fault lake / ～面 fault plane / ～作用 faulting

【断肠】heartbroken

【断炊】run out of rice and fuel; can't keep the pot boiling; go hungry

【断代】division of history into periods ◇ ～史 dynastic history

【断定】conclude; form a judgment; decide determine

【断断】absolutely: ～使不得 absolutely nothing doing

【断断续续】off and on; intermittently: ～地说 speak disjointedly / ～读过四年书 had four years of schooling off and on

【断顿】can't afford the next meal; go hungry

【断根】be completely cured; effect a permanent cure

【断后】①(走在最后的) bring up the rear; cover a retreat ②(断绝子孙) have no progeny

【断乎】[只用于否定式] absolutely: ～不可 absolutely impermissible

【断交】①(结束友谊) break off a friendship ②(断绝外交关系) sever diplomatic relations; break off diplomatic relations

【断句】①(读书时根据文义作的停顿) make pauses in reading unpunctuated ancient writings ②(按停顿加圈点) punctuate

【断绝】break off; cut off; sever: ～交通 stop traffic / ～外交关系 sever diplomatic relations ◇ ～地[军] broken terrain; broken ground

【断粮】run out of grain; run out of food

【断流器】[电] cutout: 安全～ safety cutout

【断路】[电] open circuit; broken circuit ◇ ～器 circuit breaker

【断面】[测] section ◇ ～图 sectional drawing; section

【断奶】weaning

【断气】①(停止呼吸) breathe one's last; die ②(切断煤气) cut off the gas

【断然】①(果断) absolutely; flatly; categorically: ～不能接受 absolutely inacceptable / ～否认 categorically deny / ～拒绝 flatly refuse ②(断乎; 严厉) resolute; drastic: 采取～措施 take drastic measures

【断事如神】decide matter with excellent judgement

【断送】forfeit (one's life, future, etc.); ruin

【断头】[纺] broken end ◇ ～率 end breakage rate

【断头台】guillotine: 上～ mount the guillotine; mount the scaffold

【断线风筝】a kite with a broken string; a person or thing gone beyond recall

【断崖绝壁】broken ridges and steep cliffs

【断言】say with certainty; assert categorically; affirm; state with certainty

【断语】conclusion; judgment: 遽下～ jump to conclusions

【断垣残壁】broken walls; debris

【断章取义】quote out of context; garble a statement, etc.

【断肢再植】[医] replantation of a severed limb

【断子绝孙】[骂] may you die without sons; may you be the last of your line

段 ①[量](部分) section; segment; part: 边界西～ the western sector of the boundary / 一～公路 a section of highway / 一～时间 a period of time / 这～历史 this phase of history ②(文章的段落) paragraph; passage ③[围棋] dan

【段落】①(文章的一段) paragraph ②(阶段) phase; stage

煅 ①(放在火里烧) forge: ～铁 forge iron ②(锻造) calcine

【煅烧】calcine ◇ ～炉 calciner

椴 [植] (Chinese) linden

锻 forge

【锻锤】forging hammer

【锻工】①(指工种) forging ②(指技术工人) forger; blacksmith ◇ ～车间 forging shop; forge / ～钳 band jaw tongs

【锻件】forging

【锻接】forge welding

【锻炼】①(体育锻炼) take exercise; have physical training: 坚持～ take regular exercise ②(磨炼) temper; steel; toughen

【锻炉】forge

【锻铁】wrought iron

【锻压机】forging press: 水力～ hydraulic forging press
【锻造】forging; smithing: 压力～ press forging

缎 satin
【缎带】(silk) ribbon
【缎纹】satin weave
【缎子】satin: 织锦～ brocade

duī

堆 ①(堆积) pile up; heap up; stack ②(堆积成的东西) heap; pile; stack: 草～ haystack / 粪～ manure heap; dunghill; manure dung / 土～ mound ③[量](成堆的物或人) pile; heap; crowd: 一～垃圾 a garbage heap; a rubbish heap / 一～人 a crowd of people ④(小山) hillock; mound
【堆存】store up
【堆垛机】(hay) stacker
【堆放】pile up; stack
【堆肥】[农] compost
【堆积】①(聚集成堆) pile up; heap up: ～如山 pile up like a mountain ②[地](堆积物) accumulation
【堆垒数论】[数] additive theory of numbers
【堆砌】①(使用大量华丽的词汇) load one's writing with fancy phrases ②(垒砌砖石) pile up
【堆石坝】[水] rock-fill dam
【堆栈】storehouse; warehouse

duì

兑 ①(凭票据领取) exchange; convert ②(添加) add
【兑付】cash (a cheque, etc.)
【兑换】exchange; convert: 把外币～成人民币 exchange foreign money for Renminbi ◇ ～率 rate of exchange
【兑现】①(用票据换现款) cash (a cheque, etc.) ②(诺言的实现) honor (a commitment, etc.); fulfill; make good: 说话不～ fail to keep one's promise; fail to make good one's promise / 政策～ materialize a policy

对 ①(回答) answer; reply: 无言以～ have nothing to say in reply ②(对付;对待) treat; cope with; counter: 辽宁队～陕西队 the Liaoning team versus the Shaanxi team ③(朝着) be trained on; be directed at ④(两者相对) mutual; face to face: ～饮 (two people) have a drink together / ～坐 sit facing each other ⑤(对面的;敌对的) opposite; opposing: ～岸 the opposite bank; the other side of the river ⑥(吻合;接触) bring (two things) into contact; fit one into the other: ～暗号 exchange code words ⑦(投合;适合) suit; agree; get along: ～胃口 suit one's taste / ～心眼儿 suit one down to the ground ⑧(核对) compare; check; identify: ～笔迹 identify the handwriting / ～号码 check numbers ⑨(调整) set; adjust: ～表 set one's watch; synchronize watches ⑩(正确) right; correct: 猜～了 guess right ⑪(掺和) mix; add: ～些水 dilute it with some water ⑫(均分) divide into halves: ～股劈 go halves; split fifty-fifty ⑬(对子) antithetical couplet; couplet: 喜～ wedding couplet ⑭[量](双) pair; couple: 一～夫妇 a married couple / 一～花瓶 a pair of vases
【对氨水杨酸钠】[药] sodium para-aminosalicylate (PASNa)
【对案】[外] counterproposal
【对岸】the opposite bank; the other side of the river
【对白】dialogue
【对半】①(各半) half-and-half; fifty-fifty: ～儿分 divide half-and-half; go halves ②(一倍) double: ～儿利 a double profit
【对比】①(相对比较) contrast; balance: 构成鲜明的～ form a sharp contrast / 今昔～ contrast the present with the past ②(比例) ratio: 双方人数～是一对四。 The ratio between the two sides is one to four. ◇ ～分析 comparative analysis / ～剂[医] contrast medium
【对不起】①[套](对人有愧) pardon me; I beg your pardon ②(辜负) let sb. down; be unworthy of; do a disservice to; be unfair to: ～人民 let the People down
【对策】the way to deal with a situation; countermeasure; countermove
【对策论】[数] game theory
【对唱】musical dialogue in antiphonal style; antiphonal singing
【对称】symmetry ◇ ～轴 axis of symmetry
【对答】answer; reply: ～如流 answer fluently; answer the questions without any hitch
【对打】fight each other
【对待】treat; approach; handle: 正确地～群众 adopt a correct attitude towards the masses
【对得起】not let sb. down; treat sb. fairly; be worthy of
【对等】reciprocity; equity: 在～的基础上 on the basis of reciprocity; on a reciprocal basis
【对地电阻】resistance to ground

【对调】exchange; swap: ～工作 exchange jobs / ～座位 exchange seats; swap seats
【对顶角】[数] vertical angles
【对儿童的监护】custody of children
【对方】the other side; the other party; the opposite side
【对方当事人】adversary, opposite party
【对付】①(应付) deal with; cope with; counter; tackle: ～各种复杂局面 deal with all kinds of complicate situations ②(将就) make do: 能～着看英文信 can manage to read English letters
【对歌】singing in antiphonal style
【对光】[摄] set a camera; focus a camera
【对过】opposite; across the way
【对号】①(对号数) check the number: ～入座 take one's seat according to the number on the ticket; sit in the right seat ②(相符合) fit; tally: 他说的和做的不～。His deeds don't match his words. ③(表示正确的符号) check mark(√); tick
【对话】dialogue
【对价】consideration
【对角线】[数] diagonal (line)
【对接焊】[机] butt welding
【对襟】a kind of Chinese-style jacket with buttons down the front
【对劲儿】①(称心合意) be to one's liking; suit one ②(合适) normal; right ③(合得来) get along (well)
【对进突击】[军] two-pronged assault from opposite directions
【对局】play a game of chess, etc.
【对开】①(相向开行) (of trains, buses or ships) run from opposite directions ②(对半分配) divide into two halves; go fifty-fifty ③[印](整张纸的二分之一) folio
【对抗】①(对立) antagonism; confrontation: 两国之间的～ confrontation between two states ②(抵抗) resist; oppose ◇ ～赛 [体] dual meet
【对抗性】antagonism ◇ ～矛盾 antagonistic contradiction
【对空射击】[军] antiaircraft firing
【对口】①(两人交替说唱) (of two performers) speak or sing alternately ②(相一致) be geared to the needs of the job; fit in with one's vocational training or specialty: ～训练 training geared to the needs of the job / 专业～ a job suited to one's special training ◇～唱 musical dialogue in antiphonal style; antiphonal singing / ～会谈[外] talks between representatives of similar organizations of two countries; counterpart conversations / ～赛 emulation between counterpart organizations / ～相声 cross talk; comic dialogue
【对垒】stand facing each other, ready for battle; be pitted against each other: 两军～ two armies pitted against each other
【对立】oppose; set sth. against; be antagonistic to: 唯心论和唯物论是～的。Idealism is opposite to materialism. ◇ ～关系 antagonistic relations / ～情绪 antagonism / ～物 opposite; antithesis
【对立面】[哲] opposite; antithesis: 矛盾着的～ the opposites in a contradiction
【对立统一】[哲] unity of opposites: ～规律 the law of the unity of opposites
【对联】antithetical couplet (written on scrolls, etc.)
【对流】[物] convection ◇ ～层[气] troposphere / ～雨[气] convective rain
【对路】①(合乎要求) satisfy the need ②(对劲) be to one's liking; suit one
【对骂】call each other names; abuse each other
【对门】①(大门相对) (of two houses) face each other ②(大门相对的房子) the building or room opposite
【对面】①(对过) opposite ②(正前方) right in front ③(面对面) face to face
【对内】internal; domestic; at home: ～负债 interior liabilities / ～搞活经济 invigorating the domestic economy ◇ ～政策 domestic policy
【对牛弹琴】play the lute to a cow; choose the wrong audience
【对偶】①[语](对称的字句) antithesis ②[数](双数) dual ◇ ～原理 principle of duality / ～运算 dual operations
【对瓶】[工美] twin vases
【对起诉书的答辩】plea to indictment
【对人权】right in personam
【对人之诉】action in personam
【对日照】[天] counterglow
【对生】[植] opposite ◇ ～叶 opposite leaf
【对手】①(竞赛的对方) opponent; adversary ②(劲敌) match; equal
【对数】[数] logarithm ◇ ～表 logarithmic table / ～函数 logarithmic function
【对私生子女的父亲的鉴定】affiliation
【对诉讼程序提出抗辩】demur the instance
【对损失的补偿】offset to the loss
【对损失所负责任】liability for loss
【对台戏】rival show: 唱～ put on a rival show
【对头】①(正确) correct; on the right track ②

(正常) normal ③(对劲) get on well; hit it off ④(仇敌) enemy: 死～ sworn enemy ⑤(对手) opponent; adversary
【对外】external; foreign ◇ ～工作 external work; work in the field of external relations / ～关系 external relations; foreign relations / ～经济贸易部 Ministry of Foreign Economics and Trade / ～开放 opening to the outside world / ～扩张 external expansion / ～援助 aid to foreign countries / ～政策 external policy; foreign policy / ～职能 external function
【对外贸易】foreign trade: ～逆差 foreign trade deficit; unfavorable balance of trade / ～顺差 foreign trade surplus; favorable balance of trade ◇ ～区 foreign trade zone
【对味儿】①(合口味) to one's taste; tasty ②(符合某人思想感情) seem all right: 他的发言不大～。 What he said didn't sound quite right.
【对物民事诉讼】civil process in rem
【对物权】right in rem
【对物之诉】action in rem; actions real
【对虾】prawn
【对象】①(目标) target; object: 革命～ target of the revolution / 研究～ object of study ②(恋爱的对方) boy or girl friend: 找～ look for a partner in marriage
【对消】offset; cancel each other out
【对眼】cross-eye
【对应】corresponding; homologous ◇ ～物 homologue / ～原理[物] correspondence principle
【对照】contrast; compare: ～原文修改译文 check the translation against the original and make corrections / 汉英～读本 an Chinese-English bilingual textbook / 形成鲜明的～ form a sharp contrast
【对折】50% discount: 照货价～ allow a fifty per cent discount off the price; reduce the price by one half
【对证】verify; check: ～事实 verify the facts
【对证据提出异议】demurrer to evidence
【对症下药】suit the medicine to the illness; suit the remedy to the case
【对质】confrontation (in court): 让被告与原告～ confront the accused with his accuser
【对峙】stand facing each other; confront each other: 武装～ military confrontation
【对准】①(瞄准) aim at ②(准线)[机] alignment: 轴～ shaft alignment
【对子】①(对偶的词句) a pair of antithetical phrases, etc.: 对～ supply the antithesis to a given phrase, etc. ②(对联) antithetical couplet

队 ①(行列) a row of people; line: 排成两～ fall into two lines ②(具有某种性质的集体) team; group: 国家～ national team / 军乐～ military band / 客～ visiting team / 省代表～ provincial team / 游击～ guerrilla forces; guerrillas / 种子～ seeded team / 主～ home team / 足球～ football team / 钻井～ drilling crew
【队部】the office or headquarters of a team, etc.
【队列】formation ◇ ～教练[军] (military) drill; formation drill
【队旗】[体] team pennant: 互赠～ exchange team pennants
【队伍】①(军队) troops ②(有组织的群众行列) ranks; contingent: 游行～ contingents of marchers; procession; parade
【队形】formation: 成战斗～ in battle formation / 以密集～前进 advance in close order ◇ ～变换 evolution
【队员】team member: 上场～ players in uniform / 替补～ substitute / 预备～ reserve
【队长】[体] captain: 不参加比赛的～ non-playing captain / 场上～ playing captain

dūn

敦 honest; sincere: ～请 cordially invite; earnestly request
【敦促】urge; press: ～某人做某事 urge sb. to do sth.
【敦厚】honest and sincere
【敦睦】promote friendly relations
【敦请】earnestly invite; earnestly request
【敦实】[方] stocky

墩 ①(土堆) mound: 土～ mound ②(墩子) a block of stone or wood: 桥～ pier (of a bridge) / 树～ stump ③[量](丛生或合在一起的植物) cluster
【墩布】mop; swab
【墩帽】pier cap; bridge cap
【墩子】a block of wood or stone: 菜～ chopping block
【墩座】pier base

吨 ton (t.): 长～ long ton / 短～ short ton / 公～ metric ton / 美～ short ton / 英～ long ton
【吨公里】ton kilometer
【吨海里】ton sea mile; ton nautical mile

【吨时】ton hour
【吨位】tonnage; shipping ton: 登记～ register tonnage / 净～ net tonnage / 总～ total ton; gross ton
【吨英里】ton-mile
蹲 ①(屈腿象坐的样子) squat on the heels: ～下来 squat down ②(呆着) stay: ～在家里 stay at home
【蹲膘】fatten in the shed
【蹲点】work at a selected spot (for investigation and study)
【蹲监狱】lie in prison
【蹲苗】[农] restrain the growth of seedlings (for root development)

dǔn

趸 ①(整批) wholesale ②(整批买进) buy wholesale: ～货 buy goods wholesale
【趸船】landing stage; pontoon
【趸批】wholesale: ～买进 buy wholesale / ～卖出 sell wholesale
盹 doze: 打～儿 doze off

dùn

炖 ①(用文火煮) stew: ～鸡 stewed chicken / 清～ boil sth. in its own soup without soy sauce ②(加热) warm sth. by putting the container in hot water: ～酒 warm (up) wine
顿 ①(稍停) pause ②(写字时顿笔) pause in writing in order to reinforce the beginning or ending of a stroke ③(安置) arrange; settle: 安～ arrange for; help settle down ④(叩地) touch the ground (with one's head) ⑤(跺地) stamp ⑥(立刻) suddenly; immediately: ～悟 suddenly realize the truth, etc.; attain enlightenment ⑦[量] 一天三～饭 three meals a day ⑧(疲乏) fatigued; tired: 劳～ tired out; exhausted
【顿挫】pause and transition in rhythm or melody: 抑扬～ modulation in tone
【顿改前非】immediately mend one's ways
【顿号】a slight-pause mark used to set off items in a series (、)
【顿河】the Don
【顿开茅塞】suddenly see the light
【顿时】immediately; at once; forthwith
【顿悟学习】insight learning
【顿足捶胸】stamp one's foot and beat one's breast
【顿钻钻井】churn drilling; percussion drilling; cable tool drilling
囤 a grain bin
钝 ①(不锋利) blunt; dull ②(笨拙) stupid; dull-witted: 迟～ dull-witted; slow
【钝化】[化] passivation; inactivation ◇ ～剂 passivator
【钝角】[数] obtuse angle ◇ ～三角形 obtuse triangle
【钝器伤】blunt force injury
【钝性物质】[化] inactive substance
盾 shield
【盾牌】①(防护武器) shield ②(借口) pretext; excuse
遁 escape; flee; fly
【遁词】subterfuge; quibble

duō

咄 tut-tut
【咄咄逼人】overbearing; aggressive
【咄咄怪事】monstrous absurdity
多 ①(数量大) many; much; more ②(有余;较多) more than the correct or required number; too many: 他～喝了一点儿。 He's had a drop too much. ③(过分的;不必要的) excessive; too much: ～疑 oversensitive; oversuspicious; given to suspicion ④(放在数量词后，表示有零头) more; over; odd: 四个～月 more than four months; four months and more / 五十～岁 over fifty years old ⑤(相差程度大) much more; far more: 难得～ much more difficult ⑥[副](表示程度) how: 看他～精神! Look how energetic he is!
【多半】①(大半) the greater part; most ②(大概) probably; most likely
【多倍体】[生] polyploid ◇ ～植物 polyploid plant
【多臂机】[纺] dobby
【多边】multilateral: ～会谈 multilateral talks / ～贸易 multilateral trade / ～条约 multilateral treaty / ～协议 a multilateral agreement / ～支付 multilateral payments
【多边形】[数] polygon
【多变】changeable; changeful; varied: ～的气候 a changeable climate / ～的战术 varied tactics
【多兵种合成军队】a cabined force of different arms
【多病】susceptible to diseases; constantly ill

【多才多艺】versatile; gifted in many ways
【多吃多占】eat or take more than one is entitled to; take more than one's share
【多愁善感】sentimental
【多此一举】make an unnecessary move: 何必～? Why take the trouble to do that?
【多次】many times; time and again; repeatedly; on many occasions
【多次往返有效入境出境签证】multiple entry-exit visa
【多弹头】multiple warhead ◇～导弹 multiple warhead missile / ～分导重返大气层运载工具 multiple independently targeted reentry vehicle (MIRV)
【多刀切削】[机] multiple cut; multicut
【多多少少】more or less
【多多益善】the more the better
【多发病】frequently-occurring disease
【多方】in many ways; in every way: ～设法 try all possible means; make every effort / ～协助 render all manner of help / ～阻挠 hinder in many ways; place various obstacles in the way
【多方面】many-sided; in many ways
【多福多寿】happiness and longevity; amply blessed
【多哥】Togo ◇～人 Togolese
【多股绞合线】[电] stranded wire
【多寡】number; amount: ～不等 vary in amount or number
【多管】[军] multibarrel ◇～高射机关炮 pompom / ～火箭炮 multibarrel (rocket) launcher / ～炮 multibarreled gun
【多管闲事】poke one's nose into others' business; be a busybody
【多国公司】multinational corporation
【多铧犁】multishare plough; multifurrow plough
【多会儿】①[口](什么时候) when: 你～来的? When did you come? ②(任一时间) ever; at any time: 我～有空～去。 I'll go there whenever I have time.
【多级火箭】multistage rocket
【多极】[电] multipolar ◇～发电机 multipolar generator
【多晶硅】[电子] polycrystalline silicon
【多晶体】[物] polycrystal
【多久】how long: 你来了～了? How long have you been here?
【多孔】porous ◇～动物 sponge / ～砖 porous brick; perforated brick
【多跨】[建] multispan ◇～结构 multispan structure / ～桥 multiple span bridge
【多快好省】achieve greater, faster, better and more economical results
【多亏】thanks to; luckily: ～你的帮助 thanks to your help
【多劳多得】more pay for more work
【多磷酸】[化] polyphosphoric acid
【多路通线】multichannel commnnication
【多毛症】[医] hirsutism; pilosis
【多么】[副](表示程度) how; what: ～大的变化啊! What a great change! (或: How great the change is!)
【多米尼加】Dominica ◇～人 Dominican
【多米诺骨牌】dominoes ◇～理论 the domino theory
【多面手】a many-sided person; a versatile person; an all-rounder
【多面体】[数] polyhedron
【多民族国家】multinational country
【多谋善断】resourceful and decisive; sagacious and resolute
【多幕剧】a play of many acts; a full-length drama
【多难兴邦】much distress regenerates a nation
【多瑙河】the Danube
【多年生】[植] perennial ◇～植物 perennial plant
【多年灾害】perennial scourge
【多尿症】polyuria
【多偶制】polygamy
【多普勒效应】[物] Doppler effect
【多情】full of tenderness or affection (for a person of the opposite sex)
【多刃刀具】[机] multiple-cutting-edge tool; multipoint tool
【多如牛毛】countless; innumerable
【多色】polychrome ◇～染料 polygenetic dyes / ～印刷 polychrome printing
【多少】①(数量大小) number; amount: ～不等 vary in amount or number ②(或多或少) somewhat; more or less; to some extent: ～有点失望 feel somewhat disappointed ③(问数量) how many; how much ④(不定数量) so much; as much as: 有～力, 出～力 do as much as one can
【多神教】polytheism
【多时】a long time: 等候～ have waited a long time
【多事】①(管闲事) meddlesome: 你不必～。 Don't be meddling. ②(事故多) eventful: ～之秋 an eventful period or year; troubled times

【多数】majority; most: 必要的～ the requisite majority / 简单～ simple majority / 绝大～ an overwhelming majority / 三分之二的～ a two-thirds majority / 特定～ qualified majority / 微弱的～ a small majority / 相对～ relative majority ◇ ～表决 decision by majority / ～票 majority vote
【多胎孕妊】multiple pregnancy
【多肽】[生化] polypeptide ◇ ～酶 polypeptidase
【多糖】[化] polysaccharide; polysaccharose ◇ ～酶 polysaccharase; polyase
【多头】(on the stock exchange) bull; long ◇ ～市场 Bull Market
【多头政治】polyarchy
【多退少补】refund for any overpayment or a supplemental payment for any deficiency
【多细胞生物】multicellular organism
【多相】[化] heterogeneous ◇ ～催化 heterogeneous catalysis / ～聚合 heterogeneous polymerization
【多项式】[数] multinomial; polynomial
【多谢】[套] many thanks; thanks a lot
【多心】oversensitive; suspicious
【多芯电缆】multicore cable
【多样化】diversify; make varied: ～的艺术风格 a variety of artistic styles / 使农作物～ diversify the crops
【多疑】suspicious: 你何必～? Why should you be so suspicious?
【多一事不如少一事】the less trouble the better; avoid trouble whenever possible
【多义词】[语] polysemant
【多余】unnecessary; surplus; superfluous; uncalled for: 删掉～的词语 cut out superfluous words and phrases
【多元论】[哲] pluralism ◇ ～历史观 the pluralistic concept of history / ～者 pluralist
【多元酸】[化] polybasic acid
【多云】[气] cloudy
【多灾多难】be dogged by bad luck; be plagued by frequent ills
【多种多样】varied; manifold
【多种经营】diversified economy; diversification
【多足动物】myriopod
【多嘴】speak out of turn; shoot off one's mouth: ～多舌 gossipy and meddlesome; long-tongued

哆

【哆嗦】tremble; shiver: 冷得打～ shiver with cold / 气得直～ tremble with rage

掇 pick up: 拾～ tidy up

duó

度 [书] surmise; estimate
【度德量力】estimate one's own moral and material strength; make an appraisal of one's own position

踱 pace; stroll: ～方步 walk with measured tread / ～来～去 pace to and fro; pace up and down

夺 ①(强取) take by force; seize; wrest: 从暴徒手上～下刀子 wrest a knife from a hooligan ②(争先取到) force one's way: ～门而出 force open the door and rush out; force one's way out ③(争取) contend for; compete for; strive for: ～得冠军 carry off the first prize / ～高产 strive for high yields ④(使失去) deprive: 剥～ deprive
【夺标】win the first prize
【夺佃】eviction of peasants from land leased to them by landlords or rich peasants
【夺回】recapture; retake; seize back: ～失去的时间 make up for lost time / ～一局 win a game (after losing one or more); pull up by a game / ～阵地 recapture a position
【夺目】dazzle the eyes: 光彩～ with dazzling brightness; brilliant; resplendent
【夺取】①(用武力强取) capture; seize; wrest: ～主动权 seize the initiative ②(努力争取) strive for: ～新的胜利 strive for new victories
【夺去】take away from
【夺去席位】unseat; deprive one of one's seat
【夺权】seize power

duǒ

朵 [量] 一～花 a flower / 一～云 a cloud
【朵儿】flower

垛 ①(扶垛) buttress ②(城垛) battlements
【垛口】crenel
【垛子】①(扶垛) buttress ②(城垛) battlements: 城～ battlements on a city wall

躲 ①(躲藏) hide (oneself): ～进深山老林 hide in a mountain forest ②(避) avoid; dodge: ～雨 take shelter from the rain
【躲避】①(回避) dodge; avoid; elude: ～困难 dodge difficulties ②(藏) hide

【躲藏】hide oneself; conceal oneself; go into hiding
【躲懒】shy away from work; shirk
【躲闪】dodge; evade: ～不及 be too late to dodge／躲躲闪闪 be evasive; equivocate; hedge
【躲债】avoid a creditor

duò

惰 lazy; indolent: 懒～ lazy
【惰性】inertia ◇ ～气体 inert gas／～元素 inert element

驮
【驮子】a load carried by a pack-animal; pack ◇ ～队 pack train

舵 rudder; helm: 掌～ take the helm; be at the helm
【舵柄】tiller
【舵杆】[船] rudder stock
【舵机】[船] steering engine
【舵轮】steering wheel
【舵手】steersman; helmsman
【舵索】rudder-line

垛 ①(整齐地堆) pile up neatly; stack ②(整齐地堆成的堆) pile; stack: 柴火～ a pile of faggots／麦～ a stack of wheat

剁 chop; cut: ～成两半 chop sth. in half／～肉馅 chop up meat

跺 stamp (one's foot): 气得直～脚 stamp one's foot with fury

堕 fall; sink: ～地 fall on the ground
【堕落】degenerate; sink low: 走上～、犯罪的道路 embark on the road of degeneration and crime／政治上～ be politically degenerate
【堕入】sink into; lapse into; land oneself in: ～陷阱 fall into a trap
【堕胎】①(人工流产) induced abortion ②(做人工流产) have an (induced) abortion ◇ ～药 aborticide

E

ē

阿 play up to; pander to: ～其所好 pander to sb.'s whims
【阿胶】[中药] donkey-hide gelatin
【阿弥陀佛】[佛教] ①(佛) Amitabha; Amitayus ②(祈祷用语) may Buddha preserve us; merciful Buddha
【阿魏】[植] asafoetida
【阿谀】fawn on; flatter: ～奉承 flatter and toady to; fawn upon

婀
【婀娜】(of a woman's bearing) graceful

é

额 ①(额头) forehead ②(牌匾) a horizontal tablet ③(规定的数目) a specified number or amount: 超～ above quota / 贸易～ volume of trade
【额定】specified (number or amount); rated: ～的人数 the maximum number of persons allowed; the stipulated number of personnel ◇～吨位 specified tonnage / ～功率 rated power / ～马力 rated horsepower
【额骨】[生理] frontal bone
【额角】frontal eminence
【额手称庆】put one's hand on one's forehead in jubilation; be overjoyed
【额头】forehead
【额外】extra; additional; added: ～报酬 extra pay / ～补贴 perquisite / ～负担 added burden / ～开支 extra expenses / ～收入 additional income / ～损失 extraneous loss

蛾 moth
【蛾眉】①(长而弯的眉毛) long slender eyebrows; delicate eyebrows ②(美女) beautiful woman; beauty
【蛾子】moth

锇 [化] osmium (Os)

俄 very soon; presently; suddenly
【俄罗斯】Russia ◇～人 Russian
【俄顷】in a moment; presently: ～雨止。Presently the rain let up.
【俄语】Russian (language)

鹅 goose
【鹅黄】light yellow
【鹅颈管】[机] gooseneck
【鹅口疮】[医] thrush
【鹅卵石】cobblestone; cobble
【鹅毛】goose feather
【鹅绒】goose down
【鹅行鸭步】waddle along like a duck or a goose
【鹅掌楸】[植] Chinese tulip tree

娥 pretty young woman: 宫～ palace maid; maid of honor
【娥眉】①(细而弯的眉) delicate eyebrows ②(美人) beautiful woman

讹 ①(错误) erroneous; mistaken: ～字 wrong words (in a text) / 以～传～ spread an error or a falsehood ②(讹诈) extort; blackmail; bluff: ～人 blackmail sb.; bluff sb.
【讹传】false rumor; unfounded rumor
【讹误】error (in a text)
【讹诈】extort under false pretenses; blackmail: ～钱财 extort money under false pretenses / 核～ nuclear blackmail

ě

恶
【恶心】①(要呕吐) feel like vomiting; feel nauseated; feel sick ②(使人厌恶) nauseating; disgusting

è

恶 ①(跟"善"相对) evil; vice; wickedness: 无～不作 stop at nothing in doing evil / 罪大～极 guilty of the most heinous crimes / 作～多端 commit all sorts of wickedness ②(凶恶) fierce; ferocious: ～狗 a ferocious dog a vicious dog; cur / ～骂 vicious abuse / 一场～战 a fierce battle ③(恶劣) bad; evil; wicked: ～势力 evil force / ～行 evil conduct; wicked conduct
【恶霸】local tyrant ◇～地主 despotic landlord
【恶报】retribution for evildoing; judgment
【恶病】a malignant disease
【恶病质】[医] cachexia
【恶臭】foul smell; stench
【恶毒】vicious; malicious; venomous: ～攻击 viciously attack / ～诬蔑 venomous slander
【恶感】ill feeling; malice: 对某人并无～ bear sb. no malice / 对某人有～ entertain ill feelings against sb.
【恶贯满盈】have committed countless crimes

and deserve to come to judgment; face retribution for a life of crime
【恶棍】ruffian; scoundrel; bully
【恶鬼】an evil spirit; a demon
【恶果】evil consequence; disastrous effect
【恶狠狠】fierce; ferocious: ～地瞪了他一眼 give him a ferocious stare
【恶化】worsen; deteriorate; take a turn for the worse: 病人的情况～了。The patient took a turn for the worse. / 暴风雨使燃料的缺乏状况更加～。Heavy storms worsened the fuel shortage. / 他的病情～了。His condition has worsened. / 形势～了。The situation worsened.
【恶疾】foul disease; nasty disease
【恶劣】odious; abominable; disgusting: ～环境 adverse circumstances / ～手段 mean tricks; dirty tricks / ～气候 harsh climate; vile weather / ～行径 disgusting conduct / ～作风 abominable behavior / 品质～ unprincipled; base / 影响～ make a very bad impression
【恶名】a bad reputation; infamy
【恶名昭著】notorious
【恶梦】nightmare; frightening dream; horrible dream
【恶眉恶眼】a fierce look; a ferocious look
【恶魔】demon; devil; evil spirit
【恶模恶样】a ferocious appearance; a fierce appearance
【恶念】evil intentions
【恶人】evil person; vile creature; villain: ～先告状 the villain sues his victim before he himself is prosecuted
【恶少】[旧] young ruffian
【恶事传千里】scandal spreads apace; scandal travels fast
【恶势力】vicious power; pressure groups
【恶习】bad habit; pernicious habit: 染上～ contract a bad habit; fall into evil ways
【恶性】malignant; pernicious; vicious: ～贫血 pernicious anemia / ～通货膨胀 galloping inflation; runaway inflation / ～循环 vicious circle ◇ ～肿瘤 malignant tumor; cancer
【恶意】evil intentions; ill intentions; ill will; malice: ～攻击 malicious attack / ～中伤 malicious calumniation / 并无～ bear no ill will
【恶有恶报】evil will be recompensed with evil
【恶语中伤】viciously slander; calumniate
【恶运】bad luck; ill luck
【恶战】hard fighting; a desperate fight
【恶兆】ill omen; bad omen
【恶浊】foul; filthy: ～的空气 stagnant air
【恶作剧】practical joke; prank; mischief

垩 chalk

噩 shocking; upsetting
【噩耗】sad news of the death of one's beloved
【噩梦】nightmare; frightening dream; horrible dream

厄
【厄瓜多尔】Ecuador ◇ ～人 Ecuadorian
【厄运】adversity; misfortune

苊 [化] acenaphthene

扼
【扼流圈】[电] choke
【扼杀】strangle; smother; throttle: ～在萌芽状态中 nip in the bud / ～在摇篮里 strangle in the cradle
【扼守】hold (a strategic point); guard: ～阵地 hold a position
【扼死】strangle; throttle
【扼要】to the point: 简明～ brief and to the point

轭 yoke

呃
【呃逆】[医] hiccup

遏 check; hold back: 怒不可～ be in a towering rage; be overcome with indignation; boil with anger
【遏止】check; hold back
【遏制】keep within limits; contain: ～愤怒的情绪 check one's anger ◇ ～政策 policy of containment

愕 stunned; astounded
【愕然】stunned; astounded: ～四顾 look around in astonishment

萼 [植] calyx
【萼片】[植] sepal

颚 ①(节肢动物的器官) jaw: 上～ upper jaw / 下～ lower jaw ②(腭) palate
【颚骨】jawbone
【颚针鱼】needlefish

鹗 osprey; fish hawk; sea eagle

腭 [生理] palate: 软～ soft palate／硬～ hard palate
【腭裂】[医] cleft palate
鳄 crocodile; alligator
【鳄鱼】crocodile; alligator: ～的眼泪 crocodile ears ◇ ～夹 alligator clip
饿 ①(跟"饱"相对) hungry: 挨～ go hungry ②(使受饿) starve: ～死 be starved to death
【饿饭】[方] go hungry; go without food
【饿虎扑食】like a hungry tiger pouncing on its prey
【饿殍】bodies of the starved: ～遍野 strewn with bodies of the starved everywhere

ēn

恩 kindness; favor; grace: 报～ requite a kindness; pay a debt of gratitude／施～ bestow favors
【恩爱】conjugal love ◇ ～夫妻 an affectionate couple
【恩赐】①(赏赐) bestow (favors, charity, etc.) ②(施舍) favor; charity
【恩德】favor; kindness; grace
【恩典】favor; grace
【恩惠】favor; kindness; grace; bounty
【恩将仇报】requite kindness with enmity
【恩情】loving-kindness: 党的～似海深。 The kindness of the Party is as deep as the sea.
【恩人】benefactor
【恩深义重】the spiritual debt is deep and great
【恩师】a teacher to whom one is greatly indebted; one's respected teacher
【恩同再造】a favor tantamount to giving sb. a new lease of life
【恩威并用】apply the carrot and stick judiciously
【恩义】spiritual debt; gratitude
【恩怨】①(恩惠和仇恨) feeling of gratitude or resentment ②(怨恨) resentment; grievance; old scores: 不计较个人～ not allow oneself to be swayed by personal feelings
蒽 [化] anthracene: ～酸 anthroic acid

èn

摁 press (with the hand or finger): ～电铃 ring an electric bell／～电钮 press a button; push a button／～住不放 press sth. down and hold it there
【摁钉儿】[口] drawing pin; thumbtack
【摁扣儿】[口] snap fastener

ér

而 ①[连] but; yet; while: 大～无当 large but impractical／华～不实 flashy without substance／伟大～艰巨的任务 a great and arduous task／有其名～无其实 in name but not in reality ②(表示到的意思) to: 由东～西 from east to west／自远～近 approach from a distance／一～再,再～三 again and again; time and again ③(把表示时间或方式的成分连接到动词前面) 匆匆～来 come hastening／盘旋～上 spiral up
【而后】after that; then
【而今】now; at the present time
【而且】①(表示平列) and ②(表示进一层) but also
【而已】that is all; nothing more
鸸
【鸸鹋】[动] emu
儿 ①(小孩) child: 小～ little child ②(年轻的人) youngster; youth ③(儿子) son ④(雄性的) male: ～马 stallion ⑤(后缀) 小猫～ kitten
【儿茶】[中药] catechu
【儿歌】children's song; nursery rhymes
【儿皇帝】puppet emperor
【儿科】(department of) pediatrics ◇ ～医生 pediatricians
【儿女】①(儿子和女儿) sons and daughters; children ②(青年男女) young man and woman (in love): ～情长 be immersed in love
【儿孙】children and grandchildren; descendants; posterity
【儿童】children ◇ ～保育事业 child care／～电视剧 children's TV show／～广播剧 radio play for children／～节 (International) Children's Day／～节目 children's programs／～教育 education for children／～心理 infantile psychology／～文学 children's literature; juvenile literature／～医院 children's hospital
【儿媳妇儿】daughter-in-law
【儿戏】trifling matter
【儿子】son

ěr

耳 ①(耳朵) ear: 内～ the inner ear／外～ the outer ear／中～ the middle ear ②(象耳朵的) any ear-like thing; ear of a utensil: 鼎～ ears of a tripod／银～ tremella ③(两边的) on both

sides; flanking; side: ～房 side rooms / ～门 side doors
【耳背】hard of hearing
【耳鼻喉科】①(医院部门) E.N.T. department; ear-nose-throat department; otolaryngological department ②(耳鼻喉科学) otolaryngology ◇ ～医生 E.N.T. specialist; otolaryngologist
【耳边风】unheeded advice; a puff of wind passing the ear: 当作～ let sth. in at one ear and out the other; turn a deaf ear to sth.
【耳鬓厮磨】very intimate
【耳沉】[方] hard of hearing
【耳垂】[生理] earlobe
【耳聪目明】①(听得清,看得明) (of old people) have good ears and eyes; can hear and see well ②(头脑清楚,眼光敏锐) have a thorough grasp of the situation
【耳朵】ear: ～尖 have sharp ears / ～软 credulous; easily influenced; susceptible to flattery / 揪～ pull sb. by the ear
【耳根清静】peace of mind achieved by staying away from nagging
【耳垢】earwax
【耳骨】ear bones
【耳鼓】the ear drum; tympanum
【耳光】a slap on the face; a box on the ear: 打～ slap sb.'s face; box sb.'s ear / 一记响亮的～ a ringing slap on the face
【耳环】earrings
【耳机】earphone
【耳镜】[医] otoscope
【耳科】otology ◇ ～医生 otologist; ear specialist
【耳孔】earhole
【耳廓】[生理] auricle
【耳聋】deaf
【耳轮】[生理] helix
【耳鸣】[医] tinnitus
【耳目】①(见闻) what one sees and hears; knowledge; information: ～闭塞 ill-informed / ～所及 from what one sees and hears; from what one knows ②(刺探消息的人) one who spies for sb. else: ～众多 eyes and ears everywhere; too many people around
【耳目一新】find everything fresh and new
【耳屏】[生理] tragus
【耳濡目染】be imperceptibly influenced by what one constantly sees and hears
【耳软心活】credulous and pliable
【耳塞】earplug
【耳生】strange-sounding; unfamiliar to the ear
【耳屎】[口] earwax
【耳熟】familiar to the ear
【耳提面命】pour exhortations into sb.'s ear; give earnest exhortations
【耳听八方】very alert; extremely vigilant
【耳挖子】earpick
【耳闻】hear of; hear about: ～不如目见 seeing for oneself is better than hearing from others
【耳闻目睹】what one sees and hears
【耳蜗】[生理] cochlea; acoustic labyrinth
【耳咽管】Eustachian tube; auditory tube
【耳语】whisper in sb.'s ear; whisper
【耳针疗法】[中医] auriculotherapy; ear-acupuncture therapy
【耳坠子】eardrop

铒 [化] erbium (Er)

饵 ①(糕饼) cakes; pastry: 果～ candies and cakes; confectionery ②(诱饵) bait: ～以官爵 use official title as bait; tempt sb. with official title / ～以重利 use great wealth as a bait; entice sb. with prospects of great wealth

尔

【尔代节】[伊斯兰教] Id
【尔格】[物] erg
【尔后】thereafter; subsequently: ～的战斗 the subsequent battle
【尔时】at that time
【尔虞我诈】each trying to cheat or outwit the other

迩 near: 名闻遐～ be known far and near

èr

二 ①(一加一) two: ～者必居其一 either one or the other ②(两样) different: ～心 disloyalty; half-heartedness
【二八佳人】sixteen-year-old beauty
【二把刀】[方] ①(技术不高) have a smattering of a subject ②(技术不高的人) smatterer
【二百二】[药] mercurochrome
【二百五】①[口] (做事莽撞的人) a stupid person ②[方] (半瓶醋) dabbler; smatterer
【二倍体】diploid
【二部制】two-shift system; two part-time shifts ◇ ～学校 school with two part-time shifts; two-shift school
【二重唱】[乐] (vocal) duet
【二重性】dual character; dual nature; duality
【二重奏】[乐] (instrumental) duet

【二次方程】[数] quadratic equation
【二次曲面】[数] quadratic surface
【二次投票】second ballot
【二等】second-class; second-rate: ～残废军人 disabled soldier, second class ◇ ～兵 private / ～舱 second-class cabin / ～功 Merit Citation Class II; second- class merit / ～奖 second prize / ～秘书[外] Second Secretary / ～品 goods of second quality; seconds
【二叠纪】[地] the Permian (Period)
【二分点】[天] the equinoxes
【二分法】[逻] dichotomy
【二分裂】[生] binary fission
【二分音符】[乐] minim; half note
【二副】second mate; second officer
【二锅头】a strong spirit usu. made from sorghum
【二胡】erhu fiddle
【二簧】*erhuang* melodies
【二化螟】striped rice borer
【二话】demur; objection: ～不说 without demur
【二级风】[气] force 2 wind; light breeze
【二极管】[电子] diode
【二尖瓣】[生理] mitral valve ◇ ～关闭不全 mitral insufficiency / ～狭窄 mitral stenosis
【二进制】[数] binary system ◇ ～标度 binary scale / ～数 binary number / ～数字 binary digit / ～自动计算机 binary automatic computer
【二郎腿】cross-legged: 跷起～ sit cross-legged or with ankle on knee
【二愣子】rash fellow
【二硫化物】bisulphide
【二流子】loafer; idler; bum
【二面角】[数] dihedral angle
【二名法】[生] binomial nomenclature
【二年生】[植] biennial ◇ ～根 biennial root / ～植物 biennial plant
【二全音符】[乐] breve
【二头肌】biceps; musculus
【二十八宿】[天] the lunar mansions
【二十四节气】the 24 solar terms ○立春 the Beginning of Sprint / 雨水 Rain Water / 惊蛰 the Waking of Insects / 春分 the Spring Equinox / 清明 Pure Brightness / 谷雨 Grain Rain / 立夏 the Beginning of Summer / 小满 Grain Full / 芒种 Grain in Ear / 夏至 the Summer Solstice / 小暑 Slight Heat / 大暑 Great Heat / 立秋 the Beginning of Autumn / 处暑 the Limit of Heat / 白露 White Dew / 秋分 the Autumnal Equinox / 寒露 Cold Dew / 霜降 Frost's Descent / 立冬 the Beginning of Winter / 小雪 Slight Snow / 大雪 Great Snow / 冬至 the Winter Solstice / 小寒 Slight Cold / 大寒 Great Cold
【二十四史】①(二十四部纪传体史书) the Twenty-Four Histories ②(长而复杂的故事) a long intricate story
【二四滴】[农] 2,4-D; 2,4-dichlorophenoxyacetic acid
【二踢脚】[方] double-bang firecracker
【二五眼】[方] ①(能力差) of inferior ability or quality ②(能力差的人) an incompetent person
【二项式】[数] binomial
【二象性】[物] dual property; duality: 波粒～ wave-particle duality / 物质的～ the dualistic nature of matter
【二心】disloyalty; halfheartedness
【二氧化硅】[化] silica
【二氧化碳】[化] carbon dioxide
【二氧化物】[化] dioxide
【二一添做五】go halves; go fifty-fifty
【二元】①[数] duality ②[化] binary ◇ ～酸 binary acid
【二元论】[哲] dualism
【二月】①(阳历) February ②(阴历) the second month of the lunar year; the second moon
【二者】the two; both: ～必居其一。It must be one or the other. / ～缺一不可。 Neither of the two can be dispensed.
【二至点】[天] the solstices

贰 two

【贰臣】an official who retains his position after capitulating to the new dynasty; turncoat official
【贰心】disloyalty; halfheartedness

F

fā

发 ①(送出;交付)send out; deliver; distribute; issue: ～电报 send a telegram / ～工资 pay out wages / ～货 deliver goods / ～通知 issue a notice / ～信号 give a signal ②(表达)utter; express: 两个同志在会上发了言。Two comrades spoke at the meeting. ③(发射)discharge; shoot; emit: ～光～热 emit light and heat / 万箭齐～。Ten thousand arrows shot at once. ④(扩大;开展)develop; expand ⑤(膨胀)(of foodstuffs) rise or expand when fermented or soaked: ～豆芽 raise bean sprouts; sprout beans / 面～起来了。The dough has risen. ⑥(发生;产生)come or bring into existence: 旧病复～ have an attack of a recurrent sickness; have a recurrence of an old illness; have a relapse ⑦(打开;揭露)open up; discover; expose: 揭～ expose ⑧(变;散发)get into a certain state; become: 脸色～白 lose color; become pale / 树叶开始～黄。The leaves are beginning to turn yellow. ⑨(流露)show one's feeling: ～怒 get angry / ～笑 laugh ⑩(感到)feel; have a feeling: 有点～冷 feel a bit chilly / 嘴里～苦 have a bitter taste in the mouth ⑪(起程)start; set out; begin an undertaking: 朝～夕至 set off in the morning and arrive in the evening / 车船齐～。all the boats and carts started off at the same time. ⑫[量]两～炮弹 two shells / 一百～子弹 one hundred rounds of ammunition; one hundred cartridges

【发白】turn white; become white; grow white; turn pale whitish: 气得脸都～了 turn pale with anger / 这褂子已经洗得～了。The coat has turned whitish from much washing.

【发榜】publish a list of successful candidates or applicants

【发报】transmit messages by radio, telegraphy, etc. ◇ ～机 transmitter

【发表】publish; issue: ～社论 carry an editorial / ～声明 issue a statement / ～文章 publish an article / ～演说 make a speech / ～意见 express an opinion; state one's views

【发病】(of a disease) come on

【发病率】attack; incidence of a disease; morbidity: 减少并发症～ lower the incidence of complications / 年龄～ age incidence

【发布】issue; release: ～命令 issue orders / ～新闻 release news

【发财】get rich; make a fortune; make a pile

【发潮】become damp: 衣服有点儿～。The clothes feel a bit damp.

【发车】①(派遣)dispatch (send off) a car (truck, bus) ②(开车)depart; pull out: ～时间 time of departure / 早七点半～。The train will depart at 7:30 a.m.

【发车场】[铁道] departure track

【发愁】worry; be anxious: 不要为这事～。Don't worry about it.

【发出】issue; send out; give out: ～传票 issue a summons / ～逮捕证 issue a warrant / ～紧急呼吁 send out an urgent appeal / ～警报 sound the alarm / ～警告 send out a warning / ～阵阵清香 send forth wafts of delicate fragrance / ～指示 issue a directive

【发达】developed; flourishing: 肌肉～ have well-developedmuscles / 工商业很～。Industry and commerce are flourishing. ◇ ～国家 developed country

【发呆】stare blankly; be in a trance: 她话也不说，坐在那里～。She said nothing but sat there staring blankly.

【发电】generate electricity ◇ ～量 generated energy; electricenergy production / ～能力 generating capacity / ～站 power station

【发电厂】power plant; power station: 地热～ geothermal power plant / 火力～ thermal power plant / 水力～ hydraulic power plant / 原子能～ atomic power plant ◇ ～容量 station capacity

【发电机】generator; dynamo: 汽轮～ turbo generator / 永磁～ magneto generator ◇ ～容量 generator capacity

【发电机组】generating set: 柴油～ Diesel generating set

【发动】①(使开始;使机器运转)start; launch: ～机器 start amachine; set a machine going / ～战争 launch a war ②(使行动起来) call into action; mobilize; arouse: ～群众 arouse the masses to action; mobilize the masses

【发动机】engine; motor: 柴油～ Diesel engine; Diesel motor / 煤气～ gas motor

【发抖】shiver; shake; tremble: 冷得～ shiver with cold / 吓得～ tremble with fear; shake in one's shoes

【发端】make a start

【发放】provide; grant; extend: ～救济金 deal out relief fund; dispense relief fund / ～农业贷款 grant agricultural credits

【发奋】work energetically

【发愤】make a firm resolution; make a determined effort: ～工作 put all one's energies into one's work / ～图强 work with a will to make the country strong
【发疯】go mad; go crazy; become insane; be out of one'smind: 发酒疯 be roaring drunk
【发福】grow stout; put on weight
【发糕】steamed sponge cake (usu. sweetened)
【发稿】①(发新闻专稿) distribute news dispatches ②(批准稿件付印) send manuscripts to the press: 到～时为止 at press time
【发给】issue; distribute; grant: ～复员费 issue demobilization pay / ～护照 issue a passport
【发光】① give out light; shine; be luminous: 有一分热，发一分光 give as much light as the heat can produce; do one's best, however little it may be / 群星闪闪～。 The stars twinkled. ②[物] luminescence: 场致～ electroluminescence ◇ ～度 luminosity / ～漆 luminous paint / ～强度 luminous intensity / ～体 luminous body; luminary; luminophor
【发汗】induce perspiration (as by drugs); diaphoresis ◇ ～药 sudorific; diaphoretic
【发号施令】issue orders; order people about
【发黑】①(变黑) turn black; grow black; become black; blacken; darken ②(一时眩晕) blackout: 眼睛突然～ have a sudden blackout
【发狠】①(下决心) make a determined effort ②(恼怒;动气) be angry
【发花】grow dim; see things in a blur
【发还】return sth. (usu. to one's subordinate); give back: 把作业～给学生 return the homework to the pupils
【发慌】feel nervous; get flustered; get flurried: 镇静些! 别～。 Keep calm! Don't get flurried.
【发挥】①(表现出内在的能力) bring into play; give play to; give free rein to: ～投资的效用 make the investment yield well / ～想象力 give the rein to one's imagination / ～专长 give full play tosb.'s professional knowledge or skill ②(详尽论述) develop (an idea, a theme, etc.); elaborate: 这一论点有待进一步～。 This point needs further elaboration.
【发昏】①(眩晕) feel giddy: 我的头有点儿～。 I feela bit giddy. ②(神志不清) lose one's head; becomeconfused: 你～啦! Are you out of your mind?
【发火】①(开始燃烧) catch fire; ignite ②(使爆发)detonate; go off ③(发脾气) get angry; flare up; loseone's temper ◇ ～点 [物] ignition point
【发货】send out goods; deliver goods ◇ ～单 dispatch list / ～人 consignor; shipper
【发迹】(of a poor man) gain fame and fortune; riseto power and position
【发急】become impatient
【发家】build up a family fortune
【发奖】award prizes ◇ ～仪式 prize-giving ceremony
【发酵】ferment ◇ ～粉 yeast powder; baking powder / ～饲料 fermented feed
【发窘】feel embarrassed; be ill at ease
【发觉】find; detect; discover: 错误一经～，就应改正。Mistakes should be corrected as soon as they are discovered.
【发掘】excavate; unearth; explore: ～人才 seek gifted people / ～文物 unearth cultural relics / ～祖国的医药学遗产 explore the legacy of traditional Chinese medicine and pharmacology
【发刊词】foreword to a periodical
【发狂】go mad; go crazy
【发懒】feel lazy
【发牢骚】grumble; complain; grouse
【发冷】feel cold
【发愣】[口] stare blankly; be in a daze; be in a trance
【发亮】shine
【发令枪】[体] starting gun
【发令员】starter
【发落】deal with (an offender): 从轻～ deal with sb. leniently
【发霉】go moldy; become mildewed
【发麻】tingle
【发面】①(使面发酵) leaven dough ②(已发酵的面) leavened dough ◇ ～饼 leavened pancake
【发明】①(创造) invent ②(创造出的新事物或新方法) invention ③[书](创造性地阐发) expound ◇ ～权 inventor's patentright
【发难】rise in revolt; launch an attack
【发排】send a manuscript to the compositor
【发胖】put on weight; get fat
【发脾气】lose one's temper; get angry
【发票】bill; receipt
【发起】①(倡议) initiate; sponsor: 这次会议是由二十二所院校～的。 The meeting was sponsored by 22 colleges. ②(发动) start; launch: ～反攻 launch a counterattack ◇ ～国 sponsor nation / ～人 initiator; sponsor
【发情】[动] ①(雌动物要求交配) oestrus: 同步～ synchronization of oestrus ②(动物的交尾

期) be inheat ◇ ～期 heat period; oestrus / ～周期 oestrouscycle
【发球】serve a ball: ～得分 ace / ～犯规 fault / 换～ change of service; side out / 换～! Change service! / 他～发得好。He has a very good serve. ◇ ～区 service area
【发热】①(发出热量) give out heat; generate heat ②(发烧) have a fever ③(不冷静) be hotheaded ◇～量 [化] calorific capacity
【发人深省】set people thinking; call for deep thought; provide food for thought
【发软】①(软弱无力) become limp; become flaccid ②(柔软) become soft
【发散】①(向四周散开) diffuse; (of rays, etc.) diverge ② [中医](用药把体内的热散出去) disperse the internal heat with sudorifics: 吃点药～一下 take a sudorific to sweat out a cold ◇ ～度[物] divergency / ～透镜 divergent lens; diverging lens
【发烧】have a fever; have a temperature
【发射】①(射出) launch; project; discharge; shoot; fire: ～导弹 launch a guided missile / ～炮弹 fire shells / ～人造卫星 launch a man-made satellite ②[物] transmit; emit: ～角 angle of departure ◇ ～场 launching site / ～光谱 emission spectrum / ～架 launcher / ～井 silo; launching silo / ～台 launching stand; launching pad
【发生】happen; occur; take place: 机器～故障。The machine broke down. / 那里～了强烈地震。A violent earthquake occurred there. / 这里～了巨大的变化。Tremendous changes have taken place here.
【发生炉煤气】producer gas
【发生器】[化] generator: 氨～ ammonia generator
【发声】sound production
【发誓】vow; pledge; swear: 他～要为烈士报仇。He vowed to avenge martyrs.
【发售】sell; put on sale: 新的纪念邮票将于下星期～。The new commemorative stamps will be put on sale next week.
【发水】flood
【发送】①(把无线电信号发射出去) transmit by radio: ～密码电报 transmit a coded message ②(送出) dispatch (letters, etc.)
【发送机】[无] transmitter
【发酸】①(有酸味) turn sour ②(酸痛) ache slightly: 腰有点～ have a slight backache
【发条】[机] spiral power spring; clockwork spring
【发文】outgoing message; dispatch ◇ ～簿 register of outgoing documents, letters, etc.
【发问】ask a question; raise a question
【发现】find; discover: ～问题,解决问题 discover problems and solve them
【发祥地】place of origin; birthplace: 我国古代文化的～ the birthplace of China's ancient culture
【发笑】laugh: 令人～ make one laugh; amusing
【发泄】give vent to; let off: ～不满 air one's grievances; express one's grievances
【发信】post a letter ◇ ～人 addresser
【发行】issue; publish; distribute; put on sale: ～书刊 publish books and magazines / ～影片 release a film / ～新邮票 issue new stamps / ～纸币 issue paper money / 由新华书店～ distributed by Xinhua Bookstore ◇ ～量 circulation / ～日期 issuing date / ～银行 bank of issue / ～者 publisher
【发芽】germinate; sprout: 种子还没有～。The seeds haven't sprouted yet. ◇ ～率 germination percentage / ～试验 germination test
【发言】speak; make a statement or speech; take the floor ◇ ～稿 the text of a statement or speech
【发言权】right to speak: 我们对此事当然有～。Of course we have a say in this matter.
【发言人】spokesman: 政府～ government spokesman
【发炎】[医] inflammation: 伤口～了。The wound has become inflamed.
【发扬】①(发展和提倡) develop; carry on: ～民主作风 develop a democratic style of work / ～正气 encourage healthy trends ②(发挥) make the most of; make full use of: ～火力,消灭敌人 make full use of firepower to destroy the enemy
【发扬光大】carry forward; develop; enhance
【发痒】itch; tickle: 我喉咙～。My throat tickles. / 我浑身～。I itch all over.
【发疟子】have an attack of malaria; suffer from malarial fever
【发音】pronunciation; enunciation; articulation: 这个字母不～。This letter is silent ◇ ～部位 points of articulation / ～困难 [医] dysphonia / ～器官 vocal organs
【发育】growth; development: ～健全 physically well developed ◇ ～不全 [医] hypoplasia / ～异常 [医] dysplasia
【发源】rise; originate: 黄河～于青海省。The

Huanghe River rises in Qinghai Province. ◇ ～地 place of origin; source; birthplace
【发晕】feel dizzy
【发展】①(变化)develop; expand; grow: ～生产，繁荣经济 develop production and bring about a prosperous economy ②(扩大)recruit; admit: ～新党员 admit new Party members
【发展中国家】developing country
【发胀】①(膨胀)swell: 肚子～ feel bloated / 头脑 ～ have a swelled head ②[针灸] feel distended
【发肿】swell; become swollen: 他的腿～。His legs are swelling out.
【发紫】turn blue: 嘴唇～ one's lips turn blue
【发作】①(突发)break out; show effect: 心脏病～ have a heart attack ②(发脾气) have a fit of anger; flare up: 歇斯底里大～ havea bad fit of hysterics

fá

罚 punish; penalize: 赏～分明 be fair in meting out rewards or punishments
【罚不当罪】be unduly punished
【罚出场】[体] be ordered off the field for foul play; foul out
【罚金】fine; forfeit: 处以～ fine sb.; impose a fine on sb.
【罚酒】be made to drink as a forfeit
【罚款】①(处以罚金) impose a fine or forfeit ②(被罚缴纳的钱) fine; forfeit; penalty ◇ ～条款 penalty clause
【罚球】(篮球) penalty shot; (足球) penalty kick ◇ ～区 penalty area / ～线 penalty line / ～中 convert a free throw

乏 ①(缺少) lack: 不～其人 There's no lack of such people ②(疲倦) tired; weary
【乏味】dull; insipid; drab; tasteless: 语言～ dull language

伐 ①(砍) fell; cut down ②(攻打) send an expedition against; attack: 讨～ send a punitive expedition
【伐木】lumbering; felling; cutting ◇ ～工 lumberman / ～业 lumbering
【伐区】[林] cutting area; felling area

阀 ①(有势力的人或家族) a powerful person or family: 财～ plutocrat; financial magnate / 军～ warlord ②(管道等上的控制装置) valve: 安全～ safety valve
【阀门】[机] valve ◇ ～厂 valve plant

筏 raft: 橡皮～ rubber raft
【筏道】log chute; logway
【筏子】raft

fǎ

法 ①(法律) law: 守～ observe the law / 违～ break the law / 依～惩办 be punished by law ②(方法) method; way; mode ③(仿效) follow; model after: 效～ take as model; follow ④(标准) standard; model: ～书 model calligraphy ⑤(魔法) magic arts: 戏～ conjuring tricks
【法案】proposed law; bill
【法办】deal with according to law; punish by law; bring to justice
【法宝】a magic weapon
【法场】execution ground; place of execution
【法典】code; statute book: 民法～ civil code / 刑法～ penal code; criminal code
【法定】legal; statutory ◇ ～代理人 legal representative / ～汇率 official rate (of exchange); pegged rate of exchange; pegged exchange parity / ～货币 legal tender / ～继承权 forced heirship / ～继承人 heir(s) at law / ～假日 legal holidays / ～年龄 lawful age; legal age / ～预算 legal budget
【法定人数】quorum: 达到～ have a quorum
【法官】judge; justice
【法规】laws and regulations
【法国】France ◇ ～大革命 the French Revolution (1789) / ～人 the French; Frenchman
【法纪】law and discipline: 目无～ act in utter disregard of law and discipline; flout law and discipline / 遵守～ observe law and discipline
【法警】bailiff
【法拉】[物] farad: 微～ microfarad
【法兰】[机] flange ◇ ～盘 ring flange
【法兰绒】flannel
【法郎】franc: 法国～ French Franc / 瑞士～ Swiss Franc
【法理】legal principle; theory of law ◇ ～学 jurisprudence
【法力】supernatural power
【法令】laws and decrees; decree: 政府～ government decree
【法律】law; statute: 制定～ make laws; enact laws / ～面前，人人平等。All men are equal before the law. ◇ ～保护 legal protection / ～承认 *de jure* recognition / ～地位 legal status / ～根据 legal basis / ～顾问 legal adviser / ～规定 legal provisions / ～手续 le-

gal procedure／～效力 legal effect／～制裁 legal sanction
【法螺】①[动] triton (shell) ②(螺壳)conch：自吹～ blow one's own trumpet
【法门】[佛教] initial approach to become a Buddhist believer
【法器】[宗] musical instruments used in a Buddhist or Taoist mass
【法权】right：治外～ extraterritoriality
【法人】[法] legal person; juridical person ◇～税 corporation tax／～团体 body corporate; corporation／～资格 corporate organization
【法书】①(可作书法典范的字) model calligraphy ②[敬]your calligraphy
【法术】magic arts
【法庭】court; tribunal：军事～ military tribunal; court-martial／民事～ civil court／刑事～ criminal court／仲裁～ arbitration tribunal; court of arbitration ◎庭长 chief of court／审判长 presiding judge／审判员 judge; justice／陪审员 assessor; juror; juryman／书记员 clerk of a court／公诉人 public prosecutor／当事人 litigant; party to a dispute／原告 complainant; plaintiff／被告 defendant; the accused／辩护人 advocate／律师 lawyer／开庭 the court is sitting; open a court session; call the court to order／开庭审理 hear; hold a hearing／退庭 the court is adjourned／公诉 public prosecution／公审 public trial／复审 review; retry; retrial／仲裁 arbitration／裁定 ruling／断案 settle a lawsuit／结案 wind up a case／判决书 written judgement／胜诉 win a lawsuit／错案 a misjudged case／抗诉 counter-charge／上诉 appeal to a higher court／合议庭 collegiate branch／起诉 prosecute(刑事); bring a case to court(民事)／起诉书 indictment(刑事); complaint(民事)／传讯 summon for interrogation or trial; cite／传票 (court) summons; subpoena／控告 charge; accuse; bring in an indictment against sb.／诉讼 suit; lawsuit; litigation／立案 filing a case; place a case on file for investigation and prosecution／提审 bring (a prisoner) before the court; bring to trial／原审 first instance; first hearing; first trial／终审 final instance
【法统】legally constituted authority
【法网】the net of justice; the arm of the law：落入～ be caught in the meshes of law／难逃～ can hardly escape from the meshes of law
【法西斯】fascist ◇～化 fascistization／～主义 fascism
【法学】the science of law; law ◇～家 jurist
【法眼】[佛教]Buddha's Dharma-eye
【法医】legal medical expert ◇～学 medical jurisprudence; forensic medicine
【法语】French (language)
【法院】court of justice; law court; court：地方人民～ local people's court／第一审～ court of first instance／高级人民～ Higher People's Court／区人民～ district people's court／中级人民～ intermediate people's court／最高人民～ the Supreme People's Court ◇～院长 president of court
【法则】rule; law：自然～ law of nature
【法治】rule by law
【法制】legal system; legal institutions; legality：加强～ strengthen the legal system ◇～委员会 Commission for Legal Affairs
【法子】way; method：想个～ have to think of a way to

砝

【砝码】weight (used on a balance)

fà

发 hair：理～ haircut／留短～ wear one's hair shingled; wear one's hair bobbed／染～ have one's hair dyed／烫～ have a permanent wave／洗～ have one's hair washed; have one's hair shampooed
【发辫】plait; braid
【发膏】pomade
【发髻】bun; chignon
【发夹】hairpin; bobby pin ◇卷～ hair-curler
【发蜡】pomade
【发卡】hairpin
【发刷】hairbrush
【发网】hairnet
【发型】hair style; hairdo; coiffure
【发癣】[医] ringworm of the scalp; tinea capitis
【发油】hair oil; brilliantine; hair tonic
【发指】bristle with anger; boil with anger：敌人的暴行令人～。The enemy's atrocities made one boil with anger.

珐

【珐琅】enamel
【珐琅质】[生理] enamel
【珐琅制品】enamelwork

fān

帆 sail: 大三角～ spinnaker / 三角～ jib / 主～ mainsail / 降～ lower a sail / 扬～ unfurl a sail / ～樯林立 a forest of masts

【帆布】canvas; duck ◇ ～包 canvas bag; kit bag / ～床 cot; camp bed / ～篷 canvas roof; awning / ～鞋 plimsolls

【帆船】sailing boat; sailing ship; junk; yacht; sailboat: ◇ ～船身板 planking / ～俱乐部 yacht club / ～运动 yachting; sailing; yacht racing / ～运动员 yachtsman; sailor

【帆桁】yard

【帆具】rigging

【帆面积】sail area

【帆索具】rigging

番 [量](种; 回; 次)出于一～好意 out of good intention / 听了这一～话 having heard this talk / 别有一～天地。There seems a different world. / 粮食产量翻了一～。Grain output has doubled.

【番号】designation (of a military unit)

【番荔枝】sweetsop

【番木瓜】[植] papaya

【番茄】tomato ◇ ～酱 tomato ketchup / ～汁 tomato juice

【番薯】[方] sweet potato

幡 long narrow flag; streamer

【幡状云】fall streak; virga

藩

【藩篱】hedge

【藩属】vassal state

翻 ①(移位) turn over: ～谷子 turn over the grain(to dry) / 船～了。The ship capsized. ②(越过) cross; get over: ～过山头 cross a mountaintop / ～墙 climbover a wall ③(翻找) rummage; search: ～参考书 look through reference works / 抽屉我都～遍了,还是找不到。I rummaged all the drawers, but still couldn't find it. ④(译) translate: 把中文～成西班牙文 translate the Chinese into Spanish ⑤(推翻) reverse: ～案 reverse a verdict ⑥(使成倍增长) multiply: 使炼油能力～一番 double the oil-refining capacity ⑦(翻脸)[口] fall out; break up

【翻案】reverse a (correct) verdict

【翻白眼】show the whites of one's eyes (as from emotion or illness)

【翻版】reprint; reproduction; refurbished version

【翻场】[农] turn over the grain on the threshing ground

【翻车机】[矿] tipper; dumper; tipple

【翻地】turn up the soil

【翻斗】tipping bucket; skip bucket ◇ ～车 skip car; tipcart / ～卡车 tipping lorry; tip lorry; dump truck

【翻盖】renovate (a house)

【翻改】turn; renovate: ～一件短上衣 have a jacket turned

【翻杠子】[体] do gymnastics on a horizontal bar or on parallel bars

【翻跟头】turn a somersault; loop the loop

【翻供】withdraw a confession; retract one's testimony; revoke verbal evidence

【翻滚】roll; toss; tumble: 白浪～ The waves rolled and foamed

【翻悔】back out (of a commitment, promise, etc.)

【翻江倒海】overturning rivers and seas; overwhelming; stupendous; terrific

【翻浆】[交] frost heave; frost boil

【翻看】leaf through: ～日历 leaf through the calendar

【翻来复去】①(来回翻动身体) toss and turn; toss from side to side: 他在床上～睡不着。He tossed and turned in bed, unable to sleep. ②(重复) again and again; repeatedly: 这种话,她～不知说过多少遍了。This is what she has been saying over and over again.

【翻老帐】bring up old scores again

【翻脸】fall out; suddenly turn hostile: ～不认人 turn against a friend

【翻领】turndown collar

【翻然】(change) quickly and completely: ～悔悟 quickly wake up to one's error

【翻砂】[机] founding; molding; casting ◇ ～车间 foundry shop / ～工 foundry worker; caster

【翻山越岭】cross over mountain after mountain; tramp over hill and dale

【翻身】①(躺着转动身体) turn over ②(解放出来) free oneself; stand up

【翻腾】① [跳水] tuck dive: 向内～两周半 backward tuck dive with two-and-a-half somersaults / 向前～两周半 forward tuck dive with two-and-a-half somersaults ②(上下滚动) seethe; rise; churn ③(翻动) turn sth. over and over

【翻天覆地】earth-shaking; world-shaking: ～的

变化 an earth-shaking change／～的时代 an earth-shaking era
【翻胃】gastric disorder causing nausea
【翻箱倒柜】rummage through chests and cupboards; ransack boxes and chests
【翻新】renovate; recondition; make over: 花样～(the same old thing) in a new guise
【翻修】rebuild: ～房屋 have the house rebuilt
【翻译】①(把一种语言译成另一种语言) translate; interpret: 请你帮我～一下好吗? Would you mind translating for me? ②(译员) translator; interpreter ◇～本 translation／～程序法 interpretive programming／～程序语言 language interpreter; translator language／～存储器 translation memory／～计算机 source-computer／～片 dubbed film
【翻印】reprint; reproduce
【翻阅】browse; look over; glance over; leaf through
【翻云覆雨】produce clouds with one turn of the hand and rain with another; given to playing tricks; shifty

fán

烦 ①(烦闷) be vexed; be irritated; be annoyed: 心～ feel vexed ②(厌烦) be tired of: 厌～ be fed up with ③(又多又乱) superfluous and confusing: 要言不～ giving the essentials in simple language; pithy; terse ④(烦劳) trouble: ～交某人 please forward this to so-and-so
【烦劳】trouble: ～您带几本书给他。Would you mind taking a few books to him?
【烦闷】be unhappy; be worried
【烦恼】be vexed; be worried: 自寻～ bring vexation on oneself; worry over for nothing
【烦扰】①(搅扰) bother; disturb ②(心烦) feel disturbed
【烦人】annoying; vexing; troubling: 多么～呀! How vexing! (或: How annoying!)
【烦冗】①(繁杂) (of one's affairs) diverse and complicated ②(烦琐冗长) (of speech or writing) lengthy and tedious; prolix
【烦琐】loaded down with trivial details: ～的手续 overelaborate procedure
【烦琐哲学】①(经院哲学) scholasticism ②[口] overelaboration; hairsplitting
【烦嚣】noisy and annoying
【烦躁】be fidgety; be agitated
【烦躁不安】dysphoria

樊 [书] fence
【樊篱】①(篱笆) fence ②(限制) barriers; restriction
【樊笼】bird cage; place or condition of confinement

蕃 ①(茂盛) luxuriant; growing in abundance: ～茂 luxuriant; lush ②(繁殖) multiply; proliferate: ～衍 multiply; increase gradually in number or quantity

繁 ①(繁多) in great numbers; numerous; manifold: ～星满天 a starry sky／删～就简 simplify by weeding out superfluities ②(繁殖) propagate; multiply
【繁多】various: 花样～ of all shapes and colors
【繁分数】[数] complex fraction
【繁复】heavy and complicated: 有了计算机, ～的计算几秒钟就能完成。A computer does complicated calculations in a few seconds.
【繁花似锦】flowers blooming like a piece of brocade; a flourishing scene of prosperity
【繁华】flourishing; bustling; busy: 城里最～的地区 the busiest section of town; the downtown district
【繁忙】busy: ～的生活 a busy life
【繁茂】lush; luxuriant: 枝叶～ with luxuriant foliage
【繁密】dense: 林木～ densely wooded
【繁难】hard to tackle; troublesome: ～的问题 knotty problem
【繁荣】①(蓬勃发展) flourishing; prosperous; booming: ～富强 rich, strong and prosperous ②(使昌盛) make sth. prosper: ～经济 bring about a prosperous economy; promote economic prosperity
【繁荣昌盛】thriving and prosperous: 祝贵国～, 人民幸福。We wish your country prosperity and her people happiness.／祖国日益～。Our country is thriving and prospering day by day.
【繁盛】thriving; flourishing; prosperous
【繁体字】the original complex form of a simplified Chinese character
【繁文缛节】unnecessary and overelaborate formalities; red tape
【繁细】overloaded with details; excessively detailed
【繁衍】[书] multiply; increase gradually in number or quantity
【繁育】breed: ～家畜 breed stock／～优良品种 breed good strains ◇～推广体系 a network

for breeding and popularizing
【繁杂】many and diverse; miscellaneous
【繁殖】[生] breed; reproduce; propagate ◇ ～力 reproductive capacity; fecundity; fertility / ～率 rate of reproduction; breeding rate
【繁重】heavy; strenuous; onerous: ～的体力劳动 strenuous manual labor / 任务～。 The tasks are arduous.

凡 ①(平凡) commonplace; ordinary: 非～ extraordinary ②(人世间) this mortal world; the earth: 天仙下～ a celestial beauty come down to earth ③(凡是) every; any; all: 在我国, ～年满十八岁的公民, 都有选举权和被选举权。 In China, every citizen who has reached the age of eighteen has the right to vote and stand for election. ④ [书](总共) altogether: 全书～十卷。 The set consists of 10 volumes altogether
【凡立丁呢】[纺] valitin
【凡例】notes on the use of a book, etc.; guide to the use of a book, etc.
【凡人】①(平常的人) ordinary person ②(俗人) mortal
【凡士林】vaseline; petrolatum: ～纱布 petrola
【凡事】everything: ～开头难。 It's the first step that costs. / ～要小心。 Be careful in all things.
【凡是】every; any; all
【凡庸】commonplace; ordinary

矾 [化] vitriol: 绿～ green vitriol; copperas / 明～ alum
【矾石】[矿] aluminite; websterite
【矾土】[矿] alumina ◇ ～水泥 alumina cement
【矾土肺】aluminosis

钒 [化] vanadium (V)
【钒钢】vanadium steel
【钒铅锌矿】descloizite
【钒铁】vanadium iron
【钒云母】roscoelite
【钒中毒】vanadiumism

fǎn

反 ①(翻过来;转换) turn over: ～败为胜 turn defeat into victory; turn the tide / 易如～掌 as easy as turning one's hand over ②(方向向背) in an opposite direction; in reverse; inside out: ～科学 contrary to science / ～面 the reverse side / ～其道而行之 do exactly the opposite / 适得其～。 The result is just the contrary. ③(反之) on the contrary; instead ④(回;还) return; counter: ～击 counterattack; strike back / ～问 counter with a question; ask in retort ⑤(背叛) revolt; rebel: ～叛 revolt; rebel ⑥(反对) oppose; combat: ～间谍 counterespionage ⑦(指反动派;反革命) counterrevolutionaries; reactionaries: 肃～ elimination of counterrevolutionaries
【反霸】①(反恶霸) oppose local despots ②(反霸权) opposehegemonism
【反败为胜】come from behind to win the game
【反比】[数] inverse ratio: 成～ be in inverse proportion to
【反比例】[数] inverse proportion
【反驳】retort; rebut; counterplea: ～一个论点 refute an argument; disprove an argument
【反剥削】oppose exploitation
【反差】[摄] contrast: ～指数 contrast index
【反常】unusual; abnormal; perverse; strange: 最近天气有点儿～。 The weather is a bit unusual these days.
【反超子】anti-hyperon
【反衬】set off by contrast; serve as a foil to
【反冲】[物] recoil ◇ ～核 recoil nucleus
【反刍】ruminate; chew the cud ◇ ～动物 ruminant (animal)
【反唇相讥】recriminate; answer back sarcastically
【反导弹导弹】antimissile missile
【反倒】on the contrary; instead
【反帝】anti-imperialist; against imperialism
【反动】①(逆革命潮流而动) reactionary: ～势力 reactionary forces / ～透顶 ultra-reactionary ②(相反的作用) reaction ◇ ～本质 reactionary nature / ～分子 reactionary element; reactionary / ～观点 reactionary viewpoint / ～派 reactionaries
【反对】oppose; be against; fight; combat: ～官僚主义 combat bureaucracy / ～贪污浪费 fight against corruption and waste / ～资产阶级自由化 oppose bourgeois liberalization ◇ ～党 opposition / ～派 opposition faction / ～票 dissenting vote; negative vote
【反而】[连] on the contrary; instead
【反封建】anti-feudal; against feudalism
【反复】①(重复) repeatedly; again and again; over and over again: ～较量 repeated trials of strength ②(翻悔) reversal; relapse: 思想上有～ have ideological relapses ◇ ～记号[乐] repeat

【反复无常】changeable; tickle; capricious
【反干扰】anti-disturbance
【反感】be disgusted with; be averse to; dislike; take unkindly to: 我对她的话很~。I'm disgusted with what she said
【反戈一击】turn one's weapon around and strike; turn againstthose one has wrongly sided with
【反革命】counterrevolutionary ◇ ~分子 a counterrevolutionary / ~两面派 counterrevolutionary double-dealer / ~杀人罪 counterrevolutionary murder / ~宣传煽动罪 crime of counterrevolutionary propaganda and inflammatory delusion / ~造谣罪 crime of counterrevolutionary rumormongering / ~政变 counterrevolutionary coup d'etat
【反攻】counteroffensive; counterattack
【反攻倒算】counterattack to settle old scores
【反躬自问】examine oneself; examine one's conscience
【反顾】[书] look back: 义无~ be duty-bound not to turn back
【反光】①(反射的光线) reflect light ②(光的反射) reflection of light ◇ ~灯 reflector lamp / ~镜 reflector
【反函数】[数] inverse function
【反核子】antinucleon
【反话】irony
【反悔】go back on one's word; go back on one's promise
【反击】strike back; beat back; counterattack: 自卫~ counterattack in self-defense
【反剪】①(两手背后) with one's hands behind one's back ②(两手反绑) with one's hands tied behind one's back
【反间】sow distrust or dissension among one's enemies; sow discord within the enemy camp ◇ ~计 stratagem of sowing distrust or discord among one's enemies
【反间谍】counterespionage
【反诘】ask in retort; counter with a question
【反介子】antimeson
【反抗】revolt; resist: ~精神 spirit of revolt; rebellious spirit
【反客为主】reverse the positions of the host and the guest
【反空降】[军] anti-airborne defense
【反控】countercharge; recriminate
【反馈】[电] feedback: 负~ negative feedback / 正~ positive feedback ◇ ~电路 feedback circuit / ~抑制 feedback inhibition
【反粒子】[物] antiparticle
【反面】①(与正面相反的一面) reverse side; wrong side; back: 唱片的~ the reverse side of a disc ②(坏的; 消极的一面) opposite; negative side: ~的教训 a lesson learnt from negative experience / ~教材 negative example which may serve as a lesson
【反目】fall out (esp. between husband and wife)
【反派】villain (in drama, etc.); negative character: 演~人物 play a negative role
【反叛】revolt; rebel
【反批评】counter-criticism
【反扑】pounce on sb. again after being beaten off
【反其道而行之】do exactly the opposite; act in a diametrically opposite way
【反气旋】[气] anticyclone
【反潜机】antisubmarine plane
【反潜舰艇】antisubmarine vessels
【反弱为强】convert weakness into strength
【反射】①[生](反应) reflex: 条件(非条件)~ conditioned (unconditioned) reflex ②[物](折回) reflection ◇ ~比[物] reflectance / ~波 back wave / ~测云器 reflecting nephoscope / ~弧[生理] reflex arc / ~计 reflectometer / ~角[物] angle of reflection / ~镜 reflector / ~炉[冶] reverberatory furnace / ~望远镜 reflecting telescope
【反手】[体] backhand: ~抽球 backhand drive
【反诉】[法] countercharge; counterclaim; counter charge
【反坦克炮】antitank gun
【反坦克武器】antitank weapon
【反特】anti-espionage: ~片 anti-espionage film
【反题】[哲] antithesis
【反铁电】[物] antiferroelectric ◇ ~现象 antiferroelectricity
【反围盘】[机] reverse repeater (used in steel rolling)
【反胃】gastric disorder causing nausea
【反问】①(以问代答) ask (a question) in reply ②[语](含反意的发问) rhetorical question
【反污染】antipollution
【反物质】antimatter
【反响】repercussion; echo; reverberation: 在世界上引起广泛的~ evoke worldwide repercussions
【反向】opposite direction; reverse ◇ ~铲 backhoe / ~电流 reverse current / ~卫星 retrograde satellite

【反小说】anti-novel
【反斜面】[军] reverse slope; rear slope
【反信风】[气] antitrades (winds); countertrades (winds)
【反省】introspection; self-questioning; self-examination
【反宣传】①(相反的宣传) counterpropaganda ②(造谣中伤) slander campaign
【反咬一口】trump up a countercharge against one's accuser; make a false countercharge
【反义词】[语] antonym
【反应】①(化学变化) reaction: 放热~ exothermic reaction / 核~ nuclear reaction / 化学~ chemical reaction / 碱性~ alkaline reaction / 取代~ displacement reaction / 热核~ thermo-nuclear reaction / 酸性~ acid reaction / 吸热~ endothermic reaction / 阳性~ positive reaction / 阴性~ negative reaction / 中性~ neutral reaction ②(反响) response; repercussion; reaction: ~不一。Reactions vary. ◇ ~本领 reaction capacity / ~塔 reaction tower / ~物 reactant
【反应堆】[物] reactor: 非均匀~ heterogeneous reactor / 高温~ high-temperature reactor / 核~ nuclear reactor; nuclear reaction pile / 均匀~ homogeneous reactor / 快中子~ fast neutron reactor / 浓缩铀~ enriched uranium reactor / 轻水慢化~ light-water-moderated reactor / 热中子~ thermal reactor / 石墨减速~ graphite-moderated reactor / 原子~ reactor; atomic pile / 增殖~ breeder reactor ◇ ~安全壳 containment vessel; reactor containment
【反映】①(反照) reflect; mirror: 时代的~ mirror of the times ②(报告上级) report; make known: 向上级~ report to the higher level
【反映论】[哲] theory of reflection: 唯物论的~ the materialist theory of reflection
【反语】[语] irony
【反原子】antiatom
【反照】reflection of light: 夕阳~ evening glow; sunset glow ◇ ~镜 rearview mirror
【反正】①(敌方人员投诚) come over from the enemy's side ②[副] anyway; anyhow; in any case: ~得去一个人，就让我去吧。Since someone has to go anyway, let me go.
【反证】disproof; counterevidence
【反证法】reduction to absurdity
【反之】conversely; on the contrary; otherwise
【反殖】anti-colonialist; against colonialism
【反质子】[物] antiproton
【反中子】[物] antineutron
【反种族歧视】antiracism
【反转】reverse ◇ ~片 [电影] reversal film
【反坐】sentence the accuser to the punishment facing the person he falsely accused
【反作用】counteraction; reaction ◇ ~力[物] reacting force

返 return: 流连忘~ linger on without any thought of leaving
【返潮】get damp
【返防】[军] return to stations
【返工】do poorly done work over again: 这件上衣需要~。This jacket needs to be remade.
【返航】return to base; return to port
【返回】return; come or go back
【返老还童】recover one's youthful vigor; feel rejuvenated
【返青】return green (after winter or after transplantation)
【返校】return to school after the vacation
【返盐】[农] accumulation of salt in the surface soil
【返祖现象】[生] atavism

fàn

泛 ①(漂浮) float: ~舟东湖 go boating on the East Lake ②(透出) be suffused with: 脸上~出红晕 with one's cheeks suffused with blushes ③(泛滥) flood; inundate ④(广泛) extensive; general; nonspecific: 广~ wide; extensive / 空~ containing nothing but generalities
【泛称】general term
【泛读】extensive reading
【泛泛而谈】talk in generalities
【泛光灯】floodlight
【泛函分析】[数] functional analysis
【泛滥】①(溢出) be in flood; overflow: 河水~。The river was in flood. ②(扩展) spread unchecked: ~成灾 run rampant; run wild
【泛美主义】Pan-Americanism
【泛生论】pangenesis
【泛神论】[哲] pantheism
【泛水】[建] flashing
【泛酸】[化] pantothenic acid
【泛音】[乐] overtone; harmonic ◇ ~列 harmonic series
【泛指】make a general reference

范 ①(模子) [书] pattern ②(模范) model; example; pattern: 典~ example / 示~ dem-

onstrate ③(范围) limits: 就～ submit
【范本】model for calligraphy or painting
【范畴】category
【范例】example; model
【范围】scope; limits; range: 活动～ scope of activities / 势力～ sphere of influence / 在法律许可～内 within the limits permitted by law / 在协定规定的～内 within the framework of the agreement
【范文】model essay
【范性】[物] plasticity ◇ ～形变 plastic deformation

梵 Buddhist: ～宫 Buddhist temple
【梵蒂冈】Vatican
【梵文】Sanskrit

贩 ①(商人买货) buy to resell: ～毒 traffic in narcotics ②(贩子) dealer; monger; pedlar: 小～ vendor; pedlar
【贩卖】traffic; peddle; sell: ～妇女 traffic in women / ～人口 traffic in person / ～牲口 deal in draught animals
【贩运】transport goods for sale; traffic
【贩子】dealer; monger: 鱼～ fish-monger

饭 ①(煮熟的谷类食品) cooked rice or other cereals ②(每日定时吃的食物) meal: ～前洗手。Wash your hands before meals.
【饭菜】①(饭和菜) meal; repast: ～可口,服务周到 tasty food and good service ②(下饭的菜) dishes to go with rice, steamed buns, etc.
【饭店】①(大旅馆) hotel ②[方](饭馆) restaurant
【饭馆】restaurant; luncheonette
【饭锅】pot for cooking rice; rice cooker
【饭盒】lunch-box; mess tin; dinner pail
【饭来张口,衣来伸手】have only to open one's mouth to be fed and hold out one's hands to be dressed; lead an easy life, with everything provided; be waited on hand and foot
【饭量】appetite: ～大 have a good appetite; have a good stomach / ～小 have a poor appetite; have a poor stomach
【饭票】meal ticket; mess card
【饭铺】(small) restaurant; eating house
【饭食】food
【饭厅】dining hall; dining room; mess hall
【饭桶】①(盛饭的桶) rice bucket ②(饭量大) big eater ③(无用之人) fathead; good-for-nothing
【饭碗】①rice bowl ②(指职业) job: 丢了～ lose one's job
【饭庄】(big) restaurant
【饭桌】dining table

犯 ①(违反) violate; offend (against law, etc.): ～纪律 violate discipline ②(侵犯) attack; assail; work against: 人不～我,我不～人;人若～我,我必～人。We will not attack unless we are attacked; if we are attacked, we will certainly counterattack. ③(罪犯) criminal: 重～ recidivist / 初～ first offender; first offense / 从～ accomplice / 惯～ old offender; inveterate criminal / 教唆～ abettor; abetter / 青少年～ juvenile offender; juvenile delinquent / 囚～ prisoner / 杀人～ murderer / 未决～ prisoner awaiting trial / 嫌疑～ suspect / 现行～ an offender caught red-handed / 胁从～ accomplice under duress / 诈骗～ swindler / 主～ principal offender; principal culprit / 纵火～ arsonist; incendiary ④(发作) have a recurrence of (an old illness): 她的气喘病又～了。She's got another attack of asthma. ⑤(发生) commit (a mistake, crime, etc.): ～错误 make a mistake
【犯案】be found out and brought to justice
【犯病】have an attack of one's old illness
【犯愁】worry; be anxious
【犯法】violate the law: ～的人 law-breaker; offender ◇ ～行为 offense against the law
【犯规】①(违反规定) break the rules ②[体] foul: 侵人～personal foul / 双方～ double foul
【犯忌】violate a taboo
【犯禁】violate a ban; break prohibition
【犯人】prisoner; convict
【犯上】go against one's superiors
【犯上作乱】defy one's superiors and start a rebellion
【犯嫌疑】arouse suspicion; come under suspicion
【犯疑】suspect; be suspicious
【犯罪】commit a crime: ～的中止 desistance from offense / ～和青少年～ crime and delinquency / 共同～ joint offense / 故意～ intentional offense / 过失～ offense through negligence; unpremeditated crime ◇ ～分子 offender; criminal / ～工具 guilty tools / ～集团 criminal gang / ～率 crime rate / ～未遂 attempted crime; attempted offense / ～心理学 criminal psychology / ～行为 criminal offense / ～学 criminology

fāng

方 ①(方形) square: ～桌 square table／四英尺见～ four feet square ②[数](乘方) involution; power: 二的三次～是八。 The third power of 2 is 8. ③[量](立方) short for square meter or cubicmeter: 十～土 ten cubic meter of earth／一～木材 a cubic meter of lumber ④(正直) upright; honest: 品行～正 have an upright character⑤(方向) direction: 南～ the south／前～ the front／四面八～in all directions ⑥(方面) side; party: 对～ the opposite side; other side／双～ both sides／我～ our side／有关～面 the parties concerned ⑦(地方) place; region; locality: 远～ a faraway place ⑧(方法)method; way: 教子有～ good at educating one's children／千～百计 in a hundred and one ways; by every conceivable means ⑨(药方) prescription: 处～ make out a prescription ⑩(方才) just; at the time when: 年～四十 be just forty years old

【方案】scheme; plan; program: 制定～ draw up a plan for

【方便】①(便利) convenient: ～群众 make things convenient for the people／交通～ have a good transport／为了～ for convenience ②(上厕所)[口] go to the lavatory; 你要不要～一下? Do you want to use the lavatory? ③(有富余的钱)[婉] have money to spare or lend: 手头不～ have little money to spare

【方便之门】the door is thrown wide open for: 大开～ do everything to suit sb.'s convenience

【方步】measured steps

【方才】just now: 他～回来。He has just come back. (或: He came back just now.)

【方程】[数] equation: 二次～ quadratic equation／解～ solve an equation／三次～ cubic equation／一次～ simple equation

【方程式】①[数] equation ②[化] equation: 化学～ chemical equation

【方寸】[书] heart: ～已乱 with one's heart troubled and confused; greatly agitated

【方法】method; way; means: 科学～ scientific method／学习～ method of study

【方法论】[哲] methodology

【方钢】[冶] square steel

【方格】check: ～呢 plaid／～纸 squared paper

【方根】[数] root

【方剂】[中医] prescription; recipe

【方济各会】[天主教] the Franciscan Order

【方尖塔】[建] obelisk

【方解石】[矿] calcite

【方块图】block diagram; block chart; functional diagram

【方块字】Chinese characters

【方框图】block plan; block diagram

【方括号】square brackets ([ ])

【方略】general plan

【方面】respect; aspect; side; field: 矛盾的次要～ secondary aspect of a contradiction／矛盾的主要～ principal aspect of a contradiction／在这～ in this respect

【方铅矿】galena

【方枘圆凿】like a square tenon for a round mortice; at variance with each other

【方士】①(巫师) necromancer ②(炼丹人) alchemist

【方式】way; fashion; pattern: 领导～ style of leadership／生产～ mode of production／生活～ way of life; life-style／一反过去因袭的～ depart from the formula followed in the past

【方糖】cube sugar; lump sugar

【方铁矿】[矿] wustite

【方位】position; bearing; direction; points of the compass;placement ◇ ～词[语] noun of locality／～角 angle of position; azimuth; position angle; bearing angle／～罗盘[测] azimuth compass／～天文学 positional astronomy／～图 orientation diagram; bearing diagram／～物[军] topographic marker; landmarker

【方向】direction; orientation

【方向舵】[航空] rudder

【方向盘】[汽车] steering wheel

【方向误差】[军] deflection error; error in line

【方向修正】adjustment in direction

【方兴未艾】be just unfolding; be in the ascendant

【方形】square

【方言】[语] dialect: 地理～ geographical dialect／社会～ social dialect ◇ ～学 dialectology

【方言土话】dialects of various places

【方以类聚,物以群分】like attracts like and birds of a feather flock together

【方圆】①(周围之长度) circumference ②(周围)neighborhood

【方丈】①(住持) Buddhist abbot ②(住持住房) abbot'sroom

【方针】policy; guiding principle: ～政策 general and specificpolicies／党的教育～ the Party's educational policy／文艺～ guiding principles for literature and art／战略～ strategic princi-

ple
【方正】①(书法方正) upright and foursquare: 字要写得～。In writing, make the characters square and upright. ②(正直) upright: 为人～ be an upright man
【方志】local records; local chronicles
【方子】①(药方) prescription: 开～ write out a prescription ②(配方) directions for mixing chemicals; formula
【方钻杆】[石油] kelly (bar)

芳 ①(香) sweet-smelling; fragrant: ～草 fragrant grass ②(美好的名声) good (name or reputation); virtuous: 流～百世 leave a good name to posterity
【芳草如茵】a carpet of green grass
【芳华虚度】youth passes away in vain
【芳烃】[化] aromatic hydrocarbon
【芳香】fragrant; aromatic ◇ ～剂 aromatic
【芳香馥郁】rich in fragrance
【芳香扑鼻】feel a sharp aroma; the fragrance (of...) assailed one's nostrils
【芳香四溢】sweet perfumes are diffused all around
【芳心无主】a lady not knowing what to do
【芳心已许】love a man silently
【芳族】[化] aromatics ◇ ～化合物 aromatic compound; aromatic / ～酸 aromatic acid

坊 lane (usu. as part of a street name)
【坊本】block-printed edition prepared by a bookshop
【坊间】①(街市) on the street stalls ②(书坊) in the bookshops

妨 (用于否定与疑问) harm: 不～早点动身。There's no harm in leaving a little earlier.

钫 [化] francium (Fr)

fáng

房 ①(房子) house: 平～ single-story house ②(房间) room: 病～ sickroom; ward / 客～ guest room / 书～ study ③(类似房屋之物) a house-like structure: 蜂～ beehive
【房舱】passenger's cabin in a ship
【房产】house property
【房地产】real estate
【房地契】real estate; title deeds
【房隔】atrial septum
【房顶】roof
【房东】the owner of the house one lives in; landlord or landlady
【房荒】house shortage
【房基】foundations (of a building)
【房间】room: ～号码 room number / ～钥匙 room key ◇ 单人～ single room / 双人～ double room / 一套～ suite
【房间隔缺损】[医] atrial septal defect
【房客】tenant (of a room or house); lodger
【房梁】(house) beam
【房契】title deed (for a house)
【房事】sexual intercourse (between a married couple)
【房屋】houses; buildings
【房屋纠纷】housing dispute
【房檐】eaves
【房主】house-owner
【房子】①(供人住之建筑物) house; building ②(房间) room
【房租】rent (for a house, flat, etc.)

防 ①(防备) guard against; provide against: ～病 prevent disease / 以～万一 be prepared for all contingencies; be ready for any eventuality ②(防守) defend: ～身 defend oneself / 边～ frontier defense / 国～ national defense ③(堤) dike; embankment
【防爆开关】flame-proof switch
【防备】guard against; take precautions against: ～出事 take precautions against all accidents / ～天灾 make preparations against natural disasters
【防变】be prepared for an unfavorable turn of events; get ready for any emergency
【防病治病】prevent and cure
【防波堤】breakwater; mole
【防不胜防】impossible to defend effectively; very hard to guard against: 他球路多变，对手～。His opponent couldn't stand up to his varied and fast-changing tactics.
【防潮】①(防潮湿) dampproof; moistureproof ②(防潮水) protection against the tide: ～堰堤 tidal barrage ◇ ～层[建] dampproof course; damp course / ～火药 moistureproof powder; nonhygroscopic powder / ～纸 moistureproof paper / ～砖 moistureproof brick
【防尘】dustproof ◇ ～圈[机] dust ring / ～罩 dust cover
【防磁】[物] antimagnetic
【防弹】bulletproof; shellproof ◇ ～背心 bulletproof vest / ～玻璃 bulletproof glass / ～汽车 bulletproof car

【防盗】guard against theft; take precautions against burglars
【防地】[军] defense sector; station (of a unit)
【防冻】prevent frostbite
【防冻液】anti-icing fluid
【防毒】gas defense ◇ ～面具 gas mask / ～器材 gas protection equipment / ～衣 protective clothing
【防范】be on guard; keep a lookout
【防风林】windbreak (forest): ～带 windbreak belt
【防风制沙】check winds and control sand
【防辐射】radiation protection
【防腐】antiseptic; anticorrosive ◇ ～材料 antirot material / ～剂 preservative; antiseptic / ～药 antiseptic
【防洪】prevent or control flood: ～措施 flood control measures ◇ ～工程 flood control works / ～闸门 floodgate
【防护】protect; shelter: 人体～ [军] physical protection ◇ ～堤 (protecting) embankment / ～涂层 protective coating / ～罩 protection casing; protection cover / ～装置 protective device
【防护林】shelter-forest ◇ ～带 shelterbelt / ～体系 shelterbelt network
【防滑链】[汽车] tire chain; skid chain
【防化学兵】antichemical warfare corps
【防患未然】take preventive measures; provide against possible trouble
【防火】fire prevention; fireproof ◇ ～层 fireprotection layer / ～隔离线 [林] fire lane / ～了望台 [林] fire watchtower / ～幕 fireploof curtain / ～墙 fire wall / ～区 [林] fire warden / ～装置 fireproof installations
【防空】air defense; antiaircraft ◇ ～部队 air defense forces / ～导弹 air defense missile; interceptor missile / ～洞 air raid shelter / ～火箭 anti-aircraft rocket / ～警报 air defense warning / ～炮 air defense artillery (ADA) / ～演习 air defense exercise; air-raid drill
【防涝】prevent waterlogging
【防凌】reduce the menace of ice run
【防区】defense area; garrison area
【防染剂】[纺] resist
【防沙林】sand-break
【防守】defend; guard: ～边境 keep guard on the frontier; guard the frontier
【防暑】heatstroke prevention; sunstroke prevention ◇ ～药 heatstroke preventive
【防水】waterproof: ～表 waterproof watch / ～布 waterproof cloth / ～层 [建] waterproof layer
【防缩】[纺] shrinkproof: ～整理 shrinkproof finish; shrinkage control finish
【防坦克】antitank defense ◇ ～地雷 antitank mine / ～壕 tank ditch / ～炮 antitank gun / ～阵地 antitank position
【防特】guard against enemy agents
【防微杜渐】nip an evil in the bud; check erroneous ideas at the outset
【防卫】defend: ～行为 act of defense
【防务】matters pertaining to defense; defense
【防线】line of defense
【防锈】antirust ◇ ～剂 rust inhibitor; antirusting agent / ～漆 antirust paint / ～脂 antirust grease
【防汛】flood prevention or control: 组成～大军 organize an army of flood-fighters ◇ ～指挥部 flood control headquarters
【防疫】epidemic prevention ◇ ～人员 epidemic prevention worker / ～站 epidemic prevention station / ～针 (prophylactic) inoculation
【防雨布】waterproof cloth; tarpaulin
【防御】defense: 积极～ active defense / 消极～ passive defense / 由～转入进攻 go over from the defensive to the offensive ◇ ～部队 defending force / ～地带 zone of defense; defense of zone / ～工事 fortifications; defenses / ～设施 defense installations / ～手段 means of defense / ～行为 act of defense / ～战 defensive warfare
【防灾】take precautions against natural calamities; ～抗灾 prevent and fight natural adversities
【防震】①(采取措施，预防地震) take precautions against earthquakes: ～措施 precautions against earthquakes ②(防震动) shockproof; quakeproof ◇ ～设计 aseismatic design
【防止】prevent; guard against; forestall; avoid: ～感染 protect from infection; prevent infection / ～浪费人力 avoid waste of manpower / ～煤气中毒 guard against gas poisoning
【防治】prevention and cure; prophylaxis and treatment: ～病虫害 the prevention and control of plant diseases and elimination of pests / ～血吸虫病 the prevention and cure of schistosomiasis / ～职业病 prophylaxis and treatment of occupational diseases

坊 workshop; mill: 染～ dyer's workshop / 油～ oil mill

妨 hinder; obstruct
【妨碍】hinder; hamper; impede; obstruct: ～交通 block traffic / ～生产的发展 hamper the growth of production / ～视线 obstruct the view / ～团结 hinder unity
【妨碍公务】[法] interference with public function
【妨碍司法执行】[法] obstructing justice
【妨害】impair; jeopardize; be harmful to: ～集体利益 be injurious to collective interests / ～健康 be harmful to one's health / ～商务 be injurious to trade
【妨害公共秩序罪】offense acts against public order
【妨害婚姻家庭罪】crime of interference with marriage and the family
【妨害社会秩序】blemishing the peace
【妨害私人秘密】interference with right of privacy
【妨害行为】nuisance
【妨害治安】disturbance of the peace; disorderly conduct

fǎng

访 ①(访问) visit; call on: 互～ exchange visits ②(调查;寻求) seek by inquiry or search; try to get: 暗～ investigate secretly
【访查】go about making inquiries; investigate
【访贫问苦】inquire of the poor about their past sufferings
【访求】seek; search for
【访问】visit; call on: 非正式～ an unofficial visit / 国事～ state visit / 私人～ private visit / 友好～ friendly visit; goodwill visit / 正式～ an official visit ◇ ～团 visiting mission
【访友求贤】call on friends and seek for worthies

仿 ①(仿效) imitate; copy ②(类似) resemble; be like: 相～ be very much alike; be similar
【仿单】instructions for the use of an article sold (esp. a medicine)
【仿佛】①(好象) seem; as if ②(类似) be more or less the same; be alike
【仿古】modelled after an antique: ～青铜器 an imitation of an ancient bronze
【仿生学】bionics
【仿效】imitate; follow the example of: ～机手 master-slave manipulator
【仿形】[机] profile modelling ◇ ～车床 copying lathe; repetition lathe / ～机 profiling mechanism
【仿形板】gauge finder
【仿形规】contour gauge
【仿造】copy; be modelled on; counterfeit: ～的古董 imitation curio; fake curio
【仿照】imitate; follow: ～办理 act in imitation of
【仿真】emulate; emulation
【仿制】copy; be modelled on ◇ ～品 imitation; replica; copy

舫 boat: 画～ a gaily-painted pleasure-boat

纺 ①(把纤维拧成纱) spin ②(丝织品) a thin silk cloth
【纺车】spinning wheel
【纺绸】a soft plain-weave silk fabric
【纺锤】spindle
【纺锤丝】spindle fiber
【纺纱】spinning ◇ ～工人 spinner / ～机 spinning machine
【纺丝】[化纤] spinning ◇ ～泵 spinning pump / ～罐 spinning box / ～机 spinning machine / ～浴 spinning bath
【纺织】spinning and weaving ◇ ～厂 textile mill / ～废水 textile waste / ～工业 textile industry / ～染色 [化纤] dope dyeing / ～纤维 dope-dyed fiber / ～印花 textile printing
【纺织娘】[动] katydid
【纺织品】textile ○帆布 canvas / 府绸 poplin / 灯芯绒 corduroy / 花布 cotton prints / 卡其布 khaki drills / 双面卡 reversible khaki / 阔幅平布 sheetings / 泡泡纱 seersucker; plisse crepe; blister crepe / 漂白棉布 bleached cotton cloth / 乔其纱 georgette / 绒布 flannelet / 提花床单布 broche quilts / 夏布 grass cloth / 香云纱 gambirded Guangshou gauze / 斜纹布 drills / 亚麻布 linen; linen cloth / 羽纱 camblet / 哔叽 serge / 毛哔叽 wool serge / 长毛绒 plush / 粗花呢 tweed / 大衣呢 overcoating / 凡立丁 valetin / 法兰绒 flannel / 精纺法兰绒 worsted flannel / 格子呢 tartan / 海军呢 admiralty cloth / 华达呢 gabardine / 开司米 cashmere / 麦尔登呢 melton / 派力斯 palace / 人字呢 herringbone / 茧绸 pongee / 锦缎 brocade / 罗缎 faille; bengaline; tussore / 塔夫绸 taffeta / 奥纶 orlon / 涤纶 terylene / 锦纶 nylon; jinlun / 腈纶 acrylic fibers / 卡普纶 capron; kapron / 维纶

winylon / 毛涤纶 modelon / 棉涤纶 trueran / 尼龙薄绸 nylon chiffon / 粘胶纤维 viscose acetal fiber / 人造毛 artificial wool / 人造棉 artificial cotton

fàng

放 ①(使自由) let go; set free; release: 把俘虏～了 release the captives / 把游泳池里的水～掉 let the water out of the swimming pool ②(点燃) let off; give out: ～枪 fire a gun / ～焰火 set off fireworks ③(放牧) put out to pasture: ～牛 put cattle out to pasture; pasture cattle ④(放纵) let oneself go; give way to: ～声大哭 cry loudly and bitterly ⑤(借钱给人) lend (money) for interest: ～高利贷 practice usury ⑥(扩展) let out; expand; make larger, longer, etc.: 把裤腰～宽 let out the trousers round the waist ⑦(开) blossom; open: 百花齐～ a hundred flowers in bloom ⑧(加进) put in; add: ～点盐 put in some salt ⑨(使处一定位置) put; place: 把书～在桌子上。 Put the book on the table. ⑩(搁置) leave alone; lay aside ⑪(放逐) send away: 流～ send into exile ⑫(控制行动; 达到某种分寸) readjust (attitude, behavior, etc.) to a certain extent: 脚步～轻些。 Tread softly. ⑬(放映) show: ～电影 show a film / ～幻灯 show slides

【放暗箭】stab (sb.) in the back; snipe (at sb.)

【放长线,钓大鱼】throw a long line to catch a big fish; adopt a long-term plan to secure sth. big

【放出】give out; let out; emit: ～清香 exude a delicate fragrance / ～蒸汽 let off steam

【放大】enlarge; magnify; amplify: 把照片～ have a photograph enlarged ◇ ～尺 pantograph / ～机[摄] enlarger / ～镜 magnifying glass; magnifier / ～率 magnifying power / ～纸 enlarging paper

【放大器】[电子] amplifier: 共益～ bootstrap amplifier

【放胆】act boldly and with confidence

【放诞】wild in speech and behavior

【放荡】①(淫乐的) dissolute; dissipated ②(行为不捡点) unconventional: ～不羁 unconventional and unrestrained

【放电】[物] (electric) discharge: 尖端～ point discharge

【放刁】make difficulties for sb.; act in a rascally manner

【放毒】①(投放毒物) put poison in food, water, etc.; poison ②(散布反动言论) make vicious remarks; spread poisonous ideas

【放风】①(使空气流通) let in fresh air ②(让坐牢的人到院里散步) let prisoners out for exercise or to relieve themselves ③(透露或散布消息) leak certain information; spread news or rumors

【放风筝】kiteflying

【放工】(of workers) knock off

【放过】let off; let slip: ～了一个机会 let slip an opportunity

【放虎归山】set free a tiger back to the mountains; lay by trouble for the future

【放火】①(引火烧毁房屋等) set fire to; set on fire ②(煽动) create disturbances ◇ ～犯 arsonist

【放假】have a holiday or vacation; have a day off: ～半天 have a half-day holiday / 放寒假 have winter holidays; have a winter vacation / 放暑假 have summer holidays; have a summer vacation

【放空枪】fire blank shots

【放空炮】talk big; spout hot air

【放空气】drop a hint; spread word; create an impression

【放宽】relax restrictions; relax: ～期限 extend a time limit / ～条件 soften the terms

【放款】make loans; loan; credit: 长期～ long-term loan / 短期～ short-term loan / 中期～ medium-term loan

【放浪】①(不受约束的) unrestrained ②(不道德的) dissolute

【放牧】put out to pasture; herd

【放排】[林] rafting

【放盘】sale at reduced prices; sale

【放炮】①(打炮) fire a gun ②(车胎爆裂) blowout (of a tire,etc.): 轮胎～了。 The tire's had a blowout. ③(炸开) blasting④(攻击) shoot off one's mouth

【放屁】① break wind; fart ②(骂人话) What crap! (或: Shit!)

【放弃】abandon; give up: ～表决权 abstain from voting / ～财产 relinquish one's possessions / ～原则 forsake one's principles / ～职守 desert one's post

【放弃国籍声明】declaration of alienage

【放弃继承权】renunciation of succession

【放弃诉讼】renounce an action

【放气阀】air bleeder

【放气孔】air-bleed hole

【放晴】clear up (after rain)

【放热】[化] exothermic ◇ ～反应 exothermic

reaction
【放任】①(听其自然) not interfere; let alone: ～自流 let things drift ②(不干涉) noninterference: ～政策 Laissez-faire policy / ～主义 the policy of noninterference
【放散】(of smoke, scent, etc.) diffuse; spread
【放哨】stand sentry; be on sentry go
【放射】radiate ◇ ～病 radiation sickness / ～疗法 radiotherapy / ～区域 radioactive area / ～现象[物] radioactivity / ～线 radioactive rays
【放射性】[物] radioactivity ◇ ～示踪物 radioactive tracer / ～碎片 radioactive debris / ～微尘 radioactive dust; fallout / ～污染 radioactive pollution / ～元素 radioactive element; radioelement / ～原子尘埃 radioactive fallout / ～沾染 radioactive contamination
【放射性同位素】radio isotope: 人造～ induced radio isotope
【放声歌唱】lift up one's voice and sing
【放声痛哭】utter a stifled cry of agony
【放生】①(把捉住的动物放掉) free captive animals ②(特指信佛的人把别人捉到的鱼鸟等买来放掉) buy captive fish or birds and set them free
【放手】①(松开手) let go; let go one's hold ②(解除顾虑或限制) have a free hand; go all out: 让他们～工作 give them a free hand in their work
【放水】①(开水龙头) turn on the water ②(开水闸) (of a reservoir) draw off
【放肆】unbridled; wanton: 胆敢如此～! How dare you take such liberties!
【放松】relax; slacken; loosen: ～警惕 relax one's vigilance ◇ ～活动 [体] relaxation exercise; limbering-up exercise
【放送】send out
【放下】lay down; put down: ～包袱 lay down the burden / ～架子 discard one's haughty airs; come off the high horse / ～武器 lay down arms
【放下屠刀,立地成佛】a butcher becomes a Buddha the moment hedrops his cleaver; a wrongdoer achieves salvation as soon as he gives up evil
【放线菌】[微] actinomyces
【放心】set one's mind at rest; be at ease; rest assured; feel relieved: 对他不大～ not quite trust him
【放行】let sb. pass
【放学】classes are over
【放血】bloodletting
【放眼】take a broad view; scan widely: ～世界 keep the whole world in view
【放养】put (fish, etc.) in a suitable place to breed: ～鱼秧 breed fish fry
【放映】show; project: ～电影 show a film; have a film show ◇ ～队 film projection team / ～机 (film) projector / ～室 projection room / ～员 projectionist
【放之四海而皆准】universally applicable; valid everywhere
【放置】lay up; lay aside: ～不用 lie idle
【放逐】send into exile; exile; banish
【放纵】①(纵容) let sb. have his own way; connive at; indulge: 你太～孩子了。You are too indulgent with your children. ②(不守规矩) self-indulgent; undisciplined
【放走】release; set free; let go

fēi

非 ①(错;不对) wrong; evildoing: 分清是～ distinguish between right and wrong / 为～作歹 do evil ②(不合于) not conform to; run counter to: ～分 overstepping one's bounds; assuming; presumptuous ③(不以为然) censure; blame: 未可厚～ not altogether inexcusable; excusable ④(不是) not; no: 答～所问 give an irrelevant answer ⑤(跟"不"呼应,表示"必须")难道～你去处理这件事不成? Are you really the only one who can handle the matter? ⑥[口](必须) have got to ⑦(指非洲) Africa
【非暴力反抗】civil disobedience
【非暴力主义】nonviolence ◇ ～者 exponent of nonviolence
【非比寻常】out of the ordinary
【非病原菌】nonpathogenic bacteria
【非常】①(特殊) extraordinary; unusual; special: ～措施 emergency measures / ～会议 extraordinary session / ～局面 unusual situation / ～时期 unusual times / ～事件 uncommon incident / ～支出 a special expenditure ②(十分) very; extremely; highly: ～抱歉 awfully sorry / ～精彩 simply marvellous / ～清楚 perfectly clear / ～重视 attach great importance to / ～重要 extremely important
【非常任理事国】nonpermanent member of the UN Security Council
【非常征用】[法] angary
【非但】not only
【非得】have got to; must

【非独】[书] not merely: ～无益,而且有害 not merely useless, but harmful
【非对抗性】 nonantagonistic ◇ ～矛盾 nonantagonistic contradiction
【非法】illegal; unlawful; illicit: ～合同 illegal contract / ～集会 unlawful assembly / ～监禁 detain illegally / ～交易 illegal transaction / ～侵入[法] trespass / ～收入 illicit income / ～手段 illegal means / ～同居 illicit cohabitation / ～行为 unlawful act; illegal act
【非凡】 outstanding; extraordinary; uncommon: ～的成就 outstanding achievements / ～的贡献 outstanding contribution / ～的人 man out of the common run
【非分】overstepping one's bounds; assuming: ～之想 inordinate ambitions
【非公莫入】no admittance except on business
【非官方】unofficial: ～人士 unofficial figures / ～消息 news from unofficial sources
【非婚生】 bastardy; illegitimacy ◇ ～女 bastarda / ～子 bastard
【非交战国】nonbelligerent
【非金属】nonmetal ◇ ～材料 nonmetallic materials / ～元素 nonmetallic elements
【非晶质】[矿] noncrystalline; amorphous ◇ ～体 amorphous body
【非军事化】demilitarize
【非军事区】demilitarized zone
【非军事人员】civilian personnel
【非君莫属】only you can fill the post
【非驴非马】neither ass nor horse; neither fish, flesh, nor fowl
【非轮回亲本】[农] nonrecurrent parent
【非卖品】(articles) not for sale
【非那更】[药] phenergan
【非难】(多用于否定式) blame; censure; reproach: 无可～ above criticism; not blameworthy
【非亲非故】be neither kith nor kin
【非请莫入】No Admittance Except on Invitation
【非人】①[书](不合适的人) not the right person: 所用～ choose the wrong person for a job ②(残忍的) inhuman: ～待遇 inhuman treatment
【非生产部门】nonproductive departments
【非生产劳动】nonproductive labor
【非生产性】unproductive; nonproductive: ～工程 nonproductive construction / ～开支 nonproductive expenditure / ～人员 nonproductive personnel
【非同小可】no small matter
【非刑】 brutal torture: ～拷打 torture sb. brutally; subject sb. to brutal torture
【非虚构小说】nonfiction novel
【非一日之功】cannot be done in a short time; take time and efforts
【非议】(多用于否定式) reproach; censure: 无可～ beyond reproach; irreproachable
【非约束性条款】permissive provision
【非预谋杀人】man slaughter
【非战斗人员】noncombatant
【非正规军】irregular troops; irregulars
【非正式】unofficial; informal: ～访问 unofficial visit / ～会议 informal meeting / ～声明 unofficial statement / ～译文 unofficial translation
【非正统】unorthodox
【非正义战争】unjust war
【非洲】Africa ◇ ～人 African
【非做不可】have no choice but to

扉 door leaf

【扉页】[印] title page

霏

【霏霏】[书] falling thick and fast: 雨雪～。 It was sleeting hard.

菲 ①(花草美) luxuriant ②(香味浓) rich with fragrance ③[化] phenanthrene

【菲菲】[书]①(花草美) luxuriant and beautiful ②(香味浓) richly fragrant
【菲律宾】the Philippines ◇ ～群岛 the Philippine Islands / ～人 Filipino

蜚

【蜚声】[书] make a name; become famous: ～文坛 win renown in literary circles
【蜚语】rumors; gossip: 散布流言～ spread rumors

鲱 Pacific herring

绯 red

【绯红】bright red; crimson: ～的晚霞 rosy evening clouds

飞 ①(鸟等展翅飞) fly; flit ②(空中飘浮) hover or flutter in the air: ～絮 willow catkins flying in the air / ～鸢 a hovering kite ③(极快) swiftly: ～奔 dash; tear along ④(意外的) unexpected; accidental: ～来横祸 unexpected disaster ⑤(无根据的) unfounded; groundless:

流言～语 rumors and slanders ⑥[口](挥发) disappear through volatilization
【飞车走壁】[杂技] stunt cycling, driving or motorcyclingon the inner surface of a cylindrical wall
【飞驰】speed along: 汽车～而过。A car sped by.
【飞虫】winged insect
【飞船】airship; dirigible: 太空～ spaceship; spacecraft
【飞弹】①(有自动飞行设备的炸弹) missile ②(流弹) stray bullet
【飞地】①(指在其它省或县内) land of one province or county enclosed by that of another ②(指在别国的国境内) enclave; exclave
【飞短流长】spread embroidered stories and malicious gossip
【飞蛾投火】a moth darting into a flame; bringing destruction upon oneself; seeking one's own doom
【飞飞扬扬】spread embroidered stories
【飞花】[纺] flyings; fly
【飞黄腾达】make rapid advances in one's career; have a meteoric rise
【飞蝗】migratory locust
【飞机】aircraft; airplane; plane: 不明国籍的～ unidentified aircraft／垂直降落～ vertical takeoff and landing aircraft (VTOL)／短距起落～ short takeoff and landing aircraft (STOL)／螺旋桨～ propeller-driven plane／水上～ hydroplaner; marine aircraft; seaplane ◇～场 airfield; airport; aerodrome／～出动架次 sortie／～库 hangar／～制造业 aviation industry; aircraft industry ○轰炸机 bomber／重型轰炸机 heavy bomber／中型轰炸机 medium bomber／战略轰炸机 strategic bomber／战术轰炸机 tactical bomber／攻击轰炸机 attack bomber／鱼雷轰炸机 torpedo bomber／远程轰炸机 long-distance bomber／战斗机 fighter／高空战斗机 high-altitude fighter／截击战斗机 interceptor fighter／截击机 interceptor／全天候截击机 all-weather interceptor／侦察机 reconnaissance plane／无人驾驶高空侦察机 pilotless high-altitude reconnaissance plane／运输机 transport plane; aerotransport [参见航空]
【飞溅】splash: 钢花～ sparks flying off molten steel
【飞快】①(非常迅速) very fast ②(锋利) extremely sharp
【飞轮】①[机] flywheel ②(指自行车后轮上的) free wheel
【飞沫传染】[医] infection through breathing in flying particles of the saliva or phlegm of a sick person
【飞禽走兽】birds and beasts
【飞泉】cliffside spring
【飞沙走石】sand flying about and stones hurtling through the air
【飞速】at full speed
【飞腾】fly swiftly upward; soar
【飞艇】airship; dirigible
【飞舞】dance in the air; flutter: 雪花～。Snowflakes are dancing in the air.
【飞翔】circle in the air; hover
【飞行】flight; flying: 低空～ low-altitude flying／高空～ altitude flying／水平～ level flight／特技～ stunt flying; aerobatics; aerial acrobatics／无线电遥控～[航模] radio controlled flight／线操纵圆圈～[航模] control line circular flight／自由～[航模]free flight◇～半径 flying radius／～表演 demonstration flight／～服 flying suit／～管制 air traffic control／～记录薄 flight log／～帽 aviator's helmet／～马赫数 flight Mach number／～模拟器 flight simulator／～能见度 flight visibility／～小时 pilot time／～性能 airplane characteristics
【飞行员】pilot; aviator; flyer
【飞檐】[建] upturned eaves
【飞檐走壁】leap onto roofs and vault over walls
【飞眼】make eyes; ogle
【飞扬】fly upward; rise: 尘土～ clouds of dust flying up
【飞扬跋扈】arrogant and domineering
【飞鱼】[动] flying fish
【飞跃】①(突飞猛进) leap ②[哲] leap: 认识过程的一次～ a leap in the process of cognition
【飞灾】unexpected disaster
【飞贼】①(能登墙上房的贼) a burglar who makes his way into a house over walls and roofs ②(空中进犯之敌) an intruding enemy airman; air marauder
【飞涨】(of prices, etc.) soar; skyrocket: 物价～。Prices were skyrocketing.
【飞针走线】quick work with the needle; sew quickly

妃 ①(皇帝之妾) imperial concubine ②(王、侯、太子之妻) the wife of a prince
【妃色】light pink

【妃子】imperial concubine

féi

腓 calf (of the leg)
【腓肠】calf; sura
【腓肠肌】musculus gastrocnemius
【腓骨】[生理] fibula

肥 ①(含脂肪多) fat: 这肉太～了。The meat is too fat. ②(肥沃) fertile; rich ③(使肥沃) fertilize: ～田 fertilize the soil ④(肥料) fertilizer; manure: 春～ spring top-dressing / 氮～ nitrogenous fertilizer / 冬～ winter top-dressing / 化～ chemical fertilizer / 钾～ potash fertilizer / 厩～ barnyard manure / 磷～ phosphate fertilizer / 绿～ green manure / 农家～ farmyard manure / 合理施～ apply fertilizer rationally / 积～ manure accumulation; stock up manure; store compost / 沤～ make compost / 施～ manuring' apply manure; manuring' apply fertilizer / 施底～ application of base manure to the subsoil; work manure deep into the soil / 施种～ manuring at sowing / 追～ apply additional fertilizer; additional manuring; top-dressing ⑤(肥大) loose-fitting; loose; large: 腰身～ large in girth
【肥大】①(宽又大) loose; large ②(粗大) fat; plump; corpulent: ～的鲤鱼 a fat carp ③(身体某部分体积增大)[医] hypertrophy: 扁桃体～ hypertrophy of the tonsils / 心脏～ hypertrophy of the heart
【肥大细胞】mast cell; mastocyte
【肥分】[农] (the percentage of) nutriment in a fertilizer

【肥厚】plump; fleshy: 果肉～。The pulp is full and fleshy.
【肥力】[农] fertility (of soil): 土壤～ soil fertility
【肥料】fertilizer; manure: 腐殖酸～ humic acid fertilizer / 化学～ chemical fertilizer / 抗生素～ antibiotic fertilizer / 无机～ inorganic fertilizer / 细菌～ bacterial fertilizer / 有机～ organic fertilizer
【肥美】①(肥沃) fertile; rich ②(丰美) luxuriant; plump; fat: 水草～牛羊壮 rich pastures and thriving herds
【肥胖】fat; corpulent ◇ ～病 [医] obesity
【肥缺】lucrative post
【肥实】[口] fat; stout
【肥硕】①(又大又饱满)(of fruit) big and fleshy ②(大而肥胖) (of limbs and body) large and firm-fleshed
【肥田】①(使土地肥沃) fertilize the soil ②(肥沃的土地) fertile land ◇ ～粉 ammonium sulphate
【肥沃】fertile; rich
【肥效】[农] fertilizer efficiency
【肥育】[牧] fattening ◇ ～期 stage of fattening
【肥源】[农] source of manure
【肥皂】soap: 一块～ a cake of soap; a bar of soap ◇ ～粉 soap powder / ～泡 soap bubble / ～水 soapsuds
【肥壮】stout and strong

fěi

斐
【斐济】Fiji ◇ ～人 Fijian / ～语 Fijian (language)
【斐然】[书] striking; brilliant: ～成章 show striking literary merit

悱 [书] be at a loss for words
【悱恻】[书] laden with sorrow; sad at heart: 缠绵～exceedingly sentimental; extremely sad

诽 slander
【诽谤】slander; calumniate; libel; defame; evil speaking ◇ ～案 case of libel / ～罪 crime of defamation; offense of libel
【诽谤名誉】defamatory libel
【诽谤名誉的侵权行为】character defamation
【诽谤诉讼】libel action

菲 [书] poor; humble; unworthy: ～材 my humble talent / ～仪 my small gift / ～酌 a simple meal
【菲薄】①(微薄) humble; poor: ～的礼物 a small gift ②(瞧不起) belittle; despise: 不要妄自～。Don't belittle yourself unduly.
【菲仪】my humble gift

匪 ①(强盗) bandit; brigand; robber ②[书](非) not: 获益～浅 reap no little benefit
【匪帮】bandit gang
【匪巢】bandits' lair
【匪患】the evil of banditry; banditry
【匪军】bandit troops
【匪窟】bandits' lair
【匪首】bandit chieftain
【匪徒】gangster; bandit
【匪夷所思】unimaginably queer; fantastic

榧 [植] Chinese torreya

【榧子】①[植] Chinese torreya ②(榧子树种) Chinese torreya nut

蜚

【蜚蠊】[动] cockroach; roach

翡

【翡翠】①[矿] jadeite ②[动] halcyon

fèi

吠 bark; yap

【吠犬不咬】barking dogs seldom bite; great barkers are no biters

【吠形吠声】when one dog barks at a shadow all the others join in; slavishly echo others

痱

【痱子】[医] prickly heat ◇ ～粉 prickly-heat powder

废 ①(不再使用;不再继续) give up; abandon; abolish; abrogate: 半途而～ give up halfway / 不以人～言 not reject an opinion because of the speaker ②(无用) waste; useless; disused: ～井 a disused well / ～矿 an abandoned mine / ～热 waste heat / ～油 used oil / 修旧利～ repair old equipment and make use of waste materials ③(残废) disabled; maimed: ～疾 disability

【废弛】①(失去约束作用) (of a law, custom, etc.) cease to be binding ②(松弛)(of discipline, etc.) become lax

【废除】abolish; abrogate; annul; repeal: ～烦琐的礼节 doaway with tedious formalities / ～一切陈规 annul all outdated rules; cancel all outdated rules

【废除令】abatement order

【废除债务】abatement of debts

【废黜】dethrone; depose

【废话】superfluous words; nonsense; rubbish: ～连篇 pages of nonsense; reams of rubbish

【废料】waste material; waste; scrap

【废票】①(指车船票、戏票等) invalidated ticket ②(指选票) invalidated ballot

【废品】①(不合格产品) waste product; reject ②(废旧物品) scrap; waste ◇ ～回收 waste recovery; salvage of waste material / ～率 reject rate / ～收购站 salvage station

【废气】waste gas or steam

【废弃】discard; abandon; cast aside: ～陈规旧习 discard outdated regulations and customs

【废寝忘食】(so absorbed or occupied as to) forget food and sleep

【废人】①(残废者) disabled person ②(泛指无用的人) good-for-nothing

【废水】waste water; liquid waste ◇ ～处理场 waste water processing station / ～处理池 purification tank for liquid waste / ～渗透 waste water infiltration

【废丝】[纺] waste silk

【废铁】scrap iron

【废铜烂铁】metal scraps

【废物】①(失去原有使用价值之物) waste material; trash: ～利用 make use of waste material ②(无用的人) good-for-nothing

【废液】waste liquid

【废墟】ruins; debris

【废渣】waste residue

【废止】abolish; annul; put an end to

【废止合同】avoid a contract

【废纸】waste paper: 不要乱扔～ Don't litter the place with waste paper

【废置】put aside as useless: 一口～不用的水井 a disused well

肺 lungs

【肺癌】carcinoma of the lungs; lung cancer

【肺病】pulmonary tuberculosis (TB)

【肺出血】pneumorrhagia

【肺动脉】[生理] pulmonary artery

【肺腑】the bottom of one's heart: ～之言 words from the bottom of one's heart / 出自～ straight from the heart; from the depths of one's heart / 感人～ move one deeply

【肺活量】vital capacity

【肺结核】pulmonary tuberculosis (TB)

【肺静脉】[生理] pulmonary vein

【肺痨】consumption; tuberculosis

【肺脓肿】pulmonary abscess

【肺泡】[生理] pulmonary alveolus

【肺气肿】pulmonary emphysema

【肺切除术】pneumonectomy

【肺水肿】pulmonary edema

【肺吸虫】lung fluke ◇ ～病 paragonimiasis

【肺循环】[生理] pulmonary circulation

【肺炎】pneumonia: 大叶～ lobar pneumonia / 小叶～ lobular pneumonia

【肺叶】[生理] a lobe of the lung

【肺脏】lungs

【肺蛭】lung fluke

沸 boil: ～水 boiling water

【沸点】[物] boiling point: ～温度 boiling tem-

perature
【沸沸扬扬】bubbling with noise; in a hubbub
【沸泉】[地] boiling spring
【沸石】[矿] zeolite
【沸腾】①[物](液体气化) boiling; ebullition ②(情绪高昂) seethe with excitement; boil over: 热血～ one's blood boils

费 ①(费用) fee; dues; expenses; charge: 报～ subscription for a newspaper / 车～ fare / 会～ membership dues / 军～ military expenditure / 免～ free of charge / 生活～ living expenses / 水电～ charges for water and electricity / 学～ tuition fees; tuition / 医药～ medical free ②(花费) cost; spend; expend ③(浪费) wasteful; consuming too much; expending sth. too quickly
【费工】take a lot of work; require a lot of labor
【费工夫】take time and energy; be time-consuming
【费解】hard to understand; obscure
【费尽心机】rack one's brains in scheming
【费劲】need or use great effort; be strenuous
【费口舌】require a lot of talking; require a lot of arguing
【费力】need or use great effort; be strenuous: ～不讨好 do a hard but thankless job
【费钱】cost a lot; be costly
【费神】[套] may I trouble you (to do sth.); would you mind (doing sth.)
【费时】take time; be time-consuming
【费事】give or take a lot of trouble
【费心】①(耗费心思) give a lot of care; take a lot of trouble② [套] may I trouble you (to do sth.); would you mind (doing sth.)
【费用】cost; expenses: 生产～ production cost / 生活～ cost of living; living expenses

镄 [化] fermium (Fm)

狒
【狒狒】[动] baboon

fēn

分 ①(分开) divide; separate; part: 难舍难～ cannot bear to part from each other ②(分配) distribute; assign; allot: ～田地 distribute the land ③(辨别) distinguish; differentiate: 是非不～ make no distinction between right and wrong ④(分支;部分) branch (of an organization): ～店 branch (of a shop) ⑤(表示分数) fraction: 三～之二 two-thirds ⑥(计量单位的十分之一) one tenth ⑦(货币单位) *fen*, a fractional unit of money in China (=1 / 100 of a *yuan* or 1 / 10 of a *jiao*): 四元五角六～ 4.56 *yuan*; four *yuan* fifty-six *fen* ⑧(时间单位) minute (=1 / 60 of an hour) ⑨(弧或角的计算单位) minute (=1 / 60 of a degree) ⑩(评定成绩、胜负的记数单位) point; mark ⑪(利率)年利九～ 9% interest a year
【分班】①(学校) divide into classes ②(军队) divide into squads
【分贝】[物] decibel (db)
【分崩离析】disintegrate; fall to pieces; come apart
【分辨】①(辨别) distinguish; differentiate: ～是非 distinguish between right and wrong / ～真假 distinguish truth from falsehood ② [物] resolution ◇ ～率 resolving power
【分辩】defend oneself (against a charge); offer an explanation: 不容～ allowing no explanation to be offered
【分别】①(离别) part; leave each other ②(辨别) distinguish; differentiate: ～轻重缓急 differentiate the important from the less important and the urgent from the less urgent; do things in order of importance and urgency ③(不同) difference: 毫无～ without any difference ④(各自) respectively; separately: ～和他们谈话 have a talk with them separately
【分兵】divide forces: ～把守 divide the forces for defense
【分布】be distributed (over an area); be dispersed be scattered: 人口～ population distribution
【分册】a separately published part of a book; fascicle
【分杈】[农] ① branching ② branch
【分词】[语] participle: 过去～ past participle / 现在～ present participle ◇ ～短语 participial phrase
【分寸】proper limits for speech or action; sense of propriety: 不知～ lack tact; have no sense of propriety
【分担】share responsibility for: ～责任 share the responsibility / 男女～家务劳动. Men and women are encouraged to share household duties.
【分道】[体] lane: 第一～ the first lane
【分道扬镳】separate and go different ways; part company, each going his own way
【分等】grade; classify: 产品按质～ grade prod-

ucts according to quality
【分队】a troop unit corresponding to the platoon or squad; element
【分而治之】divide and rule
【分发】distribute; hand out; issue: 给优胜者～奖品 distribute prizes to the winners／～证件 issue certificates individually
【分肥】share out ill-gotten gains; divide booty
【分赴】leave for different destinations: ～不同的工作岗位 go to take up different posts
【分割】cut apart; break up; carve up: ～别国的领土 partition territory of another country; carve up territory of another country
【分隔】separate; divide
【分工】divide the work; division of labor: ～负责 division of labor with individual responsibility／社会～ social division of labor
【分管】be assigned personal responsibility for; be put in charge of
【分光计】[物] spectrometer
【分光镜】[物] spectroscope
【分行】branch (of a bank): 国内～ home branch; domestic branch／国外～ overseas branch
【分毫】fraction; iota: 不差～ without the slightest error; just right
【分号】① [语] semicolon (；) ② (分支机构) branch (of a firm, etc.)
【分红】share out bonus; draw extra dividends
【分洪】flood diversion ◇ ～工程 flood-diversion project／～区 flood-diversion area／～闸 flood-diversion sluice
【分户帐】ledger
【分化】① (变为不同物;分裂) split up; become divided; break up ② [生] differentiation
【分化瓦解】divide and disintegrate; split and disintegrate
【分会】branch (of a society, association, etc.); chapter
【分机】extension (telephone)
【分级】grade; classify: ～管理 manage by different levels
【分级机】[农] grader; sorter: 马铃薯～ potato sorter
【分家】divide up family property and live apart; break up the family and live apart
【分解】① (整体的分割) resolve; decompose; break down: 力的～ resolution of force ② (解说,章回小说用语) recount; disclose: 欲知后事如何,且听下回～。 But as to what happened thereafter, that will be disclosed in the ensuing chapter. ◇ ～代谢 [生] catabolism／～反应 decomposition reaction／～热 decomposition heat
【分界】① (划分界线) have as the boundary; be demarcated by ② (划分的界线) dividing line; line of demarcation
【分界线】line of demarcation; boundary: 军事～ a military demarcation line
【分进合击】[军] concerted attack by converging columns
【分居】(of members of a family) live apart: ～赡养费 separate maintenance
【分开】separate; part: 把打架的人～ part the fighters
【分类】classify ◇ ～法 classification／～数字 breakdown figures／～索引 classified index／～帐 ledger
【分类学】taxology; taxonomy; systematics: 动物～ systematic zoology／植物～ systematic botany
【分离】separate; sever: 从空气中把氮～出来 separate nitrogen from air ◇ ～器 [化] separator
【分离度】separating degree
【分离段】[航空] fallaway section
【分理处】[经] a small local branch (of a bank)
【分力】[物] component (of force)
【分列式】[军] march-past
【分裂】① (分开) split; divide; break up: 使国家～ disrupt a state／制造～ create dissensions ② [生] [物] fission; division: 核～ nuclear fission／减数～ meiosis; reduction division／细胞～ cell division／有丝～ mitosis
【分裂主义】splittism: ◇ ～分子 splittist／～路线 splittist line; divisive line
【分馏】[化] fractional distillation; fractionation ◇ ～器 fractionator／～塔 fractionating tower; fractional column
【分路】① (分若干路线) along separate routes; from several directions: ～出击 attack from several directions／～前进 advance along separate routes ② [电] shunt ◇ ～电流 branch current／～电阻 shunt resistance
【分袂】leave each other; part
【分门别类】put into different categories; classify
【分米】decimeter (dm.)
【分泌】[生] secrete: ～胃液 secrete gastric juice ◇ ～物 secretion
【分娩】childbirth; parturition
【分娩台】obstetric table

【分秒必争】seize every minute and second; every second counts; not a second is to be lost
【分明】①(清楚) clearly demarcated; sharply contoured; distinct: 爱憎~ be clear about what to love and what to hate ②(明显) clearly; plainly; evidently: 这~是你的错。Evidently you are in the wrong.
【分母】[数] denominator
【分蘖】[农] tillering: 有效~ effective tillering ◇~节 tillering node / ~期 tillering stage
【分派】assign (to different persons); apportion
【分配】①(分东西) distribute; allot; assign: ~住房 allot dwelling houses ②[经] distribution ◇~律[数] distributive law / ~理论 theory of distribution
【分批】in batches; in turn: ~交货 deliver the goods in botches / ~装运 partial shipment
【分片包干】divide up the work and assign a part to each
【分期】by stages: ~付款 payment by instalments; hire purchase; instalment plan / ~交货 instalment delivery
【分期偿还】amortize
【分期摊销】amortize
【分歧】difference; divergence: 消除~ iron out differences / 原则~ a difference in principle
【分清】distinguish; draw a clear distinction between; draw a clear line of demarcation between: ~界限 draw a clear line of demarcation
【分清敌友】know a friend from an enemy
【分清是非】discern between right and wrong
【分群】(of bees) hive off
【分散】disperse; scatter; decentralize: ~兵力 dispersion of forces / ~注意力 divertone's attention; take sb.'s mind off sth. ◇~剂 [化] dispersing agent / ~染料 disperse dyes / ~指挥 decentralized command; decentralized direction / ~主义 decentralism
【分色机】[印] color scanner
【分色镜】color selective mirror; dichroic mirror
【分身】spare time from one's main work to attend to sth. else
【分神】give some attention to
【分手】part company; say good-bye
【分数】① [数] fraction: 带~ mixed number / 繁~ complex fraction / 假~ improper fraction / 真~ proper fraction ②(表示成绩等的数字) mark; grade
【分水岭】①(分水线)[地] watershed; divide ②(分界)line of demarcation; watershed
【分水闸门】[水] bifurcation gate
【分说】defend oneself (against a charge); explain matters: 不容~ allow no explanation
【分送】send; distribute
【分摊】share: ~费用 share the expenses / ~税 apportioned tax
【分庭抗礼】stand up to sb. as an equal; make rival claims as an equal
【分头】①(若干人分别进行工作) separately; severally ②(指头发式样) parted hair
【分文】a single cent; a single penny: ~不取 not take a single cent; free of charge / ~不值 not worth a farthing; be not worth straw
【分析】analyse: ~当前形势 analyse the present situation / 阶级~ class analysis ◇~化学 analytical chemistry / ~会计 analytical accounting / ~心理学 analytic psychology / ~语[语] analytical language
【分线规】dividers
【分享】share (joy, rights, etc.); partake of: ~胜利的喜悦 share the joys of victory
【分相】[电] split phase ◇~器 phase splitter
【分销店】retail shop
【分晓】①(底细;结果) outcome; solution ②(明白;清楚) seeor understand clearly: 问个~ inquire about and get to the bottom of a matter
【分心】divert one's attention
【分压器】[电] voltage divider
【分野】dividing line
【分忧】share sb.'s cares and burdens; help sb. to get over a difficulty
【分赃】divide the spoils; share the booty; share the loot
【分针】minute hand
【分支】branch: 银行的~机构 branches of a bank
【分子】① [数] numerator (in a fraction) ② [化] molecule: 克~ gram molecule ◇~病 molecular disease / ~仿生学 molecular bionics / ~结构 molecular structure / ~量 molecular weight / ~筛 molecular sieve / ~生物学 molecular biology / ~式 molecular formula / ~遗传学 molecular genetics
【分组】divide into groups: ~学习 study in groups

芬 sweet smell; fragrance
【芬芳】①(香的)sweet-smelling; fragrant ②(香气) fragrance
【芬芳扑鼻】a fragrance strike the nostrils

【芬芳郁馥】rich in fragrance; both fragrant and beautiful
【芬兰】Finland ◇ ～人 Finn; Finlander / ～蒸汽浴 sauna / ～语 Finnish (language)

酚 [化] phenol
【酚磺酞】[药] phenolsulphonphthalein
【酚醛】[化] phenolic aldehyde ◇ ～树脂 Bakelite / ～塑料 phenolic plastics
【酚酞】[化] phenolphthalein: ◇ ～试纸 phenolphthalein test paper
【酚油】carbolic oil

吩
【吩咐】tell; instruct: 照你的～去办吧。 Do as you are told.

氛 atmosphere
【氛围】atmosphere

纷 ①(杂乱) confused; tangled; disorderly ②(多) many and various; profuse; numerous: 大雪～飞。 It was snowing hard.
【纷繁】numerous and complicated: 头绪～ have too many things to take care of
【纷纷】①(接二连三) one after another; in succession ②(多而杂乱) numerous and confused
【纷乱】numerous and disorderly; helter-skelter; chaotic: ～的脚步声 hurried footsteps
【纷扰】confusion; turmoil
【纷纭】diverse and confused: 众说～。 Opinions are widely divided.
【纷争】dispute; wrangle: ～不已 endless dispute / 内部～ internal dissension
【纷至沓来】come in a continuous stream; come thick and fast; keep pouring in

fén

坟 grave; tomb: 祖～ the grave of one's ancestors
【坟地】graveyard; cemetery
【坟墓】grave; tomb
【坟头】grave mound
【坟茔】①(坟墓) grave; tomb ②(坟地) graveyard; cemetery

焚 burn: ～香 burn incense
【焚风】[气] foehn
【焚化】incinerate; cremate ◇ ～炉 incinerator; cremator
【焚毁】destroy by fire; burn down
【焚烧】burn; set on fire: ～垃圾 burn away the refuse
【焚尸灭迹】burn someone's body to cover up the crime
【焚尸扬灰】burn somebody's corpse and scatter the ashes to the winds
【焚书坑儒】burn books and bury the literatis in pits

fěn

粉 ①(粉末) powder: 合成洗衣～ synthetic detergent / 面～ flour / 奶～ powdered milk / 漂白～ bleaching powder / 去污～ cleanser / 爽身～ talcum powder ②(淀粉制食品) noodles or vermicelli made from bean or sweet potato starch ③(粉刷) white: ～墙 white-washed wall ④(粉红) pink: ～色 pink color
【粉笔】chalk: 彩色～ colored chalk ◇ ～槽 a ledge for chalk on a blackboard / ～画 chalk drawing; crayon
【粉彩】[工美] famille rose
【粉刺】[医] acne
【粉蝶】[动] white (butterfly)
【粉盒】powder box; compact
【粉红】pink
【粉剂】①[药] powder ②[农] dust
【粉瘤】[医] sebaceous cyst
【粉末】powder ◇ ～金属 powdered-metal / ～冶金 powder metallurgy
【粉墨登场】make oneself up and go or stage; embark upon a political venture
【粉皮】sheet jelly made from bean or sweet potato starch
【粉扑儿】powder puff
【粉墙】①(刷墙) whitewash a wall ②(白墙) whitewashed wall
【粉身碎骨】have one's body smashed to pieces; die the most cruel death
【粉饰】gloss over; whitewash: ～错误 whitewash mistakes / ～太平 present a false picture of peace and prosperity
【粉刷】whitewash
【粉丝】vermicelli made from bean starch, etc.
【粉碎】①(使失败、毁灭) smash; shatter; crush: ～经济封锁 smash an economic blockade ②(破碎) broken to pieces: 玻璃杯摔得～。 The glass was smashed to pieces
【粉碎机】pulverizer; grinder; kibbler: 球磨～ ball mill pulverizer / 饲料～ fodder grinder
【粉条】noodles made from bean or sweet potato starch

【粉线】tailor's chalk line
【粉妆银砌】(the landscape) gave the illusion of a silver-and-powder world

fèn

粪 ①(粪便) excrement; feces; dung; droppings ② [书](施肥) apply manure: ~田 manure the fields
【粪便】excrement and urine; night soil ◇ ~检查 stool examination
【粪便恐怖】[心理] coprophohia
【粪便污水】fecal sewage
【粪便无害处理】decontamination of feces
【粪车】dung-cart; night-soil cart
【粪池】manure pit
【粪堆】dunghill; manure pile; manure heap
【粪肥】muck; manure; dung
【粪坑】manure pit
【粪筐】manure basket
【粪桶】night-soil bucket; manure bucket
【粪土】dung and dirt; muck: 视如~ look upon as dirt

愤 indignation; anger; resentment: 公~ public indignation
【愤不欲生】would end life in a fit of bitterness
【愤愤不平】be indignant; feel aggrieved; be resentful
【愤恨】indignantly resent; detest
【愤慨】(righteous) indignation: 表示~ express one's indignation / 感到无比~ be overwhelmed with indignation
【愤懑】depressed and discontented; resentful
【愤怒】indignation; anger; wrath: ~声讨 angrily denounce
【愤世嫉俗】detest the world and its ways

奋 ①(振作) exert oneself; act vigorously: 振~ rouse oneself ②(举起) raise; lift
【奋臂高呼】raise one's hand and shout
【奋不顾身】dash ahead regardless of one's safety
【奋斗】struggle; fight; strive: ~目标 the objective of a struggle
【奋斗终身】work and fight all one's life
【奋斗到底】struggle to the very end; to make strenuous efforts in spite of difficulties
【奋发】rouse oneself; exert oneself
【奋发图强】go all out to make the country strong
【奋发有为】be promising and diligent in one's work
【奋力】do all one can; spare no effort
【奋勉】make a determined effort
【奋乃静】[药] perphenazine
【奋起】rise with force and spirit; rise: ~直追 do all one can to catch up / ~自卫 rise in self-defense
【奋勇】summon up all one's courage and energy: ~前进 forge ahead courageously / ~作战 fight bravely
【奋战】fight bravely: 日夜~ struggle hard day and night / 浴血~ fight a bloody battle

分 ①(成分) component: 盐~ salt content ②(职责或权利的限度) what is within one's rights or duty: 本~ one's duty / 过~ exceeding what is proper; going too far; excessive
【分量】weight: 给足~ give full measure / 他这话说得很有~。 What he said should not be taken lightly.
【分内】one's job; one's duty: 这是我~的事。It's something within my duty.
【分外】①(特别) particularly; especially: ~高兴 particularly happy / ~香 especially fragrant ②(本分以外) not one's job; not one's duty
【分子】member; element: 反革命~ counterrevolutionary / 积极~ active member / 先进~ advanced element

份 ①(整体中的一部) share; portion: 股~ stock; share ②[量] 一~儿礼 a gift
【份额】share; portion
【份儿饭】table d'hote; set meal
【份子】one's share of expenses for a joint undertaking, as in buying a gift for a mutual friend: 凑~ club together to present a gift to sb.

fēng

丰 ①(丰富) abundant; plentiful: ~收 a bumper harvest ②(大) great ③(美好的容貌、姿态) fine-looking; handsome
【丰碑】①(高大石碑) monument ②(杰作;功绩) monumental work
【丰采韶秀】be of a most refined and prepossessing appearance
【丰产】high yield; bumper crop
【丰登】bumper harvest: 五谷~ a bumper grain harvest
【丰登之年】a year of abundance; a prosperous year

【丰富】①(种类多、数量大) rich; abundant; plentiful: ～的经验 rich experience / 资源～ rich in natural resources ②(使丰富) enrich: ～知识 enrich one's knowledge
【丰富多彩】rich and varied; rich and colorful: ～的传统出口商品 a rich array of traditional products for export / 演出了～的节目 present a varied and interesting program
【丰功伟绩】great achievements; signal contributions
【丰厚】①(多而厚实) thick: 绒毛～ rich and thick fur②(丰富) rich and generous: ～的礼品 generous gifts
【丰满】①(充足) plentiful ②(匀称好看) full and round; well-developed; full-grown: 羽毛～ full-fledged
【丰茂】luxuriant; lush
【丰美】lush: 水草～ lush pasture
【丰年】bumper harvest year; good year
【丰年虫】fairy shrimp
【丰饶】rich and fertile
【丰润】plump and smooth-skinned
【丰盛】rich; sumptuous: ～的酒席 a sumptuous feast
【丰收】bumper harvest: 连年～ score consecutive bumper harvests; have good harvests for many years in succession / ～在望。A good harvest is in sight.
【丰硕】plentiful and substantial; rich: 取得～的成果 score great successes
【丰衣足食】have ample food and clothing; be well-fed and well-clothed
【丰盈】①(丰满) have a full figure ②(富裕) plentiful
【丰裕】well provided for; in plenty: 生活～ live in plenty; be comfortably off
【丰姿绰约】agreeable manners; her manner is graceful
【丰足】abundant; plentiful: 衣食～ have plenty of food and clothing

封
①(封闭) seal: ～门 seal up a door ②(信封) envelope: 信～ envelope ③[量] 一～信 a letter
【封闭】①(关住) seal ②(查封) seal off; close: ～机场 close an airport ◇ ～层[石油] confining bed
【封存】seal up for safekeeping
【封底】[印] back cover
【封地】fief; feud; manor
【封冻期】a period of freezing weather; freeze
【封二】[印] inside front cover
【封官许愿】[贬] offer official posts and make lavish promises
【封罐机】tin seamer; can seamer
【封焊】soldering and sealing
【封火】bank up a fire
【封建】①(一种分封的政治制度) the system of enfeoffment ②(封建主义社会形态) feudalism: 头脑～ feudal-minded ◇ ～社会 feudal society / ～主义 feudalism
【封建割据】feudal separationist rule
【封建余孽】the spawn of feudalism
【封口】①(封闭张开之处) seal ②(闭合) heal: 臂上的伤已经～了。The arm wound has healed. ③(闭口不谈) say sth. definitive so as to prevent further discussion
【封蜡】sealing wax
【封里】[印] ①(封二) inside front cover ②(封三) inside back cover
【封面】①(指线装书印有书名的一页) the title page of a thread-bound book②(指书最外面的书皮) the front and back cover of a book ③[印](封一) front cover
【封面布】[印] book cloth
【封面纸】cover paper
【封泥】[冶] lute
【封皮】①(图书的封面) paper wrapping ②(信封) envelope
【封妻荫子】get one's wife and children rewarded by heritage
【封三】[印] inside back cover
【封山】seal a mountain pass; close a mountain pass ◇ ～育林 close hillsides (to livestock grazing and fuel gathering) to facilitate afforestation
【封四】[印] back cover
【封锁】blockade; block; seal off: ～边境 close the border / ～消息 block the passage of information / 经济～ economic blockade ◇ ～线 blockade line; blockade
【封套】big envelope (for holding documents, books, etc.)
【封条】a strip of paper used for sealing (doors, drawers, etc.); paper strip seal
【封网】[排球] block
【封一】[印] front cover
【封印】[邮] seal; sealing stamp

烽
beacon
【烽火】①(古时报警烟火) beacon-fire (used to give border alarm in ancient China); beacon ②

(战火) flames of war: ～连天 flames of battle raging everywhere ◇ ～台 beacon tower
【烽火连连】continuous wars
【烽火燎原】the flames of war spread far and wide
【烽烟】beacon-fire; beacon
【烽烟滚滚】the flames of war are raging
【烽烟四起】a land beset by war

蜂 ①(黄蜂) wasp ②(蜜蜂) bee: 蜜～ honeybee / 养～ keep bees; apiculture ○养蜂业 bee-keeping; apiculture / 养蜂人 bee-keeper; apiarist / 养蜂场 apiary; bee yard; beehouse / 蜜蜂 honey-bee / 雄蜂 drone / 工蜂 working bee; worker bee / 王浆 royal jelly / 隔王板 excluder / 摇蜜机 honey extractor ③(成群地) in swarms: ～起 rise in swarms
【蜂巢】honeycomb
【蜂刺】the sting of a bee or wasp
【蜂毒】bee venom; sting poison
【蜂房】any of the six-sided wax cells in a honeycomb
【蜂糕】steamed sponge cake
【蜂花醇】melissyl alcohol; myricyl alcohol
【蜂蜡】beeswax
【蜂蜜】honey
【蜂鸣器】buzzer
【蜂鸟】hummingbird
【蜂群】(bee) colony
【蜂乳】[药] royal jelly
【蜂王】①(母蜂) queen bee ②(母的黄蜂) queen wasp
【蜂王浆】[药] royal jelly
【蜂窝】①(蜂巢) honeycomb ②(蜂巢状物) a honeycomb-like thing ◇ ～煤 honeycomb briquette / ～胃[动] reticulum; honeycomb stomach / ～织炎[医] phlegmon; cellulitis
【蜂箱】beehive; hive
【蜂音】hum; buzz
【蜂拥】swarm; flock: ～而来 come swarming; swarm forward
【蜂拥而起】like the bees rising in swarms
【蜂拥而至】come in great numbers

峰 ①(尖顶) peak; summit: 浪～ the crest of a wave / 山～ mountain peak / 攀登科学高～ scale the heights of science ②(峰状物) hump: 驼～ camel's hump
【峰环水抱】surrounded by hills and water
【峰回路转】the path winds through high peaks
【峰峦】ridges and peaks
【峰峦屏嶂】rounding ranges of hills
【峰态】[数] kurtosis
【峰形】spike: ～冰山 irregular iceberg
【峰值】[电] peak value; crest value

锋 ①(刀、剑锋利部分) the sharp point or cutting edge of a sword, etc. ②(带头的) van: 先～ vanguard ③ [气] front
【锋刚】high speed steel; rapid steel
【锋利】①(尖,面薄) sharp; keen: ～的钢刀 a sharp knife ②(文笔尖利) incisive; sharp; poignant: ～泼辣的笔调 a sharp and pungent style
【锋芒】①(刀剑的尖利部分) cutting edge; spearhead; ～所向 target of attack ②(比喻才干) talent displayed; abilities: 不露～ refrain from showing one's ability; be able but modest
【锋芒毕露】make a showy display of one's abilities
【锋芒小试】display only a small part of one's talent
【锋面】[气] frontal surface: 暖～ anaphalanx ◇～低压 frontal low / ～气旋 frontal cyclone

风 ①(空气流动) wind: 季节～ monsson / 强台～ violent typhoon / 台～ typhoon / 信～ trade wind / 旋～ whirlwind; cyclone / 阵～ gust ○蒲福～级 Beaufort scale / 无～ clam / 一级～ light air / 二级～ light breeze / 三级～ gentle breeze / 四级～ moderate breeze / 五级～ fresh breeze / 六级～ strong breeze / 七级～ moderate gale; near gale / 八级～ fresh gale / 九级～ strong gale / 十级～ whole gale / 十一级～ storm / 十二级～ hurricane ②(借风力吹干) put out to dry or air ③(借风力使纯净) winnow: 晒干～净 sun-dried and well winnowed ④(态度) style; practice; custom: 纠正不正之～ correct unhealthy tendencies / 文～ style of writing ⑤(景象) scene; view: ～景 scenery; landscape ⑥(消息) news; information: 闻～而动 act without delay upon hearing sth.
【风暴】windstorm; storm; tempest: 革命～ revolutionary storm
【风泵】air pump; air compressor
【风痹】[中医] wandering arthritis
【风标】weathercock; weather vane
【风波】disturbance: 平地起～ a storm out of nowhere
【风采】elegant demeanor; graceful bearing
【风餐露宿】eat in the wind and sleep in the

dew; endure the hardships of an arduous journey or fieldwork
【风潮】agitation; unrest
【风车】①(动力机械) windmill ②(吹净糠皮的扇车) winnower ③(玩具) pinwheel
【风尘】①(旅途劳累) travel fatigue ②(纷乱的社会) hardships or uncertainties in an unstable society: 沦落～ be driven to prostitution
【风尘仆仆】endure the hardships of a long journey; be travel-stained
【风驰电掣】swift as the wind and quick as lightning
【风传】hearsay; rumor
【风吹草动】the rustle of leaves in the wind; a sign of disturbance or trouble
【风吹浪打】be beaten by wind and waves; be battered by a storm
【风大雨狂】the wind is high and the rains pour down
【风挡】[汽车] windscreen; windshield
【风动】[机] pneumatic: ～工具 pneumatic tools
【风洞】[航空] wind tunnel
【风斗】wind scoop
【风度】demeanor; bearing: ～大方 have an easy manner
【风发】①(象风一样迅速) swift as the wind ②(奋发) energetic: 意气～ daring and energetic
【风干】air-dry
【风镐】[矿] pneumatic pick; air pick
【风格】style: 京剧的独特～ the characteristic style of Beijing opera
【风骨】①(刚强气概) strength of character ②(雄健有力的风格) vigor of style
【风光】scene; view; sight: 好～ a wonderful sight
【风害】damage caused by a windstorm; windburn
【风寒】chill; cold: 受～ have a cold; catch cold
【风和日暖】bright sunshine and gentle breeze; warm and sunny weather
【风花雪月】wind, flowers, snow and moon
【风华】elegance and talent: ～正茂 at life's full flowering; in one's prime
【风化】①(风俗教化) morals and manners; decency: 有伤～ an offense against decency ②[化] efflorescence ③[地] weathering
【风机】fan
【风级】[气] wind scale
【风纪】conduct and discipline; discipline
【风井】[矿] ventilating shaft; air shaft
【风景】scenery; landscape: 以～优美著称 famous for its scenic beauty ◇ ～画 landscape painting / ～林 scenic forest / ～区 scenic spot
【风镜】goggles
【风卷残云】a strong wind scattering the last clouds; make a clean sweep of sth.
【风口】① a place where there is a draught ②[地] wind gap ③ [冶] (blast) tuyere: 渣～ slag tuyere
【风口浪尖】where the wind and the waves are highest
【风浪】stormy waves; storm
【风雷】wind and thunder; tempest: ～激荡 a storm raging in all its fury
【风力】①(风的力量) wind-force ②(以风为动力) wind power ◇ ～发电机 wind-driven generator; windmill generator / ～发电站 wind power station
【风凉】cool
【风凉话】irresponsible and sarcastic remarks: 说～ make sarcastic comments
【风铃】aeolian bells
【风流】①(有功绩并有文采的) distinguished and admirable: ～人物 truly great men ②(有才学但不拘礼法的) talented and romantic; talented in letters and unconventional in life style: ～才子 talented and romantic scholar ③(行为放荡的) dissolute; loose
【风流云散】blown apart by the wind and scattered like the clouds; (of old companions) separated and scattered
【风马牛不相及】have absolutely nothing to do with each other; be totally unrelated
【风帽】①(挡风之帽) a cowl-like hat worn in winter ②(兜帽) hood
【风貌】①(风格和面貌) style and features: 民间艺术的～ the style and features of folk art ②(景象) view; scene;社会主义农村的新～ the new look of the socialist countryside
【风媒传粉】wind pollination
【风媒花】[植] anemophilous flower
【风门】[矿] air door; ventilation door
【风靡】fashionable: ～一时 become fashionable for a time; be all the rage at the time
【风磨】windmill
【风平浪静】the wind has subsided and the waves have calmed down; calm and tranquil
【风起云涌】like a rising wind and scudding clouds; rolling on with full force
【风气】general mood; atmosphere; common practice: 社会～ current tendencies in society;

social mode
【风琴】organ: 管～ pipe organ; organ / 簧～ reed organ; harmonium
【风情】amorous feelings; flirtatious expressions: 卖弄～play the coquette; coquette
【风趣】humor; wit: 很有～ full of humor; feel of wit
【风圈】solar or lunar halo
【风骚】① [书](泛称文学) literary excellence ②(轻佻) coquettish
【风色】how the wind blows: 看～ see which way the wind blows; see how things stand
【风沙】sand blown by the wind: ～化 the shifting of sand dunes / 这里冬天～很大。It's very windy and dusty here in winter.
【风扇】①(电扇) electric fan ② [机] fan: 散热～ radiator fan
【风尚】prevailing custom: 社会新～ new social tendency / 新的道德～ new morality and custom
【风声】rumor: ～很紧。The situation is getting tense.
【风声鹤唳】the sound of the wind and the cry of cranes; a fleeing army's suspicion of danger at the slightest sound
【风湿】[医] rheumatism ◇ ～热 rheumatic fever / ～痛 rheumatalgia / ～疹 rheumatid
【风湿性关节炎】[医] rheumarthritis
【风蚀】[地] wind erosion
【风霜】wind and frost; hardships of a journey or of one's life: 饱经～ weather-beaten; having had one's fillof hardships
【风水】the location of a house or tomb, supposed to have an influence on the fortune of a family; geomantic omen
【风俗】custom ◇ ～画 *genre*
【风速】wind speed; wind velocity ◇ ～表 anemometer / ～计 anemograph / ～器 wind gauge
【风瘫】paralysis
【风调雨顺】favorable weather; good weather for the crops
【风头】①(情势) the trend of events (as affecting a person): 避避～ lie low until sth. blow over ②(出头露面) the publicity one receives: 爱出～ be fond of the limelight
【风图】wind rose
【风土】natural conditions and social customs of a place: ～人情 local conditions and customs
【风土驯化】[农] acclimatization
【风味】special flavor; local color: 地方～ local cuisine / 家乡～ local flavor ◇ ～菜 typical local dish
【风闻】learn through hearsay; get wind of
【风险】risk; hazard: ～很大 full of hazard / 冒～ take risks
【风箱】bellows
【风向】wind direction ◇ ～标 wind vane / ～袋 wind sleeve; wind sock / ～计 registering weather vane / ～图 wind rose / ～仪 anemoscope
【风信子】[植] hyacinth
【风行】be in fashion; be popular: ～一时 be popular for a while
【风选】[农] selection by winnowing ◇ ～机 winnower
【风压】[气] wind pressure
【风雅】①(泛指诗文方面的事) literary pursuits ②(文雅) elegant; refined
【风言风语】groundless talk; slanderous gossip
【风雨】wind and rain; the elements; trials and hardships
【风雨大作】the storm rages
【风雨交加】it's raining and blowing hard; it's wet and windy
【风雨飘摇】swaying in the midst of a raging storm; precarious; tottering
【风雨如晦】wind and rain sweeping across a gloomy sky; a grim and grave situation
【风雨同舟】in the same storm-tossed boat; stand together through thick and thin
【风雨无阻】stopped by neither wind nor rain; regardless of the weather; rain or shine
【风云】wind and cloud; a stormy or unstable situation: ～变幻 a changeable situation / ～不测 unexpected misfortune
【风云人物】man of the hour
【风云突变】a sudden change in the situation
【风韵】graceful bearing; charm
【风灾】disaster caused by a windstorm
【风闸】[机] pneumatic brake
【风障】[农] windbreak
【风疹】[医] nettle rash; urticaria
【风筝】kite: 放～ fly a kite
【风烛残年】old and ailing like a candle guttering in the wind
【风姿】graceful bearing; charm
【风钻】pneumatic drill

疯 ①(神经错乱) mad; insane; crazy ②(生长旺盛,不结果实) spindle

【疯癫】insane; mad
【疯疯癫癫】be mentally deranged; act like a lunatic; be flighty
【疯狗】mad dog; rabid dog
【疯狂】①(发疯) insane ②(猖狂) frenzied; unbridled:～反扑 a desperate counterattack / ～叫嚣 frenzied clamoring
【疯人院】madhouse; lunatic asylum
【疯瘫】paralysis
【疯长】[农] overgrowth; spindling: 防止～ prevent spindling
【疯子】lunatic; madman

枫 ①(枫香树) Chinese sweet gum ②(枫树) maple: ～叶 maple leaf
【枫茅】citronella
【枫糖】maple sugar ◇～浆 maple syrup
【枫香树】Chinese sweet gum

砜 [化] sulphone

féng

逢 meet; come upon: ～人便问 ask whoever happens to come one's way
【逢场作戏】join in the fun on occasion
【逢集】market day
【逢年过节】on New Year's Day or other festivals
【逢山开路，遇水搭桥】cut paths through mountains and build bridges across rivers
【逢凶化吉】turn ill luck into good
【逢迎】make up to; fawn on; curry favor with: 阿谀～ flatter and toady

缝 stitch; sew: ～合切口 sew up the incision / (医生)给～三针 put in three stitches
【缝补】sew and mend
【缝补浆洗】mend and wash one's clothes
【缝缝补补】sew up
【缝缝连连】sewing and mending
【缝焊】[电工] seam welding
【缝合】[医] suture; sew up: ～伤口 sew up a wound ◇～针 [医] sewing needle; suture needle
【缝匠肌】musculus sartorius; tailor's muscle
【缝纫】sewing; tailoring ◇～机 sewing machine
【缝线】[医] suture: 吸收性～ absorbable suture / 羊肠～ catgut suture

fěng

讽 ①(讽刺) satirize; mock: 冷嘲热～ burning satire and freezing irony ②[书](吟诵) chant; intone
【讽刺】satirize; mock ◇～画 satirical drawing; caricature; cartoon / ～诗 satirical poem / ～小品 satirical essay
【讽诵】[书] read with intonation and expression
【讽喻】parable; allegory

fèng

奉 ①(给) give or present with respect: ～上新词典一部。I am forwarding you a new dictionary. ②(接受) receive (orders, etc.): ～上级指示 on orders from above ③(信奉) believe in: 信～伊斯兰教 believe in Islam ④(尊崇) esteem; revere: ～为典范 look upon as a model ⑤(奉养) wait upon; attend to: 侍～老人 attend to aged parents or grandparents
【奉承】flatter; fawn upon; toady ◇～话 flattery
【奉告】let sb. know; inform: 无可～。No comment.
【奉公守法】be law-abiding
【奉还】[敬] return sth. with thanks
【奉命】receive orders; act under orders: ～于危难之际 be entrusted with a mission at a critical and difficult moment
【奉命出发】be under orders to start
【奉命惟谨】receive orders respectfully
【奉陪】keep sb. company: 恕不～。Sorry, I won't be able to keep you company.
【奉陪到底】have the honor of keep sb. company until the end
【奉劝】may I offer a piece of advice
【奉若神明】worship sb. or sth.; make a fetish of sth.
【奉觞进酒】offer gifts of wine
【奉觞上寿】drink a toast of longevity
【奉送】offer as a gift; give away free
【奉为楷模】hold up as a model
【奉为至宝】value highly ...
【奉献】offer as a tribute; present with all respect
【奉行】pursue (a policy, etc.): ～不结盟政策 pursue a policy of nonalignment
【奉行故事】follow established practice
【奉养】support and wait upon (one's parents, etc.)

俸 pay; salary
【俸禄】[旧] an official's salary

凤 phoenix
【凤飞麟散】there are no wise men in the government
【凤肝龙心】rare delicacies
【凤冠】phoenix coronet: ～霞帔 a chaplet and official robes
【凤凰】phoenix ◇ ～座 [天] Phoenix
【凤梨】pineapple
【凤毛麟角】(precious and rare as) phoenix feathers and unicorn horns; rarity of rarities
【凤鸣朝阳】phoenix singing in morning sun; good omen for the country
【凤头鹦鹉】cockatoo
【凤尾鱼】anchovy
【凤尾竹】fernleaf hedge bamboo
【凤仙花】garden balsam

缝 ①(接合处) seam: 无～钢管 seamless steel tube ②(缝隙) crack; crevice; fissure: 石～ crevice in the rock
【缝隙】chink; crack; crevice

fó

佛 ①(佛;佛陀) Buddha: 拜～ worship Buddha ②(佛教) Buddhism: 信～ believe in Buddhism ③(佛像) image of Buddha: 铜～ a bronze statue of Buddha
【佛得角】Cape Verde ◇ ～人 Cape Verdean
【佛法】①(佛教教义) Buddha dharma; Buddhist doctrine ②(法力) power of Buddha
【佛法僧】①[佛教] Buddha-dharma-sangha ②(鸟) roller
【佛法无边】The powers of Buddha are unlimited
【佛教】Buddhism ◇ ～徒 Buddhist ○大乘 Mahayana doctrine; the Great Vehicle / 小乘 Hinayana doctrine; the Little Vehicle / 瑜伽宗 Yogacara Sect / 天台宗 Tiantai Sect / 禅宗 Chan Sect; Zen Sect / 喇嘛教 Lamaism
【佛经】Buddhist Scripture; Buddhist sutra
【佛龛】niche for a statue of Buddha
【佛门】Buddhism ◇ ～弟子 followers of Buddhism; Buddhists
【佛面蛇心】with a Buddha's face and the heart of snake
【佛事】Buddhist ceremony; Buddhist service: 做～ hold a Buddhist service
【佛手】[植] fingered citron; Buddha's-hand
【佛塔】pagoda
【佛堂】family hall for worshipping Buddha
【佛陀】Buddha
【佛像】figure of Buddha; image of Buddha
【佛学】Buddhism
【佛牙】tooth relic of Buddha
【佛眼相看】regard with mercy
【佛要金装,人要衣装】fine feathers make fine birds; the tailor makes the man
【佛爷】Buddha

fǒu

否 ①(否定) negate; deny: ～认 deny ②[书](不同意) nay; no ③(表示询问)先生知其事～? Do you know about that matter, Sir? ④(表示不肯定) whether; if: 我不知道他是～能来。I don't know if he'll be able to come.
【否定】①(否认事物之存在) negate; deny: ～之～ the negation of negation ②(表示否认的) negative: ～的答复 a negative answer; an answer in the negative
【否定裁决】notwithstanding
【否决】vote down; veto; overrule: 这一提案已被～。This proposal has been voted down. ◇ ～权 veto power; veto
【否认】deny; repudiate: ～罪责 deny one's guilt
【否认原告指控的抗辩】negative plea
【否则】[连] otherwise; if not; or else: 要谦虚,～就要落后。 Be modest, otherwise you will lag behind.

fū

夫 ①(丈夫) husband ②(成年男子) man: 匹～ ordinary man / 一～当关,万～莫开。 If one man guards the pass, ten thousand are unable to get through. ③[旧](体力劳动者) a person engaged in manual labor: 船～ boatman / 樵～ woodcutter
【夫唱妇随】harmony between husband and wife
【夫妇】husband and wife
【夫妻】man and wife ◇ ～感情破裂 alienation mutual affection / ～共有财产 estates by the entirety / ～关系 conjugal relation / ～同居权 conjugal right
【夫妻店】small shop run by husband and wife
【夫权】authority of the husband
【夫人】Lady; Madame; Mrs.: 第一～ the first lady / 各国使节和～ foreign diplomatic envoys and their wives
【夫荣妻贵】a woman of low birth may marry into the purple

【夫子】①(学者、老师之旧称) an ancient form of address to a Confucian scholar or to a master by his disciples ②(迂腐之人) pedant: 迂～ a pedantic old fogey
【夫子自道】the master speaks of himself; one speaks of oneself

麸

【麸皮】(wheat) bran
【麸子】(wheat) bran

呋

【呋喃】[化] furan; furfuran; tetrol
【呋喃丹啶】[药] furantoin; furadantin
【呋喃西林】[药] nitrofurazone; furacin
【呋喃树脂】furane resins

肤 skin: 切～之痛 keenly felt pain
【肤泛】superficial; shallow: ～之论 shallow views
【肤觉】[生理] dermal sensation
【肤皮潦草】cursory; casual; perfunctory
【肤浅】superficial; shallow: ～的认识 a superficial understanding of; a superficial knowledge of
【肤轻松】[药] fluocinolone acetonide
【肤色】color of skin

敷 ①(擦;涂) apply (power, ointment, etc.): 外～for external application ②(铺开) spread; lay out: ～设 lay (pipes, etc.) ③(足;够) be sufficient for: 入不～出 unable to make ends meet
【敷金属法】metallization
【敷料】[医] dressing
【敷铝】aluminize
【敷霜】[光] blooming
【敷设】lay: ～电缆 cabling / ～铁轨 lay a railway track
【敷衍】①(表面上应付) be perfunctory; go through the motions ②[书](叙述并发挥) elaborate; expound
【敷衍了事】muddle through one's work
【敷衍塞责】perform one's duty in a perfunctory manner
【敷衍搪塞】explain away; give a lame excuse

孵 hatch; brood; incubate: ～小鸡 hatch chickens
【孵化】hatching; incubation: 人工～ artificial incubation ◇ ～场 hatchery / ～期 incubation period
【孵卵】hatch; brood; incubate ◇ ～鸡 brooding hen / ～器 incubator

跗 instep
【跗骨】[生理] tarsus; tarsal bones: ～炎 tarsitis
【跗面】instep

fú

芙

【芙蓉】[植] ①(木芙蓉) cottonrose hibiscus ②(荷花) lotus
【芙蓉出水】hibiscus rising out of water
【芙蓉其面,蛇蝎其心】she has a fair face, but a foul heart
【芙蓉铀矿】furongite

扶 ①(用手支持) support with the hand; place a hand on sb. or sth. for support: ～着栏杆上楼 walk upstairs with one's hand on the banisters ②(扶起) help sb. up; straighten sth. up ③(扶助) help; relieve: 救死～伤 heal the wounded and rescue the dying
【扶壁】[土] counterfort
【扶病】in spite of illness
【扶持】help sustain; give aid to; help sb. to stand or walk; support
【扶老携幼】holding the old by the arm and the young by the hand; bringing along the old and the young
【扶弱济危】assist the weak and oppressed
【扶弱抑强】support the weak and restrain the powerful
【扶手】①(指栏杆上的横木) handrail; rail; banisters ②(指椅子的扶手) armrest ◇ ～椅 armchair
【扶疏】[书] luxuriant and well-spaced: 枝叶～。The branches and leaves are luxuriant but well-spaced.
【扶梯】staircase
【扶危济困】help the distressed and succor those in peril
【扶养】provide for; foster; bring up: (把孩子)～成人 bring up (a child)
【扶摇直上】soar on the wings of a cyclone; rise steeply; skyrocket
【扶植】foster; prop up: ～社会主义新生事物 foster the new things emerging under the socialist system
【扶助】help; assist; support: ～老弱 help the old and the weak

福 good fortune; blessing; happiness: 造～人类 promote the well-being of mankind; benefit

mankind
【福尔马林】[化] formalin
【福分】share of happiness allotted by destiny
【福气】happy lot; good fortune
【福利】material benefits; well-being; welfare ◇ ～费 welfare funds／～国家 welfare state／～基金 welfare fund／～设施 welfare facilities／～事业 welfare projects
【福如东海,寿比南山】may your fortune be as boundless as the East Sea and may you live a long and happy life
【福寿双全】enjoy both felicity and longevity
【福无双至,祸不单行】joy comes never more than once but sorrows never come singly
【福星】lucky star; mascot
【福星高照】The lucky star is in the ascendant to ride the high tide of good luck.
【福音】① [基督教] Gospel ② (好消息) glad tidings; good news ◇ ～书 the Gospels
【福至心灵】when good luck comes one has good ideas

辐 spoke: 轮～ the spoke of a wheel
【辐散】[气] divergence ◇ ～场 divergence field
【辐射】[物] radiation: 电磁～ electromagnetic radiation／受激～ stimulated radiation／自发～ spontaneous radiation ◇ ～带[天] radiation zone／～计 radiometer／～剂量 radiation dosage／～频率 radiation frequency／～容限 radiotolerance／～体 radiating body／～学 radiology／～育种[农] radioactive breeding／～源 radiant
【辐条】[口] spoke
【辐照】[物] irradiation ◇ ～度 irradiance

幅 ①(布帛等的宽度) width of cloth: 单～床单 single- width bed sheet／双～床单 double-width bed sheet ②(泛指宽度) size: 大～照片 a large-sized photo ③ [量] 一～画 a picture; a painting
【幅度】range; scope; extent: 大～调整物价 readjust the prices by a big margin
【幅面】width of cloth
【幅员】the area of a country's territory; the size of a country
【幅员辽阔】(a country) with a vast expanse; (a country) with a vast territory

蝠 [动] bat
【蝠鲼】devil ray; manta ray

孚 inspire confidence in sb.: 深～众望 enjoy great popularity; enjoy high prestige

浮 ①(停在液体表面) float: 木头～在水上。Wood floats on water.／他脸上～起了笑容。A faint smile played on his face. ② [方](游) swim ③(在表面上的) on the surface; superficial: ～土 dust on the surface ④(暂时的) temporary; provisional: ～支 expenditure not in the regular account ⑤(轻浮) flighty; unstable; superficial ⑥(空虚) hollow; inflated: ～名 bubble reputation ⑦(多余) excessive; surplus: 人～于事 be overstaffed
【浮报】give inflated figures in a report
【浮标】buoy
【浮标灯】floating lamp; floating light
【浮表】float gage
【浮冰】floating ice; (ice) floe
【浮尘】floating dust; surface dust
【浮尘子】[动] leafhopper
【浮沉】now sink, now emerge; drift along: 与世～ follow the trend; swim with the tide
【浮船坞】floating (dry) dock
【浮存】floating account
【浮荡】float in the air
【浮岛】chinampa
【浮雕】relief (sculpture): ～群像 a relief sculpture of a group of people
【浮吊】[机] floating crane
【浮动】①(漂动) float; drift ②(不稳定) be unsteady; fluctuate: 人心～ a general feeling of insecurity ③ [经] float: 货币共同～ a joint currency float ◇ ～工资 floating wage／～汇率 floating (exchange) rate／～留置权 floating lien
【浮动块】rocker piece
【浮动轴】floating axle; floating shaft
【浮而不实】giddy and insincere; an empty show
【浮泛】① [书](漂浮在水面) float about: 轻舟～ a light boat gliding past ②(流露) reveal; display ③(空泛) superficial; too abstract: 她的发言内容～。Her speech was superficial and full of generalities.
【浮光掠影】skimming over the surface; hasty and casual; cursory
【浮华】showy; ostentatious; flashy: ～的生活 a showy and luxurious life／～虚礼 empty show and pretentious ceremony
【浮记】keep a tally of a transaction before entering it in the regular accounts; keep a temporary account

【浮夸】be boastful; exaggerate: ～作风 proneness to boasting and exaggeration
【浮雷】[军] floating mine
【浮力】[物] buoyancy
【浮码头】floating pier
【浮面】surface
【浮皮蹭痒】scratching the surface; superficial
【浮皮儿】①(表皮) epidermis; outer skin ②(表面) surface
【浮萍】[植] duckweed
【浮浅】superficial; shallow
【浮桥】pontoon bridge; floating bridge
【浮生若梦】man's life is like a dream
【浮石】[矿] pumice (stone)
【浮筒】float; pontoon; buoy
【浮屠】[佛教] ①(佛) Buddha ②(和尚) Buddhist monk ③(佛塔) pagoda; stupa: 七级～ seven-stories pagoda
【浮土】dust collected on furniture, etc.; surface dust
【浮文】verbiage; padding
【浮现】appear before one's eyes
【浮现脑海】arise in one's mind; recur to one's mind
【浮现眼前】rise before one's eyes
【浮想】thoughts or recollections flashing across one's mind
【浮想联翩】thoughts thronging one's mind
【浮选】[矿] flotation ◇ ～剂 flotation agent
【浮游动物】zooplankton
【浮游生物】plankton
【浮游植物】phytoplankton
【浮游资金】floating fund
【浮云】floating clouds
【浮云翳日】floating clouds obscure the sun
【浮云朝露】floating clouds and morning dew; life is like an empty dream
【浮躁】impetuous; impulsive
【浮渣】[冶] dross
【浮肿】[医] dropsy; edema
【浮舟】pontoon
【浮子】① [渔] float ② [汽车] carburetor float

蜉

【蜉蝣】[动] mayfly

俘

①(俘获) capture; take prisoner ②(战俘) prisoner of war; captive
【俘获】① capture ② [物] capture: 裂变～ fission capture / 中子～ neutron capture
【俘虏】①(俘获) capture; take prisoner ②(被俘的敌人) captive; captured personnel; prisoner of war (P.O.W.)

伏

①(趴) bend over: ～案读书 bend over one's desk reading ②(卧) lie prostrate: ～地不动 lie still on the ground with one's face downward ③(低下去) subside; go down: 此～彼起 down here, up there ④(隐藏) hide: 昼～夜出 hide by day and come out at night ⑤(伏天) hot season; dog days ⑥(低头承认) admit (defeat or guilt) ⑦(伏特) [电] volt
【伏安】[电] volt-ampere
【伏笔】a hint foreshadowing later developments in a story, essay, etc.; foreshadowing
【伏兵】(troops in) ambush
【伏兵四起】the ambushed soldiers leapt out all round
【伏打】[电] voltaic ◇ ～电池 voltaic cell
【伏地僵卧】lying prostrate on the ground
【伏地请罪】throw oneself on the ground acknowledging one's faults
【伏法】be executed
【伏击】ambush ◇ ～圈 ambush ring
【伏几而卧】fall asleep leaning over the table
【伏流】[地] subterranean drainage; underground stream
【伏暑】hot season; dog days
【伏特】[电] volt ◇ ～计 voltmeter
【伏天】hot summer days; dog days
【伏贴】fit perfectly: 这身衣服穿着很～。This suit fits perfectly.
【伏汛】summer flood
【伏罪】plead guilty (to a crime); admit one's guilt

茯

【茯苓】[中药] fuling (poris cocos)

符

①(符节) a tally issued by a ruler to generals, envoys, etc., as credentials in ancient China ②(标记) symbol: 音～ musical notes ③(符合) tally with; accord with: 数字不～ the figures do not tally / 与事实不～ not tally with the facts ④(符咒) magic figures drawn by Taoist priests to invoke or expel spirits and bring good or ill fortune: 护身～ amulet
【符号】①(记号) symbol; mark; sign: 标点～ punctuation mark / 化学～ chemical symbol / 语音～ phonic symbol / 作～ make a sign ②(标志) insignia ◇ ～学 Semiology; Semiotics / ～义 Signifier; Signified
【符合】①(相合) accord with; tally with; con-

form to; be in keeping with: ～标准 be up to the standard / ～宪法 constitutionality / ～要求 accord with the demands / ～原则 be in conformity with the principle ② [物] coincidence: ～摆 coincidence pendulum

【符合法】coincidence method; method of coincidence

【符合门】coincidence gate

【符山石】vesuvianite

【符咒】Taoist magic figures or incantations

凫 ①(野鸭) wild duck ②(浮) swim: ～水 swim

【凫翁】[动] water cock

服 ①(衣服) clothes; dress: 常～ suit / 礼～ evening suit(男); evening dress(女) / 男～ men's wear / 女～ women's wear ②(服用) take (medicine): 必要时～ taken when necessary / 临睡时～ taken at bedtime / 日～三次 to be taken three times a day ③(承当) serve: ～刑 serve a sentence ④(信服) be convinced; obey: 以理～人 convince people by force of argument ⑤(适应) be accustomed to: 不～水土 not accustomed to the climate; not acclimatized

【服从】obey; submit (oneself) to; be subordinated to: ～命令 obey orders / ～全局利益 be subordinated to overall interest / ～真理 submit to the truth

【服从法律】amenable to law; subject to the law

【服从判决】acceptance of a judgment

【服毒】take poison: ～而死 have died of poison / ～自杀 poison oneself

【服法】[药] instructions about how to take medicine

【服气】be convinced

【服丧】be in mourning (for the death of a kinsman, etc.)

【服饰】dress and personal adornment; dress: ～华丽 elegantly attired; in elegant attire

【服侍】wait upon; attend: ～病人 attend the sick

【服输】admit defeat; acknowledge defeat

【服帖】①(顺从) docile; obedient; submissive ②(信服) be convinced: 心里很～ be positively convinced ③(妥当) fitting; well arranged

【服务】give service to; be in the service of; serve: ～周到 provide good service / 提高～质量 improve one's service / 为人民～ Serve the people ◇ ～费 service charge / ～行业 service trades; service industries / ～事业 service business / ～台 service desk / ～态度 attitude in attending to customers; attitude in waiting on guests / ～员 attendant; waiter / ～站 neighborhood service center

【服刑】serve a sentence: ～期满 complete a term of imprisonment

【服役】①(服兵役) be on active service; enlist in the army: ～期间 during one's term of military service; during the period of enlistment / ～期满 complete one's term of service ②(服苦役) do corvée labor

【服用】take (medicine)

【服装】dress; clothing; costume: 现成～ ready-made clothes ◇ ～表演 fashion show / ～厂 clothing factory / ～革 clothing leather / ～商店 clothes shop; clothing store / ～设计 dress designing; costume designing ○ 尺寸 size / 大号 large size / 中号 middle size / 小号 small size / 长度 length / 肩宽 shoulder width / 胸围 chest measurement / 腰围 waist / 臀围 buttocks / 下摆 lower hem of a gown, jacket or shirt / 领子 collar / 袖子 sleeve / 衣袋 pocket / 里子 lining / 活里子 detachable lining; detachable (zip-in) lining / 试样 fitting; try on

【服罪】plead guilty; admit one's guilt; admit that one is guilty: 不～ plead innocent; plead not guilty

弗 [书] not: 自愧～如 feel ashamed of one's inferiority

拂 ①(轻轻擦过) stroke: 春风～面 a spring breeze stroking the face ②(拂去) whisk; flick: ～去桌上的尘土 whisk the dust off a desk ③(违背) go against (sb.'s wishes)

【拂尘】horsetail whisk

【拂拭】whisk or wipe off

【拂晓】before dawn

【拂袖而去】leave with a flick of one's sleeve; go off in a huff

氟 [化] fluorine (F)

【氟化氢】[化] hydrogen-fluoride

【氟化物】[化] fluoride

【氟利昂】[化] freon

【氟美松】dexamethasone

【氟塑料】fluoroplastics

fǔ

府 ①(政府;行政区域) prefecture; seat of government; government office: 首～ capital ②

(府第) official residence; mansion: 总统～ presidential palace ③[敬](住宅) your home: 贵～ your home
【府绸】poplin
【府邸】mansion; mansion house
【府第】mansion; mansion house
【府上】[敬] ①(指住处) your home; your family ②(指老家) your native place
腐 ①(腐败) rotten; putrid; stale; corroded: ～肉 rotten meat／去～生新 remove the rotten and let fresh grow ②(豆腐) bean curd
【腐败】①(腐烂) rotten; putrid; decayed: 吃了～的食物容易生病 eat putrid food is liable to get sick ②(陈旧;堕落) corrupt; rotten: ～透顶 rotten to the core／～无能 corrupt and incompetent
【腐败堕落】become corrupt and degenerate
【腐臭】stinking; rancid
【腐化】①(思想行为变坏) degenerate; corrupt; dissolute; depraved: ～堕落 morally degenerate／生活～ lead a dissolute life ②(堕落) rot; decay ◇ ～分子 corrupt element; corruptionist／～罪 crime of corruption
【腐烂】①(某物败坏) decomposed; putrid ②(腐败) corrupt; rotten
【腐烂病】[植] rot
【腐泥煤】sapropelic coal
【腐乳】fermented bean curd
【腐生】[生] saprophytic: ～细菌 saprophytic bacteria／～植物 saprophyte
【腐蚀】①(消损破坏) corrode; etch: 金属的～ corrosion of metal ②(思想行为变质堕落) corrupt; corrode: 黄色电影～人们的灵魂 blue cinema corrupts the souls of people ◇ ～板[印] etched plate／～机[印] etching machine／～剂[化] corrosive; corrodent／～性 corrosiveness
【腐熟】[农] (of compost, etc.) become thoroughly decomposed
【腐朽】①(木料等损破) rotten; decayed: 这些木材已经～了。 The timber has rotted. ②(陈腐; 堕落; 败坏) decadent; degenerate: ～糜烂的生活 decadent and dissolute life
【腐朽没落】rotten to the core and declining; decadent and declining
【腐殖煤】humic coal
【腐殖泥】sapropel
【腐殖酸】[农] humic acid: ～类肥料 humic acid fertilizers
【腐殖土】[农] humus soil
【腐殖质】[地] humus
【腐竹】dried bean milk cream in tight rolls
俯 ①(低下) bow (one's head); ～视 overlook ②[敬](旧时公文书信中用来称对方的动作) ～就 condescend to take the post
【俯冲】[航空] dive ◇ ～轰炸机 dive bomber／～角 dive angle
【俯伏】lie prostrate
【俯角】[测] angle of depression
【俯瞰】look down at; overlook: 从山上～全城 overlook the whole town from a hill ◇ ～摄影 crane shot
【俯就人意】bend one's will
【俯射】plunging fire
【俯拾即是】can be found everywhere; be extremely common
【俯视】look down at; overlook ◇ ～图[机] vertical view
【俯首】bow one's head (in submission): ～帖耳 be docile and obedient; be all obedience; be servile／～听命 obey submissively
【俯首就范】obey; submit (oneself) to the law
【俯首受辱】docilely submit to sb.'s abuse
【俯卧撑】[体] push-up
【俯仰】a bending or lifting of the head; a simple move or action: 随人～ be at sb.'s back and call ◇ ～角 [航空] angle of pitch／～运动 [机] pitching movement
【俯仰由人】be at sb.'s back and call
【俯仰之间】in the twinkling of an eye; in an instant; in a flash
【俯泳】[体] breaststroke
拊 [书] clap
【拊背扼喉】occupy a strategic position
【拊掌】[书]clap hands
辅 assist; complement; supplement: 相～相成 complement each other
【辅币】fractional currency; fractional money: 硬～ subsidiary coin; minor coin
【辅车相依】as dependent on each other as the jowls and the jawbone; as close as the jowls and the jaws
【辅导】give guidance in study or training; coach ◇ ～报告 guidance lecture (supplementary lecture on background, study method, etc.)／～材料 guidance material
【辅导员】(political and ideological) assistant; instructor: 理论～ instructor in political theory／少先队～ Young Pioneers counsel-

lor / 校外～ after-school activities counsellor
【辅国安民】serve the state and pacify the people; protect the people and bring peace to the state
【辅课】subsidiary course
【辅酶】coenzyme
【辅音】[语] consonant
【辅助】①(从旁帮助) assist ②(非主要的) supplementary; auxiliary; subsidiary ◇ ～车间 auxiliary shop / ～惩罚措施 auxiliary penal measure / ～读物 supplementary reading material / ～工 auxiliary worker / ～机构 auxiliary body / ～舰船 auxiliary vessels / ～劳动 auxiliary labor; auxiliary jobs / ～人员 auxiliary staff members / ～授粉[农] supplementary pollination / ～仪器 supplementary instrument
【辅佐】assist a ruler in governing a country

脯

①(肉干) dried meat: 鹿～ dried venison ②(蜜饯果干) preserved fruit: 桃～ preserved peaches / 杏～ preserved apricot

抚

①(安慰;慰问) comfort; console: 安～ placate; apease ②(保养) nurture; foster: ～养 foster; raise; bring up ③(轻轻按着) stroke: ～琴 play the zither
【抚爱】caress; fondle
【抚躬自问】examine one's own conscience; hold communion with oneself
【抚今追昔】recall the past and compare it with the present; reflect on the past in the light of the present
【抚摩】stroke
【抚弄】stroke; fondle
【抚慰】comfort; console; soothe: ～死难家属 console the bereaved family
【抚恤】comfort and compensate a bereaved family ◇ ～金 pension for the disabled or for the family of the deceased
【抚养】foster; raise; bring up
【抚养费】cost of maintenance; cost of upbringing
【抚养义务】duty to rear; duty to provide support
【抚育】foster; nurture; tend: ～烈士子女 bring up the children of revolutionary martyrs
【抚掌大笑】laugh loud and clap one's hands

斧

axe; hatchet
【斧石】axinite; glass schorl
【斧头】axe; hatchet
【斧正】(please) make corrections
【斧子】axe; hatchet

釜

a kind of caldron used in ancient China
【釜底抽薪】take away the firewood from under the caldron; take a drastic measure to deal with a situation
【釜底游鱼】a fish swimming in the bottom of a caldron; a person whose fate is sealed
【釜顶蒸气】top steam
【釜脚】stillage bottoms

fù

傅

①(教导) teach; instruct ②(教导或传授技艺的人) teacher; instructor: 师～ master (worker); teacher (称呼) ③(加上) lay on; apply: ～彩 lay on colors / ～粉 put powder on; powder (the face, etc.)
【傅科摆】foucault pendulum
【傅科法】foucault method

缚

tie up; bind fast
【缚鸡之力】strength for binding a chicken
【缚手缚脚】tie sb.s hand and foot

富

rich; wealthy; abundant: ～于创造性 be highly creative / ～于养分 be rich in nutrition
【富国利民】enrich the state and bring benefits to the people; enrich the country and benefit the people
【富国强兵】make one's country rich and build up its military power
【富贵】riches and honor; wealth and rank
【富贵不能淫】neither riches nor honors can lead one astray; be impervious to the temptation of wealth and high position
【富贵浮云】regard honor and riches as floating clouds
【富贵荣华】riches; honor and splendor
【富贵勿易】no riches or honors would move someone
【富贵易同心,患难见真情】prosperity makes friends and adversity tries them
【富豪】rich and powerful people
【富矿】[矿] rich ore; high-grade ore ◇ ～脉 bonanza; pay streak / ～体 ore shoot
【富丽堂皇】sumptuous; gorgeous; splendid
【富农】rich peasant
【富强】prosperous and strong
【富饶】richly endowed; fertile; abundant: 美丽～的国家 a beautiful and richly endowed coun-

try
【富庶】rich and populous
【富翁】man of wealth
【富有】①(拥有大量财产) rich; wealthy ②(大量具有) rich in; full of: ～经验 rich in experience; very experienced / ～朝气的青年 young people full of vigor
【富裕】prosperous; well-to-do; well-off
【富余】have more than needed; have enough and to spare: 粮食有～ have a surplus of grain
【富源】natural resources
【富足】plentiful; abundant; rich

副

①(居第二位的) deputy; assistant; vice: ～博士 candidate doctor / ～部长 vice-minister / ～国务卿 undersecretary of state / ～驾驶员 co-pilot; second pilot / ～教授 associate professor / ～经理 assistant manager / ～领事 vice-consul / ～秘书长 deputy secretary general / ～司令员 assistant commanding officer / ～主席 vice-chairman / ～总理 vice-premier ②(辅助的) auxiliary; subsidiary; secondary: ～泵 auxiliary pump ③(符合) correspond to; fit: 名不～实。The name falls short of the reality. ④[量]一～手套 a pair of gloves / 一～眼镜 a pair of glasses / 装出一～笑脸 assume a smiling face
【副本】duplicate; transcript; copy
【副标题】subheading; subtitle
【副产品】by-product
【副赤道带】[气] subequatorial belt
【副词】[语] adverb
【副歌】[乐] refrain
【副官】adjutant; aide-de-camp
【副交感神经】[生理] parasympathetic nerve
【副井】[矿] auxiliary shaft
【副刊】supplement: 文学～ literary supplement
【副品】substandard goods
【副热带】subtropic
【副伤寒】[医] paratyphoid (fever)
【副神经】[生理] accessory nerve
【副食】non-staple food; non-staple foodstuffs ◇ ～加工厂 non-staple food processing factory / ～商店 grocery; grocer's
【副手】assistant
【副署】countersign: ～提单 countersigned bill of lading
【副业】sideline; side occupation
【副翼】[航空] aileron
【副油箱】[航空] auxiliary tank; drop tank
【副职】the position of a deputy to the chief of an office, department, etc.
【副轴】[机] countershaft; layshaft
【副作用】①(附带发生的不好作用) side effect; by-effect: 麻醉药物的～ side effects from the use of anesthetics / 没有～ be free from side effects ②[机] secondary action

讣

obituary
【讣告】①(报丧) announce sb.'s death ②(报丧的通知) obituary (notice)
【讣闻】obituary (notice)

赴

go to; attend: ～考 go and sit for an examination / ～宴 attend a banquet / ～英留学 go to study in England / ～约 go to fulfill an appointment
【赴难】go to the aid of one's country; go to help save the country from danger
【赴任】go to one's post; be on the way to one's post
【赴汤蹈火】go through fire and water

赋

①(交给) bestow on; endow with; vest with: ～予新的意义 give a new meaning ②(税) tax: 田～ land tax ③(我国古代文体) descriptive prose interspersed with verse ④(作诗、词) compose (a poem): ～诗一首 compose a poem
【赋格曲】[乐] fugue
【赋税】taxes
【赋闲】(of an official, etc.) be unemployed
【赋形剂】[药] excipient
【赋性】inborn nature
【赋役】taxes and corvée
【赋有】possess (naturally); be gifted (with): ～才能 possess talents; be gifted with talents

父

①(父亲) father ②(家族中的长辈男子) male relative of a senior generation: 祖～ grandfather
【父辈】elder generation
【父本】[植] male parent: ～植株 paternal plant
【父老】elders (of a country or district): ～兄弟 elders and brethren
【父母】father and mother; parents: ～见背 one's parents have passed away / ～健在 both parents are in good health ◇ ～形象 [心理] parent figure
【父母之命,媒妁之言】arrange a match by parents' order and on the matchmaker's word
【父亲】father
【父权制】patriarchy

【父系】paternal line; the father's side of the family ◇ ～亲属 relatives on the paternal side
【父兄】①(父亲和哥哥) father and elder brothers ②(泛指家长) head of a family
【父严母慈】the father is severe, the mother is indulgent
【父债子还】the son must pay his father's debts
【父子关系】set membership
复 ①(重复) compound; complex: ～光谱 complex spectrum / ～姓 compound surname; two-character surname ②(转过来,转过去) turn round; turn over: 翻来～去睡不着 toss in bed, unable to sleep ③(回答) answer; reply: 请即电～ please reply by wire immediately ④(恢复) recover; resume: ～职 resume one's post ⑤(报复) revenge ⑥(再;又) again: 一去不～返 gone never to return
【复背斜】[地] anticlinorium
【复本】duplicate
【复本位制】[经] bimetallism
【复辟】①(失位君主复位) restoration of a dethroned monarch: 阴谋～ plot to stage a comeback ②(恢复旧制度) restoration of the old order: ～活动 restorationist activities
【复波】[物] complex wave
【复查】check; reexamine: ～帐目 check the accounts / 一个星期后到医院～ come back to the hospital for a check in a week's time
【复仇】revenge; avenge: ～心理 vindictiveness; a desire for revenge ◇ ～主义 revanchism
【复电】telegram in reply (to one received)
【复调音乐】polyphony
【复发】have a relapse; recur: 旧病～ have an attack of an old illness; have a relapse
【复方】[药] medicine made of two or more ingredients; compound ◇ ～阿斯匹林 aspirin compound (APC) / ～甘草合剂 brown mixture
【复分解反应】[化] double decomposition reaction
【复根】[化] compound radical
【复工】return to work (after a strike or layoff)
【复古】restore ancient ways; return to the ancients ◇ ～思想 back-to-the-ancients ideology / ～主义 the doctrine of "back to the ancients"
【复轨器】rerailer; retracker; car replacer
【复合】compound; complex; composite ◇ ～词 compound (word) / ～电路 compound circuit / ～肥料 compound fertilizer / ～句 compound or complex sentence / ～量词 compound classifier / ～元音 compound vowel
【复核】①(核对) check: 把数字～一下 check the figures ②[法](特定司法程序) (of the Supreme People's Court) review a case in which a death sentence has been passed by a lower court
【复婚】restoration of a marriage; resume matrimonial relation
【复活】①(死而复活) bring back to life; revive: ～军国主义 revival of militarism ②[基督教](耶稣复活日) Resurrection
【复活节】Easter ◇ ～彩蛋 Easter egg
【复激】[电] compound excitation ◇ ～发电机 compound generator
【复交】[外] reestablish diplomatic relations
【复旧】restoration of old ways; return to the past
【复旧倒退】restore the old and work for retrogression
【复句】[语] a sentence of two or more clauses
【复卷机】[纸] rewinding machine; rewinder
【复刊】resume publication
【复利】[经] compound interest
【复命】report on completion of a task
【复捻】[纺] second twist
【复赛】[体] intermediary heat
【复审】①(再一次审查) reexamine ②[法](再一次审理) review a case ◇ ～请求书 [法] petition for rehearing
【复式】[簿记] double entry
【复式车床】[机] double lathe
【复述】①(重说一遍) repeat: ～命令 repeat an order ②(语言教学用语) retell (in language learning): 把故事～一遍 retell a story
【复数】①(两个以上)[语] plural (number) ②[数] complex number
【复丝】[化纤] multifilament
【复苏】①(苏醒) come back to life or consciousness; resuscitate ②(经济萧条后生产逐渐恢复) recovery: 经济～ economic recovery
【复蹈故辙】follow in the old track; return to the old rut
【复位术】[医] reduction
【复习】review; revise: ～提纲 outline for review
【复息】compound interest
【复线】[交] multiple track: ～铁路 double-tracking railways
【复向斜】[地] synclinorium
【复写】make carbon copies; duplicate ◇ ～纸

carbon paper
【复信】①(答复来信) write a letter in reply ②(回信) letter in reply; reply
【复兴】revive; resurge; rejuvenate: 文艺～ the Renaissance
【复姓】compound surname; two-character surname
【复修旧好】restore the former intimacy; become reconciled
【复学】go back to school (after prolonged absence for health reasons, etc.)
【复盐】[化] double salt
【复眼】[动] compound eye
【复业】①(恢复旧业) resume one's occupation; reestablish one's business ②(商店恢复营业) reopen
【复叶】[植] compound leaf
【复议】reconsider (a decision)
【复音】[物] complex tone
【复音词】[语] disyllabic or polysyllabic word
【复印】[印] duplicate ◇～机 duplicator; duplicating machine／～纸 duplicating paper
【复员】demobilize ◇～费 demobilization pay／～军人 demobilized soldier; ex-serviceman
【复原】①(病后康复) recover from an illness; be restored to health ②(恢复原状) restore; rehabilitate
【复圆】[天] (of an eclipse) fourth contact; last contact
【复杂】complicated; complex: ～的局面 complicated situation／～的心情 mixed feelings
【复照】[外] a note in reply
【复诊】further consultation (with a doctor); subsequent visit
【复职】resume one's post; be reinstated
【复制】duplicate; reproduce; make a copy of ◇～模型 reconstructed model／～片 duplicated film; copy of a film／～品 replica; reproduction
【复种】[农] multiple cropping ◇～面积 multiple cropping area／～指数 multiple crop index

覆 [书] ①(遮盖) cover ②(翻过来;歪倒) overturn; upset: ～舟 capsized boat
【覆巢无完卵】when the nest is overturned no egg stays unbroken; in a great disaster no one can escape unscathed
【覆盖】①(遮盖) cover ②(地面植物的保护作用) plant cover; vegetation ◇～层[地] overburden
【覆灭】destruction; complete collapse
【覆没】①(翻而沉没) capsize and sink ②(被消灭) be overwhelmed; be annihilated: 敌人全军～。The enemy's whole army was destroyed.
【覆盆之冤】dark injustice; irredeemable wrong
【覆盆子】[中药] Korean raspberry
【覆水难收】spilt water can't be gathered up; what is done can't be undone
【覆亡】fall (of an empire, nation, etc.)
【覆辙】the track of an overturned cart: 重蹈～ take the same disastrous road

蝮
【蝮蛇】Pallas pit viper

馥 [书] fragrance
【馥郁】[书] strong fragrance; heavy perfume: ～的花香 the strong scent of flowers

腹 belly; abdomen; stomach
【腹背受敌】be attacked front and rear
【腹藏机谋】one's breast conceals tactics
【腹藏奸谋】harbor a sinister design
【腹带】binder
【腹地】hinterland
【腹诽】[书] unspoken criticism
【腹稿】a draft worked out in one's mind; mental notes
【腹股沟】[生理] groin: ～疝 inguinal hernia
【腹肌】abdominal muscle
【腹鸣】borborygmus
【腹膜】[生理] peritonaeum ◇～炎 peritonitis
【腹鳍】[动] ventral fin
【腹腔】[生理] abdominal cavity ◇～穿刺术 abdominocentesis／～动脉 arteria coeliaca／～镜[医] peritoneoscope／～妊娠 abdominal gestation
【腹水】[医] ascites: 抽～ tap the abdomen
【腹痛】abdominal pain: ～如绞 have an excruciating pain in the belly
【腹泻】diarrhea
【腹胀】abdominal distension: ～如鼓 a well-to-do person; one's belly is tight as a drum; well fed and content
【腹足】[动] abdominal foot; proleg

鲋 crucian carp

付 ①(交给) hand over to; turn over to; commit to: ～表决 put to vote／～诸实施 put into effect ②(支付) pay: ～税 pay taxes／逾期未～ fail to pay on time

【付出】pay; expend: ～代价 pay a price
【付定金】pay earnest money
【付方】credit side; credit
【付款】pay a sum of money: 分期～ pay by instalments / 货到～ cash on delivery (C.O.D.); delivery against payment / 凭单～ cash against documents ◇ ～办法 methods of payment / ～保证银行 certifying bank / ～方式 type of payment / ～凭单 paying certificate / ～凭证 payment voucher / ～人 payer; drawee / ～条件 terms of payment / ～条例另议 terms of payment are to be arranged / ～通知 payment order / ～银行 paying bank
【付排】send to the compositor
【付讫】(of a bill) paid
【付清】pay in full; pay off; clear (a bill): 一次～ pay off in one lump sum
【付托】put sth. in sb.'s charge; entrust
【付托得人】have entrusted the matter to the right person
【付息】payment of interest
【付现】pay in cash; cash
【付印】①(准备出版) send to the press ②(交付印刷) turn over to the printing shop (after proofreading) ◇ ～样 pass sheet
【付邮】take to the post; post
【付帐】pay a bill
【付之一炬】commit to the flames
【付之一笑】laugh away; dismiss with a laugh
【付诸东流】thrown into the eastward flowing stream; all one's efforts wasted; irrevocably lost
【付诸公判】be submitted to public trial
【付诸行动】put into practice

附 ①(附带) add; attach; enclose: ～表 attached list or chart ②(靠近) get close to; be near ③(依从) agree to: ～议 second a motion
【附笔】postscript; additional note
【附带】①(顺便) in passing: ～说一下 by the way ②(附加) attach: 不～任何条件的援助 aid with no strings attached ③(非主要的) subsidiary; supplementary: 从事～劳动 do supplementary labor
【附带民事诉讼】incidental civil
【附带判决】incidental judgement
【附带上诉】incidental appeal
【附带原因】contributory cause; inherent cause
【附敌叛国】betray one's country and attach oneself to the enemy
【附点】[乐] dot ◇ ～音符 dotted note
【附耳】move close to sb.'s ear: ～低语 whisper in sb.'s ear
【附睾】[生理] epididymis ◇ ～炎 epididymitis
【附和】echo; chime in with: 随声～ chime in with others
【附会】draw wrong conclusions by false analogy; strain one's interpretation: 穿凿～ give strained interpretations and draw farfetched analogies / 牵强～的结论 far-fetched conclusion
【附加】①(额外加上) add; attach ②(附带的) additional; attached; appended ◇ ～费 extra charges; surcharge / ～税 surtax; additional tax; supertax / ～条款 additional article; memorandum clause / ～文件 [外] appended document / ～险 accessory risk / ～刑 accessory punishment / ～议定书 [外] additional protocol / ～指控 accessory charges
【附件】①(文件的) appendix; annex ②(书信中的) enclosure ③[机] accessories; attachment: 车床～ lathe accessories
【附近】①(靠近) nearby; neighboring: ～地区 nearby regions ②(附近的地方) close to; in the vicinity of: 住在工厂～ live close to the factory
【附录】appendix
【附上】enclosed herewith
【附设】have as an attached institution: 这个商店～了一个早晚服务部。This store has set up an after-hours department.
【附生植物】epiphyte
【附属】subsidiary; auxiliary; attached; affiliated: 医学院～医院 a hospital attached to a medical college ◇ ～国 dependency / ～机构 subsidiary body / ～品 accessory; appendage / ～中学 affiliated middle school
【附图】attached map or drawing; figure: 见～一。See Figure 1.
【附言】postscript (P.S.)
【附议】second a motion; support a proposal
【附庸】dependency; vassal; appendage
【附庸风雅】(of landlords, merchants, etc.) mingle with men of letters and pose as a lover of culture
【附则】supplementary articles (appended to a treaty, decree, etc.)
【附注】notes appended to a book, etc; annotations
【附赘悬疣】goiters — something redundant and not needed; an appendix wen and a hanging tumor — useless
【附着】adhere to; stick to ◇ ～力 [物] adhesive force; adhesion

【附子】[中药] monkshood

驸

【驸马】emperor's son-in-law

阜 [书] ①(土山) mound ②(多) abundant: 物～民丰。 Products abound and the people live in plenty.

服 [量](用于中药) dose: 一～药 a dose of medicine

负 ①(背) carry on the back or shoulder; shoulder; bear: ～经济责任 held economically responsible／ 如释重～ feel as if relieved of a heavy load; feel greatly relieved／ 身～重任 shoulder an important task ②(依仗) have at one's back; rely on: ～险固守 put up a stubborn defense by relying on one's strategic position ③(遭受) suffer: ～伤 get wounded ④(享有) enjoy: 久～盛名 have long enjoyed a good reputation ⑤(亏欠) owe: ～债 be in debt ⑥(背弃) fail in one's duty, obligation, etc.; betray: 忘恩～义 be ungrateful ⑦(失败) lose (a battle, game, etc.);be defeated: 不分胜～ end in a draw; end in a tie; break even ⑧ [数](小于零的) minus; negative: ～号 negative sign ⑨ [电] negative

【负担】①(承当) bear (a burden); shoulder ②(承受的责任) burden; load; encumbrance: 财政～ financial burden／ 家庭～ family burden (esp. financial)／ 减轻学生～ lighten the students' load／ 精神～ mental burden; load on one's mind

【负电荷】negative (electric) charge

【负电极】negative electrode; cathode

【负反馈】negative feedback

【负号】negative sign

【负荷】[电] load: 安全～ safe load／ 超～ overload／ 最大～ peak load

【负极】[电] negative pole

【负加速度】[物] negative acceleration

【负荆请罪】proffer a birch and ask for a flogging; offer a humble apology

【负疚】[书] feel apologetic; have a guilty conscience

【负咎引退】bear the consequence (of ...) and tender one's resignation

【负离子】[物] anion

【负片】[摄] negative

【负气】do sth. in a fit of pique: ～而去 leave angrily out of spite

【负伤】be wounded; be injured: 光荣～ be wounded in action

【负数】[数] negative number

【负像】[物] negative image

【负隅顽抗】(of an enemy or a robber) fight stubbornly with one's back to the wall; put up a desperate struggle

【负约】break a promise; go back on one's word

【负载】[电] load: 高峰～ peak load／ 工作～ operating load ◇ ～损失 load loss／ ～调整 load regulation／ ～效率 load efficiency／ ～作用 loading effect

【负责】①(担负责任) be responsible for; be in charge of: 对旅客的安全～ be responsible for the safety of the passengers ②(认真踏实) conscientious: 对工作很～ be very conscientious in one's work ◇ ～干部 cadre in charge／ ～人 person in charge

【负债】①(欠人钱财) be in debt; incur debts: ～累累 be heavily in debt; be up to one's eyes in debt ②(资产负债表的一方) libialities: 长期～ long-term liabilities／ 流动～ current liabilities

【负债额】amount of obligation; liabilities

【负重】bear a heavy burden: ～竞走 loaded footrace

【负重轮】bogie wheel; loading wheel

【负重致远】bear a heavy burden and cover a long distance; be able to shoulder important tasks

妇 ①(妇女) woman: ～孺 women and children ①(已婚女子) married woman: 少～ young married woman ③(妻) wife

【妇产科】(department of) gynecology and obstetrics

【妇产医院】a hospital for gynecology and obstetrics

【妇道人家】the fair sex; the womenfolk

【妇科】(department of) gynecology ◇ ～医生 gynecologist

【妇联】[简](妇女联合会) the Women's Federation

【妇女】woman ◇ ～病 gynecological disease／ ～节 Women's Day／ ～解放 emancipation of women／ ～权利 women's rights／ ～用品商店 ladies' goods store

【妇人】married woman

【妇人之仁】woman's soft nature

【妇孺皆知】even woman and children all know; It is known even to women and children

【妇幼】 women and children ◇ ～保健站 health center for women and children; maternity and child care center / ～卫生 maternity and child hygiene

G

gā

夹

【夹肢窝】armpit

嘎

【嘎巴】[方]①(沾的东西干后附在器物上) form into a crust; crust: 瞧,浆糊都~在你袖子上了。Look, the paste has crusted on your sleeve. ②(附在器物上的稀物) crust: 粥~儿 porridge crust

【嘎嘎】[象]quack

【嘎然长鸣】give a long and loud cry

【嘎吱】[象] creak

旮

【旮旯儿】[方] ①(所有的角落) nook; corner: 旮旮旯旯儿都打扫干净了。 Every nook and cranny has been swept clean. ②(狭窄偏僻处) out-of-the-way place: 山~ a mountain recess

伽

【伽马】gamma ◇ ~射线[物] gamma ray

咖

【咖喱】curry: ~牛肉 beef curry ◇ ~粉 curry powder

gá

钆 [化] gadolinium (Gd)

gāi

该 ①(应当) ought to; should: ~插秧了。It's time for transplanting. / 我们~走了。It's time we were leaving. ②(应当轮到) be sb.'s turn to do sth.: 下一个~谁发言? Who's the next speaker? / 这一回~我了吧? It's my turn now, isn't it? ③ (理应如此) deserve: 她~受到表扬。 She deserves to be commended ④(推断出必然或可能的结果) most likely; probably; ought to; should: 我们再走一小时就~到了。We ought to be able to get there in another hour. ⑤(用在感叹句中加强语气)要是水泵今天就运到,~多好哇! If only the pump could arrive today! ⑥(欠) owe: 我不~他钱。 I don't owe him any money. ⑦(多用于公文) this; that; the said; the above-mentioned: ~厂 this factory; the said factory / ~项工作 the job in question / ~校 that school

【该当】①(应受) deserve: ~何罪? What punishment do you think you deserve? ②(应当) should: 这是大家的事,我们~尽力。It's for everybody and we should do our best.

【该死】[口] ~的天气! What wretched weather!

【该帐】be in debt

赅 [书] complete; full: 言简意~ terse but comprehensive

gǎi

改 ①(改变) change; transform: ~洼地为鱼塘 transform waterlogged land into fish pond ②(修改) alter; revise: ~灶节煤 make alterations in an oven so that it will burn less coal ③(改正) correct; rectify; put right: ~作业 correct students' homework or papers ④(后接动词) switch over to (doing sth. else): ~用良种 begin to use improved varieties / 他现在~踢左后卫。 He's playing left-back now.

【改版】[印] correcting

【改编】①(根据原著重写) adapt; rearrange; revise: 这支歌已~成大合唱。 The music of the song has been rearranged for the cantata. ②(改变原编制) reorganize; redesignate: ~军队 reorganize an army / 把部队~成两个团 reorganize the troops into two regiments

【改变】change; alter; transform: ~国籍 change of nationality / ~航线 alter the course / ~主意 change one's mind

【改变符号】reindexing

【改朝换代】change of dynasty or regime; dynastic changes

【改道】①(改变旅行路线) change one's route ②(河流改变路线) change its course: 历史上,黄河曾经多次~。 The Huanghe River has changed its course many times over the centuries.

【改掉】give up; drop: ~坏习惯 give up bad habits

【改订】reformulate; rewrite: ~规章制度 draw up new rules and regulations / ~价格 revise of price / ~税则 revise the tariff

【改动】change; alter; modify: 文字上作少许~ make a few changes in wording

【改恶从善】abandon evil and do good; turn over a new leaf; mend one's ways

【改革】reform: ~不合理的规章制度 reform irrational rules and regulations / 工具~ improvement of tools / 民主~ democratic reforms / 文字~ reform of a writing system

【改观】change the appearance; get a new look; take on a new look
【改过】mend one's ways; correct one's mistakes: ～自新 correct one's errors and make a fresh start; mend one's ways; turn over a new leaf
【改行】change one's profession
【改换】change over to; change: ～包装 repacking / ～名称 rename / ～日期 change the date
【改悔】repent: 毫无～之意 show not the least sign of repentance; absolutely unrepentant
【改嫁】(of a woman) remarry
【改建】reconstruct; rebuild
【改进】improve; make better: ～耕作方法 improve farming methods / ～工作作风 improve one's work style
【改口】withdraw or modify one's previous remark; correct oneself, change ene's tone
【改良】①(去掉缺点,使之更适合要求) improve; ameliorate: ～家畜品种 improve the breed of domestic animals / ～土壤 improve the soil ②(改善) reform ◇ ～派 reformists
【改良主义】reformism ◇ ～者 reformist
【改名换姓】change one's name (and surname); disguise oneself under a alias
【改判】[法] change the original sentence; commute; amend a judgment: 由死刑～无期徒刑 commute the death sentence to life imprisonment
【改平】level off
【改平位置】level-off position
【改期】change the date: 会议～了。The date for the meeting has been changed.
【改任】change to another post
【改日】another day; some other day: 咱们～再商量吧! Let's talk it over another day.
【改容易色】one's face changed its hue
【改善】improve; ameliorate: ～伙食 improve the food / ～劳动条件 improve working conditions / 两国关系有所～。The relations between the two countries have shown some improvement.
【改天换地】transform heaven and earth; change the world; remake nature
【改头换面】[贬] change the appearance; dish up in a new form
【改弦更张】change over to new ways; make a fresh start
【改弦易辙】change one's course; strike out on a new path
【改邪归正】give up evil and return to good; turn over a new leaf
【改写】rewrite; adapt: 经过～,文章生动多了。Rewriting has livened up the article. ◇ ～本 adaptation
【改型】[工] retrofit
【改型工具】modification kit
【改性】[化] modified ◇ ～剂 modifier / ～树脂 modified resin
【改选】reelect: 每四年～一次 hold elections one every four years.
【改元】change the designation of an imperial reign; change the title of a reign
【改造】transform; reform; remold; remake: ～老企业 transforming outmoded enterprises / ～盐碱地 transform saline-alkali land / ～自然 remake nature
【改正】correct; amend; put right: ～错案 redress a misjudged case / ～错误 correct one's mistakes
【改正片】[光] corrector plate; correcting plate
【改正行为】correcting behavior
【改装】①(改变服装) change one's costume or dress ②(改变色装) repackage; repack ③(改变原装置) reequip; refit: ～一辆卡车 refit a truck ◇ ～费用 repacking charge; recondition expense
【改锥】screwdriver
【改组】reorganize; reshuffle: ～管理机构 reorganize the management / ～内阁 reshuffle the cabinet

gài

盖 ①(盖子) lid; cover: 茶壶～ teapot lid / 引擎～ bonnet of an engine / 轴承～[机] bearing cap ②(甲壳) shell (of a tortoise, crab, etc.) ③(伞状盖) canopy: 亭亭如～ (of a tree) stand towering with a canopy of leaves ④(覆盖) cover: 用塑料薄膜～住秧苗 cover the seedlings with plastic sheeting ⑤(打上) affix (a seal) ⑥(超过;压倒) surpass; top: 他的跳远成绩～过了所有的选手。He excelled all the other contestants in the long jump. ⑦(建筑) build: ～房 build a house ⑧[书](大概) approximately; about; around: 与会者～二千人。About two thousand people attended the meeting. ⑨[书](原因) for; because; in fact: 有所不知,～未学也。If there are things we do not know, it is because we haven't learnt them.
【盖板】blind flange; butt plate; spear plate: ～接合 concealed joint / ～接头 strap joint
【盖菜】[植] leaf mustard

【盖棺论定】final judgment can be passed on a person only when the lid is laid on his coffin
【盖浇饭】rice served with meat and vegetables on top
【盖然性】[逻] probability
【盖世】unparalleled; matchless; peerless: ～无双 unparalleled anywhere in the world / ～英雄 peerless hero
【盖世太保】Gestapo
【盖章】affix one's seal; seal; stamp: 由本人签字～ to be signed and sealed by the recipient or applicant
【盖子】①(器物上部遮蔽之物) lid; cover; cap; top: ②(甲壳) shell (of a tortoise, etc.)

丐 [书] ①(乞求) beg ②(乞丐) beggar

芥

【芥菜】[植] leaf mustard
【芥蓝】[植] cabbage mustard

钙 [化] calcium (Ca)

【钙化】[医] calcification
【钙化不全】hypocalcification
【钙化作用】calcification
【钙镁磷肥】calcium magnesium phosphate
【钙片】calcium tablet
【钙缺乏】calcium deficiency
【钙生植物】calcicole
【钙铀云母】autunite

概 ①(大略) general; approximate: ～而论之 generally speaking ②(无例外) without exception; categorically: ～不赊欠 no credit allowed to anybody / ～不追究 no action will be taken (against sb. for his past offenses) / ～莫能外 admit of no exception whatsoever ③(神气) the manner of carrying oneself; deportment: 气～ mettle; spirit

【概差】probable error
【概况】general situation; survey: 《亚洲～》A Survey of Asia
【概括】①(总括) summarize; generalize; epitomize: ～起来说 to sum up / 高度的艺术～ a highly artistic condensation ②(扼要) briefly; in broad outline: ～地说 to put it briefly
【概括性】generality: 这篇文章～很强。 This article is a succinct summary.
【概率】[数] probability ◇ ～分析 probability analysis / ～空间 probability space / ～流量 probability current / ～律 probability law / ～论[数] probability theory; law of probability; theory of probability / ～误差 probable erro
【概略】outline; summary
【概略图】skeleton diagram
【概论】(多用于书名) outline; introduction: 《化学～》*An Introduction to Chemistry*
【概貌】general picture
【概莫能外】admit of no exception what soever; without exception; It admits of no exception
【概念】concept; conception; notion; idea: 基本～ fundamental conception; basic concept / 形成一个～ from a concept
【概念化】deal in generalities; write or speak in abstract terms: 公式化、～的作品 literary works which tend to formularize and generalize
【概数】approximate number; round number
【概算】[经] budgetary estimate
【概要】(多用于书名) essentials; outline: 《英语语法～》 *Essentials of English Grammar*

gān

干 ①(干燥) dry: 口～ thirsty / 水塘快～了。 The pond is running dry. / 油漆未～。 Wet paint. ②(干食品) dried food: 牛肉～儿 dried beef; jerked beef ③(空虚) empty; hollow; dry: ～号 cry aloud but shed no tears; affected wailing / 外强中～ outwardly strong but inwardly weak ④(徒然) (do sth.) for nothing; futilely: 他们上午不来,咱们别～等了。 They're not coming this morning. Let's not waste time waiting for them. ⑤(涉及) have to do with; be concerned with; be implicated in: 与你何～? What has this to do with you? ⑥[书] (冒犯) offend

【干巴巴】dull and dry; insipid; dryasdust; dull as ditchwater: 文章写得～的。 The article is dull.
【干板】[摄] dry plate
【干杯】drink a toast: 为朋友们的健康～! Here's to the health of our friends — to your health!
【干贝】[食品] dried scallop (adductor)
【干瘪】shrivelled; wizened
【干冰】[化] dry ice
【干菜】dried vegetable
【干草】hay ◇ ～垛 haystack
【干柴烈火】(like) a blazing fire and dry wood, caught in a passion
【干船坞】dry dock
【干脆】①(直截了当) clear-cut; straightforward; not mince one's words: ～一点嘛! Make

it snappy! / 他办事喜欢～。He likes to be straightforward in doing things. ②(爽快) simply; just; altogether: 你～说"行"还是"不行"。Just say yes or no.

【干打雷,不下雨】thunder but no rain; much noise but no action

【干打垒】tamped-earth house; house with tamped clay walls

【干瞪眼】[口] stand by anxiously, unable to help; look on in despair

【干电池】dry cell ◇～组 dry battery

【干犯】offend; encroach upon: ～法纪 break the law and violate discipline

【干饭】cooked rice

【干纺】dry spinning ◇～纱 dry-spun yarn

【干戈】weapons of war; arms; war: 动～ take up arms; go to war

【干戈入库,偃武修文】the sword sleeps in the scabbard

【干戈四起】civil war breaks out in the country; fighting broke out all over the country

【干果】①(硬壳果) dry fruit (e. g. nuts) ②(晒干的果子) dried fruit

【干旱】arid; dry: ～地区 arid area / 战胜～ conquer drought

【干涸】dry up; run dry: 河道～。The river dried up.

【干货】dried food and nuts (as merchandise)

【干结】dry and hard: 大便～ constipated

【干净】①(清洁) clean; neat and tidy: 把院子打扫～ sweep the yard clean ②(完全) completely; totally: 忘得干干净净 have completely forgotten; clean forgot

【干净核弹】clean bomb

【干净利落】neat and tidy; neat; efficient: 她办事～。She's very efficient.

【干咳】dry cough

【干枯】dried-up; withered; shrivelled; wizened: ～的树木 withered trees / 小河～了。The stream has dried up.

【干酪】cheese ◇～素 casein

【干冷】dry and cold: 天气～。The weather is dry and cold.

【干粮】solid food (prepared for a journey); field rations; rations for a journey: 明天郊游,请自带～。Bring your own food on tomorrow's outing. ◇～袋 haversack; ration bag

【干裂】season check, weather-shack

【干馏】[化] dry distillation

【干呕】[医] retch

【干汽】net gas

【干亲】nominal kinship

【干扰】①(扰乱) disturb; interfere; obstruct: ～某人的工作 interfere with sb.'s work / 把收音机开小点儿,别～人家。Turn down the radio, or you'll disturb people. ②[电](无线电人为干扰) interference; jam ◇～台 jamming station

【干鞣法】[皮革] dry tannage

【干涉】①(不该管硬要管) interfere; intervene; meddle: 互不～内政 noninterference in each other's internal affairs / 外来～ external interference / 武装～ armed intervention ②[物](干涉现象) interference: 相消～ destructive interference / 相长～ constructive interference

【干涉婚姻自由】interference with the freedom of marriage ◇～案 case of interference with the freedom of marriage

【干涉内政者】interventionist

【干涉仪】interferometer

【干湿表】[气] psychrometer

【干瘦】skinny; bony

【干洗】dry-clean; dry cleaning

【干系】responsibility; implication: 他同这桩案子有～。He is involved in the case.

【干笑】hollow laugh

【干薪】salary drawn for a sinecure: 领～ hold a sinecure

【干选】[矿] dry separation

【干预】intervene; interpose; meddle ◇～者 meddler

【干预个人事务】[法] invasion of privacy

【干燥】①(无水分或少水分) dry; arid: 大便～ constipated; costive / 气候～ arid climate ②(无趣味) dull; uninteresting: ～无味 dryasdust; dull ◇～剂 drier; drying agent; desiccating agent / ～率 index of aridity / ～器[化] desiccator / ～箱 drier; drying oven

【干着急】be anxious but unable to do anything

【干支】the Ten Celestial Stems and Twelve Branches; the Heavenly Stems and Earthly Branches

杆

pole; staff: 电线～ pole (for telephone or electric power lines, etc.) / 旗～ flagstaff; flagpole

【杆子】pole

酐

[化] anhydride: 醋酸～ acetic anhydride / 碱～ basic anhydride

矸

【矸石】[矿] waste (rock)

竿 pole; rod: 钓鱼～ fishing rod / 竹～ bamboo pole

【竿子】bamboo pole

肝 liver: ～大可以触知 (liver) be palpable / ～大一指 (liver) be palpable one finger breath

【肝癌】[医] cancer of the liver

【肝肠寸断】liver and intestines are cut into inches; deep affliction

【肝胆】①(肝与胆) liver and gall: ～俱裂 overwhelmed by grief or terror; heart-broken or terror-stricken ②(勇气) heroic spirit; courage: ～过人 unsurpassed in valor ③(真心相见) openheartedness; sincerity: ～相照 show utter devotion to (a friend, etc.)

【肝功能】liver function ◇ ～试验 liver function test

【肝昏迷】hepatic coma

【肝火】irascibility: 动～ get worked up; fly into a rage / ～旺 hot-tempered; irascible

【肝脑涂地】(ready to) die the cruelest death

【肝气】①[中医](两肋胀痛、腹泻等症状的病) diseases with such symptoms as costal pain, vomiting, diarrhea, etc. ②(易怒的心理状态) irritability

【肝素】[药] heparin

【肝泰乐】[药] glucurolactone; glucurone

【肝吸虫】[动] liver fluke

【肝炎】[医] hepatitis

【肝硬变】[医] cirrhosis (of the liver)

【肝脏】liver

【肝蛭】liver fluke

【肝肿大】[医] hepatomegaly

甘 ①(甜) sweet; pleasant: ～泉 sweet spring water ②(乐于) willingly; of one's own accord: 不～落后 unwilling to lag behind / ～当小学生 make oneself a willing pupil

【甘拜下风】candidly admit defeat (in friendly competition, etc.)

【甘草】[中药] licorice root

【甘汞】[化] calomel; mercurous chloride ◇ ～电池 calomel cell

【甘居中游】rest content with being middling

【甘苦】①(喻美好和艰苦的处境) sweetness and bitterness; weal and woe: 同～ share the joys and sorrows ②(体会到的滋味) hardships and difficulties experienced in work: 深知其中～ know well the taste of it; know well what it is like

【甘蓝】[植] wild cabbage

【甘霖】a good rain after a long drought; timely rainfall

【甘霖普降】seasonable rain has fallen everywhere

【甘露】①(甜的汁液) sweet dew ②[医] manna

【甘露糖】mannose

【甘露子】[植] Chinese artichoke

【甘美】sweet and refreshing

【甘守清贫】glory in honest poverty

【甘薯】sweet potato ◇ ～黑斑病 sweet potato black rot / ～软腐病 sweet potato soft rot

【甘心】①(愿意) willingly; readily: ～情愿 willingly and gladly ②(称心满意) be reconciled to; resign oneself to; be content with

【甘休】willingly give up; be willing to stop: 不达目的决不～ will never give up till the aim is achieved

【甘言密语】honeyed words; one's honeyed tongue and sugary words

【甘言悦耳】tickle sb.'s ear

【甘油】[化] glycerine

【甘油醛】[化] glyceraldehyde

【甘油三酯】[化] triglyceride

【甘油酸】[化] glyceric acid

【甘油酯】[化] glyceride

【甘于】be willing to; be ready to; be happy to

【甘愿】willingly; readily: ～冒风险 be willing to take the risk / ～效劳 be glad to offer one's services; be glad to do sth. for sb.

【甘蔗】sugarcane ◇ ～板[建] cane fiber board / ～渣 bagasse / ～渣浆厂 bagasse-pulp mill

【甘之如饴】enjoy sth. bitter as if it were malt sugar; gladly endure hardships

泔

【泔水】swill; slops; hogwash; pigwash; wash

坩

【坩埚】[化] crucible: 石墨～ graphite crucible ◇ ～炉[冶] crucible furnace

【坩埚钢】pot steel

【坩埚粘土】pot clay

苷 [化] glucoside

柑 mandarin orange

【柑橘】①(橙; 红橘) oranges and tangerines ②(柑橘) citrus ◇ ～酱 marmalade

【柑子】mandarin orange

尴

【尴尬】awkward; embarrassed: 处境～ in an awkward position; in a dilemma／感到～ feel embarrassed

【尴尬异常】look very put out; be much embarrassed

gǎn

赶 ①(追) catch up with; overtake: 你追我～ emulate one another; each trying to outstrip the other ②(紧追) try to catch; make a dash for; rush for: ～头班车 catch the first bus ③(加快) hurry through: ～任务 rush through one's job ④(驾御) drive: ～大车 drive a cart ⑤(驱逐) drive away: ～下台 drive (sb.) off the stage, oust (sb.) from office ⑥(碰巧) happen to; find oneself in; avail oneself of: 得～好天把场打完。 We've got to finish the threshing while the good weather lasts. ⑦[介]咱们的婚事～春节再办吧。 Let's put off our wedding till the Spring Festival.

【赶场】①(演员赶往另一处接连演出) be on the rush to give performance elsewhere ②(赶集) go to a village fair

【赶超】catch up with and surpass: ～世界先进水平 catch up with and surpass the world's advanced levels

【赶集】go to market; go to a fair

【赶紧】lose no time; hasten: ～解释 hasten to explain／～刹车 quickly put on the brakes

【赶尽杀绝】spare none; be ruthless

【赶快】at once; quickly: ～回家 go home at once

【赶浪头】follow the trend

【赶路】hurry on with one's journey: 赶了一天的路，你累了吧？Aren't you tired after such a hard day's journey?

【赶忙】hurry; hasten; make haste: 他～道歉。 He hastened to apologize.

【赶前不赶后】It's better to hurry at the beginning than to do things in a rush at the last moment.

【赶巧】happen to; it so happened that: 我～在一家旧书店里买到这本书。 I got this book by chance at a second-hand bookshop.

【赶上】①(追上) overtake; catch up with; keep pace with: ～末班车 catch the last bus／～时代的步伐 keep abreast of the times ②(碰上) run into: 没～公共汽车 miss the bus ③(及时) be in time for: ～午饭 be in time for lunch ④(差不多一样) about as... as: 他都～他爸爸的个子了。 He is about as tall as his father now.

【赶时髦】follow the fashion; try to be in the swim

【赶鸭子上架】drive a duck onto a perch; make sb. do sth. entirely beyond him

杆 ①(器物的像棍子的细长部分) the shaft or arm of sth.: 保险～ [机] bumper bar／秤～ the arm of a steelyard／枪～ the barrel of a rifle／调整～ [机] adjusting rod ②[量]一～红旗 a red flag／一～枪 a rifle

【杆臂】lever arm

【杆菌】[微] bacillus: 结核～ tubercle bacillus ◇～载体 bacillus carrier

秆 stalk: 高粱～ sorghum stalk

【秆锈病】[农] stem rust

擀 roll (dough, etc.): ～面条 make noodles

【擀面杖】rolling pin

感 ①(觉得) feel; sense: 略～不适 not feel very well; feel a bit unwell; be under the weather; be out of sorts ②(感动) move; touch; affect: ～人 touching; moving ③(谢意) be grateful; be obliged: 请早日寄下为～。 I should be grateful if you would send it to me at an early date. ④[中医](感受风寒) be affected: 外～风寒 be affected by the cold; have a cold ⑤(感觉) sense; feeling: 安全～ feeling of security／光荣～ feeling of honor／荣誉～ sense of honor／责任～ sense of responsibility／自豪～ sense of pride

【感触】thoughts and feelings; feeling: 深有～地说 say with deep feeling

【感到】feel; sense: ～高兴 feel happy／～委曲 feel oneself wronged

【感动】move; touch: ～得流下眼泪 be moved to tears／听众深为～。 The audience was deeply affected.

【感恩】feel grateful; be thankful: ～不尽 be everlastingly grateful／～图报 be grateful to sb. and seek ways to return his kindness

【感恩戴德】be deeply grateful

【感恩非浅】esteem it a great favor

【感恩节】Thanksgiving Day

【感恩莫名】(I) do not know how to express (my) gratitude

【感奋】be moved and inspired; be fired with enthusiasm

【感官】sense organ; sensory organ: ～的印象

sense perception
【感光】[摄] sensitization ◇ ～度 (light) sensitivity / ～计 sensitometer / ～树脂版[印] photopolymer plate / ～性 photonasty; photoelectric sensitivity / ～纸 sensitive paper
【感化】help sb. to change (by education and persuasion)
【感化院】reformatory
【感怀】①(有所感触) recall with emotion: ～往事 recall past events with deep feeling ②(感伤地怀念) reflections; thoughts; recollections: 新春～ thoughts on the Spring Festival
【感激】feel grateful; be thankful; feel indebted: ～不尽 be indebted forever; be very grateful / ～涕零 shed grateful tears; be moved to tears of gratitude / ～莫铭 not to know how to express one's gratitude / 不胜～ be deeply grateful; feel very much indebted
【感觉】(对客观事物的反映) sense perception; sensation; feeling: 概念同～的区别 the difference between concepts and sense perceptions ②(觉得) feel; perceive; become aware of: 你～怎么样? How do you feel now? ◇ ～过程 sense process / ～过敏 hyperresthesia / ～减退 hypesthesia / ～论 [哲] sensualism / ～器官 sense organ / ～神经 sensory nerve / ～特性 sense quality / ～异常 paresthesia
【感觉区】sensory area
【感慨】sigh with emotion: ～流涕 be moved to tears / ～万端 all sorts of feelings well up in one's mind / ～系之 sigh with deep feeling
【感抗】inductive reactance; positive reactance
【感愧交集】feel grateful and uneasy at the same time; moved and ashamed simultaneously
【感冒】common cold: 患～ catch cold; have a cold
【感念】remember with gratitude; recall with deep emotion
【感情】①(心理反映) emotion; feeling; sentiment: 动～ be carried away by one's emotions; get worked up / 伤～ hurt sb.'s feelings ②(关切、喜爱之情) affection; attachment; love
【感情不合】incompatibility
【感情冲动】emotional impulse; act on a momentary impulse
【感情洋溢】an exuberance of feeling
【感情用事】give away to one's feeling; be swayed by one's emotions; act impetuously
【感染】①(受到传染) infect: 轻度～ light infection / 手术后～ postoperative infection / 细菌～ bacterial infection / 防止再～ avoid reinfection ②(引起相同的思想感情) influence; infect; affect: 艺术～力 artistic appeal
【感人肺腑】touch one deeply in the heart; touch the chords of one's heart
【感伤】sad; sorrowful; sentimental: ～小说 sentimental novel
【感生】[电] induced: ～电流 induced current
【感受】①(受到) be affected by: ～风寒 be affected by the cold; catch cold ②(体会) experience; feel: 这次参观中的所见所闻使我～很深。 I was deeply impressed by what I saw and heard during this visit. ◇ ～器 [生理] receptor / ～性 nyctinasty
【感叹】sigh with feeling ◇ ～词 interjection; exclamation / ～号 exclamation mark; exclamation point (!) / ～句 exclamatory sentence
【感同身受】feel indebted as if it were received in person; I shall count it as a personal favor
【感物伤怀】be deeply affected at seeing something; touched to the heart
【感想】impressions; reflections; thoughts: 你对此事有何～? What are your feelings about this matter?
【感谢】thank; be grateful: 表示衷心的～ express heartfelt thanks ◇ ～信 letter of thanks
【感性】perceptual: ～认识 perceptual knowledge ◇ ～运动 [生] nastic movement / ～知觉 [心] sense impressions
【感应】①(受外界影响而引起相应的感情或动作) response; reaction; interaction ②[生](刺激感受性) irritability ③[电](诱导) induction: 电磁～ electromagnetic induction / 静电～ electrostatic induction ◇ ～场 induction field / ～干扰 inductive interference / ～计 induction meter / ～炉 induction furnace / ～率 inductivity / ～热 induction heat / ～式话筒 inductor microphone / ～线圈 induction coil; inductor
【感召】move and inspire; impel
【感知】[哲] perception

敢

①(有勇气、胆量) bold; courageous; daring: 果～ courageous and resolute; daring ②(有胆量做某事) dare: ～想、～说、～干 dare to think, dare to speak and dare to act ③(有把握) have the confidence to; be certain: 我不～说他究竟什么时候来。 I am not sure just when he will come. ④ [书] (请求) make bold; venture: ～问 I venture to ask; may I ask
【敢怒而不敢言】be forced to keep one's resentment to oneself; suppress one's rage; choke

with silent fury
【敢说敢为】dare to speak and act
【敢死队】dare-to-die corps
【敢想敢干】have the courage to think and act
【敢于】dare to; be bold in; have the courage to: ～面对现实 have the courage to face reality
【敢在太岁头上动土】dare to leap on an earth god's head to make trouble; to provoke somebody far superior in power
【敢做敢当】be bold enough to do something and accept responsibility for it
【敢作敢为】dare to do everything; be afraid of no difficulties

橄

【橄榄】[植] ①(齐墩果; 青果) Chinese olive; the fruit of the canary tree ②(油橄榄) olive ◇ ～绿 olive green / ～石 [矿] olivine / ～岩 [地] peridotite / ～油 olive oil / ～枝 olive branch
【橄榄球】[体] Rugby (football); American football: 十三人制～ Rugby League football; Northern Union football / 十五人制～ Rugby; Rugby football; Rugby Union football

gàn

干 ①(事物的主体) trunk; main part: 骨～ backbone; hard core; mainstay / 树～ tree-trunk; trunk / ②[简](干部)cadre: ～群关系 relations between cadres and the masses; cadre-mass relations ③(做) do; work: ～得很好! Well done! / 有好多事情要～。 There's a lot to do. ④(斗争) fight; strike: ～到底 fight to the bitter end ⑤(能干的) capable; able: ～员 a capable official
【干部】cadre: ～带头 cadres stand in the van to ◇ ～政策 policy towards cadres; cadre policy
【干才】①(办事才能) ability; capability ②(有才能的人) capable person
【干掉】[口] kill; get rid of; put sb. out of the way
【干活】work; work on a job: ～去吧。 Let's get to work.
【干架】①(吵架) quarrel ②(动手打架) come to blows
【干将】capable person; go-getter
【干劲】drive; vigor; enthusiasm: ～十足 be full of vigor / 冲天～ boundless energy
【干尽坏事】do every evil; do all manner of evil; stop at no evil
【干练】capable and experienced
【干流】trunk stream; main stream; master stream
【干吗】[口] ①(为什么) Why on earth; whatever for: ～这么大规矩? Why all this formality? ②(干什么) What to do: 你想～? What are you up to?
【干渠】[水] trunk canal; main canal
【干事】a secretary in charge of sth.: 文娱～ person in charge of recreational activities / 宣传～ person in charge of propaganda work
【干线】main line; trunk line; artery; 公路～ arterial or main highway / 交通～ main lines of communication

绀 dark purple
【绀青】dark purple; prune purple

gāng

缸 vat; jar; crock: 金鱼～ goldfish bowl / 牛奶～ milk server / 奶油～ cream jug / 糖～ sugar bowl / 烟灰～ ashtray
【缸壁】casing wall
【缸管】earthen pipe
【缸径】bore; cylinder diameter
【缸盆】glazed earthen basin
【缸瓦】a compound of sand, clay, etc. for making earthenware
【缸瓦器】stoneware
【缸砖】clinker (tile); quarry tile
【缸子】mug; bowl: 茶～ tea mug

肛 anus
【肛裂】[医] anal fissure
【肛瘘】[医] anal fistula
【肛门】anus
【肛门镜】anoscope

冈 ridge (of a hill): 山～ low hill; hillock
【冈比亚】the Gambia ◇ ～人 Gambian

刚 ①(坚强;硬) firm; strong; indomitable: 她的舞蹈柔中有～。 There is strength as well as grace in her dancing. ②(恰好) just; exactly: 不大不小, ～合适 neither too big nor too small, just fit ③[副](仅仅) barely; only just ④[副](刚刚) only a short while ago; just: 他～回来。He's just come back.
【刚愎不仁】stubborn and unkind
【刚愎自用】self-willed; headstrong; opinionated
【刚才】just now; a moment ago: 他比～好一点了。 He is a little better than he was a short while ago.

【刚度准则】stiffness criterion
【刚刚】①(正好) just; only; exactly: 那时候天～亮。It was just beginning to get light. ②(刚才) a moment ago; just now: 杂志～到。The magazine came just now.
【刚果】the Congo ◇～河 the Congo (River)／～红 Congo red／～人 Congolese／～语 Congolese (language)
【刚好】①(正合适) just; exactly: 你们来得～。You've come in the nick of time. ②(正巧) happen to; it so happened that: 我～在路上碰见他。I happened to meet him on the road.
【刚架】[建] rigid frame ◇～结构 rigid-framed structure
【刚健】vigorous; energetic; robust: ～的舞姿 vigorous movements of a dancer
【刚劲】bold; vigorous; sturdy: 笔力～ write in a bold hand
【刚劲有力】(the calligraphy) is powerful and vigorous; vigorous and forcible
【刚强】firm; staunch; unyielding
【刚柔并济】couple hardness with softness; temper force with mercy
【刚体】[物] rigid body
【刚性】[物] rigidity ◇～结构 rigid structure
【刚毅】resolute and steadfast
【刚玉】[矿] corundum
【刚正】upright; honorable; principled: ～不阿 upright and never stooping to flattery
【刚直】upright and outspoken

钢

steel: 炼～ steelmaking／不锈～ stainless steel／槽～ channel steel／超高强度～ super-high strength steel／低合金高强度结构～ low-alloy high strength structural steel／低碳～ low carbon steel／高合金结构～ high alloy-structural steel／高级碳素～ high-grade carbon steel／高碳～ high carbon steel／铬～ chromium steel／工具～ tool steel／工字～ I steel beam／硅～ silicon steel／合金～ alloy steel／结构～ structural steel／锰～ manganese steel／钼～ molybdenum steel／镍～ nickel steel／弹簧～ spring steel／钨～ wolfram steel
【钢板】①(板状钢材) steel plate; plate: 薄～ steel sheet; sheet steel／成型～ shaped steel plate／锅炉～ boiler plate／厚～ steel plate／抗低温～ low temperature resistant steel plate／网眼～ perforated plate／造船～ ship plate ②(汽车上的片状弹簧) spring ③(誊写钢板的简称) stencil steel board
【钢笔】pen; fountain pen ◇～画 pen drawing
【钢材】steel products; steels; rolled steel
【钢尺】steel rule
【钢带】steel band; steel belt; steel strip
【钢刀虽快,不斩无罪之人】The innocent need have no fear of the law
【钢锭】steel ingot
【钢鼓】steel drum
【钢骨水泥】reinforced concrete
【钢管】steel tube: 焊接～ welded steel pipe／无缝～ seamless steel tube／异形～ chapped pipe
【钢轨】rail ◇～连接板 joint fastening／～铝热剂 railroad thermit／～探伤仪 rail flaw detector
【钢号】[冶] steel grade
【钢花】spray of molten steel: ～飞舞 sparks of molten steel flying
【钢化玻璃】toughened glass
【钢结构】[建] steel structure; steel construction
【钢筋】reinforcing bar: 高强度～ high strength cast iron; nodular iron ◇～混凝土 reinforced concrete
【钢筋铁胆】have an iron resolve and steel sinews
【钢筋铁骨】a body strong as iron; muscles of steel
【钢精】aluminium (as used for utensils) ◇～锅 aluminium pan
【钢锯】hacksaw ◇～架 hacksaw frame／～条 hacksaw blade
【钢卷尺】tape; steel tape; steel measuring tape
【钢盔】(steel) helmet
【钢坯】[冶] billet; 大～ bloom
【钢片琴】[乐] celesta
【钢瓶】steel cylinder
【钢钎】drill rod; drill steel
【钢琴】piano: 弹～ play the piano／竖式～ upright piano ◇～家 pianist
【钢水】[冶] molten steel ◇～包 steel ladle
【钢丝】(steel) wire: 走～ [杂技] walk the wire; walk the tightrope; high-wire walking ◇～床 spring bed／～垫子 spring mattress／～锯 fret saw; scroll saw／～钳 combination pliers; cutting pliers／～绳 steel cable; wire rope
【钢索】cable wire; steel rope; wire rope
【钢条】steel bar; steel girder
【钢铁】iron and steel; steel: ～意志 iron will／～运输线 an unbreakable transportation line ◇～厂 steelworks／～工业 iron and steel industry／～公司 iron and steel company

【钢铁联合企业】integrated iron and steel works; iron and steel complex ◯炼焦厂 coking plant / 化铁厂 iron-smelting plant / 铸铁厂 iron foundry / 炼铁厂 blast furnace plant / 炼钢厂 steel mill; steel works / 无缝钢管厂 seamless tubing mill / 大型轧钢厂 heavy steel rolling mill / 初轧厂 rough rolling mill / 中板厂 medium plate mill / 薄板厂 sheet-steel mill / 附属加工厂 subsidiary factory / 烧结 sintering / 高炉 blast furnace / 炉顶 furnace top / 炉门 furnace door / 观察孔 poking door; inspection hole / 装料门 charging door / 炉膛 hearth / 炉衬 lining / 炉衬寿命 life span of furnace lining / 风口 air hole / 出铁口 tapping hole / 风管 wind pipe / 焦比 coke ratio / 化铁炉 blast cupola / 平炉 open hearth furnace; Martin furnace / 转炉 converter; Bessemer converter / 氧气顶吹转炉 top-blown oxygen converter / 电炉 electric furnace / 电弧炉 electric arc furnace / 轧钢机 rolling mill / 冷轧机 cold rolling mill / 中板轧机 medium plate rolling machine / 板材矫直机 levelling machine / 弯板机 bending machine / 锻压 forging and pressing / 挤压 extrusion; squeezing / 冲压 stamping / 热压 hot-pressing / 热轧 hot rolling / 冷轧 cold rolling / 初轧 roughing down; breaking down / 真空铸造 vacuum pressing and casting

【钢芯铝线】steel-cored aluminium; aluminum cable steel-reinforced

【钢芯铜线】steel-cored copper conductor

【钢印】①(硬印) steel seal; embossing seal ②(印痕) embossed stamp

【钢渣】slag

【钢纸】vulcanized fiber (paper)

【钢珠】steel ball (in a ball bearing); ball bearing; ball

【钢罩】steel cage

纲 ①(提网的绳) the headrope of a fishing net ②(总纲) key link; guiding principle ③(大纲) outline; program ④ [生](动植物分类) class: 哺乳动物～ the class of mammals / 亚～ subclass

【纲纪国法】discipline, morale, law and order

【纲纪律法】law, discipline and rules of conduct

【纲纪四方】rule over the whole country

【纲举目张】once the key link is grasped, everything falls into place; once the headrope of a fishing net is pulled out, all its meshes open

【纲领】program; guiding principle: ～性文件 programmatic document

【纲目】(多用于书名) detailed outline outline

【纲要】①(提纲) outline; sketch ②(用于书名、文件名) essentials; compendium:《汉语语法～》*Essentials of Chinese Grammar*

gǎng

港 port; harbor: 避难～ port of distress / 不冻～ ice-free port / 出发～ port of departure / 船籍～ port of registry / 到达～ port of arrival / 对外贸易～ open port / 目的～ port of destination / 天然～ natural harbor / 停靠～ port of call / 卸货～ port of discharge / 装煤～ port of coaling / 自由～ free port

【港币】Hong Kong currency; Hong Kong Dollar

【港汊】branching stream

【港规】harbor regulation

【港口】port; harbor: 沿海～ coastal port ◇ ～费用 port charge / ～惯例 custom(s) of the port / ～规章 harbor regulations / ～河段 [地理] harbor reach / ～建筑线 pierhead line / ～税 port dues / ～吞吐量 traffic (of a port)

【港湾】harbor ◇ ～沉积 estuarine deposit / ～泥 estuarine muds

【港务费】harbor dues

【港务监督】harbor superintendency administration

【港务局】port office; harbor bureau

岗 ①(山岗) hillock; mound ②(岗位;岗哨) sentry; guard: 站～ be on sentry ③(狭长的隆起物) ridge

【岗警】policeman on point duty

【岗楼】watchtower

【岗峦起伏】undulating hills; the undulations of the ridge and hills

【岗哨】①(岗位) lookout post ②(放哨人) sentry; sentinel: 设置～ post sentries

【岗亭】sentry box; police box

【岗位】post; station: 坚守～ stand fast at one's post; stick to one's guns / 走上新的～ take up a new post; take on a new job

【岗位责任制】system of post responsibility (for the work done by each individual at his post)

【岗子】①(不高的山) mound; hillock ②(平面上凸起的一条长道子) ridge; wale; welt

gàng

杠 ①(较粗的棍子) thick stick; stout carrying pole ② [体](体育器械) bar: 单～ horizontal bar／双～ parallel bars ③(划线作为标记) thick line (drawn beside or under words in reading, correcting papers, etc.) ④(把错字勾出) cross out; delete: 把错字～掉 cross out the wrong word

【杠棒】stout carrying pole

【杠杆】lever; heaver; pry bar ◇～臂 lever arm／～定律 lever law／～率 leverage／～原理 lever principle／～支点 balance pivot; lever fulcrum／～作用 leverage; lever action

【杠铃】[体] barbell ◇～片 disc

【杠子】①(较粗的棍子) thick stick; stout carrying pole ②[体](体育用品) bar

钢 ①(磨) sharpen; whet; strop: ～菜刀 sharpen a kitchen knife／～剃刀 strop a razor ②(在刀口上加钢) reinforce the edge by adding steel and retempering

【钢刀布】(razor) strop

gāo

高 ①(从下向上距离大) tall; high: 大坝～六十米。 The dam is 60 meters high ②(等级在上的) of a high level or degree; above the average: ～风格 fine style／～难度动作 exceedingly difficult movements; operations of extraordinary difficulty／～年级 higher grades／～质量 high quality ③(大声) loud: ～喊 shout loudly; raise a cry ④(价高) high-priced; dear; expensive: 要价太～ ask too high a price ⑤[敬] your: ～见 your opinion

【高矮】height: ～不一 some tall and some short

【高昂】①(高高扬起) hold high (one's head, etc.) ②(向上高起) high; elated; exalted: 情绪～ be in high spirits ③(昂贵) dear; expensive; exorbitant

【高傲】supercilious; arrogant; haughty

【高不成,低不就】①(高的工作做不了,低的工作不肯做) be unfit for a higher post but unwilling to take a lower one ②(合意的得不到,不合意的不肯要) can't have one's heart's desire but won't stoop to less

【高不可攀】too high to reach; unattainable

【高才生】a brilliant student; a outstanding student

【高层云】[气] altostratus

【高产】high yield; high production: ～稳产 high and stable yield; stable high yield ◇～品种 high-yield variety／～田 high yield field; high yield plot

【高唱】①(高声歌唱) sing loudly; sing with spirit ②(大声叫喊) talk glibly about; call out loudly for

【高超】superb; excellent: 技艺～ superb skill

【高超音速】[物] hypersonic speed ◇～火箭 hypersonic rocket

【高潮】①(最高潮位) high tide; high water ②(发展的顶点) upsurge; climax; high tide: 全剧的～ the climax of the play ◇～线 [海] high-water mark

【高大】①(又高又大) tall and big; tall: 身材～ be of great stature ②(崇高的; 高尚的) lofty: 革命英雄的～形像 the lofty image of a revolutionary hero

【高蛋白】high protein

【高档】top grade; superior quality ◇～商品 high-grade goods; expensive goods

【高等】higher ◇～哺乳动物 higher mammal／～法院 high court／～教育 higher education／～数学 higher mathematics／～院校 institutions of higher learning; colleges and universities

【高低】①(高低的程度) height: 山崖的～ the height of a cliff／声调的～ the pitch of a voice ②(高下) relative superiority or inferiority: 难分～ hard to tell which is better／争个～ vie with each other to see who is better ③(深浅轻重) sense of propriety; discretion: 不知～ not know what's proper; have no sense of propriety

【高低杠】[体] uneven bars

【高低角】[军] angle of site

【高地】①(高处之地) highland; upland; elevation: ～田 an upland field ②[军] height

【高调】lofty tone; high-sounding words: 唱～ mouth high-sounding words; say fine-sounding things

【高度】①(高低程度) altitude; height: 飞行～ flying altitude／山的～ the height of a mountain ②(程度高) high; high degree; highly: ～评价 hold (sth.) in high regard／～现代化 highly modernized／～赞扬 speak highly of; praise highly ◇～表 altimeter／～差 altitude difference／～规 height gage

【高尔夫球】①(指项目) golf ②(指球) golf ball ◇～棒 golf club／～场 golf course; golf links／～俱乐部 golf club

【高尔基体】[生] golgiosome

【高分子】[化] high polymer; macromolecule ◇～化合物 macromolecular compound; high-molecular compound／～化学 (high) polymer chemistry／～聚合物 high polymer

【高风格】lofty style; high-mindedness
【高风亮节】exemplary conduct and nobility of character; have a sharp sense of integrity
【高峰】peak; summit; height: 攀登科学的～ scale the heights of science
【高浮雕】high relief
【高杆作物】long-stalked crops
【高高在上】stand high above the masses; be far removed from the masses and reality
【高歌猛进】stride forward singing militant songs; advance triumphantly
【高个儿】a tall person
【高跟鞋】high-heeled shoes
【高官厚禄】high position and handsome salary; high posts with salaries to match
【高官显爵】be honored with high official titles
【高贵】①(达到高度道德水平) noble; high: ～品质 noble quality ②(杰出的) highly privileged; elitist
【高积云】[气] altocumulus
【高级】①(级别高) senior; high-ranking; high-level; high: ～参谋 senior staff officer／～官员 high-ranking official／～将领 high-ranking general officers／最～会议 summit meeting ②(质量、水平高) high-grade; high-quality; advanced: ～墨水 high-quality ink／～染料 high-grade dyestuff ◇ ～人民法院 higher people's court／～人民检查院 high people's procuratorate; superior people's procuratorate／～中学 senior middle school
【高级神经活动】[生理] higher nervous activity
【高级神经中枢】[生理] higher nervous center
【高价】high price: ～出让 go off at high price／～勒索 demand extortionate prices／～收买 buy over at a high price ◇ ～货物 expensive goods
【高价待沽】wait for the opportunity to sell... at high price
【高架公路】flyover
【高架桥】viaduct
【高架铁道】overhead railway; elevated railway
【高见】[敬] your brilliant idea; your opinion: 有何～? What do you think about it?
【高洁】noble and unsullied
【高精尖】high-grade, precision and advanced (industrial products)
【高举】hold high; hold aloft: ～革命的红旗 hold high the revolutionary red banner
【高举远引】seclude oneself and avoid all worldly cares
【高聚物】[化] high polymer
【高踞】stand above; set oneself above; lord it over
【高峻】high and steep
【高亢】loud and sonorous; resounding: ～的歌声 sonorous singing
【高空】high altitude; upper air ◇ ～病 altitude sickness／～飞行 high-altitude flight／～核试验 high-altitude nuclear test／～气象学 aerology／～适应 high-altitude adaptation／～作业 work high above the ground
【高栏】[体] high hurdles
【高利贷】usury; usurious loan: 放～ practice usury ◇ ～者 usurer; loan shark／～资本 usurer's capital
【高利盘剥】practice usury; exploit by usury
【高粱】Chinese sorghum ◇ ～米 husked kaoliang／～饴 sweets made of sorghum syrup; sorghum candy
【高龄】advanced age; venerable age: 八十～ the advanced age of 80
【高岭石】kaolinite
【高岭土】kaolin
【高楼大厦】high buildings and large mansions
【高炉】[冶] blast furnace ◇ ～利用系数 capacity factor of a blast furnace／～煤气 blast furnace gas／～寿命 life of a blast furnace
【高氯酸】[化] perchloric acid
【高论】[敬] enlightening remarks; brilliant views
【高帽子】①(纸帽) tall paper hat (worn as a sign of humiliation) ②(恭维话) flattery
【高锰酸钾】[化] potassium permanganate
【高妙】ingenious; masterly: 手艺～ masterly craftsmanship
【高明】brilliant; wise: 另请～。Find someone better qualified (than myself).
【高能燃料】high-energy fuel
【高能物理学】high-energy physics
【高攀】make friends or claim ties of kinship with someone of a higher social position
【高朋满座】a great gathering of distinguished guests
【高频】high frequency: 超～ ultrahigh frequency (UHF)／甚～ very high frequency (VHF) ◇ ～淬火[机] high-frequency quenching／～感应电炉[冶] high-frequency induction furnace／～手表 high-frequency watch／～扬声器 tweeter
【高气压】high atmospheric pressure ◇ ～区 high-pressure area; region of high barometric pressure

【高强】excel in; be master of: 武艺～ excel in martial arts

【高强度】high strength: ～钢 high-strength steel; high-tensile steel

【高跷】stilts: 踩～ walk on stilts

【高球】[体] high ball; lob: 放～ lob

【高人逸士】one who lives secluded and does not admire wealth and high emolument; holy hermits; hermits and recluses

【高人一等】a cut above other people: 他总以为自己～。 He always thinks he's a cut above others.

【高僧】eminent monk

【高山病】mountain sickness

【高山大川】high mountains and great rivers

【高山反应】reaction to high altitudes

【高山流水】lofty mountains and flowing water; difficult to find a man who appreciates one's music

【高山仰止】admire... greatly (as one stops looking up at a peak); behold a high mountain with awe

【高山植物】alpine plant

【高尚】noble; lofty: ～的人 noble-minded person / ～的品质 noble character

【高烧】high fever: 发～ have a high fever

【高射机关枪】antiaircraft machine gun

【高射炮】antiaircraft gun

【高深】advanced; profound; recondite: ～的造诣 high attainment / 莫测～ unfathomable

【高深莫测】too profound to be understood; too high and deep to be measured

【高视阔步】carry oneself proudly; prance; strut

【高手】past master; master-hand; ace: 国际象棋～ master chess player

【高寿】①(长寿) longevity; long life ②[敬] your venerable age: 老大爷, 您今年～? May I ask how old you are, Grandpa?

【高耸】stand tall and erect; tower: ～的纪念碑 a towering monument / ～入云 reach to the sky; tower into the clouds

【高速】high speed: ～发展 develop by leaps and bounds; develop at top speed / ～转弯很危险。 It's dangerous to turn a corner at high speed. ◇ ～车 speed car / ～档 top gear; high gear / ～钢 high-speed steel; rapid steel / ～工具钢 high-speed tool steel / ～公路 expressway / ～切割 [机] high-speed cutting / ～摄影 high-speed handling

【高抬贵手】be magnanimous; be generous; not be too hard on sb.

【高谈阔论】indulge in loud and empty talk; talk volubly or bombastically; harangue

【高碳钢】[冶] high-carbon steel

【高汤】①(煮鸡、鸭、肉等的老汤) soup-stock ②(清汤) thin soup

【高文典册】order by imperial decree; great literature and classical works

【高温】high temperature ◇ ～车间 high-temperature workshop / ～淬火 quench hot / ～计 [仪表] pyrometer / ～气候 megathermal climate / ～切削 [机] high-temperature machining / ～试验 hot test / ～消毒 high-temperature sterilization / ～作业 [冶] high-temperature operation

【高屋建瓴】pour water off a steep roof; sweep down irresistibly from a commanding height; operate from a strategically advantageous position

【高效肥料】concentrated fertilizer

【高兴】①(愉快而兴奋) glad; happy; cheerful: 他听了这个好消息非常～。 He was very happy to hear the good news. ②(喜欢做某件事) be willing to; be happy to: 你不～去, 就甭去了。 You needn't go if you don't feel like it.

【高血压】[医] hypertension; high blood pressure

【高压】① [物] [气] (气压) high pressure ② [电] (电压) high tension; high voltage ③ (残酷迫害) high-handed: ～手段 high handed measure / ～政策 high-handed policy ④ (血压) maximum pressure ◇ ～泵 [机] high-pressure pump / ～电力网 high-tension network / ～风 high-pressure blast / ～釜 autoclave / ～锅 pressure cooker / ～锅炉 high-pressure boiler / ～脊[气] ridge of high pressure; pressure ridge / ～灭菌器 autoclave / ～室 hyperbaric chamber / ～线 high-tension line / ～症 [航空] hyperbarism

【高眼鲽】[动] plaice

【高音】[乐] high pitch; high-pitched voice: 男～ tenor / 女～ soprano

【高音喇叭】tweeter

【高原】plateau; highland; tableland ◇ ～病 altitude sickness

【高瞻远瞩】stand high and see far; take a broad and long-term view; show great foresight

【高涨】rise; upsurge; run high: 群众热情～。 The enthusiasm of the masses ran high. ◇ ～年 boom year

【高枕安卧】lay one's head on one's pillow and just drop off to sleep; sleep peacefully
【高枕无忧】shake up the pillow and have a good sleep; sit back and relax
【高中】[简](高级中学) senior middle school
【高姿态】lofty stance; magnanimous attitude
【高足】[敬] your brilliant disciple; your pupil
【高足弟子】one's best pupil
【高掌远跖】extend everywhere; great ambition and aspiration
【高祖】(paternal) great-great-grandfather
【高祖母】(paternal) great-great-grandmother

膏 ①(脂肪;油) fat; grease; oil: 春雨如～。Rain in spring is as precious as oil. ②(糊状物) paste; cream; ointment: 软～ ointment / 牙～ tooth paste / 硬～ plaster
【膏火自煎】fat burns and fries itself; one who has talent incurs misfortune
【膏剂】[药] medicinal extract; electuary
【膏粱】fat meat and fine grain; rich food ◇ ～子弟 good-for-nothing sons of the idle rich
【膏药】plaster: 贴～ apply a plaster to
【膏腴】[书] fertile: ～之地 fertile land

篙 punt-pole; barge-pole

羔 lamb; kid; fawn
【羔皮】lambskin; kidskin; kid
【羔羊】lamb; kid: ～痢疾 lamb dysentery

糕 cake; pudding: 蛋～ cake
【糕饼】cake; pastry
【糕点】cake; pastry
【糕干】sweetened rice flour
【糕状型芯】cake core

睾
【睾囊】[生理] scrotum
【睾丸】[生理] testis; testicle ◇ ～粘连 synorchidism / ～炎 orchitis
【睾丸甾酮】testosterone

gǎo

槁 withered
【槁木死灰】dead trees and cold ashes; complete apathy; rotten wood and cold ashes; a living corpse

搞 ①(做;干) do; carry on; be engaged in: ～副业生产 engage in sideline production / ～计划生育 carry on family planning / 他是～法律的。He is in the law. ②(弄) cause; make; produce; work out: ～一个计划 draw up a plan / 把人～糊涂了 cause people to get into confusion; make people confused ③(搞出) set up; start; organize: 我们～一点核武器完全是为了自卫。It is purely for self-defense that we have produced some nuclear weapons. ④(取得) get; get hold of; secure: ～到一批原料 get a batch of raw materials / 你去给我们～点吃的来。Go and get us something to eat. ⑤(后面接补语) produce a certain effect or result; cause to become: 把事情～糟了 make a mess of things
【搞臭】discredit; put to shame
【搞出】work out; achieve; produce: ～成果 achieve results / 把方案～来 work out a plan
【搞错】mistake: 我把他们俩～了。I mistook the one for the other.
【搞掉】get rid of; do away with: 把这堆垃圾～ get rid of this rubbish dump
【搞鬼】play tricks; be up to some mischief
【搞好】make a good job of; do well: ～关系 build 〈foster〉 good relations (with sb.) / ～团结 strengthen unity
【搞坏】damage; impair; spoil: 把身体～ impair one's health
【搞垮】disrupt; undermine
【搞乱】confuse; muddle; mess up: 把秩序～了 get the order confused
【搞糟】mess up; make a mess (of)

镐 pick; pickaxe
【镐头】pick; pickaxe

稿 ①[书](谷物的茎) stalk of grain; straw ②(草稿) rough copy; draft; sketch: 初～ first draft ③(稿子) manuscript; original text: 定～ finalize a text / 来～ contributed article; contribution / 遗～ literary remains; posthumous papers
【稿本】manuscript: ～目录 written catalogue
【稿费】payment for an article or book written; contribution fee; author's remuneration
【稿秆】straw; stalk
【稿件】manuscript; contribution
【稿约】notice to contributors
【稿纸】squared or lined paper for making drafts or copying manuscripts
【稿子】①(草稿) draft; sketch: 起个～ make a draft ②(稿件) manuscript; contribution: 给黑板报写～ write sth. for the blackboard newspaper ③(心里的计划) idea; plan: 我心里还没

个准～。I haven't got any definite plan yet.

缟 a thin white silk used in ancient China
【缟素】white mourning dress
【缟状云】[气] fumulus; velum

gào

膏 ①(加润滑油) lubricate: ～车 lubricate the axle of a cart ②(在砚台边把毛笔掭匀) dip a brush in ink and smooth it on an inkstone before writing

告 ①(陈述、解说) tell; inform; notify: 何日启程，盼～。Please inform me of your date of departure. ②(控告) accuse; go to law against; bring an action against ③(为某事而请求) ask for; request; solicit: ～假 ask for leave ④(表明) declare; announce: 不～而别 go away without taking leave; leave without saying good-bye / 自～奋勇 volunteer to do sth. ⑤(宣布或表示某种情况的实现)大功～成。The task is at last accomplished.
【告白】public notice
【告别】①(离别) leave; part from: 向亲友～ take leave of one's relatives and friends ②(辞行; 向死者诀别) bid farewell to; say good-bye to: 挥手～ wave farewell / 向遗体～ pay one's last respects to the deceased ◇ ～词 farewell speech; valediction / ～宴会 farewell banquet / ～仪式 farewell ceremony
【告成】accomplish; complete: 大功～ the important task is accomplished; be brought to successful completion
【告吹】fizzle out; fail
【告辞】take leave
【告贷】ask for a loan
【告贷无门】have no means to borrow money; nowhere to borrow money
【告发】report (an offender); inform against; lodge an accusation against; delate
【告发同案犯】split on an accomplice
【告急】①(情况紧急) be in an emergency ②(请求援救) report an emergency; ask for emergency help
【告假】ask for leave
【告捷】①(取得胜利) win victory: 首战～ win in the very first battle or game ②(报告得胜消息) report a victory
【告诫】warn; admonish; exhort: ～某人要… exhort sb. to...
【告警】report an emergency; give an alarm
【告警灯】stand by lamp
【告警阀】alarm valve
【告警系统】warning system
【告老】retire on account of age: ～还乡 retire on account of old age and return to one's native place
【告密】inform against sb. ◇ ～者 informer
【告罄】run out; be exhausted: 弹药～。Ammunition has run out.
【告饶】beg for mercy; ask pardon
【告示】official notice; bulletin; placard: 安民～ notice to reassure the people
【告示牌】billboard
【告诉】tell; let know: ～某人某事 tell sb. (about) sth.; inform sb. of sth.
【告退】ask for leave to withdraw from a meeting, etc.
【告知】inform; notify
【告终】come to an end; end up: 以失败～ end in failure
【告状】①(请求司法机关审理) go to law against sb.; bring a lawsuit against sb. ②(向某人的上级或长辈诉说) lodge a complaint against sb. with his superior

诰 imperial mandate: ～封 the conferment of honorary titles by imperial mandate
【诰命】imperial mandate

锆 [化] zirconium (Zr)
【锆鞣】zirconium tanning ◇ ～革 zirconium tanned leather
【锆石】[矿] zircon

gē

割 cut: ～草 cut grass; mow / ～麦子 cut wheat
【割爱】give up what one treasures; part with some cherished possession: 忍痛～ part reluctantly with what one treasures / ～见遗 give away one's beloved thing to another
【割草机】mower
【割除】cut off; cut out; excise: ～扁桃体 cut off the tonsil; remove the tonsil
【割地赔款】cede territory and pay indemnities
【割地求和】cede territory and ask for peace
【割断】sever; cut off: ～联系 sever relations; cut off connections (with)
【割缝】kerf
【割鸡焉用牛刀】why use an oxcleaver to kill a chicken; why break a butterfly on the wheel
【割胶】rubber tapping

【割炬】[机] cutting torch
【割据】set up a separatist regime by force of arms: 封建～ feudal separatist rule / 军阀～ separatist warlord regimes
【割尖】[工] cutting tip
【割捆机】[农] self-binder; binder
【割礼】[宗] circumcision
【割裂】cut apart; separate; isolate: 把问题～开来看 look at the problem in isolation
【割袍断义】break off all intercourse with a friend
【割让】cede: ～领土 cession of territory
【割绒刀】trevette; trivet
【割肉补疮】cutting the flesh to cure a boil; a makeshift to tide over a present difficulty
【割肉充饥】cut off one's flesh to feed oneself
【割晒机】[农] swather; windrower
【割舍】give up; part with: 难以～ find it hard to part with
【割席绝交】break up an old friendship; break off friendly relations with somebody; break off with one's friend
【割线】[数] secant

哥 (elder) brother
【哥德巴赫猜想】[数] Goldbach's conjecture
【哥哥】(elder) brother
【哥伦比亚】Colombia ◇ ～人 Colombian
【哥儿们】[口] ①(弟兄们) brothers ②(用于朋友间) buddies; pals: 穷～ we, the poor
【哥斯达黎加】Costa Rica ◇ ～人 Costa Rican
【哥特式】[建] Gothic: ～教堂 Gothic cathedral
【哥特体】[印] gothic

歌 ①(歌曲) song: 民～ folk song ②(唱) sing: 纵情高～ sing loudly and without constraint
【歌本】songbook
【歌唱】sing ◇ ～家 singer; vocalist
【歌词】words of a song
【歌功颂德】eulogize sb.'s virtues and achievements; sing the praises of sb.
【歌喉】(singer's) voice; singing voice: ～婉转 sing in a beautiful voice
【歌剧】opera: 小～ operetta ◇ ～剧本 libretto / ～团 opera troupe / ～舞剧院 opera and ballet theatre; theatre of the opera and dance drama / ～院 opera house
【歌诀】formulas or directions put into verse: 汤头～ (a handbook of) Chinese herb prescriptions in verse
【歌片儿】song sheet
【歌谱】music of a song
【歌曲】song
【歌声嘹亮】the (sound of) singing was loud and clear
【歌声盈耳】the sound of singing fills the ear
【歌手】singer; vocalist
【歌颂】sing the praises of; extol; eulogize: ～劳动英雄 sing the praises of labor heroes
【歌舞】song and dance ◇ ～剧 song and dance drama / ～片 musical film / ～团 song and dance ensemble
【歌舞升平】sing and dance to extol the good times; put on a false show of peace and prosperity
【歌榭舞场】sing song houses and dancing halls
【歌谣】ballad; folk song; nursery rhyme
【歌咏】singing ◇ ～比赛 singing contest / ～队 singing group; chorus

戈 (古兵器) dagger-axe
【戈壁】gobi; the Gobi Desert

搁 ①(放) put: 把箱子～在房里 put the trunk in the room ②(搁置) put aside; leave over; shelve: 这件事得～一～再办。We'll have to put the matter aside for the time being.
【搁浅】①(陷在浅滩上) run aground; be stranded: 船～了。The ship got stranded. ②(遇阻停顿) be held up; be at a deadlock; reach a deadlock: 对话由于工资问题而～了。The dialogue deadlocked over the wage issue. / 会谈～了。The negotiations have come to a deadlock.
【搁置】shelve; lay aside; pigeonhole: ～一项动议 shelve a motion

疙
【疙瘩】①(疣或树瘤) a swelling on the skin; pimple; lump ②(结成团或块状的东西) lump; knot: 线结成～了。The thread has got tangled ③(难解决的事) a knot in one's heart; hang-up
【疙疙瘩瘩】[口] rough; knotty; bumpy

鸽 pigeon; dove: 家～ pigeon / 野～ wild pigeon; dove / 通信～ carrier pigeon; homing pigeon
【鸽子】pigeon; dove ◇ ～笼 dovecote; pigeon house; loft

咯
【咯噔】[象] click
【咯咯】[象] ①(形容母鸡叫声) cluck; chuckle; cackle ②(形容笑声) chuckle; titter

【咯吱】[象] creak; groan

胳

【胳臂】arm

【胳膊】arm ◇ ～腕子 wrist / ～肘儿 elbow

gé

革

①(皮革) leather; hide: ～制品 leather goods / 人造～ imitation leather / 制～厂 tannery ②(改变) change; transform: 洗心～面 turn over a new leaf ③(开除) remove from office; expel

【革除】①(去掉) abolish; get rid of: ～陈规陋习 abolish outmoded regulations and irrational practices ②(开除) expel; dismiss; remove from office

【革故鼎新】discard the old ways of life in favor of the new

【革履】leather shoes

【革命】revolution: ～到底 carry the revolution through to the end ◇ ～党 revolutionary political party / ～导师 a teacher of the revolution / ～干劲 revolutionary impetus / ～公墓 cemetery for revolutionaries / ～家 revolutionary; revolutionist / ～化 revolutionize; do sth. in a revolutionary way / ～回忆录 reminiscences of earlier revolutionary times / ～军人 revolutionary armyman / ～浪漫主义 revolutionary romanticism / ～乐观主义 revolutionary optimism / ～烈士 revolutionary martyr / ～人道主义 revolutionary humanitarianism / ～首创精神 the revolutionary pioneering spirit / ～现实主义 revolutionary realism / ～性 revolutionary nature; revolutionary character / ～英雄主义 revolutionary heroism / ～者 revolutionary

【革新】innovation: 技术～ technological innovation / 传统的手工艺技术不断～。 Traditional handicraft techniques are being steadily improved. ◇ ～派 reform party; reformers

【革职】remove from office; cashier

【革职留用】be degraded and retained at a post

【革职为民】reduce an official to the ranks of the people; be dismissed from one's office and reduced to the rank of a common citizen

葛

[植] kudzu vine

【葛布】ko-hemp cloth

【葛根】[中药] the root of kudzu vine ◇ ～素 puerarin

【葛藤】①[植] kudzu; kudzu vine ②(纠缠不清的关系) tangled and involved dificulty; complications

嗝

①(饱嗝) belch ②(呃逆) hiccup

膈

diaphragm

【膈膜】diaphragm

【膈膜炎】diaphragmatitis

【膈疝】diaphragmatocele

镉

[化] cadmium (Cd)

【镉灯】cadmium lamp

【镉电池】cadmium cell

【镉中毒】cadium poisoning

隔

①(遮断) separate; partition; stand or lie between: 把一间屋～成两间 partition a room into two ②(间隔) at a distance from; after or at an interval of: ～四小时服一片 take one tablet every four hours / 相～很远 at a great distance from each other

【隔岸观火】watch a fire from the other side of the river; look on at sb.'s trouble with indifference

【隔岸相望】face each other across the river; people on either shore can see (the houses and field) on the other side

【隔壁】next door: ～邻居 next-door neighbor

【隔别教载】separated for several years

【隔断】①(阻隔;使断绝) cut off; separate; obstruct: 同外面的联系完全～ cut off completely from the outside ②[建] partition (wall, board, etc.)

【隔阂】①(彼此情意不同) estrangement; misunderstanding: 消除～ remove the misunderstanding / 制造～ foment feelings of estrangement ②(语言障碍) barrier: 语言的～ language barrier

【隔行如隔山】difference in profession makes one feel worlds apart

【隔行书写】write in alternate lines

【隔绝】completely cut off; isolated: 和外界～ be cut off from the outside world / 音信～ (with) all news cut off

【隔离】keep apart; isolate; segregate: 被～检疫 be put under quarantine / ～病人 isolate the patients / 种族～ racial segregation; apartheid ◇ ～病房 isolation ward / ～法 isolation method

【隔离体】slider

【隔膜】①(情意不相通) lack of mutual understanding: 他们之间有些～。 They are rather

estranged from each other. ②(外行) unfamiliar with: 我对那里的情况很~。 I know very little about the situation there. ③(女用避孕器具) diaphragm: 阴道~ (contraceptive) diaphragm
【隔片】[机] spacer
【隔墙】[建] partition (wall)
【隔墙有耳】 walls have ears; beware of eavesdroppers
【隔热】[建] heat insulation
【隔扇】 partition board
【隔声】[建] sound insulation ◇ ~板 sound insulating board / ~材料 sound insulator
【隔靴搔痒】 scratch an itch from outside one's boot; attempt an ineffective solution
【隔夜】 of the previous night: 把~的菜热一热 warm up last night's leftovers
【隔音】 sound insulation
【隔音板】 acoustic celotex board; acoustic septum
【隔音符号】 syllable-dividing mark
【隔音室】 soundproof room
【隔音止响】 deafening
【隔音装置】 deafening; sound arrester

蛤 clam
【蛤蜊】 clam

阁 ①(亭) pavilion ②(内阁) cabinet: 倒~ bring down a cabinet / 组~ form a cabinet
【阁楼】 attic; loft; garret
【阁下】[敬](直接称呼) Your Excellency; (间接称呼) His or Her Excellency: 总统~ Your Excellency Mr. President; His Excellency the President
【阁员】 member of the cabinet

格 ①(方格) squares formed by crossed lines; check: 在纸上打方~儿 square off the paper ②(格子) division (horizontal or otherwise): 横~纸 ruled paper / 五~儿的书架 a bookcase with five shelves / 每服二小~。 Does: two measure each time. ③(格式) standard; pattern; style: 别具一~ have a style of its own / 合~ up to standard ④[语] case: 宾~ the objective case / 主~ the nominative case
【格调】 ①(艺术特点的综合表现) (literary or artistic) style: ~豪放 a vigorous and flowing style ② [书](人的风格或品格) one's style of work as well as one's moral quality
【格斗】 grapple; wrestle; fistfight
【格格不入】 incompatible with; out of tune with; out of one's element; like a square peg in a round hole
【格局】 pattern; setup; structure: 在这场足球比赛中我队始终保持"四四二"的~。 Throughout the football match our team kept to the 4-4-2 pattern.
【格林纳达】 Grenada ◇ ~人 Grenadian
【格林效应】[航空] Glenn effect
【格林威治平时】 Greenwich mean time (GMT)
【格林威治子午线】 Greenwich meridian
【格陵兰】 Greenland
【格律】 rules and forms of classical poetic composition (with respect to tonal pattern, rhyme scheme, etc.): 诗词~ poetical meter; metrical pattern of poetry
【格杀勿论】 kill on the spot with the authority of the law
【格式】 form; pattern: 公文~ the form of an official document
【格外】[副] especially; all the more
【格物致知】 study the phenomena of nature in order to acquire knowledge; study the nature of things
【格言】 maxim; motto; aphorism
【格栅】 grille; grillage; grizzly column
【格子】 check; chequer ◇ ~布 checked fabric; check / ~窗 lattice window / ~花呢 tartan

搁 bear; stand; endure: ~不住压 cannot stand

gě

舸 barge

葛
【葛仙米】[植] nostoc

gè

个 ①[量] 十~苹果 ten apples / 一~故事 a story / 忙~不停 be as busy as a bee / 睡~好觉 have a good sleep / 洗~澡 have a bath ②(单独的) individual
【个把】 one or two
【个别】 ①(单个) individual; specific: ~辅导 individual coaching / ~照顾 special consideration for individual cases ②(极少数) very few; one or two; exceptional: 只有~人请假。 Only one or two people asked for leave
【个别差异】[心] individual differences
【个别合约】 specific contract
【个别能级】 single level

【个别项目】end item
【个儿】①(大小) size; height; stature ②(一个个的人、物) persons or things taken singly: 挨～握手 shake hands with each one / 论～卖 be sold by the piece
【个个】each and every one; all
【个个计竭】no one has any plan or suggestion to offer
【个人】①(一个人) individual (person): 用他～的名义 in his own name / ～负责 individual responsibility ②(自称) I: ～认为 in my opinion ◇ ～财产 personal property / ～得失 personal gains or losses / ～防卫 private defense / ～冠军 individual championship / ～开支 personal outlays / ～迷信 cult of the individual; personality cult / ～名誉 personal reputation / ～企业 individual enterprise / ～权利 individual rights / ～收入 personal income / ～所得税 individual income tax; personal income tax / ～卫生 personal hygiene / ～项目[体] individual events / ～信用 personal credit / ～野心 personal ambition / ～野心家 careerist / ～隐私 individual privacy / ～英雄主义 individualistic heroism / ～责任 personal liability / ～帐户 personal account / ～主义 individualism / ～资本 personal capital
【个体】individual ◇ ～工商业 private industry and commerce / ～经济 individual economy / ～经营执照 individual license / ～劳动者 a person who works on his own; self-employed laborer / ～农业经营 individual farming / ～生产者 individual producer / ～所有制 individual ownership
【个性】individual character; individuality; personality: 共性和～ the general and specific character of sth. / 她～很强。 She is a woman of strong character <personality>.
【个中】[书] therein: ～奥妙 the inside story; the secret of it ◇ ～人 a person in the know
【个中老手】an expert in a given field; a big pot; an old hand in
【个中三昧】secret known only to experts
【个中原因】the whys and wherefores of
【个子】height; stature; build: 高～ a tall person

各 each; every; various; different: ～不相同 have nothing in common with each other
【各安职守】stay at their posts; each minds his own business
【各霸一方】each one lorded it over a district
【各半】half and half; fifty-fifty: 成败的可能性～。The chances of success are fifty-fifty.
【各奔前程】each pursues his own course; each goes his own way
【各别】①(区别) distinct; different: ～对待 treat differently; treat each on its (his, etc.) own merits ②(特殊) peculiar; out of the ordinary: 式样很～ the form is rather peculiar
【各不相犯】each keeping within his sphere
【各不相关】each looks after himself
【各不相让】each refused to give in to the other
【各持己见】each sticks to his own view
【各打五十大板】blame both sides without discrimination; punish the innocent and the guilty alike
【各得其所】each is in his proper place; each is properly provided for; each has a role to play
【各个】①(每个) each; every; various ②(分别) one by one; separately: ～解决 piecemeal solution
【各怀鬼胎】each with his own axe to grind
【各行各业】the various walks of life; all trades and professions
【各级】all or different levels: ～党组织 party organizations at different levels
【各界】all walks of life; all circles: ～人士 personalities of various circles
【各尽其用】each answers the purpose intended
【各尽其责】each one does his duty
【各尽所能,按劳分配】from each according to his ability, to each according to his work
【各尽所能,按需分配】from each according to his ability, to each according to his needs
【各就位】[体] On your marks!
【各人】each one; everyone: ～自扫门前雪,不管他人瓦上霜。 Each one sweeps the snow from his own doorstep and doesn't bother about the frost on his neighbor's roof.
【各色】of all kinds; of every description; assorted: 商店里～货物,一应俱全。 The shop is well stocked with goods of all kinds.
【各抒己见】each airs his own views
【各位】①(大家) everybody: ～请注意! Attention please, everybody. ②(每个) every: ～代表 fellow delegates
【各显神通】each showing his special prowess <skill>; each trying for all one is worth
【各向同性】[物] isotropy
【各向异性】[物] anisotropy
【各行其是】each does what he thinks is right; each goes his own way

【各有千秋】each has something to recommend him; each has his strong points
【各有所长】each has his own strong points
【各有所好】each has his likes and dislikes; each follows his own bent
【各执一词】each sticks to his own version or argument
【各自】each; respective: 既要～努力,也要互相帮助。 There must be both individual effort and mutual help.
【各自为政】each does things in his own way

硌 [口] (of sth. hard or bulging)press or rub against: 鞋里有砂子,～脚。 There's some grit in the shoe, and it hurts my foot.

铬 [化] chromium (Cr)
【铬钢】[冶] chromium steel; chrome steel
【铬镍钢】chrome-nickel steel
【铬鞣】[皮革] chrome tanning ◇ ～革 chrome leather
【铬铁】ferrochrome ◇ ～矿 chromite

gěi

给 ①(使人得到) give; grant: ～某人某物 give sb. sth. / ～他一个星期的假 grant him a week's leave ②(用在动词后面,表示交付)信已经交～他了。 I've handed the letter to him. ③(表示行为的对象或有关事物) for; for the benefit of: 她～旅客送水倒茶。 She brought drinking water and tea for the passengers. ④(让) let; allow: ～我看看。 Let me have a look. ⑤(表示被动,相当于"被")我们的衣服～汗水湿透了。 Our clothes were soaked with sweat. ⑥ [助] 把纸收起来,别叫风～刮散了。 Put away all the paper. Don't let it get blown about.
【给以】give; grant: ～充分的重视 pay ample attention to / ～好评 give good opinions; accord commendation
【给以补偿】recompense

gēn

根 ①(植物的根) root: 块～ root-tuber / 树～ root of a tree / 须～ fibrous root / 支撑～ buttress root / 直～ tap root / 连～拔 pull up by the root ②[数](代数方程的解) root: 立方～ cube root / 平方～ square root ③ [化](带电的基) radical: 酸 ～acid radical ④(下部;基部) root; foot; base: 城墙～ the foot of a city wall / 舌～ the root of the tongue ⑤(本原) cause; origin; source; root: 祸～ the root of trouble or disaster ⑥(彻底) thoroughly; completely: ～除 completely do away with; eradicate ⑦[量](用于细长的东西)一～火柴 a match
【根本】①(最重要的) basic; fundamental; essential; cardinal: ～原因 basic reason; root cause / ～原则 cardinal principle ②(从来,多用于否定) at all: 我～不同意他的看法。 I entirely disagree with him. / 这～行不通。 This is utterly impracticable. ③(彻底) radically; thoroughly: 必须～改变我们这里的落后面貌。 We must thoroughly overcome our backwardness.
【根本法】fundamental law; a body of basic laws
【根除】thoroughly do away with; eradicate; root out; eliminate: ～水患 eliminate the scourge of floods / ～一切形式的殖民主义 eradicate all forms of colonialism
【根粗实累】when the root is vigorous the trees will be laden with fruit
【根底】①(基础) foundation: ～浅 have a shaky foundation / 他的中文～很好。 He has a solid foundation in Chinese. ②(底细) cause; root: 追问～ inquire into the cause of the matter
【根腐病】[农] root rot
【根冠】[植] root cap
【根号】[数] radical sign
【根基】foundation; basis: 打好～ lay a solid foundation
【根茎】[植] rhizome; rootstock
【根究】make a thorough investigation of; get to the bottom of; probe into: ～缘由 probe into the cause
【根据】①(依据) on the basis of; according to; in the light of; in line with: ～情节轻重 in accordance with the seriousness of the case / ～天气预报 according to the weather forecast ②(作为根据的事物) basis; grounds; foundation: 毫无～ utterly groundless / 说话要有～。 One should avoid making assertions without good grounds.
【根据地】base area; base
【根绝】stamp out; eradicate; exterminate: ～事故 eliminate accidents / ～血吸虫病 stamp out snail fever
【根绝弊端】do away with bad practices
【根瘤】[植] root nodule ◇ ～菌 nodule bacteria
【根毛】[植] root hair
【根苗】①(根和幼苗) root and shoot ②(事物的来由) source; root ③ [旧](子孙) offspring

【根深蒂固】deep-rooted; ingrained; inveterate: ～的偏见 deep-rooted prejudice
【根深叶茂】have deep roots and luxuriant leaves; be well established and vigorously developing
【根式】[数] radical
【根外追肥】foliage dressing; foliage spray
【根系】[植] root system
【根由】cause; origin
【根源】source; origin; root
【根指数】[数] radical exponent; index of the root
【根治】effect a radical cure; cure once and for all; bring under permanent control: ～支气管炎 effect a radical cure of bronchitis ◇ ～手术 [医] radical operation
【根轴系】[植] root system
【根子】[口] root; source; origin

跟 ①(脚跟) heel: 鞋后～ the heel of a shoe / 左脚～ heel of the left foot ② (跟随) follow: ～我来。 Come along with me. ③[介] (表示"和""同"): 有事要～群众商量。 Consult the masses when a problem crops up. ④[介](表示"向""对"): 快～大伙说说。 Tell us all about it. ⑤[介] (引进比较的对象): 今天的活儿～往常一样。 Our job today is the same as before. ⑥[连] and: 种子～农药都准备好了。 The seeds and the pesticide are both ready.
【跟班】①(随同某一集体劳动、学习) join a regular shift or class: ～劳动 go to work in a workshop for a specified period of time ②[旧] (扈从) footman
【跟脚】[口] ①(鞋大小合适) fit well ②(随即) close upon sb.'s heels
【跟前】in front of; close to; near: 桌子～靠着一支猎枪。 A shotgun leans against the table.
【跟上】keep pace with; catch up with; keep abreast of: ～形势 keep abreast of the current situation / 快～! Close up!
【跟随】follow; go after: 他～着先辈的足迹前进。 He followed in the steps of forerunners and marched onward.
【跟随左右】follow one wherever one goes
【跟头】①(摔倒) fall: 摔～ have a fall ②(身体向下翻转) somersault: 翻个～ turn somersault
【跟头虫】wiggler; wriggler
【跟踪】follow the tracks of: 雪地～ follow sb.'s tracks in the snow / ～敌舰 shadow the enemy warships / ～追击 go in hot pursuit of
【跟踪缉捕】follow a clue and put someone under arrest (as a thief)

gén

哏 [方] ①(滑稽) amusing; comical; funny: 这孩子笑的样子有点儿～。 The way the child laughs is quite funny. ②(有趣的语言或动作) clownish speech or behavior; clowning; antics: 逗～ play the fool

gěn

艮 [方] ①(坚韧不脆) (of food) tough; leathery ②(性子直; 说话生硬) straightforward; forthright; blunt: 他说的话太～。 He put it too sharply. / 这人真～! That fellow is really blunt.

gèn

亘 extend; stretch: ～古及今 from time immemorial down to the present day
【亘古未有】There was none ever since the beginning of the age.

gēng

庚 ①(天干的第七位) the seventh of the ten Heavenly Stems ②(年龄) age: 同～ of the same age

羹 a thick soup: 鸡蛋～ egg custard (usu. salty)
【羹匙】soup spoon; tablespoon

耕 plough; till: 春～ spring ploughing / 精～细作 intensive cultivation / 深～细作 deep ploughing and intensive cultivation
【耕层】[地质] topsoil
【耕畜】farm animal
【耕地】①(翻松土地) plough; till ②(种植农作物的土地) cultivated land: 可～ arable land / 未～ uncultivated land / 休～ fallow land / ～面积 area under cultivation; cultivated area
【耕具】tillage implements
【耕牛】farm cattle
【耕云播雨】command the clouds and rain
【耕耘】ploughing and weeding; cultivation: 一分～,一分收获。 The more ploughing and weeding, the better the crop.
【耕者有其田】land to the tiller
【耕种】till; cultivate
【耕作】tillage; cultivation; farming ◇ ～方法 methods of cultivation; farming methods / ～机械 tillage machinery / ～技术 farming technique / ～园田化 garden-style cultivation of

farmland; gardenization / ～制度 cropping system
【耕作土】[地质] mold
更 ①(改变;改换) change; replace: 除旧～新 replace the old with the new ②[书](经历) experience: 少不～事 young and inexperienced ③[旧](更次) one of the five two-hour periods into which the night was formerly divided; watch: 打～ beat the watches / 三～半夜 in the dead of night
【更杯易箸】bring clean cups and chopsticks; to renew the feast
【更迭】alternate; change: 内阁～ a change of cabinet
【更动】change; alter: 人事～ personnel changes
【更番】alternately; by turns
【更夫】[旧] night watchman
【更改】change; alter: 由于天气恶劣,飞机不得不～航线。 Owing to bad weather the plane had to change its course.
【更换】change; replace: ～位置 change places / ～姓名 change of name
【更年期】climacteric; climacterium; menopause ◇ ～精神病 involutional psychosis; involutional melancholia
【更仆难数】too many to count; innumerable
【更深漏残】in the dead of night
【更深人静】deep is the night and all is quiet
【更深夜阑】in the dead of night
【更生】①(重新得到生命) regenerate; revive: 自力更～ regeneration through one's own efforts; self-reliance ②(再生) renew: 可～和不可～的海洋资源 renewable and nonrenewable marine resources
【更生霉素】[药] actinomycin D
【更替】replace: 生产方式的～ the replacement of one mode of production by another
【更新】renew; replace: 设备～ renewal of equipment / 万象～。 Everything takes on a new look.
【更新伐】[林] regeneration felling
【更新改造】make replacements and technical innovations
【更新率】turnover rate
【更新世】[地] the Pleistocene (Epoch)
【更新造林】reforestation
【更衣】change one's clothes ◇ ～室 change-room; locker room
【更正】make corrections: 作必要的～ make necessary corrections / ～通知 advice to correction

gěng

耿 ①[书](光明) bright ②(专注的;耿耿) dedicated ③(耿直) honest and just; upright
【耿耿】①(忠诚) devoted; dedicated: 忠心～ loyal and faithful ②(有心事) have sth. on one's mind; be troubled: ～不寐 lose sleep over sth. / ～于怀 brood on (an injury, one's neglected duty, etc.); take sth. to heart
【耿氏模式】Gunn mode
【耿氏效应】Gunn effect
【耿直】honest and frank; upright: 秉性～ be upright by nature
埂 ①(田埂) a low bank of earth between fields ②(泥筑的堤防) an earth dike ③(地势高起的长条地方) a long, narrow mound
梗 ①(植物枝茎) stalk; stem: 茶叶～ tea stalks / 荷～ lotus stem ②(细长的木棍或金属棍) a slender piece of wood or metal: 火柴～ matchstick ③(挺直) straighten: ～着脖子 straighten up one's neck ④(阻塞;妨碍) obstruct; block: 从中作～ place obstacles in the way; put a spoke in sb.'s wheel
【梗概】broad outline; main idea; gist: 故事的～ the gist of a story; synopsis / ～表示 skeletal representation
【梗塞】①(阻塞) block; obstruct; clog: 道路～ with the road blocked up / 交通～ traffic jam ②[医](动脉堵塞) infarction: 心肌～ myocardial infarction
【梗死】infarct: ～形成 infarction
【梗直】honest and frank; upright
【梗阻】①(拦挡) block; obstruct; hamper: 横加～ unreasonably obstruct / 山川～ be separated by mountains and rivers; be far away from each other ②[医] obstruction: 不完全～ partial obstruction / 肠～ intestinal obstruction / 幽门～ pyloric obstruction
【梗阻性黄疸】obstructive jaundice
【梗阻性痛经】mechanical dysmenorrhea
哽 choke (with emotion); feel a lump in one's throat
【哽咽】choke with sobs
【哽咽难言】choke with sobs and be unable to speak
鲠 ①[书](鱼骨) fishbone ②(鱼骨等卡在喉咙里) (of a fishbone) get stuck in one's throat

【鲠直】honest and frank; upright

gèng

更 ①(更加) more; still more; even more: 这样～好。 It's better this way ②(再;又) further; furthermore; what is more: ～进一步 go a step further / ～上一层楼 climb one story higher; attain a yet higher goal; scale new heights / ～有甚者 What is more

【更加】more; still more; even more: 问题～复杂了。 The problem became even more complicated.

gōng

宫 ①(宫殿) palace ②(庙宇名称) temple: 雍和～ the Lama Temple of Peace and Harmony (in Beijing) ③(文化娱乐场所) a place for cultural activities and recreation: 工人文化～ the Worker's Cultural Palace ④[生理] womb; uterus

【宫灯】palace lantern

【宫殿】palace ◇ ～式建筑 palatial architecture

【宫调】[乐] modes of ancient Chinese music

【宫颈糜烂】cervical erosion

【宫内节育器】intrauterine device (IUD)

【宫女】a maid in an imperial palace; maid of honor

【宫阙】[书] imperial palace

【宫廷】①(帝王的住所) palace ②(帝王大臣构成的统治集团) royal or imperial court; court

【宫廷政变】palace coup

【宫外孕】[医] ectopic pregnancy; extrauterine pregnancy

【宫闱】[书] palace chambers

【宫刑】castration (a punishment in ancient China)

工 ①(工人;工人阶级) worker; workman; the working class: 矿～ miner / 女～ woman worker ②(工作) work; labor: 上～ go to work ③(工程) (construction) project: 动～ begin a project / 竣～ complete a project ④(工业) industry: 化～ chemical industry / ～交战线 the industry and communications front; industry and communications ⑤(一个劳动日的工作) man-day: 这个活需要五个～。 This job needs five man-days. ⑥(技术和技术修养) skill; craftsmanship: 唱～ (art of) singing / 做～ acting ⑦(善于) be versed in; be good at: ～诗善画 be well versed in painting and poetry

【工本】cost (of production): 不惜～ spare no expense ◇ ～费 cost of production

【工笔】[美术] fine brushwork

【工兵】engineer: 坑道～ sapper / 轻～ pioneer

【工部】the Ministry of Works in feudal China

【工厂】factory; mill; plant; works: 铁～ iron works ◇ ～交货 free at factory / ～经济 factory economy / ～区 factory district

【工场】workshop

【工潮】workers' demonstration; workers' protest movement; strike movement

【工尺】a traditional Chinese musical scale ◇ ～谱 a traditional Chinese musical notation

【工程】engineering; project: 采矿～ mining engineering / 电机～ electrical engineering / 机械～ mechanical engineering / 水利～ water conservancy project / 土木～ civil engineering / ～浩大 a gigantic project; a tremendous amount of work ◇ ～地质学 engineering geology / ～队 construction brigade / ～管理 engineering management; schedule control / ～技术人员 engineers and technicians / ～图样 engineering drawing / ～验收[建] acceptance of work

【工程兵】engineer: ～部队 engineer troops

【工程师】engineer: 总～ chief engineer

【工党】(英) the Labour Party

【工地】building site; construction site

【工段】①(施工组织) a section of a construction project ②(车间内的基层生产组织) workshop section ◇ ～长 section chief

【工蜂】worker (bee)

【工夫】①(时间) time: 需要一个月的～ will take a month's time / 明天有～再来吧。 Come again tomorrow if you have time. ②(本领;造诣) workmanship; skill; art: 练～ (of actors, athletes, etc.) practice ③(化费力气) work; labor; effort: 只要～深,铁杵磨成针。 If you work at it hard enough, you can grind an iron rod into a needle.

【工会】trade union; labor union

【工价】labor cost

【工间操】work-break exercises

【工件】workpiece; work ◇ ～夹具 workpiece holder; (work) fixture / ～架 work rest

【工匠】craftsman; artisan

【工具】tool; means; instrument; implement: 爱护～ take good care of the tools / 木工～ carpenter's tools / 生产～ implements of production / 运输～ means of transport / ～改革 improvement of tools ◇ ～车床 toolmaker lathe / ～袋 kit bag; workbag / ～房

toolhouse / ～钢 tool steel / ～箱 toolbox; tool kit; workbox
【工具书】reference book
【工科】engineering course ◇～大学 college of engineering
【工况】operating mode
【工矿企业】industrial and mining enterprises
【工力】skill; craftsmanship: ～深厚 remarkable craftsmanship / 颇见～ show the hand of a master
【工联主义】unionism
【工料】labor and materials
【工龄】length of service; standing; seniority: 一个有二十年～的工人 a worker of twenty years' standing
【工农】workers and peasants ◇～差别 the difference between industry and agriculture / ～大众 the broad masses of workers and peasants / ～联盟 worker-peasant alliance; alliance of the workers and peasants
【工农业总产值】gross output value of industry and agriculture
【工棚】①(工地上临时的简便住屋) builders' temporary shed ②(干活的简便房屋) work shed
【工期】time limit for a project
【工钱】①(作零活的报酬) money paid for odd jobs; charge for a service: 做这套衣服要多少～? How much should I pay for having the suit made? ②[口](工资) wages; pay
【工巧】exquisite; fine
【工区】work area (a grass-roots unit of an industrial enterprise)
【工人】worker; workman: 产业～ industrial worker / 农业～ farm worker ◇～贵族 labor aristocracy / ～阶级 the working class / ～俱乐部 worker's club / ～运动 labor movement
【工伤】injury suffered on the job; industrial injury: ～事故 industrial accident
【工商界】industrial and commercial circles; business circles
【工商联】[简](工商业联合会) association of industry and commerce
【工商统一税】industrial and commercial consolidated tax
【工商业】industry and commerce: 私营～ privately owned industrial and commercial enterprises ◇～者 industrialists and businessmen
【工时】man-hour
【工事】fortifications; defense works
【工头】foreman; overseer
【工团主义】syndicalism
【工效】work efficiency
【工休日】day off; holiday
【工序】working procedure; process
【工学院】engineering college
【工业】industry: 采矿～ mining industry / 电力～ power industry / 飞机制造～ aircraft industry / 服装～ clothing industry / 化学～ chemical industry / 机械制造～ machine-building industry / 基础～ basic industry / 加工～ processing industry / 煤炭～ coal industry / 汽车制造～ automobile industry / 轻～ light industry / 燃料～ fuel industry / 石油～ petroleum industry / 手～ handicraft / 塑料～ plastics industry / 冶金～ metallurgical industry / 原材料～ raw material industry / 重～ heavy industry ◇～成本制度 industrial cost system / ～大气压 technical atmosphere / ～电视 industrial television (ITV) / ～革命 the Industrial Revolution / ～国 industrialized country / ～基地 industrial base / ～酒精 industrial alcohol / ～品 industrial products; manufactured goods / ～企业 industrial enterprise / ～生产指数 industrial production index / ～体系 industrial system / ～噪声 man-made noise / ～总产值 gross value of industrial out put
【工业化】industrialization
【工艺】technology; craft: 手～ handicraft ◇～流程 technological process / ～美术 industrial art; arts and crafts / ～品 handicraft article; handiwork; handicraft / ～设计 technological design / ～水平 technological level / ～要求 technological requirements
【工友】[旧]①(同类工人) fellow worker ②(机关学校的勤杂人员) manual worker (such as janitor, cleaner, etc.)
【工欲善其事,必先利其器】A workman must first sharpen his tools if he is to do his work well.
【工贼】scab; blackleg
【工长】section chief (in a workshop, or on a building site); foreman
【工整】carefully and neatly done: 字迹～ neatly lettered
【工种】type of work in production
【工装裤】overalls: 蓝斜纹布～ blue jeans
【工资】wages; pay: 附加～ supplementary wages / 货币～ money wages / 基本～ basic wages / 计件～ piece rate wage / 计时～ payment by the hour; time wage / 名义～

nominal wage / 实际～ real wage / 给…补发～ give the back pay due to ◇ ～表 payroll; pay sheet / ～袋 pay packet / ～冻结 wage freeze / ～改革 reform of the wage system / ～级别 wage scale / ～理论 wage theory / ～率 wage rate / ～指数 index number of wage / ～制 wage system

【工字钢】I-steel

【工字形】I-shaped

【工作】work; job: 分配～ assign jobs / 努力～ work hard ◇ ～单位 an organization in which one works; place of work / ～服 work clothes; boiler suit / ～会议 working conference / ～量 amount of work; work load / ～人员 working personnel; staff member; functionary / ～日 workday; working day / ～台[机] working table; bench / ～语言 working language / ～作风 style of work / ～阻力 working resistance

【工作面】①[矿](开采矿物地点) face: 采煤～ coal face / 回采～ stope ②[机](机械加工部位) working surface ◇ ～运输机[矿] face conveyor

【工作母机】machine-tool

【工作效率】efficiency

【工作样片】[电影] rushes

【工作者】worker: 教育～ educational worker / 美术～ art worker; artist / 文艺～ literary and art workers; writers and artists / 新闻～ journalist / 音乐～ musician

【工作证】employee's card; I.D. card

攻 ①(攻打) attack; take the offensive: ～城 attack a city / 全～型选手 an all-out attack player ②(指责) accuse; charge: 群起而～之。Everyone points an accusing finger at him. ③(致力研究) study; specialize in: 他专～考古学。He specializes in archaeology.

【攻城略地】take cities and seize territory

【攻城为下,攻心为上】It is better to win the heart of the people than to capture the city; The winning of a man's friendship is a greater achievement than the capturing of a city

【攻打】attack; assault; assail

【攻读】①(勤奋学习) assiduously study; diligently study: ～法学 study law seriously ②(钻研某一门学问) specialize in: 他是～地质学的。He specializes in geology.

【攻关】①(攻打要隘) storm a strategic pass ②(解决关键性问题) tackle key problems

【攻击】①(进攻) attack; assault; launch an offensive: 发起总～ launch a general offensive ②(恶意指摘) accuse; charge; vilify: 人身～ personal attack / 无端的～ groundless charges ◇ ～点 a point chosen for attack; point of attack

【攻击机】attack plane; attacker

【攻击航线】[航空] target pattern

【攻击航向】[航空] attack heading

【攻击素】[医] aggressins

【攻坚】storm fortifications; assault fortified positions ◇ ～部队 assault troops / ～战 storming of heavily fortified positions / ～战术 tactics of storming heavily fortified points

【攻角】angle of attack

【攻讦】[书] rake up sb.'s past and attack him; expose sb.'s past misdeeds

【攻克】capture; take: ～敌军据点 capture an enemy stronghold / ～技术难关 surmount a technical difficulty; solve a difficult technical problem

【攻苦食淡】work hard and live plainly and frugally

【攻破】make a breakthrough; breach: ～敌军防线 break through the enemy defense lines

【攻其不备】strike where or when the enemy is unprepared; take sb. by surprise; catch sb. unawares

【攻其一点,不及其余】attack sb. for a single fault without considering his other aspects; seize upon one point and ignore the overall picture

【攻取】storm and capture; attack and seize

【攻势】offensive: 采取～ take the offensive / 冬季～ winter offensive / 政治～ political offensive / 客队～凌厉。The visiting team maintained a powerful offensive. ◇ ～防御 offensive defense / ～作战 offensive operation

【攻守同盟】①(结成同盟) offensive and defensive alliance; military alliance ②(相互隐瞒) an agreement between partners in crime not to give each other away; a pact to shield each other

【攻丝】[机] tapping ◇ ～机 tapping machine / ～夹头 tap holder / ～装置 chasing bar

【攻无不克】ever-victorious; all-conquering

【攻下】capture; take; overcome

【攻陷】capture; storm

【攻心】make a psychological attack; try to persuade an offender to confess

【攻占】attack and occupy; storm and capture

功 ①(功劳) meritorious service; merit; exploit: 二等～ Merit Citation Class II / 立～

perform meritorious service／ 立大～ render outstanding service／立一等～ be awarded a first class merit ②(成效) achievement; result: 好大喜～ crave greatness and success／劳而无～ work hard but to no avail／事半～倍 yield twice the result with half the effort ③(技术修养) skill: 练～ do exercises in gymnastics, acrobatics, etc.; practice one's skill ④[物] work: 机械～ mechanical work
【功败垂成】fail in a great undertaking on the verge of success; suffer defeat when victory is within one's grasp
【功标青史】cause one's fame to glow in the pages of history
【功不恕过】No merit can wipe out one's faults.
【功臣】a person who has rendered outstanding service: 不要以～自居。 Don't give yourself the airs of a hero.
【功成不居】disclaim all achievements one has made; claim no credit for one's service
【功成名遂】achieve success and win recognition; have work done and recognized
【功成身退】retire from political life after winning tremendous successes, retire from public life after a great work is done
【功成业就】be crowned with success
【功垂竹帛】one's deeds will live forever in history
【功大于过】one's achievements outweigh one's errors
【功到自然成】Constant effort yields sure success.
【功德】①(功劳和恩德) merits and virtues ②[佛教] charitable and pious deeds; benefaction; beneficence; works
【功德无量】kindness knows no bounds; great service to mankind; boundless beneficence
【功德圆满】come to a successful issue; round it off
【功过不分】no distinction is made between merits and demerits
【功过相抵】merits offset faults; deserts equal faults
【功绩】merits and achievements; contribution
【功课】schoolwork; homework: 做～ do homework
【功亏一篑】fail to build a mound for want of one final basket of earth; fall short of success for lack of a final effort
【功劳】contribution; meritorious service; credit: 立下了～ have great achievements to one's credit／躺在～簿上 rest on one's laurels
【功劳簿】record of merits
【功利】utility; material gain ◇～主义 utilitarianism／～主义者 utilitarian
【功率】[物] power ◇～计 dynamometer
【功名】scholarly honor or official rank (in feudal times)
【功名富贵】fame and fortune; fame and riches and honors
【功名利禄】high official positions and riches
【功能】function: ～锻炼 functional training／肝～正常。The liver is functioning normally. ◇～键 function key／～信号 function signal／～主义 functionalism
【功能性障碍】functional disorder
【功微德薄】(my) merit is small and (my) virtue meager
【功效】efficacy; effect
【功勋】exploit; meritorious service: 立下了不朽的～ have performed immortal feats
【功业】exploits; achievements
【功业彪炳】one's achievement is distinguished; one's feat is noteworthy and successful
【功用】function; use
【功照日月】one's achievements outshine the sun and moon

供 ①(供应;供给) supply; feed: ～不上 run out; be in short supply／～电系统 system of power supply／～水 water supply ②(提供) for (the use or convenience of): 仅～参考 for your reference only／专～矿工用的疗养所 a rest home specially for miners
【供不应求】supply falls short of demand; demand exceeds supply
【供电】power supply ◇～调度员 load dispatcher／～干线 supply main／～局 power supply bureau／～中心 center of supply
【供给】supply; provide; furnish: 原料由国家～。Raw materials are provided by the state.
【供给制】the supply system; a system of payment in kind
【供过于求】drug on 〈in〉 the market; supply exceeds demand
【供暖】[建] heating: 热水～ hot water heating／蒸气～ steam heating ◇～,通风和空气调节 heating, ventilating, and air conditioning／～系统 heating system
【供气】[机] air feed ◇～管 air supply pipe

【供求】supply and demand：～关系 the relation between supply and demand／～平衡 balance between supply and demand
【供水管】feed pipe；water service
【供销】supply and marketing：～合作社 supply and marketing cooperative
【供血者】blood donor
【供养】provide for (one's parents or elders)；support
【供应】supply：～充足 be in liberal supply／敞开～ can be bought without any restriction／计划～ planned supply／市场～ supply of commodities；market supplies／医药品～ medical supplies／商品～充足。There is an abundance of commodity supplies on the markets. ◇～点 supply center／～线 supply line／～船 tender
【供油泵】fuel feed pump
【供油箱】oil feeding reservoirs

恭 respectful；reverent：洗耳～听 listen respectfully with attentive ears
【恭贺】congratulate
【恭贺新禧】Happy New Year；with best wishes for a happy New Year
【恭候】[敬] await respectfully：～光临。We request the pleasure of your company.
【恭俭温和】be modest and retiring by nature
【恭谨】respectful and cautious
【恭敬】respectful；with great respect
【恭敬不如从命】Obedience is better than politeness.（或：Obedience is a better way of showing respect than outward reverence.）
【恭聆明诲】I will assuredly listen most reverently to your words
【恭顺】respectful and submissive
【恭桶】closestool；commode
【恭维】flatter；compliment ◇～话 flattery；compliments
【恭喜】[套] congratulations

公 ①（国家或集体的）public；state-owned；collective：交～ turn over to the authorities／出以～心 act in a public spirit ②（共同的）common；general：～分母 common denominator ③（国际间的）metric：～里 kilometer ④（使公开）make public：～之于世 make known to the world；reveal to the public ⑤（公平；公正）equitable；impartial；fair；just：秉～办理 handle affairs equitably or impartially；be evenhanded ⑥（公事）public affairs；official business：～余 after work／因～外出 be away on official business ⑦（公爵）duke ⑧（对中年以上男子的一种尊称）张～ the revered Mr. Zhang ⑨（丈夫之父）husband's father；father-in-law ⑩（雄性的）male (animal)：～鸡 cock；rooster／～牛 bull
【公安】public security ◇～部 the Ministry of public Security／～部队 public security troops／～机关 public security organs／～局 public security bureau／～人员 public security officer／～侦察机关 public security inspectorate
【公案】[旧] a complicated legal case：无头～ an intricate case without a clue
【公报】communiqué；bulletin：新闻～ press communiqué／政府～ (government) bulletin
【公报私仇】avenge a personal wrong in the name of public interests；abuse public power to retaliate on a personal enemy
【公倍数】[数] common multiple：最小～ least common multiple；lowest common multiple
【公比】[数] common ratio (of a geometric series)
【公布】promulgate；announce；publish；make public：～开会日期 announce the date of the meeting／～法令 promulgate a decree／～罪状 announce sb.'s crimes／～名单 publish a name list
【公差】① [数] common difference ② [机] tolerance：安装～ location tolerance／制造～ manufacturing tolerance
【公差】①（临时做公务）public errand；noncombatant duty：出～ go on a public errand；go on official business；perform noncombatant duty ②（差役）a person on a public errand
【公产】public property
【公称】nominal ◇～尺寸 [机] nominal dimension
【公尺】meter；metre (m.)
【公出】be away on official business
【公畜】male animal (kept for breeding)；stud
【公担】quintal (q.)
【公道】①（正义）justice：主持～ uphold justice ②（公平）fair；just；reasonable；impartial：办事～ be evenhanded；be impartial／说句～话 to be fair；in fairness to sb.／价钱～。The price is reasonable.
【公德】social morality；social ethics：有～心 be public-spirited
【公敌】public enemy
【公断】arbitration

【公断人】arbitrator; umpire
【公吨】metric ton (MT)
【公而忘私】so devoted to public service as to forget one's own interests; selfless
【公法】[法] public law
【公费】at public expense ◇ ~医疗 free medical service; public health services
【公愤】public indignation; popular anger: 引起~ arouse public indignation
【公干】business: 有何~? What important business brings you here?
【公告】announcement; proclamation
【公共】public; common; communal ◇ ~财产 public property / ~厕所 public conveniences; public latrine / ~场所 public places / ~秩序 public order / ~公证人 public surveyor / ~关系 public relations / ~积累 common accumulation; accumulation fund / ~建筑 public buildings / ~食堂 canteen; mess / ~事务 public affairs / ~卫生 public health
【公共道德】public morality: ~败坏 corruption of public moral
【公共交通】mass transit; public transit ○电车 tram; streetcar / 无轨电车 trolley bus; trackless trolley / 电车站 tram stop; streetcar stop / 公共汽车 bus / 双层公共汽车 double-decker / 长途公共汽车 inter-city bus; long-distance bus; coach / 出租汽车 taxi; taxicab / 地下铁道 underground railway; subway / 售票员 conductor / 上车 get on (a bus) / 下车 get off (a bus) / 赶上汽车 catch (a bus) / 没赶上汽车 miss (a bus) / 让乘客上车 pick up passengers / 让乘客下车 let down passengers / 乘15路汽车到… take No. 15 bus to / 在…换乘15路汽车 change to No. 15 bus at / 让座位 offer one's seat to
【公共汽车】bus ◇ ~线路 bus line / ~站 bus stop
【公公】①(丈夫之父) husband's father; father-in-law ② [方](祖父) grandfather ③[尊](尊称年老男子) grandpa; grandad
【公股】government share (in a joint state-private enterprise)
【公馆】[旧] residence (of a rich or important person); mansion
【公国】duchy; dukedom
【公海】high seas ◇ ~捕鱼权 common fishery / ~上的自由航行权 freedom of the seas / ~自由 freedom of the open sea
【公害】social effects of pollution; environmental pollution: 消除~ eliminate the public hazard
【公函】official letter
【公积金】accumulation fund
【公祭】public memorial ceremony
【公家】[口] the state; the public; the organization
【公检法】public security organs, procuratorial organs and people's courts
【公教人员】[旧] government employees and teachers
【公斤】kilogram (kg.); kilo
【公爵】duke ◇ ~夫人 duchess
【公开】①(不加隐蔽) open; overt; public: ~的秘密 an open secret ②(使秘密公开) make public; make known to the public ◇ ~裁决 public verdict / ~法庭 open court / ~论战 open polemics / ~审判 public trial; open trial / ~审讯 hold public sittings / ~挑衅 overt provocation / ~投票 disclosed ballot; open ballot / ~隐瞒 bare concealment
【公开化】come out into the open; be brought into the open
【公开毁坏他人名誉罪】public defamation
【公开市场】market overt: ~政策 open market policy
【公开投标】public bidding; public tender
【公开信】open letter
【公开招标】competitive bidding; open tender: ~价格 competitive price
【公款】public money
【公里】kilometer (km.)
【公理】①(公认的道理) generally acknowledged truth; self-evident truth ② [数] axiom
【公历】the Gregorian calendar
【公立】established and maintained by the government; public: ~学校 public school
【公粮】agricultural tax paid in grain; grain delivered to the state; public grain
【公量】[纺] conditioned weight
【公路】highway; road: 超级~ superhighway / 分道行驶~ dual highway / 干线~ arterial highway; main-line highway / 高速~ express highway; expressway / 双车道~ two-lane highway / 四车道~ four-lane highway ◇ ~工程 highway engineering / ~交通 highway communication; highway traffic / ~路面板 slab / ~立体交叉 highway grade separation / ~平面交叉 highway grade crossing / ~桥 highway bridge / ~容量 highway capacity / ~运输 highway transportation / ~支线 feeder road / ~自行车赛 road race

【公论】public opinion; verdict of the masses: 是非自有～。 Public opinion will decide which is right and which is wrong.
【公民】citizen: ～住宅不受侵犯 home(s) of citizens are inviolable ◇ ～基本义务 fundamental duties of citizens / ～权 civil rights; citizenship; citizen's rights / ～身分 citizenhood; citizenship / ～投票 referendum; plebiscite
【公亩】are (a.)
【公墓】cemetery
【公平】fair; just; impartial; equitable: ～待人 deal fairly with someone / ～的协议 an equitable agreement / ～合理 fair and reasonable / ～交易 fair deal / ～价格 fair price / 买卖～ be fair in buying and selling; buy and sell at reasonable prices
【公婆】husband's father and mother; parents-in-law
【公仆】public servant
【公切线】[数] common tangent
【公勤人员】service personnel in an office; office attendants
【公顷】hectare (ha.)
【公然】[贬] openly; undisguisedly; brazenly: ～反对 openly oppose / ～撕毁协议 brazenly tear up an agreement
【公认】generally acknowledged; (universally) accepted; established: ～的国际法准则 established principles of international law / ～的国际关系原则 generally recognized principles governing international relations / ～的领袖 acknowledged leader
【公设】[数] postulate
【公社】①(原始公社) primitive commune ②(政权形式) commune: 巴黎～ the Paris Commune / 人民～ people's commune
【公审】[法] public trial: ～大会 public trial meeting
【公使】envoy; minister ◇ ～馆 legation / ～衔参赞 counsellor with the rank of minister; minister-counsellor
【公式】formula
【公式化】①(文艺创造的固定格式) formulism (in art and literature) ②(用固定方式处理问题) formulistic; stereotyped
【公事】public affairs; official business 还是～要紧。 Public affairs should come first. ◇ ～包 briefcase; portfolio
【公事公办】do official business according to official principles; not let personal considerations interfere with one's execution of public duty
【公署】government office: (地区)专员～ prefectural commissioner's office
【公说公有理,婆说婆有理】Each says he is right. (或: Both parties claim to be in the right.)
【公司】company; corporation: 保险～ insurance company / 钢铁～ iron and steel company / 进出口～ import and export corporation ◇～法 company act; law of corporation / ～律师 corporation lawyer / ～所得税 corporation income tax / ～信贷 company credit / ～章程 corporation by-laws
【公私分明】be scrupulous in separating public from private interests
【公私兼顾】give consideration to both public and private interests
【公诉】[法] public prosecution: 对罪犯提起～ institute proceedings against a legal offender / 提起～ institute prosecution; prefer a public charge ◇ ～权 right of prosecution / ～人 public prosecutor; the prosecution / ～书 bill of indictment; bill of prosecution; public indictment
【公摊】equally shared by all
【公堂】[旧] law court; tribunal: 私设～ set up a clandestine tribunal
【公推】recommend by general acclaim
【公文】official document ◇ ～程式 forms and formulas of official documents / ～袋 document envelope / ～纸 paper for copying documents
【公务】public affairs; official business ◇ ～护照 service passport / ～人员 government functionary
【公务繁冗】be overburdened with official duties
【公务羁身】be tied down 〈up〉 by one's duties
【公务员】orderly ◇ ～法 a public servant law
【公物】public property
【公物还家】return property where it belongs
【公休】general holiday; official holiday
【公演】perform in public; give a performance
【公议】public or mass discussion: 自报～ self-assessment and public discussion
【公益】public good; public welfare: 热心～ public-spirited ◇ ～法人 public welfare legal entity / ～金 public welfare fund
【公意】public will; will of the public
【公因子】[数] common factor: 最大～ greatest common factor

【公营】publicly-owned; publicly-operated; public ◇ ～经济 the public sector of the economy; public economy / ～企业 public enterprise
【公用】for public use; public; communal ◇ ～电话 public telephone / ～事业 public utilities / ～印章 common seal
【公用权】commonage
【公有】publicly-owned; public: ～财产 public property ◇ ～化 transfer to public ownership; socialization / ～制 public ownership
【公寓】①(分成单元的住宅楼) flats; apartment house ②[旧] lodging house
【公元】the Christian era: ～前二十八年 28 B.C. / ～一九九零年 A.D. 1990
【公园】park
【公约】①(条约名称) convention; pact: 北大西洋～ the North Atlantic Treaty / 日内瓦～ the Geneva Convention ②(共同遵守的章程) joint pledge: 服务～ service pledge (given by workers in the service trades)
【公约数】[数] common divisor
【公允】just and sound; fair and equitable; evenhanded: 持论～ be just and fair in argument / 貌似～ pretend to be just and fair; put on an appearance of impartiality
【公债】(government) bonds; public debt: 经济建设～ economic construction bonds
【公章】official seal
【公正】just; fair; impartial; fair-minded: ～的裁决 true verdict / ～的判决 reasonable judgment / ～的审判 impartial trial / ～的舆论 fair-minded public opinion / ～的仲裁人 impartial arbitrator
【公证】notarization ◇ ～处 notary office; notary public office / ～会计师 certified public accountant / ～机关 notary organs / ～监督 notary supervision / ～鉴证 notary verification / ～人 notary public; notary / ～书 authentic act; notarial deed / ～委托书 letter of commitment / ～遗嘱 notary testament / ～证明 certification by a notary; notarial certification / ～证书 notarial certificate; notarial document
【公职】public office; public employment: 开除～ discharge from public employment
【公制】the metric system: 折成～ convert into the metric system ◇ ～螺纹 [机] metric thread
【公众】the public
【公诸同好】share enjoyment with those of the same taste
【公诸于众】make public
【公主】princess
【公转】[天] revolution
【公子】son of a feudal prince or high official
【公子哥儿】a pampered son of a wealthy or influential family

肱 [书] the upper arm; arm
【肱动脉】arteria brachialis
【肱二头肌】bicipital muscle of arm
【肱骨】[生理] humerus

觥 [考古] an ancient wine vessel made of horn
【觥筹交错】toast each other; wine flows freely at a dinner party; the cups go gaily round

弓 ①(发箭器械) bow: ～箭 bow and arrow ②(使弯曲) bend; arch; bow: ～着背 arch one's back; bend low / ～着腰走路 walk with one's back bent forward ③(弓子) anything bow-shaped: 小提琴的～ bow of a violin
【弓背】①(弓的背儿) back of a bow ②(曲背) hunchbacked; stooping
【弓锯】bow saw
【弓弦】bowstring ◇ ～乐器 bowed stringed instrument; bowed string instrument; bowed instrument
【弓形】①[数] segment of a circle ②(像弓的形状) bow-shaped; arched; curved
【弓子】①(发箭的弓) bow ②(弓形物) anything bow-shaped

躬 ①[书](自身;亲身) personally: ～行实践 practice what one preaches ②(弯下) bend forward; bow: ～身 bend at the waist
【躬逢其胜】be present in person on the grand occasion
【躬亲】attend to personally: 事必～ attend to everything personally
【躬亲其事】undertake an affair personally
【躬身下拜】bend the knee in obeisance

gǒng

巩 consolidate
【巩固】①(使坚固) consolidate; strengthen; solidify: ～阵地 consolidate a position ②(坚固; 不易动摇) consolidated; strong; solid; stable: ～的国防 strong national defense
【巩角膜】[生理] sclerocornea
【巩膜】[生理] sclera
【巩膜沟】[生理] scleral groove; sulcus sclerae
【巩膜膨胀】[医] sclerectasia; sclerectasis

【巩膜炎】[医] scleritis

汞 [化] mercury (Hg)

【汞弧灯】[电] mercury-arc lamp

【汞化】[化] mercuration; mercurization ◇～物 mercuride

【汞溴红】[药] mercurochrome

【汞中毒】mercurialism; mercury poisoning

珙 [书] a kind of jade

【珙桐】[植] dove tree

拱 ①(拱手) cup one hand in the other before the chest ②(环绕) surround: 众星～月。A myriad of stars surround the moon. ③(弯成弧形) hump up; arch ④ [建](弧形建筑物) arch: ～道 archway / ～式涵洞 arch culvert ⑤(推撞) push without using one's hands: 用身子把门～开 push the door open with one's body ⑥(用身子拨开土地) (of pigs, etc.) dig earth with the snout; (of earthworms, etc.) wriggle through the earth ⑦(植物从土里长出) sprout up through the earth: 苗儿～出土了。The shoots spring up from the soil.

【拱坝】[水] arch dam

【拱抱】surround: 群峰～的山坞 a cove surrounded by cliffs

【拱点】[天] apsis; apse

【拱顶】[建] vault

【拱顶支架】bow supporter

【拱肩缩背】hunch one's shoulders and bow one's back

【拱梁】arched girder

【拱门】[建] arched door

【拱桥】[建] arch bridge: 双曲～ double-curvature arch bridge

【拱曲】hog; hogging

【拱手】①(双手在胸前相抱) make an obeisance by cupping one hand in the other before one's chest ②(顺从地) submissively: ～让人 surrender sth. submissively; hand over sth. on a silver platter

【拱手称谢】join one's hands together in salute and thank

【拱手告别】bid farewell in a respectful manner

【拱手旁观】look on with folded arm

【拱手致礼】salute with joined hands; salute with the hands folded

【拱卫】surround and protect

【拱形桁架】arch truss

【拱形支护】arching

【拱券】[建] arch

【拱座】abut

gòng

贡 tribute: 进～ pay tribute (to an imperial court)

【贡品】articles of tribute; tribute

【贡税】tribute and taxes

【贡献】contribute; dedicate; devote: ～自己的一分力量 contribute one's share; do one's bit (for) / 对某事有所～ contribute to sth.

共 ①(共同) common; general: ～享 enjoy in common / ～性 general character ②(共同具有或承受) share: ～患难 share hardship / 同呼吸,～命运 share a common fate; throw in one's lot with sb. ③(一起; 一齐) doing the same thing; together ④(总计) altogether; in all; all told: ～三万字 a total of thirty thousand words; thirty thousand words altogether ⑤[简](共产党) the Communist Party: 中～ the CPC (the Communist Party of China)

【共产党】the Communist Party

【共产国际】the Communist International (1919-1943); Comintern

【共产主义】communism ◇～道德 communist morality / ～风格 communist style / ～精神 communist spirit / ～觉悟 communist consciousness / ～战士 fighter for communism

【共产主义青年团】the Communist Youth League

【共处】coexist: 和平～五项原则 the five principles of peaceful coexistence

【共存】coexist; survive together

【共电制】[讯] common-battery system ◇～电话机 common-battery telephone / ～电话局 common-battery telephone exchange

【共轭】conjugate ◇～点[数] conjugate point / ～根[数] conjugate roots / ～角[数] conjugate angles / ～面 conjugate surface / ～像[物] conjugate image / ～轴 conjugate axis

【共发射极】[电子] common emitter

【共犯】[法] accomplice

【共赴国难】call for united efforts to save the nation

【共管】[外] condominium

【共和】republicanism; republic

【共和国】republic

【共基极】[电子] common base

【共计】amount to; add up to; total: ～五千元 the total is five thousand *yuan*

【共济失调】[医] incoordination; dystaxia;

ataxia
【共价】[化] covalence ◇ ～键 covalent bond
【共聚】[化] copolymerization ◇ ～物 copolymer
【共聚一堂】gather in the same hall
【共面法】coplaner method
【共面力】coplaner force
【共勉】mutual encouragement: 愿～之。Let us encourage each other in our endeavors.
【共鸣】① [物] resonance ② (引起相同的情绪) sympathetic response: 引起～ arouse sympathy; strike a sympathetic chord ◇ ～器 resonator
【共模】[电子] common mode
【共谋】[法] collusion; conspire
【共栖】[生] commensalism
【共生】① [地] intergrowth; paragenesis: 矿物～ mineral intergrowth ② [生] symbiosis ◇ ～次序 [地] paragenesis / ～细菌 symbiotic bacteria
【共生体】homobjum
【共生星】symbiotic object
【共事】work together (at the same organization); be fellow workers: ～多年 have been colleagues for many years
【共室】coenecium; coenoecium
【共体】coenosarc
【共通】applicable to both or all
【共同】① (属于大家的) common: ～敌人 common enemy / ～关心的问题 matters of common concern; issues of common interest / ～语言 common language / 有～之处 have something in common ② (大家一起) together; jointly: ～努力 make joint efforts / ～战斗 fight side by side ◇ ～保险 coinsurance; coinsure / ～被告 joint defendants / ～财产 community property / ～担保 coinsurance; joint guarantee / ～点 common ground / ～犯罪 joint offense; fellowship in crime / ～纲领 common program / ～过失 joint negligence / ～继承 coinheritance; parcenary / ～诉讼 colitigation; joint action / ～投资 joint venture / ～原告 coplaintiff / ～愿望 common aspiration / ～债权 joint credit / ～债务 common obligation; joint debt / ～帐户 joint account
【共同市场】the Common Market
【共同体】community: 欧洲经济～ the European Economic Community
【共析】[冶] eutectoid ◇ ～钢 eutectoid steel
【共叙友情】exchange sentiments of friendship with sb.
【共襄义举】let everybody help to promote this worthy undertaking
【共享】enjoy together; share: ～甘苦 share the sweet and the bitter together
【共性】general character; generality
【共振】[物] resonance ◇ ～灯 resonance lamp / ～腔 formant; resonance cavity / ～器 resonator / ～示波器 resonoscope / ～态 resonances
【共轴】coaxial ◇ ～圆柱 [数] coaxial cylinders / ～圆锥 coaxial cone

供

① (供奉) lay (offerings) ② (供品) offerings ③ (供认) confess; own up: 据该犯～称 as was confessed by the culprit / 他～出了主犯的名字。He gave the name of the chief culprit. ④ (供词) confession; deposition: 口～ oral confession / 串～ act in collusion with others to make confessions tally
【供词】a statement made under examination; confession
【供奉】enshrine and worship; consecrate
【供具】sacrificial vessel
【供品】offerings
【供认】confess: ～不讳 confess everything; candidly confess
【供职】hold office
【供状】written confession; deposition
【供桌】altar

gōu

篝

[书] cage
【篝火】bonfire; campfire
【篝火狐鸣】a rebellion is afoot

佝

【佝偻病】rickets; rachitis

勾

① (删除) cancel; cross out; strike out; tick off: ～了这笔帐 cancel the debt / ～了这一项 strike out this item ② (描画) delineate; draw: ～出一个轮廓 draw an outline ③ (用灰泥涂抹) fill up the joints of brickwork with mortar or cement; point: ～墙缝 point a brick wall ④ (调和使粘) thicken: ～芡 thicken soup ⑤ (招引; 引起) induce; evoke; call to mind: 这件事～起了他的回忆。This incident called forth his reminiscences. ⑥ (结合) collude with; gang up with ⑦ (中国古代数学中不等腰直角三角形的短直角边) the shorter leg of a right triangle
【勾搭】① (互相串通) gang up with: 这三个坏

家伙～上了。 The three scoundrels ganged up ②(勾引) seduce
【勾股定理】[数] the Pythagorean theorem; the Pythagorean proposition
【勾股形】(中国古代数学) right triangle
【勾画】draw the outline of; delineate; sketch
【勾魂摄魄】summon spirits; have the power to make men crazy
【勾结】collude with; collaborate with; gang up with
【勾栏中人】a prostitute; the women in the prostitutes' quarter
【勾勒】①(画出轮廓) draw the outline of; sketch the contours of ②(写出大致情况) give a brief account of; outline
【勾通】collude with; work hand in glove with
【勾销】liquidate; write off; strike out: ～债务 liquidate a debt / 一笔～ write off at one stroke
【勾心斗角】intrigue against each other; jockey for position
【勾引】tempt; entice; seduce

沟 ①(沟渠) ditch; channel; trench: 排水～ drainage ditch; drain / 阴～ sewerage ②(浅槽) groove; rut; furrow: 开～播种 make furrows for sowing ③(一般水道) gully; ravine: 乱石～ boulder strewn gully / 七～八梁 seven gullies and eight ridges; full of gullies and ridges
【沟槽】[军] cannelure; furrow; plow groove
【沟槽刨】dado plane; trenching plane
【沟灌】[农] furrow irrigation
【沟壑】gully; ravine
【沟渠】irrigation canals and ditches: 田野上～纵横。 The field are crisscrossed by irrigation canals and ditches.
【沟流】[化] channet
【沟通】link up: ～两国文化 link the cultures of the two countries / ～南北的铁路 railroad connecting north and south
【沟纹】[天] rill

钩 ①(钩子) hook: 窗～ window catch / 挂衣～ clothes-hook / 鱼～ fishhook ②(汉字笔画) hook stroke (in Chinese characters) ③(钩形符号) check mark; tick ④(钩住) secure with a hook; hook: 围巾给树枝～住了。 The scarf was caught by the branches. ⑤(用针粗缝) sew with large stitches: ～贴边 sew on an edging ⑥(用带钩的针编织) crochet: ～花边 crochet lace
【钩虫】hookworm ◇ ～病 hookworm disease; ancylostomiasis
【钩端螺旋体病】leptospirosis
【钩骨】hamate bone; os hamatum
【钩键】switch hook; hook switch
【钩肩搭背】bend one's arm round somebody's shoulder
【钩头键】gib head taper stock key
【钩突】unciform process; processus uncinatus
【钩吻】[植] elegant jessamine
【钩心斗角】intrigue against each other; jockey for position
【钩形板手】claw wrench; hook key; hook spanner
【钩针】crochet hook ◇ ～编织品 crochet
【钩子】hook

gǒu

苟 ①(随便) careless; negligent; indifferent (to right or wrong): 不～言笑 be not indiscreet in speech and manner; be solemn and dignified / 不敢～同 can hardly agree / 一丝不～ be not the least bit negligent; be scrupulous about every detail; be conscientious and meticulous ②[书](假使) if: ～不努力,必将落后。 If you do not make any effort, you are bound to lag behind. / ～能坚持,必将胜利。 If you can persist, you are sure to win.
【苟安】seek momentary ease; be content with temporary ease and comfort
【苟安一时】gain some respite for oneself; seek security for a time
【苟合】illicit sexual relations
【苟活】drag out an ignoble existence; live on in degradation
【苟且】①(得过且过) drift along; be resigned to circumstances ②(敷衍了事) perfunctorily; carelessly ③(不正当的) illicit (sexual relations); improper
【苟且偷安】live in a fool's paradise; seek temporary peace
【苟且偷生】drag out an ignoble existence; keep alive without serious ambition
【苟全】preserve (one's own life) at all costs
【苟全性命】barely to manage to survive; manage to stay alive with sacrifice of principles
【苟同】(用于否定句) agree without giving serious thought; readily subscribe to (sb.'s view): 不敢～ beg to differ; cannot agree
【苟延残喘】be on one's last legs; linger on in a steadily worsening condition

枸

【枸骨】[植] Chinese holly

【枸杞子】[中药] the fruit of Chinese wolfberry

狗 ①(犬) dog: 癞皮～ mangy dog／猎～ hunting dog; hound ② [骂] damned; cursed: ～地主 the cursed landlord

【狗宝】[中药] the stone of a dog's gallbladder, kidney or bladder

【狗胆包天】monstrous audacity

【狗吠非主】The dog barks at a man who is not his master

【狗苟蝇营】shamelessly to seek personal gain

【狗獾】badger

【狗急跳墙】a cornered beast will do something desperate

【狗脊蕨】[植] chain fern

【狗恐怖】[心理] cynophobia

【狗拿耗子,多管闲事】a dog trying to catch mice; too meddlesome; poke one's nose into other people's business

【狗皮膏药】① [中药](药膏涂在狗皮上的膏药) dogskin plaster (a plaster for rheumatism, strains, contusions, etc., formerly spread on dogskin, but now usu. on cloth) ② (骗人的货色) quack medicine

【狗屁】[骂] horseshit; rubbish; nonsense: ～不通 unreadable rubbish; mere trash

【狗橇比赛】dogsled racing

【狗头军师】①(出坏主意的人) a person who offers bad advice; inept adviser ②(极坏的顾问) villainous adviser

【狗偷鼠窃】a petty theft

【狗腿子】[口] hired thug; lackey; henchman

【狗尾草】[植] green bristlegrass

【狗尾续貂】a wretched sequel to a fine work

【狗窝】kennel; doghouse

【狗熊】①(黑熊) black bear ②(懦夫; 胆怯者) coward

【狗血喷头】(多用于)骂得～ let loose a stream of abuse against sb.; pour out a flood of invective against sb.

【狗眼看人低】be damned snobbish; act like a snob

【狗牙草】Bermuda grass

【狗咬狗】dog-eat-dog; dogfight: 帝国主义～的战争 an imperialist dog-eat-dog type of war

【狗咬吕洞宾】snarl and snap at Lü Dongbin; mistake a good man for a bad one

【狗蝇】dog louse fly

【狗鱼】[动] pike

【狗崽子】① pup; puppy ② [骂] son of a bitch

【狗蚤】dog flea

【狗仗人势】[骂] like a dog threatening people on the strength of its master's power; be a bully under the protection of a powerful person

【狗彘不如】worse than a cur or a swine

【狗嘴里吐不出象牙】a dog's mouth emits no ivory; a filthy mouth can't utter decent language; what can you expect from a dog but a bark

gòu

媾 [书] ①(结为婚姻) wed: 婚～ marriage ②(友好) reach agreement ③(交配) coition: 交～ copulate

【媾和】make peace: 单独～ make peace without consulting one's allies; make a separate peace

彀 a bow drawn to the full

【彀中】[书] shooting range: 尽入～ have all come within shooting range; have all fallen into the trap

诟 [书] ①(耻辱) shame; humiliation ②(辱骂; 怒骂) revile; talk abusively

【诟病】[书] denounce; castigate: 为世～ become an object of public denunciation

【诟骂】revile; abuse; vilify

垢 ① [书](污秽) dirty; filthy: 蓬头～面 with dishevelled hair and a dirty face ②(脏东西) dirt; filth: 尘～ dust and dirt／牙～ dental calculus／油～ grease stain／藏污纳～ harboring the filth; shelter evil people and countenance evil practices ③ [书](耻辱) disgrace; humiliation: 含～忍辱 endure humiliation and insult; (be forced to) swallow insults

够 ①(足够) enough; sufficient; adequate: ～三天的粮食 provisions enough for three days ②(达到某一点或某种程度) reach; be up to (a certain standard, etc.): ～得着吗? Can you reach it? ③(确实的; 真的) quite; rather; really: 这场足球赛可真～意思。That football match was really something. ／这儿的土～肥的。The soil here is quite fertile.

【够本】make enough money to cover the cost; break even

【够格】be qualified; be up to standard: 她当教师很～。 She is highly qualified to be a teacher.
【够劲】strong enough: 这酒不～。The liquor is not strong enough.
【够朋友】deserve to be called a true friend; be a friend indeed
【够戗】[方] unbearable; terrible: 疼得～ unbearably painful / 这家伙真～! He's simply impossible.
【够瞧的】really awful; too much: 天热得真～。The weather is terribly hot.
【够受的】quite an ordeal; hard to bear: 累得～ be dog-tired
【够数】sufficient in quantity; enough: 你领的书不～。You didn't get enough books to go round.
【够条件】reach the standard; be qualified
【够味儿】just the right flavor; just the thing; quite satisfactory
【够意思】①(好极了) really something; terrific: 这场球赛可真～。That was really a terrific game. ②(慷慨;大方) generous; really kind: 不～ unfriendly; ungrateful

勾

【勾当】[贬] business; deal: 肮脏～ a dirty deal / 罪恶～ criminal activities

构

①(组合;构造) construct; form; compose: ～词 form a word ②(结成) fabricate; make up: 虚～ fabrication ③(指文艺作品) literary composition: 佳～ a good piece of writing
【构成】constitute; form; compose; make up: ～部分 component part / ～犯罪的要件 requisites to constitute a crime / ～威胁 constitute a threat
【构词法】[语] word-building; wood-formation
【构架】boom; staging; structural frame
【构架杆】truss bar
【构件】①[建] (structural) member; component ②[机](零件;刚件) component (part)
【构思】①(运用心思) (of writers or artists) work out the plot of a literary work or the composition of a painting: 影片的～相当巧妙。 The plot of the film is ingeniously conceived. ②(想象) conception: 大胆的～ boldness of conception / ～新颖 original in conception
【构图】[美术] composition (of a picture)
【构形】[教] configuration: ～成分 configurational symmetry
【构型】configuration: ～分析 conformational analysis
【构造】①(各组成部分及其相互关系) structure; construction: 句子～ sentence construction / 人体～ the structure of the human body ②[地] tectonic; structural ◇～地震 tectonic earthquake / ～地质学 structural geology / ～复合 compounding of structures / ～陆地 tectonic land / ～论 constructionism / ～体系 structural system / ～序次 structural generation / ～要素 structural element / ～运动 tectonic movement / ～转向 vergence
【构筑】construct (military works); build: ～工事 construct field works; build defenses; dig in ◇～物 [建] structures

购

purchase; buy: ～粮 purchase grain / 订～ place an order / 赊～ buy on credit; sales on credit
【购货单】order form; order
【购买】purchase; buy
【购买力】purchasing power; purchasing power ◇～平价理论 purchasing-power-parity theory
【购买欲】desire to buy; deside to purchase
【购销】purchase and sale; buying and selling: ～两旺 brisk buying and selling; both purchasing and marketing are brisk
【购置】purchase (durables): ～农具 purchase farm implements

gū

沽

①(买) buy: ～酒待客 buy wine to entertain a guest ②(卖) sell: ～酒为生 sell wine for a living / 待价而～ wait to sell at a good price; wait for the highest bid
【沽名钓誉】fish for fame and compliments

辜

guilt; crime: 无～ guiltless; innocent / 死有余～。 Even death would not expiate all his crimes.
【辜负】let down; fail to live up to; be unworthy of; disappoint: 我们决不要～老师的谆谆教导。 We must never be unworthy of our teachers' untiring and sincere teachings

轱

【轱辘】[口] wheel ◇～鞋 roller skates

咕

[象]①(母鸡等叫声) cluck ②(斑鸠等叫声) coo
【咕咚】[象] thud; splash; plump: ～一声,大石头

掉进池里去了。 Polp! The big stone dropped into the pond.
【咕嘟】①[象] bubble; gurgle: 泉水～～地往外冒。 The spring kept bubbling up. ②(长时间煮) boil for a long time: 白菜早就～烂了。The cabbage is overcooked. ③[方](撅嘴) purse (one's lips): 不高兴地～着嘴 purse one's lips in displeasure
【咕唧】① [象] squelch ②(小声说话) whisper; murmur: 他俩～了半天。 They whispered to each other for a long time.
【咕隆】[象] rumble; rattle; roll: 远处雷声～～地响。 Thunder rumbled in the distance.
【咕噜】①[象] rumble; roll: 肚子～～直响 one's stomach keeps rumbling ②(小声说话) murmur; whisper
【咕哝】murmur; mutter; grumble: 她在～些什么? What is she muttering about?

估

estimate; appraise: ～一～损失多少 assess the damages
【估产】①(估计产量) estimate the yield ②(评估) appraise the assets; assess
【估定价值】assessed valuation; appraised value
【估计】estimate; appraise; reckon: 约略的～ a rough estimate / ～今年又是一个丰收年。 It looks as if there'll be another good harvest this year. ◇ ～成本 estimated cost / ～纯利 estimated net profit / ～市场 estimated market value / ～误差 evaluated error
【估价】①(评价) appraise; evaluate: 对历史人物的～ evaluation of historical personages ② [经](估计商品价格) appraised price ◇ ～标准 basis of valuation / ～表 schedule of prices / ～单 list of cost estimate / ～调查 appraisal survey
【估量】appraise; estimate; assess: 不可～的损失 an immeasurable loss
【估量积分】weighed integral
【估算利息】imputed interest
【估算收入】imputed income
【估算数据】estimated data

姑

①(姑母) father's sister; aunt ②(丈夫的姐妹) husband's sister; sister-in-law ③(出家修行的妇女) nun: 道～ Taoist nun / 尼～ Buddhist nun ④[书](暂且) tentatively; for the time being
【姑表】the relationship between the children of a brother and a sister; cousinship: ～兄弟 cousins
【姑夫】the husband of one's father's sister; uncle
【姑姑】[口] father's sister; aunt
【姑母】father's sister (married); aunt
【姑奶奶】①(已出嫁的女儿) married daughter ②(父亲的姑母) the sister of one's paternal grandfather; grandaunt
【姑娘】①(未婚女子) girl ②[口](女儿) daughter
【姑且】tentatively; for the moment: ～不谈 leave sth. aside for the moment
【姑且不论】let us not go into the question now; let's not discuss it for the present
【姑嫂】a woman and her brother's wife; sisters-in-law
【姑妄听之】see no harm in hearing what sb. has to say
【姑妄言之】tell sb. for what it's worth
【姑息】appease; indulge; tolerate: 不应该～他的错误。 We shouldn't be indulgent towards his mistakes. ◇ ～剂 palliative; alleviating medicine / ～疗法 palliative treatment; alleviative treatment / ～手术 palliative operation
【姑息养奸】to tolerate evil is to abet it: ～无异犯罪 show mercy to an evildoer is to commit a crime oneself
【姑爷】[口] a form of address for a man used by the senior members of his wife's family

菇

mushroom

骨

【骨朵儿】[口] flower bud
【骨碌】roll: 从床上一～爬起来 roll out of bed

箍

①(箍儿) hoop; band: 铁～ hoop iron ②(捆紧) bind round; hoop: 用铁丝把桶～上 bind a bucket with wire
【箍钢】hoop iron
【箍钢带】hoop steel
【箍铁轧机】hoop mill
【箍桶匠】cooper; hooper

呱

【呱呱】[书](小儿哭声) the cry of a baby: ～坠地 come into the world with a cry; be born

孤

①(孤儿) (of a child) fatherless; orphaned ②(孤单) solitary; isolated; alone: ～岛 an isolated island / ～雁 a solitary wild goose ③(王侯自称) I
【孤傲】proud and aloof: 去掉～习气 rid oneself of aloofness and arrogance

【孤本】the only copy extant; the only existing copy
【孤单】①(单身无靠) alone: 孤孤单单的一个人 all alone; all by oneself; a lone soul ②(孤寂) lonely; friendless
【孤点】acnode
【孤独】lonely; solitary: 过着～的生活 live in solitude
【孤独癖】autism
【孤独无偶】all alone without a mate
【孤独无援】alone and with no help
【孤儿】orphan: ～寡妇 orphan and widow
【孤芳自赏】a solitary flower in love with its own fragrance; a lone soul admiring his own purity; indulge in self-admiration
【孤寡】orphans and widows
【孤寂】lonely
【孤家寡人】a person in solitary splendor; a person who has no mass support; a loner
【孤军】an isolated force: ～深入 an isolated force penetrating deep into enemy territory／～作战 fight in isolation
【孤苦伶仃】orphaned and helpless; friendless and wretched
【孤立】①(与其它事物不相联系) isolated: 处境～ find oneself in an isolated position／～无援 isolated and cut off from help ②(使得不到同情、支援) isolate: ～敌人 isolate the enemy ◇～案件 isolated case／～病灶 solitary sick point／～数据 orphan／～主义 isolationism
【孤零零】solitary; lone; all alone
【孤陋寡闻】ignorant and ill-informed
【孤僻】unsociable and eccentric: 性情～ of an uncommunicative and eccentric disposition
【孤身一人】be all on one's own
【孤行己见】follow one's bigoted course
【孤掌难鸣】it's impossible to clap with one hand; it's difficult to achieve anything without support
【孤注一掷】stake everything on a single throw; risk everything on a single venture; put all one's eggs in one basket

菰 [植] wild rice

gú

骨 bone

【骨头】①(骨骼) bone ②(喻人的品质) character; a person of a certain character: 懒～ lazybones／软～ a spineless creature
【骨头架子】①(骨架) skeleton ②(极瘦的人) a bag of bones; skin and bones: 瘦得成了一副～ become a bag of bones; be worn to a shadow; be all skin and bone

gǔ

汩

【汩汩】gurgle: 水声～ the gurgling of water／水～地流入稻田。The water gurgled into the paddy field.

鼓 ①(打击乐器) drum: 大～ base drum／～声 drumbeats ②(使发声) beat; strike; sound: ～掌 clap one's hands／～其如簧之舌 talk glibly ③(煽) blow with bellows, etc.: ～风 work a bellows ④(发动;振奋) rouse; agitate; pluck up: ～起勇气 pluck up one's courage／～实劲不～虚劲 exert genuine and not sham effort ⑤(隆起) bulge; swell: 墙上有几处～起来了。The wall bulged out at several places.

【鼓板】[乐] clappers
【鼓吹】①(宣传提倡) advocate ②[贬](吹嘘) preach; advertise; play up ◇～者 advocator; trumpeter
【鼓槌】drumstick
【鼓点子】①(鼓声节奏) drumbeats ②(戏曲中指挥其它乐器的鼓板节奏) clapper beats which set the tempo and lead the orchestra in traditional Chinese operas
【鼓动】①(激发人们行动) agitate; arouse: ～群众 arouse the masses ②(唆使) instigate; incite: 这些坏事是谁～你干的? Who put you up to all these dirty tricks?
【鼓风】[冶] (air) blast: 富氧～ oxygen-enriched (air) blast ◇～管 blast pipe／～机 air-blower; blower／～口 blast orifice／～炉 blast furnace／～容积 blast volume
【鼓鼓囊囊】bulging; swelling out; bellying
【鼓角齐鸣】beat the drums and blare the trumpets; the beating of drums and the blare of trumpets came from every side
【鼓励】encourage; urge: 这些～使我增添了克服困难的信心。These encouragements gave me greater confidence in overcoming difficulties.
【鼓楼】drum-tower
【鼓膜】[生理] tympanic membrane; eardrum ◇～穿孔[医] perforation of the tympanic membrane／～炎[医] myringitis
【鼓起】call up; muster up; pluck up: ～精神 stir up one's spirit／～勇气 pluck up one's courage; muster up one's courage

【鼓舌骨】[生理] tympanohyal
【鼓舌如簧】talk glibly; wag one's tongue with honeyed words
【鼓舌摇唇】spread rumors; flatter; gossip
【鼓室】[生理] tympanum; ear drum; tympanic ◇ ～盖 roof of tympanum / ～炎 tympanitis
【鼓手】drummer
【鼓舞】inspire; hearten: ～斗志 inspire 〈stimulate〉 the fighting spirit / ～人心 set the hearts of the people aflame / ～士气 enhance troop morale
【鼓乐】strains of music accompanied by drumbeats
【鼓乐大作】strike up the drums and trumpets
【鼓乐喧天】great din of drums and pipes; loud music fills the air
【鼓噪】make an uproar; raise a hubbub; clamor: ～一时 make a great to-do about sth. for a time
【鼓噪四起】rise up with a great clamor; rise in a hubbub
【鼓掌】clap one's hands; applaud: 热烈～ warmly applaud / ～通过 approve by acclamation
【鼓足干劲】go all out; exert the utmost effort
【鼓足勇气】call up all one's courage; muster (up) one's courage

臌

【臌胀】[中医] distension of abdomen caused by accumulation of gas or fluid due to dysfunction of liver and spleen; tympanites

古

ancient; age-old; paleo-: ～时候 in ancient times; in olden days / ～书 ancient books
【古奥】archaic and abstruse
【古巴】Cuba ◇ ～人 Cuban
【古板】old-fashioned and inflexible
【古瓷】old china
【古代】ancient times; antiquity: ～文化 ancient civilization ◇ ～史 ancient history
【古道可风】The old ways may be adopted and practiced
【古道热肠】considerate and warmhearted; sympathetic
【古典】①(典故) classical allusion ②(古代流传下来的典范) classical ◇ ～文学 classical literature / ～学派 classical school / ～艺术 classical art / ～音乐 classical music / ～主义 classicism / ～作品 classic
【古董】①(古代流传下来的器物) antique; curio ②(守旧的人) old fogey ◇ ～鉴赏家 connoisseur of curios
【古动物学】galaeozoology
【古都】ancient capital
【古尔邦节】[伊斯兰] Corban
【古风】ancient customs; antiquities
【古风遗俗】old customs
【古怪】eccentric; old; strange: ～脾气 eccentric character
【古画】ancient painting
【古话】old saying: ～说,有志者事竟成。 As the old saying goes, where there's a will there's a way.
【古迹】historic site; place of historic interest
【古籍】ancient books
【古今中外】ancient and modern, Chinese and foreign; at all times and in all countries: ～,概莫能外。There is no exception to this in modern or ancient times, in China or elsewhere.
【古井不波】A dried-up well does not have ripples; impervious to desires and passions
【古旧】antiquated; archaic: ～词语 archaic words and expressions; archaisms
【古柯】[植] coca ◇ ～碱 [药] cocaine
【古兰经】[伊斯兰] the Koran
【古老】ancient; age-old: ～的传说 legend
【古里古怪】be rather peculiar; eccentric; odd
【古朴】simple and unsophisticated; of primitive simplicity
【古人】the ancients; our forefathers
【古人类学】paleoanthropology ◇ ～家 paleoanthropologist
【古色古香】antique; quaint
【古生代】[地] the Paleozoic Era
【古生代植物】paleophyte
【古生物学】paleontology ○三叶虫 trilobite / 软骨鱼 cartilaginous fish; chondrichthy / 硬骨鱼 bony fish; osteichthy / 无颚鱼 jawless fish / 总鳍鱼 crossopterygii; lobefin fish / 矛尾鱼 latimeria / 恐龙 dinosaur / 鱼龙 ichthyosauru / 翼龙 pterosaur / 禽龙 iguanodon / 鸭嘴龙 duck-billed dinosair / 梁龙 diplodocus / 始祖鸟 archaeopteryx; ancestral bird / 猛犸 mammoth / 剑齿虎 saber-toothed tiger; smilodon / 剑齿象 stegodon / 乳齿象 mastodon / 森林古猿 dryopithecus / 南方古猿 australopithecus / 拉玛古猿 Ramapithecus / 猿人 pithecanthropus / 古人 paleanthropus / 新人 neoanthropus / 放射性

碳方法 radio carbon method／ 钾氩法 potassium-argon method／ 孢粉分析 spore-pollen analysis
【古诗】ancient poetry
【古书】ancient books
【古塔胶】[化] gutta-percha
【古体诗】a form of pre-Tang poetry
【古铜色】bronze-colored; bronze
【古玩】antique; curio
【古往今来】through the ages; of all ages; since time immemorial
【古为今用】make the past serve the present
【古文】①(文言文) ancient style prose; ancient Chinese prose ②(秦以前的字体) Chinese script before the Qin Dynasty (221—207 B.C.)
【古文字】ancient writing ◇～学 paleography
【古物】ancient objects; antiquities ◇～陈列馆 museum of antiquities
【古稀】seventy years of age: 年近～ getting on for seventy
【古雅】of classic beauty and in elegant taste; of classic elegance
【古已有之】have existed since ancient times
【古语】①(古词) archaism ②(古谚语) old saying
【古植物学】paleobotany
【古装】ancient costume: ～戏 costume piece; drama in historical costume

钴 [化] cobalt (Co)
【钴弹】cobalt bomb
【钴钢】cobalt steels
【钴线疗法】[医] cobalt therapy

牯 bull
【牯牛】bull

贾 ①(商人) merchant ②(做买卖) engage in trade ③ [书](卖) sell; afford: 余勇可～ still having plenty of fight left in one; with one's strength not exhausted

蛊 a legendary venomous insect
【蛊惑】poison and bewitch
【蛊惑人心】confuse and poison people's minds; resort to demagogy: ～的作品 demagogic writings

骨 ①(骨头) bone ②(架子) skeleton; framework: 钢～水泥 reinforced concrete ③(品质;气概) character; spirit: 傲～ lofty and unyielding character／ 媚～ obsequiousness
【骨癌】osteocarcinoma; cancer in the bones
【骨刺】[医] spur
【骨肥】fertilizer made from animal bones; bone fertilizer
【骨粉】bone meal; bone dust
【骨干】① [生理](长骨的中央部分) diaphysis ② (起主要作用的) backbone; mainstay: 科技队伍的～力量 the backbone of the scientific and technological contingents／ 起～作用 be a mainstay ◇ ～分子 core member; key member
【骨骼】[生理] skeleton ◇ ～肌 skeletal muscle
【骨鲠在喉】have a fishbone caught in one's throat: 如～，一吐方快。I feel suffocated if I don't speak out; have an opinion one cannot suppress.
【骨骺】[生理] epiphysis
【骨化】[生理] ossify
【骨灰】①(骨头烧成的灰) bone ash ②(火葬的骨灰) ashes of the dead ◇ ～盒 cinerary casket／ ～坛 cinerary urn
【骨架】skeleton; framework: 房屋的～ the framework of a house ◇ ～图 skeleton drawing
【骨胶】[化] bone glue
【骨节】[生理] joint
【骨结核】bone tuberculosis
【骨科】[医] (department of) orthopedics ◇ ～医生 orthopedist
【骨刻】[工美] bone sculpture; bone carving
【骨痨】[中医] tuberculosis of bones and joints
【骨料】[建] aggregate: 轻～ light aggregate
【骨瘤】[医] osteoma
【骨膜】[生理] periosteum ◇ ～炎 periostitis
【骨牌】dominoes
【骨盆】[生理] pelvis
【骨气】strength of character; moral integrity; back bone
【骨器】bone object; bone implement
【骨肉】flesh and blood; kindred: 亲生～ one's own flesh and blood／ ～情谊 kindred feelings; feelings of kinship／ ～团聚 a family reunion／ ～相残 fratricidal fighting／ ～相聚 be together as people of the same flesh and blood should be／ ～相连 as closely linked as flesh and blood／ ～兄弟 blood brothers; one's own brothers／ ～之亲 blood relations／ ～至亲 blood relationship; a pack of bones
【骨软化】[医] osteomalacia
【骨软筋麻】(strive until) one's bones were weak and one's muscles numbed, paralyzed; enervated

【骨瘦如柴】thin as a lath; worn to a shadow; a mere skeleton; a bag of bones
【骨瘦形销】(she) grew greatly emaciated
【骨髓】[生理] marrow ◇～炎 osteomyelitis
【骨炎】[医] osteitis; ostitis
【骨硬化】[医]osteosclerosis
【骨学】[医] osteology
【骨折】[医] fracture: 粉碎～ comminuted fracture／开放～ open fracture ◇～复位 reduction of the fracture
【骨质疏松】osteoporosis
【骨子】frame; ribs: 伞～ umbrella frame; the ribs of an umbrella／扇～ the ribs of a fan
【骨子里】in one's heart of hearts: 这人～很狡猾,虽然表面上看不出来。This man was in fact very crafty, though he didn't look so.

谷 ①(山谷) valley; gorge: 深～ a deep valley; gorge ②(谷类) cereal; grain: ～类作物 cereal crops ③(粟) millet ④[方](稻谷) unhusked rice
【谷氨酸】[药] glutamic acid
【谷氨酸钠】sodium glutamate
【谷仓】granary; barn
【谷草】①(谷秆) millet straw ②[方](稻草) rice straw
【谷蛋白】gluten
【谷蛾】[动] grain moth
【谷贱伤农】low prices for grain hurt the peasants; low grain price hurts the farmer
【谷壳】husk (of rice)
【谷物】cereal; grain
【谷象】grain beetle
【谷雨】Grain Rain (6th solar term)
【谷种】seed-grain; seed-corn
【谷子】①(粟) millet ②[方](稻子) unhusked rice ◇～白发病 downy mildew of millet
【谷值点】valley point
【谷值电流】valley point current
【谷值电压】valley point voltage

股 ①(大腿) thigh ②(组织单位) section (of an office, enterprise, etc.): 财务～ accounting section ③(绳线组成部分) strand; ply: 三～的绳子 a rope of three strands ④(股份) one of several equal parts; share in a company: 分～ divide into equal parts／优先～ preference shares; preferred stock ⑤[量](用于成条的东西)一～泉水 a stream of spring water／一～线 a skein of thread ⑥[量](用于气体、气味、力气)一～劲 a burst of energy／一～热气 a stream 〈puff〉 of hot air／一～香味 a whiff of fragrance ⑦[量][贬](用于成批的人)两～土匪 two gangs of bandits
【股本】capital stock: ～折价 discount on capital stock
【股东】shareholder; stockholder: ～产权 stockholder's equity
【股动脉】arteria femoralis
【股匪】gang of bandits
【股份】share; stock: ～转让 stock transfer／～资本 share capital
【股份公司】joint-stock company; stock company ◇～法 joint stock companies act
【股份不公开公司】close corporation
【股份有限公司】limited-liability company; limited company (Ltd.)
【股肱】[书] right-hand man
【股肱之臣】the most trustworthy ministers
【股骨】[生理] thighbone; femur: ～头 head of femur
【股金】money paid for shares (in a partnership or cooperative)
【股票】share certificate; share; stock ◇～行市 current prices of stocks; quotations on the stock exchange／～交易 buying and selling of stocks／～交易所 stock exchange／～经纪人 stockbroker; stockjobber／～市场 stock market
【股息】dividend
【股线】[纺] plied yarn
【股癣】jock itch; tinea cruris
【股长】section chief
【股子】①(股份) share ②[量](用于力量、气味等)他有一～使不完的劲。He just doesn't know what it is to be tired.

毂 hub
【毂盖】hub cap; hub cover
【毂环】nave collar
【毂键】hub key

gù

雇 hire; employ: ～车 hire a car／～船 hire a boat
【雇工】①(雇用工人) hire labor; hire hands: ～剥削 exploitation through the hiring of labor ②(受雇工人) hired laborer
【雇农】farmhand; farm laborer
【雇佣】employ; hire
【雇佣兵役制】mercenary system
【雇佣观点】hired hand mentality; the attitude

of one who will do no more than he is paid for
【雇佣军】mercenary army
【雇佣劳动】wage labor
【雇用合同】contract of employment; contract of service
【雇用劳动力】employed labor force
【雇用人】hirer
【雇员】employee
【雇主】employer

顾 ①(看;转过头看) turn round and look at; look at: 环～ look around / 回～ look back; retrospect / 相～一笑 smile at each other knowingly / 左～右盼 look to the left and to the right ②(照管;注意) attend to; take into consideration: 兼～ give consideration to both ③(拜访) visit; call on: 三～茅芦 call on sb. repeatedly (to enlist his help, etc.)
【顾此失彼】attend to one thing and lose sight of another; have too many things to take care of at the same time
【顾及】take into account; attend to; give consideration to: 无暇～ have no time to attend to the matter
【顾忌】scruple; misgiving: 不能不有所～ have to think twice (before doing sth.); be unable to overcome certain misgivings / 毫无～ without scruple; have no scruples
【顾客】customer; shopper; client
【顾虑】misgiving; apprehension; worry: 打消～ dispel one's misgivings
【顾虑重重】be full of worries; have no end of misgivings; have scruples
【顾虑周详】consider something down to the minutest details
【顾名思义】seeing the name of a thing one thinks of its function; just as its name implies; as the term suggests
【顾盼】[书] look around
【顾盼生姿】look around charmingly
【顾盼自如】gaze round as one wishes — free and easy
【顾盼自雄】look about complacently; be as proud as a peacock
【顾前不顾后】drive ahead without considering the consequences; act rashly
【顾全】show consideration for and take care to preserve: ～大局 take the interests of the whole into account / ～名誉 take into consideration one's reputation
【顾问】adviser; consultant ◇ ～班子 council / ～工程师 consulting engineer / ～律师 chamber barrister / ～团 advisory group / ～委员会 consultative committee
【顾惜】take good care of: ～劳动力 be sparing of labor power; 要～你的身体 take good care of your health
【顾影自怜】look at one's reflection and admire oneself; look at one's shadow and lament one's lot
【顾主】[旧] customer; client; patron

故 ①(事故) incident; happening: 变～ unforeseen event; misfortune / 事～ accident ②(缘故) reason; cause: 无～缺勤 be absent without reason ③(故意) on purpose; intentionally: 明知～犯 willfully violate (a law or rule) ～作惊讶 put on a show of surprise; feign surprise / ～作镇静 pretend to be calm ④(所以) hence; therefore; consequently; for this reason: 无私～能无畏。 Fearlessness stems from selflessness. ⑤(从前的;旧的) former; old: ～址 site (of an ancient monument, etc.) ⑥(朋友) friend; acquaintance: 非亲非～ neither relative nor friend; a perfect stranger ⑦(死亡;已死亡之人) die: 病～ die of illness
【故步自封】stand still and refuse to make progress; be complacent and conservative
【故地】old haunt
【故典史籍】ancient works and historical records
【故都】onetime capital
【故宫】the Imperial Palace ◇ ～博物院 the Palace Museum
【故伎】stock trick; old tactics: ～重演 play the same old trick
【故旧】old friends and acquaintances
【故居】former residence; former home
【故里】native place
【故弄玄虚】purposely turn simple things into mysteries; be deliberately mystifying
【故去】die; pass away
【故人】old friend
【故杀】[法] premeditated murder; willful murder
【故甚其辞】purposely exaggerate
【故事】①(旧日的制度;例行的事) old practice; routine: 奉行～ follow established practice mechanically ②(用作讲述的事情) story; tale: 民间～ folktale; folk story ③(故事性) plot ◇ ～会 a gathering at which stories are told; story-telling session / ～(影)片 feature

film／～员 storyteller
【故态复萌】slip back into one's old ways
【故土】native land
【故乡】native place; hometown; birthplace
【故意】intentionally; willfully; deliberately; on purpose: ～刁难 place obstacles in sb.'s way ◇ ～犯 intentional offender／～犯罪 calculated crime; intentional offense／～过失 active negligence; intentional negligence／～侵害人身 intentional interference with the person／～侵权行为 intentional tort／～杀人 intentional homicide; intent to kill／～杀人既遂罪 completed offense of intentional homicide／～使用伪造证券 uttering (of) forged documents／～损害财产 intentional interference with property／～隐瞒 active concealment; intentional concealment／～阻挠 obstruct willfully
【故障】hitch; breakdown; stoppage; trouble: 排除～ fix a breakdown; clear a stoppage／发动机出了～。 The engine has broken down. ◇ ～测定 failure terms／～率 fault rate; failure rate／～预测 failure predication
【故纸堆】a deep of musty old books or papers
【故作姿态】make a deliberate gesture; strike an attitude on purpose

固 ①(牢固) solid; firm: 加～ make sth. more solid; strengthen; reinforce／本～枝荣。When the root is firm, the branches flourish. ②(坚决地) firmly; resolutely: ～辞 resolutely refuse; firmly decline ③(使坚固) solidity; consolidate; strengthen: ～堤 strengthen the dike ④[书](原本) originally; in the first place; as a matter of course: ～当如此。 It is just as it should be. ⑤[书](固然) admittedly; no doubt: 乘车～可,乘船亦无不可。 Admittedly we can make the journey by train, but there is no harm in our travelling by boat.
【固氮菌】[微] nitrogen-fixing bacteria; azotobacter
【固氮作用】[农] nitrogen fixation; azofication
【固定】①(不变动) fixed; regular: 电视台的～节目 a regular TV program ②(使固定) fix; regularize: 把灯座～在车床上 fix the lamp stand on the lathe ◇ ～保释金 fixed bail／～工资制 fixed-wage system／～汇率 fixed (exchange) rate／～基金 fixed fund／～机库 permanent hangar／～价格 fixed price／～平价 fixed parity／～式平炉 [冶] stationary open-hearth furnace／～收入 fixed income; regular income／～摊位执照 fixed pitch licence／～职业 permanent occupation／～资本 fixed capital／～资产 fixed assets
【固化】[化] solidify ◇ ～酒精 solidified alcohol
【固件】[计算机] firmware
【固井】[石油] well cementation
【固陋不义】rustic and ignorant
【固然】no doubt; it is true; true; of course; admittedly
【固若金汤】strongly fortified; impregnable
【固沙林】sand-fixation forest; dune-fixing forest
【固守】defend tenaciously; by firmly entrenched in: ～阵地 tenaciously defend one's position
【固态】[物] solid state ◇ ～激光器 solid-state laser／～键盘 solid-state keyboard／～物理学 solid-state physics
【固体】solid body; solid ◇ ～电路 solid-state circuit／～废物 solid waste／～火箭 solid-rocket／～酱油 solidified soy sauce／～理论 theory of solids／～汽油 gasoline gel; solidified gasoline／～燃料 solid fuel／～燃料火箭发动机 solid propellant (rocket) engine; solid engine
【固有】intrinsic; inherent; innate: ～的属性 intrinsic attributes
【固执】①(不肯改变) obstinate; stubborn ②(坚持己见) persist in; cling to: ～己见 stubbornly adhere to one's opinions
【固执如驴】be as stubborn as mules

痼 chronic; inveterate
【痼疾】chronic illness
【痼癖】inveterate weakness; deep-rooted liking for
【痼习】inveterate habit

锢 ①(溶化金属堵塞) plug with molten metal; run metal into cracks ②[书](禁锢) hold in custody; imprison
【锢囚】[气] occlusion
【锢囚锋】[气] frontal occlusion; occluded front; occlusion
【锢囚气旋】[气] occluded cyclone

梏 wooden handcuffs: 桎～ fetters; shackles

guā

栝
【栝楼皮】[中药] the fruit-rind of Chinese trichosanthes

刮 ①(用刀去掉某物) scrape: ~胡子 shave the beard / ~鱼鳞 scale a fish ②(涂) smear with (paste, etc.) ③(搜刮) plunder; fleece; extort ④(风吹) blow: ~大风了。It's blowing hard.

【刮除术】[医] curettage

【刮刀】scraping cutter; scraper: 三角~ triangular scraper

【刮地皮】batten on extortions

【刮宫】[医] dilatation and curettage (D. and C.)

【刮垢磨光】scrape the dirt off an object and make it shine

【刮脸】shave (the face) ◇ ~刀 razor

【刮目相看】look at sb. with new eyes; treat sb. with increased respect

【刮痧】[中医] a popular treatment for sunstroke by scraping the patient's neck, chest or back

【刮水器】wiper

【刮削】scrape ◇ ~器 scraper

瓜 melon, gourd, etc.: 冬~ white gourd / 西~ watermelon

【瓜氨酸】[化] citrulline

【瓜分】carve up; divide up; partition: ~全世界的野心 the ambition of carving up the whole world

【瓜分豆剖】divide it like a melon; split it into two like a bean

【瓜葛】connection; implication; association: 不难看出他们之间是有~的。It was not hard to see that they got entangled with each other.

【瓜葛亲】distant relatives

【瓜果】①(水果) melon and fruit ②[植] amphisarca

【瓜枯萎病】cucurbit wilt

【瓜皮帽】a kind of skullcap resembling the rind of half a watermelon; skullcap

【瓜瓤】melon pulp

【瓜熟蒂落】when a melon is ripe it falls off its stem; things will be easily settled when conditions are ripe

【瓜田不纳履】don't do up your shoe in a melon-patch; to avoid suspicions: ~,李下不整冠 don't pull on your shoe in a melon-patch and don't adjust your cap under a plum tree; don't do anything to arouse suspicion

【瓜藤】melon vine

【瓜藤豆茎】leafy pumpkin vines and entwining bean stalks

【瓜秧】melon seedling

【瓜子儿】melon seeds ◇ ~脸 oval face

【瓜子玉】[矿] andesite

【瓜状体】legena

呱

【呱嗒】①[象] clip-clop; clack: ~~的马蹄声 clatter of horsehoofs ②[方](板起脸) ~着脸 pull a long face

【呱呱】[象](鸭子叫声) quack; (青蛙叫声) croak; (乌鸦叫声) caw

【呱呱叫】[口] tiptop; top-notch

【呱唧】[象] clap (hands)

胍 [化] guanidine

【胍基乙酸】glycocyamine

【胍呱四环素】guamecycline

【胍生】envacar; guanoxane

【胍乙啶】guanethidine; ismeline

guǎ

寡 ①(少) few; scant: 沉默~言 uncommunicative; taciturn / 失道~助。An unjust cause finds scant support. ②(淡而无味) tasteless: 清汤~水 watery soup; something insipid ③(死了丈夫之妇女) widowed: 鳏~ widowers and widows / 守~ live in widowhood

【寡不敌众】be hopelessly outnumbered

【寡妇】widow

【寡妇再醮】remarry after one's husband's death

【寡廉鲜耻】lost to shame; shameless

【寡情薄义】have no affections and be a man of shallow feelings

【寡人】I, the sovereign; we

【寡头】oligarch: 金融~ financial oligarchy; financial magnates ◇ ~垄断 oligopoly / ~政治 oligarchy

【寡言】of few words; taciturn; sparing of words

【寡欲清心】have few desires and cleanse the heart

剐 ①(割肉离骨) cut to pieces (a form of capital punishment in a ancient times); dismember: 千刀万~ be cut to pieces / 舍得一身~,敢把皇帝拉下马。He was fears not being cut to pieces dares to unhorse the emperor. ②(刺破) cut; slit: 手上~了个口子 cut one's hand

guà

褂 a Chinese-style unlined garment; gown: 大~儿 long gown / 短~儿 short gown

【褂子】a Chinese-style unlined upper garment; short gown

挂 ①(悬挂) hang; put up: 天上～着一轮明月。 A bright moon hung in the sky. ②(钩) hitch; get caught: 她的衣服给钉子～住了。Her dress got caught on a nail. ③(使电路断开) ring off: 她已经把电话～了。 She's hung up. ④(打电话) call up; put sb. through to: 我呆会儿再给他～电话。 I'll ring him up again. ⑤[方](牵挂) be concerned about: 我会把这件事～在心上的。 I shall keep the matter in mind. ⑥[方](糊) be covered with; be coated with: 瓦盆里面～一层釉子。 The earthen pot is glazed inside. ⑦(登记) register (at a hospital, etc.): ～外科 register for surgery ⑧[量](用于成套或成串的东西)十多～鞭炮 a dozen strings of firecrackers / 一～大车 a horse and cart

【挂表】pocket watch

【挂彩】①(结彩) decorate with colored silk festoons; decorate for festive occasions ②(负伤) be wounded in action

【挂车】trailer

【挂齿】mention: 区区小事,何足～。 Such a trifling matter is not worth mentioning.

【挂挡】put into gear: 挂高速挡 change to high gear

【挂钩】①[交](连接车辆装置) couple (two railway coaches); articulate ②(联系) link up with; establish contact with; get in touch with: 工厂应该与科研单位～。 Factories should establish close contact with institutes of scientific research.

【挂冠封印】resign and go home; hang up the cap and close up the seal; give up office and leave the place

【挂冠归里】retire

【挂号】①(编号登记) register (at a hospital, etc.): 请排队～。 please queue up to register. ②(挂号信) send by registered mail: 你这封信要不要～? Do you want to have this letter registered? ◇ ～处 registration office / ～费 registration fee / ～信 registered letter

【挂花】be wounded in action

【挂零】odd: 五十～ fifty odd

【挂虑】be anxious about; worry about

【挂面】fine dried noodles; vermicelli

【挂名】titular; nominal; only in name

【挂名夫妻】husband and wife in name only; a false couple

【挂念】worry about sb. who is absent; miss: 十分～ miss sb. very much

【挂牌】hang out one's shingle; put up one's brass plate

【挂牌开业】hang out one's shingle; put one's plate in (a certain street)

【挂屏】hanging panel: 立体～ relief panel

【挂失】report the loss of (identity papers, checks, etc.)

【挂帅】be in command; assume command; assume leadership: ～人物 person in command / 亲自～ take command in person

【挂锁】padlock

【挂毯】tapestry

【挂图】①(指地图) wall map ②(指图表) hanging chart

【挂线疗法】[中医] ligating method for treating anal fistula

【挂心】keep in mind; be concerned about; be anxious for

【挂羊头,卖狗肉】hang up a sheep's head and sell dogmeat; try to palm off sth. inferior to what it purports to be

【挂一漏万】for one thing cited, ten thousand may have been left out; the list is far from complete

【挂衣钩】clothes-hook

【挂钟】wall clock

【挂轴】hanging scroll (of Chinese painting or calligraphy)

卦 divinatory symbols: 算～ tell fortunes / 占～ divination

guāi

掴 slap; smack: ～耳光 box sb.'s ears; slap sb. on the face

乖 ①(听话) well-behaved (child); good: 真是个～孩子。 There's a dear. ②(伶俐) clever; shrewd; alert: 学～了 become a little wiser ③[书](违反常理) perverse; contrary to reason: 有～常理 run counter to reason

【乖乖】①(听话) well-behaved; obedient ②(对小孩的爱称) little dear; darling ③[叹] good gracious

【乖觉】alert; quick

【乖戾】perverse (behavior); disagreeable (character)

【乖谬】absurd; abnormal

【乖僻】eccentric; odd

【乖巧】①(机灵) clever ②(讨人喜欢) cute; lovely
【乖张】eccentric and unreasonable

guǎi

拐 ①(转变方向) turn: ~过墙角 turn the corner of a house / 往左~ turn to the left ②(瘸) limp: 一~一~地走 limp along; walk with a limp ③(拐杖) crutch: 走路架着双~ walk with crutches ④(绑架) abduct; kidnap ⑤(拐骗) swindle; make off with: ~款潜逃 abscond with funds
【拐脖儿】elbow (of a stove pipe)
【拐点】inflection point; knee; point of inflection
【拐法线】inflectional normal
【拐切线】inflectional tangent
【拐带儿童】child-stealing
【拐棍】walking stick
【拐角】corner; turning: 在房子的~ at the corner of the house
【拐卖人口】kidnap and sell people
【拐骗】①(用于财物) swindle: ~钱财 swindle money (out of sb.) ②(用于人) abduct: ~妇女 abduct women
【拐骗犯】kidnapper
【拐骗罪】kidnaping
【拐弯】①(行路转换方向) turn a corner; turn: ~要慢行。Slow down when turning a corner ②(思路、语言等转换方向) turn round; pursue a new course
【拐弯抹角】talk in a roundabout way; beat about the bush: 说话不要~。Get to the point. Don't beat about the bush.
【拐杖】walking stick
【拐子】①[口](跛者) cripple ②(拐骗之人) abductor ③(骗子; 诈骗犯) swindler ④(木制绕线架) I-shaped reel

guài

怪 ①(奇怪) strange; odd; queer; bewildering: 一点儿也不~ not at all strange ②(觉得奇怪) find sth. strange; wonder at: 那有什么可~的? Is that anything to be surprised at? ③(很; 非常) quite; rather: 箱子~沉的。The suitcase is rather heavy. ④(妖怪) monster; demon; evil being: 鬼~ demons, ghosts and goblins; forces of evil ⑤(责怪) blame: 不能~他们。They're not to blame.
【怪不得】①(明白原因后,不觉得奇怪) no wonder; so that's why; that explains why: 她管这么多事,~老是很忙。No wonder she's always busy, she has so many things to attend to. ②(不能责备) not to blame: 这事~他。He's not to blame for this.
【怪诞】weird; absurd; strange: ~不经 weird and uncanny; fantastic; crazy; supernatural and unreasonable
【怪话】cynical remark; grumble; complaint: 说~ make cynical remarks
【怪里怪气】eccentric; peculiar; queer: ~的人 an eccentric fellow
【怪模怪样】queer-looking; grotesque
【怪僻】eccentric: 性情~ eccentric
【怪腔怪调】speak in a queer way; a queer tune
【怪人】eccentric; a peculiar person; a strange person
【怪石嶙峋】jagged rocks of grotesque shapes
【怪声怪气】(speak in a) strange voice or affected manner
【怪物】①(妖怪) monster; monstrosity; freak ②(性情古怪的人) an eccentric person
【怪异】monstrous; strange; unusual

guān

官 ①(政府、军队中的公职人员) government official; officer; officeholder: 外交~ diplomat; diplomatic officer / ~兵 officer and soldiers ②[旧](政府的或公家的) government-owned; government-sponsored; official; public: ~办 run by the government; operated by official bodies ③(器官) organ: 感~ sense organ
【官卑职小】a petty official; be merely a petty official
【官逼民反】being oppressed by officials the masses revolt against them; misgovernment drives the people to revolt; misgovernment makes the people rebel
【官兵一致】unity between officers and men
【官场】[旧] officialdom; official circles
【官场得意】successful in one's official career
【官场积习】tradition of official circles
【官邸】official residence; official mansion: 大使~ ambassador's residence
【官定贴现率】official discount rate
【官方】of or by the government; official: ~机密 official secrecy; official secrets / ~没收 official confiscation / ~人士 official quarters / ~消息 news from government sources; official sources / 以~身分 in an official capacity
【官府】[旧] ①(地方政府) local authorities ②(官吏) feudal official

【官复原职】restore an official to his original post; be reinstated
【官官相护】bureaucrats shield one another
【官价】official price: ～汇率 official rate of exchange
【官架子】the airs of an official; bureaucratic airs
【官阶】official rank
【官吏】[旧] government officials
【官僚】bureaucrat: 封建～ feudal bureaucrat / 清除～习气 get rid of bureaucratic practices ◇ ～资本 bureaucrat capital
【官僚主义】bureaucracy ◇ ～者 bureaucrat / ～作风 bureaucratic style of work; bureaucratic way of doing things
【官能】(organic) function; sense: 视、听、嗅、味、触这五种～ the five senses of sight, hearing, smell, taste and touch ◇ ～团 [化] functional group / ～症 [医] functional disease
【官气】bureaucratic airs
【官气十足】be puffed up with self-importance
【官腔】bureaucratic tone; official jargon: 打～ speak in a bureaucratic tone; stall with official jargon
【官司】[口] lawsuit: 和人打～ go to law against sb.
【官衔】official title
【官样文章】mere formalities; officialese
【官员】official: 外交～ diplomatic official
【官职】government post; official position

棺 coffin
【棺材】coffin
【棺床】[考古] coffin platform
【棺椁】[考古] inner and outer coffins
【棺架】bier
【棺木】coffin

倌 ①(专管饲养某些家畜的人) a keeper of domestic animals; herdsman: 马～儿 groom / 羊～儿 shepherd / 猪～儿 swineherd ②[旧](被雇专做某事的人) a hired hand in certain trades: 堂～儿 waiter

关 ①(合拢) shut; close: 出门时随手～灯 turn the light off as you go out / 随手～门 please shut the door after you / 请～上门。Please shut the door. / 这扇窗～不上。The window won't shut. ②(禁闭) lock up; shut in: ～进监狱 lock up (in prison); put behind bars / 老虎～在笼子里。The tiger is shut in the cage. ③(切断) turn off: ～收音机 turn off the radio ④(倒闭;歇业) close down ⑤(关隘) pass: 把～ guard the pass; check ⑥(海关) customhouse ⑦(重要转折点) barrier; critical juncture: 技术难～ technical barriers ⑧(关系) concern; involve: 有～方面 the parties concerned
【关隘】[书] (mountain) pass
【关闭】①(合拢) close; shut: ～门窗 close the doors and windows ②(歇业) close down; shut down: 工厂～ the closing down of factories
【关闭效应】blackout effect
【关闭指令】out code
【关东糖】a kind of malt candy (originating in the Northeast)
【关防】official seal
【关怀】show loving care for; show solicitude for: ～备至 show the utmost solicitude
【关键】hinge; key; crux: 问题的～ the crux of the matter; the key to the question / ～时刻 the crucial moment ◇ ～码表 key code table / ～证人 key witness / ～字项 keyword item
【关节】①[生理](骨头连接处) joint ②(起关键作用的环节) key links; links
【关节痛】arthralgia; arthrodynia
【关节脱位】abarticulation; dislocation of joint
【关节炎】[医] arthritis: 风湿性～ rheumatic arthritis
【关禁闭】put in confinement; lock up
【关口】①(必经的处所) strategic pass ②(关头) juncture
【关联】be related; be connected: 互相～的 be interrelated
【关门】①(停业) close: 展览馆十七点半～。The exhibition center closes at 17:30. ②(无商量余地) slam the door on sth.; refuse discussion or consideration ③(不愿容纳) behind closed doors ◇ ～主义 closed-doorism
【关门打狗】shut the dogs up to beat them; block the enemy's retreat and the then destroy him
【关门大吉】put up the shutters; close down
【关门捉贼】catch the thief by closing his escape route
【关内】inside Shanhaiguan Pass
【关卡】[旧] an outpost of the tax office
【关切】be deeply concerned; show one's concern over: 表示严重～ show grave concern over

【关山迢递】the gate and mountain are far away; a long distance; be separated far apart
【关税】customs duty; tariff: 保护～ protective tariff / 特惠～ preferential tariff / ～未付 duty unpaid / ～已付 duty paid ◇ ～壁垒 tariff barrier / ～豁免 exemption from customs duties / ～配额 customs quota system / ～税率 customs tariff / ～同盟 customs union / ～优惠 tariff preference / ～自主 tariff autonomy
【关头】juncture; moment: 紧要～ a critical moment
【关外】outside Shanhaiguan Pass; northeast China
【关系】①(相互关系) relation; relationship: 外交～ diplomatic relations ②(有影响或重要性) bearing; impact; significance: 这个问题～到一个更重要的问题。This question has a bearing on a much more important one. ③(泛指原因条件)由于时间～,就谈到这里吧。Since time is limited, I'll have to stop here. ④(关联;牵涉) concern; affect; have a bearing on; have to do with: 农业～国计民生极大。Agriculture is of vital importance to the nation's economy and the people's livelihood. ⑤(组织关系) credentials showing membership in or connection with an organization
【关系人】[法] party
【关系人受牵连的过失】imputed negligence
【关心】be concerned with; show solicitude for; be interested in; care for: 双方共同～的问题 matters of interest to both sides / ～国家大事 concern oneself with state affairs
【关押】lock up; put in prison
【关于】[介] about; on; with regard to; concerning: ～保护森林的若干规定 regulations concerning the protection of forests
【关张】[方] close down
【关照】①(关心照顾) look after; keep an eye on: 感谢你的～。Thank you for the trouble you've taken on my behalf. ②(口头通知) notify by word of mouth
【关注】follow with interest; pay close attention to; show solicitude for: 多蒙～。Thank you for so much concern.

冠 ①(帽子) hat: 免～照片 bare-headed photo / 衣～整齐 be neatly dressed ②(冠状物) corona; crown: 花～ corolla / 树～ the crown of a tree / 王～ crown / 牙～ the crown of a tooth ③(鸟的冠) crest; comb: 鸡～ cock's comb; crest
【冠盖如云】many official and dignitaries
【冠盖相望】a gathering of dignitaries
【冠冕】royal crown; official hat
【冠冕堂皇】highfalutin; high-sounding: ～的理由 high-sounding excuses
【冠心病】[医] coronary heart disease
【冠周炎】[医] pericoronitis
【冠状动脉】[生理] coronary artery: ～栓塞 coronary embolism / ～硬化 coronary sclerosis; coronary arteriosclerosis
【冠状静脉】coronary vein
【冠状循环】coronary circulation
【冠子】crest; comb

鳏 wifeless; widowered
【鳏夫】[书] an old wifeless man; bachelor or widower
【鳏寡孤独】widowers, widows, orphans and the childless; those who have no kith and kin and cannot support themselves

观 ①(看) look at; watch; observe: 以～后效 see how the offender behaves in future ②(景象) sight; view: 奇～ wonderful sight / 外～ outward appearance / 壮～ impressive sight ③(认识;看法) outlook; view; concept: 世界～ world outlook / 正确的人生～ a correct outlook on life
【观测】observe: ～气象 make weather observations ◇ ～读数 observed reading / ～机 observation airplane / ～哨 observation post; observation point / ～误差 error in observation / ～值 observed value
【观察】observe; watch; survey: ～地形 survey the terrain / ～动静 watch what is going on / ～形势 observe the situation; examine the situation ◇ ～机 observation aircraft / ～家 observer / ～镜 observation mirror / ～视差 observation parallax / ～所 observation post / ～误差 error of observation / ～员 observer; inspector / ～站 observation station
【观潮派】a person who takes a wait-see attitude; onlooker; bystander
【观点】point of view; viewpoint; standpoint: ～一致 identity of views / 阐明～ explain one's position / 增强劳动～ improve one's attitude towards labor
【观风】be on the lookout; serve as a lookout
【观感】impressions
【观光】go sightseeing; visit; tour ◇ ～团 sight-seeing party; visiting group / ～者 sight-

seer
【观过知仁】understand a man by his faults; you can know a man by observing his mistakes
【观看】watch; view: ～足球比赛 watch a football match
【观礼】attend a celebration or ceremony ◇～台 reviewing stand; visitors' stand
【观摩】inspect and learn from each other's work; view and emulate ◇～教学 demonstration lecture / ～演出 performance before fellow artists for the purpose of discussion and emulation
【观念】sense; idea; concept: 私有～ private ownership mentality / 组织～ sense of organization ◇～形态 ideology
【观赏】view and admire; enjoy the sight of ◇～植物 ornamental plant
【观通站】[军] observation and communication post (of the naval service)
【观望】wait and see; look on (from the sidelines): ～不前 hesitating and stalling / 采取～态度 take a wait-and-see attitude
【观微知巨】a straw shows which way the wind blows
【观象台】[天] observatory
【观音】[佛教] Avalokitesvara; Guanyin (a Bodhisattva)
【观音竹】[植] fernleaf hedge bamboo
【观云知天】watch clouds and know signs in the sky
【观瞻】the appearance of a place and the impressions it leaves; sight; view: 以壮～ assume an imposing / 有碍～ be unsightly; be repugnant to the eye; offend the eye
【观者如堵】spectators stood round like a wall; there was a crown of spectators
【观众】spectator; viewer; audience

guǎn

管 ①(管子) tube; pipe: 试～ test tube / 输油～ oil pipeline / 水～ water pipe ②(吹奏的乐器) wind instrument: 单簧～ clarinet / 铜～乐器 brass wind ③ [量] 一～毛笔 a writing brush ④ [电子](形状象管的电器件) valve; tube: 电子～ electron tube ⑤(管理) manage; run; be in charge of: 这事我～不了。 I'm not in a position to take care of this business. ⑥(管教) subject sb. to discipline ⑦(过问) bother about; mind: 别～我! Don't bother about me. ⑧(负责供给) provide; guarantee: ～住 provide accommodation
【管保】①(保证) guarantee; assure: 我～你吃了这药就好。 I guarantee that if you take this medicine, you'll soon get well. ②(肯定) certainly; surely: 他～不知道。 I'm sure he doesn't know.
【管道】pipeline; piping; conduit; tubing: 煤气～ gas piping ◇～安装 piping erection / ～敷设 pipe laying / ～工 pipe fitter; pipelayer / ～工程 plumbing
【管风琴】[乐] pipe organ; organ
【管家】① [旧](仆人) steward; butler ②(管理财物的人) manager; housekeeper
【管见】[谦] my humble opinion; my limited understanding: 容陈～。 Let me state my humble opinion.
【管教】subject sb. to discipline
【管井】[水] tube well
【管窥】look at sth. through a bamboo tube; have a restricted view: ～所及 in my humble opinion
【管窥蠡测】look at the sky through a bamboo tube and measure the sea with a calabash; restricted in vision and shallow in understanding
【管理】manage; run; administer; supervise: ～生产 manage production / 加强企业～ strengthen the administration of enterprises ◇～处 administrative office / ～当局 administering authority / ～费 management expenses; costs of administration / ～机构 administrative organ / ～权 administrative power; right of supervision / ～人员 administrative personnel / ～制度 regulatory regime
【管路】[机] pipeline ◇～铺设 pipe laying / ～输送能力 carrying capacity of a pipeline; delivery capacity
【管区】district
【管事】①(负责) run affairs; be in charge: 这里谁～? Who's in charge here? ② [口](管用) efficacious; effective; of use: 这药很～儿。 This medicine is very effective. / 找他不～。 It's no use asking him. ③ [旧](管总务的人) manager; steward
【管束】restrain; check; control: 严加～ keep sb. under strict control
【管辖】have jurisdiction over; administer: 在～范围之内 come within the jurisdiction of ◇～范围 compass of competency; extent of jurisdiction; sphere of jurisdiction / ～区域 compass of competency; jurisdictional area / ～权 jurisdiction
【管弦乐】orchestral music ◇～队 orchestra /

～配器法 orchestration
【管心距】tube pitch
【管心针】stylet
【管押】take sb. into custody; keep in custody; detain
【管乐队】wind band; band
【管乐器】wind instrument
【管制】①(强制管理) control: 外汇～ foreign exchange control / 交通～ traffic control / 军事～ military control ②(对罪犯强制管束) put under surveillance: ～处分 punishment of control / ～劳动 labor 〈work〉 under surveillance
【管中窥豹】look at a leopard through a bamboo tube; have limited view of sth.
【管中窥豹,可见一斑】look at one spot on a leopard and you can visualize the whole animal; conjure up the whole thing through seeing a part of it
【管状花】[植] tubular flower
【管状腺】tubular gland
【管状云】pendant cloud; tornado cloud; tuba
【管子】tube; pipe ◇ ～工 plumber; pipe fitter

馆 ①(招待宾客居住的房屋) accommodation for guests: 宾～ guest house / 旅～ hotel ②(大使馆) embassy, legation or consulate: 办理建～事宜 arrange for the setting up of an embassy ③(服务性商店的名称) shop: 茶～ teahouse / 饭～ restaurant / 酒～ wineshop / 咖啡～ coffee house; cafe / 理发～ barbershop / 照相～ photo studio ④(公共文化活动场所) a place for cultural activities: 博物～ museum / 美术～ art gallery / 体育～ gymnasium / 图书～ library / 文化～ cultural center / 展览～ exhibition hall
【馆子】restaurant; eating house: 下～ eat at a restaurant

guàn

冠 ①[书](把帽子戴在头上) put on a hat ②(加名号或文字) precede; crown with ③(居第一位) first place; the best: 勇～三军 the bravest of the whole army; top the entire army in courage
【冠词】[语] article
【冠军】champion; gold medalist: 全能～ all- round champion / 上届～ defending champion / 新的世界～ new world champion ◇ ～赛 championships; tournament

灌 ①(浇) irrigate: 冬～ winter irrigation ②(倒进;装进) fill; pour: 往游泳池里～水 fill the swimming pool with water / ～醉 get sb. drunk / 满堂～ cram students; spoonfeed
【灌肠】①[医] enema; clyster ②(食品) sausage
【灌唱片】make a gramophone record; cut a disc
【灌溉】irrigate: 提水～ irrigation by pumping ◇ ～面积 irrigated area / ～渠 irrigation canal / ～网 irrigation network / ～系统 irrigation system / ～站 irrigation station
【灌浆】① [建] grouting ② [农] (of grain) be in the milk ③ [医] form a vesicle (during smallpox or after vaccination) ④ [采矿] injection ◇ ～泵 grouting pump / ～成型 slurry molding / ～帷幕 grout curtain
【灌米汤】bewitch sb. by means of flattery
【灌木】bush; shrub
【灌输】instil into; inculcate; imbue with
【灌音】have one's voice recorded
【灌注】pour into: 把铁水～到砂型里 pour molten iron into a sand mold
【灌醉】make sb. drunk; inebriate

罐 ①(罐子) jar; pot; tin: 茶叶～ tea caddy / 水～ water pitcher / 一～苹果酱 a jar of apple jam / 坛坛～～ pots and pans ② [矿](斗车) coal tub
【罐车】tank car; tank truck; tanker
【罐笼】[矿] cage ◇ ～隔间 cage compartment / ～间隙 cage clearance
【罐头】tin; can: 肉类～ canned meat / 水果～ canned fruit / 蔬菜～ canned vegetables / 鱼类～ canned fish ◇ ～牛肉 tinned beef / ～食品 tinned food
【罐子】pot; jar; pitcher; jug: 玻璃～ glass pot; glass jar

鹳 stork

观 Taoist temple

盥 [书] wash (the hands or face)
【盥漱】wash one's face and rinse one's mouth
【盥洗】wash one's hands and face ◇ ～室 washroom / ～台 washstand / ～用具 toilet articles

贯 ①(贯通) pass through; pierce: 学～古今 well versed in both ancient and modern learning ②(连贯) be linked together; follow in a continuous line: 鱼～而入 file in ③(出生地) birthplace; native place: 籍～ the place of one's birth or origin

【贯彻】carry out; implement; put into effect: ～到底 carry through to the end / ～始终 prosecute to the end
【贯穿】run through; penetrate: 这条公路～十几个城镇。This highway runs through a dozen towns. ◇ ～辐射 penetrating radiation / ～切割 through cutting / ～术 transfixion / ～针 transfixion pin
【贯串】run through; permeate
【贯通】①(透彻了解) have a thorough knowledge of; be well versed in: ～中西医学 have a thorough knowledge of both Western and traditional Chinese medicine / 豁然～ suddenly see the light ②(接通) link up; thread together 这条铁路已全线～。The whole railway line has been joined up.
【贯通伤口】penetrating wound
【贯众】[中药] the rhizome of cyrtomium
【贯注】①(集中) concentrate on; be absorbed in: 全神～ be wholly absorbed; be rapt ②(连贯;贯穿) be connected in meaning or feeling: 这五句是一气～下来的。These five sentences run together as a coherent whole

惯 ①(习惯) be used to; be in the habit of: 这是他～用的字眼。These are his customary words and expressions. ②(纵容) indulge; spoil: 孩子要管不要～。Children should be disciplined and not indulged. / 这孩子让父母～坏了。The child was spoiled by his parents.
【惯盗】common thief; habitual robber; incorrigible thief
【惯犯】habitual offender; hardened criminal; recidivist; repeater
【惯匪】hardened bandit; professional brigand
【惯技】[贬] customary tactic; old trick
【惯例】convention; usual practice: 国际～ international practice / 打破～ break away from practices 〈customs〉
【惯例法】code of practice; conventional law
【惯量】[物] inertia
【惯窃】hardened thief
【惯偷】confirmed thief; habitual thief
【惯性】[物] inertia ◇ ～导航 inertial guidance / ～导航仪 inertial navigator / ～定律 the law of inertia / ～飞行 coasting flight / ～飞行导弹 coasting missile; coaster / ～矩 moment of inertia / ～领航 inertial navigation / ～系数 inertia coefficient / ～中心 center of inertia
【惯用】①(惯于使用) habitually practice; consistently use ②(惯用的) habitual; customary: ～伎俩 customary tactics; old tricks / ～手法 habitual practice; usual tactics

掼 [方]hurl; fling
【掼纱帽】[方] throw away one's official's hat in a huff; resign in a huff; resign in resentment quit office

guāng

光 ①(光线) light; ray: 爱克斯～ X ray / 日～ sunlight ②(光亮) brightness; lustre: 两眼无～ dull-eyed ③(光荣) honor; glory: 争～ win honor for one's country; bring credit to one's country ④(景物) scenery: 春～ sights and sounds of spring; spring scene ⑤(光滑) smooth; glossy; polished: 这种纸两面～。This kind of paper is smooth on both sides. ⑥(一点不剩) used up; nothing left: 墨水用～了。The ink's used up. ⑦(露着) bare; naked: ～着头 be bareheaded ⑧(只;单) solely; only; merely; alone: ～说不做,当然不行。Only talk and no action, naturally that won't do.
【光斑】[天] facula
【光板儿】worn-out fur
【光膀子】seripped to the waist
【光泵】optical pump
【光笔】light gun; light pen
【光波】light wave
【光彩】①(颜色和光泽) lustre; splendor; radiance: ～夺目 dazzlingly brilliant ②(光荣) honorable; glorious: 扮演一个极不～的角色 play a most inglorious part
【光电】[物] photoelectricity ◇ ～导体 photoconductor / ～二极管 photodiode; photorectifier / ～发射 photoelectric emission; photoemission / ～管 photocell; phototube
【光电子】[物] photoelectron
【光度】[物] luminosity ◇ ～计 photometer
【光辐射】ray radiation ◇ ～伤害 ray radiation injury
【光复】recover: ～旧物 recover lost territory
【光复失地】regain possession of lost territory
【光杆儿】①(花叶尽落) a bare trunk or stalk ②(失去家属的人) a man who has lost his family ③(失去群众的领导) a person without a following: ～司令 a general without an army; a leader without a following
【光顾】patronize: 如蒙～,无任欢迎。Your patronage is cordially invited.
【光怪陆离】grotesque in shape and gaudy in

color; bizarre and motley: ～的广告 grotesque and gaudy advertisements
【光棍】ruffian; hoodlum
【光棍儿】unmarried man; bachelor
【光合作用】[植] photosynthesis
【光滑】smooth; glossy; sleek
【光华】brilliance; splendor
【光化】[化] ① actinic: ～射线 actinic ray ② photochemical: ～作用 photochemical action
【光环】①(行星周围的明亮环状物) a ring of light: 土星～ Saturn's ring ②[宗](神像头上的环状光圈) halo; aureole
【光辉】①(闪烁耀目的光) radiance; brilliance; glory: 太阳的～ the rays of the sun ②(光明;灿烂) brilliant; magnificent; glorious: ～榜样 a shining example / ～的一生 a glorious life / ～事迹 glorious deeds; brilliant achievements / ～形象 magnificent images
【光辉灿烂】shine with great splendor; shine with increasing splendor
【光辉篇章】brilliant chapter
【光辉耀目】glitter and glow in the sunlight
【光洁】bring and clean ◇ ～度 [机] smooth finish
【光介子】[物] photomeson
【光景】①(情景) scene: 咱俩初次见面的～我还记得很清楚。 I still remember very clearly the scene of our first meeting. ②(景况) circumstances; conditions: ～越来越好 the life is becoming better and better ③(大约) about; around: 半夜～起了风。 About midnight a wind rose.
【光刻】[物] photoetching
【光可鉴人】be brilliant enough to reflect one's image; a surface so bright that it can serve as a mirror
【光亮】bright; luminous; shiny
【光阑】[光] diaphragm; stop; light diaphragm
【光疗】[医] phototherapy
【光临】[敬] presence (of a guest, etc.): 敬请～。 Your presence is cordially requested.
【光溜溜】①(光滑) smooth; slippery: ～的大理石地面 a smooth marble floor ②(无遮盖) bare; naked: 孩子们脱得～的在河里游泳。 The children stripped off their clothes and swam naked in the river.
【光流】light stream; luminous flux
【光芒】rays of light; brilliant rays; radiance: 旭日东升,～四射。 The morning sun rises in the east, shedding its rays in all directions.
【光芒万丈】shining with boundless radiance; gloriously radiant; resplendent
【光敏】[物] photosensitive ◇ ～电阻 photo resistance / ～二极管 photodiode
【光明】①(亮光) light: 黑暗中的一线～ a streak of light in the darkness / 重见～ see the light again ②(明亮) bright; promising: ～前途 bright prospects; brilliant future ③(没有私心) openhearted; guileless: ～磊落 open and aboveboard; frank and openhearted
【光明正大】open and aboveboard; just and honorable
【光屁股】in the nude; stark naked
【光年】[天] light-year
【光谱】[物] spectrum: 暗线～ dark-line spectrum / 明线～ bright-line spectrum / 太阳～ solar spectrum ◇ ～比较仪 spectrocomparator / ～分析 spectrum analysis / ～学 spectroscopy / ～学家 spectroscopist
【光气】[化] phosgene
【光球】[天] photosphere
【光圈】[摄] diaphragm; aperture
【光荣】honor; glory; credit: ～称号 a title of honor / ～使者 an honored envoy ◇ ～榜 honor roll / ～传统 glorious tradition / ～人家 honorable family
【光润】(of skin) smooth
【光栅】[物] grating
【光渗】[物] irradiation
【光束】[物] light beam: 参考～ reference beam
【光速】[物] velocity of light
【光天化日】broad daylight; the light of day
【光通量】[物] luminous flux
【光头】①(不戴帽子) bareheaded ②(秃头) shaven head; shaven-headed: 剃～ have one's head shaved
【光秃秃】bare; bald: ～的山坡 bare hillsides / ～的树枝 naked branches
【光线】light; ray
【光效应艺术】Op art
【光行差】[天] aberration
【光学】optics: 非线性～ nonlinear optics / 几何～ geometrical optics ◇ ～玻璃 optical glass / ～工程 optical engineering / ～录音 optical recording / ～录音机 photographic sound recorder / ～谐振腔 optical resonator / ～仪器 optical instrument / ～影象 optical image
【光焰】radiance; flare

【光耀】①(闪烁耀目的光) brilliant light; brilliance ②(荣耀) glorious; honorable
【光耀门庭】win honor and distinction for one's family; bring honor to the family name
【光阴】time: ～虚度 lose one's time by trifling / ～似箭。Time flies like an arrow.
【光阴冉冉】time passes away slowly; the years roll on smoothly
【光阴荏苒】time passes very quickly; the time slipped away
【光阴如水】time passes like flowing water
【光源】[物] light source; illuminant
【光泽】lustre; gloss; sheen
【光照】[植] illumination ◇～阶段 photostage
【光照人寒】(the moon) sent forth cold rays that made sb. shudder
【光照日月】shine like the sun and the moon
【光制】[机] finishing: 最后～ final finishing ◇～品 finished product
【光质子】[物] photoproton
【光子】[物] photon ◇～火箭 photon rocket
【光宗耀祖】bring honor to one's ancestors

桄

【桄榔】[植] gomuti palm

guǎng

广 ①(宽阔) wide; vast; extensive: 地～人稀 a vast and thinly populated area / 见多识～ have wide experience and extensive knowledge ②(多) numerous: 在大庭～众之中 before a large audience; in public ③(扩大) expand; spread: 以～见闻 widen one's knowledge / 以～流传 so that it may spread far and wide
【广板】[乐] largo
【广播】broadcast; be on the air: 开始～ go on the air / 实况～ live broadcast; live transmissions over the radio or television / 停止～ go off the air ◇～电台 broadcasting station / ～稿 broadcast script / ～讲话 broadcast speech; radio talk / ～节目 broadcast program / ～剧 radio play / ～喇叭 loudspeaker / ～体操 setting-up exercises to radio music / ～网 rediffusion network / ～信道 broadcast channel / ～员 (radio) announcer; broadcaster / ～站 broadcasting station (of a factory, school, etc.); rediffusion station ○频率 frequency / 射频 radio frequency (RF) / 高频 high frequency (HF) / 中频 intermediate frequency (IF) / 低频 low frequency (LF) / 音频 audio frequency (AF) / 调幅 amplitude / 调频 module frequency (MF) / 相位 phase / 波段 waveband; range / 波长 wave length / 长波 long wave (LW) / 中波 medium wave (MW) / 短波 short wave (SW) / 超短波 ultra short wave / 微波 microwave / 千周 kilocycle (KC) / 兆周 megacycle (MC) / 千赫 kilocycle per second (KHz) / 兆赫 megacycle per second (MHz)
【广博】(of a person's knowledge) extensive; wide: 知识～ have extensive knowledge; erudite
【广场】public square; square
【广大】①(宽阔) vast; wide; extensive: ～地区 vast areas; extensive regions / ～农村 vast rural areas / 幅员～ vast in territory ②(巨大) large-scale; wide spread ③(众多) numerous: ～读者 the reading public
【广度】scope; range: 向生产的～和深度进军 develop the range and quality of production
【广而言之】speaking generally; in a general sense
【广泛】extensive; wide-ranging; widespread: ～的兴趣 wide interests / ～而深入的影响 a widespread and profound influence / ～征求意见 solicit opinions from all sides
【广柑】a kind of orange
【广告】advertisement: 分类～ classified advertisements / 整页的～ a full-page advertisement / 做～ advertise ◇～费 rate of advertising / ～画 poster / ～栏 advertisement column / ～牌 billboard / ～色 poster color / ～设计 advertising design
【广寒宫】the Moon Palace (the mythical palace in the moon)
【广见博识】have a rich experience and extensive knowledge
【广角漫射】wide-angle diffusion
【广角透镜】wide-angle lens
【广角物镜】wide-angle lens; wide-angle objective; wide-angled object glass
【广交四海】make friends extensively
【广开才路】open all avenues for people of talent
【广开言路】encourage the free airing of views; provide wide opportunities for airing views
【广开眼界】one's field of vision is vastly enlarged; widen one's horizon
【广阔】vast; wide; broad: ～天地 a wide world; a vast new world
【广漠】vast and bare: 在～的沙滩上 on the bare expanse of the beach
【广谋博采】seek advice from all sides
【广木香】[中药] costusroot

【广纳贤士】send far and wide to invite men of ability
【广延】[物] extension ◇ ～量 extensive quantity
【广延宾客】keep open house
【广义】①(范围较宽的定义) broad sense: ～地说 in a broad sense; broadly speaking ②[物] generalized: ～动量 generalized momentum／～级数 generalized series／～空间 generalized space／～坐标 generalized coordinates
【广义相对论】[物] general relativity
【广征博引】quote and prove fully
【广种薄收】extensive cultivation

犷 rustic; uncouth; boorish
【犷悍】[书] tough and intrepid

guàng

逛 stroll; ramble; roam: ～公园 stroll in the park; have a stroll in the park／～街头 stroll around the streets
【逛荡】loiter; loaf about

guī

规 ①(圆规) compasses; dividers: 一个圆～ a pair of compasses ②(规则;成例) regulation; rule: 校～ school regulations ③(劝告) admonish; advise: ～劝 admonish ④(谋划) plan; map out: ～划 plan ⑤[机] gauge: 侧隙～ feeler (gauge)／线～ wire gauge
【规避】evade; dodge; avoid: ～义务 evade obligations
【规程】rules; regulations: 操作～ rules of operation
【规定】①(所做出的决定) stipulate; provide: 法律～的措施 measures provided for by law／～纳税责任 stipulate their tax liabilities ②(所规定的内容) fix; set; formulate: ～的表格 prescribed forms／～的指标 a set quota／在～的时间内 within the fixed time ◇ ～电压 assigned voltage／～动作[体] compulsory exercise／～负荷 given load／～数额[经] quota／～载荷 ordinance load
【规度】normality
【规范】standard; norm: 合乎～ conform to the standard／用法不～ not conform with regular usage ◇ ～化 standardize
【规范变化】[电] gage transformation
【规范场论】gage theory
【规范回答】canonical answer
【规范句型】canonical sentential form
【规格】specifications; standards; norms: 统一的～ unified standards／不合～ not be up to standard; fall short of specifications ◇ ～化 standardization
【规划】program; plan: 长远～ long-term program／作出全面～ make a comprehensive program
【规矩】①(法则) rule; established practice; custom: 这里的～是这样的。The custom here is this. ②(端正老实) well-behaved; well-disciplined: 没～ have no manners; be impolite／守～ abide by the rules; behave oneself
【规矩准绳】standards; norms; criteria
【规律】law; regular pattern: 客观～ objective law／历史发展的～ law of the development of history／自然～ law of nature／生活有～ live a regular life ◇ ～性 law; regularity
【规模】scale; scope; dimensions: ～宏大 broad in scale／大～的调查 investigation of wide scope
【规劝】admonish; advise: 好意～ give well-meaning advice
【规行矩步】①(合规矩,不苟且) behave correctly and cautiously ②(墨守成规) stick to established practice; follow the beaten track
【规约】stipulations of an agreement
【规则】①(章则) rule; regulation: 交通～ traffic regulations ②(整齐) regular: 这条河流的水道原来很不～。The course of this river used to be quite irregular ◇ ～变星 regular variable star／～多面体 regular polyhedron／～组 set of rules
【规章】rules; regulations: ～制度 rules and regulations

圭 an elongated pointed tablet of jade held in the hands by ancient rulers on ceremonial occasions
【圭表】an ancient Chinese sundial consisting of an elongated dial and one or two gnomons
【圭臬】[书] criterion; standard: 奉为～ look up to as the standard
【圭亚那】Guyana ◇ ～人 Guyanese

闺 boudoir
【闺房】boudoir
【闺阁名媛】daughters of rich families
【闺阁千金】a young lady
【闺女】①(未婚女子) girl; maiden ②[口](女儿) daughter
【闺中腻友】a woman's paramour

硅 [化] silicon (Si)
【硅二极管】silicon detector; silicon diode; silicon diode rectifier
【硅肺】[医] silicosis
【硅钢】[冶] silicon steel
【硅华】[地] siliceous sinter; silica sinter
【硅胶】[化] silica gel
【硅晶体管】silicon transistor
【硅可控整流器】[电] silicon controlled rectifier; thyristor
【硅铝带】[地] sial
【硅镁带】[地] sima
【硅锰钢】silico-manganese steel
【硅石】silica
【硅酸】[化] silicic acid ◇ ～钠 sodium silicate
【硅酸盐】[化] silicate ◇ ～工业 silicate industry / ～水泥 portland cement / ～砖 silicate brick
【硅铁】ferrosilicon
【硅酮】[化] silicone
【硅藻】[植] diatom ◇ ～土 diatomaceous earth; diatomite
【硅砖】silica brick

鲑 salmon

归 ①(返回) go back to; return: ～期 date of return / 胜利～来 return victoriously / 无家可～ be homeless ②(归还) give back to; return sth. to: 物～原主 return a thing to its rightful owner ③(趋向于;集中于) converge; come together: 把性质相同的问题～为一类 group together problems of a similar nature ④(由谁负责) turn over to; put in sb.'s charge: 饭～你做。 Rice cooking goes to you. / 这件事～我。 Leave this to me. ⑤(用在重叠动词之间,表示不相干或无结果)批评～批评,他就是不改。 Despite our repeated criticisms, he simply won't mend his ways. ⑥[数] division on the abacus with a one-digit divisor
【归案】bring to justice: ～法办 bring back to court for trial and punishment / 缉拿～ arrest and bring to justice
【归并】①(合并到) incorporate into; merge into: 一家公司～到另一家公司 merge one company into another ②(合并在一起;归拢) lump together; add up: 三笔帐～起来,一共二百元。 The three entries when put together come to two hundred yuan.
【归程】return journey
【归除】[数] division on the abacus with a divisor of two or more digits
【归档】place on file; file
【归队】①(回原部队) rejoin one's unit; return to the ranks ②(重干本行) return to the profession one was trained for
【归附】submit to the authority of another: ～国法 obey the law of the state
【归根结底】in the final analysis: ～,就是如此 in the final analysis, it's like this
【归公】go to the public 〈state〉; be made a public possession: 缴获～ turn in the captured things
【归功于】give the credit to; attribute the success to: 试验的成功～群众的智慧。 Success of the experiment is due to the wisdom of the masses.
【归国】return to one's country: ～观光 return to one's homeland on a sightseeing tour ◇ ～华侨 returned overseas Chinese
【归航】[航空] homing ◇ ～飞行 homing flight / ～附加器 homing adapter / ～台 homer / ～应答器 homing transponder / ～指点标 homer
【归还】return; revert; send back; give back: 如无法投递,则～寄件人。 In case of nondelivery, return to the sender.
【归结】①(总括) sum up; put in a nutshell: 原因是多方面的,但～起来,不外乎三点。 The causes are manifold, but can be reduced to three. ②(结局) end (of a story, etc.): 故事的～ the end of the story
【归咎】impute to; attribute a fault to; put the blame on: ～于你 the blame falls entirely on you / 把失败都～于客观原因是不正确的。 It is incorrect to put all the blame on objective causes for our failure.
【归类】sort out; classify
【归零法】return-to-zero method
【归零方式】return-to-zero mode
【归拢】put together
【归谬法】reduction to absurdity
【归纳】induce; conclude; sum up: 这是她从大量事实中～出来的结论。 This is a conclusion which she has drawn from numerous facts. ◇ ～法 inductive method; induction
【归侨】[简](归国华侨) returned overseas Chinese
【归入】classify; include: 这些问题可～一类。 These questions may be included in the same category.

【归师勿掩，穷寇莫追】one should crush a retreating enemy and not pursue broken rebels
【归属】belong to; come under the jurisdiction of: 该岛的～早已确定无疑。 The ownership of the island has long been established beyond dispute.
【归顺】come over and pledge allegiance
【归宿】a home to return to: 人生～ destination of one's life voyage
【归天】[旧] pass away; die
【归途】homeward journey; one's way home
【归位】homing
【归向】turn towards (the righteous side); incline to: 人心～ the inclination of the hearts of the people
【归心似箭】with one's heart set on speeding home; impatient to get back; anxious to return
【归一化】normalize; normalization ◇ ～电流 normalized current / ～因数 normalization factor
【归于】①(属于) belong to; be attributed to: 光荣～人民 the honor belongs to the people ②(趋向于) tend to; result in; end in: 意见～一致 agreement was reached
【归约】reduced; reduction ◇ ～算法 reduction algorithm / ～证明 proof by reduction
【归真反璞】(drop all affectation and) return to original purity and simplicity
【归总】put (items, etc.) together; sum up: ～一句话 to put it in a nutshell
【归罪】put the blame on; impute to
【归罪于】impute; incriminate; inculpate

皈

【皈依】[宗] ①(佛教的入教仪式) the ceremony of proclaiming sb. a Buddhist ②(信佛教或人其它宗教组织) be converted to Buddhism or some other religion
【皈依佛法】follow the laws of Buddha
【皈依归真】become a Buddha after death
【皈依三宝】become a Buddhist

瑰 [书] rare; marvellous

【瑰宝】rarity; treasure; gem
【瑰奇绮丽】elegant; magnificent; surpassingly beautiful
【瑰丽】surpassingly beautiful; magnificent: 我们看到大海的～之景。 We had a superb view of the sea.
【瑰纬】[书] ①(奇特) remarkable ②(华丽) (of language of style) ornate
【瑰意琦形】extraordinary ideas and admirable action; praise of a man of high integrity

龟 tortoise; turtle

【龟板】[中药] tortoise plastron
【龟背】[中药] curvature of the spinal column
【龟背石】septarium; beetle stone; septarian boulder; septarian nodule; turtle stone
【龟甲】tortoise-shell
【龟缩】huddle up like a turtle drawing in its head and legs; withdraw into passive defense; hole up: 敌人～在几个孤立的据点里。 The enemy was holed up in a few isolated strongholds.
【龟纹】moire
【龟头】[生理] glans penis; balanus
【龟头包皮炎】[医] balanoposthitis
【龟头脓溢】[医] balanorrhagia
【龟头炎】[医] balanitis

guǐ

庋 ①[架子] shelf ②(保存) keep; preserve: ～藏 store up; preserve

晷 ①[书](日影) a shadow cast by the sun ②[书](时光) time: 余～ spare time ③(日晷仪) sundial

鬼 ①(迷信的人认为人死后的灵魂) ghost; spirit; apparition: ～故事 ghost stories / 不信～，不信神 believe in neither ghosts nor gods ②[骂]胆小～ coward / 酒～ drunkard / 懒～ lazy bones ③(不光明) stealthy; surreptitious ④(不可告人的勾当) sinister plot; dirty trick: 心里有～ have a guilty conscience / 这里边有～。 There's some dirty work going on here. ⑤(恶劣) terrible; damnable: ～地方 a damnable place / ～天气 terrible weather ⑥[口](机灵) clever; smart; quick: 这家伙真～。 He's an artful devil.

【鬼把戏】sinister plot; dirty trick
【鬼点子】[方] wicked idea; trick: 出～ give devilish advice; make a wicked suggestion / 他～多。 He's full of wicked ideas.
【鬼斧神工】uncanny workmanship; superlative craftsmanship
【鬼怪】ghosts and monsters; monsters of all kinds; forces of evil
【鬼鬼祟祟】sneaking; furtive; stealthy: 这家伙～的，想干什么？ What's that fellow up to, sneaking around like that?
【鬼话】lie: ～连篇 a pack of lies

【鬼混】lead an aimless or irregular existence; fool around: 和不三不四的人～ hang around with shady characters
【鬼火】will-o'-the-wisp; jack-o'-lantern
【鬼计多端】wily and mischievous
【鬼哭狼嚎】wail like ghosts and howl like wolves; set up wild shrieks and howls
【鬼脸】①(脸部表情滑稽) funny face; wry face; grimace: 做～ make a wry face; make faces; make grimaces ②(假面具) mask used as a toy
【鬼魅】[书] ghosts and goblins; forces of evil
【鬼门关】the gate of hell; danger spot; a trying moment
【鬼迷心窍】be possessed; be obsessed
【鬼模】ghosts mode
【鬼神】ghosts and gods; spirits; supernatural beings
【鬼神恐怖】[心理] demonophobia
【鬼使神差】doings of ghosts and gods; unexpected happenings; a curious coincidence
【鬼宿星团】[天] Praesepe; Beehive; Manger
【鬼胎】sinister design; ulterior motive: 心怀～ harbor sinister designs
【鬼头鬼脑】thievish; stealthy; furtive
【鬼星团】[天] Praesepe
【鬼蜮】evil spirit; demon; treacherous person: ～伎俩 devilish stratagem; evil tactics
【鬼域为灾】this calamity was cause by evil spirits
【鬼针草】[植] beggar-ticks
【鬼主意】evil plan; wicked idea
【鬼子】devil (a term of abuse for foreign invaders): 洋～ foreign devil

诡 ①(奸诈) deceitful; tricky; cunning: ～称 say deceitfully; falsely allege; pretend ②[书](奇异) weird; eerie
【诡辩】sophistry; sophism; quibbling: ～改变不了事实。 Sophistry won't alter facts. ◇～术 sophistry
【诡称】falsely allege; pretend
【诡计】crafty plot; cunning scheme; trick; ruse: ～多端 have a whole bag of tricks; be very crafty
【诡谲】[书] strange and changeful; treacherous
【诡谲多诈】be sly and shrewd
【诡雷】[军] booby mine; booby trap
【诡秘】surreptitious; secretive: 行踪～ surreptitious in one's movements
【诡诈】crafty; cunning; treacherous

轨 ①(路轨) rail; track: 单～ single track / 双～ double track / 出～ be derailed ②(规划) course; path: 常～ normal practice / 纳入正～ put on the right track / 越～ violate the rule; be against the law
【轨道】①(供火、电车行驶的铁轨) track: 地铁～ underground railway track ②(物体运行路线) orbit; trajectory: 人造卫星已进入～。The man-made satellite is now in orbit. ③(规则;范围) course; path: 工作已走上～。 The work has got onto the right track. ◇～变换[天] orbital transfer / ～车 motor-trolley / ～衡[铁道] track scale / ～火箭 orbital rocket / ～交角[天] orbit inclination / ～空间站[天] orbital space station / ～平面[天] orbit plane / ～运动[天] orbital motion
【轨范】standard; criterion
【轨迹】①[数](某点在空间移动的路线) locus ②[天](轨道) orbit
【轨迹角】track angle
【轨迹线】trajectory
【轨距】[铁道] gauge: 标准～ standard gauge
【轨辙】rut; beaten track
【轨枕】[铁道] sleeper; tie: 纵向～ longitudinal sleeper ◇～板 concrete slab sleeper

癸 the last of the ten Heavenly Stems
【癸基】[化] decyl
【癸烷】[化] decane
【癸烯】[化] decylene

guì

桂 ①(肉桂) cassibarktree ②(月桂) laurel; bay tree ③(木犀) sweet-scented osmanthus
【桂冠】laurel (as an emblem of victory or distinction)
【桂花】[植] sweet-scented osmanthus ◇～酒 wine fermented with osmanthus flowers
【桂皮】cassia bark; Chinese cinnamon: ～油 cassia oil
【桂圆】longan ◇～肉 dried longan pulp
【桂枝】[中药] cassia twig

柜 cupboard; cabinet: 书～ bookcase / 碗～ kitchen cupboard / 衣～ wardrobe
【柜房】cashier's office
【柜榴石】[矿] cinnamon stone; hessonite
【柜台】counter; bar: 玻璃～ show case / 站～ serve behind the counter

【柜子】cupboard; cabinet

贵 ①(价格高) expensive; costly; dear: 价钱很～very expensive / 春雨～如油 Spring rains are as valuable as oil. ②(值得重视) highly valued; valuable; precious: ～在坚持。The important thing is perseverance. / 人～有自知之明。 It is a good thing for one to be able to have a proper appraisal of oneself. ③(地位优越) of high rank; noble ④[敬] your: ～国 your country / (您)～姓? May I ask your name?

【贵宾】honored guest; distinguished guest ◇ ～席 seats for distinguished guests; distinguished visitors' gallery / ～休息室 reserved lounge (for honored guests)

【贵妃】highest-ranking imperial concubine

【贵金属】noble metal

【贵客盈门】The house was full of distinguished guests

【贵人多忘事】A man of your eminence has a short memory

【贵体违和】Your precious health was not all that it might be

【贵重】valuable; precious: ～物品 valuables / ～药品 costly medicines

【贵族】noble; aristocrat: 封建～ feudal nobles / 精神～ intellectual aristocrats

跪 kneel; go down on one's knees

【跪拜】worship on bended knees; kowtow

【跪倒】throw oneself on one's knees; prostrate oneself; grovel: ～脚下 throw oneself at sb.'s feet / ～在地 go down on one's knee

【跪地哀求】kneel on the ground crying for mercy

【跪乳之恩】filial piety

【跪射】[军] kneeling fire

【跪姿】kneeling position: ～射击 shoot from a kneeling position

桧 [植] Chinese juniper

刽 cut off; chop off

【刽子手】①(旧称执行死刑的人) executioner; headsman ②(屠杀人民的人) slaughterer; butcher

鳜 mandarin fish

gǔn

衮 ceremonial dress for royalty

【衮衮】[书] ①(继续不断) continual ②(众多) numerous

【衮衮诸公】[讽] ①(官僚) high-ranking officials ②(大人;阁下) Your Excellencies

滚 ①(滚动) roll; trundle: ～下山去 roll down the hill / 到处乱～ roll all over the place ②(走开) get away; beat it: ～出去! Get out of here! / 你给我～! Get away with you! ③[方](沸腾) boil: 用～水泡茶 make tea with boiling water / 水～了。 The water is boiling. ④(缝纫方法) bind; trim: 袖口上～一道花边 make an ambroidered hem on the cuff

【滚槽机】[机] channelling machine

【滚齿法】generating process

【滚齿机】gear-hobbing machine; hobber; hobbing machine

【滚存盈余】accumulated surplus

【滚存资金】deferred assets

【滚蛋】[骂] scram; get away

【滚刀】[机] hobbing cutter; hob

【滚动】roll; trundle ◇ ～摩擦[物] rolling friction / ～轴承[机] rolling bearing

【滚翻】[体] roll: 侧～ sideward roll / 后～ backward roll / 前～ forward roll

【滚瓜烂熟】(recite, etc.) fluently; (know sth.) pat: 背得～ have memorized sth. thoroughly; have sth. pat

【滚滚】roll; billow; surge: ～的浓烟 billowing smoke / ～而来 roll in; come in torrents / ～而下 (the water) rushed down in a torrent / ～入海 roll to the sea

【滚剪机】[机] slitting mill; slitting roller

【滚净筒】cleaning cage; cleaning mill

【滚雷】[军] rolling mine

【滚轮】[体] gyro wheel

【滚水坝】[水] overflow dam

【滚烫】boiling hot; burning hot

【滚筒】cylinder; roll ◇ ～板 roller plate / ～印刷机 cylinder press / ～轴 roller spindle

【滚雪球】snowball

【滚圆】round as a ball

【滚轧】[机] rolling ◇ ～机 rolling mill

【滚针轴承】needle bearing

【滚珠】[机] ball ◇ ～轴承 ball bearing

【滚柱轴承】roller bearing

【滚装船】roll-on-roll-off ship; ro-ro

【滚子链】roller chain

磙 ①(磙子) roller: 石～ stone roller ②(用滚子轧地) level (ground, etc.) with a roller: ～地 roll the ground

【磙子】①(碌碡) stone roller ②(碾轧器具)

roller

辊 [机] roller
【辊筒印花】[纺] roller printing
【辊压淬火】rolled hardening
【辊压接合】roll bonding
【辊子】roller

绲 [书] ①(带子) band; tape ②(绳索) string; cord

gùn

棍 ①(棍子) rod; stick ②(无赖) scoundrel; rascal: 赌～ gambler / 恶～ ruffian; rascal
【棍棒】①(武术器械) club; cudgel; bludgeon ②(器械体操用具) a stick or staff used in gymnastics
【棍子】rod; stick

guō

郭 the outer wall of a city: 城～ the inner and outer city walls

聒 noisy
【聒耳】grate on one's ears
【聒噪】[方] noisy; clamorous
【聒噪不休】wag one's tongue

过 [口] beyond the limit; undue; excessive: ～费 go to undue expense

蝈
【蝈蝈儿】katydid; long-horned grasshopper

锅 ①(炊事用具) pot, pan, boiler, caldron, etc.: 炒菜～ frying pan / 大～ caldron / 煎～ frying- pan / 铝～ aluminium pot / 砂～ earthenware cooking pot / 压力～ pressure cooker / 蒸～ steamer / 煮～ saucepan; stewpan ②(器物上象锅的部分) bowl (of a pipe, etc.): 烟袋～儿 the bowl of a pipe; pipe
【锅巴】crust of cooked rice; rice crust
【锅饼】(large, thick) wheat cake
【锅底朝天】no rice at all in the pot; nothing to eat in the house
【锅炉】boiler: 火管～ fire tube boiler / 水管～ water tube boiler ◇ ～防垢剂 boiler compound / ～房 boiler room / ～给水 boiler feedwater / ～垢 boiler scale / ～效率 boiler efficiency / ～性能 boiler performance / ～压力 boiler pressure
【锅台】the top of a kitchen range: 围着～转 be tied to the kitchen sink
【锅贴儿】lightly fried dumpling
【锅驼机】[机] portable steam engine; locomobile
【锅烟子】soot on the bottom of a pan
【锅子】①(器物上象锅的部分) bowl (of a pipe, etc.) ② (火锅) chafing dish

guó

国 ①(国家) country; state; nation: 收归～有 be nationalized; be taken over by the state ②(代表国家的) of the state; national: ～旗 national flag ③(指我国的) of our country; Chinese: ～画 traditional Chinese painting
【国宝】national treasure
【国宾】state guest ◇ ～馆 state guest house
【国策】the basic policy of a state; national policy
【国产】made in our country; made in China: ～品 national products; domestic products / ～远洋货轮 Chinese-built oceangoing freighter
【国耻】national humiliation
【国粹】the quintessence of Chinese culture
【国都】national capital; capital
【国度】country; state; nation
【国法】the law of the land; national law; law
【国法无情】the law is no respecter of persons
【国防】national defense ◇ ～部 the Ministry of National Defense / ～工业 national defense industry / ～建设 the building up of national defense / ～力量 defense capability / ～生产 defense production / ～现代化 modernization of national defense / ～线 national defense line / ～支出 expenditure on national defense; defense spending
【国富民殷】The people are noble and the country prosperous
【国歌】national anthem
【国故】national cultural heritage; ancient learning
【国号】title of a reigning dynasty
【国画】traditional Chinese painting
【国徽】national emblem
【国会】parliament; (美) Congress; (日) the Diet
【国货】China-made goods; Chinese goods
【国籍】nationality: 出生～ nationality by birth / 多重～ plural nationality / 双重～ dual nationality / 无～ statelessness; stateless / 保留～ retain citizenship; retain nationality / 剥夺～ deprivation of nationality;

denationalization / 根据出生地取得～ citizenship by birth / 根据血统取得～ citizenship by descent / 恢复～ restoration of nationality / 申请加入～ apply for naturalization / 退出～ renounce one's nationality / 选择～ choose one's nationality / 自动丧失～ lose one's nationality automatically / ～不明 uncertain nationality / ～不明的飞机 unidentified aircraft / ～的放弃 renunciation of nationality

【国籍法】law of nationality; nationality law

【国计民生】the national economy and the people's livelihood

【国际】international: 共产～ the Communist International / 第三～ the Third International ◇ ～标准 international standard / ～博览会 international fair / ～储备货币 international reserve currency / ～地位 international status / ～法 international law; the law of nations / ～法人 international person / ～公法 (public) international law / ～公制 the metric system / ～共管 international condominium / ～关系 international relations / ～惯例 international practice / ～航道 international waterway / ～合作 cooperation among nations / ～化 internationalization / ～货币 convertible foreign exchange; international currency / ～经济新秩序 a new international economic order / ～恐怖组织 international terrorist organization / ～礼让 comity of nations / ～列车 international train / ～贸易 international trade / ～市场 international market / ～事务 international affairs / ～水平 international standards / ～私法 private international law / ～行为准则 international code of conduct / ～形势 the international situation / ～音标[语] the International Phonetic Symbols / ～影响 international repercussions; impact abroad / ～友人 foreign friends / ～舆论 world opinion; public opinion / ～争端 international dispute / ～支付手段 medium of international payment / ～专利 international monopoly

【国际奥林匹克委员会】the International Olympic Committee (IOC)

【国际标准期刊编号】International Standard Serial Number; ISSN

【国际标准书号】International Standard Book Number; ISBN

【国际儿童节】International Children's Day (June 1)

【国际法院】the International Court of Justice; the World Court

【国际歌】*The Internationale*

【国际警察组织】Interpol

【国际共产主义运动】the international communist movement

【国际劳动妇女节】International Working Women's Day (March 8)

【国际劳动节】International Labor Day; May Day (May 1)

【国际联盟】the League of Nations (1920-1946)

【国际收支】balance of (international) payments: ～不平衡 disequilibrium of balance of payments / ～逆差 international payments deficit; unfavorable balance of payments / ～顺差 international payments surplus; favorable balance of payments / ～危机 international payments crisis

【国际象棋】chess

【国际主义】internationalism ◇ ～义务 internationalist duty / ～者 internationalist

【国际组织】international organization ◯ ①共同体 Community: 东非共同体 East African Community; EAC / 法兰西共同体 French Community / 加勒比共同市场 Caribbean Common Market / 加勒比共同体 Caribbean community; CARICOM / 经济互助委员会 Council for Mutual Economic Aid; CMEA / 欧洲共同市场 European Common Market / 欧洲经济共同体 European Economic Community; EEC / 西非国家经济共同体 Economic Community of West African States; ECOWAS / 西非经济共同体 Communaute Economique de l'Afrique de l'Ouest; CEAO / 中美洲共同市场 Central American Market; CACM ②会议 Conference: 阿拉伯首脑会议 Arab Summit Conference / 不结盟国家和政府首脑会议(不结盟国家会议) Conference of Heads of State and Government of Non-Aligned Countries / 东非和中非国家元首和政府首脑会议(东非和中非国家首脑会议) Conference of Heads of State and Government of East and Central African Countries / 发展中国家会议(七十七国集团) Conference of Developing Countries (the 77-nation group) / 非洲国家和政府会议 Assembly of the Heads of State and Government of the Organization of African Unity / 伊斯兰会议 Islamic Conference ③理事会 Council: 北欧理事会 Nordic Council / 美澳新理事会 Australia, New Zealand, United States Treaty Organization

Council; ANZUS / 世界和平理事会 World Peace Council / 亚洲太平洋地区理事会 Asian and Pacific Council; ASPAC ④联合会 Confederation: 国际自由工会联合会 International Confederation of Free Trade Union; ICFTU / 联合国协会世界联合会 World Federation of United Nations Associations; WFUNA / 世界工会联合会 World Federation of Trade Unions; WFTU / 世界劳工联合会 World Confederation of Labor; WCL ⑤联盟 Association: 阿拉伯国家联盟 League of Arab States (Arab League) / 东南亚国家联盟(东盟) Association of South East Asian Nations; ASEAN / 各国议会联盟 Inter-Parliamentary Union; IPU / 国际合作社联盟 International Cooperative Alliance; ICA / 欧洲广播联盟 European Broadcasting Union; EBU / 欧洲自由贸易联盟 European Free Trade Association; EFTA / 西欧联盟 Western European Union; WEU ⑥协会 Association: 国际笔会 International Pen / 国际新闻工作者协会 International Organization of Journalists; IOJ / 国际红十字大会 International Red Cross Conference; IRCC / 国际世界语协会 Universal Esperanto Association; UEA / 红十字会协会 League of Red Cross Societies; LRCS / 红十字国际委员会 International Committee of the Red Cross; ICRC / 欧洲作家协会 European Community of Writers / 亚非新闻工作者协会 Afro-Asian Journalists' Association / 亚非作家协会 Afro-Asian Writers' Association ⑦银行 Bank: 非洲开发银行 African Development Bank; AFDB / 国际经济合作银行 International Bank for Economic Co-operation; IBEC / 加勒比开发银行 Caribbean Development Bank; CDB / 美洲开发银行 Inter-American Development Bank; IADB / 欧洲投资银行 European Investment Bank; CDB / 世界银行 World Bank / 亚洲开发银行 Asian Development Bank; ADB / 伊斯兰开发银行 Islamic Development Bank; IDB ⑧组织 Organization: 阿拉伯石油输出国组织 Organization of Arab Petroleum Exporting Countries; OAPEC / 安第斯条约组织(安第斯集团) Andean Pact Organization (Andean group); ANCOM / 北大西洋公约组织 North Atlantic Treaty Organization; NATO / 非洲统一组织 Organization of African Unity; OAU / 国际广播电视组织 International Radio and Television Organization; OIRT / 国际通讯卫星组织 International Telecommunication Satellite Consortium; INTELSAT / 华沙条约组织(华沙友好合作互助条约) Warsaw Treaty Organization (Warsaw Treaty of Friendship, Co-operation and Mutual Assistance); WTO / 美洲国家组织 Organization of American States; OAS / 欧洲核研究组织 European Organization for Nuclear Research; CERN / 石油输出国组织 Organization of the Petroleum Exporting Countries; OPEC / 伊斯兰会议组织 Organization of the Islamic Conference; OCAS / 中美洲国家组织 Organization of central American States

【国家】country; state; nation: 第三世界~ Third World countries / 友好~ friendly country ◇ ~补助 state aid / ~财政 state revenue and expenditure / ~大事 national affairs / ~典礼 state functions / ~队 national team / ~法 constitutional law; the law of the state / ~机关 state organs; government offices / ~机关工作人员 personnel of organs of state; state personnel / ~机器 state apparatus / ~机密 state secrets / ~决算 final accounts of state revenue and expenditure; final state accounts / ~权力 force of state; power of state / ~学说 theory of the state / ~银行 state bank / ~元首 head of state / ~政权 state power / ~职能 functions and powers of the state / ~仲裁人 government arbitrator / ~主权 national sovereignty / ~资本主义 state capitalism

【国家兴亡,匹夫有责】The rise and fall of the nation is the concern of every citizen. (或: All men share a common responsibility for the fate of their country.)

【国交】diplomatic relations between nations

【国教】state religion

【国界】national boundaries

【国境】territory: ~线 boundary line

【国君】monarch

【国库】national treasury; exchequer ◇ ~债务 exchequer bond / ~证券 exchequer bills

【国库券】treasury bill (T.B.)

【国力】national power: ~雄厚 have solid national strength

【国立】state-maintained; state-run ◇ ~大学 national university

【国民】national: 本生的~ natural-born national / 归化的~ naturalized national ◇ ~教育 national education; popular education / ~经济 national economy / ~生产总值 gross national product (GNP) / ～收入 national income

【国民党】nationalist party; the Kuomintang (KMT)

【国民生计】national economy and the people's livelihood

【国难】national calamity (caused by foreign aggression)

【国内】internal; domestic; home: ～法 domestic law; national law / ～汇兑 domestic exchange / ～立法 internal legislation; national legislation / ～贸易 domestic trade / ～市场 domestic market / ～通信卫星 domestic satellite (DOMSAT) / ～新闻 home news / ～治安 internal security

【国破家亡】the country is defeated and the home lost

【国旗】national flag

【国情】the condition of a country; national conditions ◇ ～咨文(美) State of the Union Message

【国庆】National Day ◇ ～节 National Day (October 1)

【国人】[书] compatriots; fellow countrymen; countrymen

【国色天香】national beauty and heavenly fragrance peony

【国色天姿】possess surpassing beauty; woman of great beauty; celestial beauty

【国事】national affairs ◇ ～访问 state visit

【国手】national champion (in chess, etc.); grand master

【国书】letter of credence; credentials: 递交～ present credentials

【国术】traditional Chinese boxing and fencing

【国泰民安】the country is prosperous and the people live in peace

【国体】①(国家体制) state system ②(国家的体面) national prestige

【国土】territory; land: 神圣～ our sacred land

【国外】external; overseas; abroad: ～来信 letter from abroad ◇ ～代理行 foreign agency / ～共同海损 foreign general average / ～汇兑 foreign exchange / ～侨胞 fellow countrymen residing abroad / ～市场 overseas market / ～事务 external affairs / ～投资 foreign investment / ～资产 external assets

【国王】king

【国务会议】state conference: 最高～ the Supreme State Conference

【国务卿】(美) Secretary of State

【国务院】① the State Council ② (美) the State Department ○总理 Premier / 副总理 Vice-premier / 国务委员 State councillor / 国务院秘书长 Secretary-General of the State Council / 部长 minister / 外交部 Ministry of Foreign Affairs / 国防部 Ministry of National Defense / 国家计划委员会 State planning Commission / 国家经济体制改革委员会 State Commission for Restructuring Economy / 国家教育委员会 State Education Commission / 国家科学技术委员会 State Science and Technology Commission / 国防科学技术工业委员会 State Commission of Science, Technology and Industry for National Defense / 国家民族事务委员会 State Nationalities Affairs Commission / 公安部 Ministry of Public Security / 国家安全部 Ministry of State Security / 监察部 Ministry of Supervision / 民政部 Ministry of Civil Affairs / 司法部 Ministry of Justice / 财政部 Ministry of Finance / 人事部 Ministry of Personnel / 劳动部 Ministry of labor / 地质矿产部 Ministry of Geology and Mineral Resources / 建设部 Ministry of Construction / 能源部 Ministry of Energy Resources / 铁道部 Ministry of Railways / 交通部 Ministry of Communications / 机械电子工业部 Ministry of Machine-Building and Electronics Industry / 航空航天工业部 Ministry of Aeronautics and Astronautics Industry / 冶金工业部 Ministry of Metallurgical Industry / 化学工业部 Ministry of Chemical Industry / 轻工业部 Ministry of Light Industry / 纺织工业部 Ministry of Textile Industry / 邮电部 Ministry of posts and Telecommunications / 水利部 Ministry of Water Resources / 农业部 Ministry of Agriculture / 林业部 Ministry of Forestry / 商业部 Ministry of Commerce / 对外经济贸易部 Ministry of Foreign Economic Relations and Trade / 物资部 Ministry of materials / 文化部 Ministry of Culture / 广播电影电视部 Ministry of Radio, Film and Television / 卫生部 Ministry of Public Health / 国家体育运动委员会 State Physical Culture and Sports Commission / 国家计划生育委员会 State Family Planning Commis-sion / 中国人民银行 People's Bank of China / 审计署 Auditing Administration

【国玺】broad seal; great seal; seal of the state

【国宴】state banquet

【国营】state-operated; state-run ◇ ～工商业

state-operated industry and commerce / ～经济 state sector of the economy; state-owned economy / ～农场 state farm / ～牌价 state-set prices / ～企业 state enterprise
【国有国法,家有家法】a country has its laws and a family its rules
【国有化】nationalization
【国葬】state funeral
【国债】national debt
【国中之国】a state within a state
【国子监】Directorate of Imperial Academy

guǒ

果 ①(果实) fruit: 单～ simple fruit / 复合～ composite fruit / 干～ dry fruit / 核～ drupe / 荚～ pod; legume / 坚～ nut / 浆～ berry / 聚合～ aggregate fruit / 梨～ pome / 鲜～ fresh fruit / 开花后,不久就结～。 After blossoming, it will soon bear fruit. ②(结果,后果) result; consequence: 恶～ a disastrous result; dire consequences ③(果断) resolute; determined: 言必信,行必～ true to one's word and resolute in action ④(果然) really; as expected; sure enough: ～果然不出所料 just as one expected ⑤(果真) if indeed; if really: ～能如此 if things can really turn out that way; if that is so / ～有所见,不妨实说。 If (indeed) you should have any opinion, you might as well tell us frankly.
【果不其然】indeed; as expected, not unexpectedly ～,下了雨。 As expected, it rained.
【果冻】jelly
【果断】resolute; decisive: 办事～ handle affairs in a decisive manner / 行为～ resolute in action
【果脯】preserved fruit; candied fruit
【果腹】fill the stomach; satisfy one's hunger: 食不～ have not enough food to fill one's belly
【果敢】courageous and resolute: 采取～的行动 take resolute action
【果酱】jam
【果胶】pectin
【果料儿】raisins, kernels, melon seeds, etc. used in making cakes, buns, etc.
【果木】fruit tree ◇ ～园 orchard
【果农】orchard worker; orchardist; orchardman
【果皮】the skin of fruit; rind
【果品】fruit: 干鲜～ fresh and dried fruit
【果然】really; as expected; sure enough: ～名不虚传 a really well-deserved reputation
【果仁儿】kernel
【果肉】the flesh of fruit; pulp
【果如所料】It happened exactly as expected.
【果实】①(果子) fruit: ～累累 fruit growing in close clusters; fruit hanging heavy on the trees ②(收获) gains; fruits: 劳动～ fruits of labor
【果实压枝】the fruit weighed the branches down
【果树】fruit tree ◇ ～栽培 fruit growing; pomiculture
【果糖】[化] fructose; levulose
【果序】infructescence
【果芽】fruit bud
【果蝠】fruit bat
【果园】orchard
【果真】if indeed; if really: ～如此,我放心了。If this is really true, it'll take a load off my mind.
【果汁】fruit juice
【果枝】①(果树枝) fruit-bearing shoot; fruit branch ②(结棉桃的枝) boll-bearing branch (of the cotton plant)
【果子】fruit ◇ ～酒 fruit wine / ～露 fruit syrup
【果子剥离机】macerator
【果子狸】masked civet
【果子盐】[药] fruit salt

裹 bind; wrap: 把伤口～好 bind up the wound
【裹法】pack; packing
【裹脚】foot-binding
【裹腿】puttee
【裹胁】force to take part; coerce: 在～下 under duress
【裹足不前】hesitate to move forward

guò

过 ①(通过) cross; pass: 他们二人刚～去。 They two have just passed by. ②(穿过) across; past; through; over: ～桥 go over a bridge ③(度过) spend (time); pass (time): 假期～得怎样? How did you spend your holiday? ④(过去) after; past: ～了夏至,天就开始变短。 The days get shorter after the Summer Solstice. ⑤(使经过) undergo a process; go through; go over: 咱们把这篇稿子再～一遍。 Let's go over the draft once again. ⑥(超过) exceed; go beyond: 小心别坐～了站。 Be sure you don't go past your station. / 大雪深～膝盖。 The snow is more than knee-deep. ⑦ (过于;过分;太) excessively; unduly: ～长 too long; unduly long / ～早 too early; premature / 雨水～多 excessive rainfall; too much rain ⑧[化][物] per-; super-; over-: ～熔 superfusion / ～氧化物 peroxide ⑨ (过失) fault; mistake: 记～ put a person's error on record / 勇于改～ be bold in correcting one's mistakes ⑩ (用在动词后,跟"得"或"不"连用,表示胜过或通过)要说跑,咱们谁也比不～他。None of us can run as fast as he can. / 这样的大学生,我们信得～。 We have confidence in college students of this sort. ⑪[量] time: 衣服漂了三～儿了。 The clothes have been rinsed three times. ⑫ (用在动词后,表示完毕)桃花都已经开～了。 The peach blossoms are over.

【过半数】more than half; majority: 这个工厂～的职工是妇女。 More than half the workers and staff members in this mill are women.

【过磅】weigh (on the scales): 行李～了没有? Has the luggage been weighed?

【过饱和】[化] supersaturation; oversaturation ◇～溶液 supersaturated solution

【过不去】① (不能通过) cannot get through; be unable to get by; be impassable: 路挡住了,车子～。The road was blocked, no car could pass. ② (为难) be hard on; make it difficult for; embarrass: 他有意跟我～。 He deliberately made things difficult for me. ③ (过意不去) feel sorry 不必十分～。 Don't take it so much to heart.

【过场】①[剧] interlude ②[剧] cross the stage ③ (做样子敷衍过去)走～ do sth. as a mere formality; go through the motions

【过程】course; process: 缩短制作～ shorten the process of manufacture / 在讨论～中 in the course of the discussion

【过秤】weigh (on the steelyard)

【过从】[书] have friendly intercourse; associate: ～甚密 be in close association with sb.

【过冲】overshoot; overswing: ～失真 over throw distortion

【过错】fault; mistake

【过道】passageway; corridor

【过得去】① (可通过) be able to pass; can get through: 桥很宽,卡车～。 The bridge is wide enough for trucks to pass. ② (不坏) passable; tolerable; so-so; not too bad: 那里的生活条件还～。 Living conditions there are not bad. ③ (多用于反问) feel at ease: 你什么都不要,叫我们怎么～呢? How could we feel at ease while you refuse to take anything from us?

【过得硬】be able to stand all tests; become truly proficient (in sth.)

【过冬】pass the winter; winter: 这种鸟在哪儿～? Where do these birds winter?

【过度】excessive; undue; over-: ～疲劳 be overtired; be over fatigue / 饮酒～ drink to excess; over-drink

【过度感染】superinfection

【过度修正】over correction

【过渡】transition; interim: ～时期 a period of transition; a transitional period ◇～国际法 international law of transition / ～流[航空] transition flow

【过渡性条文】transitional provision

【过伐】[林] overcutting

【过分】excessive; undue; over-: ～的要求 excessive demands / ～强调 put undue stress on; overemphasize / ～自信 overconfidence in oneself / 做得太～ go too far; overdo sth.

【过关】① (通过关口) pass a barrier; go through an ordeal ② (比喻,指通过检验) pass a test; reach a standard: 这项产品的质量已经～。The quality of this product is up to standard.

【过河拆桥】remove the bridge after crossing the river; drop one's benefactor as soon as his help is not required; kick down the ladder

【过河弃舟】set adrift the boat once the river is crossed

【过后】afterwards; later: 我们～再谈。 Let's talk about it later.

【过户】[法] transfer ownership; change the name of the owner in a register: 把某物～到某人名下 have sth. transferred into sb.'s name

【过活】make a living; live: 勉强～ make a bare living; live from hand to mouth

【过火】go too far; go to extremes; overdo: ～的行动 excesses

【过激】too drastic; extremist: ～的言论 extremist opinions ◇～分子 extremist; ultra / ～主义 ultraism

【过继】① (以亲属的儿子做为自己的儿子) adopt a young relative ② (把自己的儿子给亲属做儿子) have one's child adopted by a relative

【过奖】[谦] overpraise; undeserved compliment: 您～了。 You flatter me.

【过节】celebrate a festival

【过街老鼠】a person hated by everyone

【过街老鼠,人人喊打】When a rat runs across the street everybody cries, "kill it!"

【过境】pass through the territory of a country; be in transit ◇ ～货物 transit goods / ～贸易 transit trade / ～签证 transit visa / ～权 right of passage / ～税 transit duty

【过客】passing traveller; transient guest

【过来】①(向着自己的方向) come over; come up: 快～! Come over here, quick! ②(回到正常状态) round: 他醒～了。He woke up; He came round. ③(有充分的时间、能力) can manage: 活儿不多,我一个人也忙得～。The work is not much, I can manage it alone.

【过来人】a person who has had the experience: 你是～,当然明白其中道理。You are a man that has gone through it all, naturally you know the whys and wherefores of it. / 要知水深浅,须问～。He knows the water best who has waded through it.

【过冷】[物] super-cooling

【过梁】[建] lintel

【过量】excessive; over-; over-dose; surfeit: 饮食～ excessive eating and drinking / 这种药千万不能服～。Whatever happens, never take an overdose of this medicine.

【过磷酸铵】[化] ammonium superphosphate

【过磷酸钙】[化] calcium superphosphate

【过路】pass by on one's way ◇ ～人 passerby

【过路财神】the man through whose hands pass large sums of money; those who manage financial affairs

【过虑】be over-anxious; worry overmuch; worry unnecessarily: 你何必～? Why worry unnecessarily?

【过滤】filter; filtrate ◇ ～程序 filter / ～机组 filter bank / ～器 filter / ～设备 filter plant / ～嘴 filter tip (of a cigarette)

【过门】move into one's husband's household upon marriage

【过门儿】[乐]①(指歌曲开始的乐器演奏部分) opening bars ②(指歌曲中间的乐器演奏部分) short interlude between verses

【过敏】[医] irritability; allergy ◇ ～性反应 allergic reaction

【过敏素】anaphylactin

【过敏性】anaphylaxis; hypersensitivity; hypersusceptibility

【过敏原】anaphylactogen; sensitinogen

【过目】look over (papers, lists, etc.) so as to check or approve: 名单已经排好,请您～。Here's the list for you to go over.

【过目成诵】be able to recite sth. after reading it over once; have a photographic memory

【过年】celebrate the New Year; spend the New Year: 快～了。It'll soon be New Year. / 他今年回家～。He'll be home for the New Year holiday.

【过期】exceed the time limit; be overdue: ～作废 in valid after the specified date / 你借的书已经～。The book you borrowed is overdue. ◇ ～胶卷 expired film / ～票据 pass due note; overdue bill / ～提单 stale bill of lading / ～支票 overdue check; state check / ～杂志 back number of a magazine

【过谦】too modest: 你～了。You are being too modest.

【过去】①(以前) in or of the past; formerly; previously: 现在和～不同了。Now it is different from the past. ②(趋向) go over; pass by: 我～看一看。I'll go over and take a look. ③(得逞): 骗～了 get away by trickery ④(去世) pass away: 他祖父昨天夜里～了。His grandfather passed away last night.

【过人】surpass; excel: ～的记忆力 a remarkable memory / ～之处 the things one excels in; one's forte / 精力～ surpass many others in energy / 勇气～ excel in courage

【过日子】live; get along: 勤俭～ live industriously and frugally / 挺会～ can manage to get along quite well

【过筛】sift out

【过山车】Roller Coaster

【过甚】exaggerate; overstate: ～其词 give an exaggerated account; overstate the case

【过剩】excess; surplus: 生产～ overproduction

【过失】①(因疏忽而犯的错误) fault; slip; error ②[法] unpremeditated crime; offense ◇ ～犯罪 criminal negligence; involuntary crime / ～杀人 commit manslaughter; involuntary homicide / ～伤害 corporal wound by mistake; negligent injury / ～误差 gross errors / ～行为 negligent act / ～责任 liability for fault; neglect of duty / ～罪 negligent crime

【过时】①(陈旧不合时宜) out-of-date; outmoded; obsolete; antiquated; out of fashion: ～的观念 antiquated ideas / ～的设备 outmoded equipment ②(错过规定的时间) past the appointed time ◇ ～商品 obsolete merchandise

【过时不候】no waiting after the set time

【过手】take in and give out (money, etc.); receive and distribute; handle: 银钱～,当面点清。Count the money on the spot.

【过熟林】[林] overmature forest

【过数】count
【过堂】[旧] appear in court to be tried
【过堂风】draught
【过头】go beyond the limit; overdo: 聪明～ be too clever by half
【过屠门而大嚼】pass the butcher's and start munching; feed oneself on illusions
【过往】①(来去) come and go: ～车辆 vehicular traffic / ～行人 pedestrian traffic ②(交往; 来往) have friendly intercourse with; associate with
【过问】concern oneself with; take an interest in; bother about: 亲自～ take up matter personally; take a personal interest in a matter / 无人～ not be attended to by anybody; be nobody's business
【过为已甚】overdo a thing; overstep the limits
【过细】meticulous; careful: ～地做工作 work carefully; work with meticulous care
【过眼云烟】as transient as a fleeting cloud
【过夜】pass the night; put up for the night; stay overnight
【过意不去】feel apologetic; feel sorry
【过瘾】satisfy a craving; enjoy oneself to the full; do sth. to one's heart's content: 今天我一口气游了三千米,真～。 Today I really swam to my heart's content. I did 3000 meters at a stretch.
【过硬】have a perfect mastery of sth.; be really up to the mark; be able to pass the stiffest test: ～操作技术 a perfect mastery of operational technique
【过犹不及】going too far is as bad as not going far enough
【过于】too; unduly; excessively: ～劳累 overtired / ～迁就 too accommodating; too ready to compromise / ～自信 take something too much for granted
【过载】①(转装) transship ②(超载) overload ◇ ～电平[电] overload level / ～失真 blasting / ～特性 overload characteristic / ～系数 overload factor / ～容量 overload capacity
【过帐】transfer items (as from a daybook to a ledger); post
【过重】overweight ◇ ～加费[邮] overweight charge

H

hā

哈 （哈气）blow one's breath; breathe out (with the mouth open): ～了一口气 breathe out a breath／往手上～气取暖 breath on one's hands to keep them warm ②（表示笑声）ha; ha-ha: ～～大笑 laugh heartily; roar with laughter ③[叹]～～，我猜着了。Aha, I've got 〈guessed〉 it.／～～，这回你溜不掉了。Aha, you can't slip away this time.
【哈哈镜】distorting mirror
【哈吉】[伊斯兰教] haji
【哈雷慧星】[天] Halley's Comet; the Halley Comet
【哈里发】[伊斯兰教] caliph
【哈密瓜】Hami melon (a variety of musk-melon)
【哈乃斐派】[伊斯兰教] the Hanafite school
【哈欠】yawn: 打～ give a yawn
【哈萨克】(苏联) Kazakhstan ◇ ～人 kazakh／～语 kazakh (language)
【哈萨克族】(中国) Kazak (nationality) ◇ ～人 kazak／～语 kazak (language)
【哈腰】[口] ①（弯腰）bend one's back; stoop: 我一～把钢笔掉在地上了。As I bent over; my fountain pen fell to the ground. ②（鞠躬）bow: ～曲背 humble oneself in serving (a master)／点头～ bow unctuously; bow and scrape

铪 [化] hafnium (Hf)
【铪板】hafnium plate
【铪锭】hafnium ingot

há

蛤
【蛤岸】clamshell bucket
【蛤蟆】①（青蛙）frog ②（蟾蜍）toad
【蛤珍珠】clam

hǎ

哈
【哈巴狗】①（狗的一个品种）Pekinese ②（奴才）toady; sycophant
【哈达】hada: 献～ present a hada

hà

哈
【哈什蚂】Chinese forest frog ◇ ～油[中药] the dried oviduct fat of the forest frog

hāi

嗨
【嗨哟】[叹] heave ho; yo-heave-ho; yo-ho

咳 [叹]（表示伤感、后悔或惊异）～，我怎么这么糊涂！Dammit! How stupid I was!

hái

骸 ①（骨）bones of the body; skeleton: 四肢百～ all the limbs and bones ②（指身体）body: 病～ frail body／形～ the human body／遗～ (dead) body; corpse; remains
【骸骨】human bones; skeleton

孩 child: 小女～儿 a little girl／她生了一个男～。She gave birth to a boy.
【孩儿参】[中药] caryophyllaceous ginseng
【孩提】[书] early childhood; infancy
【孩子】①（儿童）child: 男～ boy／女～ girl ②（子女）son or daughter; children: 她有两个～。She has two children.
【孩子气】childishness: 别再～啦。Don't be childish any more.

还 [副] ①（仍旧）still; yet: ～早 still early／～有一些具体问题要解决。Some specific problems have yet to be solved.／他～没有来。He has not come yet. ②（更加）even more; still more: 今天比昨天～冷。Today is even colder than yesterday.／这个问题～要困难。This problem is still more difficult. ③（另外）also; too; as well; in addition: 我们看了一个故事片，～看了一个动画片。We saw a cartoon in addition to a feature film. ④（勉强过得去）passably; fairly: 屋子不大，收拾得倒～干净。The room is small, but it's kept quite tidy. ⑤（尚且）even: 你～不知道，何况我呢？If you don't know, still less do I. ⑥（用以加强语气）这～了得！This is the limit! ⑦（表示意外）他～真有办法。He really is resourceful.
【还好】①（不坏）not bad; passable: 你今天感觉怎样？——～。How are you feeling today? — Not so bad. ②（幸运）fortunately: ～，这场大水没有把堤坝冲坏。Fortunately, the flood did not break the dike.
【还是】①[副]（仍旧）still; nevertheless; all the same: 尽管如此，～要谢谢你。Thank you all the same／他～失败了。He failed after all. ②（表示希望）had better: 要下雨了，你～带把伞吧。It looks like rain, you'd better take an

umbrella with you. ③(表示选择) or: 你早上去，～下午去? Are you going in the morning or in the afternoon? ④[副](表示意外)我没想到这事儿～真难办。I didn't expect it to be so difficult.

hǎi

海 (海洋；大湖) sea: big lake: 南～ the Nanhai Sea / 出～ put out to sea ②(多) a great number of people or things coming together: 林～ a vast stretch of forest / 人～ a sea of people; crowds of people / 红旗如～ a sea of red flags ③(大) extra large; of great capacity: ～碗 a very big bowl

【海岸】seacoast; coast; seashore: 沿～航行 sail along the coast ◇～炮 coast gun / ～炮兵 coast artillery / ～线 coastline

【海拔】height above sea level; elevation: ～四千米 4000 meters above sea level; with an elevation of 4000 meters

【海百合】[动] sea lily; crinoid

【海报】playbill: 张贴～ put up a playbill

【海豹】seal

【海宝】Hypon

【海滨】seashore; seaside: ～浴场 seaside resort / 去～度假 go to the seaside for one's holidays

【海标】seamark

【海波】[化] hypo

【海不扬波】the sea is clam; peace in the country

【海菜】edible seaweed

【海产】marine products

【海草】seaweed

【海昌蓝】[纺] hydron blue

【海潮】(sea) tide

【海程】distance travelled by sea; voyage

【海船】seagoing vessel

【海带】kelp

【海胆】[动] sea urchin

【海岛】island (in the sea): ～居民 islander / ～民兵 island militia

【海盗】pirate; sea rover ◇～船 pirate (ship); sea rover / ～行为 piracy

【海堤】sea wall

【海底】the bottom of the sea; seabed; sea floor ◇～采矿 undersea mining; offshore mining / ～电报 submarine telegraph; cablegram / ～电缆 submarine cable / ～勘察 submarine exploration / ～矿 submarine mine / ～山 seamount / ～水雷 ground mine / ～油田 offshore oilfield / ～资源 seabed resources; submarine resources

【海底捞月】try to fish out the moon from the bottom of the sea; strive for the impossible or illusory: ～一场空 be as futile as fishing for the moon in the sea

【海底捞针】fish for a needle in the ocean; look for a needle in a haystack

【海地】Haiti ◇～人 Haitian

【海防】coast defense ◇～部队 coastal defense force / ～前哨 outpost of coastal defense / ～前线 coastal front / ～艇 coastal defense boat

【海风】sea breeze; sea wind

【海港】seaport; harbor ◇～设备 harbor installations

【海沟】(oceanic) trench; submarine trench

【海狗】fur seal; ursine seal

【海关】customhouse; customs ◇～登记 customs entry / ～放行 customs clearance / ～检疫 customs quarantine control / ～人员 customs officer / ～申报单 customs declaration / ～手续 customs formalities / ～税则 customs tariff / ～通行证 customs pass / ～巡逻船 customs cruiser / ～证明书 customs certificate / ～总署 customs head office ○报关 customs declaration / 结关 customs clearance / 验关 customs examination

【海关检查】customs inspection: 通过～ go through customs ◇～站 customs inspection post

【海龟】green turtle

【海涵】[敬] be magnanimous enough to forgive or tolerate (sb.'s errors or shortcomings): 招待不周，还望～。Please forgive us if we have not looked after you well.

【海魂衫】sailor's striped shirt

【海货】marine products

【海鲫】Japanese seaperch

【海疆】coastal areas and territorial seas

【海角】cape; promontory

【海角天涯】(go to) the ends of the earth; the corner of the world; far-off regions

【海进】[地]transgression

【海禁】ban on maritime trade or intercourse with foreign countries

【海景】seascape

【海鸠】[动] guillemot

【海军】navy ◇～航空兵 naval air force / ～基地 navel base / ～陆战队 marine corps; marines / ～演习 naval maneuver; naval exercise

【海口】①(港口) seaport ②(说大话): 夸～

boast about what one can do; talk big
【海枯石烂】(even if) the seas run dry and the rocks crumble: ～,此心不变。 Seas may run dry, rocks turn to dust, but I'll always be loyal to you.
【海葵】[动] sea anemone
【海阔天空】①(大自然的广阔) as boundless as the sea and sky; unrestrained and far-ranging ②(说话无拘束、无边际) discursive; rambling
【海蓝宝石】[矿] aquamarine
【海狸】beaver
【海狸鼠】coypu; nutria
【海里】nautical mile; sea mile
【海量】①[敬](宽洪的度量) magnanimity: 对不住的地方,望您～包涵。 I hope you will be magnanimous enough to excuse any incorrect behavior on my part. ②(酒量大) great capacity for liquor: 您是～,再来一杯。Have another one. You can hold your liquor.
【海流】ocean current
【海龙】①[口](海獭) sea otter ②(鱼) pipefish
【海路】sea route; sea-lane; seaway; 走～ travel by sea
【海轮】seagoing vessel
【海螺】conch
【海洛因】heroin
【海绿石】[矿] glauconite
【海马】sea horse
【海鳗】conger pike
【海米】dried shrimps
【海绵】①(低等多细胞动物) sponge ②(橡、塑制品) foam rubber or plastic; sponge ◇～底球鞋 sponge-insoled shoes / ～垫 foam-rubber cushion / ～球拍 foam-rubber table-tennis bat
【海绵铁】sponge iron
【海难】perils of the sea ◇～例外条款 peril of the sea exception / ～条款 peril clause
【海内】within the four seas; throughout the country
【海内存知己,天涯若比邻】A bosom friend afar brings a distant land near.
【海内人望】be the cynosure of the whole country
【海内晏如】Peace reigns throughout the land; a peaceful time
【海鲇】sea catfish
【海牛】manatee; sea cow; sea slug
【海鸥】sea gull
【海盘车】[动] starfish
【海泡石】[矿] sepiolite; sea-foam
【海平面】sea level: ～气压 sea-level pressure
【海区】[军]sea area
【海群生】[药] hetrazan
【海鳃】sea pen; sea feather
【海商法】maritime law
【海商法规】commercial maritime code
【海商旗】maritime flag
【海上】at sea; on the sea: ～风暴 a storm at sea ◇～霸权 maritime hegemony / ～保险 marine insurance / ～保险诉讼 marine insurance action / ～避撞规则 rules of the road at Sea / ～补给 sealift; seaborne supple / ～封锁 naval blockade / ～过失 maritime negligence / ～交通线 sea route; sea-lane / ～救助 salvage / ～空间 air space above the sea / ～力量 sea power / ～留置权 maritime lien / ～掠夺 piracy / ～侵权行为 maritime tort / ～事故 accident at sea; maritime casualty / ～遇险信号 signal of distress; SOS / ～运输 marine transportation / ～钻探 offshore drilling / ～作业 operation on the sea
【海蛇】sea snake
【海参】sea cucumber; sea slug; trepang
【海蚀】marine abrasion; wave erosion
【海狮】sea lion
【海市蜃楼】mirage
【海事】maritime affairs ◇～保险 maritime insurance / ～裁判官 marine magistrate / ～裁判权 admiralty jurisdiction; maritime jurisdiction / ～法 marine law; maritime law / ～法庭 maritime court / ～惯例 customs of the sea / ～管辖 maritime jurisdiction / ～纠纷 admiralty dispute / ～检查员 marine inspector / ～诉讼 admiralty proceedings / ～仲裁 maritime arbitration
【海誓山盟】(make) a solemn pledge of love
【海水】seawater; brine; the sea: ～不可斗量。The sea cannot be measured with a bushel — great minds can not be fathomed. ◇～工业 marine industry / ～养殖 sea-water aquiculture / ～浴 sea-water bath; sea bathing
【海损】[商] average: 单独～ particular average / 共同～ general average ◇～货物 goods damaged by sea / ～理算 average adjustment / ～理算人 average adjuster
【海獭】sea otter
【海滩】seabeach; beach; perils of the sea
【海棠】[植]Chinese flowering crabapple
【海天一色】The sea melted into the sky. (或: The sea and the sky merged into one.)
【海桐花】tobira
【海图】sea chart ◇～室 chart room

【海退】[地] regression
【海豚】[动] dolphin ◇ ～泳[体] dolphin butterfly; dolphin fishtail; dolphin
【海外】overseas; abroad: ～华侨 overseas Chinese / ～同胞 countryman residing abroad ◇ ～版 overseas edition / ～倾销 dumping abroad
【海外奇谈】strange story from over the seas; traveller's tale; tall story
【海湾】bay; gulf: ～地区 gulf areas / ～国家 gulf states
【海王星】[天]Neptune
【海味】choice seafood: ～山珍 all sorts of delicacies
【海峡】strait; channel: 台湾～ the Taiwan Straits / 英吉利～ the English Channel
【海鲜】seafood
【海相】[地] marine facies ◇ ～沉积[地] marine deposit
【海象】walrus; morse
【海啸】tsunami; seismic sea wave
【海星】[动] starfish; sea star
【海熊】fur seal; ursine seal
【海寻】nautical fathom
【海盐】sea salt
【海蜒】[动] anchovy
【海燕】(storm) petrel
【海洋】seas and oceans; ocean ◇ ～霸权 maritime hegemony / ～动物 marine animal / ～法 law of the sea / ～工程 oceanographic engineering / ～公约 maritime convention / ～国家 maritime state / ～气象船 ocean weather ship / ～气象学 marine meteorology / ～权 maritime rights / ～生物学 marine biology / ～性气候 maritime climate / ～学 oceanography; oceanology / ～渔业 sea fishery / ～资源 marine resources / ～钻井 marine drilling / ～主权 maritime sovereignty
【海域】sea area; maritime space: 南海～ Nanhai Sea waters
【海员】seaman; sailor; mariner ◇ ～俱乐部 seamen's club / ～用语 nautical expression
【海运】sea transportation; ocean shipping; ocean carriage ◇ ～法 shipping law / ～费 ocean freight / ～货物保险 marine cargo insurance / ～提单 ocean bill of lading
【海葬】sea-burial
【海枣】[植] date palm; date
【海藻】marine alga; seaweed; kelp
【海战】sea warfare; naval battle
【海蜇】jellyfish
【海震】[地] seaquake; submarine earthquake

胲 [化] hydroxylamine

hài

害 ①(祸害; 害处) evil; harm; calamity: 为～不浅 do great harm; cause much damage / 为民除～ rid the people of a scourge / 灾～ calamity; disaster ②(有害的) harmful; destructive; injurious: ～鸟 harmful bird / ～兽 harmful animal ③(使受害) do harm to; impair; cause trouble to: ～人不浅 do people great harm ④(杀害) kill; murder: 他遇～身亡了。He was murdered. ⑤(生病) contract (an illness); suffer from: ～了一场大病 have a serious attack of illness / ～眼 suffer from eye disease ⑥(发生不安的情绪) feel (ashamed, afraid, etc.)
【害草净】lambast
【害虫】injurious insect: 除～ get rid of insects
【害处】harm: 吸烟过多对身体有～。Excessive smoking is harmful to one's health.
【害多利少】more harm than good; more disadvantages than advantages; do more harm than good
【害鸟】harmful bird
【害怕】be afraid; be scared: ～得要命 be scared to death; be mortally afraid
【害群之马】an evil member of the herd; one who brings disgrace on his group; black sheep
【害人虫】an evil creature; pest; vermin: 扫除一切～。Away with all pests.
【害人非浅】cause deep injury to someone; do people great harm; cause infinite harm to people
【害人害己】bite off one's own head; curses come home to roost
【害臊】[口] feel ashamed; be bashful: 替他～ be ashamed of him
【害羞】be bashful; be shy: 在生人面前显得～ be shy in presence of strangers
【害羞赧颜】feeling ashamed; blush a scarlet-red
【害眼】have eye trouble

亥 the last of the twelve Earthly Branches
【亥时】the period of the day from 9 p.m. to 11 p.m.

氦 [化] helium (He)
【氦层】heliosphere

【氦灯】helium lamp
【氦星】helium stars
【氦液化器】helium liquefier
【氦致冷器】helium refrigerator

骇 be astonished; be shocked
【骇怪】be shocked; be astonished
【骇然】struck dumb with amazement; gasping with astonishment
【骇人听闻】shocking; appalling: ～的暴行 horrifying atrocities / ～的事件 astounding event / ～的消息 shocking news
【骇异】be shocked; be astonished: 不胜～ be greatly shocked; be greatly astounded; be greatly surprised

hān

犴 [动] elk; moose

顸 [方] thick: 这线太～了。 This thread is too thick.

鼾 snore: 打～ snore
【鼾声】sound of snoring: ～如雷 snore thunderously
【鼾声呼吸】stertorous breathing
【鼾睡】sound, snoring sleep
【鼾音】sonorous rale

憨 ①(傻) foolish; silly: ～痴 idiotic ②(朴实; 天真) straightforward; naive; ingenuous
【憨厚】simple and honest; straightforward and good-natured
【憨态可掬】charmingly naive
【憨头憨脑】(man) with a stupid head and a dull brain
【憨笑】smile fatuously; simper
【憨直】honest and straightforward

酣 ①(饮酒尽兴) (drink, etc.) to one's heart's content: 半～ half drunk ②(泛指尽兴、畅快等) fully; heartily; heatedly; in the heat of
【酣畅】①(指饮酒) merry and lively ②(指睡眠) sound ③(指文艺作品) with ease and verve; fully: ～的笔墨 something written with ease and verve
【酣歌狂舞】sing and dance rapturously
【酣梦】sweet dream
【酣然入睡】fall into a profound sleep
【酣睡】sleep soundly; be fast asleep; sopor
【酣饮】drink to the full; carouse
【酣战】hard-fought battle: 两军～ two armies locked in fierce battle
【酣醉】be dead drunk

蚶 [动]blood clam
【蚶子】[动] blood clam

hán

寒 ①(冷) cold: 受～ catch a cold / 天～地冻。 The weather is cold and the ground is frozen. ②(害怕;畏惧) tremble (with fear): 胆～ be terrified ③(穷困) poor; needy: 贫～ in indigent circumstances; poverty-stricken ④[谦] my humble: ～舍 my humble home
【寒北风】[气] scharnitzer
【寒痹】[中医] arthritis (aggravated by cold)
【寒潮】[气] cold wave; cold-airoutbreak
【寒碜】①(难看) ugly; unsightly: 长得不～ not bad-looking ②(丢脸;不体面) shabby; disgraceful: 这有什么～。 There's nothing to be ashamed of. ③(讥笑) ridicule; put to shame: 叫人～了一顿 be ridiculed by sb.
【寒窗苦读】studying at a cold window; the life of a poor scholar
【寒带】[地] frigid zone
【寒冬腊月】severe winter; dead of winter
【寒光闪闪】(the swords) glittered like frost and snow
【寒风】cold wind: ～刺骨。 The cold wind chilled one to the bone.
【寒风指数】[气] chill wind factor
【寒花晚节】the cold flower in its ending; to preserve one's personality in old age
【寒极】[气] cold pole
【寒假】winter vacation
【寒噤】shiver (with cold or fear): 他打了个～。 A shiver ran over his body.
【寒苦】destitute; poverty-stricken
【寒来暑往】as summer goes and winter comes; with the passage of time
【寒冷】cold; frigid: ～的气候 a cold climate
【寒流】[气] cold current
【寒毛】fine hair on the human body
【寒梅斗雪】The plum trees in full bloom are braving the snow (in the depth of winter).
【寒梅吐幽】The winter plum trees exuded a fragrant smell.
【寒气】cold air; cold draught; cold: ～逼人。 There is a nip in the air.
【寒热】[中医] chills and fever ◇ ～往来 alternating spells of fever and chills
【寒暑表】thermometer

【寒酸】(of a poor scholar in the old days) miserable and shabby
【寒腿】[口] rheumatism in the legs
【寒武纪】[地] the Cambrian (Period)
【寒武系】[地] Cambrian system
【寒心】be bitterly disappointed: 令人～ bitterly disappointing
【寒暄】exchange of conventional greetings; exchange of amenities: 她同客人～了几句。She exchanged a few words of greeting with the guests.
【寒鸦】jackdaw
【寒烟衰草】chilly mists and the fading plants
【寒衣】winter clothing
【寒意】a nip in the air: 初春季节仍有～。It's spring but there's still a chill in the air.
【寒战】shiver (with cold or fear); chill; rigor
【寒症】[中医] symptoms caused by cold factors (e.g. chill, slow pulse, etc.)

含 ①(衔在口中) keep in the mouth: ～着润喉片 have a throat tablet in the mouth ②(藏在里面; 包含) contain: ～泪 with tears in one's eyes / ～沙量 silt content ③(不露出来) nurse; cherish; harbor: ～恨 nurse one's hatred / ～怒 be filled with anger
【含苞】[书] in bud: ～待放 be in bud
【含垢忍辱】endure contempt and insults; bear shame and humiliation
【含毫吮墨】moisten the tip of the writing brush with one's lips
【含恨终天】die with a deep regret
【含糊】①(不清晰;不明确) ambiguous; vague: ～不清 ambiguous and vague / ～其词 talk ambiguously ②(马虎) careless; perfunctory: ～了事 finish a job carelessly / 这事一点不能～。We'll have to handle the matter with meticulous care.
【含混】indistinct; ambiguous: 言词～,令人费解 speak so ambiguously as to be barely intelligible
【含量】content: 牛奶的乳糖～ the lactose content of the milk / 油的酯～ the ester content of oil
【含怒】in anger
【含片】lozenge: 止咳～ cough lozenge
【含情脉脉】(soft eyes) exuding tenderness and love
【含沙射影】attack by innuendo; make insinuations: 采用～的卑劣手法 resort to insinuation / ～, 恶语中伤 vilify sb. with insidious language
【含漱剂】[药] gargle
【含水】containing water; containing moisture ◇ ～层[地] water-bearing stratum; aquifer / ～率 moisture content
【含笑】have a smile on one's face: ～点头 nod with a smile
【含笑九泉】smile in the underworld; one has nothing to regret in life; One would sink happily into the nether world.
【含辛茹苦】endure all kinds of hardships; put up with hardships
【含羞】with a shy look; bashfully: ～不语 be silent with shame; go into one's shell / ～忍辱 suffer humiliations
【含羞草】[植] sensitive plant
【含羞娇嗔】pout prettily in embarrassment
【含蓄】①(包含) contain; embody ②(不直接说出的) implicit; veiled: ～的批评 implicit criticism ③(不外露) reserved
【含血喷人】make slanderous accusations
【含饴弄孙】mouth malt sugars and dally with one's grandson; an old man enjoys life with no cares
【含义】meaning; implication: 这句话～深刻。This remark has profound implications.
【含英咀华】study and relish the beauties of literature; containing the cream of the literary tradition
【含油层】[石油] oil-bearing formation
【含油废水】oily waste water
【含油树脂】oleoresin
【含油轴承】oil-impregnated bearing; oil-retaining bearing ◇ ～合金 oil-impregnated metal; oilite
【含冤】suffer a wrong: ～死去 die uncleared of a false charge
【含有】contain; have: 海水～盐分。Sea water contains salt.
【含冤负屈】suffer an unjust grievance; suffer an iniquitous wrong
【含冤莫白】suffer a grievous wrong with no hope of vengeance
【含冤终天】die with one's name uncleared
【含怨】bear a grudge; nurse a grievance
【含怨忍辱】passively to accept insults and humiliations

焓 [物] enthalpy; total heat

函 ①(匣;封套)[书] case; envelope: 镜～ a

case for a mirror ②(信件) letter: 公～ official letter / ～复 reply by letter; write a letter in reply / ～告 inform by letter
【函购】purchase by mail; mail order
【函件】letters; correspondence
【函授】teach by correspondence; give a correspondence course ◇ ～部 correspondence department (of a school) / ～学校 correspondence school
【函数】[数] function ◇ ～计算机 function calculator / ～计算器 function calculator; functional counter / ～运算 functional operation

涵 ①(包容) contain ②(涵洞) culvert: 桥～ bridges and culverts
【涵洞】culvert: 拱式～ arched culvert / 平顶～ flat topped culvert ◇ ～闸门 clough
【涵养】①(修养) ability to control oneself; self restraint: 很有～ know how to exercise self-control ②(蓄积) conserve: 用造林来～水源 conserve water through afforestation

hǎn

罕 rarely; seldom: ～闻 seldom heard of
【罕百理派】[伊斯兰教] the Hanbalite school
【罕见】seldom seen; rare: 一场～的洪水 an exceptionally serious flood
【罕言寡语】be quiet and unexpressive
【罕有】rare; unusual; exceptional: ～的机会 a rare opportunity / ～的现象 an exceptional phenomenon

喊 ①(大声叫) shout; cry out; yell: ～救命 cry "Help! Help!" / ～口号 shout slogans / 把嗓子～哑了 shout oneself hoarse ②(叫) call (a person): 你走以前～他一声。 Give him a shout before you go.
【喊话】①(对敌宣传) propaganda directed to the enemy at the front line: 向敌兵～:"缴枪不杀!" shout at the enemy soldiers, "Hands up! No harm!" ②(呼叫) communicate by tele-equipment
【喊叫】shout; cry out
【喊冤叫屈】cry out about one's grievances; complain loudly about an alleged injustice

hàn

汉 ①(汉语) Chinese (language): ～英词典 a Chinese-English dictionary ②(男子) man: 大～ a big fellow / 好～ hero; man of courage / 老～ an old man
【汉白玉】white marble
【汉奸】traitor (to China): ～卖国贼 traitor and collaborator
【汉人】the Hans; the Han people
【汉森病】[医] Hansen's disease; leprosy
【汉学】①(汉代经学) the Han school of classical philology ②(中国学) Sinology ◇ ～家 Sinologist
【汉语】Chinese (language) ◇ ～拼音方案 the Scheme for the Chinese Phonetic Alphabet / ～拼音字母 the Chinese phonetic alphabet
【汉字】Chinese character: 简化～ simplify (simplified) Chinese characters / 日语～ kanji; honji ◇ ～改革 reform of Chinese characters / ～简化方案 the Scheme for Simplifying Chinese Characters / ～输入装置 Chinese input unit / ～显示装置 Chinese display / ～注音 phonetic annotation of Chinese characters / ～字盘 Chinese selecting
【汉子】①(男子) man; fellow ②[方](丈夫) husband

汗 sweat; perspiration: 出～ sweat; perspire / 发～ induce perspiration / ～如雨下 dripping with perspiration
【汗斑】①(汗碱) sweat stain ②[医](花斑癣) tinea versicolor
【汗背心】sleeveless undershirt; vest; singlet
【汗臭】bad smell of perspiration; stink of sweat
【汗碱】sweat stain
【汗脚】feet that sweat easily; sweaty feet
【汗孔】[生理] pore of a sweat gland
【汗流浃背】streaming with sweat (from fear or physical exertion)
【汗马功劳】①(战功) distinctions won in battle; war exploits: 立下了～ perform deeds of valor in battle ②(工作中的贡献) one's contributions in work
【汗毛】fine hair on the human body
【汗牛充栋】enough books to make the ox carrying them sweat or to fill a house to the rafters — an immense number of books
【汗青】①(青竹) sweating green bamboo strips ②(史册) historical records; chronicles; annals: 光照～。 The brilliance of his name shines in the pages of history. / 人生自古谁无死,留取丹心照～。 What man was ever immune from death? Let me but leave a loyal heart shining in the pages of history.
【汗如雨下】sweat profusely; The sweat runs down like raindrops.
【汗衫】undershirt; T-shirt

【汗腺】[生理] sweat gland ◇ ～囊瘤 hydrocystoma／～腺瘤 hidradenoma; syringoma／～炎 hidradenitis; hidroadenitis
【汗颜】[书] blush with shame; feel deeply ashamed
【汗疹】heat-rash; sudamina
【汗珠子】beads of sweat

旱 ①(降水少; 无降水) dry spell; drought: 抗～ combat drought／久～的禾苗逢甘雨。A sweet rain falls on the parched seedlings. ②(非水田的) dryland: ～稻 dry rice ③(陆地交通) on land: ～路 overland route
【旱船】land boat, a model boat used as a stage prop in some folk dances
【旱稻】upland rice; dry rice
【旱地】nonirrigated farmland; dry land
【旱谷】[地质] arroyo
【旱瓜涝桃】Melon will be good in drought and peach in wet days.
【旱季】dry season
【旱金莲】[植] nasturtium
【旱井】①(防旱的井) water-retention well ②(冬季贮菜井) dry well (used to store vegetables in winter)
【旱涝保收】ensure stable yields despite drought or excessive rain
【旱涝保收,稳产高产】give stable, high yields irrespective of drought or water logging
【旱柳】dryland willow
【旱路】overland route: 走～ travel by land
【旱苗逢雨】It is like rain upon parched rice plants.
【旱年】year of drought
【旱芹】celery
【旱桥】viaduct; overpass; flyover
【旱情】damage to crops by drought: ～严重。The drought is serious.／～已趋缓和。The drought has eased up.
【旱伞】[方] parasol
【旱生动物】xerophilous animal
【旱生植物】xerophyte
【旱獭】[动] marmot: 藏～ Himalayan marmot
【旱田】dry farmland; dry land
【旱象】signs of drought
【旱烟】tobacco (smoked in a long-stemmed Chinese pipe) ◇ ～袋 long-stemmed Chinese pipe
【旱灾】drought

悍 ①(勇猛) brave; bold: 一员～将 a brave warrior／有精兵～将 have staunch fighting men and brave warriors ②(凶狠; 蛮横) fierce; ferocious: ～妇 a hot-tempered wife; a fiery woman／凶～ fierce and tough; ferocious
【悍然】outrageously; brazenly; flagrantly: ～入侵 outrageously invade／～撕毁协议 flagrantly scrap an agreement／～宣布 brazenly declare

焊 weld; solder: 气～ gas welding
【焊缝】welding seam; weld line
【焊膏】soldering paste
【焊工】①(焊接工作) welding; soldering ②(焊接工人) welder; solderer
【焊接】welding; soldering: 电弧～ (electric) arc welding ◇ ～操作 welding operation／～车间 welding shop／～钢管 welded steel pipe
【焊机】welder; welding machine
【焊炬】welding blow lamp
【焊料】solder
【焊枪】welding torch; (welding) blowpipe
【焊条】welding rod
【焊头】bonding tool; welding head
【焊锡】soldering tin; tin solder
【焊药】welding agent
【焊液】welding fluid; soldering fluid
【焊油】soldering paste

捍 defend; guard
【捍卫】defend; guard; protect: ～边疆 defend national frontier／～革命成果 safeguard the fruits of revolution／～国家主权 uphold state sovereignty

翰 [书]①(毛笔) writing brush: 挥～ wield one's writing brush; write (with a brush) ②(笔迹; 文字) writing: 华～[敬] you letter
【翰林】member of the Imperial Academy ◇ ～院 Hanlin Academy; Imperial Academy
【翰墨】[书] brush and ink — writing, painting, or calligraphy

瀚 [书] vast
【瀚海】[书] big desert

憾 regret: 引以为～ deem it regrettable
【憾事】a matter for regret

撼 shake: 蚍蜉～树 an ant trying to shake a tree — ridiculously overrating oneself／～山岳,泣鬼神 so moving as to shake the mountains and wring tears from the spirits

【撼动】shake; vibrate
【撼天动地】cause a great sensation

颔 [书] ①(下巴) chin ②(点头) nod
【颔首】[书] nod: ～会意 nod one's comprehension / ～示意 give a nod as a signal / ～微笑 nod smilingly

hāng

夯 ①(砸实地基用的工具) rammer; tamper: 木～ wooden ram ②(用夯砸) ram; tamp; pound: 把土～实 ram the earth
【夯板】tamping plate
【夯镐】tamping pick
【夯歌】rammers' work chant
【夯具】rammer; tamper
【夯实厚度】compacted thickness
【夯实压路机】tamping roller; tamping-type roller
【夯土机】rammer; tamper

háng

杭 short for Hangzhou
【杭育】[叹] heave ho; yo-heave-ho; yo-ho

吭 throat: 引～高歌 sing lustily

航 ①(船) boat; ship ②(航行) navigate (by water or air): 民～ civil aviation / 首～ maiden voyage or flight / 夜～ night navigation
【航班】scheduled flight; flight number
【航标】navigation mark
【航差】drift
【航测】[简](航空测量) aerial survey
【航程】voyage; passage; range ◇ ～记录器 odograph
【航船】boat that plies regularly between inland towns
【航次】①(出航次第) the sequence of voyages or flights; voyage or flight number ②(出航的次数) the number of voyages or flights made
【航道】channel; lane; course: 重要的国际～ an important international sea-lane / 主～ the main channel
【航海】navigation ◇ ～多项运动 navigational sports / ～法规 navigation law / ～家 navigator; voyager / ～罗盘 mariner's compass / ～日志 logbook; log / ～天文历 nautical almanac / ～天文学 nautical astronomy / ～图 nautical chart / ～仪器 nautical instrument / ～用语 nautical term
【航迹】[航空] flight path; track ◇ ～角 flight-path angle / ～轴 flight path axis
【航空】aviation: 民用～ civil aviation ◇ ～版 airmail edition / ～保险 aviation insurance / ～标塔 airway beacon / ～测量 aerial survey / ～磁测 aeromagnetic survey / ～地图 aeronautical chart; aerial map / ～电子学 avionics / ～发动机 aero-engine; aircraft engine / ～法 air law / ～港 air harbor / ～工程 aeronautical engineering / ～工业 aviation industry / ～公司 airline company; airways / ～管制 air traffic control / ～航天工业 aerospace industry / ～航天医学 aerospace medicine / ～货运 airfreight / ～机械员 aircraft mechanic / ～力学 aeromechanics / ～联运 through air transport / ～模型 model airplane / ～母舰 aircraft carrier / ～汽油 aviation gasoline / ～气象台 air weather station; aeronautical meteorological station / ～气象学 aeronautical meteorology / ～器材 air material / ～燃料 aviation fuel / ～日志 aircraft logbook / ～探矿 mineral exploration aviation; aerial prospecting / ～提单 air bill of lading; air freight bill / ～体育运动 air sports; flying sports / ～天文历 air almanac / ～通信 air communications / ～线 airline; airway / ～协定 air transport agreement / ～信 airmail letter; air letter; airmail / ～学 aeronautics; aviation / ～邮件 airmail / ～运费 air freight / ～运输 air transportation / ～站 airport / ～照相 aerial photography / ～照相机 aerocamera; aerial camera ○ 中国民航 CAAC / 中国国际航空公司 Air China / 机场 airport / 停机坪 tarmac / 跑道 runway / 机场主楼 main airport building; terminal building / 塔台 control tower / 机场灯标 airport beacon / 上飞机 board a plane; get into a plane / 下飞机 get off a plane; alight from a plane / 误飞机 miss the plane / 扶梯 ramp / 班机 airliner / 客机 passenger plane / 国内班机 domestic flight / 国际班机 international flight / 旅游旺季 on season / 旅游淡季 off season / 售票处 booking office / 问讯处 inquiry office; information counter / 预先买票 book a ticket in advance / 定座 make reservation / 退票 refund a ticket; get a refund for a ticket / 飞机票 plane ticket / 登机牌 boarding check / 头等 first class / 经济座 economy class; tourist class / 涡轮螺旋桨飞机 turboprop aircraft / 中程客机 medium haul airliner / 远程客机 long haul airliner / 远程宽机身客机 wide-body long haul airliner / 大型

喷气客机 jumbo-jet／中短程客机 short-medium haul airliner／机组 aircraft crew; air crew／机长 pilot／副机长 co-pilot; second pilot／领航员 navigator／报务员 radio operator／女乘务员（空中小姐）stewardess; hostess／男乘务员 steward／地勤人员 ground crew／客舱 passenger cabin／密封舱 pressurized cabin／行李舱 luggage compartment／驾驶舱 pilot's cockpit／机翼 wing／右翼 starboard wing／左翼 port wing／机身 fuselage; body／起落架 undercarriage／起落架轮 undercarriage wheel／方向舵 rudder／升降舵 elevator／水平尾翼 tail plane／航行灯 navigation light／起飞 take off; take-off／加油 refueling／着陆 landing／滑行 taxi along／飞行 flight; flying／直飞 direct flight; straight flight／连续飞行 non-stop flight／迫降 forced landing／高度 altitude; height／巡航速度 cruising speed／最高速度 top speed／上升限度 ceiling／盘旋 circling／爬升 climbing; gain height／降低 lose height; to fly low／迎风 face the wind／颠簸 rock; toss; bump／不平稳的飞行 bumpy flight／平稳的飞行 smooth flight／晕机 airsick／加班 extra flight／衔接航班 connecting flight／夜航 night service

【航空兵】①（指兵种）air arm ②（指飞行员）airman ◇～部队 air unit

【航空站】airport：国际～ international airport

【航路】air or sea route ◇～标志 route markings

【航速】speed of a ship or plane：～为二十节 sail at a speed of 20 knots

【航天】spaceflight; aerospace ◇～舱 space capsule／～飞机 space shuttle／～技术 space technology／～通信 space communication (SPACECOM)／～员 astronaut／～站 spaceport

【航图】chart

【航务】navigational matters

【航线】air or shipping line; route; course：传统～ customary route／内河～ inland navigation line／指定～ advertised route

【航向】course (of a ship or plane)：改变～ change course／右～ starboard tack／左～ port tack ◇～指示器 direction indicator

【航行】①（水中行驶）navigate by water; sail：内河～ inland navigation／逆风～ sail to windward／顺风～ sail before the wind; sail downwind ②（空中行驶）navigate by air; fly：空中～ aerial navigation ◇～半径 navigation radius／～灯 navigation light／～规则 navigation rules／～权 right of navigation／～事故 navigation accidents

【航运】shipping ◇～保险 shipping insurance／～法 law of navigation／～公司 shipping company／～权 navigation right／～业 shipping business

行 ①（行列）line; row：排成两～ fall into two lines／杨柳成～ lined with rows of willows ②（排行）seniority among brothers and sisters：你～几？—我～三。Where do you come among your brothers and sisters? — I'm the third. ③（行业）trade; profession; line of business：改～ change one's profession／干一～爱一～ love whatever job one takes up／各～各业 all trades and professions; different walks of life／他干哪～？What's his line? ④（营业机构）business firm：拍卖～ auctioneer's／银～ bank ⑤［量］四～诗句 four lines of verse／一～树 a row of trees

【行当】①［口］trade; profession; line of business ②［剧］type of role (in traditional Chinese operas)

【行贩】pedlar

【行行出状元】One may distinguish himself in any trade.

【行话】jargon; cant

【行会】［旧］guild ◇～制度 the guild system

【行家】expert; connoisseur：～话 expert remarks／～里手 experts and master hands

【行阶】row order

【行距】［农］row spacing

【行列】ranks：排成整齐的～ be drawn up in orderly ranks

【行列式】［数］determinant

【行频】［电视］line frequency

【行情】quotations (on the market); prices ◇～表 quotations list

【行市】quotations (on the market); prices

【行式打印】line printing ◇～机 line printer

【行式摄影】line image

【行伍】［旧］the ranks：～出身 rise from the ranks

【行业】trade; profession; industry：服务～ service trades ◇～暗语 lingo／～语 jargon; cant

【行长】president (of a bank)

绗 sew with long stitches：～被子 sew on the quilt cover with long stitches

hàng

沆

【沆瀣】evening mist
【沆瀣一气】act in collusion with; wallow in the mire with; like attracts like

巷

【巷道】[矿] tunnel ◇ ～壁 wall / ～掘进机 tunnelling machine / ～刷帮 flitching / ～刷大 slipping / ～挑顶 rip / ～腰线 gradeline

hāo

蒿

【蒿里之歌】funeral scrolls of elegies written on the death of a friend
【蒿目时艰】survey the country's situation with concern; look with anxiety at the world's ills
【蒿子】[植]wormwood; artemisia
【蒿子杆儿】crown daisy chrysanthemum (as a vegetable)

薅 pull up (weeds, etc.)

【薅草】weeding

háo

豪 ①(具有杰出才能的人) a person of extraordinary powers or endowments: 文～ a literary giant ②(气魄大;爽快) bold and unconstrained; forthright; unrestrained; ～气 heroic spirit / ～雨 torrential rain / ～饮 unrestrained drinking ③(强横) despotic; bullying: 土～ local despot / 巧取～夺 take away by force or trickery

【豪放】bold and unconstrained: ～的性格 a bold and uninhibited character
【豪放不羁】being vigorous and unrestrained
【豪富】①(有钱有势) powerful and wealthy ②(有财势的人) the powerful rich; the rich and powerful
【豪横】despotic; bullying
【豪华】luxurious; sumptuous: ～本 de luxe / ～的饭店 a luxury hotel
【豪杰】person of exceptional ability; hero
【豪举】①(有魄力之举) bold move ②(阔绰之举) munificent act
【豪迈】bold and generous; heroic: ～的气概 heroic spirit / ～的誓言 a bold pledge
【豪门】rich and powerful family; wealthy and influential clan: ～富户 the rich and powerful families / ～贵族 powerful family and honorable nobility
【豪马】a splendid horse
【豪气】heroism; heroic spirit
【豪强】①(强横) despotic; tyrannical ②(强横的人) despot; bully
【豪情】lofty sentiments: ～满怀 full of pride and enthusiasm / ～壮志 lofty sentiments and aspirations
【豪绅】despotic gentry
【豪爽】straightforward; forthright
【豪侠】[旧]①(勇敢而有义气) gallant ②(勇敢而有义气的人) gallant man
【豪兴】exuberant spirits; exhilaration; keen interest
【豪言壮语】brave words
【豪雨泼窗】Heavy rain pelted against the tightly closed windows.
【豪猪】porcupine
【豪壮】grand and heroic: ～的事业 a grand and heroic cause

壕 ①(护城河) moat ②(壕沟) trench: 防空～ air-raid shelter / 交通～ communication trench / 掘～ dig trenches; dig in

【壕沟】①[军]trench ②(沟) ditch
【壕堑战】[军]trench warfare

嚎 howl; wail: 狼～ the howl of a wolf

【嚎啕】cry loudly; wail: ～大哭 cry bitter tears; cry loudly with abandon

号 ①(号叫) howl; yell: 北风怒～。A north wind is howling. ②(大声哭) wail: 哀～ cry piteously; wail

【号哭】wail
【号叫】howl; yell
【号啕】cry loudly; wail: ～大哭 cry one's eyes out

嗥 (狼叫) howl

【嗥鸣雷动】roar like a thunderpeal

貉

【貉绒】racoon dog fur
【貉子】racoon dog

毫 ①(毛) fine long hair: 羊～笔 a writing brush made of goat's hair ②(毛笔) writing brush: 挥～ wield one's writing brush; write or draw a picture (with a brush) ③(一点儿) in the least; at all: ～不足怪 not at all surprising ④(千分之一) milli-: ～米 millimeter / ～升 milliliter ⑤(辅币) 10 cents

【毫安】[电] milliampere ◇～表 milliammeter
【毫巴】[气] millibar
【毫不动摇】 unswervingly; not waver in the least
【毫不含糊】 make a clear and definite commitment
【毫不利己,一心为公】 be dedicated to serving the public without any thought of oneself
【毫不足取】 not worth taking; not worth a fig
【毫法】[电] millifarad
【毫发】[书](多用于否定式) a hair; the least bit; the slightest: ～不爽 not deviate a hair's breadth; be perfectly accurate
【毫伏】[电] millivolt ◇～计 millivoltmeter
【毫克】 milligram (mg.)
【毫厘】 the least bit; an iota: ～不差 without the slightest error; just right
【毫毛】 soft hair on the body: 不准你动他一根～。 Don't you dare do him the slightest harm.
【毫米】 millimeter (mm.) ◇～波[无] millimeter wave
【毫秒】 millisecond
【毫升】 milliliter (ml.)
【毫微法】[电] millimicrofarad
【毫微米】 millimicron ($m\mu$)
【毫微秒】 nanosecond; millimicrosecond
【毫无道理】 utterly unjustifiable; for no reason whatsoever
【毫无二致】 without the slightest difference; just the same; identical
【毫无顾虑】 free from all inhibitions
【毫无例外】 admit of no exception
【毫无用处】 be utterly useless; good for nothing
【毫无原则】 total lack of principle
【毫无着落】 nowhere to be found
【毫子】[方] silver coin

蚝 oyster
【蚝油】 oyster sauce

hǎo

好 ①(优良的) good; fine; nice: ～看 good-looking; pleasant to the eye; beautiful / ～主意 good idea ②(友好) friendly; kind: ～朋友 great friend / 他们俩很要～。 They are on friendly terms. ③(健康;痊愈) be in good health; get well: 你～! Hello! / 我的病～了。 I'm well now. ④(用在动词后,表示完成)工具都准备～了。 The tools are ready. ⑤(表示赞许、结束、不满等语气)～,就这么办。 O.K., it's settled. / ～了,不要再说了。 All right, no need to say any more. / ～,我们走吧。 Well, let's go. ⑥(容易) be easy (to do); be convenient: 这个问题～回答。 This question is easy to answer. ⑦(便于) so as to; so that: 今儿早点睡,明儿～早起赶火车。 Let's turn in early, so as to get up early tomorrow to catch the train ⑧(表示程度深、数量多、时间长等)～大的工程! What a huge project! / ～冷啊! How cold it is? / ～漂亮! How beautiful! ⑨(用在形容词前面问数量或程度)机场离这儿～远? How far is the airport from here? ⑩[方](可以;应该) may; can; should: 时间不早了,你～走了。 It's getting late. You ought to get going. / 我～进来吗? May I come in? ⑪(用于套语)～走! Good-bye!
【好办】 easy to handle: 这件事～。 That can be easily arranged.
【好比】 can be compared to; may be likened to; be just like: 我们的工作～一场战斗。 Our work can be compared to a battle.
【好不】[副](表示程度深,并带感叹语气) how; what: 接到你的信 ～高兴! How glad I am to have heard from you! / 人来人往,～热闹! What a busy place, with so many people coming and going.
【好吃】 good to eat; tasty; delicious
【好处】①(益处) good; benefit; advantage: 戒烟～多。 To give up smoking has many advantages. / 空谈没有～。 Empty talk is no good. / 每天做操对他有～。 He benefits by daily exercises. ②(利益) gain; profit: 他没有捞到任何～。 He has gained nothing. ③(好相处) be easy to get along with: 新来的主任很～。 The new director is easy to get along with.
【好歹】①(好坏) good and bad; what's good and what's bad: 不知～ unable to tell what's good or bad for one; not appreciate a favor ②(危险) mishap; disaster: 万一她有个～,请立刻打电话给我。 If something should happen to her, please ring me up at once. ③(不论如何) in any case; at any rate; anyhow: ～我先开个头。 Let me make a start, anyhow. ④(将就) no matter in what way; anyhow: 别再做什么了,～吃点儿就得了。 Don't cook us anything more. We'll have whatever there is.
【好端端】 in perfectly good condition; when everything is all right: ～的,怎么生起气来了? Why are you angry when everything is perfectly all right?
【好多】①(许多) a good many; a good deal; a lot of: 有～事要做 have a good many things to

do ②[方](数量) how many; how much: 今天到会的人有～? How many came to the meeting today?

【好感】good opinion; favorable impression: 对他有～ be well disposed towards him; have a good opinion of him / 给人～ make a good impression on people

【好过】①(日子容易过) have an easy time; be in easy circumstances: 日子很不～ have a very hard time ②(好受) fell well: 他吃了药,觉得～一点了。He felt a bit better after taking the medicine.

【好汉】brave man; true man; hero: ～不吃眼前亏。A wise man does not fight when the odds are against him. / ～做事～当。A true man has the courage to accept the consequences of his own actions.

【好好】①(尽力;尽情) well; all out; to one's heart's content: ～想一想 think it over carefully / ～休息一下 have a good rest ②(好端端) in perfectly good condition: ～一支笔丢了,真可惜! What a pity to have lost such a nice pen!

【好好先生】one who tries not to offend anybody

【好好学习,天天向上】Study well 〈hard〉 and make progress every day.

【好话】①(有益的话; 赞扬的话) a good word; word of praise: 给他说句 ～put in a good word for him. ②(漂亮话) fine words: ～说尽,坏事做绝 say every fine word and do every foul deed

【好家伙】[叹](表示惊讶或赞叹) good god; good heavens; good gracious: ～! 你做得真快! Good Lord, you're really a deft hand.

【好景不长】good times don't last long

【好看】①(美观) good-looking; nice: 一位～的姑娘 a good-looking girl ②(使人感兴趣) interesting: 这出戏很～。It is a delightful play. ③(脸上有光彩) honored: 儿子立功,做娘的脸上也～。It's an honor for a mother like me to see my son do a meritorious deed. ④(难堪) be in a fix; in an embarrassing situation; on the spot: 等着吧,有他的～。You can be sure he'll soon find himself on the spot.

【好了疮疤忘了痛】forget the pain once the wound is healed; forget the bitter past released from one's suffering

【好评】favorable comment; high opinion: 博得读者～ be well received by the readers

【好球】well played; good shot; bravo

【好人】①(品质好的人) good person: ～好事 good people and good deeds; fine people and fine deeds ②(健康的人) a healthy person ③(老好人) a person who tries to get along with everyone ◇ ～主义 seeking good relations with all and sundry at the expense of principle

【好日子】①[旧](吉利的日子) auspicious day ②(办喜事的日子) wedding day ③(美好的生活) good days; happy life: 过～ live a happy life; live well; live in happiness

【好容易】with great difficulty; have a hard time (doing sth.): 他们～才找到我这儿。They had a hard time finding my place. (又作"好不容易")

【好生】①(多么;很) quite; exceedingly: ～奇怪! What an exceedingly strange thing! ②[方](好好儿地) carefully; properly: ～想一想。Think it over carefully.

【好声好气】[口] in a kindly manner; gently

【好使】be convenient to use; work well: 这架录音机很～。This tape recorder is very reliable.

【好事】①(好事情) good deed; good turn: 坏事变成～。Bad things can be turned to good things. ②[旧](慈善事) an act of charity; good works

【好事不出门,恶事传千里】Good deeds are never heard of outside the door, but bad deeds are proclaimed for three hundred miles. (或: Good works most frequently remain hidden, while evil deeds become known at a distance of a thousand *li*.)

【好事多磨】①(好事情常遇波折) the road to happiness is strewn with setbacks ②(真挚爱情常经历曲折) the course of true love never did run smooth

【好死不如恶活】A living ass is better than a dead lion.

【好手】good hand; past master: 做针线活儿,她可是把～。She is adept at needlework.

【好受】feel better; feel more comfortable: 我有些不～。I'm not feeling quite well.

【好说】[套]～,～! 您太夸奖了。It's very good of you to say so, but I don't deserve such praise.

【好说歹说】try every possible way to persuade sb.: 我～,他总算答应了。He agreed, but only after I had pleaded with him in every way I could.

【好说话】amiable; good-natured; easy to deal with

【好似】seem; be like: 大坝～铜墙铁壁,顶住了洪水的冲击。Like an iron bastion, the dam

withstood the rushing floodwaters.
【好天儿】fine day; lovely weather: 这些衣服～要拿出去晒。 These clothes should be put out to air on a sunny day.
【好听】pleasant to hear: 他说的比唱的还～。His glib talk sounds as sweet as a song. / 这支歌很～。This is a very pleasant song.
【好玩儿】amusing; interesting: 这可不是～的! This is no joking matter. / 这小娃娃挺～。The baby is very cute.
【好戏】①(好戏剧) good play ②[讽] great fun: 这回可有～看了! We're going to see some fun!
【好下场】good end
【好象】seem; be like: ～要下雨。 It looks like rain.
【好笑】laughable; funny; ridiculous: 又好气又～ be annoying and amusing at the same time
【好些】quite a lot; a good deal of: 她提了～建议。 She has made quite a lot of suggestions
【好心】good intention: 一片～ with the best of intentions / ～当作驴肝肺 take sb.'s goodwill for ill intent
【好心好意】have one's heart in the right; with good intentions
【好心有好报】Good-heartedness often meets with recompense. (或: Your charity would be rewarded.)
【好样儿的】example; great fellow: 是～,就站出来吧! If you are man enough, come out with what you have to say!
【好一个】what a: ～正人君子! An honorable man, indeed!
【好意】good intention; kindness: ～相劝 give well-intentioned advice
【好意思】(多用在反问句中) have the nerve: 做了这种事,亏他还～说呢!Fancy his doing that sort of thing and then having the nerve to talk about it!
【好在】fortunately; luckily: ～他伤势不重。Luckily he was not very seriously wounded.
【好自为之】conduct oneself well
【好转】take a turn for the better; take a favorable turn; improve: 病情～。 The patient is on the mend. / 目前形势正在～。The present situation is taking a turn for the better.

hào

耗 ①(消耗) consume; cost: ～电量 power consumption / ～热量 heat consumption / ～油率 fuel consumption rate ②(拖延) waste time; dawdle: 别～着了,快走吧。 Stop dawdling and get going. ③(坏消息) bad news: 噩～ the sad news of the death of one's beloved
【耗电量】power consumption
【耗费】consume; expend: ～人力物力 consume manpower and material resources / ～时间、金钱 expend time and money
【耗功】wasted work
【耗竭】exhaust; use up: 人力～ be drained of manpower ◇ ～学说 exhaustion hypothesis
【耗尽】exhaust; use up: ～体力 use up all one's strength / ～心血 exhaust all one's energies
【耗尽层】barrier layer; blocking layer
【耗尽区】depletion region; exhaustion region
【耗散】[物] dissipation: 功率～ power dissipation ◇ ～网络 dissipative network / ～尾迹 dissipation trail; distrail / ～因数 dissipation factor
【耗损】consume; waste; lose: 机器～ wear and tear of machinery / 减少水果在运输中的～ reduce the wastage of fruit in transit
【耗子】[方] mouse; rat ◇ ～药 ratsbane

号 ①(名称) name: 绰～ nickname ②(别号;字) assumed name; alternative name ③(商店) business house: 分～ branch (of a firm, etc.) / 银～ banking house ④(标志;信号) mark; sign; signal: 问～ question mark / 举火为～ light a beacon ⑤(次第) number: 编～ serial number ⑥(等级) size: 大～ large size / 小～ small size / 中～ medium size ⑦(日期) date: 今天几～? What date is it today? ⑧(号令) order: 发～施令 issue orders ⑨[乐] brass wind instrument: 军～ bugle / 小～ trumpet ⑩(号角) anything used as a horn: 螺～ conch-shell trumpet; conch ⑪(号声) bugle call; any call made on a bugle: 熄灯～ taps / 吹冲锋～ sound the call to charge ⑫[量](用于人数)一百多～人 over a hundred people
【号称】①(以某名著称) be known as ②(名义上是) claim to be: ～五十万大军 an army claiming to be half a million strong
【号角】①(喇叭一类的东西) bugle; horn ②(号角声) bugle call: 吹响了向科学技术现代化进军的～ sound the clarion call to march towards the modernization of science and technology
【号控机】number switch; numerical switch
【号令】verbal command; order
【号码】number: 电话～ telephone number ◇ ～标志 number marking / ～灯 number light / ～机 numbering machine
【号脉】[中医] feel the pulse

【号手】trumpeter; bugler
【号外】extra (of a newspaper)
【号形角】horn angle
【号衣】[旧]livery or army uniform
【号召】call; appeal: 积极响应国家的～ actively respond to the call of one's country ◇ ～书 appeal
【号子】a work song sung to synchronize movements, with one person leading

浩 great; vast; grand
【浩大】very great; huge; vast: ～的工程 a huge project
【浩荡】vast and mighty: ～的长江 the mighty Changjiang River / ～的东风 strong east wing
【浩繁】vast and numerous: ～的开支 heavy expenditure / 卷帙～ a voluminous work; a vast collection of books
【浩瀚】vast: ～的沙漠 a vast expanse of desert / 典籍～ a vast accumulation of ancient literature
【浩瀚如海】be vast as the ocean
【浩浩荡荡】go forward with great strength and vigor
【浩劫】great calamity; catastrophe; havoc: 空前～ an unheard-of calamity
【浩淼】(of water) extending into the distance; vast: 太湖之上,烟波～。On the Taihu Lake, mists and waves stretch far into the distance.
【浩气】noble spirit: ～长存 imperishable noble spirit
【浩然正气】awe-inspiring righteousness
【浩如烟海】(of data, etc.) vast as the open sea; tremendous amount of; voluminous
【浩叹】heave a deep sigh; sigh deeply

皓 ①(白)white: ～齿 white teeth ②(明亮) bright; luminous: ～月当空。A bright moon hung in the sky.
【皓齿朱唇】have pearly white teeth and crimson lips; white teeth and red lips
【皓矾】[化] zinc sulphate
【皓首】[书] hoary head
【皓首穷经】A hoary head does research in the classics — an aged person still learns.

好 ①(喜爱) like; love; be fond of: ～表现 like to show off / 虚心～学 be modest and eager to learn ②(易于) be liable to: ～晕船 be liable to seasickness; be a bad sailor
【好吃懒做】be fond of eating and averse to work; be gluttonous and lazy
【好出风头】like to get into the limelight; be fond of the limelight
【好大喜功】crave for greatness and success; have a fondness for the grandiose
【好高骛远】reach for what is beyond one's grasp; aim too high; bite off more than one can chew
【好洁成癖】excessive fondness for cleanliness
【好酒废事】be given to wine and neglect one's business
【好客】be hospitable; keep open house
【好利无信】prefer benefits at the expense of one's reputation
【好谋无断】be fond of scheming, but fall to make a decision; be fond of grandiose schemes, but unable to take a decision
【好奇】be curious; be full of curiosity ◇ ～心 curiosity
【好强】eager to do well in everything
【好胜】seek to do others down: ～心强 loving to excel others
【好事】meddlesome; officious ◇ ～之徒 busybody
【好为人师】be fond of teaching others; like to be a master to others
【好恶】likes and dislikes; taste: 翻译时不应根据自己的～改变原文的意思。In doing translation, one should not alter the meaning of the original to suit one's own taste.
【好逸恶劳】love ease and hate work
【好谀悦色】regard flattery as pleasure and cherish lust as desire
【好战】bellicose; warlike ◇ ～分子 bellicose elements; warlike elements
【好整以暇】remain calm and composed while handling pressing affairs

hē

诃 scold
【诃子】[植] myrobalan

呵 ①(呼气) breathe out (with the mouth open): ～手 breathe on one's hands (to warm them) / ～一口气 give a puff ②(呵斥) scold: ～责 scold sb. severely; give sb. a dressing down
【呵壁问天】blame god and fate; give free vent to one's griefs
【呵斥】berate; excoriate: ～孩子 scold a child in a loud voice
【呵呵】[象] ～大笑 laugh loudly; roar with

laughter

【呵欠】yawn

嗬 [叹] ah; oh: ～，真是奇迹！Oh, what a wonder!

喝 ①(饮) drink: ～茶 drink tea / ～汤 drink soup ②(饮酒) drink alcoholic liquor: ～醉了 be drunk / 爱～两盅 be fond of drinking

【喝水不忘掘井人】When you drink the water, think of those who dug the well.

【喝西北风】drink the northwest wind have nothing to eat

hé

涸 [书] dry up

【涸辙之鲋】a fish trapped in a dry rut; a person in a desperate situation

【涸井打水】drop a bucket into an empty well

阂 cut off from; not in communication with: 隔～ misunderstanding; estrangement

核 ①(果核；果仁) pit; stone: 桃～ peach-pit; peach-stone / 无～葡萄干 seedless raisins ②(核状物) nucleus: 细胞～ cell nucleus / 原子～ atomic nucleus ③(查对) examine; check: ～准 check and approve; ratify ④(核动力的) nuclear-powered: ～潜艇 nuclear-powered submarine

【核保护伞】nuclear umbrella

【核爆炸】nuclear explosion

【核尘】nuclear dust

【核打击力量】nuclear strike capability

【核大国】nuclear power

【核蛋白】[生化] nucleoprotein

【核弹头】nuclear warhead

【核导弹】nuclear missile

【核电站】nuclear power plant

【核对】check; check up

【核定】check and ratify; appraise and decide ◇～股本 authorized stock / ～预算 approved budget

【核动力】nuclear power

【核对】check: ～事实 check the facts / ～数字 check figures / ～帐单 check a bill / ～帐目 check the accounts

【核讹诈】nuclear blackmail

【核反应】nuclear reaction ◇～堆 nuclear reactor

【核辐射】nuclear radiation

【核苷】[生化] nucleoside ◇～酸 nucleotide

【核果】[植] drupe

【核黄素】[药] riboflavin; lactoflavin

【核火箭】nuclear rocket

【核计】calculate: ～成本 assess the cost

【核矩】nuclear moment

【核聚变】nuclear fusion

【核扩散】nuclear proliferation

【核裂变】nuclear fission

【核垄断】nuclear monopoly

【核模】nuclear membrane; karyolemma; karyotheca

【核能】nuclear energy

【核能源】nuclear power source

【核潜艇】nuclear-powered submarine

【核燃料】nuclear fuel

【核仁】[生] nucleolus ②(果仁) kernel (of a fruit-stone)

【核实】verify; check: ～的产量 verified output / ～的材料 verified materials / 请把这些数字～一下。Please check these figures.

【核试验】nuclear test: 大气层～ atmospheric nuclear test / 地下～ underground nuclear test / 高空～ high-altitude nuclear test

【核素】①[化] nuclein ②[物] nuclide

【核酸】[化] nucleic acid ◇～酶[生化] nuclease / ～内切酶[生化] endonuclease

【核算】business accounting: 成本～ cost accounting / 独立～ keep separate accounts

【核算单位】accounting unit: 独立～ independent accounting unit / 基本～ basic accounting unit / 经济～ economic accounting unit

【核桃】walnut

【核糖】[生化] ribose

【核糖核酸】[生化] ribonucleic acid; RNA: 脱氧～ deoxyribonucleic acid; DNA / 信息～ messenger ribonucleic acid; m-RNA

【核体】nucleome

【核威慑力量】nuclear deterrent (power)

【核威胁】nuclear threat

【核微粒沾染】contamination from nuclear fallout

【核物理】nuclear physics

【核武器】nuclear weapon ◇～储备 stockpiling of nuclear weapons; nuclear weapons stockpile

【核销】cancel after verification

【核心】nucleus; core; kernel: ～力量 force at the core / ～人物 key figure

【核心内阁】inner cabinet

【核优势】nuclear superiority

【核战争】nuclear war
【核准】examine and approve ◇ ～发行的债券 authorized bonds
【核装置】nuclear device
【核子】[物] nucleon ◇ ～学 nucleonics
劾 expose sb.'s misdeeds or crimes: 弹～ impeach
阖 [书]①(全;总共) entire; whole: ～城 the whole town / ～家 the whole family ②(关闭) shut; close: ～户 close the door / ～眼 close one's eyes
【阖第光临】The whole family is invited.
【阖家团圆】reunion of homefolks after a temporary separation; One's family was reunited.
翮 ①(鸟羽的茎) shaft of a feather; quill ②(翅膀) wing (of a bird): 振～高飞 flap the wings and soar high into the sky
河 river: 黄～ the Huanghe River
【河岸】river bank
【河边】river bank
【河叉子】a branch of a river
【河床】riverbed
【河道】river course
【河堤】dike
【河防】flood-prevention work done on rivers
【河沟】brook; stream
【河谷】river valley
【河涸鱼竭】The rivers dried up and the fish in them died.
【河口】river mouth; stream outlet ◇ ～湾 estuary
【河狸】beaver
【河狸鼠】nutria
【河流】rivers: 不通航～ non-navigable river / 多国～ multinational river / 国内～ national river / 通航～ navigable river ◇ ～沉积 fluvial deposit / ～袭夺[地] river capture; river piracy
【河马】hippopotamus; hippo; river horse
【河鳗】river eel
【河泥】river silt; river mud
【河清海晏】The Huanghe River is clear and the seas are calm. (或: The world is at peas.)
【河清难挨】It is hard to wait till the Huanghe River is clear. (或: The time would be too long to wait for something to happen.)
【河曲】bend (of a river); meander
【河渠】rivers and canals; waterways: ～纵横 be crisscrossed by rivers and canals
【河山】rivers and mountains; land; territory: 锦绣～ a land of enchanting beauty / 祖国～美丽如画。 Our native land is as pretty as a picture.
【河滩地】flood land
【河套】the bend of a river
【河豚】globefish; balloonfish; puffer
【河外星云】[天] extragalactic nebula
【河网】a network of waterways: ～化 build a network of waterways ◇ ～地带 area crisscrossed by waterways
【河系】river system
【河蟹】river crab
【河源】river head
【河运】river transport
何 [书]①(表示疑问) ～处 what place; where / ～人 who / ～时 what time; when / ～往 whither / 从～而来? Where from? ②(表示反问) ～济于事? Of what avail is it? / 有～不可? Why not?
【何必】there is no need; why: ～去哪么早。 There is no need to go so early.
【何不】why not: ～再试一下? Why not try once more?
【何尝】(用于反问) 我～不想去,只是没时间罢了。 Not that I don't want to go; I just haven't got the time.
【何啻万千】more than thousands and myriads; far more than thousands and tens of thousands
【何等】①(什么样的) what kind: ～人物 what kind of person ②(多么) how; what: 他是～伟大啊! How great he was! What a great man he was!
【何妨】why not; might as well: ～一试? Why not have a try?
【何患无辞】A pretext for ... is never wanting
【何患之有】There's no need to worry.
【何苦】why bother; is it worth the trouble: ～自寻烦恼? Why worry yourself sick?
【何况】[连] much less; let alone: 我们死都不怕, ～困难? We fear no death, let alone difficulties.
【何乐而不为】What is there against it. (或: What's the sense of not doing that?)
【何其】how; what: ～相似乃尔! What a striking likeness!
【何去何从】what course to follow: ～, 速作抉择。 What course to follow — that is a question you must quickly decide for yourselves.
【何如】①(怎么样) how about: 请君一试, ～?

How about you having a try? ②(不如)与其求人,~自力更生。It would be better to rely on oneself than on others.

【何首乌】[中药] the tuber of multiflower knot weed

【何谓】[书] what is meant by; what is the meaning of: ~自由? What is meant by freedom?

【何许】[书] what kind of; what: ~人 what sort of person

【何以】how; why: ~见得? What makes you think so?

【何在】where: 用心~? What's the motive! / 原因~? What is the reason for it?

【何止】far more than: 例子~这些。There are far more instances than we have just enumerated. / 煤的用途~那一些。The uses of coal are by no means limited to those.

【何足挂齿】not worth bothering about; not worth mentioning at all

【何足轻重】Of what consequence is it?

【何足为怪】There is nothing to be surprised at

荷 lotus

【荷包】①(随身带的小包) small bag (for carrying money and odds and ends); pouch ②(衣袋) pocket (in a garment)

【荷包蛋】fried eggs; poached eggs

【荷花】lotus

【荷兰】the Netherlands; the Holland ◇ ~人 the Dutch; Dutchman / ~语 Dutch (language)

【荷兰法】[化] Dutch process

【荷兰芹】caraway

【荷叶】lotus leaf

【荷质比】charge-mass ratio

合

①(闭;合拢) close; shut: ~上眼 close one's eyes / 请把书~上。Close your book, please. ②(结合在一起) join; combine: ~力 combined strength; joint effort ③(全) whole: ~家团聚 a reunion of the whole family ④(符合) suit; agree: ~得来 get along well / ~胃口 suit one's taste; be to one's taste / 正~我意。It suits me fine. ⑤(折合;共计) be equal to; add up to: 这件上衣连工带料~多少钱? How much will this coat cost, including material and tailoring? ⑥[书](应当;应该) proper: 理~声明。I deem it appropriate to make a statement. ⑦[天] conjunction

【合瓣】[植] sympetalous; gamopetalous ◇ ~花 sympetalous flower / ~花类 metachlamydeae; sympetalae

【合抱】(of a tree, etc.) so big that one can just get one's arms around: ~之木,生于豪末。A huge tree grows from a tiny seedling.

【合璧】(of two different things) combine harmoniously; match well: 中西~ a good combination of Chinese and Western elements

【合并】merge; amalgamate: 这三个提议~讨论。The three proposals will be discussed together. ◇ ~处罚 mixed punishment / ~管辖 amalgamated jurisdiction / ~检控 joinder of charges / ~收益 consolidated returns / ~诉讼 amalgamated action; consolidated action / ~装运 consolidated shipment

【合唱】chorus: 大~ cantata / 混声~ mixed chorus / 男声~ male chorus / 女声~ female chorus / 童声~ children's chorus ◇ ~队 chorus / ~曲 chorus / ~团 chorus / ~团指挥 chorus master

【合成】①(由部分组成整体) compose; compound: 力的~[物] composition of forces / 由两部分~ be composed of two parts ②[化] synthetise; synthetize ◇ ~氨 synthetic ammonia / ~词[语] compound word / ~结晶牛胰岛素 synthetic crystalline bovine insulin / ~军队 combined arms unit / ~酶 [生化] synzyme / ~树脂 synthetic resin / ~塔[化] synthetic tower / ~洗涤剂 synthetic detergent / ~纤维 synthetic fiber / ~橡胶 synthetic rubber; synthal / ~照片 photomontage

【合订本】one-volume edition; bound volume: 过期报刊~ back file; bound volume of back issues

【合二而一】[哲] two combine into one

【合法】legal; lawful; legitimate; rightful: 唯一~政府 the sole legal government ◇ ~不动产 legal estate / ~当局 constituted authorities; legitimate authorities / ~地位 legal status / ~斗争 legal struggle / ~化 legalize; legitimize / ~继承人 rightful heir / ~扣留 legal detention legal identity / ~贸易 fair trade / ~权利 legitimate right; lawful right / ~权益 legitimate rights and interests / ~身份 legal identify / ~收入 lawful earned income / ~途径 legal means / ~行为 lawful act / ~证据 lawful evidence / ~自卫 lawful self-defense; legitimate self-defense

【合格】qualified; up to standard: ~的司机 a qualified driver / 产品~。The product is up to standard. ◇ ~公证人 qualified notary / ~技术人员 qualified technicians / ~人员 com-

petent person / ～选举人 elector / ～证书 certificate of soundness; certificate of inspection; certificate of quality
【合股】①(几个人合资经营)pool capital; form a partnership ②[纺] plying: ～线 ply yarn
【合乎】conform with; correspond to; accord with; tally with: ～规格 up to the specifications / ～逻辑 logical / ～实际 conform to the actual situation / ～事实 tally with the facts / ～情理 reasonable; sensible / ～自然规律 in conformity with the law of nature
【合欢】[植] silk tree
【合伙】form a partnership: ～经营 run a business in partnership ◇ ～关系 copartnery; partnership / ～合同 partnership contract / ～契约 partnership agreement / ～条件 articles for partnership
【合击】make a joint attack on: 分进～ concerted attack by converging columns
【合计】①(总共) amount to; add up to; total: 把这一栏的数字～一下。Add up the figures in this column. ②(商量) consult: 大家～一下,看哪个方案比较好。Let's lay our heads together to see which of the plans is better.
【合剂】[药] mixture: 复方干草～ brown mixture / 咳嗽～ cough mixture
【合金】alloy: 二元～ binary alloy / 三元～ ternary alloy ◇ ～结构钢 structural alloy steel / ～元素 alloying element
【合金钢】alloy steel: 高～ high-alloy steel
【合刊】combined issue (of a periodical)
【合口】①(伤口愈合)(of a wound) heal up ②(合味) (of a dish) be to one's taste
【合理】rational; reasonable; equitable: ～的价格 a reasonable price / ～的上涨 justified raise / ～解决两国之间的争端 equitable settlement of the issues between the two countries / ～利用资源 put resources to rational use; make rational use of resources / ～施肥 apply fertilizer rationally ◇ ～处罚 reasonable punishment / ～分工 rational division of labor / ～负担 equitable distribution of burden / ～轮作 proper rotation of crops / ～赔款 just compensation / ～讼因 just cause
【合理化】rationalize: ～建议 rationalization proposal
【合力】①(一起出力) join forces; pool efforts: ～修建水库 pool efforts to build a reservoir ②[物] resultant of forces
【合流】①(河流汇合) flowing together; confluence ②(思想行动的一致) collaborate; work hand in glove with sb.
【合龙】①(指堤坝等) closure ②(指桥梁等) join the two sections of a bridge, etc.
【合霉素】[药] syntomycin
【合谋】①(共同策划) conspire; plot together ②[法] conspiracy
【合谋不轨】plot sedition together
【合拍】in time; in step; in harmony: 与时代潮流～ instep with the trend of the times
【合浦珠还】recover a thing which has been lost; The lost wealth is recovered.
【合情合理】fair and reasonable; fair and sensible: 这个建议～。The proposal is fair and reasonable.
【合群】①(合得来) get on well with others ②(结成团体) be gregarious
【合身】fit: 这件上衣很～。 This jacket fits well.
【合十】[佛教] put the palms together
【合十虔诚】put one's hands together in prayer
【合适】suitable; appropriate; becoming; right: 这个词用在这里不～。 This isn't the right word to use here. / 这个日期对你～么? Will the date suit you? / 这双鞋我穿着正～。These shoes fit me beautifully.
【合算】①(划得来) paying; worthwhile ②(算计) reckon up
【合题】[哲] synthesis
【合同】contract: 购货～ purchase contract / 互惠～ reciprocal contract / 签订～ sign a contract / 售货～ sales contract / 撕毁～ tear up a contract ◇ ～法 contract law / ～工 contract worker / ～期限 contract period / ～条款 contract terms / ～医院 assigned hospital (to which people from a given organization or area go for treatment)
【合围】[军] surround; encircle
【合叶】hinge
【合议庭】[法] collegiate bench
【合议制】[法] collegiate system
【合意】suit; be to one's liking: 这双鞋你～么? How do you like this pair of shoes?
【合营】jointly owned; jointly operated
【合影】group photo ～留念 have a group photo taken to mark the occasion
【合辙】①(押韵) in rhyme ②(一致) in agreement: 两人一说就～。 The moment they started talking they found themselves in complete agreement.
【合著】write in collaboration with; coauthor
【合资经营】joint venture

【合资企业】joint ventures
【合奏】instrumental ensemble: 民乐～ ensemble of national instruments
【合作】cooperate; collaborate; work together: 互相～ cooperate with each other ◇～经济 cooperative economy; cooperative sector of the economy／～经营 cooperative business operation／～企业 co-operative enterprise／～商店 cooperative shop／～社 cooperative; co-op／～生产 joint production／～医疗 cooperative medical service

貉 racoon dog

颌 [书] jaw: 上～ the upper jaw／下～ the lower jaw

盒 box; case: 铅笔～ pencil case; pencil box／一～火柴 a box of matches
【盒式磁带】cartridge tape; cassette
【盒式胶卷】cassette film
【盒子】box; case; casket

禾 standing grain (esp. rice)
【禾本科】[植] the grass family ◇～植物 grass
【禾场】threshing floor
【禾苗】seedlings of cereal crops

和 ①(平和) gentle; mild; kind: 风～日暖 bright sunshine and gentle breeze ②(和睦) harmonious; on good terms: 兄弟不～ brothers on bad terms with each other ③(和平) peace: 媾～ negotiate for peace／讲～ make peace ④[体] draw; tie: 那盘棋～了。 That game of chess ended in a draw. ⑤(连带) together with: ～衣而卧 sleep with one's clothes on; sleep in one's clothes ⑥[介](表示相关、比较等) ～这件事没有关系 have nothing to do with the matter; bear no relation to it ⑦[连] and: 你～我 you and I ⑧[数] sum: 两数之～ the sum of the two numbers
【和蔼】kindly; affable; affable; amiable: ～的教师 a kindly teacher／～可亲 affable; genial／态度～ amiable
【和畅】(of a wind) gentle and pleasant: 惠风～ a gentle and pleasant breeze
【和风】①(温和的风) soft breeze: ～拂面 a gentle breeze caressing one's ～丽日 a gentle breeze and a bright sun; fine weather ②[气] moderate breeze
【和风细雨】like a gentle breeze and a mild rain; in a gentle and mild way
【和服】kimono
【和党同尘】of the same hidden virtue and the same commonplace; drift with the current
【和好】become reconciled: ～如初 be on good terms again; restore good relations
【和缓】①(平和) gentle; mild: 水流～ gentle flow of a stream／态度～ adopt a mild attitude ②(使缓和) ease up; relax: ～一下气氛 relieve the tension a little
【和会】peace conference
【和奸】[法] adultery by consent: ～年龄 age of consent
【和解】become reconciled: 采取～的态度 adopt a conciliatory attitude ◇～书[法] reconciliation agreement
【和局】drawn game; draw; tie
【和睦】harmony; concord; amity: ～相处 live in harmony／家庭～ family harmony; domestic peace／民族～ national concord
【和暖】pleasantly warm; genial: ～的阳光 warm sunshine／～宜人 pleasantly warm／天气～ warm, genial weather
【和盘托出】reveal everything; make a clean breast of everything: 把自己的想法～ reveal everything on one's mind
【和平】①(无战争状态) peace: ～解决边界争端 peaceful settlement of a boundary dispute／～利用原子能 peaceful utilization of atomic energy; use of atomic energy for peaceful purposes ②(温和) mild: 药性～。The medicine is mild. ◇～倡议 peace proposals／～攻势 peace offensive／～共处 peaceful coexistence／～过渡 peaceful transition／～竞赛 peaceful competition／～谈判 peace negotiations／～中立政策 policy of peace and neutrality／～主义 pacifism／～主义者 pacifist
【和平共处五项原则】the Five Principles of Peaceful Coexistence ○互相尊重主权和领土完整 mutual respect for territorial integrity and sovereignty／互不侵犯 mutual non-aggression／互不干涉内政 non-interference in each other's internal affairs／平等互利 equality and mutual benefit／和平共处 peaceful coexistence
【和棋】a draw in chess or other board games
【和气】gentle; kind; polite; amiable: 和和气气 polite and amiable／伤了～ hurt sb.'s feelings／说话～ speak politely; be soft-spoken
【和气生财】Friendliness is conducive to business success. (或: An even temper brings wealth.)

【和气致祥】Good-naturedness leads to propitiousness.
【和亲】[史] attempt to cement relations with rulers of minority nationalities in the border areas by marrying daughters of the Han imperial family to them
【和善】kind and gentle; genial: ～的面容 benign faces／～的笑容 genial smiles／心地～ kind-hearted
【和尚】Buddhist monk
【和声】[乐] harmony
【和事老】peacemaker; person who is more concerned with stopping the bickering than settling the issue
【和数】summation
【和谈】peace talks
【和弦】[乐]chord
【和项】alterm
【和谐】harmonious: ～的气氛 a harmonious atmosphere／～一致 in perfect harmony
【和谐音】chord
【和煦】pleasantly warm; genial: ～的阳光 genial sunshine／春风～。The spring breeze is pleasantly warm.
【和颜悦色】with kind and pleasant countenance
【和衣而卧】sleep in one's clothes; sleep with one's clothes on; sleep all dressed
【和约】peace treaty
【和衷共济】work together with one heart (in times of difficulty)

hè

鹤 crane
【鹤发童年】white hair and ruddy complexion; healthy in old age; hale and hearty
【鹤骨松姿】like a crane's bones and a pine's appearance
【鹤唳长空】The stork sings through the sky.
【鹤立鸡群】like a crane standing among chickens; stand head and shoulders above others
【鹤鸣九皋】The crane screams in the middle marsh.
【鹤嘴锄】pick; pickaxe; mattock

赫 ①(显著)conspicuous; grand: 显～ distinguished and influential; illustrious ②[电](赫兹) hertz: 千～ kilohertz／兆～ megahertz
【赫赫】illustrious; very impressive: ～有名的人物 an illustrious personage／～战功 illustrious military exploits; brilliant military success
【赫然】①(突然呈现) impressively; awesomely ②(大怒)terribly (angry): ～震怒 get into a terrible temper; fly into a violent rage
【赫兹】[电] hertz

荷 [书] ①(扛;背) carry on one's shoulder or back: ～锄而归 come home with a hoe on one's shoulder／～枪实弹 carry a loaded rifle ②(负担) burden; responsibility: 肩负重～ shoulder heavy responsibilities ③(多用于书信, 表示感谢) grateful; obliged: 请早日示复为～。An early reply will be appreciated.／无任感～。I'll be very much obliged.
【荷载】load ◇ ～图 load diagram
【荷重】loading

壑 gully; big pool: 千山万～ innumerable mountains and valleys

吓 ①(恐吓) threaten; intimidate ②[叹](表示不满) ～, 怎么能干这种事呢? Tut-tut, how could you do that?

褐 brown
【褐斑】foxiness
【褐黄斑】[医] chloasma
【褐煤】brown coal; lignite
【褐色土】drab soil
【褐铁矿】brown iron ore; limonite
【褐云母】anomite
【褐藻】[植] brown alga

喝 shout loudly: ～问 shout a question to／大～一声 give a loud shout
【喝彩】acclaim; cheer: ～叫好 a shout of applause／齐声～ cheer in chorus; cheer with one accord
【喝倒彩】make catcalls; hoot; boo
【喝令】shout an order

和 ①(和谐地跟着唱) join in the singing: 一唱百～。When one starts singing, all the others join in. ②(和诗) compose a poem in reply: 奉～一首 write a poem in reply (to one sent by a friend, etc, using the same rhyme sequence)

贺 congratulate
【贺词】speech of congratulation; message of congratulation; congratulations; greetings: 新年～ New Year message
【贺电】message of congratulation; congratulatory telegram

【贺客盈门】Friends crowded (the) door, offering their congratulations upon the happy event.（或：The house is crowded with well-wishers.）
【贺礼】gift (as a token of congratulation)：生日～ birthday present
【贺年】extend New Year greetings or pay a New Year call ◇ ～片 New Year card
【贺喜】congratulate sb. on a happy occasion
【贺信】congratulatory letter; letter of congratulation

hēi

黑 ①（黑色）black：～发 black hair／穿一身～衣服 be dressed in black ②（黑暗）dark：天～了。It's dark. ③（秘密）secret; shady：～交易 shady deal ④（坏）wicked; sinister：～帮 sinister gang／～货 sinister stuff
【黑矮星】[天] black dwarf
【黑暗】dark：～的角落 a dark corner／～面 the seamy side; a dark aspect
【黑白】black and white; right and wrong：～分明 with black and white sharply contrasted; in sharp contrast ◇ ～电视 black-and-white television／～（电影）片 black-and-white film／～图象 monochrome picture
【黑斑病】[植] black rot
【黑板】blackboard ◇ ～报 blackboard newspaper／～擦 eraser
【黑帮】reactionary gang; sinister gang
【黑不溜秋】very black in appearance; swarthy
【黑潮】[地] Kuroshio; Japan Current
【黑貂】sable ◇ ～皮 sable fur
【黑鲷】black porgy
【黑洞】[天] black hole
【黑洞洞】pitch-dark
【黑豆】black soya bean
【黑非洲】Black Africa
【黑粉病】smut
【黑粪】[医] melaena
【黑风】[气] reshahar
【黑钙土】chernozem; black earth
【黑更半夜】[口] in the dead of night
【黑咕隆咚】very dark; pitch-dark
【黑管】[乐] clarinet
【黑光】black light
【黑海】the Black Sea
【黑糊糊】①（颜色发黑）black blackened：墙熏得～的。The wall was blackened by smoke. ②（光线昏暗）rather dark; dusky：屋子里～的。It's rather dark in the room. ③（模糊不清）indistinctly observable in the distance：远处是一片～的树林。A dark mass of trees loomed in the distance.
【黑话】①（秘密话）(bandits') argot; (thieves') cant：土匪～ bandit argot ②（反动而隐晦的话）double talk; malicious words
【黑鲩】black carp
【黑货】①（走私货）smuggled goods; contraband ②（违禁物）sinister stuff; trash
【黑胶布】[电] black tape; friction tape
【黑里康大号】[乐] helicon
【黑瘤】[医] melanoma
【黑麦】rye
【黑面包】black bread; brown bread; rye bread
【黑名单】blacklist
【黑幕】inside story of a plot, shady deal, etc.：揭穿～ expose a sinister project; tell the inside story of a plot, etc.
【黑啤酒】dark beer; stout
【黑漆漆】pitch-dark
【黑热病】[医] kala-azar
【黑人】Black people; Black; Negro：美国～ Black American; Afro-American
【黑色】black ◇ ～火药 black powder／～金属 ferrous metal／～素[生化] melanin
【黑色人种】the black race
【黑市】black market：～汇兑 black market exchange／～价 black rate／～交易 off-the-books deal／～商人 blackmarketeer／～投机买卖 black market; speculative trading
【黑手】a vicious person manipulating sb. or sth. from behind the scenes; evil backstage manipulator
【黑手党】Mafia; Maffia
【黑死病】the plague
【黑素癌】melanotic cancer; melanocarcinoma
【黑素瘤】melanoma
【黑穗病】[农] smut
【黑陶】[考古] black pottery ◇ ～文化 black-pottery culture
【黑体】①[物] blackbody ②[印] boldface ◇ ～字 boldface type
【黑土】black earth
【黑钨矿】wolframite
【黑心】black heart; evil mind
【黑信】anonymous letter from a hostile pen; poison-pen letter
【黑猩猩】chimpanzee
【黑熊】black bear
【黑压压】a dense or dark mass of：远处～的一片，看不清是些什么东西。One couldn't make

out what the dark mass was from a distance.
【黑眼镜】sunglasses
【黑曜岩】[矿] obsidian
【黑油油】jet-black; shiny black: ～的头发 shiny black hair
【黑黝黝】①(黑得发亮) shiny black: ～的脸 swarthy face ②(光线昏暗) dim; dark: 四周～的。It's dark all around
【黑鱼】snakeheaded fish; snakehead
【黑云母】[矿] black mica; biotite
【黑云压城城欲摧】dark clouds bearing down on the city threaten to overwhelm it; reactionary forces in the ascendent
【黑枣】dateplum persimmon
【黑子】①[书](黑痣) black mole ②[天](太阳的黑斑) sunspot
嘿 [叹] hey: ～,快点! Hey, be quick! / ～,下雪了! Why, it's snowing!

hén

痕 mark; trace: 刀～ a mark or scar left by a knife-cut / 泪～ tear stains / 伤～ a scar from a wound
【痕迹】mark; trace; vestige; imprint; print: 轮子的～ wheel tracks / 不露～ make no sign
【痕迹器】vestige
【痕量】[化] trace ◇ ～化学 trace chemistry / ～金属 trace metal / ～元素 trace elements / ～杂质 trace impurity / ～组分 trace components

hěn

很 [副] very; quite; awfully: 好得～ very good / ～可能 very likely / ～满意 feel quite pleased / ～有意义 highly significant

hèn

狠 ①(凶恶;残忍) ruthless; relentless: 比豺狼还～ more savage than a wolf / 凶～ ferocious and ruthless ②(控制感情) suppress (one's feelings); harden (the heart) ③(坚决) firm; resolute: ～～打击歪风邪气 take vigorous measures to counter evil trends
【狠毒】vicious; venomous: 用心～ with vicious intent
【狠心】cruel-hearted; heartless
恨 ①(仇视;怨恨) hate: ～得咬牙切齿 grind one's teeth with hatred / ～得要命 hate intensely; bitterly hate / ～之入骨 hate sb. to the very marrow of one's bones / 怀～在心 nurse hatred in one's heart ②(悔恨;不称心) regret: 遗～ eternal regret
【恨海难填】The sea of hatred is hard to fill up.
【恨海无边】a sea of eternal regrets
【恨事】a matter for regret
【恨铁不成钢】wish iron could turn into steel at once; set a high demand on somebody in the hope that he will improve

hēng

亨 ①(顺利) go smoothly ②[电] henry
【亨利】[电] henry: 微～ microhenry ◇ ～单位 Henry unit / ～定律 Henry's law / ～计 inductance meter
【亨生线】Hensen's line
【亨特氏病】dyssynergia cerebellaris myoclonica; Hunt's disease
【亨特氏综合症】mucopolysacchari dosis II; Hunter's syndrome
【亨通】go smoothly; be prosperous: 万事～。Everything is going smoothly.
哼 ①(鼻子发出的声音) snort: 蔑视地～了一声 give a snort of contempt ②(低声吟唱) hum; croon: 他一边走,一边～着曲子。He was humming a tune as he walked along. ③(发出呻吟声) groan: 伤员的～声 the groan of the wounded / 痛得直～～ groan with pain
【哼哧】[象] puff hard: ～～地直喘 be puffing and blowing
【哼儿哈儿】[象] hem and haw: 他总是～的,就是不说句痛快话。He hemmed and hawed but wouldn't say anything definite.
【哼哼哈哈】hem and ham; "You, sir" type (of person)
【哼声】[电] hum; hum note; groaning ◇ ～测量器 hum noise measuring instrument / ～电位计 hum potentiometer / ～器 hummer / ～调制 hum modulation
【哼唷】[叹] heave ho; yo-heave-ho; yo-ho

héng

恒 ①(永久;持久) permanent; lasting: 永～ eternal; everlasting ②(恒心) perseverance: 持之以～ persevere in (doing sth.) ③(平常;经常) usual; common: ～言 common saying
【恒齿】[生理] permanent tooth

【恒等】[数] identically equal; identical ◇ ～变换 identical transformation / ～式 identical equation; identity / ～网络 identical network / ～元素 identical element
【恒定流量】constant-rate of flow
【恒定振幅】uniform amplitude
【恒河沙数】as numerous as the sands of the Ganges; innumerable; countless
【恒量】[物] constant
【恒频】constant frequency
【恒温】constant temperature ◇ ～动物 homoiothermal animal / ～器 thermostat / ～炉 constant temperature furnace / ～室 thermostatic chamber; constant heat cabinet / ～箱 incubator; thermostated container
【恒心】perseverance; constancy of purpose
【恒星】[天] (fixed) star ◇ ～大气 stellar atmosphere / ～光度 stellar luminosity / ～年 sidereal year / ～日 sidereal day / ～时 sidereal time / ～视差 stellar parallax / ～天文学 stellar astronomy / ～物理学 stellar physics / ～系 stellar system; galaxy / ～月 sidereal month / ～云 star cloud
【恒压】constant voltage: ～电池 constant cell
【恒压器】barostat
【恒牙】dentes permanentes; permanent teeth: ～列 permanent dentition

横

①(与地面平行的) horizontal; transverse: 人行～道 (pedestrians') street crossing / 纵～ vertical and horizontal ②(从左到右或从右到左) across; sideways: ～过街道 across the street / ～写 write words sideways ③(使物成横向) move crosswise; traverse: ～刀跃马 gallop ahead with sword drawn ④(纵横杂乱) unrestrainedly; turbulently: 江河～溢 turbulent waters overflowing their banks ⑤(蛮横) violently; fiercely; flagrantly: ～加干涉 flagrantly interfere / ～加阻挠 willfully obstruct ⑥(汉字笔划) horizontal stroke (in Chinese characters)
【横暴】brutal violence; violent
【横波】[物] transverse wave
【横冲直撞】push one's way by shoving or humping; jostle and elbow one's way; dash around madly; barge about
【横担】[电] cross arm
【横渡】traverse; sail across
【横断层】cross fault; transverse fault
【横断面】cross section
【横队】rank; row: 排成三列～ line up three deep
【横风】beam wind
【横幅】①(横的书画) horizontal scroll of painting or calligraphy ②(横的标语) banner; streamer: 欢迎群众举着～标语。The welcoming crowd carried banners with slogans on them.
【横格纸】lined paper
【横膈膜】[生理] diaphragm
【横亘】lie across; span: 一座雄伟的大桥～在大江之上。A magnificent bridge spans the river.
【横巷】[矿] crosscut
【横谷】transverse valley; cross valley
【横加诬蔑】slander wildly
【横加指责】blame unscrupulously
【横加阻梗】obstruct wilfully; obstruct unreasonably
【横结肠】[生理] transverse colon
【横跨】stretch over or across: 一道彩虹～天际。A rainbow arched across the sky.
【横浪】beam sea
【横梁】①[建] crossbeam ②[汽车] cross member
【横流】flow over
【横眉】frown; scowl: ～怒目 face others with frowning brows and angry eyes; dart fierce looks of hate / ～冷对千夫指,俯首甘为孺子牛。Fierce-browed, I coolly defy a thousand pointing fingers, Head-bowed, like a willing ox I serve the children.
【横眉怒目】face other with frowning brows and angry eyes; dark fierce looks of hate
【横拍握法】[体] tennis grip; hand-shake grip
【横批】a horizontal scroll bearing an inscription
【横披】a horizontal wall inscription; a horizontal hanging scroll
【横剖面】cross section
【横七竖八】in disorder; at sixes and sevens; higgledy-piggledy: 过道里～地放着各种东西。The passage was cluttered up with all sorts of things.
【横切】crosscut ◇ ～锯 crosscut (saw) / ～面 cross section
【横肉】一脸～ look ugly and ferocious
【横扫】sweep away; make a clean sweep of: ～千军如席卷 rolling back the enemy as we would a mat
【横生】①(杂乱生长) grow wild: 蔓草～ be overgrown with weeds ②(层出不穷) be overflowing with; be full of: 妙趣～ be full of wit and humor ③(意外发生) happen unexpectedly

【横生事端】make trouble on every hand

【横生枝节】①(意外插进的问题) side issues or new problems unexpectedly crop up ②(提出难题) raise obstacles; deliberately complicate an issue

【横竖】anyhow; anyway; in any case: ~我要去的,不用请他来。No need to send for him, I'll be going there anyway.

【横式】horizontal type

【横挑鼻子竖挑眼】find fault in a petty manner; pick holes in sth.; nit-pick

【横尾翼】[航空] tail plane; horizontal stabilizer

【横纹肌】[生理] striated muscle

【横向】crosswise: ◇ ~进刀[机] cross feed; traverse feed

【横向经济联系】lateral economic ties

【横心】steel one's heart; become desperate: 横下一条心 resolve to do sth. in desperation

【横行】run wild; run amuck; be on a rampage: ~一时 run wild for a time

【横行霸道】ride roughshod; play the tyrant; tyrannize; domineer

【横行不法】act illegally; act against law and reason

【横行无忌】run amuck; run wild

【横征暴敛】extort excessive taxes and levies; levy exorbitant taxes

【横轴】[机] cross axle

【横坐标】[数]abscissa

珩 the top gem of a girdle-pendant (as worn by aristocrats and high officials in ancient China)

【珩床】[机] honing machine

【珩磨】[机] honing

桁 [建]purlin

【桁架】[建] truss ◇ ~桥 truss bridge

【桁梁】[航空] longeron

【桁条】[航空] stringer

衡 ①(秤杆) the graduated arm of a steelyard ②(称重量) weighing apparatus ③(衡量) weigh; measure; judge: ~情度理 considering the circumstances and judging by common sense; all things considered

【衡量】weight; measure; judge: ~得失 weigh up the gains and looses

【衡平权】equitable interest ◇ ~比率 equity ratio

【衡器】weighing apparatus

【衡情度理】considering the circumstances and judging by common sense; all things considered

【衡消】balance out

hèng

横 ①(粗暴) harsh and unreasonable; perverse: ~话 harsh, unreasonable words / 发~ act in an unreasonable way ②(意外的;不吉利的) unexpected: ~事 an untoward accident

【横暴】perverse and violent

【横财】ill-gotten wealth: 发~ get rich by foul means

【横祸】unexpected calamity; sudden misfortune: 飞来~ an unforeseen disaster

【横死】die a violent death; meet with a sudden death

【横遭不幸】suffer a sudden misfortune

hng

哼 [叹] humph: ~,你信他的! Humph, don't tell me you take him at his word!

hōng

烘 ①(使热;使干) dry or warm by the fire: ~面包 bake bread / ~手 warm one's hands at the fire / 把湿衣服~一~ dry wet clothes by the fire ②(衬托) set off: ~衬 set off by contrast; serve as a foil to

【烘焙】cure (tea or tobacco leaves)

【烘干】[化] stoving ◇ ~器 baker; drying apparatus

【烘缸】dryer

【烘烤】toast; bake

【烘漆】baking finish; stoving finish

【烘丝机】cut-tobacco drier

【烘托】①(画法) add shading around an object to make it stand out ②(使明显突出) set off by contrast; throw into sharp relief: ~出音乐的主题 set off the *leitmotiv* by contrast

【烘箱】oven

【烘相器】[摄]print drier

【烘云托月】paint clouds to set off the moon; provide a foil to set off a character or incident in a literary work: 收到了~的艺术效果 achieve the artistic effect of prominence through contrast

哄 ①[象](大笑声) roars of laughter ②(嘈杂) hubbub

【哄传】(of rumors) circulate widely: 这个消息不久就~开了。It was not long before the news was widely circulated.

【哄动】cause a sensation; make a stir
【哄然】boisterous; uproarious: ～大笑 burst into uproarious laughter / ～而散 disperse with great noise and hubbub
【哄抬】drive up (prices): ～物价 forcing up prices
【哄堂大笑】the whole room rocking with laughter: 引起～ set the room in a roar

轰 ①[象] bang; boom: 炮～声 boom of guns ①(击;鸣) rumble; bombard; explode: 雷～电闪。Thunder rumbled and lightning flashed. / 万炮齐～ ten thousand cannons booming ③(赶) shoo away; drive off: ～下台 hoot sb. off the platform; oust sb. from office or power
【轰动】cause a sensation; make a stir: ～全国 make a sensation throughout the country / ～一时 create a furore; make a great stir
【轰轰烈烈】on a grand and spectacular scale; vigorous; dynamic: ～的场面 vigorous situation
【轰击】shell; bombard: ～敌人阵地 shell enemy positions / 中子～[物] neutron bombardment
【轰雷掣电】with the force of thunder and lightning
【轰隆】[象] rumble; roll: ～～的机器声 the hum of machines
【轰鸣】thunder; roar: 雷声～。There was a peal of thunder. / 马达～。Motors roared.
【轰然】with a loud crash
【轰炸】bomb: 饱和～ saturation bombing / 定点～ pinpoint bombing / 俯冲～ dive-bomb ◇ ～机 bomber / ～瞄准具 bombsight / ～误差 bombing error

hóng

鸿 ①(鸿雁) swan goose ②[书](书信) letter: 远方来～ a letter from afar ③(大) great; grand: ～图 great plans; grand prospects
【鸿飞冥冥】The wild goose flies to the unseen world. (或: One's whereabouts is unknown.)
【鸿福齐天】One's vast happiness is as high as the sky
【鸿沟】wide gap; chasm: 不可逾越的～ an unbridgeable gap; an impassable chasm
【鸿鹄之志】lofty ambition; high aspirations
【鸿毛】[书] a goose feather; something very light or insignificant: 轻如～ as light as a feather / 轻于～ lighter than a feather
【鸿雁】swan goose
【鸿雁传书】a letter from afar
【鸿运高照】bring somebody good luck and success in life; be in luck; having good luck

虹 rainbow
【虹彩】iridescence
【虹膜】[生理] iris ◇ ～麻痹 iridoplegia / ～炎 iritis
【虹吸管】siphon
【虹吸现象】siphonage
【虹雉】monal

红 ①(红色) red: ～墙 a red ochre wall / 穿着～衣服 dressed in red ②(象征革命) revolutionary; red: 心～志坚 red in heart and firm in will / 又～又专 both red and expert; both socialist-minded and professionally proficient ③(象征喜庆的红布) red cloth, bunting, etc. used on festive occasions: 挂～ hang up red festoons / 披～ wear red sashes or cloth as a sign of honor, festivity, etc. ④(顺利; 成功) symbol of success: 开门～ get off to a good start / 他唱戏唱～了。He has won success as an actor. ⑤(红利) bonus; dividend: 分～ distribute or draw dividends
【红白喜事】weddings and funerals
【红斑】[医] erythema ◇ ～狼疮 lupus erythematosus
【红榜】honor roll
【红宝石】ruby: 晶体～ ruby crystal
【红菜头】beetroot
【红茶】black tea
【红潮】blush; flush
【红尘】[旧] the world of mortals; human society: 看破～ see through the vanity of the world; be disillusioned with this human world
【红绸】red silk ◇ ～舞 red silk dance
【红丹】[化] red lead; minium ◇ ～漆 red lead paint
【红得发紫】extremely popular
【红电气石】[矿] rubellite
【红灯】[交] red light; traffic light
【红豆】①(乔木红豆树) ormosia ②(象征相思) love pea
【红豆杉】Chinese yew
【红豆相思】red beans that inspire the memory of one's love
【红汞】[药] mercurochrome
【红骨顶】[动] moorhen
【红光满面】one's face glowing with health; in

ruddy health
【红果】[方] the fruit of large Chinese hawthorn; haw
【红鹤】ibis
【红狐】red fox
【红花】[中药] safflower
【红脚鹬】redshank
【红巨星】[天] red giant star
【红军】①[简](中国工农红军) the Chinese Workers' and Peasants' Red Army (1928—1937); the Red Army ②(红军战士) Red Army man
【红利】bonus; extra dividend ◇ ～分配 profit sharing / ～股 bonus dividend; bonus stock / ～股票 stock dividend / ～帐目 bonus account
【红脸】①(害羞) blush ②(发怒) flush with anger; get angry: 他俩从没有红过脸。There has never been a cross word between the two of them.
【红磷】red phosphorus
【红铃虫】pink bollworm
【红领巾】①(红色领巾) red scarf ②(少先队员) Young Pioneer
【红领章】red collar tab
【红绿灯】traffic light; traffic signal
【红鸾星动】The propitious star governing marriage is in the ascendant.
【红麻】[植]bluish dogbane
【红霉素】[药] erythromycin
【红焖】stew in soy sauce: ～鸡 stewed chicken
【红米】red rice
【红模子】a sheet of paper with red characters printed on it, to be traced over with a brush by children learning calligraphy
【红木】padauk; annatto; cigar-box wood
【红娘鱼】sea robin; red gurnard
【红皮书】Red Data Book
【红旗】red flag; red banner: 在～下长大 grow up in socialist society / ～招展。Red flags are fluttering gaily in the wind. ◇ ～单位 red-banner unit; advanced unit
【红人】a favorite with sb. in power; fair-haired boy
【红日高照】The red sun shone brightly.
【红润】ruddy; rosy: 脸色～ ruddy complexion; rosy cheeks
【红三叶草】red clover
【红色】①(红颜色) red ②(象征革命) revolutionary; red: ～根据地 revolutionary base; red base / ～政权 red political power
【红杉】①(中国太白红杉) Chinese larch ②(北美红杉) redwood
【红烧】braise in soy sauce: ～肉 pork braised in brown sauce
【红十字会】the Red Cross: 中国～ the Red Cross Society of China
【红薯】sweet potato
【红树】mangrove
【红松】Korean pine
【红糖】brown sugar
【红陶】red pottery; terra-cotta
【红藤】Sargent gloryvine
【红铜】red metal; pure copper; red bronze; rose copper
【红通通】bright red; glowing: ～的火苗 glowing red flames / ～的晚霞 evening glow / 他脸儿晒得～的。His face is aglow from exposure to the sun.
【红土】red soil
【红外激射】[物] iraser
【红外摄影】infrared photography
【红外探测】infrared acquisition
【红外线】[物] infrared ray ◇ ～辐射 infrared radiation / ～理疗 / infrared therapy / ～扫描装置 infrared scanner / ～探测器 infrared detector / ～照相 infrared photography
【红细胞】red blood cell; erythrocyte ◇ ～计数 red cell count
【红小豆】red bean
【红心】red heart; loyal heat; true heart
【红新月会】Red Crescent Society; the Red Crescent
【红星】red star: ～帽徽 red star cap insignia
【红血球】red blood cell; erythrocyte
【红颜薄命】A beautiful girl has (often) an unfortunate life. (或: The beautiful woman suffered a harsh life.)
【红颜易老】Beauty is a fragile good
【红眼】①(眼疾) blood-shot eye; ophthalmia ②(发怒) see red; become infuriated ③(眼红) be jealous of sb.
【红艳艳】brilliant red
【红药水】mercurochrome
【红叶】red autumnal leaves (of the maple, etc.)
【红衣主教】[天主教] cardinal
【红缨枪】red-tasselled spear
【红鱼】(red) snapper
【红运】good luck: 走～ have good luck
【红晕】blush; flush: 脸上泛出～ one's face blushing scarlet
【红藻】red alga
【红蜘蛛】red spider (mite); spider mite

【红肿】red and swollen
【红柱石】[地] andalusite
【红装】[书] ①(妇女的艳丽装束) gay feminine attire ②(青年妇女) young woman [又作"红妆"]
【红装素裹】A beauty is dressed in mourning white.(或: clad in white, adorned in red.) [又作"红妆素裹"]

洪 ①(大) big; vast: ~涛 big waves ②(洪水) flood: 防~ control or prevent flood / 分~ flood diversion; reduce flood; mitigate flood / 分~区 flood diversion are / 山~ mountain torrents; freshet
【洪波滚雪】The wind whipped the waves into a snowy foam.
【洪大】loud: ~的回声 resounding echoes
【洪都拉斯】Honduras ◇ ~人 Honduran
【洪泛区】floodplain; flooded area
【洪峰】flood peak
【洪福齐天】be supremely fortunate; have great luck
【洪积层】[地] diluvium
【洪亮】loud and clear; sonorous: 嗓音~ a sonorous voice
【洪量】①(大的气量) magnanimity; generosity ②(酒量大) great capacity for liquor
【洪流】mighty torrent; powerful current: 不可抗拒的革命~ irresistible trend of revolution / 时代的~ the powerful current of the times
【洪炉】great furnace: 在革命~中经受锻炼 be tempered in the mighty furnace of revolution
【洪脉】[中医] pulse beating like waves; full pulse
【洪水】flood; floodwater ◇ ~水位 flood level
【洪水恐怖】[心理] antlophobia
【洪水猛兽】fierce floods and savage beasts; great scourges
【洪钟】[书] large bell: 声如~ have a stentorian voice

宏 great; grand; magnificent
【宏大】grand; great: ~的志愿 great aspirations / 规模~ on a grand scale / 建设~的科学技术队伍 build a mammoth force of scientific and technical personnel
【宏代码】macrocode
【宏观】[物] macroscopic ◇ ~结构 macrostructure / ~经济学 macroeconomics / ~控制 macro-control / ~世界 macrocosm / ~指导 centralize guidance
【宏论】informed opinion; intelligent view
【宏图】great plan; grand prospect
【宏伟】magnificent; grand: ~的前景 grand prospects / 发展国民经济的~计划 the magnificent plan for developing the national economy
【宏愿】great aspirations; noble ambition
【宏旨】main theme; leading idea of an article: 无关~ insignificant

弘 ①(大) great; grand; magnificent ②(扩大; 光大) enlarge; expand
【弘大】grand

泓 ①(水深) deep ②[量] 一~清泉 a clear spring / 一~秋水 an expanse of limpid water in autumn

hǒng

哄 ①(哄骗) fool; humbug: ~人 make a fool of sb. / 不要~我。 Don't fool me. ②(哄孩子) coax; humor: ~孩子吃药 coax a child to take medicine / ~婴儿睡觉 soothe a baby to sleep
【哄骗】cheat; humbug; hoodwink

hòng

哄 uproar; horseplay: 一~而散 break up in an uproar

hóu

骺 epiphysis

侯 ①(爵位) marquis ②(达官贵人) a nobleman or a high official: ~门似海。 The mansions of the nobility were inaccessible to the common man.
【侯爵】marquis ◇ ~夫人 marquise

瘊 wart
【瘊子】[医] wart

喉 larynx; throat
【喉癌】laryngocarcinoma
【喉擦音】[语] guttural fricative
【喉结】[生理] Adam's apple
【喉镜】[医] laryngoscope
【喉咙】throat: ~痛 have a sore throat
【喉塞音】[语] glottal stop
【喉痧】[中医] scarlet fever
【喉舌】mouthpiece: 人民的~ the mouthpiece of the people
【喉头】larynx; throat

【喉炎】laryngitis: 慢性～ chronic laryngitis; clergyman's sore throat

猴 ①(猴子) monkey ②(机灵) clever boy; smart chap

【猴面包树】[植]monkey-bread free; baobab

【猴皮筋】rubber band: 跳～ skip rubber band

【猴头】[植]hedgehog hydnum

【猴戏】a show by a performing monkey; monkey show

【猴子】monkey

hǒu

吼 roar; howl: 大炮的～ the roar of the cannon / 狮～ the roar of a lion

hòu

厚 ①(与薄相对) thick: ～木板 a thick plank / ～纸 thick paper ②(深) deep; profound: 深情～谊 profound friendship ③(厚道) kind; magnanimous: 忠～ honest and kind ④(大) large; generous: ～礼 generous gifts / ～利 large profits / ～薪 high pay ⑤(浓) rich or strong in flavor: 酒味很～。 The wine tastes strong. ⑥(优待;推崇) favor; stress: ～此薄彼 favor one and be prejudiced against the other

【厚薄】thickness: ～合适。 It's just the right thickness. / 纸张～无关紧要。 It doesn't matter whether the paper is thick or thin.

【厚薄规】[机] feeler

【厚此薄彼】favor one more than another; treat with partiality

【厚道】honest and kind

【厚德载福】Great virtue carries happiness with it.

【厚度】thickness

【厚墩墩】very thick: ～的棉大衣 a heavy padded overcoat

【厚古薄今】stress the past, not the present

【厚今薄古】stress the present, not the past

【厚脸皮】thick-skinned; brazen; cheeky: 厚着脸皮说 have the nerve to say

【厚人薄己】treat others well but be frugal with oneself

【厚漆】paste paint

【厚实】thick and solid: ～的被褥 thick, heavy quilt and mattresses

【厚望】great expectations: 不负～ live up to sb.'s expectations; not let sb. down

【厚颜无耻】impudent; brazen; shameless

【厚意】kind thought; kindness: 多谢你的～。 Thank you for your kindness.

候 ①(等待) wait; await: ～复 waiting for your reply / 请稍～一会儿。 Please wait a moment. ②(问好) inquire after: 致～ send one's regards ③(时节) time; season: 季～ season / 时～ time ④(情况) condition; state: 症～ symptom

【候补】be a candidate (for a vacancy); be an alternate ◇ ～键 candidate key / ～卷宗 candidate volume / ～委员 alternate member

【候车室】waiting room (in a railway or bus station)

【候机室】airport lounge or waiting room

【候教】[敬] await your instructions

【候鸟】migratory bird; migrant

【候审】[法] await trial ◇ ～犯人 criminal for trial

【候选人】candidate: 提出～ nominate candidates ◇ ～名单 list of candidates / ～资格 qualifications for standing for election

【候诊】wait to see the doctor ◇ ～室 waiting room (in a hospital)

后 ①(靠后的部分) behind; back; rear: ～排 back row / 敌～ the enemy's rear / 请往～站! Stand back, please! ②(未来的; 较晚的) after; afterwards; later: 不久以～ soon afterwards; before long / 课～ after class ③(后代的人) offspring: 无～ without male offspring; without issue ④(君主的妻子) empress; queen

【后半】later half; second half: 十九世纪～期 in the later half of the nineteenth century ◇ ～场比赛 the second half of the game / ～生 the later half of one's life

【后半天】afternoon

【后半夜】after midnight

【后备】reserve: 留有～ keep sth. in reserve ◇ ～部队 reserve units / ～基金 reserve fund / ～力量 reserve forces

【后备军】①(预备役军人) reserves ②(补充力量) reserve force: 产业～ industrial reserve army; industrial reserve; reserve army of labor

【后辈】①(年轻的或资历浅的人) younger generation ②(后代) posterity

【后步】room for maneuver: 留～ leave sufficient room for maneuver

【后补进口手续】post entry

【后尘】[书] 步人～ follow in sb.'s footsteps

【后处理】①[化] aftertreatment ②[纺] finishing

【后代】①(某一时间以后的时代) later periods (in history); later ages ②(后代的人) later gen-

erations; descendants; posterity: 好好培养革命的～ take paints to bring up successors to the revolution / 为～着想 for the sake of future generations; in the interest of future generations ③[生] progeny
【后灯】[汽车] taillight; tail lamp
【后爹】stepfather
【后盾】backing; backup force: 坚强的～ powerful backing
【后发制人】gain mastery by striking only after the enemy has struck
【后方】rear: ～工作 rear-area work; work in the rear ◇ ～基地 rear base / ～留守处 rear headquarters / ～医院 rear hospital
【后跟】heel (of a shoe or sock)
【后顾】①(回头照顾后方) turn back (to take care of sth.): ～之忧 fear of disturbance in the rear; trouble back at home ②(回忆) look back (on the past)
【后滚翻】[体] backward roll
【后果】consequence; aftermath: 承担～ accept the consequences / 前因～ cause and effect / ～不堪设想。The consequences would be too ghastly to contemplate.
【后患】future trouble: ～无穷 no end of trouble for the future
【后悔】regret; repent: ～不已 be overcome with regret / ～莫及 too late to repent / 一点不～ have no regrets at all
【后会有期】we'll meet again some day
【后记】postscript
【后继】succeed; carry on: ～乏人 lack worthy successors / ～有人 have qualified successors; successors will come
【后襟】the back of a Chinese robe or jacket
【后进】lagging behind; less advanced; backward: ～赶先进。The advanced strive to catch with the more advanced.
【后劲】①(显露较慢的作用) delayed effect; aftereffect: 这酒～大。This wine has a strong delayed effect. ②(用在后一阶段的力量) reserve strength; stamina: 他干活有～。He has staying power when he's doing a job.
【后景】background
【后空翻】[体] backward somersault
【后来】afterwards; later: ～怎么样? What happened afterwards: ◇ ～人 successors
【后来居上】the latecomers surpass the old-timers; the new generation will surpass the old
【后浪推前浪】waves urging waves; rolling on; making progress steadily
【后路】①(退路) communication lines to the rear; route of retreat: 切断敌人～ cut off the enemy's route of retreat ②(回旋余地) room for maneuver; a way of escape: 留～ leave oneself a way of escape; leave oneself a way out
【后掠】sweepback ◇ ～角[航空] sweep angle; sweepback / ～翼[航空] swept-back wing
【后轮】rear wheel
【后妈】[口] stepmother
【后门】①(后面的门) back door: 大院的～ the back gate of a compound ②(舞弊的途径) back door influence: ～成风。The malpractice of going in for "back door" deals became common. / 走～是一种不正之风。Backdoor deals are unhealthy practice.
【后门送旧,前门迎新】The old clients are let out at the back door, the new ones admitted by the front gate.
【后面】①(位置靠后的部分) at the back; in the rear; behind: ～还有座位。There are vacant seats at the back. ②(次序靠后的部分) later: 这个问题我～还要讲。I'll come back to this question later.
【后脑】[生理] hindbrain; rhombencephalon
【后年】the year after next
【后娘】stepmother
【后怕】fear after the event
【后排】back row: ～座位 back row seats
【后期】later stage; later period: 二十世纪八十年代～ the late 1980s / 抗日战争～ the latter part of the Anti-Japanese War
【后起】(of people of talent) of new arrivals; of the younger generation: ～的青年作家 budding young writers
【后起之秀】an up-and-coming youngster; a promising young person
【后桥】[汽车] rear axle ◇ ～壳 rear axle housing
【后勤】rear service; logistics ◇ ～部 rear-service department; logistics department / ～部队 rear-service units; rear services / ～基地 logistics base; rear supply base / ～人员 rear-service personnel / ～支援 logistic support
【后人】①(后代之人) later generations ②(子孙) posterity; descendants
【后任】successor
【后三角队形】[军] V formation

【后身】①(身体后边的部分) the back of a person: 我只看见个～,认不清是谁。I couldn't make out who he was as I only saw his back. ②(衣服的后背) the back of a garment: 这件衬衫的～太长了。The back of the shirt is too long.
【后生可畏】a youth is to be regarded with respect; the younger generation will surpass the older
【后世】①(某一时代以后的时代) later ages ②(后代的人) later generations ◇～子孙 descendants; posterity
【后事】①(以后的事情) what happened afterwards: 欲知～如何,且听下回分解。If you want to know what happened afterwards, read the next chapter. ②(丧事) funeral affairs: 料理～ make arrangements for a funeral
【后视镜】[汽车] rearview mirror
【后视图】[机] back view; rearview
【后手】①(被动的形势) defensive position (in chess) ②(后路) room for maneuver; a way of escape
【后熟作用】[农] afterripening
【后送】[军] evacuation
【后膛装填】[军] breech loading
【后体】metasoma
【后台】①(舞台后部) backstage ②(背后操纵和支持的人) backstage supporter; behind-the-scenes backer: ～很硬 have very strong backing ◇～老板 backstage boss
【后天】①(明天的明天) day after tomorrow: 大～ three days from today ②(与先天相对) postnatal; acquired: 知识是～获得的,不是先天就有的。Knowledge is acquired, not innate. ◇～性免疫 acquired immunity
【后退】draw back; fall back; retreat: 迫使敌人～ force the enemy to retreat / 遇到困难决不～ never shrink from difficulties
【后卫】①[军] rear guard ②[足球] full back: 右～ right back / 左～ left back ③[篮球] guard ◇～战斗 rear-guard action
【后项】consequent
【后续部队】[军] follow-up units
【后续成本】after-cost
【后续会议】follow-up meeting
【后悬】[汽车] rear overhang
【后遗症】sequelae: 脑震荡～ sequelae of cerebral concussion / 小儿麻痹～ sequelae of infantile paralysis / 病愈没有～ recover without any after-effect
【后裔】descendant; offspring
【后影】the shape of a person or thing as seen from the back
【后元音】[语] back vowel
【后援】reinforcements; backup force; backing
【后院】backyard
【后者】the latter
【后肢】[动] hind legs
【后轴】rear axle
【后缀】[语] suffix
【后座】back-lash ◇～议员 backbencher
【后坐力】[军] recoil: 无～炮 recoilless gun

hū

糊 plaster: ～墙 plaster a wall / ～一层泥 spread a layer of mud

乎 ①[书][助](表示疑问或揣度)成败之机,其在斯～? Does not success or failure hinge on this? / 一之为甚,其可再～? Once is more than enough. How can you do it again? ②(动词后缀)超～寻常 be out of the ordinary / 出～意料 exceed one's expectations; be beyond one's expectations / 合～客观规律 conform to an objective law ③(形容词或副词后缀)确～重要 very important indeed / 巍巍～ towering; lofty

烀 stew in shallow water

呼 ①(吐气) breathe out; exhale: ～出二氧化碳 exhale carbon dioxide ②(大声喊) shout; cry out: ～口号 shout slogans ③(叫) call: 惊～ cry out in surprise / 直～其名 address sb. disrespectfully (by name) / ～之即来,挥之即去 have sb. at one's beck and call ④[象] 北风～～地吹。A north wind is whistling.
【呼哧】[象] ～～直喘 puff and blow
【呼风唤雨】①(使刮风下雨) summon wind and rain; control the forces of nature ②(煽动) stir up trouble
【呼喊】call out; shout: 大声～ raise a cry / 高兴得～起来 exclaim with delight
【呼号】①(哭叫) wail; cry out in distress: 奔走～ go around crying for help ②[讯] call sign; call letters ③(专用的口号) catchword (of an organization) ◇～机 call signal apparatus
【呼唤】call; shout to
【呼叫】①(呼喊) call out; shout ②[讯] call ◇～表 calling list / ～灯 calling lamp / ～局 calling station / ～声 ring tone / ～信号 calling signal
【呼救】call for help ◇～信号 signal for help;

SOS
【呼噜】[口] snore: 打～ snore
【呼朋引类】gang up
【呼哨】whistle: 打～ give a whistle
【呼声】cry; voice: 发出正义的～ give the cry of justice / 世界舆论的强大～ the powerful voice of world opinion
【呼天抢地】lament to heaven and knock one's head on earth; utter cries of anguish
【呼吸】breathe; respire: ～急促 be short of breath / ～困难 breathe with difficulty; lose one's breath / ～新鲜空气 have a breath of fresh air ◇ ～道 respiratory tract / ～率 respiratory rate / ～器 respirator / ～器官 respiratory organs / ～系统 respiratory system / ～运动学 respirometry
【呼啸】whistle; scream; whizz: 寒风～。 A cold wind is whistling.
【呼幺喝六】play at dice; shout for the top number to come up
【呼应】echo; work in concert with: 遥相～ echo each other over a distance; echo from afar
【呼语】[语] vocative expression; direct address
【呼吁】appeal; call on: ～团结 appeal for unity / 发出紧急～ make an urgent appeal ◇ ～书 letter of appeal; appeal
【呼之欲出】seem ready to come out at one's call (said of lifelike figures in pictures or characters in novels); be vividly portrayed

忽 ①(不注意) neglect; overlook; ignore ②(忽然) suddenly: ～发奇想 suddenly have a strange idea
【忽而】now..., now...: ～哭, ～笑 cry and laugh by turns / 天气～冷～热。 The weather is now hot, now cold.
【忽略】ignore; neglect; overlook; lose sight of: 不要～枝节问题。 Don't neglect minor issues ◇ ～码 ignore code
【忽略精度】default precision
【忽然】[副] suddenly; all of a sudden
【忽视】ignore; overlook; neglect: 不可～的力量 a force not to be ignored; a force to be reckoned with / 不应～困难。 We should not overlook the difficulties.
【忽悠】[方] flicker: 渔船上的灯火～～的。 Lights flickered on the fishing boats.

hú

壶 ①(容器) kettle; pot: 茶～ teapot / 水～ kettle / 油～ oil can ②(瓶) bottle; flask: 暖～ thermos bottle / 行军～ water bottle; canteen

胡 ①[史](旧称北方和西方少数民族) non-Han nationalities living in the north and west in ancient times ②(来自北方、西方的东西) introduced from the northern and western nationalities or from abroad: ～萝卜 carrot / ～桃 walnut ③[副](随意乱来) recklessly; wantonly; outrageously: ～编 recklessly concoct / ～吹 boast outrageously; talk big ④[书](何故) why: ～不归? Why not return? ⑤(胡子) moustache, beard or whiskers
【胡扯】①(胡说) talk nonsense; drivel ②(闲谈) chat; small talk
【胡蜂】wasp; hornet
【胡搞】①(胡来) mess things up; meddle with sth. ②(乱搞男女关系) carry on an affair with sb.; be promiscuous
【胡话】ravings; wild talk: 烧得直说～ be delirious from fever
【胡椒】pepper
【胡椒鲷】[动] grunt
【胡搅】①(扰乱) pester sb.; be mischievous ②(狡辩) argue tediously and vexatiously; wrangle
【胡搅蛮缠】harass sb. with unreasonable demands; pester sb. endlessly
【胡来】①(任意乱做) mess things up; fool with sth.: 你要是不会修就别～。 If you don't know how to repair it, don't fool with it. ②(胡闹) run wild; make trouble
【胡狼】jaokal
【胡里胡涂】confused in one's thinking; with one's mind in a haze
【胡乱】carelessly; casually; at random: ～猜测 make wild guesses / ～吃了点饭 eat a hasty meal; grab a quick bite / ～写了几行 scribble a few lines
【胡萝卜】carrot ◇ ～素[生化] carotene
【胡麻】flax ◇ ～籽 flaxseed; linseed
【胡闹】run wild; be mischievous
【胡琴】*huqin*; a general term for certain two-stringed bowed instruments ○二胡 *erhu* / 京胡 *jinghu*
【胡说】①(瞎说) talk nonsense; drivel: 别～! Don't talk rot! / 他～些什么? What's he drivelling about? ②(无道理的话) nonsense
【胡说八道】①(胡说) talk nonsense: 他的发言全是～。 His speech was all rot. ②(无价值的话) sheer nonsense; rubbish
【胡思乱想】imagine things; go off into wild

flights of fancy; let one's imagination run away with one
【胡同儿】lane; alley
【胡颓子】[植] thorny elaeagnus
【胡须】beard, moustache or whiskers
【胡言乱语】talk nonsense; rave
【胡杨】[植] diversiform-leaved poplar
【胡枝子】shrub lespedeza
【胡诌】fabricate wild tales; cook up: ～了一大堆理由 cook up a lot of excuses
【胡子】beard; moustache; whiskers
【胡子拉碴】wear a stubble of untrimmed beard; with a bristly unshaven chin
【胡作非为】act wildly in defiance of the law or public opinion; commit all kinds of outrages

湖 lake

【湖笔】writing brush produced in Huzhou
【湖滨】lakeside
【湖沉积】[地] lacustrine deposits
【湖成沉积物】[地] lacustrine sediments
【湖光掠影】The lake ripples and sparkles.
【湖光山色】a landscape of lakes and mountains; the natural beauty of lakes and mountains
【湖泊】lakes
【湖泊效应】[气象] lake effect
【湖色】light green
【湖田】land reclaimed from a lake; shoaly land
【湖心亭】a pavilion in the middle of a lake; mid-lake pavilion
【湖沼学】limnology

糊

①(浆糊) paste: 将面粉加水调成～状 mix flour and water into a paste ②(粘住) stick with paste; paste: ～窗户 paste a sheet of paper over a lattice window or seal with paper the cracks around a window ③(焦) (of food) burnt
【糊精】[化] dextrin; artificial gum
【糊口】keep body and soul together; eke out one's livelihood
【糊口谋生】make a living; eke out the barest of living
【糊墙纸】wall paper
【糊涂】muddled; confused; bewildered: ～观念 a muddled idea / 别装～。Don't play the fool. ◇ ～虫 blunderer; bungler / ～帐 chaotic accounts; a mess

煳

(of food) burnt: 饭～了。The rice is burnt.

葫

【葫芦】bottle gourd; calabash: 他的～里到底卖的是什么药? What has he got up his sleeve?

蝴

【蝴蝶】butterfly ◇ ～阀[水] butterfly valve / ～花 fringed iris / ～结 bow / ～鱼 butterfly fish

猢

【猢狲】macaque: 树倒～散。When the tree falls, the monkeys scatter. (或: When the boss falls from power, his lackeys disperse.)

核

【核儿】[口] ①(水果核) stone; pit; core: 犁～ pear core / 杏～ apricot stone ②(果核状物) sth. resembling a fruit stone: 煤～ partly-burnt coals or briquets; cinders

囫

【囫囵】whole: ～吞下 swallow sth. whole
【囫囵吞枣】swallow dates whole; lap up information without digesting it; read without understanding

鹄 swan

【鹄候】[书] await respectfully; expect: ～回音。I am awaiting your reply.
【鹄望】[书] eagerly look forward to

狐 fox

【狐臭】body odor; bromhidrosis; hircus
【狐假虎威】the fox borrows the tiger's terror (by walking in the latter's company); bully people by flaunting one's powerful connections
【狐狸】fox ◇ ～精 fox spirit; seductive woman
【狐狸尾巴】fox's tail; something that gives away a person's real character or evil intentions; cloven hoof: ～总是要露出来的。A fox cannot hide its tail.
【狐狸座】[天] little Fox; Vulpecula
【狐媚】bewitch by cajolery; entice by flattery
【狐媚淫态】(by) attractive looks and seductive manners
【狐朋狗友】evil associates
【狐裘】fox-fur robe
【狐裘羔袖】a fox-fur robe with lamb-skin sleeves; good on the whole but not perfect
【狐群狗党】a pack of rogues; a gang of scoundrels

【狐疑】doubt; suspicion: 满腹～ be full of misgivings; be very suspicious
【狐疑不决】be wavering and unable to decide; indecisive in one's mind; suspicious and undecided
【狐踪兔迹】the tracks of fox and hare

弧 [数] arc
【弧度】[数] radian; radian measure
【弧光】arc light; arc ◇～灯 arc lamp; arc light
【弧菌】vibrio
【弧面】cambered surface
【弧圈球】[乒乓球] loop drive
【弧形】arc; curve ◇～闸门[水] radial gate

槲 [植] Mongolian oak
【槲寄生】[植] mistletoe
【槲栎】[植] oriental white oak

hǔ

浒 waterside

虎 ①(老虎) tiger: 小～ a tiger cub ②(威武; 勇猛) brave; vigorous: ～将 brave general
【虎胆】as brave as a tiger: ～英雄 hero as brave as a lion
【虎毒不食子】All men, good or bad, rarely illtreat their own children.
【虎耳草】[植] saxifrage
【虎伏】[体] gyro wheel
【虎骨酒】tiger-bone liquor
【虎虎有生气】vigorous and energetic; be full of vigor
【虎劲】dauntless drive; dash: 有一股子～ be full of drive and daring; have plenty of dash
【虎鲸】killer whale
【虎口】①(老虎嘴) tiger's mouth; jaws of death: ～拔牙 pull a tooth from the tiger's mouth; dare the greatest danger; beard the lion in his den / ～余生 survive a disaster; have a narrow escape ②(拇指与食指相联接部分) part of the hand between the thumb and the index finger
【虎落平阳被犬欺】If the tiger went down to level land, he would be insulted by dogs.(或: A man who loses position and influence may be subjected to much indignity.)
【虎皮鹦鹉】budgerigar
【虎钳】vice: 台～ bench vice / 万能～ universal vice ◇～口 vice jaw
【虎鲨】bullhead shark
【虎视眈眈】glare like a tiger eyeing its prey; eye covetously
【虎头拍蝇】beat a fly on the head of a tiger
【虎头蛇尾】in like a lion, out like a lamb; fine start and poor finish
【虎威】prowers (of a general)
【虎啸】roars of a tiger
【虎啸风生】Tigers howl with the rise of winds.
【虎啸猿啼】Tigers roar and monkeys cry.
【虎穴】tiger's den: ～追踪 track the tiger to its lair

琥
【琥珀】amber
【琥珀酸】butanedioic acid; succinic acid
【琥珀油】amber oil

唬 [口] bluff: 她没被～住。She wasn't intimidated.
【唬起嘴脸】put on a solemn face

hù

糊 paste: 玉米～(cornmeal) mush
【糊弄】[方] ①(欺骗) fool; deceive; palm sth. off on: 你别～我。Don't try to fool me. ②(将就) go through the motions; be slipshod in work: 这可是细活,不能瞎～。This is a delicate job. It mustn't be done carelessly.

户 ①(门) door: ～外活动 outdoor activities / 足不出～ never step out of doors; confine oneself within doors ②(人家;住户) household; family: 挨～通知 notify the households one by one / 家家～～ each and every family ③(户头) (bank) account: 存～ (bank) depositor ④(门第) family status: 门当～对 matched in family status
【户籍】①(登记居民的册子) census register; household register ②(居民身分) registered permanent residence ◇～登记簿 census register / ～警 policeman in charge of household registration
【户口】①(住户和人口) number of households and total population ②(户籍) registered permanent residence: 报～ register one's residence; apply for residence / 查～ check residence cards; check on household occupants / 临时～ temporary resident / 迁～ report to the local authorities for change of domicile / 销～ cancel one's residence registration ◇～簿 (permanent) residence booklet / ～清册 census record / ～证件 household registry certificate

【户枢不蠹】a door-hinge is never worm-eaten
【户头】(bank) account: 开～ open an account
【户外生活】open air life
【户外天线】open antenna
【户限】[书] threshold: ～为穿 a threshold worn low by visitors; an endless flow of visitors
【户主】head of a household

戽 bail: ～水灌田 bail water to irrigate fields
【戽斗】bailing bucket ◇ ～板 bucket board / ～车 bucket car / ～式 scoop-type / ～式进料器 scoop feeded
【戽水机】scooping machine
【戽索】bucket rig

护 ①(保护) protect; guard; shield: ～厂 guard a factory / ～林 protect a forest ②(袒护) be partial to; shield from censure: 别～着自己的孩子。Don't be partial to your own child.
【护岸】[水] bank revetment ◇ ～林 protective belt (of trees) along an embankment
【护城河】city moat
【护持】shield and sustain
【护堤】protect a dike
【护短】shield a shortcoming or fault
【护耳】earflaps; earmuffs ◇ ～器 ear defenders
【护封】book jacket; jacket
【护盖】protecting cover; protecting hood
【护航】escort; convoy: 由三艘军舰～ be convoyed by three warships; have an escort of three warships ◇ ～部队 escort force / ～飞机 escort aircraft / ～舰 convoy ship
【护肩】shoulder pad
【护胫】shin guard
【护理】nurse; tend and protect: ～伤病员 nurse the sick and the wounded ◇ ～人员 nursing staff
【护路】①(巡逻和保护铁路或公路) patrol and guard a road or railway ②[交](维护) road maintenance ◇ ～队 road maintenance team / ～林 protective belt (of trees) along a road
【护面】[体] mask
【护目镜】goggles
【护目罩】eye guard; eye shield; head shield
【护坡】[水][交] slop protection
【护身符】①(符) amulet; protective talisman ②(保护自己的人和物) a person or thing that protects on from punishment or censure; shield
【护士】(hospital) nurse ◇ ～学校 nurses' school / ～长 head nurse
【护手盘】[击剑] hand guard
【护送】escort; convoy: ～救灾物资 convoy vehicles bringing relief to a disaster-stricken area
【护腿】[体] shinguard; gaiter; thigh-protector
【护卫】protect; guard ◇ ～舰 escort vessel; corvette
【护膝】[体] kneepad; kneecap
【护穴】phragmosis
【护胸】[体] chest protector
【护养】①(培育) cultivate; nurse: ～秧苗 cultivate seedlings; nurse young plants / ～羊羔 rear lambs; look after lambs ②(养护) maintain: ～公路 maintain a highway
【护照】passport: 公务～ service passport / 外交～ diplomatic passport

扈 [书] retinue
【扈从】[书] retinue; retainer

怙 [书] rely on: 失～ have nobody to rely on, one's father being dead; have lost one's father
【怙恶不悛】be steeped in evil and refuse to repent: ～的罪犯 incorrigible criminal

瓠
【瓠子】a kind of edible gourd

互 mutual; each other: ～不干涉内政 noninterference in each other's internal affairs / ～教～学 teach and learn from each other / ～派常驻使节 exchange resident envoys; mutually accredit resident envoys / ～通情报 exchange information; keep each other informed / ～为条件 mutually conditional; interdependent
【互殴】handplay
【互不干涉】mutual noninterference
【互不侵犯】mutual nonaggression
【互不侵犯条约】nonaggression treaty
【互导】[电] mutual conductance; transconductance
【互访】exchange visits: 两国体育代表团的～ exchange of sports delegations between two countries
【互感】[电] mutual inductance
【互换】exchange: ～大使 exchange of ambassadors / ～记者 exchange correspondents / ～批准书 exchange instruments of ratification
【互惠】mutually beneficial; reciprocal: 贸易～ reciprocity in trade ◇ ～待遇 reciprocal treatment / ～关税 mutually preferential tariff / ～条约 reciprocal treaty / ～政策 give-and-take policy

【互敬互爱】mutually respect and love
【互利】mutually beneficial; of mutual benefit: 在平等～的基础上 on the basis of equality and mutual benefit
【互谅互让】mutually understanding and mutual accommodation
【互谦互让】mutually making a compromise
【互通有无】each supplies what the other needs; help supply each other's needs
【互为因果】reciprocal causation
【互相】[副] mutual; each other: ～勾结 work in collusion / ～利用 each using the other for his own ends / ～排斥 be mutually exclusive / ～配合 work in coordination / ～倾心 become greatly attached to each other / ～渗透 interpenetrate / ～依存 depend on each other for existence; be interdependent / ～制约 mutually condition / ～掣肘 hold each other back / ～转化 mutual transformation
【互助】help each other: ～合作 mutual aid and cooperation ◇～组 mutual aid team

笏 a tablet held before the breast by officials when received in audience by the emperor

huā

化 spend; expend: ～工夫 spend time; take time / ～钱 spend money; cost money

花 ①(种子植物的有性繁殖器官) flower; blossom; bloom: 雌～ female flower / 两性～ hermaphrodite flower / 雄～ male flower / 摘一朵～ pluck a flower / 种～ cultivate flowers ②(象花的东西) anything resembling a flower: 火～ spark / 浪～ spray / 雪～ snowflakes ③(烟火) fireworks: 放～ let off fireworks ④(花纹) pattern; design: 这被面的～儿很大方。The design on this quilt cover is quite elegant. ⑤(花纹装饰;颜色错杂) multicolored; colored; variegated: ～蝴蝶 variegated butterfly / ～衣服 bright-colored clothes / 小～狗 spotted puppy ⑥(模糊迷乱) blurred; dim: 看书看得眼睛都～了 read until the print looks blurred ⑦(棉花) cotton: 废～ waste cotton / 轧～ gin cotton ⑧(痘) smallpox: 出～儿 get smallpox / 种～儿 vaccinate ⑨(受伤) wound: 战斗中挂了～ get wounded in action ⑩(花费) spend; expend: ～了不少钱 spend a lot of money / 很～时间 take a lot of time; be time-consuming ⑪(精华) flower: 文艺之～ flower of literature and art

【花白】grey; grizzled: 头发～ with grey hair; grey-haired
【花斑】piebald: ～马 a piebald horse
【花斑癣】[医]tinea versicolor
【花瓣】petal
【花苞】bud
【花被】[植] perianth; floral envelop
【花边】①(边缘上的花纹) decorative border: 瓶口上有一道～。There is a floral border round the mouth of the vase. ②(编织物) lace: ～装饰 lace trimmings / 镶～的衣服 dress trimmed with laces ③(印刷用语) fancy borders in printing
【花布】cotton print; print
【花草】flowers and plants ○百合花 lily / 长春藤 ivy / 翠菊 China aster / 倒挂金钟 fuchsia / 丁香 lilac; clove / 杜鹃花 azalea / 凤尾草 brake / 凤仙花 balsam / 含羞草 touch-me-not; sensitive plant / 旱金莲 bush nasturtium / 鸡冠花 cockscomb / 夹竹桃 oleander / 菊花 chrysanthemum; mum / 腊梅 wintersweet / 兰花 orchid / 铃兰 lily-of-the-valley / 玫瑰 rose / 美人蕉 canna; Indian shot / 茉莉 white jasmine / 牡丹 peony / 秋海棠 begonia / 三色堇 pansy / 山茶花 camellia / 芍药 herbaceous peony / 十样锦 gladiolus / 水仙 narcissus; daffodil / 昙花 epiphyllum / 晚香玉 tuberose / 勿忘我 forget-me-not / 西番莲 dahlia / 仙客来 cyclamen / 洋绣球 hydrangea / 一串红 scarlet sage / 郁金香 tulip / 虞美人 corn (field) poppy / 月季 Chinese rose / 栀子花 gardenia / 紫罗兰 violet / 紫茉莉 mervel of Peru; four o'clock
【花茶】scented tea: 茉莉～ jasmine tea
【花车】festooned vehicle
【花池子】flower bed
【花丛】flowering shrubs; flowers in clusters
【花大姐】[动] potato ladybird
【花旦】[戏] female role in Chinese opera
【花灯】festive lantern (as displayed on the Lantern Festival)
【花缎】figured satin; brocade
【花朵】flower: 儿童是祖国的～。Children are the flowers of our motherland.
【花萼】[植] calyx
【花房】greenhouse
【花费】①(用法) spend; expend; cost: ～时间 spend time; take time / ～心血 take pains ②(费用) money spent; expenditure; expenses
【花粉】[植] pollen ◇～管 pollen tube

【花岗岩】granite: ～脑袋 a granite-like skull; ossified thinking ◇ ～化 granitization
【花格墙】lattice wall
【花梗】pedicle; flower stalk
【花骨朵】(flower) bud
【花冠】corolla: 合瓣～ gamopetalous corolla / 离瓣～ choripetalous corolla
【花好月圆】blooming flowers and full moon; perfect conjugal bliss
【花红】①[植](沙果) Chinese pear-leaved crab apple ②[旧](红利) bonus ③(婚事喜庆礼物) gift for a wedding, etc.
【花红柳绿】bright red blossoms and green willows; a profusion of garden flowers
【花候】[植] flowering season
【花花公子】dandy; coxcomb; fop; playboy
【花花绿绿】brightly colored; colorful: 穿得～的 be colorfully dressed
【花花世界】[贬] the dazzling human world with its myriad temptations; this mortal world
【花环】garland; floral hoop
【花卉】①(花草) flowers and plants ②[美术] painting of flowers and plants in traditional Chinese style ◇ ～画 flower-and-plant painting
【花鸡】bramble finch; brambling
【花甲】a cycle of sixty years: 年逾～ over sixty years
【花剑】[体] foil
【花键】[机] spline ◇ ～轴 spline shaft / ～座 splined hub
【花匠】gardener
【花椒】Chinese prickly ash
【花轿】[旧] bridal sedan chair
【花秸】chopped straw
【花镜】presbyopic glasses
【花茎】rachis; floral axis
【花卷】steamed twisted roll
【花篮】①(献花用的) a basket of flowers ②(装饰美丽的篮子) gaily decorated basket
【花蕾】(flower) bud
【花里胡哨】①(过于鲜艳) gaudy; garish ②(华而不实) showy; without solid worth
【花脸】[戏] male character in Chinese opera with a painted face
【花鲢】variegated carp
【花柳病】venereal disease (V.D.)
【花露】(medicinal) liquid distilled from honeysuckle flowers or lotus leaves
【花露水】toilet water
【花蜜】[植] nectar
【花面狸】masked civet; gem-faced civet
【花苗】flower seedling
【花名册】register (of names); membership roster; muster roll: 造～ make a personnel roster
【花木】flowers and trees (in parks or gardens)
【花呢】fancy suiting
【花鸟】[美术] painting of flowers and birds in traditional Chinese style ◇ ～画 flower-and-bird painting
【花农】flower grower
【花盘】①[植] flower disc ②[机] disc chuck; faceplate
【花炮】fireworks and firecrackers
【花盆】flowerpot
【花瓶】flower vase; vase
【花圃】flower nursery
【花期】[植] florescence: ～已过 pass out of bloom
【花前月下】before the flowers and under the moon; ideal setting for a couple in love
【花枪】①(兵器) a short spear used in ancient times ②(花招) trickery: 要～ play tricks
【花腔】①(唱法) florid ornamentation in Chinese opera singing; *coloratura* ②(花言巧语) guileful talk: 要～ speak guilefully ◇ ～女高音 *coloratura* soprano; *coloratura*
【花墙】tracery wall
【花青】[化] cyanine ◇ ～染料 cyanine dyes / ～素 anthocyanidin
【花圈】(floral) wreath: 向革命烈士墓献～ place a wreath at the tomb of a revolutionary martyr
【花拳】flowery boxing; fancy boxing
【花蕊】[植] pistil (雌); stamen (雄)
【花色】①(花纹色彩) design and color: ～好看 be beautiful in design and color ②(品种) (of merchandise) variety of designs, sizes, colors, etc.: ～繁多 a great variety / ～品种 colors and patterns; range of goods / ～齐备 have a rich assortment of goods
【花纱布】a collective name for cotton, cotton yarn and cloth
【花哨】①(颜色鲜艳) garish; gaudy; showy: ～的衣服 showy dress ②(花样多) full of flourishes; flowery
【花生】peanut; groundnut ◇ ～饼[农] peanut cake / ～黑斑病 cercospora black spot of peanut / ～酱 peanut butter / ～壳 peanut shell / ～米 shelled peanut; peanut kernel / ～糖 peanut brittle / ～油 peanut oil
【花饰】ornamental design
【花鼠】Siberian chipmunk; chipmunk
【花束】a bunch of flowers; bouquet

【花丝】①[植] filament ②[工美] filigree ◇～工 filigree work
【花坛】(raised) flower bed; flower terrace
【花天酒地】indulge in dissipation; lead a life of debauchery
【花筒】cylindrical fireworks
【花头】①(新奇主意) fresh idea: 他的～真多。He is full of fresh ideas. ②(奥妙之处) knack: 这种游戏的～真不少。 There is quite a knack in the game.
【花团锦簇】bouquets of flowers and piles of silks — rich multicolored decorations
【花托】[植] receptacle
【花纹】decorative pattern; figure: 这些瓷盘的～很别致。These porcelain plates have rather original designs on them. ◇～玻璃 figured glass
【花线】①(彩色线) colored thread ②[电] flexible cord; flex
【花消】[口] cost; expense
【花序】[植] inflorescence
【花絮】titbits (of news); interesting sidelights: 运动会～ sidelights on the sports meet
【花押】cypher
【花芽】[植] (flower) bud
【花言巧语】sweet words; blandishments: ～的骗子 smooth-tongued humbug
【花眼】presbyopia
【花样】①(花纹样式) pattern; variety: ～翻新 the same old thing in a new guise / ～繁多 a great variety ②(花招) trick: 玩～ play tricks
【花样滑冰】[体] figure skating
【花药】[植] anther
【花椰菜】cauliflower
【花叶病】[农] mosaic (disease): 甜菜～ beet mosaic
【花容月貌】one's face is like flowers and one's features like the moon- said of a woman or maiden; fair as a flower and beautiful as the moon
【花园】flower garden; garden: ～屋顶 roof garden
【花帐】padded accounts or bills: 开～ make out a padded account; pad accounts
【花招】①(武术动作) showy movement in *wushu*: flourish ②(骗人的手段等) trick; game: 玩弄政治～ resort to political maneuver / 别耍～! None of your little tricks.
【花枝】[植] spray
【花枝招展】(of women) be gorgeously dressed
【花轴】[植] floral axis
【花柱】[植] style

哗 [象]大雨～～地下。The rain came down in torrents / 溪水～～地流。The stream went gurgling on.
【哗啦】[象]风吹得树叶～～地响。The leaves rustled in the wind. / 雨～～地下个不停。The rain kept pouring down.

huá

划 ①(拨水前进) paddle; row: ～船 paddle a boat; to boating ②(擦;割) scratch; cut the surface of: ～火柴 strike a match / 几道闪电～破长空。 Flashes of lightning streaked across the sky. / 她手～破了。Her hands were scratched. ③(合算) be to one's profit; pay: ～得来。It is worthwhile. (或: It pays.)
【划拳】finger-guessing game; a drinking game at feasts
【划算】①(盘算) calculate; weigh: ～来,～去 carefully weigh the pros and cons ②(合算) be to one's profit; pay: 做某事～。It is worthwhile doing sth.
【划艇】Canadian canoe; canoe; rowboat: 单人～ Canadian single / 双人～ Canadian pair
【划子】small rowboat

滑 ①(光滑) slippery; smooth: 路～。 The road is slippery. ②(滑动) slip; slide: ～了一跤 slip and fall / 在冰上～着走 slide on the ice ③(奸滑) cunning; crafty; slippery: 又奸又～ mean and crafty
【滑板】①[机] slide ②[乒乓球] feint play
【滑板运动】skateboarding
【滑冰】ice-skating; skating: 单人花样～ single skating; solo-skating / 花样～ figure skating / 双人花样～ pair skating; pairs / 速度～ speed skating
【滑冰场】skating rink
【滑冰航行】skage sailing
【滑冰鞋】skate: 花样～ figure skate
【滑冰运动员】skater: 花样～ figure skater
【滑车】[机] pulley; block ◇～组 block and tackle; pulley block
【滑车神经】[生理] trochlear nerve
【滑道】chute; slide
【滑动】[物] slide ◇～轴承[机] sliding bearing
【滑阀】[机] guiding valve; slide valve: ～冲程 slide valve stroke / ～装置 slide valve gear
【滑竿】a kind of litter
【滑稽】①(引人发笑) funny; amusing; comical;

～故事 funny story／ 滑天下之大稽 be the biggest joke in the world; be the object of universal ridicule ②[曲艺] comic talk ◇ ～戏 farce
【滑精】[中医] involuntary emission; spermatorrhea
【滑距】actual relative movement; slip; total displacement
【滑溜】①(烹调法) stir-fried with thick gravy; saute with starchy sauce: ～里脊 saute fillet with thick gravy. ②(光滑) smooth; slick: 这缎子摸起来很～。The satin feels smooth.
【滑轮】pulley; block ◇ ～组 pulley block
【滑脉】[中医] smooth pulse
【滑面】[机] sliding surface; slide face
【滑腻】(of the skin) satiny; velvety; creamy
【滑坡】[地] landslide; landslip
【滑润】smooth; well-lubricated
【滑升模板施工】slip-form construction technique
【滑石】talcum; talc ◇ ～粉 talcum powder
【滑水运动】water-skiing
【滑膛枪】smoothbore (gun); musket
【滑梯】(children's) slide
【滑头】①(不老实的人) slippery fellow; sly customer ②(油滑) slippery; shifty; slick: ～滑脑 crafty; artful, slick ③(不老实) foxy; cunning
【滑翔】glide
【滑翔机】glider; sailplane ◇ ～运动员 glider pilot
【滑翔运动】gliding; sailplaning
【滑行】slide; coast: ～下坡 coast down a slope／ 冰上～ slide on the ice
【滑行道灯】taxi-track lights
【滑行细菌】gliding bacteria
【滑雪】skiing: 高山～ alpine skiing／ 速度～ cross-country ski racing ◇ ～板 skis／ ～鞋 ski boots／ ～运动员 skier／ ～杖 ski pole
【滑音】①[语] glide ②[乐] *portamento*
【滑脂枪】[机] grease gun
【滑座】slide; slide carriage

猾 cunning; crafty; sly

华 ①(光辉灿烂) magnificent; splendid: ～屋 magnificent house ②(繁盛) prosperous; flourishing: 繁～ flourishing; bustling ③(精华) best part; cream: 精～ the cream; the best part ④(奢华) flashy; extravagant: 朴实无～ simple and unadorned／ 奢～ extravagant; luxurious ⑤(花白) grizzled; grey: ～发 grey hair ⑥[气] corona ⑦(中国) China: 驻～大使 ambassador to China ⑧(汉语) Chinese: 英～词典 an English-Chinese dictionary
【华表】ornamental columns erected in front of palaces, tombs, etc.
【华达呢】gabardine
【华灯】colorfully decorated lantern; light: ～初上 when the evening lights are lit
【华而不实】flashy and without substance; superficially clever
【华尔兹】waltz
【华盖】①[书] canopy (as over an imperial carriage) ②[气] aureole
【华贵】luxurious; sumptuous; costly: ～的地毯 luxurious carpet
【华丽】magnificent; resplendent; gorgeous: ～的词藻 flowery language／ ～的宫殿 a magnificent palace／ 服饰～ gorgeously dressed and richly ornamented
【华美】magnificent; resplendent; gorgeous
【华侨】overseas Chinese: 归国～ returned overseas Chinese ◇ ～投资公司 Overseas Chinese Investment Corporation／ ～委员会 Overseas Chinese Committee
【华沙条约】the Warsaw Treaty (1955) ◇ ～组织 the Warsaw Treaty Organization
【华氏温度】degree Fahrenheit; Fahrenheit
【华氏温度计】the Fahrenheit thermometer
【华夏】Hua Xia; an ancient name for China
【华夏系构造】[地] Cathaysian (structural) system
【华裔】foreign citizen of Chinese origin

哗 noise; clamor: 寂静无～ silent and still; very quiet
【哗变】mutiny
【哗然】in an uproar; in commotion: ～起哄 rising up in an uproar (of protest)／ 举座～。The audience burst into an uproar.
【哗笑】uproarious laughter
【哗众取宠】try to please the public with claptrap

铧 ploughshare: 双～犁 double-shared plough; double-furrow plough

huà

话 ①(表达思想的声音或文字) word; talk: 编者的～ editor's remark／ 留～ leave a message; leave word／ 说几句～ say a few words ②(说;谈) talk about; speak about: ～家常 chitchat; exchange small talk／ ～不投机半句多。When the conversation gets disagreeable,

to say one word more is a waste of breath.
【话本】script for story-telling (in Song and Yuan folk literature); text of a story
【话别】say a few parting words; say good-bye
【话柄】subject for ridicule; handle: 传为～ become the joke of the town
【话碴儿】[方] ①(话头) thread of discourse: 接上～ take up the thread of a conversation ②(口风) tone of one's speech: 听他的～,这件事好办。From what he says, that'll be easily done.
【话多不甜】Too much talk is unpleasant.
【话锋】thread of discourse; topic of conversation: 把～一转 switch the conversation to some other subject
【话旧】talk over old times; reminisce
【话剧】modern drama; stage play: 演出～ put on a play; perform a play ◇ ～团 modern drama troupe; theatrical company
【话里有话】the words mean more than they say; there's more to it than what is said
【话频】[讯] telephone frequency
【话题】subject of a talk; topic of conversation: 转～ change the subject
【话筒】①(发话筒) microphone; telephone transmitter ②(传声筒) megaphone
【话头】thread of discourse: 打断～ interrupt sb.; cut sb. short / 拾起～ take up the thread of a conversation
【话务员】(telephone) operator
【话匣子】[方]①(留声机) gramophone ②(收音机) radio receiving set ③(爱说话的人) chatterbox: 他打开～就没个完。Once he opens his mouth, he never stops.
【话音】①(说话的声音) one's voice in speech: ～儿未落 when one has hardly finished speaking ②[口](音外之意) tone; implication: 听他的～,他准是反对的。Judging from his tone, there's no doubt he is against it.
【话音识别】speech recognition
【话终信号】clearing signal

画 ①(绘画) draw; paint: ～素描 sketch; make a sketch / ～油画 paint in oil / 画水彩～ paint in water colors / 绘～ painting ②(成品画) drawing; painting; picture: 版～ engraving / 壁～ mural painting / 彩粉～ pastel / 插～ illustrations / 风景～ landscape painting / 钢笔～ pen drawing / 古典～ classical painting / 广告～ poster / 静物～ still life / 漫～ caricature / 木炭～ charcoal drawing / 年～ New Year picture / 铅笔～ pencil sketch / 石版～ lithograph / 水彩～ water color / 肖像～ portrait / 油～ oil painting / 中国国～ traditional Chinese painting / 中国水墨～ Chinese ink painting ◯ 木刻 woodcut / 速写 sketch / 自画像 self-portrait / 全身像 full-length portrait / 半身像 half-length portrait ③(用画装饰的) be decorated with paintings or pictures: ～栋雕梁 painted pillars and carved beams (of a magnificent building) ④(汉字的一笔) stroke (of a Chinese character): "人"字两～。The character "人" is made up of two strokes.
【画板】drawing board
【画报】illustrated magazine or newspaper; pictorial
【画笔】painting brush; brush
【画饼充饥】draw cakes to allay hunger; feed on illusions
【画布】canvas (for painting)
【画册】an album of paintings; picture album
【画地为牢】draw a circle on the ground to serve as a prison; restrict sb.'s activities to a designated area or sphere
【画点测验】[心理] dotting test
【画道】[摄] picture track
【画法】technique of painting or drawing: ～新颖 a novel technique in painting or drawing
【画舫】gaily-painted pleasure-boat
【画幅】①(图画总称) picture; painting ②(画的尺寸) size of a picture
【画稿】rough sketch (for a painting)
【画格】[电影] frame
【画虎类狗】try to draw a tiger and end up with the likeness of a dog; make a poor imitation
【画家】painter; artist
【画架】easel
【画匠】①(绘画的工匠) artisan-painter ②[旧](缺乏艺术性的画家) inferior painter
【画境】picturesque scene: 如入～ feel as though one were in a landscape painting
【画具】painter's paraphernalia
【画卷】picture scroll; scroll painting
【画绢】silk for drawing on; drawing silk
【画刊】①(报刊上的画报栏) pictorial section of a newspaper ②(画报) pictorial
【画廊】①(有图画的走廊) painted corridor ②(画展走廊)(picture) gallery
【画龙点睛】①(在画面的关键地方点明要点,使更生动) bring the painted dragon to life by putting in the pupils of its eyes; add the touch

that brings a work of art to life; add the finishing touch ②(在关键地方加一两句重要的话)add a word or two to clinch the point
【画眉】[动] babbler; chatterer
【画面】①(画幅上呈现的形象) general appearance of a picture; tableau ②[电影] frame
【画面空间】[摄] image spacing
【画皮】disguise or mask of an evildoer: 剥～ rip off sb.'s mask
【画片】a miniature reproduction of a painting
【画屏】[工美] painted screen
【画谱】a book on the art of drawing or painting
【画蛇添足】draw a snake and add feet to it; ruin the effect by adding sth. superfluous
【画师】painter
【画十字】cross; make the sign of the cross: 在胸口～ cross oneself
【画室】studio
【画帖】a book of model paintings or drawings
【画图】①(画各种图形) draw designs, maps, etc. ②(图画) picture
【画外音】[电影] offscreen voice
【画像】①(画人像) draw a portrait; portray: 让人～ sit for one's portrait ②(画成的人像) portrait; portrayal: 巨幅～ huge portrait
【画押】make one's cross sign: 在契约上～ make one's cross on a contract
【画页】page with illustrations (in a book or magazine); plate
【画展】art exhibition; exhibition of paintings
【画轴】painted scroll; scroll painting

划 ①(划分) delimit; differentiate: ～界 delimit a boundary ②(划拨) transfer; assign: ～款 transfer money ③(计划) plan: 筹～ plan and prepare ④(划线作标记) draw; mark; delineate: ～出课文要点 mark the main points in the text / ～线 draw a line ⑤(笔划) stroke (of a Chinese character)
【划变】[摄] wipe; wipe-off
【划拨】①(从一单位或户头转入另一单位或户头) appropriate; transfer: ～一笔款子 appropriate a sum ②(分出来拨给) assign; allot: ～一间屋子作阅览室 assign a room for reading
【划地绝交】sever friendship by making a mark on the ground
【划定】delimit, designate: ～边界 delimit a boundary line / 在～的区域内捕鱼 fish in the designated areas
【划分】①(分成部分) divide: ～势力范围 carve out spheres of influence / ～行政区域 divide a country into administrative areas ②(区别) differentiate
【划归】put under (sb.'s administration, etc.); incorporate into: 这个企业已～地方管理。The enterprise has been put under local administration.
【划痕器】scratcher
【划框框】set limits; place restrictions
【划清】draw a clear line of demarcation; make a clear distinction: ～界限 draw a demarcation line with sb.; draw clear distinction / ～是非界限 make a clear distinction between right and wrong
【划时代】epoch- making: ～的大事 epoch-making event / 具有～的意义 have epoch-making significance
【划一】standardized; uniform: 尺寸～ be of uniform size / 整齐～ uniform
【划一不二】①(不二价) fixed; unalterable; rigid: 价钱～ fixed price (not subject to bargaining) ②(一律;刻板) uniform; stereotyped

化 ①(变化) change; turn; transform: ～悲痛为力量 transform grief into strength / ～害为利 turn harm into good; turn a disadvantage into an advantage ②(教化) convert; influence: 潜移默～ exert a subtle influence on sb.'s character, thinking, etc. ③(融化) melt; dissolve: 雪～了。The snow has melted. ④(消化) digest: ～食 help digestion ⑤(火化) burn up: 焚～ burn up; incinerate / 火～ cremate ⑥[简](化学) chemistry ⑦(表示转变成某种性质或状态) -ize; -ify: 电气～ electrify / 工业～ industrialize / 简～ simplify / 现代～ modernize ⑧(向人求布施) beg alms: ～斋 beg a (vegetarian) meal ⑨[宗](死) die: 坐～ pass away in a sitting posture
【化肥】[简] chemical fertilizer
【化粪池】septic tank
【化废为宝】make a silk purse out of a sow's ear; turn wastes into useful thing
【化干戈为玉帛】bury the hatchet; turn swords into plowshares; turn hostility into friendship
【化工】[简] chemical industry ◇～产品 chemical products / ～厂 chemical plant / ～原料 industrial chemicals
【化合】[化] chemical combination: 原子的～与分解 combination and dissociation of atoms ◇～反应 combination reaction / ～价 valence

【化合物】chemical compound: 高分子～ high polymer; highly polymerized compound
【化剑为犁】turn swords into plowshares
【化境】sublimity; perfection: 这幅山水画已臻～。This landscape painting is a consummate work of art.
【化名】(use an) assumed name; alias
【化脓】fester; suppurate: 伤口～了。The wound is festering.
【化身】incarnation; embodiment: 智慧和勇敢的～ the embodiment of wisdom and courage
【化石】fossil: 变成～ become fossilized / 标准～ index fossil / 微体～ microfossil / 整理～ dress fossils / 指相～ facies fossil ◇ ～作用 fossilization
【化痰】reduce phlegm
【化痰生津】act as expectorant
【化铁炉】[冶] cupola furnace
【化为灰烬】turn to dust and ashes; reduced to ashes
【化为泡影】vanish like soap bubbles; melt into thin air
【化为乌有】melt into thin air; vanish; come to naught
【化纤】[简] chemical fiber
【化险为夷】turn danger into safety; head off a disaster
【化凶为吉】change the portentous into the propitious
【化学】chemistry: 放射～ radiochemistry / 分析～ analytical chemistry / 理论～ theoretical chemistry / 生物～ biochemistry / 无机～ inorganic chemistry / 物理～ physical chemistry / 应用～ applied chemistry / 有机～ organic chemistry ◇ ～变化 chemical change / ～成分 chemical composition / ～当量 chemical equivalent / ～反应 chemical reaction / ～方程式 chemical equation / ～分析 chemical analysis / ～符号 chemical symbol / ～工程 chemical engineering / ～合成 chemical synthesis; chemosynthesis / ～化 [农] extensive use of chemical fertilizers and other farm chemicals / ～鉴定 chemical analysis / ～疗法 chemotherapy / ～亲合力 chemical affinity / ～式 chemical reaction / ～试剂· chemical reagent / ～污染 chemical pollution / ～武器 chemical weapons / ～性质 chemical property / ～需氧量[环保] chemical oxygen demand / ～药品 chemicais / ～元素 chemical element / ～战争 chemical warfare / ～作用 chemical action
【化学性损伤】chemical lesion
【化验】chemical examination; laboratory test: 血常规～ ordinary blood test ◇ ～报告 analysis report / ～单 laboratory test report / ～室 laboratory / ～员 laboratory technician / ～证明书 analysis certificate; certificate of analysis
【化油器】[机] carburettor
【化缘】[宗] beg alms
【化整为零】break up the whole into parts
【化妆】put on makeup; make up
【化妆品】cosmetics: ～工业 cosmetics industry
【化装】①(装扮角色) (of actors) make up ②(假扮) disguise oneself; ～侦察 go reconnoitering in disguise ◇ ～品 makeup / ～师 makeup man / ～室 dressing room

桦 [植]birch: ～焦油 birch tar oil
【桦木】birch
【桦木酸】betulinic acid

huái

怀 ①(胸部或胸前) bosom: ～里的孩子 baby in one's arms ②(胸怀) mind: 襟怀坦白 frank and open-minded ②(心里存有) keep in mind; cherish: ～有恶意 harbor malice / ～有希望 cherish hopes ④(怀念) think of; yearn for: ～乡 yearn for one's native place; be homesick / ～友 think of a friend ⑤(腹中有胎) conceive (a child): ～了孩子 become pregnant; be with child
【怀抱】①(抱在怀里) bosom: 回到祖国的～ return to the embrace of one's homeland ②(心里存有) cherish: ～远大的理想 cherish lofty ideals
【怀璧其罪】The precious stone lands its innocent possessor in jail. (或: An innocent man gets into trouble because of his wealth.)
【怀表】pocket watch
【怀才不露】sheathe one's talent; be modest about one's talent
【怀才不遇】have unrecognized talents; have talent but no opportunity to use it
【怀古】meditate on the past; reflect on an ancient event
【怀鬼胎】have a bad conscience; with ulterior motive
【怀恨】nurse a hatred (for); harbor resentment; bear a grudge
【怀恨在心】harbor resentment in one's bosom; have resentment rankling in one's mind

【怀旧】remember past times or old acquaintances
【怀恋】think fondly of (past times, old friends, etc.); look back nostalgically
【怀念】cherish the memory of; think of: ～故乡 yearn for one's native land / ～远方的友人 think of an absent friend who is far away
【怀柔】(of feudal rulers) make a show of conciliation in order to bring other nationalities or states undercontrol ◇ ～政策 policy of control through conciliation; policy of mollification
【怀胎】be pregnant
【怀乡病】[心理] nostopathy
【怀想】think about with affection (a faraway person, place, etc.); yearn for
【怀疑】distrust; doubt; suspect: 持～态度 take a sceptical attitude / 引起～ raise doubts; arouse suspicion ◇ ～论[哲] scepticism
【怀孕】conception; in the family way; be pregnant ◇ ～期 period of pregnancy; gestation period

槐 pagoda tree; sophora
【槐角】[中药] the pod of pagoda tree

踝 ankle: ～骨 anklebone
【踝宽】ankle breadth
【踝最小厚】ankle thickness

huài

坏 ①(不好) bad: ～习惯 bad habit / ～透了 downright bad; rotten to the core / 越来越～ from bad to worse ②(变坏;有害) go bad; spoil; ruin: 机器～了。 The machinery broke down. / 鱼～了。The fish has gone bad. ③(程度深) badly; awfully; very: 乐～了 be wild with joy / 累～了 be very tired / 气～了 be beside oneself with rage / 吓～了 be badly scared ④(坏主意) evil idea; dirty trick: 使～ play a dirty trick / 一肚子～ full of tricks
【坏处】harm; disadvantage: 这对你没～。 It will do you no harm.
【坏蛋】[口] bad egg; scoundrel; bastard
【坏东西】bastard; scoundrel; rogue
【坏分子】[法] bad element; evildoer
【坏话】①(恶意的话) malicious remarks; vicious talk: 讲别人～ speak ill of others ②(不好听的话) unpleasant words: 好话～都要让人讲完。 One should let others finish what they have to say whether it sounds pleasant or unpleasant.
【坏疽】[医] gangrene
【坏人】bad person; evildoer; scoundrel
【坏人当道，好人受气】Bad elements hold sway while good people are pushed around. (或: The bad eggs wielded power while the good people were oppressed.)
【坏人坏事】evil doers and evil deeds
【坏事】①(不好的事) bad thing; evil deed: ～可以变成好事。A bad thing can be turned into a good one. ②(把事搞坏) ruin sth.; make things worse: ～了！Something terrible has happened. / 急躁只能～。Impetuosity will only make things worse.
【坏事做绝】commit terrible crimes without end; stop at no evil; stop at nothing in committing all kinds of evil
【坏死】[医] necrosis: 局部～ local necrosis / 牙～ dental necrosis
【坏心眼儿】[口] evil intention; ill will
【坏血病】scurvy; scorbutus
【坏字】[印] batter

huān

獾 [动] badger
【獾油】[药] badger fat (for treating burns)

欢 ①(快乐;高兴) joyous; merry; jubilant: ～唱 sing merrily / ～跃 jump for joy ②[方](起劲;活跃) vigorously; with great drive; in full swing: 争辩得正～ be hot over an argument / 篝火烧得正～。The bonfire is blazing.
【欢蹦乱跳】healthy-looking and vivacious: 一个～的青年人 an exuberant youngster
【欢畅】thoroughly delighted; elated
【欢度】spend (an occasion) joyfully: ～佳节 celebrate a festival with jubilation
【欢呼】hail; cheer; acclaim: ～声 shouts of joy; cheers / ～伟大的胜利 hail the great victory
【欢聚】happy get-together; happy reunion: ～一堂 happily gather under the same roof
【欢快】cheerful and light-hearted; lively: ～的曲调 a lively melody / ～的音乐 merry music
【欢乐】happy; joyous; gay: ～的歌声 merry songs / ～的景象 a scene of great joy / ～的人群 happy crowds
【欢乐主义】[心理] hedonism
【欢庆】celebrate joyously
【欢声雷动】cheers resound like rolls of thunder: 全场～。 The audience broke into deafening cheers.
【欢送】see off; send off: ～客人 see the guests off cordially ◇ ～会 farewell meeting; send-off meeting / ～仪式 seeing-off ceremony

【欢腾】great rejoicing; jubilation: 全市一片~。The whole city was a scene of jubilation
【欢天喜地】with boundless joy; wild with joy; overjoyed: 听到这好消息,大家都~。All were extremely delighted at the good news.
【欢喜】①(快乐) joyful; happy; delighted: 满心~ be filled with joy ②(喜欢) like; fond of; delight in: ~游泳 like swimming; be fond of swimming
【欢笑】laugh heartily
【欢心】favor; liking; love: 想博取~ try to win sb.'s favor
【欢欣鼓舞】be filled with exultation; be elated: 为我们的成就而~ rejoice over our achievements
【欢迎】①(高兴地迎接) welcome; greet: 夹道~ line the streets to give sb. a welcome / 到机场~贵宾 meet distinguished guests at the airport ②(乐意接受) welcome; favorably receive: ~订购。 Orders are welcomed. / 受到观众~ be warmly received by the audience ◇ ~词 welcoming speech; address of welcome / ~会 a party to welcome sb.

huán

还 ①(回) go back; come back: ~家 return home ②(归还) give back; return; repay: ~东西 return borrowed articles / ~清贷款 pay off one's loan / 到期该~的书 book due for return ③(回报别人对自己的行动) give or do sth. in return: ~手 return a blow; strike back
【还本】repayment of principal
【还本付息】repay capital with interest
【还愿】[佛教] votive; redeem a vow to a god
【还魂】①(死而复生) revive after death; return from the grave ②[方](再生) reprocessed ◇ ~纸 reprocessed paper
【还击】①(回击) fight back; return fire; counterattack: 进行自卫~ fight in self-defense ②[击剑] riposte
【还价】counter-offer; counter-bid; abate a price
【还款信用状】reimbursement letter of credit
【还礼】①(回答敬礼) return a salute ②(回赠礼物) present a gift in return
【还清】pay off: ~债务 pay off one's debts
【还手】strike back
【还俗】(of Buddhist monks and nuns or Taoist priests) resume secular life
【还我河山】restore our lost territories
【还乡】return to one's native place
【还乡团】landlords' restitution corps; home-going legion
【还原】①(恢复原状) return to the original condition or shape; restore ②[化] reduction ◇ ~剂 reducing agent; reductant / ~酶 reductase
【还愿】①(酬神) redeem a vow to a god ②(实践诺言) fulfill one's promise: 既然许愿就要~。If you make a promise, you should carry it out. (或: Promise is debt.)
【还债】pay one's debt; repay a debt: 借债容易~难。It is much easier to run into debt than to get out of it.
【还帐】pay bills (on credit sales); settle account on credit sales
【还政于民】hand state power back to the people; return the power of government to the hands of the people
【还嘴】[口] answer back; retort

环 ①(环子) ring; hoop: 耳~ earring ②(环节) link: 一~套一~ all linked with one another; wheels within wheels / 最薄弱的一~ the weakest link ③(围绕) surround; encircle; hem in: 四面~山 be surrounded by mountains ④[体] ring: 命中九~ hit the nine-point ring
【环靶】[体] round target
【环抱】surround; encircle; hem in: 群山~的村庄 a village nestling among the hills
【环比指数】chain index
【环衬】lining papers
【环城】around the city ◇ ~马路 round-the-city highway / ~赛跑 round-the-city race
【环带】[动] clitellum
【环堵萧然】in a cold, bare room
【环顾】look about: ~四周 look all round
【环礁】[地] atoll
【环节】①(许多事物中的一个) link: 生产~ links in the production chain / 主要~ a key link ②[动] segment ◇ ~动物 annelid
【环颈雉】[动] ring- necked pheasant
【环境】environment; surroundings; circumstances: ~顺利 under favorable circumstances / 换换~ have a change of environment / 自然~ natural environment ◇ ~保护 environmental protection / ~保护法 environment law; law of environment protection / ~改良 environmental improvement / ~监测系统 environmental monitoring system / ~科学 environmental science / ~空气 ambient air / ~卫生 environmental sanitation; general sanitation / ~温度 ambient humidity / ~污染 pollution of the environment / ~噪音 ambient noise / ~

证据 circumstantial evidence
【环流】[气] circulation: 大气～ atmospheric circulation ◇ ～型 [气] circulation pattern
【环球】①(围绕地球) round the world: ～航行 make a voyage around the world / ～旅行 travel round the world; a round-the- world tour ②(整个地球) the earth; the whole world
【环绕】surround; encircle; revolve around: 地球～太阳运行。 The earth revolves round the sun.
【环蛇】krait
【环食】[天]annular eclipse (of the sun)
【环视】look around
【环烃】[化] cyclic hydrocarbon
【环烷】[化] cycloalkanes; cycloparaffin; naphthene
【环行】going in a ring: ～一周 make a circuit ◇ ～公共汽车 bus with a circular route / ～公路 ring road; belt highway / ～铁路 circuit railway; belt line
【环形】 annular; ringlike ◇ ～癌 annular carcinoma / ～联合 circular integration / ～山 [天] ring structure; lunar crater
【环氧树脂】[化] epoxy resin
【环子】ring; link: 门～ knocker

寰 extensive region: 人～ the world of man
【寰球】the earth; the whole world

huǎn

缓 ①(慢;迟) slow; unhurried: ～步而行 walk slowly / ～流 flow slowly / ②(延缓;推迟) delay; postpone; put off: ～办 postpone doing sth. / ～口气 have a respite ③(缓和;不紧张) not tense; relaxed ④(恢复) recuperate; revive; come to: 她不久便～过来了。 Soon she got over.
【缓兵之计】stratagem to gain a respite; stalling tactics
【缓不济急】slow action cannot save a critical situation
【缓冲】①(使冲突缓和) buffer; cushion: 弹性～ elastic buffer ②[化] buffer ◇ ～地带 buffer zone / ～国 buffer state / ～剂[化] buffer / ～器 buffer; bumper / ～作用 cushioning effect
【缓和】①(变和缓) relax; ease up; mitigate; alleviate: ～紧张局势 relax the tension / 风势渐趋～。The wind is subsiding. ②(使和缓) détente ◇ ～剂 [物] moderator
【缓急】①(和缓与急迫) pressing or otherwise; of greater or lesser urgency: 分别轻重～ do things in order of importance and urgency ②(急迫的事) emergency: ～相助 give mutual help in an emergency; help other in case of need
【缓解】lysis
【缓慢】slow: 进展～ make slow progress / 行动～ slow in action; slowmoving
【缓坡】gentle slope
【缓期】postpone a deadline; suspend: ～付款 delay payment ◇ ～执行 [法] respite; with suspended execution of sentence
【缓气】get a breathing space; have a respite; take a breather
【缓射】[军] slow fire
【缓刑】[法] probation; imprisonment with a suspension of sentence; suspension of sentence: ～二年 two years' probation ◇ ～判决 [法] probationary sentence / ～期间 probationary period / ～释放 release on probation / ～制度 system of probationary suspension
【缓役】[军] deferment (of service)
【缓征】postpone the imposition of a tax or levy

huàn

浣 ①(洗) wash: ～衣 wash clothes ②(旬) any of the three ten-day divisions of a month: 上～ the first ten days of a month
【浣熊】racoon

宦 ①(官吏) official ②(宦官) eunuch
【宦官】eunuch
【宦海】[旧] officialdom; official circles
【宦海浮沉】vicissitudes of official life; floating and sinking in the official seas
【宦途】[旧] official career

豢
【豢养】feed; groom; keep: ～家禽 feed poultry

患 ①(祸患) trouble; peril; disaster: 防～于未然 take preventive measures; provide against possible trouble / 有备无～。 Preparedness averts peril. ②(忧虑) anxiety; worry: 何～之有。 There's no need to worry. ③(害病) contract; suffer from: ～肝炎 contract hepatitis
【患病】suffer from an illness; fall ill; be ill
【患处】affected part (of a patient's body)
【患得患失】worry about personal gains and losses; be swayed by considerations of gain and loss
【患难】trials and tribulations; adversity; trouble: ～之交 friend in adversity; tested friend
【患难见人心】Calamity is the touchstone of man.

【患难相弃】leave somebody in the lurch; fail somebody in his need
【患难相助】help someone in distress; help people in trouble
【患难与共】go through thick and thin together
【患者】sufferer; patient: 结核病～ a person suffering from tuberculosis; a TB patient / 精神病～ neuropathic
【患至呼天】cry for heaven when calamity occurs; too late

涣 melt; vanish
【涣然】melt away; disappear; vanish
【涣然冰释】melt away; thaw out; clear up: 我们之间的误会～。 The misunderstanding between us has been cleared up.
【涣散】lax; slack: ～士气 breaking up of morale; lose morale / 纪律～ be lax in discipline

焕 shining; glowing
【焕发】shine; glow; irradiate: ～革命青春 fresh with revolutionary vigor / 容光～ one's face glowing with health
【焕然一新】take on an entirely new look; look brand-new

换 ①(交换) exchange; barter; trade: 用鲜血～来的教训 a lesson paid for in blood / 用一匹马～一头牛 barter a horse (with sb.) for a cow ②(变换) change: ～衣服 change one's clothes / 旧貌～新颜。 New scenes replace the old.
【换班】①(替换) change shifts ②(交接班) relieve a person on duty ③(军) changing of the guard
【换边】[体] change side
【换步】[军] change step
【换茬】[农] change of crops
【换车】change trains or buses
【换挡】[机] shift gears
【换发球】[体] change of service
【换防】[军] relieve a garrison
【换俘】[军] exchange of prisoners ◇ ～协定 agreement for exchange of prisoners
【换岗】relieve a sentry; relieve a guard
【换工】exchange labor
【换货】exchange goods; barter ◇ ～和付款协定 goods exchange and payments agreement / ～协定 barter agreement
【换机放映】[电影] changeover
【换季】change garments according to the season; wear different clothes for a new season
【换能器】[物] transducer
【换片装置】[唱机] record changer
【换气】take a breath (in swimming)
【换钱】①(钱兑换钱) change money ②(卖) sell: 这些废品可以～。 All these waste materials can sell for money.
【换取】exchange sth. for; get in return: ～外汇 gain foreign exchange
【换人】[体] substitution (of players)
【换算】conversion ◇ ～表 conversion table
【换汤不换药】the same medicine differently prepared; the same old stuff with a different label; a change in form but not in content
【换位】[体] change of positions
【换文】change of notes: 就发展贸易关系～ exchange notes on the development of trade
【换席更酌】Wine was again served.
【换向】commutation; inverting; reversing
【换牙】(of a child) grow permanent teeth
【换羽】moulting
【换约】exchange treaties
【换装站】[铁道] transshipment station

唤 call out: 呼～ call; shout
【唤起】①(号召使奋起) arouse: ～民众 arouse the masses of the people ②(引起) call; recall: ～对往事的回忆 evoke past memories
【唤醒】wake up; awaken: ～人民 arouse the people

幻 ①(不真实) unreal; imaginary; illusory: 虚～ unreal; illusory; visionary ②(奇异地变化) magical; changeable: 变～ change irregularly; fluctuate
【幻灯】① slide show: 放～ show slides / 看～ watch a slide show ②(幻灯机) slide projector ◇ ～机 slide projector; epidiascope / ～片 (lantern) slide
【幻景】illusion; mirage
【幻境】dreamland; fairyland
【幻觉】[心] hallucination: ～症 hallucinosis
【幻梦】illusion; dream
【幻灭】vanish into thin air: 她的梦想不久就～了。 He was soon disillusioned.
【幻日】mock sun; parhelion; sun dog
【幻视】[医] photism
【幻术】magic; conjuring
【幻听】[医] phonism
【幻想】illusion; fancy; fantasy; 抱有～ cherish illusions / 丢掉～ cast away illusions / 沉湎于～ indulge in fantasy; be lost in reverie ◇ ～曲

[乐] fantasia
【幻像】mirage; phantom; phantasm
【幻影】unreal image
【幻肢】phantom limb

huāng

荒 ①(荒芜) waste: 地～了, The land lies waste. ②(荒地) wasteland; uncultivated land: 垦～ open up wasteland ③(荒凉) desolate; barren: ～村 deserted village ④(荒歉) famine; crop failure: 储粮备～ store up grain against natural disasters ⑤(荒疏) neglect; be out of practice: 别把功课～了。Don't neglect your lessons. ⑥(缺乏) shortage; scarcity; 房～ housing shortage / 水～ water shortage
【荒诞】fantastic; absurd; incredible: ～的故事 an incredible story / ～的想法 a fantastic idea
【荒诞不经】be a wild legend; unbelievable; preposterous; fantastic
【荒地】wasteland; uncultivated land: ～垦拓 reclamation; soil reclamation
【荒废】①(不耕种) leave uncultivated; lie waste: 我们村没有一亩～的土地。Not a single hectare of land is left uncultivated in our village. ②(浪费;不利用) fall into disuse: 这条渠～了。This irrigation canal has fallen into disuse. ③(荒疏) neglect; be out of practice: ～学业 neglect one's studies
【荒郊】desolate place outside a town; wilderness
【荒凉】bleak and desolate; wild: ～的景色 wild scenery / 一片～ a scene of desolation
【荒凉寥落】a lonely and pathetic sight
【荒乱】in great disorder; in turmoil
【荒谬】absurd; preposterous: ～绝伦 absolutely preposterous; utterly absurd
【荒漠】desert ◇ ～结皮 patina / ～土 desert soil
【荒年】famine year
【荒僻】desolate and out-of-the-way
【荒歉】crop failure; famine
【荒时暴月】time of dearth; lean year; hard times
【荒山】barren mountain: ～秃岭 bare hills and mountains / 绿化～ clothe barren mountains with greenery
【荒疏】out of practice; rusty
【荒唐】①(荒谬) absurd; fantastic; preposterous: ～可笑 ridiculous; absurd / ～透顶 absolutely ridiculous / ～无稽 frivolous and unfounded; mystic ②(放荡) dissipated; loose; intemperate: ～的生活 a dissipated life
【荒无人烟】desolate and uninhabited: ～的地带 a region with no sign of human habitation
【荒芜】lie waste; go out of cultivation
【荒野】wilderness; the wilds
【荒淫】dissolute; licentious; debauched: ～无耻 dissipated and unashamed / ～无度 be vicious beyond measure
【荒原】wasteland; wilderness

慌 ①(慌张) flurried; flustered; confused: ～了手脚 be flustered ②[口](表示难以忍受) awfully; unbearably: 饿得～ be awfully hungry / 心里闷得～ be bored beyond endurance
【慌不择路】He fled in any path he could without heeding which he chose.
【慌里慌张】in a hurried and confused manner; all in a fluster; lose one's head
【慌乱】flurried; alarmed and bewildered
【慌忙】in a great rush; in a flurry; hurriedly: ～赶到现场 rush to the spot / ～离去 leave hurriedly
【慌张】flurried; flustered; confused: 神色～ look flurried / 不要～! 没有危险。Don't panic! There is no danger.
【慌张步态】festinating gait; festination; propulsion
【慌作一团】be struck all of a heap; be thrown into utter confusion

huáng

黄 ①(颜色) yellow; sallow: 脸色发～ a sallow face ②[口](事情中断或失败) fizzle out; fall through: 那笔买卖～了。The deal is off.
【黄斑】[生理] macula lutea
【黄檗】①[植] cork tree ②[中药] the bark of a cork tree
【黄灿灿】bright yellow; golden: ～的稻子 golden rice
【黄刺玫】yellow rose
【黄丹】yellow lead
【黄疸】jaundice: 肝原性～ hepatogenous jaundice / 阻塞性～ obstructive jaundice ◇ ～指数 icterus index
【黄道】ecliptic ◇ ～带 zodiac / ～光 zodiacal light / ～十二宫 the 12 signs of the zodiac; zodiacal signs / ～星座 zodiacal constellation / ～座标 ecliptic coordinates
【黄道吉日】propitious date; lucky day
【黄澄澄】glistening yellow; golden: ～的麦穗 golden ears of wheat
【黄鲷】yellow porgy

【黄豆】soya bean; soybean ◇ ～芽 soybean sprouts
【黄蜂】wasp
【黄姑鱼】spotted maigre
【黄瓜】cucumber: 酸～ pickled cucumber
【黄海】the Huanghai Sea; the Yellow Sea
【黄河】the Huanghe River; the Yellow River ◇ ～流域 the Huanghe valley / ～三角洲 the Huanghe delta
【黄褐色】yellowish-brown; tawny
【黄花】①(菊花) chrysanthemum ②(黄花菜) day lily
【黄花菜】day lily; citron daylily
【黄花闺女】an untouched virgin; bread-and-butter miss; be still a virgin
【黄花苜蓿】(California) bur clover
【黄花晚节】maintain one's integrity in one's later life
【黄花鱼】yellow croaker
【黄昏】dusk
【黄酱】salted and fermented soya paste
【黄金】gold ◇ ～分割[数] golden section / ～官价 official gold price / ～价格 price of gold; gold rate / ～时代 the golden age, the most flourishing period of a man or nation; the best years of one's life / ～市场 gold market / ～外流 gold outflow; gold drain / ～总库 gold pool
【黄金储备】gold reserve: 可动用的～ "free" gold reserve
【黄精】[植] solomon's seal
【黄荆】five-leaved chaste tree
【黄晶】citrine
【黄酒】yellow rice or millet wine
【黄口孺子】a suckling babe; babes and sucklings; a callow
【黄鹂】oriole: 黑枕～ black naped oriole
【黄历】almanac
【黄连】the rhizome of Chinese goldthread
【黄连木】Chinese pistache
【黄粱美梦】Golden Millet Dream; fond dream; pipe dream
【黄磷】yellow phosphorus
【黄栌】smoke tree
【黄麻】(roundpod) jute ◇ ～袋 gunnysack; gunny-bag; gunny / ～袋布 gunny (cloth)
【黄毛丫头】a chit of a girl; a silly little girl
【黄梅季】the rainy season, usu. in April and May, in the middle and lower reaches of the Changjiang River
【黄梅雨】intermittent drizzles in the rainy season in the middle and lower reaches of the Changjiang River
【黄米】glutinous millet
【黄鸟】oriole
【黄牛】ox; cattle
【黄袍加身】be draped with the imperial yellow robe by one's supporters; be acclaimed emperor
【黄芪】[中药] the root of membranous mile vetch
【黄芩】[中药] the root of large-flowered skullcap
【黄泉】netherworld
【黄雀】siskin
【黄壤】yellow earth; yellow soil
【黄热病】yellow fever
【黄色】①(颜色) yellow ②(腐化、色情) decadent; obscene; pornographic ◇ ～电影 pornographic movie; sex film; blue film / ～工会 yellow union; scab union / ～书刊 pornographic books and periodicals / ～文学 pornography / ～小说 pornographic novel / ～新闻 yellow journalism / ～音乐 decadent music
【黄色炸药】trinitrotoluene (TNT); trinol
【黄色人种】the yellow race
【黄杉】Douglas fir
【黄鳝】ricefield eel; finless eel
【黄熟】yellow maturity
【黄鼠】ground squirrel; suslik
【黄鼠狼】yellow weasel: ～给鸡拜年没安好心. The weasel goes to pay his respects to the hen — not with the best of intentions.
【黄水疮】[医] impetigo
【黄体】[生理] corpus luteum ◇ ～酮[药] progesterone
【黄铁矿】pyrite
【黄铜】brass ◇ ～管 brass pipe / ～矿 chalcopyrite / ～丝 brass wire / ～制品 brass work
【黄土】[地] loess ◇ ～高原 loess plateau
【黄萎病】[农] verticillium wilt. 棉～ verticillium wilt of cotton

【黄癣】[医] favus
【黄羊】Mongolian gazelle
【黄杨】[植] Chinese littleleaf box
【黄莺】oriole
【黄油】①(奶油) butter ②[化] grease ◇～枪 grease gun
【黄鼬】yellow weasel
【黄鱼】yellow croaker
【黄玉】[矿] topaz
【黄种】the yellow race

磺 sulphur
【磺胺】[药]sulphanilamide (SN): sulfanilamite: 长效～ sulfamethoxypyridazine (SMP) ◇～醋酰 sulphacetamide (SA)／～胍 sulphaguanidine (SG)／～嘧啶 sulphadiazine (SD)／～噻唑 sulphathiazole (ST)
【磺化】sulphonating ◇～剂 sulphonating agent
【磺酸】sulfonic acid; mabogany acid
【磺酸盐】[化] sulphonate
【磺酰胺】[化] sulphonic acid amide

簧 ①[乐] reed ②(有弹力的机件) spring: 闹钟的～断了。The main spring of the alarm clock is broken.
【簧风琴】reed organ; harmonium
【簧片】[乐]reed
【簧乐器】reed instrument

皇 emperor; sovereign: 女～ empress
【皇带鱼】oarfish
【皇帝】emperor: ～也有草鞋亲。 Even the emperor has his poor relatives.
【皇宫】(imperial) palace
【皇冠】imperial crown
【皇后】empress
【皇家】imperial family
【皇历】[旧] almanac
【皇权】imperial power
【皇亲国戚】a kinsman of the emperor; princes and princesses of the royal family
【皇上】①(在位的皇帝) the emperor; the throne; the reigning sovereign ②(直接称呼) Your Majesty; (间接称呼) His Majesty
【皇室】imperial family: ～贵胄 a scion of royalty
【皇太后】empress dowager
【皇太子】crown prince
【皇天后土】Heaven and Earth
【皇族】people of imperial lineage; imperial kinsmen

惶 fear; anxiety; trepidation: ～悚 sudden fear; fright
【惶惶】in a state of anxiety; on tenterhooks; alarmed: 人心～～。People are disquieted (on tenterhooks).
【惶惶不安】be on tenterhooks; be greatly upset; live in terror and uncertainty
【惶惶不可终日】be on tenterhooks; be in a constant state of anxiety
【惶惶无主】be panicky and not know what to do
【惶惑】perplexed and alarmed; apprehensive: ～不安 perplexed and uneasy
【惶遽】[书]frightened; scared: 神色～ look scared
【惶恐】terrified: ～万状 be seized with fear; be frightened out of one's senses

煌 bright; brilliant: 明星～～。The stars are sparkling
【煌斑岩】[地] lamprophyre
【煌绿】brilliant green

蝗 locust: ～灾 plague of locusts
【蝗虫】locust
【蝗蝻】the nymph of a locust

篁 ①(泛指竹子) bamboo grove: 幽～ a secluded and restful bamboo grove ②(长竹子) bamboo: 修～ tall bamboos

鳇 huso sturgeon

隍 dry moat outside a city wall

huǎng

谎 lie; falsehood
【谎报】lie about sth.; give false information: ～军情 make a false report about the military situation／～年龄 lie about one's age
【谎话】lie; falsehood: ～连篇 tell a pack of lie／说～ tell a lie; lie
【谎言】lie; falsehood: ～可畏。 Lies are terrible.／～总是站不住脚的。 Lies have short legs.

恍 ①(恍然) all of a sudden; suddenly ②(与"如""若"等字连用) seem; as if: ～如梦境 as if in a dream
【恍惚】①(神志不清; 精神不集中) in a trance; absentminded: 精神～ be in a trance ②(不真

切；不清楚）dimly；faintly；seemingly：～记得 faintly remember
【恍然大悟】suddenly see the light；suddenly realize what has happened：一经指点，他～。At the hint, he suddenly saw the light.
【恍如隔世】as if being cut off from the outside world for ages；as if had been in the remote past

晃 ①（闪耀）dazzle：亮得～眼 dazzlingly bright ②（很快闪过）flash past：一～半个月过去了。A fortnight passed in a flash.

幌
【幌子】①（招牌）shop sign；signboard ②（假借的名义）pretense；cover；front：骗人的～ a facade；a front／在裁军的～下 under the cover of disarmament

huàng

晃 shake；sway：象钟摆似地～着 swing like a pendulum
【晃荡】rock；shake；sway：来回～ swing to and fro／小船在江面上～。The small boat is rocking on the river.／一瓶子不响，半瓶子～。The half-filled bottle sloshes, the full bottle remains still.（或：The dabbler in knowledge chatters away, the wise man stays silent.）
【晃动】shake；rock；sway：不要～桌子。Don't wobble the table.／车轮有点～。The wheels wobble a bit.
【晃梯】[杂技] balancing on an upright ladder
【晃悠】shake from side to side；wobble stagger：树枝在风中来回～。The branches of the trees are swaying in the wind.

huī

麾 [书]①（古代指挥军队的旗子）standard of a commander ②（指挥军队）command：～军前进 command an army to march forward
【麾下】[书]①[敬]（将帅）general；commander；your excellency ②（将帅的部下）those under one's command

挥 ①（挥舞）wave；wield：～笔 wield the brush；put pen to paper／～刀 wield a sword ②（抹掉）wipe off：～泪 wipe away tears；wipe one's eyes ③（指挥）command (an army)：～师南下 command an army to march south ④（散出）scatter；disperse
【挥动】brandish；wave：～拳头 shake one's fist
【挥发】volatilize ◇ ～性 volatility／～油 volatile oil／～作用 volatilization
【挥戈】brandish one's weapons：～东指 switch east／～返日 regain what has been lost
【挥汗如雨】drip with sweat；perspiration came down like raindrops；sweat like a trooper
【挥毫】[书] wield one's writing brush；write or draw a picture (with a brush)：～作诗 take up the brush and write a poem
【挥霍】spend freely；squander：～浪费 spend extravagantly／～无度 spend without restraint
【挥金如土】throw money about like dirt；spend money like water
【挥泪而别】part in tears；wipe one's tears and leave
【挥手】wave one's hand；wave：～告别 wave farewell；wave good-bye to sb.／～致意 wave greetings to；wave to sb. in acknowledgment
【挥舞】wave；wield：brandish：～大棒 wave the big stick at...／～花束表示欢迎 wave bouquets in welcome／～拳头 brandish one's fist

辉 ①（闪耀的光彩）brightness；splendor ②（照耀）shine：与日月同～ shine for ever like the sun and the moon
【辉安岩】[地] auganite
【辉长岩】[地] gabbro
【辉光】[电] glow ◇ ～灯 glow lamp／～放电 glow discharge／～放电管 glow discharge tube
【辉煌】brilliant；splendid；glorious：～成就 brilliant achievements／～的战果 a brilliant military victory／～的文化 splendid civilization／灯火～ brilliantly illuminated；ablaze with lights
【辉绿岩】[地] diabase
【辉钼矿】molybdenite
【辉砷钴矿】cobaltite
【辉石】[地] pyroxene；augite
【辉锑矿】stibnite
【辉铜矿】chalcocite
【辉银矿】argentite
【辉映】shine；reflect：湖光山色，交相～。The lake and the hills add radiance and beauty to each other.

灰 ①（灰烬）ash：飞～ fly ash／草木～ plant ash (as fertilizer) ②（灰尘）dust：积了厚厚的一层～ accumulate a thick layer of dust ③（石灰）lime；(lime) mortar：～墙 plastered wall／和～ mix mortar ④（颜色）grey：～马 a grey horse

⑤(消沉) disheartened; discouraged: 心～意懒 feel disheartened
【灰暗】murky grey; gloomy: ～的天空 a gloomy sky
【灰白】greyish white; ashen; pale: 脸色～ look pale
【灰尘】dust; dirt: 掸掉桌上的～ dust the table
【灰斗】ash bucket; hod
【灰分】[矿] ash content
【灰鹤】grey crane
【灰黄霉素】[药] griseofulvin
【灰浆】[建] mortar ◇ ～板 fat board / ～槽 mortar-trough
【灰烬】ashes: 化为～ be reduced to ashes
【灰口铁】[冶] grey (pig) iron
【灰溜溜】gloomy; dejected; crestfallen: 他总是～的。He always looked dejected.
【灰霉病】gray mold
【灰霉素】[药] grisein
【灰蒙蒙】dusky; overcast: ～的夜色 a dusky night scene / 天色～的。The sky was overcast.
【灰锰氧】potassium permanganate
【灰泥】[建] plaster
【灰雀】bullfinch
【灰壤】[农] podzol
【灰色】①(颜色) grey; ashy ②(颓废失望) pessimistic; gloomy: ～人生观 a pessimistic outlook on life ③(态度暧昧) obscure; ambiguous
【灰沙燕】sand martin
【灰鼠】squirrel
【灰雾】[摄] fog
【灰心】lose heart; be discouraged: ～丧气 be utterly disheartened / 不要～! Don't be discouraged! (或: Don't lose heart!)
【灰指甲】[医] ringworm of the nails; onychomycosis
【灰质】[生理] grey matter

恢

extensive; vast
【恢复】①(变成原样) resume; renew: ～邦交 resume diplomatic relations / ～正常 return to normal ②(使变成原样) recover; regain: ～健康 recover one's health / ～知觉 recover consciousness; come to ③(把失去的收回) restore; reinstate; rehabilitate: ～公民权 restitution of civil rights / ～名誉 rehabilitation (of a person's reputation) / ～原职务 reinstate sb. in his post ◇ ～法 restoring method / ～期 convalescence / ～力 restoring force / ～室 recovery room / ～载波 reinsertion of carrier
【恢恢】[书] extensive; vast: 天网～，疏而不漏。The net of Heaven has large meshes, but it lets nothing through.

诙

【诙谐】humorous; jocular ◇ ～曲[乐] scherzo; humoresque

徽

emblem; badge; insignia: 国～ national emblem / 帽～ cap insignia / 校～ school badge
【徽号】title of honor
【徽章】badge; insignia: 在外套上别一枚～ pin a badge on one's jacket

huí

回

①(曲折环绕) circle; wind: 迂～ winding; circuitous; round about / 峰～路转。The path winds along mountain ridges. ②(还;返回) return; go back: ～到原地 return to where one came from / ～国 return to one's native country ③(掉转) turn round: ～过身来 turn round ④(答复) answer; reply: ～信 send a letter in reply; write back ⑤(小说的章回) chapter ⑥[量]来过一～ have been here once / 完全是两～事 two entirely different matters
【回拜】pay a return visit
【回报】①(报告工作) report back on what has been done ②(报答) repay; requite; reciprocate: ～他的盛情 repay him for his hospitality or kindness ③(报复) retaliate; get one's own back
【回避】evade; dodge; avoid (meeting sb.): ～困难 dodge difficulties / ～要害问题 evade the crucial question ◇ ～关系 avoidance relationship / ～行为 avoidance behavior / ～证人 challenge witness / ～制度 challenge system
【回禀】[旧] report back (to one's superior)
【回波】[电] echo ◇ ～脉冲 echo pulse
【回波作用】[经] backwash effect
【回驳】refute: 当面～ refute to sb.'s face
【回采】[矿] stopping; extraction: 快速～ fast extraction ◇ ～工作面 stope / ～率 percentage of recovery; recovery / ～损失 mining loss
【回肠】①(小肠的一部分) ileum ②[书](内心焦虑) worried; agitated
【回肠荡气】(of music, poems, etc.) soulstirring; heartrending
【回肠九转】One's mind is burning with grief.
【回肠炎】ileitis
【回潮】resurgence; reversion: 思想～ an ideo-

logical relapse
【回潮率】[纺] (moisture) regain
【回程】①(返回的路程) return trip ②[机] return stroke
【回春】①(冬尽春来) return of sprig: 大地~。Spring returns to the earth. ②(起生回生) bring back to life: ~灵药 a miraculous cure; a wonderful remedy
【回答】answer; reply; response: ~问题 answer a question / 事实是对造谣者最有力的~。Facts are the most powerful rebuff to rumormongers.
【回荡】resound; reverberate: 牧羊人的歌声在山谷间~。The shepherd's songs reverberated in the valleys.
【回电】wire back: 请即~。Wire reply immediately.
【回动】[机] reverse ◇ ~机构 reversing mechanism
【回访】pay a return visit
【回风道】[矿] air return way
【回复】reply (to a letter)
【回顾】look back; review: ~过去,展望未来 review the past and look forward to the future
【回光返照】①(落日前的光) the last radiance of the setting sun ②(人死前短暂的兴奋现象) momentary recovery of consciousness just before death ③(旧事物灭亡前暂时兴旺的现象) a sudden spurt of activity prior to collapse
【回光镜】reflector
【回归】[统计] regression
【回归年】[天] tropical year
【回归热】[医] relapsing fever
【回归线】[地] tropic: 北~ the Tropic of Cancer / 南~ the Tropic of Capricorn
【回锅】cook again ◇ ~肉 twice-cooked pork often with chilli seasoning
【回国述职】return to one's country and report
【回合】round; bout: 第一个~的胜利 a first-round victory
【回话】reply; answer: 请你给他带个~。Please take a message to him by way of reply
【回火】[机] tempering ◇ ~脆性 temper brittleness / ~炉 tempering furnace
【回击】fight back; return fire; counterattack: 给以有力的~。strike a powerful counterblow; hit back hard
【回家】return home: 在~路上 on one's way home
【回见】[套] see you later; cheerio
【回交】[生] backcross
【回敬】return a compliment; do or give sth. in return: ~一杯 drink a toast in return / ~一拳 return a blow
【回绝】decline; refuse: 一口~ flatly refuse / 婉然~邀请 decline an invitation politely
【回扣】rebate; sales commission
【回来】return; come back; be back: 他马上就~。He'll be back in a minute.
【回廊】winding corridor
【回礼】①(回答敬礼) return a salute ②(回赠礼物) send a present in return; present a gift in return
【回流】reflux; back-flow ◇ ~比 reflux ratio / ~阀 reflux valve; return valve
【回笼】①(再蒸) steam again ②(货币回笼) withdrawal (of currency) from circulation
【回炉】①(再熔化) melt down: 废铁~ melt down scrap iron; use scrap iron for smelting ②(再烘烤) bake (cakes, etc.) again
【回路】[电] return circuit; return; loop ◇ ~分析 loop analysis / ~增益 loop again
【回马枪】back thrust: 杀他个~ give sb. a back thrust; swing round and catch sb. off guard
【回暖】get warm again after a cold spell
【回请】return hospitality; give a return banquet
【回去】①(用作动词) return; go back; be back ②(用在动词后) back: 带~ take back
【回扫】[电子] flyback
【回升】rise again (after a fall); pick up: 气温~。The temperature has gone up again. / 指数~。The index is picking up.
【回生】①(复活) bring back to life: 起死~ bring the dying back to life ②(荒疏) forget through lack of practice; get rusty: 几个月不用,我的法语又~了。I haven't practiced my French for months and it's getting rusty.
【回声】echo ◇ ~测深仪 echo sounder; fathometer
【回收】retrieve; recover; reclaim: ~贵重金属 retrieve rare metals / 余热~ recovery of waste heat ◇ ~率 rate of recovery / ~塔[化] recovery tower / ~站 (waste materials) collection depot
【回收期】[经] payoff period; payout period
【回手】①(转身顺手) turn round and stretch out one's hand: ~把门关上 turn round and close the door ②(还击) hit back; return a blow
【回首】①(回头) turn one's head; turn round ②[书](回忆) look back; call to mind: 不堪~ can't bear to think of the past
【回水】[水] backwater

【回溯】recall; look back upon
【回天乏术】have no plan to make improvements; unable to save the situation
【回天之力】power capable of saving a desperate situation; tremendous power
【回填】[建] backfill ◇ ~土 backfill
【回嗔作喜】One's anger at once changed to joy.
【回条】a short note acknowledging receipt of sth.; receipt
【回帖】a money order receipt to be signed and returned to the sender
【回头】①(掉转头) turn one's head; turn round ②(悔悟) repent: 及早~ repent before it is too late ③[口](等一会)later: ~见! See you later! / ~再谈。We'll talk it over later.
【回头路】the road back to one's former position; the road of retrogression: 走~ take the road back; backtrack
【回头是岸】repent and be saved: 苦海无边,~。The bitter sea has no bounds, repent and the shore is at hand.
【回味】①(食后余味) aftertaste ②(回想中体会) call sth. to mind and ponder over it: ~他说的话 ponder over what he has said
【回乡】return to one's home village: ~知识青年 an educated youth having returned to do farm work in his home village
【回响】reverberate; echo; resound
【回想】think back; recollect; recall: ~童年时代的生活 recollect the days of one's childhood
【回心转意】change one's views; come around
【回信】①(写复信)write in reply; write back: 望早日~。I'm looking forward to hearing from you soon. ②(复信) a letter in reply ③(答复的话) a verbal message in reply; reply: 事情办妥了,我给你个~儿。I'll let you know when I'm through with it.
【回形针】(paper) clip
【回旋】①(盘旋) circle round: 飞机在上空~。The airplane is circling overhead. ②(进退;周转) (room for) maneuver: 这件事还有~余地。The whole thing is not final.
【回旋加速器】[物] cyclotron
【回旋炮】flexible gun; swivel gun
【回旋曲】[乐] rondo
【回忆】call to mind; recollect; recall: ~对比 recall the past and contrast it with the present / 往事的~ recollections of the past ◇ ~录 reminiscences; memoirs; recollections
【回音】①(回声) echo ②(回信) reply: 立候~hoping for an immediate reply ③[乐] turn: 逆~ inverted turn
【回游】[动] migration: 产卵~ spawning migration / 索饵~ feeding migration
【回油器】oil scavenger
【回执】a short note acknowledging receipt of sth.; receipt
【回注】[石油] recycle
【回柱】[矿] prop drawing ◇ ~机 prop drawer; post puller
【回转】turn round ◇ ~半径[航海] radius of gyration / ~工作台[机] rotary table / ~炉[冶] rotary furnace / ~式钻床 rotary drill / ~体[数] solid of revolution / ~仪[天] gyroscope; gyro
【回嘴】answer back; retort

茴

【茴香】[植] fennel; aniseed ◇ ~豆 beans flavored with aniseed / ~油 fennel oil

蛔

【蛔虫】roundworm; ascarid
【蛔虫病】roundworm disease; ascariasis
【蛔虫感染】roundworm infection

huǐ

悔 regret; repent
【悔不当初】regret having done sth.: 早知今日,~。If I'd known then what was going to happen, I wouldn't have done as I did.
【悔改】repent and mend one's ways: 毫无~之意 have no intention of mending one's ways; show no sign of repentance / 决心~ be determined to mend one's ways ◇ ~程度[法] extent of repentance / ~态度[法] attitude toward repentance
【悔过】repent one's error; be repentant; penitence: ~书 a written statement of repentance
【悔过从善】acknowledge one's errors and become a good person
【悔过自新】repent and start anew; repent and turn over a new leaf; repent and make a fresh start
【悔恨】regret deeply; be bitterly remorseful: ~交加 mixed feeling of remorse and shame / ~终身 have a secret regret for life
【悔棋】retract a false move in a chess game
【悔愧交加】torn by self-recrimination and repentance
【悔悟】realize one's error and show repentance: 毫无~之意 show no sign of repentance

【悔之晚矣】It is now too late to repent. (或: Repentance is too late.)
【悔之无及】too late to repent; too late for regrets
【悔罪】show repentance; show penitence
毁 ①(破坏掉) destroy; ruin; damage: ~于一旦 be destroyed in a moment ②(焚烧掉) burn up: 焚~ destroy by fire; burn down ③(诽谤) defame; slander
【毁谤】slander; malign; calumniate: 这纯系~。 This is slander, pure and simple.
【毁坏】destroy; damage; break; deface ◇ ~公共财产案 case of defacing public property / ~公共财物罪 offense of defacing public property
【毁家纾难】sell family properties to relieve the distress of people; offer all one has to help in charity
【毁馏】[化] destructive distillation
【毁灭】destroy; exterminate: 给以~性打击 deal a crushing blow ◇ ~物证[法] destroy the material evidence / ~证据[法] destroy evidence; destruction of evidence
【毁弃】scrap; annul
【毁伤】injure; hurt; damage ◇ ~半径[军] radius of damage; radius of rupture
【毁尸灭迹】chop up a corpse and obliterate all traces; bury the corpse in order to destroy all traces of one's crime
【毁损】damage; impair; breakage
【毁誉】praise or blame; praise or condemnation: 不计~ be indifferent to people's praise or blame
【毁誉参半】(指人) get both praise and censure; (指书等文艺作品) have a mixed reception
【毁约】①(食言) break one's promise ②(撕毁合约、契约) scrap a contract or treaty ◇ ~诉讼 breach-of-promise suit

huì

汇 ①(汇合) converge: ~成巨流 converge into a mighty torrent ②(聚集) gather together: ~印成书 have (articles on a given subject) collected and published in book form ③(聚集而成的东西) things collected; assemblage; collection: 词~ vocabulary ④(寄) remit: 给家里~钱 remit money to one's family / 电~ telegraphic transfer / 信~ mail transfer
【汇报】report; give an account of: ~调查结果 report the findings of an investigation / ~工作 report to sb. on one's work ◇ ~会 report-back meeting / ~演出 report-back performance
【汇拨支付】payment by remittance
【汇编】compilation; collection; corpus: 文件~ a collection of documents / 语言学研究资料~ a corpus of philological data / 资料~ 工作 compilation of reference material
【汇编语言】[计算机] assembly language; assembler language
【汇编指令】[计算机] assembler directive; assembler instruction
【汇兑】remittance: 国内~ domestic remittance ◇ ~价变动 exchange movement / ~价换算表 exchange table / ~率 rate of exchange / ~损失 exchange loss / ~网 remittance network / ~银行 exchange bank
【汇费】remittance fee
【汇合】converge; join: ~成一支巨大的力量 unite to form a gigantic force
【汇集】①(聚集) collect; compile: ~材料 collect all relevant data ②(集合) come together; converge; assemble
【汇聚】syntaxes
【汇款】①(寄出款) remit money; make a remittance ②(收到或寄出的款项) remittance: 汇出~ outward remittance / 收到一笔~ receive a remittance / 邮政~ postal remittance ◇ ~单 money order / ~方式 method of remittance / ~收款人 beneficiary of remittance / ~收据 fee for remittance
【汇款人】remitter ◇ ~简短附言 sender's remarks
【汇流】converge; flow together ◇ ~点[地] confluence / ~条[电] busbar
【汇率】exchange rate: 浮动~ floating (exchange) rate / 固定~ fixed (exchange) rate / 中心~ central rate
【汇票】draft; bill of exchange; money order: 即期~ demand draft / 见期~ bill at sight; sight bill / 银行~ bank draft / 邮政~ postal money order ◇ ~贴现 discount of bill ○承兑 honor; accept / 兑现 cash / 贴现 discount / 拒付 dishonor / 托收票据 bill for collection / 票据 bill
【汇总】gather; collect; pool: 把材料~上报 collect data for the higher level; present an itemized report to the higher level
【汇总缴纳】pay on a consolidated basis
溃 festering: ~脓 suppuration

讳 ①(忌讳) avoid as taboo: 直言不～ speak bluntly; call a spade a spade ②(忌讳的事情) forbidden word; taboo: 犯了他的～了。Something was said that happened to be taboo with him. ③[旧](死去的帝王、尊长的名字) the name, regarded as taboo, of a deceased emperor or head of a family
【讳疾忌医】hide one's sickness for fear of treatment; conceal one's fault for fear of criticism
【讳莫如深】closely guard a secret; not breathe a word to a soul; not utter a single word about sth.
【讳言】dare not or would not speak up: 无可～ there's no denying the fact

彗 broom
【彗差】[光] coma
【彗尾】[天] the tail of a comet
【彗星】[天]comet
【彗形波瓣】[电]coma lobe
【彗形象差】[电子] coma

慧 intelligent; bright: 智～ wisdom; intelligence
【慧剑斩情丝】cut the thread of carnal love with the sword of wisdom
【慧心】wisdom
【慧眼】①[佛教] a mind which perceives both past and future ②(锐敏的眼力) mental discernment insight; acumen
【慧眼独具】can see what others cannot
【慧眼识英雄】Discerning eyes can tell greatness from mediocrity.

卉 (various kinds of) grass: 奇花异～ rare flowers and grasses

惠 ①(恩惠) favor; kindness; benefit: 受～ receive kindness / 小恩小～ small favors ②[敬] ～鉴 be kind enough to read (the following letter) / ～书 your letter
【惠存】[敬] please keep (this photograph, book, etc. as a souvenir); to so-and-so
【惠而不费】give sb. a pleasure which would cost one nothing; a kind act which does not cost much
【惠风和畅】A gentle breeze is freely blowing.
【惠顾】[敬] your patronage
【惠临】[敬] your gracious presence: ～增光 please honor me with your visit / 敬请～。Your presence is requested.
【惠我良多】You have been very kind to me. (或: You have conferred very much kindness on me.)
【惠训不倦】never tire of teaching

蕙
【蕙兰】a species of orchid
【蕙质兰心】a pure heart and spirit
【蕙中秀外】intelligent within and beautiful without

喙 ①(鸟兽的嘴) beak or snout ②(指人嘴) mouth: 不容置～ not allow others to butt in; brook no intervention / 百～莫辩。A hundred mouths can't explain it away.
【喙形接头】bird-head bond

秽 ①(肮脏) dirty: 污～ filthy ②(丑恶) ugly; abominable: ～行 abominable behavior
【秽土】rubbish; refuse; dirt
【秽闻】ill repute (referring to sexual behavior); reputation for immorality
【秽亵言语】aeschrolalia; coprolalia
【秽行】[书] abominable behavior; immoral conduct

贿 bribe: 受～ accept bribes / 行～ practice bribery; bribe
【贿赂】①(用财物买通别人) bribe ②(用以买通别人的财物) bribery ◇ ～金 bribery fund / ～品 boodle / ～手段 means of bribery / ～证人罪 bribing witness / ～罪 bribery crime
【贿买】buy over; suborn
【贿选】practice bribery at an election; get elected by bribery

会 ①(聚合) get together; assemble: 明晨七时在门口～齐。We'll assemble at the gate at 7 o'clock tomorrow morning. ②(会见;见到) meet; see: 昨天我没有～着他。I didn't see him yesterday. ③(会议) meeting; gathering; party: get- together; conference: 欢送～ send- off party / 欢迎～ welcoming party ④(团体) association; society; union: 帮～ secret society / 工～ trade union ⑤[旧](庙会) a temple fair: ～ go to a fair ⑥(主要城市) chief city; capital: 都～ city; metropolis / 省～ provincial capital ⑦(时机) opportunity; occasion: 适逢其～ happen to be present on the occasion ⑧(理解) understand; grasp: 误～ misunderstand ⑨(通晓;熟习) can; be able to: ～滑冰 can skate / ～

英文 know English ⑩(会做) be good at; be skillful in; be likely to; be sure to: ～修各种钟表 be skillful in repairing all kinds of clocks and watches ⑪(可能实现) be likely to; be sure to: 假使你早一些来,就～看到他了。If you had come earlier, you would have seen him. ⑫(付帐) pay a bill: 饭钱我～过了。I've paid for the meal.

【会餐】dine together; have a dinner party

【会场】meeting-place; conference hall

【会道门】superstitious sects and secret societies

【会费】membership dues

【会馆】[旧] guild hall

【会合】join; meet; converge; assemble: 两军～后继续前进。The two armies joined forces and marched on.

【会合点】[军] meeting point; rallying point

【会合周期】[天] synodic period

【会话】conversation (as in a language course)◇～方式 conversational mode / ～系统 conversational system

【会话型】interactive: ～处理 conversational processing / ～终端 interactive terminal

【会籍】membership (of an association)

【会见】meet with (esp. a foreign visitor)

【会聚】assemble; flock together: 他们～在一起讨论这个问题。They got together to discuss the question.

【会聚透镜】[物] convergent lens

【会刊】①(会议文件汇编) proceedings of a conference, etc. ②(协会的刊物) the journal of an association, society, etc.

【会客】receive a visitor: ～时间 the time for receiving visitors; visiting hours ◇ ～室 reception room

【会面】meet: 近来,我同他不常～。I have not seen much of him lately.

【会签】countersign; countersignature

【会期】①(开会日期) the time fixed for a conference; the date of a meeting: 确定～ fix a date for the meeting ②(会议持续天数) the duration of a meeting: ～定为三天。The meeting is scheduled to last three days.

【会商】hold a conference or consultation: ～解决办法 consult to find a solution

【会审】①(会同审理) joint hearing; joint trial: 举行～ conduct a joint trial ②(会同审查) make a joint checkup: ～施工图纸 have a joint checkup on the blue prints for a project

【会师】join forces; effect a junction: 胜利～ join forces triumphantly

【会谈】talks: ～中的问题 questions under discussion / 双边～ bilateral talks ◇ ～纪要 minutes of talks; notes on talks; summary of a conversation

【会堂】assembly hall; hall

【会同】(handle an affair) jointly with other organizations concerned

【会晤】meet: 两国外长定期～。The foreign ministers of the two countries meet regularly.

【会心】understanding; knowing: ～的微笑 an understanding smile

【会演】joint performance (by a number of theatrical troupes, etc.): 文艺～ theatrical festival

【会厌】[生理] epiglottis

【会议】meeting; conference: 筹备～ preparatory meeting / 非正式～ unofficial meeting / 国家元首和政府首脑～ conference of heads of states or governments / 秘密～ private session; secret session / 全体～ plenary session / 小组～ group meeting / 圆桌～ round-table conference / 正式～ official meeting / 最高级～ summit meeting; summit conference ◇ ～地点 meeting-place; venue / ～日程表 the daily agenda of a conference / ～室 meeting room; council chamber / ～厅 conference hall / ～议程 conference agenda

【会意】understanding; knowing

【会阴】[生理] perineum

【会员】member: ～人数 membership / 正式～ full member ◇ ～国 member state / ～银行 member bank / ～证 membership card / ～资格 the status of a member; membership

【会战】①[军](主力决战) meet for a decisive battle ②(集中力量突击) join in a battle; launch a mass campaign: 石油大～ a great battle for oil

【会章】①(协会的章程) the constitution of an association, society, etc. ②(协会的标志、徽章) the emblem of an association, society, etc.

【会长】the president of an association or society

【会帐】pay a bill

【会诊】[医] consultation of doctors; (group) consultation: 中西医～ hold group consultations of doctors practicing Chinese and Western medicine

【会址】①(协会的地址) the site of an association or society ②(召开会议的地方) the site of a conference or meeting

烩

①(烩菜) braise: ～虾仁 braised shrimp

meat ②(烩饭和菜) cook (rice or shredded pancakes) with meat, vegetables and water

荟 [书] luxuriant growth (of plants)

【荟萃】(of distinguished people or exquisite objects) gather together; assemble: 人才～ a galaxy of talent

【荟萃一堂】a distinguished gathering; gather together in one hall

绘 paint; draw

【绘画】drawing; painting ◇～室 studio

【绘声绘色】vivid; lively: ～的描述 a vivid description

【绘图】mapping ◇～板 drawing board／～笔 border pen／～机 draught machine／～室 drawing room／～员 draftsman

【绘制】draw (a design, etc.): ～图表 draw a diagram

诲 teach; instruct

【诲人不倦】be tireless in teaching; teach with tireless zeal

【诲淫诲盗】propagate sex and violence; stir up the base passions: ～的电影 films full of sex and violence

晦 ①(农历每月的末一天) the last day of a lunar month ②(昏暗) dark; obscure; gloomy ③(夜晚) night

【晦暗】dark and gloomy

【晦气】unlucky: 自认～ be resigned to one's bad luck

【晦涩】hard to understand; obscure: ～的语言 obscure language (in poetry, drama, etc.)

【晦朔】the last and first days of a lunar month

hūn

荤 meat or fish: ～菜 meat dishes／吃～ take meat diet

【荤腥】meat or fish

【荤油】lard

昏 ①(黄昏) dusk: 晨～ at dawn and dusk ②(昏暗) dark; dim ③(头脑迷糊) confused; muddled: 利令智～ be blinded by lust for gain ④(失去知觉) lose consciousness; faint: ～倒 fall into a swoon; go off into a faint; fall unconscious

【昏暗】dim; dusky: ～的灯光 a dim light／～的天色 gloomy skies

【昏沉】①(昏暗) murky: 暮色～ murky twilight ②(昏乱) dazed; befuddled: 睡得昏昏沉沉 be stupid with sleep

【昏呆】stupor

【昏黑】dusky; dark

【昏花】dim-sighted: 老眼～ dim-sighted from old age

【昏黄】pale yellow; faint; dim: ～的灯光 a dim light／月色～ faint moonlight

【昏昏欲睡】drowsy; sleepy

【昏厥】faint; swoon: ～过去 fall into a coma; faint away

【昏君】a fatuous and self-indulgent ruler

【昏聩】decrepit and muddleheaded

【昏聩愚昧】being stupefied and foolish

【昏乱】dazed and confused; befuddled

【昏迷】stupor; coma: ～不醒 remain unconscious／～状态 comatose state

【昏睡】lethargic sleep; lethargy

【昏天黑地】①(天色昏暗) pitch-dark; in pitch darkness: ～的,山路更不好走。It was more difficult to make our way up the mountain in pitch darkness. ②(神志不清) dizzy: 因流血过多而感到～ feel terribly dizzy from bleeding ③(生活荒唐) loose; perverted; decadent: ～的生活 a dissipated life ④(社会黑暗) dark rule and social disorder

【昏头昏脑】①(糊涂) addlebrained; addleheaded; muddleheaded ②(好忘事) absentminded; forgetful

【昏星】evening star

【昏眩】dizzy; giddy

【昏庸】fatuous; muddleheaded; stupid: ～的官吏 incompetent officials

阍 [书] ①(看门) tend or guard a gate: 司～ gatekeeper; janitor ②(宫门) palace gate

【阍者】[书] gatekeeper; janitor; doorman

婚 ①(结婚) wed; marry: 晚～ marry at a mature age／未～ unmarried／已～ married ②(婚姻) marriage; wedding: 晚～ late marriage; deferred marriage／早～ early marriage

【婚飞】[动] nuptial flight

【婚嫁】marriage

【婚礼】wedding ceremony; wedding: 参加～ attend the wedding of A and B

【婚龄】(legally) marriageable age: 她还没有到～。She is not of marriageable age.

【婚期】wedding day

【婚前检查】pre-martial check-ups

【婚神星】[天] Juno

【婚生子女】[法] children born in wedlock; legitimate children: 非～ children born out of wedlock
【婚事】marriage; wedding
【婚姻】marriage; matrimony: 包办～ arranged marriage / 合法～ legal marriage / 买卖～ mercenary marriage / 美满的～ a happy marriage ◇ ～法 law of marriage; marriage law / ～介绍所 matchmaking service / ～纠纷 matrimonial dispute / ～诉讼 matrimonial suit / ～状况 marital status / ～自由 freedom of marriage / ～自主 marry the partner of one's choice / ～自主权 marital autonomy ○配偶 spouse / 结婚登记 marriage registration / 结婚年龄 marriage age / 结婚证 marriage certificate / 已婚有偶 ever-married / 丧偶 widowed / 离婚 divorced / 分居 separated / 单身(未婚) single; unmarried; not married / 一夫一妻 monogamy / 重婚 bigamy
【婚约】marriage contract; engagement: 解除～ break off one's engagement

hún

浑 ①(浑浊) muddy; turbid: ～水 muddy water / 把水搅～ muddle the water ②(不明事理) foolish; stupid ③(天然的) simple and natural; unsophisticated ④(全;满) whole; all over
【浑蛋】[骂] blackguard; wretch; scoundrel; bastard; skunk
【浑厚】①(淳朴老实) simple and honest ②(艺术风格朴实雄厚) simple and vigorous: 笔力～ (of handwriting) bold and vigorous strokes
【浑浑噩噩】ignorant; simple-minded; muddle headed
【浑金璞玉】unrefined gold and unpolished jade
【浑然天成】like nature itself — highest quality (of art)
【浑然一体】one integrated mass; a unified entity; an integral whole
【浑身】from head to foot; all over: ～发痒 itch all over / ～解数 use all one's skill (to do sth.) / ～是胆 full of courage / ～疼痛 have pains all over / 吓得～发抖 tremble all over with fear
【浑水摸鱼】fish in troubled waters
【浑俗和光】be in harmony with the rest of the world
【浑天仪】[天] ①(浑仪) armillary sphere ②(浑象) celestial globe
【浑象】[天] celestial globe
【浑仪】[天] armillary sphere
【浑圆】perfectly round
【浑浊】muddy; turbid

魂 ①(灵魂) soul: 忠～ loyal soul ②(精神;情绪) mood; spirit: 神～不定 be distracted; have the jitters ③(国家、民族崇高的精神) the lofty spirit of a nation: 民族～ national spirit
【魂不附体】as if the soul had left the body: 吓得～ be scared out of one's wits
【魂不守舍】One's mind is somewhat unhinged. (或: The spirit has left the body.)
【魂飞魄散】be frightened out of one's wits; almost swooning with fright
【魂灵】[口]soul
【魂魄】soul
【魂神星】[天] Psyche

馄
【馄饨】dumpling soup

hùn

混 ①(搀杂) mix; confuse: ～在一起 mix things up / 两者～不起来。The two things do not mix ②(蒙混) pass for; pass off as: 鱼目～珠 pass off fish eyes as pearls; pass off the sham as genuine ③(苟且地生活) muddle along; drift along: ～日子 drift along aimlessly ④(相处) get along with sb.: 同他们～得很熟 be quite familiar with them
【混充】pass oneself off as; palm sth. off as: ～充内行 pass oneself off as an expert
【混沌】①(宇宙形成前的景象) Chaos: ～初开 when earth was first separated from heaven ②(无知的样子) innocent as a child
【混纺】[纺] blending: 棉毛～ fabric of wool and cotton ◇ ～织物 blend fabric
【混汞】[冶] amalgamate
【混合】mix; blend; mingle: 把水和酒精～起来 mingle water and alcohol / 客货～列车编队 mixed train ◇ ～保险 mixed insurance / ～编队 [军] composite formation / ～率 composite rate / ～器 [化] mixer / ～色 secondary color / ～税 mixed duties / ～授粉[农] mixed pollination / ～双打[体] mixed doubles / ～诉讼 mixed action / ～物 mixture / ～岩[地] migmatite / ～帐户 mixed account
【混交林】mixed forest
【混进】infiltrate; sneak into; worm one's way into
【混流】mixed flow; admixture
【混乱】confusion; chaos: ～不堪 in utter disor-

der / 思想～ ideological confusion
【混凝剂】[化] coagulant
【混凝土】concrete; beton ◇ ～搅拌机 concrete mixer / ～结构 concrete structure / ～振捣器 (concrete) vibrator
【混频管】[电子] mixer tube
【混频级】mixer stage; mixing stage
【混频器】first detector; heterodyne modulator
【混色纱线】melange yarn
【混世魔王】fiend in human shape; devil incarnate
【混水摸鱼】fish in troubled waters
【混同】confuse; mix up: 把"价值"和"价格"～了 confuse value with price
【混为一谈】lump together; confuse sth. with sth. else: 把正义战争与非正义战争～ lump together just wars and unjust wars
【混响】[物] reverberation
【混淆】obscure; blur; confuse; mix up: ～黑白 mix up black and white / ～是非 confuse right and wrong / ～视听 mislead the public; confuse public opinion
【混血儿】a person of mixed blood; half-breed
【混杂】mix; mingle: 不要把不同的种子～在一起。 Don't mix up different kinds of seeds.
【混战】tangled warfare: 处于～状态 be locked in a tangled fight
【混帐】[骂] scoundrel; bastard; son of a bitch ◇ ～话 impudent remark
【混浊】muddy; turbid: ～的河流 muddy waters / ～的空气 foul air / ～的水 turbid water

诨 joke; jest: 打～ make gags
【诨名】nickname

huō

豁 ①(裂开) slit; break; crack: 墙上～了一个口子。 There is a breach in the wall. ②(狠心付出某种代价) give up; sacrifice: ～出三天时间, 也要把它做好。 Even if it takes us three days, we must get the job done.
【豁出去】go ahead regardless; be ready to risk everything
【豁口】opening; break; breach: 城墙～ an opening in the city wall / 碗边上的～ notch on the edge of a bowl
【豁嘴】[口] ①(唇裂) harelip ②(唇裂的人) a harelipped person

劐 [口] ①(拉开) slit or cut with a knife: 把鱼肚子～开 slit open the fish ②(农具) hoeing

擭 shovel coal, ore, etc. from one place to another: ～煤工人 coal shoveller

锪
【锪孔钻】spot-facer; spot-facing

huó

活 ①(生存) live: 他～到九十岁。 He lived to be ninety. ②(活着的;有生命的) alive; living: ～鱼 live fish / ～字典 a walking dictionary / 在他～着的时候 during his lifetime ③(救活) save: ～人无算 (of a good doctor, etc.) save countless lives ④(灵活) vivid; lively: 描绘得很～ be depicted quite vividly / 脑子很～ have a quick mind ⑤(活动的) movable; moving: ～水 flowing water ⑥(真正) exactly; simply ⑦(工作) work: 干～儿 work / 苦～ sweated labor / 零～ odd jobs ⑧(产品) product: 这批～儿做得好。 This batch of products is well made.
【活靶】[军] maneuvering target
【活版】[印] typography; letterpress ◇ ～印刷 typographic printing; typography; letterpress printing / ～印刷机 letterpress (printing) machine
【活瓣】[生理] valve
【活宝】a bit of a clown; a funny fellow
【活报剧】living newspaper; skit; street performance
【活标本】living specimen
【活茬】[口] farm work
【活到老学到老】Live and learn.(或: It is never too old to learn.)
【活地狱】hell on earth
【活动】①(运动) move about; exercise: 站起来～～ stand up and move about ②(动摇) shaky; unsteady: 这颗牙～了。 This tooth's loose. ③(不固定) movable; mobile; flexible: ～坝 movable dam / ～房屋 prefabricated house ④(行动) activity; maneuver: 户外～ outdoor activities / 开展文体～ promote cultural and sports activities ⑤(说情) use personal influence or irregular means: 替他～～ put in a word for him; use one's influence on his behalf ⑥[心] behavior ◇ ～靶 [军] maneuvering target / ～坝[水] movable dam / ～扳手[机] adjustable spanner / ～范围 sphere of activities; scope of operation / ～家 activist; public figure / ～桥 movable bridge / ～舞台 revolving

stage / ～住房 mobile home / ～资本 liquid capital
【活度】[化] activity ◇ ～系数 activity coefficient
【活佛】[宗] Living Buddha
【活该】[口] serve sb. right: ～倒楣 be destined to come to grief
【活荷载】[交][建] live load
【活化】[化] activation ◇ ～分析 activation analysis / ～剂 activator / ～能 activation energy / ～污泥法 activated-sludge method / ～吸附 activation absorption
【活活】while still alive: ～烧死 be burnt alive
【活火山】active volcano
【活计】①(手工艺、体力劳动) handicraft work; manual labor ②(手工制品) work; handiwork
【活见鬼】it's sheer fantasy; you're imagining thing
【活结】slipknot; a knot that can be undone by a pull
【活口】①(命案可提供线索的人) a survivor of a murder attempt ②(可提供情况的俘虏) a prisoner who can furnish information
【活力】vigor; vitality; energy: 充满青春～ be brimming with youthful vigor
【活灵活现】vivid; lifelike: 说得～ give a vivid description; make it come to life
【活路】①(维持生活的办法) means of subsistence; way out: 找～ look for a means of subsistence ②(行得通的办法) workable method
【活络】[方] ①(活动) loose: 牙齿有点～。The tooth has become a bit loose. ②(话不肯定) noncommittal; indefinite: 他说得很～。He was rather noncommittal.
【活埋】bury alive
【活门】[机] valve
【活命】①(维持生命) earn a bare living; scrape along; eke out an existence: 靠卖艺～ make a living as a street entertainer ②[书](救活性命) save sb.'s life: ～之恩 indebtedness to sb. for saving one's life ③(性命) life
【活命哲学】the philosophy of survival
【活泼】①(生动自然;不呆板) lively; vivacious; vivid: 生动～的政治局面 vigorous and lively political situation / 天真～的孩子 lively children / 文字～ written in a lively style ②[化] reactive
【活期】current ◇ ～储蓄 current deposit; demand deposit / ～存款 current deposit / ～存款帐户 current account / ～放款 demand loan
【活塞】[机] piston ◇ ～杆 piston rod / ～环 packing ring; piston ring / ～圈 piston ring
【活生生】①(实际生活中的) in real life; actual; real; living: ～的例子 actual example ②(活活) while still alive: ～烧死 burnt alive
【活色生香】Lively color brings forth fragrance. (或: One's writing is lively and colorful.)
【活受罪】[口] have a hell of a life
【活水】flowing water; running water
【活体鉴定】[法] identification of the living
【活体解剖】vivisection
【活现】appear vividly; come alive
【活象】look exactly like; be the spit and image of; be an exact replica of: 这孩子长得～他父亲。 The child is the very spit of his father.
【活性】[化] active; activated ◇ ～染料 reactive dyes / ～碳[化] active carbon
【活血】[中医] invigorate the circulation of blood
【活阎王】devil incarnate; tyrannical ruler
【活页】loose-leaf ◇ ～笔记本 loose-leaf notebook / ～夹 loose-leaf binder; spring binder / ～文选 loose-leaf selections / ～纸 paper for a loose-leaf notebook
【活跃】①(活泼而积极;蓬勃热烈) brisk; active; dynamic: 市场～。 Business is brisk. / 讨论会开得很～。 The discussion was very lively. ②(使活跃) enliven; animate; invigorate: ～城乡物资交流 stimulate the interchange of urban and rural products / ～农村经济 activate rural economy / ～市场 enliven the market
【活字】[印] type; letter ◇ ～盘 type case; letter board / ～印刷 movable-type printing
【活组织检查】[医] biopsy

和 mix (powder) with water, etc.: ～点儿灰泥 prepare some plaster
【和面】knead dough ◇ ～机 flour-mixing machine

huǒ

火 ①(光焰) fire: 生～ make a fire / 着～ be on fire / 这屋里有～。 There's a fire in the room. ②(炮火) firearms; ammunition: 交～ exchange shots ③[中医](引起发炎等症状的原因) internal heat ④(红色) fiery; flaming: ～红 red as fire; flaming ⑤(紧急) urgent; pressing: ～速回电。 Cable reply immediately. ⑥(发怒) anger; temper: 把人惹～了 make sb. angry / 心头～起 flare up in anger
【火把】torch
【火坝】boiler bridge; bridge in furnace; bridge

of boiler
【火棒】lighted torch (used in acrobatics)
【火暴】[方] fiery; irritable: ～性子 a hot temper
【火并】open fight between factions
【火柴】match: 擦～ strike a match ◇ ～盒 matchbox
【火场】the scene of a fire
【火车】train: 乘～ go by train ◇ ～机车 (railway) engine; locomotive / ～轮渡 train ferry / ～票 railway ticket / ～上交货价 free on rail; F. O. B. train / ～时刻表 railway timetable; train schedule / ～司机 engine driver; (locomotive) engineer / ～站 railway station
【火成喷出岩】extrusive igneous rocks
【火成侵入岩】instrusive igneous rocks
【火成碎屑岩】pyroclastic igneous rocks
【火成岩】[地] igneous rock
【火电站】thermal power station
【火法冶金】[冶] pyrometallurgy
【火攻】fire attack (using fire as a weapon against enemy personnel and installations)
【火钩子】fire hook; poker
【火罐】[中医] cupping jar: 拔～ cupping
【火光】flame; blaze: ～冲天。The flames lit up the sky.
【火锅】chafing dish
【火海】a sea of fire
【火红】red as fire; fiery; flaming: ～的太阳 a flaming sun
【火候】①(火力大小、时间长短) duration and degree of heating, cooking, smelting, etc.: 这鸭子烤得正到～。This roast duck is done to a turn. ②(功夫深浅) level of attainment: 他的艺术到了～了。His art has matured ③(紧急时刻) a crucial moment: 正在战斗的～上 at the crucial moment of the battle
【火狐】red fox
【火花】spark: ～四溅 sparks flying off in all directions ◇ ～塞[机] sparking plug; spark plug; ignition plug
【火化】cremation
【火鸡】turkey
【火急】urgent; pressing: 十万～ most urgent
【火剪】①(生火用的火钳) fire-tongs; tongs ②(烫发用具) curling tongs; curling irons
【火碱】caustic soda
【火箭】rocket: 发射～ fire a rocket / 穿甲～ armor-piercing rocket / 单级～ one-stage rocket / 弹道～ ballistic rocket / 多级～ multi-step rocket / 二级～ two-stage rocket / 高超音速～ hypersonic rocket / 固体推进剂～ solid-propellant rocket / 机载高速～ high-velocity aircraft rocket / 三级～ three-stage rocket / 同温层～ stratospheric rocket / 信号～ signal rocket / 液体推进剂～ liquid-propellant rocket / 制动～ brake rocket ◇ ～部队 rocket troops / ～弹 rocket projectile; rocket shell / ～发射场 rocket launching site / ～发射台 rocket launching pad; rocket mount / ～技术 rocketry / ～炮 rocket gun / ～筒 rocket launcher; bazooka
【火警】fire alarm: ～灯 firewarning light
【火炬】torch ◇ ～赛跑 torch race / ～游行 torchlight parade
【火坑】fiery pit; pit of hell; abyss of suffering: 跳出～ escape from the living hell
【火筷子】fire-tongs; tongs
【火辣辣】burning: ～的太阳 a scorching sun / 心里～的 burning with anxiety
【火力】[军] firepower; fire: 发扬～ make full use of firepower ◇ ～点 firing point / ～发电厂 thermal power plant / ～控制 control of fire; fire control / ～配系 organization of fire; fire system / ～突击 fire assault / ～圈 field of fire / ～网 network of fire; fire net / ～掩护 fire cover / ～侦察 reconnaissance by firing (to observe enemy reactions) / ～支持 support fire; fire support
【火烈鸟】flamingo
【火流星】[天] bolide; fireball
【火龙】①(连成串的灯火) fiery dragon; a procession of lanterns or torches ②[方](通烟囱的孔道) an air channel from a brick kitchen stove to a chimney; flue
【火炉】(heating) stove
【火冒三丈】fly into a rage; fly into a towering passion: 使人～ make sb.'s blood boil
【火棉】guncotton; pyroxylin
【火苗】a tongue of flame; flame
【火捻】①(引火易燃物) kindling ②(导火线) fuse
【火炮】cannon; gun
【火盆】fire pan; brazier
【火漆】sealing wax
【火气】①[中医](引起发炎等的病因) internal heat ②(脾气) anger; temper: ～很大 have a bad temper
【火器】[军] firearm
【火钳】fire-tongs; tongs
【火枪】firelock
【火墙】a wall with flues for space heating
【火球】fireball

【火热】①(非常热) burning hot; fervent; fiery: ～的太阳 a burning sun / ～的心 a fervent heart ②(十分亲近的) intimate: 打得～ carry on intimately with; be as thick as thieves
【火伞高张】The sun is shining fiercely.
【火山】volcano: 活～ active volcano / 死～ extinct volcano / 休眠～ dormant volcano ◇ ～玻璃 volcanic glass / ～岛 volcanic island / ～地震 volcanic earthquake / ～灰 volcanic ash / ～口 crater / ～砾 lapillus / ～喷发 volcanic eruption / ～学 volcanology / ～渣 scoria / ～作用 volcanism
【火伤】burn (caused by fire)
【火上加油】pour oil on the fire; add fuel to the flames
【火上添薪】feed fresh fuel to the fire; give fuel to fire
【火烧】[食品] baked wheaten cake
【火烧火燎】①(热得难受) feeling terribly hot ②(焦急) restless with anxiety
【火烧眉毛】the fire is singeing the eyebrows; a desperate situation; a matter of the utmost urgency
【火烧油层】[石油] combustion (of oil) in situ
【火烧云】morning glow; evening glow; crimson clouds at sunrise <sunset>
【火舌】tongues of fire
【火绳】a rope of plaited plants burnt as a mosquito repellent
【火树石】flint
【火树银花】fiery trees and silver flowers; a display of fireworks and a sea of lanterns (on a festival night)
【火速】at top speed; posthaste: ～增援 rush up reinforcements
【火炭】live charcoal
【火头】①(火焰) flame ②(火候) duration and degree of heating, cooking, smelting, etc. ③(怒气) fit of anger; blaze of temper; flare-up; anger: 正在～上 at the height of one's anger
【火头军】(现用做戏谑的话) army cook
【火腿】ham
【火网】[军] network of fire; fire net
【火卫】[天][简](火星卫星) Martian satellite
【火险】fire insurance ◇ ～遇警仪 fire-danger meter
【火线】①(战场前沿) battle line: ～抢救 frontline first aid ②[电] live wire
【火硝】[化] nitre; saltpetre ◇ ～纸 touch paper
【火星】①(小的火) spark: ～迸发 a shower of sparks ②[天] Mars
【火眼】[医] pinkeye
【火眼金睛】piercing eye; penetrating insight
【火焰】flame ◇ ～光谱 flame spectrum / ～喷射器 flamethrower
【火药】gunpowder; powder ◇ ～库 powder magazine / ～桶 powder keg
【火药味】the smell of gunpowder: 这是一篇充满～的声明。This statement has a strong smell of gunpowder.
【火印】a mark burned on bamboo or wooden articles; brand
【火油】[方] kerosene
【火灾】fire (as a disaster); conflagration ◇ ～保险 fire insurance / ～损失评定 fire loss adjustment
【火葬】cremation ◇ ～场 crematorium; crematory
【火中取栗】pull sb.'s chestnuts out of the fire; be a cat's-paw
【火种】kindling material; kindling; tinder; live cinders kept for starting a new fire
【火烛】things that may cause a fire: 小心～! Be careful about fires!
【火主】a house where a fire started
【火砖】firerick

钬 [化] holmium (Ho)

伙 ①(伙食) mess; board; meals: 包～ get or supply meals at a fixed rate; board ②(同伴) partner; mate ③(由同伴组成的集体) partnership; company: 合～ enter into partnership ④[量] group; crowd; band: 一～强盗 a band of robbers ⑤(共同;联合) combine; join: ～买 club together to buy sth.
【伙伴】partner; companion: 小时候的～ a childhood pal
【伙房】kitchen (in a school, factory, etc.)
【伙夫】[旧] mess cook
【伙计】①(合伙人) partner ②[口](同伴) fellow; mate ③[旧](店员) salesman; salesclerk; shop assistant ④[旧](长工) farm laborer
【伙食】mess; food; meals ◇ ～补助 food allowance / ～费 money spent on meals; board expenses / ～节余 mess savings
【伙同】in league with; in collusion with; gang up with sb.

夥 [书] much; a great deal; many; numerous: 获益甚～ have derived much benefit

huò

豁 ①(通达) clear; open; open-minded; generous ②(免除) exempt; remit

【豁达大度】be generous and open-minded; open-minded and magnanimous

【豁亮】①(宽敞明亮) roomy and bright: 这屋子又干净,又～。The room is clean, bright and spacious. ②(响亮) sonorous; resonant: 嗓音～ have a sonorous voice

【豁免】exempt (from taxes or from customs inspection, etc.); remit: 民事管辖的～ exemption from civil jurisdiction / 刑事管辖的～ exemption from criminal jurisdiction / ～捐税 exempt sb. from taxes; remit taxes / 外交～权 diplomatic immunity ◇ ～权 immunity; charter; right of immunity / ～证明 negative clearance

【豁然贯通】suddenly see the whole thing in a clear light

【豁然开朗】suddenly see the light; be suddenly enlightened

祸 ①(祸事) misfortune; disaster; calamity: 车～ traffic accident; road accident / 惹～ cause trouble ②(损害) bring disaster upon; ruin

【祸不单行】misfortunes never come singly

【祸从口出】Out of the mouth comes evil.(或: Disaster emanates from careless talk.)

【祸从天降】Disaster comes from sky.(或: Misfortune dropped from Heaven and fell on somebody.)

【祸端】[书] the source of the disaster; the cause of ruin

【祸福与共】share thick and thin with...; go through thick and thin together

【祸根】the root of the trouble; the cause of ruin; bane: 铲除～ uproot the evil / 种下～ sow the seeds of misfortune

【祸国殃民】bring calamity to the country and the people

【祸害】①(祸事) disaster; curse; scourge ②(引起灾难的人或事物) scourge; curse; bone: 他真是我一生的～。He has been the bane of my life. ③(损害) bring disaster to; wreck; ruin

【祸起萧墙】trouble arises within the family; there is internal strife afoot

【祸事】disaster; calamity; mishap

【祸首】chief culprit

【祸胎】the root of the trouble; the cause of the disaster

【祸兮福所倚,福兮祸所伏】good fortune lieth within bad, bad fortune lurketh within good

【祸心】evil intent: 包藏～ harbor malicious intentions

霍 suddenly; quickly

【霍地】[副] suddenly: ～立起身来 suddenly stand up; spring to one's feet

【霍霍】①[象] ～的磨刀声 the scrape, scrape of a sword being sharpened ②(闪动) flash: 电光～。The lightning flashed.

【霍乱】①[医] cholera ②[中医] acute gastroenteritis

【霍然】①[副](突然) suddenly; quickly: ～而起 spring to one's feet / 手电筒～一亮。Suddenly somebody flashed an electric torch. ②[书](病痛迅速消除) (of an illness) be cured quickly: 数日之后,定当～。You will be restored to health in a matter of days.

【霍然痊愈】Suddenly one recovers from an illness.

【霍然云消】Hastily the clouds dispersed.

藿 [书] leaves of pulse plants

【藿香】[中药] wrinkled giant hyssop

获 ①(捉住) capture; catch: 捕～ capture ②(得到;获得) obtain; win; reap: ～救 be rescued / ～利 make a profit; reap profits / ～一等奖 win the first prize / 喜～丰收 happily reap a bumper harvest / 不劳而～ enjoy the fruits of other people's labor; reap without sowing

【获得】gain; obtain; acquire; win; achieve: ～独立 gain independence / ～好评 win acclaim; earn favorable comment / ～巨大的成绩 achieve great success / ～知识 acquire knowledge

【获得保释】[法] binding over

【获得性】[生] acquired character: ◇ ～免疫 acquired immunity

【获利年度】profit-making year

【获胜】win victory; be victorious; triumph: 甲队以四比一～。Team A won the match four to one.

【获释】get-off

【获悉】[书] learn (of an event)

或 ①[副](也许;或许) perhaps; maybe; probably: 代表团明晨～可达到。The delegation may arrive tomorrow morning. ②[连](或者)

or; either ... or ...: 这块地可以种高粱～玉米。We can grow sorghum or maize on this plot. ③[书](某人) some one; some people: ～曰 someone says; some say

【或大或小】either large or small

【或去或留】go or stay

【或然】probable

【或然率】[数] probability

【或许】[副] perhaps; maybe: 她～能来。She might be able to come.

【或者】①[副] perhaps; maybe ②[连] or; either ... or ...

惑 ①(迷惑) be puzzled; be bewildered: 大～不解 be greatly puzzled ②(使迷惑) delude; mislead: 造谣～众 fabricate rumors to mislead people

和 ①(搀和) mix; blend: 豆沙里～点儿糖 mix a little sugar into the bean paste／把水泥、沙和石子～在一起 mix cement, sand and pebbles ②[量]衣裳已经洗了三～。The clothes have been rinsed three times.

【和稀泥】try to mediate differences at the sacrifice of principle; try to smooth things over

货 ①(货物) goods; commodity: 交～ deliver goods／接～ receive goods／理～ tally goods／盘～ take stock; stock-taking／送～上门 sell goods at the customers' doors／无～ be all sold out; out of stock／卸～ discharge (cargo)／有～ (have sth.) in stock ②(钱) money: 通～ currency ③[骂]蠢～ blockhead; idiot

【货币】money; currency: 储备～ reserve currency／法定～ legal tender／可兑换～ convertible currency／周转～ vehicle currency／自由兑换～ convertible currency ◇ ～贬值 (currency) devaluation; (currency) depreciation／～单位 monetary unit／～地租[经] money rent／～发行 issuance of paper money／～工资 money wages／～供应量 money supply／～回笼 withdrawal of currency from circulation／～交换 exchange through money／～流通 currency circulation／～流通量 currency in circulation; money supply／～升值 (currency) revaluation; (currency) appreciation／～危机 monetary crisis／～信用 confidence in the currency／～增值 revaluation／～政策 monetary policy／～资本 money-capital ○硬币 coin／辅币 subsidiary note; subsidiary coin／镍币 nickel coin; nickel

【货舱】(cargo) hold; cargo bay (of a plane)

【货场】goods yard

【货车】①(货运列车) goods train; freight train ②(运货车皮) goods van; freight car ③(运货卡车) truck; lorry

【货船】freighter; cargo ship; cargo vessel: 定期～ cargo liner

【货单】manifest; waybill; shipping list

【货到付款】cash on delivery (COD)

【货到即提】delivery on arrival

【货到支付】delivery on arrival

【货好客自来】Good wine needs no bush.

【货机】[航空] cargo aircraft; air freighter

【货价】commodity price; price of goods

【货架子】goods shelves

【货款】money for buying or selling goods; payment for goods

【货郎】itinerant pedlar; street vendor ◇ ～担 street vendor's load (carried on a shoulder pole)

【货品】kinds or types of goods ◇ ～检验证 certificate of inspection

【货色】①(货物) goods: 上等～ first-class goods; quality goods／～齐全。Goods of every description are available. ②(指人、言论等,含贬义) stuff; trash; rubbish: 你知道他是什么～吗? Don't you know what stuff he is made of?

【货摊】stall; stand

【货物】goods; commodity; merchandise: ～名称 description of goods／～种类 type of merchandise ◇ ～残损检验 inspection on damaged cargo／～检查 examination of cargo／～税 commodity tax／～托运 consignment of goods／～装卸 cargo work

【货箱】packing box

【货样】sample goods; sample ◇ ～卡 sample card／～室 sample room

【货源】source of goods; supply of goods: ～充足 an ample supply of goods

【货运】freight transport ◇ ～单 waybill／～费 shipping cost; freight (charges)／～公司 transport company／～量 volume of goods transported; volume of rail freight; volume of road haulage／～列车 goods train; freight train／～业务 cargo service／～周转量 rotation volume of goods transport

【货栈】warehouse

【货真价实】①(货非冒牌,价钱实在) genuine

goods at a fair price ② (地道的) through and through; out-and-out; dyed-in-the-wool: ~的骗子 an out-and-out swindler

【货主】owner of cargo

J

jī

激 ①(水受阻或震荡而上涌) wash; surge; dash: 一石～起千层浪。 A tossed stone raises a thousand ripples. ②(冷水突然刺激身体使得病) fall ill from getting wet; catch a chill (from getting in the rain) ③(使发作; 使感情激动) arouse; stimulate; excite: ～起生产热潮 arouse a great upsurge in production ④(急剧; 强烈) sharp; fierce; violent: ～战 fierce fighting

【激昂】roused; aroused; excited and indignant; be emotionally wrought up: 群情～ public feeling was aroused 〈ran high〉

【激变】violent change; cataclysm

【激波】[物] shock wave

【激荡】agitate; surge; rage: 江流～ the river is surging / 心潮～ thoughts surging in one's mind

【激动】excite; stir; agitate: ～得流下眼泪 be moved to tears / ～地说 speak in an agitated tone; say with feeling / ～人心的讲话 a stirring speech; a rousing speech / 令人～的场面 an inspiring scene / 情绪～ be excited; get worked up

【激发】①(使奋发) arouse; stimulate; set off; stir up: ～某人的革命热情 arouse sb.s' revolutionary fervor ② [物] excitation: 热～ thermal excitation ◇ ～机制 excitation mechanism / ～能 excitation energy / ～能级 excitation level / ～态 excited state

【激奋】be roused to action; be stirred into activity

【激愤】wrathful; indignant; enraged: 心情～ be filled with indignation

【激光】[物] laser ◇ ～波束制导武器 laser beam riding weapon / ～测距仪 laser range finder / ～测云仪 laser ceilometer / ～打印机 laser printer / ～导弹跟踪系统 laser missile tracking system / ～干涉仪 laser interferometer / ～光谱学 laser spectroscopy / ～扫描器 laser scanner / ～手术刀 laser scalpel / ～束 laser beam / ～显微光谱分析仪 laser microspectral analyser / ～照明器 laser illuminator / ～准直仪 laser collimator

【激化】sharpen; intensify; become acute: 矛盾的～ intensification of contradictions / 斗争进一步～。 The struggle became more acute.

【激活】[物] activation ◇ ～剂 activator / ～媒质 active medium / ～能 activation energy

【激将法】prodding sb. into action; goading sb. into action (as by ridicule, sarcasm, etc.)

【激进】radical ◇ ～分子 radicals / ～派 radicals

【激励】①(激发鼓励) encourage; impel; urge; inspire: ～斗志 inspire one's fighting will / ～士气 give a boost to the morale of the army ② [电子] drive; excitation ◇ ～器 driver; exciter

【激烈】intense; sharp; fierce; violent; acute: ～的比赛 a closely fought game; a gruelling match / ～的冲突 sharp conflict / ～的语言 violent language / ～的战斗 a fierce battle / ～的争论 heated argument / 争吵得很～ quarrel bitterly

【激流】torrent; rapids; turbulent current: 闯过～ shoot the rapids

【激酶】[生化] kinase

【激怒】enrage; infuriate; exasperate: ～某人 provoke sb. to anger

【激起】arouse; evoke; stir up: ～公愤 arouse public indignation / ～某人的愤怒 stir sb.'s wrath / ～强烈反抗 evoke strong opposition / ～一场风波 cause a commotion

【激切】[书] impassioned; vehement: 言辞～ impassioned language

【激情】intense emotion; fervor; passion; enthusiasm: 满怀～ be filled with enthusiasm

【激素】[生理] hormone: 生长～ growth hormone / 性～ sex hormone

【激于义愤】be stirred by righteous indignation

【激越】intense; vehement; loud and strong

【激增】increase sharply; soar; shoot up: 产量～ steep rise in output

【激战】fierce fighting; fierce battle; pitched battle: ～竟日 a bitter battle was fought the whole of the day

【激浊扬清】drain awıy the mud and bring in fresh water; drive out evil and usher in good; eliminate vice and exalt virtue

【激子】[物] exciton

齑

【齑粉】[书] fine powder; broken bits: 碾成～ be ground to dust

跻 [书] ascend; mount

迹 ①(痕迹) mark; trace: 血～ bloodstain／足～ footmark; footprint ②(前人遗留的建筑或事物) remains; ruins; vestige: 陈～ a thing of the past／古城墙的遗～ the ruins of an old city wall ③(形迹) an outward sign; indication: ～近剽窃 an act verging on plagiarism
【迹地】[林] slash
【迹象】sign; indication

绩 ①(用麻纤维搓成线) twist hempen thread ②(功业;成果) achievement; accomplishment; merit: 功～ merits and achievements; contributions／战～ military achievement; military exploit

积 ①(积累) amass; store up; accumulate ②(长时间积累下来的) long-standing; long-pending; age-old ③[中医](积食) indigestion ④[数] product: 求～ find the product by multiplication
【积案】a long-pending case
【积弊】long-standing abuse; age-old malpractice ◇～难除 accumulated evil practices are difficult to remove
【积不相能】have always been at variance; have never been on good terms
【积层电池】packed cell
【积储】store up
【积存】store up; lay up; stockpile: ～的物资 goods in stock
【积德】accumulate virtue
【积肥】collect manure; farmyard manure; store compost
【积分】[数] integral: 不定～ indefinite integral／定～ definite integral ◇～变换 integral transforms／～方程 integral equation／～学 integral calculus／～仪 integrator
【积谷防饥】store up grain against famine
【积极】①(肯定的;正面的) positive: ～因素 positive factors／起～作用 play a positive role ②(进取的;热心的) active; energetic; vigorous: ～工作 work hard; work with all one's energy／采取～措施 adopt vigorous measures ◇～分子 activist; active element; enthusiast／～平衡 positive balance／～性 zeal; initiative; enthusiasm
【积久】accumulate in the course of time: ～成习 form a habit or custom through long-repeated practice
【积聚】gather; accumulate; build up: 资本～ concentration of capital
【积劳成疾】[书] break down from constant overwork; fall sick from overwork
【积累】accumulate: ～与消费 accumulation and consumption／～资金 accumulate funds
【积木】building blocks; toy bricks
【积年】[书] for many years: ～旧案 law cases which have piled up over the years
【积欠】①(累次欠下) have one's debts piling up ②(积累下的亏欠) outstanding debts; arrears: 还清～ clear up all outstanding debts／～股利 dividends in arrear
【积少成多】many a little makes a mickle; a penny saved is a penny earned
【积土成山,积水成川】Heaped-up earth becomes a mountain, accumulated water becomes a river.
【积习】old habit; long-standing practice: ～难除。 It is difficult to get rid of deep-rooted habits. (或: Old habits die hard.)／～难改。 A settle practice is hard to reform.
【积蓄】①(储蓄) put aside; save; accumulate: ～一笔钱 save a sum of money ②(存款) savings: 辛苦得来的～ hard-earned savings
【积雪】accumulated snow; accumulation of snow; snows: 高原～ highland snows／～盈尺 accumulated snow is over an inch
【积压】keep long in stock; overstock: ～物资 materials kept too long in stock／产品～ overstocking of products
【积雨云】[气] cumulonimbus
【积怨】accumulated rancor; piled-up grievances: ～甚多 have incurred widespread resentment; have many complaints against one
【积云】[气] cumulus: 层～ stratocumulus／高～ altocumulus／卷～ cirrocumulus
【积载因素】stowage factor
【积重难返】bad old practices die hard; old habits are difficult to get rid of
【积铢累寸】save every tiny bit; accumulate bit by bit

击 ①(打;敲打) beat; hit; strike: ～鼓 beat a drum／～掌 clap one's hands ②(攻打) attack; assault: 声东～西 feint in the east and attack in the west／痛～敌人 attack the enemy mercilessly ③(碰;接触) come in contact with; bump into: 撞～ collide with; ram
【击败】defeat; beat; vanquish
【击毙】shoot dead; strike dead

【击沉】bombard and sink; send (a ship) to the bottom
【击穿】[电] puncture; breakdown
【击发】[军] percussion ◇ ～装置 percussion lock; percussion mechanism
【击毁】smash; wreck; shatter; destroy
【击剑】[体] fencing ◇ ～场 fencing strip; fencing terrain; piste / ～服 fencing clothes / ～馆 fencing hall / ～技术 sword play; swordsmanship / ～裤 fencing breeches / ～上衣 fencing jacket / ～运动员 fencer ○花剑 foil; fleuret / 佩剑 saber / 面罩 wire-mesh mask; mask / 护胸 chest protector; breast protector(女用) / 刺中 hit on the target; hit received (scored) / 没刺中 hit off the target
【击溃】rout; put to flight; defeat utterly; smash: ～敌军 rout the enemy; put the enemy to flight
【击落】shoot down; bring down; down: ～敌轰炸机四架 bring down four enemy bombers
【击破】break up; destroy; rout: 各个～ destroy (enemy forces) one by one
【击球】[棒、垒球] batting ◇ ～员 batter; batsman
【击伤】wound; injure (a person); damage (a plane, tank, etc.)
【击退】beat back; repel; repulse: ～敌人的进攻 repulse the enemy attacks
【击弦乐器】hammered string instrument
【击以猛掌】give sb. a shove; give a sharp warning
【击乐器】percussion instrument
【击中】hit: ～目标 hit the target / ～痛处 hit one where it hurts / ～要害 hit sb.'s vital point; strike at the root of

基 ①(基础) base; foundation: 坝～ the base of a dam / 奠～ lay a foundation / 房～ foundations (of a building) / 路～ roadbed; bed ②(起头的; 根本的) basic; key; primary; cardinal: ～调 keynote / ～数 cardinal number ③[化] radical; base; group: 氨～ amino; amino-group / 石蜡～ paraffin base / 羟～ hydroxyl group / 自由～ free radical
【基本】①(根本的) basic; fundamental: ～观点 basic concept; fundamental view / ～规律 basic law / ～原则 basic principles ②(主要的) main; essential: ～特征 essential features / ～条件 main conditions ③(基础的) elementary; rudimentary: ～知识 elementary knowledge ④(大体上) basically; in the main; on the whole; by and large: 这部电影～是好的。This film is good on the whole. ◇ ～词汇[语] basic vocabulary; basic word-stock / ～点 main point; fundamental proposition / ～电荷 elementary charge / ～纲领 basic program / ～工资 basic wage; basic salary / ～核算单位 basic accounting unit / ～粒子[物] elementary particle / ～利率 prime rate / ～矛盾 basic contradiction / ～权利 fundamental right / ～条款 condition clause / ～职能 basic function / ～证据 basic evidence; primary evidence
【基本功】basic training; basic skill; essential technique: 苦练～ practice hard in basic skill
【基本建设】capital construction ◇ ～投资 investment in capital construction
【基本上】①(主要) mainly ②(大体上) on the whole; in the main
【基层】basic level; primary level; grass-roots unit ◇ ～单位 basic unit; grass-roots unit; unit at the grass-roots level / ～法院 grass-roots court / ～干部 cadre at the basic level / ～领导 leading body at the basic level; basic-level leadership / ～人民法院 basic-level people's court / ～人民检察院 basic-level people's procuratorate / ～选举 elections at the basic level / ～政权组织 organizations of political power at the grass-roots level / ～组织 primary organization; organization at the basic level
【基础】①(根基;起点) foundation; base; basis: 打～ lay a foundation / 经济～ economic base; economic basis / 理论～ theoretical basis / 物质～ material base / 在…～上 on the basis of ②(基本的) fundamental; basic; primary: ～工业 basic industries / ～教育 elementary education / ～科学 basic science / ～课 basic courses (of a college curriculum) / ～理论 basic theory / ～知识 rudimentary knowledge; elementary knowledge
【基础代谢】[生] basal metabolis
【基础设施】infrastructure
【基地】base: 导弹～ missile base / 工业～ industrial base / 军事～ military base / 原料～ source of raw materials
【基点】①(中心;重点;基础) basic point; starting point; center: 分析问题是解决问题的～。The analysis of a problem is the starting point for its solution. ②[测] base point (BP) ◇ ～价格 basing point pricing
【基调】[乐] fundamental key; main key ②(主要精神) keynote: 定下会议的～ set the keynote

for the conference
【基督】[宗] Christ
【基督教】Christianity; the Christian religion
【基督徒】Christian
【基肥】[农] base manure; base fertilizer
【基干】backbone; hard core ◇ ～民兵 primary militia; core members of the militia
【基极】[电子] base
【基价】base price
【基建】capital construction
【基金】fund; foundation: 大修～ funds for major overhaul / 福利～ welfare fund / 积累～ accumulation fund / 教育～ education fund / 消费～ consumption fund
【基金会】foundation
【基里巴斯】Kiribati
【基期】[统计] base period
【基石】foundation stone; cornerstone
【基数】①[数] cardinal number ②[统计] base
【基线】[测] datum line
【基因】[生] gene: 等位～ allele / 显性～ dominant gene ◇ ～流动 gene flow / ～突变 gene mutation / ～型 genotype / ～中心 gene-center
【基音】[乐] fundamental tone
【基于】because of; in view of; on account of: ～上述理由… For the above-mentioned reasons ... (或: In view of above-mentioned reasons ...)
【基准】①[测] datum ②(标准) standard; criterion ◇ ～点 datum point / ～面 datum plane / ～线 datum line
【基准兵】guide; base marker

奇 odd (number)
【奇数】[数] odd number; uneven number
【奇蹄目】[动] Perissodactyla
【奇整数】odd integer

畸 ①(偏) lopsided; unbalanced ②(不正常的; 不规则的) irregular; abnormal
【畸变】[物] distortion
【畸轻畸重】attach too much weight to this and too little weight to that; lopsided; unbalanced; now too much, now too little
【畸胎】monster; teratism
【畸形】①[医] deformity; malformation: 先天～ congenital malformation / 肢体发育～ have deformed limbs ②(不正常) lopsided; unbalanced; abnormal: ～发展 lopsided development / ～现象 abnormal phenomenon ◇ ～手 clubhand; talipomanus / ～胎 monster / ～学 teratology / ～足 clubfoot; talipes

犄
【犄角】①(角; 角落) corner: 在院子～ in a corner of the courtyard / 桌子～ the corner of a table ②(动物的角) horn: 鹿～ antler / 牛～ ox horn

唧 spurt; squirt
【唧唧】[象](形容虫叫声) chirp
【唧哝】talk in a low voice; whisper
【唧筒】pump

羁 [书] ①(马笼头) bridle; headstall: 无～之马 a horse without a bridle; unbridled horse ②(拘束) control; restrain: 放荡不～ unconventional and uninhibited; free from restrain ③(停留) stay; delay; detain: 事务～身 be detained by one's duties
【羁绊】[书] trammels; fetters; yoke: 摆脱…的～ shake off the yoke of / 挣脱…的～ break the fetters of
【羁留】①(在外地停留) stay; stop over: 短期～上海 short sojourn at Shanghai / 在京～五日 stop over in Beijing for five days ②(拘留) keep in custody; detain
【羁押】[书] detain; take into custody ◇ ～候审 committed for trial; custody for trial

稽 ①(查考) check; examine; investigate: 有案可～ be on record; be verifiable ②(停留; 拖延) delay; procrastinate: ～延时日 be considerably delayed
【稽查】①(检查) check (to prevent smuggling, tax evasion, etc.); inspect ②(检查人员) inspector; customs officer
【稽核】check; examine; audit: ～帐目 audit accounts
【稽考】[书] ascertain; verify: 无可～ be unverifiable
【稽留】[书] delay; detain: 因事～ be detained by business
【稽留热】[医] continued fever

几 ①(小桌子) a small table: 茶～儿 tea table; teapoy ②(几乎; 近乎) nearly; almost; practically
【几不欲生】almost did not desire to live
【几丁质】[生化] chitin
【几乎】nearly; almost; practically
【几近于零】practically negligible
【几率】[数] probability

【几维鸟】kiwi
【几于完备】approximate perfection
讥 ridicule; mock; satirize
【讥讽】ridicule; satirize; throw out innuendoes against
【讥笑】ridicule; jeer; sneer at; deride
机 ①(机器) machine; engine: 缝纫～ sewing machine／ 内燃～ internal-combustion engine ②(飞机) aircraft; airplane; aeroplane; plane: 滑翔～ glider／ 舰载～ carrier-borne aircraft／ 客～ passenger plane／ 僚～ wing plane／ 长～ lead plane; pilot plane ③(事情变化的枢纽; 有重要关系的环节) crucial point; pivot; key link: 转～ a turning point; a turn for the better ④(机会) chance; occasion; opportunity: 趁～ take advantage of the occasion; seize the opportunity 〈chance〉／ 见～行事 do as one sees fit; use one's discretion ⑤(生活机能) organic: 无～化学 inorganic chemistry／ 有～体 organism ⑥(机变; 灵活) flexible; quick-witted
【机不可失】can't afford to lose the opportunity; not let the opportunity slip through one's fingers: ～, 时不再来 don't let slip an opportunity, it may never come again; opportunity knocks but once
【机舱】①(船舶机器房) engine room ②(飞机客舱) passenger compartment; cabin
【机场】airport; airfield; aerodrome: 国际～ international airport／ 简易～ airstrip／ 军用～ military airfield ◇ ～标志 aerodrome markings／ ～待战 ground alert／ ～灯标 airport beacon
【机车】[铁道] locomotive; engine: 电力～ electric locomotive／ 内燃～ diesel locomotive／ 蒸汽～ steam locomotive ◇ ～车辆厂 rolling stock plant／ ～牵引力 hauling capacity of a locomotive／ ～组 locomotive crew
【机船】motor vessal
【机床】machine tool: 金属切削～ metal cutting machine tool／ 精密～ precision machine tool／ 木工～ woodworking machine tool／ 数字程序控制～ numerical controlled machine tool／ 重型～ heavy machine tool ◇ ～厂 machine tool plant; machine tool works／ ～工业 machine tool industry
【机电产品】mechanical and electrical products
【机动】①(利用机器开动的) motor-driven; motorized: ～车辆 motor-driven vehicle; motor vehicle ②(权宜; 灵活) flexible; expedient; mobile: ～处置 deal with sth. flexibly ③(准备灵活运用的) keep in reserve; for emergency use: ～力量 reserve force／ ～时间 time kept in reserve／ ～资金 a fund kept in reserve／ 留百分之二十作～开支 keep back 20% for extras ◇ ～粮 grain reserve for emergency use／ ～炮 mobile artillery／ ～税 sliding duties／ ～性 mobility; maneuverability／ ～自行车 moped
【机动车】motor vehicle ◇ ～保险 motor vehicle insurance
【机断】act on one's own judgment in an emergency: ～行事 act promptly at one's own discretion
【机帆船】motor sailboat; motorized junk
【机房】generator room; motor room; engine room of a ship
【机耕】[农] tractor-ploughing ◇ ～船 boat tractor; wet-field tractor／ ～面积 area ploughed by tractors
【机工】mechanic; machinist
【机构】①[机] mechanism: 传动～ transmission mechanism／ 分离～ disengaging mechanism ②(机关; 团体) organ; organization; institution; setup: 金融～ financial setup／ 科研～ institution for scientific research／ 行政～ administrative setup／ 宣传～ propaganda organ／ 政府～ government organization ③(机关团体内部组织) the internal structure of an organization: 精简行政～ simplify administrative structure／ 调整～ adjust the organizational structure／ ～臃肿 inflated departments; overexpansion of organizations／ ～臃肿, 层次重叠 unwieldy and over-lapping organizations; overstaffed and divided into many overlapping levels
【机关】①[机] mechanism; gear: 起动～ starting gear ②(机械控制的) machine-operated: ～布景 machine-operated stage scenery ③(办理事务的部门) office; organ; body: 党政～ Party and government organizations／ 公安～ public security organs／ 领导～ leading bodies／ 上级～ higher leading bodies／ 文化教育～ cultural and educational institutions ④(周密巧妙的计谋) stratagem; scheme; intrigue: 识破～ see through a trick ◇ ～报 organ; official newspaper／ ～干部 government functionary; office worker／ ～刊物 organ／ ～炮 machine cannon／ ～枪 machine gun
【机会】chance; opportunity: 错过～ lose a chance／ 借此～ take this occasion to; avail oneself of the opportunity to／ 抓住～ seize a

chance
【机会均等】principle of equal opportunity for all; equal opportunity
【机会主义】opportunism ◇ ～路线 opportunist line／～者 opportunist
【机件】[机] parts; works: 车床～ parts of a machine tool／钟表～ the works of a clock or watch
【机井】motor-pumped well
【机警】alert; sharp-witted; vigilant: ～的战士 a vigilant soldier
【机具】machines and tools: 农业～ farm implements
【机库】[航空] hangar
【机理】mechanism: 分娩～ [医] mechanism of labor／腐蚀～ corrosion mechanism／结晶～ crystallization mechanism
【机灵】clever; smart; sharp; intelligent: ～的孩子 a clever child; a sharp child／这人办事挺～。This chap manages things quite cleverly.
【机米】machine-processed rice
【机密】①(重要而秘密) secret; classified; confidential: ～文件 classified papers; confidential documents ②(机密的事) secret: 严守国家～ strictly guard state secrets
【机敏】alert and resourceful
【机谋】[书] stratagem; artifice; scheme
【机能】[生] function ◇ ～不全 insufficiency／～亢进 hyperfunction／～紊乱 disorder／～性肥大 functional; hypertrophy／～障碍 dysfunction; functional disturbance
【机器】machine; machinery; apparatus: 安装新～ install new machinery／国家～ state apparatus: state machine／战争～ the war machine ◇ ～保养 machine maintenance／～打包 machine press-packing／～翻译 machine translation／～负荷 machine burden／～设备 machinery equipment／～造型 machine molding／～制造 machine building
【机器人】robot ◇ ～学 robotics
【机器油】lubricating oil; lubricant
【机枪】machine gun: 多管～ multiple machine gun／高射～ antiaircraft machine gun／轻～ light machine gun／手提～ light automatic gun／圆盘～ drum-fed gun／重～ heavy machine gun◇ ～手 machine gunner
【机巧】adroit; ingenious
【机群】a group of planes: 大～ air armada; air fleet
【机身】[航空] fuselage
【机体】① [生] organism ② [航空] airframe
【机头】[航空] nose ◇ ～炮 nose gun
【机尾】[航空] tail
【机务人员】①(维修人员) maintenance personnel ② [航空] (地勤人员) ground crew
【机械】①(装置) machinery; machine; mechanism: ～故障 mechanical failure; mechanical breakdown ②(死板;刻板) mechanical; inflexible; rigid: ～地搬用外国经验 mechanically copy foreign experience ◇ ～动力学 mechanical kinetics／～工程 mechanical engineering／～工业 engineering industry／～功 mechanical work／～加工 machining／～零件 machine components／～师 machinist／～效率 mechanical efficiency／～制图 mechanical drawing
【机械厂】machinery plant: 纺织～ textile machinery plant／机车车辆～ rolling stock parts plant／建筑～ building machinery plant／矿山～ mining machinery plant／农业～ agricultural machinery plant／重型～ heavy machinery plant
【机械化】mechanize: 农业～ mechanization of agriculture; mechanization of farm work ◇ ～部队 mechanized force; mechanized troops; mechanized unit
【机械论】mechanism
【机械能】[物] mechanical energy
【机械手】[机] manipulator: 仿效～ master-slave manipulator／万能～ general-purpose manipulator
【机械唯物主义】[哲] mechanical materialism
【机械运动】[物] mechanical movement
【机心】[钟表] movement
【机型】①(飞机型号) type ②(机器型号)model
【机修车间】machine repair shop; mechanical repair shop
【机要】confidential ◇ ～部门 departments in charge of confidential or important work／～工作 confidential work／～秘书 confidential secretary
【机宜】principles of action: 面授～ brief sb. on how to act
【机翼】[航空] wing
【机油】engine oil; machine oil
【机遇】favorable circumstances; opportunity
【机缘】good luck; lucky chance: ～凑巧 as luck would have it; by chance; by a lucky coincidence
【机载导弹】air-launched missile
【机载警戒与控制系统】airborne warning and control system (AWACS)

【机长】aircraft commander; crew commander; captain of an airplane
【机罩】[航空] bonnet
【机制】①(机器制造或加工的) machine-processed; machine-made: ～糖 machine-processed sugar / ～纸 machine-made paper ②(机器的构造;生物的功能;自然现象的规律) mechanism: 分娩～ mechanism of labor / 激发～ [物] excitation mechanism / 结晶～ crystallization mechanism / 竞争～ competition
【机智】quick-witted; resourceful: ～勇敢的侦察兵 brave and resourceful scouts
【机组】① [机] unit; set: 发电～ generating unit; generating set / 液压～ hydraulic unit ②(飞机工作人员) aricrew; flight crew

叽

【叽咕】talk in a low voice; whisper mutter
【叽叽喳喳】[象] chirp; twitter
【叽里咕噜】[象] ①(形容说话别人听不清楚或听不懂) gabble; jabber; talk in an indistinct manner ②(形容滚动的声音) rumble

肌

肌 muscle; flesh: 腹～ abdominal muscle / 随意～ voluntary muscle / 胸～ pectoral muscle
【肌氨酸】methyl aminoacetic acid; methyl glycocoll; sarcosine
【肌病】[兽医] myopathies
【肌醇】inositol
【肌肤】[书] (human) skin ◇ ～之亲 blood relations; intimate relations between man and woman
【肌苷酸】inosinic acid
【肌腱】[生理] tendon
【肌理】[书] skin texture: ～细腻 fine-textured skin
【肌强直】[医] myotonia
【肌肉】muscle: ～发达 muscular ◇ ～疾病 muscle diseases / ～系统 muscle systems / ～肿瘤 muscle tumor / ～注射 intramuscular injection
【肌体】human body; organism
【肌萎缩】[医] amyotrophy; muscular atrophy
【肌炎】[医] myositis
【肌营养不良】muscular dystrophy

饥

饥 ①(饥饿) be hungry; starve; famish ②(饥荒) famine; crop failure
【饥饱劳碌】slave all day long with no assurance when the next meal will come
【饥不择食】a hungry person is not choosy about his food; all food is delicious to the starving
【饥肠】[书] empty stomach; ～辘辘 one's stomach rumbling with hunger; rumblings of an empty stomach
【饥饿】hunger; starvation; be hungry; starve
【饥寒交迫】suffer hunger and cold; live in hunger and cold; be poverty-stricken; suffer from cold and hunger; suffer the terrible hardships of hunger and cold
【饥荒】①(庄稼收成不好或无收成) famine; crop failure ②(经济困难) be hard up; be hard pressed for money; be short of money ③(债) debt
【饥馑】famine; crop failure
【饥民】famine victim; famine refugee

汲

汲 draw (water): 从井里～水 draw water from a well
【汲取】draw; derive

芨

【芨芨草】splendid achnatherum

鸡

鸡 chicken: 雏～ chick; chicken / 公～ cock; rooster / 母～ hen / 养～场 chicken farm
【鸡蛋】(hen's) egg: 炒～ scrambled eggs / 煎～ fried eggs / 煮～ hard-boiled eggs ◇ ～糕 (sponge) cake
【鸡蛋里挑骨头】look for a bone in an egg; look for a flaw where there is none; find fault; nitpick
【鸡蛋碰石头】like an egg striking a rock; attack sb. far stronger than oneself
【鸡飞蛋打】the hen has flown away and the eggs in the coop are broken; all is lost
【鸡冠】cockscomb
【鸡冠花】[植] cockscomb
【鸡冠石】[矿] realgar
【鸡霍乱】fowl cholera
【鸡奸】sodomy; buggery
【鸡口牛后】fowl's beak and ox buttocks; preferable to lead in petty position than to follow behind greater leader
【鸡肋】[书] chicken ribs; things of little value or interest: 味同～ taste like chicken ribs; be of little or no value
【鸡零狗碎】in bits and pieces; fragmentary; odds and ends
【鸡毛】chicken feather
【鸡毛掸子】feather duster

【鸡毛当令箭】make a fuss about a casual remark dropped by one's superior: 拿着～ take a chicken feather for a warrant to give commands; treat one's superior's casual remark as an order and make a big fuss about it
【鸡毛蒜皮】chicken feathers and garlic skins; trifles; trivialities
【鸡鸣而起】rise at cockcrow
【鸡鸣狗盗】(ability to) crow like a cock and snatch like a dog; small tricks: ～之徒 people who know small tricks
【鸡内金】[中药] the membrane of a chicken's gizzard
【鸡皮疙瘩】gooseflesh; gooseskin
【鸡犬不惊】even fowls and dogs are not disturbed; excellent army discipline; peace and tranquility
【鸡犬不留】even fowls and dogs are not spared; ruthless mass slaughter; utter extermination
【鸡犬不宁】even fowls and dogs are not left in peace; general turmoil
【鸡犬相闻】live nearby or in the neighborhood
【鸡肉】chicken (as food)
【鸡舍】chicken coop
【鸡食】chicken feed
【鸡汤】chicken broth
【鸡头米】[植] Gorgon fruit
【鸡尾酒】cocktail ◇ ～会 cocktail party
【鸡瘟】chicken pest; fowl plague; fowl pest
【鸡窝】chicken coop; henhouse; roost
【鸡新城疫】newcastle disease
【鸡心】①(指形状) heart-shaped ②(首饰) a heart-shaped pendant ◇ ～领 V-neck
【鸡胸】[医] pigeon breast; chicken breast
【鸡血石】bloodstone; hiliotrope; oriental jasper
【鸡血藤】[植] reticulate millettia
【鸡眼】[医] corn; clavus ◇ ～膏 corn plaster
【鸡杂】chicken giblets

姬

【姬蜂】[动] ichneumon wasp; ichneumon

缉

缉 seize; arrest
【缉捕】seize; arrest ◇ ～人员 hounds of law
【缉拿】seize; arrest; apprehend: ～归案 bring (a criminal) to justice / ～凶手 apprehend the murderer
【缉私】seize smugglers; smuggled goods; suppress smuggling; search for smuggler ◇ ～船 anti-smuggling patrol boat; coast guard vessel / ～人员 anti-contraband personnel

jí

疾

疾 ①(疾病) disease; sickness; illness: 痼～ a stubborn illness / 眼～ eye trouble ②(痛苦) suffering; pain; difficulty: ～苦 sufferings; hardships ③(痛恨) hate; abhor ④(急速; 猛烈) fast; quick: ～驶 whirl away / (马)～驰 gallop away at full speed / ～走 walk fast
【疾病】disease; illness: 防治～ prevention and treatment of disease / 减少～ reduce disease / 控制～的发生和蔓延 control the outbreak and spread of a disease / ～丛生 be infested with all diseases
【疾恶如仇】hate evil as much as one hates an enemy; hate evil like an enemy
【疾风】①(强劲的风) strong wind; gale ②[气] moderate gale
【疾风知劲草】the force of the wind tests the strength of the grass; strength of character is tested in a crisis; loyal and virtuous
【疾苦】sufferings; hardships; difficulties: 关心群众的～ be concerned about the weal and woe of the masses
【疾首蹙额】with aching head and knitted brows; with abhorrence; knit the brows with hatred
【疾言厉色】harsh words and stern looks; sudden outpourings and fierce looks

蒺

【蒺藜】[植] puncture vine: 铁～[军] caltrop

嫉

嫉 ①(妒忌) be jealous; be envious ②(憎恨) hate
【嫉妒】be jealous of; envy
【嫉恨】envy and hate; hate out of jealousy

脊

【脊梁】back (of the human body)
【脊梁骨】backbone; spine
【脊檩】[建] ridgepole; ridgepiece
【脊瓦】[建] ridge tile

瘠

瘠 ①(瘦弱) lean; thin and weak ②(瘠薄) barren; poor; lean: ～土 poor soil; barren land
【瘠薄】barren; unproductive
【瘠田】barren land

吉

吉 lucky; auspicious; propitious: 万事大～。All is well.

【吉贝树】kapok tree; silk cotton tree
【吉卜赛人】Gypsy
【吉布提】Djibouti ◇～人 Djiboutian
【吉光片羽】a fragment of a highly treasured relic
【吉利】lucky; auspicious; propitious
【吉普车】jeep
【吉庆】auspicious; propitious; happy
【吉人天相】heaven helps a good man
【吉他】[乐] guitar
【吉屋招租】house for rent; house to let
【吉祥】lucky; auspicious; propitious
【吉祥如意】good fortune as one wishes
【吉星高照】the lucky star shines bright
【吉凶】good or ill luck
【吉凶未卜】cannot predict the outcome, good or bad; no one knows how it will turn out
【吉兆】good omen; propitious sign

籍 ①(书籍) book; record: 古～ ancient books ②(籍贯) native place; home town; birthplace: 回原～ return to one's native place／祖～ the land of one's ancestors ③(组织关系) member ship: 党～ party membership／国～ nationality
【籍贯】the place of one's birth or origin native place

棘 ①(酸枣) sour jujube ②(荆棘) thorn bushes; brambles ③[动] spine; spina
【棘轮】[机] ratchet (wheel)
【棘皮动物】echinoderm
【棘手】thorny; troublesome; knotty: ～的问题 a knotty problem
【棘爪】[机] pawl; detent; pallet; cam pawl

辑 ①(编辑) collect; compile; edit: 编～ edit; compile ②(丛书的部分) part; volume; division
【辑录】compile
【辑要】summary; abstract: 会议事项～ abstract of the proceedings at the meeting

集 ①(集合; 聚集) gather; collect; assemble: 聚～ gather together ②(集市) country fair; market: 赶～ go to a country fair; go to market ③(集子) collection; anthology: 地图～ atlas／歌曲～ song book／画～ an album of paintings／诗～ a collection of poems／选～ selected works ④(书、影片的一部分) volume; part: 上～ first part; part one／下～ second part; part two／分三～出版 published in three volumes
【集材】[林] logging; skidding; yarding: 索道～ cable logging ◇～道 skid road／～绞盘机 yarder
【集尘器】[机] dust arrester; dust collector; duster
【集尘设备】dust-collecting equipment
【集成电路】[电子] integrated circuit ◇～晶体管 integrated circuit transistor／～块 integrated circuit block
【集大成】be a comprehensive expression of; be an agglomeration of; epitomize
【集电杆】trolley pole
【集电弓】bow collector; bow trolley
【集电轨】collector rail
【集电极】[电子] collecting electrode; collector
【集电靴】collector shoe
【集合】①(聚集;使集合) gather; assemble; muster; call together: ～地点 assemble place; rendezvous／紧急～ emergency muster ②(口令) Fall in! ◇～号 bugle call for fall-in; assembly／～名词[语] collective noun
【集合论】[数] set theory
【集合体】[矿] aggregate
【集会】assembly; rally; gathering; meeting: ～结社自由 freedom of assembly and association／举行群众～ hold a mass rally
【集结】mass; concentrate; build up: ～待命 assemble and await orders／～军队 mass troops; concentrate forces／～力量 build up strength ◇～地域[军] assembly area
【集锦】a collection of choice specimens: 儿童画～ outstanding examples of children's drawings／图片～ collection of choice pictures
【集聚】gather; collect; assemble: ～一堂 assemble together in one place
【集刊】collected papers (of an academic institution)
【集流环】[电] slip ring
【集权】centralization of state power: 中央～的 centralized
【集日】market day
【集散地】collecting and distributing center; distributing center
【集市】country fair; market ◇～贸易 country fair trade; village fair trade
【集少成多】many a little makes a mickle
【集思广益】draw on collective wisdom and absorb all useful ideas; pool the wisdom of the masses; put heads together so as to get better results
【集体】collective: 战斗的～ a militant col-

lective／荣立～二等功 gain a Collective Award of Merit, Second Class ◇～安全 collective security／～财产 collective property; property collectively owned／～财产所有人 owners of collective property／～福利 collective welfare／～观念 collective spirit／～交易 collective bargaining／～经济 collective economy／～利益 collective interest／～领导 collective leadership／～农庄 collective farm／～企业 collective enterprise／～生产劳动 collective productive labor／～宿舍 dormitory／～所有制 collective ownership／～谈判 colletive bargaining／～舞 group dancing／～行为 collective behavior／～英雄主义 collective heroism／～治疗 group therapy／～主义 collectivism

【集体化】collectivization: 农业～ collectivization of agriculture

【集团】group; clique; circle; bloc: 军事～ a military bloc／统治～ the ruling clique; the ruling circle／小～ a small clique ◇～犯罪 organized crime／～军 group army; army group

【集训】assemble for training ◇～队[体] team of athletes in training

【集腋成裘】the finest fragments of fox fur, sewn together, will make a robe; many a little makes a mickle

【集邮】stamp collecting; philately ◇～簿 stamp-album／～者 stamp-collector; philatelist

【集约】[农] intensive ◇～经营 intensive farming／～农业 intensive agriculture

【集镇】town; market town

【集中】concentrate; centralize; focus; amass; put together: 思想不～ be absent-minded／～反映 embody a concentrated reflection of／～各方面的正确意见 sum up correct ideas from all quarters／～火力 concentrate fire (on a target)／～精力 concentrate one's energy／～目标 concentrate on the same target／～优势兵力 muster superior forces／～注意力 concentrate on; focus one's attention on ◇～采购 centralized purchasing／～供暖 central heating／～管理 centralized management／～轰炸 mass bombing／～指挥 centralized direction; centralized command

【集中营】concentration camp

【集装箱】[交] container ◇～车 container car／～船 container ship／～堆放场 container pool; container yard／～化 containerization／～码头 container terminal／～运输 containerized traffic

【集注】①(集中) focus ②(集解; 集释) variorum ◇～本 variorum edition

【集资】raise funds; collect money; pool resources; fund raising

【集子】collection; collected works; anthology

及 ①(达到) reach; come up to: 力所能～ within one's power／目力所～ as far as the eye can reach／由此～彼 proceed form one point to another ②(赶上) in time for: ～时 timely; in time ③[连] and

【及第】pass an imperial examination

【及格】pass a test, examination, etc.; pass: 入学考试～ pass the entrance examination

【及时】①(正赶上时候) timely; in time; seasonable: ～播种 sow in good time／～雨 timely rain ②(不拖延;立即) promptly; without delay: ～汇报 report without delay／要～治疗 need prompt medical treatment

【及物动词】[语] transitive verb

【及早】at an early date; as soon as possible; before it is too late

【及至】[连] up to; until

极 ①(顶点; 尽头) the utmost point; extreme: 无所不用其～ go to every extreme; go to any extreme; stop at nothing／愚蠢之～ be the height of folly／～而言之 talk in extreme terms ②(两极) pole: 北～ the North Pole／南～ the South Pole／阳～ positive pole／阴～ negative pole ③(最) extremely; exceed- ingly; utmost: ～为重要 of the utmost importance／～困难的问题 exceedingly difficult problem／～少数 a tiny minority; only a few; a handful

【极板】[电] plate

【极地】polar region ◇～航空 polar aviation／～航行 arctic navigation; polar air navigation

【极点】the limit; the extreme; the utmost: 感动到了～ be extremely moved／混乱到了～ in a state of utmost confusion

【极度】extreme; exceeding; to the utmost: ～疲劳 be extremely tired; be overcome with fatigue／～兴奋 be elated; get extremely excited

【极端】①(达到顶点) extreme: 走～ go to extremes ②(达到极点的) extreme; exceeding: ～仇视 extremely hostile; show extreme hatred for／～腐败 rotten to the core／～负责 have a boundless sense of responsibility／～困难 exceedingly difficult／～贫困 in dire poverty; extreme poverty／～愚蠢 be exceedingly fool-

ish / 对工作～负责 have a boundless sense of responsibility in one's work ◇ ～个人主义者 out-and-out egoist / ～民主化 ultra-democracy
【极光】[天] aurora; polar lights: 北～ aurora borealis; northern lights / 南 ～ aurora australis; southern lights
【极化】[物] polarization ◇ ～强度 polarization intensity / ～张量 polarization tensor / ～子 polaron
【极乐鸟】bird of paradise
【极乐世界】[佛教] Sukhavati; Pure Land; Western Paradise; Elysian Fields
【极力】do one's utmost; spare no effort: ～帮助 do one's best to help sb. / ～吹捧 laud sb. to the skies / ～鼓吹 vigorously publicize (an erroneous theory, etc.); clamorously advocate / ～扩大 expand to the maximum / ～劝阻 try very hard to dissuade sb. from doing sth. / ～缩小 reduce to the minimum; minimize / ～宣扬 vigorously advertise
【极量】[医] maximum dose
【极目】look as far as the eye can see: ～河山 hills and rivers that can be seen by the eyes / ～四望 take panoramic view from some vantage point / ～远眺 gaze into the distance
【极谱】[物] polarogram ◇ ～分析 polarographic analysis / ～学 polarography / ～仪 polarograph
【极其】[副] most; extremely; exceedingly
【极区】[地] polar region
【极圈】[地] polar circle: 北～ the Arctic Circle / 南～ the Antarctic Circle
【极权主义】totalitarianism ◇ ～者 totalitarian
【极盛】heyday; zenith; acme: ～时期 golden age; prime
【极限】①(最高的限度) the limit; the maximum; the ultimate ②[数] limit ◇ ～负载[电] limit load / ～压力[物] limiting pressure / ～值[数] limit value
【极刑】capital punishment; the death penalty; extreme penalty
【极夜】polar night
【极右】ultra-Right
【极值】[数] extreme value
【极左】ultra-Left ◇ ～思潮 ultra-Left trend of thought / ～分子 ultra-Leftist
【极坐标】polar coordinates

岌

【岌岌】precarious: ～不可终日 live in constant fear; live precariously / ～可危 in imminent danger

级

①(等级) level; rank; grade: 高～干部 cadres of higher rank / 各～领导干部 leading cadres at all levels / 六～地震 an earthquake of magnitude 6 (on the Richter scale) / 五～风 force 5 wind (on the Beaufort scale) / 一～茶叶 grade A tea; first-class tea / 最高～会谈 top-level talks; summit talks ②(年级) course; grade; class; form: 留～ repeat the year's work; stay down / 升～ promoted to a higher grade / 升留～制度 system of promoting or holding back students ③(台阶) step: 石～ stone steps ④[量] step; stage: 多～火箭 multistage rocket / 十几～台阶 a flight of a dozen steps ⑤[语] degree: 比较～ the comparative degree / 最高～ the superlative degree
【级别】rank; level; grade; scale: 干部～ the rank of a cadre / 工资～ wage scale; grade on the wage scale / 外交～ diplomatic rank
【级差地租】differential (land) rent
【级差佣金】graded commission
【级间分离】[宇航] stage separation
【级数】[数] progression; series

急

①(着急) impatient; anxious: ～着要答复 be impatient for a reply ②(使着急) worry: 什么事使你焦～? What's worrying you? ③(容易发怒;急躁) irritated; annoyed; nettled ④(急促) fast; rapid; violent: ～病 acute disease / ～雨 pelting rain ⑤(急迫;紧急) urgent; pressing ⑥(急事) urgency; emergency: 当务之～ the urgent task at present; the urgent matter confronting us / 告～ appeal for emergency help / 应～ meet an emergency ⑦(赶快帮助) be eager to help: ～人之难 be eager to help those in need
【急板】[乐] presto
【急病】acute disease ◇ ～请三医 in case of serious illness, the patient asks for more doctors
【急不可待】too impatient to wait; extremely anxious
【急驰而过】whirl away
【急赤白脸】face red or pale with too much anxiety
【急促】①(快而短促) hurried; rapid: ～的脚步声 hurried foot-steps / ～的枪声 rapid gunfire / 脉搏～ have a short, quick pulse / 呼吸～ be short of breath ②(时间短促) short; pressing
【急电】urgent telegram; urgent cable

【急风暴雨】violent storm; hurricane; acute abdomen
【急腹症】[医] acute abdominal disease; acute abdomen
【急公好义】zealous for the common weal; public-spirited; zealous for public interests
【急功近利】eager for quick success and instant benefit; too eager to be successful that one sees only the immediate advantages
【急件】urgent document; urgent dispatch
【急进】radical
【急惊风】[中医] acute infantile convulsions
【急惊风遇到了慢郎中】deferred action taken in cases requiring prompt attention
【急救】first aid; emergency treatment: 采取～措施 take first-aid measures ◇～包 first-aid dressing / ～车 emergency ambulance; breakdown van / ～人员 first- aid personnel / ～药品 first- aid medicine / ～药箱 first- aid kit; first-aid case / ～站 first-aid station
【急就章】hurriedly-written essay; hasty work; improvisation
【急剧】rapid; sharp; sudden: ～的变化 rapid change / ～上升 steep rise / ～上涨 steep rise; sharp rise / ～下降 sudden drop; sharp decline / ～增加 rapid increase / ～转折 abrupt turn
【急遽】rapid; sharp; sudden
【急来抱佛脚】clasp Buddha's feet when in trouble; seek help at the last moment
【急流】①(湍急的水流) torrent; rapid stream; rapids: 闯过～险滩 sweep over rapids and shoals ②[气] jet stream; jet flow
【急流划艇竞赛】[体] wild-water racing
【急流勇进】forge ahead against a swift current; press on in the teeth of difficulties
【急流勇退】resolutely retire at the height of one's official career
【急忙】in a hurry; in haste; hurriedly; hastily: 他～走了。He hurried off.
【急迫】urgent; pressing; imperative: 最～的任务 the most pressing task; task of extreme urgency
【急起直追】rouse oneself to catch up; make amends while there is yet time
【急切】①(迫切) eager; impatient; urgent; imperative ②(仓猝) in a hurry; in haste
【急群众之所急】be eager to meet the needs of the masses
【急如燃眉】as urgent as if the eyebrows were burning
【急如星火】extremely pressing; most urgent; posthaste; most hurriedly
【急刹车】①(紧急制动) slam the brakes on ②(突然中止) bring to a halt
【急速】very fast; at high speed; rapidly
【急湍】swift current
【急弯】①(陡弯) sharp turn ②(急转弯) turn suddenly ③[地] elbow
【急务】urgent task: 有～在身 have some urgent task on hand
【急先锋】daring vanguard
【急行军】rapid march
【急性】acute: ～传染病 acute infectious disease / ～阑尾炎 acute appendicitis
【急性病】①[医] acute disease ②(急于求成) impetuosity: 犯～ become impetuous
【急性子】①(性情急躁) of impatient disposition; impetuous ②(性情急躁的人) an impetuous person
【急需】①(迫切需要) be badly in need of ②(紧急需要) urgent need: 以应～ meet a crying need
【急用】urgent need
【急于】eager; anxious; impatient: ～表态 impatient to state one's position / ～求成 overanxious for quick results; impatient for success / 别～下结论。Don't jump at a conclusion.
【急躁】①(遇事好激动烦躁) irritable; irascible ②(性急) impetuous; rash; impatient: 产生～情绪 give way to impatience / 防止～情绪 guard against impetuosity
【急诊】emergency call; emergency treatment ◇～病人 emergency case / ～室 emergency ward
【急智】nimbleness of mind in dealing with emergencies; quick-wittedness
【急中生智】suddenly hit upon a way out of a predicament; show resourcefulness in an emergency
【急转弯】①(急剧转弯) take a sudden turn ②(大转变) make a radical change
【急转直下】(of the march of events, etc.) take a sudden turn and then develop rapidly

即 ①(靠近; 接触) approach; reach; be near: 可望而不可～ within sight but beyond reach ②(到; 开始从事) assume; undertake: ～位 ascend the throne ③(当下; 目前) at present; in immediate future: ～日 this or that very day / 成功在～。Success is in sight. ④(就着) prompted by the occasion: ～兴 impromptu ⑤

[书](就是) be; mean; namely: 非此～彼。It must be either this or that. ⑥[书](就便) promptly; at once; immediately; in no time: 闻过～改 correct one's mistake as soon as it is pointed out / 招之～来 be on call at any hour; be able to come at the first call ⑦[书](即使) even; even if

【即便】【即或】even; even if; even though

【即将】be about to; be on the point of; soon: 会议～开始 the meeting is about to begin / 胜利～到来 victory is at hand / 展览～闭幕 the exhibition will soon be closed

【即景】[书] (of a literary or artistic work) be inspired by what one sees

【即景生情】the scene brings back memories; the scene touches a chord in one's heart

【即刻】at once; immediately; instantly; right away

【即令】even; even if; even though

【即期】[经] immediate; spot ◇ ～付现 immediate cash payment; prompt cash payment / ～汇价 spot rate / ～汇票 demand bill; damand draft / ～交货 delivery on spot / ～票据 demand note; not at sight / ～外汇 spot exchange

【即日】[书] ①(当天) this or that very day: 本条例自～起施行。The regulations come into force as of today. / 本协定自～起生效。The present agreement takes effect as from today. ②(最近几天) within the next few days: 本片～上映。The film will be shown within a few days.

【即时】immediately; forthwith ◇ ～付款 immediate payment / ～交付 pay down / ～交货 prompt delivery

【即使】even; even if; even though

【即位】ascend the throne

【即席】[书]①(在宴席或集会上) impromptu; extemporaneous; offhand: ～发言 make an extemporaneous speech; give an offhand talk; give a talk offhand ②(入席; 就位) take one's seat (at a dinner table, etc.) ◇ ～录声 instantaneous recording

【即兴】impromptu; extemporaneous ◇ ～表演 improvisation / ～曲[乐] impromptu / ～诗 extempore verse

【即以其人之道还治其人之身】pay a person back in his own coin; deal with a man as he deals with you

亟 [书] urgently; anxiously; earnestly: ～待解决 have to be settled urgently; need prompt solution / ～盼 earnestly hope / ～须纠正 must be speedily put right / ～欲 desire most ardently; want very much

jǐ

脊 ①(脊骨) spine; backbone; vertebra ②(脊状物) ridge: 山～ the ridge of a hill or mountain / 屋～ the ridge of a roof

【脊背】back

【脊鳍】[动] dorsal fin

【脊神经】spinal nerve

【脊髓】spinal cord ◇ ～灰质炎 poliomyelitis; polio / ～炎 myelitis

【脊索】[动] notochord ◇ ～动物 chordate (animal)

【脊柱】spinal column; vertebral column; backbone; spine

【脊椎】vertebra ◇ ～动物 vertebrate / ～骨 vertebra; spine

济

【济济】(of people) many; numerous: 人才～ an abundance of capable people; a galaxy of talent / ～一堂 a gathering of many people; gather together under the same roof

挤 ①(拥挤) crowd; pack; cram; throng: ～满了人 be crowd with people / ～做一团 huddle together; packed like sardines / 大厅已经～满。The hall is filled to capacity. ②(挤开) jostle; push against: ～进去 force 〈elbow; shoulder; push〉 one's way in; squeeze in / ～上公共汽车 pack into a bus / ～上前 push to the front ③(挤压) squeeze; press: ～掉水 squeeze the water out / ～牛奶 milk a cow / ～时间 find time

【挤兑】a run on a bank

【挤咕】[方] wink: ～眼儿 wink at

【挤眉弄眼】make eyes; wink

【挤奶】milk (a cow, etc.) ◇ ～机 milking machine; milker

【挤牙膏】squeeze toothpaste out of a tube; be forced to tell the truth bit by bit

【挤压】[冶] extruding ◇ ～机 extrusion press; extruder

【挤压伤】crush injuries

几 ①(询问数目) how many: ～点钟? What's the time? (或: What time is it?) / 今天～号? What's the date today? / 你～号? What's your number? ②(表示十以内的不定数) a few; several; some: ～十 tens; dozens; scores／～天 several days／二十～个人 twenty odd people／过～天 in a couple of days／十～岁的孩子 teenager／三十～ thirty-odd; over thirty; more than thirty／说～句话 say a few words／相差无～ not much difference

【几番风雨】the devastation of a few storms and gusts

【几分】a bit; somewhat; rather

【几何】①[书](多少) how much; how many ②(几何学) geometry ◇～光学 geometrical optics／～级数[数] geometric progression; geometric series／～图形[数] geometric figure

【几何学】[数] geometry: 解析～ analytic geometry／立体～ solid geometry／平面～ plane geometry

【几历沧桑】marked by vicissitudes

【几内亚】Guinea ◇～人 Guinean

【几内亚比绍】Guinea-Bissau ◇～人 Guinean (Bissau)

【几时】what time; when

麂

muntjac

【麂皮】chamois (leather); chammy

【麂子】muntjac; barking deer

虮

【虮子】the egg of a louse; nit

己

(自己) oneself: 舍～为公 make personal sacrifices for the public good／舍～为人 sacrifice oneself for others ②(自己的) one's own; personal: 引为～任 regard as one's (own) duty; deem it one's duty to do sth.／各抒～见。Each airs his own views.

【己二酸】adipic acid

【己方】one's own side

【己所不欲,勿施于人】Do not do to others what you would not have them do to you.(或: Do as you would be done by.)

给

①(供给;供应) supply; provide ②(富裕充足) ample; well provided for: 家～富足。Every household is well provided for.

【给水】①[建] water supply ②[机] feed water: 锅炉～ boiler feed water ◇～工程 water-supply engineering／～管 feed pipe／～箱 feed-tank／～器 water feeder

【给养】provisions; victuals; rations: ～不足 be short of provisions／～充足 be abundantly provisioned

【给予】[书] give; render; offer: ～补助 give subsidies to／～公正裁判 give a fair trial／～很高的评价 have a very high opinion of; appreciate highly／～同情 show sympathy for／～协助 render assistance to／～一个期限 give time／～折扣 discount granted／～支持 give support to／～正式承认 give official recognition to

jì

寂

①(寂静) quiet; still; silent ②(寂寞) lonely; lonesome; solitary: 枯～ bored and lonely

【寂静】quiet; still; silent: ～无声。It is deadly still.

【寂寞】lonely; lonesome

济

①(过河) cross a river: 同舟共～ people in the same boat help each other; pull together to tide over difficulties ②(救;救济) give relief; aid; help: ～人之急 relieve sb. in need／扶危～困 help the distressed and succor those in peril ③(对事情有益) be of help; benefit: 无～于事 not help matters; be of no help

【济事】(多用于否定) be of help; be of use

荠

【荠菜】[植] shepherd's purse

剂

①(药剂;制剂) a pharmaceutical preparation; a chemical preparation: 搽～ liniment／催醒～ analeptic／滴鼻～ nasal drops／滴眼～ eye drops／酊～ tincture／锭～ lozenge／合～ mixture／糊～ paste／浸～ infusion／麻醉～ narcotic; anesthetic／片～ tablet／乳～ emulsion／散～ powder medicine／漱口～ gargle／栓～ suppository／吐～ emetic／丸～ pill; bolus／吸入～ inhalation／洗～ lotion／洗眼～ eye lotion／消毒～ disinfectant／针～ injection／止痛～ pain-killer; analgesic／注射～ injection ②[化] agent: 催化～ catalyst; catalytic agent／防腐制～ preservative; antiseptic／干燥～ drying agent; desiccant ③[量](用于汤药) dose: 一～中药 a dose of Chinese herbal medicine

【剂量】[药] dosage; dose: 有效～ effective dose

【剂型】[药] the form of a drug (e.g. liquid, powder, pill)

鲚

long-tailed anchovy; anchovy

计

①(计算) count; compute; calculate; number: 不～其数 countless; innumerable／数以万～ by the tens of thousands; numbering tens of thousands ②(仪器) meter; gauge: 体温～ clinical thermometer / 压力～ piezometer; pressure gauge / 雨量～ rain gauge ③(主意; 策略;计划) idea; ruse; stratagem; plan: 缓兵之～ stalling tactics; stratagem to gain a respite / 为长远～ from a long-term point of view / 中～ fall into a trap

【计步器】pedometer

【计策】stratagem; plan

【计程表】(出租汽车) taximeter

【计程仪】[航海] log

【计出万全】a foolproof plan or scheme

【计划】①(预先拟定的内容和步骤) plan; project; program: 宏伟的～ a magnificent project / 制定～ make a plan; draw up a plan; work out a plan ②(做计划) map out; plan ◇ ～供应 planned supply / ～经济 planned economy / ～生产 planned production / ～体制 planning system / ～指标 plan targets

【计划生育】family planning; birth control ◎独生子女证 one-child certificate; single-child certificate / 妊娠率 pregnancy rate / 育龄妇女 women of child-bearing age / 育龄夫妇 couples of child-bearing age / 节育手术 birth control surgery / 避孕 contraception / 避孕措施 contraceptive devices / 避孕用品 contraceptives / 结扎输精管 vasectomy / 结扎输卵管 tubectomy / 宫内避孕器 IUD; intrauterine device / 上环 be fitted with a contraceptive ring / 子宫帽 diaphragm pessary / 避孕环 intrauterine ring; interuterine coil / 口服避孕药 oral contraceptives / 避孕套 condom / 安全期避孕法 rhythm method; safe period / 绝育手术 sterilization / 人工流产 induced abortion

【计价】valuation: ～过低 undervaluation / 过高 over valuation

【计件】reckon by the piece ◇ ～工资 piece rate wage / ～工资制 piecework system / ～工作 piecework

【计较】①(在乎) bother about; haggle over; fuss about: 不～个人得失 not to be concerned for personal gains or losses / ～小事 be too particular about trifles ②(争论) argue; dispute: 我现在不同你～。 I won't dispute 〈argue〉 with you now. ③(打算) think over; plan: 日后再作～。 Let's think it over later.

【计量】measure; calculate; estimate: 不可～ inestimable ◇ ～经济学 econometrics / ～学 metrology

【计谋】scheme; stratagem; plot

【计穷力竭】come to the end of one's tether; have shot one's bolt; scheme exhausted and strength worn out

【计日程功】estimate exactly how much time is needed to complete a project; have the completion of a project well in sight

【计时】reckon by time ◇ ～工资 payment by the hour; hourly wages; time wage / ～工作 timework

【计时器】hourmeter; hour-counter

【计时赛】[体] (自行车项目) time trial

【计时员】[体] timekeeper

【计数】count

【计数器】[物] counter: 盖革～ Geiger counter / 闪烁～ scintillation counter

【计算】①(求得未知数) count; compute; calculate; reckon: ～出席人数 count the number of people present / ～生产成本 calculate the cost of production / 把某事物～在内 reckon sth. in ②(考虑; 筹划) consideration; planning: 做事不能没个～。 We shouldn't do anything without a plan.

【计算尺】slide rule

【计算机】computer; calculating machine: 电子～ electronic computer / 数字控制～ digital、control computer / 微型～ personal computer (PC) / 小型～ microcomputer ◇ ～程序 computer program / ～程序设计 computer programming / ～存储器 computer storage / ～代码 computer code / ～犯罪 computer crime / ～机房 computer room / ～科学 computer science / ～软件 computer software / ～网络 computer network / ～硬件 computer hardware

【计算器】calculator

【计算数学】numerical analysis

【计算语言学】computational linguistics

【计无所出】unable to think of a way

【计议】deliberate; talk over; consult: 从长～ take one's time in coming to a decision; think sth. over carefully; give the matter further consideration and discuss it later

髻 hair worn in a bun or coil

蓟 [植] setose thistle

【蓟马】[动] thrips: 烟～ tobacco thrips

技 skill; ability; trick; technique: 绝～ unique skill / 献～ make a display of one's feats / 一～之长 what one is skilled in; skill / 黔驴～穷

at one's wits' end
【技工】[简](技术工人) skilled worker; mechanic; technician ◇ ～学校 skilled workers' school
【技能】technical ability; mastery of a skill or technique: 生产～ skill in production
【技巧】skill; technique; craftsmanship: 刺绣～ one's skill at embroidery / 写作～ writing technique / 艺术～ artistry
【技巧运动】acrobatic gymnastics; sports acrobatics; tumbling
【技师】technician
【技术】technology; skill; technique: 科学～ science and technology / 采用先进～ adopt advanced techniques / ～精良 one's art is skillful / ～熟练 be skillful at; be proficient in ◇ ～操作规程 regulations for technical operations / ～改造 technical innovation; technical transformation; technological innovation / ～革命 technological revolution / ～革新 technological innovation / ～工人 skilled worker / ～观 conceptions of technology / ～规范 technical specification; technological specification / ～合作 technological cooperation / ～鉴定 technical appraisement / ～交流 exchange of technical know-how; technical exchange / ～科学 technological sciences / ～力量 technical force; technical personnel / ～名词 technical term / ～情报 technical intelligence / ～人员 technical personnel; technical staff / ～手册 technical manual; technological manual / ～学校 technical school / ～研究所 technological research institute / ～援助 technical assistance / ～职称 titles for technical personnel / ～资料 technical data; technological data
【技术推广站】technical advice station: 农业～ agrotechnical station
【技术性】technical; of a technical nature: ～问题 technical matters
【技术员】technician
【技术指导】①(技术上的指导) technological guidance; technical guidance ②(担任指导的人) technical adviser
【技艺】skill; artistry: ～超群 one's skill in mechanical arts is above the average / ～高超 superb skills / ～精湛 highly skilled; masterly

伎

skill; ability; trick: 故～重演 be up to one's old tricks again; play the same old trick
【伎俩】trick; intrigue; maneuver; foul means: 惯用的～ one's favorite trick / 玩弄卑劣～ play a nasty trick

妓

prostitute
【妓女】prostitute; harlot; whore
【妓院】brothel

寄

①(递送) send; post; mail: ～包裹 send a parcel by post / ～费 postage / ～钱 remit money / ～信 post a letter; mail a letter ②(托付;寄托) entrust; deposit; place; park ③(依附) depend on; attach oneself to: ～食 live with a relative, etc. (because of one's straitened circumstances)
【寄存】deposit; leave with; check: ～行李 deposit one's luggage; check one's luggage / 行李～处 left-luggage office; checkroom / 自行车～处 bicycle park
【寄存器】[计算机] register: 变址～ index register / 进位～ carry storage register
【寄放】leave with; leave in the care of
【寄件人】sender
【寄居】live away from home
【寄居蟹】hermit crab
【寄卖】consign for sale on commission; put up for sale in a secondhand shop ◇ ～商店 commission store; secondhand shop
【寄人篱下】live under another's roof; depend on sb. for a living; depend on another person for support
【寄生】①[生] parasitism ②(靠别人生活) parasitic: ～生活 parasitic life ◇ ～动物 parasitic animal / ～蜂 parasitic wasp / ～甲壳动物 crustacean louse / ～菌 bacterial parasite / ～植物 parasitic plant
【寄生虫】parasite ◇ ～病 parasitic disease; parasitosis / ～学 parasitology
【寄生物学】parasitology
【寄生振荡】[电] parasitic oscillation
【寄售】consign for sale; put up for sale in a second-hand shop ◇ ～品 consignment merchandise / ～商店 commission store; secondhand shop
【寄宿】①(借宿) lodge; put up: ～在某人处 lodge with sb.; put up at sb.'s house ①(住校) board ◇ ～生 resident student; boarder / ～学校 boarding school; residential college
【寄托】①(托付) entrust to the care of sb.; leave with sb. ②(把希望、理想等放在某人某事上) place (hope, etc.) on; find sustenance in; repose: 精神有所～ have something to repose one's trust in; have spiritual sustenance

【寄信人】sender
【寄养】entrust one's child to the care of sb.; ask sb. to bring up one's child
【寄予】①(寄托) place (hope, etc.) on: ～希望 place hopes on; put one's hopes on ②(给予) show give; express: ～关怀 show concern for sb. / ～深切的同情 show heartfelt sympathy to
【寄语】[书] send word: ～同窗 send verbal message to a fellow-student
【寄主】[生] host (of a parasite)

冀 [书] hope; long for; look forward to: 希～ hope for; look forward to / ～其成功 look forward to the success of sb. or sth.

觊
【觊觎】[书] covet; cast greedy eyes on; cast one's covetous eyes on

季 ①(季节) season: 淡～ dull season / 旺～ busy season / 一年四～ the four seasons of the year; all the year round / 雨～ rainy season; wet season ②(种植次数) the yield of a product in one season; crop: 由种一～改为种两～ reap two crops a year instead of one ③(一季的最后一个月) the last month of a season: ～春 the last month of spring ④(兄弟中排行第四或最小的) the fourth or youngest among brothers: ～弟 the fourth or youngest brother
【季报】quarterly reports
【季度】quarter (of a year): 第一～ the first quarter (of a year) / ～预算 quarterly budget
【季风】[气] monsoon: 冬～ dry monsoon / 夏～ wet monsoon ◇ ～气候 monsoon climate / ～雨 monsoon rain
【季节】season: 旅游～ tourist season / 农忙～ a busy farming season / 收获～ harvest season; harvest time ◇ ～变化 seasonal variation / ～差价 seasonal variations in price / ～工 seasonal worker / ～关税 seasonal duties / ～回游 seasonal migration / ～迁徙 transhumance
【季节性】seasonal ◇ ～工作 seasonal work; seasonal jobs / ～行业 seasonal trade
【季刊】quarterly publication; quarterly

悸 [书] (of the heart) throb with terror; palpitate: 惊～ palpitate with terror / 心有余～ have a lingering fear

祭 ①(祭奠) hold a memorial ceremony for ②(祭祀) offer a sacrifice to: ～天 offer a sacrifice to Heaven; worship Heaven / ～祖 offer a sacrifices to an ancestor
【祭奠】hold a memorial ceremony for
【祭礼】①(祭祀仪式) sacrificial rites ②(祭奠仪式) memorial ceremony ③(祭品) sacrificial offerings
【祭品】sacrificial offerings; oblation
【祭祀】offer sacrifices to gods or ancestors
【祭坛】altar; sacrificial altar
【祭文】funeral oration; elegiac address
【祭献】sacrifice

鲫 crucian carp

系 tie; fasten; do up; button up: ～围裙 wear an apron / ～鞋带 tie shoe laces / ～衣服扣子 button up a jacket
【系泊】moor (a boat) ◇ ～浮筒 mooring buoy
【系船浮筒】mooring buoy
【系船索】mooring rope; mooring line
【系船柱】dolphin
【系留】moor (a balloon or airship) ◇ ～塔 mooring mast; mooring tower

既 ①(已经) already: ～得权利 vested right ② [连](既然) since; as; now that ③[连](与"且""又""也"等副词连用) both... and; as well as: ～不实用，又不美观 neither useful nor attractive / 这孩子～健康又活泼。The child is lively as well as healthy.
【既成事实】accomplished fact; fait accompli: 承认～ accept a fait accompli / 造成～ present a fait accompli; make sth. an accomplished fact
【既得利益】vested interest ◇ ～集团 vested interests
【既定】set; fixed; established: ～方案 existing plan / ～目标 set objective; fixed goal
【既而】[副] soon afterwards; later; subsequently
【既决犯】[法] convict
【既决罪犯】[法] convicted prisoner
【既来之，则安之】Since we have come, let us stay and enjoy it. (或: Since we are here, we may as well stay and make the best of it.)
【既然】[连] since; as; now that: ～如此 since it is so; such being the case; under these circumstances
【既是】[连] since; as; now that
【既遂】[法] accomplished offense ◇ ～犯 accomplished offender; consummated offender / ～罪 accomplished offense; completed offense; consummated crime

【既往不咎】forgive sb.'s past misdeeds; let bygones be bygones

记 ①(记得;记住) remember; bear in mind; commit to memory: ～不清 cannot recall exactly; remember only vaguely / ～错了 remember wrongly / 如果我没有～错的话 if I remember right; if my memory serves me / 死～硬背 learn by rote / ～得快忘得也快。Soon learnt, soon forgotten. ②(记录;记载;登记) write down; record; jot down; take down: ～笔记 take notes / ～日记 keep a diary / ～下电话号码 jot down the telephone number / ～下地址 write down the address / ～在笔记本上 write it down in a notebook / 把结果～下来 record the results ③(记载的文章等) notes; record: 大事～ a chronicle of events / 游～ travel notes ④(标志;符号) mark; sign ⑤(胎记) birthmark ⑥[量] 一～耳光 a slap in the face

【记仇】bear grudges; harbor bitter resentment

【记得】remember; recall; call back to mind; keep in memory

【记分】①(比赛用) keep the score; record the points (in a game) ②(学校用) register a student's marks ◇ ～册 (teacher's) markbook / ～员 scorekeeper; scorer; marker

【记分牌】scoreboard; score indicator: 电动～ electric scoreboard

【记功】cite sb. for meritorious service; record a merit

【记挂】be concerned about; keep thinking about; miss

【记过】record a demerit

【记号】mark; sign: 做个～ make a sign; mark out

【记恨】bear grudges

【记录】【纪录】①(写下来) take notes; keep the minutes; record: ～在案 in record; of record ②(当场记录下来的材料) minutes; notes; record: 会谈～ a transcript of talks / 会议～ the minutes of a meeting / 正式～ official record / 逐字～ verbatim record / 摘要～ summary record / 原始～ original records / 列入会议～ place on record in the minutes; minute ③(记录者) notetaker; recorder ④(最高成绩) record: 国家～ national record / 少年～ junior record / 世界～ world record / 亚洲～ Asian record / 保持～ keep a record; hold a record / 保持不败～ keep a perfect record / 创～ set a record; chalk up a record / 打破～ break a record / 平～ equal a record; match a record / 刷新～ improve a record ◇ ～保持者 record holder / ～本 minute book / ～副本 transcript / ～员 recorder

【记录片】【纪录片】documentary film: 大型～ full length documentary film / 电视～ televised documentary

【记名】put down one's name (on a check, etc. to indicate responsibility or claim); sign: 无～投票 secret ballot ◇ ～支票 order check / ～债券 registered bounds

【记谱法】[乐] musical notation

【记起】recall; recollect

【记取】remember; bear in mind: ～这个教训 bear firmly in mind this lesson / ～正反两方面的经验 draw on experience both positive and negative

【记时仪】[天] chronograph

【记事】①(把事情记录下来) keep a record of events; make a memorandum ②(记述的历史经过) account; record of events; chronicles

【记事儿】(of a child) begin to remember things

【记述】record and narrate; give an account of

【记诵】commit to memory and be able to recite; learn by heart

【记性】memory: ～好 have a good memory / ～坏 have a poor memory; have a short memory

【记叙】narrate: ～简练 concise in narrative ◇ ～体 narration; narrative / ～文 narration; narrative

【记忆】①(记住;想起) remember; recall: 就～所及 so far as I can remember; as far as I recollect; to the best of my memory ②(记住的印象) memory: ～犹新 remain fresh in one's memory; the memory is still fresh ◇ ～储存 memory storage / ～广度 memory span / ～恢复 reminiscence / ～曲线 memory curve / ～缺失 amnesia

【记忆力】the faculty of memory; memory: ～强 have a good memory / ～弱 have a poor memory / ～衰退 one's memory is failing

【记载】①(写下来) put down in writing; record ②(记事的文章) record; account: 历史～ historical records; annals

【记帐】keep accounts; keep books: ～买卖 transaction for account / 记入帐户 enter an item in an account

【记者】reporter; correspondent; newsman; journalist: 采访～ reporter / 常驻～ resident correspondent / 随军～ war correspondent / 特派～ special correspondent / 体育新闻～

sports reporter / 通讯～ correspondent / 新闻～ newspaper reporter; newsman ◇ ～席 press gallery / ～协会 journalists' association / ～招待会 press conference; news conference / ～证 press card

【记住】remember; learn by heart; bear in mind; commit to memory

忌 ①(忌妒) be jealous of; envy: ～才 be jealous of other people's talent; resent people more able than oneself / ～恨 envy and hate ②(怕) fear; dread; scruple: 横行无～ ride roughshod; run amuck ③(避免) avoid; shun; abstain from: ～生冷 avoid cold and uncooked food / 婴孩～服 not to be taken by babies ④(戒除) quit; give up: ～酒 give up alcohol; abstain from wine / ～烟 quit smoking

【忌辰】the anniversary of the death of a parent, ancestor, or anyone else held in esteem

【忌惮】dread; fear; scruple: 肆无～ stopping at nothing; unscrupulous

【忌妒】be jealous of; envy: 不～任何人 be jealous of nobody / 出于～ out of jealousy

【忌讳】①(禁忌) taboo: 犯～ violate a taboo; break a taboo ②(避免) avoid as harmful; abstain from: ～不言 be silent avoiding superstitious fear

【忌刻】jealous and mean; jealous and malicious

【忌口】avoid certain food (as when one is ill); be on a diet

纪 ①(纪律) discipline: 军～ military discipline / 违法乱～ break the law and violate discipline / 遵～守法 abide by the law and discipline ②(记录下来) put down in writing; record: ～事 chronicle ③(世纪) age; epoch: 世～ century / 中世～ the Middle Ages ④[地] period: 震旦～ the Sinian Period

【纪录】同"记录"

【纪录片】同"记录片"

【纪律】discipline: 劳动～ labor discipline; labor regulations / 加强～性 heighten one's sense of discipline / 遵守～ keep discipline; observe discipline / ～严明 highly disciplined; be strict in discipline ◇ ～检查委员会 commission for inspecting discipline / ～制裁 disciplinary sanction

【纪律处分】disciplinary award; disciplinary treatment: 给予～处分 take disciplinary measures against sb.

【纪年】①(记年代) a way of numbering the years ②(史书) chronological record of events; annals

【纪念】①(表示怀念) commemorate; mark: ～活动 commemorative activities / 举行～大会 hold a commemoration meeting ②(纪念物) souvenir; keepsake; memento: 留个～ keep sth. as a souvenir ③(用来表示纪念的物品) commemorative ④(周年; 纪念日) commemoration day; anniversary: 十周年～ the tenth anniversary

【纪念碑】monument; memorial: 人民英雄～ the Monument to the People's Heroes

【纪念册】autograph book; autograph album; commemorative album: 题词～ autograph album

【纪念馆】memorial hall; museum in memory of: 鲁迅～ the Lu Xun Museum

【纪念品】souvenir; keepsake; memento

【纪念日】commemoration day; red-letter day

【纪念塔】memorial tower; monument

【纪念堂】memorial hall; commemoration hall; mausoleum: 毛主席～ the Chairman Mao Memorial Hall / 中山～ Sun Zhongshan's 〈Sun Yat-sen's〉 Memorial Hall

【纪念邮票】commemorative stamp

【纪念章】souvenir badge

【纪实】record of actual events; on-the-spot report

【纪行】travel notes; notes on a trip

【纪要】summary of minutes; summary: 会谈～ summary of conversations; summary of talks / 座谈会～ summary of a forum or panel discussion

【纪元】①(纪年的开始) the beginning of an era (e.g. an emperor's reign) ②(时代) epoch; era: 开辟了历史的新～ usher in a new era in history

际 ①(边缘或分界处) border; boundary; edge: 水～ the edge of a body of water; waterside / 天～ horizon / 无边无～ boundless ②(彼此之间) between; among; inter-: 春夏之～ between spring and summer / 国～ international; between nations / 校～比赛 intercollegiate games; interschool matches / 星～旅行 interplanetary travel / 洲～导弹 intercontinental missile ③(中间; 里边) inside: 脑～ in one's head; in one's mind ④(时候) occasion; time: 临别之～ at the time of parting / 危急之～ at a critical moment ⑤(正当) on the

occasion of: ～此盛会 on the occasion of this grand gathering ⑥(遭遇) one's lot; circumstances: 遭～ vicissitudes in one's life; one's lot
【际遇】[书] favorable or unfavorable turns in life; spells of good or bad fortune

继 ①(接续) continue; succeed; follow: 后～有人 have qualified successor / 前赴后～ advance wave upon wave / 日以～夜 day and night ②(继而) then; afterwards
【继承】inherit; succeed; carry on: ～财产 inherit property / ～革命事业 carry on the revolutionary cause / ～衣钵 step into the shoes of; take over the mantle of / ～遗产 heritage / ～遗志 carry on the unfinished lifework of (the father or leader) / ～优良传统 carry forward the good traditions
【继承法】inheritance act; law of succession
【继承归属原则】canon of inheritance; right of succession
【继承权】right of succession; right of inheritance: 剥夺～ disinherit sb. / 长子～ primogeniture
【继承人】heir; successor; inheritor: 法定～ heir at law; legal heir / 王位～ successor to the throne / 直系～ lineal successor
【继电器】[电] relay
【继而】[副] then; afterwards
【继父】stepfather
【继母】stepmother
【继任】succeed sb. in a post: ～首相 succeed sb. as prime minister
【继室】one's second wife
【继嗣】①(过继) be adopted ②(嗣子) adopted son
【继往开来】carry forward the cause and forge ahead into the future; carry on the past and open a way for future
【继位】succeed to the throne
【继续】①(延续下去) continue; go on (with); keep on; proceed: ～工作 continue working / ～有效 remain valid; remain in force / ～执政 continue in office; remain in power ②(跟某事有连续关系的另一事) continuation ◇ ～保险 continued insurance / ～成本 continuing cost; running cost / ～年金 continuous annuity / ～审计 continuous audit

jiā

家 ①(家庭) family; household: 养～ support one's family ②(家庭住所) home: 不在～ not be in; be out / 回～ go home ③(经营某种行业的人) a person or family engaged in a certain trade: 船～ boatman / 渔～ fisherman's family ④(精通某行的人) expert; a specialist in a certain field: 画～ painter / 科学～ scientist / 文学～ a man of letters; writer / 艺术～ artist / 政治～ statesman / 作曲～ composer ⑤(学术流派) a school of thought; school: 法～ the Legalist School / 儒～ the Confucian School / 百～争鸣。 A hundred schools of thought contend. ⑥[谦] ～父 my father / ～兄 my elder brother ⑦(饲养的) domestic; tame: ～兔 rabbit / ～蝇 housefly ⑧[量] 两～人家 two families / 三～饭馆 three restaurants / 五～商店 three shops / 一～电影院 a cinema
【家蚕】silkworm
【家产】family property ◇ ～处分[法] family arrangement
【家常】the daily life of a family; domestic trivia: 拉～ engage in small talk ◇ ～话 small talk; chitchat
【家常便饭】①(普通饭菜) homely food; simple meal ②(常事) common occurrence; routine; all in the day's work; usual practice
【家丑】family scandal; the skeleton in the cupboard <closet>
【家丑不可外扬】Domestic shame should not be made public. (或: Don't wash your dirty linen in public.)
【家畜】domestic animal; livestock ◇ ～险 livestock insurance / ～运送险 livestock transit insurance
【家传】handed down from the older generations of the family ◇ ～秘方 a secret recipe handed down in the family
【家当】[口] family belongings; property
【家道】family financial situation: ～小康 be comfortably off / ～中落 the means of the family is declining
【家底】family property accumulated over a long time; resources: ～薄 without substantial resources; not financially solid
【家访】a visit to the parents of schoolchildren or young workers
【家鸽】pigeon
【家和万事兴】If the family lives in harmony, all affairs will prosper
【家伙】[口] ①(工具;武器) tool; utensil; weapon ②(人) fellow; guy

【家给人足】each family is provided for and every person is well-fed and well-clothed; all live in plenty
【家计】[书] family livelihood
【家家户户】each and every family; every household; all households without exception
【家家有本难念的经】Every family has a skeleton in the cupboard
【家教】family education; upbringing: ～严 be strict with one's children / 没有～ not properly brought up; ill-bred
【家境】family financial situation; family circumstances: ～好 come from a well-to-do family; come from a well-off family / ～清寒 with one's family in straitened circumstances
【家具】furniture: 两件～ two pieces of furniture / 一套～ a set of furniture ◇ ～工业 furniture industry
【家眷】①(家属) wife and children; one's family ②(妻子) wife
【家口】members of a family; the number of people in a family
【家累】family burden; family cares: ～重 be encumbered with a large family
【家麻雀】house sparrow
【家贫如洗】utterly destitute; penniless
【家破人亡】family ruined; with one's family broken up, some gone away, some dead
【家谱】family tree; genealogical tree; genealogy
【家禽】domestic fowl; poultry: ～饲养场 poultry farm / ～饲养员 poultryman
【家史】family history
【家什】utensils, furniture, etc.
【家书】①(给家里写的信) a letter home ②(收到的家信) a letter from home: ～抵万金 a letter from home is worth ten thousand pieces of gold
【家属】family members; (family) dependents: 工人～ families of workers / 军人～ armymen's families ◇ ～工厂 factory run by family members of workers, cadres, armymen, etc.
【家私】family property
【家庭】family; household: 大～ extended family / 核心～ nuclear family / 血亲～ con- sanguine family / 姻亲～ conjugal family ◇ ～背景 family background / ～成员 family members / ～出身 class status of one's family; family origin / ～服务 domestic service / ～负担 family responsibilities / ～妇女 housewife / ～副业 household sideline production / ～工业 household industry / ～观念 attachment to one's family / ～教师 private teacher; tutor / ～教育 family education; home education / ～纠纷 family quarrel; domestic discord; family dispute; family dissension / ～破裂 breakdown of a family / ～生活 home life; family life / ～作业 homework
【家徒四壁】have nothing but the bare walls in one's house; be utterly destitute
【家务】household duties ◇ ～劳动 housework; household chores
【家乡】hometown; native place ◇ ～话 native dialect
【家小】wife and children
【家信】a letter to or from one's family
【家业】family property; property
【家蝇】housefly
【家用】family expenses; housekeeping money ◇ ～冰箱 domestic refrigerator / ～电器 household appliances
【家喻户晓】widely known; known to all; known to every family: 做到～ make known to every household
【家园】home; homeland: 重建～ rebuild one's homeland; rebuild one's village or town
【家贼难防】A thief in the family is difficult to detect
【家长】①(一家之长) the head of a family; patriarch ②(父母或监护人) the parent or guardian of a child ◇ ～式统治 paternalism; arbitrary rule as by a patriarch; rule arbitrarily / ～制 patriarchy; patriarchal system / ～作风 a high-handed way of dealing with people; patriarchal behaviour
【家族】clan; family

镓 [化] gallium (Ga)

夹 ①(夹住) press from both sides; place in between; clip; clamp; pinch: ～着把雨伞 carry a anumbrella under one's arm / 把文件～在一起 clip papers together / 把画片～在书里 put the pictures in between the leaves of a book / 用筷子把菜～起来 pick up food with chopsticks / 用钳子把烧红的铁～住 grip a piece of red-hot iron with a pair of tongs / 敌军～起尾巴逃跑了。 The enemy troops ran away with their tails between their legs. / 门～了她的手指头。 The door squeezed her fingers. / 他～在我们两人中间。 He was sandwiched be-

tween the two of us. / 鞋子～我脚。The shoe pinches me. ② (夹杂) mix; mingle; intersperse: ～叙～议 narration interspersed with comments / ～在人群里 mingle with the crowd / 狂风～着暴雨 a violent wind accompanied by a torrential rain ③ (夹子) clip; clamp, folder, etc.: 发～ hairpin / 文件～ folder / 纸～ paper clip
【夹板】① (夹东西的木板) boards for pressing sth. or holding things together ② [医] splint: 石膏～ plaster splints / 上～ put (a limb, etc.) in splints
【夹背锯】tenon saw
【夹层】[建] mezzanine
【夹层玻璃】sandwich glass
【夹带】① (走私) carry secretly; smuggle ② (考试作弊的材料) notes smuggled into an examination hall
【夹道】① (窄的路) a narrow lane; passageway ② (排列在道路两旁) line both sides of the street: ～欢迎 line the street to welcome; welcoming crowds lining the streets
【夹缝】a narrow space between two adjacent things; crack; crevice
【夹攻】attack from both sides; converging attack; pincer attack: 内外～ attack from within and outside / 前后～ attack from the front and the rear simultaneously / 受到两面～ be under a pincer attack; be caught in a two-way squeeze / 左右～ under the crossfire from left and right
【夹击】converging attack; pincer attack
【夹剪】tongs
【夹角】[数] included angle
【夹具】[机] clamping apparatus; fixture; jig
【夹七夹八】incoherent; confused; cluttered (with irrelevant remarks)
【夹生】half-cooked; half-baked: ～饭 half-cooked rice
【夹馅】stuffed (pastry, etc.)
【夹心】with filling: 果酱～糖 sweets with jam center ◇ ～饼干 sandwich biscuits
【夹杂】be mixed up with; be mingled with
【夹竹桃】[植] (sweet-scented) oleander
【夹注】interlinear notes
【夹子】① (夹东西用的) clip; tongs: 点心～ cake tongs / 弹簧～ spring clip / 衣服～ clothes-peg; clothespin ② (夹文件、钱等用的) folder; wallet: 皮～ wallet; pocketbook / 文件～ folder; binder

佳 good; fine; beautiful: 成绩甚～ achieve very good results / 风景绝～ beautiful scenery / 身体欠～ not feel well; be indisposed
【佳话】a deed praised far and wide; a story on everybody's lips; a much-told tale: 传为～ become an interesting story on everybody's lips
【佳节】happy festival time; festival: 中秋～ the joyous Mid-Autumn Festival / 欢度国庆～ celebrate the joyous festival of National Day
【佳期】wedding day; nuptial day
【佳人】beautiful woman
【佳肴】delicacies
【佳音】welcome news; good tidings; favorable reply: 静候～。I am awaiting the news of your success.
【佳作】a fine piece of writing; an excellent work

加 ① (和在一起) add; plus: 三～五等于八。Three plus five makes eight. (或: Three and five is eight.) ② (增加) increase; augment: ～大油门 open the throttle; step on the gas / ～工资 increase sb.'s wages; raise sb.'s wages ③ (添上; 安放) put in; add; append: ～点热水 add some hot water to / ～上标点符号 put in the proper punctuation marks / ～注解 append notes to / 汤里～点盐 put some salt in the soup ④ (加以) 不～考虑 not consider at all / 大～赞扬 praise highly; lavish praise on
【加班】work overtime; work an extra shift: ～加点 work extra shifts or extra hours; put in extra hours ◇ ～车 extra bus / ～费 overtime pay
【加倍】① (加一倍) double; be twice as much ② (程度深) double; redouble: ～警惕 redouble one's vigilance / ～努力 redouble one's efforts / ～注意 redouble one's eforts; be doubly careful
【加车】(put on) extra buses 〈trains〉
【加成】[化] addition ◇ ～反应 addition reaction / ～化合物 additive compound; addition compound
【加法】[数] addition
【加感】[讯] loading
【加工】① (加工过程) process: ～中草药 process medicinal herbs / 来料～ processing of investor's raw materials; processing raw materials on client's demands / 来样～ processing according to investor's samples / 食品～ food processing / 艺术～ artistic treatment ② [机] machining; working: 机～ machining / 冷～ cold working ◇ ～厂 processing factory / ～

成本 finished cost / ～船 factory ship / ～费 processing charges / ～工业 processing industries / ～留量 allowance / ～贸易 improvement trade / ～硬化 work hardening

【加固】reinforce; consolidate: ～堤坝 reinforce dikes and dams / ～工事 improve defense works

【加官进爵】receive official promotion; advance in rank and position

【加害】injure; do harm to: ～于人 do harm to sb.; do sb. an injury ◇ ～方 injuring party; party causing the injury / ～人 author of the injury; inflictor; person causing the injury

【加号】[数] plus sign (+)

【加厚剂】[摄] intensifier

【加急电(报)】urgent telegram; urgent cable

【加紧】step up; speed up; intensify: ～生产 step up production / ～训练 intensify the training / ～准备 speed up preparation

【加劲】put more energy into; make a greater effort

【加剧】aggravate; intensify; exacerbate: ～紧张局势 aggravate tension / 忧虑使他的病情～了。His illness is being aggravated by anxieties.

【加快】quicken; speed up; accelerate; pick up speed: ～步子 quicken one's step / ～社会主义建设的速度 quicken the tempo of socialist construction

【加宽】broaden; widen: ～路面 widen the road

【加勒比海】Caribbean Sea

【加力】[航空] thrust augmentation; afterburning ◇ ～俯冲 afterburning dive

【加料】①(进料) feed in raw material: 自动～ automatic feeding ②(用优质原料或多用原料制成的) reinforced: ～药酒 reinforced tonic wine

【加仑】gallon

【加榴炮】gun-howitzer

【加码】①(提高价格) raise the price of commodities; overcharge ②(提高赌注) raise the stakes in gambling ③(提高指标) raise the quota: 层层～ raise the quota at each level

【加煤机】firing machine

【加冕】coronation; crowing ◇ ～典礼 Coronation / ～日 Coronation Day

【加拿大】Canada ◇ ～人 Canadian

【加纳】Ghana ◇ ～人 Ghanaian

【加捻】[纺] twisting

【加农炮】gun; canon

【加蓬】Gabon ◇ ～人 Gabonese

【加气】[建] air entrainment ◇ ～混凝土 aerocrete / ～水泥 air entraining cement

【加强】strengthen; enhance; augment; reinforce: ～党和群众的联系 strengthen the Party's ties with masses / ～管理 tighten up the management / ～纪律性 strengthen discipline ◇ ～连 reinforced company / ～排 reinforced platoon / ～营 reinforced battalion

【加氢裂化】[化] hydrocracking

【加氢脱硫】[化] hydrodesulfurization

【加权平均法】weighted averages method

【加权平均数】weighted averages

【加热】heating ◇ ～炉 heating furnace / ～器 heating apparatus; heater

【加入】①(加上) add; mix; put in ②(参加) join; accede to: ～共产党 join the Communist Party / ～条约 accede to a treaty ◇ ～国 acceding state / ～书 instrument of accession

【加塞儿】[口] push into a queue out of turn; jump a queue; quene-jump

【加深】deepen: ～对某人的同情 deepen one's sympathies for sb. / ～河道 deepen the channel of a river / ～理解 get a deeper understanding / ～裂痕 widen the rift

【加数】[数] addend

【加速】quicken; speed up; accelerate; expedite: ～我军的现代化建设 speed up the modernization of our army

【加速度】[物] acceleration: 重力～ acceleration of gravity ◇ ～计 accelerometer

【加速器】[物] accelerator: 回旋～ cyclotron / 静电～ electro-static accelerator / 粒子～ particle accelerator / 同步～ synchrotron / 直线～ linear accelerator

【加速运动】[物] accelerated motion

【加线】[乐] ledger line; leger line

【加线台球】balkine billards

【加刑】[法] increase of penalty; infliction

【加以】①(表示如何对待或处理前面提到的事物)有问题要及时～解决。Problems should be resolved in good time. ②(加上) in addition; moreover

【加意】with special care; with close attention: ～保护 protect with special care / ～提防 be particularly watchful / ～防范 guard with watchful care

【加油】①(加润滑油) oil; lubricate ②(加燃料油) refuel; oil: 给油灯～ pour 〈put〉 oil into a lamp / 空中～ flight refuelling; air refuelling ③(加劲) make a greater effort; make an extra effort: ～干 work harder; work with added vig-

or / 小王，～! Come on, Xiao Wang! ◇ ～车 refuelling truck; refueller / ～飞机 tanker aircraft / ～站 filling station (英); petrol station; gas station (美)
【加油添醋】exaggerate 〈give〉 embellishment to a story; add highly colored details to a story
【加重】①（增加重量）make or become heavier; increase the weight of: ～任务 add to one's tasks / ～思想负担 add to one's worries / ～语气 say sth. with emphasis ②（增加程度）make or become more serious; aggravate: ～危机 aggravate the crisis / 病情～。The patient's condition worsened.

痂 scab; crust: 结～ form a scab; crust

袈

【袈裟】kasaya; a patchwork outer vestment worn by a Buddhist monk

嘉 ①（美好）good; fine: ～宾 honored guest; welcome guest ②（夸奖）praise; commend: 精神可～ a praiseworthy spirit
【嘉奖】commend; cite: 呈请～ make a request for commendation / 传令～ cite sb. for meritorious service ◇ ～令 citation
【嘉勉】[书] praise and encourage
【嘉许】[书] praise; approve

茄

【茄克】jacket

枷 cangue
【枷锁】yoke; chains; shackles; fetters: 精神～ spiritual shackles / 摆脱…的～ shake off the yoke of

jiá

夹 double-layered; lined: ～袄 lined jacket

荚 pod: 结～ bear pods; pod / 豌豆～ pea pods
【荚果】[植] pod; legume

颊 cheek: 两～红润 with rosy cheeks
【颊骨】[生理] cheekbone
【颊囊】[动] cheek pouch

蛱

【蛱蝶】blackleg tortoiseshell

戛

【戛然】[书] ①[象]（多形容嘹亮的鸟声）～长鸣 long and loud cries ②（多用于）～而止 (of a sound) stop abruptly

jiǎ

甲 ①（第一）first: ～等 first-rate; first-class / ～级 Grade A ②（用作代称）～队和乙队 team A and team B / ～方和乙方 the first party and the second party / 某～与某乙 Mr. A and Mr. B ③（甲壳）shell; carapace: 龟～ tortoise shell ④（指甲）nail: 手指～ fingernail ⑤（装甲;盔甲等防护装备）armor: 装～车 armored car / 丢盔卸～ throw away one's helmet and coat of mail
【甲板】deck: 后～ quarterdeck; after deck / 前～ fore deck / 散步～ promenade deck / 上～ upper deck / 尾楼～ poop deck / 下～ lower deck / 中～ middle deck / 主～ main deck / 最下～ orlop deck
【甲苯】[化] toluene; methylbenzene
【甲虫】beetle
【甲醇】[化] methyl alcohol; methanol
【甲酚】[化] cresol
【甲睾酮】[药] methyltestosterone
【甲沟炎】[医] paronychia
【甲骨文】inscriptions on bones or tortoise shells
【甲基】methyl ◇ ～溴[化] methyl bromide / ～纤维素[药] methylcellulose
【甲壳】crust ◇ ～动物 crustacean
【甲硫氨酸】[生化] methionine
【甲醛】[化] formaldehyde ◇ ～水 formalin
【甲酸】[化] formic acid; methanoic acid
【甲烷】[化] methane
【甲午战争】the Sino-Japanese War of 1894–1895
【甲癣】[医] onychomycosis; ringworm of the nails
【甲氧胺】[药] methoxamine
【甲氧基氯】[农药] methoxychlor
【甲鱼】soft-shelled turtle
【甲种维生素】vitamin A
【甲状旁腺】parathyroid gland ◇ ～功能减退 hypoparathyroidism / ～机能亢进 hyperparathyroidism / ～腺瘤 parathyroid adenoma
【甲状腺】[生理] thyroid gland ◇ ～功能减退 hypothyroidism / ～机能亢进 hyperthyroidism / ～切除 thyroidectomy / ～素 thyroxine / ～腺瘤 thyroid adenoma / ～炎 thyroiditis / ～肿 goitre / ～肿瘤 thyroid tumors

【甲紫】gentian violet

岬

①（岬角）cape; promontory ②（两山之间）a narrow passage between mountains

【岬角】cape; promontory

钾

[化] potassium (K)

【钾肥】potash fertilizer

【钾碱】potash

【钾一氩年龄测定法】potassium-argon dating

【钾盐】sylvite

胛

【胛骨】[生理] shoulder blade

假

①（跟"真"相对）false; fake; sham; phoney; artificial: ～慈悲 shed crocodile tears / ～和平 phoney peace / ～检讨 insincere self-criticism / ～民主 bogus democracy; sham democracy / ～腿 artificial leg / 以～乱真 create confusion by passing off the spurious as genuine ②（借用）borrow; avail oneself of; make use of: ～革命之名 usurp the name of revolution / 久～不归 keep putting off returning sth. one has borrowed; appropriate sth. borrowed for one's own use ③（假如）if; suppose: ～令 in case

【假案】feigned case

【假扮】disguise oneself as; dress up as

【假充】pretend to be; pose as: ～内行 pretend to be an expert / ～英雄 pose as a hero

【假道】via; by way of: 代表团～欧洲去联合国. The delegation went to the United Nations via Europe.

【假定】①（姑且认定）suppose; assume; grant; presume ②（科学上的假设）hypothesis ◇～成本 assumed cost; hypothetical cost / ～负债 nominal liability / ～利息 assumed interest; hypothetical interest

【假发】wig: 戴～ wear a wig; be wigged

【假分数】[数] improper fraction

【假根】[植] rhizoid

【假公济私】use public office for private gain; practice jobbery; work for one's own ends in public affairs

【假果】[植] pseudocarp; spurious fruit

【假花】artificial flower

【假话】lie; falsehood: 说～ tell lies

【假借】make use of: ～名义 under the guise of; in the name of; under false pretenses

【假两性畸形】pseudohermaphrodism; pseudohermaphroditism

【假冒】pass oneself off as; palm off (a fake as genuine): 谨防～. Beware of imitations. ◇～签名 forged signature / ～商标 counterfeit trademark / ～他人专利 pass off the patent / ～他人专利的人 person passes off the patent of another person

【假面具】mask; false front: 揭穿某人的～ expose sb.'s hypocrisy; tear the mask off sb.

【假面舞会】masquerade

【假名】①（非真名；笔名）pseudonym ②（日语字母）kana: 片～ katakana / 平～ hiragana

【假票据】fictitious bill

【假仁假义】pretended benevolence and righteousness; hypocrisy; be a wolf in sheep's skin ◇～的人 hypocrite

【假如】if; supposing; in case

【假若】if; supposing; in case

【假嗓子】falsetto

【假山】rockery

【假设】①（假定）suppose; assume; grant; presume ②（科学上的假设）hypothesis: 科学～ a scientific hypothesis

【假声】[乐] falsetto

【假使】if; in case; in the event that

【假释】[法] release on parole; conditional release; free a prisoner on probation; parole provisional release ◇～犯 parolee

【假手】do sth. through sb. else; make a cat's-paw of sb.: ～于人 make sb. else do the work; (achieve one's end) through the instru- mentality of sb. else

【假说】hypothesis

【假死】①[医] suspended animation ②[动] play dead; feign death; play possum

【假托】①（推托）on the pretext of: ～有病 on the pretext of illness ②（用别人的名义）under sb. else's name ③（凭借）by means of; through the medium of

【假想】①（想象；假设）imagination; hypothesis; supposition ②（想象的；假定的）imaginary; hypothetical; fictitious ◇～敌[军] imaginary enemy / ～基金 imaginary fund

【假象】①（跟事实不符合的表面现象）false appearance: 识破～ see through a false appearance / 制造～ create a false impression; put up a false front ②[地] pseudomorph

【假惺惺】hypocritically; unctuously

【假牙】dental prosthesis; false tooth; denture

【假眼】ocular prosthesis; artificial eye; glass eye

【假意】①(虚假的心意) unction; insincerity; hypocrisy ②(故意表现出) pretend; put on: ～奉承 cheap flattery
【假造】①(伪造) forge; counterfeit: ～的文件 forged document / ～证件 forge a certificate ②(捏造) invent; fabricate: ～理由 invent an excuse / ～罪名 cook up a false charge against; frame up
【假肢】artificial limb
【假植】[农] heel in
【假装】pretend; feign; simulate; make believe: ～不知道 feign ignorance / ～积极 pretend to be (politically) active; simulate enthusiasm / ～有兴趣 make a show of interest

jià

稼 ①(种植) sow (grain): 耕～ ploughing and sowing; farm work ②(谷物) cereals; crops: 庄～ crops; standing grain
【稼穑】[书] sowing and reaping; farming; farm work

嫁 ①(出嫁) marry: ～女儿 marry off a daughter / ～人 get married ②(转嫁) shift; transfer
【嫁祸于人】shift the misfortune onto sb. else; put the blame on sb. else; shift the blame to sb. else
【嫁鸡随鸡】advice to be contented with the man a woman has married; a woman follows her husband no matter what his lot
【嫁接】[植] graft
【嫁娶】marriage
【嫁妆】dowry; trousseau: 给女儿一份～ dower a daughter

价 ①(价格) price: 单～ unit price / 高～ high-priced / 基～ basic price / 集市贸易～ rural fair price / 廉～ low-priced / 零售～ retail price / 牌～ price / 批发～ wholesale price / 实～ net price / 市～ market price / 议～ negotiated price / 变相涨～ price to be raised in a disguised form / 减～ reduce the price / 擅自提～ price to be raised without permission / 削～ reduced price / 要～ ask a price ②(价值) value: 等～交换 exchange of equal values / 估～ estimate the value of; evaluate ③[化] valence: 电～ electrovalence / 共～ covalence / 原子～ atomic valence / 多～元素 element of multivalence
【价带】valence band
【价电子】valence electron
【价格】price: 标明～ mark (goods) with a price tag; have goods clearly priced / 降低～ reduce a price / 提高～ raise a price / 调整～ readjust price ◇ ～标签 price tag / ～管制 price control / ～平衡 price equilibrium / ～歧视 price discrimination / ～体系 price system / ～稳定 price steadiness / ～政策 price policy
【价款】money paid for sth. purchased or received for sth. sold; cost
【价目】marked price; price ◇ ～表 price list
【价钱】price: ～公道 a fair price; a decent price; a reasonable price / 讲～ bargain / 西红柿什么～? How much are the tomatoes?
【价值】①[经] value: 交换～ exchange value / 经济～ economic value / 票面～ face value / 剩余～ surplus value / 使用～ use value ②(值多少钱) cost; be worth: ～五百万美元的设备 five million dollars worth of equipment; equipment worth five million dollars ③(重要作用) worth; value: 毫无～ completely worthless ◇ ～尺度 measure of value / ～规律 law of value / ～量 magnitude of value / ～形态 form of value / ～学 axiology
【价值连城】worth several cities; invaluable; priceless
【价值学】[哲] axiology

假 ①(假期) holiday; vacation: 暑～ summer vacation / 度～ spend one's holidays ②(休假) leave of absence; furlough: 病～ sick leave / 事～ leave of absence to attend to personal affairs / 超～ overstay one's leave of absence / 请～ ask for leave / 休～ be on leave; be on furlough
【假期】①(放假期间) vacation ②(休假期间) period of leave
【假日】holiday; day off
【假条】①(请假条) application for leave ②(准假条) leave permit: 病～ doctor's certificate (for sick leave)

架 ①(架子) frame; rack; shelf; stand: 车～ frame of a cart / 房～ the frame of a house / 钢～桥 steel-framed bridge / 工具～ tool rack / 黄瓜～ cucumber trellis / 篮球～ basketball stand / 书～ bookshelf / 行李～ luggage-rack / 衣～儿 clothes hanger / 衣帽～ clothes tree / 闭～式 closed shelves / 开～式 open shelves ②(搭起;支起) put up; erect: ～电话线 set up telephone lines / ～起机枪

mount a machine gun / ～枪 stack rifles / ～桥 put up a bridge; build a bridge ③(招架) fend off; ward off; withstand ④(扶持) support; prop; help: ～着拐走 walk on crutches ⑤(绑架) kidnap; take sb. away forcibly: 强行～走 carry sb. away by force; kidnap ⑥(殴打; 争吵) fight; quarrel: 劝～ step in and patch up a quarrel; mediate between quarrelling parties ⑦[量] 两～电视机 two TV sets / 一～收音机 a radio set

【架不住】[方]①(禁不住; 受不住) cannot sustain (the weight); cannot stand (the pressure); cannot stand up against: ～这么大的分量 cannot sustain such heavy weight ②(抵不上) be no match for; cannot compete with

【架次】sortie: 出动三批飞机共九十～ fly ninety sorties in three groups

【架空】①(离地) built on stilts; overhead; aerial: ～电缆 aerial cable / ～管道 overhead pipe / ～索道 cableway / ～铁路 overhead railway ②(没有基础) impracticable; unpractical ③(使徒具虚名) make sb. a mere figurehead; render unfeasible

【架设】set up; erect: ～电话线 set up telephone lines / ～输电线路 erect power transmission lines / 在河上～浮桥 throw a pontoon bridge across the river

【架势】[口] posture; stance; manner

【架子】①(框架; 支架; 搁置物品的架子) frame; stand; rack; shelf ②(事物的组织、结构) framework; skeleton; outline ③(傲慢作风) airs; haughty manner: 官僚～ bureaucratic airs / 摆～ put on airs / 放下～ get down from one's high horse / 没有～ be modest and unassuming; be easy of approach ④(架式) posture; stance

【架子工】[建] scaffolder

【架子猪】feeder pig

驾 ①(套牲口) harness; draw (a cart, etc.); yoke ②(驾驶) drive (a vehicle); pilot (a plane); sail (a boat) ③(原指车辆, 借用于对人的敬辞) 大～ your good self

【驾临】[敬] your arrival; your esteemed presence: 恭候～ Your presence is requested

【驾轻就熟】drive a light carriage on a familiar road; be able to handle a job with ease because one has had previous experience; do a familiar job with ease; do something with ease through experience

【驾驶】drive; pilot: ～车辆 drive a vehicle / ～飞机 pilot a plane / ～轮船 pilot a ship / ～拖拉机 drive a tractor ◇ ～舱[航空] control cabin; cockpit; pilot's compartment / ～杆[航空] control stick; control column; joystick / ～盘 steering wheel / ～室 driver's cab / ～台[航海] bridge / ～仪[航空] pilot / ～员 (车辆的) driver; (飞机的) pilot / ～执照 driving license; driver's license

【驾驭】①(驱使车马) drive: ～烈马 get control of a mettlesome horse ②(控制) control; master; dominate: ～局势 control the situation / ～形势 have the situation well in hand / ～自然 tame nature

【驾辕】pull a cart or carriage from between the shafts; be hitched up; be harnessed in the shafts

jiān

煎 ①(油煎) fry: ～鸡蛋 fried eggs ②(用文火煮) simmer in water; decoct: ～药 decoct medicinal herbs

【煎熬】suffering; torture; torment: 受尽～ suffer all kinds of torments

【煎饼】thin pancake made of millet flour, etc.

兼 ①(两倍的) double; twice: ～旬 twenty days ②(同时涉及或具有) simultaneously; concurrently: ～而有之 have both at the same time / ～管 be concurrently in charge of; also look after / 身～数职 hold several posts simultaneously / 总理～外交部长 Premier and concurrently Minister of Foreign Affairs

【兼备】have both... and...: 德才～ have both political integrity and ability; combine ability with political integrity

【兼并】annex (territory, property, etc.): 土地～ annexation of land

【兼程】travel at double speed: ～而往 travel by long stages / ～前进 advance at the double / 日夜～ travel day and night

【兼顾】give consideration to (take account of) two or more things: 公私～ give consideration to both public and private interests

【兼课】①(本身职务以外兼教课) do some teaching in addition to one's main occupation ②(兼两种或两种以上课程) hold two or more teaching jobs concurrently

【兼任】①(同时担任几个职务) hold a concurrent post ②(非专职) part-time ◇ ～教师 part-time teacher

【兼容】[电视] compatible ◇ ～制彩色电视系统 compatible color television system
【兼容并包】include in one and monopolize it; all-embracing
【兼收并蓄】incorporate things of diverse nature; take in everything
【兼听对方意见】hear both sides of the question
【兼听则明,偏听则暗】listen to both sides and you will be enlightened; heed only one side and you will be benighted
【兼之】[书] furthermore; besides; in addition; moreover
【兼职】①(同时担任几个职务) hold two on more posts concurrently: ～过多 hold too many posts at the same time ②(在本职之外兼任的职务) concurrent post; part-time job: 辞去～ resign one's concurrent job / 减少～ reduce concurrent posts ◇ ～法医 adjunct medical examiner

间 ①(中间) between; among: 同志之～ among comrades / 五点钟与六点钟之～ between five and six o'clock ②(一定空间或时间里) within a definite time or space; in; during: 假期～ during the vacation / 世～ (in) the world / 田～ (in) the fields / 晚～ (in the) evening; (at) night ③(房间) room: 里～ inner room / 外～ outer room / 衣帽～ cloakroom ④[量]三～门面 a three-bay shop front / 一～卧室 a bedroom
【间苯二酚】resorcinol
【间冰期】[地] interglacial stage; interglacial
【间不容发】not a hair's breadth in between; the situation is extremely critical
【间架】①(汉字书写的笔画) form of a Chinese character ②(文章的结构部局) structure of an essay
【间量】the area of a room; floor space: 这屋子～儿太小。This room is not spacious enough.
【间奏曲】intermezzo

肩 ①(肩膀) shoulder: 并～战斗 fight shoulder to shoulder ②(担负) take on; undertake; shoulder; bear: 身～重任 shoulder heavy responsibilities
【肩膀】shoulder
【肩负】take on; undertake; shoulder; bear: ～光荣任务 undertake a glorious task / ～重荷 bear a heavy burden; shoulder heavy loads
【肩胛骨】[生理] scapula; shoulder blade
【肩摩毂击】shoulder to shoulder and hub to hub; crowded with people and vehicles
【肩摩踵接】rubbing the shoulder and following the steps
【肩挑背负】carry on the shoulder and back
【肩章】shoulder loop; shoulder-strap; epaulet

笺 [书]①(写信、题词用纸) writing paper: 信～ letter paper ②(信札) letter ③(注解) annotation; commentary
【笺注】[书] notes and commentary on ancient texts

菅 [植] villous themeda

歼 annihilate; wipe out; destroy: ～敌三百 annihilate 300 enemy troops
【歼击机】fighter plane; fighter
【歼灭】annihilate; wipe out; destroy: ～敌人有生力量 wipe out the enemy's effective strength / 给予～性的打击 strike a crushing blow; deal a smashing blow
【歼灭战】war <battle> of annihilation

缄 seal; close
【缄口】[书] keep one's mouth shut; hold one's tongue; say nothing
【缄默】keep silent; be reticent: 保持～ remain silent

监 ①(监视) supervise; inspect; watch ②(牢狱) prison; jail: 女～ women's prison / 收～ put sb. into jail; be taken to prison
【监测器】[物] monitor: 污染～ contamination monitor
【监察】supervise; control ◇ ～委员会 control commission; supervisory committee / ～员 supervisor; controller / ～制度 supervisory system
【监督】①(察看; 监视) supervise; superintend; control: 国际～ international control / ～劳动改造 labor reform under surveillance ②(监督人) supervisor
【监督权】authority to supervise
【监犯】prisoner; convict
【监工】①(监督工作) supervise work; oversee ②(监工者) overseer; supervisor
【监护】[法] guardianship; tutelage ◇ ～权 guardianship / ～人 guardian; guarder
【监禁】take into custody; imprison; put in jail; put in prison; imprisonment ◇ ～处罚 punishment of imprisonment
【监考】invigilate ◇ ～人 invigilator

【监牢】prison; jail
【监票】scrutinize balloting ◇ ～人 scrutineer
【监视】keep watch on; keep a lookout over; guard: ～敌人的行动 keep watch on the movements of the enemy / ～某人 place sb. under surveillance ◇ ～雷达 surveillance radar equipment / ～器 monitor / ～人 guarder / ～哨[军] lookout
【监守】have custody of; guard; take care of
【监守自盗】steal what is entrusted to one's care; embezzle; defalcate; custodian turned thief
【监听】monitor ◇ ～器 monitor / ～台 monitor board / ～无线电台 monitoring station
【监外执行】executing a sentence outside of jail; serving sentence outside the prison under surveillance
【监狱】prison; jail ◇ ～看守 guard / ～长 warden
【监制】supervise the manufacture of

坚 ①(硬;坚固) hard; solid; firm; strong: ～冰 solid ice; hard ice ②(坚固的东西或阵地) a heavily fortified point; fortification; stronghold: 攻～ storm strongholds ③(坚定;坚决) firmly; steadfastly; resolutely: ～信 firmly believe
【坚壁】hide supplies to prevent the enemy from seizing them; place in a cache; cache
【坚壁清野】strengthen defense works, evacuate noncombatants, and hide provisions and livestock; strengthen the defenses and clear the fields; fortify the defense works and to leave nothing usable to the invading enemy
【坚不可摧】indestructible; impregnable
【坚持】persist in; persevere in; uphold; insist on; stick to; adhere to: ～错误 persist in one's errors / ～到底 stick it out; hold on straight to the end; carry through firmly to the end / ～己见 hold on to one's own views / ～原则 adhere to principles; uphold principles; stick to principles / ～原来的计划 adhere to the original plan / ～真理 uphold the truth; hold firmly to the truth / 再～一会儿 hold out a little longer
【坚持不懈】unremitting: 作～的努力 make unremitting efforts
【坚持不渝】persistent; persevering: ～的斗争 firm and unswerving struggle
【坚定】①(稳定坚强) firm; staunch; steadfast: ～不移 firm and steadfast; resolute and firm; firm and unshakable; unswerving / ～的步伐 firm strides / ～的立场 a firm stand / ～的意志 constancy of purpose ②(使坚定) strengthen: ～…的决心 strengthen one's resolve to / ～信念 strengthen one's conviction
【坚固】firm; solid; sturdy; strong: ～的基础 solid foundation / ～耐用 sturdy and durable
【坚果】[植] nut
【坚决】firm; resolute; determined; resolved: ～、彻底、干净、全部地消灭敌人 annihilate the enemy resolutely, thoroughly, wholly and completely / ～反对 resolutely oppose / ～支持 firmly support; stand firmly by / 意志～ be resolute in one's determination
【坚苦卓绝】showing the utmost fortitude; endure hardship being prominent above all
【坚牢度】[纺] fastness: 耐日光～ fastness to sunlight; sunfastness / 耐洗～ washfastness
【坚强】①(强固有力) strong; firm; staunch: ～的决心 strong determination / ～的领导 firm leadership / ～的意志 strong will ②(加强) strengthen
【坚忍】steadfast and persevering (in face of difficulties)
【坚忍不拔】firm and indomitable; stubborn and unyielding
【坚韧】tough and tensile; firm and tenacious
【坚韧不拔】firm and indomitable; persistent and dauntless
【坚如磐石】solid as a rock; rock-firm
【坚实】solid; substantial: 打下～的基础 lay a solid foundation / 迈出～的步子 make solid progress
【坚守】stick to; hold fast to; stand fast: ～岗位 stick to one's post / ～阵地 hold fast to one's position; hold one's ground
【坚挺】[金融] strong
【坚信】firmly believe; be firmly convinced; be fully confident of: ～共产主义一定能实现 be fully confident of the realization of communism / ～这种学说 have infinite faith in this doctrine
【坚毅】firm and persistent; with unswerving determination; with inflexible will
【坚硬】hard; solid: ～的岩石 solid rock
【坚贞】faithful; constant
【坚贞不屈】remain faithful and unyielding; stand firm and unbending

鲣 oceanic bonito; skipjack (tuna)
【鲣鸟】booby; sula

尖 ①(物体的尖儿) point; tip; top: 铅笔～ the tip of a pencil / 塔～ the pinnacle of a pagoda / 针～ the point of a needle or pin; pinpoint / 指～ fingertip ②(尖削的) pointed; tapering: ～下巴 a pointed chin / 削～铅笔 sharpen a pencil ③(声音高而细) shrill; piercing: ～叫 scream / ～声～气 in a shrill voice ④(灵敏) sharp; acute: 鼻子～ have an acute 〈sharp〉 sense of smell / 耳朵～ have sharp ears; be sharp-eared / 眼～ have sharp eyes; be sharp-eyed ⑤(出类拔萃的人或物) the best of its kind; the pick of the bunch; the cream of the crop: 拔～儿的 top-notch; the pick of the bunch

【尖兵】①[军] point ②(开创者) trailblazer; pathbreaker; pioneer; vanguard

【尖刀】sharp knife; dagger ◇ ～班[军] dagger squad

【尖端】①(尖锐的顶端) pointed end; acme; peak: 标枪的～ the point of a javelin ②(发展得最高的) most advanced; sophisticated ◇ ～产品 highly sophisticated products / ～科学 most advanced branches of science; frontiers of science / ～武器 sophisticated weapons

【尖端放电】point discharge

【尖刻】acrimonious; caustic; biting: ～的讽刺 poignant satire / 说话～ speak with biting sarcasm

【尖括号】angle brackets (〈 〉)

【尖利】①(锋利) sharp; keen; cutting: ～的钢刀 a sharp knife ②(声音高而刺耳) shrill; piercing: ～的叫声 a shrill cry

【尖脐】①(雄蟹的脐) the narrow triangular abdomen of a male crab ②(雄蟹) male crab

【尖锐】①(锋利) sharp-pointed ②(深刻; 敏锐) penetrating; incisive; sharp; keen: ～的批评 incisive criticism; sharp criticism / ～地指出 point out sharply ③(声音高而刺耳) shrill; piercing: ～的汽笛声 shrill whistle ④(激烈) intense; acute; sharp: ～的思想斗争 sharp mental conflicts / ～对立 be diametrically opposed to each other ◇ ～化 sharpen; intensify; become more acute

【尖酸】acrid; acrimonious; tart: ～刻薄 tart and mean; bitterly sarcastic

【尖塔】[建] steeple

【尖子】the best of its kind; the pick of the bunch; the cream of the crop; the pick; ace: 全班学生中的～ the cream of a class of pupils; the top student in a class

【尖嘴薄舌】have a caustic and flippant tongue

艰 difficult; hard

【艰巨】arduous; formidable; onerous: ～的任务 a arduous task / 付出～的劳动 make tremendous efforts

【艰苦】arduous; difficult; hard; tough: ～创业 pioneer an enterprise with painstaking efforts / ～的斗争 arduous struggle / ～的生活 hard life / ～朴素 hard work and plain living / 在～的条件下 under arduous conditions

【艰难】difficult; hard; arduous: ～困苦 difficulties and hardships / ～曲折 arduous and tortuous; full of difficulties and reverses / ～岁月 difficult days; arduous years / ～险阻 difficulties and obstacles; hardships and obstructions / 生活～ live in straitened circumstances / 行动～ walk with difficulty

【艰涩】involved and abstruse; intricate and obscure: 文词～ involved and abstruse writing

【艰深】difficult to understand; abstruse: ～的理论 abstruse theories

【艰危】difficulties and dangers (confronting a nation)

【艰险】hardships and dangers; perilous

【艰辛】hardships: 历尽～ experience all kinds of hardships

犍

【犍牛】bullock

奸 ①(奸诈) wicked; evil; treacherous: ～计 an evil plot ②(不忠于国家或民族的) traitor: 汉～ traitor to the Chinese nation; traitor / 内～ a secret enemy agent within one's ranks; hidden traitor / 锄～ eliminate traitors; weed out enemy agents ③(自私; 取巧) self-seeking and wily: 那个人才～哪。 That chap is just a self-seeker. ④(非法性关系) liaison; illicit sexual relations: 强～ rape / 通～ have illicit sexual relations; commit adultery / 诱～ seduce

【奸臣】treacherous court official; traitor minister

【奸夫】intrigant; paramour

【奸妇】paramour

【奸猾】treacherous; crafty; deceitful

【奸佞】[书]①(奸邪谄媚) crafty and fawning ②(奸佞的人) crafty sycophant

【奸商】unscrupulous merchant; profiteer; dishonest trader

【奸污】rape; seduce; debauchery: ～妇女 rape a woman; commit a rape on a woman ◇ ～处

女 defloration
【奸细】spy; enemy agent
【奸险】wicked and crafty; treacherous; malicious
【奸笑】sinister (villainous) smile
【奸邪】[书]①(奸诈邪恶) crafty and evil; treacherous ②(奸诈邪恶的人) a crafty and evil person
【奸雄】a person who achieves high position by unscrupulous scheming; arch-careerist
【奸淫】①(不正当的性行为) illicit sexual relations; adultery ②(奸污) rape; seduce: ～掳掠 rape and loot; rape and plunder / ～烧杀 engage in rape, arson and murder ◇ ～少女罪 crime of fornication with an underage girl / ～已遂 consummation of fornication / ～幼女 fornication with an underage girl / ～罪 carnal intercourse
【奸贼】traitor; conspirator
【奸诈】fraudulent; crafty; treacherous

jiǎn

简 ①(简单) simple; simplified; brief: ～而言之 in brief; in short; put it in a nutshell; in a word / 从～ conform to the principle of simplicity ②(竹片) bamboo slips ③(信件) letter: 书～ letters; correspondence ④[书](选择人才) select; choose ～拔 select and promote
【简报】bulletin; brief report: 会议～ conference bulletin; brief reports on conference proceedings / 新闻～ news bulletin
【简本】abridged edition
【简编】(多用于书名) short course; concise edition; abridged edition
【简便】simple and convenient; handy: ～的方法 a simple and convenient method; a handy way / 操作～ easy to operate
【简称】①(名称的简化形式) the abbreviated form of a name; abbreviation; shorter form ②(使名称简化) be called sth. for short
【简单】①(不复杂) simple; uncomplicated: ～明了 simple and clear; concise and explicit; brief and clear / 构造～ simple in structure ②(平凡)(多用否定式) commonplace; ordinary: 不～ not simple; fairly complicated; remarkable ③(草率;不细致) oversimplified; casual: ～从事 do things in a casual 〈perfunctory〉 way / ～粗暴 do things in an oversimplified and crude way; simple and crude / ～地看问题 take a naive view; oversimplify a problem / 头脑～ simple-minded; seeing things too simply ◇ ～多数 simple majority
【简单化】oversimplify
【简单再生产】simple reproduction
【简短】brief; short
【简分数】common fraction; simple fraction
【简化】simplify: ～工序 simplify working processes
【简化汉字】①(简化汉字的笔画) simplify Chinese characters ②(经过简化的汉字) simplified Chinese characters
【简洁】succinct; terse; pithy; concise
【简捷】simple and direct; forthright
【简介】brief introduction; synopsis; summarized account: 剧情～ the synopsis of a drama
【简括】brief but comprehensive; compendious; briefly; compendiously
【简历】biographical notes; curriculum vitae; resume
【简练】terse; succinct; pithy
【简陋】simple and crude: ～的房间 shabby rooms / 设备～ simple and crude equipment
【简略】simple (in content); brief; sketchy
【简码】brevity code
【简慢】negligent
【简明】simple and clear; concise: ～扼要 terse and concise; brief and to the point ◇ ～词典 a concise dictionary / ～新闻 news in brief
【简朴】simple and unadorned; plain: ～的语言 plain language / 生活～ a simple and frugal life; plain living
【简谱】[乐] numbered musical notation
【简体字】simplified Chinese character
【简图】sketch; diagram; abbreviated drawing; simplified schematic
【简谐运动】[物] simple harmonic motion
【简谐振子】[物] simple harmonic oscillator
【简写】①(简写字) write a Chinese character in simplified form ②(文字浅显) simplify a book for beginners ◇ ～本 simplified edition
【简讯】news in brief
【简要】concise and to the point; brief: ～的介绍 a brief introduction; briefing
【简仪】[天] abridged armilla
【简易】①(简单而容易) simple and easy: ～的方法 a simple and easy method ②(设备不完备) simply constructed; simply equipped; unsophisticated ◇ ～病房 simply equipped ward / ～读物 easy reader / ～公路 simply road; rough road; simply-built highway / ～机场 airstrip
【简约】brief; concise; sketchy

【简章】general regulations
【简直】[副] simply; at all; virtually: ～不堪想象 virtually unimaginable / ～是浪费时间。It's a sheer waste of time.

剪 ①(剪刀) scissors; shears; clippers ②(切断) cut (with scissors); clip; trim: ～掉 cut off; scissor off / ～断 cut off; nip; snip / ～开 cut open / ～发 have one's hair cut / ～羊毛 shear a sheep / ～指甲 trim one's nails ③(除去) wipe out; exterminate: ～除 wipe out; annihilate
【剪报】newspaper cutting; newspaper clipping
【剪裁】①(裁衣料) cut out (a garment); tailor ②(对材料取舍安排) cut out unwanted material (from a piece of writing); prune
【剪彩】cut the ribbon at an opening ceremony
【剪除】wipe out; annihilate; exterminate
【剪床】[机] shearing machine
【剪刀】scissors; shears: 一把～ a pair of scissors
【剪刀差】scissors movement of prices; scissors differential; scissors difference; price scissors: 扩大～ widen the price scissors / 缩小～ narrow the price scissors
【剪辑】①[电影] montage; film editing: 电影～机 motion-picture editing machine; movieola ②(剪裁重编) editing and rearrangement: 电影录音～ highlights of a live recording of a film
【剪接】montage; film editing
【剪毛】[牧] shearing; clipping ◇ ～机 shearing machine
【剪票】punch a ticket ◇ ～铗 conductor's punch
【剪秋萝】[植] senno campion
【剪贴】①(剪贴资料) clip and paste (sth. out of a newspaper, etc.) in a scrapbook or on cards ②(儿童手工) cutting out ◇ ～簿 scrapbook
【剪影】①(剪纸像) paper-cut silhouette ②(事物轮廓的描写) outline; sketch
【剪应力】[机] shearing stress
【剪纸】[工美] paper-cut; scissor-cut
【剪子】scissors; shears; clippers

茧 ①(蚕茧) cocoon: 蚕～ silkworm cocoon ②(老趼) callus: 老～ thick callus
【茧绸】pongee; tussah silk

柬 card; note; letter: 请～ invitation card
【柬埔寨】Kampuchea ◇ ～人 Kampuchean / ～语 Kampuchean (language)
【柬帖】note; short letter

拣 ①(挑选) choose; select; pick out ②(拾取) pick up; collect; gather
【挑选】select; choose

减 ①(去掉) subtract: 八～三得五。Eight minus three is five. (或: Three from eight is five.) ②(降低) reduce; decrease; cut: ～半 reduce by half
【减产】reduction of output; drop in production
【减充血药】decongestant drug
【减低】reduce; lower; bring down; cut: ～百分之十 be lowered by ten per cent / ～标价 mark down / ～定价 reduce prices / ～速度 lower speed; slacken speed; slow down / ～运费 freight cutting
【减法】subtraction
【减肥】reduce
【减股盈余】surplus from cancellation of stock
【减号】minus sign (−)
【减河】[水] distributary
【减缓】retard; slow down: ～进程 slow down the pace 〈progress〉
【减价】reduce the price; mark down: ～出售 sell at a reduced price / ～五元 reduce five *yuan* from the price; be marked down by five *yuan* / ～一成 be marked down by 10%
【减免】①(指刑罚) mitigate a punishment; annul a punishment ②(指捐税、债务等) reduce; remit (taxation, etc.): ～学费 reduce; tuition; remit tuition
【减摩】[机] antifriction: ～合金 antifriction alloy; antifriction metal / ～轴承 antifriction bearing
【减轻】lighten; ease; alleviate; mitigate: ～病人的痛苦 alleviate 〈ease〉 a patient's suffering / ～惩罚 commutation of punishment / ～处分 mitigate a punishment / ～国家的负担 lighten the burden on the state / ～劳动强度 reduce labor intensity / ～刑罚 commutation of penalty; reduction of punishment / ～责任 diminished responsibility / ～罪行 extenuation of a crime
【减弱】weaken; abate: 体力大大～ be much weakened physically
【减色】lose lustre; impair the excellence of; detract from the merit of: 情节松散使这篇小说大为～。Its loose plot spoils the novel.
【减色法】[电影] subtractive process

【减少】reduce; decrease; lessen; cut down: ～百分之十 cut down by ten per cent / ～非生产性开支 reduce nonproductive expenditure / ～开会时间 reduce time taken by meetings / ～危险 lessen the danger / 交通事故～了。 There has been a decrease in traffic accidents.
【减声器】[机] muffler
【减数】[数] subtrahend
【减数分裂】[生] meiosis; reduction division
【减税】abatement of tax; tax abatement; tax reduction
【减速】slow down; decelerate; retard ◇ ～度 deceleration / ～副翼[航空] deceleron / ～火箭 retro-rocket / ～剂[物] moderator / ～器[机] reduction gear; reduction device / ～伞[航空] drag parachute; deceleration parachute / ～运动[物] retarded motion
【减缩】reduce; cut down; retrench: ～开支 reduce expenditure
【减退】drop; go down; abate; subside; decrease: 热度～ one's fever has abated; one's fever has come down / 视力～ one's eyesight is failing
【减刑】[法] reduce a penalty; commute a sentence; mitigate a sentence; abatement from penalty ◇ ～特赦 commutation pardon
【减压】reduce pressure; decompress ◇ ～器 pressure reducer; decompressor / ～室 decompression chamber
【减员】depletion of numbers (in the armed forces): 战斗～ combat depletion of strength
【减震】shock absorption; damping ◇ ～器[机] shock absorber; damper
【减租减息】reduction of rent for land and of interest on loans; reduction of rent and interest
【减罪】palliate a crime

碱 ①[化] alkali ②(碳酸钠) soda: 纯～ soda; soda ash / 洗涤～ washing soda ③(盐基) base
【碱地】alkaline land
【碱化】alkalization; basification
【碱金属】alkali metal; alkaline metal
【碱石灰】soda lime
【碱式盐】[化] basic salt
【碱土金属】alkaline-earth metal
【碱性】basicity; alkalinity ◇ ～长石 alkali feldspars / ～法[冶] basic process / ～反应 alkaline reaction / ～染料 basic dyes / ～土 alkaline soil / ～岩 alkaline rock / ～氧气炼钢片 basic oxygen process
【碱液】lye
【碱中毒】alkalosis

检 ①(查) check up; inspect; examine ②(约束;检点) restrain oneself; be careful in one's conduct: 行为不～ depart from correct conduct
【检波】[电子] detection ◇ ～管 detection tube / ～器 detector
【检查】①(用心查看) check up; inspect; examine: ～工作 check up on work / ～护照 inspect sb.'s passport / ～身体 have a physical examination; have a health check; have a medical check-up / ～视力 test sb.'s eyesight / ～行李 inspect sb.'s luggage; examine sb.'s luggage / ～帐目 check the account; audit the account / ～质量 check on the quality of sth. / 钡餐～ barium meal examination / 常规～ routine examination / 全身～ general check-up / 透视～ examine by fluoroscopy / 卫生～ sanitary inspection / 新闻～ press censorship / 邮件～ postal inspection / 做诊断～ work up a patient ②(检讨) self-criticism: 写～ write a self-criticism / 作～ criticize oneself; make a self-criticism ◇ ～井[建] inspection shaft; inspection well / ～哨 checkpost / ～团 inspection party / ～员 examining officer; inspector / ～站 checkpoint; checkpost; inspection station
【检察】procuratorial work ◇ ～官 public procurator; public prosecutor; prosecuting attorney / ～机关 procuratorial organ / ～院 procuratorate / ～长 chief procurator; public procurator-general
【检点】①(查看) examine; check: ～一下工具,看有无丢失。 Check the tools and see if anything is missing. ②(注意约束) be cautious (about what one says or does): 言行有失～ be careless about one's words and acts; be indiscreet in one's speech and conduct
【检举】report (an offense) to the authorities; inform against (an offender); accuse: ～贪污分子 inform against an embezzler ◇ ～信 letter of accusation; written accusation / ～箱 a box for accusation letters
【检漏】[电] leak hunting ◇ ～器 leak detector; leak localizer
【检数单】tally sheet
【检索】retrieve; search
【检讨】self-criticism: 作～ make a self-criticism / ～自己的错误 examine one's mistakes ◇ ～书 written self-criticism
【检修】examine and repair; overhaul: ～机器

overhaul a machine / 这台发动机需要～。This engine needs an overhaul.

【检验】test; examine; inspect; verify: 商品～ commodity inspection / 实践是～真理的唯一标准。Practice is the sole criterion of truth. ◇～程序 check routine; check problem; test program; test routine / ～单 check list; inspection sheet / ～飞行 check flight / ～室 checkout room; laboratory / ～证 certificate of inspection; testing certificate

【检疫】quarantine ◇～范围 quarantine range / ～锚泊地 quarantine anchorage / ～旗 quarantine flag; yellow flag / ～员 quarantine officer / ～站 quarantine station / ～证明书 quarantine certificate; vaccination certificate

【检阅】review (troops, etc.); inspect: ～军队 inspect troops / ～仪仗队 review a guard of honor ◇～台 reviewing stand

【检字表】index of Chinese characters: 汉语拼音～ phonetic alphabetical index of Chinese characters

【检字法】indexing system for Chinese characters

捡 pick up; collect; gather: ～柴火 gather firewood / ～粪 collect manure

【捡了芝麻,丢了西瓜】pick up the sesame seeds but overlook the watermelons; mindful of small matters to the neglect of large ones; penny wise and pound foolish

【捡漏】[建] repair the leaky part of a roof; plug a leak in the roof

【捡破烂儿】pick odds and ends from refuse heaps

【捡拾压捆机】[农] pick-up bale; pick-up press

睑 eyelid

【睑结膜炎】[医] blepharoconjunctivitis

【睑内翻】entropin

【睑外翻】ectropin

【睑腺炎】[医] sty

【睑下垂】ptosis

【睑炎】blepharitis

俭 thrifty; frugal: 省吃～用 eat sparingly and spend frugally; be economical in everyday spending

【俭朴】thrifty and simple; economical: 生活～ lead a thrifty and simple life / 衣着～ dress simply

【俭省】economical; thrifty; frugal; sparing: 用钱～ be economical of one's money; spend one's money sparingly

jiàn

渐 gradually; by degrees

【渐变】①(逐渐变化) gradual change ②[进化] anamorphism; anamorphosis

【渐次】[书] gradually; one after another

【渐渐】[副] gradually; by degrees; little by little: 脚步声～消失了。The sound of footsteps gradually died away. / 雨～小了 The rain is beginning to let up.

【渐进】advance gradually; progress step by step: 循序～ advance gradually in due order

【渐显】[摄] fade in

【渐隐】[摄] fade out

间 ①(空隙) space in between; opening: 乘～ seize an opportunity / 亲密无～ closely united; on intimate terms ②(相隔) separate: 黑白相～ checkered with black and white / 晴～多云 fine with occasional clouds ③(挑拨使人不和) sow discord: 离～ sow discord; drive a wedge between ④(间苗) thin out (seedlings): ～玉米苗 thin out maize seedlings

【间谍】spy ◇～飞机 spy plane / ～活动 espionage / ～网 espionage network / ～卫星 spy satellite

【间断】be disconnected; be interrupted

【间断性】[哲] discontinuity: 不～ continuity

【间隔】interval; intermission

【间隔号】[语] separation dot

【间或】occasionally; now and then; sometimes; once in a while

【间接】indirect; secondhand ◇～宾语[语] indirect object / ～成本 indirect cost; overhead / ～犯罪 consequential crime / ～肥料 indirect fertilizer / ～费 overhead cost; indirect expense / ～负债 indirect liabilities / ～汇兑 indirect exchange / ～汇率 indirect rate / ～接触[医] mediate contacts / ～经验 indirect experience / ～贸易 indirect trade / ～输血 indirect transfusion / ～税 indirect tax; indirect duty / ～损失 indirect losses / ～投资 indirect investment / ～推理[逻] mediate inference / ～消费 indirect consumption / ～选举 indirect election / ～引语[语] indirect speech / ～原因 remote cause

【间苗】thin out seedlings

【间日疟】[医] tertian fever; tertian malaria

【间隙】①(空隙) interval; gap; space: 利用工作～学习 study in the intervals of one's work ②[机] clearance: 齿轮～ gear clearance
【间歇】intermittence; intermission ◇ ～反应[化] intermittent reaction / ～泉[地] geyser; intermittent spring / ～热[医] intermittent fever / ～性精神病 intermittent insanity; periodic insanity / ～训练 interval training
【间杂】be intermingled; be mixed
【间奏曲】intermezzo
【间作】[农] intercropping

涧
ravine; gully

践
①(踩) trample; tread ②(履行) act on; carry out
【践踏】tread on; trample underfoot: 请勿～草地。Keep off the grass.
【践约】keep a promise; keep an appointment

贱
①(价格低廉) low-priced; inexpensive; cheap: ～卖 sell cheap; sell low ②(地位低下) lowly; humble: 贫～ poor and lowly ③(卑鄙; 下贱) low-down; base; despicable; mean: 下～ low-down; base ④[谦] my: ～躯尚健 my mean body is still strong
【贱骨头】[骂] miserable wretch; contemptible wretch
【贱货贵德】despise material goods and respect virtue
【贱买贱卖】buy and sell at low price

溅
splash; spatter; sputter: ～了一身水 be spattered with water / 火花飞～。Sparks flew about.
【溅落】[宇航] splash down ◇ ～点 splash point

饯
①(饯行) give a farewell dinner ②(蜜饯) candied fruit; preserved fruit
【饯别】give a farewell dinner: ～友人 give a friend a farewell dinner
【饯行】give a farewell dinner

荐
recommend
【荐举】propose sb. for an office; recommend

鉴
①(古代铜镜) ancient bronze mirror ②(照) reflect; mirror: 光可～人 so shining and bright that it can serve as a looking glass / 水清可～。The water is so clear that you can see your reflection in it. ③(教训) warning; object lesson: 引以为～ take it as an object lesson; take warning from it / 前车之覆,后车之～。The overturning of the cart in front is a warning to the carts behind. ④(仔细看; 审察) inspect; scrutinize; examine ⑤(书信套语)某先生台～ Dear Mr. so-and-so: May I draw your attention to the following
【鉴别】distinguish; differentiate; discriminate; discern; identify: ～能力 ability to differentiate / ～文物 make an appraisal of a cultural relic / ～真伪 discern the false from the genuine
【鉴定】①(评语) appraisal (of a person's strong and weak points): 毕业～ graduation appraisal ②(评定) appraise; identify; authenticate; determine: ～产品质量 appraise the quality of a product / ～文物年代 determine the date of a cultural relic ◇ ～结论 expert conclusion / ～人 identifier; surveyor / ～书 expertise report
【鉴戒】warning; object lesson
【鉴赏】appreciate: ～家 connoisseur / ～能力 ability to appreciate painting, music, etc.; connoisseurship / 对音乐有～力 have a good ear for music
【鉴于】in view of; seeing that; in consideration of

监
an imperial office: 国子～ the Imperial College

见
①(看见) see; catch sight of: 亲眼所～ see with one's own eyes; see for oneself ②(遇到) meet with; be exposed to: 一～如故 become fast friends at the first meeting ③(看得出; 显得出) show evidence of; appear to be: 病已～轻。The patient's condition has improved. ④(参见; 参看) refer to; see; vide: ～第 234 页 see page 234 / ～后 see after; vide post / ～前 see before; vide ante / ～上 see above; vide supra / ～下 see below; vide infra ⑤(会见; 会面) meet; call on; see: 明天～。See you tomorrow! ⑥(见解) view; opinion: 依我之～ in my opinion; to my mind / 英雄所～略同。Heroes have similar views. (或: Great minds think alike.) ⑦[书][助] ～弃 be rejected; be discarded / ～责 be blamed / 即希～告。Hope to be informed immediately.
【见报】appear in the newspapers
【见不得】①(不能遇见) not to be exposed to; unable to stand: ～阳光 not to be exposed to the sunlight ②(见不得人的) not fit to be seen or revealed: ～人 shameful; scandalous; too ashamed to show up in public

【见财起意】be moved to commit crimes by sight of money
【见长】be good at; be expert in: 他以音乐～。He is good at music.
【见地】insight; judgment: 很有～ have keen insight; show sound judgment
【见多识广】experienced and knowledgeable; with rich experience and extensive knowledge
【见方】[口] square: 一米～ one meter square
【见风使舵】trim one's sails; swap horses when crossing a stream
【见缝插针】stick in a pin wherever there's room; make use of every bit of time 〈space〉
【见怪】mind; take offense: 对我讲的话请别～。I hope you will not take any offense at my words.
【见怪不怪】be inured to the unusual
【见怪不怪，其怪自败】face the fearful with no fears, and its fearfulness disappears; If one remains calm upon seeing strange things, the strangeness will do no harm.
【见鬼】①(离奇古怪) fantastic; preposterous; absurd: 种庄稼不除草不是～么? Isn't it absurd to plant crops and not weed the fields? ②(灭亡；毁灭) go to hell: 让种族歧视～去吧! To hell with racial discrimination
【见好】get better; mend: 他的病～了。He's on the mend.
【见机】as the opportunity; as befits the occasion; according to circumstances: ～行事 act according to circumstances; do as one sees fit
【见教】[套] favor me with your advice; instruct me: 有何～? Is there something you want to see me about?
【见解】view; opinion; understanding; idea: 持不同～ have different views; hold different opinions／提出新的～ put forward some new ideas
【见景生情】Memories revive at the sight of familiar places.
【见谅】[书] excuse me; forgive me: 务希～。I sincerely hope you'll excuse me.(或: I hope you would excuse me.)
【见利忘义】forget honor at sight of money; forget all moral principles at the sight of profits
【见猎心喜】thrill to see one's favorite sport and itch to have a go
【见面】①(相见) meet; see: 初次～ meet for the first time ②(接触) contact; link: 领导应和群众经常～。It's necessary for leading cadres to keep frequent contacts with the masses. ◇～礼 a present given to sb. on first meeting him
【见票即付】[商] payable at sight; payable to bearer
【见仁见智】different people, different views; opinions differ
【见世面】see the world; enrich one's experience; face the world
【见事生风】arouse trouble with very little cause
【见识】①(扩大见闻) widen one's knowledge; enrich one's experience ②(见闻知识) experience; knowledge; sensibleness: 长～ widen one's knowledge; broaden one's horizons／他是一个有～的人。He is a man of rich experience.
【见树不见林】not see the wood for the trees; fail to see the wood for the trees
【见死不救】see someone in mortal danger without lifting a finger to save him
【见所未见】see what one has never seen before; never seen before; unprecedented
【见外】regard sb. as an outsider: 在我这儿可别～。Just make yourself at home. (或: Please make yourself at home.)
【见微知著】from the first small beginnings one can see how things will develop; from one small clue one can see what is coming; a straw shows which way the wind blows
【见闻】what one sees and hears; knowledge; information: ～广 knowledgeable; well-informed／增长～ add to one's knowledge
【见物不见人】see things but not people; see only material factors to the neglect of human ones
【见习】learn on the job; be on probation: 一年～期 on one-year probation／在工厂～ work on probation ◇～技术员 technician on probation／～领事 student consul／～生 probationer／～医生 intern
【见效】become effective; produce the desired result: 这方法很～。This method worked effectively.
【见笑】laugh at (me; us): 如果做得不好，请别～。Please don't laugh at me if I can't do it well.
【见血封喉】[植] upas
【见义勇为】ready to take up the cudgels for a just cause; never hesitate to do what is righteous; gallantry rise to the occasion
【见异思迁】change one's mind the moment one sees something new; be inconstant; be irresolute; be like a rolling stone
【见影怖鬼】see a shadow and fear ghosts
【见长】grow perceptibly: 这孩子不～。The

child doesn't seem to be growing.
【见证】witness; testimony ◇ ～人 eyewitness; witness

舰

舰 warship; naval vessel; man-of-war: 导弹驱逐～ guided missile destroyer / 导弹巡洋～ guided missile cruiser / 登陆～ landing ship / 供应～ tender / 航空母～ aircraft carrier; carrier / 护航～ convoy ship / 护卫～ escort vessel / 军～ warship / 旗～ flagship / 驱逐～ destroyer / 巡洋～ cruiser / 直升机航空母～ helicopter aircraft carrier / 指挥～ control vessel / 主力～ capital ship ○靶船 target ship / 布雷艇 mine layer / 登陆艇 landing boat / 核潜艇 nuclear-powered submarine / 救生船 life boat / 救生艇 life craft / 救援艇 crash rescue boat / 炮艇 gunboat / 潜水艇 submarine; U-boat / 扫雷艇 mine sweeper / 巡逻艇 patrol boat / 鱼雷快艇 torpedo craft
【舰船中间补给基地】staging base
【舰队】fleet; naval force: 南海～ the Nanhai Sea Fleet / 特混～ task force; task fleet ◇ ～司令 admiral
【舰队岸转移】ship-to-shore movement
【舰控进场系统】carrier-controlled approach
【舰首】bow ◇ ～炮 bow chaser
【舰艇】naval ships and boats; naval vessels: 两栖作战～ amphibious warfare vessel / 作战～ combat ship
【舰尾】stern ◇ ～炮 stern chaser
【舰位】berth
【舰载】carrier-borne; carrier-based; ship-based ◇ ～导弹 ship-based missile / ～飞机 shipboard aircraft; deck-landing aircraft; ship plane
【舰长】captain (of a warship)
【舰只】warships; naval vessels: 海军～ naval vessels

剑

剑 sword: 花～ epee / 轻～ foil / 重～ saber
【剑拔弩张】with swords drawn and bows bent; at daggers drawn; saber-ratting
【剑柄】handle of a sword; hilt
【剑道】[日] kendo
【剑杆】arrow shaft
【剑客】swordsman
【剑麻】[植] sisal hemp
【剑眉】straight eyebrows slanting upwards and outwards; dashing eyebrows
【剑鞘】scabbard; sheath
【剑术】art of fencing; swordsmanship
【剑突】[生理] ensiform process; xiphoidprocess
【剑舞】sword dance
【剑鱼】swordfish

箭

箭 arrow: 射～ shoot an arrow
【箭靶子】target for archery
【箭步】a sudden big stride forward
【箭毒】curare; curari
【箭杆】arrow shaft
【箭楼】an embrasured watchtower over a city gate
【箭筒】quiver
【箭头】①(箭的尖头) arrowhead; arrow tip; arrow point; head of arrow ②(箭头形符号) arrow
【箭在弦上】like an arrow on the bowstring; there can be no turning back
【箭猪】[动] porcupine
【箭镞】metal arrowhead

僭

僭 [书] overstep one's authority: ～越 overstep one's authority

件

件 ① [量] piece: 一～衬衫 a shirt / 一～工作 a piece of work; a job / 一～事 a matter; a thing ②(指可以一一计算的事物): 案～ (law) case / 锻～ forged piece; forging / 工～ workpiece ③(文件;信件) letter; correspondence; paper; document: 来～ a communication, document, etc. received / 密～ confidential documents; classified documents; secret papers

建

建 ①(建设) build; construct; erect: ～电站 build a power station / 新～的工厂 newly-built factory ②(建立) establish; set up; found: ～立一个国家 establish a state
【建党】①(指建立) found a party ②(指建设) Party building ◇ ～路线 line for Party building
【建都】found a capital; make (a place) the capital
【建国】①(建立国家) found a state; establish a state ②(建设国家) build up a country: ～方略 constructive scheme for our country / ～宏图 a grand project for national reconstruction / 勤俭～ build up our country through diligence and frugality
【建交】establish diplomatic relations
【建军】①(建立军队) found an army ②(建设军队) army building: ～原则 principles for army building ◇ ～节 Army Day
【建兰】[植] sword-leaved cymbidium

【建立】build; establish; set up; found: ～功勋 perform meritorious deeds / ～统一战线 form a united front / ～外交关系 establish diplomatic relations / ～信心 build up one's confidence
【建设】build; construct: ～社会主义 build socialism / 加强党的思想～和组织～ strengthen the Party ideologically and organizationally / 社会主义～ socialist construction
【建设性】constructive: ～的意见 constructive suggestions
【建树】[书] make a contribution; contribute; score an achievement
【建议】①(提建议) propose; suggest; recommend: 我～再试一次。I suggest trying once more. ②(所提出的建议) proposal; suggestion; recommendation: 反～ counterproposal / 合理化～ rationalization proposal
【建造】build; construct; make
【建制】organizational system: 部队～ the organizational system of the army ◇ ～部队 organic unit
【建筑】①(建造) build; construct; erect: ～高楼 erect a tall building / ～桥梁 construct a bridge / ～铁路 build a railway ②(建筑物) building; structure; edifice: 古老的～ an ancient building / 宏伟的～ a magnificent structure ③(建筑学) architecture: 现代～学 modern architecture ◇ ～材料 building materials / ～法规 building laws / ～工程学 architectural engineering / ～工地 building site; construction site / ～工人 building worker; builder / ～红线 property line / ～群 architectural complex / ～设计 architectural design / ～师 architect / ～物 building; structure

键

①[机] key: 轴～ shaft key ②(按键) key (of a typewriter, piano, etc.) ③[化] bond: 共价～ covalent bond
【键槽】[机] keyway; key slot; key seat
【键盘】keyboard; fingerboard
【键盘乐器】keyboard instrument; clavier

毽

shuttlecock
【毽子】shuttlecock: 踢～ kick the shuttlecock

健

①(强健) healthy; strong ②(使强健) strengthen; toughen; invigorate: ～脾 invigorate the function of the spleen / ～胃 be good for the stomach③(善于) be strong in; be good at: ～谈 be a good talker
【健步】walk with vigorous strides: ～如飞 walk as if on wings; walk fast and vigorously
【健儿】①(健儿) valiant fighter ②(运动员) good athlete; skilled athlete
【健将】master sportsman; top-notch player: 运动～ master sportsman / 足球～ top-notch footballer
【健康】①(指身体) in good health; sound; fit: ～成长 grow up healthy and sound / 保持～ keep fit / 身体～ be in good health; be healthy / 提高～水平 improve health conditions / 祝你～。I wish you good health. ②(情况正常) healthy; wholesome; sound: ～的思想 wholesome thought; sound ideas ◇ ～带菌者 healthy carrier / ～水平 general level of the health / ～险 health insurance / ～证明书 health certificate / ～状况 state of health; physical condition
【健美】strong and handsome; vigorous and graceful
【健美运动】body building
【健全】①(强健而无缺陷) sound; perfect: 身心～ sound in mind and body / 头脑～的人 a person in his or her right mind ②(完善; 完备) regular; perfect: ～合理的规章制度 amplify necessary rules and regulations ③(使完备) perfect; improve; strengthen: ～法制 perfect legal system / ～管理制度 improve the management system / ～民主集中制 strengthen democratic centralism
【健身房】gymnasium; gym
【健身运动】body-building
【健谈】be a good talker; be a brilliant conversationalist
【健忘】forgetful; having a bad memory; have a poor memory
【健忘症】[医] amnesia
【健旺】healthy and vigorous
【健胃药】stomachic tonic
【健在】[书] (of a person of advanced age) be still living and in good health
【健壮】healthy and strong; robust

腱

[生理] tendon
【腱鞘】[生理] tendon sheath
【腱鞘囊肿】ganglion
【腱鞘炎】tenosynovitis; tendovaginitis

jiāng

将

①(将要) be going to; be about to; will; shall: 会议～开始。The meeting is about to begin. ②[棋] check: ～死 checkmate ③(用言语

刺激) incite sb. to action; challenge; prod: 他只需几句话一～，就会干。Just a few words will incite him into action. ④ (拿;用) with; by means of; by: ～功补过 make amends for one's faults by good service ⑤ (把)～门关上 shut the door / ～某人捉拿归案 bring sb. to justice / ～书合上 close the book / ～他请来 invite him to come over ⑥[助](用在动词和表示趋向的补语之间)唱～起来 start to sing / 赶～上去 hurry to catch up / 走～进去 get into the room

【将错就错】leave a mistake uncorrected and make the best of it; make the best of a mistake without correcting it

【将到货物】goods to arrive

【将到期负债】maturing liabilities

【将功补过】make amends for one's faults by good deeds

【将功赎罪】atone for a crime by good deeds; expiate one's crime by good deeds; redeem sins by good deeds

【将功折罪】expiate one's crime by good deeds; lenient consideration for person's mistakes in view of past or future achievements

【将计就计】turn sb.'s trick against him; beat sb. at his own game

【将近】close to; nearly; almost: ～六十岁 be close on sixty / ～半夜了。It was close upon midnight.

【将就】make do with; make the best of; put up with

【将军】①[军] general ②[象棋] check ③(使人为难)put sb. on the spot; embarrass; challenge

【将来】future: 美好的～ happy future / 在不远的～ in the not too distant future; before long; in the near future

【将息】rest; recuperate

【将心比心】feel for others; judge other people's feelings by one's own; compare one's feeling with other's

【将信将疑】half believing, half doubting; take with a grain of salt; half believe and half doubt

【将养】rest; recuperate

【将要】be going to; will; shall; be about to

【将欲取之，必先与之】in order to take, one must first give; give in order to take

浆

① (浓的液体) thick liquid: 豆～ bean milk / 糖～ syrup / 纸～ pulp ② (上浆) starch: ～衣服 starch clothes

【浆果】[植] berry

【浆膜】serosa; serous coat

【浆洗】wash and starch: ～衣服 wash and starch clothes

【浆纸机】[纸] coating machine

姜

ginger

【姜黄】[植] turmeric

【姜片虫】fasciolopsis ◇～病 fasciolopsiasis

【姜汁啤酒】ginger beer

江

① (河流) river: 在～心 in the middle of a river ② (长江) the Changjiang River: ～北 the north of the Changjiang River / ～南 the south of the Changjiang River

【江岸】river bank

【江防】defense works along the Changjiang River

【江河日下】go from bad to worse; be on the decline; decline steadily

【江湖】① (河流;湖泊) rivers and lakes ② (四方各地) all corners of the country: 流落～ live a vagrant life / 走～ tramp from place to place to make a living ③ (四处流浪卖艺卖药) itinerant entertainers, quacks, etc.: 讲～义气 be loyal to brotherhood ◇ ～骗子 swindler; charlatan / ～医生 quack; mountebank / ～艺人 itinerant entertainer

【江郎才尽】have used up one's literary talent or energy; a writer whose creative powers are exhausted

【江轮】river steamer

【江米】polished glutinous rice ◇ ～酒 fermented glutinous rice

【江畔】river bank; beside the river

【江山】① (江河山川) rivers and mountains; land; landscape: ～如此多娇。The land is so rich in beauty. ② (国家; 国家政权) country; state power: 打～ fight to win state power / 坐～ rule the country

【江山如画】a picturesque landscape; beautiful scenery

【江山易改,本性难移】A leopard never changes his spots. (或: It's easy to change rivers and mountains but hard to change a person's nature.)

【江豚】black finless porpoise

【江洋大盗】an infamous robber ⟨pirate⟩; a notorious bandit leader

【江珧】[动] pen shell ◇～柱 the dried adductor of a pen shell

豇

【豇豆】cowpea

僵 ①（僵硬）stiff; numb: ～直 stiff and rigid／冻～ stiff with cold ②（难于处理;停滞不前）deadlocked: 他把事情搞～了。He's brought things to a deadlock.

【僵持】(of both parties) refuse to budge; be stalemated; refuse to give in

【僵化】become rigid; ossify: 思想～ a rigid way of thingking; a ossified way of thinking

【僵局】deadlock; impasse; stalemate: 打破～ break a deadlock／谈判陷入～。The negotiations have reached an impasse.

【僵尸】corpse: 政治～ a political mummy

【僵死】dead; ossified

【僵硬】①（不能活动）stiff: 觉得两腿～ feel stiff in the legs ②（呆板的）rigid; inflexible: ～的公式 a rigid for formula／～态度 rigid attitude

疆 boundary; border: 边～ frontier; borderland

【疆场】battlefield

【疆界】boundary; border; frontier

【疆土】territory

【疆域】territory; domain

缰 reins; halter: 收～ draw rein; rein up

【缰绳】reins; halter

jiǎng

桨 oar; (短桨) scull; (阔桨)paddle

【桨板】paddle board

【桨叶】paddle

奖 ①（奖赏）encourage; praise; reward; award: 有功者～。Those who have gained merit will be rewarded. ②（奖品;奖状;奖金）award; prize; reward: 得～ win a prize／得～人 prize-winner; awardee／发～ give awards; give prizes

【奖杯】cup (as a prize)

【奖惩】rewards and punishments; bonus-penalty ◇ ～条例 regulations concerning rewards and disciplinary sanctions／～制度 system of rewards and penalties

【奖金】money award; bonus; premium: 年终～ year-end bonus ◇ ～制度 bonus system

【奖励】encourage and reward; award; reward: ～先进生产者 give awards to advanced workers／精神～ spiritual encouragement／物质～ material reward ◇ ～工资 premium wages

【奖牌】medal

【奖品】prize; award; trophy

【奖券】lottery ticket

【奖赏】award; reward

【奖学金】scholarship; exhibition

【奖掖】[书]reward and promote; encourage by promoting ad rewarding: ～后进 exhort and promote a new comer

【奖章】medal; decoration: 纪念～ commemorative medal／金质～ gold medal／铜质～ bronze medal／银质～ silver medal ◇ ～获得者 medalist

【奖状】certificate of merit; certificate of award; citation; honorary credential; diploma

讲 ①（说）speak; say; tell; talk about: ～的是一套，做的是另一套 say one thing and do another／～故事 tell stories／～几点意见 make a few remarks／～几句话 say a few words／～英语 speak English ②（解释;说明）explain; make clear; interpret ③（商量）discuss; negotiate: 不～条件 unconditionally／～条件 negotiate the terms; insist on the fulfillment of certain conditions ④（讲求；重视）stress; lay emphasis on; pay special attention to; be particular about: ～吃～穿 be particular about food and clothing／～排场 go in for ostentation and extravagance; go in for showy display; be ostentatious／～团结 stress unity／～卫生 pay attention to hygiene／～质量 stress quality／不～情面 have no consideration for anyone's sensibilities／不～政策 without any sense of policy ⑤（论；就某方面说）as far as sth. is concerned; when it comes to; as to; with regard to; as regards: 从效果来～ with regard to effects／～能力，我不如你。As to ability, I am not your match.

【讲稿】①(讲话稿) the draft or text of a speech ②(讲课稿)lecture notes

【讲和】make peace; settle a dispute; become reconciled

【讲话】①（说话）speak; talk; address ②（讲演）speech; talk: 鼓舞人心的～ an inspiring speech ③（用作书名）guide; introduction:《语法～》*An Introduction to Grammar*

【讲价】bargain; haggle over the price

【讲解】explain; interpret; expound: ～难句 explain difficult sentences

【讲解员】guide; interpreter; (体育比赛的) announcer

【讲经说法】expound the texts of Buddhism

【讲究】①（讲求；重视）be particular about; pay attention to; stress; strive for: ～吃穿 be too fastidious about one's food or clothing／～实际效果 stress practical results／～卫生 pay attention to hygiene ②（精美）exquisite; tasteful; elegant: ～的服装 elegant dress／～的早餐 fine breakfast ③（推敲）careful study: 教学法大有～。Teaching methods need careful study.

【讲课】teach; lecture ◇ ～时数 teaching hours

【讲理】①（评是非曲直）reason with sb.; argue: 你最好心平气和地同他～。You'd better reason thing out with him calmly. ②（服从道理）listen to reason; be amenable to reason; be reasonable; be sensible: 蛮不～ be utterly unreasonable; be impervious to reason

【讲明】explain; make clear; state explicitly

【讲排场】go in for pomp; hanker after vainglory

【讲评】comment on and appraise: ～学生的作业 comment on the students' work

【讲情】intercede; plead for sb.

【讲求】be particular about; pay attention to; stress; strive for: ～效率 strive for efficiency

【讲师】lecturer

【讲授】lecture; instruct; teach; give a lecture: ～提纲 an outline for a lecture; teaching notes

【讲述】tell about; give an account of; narrate; relate: ～解放军的英雄故事 tell about the heroic exploits of the PLA

【讲台】platform; dais; rostrum

【讲坛】①（讲台）platform; rostrum ②（讨论的场所）forum

【讲堂】lecture room; classroom

【讲题】subject of a lecture; topic of a speech

【讲习】lecture and study ◇ ～班 study group／～所 institute (for instruction or training)

【讲笑话】tell funny stories; break a jest

【讲学】give lectures; discourse on an academic subject

【讲演】lecture; speech

【讲义】(mimeographed or printed) teaching materials

【讲座】a course of lectures: 英语广播～ English lessons over the radio; English by radio

jiàng

酱 ①（豆酱；面酱）a thick sauce made from soya beans, flour, etc. ②（用酱或酱油腌、煮）cooked or pickled in soy sauce: ～黄瓜 pickled cucumber／～肉 pork cooked in soy sauce; braised pork seasoned with soy sauce ③（象酱的糊状食品）sauce; paste; jam: 豆瓣～ soybean paste／番茄～ tomato sauce; ketchup／花生～ peanut butter／辣～ hot sauce／辣椒～ chili sauce／苹果～ apple jam／甜面～ sweet brown sauce／虾～ shrimp sauce／芝麻～ sesame butter; sesame paste

【酱菜】vegetables pickled in soy sauce; pickles

【酱豆腐】fermented bean curd

【酱色】dark reddish brown

【酱油】soy sauce; soy: 辣～ pungent sauce

【酱园】a shop making and selling sauce, pickles, etc.; sauce and pickle shop

【酱紫】dark reddish purple

将 ①（将军）general ②（统兵者）commander in chief ③［书］（带兵）command; lead: ～兵 command troops

【将官】［口］high-ranking military officer; general

【将领】high-ranking military officer; general

【将士】［书］officers and men

【将相】generals and ministers of state

【将遇良才】a Roland for an Oliver

【将在外君命有所不受】A field commander must decide even against king's orders.

匠 craftsman; artisan: 木～ carpenter／能工巧～ skilled craftsmen／石～ stonemason／铁～ blacksmith／瓦～ bricklayer; tiler／鞋～ shoemaker; cobbler

【匠人】artisan; craftsman

【匠心】［书］ingenuity; craftsmanship: 独具～ show ingenuity; have great originality

降 fall; drop; lower: ～价 lower prices／～雨 a fall of rain; rainfall／温度～到零摄氏度以下。The temperature fell below zero degrees Centigrade.

【降低】reduce; cut down; drop; lower: ～定额 cut down the norm for／～生产成本 reduce production costs／～原料消耗 cut down the consumption of raw materials

【降格】［书］lower one's standard or status: ～以求 fall back on sth. inferior to what one originally wanted; accept a second best

【降号】［乐］flat

【降级】①（降低级别）reduce to a lower rank; demote ②（留级）send to a lower grade

【降结肠】colon descendens

【降临】[书] befall; arrive; come: 夜色～。Night fell.
【降落】descend; land; touch down: 安全～ land safely / 垂直～ vertical landing / 紧急～ make an emergency landing / 强迫～ forced landing / 自由～ free fall ◇ ～场 landing field / ～辅助设备 landing aid / ～区 dropping zone / ～伞 parachute / ～设备 landing equipment
【降旗】lower a flag: 降半旗 fly a flag at half-mast
【降生】be born
【降水】[气] precipitation: 人工～ artificial precipitation ◇ ～量 precipitation / ～区 precipitation area
【降温】①(降低温度) lower the temperature (as in a workshop) ②[气] drop in temperature; cool down
【降压】① [电] step-down ② [医] bring high blood pressure ◇ ～变电站 step-down sub-station / ～变压器 step-down transformer
【降压片】[药] hypertension pill
【降压药】depressor
【降雨量】rainfall: 年～ annual rainfall

绛

绛 deep red; crimson
【绛紫】dark reddish purple

强

强 stubborn; unyielding: 倔～ unbending; unyielding
【强嘴】reply defiantly; answer back; talk back

糨

糨 thick
【糨糊】paste: 打～ make paste

犟

犟 obstinate; stubborn; self-willed; bullheaded: ～脾气 be obstinate in disposition / 你这人真～! How stubborn you are!

jiāo

浇

浇 ①(洒) pour liquid on; sprinkle: 把火～灭 put out a fire by water / ～水 sprinkle water on ②(浇灌) irrigate; water: ～地 irrigate the fields / ～花 water flowers ③[印](浇铸) cast: ～铅字 type casting; type founding
【浇版】[印] casting ◇ ～机 casting machine
【浇灌】①(灌溉) water; irrigate ②(灌进模子内) pour: ～混凝土 pour concrete
【浇口】[冶] runner
【浇注】[冶] pouring; teeming ◇ ～包 pouring ladle / ～口 sprue gate
【浇铸】[冶] casting; pouring: 金属～ metal founding / 砂型～ sand casting ◇ ～版[印] cast / ～机 casting machine

交

交 ①(交出;交付;交纳) hand in; hand over; give up; deliver: ～党费 pay Party membership dues / ～公粮 deliver tax grain to the state / ～还 give back; return / ～活 turn over a finished item ②(到某一时辰或季节) reach; set in; come: 今年～春早。Spring set in early this year. ③(时间、地区相连接) meet; join: 春夏之～ when spring is changing into summer / 夏秋之～ when summer is changing into autumn ④(交叉) cross; intersect: AB 和 CD 两线相～于 E 点。Line AB intersects line CD at E. ⑤(结交) associate with: ～朋友 make friends ⑥(交往;交谊) friend; acquaintance; friendship; relationship; relation: 点头之～ nodding acquaintance / 多年之～ a friendship of many years / 患难之～ tested friend / 建～ establish diplomatic relations / 绝～ sever relations; break off relations / 一面之～ a casual acquaintance ⑦(性交) have sexual intercourse ⑧(交配) mate; breed: 杂～ crossbreed ⑨(互相) mutual; reciprocal; each other: ～换 exchange ⑩(一齐;同进) together; simultaneous: 内外～困 be beset with difficulties at home and abroad ⑪(交易) deal; bargain; business transaction: 成～ strike a bargain; conclude a transaction; make a deal ⑫(跟头) fall
【交白卷】①(不能回答试题) hand in a blank examination paper; hand in an unanswered examination paper ②(没有完成任务) fail to carry out an assignment; completely fail to accomplish a task
【交班】hand over to the next shift; turn over one's duty
【交臂失之】fail to meet someone by a narrow chance
【交变电流】[电]alternating current (AC)
【交兵】[书] (of two or more parties) be at war; wage war
【交叉】①(互相穿过) intersect; cross; crisscross: 两条铁路在此～。The two railways cross here. ②(相重的) overlapping: 两个方案中～的部分 the overlapping parts of the two plans ③(间隔穿插) alternate; stagger: ～进行 do alternately ◇ ～点 intersect / ～火力 [军] cross fire
【交差】report to the leadership after accomplishing a task; report on one's mission; report on the fulfillment of one's duty

【交出】surrender; hand over: ～武器 surrender one's weapons
【交存】deposit; hand in for safekeeping: ～批准书 deposit instruments of ratification
【交错】①(交叉; 错杂) interlock; crisscross; interlace: 管道纵横～ ducts and pipes crisscross ②[机] staggered ◇ ～气缸 staggered cylinder
【交代】【交待】①(移交) turn over; hand over; transfer: ～工作 hand over work to one's successor; brief one's successor on handing over work ②(嘱咐) tell; leave words; order: 已～得清清楚楚 have given clear-cut orders ③(说明) explain; make clear; brief: ～任务 assign and explain a task; brief sb. on his task / ～政策 explain policy ④(说清楚) account for; justify oneself: ～不过去 be unable to justify an action ⑤(坦白) confess: ～罪行 confess a crime / 彻底～ make a clean breast of one's guilt
【交底】tell sb. what one's real intentions are; put all one's cards on the table
【交点】①[数] point of intersection ②[天] node: ～月 nodical month
【交锋】cross swords; engage in a battle or contest; fight with; struggle with: 思想～ confrontation of ideas; confrontation between differing views / 与敌人～ fight a battle with the enemy; engage the enemy in battle
【交付】①(付钱) pay: ～定金 down payment / ～现金 pay cash; payment in cash / ～租金 pay rent ②(交给) turn over; hand over; deliver; consign: ～表决 put to the vote / ～审查 hand over for investigation / ～审判 committal for trial; submit to trial / ～日期 due date; date of delivery / ～宣判 render a verdict
【交感神经】[生理] sympathetic nerve
【交割】complete a business transaction
【交工】hand over a completed project
【交公】hand over to the collective or the state
【交媾】sexual intercourse; copulation; coitus
【交好】(of people or states) be on friendly terms; be friendly with
【交互】①(互相) each other; mutual: ～订正课外作业 correct each other's homework ②(替换着) alternately; in turn: 两种方法可～使用。The two methods can be used alternately.
【交换】exchange; swap; interchange: ～场地[体] change of courts, goals or ends / ～意见 interchange opinions; exchange views; compare notes / ～战俘 exchange of POW (prisoner of war) / ～照会 exchange of notes / 商品～ exchange of commodities / 用钢材～石油 swap steels for oil / 用小麦～大米 barter wheat for rice ◇ ～齿轮[机] change gear / ～机[电话] switchboard; exchange / ～价值[经] exchange value / ～器[电] converter / ～条件 give-and-take conditions
【交火】fight
【交货】delivery: 仓库～ ex warehouse / 船上～ ex ship / 分批～ partial delivery / 铁路旁～ ex rail / 即期～ prompt delivery / 近期～ near delivery / 远期～ forward delivery ◇ ～不足 short delivery / ～单 delivery order / ～港 port of delivery / ～期 date of delivery / ～收据 delivery receipt / ～收款 cash on delivery
【交集】(of different feelings) be mixed; occur simultaneously: 悲喜～ mixed feelings of grief and joy / 惊喜～ surprise and joy inter- mingled
【交际】social intercourse; communication: ～很广 have a large circle of acquaintances / ～应酬 business entertainment / 忙于～ be busy with social activities / 善于～ be a good mixer ◇ ～费 allowance for entertainment / ～花 social beauty; social butterfly / ～舞 ballroom dancing; social dancing
【交加】[书] (of two things) accompany each other; occur simultaneously: 风雪～ a raging snowstorm / 悔恨～ regret mingled with self-reproach / 雷电～ lightning accompanied by peals of thunder; there was thunder and lightning / 贫病～ be plagued by both poverty and illness
【交接】①(连接) join; connect ②(移交;接替) hand over and take over: ～班 relief of a shift ③(结交) associate with; make friends with: 他所～的朋友 the friends he has made; the people he associated with
【交界】have a common border; have a common boundary; border on; juncture of: 福建北面与浙江～。Fujian is bounded on the north by Zhejiang.
【交卷】①(交考卷) hand in an examination paper ②(完成任务) fulfill one's task; carry out an assignment; finish up one's job
【交口称誉】praise sb. <sth.> unanimously
【交流】①exchange; interflow; interchange: ～经验 exchange experience; draw on each other's experience / 城乡物资～ flow of goods

and materials between city and country / 经济和技术～ economic and technical interchange / 文化～ cultural exchange②[电] alternating ◇～电 alternating current / ～发电机 alternating current generator; alternator / ～声 hum

【交纳】pay (to the state or an organization); hand in: ～会费 pay membership dues / ～所得税 pay income tax

【交配】mating; copulation ◇～期 mating season / ～行为 mating behavior

【交情】friendship; friendly relation: 讲～ do things for the sake of friendship / 老～ long-standing friendship / 有～ be on friendly terms with sb.

【交融】blend; mingle: 水乳～ blend as well as milk and water; be in perfect harmony

【交涉】negotiate; make representations; take up with: 办～ carry on negotiations with; take up a matter with / 与某人～ negotiate with sb. on sth.

【交手】fight hand to hand; be engaged in a hand-to-hand fight; come to grips

【交售】sell (to the state): 向国家多～粮棉 sell more grain and cotton to the state

【交谈】talk with each other; converse; chat; have a conversation: 亲切～ have a heart-to-heart talk

【交替】①(接替) supersede; replace: 新老～ the new replace the old; the old gives place to the new ②(替换看) alternately; take place by turn; in turn: 昼夜～ day alternates with night

【交替进行】[乐] hocket

【交通】①(往来通达; 运输) traffic; communications: ～便利 have transport facilities / ～繁忙 heavy traffic / 妨碍～ interfere with the traffic / 公路～ highway traffic / 陆上～ land traffic / 市区～ urban traffic / 违反～规则 contravene the traffic regulations ②(解放前指通讯和联络工作) liaison; liaison man: ～员 liaison man; underground messenger / 跑～ do liaison work ◇～安全 traffic safety / ～安全法规 traffic safety code / ～标线 traffic marking / ～标志 traffic sign / ～部 the Ministry of Communications / ～干线 main line of communication; main road; artery of traffic / ～高峰 traffic peak / ～管理 traffic control / ～管理色灯 traffic lights / ～规则 traffic regulations / ～壕[军] communication trench / ～警 traffic police / ～量 volume of traffic / ～事故 traffic accident; road accident / ～事故现场 scene of a traffic accident / ～网 network of communication lines / ～信号 traffic signal / ～要道 vital communication line / ～运输 communications and transportation / ～指挥台 police stand / ～阻塞 traffic jam; traffic block ○超速行驶 driving over the speed limit / 酒后驾车 drive when intoxicated / 违章肇事人 delinquent / 超车 overtake another car / 加速 speed up (the car) / 减速 slow down (the car) / 靠右边走 keep to the right / 车辆可以通行 open to traffic / 人行道 pavement; sidewalk / 人行横道 curb; kerb / 快车道 fast traffic lane / 慢车道 slow traffic lane / 停车线 stop line / 安全岛 safety island / 单行线 one-way traffic / 红绿灯 traffic lights / 红灯 red light / 绿灯 green light / 禁止超车 No Overtaking / 不要超车 Do Not Overtake / 禁止鸣笛 No Tooting / 慢 Slow / 此路不通 No Thoroughfare / 此处不许停车 No Parking Here / 时速限制: 30公里/小时 Speed Limit: 30 Km/h / 单行道 One Way Traffic / 向右绕行 Turn to the Right / 连续弯路 Consecutive Curves / 中速行驶,安全礼让 Drive at Moderate Speed; Give Another Vehicle the Right of Way for the Sake of Safety / 一慢,二看,三通过 Slow own, Look Around, and Cross

【交头接耳】speak in each other's ears; whisper to each other; to bill and coo

【交往】association; contact; associate with; be in contact with: ～甚密 have an intimate association with; have close contact with ◇～自由 freedom of communication

【交尾】mating; pairing; coupling

【交恶】fall foul of each other; become enemies

【交响诗】[乐] symphonic poem; tone poem

【交响乐】[乐] symphony; symphonic music ◇～队 symphony orchestra; philharmonic orchestra

【交心】lay one's heart bare; open one's heart to: 互相～ have heart-to-heart talk

【交易】business; deal; trade; transaction: 肮脏的政治～ a dirty political deal / 黑市～ black market bargain / 决不拿原则做～ never barter away principles / 赊帐～ credit transaction / 现款～ cash transaction / 做成一笔～ make a deal ◇～额 volume of trade / ～券 trading stamp

【交易会】trade fair: 商品～ trade fair; commodities fair

【交易所】exchange: 商品～ commodity exchange / 证券～ stock exchange

【交谊】friendship; friendly relations
【交游】make friends: ～甚广 have a large circle of friends
【交战】be at war; fight; wage war: ～的一方 a belligerent / ～双方 the two belligerent parties / ～状态 state of war; belligerency ◇～国 belligerent countries (states, nations)
【交帐】①(移交帐务) hand over the accounts ②(说明原因) account for
【交织】interweave; intertwine; mingle: 惊异和喜悦～在一起。 Joy mingled with surprise.

茭

【茭白】[植] wild rice stem

胶

①(粘性物) glue; gum: 树～ gum / 鱼～ fish glue ②(用胶粘) stick with glue; glue ③(象胶一样粘的) gluey; sticky; gummy ④(橡胶) rubber
【胶版】offset plate ◇～复印法 hectograph / ～印刷 offset printing; offset lithography; offset / ～印刷机 offset press / ～纸 offset paper
【胶布】①(涂上胶的布) rubberized fabric ②[口](橡皮膏) adhesive plaster ◇～带[电] rubberized tape; adhesive tape
【胶带】adhesive tape; gummed tape
【胶冻】jelly
【胶合】glue together; veneer ◇～板 plywood; veneer board
【胶结】glued; cemented ◇～材料 cementing material / ～剂 cementing agent
【胶卷】roll film; film: 彩色～ color film / 全色～ panchromatic film / 冲～ have one's film developed ◇～暗盒 cassette; film cassette
【胶粒】micelle; colloidal particle
【胶轮】rubber tire ◇～车 rubber-tired cart
【胶木】bakelite
【胶囊】capsule
【胶泥】clay
【胶粘剂】adhesive
【胶凝作用】[化] gelation
【胶皮】(vulcanized) rubber
【胶片】film: 彩色～ color film / 电影～ cinefilm / 全色～ panchromatic film
【胶溶剂】peptizing agent
【胶乳】[化] latex: 硫化～ vulcanized latex; vultex
【胶水】mucilage; glue
【胶体】[化] colloid ◇～化学 colloid chemistry
【胶鞋】rubber overshoes; galoshes; rubbers rubber-soled shoes; tennis shoes
【胶靴】high rubber overshoes; galoshes
【胶印】offset printing; offset lithography; offset ◇～机 offset press; offset (printing) machine
【胶原】collagen
【胶柱鼓瑟】stubbornly stick to old ways in the face of changed circumstances
【胶着】deadlocked; stalemated: ～状态 deadlock; stalemate; impasse

郊

suburbs; outskirts: 市～ outskirts of a city / 四～ the suburbs / 西～ the western suburbs / 远～ the outer suburbs; the remoter outskirts of a city
【郊区】suburban district; suburbs; outskirts ◇～居民 suburban; suburbanite
【郊外】the countryside around a city; outskirts
【郊游】outing; excursion

教

teach; instruct: ～英语 teach English
【教书】teach school; teach: 在中学～ teach in a middle school; be a middle school teacher
【教书育人】impart knowledge and educate people

椒

hot spice plants: 胡～ pepper / 辣～ chili; red pepper
【椒盐】spiced salt

娇

①(柔嫩;美丽可爱) tender; lovely; charming: 嫩红～绿 tender blossoms and delicate leaves ②(虚弱) fragile; frail; delicate ③(娇气) squeamish ④(过度爱护) pamper; spoil
【娇滴滴】delicately pretty; affectedly sweet
【娇惯】pamper; coddle; spoil: ～孩子 pamper a child
【娇贵】enervated (by good living); pampered
【娇憨】young and ignorant
【娇媚】①(撒娇献媚) coquettish ②(妩媚) sweet and charming
【娇嫩】①(柔嫩) tender and lovely ②(柔弱) fragile; delicate: ～的身子 delicate health / ～的幼苗 delicate seedlings / ～细腻 pretty, delicate and glossy
【娇气】squeamish; finicky: ～十足 be full of finicky airs / 去掉～ get rid of squeamishness
【娇娆】enchantingly beautiful
【娇生惯养】pampered since childhood; pampered by doting parents; brought up in clover
【娇声娇气】speak in a seductive tone
【娇小玲珑】delicate and exquisite; dainty and cute
【娇艳】delicate and charming; tender and beau-

tiful: ～的桃花 delicate and charming peach blossoms / ～无比 being gay to excel all
【娇纵】indulge (a child); pamper; spoil

骄 proud; arrogant; conceited: 胜不～,败不馁 not be dizzy with success, nor discouraged by failure
【骄傲】①(自大) arrogant; conceited: ～自大 swollen with pride; conceited and arrogant / ～自满 conceited and self-satisfied; arrogant and complacent ②(自豪) be proud; take pride in ③(值得自豪的人或事物) pride: 民族的～ the pride of the nation
【骄兵必败】An army puffed up with pride is bound to lose
【骄横】arrogant and imperious; overbearing: ～不法 presumptuous and unlawful
【骄矜】[书] self-important; proud; haughty
【骄气】overbearing airs; arrogance
【骄奢淫逸】lordly, luxury-loving, loose-living and idle; wallowing in luxury and pleasure; extravagant and dissipated
【骄阳】[书] blazing sun: ～似火 scorching sun
【骄纵】arrogant and willful: ～放恣 overbearing and de-bauchery

焦 ①(烧焦) burnt; scorched; charred: ～味 smell of burning / 饭烧～了。The rice has burnt. ②(焦炭) coke: 炼～ coking ③(着急) worried; anxious: 心～ worried
【焦比】[冶] coke ratio
【焦点】①[物] focal point; focus: 虚～ virtual focus / 主～ principal focus ②(引人注意的集中点) central issue; point at issue: 斗争的～ the focus of the struggle / 矛盾的～ focal points of contradictions / 问题的～ the heart of the matter / 争论的～ the points at issue
【焦耳】[物] joule ◇～定律 Joule's law
【焦黑】burned black
【焦化】[化] coking: 延迟～ delayed coking
【焦黄】sallow; brown: 脸色～ sallow face
【焦急】anxious; worried: 大家都在～地等着医生。Everybody is waiting anxiously for the doctor.
【焦痂】[医] eschar
【焦距】[物] focal distance; focal length
【焦渴】terribly thirsty; parched; dying of thirst
【焦枯】shrivelled; dried up; withered
【焦沥青】pyrobitumen
【焦虑】feel anxious; have worries and misgivings ◇～反应[心理] anxiety reaction
【焦煤】coking coal
【焦炭】coke: 沥青～ pitch coke
【焦糖】caramel
【焦头烂额】badly battered; in a terrible fix; be bruised and battered
【焦土】scorched earth; ravages of war 一片～ with everything burned down and lying in ruins ◇～政策 scorched earth policy
【焦油】[化] tar: 煤～ coal tar / 木～ wood tar
【焦躁】restless with anxiety; impatient: 克服～情绪 curb one's impatience

蕉 any of several broadleaf plants: 美人～ canna / 香～ banana
【蕉麻】[植] abaca; Manila hemp

礁 reef: 触～ strike a reef; run up on a rock
【礁石】reef; rock

鹪
【鹪鹩】wren
【鹪莺】wren warbler

jiáo

嚼 masticate; chew; munch: 细～慢咽 chew carefully and swallow slowly; chew one's food well before swallowing it
【嚼舌】①(信口胡说;搬弄是非) wag one's tongue; chatter away; gossip ②(无谓争论) argue senselessly; squabble
【嚼烟】chewing tobacco
【嚼子】bit (of a bridle)

jiǎo

铰 [口] ①(剪) cut with scissors: ～成两半 cut in two; cut into halves; cut in half ②(用铰刀切削) bore with a reamer; ream: ～孔 ream a hole
【铰刀】[机] reamer
【铰接】[机] join with a hinge; articulate ◇～式公共汽车 articulated bus
【铰链】hinge: ～接合 hinge joint

佼 [书] handsome; beautiful
【佼佼】[书] above average; outstanding

皎 clear and bright: ～月当空。A bright moon hung in the sky.
【皎皎】very clear and bright; glistening white: ～者易污。The immaculate stains easily. (或: The immaculate is easily sullied.)
【皎洁】(of moonlight) bright and clear

狡 crafty; foxy; cunning: ～计 crafty trick;

ruse
【狡辩】quibble; indulge in sophistry
【狡猾】【狡滑】sly; crafty; cunning; tricky
【狡赖】deny (by resorting to sophistry): 证据确凿,不容～。 It's no use denying it, the evidence is conclusive.
【狡兔三窟】a wily hare has three burrows; a crafty person has more than one hideout; many provisions for cunning escape
【狡兔死,走狗烹】After the cunning hare is killed, the hound is boiled.
【狡黠】[书] sly; crafty; cunning
【狡诈】deceitful; crafty; cunning

饺 dumpling: 蒸～ steamed dumplings
【饺子】dumpling (with meat and vegetable stuffing) ◇ ～皮 dumpling wrapper / ～馅 filling for dumplings; stuffing

绞 ①(拧;扭在一起) twist; wring; entangle: ～尽脑汁 rack one's brains; beat one's brains / 心如刀～ feel as if a knife were being twisted in one's heart ②(转动轮轴) wind: ～动辘辘 wind a windlass ③(绞刑) hang by the neck ④[机] reaming ⑤[量] skein; hank: 一～毛线 a skein of woolen yarn
【绞肠痧】[中医] dry cholera
【绞车】winch; windlass
【绞架】gallows
【绞盘】capstan: 推杆～ bar capstan
【绞肉机】meat mincer; mincing machine
【绞杀】strangle; hang; garrotte
【绞死】gibbet
【绞索】(the hangman's) noose
【绞痛】[医] angina: 肚子～ abdominal angina; colic / 心～ angina pectoris
【绞刑】hanging; sentence to be hung; sentence to the gallows
【绞刑架】gallows

搅 ①(搅拌) stir; mix ②(扰乱;打扰) disturb; annoy
【搅拌】stir; agitate; mix
【搅拌机】mixer; agitator; stirring machine
【搅拌器】stirrer; agitator; stirring apparatus
【搅动】mix; stir: ～灰浆 stir the plaster
【搅混】[口] mix; blend; mingle
【搅和】①(混合) mix; blend; mingle: 把水泥和沙石～在一起 mix cement, sand and pebbles ②(扰乱) mess up; spoil
【搅乱】confuse; throw into disorder; mess up: ～了计划 mess up the plan
【搅扰】disturb; annoy; cause trouble
【搅乳器】churn

矫 ①(矫正) rectify; straighten out; correct ②(强壮;勇武) strong; brave: ～若游龙 as powerful as a flying dragon; as strong and brave as a lion ③(假托) pretend; feign; dissemble: ～命 counterfeit an order; issue false orders
【矫健】strong and vigorous: ～的步伐 vigorous strides
【矫捷】vigorous and nimble; brisk
【矫揉造作】affected; artificial; the affectedness of one's manner
【矫饰】feign in order to conceal sth.; dissemble
【矫枉过正】exceed the proper limits in righting a wrong; overcorrect; overdo in righting a wrong
【矫形】[医] orthopedic ◇ ～术 orthopedics / ～外科 orthopedic surgery / ～医生 orthopedist
【矫正】correct; put right; rectify: ～发音 correct sb.'s pronunciation mistakes / ～口吃 correct a stammer / ～偏差 correct a deviation / ～视力 correct defects of vision; correct faulty vision
【矫直机】[冶] straightening machine; straightener

侥
【侥幸】lucky; by luck; by a fluke: ～得免 escape by good fortune / ～取胜 gain victory by sheer good luck; win by a fluke / ～心理 the idea of leaving things to chance; trusting to luck

缴 ①(交出) pay; hand over; hand in: ～税 pay taxes / 上～ turn over 〈in〉 to the higher authorities ②(迫使交出) capture: ～枪不杀! Lay down your arms and we'll spare your lives!
【缴获】capture; seize: 一切～要归公。 Hand in everything captured.
【缴款通知】payment notice
【缴入盈余】paid-in surplus
【缴入资本】paid-in capital
【缴销】hand in for cancellation
【缴械】①(解除武装) disarm ②(被迫交出武器) surrender one's weapons; lay down one's arms: ～投降 lay down one's arms and surrender

脚 ①foot: 赤～ barefoot ②(东西的最下部) foot; base: 墙～ the foot of a wall / 山～ the

foot of a hill
【脚背】instep
【脚本】script; scenario: 电影～ film script
【脚脖子】ankle: 扭了～ sprain one's ankle
【脚步】step; pace: 加快～ quicken one's pace ◇ ～声 footfall; footsteps
【脚灯】[剧] footlights
【脚蹬子】pedal; treadle
【脚夫】porter
【脚跟】heel: 站稳～ stand firm; gain a firm foothold
【脚尖】the tip of a toe; tiptoe: 踮着～走 walk on tiptoe
【脚扣】climbers; climbing iron; grapnel
【脚镣】fetters; shackles
【脚炉】foot warmer; foot stove
【脚面】instep
【脚气】①[医] beriberi ②(脚癣) athlete's foot
【脚钱】payment to a porter
【脚刹车】foot brake
【脚手架】[建] scaffold; falsework
【脚踏两头便落空】Between two stools a person falls to the ground.
【脚踏两只船】straddle two boats, have a foot in either camp; sit on the fence
【脚踏实地】have one's feet planted on solid ground; earnest and down-to-earth; stand on solid ground: ～的工作作风 a down-to-earth style of work
【脚踏脱粒机】pedal thresher
【脚踏制动】pedal brake
【脚腕子】ankle
【脚心】arch (of the foot)
【脚癣】[医] ringworm of the foot; tinea pedis; athlete's foot
【脚印】footprint; footmark; track
【脚掌】sole (of the foot)
【脚指甲】toenail
【脚趾】toe
【脚注】footnote

角 ①(动物的角) horn: 鹿～ antler / 牛～ ox horn ②(号角) bugle; horn: 号～ bugle ③(岬角) cape; promontory; headland: 好望～ the Cape of Good Hope ④(角落) corner: 墙～ corner (of a wall) / 眼～ corner of the eye ⑤[数] angle: 补～ complementary angle / 对～ opposite angle / 钝～ obtuse angle / 多面～ polyhedral angle / 反射～ reflex angle / 邻～ adjacent angle / 内～ interior angle / 锐～ acute angle / 同位～ corresponding angle / 外～ exterior angle / 直～ right angle ⑥(中国辅币) *jiao* (=1/10 of a *yuan* or 10 *fen*)
【角尺】angle square
【角蛋白】keratin
【角动量】[物] angular momentum
【角豆树】carob
【角度】①[数] angle ②(看问题的出发点) point of view; angle
【角度计】goniometer; angle gauge
【角钢】[冶] angle steel; angle bar; angle iron
【角弓反张】[医] opisthotonos
【角规】angle gauge
【角化病】keratosis
【角砾岩】[地] breccia
【角楼】a watchtower at a corner of a city wall; corner tower; turret
【角落】corner; nook: 找遍每一个～ search every nook and corner
【角马】gnu
【角门】small side door
【角膜】[生理] cornea ◇ ～混浊[医] opacity of the cornea / ～溃疡 the ulcer of the cornea / ～炎 keratitis / ～移植术 corneal trans-plantation; keratoplasty / ～云翳 nebula
【角球】[足球] corner (kick)
【角鲨】spiny dogfish
【角闪石】[矿] hornblende; amphibole
【角速度】[物] angular velocity
【角铁】[冶] angle iron
【角宿一】[天] Spica
【角质】[生] cutin ◇ ～层[植] cuticle
【角柱体】prism
【角锥体】pyramid

剿 send armed forces to suppress; put down: ～匪 suppress bandits; put down bandits
【剿灭】exterminate; wipe out

jiào

窖 ①(地窖) cellar or pit for storing things: 菜～ vegetable cellar ②(收藏在窖里) store sth. in a cellar or pit

觉 sleep: 睡一～ have a sleep / 午～ midday nap

校 check; proofread; collate: ～长条样 read galley proofs / 二～ the second proof; proofread for the second time
【校订】check against the authoritative text
【校对】①(核对) proofread; proof ②(校对人) proofreader ③(核对是否符合标准) check

against a standard; calibrate
【校对符号】proofreader's mark
【校改】read and correct proofs
【校勘】collate ◇ ～学 textual criticism
【校样】[印] proof sheet; proof: 长条～ galley proof / 二～ the second proof / 付印～ final proof
【校阅】read and revise
【校正】proofread and correct; rectify: ～错字 correct misprints / ～仪器 rectify an instrument
【校准】[机] calibration: 方位～ bearing calibration ◇ ～器 calibrator

较 ①(比较) compare; as compare with; in comparison with: 产量～去年增加百分之二十。 The output has increased by twenty per cent as compared with last year. ②(一定程度的) comparatively; relatively; fairly; quite; rather: ～差 relatively poor / ～好 fairly good; quite good / 取得～大的进步 make considerable progress ③(明显)clear; obvious; marked
【较比】[方] comparatively; relatively; fairly; quite
【较量】①(比高低) measure one's strength with; have a contest; have a trial 〈test〉 of strength ②(计较) haggle; argue; dispute

教 ①(教导) teach; instruct: 请～ ask for advice; consult / 言传身～ teach by precept and example ②(宗教) religion: 佛～ Buddhism / 基督～ Christianity / 罗马天主～ Roman Catholic Church / 信～ believe in a religion; be religious
【教案】teaching plan; lesson plan
【教鞭】(teacher's) pointer
【教材】teaching material
【教程】course (of study): 近代史～ A Course in Modern History
【教导】①(教育指导) instruct; teach; give guidance ②(学说; 指引) teaching; guidance
【教导队】training unit
【教导员】political instructor
【教范】[军] manual: 兵器～ a manual of arms; manual
【教父】god-father
【教父教母】godparent
【教改】[简](教学改革)educational reform
【教工】teaching and administrative staff (of a school)
【教官】drillmaster; instructor
【教规】[宗] canon
【教皇】[宗] pope; pontiff ◇ ～通谕 Papal Encyclical / ～诏书 bull / ～自动诏书 motu proprio
【教会】[宗] church: 东正～ Greek Orthodox Church ◇ ～法规 canon law / ～教育 parochial education / ～学校 missionary school
【教诲】[书] teaching; instruction: 谆谆～ earnest teachings
【教具】teaching aid: 直观～ aids to object teaching; audio-visual aids
【教科书】textbook
【教练】①(训练) train; drill; coach ②(教练员) coach; instructor: 足球～ football coach ◇ ～车 learner-driven vehicle / ～船 training ship / ～弹 dummy shell (bullet) practice projectile; dummy projectile; dummy / ～机 trainer aircraft; trainer / ～员 coach; instructor; trainer
【教母】god-maother
【教女】god-daughter
【教派】[宗] religious sect; denomination
【教区】parish
【教师】teacher; schoolteacher: 兼职～ part-time teacher / 专职～ full-time teacher
【教士】[宗] priest; clergyman; Christian missionary
【教室】classroom; schoolroom: 大～ lecture hall / 阶梯～ lecture theatre
【教授】①(指职称) professor: 副～ associate professor / 客座～ visiting professor; guest professor ②(讲解说明教材内容) instruct; teach: ～得法 have tact in teaching / ～历史 teach history ◇ ～法 teaching methods; pedagogics
【教唆】instigate; abet; put sb. up to sth.: ～犯罪(～他人犯罪) instigate to crime ◇ ～犯 instigator; abetter / ～犯罪(被～犯罪) subornation / ～者 abetter / ～罪 guilt of instigation; solicitation
【教堂】church; cathedral: 大～ abbey; cathedral / 威斯敏斯特大～ Westminster Abbey ◇ ～长凳 pew
【教条】dogma; doctrine; creed; tenet
【教条主义】dogmatism; doctrinairism ◇ ～者 dogmatist; doctrinaire
【教廷】the Vatican; the Holy See ◇ ～大使 nuncio / ～公使 internuncio
【教徒】believer (follower) of a religion
【教务】educational administration ◇ ～处 Dean's Office / ～长 Dean of Studies
【教学】teaching; education ◇ ～大纲 teaching

program; syllabus / ～方法 teaching method / ～方针 principles of teaching / ～改革 transformation of education; reform in education / ～内容 content of courses / ～人员 faculty; teaching staff
【教学法】pedagogy
【教学相长】teaching benefits teacher and student alike; teaching benefits teachers as well as students; teaching and learning promote and enhance each other
【教训】①(鉴戒) lesson; moral: 吸取～ draw a lesson from sth.; take warning from sth. / 血的～ a lesson paid for with blood; lesson written in blood ②(教育训诫) teach sb. a lesson; give sb. a talking-to; chide
【教研室】teaching and research section
【教研组】teaching and research group
【教养】①(教育和培养) bring up; train; educate ②(文化品德修养) upbringing; education; culture; breeding: 有～的 cultured; accomplished ③[法] correction
【教养员】kindergarten teacher
【教养院】correctional; indoctrination center; reformatory; workhouse
【教义】[宗] religious doctrine; creed
【教益】[书] benefit gained from sb.'s wisdom; enlightenment
【教育】① education ②(教导;启发) teach; educate; inculcate ◇ ～程度 level of education / ～方针 policy for education; educational policy / ～家 educationist; educator / ～界 educational circles; educational world / ～经济学 economics of education / ～心理学 educational psychology / ～学 pedagogy; pedagogics; education / ～哲学 philosophy of education / ～制度 system of education
【教员】teacher; instructor
【教长】[宗] iman; dean: 坎特伯雷～ Dean of Canterbury
【教长国】[伊斯兰教] imamate
【教职员】teaching and administrative staff
【教主】the founder of a religion
【教子】god-son

酵
ferment; leaven
【酵母】yeast: 鲜～ yeast cake ◇ ～菌 yeast; saccharomycete
【酵素】[化] ferment; enzyme

叫
①(发出声音) cry; shout: 大～一声 give a loud cry; shout; cry out loudly / 狗～ bark / 鸡～ crow / 驴～ bray / 羊～ bleat ②(招呼;招唤) call; greet ③(雇车;点菜;买煤) hire; order: ～菜 order dishes / ～出租汽车 hire a taxi; call a taxi / ～两吨煤 order two ton(s) of coal ④(称为) name; call: 我们～他小王。We call him Xiao Wang. ⑤(吩咐) ask; order; make: 医生～我好好休息。The doctor ordered me to have a good rest. ⑥(被)书～谁拿走了? Who took the book?
【叫喊】shout; yell; howl
【叫好】applaud; shout "Bravo!"; shout "Well done!"
【叫花子】beggar
【叫唤】cry out; call out: 疼得直～ cry out with pain
【叫苦】complain of hardship or suffering; moan and groan: ～不迭 pour out endless grievances / ～叫累 complain of hardship or fatigue / ～连天 one's cry for bitterness is heavenly high; ventilate one's endless grievances / 暗暗～ groan inwardly
【叫骂】shout curses
【叫卖】cry one's wares; peddle; hawk: 沿街～ hawk one's wares in the streets
【叫门】call at the door to be let in
【叫屈】complain of being wronged; protest against an injustice
【叫嚷】shout; howl; clamor; raise a hue and cry
【叫嚣】clamor; raise a hue and cry: 发出战争～ clamor for war
【叫醒】wake up; awaken: 到时～我。Wake me up in time.
【叫座】draw a large audience; draw well; appeal to the audience: 这部电影很～。The film draws well.
【叫做】be called; be known as: 这种机器～起重机。This machine is called a crane.

轿
sedan (chair)
【轿车】①[旧] (horse-drawn) carriage ②(汽车) bus; car: 大～ bus; coach / 小～ car; sedan / (前后排座隔开的)三排座小～ limousine
【轿子】sedan (chair)

jiē

秸
stalks; straw: 豆～ bean stalks / 麦～ wheat straw / 秫～ sorghum stalks
【秸秆】straw

结
bear (fruit); form (seed); produce: 开花～果 blossom and bear fruit
【结巴】①(口吃) stammer; stutter ②(口吃的

人) stammerer; stutterer

【结实】①(结出果实) bear fruit; fructify; fruit: 这些树每年～。These trees fruit annually. ②(坚固耐用) solid; sturdy; durable: ～的材料 durable stuff ③(健壮) strong; sturdy; tough; robust: 身体～的人 sturdily built man

接 ①(靠近;接触) come into contact with; come close to ②(连接;使连接) connect; join; put together: ～电线 connect wires / ～管子 join two pipes together / 请～2175分机。Extension 2175, please. ③(托住;承受) catch; take hold of: ～球 catch a ball ④(接受;接到) receive; take: ～到信 receive a letter / ～到上级指示 receive instructions from higher authorities / ～电话 answer the phone; receive a phone call; take a phone call ⑤(迎接) meet; welcome: 到机场～人 go to the airport to meet sb. ⑥(接替) take over: ～工作 take over a job

【接班】take one's turn on duty; take over from; succeed; carry on: 我们下午一点～。We'll take our turn on duty at 1 p.m.

【接班人】successor: 培养和造就千百万社会主义现代化事业的～ train and bring up millions of successors to carry on the cause of socialist modernization

【接触】①(交往) come into contact with; get in touch with ②(冲突) engage: 双方武装力量脱离～ disengage the armed forces of the two sides ③(挨上;碰着) contact: ～不良 loose contact; poor contact ◇ ～传染 contagion / ～法[化] contact process / ～故障[电] contact fault / ～炉[化] contact furnace / ～性皮炎 contact dermatitis / ～眼镜 contact lenses

【接待】receive; admit: 受到亲切～ be accorded a cordial reception / 博物馆从上午九点～观众。The museum is open from 9 a.m. ◇ ～单位 host organization / ～人员 reception personnel / ～室 reception room / ～站 reception center

【接地】①[电] ground connection; grounding; earthing ②[航空] touchdown; ground contact ◇ ～线[电] ground wire; earth lead / ～迎角[航空] landing angle

【接点】[电] contact

【接二连三】one after another; in quick succession; repeatedly; continuously

【接防】relieve a garrison; relieve; take over the defense

【接风】give a dinner for a visitor from afar; give a reception in honor of a guest from afar ◇ ～洗尘 give a dinner of welcome

【接羔】[牧] deliver lambs ◇ ～房 lamb-delivery room / ～季节 lambing season

【接骨】[医] set a (broken) bone; set a fracture

【接管】take over control; take over: ～官僚资本企业 take over the enterprises run by bureaucrat-capital

【接合】joint; join; link: 气密～ airtight joint / 水密～ watertight joint ◇ ～点 junction; junction point

【接火】[口] ①(互相射击) start to exchange fire ②(接通电源) energize

【接济】give material assistance to; give financial help to; offer pecuniary aid to

【接见】receive sb.; grant an interview to: ～外宾 receive foreign guests

【接近】be close to; near; approach: ～世界先进水平 approach the advanced world level / ～最高水平 approach the topmost level / 易于～ be easy of approach

【接颈交臂】be very intimate with

【接口】interface

【接力】relay: 四乘一百米～ 4×100-meter relay ◇ ～棒 relay baton / ～区 tak-over zone / ～赛跑 relay race; relay

【接连】on end; in a row; in succession; running: ～获得丰收 reap several bumper harvests running ◇ ～不断 continuously; incessantly; in rapid succession

【接目镜】eyepiece; ocular

【接纳】admit (into); take in: 她被～为工会会员。She was admitted as a member of the trade union.

【接片】[电影] splicing ◇ ～机 splicer

【接气】coherent; consistent: 这两段文章不太～。These two paragraphs are not coherent enough.

【接洽】take up a matter with; arrange with; consult with: 与有关单位～ take up the matter with departments concerned; consult with departments concerned

【接壤】border on; be contiguous to; be bounded by ◇ ～地区 contiguous areas

【接任】take over a job; replace; succeed: ～主席 take over the chairmanship

【接生】deliver a child; practice midwifery ◇ ～员 midwife

【接收】①(收受) receive: ～无线电信号 receive radio signals ②(接管) take over; expropriate ③(接纳) admit ◇ ～机 receiver / ～天线

receiving antenna; receiving aerial / ～仪式 take-over ceremony

【接手】①（接管）take over (duties, etc.); take up matters ②[棒、垒球] catcher

【接受】accept; take; embrace: ～定货 execute (accept) an order / ～教训 learn a lesson / ～贿赂 on the take / ～考验 face up to a test / ～马克思主义 embrace Marxism / ～群众监督 subject oneself to supervision by the masses / ～劝告 take sb.'s advice / ～任务 accept an assignment / ～邀请 accept an invitation / ～意见 take sb.'s advice / 容易～新思想 be readily receptive to new ideas ◇ ～保证人 guarantee / ～承诺人 acceptee / ～处罚 acceptance of punishment / ～抗诉 acceptance of protest / ～判决 acceptance of a judgment / ～赔偿人 indemnitee / ～书 instrument of acceptance

【接穗】[植] scion

【接替】take over; replace; succeed; take the place of: ～某人当大使 succeed sb. as ambassador

【接通】put through: 电话～了吗? Have you got through?

【接头】①(连接起来) connect; join; joint ②[纺] (纱条) piecing; (经纱) tying-in ③(找人联系) contact; get in touch with ④(了解情况) know about; have knowledge of: 这事她不～。 She doesn't know anything about it. ⑤[机] connection; joint; junction: 四通～ four-way connection / 万向～ universal joint

【接吻】kiss

【接物待人】attend a matter and to receive a person

【接物镜】objective lens; objective

【接线】①[电] wiring ②(电话员接通线路) work a telephone switchboard ◇ ～图 wiring diagram; connection diagram / ～箱 junction box / ～员 (telephone) operator / ～柱 terminal; binding post

【接续】continue; follow

【接应】①(配合) come to sb.'s aid; coordinate with; reinforce ②(供应) supply

【接着】①(用手接) catch: 给你一个桃子,～! Here's a peach for you. Catch! ②(连着;紧跟着) follow; carry on; go on (with); proceed: ～干吧。 Carry on with your work.

【接踵】[书] following on sb.'s heels: ～而至 come one after another; come hot on one another's heels; follow hard at heel / 摩肩～ jostle each other in a crowd

【接种】[医] have an inoculation; inoculate: ～防霍乱疫苗 inoculate sb. against cholera / ～防伤寒疫苗 be inoculated against typhoid / ～牛痘疫苗 be vaccinated

揭 ①(把粘贴的片状物取下) tear off; take off ②(把盖在上面的东西拿起) uncover; lift (the lid, etc.) ③(揭露) expose; show up; bring to light

【揭穿】expose; lay bare; show up: ～谎言 expose a lie / ～阴谋 show up an evil plot

【揭底】reveal the inside story; lay bare the inside story

【揭短】rake up sb.'s faults

【揭发】expose; unmask; bring to light; lay open: ～罪行 expose a crime / 假～ sham exposure

【揭竿而起】raise the standard of revolt; start an uprising; rise in rebellion

【揭开】uncover; reveal; open; disclose: ～奥秘 reveal the secrets of / ～内幕 reveal the inside story of; get to the bottom of / ～序幕 raise the curtain on; usher in

【揭露】expose; unmask; ferret out; uncover; disclose; show up; lay bare; bring to light: ～其真面目 expose sb.'s true colors; show sb. up for what he is; unveil the actual role of / ～事物的本质 uncover 〈unveil〉 the essence of things / ～无余 give the plain truth; completely unmask / ～阴谋 expose one's plot / ～真相 betray the fact

【揭幕】unveil (a monument, etc.); inaugurate ◇ ～式 unveiling ceremony

【揭示】①(公布) announce; promulgate; proclaim ②(使人看到) reveal; bring to light: ～了社会发展的客观规律 reveal the objective laws governing the development of society

【揭晓】announce; make known; publish: 足球赛的结果已～。 The result of the football match has been announced.

皆 all; each and every: 人人～知 it is known to all; it is public knowledge

【皆大欢喜】everybody is happy; to the satisfaction of all; everybody is satisfied

阶 ①(台阶) steps; stairs: 台～ a flight of steps ②(等级) rank: 军～ military rank

【阶层】(social) stratum: 社会～ social stratum / 特权～ the privileged stratum / 中间～ intermediate stratum

【阶地】[地] terrace
【阶段】stage; phase; period: 分～by stages / 过渡～ transitional stage / 历史～ historical period
【阶级】(social) class: 剥削～ exploiting class / 被剥削～ exploited class / 官僚资产～ bureaucrat bourgeoisie / 垄断资产～ monopoly bourgeoisie / 买办资产～ comprador bourgeoisie / 民族资产～ national bourgeoisie / 无产～ proletariat / 小资产～ petty bourgeoisie / 有产～ propertied class / 资产～ bourgeoisie / 自为～ class in itself / 自在～ class for itself ◇～本能 class instinct / ～本质 class nature / ～成分 class status / ～冲突 class conflict / ～斗争 class struggle / ～对抗 class antagonism / ～队伍 class ranks / ～分化 class polarization / ～分析 class analysis / ～感情 class feeling / ～根源 class origin / ～基础 class basis / ～较量 trials of class strength / ～警觉性 class vigilance / ～觉悟 class consciousness / ～烙印～ brand of a class / ～立场 class stand / ～力量的对比 balance of class forces / ～利益 class interests / ～路线 class line / ～矛盾 class contradictions / ～内容 class content / ～性 class character; class nature / ～友爱 class love; class solidarity / ～阵线 class alignment
【阶梯】a flight of stairs; ladder: 进身的～ stepping stone ◇～教室 lecture theatre
【阶下囚】prisoner; captive

嗟 [书] sigh; lament: ～悔无及 too late for regrets and lamentations / 悲～ sigh in sorrow
【嗟来之食】food handed out in contempt; a handout

街 street
【街道】① street ②(街道地区) residential district; neighborhood ◇～办事处 subdistrict office / ～服务站 neighborhood service center / ～工厂 neighborhood factory / ～居民委员会 neighborhood committee
【街坊】neighbor
【街垒】street barricade
【街面儿上】①(市面) (activities, etc.) in the street ②(附近街巷) neighborhood
【街市】downtown streets
【街谈巷议】street gossip; rumors; common talk
【街头】street corner; street: 十字～ (at the) crossroads / 流落～ tramp the streets; be down and out in a city / 涌上～ pour into the streets ◇～叫卖声 street cry / ～剧 street-corner skit; street performance
【街头巷尾】streets and lanes; street corners and alleys

节
【节骨眼】①(关键时刻) critical juncture; at the critical moment; in the nick of time ②(紧要环节) vital link

疖
【疖子】① [医] furuncle; boil ② [植] knot (in wood)

jié

洁 clean: 清～ clean / 整～ clean and tidy; clean and neat
【洁白】spotlessly white; pure white: ～无暇 spotless and flawless
【洁净】clean; spotless
【洁身自好】①(独善其身) refuse to be contaminated by evil influence; preserve one's purity; exercise self-control so as to protect oneself from immorality ②(不招惹是非) keep to one's own business; mind one's own business in order to keep out of trouble
【洁牙剂】dentifrice
【洁治】scaling
【洁治钩】tooth scaler

诘 [书] closely question; interrogate
【诘问】closely question; interrogate; cross-examine

拮
【拮据】short of money; hard up; in straitened circumstances

桔
【桔梗】[中药] the root of balloonflower

结 ①(编织) tie; knit; knot; weave: ～毛衣 knit wool into a sweater; knit a sweater out of wool / ～网 weave a net; weave a web (蛛网) ②(结子) knot: 打～ tie a knot / 蝴蝶～ bowknot / 解～ untie a knot; undo a knot / 活～ slipknot / 死～ fast knot ③(结合;结成) tie; form: ～为夫妇 be tied in wedlock ④(凝结) congeal; form: ～痂 form a scab; scab ⑤(结束;了结) settle; conclude: ～帐 settle accounts ⑥(保证负责的字据) written guarantee; affidavit: 具～ give a written guarantee ⑦ [电子] junction ⑧[生] node: 淋巴～ lymph node

【结案】wind up a case; close a case; settle a lawsuit
【结疤】①[冶] scab ②[医] become scarred
【结拜】[旧] become sworn brothers or sisters
【结伴】go with: ～而行 go or travel in a group
【结冰】freeze; ice up; ice over
【结彩】adorn with festoons; be decorated with festoons: 张灯～ decorate with lanterns and festoons
【结肠】[生理] colon ◇ ～下垂 coloptosis／～炎 colitis／～造口术 colostomy
【结成】form: ～深厚的友谊 form〈forge〉a profound friendship／～同盟 form an alliance; become allies／～一伙 gang up; band together
【结仇】start a feud; become enemies; breed enmity with; incur hatred of
【结存】①(指款项) cash on hand; balance ②(指货物) goods on hand; inventory
【结党营私】form a clique to pursue selfish interests; gang together for clandestine and illegal activities
【结缔组织】[生理] connective tissue
【结发夫妻】husband and wife by the first marriage
【结构】①(各个部分的配合;组织) structure; composition; construction: 经济～ economic structure／人体～ structure of the human body／土壤～ soil structure／原子～ atomic structure／组织～ framework of organization／～严谨 of a tightly knit structure; be compact and well organized ②[建] structure; construction: 钢～ steel structure／钢筋混凝土～ reinforced concrete structure／钢屋架～ steel roof truss structure／焊接～ welded construction／铆合～ riveted construction／轻质混凝土～ light-concrete structure／～牢固 be a very solid construction ③[地] texture: 斑状～ porphyritic texture／致密～ compact texture ◇ ～材料 material of construction／～钢 structural steel／～力学 structural mechanics／～式[化] structural formula／～图 structural drawing／～异构[化] structural isomerism／～语言学 structural linguistics
【结关】customs clearance
【结果】①(结局) result; outcome; ending: 必然～ inevitable result ②(最终) finally; at last: ～,他们打赢了。 Finally, they won the game. ③(杀死) kill; finish of
【结合】①(发生密切联系) combine; unite; integrate; link: 与群众相～ integrate oneself with the masses; become one with the masses／把理论与实践～起来 combine〈link〉theory with practice ②(结为夫妻) marry; be united in wedlock; be tied in wedlock
【结合能】[物] binding energy
【结核】①[医] tuberculosis: 肺～ pulmonary tuberculosis／骨～ bone tuberculosis ②[矿] nodule: 锰～ manganese nodule ◇ ～病 tuberculosis／～病院 tuberculosis hospital or sanatorium／～杆菌 tubercle bacillus; bacillus tubercle／～菌素 tuberculin／～菌素试验 tuberculin test
【结喉】[生理] Adam's apple
【结汇】settlement of exchange ◇ ～证 marginal receipt
【结婚】marry; get married ◇ ～登记 marriage registration／～年龄 age for marriage; matrimonial age／～证书 marriage certificate; marriage lines
【结伙】gang: ～盗窃 gang stealing／～斗殴 gang fighting／～抢劫 gang robbery
【结集】concentrate; mass: ～军队 concentrate troops; mass forces
【结痂】eschar; scab
【结交】make friends with; associate with
【结节】[生] tubercle; node
【结节虫】nodular worm
【结晶】①(析出晶体) crystallize ②(晶体) crystal: 盐～ salt crystals ③(成果) crystallization; fruit; product; quintessence: 劳动的～ the fruit of labor／知识的～ quintessence of knowledge／智慧的～ a crystallization of wisdom ◇ ～度 crystallinity／～断面[化] crystalline fracture／～固体 crystalline solid／～化学 crystal chemistry／～水 water of crystallization; crystal water／～学 crystallography／～岩石 crystalline rock
【结局】final result; outcome; ending; upshot: 悲惨的～ tragic result; sad denouement
【结论】①[逻] conclusion (of a syllogism) ②(最后论断) conclusion; verdict: 匆忙下～ jump to a conclusion／得出～ draw a conclusion; come to a conclusion
【结盟】form an alliance; ally; align: 不～国家 nonaligned countries／不～政策 nonalignment policy
【结膜】[生理] conjunctiva ◇ ～炎 conjunctivitis
【结欠】balance due
【结亲】①(结婚) marry; get married ②(结成姻亲) become related by marriage
【结秦晋之好】marriage between two families

【结清】settle; square up: ～债务 settle a debt / ～帐目 square accounts (with sb.)
【结球甘蓝】cabbage
【结社】form an association ◇ ～自由 freedom of association
【结石】[医] stone; calculus: 胆～ gall stone / 肾～ kidney stone / 排出～ discharge of stones; pass stones
【结识】get acquainted with sb.; get to know sb.
【结束】end; finish; conclude; wind up; close: ～讲话 wind up a speech / ～战争状态 terminate the state of war / 演出到此～。That's the end of our performance. ◇ ～语 concluding remarks
【结算】settle accounts; close an account; wind up an account: 用瑞士法郎计价～ use Swiss Franc for quoting prices and settling accounts ◇ ～贷款 loan for the settlement of accounts / ～单据 document of settlement
【结为秦晋】united in matrimony
【结尾】①(末尾;结束的阶段) ending; winding-up stage ②[乐] coda
【结业】complete a course; wind up one's studies
【结余】cash surplus; surplus; balance
【结缘】form ties (of affection, friendship, etc.); become attached to; take a fancy to
【结怨】contract enmity; incur hatred
【结扎】[医] ligation; ligature: ～血管 ligature blood vessels / 输精管～术 vasoligation / 输卵管～术 ligation of oviduct
【结帐】settle accounts; square accounts; balance the books; closing accounts ◇ ～记录 closing entries
【结组】[登山] roped party

截 ①(截断) cut; sever: ～成两段 cut in two ②[量] section; chunk; length ③(阻拦) stop; check; stem; intercept: ～住惊马 stop a bolting horse / ～流 dam a river / ～球 intercept a pass ④(截止) by; up to: ～至二月底 up to the end of February
【截长补短】take from the long to add to the short; draw on the strength of each to offset the weakness of the other; even up scarcity and superabundance
【截断】①(切断) cut off; block: ～河流 dam a river ②(打断;拦住) cut short; interrupt: ～别人的讲话 interrupt sb.'s speech
【截获】intercept and capture: ～情报 intercept (enemy) information
【截击】intercept: ～敌人的轰炸机 intercept the enemy's bombers ◇ ～导弹 interceptor missile; interception missile / ～机 interceptor
【截留】hold back: ～国家财政收入 hold back state revenue
【截煤机】[矿] coalcutter; cutter: 滚筒式～ drum cutter
【截面】①[数] section: ～图 sectional drawing / 横～cross section / 正～ normal section ②[物] cross section
【截取】cut out: ～一段铁丝 cut out a section of wire
【截然】sharply; completely; entirely: ～不同 poles apart; completely different; different as black and white; entirely different / ～对立 be diametrically opposed / ～相反 completely contradict / 二者不能～分开。No hard and fast line can be drawn between the two.
【截瘫】[医] paraplegia
【截肢】[医] amputation
【截止】①(停止) end; close: 付款～期 deadline for payment / 注册已经～。The enrollment has closed. ②[电] cut-off: ～电平 cut-off level ◇ ～阀 stop valve
【截至】by (a specified time); up to: ～今年年底 by the end of this year / ～目前为止 up to now

劫 [书] ①(抢劫) rob; plunder; raid: 打～ rob; loot ②(灾难) calamity; disaster; misfortune: 浩～ a great calamity; dire disaster / 遭～ meet a disaster; suffer a disaster ③[佛教] kalpa
【劫持】kidnap; hold under duress; hijack; abduction: ～船只 hijack a ship / ～飞机 hijack an airplane; air piracy ◇ ～者 hijacker; abductor
【劫夺】seize (a person or his property) by force
【劫富济贫】rob the rich and help the poor
【劫后余生】be a survivor of a disaster; a lucky survivor from a holocaust; outlive a calamity
【劫掠】plunder; loot
【劫数】[佛教] inexorable doom; predestined fate ◇ ～难逃。A fatal lot is unavoidable.
【劫狱】break into a jail and rescue a prisoner; jail delivery

节 ①(各段之间相连处) joint; node; knot: 骨～ joint (of bones) / 脱～ out of joint / 竹～ bamboo joint ②(段落) division; part: 音～ syllable ③[量] section; length: 第三章第二～ Chapter Three, Section Two / 十八～车厢

eighteen railway coaches／四～课 four periods; four classes／一～钢管 a length of steel tube ④(节日) festival; red-letter day; holiday: 春～ the Spring Festival／国庆～ National Day／过～ celebrate a festival; observe a festival ⑤(删节) abridge: ～译 abridged translation ⑥(节约;节制) economize; save: ～煤 economize on coal; save coal ⑦(事项) item: 生活小～ trifling personal matters／细～ details ⑧(节操) moral integrity; chastity: 气～ moral integrity／晚～ integrity in one's later year ⑨(船速单位; 1节＝每小时1海里) knot

【节哀】restrain one's grief

【节本】abridged edition; abbreviated version

【节操】[书] high moral principle; moral integrity

【节俭】thrifty; frugal: 提倡～ encourage frugality

【节节】successively; steadily: ～败退 retreat in defeat again and again; keep on retreating／～胜利 win many victories in succession; go from victory to victory; gain victory after victory

【节理】[地] joint: 倾向～ dip joint／走向～ strike joint

【节令】climate and other natural phenomena of a season

【节流】① reduce expenditure: 开源～ broaden sources of income and reduce expenditure ② [机] throttle: 全～ full throttle

【节录】extract; excerpt

【节目】program; item (on a program); number: 广播～ broadcasting program ◇ ～单 program; playbill

【节拍】[乐] meter ◇ ～器 metronome

【节气】solar terms

【节日】festival; red-letter day; holiday: ～气氛 festive air／致以～的祝贺 extend holiday greetings

【节省】economize; save; use sparingly; cut down on: ～开支 cut down expenses; save unnecessary expenses／～篇幅 save space／～人力物力 use manpower and material resources sparingly／～时间 save time／～原材料 economize raw materials

【节外生枝】side issues or new problems crop us unexpectedly; raise obstacles; deliberately complicate an issue; cause complications

【节衣缩食】economize on food and clothing; live frugally

【节油器】economizer; gasoline economizer

【节余】surplus (as a result of economizing)

【节育】birth control ◇ ～环 intrauterine device (IUD); the loop

【节约】practice thrift; economize; save: ～费用 reduce expenses／～开支 cut down expenses; retrench (expenditure)／～用水 economize on water; save on water／～用电 economize on electricity; save on electricity／厉行～ practice strict economy

【节肢动物】arthropod

【节制】①(限制;控制) control; check; be moderate in: ～饮食 be moderate in eating and drinking ②(克制) temperance; abstinence ◇ ～闸[水] check gate

【节奏】rhythm: ～明快 lively rhythm／～轻快 play in quick rhythm

捷

①(快) prompt; nimble; quick: 敏～ quick; nimble; agile ②(战胜) victory; triumph: 报～ announce a victory／大～ a great vic- tory; major victory／首战告～ be victorious in the first battle; win the first battle

【捷报】news of victory; report of a success

【捷径】shortcut: 走～ take a shortcut

【捷克斯洛伐克】Czechoslovakia ◇ ～人 Czechoslovak; Czechoslovakian

【捷克语】Czech (language)

【捷足先登】the swift-footed arrive first; the race is to the swiftest; the early bird catches the worms

睫

eyelash; lash

【睫毛】eyelash; lash

【睫状肌】ciliary muscle

竭

exhaust; use up

【竭诚】wholeheartedly; with all one's heart: ～拥护 give wholehearted support

【竭尽】use up; exhaust: ～全力 spare no effort; do one's utmost; do all one can; make all-out efforts／～造谣诬蔑之能事 stop at nothing in spreading lies and slanders; exhaust every means to spread lies and slanders

【竭蹶】[书] destitute; impoverished: 艰难～ hardship and destitution

【竭力】do one's utmost; use every ounce of one's energy; spare no efforts: ～反对 oppose strongly; actively oppose／～鼓吹 boost with all one's might; energetically advocate／～宣传 assiduously propagate／～支持 give all-out support

【竭泽而渔】drain the pond to get all the fish;

kill the goose that lays the golden eggs

羯

【羯羊】wether

碣 stone tablet: 墓～ tombstone

桀

【桀骜不驯】[书]stubborn and intractable; obstinate and unruly

【桀犬吠尧】the tyrant Jie's cur yapping at the sage-king Yao; utterly unscrupulous in its zeal to please its master

杰 ①(杰出) outstanding; prominent; distinguished ②(才能出众的人) outstanding person; hero

【杰出】outstanding; remarkable; prominent: ～的共产主义战士 outstanding communist fighter／～贡献 a brilliant contribution

【杰作】masterpiece

孑 [书]lonely; all alone

【孑孓】wiggler; wriggler

【孑然】[书] solitary; lonely; alone: ～一身 all alone in the world

jiě

解 ①(分开) separate; divide: ～剖 dissect／溶～ dissolve／瓦～ disintegrate ②(把束缚着或系着的东西打开) untie; undo: ～扣儿 unbutton／～缆 untie the mooring rope／～鞋带 undo shoelaces③(解除) allay; alleviate; dispel; dismiss: ～惑 dispel sb.'s doubts／～热 allay a fever／～痛 alleviate pain ④(解释) explain; interpret; solve: ～题 solve a (mathematical, etc.) problem／详～ explain in detail; detailed explanation／新～ a new interpretation／注～ (explanatory) notes; annotation ⑤(了解;明白) understand; comprehend: 费～ hard to understand; obscure／令人不～ puzzling; incomprehensible ⑥(解手;便溺) relieve oneself: 大～ go to the lavatory (to defecate)／小～ go to the lavatory (to urinate) ⑦[数] solution: 求～ find the solution

【解馋】satisfy a craving for good food; satisfy an appetite for good food

【解嘲】try to explain things away when ridiculed; try to get out of a scrape when ridiculed: 自我～ console oneself with soothing remarks; find excuses to console oneself

【解除】remove; relieve; get rid of: ～顾虑 free one's mind of apprehensions／～合同 terminate a contract; dissolution of contract／～婚姻关系 dissolution of marriage／～婚约 renounce an engagement; dissolution of engagement／～禁令 lift a ban／～警报 sound the all clear／～扣押 release of distress／～武装 disarm／～宵禁 lift the curfew／～职务 remove sb. from his post; relieve sb. of his office／旱象已经～。The dry spell is over.

【解答】answer; explain: ～疑难问题 answer difficult questions／问题～ answers to questions; solutions to problems

【解冻】①(融化) thaw; unfreeze: ～季节 thawing season ②(解除冻结) unfreeze (funds, assets, etc.) ③(关系缓和) thaw: 两国关系的～ a thawing of the relations between the two countries

【解毒】①[医] detoxify; detoxicate ②[中医] relieve internal heat or fever ◇～药 antidote

【解饿】satisfy 〈appease〉 one's hunger; stay one's stomach

【解乏】recover from fatigue; refresh oneself

【解法】[数] solution

【解放】liberate; emancipate: ～后 after liberation／～生产力 liberate the productive forces; emancipate the productive forces／～思想 emancipate the mind; free oneself from old ideas／～前 before liberation

【解放军】①(为取得解放而战的军队) liberation army ②[简](中国人民解放军) the Chinese People's Liberation Army; the PLA ③(解放军战士) PLA man

【解放区】liberated area

【解放战争】①(争取解放的战争) war of liberation ②(指第三次国内革命战争) China's War of Liberation

【解雇】discharge; dismiss; fire; give the sack

【解恨】vent one's hatred; have one's hatred slaked

【解甲归田】take off one's armor and return to one's native place; be demobilized; quit military service and resume civilian life; retire from military service

【解禁】lift a ban: ～期间 open season; open time

【解救】save; rescue; deliver: ～某人脱险 rescue 〈save, deliver〉 sb. from danger

【解决】①(处理问题) solve; resolve; settle: ～困难 overcome a difficulty; find a way out of a difficulty／～赔偿 settle a claim; settle a bill／

～问题 solve a problem; settle a question; work out a solution／～争端 settle a dispute ②(消灭) finish off; dispose of: ～了敌人一个连 have finished off (wiped out) a company of enemy troops; have wiped out a company of enemy troops

【解开】untie; undo; unfasten: ～包裹 untie a package／～钮扣 undo a button／～上衣 unbutton one's jacket／～头巾 untie a kerchief／～鞋带 unlace the shoes

【解渴】quench one's thirst

【解款】transfer of fund ◇ ～单 cash remittance note

【解铃系铃】let him who tied the bell on the tiger take it off; whoever started the trouble should end it; let the mischief-maker undo the mischief

【解闷】divert oneself (from boredom)

【解囊】[书] open one's purse: ～相助 help sb. generously with money

【解聘】dismiss an employee (usu. at the expiration of a contract)

【解剖】dissect: 活体～ vivisection／尸体～ autopsy; postmortem examination ◇ ～刀 scalpel／～学 anatomy

【解气】vent one's spleen; work off one's anger; vent one's spite upon sb.

【解劝】soothe; mollify; comfort

【解散】①(集合的人分散开) dismiss ②(口令) Dismiss! ③(取消) dissolve; disband: ～法人团体 disincorporate／～国会 dissolve a parliament／～协会 disband an association／～组织 disband an organization

【解释】explain; expound; interpret: ～法律 interpret laws／～新词 explain a new word／权威～ authentic interpretation ◇ ～权 power of interpretation; right to interpret／～性备忘录 explanatory memorandum／～性法规 declaratory statute／～性发言 explanatory statement

【解手】relieve oneself; go to the toilet

【解数】skill; art: 使出浑身～ bring all one's skill into play

【解说】explain orally; comment ①(口头的) commentary ②(书面的) caption: ～员 announcer; narrator; commentator

【解体】disintegrate; break up: 封建制度的～ the disintegration 〈breakup〉 of the feudal system

【解脱】①(摆脱) free oneself from; get rid of; extricate oneself: 从困境中～出来 extricate oneself from a predicament ②(开脱) exonerate; absolve: 为某人～ absolve sb. from responsibility; plead for sb. ③ [佛教] mukti; vimukta

【解围】①(解除包围) force an enemy to raise a siege; rescue sb. from a siege; come to the rescue of the besieged ②(摆脱窘境) help sb. out of a predicament; save sb. from embarrassment; get sb. out of a fix

【解析几何】analytic geometry

【解严】declare martial law ended; lift a curfew

【解约】terminate an agreement; cancel a contract; annul an agreement: ～条款 cancelling clause

【解职】dismiss from office; discharge; relieve sb. of his post

姐

姐 elder sister; sister

【姐夫】elder sister's husband; brother-in-law

【姐姐】elder sister; sister

【姐妹】①(姐和妹) sisters: 同胞～ sister of the whole blood; full sister／异父～ half sister／异母～ half sister ②(兄弟姐妹) brothers and sisters ◇ ～之情 the affection of elder and younger sisters

jiè

戒 ①(防备;警惕) guard against; avoid ②(警告) exhort; admonish; warn: 引以为～ take warning from sth.; take sth. as an object lesson ③(戒除) give up; drop; stop: ～荤腥 go on a vegetarian diet／～酒 stop drinking／～烟 give up smoking ④(佛教戒律) Buddhist monastic discipline: 受～ attain the full status of a monk or nun ⑤(戒指) ring: 钻～ diamond ring

【戒备】guard; take precautions; be on the alert; keep a sharp lookout: ～森严 be heavily guarded／处于～状态 be on the alert

【戒除】give up; drop; leave off; stop: ～恶习 give up a bad habit

【戒骄戒躁】guard against arrogance and rashness; be on guard against conceit and impetuosity; guard against pride and haste

【戒律】[宗] religious discipline; commandment; precept

【戒奢崇俭】refrain from luxury and uphold frugality

【戒心】vigilance; wariness; alertness: 对某人怀有～ be on one's guard against someone; keep a wary eye on someone; be vigilant against sb. / 对某事怀有～ be wary of sth. / 怀有～ keep a sharp look out; have one's eyes open

【戒严】enforce martial law; impose a curfew; cordon off an area: 宣布～ proclaim martial law ◇ ～地区 district under martial law / ～法 martial law / ～令 order of martial law

【戒指】(finger) ring: 订婚～ engagement ring / 结婚～ wedding ring

诫 ① warn; admonish: 告～ give warning; admonish ②[宗] commandment: 十～ the Ten Commandments

介 ①(在两者中间) be situated between; interpose ②(介意) take seriously; take to heart; mind: ～意 take offense; mind

【介词】[语] preposition

【介电常数】dielectric constant

【介电塑料】plastic dielectric

【介壳】shell (of oysters, snails, etc.)

【介壳虫】scale insect

【介入】intervene; interpose; get involved ◇ ～诉讼当事人 intervenient party; intervening party / ～行为 intervening act

【介绍】①(使双方相识) introduce; present: ～对象 introduce sb. to a potential marriage partner; find sb. a boy or girl friend / 作自我～ introduce oneself ②(推荐) recommend; suggest: ～某人入党 recommend sb. for Party membership / ～新的工作方法 introduce sb. to a new method of work / 为某人～工作 recommend sb. for a position ③(使了解) let know; brief; give information; provide information: ～经验 pass on experience / ～情况 brief sb. on the situation; put sb. in the picture; fill sb. in ◇ ～费 middleman fee; procuration

【介绍人】①(组织介绍者) introducer; sponsor: 他是我的入党～。He was my sponsor when applied for party membership. ②(婚姻介绍人) matchmaker

【介绍信】①(介绍认识的书信) letter of introduction ②(推荐信)letter of recommendation

【介形虫】mussekshrimp

【介意】(多用于否定词后) take offense; mind; get annoyed; care about: 区区小事,何必～。Never mind such trifles.

【介原子】[物] mesic atom

【介质】[物] medium: 工作～ actuating medium

【介子】[物] meson; mesotron

疥 scabies

【疥虫】[医] sarcoptic mite

【疥疮】[医] scabies; psora

【疥螨】itch mite; sarcoptic mite

【疥螨病】psoroptic mange; sarcoptic mange

【疥癣】[牧] mange: 羊～ sheep scab

芥 mustard

【芥菜】leaf mustard ◇ ～疙瘩 rutabaga

【芥蒂】[书] ill feeling; unpleasantness; grudge: 不存～ harbor no grudge; bear no grudge / 心存～ bear a grudge

【芥末】mustard

【芥子】mustard seed

【芥子气】[化] mustard gas ◇ ～中毒 mustard gas poisoning

【芥子油】mustard oil

界 ①(界线) boundary; border: 国～ the boundary of a country; national boundary / 越～ cross the border ②(一定的范围) bounds; scope; extend: 外～ external world / 眼～ field of vision ③(社会的各界) circles; world: 各～人士 people of all walks of life; all sections of the people / 文艺～ world 〈circles〉 of literature and art / 新闻～ press circles / 学术～ academic circles ④(自然界类别) primary division; kingdom: 动物～ the animal kingdom / 矿物～ the mineral kingdom / 植物～ the vegetable kingdom ⑤[地] group: 古生～ the Paleozoic group ⑥[数] bound: 上～ upper bound / 下～ lower bound

【界碑】boundary tablet; boundary marker

【界尺】ungraduated ruler

【界河】boundary river

【界面】interface

【界内球】[体]in bounds; in

【界石】boundary stone or tablet

【界外球】out-of-bounds

【界限】①(分界) demarcation line; dividing line; limits; bounds: 打破行业～ break the bounds of different trades ②(限度)limit; end

【界限塞规】limit plug gauge

【界线】①(分界线) boundary line ②(界限) demarcation line; dividing line; limits

【界桩】 boundary stone; boundary marker; boundary post

借 ①(借进) borrow: ～钱 borrow money from sb. / 从图书馆～书 borrow a book from the library ②(借出) lend: ～书给某人 lend sb. a book; lend a book to sb. ③(凭借) make use of; take advantage of: ～此机会 take this op-

portunity to ④(假托)use as a pretext ⑤(借阅图书)～出 in circulation; out / 不外～ not for circulation / 续～ renewal
【借词】[语] loanword; loan
【借贷】①(借钱) borrow or lend money ②(借方和贷方) debit and credit sides ◇ ～抵押证书 bill of goods adventure / ～利息 loan interest / ～资本 loan capital
【借刀杀人】murder with a borrowed knife; kill sb. by another's hand; make use of another person to get rid of an adversary
【借调】temporarily transfer; be on loan: 他～到外文出版社工作去了。 He is on loan to the Foreign Languages Press.
【借读】study at a school on a temporary basis
【借端】use as a pretext; find a pretext; invent an excuse: ～讹诈 make use of anything as a pretext for blackmailing / ～生事 find an excuse to make trouble; avail oneself of a pretext to stir up trouble / ～寻衅 make use of anything as a pretext for seeking a quarrel
【借方】[簿记] debit; debit side ◇ ～差额 debit balance
【借风使船】sail with the wind
【借公济私】seek for private interest through public affairs
【借古讽今】use the past to disparage the present
【借故】find an excuse: ～生端 make use of anything as a pretext for provocation / ～推托 find an excuse to refuse
【借光】[套] excuse me
【借花献佛】present Buddha with borrowed flowers; borrow sth. to make a gift of it; make a present provided by sb. else
【借火】ask for a light
【借鉴】use for reference; draw lessons from; draw on the experience of; have successful experiences of others to go by: ～外国的经验 use the experience of other countries for reference
【借酒浇愁】drink sorrow down; drown care in the wine bowl; drown one's worries in drink
【借据】receipt for a loan (IOU); certificate of indebtedness; debt on bond; evidence of debt
【借口】①(以某事为理由) use as an excuse; on the pretext of; on the excuse of: 他～头疼走了。 He went away on the pretext (excuse) of a headache. ②(假托的理由) excuse; pretext: 找～ find an excuse; find a pretext / 制造～ make an excuse; invent an excuse; cook up a pretext
【借款】①(借入) borrow money; ask for a loan ②(贷出) lend money; offer a loan ③(借用的钱) loan ◇ ～股份 debenture stock / ～企业 borrowing enterprise; borrowing venture / ～契约 loan agreement; loan contract / ～人 borrower
【借入资本】borrowed capital; debenture capital
【借尸还魂】(of sth. evil) revive in a new guise; (of a dead person's soul) find reincarnation in another's corpse
【借是友,讨是敌】He that does lend will lose his friend.
【借书处】loan desk (of a library)
【借书证】library card; reader's card; admission card
【借水行舟】borrow water to sail a boat
【借宿】stay overnight at sb. else's place; put up for the night: 到朋友家～ find a lodging in a friend's house
【借题发挥】make use of the subject under discussion to put over one's own ideas; seize on an incident to exaggerate matters; make use of a subject as a pretext for one's flown talk
【借条】receipt for a loan (IOU)
【借位】[数] borrow
【借问】[敬] may I ask
【借项】debit; charges ◇ ～凭单 debit memo / ～清单 debit note
【借以】so as to; for the purpose of; by way of
【借用】①(借来使用) borrow; have the loan of ②(用于别种用途) use sth. for another purpose
【借债】borrow money; raise (contract) a loan: ～度日 live by borrowing; pass day by borrowing ◇ ～抵押品 security for a loan
【借支】obtain an advance on one's salary; ask for an advance on one's pay
【借重】rely on for support; enlist sb's help; count on for support
【借助】have the aid of; draw support from; with the help of: ～望远镜观察月球 observe the moon with the aid of a telescope

解 send under guard

【解送】send under guard: ～犯人 send criminals under guard / ～款项 send money (to a bank)

届 ①(到) fall due: ～期 when the day comes; on the appointed date ②[量] session: 八八～毕业生 graduates of 1988 / 本～毕业生 this

year's graduates / 本～联大 the present session of the U.N. General Assembly / 第七～全国人民代表大会 the Seventh National People's Congress / 上～毕业生 last year's graduates

【届满】expiration; at the expiration of one's term of office: 任期～。 The term of office has expired.

【届时】when the time comes; at the appointed time; on the occasion: ～务请出席。 Your presence is requested for the occasion.

jīn

津 ①(唾液) saliva: 生～止渴 help produce saliva and slake thirst ②(汗) sweat: 遍体生～ perspire all over ③(渡口) ferry crossing; ford: ～渡 a ferry crossing

【津巴布韦】Zimbabwe ◇～人 Zimbabwean

【津津乐道】take delight in talking about; dwell upon with great relish; talk with great relish

【津津有味】with relish; with gusto; with keen pleasure: 吃得～ eat with great relish / 讲得～ talk with gusto / 听得～ listen to sth. with keen interest

【津贴】subsidy; allowance: 残废～ disablement allowance / 给予～ grant an allowance / 生活～ subsistence allowance

【津液】①[中医] body fluid ②(唾液) saliva

禁 ①(禁受; 耐) bear; stand; endure: 涤纶～洗。 Polyester bears 〈stands〉 lot of washing. ②(忍住) restrain oneself; contain oneself: 不～流下眼泪 cannot hold back one's tears / 不～失笑 cannot help giggling

【禁不起】be unable to stand (tests, trials, etc.); cannot withstand; cannot bear: ～考验 fail to stand tests; cannot pass tests

【禁不住】①(承受不住) be unable to bear or endure: ～折磨 can't endure torture / 晚稻～霜冻。 The late rice cannot stand frost. ②(抑制不住;不由得) can't help (doing sth.); can't refrain from: ～笑了起来 can't help laughing; burst out laughing

【禁得起】be able to stand (tests, trials, etc.): ～各种考验 be able to stand all tests

【禁得住】be able to bear or endure

【禁受】bear; stand; endure

襟 ①(指衣服前襟) front of a garment ②(连襟) brothers-in-law whose wives are sisters: ～兄 husband of one's wife's elder sister; brother-in-law

【襟怀】[书] bosom; (breadth of) mind: ～坦白 openhearted and above board; unselfish and magnanimous; have an open heart

【襟翼】[航空] (wing) flap

巾 a piece of cloth (as used for a towel, scarf, kerchief, etc.): 餐～ napkin / 领～ scarf / 毛～ towel / 头～ kerchief / 围～ scarf

【巾帼】woman ◇～英雄 heroine

今 ①(现在;现代) modern; present-day; nowadays; now: ～人 moderns; contemporaries; people of our era / 从～以后 from now on; henceforth / 至～ until now; up to now; to date; up to the present ②(今天) today: ～晨 this morning / ～明两天 today and tomorrow / ～晚 tonight; this evening ③(当前的) this (year); of this year: ～冬 this winter

【今非昔比】The present cannot compare with the past. (或: The present is superior to the past.)

【今古奇谈】modern and ancient strange talks

【今后】from now on; in the days to come; henceforth; hereafter; in future

【今年】this year: ～夏天 this summer (用于春夏季时); last summer (用于秋冬季时)

【今日】①(今天) today: ～事,～毕。 Never put off till tomorrow what can be done today. ②(当前的) present; now: ～的社会制度 the present social system

【今生】this life

【今胜于昔】The present is superior to the past

【今是昨非】present are right and past are wrong

【今世】①(今生) this life ②(当代) this age; the contemporary age

【今天】①today: 一年前的～ a year ago today ②(现在;目前) today; present; now: ～的年轻人 young men of today

【今昔】the present and the past; today and yesterday: ～对比 contrast the past with the present

【今译】modern translation; modern-language version: 古诗～ ancient poems rendered into modern Chinese

【今朝】[书] today; the present; now: ～有酒～醉 enjoy while one can; today's wine I drink today

矜 ①(怜悯;怜惜) pity; sympathize with: ～恤 show sympathy and consideration for ②(自尊自大;自夸) self-important; conceited: 骄～之

气 arrogant airs ③(慎重;拘谨) restrained; reserved: ～重 reserved and dignified
【矜持】restrained; reserved: 举止～ have a reserved manner
【矜夸】conceited and boastful
金 ①(金属) metals: 合～ alloy／五～店 hardware store ②(钱) money: 现～ cash; ready money ③(古时指锣等金属打击乐器) gong: ～鼓齐鸣。All the gongs and drums are beating. ④[化] gold (Au): ～戒指 gold ring／～块 gold bullion／纯～ pure gold／镀～ gild ⑤(金色的) golden: ～发 golden hair／红底～字 golden characters on a red background
【金榜题名】succeed in a government examination
【金镑汇兑】sterling exchange
【金杯】golden cup
【金本位制】[经] gold standard
【金笔】fountain pen; fountain pen with a gold nib
【金币】gold coin ◇ ～汇兑 gold exchange
【金碧辉煌】looking splendid in green and gold; resplendent and magnificent; magnificent and splendid
【金箔】goldleaf; gold foil ◇ ～加工 goldbeating
【金不换】not to be exchanged even for gold; invaluable; priceless; be more valuable than gold: 浪子回头～。A prodigal who returns is more precious than gold.
【金蝉脱壳】slip out of a predicament like a cicada sloughing its skin; escape by cunning maneuvering; a disappearance act from an entangled situation
【金城汤池】ramparts of metal and a moat of boiling water; impregnable fortress; a strongly fortified city
【金翅雀】greenfinch
【金疮】[中医] metal-inflicted wound; incised wound
【金雕】golden eagle
【金额】[书] amount of money
【金刚】Buddha's warrior attendant: 四大～ four guardian warriors
【金刚经】[佛教] Vajracchedika-sutra
【金刚努目】glare like a temple door god; be fierce of visage
【金刚砂】[机] emery; corundum; carborundum ◇ ～磨床 emery grinder
【金刚石】diamond
【金刚钻】diamond ◇ ～钻头 diamond bit
【金戈铁马】shining spears and armored horses
【金工】metalworking; metal processing ◇ ～车间 metalworking workshop／～机械 metalworking machinery
【金光】golden light: ～大道 golden road; bright broad highway／～闪闪 glittering; glistening／万道～ myriad golden rays
【金龟】tortoise
【金龟子】[动] scarab; cockchafer; dung beetles; may beetles; may bug
【金合欢】[植] acacia; sponge tree
【金衡】troy (weight) ◇ ～制 troy weight; troy
【金红石】[矿] rutile
【金花菜】(California) bur clover
【金黄】golden yellow; golden
【金汇兑本位制】[经] gold exchange standard
【金婚】golden wedding
【金鸡】golden pheasant
【金鸡独立】standing on one foot as the cock does
【金鸡纳树】cinchona
【金鸡纳霜】[药] quinine
【金鸡纳素】cinchonine
【金橘】[植] kumquat
【金科玉律】golden rule and precious precept; laws and regulations: 奉为～ accept as infallible law
【金口玉言】oracular words
【金库】national (state) treasury; exchequer
【金块】gold bullion
【金兰之契】sworn brothers
【金绿宝石】[矿] chrysoberyl
【金銮殿】emperor's audience hall; throne room
【金霉素】[药] aureomycin
【金迷纸醉】living an extravagant life; given to sensual pleasures
【金牛座】[天] Taurus
【金瓯无缺】unimpaired territorial integrity
【金牌】[体] gold medal
【金漆】[工美] gold lacquer; ningpo lac; Ningpo varnish ◇ ～镶嵌制品 inlaid gold lacquerware
【金器】gold vessel
【金钱】money ◇ ～万能 money is almighty; money talks／～至上 exclusive concern with money／～主义 money grabbing spirit; venality
【金钱豹】leopard
【金枪鱼】tuna
【金融】finance; banking ◇ ～呆滞 financial stringency／～公司 finance company／～寡

头 financial oligarch; financial magnate / ～机构 financial institution; banking institution / ～界 financial circles / ～紧缩 tight money policy; deflationary policy / ～市场 money (financial) market / ～体制 monetary system / ～投机 monetary speculation / ～危机 financial crisis / ～制裁 financial sanction / ～中心 financial (banking) center / ～资本 financial capital
【金色】golden: ～的朝阳 golden rays of the morning sun; golden dawn
【金石】① [书](坚硬的东西) metal and stone; a symbol of hardness and strength ② (铜器、石碑上的铭刻) inscriptions on ancient bronzes and stone tablets ◇ ～同盟 alliance of perpetuity / ～为开 sincerity can make metal and stone crack / ～学 the study of inscriptions on ancient bronzes and stone tablets; epigraphy / ～之交 close and intimate friendship
【金石丝竹】musical instruments made of metals, stone, strings and bamboo
【金属】metal: 粉末～ powdered metal / 黑色～ ferrous metal / 稀有～ rare metals / 有色～ nonferrous metal ◇ ～废料 scrap metal / ～互化物 intermetallic compound / ～键 metallic bond / ～加工 metal processing; metalworking / ～结构 metal structure / ～结构厂 metal structures plant / ～理论 theory of metals / ～模 metal pattern / ～疲劳 metal fatigue / ～切削机床 metal-cutting machine tool / ～探伤器 flaw detector / ～陶瓷 cermet / ～涂料 metallic paint / ～纤维 metallic fiber / ～性 metallicity / ～制品 metalwork
【金属分币】coin; small denomination coin
【金丝猴】golden monkey; snub-nosed monkey
【金丝雀】canary
【金丝镶嵌】[工美] gold filigree
【金丝燕】[动] esculent swift
【金条】gold bar
【金童玉女】boy and girl attendants of fairies
【金屋藏娇】live with one's young wife in a plush apartment; keep a mistress in a love nest
【金线鱼】[动] red coat; golden thread
【金相学】metallography
【金小蜂】tiny golden wasp; pteramalid
【金星】① [天] Venus ② (金黄色五角星) golden star ③ (象星的小点) spark; star: 两眼冒～ see stars
【金星玻璃】aventurine glass
【金要足赤,人要完人】gold must pure and man must be perfect; perfectionism
【金银财宝】treasures
【金银花】honeysuckle ◇ ～露[中药] distilled liquid of honeysuckle
【金樱子】[中药] the fruit of Cherokee rose
【金鱼】goldfish
【金玉】[书] gold and jade; precious stone and metals; treasures: ～良言 golden saying; invaluable advice; good counsels / ～之言 precious and valued advice / ～其外,败絮其中 rubbish coated in gold and jade; fair without, foul within; a rotten interior beneath a fine exterior
【金云母】[矿] phlogopite
【金盏花】[植] pot marigold
【金针菜】[植] day lily
【金针虫】[动] wireworm
【金枝玉叶】descendants of royal families
【金字塔】pyramid
【金字招牌】① (商店招牌) gold-lettered signboard ② (炫耀于人的称号) vainglorious title
【金子】gold

筋

①(肌的旧称) muscle ② (韧带) tendon; sinew: 扭了～ get one's sinew 〈tendon〉 sprained ③ (可以看见的皮下静脉) vein: 青～ blue veins ④ (象筋的东西) anything resembling a tendon or vein: 钢～ reinforcing steel; steel reinforcement / 叶～ ribs of a leaf
【筋斗】① (跟斗) somersault: 翻～ turn a somersault ② (跌交) fall; tumble (over): 摔了个～ fall; have a fall; tumble over; have a tumble
【筋骨】muscles and bones; physique: 锻炼～ develop one's muscles; strengthen the physique
【筋疲力竭】be completely exhausted; have used up all energy
【筋疲力尽】utterly exhausted; played out; worn out; tired out; dead tired
【筋肉】muscles

斤

jin (= 1 / 2 kilogram)
【斤斤】fuss about; haggle over: ～计较 haggle over every ounce; be calculating / ～计较个人得失 be preoccupied with one's personal gains and losses
【斤两】weight: ～不足 short weight; underweight

jǐn

堇

【堇菜】[植] violet
【堇青石】[矿] cordierite

【堇色】violet

谨 ①(谨慎;小心) careful; cautious; circumspect: ～记在心 bear in mind / ～守规则 strictly adhere to the rules / ～守诺言 carefully keep one's promise ②(郑重) solemnly; sincerely: ～启 yours respectfully / ～致谢意 thank you earnestly; please accept my sincere thanks

【谨防】guard against; beware of; be alert to; be cautious of: ～暗箭。Guard against a hidden arrow. / ～扒手。Beware of pickpockets. / ～恶犬。Beware of vicious dog. / ～假冒。Beware of imitations. / ～有误 caution oneself against error

【谨上】[套](用于书信具名后) sincerely yours

【谨慎】prudent; careful; cautious; circumspect: ～从事 act with caution / 谦虚～ modest and prudent / 说话～ be guarded in one's speech / 言行～ be prudent in one's words and deeds

【谨小慎微】overcautious; timorous and punctilious; cautious and meticulous: ～的君子 well-mannered man

【谨严】careful and precise: 治学～ careful and exact scholarship / 文章结构～。The article is compact and carefully constructed.

【谨言慎行】speak and act cautiously; be discreet in word and deed; be prudent in making statements and careful in personal conducts

紧 ①(不松) tight; taut; close: 拉～绳子 pull the rope taut / 螺丝拧～ tighten the screw / 抓～时间 make the full use of one's time / 这鞋子太～。The shoes are too tight. (或: The shoes pinch.) ②(牢固) fast; firm: 把门关～ make the door fast ③(贴近) close: ～靠墙站着 stand close against the wall ④(厉害) hard; heavily; violently: 风刮得很～。The wind blow hard. ⑤(紧急) urgent; pressing; tense: 任务～ the task is urgent / 时间很～。Time is pressing. ⑥(拮据) short of money; hard up: 手头～ be short of money; be hard up / 银根～ money is tight

【紧巴巴】①(不松) tight; taut ②(拮据) hard up

【紧绷绷】①(不松) tight; taut ②(表情不自然) strained; stiffened; sullen: 脸上～的 look strained; look sullen; with a serious face

【紧逼】press hard; close in on: 步步～ press on at every stage / 全场～[篮球] full-court press

【紧凑】compact; terse; well-knit; tight: 情节～ have a well-knit plot

【紧跟】follow closely; keep in step with; keep up with; tread on the heels of: ～时代的步伐 keep in step with the times / ～形势 keep abreast of the situation

【紧箍咒】inhibition; trammels

【紧固件】fastener; fastening piece

【紧急】urgent; pressing; critical; emergent ◇～避难 act of rescue / ～措施 emergency measures / ～法令 emergency act / ～关头 critical moment / ～会议 emergency meeting; emergency session / ～呼叫 urgent call / ～呼吁 urgent appeal / ～集合 emergency muster / ～空袭警报 emergency air-raid alarm / ～起飞 scramble / ～情况 critical situation / ～任务 urgent task / ～刹车 snub / ～信号 emergency signal; distress signal / ～状态 state of emergency / ～追捕 hot pursuit / ～着陆 emergency landing

【紧紧】closely; firmly; tightly: ～盯着 watch closely; stare fixedly; gaze steadfastly / ～相连 closely linked

【紧邻】close neighbor

【紧锣密鼓】wildly beating gongs and drums; intense publicity campaign in preparation for some sinister undertaking, etc.

【紧密】①(密切) close together; inseparable: ～地团结在党中央周围 rally closely round the Party Central Committee ②(密集) thick and fast; rapid and intense

【紧迫】pressing; urgent; imminent: 任务～ the task is urgent / 时间～ be pressed for time ◇～感 sense of urgency

【紧身】close-fitting vest; undershirt; close-fitting undergarment ◇～健美服 corset

【紧缩】reduce; retrench; tighten; cut down: ～包围圈 tighten the ring of encirclement / ～编制 reduce staff / ～开支 cut down expenses; retrench; curtail outlay / ～生产 curtail production ◇～时期 period of contraction

【紧要】critical; crucial; vital: ～关头 critical moment; crucial moment; critical point; critical juncture / 无关～ of no consequence; of no importance

【紧张】①(兴奋不安) nervous; keyed up: 神情～ look nervous ②(激烈或紧迫) ～而有秩序的工作 intense but orderly work / tense; intense; strained: ～局势 a tense situation / ～气氛 a tense atmosphere / 缓和国际～局势 ease international tension ③(供应不足) in short supply; tight: 这些货物供应～。These goods are in short supply

锦 ①(丝织品) brocade ②(色彩鲜明华丽) bright and beautiful: ～霞 rose-tinted clouds / 前程似～ splendid prospects; glorious future

【锦标】prize; trophy; title ◇ ～主义 cups and medals mania

【锦标赛】championship contest; championships: 世界游泳～ the World Swimming Championships

【锦缎】brocade

【锦鸡】golden pheasant

【锦葵】[植] high mallow

【锦纶】[纺] polyamide fiber; jinlun

【锦囊妙计】instructions for dealing with an emergency; wise counsel; a secret master plan

【锦旗】silk banner (as an award or a gift)

【锦上添花】add flowers to the brocade; make perfection still more perfect; gild refined gold

【锦绣】as beautiful as brocade; beautiful; splendid: ～前程 glorious future / ～山河 a land of charm and beauty; a beautiful land; a beautiful landscape / ～文章 an embroidered piece of literature

仅 [副] only; merely; barely; simply; solely: ～次于 second only to; next to / 世所～见 have no parallel anywhere

【仅仅】[副] only; merely; barely; simply; solely; just; alone: 这～是开始。 This is only the beginning. / 这～是时间问题。 It is merely 〈simply〉 a matter of time.

尽 ①(尽量) to the greatest extent: ～早 as early as possible; at the earliest possible date ②(不得超过) within the limits of; no more than: ～着五十元花。 Don't spend more than fifty *yuan*. ③(尽先) give priority to; first: ～着女同志先坐。 Let women comrades sit down first. ④(最,用在方位词前面) at the furthest end of; most: ～北边 the northernmost end / ～底下 at the very bottom / ～后头 rearmost

【尽管】①[副](不必考虑别的) feel free to; not hesitate to: 你～直说。Just speak out. / 有意见～提好了。 Don't hesitate to make comments or suggestions if you have any. ②[连](虽然) though; even though; in spite of; despite; notwithstanding: ～任务困难,他们仍按时完成了。Though the task was difficult, they managed to accomplish it in time.

【尽可能】as far as possible; to the best of one's ability: ～早点儿来 come as early as possible

【尽快】as quickly 〈soon; early〉 as possible: ～投入生产 put into production as soon 〈early〉 as possible

【尽量】to the best of one's ability; as far as possible: ～采用先进技术 make the widest possible use of advanced technology / ～利用时间 make the best of one's time

【尽先】[副] give priority to; first: ～考虑住房问题 give first priority to the housing problem

jìn

进 ①(向前移动) advance; move forward; move ahead: 不～则退。 Move forward, or you'll fall behind. ②(从外面到里面) enter; come into; go into; get into: ～大学 enter college / ～屋 enter a house; go into the room / ～医院 be sent to hospital; be hospitalized / 请～! Come in! ③(收入) receive: ～货 lay in a new stock of goods / ～款 income ④(进食) eat; drink; take: 共～晚餐 have supper together ⑤(呈上) submit; present: ～一言 give a word of advice ⑥(用在动词后,表示到里面) into; in: 走～客厅 walk into the hall ⑦[体](进球) score a goal

【进逼】close in on; advance on; press on towards: 步步～ steadily close in

【进步】①(向前发展) advance; progress; improve: 世界是在～的。 The world is progressing / 他作文有～。 His composition has improved. ②(先进的) (politically) progressive: ～人士 progressive personages / ～势力 progressive forces / 思想～ have progressive ideas

【进场】①(进入场地) march into the arena ②[航空] approach: ～失败 missed approach

【进城】①(到城里去) go into town; go to town ②(到城市生活和工作) enter the bit cities

【进程】course; process; progress: 历史～ the course of history

【进尺】[矿] footage: 掘进～ drifting footage / 开拓～ tunnelling footage / 凿岩～ drilling footage

【进出】①(进来和出去) pass in and out; get in and out ②(收入和支出) receipts and payments; turnover: 这个商店每天有好几千元的～。 This store has a daily turnover of several thousand *yuan*.

【进出口】①(商品进出口) imports and exports ②(出入门口) exits and entrances; exit ◇ ～公司 import and export corporation / ～货物 cargoes imported and exported / ～货物报关单 customs declaration for imports and

exports / ～贸易 import and export trade; foreign trade / ～申报 import and export declaration / ～业务 imports and exports
【进刀】[机] feed ◇ ～装置 feed arrangement; feed gear; feeder
【进度】①(工作进行的速度) rate of progress; rate of advance: 加快～ quicken the pace; quickent the tempo ②(工作进行的计划) planned speed; schedule ◇ ～报告 progress report / ～表 progress chart
【进而】proceed to the next step; and then: 先搞试点,～逐步推广 first make experiments at chosen units and then extend its application step by step
【进发】set out; start: 向南京～ head for Nanjing
【进犯】intrude into; invade: 打败～之敌 beat back the invading enemy
【进风井】[矿] downcast (shaft)
【进攻】attack; assault; offensive: 发起全面～ launch all-out offensive ◇ ～队形 attack formation / ～命令 order to attack / ～性武器 offensive weapon / ～正面 frontage in attack; front of attack
【进贡】pay tribute (to a suzerain or emperor)
【进化】evolution ◇ ～论 the theory of evolution; evolutionism
【进货】stock (a shop) with goods; lay in a stock of merchandise; replenish one's stock ◇ ～成本 purchasing cost / ～分类帐 purchase ledger; bought ledger / ～折让 purchase discounts and allowances
【进见】call on (sb. holding high office); have an audience with
【进军】march; advance; move: 向边界～ march 〈advance〉 towards the boundary / 向四个现代化～ march towards the four modernization
【进口】①(船只进港) enter port; sail into a port ②(外贸进口) import ③(入口) entrance ◇ ～补贴 import subsidy / ～报单 import declaration / ～代理商 import agent / ～港 port of entry / ～货 imported goods; imports / ～商 importer / ～税 import duty; customs import duty / ～托收 inward collection / ～限额 import quota / ～许可证 import license; import authorization; import permit
【进款】income; receipts
【进来】①(到里面来) come in; get in; enter ②(用在动词后,表示向里) in: 搬～ move in / 走～ walk in
【进取】keep forging ahead; be eager to make progress; be enterprising; be up-and -coming ◇ ～心 enterprising spirit; initiative; gumption; push
【进去】①(到里面去) go in; get in; enter ②(用在动词后,表示向里) in: 冲～ rush in / 闯～ break in
【进入】get into; enter: ～决赛阶段 enter the finals / ～角色 enter into the spirit of a character; live one's part / ～战斗 go into action / ～阵地 get into position
【进身之阶】stepping-stone (in one's official career)
【进食】take food; have one's meal
【进水闸】intake work; intake; head gate; intake gate
【进退】①(进和退) advance and retreat: ～两难 difficult to advance or to retreat; equally difficult to go on or retreat; be in a dilemma / ～维谷 caught in a dilemma; to be in a cleft stick; between the devil and the deep blue sea / ～自如 free to advance or retreat (in a battle or game); have room for maneuver ②(分寸) sense of propriety: 不知～ have no sense of propriety
【进位】[数] carry (a number, as in adding)
【进项】income; receipts
【进行】①(开展) be in progress; be underway; go on: 会议的准备工作正在～。 Preparations for the meeting are in progress. ②(从事) carry on; carry out; conduct; make: ～报复 make reprisals / ～表决 put a question to the vote / ～抵抗 put up a resistance / ～动员 mobilize; make a mobilization speech / ～核试验 conduct a nuclear test / ～科学实验 engage in scientific experiment / ～民事诉讼 sue in civil action / ～亲切的谈话 have a cordial conversation / ～侵略 commit aggression / ～社会调查 make social investigations / ～诉讼 engage in a lawsuit / ～讨论 hold discussions / ～投机倒把 engage in speculation and profiteering / ～刑事诉讼 institute penal proceedings / ～一场激烈的争论 carry on a spirited debate / ～英勇斗争 wage a heroic struggle / ～自我批评 undertake self-criticism / 将革命～到底 carry the revolution through to the end ③(前进) be on the march; march; advance
【进行曲】march: 义勇军～ March of the Volunteers
【进修】engage in advanced studies; take a re-

fresher course: 教师的业务～ teachers' vocational studies / 在职～ in-service training; on-the-job training ◇ ～班 class for advanced studies / ～生 graduate student

【进一步】go a step further; further: ～提高质量 make further improvement on the quality / 作～调查 make further investigation

【进展】make progress; make headway: ～神速 advance at a miraculous pace

【进占】attack and occupy; attack and capture; capture

【进站】get in; pull in; draw up at a station: 准时～ drew up at the station on time

【进驻】enter and be stationed in; enter and garrison; march into (a place) and station there

晋 ①(进) enter; advance: ～见 have an audience with ②(提高) promote: 加官～爵 be promoted to a higher office and rank

【晋级】rise in rank; be promoted

【晋见】call on (sb. holding high office); have an audience with

【晋升】promote to a higher office

【晋谒】[书] call on (sb. holding high office); have an audience with

觐 ①(朝见) present oneself before ②(朝拜) go on a pilgrimage

【觐见】present oneself before (a monarch); go to court; have an audience with

禁 ①(禁止) prohibit; forbid; ban: 解～ lift a ban; remove a ban; withdraw a prohibition / 严～烟火。Smoking and lighting fires strictly prohibited. ②(监禁) imprison; detain: 监～ imprison ③(法令或习俗所不允许的事项) what is forbidden by law or custom; a taboo: 入国问～ on entering a country ask about its taboos / 违～品 contraband (goods)

【禁闭】confinement (as a punishment): 关～ be placed in confinement ◇ ～室 guard room

【禁地】forbidden area; restricted area; out-of-bounds area

【禁赌】prohibition of gambling; suppress gambling

【禁伐林】forest prohibiting on cutting and chopping timber

【禁锢】①(禁止异已做官) debar from holding office (in feudal times) ②(监禁) keep in custody; imprison; put in jail ③(强制束缚) confine; shackle: 被旧习惯所～ be confined by old customs

【禁忌】①(忌讳的话和行动) taboo: 属～之列 under taboo ②(戒) avoid; abstain from: ～辛辣油腻 abstain from peppery or greasy food ③[医] contraindication

【禁酒】prohibition against alcoholic drinks

【禁例】prohibitory regulations; prohibitions

【禁令】prohibition; ban: 解除～ lift a ban / 取消～ nullify a prohibition

【禁区】①(禁止进入地区) forbidden zone; restricted zone; out-of-bounds area: 空中～ restricted airspace ②(自然保护区) preserve; reserve; natural park ③[足球] penalty area ④[篮球] restricted area

【禁食】fasting

【禁食疗法】fasting treatment; starvation cure

【禁书】official banned book; banned book

【禁烟】ban on opium-smoking and the opium trade

【禁欲主义】asceticism ◇ ～者 ascetic

【禁运】embargo ◇ ～货物 contraband goods / ～品 contraband

【禁止】prohibit; ban; forbid: ～保释 forbid bail / ～出口 embargo / ～记者采访 be barred to the press / ～进口 nonimportation / ～流通 demonetize / ～某人做某事 forbid sb. to do sth.; prohibit sb. from doing sth. / ～旁听 clear the court / ～邮寄的物品 goods prohibited from mail / ～拍照。Cameras are forbidden. / ～入内。No admittance. / ～践踏草地。Keep off the grass. / ～停车。No parking. / ～通行。No thoroughfare. (或: Closed to traffic.) / ～招贴。Post no bills.

【禁治产】interdiction ◇ ～者 imbecile; interdicted person

【禁制品】articles the manufacture of which is prohibited except by special permit; banned products

噤 ①(闭口不做声) keep silent ②(因寒冷而哆嗦) shiver: 寒～ shiver with cold

【噤若寒蝉】as silent as a cicada in cold weather; keep quiet out of fear

近 ①(距离或时间短) near; close; immediate: ～几年来 in recent years / ～在眼前 right before one's eyes; near at hand / ～在咫尺 be close at hand; be well within one's reach; round the corner ②(接近) approaching; approximately; come near to; close to: 年～八十 approaching eighty; getting on for eighty / 观众～七万人。There were nearly 70000 spec-

tators./时～午夜。It was close upon midnight.③(亲密) intimate; closely related: 两家关系很～。The two families are on intimate terms. ④(浅显) easy to understand: 浅～ simple and easy to understand/言～旨远 simple in language but profound in meaning
【近便】close and convenient; close at hand
【近程】short range: ～雷达 short-range radar
【近代】modern times ◇～史 modern history
【近道】shortcut
【近地点】[天] perigee
【近东】the Near East
【近海】coastal waters; inshore; offshore ◇～航运 shipping in coastal waters/～渔业 inshore fishing
【近乎】①(接近于) close to; little short of; near; almost; approximately: ～不可能 almost impossible/～黑色的深棕色 dark brown coming near black ②(关系的亲密) intimate; friendly: 套～ try to be friendly with; try to chum <pal> up with; cotton up to
【近郊】outskirts of a city; suburbs; environs
【近景】[摄] close shot
【近距离】close quarter; at short range
【近况】recent developments; how things stand: ～日趋紧张。The current situation is growing increasingly tense./你～如何? How are you getting along?
【近来】recently; of late; lately
【近邻】near neighbor
【近路】shortcut: 走～ take a shortcut
【近旁】nearby; near
【近期】in the near future; short term ◇～预报 short-term forecast
【近亲】close relative; near relation ◇～繁殖[生] inbreeding; close breeding/～结婚 consanguineous marriage/～收养 adoption by close relative/～属 close relative
【近日】①(指过去) recently; in the past few days ②(指将来) within the next few days
【近日点】[天] perihelion
【近视】myopia; nearsightedness; shortsightedness: 他是～眼。He is shortsighted 〈nearsighted〉 ◇～散光 myopic astigmatism/～眼镜 spectacles for nearsighted persons
【近水楼台】waterside pavilion; a favorable position
【近水楼台先得月】a waterfront pavilion gets the moonlight first; the advantage of being in a favored position; enjoy the benefits of a favorable position
【近似】approximate; similar ◇～读数 approximate reading/～计算 approximate calculation/～值[数] approximate value
【近因】immediate cause
【近于】bordering on; little short of: ～粗暴 border on rudeness/～荒唐 little short of preposterous
【近月点】[宇航] perilune
【近战】fighting at close quarters; close combat ◇～武器 close-in weapons
【近朱者赤,近墨者黑】One who mixes with vermilion will turn red, one who touches pitch shall be defiled therewith. (或: He who stays near vermilion gets stained red, and he who stays near ink gets stained black/或: Evil communications corrupt good manners/或: One takes on the color of one's company)

劲 ①(力气) strength; energy: 用～ put forth strength/身上没～ feel weak ②(精神;情绪) vigor; spirit; drive; zeal: 干得非常起～ work with unusual vigor; work with great zeal ③(神情;态度) air; manner; expression: 他们都显出高兴～。They all have a look of happiness. ④(趣味) interest; relish; gusto; mood: 打扑克没～。Playing cards is no fun.
【劲头】①(力气) strength; energy ②(精神;情绪) vigor; spirit; drive; zeal

浸 soak; steep; immerse
【浸镀】immersion plating
【浸膏】[药] extract
【浸焊】dip soldering; solder dipping
【浸剂】[药] infusion
【浸礼】[基督教] baptism; immersion
【浸没】immersion ◇～透镜[物] immersion lens/～折射计[物] immersion refractometer
【浸泡】soak; immerse
【浸染】①(针织物染色) dip-dye ②(逐渐沾染) be gradually tainted with; be contaminated; be gradually influenced
【浸润】①(渐渐渗入) soak; infiltrate: 雨水～了表土。The rain has soaked into the subsoil. ②[医] infiltration
【浸透】soak; saturate; steep; infuse
【浸种】[农] seed soaking (in water); water treatment
【浸渍】soak; ret; macerate: 亚麻～ flax retting ◇～槽 macerate tank/～剂 soaker/～液 maceration extract

尽 ①(完) exhausted; finished: 取之不～ inexhaustible / 无穷无～ endless; inexhaustible / 知无不言,言无不～ say all you know and say it without reserve ②(达到极端) to the utmost; to the limit; to the full: 用～气力 exert oneself to the utmost ③(全部用完) use up; exhaust: 一言难～。 It can't be expressed in a few words. (或: It's a long story.) / 一饮而～ empty a glass at one gulp; drain the cup with one gulp ④(全部使出) try one's best; put to the best use; exert oneself; do one's best: ～全力 do one's best; exert one's utmost; do all in one's power / ～一切办法 seek in every way to; by every means / ～责任 do one's duty; discharge one's responsibility / 人～其才,物～其用 make the best possible use of men and material ⑤(全;所有的) all; completely; exhaustive: 不可～信 not to be believed word for word; to be taken with a grain of salt / ～收眼底 have a panoramic view

【尽瘁而死】fag oneself to death

【尽瘁至死】slave oneself to death

【尽欢而散】leave only after each has enjoyed himself to the utmost

【尽捐前嫌】forget all the past quarrel, ill will or enemy

【尽力】do all one can; try one's best; do sth. to the best of one's ability: ～而为 do one's best; do everything in one's power; shoot one's bolt

【尽量】(drink or eat) to the full

【尽其本分】exhaust one's obligation; do one's part; do one's bit

【尽其所长】work or endeavor to the best of one's ability

【尽情】to one's heart's content; as much as one likes: ～歌唱 sing to one's heart's content / ～欢呼 cheer heartily / ～款待 treat with the utmost kindness

【尽人皆知】be known to all; be common knowledge

【尽人事】do what one can (to save a dying person, etc.); do all that is humanly possible (though with little hope of success)

【尽善尽美】the acme of perfection; perfect; perfectly satisfactory

【尽是】full of; all; without exception: 这里展销的～新产品。 All the exhibits here are for sale new products.

【尽头】end: 路的～ the end of the road ◇ ～路 cul-de-sac

【尽心】with all one's heart; put one's heart and soul into: ～竭力 (do sth.) with all one's heart and all one's might; with heart and soul and with might and main

【尽兴】to one's heart's content; enjoy oneself to the full: ～而归 return after thoroughly enjoying oneself

【尽义务】①(尽责任) do one's duty; fulfill one's obligation: 尽国际主义义务 fulfill one's internationalist duty ②(无报酬劳动) be a volunteer; work for no reward; work without reward

【尽早】with the least delay; at one's earliest convenience; as soon as possible

【尽职】fulfill one's duty; discharge one's duty

【尽忠】①(竭尽忠诚) be loyal to ②(牺牲) sacrifice one's life for; lay down one's life for ◇ ～报国 be loyal and patriotic

烬 cinder: 灰～ ashes; cinders

荩

【荩草】[植] hispid arthraxon

jīng

京 ①(首都) the capital of a country: 进～ go to the capital ②(北京) Beijing: ～津唐 Beijing-Tianjin-Tangshan

【京城】the capital of a country

【京都】the capital of a country

【京胡】*jinghu*, a two-stringed bowed instrument with a high register; Beijing opera fiddle

【京畿】[书] the capital city and its environs

【京剧】Beijing opera

惊 ①(惊恐) start; be frightened: ～呆了 be stupefied / ～叫 give a cry of alarm ②(惊动) surprise; shock; alarm: 一声～雷 a sudden clap of thunder ③(动物受惊) shy; stampede: 马～了。 The horse shied.

【惊诧】[书] surprised; amazed; astonished

【惊动】①(使吃惊) startle; alarm; shock: 这消息～了全城。 The whole town was startled by the news. ②(使警惕) alert: 不要～敌人 don't alert the enemy ③(惊扰) disturb; bother: 不要～病人 don't disturb the patient

【惊愕】[书] stunned; stupefied: ～不已 extremely surprised

【惊风】[中医] infantile convulsions: 急～ acute infantile convulsions

【惊弓之鸟】a bird startled by the mere twang of a bow-string; a badly frightened person; a panic-stricken person
【惊骇】[书] frightened; panic-stricken
【惊呼】cry out in alarm
【惊慌】alarmed; scared; panic-stricken: ～不安 jittery; nervy / ～失措 frightened out of one's wits; be thrown into panic and confusion
【惊魂未定】not yet recovered from a fright; still badly shaken
【惊叫】cry in fear; scream
【惊厥】①(因害怕而晕过去) faint from fear ② [医] convulsions
【惊恐】alarmed and panicky; terrified; seized with terror: ～失色 pale with fear / ～万状 be in a great panic; be terribly frightened; convulsed with fear
【惊奇】wonder; be surprised; be amazed
【惊扰】alarm; agitate: 自相～ raise a false alarm
【惊人】astonishing; amazing; alarming: ～的成就 amazing achievements / ～的速度 surprising rapidity / ～的毅力 amazing willpower / ～之举 shocking action
【惊叹】wonder at; marvel at; exclaim (with admiration) ◇ ～号 exclamation mark (!)
【惊涛骇浪】①(大风浪) terrifying waves; stormy sea; fearful storm ②(险恶环境或遭遇) a situation or life full of perils; perilous situation
【惊天地,泣鬼神】startle the universe and move the gods
【惊天动地】shaking heaven and earth; earthshaking; world-shaking; sky-rocking and earth shaking: ～的奇迹 incredible wonders / ～的事业 earthshaking undertaking
【惊跳反应】startle reaction
【惊悉】be shocked to learn: ～某人不幸逝世 be distressed to learn of the passing away of sb.
【惊喜】pleasantly surprised: ～交集 be filled with elation and amazement
【惊吓】frighten; scare; shock
【惊险】alarmingly dangerous; breathtaking; thrilling: ～的表演 breathtaking performance / ～的场面 thrilling scene / ～动作 astounding feat ◇ ～飞行表演 stunt flying / ～小说 thriller
【惊心动魄】soul-stirring; profoundly affecting; struck with fright or horror
【惊心怵目】be thoroughly frightened
【惊醒】①(受惊醒来) wake up with a start ②(使醒来) cause to wake up; rouse suddenly from sleep; awaken ③(容易醒来) sleep lightly; easily roused from sleep: 她睡觉很～。She sleeps lightly. (或: She is easily roused from sleep.)
【惊讶】surprised; amazed; astonished; astounded
【惊疑】surprised and bewildered
【惊异】surprised; amazed; astonished; astounded
【惊蛰】the Waking of Insects (3rd solar term)

鲸 whale: 雌～ cow whale / 蓝～ blue whale / 抹香～ sperm whale / 雄～ bull whale / 幼～ whalecalf ○捕～船 whale catcher; whaleship / 捕～炮 whaling-gun / 捕～者 whaler
【鲸蜡】spermaceti; spermaceti wax
【鲸肉】whalemeat
【鲸鲨】whale shark
【鲸吞】swallow like a whale; annex (territory): ～蚕食 take by force and encroach upon land or territory
【鲸须】baleen; whalebone
【鲸油】whale oil; blubber
【鲸鱼】whale
【鲸仔】whale calf

旌 an ancient type of banner hoisted on a feather-decked mast
【旌旗】banners and flags: ～招展 banners and flags are fluttering about

粳
【粳稻】round-grained rice
【粳米】polished round-grained rice

精 ①(经提炼或挑选的) refined; picked; choice: ～白米 polished white rice / ～金 fine gold / ～盐 refined salt ②(精华) essence; extract: 去粗取～ discard the dross and select the essence / 柠檬～ lemon extract ③(完美;最好) perfect; excellent: ～良 excellent; superior; of the best quality ④(细) meticulous; fine; precise ⑤(机灵心细) smart; sharp; clever; shrewd: 小算盘打得～ be selfish and calculating ⑥(精通) skilled; conversant; proficient: ～于此道 be proficient in the knowledge of / ～于绘画 be skilled in painting / ～于武术 be proficient at wushu ⑦(精神;精力) energy; spirit: 聚～会神 concentrate one's attention; be all attention ⑧

(精液;精子) sperm; semen; seed: 受～ fertilization ⑨[方](非常;十分) extremely; very: ～瘦 very lean; all skin and bone ⑩(妖精) goblin; spirit; demon
【精兵】picked troops; crack troops
【精兵简政】better troops and simpler administration; streamlined administration
【精彩】brilliant; splendid; wonderful; marvelous
【精巢】[生理] spermary; testis; testicle
【精诚】[书] absolute sincerity; good faith
【精诚所至,金石为开】complete sincerity can affect even metal and stone
【精粹】succinct; pithy; terse
【精打细算】careful calculation and strict budgeting; use one's calculating funds with meticulous care
【精当】precise and appropriate: 用词～ precise and appropriate wording; masterly choice of words
【精雕细刻】work at sth. with the care and precision of a sculptor; work at sth. with great care
【精读】read carefully and thoroughly; intensive reading
【精度】precision: 高～ high precision
【精干】①(指人少而精) small in number but highly trained; crack: 一支～的小分队 a small detachment of picked troops ②(精明强干) keen-witted and capable
【精耕细作】intensive and meticulous farming; intensive cultivation
【精光】①(一无所有) with nothing left: 钱用得～。Not a penny is left. ②(光洁) shiny; bright
【精悍】①(精明能干) capable and vigorous ②(精练犀利) pithy and poignant
【精华】cream; essence; quintessence: 中国古典文学的～ the cream of the Chinese classical literature / 去其糟粕,取其～ discard the dross and select the essence
【精加工】[机] finish machining; precision work; finely process
【精简】retrench; simplify; cut; reduce: ～报表 reduce the number of forms; cut down paper work / ～编制 reduce the staff / ～非生产人员 reduce nonproductive personnel / ～会议 cut (the number of) meetings to a minimum / ～节约 simplify administration and practice economy / ～开支 cut expenses; retrench / ～课程 cut down curriculum / ～整编 streamline and reorganize the structure
【精矿】concentrate
【精力】energy; vigor; vim: ～不足 be deficient in energy / ～充沛 very energetic; full of vigor; full of vitality / ～过人 exceptional vitality / ～旺盛 be full of vitality
【精练】concise; succinct; terse: 语言～ succinct language
【精炼】①[冶] refine; purify: 火法～ fire refining / 真空～ vacuum refining ②(精练) concise; succinct; terse ◇ ～炉 refining furnace / ～期 refining period
【精良】excellent; superior; of the best quality: 制作～ of excellent workmanship / 装备～ well-equipped
【精灵】①(鬼怪) spirit; demon ②(机灵) clever; smart; intelligent
【精馏】[化] rectification ◇ ～塔 rectifying tower; fractionating tower
【精美】exquisite; elegant: 制作～ exquisitely made / 包装～ beautifully packaged
【精密】precise; accurate: ～的观察 accurate observation; close observation ◇ ～度 precision / ～机床 precision machine tool / ～机械 precision optical machinery / ～仪器 precision instrument / ～铸造 precision casting
【精明】astute; shrewd; sagacious: ～能干 know one's own business / ～强干 intelligent and capable; able and efficient / 为人～ be a man of sagacity
【精囊】[生理] seminal vesicle ◇ ～炎 vesiculitis
【精疲力竭】exhausted; worn out; tired out; spent
【精疲力尽】spirit and strength totally used up; extremely fatigued
【精辟】penetrating; incisive: ～的论述 a brilliant exposition / 进行～的分析 make a penetrating analysis
【精巧】exquisite; ingenious: ～的技艺 exquisite craftsmanship / 构造～ ingeniously constructed
【精确】accurate; exact; precise: ～的统计 accurate statistics / 下一个～的定义 give a precise definition ◇ ～文本 perfect copy / ～计量 accurate measurement
【精锐】crack; picked: ～部队 crack troops; picked troops
【精深】profound: 博大～ have both extensive knowledge and profound scholarship / ～的理论 a comprehensive and profound theory
【精神】①(心理状态) spirit; mind; consciousness: 国际主义～ the spirit of internationalism / 给予～上的支持 give moral sup

port ②(宗旨;主要意义) essence; gist; spirit; substance: 传达文件的～ pass on the gist of a document ③(活力) vigor; vitality; drive: 振作～ bestir oneself; summon up one's energy; get up steam / 没有～ listless; languid ④(活跃;有生气) lively; spirited; vigorous: 那孩子怪～的。The child is full of life. ◇ ～饱满 energetic; full of energy; full of vitality / ～不死 one's spirit will not vanish / ～财富 spiritual values / ～错乱 have a slate loose / ～抖擞 brace up; vigorous and energetic / ～反常 lose one's mental balance / ～鼓励 moral encouragement / ～贯注 concentrate attention / ～贵族 intellectual aristocrats / ～焕发 in high spirits, fresh with energy / ～恍惚 absent-minded / ～枷锁 spiritual shackles / ～空虚 be spiritually barren / ～面貌 spiritual outlook; mental attitude / ～生活 cultural life / ～失常 out of one's mind / ～食粮 nourishment for the mind / ～世界 inner world; mental world / ～振奋 be inspired with enthusiasm / ～支柱 spiritual prop; ideological prop / ～状态 state of mind

【精神病】 mental disease; mental disorder; psychosis ◇ ～患者 mental patient / ～学 psychiatry / ～医生 psychiatrist / ～院 psychiatric hospital; mental home; mental hospital; mental institution

【精神产品】intellectual products

【精神错乱】①[医] amentia; alienism ②[心理] aballenation ③[法] insanity

【精神发育迟缓】[医] mental retardation

【精神分裂症】[医] schizophrenia

【精神分析】psychoanalysis

【精神衰弱】psychasthenia

【精神外科】[医] psychosurgery

【精神文明】cultural and ideological progress; culture and ideology; spiritual civilization ◇ ～建设 cultural and ideological progress

【精神治疗】psychotherapy

【精收细打】careful reaping and threshing

【精梳】[纺]combing ◇ ～机 comber / ～毛纺 worsted spinning / ～纱 combed yarn

【精饲料】concentrated feed; concentrate

【精髓】marrow; pith; quintessence

【精通】be proficient in; have a good command of; master: ～业务 be proficient in professional work / ～英语 have a good command of English

【精细】 meticulous; fine; careful: ～的计算 careful calculation / 一件～的绣品 a fine piece of embroidery

【精心】 meticulously; painstakingly; elaborately: ～护理 nurse with the best of care / ～炮制 elaborately cook up / ～培育 take pains to foster

【精选】①[矿] concentration ②(精心挑选) carefully chosen; choice ◇ ～品 select items

【精盐】refined salt; table salt

【精液】[生理] seminal fluid; semen ◇ ～分析 semen analysis

【精益求精】constantly improve sth.; keep improving; always endeavor to do still better; strive for perfection

【精轧】[冶] finish rolling

【精湛】consummate; exquisite: ～的技巧 consummate skill; superb technique / 工艺～ exquisite workmanship; perfect craftsmanship

【精制】 make with extra care; refine ◇ ～品 highly finished products; superfines

【精致】 fine; exquisite; delicate: ～的花边 exquisite lace / 做得～ be delicately made

【精装】 (of books) clothbound; hardback; hardcover ◇ ～本 de luxe edition

【精壮】able-bodied; strong

【精子】[生理] sperm; spermatozoon ◇ ～发生 spermatogenesis / ～过少 spermacrasia

睛

eyeball: 目不转～ gaze fixedly; look attentively

腈

[化] nitrile

【腈纶】acrylic fibers

荆

chaste tree; vitex

【荆棘】thistles and thorns; brambles; thorny under-growth: ～丛生 overgrown with brambles

【荆棘载途】a path overgrown with brambles a path beset with difficulties

【荆芥】schizonepeta; fineleaf schizonepeta

【荆条】twigs of the chaste tree

兢

【兢兢业业】cautious and conscientious: 为革命～地工作 work conscientiously for the revolution

晶

①(光亮) brilliant; glittering: 亮～～ shining; glittering ②(水晶) quartz; (rock) crystal ③(晶体) any crystalline substance

【晶格】[物] (crystal) lattice: 面心～ face-centered lattice / 体心～ body-centered lattice

【晶粒】[物] crystalline grain; grain
【晶石】spar
【晶体】crystal: 多～ polycrystal ◇ ～点阵 crystal lattice / ～发生学 crystallogeny / ～结构 crystal structure / ～生长 crystal growth / ～学 crystallography
【晶体管】transistor: 硅～ silicon transistor / 锗～ germanium transistor ◇ ～收音机 transistor radio
【晶体钟】crystal clock
【晶莹】sparkling and crystal-clear; glittering and translucent: ～的露珠 sparkling dew
【晶状体】[生理] crystalline lens ◇ ～脱位 lens dislocation

泾

【泾渭不分】unable to distinguish between the clear and the muddy
【泾渭分明】be quite distinct from each other; make a clear distinction between purity and impurity; entirely different

茎

茎 stem (of a plant); stalk: 地下～ subterranean stem / 根～ rhizoma / 块～ tuber / 鳞～ bulb / 球～ corn
【茎稿】straw

经

经 ①[纺] warp: ～络 channels and subsidiary channels ②[中医] channels ③[地] longitude: 东～ east longitude / 西～ west longitude ④(经营;治理) manage; deal in; engage in: ～商 engage in trade ⑤(历久不变的) constant; regular: ～常 regular; frequent / 不～之谈 preposterous statement; cock-and-bull story ⑥(经典) scripture; canon; classics: 法华～ Saddharmapundarika Sutra / 佛～ Buddhist sutra; Buddhist scripture / 金刚～ Diamond Sutra / 三藏～ Tripataka / 圣～ the Holy Bible / 藏～ the Tripitaka; the whole collection of Buddhist texts / 取～ go on a pilgrimage for Buddhist scriptures / 各有一本难念的～。Each has his own hard nut to crack. (或: Each has his own trouble.) ⑦(月经) menses; menstruation ⑧(经过) pass through; via; by way of; undergo: ～东京回国 return home via Tokyo / 身～百战 have fought many battles; be a veteran of many wars / 途～广州 pass through Guangzhou ⑨(通过) as a result of; after; through: ～法律许可 authorized by law / ～法院判定 award by court / ～某人建议 upon sb.'s proposal / ～商定 it has been decided through consultation that ⑩(经受) stand; bear; endure: ～得起时间的考验 can stand the test of time
【经闭】[中医] amenorrhea
【经编】[纺] warp knitting ◇ ～机 tricot machine / ～针织物 warp-knitted fabric
【经藏】the Sutra-Pitaka; collection of sutras
【经产】multiparity ◇ ～孕妇 multigravida
【经常】①(平常;日常) day-to-day; everyday; daily: ～工作 day-to-day work / ～开支 running expenses / ～收入 regular income / ～帐户 current account ②(常常) often; frequently; constantly; regularly: 她～去图书馆。She goes to the library regularly. ◇ ～化 become a regular practice
【经典】①(具有权威性的著作) classics: ～作家 writer of classics; author of classics ②(宗教教义著作) scriptures ③(著作具有权威性的) classical: 马克思主义～著作 Marxist classics; classical works of Marxism ◇ ～力学[物] classical mechanics
【经度】longitude
【经费】funds; outlay: 行政～ administrative expenditure / 需要大量～ require a heavy outlay ◇ ～转移 transfer of appropriation
【经管】be in charge of: ～财务 be in charge of financial affairs
【经过】①(从某处过) pass; go through; go by: 这汽车～北海公园吗? Does the bus pass the Beihai Park? ②(通过) as a result of; after; through: 她～治疗病情好转。She is getting better after treatment. ③(过程;经历) process; course: 事情的全部～ the whole course of the incident
【经纪】① (经营) manage (a business) ②(经纪人) manager; broker
【经纪人】broker; middleman; agent: 房地产～ estate agent / 外汇～ foreign exchange broker
【经济】①[经] economy: 国民～ national economy / 国营～ the state sector of the economy / 繁荣～ promote economic prosperity ②(对国民经济有利的) economic; of industrial or economic value: ～植物 economic plants / ～作物 industrial crops; cash crops ③(个人生活用度) financial condition; income: ～负担 financial burden / ～拮据 be hard up / ～宽裕 well-off; well-to-do / ～困难 be in financial difficulties 〈straits〉 ④(节省的) economical; thrifty: 不～ costing too much; uneconomical / ～地使用 use economically / ～实惠 economical and practical ◇ ～崩溃

economic collapse / ～部门 branches of the economy; economic departments / ～成分 sector of the economy; economic sector / ～成效 economic effects / ～地理学 economic geography / ～地位 economic status; economic position / ～地质学 economic geology / ～发展 economic development / ～封锁 economic blockade / ～改革 economic reform / ～杠杆 economic levers / ～合作 economic cooperation / ～核算 economic accounting; business accounting / ～核算单位 business accounting unit / ～基础 economic base; economic basis / ～计划 economic planning / ～结构 economic structure / ～恐慌 economic depression / ～林 economic forest / ～命脉 economic lifeline; economic arteries; key branches of the economy / ～赔偿 economic compensation; financial reimbursement / ～平衡 economic equilibrium / ～失调 dislocation of the economy / ～衰退 recession / ～损失 pecuniary loss / ～特区 special economic zone / ～体系 economic system / ～萎缩 economic contraction / ～危机 economic crisis / ～学 economics / ～学家 economist / ～一体化 economic integration / ～预测 economic forecasting / ～援助 economic aid / ～杂交 [牧] commercial crossbreeding / ～增长 economic growth / ～指标 economic norms / ～制裁 economic sanctions

【经济体制】economic system: 加快～改革 speed up the restructuring of economic systems

【经济效益】economic benefits; economic results; economic effect: 讲究～ take into consideration the economic benefits / 以最少的消耗取得最大的～ obtain the maximum economic results at minimum costs

【经济主义】economism

【经久】①(经过很长时间) prolonged: ～不息的掌声 prolonged applause; prolonged clapping ②(耐久) durable: ～耐用 durable; able to stand wear and tear

【经理】①(经营管理) handle; manage ②(企业负责人) manager; director

【经历】①(亲身体验) go through; undergo; experience: ～长期的磨炼 undergo a long process of tempering / ～两个阶段 go through two stages ②(经历之事) experience: 共同的～ a common experience / 他这人～多,见识广。He is a man of wide knowledge and experience.

【经年累月】for years; year in year out; for months and years on end

【经期】(menstrual) period

【经纱】[纺] warp; end

【经商】engage in trade; be in business; go into business

【经手】handle; deal with ◇ ～费 brokerage / ～人 person handling a transaction, particular job, etc.

【经受】undergo; experience; withstand; stand; weather: ～各种考验 experience all sorts of trials; stand up to all tests; withstand all trials and tribulations / 在斗争中～锻炼 be tempered in the struggle

【经售】sell on commission; deal in; distribute; sell

【经纬度】longitude and latitude

【经纬仪】theodolite; transit ◇ ～测量 transit survey

【经线】①[纺] warp ②[地] meridian (line)

【经销】sell on commission; deal in; distribute; sell ◇ ～处 agency

【经心】careful; mindful; conscientious: 漫不～ careless; casual; negligent

【经验】①(由实践得来的知识或技能) experience: ～不足 lack experience; not be sufficiently experienced / ～丰富 have rich experience; be very experienced / ～之谈 remark made by one who has had experience; the wise remark of an experienced person / 间接～ indirect experience / 交流～ exchange experience / 介绍～ pass on one's experience / 直接～ direct experience / 总结～ sum up experience ②(经历) go through; experience

【经验主义】empiricism ◇ ～者 empiricist

【经一事,长一智】By every experience a person increases his knowledge

【经营】manage; run; engage in: 发展多种～ promote a diversified economy / 改善～管理 improve management and administration / 苦心～ take great pains to build up (an enterprise, etc.) ◇ ～比率 operating ratio / ～成本 operating cost / ～周期 period of operating cycle / ～资金 floating capital

【经由】via; by way of: ～西安去重庆 be bound for Chongqing via Xi'an

【经院哲学】scholasticism

【经轴】[纺] warp beam

【经传】①(经典或古人的解释) Confucian classics and commentaries on them; Confucian canon ②(重要的古书) classical works; classics: 名不见～ not well-known; a mere nobody

jing

井 ①well: 打～ sink a well; drill a well ②(形状象井的) well; sth. in the shape of a well: 风～ air shaft / 矿～ pit; mine / 排水～ pumping shaft / 油～ oil well ③(整齐) neat; orderly
【井场】[石油] well site
【井底之蛙】a frog in a well; a person with a very limited outlook
【井灌】[水] well irrigation
【井架】①[石油] derrick: 轻便～ portable derrick ②[矿] headframe; headgear; pitheadframe ③(水井架) well head
【井井有条】in perfect order; shipshape; methodical; well-arranged
【井口】①(水井口) the mouth of a well ②[矿] pithead ③[石油] wellhead
【井喷】[石油] blowout
【井然】[书] orderly; neat and tidy; shipshape; methodical: 秩序～ in good order
【井水不犯河水】well water does not intrude into river water; I'll mind my own business, you mind yours; none may encroach upon the precincts of another
【井筒】[矿] pit shaft
【井下】in the pit; under the shaft: ～作业 operation in the pit; underpit operation
【井斜】[石油] well deflection; well deviation
【井盐】well salt

肼 [化] hydrazine

阱 trap; pitfall; pit

警 ①(戒备) alert; vigilant: ～醒 be a light sleeper ②(使人注意) warn; alarm: ～告 warn ③(危险紧急的情况) alarm: 火～ fire alarm ④(警察) police: ～亭 police box
【警报】alarm; warning; alert: 发～ sound the alarm; sound the siren / 解除空袭～ all clear / 空袭～ air-raid alarm; air-raid warning / 台风～ a typhoon warning / 战斗～ combat alert ◇ ～器 siren; alarm / ～系统 warning system / ～信号 alarm signal
【警备】guard; garrison ◇ ～区 garrison command / ～司令 garrison commander / ～司令部 garrison headquarters
【警察】police; policeman: 便衣～ plainclothes policeman / 女～ policewoman / 人民～ a people's policeman ◇ ～分局 police station / ～局 police office
【警车】patrol wagon
【警笛】①(警哨) police whistle ②(警报汽笛) siren
【警方调查】police investigation
【警告】①(告诫) warn; caution; admonish: 提出严重～ issue a serious warning to ②(处分) warning: ～处分 punishment of warning / 给予～处分 give sb. a disciplinary warning
【警告机制】[动] aposematic mechanism
【警告信号】[交] warning signal
【警棍】baton; truncheon
【警戒】①(告诫) warn; admonish ②[军] be on the alert against; guard against; keep a close watch on: ～森严 the caution is stern / 采取～措施 take precautionary measures; take deterrent measures ◇ ～部队 outpost troops; security force; security detachment / ～地带 outpost area / ～色 [动] warning coloration; aposematic coloration / ～水位 warning line; warning water level; warning stage / ～艇 guard boat / ～线 cordon; security line / ～状态 state of alert
【警句】aphorism; epigram
【警觉】vigilance; alertness: 引起～ arouse vigilance ◇ ～性 alertness
【警铃】alarm bell
【警犬】police dog; patrol dog
【警惕】be on guard against; watch out for; be vigilant; be on the alert: 保持高度～ maintain sharp vigilance / 放松～ relax vigilance / 丧失～ drop one's guard; be off one's guard / 提高～ heighten one's vigilance ◇ ～性 vigilance
【警卫】(security) guard: 担负～任务 be on guard duty ◇ ～室 guardroom / ～团 guards regiment / ～员 bodyguard
【警钟】alarm bell; tocsin: 敲～ sound the alarm for

儆 warn; admonish: 惩一～百 punish one to warn a hundred; make an example of sb. / 以～效尤 so as to deter anyone from committing the same mistake

景 ①(景致;风景) view; scenery; scene: 外～ exterior view / 雪～ a snow scene / 夜～ night view ②(景况) situation; condition: 好～不长 Good times do not last long ③[戏] scenery; scene: 第三幕第二～ Act III, scene 2 / 换～ change of scenery ④(尊敬;佩服) admire; revere; respect: ～慕 esteem; revere
【景观】[地] landscape: 岩溶～ karst landscape / 自然～ natural landscape

【景况】situation; circumstances: ～不佳 circumstances are no good / ～丰裕 be in affluent circumstance
【景片】a piece of (stage) scenery; flat
【景气】prosperity; boom: 不～ depression; slump
【景色】scenery; view; scene; landscape: ～宜人 attractive scenery / 南方～ southern landscape / 深秋～ a late autumn scene / 自然～ natural scenery
【景深】[摄] depth of field
【景泰蓝】cloisonne enamel; cloisonne
【景天】[植] red-spotted stonecrop
【景物】scenery: ～依然 the view is just as before / ～宜人 delightful scenery
【景象】scene; sight; picture: 呈现一派繁荣～ present a scene of prosperity
【景仰】respect and admire; hold in deep respect: ～大名 look up to your great name
【景遇】[书] circumstances; one's lot
【景致】view; scenery; scene: 西湖的优美～ a fine view of the West Lake

颈 neck

【颈动脉】arteria carotis
【颈项】neck
【颈椎】[生理] cervical vertebra

jìng

竟 ①(完毕) finish; complete: 未～之业 unaccomplished cause; unfinished task ②(从头到尾) throughout; whole: ～日 throughout the day; all day long / ～夜 the whole night; throughout the night ③(终于) in the end; eventually: 有志者事～成。 Where there's a way. ④(竟然; 出于意料之外) go so far as to; unexpectedly; actually

【竟敢】have the audacity; have the impertinence; dare: 你～讲这种话! How dare you say such a thing!
【竟然】[副] ①(表示有点出乎意外) to one's surprise; unexpectedly; actually: 这任务～在一周内就完成了。 To my surprise, the task was finished in only one week. ②(居然) go so far as to; go to the length of; have the impudence 〈effrontery〉 to: ～不顾事实 go so far as to disregard the facts

境 ①(疆界) border; boundary: 国～ national boundary / 迁移出～ emigrate; emigration / 驱逐出～ deport / 移居入～ immigrate; immigration / 越～ cross the border illegally / 在本省～内 within the boundaries of this province ②(地方) place; area; territory; land: 如入无人之～ like entering an unpeopled land; meeting no resistance ③(境况) condition; situation; circumstances: 困～ difficult position; predicament / 在逆～中 in adverse circumstances / 在顺～中 in favorable circumstances

【境地】condition; circumstances: 处于贫困的～ be in straitened circumstances
【境界】①(地界) boundary ②(情况;程度) extent reached; plane attained; state; realm: 崇高的思想～ realm of lofty thought / 理想～ ideal state; ideal
【境况】condition; circumstances: ～不佳 in straitened circumstances
【境域】①(境地) condition; circumstances ②(境界) area; realm: 大同～ the realm of Great Harmony
【境遇】circumstances; one's lot: 悲惨～ miserable lot / 极困难的～ extremely adverse circumstances

镜 ①(镜子) looking glass; mirror: 照～子 look at oneself in the glass / 湖平如～。 The lake is as smooth as a mirror. ②(眼镜或其他光学用器具) lens; glass: 放大～ magnifying glass; magnifier / 墨～ sunglasses

【镜花水月】flowers in a mirror or the moon in the water; an illusion
【镜框】①(装相片等的框子) picture frame ②(眼镜架) spectacles frame
【镜片】lens
【镜台】dressing table
【镜头】①(光学装置) camera lens: 广角～ wide-angle lens / 可变焦距～ zoom lens / 远摄～ telephoto lens ②(画面) shot; scene: 电影～ cinema scene / 特技～ special effect shot; trick shot / 特写～ close-up ◇ ～遮光罩 lens hood
【镜匣】dressing-case; toilet-box
【镜象】[物] mirror image
【镜子】①(各种面镜) mirror; looking glass ②(眼镜) glasses; spectacles

竞 compete; contest; vie

【竞渡】①(划船比赛) boat race ②(游泳比赛) swimming race
【竞技】sports; athletics ◇ ～场 arena
【竞技状态】form (of an athlete): ～不好 not in good form; out of form; off one's game / ～好 in good form; in top form

【竞赛】contest; competition; emulation; race: 军备～ arms race; armament race / 体育～ athletic contest; athletic competition ◇ ～规则 rules of a contest; rules of a competition / ～艇 wager boat

【竞选】enter into an election contest; campaign for (office); stand for; run for: ～总统 run for the presidency ◇ ～纲领 election program / ～活动 electioneering / ～伙伴 running mate / ～演说 stumping speech; campaign speech / ～运动 election campaign

【竞争】compete; vie; contend: 公开～ open competition / 激烈的～ keen competition; fierce rivalry / 自由～ free competition / 在价格上经得起～的商品 goods of competitive price ◇ ～地带 zone of competition / ～机制 competition / ～价格 competitive price / ～商品 rival commodities / ～市场 competitive market / ～投资 competitive investment / ～性 competitiveness

【竞走】heel-and-toe walking race

净 ①(洁净) clean: ～水 clean water / 擦～ wipe sth. clean ②(没有剩余) completely; up: 吃～ eat up / 烧～ burn up / 用～ use up ③(纯) net: ～出口 net export / ～进口 net import / ～利益 net advantage / ～收入 net income ④(没有别的;只) only; merely; nothing but

【净吨位】net tonnage

【净额】net amount

【净高】[建] clear height

【净荷载】net load

【净化】purify: 水的～ purification of water ◇ ～塔 purifying column

【净尽】completely; utterly: 消灭～ utterly annihilate

【净空】[建] headroom

【净跨】[建] clear span

【净宽】[建] clear width

【净亏】net loss; net deficiency

【净利】net profit

【净流】net flow

【净水厂】water treatment plant

【净土】[佛教] Sukhavati; Pure Land; Paradisee of the West

【净销价法】net selling price method

【净盈余说】all inclusive theory; clean surplus theory

【净余】remainder; surplus

【净值】net worth; net value: 出口～ net export value / 进口～ net import value

【净重】net weight: ◇ ～条件 net weight term

【净赚】net earnings

静 ①(安定不动) still; calm; motionless: ～立 stand still / ～卧 lie motionless / 风平浪～ calm and tranquil / 夜深人～ in the still of the night; at the dead of night ②(没有声响) silent; quiet: 请～一～。Please be quiet.

【静电】[物] static electricity ◇ ～除尘器 electrostatic precipitator / ～打印 static dump / ～纺纱 electrostatic spinning / ～感应 electrostatic induction / ～荷 electrostatic charge / ～计 electrometer / ～加速器 electrostatic accelerator / ～扫描 electrostatic scanning / ～学 electrostatics / ～印刷 xerography

【静观情势】take sounding

【静荷载】[建] dead load

【静候调遣】quietly awaiting for transference of post

【静力学】[物] statics: 气体～ aerostatics

【静脉】[生理] vein ◇ ～滴注法 intravenous drip / ～瘤 phlebangioma; venous aneurysm / ～曲张[医] varix; varicosity / ～输血法 venous transfusion / ～输液法 phleboclysis; venous transfusion / ～炎 phlebitis / ～注射 intravenous injection

【静谧】[书] quiet; still; tranquil

【静摩擦力】stiction

【静默】①(不出声) become silent: ～无声 absolute quietness ②(肃立表示悼念) mourn in silence; observe silence: ～致哀三分钟 observe three minutes' silence / 为…～致哀 mourn in silence for

【静穆】solemn and quiet

【静悄悄】very quiet

【静态】[物] static state ◇ ～电阻 static resistance / ～平衡 static equilibrium / ～特性 static characteristic

【静听】listen quietly

【静物】still life ◇ ～画 still life / ～写生 paint still life

【静养】rest quietly to recuperate; convalesce; have a rest-cure; have a good rest to recuperate

【静止】static; motionless; at a standstill: 不要～地孤立地看待事物。Don't view things as static and isolated.

【静坐】①(疗养办法) sit quietly ②(示威) sit still as a form of therapy ◇ ～罢工 sit-down (strike) / ～示威 sit-in (demonstration); sit-down (protest)

靖 ①(平安) peace; tranquillity ②(使秩序安定) pacify: ～边 pacify the border regions / ～乱 put down a rebellion

敬 ①(尊敬) respect; honor: 致～ pay one's respects; salute / 尊～ respect; esteem; honor ②(恭敬) respectfully: ～请斧正 will you be kind enough to prune / ～请光临 request the honor or your presence; May I have the honor of your presence at...～请指教 humbly request your advice ③(有礼貌地送上) serve; offer politely: ～茶 serve tea; offer a cup of tea / ～烟 offer a cigarette

【敬爱】respect and love: ～的 esteemed and beloved; respected and beloved

【敬辞】term of respect; polite expression

【敬而远之】stay at a respectful distance from sb.; keep sb. at a respectful distance

【敬奉】①(虔诚地供奉) piously worship ②(敬献) offer respectfully; present politely

【敬鬼神而远之】keep a person at a distance

【敬贺】congratulate sb. on sth.; offer congratulations

【敬酒】propose a toast; toast

【敬酒不吃吃罚酒】refuse a toast only to drink a forfeit submit to sb.'s pressure after first turning down his request; be constrained to do what one at first declined

【敬礼】①(行礼) salute; give a salute ②(致敬意) extend one's greetings ③[敬](用于书信结尾)此致～ with high respect; with best wishes

【敬佩】esteem; admire; express admiration for; have a great esteem for

【敬上】(用于书信结尾) truly yours; yours truly

【敬挽】(用于挽联、花圈等的落款) with deep condolences from sb.

【敬畏】hold in awe and veneration; revere

【敬仰】revere; venerate; admire; look up to

【敬意】respect; tribute; regards: 表示衷心的～ extend one's heartfelt respects; pay sincere tribute

【敬赠】respectfully presented by

【敬重】highly esteem; look up to with great respect; deeply respect; revere; honor: 我们都很～这位老教师。 We all have great respect for the old teacher.

痉

【痉挛】convulsion; spasm; cramp: 食管～ spasm of the esophagus; esophagospasm / 胃～ spasm of the stomach

径 ①(小路) footpath; path; track: 曲～ a winding path / 山～ mountain path ②(达到目的的办法) way; means: 捷～ an easy way; shortcut ③(径直) directly; straightaway: ～回上海 go straight back to Shanghai / ～行办理 deal with the matter straightaway; handle sth. straight ④(直径) diameter: 半～ radius

【径流】[水] runoff: 地表～ surface runoff / 地下～ groundwater runoff

【径情直遂】as smoothly as one would wish

【径赛】[体] track ◇ ～项目 track events

【径庭】[书] very unlike: 大相～ entirely different; poles apart

【径向】[物] radial ◇ ～间隙 radial clearance / ～轴承 radial bearing

【径直】[副] straight; directly; straightaway: 客机～飞往南京。 The liner flew straight 〈directly〉 to Nanjing.

【径自】[副] without leave; without consulting anyone: 她没打招呼，～走了。 She turned away without leave.

胫 [生理] shin

【胫腓骨】tibiofibula

【胫骨】shin bone; tibia

【胫骨肌】tibialis

劲 strong; powerful; sturdy: ～松 sturdy pines

【劲敌】①(指敌人) powerful enemy ②(指对手) formidable rival; strong opponent

【劲旅】strong contingent; crack force

经 [纺] warping

jiǒng

窘 ①(穷困) in straitened circumstances; hard up; poorly off; in financial straits ②(为难) awkward embarrassed; ill at ease: 露出～态 show signs of embarrassment ③(使为难) embarrass; disconcert

【窘境】awkward situation; plight; predicament: 处于～ be landed in an awkward predicament

【窘口无言】distressed mouth said nothing

【窘困潦倒】be miserably poor and greatly disappointed

【窘迫】①(非常困难) poverty-stricken; very poor: 生活～ live in poverty ②(十分为难)

hard pressed; embarrassed; in a predicament: 处境～ find oneself in a predicament

【窘态】embarrassed look: 露出一副～ show embarrassment; look embarrassed

炯 bright; shining

【炯炯】[书] (of eyes) bright; shining: 两眼～有神 have a pair of bright piercing eyes / 目光～ sparkling eyes; piercing eyes

迥 [书] widely different

【迥然】far apart; widely different: ～不同 utterly different; not in the least alike; diametrically different

jiū

阄 lot: 抓～ draw lots / 抓～决定 decide by lot

揪 ①(紧紧地抓) hold tight; seize: ～住不放 hold in a tight grip / ～住一个小偷 grab a thief ②(抓住并拉) pull; tug; drag; give a hard tug: 别那么使劲～绳子。Don't pull so hard at the rope.

【揪辫子】seize sb.'s queue; seize upon sb.'s mistakes or shortcomings

【揪出】uncover; ferret out

【揪痧】[中医] pinching the patient's neck, etc. to achieve congestion

【揪心】[方] ①(放不下心) anxious; worried ②(疼痛难忍) heartrending; agonizing; gnawing: 伤口疼得～。 There was a gnawing pain from the wound.

究 ①(仔细推求;追查) study careful; go into; investigate: 深～ go deeply into a matter; get to the bottom of a matter / ～其根源 trace sth. to its source; get to the bottom of sth.; probe to the roots of sth. / ～其真相 get to the bottom of a matter ②(用在问句中,追究) actually; really; after all: ～系何因,尚待深查。 The actual cause awaits further investigation. / ～应如何办理? How should this really be dealt with?

【究办】investigate and deal with: 依法～ investigate and deal with according to law

【究竟】①(结果;原委) outcome; what actually happened: 我们都想知道个～。 We all want to know what actually happened. ②(用于问句,表示追究) actually; exactly: 你们～要什么? What exactly do you want? ③(毕竟;到底) after all; in the end: 他～是生手。 He is new to the job after all.

鸠 [动] turtledove

赳

【赳赳】valiant; gallant: 雄～ valiant; gallant / ～武夫 a soldier of dauntless courage

纠 ①(缠绕) entangle: ～缠 get entangled; get bogged down ②(集合) gather together: ～集一伙流氓 get together a bunch of hoodlums ③(纠正) correct; rectify; put right: ～偏 rectify a deviation / 有错必～ mistakes must be corrected whenever discovered

【纠察】①(维持秩序) maintain order at a public gathering ②(维持秩序者) picket ◇ ～队 pickets / ～线 picket line

【纠缠】①(绕在一起) get entangled; be in a tangle: ～不清 too tangled up to unravel ②(捣麻烦) nag; worry; pester: 他还有事,别～他。He is busy. Stop nagging at him.

【纠纷】dispute; issue: 调解～ mediate an issue / 挑起～ stir up troubles / 无原则～ an unprincipled dispute

【纠葛】entanglement; dispute; complication

【纠集】[贬] get together; muster; draw together: ～多数 muster a majority / ～残部 muster the remaining forces

【纠结】be entangled with; be intertwined with

【纠偏】rectify a deviation; correct an error

【纠正】correct; put right; redress; rectify; set right: ～不正之风 check unhealthy tendencies; amend unwholesome ways / ～错误 correct a mistake; redress an error / ～冤案 redress a wrong; correct a wrong; remedy an injustice / ～姿势 correct sb.'s posture

jiǔ

酒 alcoholic drink; wine; liquor; spirits: 白兰地～ brandy / 白葡萄～ white wine / 不甜～ dry sweet wine / 伏特卡～ vodka / 红葡萄～ red wine / 黄～ yellow wine; rice wine / 鸡尾～ cocktail / 金～ gin; dry gin / 莲花白～ lotus white wine / 烈性～ liquor; spirit / 罗木～ rum / 啤～ beer / 葡萄～ wine / 汽～ bubbling wine / 甜～ sweet wine / 威士忌～ whisky / 味美思～ vermouth / 香槟～ champagne / 雪利～ sherry

【酒吧间】bar; barroom

【酒杯】wineglass; wine bowl

【酒不醉人人自醉】It's not the wine that intoxicates but the drinker who gets himself drunk.

【酒菜】food and drink; food to go with wine or liquor
【酒厂】brewery; winery; distillery
【酒店】wineshop; public house
【酒逢知己千杯少】When drinking with a bosom friend, a thousand cups will still be too little.
【酒馆】public house; tavern
【酒鬼】wine bibber; tippler
【酒酣耳热】warmed with wine; mellow with drink
【酒后失态】act ludicrously when drunk
【酒后失言】say something wrong when drunk
【酒后失仪】After drinking one is wanting politeness
【酒后吐真言】In wine there is truth. (或: When wine is in truth is out.)
【酒壶】wine pot; flagon
【酒花】[植]hops
【酒会】cocktail party
【酒家】wineshop; restaurant
【酒窖】wine cellar
【酒精】ethyl alcohol; alcohol ◇ ～比重计 alcoholimeter; spirit gauge / ～灯 spirit lamp; alcohol burner / ～炉 alcohol heater / ～温度计 spirit thermometer; alcohol thermometer / ～中毒 alcoholism
【酒量】capacity for liquor; one's drinking capacity
【酒囊饭袋】wine skin and rice bag; a good-for-nothing; useless person good only for feasting and drinking
【酒酿】fermented glutinous rice
【酒曲】distiller's yeast
【酒肉朋友】wine-and -meat friends; fair-weather friends
【酒石酸】tartaric acid ◇ ～锑钾[药] antimony potassium tartrate
【酒色财气】wine, women, avarice and pride; the four cardinal vices
【酒徒】wine bibber
【酒窝】dimple
【酒席】feast: 大摆～ entertain guest to a sumptuous banquet
【酒药】yeast for brewing rice wine or fermenting glutinous rice
【酒意】a tipsy feeling: 已有几分～ be slightly tipsy; be mellow
【酒糟】distiller's grains
【酒糟鼻】acne rosacea; brandy nose
【酒盅】a small handleless wine cup
【酒醉】drunkenness

韭 fragrant-flowered garlic; (Chinese) chives: 青～ young chives; chive seedlings
【韭菜】fragrant- flowered garlic; (Chinese) chives
【韭黄】hotbed chives

九 ①(数目) nine: 第～ the ninth / ～成新 ninety per cen new ②(从冬至起九天为一个九) each of the nine-day periods after the Winter Solstice: 三～ the third nine-day period after the Winter Solstice; the coldest days of winter ③(多次;多数) many; numerous: ～曲桥 a zigzag bridge
【九级风】[气] force 9 wind; strong gale
【九节狸】[动] zibet; large Indian civet
【九九表】multiplication table
【九九归一】when all is said and done; in the last analysis; after all; in the final analysis
【九牛二虎之力】the strength of nine bulls and two tigers; tremendous effort
【九牛一毛】a single hair out of nine ox hides; a drop in the ocean
【九品】the nine grades of rank in the feudal regimes
【九泉】[书] grave; the nether world: ～之下 in the nether regions; after death
【九死一生】a narrow escape from death; survival after many hazards
【九天】the Ninth Heaven; the highest of heavens
【九霄云外】beyond the highest heavens; beyond the nine clouds; far, far away
【九月】①(阳历) September ②(阴历) the ninth month of the lunar year; the ninth moon
【九折】ten per cent discount

久 ①(时间长) for a long time; long: 很～以前 long ago ②(时间的长短) of a specified duration: 三周之～ for as long as three weeks / 你来了有多～? How long have you been here?
【久别重逢】meet again after a long separation
【久病成良医】Prolonged illness makes the patient a good doctor
【久而久之】in the course of time; as time passes; in the long run
【久旱逢甘雨】have a welcome rain after a long drought; have a long-felt need satisfied
【久候不至】didn't turn up after a long waiting
【久假不归】put off indefinitely returning sth.

one has borrowed; appropriate sth. borrowed
【久经锻炼】well-steeled; long-tested
【久经风霜】have experienced all sorts of hardships
【久经考验】a long-tested; seasoned
【久久】for a long, long time
【久违】[套] how long it is since we last met; I haven't seen you for ages
【久闻大名】I've long heard of your great name.
【久仰】[套] I've long been looking forward to meeting you; I'm very pleased to meet you; I have long desired to know your
【久远】far back; ages ago; remote: 年代～ of the remote past; age-old; time-honored

灸 [中医] moxibustion; moxa treatment

jiù

就 ①(凑近) come near; move towards ②(开始从事) go to; take up; undertake; engage in; enter upon: ～席 take one's seat; be seated at the table / ～学 go to school ③(完成) accomplish; make: 功成业～ (of a person's career) be crowned with success ④(将就) accommodate oneself to; suit; fit: ～便 at sb.'s convenience / ～你的时间吧。Make it anytime that suits you. ⑤(搭着吃喝) go with: 花生仁～酒 have peanuts to go with liquor ⑥(按照) with regard to; concerning: on; in the light of; as far as: ～共同关心的问题进行会谈 hold talks on questions of common interest / ～目前情况看来 in the light of present situation / ～我所知 so far as I know ⑦(表示在很短时间内) at once; right away; in a moment: 我这～来。I'm coming right away. ⑧(表示发生或结束得早) as early as; already: 大风早晨～停了。The wind has already subsided in the morning. ⑨(表示数量大、能力强等) as much as; as many as: 一个月～节约了十吨煤 save as much as ten tons of coal in onc month ⑩(表示在时间上紧接着) as soon as; no sooner... than; right after: 说干～干 act without delay ⑪(仅仅) only; merely; just: ～等你一个了。You're the only one we're waiting for. ⑫(表示强调) exactly; precisely: 我要的～是这一个。What I want is exactly this one. ⑬(就是,即使) even if; even: 你～送来,我也不要。Even if you bring it to me, I won't take it.
【就伴】accompany sb. (on a journey); travel together
【就便】at sb.'s convenience; while you're at it
【就此】at this point; here and now; thus: 讨论～结束。The discussion was thus brought to a close.
【就地】on the spot: ～解决 settle〈solve〉a problem right on the spot / ～枪决 shoot on the spot / ～取材 use locally available materials; make use of indigenous materials; obtain raw materials on the spot / ～正法 carry out the execution on the spot ◇ ～采购[军] local procurement / ～审判 on-the-spot trial / ～视察[军] on-site -inspection / ～调解 on-the-spot mediation / ～组装 site-assembly
【就范】submit; give in: 不肯～ refuse to submit to control; refuse to give in / 迫使～ compel sb. to submit
【就近】(do or get sth.) nearby; in the neighborhood; without having to go far: ～找个住处 find accommodation in the neighborhood
【就寝】[书] retire for the night; go to bed
【就让】even if: ～她来,也晚了。Even if she comes it will be too late.
【就任】take up one's post; take office
【就势】making use of momentum
【就事论事】consider sth. as it stands; take the matter as it stands; confine oneself merely to facts as they are
【就是】①(表示同意) quite right; exactly; precisely ②(即使) even if; even: ～小学生也知道这一点。Even school children know this.
【就是说】that is to say; in other words; namely
【就手】while you're at it: ～把门关上。Close the door behind you.
【就算】[口] even if; granted that: ～有困难,也不会太大。There won't be much difficulty, if any.
【就位】take one's place
【就绪】be in order; be ready: 一切布置～。Everything is arranged〈ready〉.
【就要】be about to; be going to; be on the point of: 飞机～起飞了。The plane is about to take off.
【就业】obtain employment; take up an occupation; get a job: 充分～ full employment / ～不足 underemployment ◇ ～登记 employment registration
【就医】seek medical advice; go to a doctor
【就义】be executed for championing a just cause; die a martyr: 英勇～ face execution bravely; die a hero's death
【就正】solicit comments (on one's writing : ～

于读者 request 〈invite〉 the readers to offer their criticisms
【就职】assume office: 宣誓～ take the oath of office; be sworn in ◇ ～典礼 inaugural ceremony; inauguration / ～演说 inaugural speech
【就座】take one's seat; be seated: 在主席台前列～的有… seated in the front row on the rostrum were
鹫 vulture
厩 stable; cattle-shed; pen
【厩肥】[农] barnyard manure
救 ①(使脱离灾难或危险) rescue; save; salvage: 呼～ call out for help; send out SOS signals / 病人得～了。 The patient was saved. ②(援助) help; relieve; succor: 生产自～ tide over a disaster by production
【救兵】relief troops; reinforcements
【救国】save the nation
【救护】relieve a sick or injured person; give first-aid; rescue: ～伤员 give first-aid to the wounded ◇ ～车 ambulance / ～船 ambulance ship / ～队 ambulance corps / ～飞机 ambulance aircraft / ～所 medical aid station / ～条款 suing and laboring clause / ～站 first-aid station
【救荒】send relief to a famine area; help to tide over a crop failure
【救活】bring sb. back to life
【救火】fire fighting ◇ ～车 fire engine; fire truck; fire brigade wagon / ～队 fire brigade / ～队员 fireman; fire fighter
【救急】help sb. to cope with an emergency; help meet an urgent need: ～不救穷。 One may give financial aid to others in an emergency but should not do so if they are perennially in need of money.
【救济】relieve; succor: ～灾区人民 provide relief to the people in a disaster area / 社会～事业 social relief facilities ◇ ～金 relief payment; relief fund / ～粮 relief grain; relief food / ～品 relief
【救苦救难】help the needy and relieve the distressed
【救命】①(救人性命) save sb.'s life ②(呼救声) Help! ◇ ～稻草 a straw to clutch at / ～恩人 savior
【救人一命胜造七级浮屠】Better save one life than build a sevenstory pagoda.
【救生】lifesaving ◇ ～带 life belt / ～筏 life raft / ～圈 life buoy / ～设备 lifesaving appliance; life preserver / ～艇 lifeboat / ～衣 life jacket / ～员 lifeguard; lifesaver
【救世主】[基督教] the Savior; the Redeemer
【救死扶伤】heal the wounded and rescue the dying
【救亡】save the nation from extinction: ～图存 save the nation from subjugation and ensure its survival; save one's country so that it may survive ◇ ～运动 national salvation movement
【救险车】wrecking truck; wrecking car
【救星】liberator; emancipator; savior; deliverer
【救援】rescue; come to sb.'s help ◇ ～车 rescue car
【救灾】provide disaster relief; send relief to a disaster area; help the people tide over a natural disaster; relieve the victims of a disaster
【救治】bring a patient out of danger; treat and cure: ～伤病员 give treatment to the sick and wounded
【救助】help sb. in danger or difficulty; succor ◇ ～费用 salvage charges / ～基金 relief fund
旧 ①(过去的;过时的) past; bygone; old: ～的传统观念 outdated conventional ideas / ～风俗 old customs / ～社会 the old society / ～时代 past ages / ～思想 old way of thinking; timeworn ideas / ～习惯 old habits / ～杂志 outdated magazine ②(用过的) used; worn; old; secondhand: ～家具 timeworn furniture / ～衣服 used clothes / 买～的 buy sth. secondhand ③(以前的) former; onetime: ～址 former ④(老交情;老朋友) old friendship; old friend: 故～ old acquaintances
【旧案】①(过去的案件) a court case of long standing ②(过去的条例或事例) old regulations; former practice
【旧病复发】have a recurrence of an old illness; have a relapse
【旧地重游】revisit a once familiar place; revisit a place
【旧调重弹】harp on the same old tune; repeat the shop-worn stuff
【旧都】former capital
【旧恶】old grievance; old wrong: 不念～ forgive an old wrong
【旧观】former appearance; old look: 恢复～ be restored to original form / 迥非～ entirely different from what it used to be
【旧恨新仇】new hatred piled on old; all the old and recent sorrows

【旧货】secondhand goods; junk ◇～店 secondhand shop; junk shop／～市场 flea market
【旧交】old acquaintance
【旧居】former residence; old home
【旧框框】convention
【旧历】the old Chinese calendar; the lunar calendar
【旧梦重温】renew a sweet experience of bygone days
【旧民主主义革命】democratic revolution of the old type
【旧瓶装新酒】new wine in an old bottle; new concepts in an old framework
【旧日】former days; old days
【旧诗】old-style poetry; classical poetry
【旧石器时代】the Old Stone Age; the Paleolithic Period
【旧时】old times; old days
【旧式】old type: ～文人 old-type scholars ◇～婚姻 customary marriage
【旧事重提】repetition of the old tale; bring up a matter of the past
【旧书】①(破旧的书) secondhand book; used book; old book ②(古书) books by ancient writers ◇～店 secondhand bookstore
【旧约】[基督教] the Old Testament
【旧址】former site: 农会的～ the site of the former peasant association

臼 ①(石制的舂米器具) mortar: 石～ stone mortar ②(形状象臼的) any mortarshaped thing ③(关节) joint (of bones): 脱～ dislocation (of joints)
【臼齿】molar; molar tooth

桕 [植] Chinese tallow tree

舅 ①(舅父) mother's brother; uncle ②(妻子的弟兄) wife's brother; brother-in-law
【舅父】mother's brother; uncle
【舅母】wife of mother's brother; aunt
【舅子】[口] wife's brother; brother-in-law

疚 [书] remorse: 感到内～ have a guilty conscience

柩 a coffin with a corpse in it
【柩车】hearse

咎 ①(过失) fault; blame: 归～于人 lay the blame on sb. else ②(责备) censure; punish; blame: 既往不～ forgive sb.'s past misdeeds; let bygones be bygones
【咎由自取】have only oneself to blame; that serves him right; he asked for it; deserving reproof

jū

车 ①(中国象棋的) chariot ②(国际象棋的) castle; rook

疽 [中医] subcutaneous ulcer; deep-rooted ulcer

狙
【狙击】snipe ◇～手 sniper／～战 sniping action

鞠 rear; bring up: ～养 bring up
【鞠躬】bow: ～致谢 bow one's thanks／深深地鞠一个躬 make a deep bow; bow low
【鞠躬尽瘁】bend oneself to a task and exert oneself to the utmost; spare no effort in the performance of one's duty; fag oneself to death
【鞠躬尽瘁,死而后已】bend one's back to the task until one's dying day; give one's all till one's heart stops beating; have dedicated one's life to a cause

掬 hold with both hands: 憨态可～ charmingly naive／笑容可～ radiant with smiles／以手～水 scoop up some water with one's hands

拘 ①(逮捕或拘留) arrest; detain ②(限制) restrain; restrict; limit; constrain: 长短不～ with no limit on the length／大小不～ regardless of size／无～无束 unconstrained; free and easy ③(不变通) inflexible: ～泥 be a stickler for (form, etc.); rigidly adhere to (formalities, etc.)
【拘捕】arrest; capture ◇～权 power of arrest
【拘谨】overcautious; reserved
【拘禁】take into custody; put under arrest
【拘礼】be punctilious; stand on ceremony: 熟不～ be too familiar with each other to stand on ceremony
【拘留】detain; hold in custody; intern; detention; provisional apprehension ◇～查讯 detention for questioning／～处罚 punishment of detention／～方式 form of detention／～期间 duration of detention／～所 bridewell; house of detention; lockup

【拘泥】be a stickler for (form, etc.); rigidly adhere to (formalities, etc.): ～于细节 be very punctilious; scrupulous about minor details; stand on points / ～于小节 be tied down by trifles or petty conventions / ～于形式 rigidly adhere to form; be formalistic
【拘票】arrest warrant; warrant
【拘守绳墨】stick to the rules
【拘束】①(限制) restrain; restrict: 不要～她的正当活动。 Don't restrict her proper activities. ②(过分约束自己) constrained; awkward; ill at ease: 感到～ feel ill at ease / 显得～ look ill at ease
【拘押】take into custody
驹 ①(少壮的马) colt ②(驹子) foal: 怀～ be in foal; be with foal
【驹子】foal
居 ①(住) reside; dwell; live ②(住所) residence; house: 故～ former residence / 迁～ move house; change one's residence ③(处于; 在) be (in a certain position); occupy (a place): ～世界首位 occupy first place in the world; rank first in the world ～中 be in the middle / 身～要职 hold an important post ④(当;任) claim; assert: 以专家自～ claim to be an expert ⑤(积蓄) store up; lay by: 囤积～奇 hoarding and profiteering ⑥(停留;固定) stay put; be at a standstill: 岁月不～。 Time marches on.
【居安思危】be prepared for danger in times of peace; be vigilant in peace time; provide against danger while living in peace
【居多】be in the majority
【居高临下】occupy a commanding position; occupy a high position and descend down; have a commanding view from a vantage ground
【居功】claim credit for oneself: ～自傲 claim credit for oneself and become arrogant
【居间】(mediate) between two parties: ～调停 mediate between two parties; act as mediator ◇ ～贸易 intermediary trade / ～人 intermediary; mediator
【居留】reside: 长期～ permanent residence ◇ ～权 right of residence / ～证 residence permit
【居民】resident; inhabitant ◇ ～点 residential area
【居然】[副]①(出乎意料) unexpectedly; to one's surprise: 谁会想到～有这种事! Who would have thought of such a thing! / 他～会相信这件事! Fancy his believing it! ②(甚至于) go so far as to; have the impudence to; have the effrontery to: 敌人～使用了毒气。 The enemy went so far as to use poison gas.
【居心】harbor (evil) intentions: ～不良 harbor evil intentions; the mind is bent on evil / ～叵测 with hidden intent; with ulterior motives / ～何在? What is the motive (behind all this)?
【居中】①(在当中) (mediate) between two parties: ～斡旋 mediate between disputants ②[印] be placed in the middle
【居住】live; reside; dwell ◇ ～国 country of residence / ～建筑 residential architecture / ～面积 living space; floor space / ～期限 length of residence / ～条件 housing conditions
锔 mend (crockery) with cramps
【锔子】a cramp used in mending crockery

jú

菊 chrysanthemum
【菊花】chrysanthemum
【菊科】[植] the composite family
【菊展】chrysanthemum show
橘 tangerine
【橘红】①(颜色) tangerine (color); reddish orange ②[中药] dried tangerine peel
【橘黄】orange (color)
【橘汁】orange juice
【橘子】tangerine
局 ①(机关) office; bureau: 电话～ telephone exchange / 邮～ post office / 政治～ Political Bureau ②(比赛的一次) game; set; innings: 第一～(乒乓球等) the first game; the first set; (板球、棒球、垒球) the first innings ③(形势;情况) situation; state of affairs: 全～ the overall situation; the situation as a whole / 战～ the war situation ④(聚会) gathering: 饭～ a dinner party; a banquet ⑤(圈套) ruse; trap: 骗～ fraud; trap; swindle ⑥(部分) part; portion
【局部】part ◇ ～地区 some areas; parts of an area / ～交货 partial delivery / ～利益 partial and local interests / ～麻醉 local anesthesia / ～认付 partial acceptance / ～战争 local war; partial war
【局促】①(狭小) narrow; cramped: 这地方太～了。 This place is rather cramped. ②(短促) short: 要两天完成这活，太～了。 Two days are too short for this job. ③(拘谨) feel or show

constraint: ～不安 ill at ease
【局面】aspect; phase; situation; prospects: 处于尴尬的～ be in an embarrassing situation / 打开～ open up a new prospect; make a breakthrough / ～一新 enter upon a new phase
【局势】situation: 国际～ the international situation / 紧张～ a tense situation; tension
【局外人】outsider
【局限】limit; confine ◇ ～性 limitations

锔[化] curium (Cm)

jǔ

举 ①(往上托) lift; raise; hold up: ～杯 raise one's glass (to propose a toast) / 高～红旗 hold high the red banner ②(举动) act; deed; move: 善～ good deed / 壮～ a heroic undertaking / 一～一动 every act and every move; every action ③(兴起;起) start: ～义 rise in revolt ④(推选;推举) elect; choose: 我们公～他作代表。 We chose him for our representative. ⑤(提出;列举) cite; enumerate; take; give: ～不胜～ too numerous to mention ⑥(全) whole; entire: ～座 all those present
【举哀】go into mourning: ～追悼 observe mourning by holding a memorial service / 全国～ observe a national mourning
【举案齐眉】husband and wife treating each other with courtesy; A married couple love and respect each other for life.
【举办】conduct; hold; run: ～训练班 conduct a training course / ～音乐会 give a concert / ～展览会 put on an exhibition; hold an exhibition
【举步】step forward; stride forward
【举措】behave; move; act: ～失当 make an ill-advised move
【举动】movement; move; act; activity: ～灵活 be nimble in movement / ～缓慢 be slow in movement / 轻率的～ a rash act
【举国】whole nation; entire nation; throughout the country: ～欢腾。The whole nation is jubilant. / ～上下团结一致。 There is solid unity throughout the nation.
【举荐】recommend (a person)
【举例】give an example; cite an instance: ～来说 for example / ～说明 illustrate with examples
【举目】raise the eyes; look: ～四望 look round / ～无亲 have no one to turn to (for help); be a stranger in a strange land; find oneself in a forlorn position / ～远眺 look into the distance
【举棋不定】hesitate about 〈over〉 what move to make; be unable to make up one's mind; vacillate; shilly-shally
【举世】all over the world; the world over; throughout the world; universally: ～公认 universally acknowledged / ～皆知 known to all / ～闻名 of world renown; world-famous / ～无双 unrivaled; matchless / ～瞩目 attract worldwide attention; become the focus of world attention
【举手】raise 〈put up〉 one's hand or hands: ～表决 vote by raising hands; vote by a show of hands ◇ ～礼 hand salute / ～之劳 lift a finger
【举行】hold; stage: ～罢工 stage a strike / ～会谈 hold talks / ～宴会 give a banquet; host a banquet
【举一反三】draw inferences about other cases from one instance; from a part we may judge the whole; learn by analogy.
【举证】put to the proof
【举止】bearing manner; mien: ～大方 have poise; have an easy manner; be gentle of mien; have a dignified air / ～娴雅 deport oneself gracefully / ～庄重 deport oneself in a dignified manner; carry oneself with dignity; respectfulness of deportment
【举重】weight lifting ◇ ～台 platform / ～运动员 weight lifter; lifter; strong man ○ 52 公斤级 flyweight / 56 公斤级 bantamweight / 60 公斤级 featherweight / 67.5 公斤级 lightweight / 75 公斤级 middleweight / 82.5 公斤级 light heavyweight / 90 公斤级 middle heavyweight / 110 公斤级 heavyweight / 110 以上公斤级 super heavyweight / 试举成功 finish a lift; good lift / 试举失败 no lift; not good / 杠铃 barbell / 铃片 disc; disk; plate
【举足轻重】hold the balance; prove decisive; play a decisive role

矩 ①(曲尺) carpenter's square; square ②(法度;规则) rules; regulations ③[物] moment: 动量～ moment of momentum / 力～ moment of force
【矩臂】[物] moment arm
【矩形】rectangle ◇ ～线圈 square coil
【矩阵】[数] matrix

沮

【沮丧】dejected; depressed; dispirited; disheartened

龃

【龃龉】the upper and lower teeth not meeting properly; disagreement; discord

咀 chew

【咀嚼】①(用牙齿嚼食物) masticate; chew ②(反复体会) mull over; ruminate; chew the cud

枸

【枸橼】[植] citron ◇ ～酸[化] citric acid / ～酸钠[药] sodium citrate

jù

聚 assemble; gather; get together

【聚氨酯】[化] polyurethane
【聚宝盆】treasure bowl; a place rich in natural resources; cornucopia
【聚苯】[化] polyphenyl
【聚苯乙烯】[化] polystyrene
【聚变】[物] fusion: 核～ nuclear fusion ◇ ～反应堆 fusion reactor
【聚丙烯】[化] polypropylene
【聚丙烯腈】[化] polyacrylonitrile
【聚餐】dine together; have a dinner party
【聚丁橡胶】butadiene rubber
【聚赌】gambling in group; group gambling
【聚砜】[化] polysulfone
【聚氟乙烯】[化] polyvinyl fluoride
【聚光灯】spotlight
【聚光镜】condensing lens; condenser
【聚合】①(聚集到一起) get together ②[化] polymerization ◇ ～反应(作用) polymerization
【聚合物】[化] polymer: 高分子～ high polymers / 工程～ engineering polymers
【聚会】get together; meet
【聚积】accumulate; collect; build up: ～力量 build up strength; gather forces
【聚集】gather; assemble; collect: 老师把学生～在她周围。 The teacher gathered her pupils round her.
【聚甲基丙烯酸甲酯】polymethyl methacrylate
【聚甲醛】[化] polyformaldehyde
【聚歼】round up and annihilate; annihilate en masse
【聚焦】[物] focusing
【聚精会神】concentrate one's attention; be all attention: ～地工作 concentrate on one's work; be intent on one's work / ～地听讲 listen with rapt attention
【聚居】inhabit a region (as an ethnic group); live in a compact community: 维吾尔族～地区 regions where the Uygur nationality live in compact communities.
【聚敛】amass wealth by heavy taxation; levy heavy taxes
【聚拢】gather together
【聚氯乙烯】[化] polyvinyl chloride (PVC)
【聚醛树脂】[化] aldehyde resin
【聚伞花序】[植] cyme
【聚沙成塔】many grains of sand piled up will make a pagoda; many a little makes a mickle; a pin a day is a great a year
【聚四氟乙烯】[化] polytetrafluoroethylene (PTFE)
【聚讼纷纭】argue back and forth without coming to an agreement; opinions differ widely
【聚碳酸脂】[化] polycarbonate
【聚烯烃】[化] polyolefin
【聚酰胺】[化] polyamide
【聚星】[天] multiple star
【聚乙烯】[化] polyethylene; polythene
【聚乙烯醇】[化] polyvinyl alcohol
【聚酯】[化] polyester ◇ ～薄膜 polyester film / ～塑料 polyester plastics / ～纤维 polyester fiber
【聚众】gather a crowd ◇ ～斗殴 gather a crowd to engage in an affray / ～闹事 mob

巨 huge; tremendous; gigantic: ～款 a huge sum of money

【巨变】great change; radical change: 山乡～ tremendous changes in a mountain village
【巨擘】①(拇指) thumb ②(某方面居于首位者) authority; leading figure: 诗坛～ prince of poets / 医界～ the authority in medical circles
【巨大】huge; tremendous; enormous; gigantic; immense: ～的工程 a giant project / ～的物质力量 tremendous material force / ～的知识宝库 an immense treasury of knowledge
【巨盗】major thief
【巨额】a huge sum; a big amount: ～财富 immense amount of treasure / ～赤字 huge financial deficits / ～交易 extensive trans- action / ～利润 enormous profits / ～投资 huge investments
【巨匠】[书] great master; consummate craftsman

【巨浪】billow; surge; mountainous waves
【巨流】a mighty current
【巨人】giant; colossus ◇ ～症[医] gigantism
【巨头】magnate; tycoon: 钢铁～ steel magnate / 金融～ financial magnate
【巨蜥】[动] monitor
【巨细】big and small: 事无～ all matters, big and small
【巨星】[天] giant star; giant
【巨型】giant; heavy; mammoth; colossal: ～客机 a giant liner / ～喷气客机 a jumbo jet / ～运输机 a giant transport plane
【巨灾】catastrophe: ～危险 catastrophic hazard
【巨著】monumental work; great work: 历史～ a magnum opus of historic significance

炬 ①(火把) torch ②(烧掉) fire: 付之一～ be burnt down; be committed to the flames

拒 ①(抵抗) resist; repel: ～敌 resist the enemy; keep the enemy at bay ②(拒绝) refuse; reject: ～不接受 refuse to accept / ～不收礼 refuse a gift
【拒捕】resist arrest
【拒腐蚀】repel the corrupting influence; ward off the corrosive influence
【拒付】refuse payment; dishonor (a check)
【拒谏饰非】reject representations and gloss over errors; reject criticisms and whitewash one's mistakes
【拒绝】refuse; reject; turn down; decline: ～参加 refuse to participate / ～发表意见 refuse to comment / ～批评 reject other people's criticism
【拒人于千里之外】keep people a thousand miles away

距 ①(距离) distance: 行～ the distance between rows of plants ②(相距) be apart from; away from; be at a distance from: 相～不远 not far from each other / ～今已有二十年。That was twenty years ago. / 两城相～五公里。The two towns are about five kilometer apart. ③(鸡距) spur
【距离】①(相隔的长度) distance; range; gap: 保持一定～ keep a certain distance / 扩大～ increase the distance; widen the gap / 缩小～ reduce the distance; lessen the gap ②(相距) be apart from; away from; be at a distance from: ～约 300 公里 be about 300 kilometers apart from each other

遽 ①(匆忙;急) hurriedly; hastily: ～下结论 pass judgment hastily ②(惊慌) frightened; alarmed
【遽然】[书] suddenly; abruptly: ～变色 suddenly change countenance

具 ①(用具) utensil; tool; implement: 农～ farm tool; farm implement; agricultural implement / 文～ writing materials; stationery ②(具有) possess; have: 初～规模 have begun to take shape ③[量] 一～尸体 a corpse
【具备】possess; have; be provided with: ～必要条件 satisfy the essential requirements / ～申请资格 have qualifications for application / ～一切条件 there is every requisite
【具结】sign an undertaking; binding over; enter into recognizances: ～领回失物 sign a receipt for restored lost property / ～释放 enter into a bond and to be released / 责令～悔过 instructed to write a statement of repentance
【具名】put one's name to a document, etc.; affix one's signature
【具体】concrete; specific; particular: ～计划 practical plans / ～情况 concrete conditions / ～日期 exact date / ～政策 specific policies
【具体而微】small but complete; miniature
【具体劳动】concrete labor
【具文】mere formality; dead letter: 一纸～ a mere scrap of paper
【具有】possess; have; be provided with: ～高度责任心 possess a high sense of responsibility / ～历史意义 have profound historical significance / ～约束力 bind / ～证人资格 qualified as a witness / ～重大意义 be of great importance

惧 fear; dread: 毫无所～ fear nothing; be fearless; not cowed in the least
【惧怕】fear; dread
【惧色】a look of fear: 面无～ look undaunted / 面有～ look scared

俱 all; complete; entirely
【俱乐部】club: 工人～ worker's club / 海员～ seamen's club / 体育～ athletic club ◇ ～会员 clubber
【俱全】complete in all varieties: 一应～ be available in all varieties

飓
【飓风】hurricane

句①(句子) sentence ②[量]两～诗 two lines of verse／一～话也没说 not utter a word
【句法】①(句子的结构方式) sentence structure ②[语] syntax
【句号】full stop; full point; period (。)(.)
【句型】sentence pattern: ～练习 pattern drills
【句子】sentence ◇～成份 sentence element; member of a sentence

据①(占据) occupy; seize; take possession of; lay hold of: ～为己有 take forcible possession of; appropriate; rob for oneself ②(凭借) rely on; depend on: ～险固守 take advantage of a natural barrier to put up a strong defense ③(按照;依据) according to; on the grounds of: ～报道 according to (press) reports; it is reported that／～理力争 argue strongly on sound grounds; argue on the basis of reason／～实报告 report the facts; make a factual report／～实相告 tell according to facts／～实招供 tell the facts in court／～实招认 factual admission／～我看 as I see it; in my opinion／～我所知 as far as I know ④(证据;凭据) evidence; proof; grounds: 查无实～ be unverified upon investigation／言之有～ have good grounds in what one says
【据传】a story is going around that; rumor has it that
【据此】on these grounds; in view of the above; accordingly
【据点】strongpoint; fortified point; stronghold
【据守】guard; be entrenched in
【据说】it is said; they say; allegedly: ～情况如此。This is allegedly the case.／～她很有成功的把握。It is said she is quite sure of success.
【据悉】it is reported

踞①(蹲或坐) crouch; squat ②(盘踞) sit

剧①(戏剧) theatrical work; drama; play; opera: 悲～ tragedy／雕塑～ tableau vivant／独幕～ one-act play／歌～ opera／广播～ radio play／话～ play; stage play／活报～ skit／京～ Beijing opera／历史～ historical play／五幕～ a play of five acts／喜～ comedy／戏～ drama／小歌～ operetta／哑～ pantomime ○三幕五场 (a play) in three acts and five scenes／滑稽戏 farce／木偶戏 puppet show／地方戏 local opera／皮影戏 shadow show ②(剧烈) acute; severe; intense: ～变 a violent change; a drastic change／～痛 a severe pain／产量～增 a sharp increase in output
【剧本】play; drama: ～创作 playwriting／电影～ scenario／分镜头～ shooting script／歌剧～ libretto／话剧～ drama
【剧场】theatre: 露天～ open- air theatre; amphitheater／木偶～ puppet show theatre
【剧烈】violent; acute; severe; fierce: ～的社会变动 radical social changes／～地震动 quake violently／～运动 strenuous exercise
【剧目】a list of plays or operas: 保留～ repertoire
【剧评】a review of a play or opera; dramatic criticism
【剧情】the story〈plot〉of a play or opera: ～复杂 intricate plot／～简单 intricate plot ◇～简介 synopsis
【剧团】theatrical company; opera troupe; troupe: 芭蕾舞～ ballet troupe／实验～ experimental theatre／业余～ amateur troupe; amateur dramatic group; amateur theatrical group
【剧务】①(有关排演和演出等事务) stage management ②(担任剧务的人) stage manager
【剧院】①(剧场) theatre ②(演出团体) theatre: 北京人民艺术～ The People's Arts Theatre of Beijing
【剧照】stage photo; still: 电影～ still
【剧中人】characters in a play or opera; dramatis personae
【剧终】the end; curtain
【剧种】type of drame; genre of drama
【剧作家】playwright; dramatist

锯①(工具) saw: 手～ handsaw／圆～ circular saw／油～[林]motor chain saw ②(用锯锯开) cut with a saw; saw: ～木头 saw wood
【锯齿】sawtooth
【锯床】[机] sawing machine
【锯缝】saw kerf
【锯框】saw frame
【锯末】sawdust
【锯木厂】sawmill; lumber-mill
【锯条】saw blade

juān

圈①(围起来) shut in a pen; pen in: 把羊群～起来 herd the sheep into the pens ②[口](关押) lock up; put in jail

涓[书] a tiny stream
【涓埃】[书] insignificant; negligible: 略尽～之力 make what little contribution one can; do one's

bit
【涓滴】[书] a tiny drop; dribble; driblet: ～归公 turn in every cent of public money
【涓涓】[书] trickling sluggishly

捐 ①(舍弃) relinquish; abandon ②(捐助) contribute; donate; subscribe: ～钱 contribute money / 募～ solicit contributions; appeal for donations ③(税收) tax: 房～ housing tax / 上～ pay a tax
【捐款】①(捐助款项) contribute money ②(所捐之款) contribution; donation; subscription
【捐弃】[书] relinquish; abandon ◇ ～前嫌 throw away the past resentment
【捐躯】sacrifice one's life; lay down one's life: 为国～ lay down one's life for one's country
【捐税】taxes and levies
【捐献】contribute (to an organization); donate; present: ～一笔巨款 contribute a big sum of money
【捐赠】contribute (as a gift); donate; present ◇ ～地产 donated land / ～股份 donated stock
【捐助】offer (financial or material assistance); contribute; donate

娟[书] beautiful; graceful
【娟秀】[书] beautiful; graceful: 字迹～ beautiful handwriting; a graceful hand

镌[书] engrave
【镌刻】[书] engrave

juǎn

卷 ①(裹成圆筒形) roll up; coil; curl; furl: ～地图 roll up a map / ～起袖子就干 roll up one's sleeves and pitch in / ～支烟 roll a cigarette ②(猛力撮起或裹走) sweep off; carry along; swirl: 一个大浪把小船～走了。A huge wave swept the boat away. ③(裹成圆筒形的东西) cylindrical mass of sth.; roll: 花～儿 fancy-shaped 〈plaited; twisted〉 steamed roll / 铺盖～儿 bedding roll ④[量] roll; spool; reel: 一～胶卷 a roll of film / 一～卫生纸 a roll of toilet paper
【卷笔刀】pencil sharpener
【卷层云】[气] cirrostratus
【卷尺】tape measure; band tape: 钢～ steel tape
【卷发】curly hair; wavy hair; crimped hair
【卷积云】[气] cirrocumulus
【卷铺盖】①(把铺盖卷起来) pack up and quit ②(被解雇) get the sack
【卷刃】(of a knife blade) be turned
【卷入】be drawn into; be involved in: ～漩涡 be drawn into a whirlpool / ～一场纠纷 be involved in a dispute
【卷舌音】[语] retroflexion
【卷逃】abscond with valuables
【卷筒】reel
【卷筒纸】web ◇ ～纸印刷机 web press
【卷土重来】stage a comeback; bounce back; bob up
【卷尾猴】(weeping) capuchin; weeping monkey
【卷席而逃】roll up one's mat and run away
【卷心菜】cabbage
【卷须】[植] tendril
【卷烟】cigarette: 带过滤嘴的～ filter-tipped cigarette ◇ ～包装机 cigarette packer / ～工业 cigarette industry / ～机 cigarette (making) machine / ～纸 cigarette paper
【卷扬机】hoist; hoister
【卷叶蛾】[动] leaf roller
【卷云】[气] cirrus
【卷轴】reel: 天线～ aerial reel
【卷装货】cargo in coil; cargo in roll

juàn

眷 ①(亲属) family dependant: 女～ female members of a family ②[书](关心;怀念) have tender feeling for
【眷恋】[书] be sentimentally attached to (a person or place)
【眷念】[书] think fondly of; feel nostalgic about
【眷属】family dependants

卷 ①(书本) book: 画～ scroll of painting / 手不释～ always have a book in one's hand; be a diligent reader ②(书籍册数) volume: 第一～ Volume One / 这个图书馆藏书百万～。This is a library of a million volumes. ③(卷子) examination paper: 交～ hand in an examination paper; finish a written work; complete a task (指完成任务) ④(机关里保存的文件) file; dossier: 查～ look through the files
【卷轴】[书] scroll
【卷子】examination paper: 看～ mark examination papers
【卷宗】①(纸夹子) folder ②(文件) file; dossier

圈 pen; fold; sty: 羊～ sheepfold; sheep pen / 猪～ pigsty
【圈肥】[农] barnyard manure
【圈养】rear livestock in pens

倦 weary; tired: 面有～容 look tired
【倦意】tiredness; weariness; sleepiness: 毫无～意 not feel in the least tired
【倦游】weary of wondering and sight-seeing

绢 thin, tough silk
【绢本】silk scroll
【绢纺】silk spinning
【绢花】[工美] silk flower
【绢画】classical Chinese painting on silk
【绢丝】spun silk (yarn) ◇ ～纺绸 spun silk pongee / ～织物 spun silk fabric
【绢网印花】[纺] screen printing ◇ ～法 silk-screen process

隽
【隽永】[书] meaningful: 语颇～,耐人寻味 The remarks are meaningful and thought-provoking.

juē

撅 ①(翘起) stick up: ～着尾巴 sticking up the tail / ～嘴 pout (one's lips) ②(折) break (sth. long and narrow); snap: 把树枝～成两段 break the twig in two

jué

觉 ①(对刺激的感受和辨别) sense; feel: 触～ touch; sense of touch; tactile sensation / 错～ illusion / 光～ light sensation / 色～ color sensation / 视～ sight; sense of sight / 听～ hearing; sense of hearing; auditory sensation / 味～ taste; sense of taste; taste sensation / 嗅～ smell; sense of smell; smell perception / 知～ perception / 几乎～不出来 can hardly sense it ②(睡醒) wake (up); awake: 如梦初～ as if waking from a dream ③(觉悟) become aware; become awakened
【觉察】detect; become aware of; perceive: ～到态度有变化 perceive a change in attitude / ～到有危险 become aware of the danger
【觉得】①(发生某种感觉) feel: 一点不～累 not feel tired at all ②(认为) think; feel; find: 你～这部电影怎样? How do you think of the film? (或: How do you like the film?) / 他～这个计划不妥当。 He felt the plan to be unwise.
【觉悟】①(认识程度) consciousness; awareness; understanding: 阶级～ class consciousness / 政治～ political consciousness; political understanding ②(醒悟) come to understand; become aware of; become politically awakened: ～了的人民 an awakened people
【觉醒】awaken; awake: 世界人民的新～ a new awakening of the world people

厥 faint; lose consciousness; fall into a coma: 昏～ fall to the ground in a faint

蕨 [植] brake (fern)
【蕨类植物】pteridophyte

橛 a short wooden stake; wooden pin; peg
【橛子】a short wooden stake; wooden pin; peg

蹶 ①(摔倒) fall ②(失败;挫折) suffer a setback: 一～不振 collapse after one setback; never recover from a setback

矍
【矍铄】[书] hale and hearty

攫 seize; grab: ～为己有 seize possession of; appropriate
【攫取】seize; grab

爵 the rank of nobility; peerage: 封～ confer a title (of nobility) upon ◯公～ duke / 侯～ marquis / 伯～ earl; count / 子～ viscount / 男～ baron
【爵士】①(欧洲君主国的最低封号) knight ②(放在姓名前,用于称呼) Sir
【爵士音乐】jazz
【爵位】the rank of nobility; title of nobility

角 ①(角色) role; part; character: 主～ leading role; principal role; main character ②(演员) actor; actress: 名～ a famous actor; a famous actress ③(竞赛;斗争) contend; wrestle: ～斗 wrestle / 口～ quarrel; bicker
【角斗】wrestle ◇ ～场 wrestling ring
【角力】have a trial of strength; wrestle: 与某人～ wrestle with sb.
【角色】role; part: 反面～ negative character / 正面～ positive character / 你扮演哪个～? What role 〈part〉 do you play?
【角逐】contend; tussle; enter into rivalry: 进行～ enter into rivalry / 超级大国之间的～ fierce rivalry between the superpowers

谲 [书] cheat; swindle

【谲诈】cunning; crafty

决 ①(决定) decide; determine: ～一胜负 fight it out / 悬而未～ be undecided; be awaiting decision / 犹豫不～ hesitate; be unable to reach a decision; be in a state of indecision ②(一定;用在否定词前面) definitely; certainly; under any circumstances: ～不退让 will under no circumstances give in / ～非偶然 by no means accidental / ～无此事 no such things ever happened ③(执行死刑) execute a person: 枪～ execute by shooting ④(决口) be breached; burst: 全堤溃～。The bank burst and totally collapsed.

【决策】①(决定策略或办法) make policy; make a strategic decision; decide a policy ②(决定的策略或办法) policy decision; decision of strategic importance: 战略～ strategic decision ◇～机构 policy-making body / ～理论 decision theory / ～人 policymaker

【决定】①(做出主张) decide; resolve; make up one's mind: ～总政策 decide the general policy / 理事会～下次比赛在上海举行。The Council resolved that the next tournament should be held in Shanghai. ②(决定的事项) decision; resolution: 通过一项～ pass a resolution ③(起决定作用) determine; decide: ～性胜利 a decisive victory / ～因素 decisive factor; determinant / 存在～意识。Man's social being determines his consciousness.

【决定论】[哲] determinism

【决定权】power to make decisions: 有最后～ have the final say

【决斗】①(用武器格斗) duel ②(你死我活的斗争) decisive struggle

【决断】①(做决定) make a decision ②(有魄力) resolution; resolve; decisiveness: 有～的人 man of determination (great resolution)

【决计】①(主意已定) have decided; have made up one's mind: 我～明天就走。I've decided to go tomorrow. ②(一定) definitely; certainly: ～不会错 definitely can't go wrong

【决口】①(出现裂口) burst; be breached ②(裂口) breach; break: 堵住～ close up a breach

【决裂】break with; rupture: 与旧的传统观念彻底～ break completely with outdated conventional ideas

【决然】[书]①(坚决) resolutely; determinedly; decidedly; firmly: 毅然～地 determinedly; decidedly ②(必然) definitely; unquestionable; undoubtedly

【决赛】[体] finals: 半～ semifinals / 进入～ enter the finals

【决胜】decide the issue of the battle; determine the victory ◇～局[体] deciding game; deciding set

【决死】life-and-death: ～的斗争 a life-and-death struggle; a last-ditch fight

【决算】final accounts; final accounting of revenue and expenditure: 国家的～ final state accounts ◇～表 final statement / ～审计 audit of returns / ～书 final report / ～帐户 final account

【决心】determination; resolution: ～改正错误 be determined to correct one's mistake / 下定～ make up one's mind; be resolute / 向党表～ pledge one's determination to the Party ◇～书 written pledge; statement of one's determination

【决一雌雄】fight to see who is the stronger; fight it out

【决一死战】wage a life-and-death battle

【决议】resolution: 撤销～ cancel a resolution; revoke a resolution / 通过～ pass a resolution; adopt a resolution ◇～(草)案 draft resolution

【决意】have one's mind made up; be determined

【决战】decisive battle; decisive engagement

诀 ①(口诀) rhymed formula ②(诀窍) knack; tricks of the trade: 秘～ secret of success; key to success ③(分别) bid farewell; part: 永～ part never to meet again; part for ever

【诀别】bid farewell; part

【诀窍】secret of success; tricks of the trade; knack: 成功的～ the secret of success / 生意的～ the tricks of the trade

抉 [书] pick out; single out

【抉择】choose: 作出～ make one's choice

掘 dig: ～地道 dig a tunnel / ～井 dig a well / 自～坟墓 dig one's own grave

【掘进】[矿] driving; tunnelling

【掘墓人】gravedigger

【掘土机】excavator; power shovel

崛 rise abruptly

【崛起】①(突起) rise abruptly; rise sharply; suddenly appear on the horizon: 群山～。Mountains rise abruptly (over the plain). ②(兴起) rise; spring up

倔

【倔强】stubborn; unbending; unyielding

绝 ①(断绝) cut off; sever: ~其后路 cut off his retreat / 掌声不~ prolonged applause ②(穷尽) exhausted; used up; finished: 弹尽粮~ have run out of ammunition and provisions ③(走不通的;没有出路的) desperate; hopeless: ~境 hopeless situation; impasse ④(独一无二;极好) unique; superb; matchless: 她的书画可称双~。 Her calligraphy as well as her painting can be rated as superb works of art. ⑤(极;最) extremely; most: ~大多数 most; the overwhelming majority / ~早 extremely early ⑥(绝对,用在否定词前) absolutely; in the least; by any means; on any account: ~不可能 absolutely impossible / ~非偶然 by no means fortuitous / ~无此意 have absolutely no such intentions ⑦(肯定) leaving no leeway; making no allowance; uncompromising; definitive: 不要把话说~ don't be too definitive in what you say / 没有把话说~ didn't say anything definitive

【绝版】out of print: ~书籍 out-of-print publications

【绝笔】①(遗书) last words written before one's death ②(最后的作品) the last work of an author or painter

【绝壁】precipice

【绝唱】the peak of poetic perfection

【绝处逢生】be unexpectedly rescued from a desperate situation; find one's way out from an impasse

【绝代】[书] unique among one's contemporaries; peerless: ~佳人 beauty of beauties; an incomparable beauty; a beautiful woman of unsurpassed beauty / 才华~ unrivaled talent

【绝顶】extremely; utterly: ~聪明 extremely intelligent / ~荒谬 utterly ridiculous / ~愚蠢的行为 the height of folly

【绝对】①(没有任何条件;不受任何限制) absolute: ~优势 absolute predominance; overwhelming superiority; absolute superiority ②(完全;一定) absolutely; perfectly; definitely: ~可靠 absolutely reliable ◇ ~担保 absolute guaranty / ~地租 absolute rent / ~多数 absolute majority; overwhelming majority / ~观念[哲] absolute idea / ~豁免权 absolute immunity / ~价值 absolute value / ~禁制 absolute prohibition / ~禁制品 absolute contraband / ~利益 absolute advantage / ~量度[物] absolute measurement / ~零度[物] absolute zero / ~平均主义 absolute equalitarianism / ~湿度[气] absolute humidity / ~温度[物] absolute temperature / ~音乐 absolute music / ~真理[哲] absolute truth / ~值[数] absolute value / ~转让 absolute assignment

【绝对主义】[哲] absolutism

【绝后】①(没有后代) without offspring; without issue ②(今后不会再有) never to be seen again: 空前~ never known before and never to occur again; unique

【绝户】①(没有后代) without offspring; without issue ②(没有后代的人) a childless person

【绝迹】disappear; vanish; be stamped out: 那种传染病早已~。 That kind of epidemic disease has long been stamped out.

【绝技】unique skill; consummate skill

【绝交】break off relations (as between friends or countries)

【绝经】[生理] menopause ◇ ~期 climacteric; menopause

【绝境】hopeless situation; impasse; blind alley; cul-de-sac: 濒于~ face an impasse

【绝口】①(住口,只用在"不"字后) stop talking: 骂不~ heap endless abuse upon; pour out unceasing abuse / 赞不~ give unstinted praise; praise profusely ②(因回避而不开口) keep one's mouth shut: ~不提 never say a single word about; avoid all mention of

【绝路】road to ruin; blind alley; impasse; disaster: ~逢生 be alive in desperation / 引上~ lead sb. into a blind alley / 自寻~ court self-destruction; bring ruin upon oneself / 走上~ take the road to ruin; head for one's doom

【绝伦】unsurpassed; unequalled; peerless; matchless: 聪颖~ incomparably intelligent / 荒谬~ utterly absurd; utterly preposterous / 精美~ exquisite beyond compare; superb

【绝密】top-secret; most confidential: ~文件 top-secret papers

【绝妙】extremely clever; ingenious; excellent; perfect: ~的讽刺 perfect irony / ~的一招 a masterstroke

【绝命书】①(临自杀前的遗书) suicide note ②(临死刑前的遗书) note written on the eve of one's execution

【绝热】[物] heat insulation

【绝食】fast; go on a hunger strike

【绝世之姿】a paragon of beauty

【绝望】give up all hope; despair; hopelessness; lose all hope of
【绝无仅有】the only one of its kind; unique; singular of its kind
【绝育】[医] sterilization: ～手术 sterilization operation
【绝缘】①[电] insulation ②(跟外界不发生接触) be cut off from; be isolated from ◇～材料 insulating material; insulant／～套管 insulating sleeve; spaghetti (tubing)／～体 insulator／～子 insulator
【绝招】①(绝技) unique skill ②(最后一招) unexpected tricky move (as a last resort)
【绝症】incurable disease; fatal illness
【绝种】(of a species) become extinct; die out: 已～的动物 extinct animal

jué

蹶

【蹶子】尥～ (of horses, donkeys, etc.) kick

juè

倔 gruff; surly; blunt: 说话～ be blunt of speech／他脾气够～的。He's surly enough.
【倔头倔脑】blunt of manner and gruff of speech

jūn

军 ①(军队) armed forces; army; troops: 参～ join the army ①(军队编制) corps: ～部 corps headquarters ③(军人的;军队的) military: ～操 military drill ④(队伍) army; contingent: 劳动～ labor army; contingent of labor forces
【军备】armament; arms: 扩充～ engage in arms expansion ◇ ～费 military expenditures／～竞赛 armament race; arms race／～控制 arms control
【军车】military vehicle
【军刀】soldier's sword; saber; sabre
【军队】armed forces; army; troops: ～建制 military organization ○中央军事委员会 Central Military Commission／国防部 Ministry of National Defense／总参谋部 Headquarters of the General Staff／总政治部 General Political Department／总后勤部 General Logistics Department／军 army／师 division／旅 brigade／团 regiment／营 battalion／连 company／排 platoon／班 squad／兵团 army; corps; formation／大部队 large unit／纵队 column／(空军)大队 group／(海、空军)中队 squadron／小队 flight; unit; team; detachment／分队 element／支队 detachment／预备役 reserve service／现役 active service／义务兵役制 compulsory service system／仪仗队 guard of honor／兵种 arms／番号编制 footing; organization
【军阀】warlord ◇～战争 war among warlords／～主义 warlordism／～作风 warlord ways; warlord style
【军法】military criminal code; military law: ～从事 punish by military law ◇ ～审判 court-martial
【军费】military expenditure
【军分区】military subarea; military sub-command
【军风纪】soldier's bearing and discipline
【军服】military uniform; uniform
【军港】naval port
【军工】①(军事工业) war industry ②(军事工程) military project ◇～生产 war production
【军功】military exploit
【军官】officer ○军长 army commander／师长 division commander／旅长 brigade commander／团长 regiment commander／营长 battalion commander／连长 company commander／排长 platoon leader／班长 squad leader; section leader／总司令 commander-in-chief／总参谋长 chief of general staff／司令员 commander; commanding officer／政委 political commissar／政治部主任 director〈chief〉of the political depart- ment／教导员 battalion political instructor／指导员 political instructor／参谋 staff officer／作战参谋 operational staff／司务长 supply chief; quartermaster／文书 clerk
【军管】military control: 实行～ exercise military control ◇～会[简](军事管制委员会) military control commission
【军国主义】militarism ◇～化 militarization／～者 militarist
【军号】bugle
【军徽】army emblem
【军火】munitions; arms and ammunition ◇～工业 munitions industry; armament industry／～库 arsenal／～商 munitions merchant; arms dealer; merchant of death
【军机】①(军事机宜) military plan: 贻误～ delay or frustrate the fulfillment of a military plan ②(军事秘密) military secret: 泄漏～ leak a military secret
【军籍】military status; one's name on the army

roll: 保留～ retain one's military status / 开除～ strike sb.'s name off the army roll; discharge sb. from the army
【军纪】military discipline: 违犯～ breach of military discipline
【军舰】warship; naval vessel
【军阶】(military) rank; grade
【军界】military circles; the military
【军垦】reclamation of wasteland by an army unit ◇ ～农场 army reclamation farm; army farm
【军礼】military salute: 行～ make a salute; give a salute
【军力】military strength
【军粮】army provisions; grain for the army ◇ ～库 military grain depot; army granary
【军龄】length of military service
【军令】military orders: 颁布～ issue a military order ◇ ～如山 military orders cannot be disobeyed or revoked
【军马】army horse ◇ ～场 army horse-breeding farm; army horse ranch
【军帽】army cap; service cap
【军旗】army flag; colors; ensign
【军情】military situation; war situation: 刺探～ spy on the military movements; collect military information
【军区】military region; (military) area command: 北京～ Beijing Military Command / 省～ provincial military command ◇ ～司令部 the headquarters of a military area command
【军人】soldier; serviceman; armyman: 复员～ demobilized soldier / 现役～ persons in active service ◇ ～大会 soldiers' conference (of a company) / ～家属 soldier's dependants; armyman's family members
【军容】soldier's discipline, appearance and bearing: 整饬～ strengthen army discipline and maintain required standards for appearance and bearing
【军师】military counsellor; army adviser
【军士】noncommissioned officer (NCO)
【军事】military affairs ◇ ～表演 display of military skills / ～部署 military deployment; disposition of military forces / ～法院 military tribunal; military court; martial court / ～分界线 military demarcation line / ～工程学 military engineering / ～工业 war industry / ～管制 military control / ～管制委员会 military control commission / ～基地 military base / ～家 strategist / ～科学 military science / ～路线 military line / ～民主 military democracy / ～设施 military installations / ～素质 military qualities; fighting capability / ～体育 military sports / ～条令 military manuals / ～学 military science / ～学家 military scientist / ～学院 military academy; military institute / ～训练 military training / ～演习 military maneuver; war exercise / ～优势 military superiority / ～原则 principles of operation; military principles / ～政变 military coup d'etat
【军事化】militarize; place on a war footing: 过～的生活 follow a military routine
【军属】soldier's dependants; armyman's family
【军团】army group
【军委】①(中央军委简称) the Central Military Commission ②(中共中央军事委员会) the Military Commission of the Central Committee of the Communist Party of China
【军务】military affairs; military task
【军衔】military rank ◇ ～制度 system of military ranks ○元帅 Field-Marshal; (海军) Admiral of the Fleet; (空军) Marshal of the RAF / 五星上将 General of the Army; (海军) Fleet Admiral; (空军) General of the Air Force / 上将 General; (海军) Admiral; (英国空军) Air Chief Marshal / 中将 Lieutenant General; (海军) Vice Admiral; (英国空军) Air Marshal / 少将 Major General; (海军) Rear Admiral; (英国空军) Air Vice Marshal / 准将 (英国) Brigadier; (美国) Brigadier General; (海军) Commodore; (英国空军) Air Commodore / 上校 Colonel; (海军) Captain; (英国空军) Group Captain / 中校 Lieutenant Colonel; (海军) Commander; (英国空军) Wing Commander / 少校 Major; (海军) Lieutenant; (英国空军) Squadron Leader / 上尉 Captain; (海军) Lieutenant; (英国空军) Flight Lieutenant / 中尉 Lieutenant; (英国海军) Sub-Lieutenant; (美国海军) Lieutenant Junior Grade; (英国空军) Flying Officer; (美国空军, 美国海军陆战队) First Lieutenant / 少尉 Second Lieutenant; (英国海军) Acting Sub-Lieutenant; (美国海军) Ensign; (英国空军) Pilot Officer / 准尉 Warrant Officer; (美国陆军, 美国空军) Chief Warrant Officer / 军士长 Master Sergeant; (海军) Master Chief Petty Officer / 上士 (英国陆军) Staff-Sergeant; (美国陆军) Sergeant First Class; (英国海军) Chief Petty Officer; (美国海军) Petty Officer First Class; (英国空军) Flying Sergeant; (英国

海军陆战队)Color Sergeant / 中士 Sergeant; (英国海军) Petty Officer First Class; (美国海军) Petty Officer Second Class / 下士 Corporal; (英国海军) Petty Officer Second Class; (美国海军) Petty Officer Third Class / 技术军士 Technical sergeant / 参谋军士 Staff Sergeant / 一等兵 (英国陆军) Lance-Corporal; (美国陆军, 美国海军陆战队) Private First Class; (英国海军) Leading Seaman; (美国海军) Seaman First Class; (英国空军) Senior Aircraftman; (美国空军) Airman First Class; (英国海军陆战队) Marine First Class / 二等兵 Private; (英国海军) Able Seaman; (美国海军) Seaman Second Class; (英国空军) Leading Aircraftman; (美国空军) Airman Second Class; (英国海军陆战队) Marine Second Class / 三等兵 Basic Private; (海军) Apprentice Seaman; (空军) Airman Third Class / 新兵 Recruit; (海军) Ordinary Seaman; (空军) Aircraftman

【军饷】soldier's pay and provisions

【军校】military school; military academy

【军械】ordnance; armament ◇ ～处 ordnance department / ～库 ordnance depot; arms depot; armory / ～员 armorer

【军心】soldiers' morale; morale of the troops: 振奋～ raise 〈heighten〉 the morale of the troops

【军需】①(军用物资) military supplies ②[旧] (军需官) quartermaster

【军训】military training

【军医】medical officer; military surgeon ◇ ～大学 army medical college

【军营】military camp; barracks

【军用】for military use: military ◇ ～地图 military map / ～飞机 warplane; military aircraft / ～警报与探测系统 military warning and detection systems / ～列车 military train / ～桥 miliatry bridge / ～物资 military supplies

【军邮】army postal service; army post; army mail

【军援】military aid

【军乐】martial music; military music ◇ ～队 military band

【军长】army commander

【军政】①(军事和政治) military-political: ～大学 military and political college ②(军队和政府) army and government: ～当局 civil and military authorities / ～费用 military and administrative spending / ～人员 military and administrative personnel

【军政府】military government

【军职】official post in the army; military appointment

【军中无戏言】There are no jokes in armed forces

【军种】(armed) services: 各～兵种 all services and arms

【军装】military uniform; army uniform; uniform

皲

【皲裂】[书] (of skin) chap

均

①(均匀) equal; even: 分得不～ not evenly divided ②(全部) without exception; all: 诸事～已办妥。 All things have been properly arranged.

【均等】equal; impartial; fair: 机会～ equal opportunity / 势力～ balance of power

【均分】divide equally; share out equally

【均衡】balanced; proportionate; harmonious; even: ～发展 balanced development; harmonious development / 保持～ keep balance ◇ ～汇率 equilibrium rate of exchange

【均衡论】[哲] the theory of equilibrium

【均势】balance of power; equilibrium of forces; equilibrium; parity: 军事～ military parity

【均摊】share equally; share alike: ～费用 share the expenses equally

【均匀】even; well-distributed; uniform; regular: ～的呼吸 even breathing ◇ ～负载 uniform load / ～速度 even speed; uniform velocity

【均匀分裂】[化] homolysis

菌

①(真菌) fungus: 食用～ mushroom ②(细菌) bacterium

【菌柄】stem

【菌肥】[简] (细菌肥料) bacterial manure

【菌苗】[医] vaccine

【菌伞】cap; pileus

【菌丝】hypha

【菌丝体】mycelium

【菌血症】[医] bacteriemia

龟

【龟裂】①(田地干裂) be full of cracks ②(皲裂) chap

君 ①（君主）monarch; sovereign; supreme ruler ②（尊称）gentleman; Mr.

【君临天下】the sovereign descends the world

【君权】monarchical power ◇ ～神授 divine right of kings

【君王】king; lord

【君无戏言】the king's words are to be taken seriously

【君主】monarch; sovereign: 被废黜的～ deposed monarch / 退位～ abdicated mon- arch ◇ ～国 monarchical state; monarchy / ～立宪 constitutional monarchy / ～制 monarchy / ～专制 autocratic monarchy; absolute monarchy

【君子】a man of noble character; gentleman: 伪～ hypocrite / 正人～ a man of moral integrity / ～成人之美 a gentleman is always ready to help others attain their aims / ～一言，快马一鞭 a word from the princely man is a sufficient incentive for action / 以小人之心度～之腹 gauge the heart of a gentleman with one's own mean measure / ～忧道不忧贫。You are not acting for your interests; Your are acting against your own interests. / ～之交淡如水。The friendship between gentleman appears indifferent but is pure like water.

【君子国】the (imaginary) land of the virtuous

【君子协定】gentlemen's agreement

jùn

菌 mushroom

【菌子】mushroom

浚 dredge: ～河 dredge a river

【浚泥船】dredger

竣 complete; finish: 告～ have been completed

【竣工】(of a project) be completed

峻 ①（高大）high: 高山～岭 high mountains / 险～ precipitous ②（严厉）harsh; severe; stern: 严刑～法 harsh law and severe punishment

【峻峭】high and steep

俊 ①（相貌清秀好看）handsome; pretty ②（才智出众的人）a person of outstanding talent

【俊杰】a person of outstanding talent; hero

【俊美】pretty; handsome

【俊俏】pretty and charming

【俊秀】pretty; of delicate beauty

骏 fine horse; steed

【骏马】fine horse; steed; gallant horse

K

kā

咯 noise made in coughing or vomiting
【咯嚓】[象] crack; snap: 扁担～一声断了。The carrying-pole broke with a crack.
【咯麦隆】Cameroon ◇ ～人 Cameroonian
【咯斯特】[地] karst ◇ ～地形 karst topography

咖
【咖啡】coffee: 牛奶～ coffee with milk / 速溶～ instant coffee / 煮～ make coffee
【咖啡馆】cafe
【咖啡壶】coffee pot: 带过滤网的～ percolator
【咖啡磨】coffee-mill
【咖啡色】coffee (color)
【咖啡因】[药] caffeine; caffeinum

kǎ

卡 ①(扣住; 阻挡) block; check: ～住所有的通道 block all the passageways ②(卡路里) calorie ③(卡片) card: 分类～ classified card
【卡巴胂】[药] carbarsone
【卡宾枪】carbine
【卡车】truck (美); lorry (英) ◇ ～上交货价 free on truck
【卡尺】[机] (sliding) callipers: 游标～ vernier callipers
【卡介苗】Bacille Calmette-Guerin; BCG vaccine
【卡路里】[物] calorie
【卡那霉素】[药] kanamycin
【卡片】card ◇ ～柜 card cabinet / ～目录 card catalogue / ～式分类帐 card ledger / ～索引 card index
【卡其】[纺] khaki
【卡钳】callipers: 内～ inside callipers / 内外～ combination callipers / 外～ outside callipers
【卡塔尔】Qatar ◇ ～人 Qatari
【卡特尔】[经] cartel
【卡通】(animated) cartoon

咔
【咔叽】[纺] khaki
【咔唑】[化] carbazole

胩 [化] carbylamine; isocyanide

咯 cough up
【咯痰】cough up phlegm
【咯血】spit blood; hemoptysis

kāi

开 ①(打开) open; turn on; be on: ～灯 turn on the light / ～电视机 turn on the TV / ～门 open the door / ～锁 open a lock; unlock ②(打通; 开辟) make an opening; open up; reclaim: 在墙上～个窗口 make a window in the wall ③(舒张; 分离) open out; come loose: 花～了 the flowers are out; the flowers are open / 扣儿～了 the knot has come untied ④(河流解冻) thaw; become navigable: 河～了 the river is navigable now ⑤(解除封锁、禁令、限制等) lift: ～禁 lift a ban ⑥(发动; 操纵) start; operate; drive; pilot; run; work: ～飞机 pilot an airplane / ～机器 operate a machine / ～汽车 drive a car / ～拖拉机 drive a tractor ⑦(军队开拔) set out; move: 军队正～往前线。 The troops are moving to the front. ⑧(开办) set up; run: ～茶馆 run a teahouse / ～工厂 set up a factory ⑨(开始) begin; start: ～拍 start shooting (a film) ⑩(举行) hold; run: ～运动会 hold an athletic meet; hold a sports-meet ⑪(写出) write out; make out; draw: ～介绍信 write a letter of introduction / ～收条 make out a receipt / ～药方 write out a prescription / ～支票 make out a check ⑫(支付; 开销) pay: ～销 pay expenses ⑬(沸腾) boil: 水～了 the water is boiling ⑭(十分之几的比例) percentage; proportion: 三七～ in the proportion of three to seven ⑮[印] division of standard size printing paper: 八～ octavo / 四～ quarto ⑯(黄金中的含纯金量) carat: 二十四～金 24-carat gold
【开拔】move; set out
【开办】open; set up; start; found; establish: ～工厂 start a factory / ～一个新企业 establish a new enterprise / ～一所学校 start a school ◇ ～费 organization expenses; preliminery expenses / ～税 organization tax
【开本】[印] format; book size: 八～ octavo / 三十二～ 32 mo / 十六～ 16 mo
【开标】open sealed tenders
【开采】mine; extract; exploit: ～煤矿 mine coal / ～石油 recover petroleum / ～天然气 tap natural gas; extract natural gas
【开场】begin; opening of a show; open; start: 戏已经～了。The play has already started.

【开场白】① [戏] prologue ② (讲话或文章的开始部分) opening speech; introductory remarks
【开车】① (开动车辆) drive; start ② (发动机器) set a machine going
【开诚布公】frankly and sincerely; open-heartedly: ～地谈一谈 speak frankly and sincerely
【开诚相见】deal with each other in all sincerity; treat sb. open-heartedly
【开除】expel; discharge; dismiss; fire; sack: ～公职 discharge sb. from public employment; take sb.'s name off the books / ～某人 give sb. the sack / ～学籍 expel from school / 被～ get the sack
【开锄】start hoeing
【开船】set sail; sail ◇ ～时间 sailing time; hour of sailing
【开创】start; initiate; found; set up; pioneer; open: ～社会主义新风尚 initiate a new socialist custom / ～新纪元 open a new epoch / ～新局面 create a new situation in
【开春】beginning of spring
【开大肌】[生理] dilator muscles
【开裆裤】open-seat pants; split pants
【开刀】① (动手术) perform or have an operation; operate or be operated on: 给病人～ operate on a patient ② (先从某个方面下手) make sb. the first target of attack: 拿某人～ make an example of sb.; punish sb. first as a warning to others
【开导】help sb. to see what is right or sensible; help sb. to straighten out his wrong or muddled thinking; enlighten; give guidance to
【开倒车】turn the clock back; turn back the wheel of history; return to the past
【开道】clear the away
【开动】① (开行运转) start; set in motion: ～机器 start a machine / ～脑筋 use one's brains; think hard / ～宣传机器 set the propaganda machine in motion ② (开拔前进) move; march; be on the move
【开冻】thaw: 趁还未～，我们去溜冰吧。Let's go skating before the thaw sets in.
【开端】beginning; start: 良好的～ a good beginning
【开恩】show mercy; bestow favors
【开发】develop; open up; exploit: ～山区 develop mountain areas / ～实业 open up industries / ～油田 open up oilfields / ～自然资源 exploit natural resources ◇ ～成本 development cost / ～规划 development project / ～银行 development bank / ～周期 construction cycle
【开饭】serve a meal
【开方】extraction of a root; evolution: 开立方 extract the cube root / 开平方 extract the square root
【开放】① (开花) come into bloom ② (解除封锁、限制等) lift a ban; lift a restriction ③ (道路允许通行) open to traffic; open to public use ④ (接待游人、参观者等) be open to the public: 展览会将于 10 月 15 日起～。The exhibition is going to open to the public on October 15. ◇ ～城市 the open cities / ～地带 the open regions / ～港 open port / ～经济 open economy / ～政策 the policy of opening to the world
【开赴】march to; be bound for: ～前线 march to the front
【开工】① (工厂开始生产) go into operation; start operation; be put into operation: ～不足 be operating under capacity ② (土木工程开始修建) start; begin construction; begin building: 水库工程～了。Construction of the reservoir has started. ◇ ～率 utilization of capacity
【开沟机】ditching machine; trench digger
【开关】[电] switch; button: 分档～ step switch / 双向～ a two-way switch / 通断～ on-off switch / 闸刀～ knife switch ◇ ～厂 switchgear plant / ～屏 switchboard
【开锅】(of a pot) boil
【开国】found a state ◇ ～大典 founding ceremony (of a state) / ～元勋 a founding father of a country
【开航】① (开始通航) become open for navigation; be open to navigation ② (船只启程) set sail
【开河】① (开辟河道) construct a canal; open up a waterway ② (河流解冻) thaw; breakup of the ice in a river
【开合桥】bascule bridge; folding bridge
【开户】open an account; establish an account: 在银行～ open an account with the bank
【开花】① (花朵开放) flower; bloom; blossom; be in flower; come into flower: 心里乐～ burst with joy; feel elated / 玫瑰要～了。The roses are blooming. ② (裂开; 炸开) explode ◇ ～期 [植] florescence
【开花结果】blossom and bear fruit; yield positive results
【开化】become civilized: 未～的 uncivilized
【开怀】to one's heart's content: ～畅饮 drink

(alcohol) to one's heart's content; go on a drinking spree
【开荒】open up wasteland; reclaim wasteland
【开会】hold a meeting; have a meeting; meet; attend a meeting; go to a meeting: 他们下午～。They'll have a meeting this afternoon. / 她正在～。She is at a meeting.
【开荤】begin or resume a meat diet; end a meatless diet
【开火】open fire; fire: 向敌人～ fire at the enemy
【开豁】①(宽阔;爽朗) open and clear ②(思想开阔) with one's mental outlook broadened
【开价】opening price; state a price; make a quotation
【开架】open-shelf ◇ ～阅览室 open-shelf reading room
【开讲】begin lecturing or story-telling
【开戒】break an abstinence (from smoking, drinking, etc.)
【开禁】lift a ban
【开卷】open a book; read: ～有益 Reading is always profitable ◇ ～考试 open-book examination
【开掘】dig: ～运河 dig a canal
【开课】①(学校开始上课) school begins ②(担任某一课程的教学) give a course; teach a subject; deliver a course
【开垦】open up wasteland; reclaim wasteland; bring under cultivation: ～荒山 bring barren hills under cultivation
【开口】①(开口说话) open one's mouth; start to talk; begin to speak: 不～ hold one's tongue / 难以～ find it difficult to bring the matter up ②(开刃) put the first edge on a knife
【开口销】[机] split pin; cotter pin
【开口子】①(堤岸决口) break; burst ②(皮肤裂口) chap
【开快车】①(加快车速) drive a high speed; speed up; step on the gas ②(加快机器速度) speed up ③(加快工作速度) hurry through one's work; make short work of a job
【开矿】open up a mine; exploit a mine
【开阔】①(宽广) open; wide: ～的广场 an open square ②(乐观、畅快,不阴郁低沉) tolerant: 心胸～ broad-minded; unprejudiced ③(使开阔) widen: ～眼界 broaden one's outlook ◇ ～地 [军] open terrain; open ground; unenclosed ground
【开朗】①(地方开阔;光线充足) open and clear: 豁然～ suddenly see the light ②(乐观、畅快) sanguine; optimistic: 性情～ of a sanguine disposition; always cheerful
【开犁】start the year's ploughing
【开例】create a precedent
【开镰】start harvesting
【开列】draw up (a list); list: ～如下 as listed below / ～名单 make a list of names / ～清单 draw up a list; make out a list; make an inventory
【开路】①(开辟道路) open a way; blaze a trail; open a new road; break a fresh path: 逢山～, 遇水搭桥 cut paths through the mountains and build bridges across the rivers ② [电] open circuit ◇ ～先锋 pathbreaker; trailblazer; pioneer
【开门】open the door
【开门红】make a good beginning; get off to a good start: 来一个新年～。Let the beginning of the year be crowned with achievements.
【开门见山】come straight to the point; declare one's intention right at the outset
【开门揖盗】open the door to robbers; invite disaster by letting in evildoers
【开明】enlightened; liberal: ～的思想 liberal ideas / ～人士 enlightened persons / ～绅士 enlightened gentry
【开幕】①(演出开始) the curtain rises; begin the performance ②(会议、展览等开始) open; inaugurate ◇ ～词 opening speech; inaugural address / ～式 opening ceremony; inaugural ceremony; inauguration
【开盘】[经] opening quotation (on the exchange) ◇ ～汇率 opening rate / ～价格 opening price
【开炮】①(发射炮弹) open fire with artillery; fire; bombard ②(严厉批评) fire criticism at sb.; severely criticize
【开辟】①(打开道路) open up; hew out; break: ～航线 open an air route; open a sea route / ～一条上山的路 cut a path up the hill; hew out a path up the hill ②(创立;建设) open up; set up; establish: ～专栏 start a special column; set up a special column ③(开发) open up; develop: ～财源 tap new financial resources
【开票】①(选举) open the ballot box and count the ballots ②(开发票) make out an invoice
【开屏】(of a peacock) spread its tail; display its fine tail feathers
【开启】open: 自动～ open automatically
【开枪】fire with a rifle, pistol, etc.; shoot: ～还

击 return fire / ～射击 open fire / 向某人～fire at sb.; shoot at sb.
【开腔】begin to speak; open one's mouth
【开窍】have one's ideas straightened out; enlighten: 他的一番话使我开了窍。His words have enlightened me.
【开球】[足球] kick off
【开山】cut into a mountain (for quarrying, etc.): ～筑路 cut the mountain for a road; build roads by blasting mountains
【开山祖师】the founder (of a religious sect or a school of thought); originator
【开设】①(设立) open; set up; found: ～商店 open a shop / ～医院 establish a hospital ②(设置) offer (a course in college)
【开审】sit at session ◇ ～日期 hearing time
【开始】①(从头起) begin; start; commence: ～生效 take effect; come into effect / 从头～start from the very beginning ②(开始阶段) initial stage; beginning; outset
【开市】①(开始营业) reopen after a cessation of business ②(每天第一次成交) the first transaction of a day's business ◇ ～行情 opening quotation
【开释】release (a prisoner); acquit; acquittal
【开水】①(正在开着的水) boiling water ②(开过的水) boiled water: 用～沏茶 make tea with boiling water
【开司米】cashmere
【开台】begin a theatrical performance
【开天窗】put in a skylight; leave a blank in a publication to show that sth. has been censored
【开天辟地】①(开辟天地) the creation of the world; genesis ②(有史以来) since the dawn of history; since the beginning of history
【开庭】open a court session; call the court to order; hold a court ◇ ～审理 hold hearings / ～通知 notice of trial
【开通】①(打开通路) remove obstacles from; dredge clear: ～河道 dredge a river ②(使不闭塞) enlighten ③(不守旧) open-minded; liberal; enlightened; broad-minded
【开头】begin; start: 请你先开个头。Would you make a start? / 万事～难。The first step is always difficult.
【开脱】absolve; exonerate; exculpate: ～罪责 absolve sb. from guilt or blame / 替某人～plead for sb.
【开拓】①(开辟;扩展) open up; reclaim: ～道路 open up a path / ～荒地 reclaim wasteland ② [矿] developing; opening ◇ ～精神 pioneering spirit
【开外】over; above; more than: 离城五公里～over five kilometers from the town / 他有六十～了。He is over sixty.
【开玩笑】crack a joke; joke; make fun of
【开往】leave for; be bound for: ～上海的特别快车 the Shanghai express
【开胃】whet the appetite; stimulate one's appetite ◇ ～酒 aperitif / ～食品 appetizer
【开小差】①(当逃兵) desert ②(思想不集中) be absentminded: 思想～ be woolgathering
【开销】①(支付) pay expenses: 一切费用都已经～了。All the expenses have been paid up. ②(费用) expense: 日常的～ daily expenses; running expenses
【开心】①(快乐;舒畅) feel happy; rejoice; joyful; be delighted ②(戏弄别人) amuse oneself at sb.'s expense; make fun of sb.
【开学】school opens; term begins
【开眼】open one's eyes; widen one's view; widen one's horizons; broaden one's mind
【开演】begin; start: 戏今晚七点～。The play begins at 7 p.m. this evening.
【开业】①(商店开业) start business; open for business ②(律师、医生等开业) open a private practice: ～行医 practice medicine
【开夜车】work late into the night; put in extra time at night; burn the midnight oil
【开源节流】broaden sources of income and reduce expenditure; increase income and decrease expenditure; tap new resources and economize on expense
【开凿】dig; cut: ～隧道 cut a tunnel / ～运河 dig a canal
【开斋】①(恢复吃荤) resume a meat diet ② [伊斯兰教] come to the end of Ramadam ◇ ～节 Lasser Bairam; the Festival of Fast-breaking
【开展】①(使展开) develop; launch; unfold; carry out: ～课外活动 develop extracurricular activities / ～批评自我批评 carry out criticism and self-criticism ②(开朗;开豁) open-minded; politically progressive
【开战】make war; open hostilities
【开绽】come unsewn: 鞋子～了。The shoe has split at the seams.
【开张】①(开始营业) open a business; begin doing business: 重新打鼓另～ reopen a business to the beating of gongs and drums; start all over again ②(做第一笔交易) the first transaction of a day's business ◇ ～大吉 auspicious beginning of a new enterprise

【开仗】make war; open hostilities

【开帐】①(开列帐单) make out a bill ②(支付帐款) pay the bill

【开支】①(支付) pay (expenses); spend ②(费用) expenses; expenditure; spending: 节省～ cut down expenses; retrench / 军费～ military spending ③ [方](发工资) pay wages; pay salaries; get the pay: 我们每月 6 日～。 We got our pay on the 6th of every month.

【开宗明义】make clear the purpose and main theme from the very beginning; state the purpose from the very beginning

【开足马力】put into high gear; go full steam ahead; open the throttle; switch to top gear

锎 [化] californium (Cf)

揩 wipe: ～汗 wipe sweat / ～眼泪 wipe out one's tears

【揩油】get petty advantages at the expense of other people or the state; scrounge

kǎi

慨 ①(愤激) indignant: 对某事感到愤～ be indignant at sth. ②(感慨) deeply; touched: 感～ sign with emotion ③(慷慨) generous: ～允 consent readily; kindly promise

【慨然】①(感慨地) with emotion; with deep feeling: ～长叹 heave a sigh of regret ②(慷慨地) generously; without stint: ～相赠 give generously / ～相助 help without stint / ～允许 generously promise

【慨叹】sigh with regret

楷 ①(法式; 模范) model; pattern ②(楷书, 汉字手写正体) regular script: 大～ regular script in big characters / 小～ regular script in small characters

【楷模】model; pattern; good example

【楷书】regular script

【楷体】①(楷书, 汉字手写正体) regular script ②(拼音字母的印刷体) block letter

铠

【铠甲】(a suit of) armor

【铠装】[电] armor ◇ ～电缆 armored cable

凯 ①(胜利的乐歌) triumphant strains ②(胜利的) triumphant; victorious

【凯歌】a song of triumph; paean: 一曲壮丽的～ a magnificent paean

【凯门鳄】[动] caiman

【凯旋】triumphant return: ～归来 return in triumph ◇ ～门 triumphal arch

kān

刊 ①(排印出版) print; publish: 出～ publish / 停～ suspend publication; stop publication (of a newspaper, etc) ②(定期或不定期出版物) periodical; publication: 半月～ fortnightly / 报～ newspapers and magazines / 季～ quarterly / 纪念～ memorial volume / 年～ annual; yearbook / 期～ journal; periodical 双月～ bimonthly / 双周～ fortnightly; biweekly / 特～ special issue / 月～ monthly / 周～ weekly (publication) / 增～ supplement ③(削除; 修改) delete; correct: ～误 correct errors in printing

【刊登】publish in a newspaper 〈magazine〉; carry: ～广告 print an advertisement; advertise

【刊头】masthead of newspaper 〈magazine〉

【刊物】publication: 定期～ periodical (publication) / 科技～ publications on science and technology

【刊行】print and publish

【刊载】publish; carry

堪 ①(可以; 能够) may; can: ～称佳作 may be rated as a good piece of writing or a fine work of art / ～当重任 be capable of shouldering important tasks; can fill a position of great responsibility ②(能忍受) bear; endure: 不～回首 cannot bear to look back / 不～一击 cannot withstand a single blow; collapse at the first blow

戡 suppress

【戡乱】suppress a rebellion; put down a rebellion

勘 ①(校订; 核对) read and correct the text of; collate ②(实地查看; 探测) investigate; survey

【勘测】survey

【勘察】①(实地调查) reconnaissance ② [地] prospecting

【勘探】exploration; prospecting: 磁法～ magnetic prospecting / 地震～ seismic prospecting ◇ ～地震学 exploration seismology / ～队 prospecting team

【勘误】correct errors in printing ◇ ～表 errata; corrigenda

龛 niche; shrine

看 ①(照料) look after; take care of; tend: ~孩子 look after children / ~一群羊 tend a flock of sheep ②(守护) guard; keep watch on: ~仓库 guard the warehouse / ~瓜 keep watch in the melon fields ③(监视; 看押) keep under surveillance

【看场】guard the threshing floor (during the harvest season)

【看管】①(照管) look after; attend to: ~行李 look after the luggage ②(看守) guard; watch: ~犯人 guard prisoners

【看护】①(护理) nurse; look after; take care of; tend; attend on: ~病人 look after the patient ②(护士的旧称) nurse

【看家】①(看守; 照看门户) look after the house; mind the house ②(胜过别人的绝技等) outstanding (ability); special (skill) ◇ ~本领 one's special skill / ~狗 watchdog

【看门】①(看管大门) guard the entrance; act as doorkeeper ②(看家) look after the house

【看青】keep watch over the ripening crops

【看守】①(守卫; 照料) watch; guard: ~仓库 guard a storehouse ②(监视管理犯人) guard: ~犯人 guard prisoners ③(监狱看守人员) gaoler; jailor(美); warder; guard: 女~ wardress; matron ◇ ~所 lockup for prisoners awaiting trial; detention house

【看守内阁】caretaker cabinet

【看押】take into custody; detain

kǎn

槛 threshold

侃

【侃侃而谈】speak with fervor and assurance

坎 bank; ridge: 田~儿 a raised path through fields

【坎肩儿】sleeveless jacket; waistcoat

【坎坷】①(坑坑洼洼) bumpy; rough: ~不平的道路 a rough and bumpy road ②(不得志) full of frustrations: ~一生 a lifetime of frustrations

【坎儿井】karez; an irrigation system of wells connected by underground channels used in Xinjiang

莰 [化] camphane; bornane

砍 ①(用刀斧砍) cut; chop; hack; hew: ~柴 cut firewood / ~倒 cut down; chop down ②(去掉) cut (down): 把这篇文章~去一半。Cut the article down to half its length.

【砍刀】chopper

【砍伐】fell (trees)

【砍头】chop off the head; behead

kàn

瞰 look down from a height; overlook: 鸟~ get a bird's-eye view

看 ①(观看) see; look at; watch: ~电视 watch TV / ~电影 see a film; go to the movies / ~球赛 watch a ball game / ~戏 go to the theatre; see a play, an opera, etc. ②(阅读) read: ~报 read a newspaper / ~书 read (a book) ③(观察并加以判断) think; consider; view; observe; judge: 从实质上~ judging by essentials / 你对这件事怎么~? What's your view on this matter? ④(访问) call on; visit; go to see: ~朋友 visit a friend; call on a friend ⑤(诊治) treat: 她正在~病。She is under medical treatment. ⑥(照料) look after: ~孩子 look after children ⑦(取决于) depend on: 明天打不打场,得~天气。Whether we'll do the threshing tomorrow will depend on the weather. ⑧(提醒对方可能发生的不好情况) mind; watch out: 别跑这么快! ~摔着! Don't run so fast! Mind you don't fall! ⑨(表示试一试) 尝尝~ just taste this / 等等~ wait and see / 摸摸~ just feel it / 试试~ have a try; try and see / 让我想想~。Let me see.

【看病】①(医生治病) see a patient; treat; attend; give medical advice: 赵医生~很认真。Dr. Zhao handles his cases with great care. ②(病人就医) see a doctor; consult a doctor; go to a doctor; receive medical; advice: 带病人去~ take sb. to a doctor / 请医生来~ send for a doctor / 今天下午我去~。I'm going to see a doctor this afternoon.

【看不惯】cannot bear the sight of; frown upon; hate to see: 我们~这种浪费现象。We hate to see such waste.

【看不起】look down upon; scorn; despise; disdain; hold in contempt: 懒汉总叫人~。The lazybones are always held in contempt.

【看不上眼】spurn; disdain; hold in contempt

【看不顺眼】things are disgusting

【看成】look upon as; regard as; treat as; consider as: 不要把幻想~事实。Don't regard fantasies as truth. / 她被~全公司的标兵。She is considered a pace-setter in the corporation.

【看出】make out; see; perceive; find out; be aware of: ～问题的所在 see where the trouble is／～形势的严重性 be aware of the gravity of the situation／看不出真假 cannot tell whether it is genuine or fake

【看穿】see through; penetrate: ～敌人的阴谋 see through the enemy's plots

【看待】look upon; regard; treat: 他从不把雇工当人～。 He never regarded the farmhands as human beings.

【看到】①(看见) see; catch sight of: 看不到 fail to see; lose sight of ②(注意到) notice; note; be aware of: ～形势的变化 notice the change in situation

【看得起】have a good opinion of; think highly of; think much of

【看跌】(of market prices) be expected to fall

【看法】a way of looking at a thing; view: 对那件事我们的～有分歧。 Our opinions were divided on the matter.

【看风使舵】trim one's sails; adapt oneself to circumstances

【看见】catch sight of; see

【看来】it seems; it appears; it looks as if: 她～没有那么大年纪。 She does not look her age.／这活儿～今天可以做完。 It looks as if we'll be able to finish this job today.

【看破】see through; ～红尘 be disillusioned with the mortal world; use through the vanity of life

【看齐】①(口令) dress: 向右～! Dress right, dress!／向左～! Dress left, dress! ②(作为学习榜样) keep abreast with; keep up with; emulate: 向先进工作者～ emulate the advanced workers

【看轻】underestimate; look down upon

【看清】①(看清楚) see clearly ②(认识清楚) realize: 她～了问题的性质。 She realized the nature of the problem.

【看热闹】watch the scene of bustle; be a looker-on

【看人嘴脸】live on another's favor

【看上】like; take a fancy to; settle on: ～那个地方 take a liking to the place／～一位姑娘 take a fancy to a girl／我～了这个式样。 I like this style.

【看台】[体] bleachers; stand; grandstand (指正面看台)

【看透】①(透彻地了解) understand thoroughly; gain an insight into: ～某人的心思 gain an insight into sb.'s mind ②(透彻地认识) see through

【看头】[口] sth. worth seeing or reading: 这部电影很有～。 This film is well worth seeing.

【看图识字】learn to read with the aid of pictures

【看望】call on; visit; see: ～伤病员 go and see the wounded and sick

【看眼色】be ready to take hint: ～行事 take one's cue from sb.

【看涨】(of market prices) be expected to rise

【看中】take a fancy to; settle on

【看重】①(重视) regard as important; value; set store by: ～老师的忠告 value the teacher's advice ②(估计过高) overestimate: 别把问题～了。 Don't overestimate the seriousness of the problem.

【看准】be certain (about sth.)

【看做】look upon as; regard as

kāng

康 well-being; health

【康拜因】combine (harvester)

【康采恩】[经] concern

【康复】restored to health; recovered; recuperate one's health

【康健】healthy; in good health

【康乐】peace and happiness

【康乐球】caroms

【康铜】constantan

【康庄大道】broad road; main road; wide free road: 通向共产主义的～ the broad road to communism

慷

【慷慨】①(激昂) vehement; fervent: ～陈词 present one's views vehemently ②(不吝惜) generous; liberal: ～解囊 make generous contributions (of funds); help sb. generously with money／慷他人之慨 be generous at other's expense

【慷慨悲歌】chant in a heroic but mournful tone

【慷慨激昂】impassioned; vehement: ～的演说 vehement speech

【慷慨就义】go to one's death like a hero; die a martyr's death; die a hero

糠 ①(麦子、谷子的皮壳) chaff; bran; husk ②(萝卜变空) spongy: 萝卜～了。 The radish has turned spongy.

【糠秕】chaff
【糠醛】[化] furfural ◇ ～树脂 furfural resin

káng

扛 carry on the shoulder; shoulder: ～枪 shoulder a gun; bear arms
【扛长活】work as a farm laborer on a yearly basis
【扛活】work as a farm laborer

kàng

亢 ①(高) high; haughty: 不～不卑 neither supercilious nor obsequious／高～ loud and sonorous; resounding ②(过度; 极; 很) excessive; extreme
【亢奋】stimulated; excited
【亢旱】severe drought
【亢进】[医] hyperfunction: 甲状腺机能～ hyperthyroidism
炕 *kang*; a heatable brick bed
【炕席】*kang* mat
【炕桌】*kang* table; a small, short-legged table for use on a *kang*
抗 ①(抵抗; 抵挡) resist; combat; fight: ～暴斗争 struggle against violent repression／～敌 fight the enemy／～灾 fight natural calamities ②(拒绝; 抗拒) refuse; defy: ～捐～税 refuse to pay levies and taxes ③(对等) contend with; be a match for: 分庭～礼 stand up to sb. as an equal
【抗癌药】anticarcinogen; anticancer drugs
【抗爆】[化] antiknock ◇ ～剂 antiknock (agent); antidetonant／～汽油 antiknock gasoline
【抗辩】①(拒绝责难进行辩护) contradict ② [法] counterplea; demurer; counterargument ◇ ～者 opposing counsel
【抗病】[农] disease-resistant ◇ ～性 disease resistance
【抗磁性】[物] diamagnetism
【抗代谢物】antimetabolite
【抗倒伏】[农] lodging-resistant; resistant to lodging
【抗敌素】colistin; polymyxin E
【抗毒素】antitoxin
【抗毒血清】[医] antitoxic serum
【抗旱】fight a drought; fight against drought: ～措施 drought-relief measures ◇ ～品种 drought-resistant variety／～作物 drought-resistant crops
【抗衡】contend with; match; rival be evenly matched with
【抗洪】fight a flood; combat a flood: ～排涝 fight the flood and drain water-logged areas／～抢险 combat a flood and go to the rescue hurridy
【抗坏血酸】ascorbic acid; vitamin C
【抗击】resist; beat back; fight: ～侵略者 resist the aggressors
【抗拒】resist; defy
【抗菌素】antibiotic
【抗拉强度】[机] tensile strength
【抗美援朝战争】the War to Resist US Aggression and Aid Korea (1950-1953)
【抗命】defy orders; disobey
【抗凝血药】anticoagulant
【抗日战争】the War of Resistance Against Japan (1937-1945)
【抗渗】impervious ◇ ～试验 impermeability test
【抗生素】antibiotic
【抗霜】[农] frost-resistant
【抗水性】water-resistance; water-resisting property
【抗诉】counterappeal; lodge a protest against court judgments; oppose an action
【抗体】antibody
【抗弯强度】[机] bending strength
【抗血清】antiserum
【抗压强度】compressive strength; compressive resistance ◇ ～试验 compressive strength test
【抗药性】resistance to the action of a drug: 产生～ become drug-fast
【抗氧化剂】anti-oxidant
【抗议】protest; remonstrate; [法] object: 提出～ lodge a protest ◇ ～集会 protest rally／～照会 note of protest
【抗原】[医] antigen ◇ ～抗体反应 antigen-antibody reaction
【抗战】①(抵抗外国侵略的战争) war of resistance against aggression: ～到底 fight to the bitter end ②[简](抗日战争) Anti-Japanese War (1937-1945): ～时期 period of the War of Resistence Against Japan
【抗张强度】tensile strength
【抗震】anti-seismic: ～结构 anti-seismic structure／～救灾工作 earthquake relief work
【抗争】make a stand against; resist

钪 scandium (Sc)

伉

【伉俪】married couple; husband and wife

kǎo

考 ①(考试) examine; give an examination; take an examination; test: ～满分 get full marks／～上大学 be admitted to a university／～物理 have a test in physics／～学校 take entrance examination／补～ a make-up exam／统～ the unified state examination／应～ sit for an examination; take an examination ②(检查) check; inspect ③(推求; 研究) study; investigate; verify: 待～ remain to be verified

【考查】examîne; test; check: ～学生成绩 check students' work

【考察】①(实地调查) inspect; make an on-the-spot investigation; investigate: ～水利工程 investigate water conservancy projects／出国～ go abroad on a tour of investigation ②(细致深刻地观察) observe and study ◇～报告 report on an investigation／～团 investigation group; inspection delegation; observation delegation／～组 study group

【考场】examination hall; examination room

【考的松】[医] cortisone

【考古】①(研究古代历史) engage in archaeological studies ②(考古学) archeology ◇～学 archaeology／～学家 archaeologist

【考核】examine; check; assess (sb.'s proficiency): ～干部 check on cadres／定期～ routine check／建立～制度 set up a check-up system

【考绩制度】merit system

【考究】①(查考; 研究) observe and study; investigate; examine closely: 这个问题很值得～。 We must make a careful study of the matter. ②(讲究) care about; fastidious; particular: 穿衣服过于～ be too particular about dress ③(精美) exquisite; fine: 这本画册装订得很～。 This album is beautifully bound.

【考据】textual criticism; textual research

【考卷】examination paper

【考虑】think over; take into account; consider: ～不周的 ill considered; inadequately considered／～周到的 long thought out preparations; well-considered／不～个人得失 disregard personal gains and losses／首先～ first consideration／正在～的问题 problems under consideration

【考勤】check on work attendance ◇～簿 attendance record

【考取】pass (an entrance) examination; be admitted to school or college

【考生】candidate for an entrance examination; examinee

【考试】examination; test; exam: 大学入学～ college entrance examination／期末～ end-of-term examination／期中～ mid-term examination／入学～ entrance examination／学年～ year-end examination／参加～ take an examination／举行～ hold an examination／～不及格 fail (in) an examination／～及格 pass an examination ◇～制度 examination system

【考题】examination questions; examination paper: 出～ set an examination paper; set examination questions

【考问】examine orally; question

【考验】test; trial; ordeal: 久经～的 long-tested／经过时间～的 time-tested

【考证】textual criticism; textual research: 经～ as a result of textual research

烤 ①(用火烤熟或烤干) bake; roast; toast: ～面包 toast bread／～肉 roast meat／～羊肉串 roast mutton cubes on a skewer／把衣服～干 dry clothing near fire ②(太热) scorching: 这个炉子太～人。 This stove is really scorching

【烤电】[医] diathermy

【烤火】warm oneself by a fire

【烤炉】oven

【烤面包】toast: 一片～ a slice of toast ◇～夹 toast rack／～架 toaster

【烤肉】roast meat; roast ◇～叉 spit; skewer

【烤箱】coal-scuttle

【烤鸭】roast duck: 北京～ roast Beijing duck

【烤烟】flue-cured tobacco

栲 [植] evergreen chinquapin

【栲胶】tannin extract

拷 flog; beat; torture

【拷贝】[电影] copy: 现在上映的电影是个新～。 The picture now showing is a new copy.

【拷贝纸】copy paper; coping paper

【拷打】flog; beat; torture: 严刑～ subject sb. to severe torture

【拷花】[纺] embossing ◇～布 embossed cloth

【拷问】torture sb. during interrogation; interrogate with torture

kào

铐 ①(手铐) handcuffs ②(戴手铐) put handcuffs on; handcuff: 把罪犯～起来 handcuff the criminal

靠 ①(依着) lean against; lean on; rest against: ～在墙上 lean against a wall / ～在桌子上 lean upon a desk ②(沿着) keep to; get near; come up to: 火车都～左走。 All trains should keep to the left. ③(挨近) near; by: 疗养院～海。 The sanatorium stands by the sea. ④(依靠) depend on; rely on: ～自己的努力 rely on one's own efforts ⑤(信赖) trust: 可～ reliable; trustworthy

【靠岸】pull in to shore; draw alongside
【靠背】back (of a chair) ◇ ～椅 chair
【靠边】keep to the side
【靠不住】unreliable; undependable; untrustworthy: 她的话～。 What she said was unreliable.
【靠得住】reliable; dependable; trustworthy: 这消息～。 The news is reliable.
【靠垫】(backrest) cushion
【靠近】①(挨近) near; close to; by: 车站～我们学校。 The station is near our school. ②(靠拢) draw near; approach: 客轮慢慢地～码头。 The passenger boat is nearing the dock.
【靠拢】draw close; close up
【靠山】backer; patron; backing
【靠山吃山,靠水吃水】those living on a mountain live off the mountain, those living near the water live off the water; one has to make use of whatever resources available; make use of local resources
【靠手】armrest
【靠枕】back cushion

犒 reward with food and drink

【犒劳】reward with food and drink; give food and drink for meritorious service: ～三军 feast the army
【犒赏】reward a victorious army, etc. with bounties

kē

颏 chin

磕 ①(碰在硬东西上) knock ②(磕打) rap

【磕打】knock sth. out of a vessel, container, etc.; knock out
【磕磕绊绊】stumble; walk with difficulty
【磕磕撞撞】stagger along; reel; walk unsteadily
【磕碰】knock against; collide with; bump against
【磕头】kowtow: 给某人～ make a kowtou to sb.; give sb. a kowtou
【磕头碰脑】①(人和东西相碰) bump against things on every side ②(人和人相碰) push and bump against one another

瞌

【瞌睡】sleepy; drowsy: 打～ doze off; nod; fall into a doze

珂

【珂罗版】collotype ◇ ～印刷 collotype printing

苛 severe; exacting

【苛待】treat harshly; be hard upon
【苛捐杂税】exorbitant taxes and levies; multifarious taxes
【苛刻】harsh: ～待人 treat others harshly / ～的条件 harsh terms
【苛求】make excessive demands; be overcritical
【苛性】[化] causticity ◇ ～钾 caustic potash / ～钠 caustic soda
【苛责】criticize severely; excoriate
【苛政】harsh government; oppressive government; tyrannical government; tyranny

窠 nest; burrow: 鸟～ a bird's nest

【窠臼】set pattern: 不落～ show originality; be unconventional

颗

【颗粒】①(小而圆的东西) pellet; anything small and roundish (as a bean, pearl, etc.) ②(粮食) grain: ～归仓 get in every single grain; every grain to the granary ◇ ～肥料 granulated fertilizer / ～饲料 pellet / ～物质 particulate matter

科 ①(学科) branch of (academic) study: 理～ the sciences / 文～ the humanities; the liberal arts / 牙～ department of dentistry / 眼～ department of ophthalmology ②(机关单位) department; section: 财务～ finance section / 销售～ sales section / ～长 section chief ③[生] family: 豆～植物 bean family; legume / 犬～动物 animals of the dog family ④(判定) pass a

sentence: ～处徒刑 sentence sb. to imprisonment / ～以罚金 impose a fine on sb.; fine
【科班】regular professional training: ～出身 be a professional by training / ～出身的人 a man with professional training
【科尔夫球】[体] korfball
【科技】science and technology ◇ ～大学 university of science and technology / ～工作者 scientific and technical worker / ～规划 program for the development of science and technology / ～界 scientific and technological circles / ～情报 scientific and technical information / ～人才 skilled personnel / ～术语 scientific and technical terminology
【科教片】[简](科学教育影片) popular science film; science and educational film
【科举】imperial examination ◇ ～制度 imperial examination system
【科摩罗】Comoros
【科目】①(学科) subject (in a curriculum); course ②(会计帐目) headings in an account book
【科室】administrative or technical offices ◇ ～人员 office staff; office personnel
【科特迪瓦】the Ivory Coast
【科威特】Kuwait ◇ ～人 Kuwaiti
【科刑】sentence ◇ ～的法律条文 vindicative parts of laws / ～判决 judgment of sentence
【科学】science; scientific knowledge: 理论～ pure science / 社会～ social sciences / 应用～ applied science / 自然～ natural science ◇ ～工作者 scientific worker; scientist / ～管理 scientific management / ～幻想小说 science fiction; (美) sci-fi / ～技术 science and technology / ～家 scientist / ～理论 scientific theory / ～普及读物 popular science books; popular science / ～社会主义 scientific socialism / ～实验 scientific experiment / ～史 history of science / ～文献 scientific literature / ～研究 scientific research / ～仪器 scientific instruments; scientific apparatus / ～院 academy of sciences / ～哲学 philosophy of science
【科研】scientific research ◇ ～机构 scientific research institution / ～人员 scientific research personnel / ～项目 scientific research item

蝌

【蝌蚪】tadpole; polliwog

ké

咳

cough: 干～ dry cough / ～得利害 have a bad cough
【咳必清】carbetapentane citrate; toclase ◇ ～糖浆 toclase syrup
【咳嗽】cough: ～有痰 bring up phlegm when one coughs ◇ ～糖浆 cough syrup
【咳痰】expectoration

壳

shell: 核桃～ walnut shell / 鸡蛋～ egg shell

kě

渴

①(口干想喝水) thirsty: 解～ quench one's thirst ②(迫切地) yearningly; thirstily; eagerly: ～念 yearn for
【渴望】thirst for; long for; yearn for: ～回到战斗岗位 long to go back to one's fighting post / ～学习 thirst to learn / ～祖国早日统一 long for the early unification of one's motherland

可

①(同意) approve: 不置～否 decline to comment; be noncommittal ②(许可;可能) can; may: ～供效法 it may serve as a model / 由此～见 thus it can be seen that; this proves ③(值得) need (doing); be worth (doing): ～爱 lovable / ～悲 lamentable / ～靠 reliable ④(适合) fit; suit: 这回倒～了她的心。It suited her perfectly this time. / 这帽子你～心吗? Does the hat suit you? ⑤(表示转折,同"可是") but; yet
【可爱】lovable; likable; lovely: ～的小女孩 a lovely little girl / ～的祖国 my beloved country
【可保】insurable: ～财产 insurable property / ～价值 insurable value / ～权益 insurable interest
【可悲】sad; lamentable
【可比价格】[经] fixed price; constant price
【可鄙】contemptible; despicable; mean: ～的动机 a mean motive / 行为～ act contemptibly
【可变】variable ◇ ～成本 variable cost / ～电容器 variable condenser; variable capacitor / ～电阻器 rheostat / ～反差[摄] variable contrast / ～预算 variable budget / ～资本 [经] variable capital
【可采储量】recoverable reserves
【可操左券】be sure to succeed; be certain of success; have a winning hang
【可拆】removable; detachable
【可长可短】the length is changeable
【可乘之机】an opportunity that can be exploited to sb.'s advantage
【可耻】shameful; disgraceful; ignominious: ～

的行为 shameful conduct; disgraceful behavior

【可大可小】the size is changeable

【可待因】[药] codeine

【可的松】[药] cortisone

【可锻性】[冶] malleability; forgeability

【可锻铸铁】[冶] malleable (cast) iron

【可兑换证券】convertible bond

【可多可少】the amount doesn't matter

【可歌可泣】move one to song and tears: ～的英雄事迹 heroic and moving deeds

【可耕地】arable land; cultivable land

【可攻可守】equally valuable as a stepping-stone for offense or a strong point for defense

【可观】considerable; impressive; sizable: 损失～。 The losses are considerable.

【可贵】valuable; praiseworthy; commendable: ～品质 fine qualities

【可恨】hateful; detestable; abominable

【可换股份】convertible stock

【可加工性】[机] machinability

【可见】it is thus clear that; it is thus evident that

【可见度】visibility: ～差 poor visibility

【可见光】[物] visible light

【可进可退】be free to go forward or back out

【可惊】surprising; startling

【可敬】worthy of respect; respected

【可卡因】[药] cocaine

【可抗辩条款】contestable clause

【可靠】reliable; dependable; trustworthy: 革命事业的～接班人 reliable successors to the revolutionary cause ◇ ～性 reliability / ～帐户 reliable account

【可可】cocoa ◇ ～粉 cocoa powder / ～碱 theobromine / ～生产 cocoa production / ～脂 cocoa butter; theobrama oil

【可控硅】[电子] silicon controlled rectifier (SCR); thyristor

【可口】good to eat; nice; tasty; palatable: ～的菜肴 tasty dishe; delicious dishes

【可口可乐】Coca Cola

【可乐】Coke; Coca Cola: 百事～ Pepsi Cola

【可怜】①(值得怜悯) pitiful; pitiable; poor; pathetic: ～的孩子 a poor boy / 他露出一副～相。 He looks pitiable. ②(怜悯) pity; have pity on; take compassion on: 我～她。 I pity her. ③(数量少质量坏得不值一提) meager; wretched; miserable; pitiful: 穷得～ miserably poor / 少得～ pathetically meager

【可怜虫】pitiful creature; wretch

【可裂变物质】[物] fissile material

【可能】①(可以实现) possible; probable; can; may: 团结一切～团结的力量 unite all the forces that can be united ②(或许) probably; maybe: 她～在教室里。 Maybe she is in the classroom. ③(可能性) possibility: 甲队没有获胜的～。 Team A has no chance of winning. ◇～性 possibility

【可逆】reversible ◇ ～反应 [化] reversible reaction

【可怕】fearful; frightful; terrible; terrifying: ～的灾难 a dreadful disaster

【可欺】①(可欺侮) easily cowed; can be bullied; browbeaten ②(可欺骗) gullible; easily duped

【可气】annoying; exasperating

【可巧】as luck would have it; by a happy coincidence; it happened that: ～我那天身边没带钱。 It so happened that I had no money with me that day.

【可取】desirable; advisable; recommendable; commendable: 无～ nothing to recommend / 她的建议有～之处。 Her proposal has something to recommend it.

【可燃性】[化] combustibility; flammability

【可溶性】solubility

【可是】[连] but; yet; however

【可塑性】plasticity

【可望而不可即】within sight but beyond reach; unattainable; inaccessible

【可谓】one may well say; it may be said; it may be called

【可恶】hateful; abominable; detestable; loathsome: ～之极 utterly detestable

【可惜】it's a pity; it's too bad; unfortunately; it is to be regretted: ～我帮不了你的忙。 It is to be regretted that I can't help you. / 错过这个机会真～! What a pity to miss the chance!

【可喜】gratifying; heartening; ～的成就 gratifying achievements / ～可贺 you are to be congratulated / 取得了～的进展 have made encouraging progress

【可想而知】you can imagine

【可笑】①(滑稽可笑) laughable; ridiculous; ludicrous; funny: 滑稽～的故事 funny story ②(荒唐可笑) ridiculous; absurd: ～的想法 ridiculous idea

【可心】satisfying; to the satisfaction of; to the liking of

【可行】feasible; practicable; workable: 切实～的计划 practicable plan / 唯一～的办法 the only practical thing to do ◇ ～性 feasibility

【可疑】suspicious; dubious; questionable: 形迹～ look suspicious ◇ ～分子 a suspect; a suspicious character

【可以】①(表示可能) can; may: 我～在这里吸烟吗? May I smoke here? / 星星之火, ～燎原。A single spark can start a prairie fire. ②(好;不坏) passable; pretty good; not bad: 她的英语还～。Her English is pretty good. ③(利害) awful: 她那张嘴真～。What a sharp tongue she has got! ◇ ～断言 it can be asserted that; one may predict that / ～接受 acceptable; agreeable / ～看出 it can be seen; it can be perceived / ～理解 understandable; explicit / ～允许 permissible; allowable / ～证明 demonstrable; able to stand the proof; that can be proved

【可意】gratifying; satisfactory

【可意会不可言传】can be understood but can not be described

【可有可无】not essential; not indispensable; may or may not be needed

【可着】manage to make do: 你就～这块木料做吧。You'll have to make do with this timber.

【可支配收入】disposable income

【可知性】[哲] knowability

【可转让提单】negotiable bill of lading

【可转让证券】negotiable instruments

kè

刻 ①(雕刻) carve; engrave; cut: ～蜡版 cut stencils / ～图章 engrave a seal / 木～ woodcut ②(十五分钟) quarter: 四点一～ a quarter past four; four fifteen ③(时间) moment: 此～ at the moment / 稍等片～。Wait a moment. ④(形容程度极深) in the highest degree: 深～ penetrating; profound ⑤(刻薄) cutting; penetrating: 尖～ acrimonious; biting; sarcastic

【刻板】①(刻底版) cut blocks for printing ②(呆板) mechanical; stiff; inflexible: ～地照抄 copy mechanically

【刻版】cut blocks for printing ◇ ～印刷 block printing

【刻本】block-printed edition

【刻薄】unkind; harsh; mean: 待人～ be harsh towards others; treat people meanly / 说～话 speak unkindly; make caustic remarks

【刻不容缓】brook no delay; demand immediate attention; be of great urgency: ～的任务 urgent task; pressing task

【刻刀】burin; graver

【刻毒】venomous; spiteful; malignant; acrid: 对人～ be malignant towards sb. / 言语～ venomed remarks; have an acrid tongue

【刻度】graduation ◇ ～盘 graduated disc; dial / ～瓶 graduated bottle

【刻骨】deeply ingrained; deep-rooted: ～仇恨 inveterate hatred; deep-seated hatred

【刻骨铭心】be engraved on one's bones and heart; remember with gratitude to the end of one's life; inscribe debt of gratitude on one's mind

【刻花】engraved designs; carved designs ◇ ～玻璃 cut glass

【刻画】depict; portray: ～入微 portray to the life; vivid portrayal of details

【刻苦】①(很能吃苦) assiduous; hardworking; painstaking: ～钻研 study assiduously ②(俭朴) simple and frugal: 生活～ lead a simple and frugal life

【刻意】painstakingly; sedulously: ～求工 sedulously strive for perfection / ～求精 perfect one's skill assiduously; do one's every best to achieve perfection

【刻舟求剑】cut a mark on the side of one's boat to indicate the place where one's sword has dropped into the river; take measures without regard to changes in circumstances

克 ①(能) can; be able to: 不～分身 be unable to leave what one is doing at the moment; can't get away ②(克服;克制) restrain: ～制 exercise restraint ③(攻下;战胜) overcome; subdue; capture: 迭～名城 capture one important city after another / 攻无不～ be invincible ④(消化) digest: ～食 help one's digestion ⑤(严格限定期限) set a time limit: ～期完工 set a date for completing the work ⑥(重量;质量) gram (g.)

【克当量】[化] gram equivalent

【克敌制胜】vanquish the enemy; conquer the enemy

【克分子】[化] gram molecule ◇ ～浓度 molarity

【克服】①(战胜) surmount; overcome; conquer: ～官僚主义 get rid of bureaucracy / ～困难 surmount a difficulty / ～浪费现象 eliminate waste / ～片面性 eliminate one-sidedness / ～千难万险 surmount numerous difficulties and dangers / ～缺点 overcome one's shortcomings / ～私心杂念 overcome selfish considerations ②(克制;忍受困难) put up with

【克复】retake; recapture; recover: ～失地 recover lost territory
【克己奉公】wholehearted devotion to public duty; work selflessly for the public interest
【克己复礼】deny self and return to propriety
【克扣】embezzle part of what should be issued: ～分量 give short measure
【克拉】carat
【克朗】①(瑞典、冰岛本币) Krona ②(挪威、丹麦本币) Krone ③(捷克斯洛伐克本币) Koruna
【克郎球】caroms
【克厘米】gram-centimeter
【克勤克俭】be industrious and frugal; be able to practice diligence and frugality
【克丝钳】combination pliers; cutting pliers
【克原子】[化] gram atom
【克制】restrain; exercise restraint: ～感情 restrain one's passion / ～急躁情绪 control one's temper

氪
[化] krypton (Kr)

可
【可汗】[史] khan

课
①(教学科目) subject; course: 必修～ required courses / 基础～ basic course / 选修～ option courses / 主～ the main subject / 专业～ specialized courses ②(教学时间) class: 公开～ open class / 讲～ give a lecture / 上～ go to class / 上四节～ have four classes; have four periods / 听～ visit a class; sit in on a class; attend a lecture / 下～ finish class; get out of class / 一节物理～ a class in physics ③[量] lesson: 补～ make up missed lessons / 第二～ Lesson Two / 这本教科书共有十六～。This textbook contain 16 lessons. ④(征收) levy: ～以重税 levy heavy taxes
【课本】textbook: 化学～ a textbook on chemistry
【课表】school timetable
【课程】course; curriculum ◇ ～表 school timetable / ～设置 course offered
【课间操】setting-up exercises during the break
【课间休息】break; interval; recess: 十分钟的～ a ten-minute recess
【课时】class hour; period: 每周授课十二～ teach 12 period a week
【课税】levy duty; charge duty: ～基准 basis of assessment / ～价值 taxable value / ～年度 taxable year
【课堂】classroom; schoolroom ◇ ～教学 classroom instruction; classroom teaching / ～讨论 classroom discussion / ～作业 classwork
【课题】①(研究或讨论的问题) a question for study or discussion ②(急待解决的问题) problem; task: 提出新的～ pose a new problem; set a new task
【课外】extracurricular; outside class; after school ◇ ～辅导 instruction after class / ～活动 extracurricular activities / ～阅读 outside reading / ～作业 homework
【课文】text
【课椅】tablet chair
【课业】lessons; schoolwork: 荒废～ neglect one's studies
【课余】after school; after class: ～进行义务劳动 do voluntary labor after school
【课桌】(school) desk

骒
【骒马】mare

客
①(客人) visitor; guest; caller: 常～ a frequenter; a frequent visitor / 贵～ a distinguished guest / 稀～ a rare visitor / 远～ a visitor from afar / 会～ receive a visitor / 送～ see a visitor off ②(旅客) traveller; passenger: ～舱 passenger cabin ③(寄居或迁居外地) settle or live a strange place; be a stranger: ～死他乡 die abroad / 作～他乡 live in a strange land ④(客商) travelling merchant ⑤(顾客) customer: 房～ boarder; lodger ⑥(指奔走各地从事某种活动的人) a person engaged in some particular pursuit: 刺～ assassin / 政～ politician ⑦(客观) objective: ～观 objective
【客车】①[铁路] coach; passenger train ②(大轿车) bus; coach
【客船】passenger ship; passenger boat: 定期～ liner
【客串】play a part in a professional performance; be a guest performer
【客店】inn; tavern
【客队】[体] visiting team
【客饭】①(供客人的饭) a meal specially prepared for visitors at a canteen ②(份儿饭) set meal
【客房】guest room
【客观】objective ◇ ～必然性 objective necessity / ～存在 objective reality / ～价值 objective value / ～实在 objective reality / ～世界 objective world / ～事实 objective fact / ～事

物 objective things; objective reality / ～条件 objective condition / ～唯心主义 objective idealism / ～性 objectivity; objectiveness / ～因素 objective factor / ～原因 objective cause / ～真理 objective truth / ～主义 objectivism

【客观规律】objective law: 不以人们意志为转移的～ objective law independent of man's own will

【客货船】passenger-cargo vessel

【客机】passenger plane; airliner

【客籍】①(寄居本地的外地人) a settler from another province ②(寄居的籍贯) the province into which settlers move

【客满】sold out; full house: 剧场经常～。The theatre is always filled to capacity.

【客票】passenger ticket

【客气】①(说客气话;做出客气的动作) polite; courtequs: 请不要～。(对客人说时用) Please don't stand on ceremony.(或: Make yourself at home.); (对主人说时用) Please don't bother. ②(对人谦让) modest: 您太～了。You are being too modest.

【客人】①(宾客) visitor; guest ②(旅客) passenger(车船的); guest(旅馆的)

【客商】travelling trader; travelling merchant

【客套】polite formula; civilities: ～话 polite greetings / 不讲～ do away with formalities; not stand on ceremony

【客体】[哲] object

【客厅】drawing room; parlor

【客运】passenger transport; passenger traffic ◇～列车 passenger train

【客栈】inn

恪 scrupulously and respectfully

【恪守】scrupulously abide by (a treaty, promise, etc.)

kěn

肯 ①(同意) agree; consent: 首～ nod assent ②(乐意; 愿意) be willing to; be ready to: ～帮助人 be willing to help others / ～动脑筋 be ready to beat one's brains / ～干 be willing to do hard work / ～虚心接受批评 be ready to listen to criticism with an open mind

【肯定】①(承认事物的存在或真实性) affirm; confirm; approve; regard as positive: ～成绩 affirm the achievements ②(正面的) positive; affirmative: 她的回答是～的。Her answer is in the affirmative. ③(明确的; 确定的) definite; sure: ～的答复 a definite answer ④(一定; 无疑问) certainly; undoubtedly; definitely: 我们～能按时完成任务。We can certainly have the job finished on time.

【肯尼亚】Kenya ◇～人 Kenyan

啃 gnaw; nibble; bite: ～骨头 gnaw a bone ◇～书本 delve into books

恳 ①(诚恳) earnestly; sincerely: ～谈 talk earnestly ②(请求) request; beseech; entreat: 敬～ respectfully request

【恳切】earnest; sincere: ～陈词 make a statement earnestly / ～希望 earnestly hope; sincerely hope / 言词～ speak in an earnest tone

【恳请】earnestly request; cordially invite: ～光临。You are earnestly requested to be present at the party.(或: You are cordially invited to the party.)

【恳求】implore; entreat; beseech: ～体恤 implore for sympathy / ～支持 solicit sb.'s support

【恳挚】[书] earnest; sincere: 词意～ express one- self earnestly / 情意～ show sincere feeling

垦 cultivate (land); reclaim (wasteland): 军～农场 army reclamation farm

【垦荒】reclaim wasteland; bring wasteland under cultivation; open up virgin soil

【垦区】reclamation area

【垦殖】reclaim and cultivate wasteland

kēng

坑 ①(洼下去的地方) hole; pit; hollow: 粪～ manure pit / 泥～ mud puddle / 沙～[体] jumping pit / 水～ puddle / 土～ (sunken) pit / 一个萝卜一个～ one radish, one hole; each has his own task, and there is nobody to spare ②(地洞;地道) tunnel; pit ③(坑害) entrap; cheat: ～人 trap sb.; set a trap for sb.

【坑道】①[矿]gallery ②[军]tunnel ◇～工事[军] tunnel defenses; tunnel fortifications / ～战 tunnel warfare / ～作业 tunnelling

【坑害】lead into a trap; entrap; scheme to do harm; make false accusation against sb.

【坑坑洼洼】full of bumps and hollows; bumpy; rough

【坑木】[矿] pit prop; mine timber

吭 utter a sound or a word: 一声不～ with-

out saying a word
【吭哧】①(因用力而发出声音) puff and blow ②(吞吞吐吐) hum and haw ③(费力) work hard; toil
【吭气】【吭声】utter a sound or a word
铿 [象] clang; clatter
【铿锵】ring; clang; jingle: ～有力 be sonorous and forceful
【铿然】[书] loud and clear

kōng

空 ①(不包含什么; 不切实际的) empty; hollow; void; vacant: ～瓶子 an empty bottle / ～屋 an empty room / 一场～欢喜 a hollow joy ②(天空) sky; air: 低～ low altitude / 高～ upper air; high altitude / 领～ territorial sky; territorial air / 晴～ a clear sky; blue sky / 夜～ the night sky ③(白白地; 没有结果的) for nothing; in vain: ～忙 make fruitless efforts / ～跑一趟 make a journey for nothing ④[佛教] Emptiness; Void of the world of senses: 四大皆～ all space- directions are void
【空靶】air target; aerial target; airborne target
【空包弹】[军] blank cartridge
【空舱费】[交] dead freight
【空肠】[生理] jejunum
【空挡】[机] neutral (gear)
【空荡荡】empty; deserted: 肚子里～ on an empty stomach
【空洞】①(物体内的窟窿) cavity: 肺～ pulmonary cavity ②(无内容) empty; hollow; devoid of content; vague and general: ～的词句 mere phrases; empty phraseology / ～的建议 empty proposal / ～的理论 empty theory / ～的议论 vague and general proposal / ～无物 utter lack of substance; devoid of content
【空翻】[体] somersault; flip: 后～ backward somersault; backflip
【空泛】vague and general; not specific: ～的议论 vague and general opinions; generalities
【空防】air defense
【空腹】on an empty stomach
【空话】empty talk; empty words; idle talk; hollow words: ～连篇 pages and pages of empty verbiage; fill endless pages with empty talk / 说～ indulge in idle talk
【空怀】[牧] nonpregnant; barren
【空欢喜】rejoice too soon; be or feel let down
【空幻】visionary; illusory
【空架子】a mere skeleton; a bare outline
【空间】space: 外层～ outer space ◇ ～点阵[物] space lattice / ～电荷 space charge / ～定向障碍 spatial disorientation / ～防御 space defense / ～构架 space frame / ～技术 space technology / ～科学 space science / ～站 space station / ～知觉 space perception
【空降】airborne ◇ ～兵 airborne force; parachute landing force / ～部队 air-borne troop / ～地点 landing area / ～师 airborne division
【空军】air force ◇ ～部队 air (force) unit / ～基地 air base / ～司令部 general headquarters of the air force; air command / ～司令员 commander of the air force / ～武官 air attaché
【空空如也】absolutely empty; all empty; having nothing in it
【空口吃】eat dishes without rice or wine; eat rice or drink wine with nothing to go with it
【空口说白话】make empty promises; speak without taking action
【空口无凭】a mere verbal statement is no guarantee; mere verbal statement has no binding force: ～, 立字为证。 Words of mouth being no guarantee, a written statement is hereby given.
【空旷】open; spacious: ～的原野 an expanse of open country; champaign
【空阔】open; spacious: 水天～ a vast expanse of water and sky
【空廓】open; spacious: 四望～ spacious and open on all sides
【空论】empty talk
【空门】[佛教] Buddhism: 遁入～ become a Buddhist monk 〈nun〉
【空气】①(大气) air; atmosphere: 呼吸新鲜～ breathe fresh air / 湿～ moist air; moist atmosphere ②(气氛) atmosphere: ～紧张 a tense atmosphere ◇ ～弹道 aeroballistic trajectory; atmospheric trajectory / ～动力学 aerodynamics / ～净化器 air purifier / ～净化物 air purifier / ～冷却 air-cooling / ～力学 aeromechanics / ～栓塞 air embolism / ～弹簧 air spring / ～调节器 air conditioner / ～污染 air pollution / ～压缩机 air compressor / ～制动装置 air brake
【空前】unprecedented; unparalleled: ～的低落 be at an alltime low / ～的高涨 be at an alltime high / ～的孤立 be more isolated than ever before / ～的规模 on an unprecedented scale / ～的盛况 an unprecedentedly grand occasion; an unparalleled spectacular event /

处于～的困境 be in an unprecedented predicament／ 以～的速度发展 be developing at an unprecedented rate

【空前绝后】unprecedented and unrepeatable; without both precedent and following up; unique

【空勤】air duty ◇ ～人员 aircrew; aircraft crew; flight crew

【空手】empty-handed: ～而归 return empty

【空手道】[体] karate

【空谈】①(只说不做) indulge in empty talk ②(不切合实际的言论) empty talk; idle talk; prattle ◇ ～主义 phrase-mongering

【空调】air-condition: 带～的卧室 an air-conditioned bedroom ◇ ～机 air conditioner／ ～设备 air conditioning

【空头】①[经] bear; shortseller: ～市场 bear market ②(有名无实的) nominal; phony: ～文学家 phony writer／ ～政治家 armchair politician

【空头支票】①(不能兑现的支票) dud check; rubber check; ②(不实践的诺言) empty promise; lip service

【空投】air-drop; paradrop: ～救灾物资 air-drop relief supplies (to a stricken area) ◇ ～包 parapack／ ～场 dropping ground／ ～伞 aerial delivery parachute／ ～特务 an air-dropped spy／ ～鱼雷 aerial torpedo

【空文】ineffective law, rule, etc.: 一纸～ a mere scrap of paper

【空吸】[物] suction

【空袭】air raid; air attack ◇ ～警报 air raid alarm／ ～警报器 air raid siren

【空想】idle dream; fantasy

【空想家】dreamer; visionary

【空想社会主义】utopian socialism ◇ ～者 utopian socialist

【空心】hollow ◇ ～长丝[化纤] hollow filament／ ～砖 hollow brick

【空虚】hollow; void: 生活～ lead a life devoid of meaning／ 思想～ lack mental or spiritual ballast; be impractical in one's thinking

【空穴】[电子] hole

【空穴来风】an empty hole invites the wind; weakness lends wings to rumors

【空域】airspace: 战斗～ combat airspace

【空运】air transport; air freight; airlift: ～救灾物资 airlift relief supplies (to a stricken area) ◇ ～保险 air transportation insurance／ ～单 airway bill (of lading)／ ～货物 airfreight; air cargo

【空载】weight empty: ～运费 dead freight

【空战】air battle; air action; aerial combat

【空中】in the sky; in the air; aerial; overhead ◇ ～爆炸 air burst／ ～补给 air-supply; air-resupply／ ～发射 air-launch／ ～加油 air refueling; inflight refueling／ ～劫持 aerial hijacking; air piracy／ ～劫持者 air pirate／ ～劫机 sky jacking／ ～禁区 restricted airspace; airspace reservation／ ～警戒 air alert／ ～力量 air power／ ～摄影 aerophotography／ ～掩护 air umbrella; air cover／ ～运载工具 air vehicle／ ～侦察 aerial reconnaissance／ ～支援 air support／ ～走廊 air corridor; air lane

【空中飞人】[杂技] flying trapeze

【空中楼阁】castles in the air; realm of fancy

【空中小姐】airline hostess; airplane stewardess; air hostess

【空中转】[芭蕾舞] tour en l'air

【空重】[交] empty weight

【空竹】diabolo: 抖～ play diabolo

【空转】①(机器无负荷运转) race; idle ②(轮子打滑) spin; turn without moving forward

kǒng

恐 ①(害怕) fear; dread: 惊～ be alarmed ②(使害怕) terrify; intimidate: ～吓 threaten; intimidate ③(恐怕) I'm afraid: ～非原意。I'm afraid it was not the original intent.／ ～另有原因。There may be some other reason for it.

【恐怖】terror: 白色～ White terror ◇ ～分子 terrorist／ ～集团 gang of terrorist／ ～片 horror film／ ～手段 terroristic means／ ～统治 reign of terror／ ～行为 act of terrorism／ ～症 phobia／ ～主义 terrorism／ ～组织 terroristic organization

【恐吓】threaten; intimidate; cow; menace; frighten: 用手枪～某人 menace sb. with a pistol ◇ ～信 letter of intimidation; blackmailing letter; threatening letter

【恐慌】panic; scare; fright: ～万状 panic-stricken／ 战争～ a war scare

【恐惧】fear; dread; be afraid of: ～不安 be frightened and restless／ 无所～ fear nothing; feel no fear

【恐龙】[古生物] dinosaur

【恐怕】[副] ①(表示估计兼担心) I'm afraid; fear: ～不成。I'm afraid it won't do.／ 我～不能来。I'm afraid I can't come. ②(表示估计) perhaps; I think; probably: ～要下雨。It looks like rain.

【恐水病】hydrophobia; rabies

倥

【倥偬】[书] ①(急迫匆忙) pressing; urgent: 戎马～ burdened with pressing military duties ②(穷困) poverty-stricken; destitute

孔

孔 hole; opening; aperture: 穿个～ perforate a hole / 十七～桥 a seventeen-arched bridge / 钥匙～ keyhole
【孔道】duct; drill way; passage; pass
【孔洞】opening or hole in a utensil, etc.
【孔方兄】money
【孔径】①[物] aperture ②[机] bore diameter
【孔孟之道】the doctrine of Confucius and Mencius
【孔庙】Confucian temple
【孔雀】peacock; peafowl; peahen (雌性)
【孔雀蓝】peacock blue
【孔雀绿】peacock green; malachite green
【孔雀石】[矿] malachite
【孔隙】small opening; hole
【孔型】[冶] pass
【孔穴】hole; cavity

kòng

空

空 ①(腾出来;使空) leave empty; leave blank; vacate: ～出两格来 leave two blank spaces / ～出一个抽屉 empty a drawer / ～出一个座位 vacate a seat ②(没有被利用的东西) unoccupied; vacant; blank: ～房 a vacant room / ～行 blank line / ～座位 unoccupied seat ③(尚未占用的地点) empty space; room ④(尚未占用的时间) free time; spare time; leisure; free hours: 你有～吗? Are you free? / 整个星期都没～。The whole week is engaged.
【空白】blank space; gap; margin: 填补科学技术上的～ fill the gaps in science and technology / 在左边留～ leave a margin on the left hand side of the page ◇ ～表格 blank form / ～票据 blank bill / ～支票 blank check
【空白点】blank spot; gap; blank
【空当儿】①(时间上的) interval; break: 我们趁这～唱支歌。Let's sing a song during the break. ②(空间上的) gap: 从一个～挤过去 squeeze through a gap
【空地】vacant lot; open ground; open space
【空额】vacancy: 补三个～ fill three vacancies
【空格】blank space (on a form)
【空格键】space (key)
【空缺】vacant position; vacancy
【空隙】①(时间上的) interval: 战斗～ intervals of fighting; interval between battles ②(空间上的) gap; space: 填补～ fill up a gap
【空暇】free time; spare time; leisure
【空闲】①(事情或活动停下) idle; free ②(空着的时间; 闲暇) free time; spare time; leisure
【空子】①(空隙) gap; opening: 小女孩找了个～挤了进去。The girl found a gap and squeezed in. ②(可乘的机会) chance; opportunity: 不让坏人钻～ leave no loop-hole for bad people to exploit

控

控 ①(控告) accuse; charge: 指～ accuse ②(控制) control; dominate: 遥～ remote control; telecontrol ③(使容器里的水流出) turn (a container) upside down to let the liquid trickle out: 把瓶子～一～。Turn the bottle upside down to empty it.
【控方证人】[法] prosecuting witness
【控告】charge; accuse; complain: ～某人 bring a charge against sb.; bring a suit against sb.; bring an accusation against sb. / ～某人犯谋杀罪 charge with murder / 向法院提出～ file charges in court ◇ ～权 right of complaint / ～人 accusant; accuser; accusing party / ～者 charger; delator
【控股公司】holding company
【控诉】accuse; denounce; make a complaint against ◇ ～方 accusing party; complainant / ～会 accusation meeting / ～人 accuser / ～要点 gravemen; heads of a charge / ～状 bill of complaint
【控制】control; dominate; command: ～不住自己的感情 lose control of one's feelings; cannot contain one's feelings / ～局面 have the situation under control / ～险要 command a strategic position ◇ ～经济 command economy / ～理论 control theory / ～权 control power; mastery / ～数字[经] control figure / ～塔 control tower (of an airport) / ～台 control power; mastery / ～系统 control system
【控制论】cybernetics; kybernetics

kōu

抠

抠 ①(挖) dig or dig out with a finger or sth. pointed; scratch; pick; scrape ②(雕刻) carve; cut ③(不必要的深究) delve into; study meticulously ④(吝啬) stingy; miserly
【抠门儿】stingy; miserly
【抠字眼儿】pay too much attention to the shades of meaning of words; find fault with the choice of words; be fastidious about wording

眍 (of the eyes) sink in; become sunken

kǒu

口 ①(嘴) mouth ②(容器口) mouth; rim: 瓶子～ the mouth of a bottle / 碗～the rim of a bowl / 信箱～the slit of a letter box ③(出入通过的地方) opening; entrance; inlet; outlet; exit: 出～ entrance; outlet; exit / 洞～ the mouth of a cave / 河～ the mouth of a river; river mouth; estuary / 胡同～儿 the entrance of an alley / 入～ entrance; inlet ④(口子) cut; hole; opening: 伤～ wound; cut / 衣服撕了个～子 tear a hole in one's jacket ⑤(刀刃) edge; blade: 刀～ the edge of a knife ⑥(驴、马等的年龄) the age of a draft animal: 这匹马～还轻。This horse is still young.

【口岸】port; seaport; river port: ～检查机关 inspection office at the port / 通商～ trading port; commercial port

【口碑】public praise: ～载道 be praised everywhere

【口才】eloquence: 有～ be eloquent; be endowed with eloquence

【口吃】stutter; stammer: 她有些～。She is troubled with stammer.

【口齿】①(发音) enunciation: ～清楚 have clear enunciation ②(说话的本领) ability to speak: ～伶俐 be clever and fluent

【口臭】halitosis; bad breath; ozostomia

【口传】oral instruction; from mouth to mouth

【口疮】aphtha

【口唇】lip

【口袋】①(衣兜) pocket ②(用具) bag; sack: 纸～ paper bag

【口对口呼吸】mouth to mouth breathing

【口风】one's intention or view as revealed in what one says: 不露～ give no hint / 探探某人的～ sound sb. out

【口服】①(口头上表示信服) profess to be convinced: ～心不服 pretend to be convinced / 心服～be sincerely convinced; be truly convinced ②(内服) take orally: 不得～ not to be taken orally ◇ ～避孕药 oral contraceptive; the pill

【口福】gourmet's luck; the luck to get sth. very nice to eat

【口腹】food: ～之欲 the desire for good food / 不贪～not indulge one's appetite

【口供】a statement made by the accused under examination; oral confession; verbal confession; testimony: ～记录 record of testimony / ～书 affidavit / ～证据 testimonial proof

【口号】slogan; watchword: 呼～ shout slogans

【口红】lipstick

【口惠】lip service; empty promise: ～而实不至 make a promise and not keep it; pay lip service

【口技】vocal mimicry; vocal imitation

【口角】corner of the mouth ◇ ～炎 perleche

【口紧】closemouthed; tight-lipped; secretive

【口径】①(器物圆口的直径) bore; caliber: ～203 毫米的大炮 203 mm. gun / 大～机枪 heavy-caliber machine gun / 小～步枪 small-bore rifle ②(说法) statement: 对～ arrange to give the same story; give the same account by arrangement / 统一～ unify statement

【口诀】a pithy formula

【口角】quarrel; bicker; wrangle; spat: 因某事与某人～ quarrel with sb. about sth.; have a quarrel with sb. over sth.

【口渴】thirsty

【口口声声】say again and again; keep on saying; repeatedly declare: 她～说不知道。She kept on pleading ignorance.

【口粮】grain ration; provisions

【口令】①(口头命令) word of command; word; command ②(识别敌我的口头暗号) password; watchword; countersign: ～问答 challenge and reply

【口蜜腹剑】honey-mouthed and dagger-hearted; honey on one's lips and murder in one's heart; hypocritical and malignant

【口气】①(说话的感情色彩) tone; note: 改变～ change one's tone / 埋怨的～ a tone of complaint / 严肃的～ a serious tone / 自满的～ a tone of conceit ②(言外之意) what is actually meant; implication: 听他的～ judging by the way he spoke; from the implication of his statement ③(说话的气势) manner or speaking: 好大的～! What high-sounding sentiments!

【口腔】[生理] oral cavity ◇ ～科 department of stomatology / ～外科学 oral surgery / ～卫生 oral hygiene / ～学 stomatology / ～医院 stomatological hospital

【口琴】mouth organ; harmonica

【口轻】①(味道不咸) not salty ②(爱吃淡一点的味道) be fond of food that is not salty ③(驴、马等年龄小) young

【口若悬河】let loose a flood of eloquence; be eloquent; glibly; talk volubly

【口哨儿】whistling sound through rounded

lips: 吹～ whistle (through rounded lips)
【口舌】①(争吵) quarrel; dispute; exchange of words: 与某人有～ quarrel with sb. ②(言语) talking round; words; talking: 费很大～说服某人 take a lot of talking to convince sb.
【口实】a cause for gossip; handle: 贻人～ provide one's critics with a handle
【口试】oral examination; oral test; oral quiz: 参加～ take oral examination
【口是心非】say yes and mean no; say one thing and mean another; affirm with one's lips but deny in one's heart
【口授】①(口头传授) oral instruction; pass on (sth.) through oral instruction ②(口说而由别人代写) dictate: 向秘书～信稿 dictate a letter to a secretary ◇ ～遗嘱 oral testament
【口述】oral account; nuncupate ◇ ～遗嘱 nuncupative will
【口水】saliva: 流～ slobber
【口蹄疫】[牧] foot-and-mouth disease
【口条】pig's or ox's tongue (as food)
【口头】①(用说话的方式) oral; verbal: ～练习 oral practice / ～通知 notify orally; verbal notification / ～选举 elect by acclamation ②(口头上) in words; in speech; on one's lips; verbal: ～上赞成 agree in words / ～拥护 give verbal support to ◇ ～表决 voice vote; vote by "yes" and "no" / ～革命派 a revolutionary in words / ～汇报 oral report / ～声明 oral statement / ～文学 folk tales, ballads, etc. handed down orally / ～协议 oral contract
【口头禅】pet phrase
【口味】①(各人对味道的爱好) a person's taste: 不合～ not be to one's taste / 合～ suit one's taste / 那种小说不合我的～。That sort of novel is not to my taste. ②(食品的滋味) flavor; taste of food: 这菜～好。The dish tasted nice.
【口吻】①[动] muzzle; snout ②(口气) tone; note: 玩笑的～ jocular tone / 责备的～ a note of reproach
【口香糖】chewing gum
【口信】oral message; message; word: 请给我妈妈带个～。 Will you give a message to my mother.
【口形】[语] degree of lip-rounding
【口炎】[医] stomatitis
【口眼歪斜】[中医] facial paralysis
【口译】①(口头翻译) oral interpretation; interpret ②(口译译员) interpreter
【口音】①(声调) voice ②(方音) accent: 她讲英语带法国～。 She speaks English with an French accent.
【口语】①(口头语言) colloquial language; colloquialism: ～体 colloquial style ②(与"书面语言"相对) spoken language
【口罩】gauze mask (worn over nose and mouth)
【口重】①(味道咸) salty ②(爱吃咸一点的味道) be fond of salty food
【口诛笔伐】condemn both in speech and in writing; denounce by tongue and pen
【口子】opening; hole; cut; tear: 大渠决了～。The canal has burst its banks. / 袖子撕了一个大～。 There is a tear in the sleeve.

kòu

寇 ①(强盗) bandit; robber 海～ pirate ②(侵略者) invader; enemy: 敌～ the invading enemy ③(侵犯) invade: ～边 harass border area / 入～ invade (a country)
【寇仇】enemy; foe

扣 ①(套住; 搭住) button up; buckle; fasten: ～钮扣 do up the buttons / 把门～上 bolt the door / 把皮带～上 buckle a belt / 把衣服～好 button (up) one's coat ②(器物口朝下) place a cup, bowl, etc. upside down; cover with an inverted cup, bowl, etc.: 把碗～在盘子上 turn the bowl bottom upward on the plate ③(扣留; 扣押) detain; take into custody; arrest ④(从原数额中减去一部分) deduct; discount; take off: ～工资 deduct a part of sb.'s pay; deduct part of wages / 打八～ give a 20 per cent discount; allow 20% discount ⑤(条状物系成的扣儿) knot; button; buckle: 系个～儿 make a knot ⑥(扳动) press; pull: ～动扳机 press the trigger ⑦(用力猛击) smash (the ball): 猛～一球 smash the ball forcefully
【扣除】deduct; take off: ～费用的百分之二十 deduct 20% from the expenses / 从工资中～房租 deduct rent from wages
【扣缴】withhold ◇ ～所得税报告表 withholding income tax return / ～义务人 withholding agent
【扣留】detain arrest; hold in custody: ～船只 detain the ship / ～盗窃嫌疑犯 detain the suspected thief / ～货物 detain the cargo / ～行车执照 suspend a driving licence / ～走私犯 detain the smuggler
【扣帽子】put a label on sb.: 给某人扣上…的帽子 label sb. as

【扣球】smash; spike: 斜线～ cross-court smash; cross smash; cross spike／直线～ straight spike; line spike
【扣人心弦】exciting; thrilling; soul-stirring; very touching: ～的比赛 an exciting match; an exciting game／～的歌 a soul-stirring song
【扣杀】[体] smash (the ball): 大板～ overpowering smashes (in a table tennis game)／大力～ powerful smash／连续～ repeated smashes／闪电般的～ swift killing smashes／一次性～ direct smash
【扣压】withhold; pigeonhole: ～稿件 withhold a manuscript from publication／～信件 withhold letters／～议案 smother up the motion
【扣押】①(扣留) detain; hold in custody ②[法] distrain; detention; levy on ◇ ～财产通知 garnishment／～财物 levy; distrainment; distraint／～令 writ of detention; attachment／～权 power of detention; right of detention
【扣子】①(纽扣) button ②(皮带扣) buckle ③(结) knot: 把绳子打个～ tie a knot in a rope

筘
[纺] reed

叩
①(敲; 打) knock: ～门 knock at a door ②(磕头) kowtow
【叩甲】[动] click beetle; spring beetle
【叩门求见】knock the door and asking for interview
【叩头】kowtow
【叩头虫】click beetle
【叩问】[书] make inquiries: ～缘由 make inquiries for the causes
【叩诊】percussion ◇ ～锤 percussion hammer

kū

窟
①(洞穴) hole; cave: 石～ cave; grotto ②(坏人聚集场所) den: 赌～ a gambling-den／匪～ robbers' den; bandit's den
【窟窿】①(洞) hole; cavity: 老鼠～ rat-hole／戳个～ bore a hole／磨了个～ have worn a hole ②(亏空) deficit; debt
【窟窿眼儿】small hole

枯
①(枯萎) withered: ～草 withered grass／～树 a withered tree／～叶 dead leaves／～枝 a dead branch ②(没有水) dried up: ～井 a dry well ③(枯燥) dull; uninteresting: ～坐 sit in boredom
【枯草热】[医] hay fever
【枯肠】[书] impoverished mind: 搜索～ rack one's brains (for ideas or expressions)
【枯槁】①(枯萎) withered: 草木～ trees and grass withered up ②(憔悴) haggard; wizened: 面容～ look haggard
【枯黄】withered and yellow
【枯寂】dull and lonely
【枯竭】dried up; exhausted; drain: 财源～ financial resources were exhausted／水源～ the source has dried up／他的想象力好象已经～了。His imagination seems to have dried up.
【枯窘】dried up: 文思～ the source of one's inspiration has dried up; be devoid of inspiration; run out of ideas to write about
【枯木逢春】spring comes to the withered tree; good fortune that comes after a long spell of bad luck; get a new lease of lift
【枯涩】dull and heavy: 文字～ a dull and heavy style
【枯瘦】emaciated; skinny
【枯水】low water ◇ ～期 dry season／～位 the lowest water level
【枯萎】withered
【枯叶蛾】lappet moth
【枯燥】dull and dry; uninteresting: ～的谈话 insipid conversation／～无味 dry as dust; as dry as a chip

骷
【骷髅】①(人的尸骨) human skeleton ②(死人头骨) human skull; death's-head

哭
cry weep; sob; blubber; wail: ～得死去活来 sob one's heart out／～了起来 burst into tears／～肿眼睛 cry one's eyes out／放声大～ cry loudly; cry unrestrainedly
【哭鼻子】snivel
【哭哭啼啼】endlessly weep and wail; weep and sniffle; blubber
【哭泣】cry; weep; sob
【哭穷】go about telling people how hard up one is; complain of being hard up
【哭丧着脸】put on a long face; wear a long face; go around with a long face; one looks mournful as if in bereavement
【哭声震天】noise of grief rises to heaven
【哭诉】complain tearfully
【哭天抹泪】wail and whine
【哭笑不得】not know whether to laugh or to cry; find sth. both funny and annoying

kǔ

苦 ①(味苦) bitter: 嘴发～ have a bad taste in one's mouth / 这药～得很。This medicine tastes very bitter. ②(难受; 痛苦) hardship; suffering; pain; bitterness: 他在旧社会受尽了～。He went through all kinds of hardships in the old society. ③(使痛苦; 使难受) cause sb. suffering; give sb. a hard time: 这事可～了他了。This matter really give him a hard time. ④(苦于) suffer from; be troubled by ⑤(有耐心地; 尽力地) painstakingly; doing one's utmost: ～～哀求 entreat piteously; implore urgently; implore bitterly / ～留 try hard to ask sb. to stay / ～劝 earnestly advise; earnestly exhort / 勤学～练 study and train hard

【苦差】hard and unprofitable job

【苦楚】suffering; misery; distress

【苦处】①(苦难) suffering; hardship ②(困难) difficulty

【苦胆】gall bladder

【苦干】work hard: ～加巧干 work hard and skillfully; combine hard work with ingenuity / ～精神 hard-working spirit

【苦工】hard work; manual work; hard labor: 做～ do sweated labor; do backbreaking work

【苦功】hard work; painstaking effort: 下～ take pains; make painstaking efforts / 下～学习 study hard

【苦瓜】[植] balsam pear

【苦海】sea of bitterness; abyss of misery: 脱离～ get out of the abyss of misery / ～无边 boundless sea of hardship / ～无边，回头是岸。The sea of bitterness has no bounds, repent and the shore is at hand.

【苦寒】bitter cold

【苦尽甘来】when bitterness if finished, sweetness begins; after suffering comes happiness; sweet are the fruits of labor, luck turns after hardship

【苦口】①(反复恳切地说) (admonish) in earnest: ～相劝 earnestly advise; earnestly exhort ②(引起苦的味觉) bitter to the taste

【苦口良言】a bitter mouth utters fine words

【苦口婆心】urge sb. time and again with good intentions; advice in earnest words and with good intention

【苦力】[旧] coolie

【苦练】practice hard; drill diligently: ～杀敌本领 practice hard to master the military skill / 勤学～ study diligently and practice hard

【苦闷】depressed; dejected; feeling low; gloomy: 感到～ feel depressed / 精神上很～ suffered greatly from the spiritual depression

【苦难】suffering; misery; distress; tribulation: ～家史 one's bitter family history / ～深渊 the abyss of misery / ～深重的人民 the long suffering people

【苦恼】vexed; worried; distressed; tormented; troubled: 为某事～ be distressed about sth.; be tormented with sth. / 自寻～ torment oneself

【苦肉计】the ruse of inflicting an injury on oneself to win the confidence of the enemy; a trick of having oneself tortured to win the confidence of the enemy

【苦涩】①(又苦又涩) bitter and astringent ②(形容内心痛苦) pained; agonized; anguished: ～的表情 a pained look

【苦水】①(味道苦的水) bitter water ②(口中吐出的苦的液体) gastric secretion, etc. rising to the mouth ③(内心痛苦) suffering: 在～中长大 grow up amidst sufferings

【苦思】think hard; cudgel one s brains

【苦思冥想】cudgel one's brains (to evolve an idea)

【苦痛】pain; suffering

【苦头】①(稍苦的味道) bitter taste: 这个井里的水带～儿。Water from this well has a slightly bitter taste. ②(苦痛; 磨难; 不幸) suffering; hardship: 吃尽了～ go through all sorts of hardships

【苦味酸】picric acid ◇ ～盐 picrate

【苦夏】loss of appetite and weight in summer

【苦笑】forced smile; wry smile

【苦心】trouble taken; pains: ～经营 painstakingly build up (an enterprise, etc.) / 煞费～ take great pains

【苦心孤诣】make extraordinarily painstaking efforts; do sth. with painstaking efforts

【苦行】ascetic practices ◇ ～主义 asceticism

【苦役】hard labor; penal servitude

【苦于】suffer from (a disadvantage); trouble: ～没有时间。The trouble is that there is no time.

【苦雨凄风】bitterly cold winds and rain

【苦战】wage an arduous struggle; struggle hard

【苦衷】difficulties: 难言的～ feelings of pain or embarrassment which are hard to mention / 她也许有她的～。She might have her own difficulties.

【苦竹】bitter bamboo

【苦主】the family of the victim in a murder

case

kù

库 ①（仓库；车库）warehouse; storehouse; depository; depot: 军械～ armory / 粮～ granary / 汽车～ garage ②[计算机] library
【库藏】have in storage; have a storage of; have in store
【库存】stock; reserve: ～充足 keep a large stock of goods / ～物资 goods kept in stock; reserve of materials / ～现金 cash in hand; cash on hand / ～债券 treasury bonds / 商品～ commodity stocks / 有大量～ have 〈keep〉 a large stock of goods ◇～量 storage
【库房】storehouse; storeroom: ～重地，闲人免进! Storage Room, No Admittance!
【库仑】[电] coulomb ◇～定律 Coulomb's law
【库容】[水] storage capacity

裤 trousers; pants: 衬～ panty / 短～ shorts / 工作～ work pants / 棉～ cotton-padded trousers / 男短～ shorts / 内衣～ underwear / 牛仔～ jeans / 女～ slacks / 女三角～ brief / 女用袜～ panty hose / 睡～ pyjama pant; panties（女用）/ 游泳～ swimming trunks
【裤衩】underpants; undershorts: 比基尼三角～ bikini briefs / 三角～ briefs
【裤裆】crotch (of trousers)
【裤兜】trouser pocket
【裤缝】seams of a trouser leg
【裤脚】bottom of a trouser leg
【裤料】trousering: 一块～ a piece of trousering
【裤腿】trouser legs
【裤线】creases (of trousers)
【裤形救生圈】breeches buoy
【裤腰】waist of trousers
【裤子】trousers; pants: 一条～ a pair of pants

酷 ①（残酷）cruel; brutal; oppressive ②（程度深的）very; extremely; exceedingly
【酷爱】ardently love; be very fond of
【酷寒】bitter cold
【酷烈】cruel; fierce: ～的太阳 the scorching sun
【酷热】extremely hot (weather): 天气～ a sweltering hot day
【酷暑】the intense heat of summer
【酷似】be the very image of; be exactly like; bear a strong resemblance to
【酷刑】cruel torture; savage torture; brutal corporal punishment; excruciation

kuā

夸 ①（夸大）exaggerate; overstate; boast: ～口 boast; brag ②（夸讲）praise: 自～ sing one's own praises
【夸大】exaggerate; overstate; magnify; aggrandize: ～困难 exaggerate the difficulties / ～缺点 exaggerate the short-comings
【夸大其词】make an overstatement; exaggerate: ～的报告 an exaggerated report
【夸奖】praise; commend; extol; compliment; speak well of: ～某人 sing sb.'s praises / ～某人进步快 praise sb. for his rapid progress / ～某人勇敢 extol sb.'s bravery; compliment sb. on his courage
【夸克】[物] quark
【夸口】boast; brag; talk big
【夸夸其谈】indulge in exaggerations; talk big; a big screed full of bombast
【夸耀】brag about; show off; flaunt; boast of: ～自己 brag about oneself / ～自己的能力 show off one's ability / ～自己的学识 boast of one's learning
【夸赞】speak highly of; commend; praise
【夸张】①（夸大）exaggerate; overstate: ～的语言 inflated language; exaggerations / 艺术～ artistic exaggeration / 可以毫不～地说… It is no exaggeration to say that... ②[语] hyperbole

kuǎ

垮 collapse; fall; break down; crack up: 打～敌人 put the enemy to rout / 她身体～了。Her health has broken down. / 这面墙要～了。The wall is going to collapse.
【垮台】collapse; fall from power: 敌人彻底～了。The enemy has completely collapsed.

kuà

挎 ①（挂在胳膊上）carry on the arm; sling; hang: ～着胳膊 arm in arm / ～着一个篮子 with a basket on one's arm ②（挂在肩头、脖颈或腰里）carry sth. over one's shoulder or at one's side: ～着照相机 have a camera slung over one's shoulder
【挎包】satchel

跨 ①（迈步）step; stride: ～进大门 step into a

doorway／～上公共汽车 step into the bus／向前～一步 take a step forward ②(跨骑) bestride; straddle; ride astride: ～上马 mount a horse; bestride a horse／～在马上 ride a horse astride; straddle a horse ③(时空上的跨越) cut across; go beyond: ～地区 transregional／～过小河 cross the creek

【跨度】[建] span; fly-past; strech

【跨国公司】transnational corporation

【跨行业公司】conglomerate

【跨栏跑】[体] hurdle race; the hurdles: 400 米～ 400m hurdles

【跨年度】go beyond the year ◇～预算 a budget to be carried over to the next year; a budget which spans two years

【跨线桥】flyover; overpass

【跨音速】transonic speed: ～飞行 transonic flight

【跨越】stride across; leap over; cut across: ～几个历史阶段 leap over several historical stages of development

胯 hip

【胯骨】hipbone; innominate bone

kuài

会

【会计】①(会计工作) accounting; accountancy: 财务～ financial accounting／成本～ cost accounting／工业～ industrial accounting／税务～ tax accounting／②(会计人员) accountant; bookkeeper: ～师 a chartered accountant; a certified public accountant (美) ◇～报表 accounting statement／～报告 accounting report／～程序 accounting procedure／～决算报告 statement of final accounts／～科目 account title; accounting item／～年度 financial year; fiscal year／～凭证 accounting document; accounting voucher／～原理 principles of accounting／～制度 accounting system ○审计 auditing／审计师 auditor／出纳员 cashier／簿计 bookkeeping／簿计员 bookkeeper／帐簿 a book of accounts／复式记帐 double-entry／单式记帐 single-entry／分录日记帐 journal day book／普通日记帐 general journal／双栏式日记帐 two-column journal／多栏式日记帐 columnar journal／现金收入日记帐 cash receipts journal／现金支出日记帐 cash disbursement journal／销售日记帐 sales journal／购货日记帐 purchases journal／借方 debit／贷方 credit／成本、数量、利润分析 cost-volume-profit analysis; CVP analysis／负债 liabilities／资金 funds／资本 capital／资产 assets／固定资产 fixed assets／工资 wages; salary／奖金 incentive pay／折旧 depreciation／租金 rent／利息 interest／现金 cash／库存现金 cash on hand／银行存款 deposit in banks／费用 expense／成本 cost／产品 product／商品 merchandise／利润 profit／税金 tax／公司所得税 corporation income tax／总利润 gross profit／净收益 net income／财务报表 financial statements／资产负债表 balance sheet／损益表 statement of loss and profit

【会计师】accountant: 总～ chief accountant

【会计学】accounting

侩 middleman: 市～ philistine; sordid merchant

脍

【脍炙人口】win universal praise; enjoy great popularity; be on everybody's lips

块 ①(成块的东西) piece; lump; chunk; block: 木～ wood block／糖～ lumps of sugar ②[量] 一～肥皂 a cake of soap／一～面包 a piece of bread／一～手表 a wrist watch

【块根】[植] root tuber

【块规】[机] slip gauge; gauge block

【块茎】[植] stem tuber; root tuber

【块煤】lump coal; torbanite; bitumenite

快 ①(速度高) fast; quick; rapid; swift; speedy: 动作～ be quick in action／她进步很～。 She has made rapid progress. ②(速度) speed: 这公共汽车能跑多～? How fast can this bus go? ③(赶快; 从速) hurry up; make haste ④(快要; 将要) soon; before long; be about to: 春天～来了。 Spring is coming.／她～回来了。 She will soon be back. ⑤(灵敏) quick-witted; ingenious: 她脑子～。 She understands things quickly. ⑥(锋利) sharp; keen: ～刀 a sharp knife ⑦(爽快; 痛快) straightforward; forthright; plainspoken: 心直口～ straightforward and outspoken ⑧(愉快; 高兴; 舒服) pleased; happy; gratified: 大～人心 to the immense satisfaction of the people／拍手称～ clap and cheer／心中不～ feel unhappy

【快板】[乐] *allegro*

【快板儿】*kuaibanr*; clapper talk; ballad: 说～ perform a *kuaibanr* ballad ◇～书 story re-

cited to the rhythm of bamboo clappers
【快报】wall bulletin; bulletin; stop-press news; newsflash
【快餐】fast food; quick meal ◇～部 quick-lunch counter／～馆 fast food restaurant
【快车】express train; express bus; fast train: 特别～ special express／直达～ through express
【快车道】fast traffic lane
【快当】quick; prompt: 办事～ be quick at work; do everything promptly
【快刀斩乱麻】cut a tangled skein of jute with a sharp knife; cut the Gordian knot
【快递】express delivery ◇～费 express fee／～邮件 express mail
【快干】quick-drying ◇～漆 quick-drying paint
【快感】pleasant sensation; delight
【快活】happy; merry; cheerful; joyful; joyous
【快乐】happy; joyful; cheerful; gay: ～的假日 a delightful holiday／祝您新年～! I wish you a happy New Year.
【快马加鞭】spur on the flying horse; spur the flying horse to full speed
【快慢】speed ◇～针 index lever; regulator pin
【快门】[摄] shutter: 焦点平面～ focal plane shutter／中心～ between-lens shutter ◇～开关 shutter release／～速度 shutter speed
【快凝水泥】fast-setting cement
【快人】straightforward man: ～快事 a heroic deed performed by a straightforward man／～快语 straight talk from an honest man／～一言 an intelligent man only needs a hint
【快事】pleasure; delightful event; delight: 生平一大～ one of the most delightful experiences in one's life／引为～ recall an event with great satisfaction
【快手】quick worker; deft hand; nimbled-handed person ◇～快脚 nimble of hands and fast of feet; do things quickly
【快书】quick-patter: 山东～ Shandong clapper ballad
【快速】fast; quick; high-speed; speedy ◇～部队 mobile force; mobile troops; mobile units／～切削 high-speed cutting／～走带 fast feed
【快艇】speedboat; motor boat; mosquito boat: 鱼雷～ torpedo boat
【快慰】glad; pleased
【快信】express letter
【快要】be about to; be going to; be on the verge of: 他们～出发了. They are about to start.／她～生孩子了. She is going to have a baby.
【快硬水泥】[建] quick-hardening cement
【快意】pleased; satisfied; comfortable; elated
【快鱼】Chinese herring
【快照】snapshot
【快中子】[物] fast neutron; high-speed neutron
【快子】[物] tachyon
【快嘴】careless tattler; have a loose tongue

筷 chopsticks
【筷子】chopsticks

kuān

宽 ①(跟"窄"相对) wide; broad: ～肩膀 broad shoulders／～脸 a broad face ②(宽度) width; breadth: 四米～ four meters in breadth／一手～ a hand's breadth ③(放宽;使松缓) relax; relieve: 心里～多了 be greatly relieved ④(放宽限期) extend: 期限放～两天。The deadline will be extended two days. ⑤(宽大;不严厉;不苛求) generous; lenient: ～以待人 be lenient with others／从～处理 treat with leniency ⑥(宽裕;宽绰) comfortably off; well-off: 手头～多了 have more money than before; be better off than before
【宽畅】free from worry; happy; cheerful; carefree: 胸怀～ be cheerful in one's mind
【宽敞】spacious; roomy; commodious: ～的房子 a commodious house
【宽绰】①(宽阔) spacious; commodious: 屋子～ the room is spacious ②(松缓) relax; relieve: 心里～多了 feel greatly relieved ③(富余) comfortably off; well-off
【宽打窄用】budget liberally and spend sparingly
【宽大】①(面积或容积大) roomy; spacious; wide: 一间～的客厅 a spacious sitting room ②(从宽处理) lenient; magnanimous: ～处理 be dealt with leniently; be accorded lenient treatment; receive clemency; be afforded lenient treatment; be magnanimously treated／～为怀 be magnanimous with an offender; be lenient with an offender; open-hearted; benignant／～与惩办相结合 combination of leniency with punishment ◇～政策 lenient policy
【宽待】treat with leniency; be lenient in dealing with; treat liberally; accord lenient treatment: ～战俘 treat the POWs liberally; give lenient treatment to prisoners of war; treat prisoners of war leniently

【宽贷】pardon; forgive
【宽度】width; breadth: ～十米 ten meters width / 领海～ the extent of the territorial sea
【宽广】broad; extensive; vast: ～的田野 a broad expanse of country / 心胸～ broad-minded
【宽轨】broad gauge; wide gauge ◇～铁路 broad-gauge railway
【宽宏大量】large-minded; magnanimous; generous
【宽厚】generous; lenient; kind: 待人～ be generous to people
【宽解】ease sb.'s anxiety; ease sb. of his trouble
【宽旷】extensive; vast: ～的草原 extensive grasslands
【宽阔】broad; wide: ～的河面 a broad river / ～的林荫道 a broad avenue / ～的人行道 a wide sidewalk / ～的胸怀 broad-mindedness
【宽频带】wide band
【宽饶】forgive; show mercy; give quarter
【宽容】tolerant; lenient; bear with ◇～待人 be lenient in treating a person / ～条款 allowance clause
【宽恕】forgive; pardon; excuse: ～某人 pardon sb. for / 得到～ obtain clemency from / 请求～ ask for forgiveness
【宽慰】comfort; console
【宽限】extend a time limit; grace: ～两星期 give two week's grace / 请～几天。 Please extend the deadline a few days. ◇～期 grace period; period of grace / ～日期 days of grace
【宽心】feel relieved; find relief; be relaxed; be at ease: 说几句～话 say a few reassuring words ◇～丸儿 story told to make people relax
【宽严并举】temper justice with mercy
【宽衣】[敬] take off your coat: 请～ Do take off your coat
【宽银幕】wide screen
【宽银幕电影】wide-screen film: 70毫米～ Panavision / 变形镜头式～ cinemascope / 全景～ cinerama / 遮幅式～ superscope
【宽影片】wide film
【宽裕】well-to-do; comfortably off; ample; well-off; plenty: 经济～ in easy circumstances; well-off / 时间～ there is plenty of time yet; there is enough time to spare
【宽窄】width; breadth; size

髋 hip
【髋骨】hipbone; innominate bone
【髋关节】hip joint; articulatio coxae
【髋关节炎】coxitis

kuǎn

款 ①(诚恳) sincere: ～曲 heartfelt feelings ②(招待; 款待) receive with hospitality; entertain: ～待客人 entertain guests ③(条款) section; paragraph; item: 第五条第三～ Article 5, Section 3 ④(款项) a sum of money; fund; money: 拨～ set aside a sum / 筹～ procure money; raise funds / 公～ public funds / 汇～ remit money; transfer money; make remittance / 捐～ contribute money ⑤(书画的上下款)上～ the name of recipient / 下～ the signature ⑥(缓; 慢) leisurely; slow: ～步 with deliberate steps
【款待】treat cordially; entertain; receive cordially: ～客人 entertain guests; receive guests cordially / 感谢对我们的盛情～。 Thank you for the hospitality you have shown us.
【款留】cordially urge (a guest) to stay: ～殷殷 detain a visitor with great enthusiasm
【款洽】cordial and harmonious
【款曲】heartfelt feelings: 互通～ express feelings of mutual affection or friendship
【款式】pattern; style; design; form: 时新～ up-to-date style; the vogue of the day
【款项】a sum of money; fund; item of expenditure (指支出)
【款子】a sum of money: 一大笔～ a large sum of money; a great sum of money / 汇一笔～ remit a sum of money

kuāng

匡 ①(纠正) rectify; correct: ～谬 correct mistakes ②(救; 帮助) assist; save: ～我不逮 help me to overcome my shortcomings
【匡正】rectify; correct

诓 deceive; hoax: 她不会～你的。 She won't deceive you.
【诓骗】deceive; hoax; dupe; cheat; swindle: ～某人钱财 cheat sb. of his money; swindle sb. out of his money

框 ①(框框) frame; circle ②(在周围加上线条) circumscribe; draw a frame round ③(约束; 限制) put in a strait-jacket
【框框】①(周围的圈) frame; circle ②(固有格式; 传统做法) convention; restriction; set pattern; set of rules: 打破旧～ break through the conventions; throw convention to the winds / 划～ draw up a set of rules / 老～ the old

ways; rut; groove / 条条～ regulations and restrictions

哐 [象] crash; bang: ～的一声掉在地上 fall with a crash

【哐啷】[象] crash: ～一声关上门 bang the door shut

筐 ①(竹柳等编的容器) basket; crate ②[量] basketful: 两～土 two basketfuls of earth

【筐子】small basket; crate: 编～ weave a small basket

kuáng

狂 ①(精神失常；疯狂) mad; crazy; mania; insanity: 发～ go mad; become mad; run mad ②(猛裂；声势大) violent; wild: ～饮 drink heavily / 股票价格～跌。The stocks slumped. ③(纵情地、无拘束地) wild; delirious; raving; unrestrained: ～怒 raving with fury / ～喜 be wild with joy; be delirious with delight / 欣喜若～ be wild with joy; beside oneself with joy ④(狂妄) arrogant; overbearing

【狂暴】violent; wild; frantic; rabid: ～的山洪 raging mountain torrents

【狂飙】hurricane; wild whirlwind

【狂吠】bark furiously; howl

【狂风】①[气] whole gale ②(猛烈的风) fierce wind; wild wind: ～暴雨 a violent storm; tempest / ～巨浪 wild winds and huge waves / ～呼啸。The wind howled.

【狂轰滥炸】wanton and indiscriminate bombing; bomb wantonly and indiscriminately

【狂欢】revelry; carnival

【狂澜】raging waves; roaring waves: 力挽～ do one's utmost to stem a raging tide or save a desperate situation

【狂犬病】hydrophobia; rabies

【狂热】fanaticism; fanatical; feverish; rabid: ～的信徒 a fanatical follower; fanatic; zealot ◇ ～性 fanaticism

【狂人】madman; maniac: ～呓语 ravings of a madman / 战争～ war maniacs

【狂妄】wildly arrogant; frantic; unbridled; wanton; presumptuous: ～的野心 a wild ambition / ～无知 conceited and ignorant / ～凶暴 arrogant and violent / ～自大 arrogant and conceited

【狂喜】wild with joy

【狂想曲】[乐] rhapsody

【狂笑】laugh wildly; laugh boisterously; wild laugh; roars of laughter

【狂言】ravings; wild language; crazy talk; crazy remarks: 口出～ talk wildly

诳

【诳语】lies; falsehood

kuàng

矿 ①(矿床) ore deposit; mineral deposit: 报～ report where deposits are found ②(开采场所) mine: 露天～ an open pit / 煤～ coal mine / 铁～ iron mine ③(矿石) ore: 铁～ iron ore / 选～ ore dressing

【矿藏】mineral resources; mineral reserves; ore deposits: ～丰富 rich in mineral resources ◇ ～量 ore reserves

【矿层】ore bed; ore horizon; seam; stratum of ores: 可采～ a workable seam

【矿产】mineral products; minerals: ～分布 the distribution of mineral deposits

【矿车】mine car; tub; tram; bogie; bogie truck

【矿尘】mine dust

【矿床】mineral deposit; ore deposit; deposit: 层状～ bedded deposit / 海底～ submarine deposit / 金属～ metalliferous deposit

【矿灯】miner's lamp; cap-lamp: 手提安全～ an electric hand lamp

【矿工】miner; pitman (煤矿工人) ◇ ～帽 miner's helmet

【矿井】shaft; mine; pit ◇ ～火灾 mine fire / ～通风 mine ventilation / ～瓦斯 mine gas

【矿坑】pit

【矿口发电厂】mine mouth power plant

【矿脉】mineral ore; mineral vein; lode

【矿棉】mineral wool

【矿苗】outcropping; outcrop; crop

【矿泥】sludge; slime; slurry

【矿区】mining area ◇ ～铁路 mine railway

【矿泉】mineral spring ◇ ～疗养地 spa / ～水 mineral water

【矿山】mine; mining area ◇ ～地压 rock pressure / ～工程图 mine map / ～机械 mining machinery / ～机械厂 mining machinery plant / ～机械化 mining mechanization / ～救护 mine rescue / ～运输 mine haul; mine haulage; pit haulage

【矿石】ore; mineral: 富～ high-grade ore / 贫～ low-grade ore

【矿石收音机】crystal receiver

【矿物】mineral: 伴生～ associated mineral ◇ ～肥料 mineral fertilizer / ～界 mineral kingdom / ～燃料 fossil fuel / ～纤维 mineral fi-

ber / ～学 mineralogy / ～油 mineral oil
【矿样】sample ore
【矿业】mining industry ◇ ～学院 mining institute
【矿渣】slag ◇ ～水泥 slag cement / ～砖 slag brick
【矿柱】(ore) pillar

旷 ①(空阔宽广) vast; spacious: 地～人稀 a vast territory with a sparse population ②(心境开阔) free from worries and petty ideas: 心～神怡 carefree and happy ③(荒废) neglect; waste ④(过于肥大; 间隙过大) loose-fitting
【旷达】broad-minded; bighearted; open-minded: ～之士 a profound scholar; one with an open mind
【旷废】neglect: ～学业 neglect one's studies; neglect school work
【旷费】waste: ～时间 waste one's time
【旷工】stay away from work without leave or good reason
【旷古】from time immemorial: ～奇闻 unprecedented story / ～未闻 never heard of in history; unprecedented / ～未有 never seen in past history
【旷课】be absent from school without leave; cut school; cut classes; skip school work
【旷日持久】long-drawn-out; protracted; prolonged; procrastinating: ～的谈判 long-drawn-out negotiations
【旷世】outstanding; without peer in one's generation: ～功勋 outstanding deeds / ～奇才 a remarkable talent of many ages; a genius without peer in one's generation / ～无双 stand without peer in one's generation / ～之才 a man of brilliance unequaled by contemporaries
【旷野】wilderness; open field; open country
【旷职】be absent from duty without leave

框 frame; case: 窗～ window frame; window case / 镜～儿 picture frame / 门～ door frame / 眼镜～儿 rims of spectacles
【框架】frame: 钢筋混凝土～ reinforced concrete frame ◇ ～建筑 framed building

眶 orbit; the socket of the eye; eye socket: 热泪盈～ one's eyes filling with tears / 眼泪夺～而出 tears be rolling down one's cheeks

况 ①(情形) condition; situation: 近～ recent condition / 窘～ predicament; difficult situation / 近～如何? How have you been recently? ②(比较) compare: 每～愈下 from bad to worse / 以古～今 draw parallels from history; compare the present with the past
【况且】moreover; besides; in addition; furthermore

kuī

窥 peep; spy
【窥测】spy out: ～孔 spy-hole / ～时机 bide one's time
【窥测方向, 以求一逞】spy out the land in order to accomplish one's schemes; see which way the wind blows in order to achieve one's evil ends
【窥见】get a glimpse of; catch a glimpse of; detect: ～一斑 see segment of a whole
【窥器】[医] speculum
【窥视】peep at; spy on; watch stealthily; take a furtive glance: 从锁孔～ peep through a keyhole
【窥伺】lie in wait for; be on watch for
【窥探】spy upon; pry about; pry into; poke one's nose into: ～敌情 spy on the enemy situation / ～秘密 pry into sb.'s secret / ～形势 send up a kite

亏 ①(受损失; 亏折) lose (money, etc.); have a deficit; 吃大～ suffer heavy losses / 盈～ profit and loss ②(欠缺; 短少) short of; deficient: 理～ be in the wrong / 血～ blood deficiency / 功～一篑 just one step short of success; just fall short of final completion; fail to build a mound for want of the last basket of earth ③(亏负) treat unfairly: 放心吧, ～不了你。 Don't worry, we won't be unfair to you. ④(多亏; 幸亏) fortunately; luckily; thanks to: ～得是你。 Luckily it is you. ⑤(月蚀初亏) wane
【亏本】lose money in business; lose one's capital: ～卖出 sell one's hen on a rainy day / ～生意 a losing proposition
【亏待】①(不公平) treat unfairly; treat shabbily ②(不尽心) treat unobligingly
【亏得】fortunately; luckily; thanks to
【亏负】let sb. suffer; let sb. down
【亏耗】loss by a natural process
【亏空】①(欠人财物) be in debt; be in the red; can't make both ends meet ②(所欠的财物) debt; deficit: 巨额～ a great deficit / 拉～ get into debt / 弥补～ meet a deficit; make up a deficit; make up (for) a loss ◇ ～公款

embezzle / ～公款者 defaulter
【亏欠】have a deficit; be in arrears
【亏蚀】①(日蚀; 月蚀) eclipse of the sun; eclipse of the moon ②(亏本) lose money in business
【亏损】①(支出超过收入) loss; deficit: 企业～ loss incurred in an enterprise ②(身体虚弱) general debility ◇～总额 total loss
【亏心】have a bad 〈guilty〉 conscience ◇～事 a deed that troubles one's conscience; discreditable affair / 做～事 do sth. with a bad conscience
【亏月】waning moon

盔 helmet; headpiece: 钢～ steel helmet
【盔甲】a suit of armor; helmet and armor
【盔云】[气] crest cloud; cloud crest

岿
【岿然】towering; lofty: ～不动 steadfastly stand one's ground; remain firm; remain unmoved / ～屹立 tower; stand towering

kuí

奎
【奎纳克林】[药] quinacrine
【奎宁】[药] quinine

喹
【喹啉】[化] quinoline

蝰
【蝰蛇】viper

魁 ①(为首的; 领头的) chief; head; chieftain (指土匪首领): 罪～ chief criminal; arch-criminal ②(身体高大) of stalwart build
【魁首】a person who is head and shoulders above others; the brightest and best: 文章～ outstanding writer of the day
【魁伟】big and tall
【魁梧】big and tall; stalwart; tall and strong; strong-built

葵 certain herbaceous plants with big flowers: 锦～ high mallow / 蒲～ palm / 向日～ sunflower / 蜀～ hollyhock
【葵花】sunflower ◇～油 sunflower oil / ～子 sunflower seeds
【葵扇】palm-leaf fan

睽
【睽睽】stare; gaze: 众目～之下 in the public eye

kuǐ

傀
【傀儡】puppet ◇～戏 puppet show; puppet play / ～政府 puppet government / ～政权 puppet regime

kuì

愧 ①(惭愧) ashamed ②(不安) embarrassed; uneasy: ～不敢当 embarrassed by undeserved praise / 问心无～ have a clear conscience; have nothing on one's conscience / 于心有～ have a guilty conscience; have sth. on one's conscience
【愧恨】ashamed and remorseful; remorseful
【愧色】ashamed look; sign of shame: 毫无～ look unashamed / 面有～ look ashamed

溃 ①(河水冲破堤坝) burst: 千里之堤,～于蚁穴。 One ant hole may cause the cause the collapse of a thousand-*li* dike. ②(突破包围) break through: ～围西奔 break through the encirclement and head west ③(溃败) be defeated; be routed: ～不成军 be utterly routed; the army has collapsed / 一触即～ be defeated at the first encounter ④(肌肉组织腐烂) fester; ulcerate
【溃败】be defeated; be routed
【溃决】burst; break: 堤坝～ the dam burst; the dike burst
【溃烂】fester; ulcerate: 伤口未～。 The wound did not fester.
【溃灭】crumble and fall; collapse and perish
【溃散】be defeated and dispersed; collapse in disorder; scatter
【溃逃】escape in disorder; fly pell-mell; flee helter-skelter; break and flee
【溃退】beat a precipitate retreat; retreat in disorder
【溃疡】[医] ulcer: ～穿孔 perforated ulcer / 十二指肠～ duodenal ulcer / 胃～ gastric ulcer

愦 muddleheaded: 昏～ muddleheaded

聩 deaf; hard of hearing: 振聋发～ rouse the deaf and awaken the unhearing

匮 deficient
【匮乏】short (of supplies); deficient: 劳动力～ a scarcity of labor / 资源～ want of natural resources

馈 make a present of: ～送 present (a gift); make a present of sth.
【馈电】[电] feed: 交叉～ cross feed ◇ ～线 feed line; feeder
【馈赠】present (a gift); make a present of sth.: ～礼品 make a present / 接受～ accept a present; receive sth. as a present

kūn

坤 female; feminine: ～表 woman's watch
昆
【昆布】①[中药] kelp ②[植] laminaria
【昆虫】insect: 传病～ insect vector ◇ ～防治 insect control / ～学 entomology; insectology / ～学家 entomologist
【昆仲】elder and younger brothers; brothers
醌 quinone

kǔn

捆 ①(用绳子缠绕打结) tie; bind; bundle up; truss: ～谷草 bundle up millet stalks / ～干草 truss hay / ～行李 tie up one's baggage / ～住手脚 bound hand and foot / 把他～起来 tie him up ②[量] bundle; sheaf; truss: 两～柴 two bundles of firewood
【捆绑】truss up; bind; tie up
【捆扎】tie up; bundle up

kùn

困 ①(困扰) be stranded; be hard pressed; be distressed; be beset: 内外交～ be beset with troubles both at home and abroad / 贫病交～ be distressed with poverty and sickness / 为病所～ be afflicted with illness ②(围困) surround; pin down; besiege; hem in; encircle ③(疲乏) tired; weary; fatigued: ～乏 tired; fatigued ④(疲乏想睡) sleepy
【困顿】①(劳累不能支持) tired out; exhausted; fatigued; weary; worn-out ②(生活艰难窘迫) in financial straits; hard up; in straitened circumstances
【困乏】tired; fatigued
【困惑】perplexed; puzzled; bewildered; at a loss: ～不解 feel puzzled
【困境】difficult position; predicament; straits; dilemma; plight: 摆脱～extricate oneself from a difficult position / 陷于～ find oneself in a tight corner; land oneself in a fix; find oneself in a woeful predicament
【困窘】in straitened circumstances; in a difficult position; embarrassed
【困倦】sleepy; drowsy: 感到～ feel sleepy / 使人～ make one drowsy
【困苦】hardship; deep distress; deep poverty; tribulation; privation: ～备尝 suffer many privations / ～的日子 hard times / 艰难～ difficulties and hardships / 生活～ live in privation; live in dire poverty
【困难】①(事情复杂,有阻碍) difficulty: ～的工作 a difficult work; a hard work / ～重重 be beset with difficulties ②(生活穷困) financial difficulties; straitened circumstances: 生活～ live in straitened circumstances ◇ ～户 families with material difficulties
【困扰】perplex; puzzle: 为…所～ be puzzled by
【困守】defend against a siege; stand a siege: ～孤城 be entrenched in a beseiged city / ～一隅 be hemmed in a corner
【困兽犹斗】cornered beasts will still fight; beasts at bay will put up a desperate fight

kuò

廓 ①(广阔) wide; extensive ②(外部的周围) outline
扩 expand; enlarge; extend
【扩充】expand; strengthen; augment; enlarge; extend: ～军备 arms expansion; armaments expansion / ～设备 augment the equipment / ～实力 expand (military or political) forces ◇ ～接口 extended interface / ～解释 amplified interpretation; extensive interpretation
【扩大】enlarge; expand; extend; widen; broaden: ～统一战线 broaden the united front / ～眼界 widen one's outlook; broaden one's horizons / ～营业 extend one's business / ～再生产 expanded reproduction / ～战果 exploit the victory / ～政治影响 extend political influence
【扩大化】broaden the scope; magnify
【扩大会议】enlarged meeting; enlarged session; enlarged conference
【扩建】extend; expand: ～一个车间 extend a workshop ◇ ～工程 extension (project)
【扩军】arms expansion; armaments expansion: ～备战 arms expansion and war preparations
【扩孔】[机] reaming ◇ ～钻头 reaming bit; reamer bit
【扩散】spread; diffuse; proliferate: ～光 diffuse light / 癌～ proliferation of cancer; spread of cancer / 病菌～ proliferation of germs / 核～

nuclear proliferation
【扩胸器】chest expander; chest developer
【扩音机】amplifier
【扩音器】①(话筒) microphone; megaphone ②(扬声器) loud-speaker
【扩音系统】public-address system
【扩展】expand; spread; extend; develop
【扩张】①(扩大) expand; enlarge; extend; spread; aggrandize: 对外～ expansionism; foreign aggrandizement / 领土～ territorial expansion; territorial aggrandizement ②[医] dilate: 血管～ blood vessel dilatation ◇ ～军备 arms drive / ～时期 period of expansion / ～野心 expansionist ambitions / ～战果[军] exploitation of success / ～主义 expansionism
【扩张器】[医] dilator

阔

①(宽广) wide; broad; vast ②(有钱) rich; wealthy
【阔别】long separated; long parted; have not seen each other for a long time
【阔步】take big strides; make great strides: ～前进 advance in giant strides; stride forward / 昂首～ stride forward with one's chin up; stride proudly ahead
【阔绰】ostentatious; liberal with money: ～的生活 an extravagant life
【阔幅平布】sheeting: 本色～ grey sheeting
【阔老】rich man
【阔气】luxurious; extravagant; lavish; showily rich: 摆～ go in for extravagance; display one's wealth / 花钱～ spend lavishly
【阔少】young master of a rich family: ～作风 style of a rich youth
【阔叶树】broadleaf tree

括

①(扎;束) draw together; contract ②(包括) include
【括号】brackets: 大～ braces / 方～ brackets / 圆～ parentheses
【括弧】parentheses
【括约肌】sphincter

蛞

【蛞蝓】slug

L

lā

垃

【垃圾】rubbish; garbage; refuse (metter): 焚化～ refuse incineration／ 清除～ remove refuse ◇～处理 garbage disposal
【垃圾车】garbage truck; dust-cart
【垃圾堆】rubbish heap; refuse dump; garbage heap: 被扫进历史的～ be swept onto the rubbish heap of history
【垃圾发电】garbage power
【垃圾箱】garbage can; dustbin; ash can

拉

①(用力使朝自己所在的方向移动) pull; draw; tug; drag: ～车 pull a cart／～弓 draw a bow／ ～开门 pull the door open／～起窗帘 draw back the curtain／ ～上门 pull the door shut／ 把船～到岸边 tug the boat in to shore／ 把窗帘～过来 pull the curtains across／ 把窗帘～开 draw the curtain aside／ 这抽屉～不开。 The drawer won't pull out. ②(用车运输) transport by vehicle; haul: 去～肥料 haul back the fertilizer ③(带领转移) move troops to a place: 把一连～到前沿阵地 move Company Une to the forward position ④(牵累) drag in; implicate: 这事与他无关, 不要～上他。 He has nothing to do with it, don't drag him into trouble. ⑤(拉拢;联络) draw in; win over; canvass: ～买卖 tout; canvass orders; push sales／ 为某人～选票 canvass votes for sb. ⑥(演奏) play (certain musical instruments): ～二胡 play the *erhu*／ ～手风琴 play the accordion／ ～小提琴 play the violin ⑦(拖长) drag out; draw out; space out: ～长声音说话 drawl／ 成单行～开距离 form into single file and space out ⑧(帮助) give a helping hand; help: 我们应当～他一把。 We should help him out. ⑨(排泄) empty the bowels: 又～又吐 suffer from diarrhea and vomiting ⑩(拉丁美洲的简称) Latin America
【拉拔】[机] drawing
【拉不下脸来】cannot do sth. for fear of hurting another person's feelings
【拉扯】①(拉) drag; pull ②(辛勤抚养) take great pains to bring up ③(牵涉;牵扯) implicate; drag in: 干么把我～进去? Why drag me in? ④(闲谈) chat
【拉出去】pull out; drag out
【拉床】[机] broaching machine
【拉大旗作虎皮】 use the great banner as a tiger-skin; drape oneself in the flag to impress people
【拉刀】[机] broach
【拉倒】drop it; forget about it
【拉丁美洲】Latin America
【拉丁文】Latin (language)
【拉丁字母】the Latin alphabet; the Roman alphabet
【拉肚子】 suffer from diarrhea; have loose bowels
【拉队伍】 raise a force or contingent; form a band
【拉夫】press-gang; press people into service
【拉幅机】[纺] stenter; tenter
【拉杆】[机] pull rod; drag link; draw bar; tension link ◇ ～天线 telescopic antenna
【拉关系】 try to establish a relationship with sb.; cotton up to: 拉亲戚关系 claim kinship
【拉后腿】hold sb. back; be a drag on sb.
【拉花】[工美] garland: 纸～ festoon; paper garland
【拉簧】[机] extension spring
【拉火绳】[军] lanyard
【拉家带口】have family burden
【拉脚】transport persons or goods by cart at a charge
【拉锯】①(拉大锯) work a two-handed saw ②(来回往复的) be locked in a seesaw struggle ◇～地带 area which frequently changes hands in a war; scene of a seesaw battle／ ～战 seesaw battle
【拉开】①(打开) pull open; draw back: ～抽屉 open the drawer／ ～窗帘 draw back the curtain／ ～枪栓 pull back the bolt (of a rifle) ②(增加间隔距离) increase the distance between; space out: 把比分～到四十八比二十七 pull away to 48-27; increase the lead to 48-27
【拉拉扯扯】①(动手动脚) pull sb. about; drag sb. about: 别～的! Take your hands off me! ②(拉拢吹拍) exchange flattery and favors
【拉拉队】 cheering squad; rooters ◇ ～队员 rooter／ ～队长 cheerleader
【拉拉杂杂】 not well organized and without a central theme
【拉力】[物] pulling force ◇～器[体] chest-developer; chest-expander／ ～试验[机] pull test;

tension test
【拉链】zip-fastener; zipper: 拉上～ zip up
【拉拢】draw sb. over to one's side; rope in: 不要受坏人～。 Don't get roped in by bad people.
【拉马克学说】[生] Lamarckism
【拉买卖】act as a broker; solicit business
【拉模】[机] drawing die
【拉皮条】act as a procurer; act as a pimp
【拉平】bring to the same level; even up: 把比分～ even the score up
【拉纤】①(在岸上用绳子拉船) tow a boat ②(介绍买卖或租赁房屋等并从中谋利) act as go-between
【拉橇狗】sled dog
【拉山头】form a faction
【拉伸】[纺] drawing; stretch ◇ ～加捻机 stretch twister / ～络丝机 draw winder / ～试验 tensile test / ～应变[机] tensile strain
【拉生意】solicit trade; canvass trade
【拉屎】[口] empty the bowels; shit
【拉手】①[口](握手) shake hands ②(把手) handle: 门～ doorknob
【拉丝】[冶] wiredrawing ◇ ～机 wiredrawing machine
【拉锁儿】zip fastener; zipper
【拉条】[机] brace; stay: 链～ chain stay / 斜～ batter brace
【拉稀】[口] have loose bowels; have diarrhea
【拉下脸】①(不高兴的样子) look displeased; pull a long face; put on a stern expression ②(打破情面) not spare sb.'s sensibilities
【拉下水】drag sb. into the mire; make an accomplice of sb.; corrupt sb.
【拉线开关】[电] pullswitch
【拉削】[机] broaching
【拉秧】uproot plants after their edible portions have been harvested
【拉杂】rambling; jumbled; ill-organized: 这篇文章写得很～。 The article is very badly organized.
【拉主顾】attract customers; solicit customers

邋

【邋遢】[口] slovenly; sloppy: 衣着～ be slovenly dressed

lá

拉

①(割) slash; slit; cut; make a gash in: ～成小块 cut it into pieces / 把这块皮子～开 slit the leather / 手上～了个口子 cut one's hand; get a cut in the hand ②(闲谈) chat: ～家常 have a chat

lǎ

喇

【喇叭】①(唢呐) *suona*, a woodwind instrument ②(管乐器) brass-wind instruments in general: 吹～ blow a trumpet ③(扬声器) loudspeaker ◇ ～口 bell (of a wind instrument) / ～筒 megaphone
【喇叭花】(white-edged) morning glory
【喇叭裤】flared trousers; bell-bottoms
【喇嘛】[宗] lama ◇ ～庙 lamasery
【喇嘛教】Lamaism ◇ ～徒 lamaist; lamaite ○ 白教 White Sect of Lamaism / 红教 Red Sect of Lamaism / 黄教 Yellow Sect of Lamaism

là

落

①(遗漏) leave out; be missing: ～了几个字。 A few words are left out. ②(遗忘) leave behind; forget to bring: 把眼镜～在家里 leave one's spectacles at home ③(跟不上) lag behind; fall behind; drop behind: ～下很远 fall far behind / ～在某人后面 fall behind sb.

蜡

①(动物、矿物或植物的油质) wax: 地板～ floor wax / 给地板打～ wax the floor ②(蜡烛) candle: 点一支～ light a candle
【蜡版】mimeograph stencil ◇ ～术 cerography
【蜡笔】wax crayon ◇ ～画 crayon drawing
【蜡虫】wax insect
【蜡光纸】glazed paper
【蜡果】[工美] wax fruit
【蜡黄】wax yellow; waxen; sallow: ～的脸 a sallow face
【蜡染】[纺] wax printing
【蜡人】waxwork ◇ ～馆 waxworks
【蜡塑】wax sculpture
【蜡台】candlestick
【蜡丸】wax-wrapped pill
【蜡像】waxen imagen
【蜡纸】①(包装用) wax paper ②(油印用) stencil paper; stencil: 刻～ cut a stencil
【蜡烛】candle
【蜡嘴雀】[动] hawfinch

腊

【腊肠】sausage
【腊梅】[植] wintersweet
【腊肉】cured meat; bacon

【腊味】cured meat, fish, etc.
【腊月】the twelfth month of the lunar year; the twelfth moon

辣 ①(辣味) hot; peppery; pungent: 闻到一股～味 notice a pungent smell ②(狠毒) vicious; ruthless: 心毒手～ vicious and ruthless
【辣不可言】too hot to be told
【辣根】[植] horseradish
【辣酱】thick chilli sauce
【辣酱油】pungent sauce
【辣椒】hot pepper; chili ◇～粉 chili powder／～油 chili oil
【辣手】①(毒辣的手段) ruthless method; vicious device ②[方](手段厉害或毒辣) vicious; ruthless ③[口](棘手; 难办) thorny; troublesome; knotty
【辣子】[口] hot pepper; cayenne pepper; chili
【辣味】piquancy; pungency; peppery taste

瘌
【瘌痢】[方] favus of the scalp
【瘌痢头】①(长黄癣的人) a person affected with favus on the head ②(长黄癣的头) affected with favus on the head

lái

来 ①(来到) come; arrive: ～稿 incoming manuscripts; a contribution received by an editor／～信 incoming letter; your letter／客人还没有～。The guests haven't arrived yet. ②(发生) crop up; take place: 问题一～就设法解决 try to solve a problem as soon as it crops up ③(代替意义更具体的动词)我自己～吧。Let me do it myself. (指做事); I'll help myself. (指吃东西). ④(跟"得"或"不"连用)车子进得～么? Can the car get in?／他们俩很合得～。The two of them get along very well. ⑤(表示要做某件事)大家～想办法。Let's pool out ideas and see what to do. ⑥(表示来做某件事)他回学校看望大家～了。He's come back to the college to see us folks. ⑦(表示后面部分是目的)怎么～帮助他呢? How are you going to help him? ⑧(未来的) future; coming; next: ～年 the coming year; next year ⑨(以来) ever since: 三千年～ over the past 3000 years／十几天～ for the last ten days and more ⑩(表示概数) about; around: 八十～个 around eighty／二十～岁 about twenty (years old)／三米～高 about three meters high
【来宾】guest; visitor ◇～席 seats for guests
【来不得】won't do; be impermissible: ～半点虚假 permit no dishonesty
【来不及】there's not enough time (to do sth.); it's too late
【来到】arrive; come
【来得及】there's still time; be able to do sth. in time; be able to make it: 你赶末班车还～。You'll be in time for the last bus. (或: There is still time for you to catch the last bus.)
【来得容易去时快】soon got, soon gone
【来电】①(发来的电报) incoming telegram; your telegram; your message: ～收到。Your telegram received. ②(发电报来) send a telegram here
【来而不往非礼也】it is impolite not to reciprocate; one should return as good as one receives
【来犯】come to attack us; invade our territory
【来访】come to visit; come to call
【来复枪】rifle
【来复线】rifling
【来亨鸡】Leghorn
【来回】①(往复一次) make a round trip; make a return journey; go to a place and come back: 打～儿 make a round trip ②(往复多次) back and forth; to and fro: ～摇摆 oscillate; vacillate／在屋子里～走动 pace up and down the room ◇～飞行 round-trip flight／～票 return ticket; round-trip ticket
【来回来去】[方] back and forth; over and over again: ～地跑了好多趟 run back and forth many times／～地说 say sth. over and over again; repeat again and again
【来件】communication or parcel received
【来劲】①[方](有劲头) full of enthusiasm; in high spirits: 她越干越～。The longer she worked at it, the more enthusiastic she became. ②(嘲弄; 使烦恼) jest with; annoy; offend: 你别跟我～。I won't stand any nonsense from you.
【来客】guest; visitor
【来来往往】coming and going in great number
【来历】origin; source; antecedents; background; past history: ～不明 of unknown origin (指事物); of dubious background or of questionable antecedents (指人)／查明～ trace to the source; ascertain a person's antecedents
【来临】arrive; come; approach: 国庆即将～。National Day is coming.
【来龙去脉】origin and development; cause and effect: 弄清事情的～ find out the cause and ef-

fect of the incident
【来路】①(向这里来的路) incoming road; approach ②(来历) origin; antecedents: ~不正 of questionable origin(指物); of dubious background (指人)
【来年】the coming year; next year
【…来…去】(表示动作的不断反复) back and forth; over and over again: 翻来复去睡不着 toss and turn in bed / 飞来飞去 fly back and forth / 考虑来考虑去 turn sth. over and over again in one's mind / 挑来挑去 pick and choose
【来去自如】come and go freely
【来人】bearer; messenger
【来日方长】there will be ample time; there is a long time ahead
【来生】next life
【来世】[佛教] future world; next world ◇ ~报应说 retribution
【来势】the force with which sth. breaks out; oncoming force: ~汹汹 bear down menacingly
【来苏】[药] lysol
【来头】①(来历;背景) connections; background; backing: ~不小 have powerful backing / ~大 of impressive background ②(原因;来由) the motive behind cause: 他说这些话是有~的。 He didn't say these words without a motive. ③[口](做某事的兴趣) interest; fun: 下棋没有~。I have no interest in playing chess.
【来往】①(来和去) come and go: ~的信件 correspondence / ~于京津之间 travel between Beijing and Tianjin ②(交际往来) dealings; contact; intercourse: 促进商业~ promote commercial intercourse
【来往不绝】ceaseless coming and going
【来文】document received
【来信】①(寄信来) send a letter here ②(寄来的信) incoming letter: 读者~ letters from the readers / 人民~ letters from the people
【来意】one's purpose in coming: 说明~ make clear what one has come for
【来由】reason; cause: 没~ without rhyme or reason
【来源】①(事物来的地方) source; origin: 经济~ source of income / 原料~ source of raw materials ②(起源;发生) originate; stem from: 知识~于实践 knowledge stems from practice
【来源可靠】from the horse's mouth
【来者不拒】refuse nobody; refuse nobody's request or offer
【来者不善,善者不来】①(来的是强者) he who has come is surely strong or he'd never have come along ②(来人动机不善) he who has come, comes with ill intent, certainly not on virtue bent
【来者犹可追】the future is yet for oneself to shape
【来之不易】it has not come easily; hard-earned
【来踪去迹】traces of sth.; traces of sb.'s whereabouts

莱

【莱菔】[植] radish ◇ ~子[中药] radish seed
【莱诺铸排机】[印] linotype
【莱塞】[物] laser
【莱索托】Lesotho ◇ ~人 Mosotho (单数); Basotho (复数)

铼

[化] rhenium (Re)

lài

赖

①(依靠) rely; depend: ~以生存的条件 conditions on which persons rely for existence; conditions on which things depend for existence ②(留着不肯走) hang on in a place; drag out one's stay in a place; hold on to a place: ~着不走 hang on and refuse to clear out ③(抵赖) deny one's error or responsibility; go back on one's word ④(诬赖) blame sb. wrongly; put the blame on sb. else ⑤(责怪) blame: 这件事全~我。 I'm entirely to blame for that. ⑥(不好) not good, poor
【赖氨酸】[生化] lysine
【赖皮】[口] rascally; shameless; unreasonable: 耍~ act shamelessly
【赖帐】①(赖债) repudiate a debt ②(食言) go back on one's word

癞

①[医](麻风) leprosy ②[方](黄癣) favus of the scalp
【癞蛤蟆】toad: ~想吃天鹅肉 a toad lusting after a swan's flesh; aspiring after sth. one is not worthy of
【癞皮狗】①(患疥癣的狗) mangy dog ②(令人讨厌的人) loathsome creature
【癞子】a person affected with favus on the head

籁

sound; noise: 万~具寂。 Silence reigns supreme. (或: Everything quieted down.)

lán

阑

①(将尽) late: 夜~人静 in the stillness of

the night ②(栏杆) railing; balustrade
【阑珊】[书] coming to an end; waning
【阑尾】[生理] appendix ◇～穿孔 appendicular perforation / ～切除术 appendectomy / ～炎 appendicitis

谰 calumniate; slander
【谰言】calumny; slander: 无耻～ a shameless slander

镧 [化] lanthanum (La)
【镧系元素】lanthanide series; lanthanon

兰 orchid
【兰草】fragrant thoroughwort
【兰花】cymbidium; orchid
【兰科】the orchid family

栏 ①(栏杆) fence; railing; balustrade; hurdle: 跨～赛跑 hurdle race; the hurdles / 凭～ lean on a railing ②(家畜圈) pen; shed: 牛～ cowshed / 羊～ sheep pen; sheepfold ③(部分版面) column: 备注～ remarks column / 布告～ bulletin board; notice board
【栏杆】railing; banisters; balustrade
【栏杆柱】baluster

拦 bar; block; hold back: ～住去路 block the way
【拦挡】block; obstruct
【拦河坝】a dam across a river; dam
【拦洪坝】regulating dam; flood-control dam
【拦劫】intercept and rob
【拦截】intercept: ～敌机 intercept an enemy plane
【拦路】block the way: ～抢劫 waylay; hold up
【拦路虎】obstacle; stumbling block
【拦网】[排球] block: 单人～ one-man block / 双人～ two-man block / ～成功 shut out ◇～队员 blocker
【拦蓄】retain: ～山洪 retain the mountain flood
【拦腰】by the waist; round the middle: ～抱住 seize round the middle; clasp sb. by the waist
【拦鱼栅】fish screen
【拦住】hold up; keep away
【拦阻】block; hold back; obstruct

襤
【襤褛】ragged; shabby: 衣衫～ be dressed in rags; shabbily dressed; out at elbows

蓝 ①(蓝颜色) blue ②(蓼蓝) indigo plant
【蓝宝石】sapphire
【蓝本】①(主要原始资料) writing upon which later work is based; chief source ②(指底本) original version (of a literary work)
【蓝布】blue cloth
【蓝靛】indigo
【蓝矾】[化] blue vitriol; cupric sulphate
【蓝花参】[植] tuftybell
【蓝晶石】[矿] kyanite; disthene
【蓝皮书】blue book
【蓝铜矿】azurite; chessylite
【蓝图】blueprint
【蓝印花布】blue cloth with design in white
【蓝藻】[植] blue green alga

篮 ①(篮子) basket: 提～ hand basket; hamper / 网～ net-covered basket ②(篮球球篮) goal; basket: 补～ tap in; rebound shot; tip-in; follow-up shot / 扣～ dunk shot / 跨步上～ stride lay-up / 上～ lay up / 投～ shoot a basket; shoot; shooting
【篮板】[篮球] backboard; bank
【篮板球】[篮球] rebound: 控制～ control the rebounds / 抢～ rebound; back-board recovery / 抓住～投篮入网 grab the rebound and sink a basket
【篮球】basketball: 打～ play basketball ◇～场 basketball court / ～队 basketball team; quintet / ～队员 basketball player; basketballer / ～架 basketball stands / ～赛 basketball match
【篮架】[篮球] basket support
【篮圈】[篮球] ring; hoop; basket
【篮子】basket

岚 haze; vapor; mist

lǎn

懒 ①(懒惰) lazy; indolent; slothful: 腿～ disinclined to move about; lazy about paying visits ②(疲倦; 没力气) sluggish; languid: 身上发～ feel sluggish
【懒得】not feel like; not be in the mood to; be disinclined to: 怕有雨，我～上街。It threatens to rain. I don't feel like going out.
【懒惰】lazy: ～成性 have laziness as one's second nature
【懒汉】sluggard; idler; lazybones: ～思想 the way of thinking of the sluggard
【懒散】sluggish; negligent; indolent
【懒洋洋】languid; listless: ～地躺在沙发上

lounge on a sofa
【懒于】too lazy to do sth.; not enthusiastic about sth.

览 ①(看) look at; see; view: 游～ tour; go sightseeing／一～无余 take in everything at a glance ②(阅读) read: 博～ read extensively／浏～ glance over; skim through; skim over／阅～ read／展～ exhibit; show

揽 ①(用胳膊围住) pull sb. into one's arms; take into one's arms ②(拢住) fasten with a rope, etc.: 用绳子～上 put a rope around sth. ③(拉到自己方面) take on; take upon oneself; canvass: ～买卖 canvass business orders／～责任 take the responsability upon oneself ④(把持) grasp; monopolize: ～权 arrogate power to oneself／包～ monopolize; undertake the whole thing

缆 ①(缆绳) hawser; mooring rope; cable: 解～ cast off; set sail ②(象缆的东西) thick rope; cable: 电～ power cable; cable／钢～ steel cable
【缆车】cable car ◇ ～铁道 cable railway
【缆道】cableway
【缆索】thick rope; cable ◇ ～铁道 funicular (railway)

罱 ①(捕鱼或捞水草、河泥的工具) a kind of net used for fishing or for dredging up river sludge, etc. ②(用罱捞) dredge up: ～河泥 dredge up sludge from a river
【罱泥船】a boat used in collecting river sludge for fertilizer

làn

滥 ①(泛滥) overflow; flood ②(过度; 无限制) excessive; indiscriminate: ～发钞票 reckless issuing of bank notes／～花钱 spend money lavishly／～施轰炸 indiscriminate bombing; wanton bombing／～收费 charge excessively
【滥调】hackneyed tune; worn-out theme: 陈词～ hackneyed and stereotyped expressions; clichés
【滥伐】[林] denudation
【滥用】abuse; misuse; use indiscriminately: ～公款 irregularities in use of public funds／～经费 squander funds／～无度 use without limit／～职权 abuse one's power; misuse one's authority
【滥竽充数】pass oneself off as one of the players in an ensemble; be there just to make up the number

烂 ①(松软) sodden; mashed; pappy: 雨后地上都是～泥。The ground was sodden after the rain. ②(腐烂) rot; spoil; fester; decay; putrefy: 伤口～了。The wound is festering. ③(破烂) worn-out ④(头绪乱) messy: 真是一本～帐。The accounts are all in a mess.
【烂花花边】[纺] burnt-out lace
【烂糊】(of food) mashed; pulpy
【烂漫】①(鲜明而美丽) bright-colored; brilliant: 山花～ bright mountain flowers in full bloom ②(坦率自然) unaffected: 天真～ naive; innocent
【烂泥】mud; slush ◇ ～塘 a muddy pond
【烂熟】①(肉菜十分熟) thoroughly cooked ②(十分熟悉) know sth. thoroughly: 台词背得～ learn one's lines thoroughly
【烂摊子】a shambles; an awful mess
【烂帐】①(头绪乱的帐目) accounts in a mess ②(收不回来的帐) bad debt; bad accounts
【烂纸】waste paper
【烂醉】dead drunk: ～如泥 be dead drunk; be as drunk as a lord

láng

郎 ①(古代官名) an ancient official title ②(用于对男子的称呼)货～ street vendor／令～ your son／新～ bridegroom ③(女子称丈夫或情人) my darling
【郎才女貌】a perfect match between a man and a girl
【郎舅】a man and his wife's brother
【郎中】[方] a physician trained in herbal medicine; doctor

廊 porch; corridor; veranda: 长～ the Long Corridor (in the Summer Palace, Beijing)／画～ picture gallery／回～ winding corridor
【廊檐】the eaves of a veranda
【廊子】veranda; corridor

榔
【榔头】hammer

琅
【琅琅】[象] ～的读书声 the sound of reading aloud

锒

【锒铛】①[书](铁锁链) iron chains: ～入狱 be chained and thrown into prison ②(金属撞击的声音) chank; clang

狼 wolf

【狼狈】in a difficult position; in a tight corner: ～不堪 in an extremely awkward position; in a sorry plight; in sore straits／～逃窜 flee in panic; flee helter-skelter／显出一副～相 cut a sorry figure／陷于～境地 find oneself in a fix; be caught in a dilemma
【狼狈为奸】act in collusion with each other
【狼奔豕突】run like a wolf and rush like a boar; tear about like wild beasts
【狼疮】[医] lupus
【狼多肉少】there is too little meat for so many wolves; too many looters for the limited wealth
【狼狗】wolf dog; wolfhound
【狼孩】wolf child
【狼毫】a writing brush made of weasel's hair
【狼獾】[动] glutton
【狼藉】[书] in disorder; scattered about in a mess: 杯盘～ wine cups and dishes lying about in disorder after a feast／声名～ notorious; in disrepute; discredited
【狼皮】wolf-skin
【狼群】wolf pack
【狼贪虎视】insatiably greedy like wolves and tigers
【狼吞虎咽】wolf down; gobble up; devour ravenously
【狼尾草】[植] Chinese pennisetum
【狼心狗肺】①(心肠狠毒) rapacious as a wolf and savage as a cur; cruel and unscrupulous; brutal and cold-blooded ②(忘恩负义) ungrateful
【狼烟四起】war alarms raised at all border posts
【狼子野心】wolfish ambition; wicked intention

lǎng

朗 ①(光线充足;明亮) light; bright ②(声音清晰响亮) loud and clear

【朗读】read aloud; read loudly and clearly
【朗朗】①[象](读书的声音) the sound of reading aloud ②(明亮) bright; light: ～乾坤 as bright as sun and moon
【朗诵】recite; deliver a recitation
【朗月明星】a bright moon and illuminating stars

làng

浪 ①(波浪) wave; breaker; ripple: 大～ rough sea／巨～ very rough sea／轻～ slight sea／微～ very smooth sea／无～ calm／小～ smooth sea／中～ moderate sea／白～滔天 white breakers leaping skyward／麦～起伏 wheat rippling in the wind ②(放纵; 无约束) unrestrained; dissolute: 放～ dissolute; dissipated

【浪潮】tide; wave: 罢工～ a wave of strikes／抗议的～ tidal waves of protest
【浪船】swingboat
【浪荡】①(游荡) loiter about; loaf about ②(行为不检点) dissolute; dissipated ◇～不羁 dissipated and unrestrained
【浪费】waste; squander; be extravagant: ～青春 waste one's youth／～时间 waste time; fritter away one's time／～无度 lavish profusely without limit／反对～ combat waste
【浪花】spray; spindrift
【浪迹天涯】roaming freely all over the world
【浪漫】romantic ◇～主义 romanticism
【浪人】[日本] ronin
【浪头】①[口](波浪) wave ②(潮流) trend: 赶～ follow the trend
【浪涌】[电] surge ◇～放电器 surge arrester
【浪子】prodigal; loafer; wastrel: ～回头 return of the prodigal son
【浪子回头金不换】A prodigal who returns is more precious than gold.

莨

【莨菪】[植] henbane

lāo

捞 ①(打捞) drag for; dredge up; fish for; scoop up from the water: ～鱼 net fish; catch fish／在河里～水草 dredge up water plants from the river ②(攫取) get by improper means; gain ③(得到机会) get the opportunity

【捞本】win back lost wagers; recover one's losses; recoup oneself
【捞稻草】①(捞好处) take advantage of sth. ②(在绝境中做徒劳无益的挣扎) clutch at a straw
【捞饭】rice boiled, strained and then steamed
【捞取】fish for; gain: ～水月 salvage moon in the river／～政治资本 fish for political advantage

【捞一把】reap some profit; profiteer
【捞着】get the opportunity: 那天的电影，我没～看。I missed the film the other day.

láo

牢 ①(监狱) jail; prison: 坐～ be in prison ②(牢固) firm: ～～记住 firmly bear in mind / 把犯人看～ keep a prisoner secure / 把绳拴～ tie the rope fast
【牢不可破】unbreakable; indestructible: ～的友谊 unbreakable friendship
【牢愁莫遣】worried not knowing how to drive away melancholy
【牢房】cell; ward
【牢固】firm; secure: ～的基础 solid foundation
【牢记】keep firmly in mind; remember well
【牢靠】①(坚固;稳固) firm; strong; sturdy ②(稳妥可靠) dependable; reliable: 办事～ dependable in handling matters
【牢笼】①(鸟兽笼) cage ②(束缚) bonds: 冲破旧思想的～ shake off the bonds of old ideas ③(圈套) trap; snare: 陷入～ fall into a trap; be caught in a trap
【牢骚】discontent; grievance; complaint: 发～ grumble / 满腹～ be querulous; be full of grievances / ～抑郁 grieved and depressed
【牢稳】[口] stable; safe; secure
【牢狱】prison; jail

劳 ①(劳动) work; labor: 多～多得 more pay for more work ②(烦劳) put sb. to the trouble of ③(劳苦) fatigue; toil: 积～成疾 break down from constant overwork ④(功劳) meritorious deed; service: 汗马之～ distinctions won in battle; war exploits ⑤(慰劳) express one's appreciation reward: ～军 bring greetings and gifts to army units
【劳保】[简]①(劳动保险) labor insurance ②(劳动保护) labor protection ◇ ～条例 labor insurance regulations / ～用品 appliances for labor protection
【劳动】①(人类创造性的活动) work; labor: 抽象～ abstract labor / 具体～ concrete labor / 社会～ social labor / 熟练～ skilled labor / 靠自己～生活 live by one's hands / ～创造世界。Labor creates the world. ②(体力劳动) physical labor; manual labor: ～锻炼 temper oneself through manual labor / 参加体力～ take part in manual labor ◇ ～保护 labor protection / ～保护设施 labor safety devices / ～保险 labor insurance / ～保险条例 labor insurance regulations / ～报酬 payment for labor / ～布 denim / ～产品 products of labor / ～定额 work norm; work quota; production quota / ～对象 subject of labor / ～二重性 the twofold character of labor / ～法 labor law / ～改造 reform through labor / ～号子 work song / ～纪律 labor discipline / ～教养 reeducation through labor / ～节 International Labor Day / ～竞赛 labor emulation; emulation drive; emulation campaign / ～量 amount of labor / ～模范 model worker / ～强度 labor intensity / ～权 right to work / ～群众 working people; laboring masses / ～人民 laboring people; working people / ～日 workday; working day / ～生产率 labor productivity; productivity / ～时间 hours of labor / ～市场 labor market / ～收入 income from work / ～手段 means of labor / ～效率 labor efficiency / ～英雄 labor hero / ～者 laborer; worker / ～资料 means of labor; instruments of labor
【劳动观点】attitude to labor: 树立～ form a correct attitude towards labor / 增强～ improve one's attitude to labor
【劳动力】①(人力) labor force; labor: ～不足 short of manpower; shorthanded / ～调配 allocation of the labor force / 调剂～ adjust the use of the labor force ②(劳动能力) capacity for physical labor: 丧失～ lose one's ability to work; be rendered unfit for physical labor; be incapacitated; be disabled ③(有劳动力的人) able-bodied person: 全～和半～ able-bodied and semi-ablebodied (farm) workers
【劳顿】[书] fatigued; wearied: 旅途～ fatigued by a journey; travel-worn
【劳而无功】work hard but to no avail; work fruitlessly
【劳而无怨】one lays tasks without repining
【劳方】labor: ～与资方 labor and capital
【劳改】[简](劳动改造) reform through labor ◇ ～队 group sentenced to reform through labor / ～农场 reform through labor farm
【劳工】[旧] laborer; worker ◇ ～运动 labor movement
【劳驾】①[套](要求让路等) Excuse me. ②(要求别人做事) May I trouble you?
【劳苦】toil; hard work: ～大众 toiling masses; laboring people / ～功高 have worked hard and achieved great things / ～与共 a trouble shared is a trouble halved / 不辞～ spare no pains

【劳累】tired; run-down; overworked
【劳力】labor; labor force: 合理安排～ rational allocation of labor
【劳碌】work hard; toil
【劳民伤财】tire the people and drain the treasury; waste money and manpower
【劳模】[简] model worker
【劳其筋骨】toil flesh and bone
【劳神】be a tax on (one's mind); bother; trouble: ～费力 weary mind and use strength
【劳师】tire troops: ～动众 mobilize too many troops; drag in lots of people / ～远征 tire the troops on a long expedition
【劳什子】[方] nuisance
【劳损】[医] strain: 肌腱～ muscular strain
【劳务】services ◇ ～出口 the export of labor services / ～合同 contract of service
【劳心】work with one's mind or brains
【劳燕分飞】be like birds flying in different directions; part; separate
【劳逸】labor and rest ◇ ～不均 uneven allocation of work / ～结合 strike a proper balance between work and rest; alternate work with rest and recreation
【劳资】labor and capital ◇ ～关系 relations between labor and capital; labor-capital relations / ～纠纷 trouble between labor and management / ～两利 benefit both labor and capital / ～争议 labor dispute

痨 consumptive disease; tuberculosis; consumption: 肺～ pulmonary tuberculosis

唠
【唠叨】chatter; be garrulous: 唠唠叨叨说个没完 chatter interminably

铹 [化] lawrencium (Lw)

醪 ①(浊酒) wine with dregs; undecanted wine ②(醇酒) mellow wine
【醪糟】fermented glutinous rice

lǎo

老 ①(年岁大) old; aged: 他不显～。He doesn't look old. ②(老年人) old people: 扶～携幼 bringing along the old and the young ③(历史久) of long standing; old: ～部下 a former subordinate / ～干部 a veteran cadre / ～朋友 an old friend ④(陈旧) outdated: ～机器 outmoded machine / ～式 old-fashioned; outmoded; outdated / ～习惯 outdated customs ⑤(火候大) overdone; tough ⑥(蔬菜长过了时) overgrown ⑦(颜色深) dark ⑧(经常) always; constantly: 她～是在家里。She always stays at home. ⑨(很) very: ～远 far away / ～早 very early; long ago ⑩[口](排行最末的) the youngest: ～姑娘 the youngest daughter
【老把戏】an old trick
【老百姓】[口] common people; ordinary people; civilians
【老板】boss; proprietor; shopkeeper: 后台～ backstage boss ◇ ～娘 shopkeeper's wife; proprietress
【老伴儿】[口] husband or wife: 我的～ my old man or woman
【老鸨】a woman running a brothel; procuress; madam
【老辈】one's elders; old folks
【老本】principal; capital: 把～输光 lose one's last stakes
【老兵】old soldier; veteran
【老伯】[尊] uncle
【老财】[方] moneybags; landlord
【老巢】nest; den; lair
【老成】experienced; steady: 少年～ young but steady; old head on young shoulders / ～持重 experienced and prudent / ～练达 experienced and versed in one's work
【老处女】old maid; spinster
【老粗】uneducated person; rough and ready chap
【老搭档】old partner; old workmate
【老大】①(排行第一的人) eldest child ②(木船上的船夫) master of a sailing vessel ③(很) greatly; very: 心里～不高兴 feel very annoyed
【老大不小】have grown up
【老大哥】[尊] elder brother
【老大难】long-standing, big and difficult (problem): ～单位 a unit with serious and long-standing problems / ～问题 a knotty problem of long-standing
【老大娘】[尊] aunty; granny
【老大徒伤悲】vainly regret in old age one's laziness in youth
【老大爷】[尊] uncle; grandpa
【老当益壮】old but vigorous
【老道】[口] Taoist priest
【老底】sb.'s past; sb.'s unsavory background: 揭～ dig up sb.'s unsavory past; drag the skeleton out of sb.'s closet
【老弟】young man; young fellow; my boy

【老调】hackneyed theme; platitude: ~重弹 harp on the same string; play the same old tune
【老掉牙】very old; out of date; obsolete; antediluvian: 这架打字机~了. The typewriter is obsolete.
【老而弥坚】become more firm as one grows old
【老夫老妻】an old couple
【老夫少妻】an old man with a young wife
【老干部】veteran cadre
【老公公】①[方](小孩称年老的男人) grandpa ②(丈夫的父亲) husband's father; father-in-law
【老姑娘】old spinster; old maid
【老古董】①(陈旧过时的东西) old-fashioned article; antique ②(思想或生活习惯过时的人) old fogey
【老规矩】old rules and regulations; convention; established custom or practice
【老汉】①(老年男人) old man ②(用于自称) an old fellow like me
【老好人】a benign and uncontentious person who is indifferent to matters of principle; one who tries never to offend anybody
【老狐狸】①(老的狐狸) old fox ②(非常狡猾的人) crafty scoundrel
【老虎】tiger
【老虎凳】rack used as an instrument of torture
【老虎屁股摸不得】like a tiger whose backside no one dares to touch
【老虎钳】①(台钳) vice ②(手工工具) pincer pliers
【老花镜】presbyopic glasses
【老花眼】presbyopia
【老化】[化] ageing
【老话】①(流传已久的话) old saying; saying; adage: 正如中国~说的 as the Chinese saying goes ②(谈论过去事情的话) remarks about the days
【老皇历】calendar of the past; old history; obsolete practice
【老黄牛】①(老的黄牛) willing ox ②(老实勤恳为人民服务者) a person who serves the people wholeheartedly
【老几】①(排行第几) order of seniority among brothers or sisters ②(用于反问)你算~? Who do you think you are?/ 我算~. I'm a nobody.
【老骥伏枥,志在千里】an old steed in the stable still aspires to gallop a thousand *li*; old people may still cherish high aspirations
【老家】native place; old home
【老奸巨猾】a past master of machination and maneuver; a crafty old scoundrel; a wily old fox
【老茧】callosity; callus
【老将】veteran; old-timer
【老将出马,一个顶俩】when a veteran takes the field, he can do the job of two
【老交情】long-standing friendship; an old friend
【老街坊】[口] old neighbor
【老境凄凉】a lonely, dreary life in old age
【老老实实】honestly; conscientiously; in earnest
【老泪纵横】old people weep unashamedly
【老脸皮】thick-skinned; brazen-faced
【老练】seasoned; experienced: 她办事~. She is experienced and works with a sure hand.
【老路】old road; beaten track: 走~ follow the beaten track; slip back into the old rut
【老妈子】[旧] amah; maidservant
【老马识途】an old horse knows the way; an old hand is a good guide
【老迈】aged; senile
【老毛病】old trouble; old weakness
【老谋深算】circumspect and farseeing; experienced and astute
【老脑筋】old way of thinking
【老年】old age ◇ ~斑 senile plaque / ~人 old people; the aged / ~性痴呆 senile dementia / ~医学 gerontology
【老牛破车】an old ox pulling a rickety cart; making slow progress
【老牛舐犊】dote on one's children
【老农】old farmer; experienced peasant
【老牌】old brand
【老婆子】①(年老的妇人) old biddy ②(丈夫称妻子) my old woman
【老气】①[方](老成的样子) old mannish ②(服装颜色深;样式旧) dark and old-fashioned
【老气横秋】①(摆老资格) arrogant on account of one's seniority ②(没有朝气) lacking in youthful vigor
【老前辈】one's senior; one's elder: 革命~ a veteran of the revolution
【老区】old liberated area
【老人】①(老年人) old man or woman; the aged; the old ②(上了年纪的父母或祖父母) one's aged parents or grandparents
【老人星】[天] Canopus
【老弱】the old and weak

【老弱病残】the old, weak, sick and disabled
【老弱残兵】remaining troops made up of the old and weak; those who on account of old age, illness, etc. are no longer active or efficient in work
【老少无欺】cheat neither the old nor the young
【老生常谈】commonplace; platitude
【老师】teacher
【老师傅】[尊] master craftsman; experienced worker
【老实】①(诚实) honest; frank: ～交待 come clean; own up; make a clean breast of / 做～人,说～话,办～事 be an honest person, honest in word and honest in deed ②(规规矩矩) well-behaved; good: 你～点! Behave yourself! ③[婉](不聪明) simpleminded; naive; easily taken in
【老视】[医] presbyopia
【老手】old hand; old stager; veteran: 她是个滑冰的～。She is a good hand at skating.
【老鼠】mouse; rat
【老死不相往来】not visit each other all their lives; never be in contact with each other
【老太太】[尊]①(年老的妇女) old lady ②(对人称自己的母亲或婆婆) my mother
【老太爷】[尊]①(年老的男子) elderly gentleman ②(对人称自己的父亲或公公) my father
【老态龙钟】senile; doddering
【老天爷】God; Heavens
【老头子】①(年老的男子) old fogey; old codger ②[口](妻子称丈夫) my old man
【老顽固】old stick-in-the-mud; old diehard; old fogey
【老王卖瓜,自卖自夸】Lao Wang selling melons praises his own goods; praise one's own work or wares
【老翁】old man; greybeard
【老挝】Laos ◇ ～人 Laotian; Lao / ～语 Lao (language)
【老乡】fellow-townsman; fellow-villager
【老小】grown-ups and children; one's family: 一家～ the whole family
【老兄】brother; man; old chap
【老羞成怒】fly into a rage out of shame; be shamed into anger
【老朽】decrepit and behind the times; old and useless: ～昏庸 old, worthless, dull and mean
【老学究】old pedant
【老眼光】old ways of looking at things; old views: 拿～看人 judge people from old impressions
【老眼昏花】can't see clearly; dim-sighted from old age
【老爷爷】①(曾祖父) great grandfather ②[尊](小孩尊称年老的男子) grandpa
【老爷】①(官僚) master; bureaucrat; lord: 采取～式的态度 adopt a bureaucratic attitude / 做官当～ act as lords and masters ②[方](外祖父) grandfather; grandpa ◇ ～兵 pampered soldier
【老一辈】older generation: ～无产阶级革命家 proletarian revolutionaries of the older generation; veteran proletarian revolutionaries
【老一套】the same old stuff; the same old story: 改变～的做法 change outmoded methods
【老鹰】black-eared kite; hawk; eagle
【老油子】wily old bird; old campaigner
【老于世故】versed in the ways of the world; worldly-wise
【老玉米】corn
【老丈人】a man's father-in-law
【老帐】old debts; long-standing debts
【老着脸皮】unabashedly; unblushingly
【老主顾】regular customer
【老资格】old-timer; veteran
【老子】①(父亲) father ②(骄傲的人自称): ～天下第一 regard oneself as No. 1 authority under heaven; think oneself the wisest person in the world
【老祖宗】ancestor; forefather

铑 [化] rhodium (Rh)

佬 [贬] man; guy; fellow: 阔～ a rich guy / 美国～ Yankee

姥

【姥姥】[方] grandmother; grandma

lào

涝 waterlogging: 防～ prevent waterlogging / 排～ drain waterlogged areas
【涝洼地】waterlogged lowland
【涝灾】damage or crop failure caused by waterlogging

烙 ①(烫;熨) brand; iron: ～衣服 iron clothes / 给马～上印记 brand a horse ②(烤热) bake in a pan: ～两张饼 bake a couple of cakes
【烙饼】a kind of pancake
【烙铁】①(烫衣服的) flatiron; iron ②(焊接的) soldering iron

【烙印】brand

落

【落色】discolor; fade
【落枕】have a stiff neck

酪

酪 ①(奶酪) junket ②(果酱) fruit jelly ③(干果酪) sweet paste made from crushed nuts; sweet nut paste: 核桃～ walnut cream
【酪氨酸】[化] tyrosine
【酪乳】buttermilk
【酪素】[化] casein ◇～胶 casein glue

lè

乐

乐 ①(快乐) happy; cheerful; joyful: ～得跳起来 jump with delight / 助人为～ find pleasure in helping others ②(乐于) be glad to; find pleasure in; enjoy ③(笑) laugh; be amused
【乐不可支】overwhelmed with joy; be beside oneself with hapiness
【乐不思蜀】indulge in pleasure and forget home and duty
【乐此不疲】always enjoy it; never be bored with it
【乐得】readily take the opportunity to; be only too glad to
【乐而忘返】be a slave of pleasure
【乐观】optimistic; hopeful; sanguine: ～的报道 a sanguine report / ～的看法 an optimistic view / ～的性格 optimistic nature / 对前途很～ be optimistic about the future; be sanguine about the future ◇ ～主义 optimism / ～主义者 optimist
【乐果】[农] Rogor; dimethoate
【乐呵呵】buoyant; happy and gay
【乐极生悲】extreme joy begets sorrow
【乐趣】delight; pleasure; joy: 工作中的～ delight in work / 生活中的～ joys of life
【乐善好施】love to do philanthropic work
【乐事】pleasure; delight: 以助人为～ find pleasure in helping others
【乐陶陶】[书] cheerful; happy; joyful
【乐天】carefree; happy-go-luck: ～知命 be content with what one is ◇ ～派 an easy going person
【乐土】land of happiness; paradise
【乐以忘忧】seek pleasure in order to free oneself from care
【乐意】①(甘心愿意) be willing to; be ready to: ～帮忙 be willing to help ②(满意; 高兴) pleased; happy
【乐于】be happy to; take delight in: ～从命 be happy to obey / ～接受任务 be ready to accept a task / ～助人 be happy to render helps to other
【乐园】paradise; Eden: 儿童～ children's playground / 冒险家的～ paradise of the adventures / 人间～ earthly paradise; paradise on earth
【乐滋滋】[口] contented; pleased

勒

勒 ①(收住缰绳) rein in: ～马 rein in a horse ②(强制;逼迫) force; coerce
【勒逼】force; coerce
【勒克司】[物] lux; meter-candle
【勒令】compel (by legal authority); order: ～停业 be closed down by order
【勒派】force sb. to pay levies; levy on sb.: ～税款 levy a tax on sb.
【勒索】extort; blackmail: ～钱财 extort money from sb.

鳓

鳓 Chinese herring

lēi

勒

勒 tie or strap sth. tight: ～紧裤带 tighten one's belt
【勒脚】[建] plinth

擂

擂 hit; beat: ～了一拳 give sb. a punch

léi

雷

雷 ①(随闪电的响声) thunder ②(爆炸武器) mine: 地～ mine; land mine / 饵～ body trap; trap mine / 浮～ buoyant mine / 锚～ moored mine / 石～ stone mine / 水～ mine / 信号～ signal mine / 鱼～ torpedo / 布～ lay mines / 扫～ sweep mines / 探～ detect mines ○布雷艇 mine layer / 探雷器 mine detector / 扫雷艇 mine sweeper
【雷暴】[气] thunderstorm ◇ ～雨 thunderstorm rain
【雷达】radar: 全景～ panoramic radar ◇ ～兵 radar operator; radarman / ～测距 radar ranging / ～干扰 radar jamming / ～跟踪 radar tracking / ～领航 radar navigation / ～探测区 radar coverage / ～网 radar fence / ～信标 racon; radar beacon / ～荧光屏 radar screen / ～预警机 radar picket / ～员 radar operator; radarman / ～站 radar station
【雷电】thunder and lightning: ～交作 lightning accompanied by peals of thunder ◇ ～计 ceraunograph

【雷动】thunderous: 欢声～ thunderous cheers
【雷公】Thunder God: ～打豆腐，拣软的欺。The God of Thunder strikes the beancurd; bullies pick on the soft and weak.
【雷汞】[化] mercury fulminate
【雷管】detonator; detonating cap; blasting cap; primer: 电～ electric detonator
【雷击】be struck by lightning: 遭～ be struck by thunderbolt
【雷厉风行】with the power of a thunderbolt and the speed of lightning: vigorously and speedily; resolutely
【雷米封】[药] rimifon
【雷鸣】thunderous; thundery: ～般的掌声 thunderous applause
【雷鸟】white partridge
【雷劈火烧】struck by lightning and caught fire
【雷区】[军] mine field
【雷神式导弹】Thor missile
【雷声】thunderclap; thunder: ～隆隆 the rumble of thunder / ～震耳 deafening thunder
【雷声大，雨点小】loud thunder but small raindrops; much said but little done
【雷霆】①(霹雳) thunderclap; thunderbolt ②(威力或怒气) thunder-like power or rage; wrath: 大发～ fly into a rage
【雷霆万钧】as powerful as a thunderbolt: 以～之力 with the force of a thunderbolt
【雷同】①(随声附和) echoing what others have said ②(不该相同而相同) duplicate; identical
【雷雨交加】storm accompanied by peals of thunder
【雷雨云】[气] thundercloud
【雷阵雨】thunder shower

镭

[化] radium (Ra)
【镭疗】[医] radium therapy
【镭射气】[化] radium emanation

累

【累累】①(连接成串) clusters of; heaps of: 果实～ fruit hanging in clusters; fruit hanging heavy ②(憔悴颓丧的样子) haggard; gaunt: ～若丧家之犬 wretched as a stray cur
【累赘】①(多余；麻烦) burdensome; cumbersome ②(不简洁) wordy; verbose ③(使人感到多余、麻烦的事物) encumbrance; burden; nuisance

lěi

蕾

flower bud; bud: 花～ flower bud
【蕾铃】cotton buds and bolls

磊

【磊落】open and upright: 光明～ open and aboveboard / 胸怀～ openhearted and upright

累

①(积累) pile up; accumulate: 成千～万 thousands upon thousands / 日积月～ accumulate day by day and month by month ②(屡次；连续) continuous; repeated; running: 奋战～日 carry on the fight for several days running ③(牵扯) involve: 连～ involve; implicate; get sb. into trouble
【累牍连篇】long and tedious writings
【累犯】①(指行为) recidivism ②(指人) recidivist
【累积】accumulate ◇ ～盈余 accumulated surplus / ～支出 accumulated outlay
【累及】implicate; involve; drag in: ～无辜 involve the innocent
【累计】①(加算) add up ②(合计) accumulative total; grand total
【累戒不改】refuse to mend one's ways despite repeated warnings
【累进】progression ◇ ～率 graduated rates / ～税 progressive tax; progressive taxation
【累累】①(屡屡) again and again; many times ②(累积得多) innumerable; countless: 罪行～ have a long criminal record; commit countless crimes
【累卵】a stack of eggs: liable to collapse any moment; precarious: 危如～ as precarious as a stack of eggs; in an extremely precarious situation
【累年】for years in succession; year after year
【累人累己】implicate others and self
【累世】for many generations; generation after generation
【累退率】regressive rate
【累退税】regressive taxation
【累战皆北】defeated in successive battles
【累战皆捷】one victory after another

垒

①(砌；筑) build by piling up bricks, stones, earth, etc.: ～一道墙 build a wall ②(壁垒) rampart ③[棒、垒球](起点) base: 安全上～ safe / 本～ home base / 滑～ sliding / 跑～ base running / 偷～ stealing
【垒垫】bag
【垒间线】base line
【垒球】softball: 打～ play softball ○后摆投球法 sling shot / 绕环投球法 wind mill / 8字投

球法 8 figure [参见棒球]
【垒球棒】softball bat

lèi

泪 tear; teardrop: 流～ shed tear
【泪管】tear duct; lachrymal duct
【泪痕】tear stains: ～斑斑 tear-stained / 满脸～ a face bathed in tears
【泪花】tears in one's eye
【泪流满面】tears cover one's face
【泪囊炎】[医] dacryocystitis
【泪如泉涌】tears welling up like a fountain
【泪水】tear; teardrop
【泪汪汪】with watery eyes; brimming with tears
【泪腺】[生理] lachrymal gland; tear gland ◇ ～炎[医] dacryoadenitis
【泪眼】tearful eyes: ～模糊 eyes blurred by tears
【泪沾襟】wet the front part of one's garment with tears
【泪珠】teardrop

类 ①(种类) kind; type; class; category: 各～ all sorts of / 同～ be of a kind; belong to the same category / 诸如此～ things like that; and suchlike; and what not ②(类似) resemble; be similar to: ～乎神话 sound like a fairy tale / 画虎不成反～犬 try to draw a tiger but end up with the likeness of a dog; attempt something too ambitious and end in failure
【类比】[逻] analogy: 作历史的～ draw a historical analogy
【类别】classification; category: 属于不同的～ belong to different categories / 土壤的～ classification of soil
【类病毒】[医] viroid
【类地行星】[天] terrestrial planet
【类毒素】[医] toxoid
【类风湿性关节炎】[医] rheumatoid arthritis
【类固醇】[生] steroids ◇ ～激素 steroid hormone
【类胡萝卜素】carotenoid
【类聚】birds of the same feather flock together
【类木行星】[天] Jovian planet
【类偏执反应】paranoid reactions
【类人猿】anthropoid
【类似】similar; analogous: 保证不再发生～事件 guarantee against the occurrence of similar incidents
【类推】analogize; reason by analogy: 依此～ on the analogy of this
【类星体】[天] quasar; quasi-stellar object
【类型】type ◇ ～论 theory of types / ～学 typology

擂 beat
【擂台】ring; arena: 摆～ give an open challenge; invite an emulation / 打～ take up the challenge

累 ①(疲劳) tired; fatigued; weary: ～坏了 tired out; worn out; exhausted / 不怕苦,不怕～ fear neither hardship nor fatigue ②(使疲劳) tire; strain; wear out: ～活 tiring work; heavy work ③(操劳) work hard; toil: 从早到晚～了一天 have been working hard all day

肋 ①(肋条) rib ②(胸部的侧面) costal region: 两～ both sides of the chest
【肋骨】rib ◇ ～切除术 costectomy
【肋间肌】[生理] intercostal muscle
【肋膜】[生理] pleura ◇ ～炎 pleurisy
【肋木】[体] stall bars
【肋条】[方] ①(肋骨) rib ②(作为食品的带肉的肋骨) pork ribs

léng

棱 ①(两个平面连接部分) arris; edge: 见～见角 angular / 桌子～儿 edges of a table ②(条状凸起部分) corrugation; ridge: 搓板的～儿 ridges of a washboard
【棱角】①(棱和角) edges and corners ②(锋芒) edge; pointedness: 不要把～磨掉。Don't draw in your horns.
【棱镜】[物] prism: 三～ triangular prism ◇ ～分光 prismatic decomposition
【棱线】[军] crest line
【棱柱体】[数] prism: 斜～ oblique prism / 正～ regular prism / 直～ right prism
【棱锥体】[数] pyramid

lěng

冷 ①(温度低) cold: ～天 the cold season; cold days / ～得发抖 shiver with cold / 觉得～ feel cold / 他～得发抖。He was shivering with cold. / 她怕～。She is afraid of cold. ②(不热情) cold in manner; frosty: ～若冰霜 frosty in manner ③[方](使冷) cool: ～一下再吃。Let it cool off before you eat it. ④(寂静) unfrequented; deserted: 屋里～清清的。The house looked deserted. ⑤(乘人不备) shot from hiding: ～枪 a sniper's shot

【冷拔】[机] cold-drawing
【冷板凳】①(清闲冷落的职务) an indifferent post: 坐～ hold a title without any obligations of office ②(冷遇) a cold reception: 坐～ get frosty reception; meet with cold reception
【冷冰冰】ice cold; icy; frosty: ～的脸色 cold expression; frosty looks / ～的态度 icy manners
【冷布】(cotton) gauze
【冷不防】unawares; suddenly; by surprise: 打他一个～ take him unawares; catch him off guard
【冷餐】buffet ◇ ～招待会 buffet reception
【冷藏】refrigeration; cold storage ◇ ～车 refrigerator car; refrigerator van / ～船 refrigerated carrier / ～工业 refrigeration industry / ～库 cold storage; freezer / ～汽车 cold storage truck / ～设备 refrigerating equipment / ～箱 refrigerator; fridge
【冷场】①(演员失误的局面) awkward silence on the stage when an actor enters late or forgets his lines ②(开会没人发言的情景) awkward silence at a meeting
【冷嘲热讽】freezing irony and burning satire
【冷处理】[机] cold treatment
【冷床】[农] cold bed; cold frame
【冷淡】①(不热闹) cheerless; desolate ②(不热情) cold; indifferent: ～的答复 a cold reply / ～的接待 a cold reception / ～的态度 a frigid manner / 对倡议表示～ show indifference towards a proposal / 反映～ a cold response ③(招待不热情) treat coldly; cold-shoulder; slight
【冷冻】freezing ◇ ～厂 cold storage plant / ～干燥 freeze drying / ～货 frozen cargo / ～机 refrigerator; freezer / ～剂 refrigerant / ～精液[牧] frozen semen
【冷冻手术】[医] cryosurgery
【冷锻】[机] cold forging; cold hammering
【冷堆】[气] cold dome
【冷锋】[气] cold front
【冷敷】[医] cold compress
【冷宫】cold palace: 被打入～ be consigned to limbo
【冷光】[物] cold light
【冷汗】cold sweat: 出～ be in a cold sweat; break out in a cold sweat
【冷焊】[机] cold welding
【冷货】goods not much in demand; dull goods
【冷加工】[机] cold working
【冷箭】an arrow shot from hiding; sniper's shot: ～伤人 injure a person by hidden arrows / 放～ make a sneak attack
【冷静】sober; calm; in cold blood: 保持～ keep calm / 头脑～ sober-minded; level-headed; cool-headed
【冷觉】[生理] sensation of cold; sense of cold
【冷酷】unfeeling; callous; grim: ～的人 a hard-hearted person / ～的现实 grim reality / ～无情 unfeeling; cold-blooded
【冷冷清清】cold and cheerless; desolate: ～的局面 dreary situation / 会议开得～。 The meeting was very dull.
【冷落】①(不热闹) unfrequented; desolate: 狭窄～的小巷 an unfrequented narrow alley ②(使受冷淡待遇) treat coldly; cold-shoulder; leave out in the cold: ～了客人 leave a guest out in the cold
【冷铆】[机] cold riveting
【冷门】①(不时兴) a profession, trade or branch of learning that receives little attention ②(意想不到的) an unexpected winner; dark horse: 爆了个～。 That was a surprise hit. ◇ ～货 goods not much in demand; dull goods
【冷漠】cold and detached; unconcerned; indifferent: ～无情 sternly cool and unmoved
【冷凝】[物] condensation ◇ ～点 condensation point / ～器 condenser / ～物 condensate
【冷暖】changes in temperature: ～自知 know whether it is cold or warm by oneself / 注意～ be careful about changes of temperature; take care of oneself
【冷盘】cold dish; hors d'oeuvre
【冷僻】①(冷落偏僻) deserted; out-of-the-way ②(不常见的) rare; unfamiliar: ～的典故 unfamiliar allusions / ～的字眼 rarely used words
【冷气】air conditioning ◇ ～机 air conditioner / ～设备 air-cooling system
【冷枪】sniper's shot
【冷清】cold and cheerless; desolate; lonely; deserted
【冷却】cooling ◇ ～剂 coolant; cooler / ～器 chiller / ～塔 cooling tower / ～旋管 cooling worm; cooling coil
【冷却红外探测器】cooled infrared detector
【冷热病】①[方](疟疾) malaria ②(情绪忽高忽低) capricious changes in mood; sudden waxing and waning of enthusiasm
【冷若冰霜】as cold as ice in manner; frosty in manner
【冷色】[美术] cool color
【冷杉】[植] fir

【冷食】cold drinks and snacks ◇ ～部 cold drink and snack counter
【冷霜】cold cream
【冷水】①(凉水) cold water: ～浇头 throw cold water on／泼～ pour cold water on; dampen sb.'s enthusiasm ②(生水) unboiled water ◇ ～浴 cold bath
【冷飕飕】chilling; chilly
【冷橡胶】cold rubber
【冷笑】sneer; laugh grimly; grin with dissatisfaction, helplessness, bitterness, etc.
【冷血动物】①(变温动物) cold-blooded animal; poikilothermal animal ②(没有感情的人) an unfeeling person; a coldhearted person
【冷言冷语】sarcastic comments; ironical remarks
【冷眼】①(冷静客观的态度) cool detachment: ～相看 look at coldly ②(冷淡的待遇) cold-shoulder
【冷眼旁观】①(冷静观察但不参与) look on coldly; stay aloof ②(用挑剔的态度对待) look on with a critical eye
【冷饮】cold drink
【冷遇】cold reception; cold shoulder: 遭到～ be given the cold shoulder; be left out in the cold
【冷轧】[冶] cold rolling ◇ ～钢 cold-rolled steel／～机 cold-rolling mill
【冷战】①(不使用武器的斗争) cold war ②[口](发抖) shiver: 打～ shiver with cold
【冷铸】[冶] chill casting

lèng

愣 ①(失神) distracted; stupefied; blank: 发～ stare blankly; look distracted ②(鲁莽) rash; reckless; foolhardy: ～小子 rash young fellow; young hothead
【愣干】[口] do things recklessly; persist in going one's own way
【愣劲儿】[方] dash; pep; vigor
【愣神儿】stare blankly; be in a daze
【愣头愣脑】rash; impetuous; reckless
【愣头儿青】[方] rash fellow; hothead
【愣住】become speechless because of astonishment, an unexpected question, etc.; be taken aback

lī

哩
【哩哩啦啦】[口] scattered; sporadic
【哩哩罗罗】[口] verbose and unclear in speech; rambling and indistinct

lí

离 ①(分离) leave; part from; be away from: ～沈赴京 leave Shenyang for Beijing ②(相距) off; away; from: 车站～我家有五公里路。The station is five kilometer from my house. ③(缺少) without; independent of: 植物生长～不了阳光。Plants generally can't grow without sunlight.
【离岸价格】[经] free on board (FOB)
【离瓣】[植] polypetalous; choripetalous ◇ ～花类 choripetalae
【离别】part (for a longish period); leave; bid farewell: ～故园 part from home district
【离不开】①(不可缺少) can't do without ②(太忙走不开) too busy to get away
【离愁别恨】grief of parting
【离格儿】[口] go beyond what is proper; be out of place
【离合悲欢】life is intermingled with joy and sorrow
【离合器】[机] clutch
【离婚】divorce; break a marriage: 判准甲和乙～ divorce A and B／是他要和他妻子～,还是他妻子要和他～? Did he divorce his wife or did she divorce him? ◇ ～的必要条件 divorce requirement／～夫妇的子女 child of divorce／～理由 grounds for divorce／～判决 divorce decree／～申请书 divorce petition／～诉讼 divorce proceedings; matrimonial action／～调解 divorce conciliation; divorce mediation／～者 divorcee; divorcé (男); divorcée (女)／～证书 bill of divorcement／～自由 freedom to divorce
【离间】sow discord; drive a wedge between; set one party against another: ～甲乙二人 drive a wedge between A and B ◇ ～之计 the scheme of sowing dissension
【离解】[化] dissociation
【离经叛道】depart from the classics and rebel against orthodoxy
【离境】leave a country or place ◇ ～签证 exit visa／～许可证 exit permit
【离久情疏】out of sight, out of mind
【离开】leave; depart from; deviate from: ～本题 stray from the subject; digress／离不开手儿 be too busy with the job on hand; have one's hands full
【离谱】far away from what is normal; far off the beam

【离奇】odd; fantastic; bizarre: ～的谎言 a fantastic lie／～怪诞 eccentric and wild
【离弃】abandon; desert
【离群索居】live in solitude
【离任】leave one's post: 即将～的大使 the outgoing ambassador／～回国 leave one's post for home
【离散】dispersed; scattered about; separated from one another: 骨肉～。The family were scattered in different places.
【离题】digress from the subject; stray from the point: 发言不要～。Don't diverge in your speech. (或: Please keep to the subject.)
【离心】①(不是一条心) be at odds with the community or the leadership ②(离开中心) centrifugal ◇～泵[机] centrifugal pump／～机 centrifugal machine; centrifuge／～力[物] centrifugal force／～调节器[机] centrifugal governor／～作用 centrifugation
【离心离德】dissension and discord; disunity
【离乡背井】tear oneself away from one's native place
【离异】separate; divorce: ～,无子女 divorced, without child
【离职】①(暂时离职) leave one's job temporarily ②(离开工作岗位) leave office
【离中趋势】[统计] dispersion
【离子】[物] ion: 氩～ argonion／阳～ cation／阴～ anion ◇～泵 ionic pump／～键 ionic bond／～晶体 ionic crystals／～偶 ion pair／～束 ion beam／～淌度 ionic mobility／～雾 ion-atmosphere
【离子交换】[化] ion exchange ◇～反应 ion-exchange reactions／～剂 ion exchanger／～容量 ion exchange capacity／～树脂 ion exchange resin

篱

hedge; fence: 树～ hedge; hedgerow／竹～茅舍 thatched cottage with bamboo fence
【篱笆】bamboo or twig fence: ～墙 wattled wall

厘

①(利率) a unit of monthly interest rate (= 0.1%; 1‰): 月利九～ a montly interest of 9‰ ②(一点儿) a fraction; the least: 分～不差 without the slightest error; just right
【厘米】centimeter ◇～波[电] centimeter wave／～、克、秒单位 centimeter-gram-second unit (C.G.S unit)

狸

racoon dog
【狸猫】leopard cat
【狸藻】[植] bladderwort

罹

【罹难】①(遇灾;遇险) die in a disaster or an accident ②(被害) be murdered

梨

pear
【梨膏】[中药] pear syrup (for the relief of coughs)
【梨果】[植] pome
【梨园】the operatic circle ◇～子弟 operatic players

犁

①(翻土的农具) plough ②(用犁耕地) work with a plough; plough
【犁壁】[农] mouldboard
【犁底层】[农] plough sole; plough pan
【犁铧】ploughshare; share

黎

【黎巴嫩】Lebanon ◇～人 Lebanese
【黎民】[书] the common people; the multitude
【黎明】dawn; daybreak

藜

[植] lamb's-quarters
【藜芦】[植] black false hellebore

lǐ

礼

①(仪式) ceremony; rite: 婚～ wedding／丧～ funeral ceremony; funeral ②(表示尊敬的动作) courtesy; etiquette; manners: 彬彬有～ be refined and courteous／失～行为 breach of etiquette; discourtesy／行～ (give a) salute ③(礼物) gift; present: 寿～ birthday present／送～ give a present; send a gift
【礼拜】①[宗](向神行礼) religious service: 做～ go to church; be at church ②[口](星期) week: 下～ next week ③[口](星期中的某天) day of the week: ～天 Sunday／今天～儿? What day is it today? ④(星期天) Sunday ◇～寺[伊斯兰教] mosque／～堂[基督教] church
【礼宾司】the Department of Protocol; the Protocol Department ◇～司长 Director of the Protocol Department; Chief of Protocol
【礼单】list of presents
【礼多人不怪】civility costs nothing
【礼服】ceremonial robe or dress; full dress; formal attire: 大～ full dress／男～ evening suit; dress suit; dinner suit／女～ evening dress; evening gown／晚～ evening dress
【礼花】fireworks display
【礼节】courtesy; etiquette; protocol; ceremony:

～性拜访 a courtesy call / 社交～ social etiquette
【礼金】cash gift
【礼帽】a hat that goes with formal dress: 大～ top hat
【礼貌】courtesy; politeness; manners: 有～ courteous; polite / 没～ have no manners; be impolite
【礼炮】salvo; (gun) salute: 二十一响～ 21 gun salute; a salvo of 21 guns
【礼品】gift; present: 互赠～ exchange presents ◇～部 gift and souvenir department or counter
【礼聘】cordially invite the service of
【礼轻人意重】the gift is trifling but the feeling is profound; it's nothing much, but it's the thought that counts
【礼让】give precedence to sb. out of courtesy or thoughtfulness; comity: 安全～ yield right of way for safety's sake / 国际～ the comity of nations
【礼尚往来】①(礼节上的往来) courtesy demands reciprocity ②(以其人之道还治其人之身) deal with a man as he deals with you; pay a man back in his own coin
【礼所当然】etiquette requires it
【礼堂】assembly hall; auditorium
【礼物】gift; present
【礼贤下士】courteous to the wise and condescending to the scholarly
【礼仪】etiquette; rite; protocol
【礼仪廉耻】sense of propriety, justice, honesty and honor
【礼仪之邦】a state of ceremonies
【礼遇】courteous reception: 受到～ be accorded courteous reception

李 plum

【李代桃僵】①(用一种事物取代另一种) substitute one thing for another; substitute this for that ②(代替别人受过) sacrifice oneself for another person
【李子】plum

里

①(衣服里子) lining; inside: 衣服～儿 the lining of a garment ②(里边的) inner: ～屋 inner room ③(街坊) neighborhood: 邻～ people of the neighborhood ④[书](家乡) hometown; native place: 故～ native place ⑤(长度单位) *li*; 0.5 kilometer ⑥(里面) in; inside: 手～ in one's hands / 小提箱～ in the suitcase; inside the suitcase ⑦(附在"这""那""哪"等字后边表示地点)那～ there / 这～ here / 省～发的通知 a circular issued by the province authorities
【里边】inside; in; within: 壁橱～ inside the cupboard
【里程】①(路程) mileage ②(发展的过程) course of development; course ◇～标 milepost / ～表 odometer
【里程碑】milestone: 历史的～ a milestone in history
【里出外进】①(不平整;参差不齐) uneven; not neat ②(人出来进去) a motley crowd coming and going
【里脊】tenderloin ◇～肉 lean pork taken from under the spinal column of a hog
【里里外外】inside and outside: ～一把手 competent in all one does, both inside and outside the house
【里弄】[方] lanes and alleys; neighborhood: ～工作 work on the neighborhood committee
【里面】inside; interior
【里圈】[体] inner lane
【里手】①(左边) the left-hand side ②[方](内行;专家) expert; old hand
【里通外国】have 〈maintain〉 illicit relations with a foreign country
【里外受敌】encounter enemy within and without; face opposition inside and outside
【里屋】inner room
【里应外合】act from inside in coordination with forces attacking from outside; collaborate from within with forces from without
【里子】lining: 衣服～ lining of a coat

理

①(条纹) texture; grain (in wood, skin, etc.): 肌～ skin texture / 纹～ texture; grain ②(道理) reason; logic; truth: 不可～喻 will not listen to reason; be impervious to reason ③(自然科学,有时特指物理学) natural science, esp. physics: ～工科 science and engineering / 数～化 mathematics, physics and chemistry ④(管理;办理) manage; run: ～家 keep house; manage family affairs ⑤(整理;使整齐) put in order; tidy up: ～东西 put things in order ⑥(理睬) pay attention to; acknowledge: 爱～不～ look cold and indifferent; be standoffish / 置之不～ pay no attention to sth.; brush sth. aside
【理财】manage money matters; conduct financial transactions: ～之道 the way of managing financial affair

【理睬】pay attention to; show interest in: 不予～ ignore; turn a deaf ear to; pay no heed to
【理舱费】[航运] stowage charges
【理当如此】that's just as it should be
【理短】be on the wrong side; have no justification
【理发】haircut; hairdressing ◇ ～馆 barbershop; barber's; hairdresser's / ～师 barber; hairdresser; woman barber ○ 做头发 have one's hair done; set one's hair / 留辫子 wear braids; wear plaits / 剪短 cut one's hair short / 擦油 oil one's hair / 吹风 dry one's hair / 刮脸 have a shave / 剪须 trim one's beard; trim one's moustache / 吹风机 hair drier / 电推子 electric clippers; electric shears
【理工科大学】college 〈university〉 of science and engineering
【理会】①(懂;了解) understand; comprehend: 不难～ not difficult to understand ②(理睬;过问) take notice of; pay attention to
【理货】[航运] tally ◇ ～单 tally sheet / ～员 tallyman; tall clerk
【理解】understand; comprehend: 不可～ incomprehensible; beyond one's comprehension / 不难～ not difficult to understand / 加深～ deepen one's comprehension; acquire a better understanding
【理解力】faculty of understanding; understanding; comprehension: ～差的人 man of weak apprehension / ～强 have good understanding
【理科】①(教学上所设的学科) science department in a college ②(自然科学) science ◇ ～博士 Doctor of Science (D.Sc) / ～硕士 Master of Science (M.Sc) / ～学士 Bachelor of Science (B.Sc)
【理亏】be in the wrong: 自知～ know that one is in the wrong; realize that justice is not on one's side
【理亏心虚】feel apprehensive because one is not on solid ground
【理亏语塞】principle deficient and words blocked
【理疗】[医] physiotherapy ◇ ～科医生 physiotherapist
【理乱解纷】manage and settle confusion
【理论】theory: ～学习 study of the theory / ～联系实际 integration of theory with practice / ～脱离实际 theory divorced from practice / 在～上 in terms of theory; on the theoretical plane; theortically ◇ ～化 theorize; raise to a theoretical plane / ～家 theoretician; theorist / ～水平 theoretical level
【理屈词穷】fall silent on finding oneself bested in argument; be unable to advance any further arguments to justify oneself
【理事】member of a council; director: 常任～国 permanent member state of a council ◇ ～会 council; board of directors / ～长 board chairman; chairman of the board of directors
【理算】adjustment ◇ ～人 adjuster
【理所当然】of course; naturally
【理想】ideal: 实现共产主义是我们的崇高～。To realize communism is our lofty ideal. ◇ ～国 utopia / ～化 idealize / ～家 idealists; dreamers / ～气体[物] perfect gas; ideal gas / ～语言 ideal language / ～主义 idealism
【理性】reason: 感性和～ the perceptual and the rational / 恢复～ come to one's senses / 失去～ lose one's reason ◇ ～认识[哲] rational knowledge
【理学】[哲] a Confucian school of idealist philosophy of the Song and Ming Dynasties
【理应】ought to; should: ～归公 ought to be handed over to the state or collective
【理由】reason; ground; argument: ～不足的答辩 bad plea / 没有～抱怨 have no grounds for complaint / 有充分～相信 have every reason to believe
【理直气壮】with justice on one's side, one is bold and assured: ～地回答 reply with perfect assurance / ～地予以驳斥 justly and forcefully refute
【理智】reason; intellect: 丧失～ lose one's reason; lose one's senses

锂 [化] lithium (Li)
【锂云母】[矿]lepidolite; lithia mica

俚 vulgar
【俚俗】vulgar; unrefined; uncultured
【俚语】slang

鲤 carp
【鲤鱼】carp
【鲤鱼钳】slip-joint pliers

lì

立 ①(站) stand: 起～ stand up ②(使竖立) erect; set up: ～纪念碑 erect a monument / ～界桩 erect boundary markers / 把梯子～起来 set up the ladder ③(直立的) upright; erect;

vertical ④(建立;制定) found; establish; set up: ～国 found a state / ～合同 sign a contract ⑤(自立) exist; live: 自～ be on one's feet ⑥(立刻) immediate; instantaneous

【立案】①(备案) register; put on record ②[法](设立专案) place a case on file for investigation and prosecution ◇ ～注册 register and make a record of

【立场】position; stand; standpoint: ～坚定 be steadfast in one's stand; take a firm stand / 阐明我们对这一问题的～ make clear our position on this question / 马克思列宁主义的～、观点和方法 the Marxist-Leninist stand, viewpoint and method / 丧失～ depart from the correct stand

【立春】the beginning of Spring

【立德粉】[化] lithopone

【立定】halt

【立定跳远】[体]standing long jump

【立冬】the Beginning of Winter

【立法】legislation ◇ ～机关 legislative body; legislature / ～机关的职能 legislative function / ～权 legislative power / ～权限 legislative competence / ～手续 legislative process / ～委员会 legislation committee; legislative council

【立方】①[数] cube: 四的～是六十四。The cube of four is sixty four. ②[简](立方体) cube ③[量] cubic meter; stere: 一～土 one cubic meter of earth ◇ ～根[数] cube root / ～晶系 isometric system / ～厘米 cubic centimeter / ～米 cubic meter / ～体 cube / ～英尺 cubic foot

【立竿见影】set up a pole and see its shadow; get instant results

【立功】render meritorious service; do a deed of merit; win honor; make contributions: 立大功 render outstanding service / 立集体三等功 be awarded a class three collective commendation / 立一等功 win a first class merit / ～赎罪 perform meritorious services to atone for one's crimes / ～受奖。Those who render meritorious service receive awards. ◇ ～奖状 certificate for meritorious service; certificate of merit

【立柜】clothes closet; wardrobe; hanging cupboard

【立合同】conclude a contract

【立候回音】Your immediate reply is awaited.

【立户】①(立户口) register for a household residence card; register for permanent residence ②(建立户头) open an account with the bank

【立即】[副] immediately; at once; promptly: ～回复有效 subject to immediate reply / ～交货 prompt delivery / ～照办 carry out promptly / 判处死刑,～执行 be sentenced to death and executed immediately

【立见功效】produce immediate results; feel the effect immediately

【立据】deed

【立克次氏体】[医] rickettsia

【立刻】[副] immediately; at once; right away

【立论】①(提出自己的看法) set forth one's views; present one's argument ②(表示自己的意见) argument; position; line of reasoning

【立面图】[建] elevation

【立秋】the Beginning of Autumn

【立射】[军] fire from a standing position ◇ ～射准[体] target archery

【立身处世】ways of conducting oneself in society

【立室成家】take a wife and establish a family

【立式】[机] vertical; upright: ～车床 vertical lathe / ～拉门 vertically sliding door / ～钻床 upright drill; vertical drill

【立誓】take an oath; vow

【立体】①(三维的) three-dimensional; stereoscopic ②[数](几何体) solid ◇ ～电影 stereoscopic film / ～化学 stereochemistry / ～几何学 solid geometry / ～交叉[交] grade separation / ～角[数] solid angle / ～模型 space model / ～派 cubism / ～显微镜 stereoscopic microscope; stereomicroscope / ～战争 three-dimensional warfare; triphibious warfare / ～照相机 stereopscopic camera; stereo camera

【立体声】stereophony; stereo ◇ ～系统 stereophonic sound system

【立下规矩】draw up a set of rule

【立夏】the Beginning of Summer

【立宪】constitutionalism: 君主～ constitutional monarchy / ～政体 constitutional government; constitutionalism

【立言】expound one's ideas in writing; achieve glory by writing

【立遗嘱】make a will ◇ ～处理财产的自由 freedom of testamentary disposition / ～的资格 capacity to make will

【立意】①(打定主意) be determined; make up one's mind ②(命意) conception; approach

【立于不败之地】establish oneself in an

unassailable position; remain invincible; be in an impregnable position
【立约】contract ◇ ～当事人 contracting party／～的权利 legal capacity to make contracts
【立正】stand at attention
【立证书人】drawer of document
【立志】resolve; be determined: ～改革 be determined to carry out reforms; be resolved to institute reforms
【立轴】①(长条字画) vertical scroll of painting or calligraphy ②[机] vertical shaft; upright shaft
【立住脚跟】establish oneself firmly
【立锥之地】a place to stick an awl; a tiny bit of land: 无～ not possess a speck of land
【立姿】[军] standing position
【立足】①(站住脚) have a foothold; keep a foodhold: 获得～之地 gain a foothold／(某人)难于～ be difficult to keep a foothold for sb. ②(处于某种立场) base oneself upon
【立足点】①(立脚点) foothold; footing: 找不到～ be unable to find a foothold ②(立场) standpoint; stand: ～不同 stand in different position

粒 ①(颗粒) grain; granule; pellet: 砂～儿 grains of sand／米～儿 grains of rice ②[量] 八～子弹 eight bullets／每服三～ dosage: 3 pills each time
【粒度】[矿] size
【粒肥】[简](颗粒肥料) granulated fertilizer
【粒级标准】grade scale
【粒选】[农] grain-by-grain seed selection
【粒状】granular
【粒子】[物] particle: 带电～ charged particle／高能～ energetic particle ◇ ～加速器 particle accelerator
【粒子透镜】[物] particle lens

莅 [书] arrive; be present: ～场 be present on the occasion／～会 be present at a meeting
【莅临】[书] arrive; be present: 敬请～指导。Your presence and guidance are requested.

丽 beautiful: ～人 a beauty／佳～ beautiful women／～人如云 beauties are numerous like clouds／～日和风 bright sun and pleasant breezes

俪 ①(成双的) pair; couple ②(指夫妻) husband and wife; married couple: ～影 heart-warming sight of a couple in love

栗 ①[植](栗子树) chestnut ②(发抖) tremble; shudder: 不寒而～ tremble with fear
【栗钙土】chestnut soil
【栗色】chestnut color; maroon
【栗子】chestnut

厉 ①(严格) strict; rigorous: ～禁 strictly forbid ②(严肃) stern; severe: ～声 in a stern voice
【厉兵秣马】sharpen the weapons and feed the horses; get ready for battle
【厉害】①(剧烈的、严厉的) severe; sharp; fierce: 被批评得很～ be severely criticized ②(极度的) terrible: 病得～ be seriously ill／这里热得～。It's terribly hot here.
【厉目而视】look with severe glare
【厉色正言】speak with stern countenance
【厉声】talk harshly; shout angrily
【厉声斥责】scold with an irritating voice
【厉行】strictly enforce; rigorously enforce; make great efforts to carry out: ～节约 practice strict economy

励 encourage
【励磁机】[电] exciter
【励精图治】rouse oneself for vigorous efforts to make the country prosperous

利 ①(锋利) sharp: ～刃 a sharp sword or blade／～爪 sharp claws ②(顺利) favorable: 不顺～ be unfavorable／成败～钝 advantages and disavantages; successes and failures ③(利益) advantage; benefit ④(利润或利息) profit; interest: 借款付～六分 pay 6% interest on a loan／连本带～ both principal and interest; profit as well as capital ⑤(使有利) do good to; benefit: 毫不～己，专门～人 be utterly devoted to others without any thought of self
【利比里亚】Liberia ◇ ～人 Liberian
【利比亚】Libya ◇ ～人 Libyan
【利弊】advantages and disadvantages; pros and cons: ～得失 advantages and disadvantages／权衡～ weigh the advantages and disadvantages
【利多卡因】[药] lidocaine
【利福平】[药] rifampin
【利国利民】benefit the nation and the people
【利害】①(得失) advantages and disadvantages; gains and losses: 有共同的～关系 have common interests／与…～攸关 have a stake in ②(厉害) terrible; formidable ◇ ～冲突 con-

flict of interest / ～关系人 interested party; interested person
【利己害人】benefit oneself at the expense of others
【利己利人】benefit other people as well as oneself
【利己主义】egoism ◇ ～者 egoist
【利令智昏】be blinded by lust for gain
【利禄】wealth and position
【利率】[经] rate of interest; interest rate
【利落】①(灵活敏捷) agile; nimble; dexterous: 动作～ agile movements / 手脚～ dexterous; deft / 说话不～ speak slowly and indistinctly ②(整齐有条理) neat; orderly: 房间干净～。The room is neat and tidy. ③(完毕) settled; finished: 事情总算办～了。The matter has been all settled.
【利眠宁】[药] librium
【利尿】[医] diuresis ◇ ～剂 diuretic
【利器】①(锐利武器) sharp weapon ②(有效的工具) good tool; efficient instrument
【利钱】interest
【利权】①(经济权力) economic rights ②(财政权利) financial power: ～外溢 vested rights overflow outwardly
【利润】profit: 超额～ superprofit / 企业的～留成收入 portion of the profits kept by the enterprises for their own use / 上交～ profit turned in to the state; that part of the profit turned over to the state / 这个行业～不大。The profits in this business are not large. ◇ ～税 profits tax
【利润率】profit margin; profit rate: 平均～ average rate of profit ◇ ～递减律 the law of the falling tendency of the rate of profit
【利他主义】altruism
【利息】interest ◇ ～回扣 interest rebate / ～收益 interest income
【利血平】[药] reserpine
【利益】interest; benefit; profit: ～均沾 share equal profit / 使人民群众得到～ benefit the masses of the people / 为大多数人谋～ work for the interests of the vast majority of people
【利用】①(使发挥效用) use; utilize; make use of: ～废料 make use of scrap material; turn scrap material to good account / 充分～最新科学技术成就 make full use of the latest achievements in science and technology ②(利用手段使为自己服务) take advantage of; exploit: ～矛盾 utilize contradictions / ～某人 make use of sb. / ～某人的弱点 take advantage of sb.'s weaknesses / ～逆境 make the best of a bad bargain / ～时机 take advantage of opportunities / ～职权 take advantage of one's position and power; exploit one's office / 受人～ be made use of; be a cat's-paw ◇ ～率 utilization ratio / ～系数 utilization coefficient; utilization factor
【利诱】lure by promise of gain
【利欲熏心】be blinded by greed; be obsessed with the desire for gain; be overcome by covetousness

痢 dysentery
【痢疾】dysentery: 阿米巴～ amoebic dysentery / 恶性～ malignant dysentery / 细菌性～ bacillary dysentery
【痢特灵】[药] furazolidone

例 ①(例子) example; instance: 举～ give an example; cite an instance ②(先例) precedent: 破～ break all precedents; make an exception / 援～ quote a precedent; follow a precedent ③(事例) case; instance ④(规则) rule; regulation: 旧～ an old rule ⑤(照成规办的) regular; routine
【例会】regular meeting
【例假】①(规定假日) official holiday; legal holiday ②(月经期) menstrual period; period
【例句】illustrative sentence; example sentence
【例如】for instance; for example (e.g.); such as
【例题】example
【例外】exception: 毫无～ without exception
【例外法规】exception law
【例行公事】①(按常规办的公事) routine; routine business ②(形式主义的工作) mere formality
【例言】introductory remarks; notes on the use of a book
【例证】illustration; example; case in point
【例子】example; case; instance

栎 [植] oak

砾 gravel; shingle
【砾石】gravel ◇ ～混凝土 gravel concrete / ～路 gravel road
【砾岩】[地] conglomerate

隶 ①(附属) be subordinate to; be under ②(被奴役的) a person in servitude: 奴～ slave
【隶书】official script, an ancient style of calligraphy current in the Han Dynasty (206 B.C. —

A.D. 220)
【隶属】be subordinate to; be under the jurisdiction or command of: 常州～江苏省。Changzhou is under the jurisdiction of Jiangsu Province.

力 ①(力量) power; strength; ability: 创造～ creative power / 兵～ military strength; military capabilities / 电～ electric power / 记忆～ strength of memory / 能～ ability; capability / 魄～ drive and decisiveness; boldness / 人～ manpower / 视～ power of vision / 水～ water power / 物～ material resources / 自然～ natural power ②[物](作用) force: 磁～ magnetic force / 分～ component force / 浮～ buoyancy force / 附着～ adhesive force / 合～ composite force / 拉～ tractive force / 离心～ centrifugal force / 内聚～ cohesive force / 扭～ torsional force / 弹～ elastic force / 向心～ centripetal force / 引～ attraction force / 轴向～ axial force / 张～ tension force / 阻～ resisting force ③(体力) physical strength: 大～士 a man of great strength / ～不能支 unable to stand the strain any longer; too weak to stay on one's feet ④(努力;尽力) do all one can; make every effort: 办事不～ not do one's best in one's work
【力臂】[物] arm of force
【力薄能鲜】be deficient in strength and ability
【力不从心】ability falling short of one's wishes; ability not equal to one's ambition
【力不胜任】be unequal to one's task
【力场】[物] field of force
【力度】[乐] dynamics
【力疾奔走】busy oneself in running about
【力疾从公】attend to one's duties in spite of illness
【力竭声嘶】exhausted from effort
【力戒】strictly avoid; do everything possible to avoid; guard against: ～骄傲 guard against arrogance / ～浪费 do everything possible to avoid waste
【力矩】[物] moment of force; moment: 俯仰～ pitching moment / 合～ resultant moment
【力量】①(力气) physical strength ②(能力) power; force; strength: 国防～ defense capability / 新兴～ new emerging forces / 有生～ effective strength / ～对比 balance of force; relative strength / ～悬殊 disparity in strength / 依靠群众的～ rely on the strength of the masses / 团结就是～。Union is strength. / 知识就是～。Knowledge is power.
【力偶】[物] couple
【力排众议】prevail over all dissenting views
【力气】physical strength; effort: ～活儿 heavy work; strenuous work
【力求】make every effort to; do one's best to; strive to
【力所不及】beyond one's power; out of one's ability: 这是我～的。This is beyond my power. (或: This is not within my ability.)
【力所能及】in one's power: 在～的范围内 within one's power
【力图】try hard to; strive to: ～摆脱困境 strive to get out of a predicament / ～否认 try hard to deny
【力挽狂澜】make vigorous efforts to turn the tide
【力学】mechanics: 波动～ wave mechanics / 地质～ geologic mechanics / 动～ dynamics / 断裂～ fracture mechanics / 流体～ fluid mechanics / 静～ statics / 生物～ biomechanics
【力争】①(极力争取) work hard for; do all one can to: ～上游 strive for the best / ～主动 do all one can to gain the initiative ②(极力争辩) argue strongly; contend vigorously: 据理～ argue strongly on just grounds

历 ①(经历) go through; undergo; experience: ～尽艰辛 have gone through all kinds of hardships and difficulties ②(统指过去的各个或各次) all previous ③(遍) covering all; one by one: ～访各有关部门 have visited the departments concerned one by one ④(历法) calendar: 阳～ solar calendar / 阴～ lunar calendar
【历程】course: 难忘的～ unforgettable experience / 人生的～ life's journey / 回顾战斗的～ look back on the course of the struggle
【历次】all previous: ～代表大会 the various congresses
【历代】successive dynasties; past dynasties: ～封建王朝 the feudal dynasties of past ages / ～名画 famous paintings through the ages
【历法】[天] calendar
【历届】all previous: ～毕业生 graduates of all previous years / ～内阁 successive cabinets / ～全国人民代表大会 all the previous National People's Congresses
【历经沧桑】go through all the vicissitudes of life
【历久不衰】long lasting

【历久弥坚】remain unshakable and become even firmer as time goes by
【历来】always; constantly; all long: ～认为 have invariably insisted; have consistently held; have always maintained／～如此。It has always been so.
【历历】distinctly; clearly: ～可数 can see each and every one of them distinctly／～在目 come clearly into view; leap up vividly before the eyes
【历年】①(过去多少年) over the years: ～的积蓄 savings over the years ②[天](历法上的年) calendar year
【历任】①(相继担任) have successively held the posts of; have served successively as: 他～旅长、军长等职。He successively held the posts of brigade and army commander. ②(过去各任) successive: 这所大学的～校长 all the successive presidents of this university
【历时】last; take: 邀请赛～五天。The tournament took five days.
【历史】history; past records: ～清白 have a clean record／隐瞒自己的～ conceal one's past record／用～观点看问题 look at the problem from a historical point of view ◇ ～博物馆 history museum; historical museum／～潮流 the tide of history; historical trend／～地图 historical map or atlas／～观 conception of history／～记载 historical records／～阶段 historical stage／～剧 historical play／～人物 historical personage; historical figure／～唯物主义 historical materialism／～唯心主义 historical idealism／～文物 historical relics／～问题 question of a political nature in sb.'s history／～小说 historical novel／～学家 historian／～学派 historical school／～循环论 historicism／～遗产 legacy of history; historical heritage
【历史性】historic; of historic significance: ～胜利 a historic victory
【历书】almanac
【历数】count one by one; enumerate: ～侵略者的罪行 enumerate the crimes of the aggressors
【历月】[天] calendar month

沥

①(滴落) drip; trickle: 滴～ patter ②(液体的点滴) drop: 余～ last drops
【沥涝】waterlogging
【沥青】pitch; asphalt; bitumen: 天然～ natural asphalt; natural bitumen ◇ ～混凝土 bituminous concrete; asphalt concrete／～基原油 asphalt-base crude oil／～路 bituminous road; asphalt road／～煤 pitch coal／～油毡 asphalt felt／～油纸 asphalt paper／～铀矿 uraninite
【沥水】waterlogging caused by excessive rainfall

呖

【呖呖】[象] 莺声～ warbling of the oriole

荔

【荔枝】litchi

liǎ

俩

①(两个) two: 咱～ we two; both of us; the two of us ②(不多; 几个) some; several: 接济他～钱儿。Help him with some money.

lián

帘

①(布做的望子) flag as shop sign: 酒～ wineshop sign ②(帘子) curtain: 窗～ window curtain
【帘布】cord fabric
【帘栅管】[电] screen-grid tube
【帘栅极】[电] screen grid
【帘子】[口] (hanging) screen; curtain
【帘子布】cord fabric

廉

①(廉洁) honest and clean ②(便宜) low-priced; inexpensive; cheap: 价～物美 good and cheap
【廉耻】sense of honor; sense of shame
【廉价】low-priced; cheap: ～出售 sell at a low price; sell cheap／～买进 buy cheap ◇ ～部 bargain counter／～劳动力 cheap labor／～品 cheap goods; bargain／～书 a cheap book
【廉洁】honest: ～奉公 be honest in performing one's official duties／～清正 keep the hands clean
【廉静寡欲】pure and few desires
【廉明】incorruptible and intelligent; clean-handed and clearheaded

镰

sickle
【镰刀】sickle
【镰鱼】[动] Moorish idol

臁

[生理] shank

怜

sympathize with; pity: 同病相～ fellow sufferers sympathize with each other
【怜爱】love tenderly; have tender affection for
【怜才】have sympathy for talented persons

【怜悯】pity; take pity on; have compassion for
【怜贫】pity or commiserate the poor
【怜惜】take pity on; have pity for

联 ①(联合) ally oneself with; unite; join ②(对联) antithetical couplet: 春～ Spring Festival couplets
【联邦】federation; union; commonwealth: 英～ the British Commonwealth of Nations ◇ ～调查局 the (U.S.) Federal Bureau of Investigation (FBI) / ～共和国 federal republic; federated republic / ～议会 federal parliament / ～制 federal system; federalism
【联苯胺】[化] benzidine
【联播】radio hookup; broadcast over a radio network ◇ ～节目时间 network time
【联产承包责任制】the contracted responsibility system with remuneration linked to output
【联产计酬】method of paying remuneration according to output
【联大】[简](联合国大会) the United Nations General Assembly
【联动机】[机] gear
【联队】[军] wing (of an air force)
【联防】joint defense; joint command of defense forces: 军民～ joint defense by army and militia; armycivilian defense
【联管节】[机] pipe union; pipe coupling; union joint
【联管箱】[机] header: 汽锅～ boiler header
【联合】①(结合) unite; ally: ～一切可能的力量 ally oneself with all forces that can be allied with ②(结合在一起的) alliance; union; coalition ③(共同) joint; combined: ～对外 concerted action towards our foreign counterparts / ～进攻 combined attack; concerted attack / ～举办 jointly organize or sponsor ④[生理] symphysis: 耻骨～ symphysis pubis ◇ ～兵种 combined arms / ～采煤机 cutter-loader; combine / ～词组 coordinative word group / ～公报 joint communiqué / ～公司 allied company / ～经营 coordinated management / ～企业 integrated complex / ～声明 joint statement / ～收割机 combine (harvester) / ～诉状 joint indictment / ～行动 joint action; concerted action / ～宣言 joint declaration / ～演习 [军] joint maneuverer; joint exercise / ～政府 coalition government / ～作战 combined operation
【联合国】the United Nations (U.N.) ◇ ～大会 the United Nations General Assembly / ～秘书处 the United Nations Secretariat / ～宪章 the Charter of the United Nations ○会员国; 成员国 Members; Member States / 总务委员会 General Committee / 特别政治委员会 Special Political Committee / 全权证书委员会 Credentials Committee / 行政和预算问题咨询委员会 Advisory Committee on Administrative and Budgetary Questions / 会费委员会 Committee on Contributions / 安(全)理(事)会 Security Council (SC) / 经(济及)社(会)理事会 Economic and Social Council (ECOSOC) / 托管理事会 Trusteeship Council / 国际法院 International Court of Justice / 非洲经济委员会 Economic Commission for Africa (ECA) / 亚洲及远东经济委员会 Economic Commission for Asia and the Far East (ECAFE) / 欧洲经济委员会 Economic Commission for Europe (ECE) / 西亚经济委员会 Economic Commission for Western Asia (ECWA) / 拉丁美洲经济委员会 Economic Commission for Latin America (ECLA) / 亚洲及太平洋经济社会委员会 Economic and Social Commission for Asia and the Pacific (ESCAP) / 联合国粮食及农业组织 Food and Agriculture Organization (FAO) / 关税及贸易总协定 General Agreement on Tariffs and Trade (GATT) / 国际复兴开发银行 International Bank for Reconstruction and Development / 国际民用航空组织 International Civil Aviation Organization (ICAO) / 国际开发协会 International Development Association (IDA) / 国际金融公司 International Finance Corporation (IFC) / 国际劳工组织 International Labor Organization (IMCO) / 政府间海事协商组织 Inter-Governmental Maritime Consultative Organization (IMCO) / 国际货币基金组织 International Monetary Fund (IMF) / 联合国教育科学及文化组织 United Nations Educational, Scientific and Cultural Organization (UNESCO) / 万国邮政联盟 Universal Postal Union (UPU) / 世界卫生组织 World Health Organization (WHO) / 世界知识产权组织 World Intellectual Property Organization (WIPO) / 世界气象组织 World Meteorological Organization (WMO) / 裁军委员会 Disarmament Commission / 国际原子能机构 International Atomic Energy Agency (IAEA) / 联合国贸易和发展会议 United Nations Conference on Trade and Development (UNCTAD) / 联合国开发计划署 United Nations Development Program (UNDP) / 联合

国救灾协调专员办事处 United Nations Disaster Relief Coordinator Office (UNDRO) / 联合国紧急部队 United Nations Emergency Force (UNEF) / 联合国环境规划署 United Nations Environment Program (UNEP) / 联合国人口活动基金 United Nations Fund for Population Activities (UNFPA) / 联合国难民事务高级专员办事处 United Nations High Commissioner for Refugees (UNHCR) / 联合国儿童基金 United Nations Children's Fund / 联合国工业发展组织 United Nations Industrial Development Organization (UNIDO) / 联合国社会发展研究所 United Nations Research Institute for Social Development (UNRISD) / 联合国特别基金 United Nations Special Fund (UNSF) / 世界粮食理事会 World Food Council (WFC) / 世界粮食计划署 World Food Program (WFP)

【联合会】federation; union: 妇女～ women's federation / 学生～ students' union

【联合王国】the United Kingdom

【联欢】have a get-together: 节日～ gala celebrations ◇ ～会 get-together / ～节 festival / ～晚会 (evening) party

【联接】[宇航] mate

【联结】bind; tie; join: ～两国人民的友谊纽带 the ties of friendship that join the two peoples

【联军】allied forces; united army

【联立方程】[数] simultaneous equations

【联络】①(接触;联系) get in touch with; come into contact with: ～感情 make friendly contacts / 用电话～ get in touch by telephone ②(接上关系) contact; liaison ◇ ～部 liaison department / ～处 liaison office / ～点 contact point / ～官 liaison officer / ～网 liaison net / ～员 liaison man

【联盟】alliance; coalition; league; union: 工农～ alliance of the workers and peasants

【联名】jointly signed; jointly: ～发起 jointly initiate; jointly sponsor / ～上书 submit a joint letter

【联翩】in close succession; together: 浮想～ thoughts thronging one's mind

【联赛】[体] league matches: 排球～ league volleyball tournament / 足球～ league football matches

【联锁反应】chain reaction

【联锁机构】[机] interlocking mechanism

【联席会议】joint conference; joint meeting

【联系】①(接上关系) contact; touch; connection; relation: 保持～ keep in contact with / 取得～ get in touch with; establish contact with / 事物的内部～ the internal relations of things / 有广泛的社会～ have wide social connections ②(使结合) integrate; relate; link; get in touch with: 理论～实际 integrate theory with practice; apply theory to reality / 密切～群众 maintain close links with the masses

【联想】associate; connect in the mind: ～测验 association test

【联姻】connections through marriages; unite by marriage

【联营】joint operation

【联运】[交] through transport; through traffic: 国际铁路～ international railway through transport / 火车汽车～ train- and- bus coordinated transport / 水陆～ land-and-water coordinated transport; through transport by land and water ◇ ～票 through ticket / ～提单 through bill of lading

【联轴器】[机] shaft coupling; coupling: 刚性～ rigid coupling / 挠性～ flexible coupling / 万向～ universal coupling

连 ①(连接) link; join; connect: 把零散的土地～成一片 join together scattered pieces of land ②(连续) in succession; one after another; repeatedly: ～挫强手 defeat strong opponents one after another / ～战皆捷 win a series of victories; win battle after battle ③(包括在内) include: ～皮十公斤。 It weighs 10 kilos, including packing. ④[军](连队) company ⑤(甚至) even: 这个我～想都没想过。 I haven't even thought of it.

【连本带利】both the principal and the interest

【连鬓胡子】full beard

【连茬】[农] continuous cropping

【连词】[语] conjunction

【连带】related ◇ ～保证 joint suretyship / ～原因 contributory cause / ～责任 joint liability / ～责任的债务 obligation establishing joint and several / ～债务人 joint and several debtor

【连…带…】①(表示包括前后两项) and; as well as: ～工～料，一共多少钱? How much is it, with material and labor all together? ②(表示两种动作同时发生) and while: ～说～比划 talking and gesticulating / ～蹦～跳 hopping and skipping

【连裆裤】①(裆里不开口的裤子) child's pants with no slit in the seat ②[方](相互勾结)穿～ band together; collude; gang up
【连队】[军] company
【连发】[军] running fire ◇～枪 repeating rifle; magazine gun／～射击 burst (of fire)／～武器 repeating firearms
【连番喝采】round after round of cheers
【连杆】[机] connecting rod
【连根拔起】tear up by the roots; uproot
【连亘】[书] continuous: 山岭～ a continuous stretch of mountains
【连拱坝】[水] multiple-arch dam; multi-arch dam
【连拱桥】multiple-arch bridge
【连贯】①(连接贯通) link up; piece together; hang together: 各种材料～起来考虑 piece together various kinds of data and ponder over them ②(首尾一致) coherent; consistent ◇～性 coherence; continuity
【连锅端】remove or destroy lock, stock and barrel
【连环】chain of rings ◇～画 a book with a story told in pictures; picture-story book／～计 a set of interlocking stratagems; series of stratagems
【连击】[体] double hit
【连枷】[农] flail
【连接】join; link: 把两条铁路线～起来 link up the two railway lines ◇～号 the mark (–)／～线[乐] tie
【连襟】husbands of sisters
【连累】implicate; involve; get sb. into trouble
【连理】①(树枝交错生在一起) trees whose branches interlock or join together ②(比喻恩爱夫妻) a couple very much in love
【连连】[口] repeatedly; again and again: ～点头 nod again and again
【连忙】promptly; at once
【连袂而起】rise side by side
【连绵】continuous; unbroken; uninterrupted: ～起伏的山峦 rolling hills／阴雨～ a succession of wet and overcast days
【连年】in successive years; in consecutive years; for years running; for years on end: ～丰收 reap rich harvests for many years running／～干旱 successive years of drought
【连皮】gross weight; including the packing: ～二十三公斤。 The gross weight is 23 kilos.
【连篇】①(充满篇幅) throughout a piece of writing; page after page: 空话～ pages and pages of empty verbiage ②(一篇接一篇) one article after another; a multitude of articles
【连篇累牍】lengthy and tedious; at great length: ～地发表文章 publish one article after another
【连谱号】[乐] accolade; brace
【连任】be reappointed or reelected consecutively; renew one's term of office: ～部长 be reappointed minister／～总统 be reelected President ◇～受托人 continuing trustee
【连日】for days on end; day after day: ～来 for the last few days
【连射】[军] running fire
【连锁反应】[物] chain reaction
【连体双生】Siamese twins
【连天】①(形容与天空相接) reaching the sky: 高峰～ skyscraping peaks ②(连续不间断) incessantly: 叫苦～ incessantly complain to high heaven ③(接连多天) sky-rending: 杀声～ air-rending battle cries
【连同】together with; along with: 把信～报纸一起带去。 Take the letters along with the newspapers.
【连续】continuous; successive; in a row; running: ～作战 continuous fighting; successive battles; consecutive operations／钞票上有～的号码。 Currency notes bear serial numbers. ◇～爆破 continuous demolition／～犯 continuing offense／～航次 consecutive voyages／～剧 serial／～谱[物] continuous spectrum／～性 continuity; continuance／～铸锭机 continuous casting machine
【连夜】the same night; that very night: 她～赶进城去。 She rushed to town that very night.
【连衣裙】a woman's dress
【连阴天】cloudy or rainy weather for several days running
【连载】publish in instalments; serialize: 长篇～ serial (of a novel, etc.)
【连战皆北】suffer one defeat after another
【连战皆捷】score one victory after another
【连长】company commander
【连真带假】altogether, genuine or fake
【连珠】like a chain of pearls or a string of beads; in rapid succession: ～似的机枪声 a continuous rattle of machine-gun fire
【连珠炮】continuous firing; drumfire: ～似地向他提问 bombard him with questions; fire questions at him／说话象～ chatter away like a machine gun
【连缀】①(联结) join together; put together ②

[语](辅音或元音群) cluster: 辅音～ consonant cluster
【连字号】hyphen (-)
【连奏】[乐] legato

涟
【涟漪】[书] ripples

莲 [植] lotus
【莲花】lotus flower; lotus ◇ ～纹 lotus design
【莲蓬】seedpod of the lotus
【莲蓬头】shower nozzle ◇ ～式喷头 shower head injector
【莲台】lotus throne; a Buddha's seat in the form of a lotus flower
【莲子】lotus seed

鲢 silver carp

liǎn

敛 ①(收起) hold back; restrain: ～足 hold back from going; check one's steps ②(收集) collect: 横征暴～ extort heavy taxes and levies
【敛步】slow down one's steps; hesitate to advance further
【敛财】accumulate wealth by unfair means
【敛迹】temporarily desist from one's evil ways; lie low
【敛容】[书] assume a serious expression

脸 ①(面部) face; countenance: 不要～ shameless / 丢～ lose face / 拉长了～ put on 〈pull〉 a long face / 没～见人 too ashamed to face anyone / 撕破了～ put aside all considerations of face; not spare sb.'s sensibilities / 笑～ a smiling face ②[方](物体的前部) front: 门～儿 the vicinity of a city gate; the front of a shop
【脸蛋儿】cheeks; face
【脸憨皮厚】shameless
【脸红】①(由于害羞等) blush with shame; blush ②(由于愤怒等) flush with anger; get excited; get worked up
【脸红脖子粗】get red in the face from anger or excitement; flush with agitation: 争得～ argue excitedly
【脸面】face; self-respect; sb.'s feelings
【脸盘儿】the cast of one's face
【脸盆】washbasin; washbowl ◇ ～架 washstand
【脸皮】face; cheek: ～薄 thin-skinned; very sensitive / ～厚 thick-skinned; shameless
【脸谱】types of facial makeup in operas
【脸色】①(气色) complexion; look: ～红润 a ruddy complexion ②(脸上的表情) facial expression
【脸上无光】lose face

liàn

恋 ①(恋爱) love: 初～ be in love for the first time; first love / 热～ be passionately in love ②(想念;不忍分离) long for; feel attached to: ～家 reluctant to be away from home
【恋爱】love: 谈～ be in love; have a love affair / 甲与乙～。A and B are in love with each other. (或: A fell in love with B.) / 他们在～。They are in love. ◇ ～场面 love scene / ～结婚 love match
【恋父情结】[心理] Electra complex
【恋歌】a love song
【恋旧】①(留恋过去) year for the past; to long for the good old days ②(怀念老朋友) year for old friends
【恋恋不舍】be reluctant to part with; hate to see sb. go
【恋母情结】[心理] Oedipus complex
【恋慕】have a tender feeling towards
【恋人】a sweetheart; a lover

楝
【楝树】[植] chinaberry

炼 ①(熔炼;精炼) smelt; refine: ～铅 smelt lead / ～糖 refine sugar ②(锤炼) temper with fire ③(琢磨) polish
【炼丹】make pills of immortality ◇ ～术[化] alchemy
【炼钢】steelmaking; steel-smelting ◇ ～厂 steel mill; steelworks / ～工人 steelworker / ～炉 steelmaking furnace; steel-smelting furnace
【炼焦】coking ◇ ～厂 coking plant; cokery / ～炉 coke oven / ～炉煤气 coke-oven gas / ～煤 coking coal
【炼金术】alchemy
【炼乳】condensed milk
【炼铁】iron-smelting ◇ ～厂 ironworks / ～炉 iron-smelting furnace; blast furnace
【炼油】①(分馏石油) oil refining ②(提炼动、植物油) extract oil by heat ③(加热食用油) heat edible oil ◇ ～厂[石油] (oil) refinery
【炼制】[化] refine: 石油～ petroleum refining

练 ①(白绢) white silk ②(把生丝煮熟) boil and scour raw silk: ～漂 [纺] scouring and

bleaching ③(练习) practice; train; drill: ～单杠 train 〈practice〉 on the horizontal bar / ～好本领 perfect one's skill / ～好身体 do exercises to build up one's physique / ～节目 rehearse / ～跑 practice running / ～嗓子 cultivate the voice ④(经验多) experienced; skilled; seasoned: 老～ experienced and assured
【练兵】troop training; training ◇～场 drill ground; parade ground / ～项目 training courses
【练操】drill
【练队】drill in formation; drill for a parade
【练功】do exercises in gymnastics, *wushu*, acrobatics, etc.; practice one's skill
【练气功】do breathing exercises
【练球】practice a ball game: 赛前～ warm-up (before a match); knockup
【练鹊】[动] long-tailed flycatcher
【练武】do weapon practice; practice martial arts
【练习】①(反复学习) practice: ～讲英语 practice spoken English / ～射击 practice marksmanship / ～写文章 practice writing ②(习题或作业等) exercise: ～算术 arithmetic exercises / 做～ do exercises ◇～簿 exercise-book / ～曲 [乐] étude / ～题 exercises; problems of an exercise
【练字】practice calligraphy

潋

【潋滟】①(水势浩大) overflowing ②(水波流动) continuing joining

殓

put a body into a coffin; encoffin

链

①(链子) chain: 表～ watch chain / 铁～ iron chain ②(计量海洋上距离的长度单位) cable's length (=185.2 m.)
【链钩】[机] chain hook; sling
【链轨】caterpillar track (of a tractor)
【链接】[化] catenation
【链轮】[机] sprocket; chain wheel
【链霉素】[药] streptomycin
【链球】[体] hammer: 掷～ hammer throw
【链球菌】[微] streptococcus
【链式反应】[化] chain reaction
【链式核反应】nuclear chain reaction
【链式磨木机】[纸] caterpillar grinder; chain grinder
【链套】[自行车] chain case
【链条】①(机械上传动用的链子) chain ②[自行车](链子) roller chain; chain
【链烃】[化] chain hydrocarbon
【链罩】[自行车] chain guard; chain cover
【链子】chain

liáng

梁

①(房梁) roof beam: 横～ cross beam / 架～ set a roof beam in place ②(桥) bridge: 桥～ bridge ③(物体中间隆起的长条部分) ridge: 山～ mountain ridge
【梁桥】beam bridge
【梁上君子】gentleman on the beam; burglar; thief
【梁柱体系】post and lintel system

凉

①(冷) cool; cold: ～风 cool breeze ②(灰心) discouraged; disappointed
【凉拌】(of food) cold and dressed with sauce ◇～面 cold noodles in sauce / ～生菜 tossed salad
【凉菜】cold dish
【凉粉】bean jelly
【凉风透骨】chilled to the bone
【凉糕】cake made of glutinous rice served cold in summer
【凉快】①(天气清凉爽快) nice and cool; pleasantly cool ②(纳凉) cool oneself; cool off
【凉了半截】heart chills with disappointment
【凉棚】mat-awning; mat shed
【凉爽】nice and cool; pleasantly cool: ～的秋天 pleasantly cool autumn days
【凉水】①(温度低的水) cold water ②(生水) unboiled water
【凉丝丝】coolish; rather cool; a bit cool
【凉飕飕】chilly; chill
【凉台】balcony; veranda
【凉亭】wayside pavilion; summer house; kiosk
【凉席】bed-mat for summer; summer sleeping mat
【凉鞋】sandals

椋

【椋鸟】starling

良

①(好) good; fine: ～将 a good general; an able general / ～马 a fine horse ②(善良的人) good people: 除暴安～ get rid of bullies and bring peace to good people
【良材】①(好木头) good timber ②(有能力的人) able person
【良策】good plan; sound strategy
【良辰美景】beautiful scene on a bright day

【良导体】[物] good conductor
【良方】①(好药方) effective prescription; good recipe ②(好计划) good plan; sound strategy
【良港】good harbor
【良好】good; well: ～的比赛风格 fine sportsmanship / ～的开端 good beginning; a good start / ～的愿望 good intentions / ～的祝愿 good wishes / 打下～的基础 lay a sound foundation / 为双方会谈创造～的气氛 create a favorable atmosphere for bilateral talks / 自我感觉～ feel fine
【良机】[书] good opportunity: ～勿失。 Don't let the good chance slip.
【良久】[书] a good while; a long time
【良能】[哲] intuitive ability
【良禽择木】good birds select their roosts
【良师益友】good teacher and helpful friend
【良书乃益友】a good book is a great friend
【良田】good field; fertile farmland
【良心】conscience: ～不安 have an uneasy conscience / ～发现 stung by conscience / ～丧尽 utterly conscienceless / ～有愧 guilty conscience / 没～ ungrateful; heartless / 说句～话 to be fair; in all fairness / 违背～ against one's conscience / 有～的人 people with a conscience; good-hearted people
【良药苦口】good medicine tastes bitter: ～利于病，忠言逆耳利于行 just as bitter medicine cures sickness, so unpalatable advice benefits conduct
【良医】good doctor
【良友】beneficial friend
【良莠不齐】the good and the bad are intermingled
【良缘】happy match; harmonious union
【良知】[哲] intuitive knowledge ◇ ～良能 innate knowledge and ability instinct
【良种】① [农](优良作物品种) improved variety: 水稻～ improved varieties of rice ②(优良牲畜品种) fine breed: ～马 a horse of fine breed ◇ ～场 seed multiplication farm

粮 ①(粮食) grain; food; provisions: 粗～ coarse grain / 商品～ marketable grain / 细～ refined grain / ～棉双丰收 a bumper harvest of both grain and cotton / 弹尽～绝 run out of ammunition and food ②(作为农业税的粮食) grain tax paid in kind: 交公～ pay grain tax to the state
【粮仓】granary; barn
【粮草】army provisions; rations and forage
【粮店】grain shop
【粮管所】staple food control office
【粮库】grain depot
【粮秣】army provisions; rations and forage; grain and fodder: ～被服 grain, fodder, bedding and clothing ◇ ～库 ration depot
【粮票】food coupon; grain coupon
【粮食】grain; cereals; food: 主要～ staple food ◇ ～产量 grain yield / ～储备 grain reserves; grain stock / ～定量 monthly quota of food grain for an individual / ～供应 staple food supply / ～加工 grain processing / ～加工厂 grain processing plant / ～局 grain bureau / ～作物 cereal crops; grain crops ○稻 rice / 谷子 millet / 玉米 maize / 高粱 Chinese sorghum / 大麦 barley / 小麦 wheat
【粮栈】wholesale grain store; grain depot
【粮站】grain distribution station; grain supply center

莨
【莨绸】gambiered Guangdong silk

量 measure: ～尺寸 take sb.'s measurements / ～地 measure land; measure a piece of ground / ～身材 take sb.'s measurements / ～体温 take sb.'s temperature / ～体重 weigh oneself
【量杯】measuring glass; graduate
【量度】measurement
【量角器】protractor
【量具】measuring tool ◇ ～刃具厂 measuring and cutting tools plant
【量热器】calorimeter
【量日仪】[天] heliometer
【量筒】graduated cylinder; graduate
【量图仪】map measurer
【量雪尺】[气] snow scale
【量雪器】[气] snow gauge
【量油尺】[机] oil dip rod; dipstick

liǎng

两 ①(二) two: ～本书 two books / ～层楼房 a two-story house / ～个半星期 two and a half weeks / ～匹马 two horses / ～万元 twenty thousand *yuan* / 这里有～个。 Two are here. ②(双方) both (sides); either (side): ～者缺一不可 neither is dispensable / 势不～立 irreconcilably hostile to each other; mutually exclusive ③(不定数目) a few; some: 我过～天再来。 I'll come again in a few days. ④(重量

单位) *liang*, a unit of weight (= 50 grams)
【两败俱伤】both sides suffer neither side gains
【两半儿】two halves; in half; in two: 把西瓜切成～ cut a watermelon in half
【两倍】twofold; double; twice as much: 产量是十年前的～。The output is twice as much as that of ten years ago.
【两边】①(两个边) both sides ②(两处) both places ③(两个方向) both direction ④(双方) both parties; both sides: ～讨好 try to please both sides
【两边倒】lean now to one side, now to the other; waver
【两便】be convenient to both; make things easy for both
【两鬓斑白】greying at the temples
【两重】double; dual; twofold: ～任务 a twofold task / ～意义 double meaning ◇～性 [哲] dual nature; duality
【两侧对称】bilateral symmetry
【两次运球】[篮球] double dribble
【两党制】two-party system; bipartisan system
【两抵】balance or cancel each other: 收支～ the account balances; income and expenditure balance each other
【两点论】[哲] the doctrine that everything has two aspects
【两耳不闻窗外事】not care what is going on outside one's window; be oblivious of the outside world: ～,一心只读圣贤书 busy oneself in the classics and ignore what is going on beyond one's immediate surroundings
【两分法】law of one dividing into two
【两害相权取其轻】accept the lesser of two evils
【两虎相争】fight between the two biggest
【两回事】two entirely different things; two different matters
【两极】①(地球的两极) the two poles of the earth: ～地区 polar regions ②[物](阴、阳极) the two poles
【两极分化】①(两个极端;两个对立面) polarization; division into two opposing extremes ②(分成两极) polarize; produce a polarization of
【两件套】two-piece dress
【两脚规】①(圆规) compasses ②(分线规) dividers
【两可】both will do; either will do: ～之间 not knowing which to choose; maybe, maybe not
【两可阶梯】reversible staircase
【两口儿】husband and wife; married couple: 老～ the old couple / 小～ the young couple
【两口子】[口] husband and wife; couple
【两利】be good for both sides; benefit both
【两面】①(正反面) two sides; both sides; two aspects: both aspects ②(对立的两面) having a dual character; dual; double: ～手法 double- faced tactics; double- dealing; double game / ～性 dual character
【两面光】please both parties
【两面夹攻】make a pincer attack: 受到～ be caught in cross fire; be caught in a pincer attack
【两面派】double-dealer: ～的行为 act of duplicity; double-faced behavior; double-dealing
【两面三刀】double-dealing
【两面受敌】between two fires
【两面外交】two-sided diplomacy
【两难】face a difficult choice; be in a dilemma: 进退～ can neither advance nor retreat; be in a dilemma
【两旁】both sides; either side
【两栖】[军] amphibious ◇～部队 amphibious forces; amphibious units / ～动物 amphibious animal; amphibian / ～植物 amphibious plant; amphibian / ～作战 amphibious warfare; amphibious operations / ～作战舰艇 amphibious vessel
【两讫】[商] the goods are delivered and the bill is paid
【两全】be satisfactory to both parties; have regard for both sides: ～的办法 measures satisfactory to both sides
【两全其美】satisfy both sides; satisfy rival claims
【两审终审制】[法] the system of the court of second instance being the court of last instance
【两世为人】barely escape with one's life; be lucky to have escaped death
【两手】dual tactics: 作～准备 prepare oneself for both eventualities
【两条心】in fundamental disagreement; not of one mind
【两头】①(两端) both ends; either end: ～儿跑 go back and forth between two places / ～尖 pointed at both ends ②(双方) both parties; both sides: ～敲诈 extort money from both parties / ～落空 fall between two stools / ～说情 intercede between two parties / ～为难 find it hard to please either party; find it difficult to satisfy two conflicting demands
【两头小,中间大】small at both ends and big in

the middle; a few at each extreme and many in between
【两下子】a few tricks of the trade: 你真有～! You really are smart!
【两相情愿】both parties are willing
【两厢】①(两面的厢房) wing-rooms on either side of a one-story house ②(两旁) both sides: 站立～ stand on either side
【两相】[电] two-phase ◇ ～电动机 two-phase motor
【两项运动】biathlon
【两小无猜】living and playing together in childhood innocence
【两心相许】tacit permission to each other; marry no man but him and vice versa
【两性】①(两种性别) both sexes ②[化](两种性质) amphiprotic; amphoteric ◇ ～关系 sexual relations／～花 hermaphrodite flower／～胶体[化] amphoteric colloid; ampholytoid／～人 bisexual person; hermaphrodite
【两袖清风】remain uncorrupted; have clean hands
【两翼】[军] both wings; both flanks ◇ ～包抄 double envelopment
【两用】dual purpose ◇ ～炉 dual-purpose stove／～衫 reversible jacket／～雨衣 reversible raincoat
【两院制】two-chamber system; bicameral system; bicameralism
【两愿离婚】divorce by mutual consent
【两造】①(诉讼的双方) both parties in a lawsuit; both plaintiff and defendant ②[方](二季庄稼) two crops ◇ ～诉讼 bilateral action

liàng

凉 make cool; become cool

谅 ①(原谅) forgive; understand: 本着互～互让的精神 in the spirit of mutual understanding and mutual accommodation／尚希见～。I hope you will excuse me. ②(料想) I think; I suppose; I expect
【谅必】most likely; probably
【谅解】understand; make allowance for: 达成～ reach an understanding／得到群众的～ gain the forgiveness of the masses／互相～ mutual understanding

晾 ①(在风里晾) dry in the air; air ②(晒) dry in the sun; sun: ～衣服 sun clothes; hang out the washing to dry
【晾干】dry by airing; dry in the air
【晾烟】①(在风里晾烟叶) air-curing of tobacco leaves ②(风干的烟草) air-cured tobacco
【晾衣绳】clothesline

亮 ①(明亮) bright; light: 天开始～了。It's beginning to get light. ②(使响亮) loud and clear: ～起嗓子 lift one's voice ③(开朗) enlightened: 心里～了 be enlightened ④ (发光) shine ⑤(显示) show: ～出入证 show one's pass／～观点 declare one's position; air one's view／～思想 lay bare one's innermost thoughts
【亮底】put one's cards on the table; disclose one's plan, stand, views, etc.
【亮度】[物] brightness; brilliance: 星的～ the brightness of a star／萤光屏～ screen brilliance
【亮光】light: 一道～ a shaft of light ◇ ～漆 polish lacquer
【亮节】integrity; uprightness
【亮晶晶】glittering; sparkling; glistening: ～的星星 glittering stars／～的眼睛 sparkling eyes／～的钻石 glittering diamond
【亮堂】①(敞亮;明朗) light; bright: 这屋子很～。The room is well lighted. ②(开朗;清楚) clear; enlightened: 她一讲,大家心里都～了。What she said enlightened all of us.
【亮相】①(演员的一种表演方式) strike a pose on the stage ②(表明态度;观点) declare one's position; state one's views

踉
【踉跄】stagger: ～而行 stagger along

辆 [量] 三～公共汽车 three bus／一～面包车 a mini-bus

量 ①(容纳限度) capacity: 酒～ capacity for liquor／载重～ loading capacity ②(数量; 数目) quantity; amount; volume: 保质保～ guarantee both quantity and quality／工业产～ the volume of industrial output ③(估计; 衡量) estimate; measure: ～力 estimate one's own strength or ability
【量变】[哲] quantitative change
【量才录用】give sb. work suited to his abilities; assign jobs to people according to their abilities
【量词】[语] classifier measure word
【量纲】[物] dimension ◇ ～分析 dimensional analysis
【量化】[逻] quantification

【量力】estimate one's own strength or ability (and act accordingly): 不自～ overrate one's ability; overreach oneself
【量力而为】estimate one's strength before acting
【量力而行】do what one is capable of; act according to one's capability
【量入为出】keep expenditures within the limits of income; live within one's mean
【量体裁衣】cut the garment according to the figure; act according to actual circumstances
【量刑】[法] measurement of penalty ◇ ～标准 criterion for imposing penalty; criterion of imposing penalty / ～幅度 extent for measurement of punishment
【量子】[物] quantum: 光～ light quantum ◇ ～场论 quantum field theory / ～电动力学 quantum electrodynamics / ～化 quantization / ～化学 quantum chemistry / ～力学 quantum mechanics / ～论 quantum theory / ～生物学 quantum biology / ～数 quantum number

liāo

撩 ①(掀) hold up: ～开窗帘 draw aside the curtains / ～起裙子 hold up the skirt ②(用手洒) sprinkle: ～水 sprinkle water

liáo

聊 ①(姑且) merely; just: ～表谢意 just a token of gratitude; just to show my appreciation ②(略微) a little; slightly ③[口](闲谈) chat: ～一～ have a chat
【聊备一格】may serve as a specimen
【聊可解忧】crumb of comfort
【聊赖】something to live for; something to rely upon
【聊且】tentatively; for the moment
【聊胜于无】It's better than nothing. (或: Half a loaf is better than no bread.)
【聊天儿】[口] chat
【聊以解嘲】manage somehow to relieve embarrassment
【聊以塞责】merely to avoid the charge of dereliction of duty
【聊以助兴】just for entertainment
【聊以自慰】just to console oneself
【聊以卒岁】just to tide over the year

寮 small house; hut: 茶～酒肆 teahouses and wineshops / 僧～ a monk's cell; a monk's hut
【寮棚】shed; hut

燎 burn
【燎泡】blister raised by a burn or scald
【燎原】set the prairie ablaze: ～大火 wildland fire / ～烈火 a blazing prairie fire / 星星之火，可以～。A single spark can start a prairie fire.

撩 ①(逗引) tease; tantalize ②(拨动) provoke; stir up
【撩拨】①(挑逗) tease; banter ②(招惹) incite; provoke
【撩动肝火】stir up anger
【撩逗】provoke; entice

嘹
【嘹亮】resonant; loud and clear: ～的号角 a clarion call

僚 ①(官吏) official: 官～ official; bureaucrat ②(同一官署的官吏) an associate in office: 同～ colleague
【僚机】①(跟随长机的飞机) wing plane; wingman ②(僚机飞行员) wingman
【僚舰】[军] consort
【僚属】officials under someone in authority; subordinates; staff: ～星散 official associates were scattered like stars

鹩
【鹩哥】[动] hill myna

獠
【獠牙】long, sharp, protruding teeth: 青面～ be green-faced and long-toothed; have fiendish features

缭 ①(缠绕) entangled ②(缝纫方法) sew with slanting stitches: ～贴边 stitch a hem; hem
【缭乱】confused; in a turmoil: 心绪～ in a confused state of mind / 眼花～ be dazzled
【缭绕】curl up; wind around: 炊烟～ smoke curling up from kitchen chimneys / 歌声～。The song lingered in the air.

寥 ①(稀少) few; scanty: ～～可数 just a sprinkling ②(静寂) silent; deserted: 寂～ deserted and lonely
【寥廓】boundless; vast: ～的天空 the boundless sky
【寥寥数行】just a few lines
【寥寥无伴】lonely
【寥寥无几】very few

【寥落】few and far between; sparse; scattered: 疏星～ only a few solitary stars twinkling in the sky

【寥若晨星】as sparse as the morning stars; few and far between

疗 treat; cure: 电～ electrotherapy; electrical treatment / 泥～ mud-bath treatment / 水～ hydrotherapy; hydropathic treatment / 诊～ make a diagnosis and give treatment / 治～ treat (a patient); give medical care to

【疗程】course of treatment; period of treatment

【疗法】therapy; treatment: 按摩～ massotherapy / 保守～ conservative treatment / 催眠～ hypnotherapy / 放射～ radiotherapy / 非手术方法～ non-operative methods of treatment; non-operative treatment / 封闭～ block therapy / 化学～ chemotherapy / 药物～ medicinal treatment / 放射～ radiotherapy / 物理～ physiotherapy; physical therapy / 饮食～ dietotherapy / 针刺～ acupuncture treatment / 综合～ composite treatment

【疗效】curative effect: 具有良好的～ produce 〈have〉 a good curative effect

【疗养】recuperate; convalesce: 到海滨去～ go to the seaside to recuperate ◇ ～院 sanatorium; convalescent hospital

辽 ①(远) distant; faraway ②(辽代) the Liao Dynasty (916–1125)

【辽阔】vast; extensive: ～的海洋 a vast expanse of ocean / ～的平原 vast plains / ～的土地 a vast expanse of land; vast territory

【辽落】open and spacious

【辽远】distant; faraway: ～的边疆 distant frontier regions

liǎo

蓼 [植] knotweed

【蓼蓝】[植] indigo plant

潦

【潦草】①(不工整) hasty and careless; illegible: 字迹～ careless handwriting ②(不仔细) sloppy; slovenly: 干活儿～ work in a slipshod way

【潦倒】be frustrated: 穷愁～ be penniless and frustrated; be down and out

了 ①(明白) know clearly; understand: 明～ understand ②(完毕; 结束) end; finish; settle; dispose of: 没完没～ endless / 未～之事 unfinished task; an unsettled matter ③(放在动词之后,表示可能) can: 办得～ can manage it / 来不～ could not come / 受不～ cannot stand sth.

【了案】conclude a case; close a case

【了不得】①(非常) terrific; extraordinary: 多得～ in terrific number / 高兴得～ extremely happy / 一件～的大事 a matter of the utmost importance ②(情况严重) terrible; awful: 真是～的大祸啊! What a terrible disaster!

【了不起】amazing; terrific; extraordinary: ～的成就 an amazing achievement / ～的进步 extraordinary progress / 没什么～ nothing extraordinary / 自以为～ be swell with pride

【了此残生】end this miserable life

【了此心愿】able to fulfill this wish

【了此一生】to end this life

【了得】(用于句尾,常跟在"还"字后面,表示情况严重) terrible; horrible: 那还～! How horrible!

【了结】finish; settle; wind up; bring to an end: ～一场争端 settle a dispute; end a conflict

【了解】①(知道得清楚) understand; comprehend: ～事物发展的规律 understand the laws of development of things / 增进两国人民之间的～ promote understanding between the two peoples ②(打听; 调查) find out; acquaint oneself with: ～技术发展状况 keep abreast of current developments in technology / ～群众的思想动态 find out what the masses are thinking

【了局】①(结局) end ②(解决办法) solution; settlement

【了却】settle; solve: ～心愿 have his wish fulfilled / ～一桩心事 take a load off one's mind

【了然】understand; be clear: 一目～ be clear at a glance

【了如指掌】know sth. like the palm of one's hand; have sth. at one's fingertips

【了事】dispose of a matter; get sth. over: 草草～ get through sth. in a careless or perfunctory way; rush through sth.

【了无长进】having made no progress in the least

钌 [化] ruthenium (Ru)

liào

料 ①(预料) expect; anticipate: 不出所～ as was expected / ～定敌军会有行动 anticipate

movements on the part of the enemy ②(原料) material; stuff: 燃～ fuel / 原～ raw material ③(精饲料) grain meal ④(适合做某事的人) makings; stuff: 他不是跳舞的～。He hasn't got the makings of a dancer.
【料车】[冶] skip; skip car
【料斗】[冶] (charging) hopper
【料酒】cooking wine
【料理】arrange; manage; attend to; take care of: ～后事 make arrangements for a funeral / ～家务 manage household affairs
【料器】[工美] glassware
【料峭】[书] chilly: 春寒～。There is a chill in the air in early spring.
【料事如神】predict like a prophet; foretell with miraculous accuracy
【料想】expect; think; presume
【料子】①(布料) material for making clothes ② [方](毛料) woolen fabric: ～裤 trousers made of woolen fabric

撂 ①(放;搁) put down; leave behind: ～下饭碗 put down one's rice bowl ②(弄倒) throw down; knock down; shoot down
【撂挑子】throw up one's job

镣 fetters
【镣铐】fetters and handcuffs; shackles; irons; chains: 戴上～ be shackled; be in chains

了 watch from a height or a distance
【了望】watch from a height or a distance; keep a lookout ◇ ～台 observation tower; lookout tower

钌
【钌铞儿】hasp and staple

liē

咧
【咧咧】[方]①(乱说) babble ②(小儿哭) baby's crying sound

liě

咧
【咧嘴】grin: 疼得直～ grin with pain

liè

列 ①(排列) arrange; line up: ～表 arrange in tables or columns; tabulate / ～出理由 set out one's reasons / ～队欢迎 line up to welcome sb. / ～为甲等 be classified as first-rate; be rated as class A ②(列入) list; enter in a list: ～入议程 be placed on the agenda ③(行列) row; file; rank ④[量] 一～火车 a train ⑤(类) kind; sort: 不在讨论之～ not among the subjects to be discussed ⑥(各) various; each and every: ～国 various countries
【列兵】[军] private
【列车】train: 国际～ international train / 上行～ up train / 下行～ down train / 下一班到上海的～ the next train for Shanghai / 直达～ through train ◇ ～调度员 train dispatcher / ～时刻表 train schedule; timetable / ～员 attendant / ～长 head of a train crew; conductor (美); guard (英)
【列当】[植] broomrape
【列岛】a chain of islands; archipelago: 澎湖～ the Penghu Islands
【列国】the various states or nations
【列举】enumerate; list: ～大量事实 cite numerous facts
【列宁主义】Leninism
【列强】big powers
【列氏温度计】[物] the Réaumur thermometer
【列位】all the ladies and gentlemen present
【列席】attend as a nonvoting delegate ◇ ～代表 delegate without the right to vote; nonvoting delegate / ～旁听 attend the meeting as an observer
【列支敦士登】Liechtenstein ◇ ～人 Liechtensteiner
【列传】collected biographies

烈 ①(强烈) strong; violent; intense: ～酒 strong drink / ～日 scorching sun / ～焰 a roaring blaze; raging flames ②(暴烈) fiery: ～马 fiery steed ③(为正义而死的) sacrificing oneself for a just cause: 先～ martyr / 壮～牺牲 die heroically; die a heroic death
【烈度】intensity: 地震～ earthquake intensity
【烈风】[气] strong gale
【烈火】raging fire; raging flames
【烈火见真金】pure gold proves its worth in a blazing fire; people of worth show their mettle during trials and tribulations
【烈日】burning sun; scorching sun: ～当空 with the scorching sun directly overhead
【烈士】①(为正义事业而牺牲的人) martyr: 革命～ revolutionary martyrs ②(古代指有志于建立功业的人) a person of high endeavor ◇ ～纪念碑 a monument to revolutionary martyrs / ～陵园 cemetery of revolutionary mar-

tyrs／～墓 the grave of a revolutionary martyr
【烈暑】summer at its hottest
【烈属】members of a revolutionary martyr's family
【烈性】①(性情刚烈) spirited：～汉子 a man of character ②(性质猛烈) strong; violent：～毒药 deadly poison／～酒 spirit／～炸药 high explosives

裂

split; crack; rend：～成两半 be rent in two／分～ split; break up
【裂变】[原] fission：核～ nuclear fission／自发～ spontaneous fission ◇～产物 fission product／～武器 the fission type of weapon／～物质 fissile material
【裂缝】rift; crevice; crack; fissure：墙上的～ crevices in a wall
【裂果】[植] dehiscent fruit
【裂痕】rift; crack; fissure
【裂化】[石油] cracking：催化～ catalytic cracking ◇～炉 cracking still／～气 cracked gas
【裂解】[化] splitting decomposition; splitting ◇～作用 splitting action
【裂开】split open; rend
【裂口】①(裂开的口) breach; gap; split ②[地](火山口) vent ◇～火山锥 breached cone
【裂纹】crackle (on pottery, porcelain, etc.) ◇～探测仪[冶] crack detector
【裂隙】crack; crevice; fracture ◇～水 crevice water
【裂殖菌】[微] schizomycete

趔

【趔趄】stagger; reel; 那醉汉～着走在街上。The drunkard staggered along the street.

劣

bad; inferior; of low quality：难分优～ very hard to tell which is better
【劣等】of inferior quality; low-grade; poor ◇～货 goods of inferior quality／～生 a dull student
【劣根性】deep-rooted bad habits
【劣弧】[数] minor arc
【劣迹】misdeed; evil doing ◇～昭著 notorious
【劣马】①(不好的马) inferior horse; nag ②(性情暴躁的马) vicious horse; fiery steed
【劣绅】evil gentry：土豪～ local tyrants and evil gentry
【劣势】inferior strength or position
【劣质】of poor quality; of low quality; inferior：～煤 inferior coal; faulty coal
【劣种】inferior strain

猎

hunt：～虎 tiger hunting／从事渔～ engage in fishing and hunting／去打～ go hunting
【猎豹】[动] cheetah
【猎场】hunting ground; hunting field
【猎刀】hunting knife
【猎狗】hunting dog; hound
【猎户】hunter; huntsman
【猎户座】[天] Orion
【猎获】capture or kill in hunting; bag：～两三只野兔 bag a couple of hares／～一头幼狮 trap a young lion ◇～物 bag
【猎奇】hunt for novelty; seek novelty
【猎潜舰】submarine chaser
【猎枪】shotgun; fowling piece; hunting rifle
【猎取】①(打猎获得) hunt ②(夺取) pursue; seek; hunt for：～个人名利 pursue personal fame and gain／～廉价的声誉 make a bid for cheap popularity
【猎犬】hunting dog
【猎犬座】[天] Canes Venatici; Hunting Dogs
【猎人】hunter; huntsman
【猎涉不精】read widely without intensive studies
【猎手】hunter
【猎物】prey; quarry; game
【猎鹰】falcon
【猎装】a hunting dress

鬣

mane
【鬣狗】[动] hyena; striped hyena
【鬣羚】[动] serow
【鬣蜥】[动] agama

lín

麟

【麟凤】rare treasures
【麟角】rare things

遴

【遴选】[书] select sb. for a post; select; choose

磷

[化] phosphorus (P)
【磷肥】[农] phosphate fertilizer
【磷光】[物] phosphorescence ◇～体 phosphor
【磷灰石】[矿] apatite
【磷火】will-o'-the wisp; phosphorescent light
【磷矿粉】[农] ground phosphate rock
【磷酸】[化] phosphoric acid ◇～铵 ammonium phosphate／～钙 calcium phosphate／

～盐 phosphate
【磷虾】[无脊椎] krill
【磷脂】[化] phosphatide ◇ ～酸 phosphatidic acid

辚
【辚辚】[象] rattle: 车～，马萧萧 chariots rattling and horse neighing

嶙
【嶙峋】[书] ①(山石突兀、重叠) jagged; rugged; craggy: 怪石～ jagged rocks of grotesque shapes ②(人消瘦露骨) bony; thin

鳞 ①(鱼鳞) scale: 刮鱼～ scrape the scale off the fish; scale the fish ②(象鱼鳞的) like the scales of a fish: 遍体～伤 be covered with bruises <injuries>; be a mass of bruises
【鳞次栉比】row upon row of (houses)
【鳞甲】scale and shell
【鳞茎】[植] bulb
【鳞片】①(鱼鳞) scale ②[植](覆盖在芽上的象鳞片的东西) bud scale
【鳞屑癣】ringworm
【鳞爪】①[书](鳞和爪) scales and nails ②(事情的片断) small bits; fragments; odd scraps

林 ①(大片的树木或竹子) forest; woods; grove: 保护～ forest reserve / 防风～ windbreaks; windbreak forest / 防沙～ sand-breaks / 风景～ ornamental plantation / 松～ pine forest / 天然～ natural forest / 用材～ timber forest; timberland / 竹～ bamboo groves / 伐～ cut down a forest / 护～ protect forests; protect a forest / 造～ afforestation ②(聚集在一起的同类人或事物) circles: 艺～ art circles ③(林业) forestry
【林产品】forest products
【林场】forestry center; tree farm; forest farm
【林带】forest belt: 防护～ shelter belt / 辅助～ auxiliary forest belt / 主～ main forest belt
【林地】forest land; woodland; timberland
【林管区】district of a forest warden
【林冠】[林] crown canopy; crown cover
【林海】immense forest
【林垦】forestry and land reclamation
【林立】stand in great numbers: 港口内～的桅墙 a forest of masts in a harbor / 烟囱～ a forest of chimneys
【林龄】[林] age of stand
【林木】①(树林) forest; woods: ～葱郁 densely wooded / 经济～ trees of economic value / 油料～ oil-yielding shrubs ②[林](生长在森林中的树木) forest tree ◇ ～线 timberline
【林牧业山区】hilly afforested and livestock breeding area
【林檎】[植] Chinese pear-leaved crabapple
【林区】forest zone; forest region; forest: 禁止滥伐～ ban the destructive lumbering of a forest / 营造～ afforested area
【林鸮】[动] wood owl
【林业】forestry ◇ ～工人 forest worker; forester / ～学院 forestry institute
【林荫道】boulevard; avenue
【林园】wooden land; a park of trees and vegetation
【林子】[口] woods; grove; forest

淋 pour; drench: 浑身都～湿了 be drenched from head to foot / 日晒雨～ sun-scorched and rain-drenched; exposed to the elements
【淋巴】[生理] lymph ◇ ～管 lymphatic vessel / ～球 lymph corpuscle / ～肉瘤 lymphosarcoma / ～细胞 lymphocyte / ～腺 lymphatic gland
【淋巴结】[生理] lymph node ◇ ～结核 scrofula; tuberculous lymphadenitis / ～炎 lymphnoditis / ～增大 lymphadenovarix / ～肿大 lymph node enlargement
【淋漓】①(往下滴) dripping wet: 大汗～ dripping with sweat / 鲜血～ dripping with blood ②(形容畅快) free from inhibition: 痛快～ impassioned and forceful
【淋漓尽致】incisively and vividly; thoroughly: 揭露得～ make a most telling exposure / 刻画得～ portray most vividly
【淋淋】dripping: 湿～的衣服 dripping clothes
【淋湿】be soaked; splashed wet
【淋洗】[化纤] drip washing
【淋雨】get wet in the rain; be exposed to the rain
【淋浴】shower bath; shower: 洗个～ take a shower bath

霖 continuous heavy rain: 甘～ good soaking rain; timely rain
【霖雨】continuous heavy rain
【霖雨为灾】continuous heavy rains bring disaster

琳 [书] beautiful jade
【琳琅】beautiful jade; gem
【琳琅满目】a superb collection of beautiful things; a feast for the eyes

临 ①(靠近) face; overlook: ～街的阳台 a veranda overlooking the street / 东～大海 border on the sea in the east / 如～大敌 as if confronted with a formidable enemy ②(来到) arrive; be present: 亲～指导 come personally to give guidance ③(将要) on the point of; just before; be about to: ～睡 just before going to bed; at bedtime / ～行 on the point of leaving; on the eve of departure ④(摹仿) copy: ～画 copy a painting / ～帖 practice calligraphy after a model

【临本】copy (of a painting)

【临别】at parting; just before parting: ～赠言 words of advice at parting; parting advice / 作为～纪念 as a parting souvenir

【临产】about to give birth; parturient ◇～阵痛 labor pains; birth pangs

【临床】[医] clinical: 有丰富的～经验 have rich clinical experience ◇～表现 clinical manifestation / ～观察 clinical observation / ～检查 clinical examination / ～心理学 clinical psychology / ～学 clinical medicine / ～医生 clinician / ～应用 clinical practice / ～诊断 clinical diagnosis

【临到】①(接近) be about to; just before; on the point of: ～开会 when the meeting was about to begin / ～收获 as the time for harvest draws near ②(落到) befall; happen to: 万一这种不幸～他头上 in case such misfortune should happen to him

【临机】[书] as the occasion requires: ～应变 adapt to changing circumstances; cope with any contingency

【临界】[物] critical ◇～点 critical point / ～角 critical angle / ～体积 critical size / ～态 critical state / ～温度 critical temperature

【临近】close to; close on: ～半夜 close on midnight / ～车站的饭店 a restaurant close to the station / ～黎明 close on daybreak

【临渴掘井】not dig a well until one is thirsty; not make timely preparations

【临了】finally; in the end

【临摹】copy; imitate

【临盆】be giving birth to a child; be confined; be in labor

【临深履薄】wading in deep water and treading on thin ice; with caution and care

【临时】①(事到临头) at the time when sth. happens ②(暂时的) temporary; provisional; for a short time: ～办法 a temporary arrangement; makeshift measures / ～凑合 make do for the moment / ～工作人员 a temporary member of the staff / ～起作用的因素 a factor which could play only a temporary role ◇～代办[外] chargé d'affaires ad interim / ～代理人 acting agent / ～动议 extempore motion / ～法庭 provisional court / ～费用 incidental expenses / ～工 casual laborer; temporary worker / ～户口 temporary residence permit / ～拘押 interim custody / ～拘留 temporary detention / ～扣押 interim attachment / ～停火 suspension of arms / ～舞台 makeshift stage / ～协议 interim agreement; provisional agreement / ～议程 provisional agenda / ～证书[外] temporary credentials; temporary papers / ～政府 provisional government; interim government / ～执照 provisional certificate; temporary licence / ～主席 interim chairman

【临时抱佛脚】embrace Buddha's feet in one's hour of need; seek help at the last moment; make a frantic last-minute effort

【临事仓惶】be startled at a crisis

【临事踌躇】hesitate at a step

【临死】on one's deathbed

【临头】befall; happen: 大难～ be faced with imminent disaster

【临危】①(病重将死) be dying ②(面临危险) facing death or deadly peril; in the hour of danger: ～不惧 face danger fearlessly; betray no fear in an hour of danger / ～授命 give a very important assignment in time of a national emergency

【临刑】just before execution

【临渊羡鱼】stand on the edge of a pool and idly long for fish: ～，不如退而结网 it's better to go back and make a net than to stand by the pond and long for fish; one should take practical steps to achieve one's aims

【临阵磨枪】sharpen one's spear only before going into battle; start to prepare only at the last moment

【临阵脱逃】desert on the eve of a battle; sneak away at a critical juncture

【临终】approaching one's end; immediately before one's death; on one's deathbed: ～遗言 deathbed will; deathbed words

邻 ①(邻居) neighbor: 近～ a close neighbor / 睦～政策 good-neighbor policy ②(邻近的) neighboring; near; adjacent: ～村 neighboring village / ～国 neighboring coun-

try / ～室 adjacent room / ～县 a neighboring county / ～座 an adjacent seat
【邻邦】neighboring country
【邻国为壑】bring evil upon another in malice
【邻角】[数] adjacent angles
【邻接】border on; be next to; be contiguous to; adjoin
【邻近】near; close to; adjacent to
【邻居】neighbor: 隔壁～ a next-door neighbor
【邻里】①(乡里) neighborhood ②(同一乡里的人) people of the neighborhood; neighbors
【邻位】[化] ortho-position ◇ ～化合物 ortho-compound

lǐn

凛 ①(冷) cold ②(严肃) strict; stern; severe: ～遵 strictly abide by ③(畏惧) afraid; apprehensive: ～于远行 be afraid of going on a long journey
【凛冽】piercingly cold
【凛凛】①(寒冷) cold: 寒风～ a piercing wind ②(严肃;严厉) stern; awe-inspiring: 威风～ majestic-looking; with an awe-inspiring
【凛凛如生】it seems alive
【凛然】stern; awe-inspiring: 态度～ stern in manner / 正气～ awe-inspiring righteousness
【凛若冰霜】cold as ice and dew
檩 [建] purlin
【檩条】[建] purlin

lìn

淋 strain; filter: 用纱布把药～一下 strain the herbal medicine with a piece of gauze
【淋病】[医] gonorrhea
吝 stingy; mean; closefisted
【吝啬】stingy; niggardly; miserly; mean ◇ ～鬼 miser; niggard; skinflint
【吝惜】grudge; stint: ～财物 be sparing with money and goods, not willing to spend / 不～自己的力量 spare no effort; stint no effort
赁 rent; hire: ～费 rent; rental / ～借 hire; borrow / ～书 borrow books for reading / ～屋 rent a house / ～租 rent / 房屋出～ house to let
膦 [化] phosphine

līng

拎 [方] carry; lift: 她～着一个篮子。She was carrying a basket.

líng

凌 ①(欺侮) insult: 盛气～人 arrogant and aggressive ②(逼近) approach: ～晨 before dawn ③(升高) rise high; tower aloft: ～霄 reach the clouds ④[方](冰) ice: 冰～ icicle
【凌波】ride the waves
【凌晨】in the small hours; before dawn: 九月十五日～ in the small hours of September 15
【凌迟】put to death by dismembering the body
【凌驾】place oneself above; override: ～他人 oppress another in order to advance oneself / ～一切 overriding; predominant
【凌空】be high up in the air; soar or tower aloft: ～高飞的鸟 bird soaring to the skies / ～高阁 towering pavilion / ～盘旋的飞机 plane hovering over in the sky
【凌厉】swift and fierce: 攻势～ a swift and fierce attack
【凌厉无前】be energetic without precedent
【凌乱】in disorder; in a mess: ～不堪 in a fearful mess; in a state of utter confusion / ～的头发 unkempt hair
【凌乱无序】confused and disordered
【凌日】[天] transit: 金星～ transit of Venus
【凌辱】insult; humiliate: 受到～ be humiliated; suffer humiliation / 忍受～ pocket an insult
【凌霄花】[植] Chinese trumpet creeper
【凌汛】ice run
【凌云】[书] reach the clouds; soar to the skies: ～壮志 lofty aspiration / 壮志～ cherish high aspirations
菱 [植] ling; water chestnut; water caltrop
【菱角】ling; water chestnut; water caltrop
【菱镁矿】magnesite
【菱铁矿】siderite
【菱锌矿】smithsonite
【菱形】rhombus; lozenge ◇ ～队形[军] diamond formation / ～六面体 rhombohedron
鲮
【鲮鲤】pangolin
【鲮鱼】dace
陵 ①(丘陵) hill; mound: ～谷 hills and valleys ②(陵墓) imperial tomb; mausoleum: 十三～ the tombs of 13 Ming emperors; the Ming Tombs / 中山～ the Sun Yat-sen Mausoleum
【陵谷变迁】Mountains and valleys change.
【陵居】live on the highland

【陵墓】mausoleum; tomb
【陵寝】[书] emperor's or king's resting place; mausoleum
【陵园】tombs surrounded by a park; cemetery: 烈士～ cemetery of revolutionary martyrs
绫 a silk fabric resembling satin but thinner; damask silk: ～罗绸缎 silks and satins / ～罗锦绣 expensive clothes
【绫子】thin satin; damask silk
羚 antelope
【羚牛】takin
【羚羊】antelope; gazelle: 飞～ oryx ◇ ～角[中药] antelope's horn
零 ①(数字) zero sign (0); nought: 三二～五号 No. 3205 / ～点～一五 0.015 ①(放在两个数量之间) 两年～十天 two years and ten days / 三十二元～七分 thirty *yuan* and seven *fen* ③(零头) odd; with a little extra: 年纪八十有～ a little more than eighty years old / 四千挂～ four thousand odd ④(没有数量) nought; zero; nil: 他们的努力结果等于～。Their effort came to nought. ⑤(温度表上的零度) zero: ～下二十摄氏度 20 degrees below zero centigrade; minus twenty degrees centigrade ⑥(零碎) fractional; part: 化整为～ break up the whole into parts ⑦(枯萎而落下) wither and fall: 凋～ withered, fallen and scattered about ⑧[体](零分) nil; love: ～比～ no score; love all
【零吃】[口] between-meal nibble
【零存整取】deposit small sums of money every month and draw out both the principal and interest in a lump sum when the specified time comes up
【零点】①[物](零度) zero point ②(零时) zero hour; 0:00 a.m.
【零度】zero: ～以下 below zero; sub-zero / 气温降到～。The temperature has fallen to zero.
【零分】zero; no marks; scoreless
【零工】①(指短工) odd job; short-term hired labor: 打～ do odd jobs ②(做零工的人) odd-job man; casual laborer
【零花】①(零碎地花钱) incidental expenses ②(零用钱) pocket money
【零活儿】odd jobs: 靠做～生活 make living by doing odd jobs
【零件】spare parts; spares
【零落】①(脱落) withered and fallen: 草木～ bare trees and withered grass ②(衰败) decayed: 凄凉～的景象 a desolate scene ③(稀疏) scattered; sporadic: ～的来访者 a scattering of visitors / ～的枪声 sporadic shooting; scattered reports of gunfire
【零卖】①(不成批地卖) retail; sell retail ②(零散地卖) sell by the piece or in small quantities
【零七八碎】①(零碎而纷乱) scattered and disorderly ②(零散没系统的事或没有大用的东西) miscellaneous and trifling things; odds and ends: 整天忙些个～儿 fuss over trifles all day long
【零钱】①(币值小的钱) small change; coins ②(零用钱) pocket money
【零敲碎打】do sth. bit by bit, off and on; adopt a piecemeal approach
【零散】scattered: 把～的情况凑到一块儿 piece together scraps of information
【零时】zero hour; 0:00 a.m.
【零食】between-meal nibbles; snacks: 吃～ nibble between meals
【零售】retail; sell retail ◇ ～店 retail shop; retail store / ～额 turnover / ～货物 odd lot / ～价格 retail price / ～商 retail dealer; retail trader / ～市场 retail market / ～网 retail network / ～总额 total volume of retail sales
【零碎】①(琐碎) scrappy; fragmentary; piecemeal: ～东西 odds and ends / ～活儿 odd jobs ②(零碎的事物) odds and ends; oddments; bits and pieces
【零头】①(不够一定单位的零碎数量) odd: 三米～ the odd three meters ②(材料用后剩下的零碎部分) remnant: 一块～布 a remnant
【零星】①(零碎的) fragmentary; odd; piecemeal: ～材料 fragmentary material / ～土地 odd pieces of land / 一些零零星星的消息 some odd scraps of news ②(零散) sporadic: ～小雨 occasional drizzles; scattered showers / ～战斗 sporadic fighting
【零用】①(小额费用) small incidental expenses ②(零用钱) pocket money ◇ ～费 petty cash / ～钱 pocket money / ～现金 petty cash fund / ～帐 petty cash book; petty cash account
【零指数】[数] zero exponent
玲
【玲琅】tinkling of jades
【玲珑】①(精巧细致) ingeniously and delicately wrought; exquisite: 小巧～ small and exquisite ②(灵活敏捷) clever and nimble: 娇小～ petite and dainty

【玲珑剔透】exquisitely carved; beautifully wrought: ～的玉石雕刻 exquisitely wrought jade carvings

聆 [书] listen; hear: ～教 listen to one's instructions; hear one's instructions

【聆听】listen

【聆悉】learn; hear

龄 ①(岁数) age; years: 高～ advanced in years / 年～ age / 学～儿童 school-age children ②(年限) length of time; duration: 党～ length of Party membership; Party standing / 工～ length of service; number of years worked; years of service

囹

【囹圄】[书] jail; prison: 身入～ be behind prison bars; be thrown into prison

铃 ①(响器) bell: 门～ door bell / 摇～ ring the bell ②(铃状物) anything in the shape of a bell: 哑～ dumbbell ③(蕾铃) boll; bud: 落～ shedding of cotton boll / 棉～ cotton boll

【铃铛】small bell

【铃鼓】[乐] tambourine

【铃兰】[植] lily of the valley

【铃声】the tinkle of bells: ～震耳 the sound of the bell shakes the ears

伶 [旧] actor; actress

【伶仃】left alone without help; lonely: 孤苦～ alone and uncared for

【伶俐】clever; bright; quick-witted: ～乖巧 clever and tricky / ～机警 smart and alert

【伶牙俐齿】have the gift of the gab; have a glib tongue

翎 feather; plume: 孔雀～ peacock plumes; peacock feathers

【翎毛】①(羽毛) feathers ②(以鸟类为题材的中国画) a type of classical Chinese painting featuring birds and animals

灵 ①(灵活) quick; clever; sharp: 耳朵很～ have sharp ears / 脑筋不～ one's brain doesn't seem to be working well / 心～手巧 quick-witted and nimble-fingered; clever and deft ②(灵验) efficacious; effective: ～药 an effective remedy / 这个办法很～。The method worked like a charm. ③(精神;灵魂) spirit; intelligence: 心～ the mind; the soul / 英～ the spirit of the brave departed ④ (关于神仙的) fairy; sprite; elf: ～怪 elf; goblin ⑤(关于死人的) of the deceased; bier: 守～ stand as guards at the bier; keep vigil beside the bier

【灵便】①(四肢五官灵活) nimble; agile: ～的手指 agile fingers / 耳朵不～ be hard of hearing ②(工具轻巧) easy to handle; handy: 这辆车开起来很～。The car handles well.

【灵车】hearse

【灵床】bier

【灵丹妙药】miraculous cure; panacea

【灵感】inspiration

【灵魂】soul; spirit: ～深处 in one's innermost soul; in the depth of one's soul / 出卖～ sell one's soul

【灵活】①(敏捷) nimble; agile; quick: 脑筋～ be quick-witted; have a supple mind / 手脚～ dexterous and quick in action ②(善于随机应变) flexible; elastic: ～机动的战略战术 flexible strategy and tactics ◇ ～措施 flexible measures / ～性 flexibility; adaptability; mobility

【灵活敏捷】active and intelligent; vivacious

【灵机】sudden inspiration; brain wave: ～一动 have a brain wave; a sudden inspiration

【灵柩】a coffin containing a corpse; bier

【灵猫】[动] civet: 大～ zibet

【灵敏】sensitive; keen; agile; acute: ～的头脑 quick-witted / ～的嗅觉 an acute sense of smell ◇ ～度 sensitivity

【灵巧】dexterous; nimble; skillful; ingenious: 一双～的手 a pair of clever hands

【灵堂】mourning hall

【灵通】well-informed; having quick access to information: 消息～人士 well-informed sources

【灵犀相通】extremes meet

【灵犀一点通】meeting of minds

【灵性】intelligence: 有～ have sagacity; be very intelligent

【灵验】①(有奇效) efficacious; effective: 这药很～。This medicine is highly efficacious. ②(应验) accurate; right: 她说的话果然～。What she said is proved to be correct.

【灵长目动物】primate

lǐng

令 [量] ream (of paper): ～重 ream weight

领 ①(颈) neck: 引～而望 crane one's neck for a look; eagerly look forward to ②(衣领) collar; neckband: 鸡心～ V-neck / 尖～儿 V-shaped collar / 把大衣～儿翻起来 turn up one's coat collar ③(大纲;要点) outline; main point: 要～ main points; essentials ④[量]一～席 a mat ⑤(带;引) lead; usher: ～兵打仗 lead troops into battle / ～我们参观工厂 show us round the factory / 把客人～到会客室去 usher the guests into the reception room ⑥(领有;领有的) have jurisdiction over; be in possession of: ～土 territory ⑦(领取) receive; draw; get: ～工资 take one's wages / ～奖 receive a prize / ～养老金 draw one's pension ⑧(了解) understand; comprehend; grasp: 心～神会 understand tacitly; readily take a hint

【领班】foreman; gaffer

【领兵】lead troops; a military officer

【领唱】①(在合唱时带头唱或独唱) lead a chorus ②(领唱者) leading singer

【领带】necktie; tie ◇ ～夹 a tie clip / ～扣针 tiepin

【领导】①(率领引导前进) lead; exercise leadership: ～我们从胜利走向胜利 lead us from victory to victory / ～现代化建设工作 exercise leadership in modernization program / 担任～工作 shoulder the responsibility of leadership; hold a leading position ②(领导者) leadership; leader ◇ ～班子 leading group / ～方法 method of leadership / ～方针 guiding policy / ～干部 leading cadre / ～骨干 the backbone of the leadership / ～核心 leading nucleus; the core of leadership / ～机关 leading body / ～集团 leading group / ～权 leadership; authority; overall control / ～人 leader / ～小组 leading group / ～艺术 the art of leadership / ～职务终身制 life tenure of leading posts / ～作风 work style of the leadership

【领地】①(封建主的领地) manor ②(领土) territory

【领队】①(率领队伍) lead a group ②(领队的人) the leader of a group, sports team, etc.

【领港】①(引导船舶进出港口) pilot a ship into or out of a harbor ②(领港员) pilot ◇ ～费 pilotage

【领海】territorial waters; territorial sea ◇ ～范围 extent of territorial waters / ～管辖权 jurisdiction within territorial water / ～宽度 breadth of the territorial sea / ～权 maritime domain / ～线 boundary line of territorial waters

【领航】①(引导船舶或飞机航行) navigate; pilot ②(领航员) navigator; pilot ◇ ～飞机 pathfinder aircraft / ～权 pilotage / ～设备 navigation equipment / ～图 pilotage chart / ～员 navigator

【领回】get back; take back

【领会】understand; comprehend; grasp: ～讲话的内容 follow a speech / ～文件的精神 grasp the essence of a document

【领江】①(在江河上引导船舶航行) navigate a ship on a river ②(领江人员) river pilot

【领教】①[套](用于接受人的教益或欣赏人的表演时) thanks; much obliged ②(请教) ask advice ③[讽](经历过) experience; encounter

【领结】bow tie

【领巾】scarf; neckerchief: 红～ red scarf

【领进】usher into; lead to; introduce into

【领空】territorial sky; territorial air space ◇ ～管辖权 jurisdiction within space

【领口】①(衣服的领子口) collarband; neckband: ～太小了。 The neckband is too small. ②(领子两头相合处) the place where the two ends of a collar meet

【领扣】collar button; collar stud

【领款】draw money ◇ ～人 payee

【领路】lead the way

【领略】have a taste of; realize; appreciate: ～词意 comprehend the meaning of a phrase / ～了塞北的大好风光 get some idea of how splendid are the sights north of the Great Wall / ～粤菜风味 taste Guangdong dishes

【领情】feel grateful to sb.; appreciate the kindness

【领取】draw; receive: ～办公用品 get stationery for use in the office / ～出入证 receive one's pass / ～工资 draw one's pay / ～驾驶执照 take out a driving licence

【领事】[外] consul: 代理～ pro-consul / 副～ vice-consul / 总～ consul general ◇ ～裁判官 judge consul / ～裁判权 consular jurisdiction / ～处 consular section / ～馆 consulate / ～豁免权 consular immunity / ～签证 consular invoice / ～特权 consular privileges / ～条例 consular act / ～团 consular corps / ～委任书 certificate of appointment of consul; consular commission / ～证书 exequatur

【领受】accept (kindness, etc.); receive

【领水】①(国内河流等) inland waters ②(领海) territorial waters ◇ ～员 pilot; navigator

【领头】take the lead; be the first to do sth.

【领土】territory: 保卫国家的～完整 safeguard a country's territorial integrity ◇ ～不可侵犯性 territorial inviolability / ～管辖权 jurisdic-

tion within territory / ～扩张 territorial expansion; territorial aggrandizement / ～完整 territorial integrity / ～要求 territorial claim
【领悟】comprehend; grasp
【领先】be in the lead; lead: 开始就～ have a good start / 遥遥～ hold a safe lead / 以七分～ lead by seven points
【领衔】head the list of signers (of a document)
【领袖】leader
【领养】adopt: ～一个孩子 adopt a child ◇ ～人 adopter / ～子女 adopted children
【领有】possess; own
【领域】①(一个国家行使主权的区域) territory; domain; realm: ～广大 the domain is vast ②(范围) field; sphere; domain; realm: 科学～ the field of science / 上层建筑～ the realm of the superstructure / 社会科学～ the domain of the social sciences / 文学～ the region of literature / 意识形态～ the ideological sphere / 艺术～ the world of art
【领章】collar badge; collar insignia
【领主】feudal lord; suzerain
【领子】collar
岭 ①(高大的山脉) mountain range: 秦～ the Qinling Mountains ②(顶上有路可通行的山) mountain; ridge: 崇山峻～ high mountain ridges / 翻山越～ cross over mountain after mountain
【岭南】south of the Five Ridges

lìng

另 other; another; separate: ～想办法 try to find some other way / 从一个极端跳到～一个极端 jump from one extreme to another
【另案办理】be handled as a separate case
【另册】the other register
【另打主意】make some other plans
【另订】order sth. separately; arrange sth. separately
【另搞一套】do what suits oneself; go one's own way
【另函】a separate letter; write another letter
【另寄】post separately; post under separate cover
【另立户头】open another bank account
【另谋生路】find another way of living
【另起炉灶】set up a separate kitchen; make a fresh start; start all over again
【另请高明】find someone better qualified
【另外】in addition; moreover; besides: ～你还要些什么? Would you like anything else? / 我们～找时间再谈吧。Let's talk it over again some other time.
【另行】separately: ～安排 be arranged separately; make separate arrangement / ～通知 be notified later; till further notice
【另眼相看】①(指不同于一般) regard sb. with special respect ②(指不同于过去) view sb. in a new, more favorable light; see sb. in a new light
【另议】be discussed separately
【另有打算】have other plans
【另有企图】have other intentions
令 ①(命令) command; order; decree: 法～ laws and decrees / 下～ issue an order / 遵～ obey orders ②(使) make; cause: ～人鼓舞 heartening; inspiring; encouraging / ～人深思 make one ponder; provide food for thought / ～人作呕 make one sick; nauseating; revolting ③(时节) season: 当～ in season / 夏～时间 summer time ④[敬] your: ～爱 your daughter / ～郎 your son / ～堂 your mother / ～尊 your father ⑤(酒令) drinking game: 行酒～ play a drinking game ⑥(美好) good: ～德 good character; high virtue / ～誉 good reputation
【令箭】an arrow used as a token of authority
【令箭荷花】[植] nopalxochia
【令人】make one: ～发指 make one's hair stand on end / ～费解 elude the understanding / ～鼓舞 inspiring / ～怀疑 open to suspicion / ～泪下 bring the tears to one's eyes / ～满意 satisfactory; satisfying / ～钦佩 admirable / ～深省 make one deep in thought / ～痛心 cut one to the heart / ～厌烦 bore one to tears / ～作呕 make one sick
【令行禁止】strict execution of orders and prohibitions

liū

溜 ①(往下滑) slide; glide: 从滑梯上～下来 slide down a children's slide ②(光滑; 平滑) smooth: ～光 very smooth / 滑～ slippery ③(偷偷地走开) sneak off; slip away; glide: ～掉 sneak off; slip away / ～进房间 sneak into the room
【溜边】[口] keep to the edge
【溜冰】①(滑冰) slide on the ice; skating ②[方](溜旱冰) roller-skating ◇ ～场 skating rink / ～鞋 skating shoes; roller skates
【溜达】[口] stroll; saunter; go for a walk

【溜光】[方] very smooth; sleek; glossy
【溜号】[方] sneak away; slink off
【溜肩膀】①(双肩下垂) sloping shoulders ②(不负责任) lacking a proper sense of responsibility; irresponsible
【溜须拍马】[口] fawn on; toady to; shamelessly flatter
【溜之大吉】make oneself scarce; sneak away; slink off
【溜走】leave stealthily; slip away

熘 sauté; quick-fry: ～肝尖 liver sauté / ～鱼片 fish slices sauté

liú

刘
【刘海儿】bang; fringe

浏
【浏览】glance over; skim through; browse: ～报纸上的大标题 take a glance at the newspaper headlines / 把稿子～一遍 glance over a manuscript / 在阅览室～图书 browse among books in the reading room

流 ①(流动; 液体移动) flow; run: ～鼻涕 have a running nose / ～汗 perspire; sweat / ～口水 slobber; slaver / ～泪 shed tears / 农村人口～入城市 flow of rural population into urban areas ②(移动不定) moving from place to place; drifting; wandering: ～民 refugees ③(流传) spread; circulate: ～传甚广 spread far and wide ④(向坏的方向转变) change for the worse; degenerate: ～于形式 become a mere formality ⑤(流放) banish; send into exile ⑥(江河流水) stream of water: 河～ river / 逆～而上 sail against the current / 顺～而下 down with the current / 中～ midstream ⑦(似水流的) sth. resembling a stream of water; current: 电～ electric current / 气～ air current ⑧(品类; 等级) class; rate; grade: 第一～的旅馆 first-class hotel / 第一～的作家 a first-rate writer / 二～演员 second-rate actor
【流弊】corrupt practices; abuses
【流产】①(胎儿不足月产出) abortion; miscarriage: 人工～ induced abortion / 习惯性～ habitual abortion ②(计划的事情没有实现) miscarry; fall through
【流畅】easy and smooth: 文笔～ write with ease and grace
【流程】①(工艺程序) technological process ②[矿] circuit: 浮选～ flotation circuit / 破碎～ crushing circuit ◇ ～图 flow chart; flow diagram
【流传】spread; circulate; hand down: ～后世 hand down to generations / 从祖先～下来的风俗 the custom handed down to us from our ancestors / 古代～下来的寓言 fables handed down from ancient times
【流窜】flee hither and thither: ～作案 flee thither and tither to commit offenses
【流弹】stray bullet
【流动】①(液体或气体移动) flow ②(经常变换位置) going from place to place; on the move; mobile ◇ ～电影放映队 mobile film projection team; mobile cinema team / ～红旗 mobile red banner / ～货车 shop-on-wheels / ～汇率 floating rate of exchange / ～基金 circulating fund / ～劳工 migrant labor / ～人口 floating population / ～商店 mobile shop / ～哨 soldier on patrol duties; patrol / ～售书站 mobile bookshop / ～图表 flow chart / ～图书馆 travelling library / ～性 mobility; fluidity / ～资本 circulating capital; floating capital / ～资产 floating assets; current assets
【流毒】①(毒害的流传) exert a pernicious influence: ～甚广 exert a widespread pernicious influence ②(流传的毒害) pernicious influence; baneful influence
【流芳百世】leave a good name for a hundred generations; leave a reputation which will go down to posterity
【流放】①(流放犯人) banish; send into exile ②(把木头放在河里运输) float (logs) downstream ◇ ～地 penal colony
【流感】[简](流行性感冒) flu; influenza
【流浸膏】[药] liquid extract
【流寇】roving bandits; roving rebel bands
【流浪】roam about; lead a vagrant life: ～街头 roam the streets / 到处～ wander here and there ◇ ～儿 waif; street urchin / ～汉 tramp; vagrant
【流离失所】become destitute and homeless; be forced to leave home and wander about
【流离迁徙】wander about and scatter every where
【流利】fluent; smooth: 讲一口～的英语 speak fluent English
【流里流气】rascally
【流连忘返】enjoy oneself so much as to forget to go home; linger on, forgetting to return
【流量】rate of flow; flow; discharge: 管道～

flow of a pipe / 河道～ discharge of a river / 平均～ average discharge / 总～ total flow / 最低～ minimum flow / 最高～ maximum flow ◇ ～计 flowmeter

【流露】reveal; betray; show unintentionally: 真情的～ a revelation of one's true feelings

【流落】wander about destitute: ～他乡 wander destitute far from home

【流氓】①(为非作歹的人) rogue; hoodlum; hooligan; gangster ②(施展下流手段) immoral behavior; hooliganism; indecency: 耍～ behave like a hoodlum; take liberties with women; act indecently ◇ ～成性者 incorrigible rogue / ～活动罪 offense of indecent activities / ～集团 gang of hooligans; criminal gang / ～团伙 rogues and vagabonds / ～无产者 lumpen-proletariat / ～习气 hooliganism / ～行为 indecent behavior; hooliganism

【流明】[物] lumen

【流年】①(光阴) fleeting time: 似水～ time passing swiftly like flowing water ②(一年中的运道) prediction of a person's luck in a given year: ～不利 an unlucky year

【流派】school; sect: 学术～ schools of thought

【流气】hooliganism; rascally behavior

【流入俗套】drift into inanities

【流沙】drift sand; quicksand; shifting sand

【流失】run off; be washed away: 黄金储备～ drain on gold reserves / 水土～ loss of water and erosion of soil; soil erosion

【流逝】pass; elapse: 光阴～ time flows away / 随着时间的～ with the passage of time

【流水】①(流动的水) running water: ～不腐，户枢不蠹 running water is never stale and a door-hinge never gets worm-eaten / ～潺潺 the murmuring of the running water ②(商店销售额) turnover ◇ ～号 serial number / ～线 assembly line / ～帐 day-to-day account; current account / ～作业 flow process; assembly line method; conveyer system

【流苏】tassels

【流速】①[机] velocity of flow ②[水] current velocity ◇ ～计[水] current meter

【流体】[物] fluid ◇ ～动力学 hydrokinetics; hydrodynamics / ～静力学 hydrostatics / ～力学 hydromechanics; fluid mechanics / ～压力计 manometer

【流铁槽】[冶] iron runner

【流通】circulate: 货币～ circulation of money / 空气～ circulation of air / 商品～ circulation of commodities / ～无阻 circulate without hindrance ◇ ～费用 circulation costs / ～管[机] runner pipe / ～货币 currency / ～领域 the field of circulation / ～票据 negotiable bill / ～手段 medium of circulation / ～证券 negotiable instrument

【流亡】be forced to leave one's native land; go into exile ◇ ～政府 government-in-exile

【流网】[渔] drift net

【流纹岩】[地] rhyolite

【流线型】streamline: ～汽车 streamlined car

【流星】①[天](流星体) meteor; shooting star ②[杂技] meteors: 火～ fire-meteors / 水～ water-meteors ◇ ～尘 meteoric dust / ～防护[宇航] meteoroid protection / ～群 meteor stream; meteor swarm / ～学 meteoritics / ～雨 meteor shower

【流行】prevalent; popular; fashionable; in vogue: 开始～ come into fashion / 骑自行车非常～ cycling is much in vogue / 一种～的式样 a prevailing style

【流行病】epidemic disease ◇ ～学 epidemiology

【流行歌曲】popular song

【流行性】[医] epidemic ◇ ～感冒 influenza; flu / ～脑脊髓膜炎 epidemic cerebrospinal meningitis / ～腮腺炎 mumps / ～乙型脑炎 Japanese Type-B encephalitis

【流血】bleed; shed blood: ～斗争 a sanguinary struggle / 为祖国～牺牲 shed blood for one's country

【流言】rumor; gossip: 散布～ spread rumors / ～飞语 rumors and slanders

【流域】valley; river basin; drainage area: 黄河～ the Huanghe River valley ◇ ～面积 drainage area

【流质膳食】[医] liquid diet: 半～ semiliquid diet

【流转】①(流动转移) wander about; roam; be on the move: ～四方 wander up and down the country ②(商品或资金周转) circulation (of goods or capital)

琉

【琉璃】colored glaze ◇ ～塔 glazed pagoda / ～瓦 glazed tile

硫

硫 [化] sulphur (S)

【硫代硫酸钠】[化] sodium thiosulphate

【硫华】[化] sublimed sulphur

【硫化】[化] vulcanization ◇ ～汞 mercuric sulphide／～剂 vulcanized agent; curing agent／～染料 sulphur dyes／～物 sulphide／～橡胶 vulcanized rubber; vulcanizate
【硫磺】[化] sulphur ◇ ～泉[地] sulphur spring
【硫塑料】[化] thioplast
【硫酸】[化] sulphuric acid
【硫酸钙】calcium sulfate
【硫酸铜】copper sulphate; cupric sulphate
【硫酸亚铁】ferrisulphas; green vitriol
【硫酸盐】[化] sulphate

留 ①(不离去) stay; remain: ～在亲戚家里 stay with a relative ②(使留) ask sb. to stay; keep sb. where he is: ～朋友吃饭 ask one's friend to stay for dinner ③(保留) reserve; keep; save: ～饭 save food for sb.／～座位 reserve a seat for sb.／把其余的吃的～到明天。Save the rest of the food for tomorrow.／你自己～起来吧。Keep it for yourself.／我要把这个～作将来用。I will set this aside for future use. ④(留着长) let grow; grow; wear: ～短头发 wear one's hair short; have short hair; have bobbed hair／～胡子 grow a beard／～小辫儿 wear plaits; wear one's hair in plaits ⑤(收下) accept; take: 把礼物～下 accept a present ⑥(遗留) leave: 笔记本～在宿舍里了 have left one's notebook in the dormitory／给她～个条 leave a note for her
【留步】[套] don't bother to see me out; don't bother to come any further
【留成外汇】retain a portion of one's foreign exchange
【留存】①(保存) preserve; keep: 此稿～ keep this copy on file ②(存在) remain; be extant
【留待】wait until
【留党察看】be placed on probation within the Party
【留得青山在,不愁没柴烧】as long as the green mountains are there, one need not worry about firewood
【留点】[天] stationary point
【留芳百世】have a niche in the temple of fame
【留后路】keep a way open for retreat; leave a way out: 给自己留条后路 leave oneself a way out; leave oneself an option
【留后手】leave room for maneuver
【留话】leave a message; leave word
【留级】fail to go up to the next grade; repeat the year's work
【留局候领】[邮] post restante; general delivery
【留客】detain a guest; ask a guest to stay
【留空】leave a blank; leave a space in writing
【留兰香】[植] spearmint
【留连忘返】so enchanted as to forget about home
【留恋】①(不忍离开) be reluctant to leave (a place); can't bear to part: ～家乡 be unwilling to leave one's hometown ②(缅怀过去) recall with nostalgia: ～过去 yearn for the past
【留量】[机] allowance: 机械加工～ stock allowance
【留门】leave a door unlocked 〈unbolted〉
【留名】leave behind a good reputation
【留难】make things difficult for sb.; put obstacles in sb.'s way
【留念】accept or keep as a souvenir: 照相～ have a photo taken as a memento／某某同志～。To Comrade so-and-so.
【留鸟】[动] resident (bird)
【留情】show mercy or forgiveness: 对敌人毫不～ show the enemy no mercy; give the enemy no quarter
【留取丹心照汗青】my loyalty may leave a page in the annals
【留任】retain a post; remain in office
【留神】be careful; take care: ～别碰头! Mind your head!／～,汽车来了! Mind the car!
【留声机】gramophone; phonograph
【留守】stay behind to take care of things; stay behind for garrison or liaison duty ◇ ～处 rear office／～人员 rear personnel
【留宿】①(留客人住宿) put up a guest for the night ②(停留下来住宿) stay over night; put up for the night
【留头发】allow the hair to grow
【留退路】leave ground for retreat
【留下后果】entail consequences
【留校察看】be kept in school but placed under surveillance
【留校工作】be assigned to work at Alma Mater after graduation
【留心】be careful; take care: ～听讲 listen attentively to a lecture／～路滑! Watch your step!／～那条狗! Beware of that dog!
【留学】study abroad
【留学生】student studying abroad; returned student: 外国～ foreign student
【留言】leave one's comments; leave a message ◇ ～簿 visitors' book
【留一手】hold back a trick or two

【留意】be careful; look out; keep one's eyes open
【留影】take a photo as a memento; have a picture taken as a souvenir
【留用】continue to employ; keep on ◇ ～人员 personnel who were kept on after liberation
【留余地】allow for unforeseen circumstances; leave some leeway
【留置权】[法] lien
【留种】[农] reserve seed for planting; have seed stock ◇ ～地 seed-plot; seedbed

瘤 tumor: 毒～ malignant tumor / 良性～ benign tumor / 肉～ sarcoma / 纤维～ fibroma / 腺～ adenoma
【瘤胃】[动] rumen ◇ ～臌胀[牧] bloat
【瘤子】[口] tumor

榴
【榴弹】[军] high explosive shell
【榴弹炮】howitzer: 122 毫米～ 122mm howitzer
【榴弹炮兵连】howitzer battery
【榴莲】[植] durian
【榴霰弹】shrapnel; canister

镏
【镏金】gold-plating: ～银器 gilded silverware

馏
【馏出燃料】distillate fuel
【馏出燃料油】distillate fuel oil
【馏出油】[石油] distillate oil
【馏份】[石油] fraction; cut: 轻～ light fraction / 重～ heavy fraction

liǔ

柳 willow: 垂～ weeping willow
【柳暗花明】dense willow trees and bright flowers; enchanting sight in spring time
【柳眉】the eyebrows of a beautiful woman
【柳琴】*liuqin*, a plucked stringed instrument
【柳杉】[植] cryptomeria
【柳丝】fine willow branches
【柳条】willow twig; osier; wicker ◇ ～筐 wicker basket / ～箱 wicker suitcase / ～制品 wicker; wickerwork
【柳巷花街】the red-light district
【柳絮】catkin
【柳芽】willow buddings; willow sprouts
【柳荫】shade of willow trees
【柳莺】[动] willow warbler
【柳枝】withy; willow branch

绺 [量] tuft; lock; skein: 一～丝线 a skein of silk thread / 一～头发 a lock of hair

liù

六 six: ～年 six years / 排成～个一组 arranged by sixes / 她穿～号衣服。 She wears a six. / 找到了～个。 Six were found. / 这里有～个。 Six are there.
【六边形】hexagon
【六畜兴旺】the domestic animals are all thriving
【六发左轮手枪】six-shooter
【六分仪】[天] sextant
【六根清净】free from human desires and passions
【六级风】[气] force 6 wind; strong breeze
【六极管】[无] hexode
【六角车床】[机] turret lathe
【六零六】[医] six-o-six; aesphenamine
【六六六】[农] BHC
【六轮汽车】six-wheeled vehicle; six-wheeler; three-axle truck
【六面体】hexahedron
【六亲】the six relations; one's kin: ～不认 refuse to have anything to do with all one's relatives and friends / ～无靠 nobody to turn to
【六神无主】all six vital organs failing to function; in a state of utter stupefaction
【六十】sixty: ～倍的增长 a sixtyfold increase / ～几岁的人 man in his sixties / ～年 sixty years / ～年代 the sixties / ～周年 the sixtieth anniversary / 增长～倍 increase sixtyfold
【六三三制】[教育] sixty-three-three
【六十四开】[印] sixty-fourmo; 64mo
【六弦琴】guitar
【六一国际儿童节】International Children's Day
【六月】①(阳历的) June ②(阴历的) the sixth month of the lunar year; the sixth moon
【六宅不安】successive family misfortunes

碌
【碌碡】[农] stone roller

溜 ①(湍流) swift current; turbulent flow ②(房顶上流下来的水) rainwater from the roof ③(檐沟) roof gutter ④(排) row: 一～房屋 a row of houses ⑤(附近) surroundings neighborhood: 这一～有许多工厂。 There are many factories in the neighborhood. ⑥(填塞) fill

【溜子】[矿] scraper-trough conveyer

遛 ①(散步) saunter; stroll: 出去～～。 Let's go for a stroll. ②(遛牲口) walk (a horse, etc.)

【遛遛】take a walk

【遛马】walk a horse

【遛弯儿】[方] take a walk; go for a stroll

馏 heat up in a steamer: 把馒头～一～ heat up the cold steamed bread

鹨 [动] pipit: 树～ tree pipit / 田～ paddy-field pipit

陆 six

lóng

龙 ①(传说中的动物) dragon ②(帝王的) imperial: ～袍 imperial robe ③(古生代一些大型爬行动物) a huge extinct reptile: 恐～ dinosaur

【龙船】dragon-boat: 赛～ a dragon-boat race

【龙胆】[植] rough gentian ◇ ～紫[药] gentian violet

【龙灯】dragon lantern

【龙飞凤舞】like dragons flying and phoenixes dancing; lively and vigorous flourishes in calligraphy

【龙凤呈祥】prosperity brought by the dragon and the phoenix; in extremely good fortune

【龙骨】①(鸟类的胸骨) a bird's sternum ②(船艇结构) keel ③(中药) fossil fragments

【龙骨水车】dragon-bone water lift; square-pallet chain-pump

【龙睛鱼】[动] dragon-eyes

【龙井】*longjing* tea; Dragon Well tea

【龙驹凤雏】a young, talented scholar

【龙卷】[气] spout: 海～ waterspouts

【龙卷风】tornado

【龙口夺粮】snatch food from the dragon's mouth; speed up the summer harvesting before the storm breaks

【龙葵】[植] black nightshade

【龙门刨床】[机] double housing planer

【龙门起重机】[机] gantry crane

【龙门铣床】[机] planer-type milling machine

【龙脑】[化] borneol; borneo camphor

【龙盘虎踞】like a coiling dragon and crouching tiger; a forbidding strategic point

【龙蛇混杂】the wise and the unwise huddled together

【龙舌兰】[植] century plant

【龙生龙,凤生凤】like father, like son

【龙潭虎穴】dragon's pool and tiger's den; a danger spot

【龙套】actor playing a walk-on part in old-style opera; utility man: 跑～ be a utility man in a theatrical show; play a bit role

【龙腾虎跃】dragons rising and tigers leaping; a scene of bustling activity

【龙头】①(旋塞) faucet; tap; cock: 打开～ turn on the tap / 关上～ turn off the tap ②(自行车把) handlebar

【龙头鱼】[动] Bombay duck

【龙王】the Dragon King

【龙虾】lobster

【龙涎香】ambergris

【龙须菜】asparagus

【龙须草】[植] Chinese alpine rush

【龙牙草】[植] hairyvein agrimony

【龙眼】[植] longan

【龙章凤姿】great handsome appearance

【龙爪槐】Chinese pagoda tree

【龙争虎斗】a fierce struggle between two evenly-matched opponents

【龙钟】[书] decrepit; senile: ～老者 a decrepit; old man / ～之年 the period of senility / 老态～ senile; doddering

【龙舟】dragon boat: ～竞渡 dragon-boat regatta; dragon-boat race

茏

【茏葱】verdant; luxuriantly green

聋 deaf; hard of hearing

【聋哑】deaf and dumb; deaf-mute ◇ ～人 deaf-mute / ～学校 school for deaf-mutes / ～症 deaf-mutism

【聋子】a deaf person

砻 ①(木磨) rice huller ②(去稻壳) hull

【砻谷机】[农] rice huller

【砻糠】rice chaff

笼 ①(笼子) cage; coop: 鸡～ chicken coop / 鸟～ birdcage ②(笼屉) steamer; steambox ③[方](把手放在袖筒里) put each hand in the opposite sleeve

【笼火】[方] light a coal fire with firewood; make a fire

【笼鸟】cage bird

【笼屉】bamboo or wooden utensil for steaming food; food steamer

【笼头】headstall; halter

【笼子】①(鸡笼,鸟笼) cage; coop ②(篮子) basket; container

隆 ①(盛大) grand ②(兴盛) prosperous; thriving ③(深厚) intense; deep: ～情厚谊 profound sentiments of friendship ④(凸出) swell; bulge

【隆冬】midwinter; the depth of winter: ～时节 at the time of bitter winter

【隆隆】[象] rumble: 火车～而过 the train rumbled past / 雷声～ the rumble of thunder / 炮声～ the rumble of gunfire

【隆起】rise; swell; bulge: 火山性～ volcanic upheaval / 地面渐渐～。 The ground rises gradually.

【隆头鱼】[动] wrasse

【隆重】grand; solemn; ceremonious: ～的典礼 a grand ceremony / ～开幕 open ceremoniously / 受到～的接待 be accorded a grand reception; be given a red carpet reception

lǒng

垄 ①(耕地上培成的土埂) ridge: 做～ make into ridges ②(田地分界的稍稍高起的小路) raised path between fields

【垄断】monopolize: ～市场 monopolize the market / ～市价 corner the market ◇ ～集团 monopoly group / ～价格 monopoly price / ～利润 monopolist profits / ～资本 monopoly capital / ～资本主义 monopoly capitalism / ～资产阶级 monopoly capitalist class

【垄沟】field ditch; furrow: 开～ make furrows / 新犁的～ newly turned furrows

【垄作】[农] ridge culture

拢 ①(靠近) approach; reach: ～岸 come alongside the shore ②(总共) add up; sum up: ～一～帐 sum up the accounts ③(合上) hold together: 用绳子把竹竿～住 tie bamboo poles in a bundle ④(梳) comb: ～头 comb hair

【拢共】altogether; all told; in all

【拢子】a fine-toothed comb

笼 ①(笼罩) envelop; cover: 烟～雾罩 be enveloped in mist ②(笼子) a large box or chest; trunk

【笼括】encompass; seize all: ～一切 encompass everything

【笼络】win sb. over by any means; draw over; rope in: ～人心 try to win people's support by hook or by crook

【笼统】general; sweeping: ～地解释一下 explain in general terms / ～地说 generally speaking

【笼罩】envelop; shroud: ～心头 cast shadow over one's heart / ～着阴影 overshadowed

【笼子】a large box or chest; trunk

lòng

弄 [方] lane; alley; alleyway: 里～ lanes and alleys

【弄堂】[方] lane; alley; alleyway

lōu

搂 ①(用手或工具把东西聚集到面前) gather up; rake together: ～草 rake up hay ②(用手拢着提起) hold up; tuck up: ～起袖子 tuck up one's sleeves ③(搜刮) squeeze; extort: tort: ～钱 extort money ④[方](向自己方向拨) pull: ～扳机 pull a trigger

【搂草机】[农] rake

lóu

娄

【娄子】trouble; blunder: 捅～ make a blunder; get into trouble

耧 an animal-drawn seed plough; drill barrow; drill

楼 ①(楼房) a storied building: 办公～ office building / 教室～ classroom building ②(楼房的每一层) story; floor; storey: 二～ first floor; first story (英); second story; second floor (美) / 一～ ground floor (英); first story; first floor (美) ③(建筑物上加盖的一层房子) superstructure: 城～ city-gate tower / 鼓～ drum-tower / 角～ watchtower / 钟～ clock-tower

【楼板】floor; floorslab

【楼道】corridor; passageway

【楼房】a building of two or more stories

【楼面】[建] floor ◇ ～面积 floor area

【楼上】upstairs: 睡在～ sleep upstairs

【楼梯】stairs; staircase: 太平～ fire-escape ◇ ～平台 landing

【楼下】downstairs: ～的房间 a downstairs room; a room on the floor below

喽

【喽罗】①(强盗的部下) the rank and file of a band of outlaws ②(仆从) underling; lackey

蝼

【蝼蛄】mole cricket
【蝼蚁】mole crickets and ants; nobodies; nonentities: 视同～ look upon sb. as mole cricket and ant

lǒu

搂

hold in one's arms; hug; embrace: 小女孩紧～着妈妈。The little girl gave her mother a big hug.
【搂抱】hug; embrace; cuddle

篓

basket: 竹～ bamboo basket / 字纸～ wastepaper basket; wastebasket
【篓子】basket

lòu

漏

①(渗漏) leak: ～雨了。The rain is leaking in. / 船～得利害。The ship was leaking badly. / 管子～气。The pipe leaks gas. ②(泄露) divulge; leak: 走～消息 leak information ③(遗漏) missing; leave out: ～了一行。A line is missing.
【漏报】fail to report sth.; fail to declare
【漏疮】anal fistula
【漏电】leakage of electricity
【漏洞】①(空隙;小孔) leak: 房顶上的～ a leak in the roof / 检查一下贮气罐有没有～ check and see if there is any leak in the gas tank ②(破绽) flaw; hole; loophole: ～百出 be full of loopholes / 堵塞～ stop up all loopholes
【漏斗】funnel
【漏风】①(走气) air leak; not airtight: 窗子～。There are some leaks in the window. ②(走漏风声) leak out
【漏光】①(因封闭不严而走光) light leak ②(漏完)水～了。All the water has leaked out.
【漏壶】water clock; clepsydra; hourglass
【漏尽更残】the night is waning
【漏孔】a small hole through which air or water leaks out
【漏了老底】leak out one's personal secrets
【漏排】omit some words or passages in printing due to carelessness
【漏勺】strainer; colander
【漏税】evade payment of a tax; evade taxation
【漏损条款】leakage clause
【漏网】escape unpunished; slip through the net: ～之鱼 fish that has escape the net
【漏雨】a roof leaking
【漏子】①[口](漏斗) funnel ②(漏洞) flaw; hole; loophole

瘘

fistula: 肛～ ana fistula
【瘘管】[医] fistula

镂

engrave; carve
【镂花】[工美] ornamental engraving
【镂刻】engrave; carve
【镂空】[工美] hollow out: ～的象牙球 hollowed-out ivory ball

露

[口] reveal; show
【露出破绽】show one's slip
【露底】betray a confidence; let out the whole story
【露脸】look good as a result of receiving honor or praise
【露马脚】give oneself away; let the cat out of the bag
【露面】show one's face; appear 〈reappear〉 on public occasions: 初次～ make one's first appearance
【露苗儿】sprouting; budding
【露怯】[方] display one's ignorance; make a fool of oneself
【露头】①(露出头部) show one's head ②(刚出现) appear; emerge
【露馅儿】let the cat out of the bag; give the game away; spill the beans
【露一手】make an exhibition of one's abilities or skills; show off: 一有机会总要～ seize every opportunity to show off
【露原形】reveal one's true color

陋

①(不好看) plain; ugly: 丑～ ugly ②(狭小) humble; mean: ～室 a humble room / ～巷 a mean alley ③(不文明) vulgar; corrupt; undesirable: ～习 corrupt customs; bad habits ④(见闻少) scanty; limited; shallow: 浅～ shallow; superficial
【陋规】objectionable practices
【陋室】a humble room; a room totally without decoration
【陋俗】undesirable customs
【陋习】corrupt customs; bad habits

lú

庐

hut; cottage
【庐山真面目】what Lushan Mountain really looks like; the truth about a person or a matter
【庐舍】[书] house; farmhouse

【庐帐】tent used as a dwelling

炉 ①(炉子) stove; furnace: 壁～ wall oven / 柴～ wood stove / 电～ electric stove / 火～ household stove / 鼓风～ blast furnace / 炼焦～ coke oven / 煤～ coal stove / 煤气～ gas heater; gas stove / 煤油～ kerosene stove ②[量](炼钢炉次) heat: 一～钢 a heat of steel

【炉箅子】grate

【炉衬】[冶] lining ◇ ～寿命 lining durability

【炉顶】[冶] furnace top; furnace roof

【炉灰】ashes of a stove

【炉火纯青】pure blue flame; high degree of technical or professional proficiency

【炉料】[冶] furnace charge; furnace burden

【炉龄】[冶] furnace life: 延长～ lengthen the life of a furnace

【炉门】furnace gate; stove gate

【炉盘】stone or metal plate for standing a stove on as a precaution against fire

【炉前工】[冶] blast-furnace man; furnaceman

【炉身】[冶] shaft; furnace stack

【炉台儿】mantel

【炉膛】the chamber of a stove or furnace

【炉条】fire bars; grate

【炉温】[冶] furnace temperature

【炉灶】kitchen range; cooking range: 另起～ make a fresh start

【炉渣】slag; cinder

【炉子】stove; furnace: 围着～烤火 sit round a fire to get warm

芦 [植] reed

【芦丁】[药] rutin

【芦管】reed pipe

【芦花】reed catkins

【芦荟】[植] aloe

【芦笙】a reed-pipe wind instrument

【芦笋】[植] asparagus

【芦田】sandy fields where reeds grow

【芦苇】reed ◇ ～荡 reed marshes

【芦席】reed mat

卢

【卢比】rupee

【卢布】rouble

【卢森堡】Luxembourg ◇ ～人 Luxembourger

【卢旺达】Rwanda ◇ ～人 Rwandese

颅 cranium; skull

【颅骨】[生理] skull ◇～切开术 craniotomy

【颅检查术】[医] cranioscopy

【颅腔】[生理] cranial cavity

鸬

【鸬鹚】cormorant

鲈 [动] perch

lǔ

卤 ①(盐卤) bittern ②[化](卤素) halogen ～族 halogen family ③(卤制) stew in soy sauce: ～鸡 pot-stewed chicken ④(卤汁) thick gravy used as a sauce for noodles, etc.: 打～面 noodles served with thick gravy

【卤化】[化] halogenate ◇ ～物 halogenide; halide

【卤水】①(卤) bittern ②(盐水) brine

【卤素】[化] halogen

【卤味】pot-stewed fowl, meat, etc. served cold

【卤族】[化] halogen family

虏 ①(打仗时捉住) take prisoner ②(打仗时捉住的敌人) captive; prisoner of war

【虏获】①(俘虏敌人) capture ②(俘虏的敌人和缴获武器) men and arms captured

掳 carry off; capture

【掳掠】pillage; loot: 奸淫～, 无恶不作 rape and pillage and commit all kinds of atrocities

鲁 ①(迟钝) stupid; dull ②(莽撞) rash; rough; rude

【鲁钝】dull-witted; obtuse; stupid

【鲁莽】crude and rash; rash: ～行事 act rashly; act without thought

【鲁米那】[药] luminal

橹 scull; sweep

镥 [化] lutecium; lutetium (Lu)

lù

鹿 deer: 公～ stag; buck / 梅花～ sika deer / 麋～ mi-lu; David's deer / 母～ doe / 小～ fawn

【鹿角】①(鹿的角) deerhorn antler ②(鹿砦) abatis ◇ ～胶[中药] deerhorn glue

【鹿角菜】[植] siliquose pelvetia

【鹿圈】deer enclosure; deer pen

【鹿皮】deerskin

【鹿茸】[中药] pilose antler

【鹿肉】venison

【鹿死不择荫】a desperate man will resort to anything

【鹿死谁手】at whose hand will the deer die; who will win the prize; who will gain supremacy: ～，尚难逆料 it's still hard to tell who will emerge victorious
【鹿尾】deer's tail
【鹿苑】deer park
【鹿砦】[军] abatis

漉 seep through; filter
【漉网】[纸] vat-net

麓 [书] the foot of a hill or mountain: 天山～ the foot of the Tianshan mountains／喜马拉雅山北～ at the northern foot of Himalaya

辘
【辘轳】windlass; winch
【辘辘】[象] rumble: 车轮的～声 the rumbling of cart wheels／饥肠～ so hungry that one's stomach rumbles; one's stomach growling from hunger; famished

路 ①(道路) road; path; way: 大～ broad road; highway／公～ highway／混凝土～ concrete road／沥青～ asphalt road／汽车～ motor road／碎石～ macadam road／土～ dirt road／小～ path; trail／迷～ lose one's way／引～ lead the way ②(路程) journey; distance: 一小时走六十二公里～ cover 62 kilometers an hour／走很远的～ walk a long distance; make a long journey ③(途径) way; means: 生～ means of livelihood; a way out／走邪～ go the wrong way／走正～ go the right way ④(条理) sequence; line; logic: 理～ line of reasoning／思～ train of thought ⑤(地区) region; district: 南～货 southern products／外～人 nonlocal people ⑥(路线) route: 三二三～公共汽车 No. 323 bus／三～进军 advance along three routes ⑦(种类) sort; grade; class: 头～货 top-notch goods／一～货 the same sort; birds of a feather
【路拌】[交] road mix ◇ ～路面 road-mixed pavement
【路标】①(交通标志) road sign ②[军](沿途的联络标志) route marking; route sign
【路不拾遗】no one picks up and pockets anything lost on the road; descriptive of a high moral standard in society
【路程】distance travelled; journey: 十天～ a ten days' journey／遥远的～ long journey／走了二百公里的～ have covered a distance of 200 kilometers
【路道】[方]①(途径) way; approach ②(人的行径) behavior
【路灯】street lamp; road lamp
【路堤】[交] embankment
【路段】a section of a highway or railway
【路费】travelling expenses
【路轨】①(铺轨的钢材) rail ②(轨道) track
【路过】pass by or through: 从北京到上海，～天津 pass through Tianjin en route from Beijing to Shanghai
【路基】roadbed; bed
【路见不平，拔刀相助】see injustice on the road and draw one's sword to help the victim; take up the cudgels for the injured party
【路劫】highway robbery; holdup; mugging
【路警】railway police
【路径】①(道路) route; way: ～不熟 not know one's way around／不知道最短的～怎样走 not know how to go by the shortest route ②(门路) method; ways and means: 成功的～ way to success
【路口】crossing; intersection: 三岔～ a fork in a road／十字～ crossroads
【路面】road surface; pavement: 刚性～ rigid pavement／混凝土～ concrete road surface／柔性～ flexible pavement ◇ ～基层 sub-surface／～宽度 width of roadway
【路牌】street nameplate
【路堑】[交] cutting
【路人】passerby; stranger: ～皆知 known by everybody／视若～ treat sb. like a stranger
【路上】①(道路上面) on the road: ～停着一辆公共汽车。A bus is parking on the road. ②(在途中) on the way: 在我回家的～ on my way home
【路数】①(途径) way; approach ②(着数) a movement in martial arts: 击剑的～ thrusts in fencing ③(底细) exact details; inside story
【路透社】Reuters; Reuter's News Agency
【路途】①(路) road; path ②(路程) way; journey: ～遥远 a long way to go; far away
【路线】①(经过的道路) route; itinerary: 旅行的～ the route of a journey ②(思想上、政治上所遵循的途径) line: 基本～ basic line
【路线图】route chart; line map: 参观～ visitors' itinerary
【路遥知马力】distance tests a horse's stamina: ～，日久见人心 as distance tests a horse's strength, so time reveals a person's heart／～，事久见人心 as a long road tests a horse's strength, so a long task proves a person's heart

【路远】great distance
【路障】roadblock
【路子】way; approach: ～不对等于白费劲儿 a wrong approach means a waste of effort

露 ①(露水) dew: ～珠 dew drop ②(饮料) beverage distilled from flowers, fruit or leaves; syrup: 果子～ fruit syrup / 玫瑰～ rose flavored juice ③(显露) show; reveal betray: 不～声色 not betray one's feelings 〈intentions〉/ 刚～芽 the buds are just showing
【露出】show; reveal: ～马脚 give oneself away; let the cat out of the bag / ～笑容 reveal a smile / ～原形 reveal one's true colors; betray oneself
【露点】[气] dew point: 温度～差 dew-point deficit ◇ ～湿度表 dew-point hygrometer
【露骨】thinly veiled; undisguised; barefaced: ～地干涉别国内政 flagrantly interfere in the internal affairs of another country / 说得十分～ speak undisguisedly; speak in no equivocal terms
【露光计】[摄] exposure meter
【露脊鲸】[动] right whale
【露酒】alcoholic drink mixed with fruit juice
【露水】dew: ～夫妻 man and woman not married, meeting secretly
【露宿】sleep in the open: ～风餐 sleep in open air and eat in the wind; the hardships of a traveler
【露天】in the open; outdoors: 今晚电影在～演 the film will be shown in the open air tonight ◇ ～采矿 opencast mining / ～电影院 open-air cinema / ～堆栈 open-air repository; open-air depot / ～剧场 open-air theatre / ～开采 surface mining / ～矿 opencut; opencast; open- pit; strip mine / ～煤矿 opencut coal mine
【露头】[矿] outcrop; outcropping: 断层～ the outcrop of the fault
【露头角】beginning to show ability or talent; budding
【露一手】make an exhibition of one's abilities or skills; show off
【露营】camp; encamp; bivouac

鹭 egret; heron: 池～ pond heron / 牛背～ cattle egret
【鹭鹭】egret

戮 ①(杀) kill; slay: 杀～ slaughter ②[书](并; 合) uitite; join: ～力 join hands
【戮力同心】[书] unite in a concerted effort; make concerted efforts

录 ①(记录) record; write down; copy: 抄～ copy down / 记～在案 put on record ②(采取; 任用) employ; hire: 收～ employ; take sb. on the staff ③(录音) tape-record: 把演讲～下来 record a speech ④(用做记载物的名称) record; register; collection: 回忆～ memoirs; reminiscences / 语～ quotation; a book of quotations
【录供】[法] take down a confession or testimony during an interrogation
【录取】enroll; recruit; admit: ～口供 take down verbal evidence / ～新生四百名 enroll 400 students ◇ ～名单 list of enrollees / ～名额 number of enrollees / ～通知书 admission notice
【录像机】videocorder: 磁带～ video tape recorder
【录像盘】video disk
【录音】sound recording: 磁带～ tape recording / 放～ play back the recording / 实况～ on-the-spot recording; live-recording ◇ ～报告 tape- recorded speech / ～磁头 magnetic recording head / ～带 magnetic tape; tape / ～电平 recording level / ～广播 electrical transcription / ～胶片 recording film / ～摄影机 sound camera / ～室 recording room
【录音机】recorder: 盒式磁带～ cassette tape recorder
【录用】employ; take sb. on the staff: 量才～ give a person employment commensurate with his abilities

禄 official's salary in feudal China; emolument: 高官厚～ high position and handsome salary
【禄位高升】salary and rank rise high

碌 ①(平凡) commonplace; mediocre ②(事务繁杂) busy
【碌碌】①(平庸) mediocre; commonplace: ～无能 incompetent; devoid of ability ②(辛苦) busy with miscellaneous work: ～半生 toil half a lifetime / 忙忙～ busy going about one's work; as busy as a bee ◇ ～无奇 rough and not wonderful / ～庸才 very mediocre person / ～庸人 a rough and commonplace person

绿
【绿林好汉】①(绿林英雄) heroes of the

greenwood; forest outlaws ②(聚集山林的强盗集团) a band of bandits entrenched in a mountain stronghold; brigands

陆 land: 水～交通 land and water communications

【陆半球】[地] the continental hemisphere; the land hemisphere

【陆地】dry land; land ◇ ～棉 upland cotton／～行舟 sail a boat on the land; impossibility

【陆风】[气] land breeze

【陆海空三军】the army, navy and air force

【陆军】ground force; land force; army

【陆空运输保险】insurance of transport by inland or by air

【陆连岛】land-tied island; tombolo

【陆龙卷】[气] tornado; landspout

【陆路】land route: ～交通 overland communication; land communication／走～ travel by land

【陆桥】land bridge: ～运输 land-bridge service

【陆生动物】terrestrial animal

【陆相】[地] land facies ◇ ～沉积 continental deposit

【陆续】one after another; in succession: 客人们～来了。The guests came one after another.

【陆运】land transportation

【陆战队】[军] marine corps; marines

lǘ

驴 donkey; ass: 公～ jackass; donkey／母～ jenny ass

【驴车】donkey cart

【驴唇不对马嘴】donkeys' lips don't match horses' jaws; incongruous; irrelevant: 这个例子有点～。The example is rather farfetched.

【驴打滚】snowballing usury

【驴叫】a donkey's bray; loud and unpleasant voice

【驴脸】a donkey's face; a long face

【驴骡】hinny

【驴鸣狗吠】asses braying and dogs barking; poor style of writing

【驴年马月】impossible date; a time will never come

【驴子】[方] donkey; ass

lǚ

旅 ①(旅行) travel; stay away from home ② [军](军队编制单位) brigade ③(泛指军队) troops; force: 劲～ a powerful army; a crack force／军～之事 military affairs

【旅伴】travelling companion; fellow traveller

【旅程】route; itinerary

【旅店】inn

【旅费】travelling expenses

【旅馆】hotel: 汽车游客～ motel／(结帐后)离开～ check out／住～ check in (at a hotel) ○ 宾馆 guest house／大门 main entrance／门厅 entrance hall／楼梯 staircase; stairs; stair-way／过道 corridor／外廊 verandah／走廊 lobby／电梯 lift; elevator／问询处 information desk／接待室 reception office／旅客登记簿 hotel register／登记表 registration form／邮局服务处 postal service／小卖部 shop／酒巴间 bar／休息厅 lounge／屋顶花园 roof garden／台球房 billiard-room／餐厅 dining-room; dining-hall; hotel restaurant／男盥洗室 men's room／女盥洗室 ladies' room／存衣处 cloak room／房间号码 room number／房间钥匙 room key／一套房间 suite／单人房间 single room／双人房间 double room／擦鞋棕垫 door-mat／经理 manager／服务员 attendant／值班服务员 desk clerk／餐厅服务员 waiter(男); waitress(女)／租金 rent／帐单 bill／收据 receipt

【旅进旅退】[书] always follow the steps of others, forward or backward; have no definite views of one's own

【旅居】reside abroad; sojourn: ～海外的侨胞 Chineses nationals residing abroad

【旅客】hotel guest; traveller; passenger: 出境～ outgoing passenger／过境～ stopover passenger; transit passenger／过往～ travellers passing through; transients／入境～ incoming passenger ◇ ～舱 passenger cabin／～登记簿 hotel register／～登记处 passenger registration／～责任保险 passenger liability insurance

【旅力方刚】while one's backbone retains its strength

【旅社】hotel

【旅食维艰】the difficulties in lodging and boarding

【旅途】journey; trip: ～见闻 what one sees and hears during a trip; traveller's notes／～随笔 sketches on a journey; traveler's notes／踏上～ set forth on a journey

【旅行】travel; journey; tour: 作长途～ make a long journey ◇ ～包 travelling bag／～车 wagon car; station wagon／～袋 valise／～家具 campaign furniture／～闹钟 travelling clock／～社 travel service／～团 touring party / ～平安险 travel accident insurance / ～信用证 travellers letter of credit / ～证 travel certificate / 支票 traveller's check / ～指南 guidebook

【旅游】tour; tourism: ～事业 tourist trade; tourism / ～者 tourist

【旅杂费外汇】travelling and miscellaneous expenses

【旅长】brigade commander

膂

【膂力】muscular strength; physical strength; brawn: ～过人 possessing extraordinary physical strength

偻

crooked: 伛～ humpback(ed); hunchback(ed)

屡

repeatedly; time and again: ～建奇功 make unusual contributions repeatedly / ～战～胜 have fought many battles and won every one of them; score one victory after another

【屡次】time and again; repeatedly: ～打破全国纪录 repeatedly break the national record

【屡次三番】again and again; over and over again; many times

【屡告不听】wouldn't listen to repeated advice

【屡见不鲜】common occurrence; nothing new

【屡教不改】refuse to mend one's ways despite repeated admonition

【屡经挫折】repeatedly met with setbacks

【屡屡】[书] time and again; repeatedly: ～吃亏 get the worst of it again and again

【屡试不爽】put to repeated tests and proved right; time-tested

缕

①(线) thread: 金～ gold thread ②[量](用于细的东西) wisp; strand; lock: 一～麻 a strand of hemp / 一～烟 a wisp of smoke ③(详详细细) detailed; in detail: ～陈 state in detail

【缕解】explain in detail; go into particulars

【缕缕】continuously: ～陈述 narrate in detail / 山村炊烟～上升。 Wisps of smoke rose continuously from the mountain village chimneys.

【缕述】state in detail; give all the details; go into particulars: ～条款 go into particulars of the terms

【缕析】make a detailed analysis

捋

smooth out with the fingers; stroke: ～胡子 stroke one's beard / 把纸～平 smooth out a piece of paper

铝

[化] aluminum (Al)

【铝箔】aluminum foil

【铝胶】alumina gel

【铝热剂】[化] thermite ◇ ～焊接 aluminithermic weld; thermite joint / ～燃烧弹 thermite bomb

【铝土矿】bauxite

【铝线】aluminum steel

【铝制品】aluminum products

侣

a companion; a mate: 伴～ companion; partner / 情～ lovers

履

①(鞋) shoe: 革～ leather shoes ②(踩;踏) tread on; walk on: 如～薄冰 as if walking on thin ice ③(脚步) footstep: 步～艰难 walk with difficulty; hobble along ④(履行) carry out; honor; fulfill: ～约 honor an agreement; keep an appointment

【履带】[机] caterpillar tread; track: ～式推土机 crawler dozer / ～式拖拉机 caterpillar tractor; truck tractor

【履带车】creeper truck

【履历】personal details; antecedents ◇ ～表 a biographic sketch; curriculum vitae

【履霜坚冰】walk on hoar-frost and later on solid ice

【履险如夷】cross a dangerous pass as easily as walking on level ground; handle a crisis without difficulty

【履新】take or assume one's new office or post

【履行】perform; fulfill; carry out: ～国际主义义务 fulfill internationalist obligations / ～合同 carry a contract; discharge a contract / ～合同责任 burden of contract / ～合约 meet one's engagements / ～诺言 keep one's word; fulfill one's promise / ～契约 execute one's promises / ～期限 deadline for performance / ～入党手续 go through the procedure for admission to the Party / ～诉讼手续 discharge of proceedings / ～条约 execution of treaty / ～义务 duty of performance / ～债务 fulfill obligation / ～职责 do one's duty

【履约保证】performance bond; performance guarantee

【履约而来】come keeping an appointment

lǜ

率

rate; proportion; ratio: 成活～ survival

rate／(外汇)兑换～ rate of exchange／废品～ the rate of rejects／回流～[物] reflux ratio／人口增长～ the rate of population increase

虑 ①(思考) consider; ponder; think over: 深思熟～ careful consideration ②(担忧;发愁) concern; anxiety; worry: 不必过～ need not worry overmuch; need not be over-anxious／不足为～ give no cause for anxiety／无忧无～ carefree

滤 strain; filter: 过～ filter

【滤波器】[电] wave filter: 带通～ band-pass filter／高通～ high-pass filter

【滤过性病毒】[医] filterable virus

【滤器】filter: 粗～ strainer

【滤色镜】[摄] (color) filter

【滤液】[化] filtrate

【滤纸】[化] filter paper: 定量～ quantitative filter paper／定性～ qualitative filter paper

律 ①(法律) law: ～令 laws and decrees／按～处治 punish according to law ②(规律) law: 因果～ law of causation ③(音律)十二～ twelve-tone ④(约束) discipline; restrain; keep under control: 严以～己 be strict with oneself; exercise strict self-discipline

【律赋】a form of literary style

【律己】discipline oneself; self-restraint: 严于～ be strict with oneself; exercise strict self-discipline

【律例】laws, statutes and precedents

【律例繁杂】statutes and laws are many and confused

【律师】lawyer; barrister (英); solicitor (英); attorney (美): 当～ practice as a lawyer／请～ engage a lawyer ◇ ～费 counsel fee／～留置权 lawyer's lien／～事务所 law office／～协会 bar association

【律诗】*lushi*, a poem of eight lines, each containing five or seven characters, with a strict tonal pattern and rhyme scheme

氯 [化] chlorine (Cl)

【氯丙嗪】[药] chlorpromazine; wintermine

【氯丁橡胶】[化] chloroprene rubber

【氯仿】[化] chloroform

【氯化纳】[化] common salt; sodium chloride

【氯化氰】[化] cyanogen chloride

【氯化物】[化] chloride

【氯喹】[药] chloroquine

【氯磷定】[药] pyraloxime methylchloride

【氯纶】[纺] polyvinyl chloride fiber

【氯霉素】[药] chloromycetin; chloramphenicol

【氯气】[化] chlorine

【氯噻酮】[药] chlorthalidone

【氯酸】[化] chloric acid ◇ ～钾 potassium chlorate

绿 green: ～叶 green leaves／～油油的秧苗 green and lush seedlings／碧～ bright green／嫩～ vivid green

【绿宝石】emerald

【绿草如茵】the green grass is like a cushion

【绿茶】green tea

【绿灯】①[交](可通行的绿色灯光) green light ②(准许前进) permission to go ahead with some project; green light: 开～ give the green light to

【绿豆】mung bean; green gram: ～糕 small cakes made with green bean flour／～稀饭 congee or rice porridge cooked with green beans／～芽 mung bean sprouts

【绿矾】[化] green vitriol

【绿肥】green manure ◇ ～作物 green manure crop

【绿化】make green by planting trees, flowers, etc.; afforest: ～城市 plant trees in and around the city／～地带 greenbelt／～规划 afforestation plan／～山区 afforest the mountain district／～祖国 cover the country with trees

【绿篱】[林] hedgerow; hedge

【绿泥石】[矿] chlorite

【绿萍】duckweed

【绿铅矿】pyromorphite

【绿松石】[矿] turquoise

【绿头鸭】[动] mallard

【绿野】the green field

【绿油油】glossy and green; shiny green: ～的麦苗 glossy and green wheat-shoots

【绿藻】[植] green alga

【绿洲】oasis

【绿柱石】[地] beryl

luán

栾 [植] goldenrain tree

峦 ①(陡峭的山) low but steep and pointed hill ②(连绵的山) mountains in a range

銮 a small tinkling bell

挛 contraction: 痉～ spasm; convulsions / 拘～ contraction

【挛缩】contracture

鸾 a mythical bird like the phoenix

【鸾凤】husband and wife: ～和鸣 be blessed with conjugal felicity; be a happy couple

【鸾交凤友】a couple deeply in love

【鸾飘凤泊】separation of husband and wife

孪 twin

【孪生】twin: ～姐妹 twin sisters

luǎn

卵 ovum; egg; spawn

【卵白】[动] white of an egg; albumen

【卵巢】[生理] ovary

【卵黄】[动] yolk

【卵磷脂】[生理] lecithin

【卵生】[动] oviparity ◇ ～动物 oviparous animal; ovipara

【卵石】cobble; pebble; shingle

【卵胎生】[动] ovoviviparity ◇ ～动物 ovoviviparous animal; ovovivipara

【卵细胞】egg cell; ovum

【卵翼】cover with wings as in brooding; shield: ～之下 under the wings of; under the shield of

【卵用鸡】[农] laying fowl; layer

【卵子】[生] ovum; egg: ～着床 embedding of ovum

luàn

乱 ①(无秩序) in disorder; in a mess; in confusion: ～跑 run about / ～嚷～叫 shout and scream madly / 屋子里～作一团。 There was great confusion in the house. ②(心绪不宁) be confused; be upset; be disturbed: 心烦意～ be confused in mind ③(使混乱) confuse; upset; throw sb. into chaos: ～了手脚 be thrown into chaos ④(任意;随便) indiscriminate; random; arbitrary: 给人～扣帽子 slap political labels on people right and left / ～花钱 spend money extravagantly / ～来 act recklessly / ～作决定 make an arbitrary decision ⑤(不正当的男女关系) promiscuous sexual behavior; promiscuity ⑥(战乱) disorder; upheaval; chaos; riot; unrest; turmoil: 内～ internal unrest / 叛～ armed rebellion; mutiny

【乱兵】①(叛乱或溃逃的兵) mutinous soldiers ②(完全没有纪律约束的军队) totally undisciplined troops

【乱成一团】in great confusion

【乱点鸳鸯】cause an exchange of partners by mistake between two couples engaged to marry

【乱罚】mete out unjustified punishment

【乱纷纷】disorderly; confused; chaotic: ～的人群 a tumultuous crowd

【乱坟岗】unmarked common graves; unmarked burial-mounds

【乱哄哄】in noisy disorder; in a hubbub; tumultuous; in an uproar: 街上～的 the street being in a hubbub / 她心里～的。 Her mind is in a tumult.

【乱花乱用】spend money recklessly

【乱了营】[方] be thrown into confusion; be in disarray

【乱伦】commit incest

【乱骂】verbal garbage; foul abuse

【乱蓬蓬】dishevelled; tangled; jumbled: ～的胡子 unkempt beard / ～的茅草 a jumbled mass of reeds / ～的头发 dishevelled hair; tangled hair

【乱七八糟】at sixes and sevens; in a mess; in a muddle

【乱如丝麻】tangled like silk and hemp

【乱杀】kill without discrimination

【乱世】troubled times; turbulent days

【乱说】speak carelessly; make irresponsible remarks; gossip: ～一顿 idle chatter; talk through one's hat / 当面不说,背后～ gossip behind people's backs but say nothing to their faces

【乱弹琴】[口] act or talk like a fool; talk nonsense

【乱腾】confusion; disorder; unrest

【乱糟糟】①(事物杂乱) hubbub; cluttered: 桌上～地堆满了书报。 The desk is cluttered with books and papers. ②(心里烦乱) confused; troubled; perturbed: 心里～的 feel very perturbed

【乱真】①(模仿得很象) look genuine: 以假～ pass off a fake as genuine ②[物] spurious: ～放电 spurious discharge / ～脉冲 spurious pulse

【乱子】disturbance; trouble; disorder: 闹～ create a disturbance; cause trouble

lüè

掠 ①(掠夺) plunder; pillage; sack ②(掠过) sweep past; brush past; graze; skim over: 海鸥～水而过。 The seagull skimmed over the water.

【掠地飞行】minimum-altitude flight; treetop

flight; hedgehopping
【掠夺】plunder; rob; pillage ◇ ～者 plunderer; looter
【掠过】sideswipe; fly past: 飞机从上空～过。Planes were sweeping past.
【掠美】claim credit due to others
【掠取】seize; grab; plunder: ～别国的资源 plunder the resources of other countries
【掠人之美】rob other's good point

略 ①(简单) brief; sketchy: 简～ sketchy; simple / ～述大意 give a brief account ②(略微) slightly; a little; somewhat: ～加修改 make some slight changes; edit slightly / ～有出入 vary slightly; there's a slight discrepancy / ～有错误 with a few mistakes / ～有所闻 have heard a little about the matter ③(简叙) summary; brief account; outline: 史～ outline history; brief history / 事～ a short biographical account ④(省去) omit; delete; leave out: 从～ be omitted / ～去不提 make no mention of; leave out altogether ⑤(计划;计谋) strategy; plan; scheme: 策～ tactics / 雄才大～ great talent and bold vision; rare gifts and bold strategy ⑥(夺取) capture; seize: 攻城～地 attack cities and seize territories
【略表寸心】just to show my gratitude
【略而不谈】omit to mention; skip over without reference to
【略见一斑】catch a glimpse of; get a rough idea of
【略略】slightly; briefly
【略去】omit; leave out
【略胜一筹】a notch above; slightly better
【略图】sketch map; sketch
【略微】slightly; a little; somewhat: ～懂一点英语 know a little English / ～有点感冒 have a slight cold; have a touch of flu
【略语】[语] abbreviation; shortening
【略知一二】have a smattering of; know sth. about

lūn

抡 brandish; swing: ～刀 brandish a sword / ～起大铁锤 swing a sledgehammer

lún

沦 ①(沉没) sink: 沉～ sink into depravity, etc. ②(陷入) fall; be reduced to: ～为殖民地 be reduced to the status of a colony / ～于敌手 fall into enemy hands
【沦落】fall low; come down in the world; be reduced to poverty: ～街头 be driven onto the streets / ～异乡 sinking in a strange district
【沦入风尘】fall into professions not socially respectable
【沦丧】be lost or ruined
【沦亡】be annexed
【沦陷】be occupied by the enemy; fall into enemy hands ◇ ～区 enemy-occupied area

轮 ①(轮子) wheel: 齿～ gear wheel / 三～摩托 motor tricycle ②(象轮子的东西) sth. resembling a wheel; disc; ring: 光～ halo / 年～ [植] annual ring / 月～ the moon ③(轮船) steamboat; steamer: 江～ river steamer ④(轮流) take turns; in turn: ～着发言 speak in turn / ～着看护病人 take turns to watch the sick / ～着值日 be on duty by turns ⑤[量] 一～红日 a red sun / 一～明月 a bright moon ⑥[量](循环的事物或动作) round: 第二～比赛 the second round of the match / 举行新的一～会谈 hold a new round of talks
【轮班】in shifts; in relays; in rotation: ～工作 work by relays; work in relays / ～看守 keep watch in turn ◇ ～制 rotation system; relay system
【轮唱】[乐] round
【轮齿】[机] teeth of a cogwheel
【轮虫】[动] wheel animalcule; rotifer
【轮船】steamer; steamship; steamboat
【轮渡】ferry: 火车～ train ferry
【轮番】take turns: ～轰炸 bomb in waves
【轮换】rotate; take turns
【轮机】①(涡轮机) turbine: 冲压空气～ ram-air turbine / 燃气～ combustion gas turbine ②(轮船发动机) motorship engine; engine ◇ ～室 engine room / ～员 engineer / ～长 chief engineer
【轮奸】violation of a woman by several men in turn
【轮距】track; tread
【轮空】[体] bye: 在第一轮比赛中～ be a bye in the first round of the tournament
【轮廓】outline; contour; rough sketch: 大概的～ broad outline; rough sketch / 人体的～图 outline sketch of a body / 山峰的～ outline of a mountain
【轮流】take turns; do sth. in turn: ～交替 one after the other / ～值班 take turns on duty; work on shifts in turn
【轮牧】[牧] rotation grazing

【轮生】[植] verticillate ◇ ～叶 verticillate leaves
【轮式拖拉机】[农] wheeled tractor
【轮胎】tire: 翻制～ retreaded tire / 防滑～ antiskid tire; nonskid tire / 双层～ two ply tire ◇ ～帘子线 tire cord / ～压力计 tire pressure gauge
【轮辋】rim; felloe
【轮休】have holidays by turns; rotate days off; stagger holidays
【轮训】training in rotation
【轮椅】wheelchair
【轮轴】①[物](差动滑轮) wheel and axle ②(轮子的轴) wheel axle; wheel spindle
【轮转】rotate ◇ ～印刷机 rotary press
【轮子】wheel
【轮作】[农] crop rotation: 合理～ proper rotation of crops / 粮棉～ rotation of cereal crops and cotton ◇ ～制 rotation system

伦 ①(人伦) human relations ②(条理) logic; order ③(同类; 同等) peer; match: 绝～ peerless; matchless
【伦比】[书] rival; equal: 无与～ unrivaled; unequalled; peerless
【伦常】feudal order of importance or seniority in human relationships
【伦常纲纪】the human relationship and principle of the feudal society
【伦次】coherence; logical sequence: 语无～ speak incoherently; babble like an idiot
【伦理】ethics; moral principles ◇ ～学 ethics
【伦琴射线】[物] roentgen rays

纶 ①(青丝带子) black silk ribbon ②(钓丝) fishing line ③(合成纤维) synthetic fiber: 涤～ polyester fiber / 锦～ polyamide fiber

lùn

论 ①(分析和说明事理) discuss; talk about; discourse: 就事～事 talk about a matter in isolation; deal with a matter on its merits / 讨～ discuss ②(分析和说明事理的话或文章) view; opinion; statement: 高～ your brilliant views; your wise counsel / 舆～ public opinion ③(文章等题目用语) dissertation; essay: 《实践～》 *On Practice* ④(学说) theory: 进化～ the theory of evolution / 唯物～ materialism ⑤(说;看待) mention; regard; consider: 相提并～ mention in the same breath / 又当别～ should be regarded as a different matter ⑥(衡量) decide on; determine: 按质～价 determine the price according to the quality ⑦(按照某种单位或类别说) by; in terms of: ～件 by piece / ～天 by day / ～月发工资 pay wages by the month
【论处】decide on sb.'s punishment; punish: 以违反纪律～ be punished for a breach of discipline
【论点】argument; thesis: 提出～ put forward an argument / 鲜明的～ clear-cut proposition
【论调】view; argument
【论断】inference; judgment; thesis: 符合逻辑的～ logical conclusion / 科学～ scientific thesis / 著名的～ celebrated thesis / 作出～ draw an inference
【论功行赏】dispense rewards or honors according to merit; award people according to their contributions
【论及】touch upon
【论件计酬】payment by the piece
【论据】grounds of argument; argument: ～不足 insufficient grounds / 有力的～ strong argument; valid reasons
【论理】①(按理说) normally; as things should be: ～他早该回家了。 He should have come home long ago. ②(逻辑) logic: 合乎～ be logical; stand to reason ◇ ～学 logic
【论述】discuss; expound: 精辟的～ brilliant exposition / ～国际形势 expound the international situation
【论说】①(议论) exposition and argumentation: ～文 argumentation ②[口](按理论) normally
【论坛】forum; tribune: 文艺～ forum on literature
【论题】[逻] proposition
【论文】thesis; dissertation; treatise; paper: 毕业～ graduation thesis / 博士～ thesis for the Doctorate; doctoral dissertation / 科学～ a scientific treatise / 学术～ an academic thesis
【论战】polemic; debate
【论证】①(论述过程) demonstration; proof: 无可辩驳的～ irrefutable proof ②(论述并证明) expound and prove
【论著】treatise; work; book
【论罪】decide on the nature of the guilt: 按贪污～ be found guilty of corruption

luō

捋 rub one's palm along (sth. long): ～掉树枝上的叶子 strip a twig of its leaves / ～起袖子 push up one's sleeve
【捋虎须】stroke a tiger's whiskers; do sth. very

daring; run great risks

罗

【罗嗦】①(繁复) long-winded; wordy: 他说话太～。 He's far too long-winded. ②(琐碎;麻烦) overelaborate; trouble some: 这件事真～。 The matter is really troublesome.

luó

螺 ①(软体动物) spiral shell; snail: 马蹄～ top shell / 田～ field snail ②(螺旋形的指纹) whorl

【螺贝】spiral shell

【螺钿】[工美] mother-of-pearl inlay: ～漆盘 lacquer tray inlaid with mother-in-pearl

【螺钉】screw: 木～ wood screw; screwnail

【螺号】conch; shell trumpet

【螺距】[机] pitch; thread pitch

【螺母】[机] nut ◇ ～垫圈 nut collar

【螺栓】[机] bolt: 地脚～ foundation bolt / 连接～ binder bolt; connecting bolt

【螺丝】[口] screw ◇ ～板牙 screw die; threading die / ～刀 screwdriver / ～钉 screw / ～扣 thread / ～母 nut / ～起子 screwdriver / ～钳 wrench

【螺蛳】spiral shell; snail

【螺纹】①(手指上的纹理) whorl; spiral ②[机] screw thread: 公制～ metric thread / 惠氏～ Whitworth thread ◇ ～刀具 threading tool; screw tool

【螺线】spiral

【螺线管】[物] solenoid

【螺旋】①(象螺蛳壳纹理的曲线形) spiral; helix: ～式发展 spiral development; developing in spirals ②[物] screw ◇ ～扳手 spanner / ～菌 spirillum / ～推进机 screw propeller; propeller / ～线 helix; helical line; spiral / ～钻 spiral drill; auger

【螺旋桨】[机] propeller; screw: 飞机～ airscrew; aircraft propeller / 涡轮～ turbo-propeller ◇ ～调速器 propeller governor / ～叶 propeller blade / ～直升机 propcopter

【螺旋体】[微] spirochaeta

骡 mule

【骡马店】an inn with sheds for carts and animals

【骡子】mule

罗 ①(捕鸟的网) a net for catching birds: ～网 net; trap ②(张网捕) catch birds with a net: 门可～雀 you can catch sparrows on the doorstep; visitors are few and far between ③(搜集) collect; gather together ④(陈列) display; spread out: 星～棋布 spread out like stars in the sky or chessmen on the chessboard ⑤(细筛) sieve; sift: ～面 sift flour ⑥(丝织品) a kind of silk gauze: ～扇 silk gauze fan ⑦[量] twelve dozen; a gross

【罗布麻】[植] bluish dogbane

【罗锅】arched: ～桥 arch bridge

【罗锅儿】①(驼背) hunchbacked; humpbacked ②(驼背的人) hunchback; humpback

【罗汉】[佛教] *arhat*: 十八～ the eighteen disciples of Buddha

【罗汉松】[植] yew podocarpus

【罗经】[航海] compass: 磁～ magnetic compass / 电～ gyrocompass / 航海～ mariner's compass

【罗口】[纺] rib cuff or rib collar; rib top

【罗口灯泡】screw socket bulb

【罗口灯头】screw socket

【罗列】①(分布;陈列) spread out; set out: ～珍品 set out precious things in order ②(列举) enumerate: ～事实 enumerate the facts / 光是～现象 merely list the phenomena

【罗马教皇】the Pope

【罗马教会】the Roman Catholic Church

【罗马尼亚】Romania ◇ ～人 Romanian / ～语 Romanian (language)

【罗马数字】Roman numerals

【罗马语族】[语] the Romance group of languages; Romance languages

【罗盘】compass

【罗圈腿】bowlegs; bandy legs

【罗裙】skirt of thin silk

【罗网】net; trap: 布下～ spread a net / 陷入～ fall into a trap / 自投～ walk right into the trap

【罗帷】a gauze curtain

【罗纹机】rib knitting machine; ribber

【罗音】[医] rale

【罗帐】curtain of thin silk

【罗织】frame up: ～成狱 frame up a crime / ～诬陷 frame sb. up / ～罪名 cook up charges

【罗致】enlist the services of; secure sb. in one's employment; collect; gather together: ～人材 enlist the services of talents

逻 patrol: 巡～ patrol

【逻辑】logic: 数理～ mathematical logic / 形式～ formal logic / ～上的错误 an error in logic / 按照这种～ according to that kind of reasoning / 合乎～ logical ◇ ～电路 logical circuit / ～解释 logical interpretation / ～思维 logical thinking / ～推理 logical reasoning / ～性 logicality / ～学 logic / ～学家 logician / ～哲学 philosophy of logic / ～主语[语] logical subject

萝 trailing plants: 茑～ cypress vine / 藤～ Chinese wistaria

【萝卜】radish: ～干 dried radish

【萝芙木】[植] devilpepper

箩 a square-bottomed bamboo basket

【箩筐】a large bamboo basket; a large wicker basket

锣 gong: 打铜～ beat a brass gong / 敲～ beat a gong

【锣槌】hammer

【锣鼓】①(锣和鼓) gong and drum: ～喧天 a deafening sound of gongs and drums ②(传统的打击乐器) traditional percussion instruments

luǒ

瘰

【瘰疬】[医] scrofula; struma

裸 bare; naked; exposed: 赤～～ stark-naked; undisguised

【裸鲤】naked carp

【裸露】uncovered; exposed: ～的电线 exposed electric wire / ～的煤层 exposed coal seam

【裸麦】[植] naked barley; highland barley

【裸袒】bare; naked

【裸体】naked; nude: ～画 a painting of a nude / ～像 a nude figure; a nude statue / 许多名画家都画过～模特儿。 Many famous painters have painted nude models.

【裸线】[电] bare wire

【裸装货】nude cargo

【裸子植物】[植] gymnosperm

luò

荦

【荦荦】[书] conspicuous; apparent; obvious: ～大端 salient points / ～大者 the most essential points

摞 ①(堆积) pile up; stack up: 把砖～起来 stack up the bricks ②[量] pile; stack: 一～碟子 a pile of plates / 一～书 a stack of books / 一～砖 a pile of bricks

洛

【洛氏硬度】[物] Rockwell hardness

【洛阳纸贵】become a bestseller

落 ①(掉下) fall; drop: 西红柿价格～了。The price of tomatoes has dropped. ②(下降) go down; set: 潮水～了。 The tide is low. ③(使下降) lower: ～帆 lower a sail / 把帘子～下来 lower the blinds ④(衰败) decline; come down; sink: ～到这步田地 come to such a pass / 没～ be on the downgrade / 衰～ decline; go downhill ⑤(遗留在后边) lag behind; fall behind ⑥(留下) leave behind; stay behind: 不～痕迹 leave no trace / ～一个好名声 leave a good name ⑦(停留的地方) whereabouts: 下～ whereabouts ⑧(聚居的地方) settlement: 村～ a small village; hamlet ⑨(归属) fall onto; rest with: 这个任务就～在我们肩上了。The task rested squarely on our shoulders. ⑩(得到) get; have; receive: ～褒贬 be criticized; lay oneself open to censure

【落榜】flunk a competitive examination for a job or school admission

【落笔】start to write or draw; put pen to paper: ～成文 put pen to paper and finish an essay

【落泊】【落魄】be in dire straits; be down and out

【落草】take to the greenwood; take to the heather; become an outlaw: ～为寇 turning to banditry

【落差】[水] ①(水位的差数) drop ②(蓄水高度) head

【落潮】ebb tide

【落成】completion ◇ ～典礼 inauguration ceremony

【落锤】[机] drop hammer

【落得】get; end in: ～个不是 come in for censure / ～一场空 come to nothing; end up in smoke / 将会～一个可悲的结局 head for a miserable end

【落地】①(落在地上) fall to the ground: 人头～ be killed or beheaded ②(婴儿出生) be born: 呱呱～ come into the world with a cry; be born ◇ ～窗 French window / ～灯 floor lamp; standard lamp / ～电扇 electric fan with adjustable stand / ～式收音机 console (radio) set

【落地生根】[植] air plant; life plant

【落点】[体] placement (of a ball): ～准 accuracy in placement ②[军] (火炮的落点) point of fall

【落果】[农] premature drop
【落红狼藉】falling flowers scattered all about
【落后】①(落在后面) fall behind; lag behind: ～于时代 behind the times ②(不先进) backward: 技术～ backward in technique ◇ ～地区 backward areas; less developed areas / ～分子 backward element
【落户】settle: 安家～ make a place one's home and settle down there / 在农场～ settle in a farm
【落花流水】like fallen flowers carried away by the flowing water; utterly routed: 把敌人打得～ put the enemy to rout
【落花有意，流水无情】shedding petals, the waterside flower pines for love, while the heartless brook babbles on; unrequited love
【落花生】[植] peanut; groundnut
【落荒而逃】take to the wilds; be defeated and flee the battlefield; take to flight
【落价】fall in price; drop in price
【落脚】stay; stop over; put up: 在客店～ put up at an inn / 在朋友家临时～ stay with friends for the time being ◇ ～处 temporary lodging
【落井下石】drop stones on someone who has fallen into a well; hit a person when he's down
【落空】come to nothing; fail; fall through: 两头～ fall between two stools / 希望～ fail to attain one's hope
【落款】write the names of the sender and the recipient on a painting, gift or letter; inscribe
【落泪】shed tears; weep
【落铃】[农] shedding of cotton bolls
【落落大方】natural and graceful
【落落寡合】standoffish; unsociable; aloof
【落难】meet with misfortune; be in distress
【落魄潦倒】fall on evil days
【落日】setting sun: ～余辉 light of the setting sun
【落入圈套】fall into a snare
【落入陷井】fall into a trap; be caught in a pitfall; be ensnared
【落纱机】[纺] doffer
【落实】①(切合实际) practicable; workable: 采取～的措施 adopt workable measures ②(确定) fix in advance; ascertain; make sure ③(实现) carry out; fulfill; implement; put into effect: 认真～党的政策 earnestly implement the policies of the Party
【落水】①(掉在水里) fall into water ②(堕落) go astray; degenerate: 拖人～ lead sb. astray
【落水洞】sinkhole
【落水狗】dog in the water: 痛打～ flog the cur that's fallen into the water; be merciless with bad people even if they're down
【落汤鸡】like a drenched chicken; like a drowned rat; soaked through; drenched and bedraggled
【落体】[物] falling body: 自由～ freely falling body
【落拓不羁】unconventional and uninhibited
【落网】fall into the net; be caught; be captured
【落伍】fall behind the ranks; straggle; drop behind; drop out
【落选】fail to be chosen; lose an election
【落叶】①(落下的树叶) fallen leaves ②[植](每年落叶的) deciduous leaf ◇ ～林 deciduous forest / ～树 deciduous tree / ～松 larch
【落叶归根】fallen leaves return to the roots
【落英缤纷】fallen petals lies in profusion

骆

【骆驼】camel: 单峰～ dromedary; one-humped camel / 双峰～ Bactrian camel; two-humped camel
【骆驼刺】[植] camel thorn
【骆驼队】camel train; caravan
【骆驼毛】camel hair
【骆驼绒】camel hair cloth

络

络 ①(网状物) sth. resembling a net: 橘～ tangerine pith / 丝瓜～ loofah ②(用网状物兜住) hold sth. in place with a net ③(缠绕) twine; wind: ～纱 winding yarn; spooling
【络合】[化] complexing ◇ ～物 complex compound
【络离子】[化] complex ion
【络腮胡子】whiskers; full beard
【络筒机】[纺] (high speed) cone winder; winding machine; winder
【络盐】[化] complex salt
【络绎不绝】in an endless stream

M

mā

摩

【摩挲】gently stroke

抹 ①(擦) wipe: ～一把脸 wipe one's face / ～桌子 wipe a table clean ②(用手按着向下移动) rub sth. down; slip sth. off: 把帽子～下来 slip one's cap off

【抹布】rag: 用～擦桌椅 wipe tables and chairs with a rag

【抹脸】straighten one's face; put on a stern expression: 抹不下脸来 find it difficult to be strict with sb. / 抹下脸来 show anger or displeasure suddenly

妈 [口](母亲) ma; mum; mummy; mother ②(称呼长一辈已婚妇女) aunt: 大～ aunt / 姑～ aunt / 舅～ aunt / 姨～ aunt

【妈妈】[口] ma; mum; mummy; mother

má

麻 ①(麻类植物) fiber crops: 大～ hemp / 黄～ jute / 剑～ sisal-hemp / 罗布～ bluish dogbane / 亚～ flax / 苎～ ramie; ramee ②(芝麻) sesame: ～糖 sesame candy ③(不光滑) rough; coarse ④(麻子) pocked; pockmarked; pitted; spotty: ～脸 a pockmarked face ⑤(麻木) have pins and needles; tingle: 腿发～ have pins and needles in one's legs ⑥(麻醉) anesthesia: 药～ drug anesthesia / 针～ acupuncture anesthesia

【麻包】gunny-bag; gunnysack; sack

【麻痹】①[医] paralysis: 面部神经～ facial paralysis / 全身～ general paralysis / 小儿～ infantile paralysis; poliomyelitis; polio ②(使丧失警惕) benumb; lull; blunt: ～人们的斗志 lull people's fighting will ③(疏忽) lower one's guard; slacken one's vigilance: ～大意 lower one's guard and become careless; be off one's guard

【麻布】①(粗麻布) gunny (cloth); sackcloth; burlap; hessian ②(夏布) linen

【麻袋】gunny-bag; gunnysack; sack ◇ ～片 a piece of gunnysacking

【麻刀】[建] hemp; hair: ～灰泥 hemp-fibered plaster

【麻烦】①(费事) troublesome; inconvenient: 自找～ ask for trouble / 这事要是太～，你就别管了。 Don't bother if it's too much trouble. ②(使费事) put sb. to trouble; trouble sb.; bother sb.: 对不起，～你了。 I am sorry to trouble you so much. (或: Sorry to have put you to so much trouble.)

【麻纺厂】flax mill

【麻风】[医] leprosy ◇ ～病 lepra; leprosy / ～病人 leper

【麻花】fried dough twist

【麻花钻】[机] twist drill

【麻黄】[植] Chinese ephedra ◇ ～碱[药] ephedrine

【麻将】mahjong: 打～ play mahjong ◇ ～牌 mahjong pieces; mahjong tiles

【麻酱】sesame paste

【麻胶版画】linonut

【麻利】quick and neat; dexterous; deft: 干活～ work dexterously; be a quick and neat worker / 干事～的人 quick and neat worker / 做事～ smart at work

【麻木】①(失去知觉) numb: 两腿～得一时站不起身来 one's legs are too numb to get up at once ②(麻木的感觉) apathetic; insensitive

【麻木不仁】apathetic; insensitive; unfeeling

【麻雀】sparrow

【麻雀虽小，五脏俱全】the sparrow may be small but it has all the vital organs; small but complete

【麻纱】①(麻纤维) yarn of ramie, flax, etc. ②(棉麻织品) cambric; hair cords

【麻绳】rope made of hemp, flax, jute, etc.

【麻线】flaxen thread; linen thread

【麻药】anesthetic

【麻油】sesame oil

【麻疹】[医] measles

【麻织品】linen fabrics

【麻子】①(天花疤痕) pockmarks ②(指有天花疤痕的人) a person with a pockmarked face

【麻醉】①[医] anesthesia; narcosis: 低温～ hypothermic anesthesia / 脊髓～ spinal anesthesia / 局部～ local anesthesia / 全身～ general anesthesia / 药物～ drug anesthesia / 针刺～ acupuncture anesthesia / 中药～ herbal anesthesia ②(使人认识模糊) anesthetize; poison: ～自己 hypnotize oneself; give oneself over to / 用诲淫诲盗的电影～青年人 poison young people with films full of sex and violence ◇ ～剂 anesthetic; narcotic / ～师 anesthetist / ～药 narcotic; anesthetics / ～针

anesthetic needle
【麻醉品】narcotic; drug: ～管制 narcotic control

mǎ

马 ①（哺乳动物一种）horse: 纯种～ purebred; blood-stock／母～ mare／小～ pony／种～ stallion; stud ②（中国象棋里的一个子）horse; one of the pieces in Chinese chess
【马鞍】saddle
【马鞍形】the shape of a saddle; a falling-off between two peak periods
【马帮】a train of horses carrying goods; caravan
【马鼻疽】[牧] glanders
【马鞭】horsewhip
【马表】stopwatch
【马鳖】leech
【马不停蹄】without a stop; nonstop
【马槽】manger
【马车】①（马拉的载人车）carriage ②（大车）cart
【马齿苋】[植] purslane
【马刺】spur
【马达】motor
【马达加斯加】Madagascar ◇～人 Madagascan
【马大哈】①（粗心大意）careless; forgetful ②（粗心大意的人）a careless person; scatterbrain
【马刀】saber
【马到成功】win success immediately upon arrival; gain an immediate victory; win instant success
【马灯】barn lantern; lantern
【马镫】stirrup
【马兜铃】[植] birthwort
【马肚带】belly band
【马队】①（成队的运货的马）a train of horses carrying goods; caravan ②（骑兵部队）a contingent of mounted troops; cavalry
【马尔代夫】Maldives ◇～人 Maldivian
【马尔加什语】Malagasy (language)
【马尔萨斯主义】Malthusianism: 新～ Neo-Malthusianism
【马耳他】Malta ◇～人 Maltese／～语 Maltese (language)
【马粪纸】strawboard
【马蜂】hornet; wasp
【马蜂窝】hornet's nest: 捅～ stir up a hornet's nest
【马革裹尸】be wrapped in a horse's hide after death; die on the battlefield
【马褂】mandarin jacket
【马海呢】[纺] mohair
【马号】①（马房）stable ②（号角）cavalry bugle
【马赫数】[物] Mach number
【马赫主义】[哲] Machism ◇～者 Machist
【马后炮】belated action or advice; belated effort: ～的声明 a belated statement
【马虎】careless; casual: 工作～ be careless in one's work
【马鲛鱼】Spanish mackerel
【马脚】sth. that gives the game away: 露出～ show the cloven hoof; give oneself away
【马厩】stable
【马驹子】[口] colt; foal; pony
【马克】①（德国货币）mark: 民主德国～ mark der D.D.R／西德～ Deutsche mark ②（芬兰货币）markka
【马克思列宁主义】Marxism-Leninism ◇～者 Marxist-Leninist
【马克思主义】Marxism ◇～哲学 Marxist philosophy／～者 Marxist／～政治经济学 Marxist political economy
【马口铁】tinplate; galvanized iron
【马裤】riding breeches ◇～呢 whipcord
【马拉犁】[农] horse-drawn plough
【马拉松】marathon ◇～赛跑 marathon race; marathon
【马拉维】Malawi ◇～人 Malawian
【马来人】Malay
【马来西亚】Malaysia ◇～人 Malaysian
【马来语】Malay (language)
【马蓝】[植] acanthaceous indigo
【马里】Mali ◇～人 Malian
【马力】[物] horsepower (h.p.): 开足～ at full speed; at full steam ◇～小时 horsepower-hour; hp-hr
【马列主义】[简]（马克思列宁主义）Marxism-Leninism
【马蔺】[植] Chinese small iris
【马铃薯】potato ◇～晚疫病 late blight of potato
【马鹿】[动] red deer
【马路】road; street; avenue
【马骡】[动] mule
【马马虎虎】①（不仔细）careless; casual ②（勉强可以）not so bad; just passable; so-so; fair
【马奶】mare's milk
【马棚】stable
【马匹】horses
【马钱子】[植] vomiting nut; nux vomica

【马前卒】pawn; cat's-paw
【马枪】carbine
【马球】[体] polo
【马赛克】[建] mosaic ◇ ～铺面 mosaic pavement
【马上】at once; immediately; straight away; right away: 你～就走吗? Are you leaving right away? / 她～就来。She'll come in a moment.
【马勺】ladle
【马首是瞻】take the head of the general's horse as guide; follow sb.'s lead
【马术】horsemanship
【马蹄】horse's hoof ◇ ～表 round or hoof-shaped desk clock; alarm clock / ～莲 common calla / ～声 hoofbeat; clatter of a horse's hoofs; clip-clop / ～形 the shape of a hoof; U-shaped
【马蹄铁】① (铁掌) horseshoe ② (马蹄形的磁铁) U-shaped magnet; horseshoe magnet
【马桶】nightstool; closestool; commode
【马头琴】[乐] *matouqin*, 4-stringed instrument of the Mongol nationality
【马尾松】[植] masson pine
【马戏】circus ◇ ～团 circus troupe
【马熊】brown bear
【马靴】riding boots
【马仰人翻】the rider falls as the horse rears in fright; there is a panic
【马缨丹】[植] lantana
【马蝇】horse botfly
【马扎】campstool; folding stool
【马掌】horseshoe
【马桩】hitching post
【马鬃】horse's mane

玛

【玛瑙】agate: ～念珠 agate beads

码

码 ① (数目符号) a sign or thing indicating number: 尺～ size / 价～ marked price / 页～ page number ② (堆叠) pile up; stack: ～砖 stack bricks ③ (指一件事或一类的事) 两～事 two different things / 一～事 the same thing ④ (英美制长度单位) yard (yd.)
【码头】① (停船的地方) wharf; dock; quay; pier: 储油～ oil wharf; petrol wharf / 浮～ pontoon ② [方] (商业城市) port city; commercial and transportation center: 跑～ travel from port to port as a trader; be a travelling merchant ◇ ～费 wharfage; dockage / ～工人 docker; stevedore; longshoreman / ～交货 ex wharf; ex pier; ex dock / ～交货价格 ex dock / ～税 pierage
【码子】① (数目符号) numeral: 苏州～ Suzhou numerals ② (筹码) counter; chip

吗

【吗啡】[药] morphine

蚂

【蚂蟥】[动] leech
【蚂蚁】ant
【蚂蚁搬泰山】ants can move Mount Taishan; the united efforts of the masses can accomplish mighty projects
【蚂蚁啃骨头】ants gnawing at a bone; a concentration of small machines on a big job; plod away at a big job bit by bit

mà

蚂

【蚂蚱】[方] locust

骂

骂 ① (咒骂) abuse; curse; swear; call names: ～不绝口 pour out a stream of abuse; curse unceasingly / 不～人 never swear at people ② (斥责) condemn; rebuke; reprove; scold: 把孩子～了一顿 give one's child a scolding / 小～大帮忙 scold a little but help a great deal
【骂街】shout abuses in the street; call people names in public: 泼妇～ like a shrew shouting abuses in the street
【骂骂咧咧】intersperse one's talk with curses; be foul-mouthed
【骂名】bad name; infamy: 留下千古～ earn oneself eternal infamy
【骂人话】abusive language; swearword; curse
【骂人取乐】criticize others as a pastime

mái

霾

霾 [气] haze
【霾层】[气] haze layer
【霾线】[气] haze line; haze level

埋

埋 cover up; bury: ～地雷 lay a mine / 把树根用土～起来 cover up the roots of the tree with earth
【埋藏】① (藏在土中) lie hidden in the earth; bury: ～的财物 hidden wealth ② (隐藏) hide; conceal: 把秘密～在心底 hide the secret deep in one's heart ◇ ～物 treasure trove
【埋伏】① (秘密布置兵力) ambush; lie in ambush: ～兵马 place soldiers in ambush / ～下

来 make an ambush; lay an ambush／中～ fall into an ambush ②(潜伏) hide; lie low: ～着失败的因素 bear the seed of defeat ◇ ～圈 ambuscade
【埋名】conceal one's name; live incognito
【埋没】①(埋起来) bury; cover up ②(使显不出来) neglect; stifle: ～人材 stifle real talents／～英雄 let a hero or genius lie unknown
【埋头】immerse oneself in; be engrossed in: ～读书 bury oneself in books／～苦干 quietly immerse oneself in hard work; quietly put one's shoulder to the wheel
【埋头铆钉】countersunk rivet
【埋怨】complain of; murmur at: ～他人 blame others
【埋葬】bury: 他昨天被～。 He was buried yesterday.
【埋葬虫】carrion beetle

mǎi

买 buy; purchase: ～不起 cannot afford／～得起 can afford／～东西 buy things; go shopping
【买办】comprador ◇ ～封建制度 comprador-feudal system／～资产阶级 comprador bourgeoisie
【买椟还珠】keep the glittering casket and give back the pearls to the seller; show lack of judgment
【买方】the buying party: ～负担风险 let the buyer beware ◇ ～市场 buyer's market／～样品 buyer's sample
【买好】try to win sb.'s favor; ingratiate oneself with; play up to
【买进】buy in
【买价】buying price
【买空卖空】fictitious transaction; cross trade
【买麻藤】[植] sweetberry jointfir
【买卖】①(生意) buying and selling; business; deal; transaction: ～公平 buying and selling at reasonable prices／做成一笔～ make a deal ②(商店) shop ◇ ～合同 contract note／～黄金者 gold trafficker／～婚姻 mercenary marriage／～人 businessman; trader; merchant／～人口 deal in human beings
【买入汇率】buying rate
【买通】bribe; buy over; buy off
【买一送一】buy one item with another similar item presented free
【买帐】acknowledge the superiority or seniority of; show respect for
【买主】buyer; customer ◇ ～垄断市场 monopsony

mài

麦 ①(麦类统称) a general term for wheat, barley, etc.: 春小～ spring wheat／大～ barley／黑～ rye／荞～ buckwheat／小～ wheat／燕～ oats／莜～ naked oats／割～ reap wheat／种～ raise wheat ②(专指小麦) wheat
【麦草】wheatgrass
【麦茬】[农] wheat stubble ◇ ～白薯 sweet potatoes grown after the wheat harvest／～地 a field from which wheat has been reaped
【麦蛾】gelechiid (moth)
【麦尔登呢】[纺] melton
【麦饭豆羹】coarse meals of a farming family
【麦粉】flour
【麦麸】wheat bran
【麦秆虫】skeleton shrimp
【麦红吸浆虫】wheat midge
【麦角】[药] ergot ◇ ～病 ergot／～中毒[医] ergot poisoning; ergotism
【麦秸】wheat straw ◇ ～画 straw patchwork
【麦精】malt extract ◇ ～鱼肝油 cod-liver oil with malt extract
【麦克风】microphone; mike
【麦浪】rippling wheat; billowing wheat fields
【麦粒肿】[医] sty
【麦苗】wheat seedling
【麦片】oatmeal ◇ ～粥 oatmeal porridge; oatmeal
【麦秋】wheat harvest season ◇ ～假 wheat harvest vacation
【麦乳精】malted milk; extract of malt and milk
【麦收】wheat harvest
【麦穗】ear of wheat; wheat head
【麦芒】awn of wheat
【麦芽】malt ◇ ～酵素 maltase／～糖 malt sugar; maltose
【麦蚜】[动] wheat aphid
【麦子】wheat

卖 ①(拿东西换钱) sell: ～不出去 not sell well／～得快 sell well／票全～完了。 All tickets were sold out. ②(出卖) betray: 出～朋友 betray one's friend ③(尽量用出来) exert to the utmost; not spare: ～劲儿 exert all one's strength; spare no effort ④(故意表现出来) show off: ～乖 show off one's cleverness
【卖唱】sing for a living: ～过活 maintain life

by singing
【卖方】the selling party; seller: ～负担风险 let the seller beware◇ ～市场 seller's market / ～样品 seller's sample
【卖狗皮膏药】sell quack remedies; palm things off on people
【卖瓜的说瓜甜】every peddler praises his own needles
【卖乖】show off one's cleverness
【卖关子】stop a story at a climax to keep the listeners in suspense; keep people guessing
【卖国】betray one's country; turn traitor to one's country: ～求荣 seek power and wealth by betraying one's country; turn traitor for personal gain / ～通敌 sell the country and communicate with the enemy ◇ ～集团 traitorous clique / ～条约 traitorous treaty / ～行为 treasonable act / ～贼 traitor (to one's country) / ～主义 national betrayal
【卖好】curry favor with; ingratiate oneself with; play up to
【卖价】selling price
【卖劲儿】exert all one's strength; spare no effort
【卖空】short sales
【卖力】exert all one's strength; spare no effort; do all one can
【卖力气】①(尽量用力) exert all one's strength; exert oneself to the utmost; do one's very best ②(出卖劳动力) live by the sweat of one's brow; make a living by manual labor
【卖命】①(拼命干活) work oneself to the bone for sb. ②(送死) die unworthy for
【卖弄】show off; parade: ～小聪明 show off one's smartness / ～学问 show off one's learning; parade one's knowledge
【卖弄风情】flirt and coquet
【卖契】contract of sale
【卖俏】play the coquette; coquette; flirt
【卖身】①(卖自己或家人) sell oneself or a member of one's family ②(出卖灵魂或肉体) sell one's body; sell one's soul ◇～契 an indenture by which one sells oneself or a member of one's family
【卖身投靠】barter away one's honor for sb.'s patronage; basely offer to serve some reactionary bigwig
【卖笑生涯】make a living by prostitution
【卖艺】make a living as a performer: 在街头～ be a street-performer
【卖淫】prostitution
【卖友求荣】betray friends to obtain promotion
【卖主】bargainer; seller
【卖座】draw large audiences; attract large numbers of customers: 这部影片很～。The film draws well.

迈 ①(跨;向前走) step; stride: ～过钢轨 step over the rail / ～着矫健的步伐 walk with vigorous strides ②(老) advanced in years; old: 年～ aged / 在他老～之年 in his old age
【迈步】take a step; make a step; step forward: ～走向主席台 step up to the platform / 迈出第一步 make the first step
【迈进】stride forward; forge ahead; advance with big strides: ～一大步 take a big stride forward / 向着四个现代化的宏伟目标～ forge ahead toward the great goal of the four modernization

脉 ①(动脉和静脉) arteries and veins ②(脉搏) pulse: 号～ feel sb.'s pulse ③(象血管的东西) vein: 金矿～ vein of gold / 矿～ ore vein; mineral vein / 叶～ veins in a leaf
【脉搏】pulse: 时代的～ the pulse of present era / ～微弱 have a weak pulse / 数～ count sb.'s pulse ◇ ～计 sphygmometer
【脉冲】[物] pulse ◇ ～发生器 pulser / ～计数器 pulse counter / ～雷达 pulse radar / ～信号 pulse signal / ～星[天] pulsar
【脉动】[物][天] pulsation ◇ ～电流 pulsating current / ～式喷气发动机 pulse-jet engine / ～星 pulsating star
【脉管炎】[医] vasculitis
【脉络】①[中医](对静脉和动脉的统称) a general name for arteries and veins ②(条理) thread of thought; sequence of ideas: ～分明 clear line of thought
【脉弱】[医] weak pulse
【脉石】[矿] gangue; veinstone ◇ ～矿物 gangue mineral
【脉速】[医] rapid pulse
【脉息】pulse: ～微弱 have a weak pulse
【脉泽】[物] maser
【脉诊】[中医] diagnosis by feeling the pulse

mān

颟
【颟顸】muddleheaded and careless

mán

蛮 ①(野蛮) rough; fierce; reckless; unrea-

soning: ～劲 sheer animal strength / 野～ savage ②[方](很; 挺) quite; pretty: ～好 pretty good
【蛮不讲理】be impervious to reason; persist in being unreasonable
【蛮干】act rashly; act recklessly; be foolhardy
【蛮横】rude and unreasonable; arbitrary; peremptory: ～无理 be impervious to reason / ～无理的要求 peremptory demands

埋

【埋怨】complain; blame; grumble: 过多的～ a surfeit of complaints / ～他人 blame others

瞒 hide the truth from: ～不了人 can't deceive others / 别～我 don't hide it from me / 不～你说 to tell you the truth
【瞒哄】deceive; pull the wool over sb.'s eyes
【瞒上欺下】deceive those above and bully those below
【瞒天过海】cross the sea by a trick; practice deception

蔓

【蔓青】[植] turnip

鳗 eel
【鳗鲡】eel

馒

【馒头】steamed bun; steamed bread

mǎn

满 ①(全部充实) full; filled; packed: ～～一卡车煤 a full truckload of coal / ～头大汗 one's face streaming with sweat ②(使满) fill: 把杯子斟～啤酒 fill a glass with beer ③(达到一定期限) expire; reach the limit: 不～一年 in less than one year / 年～三十五岁的公民 critizen who have reached the age of 35 ④(全) completely; entirely; perfectly ⑤(满足) satisfied: 不～ dissatisfied; discontented ⑥(骄傲) complacent; conceited: 反骄破～ combat arrogance and complacence
【满不在乎】not worry at all; not care in the least; give no heed; take no heed: 对某人的劝告～ pay no heed to sb.'s advice
【满城风雨】the talk of the town: 闹得～ create a sensation; create a scandal
【满打满算】reckoning in every item; at the very most
【满额】fulfill the quota
【满而不溢】full without flowing over
【满分】full marks
【满腹】be full of; have one's mind filled with: ～狐疑 filled with suspicion; extremely suspicious / ～经纶 an encyclopedic mind / ～牢骚 full of grievances; full of resentment
【满怀】①(心中充满) have one's heart filled with; be imbued with: ～豪情 be filled with pride and enthusiasm / ～热情 be full of enthusiasm / ～深情 be imbued with deep love / ～信心 with full confidence ②(整个前胸部分) 撞了个～ bump right into sb. ③(母畜全部怀孕) (of sheep, cattle, etc.) all with young
【满口】speak profusely: ～称赞 praise unreservedly / ～答应 readily promise / ～胡说 talk nonsense
【满脸】all over the face: ～风尘 a faceful of traveling dust / ～晦气 look very depressed / ～雀斑 freckled face / ～俗气 look very vulgar / ～通红 face reddens all over / ～笑容 be all smiles / ～皱纹 full of wrinkles
【满满当当】[口] full to the brim: 挑着～的两桶水 carry two brimming buckets of water
【满门】the whole family
【满面】have one's face covered with: ～春风 beaming with satisfaction; radiant with happiness and kindness / ～红光 glowing with health / ～怒容 red with anger / ～笑容 grinning from ear to ear; be all smiles / ～忧容 full of sorrow on one's face / 泪流～ tears streaming down one's cheeks
【满目】meet the eye on every side: ～荒凉。A scene of desolation met the eye on every side.
【满目疮痍】misery and suffering greets the eyes everywhere
【满目萧索】a melancholy and solitary aspect as far as the eyes stretch
【满脑子】have one's mind stuffed with
【满期】expire; come to deadline ◇ ～通知书 expiration notice
【满腔】have one's bosom filled with: ～仇恨 burning with hatred / ～怒火 filled with rage / ～热忱 filled with ardor and sincerity / ～热血 full of patriotic fervor / ～同情 heart-felt sympathy
【满身】have one's body covered with; be covered all over with: ～是汗 sweat all over / ～油泥 covered all over with grime
【满师】finish serving one's time; serve out one's apprenticeship: 他尚未～。He is not yet out of his apprenticeship.

【满堂喝采】universal applause
【满堂红】all-round victory; success in every field
【满天】all over the sky: ～星斗 a star-studded sky
【满头大汗】head covered with big drops of perspiration
【满心】have one's heart filled with sth.: ～欢喜 filled with joy
【满眼】①(充满眼睛) have one's eyes filled with: ～红丝 with blood shot eyes ②(充满视野) meet the eyes on every side: ～的山花 mountain flowers greeting the eye everywhere
【满意】satisfied; pleased: ～之至 satisfied to the utmost degree / 她的答复令人十分～。Her answer is most satisfactory.
【满应满许】promise anything and everything
【满园春色】the garden is filled with the brightness of spring
【满员】①[军](达到规定人数) at full strength: 保证主力部队经常～ ensure that the main forces are always kept at full strength ②(乘客满了) all seats taken
【满月】①[天](望月) full moon ②(小孩满月) a baby's completion of its first month of life
【满载】loaded to capacity; fully loaded; laden with: 一辆～木材的卡车 a truck fully loaded with timber / 一列～煤炭的火车 a train laden with coal
【满载而归】come back with fruitful results; return from a rewarding journey
【满招损,谦受益】one loses by pride and gains by modesty
【满足】①(觉得足够) satisfied; content; contented: ～于现状 be satisfied with the existing state of affairs; be content with things as they are / 不～于已经取得的成绩 not rest content <satisfied>with one's achievements ②(使满足) satisfy; meet: ～读者的需要 meet the reader's demands / ～人民的生活需要 meet the needs of the people's everyday life / ～这里的急需 satisfy the urgent needs here
【满座】capacity audience; capacity house; full house: 场场～ have a capacity audience for every show

螨 [动] mite

màn

曼 ①(柔和) graceful: 轻歌～舞 soft music and graceful dances ②(长) prolonged; long-drawn-out: ～延 draw out
【曼多林】[乐] mandolin
【曼声】lengthened sounds: ～而歌 drawl out a song / ～吟诵 recite in slow, measured tones
【曼陀罗】[植] datura
【曼延】draw out (in length); stretch: ～曲折的羊肠小道 a winding footpath stretching into the distance

漫 ①(溢出) overflow; brim over; flood; inundate: 啤酒～出了杯子。Beer overflows the glass. ②(到处都是) all over the place; everywhere ③(自由;随便) free; casual; unrestrained: ～无目标 aimless; at random / ～无止境 know no bounds; be without limit
【漫笔】informal essay; literary notes
【漫不经心】careless; casual; negligent
【漫步】stroll; ramble; roam
【漫长】very long; endless: ～的海岸线 a long coastline / ～的岁月 long period of time; long years
【漫反射】[物] diffuse reflection
【漫灌】flood irrigation
【漫画】caricature; cartoon ◇ ～家 cartoonist; caricaturist
【漫江碧透】the whole stream was emerald green
【漫骂】abuse wildly; slander with
【漫漫】very long; boundless: ～长夜 endless night / 白雪～ a boundless expanse of snow
【漫山遍野】all over the mountains and plains; over hill and dale
【漫射】[物] diffusion ◇ ～光 diffused light / ～体 diffuser
【漫谈】random talk; informal discussion: ～英国文学 random talk on English literature / 组织一次～ organize an informal discussion
【漫天】①(布满天空) filling the whole sky; all over the sky: ～大雾 a dense fog obscuring the sky / ～大雪 whirling snow ②(无限度的) boundless; limitless: ～大谎 a monstrous lie / ～要价 ask an exorbitant price
【漫无边际】①(无边际的) boundless; limitless ②(离题甚远) straying far from the subject; rambling; discursive
【漫游】go on a pleasure trip; roam; wander: ～镜泊湖 go boating on or roam around the Jingbo Lake / ～世界 roam about the world

慢 ①(低速度) slow: ～下来 slow down / 把汽车开～ slow down a car / 反应～ be slow to

react / 行动～ slow to act ②(从缓) postpone; defer ③(态度冷淡) supercilious; rude: 傲～ arrogant; haughty / 言词骄～ use arrogant language
【慢车】slow train: 乘～ take the slow train ◇ ～道 slow-traffic lanes
【慢待】treat rudely or discourteously
【慢动作】slow motion
【慢工出巧匠】fine products come from slow work
【慢工出细活】slow work yields fine products
【慢镜头】[电影] slow motion
【慢慢】①(速度慢) slowly; leisurely: ～讲 speak slowly ②(逐渐地) gradually: ～积累 accumulate gradually
【慢坡】gentle slope
【慢手慢脚】slow in doing things; slow moving
【慢说】let alone; to say nothing of: ～是你, 连他都不会。 Even he can't do it, to say nothing of you.
【慢速摄影】memomotion
【慢腾腾】at a leisurely pace; unhurriedly; sluggishly: ～地念 read unhurriedly
【慢条斯理】leisurely; unhurriedly
【慢吞吞】irritatingly slow; exasperatingly slow: ～地走 slow-poke
【慢性】①(发作得缓慢) slow in taking effect: ～毒药 slow poison ②(拖得长久) chronic: ～病 chronic disease / ～阑尾炎 chronic appendicitis / ～中毒 chronic poisoning
【慢性子】①(性情迟缓) phlegmatic temperament ②(性情迟缓的人) slow poke; slow coach
【慢悠悠】unhurriedly; without haste
【慢中子】[物] slow neutron; low-speed neutron
【慢走】①(不要着急走) don't go yet; stay; wait a minute ②[套] (用于送别时) good-bye; take care

谩 disrespectful; rude
【谩骂】hurl invectives; fling abuses; rail

蔓
【蔓草】creeping weed
【蔓生植物】trailing plant
【蔓延】creep; spread: 火势～很快。 The fire spread quickly.

幔 curtain; screen: 布～ cotton curtain
【幔帐】curtain; screen; canopy

镘 trowel
【镘刀】trowel
【镘板】patter

máng

忙 ①(事情多) busy; fully occupied: ～得不可开交 be swamped with work / ～于工作 be busy with ⟨at⟩ one's work / ～于准备考试 be busy preparing for the examination ②(匆忙) hurry; hasten; make haste: ～从里屋出来 come hurrying out of the inner room; hasten out of the inner room
【忙合】[口] be busy; bustle about
【忙里偷闲】snatch a little leisure from a busy life
【忙碌】be busy; bustle about: 一天到晚忙忙碌碌 be busy all day long
【忙乱】be in a rush and a muddle; tackle a job in a hasty and disorderly manner
【忙人】busy person: 他是个～。 He is a busy man.
【忙中有错】haste makes waste

芒 awn; beard; arista
【芒刺在背】feel prickles down one's back; feel nervous and uneasy
【芒果】[植] mango
【芒硝】[化] mirabilite; Glauber's salt
【芒种】Grain in Ear

茫 ①(形容水或其他事物没有边际) boundless and indistinct ②(无所知) ignorant; in the dark
【茫茫】boundless and indistinct; vast: ～草原 the boundless grasslands / ～大海 a vast sea
【茫然】ignorant; in the dark; at a loss: ～不知所措 be at a loss what to do; be at sea / ～无知 be utterly ignorant; be in the dark
【茫无头绪】be confused like a tangle of flax; not know where to begin

盲 blind: 偏～ hemianopsia / 色～ color-blindness / 夜～ night blindness
【盲肠】[生理] cecum ◇ ～炎 appendicitis; cecitis
【盲椿象】[动] plant bug
【盲从】follow blindly
【盲点】[生理] blind spot; scotoma
【盲动】act blindly; act rashly ◇ ～主义 putschism
【盲鳗】hagfish
【盲目】blind: ～抄袭 blind imitation / ～崇拜 worship blindly / ～乐观 be unrealistically op-

timistic／～追随(某人) follow sb. blindly ◇～飞行 blind flight; instrument flying／～轰炸 blind bombing／～着陆 blind landing
【盲目性】blindness: 克服～ curb blindness in action; refrain from blindfold action
【盲区】[无] blind area
【盲人】blind person
【盲人摸象】like the blind men trying to size up the elephant; take a part for the whole
【盲人瞎马】a blind man on a blind horse; rushing headlong to disaster
【盲人院】blind asylum
【盲文】braille
【盲哑教育】education for the blind and the deaf-mute
【盲障】[军] blindage

mǎng

莽 ①(密生的草) rank grass ②(鲁莽) rash
【莽苍】①(指原野) open country ②(景色迷茫) misty; hazy: 烟雨～ a vast blur of mist and rain
【莽汉】a boorish fellow; a boor
【莽莽】①(草木茂盛) luxuriant; rank ②(无边无际) vast; boundless
【莽原】wilderness overgrown with grass
【莽撞】crude and impetuous; rash: ～的小伙子 a young harum-scarum

蟒 boa; python
【蟒蛇】boa; python

māo

猫 cat: 波斯～ Persian cat／小～ kitten／雄～ tomcat／～叫 mewing; purring
【猫哭老鼠】the cat weeping over the dead mouse; shed crocodile tears
【猫头鹰】owl
【猫熊】panda; giant panda
【猫眼】[电] cat eye
【猫眼石】[矿] cat's eye

máo

毛 ①(皮上的毛) hair; feather; down: 桃子上的～ the down of a peach／羽～ feather／他胸部有～。 He has hair on his breast. ②(羊毛) wool: ～裤 long woolen underwear／～毯 woolen blanket／～袜 woolen stockings ③(霉) mildew: 长～ become mildewed; be covered with mildew ④(粗糙的；未加工的) semifinished: ～坯 semifinished product ⑤(不纯净的) gross: ～利 gross profit／～重 gross weight ⑥(小) little; small: ～孩子 a small child ⑦(粗心) careless; crude; rash: ～头～脑 rash; impetuous ⑧(惊慌) panicky; scared; flurried: 吓～了 be in a flurry of alarm／心里直发～ feel scared; be panic-stricken ⑨[口](货币贬值) depreciate; be no longer worth its face value ⑩[口](角;10分) *mao*(=10 *fen*)
【毛白杨】[植] Chinese white poplar
【毛笔】writing brush ◇～画 brush drawing
【毛边纸】writing paper made from bamboo
【毛病】①(疾病) illness; disease: 老～ inveterate disease ②(故障) trouble: 这台机器有点～。The machine does not work properly. ③(缺点) shortcoming; defect: 犯主观主义的～ commit the error of subjectivism／思想上的～ ideological malady／作文中的～ faults in a composition
【毛玻璃】frosted glass
【毛糙】crude; coarse; careless
【毛虫】caterpillar
【毛刺】[机] burr
【毛地黄】[药] digitalis
【毛豆】young soya bean
【毛发】hair ◇～鉴定 identification by hair
【毛纺】wool spinning: 粗梳～ woolen spinning／精梳～ worsted spinning ◇～厂 woolen mill
【毛茛】[植] buttercup
【毛骨悚然】with one's hair standing on end; absolutely terrified: 令人～ send cold shivers down one's spine; make sb.'s hair stand on end; be bloodcurdling
【毛巾】towel ◇～被 towelling coverlet／～布 towelling／～架 towel rail or rack
【毛孔】[生理] pore
【毛蓝】darkish blue ◇～土布 dyed nankeen
【毛里求斯】Mauritius ◇～人 Mauritian
【毛里塔尼亚】Mauritania ◇～人 Mauritanian
【毛利】gross profit ◇～率 rate of margin; rate of gross profit
【毛料】woolen cloth; woolens
【毛驴】donkey
【毛毛雨】drizzle
【毛囊炎】folliculitis
【毛坯】①(砖瓦等的半成品) semifinished product ②[机](坯料) blank
【毛皮】fur; pelt ◇～兽 fur-bearing animal
【毛票】[口] banknotes of one, two or five *jiao* denominations
【毛渠】sublateral canal; sublateral

【毛茸茸】hairy; downy
【毛瑟枪】Mauser
【毛纱】[纺] wool yarn: 粗纺～ woolen yarn／精纺～ worsted yarn
【毛石】[建] rubble ◇～混凝土 rubble concrete
【毛收入】gross income
【毛手毛脚】careless
【毛丝】[纺] broken filament
【毛遂自荐】offer one's service as Mao Sui did; volunteer one's services
【毛损】gross loss
【毛笋】the shoot of *mao* bamboo
【毛毯】woolen blanket
【毛桃】wild peach
【毛细管】capillary ◇～水[农] capillary water
【毛细现象】[物] capillarity
【毛细血管】[生理] blood capillary
【毛虾】shrimp
【毛线】knitting wool ◇～针 knitting needle
【毛象】[古生物] mammoth
【毛丫头】a little girl
【毛样】[印] galley proof
【毛衣】woolen sweater; sweater; woolly
【毛蚴】[动] miracidium
【毛躁】①(急躁) short-tempered; irritable ②(不细心) rush and careless
【毛泽东思想】Mao Zedong Thought
【毛毡】felt
【毛织品】①(毛料) wool fabric; woolens ②(编织物) woolen knitwear
【毛竹】*mao* bamboo

牦

【牦牛】yak

锚

anchor: ～链 anchor chain／抛～ drop anchor; cast anchor／起～ weigh anchor
【锚地】anchorage
【锚链】anchor chain; anchor cable ◇～孔 hawsehole
【锚雷】[军] mooring mine; moored buoyant mine
【锚爪】fluke

猫

【猫腰】[方] arch one's back

矛

lance; pike; spear
【矛盾】①(自相抵触) contradictory: ～百出 full of contradictions／～上交 pass on problems to a higher level instead of solving them oneself／相互～的因素 contradictory elements／自相～ self-contradictory／自相～的推论 inconsequent reasoning ②[哲][逻] contradiction: ～的次要方面 secondary aspect of a contradiction／～的斗争性 the struggle of opposite／～的普遍性 the universality of contradiction／～的特殊性 the particularity of contradiction／～的同一性 the identity of opposites／～的主要方面 the principal aspect of a contradiction／～的转化 the transformation of a contradiction／对抗性～ antagonistic contradiction／非对抗性～ nonantagonistic contradiction／非主要～ nonprincipal contradiction／主要～ principal contradiction
【矛盾律】[逻] the law of contradiction
【矛头】spearhead: ～所向 the target of attack／～针对 be spearheaded at／～指向某人 direct the spearhead at sb.

茅

[植] cogongrass
【茅草】[植] cogongrass ◇～棚 thatched shed; thatched shack
【茅房】[口] latrine
【茅膏菜】[植] sundew
【茅坑】①[口](厕所里的粪坑) latrine pit ②[方](厕所) latrine
【茅庐】thatched cottage
【茅塞顿开】suddenly see the light
【茅舍】[书] thatched cottage
【茅厕】[口] latrine
【茅屋】thatched cottage

蟊

an insect destructive of the roots of seedlings
【蟊贼】a person harmful to the country and people; pest

mǎo

卯

【卯劲儿】[口] make a sudden all-out effort
【卯时】the period of the day from 5 a.m. to 7 a.m.
【卯榫】mortise and tenon
【卯眼】mortise

铆

[机] riveting: 搭接～ lap riveting／对接～ butt riveting／风动～ pneumatic riveting
【铆钉】rivet ◇～距 rivet pitch／～枪 riveting gun
【铆工】①(指铆接的操作) riveting ②(指操作的人) riveter
【铆机】riveter: 风动～ pneumatic riveter／水力～ hydraulic riveter

【铆接】riveting; rivet joint

mào

茂 ①(茂盛) luxuriant; exuberant; profuse: 根深叶～ deep roots and exuberant foliage ②(丰富精美) rich and splendid: 图文并～ be excellent in both pictures and literary compositions ③[化] cyclopentadiene

【茂林修竹】thick forest and tall bamboos

【茂密】dense; thick: ～的森林 a dense forest / 叶子～的树 trees thick with leaves

【茂盛】luxuriant; exuberant; flourishing: 草木～ luxuriant vegetation / 园中花木～ garden flourishing with flowers

冒 ①(向外透;往上升) emit; send out; give off: ～泡 send up bubbles; be bubbling / ～气 give off steam; be steaming / 眼前～金星 see stars / 伤口还在～血。Blood was still oozing from the wound. ②(不顾) risk; brave: ～着风雨 brave the storm / ～着风浪出海 put to sea in spite of wind and wave; venture out on a stormy sea / ～着生命危险救人 risk one's life to save another ③(冒失;冒昧) boldly; rashly: ～猜一下 make a bold guess; venture a guess ④(冒充) falsely; fraudulently: ～领 falsely claim as one's own

【冒充】pretend to be; pass sb. or sth. off as: ～内行 pretend to be an expert; pose as an expert ◇ ～者 jactitator

【冒顶】[矿] roof fall: 大～ bulk caving / 工作面～ face fall

【冒犯】offend; affront: ～禁令 violate a prohibition / ～尊严 violate the sanctity

【冒风险】run risks

【冒号】[语] colon

【冒火】burn with anger; get angry; flare up: 他心里直～。He was burning with anger in his heart.

【冒尖儿】①(装得高出容器) piled high above the brim ②(超过一点) a little over; a little more than: 十公斤刚～ a little over ten kilos ③(突出) stand out; be conspicuous: ～的人物 a conspicuous figure / 怕～ be afraid of becoming too conspicuous ④(露出苗头) begin to crop up

【冒进】premature advance; rash advance

【冒口】[机] rising head; riser

【冒领】falsely claim as one's own: 虚报～ fraudulent applications and claims

【冒昧】[谦] make bold; venture; take the liberty: ～陈辞 make bold to express my views; venture an opinion / 不揣～ may I take the liberty to; I venture to

【冒名】go under sb. else's name; assume another's name: ～顶替 take another's place by assuming his name ◇ ～顶替者 imposter / ～者 personator

【冒牌】a counterfeit of a well-known trade mark; imitation; fake ◇ ～货 imitation; fake / ～医生 fake doctor

【冒然闯入】butt in

【冒失】rash; abrupt: ～的行为 rash action / ～从事 act abruptly / 说话～ speak without due consideration ◇ ～鬼 harumscarum

【冒天下之大不韪】defy world opinion; risk universal condemnation

【冒头】begin to crop up

【冒险】take a risk; run a risk: 军事～ military adventure / 不要～去做某事 not run risks to do sth. ◇ ～家 adventurer / ～借款 respondentia / ～政策 adventurist policy

【冒险主义】adventurism ◇ ～者 adventurist

【冒用商标】infringement

帽 ①(帽子) headgear; hat; cap: 安全～ safety helmet / 贝雷～ beret / 便～ cap; sports cap / 草～ straw hat / 貂皮～ mink hat / 军～ service cap / 呢～ woolen cap 女～ bonnet / 皮～ fur cap / 绒线～ woolen hat / 毡～ felt hat / 毡礼～ trilby hat ②(帽状物) cap-like cover for sth.: 笔～儿 the cap of a pen / 螺钉～ screw cap

【帽带】hatband

【帽顶】crown

【帽徽】insignia on a cap

【帽盔儿】skullcap

【帽舌】peak (of a cap); visor

【帽檐】the brim of a hat

【帽子】①(头上用品) headgear; hat; cap ②(罪名或名义) label; tag; brand: 扣～ put a label on sb.; hurl an epithet at sb. / 摘～ rid of the label; remove the label

耄 ①(八九十岁的年纪) octogenarian ②(老年) advanced in years

【耄年犹勤】aged yet industrious

貌 looks; appearance: 古城新～ the new look of an old city / 美～ good looks

【貌不惊人】look mediocre

【貌合神离】seemingly in harmony but actually at variance; be friendly in name only

【貌似】seemingly; in appearance: ～公正 seemingly impartial / ～强大 seemingly powerful; outwardly strong

【貌形古怪】feature is strange

贸 trade: 外～ foreign trade

【贸然】rashly; hastily; without careful consideration: ～说出 speak out precipitately / ～下结论 draw a hasty conclusion; jump to a conclusion

【贸易】trade: 补偿～ compensation trade / 对外～ foreign trade / 国际～ international trade / 国内～ domestic trade / 集市～ village fair / 沿海～ cabotage; coasting trade / 直接～ direct trade / 和别国进行～ trade with foreign countries; do business with other countries / 建立经常～关系 establish regular trade relations ◇ ～差额 balance of trade / ～额 volume of trade; turnover / ～法 law of trade / ～港 a trading port; a commercial port / ～公司 a trading company; a trading firm / ～货栈 trade warehouse / ～禁运 commercial embargo / ～逆差 unfavorable balance of trade / ～顺差 favorable balance of trade / ～条件 term of trade / ～条约 commercial treaty / ～协定 trade agreement / ～议定书 trade protocol / ～限制 restriction of trade / ～中心 trade center / ～总额 total volume of trade / ～自由 freedom of trade ○ 对外经济贸易部 Ministry of Foreign Economic Relations and Trade / 成交 strike a bargain; conclude a transaction / 交易额 volume of trade / 出口 export / 进口 import / 倾销 dumping / 成交确认书 sales confirmation / 客户 client; customer / 合同 contract / 互惠 reciprocity / 最惠国待遇 most-favored nation treatment / 发盘 make an offer / 还盘 counter offer / 入超 excess of imports; import surplus / 出超 excess of exports; export surplus / 合作生产 joint production / 合资经营 joint venture / 返销 buy-back / 互购 counter-purchase / 来料加工 processing of investor's raw materials; processing raw materials on client's demands / 来样加工 processing according to investor's samples / 估价 appraise; evaluate / 运价 freight rate / 付款方式 type of payment / 付讫 paid; pd. / 收讫 fully paid (F.P.); received (Rd) / 佣金 commission / 折扣 discount / 回扣 rebate / 交货 delivery / 提货 take delivery of goods / 报关 clearance of goods / 提货单 bill of lading / 理赔 settle a claim / 索赔 claims / 滞纳金 fine for delayed payment / 滞期费 demurrage charges

【贸易风】[气] trade wind

méi

没

【没出息】not promising; good for nothing

【没词儿】①(没有说的) can find nothing to say ②(不知用什么词合适) be at a loss for words; be stuck for an answer

【没错儿】①(我确信) I'm quite sure; you can rest assured ②(不会错) can't go wrong

【没法子】can do nothing about it; can't help it

【没关系】it doesn't matter; it's nothing; that's all right; never mind

【没规矩】not observing proper rules or manners; improper; inappropria

【没见过世面】green and inexperienced

【没见识】inexperienced and ignorant; unlearned and provincial

【没精打采】listless; in low spirits; out of sorts; lackadaisical

【没良心】without conscience; unconscionable; ungrateful

【没…没…】①(用在两个同义词前面，强调没有) ～皮～脸 shamelessly / ～亲～故 without any relatives or friends / ～完～了 endless; without end / ～心～肺 inattentive / ～羞～臊 shameless; have no sense of shame ②(用在两个反义词前面，表示应区别而未区别) ～大～小 impolite impertinent; impudent / ～日～夜 day and night / ～早～晚 without regard to time of day

【没门儿】①(没有门路) have no access to sth.; have no means of doing sth. ②(表示不同意) no go; nothing doing

【没命】①(死亡) lose one's life; die ②(拼命地) recklessly; desperately; like mad; for all one's worth

【没谱儿】[方] be unsure; have no idea

【没趣】feel put out; feel snubbed: 自讨～ ask for a snub

【没什么】it doesn't matter; it's nothing; that's all right; never mind

【没事】①(闲着) have nothing to do; be free; be at leisure ②(没关系) it doesn't matter; it's nothing; that's all right; never mind

【没事找事】①(自找麻烦) ask for trouble ②(找碴) find fault with

【没头案子】a criminal case without a clue for lawenforcement officers to work on

【没头没脑】completely without clue
【没心眼儿】careless; frank
【没羞】unabashed; unblushing
【没有】①(无) not have; there is not: ～共产党就～新中国。 Without the Communist Party there would be no New China. ②(不及;不如) not so ... as: 你～他高。 You are not as tall as he. ③(不到) less than ④[副](未)我今早～见到他。 I didn't see him this morning.
【没有把握】not sure of; not confident
【没有出路】find oneself in a blind alley; without a way out
【没有结果】come to nothing; with no result
【没有解决的问题】open question; suspended problem
【没有什么了不起】not enough to; not so great; nothing to be impressed by
【没有说的】①(没有缺点) really good; excellent ②(不成问题) without question; needless to say; it goes without saying
【没有依靠】with no support; have no backing
【没辙】[方] can find no way out; be at the end of one's rope
【没主意】lose one's head; cannot make up one's mind

糜

【糜子】[植] broom corn millet

煤

煤 coal: 粉～ fine coal / 块～ lump coal / 褐～ lignite / 焦～ coking coal / 泥～ peat / 无烟～ anthracite / 烟～ bituminous coal / 原～ raw coal ○回采 stopping / 开拓 opening up / 露天采煤法 opencast coal mining / 采煤工作面 coal face / 掘进 tunnelling / 割进 undercutting / 回采率 rate of extraction / 放顶 blasting down the roof / 支柱 prop setting / 顶板 roof / 冒顶 roof fall; roof caving / 气化 gasification / 井下运煤 underground hauling of coal / 大巷运输 main road haulage / 联合采煤机 combine / 截煤机 cutter / 滚筒式截煤机 drum cutter / 装载机 loader / 刨煤机 coal plough / 水采水枪 hydraulic giant / 矿灯 miner's lamp
【煤仓】coal bunker
【煤藏】coal deposits; coal reserves
【煤层】coal seam; coal bed: 薄～ thin seam / 厚～ thick seam
【煤场】coal yard
【煤尘】coal dust ◇ ～爆炸 coal-dust explosion
【煤斗】coal scuttle; scuttle ◇ ～车 hopper car
【煤矸石】gangue
【煤耗】coal consumption
【煤核儿】partly-burnt briquet; coal cinder
【煤灰】coal ash
【煤焦油】coal tar
【煤精】jet; black amber
【煤坑】coalpit
【煤矿】coal mine; colliery ◇ ～工人 coal miner
【煤气】coal gas; gas ◇ ～厂 gasworks; gashouse / ～灯 gas lamp; gas light / ～管 gas pipe / ～管线 gas-pipe line / ～罐 gas tank / ～机 gas engine / ～炉 gas stove; gas furnace / ～设备 gas fittings / ～灶 gas range; gas cooker / ～中毒 carbon monoxide poisoning; gas poisoning
【煤球】briquet
【煤炭】coal ◇ ～工业 coal industry
【煤田】coal field ◇ ～地质学 coal geology
【煤系】[地] coal measures
【煤屑】nickings
【煤烟】①(烧煤产生的烟) smoke form burning coal ②(烟灰) soot ◇ ～污染 smoke pollution
【煤窑】coalpit
【煤油】kerosene; paraffin ◇ ～灯 kerosene lamp / ～炉 kerosene stove
【煤渣】coal cinder ◇ ～路 cinder road / ～跑道 cinder track
【煤砟子】small piece of coal
【煤砖】briquet

媒

媒 ①(媒人) matchmaker; go-between: 做～ act as a matchmaker ②(媒介) intermediary: 虫～ insect-pollination; entomophily / 风～ wind-pollination
【媒介】intermediary; medium; vehicle: ～之言 the words of a match-maker / 传染疾病的～ vehicle of disease; vector / 授粉～ pollination medium
【媒婆】[旧] woman matchmaker
【媒染】mordant dyeing ◇ ～剂[化] mordant / ～染料 mordant dye
【媒人】matchmaker; go-between
【媒质】[物] medium: 吸收～ absorbing medium

玫

【玫瑰】[植] rugosa rose; rose: ～多刺 no rose without an thorn

枚

枚 [量] 二十～古币 twenty ancient coin / 两～纪念章 two badges / 不胜～举 too numerous to enumerate

霉 mold; milew: 发～ go moldy; mildew
【霉病】[农] mildew
【霉菌】[微] mold ◇ ～病 mycosis
【霉烂】mildew and rot
【霉天】early summer rains

莓 certain kinds of berries: 草～ strawberry

梅 plum
【梅毒】[医] syphilis
【梅红色】plum color
【梅花】①(梅树的花) plum blossom ②[方](腊梅) wintersweet
【梅花鹿】sika; spotted deer
【梅林止渴】when one visits the orchard of plums, he never feels thirsty
【梅雨】[气] plum rains
【梅子】plum

酶 [生化] enzyme; ferment: 消化～ digestive ferment / 诱导～ induced enzyme
【酶原】[生化] zymogen; fermentogen

眉 ①(眉毛) eyebrow; brow: 浓～ bushy eyebrows ②(书眉) the top margin of a page
【眉笔】eyebrow pencil
【眉飞色舞】with dancing eyebrows and radiant face; enraptured; exultant
【眉睫】①(眉毛和眼睫毛) the eyebrows and eyelashes ②(比喻近在眼前) urgent; imminent: 迫在～ extremely urgent
【眉开眼笑】be all smiles; beam with joy
【眉来眼去】make eyes at each other; flirt with each other
【眉棱骨】superciliary ridge
【眉毛】eyebrow; brow
【眉毛胡子一把抓】try to grasp the eyebrows and the beard all at once; try to attend to big and small matters all at once
【眉目】①(面貌) features; looks: ～清秀 have delicate feature ②(条理) logic; sequence of ideas: ～不清 not well organized ③(纲要) essential ④(事情的头绪) sign of a positive outcome; prospect of a solution: 那件事渐渐有了～。 That matter is settling into shape.
【眉目传情】flash amorous glances; make eyes at sb.
【眉批】notes and commentary at the top of a page
【眉梢】the tip of the brow: 喜上～ look very happy
【眉头】brows: ～不展 with knitted brows / ～一皱,计上心来 knit the brows and a stratagem comes to mind / 皱～ knit the brows; frown
【眉心】between the eyebrows
【眉宇】[书] forehead

楣 lintel

镅 [化] americium (Am)

měi

美 ①(美丽) beautiful; pretty: 大自然的～ beauty of nature / 真、善、～ the true, the good and the beautiful ②(好) very satisfactory; good: ～酒 good wine / ～意 good will / 价廉物～ good and inexpensive / 日子过得挺～ live quite happily / 味道～ good taste ③(得意) be pleased with oneself: ～滋滋 self-satisfied; complacent
【美不胜收】so many beautiful things that one simply can't take them all in
【美差】cushy job
【美称】laudatory title; good name
【美德】virtue; moral excellence: ～无价 virtue is beyond price
【美吨】short ton
【美感】aesthetic feeling; aesthetic perception; sense of beauty
【美工】①(电影等的美术工作) art designing ②(担任美术工作的人) art designer
【美观】pleasing to the eye; beautiful to look at: ～大方 beautiful and dignified; elegant and in good taste
【美国】the United States of America (U.S.A.) ◇ ～人 American / ～生活方式 American way of life / ～之音 Voice of America (VOA)
【美好】fine; happy; glorious: ～的回忆 happy memories / ～的景色 fine view / ～的理想 glorious ideals / ～的日子 happy days; a happy life / ～的远景 magnificent prospects
【美化】beautify; prettify; embellish: ～环境 beautify the environment / 竭力～自己 try hard to prettify oneself
【美景】beautiful scenery: ～良辰 beautiful scenery and pleasant morning; an enjoyable situation
【美酒良肴】good wine and delicious dishes
【美利奴羊】Merino ◇ ～毛[纺] Merino wool
【美丽】beautiful: ～的花朵 beautiful flowers / ～富饶的国家 a beautiful and richly-endowed country
【美满】happy; perfectly satisfactory: ～的婚姻

a happy marriage; conjugal happiness / ～的家庭 a good and perfect family / ～的结果 very satisfactory result / ～的生活 a happy life
【美梦】fond dream
【美妙】beautiful; splendid; wonderful: ～的歌喉 sweet voice / ～的青春 the wonderful days of one's youth / ～的诗句 beautiful verse / ～的音乐 beautiful melody; exquisite melody
【美名】good name; good reputation: ～胜于财富 a good name is better than riches / 博得～ win a good reputation
【美男子】handsome man; very good-looking man
【美尼尔氏症】Méniére's syndrome
【美女】beautiful woman; beauty
【美其名曰】call it by the fine-sounding name of
【美人】beautiful woman; beauty ◇ ～计 use of a woman to ensnare a man; sex-trap
【美人蕉】[植] canna; Indian shot
【美容】①(使容貌美丽) improve one's looks ②(整容) cosmetology ◇ ～术 cosmetology / ～院 beauty parlor; beauty shop / ～专家 cosmetologist
【美术】①(造型艺术) the fine arts; art: 工艺～ industrial arts; arts and crafts ②(绘画) painting ◇ ～爱好者 lover of arts / ～革 fancy leather / ～工作者 art worker; artist / ～馆 art gallery / ～家 artist / ～明信片 picture postcard / ～片 [电影] cartoons; puppet films / ～品 artistic products / ～设计 artistic designing / ～学校 art school / ～学院 academy of fine arts / ～展览会 art exhibition / ～字 artistic calligraphy; art lettering
【美谈】a story passed on with approval: 传为～ be told from mouth to mouth with general approval
【美味】①(佳肴) delicious food; delicacy ②(味美) delicious dainty: ～清香 delicious and fragrant / ～小吃 dainty snacks
【美学】aesthetics
【美言】put in a good word for sb.
【美育】aesthetic education; art education
【美元】American dollar; U.S. dollar: 欧洲～ Eurodollar
【美中不足】a blemish in an otherwise perfect thing; a fly in the ointment
【美洲】America ◇ ～人 American
【美洲虎】jaguar
【美洲狮】cougar; puma

镁 [化] magnesium (Mg)
【镁光】magnesium light ◇ ～灯 flash lamp magnesium lamp / ～照明弹 magnesium flare
【镁砂】[冶] magnesia; magnesite
【镁砖】[冶] magnesia brick

每 ①(各个) every; each; per: ～个人 every one; each one / ～公斤五元 five *yuan* a kilo / ～隔一段时间 at set intervals / ～年的平均产量 average yearly yield; average output per annum / ～四小时服一次 to be taken once every four hours / ～人一架录音机 a recorder for each person / ～周两次 twice a week / 节约～一分钱 save every penny / 以～小时七十公里的速度行驶 drive at seventy kilometers an hour ②(反复) every; often: ～逢星期天 on Sundays / ～会必到 be present at every meeting
【每当】whenever; every time
【每逢佳节倍思亲】on festive occasions more than ever we think of our dear ones far away
【每个角落】every nook and cranny
【每况愈下】steadily deteriorate; go from bad to worse
【每每】often: ～推故 often make some excuse / 他们～谈到深夜。They often talked late into the night.
【每谋辄败】fail in every scheme
【每时每刻】all the time; at all times

mèi

寐 [书] sleep
【寐而不睡】lie down but not sleep

昧 ①(糊涂) have hazy notions about; be ignorant of: 素～平生 have never made sb.'s acquaintance ②(隐藏) hide; conceal: ～着良心 against one's conscience / 拾金不～ not pocket the money one has picked up
【昧心】against one's conscience: 不说～话 never say anything against one's conscience

魅 evil spirit; demon
【魅力】glamour; charm; enchantment; fascination: 艺术～ artistic charm / 自然～ charms of nature

妹 younger sister; sister
【妹夫】younger sister's husband; brother-in-law
【妹妹】younger sister; sister

媚 ①(有意讨人喜欢) fawn on; curry favor

with; flatter; toady to: ～敌 curry favor with the enemy ②(美好) charming; fascinating; enchanting: ～人的景色 enchanting scenery
【媚骨】obsequiousness
【媚上骄下】fawn on and please superiors and to be proudly contemptuous to inferiors
【媚态】coquetry; subservience: ～娇容 a seductive appearance and attractive manner
【媚外】fawn on foreign powers: 崇洋～ worship foreign things and fawn on foreign powers

mēn

闷 ①(气闷) stuffy; stifling; close: 房间里很～。It is stifling <stuffy> in the room. ②(使不透气) cover closely; cover tightly ③(不响亮) muffled: 说话～声～气的 speak in a muffled voice ④(呆在家) shut oneself or sb. indoors: ～在家里 confine oneself indoors
【闷气】stuffy; close
【闷热】hot and suffocating; sultry; muggy: ～的房间 sweltering room／～的天气 sultry weather
【闷声不响】remain silent; be dumb: ～地听人说话 listen to sb. in silence
【闷头儿】quietly; silently: ～干 work quietly; plod away silently

mén

门 ①(出入口) entrance; door; gate: 单扇～ univalve door／后～ back door／拉～ sliding door／炉～ stove door／双扇～ bivalve door／校～ school gate／旋转～ revolving door／正～ front entrance ②(开关) valve; switch: 电～ switch／气～ air valve／水～ water valve ③(门径) way to do sth.; knack ④(家族) family: 豪～ wealthy and influential family ⑤(派别) sect; school: 佛～子弟 Buddhist follower／会道～ superstitious sects ⑥(类别) field; sphere; branch: ～～精通 be proficient in all subjects／分～别类 classify according to categories／一～知识 a field of knowledge ⑦[生] phylum: 脊椎动物～ vertebrata／亚～ subphylum ⑧[量]两～功课 two subjects; two courses／一～大炮 a piece of artillery; a cannon; a gun
【门把】door knob; door handle
【门板】①(木板门) door plank ②(店门板) shutter: 上～儿 put up the shutters
【门齿】front tooth; incisor
【门当户对】be well-matched in social and economic status
【门道】[口] way to do sth.; knack: 技术革新的～很多。There are all sorts of possibilities for technical innovation.
【门第】family status: ～相称 families related by marriage equal in social status
【门吊】[机] gantry crane
【门洞儿】gateway; doorway
【门房】①(看门人的房子) gate house; janitor's room; porter's lodge ②(看门人) gatekeeper; doorman; janitor; porter
【门缝】a crack between a door and its frame
【门岗】gate sentry
【门户】①(门) door: ～紧闭 with the doors tightly shut／～开放政策 "Open Door" policy ②(必经地) gateway; important passageway ③(派别) faction; sect: ～之见 sectarian bias; sectarianism ④(门第) family status
【门环子】knocker
【门捷列夫元素周期律】[化] Mendeleev's law
【门禁】entrance guard: ～森严 with the entrances heavily guarded
【门警】police guard at an entrance
【门静脉】[生理] portal vein
【门径】access; key; way
【门槛】threshold
【门可罗雀】you can catch sparrows on the doorstep; where visitors are few and far between
【门客】a hanger-on of an aristocrat
【门口】entrance; doorway: 在～等候 wait at the door／不要站在～。Don't stand in the doorway.
【门框】doorframe
【门廊】[建] porch; portico
【门类】class; kind; category: 基础科学和技术科学这两大～ the two major departments of basic and technical sciences
【门帘】door curtain; portiére
【门脸】①[方](城门附近的地方) the vicinity of a city gate ②(商店的门面) the facade of a shop; shop front
【门铃】the doorbell
【门楼】an arch over a gateway
【门路】①(达到个人目的的途径) pull; social connections: 找～ solicit help from potential backers ②(做事的诀窍) knack; way; know-how: 摸到一些～ have learned the ropes; know one's way around／学到做生意的～ acquire the knack for business; learn the tricks of the trade
【门面】①(店面) the facade of a shop; shop

front: 三间～ a three-bay shop front ②(外表) appearance; facade: 为了装～ for appearance's sake／装点～ keep up appearances; put up a facade; put on a front ◇ ～话 formal and insincere remarks; lip service
【门牌】house number
【门碰球】ball catch; bullet catch
【门票】entrance ticket; admission ticket: 不收～ admission free
【门桥】[军] raft of pontoons; boat raft
【门上雨罩】dorsal; dorse
【门神】door-god
【门生】pupil; disciple: 得意～ one's favorite student
【门市】retail sales ◇ ～部 retail department; sales department; salesroom／～价格 retail price
【门式起重机】gantry; portal crane
【门闩】bolt; bar
【门厅】[建] entrance hall; vestibule
【门庭若市】the courtyard is as crowded as a marketplace; a much visited house
【门徒】disciple; follower; adherent
【门外汉】layman; the uninitiated
【门卫】entrance guard
【门牙】front tooth; incisor
【门诊】outpatient service ◇ ～病人 outpatient; clinic patient／～部 clinic; outpatient department／～时间 consulting hours／～所 ambulatorium; clinic

扪 [书] touch; stroke
【扪心自问】examine one's conscience
【扪诊】[医] palpation

钔 [化] mendelevium (Md)

mèn

闷 ①(心情不舒畅) bored; depressed; in low spirits: ～～不乐 be very much bored; feel depressed ②(密闭) tightly closed; sealed
【闷棍】staggering blow
【闷葫芦】enigma; puzzle; riddle
【闷倦】bored and listless
【闷雷】①(低沉的雷) dull thunder ②(精神上的打击) unpleasant surprise; shock
【闷闷不乐】depressed; in low spirits
【闷气】the sulks: 生～ be sulky; be in the sulks
【闷子车】boxcar

焖 boil in a covered pot over a slow fire; braise: ～饭 cook rice over a slow fire／～牛肉 braised beef

mēng

蒙 ①(欺骗) cheat; deceive; hoodwink; swindle: 你～人。You are cheating. ②(瞎猜) make a wild guess: ～对了 make a lucky guess ③(昏迷) unconscious; senseless: 给打～了 be knocked senseless; be stunned by a blow
【蒙蒙亮】first glimmer of dawn; daybreak: 天～就起床 get up at daybreak
【蒙骗】deceive; cheat; hoodwink; delude
【蒙头转向】lose one's bearings; be utterly confused: 使某人～ make sb. utterly confused

méng

虻 horsefly; gadfly: 牛～ gadfly

蒙 ①(遮盖) cover: ～上一层灰尘 be covered with a layer of dust／～头睡大觉 tuck oneself in and sleep like a log／～住眼睛 be blindfolded ②(受) receive; meet with: ～不白之冤 suffer underdressed injustice ③(蒙昧) ignorant; illiterate: 启～ enlighten
【蒙蔽】hoodwink; deceive; hide the truth from; pull the wool over sb.'s eyes: ～某人 pull wool over sb.'s eye／～一部分群众 hoodwink part of the masses／一时受了～ be hoodwinked for the moment
【蒙汗药】knockout drops; a narcotic believed to have been used by highwaymen, etc. to drug their victims
【蒙哄】deceive; hoodwink; swindle; cheat
【蒙混】deceive or mislead people: ～过关 get by under false pretenses
【蒙胧】drowsy; sleepy; half asleep; somnolent: ～睡去 doze off／睡眼～ drowsy eyes; eyes heavy with sleep
【蒙昧】①(未开化的) barbaric; uncivilized; uncultured: ～时代 age of barbarism ②(无知) ignorant; benighted; unenlightened: ～无知 unenlightened; childishly ignorant; illiterate ◇ ～主义 obscurantism
【蒙蒙】drizzly; misty: ～细雨 a fine drizzle／烟雾～ misty
【蒙面盗】a masked bandit; a masked burglar
【蒙难】be confronted by danger; fall into the clutches of the enemy
【蒙受】suffer; sustain: ～耻辱 be subjected to humiliation; be humiliated／～损失 sustain a loss

【蒙太奇】[电影] montage
【蒙头盖脸】cover one's head and face
【蒙羞而亡】die an ignominious death
【蒙在鼓里】be kept inside a drum; be kept in the dark

朦

【朦胧】①(月光不明) dim moonlight; hazy moonlight ②(不清楚) obscure; dim; hazy: ～不清 not bright and not clear／～的景色 a hazy view

萌 sprout; shoot forth; bud; germinate

【萌动】begin or start an action; bud: ～待时 waiting for the budding／春意～。Spring awakens.
【萌发】[植] sprout; germinate; shoot; bud: 玫瑰～新枝。The rose sprouted new buds.
【萌生】produce; conceive
【萌芽】①(植物生芽) sprout; germinate; shoot; bud ②(新生未长成的事物) rudiment; shoot; seed; germ: ～时期 in embryo; in the rudimentary stage／～生根 sprout and grow roots／处于～状态 in the embryonic stage; in the bud／资本主义的～ the seeds of capitalism

盟 ①(同盟) alliance: 与…结～ enter into alliance with ②(内蒙古自治区的行政区域) league ③(结拜兄弟) sworn brothers

【盟邦】allied country; ally
【盟国】allied country; ally: ～军队 troops of the allied states
【盟军】allied forces
【盟誓】oath of alliance
【盟兄弟】sworn brothers
【盟友】ally
【盟员】a member of an alliance
【盟约】oath of alliance; treaty of alliance: ～条件 terms of an allied treaty
【盟主】the leader of an alliance

měng

懵 muddled; ignorant

【懵懂】muddled; ignorant; muddleheaded: ～无知 ignorant and dull

蒙

【蒙古】Mongolia ◇ ～人 Mongolian／～语 Mongol
【蒙古包】yurt
【蒙古人种】the Mongolian stock
【蒙古族】Monggol nationality
【蒙栎】[植] Mongolian oak

蠓 midge; biting midge

锰 [化] manganese (Mn)

【锰钢】manganese steel
【锰结核】manganese nodule
【锰矿】manganese mine
【锰铁】ferromanganese

猛 ①(猛烈) fierce; violent; energetic; vigorous: ～虎 a fierce tiger／～将 a valiant general／产量～增 a sharp increase in output／价格～跌 a sharp drop in the price／穷追～打 hotly pursue and fiercely attack／越干越～ work with ever growing energy ②(猛然) suddenly; abruptly: ～吃一惊 be startled／～地往前一跳 suddenly jump forward／～一拐弯 turn a corner suddenly／～一刹车 apply the brake suddenly／从梦中～地惊醒 be startled out of dream

【猛不防】by surprise; unexpectedly; unawares: ～被车撞倒 be run over by a car unexpectedly
【猛打猛冲】go full blast ahead
【猛攻】onslaught; fierce attack
【猛将如云】a great many brave warriors
【猛进】push ahead vigorously: 突飞～ advance by leaps and bounds
【猛劲儿】①(集中用力气) a spurt of energy; dash ②(爆发力) great vigor
【猛力】vigorously; with sudden force: ～冲刺 make a dash with all one's strength／～扣杀 smash with all one's strength／把手榴弹～一甩 throw a grenade with all one's might
【猛烈】fierce; vigorous; violent: ～的风暴 a violent storm／～的风势 a fierce wind／～的进攻 a vigorous offensive／～的炮火 heavy shellfire／发动～的进攻 wage a vigorous offensive
【猛犸】[古生物] mammoth
【猛扑】swoop down on: ～敌堡 swoop down on the enemy fort
【猛禽】bird of prey
【猛犬】fierce dogs; vicious dogs; bull dogs
【猛然】suddenly; abruptly: ～想起 recall sth. suddenly／～一拉 pull with a jerk
【猛士】brave warrior
【猛兽】beast of prey
【猛醒】suddenly wake up
【猛追】give a hot pursuit

mèng

梦 dream：从～中醒来 awake from a dream／做了一个～ had a dream

【梦话】①(睡梦中说的话) words uttered in one's sleep; somniloquy ②(胡言乱语) daydream; nonsense

【梦幻】illusion; dream; reverie：～般的境界 a dreamlike world; dreamland

【梦幻泡影】pipe dream; bubble; illusion

【梦境】dreamland; dreamworld; dream：如入～as if in a dream

【梦寐】dream; sleep：～难忘 hard to forget even in one's dreams／～以求 crave for; try to find even in sleep／～之事 matters in dream and sleep

【梦乡】dreamland：进入～ go off to dreamland; fall asleep

【梦想】①(妄想) dream of; vainly hope ②(渴望) fond dream; earnest wish：他的～实现了。His dream has come true.

【梦魇】[医] nightmare

【梦遗】[医] nocturnal emission; wet dream

【梦呓】①(梦话) somniloquy ②(胡言乱语) rigmarole

【梦游症】somnambulism; sleepwalking

孟 ①(农历一季的第一个月) the first month of a season ②(在兄弟里排行第一的人) eldest brother

【孟春之秋】the beginning of the first month of spring

【孟德尔主义】[生] Mendelism

【孟加拉】Bengal ◇～国 Bangladesh／～人 Bengalese; Bengali／～湾 the Bay of Bengal／～语 Bengali (language)

【孟浪】rash; impetuous; impulsive：～之谈 a rough and wasteful talk

【孟什维克】Menshevik

【孟子】Mencius

mī

咪

【咪咪】①[象](猫叫的声音) mew; miaow ②(笑貌) smilingly：笑～ be all smiles; be wreathed in smiles

眯 ①(眯着眼) narrow one's eyes：～着眼瞧 squint at／～着眼睛笑 narrow one's eyes into a smile ②[方](小睡) nap; take a nap：～一会儿 take a short nap; have forty winks

【眯缝】narrow：～着眼 with eye's narrowed into slits

mí

靡 waste：奢～ wasteful; extravagant

【靡费】waste; spend extravagantly

糜 ①(粥) gruel ②(烂) rotten ③(浪费) wasteful; extravagant

【糜费】waste; dissipation：～精力 dissipate energy／～钱财 waste money／防止～人力 avoid waste of man power

【糜烂】rotten to the corn; corrupted; profligate; debauched：生活～ lead a fast life ◇～性毒剂[军] vesicant agent; blister agent

麋 elk

【麋羚】hartebeest

【麋鹿】mi-lu; Bavid's deer

迷 ①(迷失) be confused; be lost：～了方向 lose one's bearings; get lost／～了路 lose one's way ②(沉醉) be fascinated by; be crazy about：看电影入了～ be crazy about movies ③(爱好者) fan; enthusiast; fiend：电影～ film fan／革新～ innovation enthusiast／官～ a person who craves office／乒乓球～ a table tennis fan／棋～ a chess fiend ④(使看不清) confuse; perplex; fascinate; enchant：～人的景色 scenery of enchanting beauty／被音乐～住 be bewitched by music／财～心窍 be befuddled by a craving for wealth; be obsessed by lust for money

【迷不知返】go astray not knowing how to return

【迷宫】labyrinth; maze

【迷航】drift off course; lose one's course; get lost

【迷糊】①(看不清) misted; blurred; dimmed ②(神志不清) dazed; confused; muddled：睡～了 dazed with sleep

【迷魂汤】sth. intended to turn sb.'s head; magic potion：灌～ try to ensnare sb. with honeyed words

【迷魂阵】a scheme for confusing or bewildering sb.; maze; trap：摆～ lay out a scheme to bewitch sb.; set a trap

【迷惑】puzzle; confuse; perplex; baffle：～敌人 confuse the enemy／～舆论 throw dust in the eyes of the public／感到～不解 feel puzzled; feel perplexed

【迷离】blurred; misted：～恍惚 be in a

stupor / 睡眼～ eyes dim with sleep
【迷恋】be infatuated with; madly cling to
【迷路】①(迷失道路) lose one's way; get lost ②[生理](内耳) inner ear; labyrinth ◇～炎 labyrinthitis
【迷漫】vague; hazy; indistinct
【迷茫】①(广阔而看不清) vast and hazy: 远处一片～。It's vast and hazy in the distance. ②(迷离恍惚) confused; perplexed; dazed: ～的神情 a confused look
【迷梦】pipe dream; fond illusion
【迷迷糊糊】in a daze, difficult to make out
【迷你裙】miniskirt
【迷人】charming; fascinating; enchanting; bewitching
【迷失】lose: ～方向 lose one's bearings; get lost
【迷途】①(迷失道路) lose one's way ②(错误的道路) wrong path: ～羔羊 astray lamb / 走入～ go astray
【迷途知返】recover one's bearings and return to the fold; realize one's errors and mend one's ways
【迷惘】be perplexed; be at a loss
【迷雾】①(浓厚的雾) dense fog ②(使人迷失方向的事物) anything that misleads people: 妖风～ evil wind and miasma
【迷信】①(迷信鬼神) superstition; superstitious belief; blind faith; blind worship ②(盲目迷信) have blind faith in; make a fetish of: 破除～ do away with fetishes and superstitions / 个人～ cult of the individual; personality cult
【迷走神经】[生理] vagus (nerve)

谜

①(谜语) riddle; conundrum: 猜～ guess a riddle ②(没有弄清楚的事物) enigma; mystery; puzzle: 不解之～ unfathomable enigma; insoluble mystery
【谜底】①(谜语的答案) answer to a riddle ②(真相) truth
【谜语】riddle; conundrum: ～猜释 guess and explain riddles

醚

[化] ether

弥

①(遍; 满) full; overflowing: ～漫 fill the air ②(填满) cover; fill: ～缝 plug up holes; gloss over faults ③(更加) more: 欲盖～彰 try to cover sth. up only to make it more conspicuous
【弥补】make up; remedy; make good: ～赤字 make up a deficit / ～过失 smooth over a fault / ～缺陷 remedy a defect / ～损失 make up for a loss
【弥缝】fill cracks; cover up mistakes; ～差错 patch up mistakes
【弥合】close; bridge: ～裂痕 close a rift
【弥勒】[佛教] Maitreya
【弥留】[书] be dying: ～遗言 words uttered when there is no hope of recovery / ～之际 on one's deathbed
【弥漫】fill the air; spread all over the place: ～浊气 a boundless and foul air / 烟雾～ heavy with smoke; smoke-laden; be enveloped in mist
【弥撒】[天主教] Mass
【弥天大谎】monstrous lie
【弥天大罪】monstrous crime; heinous crime
【弥望无际】boundless horizon
【弥月之敬】present for a baby just one month old

猕

【猕猴】macaque; rhesus monkey
【猕猴桃】[植] yangtao

mǐ

靡

①(顺风倒下) blown away by the wind: 所向披～ send the enemy fleeing helter-skelter; carry all before one ②[书](无) no; not: ～日不思 not a day passed without one's thinking of sth. or sb.
【靡丽】extravagant
【靡靡之音】soft, effeminate music; decadent music
【靡然从风】go with the fashion

米

①(稻米) rice: ～粒 grain of rice / 精白～ polished rice ②(去壳的种子) shelled seed; husked seed: 花生～ peanut seed; peanut kernel / 高粱～ grains of *gaoliang* ③(长度单位) meter; metre
【米波】[无] metric wave
【米饭】rice
【米粉】①(大米粉) ground rice; rice flour: ～肉 pork steamed with ground glutinous rice ②(米粉条) rice-flour noodles
【米糕】rice cake; rice pudding
【米黄】cream-colored
【米价】the price of rice
【米酒】rice wine
【米糠】rice bran ◇～油 rice bran oil
【米粮川】rich rice-producing area: 荒滩变成～。Large tracts of wasteland have become a granary.

【米色】cream-colored
【米汤】①(煮饭时取出的汤) water in which rice has been cooked ②(薄稀饭) thin rice or millet gruel; rice water
【米象】[动] rice weevil
【米制】the metric system
【米粥】congee; rice gruel
【米珠薪桂】rice is as precious as pearls and firewood as costly as cassia; exorbitantly high cost of living
【米烛光】[物] meter-candle; lux

眯 get into one's eye: 尘土～了我的眼。Dust has got into my eyes

脒 [化] amidine

mì

泌 secrete
【泌尿科】[医] urological department
【泌尿器官】[生理] urinary organs
【泌尿系统】[解] urinary system
【泌尿学】[医] urology

秘 ①(秘密的) secret; confidential: ～方 secret recipe / ～事 a secret ②(保守秘密) keep sth. secret; hold sth. back: ～而不宣 keep sth. secret; not let anyone into a secret
【秘本】treasured private copy of a rare book
【秘方】secret recipe: 祖传～ a secret recipe handed down from generation to generation
【秘诀】secret (of success): 成功的～ the secret of one's success
【秘密】secret; clandestine; confidential: ～会议 secret meeting; closed-door session / ～活动 clandestine activities / ～结社 form a secret society / ～条约 secret treaty / ～投票 secret ballot; Australian ballot / ～外交 secret diplomacy / ～文件 secret papers; confidential document / ～消息 classified information / 揭开宇宙的～ unravel the secrets of the universe
【秘史】secret history (as of a feudal dynasty); inside story
【秘书】secretary: 机要～ confidential secretary / 私人～ private secretary ◇ ～处 secretariat / ～长 secretary general

蜜 ①(蜂蜜) honey ②(甜美) honeyed; sweet: 甜如～ honey-sweet; as sweet as honey
【蜜蜂】honeybee; bee ◇ ～窝 honeycomb; beehive
【蜜柑】mandarin orange; tangerine orange
【蜜饯】candied fruit; preserved fruit
【蜜饯砒霜】sugar-coated arsenic; deceptive sweet words
【蜜橘】tangerine
【蜜蜡】beeswax
【蜜腺】[植] nectary
【蜜源】nectar source ◇ ～区 pasture / ～植物 nectariferous plant; bee plant; honey plant
【蜜月】honeymoon: ～旅行 honeymoon trip / 他们在度～。They are on their honeymoon.
【蜜枣】candied date; jujube
【蜜渍】candied; preserved in sugar

密 ①(距离近;空隙小) close; dense; thick: ～不透风 airtight / ～林 thick forest / 人口很～ densely populated / 头发很～ thick hair ②(亲密) intimate; close: ～友 close friend; bosom friend / 与某人来往很～ be on intimate terms with sb. ③(精致;细致) fine; meticulous: 周～ carefully considered; meticulous ④(秘密) secret: ～件 confidential document; classified material / 绝～ top secret; strictly confidential ⑤[纺] density: 经～ warp density / 纬～ weft density
【密闭】airtight; hermetic
【密布】densely covered: 礁石～ thick with reefs
【密陈利害】send a secret memorial concerning the profit and loss
【密电】①(密码电报) cipher telegram ②(秘密电告) secretly telegraph sb. ◇ ～码 cipher code
【密度】①(密的程度) density; thickness: 火力～ density of fire / 人口～ population density ②[物](物质密度) density: 电流～ current density ◇ ～计 densimeter
【密访】pay a secret visit; make investigation by traveling incognito
【密封】seal up; seal airtight; seal hermetically: ～的罐头 sealed tin / ～的容器 hermetically sealed chamber / ～的文件 sealed documents ◇ ～舱 sealed cabin; airtight cabin / ～垫圈 [机] sealing washer / ～机身 [航空] closed fuselage / ～胶 fluid sealant; gasket cement / ～压盖 [机] sealing gland
【密函告知】inform one by means of confidential letters
【密集】concentrated; crowded together: 人口～ densely populated; thickly populated ◇ ～队形 close formation; tight formation / ～轰炸 mass bombing / ～炮火 intensive bombardment; concentrated fire; massed fire; drumfire
【密件】a confidential paper or letter; classified

matter; classified material
【密令】a secret order
【密码】cipher; cipher code; secret code ◇ ～电报 cipher telegram / ～机 cipher machine; cryptograph / ～通讯 communicate with code telegram / ～信 message in cipher / ～员 cryptographer
【密密层层】packed closely layer upon layer; dense; thick: ～的人群 a dense crowd
【密密麻麻】close and numerous; thickly dotted
【密谋】conspire; plot; scheme: ～策划 conspire / ～犯罪 conspiracy to commit crime ◇ ～者 intriguer
【密切】①(关系近) close; intimate: ～联系群众 maintain close ties with the masses / ～两国关系 build closer relations between the two countries / ～配合 act in close coordination / ～相关 be closely related ②(仔细) carefully; intently; closely: ～注视 pay close attention to; watch closely
【密如蛛网】as fine as a spider's web
【密商】hold private counsel; hold secret talks
【密使】secret emissary; secret envoy
【密室】a room used for secret purposes: 策划于～ plot behind closed doors
【密实】closely knit; dense; thick
【密谈】secret talk; talk behind closed doors
【密探】secret agent; spy
【密通信息】secretly communicate with each other
【密陀僧】[化] litharge; yellow lead
【密纹唱片】long-playing record; micro-groove record
【密写情报】intelligence written in invisible ink, etc.
【密友】friend; bosom friend
【密约】secret agreement; secret treaty
【密云不雨】dense clouds but no rain; trouble is brewing
【密植】[农] close planting: 合理～ rational close planting

嘧

【嘧啶】[化] pyrimidine

幂

幂 ①[数] power: 四次～ the fourth power ②(覆盖东西的布) cloth cover
【幂级数】[数] power series

觅

觅 look for; hunt for; seek: ～食 look for food
【觅索词句】strive for phrases and sentences
【觅致人才】on the lookout for proper personnel

mián

眠

眠 ①(睡眠) sleep: 不～之夜 a sleepless night; a white night ②(某些动物的生理现象) dormancy: 冬～ hibernate
【眠尔通】[药] miltown; meprobamate
【眠目静思】closing the eyes and meditating

棉

棉 ①(棉类统称) a general term for cotton and kapok ②(棉花) cotton: 长绒～ long staple cotton / 海岛～ sea island cotton / 陆地～ upland cotton / 木～ silk cotton kapok / 皮～ ginned cotton / 脱脂～ absorbent cotton / 原～ raw cotton / 籽～ unginned cotton ③(衬棉花的) cotton-padded; quilted: ～大衣 cotton-padded overcoat
【棉袄】cotton-padded jacket
【棉被】a quilt with cotton wadding
【棉布】cotton cloth; cotton
【棉纺】cotton spinning ◇ ～厂 cotton mill
【棉纺织品】cotton textiles
【棉红铃虫】pink bollworm
【棉红蜘蛛】two-spotted spider mite
【棉猴儿】hooded cotton-padded coat; parka; anorak
【棉花】cotton ◇ ～签 (cotton) swab / ～胎 cotton batting
【棉卷】[纺] lap
【棉枯萎病】fusarium wilt of cotton
【棉裤】cotton-padded trousers
【棉铃】cotton boil ◇ ～虫 bollworm / ～象虫 boll weevil
【棉毛机】interlock (knitting) machine
【棉毛裤】cotton (interlock) trousers
【棉毛衫】cotton (interlock) jersey ◇ ～布 interlock
【棉帽】cotton-padded cap
【棉农】cotton grower
【棉绒】cotton velvet ◇ ～布 [纺] winsey
【棉纱】cotton yarn ◇ ～头 (cotton) waste
【棉毯】cotton blanket
【棉桃】cotton boll
【棉套】a cotton-padded covering for keeping sth. warm
【棉田】cotton field
【棉条】[纺] sliver ◇ ～桶 sliver can
【棉线】cotton thread; cotton
【棉鞋】cotton-padded shoes
【棉絮】①(棉花纤维) cotton fiber ②(棉被胎) a

cotton wadding
【棉蚜虫】cotton aphid
【棉衣】cotton-padded clothes
【棉织厂】cotton textile mill
【棉织品】cotton goods; cotton textiles; cotton fabrics
【棉籽】cottonseed ◇ ～饼 cottonseed cake／～绒 (cotton) linters／～油 cottonseed oil

绵 ①(丝棉) silk floss ②(绵延) continuous: ～长 long; lengthy ③(柔软) soft
【绵薄】[谦] meager strength; humble effort: 愿尽～。I'll do what little I can.
【绵绸】fabric made from waste silk
【绵亘】stretch in an unbroken chain: ～数千里 stretching for thousands of miles
【绵里藏针】a needle hidden in silk floss; a ruthless character behind a gentle appearance; an iron hand in a velvet glove
【绵力】my limited power
【绵绵】continuous; unbroken: ～不绝 continuing forever／～絮语 whisper continually
【绵软】①(柔软) soft: ～的羊毛 soft wool ②(身体无力) weak: 觉得浑身～ feel weak all over
【绵延】be continuous; stretch long and unbroken: ～千里的山脉 mountains extending a thousand *li*
【绵羊】sheep
【绵纸】tissue paper

miǎn

湎
【湎于酒色】indulging in wine and lewdness

腼
【腼腆】shy; bashful: 见生人有点～ feel shy with strangers

缅 remote; far back
【缅甸】Burma ◇ ～人 Burmese／～语 Burmese (language)
【缅怀】cherish the memory of; recall: ～革命先烈 cherish the memory of our revolutionary martyrs／～往事 recall past events
【缅想】think of (past events); recall: ～前尘 think of the past affairs

免 ①(去掉) excuse sb. from sth.; exempt; dispense with: ～服兵役 be exempt from military service／～试 be excused from an examination／～受攻击 immune against attacks／互相～办签证协议 mutual exemption of visas agreement ②(避免) avoid; avert; escape: ～了一场灾祸 avert an accident 以～误会 avoid misunderstanding ③(不可) not allowed; forbidden: 闲人～进。No admittance except on business.
【免不了】be unavoidable; be bound to be
【免除】①(免去) prevent; avoid: ～某人职务 remove sb. from office／兴修水利,～水旱灾害 build irrigation works to prevent droughts and floods ②(免掉) remit; excuse; exempt; relieve: ～罚金 relief against forfeiture／～劳役 dispensing with service／～刑罚 abatement and exemption from penalty／～一切捐税 exempt from obligation／～一项任务 excuse sb. from a task; relieve sb. of a task／～责任 dissolution of responsibility／～债务 remit a debt
【免得】so as not to; avoid: ～引起怀疑 so as not to arouse suspicion／～引起误会 avoid misunderstanding
【免费】free of charge; free; gratis: ～入场 admission free; be admitted gratis／～医疗 free medical care
【免耕农业】till-less agriculture
【免冠】①(脱帽) take one's hat off ②(不戴帽子) without a hat on; bareheaded: 半身～正面相片 a half-length, bareheaded, full-faced photo
【免开尊口】you had better shut up
【免赔条款】franchise clause
【免票】①(不收费的票) free ticket ②(免费) free of charge
【免税】exempt from taxation; tax-free; duty free ◇ ～单 duty free slips／～商品 free commodities／～物品 free goods／～执照 duty free certificate
【免刑】[法] exempt from punishment
【免验】exempt from customs examination: ～放行 pass without examination (P.W.E.) ◇ ～证 laissez-passer
【免役】exempt from service
【免疫】[医] immunity: 后天～ acquired immunity／先天～ congenital immunity ◇ ～球蛋白 immunoglobulin／～体 immune body／～治疗 immunization therapy／～证书 bill of health
【免于公诉】immunity from prosecution
【免于起诉】exemption from prosecution; immunity from suit
【免予处罚】remit
【免战牌】a sign used in ancient times to show

refusal to fight: 挂～ refuse battle
【免职】remove sb. from office; relieve sb. of his post
【免致后患】avoid causing future trouble
【免罪】exempt from punishment
冕 crown: 加～礼 coronation
勉 ①(努力) exert oneself; strive: ～力为之 exert oneself to the utmost; do one's best ②(勉励) encourage; urge; exhort: 互～ encourage one another / 自～ spur oneself on ③(力量不够而尽力做) strive to do what is beyond one's power: ～为其难 undertake to do a difficult job as best one can
【勉尽力量】do one's best
【勉励】encourage; urge: ～上进 incite one to make progress / ～学生努力学习 encourage the students to study with still greater efforts
【勉强】①(尽力) manage with an effort; do with difficulty ②(不心甘情愿) reluctantly; grudgingly: ～地笑了笑 force a smile / 毫不～地同意 agree without reluctance ③(强迫) force sb. to do sth.: 要是她不答应, 不要～她。 In case she doesn't agree, don't try to force her to. ④(牵强) inadequate; unconvincing; strained; farfetched: 你的理由很～。 The reason you give is rather unconvincing. ⑤(刚刚够) barely enough: ～糊口 eke out a living / 他～赶上火车。 He barely caught the train.
【勉为其难】try to do a difficult job as best one can

miàn

面 ①(脸) face: ～带笑容 with a smile on one's face / ～无惧色 not look at all afraid / ～有难色 wear an embarrassed look ②(向着) face ③(表面) surface; top; face: 地～ the surface of the earth / 路～ road surface / 水～ the surface of the water / 桌～ the top of a table; tabletop ④(当面) personally; directly; face to face: ～告 tell sb. personally / ～交 deliver personally / ～谈 speak to sb. face to face / ～谢 thank sb. in person ⑤(皮; 面) the right side; cover; outside: 被～儿 the top covering of a quilt / 夹袄的～儿 the outside of a lined jacket ⑥ [数] surface ⑦(部位; 方面) side; aspect: 右～ right side / 左～ left side / 四～包围敌人 surround the enemy on all sides / 向四～观望 look around ⑧(全面) entire area ⑨(范围) extend; range; scale; scope: 知识～广 have a wide range of knowledge ⑩(方位词后缀) 前～ in front / 外～ outside / 左～ on the left ⑪ [量] 两～旗子 two flags / 一～镜子 a mirror ⑫(粉) powder; wheat flour; flour: 白～ wheat flour / 大米～ rice flour / 胡椒～ ground pepper / 药～ medicinal powder / 玉米～ corn flour ⑬(面条) noodles: 一碗～ a bowl of noodles / 中午吃～ have noodles for lunch
【面包】bread ◇ ～房 bakery / ～干 rusk / ～渣儿 breadcrumbs; crumbs
【面包果】[植] breadfruit
【面不改色】not change color; remain calm; without turning a hair; without batting an eyelid
【面茶】seasoned millet mush
【面陈】report in person; deliver sth. to a superior in person
【面辞】go to say good-bye to sb.; take leave of sb.
【面带病容】one's face shows sickly countenance
【面带愁容】with a sad air
【面带笑意】one's face shows smiling mood
【面对】face; confront: ～现实 face reality; be realistic / 我们～困难决不退缩。 We will never flinch from difficulties.
【面对面】facing each other; face-to-face; vis-a-vis: ～的斗争 a face-to-face struggle; direct confrontation / ～地坐着 sit face-to-face; sit vis-a-vis
【面额】[经] denomination: ～为 100 元的人民币 Renminbi in 100-*yuan*notes / 各种～的纸币 banknotes of different denominations
【面肥】leavening dough; leaven
【面粉】wheat flour; flour ◇ ～厂 flour mill
【面和心不和】remain friendly in appearance but estranged at heart
【面红耳赤】be red in the face; be flushed: 羞得～ flush with shame or shyness / 争得～ argue until everyone is red in the face; have a heated argument
【面糊】[方] soft and floury
【面黄肌瘦】sallow and emaciated; lean and haggard
【面积】area: 建筑～ built-up area / 居住～ dwelling area; floor space / 大～丰收 bumper harvest over an extensive area
【面颊】cheek
【面巾纸】face tissues
【面筋】gluten
【面具】mask: 防毒～ gas mask / 剥去假～

unmask
【面孔】face: 和蔼的～ amiable countenance / 讨人喜欢的～ likable face / 严肃的～ a stern face / 板起～ put on a stern expression
【面临】be faced with; be confronted with; be up against: ～重重困难 be confronted with numerous difficulties / ～一场严重的危机 be faced with a serious crisis
【面聆教益】benefit by your advice
【面貌】①(相貌) face; features; looks: ～清秀 have delicate looks / ②(面目) appearance; look; aspect: 城市～ the appearance of a city / 精神～ mental outlook / ～一新 take on a now look
【面面俱到】attend to each and every aspect of a matter
【面面相觑】look at each other in blank dismay; gaze at each other in speechless despair
【面目】①(相貌) face; features; visage: ～可憎 repulsive in appearance / ～清秀 have delicate looks / ～狰狞 sinister in appearance ②(面貌) appearance; look; aspect: ～全非 be changed or distorted beyond recognition / 还其本来～ reveal sth. in its true colors / 显出庐山真～ show one's true colors / 政治～不清 of dubious political background ③(面子) self-respect; honor; sense of shame; face: 愧无～见人 feel too ashamed to face people
【面目一新】take on an entirely new look; present a completely new appearance; assume a new aspect
【面嫩】timid; sensitive
【面庞】contours of the face; face: 圆圆的～ a round face
【面前】face of; in front of; before: 困难～不动摇 not waver in the face of difficulties / 在敌人～无所畏惧 be dauntless before the enemy
【面人儿】dough figurine
【面容】facial features; face: ～消瘦 look emaciated
【面如菜色】the pinched look of a hungry person
【面如土色】look ashen; look pale: 吓得～ turn pale with fright
【面色】①(面上气色) complexion: ～苍白 look pale / ～红润 have rosy cheeks; be ruddy-cheeked / ～黝黑 have a dark complexion ②(面部神色) facial expression: ～忧郁 have a melancholy look; look worried / ～骤变 one's countenance suddenly changed
【面纱】veil
【面商】discuss with sb. face to face; consult personally
【面上无光】loss of prestige
【面神经】[生理] facial nerve
【面生】look unfamiliar
【面食】cooked wheaten food
【面试】an oral quiz; an audition
【面授机宜】personally instruct sb. on the line of action to pursue; give confidential briefing
【面熟】look familiar
【面谈】speak to sb. face to face; take up a matter with sb. personally
【面汤】water in which noodles have been boiled
【面条】noodles
【面团】dough
【面无人色】look ghastly pale
【面向】①(朝着) turn one's face to; turn in the direction of; face: ～国旗庄严敬礼 stand facing the national flag and salute solemnly ②(使适合) be geared to the needs of; cater to: ～广大读者 be geared to the needs of reading public
【面谢】thank sb. in person
【面有难色】appear to be reluctant; wear an embarrassed look
【面誉背毁】praise sb. to his face and abuse him behind his back
【面罩】face guard: ～寒霜 a cloud passes over one's face
【面值】①(票面价值) par value; face value; nominal value ②(纸币单位) denomination
【面砖】[建] face brick
【面子】①(物体表面) outer part; outside; face: 大衣的～ the outside of an overcoat ②(体面) reputation; prestige; face: ～问题 questions of one's reputation / 爱～ be concerned about face-saving / 不讲～ have no consideration for sb.'s sensibilities / 丢～ lose face / 给～ show due respect for sb.'s feelings / 顾～ save face / 撕破～ cast aside all considerations of face; not spare sb.'s sensibilities

miāo

喵 [象] mew; miaow

miáo

苗 ①(初生的植物) young plant; seedling: ～期 seedling stage / 麦～儿 wheat seedling ②(初生的动物) the young of some animals: 鱼～ fry ③(疫苗) vaccine: 牛痘～ (bovine) vaccine ④(形状象苗的) sth. resembling a young plant:

火～儿 flame
【苗床】seedbed
【苗木】[林] nursery stock
【苗圃】nursery (of young plants)
【苗期】[农] seedling stage
【苗条】slender; slim: ～身材 slender and graceful in stature
【苗头】symptom of a trend; suggestion of a new development
【苗子】[农]①(苗) young plant; seedling ②(比喻继承某种事业的年轻人) young successor

描 ①(照底样画) trace; copy: ～图样 trace designs; copy designs ②(重复地涂抹) touch up; retouch: ～红 trace over the red printed characters with a writing brush in black
【描画】draw; paint; depict; describe: ～出美好的前景 paint a bright future / ～出幸福的未来 picture a happy future / 生动地～ depict vividly / 详细地～ describe in detail
【描绘】depict; describe; portray
【描龙绣凤】do fine needlework
【描摹】depict; portray; delineate: ～一个人的性格 depict a person's character
【描述】describe: 详细～事情的经过 describe what happened in great detail
【描图】tracing ◇ ～员 tracer / ～纸 tracing paper
【描写】describe; depict; portray: 生动地～音乐会的情况 give a vivid description of the concert

瞄 concentrate one's gaze on; take aim: ～得准 take good aim
【瞄准】take aim; aim; train on; lay; sight: ～靶心 aim at the bull's-eye / 练习～ practice aiming ◇ ～环 ring sight / ～具 sighting device; (gun) sight / ～手 layer; pointer / ～仪 aiming sight / ～装置 sighting device

miǎo

秒 second
【秒表】stopwatch; chronograph
【秒差距】[天] parsec
【秒立方米】[水] cubic meter per second
【秒针】second hand (of a clock or watch)

渺 ①(渺茫) vast; distant and indistinct; vague: ～无人烟 wild and uninhabited / ～无希望 have slim hopes / ～无踪迹 without a trace ②(渺小) tiny; insignificant: ～不足道 insignificant; negligible
【渺茫】①(模糊不清) distant and indistinct; vague ②(难以预期) uncertain: ～得很 completely at sea; uncertain / 前途～ have an uncertain future / 希望～ have slim hopes
【渺若烟云】as vague as mist
【渺小】tiny; negligible; insignificant; paltry

藐 ①(小) small; petty ②(轻视) slight; despise: 言者谆谆，听者～～。 The words were earnest but they fell on deaf ears.
【藐视】despise; look down upon: ～法律 defiance of law / ～法庭 contempt of court
【藐小】tiny; negligible; insignificant; paltry

淼 vast
【淼茫】stretch as far as the eye can see

miào

庙 ①(供神佛或历史名人的处所) temple; shrine ②(庙会) temple fair: 赶～ go to the fair
【庙号】posthumous title of an emperor
【庙会】temple fair; fair
【庙宇】temple
【庙主】head priest of a temple

妙 ①(好;美妙) wonderful; excellent; fine: 绝～的讽刺 a supreme irony / ～极了! It is simply wonderful! / 不去为～。 Better not go there. ②(神奇;巧妙) ingenious; clever; subtle: ～语 clever remarks / 他回答得很～。 He make a clever answer.
【妙不可言】too wonderful for words; most intriguing
【妙计】excellent plan; brilliant scheme: ～奇谋 an excellent scheme and clever plan
【妙句】a quotable quote
【妙诀】a clever way of doing sth.; knack
【妙绝古今】an unparalleled wonder both in ancient and modern times
【妙龄】young; youthful
【妙论】extraordinary argument; convincing discourse; sparkling discourse
【妙品】①(高质商品) fine quality goods: 调味～ best-quality condiment ②(精致的工艺品) fine work of art
【妙趣横生】full of wit and humor; very witty
【妙手回春】effect a miraculous cure and bring the dying back to life
【妙语】witty remark; witticism: ～如珠 pearls of wisdom, sparkling sayings / ～双关 a clever double-extender

miē

咩 [象] baa; bleat

乜

【乜斜】①(眼睛微眯而斜着看) squint ②(眼睛因困倦而眯成一条缝) half-closed: ～的睡眼 half-closed eyes heavy with sleep

miè

灭 ①(熄灭) go out: 灯～了。The light goes 〈is〉 out. ②(使熄灭) extinguish; put out; turn off: ～火 put out a fire; extinguish a fire ③(淹没) submerge; drown: ～顶 be drowned ④(消灭) destroy; exterminate; wipe out: ～蝇 kill flies

【灭茬机】[农] stubble cleaner

【灭此朝食】will not have breakfast until the enemy is wiped out; be anxious to finish off the enemy immediately

【灭顶】be drowned: ～之灾 the woe of falling into river; disaster of being drowned

【灭火】①(把火弄灭) put out a fire: extinguish a fire ②(使发动机熄火) cut out an engine ◇ ～剂 fire-extinguishing chemical / ～器 fire extinguisher / ～沙 sand for extinguishing fire

【灭迹】destroy the evidence: ～焚尸 destroy traces by burning the body

【灭尽天良】destroy utterly one's conscience

【灭绝】become extinct: ～净尽 totally exterminated / 现已～的动物 extinct animals

【灭绝人性】inhuman; savage; cannibalistic

【灭绝种族】genocide

【灭口】do away with a witness or accomplice

【灭门之祸】the calamity of exterminating a family

【灭亡】be destroyed; become extinct; die out: 注定要～ be doomed to perish / 自取～ court destruction

【灭音器】muffler

【灭种之虑】anxiety for the extinction of a race

【灭族】extermination of an entire family

蔑 [书]①(小) slight; disdain: 轻～ disdain ②(无) nothing; none: ～以复加 could not be surpassed; reach the limit ③(诽谤) smear: 诬～ slander; vilify

【蔑弃】despise and cast away: ～礼仪 cast away rites

【蔑然无言】not uttering a word

【蔑视】despise; show contempt for; scorn: ～世俗 have a contempt for custom

篾 ①(薄竹片) thin bamboo strip ②(苇子或高粱杆上劈下的皮) the rind of reed or sorghum

【篾黄】the inner skin of a bamboo stem

【篾匠】a craftsman who makes articles from bamboo strips

【篾片】thin bamboo strip

【篾青】the outer cuticle of a bamboo stem

【篾席】a mat made of thin bamboo strips

mín

民 ①(人民) the people: 为国为～ for the country and the people / 为～除害 rid the people of a scourge ②(从事某种职业的人) a person of a certain occupation: 牧～ herdsman / 渔～ fisherman ③(某族的人) a member of a nationality: 回～ a Hui ④(民间的) of the people; folk: ～歌 folk song / ～谣 folk rhyme ⑤(非军人) civilian: ～船 a junk or small boat for civilian use

【民安国泰】the masses are in peace and the country is prosperous

【民办】run by the local people: ～公助 run by the local people and subsidized by the state ◇ ～小学 a primary school run by the local people

【民兵】people's militia; militia

【民变】mass uprising; popular revolt

【民不聊生】the people have no means of livelihood; the masses live in dire poverty

【民粹主义】populism

【民法】civil law ◇ ～规范 norm of the civil law / ～上的违法行为 civil offense / ～通则 general rule of the civil law / ～效力 validity of civil law / ～学 science of civil law / ～学家 civil jurist; civilian

【民房】a house owned by a citizen

【民愤】the people's wrath; popular indignation: ～极大 have earned the bitter hatred of the people; have incurred the greatest popular indignation

【民风】folkway

【民风淳朴】the people are simple, honest and unspoiled

【民歌】folk song

【民工】a laborer working on a public project

【民国】the Republic of China (1912–1949)

【民航】[简](民用航空) civil aviation ◇ ～机 civil aircraft; civil airplane

【民间】①(人民之间的) among the people;

popular; folk: ～疾苦 hardships of the people ② (非官方的) nongovernmental; people-to-people: ～来往 nongovernmental contact; people-to-people exchange ◇ ～传说 popular legend; folk legend; folklore / ～故事 folktale; folk story / ～纠纷 dispute among the people / ～文学 folk literature / ～舞蹈 folk dance / ～协定 nongovernmental agreement / ～验方 folk remedy; folk recipe / ～艺术 folk art / ～音乐 folk music

【民警】people's police; people's policeman: 女～ people's policewoman

【民力】financial resources of the people

【民气】the people's morale; popular morale

【民情】① (人民的情况) condition of the people: 熟悉地理～ be familiar with the place and the people ② (人民的心情,愿望) feeling of the people; public feeling

【民穷财尽】the means of the people have been used up

【民权】civil rights; civil liberties; democratic rights

【民生】the people's livelihood: 国计～ the national economy and the people's livelihood

【民事】[法] relating to civil law; civil ◇ ～案件 civil case / ～法律关系 civil legal relationship / ～诽谤罪 civil libel / ～管辖权 civil jurisdiction / ～过失 civil negligence / ～纠纷 civil dispute / ～立法 civil legislation / ～赔偿 civil compensation / ～侵权法 law of tort / ～上诉 civil appeal / ～审判庭 the civil division of a people's court; civil court / ～诉讼 civil action / ～诉讼法 civil procedure act / ～调解 civil mediation / ～责任 civil liability

【民俗】folk custom; folkways ◇ ～学 folklore

【民心】popular feelings; common aspiration of the people: ～所向 where the popular will inclines; the common aspiration of the people / 深得～ enjoy the ardent support of the people

【民谣】folk rhyme

【民以食为天】hunger breeds discontentment

【民意】the will of the people; popular will ◇ ～测验 public opinion poll; poll

【民用】for civil use; civil ◇ ～波段无线电通信 citizens band radio / ～航空 civil aviation / ～机场 civil airport

【民怨沸腾】the people are boiling with resentment; seething popular discontent

【民乐】music, esp. folk music, for traditional instruments ◇ ～队 traditional instruments orchestra / ～合奏 ensemble of traditional instruments ○胡琴 *huqin*, Chinese violin / 琵琶 *pipa*, 4-stringed Chinese lute / 三弦 *sanxian*, 3-stringed Chinese guitar / 月琴 *yueqin*, 4-stringed full-moon-shaped Chinese mandolin / 扬琴 *yangqin*, dulcimer / 笛子 *dizi*, Chinese flute; 8-holed bamboo flute / 箫 *xiao*, Chinese vertical bamboo flute / 唢呐 *suona*, Chinese cornet / 笙 *sheng*, Chinese wind pipe / 古琴 *guqin*, Chinese zither / 腰鼓 waist drum

【民贼】traitor to the people

【民政】civil administration ◇ ～管理 civil administration / ～机关 civil administration organ / ～事务 civil affairs

【民脂民膏】flesh and blood of the people: 搜刮～ fleece the people

【民众】the masses of the people; the common people; the populace: 唤起～ arouse the masses ◇ ～团体 people's organization; mass organization

【民主】① (人民的民主权利) democracy; democratic rights: 党内～ inner-party democracy / 社会主义～ socialist democracy ② (合于民主原则) democratic: 他的作风～。 He has a democratic work-style. ◇ ～改革 democratic reform / ～革命 democratic revolution / ～共和国 democratic republic / ～管理 democratic management / ～国家 democratic state / ～权利 democratic rights / ～人士 democratic personages / ～协商 democratic consultation / ～制度 democratic system / ～专政 democratic dictatorship

【民主党】[美国] the Democratic Party

【民主党派】democratic party ○中国国民党革命委员会(民革) Revolutionary Committee of the Kuomintang / 中国民主同盟(民盟) China Democratic League / 中国民主建国会(民建) China Democratic National Construction Association / 中国民主促进会(民进) China Association Promoting Democracy / 中国农工民主党 Chinese Peasants' and Workers' Democratic Party / 中国致公党 China Zhi Gong Dang / 九三学社 Jiu San Society / 台湾民主自治同盟(台盟) Taiwan Democratic Self-Government

【民主集中制】democratic centralism: ～的原则 principles of democratic centralism

【民主生活】democratic life: 坚持正常的～ maintain the normal practice of democracy

【民族】nation; nationality: 被压迫～ oppressed nations / 少数～ minority nationality; national minority／中华～ the Chinese nation ◇～败类 scum of a nation／～大家庭 the great family of nationalities／～独立 national independence／～对立 national antagonism／～复兴 revival of nationhood; national rejuvenation／～感情 national sentiments／～隔阂 national estrangement／～革命 national revolution／～解放运动 national liberation movement／～利己主义 national egoism／～民主革命 national-democratic revolution／～区域自治 regional autonomy of minority nationalities; regional national autonomy／～色彩 national in color／～统一战线 national united front／～团结 national unity／～文化宫 the Cultural Palace of the Nationalities／～形式 national style; national form／～学 ethnology／～压迫 national oppression／～遗产 national heritage／～意识 national consciousness／～英雄 national hero／～杂居地区 multi-national area／～政策 policy towards nationalities／～主义 nationalism／～资产阶级 national bourgeoisie／～自决 national self-determination／～自信心 national confidence／～自尊心 national pride; national self-respect／～组成 national composition ◯阿昌 Achang／白族 Baizu／崩龙 Benglong／布朗 Blang／保安 Bonan／布依 Bouyei／朝鲜 Chaoxian／傣族 Daizu／达斡尔 Daur／独龙 Derung／东乡 Dongxiang／侗族 Dongzu／鄂温克 Ewenki／高山 Gaoshan／仡佬 Gelao／京族 Ginzu／哈尼 Hani／汉族 Hanzu／赫哲 Hezhen／回族 Huizu／景颇 Jingpo／基诺 Jino／哈萨克 Kazak／柯尔克孜 Kirgiz／拉祜 Lahu／珞巴 Lhoba／傈僳 Lisu／黎族 Lizu／满族 Manzu／毛难 Maonan／苗族 Miaozu／门巴 Monba／蒙古 Mongol／仫佬 Mulam／纳西 Naxi／怒族 Nuzu／鄂伦春 Oroqen／普米 Primi／羌族 Qiangzu／俄罗斯 Russ／撒拉 Salar／畲族 Shezu／水族 Suizu／塔吉克 Tajik／塔塔尔 Tatar／土家 Tujia／土族 Tuzu／维吾尔 Uygur／乌孜别克 Uzbek／佤族 Vazu／锡伯 Xibe／瑶族 Yaozu／彝族 Yizu／裕固 Yugur／藏族 Zangzu／壮族 Zhuangzu

mǐn

悯 commiserate; pity: ～恻于怀 pity in the heart

敏 quick; nimble; agile

【敏感】sensitive; susceptible: ～的皮肤 sensitive skin／政治～ political sensitivity ◇～度 susceptibility／～区 sensitizing range／～元件 [无] sensitive element; sensor

【敏化】[物] sensibilization; sensitization ◇～剂 sensitizer／～纸 sensitized paper

【敏慧】clever; sharp-witted; keen

【敏捷】quick; nimble; agile: ～机智 quick-witted and sagacious／才思～ quick in thought／动作～ be quick in movement／举动～ as nimble as a squirrel

【敏锐】sharp; acute; keen: ～的见解 a sagacious perception／～的政治眼光 keen political insight／目光～ have sharp eyes; be sharp-eyed／听觉～ have good ears; have sharp ears／嗅觉～ have a keen sense of smell

【敏于事而慎于言】speedy as a worker and cautious as a speaker

泯 vanish; die out: 永存不～ be everlasting; be immortal

【泯灭】die out; disappear; vanish: 难以～的印象 an indelible impression

【泯没】vanish; sink into oblivion; become lost: ～无闻 dead and forgotten

【泯弃宿怨】disregard old grievances

抿 ①(用小刷子蘸水或油抹头) smooth with a wet brush ②(稍稍合拢) close lightly; furl; tuck: ～着嘴笑 smile with closed lips; compress one's lips to smile／～翅 furl the wings ③(稍沾) sip: ～一口酒 take a sip of the wine

【抿子】small hairbrush

míng

冥 ①(昏暗) dark; obscure: 幽～ dark hell; the nether world ②(深沉) deep; profound: ～想 be deep in thought ③(阴间) underworld; the neither world: ～府 the nether world ④(糊涂) dull; stupid: ～顽 thickheaded; stupid

【冥冥之中】in the unseen world

【冥器】funerary object; burial objects

【冥思出神】be in a brown study

【冥思苦想】think long and hard; cudgel one's brains

【冥顽】[书] thickheaded; stupid: ～不灵 impenetrably thickheaded

【冥王星】[天] Pluto

【冥想】deep thought; meditation: 苦思～ think long and hard; cudgel one's brains

瞑

【瞑目】close one's eyes in death; die content: ～

九泉 one's soul may rest at peace / 死不～ die discontent; die with everlasting regret

螟 snout moth's larva

【螟虫】snout moth's larva

【螟蛾】snout moth

【螟害】borer pest

【螟蛉】①（小虫）corn earworm ②（比喻义子）adopted son

明 ①（亮）bright; brilliant; light: ～月 a bright moon / 灯火通～ be brightly lit; be brilliantly illuminated ②（明白；清楚）clear; distinct: 去向不～ whereabouts unknown / 说～自己的意思 make one's meaning clear / 指～出路 point the way out ③（公开的）open; overt; explicit: ～里暗里支持 give both overt and covert support ④（眼力好）sharp-eyed; clear-sighted: 耳聪目～ have sharp ears and eyes / 眼～心亮 be sharp-eyed and clear-headed ⑤（心地光明）aboveboard; honest ～人不做暗事 an honest man doesn't do anything underhand ⑥（视觉）sight: 复～ regain one's sight / 双目失～ go blind in both eyes ⑦（懂得）understand; know: 不～真相 not know the facts; be ignorant of the actual situation / 深～大义 understand perfectly where righteousness lies ⑧（次于今年或今天的）immediately following in time: ～晨 tomorrow morning / ～年 next year / ～晚 tomorrow evening

【明暗】light and shade ◇ ～对照法 [美术] chiaroscuro

【明摆着】obvious; clear; plain: 这是～的事实。It is an obvious fact.

【明白】①（清楚；明确）clear; obvious; plain ②（懂得）know; understand: 把真相讲～ explain the actual situation in explicit terms ③（公开的；不含糊的）frank; unequivocal; explicit ④（懂道理）sensible; reasonable: 他是一个～人。He is a sensible person.

【明白表示】clear expression

【明辨是非】make a clear distinction between right and wrong

【明察暗访】observe publicly and investigate privately conduct a thorough investigation

【明察秋毫】have eyes sharp enough to perceive an animal's autumn hair; be perceptive of the minutest detail

【明畅】lucid and smooth

【明澈】bright and limpid; transparent: ～的眼睛 clear and shining eyes

【明处】①（明亮的地方）where there is light ②（公开场合）in the open; in public: 有话说在～。If you have got anything to say, say it openly.

【明灯】bright lamp; beacon: 指路～ a beacon lighting up one's way forward

【明矾】alum ◇ ～石 alumstone; alunite

【明沟】open drain

【明晃晃】gleaming; shining: ～的刺刀 gleaming bayonets

【明火执仗】carry torches and weapons in a robbery; conduct evil activities openly

【明胶】gelatin

【明净】bright and clean; clear and bright: ～的卧室 a bright and clean bedroom

【明镜】bright mirror

【明快】①（明白通畅）lucid and lively; sprightly: ～的笔调 a lucid and lively style / ～的节奏 sprightly rhythm ②（性格爽朗）straightforward; forthright: ～的性格 a forthright character

【明来暗往】have overt and covert contacts with sb.

【明朗】①（光亮）bright and clear: ～的天空 a clear sky / ～的早晨 a bright morning ②（明显）clear; obvious: 态度～ take a clear-cut position; adopt an unequivocal attitude ③（开朗爽快）forthright; bright and cheerful: ～的性格 an open and forthright character / 色彩～ in bright and gay colors

【明亮】①（光线足）light; well-lit; bright: 宽畅而～的房间 a bright and spacious room / 在～的灯光下 under a bright lamp ②（发亮的）bright; shining: ～的眼睛 bright eyes ③（明白）become clear

【明了】①（清楚地懂得）clearly understand; be clear about: ～党的政策 understand clearly the Party's policies ②（清晰）clear; plain: 简单～ simple and clear

【明令】explicit order; formal decree; public proclamation: ～嘉奖 issue a commendation; mention in a citation / ～禁止 prohibited by official order / ～取缔 proscribe by formal decree

【明码】plain code: ～电报 plain code telegram / 用～发报 send a telegram in plain code

【明媒正娶】formal wedding

【明媚】bright and beautiful; radiant and enchanting: ～的春光 a radiant and enchanting spring scene

【明明】[副] obviously; plainly: ～是 it is obvi-

ously that...／这话～是她说的。It is undoubtedly she who has said that.
【明目张胆】brazenly; flagrantly: ～地进行武装干涉 brazenly commit an act of armed intervention
【明年】next year
【明枪易躲,暗箭难防】it is easy to dodge a spear in the open, but hard to guard against an arrow shot from hiding
【明确】①(明白确定) clear and definite; clear-cut; explicit; unequivocal: ～的承诺 definite undertaking／～的答复 a definite answer／～的分工 clear-cut division of labor／～的立场 a clear-cut stand／～的目标 a clear aim／我们的态度是～的。Our attitude is clear. ②(使明白确定) make clear and definite: ～讨论的目的 make clear the purpose of the discussion
【明儿】[口] ①(明天) tomorrow ②(将来) one of these days; some day
【明日】①(明天) tomorrow ②(不久的将来) the near future
【明日黄花】overblown blossoms; things that are stale and no longer of interest
【明升暗降】kick upstairs
【明示】express: ～保释 express bailment／～承认 express recognition／～担保 express guaranteeship／～放弃 express waiver／～和解 express arrangement／～合同 express contract／～交付 express delivery／～拒绝履行 express renunciation／～诺言 express promise／～条件 express condition／～条款 express terms／～同意 express consent／～信托 express trust／～引渡 express extradition
【明太鱼】walleye pollack
【明天】①(明日) tomorrow ②(不远的将来) the near future: 展望美好的～ look forward to a bright future
【明文】proclaimed in writing: ～规定 stipulate in explicit terms; expressly provide
【明晰】distinct; clear: 有一个～的印象 have a clear impression
【明细帐簿】subsidiary book
【明虾】prawn
【明显】clear; obvious; evident; distinct: ～的成效 tangible result／～的错误 evident mistake／～的恶意 express malice／～的改进 distinct improvement／～的优势 clear superiority／～的证据 clear evidence
【明线】[电子] open-wire line; open wire ◇ ～载波设备 open-wire carrier equipment
【明效大验】clinching proof of effectiveness; outstanding effect
【明信片】postcard: 美术～ picture postcard
【明星】star: 电影～ film star; movie star／足球队的～球员 star players of a soccer team
【明修栈道,暗渡陈仓】pretend to prepare to advance along one path while secretly going along another; do one thing under cover of another
【明眼人】a person with a discerning eye; a person of good sense
【明一套暗一套】act one way in the open and another way in secret
【明于观人,暗于察己】good at knowing others but poor at knowing oneself
【明喻】simile
【明哲保身】be worldly wise and play safe
【明争暗斗】both open strife and veiled struggle
【明证】clear proof
【明知】know perfectly well; be fully aware: ～山有虎,偏向虎山行 go deep into the mountains, knowing well that there are tigers there; go on undeterred by the dangers ahead
【明知故犯】knowingly violate deliberately break (a rule, etc.); do sth. one knows is wrong
【明知故问】ask while knowing the answer
【明智】sensible; sagacious; wise: ～之举 a wise move／表现出～的态度 show a sensible attitude
【明珠】bright pearl; jewel
【明珠暗投】①(怀才不遇) cast pearls before swine; find one's ability unrecognized ②(好人失足) a good person fallen among bad company
【明子】pine torch

盟

【盟誓】take an oath; make a pledge

鸣

①(鸟兽或昆虫叫) the cry of birds, animals or insects: 鸟～ chirping of birds／蛙～ croaking of a frog ②(发声) ring; sound: 耳～ ringing in the ears／雷～ pealing of the thunder ③(表达) express; voice; air: ～不平 make complaints／自～得意 be very pleased with oneself; preen oneself
【鸣笛】whistle: 禁止～。No Tooting.
【鸣号】trumpet; sound the bugle
【鸣礼炮】fire a salute: ～二十一响 fire a salute of 21 guns
【鸣锣开道】prepare the public for a coming

event
【鸣枪】fire rifles into the air; fire a shot: ～示警 fire a warning shot
【鸣禽】singing bird; songbird
【鸣谢】express gratitude; express one's thanks formally
【鸣冤叫屈】complain and call for redress; voice grievances

名 ①(名字;名称) name; title: 地～ place name / 人～ name of a person / 书～ title of a book ②(不带姓的名字) given name ③(名声) fame; reputation; renown: 不为～, 不为利 seek neither fame nor gain ④(有名声的) famous; well-known; celebrated; noted: ～城 well-known city / ～画家 celebrated painter; noted painter / ～医 famous physician ⑤[量] 五～教师 five teachers / 第一～ the first place
【名不副实】the name falls short of the reality; be sth. more in name than in reality; be unworthy of the name or title
【名不虚传】have a well-deserved reputation; deserve the reputation one enjoys; live up to one's reputation
【名册】register; roll: 学生～ students' register; students' roll / 部队～ muster roll / 工作人员～ personnel roll
【名产】famous product; specialty goods
【名场失意】be disappointed in examination for degrees
【名称】name (of a thing or organization): ～代表什么? What's in a name?
【名垂青史】go down in history; be crowned with eternal glory
【名词】①[语] noun; substantive ②(术语) term; phrase: 科学～ scientific terms / 新～ new terms
【名次】position in a name list; place in a competition
【名存实亡】cease to exist except in name; exist in name only
【名单】name list: 候选人～ list of candidates / 主席团～ list of the members of the presidium
【名额】the number of people assigned or allowed; quota of people: 代表～ the number of deputies to be elected or sent / 招生～ the number of students to be enrolled; planned enrollment figure
【名孚众望】prestige commands public confidence
【名副其实】the name matches the reality; be sth. in reality as well as in name; be worthy of the name: ～的专家 be an expert worthy of the name
【名贵】famous and precious; rare: ～的古玩 rare curious / ～的药材 rare medicinal herbs / ～的字画 priceless scrolls of calligraphy and painting
【名家】①(古代思想流派) the School of Logicians ②(著名专家) a person of academic or artistic distinction; famous expert; master: ～手笔 the handwritings of celebrities
【名将】famous general; great soldier: 足球～ a soccer hero; a soccer star
【名利】fame and gain; fame and wealth ◇～思想 desire for personal fame and gain
【名利双收】gain both honor and money; gain both fame and wealth
【名列前茅】be among the best of the successful candidates
【名流】distinguished personages; celebrities
【名落孙山】fall in a competitive examination
【名门闺秀】daughter of an illustrious family
【名门之后】descendant of notable family
【名目】names of things; items: ～繁多 a multitude of names 〈items〉; names of every description / 巧立～ invent all kinds of names
【名牌】①(出名的牌子) famous brand: ～手表 watch of famous brand ②(写着人名的牌子) nameplate; name tag
【名片】visiting card; calling card: 留下～ leave one's card 肖像～ carte-de-visite
【名气】reputation; fame; name: ～大的人 man of great reputation / 有点～ enjoy some reputation; be quite well-known; have made a name for oneself
【名曲】a great musical composition; a masterpiece in music
【名人】famous person; eminent person; celebrity; notable: 历史～ great names in history
【名山大川】famous mountains and great rivers
【名声】reputation; repute; renown: ～不好 have a bad reputation / ～很坏 have an unsavory reputation; be held in ill repute; be notorious / ～狼籍 having a bad reputation / 享有好～ enjoy a good reputation; be held in high repute
【名胜】a place famous for its scenery or historical relics; scenic spot ◇～古迹 places of historic interest and scenic beauty; scenic spots and historical sites
【名师出高徒】an accomplished disciple owns his accomplishments to his great teacher

【名手】a famous artist, player, etc.: 象棋～ famous chess player
【名数】[数] concrete number
【名堂】①（成就;成果）result; achievement: 他们会搞出～来的。They'll certainly accomplish something. ②（道理;内容）reason ③（花样）item; variety: 搞不出～来 accomplish nothing remarkable
【名望】fame and prestige; good reputation; renown: ～颇高 one's reputation is fairly high / 有～的大夫 a famous doctor
【名位】fame and position
【名闻中外】well-known both at home and abroad
【名下】under sb.'s name; belonging or related to sb.
【名言】well-known saying; celebrated dictum; famous remark
【名扬四海】well-known in the world; world-renowned; become famous all over the world
【名义】①（名称或称号）name: 盗用…的～ usurp the name of... / 假借…的～ on the false pretenses of... / 以革新的～ in the name of innovation ②（表面上）nominal; titular; in name: ～上裁军，实际上扩军 disarmament in name, armament in reality ◇ ～被告 nominal defendant / ～代理 ostensible agency / ～代理权 ostensible authority / ～当事人 nominal party / ～负债 nominal liability / ～工资[经] nominal wages / ～合伙人 ostensible partner / ～汇价[经] nominal rate / ～所得 nominal income / ～原告 nominal plaintiff
【名誉】①（名声）fame; reputation: ～扫地 be discredited / ～攸关 affects one's reputation / 破坏某人～ damage sb.'s reputation / 追求～地位 seek after fame and position ②（名义上的）honorary ◇ ～顾问 honorary advisor / ～会员 honorary member / ～损害赔偿 indemnity for defamation / ～主席 honorary chairman; honorary president
【名噪一时】gain considerable fame among one's contemporaries
【名正言顺】come within one's jurisdiction; be perfectly justifiable
【名著】famous book; famous work: 科学～ famous science book / 文学～ a famous literary work / 英国文学～ masterpieces in English literature
【名字】①（人名）name: 那人的～叫什么? What is the name of that man? ②（事物名称）name; title: 一个学校的～ name of a school / 影片的～ title of a film

茗 tea: 品～ sip tea; sample tea

铭 ①（器物上刻的文字）inscription: 墓志～ inscription on the memorial tablet within a tomb / 座右～ motto ②（永记不忘;纪念）engrave: ～诸肺腑 bear firmly in mind
【铭感】be deeply grateful: ～终身 remain deeply grateful for the rest of one's life
【铭肌镂骨】remember with gratitude to the end of one's life; be engraved on one's mind forever
【铭记】engrave on one's mind; always remember: ～心头 be engraved on one's mind
【铭刻】①（刻在器物上的文字）inscription ②（铭记）engrave on one's mind ◇ ～学 epigraphy
【铭牌】[机] data plate; nameplate
【铭文】inscription; epigraph
【铭谢】show gratefulness
【铭心】imprint on one's mind

mǐng

酩
【酩酊大醉】be dead drunk

mìng

命 ①（生命）life: 救～ save sb.'s life / 丧～ lose one's life / 逃～ run for one's life ②（命运）lot; fate; destiny: 苦～ hard lot / 听天由～ resign to fate; submit oneself to fate ③（命令）order; command: 不肯听～于人 refuse to take orders from anybody / 待～ await orders ④（给与）assign ～题 assign a topic; set a question
【命案】a case involving the killing of a person; homicide case
【命笔】[书] take up one's pen; set pen to paper: 欣然～ gladly set pen to paper; be happy to start writing
【命根子】one's very life; lifeblood: 这女孩子是她母亲的～。The girl is the lifeblood of her mother.
【命令】order; command: ～式的口气 a commanding tone / 服从～ obey orders / 下～ issue an order ◇ ～句 [语] imperative sentence / ～主义 commandism
【命脉】lifeblood; lifeline: 经济～ economic lifelines
【命名】name: 这个海峡是以它的发现者～的。

The strait was named after its discoverer. ◇ ～大会 naming ceremony / ～法 nomenclature
【命若悬丝】life as if hanging by a thread
【命数法】[数] numeration
【命题】①（出题目）assign a topic; set a question: ～作文 assign a subject for composition ② [逻][数] proposition
【命途多舛】suffer many a setback during one's life
【命意何在】where is the meaning
【命运】destiny; fate; lot: 关心国家的～ be concerned about the destiny of the state / 掌握自己的～ grasp one's destiny in one's own hands
【命在旦夕】death may come any minute; be dying
【命之所招】be caused by fate
【命中】hit the target; score a hit: ～目标 hit the mark ◇ ～率 percentage of hits / ～偏差 deviation of impact

miù

谬 wrong; false; erroneous; mistaken: ～传 a false report / ～见 a wrong view / 大～不然 be grossly mistaken
【谬奖】[谦] overpraise
【谬论】fallacy; false theory; falsehood: ～连篇 a link of fallacious discussions / 驳斥～ refute the fallacy
【谬误】falsehood; error; mistake: ～百出 there are hundreds of errors
【谬种】error; fallacy: ～流传 the dissemination of error

mō

摸 ①（触摩）touch; feel; stroke: ～起来很平滑 smooth to the touch / ～～看水够不够热 feel whether the water is warm enough yet / 你～～看。Just feel it. ②（摸索）feel for; grope for; fumble: 在口袋里～钥匙 fumble in one's pocket for the key / 他在衣袋里～着找了一个硬币。He felt in his pockets for a coin. ③（试着了解）get to know; find out; sound out: ～不着头脑 can't make head or tail of sth. / ～清敌情 find out about the enemy's situation / ～透了他的脾气 get to know him well / ～一下他对这个问题的意见 sound her out about this matter
【摸不透】wonder; be puzzled
【摸彩】draw lot to determine the prize winners in a raffle or lottery
【摸到门路】have learned the ways of the trade
【摸底】①（了解底细）know the real situation ②（打听底细）try to find out the real intention or situation; sound sb. out
【摸黑儿】[口] grope one's way on a dark night: ～赶路 press on with the journey at night / 起早～地干 work from morning till night
【摸清底细】get to the bottom of the story; ascertain the actual situation
【摸索】①（试探着）grope; feel about; fumble: ～而行 feel one's way / 在黑暗中～ grope in the dark ②（寻找）try to find out: 不断～,掌握技能 learn technical skills through trial and error

mó

磨 ①（摩擦）rub; wear: 裤子上～出一个洞 wear a hole in one's trousers / 手上～出泡来。The rubbing raised a blister on the palm. ②（磨东西）grind; polish sharpen: ～刀 whet a knife / ～斧子 grind an axe / ～墨 rub an ink stick ③（折磨）wear down; wear out: 她给病～垮了。She was totally worn down by the disease. ④（纠缠）trouble; pester; worry: ～着某人要某物 worry sb. for sth. ⑤（拖延；耗时间）dawdle; waste time: ～时间 dawdle away one's time ⑥（消灭;磨灭）obliterate; die out: 百世不～ will endure for centuries
【磨版机】[印] graining machine
【磨蹭】①（轻微摩擦）lightly rub ②（行动缓慢）dawdle; move slowly
【磨杵成针】grind mortar into needle
【磨穿铁砚】grind through an inkstone; long years of study
【磨床】[机] grinding machine; grinder: 高精度～ high-precision grinder / 内圆～ internal grinder / 外圆～ cylindrical grinder
【磨刀霍霍】sharpen one's swords for battles
【磨刀器】knife-grinder
【磨刀石】whetstone; grindstone
【磨革】[皮革] buffing ◇ ～机 buffing machine
【磨工】[机] ①（磨工活）grinding work ②（磨工工人）grinder ◇ ～车间 grindery
【磨光】polish ◇ ～玻璃 polished glass / ～机 polishing machine; glazing machine
【磨耗】wear and tear
【磨练】put oneself through the mill; temper oneself; steel oneself: 经受艰难困苦的～ go through hardships and tribulations / 在实际工作中～自己 steel oneself in practical work
【磨料】abrasive; abradant

【磨灭】wear away; efface; obliterate: 建立不可～的功勋 perform meritorious deeds never to be obliterated / 留下不可～的印象 leave an indelible impression
【磨墨挥毫】grind the ink and flourish the brush to write
【磨木机】[纸] (wood) grinder: 袋式～ pocket grinder / 链式～ caterpillar grinder
【磨难】tribulation; hardship; suffering
【磨拳擦掌】ready for fight
【磨砂玻璃】ground glass; frosted glass
【磨损】wear and tear ◇ ～留量 [机] wear allowance
【磨削】[机] grinding ◇ ～裕量 grinding tolerance
【磨牙】①(睡觉时咬牙) grind one's teeth in sleep ②(多费口舌) indulge in idle talk; argue pointlessly ③ [解] molar (teeth): 前～ premolar
【磨洋工】loaf on the job; dawdle along
【磨琢成器】grinding and polishing to become something
【磨嘴皮子】①(磨牙) jabber; blah-blah ②(说废话) do a lot of talking

蘑 mushroom
【蘑菇】①(食用蕈类) mushroom ②(故意纠缠) worry; pester; keep on at ③(拖延时间) dawdle; dillydally ◇ ～云 mushroom cloud / ～战术 the tactics of "wear and tear"

摩 ①(摩擦; 接触) rub; scrape; touch: 峻岭～天 the high mountains seem to scrape the sky ②(研究切磋) mull over; study
【摩擦】①(摩动) rub ② [物] friction 滚动～ rolling friction / 滑动～ sliding friction ③(冲突) friction; conflict; clash: 内部～ internal friction ◇ ～力 [物] frictional force; friction / ～抛光 [机] burnishing / ～音 [语] fricative / ～桩 [建] friction pile
【摩登】modern ; fashionable
【摩电灯】dynamo-powered lamp
【摩顶放踵】delicate oneself completely to the welfare of mankind
【摩尔根主义】[生] Morganism
【摩肩接踵】jostle each other in a crowd
【摩羯座】[天] Capricornus; Sea Goat
【摩洛哥】Morocco ◇ ～人 Moroccan
【摩纳哥】Monaco ◇ ～人 Monacan
【摩拳擦掌】rub one's fists and wipe one's palms; be eager for a fight; itch to have a go
【摩天】sky scraping ◇ ～楼 skyscraper
【摩托】motor ◇ ～化部队 motorized troops / ～运动 motorcycling
【摩托车】motorcycle; motor bicycle; motorbike: 单人～ solo; solomachine / 三轮～ motor tricycle; motor tri-wheeler / 三轮运货～ three-wheel deliver motor lorry ◇ ～领先自行车赛 motor-paced race / ～选拔赛 motocycle trial / ～越野赛 motor-cross / ～运动 motorcycle sport
【摩托艇】motorboat: ～运动 motorboating; powerboating

魔 ①(魔鬼) devil; demon; evil spirit ②(神奇) magic; mystic: ～力 magic power
【魔法】magic; wizardry; sorcery; witchcraft
【魔方】magic square
【魔怪】demons and monsters; fiends
【魔鬼】devil; demon; monster
【魔窟】den of monsters
【魔力】magic power; magic; charm
【魔术】magic; conjuring; sleight of hand ◇ ～演员 magician; conjurer
【魔王】①(恶鬼) Prince of the Devils ②(凶暴的恶人) tyrant; despot; fiend
【魔掌】devil's clutches; evil hands: 逃出敌人的～ escape from the clutches of the enemy
【魔杖】magic wand
【魔爪】devil's talons; claws; tentacles: 斩断侵略者的～ cut off the tentacles of the aggressors

模 ①(法式; 规范) pattern; standard: 楷～ model; paragon ②(仿效) imitate ②(模范) model: 劳～ model worker
【模本】calligraphy or painting model
【模范】an exemplary person or thing; model; fine example: ～工作者 model worker / ～共产党员 model member of the Communist Party / ～事迹 exemplary deeds / ～行为 exemplary behavior / ～战士 exemplary fighter / ～作用 exemplary role
【模仿】imitate; copy; model oneself on: ～动物的叫声 imitate the cries of animals / ～某人 copy sb.'s example / ～鸟 mockingbird
【模糊】①(不清楚) dim; vague; indistinct; obscure: ～的景象 obscure view / ～的印象 vague impression / 童年时代的～记忆 dim memories of one's childhood / 认识～ have but a hazy understanding ②(混淆) blur; obscure; confuse; mix up: ～二者的界限 confuse the distinction between the two ◇ ～数学 fuzzy mathematics / ～信息 fuzzy message
【模棱两可】equivocal; ambiguous: ～的话

equivocality; hedge / ～的回答 equivocal reply / ～的提法 an ambiguous formulation / 采取～的态度 take an equivocal attitude
【模模糊糊】unintelligible
【模拟】imitate; simulate ◇ ～飞行[军] simulated flight / ～计算机 analogue computer / ～器 emulator; imitator; simulator / ～人像 effigy / ～试验 simulated test / ～战 mock battle; mimic warfare
【模数】[物] modulus: 弹性～ modulus of elasticity
【模特儿】model: 画～ paint model / 裸体～ nude model / 时装～ dress model / 做～ stand model
【模型】①(仿制实物) model; pattern: ～试验 model test / ～展品 scale model / 飞机～ model of an airplane / 复制～ replica / 机器的～ pattern of a machine ②(制砂型的工具) mold; pattern: ～板 mold plate

摹 copy; trace: 临～ copy a model of calligraphy or painting
【摹本】facsimile; copy
【摹古】model or pattern after ancient style: ～之作 a production copied from the old style
【摹刻】①(摹写并雕刻) copy by carving ②(摹刻的成品) carved reproduction
【摹拟】mimic; imitate: ～某人声音、姿势和举止 imitate sb.'s voice, gestures and manners
【摹写】①(照样子写) copy; imitate the calligraphy of ②(描写) describe; depict
【摹印】①(古印玺字体) ancient imperial seal script ②(摹写并印刷) copy and print

膜 ①(皮组织) membrane: 粘～ mucous membrane ②(象膜样的薄皮) film: 塑料薄～ plastic film / 结成一层薄～ form a thin film
【膜拜】prostrate oneself worship: 顶礼～ prostrate oneself in worship; pay homage to
【膜翅目】[动] Hymenoptera
【膜法】[环保] membrane method
【膜片】[仪表] diaphragm

mǒ

抹 ①(涂抹) put on; apply; smear; plaster: ～粉 put on powder / 给伤口～上药膏 apply ointment to the wound / 往面包上～黄油 spread butter on bread ②(擦去) wipe off: ～去额头的汗水 wipe the sweat off one's forehead ③(勾掉) erase; strike out; delete; blot out; cross out: ～掉磁盘上的文件 erase the file from a disc
【抹脖子】cut one's own throat; commit suicide
【抹掉】erase; wipe away: ～这个短语。Cross out this phrase.
【抹黑】blacken sb.'s name; discredit; throw mud at; bring shame on
【抹杀】blot out; obliterate; write off: ～是非界线 blur out distinctions between right and wrong / ～事实 deny facts / ～一切 wipe out everything / 一笔～ write off at one stroke; deny completely
【抹香鲸】sperm whale
【抹一鼻子灰】suffer a snub; meet with a rebuff
【抹子】[建] trowel

mò

磨 ①(磨子) mill; millstones: 电～ electric mill / 盘～ a mill / 推～ turn a millstone ②(磨碎) mill; grind: ～豆腐 grind soybeans to make bean curd / ～麦子 grind wheat ③(掉转) turn about: 把汽车～过来 turn about the car
【磨不开】①(脸上下不来) feel embarrassed ②(不好意思) for fear of hurting sb.'s feelings
【磨坊】mill
【磨面机】flour-milling machine
【磨盘】①(底盘) nether millstone ②(磨) mill; millstone

末 ①(东西的梢) tip; end: 秋毫之～ the tip of an animal's autumn hair ②(不是根本的) nonessentials; minor details: 本～倒置 take the branch for the root; put the nonessentials before the essentials; put the cart before the horse ③(最后) end; last stage: 月～ at the end of the month / 周～ weekend ④(粉末) powder; dust: 茶叶～ tea dust / 研成～ grind to powder; pulverize
【末班车】last bus
【末代】the last reign of a dynasty: ～皇帝 the last emperor of a dynasty
【末节】minor details; nonessentials: 细枝～ minor details
【末了】last; finally; in the end: ～的一个 the last one
【末路】dead end; impasse: ～途穷 reach the end of the rope
【末年】last years of a dynasty or reign
【末期】last phase; final phase; last stage: 春秋～ end of the Spring and Autumn Period / 十九世纪～ towards the end of the nineteenth century

【末日】①[基督教] doomsday; Day of Judgment; Judgment Day: ～审判 Last Judgment ②(死亡或灭亡的日子) end; doom: ～将至 one's last day is soon reached / ～可数 one's days are numbered
【末梢】tip; end: 鞭子的～ the tip of a whip ◇ ～神经 nerve ending
【末尾】①(最后的部分) end: 排在～ stand at the end of a line / 在信的～ at the end of the letter ②[乐] fine; end
【末叶】last years: 二十世纪～ the end of the 20th century

沫 foam; froth: 肥皂～ soapsuds; lather / 啤酒～ froth on beer; the head on a glass of beer / 吐白～ foam at the mouth
【沫子】foam; froth

茉

【茉莉】[植] jasmine ◇ ～花茶 jasmine tea

抹 ①(把和好了的灰或泥涂上再抹平) daub; plaster: ～墙 plaster a wall; daub plaster on a wall ②(紧挨着绕过) skirt; bypass
【抹灰】[建] plastering ◇ ～工 plasterer

秣 ①(牲口的饲料) fodder ②(喂牲口) feed animals
【秣马厉兵】feed the horses and sharpen the weapons; make active preparations for war; prepare for battle

莫 ①(不) no; not ～知所措 not know what to do; be at a loss ②(不要) don't: ～管闲事。Don't poke your nose into other's business. / ～慌张。No hurry! (或: Take it easy.) / 非公～入。No admittance except on business.
【莫不】there's no one who doesn't or isn't: ～发指。There were none who were not angry with him. / ～为之感动。There was no one who was unmoved.
【莫测高深】unfathomable; enigmatic
【莫大】greatest; utmost: ～的愤慨 the utmost indignation / ～的侮辱 a gross insult / ～的幸福 the greatest happiness / ～的愉快 utmost pleasure / 感到～的光荣 feel greatly honored
【莫非】[副] can it be that; is it possible that: ～写错了? Could it be that it was written wrongly?
【莫怪】no wonder that...
【莫过于】nothing is more ... than; nothing is better than
【莫可名状】that cannot be described
【莫名其妙】①(无人能说明其奥妙) be unable to make head or tail of sth.; be baffled ②(奇怪的事) without rhyme or reason; inexplicable; odd: 她～地生起气来。She got angry without rhyme or reason.
【莫逆】very friendly; intimate: ～之交 bosom friends
【莫如】would be better; might as well
【莫桑比克】Mozambique ◇ ～人 Mozambican
【莫为儿孙作牛马】do no slave for your children
【莫须有】unwarranted; groundless; fabricated; trumped-up: ～的罪名 a fabricated charge; an unwarranted charge
【莫衷一是】unable to agree or decide which is right

漠 ①(沙漠) desert ②(冷淡地) indifferent; unconcerned
【漠不关心】indifferent; unconcerned: 对别人的痛苦～ show indifference to the pain of others
【漠不相关】entirely unrelated
【漠漠】①(云烟密布的样子) misty; foggy ②(广漠而沉寂) vast and lonely: 黄沙～ a vast stretch of yellow sand
【漠然】indifferently; apathetically; with unconcern: ～不动 completely indifferent / ～置之 remain indifferent towards sth.; look on with unconcern
【漠视】treat with indifference; ignore; overlook; pay no attention to: ～群众的意见和要求是错误的。It is wrong to have no regard for the opinions and demands of the masses.

寞 lonely; deserted: 寂～ lonely

貘 [动] tapir

蓦 suddenly
【蓦地】suddenly; unexpectedly; all of a sudden: ～惊叫 give a sudden scream
【蓦然】suddenly; ～想起 suddenly remember

陌 ①(田间小路) a path between fields ②(泛指道路) road: ～头杨柳 roadside willows
【陌路】stranger: 视同～ regard as a stranger
【陌生】strange; unfamiliar: ～面孔 unfamiliar face / ～人 stranger

墨 ①(写字或绘画用的墨) China ink; ink stick: 一块～ an ink stick ②(写字或绘画等用

的颜料) ink ③(字画) handwriting or painting: 遗～ writing or painting by the deceased ④(学识) learning: 胸无点～ without any learning; unlettered ⑤(黑) black; dark: ～黑 pitch-dark
【墨宝】①(贵重字画) treasured scrolls of calligraphy or painting; valued pieces of calligraphy or painting ②(尊称) your beautiful handwriting or painting
【墨斗】carpenter's ink marker
【墨斗鱼】inkfish; cuttlefish
【墨盒】ink box
【墨迹】①(墨的痕迹) ink mark: ～未干 before the ink is dry / ～犹新 the trace of ink is still fresh ②(亲笔字画) sb.'s writing or painting
【墨家】Mohist School (770-221 B.C.): ～学说 Mohism
【墨晶】smoky quartz
【墨镜】sunglasses
【墨卡托投影】Mercator projection
【墨绿】blackish green
【墨囊】[动] ink sac
【墨色不均】[印] bad color
【墨守成规】stick to conventions; stay in a rut
【墨守旧习】adhere to old customs
【墨水】①(墨汁) prepared Chinese ink ②(钢笔用的水) ink ③(学问) book learning ◇ ～池 inkwell / ～瓶 ink bottle / ～台 inkstand
【墨西哥】Mexico ◇ ～人 Mexican / ～湾 the Gulf of Mexico / ～湾流 the Gulf Stream
【墨线】①(墨斗上的线绳) the line in a carpenter's ink marker ②(用墨线打出的直线) line made by a carpenter's ink marker
【墨汁】prepared Chinese ink

默 ①(不说话) silent; tacit: ～不作声 keep silent / ～坐 sit in silence ②(默写) write from memory: ～生字 write the new words from memory
【默哀】stand in silent tribute: ～三分钟 observe three minutes' silence
【默祷】pray in silence; say a silent prayer
【默读】read silently
【默默】quietly; silently: ～无言 without saying a word; silently
【默默无闻】unknown to the public; without attracting public attention: 一生～ remain obscure all one's life
【默契】①(心照不宣) tacit agreement; tacit understanding ②(秘密协定) secret agreement: 双方早有～。 The two parties reached a secret agreement long ago.
【默然】silent; speechless: ～无语 fall silent; be speechless
【默认】give tacit consent to; tacitly approve; acquiesce in: ～既成事实 acquiesce in the fait accompli ◇ ～契约 implied contract / ～作废 repeal by implication
【默示】imply ◇ ～承诺 implied promise / ～承认 implied recognition / ～担保 implied warrant / ～放弃 implied waiver / ～合同 implied contrast / ～批准 implied ratification / ～特约条款 warranty implied / ～条件 implied term / ～同意 implied consent / ～协定 implied agreement / ～信托 implied trust / ～异议 implied objection
【默诵】read silently
【默写】write from memory
【默许】tacitly consent to; acquiesce in: ～所求 tacit permission of a request / ～条件 implied condition

脉
【脉脉】affectionately; lovingly; amorously: ～含情 quietly sending the message of love / 温情～ full of tender affection; sentimental

没 ①(沉下或沉没) sink; submerge: ～入水中 sink into water / 船沉～。 The ship sank. ②(漫过或高过) overflow; rise beyond: 雪深～膝。 The snow is knee-deep. ③(隐没) disappear; hide ④(没收) confiscate; take possession of ⑤(尽;终) till the end: ～世 till the end of one's life
【没齿不忘】will never forget to the end of one's days; remember for the rest of one's life
【没顶】be drowned: 水深～。 The water goes above a man's head.
【没落】decline; wane: ～贵族 declining aristocrat
【没奈何】be utterly helpless; have no way out; have no alternative
【没世】through one's life-time; till the end of one's life: ～不忘 I shall not forget it in my life
【没收】confiscate; expropriate ◇ ～财产 expropriation; confiscation of property / ～公告 declaration of forfeiture / ～货物 confiscated goods / ～者 expropriator

mōu

哞 [象] moo; low; bellow

móu

谋 ①(主意) stratagem; plan; scheme: 有勇无～ brave but not astute / 足智多～ wise and full of stratagems; resourceful ②(谋求) work for; seek; plot: ～私利 pursue private ends / 另～出路 seek another way out / 为人民～福利 work for the welfare of the people ③(商议) consult: 不～而合 be in full agreement without previous consultation
【谋财害命】murder sb. for his money
【谋刺】plot to assassinate; make an attempt on sb.'s life
【谋反】conspire against the state; plot a rebellion：～叛逆 plot and rebel
【谋害】①(阴谋杀害) plot to murder ②(阴谋陷害) plot a frame-up against
【谋划】plan; scheme; try to find a solution
【谋利】profit; turn something to profit
【谋略】astuteness and resourcefulness; strategy
【谋求】seek; strive for; be in question of: ～霸权 seek hegemony / ～和平解决 bring a peaceful solution; try to solve by peaceful means / ～计划早日实现 strive for the early accomplishment of the plan
【谋取】try to gain; seek; obtain: ～长远利益 seek long-term interests / ～私利 play one's own game; trying to abstain profit for oneself
【谋杀】murder: ～未遂 make a vain attempt to murder ◇ ～案 case of murder / ～罪 offense of murder
【谋生】seek a livelihood; make a living: ～的手段 a means of life / ～乏术 having no means of getting a livelihood
【谋士】adviser; counsellor
【谋事】①(计划事情) plan matters ②(找职业) look for a job
【谋事在人，成事在天】man proposes, God disposes

牟 try to gain; seek; obtain
【牟利】seek profit
【牟取】try to gain; seek; obtain: ～暴利 seek exorbitant profits

眸 pupil; eye: 明～皓齿 have shining eyes and white teeth; be comely / 凝～ fix one's eyes on
【眸子】pupil (of the eye); eye

mǒu

某 ①(指一定的人或事) certain: ～日 at a certain date / ～些产品 certain products ②(指不定的人或事) some: ～地 some place / ～人 somebody
【某某】so-and-so: ～先生 Mr. so-and-so / ～学校 a certain school
【某些】certain; a few

mú

模 mold; pattern; matrix: 铜～[印] matrix; (copper) mold / 制～工 molder
【模板】①[建] shuttering; formwork ②[机](模型板) pattern plate
【模具】mold; matrix; pattern; die
【模压】mold pressing ◇ ～机 molding press / ～胶底皮鞋 leather shoes with molded-on rubber soles
【模样】①(长相) appearance; look: 这女孩子的～很讨人喜欢。 The girl has a pleasing appearance. ②(表示约略的情况) approximately; about; around: 这女孩有十岁～。The girl is around ten.
【模子】mold; matrix; pattern; die: 一个～里铸出来的 made out of the same mold; as like as two peas

mǔ

亩 *mu*：～产 per *mu* yield

牡 male: ～牛 bull
【牡丹】peony; tree peony
【牡蛎】[动] oyster

母 ①(母亲) mother: ～女俩 the mother and the daughter ②(家族或亲戚中的长辈女性) one's female elders: 姨～ aunt / 祖～ grandmother ③(雌性的) female: ～狗 bitch / ～鸡 hen / ～山羊 she-goat / ～猪 sow ④(有产生出其他事物的能力或作用的) origin; parent
【母爱】mother love; maternal love
【母本】[植] female parent: ～植株 maternal plant
【母畜】dam
【母蜂】queen bee
【母公司】parent company
【母机】①(工作母机) machine tool ②[航空] mother aircraft; launching aircraft
【母老虎】①(雌老虎) tigress ②(泼妇) vixen; shrew; termagant
【母亲】mother ◇ ～节 Mother's Day on the second Sunday in May
【母权制】matriarchy

【母体】[动] the mother's body; the (female) parent

【母系】①(在血统上属于母亲的) maternal side ②(母女相承的) matriarchal ◇～亲属 maternal relatives／～社会 matriarchal society／～氏族公社 matrilineal commune／～氏族制 matriarchy

【母线】①[电] bus; bus bar ②[数] generatrix; generator

【母校】one's old school; Alma Mater

【母性】maternal instinct

【母液】[化] mother liquor; mother solution

【母语】①(本族语) mother tongue ②(某些语言的共同来源) parent language; linguistic parent

【母株】[植] maternal plant; mother plant

【母子之情】love between mother and son

拇

【拇指】①(手的第一个指头) thumb ②(脚的第一个指头) big toe

姆

【姆夫蒂】[伊斯兰教] mufti

【姆欧】[电] mho

mù

墓 grave; tomb; mausoleum: 列宁～ the Lenin Mausoleum／烈士～ tombs of revolutionary martyrs／马克思～ Marx's grave

【墓碑】tombstone; gravestone

【墓道】①(墓外甬道) path leading to a grave; tomb passage ②(墓内甬道) aisle leading to the coffin chamber of an ancient tomb

【墓地】graveyard; burial ground; cemetery

【墓室】coffin chamber

【墓穴】coffin pit; open grave

【墓葬】[考古] grave ◇～群 graves

【墓志】inscription on the memorial tablet within a tomb

【墓志铭】inscription on the memorial tablet within a tomb; epitaph

暮 ①(傍晚) dusk; evening; sunset: 薄～ dusk; gloaming ②(时间将尽) towards the end; late: ～春 late spring／～年 declining years; the evening of one's life

【暮霭】evening mist

【暮鼓晨钟】evening drum and morning bell in a monastery; timely exhortations to virtue and purity

【暮年】declining years; old age; evening of one's life

【暮气】lethargy; apathy: ～沉沉 lethargic; apathetic; lifeless

【暮色】dusk; twilight; gloaming: ～苍茫 deepening dusk; spreading shades of dusk／～朦胧 glimmering twilight

【暮生儿】a posthumous child

幕 ①(幕布) curtain; screen: 在夜～的掩护下 under screen of night ②[戏剧] act: 五～悲剧 a tragedy in five acts／生活的一～ a scene of life

【幕布】curtain; screen

【幕后】behind the scenes; backstage: ～操纵 pull strings <wires> behind the scenes／～操纵者 wire-puller; backstage manipulator／～活动 behind-the-scenes activities; backstage maneuvering／～交易 behind-the-scenes deal; backstage deal／～新闻 behind-the-scenes news／退居～ retire backstage

【幕间休息】interval; intermission

【幕帘快门】focal shutter

【幕燕釜鱼】in dangerous spot

募 raise; collect; enlist; recruit: ～兵 recruit soldiers／～款 raise money

【募兵制】mercenary system

【募化】collect alms

【募集】raise; collect: ～资金 raise a fund

【募捐】solicit contributions; collect donations

慕 admire; yearn for: ～名 out of admiration for a famous person／爱～ adore love／仰～ look up to with admiration

【慕名而来】be attracted to a place by its reputation as a scenic spot

【慕名求见】have respect for one's name and ask for an interview

木 ①(树木) tree: 果～ fruit tree／花～ flowers and trees ②(木头) timber; wood; log ③(木制的) made of wood; wooden: ～梳 wooden comb／～箱 wooden box; chest made of wood ④(棺材) coffin: 行将就～ have one foot in the grave ⑤(麻木) numb; wooden: ～头～脑 wooden-headed; dull-witted

【木板】plank; board ◇～床 plank bed

【木版】[印] block ◇～画 woodcut; wood engraving／～印花[纺] block printing／～印刷 block printing

【木本水源】a tree has its root, a stream has its source; the root of a matter

【木本植物】[植] xylophyta; woody plant

【木菠萝】[植] jackfruit
【木材】wood; timber; lumber ◇ ～厂 timber mill / ～防腐 wood preservation / ～工业 timber industry
【木柴】firewood
【木船】wooden boat
【木醋酸】[化] pyroligneous acid
【木雕泥塑】like an idol carved in wood or molded in clay; as wooden as a dummy
【木耳】an edible fungus
【木筏】raft
【木芙蓉】[植] cotton rose
【木工】①(木工工作) woodwork; carpentry ②(做木工的人) woodworker; carpenter ◇ ～机械 woodworking machinery
【木瓜】①(木瓜树) Chinese flowering quince ②(番木瓜) papaya
【木管乐器】woodwind instrument; woodwind
【木屐】clogs
【木简】[考古] inscribed wooden slip
【木浆】[纸] wood pulp: 化学～ chemical wood pulp
【木匠】carpenter
【木焦油】[化] wood tar
【木结构】[建] timber structure; wood construction
【木刻】woodcut; wood engraving ◇ ～术 xylography
【木兰】[植] lily magnolia
【木料】timber; lumber
【木马】①[体] vaulting horse; pommelled horse ②(象马的儿童玩具) hobbyhorse; rocking horse
【木马计】stratagem of the Trojan horse; Trojan horse
【木棉】silk cotton; kapok
【木模】wooden model
【木乃伊】mummy
【木偶】①(木头偶像) wooden image; carved figure: 象～似地站着 stand as still as a carved figure ②(做戏的木人) puppet; marionette ◇ ～剧 puppet show; puppet play / ～片 puppet film
【木排】raft
【木盒】tub
【木器】wooden furniture; wooden articles
【木琴】[乐] xylophone
【木然】stupefied
【木梳】wooden comb
【木薯】[植] cassava ◇ ～淀粉 tapioca
【木丝】wood wool ◇ ～板[建] wood wool board
【木炭】charcoal ◇ ～画 charcoal drawing
【木头】wood; log; timber
【木头人儿】woodenhead; blockhead; slow coach
【木屋】log cabin
【木犀】①[植](桂花) sweet-scented osmanthus ②(经烹调的打碎的鸡蛋) egg beaten and then cooked ◇ ～饭 fried rice with scrambled eggs / ～肉 pork fried with scrambled eggs / ～汤 eggdrop soup
【木锨】wooden winnowing spade
【木星】[天] Jupiter
【木已成舟】the wood is already made into a boat; what is done cannot be undone
【木俑】[考古] wooden figurine
【木鱼】wooden fish; wooden clapper
【木贼】[植] scouring rush
【木质部】[植] xylem
【木质素】lignin

沐 wash one's hair

【沐猴而冠】a monkey with a hat on; a worthless person in imposing attire
【沐雨栉风】work very hard regardless of weather
【沐浴】①(洗澡) have a bath ②(沉浸在某种环境中) bathe; immerse: ～在欢乐之中 be immersed in joy / ～在阳光中 be bathed in the sunshine

目 ①(眼睛) eye: ～语 communicate with the eyes / 双～失明 be blind in both eyes ②(看) look; regard: ～为奇事 regard as something strange ③(项目) item: 细～ detailed items ④[生物] order: 亚～ suborder ⑤(目录) list; catalogue: 书～ list of books

【目标】①(对象) target; objective: 打不中～ miss one's aim / 发现～ detect the objective / 击中～ hit the target / 军事～ military objective ②(目的) goal; aim; objective: 达到～ achieve one's aim / 怀着崇高的～ with a lofty end in mind
【目不见睫】the eye cannot see its lashes; lack self-knowledge
【目不交睫】not sleep a wink
【目不识丁】not know one's ABC; be totally illiterate
【目不暇接】the eye cannot take it all in; there are too many things for the eye to take in
【目不斜视】look steadily forward

【目不转睛】look with fixed eyes; watch with the utmost concentration
【目测】[军] range estimation
【目次】table of contents; contents
【目瞪口呆】gaping; stupefied; dumbstruck: 吓得～ be struck dumb with fear
【目的】purpose; aim; goal; objective; end: ～明确 have a definite purpose / ～与手段 ends and means / 达到～ attain one's objective / 怀着不可告人的～ harbor evil intentions; have ulterior motives ◇ ～地 destination / ～港 [航海] port of destination / ～论[哲] teleology
【目光】①(眼光) sight; vision; view: ～短浅 shortsighted / ～锐利 sharp-eyed; sharp-sighted / ～远大 farsighted; farseeing ②(眼睛的神采) gaze; look: ～呆滞 one's eye-sight is restrained / ～无神 dull look
【目光如豆】of narrow vision; shortsighted
【目光如炬】①(眼光象火炬那样亮) eyes blazing like torches; blazing with anger ②(见识远大) look ahead with wisdom
【目击】see with one's own eyes; witness: ～的报告 on-the-scene report; on-the-spot account
【目击者】eyewitness; witness
【目见】see for oneself: 耳闻不如～。 Seeing a thing for oneself is better than hearing about it.
【目镜】[物] eyepiece; ocular
【目空一切】consider everybody and everything beneath one's notice; be supercilious
【目力】eyesight; vision: ～不好 have poor eyesight / ～好 have good eyesight
【目录】①(事物名目) catalogue; list: 分类～ classified catalogue / 书名～ title catalogue / 图书～ library catalogue / 主题～ subject catalogue / 著者～ author catalogue ②(篇章名目) table of contents; contents ◇ ～室 catalogue room / ～学 bibliography
【目录柜】catalogue cabinet: 卡片～ card catalogue cabinet
【目迷五色】①(颜色多看不清) dazzled by a riot of color ②(形势复杂分辨不清) bewildered by a complicated situation
【目前】at present; at the moment: ～的生产能力 existing production capacity / ～形势 the present situation / 到～为止 up till the present moment; up till now; so far; to date / 就～情况看 as matters stand; as things are at the moment
【目清眉秀】the eyes are clear and the eyebrows refined
【目视飞行】[航空] visual flight
【目送】follow sb. with one's eyes; watch sb. go; gaze after
【目无法纪】disregard law and discipline
【目无全牛】be supremely skilled
【目无组织】disregard organizational discipline; defy the leadership of one's organization
【目眩】dizzy; dazzled
【目中无人】consider everyone beneath one's notice; be supercilious; be overweening

苜

【苜蓿】[植] lucerne; alfalfa

钼

[化] molybdenum (Mo)
【钼钢】molybdenum steel
【钼酸】molybdic acid ◇ ～铵 ammonium molybdate

睦

peaceful; harmonious
【睦邻】good-neighborliness ◇ ～关系 good-neighborly relations / ～政策 good-neighbor policy

牧

herd; tend: ～马 herd horses / ～羊 tend sheep
【牧草】herbage; forage grass
【牧场】grazing land; pastureland; pasture; livestock farm: 天然～ natural grazing grounds
【牧笛】reed pipe
【牧地千里】grazing area extends to a thousand *li*
【牧放】herd; tend; put out to pasture
【牧歌】①(以农村生活为题材的诗歌，乐曲) pastoral song; pastoral ②[乐](情歌；小曲) madrigal
【牧工】hired herdsman
【牧马中原】become master of the country
【牧民】herdsman
【牧区】pastoral area
【牧师】[基督教] pastor; minister; clergyman
【牧童】shepherd boy; buffalo boy
【牧畜】livestock breeding; animal husbandry
【牧羊犬】shepherd dog; collie
【牧羊人】shepherd
【牧业】animal husbandry; stock raising
【牧主】herd owner

穆

solemn; reverent: 肃～ solemn
【穆民】believers in Islam
【穆斯林】Moslem; Muslim

N

ná

拿 ①(用手取物或抓住) hold; take; bring; fetch; get: ～起武器 take up arms / 把行李～上楼 take the luggage upstairs ②(用强力夺取) seize; capture: ～下敌人阵地 take the enemy's positions / ～下要塞 capture a fort ③(掌握; 把握) be sure of; have a firm grasp of; be able to do ④(刁难) put sb. in a difficult position; make things difficult for sb. ⑤(用) with: ～笔写字 write with a pen / ～拐杖走路 walk with a crutch / ～事实证明 prove with facts ⑥(引进所处置的对象) ～朋友当敌人 take a friend for foe / ～他开刀 the first to be punished / ～他没办法 cannot do anything with him

【拿办】arrest and deal with according to law; bring to justice

【拿不出手】not be presentable

【拿大顶】[体] handstand

【拿得起,放得下】be able to advance or retreat; flexible; adaptable

【拿得起来】be able to afford it; competent

【拿获】apprehend

【拿架子】put on airs

【拿开】take away

【拿腔做势】act affectedly

【拿权】wield power; be in the saddle

【拿人】make things difficult for others; raise difficulties

【拿手】adept; expert; good at: ～好戏 a game or trick one is good at

【拿稳】hold steadily; predict with confidence

【拿主意】make a decision; make up one's mind

【拿住】hold firmly; put under arrest

镎 [化] neptunium (Np)

nǎ

哪 ①(表示疑问) which; what: 你们中间～一位是王先生? Which one of you is Mr. Wang? / 你最喜欢～种音乐? What is your favorite kind of music? ②(泛指) any: 今年的收成比以往～一年都好。 This year's crop is better than any other year's. ③(表示反问) ～会有这种事? How can there be things as such?

【哪个】①(哪一个) which ②(谁) who: 这话你是听～讲的? Who told you so?

【哪会儿】①(问时间) when ②(泛指时间) whenever; any time: 你～方便就～来吧。 Come at any time that is convenient to you.

【哪里】①(问什么处所) where: 你到～去? Where are you going? / 喂,～?(打电话用语) Are you there? ②(泛指任何处所) wherever; where ③(表示否定) ～,～,你过奖了。 Oh, it's nothing! You are flattering me.

【哪能】how can: 她～说谎? How can she tell lies?

【哪怕】[连] even; even if; even though; no matter how: ～失败,我也要试一下。 I will have a try even though I should fail.

【哪儿】[口] ①(哪里) where ②(泛指任何地方) wherever; anywhere

【哪些】which; who; what: ～人是你的亲戚? Who are your relatives? / 你要～词典? Which dictionaries do you want?

【哪样】①(用于问话) what kind of; which ②(泛指) any kind of: 你挑选～颜色都行。 You can choose any color you like.

nà

捺 ①(抑制) press down; restrain; keep under control: ～着性子 keep one's temper; control one's temper ②(汉字笔画) right-falling stroke in Chinese characters

衲 ①(补缀) patch up ②(和尚法衣) patchwork vestment worn by a Buddhist monk

呐

【呐喊】shout loudly; cry out: ～助威 shout encouragement; cheer

钠 [化] sodium (Na)

【钠长石】[矿] albite

【钠钙玻璃】soda-lime glass

【钠汽灯】sodium lamp; sodium vapor

【钠缺乏】sodium deficiency

纳 ①(放进来) receive; admit ②(接受) accept; take in: 不～忠言 not accept any advice; refuse to take in admonishment / 采～adopt ③(享受) enjoy; take delight or pleasure in life ④(交付) pay; offer: ～粮 pay taxes in grain ⑤(缝纫) sew close stitches: ～鞋底 stitch soles of cloth shoes

【纳粹】Nazi ◇ ～分子 Nazi / ～主义 Nazism

【纳福】enjoy a life of ease and comfort: ～之人 a man who enjoys himself

【纳罕】be surprised; marvel: 她很～。She was much surprised.
【纳贿】①(受贿) take bribes ②(行贿) offer bribes
【纳凉】enjoy the cool
【纳赂被控】be accused of receiving bribes
【纳闷儿】[口] feel puzzled; be perplexed; wonder
【纳米比亚】Namibia
【纳入】bring into; fit into: ～国家计划 bring sth. into line with the state plan / ～正轨 put sth. on the right course
【纳税】pay taxes ◇ ～年度 tax year / ～凭证 tax payment receipt / ～人 taxpayer / ～收据 duty receipt / ～义务 duty to pay taxes; obligation to pay tax
【纳降】accept the enemy's surrender

那 ①(指较远的人或事物) that ②[连](那么) then; in that case
【那边】there; over there: 孩子们在～玩。The children are playing over there.
【那达慕】Nadam Fair (a Mongolian traditional fair)
【那儿】①(那里) that place; there ②(那时候) that time; then
【那个】①(那一个) that ②[口](用在动词,形容词之前表示夸张)瞧他们干得～欢哪! See how they're throwing themselves into their work! ③[口](代替不便直说的话)我说他的所作所为也太～了。I say his behavior is somewhat — you know what I mean.
【那会儿】[口] at that time; then
【那里】that place; there: ～气候怎么样? What's the weather like there? / 谁在～? Who's there? / 他将在～住到六月。He will stay there till June.
【那么】①(指示性质、状态、方式和程度等) like that; in that way: 你不该～做。You shouldn't have behaved like that. ②(放在数量词前,表示估计) about; or so: 再有～五六元钱就够了。Another five or six *yuan* will probably be enough. ③[连](申说应有的结果) then; in that case; such being the case
【那么点儿】so little; so few
【那么些】so many; so much
【那么着】do that; do so: ～也许会好一些。Perhaps it'll be better that way.
【那时】at that time; then; in those days: 从～以来 since then; from that time on / ～我在上海。I was in Shanghai then.
【那些】those
【那样】of that kind; like that; such; so: 象雪～白 as white as snow / 我没有说过～的话。I said no such thing as that. / 真是～的吗? Is that really so?

na

哪 [助] 加油干～! Speed up! (或: Come on!) / 谢谢您～! Thank you!

nǎi

乃 ①(是) be: 失败～成功之母。Failure is the mother of success. ②(于是) then; hence; therefore: ～有此名 hence comes the name ③(才) only then: 今～知之。I didn't know it until now. ④(你; 你的) your: ～父 your father
【乃尔】like this; to such an extent: 何其相似～。What a striking similarity.
【乃至】and even

氖 [化] neon (Ne)
【氖灯】neon lamp; neon light; neon
【氖管】neon tube

奶 ①(乳房) breasts ②(乳汁) milk: 全脂～ whole milk / 脱脂～ skimmed milk ③(哺奶) suckle; breast-feed: 给婴儿喂～ feed the baby at the breast
【奶茶】tea with milk
【奶疮】mastitis
【奶粉】milk powder; powdered milk; dried milk: 全脂～ whole milk powder / 脱脂～ skimmed milk powder
【奶糕】a baby food made of rice-flour, sugar, etc.
【奶酪】cheese
【奶妈】wet nurse
【奶名】a child's pet name; infant name
【奶奶】①[口](祖母) grandmother; grandma ②(称老年妇女) grandma; granny
【奶牛】milch cow; milk cow; cow
【奶品】milk products; dairy products
【奶瓶】①(喂婴儿用) feeding bottle; nursing bottle; baby's bottle ②(盛奶用) milk bottle
【奶水】[口] milk
【奶糖】toffee; toffy
【奶头】[口] ①(乳头) nipple; teat ②(奶嘴) nipple (of a feeding bottle)
【奶牙】milch tooth
【奶羊】milch goat
【奶油】cream ◇ ～蛋糕 cream cake / ～分离器 cream separator / ～色 cream-colored

【奶油泡夫】cream puff
【奶罩】brassiere; bra
【奶嘴】nipple (of a feeding bottle)

nài

奈
【奈何】①(反问) how; to no avail: 徒唤～ utter bootless cries / 无可～ be utterly helpless ②(拿某人怎么办) do sth. to a person: 你～不了他。 You can do nothing to him.

萘 [化] naphthalene
【萘酚】[化] naphthol
【萘乙酸】[农] methyl α-naphthyl acetate

耐 be able to bear or endure: ～穿 can stand wear and tear; be endurable / 吃苦～劳 bear hardships and stand hard work
【耐波力】seakeeping qualities
【耐不住】unable to bear; unable to stand
【耐烦】patient: 显出不～的样子 show signs of impatience
【耐寒】cold-resistant: 耐严寒 resistant to low temperature ◇ ～性 cold resistance; winter-hardiness / ～作物 cold-resistant crop
【耐旱植物】drought-enduring plant
【耐航包装】seaworthy packing
【耐火】fire-resistant; refractory ◇ ～材料 refractory, fireproof material / ～衬砌 refractory lining / ～水泥 refractory cement / ～砖 refractory brick; firebrick
【耐久】lasting long ; durable: ～之朋 friendship of long duration ◇ ～力 durability; endurance
【耐劳】able to endure hard work
【耐力】endurance; staying power; stamina
【耐磨】wear-resisting; wearproof ◇ ～合金钢 wear-resisting alloy steel / ～性 wearability; wear resistance / ～硬度 abrasion hardness
【耐热】heat-resisting; heatproof ◇ ～合金 heat-resisting alloy / ～性 heat resistance
【耐人寻味】afford food for thought
【耐蚀钢】[冶] corrosion-resisting steel
【耐水作物】water-tolerant crop
【耐酸】acidproof; acid-resisting ◇ ～缸器 acidproof stoneware / ～混凝土 acid-resisting concrete / ～漆 acid-resisting paint; etching ink
【耐心】patient: ～等待 be patient and wait for something; wait patiently / ～说服教育 patient persuasion and education / ～听取意见 listen with patience to the opinions
【耐性】patience; endurance: 我没～再听你的抱怨。 I haven't the patience to hear your complaints again.
【耐印力】[印] pressrun
【耐用】durable: ～物品 durable good; durables / ～消费品 durable consumer goods

nán

南 south: ～岸 south coast / ～风 a south wind / ～国风光 southern scenery / ～屋 a room with a northern exposure / 城～ south of the city / 华～ south China / 正～偏东 south by east
【南半球】the Southern Hemisphere
【南北】①(南边和北边) north and south ②(从南到北) from north to south: 大江～ on both sides of the Changjiang River
【南北朝】the Northern and Southern Dynasties (420–589)
【南部】southern part; south
【南船座】[天] Argo
【南方】①(南) south ②(南部地区) the southern part of the country: ～风味 southern style; southern flavor ◇ ～话 southern dialect / ～人 southerner
【南风】south wind
【南瓜】pumpkin; cushaw ◇ ～子 pumpkin seeds
【南海】the Nanhai Sea; the South China Sea ◇～诸岛 South China Sea Islands
【南寒带】the south frigid zone
【南回归线】[地] Tropic of Capricorn
【南货】delicacies from south China
【南箕北斗】something which enjoys empty name but serves no practical purposes
【南极】①(地轴的南端) the South Pole; the Antarctic Pole ②(南磁极) the south magnetic pole ◇ ～光 [天] southern lights; aurora australis / ～海 Antarctic Ocean / ～圈 Antarctic Circle / ～洲 Antarctic Continent; Antarctica
【南柯一梦】illusory joy; an imaginary dream; an empty dream
【南来北往】be always on the move
【南美洲】South America
【南冕座】[天] Corona Australis
【南腔北调】a mixed accent
【南三角座】[天] Southern Triangle
【南十字座】[天] Cross; Crux; South Cross
【南斯拉夫】Yugoslavia ◇ ～人 Yugoslav
【南天竹】[植] nandina

【南纬】south latitude
【南温带】the south temperate zone
【南亚】South Asia ◇ ～次大陆 the South Asian Subcontinent
【南辕北辙】try to go south by driving the chariot north; act in a way that defeats one's purpose
【南针】①(指南针) compass ②(行动的指南) guide; guiding principle
【南征北战】fight north and south on many fronts

喃

【喃喃】[象] mutter; murmur: ～自语 mutter to oneself

男

男 ①(男性) man; male: ～扮女装 a man disguised as a woman / ～服 men's clothing ②(儿子) son; boy: 没～没女 has neither sons nor daughters
【男才女貌】the man is able and the woman is beautiful; an ideal couple
【男厕所】men's lavatory; Gentlemen; Men; Gents
【男大当婚,女大当嫁】Upon growing up, every male should take a wife and every female should take a husband.
【男盗女娼】behave like thieves and whores; be out-and-out scoundrels
【男低音】[乐] bass
【男儿】man: ～本色 the manliness of a man / 好～ a fine man
【男方】the bridegroom's or husband's side
【男高音】[乐] tenor
【男孩】boy
【男婚女嫁】a man should take a wife and a woman should take a husband
【男家】the bridegroom's or husband's family
【男爵】baron ◇ ～夫人 baroness
【男女】men and women ◇ ～混合双打 mixed double / ～老少 men and women, old and young / ～青年 young men and women / 不分～ irrespective of sex
【男女平等】equality of men and women; equality of the sexes ◇ ～权 equal rights for both sexes
【男女有别】Males and females should be treated differently.
【男朋友】boyfriend; boy friend
【男人】①(男子) man ②(丈夫) husband
【男生】man student; boy student; schoolboy
【男声】[乐] male voice ◇ ～合唱 men's chorus; male chorus
【男性】①(性别) the male sex ②(男人) man ◇ ～不育症 [生理] male sterility
【男中音】[乐] baritone
【男主角】leading man; male title role; chief actor; hero
【男装】men's clothing: 女扮～ a woman disguised as a man
【男子】man; male ◇ ～单打 men's singles / ～双打 men's doubles / ～团体赛 men's team event
【男子汉】man: 不象个～ not manly; not man enough

难

难 ①(不易) difficult; hard; troublesome: 我们很～下结论。It's very hard for us to come to conclusion. / 这山很～爬。This hill is hard to climb. ②(使感到困难) put sb. into a difficult position ③(不大可能) hardly possible ④(不好) bad; unpleasant: ～吃 taste bad; be unpalatable / ～闻 smell bad
【难办之事】have a long row to hoe
【难保】cannot say for sure: 明天～不下雨。You can't say for sure that it won't rain tomorrow.
【难辨是非】difficult to discriminate between right and wrong
【难辨真伪】hard to distinguish between the true and false
【难产】①[医] difficult labor; dystocia ②(不易完成) be difficult of fulfillment; be slow in coming
【难处】①(不易相处) hard to get along with ②(为难之处) difficulty; trouble: 各家有各家的～。Each family has its own difficulties.
【难打交道】hard to deal with
【难倒】daunt; baffle; beat: 她的问题把我～了。Her question has baffled me.
【难道】surely it doesn't mean that...; could it be said that...: ～这是办不到的吗? Could it be said that it is impossible of attainment?
【难得】①(不易得到) hard to come by; rare: ～的好机会 a rare chance ②(不常发生) seldom; rarely: 她～到这里来。She seldom comes here.
【难点】difficult point; difficulty
【难懂】difficult to comprehend; hard to understand
【难度】degree of difficulty; difficulty
【难怪】①(怪不得) no wonder ②(不应当责

备) understandable; pardonable
【难关】difficulty; crisis: 渡过～ tide over a difficulty / 攻克技术～ break down a technical barrier; resolve key technical problems
【难管】difficult to govern; hard to rule
【难过】①(不易过活) have a hard time ②(难受) feel sorry; feel bad; be grieved
【难乎为继】hard to keep up
【难交益友】having difficulty in getting beneficial friends
【难解难分】①(双方相持不下) neither would give in; neither can get the upper hand ②(关系密切) be sentimentally attached to each other
【难堪】①(难以忍受) intolerable; unbearable: ～之事 a bitter pill to swallow ②(窘) embarrassed: 感到～ feel very much embarrassed
【难看】①(不好看) ugly; unsightly: ～的帽子 an ugly hat ②(不体面) shameful; embarrassing; disgraceful
【难免】hard to avoid: ～失败 unavoidable failure / 犯错误是～的。 It's hard to avoid mistakes.
【难能可贵】praiseworthy for one's excellent conduct; deserving praise for one's excellent performance or behavior; estimable; rare and commendable
【难人】①(使人为难) difficulty; delicate; ticklish: ～的事情 delicate matter / ～的问题 ticklish problem ②(担当为难事情的人) a person handling a delicate matter
【难色】appear to be reluctant or embarrassed: 面有～ show signs of reluctance or embarrassment
【难上加难】extremely difficult
【难舍难分】loath to part from each other
【难受】①(身体不舒服) feel unwell; feel ill; suffer pain ②(心里不痛快) feel unhappy; feel bad
【难说】it's hard to say; you never can tell
【难逃法网】be unable to escape the net of justice
【难题】difficult problem; a hard nut to crack; poser: 出～ set difficult questions
【难听】①(不悦耳) unpleasant to hear: ～的声音 unpleasant sound ②(粗俗刺耳) offensive; coarse: ～的话 offensive language; profane language ③(不体面) scandalous: 这件事情说出去多～。 The matter will create a scandal once it gets out.
【难忘】unforgettable; memorable: ～的岁月 memorable years / ～的一课 an unforgettable lesson / ～的印象 indelible impression; memorable impression
【难为】①(使人为难) embarrass; press ②(多亏) be a tough job to ③[套](感谢别人代自己做事) ～你关上门。 Would you mind shutting the door?
【难为情】①(害羞) ashamed; embarrassed; shy ②(为难) find it difficult; disconcerting
【难闻】smell unpleasant; smell bad
【难兄难弟】[讽] two of a kind
【难言之隐】sth. which it would be awkward to disclose; sth. embarrassing to mention; a painful topic
【难以】difficult to: ～比拟 hardly match; hardly equal / ～否认 undeniable; irrefutable / ～估计 be difficult to estimate; inestimable / ～挽回 hard to retrieve / ～相处 be hard to get along with / ～相信 incredible / ～想象 unimaginable / ～形容 indescribable; beyond description / ～约束 beyond one's control / ～置信 hard to believe / ～捉摸 difficult to pin down; elusive; unintelligible
【难于启齿】have a bone in the throat; difficult to speak out one's mind

nǎn

赧 blushing
【赧然】[书] blushing
【赧颜】[书] blush; be shamefaced

蝻 the nymph of a locust
【蝻子】the nymph of a locust

nàn

难 ①(灾难) catastrophe; disaster; calamity: 大～临头 be faced with imminent disaster / 逃～ be a refugee ②(质问) blame; reproach: 非～ blame
【难民】refugee ◇ ～营 refugee camp
【难兄难弟】fellow sufferers
【难友】fellow sufferer

nāng

囔
【囔囔】speak in a low voice; murmur

náng

囊 ①(口袋) bag; pocket: 胶～ capsule / 药～ medicine bag ②(象口袋的东西) anything shaped like a bag: 胆～ gall bladder
【囊虫】cysticercus ◇ ～病 cysticercosis
【囊空如洗】with empty pockets; penniless;

broke
【囊括】include; embrace: ～四海 bring the whole country under imperial rule／～一切 sweep up everything
【囊中物】sth. which is in the bag; sth. certain of attainment
【囊肿】[医] cyst ◇～病 cystic disease
【囊状果】[植] saccate fruit

nǎng

攮 stab
【攮子】dagger

nāo

孬 ①(坏) bad ②(怯懦) cowardly
【孬种】coward

náo

硇
【硇砂】[化][矿] sal ammoniac
挠 ①(轻轻地抓) scratch: ～痒 scratch an itch ②(阻止) hinder: 阻～ obstruct ③(屈服) yield; flinch: 不屈不～ indomitable; unyielding
【挠度】[建] deflection
【挠钩】long-handled hook
【挠人清梦】disturb one's dream
【挠头】①(用手抓头) scratch one's head ②(事情麻烦) difficult to tackle: 令人～的事 a knotty problem
【挠性】[物] flexibility
【挠秧】[农] weed rice fields and loosen the soil around the seedlings
蛲
【蛲虫】pinworm ◇～病 enterobiasis

nǎo

恼 ①(生气) angry; irritated; annoyed: ～恨 resent ②(烦闷) unhappy; worried: 烦～ vexed; worried
【恼恨】resent; hate
【恼火】annoyed; irritated; vexed: 别～。Don't lose your temper.
【恼怒】angry; indignant; furious: 对某人因某事而～ be angry with sb. about sth.／因某人做某事而～ be angry at what sb. has done
【恼人】irritating; annoying
【恼羞成怒】fly into a rage from shame; be shamed into anger
脑 [生理] brain: 大～ cerebrum／后～ hindbrain／小～ cerebellum／中～ midbrain／用～过度 overtax one's brain
【脑充血】[医] encephalemia
【脑出血】cerebral hemorrhage
【脑袋】[口] head
【脑电波】[生理] brain wave
【脑电图】[医] electroencephalogram (EEG)
【脑动脉】cerebral artery
【脑海】brain; mind: 往事又浮上～。Past events flashed across my mind again.
【脑积水】[医] hydrocephalus
【脑脊膜】[生理] meninges ◇～炎 meningitis
【脑脊髓炎】[医] encephalomyelitis
【脑脊液】[生理] cerebrospinal fluid (CSF)
【脑浆】brains
【脑筋】①(指思考、记忆等能力) brains; mind; head: ～昏乱 be wrong in the upper storey／～简单 simple-minded／～灵敏 keen and sharp in thinking ②(指意识) ideas: 旧～ outdated conventional ideas; an old fogey
【脑壳】①(头颅) skull ②[方](头) head
【脑力劳动】mental work ◇～者 mental worker; brain worker
【脑瘤】brain tumor: 切除～ remove a brain tumor
【脑满肠肥】heavy-jowled and potbellied
【脑门子】[方] forehead; brow
【脑膜】[生理] meninx ◇～炎 meningitis
【脑贫血】cerebral anemia
【脑神经】[生理] cranial nerve
【脑室】[生理] ventricles of the brain ◇～造影 [医] ventriculography
【脑髓】brains
【脑萎缩】encephalatrophy
【脑血管造影】[医] cerebral angiography
【脑炎】encephalitis; cerebritis: 流行性乙型～ epidemic encephalitis B
【脑溢血】[医] cerebral hemorrhage
【脑震荡】[医] cerebral concussion; concussion of the brain
【脑汁】brains: 绞尽～ rack one's brains; cudgel one's brains
【脑子】①[口](脑) brain ②(脑筋) brains; mind; head
瑙
【瑙鲁】Nauru ◇～人 Nauruan

nào

闹 ①(喧哗) noisy: ～中取静 a quite spot in a noisy neighborhood ②(吵) make a noise; stir up trouble: ～翻了天 make a hell of a fuss / ～乱子 make trouble / ～名利 be out for fame and gain ③(发泄) give vent: ～脾气 lose one's temper; vent one's spleen ④(害;发生) suffer from; be troubled by: ～肚子 have diarrhea / ～事 make trouble / ～水灾 suffer from flood ⑤(干;搞) go in for; do; make: ～技术革新 make a technical innovation

【闹别扭】be difficult with sb.; be at odds with sb.

【闹病】fall ill; be ill

【闹虫灾】suffer from insect pests

【闹得鸡犬不宁】cause such utter confusion as to make everybody nervous

【闹得满城风雨】cause a big scandal

【闹得头昏脑胀】cause such utter confusion as to crave one crazy

【闹独立性】assert one's independence; refuse to obey the leadership

【闹翻】fall out with sb.

【闹翻天】raise hell; raise a rumpus

【闹风潮】carry on agitation; stage strikes, demonstrations, etc.

【闹革命】carry out revolution; make revolution; rise in revolution

【闹鬼】①(鬼怪作祟) be haunted ②(在背后做坏事) play tricks behind sb.'s back; use underhand means

【闹哄哄】clamorous; noisy

【闹饥荒】①(遇荒年) suffer from famine ②(指经济困难) be hard up

【闹剧】farce

【闹乱子】cause trouble

【闹情绪】be disgruntled; be in low spirits

【闹市】busy shopping center; downtown area; busy streets

【闹事】create a disturbance; make trouble

【闹笑话】make a fool of oneself; make a stupid mistake

【闹意见】be on bad terms because of a difference of opinion

【闹意气】feel resentful because something is not to one's liking; sulk

【闹着玩儿】joke: 这事可不能～。 This is no joke (joking matter).

【闹钟】alarm clock: 把～拨到明晨六点 set the alarm clock at six tomorrow morning

nè

讷 [书] slow (of speech): ～于言而敏于行 slow of speech but quick in action

něi

馁 ①(饥饿) hungry; famished ②(失掉勇气) disheartened; dispirited: 气～ lose heart; be disheartened / 胜不骄，败不～ not become dizzy with success, nor be discouraged by failure

nèi

内 ①(里头) inner; inside; within: ～销 sold inside the country; home trade / ～宅 inner chambers ②(指妻或妻的亲属) one's wife or her relatives: ～弟 wife's younger brother / ～人 my wife / ～兄 wife's older brother

【内白】[剧] words spoken by an actor from offstage

【内部】inside; internal; interior: ～联系 internal relations / 人民～矛盾 contradictions among the people ◇ ～比率 internal ratio / ～结构 internal structure / ～刊物 restricted publication / ～审计 internal audit

【内场】[棒、垒球] infield ◇ ～手 infielder

【内出血】[医] internal hemorrhage

【内地】inland; interior; hinterland: ～城市 inland city / 发展～工业 promote the inland industry ◇ ～税 inland duty / ～运费 inland forwarding expenses

【内电阻】[电] internal resistance

【内定】decided at the higher level but not officially announced

【内毒素】[医] endotoxin

【内耳】[生理] inner ear

【内方外圆】square internally and round externally

【内分泌】[生理] endocrine; internal secretion ◇ ～器官 endocrinal organ / ～失调 endocrinopathy / ～系统 internal system / ～腺 endocrine glands

【内锋】[足球] inside forward

【内服】[医] to be taken orally

【内阁】cabinet: 影子～ shadow cabinet ◇ ～大臣 cabinet minister / ～改组 cabinet reshuffle / ～会议 cabinet council

【内功】exercises to benefit the internal organs

【内骨骼】[动] endoskeleton

【内顾之忧】worries for trouble at home

【内果皮】[植] endocarp

【内海】①(内陆海) continental sea ②(属于一国的海) inland sea

【内涵】[逻] intension; connotation

【内行】expert; adept: 向～人学习 learn from the one who knows how
【内河】inland river ◇ ～航行权 inland navigation rights / ～运输 inland water transport / ～运送保险 inland marine insurance
【内讧】internal conflict; internal strife; internal dissension
【内寄生物】[生] endoparasite
【内奸】a secret enemy agent within one's ranks; hidden traitor
【内角】[数] interior angle
【内接形】[数] inscribed figure
【内景】indoor setting; indoor scene; interior
【内径】[机] internal diameter; inside diameter (ID) ◇ ～规 internal gauge / ～千分尺 inside micrometer
【内疚】compunction; guilty conscience
【内聚力】[物] cohesive force; cohesion
【内眷】female members of a family
【内科】[医] internal medicine ◇ ～病房 medical ward / ～医生 physician
【内裤】briefs; knickers: 男～ briefs; short pants; underpants / 女～ women's panties; knickers
【内窥镜】[医] endoscope ◇ ～检查 endoscopy
【内涝】waterlogging
【内力】[物] internal force
【内陆】inland; interior; landlocked ◇ ～国 landlocked country / ～河 continental river / ～盆地 interior basin
【内乱】civil strife; internal disorder
【内蒙古】Inner Mongolia
【内幕】what goes on behind the scenes; inside story
【内能】[物] internal energy; intrinsic energy
【内切圆】[数] inscribed circle
【内勤】①(内部工作) office work ②(内勤人员) office staff
【内情】inside information: 了解～ be an insider; be in the know
【内燃机】[机] internal-combustion engine ◇ ～船 motor vessel
【内燃机车】diesel locomotive
【内容】content; substance: 有丰富的～ have substantial content ◇ ～提要 synopsis; résumé
【内伤】[医] internal injury
【内胎】the inner tube of a tire
【内外】①(内部和外部) inside and outside; domestic and foreign: ～交困 beset with difficulties both at home and abroad / 国～市场 domestic and foreign market ②(表示概数) around; about: 二十年～ in about twenty years
【内务】①(国内事务) internal affairs: ～部 Ministry of Internal Affairs ②(日常事务) daily routine of sanitation tasks ◇ ～条令 [军] interior service regulations
【内线】①(在对方内部活动的人) planted agent ② [军] interior lines: ～作战 carry out interior-line operations; flight on interior lines ③(电话内线) inside telephone connections ◇ ～自动电话机 interphone
【内详】name and address of sender enclosed
【内向】[心] introversion
【内斜视】[医] esotropia; cross-eye
【内心】heart; innermost being: ～活动 one's inner life / ～深处 in one's heart of hearts / ～世界 one's inner world
【内省】[心] introspection; self-examination: ～不疚 find no fault in examining one's heart ◇ ～心理学 introspective psychology
【内秀】be intelligent without seeming so
【内衣】underwear; underclothes
【内因】[哲] internal cause
【内应】a person operating from within in coordination with outside forces; a planted agent; a plant
【内应力】[机] internal stress
【内忧外患】domestic trouble and foreign invasion
【内在】inherent; intrinsic ◇ ～规律 inherent law / ～联系 inner link; internal relations / ～论 [哲] immanentism / ～矛盾 inner contradictions / ～美 inner beauty / ～因素 internal factor
【内脏】internal organs; viscera
【内债】internal debt
【内战】civil war
【内政】internal affairs: 互不干涉～ noninterference in each other's internal affairs
【内痔】[医] internal piles
【内助】[书] wife

nèn

嫩 ①(娇嫩) tender; delicate: ～的皮肤 delicate skin / ～黄瓜 young cucumber / ～芽 tender shoots / 脸皮～ shy; bashful ②(烹调时间短) underdone; tender: 这些鸡蛋煮得太～了。These eggs boiled too tender. ③(色浅) light; delicate: ～色 light color ④(阅历浅; 经验少) inexperienced; immature; green: ～手 raw hand; new hand / 他做这件工作还太～。He is still green at this job.

néng

能 ①(能力) ability; capability; skill: 无～的人 a man of small caliber / 一专多～ good at many things and expert in one ② [物](能量) energy: 动～ kinetic energy / 太阳～ solar energy / 位～ potential energy ③ (有能力的) able; capable: ～人 able person; man of large caliber ④ (能够; 会) can; be able to; be capable of: ～吃苦耐劳 able to bear hardship

【能动】active; dynamic; vigorous: ～的飞跃 active leap / ～地改造世界 play a dynamic role in reforming the world

【能动性】dynamic role; activity; initiative: 主观～ subjective activity / 自觉的～ (man's) conscious dynamic role

【能干】able; capable; competent

【能歌善舞】good at both singing and dancing

【能工巧匠】skillful craftsman; dab hand

【能够】can; be able to; be capable of: 你～开拖拉机吗? Can you drive a tractor?

【能级】[物] energy level: 费密～ Fermi level / 基态～ ground state level

【能见度】visibility: 地面～ ground visibility

【能力】ability; capacity; capability: ～所及 reach within one's capacity / 分析问题和解决问题的～ ability to analyse and solve problems / 生产～ production capacity / 阅读～ reading ability ◇ ～倾向 aptitude / ～水平 ability level

【能量】① [物] energy: ～转化 energy conversion ② (能力) capabilities ◇ ～交换 energy exchange / ～均分 equipartition of energy / ～守恒 conservation of energy

【能耐】[口] ability; capability; skill

【能…能…】can not only... but also: ～吃～喝 be healthy enough to enjoy food and drinks / ～攻～守 be good at offense and defense; be able to take the offensive or hold one's ground / ～官～民 be ready to be an official or one of the common people / ～屈～伸 be able to stoop or to stand; can take temporary setbacks / ～上～下 be ready to accept a higher or a lower post / ～文～武 be able to wield both the pen and the gun

【能骑善射】expert at horseback riding and shooting arrow

【能人背后有能人】for every able person there is always one still abler

【能诗善画】having superior abilities to write poetry and good in painting

【能事】what one is particularly good at: 竭尽挑拨离间之～ stop at nothing to sow discord

【能手】dab; expert; crackajack: 游泳～ dab at swimming

【能说会道】have the gift of the gab; have a glib tongue

【能态】[物] energy state

【能言快语】eloquent and frank in speech

【能言善辩】eloquent, glib and quick-tongued in argument

【能源】the sources of energy; energy resources; energy: ～危机 energy crisis

【能者多劳】able people should do more work

【能者为师】let those who know teach

ní

霓 [气] secondary rainbow

【霓虹灯】neon lamp; neon light; neon: ～广告 neon sign

鲵 salamander

尼 Buddhist nun

【尼庵】Buddhist nunnery

【尼泊尔】Nepal ◇ ～人 Nepalese / ～语 Nepali

【尼姑】Buddhist nun

【尼古丁】nicotine

【尼加拉瓜】Nicaragua ◇ ～人 Nicaraguan

【尼龙】[纺] nylon ◇ ～丝 nylon yarn / ～袜 nylon socks

【尼日尔】the Niger ◇ ～人 Nigerois

【尼日利亚】Nigeria ◇ ～人 Nigerian

泥 ①(泥巴) mud; mire: ～墙 mud wall ②(泥状物) mash: 蒜～ mashed garlic / 土豆～ mashed potato

【泥刀】trowel

【泥点儿】droplets of mud

【泥肥】[农] sludge; silt; mud fertilizer

【泥敷剂】cataplasm; poultice

【泥垢】dirt; grime

【泥灰岩】[地] marl

【泥浆】slurry; mud: 钻井～ drilling mud ◇ ～泵 slurry pump / ～工 mudman

【泥坑】mud pit; mire; morass: 陷在～里 get stuck in the mud

【泥疗】[医] mud therapy

【泥煤】peat

【泥泞】muddy; miry; sloppy: ～的道路 a muddy road

【泥牛入海】like a clay ox entering the sea; nev-

er to be heard of again; gone forever
【泥菩萨】clay idol: ～过河，自身难保 like a clay idol fording a river; hardly able to save oneself
【泥鳅】loach
【泥人】[工美] clay figurine: 彩塑～ painted clay figurine
【泥沙】[地] silt
【泥沙俱下】mud and sand are carried along; there is a mingling of good and bad
【泥石流】[地] mud-rock flow
【泥水匠】bricklayer; tiler; plasterer
【泥塑】clay sculpture
【泥胎】①(尚未着油彩的泥像) unpainted clay idol ②(陶坯) unfired pottery
【泥潭】mire; morass; quagmire
【泥土】①(土壤) earth; soil ②(粘土) clay
【泥腿子】bumpkin; clodhopper
【泥瓦匠】bricklayer; tiler; plasterer
【泥岩】[地] mudstone
【泥俑】[考古] clay figures buried with the dead; funerary clay figures; earthen figurines
【泥沼】mire; swamp; morass; slough
【泥足巨人】colossus with feet of clay
【泥足深陷】get into real trouble

呢 wool; woolen cloth; heavy woolen cloth; wool coating or suiting: 大衣～ heavy woolen cloth for overcoat / 格子～ woolen check / 海军～ navy cloth / 制服～ uniform coating
【呢喃】twittering
【呢绒】woolen goods; wool fabric
【呢子】woolen cloth; heavy woolen cloth; wool coating or suiting

铌 [化] niobium (Nb)
【铌铁矿】columbite

nǐ

拟 ①(设计; 起草) draw up; draft: ～电稿 draft a telegram / ～一个方案 draw up a plan ②(打算) intend; plan ③(仿效) imitate: 模～ imitate; copy
【拟订】draw up; draft; work out: ～计划 draw up a plan; draft a plan / ～具体办法 work out specific measures / ～一份决议草案 work out a draft resolution / ～一份提纲 draw up an outline
【拟稿】make a draft
【拟古】model one's literay or artistic style on that of the ancients: ～之作 a work modelled after the ancients
【拟人】[语] personification
【拟态】[生] mimicry; imitation
【拟议】①(事先的考虑) proposal; recommendation ②(草拟) draw up; draft
【拟作】a work done in the manner of a certain author

你 ①(指对方) you ②(泛指任何人) you; one: ～走你的阳关道，我走我的独木桥 you look after your own concern and leave me to my own affairs ③(第二人称复数) you: ～方 your side / ～校 your school
【你好】how do you do; how are you; hello
【你敬我一尺，我还你一丈】kindness is always returned tenfold
【你们】you
【你死我活】life-and-death; mortal: ～的斗争 a life-and-death struggle / 拼个～ fight to the bitter end
【你追我赶】try to overtake each other in friendly emulation
【你自己】yourself

nì

溺 ①(淹没在水里) drown ②(沉迷不悟) be addicted to: ～于利欲 indulge in profits and lust
【溺爱】spoil; dote on: ～子女 dote on children
【溺婴】drowning of infants; infanticide: ～罪 infanticide by drowning
【溺于酒色】given over to wine and women
【溺职】neglect of duty; dereliction

逆 ①(方向相反) contrary; counter ②(抵触) go against; disobey; defy: ～时代潮流而动 go against the trend of the times ③(背叛者的) traitor
【逆变换】[数] inverse transformation
【逆差】[商] adverse balance of trade; trade deficit: 国际收支～ an adverse balance of international payments
【逆产】traitor's property
【逆定理】[数] converse theorem
【逆耳】grate on the ear; be unpleasant to the ear: ～的话 words or advice unpleasant to hear
【逆反应】back reaction
【逆风】①(迎面对着风) against the wind; in the teeth of the wind ②(跟车船等行进方向相反的风) contrary wind; head wind; adverse wind: ～飞行 head-wind flight
【逆戟鲸】killer whale

【逆迹昭彰】signs of rebellion are evident
【逆境】adverse circumstances; adversity
【逆来顺受】meekly submit to oppression, maltreatment, etc.; resign oneself to adversity
【逆料】anticipate; foresee: 事态的发展不难～。The course of events can be foreseen. ◇～喷射 upstream injection
【逆流】adverse current; countercurrent: ～而上 go up stream / ～而行 sail against the stream
【逆时针方向旋转】anti-clockwise
【逆水】against the current: ～行舟，不进则退 a boat sailing against the current must forge ahead or it will be driven back
【逆算子】inverse operator
【逆温】[气] inversion ◇～层 [气][地] inversion layer
【逆行】go in a direction not allowed by traffic regulations; go in the wrong direction: 单行线，车辆不得～ one-way street; one-way traffic
【逆运算】[数] inverse operation
【逆转】take a turn for the worse; reverse; become worse; deteriorate: ～过敏反应 reversed anaphylaxis
【逆子】unfilial son

匿 hide; conceal: 逃～ escape and hide / 隐～ go into hiding; hide
【匿伏】be in hiding; lurk
【匿迹】go into hiding; stay in concealment: 销声～ be in hiding; disappear from the scene
【匿名】anonymous ◇～控告信 anonymous letter of accusation / ～信 anonymous letter
【匿影藏形】hide from public notice; conceal one's identity; lie low

腻 ①(食品中油脂过多) greasy; oily: 油～的汤 oily soup ②(厌烦) be bored with; be tired of: 他实在叫人～。He is so tiresome. ③(细致) meticulous: 细～的描写 a minute description
【腻虫】aphid
【腻烦】[口] ①(厌烦) be bored; be fed up: 这些话我都听～了。I'm tired of listening to all this. ②(厌恶) loathe; hate: ～油味 loathe the smell of grease
【腻味】[方] get fed up

泥 ①(用土、灰等涂抹墙壁或器物) cover with plaster; daub; plaster: ～墙 plaster a wall ②(固执) stubborn; obstinate
【泥古】obstinately follow tradition; blindly stick to old conventions
【泥子】[建] putty

昵 close; intimate: 亲～ very intimate

niān

拈 take; pick up: 信手～来 pick up at random
【拈花惹草】have many affairs
【拈阄儿】draw lots
【拈轻怕重】prefer the light to the heavy; pick easy jobs and shirk hard ones

蔫 ①(萎缩) shrivel up; wither; wilt; fade: 花儿晒～了。The flowers drooped in the heat of the sun. ②(精神不振) droopy; listless; spiritless; run-down: 这女孩看上去有点～。The girl seems droopy.

nián

黏(粘) sticky; glutinous
【黏不住】fail to stick; fail to adhere
【黏虫】armyworm
【黏度】[化] viscosity: 恩氏～ Engler viscosity ◇～计 viscosimeter
【黏附】adhere ◇～力 adhesion / ～体 adherend
【黏合】[化] bind; bond; adhere ◇～剂 binder; adhesive; bonding agent
【黏糊】①(黏) sticky; glutinous ②(行动缓慢) dull; languid; slow-moving: 这人真～。He is very dull.
【黏胶】[化] viscose ◇～长丝 viscose filament yarn / ～短纤维 viscose staple fiber / ～丝 viscose
【黏结】cohere ◇～力 cohesion; cohesive force / ～性 cohesiveness
【黏米】glutinous rice
【黏膜】[生理] mucous membrane; mucosa ◇～炎 mucositis
【黏土】clay: 耐火～ refractory clay ◇～矿物 clay mineral / ～岩 clay rock
【黏性】stickiness; viscidity; viscosity ◇～油 viscous oil
【黏液】①[生理] mucus ②[植] mucilage
【黏住】stick; adhere
【黏着】stick together; adhere
【黏着语】[语] agglutinative language

鲇 catfish
【鲇鱼】catfish

年 ①(时间单位) year: 今～ this year / 明～

next year／前～ year before／去～ last year／学～ school year; academic year／～初 at the beginning of the year／～复一～ year after year; year in year out ②(每年的) annual; yearly: 逐～ annually ③(岁数) age; year: ～高德劭 advanced in years and highly esteemed ④(时期) a period of time: 近～来 in recent years／童～ childhood ⑤(年成) harvest: 丰～ rich harvest ⑥(新年) New Year: 拜～ pay a New Year visit
【年报】annual report; annual; annals
【年表】chronological table: 大事～ chronicle of events
【年成】the year's harvest: ～不好 a lean year／好～ a good harvest
【年初】the beginning of the year: 去年～ at the beginning of last year
【年代】①(时代) age; years; time: 战争～ during the war years ②(十年的时期) a decade of a century: 二十世纪九十～ 1990s ◇ ～测定 dating／～学 chronology
【年底】the end of the year; year-end
【年度】year: 财政～ financial year; fiscal year ◇ ～报告 annual report／～计划 annual plan／～结算 annual account／～决算表 annual statement／～收入 annual revenue／～统计表 annual returns
【年份】①(指某一年) a particular year ②(所经历的年代) age; time: ～很久的瓷器 old porcelain
【年富力强】in the prime of life; in one's prime
【年高望重】aged and in high standing
【年糕】New Year cake (made of glutinous rice flour)
【年庚】the time of a person's birth; date of birth
【年关】the end of the year
【年光荏苒】the quick passing of time
【年华】time; years: ～如矢 the years glide swiftly along／虚度～ idle away one's time; waste one's life
【年号】reigh title
【年会】annual meeting
【年货】special purchases for the Spring Festival: 办～ do Spring Festival shopping
【年金】annuity
【年级】grade; year: 大学三～学生 third year university student／小学一～学生 first grade primary school pupil／他在大学与我同～。He was in my year at college.
【年纪】age: ～轻 young／上了～ old; advanced in years／她多大～了? How old is she?
【年鉴】yearbook; almanac
【年景】①(年成) the year's harvest ②(过年景象) holiday atmosphere of the Spring Festival
【年久失修】worn down by the years without repair
【年历】a calendar with the whole year printed on one sheet; single-page calendar: ～片 calendar card
【年利】annual interest ◇ ～率 annual interest rate
【年龄】age: 超过规定～ over age／未达到规定～ under age／应征～ age for enlistment／他的～超过四十岁。He is over forty years of age.／他们同～。They are of the same age. ◇ ～结构 age structure／～群 age set／～性别组成 age-sex structure
【年龄组】age group: 同～的人 cohorts
【年率】annual rate; rate per annum
【年轮】[植] annual ring; growth ring
【年迈】old; aged: ～的双亲 old parents／～力衰 old and infirm; senile
【年年】every year; year after year; year by year: ～如意 New Year greetings
【年谱】a chronicle of sb.'s life
【年轻】young: ～的一代 the young generation／～有为 able and young; young and promising
【年少气盛】young and impetuous
【年深日久】with the passage of time; as the years go by
【年岁】①(年纪) age: 上了～的人 a person who is advanced in age ②(年代) years
【年头】①(年代) days; times: 那～ in those days ②(指年份) year: 十个～ ten years ③(多年的时间) long time; years: 她从事教学有～了。She has been teaching for years. ④(收成) harvest: 今年～真好。This year we had a bumper harvest.
【年息】annual interest
【年限】fixed number of years: 工具使用～ the service life of a tool／学习～ the number of years set for a course
【年薪】yearly salary
【年夜】the eve of the lunar New Year
【年幼无知】young and ignorant; ignorance for being young of age
【年愈不惑】have passed 40
【年月】days; years
【年增长系数】annual improvement factor
【年终】the end of the year; year-end: ～报告

year-end report / ～分配 year-end distribution / ～加薪 year-end bonus / ～结存 annual balance / ～评比 year-end appraisal of work

niǎn

撵 ①(驱逐) drive out; oust: 把人～走 drive sb. away ②[方](追赶) catch up: 我会～上你的。I'll catch up with you.

捻 ①(用手指搓) twist with the finger: ～胡子 finger the beard / ～线 twist thread ②(捻子) sth. made by twisting: 灯～儿 lampwick / 纸～儿 a paper spill

【捻度】[纺] number of turns; twist

【捻线机】[纺] twisting frame

【捻子】①(用纸搓成的捻) spill ②(灯芯) wick; lampwick

碾 ①(碾子) roller ②(去掉谷物的外皮) grind or husk with a roller: ～米 husk rice ③(压碎) crush; pulverize; grind: 把玉米～成粉 grind corn into flour ④(碾平) flatten

【碾坊】grain mill

【碾磙子】stone roller

【碾米机】rice mill

【碾碎】pulverize

【碾转反侧】be sleepless

【碾子】roller: 石～ stone roller

niàn

念 ①(想念) think of; miss ②(念头) thought; idea: 心无杂～ have only one thing in mind ③(读) read: 请～给我听。 Read to me, please. ④(求学) study; attend school: ～书 read; study

【念白】spoken parts of a Chinese opera

【念叨】①(不停地说) talk about again and again in recollection or anticipation; be always talking about ②(谈论) talk over; discuss

【念佛】chant the name of Buddha; pray to Buddha

【念经】recite or chant scriptures

【念旧】①(不忘旧日的友谊) keep old friendships in mind ②(怀念旧时光) for old time's sake

【念念不忘】bear in mind constantly

【念念有词】①(念咒) mutter incantations ②(咕哝) mumble

【念诵】recite; read aloud

【念头】thought; idea; intention: 忽然想到一个～ hit upon an idea

【念咒】chant or intone chant

【念珠】beads; rosary

埝 a low bank between fields: 打～ build banks between fields

niáng

娘 ①(母亲) ma; mum; mother ②(称长一辈的妇女) a form of address for an elderly married woman: 婶～ aunt ③(年轻妇女) a young woman: 新～ bride

【娘家】a married woman's parents' home

【娘娘】①(皇后) empress ②(妃子) imperial concubine ③(女神) goddess ◇ ～庙 Temple of Goddess

【娘娘腔】sissy; womanish

【娘胎】mother's womb: 出了～ be born

【娘子军】detachment of women

niàng

酿 ①(酿造) make wine; brew beer ②(蜜蜂做蜜) make honey: 蜜蜂～蜜。 Bees make honey. ③(逐渐形成) lead to; result in ④(酒) wine: 佳～ good wine

【酿成】lead to; bring on; breed: ～巨祸 bring about a great calamity / ～争端 breed strife

【酿祸】create trouble; ferment disturbances

【酿酒】make wine; brew beer ◇ ～厂 winery; brewery / ～业 wine-making industry

【酿酶】[化] zymase

【酿热物】[农] ferment material

【酿造】make; brew

niǎo

袅 slender and delicate

【袅袅】①(缭绕上升) curl upwards: 炊烟～。 Smoke is curing upward from kitchen chimneys. ②(随风摆动) wave in the wind: 晨风中垂柳～。 The drooping willows are waving gently in the morning breeze. ③(延长不绝) linger: 余音～。 The melody is still lingering about long after the performance.

【袅袅婷婷】curvaceous and soft; lissome and graceful

【袅娜】slender and graceful; willowy

鸟 bird: 候～ migratory birds; birds of passage / 留～ resident birds / 雏～ young bird; nestling ○鹌鹑 quail / 八哥 starling; mynah / 白嘴鸦 rook / 百灵 (sky) lark / 斑鸠 turtle-doge / 布谷鸟 cuckoo / 苍鹭 (gray) heron / 苍鹰 goshawk / 长尾小鹦鹉

parakeet / 翠鸟 kingfisher; halcyon / 大葵花鹦鹉 cockatoo / 大乌鸦 raven / 大雁 wild goose / 杜鹃 cuckoo / 鸸鹋 emu / 海鸥 sea gull / 海燕 petrel / 鹤 crane / 鹳 stork / 画眉 thrush / 红尾鸲 redstart / 黄莺 oriole / 火烈鸟 flamingo / 极乐鸟 bird of paradise / 金翅雀 gold-finch / 金顶鹪鹩 golden-crested wren / 金刚鹦鹉 macaw / 金丝雀 canary / 鸨 bustard / 孔雀 peacock / 鸬鹚 cornorant / 麻雀 sparrow / 猫头鹰 owl / 企鹅 penguin / 山鸡 black grouse / 山雀 tit; titmouse / 隼 falcon; hawk / 鹈鹕 pelican / 天鹅 swan / 秃鹰 vulture / 鸵鸟 ostrich / 乌鸦 crow / 喜鹊 magpie / 燕鸥 tern / 燕子 swallow / 夜莺 nightingale / 鹰 eagle / 鹦鹉 parrot / 鹬 snipe / 鸳鸯 mandarin duck / 云雀 (sky) lark / 鹧鸪 partridge / 知更鸟 robin / 雉 pheasant / 啄木鸟 woodpecker

【鸟粪】① birds' droppings ② (海鸟粪) guano

【鸟尽弓藏】cast aside the bow once the birds are gone; cast sb. aside when he has served his purpose

【鸟瞰】① (从高处往下看) get a bird's-eye view ② (概括描写) general survey of a subject; bird's-eye view ◇ ～图 bird's-eye view

【鸟类】birds ◇ ～学 ornithology

【鸟笼】birdcage

【鸟枪】① (火枪) fowling piece; birding piece ② (汽枪) air gun

【鸟兽】birds and beasts; fur and feather: 作～散 scatter like birds and beasts; flee helter-skelter; stampede

【鸟为食亡,人为财死】birds die in pursuit of food and human beings die in pursuit of wealth

【鸟语花香】birds sing and flowers give forth their fragrance; characterizing a fine spring day

【鸟嘴】beak; bill

茑

【茑萝】[植] cypress vine

niào

尿

① (人或动物尿) urine ② (撒尿) urinate; make water; pass water

【尿崩症】[医] diabetes insipidus

【尿闭】[医] anuria

【尿布】diaper; napkin; nappy

【尿床】wet the bed; bed-wetting

【尿胆素】[医] urobilin ◇ ～原 urobilinogen

【尿道】[生理] urethra ◇ ～炎 urethritis / ～造影 urethrography

【尿毒症】[医] uremia

【尿肥】[农] urine

【尿分析】urinalysis: 常规～ routine urinalysis

【尿盆】chamber pot; urinal

【尿频】[医] frequent micturition

【尿失禁】[医] urinary incontinence; incontinence of urine

【尿素】[化] urea; carbamide

【尿酸】[化] uric acid

【尿桶】wooden pail for urine

【尿血】[医] hematuria

【尿潴留】[医] retention of urine

脲

[化] urea; carbamide

【脲醛塑料】[化] urea-formaldehyde plastics

niē

捏

① (握; 捉) hold between the fingers; pinch: ～住一枚别针 hold a pin with one's fingers ② (揉捏) mold; knead: ～泥人 mold clay figurines ③ (编造事实) fabricate; make up: ～报 fake a report

【捏合】mediate; act as go-between

【捏合机】[化纤] kneading machine

【捏一把汗】be breathless with anxiety or tension

【捏造】fabricate; concoct; fake; trump up: ～案情陷害某人 frame a case against sb. / ～口实 trump-up excuse / ～事实 invent a story; make up a story / ～数字 conjure up figures / ～谣言 fabricate a rumor / ～证据 fabricate an evidence / ～罪名 trump up charges

niè

颞

【颞骨】[生理] temporal bone

【颞颥】[生理] temple

蹑

① (放轻脚步) lighten one's step; walk on tiptoe ② (追随) tread; step on; walk with

【蹑手蹑脚】walk gingerly; walk on tiptoe

【蹑踪】follow along behind sb.; track

【蹑足不前】not to move a step forward

【蹑足其间】join a trade; associate with

【蹑足潜踪】walking stealthily

镊

【镊子】tweezers

蘖

[植] tiller

孽 evil; sin: 妖～ evildoer; monster / 造(作)～ do evil; commit a sin
【孽因】[佛教] sinful cause
【孽障】evil creature; vile spawn
涅
【涅白】opaque white
【涅槃】[佛教] nirvana
啮 [书] gnaw
【啮齿动物】rodent
【啮合】①(上下牙咬紧) clench the teeth ②(象牙齿那样咬紧) mesh; engage: 这两个齿轮～在一起。 The two cogwheels are engaged.
镍 [化] nickel: ～钢 nickel steel
【镍币】nickel coin; nickel

nín

您 [敬] you

níng

宁 peaceful; tranquil
【宁静】peaceful; tranquil; quiet: ～的小镇 a quiet little town / 拂晓前的～ the still hours before dawn
【宁静致远】still water run deep; leading a quiet life
【宁人息事】pacify people and settle matters
【宁日】peaceful day
柠
【柠檬】lemon
【柠檬茶】lemon tea
【柠檬色】lemon yellow
【柠檬水】lemonade; lemon squash
【柠檬素】[化] citrin; vitamin P
【柠檬酸】[化] citric acid
【柠檬糖】lemon drops
【柠檬油】lemon oil
【柠檬榨汁器】lemon squeezer
【柠檬汁】lemon juice
拧 ①(扭绞) twist; wring: ～毛巾 twist a towel / 把稻草～成绳 twist straw into a rope ②(捏) pinch; tweak: ～了她一下 give her a pinch
【拧眉瞪眼】raise one's eyebrows and stare in anger
狞 ferocious; hideous
【狞恶可怖】fierce and terrifying
【狞笑】grin hideously
凝 ①(凝结) congeal; curdle; coagulate ②(注意力集中) with fixed attention
【凝点】[物] condensation point
【凝固】solidify ◇ ～点[物] solidifying point; freezing point / ～汽油 gelatinized gasoline / ～汽油弹[军] napalm bomb
【凝华】[气] sublimate ◇ ～核 sublimation nucleus
【凝灰岩】[地] tuff
【凝集】[化] agglutinate ◇ ～素 agglutinin
【凝胶】gel
【凝结】①(气体变为液体) condense ②(液体变为固体) coagulate; congeal ③(结合) cement: ～成深厚的友谊 cement a profound friendship ◇ ～剂[化] coagulant / ～力 coagulability / ～器 condenser / ～物 coagulum
【凝聚】①(凝结) condense ② [化] coacervation ◇～层 coacervate / ～力 cohesion; cohesive force
【凝练】concise; condensed; compact
【凝眸而视】behold with fixed gazing
【凝神】with fixed attention: ～思索 gaze fixedly and ponder / ～谛听 listen with rapt attention; listen attentively / ～远视 look into distance with fixed gazing
【凝视】gaze fixedly; stare
【凝思】be buried in thought; be deep in contemplation: ～默想 profounding and meditating
【凝析油】[石油] condensate
【凝血酶】fibrin ferment; thrombin; thrombase
【凝血药】[医] coagulant
【凝脂】congealed fat
【凝滞】stagnate; move sluggishly: ～的目光 dull, staring eyes

nǐng

拧 ①(旋转) twist; screw; wrench: ～断铁丝 twist off a piece of wire / ～上盖子 screw a lid on / ～下笔帽 twist off the cap of a fountain pen ②(颠倒) wrong; mistaken: 全给弄～了。 All is wrong. ③(别扭) differ; disagree; be at cross-purposes: 两人越谈越～。 The more they talked, the more they disagreed.

nìng

宁 ① rather; would rather; better: ～早勿晚。 Better be early than late. / ～折不弯。 It may break, but it will not bend.

【宁可】would rather; better: ～站着死,绝不跪着生 would rather die on one's feet than live on one's knees
【宁肯】would rather
【宁缺毋滥】rather go without than have something shoddy; put quality before quantity
【宁死不屈】rather die than submit
【宁为玉碎,不为瓦全】rather be a shattered vessel of jade than an unbroken piece of pottery; better to die in glory than live in dishonor
【宁愿】would rather; better: 我～待在家里。I would rather stay at home.

佞 given to flattery: ～人 sycophant; toady

niú

牛 ox; cattle (总称): 菜～ beef cattle／纯种公～ pedigree bull／公～ bull／母～ cow／奶～ milk cow; dairy cattle／水～ water buffalo／小～ calf
【牛蒡】[植] great burdock
【牛鼻子】the nose of an ox
【牛车】ox cart; bullock cart
【牛刀小试】a master hand's first small display
【牛痘】① [医](牛的一种急性传染病) cowpox; vaccinia ② (豆状疱疹) smallpox pustule; vaccine pustule: 种～ give or get smallpox vaccination ◇ ～苗 bovine vaccine
【牛犊】calf
【牛顿】[物] newton ◇ ～环 Newton's rings／～万有引力定律 Newton's law of gravitation／～运动定律 Newton's law of motion
【牛轭湖】oxbow lake
【牛肺疫】pleuropneumonia
【牛粪】cow dung
【牛感冒】ox influenza
【牛黄】[中药] bezoar
【牛角】ox horn ◇ ～画 horn mosaic／～制品 horn-ware
【牛角尖】the tip of a horn; an insignificant or insoluble problem
【牛劲】① (大力气) great strength; tremendous effort ② (牛脾气) stubbornness; obstinacy; tenacity
【牛郎星】[天] Altair
【牛马】oxen and horses; beasts of burden: 当～ be a slave／过着～不如的生活 live a life worse than that of beasts of burden
【牛毛】ox hair: 多如～ as many as the hairs on an ox; countless; innumerable
【牛毛细雨】drizzle; fine drizzling rain
【牛虻】gadfly
【牛奶】milk: 鲜～ fresh milk ◇ ～场 dairy／～咖啡 white coffee／～糖 toffee
【牛排】beefsteak
【牛棚】cowshed; ox fence
【牛皮】① (牛的皮) oxhide; cowhide; cattlehide ② (说大话) brag: ～大王 braggart／吹～ talk big
【牛皮癣】[医] psoriasis
【牛皮纸】kraft paper
【牛脾气】stubbornness; obstinacy; pigheadedness
【牛肉】beef: 小～ veal
【牛虱】ox louse
【牛溲马勃】sth. cheap but useful
【牛头刨床】[机] shaping machine; shaper
【牛头不对马嘴】horses' jaws don't match cows' heads; incongruous; irrelevant
【牛蛙】bullfrog
【牛尾】oxtail ◇ ～汤 oxtail soup
【牛瘟】rinderpest; cattle plague
【牛眼窗】oeil-de-boeuf window
【牛仔裤】close-fitting pants; jeans

niǔ

忸
【忸怩】blushing; bashful: ～作态 behave coyly; be affectedly shy

扭 ① (扭转) turn round: ～过头去 turn about; turn round ② (拧) twist; wrench: ～开门 wrench the door open ③ (拧伤) sprain; wrench: ～伤了手腕 sprain one's wrist ④ (走路身子左右摆动) roll; swing: 走起路来一～一～的 have a rolling gait ⑤ (揪住) grapple with; grab; seize: 两个摔交的～在一起。The two wrestlers grappled together.
【扭摆舞】twist
【扭秤】[物] torsion balance
【扭打】wrestle; grapple
【扭干】wring dry
【扭角羚】[动] takin
【扭结】twist together; tangle up
【扭亏为盈】turn loss into gain
【扭亏增盈】make up deficits and increase surpluses
【扭力】[物] twisting force
【扭扭捏捏】unmanly in handling business
【扭伤】sprain; wrench: ～了筋 wrench a tendon; sprain a muscle／～了腰 sprain one's back

【扭送】seize and turn over; drag sb. off to: 小偷被～到派出所了。The thief was seized and turned over to a police station.
【扭转】①(掉转) turn round; turn about ②(纠正) turn back; reverse: ～乾坤 bring about a radical change in the situation; reverse the course of events / ～时势 turn the tide; reverse the trend

狃 be bound by; be constrained by: ～于习俗 be bound by custom

纽 ①(器物上供抓住的部分) handle; knob: 门～ doorknob ②(纽扣) button ③(枢纽) bond; tie
【纽带】link; tie; bond: 友谊的～ ties of friendship
【纽扣】button
【纽襻】button loop

niù

拗 stubborn; obstinate; difficult: ～不过某人 unable to dissuade sb.; unable to turn sb. away / 脾气～ be of a stubborn temper

nóng

农 ①(农业) agriculture; farming: 务～ go in for agriculture ②(农民) peasant; farmer: ～户 peasant household / 菜～ vegetable grower
【农产品】agricultural products; farm produce: ～加工工业 farm products processing industry
【农场】farm: 国营～ state farm
【农村】rural area; countryside; village ◇ ～城市化 urbanization of villages / ～电气化 electrification of the countryside / ～集市 village fair; rural market / ～生活 rural life
【农贷】agricultural loans
【农工商联合企业】integrated farm-industry-commerce enterprise
【农活】farm work
【农机】agricultural machinery; farm machinery
【农家】peasant family ◇ ～肥 farm manure; farmyard manure
【农具】farm implements; farm tools: 半机械化～ semi-mechanized farm implements
【农历】the traditional Chinese calendar; the lunar calendar
【农林牧副渔】farming, forestry, animal husbandry, side-line production and fishery
【农忙】busy season
【农民】peasant; peasantry: ～起义 peasant uprising / ～意识 peasant mentality / ～协会 peasant association / ～战争 peasant war
【农牧区】farming and stockbreeding areas; agricultural and pastoral areas
【农奴】serf ◇ ～制度 serf system; serfdom / ～主 serf owner
【农时】farming season: 不违～ do farm work in the right season
【农事】farm work; farming
【农田】farmland; cropland; cultivated land ◇ ～基本建设 capital construction on farmland; farmland capital construction / ～水利 irrigation and water conservancy / ～水利建设 construction of water conservancy works
【农闲】slack season
【农学】agronomy; agriculture ◇ ～家 agronomist
【农谚】farmer's proverb; farmer's saying
【农药】agricultural chemical; farm chemical; pesticide ◇ ～残留 pesticide residue / ～污染 pesticide pollution / ～中毒 pesticide poisoning
【农业】agriculture; farming ◇ ～保险 agricultural insurance / ～地质学 agrogeology / ～工程学 agricultural engineering / ～工人 agricultural laborer; farm laborer; farm worker / ～国 an agricultural country / ～化学 agricultural chemistry; agrochemistry / ～机械 agricultural machinery; farm machinery / ～技术 agricultural technology / ～经济 agricultural economy / ～科学 agricultural sciences / ～气象学 agricultural meteorology; agrometeorology / ～人口 agricultural population / ～生物学 agrobiology / ～税 agricultural tax / ～土壤学 agrology / ～现代化 modernization of agriculture; agricultural modernization
【农艺师】agronomist
【农艺学】agronomy
【农用飞机】agricultural aircraft
【农作物】crops

浓 ①(稠密) dense; thick; concentrated: ～雾 dense fog / ～烟 dense smoke; thick smoke ②(味道厚) strong; rich: ～茶 strong tea ③(程度深) great; keen: 对此兴趣不～ take not much interest in it
【浓度】consistency; concentration; density: 当量～ [化] equivalent concentration / 高～ high concentration / 矿浆～ pulp density
【浓厚】①(很浓) dense; thick: ～的烟雾 thick smog ②(色彩浓重) deep; strong; pronounced: ～的民族色彩 strong national color ③(程度

深）strong: 兴趣很～ with keen interest; take great interest in
【浓积云】[气] cumulus congestus
【浓绿】dark green
【浓眉】heavy eyebrows; thick eyebrows: ～大眼 heavy features
【浓密】dense; thick: ～的森林 a dense forest / ～相宜 proper distribution (of trees in painting)
【浓缩】[化] concentrate; enrich ◇ ～剂 concentrating agents / ～物 concentrate / ～铀 enriched uranium
【浓汤】thick soup
【浓艳】rich and gaudy: 色彩～ in gaudy colors
【浓荫蔽空】the thick branches and leaves seem to blot out the sky
【浓郁】rich; strong: ～的香味 rich fragrance
【浓云密布】overcast with heavy clouds
【浓重】dense; thick; strong
【浓妆艳抹】wear heavy make up

脓 pus: 化～ suppurate; suppuration / 流～ run with pus
【脓包】① [医](脓液形成的隆起) pustule ②(无用的人) worthless fellow; good-for-nothing
【脓疮】running sore
【脓尿】[医] pyuria
【脓胸】[医] pyothorax
【脓肿】[医] abscess: 齿槽～alveolar abscess / 肝～ liver abscess / 阑尾～ appendicular abscess ◇ ～切开 incision of abscess

nòng

弄 ①(戏耍) play with; fool with; trifle with: 小孩爱～水。 Children like to play with water. ②(做;干) do; make; handle: 别把孩子～哭了。 Don't make the child cry. ③(设法取得) get; fetch: 去～点水来。 Go and get some water. ④(玩弄) play
【弄笔】distort facts; exaggerate in writing
【弄潮儿】a seaman; a beach swimmer
【弄错】make a mistake; misunderstand
【弄饭】prepare a meal
【弄鬼】hatch plots; play tricks behind the scene
【弄好】get sth. done: 你能把我的自行车～吗? Can you get my bicycle fixed?
【弄坏】ruin; put out of order; make a mess of: 把事情～ make a mess of things
【弄假成真】what was make-believe has become reality
【弄僵】bring to a deadlock; deadlock
【弄巧成拙】try to be clever only to end up with a blunder; outsmart oneself
【弄清】make clear; clarify; gain a clear idea of; understand fully: ～情况 find out the real situation / ～事实真相 clarify the facts
【弄权】manipulate power for personal ends
【弄死】put to death; kill
【弄通】get a good grasp of
【弄虚作假】practice fraud; employ trickery; resort to deception
【弄脏】stain; soil; pollute; smudge; smear
【弄糟】make a mess of; mess up; bungle; spoil: 你把事情～了。 You have spoiled the whole thing. (或: You have made a mess of the job.)

nú

奴 ①(受奴役的人) slave; bondservant ②(象对待奴隶那样的使用) enslave
【奴才】flunkey; lackey: 帝国主义的～ a lackey of imperialism ◇ ～相 servile behavior; servility; shameless fawning
【奴化】enslave: ～教育 enslaving education / ～思想 slave ideology / ～政策 policy of enslavement
【奴隶】slave ◇ ～起义 slave uprising / ～社会 slave society / ～占有制度 slave-owning system / ～主 slave owner; slaveholder / ～主义 slavishness; slavish mentality
【奴仆】servant; lackey
【奴性】servility; slavishness
【奴颜婢膝】subservient; servile
【奴役】enslave; keep in bondage

驽
【驽钝】[书] dull; stupid
【驽马】inferior horse; jade: ～千里，功在不舍。 If a jade travels a thousand *li*, it's only through perseverance

nǔ

努 ①(使出) put forth; exert: ～劲儿 put forth all one's strength; make an effort ②(凸出) protrude; bulge: ～着嘴 protrude one's lips ③(用力太过身体受伤) injure oneself through overexertion
【努力】make great efforts; try hard; exert oneself: ～不懈 strive without cease / ～工作 work hard / ～完成任务 exert oneself to accomplish one's assignment / 尽一番～ make an effort
【努嘴】pout one's lips as a signal

nù

怒 ①(愤怒) anger; rage; fury: 大～ in great anger ②(形容气势很盛) in profusion; burst: 心花～放 be beside oneself with joy; be highly delighted; be exceedingly happy

【怒不可遏】be beside oneself with anger; boil with rage

【怒潮】①(汹涌澎湃的浪潮) angry tide; raging tide ②[地] bore

【怒斥】angrily rebuke; indignantly denounce

【怒冲冲】in a rage; furiously: 他～地离开了房间。He left the room in a huff.

【怒发冲冠】bristle with anger; be in a towering rage

【怒放】in full bloom: 百花～。 Hundreds of flowers are blooming in profusion.

【怒号】howl; roar

【怒吼】roar; howl

【怒火】flames of fury; fury: ～冲天 a surge of great fury / ～中烧 be burning with anger / 满腔～ be filled with fury

【怒骂】curse in rage: ～不休 cursing angrily without stopping

【怒目】glaring eyes; fierce stare: ～而视 stare angrily; look daggers at; glare at; glower at

【怒气】anger; rage; fury: ～冲冲 in a great rage / ～冲天 be in a towering rag; give way to unbridled fury

【怒容】an angry look: ～满面 a face contorted with anger; look very angry

【怒视】glare at; glower at; scowl at: ～不语 giving black look and saying nothing

【怒涛】furious billows; mountainous sea: ～澎湃 billows raging with great fury

【怒形于色】betray one's anger; look angry

nǚ

女 ①(女子) woman; female: ～歌手 songstress / ～医生 woman doctor ②(女儿) daughter; girl: 独生～ one's only daughter

【女扮男装】a woman disguised as a man

【女厕所】Ladies; Women; ladies room; women's lavatory

【女车】woman's bicycle; lady's bicycle

【女低音】[乐] alto

【女儿】daughter; girl

【女儿墙】[建] parapet

【女方】the bride's side; the wife's side

【女飞行员】female pilot; aviatrix

【女服务员】①(飞机上) air hostess; stewardess ②(餐馆的) waitress

【女高音】[乐] soprano

【女工】woman worker

【女皇】empress

【女眷】the womenfolk of a family

【女郎】young woman; maiden; girl

【女朋友】girl friend

【女色】woman's charms: 好～ be fond of women

【女神】goddess

【女生】woman student; girl student; schoolgirl

【女声】[乐] female voice ◇ ～合唱 women's chorus; female chorus

【女士】lady

【女式便袍】negligee

【女售货员】shopgirl; saleswoman

【女王】queen

【女巫】witch; sorceress

【女性】①(性别) the female sex ②(妇女) woman

【女修道院】convent

【女演员】actress

【女英雄】heroine

【女中音】[乐] *mezzo-soprano*

【女中丈夫】as a man amongst the woman folks

【女主角】feminine lead; leading lady

【女主人】hostess; woman of the house

【女子】woman; female: 独身～ a single woman; feme sole ◇ ～单打 women's singles / ～双打 women's doubles / ～团体赛 women's team event / ～学校 girl's school

钕 [化] neodymium (Nd)

nuǎn

暖 ①(暖和) warm; genial ②(使温暖) warm up: 烤火取～ warm oneself at fire

【暖低压】[气] warm-core cyclone; warm-core low

【暖调】[美术] warm color tone; warm tone

【暖房】greenhouse; hothouse

【暖锋】[气] warm front

【暖高压】[气] warm anticyclone; warmcore anticyclone

【暖烘烘】nice and warm

【暖壶】①(热水瓶) thermos flask; thermos bottle ②(汤壶) hotwater bottle

【暖和】①(气候不冷) warm; nice and warm: 天气～了。It's getting warm. ②(使暖和) warm up: 进屋烤烤～～吧。 Come in and warm

yourself up.
【暖流】[地][气] warm current
【暖气】central heating ◇ ～片 (heating) radiator
【暖气团】[气] warm air mass
【暖色】[美术] warm color
【暖水瓶】thermos flask; thermos bottle

nüè

疟 malaria
【疟疾】malaria; ague: 恶性～ pernicious malaria
【疟蚊】malarial mosquito
【疟原虫】plasmodium; malarial parasite

虐 cruel; tyrannical
【虐待】maltreat; ill-treat; tyrannize
【虐杀】cause sb.'s death by maltreating him; kill sb. with maltreatment
【虐政】tyrannical government; tyranny

nuó

挪 move; shift
【挪东补西】make up deficiency at one place by drawing upon the surplus at the other
【挪动】move; shift: 往前～几步 move a few steps forward
【挪借】borrow money for a short time; get a short-term loan
【挪开】move away
【挪威】Norway ◇ ～人 Norwegian / ～语 Norwegian (language)
【挪窝儿】①(离开原来所在的地方) move to another place ②(搬家) move
【挪用】①(把钱移作别用) divert ②(私自动用) misappropriate; embezzle: ～公款 misappropriate public funds

nuò

懦 cowardly; weak
【懦钝】weak and dull
【懦夫】coward; craven; weakling: ～懒汉思想 the coward's and sluggard's way of thinking
【懦怯】cowardly
【懦弱】cowardly; weak: 她性格～。 She is weak in character.

糯 glutinous
【糯稻】glutinous rice
【糯米】polished glutinous rice

诺 ①(答应) promise; assent; nod: 许～ promise ②(答应的声音) yes; yep
【诺贝尔奖金】Nobel Prize
【诺言】promise: 履行～ fulfill one's promise; keep one's word

锘 [化] nobelium (No)

O

ōu

讴 ①(歌唱) sing ②(民歌) folk songs; ballads

【讴歌】sing the praises of; celebrate in song; eulogize

鸥 gull: 海～ sea gull

欧 short for Europe

【欧风美雨】the influences of Western culture and civilization

【欧化】Europeanize; westernize

【欧椋鸟】starling

【欧美】Europe and America; Western; the West

【欧姆】[物] ohm ◇ ～表 ohmmeter / ～定律 Ohm's law

【欧鸲】[动] robin; redbreast

【欧亚大陆】Eurasia

【欧洲】Europe ◇ ～经济共同体 the European Economic Community (E.E.C.) / ～美元 Eurodollar / ～人 European

殴 beat up; hit: ～伤 beat and injure

【殴打】beat up; hit: 互相～ come to blows; exchange blows

ǒu

耦

【耦合】[物] coupling: 机械～ mechanical coupling ◇ ～电路 coupled circuit; coupling circuit / ～系数 coupling coefficient

藕 lotus root

【藕断丝连】the lotus root snaps but its fibres stay joined; apparently severed, actually still connected

【藕粉】lotus root starch

【藕荷】pale pinkish purple

【藕节儿】joints of a lotus root

【藕色】pale pinkish grey

偶 ①(木雕或泥塑人像) image; idol: 木～ wooden image; puppet ②(双数) even; in pairs: 无独有～。 It is not a unique instance, but has its counterpart. ③(配偶) mate; spouse: 配～ spouse ④(偶然) by chance; by accident; once in a while; occasionally: ～一为之 do sth. once in a while; do it by chance / ～遇 meet by chance

【偶成之得】a verse written by chance

【偶氮染料】azo dyes

【偶尔】once in a while; occasionally: ～为之 do it occasionally

【偶发】accidental; chance; fortuitous: ～事件 a chance occurrence ◇ ～性倾销 sporadic dumping

【偶合】coincidence

【偶然】accidental; fortuitous; chance: ～现象 accidental phenomena / ～遇见一个老朋友 run into an old acquaintance; meet an old friend by chance; come across an old friend ◇ ～误差 accidental error / ～性[哲] contingency; fortuity; chance

【偶生成本】non-recurring cost

【偶生收益】non-recurring income

【偶数】[数] even number ◇ ～页[印] even page

【偶蹄动物】an artiodactyl

【偶像】image; idol ◇ ～崇拜 idolatry / ～崇拜者 idolater / ～化 idolize

呕 vomit; throw up

【呕吐】vomit; throw up; be sick: ～不止 keep vomiting

【呕心】exert one's utmost effort: ～之作 a work embodying one's utmost effort

【呕心沥血】shed one's heart's blood; take infinite pains; work one's heart out

【呕血】[医] haematemesis; spitting blood

òu

沤 soak; steep; macerate: ～麻 ret flax or hemp

【沤肥】①(使粪肥起变化) make compost ②(用水泡肥) wet compost; waterlogged compost

怄 ①(使怄气) irritate; annoy ②(怄气) be irritated

【怄气】be difficult and sulky: 怄了一肚子气 have a bellyful of repressed grievances

P

pā

趴 ①(胸腹朝下卧倒) lie on one's stomach; lie prone; lie prostrate: ～下 lie down / ～在地上 lie on the ground ②(伏;身体向前靠) bend over; lean over; lean on: ～在桌上 lean over a desk

pá

扒 ①(用手或耙子等使东西聚拢或散开) gather up; rake up; rake together: 把干草～在一起 rake together the hay ②(煨烂) stew; braise: ～鸡 braised chicken / ～羊肉 stewed mutton

【扒手】pickpocket; shoplifter: 谨防～! Beware of pickpockets!

耙 ①(耙子) rake: 钉～ an iron rake / 木～ wooden rake ②(扒拢;扒平) make smooth with a rake; rake: 把地～平 rake the soil level

【耙子】rake

琶

【琶音】[音] *arpeggio*

筢

【筢子】bamboo rake

爬 ①(爬行) crawl; creep ②(攀登) climb; clamber; scramble: ～山 climb a mountain / ～绳 climb a rope / ～树 climb a tree / 飞机～到一万米的高度。 The plane climbed to a height of ten thousand meters.

【爬虫】reptile

【爬得高跌得重】The higher one climbs, the harder he falls.

【爬竿】[体] ①(指项目) pole-climbing ②(指竿) climbing pole

【爬犁】sledge; sleigh

【爬坡能力】climbing capacity

【爬山虎】[植] Boston ivy; Virginia creeper

【爬山赛】[体] hill climb

【爬绳】[体] rope climbing

【爬行】crawl; creep: 跟在别人后面一步一步地～ trail behind others at a snail's pace

【爬行动物】reptile; creeper

【爬泳】[体] the crawl; crawl stroke

pà

怕 ①(害怕) fear; dread; be afraid of: ～某人 be in dread of sb. / ～死 be afraid to die / 不～疲劳 not be afraid of fatigue / 不～任何困难 brave all difficulties / 她～蛇。 She is afraid of snakes. ②(估计) I'm afraid; I suppose; perhaps: 会议～要延期了。I'm afraid the meeting will be postponed. ③(担心) be afraid of; for fear; fear lest: 她～把孩子吵醒。She was afraid of waking the child.

【怕前怕后】timid and apprehensive of everything

【怕人】①(见人害怕) shy; timid ②(使人害怕) terrible; terrific; awful

【怕生】(of a child) be shy with strangers

【怕事】be afraid of getting into trouble; be afraid of being involved: 胆小～ timid and overcautious

【怕死】fear death; be afraid of death ◇～鬼 coward

【怕羞】coy; shy; bashful: 这女孩～。The girl is shy.

帕 handkerchief: 手～ handkerchief

pāi

拍 ①(用手掌打) clap; pat; beat; slap: ～掉身上的土 pat one's clothes to get the dust off / ～球 bounce a ball / ～手 clap one's hands / ～桌子 strike the table ②(拍子) bat; racket: 苍蝇～儿 fly swatter / 乒乓球～ ping-pong bat; table-tennis bat / 网球～ tennis racket ③[乐] beat; time: 一小节四～ four beats in ⟨to⟩ a bar ④(拍摄) take (a picture); shoot: ～电影 shoot a film; make a film / ～照 take a picture / 把小说～成电影 film the novel ⑤(发出) send (a telegram, etc.): ～电报 send a telegram ⑥(拍马屁) flatter; fawn on: ～马 lick sb.'s boots; flatter / 能吹会～ be good at boasting and toadying

【拍案】strike the table (in anger, surprise, admiration, etc.): ～而起 smite the table and rise to one's feet

【拍案叫绝】express admiration by thumping the table; thump the table and shout "bravo!"

【拍板】① [音] clappers ②(打拍板) beat time with clappers ③ [商] rap the gavel: ～成交 strike a bargain; clinch a deal ④(做出决定) have the final say; give the final verdict

【拍打】pat; beat; slap: ～身上的雪 pat the snow off one's clothes

【拍发】send (a telegram): ～消息 cable a dispatch or report
【拍马屁】lick sb.'s boots; flatter; soft-soap; fawn on
【拍卖】①(当众出售,由顾客争价,以最高价成交) auction; sell sth. at auction ②(减价抛售;甩卖) selling off goods at reduced prices; sale ◇～佣金 lot money / ～者 auctioneer
【拍摄】take (a picture); shoot: ～特写镜头 shoot a close-up / ～照片 take pictures; take photos / 在～外景 be on location
【拍手】clap one's hands; applaud: ～称快 clap and cheer; clap hands for joy / ～叫好 clap and shout "bravo!"
【拍照】take a picture; photograph: 为别人～ take a picture of sb. / 请别人为自己～ have 〈get〉 one's photo taken
【拍子】①(球拍) bat; racket: 网球～ tennis racket / 羽毛球～ badminton racket ②[乐] beat; time: 打～ beat time / 单～ simple time / 二～ duple time / 复～ compound time / 三～ triple time / 四～ quadruple time

pái

排 ①(按次序摆) arrange; put in order: ～节目单 arrange the program / ～座位 arrange seats ②(排成的行列) row; line; rank: 坐在后～ sit in the back row / 坐在前～ sit in the front row ③ [军] platoon: 民兵～ militiaman platoon ④(排演) rehearse: ～戏 rehearse a play ⑤(竹排;木排) raft: 木～ timber raft / 竹～ bamboo raft ⑥(用力除去) exclude; eject; discharge: ～脓 discharge pus / 把水～出去 drain the water away
【排版】[印] composing; typesetting: 机器～ machine composition / 照相～ photocomposition ◇～工人 compositor; typesetter
【排比】parallelism
【排笔】broad brush comprising a row of pen-shaped brushes
【排场】ostentation and extravagance: 讲～ go in for ostentation and extravagance
【排斥】repel; exclude; reject: ～异己 exclude outsiders; discriminate against those who hold different views
【排除】get rid of; remove; eliminate: ～故障 fix a breakdown / ～万难 surmount 〈overcome〉 all difficulties / ～异己 get rid of those who hold a view different from one's own or who do not conform with one's ideas / ～障碍 remove an obstacle; get over an obstacle / 不～这种可能 cannot rule out this possibility ◇～法 exclusive method / ～性采购 preclusive buying
【排挡】gear
【排队】form a line; line up; queue up: ～买票 line up for tickets / ～上车 queue up for a bus / 把问题分类～ arrange the problems in order of importance and urgency ◇～论 queuing theory
【排粪】defecation; dejection; bowl movement
【排放】emission
【排骨】spareribs; ribs: 糖醋～ spareribs in sweet-sour sauce
【排灌】irrigation and drainage ◇～设备 irrigation and drainage equipment / ～网 irrigation and drainage network / ～站 irrigation and drainage pumping station
【排行】seniority among brothers and sisters: 她～第二。She is the second child.
【排挤】push aside; push out; squeeze out; elbow out: ～某人 push sb. aside / 互相～ each trying to squeeze the other out
【排解】①(调解) mediate; reconcile; intervene: ～纠纷 mediate a dispute; reconcile a quarrel ②(排遣) divert oneself from; dispel
【排涝】drain flooded fields; drain waterlogged land
【排雷】[军]removal of mines; mine clearance
【排练】rehearse: ～节目 have a rehearsal
【排列】①(顺次序放) arrange; range; put in order: ～成行 arrange in a row 〈line; column〉 / 按字母顺序～ arrange in alphabetical order ②[数] permutation: ～和组合 permutations and combinations
【排卵】ovulate; ovulation ◇～期 period of ovulation
【排难解纷】mediate a dispute; pour oil on troubled waters
【排尿】urinate; micturate ◇～困难 [医] dysuria / ～作用 excretion of urine
【排炮】(artillery) salvo; volley of guns
【排气】[机] exhaust ◇～阀 exhaust valve / ～管 exhaust pipe
【排遣】divert oneself from (loneliness or boredom); dispel; distract one's mind form: ～烦闷 dispel boredom
【排枪】volley of rifle fire
【排球】volleyball: 打～ play volleyball ◇～场 volleyball court / ～赛 volleyball match / ～网 volleyball net
【排山倒海】topple the mountains and overturn

the seas; overwhelming or sweeping: 以～之势 with the momentum of an avalanche
【排水】drain off water; drain away water; dewatering: 开沟～ dig trenches to drain water off ◇ ～工程 drainage works / ～沟渠 escape canal / ～管 drain pipe / ～管道 drainage pipeline
【排水量】①(船舶的排水量) displacement: 满载～ load displacement / 一艘～为五万吨的船 a ship of 50000 tons displacement ②(河道、渠道等的排水量) discharge capacity (of a spillway, etc.)
【排他性】exclusiveness ◇ ～集团 exclusive bloc
【排闼】push: ～直入 push the door open and go straight in; break into room unceremoniously
【排头】the person at the head of a procession; file leader; the person at the head of a row
【排外】exclusive; antiforeign: 盲目～ blind opposition to everything foreign ◇ ～主义 exclusivism; exclusionism; xenophobia; antiforeignism
【排尾】the last person in a row; the person at the end of a row
【排泄】①(使雨水、污水等流走) drain: ～不畅 drainage difficulty ②(把体内废物排出) excrete; excretion ◇ ～器官 excretory organ / ～物 excreta; excrement; evacuation / ～系统 excretory system
【排演】rehearse
【排印】typesetting and printing: 用大号字～ print in large type
【排长】platoon leader
【排中律】[逻] the law of excluded middle
【排字】composing; typesetting: 给一篇杂志上的文章～ typeset a magazine article / 为报纸第一版～ compose the front page of a newspaper ◇ ～车间 composing room / ～工人 typesetter; compositor / ～机 typesetting machine; composing machine / ～架 composing frame / ～手托 composing stick

徘

【徘徊】①(来回走) pace up and down; walk back and forth ②(犹豫不决) hesitate; waver; hover ③[经] fluctuate
【徘徊不前】hesitate to press forward
【徘徊歧路】hesitate at the crossroads

牌

①(标志牌子) plate; tablet; board; placard: 布告～ notice board / 车～ number plate on a vehicle / 路～ signpost / 门～儿 doorplate / 招～ shop sign; signboard ②(商标) brand: 长城～ Great Wall brand / 名～货 goods of a well-known brand ③(娱乐用品) cards; dominoes: 扑克～ playing cards
【牌匾】board; tablet
【牌坊】memorial archway; memorial gateway; honorific arch
【牌号】①(商店字号) the name of a shop; shop sign ②(商标) trademark
【牌价】①(标出价格) list price ②(市价) market quotation; prevailing price
【牌楼】①(指装饰用建筑) *pailou*,dccorated archway ②(指喜庆日临时搭成的) temporary ceremonial gateway
【牌位】memorial tablet
【牌照】license plate; license tag: 自行车～ bicycle number plate ◇ ～税 licence tax
【牌子】①(标志板) plate; sign: 存车～ tally ②(商标) brand; trademark: 老～ old brand; well-known brand / 我们有各种最好～的酒。We have the best brands of wines.

pǎi

排

【排子车】large handcart

迫

【迫击炮】mortar: 敌人正在用～轰击山头儿。The enemy was mortarring a hilltop. ◇ ～弹 mortar projectile; mortar shell

pài

派

①(派别; 派系) group; school; faction; clique: 党～ political parties and groups / 学～ school of thought / 左～ leftists ②(作风; 风度) style; manner and air: 气～ bearing ③(分配; 派遣; 委派) send; dispatch; assign; appoint: ～出代表团 send a delegation to / ～她担任秘书 appoint her to be secretary
【派别】group; school; faction: ～斗争 factional strife
【派兵遣将】dispatch troops and send generals
【派不是】put <lay> the blame on sb.
【派出机构】agency
【派出所】local police station; police substation
【派款】impose levies of money
【派力司】[纺] palace
【派遣】send; dispatch: ～代表团 send a delegation / ～武装部队 dispatch armed forces / ～驻外全权代表 dispatch a plenipotentiary (en-

voy) to a foreign country ◇ ～军 expeditionary forces
【派生】derive ◇ ～词[语] derivative
【派头】style; manner: 他～真不小! He certainly puts on quite a show!
【派系】factions
【派驻】accredit: ～联合国的代表 a representative accredited to the United Nations / 大使是本国～外国的代表。 An ambassador is accredited as the representative of his own country in a foreign land. / 他被～到华盛顿。 He was accredited at Washington.

哌
【哌嗪】[化] piperazine

pān

攀 ①(攀登) climb; clamber: ～着铁索往上爬 climb up a cable hand over hand ②(高攀) seek connections in high places ③(牵扯) involve; implicate: 乱咬乱～ make wild charges, while under interrogation, to implicate others
【攀扯】implicate (sb. in a crime)
【攀登】climb; clamber; scale: ～高峰 scale the heights / ～峭壁 climb up a cliff / ～珠穆朗玛峰 climb up Mt. Qomolangma / 单人～ [体] solo climb
【攀龙附凤】play up to people of power and influence; put oneself under the patronage of a bigwig; fawn upon the influential people
【攀亲】①(拉亲戚关系) claim kinship: ～道故 claim ties of blood or friendship ②(议婚) arrange a match; seek a match; ask for a match
【攀谈】engage in small talk; chitchat: 跟朋友～ have a chat with a friend
【攀缘】①(抓着东西往上爬) climb; clamber: ～而上 climb up / 由树上～而下 climb down a tree ②(攀附投靠) climb the social ladder through pull ◇ ～茎 climbing stem / ～植物 climber
【攀折】pull down and break off (twigs, etc.): 请勿～花木。 Please don't pick the flowers.

pán

蹒
【蹒跚】walk haltingly; limp; hobble; stagger; dodder: ～着向门口走去 limp to the door

盘 ①(盘子) tray; plate; dish: 茶～儿 tea tray / 古瓷～ old china plate ②(形状或功用象盘子的东西) sth. shaped like or used as a tray, plate, etc.: 磨～ millstone / 棋～ chessboard ③(回旋地绕) coil; wind; twist: 把绳子～起来 coil up the rope ④(垒、砌、搭炕或灶) build: ～灶 build a brick cooking range ⑤(仔细查问或清点) check; examine; interrogate: ～根究底 try to get to the heart of a matter ⑥(转让工商企业) transfer: ～店 transfer the ownership of a shop ⑦[体] game; set: 下一～棋 play a game of chess
【盘剥】practice usury; exploit: ～取利 be a Shylock / 重利～ lend money at usurious rates; exploit by lending money at exorbitant rates of interest
【盘查】interrogate and examine; question; examine thoroughly: ～行人 question the passers-by
【盘缠】money for the journey; travelling expenses
【盘秤】a steelyard with a pan
【盘存】take inventory: ～折旧 depreciation inventory
【盘点】check; make an inventory of: ～存货 take stock / 今日～,明日照常营业。 Stocktaking today. Business as usual tomorrow.
【盘根错节】with twisted roots and gnarled branches; complicated and difficult to deal with; deep-rooted
【盘管】coil (pipe)
【盘桓】stay; linger: ～终日 linger about all day long
【盘簧】[机] coil spring
【盘货】make an inventory of stock on hand; take stock
【盘诘】cross-examine; question
【盘踞】illegally or forcibly occupy; be entrenched; settle in: ～要津 hold a place of importance
【盘库】make an inventory of goods in a warehouse
【盘弄】play with; fiddle with; fondle
【盘儿菜】ready-to-cook dish (sold at market)
【盘绕】twine; coil; wreathe
【盘算】calculate; figure; plan
【盘损】adjustment debit
【盘梯】winding staircase; spiral staircase
【盘腿】cross one's legs: ～坐在床上 sit cross-legged on a bed
【盘问】cross-examine; interrogate: ～个可疑的人 interrogate a suspicious person
【盘香】incense coil
【盘旋】①(环绕着飞或走) spiral; circle; wheel;

hover: 飞机降落前在机场上空～。The plane spiraled the airport before. ②(徘徊;逗留) linger; stay
【盘羊】[动] argali
【盘盈】inventory profit
【盘帐】check accounts; audit accounts; examine accounts
【盘子】tray; plate; dish

磐

【磐石】huge rock; bedrock: ～般的团结 rocklike unity; monolithic unity / 坚如～ as solid as a rock; as firm as a rock

pàn

判 ①(分开;分辨) distinguish; discriminate ②(显然有区别) obviously (different): 前后～若两人 be quite a different person; be no longer one's old self ③(评定) appraise; give a mark: ～卷子 mark examination papers ④(判决) judge; decide: ～案 decide a case ⑤(判处) sentence; condemn: ～三年徒刑 be sentenced to three years' imprisonment
【判别】differentiate; distinguish: ～真假 distinguish the true from the false ◇ ～式 [数] discriminant
【判处】sentence; condemn: ～死刑 sentence sb. to death
【判词】[法] court verdict
【判定】judge; decide; determine
【判断】①(断定) judge; decide determine: ～错误 back the wrong horse / ～情况 assess the situation; size up the situation / ～是非 judge 〈decide〉 what is right and what is wrong ②[逻] judgment ◇ ～力 judgment
【判决】[法] court decision; judgment: ～无罪 pronounce sb. not guilty / ～有罪 pronounce sb. guilty ◇ ～书 court verdict; written judgment
【判例】[法] legal precedent; judicial precedent; case law
【判明】distinguish; ascertain; draw a clear distinction: ～是非 distinguish between right and wrong / ～真相 ascertain the facts / ～责任 establish responsibility (for what has happened)
【判若鸿沟】the difference is as far-reaching as between the water-course
【判若两人】different as if he weren't the same person; no longer be one's old self
【判若云泥】as far removed as heaven is from earth; poles apart
【判罪】declare guilty; convict: 判某人犯盗窃罪 convict sb. of theft

畔 ①(江湖道路旁边) side; bank: 河～ river bank; riverside / 湖～ the shore of a lake; by the lake ②(田地的边界) boundary; border: 田～ the border of a field

叛 betray; rebel against
【叛变】betray one's country, party, etc.; turn traitor; turn renegade; defect: ～投敌 turn traitor and go over to the enemy
【叛匪】rebel bandit; bandit rebels
【叛国】betray one's country; commit treason ◇ ～分子 traitor to one's country / ～罪 treason
【叛军】rebel army; rebel forces; insurgent troops
【叛离】betray; desert
【叛乱】armed rebellion; insurrection; revolt; mutiny: 煽动～ incite people to rise in rebellion / 镇压～ suppress a rebellion; put down a rebellion
【叛卖】betray; sell: ～祖国 betray one's country ◇ ～活动 traitorous activity; acts of treason
【叛逆】①(背叛) rebel against; revolt against ②(有背叛行为的人) rebel: 封建礼教的～ a rebel against feudal ethics
【叛徒】traitor; renegade; turncoat; rebel

襻 ①(扣住纽扣的套) a loop for fastening a button: 纽～儿 button loop ②(象纽襻的东西) sth. shaped like a button loop or used for a similar purpose: 鞋～儿 shoe strap ③(用绳子、线等绕住,使连在一起) tie; fasten with a rope, string, etc.: ～上几针 put in a few stitches / 用绳子～上 fasten with a rope

盼 ①(盼望) hope for; long for; expect; yearn for; look forward to: ～即示复 hoping for your immediate reply / ～家信 look forward to hearing from one's family ②(看) look: 左顾右～ glance right and left; look round
【盼头】sth. hoped for and likely to happen; good prospects: 这下子可有～了。There is hope then.
【盼望】hope for; long for; look forward to: ～佳音 long for your good voice

pāng

滂

【滂湃】(of water) roaring and rushing: ～的急流 rushing torrent

【滂沱】torrential; pouring: ～大雨 torrential rain; pouring rain / 涕泗～ let loose a flood of tears; be in a flood of tears / 大雨～。The rain poured down in torrents. (或: It's raining in torrents.)

膀 swell: ～肿 swollen; bloated

乓 [象] bang: 门～地一声关上了。The door banged shut. / 他～的一声关上了门。He closed the door with a bang.

páng

旁 ①(旁边) side: 马路两～ both sides of the street / 在一～ on one side / 站在路～ stand by the roadside ②(其他;另外) other; else: ～的事 other thing

【旁白】aside (in a play)

【旁边】side: 请坐在我～。Please sit by my side.

【旁观】look on; be an onlooker: 袖手～ look on with folded arms ◇ ～者 onlooker; bystander; spectator

【旁观者清】the spectator sees most clearly; the onlooker sees the game best; the onlooker is clear-headed

【旁路】[电] bypass ◇ ～电容器 bypass capacitor

【旁门】side door

【旁门左道】heresy; heterodox school; unorthodox ways

【旁敲侧击】attack by innuendo; make oblique thrusts

【旁切圆】[数] escribed circle

【旁人】other people

【旁若无人】act as if there was no one else present; self-assured; supercilious; act as if there was no one else present

【旁听】be a visitor at a meeting, in a school class, etc.; audit ◇ ～生 auditor / ～席 visitors' seats; public gallery

【旁通阀】[机] bypass valve

【旁通管】[机] bypass pipe

【旁系】collateral line ◇ ～继承人 collateral heir / ～亲属 collateral; collateral consanguinity; collateral relative / ～血亲 collateral relatives by blood

【旁压力】[物] lateral pressure

【旁征博引】quote copiously from many sources

【旁证】circumstantial evidence; collateral evidence; side witness; extraneous evidence

【旁支】collateral branch (of a family)

磅

【磅礴】①(广大) boundless; majestic; vast; widespread; tremendous: 气势～ of tremendous momentum ②(充满) fill; permeate

螃

【螃蟹】crab

膀

【膀胱】(urinary) bladder ◇ ～结石 cystolith; vesical calculus / ～镜 cystoscope / ～炎 cystitis; urocystitis / ～造影 cystography

彷

【彷徨】walk back and forth, not knowing which way to go; hesitate about which way to go; hesitate: ～歧途 hesitate at the crossroads

庞 ①(庞大) huge; tremendous; immense ②(多而杂乱) innumerable and disordered ③(脸盘) face: 面～ face

【庞大】huge; enormous; colossal; gigantic; big; immense: ～的代表团 a large delegation / ～的预算 huge budget / ～的正规军 a massive regular army / 机构～ an unwieldy organization / 开支～ an enormous expenditure

【庞然大物】huge monster; colossus; formidable giant; monstrous creature

【庞杂】numerous and jumbled; multifarious and disorderly; in a cumbersome jumble: 机构～ cumbersome administrative structure / 内容～ the contents are multifarious and disorderly / 议论～ numerous and jumbled views

pǎng

耪 loosen soil with a hoe: ～地 hoe the soil

pàng

胖 fat; stout; plump: 矮～的人 a man of stout build / 长～ get fat; put on weight

【胖大海】the seed of boat-fruited sterculia

【胖乎乎】plump; chubby; pudgy: ～的脸蛋 plump cheeks; chubby cheeks

【胖头鱼】bighead; variegated carp

【胖子】fat person; fatty

pāo

泡 ①(鼓起而松软的东西) sth. puffy and soft: 豆腐～儿 beancurd puff ②(虚而松软) spongy: 这木料发～。This wood is spongy.

【泡货】bulky cargo

【泡桐】paulownia

抛 ①(扔) throw; toss; fling: ～救生圈 throw a life buoy／他把球～给了我。He threw the ball to me. ②(丢下) leave behind; cast aside: 把某人远远～在后面 leave sb. far behind／被时代～在后面 be tossed to the rear by the times

【抛光】[机] polishing; buffing: ～零件 polish parts ◇～剂 polishing compound; polish／～轮 polishing wheel; buff

【抛锚】①(下锚使船停稳) drop anchor; cast anchor: 这船在岸边抛下了锚。The ship anchored along the shore. ②(车辆发生故障) break down; be out of order: 公共汽车在路上～了。The bus broke down on the way.

【抛弃】abandon; forsake; cast away; cast aside; discard: ～坏习惯 forsake bad habits; give up bad habits／～旧的传统观念 discard old traditional ideas

【抛射体】[物] projectile

【抛售】undersell; dump; sell sth. in big quantities

【抛头颅，洒热血】shed one's blood and lay down one's life (for a just cause)

【抛头露面】show one's face in public; go out and be seen in public

【抛物面】[数] paraboloid

【抛物面天线】parabolic antenna

【抛物线】[数] parabola

【抛物线飞行】[航] parabolic flight

【抛掷】throw; cast; toss

【抛砖引玉】cast a brick to attract jade; throw out a minnow to catch a whale; offer a few commonplace remarks by way of introduction so that others may come up with valuable opinions

páo

炮

【炮制】①[中医] prepare (Chinese medicine) ②(搞出) concoct; cook up: ～反动纲领 concoct a reactionary program／如法～ act after the same fashion; follow suit

袍 robe; gown: 棉～ cotton-padded robe

【袍笏登场】dress up and go on stage; said of a puppet upon his take-over

【袍子】robe; gown: 皮～ fur robe

咆

【咆哮】roar; thunder: ～如雷 be in a thundering rage; roar with rage

刨 ①(挖掘) dig; excavate: ～白薯 dig (up) sweet potatoes／～地 dig the ground／～坑儿 dig a hole; dig a pit ②(减去) excluding; not counting; minus

【刨根问底】get to the root of the matter; get to the bottom of things

【刨煤机】coal plough: 动力～ activated plough

狍 roe deer

【狍子】roe deer

pǎo

跑 ①(迅速前进) run: ～八百米 run the 800-meter dash／火车通常一小时～120公里。The train usually goes 120 kilometers in one hour. ②(逃走) run away; escape; flee: 别让坏蛋～了。Don't let the rascal escape.／车带～气了。Air is escaping from the tire. ③(为某种事物而奔走) run about doing sth.; run errands: ～材料 run about collecting material or making inquiries／～买卖 be a commercial traveller ④(液体挥发) evaporate: 瓶里的汽油～了。The gasoline in the bottle has evaporated. ⑤(接在动词后面，表示离开原有位置) away; off: 吓～ frighten away／桌上的纸叫风给刮～了。The paper blew off the table.

【跑遍】go around; travel all over

【跑表】[体] stopwatch

【跑步】run; march at the double: 去～ go for a run

【跑步前进】[口令] Double time!

【跑步】[口令] At the double, quick march!

【跑车】racing bike

【跑单帮】travel around trading on one's own

【跑道】①[航空] runway ②[体] track; athletic track: ～里圈 inside lane; inner lane／～外圈 outside lane; outer lane／雷考坦塑料～ Rekortan track／全天候～ all-weather track／塑料～ plastic track

【跑电】[电] leakage of electricity; electric leakage

【跑江湖】wander about, making a living as an acrobat, fortuneteller, physiognomist, etc.

【跑警报】run for shelter during an air raid; run for air-raid shelter

【跑来跑去】run back and forth
【跑了和尚跑不了庙】The runaway monk can't run away with the temple. (或: A fugitive must belong to some place that can provide clues.)
【跑垒】[棒、垒球] baserunning ◇ ～员 base runner
【跑龙套】play a bit role; be a utility man
【跑马】①(骑着马跑) have a ride on a horse; run a horse ②(赛马) horse race ◇ ～场 racecourse; the turf
【跑跑颠颠】bustle about; be on the go
【跑墒】evaporation of water in soil
【跑堂儿的】waiter (in a restaurant)
【跑腿儿】run errands; do legwork
【跑鞋】[体] running shoes; track shoes

pào

泡 ①(气泡) bubble: 肥皂～儿 soap bubbles / 冒～儿 send up bubbles; rise in bubbles ②(泡状物) sth. shaped like a bubble: 电灯～ electric light bulb / 脚上起了～ get 〈raise〉 blisters on one's feet ③(浸泡) steep; soak ④(故意消磨时间) dawdle; dillydally
【泡菜】pickled vegetables; pickles; sauerkraut: 朝鲜～ kimchi
【泡茶】make tea
【泡饭】①(把汤或水加在饭里) soak cooked rice in soup or water ②(加水重煮的饭) gruel
【泡蘑菇】①(拖延时间) use delaying tactics; play for time; play a game of stalling; stall for time ②(纠缠) importune; pester
【泡沫】foam; froth: 啤酒～ the head on a glass of beer ◇ ～玻璃 foam glass / ～混凝土 foam concrete; cellular concrete / ～剂 foaming agent / ～灭火器 foam extinguisher / ～塑料 foamed plastics; aerated plastics / ～橡胶 foam rubber; air foam rubber
【泡泡纱】seersucker; crimp cloth
【泡泡糖】bubble gum
【泡影】visionary hope, plan, scheme, etc.; bubble: 化为～ vanish like soap bubbles; melt into thin air; go up in smoke; come to nothing; end up in smoke

疱 blister; bleb
【疱疹】①(天花、水痘等的症状) bleb ②(皮肤病) herpes: 带状～ herpes zoster; zoster ◇ ～病毒 herpesvirus / ～净[药] idoxuridine

炮 ①(火炮) big gun; cannon; artillery piece: 多管火箭～ multibarrel rocket launcher / 反坦克～ antitank artillery; AT gun; tank destroyer / 高射～ antiaircraft artillery; antiaircraft gun / 海岸～ coast gun / 火箭～ rocket launcher / 加农～ cannon / 榴弹～ howitzer / 迫击～ mortar / 轻型火～ light artillery / 山～ mountain artillery / 无后座力～ recoilles gun / 要塞～ garrison gun; garrison ordnance / 野战～ field artillery / 远射程～ long-range gun / 重～ heavy artillery ② [棋] cannon
【炮兵】artillery; artillerymen ◇ ～部队 artillery (troops) / ～连 battery / ～阵地 artillery position; gun emplacement
【炮弹】(artillery) shell: 毒气～ gas shell / 空～ blank shell / 迫击～ mortar shell / 杀伤～ fragmentation shell / 未爆～ dud shell
【炮轰】bombard; shell: ～敌人阵地 bombard the enemy's position
【炮灰】cannon fodder
【炮火】artillery fire; gunfire: ～连天 gunfire licks the heavens
【炮击】bombard; shell
【炮架】gun carriage; gun mount
【炮舰】gunboat ◇ ～外交 gunboat diplomacy / ～政策 gunboat policy
【炮楼】blockhouse
【炮身】barrel ◇ ～托架 gun stay
【炮声】thunder of guns; report of artillery; roar of guns: ～沉寂 the guns are quiet / ～隆隆 boom of guns
【炮手】gunner; artilleryman
【炮闩】breech block
【炮塔】gun turret; turret
【炮台】fort; battery
【炮艇】gunboat
【炮筒】barrel (of a gun)
【炮筒子】a person who shoots off his mouth
【炮位】emplacement
【炮眼】① [军] porthole; embrasure ② [矿] blasthole; dynamite hole; borehole
【炮衣】gun cover
【炮战】artillery action; artillery engagement
【炮仗】firecracker; cracker
【炮座】gun platform

pēi

呸 [叹] pah; bah; pooh

胚 [生] embryo
【胚层】[生] germinal layer: 内～ entoderm / 外～ ectoderm / 中～ mesoderm
【胚根】[植] radicle

【胚盘】[动] blastodisc; germinal disc
【胚乳】[植] endosperm
【胚胎】[生] embryo; embryon ◇ ～学 embryology / ～移植[牧] embryo transfer; embryonic implantation
【胚芽】[植] plumule
【胚轴】[植] plumular axis
【胚珠】[植] ovule

péi

培 ①(在根基部堆上土) bank up with earth; earth up ②(培养) cultivate; foster; train
【培土】hill up; earth up; bank up with earth; ridge: 给植物或它的根～ earth up a plant or its root
【培训】cultivate; train ◇ ～班 training class
【培养】①(教育和训练) foster; train; develop; educate; cultivate: ～接班人 train successors / ～人才 train men for profession ②[生] culture; cultivate: ～细菌 culture of bacteria ◇ ～基[生] culture medium / ～皿 culture dish; culture plate / ～瓶 culture bottle / ～液 cultivation liquid; culture solution
【培育】cultivate; foster; breed; nurture; rear: ～树苗 grow saplings / ～水稻新品种 breed new varieties of rice
【培植】cultivate; foster; train; educate: ～果树 cultivate fruit trees / ～私人势力 build up one's personal influence; foster one's personal influence

赔 ①(赔偿) compensate; pay for: 损坏东西要～。 Pay for anything you damage. ②(亏本) stand a loss: ～钱 lose money in business transactions
【赔本】sustain losses in business; run a business at a loss; lose one's outlay; lose one's capital: ～的买卖 losing business
【赔不是】apologize
【赔偿】compensate; pay for: ～名誉 indemnity for defamation / ～损害 make a reparation for an injury; repair the injury / ～损失 compensate for a loss; make good a loss / 战争～ war reparations / 照价～ compensate according to the cost / 保留要求～的权利 reserve the right to demand compensation for losses ◇ ～保证书 letter of indemnity / ～代理人 claim settling agent / ～范围 extent of compensation / ～费 damages / ～合同 indemnity contract / ～条款 indemnity clause / ～协定 reparations agreement / ～责任 liability to pay compensation
【赔款】①(赔偿损失) pay an indemnity; pay reparations ②(赔偿费) indemnity; reparations; amende; amends
【赔礼】offer 〈make〉 an apology; apologize: ～道歉 offer an apology
【赔了夫人又折兵】throw the helve after the hatchet; pay a double penalty
【赔笑】smile obsequiously or apologetically
【赔帐】pay for the loss of cash or goods entrusted to one
【赔罪】apologize

锫 [化] berkelium (Bk)

陪 accompany; keep sb. company: ～外宾参观农场 show foreign visitors round the farm / 请～我散步。 Please accompany me on my walk. / 我～你回家。 I'll accompany you home.
【陪伴】for company; keep sb. company: 我要～你一直到车站。 I'll go with you as far as the station for company.
【陪绑】①(指与将处决的犯人一起被绑赴刑场) be taken to the execution ground together with those to be executed as a form of intimidation ②(指无过错的人与有过错的人一起被批评或处分) be criticized or punished together with the guilty
【陪衬】①(衬托) serve as a contrast or foil; set off ②(陪衬物) foil; setoff
【陪客】①(主人邀来陪伴客人的人) a guest invited to help entertain the guest of honor at a dinner party ②(陪伴客人) accompany a guest; with a guest for company
【陪审】act as an assessor; serve as an assessor; serve on a jury ◇ ～团 jury / ～席 box / ～员 juror; juryman / ～制 jury system
【陪同】accompany; be in the company of: 在某人的～下 accompanied by sb.; in the company of sb.
【陪葬】be buried with the dead

pèi

沛 copious; abundant: 精力充～ be full of energy

配 ①(两性结合) join in marriage; match sb. with sb.; pair off with sb.: 婚～ marry ②(使动物交配) mate: ～猪 mate pigs ③(按适当比例调和或凑在一起) blend; compound; mix: ～颜色 mix colors / ～药 make up a prescription ④

(有计划地分派) distribute according to plan; apportion: ～售 ration ⑤ (补足;配齐) fit; complete: ～成一套 complete the set / ～零件 replace parts / ～钥匙 have a key made to fit a lock ⑥ (衬托;陪衬) match: 这顶帽子与外衣很相～。 The hat is a match for the coat. ⑦ (够得上;符合;相当) deserve; be worthy of; be qualified: 他不～做一名教师。 He is not qualified to be a teacher.

【配备】① (根据需要分配) allocate; provide; fit out: ～拖拉机 allocate tractors / ～助手 provide assistants ② (布置兵力) dispose (troops, etc.); deploy ③ (成套的器物等) outfit; equipment: 现代化的～ modern equipment

【配电】(power) distribution ◇ ～盘 distributor / ～网 distribution network / ～线路 distribution line / ～站 power distribution station

【配殿】side hall in a palace or temple

【配对】① (配合成双) pair; make a pair ② (动物交配) mate

【配额】quota ◇ ～限制 quota restrictions / ～制 quota system

【配方】① (配药) make up a prescription; fill a prescription ② (配制方法) formula: 止咳合剂的～ a formula for cough mixture

【配合】coordinate; cooperate; concert: ～得宜 be in harmony with one another / ～行动 take concerted action / ～作战 coordination of military operations / 起～作用 play a supporting role

【配给】ration; allotment; allocate ◇ ～品 rationed goods / ～证 ration card / ～制 ration system; allotment system

【配货提单】order blank

【配件】① fittings (of a machine, etc.): 窗～ window fittings / 管子～ pipe fittings ② (损坏后重新安上的零件) a replacement

【配角】① (共同主演;合演主角) appear with another leading player; costar ② (次要角色) supporting role; minor role: 男～ a supporting actor / 女～ a supporting actress

【配料】[冶] burden: 高炉～ blast-furnace burden ◇ ～表 burden sheet / ～计算 burden calculation

【配偶】spouse; consort

【配色】match colors; harmonize colors

【配套】form a complete set ◇ ～工程[水] conveyance system / ～器材 necessary accessories / ～饰物 parure

【配伍】[药] compatibility of medicines ◇ ～禁忌 incompatibility

【配戏】support a leading actor; play a supporting role

【配药】make up a prescription

【配页】[印] gathering (leaves of a book)

【配音】dub (a film, etc.): 给外国电影～ dub foreign films

【配乐】dub in background music

【配制】compound; make up: ～药剂 compound medicines

【配置】dispose (troops, etc.); deploy: ～兵力 dispose forces / 纵深～ disposition in depth

【配种】breeding; mating ◇ ～率 breeding rate / ～站 breeding station

【配子】[生] gamete ◇ ～体 gametophyte

辔 bridle: 鞍～ saddle and bridle

【辔头】bridle

佩 ① (佩带) wear (at the waist, etc.): ～刀 wear a sword / 腰～手枪 carry a pistol in one's belt ② (佩服) admire: 他的精神可～。 His spirit is admirable.

【佩带】wear; bear; carry: ～徽章 wear a badge

【佩服】admire; have admiration for: ～他的工作能力 admire his capacity for work

pēn

喷 ① (喷出) spurt; spout; gush; jet: 伤口～血。 Blood spouts from wound. ② (喷洒) spray; sprinkle: 给玫瑰花～点水 sprinkle some water on the roses

【喷薄】gush; spurt: 一轮～欲出的红日 the emerging sun with all its shimmering rays

【喷出岩】[地] extrusive rock

【喷灯】blowtorch; blowlamp

【喷粉器】duster

【喷灌】sprinkling irrigation; spray irrigation: 机械化～ mechanized spray irrigation ◇ ～器 sprinkler

【喷壶】watering can; sprinkling can

【喷火器】flamethrower; flame projector

【喷浆】[建] ① (粉刷) whitewashing ② (灌混凝土) guniting

【喷漆】spray paint; spray lacquer ◇ ～枪 paint (spraying) gun; varnish spray gun

【喷气发动机】jet engine

【喷气燃料】jet fuel

【喷气式】jet-propelled ◇ ～飞机 jet plane; jet aircraft; jet / ～轰炸机 jet bomber / ～客机 jet airliner; jetliner / ～空中加油机 jet

tanker／～运输机 jet transport／～战斗机 jet fighter
【喷气织机】air-jet loom
【喷枪】spray gun
【喷泉】fountain
【喷洒】spray; sprinkle: ～农药 spray insecticide
【喷射】spray; spurt; jet: ～火焰 spurt flames ◇ ～泵 jet pump／～剂 propellant／～器 sprayer; ejector; thrower／～推进 jet propulsion
【喷水池】fountain
【喷丝头】[纺] spinning jet; spinning nozzle
【喷嚏】sneeze: 打～ sneeze
【喷头】①(淋浴喷头) shower nozzle ②(洒水车喷头) sprinkler head
【喷雾】spraying ◇ ～器 sprayer; atomlser
【喷子】sprayer; spraying apparatus
【喷嘴】spray nozzle; spray head

pén

盆

basin; tub; pot: 花～ flowerpot／脸～ washbasin／澡～ bathtub
【盆地】basin
【盆花】potted flower
【盆景】potted landscape; miniature trees and rockery: 日本～ bonsai
【盆腔】[生理] pelvic cavity ◇ ～炎 pelvic infection
【盆浴】bath in a tub; tub

pèn

喷

【喷香】fragrant; delicious

pēng

烹

①(煮) boil; cook: ～茶 brew tea; make tea ②(热油略炒加作料搅拌) fry quickly in hot oil and stir in sauce: ～对虾 quick-fried prawns in brown sauce
【烹饪】cooking; culinary art: 擅长～ be good at cooking; be a good cook ◇ ～法 cookery; cuisine; recipe
【烹调】cook (dishes): 他的妻子精于～。His wife is a good cook. ◇ ～法 gastronomy

抨

【抨击】attack (in speech or writing); assail; lash out at; flay: ～某人 assail sb.

澎

splash; spatter
【澎湃】surge; be in an upsurge: 波涛～ waves surge high／心潮～ feel an upsurge of emotion

péng

澎

【澎湖列岛】the Penghu Islands

膨

【膨大】expand; inflate
【膨松剂】leavening agent
【膨体纱】[纺] bulk yarn
【膨胀】expand; swell; dilate; inflate: 通货～ inflation ◇ ～计[物] dilatometer／～曲线 cut-off curve; expansion curve／～系数[物] coefficient of expansion; swell factor／～性 expansibility

蓬

①[植] bitter fleabane ②(蓬乱) fluffy; dishevelled: ～着头 with dishevelled hair
【蓬勃】vigorous; flourishing; full of vitality: ～发展 grow vigorously; grow with vigor／～兴起 spring up energetically; spring up exuberantly
【蓬莱】a fabled abode of immortals
【蓬松】fluffy; puffy: ～的头发 fluffy hair
【蓬头垢面】with dishevelled hair and a dirty face; unkempt

篷

①(遮蔽物) covering; awning: 车～ awning on a car ②(船帆) sail: 扯～ hoist the sails／落～ drop the sails
【篷布】tarpaulin

朋

friend: 良～ good friend／宾～满座。There was a houseful of guests. (或: Visitors filled all the seats.)
【朋比为奸】act in collusion with; conspire; collude; gang up; associate with for treasonable purpose
【朋党】clique; cabal
【朋友】①(有交情的人) friend: 与…交～ make friends with／他有很多～。He has a large number of friends.／我们是好～。We are great friends. ②(恋爱对象) boy friend; girl friend

棚

①(用竹木等搭成的小棚) booth; awning; canopy: 茶～ tea booth／凉～ awning ③(简陋棚屋) shed; shack: 草～ straw mat shed／牲口～ livestock shed／自行车～ bicycle shed
【棚车】①[铁道] box wagon; boxcar ②(带棚的卡车) covered truck
【棚户】slum-dwellers; shack-dwellers ◇ ～区 shanty town

【棚子】shed; shack: 草～ straw mat shed

硼 boron (B)

【硼钢】boron steel

【硼砂】borax; sodium borate ◇ ～玻璃 borax glass

【硼酸】boric acid ◇ ～盐 borate

鹏 roc

【鹏程万里】(make) a roc's flight of 10000 *li*; have a bright future

pěng

捧 ①(用双手托) hold or carry in both hands: ～着一个大碗 hold a big bowl in both hands ②(奉承;吹嘘) boost; exalt; extol; flatter: 把某人～上天 praise sb. to the skies

【捧场】①(在剧场喝彩) be a member of a *claque* ②(吹嘘) boost; sing the praises of; flatter: 无原则的～ unprincipled praise

【捧腹】split 〈shake; burst〉 one's sides with laughter: ～大笑 be convulsed with laughter / 令人～ set people roaring with laughter; make one burst out laughing

pèng

碰 ①(接触;碰击) touch; bump: ～得头破血流 knock one's head against a brick wall / ～翻墨水瓶 knock the ink-bottle over / 头～到门上 bump one's head against the door ②(碰见;遇到) meet; run into ③(试探) take one's chance: ～～机会 take a chance

【碰杯】clink glasses

【碰壁】run up against a stone wall; meet a rebuff; be rebuffed; have one's nose put out of joint: 到处～ run into snags and be foiled everywhere

【碰钉子】meet with a rebuff; run into snags: 碰了个软钉子 be tactfully rebuked; be mildly rebuffed

【碰簧锁】spring lock

【碰见】meet unexpectedly; run into; encounter; come across: 在街上～一个熟人 run into an acquaintance in the street

【碰碰车】bumper car; Scooter car

【碰巧】by chance; by coincidence; happen to: 她来访时我～不在家。 I happened to be out when she called.

【碰锁】spring lock

【碰头】①(开简短的会) meet and discuss; put (our, your, their) heads together ②(见面) see each other ◇ ～会 brief meeting (mainly to exchange information)

【碰一鼻子灰】be snubbed; meet with a rebuff; meet rejection

【碰运气】try one's luck; take a chance; depend upon luck

【碰撞】①(猛然碰上) collide; run into; knock against; run foul of; crash: 互相～ ran foul of each other / 公共汽车～了一辆停着的小汽车。 A bus ran into a parked car. ②[物] collision; impact: 核～ nuclear collision ◇ ～负载 impact load / ～理论 collision theory / ～伤 impact injury / ～险 collision insurance / ～责任 collision liability

pī

坯 ①(坯件) base; semifinished product; blank: 铜～ cooper base; cooper blank ②(砖坯;土坯) unburned brick; earthen brick; adobe

【坯布】grey (cloth)

【坯件】blank: 螺栓～ bolt blank

【坯模】mold

【坯子】semifinished product; base; blank

砒 arsenic

【砒霜】(white) arsenic

批 ①(用手掌打) slap ②(批判) criticize; refute ③(批示) write instructions or comments on (a report from a subordinate, etc.): ～文件 write instructions on documents ④(大量买卖) wholesale: ～购 buy goods wholesale ⑤[量] batch; lot; group: 分～ in separate batches / 一大～ large numbers / 一～货物 a batch of goods ⑥(线批;麻批) fibers of cotton, flax, etc. ready to be drawn and twisted

【批驳】①(否决别人的意见或请求) veto an opinion or a request from a subordinate body ②(批评) refute; criticize; rebut: 逐点予以～ refute point by point

【批次】batch (of aircraft, etc.)

【批发】①(成批销售)wholesale ②(批准发布) be authorized for dispatch ◇ ～部 wholesale department / ～价格 wholesale price / ～价格指数 wholesale price index / ～商 wholesaler / ～市场 terminal market / ～业 wholesale business

【批复】give an official, written reply to a subordinate body

【批改】correct: ～作业 correct students' papers

【批号】lot number; batch number

【批量】batch; lot: ～生产 batch process

【批判】criticize; repudiate ◇ ～现实主义 critical realism
【批判地】critically; discriminatingly: ～继承 critically inherit / ～吸收 critically assimilate; assimilate with discrimination
【批评】criticize: ～某人 criticize sb. / ～缺点和错误 criticize shortcomings and mistakes ◇ ～与自我～ criticism and self-criticism
【批示】①(批示的话) written instructions or comments ②(写意见和指示) make comments and instructions
【批语】①(对文章的评语) remarks on a piece of writing ②(批示) comments and instructions
【批阅】read over; read and amend or comment on; read over and give remarks: ～文件 read and comment on documents
【批注】①(加批语和注释) annotate and comment on ②(批语和注释) annotations and commentaries; marginalia; head-note
【批准】ratify; approve; sanction; authorize: ～条约 ratify a treaty / ～一项工程 authorize a project / ～一项决议 approve a resolution
【批准书】instrument of ratification: 互换～的证书 protocol on the exchange of instruments of ratification / 交存～ deposit the instruments of ratification

纰 (of cloth, thread, etc.) become unwoven or untwisted; be spoilt
【纰漏】careless mistake; small accident; slip: 出了～ make a small error; make a slip
【纰缪】error; mistake

披 ①(覆盖或搭在肩背上) drape over; wrap around; throw on; spread: ～上节日的盛装 be colorfully decorated for the festival / ～着大衣 have an overcoat draped over one's shoulders / ～着…的外衣 under the cloak of; be clothed in; disguised as / 一只～着羊皮的狼 a wolf in sheep's clothing ②(打开) open; unroll; spread out: ～卷 open a book ③(竹木等裂开) split open; crack; break
【披风】cloak; mantle
【披肝沥胆】①(开诚相见) open up one's heart; speak without reserve; be open and sincere ②(忠诚) be loyal and faithful; show a loyal heart
【披红】drape a band of red silk over sb.'s shoulders: ～戴花 have red silk draped over one's shoulders and flowers pinned on one's breast
【披坚执锐】wear armor and carry weapons; be a warrior; be warrior-like
【披肩】①(无袖短外衣) cape ②(女用披巾) shawl
【披巾】shawl
【披荆斩棘】break through brambles and thorns; hack one's way through difficulties; clear away obstacles in one's way
【披露】①(发表;公布) publish; announce; make public ②(表露) reveal; show; disclose
【披麻带孝】put on mourning apparel
【披靡】①(草木随风散乱倒下) be swept by the wind; be blown about by the wind ②(军队溃散) be routed; flee; put to rout: 望风～ flee pell-mell at the sight of the opponent army / 大军所向～。The main forces put all to rout wherever they went. (或: The main forces carried all before them.)
【披散】(of hair, etc.) hang down loosely
【披沙拣金】sort out the fine gold from the sand; get essentials from a large mass of material
【披头散发】with hair dishevelled; with hair in disarray; wear hair dishevelled
【披星戴月】under the canopy of the moon and the stars; work or travel night and day
【披阅】open and read (a book); peruse

霹
【霹雷】thunderbolt; thunderclap
【霹雳】thunderbolt; thunderclap: 晴天～ a bolt from the blue
【霹雳舞】break dance

劈 ①(用刀斧破开) split; chop; cleave: ～成两半 cleave sth. in two / ～木柴 chop wood; split logs ②(正对着;冲着) right against: 大浪朝我们～面打来。Huge waves came crashing almost on top of us. ③(雷电击毁或击毙) strike ④[物] (尖劈) wedge
【劈波斩浪】cleave through the waves
【劈刺】[军] saber or bayonet fighting ◇ ～训练 bayonet drill
【劈刀】①(刀背较厚的刀) chopper ②[军] saber fighting
【劈理】[矿] cleavage
【劈脸】right in the face: 一个球～向她打来。A ball came hurtling towards her face.
【劈山】level off hilltops; blast cliffs: ～引水 cleave hills and lead in water; cut through mountains to bring in water / ～筑路 blast cliffs to build highways <railways>
【劈手】make a sudden snatch: ～抢来 grab sth.

in a flash
【劈头】①(迎头) straight on the head; right in the face: ～一拳 hit sb. right on the head ②(开头) at the very start; right at the beginning
【劈头盖脸】right in the face; direct to one's head and face: ～地提了许多问题 shower a lot of questions upon sb.; fire a volley of questions at sb.
【劈胸】right against the chest; straight on the breast: ～一把抓住 grasp sb. by the front of his coat

pí

琵
【琵琶】*pipa*, a plucked string instrument with a fretted fingerboard
【琵琶桶】barrel

枇
【枇杷】loquat

毗
adjoin; be adjacent to
【毗连】【毗邻】adjoin; border on; be adjacent to: ～地区 contiguous zone / 墨西哥与美国毗邻。 Mexico adjoins United States. (或: Mexico and the United States adjoin.)

蚍
【蚍蜉】ant
【蚍蜉撼大树】an ant trying to topple a giant tree; an ant trying to shake a big tree; ridiculously overrating one's own strength

啤
【啤酒】beer: 淡色～ ale / 黑～ porter; brown ale; stout; dark beer / 上面～ ale / 生～ draught beer ◇ ～厂 brewery
【啤酒花】[植] hops

蜱
tick

脾
spleen
【脾破裂】lienal rupture; rupture of spleen
【脾气】①(性情) temperament; disposition: ～很好 have a good temper / ～急躁 have a hot temper ②(易怒) bad temper: ～大 hot-tempered / 发～ lose one's temper; flare up ③(事物的特性) behavior; characteristic: 摸熟机器的～ get to know the characteristics of a machine; get to know how the machine behaves
【脾切除】splenectomy
【脾胃】taste: ～相投 have similar likes and dislikes / 不合～ not suit one's taste; not be to one's liking
【脾脏】spleen: ～切除 splenectomy
【脾肿大】splenomegaly

皮
①(外皮) skin: 擦破一块～ scrape a bit of skin off / 土豆～ potato peel / 树～ bark / 香蕉～ banana skin / 西瓜～ watermelon rind / 猪～ pigskin ②(皮革) leather; hide: ～靴 leather boots ③(毛皮) fur: ～大衣 fur coat / ～领 fur collar / 上等的狐～ a fine fox fur ④(书的封面;包装物) cover; wrapper: 包袱～儿 cloth-wrapper / 书～儿 book cover; jacket ⑤(表面) surface: 水～儿 surface of the water ⑥(薄片状物) a broad, flat piece (of some thin material); sheet: 奶～儿 skin (on boiled milk) / 铁～ iron sheet ⑦(受潮而不再酥脆) become soft and soggy ⑧(顽皮;调皮) naughty: 这孩子真～! What a naughty boy! ⑨(感到无所谓) case-hardened: 他老挨罚，都～了。 He gets punished so often that he no longer cares.
【皮袄】fur-lined jacket
【皮包】leather handbag; briefcase; portfolio
【皮包骨头】skinny: 瘦得～ be only skin and bone
【皮鞭子】leather-thonged whip
【皮层】①[生] cortex ②[生理] cerebral cortex
【皮尺】tape measure; tape
【皮带】①(腰带) leather belt ②(传动带) belt: 交叉～ cross belt / 三角～ triangle belt ◇ ～车床 belt-driven lathe / ～传动 belt transmission / ～扣 belt fastener / ～轮 (belt) pulley / ～油 belt dressing; belt filler / ～运输机 belt conveyer
【皮垫圈】leather washer; leather packing collar
【皮筏】skin raft
【皮肤】skin
【皮肤病】skin disease; dermatosis; dermatopathy; dermopathy ◇ ～学 dermatology
【皮肤科】dermatological department; dermatology ◇ ～医生 dermatologist
【皮肤真菌病】dermatomycosis
【皮革】leather; hide
【皮猴儿】hooded fur overcoat; fur parka; fur anorak
【皮划艇运动】canoeing
【皮货】fur; pelt ◇ ～店 fur shop / ～商 furrier; fur trader ○白狐 white fox / 豹皮 leopard fur / 貂皮 mink / 羔皮 lamb skin / 海狸皮 beaver / 黑貂皮 sable fur / 黑狐 black

fox / 狐皮 fox fur / 虎皮 tiger fur / 黄鼠狼皮 yellow weasel / 灰鼠皮 squirrel / 山羊皮 goat skin; kid-skin / 猞猁皮 lynx / 水獭皮 otter / 兔皮 rabbit / 小山羊皮 kid / 雪貂皮 ferret fur / 羊皮 sheep skin / 银狐 silver fox / 银鼠皮 white squirrel / 鼬鼠皮 weasel

【皮夹子】wallet; pocketbook

【皮匠】①(制鞋的) cobbler; shoemaker ②(制革或鞣皮的) tanner

【皮筋儿】rubber band; elastic band

【皮开肉绽】the skin is torn and the flesh gapes open; badly bruised from flogging: 打得～ be bruised and lacerated (from flogging)

【皮毛】①(带毛兽皮) fur ②(表面的知识) smattering; superficial knowledge: 略知～ have only a superficial knowledge

【皮毛兽】fur-bearer

【皮棉】ginned cotton; lint (cotton)

【皮袍】furred robe

【皮破血流】wounded and bleeding

【皮球】rubber ball; ball

【皮实】①(不易得病) sturdy ②(不易破损) durable

【皮艇】kayak: 单人～ single kayak; kayak single; one-seater kayak / 双人～ double kayak; kayak pair; two-seater kayak; tandem kayak / 四人～ kayak four

【皮桶子】fur lining

【皮下注射】subcutaneous injection; hypodermic injection

【皮下组织】subcutaneous tissue

【皮线】[电] rubber-insulated wire; rubber-covered wire

【皮箱】leather suitcase; leather trunk

【皮笑肉不笑】put on a false smile; smile hypocritically

【皮鞋】leather shoes ◇ ～油 shoe polish

【皮炎】dermatitis; cutitis; dermitis: 神经性～ neurodermatitis

【皮衣】①(毛皮的) fur clothing: 她穿着非常昂贵的～。 She was wearing very expensive furs. ②(皮革的) leather clothing

【皮影戏】leather-silhouette show; shadow play

【皮张】hide; pelt

【皮疹】rash

【皮之不存,毛将焉附】with the skin gone, what can the hair adhere to; a thing cannot exist without its basis

【皮脂腺】[生理]sebaceous glands

【皮质】[生理] cortex: 大脑～ cerebral cortex

【皮重】tare

【皮子】①(皮革) leather; hide ②(毛皮) fur

疲

tired; weary; exhausted: 精～力尽 completely exhausted; be tired out

【疲惫】tired out; exhausted: ～不堪 be in a state of utter exhaustion; extremely tired

【疲乏】weary; tired: 感到～ feel weary

【疲倦】tired; weary; fatigued: 觉得～ feel tired

【疲劳】①(疲乏劳累) tired; fatigued; weary: ～过度 excessive fatigue / 身心～ be weary in body and mind / 使人忘掉～ relieve sb. of his fatigue ②(机能或反应能力减弱) fatigue; strain: 肌肉～ muscular fatigue / 听觉～ auditory fatigue ③[物] fatigue: 金属～ metal fatigue / 弹性～ elastic fatigue ◇ ～断裂 fatigue failure / ～轰炸 long and tedious harangue / ～强度 fatigue strength / ～审讯 grueling trial; persistent cross- examination aimed at blunting prisoner's senses and mind / ～试验 fatigue test / ～战 harassing warfare

【疲软】①(疲乏无力) fatigued and weak: 两腿～ be weak in the legs ②[金融] weaken; slump

【疲塌】slack; negligent: 工作～ be slack at one's work

【疲于奔命】be kept constantly on the run; be tired from running around; be weighed down with work: 使之～ tire sb. out by keeping him on the run

铍

[化] beryllium (Be)

pǐ

否

【否极泰来】out of the depth of misfortune comes bliss; When misfortune reaches the limit, good fortune is at hand.

痞

①(痞块) a lump in the abdomen ②(恶棍;流氓) ruffian; riffraff: 地～ local ruffian

【痞子】ruffian; riffraff

匹

①(比得上) be equal to; be a match for: 世无其～ matchless; peerless ②[量]两～马 two horses / 一～布 a bolt of cloth

【匹敌】be equal to; be well matched: 无可～ be withou a rival / 双方实力～。 The two sides are well matched.

【匹夫】①(平常人) ordinary man: 国家兴亡,～有责。 Every man has a share of responsibility for the fate of his country. ②(无学识、无智谋者) an ignorant person: ～之勇 reckless courage; foolhardiness

【匹配】①[书] mate; marry ②[电] matching: 阻抗～ impedance matching ◇ ～变压器 matching transformer

癖 addiction; weakness for: 嗜酒成～ be addicted to drinking

【癖好】favorite hobby; fondness for; weakness for; partiality for: 种玫瑰花是她的～。Growing roses is her hobby.

【癖性】natural inclination; proclivity; propensity

劈 ①(分开) divide; split: 把绳子～成两股 split the rope into two strands ②(剥;分裂) break off; strip off: ～莴苣叶 strip the outer leaves off lettuces

【劈叉】do the splits

【劈柴】kindling; firewood

pì

屁 wind (from bowels); fart: 放～ break wind; fart

【屁股】①(臀部) buttocks; bottom; behind; backside: 摔了个～蹲儿 fall on one's bottom / 在右～上打一针 have an injection in the right buttock ②(动物的臀部) rump; haunch; hindquarters ③(物体的末尾部分) end; butt: 香烟～ cigarette butt

【屁滚尿流】吓得～ scare the shit out of sb.; wet one's pants in terror; be frightened out of one's wits

【屁话】shit; nonsense; rubbish

媲

【媲美】compare favorably with; rival

辟 ①(开辟) open up (territory, land, etc.); break (ground): 开～果园 lay out an orchard / 另～专栏 start a new column (in a newspaper, etc.) ②(透彻) penetrating; incisive: 精～ profound; incisive ③(驳斥) refute; repudiate: ～谣 refute a rumor

【辟谣】refute a rumor

譬 example; analogy

【譬如】for example; for instance; such as

【譬喻】metaphor; simile; figure of speech

僻 ①(偏僻) out-of-the-way; secluded: ～处一隅 live in a remote corner / ～巷 side lane ②(性情古怪) eccentric: 怪～ eccentric ③(不常见的) rare: ～字 rare word

【僻地港】out port ◇ ～附加费 out port surcharge

【僻静】secluded; lonely: ～的地方 a secluded place

【僻壤】an out-of-the-way place; little known region: 穷乡～ remote and undeveloped place

piān

篇 ①(篇章) a piece of writing: 不朽的诗～ an immortal poem ②(写着或印着文字的单张纸) sheet (of paper): 单～儿油印材料 mimeographed sheets / 歌～儿 song sheet ③[量] piece; sheet: 一～文章 a piece of writing / 一～序文 a preface

【篇幅】①(文章的长短) length (of a piece of writing): 我们不需要～太长的文章。We don't need articles of great length. ②(指书刊的篇幅) space: ～有限 have limited space / 限于～ due to limited space

【篇目】table of contents; contents; list of articles

【篇章】sections and chapters; writings: 人类历史的新～ a new chapter of human history

偏 ①(倾斜) inclined to one side; slanting; leaning: 东南～东 southeast by east / 方向略～ in a slight slanting direction / 太阳～西了。The sun is to the west. / 这个指标～低。The target is on the low side. ②(不公正) partial; prejudiced: ～爱 have partiality for sth.; show favoritism to sb. / 这个意见太～。This opinion is rather one-sided.

【偏爱】have partiality for sth.; show favoritism to sb.; have a preference for; be partial to; love one more than another

【偏安】be content to retain sovereignty over a part of the country: ～一隅 content to exercise sovereignty over a part of the country

【偏差】①(离开确定方向的角度) deviation; declination ②(工作差错) deviation; error: 纠正执行政策中的～ correct deviations made in implementing a policy

【偏方】folk prescription

【偏废】do one thing and neglect another; emphasize one thing at the expense of another: 二者不可～。Neither should be overemphasized at the expense of the other.

【偏航】going off course; off-course; yaw

【偏护】be partial to and side with; show partiality for; take sides with

【偏激】extreme: 意见～ hold extreme views / 她这个人比较～。She tends to go to extremes.

【偏见】prejudice; bias; preconception; partial opinion: 对某人有～ have a prejudice against sb.; be biased against sb. / 毫无～ be free from prejudice / 消除～ dispel prejudices
【偏口鱼】flatfish
【偏离】deviate; diverge: ～航线 drift off the course
【偏旁】Chinese character component; a radical on one side of a character
【偏僻】remote; out-of-the-way; far-off: ～的山区 a remote mountainous district / 这是个～的地方。 It is an out-of-the-way place.
【偏巧】it so happened that; as luck would have it; by chance; fortunately
【偏食】①[天] partial eclipse: 日～ partial solar eclipse / 月～ partial lunar eclipse ②[医] partiality for a particular kind of food
【偏瘫】[医] hemiplegia; hemiparalysis
【偏袒】be partial to and side with; show partiality for; take sides with: ～某人 bias for sb. / ～一方 be partial to one side / 不～任何一方 show no partiality to either side; be impartial; be unbiased
【偏题】a catch question; a tricky question
【偏听偏信】heed and trust only one side; listen only to one side; believe in one-side story
【偏头痛】migraine; hemicrania
【偏向】①(不正确的倾向) erroneous tendency; deviation: 纠正～ correct a deviation ②(袒护) be partial to: ～某人 be partial to sb.
【偏心】①(不公正) partiality; bias ②[机] eccentric ◇ ～轮 eccentric wheel / ～凸轮 eccentric cam
【偏压】[电] bias voltage; bias: 截止～ cut-off bias ◇ ～电池 bias battery
【偏远】remote; faraway: ～的村庄 a remote village / ～地区 remote districts
【偏振】[物] polarization: 光的～ polarization of light
【偏振光】[物] polarized light ◇ ～镜 Polariscope / ～显微镜 polarizing microscope
【偏执己见】one is bigoted in one's opinions
【偏重】lay particular stress on; one-sidedly emphasize: ～于理论 lay particular stress on theory
【偏转】[物] deflection ◇ ～系统 deflection system

翩

【翩翩】①(轻快飞舞) dance lightly: ～飞舞 flutter / ～起舞 dance trippingly ②(举止洒脱) elegant; smart: ～少年 an elegant young man
【翩然】lightly; trippingly: ～而至 come tripping down
【翩跹】lightly; trippingly: ～起舞 dance with quick, light steps; dance trippingly

片

film; movie: 故事～ feature film
【片长】[电影] length of a film (in term of showing time)
【片酬】[电影] remuneration for a movie actor 〈actress〉
【片盒】[电影] film magazine
【片孔】[电影] (film) perforation
【片盘】[电影] film spool; bobbin
【片头】[电影] titles (of a film)
【片子】①(电影胶片) a roll of film ②(影片) film; movie ③(唱片) gramophone record; disc

pián

胼

【胼胝】callosity; callus
【胼胝体】[生理] corpus callosum

蹁

【蹁跹】whirling about (in dancing)

便

【便宜】①(价钱低) cheap; inexpensive: ～货 goods sold at bargain prices / 东西买得～ buy a thing cheap / 价格～的商店 a very cheap store / 并不～。 It is by no means cheap. ②(不应得的利益) small advantages; petty gain: 贪小～ out for small advantages; on the fiddle ③(使得到便宜) let sb. off lightly: 这次～了他。 This time we have let him off lightly.

piàn

片

①(平而薄的小东西) a flat, thin piece; slice; flake: 布～儿 small pieces of cloth / 面包～ slices of bread; sliced break / 牛肉～ slices of beef / 碎纸～儿 scraps of paper / 雪～ snowflakes ②(较大地区内划分的较小地区) part of a place: 分～包干 divide up the work and assign a part to each individual or group ③(切成薄片) cut into slices: ～肉片儿 slice meat / ～鱼片儿 flake a fish ④(不全的; 零星的; 简短的) incomplete; fragmentary; partial; brief: ～言 a few words ⑤[量] 两～药 two tablets of medicine / 一～草地 a tract of meadow / 一～柠檬 a slice of lemon / 一～欢腾 a scene of great rejoicing / 一～真心 in all

sincerity
【片窗】[摄] gate
【片段】part; passage; extract; fragment: ～的回忆 fragments of sb.'s reminiscences / 生活的～ a slice of life; an episode of sb.'s life
【片儿汤】flat pieces of dough served with soup
【片簧】[机] leaf spring: 多片～ multiple leaf spring
【片基】[摄] film base: ～密度 base density
【片剂】tablet
【片甲不存】not a single armored warrior remains; the army is completely wiped out; have not even a fragment of armor remaining
【片刻】a short while; an instant; a moment; a little while; a minute
【片麻岩】[地] gneiss
【片面】①(单方面的) unilateral; one-sided: ～撕毁协议 unilaterally tear up an agreement / ～之词 an account given by one party only; one party's version of an event, etc.; one person's word against another's ②(不全面) one-sided; lopsided: ～地看问题 take a one-sided approach to problems / ～夸大 one-sided exaggeration; exaggerate certain aspects of / ～强调 put undue emphasis on; one-sided stress; unbalanced stress / 对一场争论的～叙述 a one-sided account of a quarrel ◇～观点 lopsided view / ～性 one-sidedness
【片瓦无存】not a single tile remains; be razed to the ground
【片言】a few words; a phrase or two: ～可决 can be settled in a few words / ～只辞 a few words and phrases / ～只语 just a few words / ～折狱。A single word uttered by a wise man can decide a legal case.
【片岩】[地] schist
【片纸只字】fragments of writing; a brief note or letter
【片子】①(平而薄的东西) flat, thin piece; slice; flake; scrap: 铁～ small pieces of sheetiron ②(名片) visiting card

骗 ①(欺骗) deceive; fool; hoodwink: ～某人去做某事 deceive sb. into doing sth. / 受～ be taken in; be deceived / 他说谎～老师。He deceived the teacher by lying. ②(骗取) cheat; swindle: ～钱 cheat sb. out of his money; swindle money out of sb.
【骗局】fraud; hoax; swindle: 和谈～ a peace talk swindle / 政治～ a political fraud
【骗取】gain sth. by cheating; cheat sb. out of sth.; swindle; defraud: ～财物 defraud sb. of his money and belongings / ～信任 worm one's way into sb.'s confidence / 弄虚作假，～荣誉 seek honor through fraud and deception
【骗人】deceive people: ～的勾当 a fraudulent practice / ～的空话 deceitful empty talk
【骗术】deceitful trick; ruse; hoax: 施行～ perpetrate a fraud
【骗子】swindler; cheat; trickster; fraudulent person

piāo

漂 float; drift: ～洋过海 travel far away across the sea / 船～到海里去了。The boat drifted out to sea. / 木头～在水上。Wood floats on water.
【漂泊】lead a wandering life; rove; wander; drift: ～江湖 lead a wandering life all over the country / ～异乡 wander aimlessly in a strange land
【漂浮】①(漂) float: 树叶～在水上。Leaves were floating on the water. ②(工作不踏实) showy; superficial
【漂浮生物】neuston
【漂砾】erratic
【漂流】①(漂在水面随着水流浮动) be driven by the current; drift about: 随波～ drift with the tide / 我们顺河～而下。We drifted down the stream. ②(漂泊) lead a wandering life; rove; drift ③[海] drift current; wind-driven current ◇～瓶 drift bottle; floater
【漂移】[电子] drift ◇～晶体管 drift transistor

剽 ①(抢劫;掠夺) rob: ～掠 plunder; loot ②(动作敏捷) nimble; swift
【剽悍】agile and brave; quick and fierce
【剽窃】plagiarize; lift: 头两章是从那部书上～来的。The first two chapters lift two sections from the book.

飘 wave to and fro; float (in the air); flutter: 灰尘～在空中。Dust floats in the air.
【飘尘】floating dust
【飘带】streamer; ribbon
【飘荡】drift; wave; flutter: 几只小船随波～。A few boats were drifting with the tide.
【飘忽】①(轻轻地移动) move swiftly; fleet ②(摇摆) mobile; uncertain: ～不定 drift from place to place
【飘零】①(凋谢) faded and fallen ②(失去依靠) wandering; adrift; homeless; forsaken

【飘飘然】smug; self-satisfied; complacent: 心中～ feel quite self-satisfied
【飘然】floating in the air
【飘洒】①(飘扬) float; drift ②(姿态自然) suave; (书法) facile and graceful
【飘扬】wave; flutter; fly: 迎风～ flying 〈fluttering〉 in the wind
【飘摇】sway; shake; totter: 风雨～ buffeted by wind and rain; precarious; tottering
【飘逸】possessing natural grace; elegant: 神采～ have an elegant bearing

缥
【缥缈】dimly discernible; misty: 虚无～ visionary; illusory

piáo

瓢 gourd ladle; wooden dipper
【瓢虫】ladybug; ladybird
【瓢泼大雨】heavy rain; torrential rain; downpour

嫖 visit prostitutes; go whoring

piǎo

漂 ①(漂白) bleach ②(用水冲干净) rinse: 把衣服～干净 give the clothes a good rinse
【漂白】bleach: ～棉布 bleached cotton cloth ◇～粉 bleaching powder / ～机 bleaching machine; bleacher / ～剂 bleaching agent / ～率 bleachability
【漂洗槽】[化] potcher

瞟 look sidelong at; look askance at; glance sideways at: ～了他一眼 cast a sidelong glance at him

piào

票 ①(车票; 戏票等) ticket: 火车～ train ticket; railway ticket / 来回～ return ticket; round trip ticket / 买～ get a ticket; buy a ticket ②(选票) ballot: 反对～ negative vote / 废～ spoilt vote / 投～ cast a ballot; vote / 投～否决 vote down / 投～通过 vote through / 投某人的～ give one's vote to sb. / 赞成～ affirmative vote ③(钞票) bill; bank note: 零～儿 notes of small denominations; change
【票额】the sum stated on a check or bill; denomination; face value
【票房】①(车站等的) booking office ②(戏院等的) box office ◇～价值 box-office value
【票根】counterfoil; stub
【票价】the price of a ticket; admission fee; entrance fee
【票据】①(写有支付金额义务的证件) bill; note: ～托收 collection on bill / 到期未付～ overdue bill / 即期～ a demand note / 流通～ negotiable instruments; negotiable papers / 应付～ bills payable / 应收～ bills receivable ②(出纳或货运凭证) voucher; receipt ◇～交换 bank clearing / ～交换所 clearinghouse / ～贴现 bill discounted; discounted notes
【票面】face value; par value; nominal value: 各种～的邮票 stamps of various denominations ◇～价值 face value; par (value)
【票面值】denomination
【票期】usance
【票箱】ballot box
【票子】bank note; paper money; bill

漂
【漂亮】①(好看;美观) handsome; good-looking; pretty; beautiful: ～的小伙子 handsome fellow / ～的小姑娘 pretty little girl / ～的衣服 pretty dress; fine clothes / 把自己打扮得漂漂亮亮 trim oneself up ②(出色) smart; remarkable; brilliant; splendid; beautiful: 法语讲得很～ speak French beautifully ◇～话 fine words; high-sounding words

嘌
【嘌呤】purine

piē

撇 ①(弃置不顾;抛弃) cast aside; throw overboard; neglect: 我们不能～下他不管。We should not ignore him. ②(从液面上轻轻地舀) skim: ～沫儿 skim off the scum / ～油 skim off the grease
【撇开】leave aside; bypass: ～次要问题 bypass questions of minor importance
【撇弃】cast away; abandon; desert

瞥 shoot a glance at; dart a look at
【瞥见】get a glimpse of; catch sight of

氕 protium (H^1)

piě

苤
【苤蓝】[植] kohlrabi

撇 throw; fling; cast: ～手榴弹 throw hand grenades

【撇嘴】curl one's lip; twitch one's mouth

pīn

拼 ①(合在一起;连合) put together; piece together: 把两件～成一件 put the two together to form one piece ②(不顾一切地干;豁出去) be ready to risk one's life (in fighting, work, etc.); go all out in work: ～到底 fight to the bitter end

【拼板游戏】jigsaw puzzle

【拼版】[印] makeup

【拼刺】①(军事训练) bayonet drill; bayonet practice ②(枪刺格斗) bayonet charge: 和敌人～ fight it out with the enemy with bayonets

【拼凑】piece together; knock together; rig up

【拼命】①(把性命豁出去;舍命) risk one's life; defy death; go all out regardless of danger to one's life: ～抵抗 risk the life in resistance / ～精神 the death-defying spirit ②(尽最大的力量;极度地) exerting the utmost strength; for all one is worth; with all one's might; desperately: ～奔跑 run for all one is worth / ～工作 work with all one's might

【拼盘】assorted cold dishes; hors d'oeuvres

【拼死】risk one's life; defy death; fight desperately: ～挣扎 wage a desperate struggle

【拼写】spell; transliterate: 用汉语拼音字母～汉字 transliterate Chinese characters into the Chinese phonetic alphabet / 你的名字怎样～? How do you spell your name? ◇ ～法 spelling; orthography

【拼音】combine sounds into syllables ◇ ～文字 alphabetic (system of) writing / ～字母 phonetic alphabet; phonetic letters

姘 have illicit relations with

【姘居】cohabit; live illicitly as husband and wife

【姘头】paramour; mistress

pín

频 ①(屡次;连续几次) frequently; repeatedly ②[物] frequency: 音～ audio frequency

【频传】keep pouring in: 捷报～ reports of new victories keep pouring in

【频带】[物] frequency band ◇ ～宽度 bandwidth

【频道】frequency channel: 十五～ channel 15; channel fifteen

【频繁】frequently; often: ～交往 frequent contacts

【频率】frequency: ～范围 frequency range / 低～ low frequency / 高～ high frequency

【频频】again and again; repeatedly: ～举杯 propose repeated toasts / ～招手 wave one's hand again and again

贫 ①(穷) poor; impoverished ②(缺少) be deficient in; be poor in; be scanty of: ～油国 oil-poor country; country poor in oil ③(絮叨可厌) garrulous; loquacious

【贫病交集】sick and in straits

【贫病交迫】suffering from both poverty and sickness; sick as well as poor; be beset by poverty and illness

【贫病相连】poverty and sickness are closely connected with

【贫乏】poor; short; lacking: 经验～ lack experience / 内容～ devoid of matter / 语言～ flat, monotonous language / 资源～ poor in natural resources; short of natural resources

【贫富不均】too much difference between the rich and the poor

【贫富悬殊】extreme disparity between the rich and the poor

【贫雇农】poor peasants and farm laborers

【贫寒】poor; poverty-stricken: ～人家 an impoverished family / 出身～ be born poor; came from a poor family

【贫化】[矿] dilution: 矿石～ ore dilution

【贫瘠】barren; infertile; poor: ～的土壤 poor soil; impoverished soil

【贫贱】poor and lowly; in straitened and humble circumstances: ～不能移 not to be shaken or modified by one's poverty or destitution

【贫苦】poor; poverty-stricken; badly off; impoverished

【贫矿】lean ore; low grade ore

【贫困】poor; impoverished; in straitened circumstances: 处在～之中 be in dire necessity / 生活～ live in poverty ◇ ～化 pauperization

【贫民】poor people; pauper: 城市～ the urban poor ◇ ～窟 slum / ～区 slum area; slum district

【贫农】poor peasant

【贫气】①(小气) stingy; miserly; mean ②(絮叨可厌) garrulous; loquacious

【贫穷】poor; needy; impoverished

【贫弱】poor and weak

【贫下中农】poor and lower-middle peasants

【贫血】anemia: 恶性～ pernicious anemia／脑～ cerebral anemia／再生障碍性～ aplastic anemia
【贫嘴】garrulous; loquacious
【贫嘴薄舌】garrulous and sharp-tongued; light and airy utterance

pǐn

品 ①(物品) article; product: 工业～ industrial products／农产～ farm produce／商～ commodity; merchandise／生活必需～ daily necessities ②(等级;品级) grade; class; rank: 上～ highest grade; top grade ③(品质) character; quality: ～学兼优 of good character and scholarship／人～ moral quality; character④(品评; 辨别好坏) taste sth. with discrimination; sample; savor: ～茶 sample tea／～～味儿 savor the flavor
【品尝】taste; sample; savor
【品德】moral character; morality
【品格】①(品性;品行) one's character and morals: ～高雅 high character ②(文艺作品质量和风格) quality and style
【品红】①(颜色) pinkish red ②(染料) magenta; fuchsin
【品级】grade; class
【品蓝】reddish blue
【品绿】light green; malachite green
【品貌】①(相貌) looks; appearance: ～端正 well-shaped figure and decorous appearance ②(人品和相貌) character and looks; conduct and appearance: ～端庄 one's bearing is respectable
【品名】the name of an article; the name or description of a commodity
【品目】names of things; items: ～繁多 numerous names; names and descriptions of articles are numerous
【品评】judge; comment on
【品头论足】①(评人外表) make frivolous remarks about a woman's appearance ②(挑剔) find fault with; be overcritical
【品脱】[量] pint
【品位】[矿] grade
【品味】taste; savor
【品系】[生] strain
【品行】conduct; behavior: ～不良 having bad conduct; illbehaved／～端正 behave oneself well／～优良 having good conduct; well-behaved
【品性】moral character
【品质】①(人的本质) character; quality: 道德～ moral character ②(质量) quality: ～优良 of the best quality ◇ ～管制 quality control／～检验 quality restriction／～证明书 certificate of quality
【品种】①[生] breed; strain; variety: 优良～ improved breeds／～纯度 varietal purity／～间杂交 interbreed ②(产品种类) variety; assortment: 货物～齐全 have a good assortment of goods／增加花色～ increase the variety of colors and designs

pìn

聘 ①(聘请) engage: ～某人为教授 engage sb. as a professor／被～为技术顾问 be employed 〈invited〉 to be technical adviser ②(定亲) betroth
【聘礼】betrothal presents; bride-price
【聘请】engage; invite; employ: ～律师 brief a barrister
【聘任】engage; appoint to a position: ～她为经理 appoint her to be manager
【聘书】letter of appointment; contract

牝 female: ～鸡 hen／～马 mare／～牛 cow

pīng

娉
【娉婷】have a graceful demeanor

乒 table tennis; ping-pong: ～坛 table tennis circles
【乒乓球】①[体] table tennis; ping-pong: 打～ play ping-pong ②(乒乓球运动用的球) table tennis ball; ping-pong ball ◇ ～拍 table tennis bat／～台 table tennis table／～网 table tennis net ○防守型选手 defensive player／攻击型选手 attacking player／上旋球 topspin／下旋球 under spin; back spin／侧旋球 side spin／弧圈球 loop (drive)／抽球 drive; smash／正手抽球 forehand smash／反手抽球 backhand smash／近台快攻 close-to-table fast attack／远台长抽 off the table long drive／长抽短吊 combine long drives with drop shots／海绵拍 sponge bat／橡胶拍 rubber bat

píng

瓶 bottle; vase; jar; flask: 醋～ vinegar cruet／胡椒～ pepper-box; pepper pot; pepper castor／花～ flower vase／酱油～ sauce cruet／芥末～ mustard pot／热水～ thermos

flask / 盐～ saltcellar; saltshaker / 两～牛奶 two bottles of milk / 装～ bottling
【瓶胆】glass liner
【瓶盖】cap
【瓶颈】bottleneck
【瓶塞】bottle stopper: 软木～ cork
【瓶装】bottled; in bottles: ～啤酒 beer in bottle / ～液化气 bottled gas; bugas
【瓶子】bottle

屏 ①(屏风) screen: 画～ painted screen ②(遮挡) shield; screen
【屏蔽】shield; screen: ～着这个地区 provide a protective screen for this region ◇ ～电缆 shielded cable; screened cable / ～天线 screened antenna; shielded antenna
【屏风】screen
【屏极】[电] plate ◇ ～电路 plate circuit
【屏幕】screen: 电视～ telescreen; screen
【屏条】a set of hanging scrolls
【屏障】protective screen; shield; shelter; barrier: 天然～ natural barrier; natural defense

平 ①(无凹凸；不倾斜) flat; level; even; smooth: 把衣服弄～ smooth one's dress / 躺～ lie stretched out; lie flat / 这块地相当～。This piece of ground is flat enough. / 这张桌子的桌面不～。 The top of this table is not even. ②(使平) level; even: ～一～地面 level the ground; even the land ③(高度相同) be on the same level; be on a par; equal: ～世界纪录 equal a world record / 积雪与窗子相～。The snow is even with the window. / 水已经～了河岸。The water has been on the same level with the banks. ④[体](平局) make the same score; tie; draw: 场上比分是五～。The score is now five all. / 双方踢成二～。The two teams tied at 2-2. / 这场足球赛最后踢～了。The football game ended in a draw ⑤(平均;公平) equal; fair; impartial: ～分 divide equally / 持～之论 a fair argument; an unbiased view ⑥(安定;平定) calm; peaceful; quiet: ～民愤 assuage popular indignation / 风～浪静 the sea was calm / 为民～愤 redress the grievances of the people ⑦(镇压;平定) put down; suppress: ～乱 suppress a riot / ～叛 put down a rebellion; put down a revolt ⑧(经常的;普通的) average; common: ～日 on ordinary days
【平安】safe and sound; without mishap; well: ～到达 arrive safe and sound; arrive without mishap / ～康泰 peaceful and well-being / ～无事。All is well. / 全家～。The whole family is well. / 一路～! Have a good trip! (或: Bon voyage!) ◇ ～险 [商] free of particular average (F.P.A.)
【平白】for no reason; gratuitously: ～挨一顿骂 get a scolding for no reason at all
【平板】dull and stereotyped; flat; monotonous: 写得～ written in a dull style
【平板玻璃】plate glass
【平板车】flatbed tricycle; flatbed
【平版】lithographic plate ◇ ～印刷 lithographic printing; planographic printing
【平辈】of the same generation
【平布】plain cloth
【平步青云】rapidly go up in the world; have a meteoric rise
【平槽】the water has been on the same level with the banks
【平产】be equal in output; have the same output
【平常】①(普通) ordinary; common: 她穿着～的衣服。 She is in ordinary dress. ②(平时) generally; usually; ordinarily; as a rule: 到得比～晚 arrive later than usual ③(不出色) average; mediocre: 他的才能很～。 He is a man of average ability.
【平车】[铁道] flatcar; platform wagon; platform car
【平川】level land; flat, open country; plain: 一马～ a vast stretch of flat land
【平淡】flat; insipid; prosaic; pedestrian; dull: ～无奇 prosaic; pedestrian; nothing exciting / ～无味 insipid; dull; without flavor
【平等】equality: 男女～ equality between the sexes / ～待人 treat others as equals / 在～的基础上 on an equal footing ◇ ～待遇 equal treatment / ～关系 relations on an equal basis / ～互惠 reciprocal favored treatment / ～互利 equality and mutual benefit / ～权利 equal rights / ～协商 consultation on the basis of equality
【平地】①(平整土地) level the land; level the ground; rake the soil smooth ②(平坦的土地) level ground; flat ground
【平地风波】a sudden storm on a calm sea; a sudden, unexpected turn of events; unforeseen trouble
【平地机】①[农] land leveller; grader ②[交] road grader
【平地楼台】high buildings rise from the ground; start from scratch

【平地一声雷】a sudden clap of thunder; a sudden big change, e. g. a sudden rise in fame and position; an unexpected happy event; a bolt from the blue
【平定】①(平稳安定) calm down: 他的情绪终于～下来。He calmed down in the end. ②(平息) suppress; put down: ～叛乱 put down a rebellion
【平凡】ordinary; common; commonplace: ～的岗位 ordinary post / ～的工作 ordinary work
【平反】redress (a mishandled case); rehabilitate: ～错案 redress a mishandled case / ～冤案 redress of a grievance / 宣布给某人～ announce sb.'s rehabilitation
【平方】①square: 九是三的～。Nine is the square of three. ②(平方米) square meter: 十五～米的房间 a room of fifteen square meters ◇ ～根 square root / ～公里 square kilometer / ～数 square numbers / ～英尺 square foot
【平房】single-story house; one-story house
【平分】divide equally; share and share alike; go halves; go fifty-fifth: ～土地 equal distribution of land ◇ ～线[数] bisector
【平分秋色】have equal shares; share on a fifty-fifty basis; divide the cheerful countenance equally
【平复】①(恢复平静) calm down; subside; be pacified: 事态～。The situation has quieted. ②(痊愈) be cured; be healed: 伤口～了。The wound is healed.
【平光】zero diopter; plain glass ◇ ～眼镜 plain glass spectacles
【平巷】[矿]drift; level
【平和】gentle; mild; moderate; placid: ～的语气 mild tone / 性情～ be of gentle 〈mild〉 disposition / 这药药性～。This medicine is quite mild.
【平衡】balance; equilibrium: 保持～ maintain one's equilibrium; keep one's balance / 发展不～ unbalanced development; uneven development / 供求～ equilibrium of supply and demand / 破坏心理上的～ upset one's equilibrium / 失去～ lose one's balance; be in a state of imbalance / 收支～ balance between income and expenditure ◇ ～常数 equilibrium constant / ～汇率 equilibrium rate of exchange / ～价格 equilibrium price / ～觉[生理] sense of equilibrium / ～力 equilibrant / ～预算 balanced budget
【平衡木】[体] balance beam
【平滑】level and smooth; smooth ◇ ～肌[生理] smooth muscle
【平缓】①(平坦;倾斜度小) gently: 地势～。The terrain slopes gently. / 水流～。The water flows gently ②(缓和;平和) mild; placid; gentle: ～的语调 a mild tone
【平价】par; parity: 固定～ fixed parity / 汇兑～ par of exchange / 铸币～ specie par
【平角】[数] straight angle
【平静】calm; quiet; tranquil: ～的海面 a calm sea / ～的生活 a quiet life / ～的夜晚 a quiet night
【平局】draw; tie: 扳成～ equalize the score / 打成～ level the score; play even / 比分屡次出现～。The score was tied several times. / 比赛以～结束。The match ended in a draw.
【平均】①(按份均匀计算) average; mean: ～每年增长百分之七 increase by an average of 7% a year / 按人口～计算收入 per capita income / 我们～每天走 150 公里。We averaged 150 km. a day. ②(没有轻重多少的区别) equally; share and share alike: ～分摊 share out equally ◇ ～高度 average height / ～价格 average price / ～利润 average profit / ～年龄 composite life; average age / ～寿命 average life span; life expectancy / ～速度 average speed; mean velocity
【平均数】average; mean: 取其～ take the mean
【平均值】average value; mean: 取其～ take the mean
【平均主义】equalitarianism; egalitarianism
【平口钳】flat-nose pliers
【平列】place side by side; place on a par with each other; put on a par with
【平流】[气] advection ◇ ～层 stratosphere
【平炉】open-hearth furnace; open hearth ◇ ～钢 open-hearth steel / ～炼钢法 open-hearth process
【平路机】grader
【平面】plane ◇ ～波 plane wave / ～几何 plane geometry / ～交叉[交] grade crossing; level crossing / ～镜 plane mirror / ～磨床 surface grinding machine
【平面图】①(平面图形) plane ②(垂直投影图形) plane figure
【平民】common people; common man; the populace
【平年】①[天] non-leap year; common year ②(平常年景) average year
【平平】average; mediocre; indifferent: ～无奇 be average and not clever / 成绩～。The results are about up to the average. / 年成～。

The crop is just so-so.
【平铺直叙】①(简单直接地叙述) tell in a simple, straightforward way; narrate in a simple direct way ②(不讲求修辞) speak or write in a dull, flat style
【平起平坐】sit as equals at the same table; be on an equal footing
【平权】(enjoy) equal rights: 男女～ equal rights for men and women
【平绒】velveteen
【平射】flat (trajectory) fire ◇ ～炮 flat fire gun; flat trajectory gun
【平生】all one's life; one's whole life: ～大事 the biggest events in all one's life / ～夙愿 one's lifelong aspiration / ～唯一的机会 chance of a lifetime
【平时】①(通常;日常) at ordinary times; in normal times; usually ②(和平时期;非战时) in peacetime: ◇～兵力 peacetime strength / ～编制[军] peacetime establishment; peace organization; peace footing
【平时不烧香,临时抱佛脚】Neglect one's prayers in times of peace, then embrace the Buddha's feet in a crisis.
【平手】draw: 两队打了个～。The two teams drew.
【平顺】smooth-going; plain sailing
【平素】usually; customarily
【平台】terrace; platform
【平台印刷机】flatbed press
【平坦】level; even; smooth: ～的公路 a smooth highway / 地势～ smooth terrain
【平头】closely cropped hair; crop; crew cut: 留着～ have closely cropped hair
【平头百姓】common people
【平纹】plain weave ◇ ～织物 plain cloth
【平稳】smooth and steady; smooth; stable: 飞行～ have a smooth flight / 物价～ prices are stable; commodity prices are stable
【平息】①(平静) calm down; quiet down; subside ②(平定) put down; suppress: ～风波 [海] pour oil on troubled water
【平心而论】in all fairness; to give sb. his due; objectively speaking
【平心静气】calmly; dispassionately; cool-headed: ～地讨论 calmly discuss
【平信】①(非挂号信) ordinary mail ②(非航空信) surface mail
【平行】①(等级相同) of equal rank; on an equal footing; parallel; of the same level: ～机关 organizations of a level; units of equal rank; parallel organizations ②(同时进行的) simultaneous; parallel: ～作业 parallel operations / 就各种问题举行～的会谈 hold simultaneous talks on different subjects ③[数] parallel: 相互～ be parallel to each other ◇ ～六面体[数] parallelepiped / ～脉[植] parallel veins / ～四边形[数] parallelogram / ～线 parallel lines
【平易】①(谦逊和蔼) unassuming; amiable: ～近人 amiable and easy of approach; easy to approach; unassuming and approachable
【平印】lithography
【平庸】mediocre; indifferent; commonplace: ～的作家 a mediocre writer / 才能～ of limited ability
【平原】plain: 冲积～ alluvial plain
【平展】(of land, etc.) open and flat
【平针组织】plain stitch
【平整】①(整地) level: ～土地 level the land ②(平正整齐) neat; smooth
【平装】paperback; paper-cover: ～本 paperback (book)
【平足】flatfoot

评

①(评论;批评) comment; criticize; review: 博得好～ receive favorable comments; be well received / 短～ brief commentary / 书～ book review ②(评判) judge; appraise
【评比】appraise through comparison; compare and assess: ～产品质量 compare and appraise the quality of different products; make a public appraisal of the quality of different products
【评定】pass judgment on; evaluate; assess; rate ◇ ～人 adjuster
【评断】judge; arbitrate: ～是非 judge between right and wrong; arbitrate a dispute
【评分】give a mark; mark (students' papers, etc.); grade; score: ～标准 standards of grading / 给测验～ score a test / 给试卷～ mark papers
【评功】appraise sb.'s merits; evaluate achievements: ～授奖 announce commendations and issue awards
【评级】grade; rate
【评价】appraise; evaluate; assess: ～过低 rate sth. unreasonably low / ～历史人物 appraise historical figures / 高度～ set a high value on; speak highly of; highly appraise
【评奖】decide on awards through discussion
【评介】review (a new book, etc.): 新书～ book review
【评理】judge between right and wrong; decide

which side is right
【评论】①(批评或议论) comment on; discuss: ～是非 discuss right and wrong / 他对这个问题未作～。 He made no comment on the subject. ②(批评或议论的文章) comment; commentary; review: 时事～ comments on current events ◇ ～家 critic; reviewer / ～员 commentator
【评判】pass judgment on; judge: ～胜负 decide who is the winner; judge between contestants / ～优劣 judge which is superior / 我不能～他是对是错。I can't judge whether he was right or wrong. ◇ ～员 judge(体育、演讲等的); adjudicator (音乐等的)
【评头论足】find fault with; criticize from head to feet; comment from head to feet
【评选】choose through public appraisal; discuss and elect: 被～为劳动模范 be chosen as a model worker
【评议】appraise sth. through discussion; deliberate: ～确定 be discussed and determined by the masses ◇ ～会 appraisal meeting
【评语】comment; remark
【评阅】read and appraise
【评注】①(评论并注释) make commentary and annotation ②(作出的评论和注释) notes and commentary
【评传】critical biography

坪 level ground: 草～ lawn; grassplot / 停机～ aircraft park; apron

苹

【苹果】apple: ～干 dried apple slices / ～脯 preserved apple
【苹果酱】apple jam
【苹果酒】cider; applejack
【苹果绿】apple green
【苹果树】apple tree
【苹果园】apple orchard

萍 duckweed

【萍水相逢】meet by chance like patches of drifting duckweed

鲆 left-eyed flounder

凭 ①(靠看) lean on; lean against: ～栏远眺 lean on a railing and look into the distance ②(依靠) rely on; depend on: ～险抵抗 make use of a strategic vantage point to fight back ③(证据) certification; evidence; proof: 真～实据 ironclad evidence / 口说无～。 Verbal statements are no guarantee. / 以此为～。 This will serve as certification. ④(根据) go by; base on; take as the basis: ～表面现象判断 on the face of it / ～良心说 in all fairness / ～票付款 payable to bearer / ～票入场。 Admission by ticket only. ⑤(任凭;不管) no matter (what, how, etc.): ～你是谁 no matter who you may be
【凭单】voucher; indenture; a certificate for drawing money, goods, etc. ◇ ～索引 vouchers index / ～制度 voucher system
【凭吊】visit (a historical site, etc.) and ponder on the past: ～古战场 pay a visit to an ancient battle-ground
【凭借】rely on; depend on: ～暴力 by force; relay on violence / ～经验 by virtue of experience / ～想象力 draw on one's imagination / ～自己的力量 rely on one's own strength
【凭据】evidence; proof
【凭空】out of the void; out of thin air; without foundation; groundless: ～杜撰 create groundless rumors / ～捏造 fabrication founded upon nothing; make sth. out of nothing
【凭眺】enjoy a distant view from a height; gaze from a high place into the distance
【凭信】trust; believe: 他的话不足～。 His words are not to be trusted.
【凭依】base oneself on, rely on; have something to go by: 无所～ have nothing to go by
【凭仗】rely on; depend on
【凭证】proof; evidence; certificate; voucher: ～记录 evidence record / 付款～ paying voucher / 收款～ receiving voucher / 完税～ tax payment receipt

pō

泊 lake: 血～ pool of blood

钋 [化] polonium (Po)

坡 ①(倾斜的土地) slope: 陡～ a steep slope / 山～ a mountain slope; hillside ②(倾斜) sloping; slanting: 把梯子放～一点。 Slope the ladder a bit.
【坡地】hillside fields; sloping fields; land on the slopes
【坡度】slope; gradient: 三十度的～ a slope of 30 degrees / 这电影院的地板从后面座位到前面有两米的～。 The floor of the cinema has a slope of two meters from the back seats to the front.

颇 ①[书]（偏;不正）inclined to one side; oblique: 偏～ biased; partial ②（很;相当地）quite; rather; considerably: ～不以为然 highly disapprove of sth. / ～感兴趣 have considerable interest / ～佳 quite good / ～为费解 rather difficult to understand / 影响～大 exert a considerable influence

泼 ①（倒出;洒出）sprinkle; splash; spill: ～点儿水 sprinkle some water / 用水～地 splash the floor with water ②（蛮不讲理）rude and unreasonable; shrewish: 撒～ act hysterically and refuse to see reason

【泼妇】shrew; vixen: ～骂街 like a shrew shouting abuse in the street

【泼辣】①（凶悍而不讲理）rude and unreasonable; shrewish; vixenish; fierce and unreasonable ②（文章有力）pungent; forceful: 文章写得～。The article is written in a pungent style ③（有魄力）bold and vigorous: 工作～ be bold and vigorous in one's work

【泼冷水】pour 〈throw〉 cold water on; dampen the enthusiasm 〈spirits〉 of; put a dumper on; discourage

pó

婆 ①（老年妇女）old woman ②（丈夫的母亲）husband's mother; mother-in-law

【婆家】husband's family

【婆罗门】Brahman ◇～教 Brahmanism

【婆婆】①（丈夫的母亲）husband's mother; mother-in-law ②（祖母;外祖母）grandmother

【婆婆妈妈】①（行动缓慢;言语罗唆）womanishly fussy; garrulous ②（感情脆弱）mawkish; maudlin; foolishly sentimental; sentimentally silly

【婆娑】whirling; dancing: ～起舞 start dancing / 杨柳～。The willows dance in the breeze.

pǒ

叵 impossible

【叵测】unfathomable; unpredictable: 居心～ with hidden intent / 心怀～ harbor dark designs; nurse evil intentions

钷 [化] promethium (Pm)

笸

【笸箩】shallow basket

pò

朴 [植] Chinese hackberry

破 ①（破旧）broken; damaged; torn; worn-out: ～布 rags / ～家具 dilapidated furniture / ～碗 a broken bowl / ～衣服 worn-out clothes; ragged clothes ②（破开）break; split; cleave; cut: 一～两半 break into two; split into two / ～浪前进 cleave 〈cut; plough〉 through the waves / 把百元的票子～开 break a hundred yuan note ③（使损坏）break; break down; damage: 小心别打～杯子。Be careful not to break the cup. ④（突破）break; break through: ～世界纪录 break the world record / 突～重围 break through the heavy encirclement ⑤（破除）get rid of; destroy; break with: ～旧立新 destroy the old and establish the new ⑥（打败;攻克）defeat; capture: 大～敌军 inflict a crushing defeat on the enemy ⑦（揭穿）expose the truth of; lay bare: 看～ see through / 一语道～ get to the heart of the matter in a few words; puncture a fallacy with one remark ⑧（讥讽质量不好）poor; paltry; lousy; shabby: 谁看那个～电影! Who wants to see that poor film!

【破案】solve a case; clear up a case; crack a criminal case; track down the criminal: 警察只用了三天就～了。The police needed only three days to break a case. ◇～率 detection rate

【破败】ruined; dilapidated; tumble-down: 这楼已～不堪。The building is dilapidated.

【破冰船】icebreaker: 原子～ atomic icebreaker

【破财】suffer unexpected personal financial losses

【破产】①（丧失全部财产）go bankrupt; become insolvent; become impoverished: ～地主 bankrupt landlords / ～农民 impoverished peasants / 宣告～ declare bankruptcy / 银行～ bank failure / 公司已经～。The company is bankrupt. ②（失败）go bankruptcy; fall through; come to naught: 阴谋～了。The plot has fallen through. ◇～当事人 party in bankruptcy / ～法 insolvent law; law of bankruptcy / ～债务人 insolvent debtor / ～者 lame duck

【破除】do away with; get rid of; eradicate; break with: ～恶习 break with foul habits / ～迷信 do away with superstitions or blind faith; topple old idols / ～情面 not spare anybody's

feelings
【破费】[套] spend money; go to some expense: 何必这么～? Why must you go to this expense?
【破釜沉舟】break the caldrons and sink the boats (after crossing); cut off all means of retreat; burn one's boats
【破格】break a rule; make an exception: ～接待 break protocol to honor sb. / ～录用 break a rule to engage sb. / ～提升 break a rule to promote sb.
【破罐破摔】smash a pot to pieces just because it's cracked; write oneself off as hopeless and act recklessly; one who acts recklessly because he thinks his faults are irremediable
【破坏】①(损毁) destroy; wreck: ～桥梁 destroy a bridge / ～铁路 wreck railways ②(使受到损害) do great damage to; damage; disrupt: ～边界现状 disrupt the status quo along the boundary line / ～名誉 damage sb.'s reputation / ～社会秩序 disrupt public saboteur / ～生产 sabotage production / ～团结 disrupt unity; undermine unity ③(变革) destroy; demolish; change completely: ～旧秩序 demolish the old order ④(违反) violate; break; breach: ～合同 breach of contract / ～纪律 breach of discipline / ～停战协定 violate an armistice agreement / ～协议 break an agreement ⑤(物体的组织损坏) decompose; destroy: ～维生素 C destroy vitamin C ◇ ～分子 saboteur / ～活动 sabotage / ～力 destructive power / ～社会主义经济秩序罪 offense against the socialist economic order / ～交通罪 crime of sabotaging communications / ～婚姻家庭罪 offense against marriage and the family / ～试验 breaking test; destruction test / ～性 destructiveness / ～性试验 destructive testing; breakdown test
【破获】unearth; uncover; ferret out: ～一个间谍组织 unearth a spy ring / ～一起反革命案件 crack a counterrevolutionary case
【破戒】①(违反宗教戒律) break a religious precept ②(违反个人戒规) break one's vow of abstinence
【破镜重圆】a broken mirror joined together; reunion of husband and wife after an enforced separation or rupture
【破旧】old and shabby; worn-out; dilapidated: ～的房子 an old, dilapidated house
【破口大骂】shout abuse; let loose a torrent of abuse; hurl all kinds of abuse
【破烂】①(残破的) tattered; ragged; worn-out: ～东西 worm-out stuff ②(破烂的东西;废品) junk; scrap ◇ ～货 worthless stuff; rubbish; trash
【破例】break a rule; make an exception
【破脸】turn against sb.; fall out
【破裂】burst; split; rupture; crack: 血管～ rupture of a blood vessel / 他们夫妇感情～了。Their marriage has broken up. / 谈判～了。The negotiations broke down.
【破落】decline (in wealth and position); fall into reduced circumstances; be reduced to poverty: ～地主家庭 an impoverished landlord family ◇ ～户 a family that has gone down in the world; impoverished family
【破门】①(砸开门) burst 〈force〉 open the door: ～而入 break into; force open a door ②[宗] excommunicate
【破灭】be shattered; fall through; evaporate; be disillusioned: 幻想～ ruin of an illusion / 希望～ one's hopes were shattered / 他的幻想～了。He was disillusioned.
【破伤风】tetanus; lockjaw
【破碎】①(破成碎块的) tattered; broken: ～的玻璃 broken glass / ～的纸 paper in tatters ②(使破成碎块) smash sth. to pieces; break into pieces; crush ◇ ～机 crusher; breaker / ～险 [商] risk of breakage
【破损】damaged; worn; torn; damage; breakage: 避免～ avoid breakage ◇ ～险 risk of breakage
【破题儿第一遭】the first time one ever does sth.; the first time ever
【破涕为笑】smile through tears; turn tears into smiles: 她不禁～。She couldn't refrain from smiling through tears.
【破天荒】occur for the first time; be unprecedented; without precedent: ～的大事 the biggest event without any precedent
【破土】①(挖地动工) break ground: ～建新的图书馆。The ground was broken for a new library. ②(开始耕种) start spring ploughing ③(种子发芽出土) break through the soil
【破相】(of facial features) be marred by a scar, etc.
【破晓】dawn; daybreak: ～时分 at dawn; at the peep of day / 天将～。Day is breaking.
【破鞋】loose woman
【破颜】break into a smile
【破约】break one's promise
【破绽】①(衣服开线) a burst seam ②(漏洞)

flaw; weak point: ～百出 be full of flaws / 看出～ spot sb.'s weak point
【破折号】dash (—)

迫 ①(逼迫;强迫) compel; force; press: ～敌投降 force the enemy to surrender / ～于形势 be compelled by circumstances; under the pressure of events / 被～撤退 be forced to retreat / 被～拿起武器 be compelled to take up arms / 为饥寒所～ be driven (to do sth.) by cold and hunger ②(急促) urgent; pressing: 从容不～ calm and unhurried ③(迫近) approach; go towards; go near: ～近 get close to
【迫不得已】have no alternative (but to); be forced to; be compelled to; (do sth.) against one's will; compelled by circumstances
【迫不及待】unable to hold oneself back; too impatient to wait; brook no delay
【迫害】persecute: 遭受～ suffer persecution; be subjected to persecution / 政治～ political persecution
【迫降】[航空] forced landing; distress landing
【迫近】approach; get close to; draw near: 行期～。The day of departure is drawing near.
【迫切】urgent; pressing; imperative: ～陈词 urge to give a statement / ～的任务 urgent task / ～的心情 eager desire; eagerness / ～的需要 an urgent need; a crying need ◇ ～性 urgency
【迫使】force; compel; coerce: ～某人做某事 force sb. to do sth.; force sb. into doing sth.
【迫在眉睫】①(紧迫) extremely urgent; extremely urgent and near ②(临近眼前) imminent; stare sb. in the face

魄 ①(魂魄) soul: 魂飞～散 (be frightened) out of one's wits ②(精力;魄力) vigor; spirit: 气～ boldness of vision; spiritedness
【魄力】daring and resolution; boldness: 有～ be bold and resolute in action / 很有～的人 man with plenty of guts

pōu

剖 ①(破开) cut open; rip open: 把鸡肚子～开 cut open the belly of a chicken ②(分辨;分析) analyse; examine; dissect: ～明事理 analyse the whys and wherefores
【剖白】explain oneself; vindicate oneself: ～心迹 lay one's heart bare
【剖腹产】Caesarean birth ◇ ～术 Caesarean section; Caesarean operation
【剖腹自杀】lay open the bowel and commit suicide; hara-kiri
【剖肝泣血】bare one's heart in all sincerity
【剖解】analyse; dissect: ～细密 make a minute analysis
【剖面】section: 横～ cross section / 纵～ longitudinal section ◇ ～图 sectional drawing; section
【剖面符号】section symbols
【剖面线】thalweg
【剖视图】cutaway view
【剖析】analyse; dissect: ～失败的原因 analyse the cause of failure / ～问题的实质 analyse the essence of the problem

pū

扑 ①(冲向) throw oneself on; pounce on: 老虎向山羊～去。The tiger sprang on the goat. ②(全力以赴) devote: 一心～在癌症的研究上 devote oneself heart and soul to the research on cancer ③(扑打;进攻) rush at; attack: ～蝴蝶 catch butterflies / 直～匪巢 swoop down on the bandits' lair ④(拍打) flap; flutter: 鸭子～着翅膀。The duck flapped its wings.
【扑鼻】assail the nostrils: 臭气～。An offensive smell greeted us.
【扑打】①(猛然打) swat: ～苍蝇 swat a fly ②(轻轻地拍) beat; pat: ～身上的尘土 dust off one's clothes
【扑尔敏】[药] chlorpheniramine
【扑粉】①(化妆用香粉) face powder ②(爽身粉) talcum powder ③(搽粉) apply powder; put on powder
【扑救】put out a fire to save life and property
【扑克】playing cards; poker: 打～ play cards
【扑空】fail to get or achieve what one wants; come away empty-handed: 昨天我去找他,又～了。Yesterday I went to see him, but again he wasn't home.
【扑面】blow on one's face; against one's face: 春风～。The spring wind caressed our faces.
【扑灭】①(扑打消灭) stamp out; put out; extinguish: ～火灾 put out a fire ②(消灭;灭亡) exterminate; wipe out: ～虫害 exterminate insect pests / ～蚊蝇 wipe out mosquitoes and flies
【扑热息痛】[药] paracetamol
【扑朔迷离】complicated and confusing; puzzling
【扑簌】(of tears) trickling down: 眼泪～地往下

掉 tears trickled down one's cheeks
【扑腾】①[象] thupm; thud ②(跳动) move up and down; throb; palpitate
【扑通】[象] flop; thump; splash; pit-a-pat: ～一声,跌倒在地上 fall with a flop on the ground / ～一声,掉进水里 fall into the water with a splash
【扑翼】flapping wing ◇ ～机 flapping-wing aircraft; ornithopter

仆 fall forward; fall prostrate: 前～后继 one stepping into the breach as another falls

噗 [象] puff: ～,一口气吹灭了蜡烛 blow out a candle with one puff

铺 ①(展开;摊平) spread; extend; unfold: ～桌布 spread a table-cloth ②(铺设) pave; lay: ～路面 surface a road / ～平道路 pave the way / ～铁轨 lay a railway track / ～一条街 pave a street
【铺衬】small pieces of cloth used for patches
【铺床】make the bed
【铺地砖】floor tile; paving tile
【铺盖】bedding; bedclothes ◇ ～卷儿 bedding roll; bedroll; luggage roll
【铺轨】lay a railway track ◇ ～机 track-laying machine; tracklayer
【铺路】①(铺设道路) pave a road; pave the way; surface a road ②(预作安排;排除障碍) pave the way for: 那个条约是为进一步的合作～。The treaty paved the way to still closer forms of joint action.
【铺路机】paver
【铺平】pave nicely; smooth out: ～道路 pave the way for / 把床单～ smooth out the bed sheet
【铺砌】pave: 用卵石～的小路 a path paved with pebbles
【铺设】lay; build: ～双轨 lay a double-track
【铺天盖地】blot out the sky and cover up the earth; in an overwhelming manner
【铺叙】narrate in detail; elaborate
【铺展】spread out; sprawl
【铺张】extravagant: ～浪费 extravagance and waste

pú

菩
【菩萨】①[宗] Bodhisattva ②(佛;神) Buddha; Buddhist idol ③(心肠慈善的人) a term applied to a kindhearted person: ～心肠 kindhearted and merciful
【菩提】[佛教] bodhi; supreme
【菩提树】pipal; bo tree; bodhi tree

蒲 [植] cattail
【蒲包】cattail bag; rush bag
【蒲草】the stem or leaf of cattail
【蒲福风级】Beaufort scale
【蒲公英】dandelion
【蒲葵】Chinese fan palm
【蒲柳】big catkin willow
【蒲绒】cattail wool, used for stuffing pillows
【蒲扇】cattail leaf fan
【蒲式耳】bushel
【蒲团】cattail hassock; rush cushion
【蒲席】cattail mat; rush mat

葡
【葡萄】grape: 一串～ a bunch of grapes; a cluster of grapes
【葡萄干】raisin: 无核～ currant
【葡萄架】grape trellis
【葡萄酒】(grape) wine
【葡萄球菌】staphylococcus
【葡萄胎】hydatidiform mole; vesicular mole
【葡萄糖】glucose; grape sugar; dextrose
【葡萄藤】grapevine
【葡萄牙】Portugal ◇ ～人 Portuguese / ～语 Portuguese (language)
【葡萄园】vineyard; grapery

匍
【匍匐】①(爬行) crawl; creep: ～前进 crawl forward ②(趴) lie prostrate ◇ ～茎 stolon / ～植物 creeper

璞 uncut jade
【璞玉浑金】uncut jade and unrefined gold unadorned beauty

镤 [化] protactinium (Pa)

仆 servant: 公～ public servant / 男～ manservant / 女～ maidservant; servant girl
【仆从】footman; retainer; henchman ◇ ～国 vassal country
【仆人】(domestic) servant

pǔ

普 general; universal: ～天下 all over the world; everywhere in the world
【普遍】universal; general; widespread; com-

mon：有～意义 be of universal significance／～的要求 universal demand ◇～裁军 universal disarmament／～规律 universal law／～性 universality／～优惠制 generalized preferential system／～真理 universal truth
【普查】general investigation; general survey：常见病～ general survey of common diseases／地质～ reconnaissance survey／人口～ census ◇～人员 census enumerator
【普及】①(普遍推广) popularize; disseminate; spread：～文化科学知识 spread cultural and scientific knowledge among the people／～与提高相结合 combine popularization with the raising of standards／～中等教育 make secondary education universal ②(普遍地传到) universal; popular ◇～本 popular edition／～教育 universal education
【普鲁本辛】[药] propantheline (bromide); probanthine
【普鲁卡因】[药] procaine
【普天同庆】the whole world or nation joins in the jubilation; universal or national celebration
【普天之下】in all the world; all over the world
【普通】ordinary; common; average：～一兵 an ordinary soldier; a soldier in the ranks; a rank-and-filer／～人 the average person; the man in the street ◇～法[法] common law／～股 common stock; ordinary share／～基金 general fund／～会计学 general accounting／～劳动者 ordinary laborer／～税则 general tariff／～条款 general clause／～心理学 general psychology／～照会 verbal note
【普通话】common spoken Chinese：推广 popularize the common spoken Chinese
【普选】general election ◇～权 universal suffrage
【普照】illuminate all things

谱

①(按类别或系统编的参考书) table; chart; register：家～ family tree; genealogy／食～ cookbook; menu ②(指导练习用的格式或图形) manual; guide：棋～ chess manual ③(乐谱；曲谱) music score; music：歌～ music of a song／乐～ music score; music ④(谱曲) set to music; compose：把一首诗～成歌曲 set a poem to music ⑤(把握) sth. to count on; a fair amount of confidence：心里没个～儿 have nothing definite in mind／做事有～儿 do things with confidence; know what one is doing
【谱斑】[天] flocculus
【谱表】[乐] stave; staff：大～ great stave
【谱号】[乐] clef：低音～ bass clef; F clef／高音～ treble clef; G clef／中音～ tenor clef; alto clef; C clef
【谱系】[生] pedigree
【谱写】compose (music)
【谱子】music score; music

镨

[化] praseodymium (Pr)

朴

simple; plain
【朴实】①(朴素) simple; plain：～无华 simple and unadorned／文风～ simple style of writing ②(踏实) down-to-earth; sincere and honest; guileless：～的工作作风 a down-to-earth style of work
【朴素】simple; plain：～的语言 unaffected language／艰苦～的生活作风 a habit of simple and plain living／衣著～ simple dressed
【朴直】honest and straightforward：文笔～ simple and straightforward writing
【朴质】simple and unadorned; natural

圃

garden：菜～ vegetable plot／花～ flower nursery／苗～ seed plot; (seedling) nursery

蹼

web
【蹼趾】webbed toe
【蹼足】webfoot; palmate foot

pù

铺

①(商店) shop; store ②(用板子搭的床) plank bed
【铺板】bed board; bed plank
【铺面】shop front
【铺位】bunk; berth：上～ upper berth／下～ lower berth
【铺子】shop; store

瀑

waterfall
【瀑布】waterfall; falls; cataract

曝

expose to the sun
【曝光】exposure：～不足 underexposure; underexposed／～过度 overexposure; over-exposed ◇～表 exposure meter／～宽容度 exposure latitude

Q

qī

期 ①(一段时间) a period of time; phase; stage: 第一～工程 the first phase of the project / 服役～ term of military service / 假～ vacation / 潜伏～ incubation period / 危险～ critical phase / 学～ school term ②(预定的时日) designated time; scheduled time: 到～ fall due / 过～ be overdue; become overdue / 限～ set a time limit ③[量] issue; number; term: 最近一～《英语学习》 the current ⟨latest⟩ issue of English Language Learning / 短训班办了两～。The short-term training class has been run two times. ④(约定时日) make an appointment: 不～而遇 meet unexpectedly; meet by chance / 约～会晤 appoint a date for interview ⑤(等待;盼望) expect: ～待 expect; await

【期待】anticipate; await; expect; look forward to: ～的权益 interest in expectancy / ～获得的财产 expectation of property / ～继承人 expectant heir / 我们殷切地～你早日光临。We eagerly await your early arrival.

【期货】[经] futures: 做～交易 deal in futures ◇ ～价格 forward price / ～交易 forward business / ～合同 forward contract; futures contract / ～汇率 forward exchange rate / ～市场 option market

【期间】time; period; course; duration: 会议～ in the course of the conference; during the conference / 就在这～ during this time; in this very period ◇ ～保单 time policy

【期刊】periodical ◇ ～阅览室 periodical reading room ○国际标准期刊编号 ISSN (International Standard Serial Number)

【期考】end-of-term examination; terminal examination

【期满】expire; run out; come to an end: 服役～ complete one's term of (military) service / 合同～ when the contract expires; on the expiration of the contract ◇ ～通知书 notice of expiry

【期票】promissory note; term bill ◇ ～承兑人 acceptance house / ～附件 additional part of a bill / ～推销人 bill broker

【期望】hope; expectation: ～成功 hope to succeed / 辜负某人的～ disappoint one's expectations / 寄予很大的～ place high hopes on

【期限】allotted time; time limit; deadline: 超过～ go beyond the time limit; exceed the deadline / 规定一个～ set a deadline; fix a target date / 延长～ extend the time limit / 在规定的～内 in the allotted time

【期中报告】interim report

【期中考试】midterm examination

【期中审计】interim audit

【期中决算表】interim statements

【期终考试】term examination; end-of-term examination; final examination

欺 ①(欺骗) deceive: 自～～人 deceive oneself as well as others / ～人之谈 deceitful words; deceptive talk ②(欺负) bully; take advantage of: 可～ can be taken advantage of / ～人太甚。That's going too far. (或: That's too much of a bully.)

【欺负】bully; treat sb. high-handedly; browbeat; treat sb. rough

【欺凌】bully and humiliate: 任人～ allow oneself to be trodden upon; allow others to tread on one's neck / 受尽了～ be subjected to endless bullying and humiliation

【欺瞒】hoodwink; dupe; pull the wool over sb.'s eyes; cover up the truth to deceive

【欺蒙】hoodwink; dupe; pull wool over sb.'s eyes; befool

【欺骗】deceive; cheat; dupe; swindle; defraud; take in: ～不明真相的人 deceive those who do not know the truth; mislead those who are not aware of the facts / ～世界舆论 befuddle world opinion / ～无知的人 dupe the ignorant people ◇ ～案 case of victimization / ～行为 act of swindling; fraudulent act / ～性证明 fraudulent proof

【欺软怕硬】bully the weak and fear the strong; meek towards the brutal and brutal towards the meet

【欺上瞒下】deceive one's superiors and delude one's subordinates

【欺生】①(欺负或欺骗新来的人) bully or cheat strangers ②(马驴等对生人不驯服) be ungovernable by strangers

【欺世盗名】gain fame by deceiving the public; angle for undeserved fame

【欺侮】bully; treat sb. high-handedly

【欺压】bully and oppress; ride roughshod over

【欺诈】cheat; swindle ◇ ～行为 fraudulent conduct / ～意图 fraudulent intention / ～罪

crime of false pretenses

栖 ①(鸟停在树上) perch ②(居住;停留) dwell; stay

【栖身】stay; sojourn: 无处～ have no place to stay; have no roof to live under／暂作～之计 plan to make a stay for a time

【栖息】perch; rest: 在树上～ perch on the tree ◇～地 habitat

桤

【桤木】[植] alder

漆 ①(涂料) lacquer; paint: ～盘 lacquer tray ②(涂漆) paint; lacquer; varnish; coat with paint 〈lacquer; varnish〉: 把门～成红色 paint the door red

【漆包线】enamel-insulated wire; enameled wire

【漆布】varnished cloth

【漆革】patent leather

【漆工】①(指工作) lacquering; painting ②(指人) lacquerer; lacquer man; painter

【漆黑】①(光线很暗) pitch-dark ②(非常黑) pitch-black; jet-black

【漆黑一团】①(非常黑暗) pitch-dark; complete darkness: 把…说得～ paint a dark picture of ②(一无所知) be entirely ignorant of; be in the dark: 这个问题在他心中还是～。He is still completely in the dark about the matter.

【漆画】lacquer painting

【漆皮】①(一层漆) coat of paint ②(虫胶清漆) shellac

【漆器】lacquerware; lacquerwork: 脱胎～ bodiless lacquerware

【漆树】lacquer tree; varnish tree

戚 ①(亲戚) relative: 皇亲国～ relatives of an emperor ②(悲哀;忧愁) sorrow; woe: 休～相关 share joys and sorrows; share weal and woe

槭

【槭树】maple

嘁

【嘁嘁喳喳】chatter away; jabber

七 seven: 第七 seventh／第十～ seventeenth／十～ seventeen／乱～八糟 at sixes and sevens

【七边形】heptagon; septilateral

【七颠八倒】at sixes and sevens; all upside down; topsy-turvy

【七级风】force 7 wind; moderate gale

【七极管】heptode

【七绝】a four-line poem with seven characters to a line

【七零八落】scattered here and there; in disorder: ～的几间草房 a few ramshackle huts scattered here and there

【七律】an eight-line poem with seven characters to a line

【七扭八歪】crooked; uneven; irregular; in a state of great disorder

【七拼八凑】piece together; knock together; rig up; scrape together

【七巧板】seven-piece puzzle; tangram

【七窍】the seven apertures in the human head ○眼 eyes／耳 ears／鼻孔 nostrils／口 mouth

【七窍流血】bleeding from nose and mouth

【七窍生烟】fume with anger; foam with rage; fumigate with anger

【七情】the seven human emotions ○喜 joy／怒 anger／哀 sorrow／惧 fear／爱 love／恶 hate／欲 desire

【七情六欲】the seven emotions and the six sensory pleasures

【七鳃鳗】lamprey

【七上八下】an unsettled state of mind; be agitated; be perturbed

【七十】seventy: ～年代 seventies／第～ seventieth

【七十二变】countless changes of tactics

【七十二行】all sorts of occupations; in every conceivable line of work

【七手八脚】with everybody lending a hand; in a bustle

【七夕】the seventh evening of the seventh moon

【七月】①(阳历) July ②(阴历) the seventh month of the lunar year; the seventh moon

【七折八扣】various deductions; not pay up full amount

【七嘴八舌】all talking at once; lively discussion with everybody trying to get a word in

沏 infuse: ～茶 infuse tea; make tea

妻 wife: 夫～ husband and wife／娶某人为～ take sb. to wife

【妻儿老小】a married man's entire family (parents, wife and children)

【妻离子散】be separated from one's wife and children; breaking up or scattering of one's

family; family broken up
【妻子】①(男子的配偶) wife ②(妻子和儿女) wife and children

凄 ①(寒冷) chilly; cold ②(冷落萧条) bleak and desolate: ～清 lonely and sad ③(悲伤难过) sad; wretched; miserable: ～楚 miserable
【凄惨】wretched; miserable; tragic
【凄风苦雨】miserable conditions; wailing wind and weeping rain; wretched circumstances
【凄厉】sad and shrill: ～的叫声 sad, shrill cries / 风声～ with the wind moaning
【凄凉】dreary; desolate; miserable: ～的地方 desolate place / ～的前景 bleak prospects / ～的生活 forlorn life / 满目～ desolation all round / 晚景～ lead a miserable and dreary life in old age
【凄切】plaintive; mournful
【凄然】sad; mournful: ～泪下 shed tears in sadness

萋
【萋萋】luxuriant: 芳草～ a luxuriant growth of grass

蹊
【蹊跷】odd; queer; fishy

qí

齐 ①(整齐) neat; even; uniform; well arranged; in good order: 把桌子摆～ arrange the tables in an orderly way / 长短不～ not of uniform length / 剪得很～ be evenly trimmed / 整～ neat and tidy ②(达到同样高度) on a level with; be flush with; on the same plane with: 河水～岸。The river is flush with its banks. / 水深～腰。The water comes up to my waist. (或: The water is waist-deep.) ③(同样; 一致) alike; similar: 心～ think alike; think in similar way ④(一块儿; 同时) together; simultaneously: 师生～动手。Teachers and students all pitched in. ⑤(完备;全) all ready; all present: 人都～了。All are present. / 一切都～了。Everything is ready. (或: Everything is arranged.) ⑥(取齐) along: ～根切 cut to the roots / ～着边儿画线 draw a line along the edge
【齐备】all ready; complete: 商店里货色～。The store carries a complete stock.
【齐步走】[军] ①(以整齐的步伐前进) quick march ②(口令) Quick time, march!
【齐唱】singing in unison; unison
【齐楚】[书] neat and smart: 衣冠～ be smartly dressed
【齐东野语】what folks say; popular report
【齐集】assemble; gather; collect
【齐家治国】regulate the family and rule the state
【齐名】enjoy equal popularity; be equally famous
【齐全】complete; all in readiness: 尺码～ have a complete range of sizes / 货物～ have a satisfactory variety of goods / 设备～ have all the necessary fittings / 装备～ be fully equipped
【齐射】[军] salvo; volley
【齐声】in chorus; in unison: ～附合 second with one voice / ～歌唱 sing in unison / ～欢呼 cheer in unison / ～回答 answer in chorus
【齐头并进】advance side by side; do two or more things at once; go forward together
【齐心】be of one mind 〈heart〉: ～协力 work as one; make concerted efforts; pull together
【齐整】neat; uniform
【齐奏】[乐] playing (instruments) in unison; unison

蛴
【蛴螬】[动] grub

脐 ①[生理] navel; umbilicus ②(蟹脐) the abdomen of a crab
【脐带】umbilical cord

鳍 [动] fin: 背～ dorsal fin / 腹～ ventral fin; pelvic fin / 尾～ caudal fin / 胸～ pectoral fin
【鳍足】[动] clasper ◇ ～动物 Pinnipedia; pinniped

其 ①(他的; 她的; 它的; 他们的; 她们的; 它们的) his; her; its; their: ～母 his mother / 各得～所。Each is in his proper place. (或: Everyone is properly provided for.) / 各尽～力。Everyone does his best. ②(他;她;它;他们;她们;它们) he; she; it; they: 不要任～自流 don't let things slide / 促～早日实现 help bring it about at an early date / 听～自然 let it alone ③(那个; 那样) that; such: 正当～时 just at that time; at the opportune moment; at the right moment / 不乏～人。There is no lack of such people. / 查无～事。It is found that there was no such thing. ④(与"大"连用,虚指) 大吹～牛 greatly exaggerate; talk big without restraint / 大请～客 feast many guests; entertain lavishly; invite many guests to dinner

【其次】①(次第较后) next; secondly; then: 他首先发言，～轮到我。He will speak first, and I next. ②(次要的地位) secondary: 主要是内容，～才是形式。Content comes first, form second.

【其间】between them; between which; among them; among which

【其实】[副] actually; in fact; as a matter of fact; really; in reality: ～他是个很聪明的人。He's really a very clever man.

【其他】other; else: ～事情 other things / 到～地方去 go to some other place

【其余】the others; the rest; the remainder: ～的不用说了。The rest needs no telling. / 班里只有十名女孩，～都是男孩。There are only ten girls in the class, the rest are boys.

【其中】among (which, them, etc.); in (which, it, etc.) 乐在～ find pleasure in it / 他爸爸给他六个苹果,～有一个是坏的。His father gave him six apples. There was a bad one among them.

麒

【麒麟】*kylin*; (Chinese unicorn

【麒麟座】[天] Monoceros

旗

flag; banner; standard: 白～ white flag / 队～ team pennant / 国～ national flag / 红～ red flag / 锦～ brocade banner / 升～ raise a flag / 下半～致哀 hoist a flag half-mast high

【旗杆】flagpole; flag post

【旗鼓相当】be well-matched; be equal in match or contest strength

【旗号】[贬] banner; flag: 打着…的～ flaunt the banner of; under the signboard of

【旗舰】flagship ◇ ～舰长 flag captain

【旗开得胜】win victory the moment one raises one's standard; win victory in the first battle; win speedy success; succeed from the very start

【旗袍】a close-fitting woman's dress with high neck and slit skirt; cheongsam; a sheath with a slit skirt

【旗人】bannerman

【旗手】standard-bearer

【旗塔】[航海] flag tower

【旗鱼】sailfish

【旗语】semaphore; flag signal: 打～ signal by semaphore; semaphore / 用～通信 flag a message

【旗帜】①(旗子) banner; flag ②(有代表性的思想、学说或政治力量等) stand; colors: ～鲜明 have a clear-cut stand

【旗状云】[气] banner cloud; cloud banner

【旗子】flag; banner; pennant; standard

棋

chess or any board game: 下～ play chess / 下一盘～ play a game of chess / 走～ move ◯(中国)象棋 Chinese chess / 国际象棋 (international) chess / 象棋大师 master / 特级大师 grandmaster / 王 king / 后 queen / 车 chariot / 象 elephant; bishop / 马 horse; knight / 兵 pawn / 炮 cannon / 象〈相〉elephant / 士 guard / 将〈帅〉commander / 卒〈兵〉soldier / 出子 develop / 吃子 capture / 将军 check / 将死 checkmate

【棋逢对手】meet one's match in a chess tournament; be well-matched in a contest; diamond cut diamond

【棋迷】chess fan; chess enthusiast

【棋盘】chessboard; checkerboard

【棋谱】chess manual

【棋手】chess player

【棋子】piece (in a board game); chessman

祈

①(祈祷) pray: ～年 pray for a good harvest ②(请求;希望) entreat

【祈祷】pray; say one's prayers

【祈年殿】Hall of Prayer for Good Harvest

【祈求】earnestly hope; pray for

【祈使句】[语] imperative sentence

【祈望】hope; wish

歧

①(岔道;大路分出的路) fork; branch ②(不相同;不一致) divergent; different

【歧管】[机] manifold

【歧路】branch road; forked road

【歧路亡羊】a lamb going astray at a fork in the road; go astray in a complex situation

【歧视】discriminate against: 种族～ racial discrimination

【歧途】wrong road: 被引入～ be led astray / 误入～ take the wrong road by mistake; go on the wrong track

【歧义】different meanings; various interpretations: 这段文章有～。This passage is open to different interpretations.

奇

①(罕见的;特殊的;非常的) strange; queer; rare; uncommon; unusual: ～才 prodigy; unusual talent; remarkable talent / ～事 a strange affair; an unusual phenomenon ②(出人意料;令人难测) unexpected; unpredictable: 出～制胜 win victory through unexpected moves ③(惊异) surprise; wonder; astonish: 不

足为～ be nothing surprising; nothing to be surprised at ④(非常) extremely; exceeding: ～冷 extremely cold; exceedingly cold／～痛 unbearably painful
【奇兵】an army suddenly appearing from nowhere; an ingenious military move
【奇耻大辱】galling shame and humiliation; deep disgrace; burning shame
【奇风异俗】exotic customs; strange customs
【奇功】outstanding service: 屡建～ repeatedly perform outstanding service
【奇怪】odd; queer; strange; surprising: 多～ How odd!／这事情有些～。There's sth. fishy about the matter.／真～! That's queer, indeed!
【奇观】marvellous spectacle; wonder; wonderful sight: 大自然的～ a marvellous natural phenomenon
【奇花异卉】exotic flowers and rare herbs
【奇货可居】hoard as a rare commodity; a rare commodity worth boarding
【奇迹】miracle; wonder; marvel: 创造～ work wonders; accomplish wonders; perform miracles／科学上的～ a marvel of science
【奇景】wonderful view; extraordinary sight: 冰峰～ a wonderful view of ice-capped peaks
【奇妙】marvellous; wonderful; intriguing
【奇谈】strange tale; absurd argument: 海外～ strange tales from over the seas／～怪论 strange tales and absurd arguments
【奇特】peculiar; queer; singular: ～的服装 peculiar dress／～的习惯 queer habits
【奇文】①(精美的文章) a remarkable piece of writing: ～共欣赏,疑义相与析。A remarkable work should be shared and its subtleties discussed ②(荒谬文章) preposterous piece of writing; queer writing
【奇闻】sth. unheard-of; a thrilling, fantastic story: ～逸事 a strange news and an extraordinary affair／千古～ an unheard-of fantastic story
【奇袭】surprise attack; raid
【奇形怪状】grotesque or fantastic in shape or appearance; of strange or grotesque shapes and sizes
【奇异】①(奇怪) queer; strange; bizarre; odd ②(惊异) surprising; curious: 用～的眼光看人 look at sb. in surprise; look at sb. with curious eyes
【奇遇】①(意外相逢) happy encounter; fortuitous meeting ②(奇特遭遇) adventure
【奇珍异宝】rare treasures
【奇志】high aspirations; lofty ideal; noble ambition
【奇装异服】exotic costume; bizarre dress; outlandish clothes

崎

【崎岖】rugged; rough: ～不平 rugged and rough／～的山路 a rugged mountain path

骑

①(两腿跨坐) ride; sit on the back of: ～车回家 go home by bicycle／～马 ride a horse; be on horseback ②(骑兵) cavalryman; cavalry: 铁～ cavalry
【骑兵】cavalryman; cavalry ◇ ～部队 mounted troops; cavalry unit
【骑缝】a junction of the edges of two sheets of paper; perforation: ～章 seal on the perforation
【骑虎难下】ride a tiger and find it hard to get off; have no way to back down; irrevocably committed
【骑马订】[印] saddle stitching
【骑马找马】look for a horse while sitting on one; hold on to one job while seeking a better one
【骑墙】sit on the fence; straddle: 在争论中采取～态度 sit on the fence in the argument ◇ ～派 fence-sitter
【骑士】knight; cavalier
【骑术】horsemanship; equestrian skill
【骑无鞍劣马】bareback bronc-riding
【骑在头上】ride roughshod over; sit on the backs of; lord it over

畦

plot of land: 菜～ a vegetable bed
【畦灌】border method of irrigation
【畦田】ridge-bordered plots

qǐ

启

①(开) open: ～门 open the door／～幕 raise the curtain／幕～。The curtain rises. ②(开始) start; initiate: ～行 start on a journey; start going; set off ③(开导) enlighten; awaken: ～发 arouse; inspire; enlighten ④(陈述) state; inform: 敬～者 I beg to state; I wish to inform you
【启闭机】headstock gear
【启程】set out; start on a journey
【启齿】open one's mouth; start to talk about sth.: 难以～ find it difficult to bring the matter up
【启迪】open and enlighten

【启动】start (a machine, etc.); switch on
【启发】arouse; inspire; enlighten: 从…中得到～ draw inspiration from / 他的谈话给了我们很多～。His talk has greatly inspired us.
【启发过程研究】[心理] heuristics
【启发式】elicitation method (of teaching); heuristic method
【启封】①(除去封印、封条) unseal; break the seal; remove the seal ②(拆信;拆包) open an envelop or wrapper
【启航】set sail; weigh anchor: 李先生昨日乘船～去日本。Mr. Li set sail for Japan yesterday.
【启蒙】①(使初学者得到基本的入门知识) impart rudimentary knowledge to beginners; initiate: ～课本 children's primer / ～老师 the teacher who introduces one to a certain field of study ②(使摆脱愚昧和迷信) enlighten; free sb. from prejudice or superstition ◇ ～时期 period of enlightenment / ～运动 the Enlightenment; enlightenment campaign
【启明星】[天] Venus
【启示】enlightenment; inspiration; revelation: 从…得到～ draw inspiration from; gain enlightenment from
【启事】notice; announcement: 张贴～ put up a notice / 征稿～ a notice inviting contributions (to a magazine, newspaper, etc.) / 他们的结婚～已见报了。The announcement of their marriage appeared in the newspaper.
【启通脉冲】[电子] unblanking pulse
【启衅】start a quarrel; provoke discord; provoke dispute; pick a quarrel
【启用】start using
【启用前检查】[计算机] readiness review
【启运】start shipment

企

①(抬起脚跟站着) stand on tiptoe ②(盼望) anxiously expect sth.; look forward to
【企鹅】penguin
【企口】[建] tongue-and-groove: ～接合 tongue-and-groove joint ◇ ～板 matched boards
【企求】desire to gain; seek for; hanker after
【企图】attempt; seek; try: ～复辟 attempt to stage a comeback; try to stage a comeback
【企望】hope for; look forward to
【企业】enterprise; business: ～管理 business management / 工矿～ factories, mines and other enterprises ◇ ～发展基金 venture expansion fund / ～法 law of enterprises; enterprise law / ～家 entrepreneur; enterpriser / ～自主权 decision-making power of enterprises; right of autonomy for enterprises

乞

beg (for alms, etc.); supplicate: ～哀告怜 piteously beg for help / ～食 beg for food / 行～ go begging
【乞丐】beggar
【乞和】sue for pity
【乞怜】beg for pity 〈mercy〉: 摇尾～ be like a dog wagging its tail pitifully; abjectly beg for mercy
【乞灵】resort to; seek help from
【乞求】beg for; supplicate; implore: ～宽恕 beg for mercy 〈pardon〉
【乞讨】beg; go begging: 沿街～ go begging from door to door
【乞降】beg to surrender
【乞援】ask for assistance; beg for aid

起

①(站起;坐起) rise; get up; stand up: 早睡早～ early to bed early to rise ②(取出;取走) draw out; remove; extract; pull: ～地雷 clear mines / ～钉子 draw out a nail / ～瓶塞 pull the cork from a bottle ③(长出) appear; raise: 背上～痱子 heat rash rises on one's back / 脚上～水泡 get blisters on one's feet ④(发生) rise; grow: ～风 the wind is rising / ～疑心 become suspicious / ～作用 take effect ⑤(拟定) draft; work out: ～草 draft / ～稿子 work out a draft; make a draft ⑥(建立) build; set up: ～伙 set up a mess / ～一堵墙 build a wall ⑦(开始) start; begin: 从今天～ starting from today ⑧[量](件;次) case; instance; (批) batch; group; lot: 两～火警 two cases of fire alarm / 分两～ in two groups
【起岸】bring (cargo, etc. from a ship) to land; unship a cargo; unload a ship
【起爆】detonate: 准时～ detonate at the designated time ◇ ～剂 detonating agent; primer / ～帽 detonating cap
【起笔】①(写字的第一笔) the first stroke of a Chinese character ②(每一笔的开始) the start of each stroke in writing a Chinese character
【起搏器】pacemaker
【起草】draft; draw up: ～文件 draft a document; draw up a document ◇ ～人 draftsman / ～委员会 drafting committee
【起承转合】introduction, elucidation of the theme, transition to another viewpoint and summing up; the four steps in the composition of an essay

【起程】leave; set out; start on a journey
【起初】originally; at first; at the outset; in the beginning
【起床】get up; get out of bed; rise
【起床号】reveille
【起点】starting point: 赛跑的～ the starting mark 〈line〉 of a race
【起电】electrification; charge ◇ ～盘 electrophorus
【起动】start (a machine, etc.) ◇ ～电动机 starting motor / ～机 starter
【起飞】(of aircraft) take off: 班机准时～。The airliner took off on time.
【起伏】rise and fall; undulate
【起稿】make a draft; draft
【起航】set sail
【起哄】①(胡闹) gather together to create a disturbance; gather together to stir up trouble ②(开玩笑) jeer; boo and hoot
【起火】①(失火) fire breaking out; outbreak of a fire ②(做饭) cook meals; prepare meals; do cooking
【起货】unload; landing ◇ ～单 landing permit / ～机 winch / ～码头 landing pier
【起家】build up; grow and thrive; make one's fortune, name, etc.: 白手～ build up from nothing; start from scratch
【起见】for the purpose of; in order to: 为醒目～ in order to make it stand out clearly
【起劲】vigorously; energetically; enthusiastically: 干得很～ work very energetically
【起居】daily life: ～有恒 lead a regular life ◇ ～室 living room
【起圈】[农] remove manure from a pigsty, sheepfold, etc.
【起来】①(站起; 坐起) stand up; sit up; rise to one's feet ②(起床) get up; get out of bed ③(奋起) rise; arise; revolt: ～反抗 rise in revolt ④(表示动作向上)把孩子抱～ take a child up in one's arms ⑤(表示动作开始并继续)唱～ start to sing ⑥(表示印象或看法)听～没什么问题。It sounds quite all right.
【起立】stand up; rise to one's feet: ～欢迎 rise to welcome sb. / 司令进来时,全体～。Everyone stood when the commander come in.
【起垄】[农] ridging
【起落】rise and fall; up and down
【起落架】[航空] landing gear; alighting gear; undercarriage: ～放下 gear down; landing gear lowering / ～收上 gear up; landing gear raising
【起码】①(最低的;初步的) minimum; rudimentary; elementary: ～的要求 minimum requirements / ～的知识 rudimentary knowledge; elementary knowledge / ～资格 minimum qualification ②(最低限度) at least: 最～的酬劳 the least reward
【起毛】[纸] fluff
【起锚】weigh anchor; set sail
【起名儿】give a name; name
【起跑】[体] start of a race: 集体～ mass start / 梯形～ staggered start / 站立～ standing start / 在跑道上练～ practice starts on a running track ◇ ～信号 starting signal; starting mark / ～线 starting line (for a race); scratch line (for a relay race); starting mark; balkline
【起泡】①(起水泡) blister; bubble ②(起泡沫) foam; bead ◇ ～剂 blowing agent; foaming agent
【起讫】the beginning and the end
【起绒】gigging; raising ◇ ～织物 pile
【起色】improvement; pickup: 工作有～ there's some improvement in the work / 健康有～ one's beginning to pick up now
【起身】①(起床) get up; get out of bed ②(动身) leave; set out; get off
【起事】start armed struggle; rise in rebellion
【起誓】take an oath; swear
【起死回生】(of a doctor's skill) bring the dying back to life; snatch a patient from the jaws of death; restore the dying to life
【起诉】bring a suit against sb.; bring an action against sb.; sue; prosecute: ～和辩护的能力 capacity to sue and defend ◇ ～人 suitor; prosecutor / ～书 indictment; bill of complaint; statement of charges / ～资格 standing to sue
【起跳】[体] take off ◇ ～板 take-off board / ～线 take-off line; take-off mark
【起头】①(开始;发起) start; originate; initiate: 这件事都是他～的。It was he who started all this. ②(开始的时候) at first; in the beginning; originally: ～她是同意的,后来变了卦。At first she did agree but then she changed her mind. ③(开始;开端) beginning: 万事～难。Everything is hard in the beginning.
【起网】(net) hauling ◇ ～机 net hauler
【起息期】date of value
【起先】at first; in the beginning
【起夜】get up in the night to urinate
【起义】uprising; insurrection; revolt: 农民～ a peasant uprising / ～投诚 come over to our

side; revolt and cross over ◇～军 insurrectionary army
【起意】[贬] conceive a design
【起因】cause; origin: ～于一件小事 arise from a mere trifle / 调查事故的～ investigate the cause of the accident / 争执的～ origin of the dispute
【起用】reinstate a retired or dismissed official
【起源】①(根源) origin: 生命的～ the origin of life / 文明的～ the origin of civilization ②(开始发生) originate from; stem from; come from: 舞蹈～于劳动。 Dancing originates from labor.
【起运】start shipment: 货物业已～。 The goods are on their way. ◇～地点 starting place for shipping; place of dispatch
【起赃】track down and recover stolen goods
【起早贪黑】start work early and knock off late; work from down to dusk
【起止系统】start-stop system
【起重车】derrick car
【起重船】crane ship
【起重磁铁】lifting magnet
【起重滑车】lifting block: 船用～ gin block
【起重机】hoist; crane; derrick: ～的起重能力 the lifting capacity of a crane / 龙门～ gantry crane / 门式～ portal crane / 塔式～ tower crane
【起绉】wrinkle; crumple ◇～工艺 [纺] creping
【起子】①(开瓶盖的) bottle opener ②(螺丝刀) screwdriver ③(发酵粉) baking powder
【起作用】①(起影响) play a part (in) ②(产生效果) take effect

杞

【杞人忧天】like the man of Qi who was haunted by the fear that the sky might fall; entertain imaginary or groundless fears; with unwarranted anxiety

岂

【岂不】(表示反问) Isn't that; Doesn't that; Hasn't that; Won't that: 这样～更好些? Wouldn't that be better?
【岂但】not only: ～你不知道，就连她的父母也不知道。 Not only are you ignorant of that, even her parents are in the dark.
【岂敢】you flatter me; I don't deserve such praise or honor
【岂能】how could; how is it possible
【岂有此理】preposterous; outrageous: 真是～! It's really outrageous!

绮

①(丝织品) figured woven silk material; damask ②(美丽) beautiful; gorgeous
【绮丽】beautiful; gorgeous: 风景～ beautiful scenery

qì

泣

①(哭) weep; sob: ～不成声 choke with sobs / ～诉 accuse while weeping; accuse amid tears ②(泪) tears: ～下如雨 shed tears like rain; weep copious tears

弃

throw away; discard; abandon; give up: ～城而逃 abandon the city and flee / ～之可惜 hesitate to discard sth.; be unwilling to throw away
【弃暗投明】forsake darkness for light; leave the reactionary side and cross over to the side of progress; give oneself up to the government
【弃本逐末】run after the less important things, forgetting the important
【弃短取长】forget someone's shortcomings and make use of his strong points
【弃甲曳兵】(of troops) throw away their armor and trail their weapons behind them; be routed; flee pell-mell
【弃旧图新】turn over a new leaf; change new for the old
【弃旧迎新】replace the old with the new
【弃权】①(放弃权利) abstain from voting: 四票～ four abstentions ②[体] waive the right (to play); forfeit
【弃世】pass away; die
【弃置】discard; throw aside: ～不用 be discarded; lie idle

契

①(文书;凭证) contract; deed: 地～ title deed for land; land deed / 房地～ title deed for real property ②(投合) agree; get along well: 默～ tacit agreement; tacit understanding
【契合】agree with; tally with; correspond to: 与进化论相～ agree with the theory of evolution
【契机】①[哲] moment ②(事物转化的关键) turning point; juncture
【契据】deed; contract; receipt
【契友】close friend; bosom friend
【契约】contract; deed; charter: 租船～ contract of affreightment; charter party ◇～当事人 contracting parties / ～法 contract law / ～生效 validity of contract; execution of contract / ～条款 contract terms

葺 ①(用茅草覆盖房顶) cover a roof with straw; thatch ②(修理房屋) repair; mend: 修～ repair (a house); make repairs

碛 ①(砂石浅滩) moraine ②(沙漠) desert

砌 build by laying bricks or stones: ～墙 build a wall (with bricks, stones, etc.) / ～烟囱 build a chimney / ～砖 lay bricks

器 ①(器具) implement; utensil; ware: 瓷～ chinaware; china; porcelain / 漆～ lacquerware / 玉～ jade article / 乐～ musical instrument ②(器官) organ: 生殖～ reproductive organs; generative organs; genitals ③(度量;才能) capacity; talent: ～识 capability and judgment

【器材】equipment; material: 线路～ line materials / 照相～ photographic equipment

【器度不凡】uncommon personality

【器官】organ; apparatus: 发音～ organs of speech / 呼吸～ respiratory apparatus / 消化～ digestive organs

【器件】parts of an apparatus or appliance: 电子～ electronic device

【器具】utensil; implement; appliance: 日用～ household utensils; articles of daily use

【器量】tolerance: ～大 broad-minded / ～小 narrow-minded

【器皿】household utensils; containers esp. for use in the house

【器物】implements; utensils

【器械】①(器具) apparatus; appliance; instrument: 光学～ optical instrument / 体育～ sports apparatus / 医疗～ medical appliances ②(武器) weapon; weaponry (总称)

【器械体操】gymnastics on apparatus

【器宇】bearing; deportment: ～不凡 with unusual deportment / ～高雅 a man's demeanor is high and elegant / ～轩昂 of dignified bearing; one's deportment is dignified

【器乐】instrumental music ◇ ～曲 composition for an instrument

【器质性精神病】organic psychosis

【器重】think highly of; regard highly; have a high opinion of

气 ①(气体) gas: 毒～ poisonous gas; poison gas / 沼～ marsh gas; methane ②(空气) air: 到室外透透～ go out and take some fresh air ③(气息) breath: 喘不过～来 lose one's breath; gasp for breath / 呼～ breathing out; exhale / 换～ breathing; change breath / 上～不接下～ be out of breath; gasp for breath / 吸～ breathing in; inhale / 歇口～ catch one's breath ④(天气) weather: 秋高～爽 fine autumn weather ⑤(味儿) smell; odor: 臭～ bad odor; foul smell / 香～ sweet smell ⑥(精神状态) spirit; morale: 垂头丧～ in low spirits / 打～ boost the morale; cheer on / 朝～勃勃 vigorous; full of youthful vigor ⑦(作风;习气) airs; manner; style: 官～ bureaucratic airs / 孩子～ childish / 书生～ bookish; pedantic ⑧(使人生气) make angry; enrage ⑨(生气;发怒) get angry; be enraged: ～得发抖 tremble with anger ⑩(欺压) bully; insult: 挨打受～ be bullied and beaten

【气昂昂】full of mettle; full of dash

【气泵】air pump

【气冲冲】furious; beside oneself with rage

【气喘】asthma: 阵发性～ spasmodic asthma

【气窗】transom (window); fanlight

【气锤】pneumatic hammer; air hammer

【气垫】air cushion ◇ ～车 air cushion car; hovercraft / ～船 hovercraft / ～运载器 air cushion machine

【气动】pneumatic ◇ ～工具 pneumatic tool

【气度】tolerance; bearing: ～不凡 in a laudable tolerant spirit

【气短】①(呼吸短促) breathe hard; be short of breath; pant ②(泄气) lose heart; be discouraged

【气氛】atmosphere; air: 节日～ festive air / 亲切友好的～ cordial and friendly atmosphere / 热烈的～ lively atmosphere

【气愤】indignant; furious: 感到～ be filled with indignation

【气腹】[医] pneumoperitoneum: 人工～ artificial pneumoperitoneum

【气概】lofty quality; mettle; spirit: 革命～ revolutionary spirit / 英雄～ heroic mettle

【气缸】[机] air cylinder; cylinder

【气割】[机] gas cutting; blowpipe

【气根】[植] aerial root

【气鼓】[医] tympanites

【气管】windpipe; trachea ◇ ～切开术 tracheotomy / ～炎 tracheitis

【气贯长虹】imbued-with a spirit as lofty as the rainbow spanning the sky; full of noble aspiration and daring

【气焊】[机] gas welding

【气候】①(气象情况) climate: 赤道～ equatorial climate / 大陆性～ continental climate / 高原～ plateau climate / 海岸～ coastal climate / 海洋性～ oceanic climate / 改良～条件 improve weather conditions / 调节～ regulate the climate ②（局势）climate; situation: 政治～ political climate ③（结果；成就）successful development: 成不了～ will not get anywhere; will not amount to anything ◇～带 climatic zone / ～适应 climatic adaptation / ～图 climatic chart; climatic map / ～志 climatography

【气侯学】climatology: 生物～ bioclimatology

【气呼呼】in a huff; panting with rage

【气化】gasification

【气急败坏】flustered and exasperated; utterly discomfited; be in exasperation

【气节】integrity; moral courage: 民族～ national integrity / 有～的人 man of moral courage

【气井】[石油] gas well

【气阱】[航空] air pocket

【气可鼓而不可泄】morale should be boosted, not dampened

【气孔】①[植] stoma ②[动] spiracle ③[冶] gas hole ④[建] air hole

【气浪】blast (of an explosion)

【气冷】[机] air cooling ◇～式发动机 air-cooled engine

【气力】effort; energy; strength: 费很大～ exert great efforts / 用尽～ with all one's energy

【气量】tolerance: ～大 large-minded; magnanimous / ～小 narrow-minded; petty

【气量表】gas meter

【气流】①[气] air current; airflow; airstream ②[语] breath ◇～纺纱 open-end spinning / ～干扰 interference in airflow / ～畸变 flow distortion

【气门】①（轮胎的充气活门）(air) valve of a tire ②[动] spiracle; stigma

【气门心】valve inside

【气密】airtight; gastight; gasproof ◇～接合 [机] airtight joint / ～试验 [航空] air seal test; leakage test

【气囊】①（鸟的）air sac ②（高空气球的）gasbag

【气恼】get angry; take offense; be ruffled; be peeved: 因某事～ be peeved over sth.

【气馁】become dejected; be discouraged; lose heart

【气派】manner; style; air: 东方～ oriental style / 学者～ scholarly manner

【气泡】air bubble; bubble

【气瓶】gas cylinder

【气魄】boldness of vision; breadth of spirit; daring: 有扭转乾坤的～ have the daring to reverse the course of events

【气枪】air gun; pneumatic gun: 玩具～ popgun

【气壳星】[天] shell star

【气球】balloon: 彩色～ colored balloon / 测风～ pilot balloon / 定高～ constant-level balloon ◇～飞翔运动 ballooning

【气色】complexion; color: ～不好 look pale; be off color / ～很好 have a rosy complexion; have a good color

【气势】momentum; imposing manner: ～磅礴 of great momentum; powerful; with a tremendous momentum / ～雄伟的建筑 imposing building

【气势汹汹】fierce; truculent; overbearing; in a threatening manner

【气数已尽】spell of good fortune has run out; his days are numbered

【气态】gaseous state

【气体】gas ◇～动力学 aerodynamics / ～发生器 gas generator / ～分离器 gas separator / ～力学 pneumatics / ～燃料 gaseous fuel / ～燃烧器 gas burner

【气田】[石油] gas field

【气筒】inflator; bicycle pump

【气头上】in a fit of anger

【气团】[气] air mass: 冷～ cold air mass ◇～变性 air-mass modification

【气吞山河】imbued with a spirit that can conquer mountains and rivers; full of daring

【气味】① smell; odor; flavor: ～难闻。The smell is awful. ②（性格；志趣）smack; taste: ～相投 congenial to each other; have the same tastes and temperament; be two of a kind

【气温】air temperature; atmospheric temperature

【气雾剂】aerosol

【气息】①（呼吸）breath: ～奄奄 at one's last gasp; at the point of death; like a person who is sinking fast ②（气味；色彩）flavor; smell: 具有强烈的生活～ have the rich flavor of life / 时代～ the spirit of the times

【气象】①（大气现象）meteorological phenomena ②（气象学）meteorology ③（情景）atmosphere; scene: 生气勃勃的新～ a new and dynamic atmosphere ◇～部门 meteorological department / ～工作者 a worker in meteorology / ～观测 meteorological observation / ～

火箭 meteorological rocket／～哨 weather post／～台 meteorological observatory／～图 meteorological map／～卫星 meteorological satellite; weather satellite／～学 meteorology／～预报 weather forecast／～员 weatherman／～站 weather station; climatological station

【气象万千】majestic in all its variety; things change in countless ways; spectacular

【气性】temperament; disposition

【气胸】[医] pneumothorax: 人工～ artificial pneumothorax

【气吁吁】panting; gasping for breath

【气虚】deficiency of vital energy

【气旋】cyclone: 反～ anticyclone

【气压】atmospheric pressure; barometric pressure: 高～ high pressure ◇～表 barometer

【气眼】① [建] air hole ② [冶] gas hole

【气焰】arrogance; bluster: ～嚣张 be swollen with arrogance

【气闸】air brake

【气质】①(个性特点) temperament; disposition ②(风格) qualities; makings: 革命英雄的～ qualities of a revolutionary hero

【气壮如牛】fierce as a bull

【气壮山河】full of power and grandeur; magnificent: 一篇～的宣言 a magnificent manifesto

汽

vapor; steam

【汽车】automobile; motor vehicle; car: 公共～ bus; autobus／微型～ baby car; midget car／小～ car; sedan／小型公共～ minibus／越野～ cross-country car／载重～ truck; heavy-duty truck／自卸～ dump truck／她乘～去上班。She goes to work by car.／他正在学开～。He is learning to drive. ○轿车 sedan; limousine／敞篷车 convertible; open car／活动车顶轿车 sliding roof limousine／吉普车 jeep／大客车 town bus／卡车 truck; lorry; wagon／跑车 roadster／自动装货车 automatic freight handling car／自动卸货车 self-discharging truck; tip lorry; tipper; dumper／邮政车 postal automobile; mail truck／冷藏车 refrigeration truck／救护车 ambulance／消防车 fire engine; fire brigade wagon; fire fighting truck／洒水车 road-sprinkler; sprinkling truck; watering lorry／扫街车 road-sweeper／扫雪车 snow-sweeper／救险车 wrecking truck; trouble car: breakdown lorry／修理车 shop truck／紧急修理车 emergency repair truck／油罐车 tank truck／牵引车 tractor／(半)拖(挂)车 (semi-)trailer／司机室 driver's cab／司机座 driver's seat／前座 front seat／后座 back seat／折座 folding seat／顶灯 dome lamp／大灯 headlight; headlamp／(车前)方向灯 front blinker; front direction indicator／(车后)方向灯 blinker; trafficator; direction light／停车灯 stop light; stop lamp／倒车灯 reversing light／离合器 clutch／头档 first gear; first speed gear／空档 neutral point; neutral position／倒车档 reverse gear／蓄电池 battery／排气管 exhaust pipe／减震器 shock absorber; vibration-damper; snubber／消声器 silencer; muffler／防滑链 non-skid chain; anti-skid chain／发动机 engine／汽化器式发动机 carburetor engine／四冲程发动机 four-stroke engine; four cycle engine／火花塞 sparking plug; spark plug／加油站 filling station; petrol station; service station／汽油泵 petrol pump; gasoline pump／油箱 oil box／柴油机 Diesel engine／车身 body／底盘 chassis／保险杠 bumper／挡泥板 mudguard; wing; fender／车门 car door／车门柄 car door handle／车门锁 car door lock／(车侧)脚踏板 running-board／带手摇曲柄的车窗 door window with crank handle／后窗 rear window／前三角玻璃通风窗 front quarter vent／前轮 front wheels／后轮 rear wheels／轮圈 wheel rim／轮胎 tire／内胎 inner tube／行李箱 boot; baggage compartment; luggage hold; trunk／备用轮 spare wheel／备用胎 spare tire／车号牌 number plate／油箱 (petrol, fuel) tank; gasoline tank／加油孔 tank filler sleeve (with cap)／风挡 windscreen; windshield／风挡刮水器 windscreen wiper／后视镜 driving mirror; rear vision mirror; back view mirror／通风器 ventilator／方向盘 steering wheel／变速杆 gear shift lever; gear lever／自动换挡;自动变速 fluid drive／手闸杆 hand-brake lever／喇叭 horn; hooter／加速踏板 accelerator pedal／制动踏板 brake pedal; foot brake; service brake／仪表板 dashboard; instrument panel; instrument board／仪表板灯 dash-light／起动机 starter／发火开关;电门 ignition switch／速度计附里程表〈路码表〉speedometer with mileage recorder／汽油表 petrol gauge; gasoline gauge

【汽车保险】automobile insurance

【汽车吊】truck crane

【汽车队】motor transport corps; fleet of cars; fleet of trucks

【汽车工业】auto industry; automobile industry; automotive industry
【汽车混凝土搅拌机】motor-truck concrete mixer
【汽车起重机】automobile crane
【汽车修配厂】motor repair shop
【汽车拉力赛】rally
【汽车驾驶赛】gymkhana
【汽车驾驶执照】driver's licence
【汽车俱乐部】automobile club
【汽车库】garage
【汽车旅馆】motel
【汽车赛】automobile racing
【汽车制造厂】automobile factory; motor works: 长春第一～ Changchun No. 1 Motor Vehicle Plant
【汽船】steamship; steamer
【汽锤】steam hammer
【汽灯】gas lamp
【汽笛】steam whistle; siren; hooter: 鸣～ sound a siren
【汽缸】cylinder ◇ ～体 cylinder body / ～组 cylinder block
【汽化】[物] vaporization
【汽化器】①[机] carburettor ②[化] vaporizer
【汽酒】light sparkling wine
【汽轮发电机】turbogenerator: 双水内冷～ turbogenerator with inner water-cooled stator and rotor
【汽轮机】steam turbine
【汽水】aerated water; soft drink; soda water
【汽艇】motorboat
【汽油】petrol; gasoline; gas: 航空～ aviation gasoline / 加氧～ oxygenated gasoline / 抗爆～ anti-knock gasoline / 凝固～ napalm / 人造～ synthetic petroleum ◇ ～弹 naplam / ～(发动)机 gasoline engine / ～加油站 gasoline filling station; filling station; petrol station / ～添加剂 gasoline additive

憩 rest: ～息 rest; have a rest
【憩室】diverticulum

讫 ①(完结) settled; completed: 付～ paid / 收～ received in full / 验～ checked; examined / 银货两～ delivered and paid ②(截止) end: 起～ the beginning and the end

迄 ①(直到) up to; till: ～今 up to now; to this day; so far ②(用于"未"或"无"前) so far; all along: ～无音信。We have received no information so far. / ～未见效。So far there hasn't been any result.
【迄今】up to now; to this day; to date; so far

qiā

掐 ①(用手指捏或截断) pinch; nip: ～花 nip off the flowers / 把香烟～了 stub out the cigarette ②(用手紧卡住) clutch: ～脖子 seize sb. by the throat / ～死 choke to death; throttle
【掐断】nip off; cut off: ～电线 disconnect the wire / ～水源 cut off the water supply
【掐算】count sth. no one's fingers
【掐头去尾】break off both ends; leave out the beginning and the end; remove the superfluous part

qiǎ

卡 ①(夹在中间不能活动) wedge; get stuck; be jammed; become tightly wedged: ～在冰中间的船 a ship jammed in the ice / 鱼刺～在他嗓子里。A fish bone sticks in his throat. ②(卡子) clip; fastener: 发～ hairpin ③(关卡) checkpost: 关～ checkpost
【卡刀】[机] swaging clamp
【卡具】[机] clamping apparatus; fixture
【卡壳】①[军] jamming of cartridge or shell case ②(遇到障碍) get stuck; be held up; have a temporary stoppage
【卡口灯泡】bayonet-socket bulb
【卡口灯头】bayonet socket
【卡盘】[机] chuck
【卡子】①(夹东西的器具) clip; fastener ②(关卡) checkpost

qià

髂
【髂动脉】common iliac artery; iliac artery
【髂骨】ilium; iliac bone
【髂窝】fossa iliaca

洽 ①(融洽) be in harmony; agree: 融～ be in harmony / 意见不～ have different opinions; not see eye to eye ②(接洽) consult; arrange with
【洽谈】make arrangements with; talk over with: ～贸易事宜 hold trade talks / 与某人～某事 talk sth. over with sb.

恰 ①(恰当) appropriate; proper ②(恰恰) just; exactly: ～到好处 just right / ～似 exactly like
【恰当】proper; suitable; fitting; appropriate: 采

取～的措施 adopt appropriate measures / 用词～ use proper words / 找～的人担任这项工作 try to get a suitable man for the job
【恰好】just right; as luck would have it: 这里的气候～宜于种植水稻。 The climate here is just right for planting rice.
【恰恰】just; exactly; precisely: ～相反 just the opposite; exactly the reverse; quite the contrary
【恰巧】by chance; fortunately; as chance would have it; happen to: 我～在路上遇见了她。 I met her by chance on the way.
【恰如其分】apt; appropriate; just right: ～的评价 an apt appraisal / ～地估计成绩和缺点 make an appropriate estimate of the achievements and shortcomings / 给予～的批评 give a balanced criticism

qiān

悭
【悭吝】stingy; miserly: ～人 miser

谦 modest: ～和 modest and amiable
【谦卑】humble; modest
【谦辞】self-depreciatory expression
【谦恭】modest and courteous; modest and polite: ～有礼 respectful and polite
【谦谦君子】①(旧指谦虚的人) a modest, self-disciplined gentleman ②(故作谦虚的人) a hypocritically modest person
【谦让】modestly decline
【谦受益,满招损】the modest receive benefit, while the conceited reap failure
【谦虚】①(虚心) modest; self-effacing: ～谨慎 modest and prudent ②(说谦虚的话) make modest remarks
【谦逊】modest; unassuming

牵 ①(拉) lead along; pull: ～马 lead a horse / ～着狗散步 take one's dog for a walk / 手～手 hand in hand ②(牵连) involve
【牵肠挂肚】feel deep anxiety about; be very worried about; be deeply concerned
【牵扯】involve; implicate; drag in: 你犯了错误不要～别人。 Don't involve other people in your mistake.
【牵掣】①(阻碍) hold up; impede: 互相～ hold each other up ②(牵制) pin down; check; contain
【牵动】affect; influence: ～全局 affect the situation as a whole
【牵挂】worry; care: 没有～ free from care
【牵累】①(因牵制受累) tie down: 受家务～ be tied down by household chores ②(连累) implicate; involve (in trouble)
【牵连】involve (in trouble); implicate; tie up with ◇～犯 implicated offender; act related to a crime
【牵牛花】(while-edged) morning glory
【牵牛星】[天] Altair
【牵强】forced (interpretation, etc.); farfetched: ～附会 draw a forced analogy; make a farfetched 〈irrelevant〉 comparison; give a strained interpretation; stretch of language
【牵涉】concern; drag in; involve: 营业上的这些变动～到所有股东的利益。 These changes in the business involve the interests of all owners. / 这项计划～到许多部门。 This project concerns many department.
【牵线】①(幕后操纵) pull strings; pull wires; control from behind the scenes: ～人 wirepuller ②(充当介绍人) act as go-between
【牵一发而动全身】pull one hair and the whole body is affected; a slight move in one part may affect the situation as a whole
【牵引】tow; draw; haul: 由内燃机车～ drawn by diesel engine ◇～车 tractor; tractor truck / ～犁 trailed plough / ～力 [物] traction force; traction; pulling force / ～炮 towed artillery / ～器[医] tractor / ～式滑翔机 towed glider
【牵制】pin down; tie up; check; contain: ～敌人 pin down the enemy ◇ ～行动 containing action / ～性攻击 diversionary attack

签 ①(签名) sign; autograph ②(签署意见) make brief comments on a document ③(竹签) bamboo slips used for divination or drawing lots: 抽～ draw lots ④(作为标志的小纸条) label; sticker: 标～ label; sticker / 航空邮～ air mail sticker / 书～ bookmarker ⑤(竹木制小细棍) a slender pointed piece of bamboo or wood: 牙～ tooth pick ⑥(粗粗地缝) tack: ～袖口 tack on a cuff
【签到】register one's attendance at a meeting or at an office; sign in ◇ ～簿 attendance book / ～处 sign-in desk
【签订】conclude and sign: ～合同 sign a contract / ～条约 sign a treaty / ～协定的各方 the parties signatory to the agreement / 为某事与某人～合同 sign a contract with sb. for sth.
【签发】sign and issue (a document, certificate, etc.)

【签名】sign one's name; autograph: ～盖章 sign and affix one's seal; set one's hand seal to / 亲笔～的照片 an autographed picture / 来宾～簿 visitors' book / 他忘了～。He's forgotten to sign his name.
【签收】sign after receiving sth.; sign to acknowledge the receipt of sth.; sign for: ～挂号信 sign for a registered letter
【签署】sign; affix; subscribe: ～联合公报 sign a joint *communiqué* / ～意见 write comments and sign one's name ◇ ～人 under-signed
【签约】sign a contract: 上星期～雇用了一百名工人。The firm signed on a hundred workmen last week. ◇ ～人 parties to a contract / ～日期 date of contract
【签证】visa; vise: 出境～ exit visa / 过境～ transit visa / 入境～ entry visa / 办理护照～手续 get one's passport visaed ◇ ～机关 visa-granting office / ～费 certificate fee
【签注】attach a slip of paper to a document with comments on it; write comments on a document
【签字】sign; affix one's signature: ～后立即生效 come into force upon signature / 中转～ sign a transfer / 他～把财产移交给他兄弟。He signed over the property to his brother. ◇ ～国 signatory state; signatory / ～认证 authentication / ～仪式 signing ceremony

铅 ①[化] lead (Pb): 重得象～。It's as heavy as lead. ②(铅笔心) lead; black lead
【铅白】[化] white lead
【铅版】[印] stereotype
【铅笔】pencil: 半硬～ medium pencil / 彩色～ colored pencil; crayon / 软～ soft pencil / 硬～ hard pencil ◇ ～刀 small knife for sharpening pencils; pen-knife / ～盒 pencil-case / ～画 pencil drawing / ～心 lead (in a pencil); black lead
【铅玻璃】lead glass
【铅垂线】[建] plumb line
【铅锤】[建] plummet; plumb (bob)
【铅封】lead sealing
【铅灰色】leaden (color): ～的云 leaden clouds
【铅球】[体] shot: 推～ shot put; putting the shot ◇ ～运动员 shot-putter
【铅丝】①(镀锌铁丝) galvanized wire ②[电] lead wire
【铅条】①[印] slug; lead ②(自动铅笔心) lead
【铅印】letterpress printing; relief printing; typographic printing; stereotype
【铅制品】leadwork
【铅中毒】lead poisoning; saturnism
【铅字】type; letter: 小号～ small type / 这本书用大号～印刷。The book is printed in large type. ◇ ～合金 type metal / ～面 typeface / ～盘 type case; letter board

千 ①(十个百) thousand: ～分之一 one thousandth; a thousandth / ～～万万 thousands upon thousands / 成～上万 by the thousands and tens of thousands ②(很多) a great amount of; a great number of; a thousand
【千变万化】ever changing; change in thousands of ways; volatile
【千不该万不该】really should not
【千层饼】multi-layer steamed bread
【千差万别】differ in thousands of ways; be in endless variety
【千疮百孔】one thousand boils and a hundred holes
【千锤百炼】①(多次斗争和考验) thoroughly tempered; well-seasoned; much-steeled ②(多次精细修改) be polished again and again; be revised and rewritten many times; be highly finished
【千刀万剐】thousand cuts and myriad pieces; hack sb. to pieces
【千叮咛万嘱咐】exhort sb. repeatedly
【千恩万谢】many thanks
【千儿八百】a thousand or slightly less
【千方百计】in a thousand and one ways; by every possible means; by hook or by crook; use every conceivable stratagem
【千分表】dial gauge; dial indicator
【千分尺】micrometer: 内径～ inside micrometer / 外径～ outside micrometer / 游标～ vernier micrometer
【千夫】numerous people: ～所指 be universally condemned; face a thousand accusing fingers
【千伏】kilovolt (Kv.) ◇ ～安 kilovolt-ampere (KVA)
【千古】①(长远的年代) through the ages; eternal; for all time: ～奇谈 strange stories of the ages / ～奇闻 a fantastic story; a forever strange tale / ～奇冤 the most appalling injustice through the ages / ～遗恨 eternal regret ②(用于挽联、花圈的上款)李先生～! Eternal repose to Mr. Li!
【千赫】kilohertz
【千斤】①(千斤顶) hoisting jack; jack ②(防止

齿轮倒转的装置) pawl ③(责任重) very heavy; weighty: ～重担 an exceptionally heavy load or responsibility

【千斤顶】[机] hoisting jack; jack: 油压～ hydraulic jack

【千金】①(很多的钱) a thousand pieces of gold; a lot of money: ～难买 not to be had even for 1000 pieces of gold; not to be bought with money ②[敬](女儿) daughter

【千军万马】like the charge of a powerful army which no force can stop; a large number of mounted and foot soldiers

【千钧一发】a hundredweight hanging by a hair: be in an extremely critical situation

【千卡】[物] kilocalorie (Kcal.)

【千克】kilogram (kg.)

【千里】a thousand *li*; a long distance: ～迢迢 a far distance; from afar; from a thousand *li* away / 沃野～ a vast expanse of fertile farmland ◇ ～马 a winged steed; a horse that covers a thousand *li* a day / ～眼 farsighted person

【千里光】[植] climbing groundsel

【千里送鹅毛】a goose feather sent from a thousand *li* away; a small gift that conveys great affection

【千里之堤,溃于蚁穴】one ant-hole may cause the collapse of a thousand *li* dike; slight negligence may lead to great disaster; a small leak will sink a great ship

【千里之行,始于足下】a thousand-*li* journey is started by taking the first step

【千米】kilometer (km.)

【千难万险】innumerable hazards and hardships

【千篇一律】stereotyped; following the same pattern; thousand pieces of the same tune

【千奇百怪】all kinds of strange things; an infinite variety of fantastic phenomena; a great variety of fantasies

【千秋】a thousand years; centuries: ～万代 throughout the ages; for all generations to come

【千日红】[植] globe amaranth

【千山万水】numerous mountains and rivers; a long and arduous journey

【千丝万缕】countless ties; a thousand and one links; interrelated in innumerable ways: 有着～的联系 have a thousand and one links; be tied in a hundred and one ways

【千头万绪】thousands of strands and loose ends; a multitude of things; a thousand and one things to attend to

【千瓦】kilowatt (KW) ◇ ～小时 kilowatt-hour (KWh)

【千万】①(一千个万) ten million; millions upon millions ②(务必) be sure so; do: 这事你～记着。Be sure to bear this in mind.

【千辛万苦】innumerable trials and tribulations; untold hardships: 历尽～ after innumerable hardships

【千言万语】thousands and thousands of words; innumerable words: ～说不尽 no words can express

【千载难逢】occurring only once in a thousand years; very rare; once in a blue moon: ～的机会 a golden opportunity; the chance of a lifetime

【千载一时】only once in a thousand years; the chance of a lifetime; a rare opportunity

【千真万确】absolutely true; very real

【千周】kilocycle (KC)

迁 ①(迁移) move: ～出 move out / ～人 move in / ～往郊区 move to the suburbs ②(转变) change: 事过境～。The matter is all over, and the situation has changed.

【迁都】move the capital to another place

【迁就】accommodate oneself to; yield to: ～姑息 excessively accommodating; overlenient / 过于～ excessively accommodating / 无原则的～ unprincipled accommodation

【迁居】move (house); change residence: ～外地 move away to another place

【迁怒】vent one's anger on sb. who's not to blame; take it out on sb.: ～他人 transfer one's anger to others

【迁徙】move; migrate; change one's residence: 有些鸟随季节～。Some birds migrate.

【迁延】delay; defer; procrastinate: ～时日 cause a long delay; become long-drawn-out

【迁移】move; remove; migrate

扦 a short slender pointed piece of metal, bamboo, etc.: 蜡～儿 candlestick / 竹～ bamboo spike

【扦插】[农] cuttage

【扦子】①(竹扦;金属扦) a slender pointed piece of metal, bamboo, etc. ②(取样器) a sharp-pointed metal tube used to extract samples of grains, etc. from sacks

钎 drill rod; drill steel; borer
【钎子】hammer drill; rock drill
阡 a footpath between fields, running north and south
【阡陌】crisscross footpaths between fields: ~纵横。 The crisscross paths on a farmland.
愆 fault; transgression: 前~ past faults
【愆期】pass the appointed time; delay

qián

潜 ①(隐藏的) latent; hidden: ~能 latent energy ②(秘密的) seeretly; stealthily; on the sly
【潜藏】hide; go into hiding
【潜伏】hide; conceal; lie low: ~的疾病 an insidious disease / ~特务 hidden enemy agent / ~着的危机 a latent crisis ◇ ~期 [医] incubation period / ~性病毒感染 latent-virus infection
【潜航】submerge ◇ ~深度 submerged depth / ~速度 submerged speed
【潜力】latent capacity; potential; potentiality: 充分发挥~ fully bring out latent potentialities; bring the potential into full play / 挖掘~ exploit potentialities; tap potentials / 有很大~ have great potentialities
【潜流】[地] undercurrent; underflow
【潜热】[物] latent heat
【潜入】①(偷偷地进入) slip into; sneak into; steal in ②(钻进水中) dive into; go under (water); submerge: 他~水中,救出溺水的小孩。 He dived into the water and rescued the drawning boy.
【潜水】①(在水下潜游) go under water; dive ② [地] phreatic water ◇ ~器 scuba / ~衣 diving suit / ~运动 underwater swimming / ~钟 diving bell
【潜水泵】submersible pump; immersible pump
【潜水员】diver; frogman ◇ ~病 caisson disease; compressed- air illness; decompression sickness
【潜水艇】submarine; U-boat
【潜台词】[剧] unspoken words in a play left to the understanding of the audience or reader
【潜逃】abscond: 携公款~ abscond with public money
【潜艇】submarine: 核~ nuclear-powered submarine ◇ ~探测器 submarine detector
【潜望镜】periscope
【潜心】(do sth.) with great concentration; devote oneself to sth.: ~研究 concentrate on studies
【潜行】①(在水面下行动) move under water ②(秘密行走) move stealthily; slink
【潜血】[医] occult blood ◇ ~试验 occult blood test
【潜移默化】exert a subtle influence on sb.'s character, thinking, etc.; imperceptibly influence; a silent transforming influence
【潜意识】the subconscious; subconsciousness
【潜影】[摄] latent image
【潜泳】underwater swimming: 屏气~ apnea swimming
【潜游运动】skin-diving
【潜在】latent; potential: ~的力量 latent power; potentially ◇ ~竞争 potential competition / ~效用 potential utility / ~需要 potential demand / ~资本 latent capital
前 ①(在正面的) front: ~部 front part / ~窗 front window / ~门 front door / ~院 front courtyard / 楼~ in front of the building / 舞台~ in front of the stage ②(向前) forward; ahead: 向~看 look forward / 勇往直~ go bravely forward; forge ahead dauntlessly ③(以前) ago; before; preceding: ~一阶段 the preceding stage / 解放~ before liberation / 日~ a few days ago; the other day / 三十年~ thirty years ago / 晚饭~ before supper / 战~ prewar ④(从前的) former; formerly: ~市长 former mayor; ex-mayor ⑤(次序在先的) first: ~六排座位 the first six rows of seats / ~三名 the first three places
【前半天】forenoon; morning
【前半夜】the first half of the night; from nightfall to midnight; from dusk till midnight
【前辈】senior (person); elder; the older generation: 革命~ revolutionaries of the older generation
【前臂】forearm
【前边】①(在前面) in front; ahead ②(从前) above; preceding
【前叉】(自行车的) front fork
【前车之覆,后车之鉴】the overturned cart ahead is a warning to the carts behind; take warning from previous failure
【前车之鉴】warning taken from the over turned cart ahead; lessons drawn from other's mistakes; a lesson from the failure of one's predecessor
【前程】future; prospect: ~远大 have brilliant

prospects; have excellent prospects / 锦绣～ a bright future; a rosy future

【前导】①(在前面引路) lead the way; march in front; precede ②(在前面引路的人) a person who leads the way; guide: 以仪仗队为～ with the guard of honor marching at the head

【前敌】front line ◇ ～委员会 front committee / ～总指挥 frontline commander-in-chief

【前度刘郎】a person who returns to a place he once abandoned

【前额】forehead

【前方】①(前面) ahead: 注视着～ look ahead; gaze ahead ②(前线) the front: 开赴～ be dispatched to the front / 支援～ support the front

【前锋】①(先头部队) vanguard ②[体] forward

【前夫】①(离了婚的) former husband; ex-husband ②(死去的) late husband

【前俯后仰】bend forward and backward; bowing forward and leaning backward

【前赴后继】advance wave upon wave

【前功尽弃】 all that has been achieved is spoiled; all one's previous efforts are wasted; all the former tasks have been thrown away

【前后】①(时间接近) around; about: 九点～ around 9 o'clock / 在 1989 年～ round about 1989 ②(自始至终) from start to finish; from beginning to end; altogether ③(前面和后面) in front and behind: ～受敌 be attacked by the enemy both front and back; be caught between two fires / ～左右 on all sides; all around

【前呼后拥】with many attendants crowding round; with a large retinue

【前脚】①(踏在前面的脚) the forward foot in a step ②(与"后脚"连用,表示时间靠近) no sooner ... than; the moment (when); hardly ... when: 你～走,她后脚就来了。 She arrived the moment you had left. (或: No sooner had you left than she arrived.)

【前襟】 the front part of a Chinese robe or jacket

【前进】 advance; go forward; forge ahead; make progress: 不断地～ make constant progress / 大踏步～ make big strides forward / 继续～ continue to make progress

【前景】①[摄] foreground ②(将要出现的景象) prospect; vista; perspective: 美好的～ good prospects; a bright future

【前臼齿】premolar teeth

【前倨后恭】first supercilious and then deferential; change from arrogance to humility

【前空翻】[杂技] forward somersault in the air

【前科】criminal record; pedigree: ～的证据 evidence of previous convictions ◇ ～犯 criminal with previous conviction; ex-convict; ex-prisoner; pedigree-man / ～记录 record of previous crime / ～罪 previous crime

【前例】precedent: 有～可援 have precedents to go by / 史无～ without precedent in history; unprecedented

【前列】front row; front rank; forefront; van: 站在斗争的～ stand in the forefront of the struggle

【前列腺】[生理] prostate (gland) ◇ ～肥大 hypertrophy of the prostate / ～疾病 prostatic disorders / ～素[药] prostaglandin / ～炎 prostatitis

【前轮】①(车辆的) front wheel ②(飞机的) nosewheel

【前门】front door

【前门拒虎,后门进狼】drive the tiger away from the front door and let a wolf in at the back; fend off one danger only to fall prey to another

【前面】①(位置靠前) in front; at the head; ahead: 在房子～ in front of the house / 走在队列～ march at the head of the column / 它正在你的～。 It's just in front of you. / 王先生在～。 Mr. Wang was in front. ②(次序靠前) above; preceding: ～的一章 the preceding chapter / ～提到的原则 the above-mentioned principle

【前磨牙】dentes premolares; premolar teeth

【前脑】[生理] forebrain

【前年】the year before last

【前怕狼,后怕虎】fear wolves ahead and tigers behind; be full of fears

【前排】front row: ～座位 front-row seats / 在主席台的～就座 be seated in the front row on the rostrum

【前判】former adjudication

【前仆后继】no sooner has one fallen than another steps into the breach

【前期】earlier stage; early days ◇ ～滚结帐目 brought forward account / ～损益 profit and loss for the previous period

【前妻】①(离了婚的) former wife; ex-wife ②(死去的) late wife

【前前后后】the whole story; the ins and outs; from front to rear: 一件事情的～ the ins and outs of a matter

【前桥】[汽车] front axle

【前驱】forerunner; precursor; pioneer
【前驱期】[医] prodromal stage
【前人】forefathers; predecessors: ～开路后人行。One generation opens the road upon which another generation travels. / ～栽树后人乘凉。One generation plants the trees under whose shade another generation rests.
【前任】predecessor: ～书记 former secretary / ～总统 ex-president / 他的～ his predecessor
【前日】the day before yesterday
【前哨】outpost; advance guard ◇ ～战 skirmish / ～阵地 outpost position
【前身】predecessor
【前审】former trial
【前世】previous existence; previous life
【前事不忘,后事之师】past experience, if not forgotten, is a guide for the future; the past is the prologue
【前思后想】think over again and again; turn over in one's mind
【前松后紧】be slack at the beginning and have to speed up towards the end
【前所未闻】never heard before
【前所未有】hitherto unknown; unprecedented: ～的盛况 an unprecedentedly grand occasion / 工业的～的发展 an unprecedented expansion in industry / 遇到～的困难 encounter greater difficulties than ever
【前台】①(舞台) proscenium ②(公开) (on) the stage
【前提】①[逻] premise: 大～ major premise / 小～ minor premise ②(必要的条件) prerequisite; presupposition: 必要的～ essential prerequisite / 安定团结是实现四个现代化的～。Stability and unity are a prerequisite to the four modernizations.
【前天】the day before yesterday: ～早晨 the morning before last
【前厅】antechamber; vestibule
【前庭】[生理] vestibule ◇ ～大腺 greater vestibular gland / ～炎 vestibulitis
【前条】preceding article
【前途】future; prospect: ～暗淡 prospects are dim / ～茫茫 have a bleak future; have gloomy prospects / ～无量 have boundless prospects / 他们的～似难预测。Their future seems very uncertain.
【前往】go to; leave for; proceed to: 经上海～东京 go to Tokyo via Shanghai / 已动身～罗马 have left for Rome
【前委】[简](前敌委员会) the front committee
【前卫】①[军] advance guard; vanguard ②[体] halfback: 左～ left halfback; left half
【前无古人】without parallel in history; unprecedented; have no predecessors
【前夕】eve: 第二次世界大战～ on the eve of the Second World War / 解放～ on the eve of liberation; shortly before liberation
【前线】front; frontline: 上～ go to the front / 远离～ far from the frontline
【前项】preceding paragraph; former term
【前言】preface; foreword; introduction
【前言不搭后语】utter words that do not hang together; talk incoherently; self-contradictory
【前沿】forward position ◇ ～阵地 forward position / ～指挥所 forward command post
【前仰后合】rock (with laughter): 笑得～ rock with laughter; shake with laughter
【前夜】eve
【前意识】preconscious
【前因后果】cause and effect; the entire process; the ways and wherefores
【前院】front courtyard
【前兆】omen; forewarning; premonition: 地震的～ warning signs 〈indications〉 of an earthquake
【前者】the former
【前震】foreshock
【前肢】fore limb; foreleg
【前置词】[语法] preposition ◇ ～宾语 prepositional object / ～短语 prepositional phrase
【前缀】prefix
【前奏】prelude ◇ ～曲[乐] prelude

乾

【乾坤】heaven and earth; the universe: 扭转～ bring about a radical change in the existing state of affairs; reverse the course of events

荨

【荨麻】nettle
【荨麻疹】nettle rash; urticaria

掮

【掮客】broker: 房地产～ real estate broker / 政治～ political broker ◇ ～佣金 brokerage; commission

虔

pious; sincere
【虔诚】pious; devout: ～的祷告 a devout prayer / ～的基督徒 a devout Christian / 她很～。She is pious.
【虔敬】reverent

黔

【黔驴技穷】the proverbial donkey in ancient Guizhou has exhausted its tricks; at one's wit's end

【黔驴之技】tricks not to be feared; cheap tricks; one's low skill is disclosed

钱

①(硬币) coin; cash: 古～ ancient coins ②(钱财) money; wealth; riches: 很多～ a lot of money / 缺～ be short of cash / 有～人家 wealthy family / 有～有势 both rich and influential / 挣～ make money / ～能通神。Money can move the gods. / 据说他很有～。They say he is made of money. / 我把～留在家里了。I've left my money at home. / 有～能使鬼推磨。Money makes the mare go. / 这个多少～? How much is this? ③(款项) fund; sum

【钱包】wallet; purse; moneybag

【钱币】coin

【钱财】wealth; money: 浪费～ waste of money / 拥有～ possess wealth / ～身外物。Money is not an inherent part of the human being.

【钱庄】private bank

钳

①(钳子) pincers; pliers; tongs: 火～ fire tongs; coal tongs / 克丝～ combination pliers / 老虎～ pincer pliers / 手～ hand vice ②(用钳子夹) grip; clamp

【钳工】①(工种) benchwork ②(工人) fitter

【钳形】pincerlike: ～攻势 a pincer movement; a two-pronged offensive / 形成～包围 form a pincerlike encirclement

【钳制】clamp down on; suppress: ～舆论 muzzle <gag> public opinion

【钳子】pliers; pincers; forceps

qiǎn

浅

①(跟"深"相对) shallow: ～盘 shallow dish / ～水 shallow water; shoal water / ～种 shallow sowing ②(浅显) simple; easy: 这课书很～。This text is very easy. ③(浅薄) superficial: ～薄的知识 superficial knowledge ④(感情不深厚) not intimate; not close: 交情很～ not on familiar terms ⑤(颜色淡) light: ～红 pale red / ～黄 pale yellow / ～蓝 light blue / ～绿 light green ⑥(时间短) not long in time: 相处的日子还～ have not been together for long

【浅薄】shallow; superficial; meager; limited: ～的体会 limited experience / ～的议论 shallow argument

【浅尝辄止】stop after getting a little knowledge of a subject or about sth.; be satisfied with a smattering of a subject; do not study further or deeper

【浅成岩】[地] hypabyssal rock

【浅而易见】easily understood

【浅耕】shallow ploughing

【浅海】[地] shallow sea; epeiric sea; epicontinental sea ◇ ～带 neritic zone / ～水域 the shallow waters along the coast / ～养殖 fish-farming in shallow marine water

【浅见】superficial view; humble opinion: 依我～ in my humble opinion

【浅近】simple; plain; easy to understand: ～的文字 simple language

【浅陋】shallow; meager; mean: ～不堪 very shallow and detestable / 学识～ have meager knowledge

【浅色】light color

【浅释】simple explanation

【浅水池】the shallow end of a swimming pool; shallow pool

【浅说】elementary introduction

【浅滩】shoal; shallows

【浅显】plain; easy to read and understand: ～的道理 a plain truth / ～的科普读物 simple popular scientific literature / ～易明 so shallow and clear that it is easy to understand

【浅易】simple and easy: ～读物 easy readings

遣

①(派遣;打发) send; dispatch; designate: 派～ dispatch / 调兵～将 dispatch officers and men; move troops; deploy forces ②(消除;发泄) dispel; expel: ～闷 dispel boredom / 无以自～ have no diversion / 消～ diversion; pastime

【遣愁】drove away melancholy

【遣返】repatriate: ～战俘 repatriate prisoners of war / 强迫～ forced repatriation / 自愿～ repatriation of one's own accord

【遣散】disband; dismiss; send away: ～费 termination pay; release pay

【遣送】send back; repatriate: ～出境 deport / ～回国 repatriate

谴

【谴责】condemn; denounce; censure; reprimand

qiàn

歉 ①(对不住人的心情) apology; sorry; ill at ease: 抱～ be sorry / 道～ offer an apology; make an apology; apologize / 迟复为～。I feel sorry for not giving you a quick reply.
【歉年】lean year
【歉收】crop failure; poor harvest: 因遭水灾而～ have a bad harvest due to natural inundation
【歉意】apology; regret: 表示～ offer an apology; express one's regret / 谨致～。Please accept my apologies.

茜
【茜草】[植] madder
【茜素染料】[化] alizarin dyes

堑 moat; chasm: 天～ natural chasm
【堑壕】trench; entrenchment: ～工事 entrenchment works ◇ ～战 trench warfare

欠 ①(未还) owe; be behind with: ～情 owe sb. a debt of gratitude; be indebted to sb. / ～条 IOU / ～债 owe a debt; run into debt; get into debt / ～租 be behind with the rent ②(不够;缺少) insufficient; not enough; lacking; wanting: ～佳 not good enough; not up to the mark / ～慎重 lack of prudence / 说话～考虑 speak without due consideration / 文字～通顺 The writing is not altogether grammatical ③(身体一部分稍微向上移动) raise slightly: ～脚儿 slightly raise one's heels ④(呵欠) yawn
【欠产】shortfall in output
【欠款】money that is owing; arrears; balance due; debt
【欠缺】①(不够) be deficient in; be short of ②(不够的地方) shortcoming; deficiency
【欠伸】stretch oneself and yawn
【欠身】raise oneself slightly; half rise from one's seat: ～倾听 bend one's body slightly forward to listen
【欠条】IOU; a bill signed in acknowledgement of debt
【欠妥】not proper: 措词～ not properly worded
【欠债】①(欠别人的钱) be in debt; run into debt ②(所欠的债) debt due; outstanding accounts: 还清～ clear off all one's debt
【欠帐】bills due; outstanding accounts
【欠资】[邮] postage due ◇ ～通知单 notice for postage due / ～信 postage-due letter / ～邮件 postage-due mails

芡 [植] Gorgan euryale
【芡粉】①(芡实的粉) the seed powder of Gorgon euryale ②(勾芡用的淀粉) any starch used in cooking
【芡实】Gorgon fruit

嵌 inlay; embed; set: ～花地面 a mosaic pavement
【嵌甲】ingrowing nail

纤 a rope for towing a boat; tow line: 拉～ track (a boat)
【纤夫】boat tracker
【纤路】towpath; towing path; track road
【纤绳】towline; towrope

qiāng

锵 [象] clang; gong

羌
【羌活】[植] notopterygium

蜣
【蜣螂】dung beetle

枪 ①(兵器) rifle; gun; firearm: 步～ rifle / 冲锋～ sub-machine gun / 机～ machine gun / 卡宾～ carbine; machine carbine / 手～ pistol / 信号～ flare pistol; ground signal projector / 开～ open fire; fire with rifle / 开了一～ fire a shot②(旧式兵器) spear: 红缨～ red-tasselled spear
【枪靶】target; mark
【枪把】the small of the stock; pistol grip
【枪毙】execute by shooting
【枪刺】bayonet
【枪弹】cartridge; bullet
【枪法】marksmanship: 这个战士～高明。This soldier is a crack shot.
【枪放下】[军](口令)Order arms!
【枪杆子】the barrel of a gun; gun; arms
【枪管】barrel (of a gun)
【枪机】rifle bolt
【枪架】rifle rack
【枪决】execute by shooting
【枪口】muzzle
【枪林弹雨】a hail of bullets; storm of shots and shells: 冒着～ braving the storm of shots and shells; under enemy's heavy fire / 在～中 amid hails of bullets; on the raging battlefield

【枪榴弹】rifle grenade
【枪炮】firearms; arms; guns
【枪杀】shoot dead; gun killing
【枪伤】bullet wound
【枪声】report of a gun; shot; crack: 听到～ heard gunshots
【枪手】①(射击手) marksman; gunner: 神～ an expert marksman; a crack shot ②[旧] spearman
【枪栓】rifle bolt
【枪膛】bore (of a gun)
【枪托】(rifle) butt; buttstock
【枪乌贼】[动] squid
【枪械】firearms
【枪眼】①(射击孔) embrasure; loophole ②(枪弹打的洞) bullet hole
【枪战】gunplay
【枪支】firearms ◇ ～弹药 firearms and ammunition

戗 ①(方向相对) in an opposite direction: ～风行船 sail against the wind ②(言语冲突) clash; be at loggerheads with

呛 choke: 吃饭吃～了 choke over one's food

镪

【镪水】strong acid: 硝～ nitric acid

腔 ①(动物身体内部空的部分) cavity: 鼻～ the nasal cavity / 口～ the oral cavity / 胸～ thoracic cavity / 满～热情 full of enthusiasm ②(话) speech: 不开～ keep mum / 答～ answer ③(乐曲的调子) tune; pitch: 高～ high pitched tune / 唱走了～儿 sing out of tune ④(说话的腔调) accent: 学生～ schoolboy talk; classroom tone of a schoolboy / 说话南～北调 speak with a mixture of accents
【腔肠动物】coelenterate
【腔调】①(曲调) tune ②(口音、语调) accent; intonation
【腔静脉】[生理] vena cava

qiáng

墙 wall: 围～ surrounding wall / 砖～ brick wall / 把画挂到～上。 Hang the picture on the wall. / 隔～有耳。 Walls have ear.
【墙板】wallboard
【墙报】wall newspaper
【墙壁】wall
【墙倒众人推】when a wall is about to collapse, everybody gives it a push; everybody hits a man who is down
【墙根】the foot of a wall
【墙角】a corner formed by two walls ◇ ～家具 corner furniture / ～石 cornerstone
【墙脚】①(墙根) the foot of a wall ②(基础) foundation: 挖～ cut the ground (from under sb.'s feet); undermine the foundation of
【墙裙】[建] dado
【墙手球】[体] handball
【墙头】the top of a wall ◇ ～草 grass on the top of a wall which sways with every wind; a person who bends with the wind / ～诗 wall poems
【墙网球】[体] rackets: 单打～ squash tennis
【墙纸】[建] wall paper

蔷

【蔷薇】rose ◇ ～科 the rose family

樯 mast: 帆～如林 a forest of masts

强 ①(力量大;程度高) strong; powerful: ～国 powerful nation; power / 能力很～ very capable / 身～体壮 strong and healthy / 责任心～ have a strong sense of responsibility ②(强迫) by force: ～取 take by force / ～令执行 arbitrarily give orders to carry out sth. ③(优越;好) better than; surpass; excel: 我们的生活条件一年比一年～。 Our living conditions are getting better each year. / 一代更比一代～。 Each generation surpasses the preceding one. ④(略多些) slightly more than; plus: 四分之一～ slightly more than one fourth
【强暴】①(强横凶暴) violent; brutal: ～的行为 act of violence ②(强暴的势力) ferocious adversary: 不畏～ defy brute force
【强大】big and powerful; powerful; formidable: ～动力 powerful motive force
【强盗】robber; bandit; pirate; highwayman: ～行为 banditry; robbery; piratical act ◇ ～逻辑 gangster logic / ～头子 gang boss; bandit chieftain / ～遇着贼打劫 robbers are plundered by thieves
【强的松】[药] prednisone
【强调】stress; emphasize; underline: ～指出 point out emphatically / 不适当地～情况特殊 lay undue stress on special circumstances
【强度】intensity; strength: 辐射～ radiation intensity / 钢的～ the strength of the steel / 抗震～ shock strength / 劳动～ the intensity of labor
【强渡】fight one's way across a river; force a

river
【强夺】grab
【强风】strong breeze
【强攻】take by storm; storm: ～敌人阵地 storm the enemy position
【强固】strong; solid: ～的工事 strong fortifications / ～的基础 a solid foundation
【强悍】intrepid; doughty; valiant
【强横】brutal and unreasonable; tyrannical; insolent; overbearing
【强化】strengthen; intensify; consolidate: ～人民的国家机器 strengthen the people's state apparatus
【强击机】attack plane
【强加】impose; force: 不要～于人。 Don't force your views on others.
【强奸】rape; violate: ～妇女 indecent assault on women / ～民意 defile public opinion ◇ ～犯 raper; rapist / ～罪 forcible rape; offense of rape; rape
【强碱】[化] alkali; strong base
【强健】strong and healthy: 体魄～ be physically strong; have a strong constitution
【强将手下无弱兵】there are no poor soldiers under a good general; there are no weak troops under a strong general
【强劲】powerful; forceful: ～的海风 a strong wind blowing from the sea
【强劳动力】able-bodied laborer
【强力霉素】[药] doxycycline
【强烈】strong; intense; violent: ～的仇恨 intense hatred / ～的对比 a striking contrast / ～的反应 strong reaction / ～的光线 forceful rays / ～的愿望 a strong desire / ～反对 strongly oppose / ～遣责 vehemently condemn; vigorously denounce / 味道～的调味品 strong seasoning / 他有～的责任感。 He has a strong sense of duty.
【强龙难压地头蛇】the mighty dragon is no match for the native serpent
【强买强卖】buy or sell under coercion
【强弩之末】an arrow at the end of its fight; a spent force
【强拍】[乐] strong beat; accented beat: 次～ subsidiary strong beat
【强取豪夺】rapacity
【强权】power; might: ～即公理 might is right / ～政治 power politics
【强盛】(of a country) powerful and prosperous
【强似】be better than; be superior to
【强酸】[化] strong acid
【强心剂】cardiac stimulant; cardiotonic
【强行】force: ～闯入 force one's way in / ～登陆 force a landing / ～通过一项议案 force through a bill / ～越狱 breakout
【强行军】forced march
【强硬】strong; tough; unyielding: 采取～态度 take an intransigent attitude / 措词～的声明 a strongly worded statement / 提出～抗议 lodge a strong protest / 用～的手段 with a high hand ◇ ～路线 tough line; hard line / ～派 hardliner
【强有力】strong; vigorous; forceful: 采取～的行动 take vigorous action
【强占】forcibly occupy; seize
【强震】strong shock ◇ ～区 meizoseismal area
【强制】force; compel; coerce: ～保险 compulsory insurance / ～措施 coercive measures / ～机关 institutions of coercion / ～接受业务 assigned risk / ～劳动 forced labor / ～手段 compulsory means; coercive measure / ～执行 enforce / ～性的命令 mandatory order / ～性法规 mandatory rules of law / ～性规定 mandatory provision / ～性条款 mandatory term / ～性制裁 mandatory sanction
【强中自有强中手】however strong you are, there's always someone stronger; diamond cut diamond
【强壮】strong; sturdy; robust: 他已六十多岁了,可身体还很～。 He is now well over sixty, but is still going strong. ◇ ～剂[药] roborant; tonic
【强子】[物] hadron

qiǎng

羟
【羟甲基】[化] hydroxyl (group) ◇ ～化物 hydroxylate

抢
① (抢劫) rob; loot: ～了某人的钱 rob a man of his money ② (抢夺) snatch; grab: 从某人手中把信～了过去 snatch away the letter from a person's hand ③ (抢先;争先) vie for; scramble for: ～球 scramble for the ball / ～干重活 vie with each other for the hardest job ④ (赶紧;突击) rush: ～收 rush in the harvest / ～运 rush-transport sth. ⑤ (刮掉或擦掉物体表面的一层) scrape; scratch: 磨剪子～菜刀 sharpen scissors and kitchen knives
【抢白】reprove or satirize sb. to his face
【抢地呼天】strike the head on the ground and call on Heaven

【抢渡】speedily cross (a river)
【抢夺】snatch; wrest; seize: ～胜利果实 seize the fruits of victory
【抢风行驶】tack
【抢购】rush to purchase: ～风潮 panic purchasing
【抢婚】marriage by capture
【抢劫】rob; loot; plunder: ～财物 plunder everything valuable; rob the people of their valuables / ～银行 rob the bank / ～一空 loot or rob to the last pin ◇ ～罪 crime of pillage; offense of robbery
【抢救】rescue; save; salvage: ～病人 give emergency treatment to a patient; rescue a patient / ～国家财产 save state property / ～遇险矿工 rescue trapped miners / ～无效。 All rescue measures proved ineffectual. ◇ ～费 salvage / ～工作 rescue work / ～组 rescue party
【抢时间】race against time
【抢收】rush in the harvest; get the harvest in quickly
【抢先】try to be the first to do sth.; anticipate; forestall; do sth. before others have a chance to
【抢险】rush to deal with an emergency ◇ ～队 emergency squad
【抢修】rush to repair; do rush repairs; make urgent repair on: ～公路和桥梁 make urgent repair on highway and bridges
【抢占】race to control; seize; grab: ～制高点 race to control a commanding point
【抢种】rush-planting

强 make an effort; strive: ～不知以为知 pretend to know what one does not know / ～作镇静 make an effort to appear composed; try hard to keep one's composure
【强逼】compel; force
【强辩】defend oneself by sophistry
【强词夺理】use lame arguments; resort to sophistry; reason fallaciously; argue irrationally
【强记】strain to memorize sth.; cram sth. into one's memory
【强迫】force; compel; coerce: ～包办婚姻 compulsory marriage on account of an arrangement / ～敌机降落 compel the enemy plane to land / ～卖淫 force harlotry / ～命令 resort to coercion and commandism / 不得不使用～手段 have to use force
【强求】insist on; impose: ～一律 insist on identity with; unify by force; forcibly unified
【强人所难】try to make sb. do sth. which he won't or can't
【强使】force; compel
【强笑】forced smile
【强颜欢笑】put on an air of cheerfulness; try to look happy when one is sad; a forced countenance with a smile

襁
【襁褓】swaddling clothes: ～中 be in one's infancy

qiàng

呛 irritate (respiratory organs): 辣椒味～鼻子 the smell of red pepper irritates the nose

戗 ①[建] prop ②(支撑) prop up; shore up: ～脊瓦挂钩 hip hook

qiāo

敲 knock; beat; strike: ～警钟 sound the alarm / ～门 knock at the door / 谁～门? Who is knocking?
【敲边鼓】speak or act to assist sb.; back sb. up
【敲打】①(在物体上面打) beat; rap; tap ②(用言语刺激) say sth. to irritate sb.: 冷言冷语～人 irritate people with sarcastic remarks
【敲骨吸髓】break the bones and suck the marrow; cruel, bloodsucking exploitation; suck the lifeblood
【敲锣打鼓】beat drums and strike gongs
【敲门砖】a brick picked up to knock on the door and thrown away when it has served its purpose; a stepping-stone to success
【敲诈】extort; blackmail; racketeer: ～勒索 blackmail and impose exactions on; extort and racketeer / ～钱财 extort money ◇ ～勒索者 racketeer / ～者 blackmailer
【敲竹杠】take advantage of sb.'s being in a weak position to overcharge him; fleece

悄
【悄悄】①(低声或无声) quietly; on the quiet; with as little noise as possible: ～地告诉 tell sb. quietly / ～离开 leave quietly ②(不让人知道) without being noticed: ～地拿走 take away sth. without being noticed

橇 sledge; sled; sleigh

跷 ①(抬起) lift up (a leg); hold up (a finger): ～着腿坐着 sit with one's legs crossed ②(踮

起) on tiptoe: ～着脚走路 walk on tiptoe ③(高跷) stilts
【跷蹊】fishy; dubious
【跷板】seesaw: 玩～ play on a seesaw
缲 hem with invisible stitches: 给手绢儿～边 hem a handkerchief
锹 spade: 每一～煤 each shovelful of coal / 挖一～深 dig a spade's depth; dig a spit deep
劁 geld; castrate: ～猪 castrate a pig

qiáo

翘 ①(抬起) raise (one's head) ②(翘棱) become warped: 木板～了。The board has warped.
【翘首】raise one's head and look: ～星空 look up at the starry sky / ～以待 raise one's head and look forward to
乔 ①(高大) tall ②(假扮) disguise
【乔木】arbor; tree
【乔其纱】[纺] georgette
【乔迁】(多用于祝贺) move to a better place; have a promotion: ～之喜。Best wishes for your new home.
【乔装】disguise; dress up: ～男人 disguise oneself as a man
荞
【荞麦】buckwheat: ～面 buckwheat flour
桥 bridge: 高架～ viaduct / 公路～ highway bridge / 弓弦式桁架～ bow-string girder bridge / 拱～ arch bridge / 桁架～ truss bridge / 跨线～ overline bridge / 平～ level bridge; flat bridge / 人行～ foot bridge; pedestrian bridge / 双层～ double-deck bridge / 双线双层铁路公路两用～ double-deck, double-track rail and highway bridge / 铁路～ railway bridge / 铁路公路两用～ combined bridge / 悬索～ suspension bridge / 引～ approach / 预应力混凝土～ pre-stressed concrete bridge / 在河上架～ bridge the river; put up a bridge across the river
【桥边人行道】footpath; sidewalk
【桥洞】bridge opening
【桥墩】(bridge) pier: 水下～ underwater foundation ◇ ～破冰构造 ice guard
【桥拱】bridge arch
【桥孔】bridge opening
【桥栏杆】bridge railings
【桥梁】bridge: ～工程 bridge work / ～工程队 bridge-gang / ～工事 bridge work / ～构架 bridge framework / ～跨度 bridge span / 起～作用 play the role of a bridge; serve as a link
【桥门】portal
【桥面】road of bridge; bridge floor
【桥牌】bridge (a card game): 打～ play bridge
【桥式起重机】bridge crane; overhead travelling crane
【桥塔】bridge tower
【桥台】abutment
【桥头】either end of a bridge
【桥头堡】①[军] bridgehead ②[建] bridge tower
侨 ①(侨居) live abroad ②(侨民) a person living abroad: 华～ overseas Chinese; Chinese nationals residing abroad / 外～ foreign residents; aliens
【侨胞】countrymen residing abroad
【侨汇】overseas remittance
【侨居】live abroad; emigration ◇ ～国 country of residence
【侨眷】relatives of nationals living abroad: 华侨及～ overseas Chinese and their relatives
【侨民】a national of a particular country residing abroad ◇ ～法 alien act
【侨务】affairs concerning nationals living abroad
憔
【憔悴】①(瘦弱;面色不好看) wan and sallow; thin and pallid: 面容～ haggard face ②(用于花木) withered: 花已显得～。The flowers are withering away.
瞧 look; see; glance at: ～亲戚 visit 〈see〉 a relative; call on a relative / ～书 read a book / ～一～ take a look at; have a look at / 东～西～ look about
【瞧病】①(就医) see a doctor; consult a doctor ②(诊病) see a patient; examine a patient
【瞧不起】look down upon; hold in contempt; look down one's nose at
【瞧不上眼】consider beneath one's notice; turn one's nose up at; not worth so much a look at; not at all appealing to the eye; beneath notice
【瞧得起】think much of; look up to; see much in
【瞧见】see; catch sight of

qiǎo

悄 quiet; silent
【悄然】①(忧愁) sorrowfully; sadly: ～泪下 shed tears in sorrow; shed sad tears ②(寂静无声) quietly; softly: ～离去 leave quietly
【悄声】quietly; in a low voice

巧 ①(灵巧;技术高明) skillful; ingenious; clever: 手～ a dab hand; a clever hand; (指人) clever with one's hands; dexterous / 嘴～ silver-tongued; have a nimble tongue ②(虚浮不实) cunning; deceitful; artful: ～言 cunning words; deceitful talk ③(恰好;恰巧) opportunely; coincidentally; as it happens; as luck would have it: 来得真～ have come at a most opportune moment / 偏～ as luck would have it
【巧夺天工】wonderful workmanship excelling nature; superb craftsmanship excelling nature
【巧妇难为无米之炊】the cleverest housewife can't cook a meal without rice; one can't make bricks without straw; you cannot make an omelet without breaking egg
【巧干】work ingeniously; do sth. in a clever way
【巧合】coincidence: 意外的～ curious coincidence / 真是～! What a happy coincidence!
【巧计】clever device; artful scheme
【巧匠】clever artisan; skilled workman
【巧克力】chocolate: 果仁～ nut chocolate / 酒心～ whisky heart chocolate / 一杯～饮料 a cup of chocolate / 一盒～糖 a box of chocolates ◇ ～饼干 chocolate cream biscuit / ～豆 marble chocolate / ～壶 chocolate pot
【巧立名目】concoct various pretexts; invent all sorts of names
【巧妙】ingenious; clever: ～的手段 a clever move / ～的战术 ingenious tactics
【巧取豪夺】extort by trick or by force; grab through deceit or by force; take by art and plunder by force
【巧舌如簧】have a glib tongue
【巧事】coincidence
【巧言令色】clever talk and an ingratiating manner
【巧遇】chance encounter

qiào

窍 ①(孔洞) aperture ②(窍门) a key to sth.: 诀～ knack; trick of a trade
【窍门】key (to a problem); knack: 找～ try to find the key to a problem; try to get the knack of doing sth.

壳 shell; hard surface
【壳菜】mussel

撬 prize; pry: ～开箱子 prize open a box / ～锁 pick a lock
【撬杆】pinch bar
【撬杠】crowbar

翘 stick up; hold up; bend upwards; turn upwards
【翘尾巴】be cocky; get stuck-up

鞘 sheath; scabbard
【鞘翅】[动] elytrum ◇ ～目 coleoptera

峭 ①(山势又高又陡) high and steep; precipitous ②(严厉) severe; stern
【峭拔】①(高而陡) high and steep ②(文笔雄健) vigorous: 笔锋～ have a vigorous style of writing
【峭壁】cliff; precipice; steep
【峭立】rise steeply

俏 ①(俊俏) pretty; smart; good-looking; handsome: 打扮得真～ be smartly dressed ②(销路好) sell well; be in great demand: ～货 goods in great demand / 这些商品很走～。These goods are selling well.
【俏皮】①(俊俏) good-looking; smart ②(活泼风趣) lively and delightful; witty
【俏皮话】①(歇后语) witty remark; witticism; wisecrack ②(讽刺的话) sarcastic remark

qiē

切 ①(用刀分开) cut; slice: ～菜 cut up vegetables / ～成两半 cut into halves; cut in two / ～成许多块 cut in pieces; cut into pieces / 用快刀～肉 cut meat with a sharp knifc ②[数] tangency
【切变】[物] shear: 风～ [气] wind shear
【切除】[医] excision; resection: ～扁桃腺 removal of tonsils / ～脂肪瘤 the resection 〈removal〉 of a lipoma / 部分～ partial excision / 全～ total excision
【切磋】learn from each other by exchanging views; compare notes
【切点】[数] point of tangency; point of contact
【切断】cut off: ～敌人后路 cut off the enemy's

retreat / ～电源 cut off the electricity supply / ～水源 cut off the water supply
【切分音】syncopation
【切割】cut: ～金属 cut through metal◇ ～机 cutting machine; cutter / ～线 line of cut
【切开】①(用刀子切割开) cut open; cut apart: ～西瓜 cut open a water melon ②[医] incision: ～引流 incision and drainage
【切块】[食品] stripping and slicing
【切面】①[数] tangent plane ②(剖面) section ③(面条) cut noodles; machine-made noodles
【切片】①(切成片) cut into slices ②[医] section: ～检查 cut sections (of organic tissues) for microscopic examination
【切片机】①(切片的机器) slicer ②[医] microtome
【切丝】cut into shreds; shred
【切线】[数] tangent (line)
【切碎】cut up: 厨师把肉～做肉饼。The cook cut up the meat for the pie.
【切削】[机] cutting: 粗～ rough cut / 高速～ high-speed cutting / 金属～ metal cutting
【切牙】incisor; incisor teeth
【切纸机】paper cutting machine; paper cutter

qié

茄 eggplant; aubergine
【茄子】eggplant; aubergine

qiě

且 ①(暂且;姑且) just; for the time being: ～等一会儿 just wait a while ②(经久) for a long time: 这种鞋～穿呢。These shoes will last long. ③(尚且) even: 死～不惧,何畏困难! Even death holds no fears for us, to say nothing of difficulties. ④(并且) both ... and ...; as well (as): 既高～大 both tall and heavy set; both high and wide / 既智～勇 be both brave and clever
【且慢】wait a moment; not go or do so soon: ～高兴! Don't rejoice too soon! / ～,听我把话说完 Wait a minute, let me finish what I have to say.
【且…且…】while; as: ～战～退 carry on the fight while beating a retreat / 他们～说～笑。They laughed as they talked.

qiè

妾 concubine
锲 carve; engrave
【锲而不舍】keep on carving unflaggingly work with perseverance
挈 ①(带领) take along: ～眷 take one's family along ②(举;提) lift; raise; take up: 提纲～领 hold a net by the headrope or a coat by the collar; concentrate on the main points
怯 timid; cowardly; nervous
【怯场】have stage fright
【怯懦】timid and overcautious
【怯弱】timid and weak-willed
【怯生】shy with strangers
【怯阵】①(临阵胆怯) feel nervous when going into battle; be battle-shy ②(怯场) have stage fright
惬 be satisfied
【惬意】be pleased; be satisfied
切 ①(合;符合) correspond to; be close to: 不～实际 not correspond to reality; unrealistic; impractical ②(迫切;急切) eager; anxious: 回国心～ be anxious to return to one's country ③(切实;千万要) be sure to: ～不可自以为是 one should never be presumptuous and opinionated / ～勿悲观 hope for the best / ～勿迟延 be sure not to delay
【切齿】gnash one's teeth: ～痛恨 gnash one's teeth in hatred / ～之仇 bitter hatred that gnashes the teeth
【切肤之痛】pain of cutting one's body / keenly-felt pain
【切合】suit; fit in with; correspond with: ～实际 be geared to actual circumstances / ～需要 fit in with the needs
【切记】be sure to keep in mind; must always remember
【切忌】must guard against; avoid by all means: ～生冷 cold and raw food strictly forbidden / ～饮酒过度 by all means avoid excessive drinking / ～主观片面。Be careful to avoid being subjective and one-sided.
【切脉】[中医] feel the pulse
【切切】①(千万;务必) be absolutely sure to: ～不可过份 be absolutely sure not go too far / ～不可骄傲。Be sure not to become conceited. (或: Guard against arrogance by every means.) ②(用于布告、条令等末尾)～此布。This proclamation is hereby issued in all sincerity and earnestness. / ～此令。This order is to be observed. ③(声音细小) in small voice: ～私语 a

private talk in small voice
【切身】①(跟自己有密切关系的) of immediate concern to oneself: ～利益 one's immediate or vital interests ②(亲身) personal: ～体会 personal understanding; intimate knowledge / ～体验 one's own experience
【切实】①(切合实际) feasible; practical; realistic: ～可行的计划 a feasible plan; a realistic plan / ～有效的办法 practical and effective measures ②(实实在在) conscientiously; earnestly: ～改正错误 correct one's mistakes in real earnest / 切切实实地工作 do one's job conscientiously
【切题】keep to the point; be relevant to the subject; be pertinent to the subject
【切勿】be sure not to: ～倒置 do not turn over; keep upright / ～干燥 keep wet / ～平放 not to be laid flat / ～受潮 keep dry
【切中要害】hit the mark; strike home

窃 ①(偷) steal; pilfer: 扒～ do picking and stealing / 失～ be stolen / 行～ steal; practice theft ②(偷偷地) secretly; surreptitiously; furtively: ～笑 laugh secretly; laugh up one's sleeve / ～～私议 exchange whispered comments / ～～私语 whisper to each other
【窃案】larceny; burglary
【窃盗险】burglary insurance
【窃国】usurp state power ◇ ～大盗 arch usurper of state power
【窃据】usurp; unjustly occupy: ～要职 usurp a high post; unjustly occupy a high post
【窃取】usurp; steal; grab: ～别人的劳动果实 grab the fruits of other people's labor / ～机密情报 steal secret information / ～某人名义 usurp one's name / ～胜利果实 steal the fruits of victory
【窃听】eavesdrop; wiretap; bug ◇ ～电话 tap phone / ～器 tapping device; listening-in device; bug
【窃贼】thief; burglar; pilferer

qīn

亲 ①(父母) parent: 双～ parents ②(血缘最接近) blood relation; next of kin: ～兄弟 blood brother ③(有亲属关系) relative; kinsfolk: 近～ close relative; near kin / 远～ distant relative ④(婚姻) marriage; match: 成～ be married / 定～ be engaged; be betrothed to / 说～ act as a matchmaker ⑤(新娘) bride: 迎～ (of the groom's family) send a party to escort the bride to the groom's home ⑥(关系近;感情好) close; intimate; dear: ～如手足 as dear to each other as brothers / ～如一家 as dear to each other as members of one family / 不分～疏 regardless of close or distant relationship ⑦(亲自) in person; oneself ～临出事现场 visit the scene of the accident in person ⑧(用嘴唇接触) kiss ⑨(亲善;赞成) pro-: ～华 pro- Chinese / ～美 pro-American / ～日 pro-Jap- anese / ～苏 pro-Soviet
【亲爱】dear; beloved: ～的同志们 dear comrades / ～的祖国 one's beloved motherland
【亲本】[生] parent: 轮回～ recurrent parent
【亲笔】①(亲笔写) in one's own handwriting; in one's own hand ②(亲笔写的字) one's own handwriting ◇ ～签名 one's own signature; autograph / ～信 a personal, handwritten message; an autograph letter / ～遗嘱 holographic will
【亲代】[生] parental generation
【亲等】degree; degree of kinship; degree of relationship
【亲和力】[化] affinity
【亲近】be close to; be on intimate terms with: 我们之间很～。We are on intimate terms.
【亲眷】one's relatives
【亲口】(say sth.) personally: ～答应 make a promise personally
【亲密】close; intimate: ～无间 be on very intimate terms with each other
【亲昵】very intimate: ～的称呼 an affectionate form of address
【亲戚】relative; kinsman; kinswoman; kinsfolk
【亲切】cordial; kind: ～的关怀 kind attention; loving care / ～的教导 kind guidance / ～的谈话 a cordial conversation
【亲热】affectionate; intimate; warmhearted
【亲人】①(直系亲属或配偶) one's parents, spouse, children, etc.; one's family members ②(关系密切、感情深厚的人) dear ones; those dear to one
【亲如手足】as close as brothers
【亲善】goodwill (between countries)
【亲上加亲】cement old ties by marriage as marriage between cousins, etc.
【亲身】personal; firsthand: ～经历 personal experience; firsthand experience
【亲生】one's own (children, parents): ～父母 one's own parents
【亲事】marriage
【亲手】with one's own hands; personally; one-

self: ～交付 deliver sth. personally / 你～做一做。Do it yourself.

【亲属】kinsfolk; relatives ◇～关系 kinship

【亲嗣关系】filiation

【亲痛仇快】sadden one's own people and gladden the enemy; grieve one's friends and gladden one's enemies

【亲王】prince

【亲信】trusted follower

【亲兄弟,明算帐】even brothers keep careful accounts

【亲眼】with one's own eyes; personally: 我～看见的。I saw it with my own eyes. (或: I was an eyewitness to that.)

【亲友】relatives and friends; kith and kin

【亲自】personally; in person; oneself: ～拜访 make a personal call / ～参加 take part in person / ～动手 personally take a hand in the work; do the job oneself / 她～操持家务。She runs the household herself.

【亲族】members of the same clan

【亲嘴】kiss

钦 ①(敬重) admire; respect ②(皇帝亲自) by the emperor himself: ～赐 granted by an emperor / ～定 (of a book, etc.) made by imperial order

【钦差】imperial envoy; imperial commissioner

【钦敬】admire and respect

【钦佩】admire; esteem: 表示～ express admiration for / 使我们十分～ command our great admiration

侵 ①(侵入) invade; intrude into; infringe upon: 入～之敌 invading enemy; invaders ②(接近) approaching: ～晓 approaching daybreak

【侵犯】encroach on; infringe upon; violate: ～版权 infringe a copyright / ～领土和主权 violate a country's territorial integrity and sovereignty / ～人权 infringe upon human rights / ～商标权 infringement of trade marks / ～所有权 infringement of title / ～专利权 infringement of patent; piracy

【侵害】encroach on; make inroads on: ～公民权利的行为 infringement of the rights of citizens / ～某人的权利 aggression upon one's right / 防止蝗虫～农作物 prevent the inroads of locusts on the crops / 减少风沙的～ reduce encroachments by sandstorms

【侵略】aggression; invasion: ～别国 commit aggression against another country / ～成性 be aggressive by nature ◇～国 aggressor (nation) / ～军 aggressor troops; invading army / ～行为 act of aggression / ～战争 war of aggression / ～者 aggressor; invader / ～罪 crime of aggressive war

【侵权】tort ◇～人 infringer / ～行为 act of tort; infringement act / ～责任 liability for tort

【侵扰】invade and harass: ～边境 harass a country's frontiers; make border raids

【侵入】invade; intrude into; made incursions into: ～领海 intrude into a country's territorial waters / ～领空 intrude into a country's air space ◇～住宅 house-breaking

【侵蚀】corrode; erode: 风雨的～ erosion by wind and rain / 土壤～ soil erosion / 抵制资产阶级思想的～ resist the corrosive influence of bourgeois ideology

【侵吞】①(暗中非法占有) embezzle; peculate: ～财产 annexation of property / ～公款 embezzle public funds / ～社会财富 appropriate social property / ～税款 embezzlement of tax funds ②(用武力吞并) swallow up; annex: ～别国领土 annex another country's territory

【侵袭】make inroads on; invade and attack; hit: 台风～沿海地区。The typhoon hit the coastal areas.

【侵占】invade and occupy; seize: ～别国领土 invade and occupy another country's territory / ～财产 conversion of property / ～公产 purpresture / ～他人的土地 encroach upon another man's land

qín

秦

【秦艽】[植] large-leaved gentian

【秦晋之好】inter marriage

【秦晋之交】relation of two states; to form a marriage alliance

勤 ①(勤劳) diligent; industrious; hardworking: ～学苦练 study diligently and train hard ②(次数多;经常) often; frequently; regularly: 她来得最～。She comes here more often than any other. / 夏季雨水～。Rain is frequent in summer ③(勤务) duty; attendance: 考～ check on work attendance / 内～ office work / 外～ field work / 值～ be on duty

【勤奋】diligent; assiduous; industrious: 学习～ be diligent in one's studies

【勤工俭学】part-work and part-study system;

work-study program
【勤俭】hardworking and thrifty: ～持家 be industrious and thrifty in managing a household／～建国 build up the country through thrift and hard work／办任何事情都要～。Be thrifty and hardworking in whatever undertaking you are to start on.
【勤恳】diligent and conscientious; earnestly and assiduously: 一个非常～的工人 a most painstaking worker
【勤快】diligent; hardworking
【勤劳】diligent; industrious; hardworking; industrious and courageous: ～勇敢 brave and industrious／～致富 achieving prosperity through industrious work
【勤勉】diligent; assiduous: ～好学 diligent and eager to learn
【勤务】duty; service ◇ ～兵 orderly／～员 odd-jobman; servant
【勤有功,嬉无益】Reward lies ahead of diligence, but nothing is gained by indolence.
【勤杂工】odd-jobman; handyman
【勤杂人员】personnel regularly doing certain odd jobs; odd-jobmen

芹
【芹菜】celery

禽 birds: 家～ (domestic) fowls; poultry／鸣～ song birds
【禽兽】birds and beasts: 衣冠～ a beast in human clothing／～行为 brutish acts; bestial acts

擒 capture; catch; seize: 生～ capture alive／～贼先～王。To catch bandits, first catch the ringleader.
【擒拿】arrest; catch
【擒纵机】[机] escapement
【擒纵轮】[机] escape wheel
【擒纵自如】arrest and release at will; in perfect control of situation

噙 hold in the mouth or the eyes: ～着烟袋 hold a pipe between one's lips／～着眼泪 eyes brimming with tears

琴 a general name for certain musical instruments: 钢～ piano／口～ harmonica／小提～ violin
【琴凳】music stool
【琴键】key (on a musical instrument)
【琴马】[乐] bridge (of a stringed instrument)
【琴鸟】lyrebird
【琴师】music master; accompanist
【琴弦】string (of a musical instrument)

qǐn

寝 ①(睡) sleep: 废～忘食 (so absorbed or occupied as to) forget about eating and sleeping／入～ fall asleep ②(卧室) bedroom: 就～ go to bed; turn in; retire to bed ③(寝宫) coffin chamber: 陵～ imperial burial place; mausoleum
【寝具】bedding
【寝食】sleeping and eating: ～不安 feel uneasy even when eating and sleeping; be worried waking or sleeping
【寝室】bedroom; dormitory

qìn

沁 ooze; seep; exude: 汗珠～了出来。Beads of perspiration oozed out.
【沁人心脾】gladdening the heart and refreshing the mind; mentally refreshing; refreshing

qīng

青 ①(青色) blue or green: ～椒 green pepper／～天 blue sky ②(黑色) black: ～布 black cloth ③(青草) green grass: 踏～ walk on the green grass; go for an outing in early spring ④(未成熟的庄稼) young crops: 看～ keep watch on the ripening crops ⑤(年轻) young (people): ～工 young workers
【青菜】①(蔬菜) green vegetables; greens ②(小白菜) Chinese cabbage
【青草】green grass
【青出于蓝而胜于蓝】indigo blue is extracted from the indigo plant but is bluer than the plant it comes from; the pupil surpasses the master; the pupil excels the teacher
【青春】youth; youthfulness: ～的活力 youthful vigor; youthful energy／充满着～的活力 full of youth and vigor／焕发了革命～ regain one's revolutionary vigor
【青春期】puberty
【青瓷】celadon (ware)
【青葱】verdant; fresh green: ～的山林 green wooded mountain
【青翠】verdant; fresh and green
【青豆】green soya bean
【青光眼】glaucoma
【青果】white canarytree fruit
【青红皂白】wheat and chaff; right and wrong:

不分～ make no distinction between right and wrong
【青花瓷】blue and white porcelain
【青黄不接】when the new crop is still in the blade and the old one is all consumed; temporary shortage
【青筋】blue veins
【青金石】lapis lazuli
【青稞】highland barley
【青睐】favor; good graces: 获得某人的～ find favor in sb.'s eyes; be in sb.'s good graces
【青绿】dark green ◇ ～山水 traditional landscape painting characterized by the prominence of blue and green color
【青梅】green plum
【青梅竹马】the games of childhood; the period when a boy and a girl grew up together
【青霉素】penicillin: ～眼药水 penicillin eye drops
【青面獠牙】green-faced and long-toothed; terrifying in appearance: 露出～的凶相 reveal the ferocious features of an ogre
【青苗】young crops; green shoots of (food) grains
【青年】youth; young people: ～人 young people; youth / ～时代 one's youth / ～学生 young students; student youth ◇ ～工作 youth work / ～节 Youth Day (May 4) / ～运动 youth movement
【青年会】[基督教] Young Men's Christian Association ◇ 女～ Young Women's Christian Association
【青纱帐】the green curtain of tall crops
【青山】green hill: 留得～在, 不愁没柴烧。As long as the green hills are there, one need not worry about firewood.
【青少年】young boys and girls; teen-agers; youngsters ◇ ～犯 young-offender / ～感化院 borstal
【青史】annals of history: 永垂～ go down in the annals of history
【青饲料】greenfeed; green fodder
【青松】pine
【青蒜】garlic sprouts
【青苔】moss
【青天】①(蓝天) blue sky ②[旧](清官) a just judge; an upright magistrate
【青天霹雳】a bolt from the blue
【青铜】bronze ◇ ～器 bronze ware / ～时代 the Bronze Age
【青蛙】frog
【青虾】freshwater shrimp
【青葙】[植] feather cockscomb ◇ ～子[中药] the seed of feather cockscomb
【青杨】Cathay poplar
【青鱼】black carp
【青玉】gray jade
【青云】high official position: ～直上 rapid advancement in one's career; meteoric rise; soar higher and higher
【青砖】blue bricks
【青贮】[农] ensiling ◇ ～窖 silo / ～饲料 ensilage; silage / ～作物 silo crop; silage crop

清 ①(纯净没有混杂的东西) unmixed; clear: ～水 clear water / ～汤 clear soup / 这水很～。The water is quite clear. ②(寂静) quiet: ～静 quiet ③(清楚) distinct; clarified: 查～事实 check up on the facts / 分～ make a clear distinction / 数不～ countless / 说不～ hard to explain / 问～底细 make sure of every detail; get to the bottom of the matter ④(一点不留) completely; thoroughly: 把帐还～ pay up what one owes / 扫～障碍 sweep away the obstacles completely ⑤(使组织纯洁) clean up ⑥(还清;结清) settle; clear up: ～帐 clear up an account ⑦(点验) count: ～一～行李的件数 count the pieces of luggage and see how many there are
【清白】pure; clean; stainless: ～无辜 innocent / 历史～ have a clean personal record ◇ ～人家 a decent family
【清仓】make an inventory of warehouses; make an inventory of the stock in the storehouse
【清册】inventory; detailed list: 材料～ detailed list of materials / 固定资产～ an inventory of fixed assets
【清茶】①(绿茶) green tea ②(只有茶水,没有点心) tea served without refreshments
【清查】①(查对) check: ～户口 check on residents; check residence cards / ～帐目 check the acounts ②(查出) ferret out; uncover; comb out: ～反革命 ferret out counter- revolutionaries
【清偿】pay off; clear off: ～债务 pay off debts; clear off debts ◇ ～损失额 liquidated damages / ～责任 liability for satisfaction
【清唱】sing opera arias (without makeup and acting) ◇ ～剧[乐] oratorio
【清澈】limpid; clear: ～的池塘 a limpid pool / ～透明 as clear as crystal

【清晨】early morning
【清除】clear away; eliminate; get rid of; sweep away: ～不良工作作风 get rid of the bad working style / ～垃圾 clear away the rubbish / ～流毒 eliminate the pernicious influence / ～水沟 clear out drain / ～障碍 remove obstacles; clear away obstacles
【清楚】①(容易让人了解、辨认) clear; distinct: 把工作交代～ explain one's job clearly on handing it over / 发音～ a clear pronunciation / 一清二楚 (as) clear as daylight / 字迹～ written in a clear hand / 你～了吗? Are you clear about it? ②(不糊涂) clear; lucid: 头脑～ a clear head; clear-headed ③(了解) be clear about; understand
【清脆】clear and melodious: ～的歌声 clear and melodious singing
【清单】detailed list; detailed account: 货物～ a detailed list of goods; inventory
【清淡】①(不浓烈) light; weak; delicate: ～的花香 the delicate fragrance of flowers / 一杯～的绿茶 a cup of weak green tea ②(不油腻) not greasy or strongly flavored; light: ～的食物 light food ③(萧条) dull; slack: ～季 slack season / 生意～. Business is slack.
【清点】check; make an inventory; sort and count: ～货物 take stock / ～人数 check the number of people / ～物资 make an inventory of equipment and materials / ～战利品 check and sort out spoils of war
【清炖】boiled in clear soup: ～鸡 stewed chicken in clear soup
【清风】cool breeze; refreshing breeze: ～徐来。A cool breeze blows gently
【清高】aloof from politics and material pursuits; above politics and worldly interests: 自鸣～ profess to be above politics and worldly considerations
【清稿】fair copy; clean copy
【清官】honest and upright official: ～难断家务事。Even an upright official finds it hard to settle a family quarrel.
【清规】[佛教] monastic rules for Buddhists
【清规戒律】① [宗] regulations, taboos and commandments ②(规章制度) restrictions and fetters
【清寒】①(清贫) poor; in straitened circumstances: 家境～ come of an impoverished 〈poor〉 family ②(清朗而有寒意) cold and clear: 月色～ clear, cold moonlight
【清剿】clean up; suppress; eliminate: ～土匪 clean up bandits; suppress bandits
【清教徒】Puritan
【清洁】clean: 整齐～ clean and tidy / 注意～卫生 pay attention to sanitation and hygiene. ◇～队 cleaning squad / ～工人 sanitation worker; street cleaner
【清劲风】[气] fresh breeze
【清净】peace and quiet: 图～, 怕麻烦 seek peace and quiet and shirk taking any trouble
【清静】quiet; secluded: ～的地方 a quiet spot
【清朗】①(清澈晴朗) clear and bright: ～的天气 clear and bright weather ②(清楚响亮) clear; ringing; clear and resounding: ～的声音 clear voice
【清冷】①(微寒) chilly: 一个～的秋夜 a chilly autumn night ②(冷清) deserted; desolate
【清理】put in order; check up; clear; sort out: ～仓库 take stock; make an inventory of warehouse stocks / ～档案 put the archives in order; sort out documents / ～东西 put things in order / ～积案 clear the docket / ～物资 check up on equipment and materials / ～债权债务 settle claims and debts / ～债务 clear up debts / ～帐户 liquidation account / ～资产 liquidation assets / 把房间～～ put the room in order; clean up the room
【清廉】honest and upright; free from corruption
【清凉】cool and refreshing: ～饮料 cold drink; cooler ◇～油 cooling ointment; essential balm
【清零】zero clearing; minimum clearing
【清明】①(清澈明朗) clear and bright: 月色～ clear and bright moonlight ②(心里清楚而镇静) sober and calm: 神志～ having a clear mind ③(节气) Clear and Bright
【清贫】be poor: 家境～ be a person of scanty means
【清平世界】a peaceful and orderly world
【清漆】varnish: 透明～ clear varnish / 皱纹～ shrivel varnish
【清热药】antipyretic
【清嗓子】clear one's throat
【清扫战场】mop up enemy remnants on battle field
【清瘦】thin; lean; spare: 面容～ meager face
【清爽】①(清洁凉爽) fresh and cool; brisk; refreshing: 空气～ fresh air ②(轻松爽快) easy; light; relieved; relaxed: 心里～了一些 feel relaxed; be in an easier state of mind
【清水墙】[建] dry wall
【清算】①(彻底地计算) clear (accounts);

square ②(列举全部罪恶并做出处理) settle accounts; expose and criticize: ～罪行 expose and criticize the crimes ◇ ～人 receiver; liquidator / ～协定 clearing agreement / ～银行 clearing bank / ～帐户 clearing account / ～债务 settle a claim

【清谈】idle talk; empty talk: ～误国 pure theories and talks without action will get the nation into trouble

【清汤】clear soup; light soup

【清晰】distinct; clear: 发音～ pronounce distinctly / 轮廓～可见 the outlines of sth. are clearly discernible

【清晰度】①(电视的) clarity ②(传声的) articulation

【清洗】①(洗净) rinse; wash; clean: ～炊具 clean cooking utensils / ～口腔 rinse the mouth ②(清除) purge; comb out: ～坏分子 comb out the bad elements

【清闲】at leisure; idle: 过～的生活 live a leisurely life; enjoy leisure

【清香】delicate fragrance; faint scent; refreshing fragrance: ～扑鼻 a sweet scent assails the nostrils

【清新】pure and fresh; fresh: 空气～ pure and fresh air; crisp air

【清醒】①(清楚;明白) clear-headed; sober: ～的估计 clear-headed appraisal / 保持～的头脑 keep a clear 〈cool〉 head; keep sober-minded ②(神志恢复正常) come to; come round; regain consciousness: 病人～过来了。 The patient has come to.

【清秀】delicate and pretty: 面貌～ of fine, delicate features / 山清水秀 beautiful landscape

【清雅】elegant; refined: 风格～ in an elegant style

【清样】[印] final proof; foundry proof

【清夜扪心】examine one's conscience in the stillness of night

【清一色】①(同一颜色) all of the same color; uniform; homogeneous: 穿着～的白衣服 all dressed alike in white ②(完全一致) uniform; identical: ～的看法 identical views ③(牌戏中的同一花色) all of the same suit

【清音】[语] voiceless sound

【清幽】(of a landscape) quiet and beautiful

【清早】early in the morning; early morning

【清帐】square an account; clear an account

【清真】Islamic; Muslim ◇ ～食堂 Muslims canteen / ～寺 mosque

【清蒸】steamed in clear soup: ～鸡 steamed chicken / ～鱼 steamed fish

蜻

【蜻蜓】dragonfly

【蜻蜓点水】like a dragonfly skimming the surface of the water; touch on sth. without going into it deeply

鲭
mackerel

倾
①(歪;斜) incline; lean; bend: 身子向前～ bend forward; lean forward / 向左～ incline to the left ②(倾向) deviation; tendency: 右～机会主义 right opportunism / "左"～盲动主义 "Left" putschism ③(倒塌) collapse ④(尽数倒出) overturn and pour out; empty ⑤(用尽力量) do all one can; use up all one's resources: ～全力 exert oneself to the utmost to do sth.

【倾巢】(of the enemy or bandits) turn out in full force: ～出动 turn out in full strength

【倾城倾国】so beautiful as to overrun cities and ruin states; exceedingly beautiful

【倾倒】①(由歪斜而倒下) topple and fall; topple over ②(十分佩服) greatly admire: 为之～ be infatuated with sb.; be overwhelmed with admiration for sb. ③(倒出) tip; dump; empty; pour out: ～垃圾 dump rubbish

【倾覆】overturn; topple; capsize

【倾家荡产】lose a family fortune; be reduced to poverty and ruin; be brought to total ruin

【倾角】①[物] dip ②[数] inclination ③[地] dip angle ◇ ～测量仪 dipmeter

【倾慕】have a strong admiration for; adore

【倾囊相助】empty one's purse to help; give generous financial support; give one's all to help

【倾盆大雨】heavy downpour; torrential rain; cloudburst: 遇上一场～ be caught in a downpour / 下起了～。 The rain was pelting down. (或: It was raining cats and dogs.)

【倾诉】pour out; pour forth: ～衷肠 pour out one's heart; reveal one's innermost feelings

【倾听】listen attentively to; lend an attentive ear to: ～群众意见 listen attentively to the views of the masses

【倾吐】say what is on one's mind without reservation: ～衷情 unbosom oneself

【倾箱倒箧】turn out all one's boxes and suitcases; give away all one has

【倾向】①(趋势) tendency; trend; inclination; deviation: 有害～ harmful tendency / 政治～

political inclination ②(偏于赞成) be inclined to; prefer: ～于乐观 tend to be optimistic; tend to optimism / 我～于她的方案。I prefer her plan.

【倾向性】tendentiousness: 她发表了有～的讲话。She made a tendentious speech.

【倾销】dump: ～货物 dump goods / ～剩余农产品 dump surplus farm produce

【倾斜】tilt; incline; slope; slant: 地势向海面～。The ground slopes down to the sea. ◇～度 gradient / ～角 angle of inclination; [航空] bank angle / ～面 inclined plane

【倾泻】come down in torrents

【倾卸汽车】dump truck; tipper

【倾心】①(爱慕) admire; fall in love with; lose one's heart to: 一见～ fall in love at first sight ②(真诚) cordial; heart-to-heart: ～交谈 have a heart-to-heart talk

【倾轧】engage in internal strife; jostle against one another

【倾注】①(由上而下流入) pour into; stream down into ②(精力、感情等集中于一个目标) devote to; direct to; throw into: 把精力～在工作上 devote all one's energies to work; throw all one's energy into one's work

轻 ①(重量小;比重小) light: ～如鸿毛 as light as a feather / 这个箱子比那个～。This box is lighter than that. ②(数量少;程度浅) small in number, degree, etc., slight; not serious: 年纪很～ be very young / 某人的病很～ one's illness in not at all serious ③(不重要) not important: 责任～ carry a light responsibility ④(轻松) light; relaxed: ～音乐 light music ⑤(用力不猛) gently; softly; lightly: ～拿～放 handle with care; handle gently ⑥(轻率) rashly: ～信 readily believe ⑦(轻视) belittle; make light of: 掉以～心 take sth. lightly / 文人相～。Men of letters tend to despise one another.

【轻便】light; portable: ～帆船 handy sailing-boat / ～放映机 portable projector / ～桥 portable bridge / ～铁道 light railway / 一双～鞋 a pair of light shoes

【轻薄】given to philandering; frivolous: 举止～ frivolous behavior; philandering act

【轻车简从】travel with light luggage and few attendants for quick movements

【轻车熟路】(drive in) a light carriage on a familiar road; (do) something one knows well and can manage with ease

【轻敌】take the enemy lightly; underestimate the enemy: ～思想 tendency to take the enemy lightly

【轻而易举】easy to do; easy to accomplish: 这不是～的事。This is no easy job.

【轻放】put down gently: 小心～! Handle with care!

【轻浮】frivolous; flighty; light: ～的女人 a light woman / ～态度 frivolous manner / ～的行为 frivolous conduct / 举止～ behave frivolously

【轻歌曼舞】sing merrily and dance gracefully; soft music and graceful dances

【轻工业】light industry

【轻混凝土】lightweight concrete

【轻活】light work; soft job

【轻机枪】light machine gun

【轻捷】spry and light; nimble: ～的脚步 brisk steps

【轻驾车赛马】harness racing

【轻金属】light metal

【轻举妄动】act rashly; take reckless action: 不可～ make no move without careful thought

【轻快】①(不费力) brisk; spry: 迈着～的步子 walk at a brisk pace ②(轻松愉快) relaxed; lively: ～的曲调 lively tune

【轻狂】extremely frivolous: ～的举动 extremely frivolous behavior

【轻量货品】light cargo

【轻量级】[体]lightweight

【轻慢】treat sb. without proper respect; slight: ～客人 treat a guest with impropriety

【轻描淡写】touch on lightly; mention casually; describe with a delicate touch

【轻蔑】scornful; disdainful; contemptuous: ～的眼光 a disdainful look

【轻诺寡信】make promises easily but seldom keep them

【轻泡货】light cargo

【轻飘飘】light; buoyant

【轻骑兵】light cavalry

【轻巧】①(灵巧;轻便) light and handy ②(操作轻松灵巧) deft; dexterous: ～的双手 dexterous hands ③(简单容易) easy; simple: 说起来～,做起来难。Easier said than done. (或: It's easy to talk, but difficult to put into practice.)

【轻轻】lightly; gently; softly: ～地说 speak softly / ～地走 walk with light steps

【轻取】beat easily; win an easy victory; win hands down: ～第一局 win the first game easily

【轻柔】soft; gentle: ～的声音 a gentle voice /

～的枝条 pliable twigs
【轻伤】slight wound; minor wound; flesh wound ◇ ～员 ambulant patient; ambulant case; walking wounded
【轻生】make light of one's life; commit suicide
【轻声】in a soft voice; softly: ～低语 speak softly; whisper
【轻视】despise; look down on; underestimate; make light of; take things lightly: ～困难 make light of difficulties / ～某人 look down upon sb.
【轻手轻脚】gently; softly: ～地走动 move about softly
【轻率】rash; hasty; indiscreet: ～从事 act rashly / ～的态度 reckless attitude / ～的行动 indiscreet act
【轻松】light; relaxed: ～的工作 light work; soft job; cushy job / ～地打败了对手 beat one's opponent with ease / ～愉快 happy and relaxed
【轻佻】frivolous; skittish; giddy: ～的女子 coquettish woman / 举止～ skittish behavior
【轻微】light; slight; trifling; to a small extent: ～的伤亡 light casualties / ～的损失 a trifling loss / ～的头痛 a slight headache
【轻武器】light arms; small arms
【轻雾】[气] mist
【轻泻剂】laxative
【轻信】be credulous; readily place trust in; readily believe
【轻型】light-duty; light: ～飞机 light aircraft / ～机械 light-duty machinery / ～载重汽车 light truck; light-duty truck
【轻型汽车赛】sports-car racing
【轻易】①(简单;容易) easily; readily: 这些资料不是～能得到的。These data are not readily available. ②(随随便便) lightly; rashly: ～地下结论 draw hasty conclusions
【轻音乐】light music
【轻盈】slim and graceful; lithe; lissom: 动作～ lissom movements; quick and graceful movements / 体态～ with a slim and graceful figure / 笑语～ talk and laugh merrily and lightheartedly
【轻油】light oil; naphtha
【轻于鸿毛】lighter than a goose feather
【轻重】①(重量大小) weight: ～相等 equal in weight ②(程度的深浅;事情的主次) degree of seriousness; relative importance: 无足～ a matter of no consequence / 工作应分～缓急。Work should be done in order of importance and urgency. ③(适当的程度) propriety: 做事不知～ not know the proper way to act
【轻重倒置】put the trivial above the important
【轻装】light; with light packs: ～前进 march with light packs / ～上阵 go into battle with a light pack; take part in a political movement with nothing on one's conscience
【轻子】[物] lepton
【轻罪】[法] misdemeanor: minor offense; minor crime

氢 hydrogen (H)

【氢弹】hydrogen bomb ◇ ～起爆弹 trigger for H-bomb / ～头 hydrogen warhead
【氢氟酸】hydrofluoric acid
【氢化】hydrogenation ◇ ～可的松 cortisol / ～物 hydride
【氢气】hydrogen ◇ ～球 hydrogen balloon
【氢氰酸】hydrocyanic acid
【氢氧】oxyhydrogen ◇ ～吹管 oxyhydrogen blowpipe / ～焰 oxyhydrogen flame
【氢氧化铵】[化] ammonium hydroxide
【氢氧化钙】[化] calcium hydroxide
【氢氧化钾】[化] potassium hydroxide
【氢氧化钠】[化] sodium hydroxide
【氢氧化物】[化] hydroxide

qíng

情 ①(感情) feeling; affection; sentiment: 热～ enthusiasm / 温～ tender sentiments ②(情面) favor; kindness; sensibilities; feelings: 求～ ask for a favor; plead with sb.; beg for leniency ③(爱情) love; passion: ～欲 sexual passion / 谈～说爱 be courting; talk love ④(情形;情况) situation; circumstances; condition: 病～ patient's condition / 军～ military situation / 实～ actual state of affairs

【情报】intelligence; information: 互通～ exchange information / 科技～ scientific and technological information / 秘密～ secret intelligence / 搜集～ collect intelligence ◇ ～机关 intelligence agency / ～局〈处〉 intelligence bureau / ～人员 intelligence personnel; intelligence agent / ～系统 intelligence channel / ～资料 intelligence data
【情不可却】can hardly decline sb.'s kind offer
【情不自禁】cannot refrain from; cannot help (doing sth.); be seized with a sudden impulse to: ～地流下泪来 cannot refrain from tears / ～地笑起来 cannot refrain from laughing
【情操】sentiment: 革命～ revolutionary senti-

ment
【情长纸短】the paper is too short to contain what I have to say
【情场失意】frustrated in love; unlucky in love
【情敌】rival in love
【情调】sentiment; emotional appeal: 不健康的～ unhealthy sentimentalism
【情窦初开】(of a young girl) first awakening 〈dawning〉 of love; reach puberty
【情分】mutual affection: 夫妻～ marital affection; marital relationship / 朋友～ friendship / 兄弟～ fraternity; brotherhood
【情夫】lover
【情妇】mistress
【情感】emotion; feeling: 易动～ emotional; sentimental ◇～曲线 feeling curve
【情歌】love song
【情话】lovers' prattle; lover's honeyed words
【情怀】feelings: 抒发革命～ express the revolutionary thoughts and feelings
【情急智生】hit on a good idea in a moment of desperation; emergency fillips one's wits
【情节】①(内容) plot; story: 动人的～ fascinating story / ～紧凑 a tightknit plot / ～曲折 the details are intricate ②(事实经过) circumstances: ～相同 with similar details
【情景】scene; sight; circumstances: 感人的～ a moving sight / 兴奋热烈的～ an exhilarating scene
【情景交融】(of literary work) feeling and setting happily blended
【情况】①(情形) circumstances; situation; condition; state of affairs: 健康～ health condition / 特殊～ special circumstances / 根据具体～ in accordance with specific conditions / 在许多～下 in many cases / 在这种～下 under these circumstances; such being the case / ～怎么样? How do matter stand? ②(军事上的变化) military situation: 前线有什么～? How is the situation at the front?
【情理】reason; sense: ～难容 incompatible with the accepted code of human conduct; contrary to reason or common sense / 不近～ unreasonable; irrational / 合乎～ be reasonable; stand to reason
【情侣】sweethearts; lovers
【情面】feelings; sensibilities: 不顾～ have no consideration for sb.'s feelings / 留～ spare sb.'s feelings
【情趣】①(性情志趣) temperament and interest: 我们～相投。 We are temperamentally compatible. ②(情调趣味) interest; appeal; delight: 生活～ delight of life
【情人】lover; sweetheart: ～眼里出西施。 Love is blind. (或: In the eye of the lover, his beloved is a beauty.)
【情杀】crime passionnel
【情深似海】love is as deep as the sea
【情势】situation; trend of events; circumstances: ～不妙 the trend of events being unfavorable / ～所迫 under the force of circumstances / 估计～ size up the situation
【情书】love letter
【情随事迁】people's feelings change with the circumstances
【情态】spirit; mood ◇～动词 modal verb
【情同手足】like brothers; with brotherly love for each other; be attached to each other like brothers
【情投意合】find each other congenial; hit it off perfectly; be congenial
【情文并茂】excellent in both content and language
【情形】circumstances; situation; condition; state of affairs: 各地～大不相同。 Conditions vary somehow in different places.
【情绪】①(心理状态) morale; feeling; mood; sentiments: 对立～ antagonistic sentiments / 急躁～ rashness / ～不高 be in low spirits / ～低落 be in low spirits / ～高涨 be in high spirits ②(不愉快的情绪) depression; moodiness; the sulks: 闹～ be in a fit of depression; be in low spirits; have a fit of the sulks / 有点儿～ rather sulky ◇～不稳定 emotional instability / ～紧张 emotional stress / ～适应 emotional adjustment / ～阻滞 emotional blocking
【情义】ties of friendship; tie of comradeship: 夫妻～ true feelings between husband and wife
【情谊】friendly feelings; friendly sentiments: 兄弟～ brotherly affection / 战斗～ militant bonds of friendship
【情意】tender regards; affection; goodwill: 深厚的～ deep affection / ～绵绵 lingering sentiments; everlasting love
【情由】the hows and whys: 不问～ without asking about the circumstances and causes; without going into the hows and whys
【情有可原】excusable; pardonable
【情欲】sexual passion; lust; erotic feeling
【情愿】①(愿意) be willing to: 两相～ by mutual consent; both parties being willing ②(宁

愿) prefer; would rather; had rather

晴 fine; clear: ～天 fine day／～转多云 change from fine to cloudy

【晴和】warm and fine: 天气～。It's a fine, warm day.

【晴空】clear sky; cloudless sky: 万里～ a clear and boundless sky

【晴朗】fine; sunny: ～的天 bright day／天气～。It's a sunny day.

【晴天】fine day; sunny day

【晴天霹雳】a bolt from the blue

【晴雨表】weatherglass; barometer

氰 [化] cyanogen; dicyanogen

【氰钴胺】cyanocobalamin; vitamin B_{12}

【氰化】cyaniding ◇～法 cyanidation／～钾 potassium cyanide／～氢 hydrogen cyanide／～物 cyanide

【氰酸】cyanic acid

擎 prop up; hold up; lift up: 高～火炬 hold aloft the torch

qǐng

请 ①(请求) request; ask: ～勿动手。Keep your hands off. ②(邀请) invite; engage: ～某人吃糖果 treat sb. to some candy／～医生 send for a doctor ③[敬] please: ～安静。Be quiet, please.／～坐。Sit down, please. (或: Please be seated.)

【请安】pay respects to sb.; wish sb. good health

【请便】do as you wish; please yourself: 你一定要走,那就～吧。If you insist on leaving now, please go ahead.

【请功】ask the higher level to record sb.'s meritorious deeds

【请假】ask for leave: 请两天病假 ask for two days' sick leave ◇～条 written request for leave (of absence)

【请柬】invitation card: 发～ send out invitations

【请教】ask for advice; consult: 向某人～ ask for sb.'s advice; ask advice of sb.／虚心向内行～ learn modestly from experts

【请君入瓮】kindly step into the vat: try what you have devised against others

【请客】stand treat; entertain guests; invite sb to dinner; give a dinner party: 今天我～。I'll stand treat today.

【请命】plead on sb.'s behalf: 为民～ plead on behalf of the people

【请求】ask; request: ～宽恕 ask for forgiveness／～某人帮忙 ask a favor of sb.／～一项贷款 request a loan

【请示】request 〈ask for〉 instructions: 向上级～ ask for instructions from one's superior; ask one's superior for instructions

【请帖】invitation card; invitation: 发～ send out invitations

【请问】①[敬](用于请对方回答问题) excuse me; please: ～,到机场怎么走? Excuse me, but could you tell me how to get to the airport? ②(向对方提出问题时用) we should like to ask; it may be asked; one may ask

【请勿】please don't: ～迟到。You are requested to come on time.／～践踏草地。Keep off the lawn 〈grass〉.／～攀折花木。Please don't pick the flowers.／～入内。No admittance.／～随地吐痰。No spitting.／～吸烟。No smoking.

【请降】beg to surrender

【请愿】present a petition; petition: 为某事向某人～ petition sb. for sth. ◇～书 petition

【请战】ask for a battle assignment ◇～书 written request for a battle assignment

【请罪】①(请求处分) admit one's error and ask for punishment ②(道歉) apologize; humbly apologize

苘

【苘麻】piemarker

顷 ①(公顷) hectare: 碧波万～ a boundless expanse of blue water ②(顷刻;短时间) moment; instant; a little while: 少～ after a while; in a moment ③(不久以前) just; just now: ～接来函。I have just received your letter.

【顷刻】in a moment; in an instant; instantly: ～瓦解 collapse instantly／～消失 vanish in an instant／～之间 in a twinkling; in no time

qìng

亲

【亲家】①(儿子的岳父母或女儿的公婆) parents of one's daughter-in-law or son-in-law ②(婚配后的亲戚关系) relatives by marriage

庆 ①(庆祝;庆贺) celebrate; congratulate: ～丰收 celebrate a good harvest ②(值得庆祝的周年纪念日) occasion for celebration: 国～ National Day／校～ anniversary of the founding of a school

【庆大霉素】[药] gentamicin
【庆典】celebration; a ceremony to celebrate: 盛大～ grand celebrations
【庆父不死,鲁难未已】until Qing Fu is done away with, the crisis in the state of Lu will not be over; there will always be trouble until he who stirs it up is removed
【庆功会】victory meeting
【庆功论赏】confer honors according to merits in service while celebrating a success or victory
【庆贺】congratulate; celebrate
【庆幸】rejoice: 值得～的事 a matter for rejoicing／可～的是,大家都平安地回到家里。Happily, all have returned safe and sound.
【庆祝】celebrate: ～国庆 celebrate National Day／～生日 celebrate one's birthday ◇ ～大会 celebration meeting

磬 ①(古代打击乐器) chime stone ②(佛教打击乐器) inverted bell

罄 use up; exhaust; empty: ～其所有 empty one's purse; offer all one has／告～ be all used up; be exhausted; run out
【罄尽】with nothing left; all used up
【罄竹难书】too numerous to record: 这个歹徒罪行累累,～。The scoundrel's crimes were too numerous to record.

qióng

穹 ①(穹隆) vault; dome ②(天空) the sky
【穹苍】the vault of heaven; the firmament; the sky; the heavens
【穹顶】[建] dome
【穹隆】vault; arched roof
【穹形】vaulted; arched: ～的屋顶 a vaulted roof

穷 ①(贫穷) poor; poverty-stricken; badly off: 人～志不～。One may be poor but never ceases to be ambitious. ②(穷尽) limit; end: 技～ exhaust one's whole bag of tricks／无～无尽 endless; inexhaustible／智～ at one's wit's end ③(彻底) thoroughly: ～究 make a thorough inquiry ④(极端) extremely
【穷棒子】pauper; the destitute
【穷兵黩武】use all one's armed might to indulge in wars of aggression; wantonly engage in military aggression; wage war frequently
【穷愁潦倒】be penniless and frustrated
【穷光蛋】pauper; poor wretch; a penniless vagrant
【穷极思变】one will start thinking about changes when he is in extreme poverty
【穷极无聊】①(感到厌烦) be utterly bored ②(令人讨厌) absolutely senseless; disgusting
【穷家富路】practice thrift at home but be amply provided while traveling
【穷尽】limit; end: 群众的创造力是无穷无尽的。The creative power of the masses knows no limits.／知识是无～的。Knowledge knows no bounds.
【穷寇】hard-pressed enemy; tottering foe: 宜将剩勇追～。With power and to spare we must pursue the tottering foe.／～勿追。The stag at bay is a dangerous foe. (或: Don't pursue a beaten enemy.)
【穷苦】poverty-stricken; impoverished
【穷困】poverty-stricken; destitute; in straitened circumstances
【穷年累月】for years on end; year after year
【穷期】termination; end: 战斗正未有～ The struggle will go on and on
【穷人】poor people; the poor
【穷日子】days of poverty; straitened circumstances
【穷山恶水】barren mountains and unruly rivers; barren and unwholesome view
【穷奢极欲】extremely extravagant and luxurious; go to the extremes of extravagance; (live a life of) wanton extravagance: 过着～的生活 wallow in luxury
【穷酸】poor and pedantic
【穷途末路】have come to a dead end; be driven into an impasse
【穷乡僻壤】a remote, backward place; obscure village
【穷形尽相】describe in minute, vivid detail; appear in all one's ugliness
【穷凶极恶】extremely violent and wicked; extremely vicious; utterly evil; atrocious; diabolical: ～的敌人 most vicious enemy／一副～的样子 with the look of a fiendish brute
【穷于应付】be hard put to cope with
【穷原竟委】get to the bottom of the matter; make a thorough inquiry into sth.
【穷则变,变则通】When all means are exhausted, changes become necessary, once changed, a solution emerges.
【穷则思变】poverty gives rise to a desire for change
【穷追】go in hot pursuit: ～不舍 run down a

convict / ～猛打 vigorously pursue and fiercely maul

琼 fine jade: ～阁 a jewelled palace

【琼浆玉液】nectar

【琼楼玉宇】a richly decorated jade palace; a magnificent building; a splendid building

【琼筵】luxurious dinner; sumptuous banquet

【琼脂】agar-agar; agar

茕 ①(孤单;孤独) solitary; alone ②(忧愁) dejected

【茕茕】all alone; lonely: ～孑立,形影相吊 standing all alone, body and shadow comforting each other

qiū

秋 ①(秋季) autumn; fall (美): 深～ late autumn / ～风 autumn wind / ～高气爽 (a fine day) with clear autumn sky and crisp air; the clear and crisp autumn climate / ～去冬来。Winter comes after autumn. ②(庄稼成熟季节) harvest time: 麦～ time for the wheat harvest ③(年) year: 千～万代 for thousands of years ④(多指某个不好的时期) time; period; a period of time: 多事之～ an eventful period; troubled times / 危急存亡之～ at a critical time of survival or extinction

【秋波】bright eyes of a beautiful woman: 送～ (of a woman) make eyes; ogle; cast amorous glances

【秋播】autumn sowing

【秋分】the Autumnal Equinox (16th solar term)

【秋耕】autumn ploughing

【秋海棠】begonia

【秋毫无犯】not commit the slightest offense against the civilians; not encroach on the interests of the people to the slightest degree

【秋后的蚂蚱】a grasshopper at the end of autumn; nearing its end

【秋后算帐】square accounts after the autumn harvest; wait one's opportunity to settle accounts with sb. ◇ ～派 people who bide their time to take revenge

【秋季】autumn; fall (美) ◇ ～作物 autumn crops

【秋景】①(景色) autumnal scenery ②(收成) prospects for the autumn harvest

【秋老虎】[气] a spell of hot weather after the Beginning of Autumn; old wives' summer

【秋凉】cool autumn days

【秋令】①(秋季) autumn ②(秋季气候) autumn weather

【秋千】swing: 打～ have a swing

【秋色】autumn scenery: ～宜人 charming autumn scenery

【秋沙鸭】[动] merganser

【秋收】autumn harvest

【秋水】autumn waters; limpid eyes (of a woman): ～伊人 thinking of an old acquaintance on seeing a familiar scene / 望穿～ gaze anxiously till one's eyes are worn out; eagerly look forward (to seeing a dear one)

【秋水仙】[植] meadow saffron; autumn crocus

【秋天】autumn; fall(美)

【秋雨】autumn rains

【秋种】autumn planting

【秋庄稼】autumn crops

楸 [植] Chinese catalpa

丘 ①(小土山;土堆) mound; hillock: 荒～ a barren hillock / 沙～ a sand dune ②(坟) grave: 坟～ grave; tomb

【丘陵】hills: ～起伏 a chain of undulating hills ◇ ～地带 hilly country; hilly land

【丘脑】thalamus; cerebral ganglion

【丘疹】[医] papule

蚯

【蚯蚓】earthworm

qiú

酋 ①(酋长) chief of a tribe ②(首领) chieftain: 敌～ enemy chieftain / 匪～ bandit chief

【酋长】①(部落首领) chief of a tribe ②(酋长国首领) emir; sheik(h)

【酋长国】emirate; sheikhdom

求 ①(请求) ask; beg; request; entreat; beseech: ～人帮忙 ask sb. a favor; ask a favor of sb. / 有～于人 have to look to others for help / 他恳～我帮忙。He entreated a favor of me. / 我～你一件事。I must ask you a favor. ②(追求;探求;寻求) strive for; seek; try: ～进步 strive for further progress / ～学问 seek knowledge / 力～改进 strive for improvement / 实事～是 seek truth from facts ③(需求;需要) demand: 供不应～。Supply falls short of demand.

【求爱】pay court to; woo; court

【求成】hope for success: 急于～ be impatient for success; hope to achieve quick results
【求大同,存小异】seek common ground on major issues while reserving differences on minor ones
【求告】①(求得帮助、支持) solicit; petition ②(求情、求恕) implore; entreat; supplicate
【求根】[数] extract a root
【求和】①(请求停战) sue for peace ②(设法作成平局) try to draw a match 〈game〉
【求婚】make an offer of marriage; propose
【求积仪】[数] planimeter
【求见】ask to see; request an interview; beg for an audience
【求教】ask for advice: 登门～ call on sb. for counsel; come to seek advice
【求救】ask sb. to come to the rescue; cry for help: 发出～的信号 signal an SOS; send an SOS ◇ ～呼号 distress call; SOS
【求偶】courtship
【求乞】beg: 沿门～ go begging from door to door
【求签问卜】ask the gods for an oracle
【求亲】seek a marriage alliance
【求亲靠友】ask favors of relatives and friends
【求情】plead; intercede; ask for a favor; beg for leniency: 为某人～ intercede for sb.; beg (for mercy) on sb.'s behalf / 向某人～ plead with sb.
【求全】①(要求完美无缺) demand perfection: 不要～责备。 We shouldn't demand perfection. (或: Don't nitpick.) ②(希望事情成全) try to round sth. off; try to bring sth. to a satisfactory conclusion: 委曲～ make concessions to achieve one's purpose
【求饶】beg for mercy; ask for pardon
【求人】ask for help
【求人不如求己】Self-help is better than help from others. (或: God helps those that help themselves.)
【求生】seek to live on
【求胜】strive for victory: ～心切 be anxious to gain victory
【求实】be realistic; be practical-minded ◇ ～精神 matter-of-fact attitude; realistic approach; down-to-earth approach
【求同存异】put aside minor differences so as to seek common ground; seek common ground while reserving differences
【求降】beg to surrender; hang out the white flag
【求学】①(在学校学习) go to school; attend school ②(探求学问) pursue one's studies; seek knowledge
【求援】ask for help; request reinforcements
【求战】①(寻求战斗) seek battle ②(要求参加战斗) ask to take part in a battle: 战士们～心切。The men are itching to fight.
【求之不得】all that one could wish for; most welcome: ～的事情 just what one wants / 这是～的好机会。This is a most welcome opportunity.
【求知】seek knowledge ◇ ～欲 thirst for knowledge
【求职】job wanted
【求值】[数] evaluation
【求助】turn to sb. for help; seek help ◇ ～程序 help program

球 ①[数] sphere ②[体] ball: 传～ pass the ball / 高尔夫～ golf / 篮～ basketball / 排～ volleyball / 网～ tennis / 足～ football / 我把～传给他。I passed the ball to him. ③(球形物) ball; anything shaped like a ball: 雪～ snowball / 线已经绕成～了。 The threads are wound into a ball. ④(地球) the globe; the earth: 东半～ the Eastern Hemisphere / 全～战略 global strategy / 西半～ the Western Hemisphere
【球场】①(打球的场地) a ground where ball games are played ②(排、篮、网、羽毛球场) court ③(足、棒、垒球场) field
【球胆】bladder (of a ball)
【球蛋白】globulin
【球队】(ball game) team
【球罐】[石油] sphere
【球果】[植] cone
【球茎】[植] corm
【球茎甘蓝】[植] kohlrabi
【球菌】[微] coccus
【球类运动】ball games
【球门】[体] goal ◇ ～区 goal area / ～网 goal-net / ～柱 goalpost
【球迷】(ball game) fan: 乒乓～ ping-pong fan / 足～ football fan
【球面】spherical surface; sphere ◇ ～车床[机] spherical turning lathe / ～几何学 spherical geometry / ～镜 [物] spherical mirror / ～天文学 spherical astronomy
【球磨床】[机] ball grinder
【球磨机】ball mill
【球墨铸铁】nodular cast iron; spherulitic iron

【球拍】①(网球、羽毛球拍) racket ②(乒乓球拍) bat
【球赛】ball game; match
【球坛】the ball-playing world; ball-playing circles; ball-players: ～盛会 a grand gathering of players / ～新手 a new player; a newcomer to the tournament
【球体】spheroid
【球网】net (for ball games)
【球窝节】[机] ball-and-socket joint
【球鞋】gym shoes; tennis shoes; sneakers
【球形】spherical; globular; round ◇ ～安全阀 ball relief valve / ～阀 spherical valve / ～闪电 ball lightning
【球艺】skills in playing a ball game; ball game skills

裘 fur coat

囚 ①(囚禁) imprison: 被～ be thrown into prison ②(囚犯) prisoner; convict: 死～ a convict sentenced to death
【囚车】prison van; prisoners' van; patrol wagon; black maria
【囚犯】prisoner; convict
【囚禁】imprison; put in jail; keep in captivity
【囚牢】prison; jail
【囚室】prison cell
【囚首垢面】with unkempt hair and dirty face
【囚徒】convict; prisoner

泅 swim
【泅渡】swim across: 武装～ swim across with one's weapons; swim across fully armed
【泅水】swim

犰
【犰狳】[动] armadillo

qū

趋 ①(快走) hasten; hurry along: ～步飞跑 quicken pace and run fast / ～前 hasten forward / 疾～而过 hurry past ②(趋向) tend towards; tend to become: ～于稳定 be tending towards stability / ～于一致 be reaching unanimity / 大势所～ irresistible general trend / 日～繁荣 be getting prosperous with every passing day
【趋奉】toady to; fawn on: ～权贵 dance attendance on a nobleman
【趋附】ingratiate oneself with: curry favor with: ～权贵 be a hanger-on of high officials
【趋光性】[生] phototaxis
【趋热性】[生] thermotaxis
【趋势】trend; tendency: 营业有改善的～。Business is showing a tendency to improve.
【趋向】①(朝某个方向发展) tend to; incline to: ～灭亡 head for extinction / ～完善 tend to reach perfection / 日益～好转 tend to improve with each passing day ②(趋势) trend; direction: 物价的～是仍在上涨。The trend of prices is still upwards.
【趋性】[生] taxis
【趋炎附势】curry favor with the powerful; play up to those in power; be follower of the rich and powerful
【趋药性】[生] chemotaxis
【趋之若鹜】go after sth. like a flock of ducks; scramble for sth.; go after in a swarm

祛 dispel; remove; drive away: ～暑 drive away summer heat
【祛除】dispel; get rid of; drive out: ～邪魔 exorcise evil spirits; drive out evil spirits / ～疑虑 dispel one's misgivings
【祛风】[中医] dispel the wind; relieve rheumatic pains, colds, etc.
【祛风湿药】medicine for rheumatism
【祛痰】make expectoration easy ◇ ～剂 expectorant
【祛疑】remove suspicion or doubts
【祛瘀活血】[中医] remove blood stasis and promote blood circulation

区 ①(地区;区域) area; district; region: 风景～ scenic spot / 工业～ industrial area / 林～ wooded district; forest / 贫民～ slum district / 山～ mountainous district / 商业～ shoping district; business section (of a city) / 邮～ postal district / 住宅～ residential district ②(行政区划单位) district; region; division: 郊～ suburban district / 市～ urban district / 行政～ administrative division / 自治～ autonomous region ③(区别;划分) distinguish; classify; subdivide ④[体](球类比赛场地上的特定区域) area; zone: 端～ end zone / 攻～ attacking zone / 球门～ goal area; goal crease / 守～ defending zone
【区别】①(比较;分别) distinguish; differentiate; make a distinction between: ～不同情况 distinguish among differing cases / ～对待 deal with each case on its merits; deal with different things or people in different ways / ～好

坏 distinguish between good and bad / ～罪与非罪 distinguishment of crime from noncrime / 把两者～开来 differentiate one from the other ②（彼此不同的地方）difference: 主要～ the main distinction
【区分】differentiate; distinguish: ～两个历史时期 mark off two historical periods
【区划】division into districts: 行政～ administrative divisions
【区间车】shuttle bus; a train or bus travelling only part of its normal route
【区间贸易】inter-regional trade
【区区】trivial; trifling: ～之数 this is a petty amount / ～小事,何足挂齿。Such a trifling thing is hardly worth mentioning.
【区时】zone time
【区域】region; area; district: ～间合作 inter-regional cooperation ◇ ～会议 regional conference; local conference / ～配额制 regional quota / ～自治 regional autonomy
【区域性】regional ◇ ～公约 regional convention / ～同盟 regional alliance / ～问题 a matter of regional significance / ～战争 regional war

躯 the human body: 为国捐～ lay down one's life for one's country / 血肉之～ mortal flesh and blood
【躯干】trunk; torso
【躯壳】the body; outer form: 失去精神实质,只留下～ have lost the essential spirit leaving only the outer form
【躯体】body

驱 ①（赶牲口;驾车）drive: ～车前往 drive (in a vehicle) to a place / ～马前进 urge a horse on ②（驱除;驱散）expel; disperse: ～云防雹 disperse clouds to prevent a hailstorm ③（快跑）run quickly: 并驾齐～ run neck and neck / 长～直入 drive deep into; drive straight in / 驰～ gallop
【驱策】①（用鞭子赶）drive; whip on ②（驱使）order about: 任人～ allow oneself to be ordered about
【驱虫净】tetramisole; tetramizole
【驱虫药】anthelmintic; vermifuge
【驱除】drive out; get rid of
【驱动】[机] drive: 前轮～ front-wheel drive / 四轮～ four-wheel drive ◇ ～齿轮 driving gear
【驱蛔灵】[药] piperazine citrate
【驱散】disperse; dispel; break up: 风不久就把雾～了。The wind soon dispelled the fog.
【驱使】①（迫使;使唤）order about: 供～ be ordered about; be at sb.'s beck and call ②（推动）prompt; urge; spur on: 为好奇心所～ be prompted by curiosity
【驱逐】drive out; expel; banish: ～出境 deport; expel ◇ ～机 pursuit plane / ～舰 destroyer

蛆 maggot

曲 ①（弯曲）bent; crooked: 弯腰～背 with one's back bent ②（弯曲处）bend: 河～ bend of a river ③（理亏）wrong; unjustifiable: 是非～直 the rights and wrongs of a matter ④（发酵剂）leaven; yeast
【曲别针】paper clip
【曲柄】[机] crank ◇ ～销 wrist pins / ～轴 crank axle / ～钻 brace drill
【曲尺】carpenter's square; zigzag rule
【曲从众意】bend to the public opinion
【曲高和寡】caviar to the general
【曲拱】arched: ～石桥 arched stone bridge
【曲古霉素】[药] trichomycin
【曲棍球】①（指运动）field hockey ②（曲棍球用球）hockey ball ◇ ～场 field; hockey field / ～球棍 hockey stick; stick
【曲解】(deliberately) misinterpret; twist: 他～了我的话。He twisted what I said.
【曲尽其妙】(describe) in a subtle and skilful way
【曲径通幽】a winding path leading to a secluded spot
【曲里拐弯】tortuous; zigzag: ～的小道 a tortuous path / 这条小道～地通往山上。The path ran zigzag up the hill.
【曲流】[地] meander
【曲率】[数] curvature ◇ ～计 flexometer
【曲霉】aspergillus
【曲面】curved surface; camber: 内～ negative camber / 外～ positive camber
【曲曲弯弯】full of twists and turns: ～的小路 winding path
【曲射】[军] curved fire ◇ ～弹道 curved trajectory / ～炮 curved-fire gun
【曲突徙薪】bend the chimney and remove the fuel (to prevent a possible fire); take precautions against a possible danger
【曲线】[数] curve ◇ ～球 [棒、垒球] curve ball / ～图 diagram (of curves) / ～运动 [物] curvilinear motion
【曲线板】French curve; irregular curve; curve board

【曲意逢迎】go out of one's way to curry favor
【曲折】①(弯曲) tortuous; winding: 长期～的斗争 protracted tortuous struggle ②(复杂的、不顺当的情节) complications: 故事的情节是～的。The story has a complicated plot.
【曲直】right and wrong: ～不分 not distinguish between right and wrong
【曲轴】[机] crankshaft; bent axle ◇ ～磨床 crankshaft grinding machine / ～箱 crankcase

黢 black; dark: 黑～～ pitch-black; pitch-dark
【黢黑】pitch-black; pitch-dark

屈 ①(弯曲) bend; bow; crook: ～臂 crook one's arm / ～伸自如 bend and stretch the limbs freely ②(屈服) subdue; submit: 不～不挠 indomitable; dauntless; unyielding / 宁死不～ would rather die than yield ③(委曲) wrong; injustice: 叫～ complain about an injustice / 受～ be wronged ④(理亏) in the wrong: 理～ have a weak case
【屈才】do work unworthy of one's talents; put sb. on a job unworthy of his talents
【屈从】submit to; yield to; knuckle under to
【屈打成招】confess to false charges under torture; beat a man in order to extort a confession
【屈服】surrender; yield; knuckle under: ～于外界的压力 yield to pressure from outside / 迫使敌人～ force the enemy to give in
【屈光度】diopter
【屈光透镜】diopteric lens
【屈就】condescend to take a post offered
【屈居人下】reluctant to be placed under others
【屈辱】humiliation; mortification: 感到～ feel mortified at <by> / 使某人蒙受～ bring humiliation on sb.
【屈死】be wronged and driven to death; be persecuted to death
【屈体】[体] picked
【屈膝】go down on one's knees; bend one's knees: ～求和 bow the knees and sue for peace / ～谈心 have heart-to-heart talk with knees together / ～投降 go down on one's knees in surrender; knuckle under
【屈指】count on one's fingers: ～一算 count on one's fingers
【屈指可数】can be counted on one's fingers; very few: 他的日子～了。His days are numbered.
【屈尊】condescend

qú

渠 canal; ditch; channel: 干～ main canal / 沟～ ditches / 灌溉～ irrigation canal / 排水～ drainage ditch / 引水～ inlet channel
【渠道】①(引水灌溉的水道) irrigation ditch; canal; channel ②(途径) medium of communication; channel: 通过各种～ through various channels / 通过外交～ through diplomatic channels
【渠灌】canal irrigation

蠼
【蠼螋】earwig

qǔ

取 ①(拿到身边) take; get; fetch: ～款 draw on an account; draw money / 回去～钥匙 go back to fetch one's key ②(得到;招致) aim at; seek: ～乐 seek pleasure / ～信于人 win the people's confidence / 自～灭亡 court destruction ③(采取;选取) adopt; assume; choose: ～慎重态度 adopt a cautious attitude / ～中立态度 adopt a neutral attitude / 不足～ inadvisable; undesirable / 给孩子～个名儿 choose a name for a child; give a name to a child / 可～ advisable; desirable
【取保】get sb. to go bail for one: ～释放 be released on bail; be bailed out
【取材】draw materials: ～于 have drawn sth. from / 就地～ make use of local materials; draw on local materials
【取长补短】learn from others' strong points to offset one's weaknesses; overcome one's weaknesses by acquiring other's strong points
【取长补短,共同提高】draw on each other's merits and raise the level together, learn each other's good points for common progress
【取代】replace; substitute for; supersede; supplant: 电灯已经～了煤气灯。Electric light has superseded gas light. / 公共汽车已经～了电车。Buses have replaced streetcar <trams>. ◇ ～反应 substitution reaction
【取道】by way of; via: ～上海前往东京 go to Tokyo via Shanghai
【取得】gain; acquire; obtain: ～律师资格 call to the bar / ～相当大的进展 make considerable headway / ～一致意见 reach complete identity of views / ～圆满成功 be crowned with success; achieve complete success
【取缔】outlaw; ban; suppress: ～非法组织 ban

illegal organization / ～倾销 anti-dumping
【取而代之】replace sb.; supersede sb.; take it over
【取法】take as one's model; follow the example of
【取经】①(求佛取经) go on a pilgrimage for Buddhist scriptures ②(吸取经验) learn from sb. else's experience
【取精用弘】select the finest from a vast quantity
【取景】find a view: 摄影～ find a view to photograph ◇～器[摄] viewfinder
【取决】be decide by; depend on; hinge on
【取乐】seek pleasure; find amusement; amuse oneself; make merry: 饮酒～ drink and make merry
【取暖】warm oneself (by a fire, etc.): 烤火～ warm oneself by the fire; keep warm by the fire / 用热水袋～ keep warm with hot-water bag
【取齐】①(使数量、长度或高度相等) make even; even up ②(聚齐;集合) assemble; meet each other
【取巧】resort to trickery to serve oneself: 投机～ resort to dubious shifts to further one's interests; be opportunistic
【取人之长】suck a person's brains
【取舍】accept or reject; make one's choice: 决定～decide which to choose ◇～权 option
【取胜】win victory; score a success: 侥幸～gain a victory by sheer luck / 以多～ win victory through numerical superiority
【取消】cancel; call off; abolish: ～定货 cancel an order / ～会员资格 deprive sb. of his membership / ～禁令 lift a ban / ～律师资格 disbar / ～一场比赛 call off a match / ～一次会议 cancel a meeting; call off a meeting / ～一项决议 revoke a decision / ～证人资格 recall of witness / ～资格 disqualify
【取消主义】liquidationism
【取消符】[计算机] ignore character
【取笑】ridicule; make fun of; poke fun at
【取信于民】win the people's confidence
【取样】sampling: ～办法 sampling method / 检查 take a sample to check ◇～管 probe tube / ～勺 sample spoon
【取悦】try to please; ingratiate oneself with sb.
【取证】obtain evidence
【取之不尽,用之不竭】inexhaustible; inexhaustible in supply and always available for use
【取之于民,用之于民】what is taken from the people is used in the interests of the people

娶
marry (a woman); take to wife
【娶亲】(of a man) get married

龋
【龋齿】①(病) dental caries ②(蛀齿) decayed tooth

曲
①(歌曲) song; tune; melody: 高歌一～ lustily sing a song / 小～儿 ditty ②(歌谱;乐谱) music: 芭蕾舞组～ ballet suite / 变奏～ variation / 钢琴～ piano music / 进行～ march / 练习～ etude / 圆舞～ waltz / 协奏～ concerto / 序～ overture / 叙事～ ballad / 奏鸣～ sonata / 作～ compose music
【曲调】tune (of a song); melody: 古老～ old melody / 流行～ popular tune
【曲高和寡】highbrow songs find few singers; too highbrow to be popular; melodies of superior taste find few to join in chorus
【曲式】[乐] musical form
【曲艺】*quyi*, Chinese folk art forms
【曲子】song; tune; melody

qu

阒
quiet; still: ～然无声 very quiet; absolutely still / ～无一人 all was quiet and not a soul was to be seen

趣
①(趣味;兴味) interest; delight: 有～ interesting; delightful; amusing ②(有趣味的) interesting: ～事 an interesting episode
【趣味】①(使人愉快、有吸引力的特性) interest; delight: ～无穷 be of infinite interest; afford the greatest delight; be fascinating ②(爱好) taste; liking; preference: 迎合低级～ cater to vulgar tastes
【趣闻】interesting news

去
①(离开某地到某地) go; leave: 从北京～上海 leave Beijing for Shanghai / 你～过广州没有? Have you ever been to Guangzhou? ②(除去;除掉) remove; get rid of: ～掉几个字 take off some words / ～掉思想上的负担 get a load off one's mind ③(距离) be apart from: ～今三十多年 more than thirty years ago / 两地相～二十公里 the two places are 20km. apart ④(去年的) of last year: ～秋 last autumn
【去臭】deodorizing
【去处】①(去的地方) place to go; whereabouts: 他的～不明。 His present

whereabouts is unknown. ②(场所;地方) place; site: 这是一个极好的避暑～。 This is a very nice place for summer.
【去粗取精】discard the dross and select the essential
【去恶务尽】do away with evil and must do it wholly
【去垢剂】detergent
【去火】[中医] reduce internal heat; relieve inflammation or fever
【去旧更新】do away the old and change it for new
【去留未决】dismiss or return is not yet decided; to go or to stay is yet to be decided
【去留悉听尊便】you are liberty to go or stay
【去路】the way along which one is going; outlet: 挡住某人的～ block one's way / 给大水找到～ find an outlet for the flood
【去敏灵】[药] tripelennamine
【去年】last year: ～此时 this time last year / ～九月 last September
【去皮】①(去掉皮) remove the peel ②(净重) net weight: ～四十公斤 net weight forty kilograms ◇ ～机 decorticator; sheller
【去取之间】between taking and leaving
【去如黄鹤】gone as the yellow crane
【去世】die; pass away
【去势】[兽医] castrate
【去痛定】[药] piminodine esylate
【去邪归正】return to orthodox path
【去伪存真】eliminate the false and retain the true
【去污】decontamination
【去污粉】household cleanser; cleanser
【去芜存菁】get rid of the weed and keep the flower of the leek
【去向】the direction in which sb. or sth. has gone: ～不明 whereabouts unknown / 不知～ be nowhere to be found
【去雄】[植] emasculate; castrate
【去杂去劣】[农] roguing
【去职】no longer hold the post

觑 look; gaze: 面面相～ gaze at each other in speechless despair

quān

悛 repent; make amends: 怙恶不～ be steeped in evil and refuse to repent

圈 ①(环形) circle; ring: 包围～ ring of encirclement; encirclement / 画～儿 draw a circle / 绕跑道跑两～ run around the track twice / 围成一～ form a ring / 钥匙～儿 key ring ②(范围) circle; group: 不是～里人 be not on the inside; do not belong to the inner circle ③(围) enclose; encircle: 把菜园～起来 enclose the vegetable garden ④(画圈儿) mark with a circle
【圈点】①(加标点作为句读的记号) punctuate ②(标出值得注意的语句) mark words and phrases for special attention
【圈套】snare; trap: 落入～ fall into a trap; play into sb.'s hands / 设～ lay a snare
【圈椅】round-backed armchair
【圈阅】make a circle round one's name listed on a document after reading it
【圈子】①(环形) circle; ring: 说话绕～ speak in a roundabout way; beat about the bush / 站成一个～ stand in a circle ②(集体或生活范围) circle; group: 搞小～不好。 A few banding together is no good. / 她的活动～很小。 She moves in a very small circle.

quán

拳 ①(拳头) fist: 挥～ shake one's fist / 握～ clench one's fist; close a fist ②(拳术) boxing; pugilism: 打～ do Chinese boxing / 练～ practice shadow boxing
【拳打脚踢】cuff and kick; beat up; strike and kick
【拳击】boxing; pugilism ◇ ～比赛 boxing match; bout; fight / ～等级 boxing weights / ～手套 boxing glove / ～台 boxing ring / ～鞋 boxing shoe / ～运动员 boxer; pugilist ○击倒 knock down / 击倒对方获胜 knock out; KO / 被击倒失败 count out / 48 公斤级 light flyweight / 51 公斤级 flyweight / 54 公斤级 bantamweight / 57 公斤级 featherweight / 60 公斤级 lightweight / 63.5 公斤级 light welterweight / 67 公斤级 welterweight / 71 公斤级 light middleweight / 75 公斤级 middleweight / 81 公斤级 light heavyweight / 81 以上公斤级 heavyweight
【拳曲】curl; twist; bend: ～的头发 curly hair / ～着身子 curl oneself up
【拳师】boxing coach; pugilist
【拳术】Chinese boxing
【拳头】fist

鬈 curly; wavy: ～发 curly hair
【鬈曲】[纺] crimp; crinkle; curl: ～的羊毛 crimpy wool; crinkled wool

颧 cheekbone
【颧骨】cheekbone: ～突起 have prominent cheekbones
蜷 curl up; huddle up
【蜷伏】curl up; huddle up; lie with the knees drawn up
【蜷曲】curl; coil; twist: 一条蛇～在树枝上。A snake coiled round the branch.
【蜷缩】roll up; huddle up; curl up: 把身体～成一团 huddle oneself up
权 ①(权利) right: 表决～ right to vote / 特～ special right and privilege / 选举～和被选举～ the right to vote and stand for election / 在这个问题上没有发言～ not be entitled to speak on the matter; have no say in the matter ②(权力) power; authority: 行政～ administrative power / 当～ in power / 受～ be authorized (to do sth.) / 越～ overstep one's authority ③(有利的形势) advantageous position: 霸～ hegemony / 制空～ mastery of the air; air supremacy / 主动～ initiative ④(权衡) weigh: ～其轻重 weigh up one thing against another; weigh up the matter carefully ⑤(权且;姑且) tentatively; for the time being: ～充 act temporarily as; serve as a stopgap for ⑥(权变;权宜) expediency: 通～达变 adapt oneself to circumstances
【权变】adaptability in tactics; tact; adaptation to circumstances
【权柄】power; authority: 掌握～ be in power; be in the saddle
【权贵】influential officials; bigwigs
【权衡】weigh; balance: ～得失 weigh gains and losses / ～利弊 weigh the advantages and disadvantages; weigh the pros and cons
【权力】power; authority: ～机构 organ of power / ～下放 delegate power to the lower levels / 行使会议主席的～ exercise the functions of chairman of a conference; invoke the authority of chairman of a conference / 有做某事的～ have the authority to do sth. / 这是我们～范围内的事。This matter comes within our jurisdiction.
【权利】right: 不可剥夺的～ inalienable rights / 公民的～和义务 rights and duties of citizens / 劳动的～ the right to work / 受教育的～ the right to education / 政治～ political rights ◇ ～人 obligee / ～证书 document of title; letter of authorization
【权谋】(political) tactics; trickery
【权能】powers and functions
【权且】for the time being; as a temporary measure
【权势】power and influence: 很有～的人 man of great influence
【权术】political trickery; shifts in politics: 玩弄～ play politics
【权威】authority; authoritativeness; a person of authority: 学术～ an academic authority / 语言学～ an authority on phonetics ◇ ～解释 authentic interpretation / ～人士 authoritative person; authoritative sources
【权限】limits of authority; jurisdiction; competence; extent of authority; extent of power: ～范围 extent of competence / 超过某人的～ exceed sb.'s authority / 在…～之内 within the power 〈jurisdiction〉 of
【权宜】expedient: ～之计 an expedient measure; makeshift (device)
【权益】rights and interests: 合法～ lawful rights and interests / 维护合法～ safeguard lawful rights and interests
【权诈】trickery; craftiness
全 ①(完备;齐全) complete: 获得～胜 win complete victory ②(保全;使完整不缺) make perfect or complete; keep intact: 两～其美 satisfy both sides ③(整个) whole; entire; full; total: ～称 full name / ～国 the whole nation; the entire nation / ～世界 the whole world; all over the world / ～中国 the whole of China; all over China ④(完全;都) entirely; completely: ～错 completely wrong / ～对了 entirely correct
【全豹】whole picture; overall situation: 窥其～ see the whole picture / 未窥～ fail to see the whole picture; fail to grasp the overall situation
【全波整流】full-wave rectification ◇ ～器 full-wave rectifier
【全部】whole; complete; total; all: ～保险 full insurance / ～付讫 payment in full / ～过程 the whole process / ～精力 all one's efforts / ～开支 total expenditure / 为…贡献自己的～力量 contribute one's all to / 要求赔偿～损失 demand full compensation for the loss incurred
【全才】a versatile person; all-rounder: 文武～ be versed in both civil and military affairs
【全场】①(全部在场者) the whole audience; all those present: 博得～喝采 bring down the house / ～欢声雷动。The audience broke out

into thunderous cheers. ② [体] full-court; all-court: ～紧逼 all-court press; full-court press
【全程】whole journey; whole course
【全等】[数] congruent ◇～形[数] congruent figures
【全都】all; without exception: 你们这些男孩～需要更加用功。 All you boys need to work harder.
【全反射】[物] total reflection
【全方位】omnibearing
【全份】complete set: ～表册 a complete set of lists and forms
【全副】complete: ～武装 fully armed; in full battle array
【全国】nation-wide; country-wide; the whole nation; the whole country; throughout the country: ～规模 nationwide scale / ～人民 the people of the whole country; the people throughout the country; the whole nation / ～上下 the whole nation from the leadership to the masses ◇～冠军 national champion / ～记录 national record / ～人口普查 nationwide census / ～运动会 the national games
【全国人民代表大会】National People's Congress ◇～常务委员会 Standing Committee of the National People's Congress
【全国性】nation-wide; country-wide; national: ～报纸 a national newspaper; a newspaper with a nation-wide circulation
【全国一盘棋】coordinate all the activities of the nation like as a whole
【全会】plenary meeting; plenary session: 十一届三中～ the Third Plenary Session of the Eleventh Central Committee
【全集】complete works; collected works:《列宁～》The Collected Works of Lenin
【全景】panorama; full view; whole scene ◇～电影 cinerama / ～画 panorama / ～摄影机 panoramic camera
【全局】overall situation; situation as a whole: ～利益 interests of the whole; general interests / ～性问题 a matter of overall importance / 从～出发 proceed from the situation taken as a whole / 树立～观点 adopt an overall point of view / 通观～ take a comprehensive view of the general situation / 影响～ affect the overall situation
【全军】the whole army; the entire army: ～指战员 the officers and men of the whole army / ～运动会 army-wide sports meet
【全军覆没】completely annihilated; total destruction of an army; the whole army was annihilated
【全开】[印] a standard-sized sheet
【全劳动力】able-bodied farm worker
【全力】with all one's strength; all-out; sparing no effort: ～以赴 go all out; spare no effort / ～支持 support with all one's strength; spare no effort to support; give all-out support / 竭尽～ exert all one's strength; move heaven and earth; throw in one's whole might
【全貌】complete picture; full view: 窥其～ see the whole picture / 弄清问题的～ try to get a complete picture of the problem
【全面】overall; comprehensive; all-round: ～崩溃 total collapse / ～对抗 all-out confrontation / ～发展 all-round development / ～规划 overall planning / ～检查 comprehensive review; overall checkup / ～进攻 an all-out attack / ～禁止和彻底销毁核武器 complete prohibition and thorough destruction of nuclear weapons / ～战争 a full-scale war / ～总结 comprehensive summing-up
【全民】the whole people; the entire people; all the people: ～皆兵 an entire nation in arms; every citizen a soldier / ～总动员 general mobilization of the nation ◇～所有制 ownership by the whole people
【全名】full name
【全能】[体] all-round: ～冠军 all-round champion / ～体操比赛 combined exercises / ～运动员 all-round athlete; all-rounder / 十项～运动 decathlon / 五项～运动 pentathlon / 获得女子～冠军 win the women's individual all-round title
【全年】annual; yearly: ～的产量 annual production / ～的收入 annual income / ～平均温度 mean annual temperature / ～雨量 yearly rainfall
【全盘】overall; comprehensive; wholesale: ～否定 total repudiation / ～考虑 give overall consideration to / ～接受 total and uncritical acceptance / ～西化 wholesale Westernization
【全球】the whole world: ～战略 global strategy / 遍及～的影响 worldwide influence / 在～范围内 on a global scale
【全权】full powers; plenary powers: ～代表 plenipotentiary / ～证书 full powers / 特命～大使 ambassador plenipotentiary and extraordinary / 特命～公使 envoy extraordinary and minister plenipotentiary
【全然】completely; entirely: ～不顾个人安危

give no thought to one's own safety / ～不计后果 in utter disregard of the consequences / ～不了解 be completely ignorant of sth. / ～不同 entirely different / ～不知 not to know beans

【全日制】full-time ◇ ～教育 full-time schooling / ～学校 full-time school

【全色】[摄] panchromatic ◇ ～胶片 panchromatic film

【全色盲】monochromasia; monochromatism

【全身】the whole body; all over (the body): ～发抖 shake all over / ～检查 a general physical checkup / ～麻醉 general anesthesia / ～湿透 be soaked to the skin / ～是伤 be covered with cuts and bruises / ～像 full-length picture

【全神贯注】be absorbed in; be engrossed in; be preoccupied with: ～地考虑问题 be preoccupied with a problem / ～地听着 be all ears

【全盛】flourishing; in full bloom: ～时期 period of full bloom; prime; heyday

【全食】[天] total eclipse ◇ ～带 path of total eclipse; belt of totality; zone of totality

【全始全终】see sth. through; stick to sth. to the very end

【全视图】full view; general view

【全数】total number; whole amount: ～付讫 have paid the whole amount; paid in full

【全速】full speed; maximum speed; top speed: ～前进! Full speed ahead!

【全损】total loss ◇ ～险[经] total loss only (T.L.O.)

【全套】complete set ◇ ～设备 a complete set of equipment

【全体】all; entire; whole: ～出席 full attendance / ～船员 the crew (of a ship); the ship's complement / ～反对 unanimous opposition / ～工作人员 the whole staff / ～会议 plenary session / ～演员 the entire cast / 开～会 meet in full session; hold a plenary session

【全天候】all-weather ◇ ～飞机 all-weather aircraft / ～公路 all-weather road

【全托】put one's child in a boarding nursery ◇ ～托儿所 boarding nursery

【全文】full text: ～发表 publish in full / ～记录 verbatim record / ～如下。 The full text follows.

【全无头绪】make neither head nor tail of it

【全无章法】utterly without a literary style

【全息电影】holographic movie

【全息摄影】holography

【全息照相】hologram: 激光～ laser hologram ◇ ～存储器 holographic memory / ～术 holography

【全线】all fronts; the whole line; the entire length: 边界～ the entire length of the boundary / ～出击 launch an attack on all fronts / ～反攻 counteroffensive on all fronts / 敌军已～崩溃。 The enemy troops were put to rout all along the line. / ～已通车。 The whole line has been opened to traffic.

【全向导航】omnirange; omnidirectional range

【全向天线】omnidirectional antenna

【全向无线电信标】[航空] omnidirectional radio beacon (ORB)

【全心全意】wholeheartedly; heart and soul: ～地为人民服务。 Serve the people wholeheartedly.

【全休】complete rest

【全音】[乐] whole tone

【全音符】[乐] whole note; semibreve

【全音阶】[乐] diatonic scale

【全运会】[简](全国运动会) the national games

【全知全能】omniscient and omnipotent

【全脂奶粉】whole milk powder

痊 recover from an illness

【痊愈】fully recover from an illness; be fully recovered: 祝你早日～。 I wish you a speedy recovery.

醛 [化] aldehyde

【醛氨】aldehyde ammonia

【醛酸】aldehydic acid

【醛糖】aldose

【醛酯】aldehydo-ester

诠

【诠释】annotation; explanatory notes

【诠注】notes and commentary

泉 spring: 矿～ mineral spring / 喷～ fountain / 温～ hot spring

【泉华】[地] sinter

【泉水】spring water; spring

【泉眼】the mouth of a spring; spring

【泉源】①(水源) fountainhead; springhead; wellspring ②(来源) source: 力量的～ source of strength / 知识的～ source of knowledge / 智慧的～ source of wisdom

quǎn

犬 dog: 猎～ hunting dog; hound / 警～ po-

lice dog／牧～ shepherd dog; sheep dog／鸡鸣～吠 the crowing of cocks and the barking of dogs; country sounds／丧家之～ a stray cur
【犬齿】canine tooth
【犬马之劳】serve like a dog or a horse: 效～ serve one's master faithfully; be at sb.'s beck and call
【犬热病】(canine) distemper
【犬儒】cynic ◇ ～主义 cynicism
【犬牙】①(犬齿) canine tooth ②(狗牙) fang
【犬牙交错】jigsaw-like; interlocking: ～的战争 jigsaw pattern warfare／形成～的状态 form a jagged, interlocking pattern

quàn

券 certificate; ticket: 公债～ government bond／奖～ lottery ticket／入场～ admission ticket／外汇兑换～ foreign exchange certificate／赠～ free ticket

劝 ①(说服;使人听从) advise; urge; try to persuade: ～某人不要做某事 try to persuade sb. not to do sth.; try to dissuade sb. from doing sth.／～某人做某事 try to persuade sb. to do sth.／～他戒烟 advise him to give up smoking／～他休息 urge him to take a rest ②(勉励) encourage ～学 encourage learning
【劝导】try to persuade; advise; induce: 耐心～ try patiently to talk sb. round
【劝告】advise; urge; exhort: 经再三～ after repeated exhortations／～他不要多喝酒 advise him not to drink exccessively／～她要勤俭 exhort her to be diligent and thrifty／医生～她好好休息。 The doctor advised her to have a good rest.
【劝和】mediate; try to make peace
【劝架】try to reconcile parties to a quarrel; try to stop people from fighting each other; mediate between two quarrelling parties; mediate
【劝解】①(劝导宽解) help sb. to get over his worries, etc. ②(劝架) mediate; make peace between; bring people together
【劝戒】admonish; expostulate
【劝酒】urge sb. to drink (at a banquet)
【劝勉】advise and encourage: 互相～ help and encourage each other
【劝说】persuade; advise; talk (sb.) into
【劝慰】console; soothe
【劝降】induce to capitulate
【劝诱】induce; prevail upon
【劝阻】dissuade sb. from; advise sb. not to; warn sb. against; talk sb. out of: ～无效 try in vain to talk sb. out of doing sth.／你最好～她别这样早结婚。 You'd better dissuade her from marrying so early.

quē

阙 fault; error
【阙如】be wanting
【阙疑】leave the question open

炔 [化] alkyne: 乙～ acetylene
【炔雌醇】ethinyloestradiol
【炔诺酮】norethindrone

缺 ①(缺乏;缺少) be short of; lack: ～吃～穿 have insufficient food and clothing／～人手 be short of hands／什么也不～。 Nothing is lacking. ②(残破;残缺) incomplete; imperfect: 残～不全 incomplete; fragmentary／完美无～ flawless; perfect; impeccable ③(该到而未到) be absent: 人齐了,一个不～。 No one is absent. (或: Everybody's here.) ④(空额) vacancy: 补～ supply a vacancy; fill a vacancy
【缺德】mean; wicked; villainous: 做～事 do sth. mean; play a mean trick／他这样做可真～。 It's wicked of him to act like that.
【缺点】shortcoming; defect; weakness; drawback: 克服～ overcome one's shortcomings／不守时间是他的最大～。 Not being punctual is his greatest shortcoming.／每个人都有～。 No person is without defects.
【缺额】vacancy
【缺乏】be short of; lack; be wanting in: ～锻炼 lack of exercise (体育的); need tempering (思想的)／～经验 lack experience／～劳动力 be short of labor power／～战斗力 have poor fighting capacity／～证据 want of proof; absence of proof／～资源 be deficient in resources
【缺环】missing link
【缺货】be in short supply; be out of stock ◇ ～单 want slips
【缺课】be absent from school; miss a class: ～两天 be absent from school for two days／缺了两课 miss two lessons
【缺口】①(物体边缘缺掉一块) breach; gap: 打开～ make a breach; open a breach ②(不足;缺额) insufficiency ③[机] notch
【缺门】gap (in a branch of learning, etc.): 填补工业中的一个～ fill a gap in industry
【缺勤】absence from duty 〈work〉 ◇ ～率 absence rate

【缺少】lack; be short of: ～零件 lack spare parts / ～人手 be short of hands; be short-handed / 不可～的条件 indispensable conditions
【缺铁性贫血】iron-deficiency anemia
【缺席】absent (from a meeting, etc.): 无故～ be absent without excuse / 因事～ be absent on business ◇ ～判决[法] judgment by default / ～审判[法] trial by default / ～投票人 absentee voter
【缺陷】defect; drawback; flaw; blemish: 生理～ physical defect
【缺医少药】be short of medical services and supplies

qué

瘸 be lame; limp: 一步一～ walk with a limp / 右腿～了 be lame in the right leg
【瘸腿】lame: ～的人 a lame person
【瘸子】a lame person; cripple

què

阙 ①(宫门外望楼) watchtower on either side of a palace gate ②(帝王的宫殿) imperial palace: 宫～ imperial palace
却 ①(后退) step back; fall back: 退～ go back; retreat ②(使退却) drive back; repulse: ～敌 repulse the enemy ③(推辞;拒绝) decline; refuse; reject; turn down: 推～ decline; refuse
【却病】prevent or cure a disease: ～延年 prevent disease and prolong life; banish illness and increase long life
【却步】step back (from); hang back (from): 望而～ shrink back at the sight / ～而行 walk stepping backwards
【却而不受】refuse to accept
【却之不恭】it would be impolite to decline: ～, 受之有愧。 To decline would be disrespectful but to accept is embarrassing.
鹊 magpie
【鹊巢鸠占】the magpie's nest is occupied by the turtledove: one person seizes another person's place, land, etc.
榷 discuss: 商～ discuss; deliberate over
雀 sparrow
【雀斑】freckle
【雀麦】bromegrass; brome
【雀鹰】sparrow hawk
【雀跃】jump for joy: 欢呼～ shout and jump for joy
确 ①(符合事实;真实) true; reliable; authentic: ～有其事。 It's a fact. (或: It really happened.) ②(坚固;坚定) firmly: ～信 firmly believe
【确保】ensure; guarantee: ～安全生产 ensure safety in production / ～质量 guarantee quality
【确定】①(使确定) define; fix; determine; ascertain: ～犯罪性质 determine the nature of an offense / ～会议的日期和地点 determine 〈fix〉 the time and place for a meeting / ～会议宗旨 define the aims of the conference / ～领海宽度 delimit the extent of territorial waters / ～任务 set the tasks ②(明确而肯定) definite; certain; for sure: ～不移的结论 an incontestable conclusion / ～的答复 a definite reply
【确乎】really; indeed: ～有效 really effective
【确立】establish: ～共产主义世界观 form a communist world outlook
【确切】definite; exact; precise: ～的解释 a clear and unambiguous explanation / ～的日期 an exact date / ～地说 to be exact / 下个～的定义 give a precise definition
【确认】affirm; confirm; acknowledge: ～如下原则 affirm the following principles / ～所陈属实 confirm the statement
【确实】①(真实可靠) true; reliable: ～的消息 reliable information ②(的确) really; indeed: ～太晚了。 Indeed, it's too late.
【确信】firmly believe; be convinced; be sure: ～无疑 firmly believe sth. to be true / 我们～他会成功。 We firmly believed that he would succeed.
【确诊】make a definite diagnosis; diagnose
【确证】ironclad proof; conclusive evidence
【确凿】conclusive; authentic; irrefutable: ～的事实 irrefutable facts / ～的证据 conclusive evidence; absolute proof / 证据～ irrefutable evidence; verified evidence; valid evidence

qún

麇 flock together
【麇集】swarm; flock together
群 ①(聚在一起) crowd; group: 建筑～ a building complex; a cluster of buildings / 人～ crowd / 鱼～ shoals of fish / ～山环抱 surrounded by hills / 成～结队 in crowds, in

flocks; in groups / 人们三五成～地散立着。People were standing about in small groups. ②[量] group; herd; flock: 一～匪徒 a band of gangsters / 一～狼 a pack of wolves / 一～鹿 a herd of deer / 一～马 a drove of horses / 一～蜜蜂 a swarm of bees / 一～牛 a herd of cattle / 一～人 a crowd of people / 一～小孩 a group of children / 一～羊 a flock of sheep / 一～野鸭 a flock of wild ducks

【群策群力】pool the wisdom and efforts of everyone; work with collective wisdom and concerted efforts

【群岛】archipelago; islands

【群而不党】be social, but not clannish

【群芳】beautiful and fragrant flowers; all kinds of flowers: ～竞艳 flowers vying with each other in beauty; a host of beautiful women compete for attention

【群婚】group marriage; communal marriage

【群居】living in groups; gregarious; social ◇～动物 social animal / ～昆虫 social insect

【群控制】group control

【群龙无首】a host of dragons without a head; a group without a leader

【群落】[生] community ◇ ～交错区 ecotone

【群魔乱舞】a host of demons dancing in riotous revelry; rogues of all kinds running wild

【群起而攻之】all rise against sb.; rally together to attack sb.

【群青】[化] ultramarine

【群情】public sentiment; feelings of the masses: ～激昂。Popular feeling ran high. / ～振奋。Everyone is exhilarated.

【群体】①[生] colony ②[社会学] group: 次属～ secondary group / 首属～ primary group / 小～ small group / 指涉～ reference group ◇～生态学 synecology

【群威群胆】mass heroism and daring

【群言堂】allowing everyone to have his say

【群众】the masses: 广大人民～ the broad masses of the people / 遵守～纪律 maintain discipline in relations with the masses ◇ ～大会 mass rally / ～工作 mass work / ～关系 one's relations ⟨ties⟩ with the masses / ～观点 the mass viewpoint / ～监督 surveillance by the masses / ～团体 mass organization / ～运动 mass movement

【群众路线】the mass line: 走～ follow the mass line

【群众性】of a mass character: ～体育活动 mass sports activities

裙 skirt: 百褶～ pleated skirt / 超短～ miniskirt / 衬～ slip; petticoat / 绸～ silk skirt / 围～ apron

【裙带】connected through one's female relatives: 通过～关系 with the help of one's female relatives; through petticoat influence

【裙子】skirt

R

rán

髯 whiskers; beard

【髯口】artificial whiskers worn by actors in Beijing opera

然 ①(对) right; correct: 不以为～ object to; not approve／大谬不～ entirely wrong; absurd ②(如此;这样) so; like that: 不尽～ not exactly so; not exactly the case／知其～,不知其所以～ know the hows but not the whys ③[书][连] but; nevertheless; however ④(副词或形容词后缀)忽～ suddenly; all of a sudden／巍～屹立 tower majestically／显～ obviously

【然而】[连] yet; but; however: 他已年老,～记忆力还好。 He is aged, but his memory is still good.

【然后】[副] then; after that; afterwards: 我们先研究一下,～再决定。 We'll look into the matter first before coming to a decision／学～知不足。 One discovers his ignorance only through learning.

【然诺】[书] promise; pledge: 重～ be serious about making and keeping a promise

燃 burn; ignite; light: ～起一堆篝火 light a bonfire／易～物品 combustibles; inflammables

【燃点】①(点着) ignite; kindle; set fire to; light ② [化] ignition point; burning point; kindling point

【燃放】set off: ～爆竹 set off firecrackers／～烟火 set off fireworks

【燃料】fuel: 标准～ ideal fuels／低热值～ low calorie fuels／粉状～ dusty fuels; pulverized fuels ◇～比[冶] fuel ratio／～电池 fuel cell／～附加费 bunker surcharge／～库 fuel depot; fuel reservoir／～油 fuel oil

【燃眉之急】as pressing as a fire singeing one's eyebrows; a matter of extreme urgency; a pressing need

【燃气轮机】gas turbine ◇～发电厂 gas turbine power station

【燃烧】①(烧) burn; kindle: 怒火～ burning with rage／干柴容易～。 Dry wood burns easily. ②[化] combustion; inflammation ◇～弹 incendiary bomb／～剂 incendiary agent／～室 [机] combustion chamber; blast chamber; combustor／～性能 combustibility

【燃油泵】fuel pump

【燃油炉】oil burner

rǎn

染 ①(着色) dye: ～发 dye hair／～指甲 paint fingernails／把白衣服～成蓝色 dye a white dress blue ②(感染) catch (a disease): ～上了流感 have caught flu; become infected with influenza ③(沾染) acquire (a bad habit, etc.); soil; contaminate: ～上恶习 acquire a bad habit／出污泥而不～ emerge unstained from the filth／污～ pollution／一尘不～ not soiled by a speck of dust; spotless

【染病】catch an illness; be infected with a disease

【染毒】[军] contamination

【染坊】dyehouse; dye-works

【染缸】dye vat; dyejigger

【染料】dyestuff; dye: 活性～ reactive dye ◇～敏化 dye sensitization／～调色法[摄] dye toning

【染漂法】[摄] dye bleaching process

【染色】dyeing; coloring ◇～法 decoration method; staining／～剂 coloring agent／～性 dyeability

【染色体】[生] chromosome ◇～疾病 chromosomal disorders

【染色质】[生化] chromatin

【染液】dye liquor

【染印法】[电影] dye transfer process ◇～彩色电影 color film made by the dye transfer process; technicolor

【染指】take a share of sth. one is not entitled to; encroach on: 妄图～别国资源 attempt to encroach on the resources of other countries

【染指择肥】dip one's finger in the pie and claim the lion's share

冉 [书] slowly

【冉冉】[书] slowly; gradually: ～而上 go up slowly／～上升 rise slowly

rāng

嚷

【嚷嚷】[口] ①(喧哗) shout; yell; make an uproar: 外面人们在～。 People are shouting outside. ②(声张) make widely known: 这事暂时不要～出去。 Don't breathe a word about this for the time being.

ráng

禳 [书] avert (a misfortune or disaster) by prayers

瓤 ①(瓤子) pulp; flesh; pith: 南瓜～ the pulp of a pumpkin ②(皮、壳里的东西) the interior part of certain things: 光剩了个空信封,里头没有信～儿。There's only an empty envelope with nothing in it.

【瓤子】pulp; flesh; pith

rǎng

壤 ①(土壤) soil: 沃～ fertile soil; rich soil ②(地) earth: 天～之别 be as far removed as heaven from earth; be vastly different ③(地区) area: 接～ have a common border; be adjacent to each other / 穷乡僻～ a remote, backward place

【壤土】[农] loam

攘 [书] ①(排斥) reject; resist: ～外 resist foreign aggression ②(抢) seize; grab ③(捋起) push up one's sleeves

【攘臂】[书] push up one's sleeves and bare one's arms (in excitement or agitation): ～高呼 raise one's hands and shout

【攘除】[书] get rid of; weed out; reject: ～奸邪 get rid of the wicked

【攘夺】[书] seize; grab

嚷 shout; yell; make an uproar: 别～了! Stop yelling. / 她在～什么? What's she shouting about?

ràng

让 ①(把好处给别人) give way; give ground; yield; give up: 寸步不～ refuse to yield an inch; not budge an inch / 各不相～。Neither is willing to give ground. ②(请人接受招待) invite; offer: ～茶 offer sb. tea ③(听任) let; allow; make: ～我试一试。Let me have a try. / 对不起,～你久等了。Sorry to have kept you waiting. ④(出让;转让) let sb. have sth. at a fair price: 我们可以按半价把这本词典～给你。We can let you have this dictionary at half its price. ⑤[介] 我～雨淋着了。I was caught in the rain. / 行李～雨淋湿了。The luggage got wet in the rain.

【让步】make a concession; give in; give way; yield: 不向无理要求～ not yield to any unreasonable demand / 在原则问题上不～ make no concession in matters of principle

【让步比】odds ratio

【让车线】passing track

【让开】get out of the way; step aside; make way

【让路】make way for sb. or sth.; give way; give sb. the right of way

【让球】concede points: 他～她五个球。He conceded her five points.

【让位】①(让出统治地位) resign sovereign authority; abdicate ②(让座) offer one's seat to sb. ③(转向) yield to; give way to; change into

【让贤与能】retire and give room to better men

【让座】①(让座位) offer one's seat to sb. ②(请客人入座) invite guests to be seated

ráo

荛 [书] firewood; faggot

【荛花】canescent wikstroemia

桡 oar

【桡动脉】[生理] radial artery

【桡骨】[生理] radius

【桡神经】nervus radialis

饶 ①(丰富) rich; plentiful: ～有风趣 full of wit and humor ②(饶恕) have mercy on; let sb. off; forgive: 求～ beg for mercy / ～他这回吧。Let him off this time. ③(另外添) give sth. extra; let sb. have sth. into the bargain: 给你～上一个。I'll let you have one more. ④[口][连] although; in spite of the fact that: ～这么细心核对,还是有遗漏。In spite of careful checking there are still omission.

【饶命】spare sb.'s life

【饶舌】①(唠叨) too talkative; garrulous ②(多嘴) say more than is proper; shoot off one's mouth

【饶恕】forgive; pardon

【饶沃】(of soil) fertile; rich

rǎo

扰 ①(扰乱) harass; trouble: ～人清梦 disturb one's sweet dream / 纷～ tumult; turmoil ②(客套话) trespass on sb.'s hospitality: 叨～, 叨～。Thank you for your hospitality.

【扰动】perturbation motion; disturb ◇ ～气流 rough air / ～速度 disturbance velocity / ～中心 center of disturbance

【扰流器】spoiler; vortex generator

【扰乱】harass; disturb; create confusion: ～边境 harass the border of a country / ～军心 undermine the morale of an army / ～市场 disrupt the market / ～视线 interfere with sb.'s view / ～治安 disturb public order ◇ ～反射体 confusion reflector / ～性射击 harassing fire / ～性药剂 harassing agent

【扰攘】hustle and bustle; noisy confusion; tumult: 干戈～ in the tumult of a raging war

rào

绕 ①(缠绕) wind; coil: ～线 wind thread ②(围着转动) move round; circle; revolve: 月球～地球运行。The moon revolves round the earth. ③(迂回过去) make a detour; bypass; go round: 道路施工,车辆～行。Detour. Road under repair. ④(纠缠) confuse; baffle; befuddle: 你的话把他～住了。What you said confused him.

【绕脖子】[方] ①(不直截了当) beat about the bush; speak or act in a roundabout way ②(言语曲折费思索) involved; knotty; tricky: 这句话太～了。This sentence is too involved.

【绕场一周】go round the stadium

【绕道】make a detour; go by a roundabout route: ～而行 take a devious route; go in a roundabout way / ～迂回 make a detour

【绕焊】boxing; end turning

【绕接】solderless wrapped connection; wire-wrap connection

【绕口令】tongue twister

【绕圈子】①(走迂回曲折的路) circle; go round and round ②(不直说) take a circuitous route; make a detour

【绕弯儿】①(散步) go for a stroll ②(不照直说) talk in a roundabout way; beat about the bush: 你直说吧,别～。You might come straight to the point. Don't beat about the bush.

【绕远儿】go the long way round

【绕组】[电] winding: 双线～ bifilar winding ◇ ～线 winding wire

【绕嘴】(of a sentence, etc.) not be smooth; be difficult to articulate: 这句话很～。This sentence is a tongue twister.

rě

惹 ①(引起) invite or ask for (sth. undesirable): ～麻烦 ask for trouble; invite trouble / ～是非 provoke a dispute; stir up trouble ②(触动对方) offend; provoke; tease: 别～他。Don't offend him. / 他不是好～的。He's not a man to be trifled with. ③(引起爱憎的反应) attract; cause: ～人讨厌 make a nuisance of oneself / ～人注意 attract attention

【惹草拈花】stir bushes and pick flowers; have a fancy for prostitutes

【惹火烧身】stir a fire only to burn oneself; court disaster; ask for trouble

【惹祸】court disaster; stir up trouble

【惹气】get angry

【惹事】stir up trouble

【惹是生非】provoke a dispute; stir up trouble

rè

热 ① [物] heat: 传～ transmission of heat; heat transfer; conduct heat / 导～性 heat conductivity / 凝固～ heat of solidification / 汽化～ heat of vaporization ②(温度高) hot: ～水 hot water / ～天 hot weather ③(加热) heat up; warm up; warm: 把牛奶～一下 heat up the milk ④(体温高) fever; temperature: 发～ have a fever; run a fever ⑤(情意深厚) ardent; warmhearted: ～望 ardently wish; fervently hope / ～心肠 warmheartedness; ardor ⑥(受欢迎的) in great demand; popular: ～货 goods in great demand; goods which sell well ⑦(热潮) craze; fad: 乒乓～ intense popular interest in table tennis; ping-pong craze ⑧(羡慕) envious; eager: 眼～ feel envious at the sight of sth. ⑨(放射性强) thermal; thermo-: ～磁 thermomagnetic / ～中子 thermal neutron

【热爱】ardently love; have deep love for: ～科学 take a keen interest in science / ～祖国 have ardent love for the motherland

【热病】pyreticosis

【热补】vulcanize (tire, etc.)

【热潮】great mass fervor; upsurge: 生产～ a great upsurge in production

【热忱】zeal; warmheartedness; enthusiasm and devotion: 革命～ revolutionary zeal

【热诚】warm and sincere; cordial: ～地希望 sincerely hope / ～欢迎 cordially welcome

【热处理】[机] heat treatment; thermal treatment ◇ ～钢 heat-treated steel / ～炉 heat-treatment furnace

【热磁】thermomagnetic

【热脆性】[冶] hot-shortness; red-shortness

【热带】the torrid zone; the tropics ◇ ～草原 savanna / ～风暴 tropical storm / ～雨林 selva; tropical rainforest / ～植物 tropical plants / ～作物 tropical crops

【热当量】heat equivalent
【热导体】[物] heat conductor
【热电】[物] pyroelectricity; thermoelectricity ◇ ～厂 heat and power plant / ～偶 thermocouple / ～体 pyroelectrics / ～效应 pyroelectric effect / ～学 pyroelectricity / ～站 heat and power station / ～阻 thermal resistance
【热度】①(冷热程度) degree of heat; heat ②(高于正常体温) fever; temperature: 他的～退了吗? Has his temperature come down?
【热风】[气] hot wind; hot-blas-air
【热风炉】[冶] hot-blast stove
【热敷】[医] hot compress
【热辐射】[物] heat radiation; thermal radiation
【热功当量】[物] mechanical equivalent of heat
【热功率】thermal power
【热固塑料】[化] thermosetting plastic
【热锅上的蚂蚁】ants on a hot pan: 急得象～一样 as restless as ants on a hot pan
【热耗】heat rate
【热核】thermonuclear ◇ ～爆炸 thermonuclear explosion / ～弹头 thermonuclear warhead / ～技术 thermonucleonics / ～武器 thermonuclear weapon
【热核反应】[物] thermonuclear reaction: 受控～ controlled thermonuclear reaction ◇ ～堆 thermonuclear reactor
【热烘烘】very warm: 屋里～的。It's very warm in the room
【热乎】①(热) nice and warm; warm: 汤还～。The soup is still warm. ②(亲热) warm and friendly; pally; chummy; thick
【热乎乎】warm: 她的话使我心里感到～的。Her words warmed my heart.
【热火朝天】buzzing with activity; in full swing
【热机】[机] heat engine
【热寂】[物] heat death
【热加工】[冶] hot-working; hot work
【热扩散】[物] thermal diffusion
【热辣辣】burning hot; scorching: 觉得脸上～的 feel one's cheeks burning
【热浪】[气] heat wave; hot wave
【热泪滚滚】warm tears streaming down one's face; hot tears rolled down one's cheeks
【热泪盈眶】one's eyes brimming with tears
【热离子】[物] thermion ◇ ～电流 thermionic current / ～发射 thermionic emission / ～装置 thermionic device
【热力】[机] heating power ◇ ～学 thermodynamics
【热恋】be passionately in love; be head over heels in love
【热量】[物] quantity of heat ◇ ～单位 thermal unit / ～计 calorimeter
【热烈】warm; enthusiastic; ardent: ～的祝贺 warm congratulations / ～的欢迎 warm welcome / ～欢呼 warmly hail / ～欢送 give sb. a warm send-off / 反应～ respond enthusiastically
【热裂化】[石油] thermal cracking
【热流】①[气] thermal current ②(激动振奋的感受) warm current: 我握着老师的手,一股～传遍全身。I clasped my teacher's hand and felt a warm current coursing through my body.
【热门】in great demand; popular: ～货 goods in great demand; goods which sell well / 赶～ follow a craze
【热敏电阻】[电] thermal resistor; thermistor
【热闹】①(繁盛活跃) lively; bustling with noise and excitement: ～的菜市场 a food market bustling with activity; a busy food market ②(使欢跃愉快) liven up; have a jolly time: 晚会开始～起来了。The evening party is beginning to liven up. ③(热闹的景象) a scene of bustle and excitement; a thrilling sight: 看～ watch the excitement; watch the fun
【热能】[物] heat energy
【热膨胀】thermal energy
【热气】steam; heat: ～腾腾的馒头 steaming hot buns
【热切】fervent; earnest: ～的愿望 earnest wish; fervent hope / ～地希望 earnestly hope
【热情】①(热烈的感情) enthusiasm; zeal; warmth: 一封～洋溢的感谢信 an ebullient letter of thanks ②(有热情) warm; fervent; enthusiastic; warmhearted: ～接待 warmly receive; give sb. a warm reception
【热情奔放】an outburst of enthusiasm; bubbling with enthusiasm
【热情洋溢】glowing with enthusiasm; brimming over with enthusiasm: ～的讲话 a speech brimming with warm feeling
【热身运动】warm-up: 比赛开始前先做～ warm up before entering a game
【热水袋】hot-water bottle
【热水瓶】[口] thermos bottle; thermos; vacuum bottle
【热塑塑料】[化] thermoplastic
【热腾腾】steaming hot: ～的汤面 steaming hot noodles in soup
【热天】hot weather; hot season; hot days

【热望】fervently hope; ardently wish
【热线】①[物] heat ray ②(两国首脑之间的直通电话线路) hot line
【热象仪】[电子] thermal imaging system
【热效发光】thermoluminescence
【热效率】thermal efficiency; fuel efficiency; thermal flow
【热效应】fuel factor; heat effect; thermal results
【热心】enthusiastic; ardent; earnest; warmhearted: ~科学 eager to promote science / ~为顾客服务 warmheartedly serve the customers
【热心肠】[口] warm heart; warmheartedness
【热性肥料】[农] hot manure
【热学】[物] heat (a branch of physics)
【热血】warm blood; righteous ardor: ~沸腾 burning with righteous indignation / ~青年 ardent youth ◇ ~动物 warm-blooded animal; warm blood
【热压】[化] hot pressing
【热药】[中医] medicines of a hot or warm nature; tonics and stimulants
【热源】[物] heat source
【热轧】[冶] hot-rolling ◇ ~机 hot-rolling mill
【热战】hot war; shooting war
【热胀冷缩】expand with heat and contract with cold; expansion caused by heat and contraction caused by cold; heat makes something expand and cold makes it contract
【热障】[物] heat barrier
【热证】[中医] heat symptom-complex; febrile symptoms
【热值】[物] calorific value
【热中】①(急切想望) hanker after; crave: ~于个人名利 hanker after personal fame and gain ②(十分爱好) be fond of; be keen on: ~于溜冰 be very fond of skating
【热中子】[物] thermal neutron

rén

人 ①(泛指人) man; person; people; human being: 城里~ townman / 好~ good person / 坏~ bad person / 黄种~ the yellow race / 男~ man / 女~ woman / 外国~ foreigner; foreign national / 乡下~ countryman / 我们一共是六十~。We are sixty in number. ②(成年人) adult; grown-up: 长大成~ become a grown-up ③(某种人) a person engaged in a particular activity: 工~ worker / 军~ soldier / 主~ host ④(别人) other people; people: 助~为乐 take pleasure in helping people ⑤(人品; 性格) personality; character: 为~公正 upright in character / 为~谦恭 modest in character ⑥(人的身体或意识) state of one's health; how one feels: 送到医院,~已经死了。When the patient was taken to hospital, he had already died. ⑦(每人) everybody; each; all: ~所共知 be know to all ⑧(人手;人才) manpower; hand: 我们这里还缺~。We are still short of hands.
【人本主义】[哲] humanism
【人不可貌相】Men cannot be judged by their looks. (或: You cannot judge a tree by its bark.)
【人不为已,天诛地灭】unless a man looks out for himself, Heaven and Earth will destroy him; everyone for himself and the devil take the hindmost
【人不知,鬼不觉】without a soul knowing anything about it
【人才】①(德才兼备的人) a person of ability; a talented person; talent; qualified personnel: 科技~ qualified scientists and technicians / 难得的~ a person of extraordinary ability ②(美丽端庄的相貌) handsome appearance: 一表~ a man of striking appearance
【人才辈出】people of talent coming forth in large numbers
【人才出众】a person of exceptional ability or striking appearance
【人才济济】a galaxy of talent
【人才培养】personnel training
【人才外流】the brain drain
【人财两旺】prosperous both in family and purse
【人称】[语] person: 不定~ indefinite person / 第一~ the first person ◇ ~代词 personal pronoun
【人次】person-time; man-time
【人大】[简](全国人民代表大会) the National People's Congress ◇ ~常委会 the Standing Committee of the NPC / ~代表 deputy to the NPC
【人道】①(爱护、尊重、关怀人的道德) humanity; human sympathy ②(仁慈的) human; humane: ~的待遇 humane treatment / 不~ inhuman ◇ ~主义 humanitarianism / ~主义者 humanitarian
【人地生疏】be unfamiliar with the place and the people; be a complete stranger
【人丁】population; family number: ~兴旺 have a growing family; have a flourishing pop-

ulation ◇～税 head tax
【人定胜天】man can conquer nature; man will triumph over nature
【人多势众】overwhelm with numerical strength; many hands provide great strength
【人多智广】More people means more ideas. (或: Two heads are better than one.)
【人多嘴杂】Agreement is difficult if there are too many people.
【人贩子】trader in human beings
【人防】[简](人民防空) people's air defense; civil air defense
【人粪尿】[农] night soil; human wastes
【人浮于事】have more hands than needed; be overstaffed
【人格】①(品性) personality; character; moral quality: ～高尚 have a noble character; have moral integrity ②(尊严) human dignity ◇～化 personification
【人工】①(人为的) man-made; artificial ②(人力) manual work; work done by hand: 用机器代替～ install machines to replace manual labor ③(工作量计算单位) manpower; man-day: 修这条渠道要多少～? How many man-days will be needed to construct this irrigation canal? ◇～草场 artificially sown pastures／～繁殖 [农] artificial propagation／～孵化 artificial incubation／～更新[林] artificial regeneration／～合成蛋白质 synthetic protein／～合成结晶胰岛素 synthetic crystalline insulin／～呼吸 artificial respiration／～湖 man-made lake／～降水 artificial rainmaking; artificial precipitation／～降雨 artificial rainfall／～降雨装置 artificial rain device; sprinkler／～流产 induced abortion／～器官 artificial organs／～肾 artificial kidney／～授粉[农] artificial pollination／～授精 [牧] artificial insemination／～心肺机 heart-lung machine; extra-corporeal circulation apparatus／～选择 [生] artificial selection／～语言 synthetic language; artificial language／～智能 artificial intelligence
【人公里】[交] passenger-kilometer
【人海】a sea of faces; a huge crowd (of people)
【人和】unity and coordination within one's own ranks; support of the people
【人欢马叫】people bustling and horses neighing; a busy, prosperous country scene
【人寰】[书] man's world; the world
【人机对话】man-machine interaction; interactive
【人机联系】man-machine interface
【人机模拟】man-machine simulation
【人迹】human footmarks traces of human presence: ～罕至的地区 an untraversed region
【人家】①(住户) household: 十户～ ten households ②(家庭) family: 勤俭～ an industrious and frugal family ③(未来的丈夫家) fiancé's family: 她已经有了～儿了。She is engaged to be married. ④(别人) other; another: ～都这么说。That's what everybody says.／～能做到的,我们也能做到。If other people can do it, so can we. (或: What other people can do we can do, too.) ⑤(指某个人或某些人) he; she; they: 把信给～送去。Take the letter to him 〈her; them〉. ⑥(指说话者本人) I: ～等你半天了。I've been waiting for you for quite a while.／～这会儿忙着呢! I am busy just now!
【人间】man's world; the world: ～地狱 a hell on earth／～乐园 earthly paradise／～奇迹 a miracle／～天堂 heaven on earth
【人杰】an outstanding personality: ～地灵 a place propitious for giving birth to great men; the greatness of a man lends glory to a place
【人尽其才】make the best possible use of men
【人口】①(居住某地人的总数) population: ～稀少 have a sparse population／～众多 have a very large population／标准～ standard population／常住～ resident population; permanent population／城市～ urban population／待业～ population waiting for employment／非农业～ non-agricultural population／流动～ transient population／流动过剩～ floating overpopulation／盲流～ aimlessly drifting population／农业～ agricultural population／适度～ optimum population／现住～ de facto population／乡村～ rural population／在业～ working population ②(一户人家的人数) number of people in a family: 他家～不多。There aren't many people in his family. ◇～爆炸 population explosion／～不足 underpopulation／～出生率 human fertility／～分布 population distribution／～负增长 NPG; negative population growth／～更替 population replacement／～过剩 overpopulation／～基数 population base／～减少 population decline; population decrease／～结构 population structure／～老化 aging of population／～理论 population theory／～零点增长 ZPG; zero population growth／～密度 density of

population; population density / ～年龄金字塔 population pyramid / ～年轻化 rejuvenation of population / ～普查 census / ～生育力 human fecundity / ～死亡率 human mortality / ～特性 characteristics of the population / ～统计 vital statistics; population statistics / ～预测 population projection / ～再生产 population reproduction / ～增长 population growth / ～自然增长 natural growth of the population / ～质量 population quality

【人口学】demography ◇ ～家 demographer ○马尔萨斯主义 Malthusianism / 新马尔萨斯主义 neo-Malthusianism

【人困马乏】the men weary, their steeds spent; the entire force was exhausted

【人老识广】The devil knows many things because he is old.

【人老心不老】be old in age, but young in mind; though old, still young in heart

【人老珠黄】one getting old like the pearl becoming yellow; lose one's looks; no longer held in esteem

【人类】mankind; humanity: ～起源 the origin of mankind; the origin of the human species ◇ ～环境 human environment / ～生态学 human ecology / ～行为 human behavior / ～学 anthropology

【人力】manpower; labor power: ～资源 human resources / 爱惜～物力 treasure manpower and material resources

【人力车】①(人力拉或推的车) a two-wheeled vehicle drawn by man ②[旧](人力拉的载客的车) rickshaw

【人流】stream of people

【人伦】human relations (according to feudal ethics)

【人马】forces; troops: 原班～ the same batch of people; the old cast

【人马座】[天] Sagittarius

【人满为患】The house is crowded in every part. (或: The place is packed.)

【人们】people; men; the public

【人面兽心】have the face a man but the heart of a beast; a beast in human shape

【人民】the people ◇ ～代表大会 people's congress / ～法院 people's court / ～法院院长 president of the people's court / ～防空 people's air defense; civil air defense / ～公社 people's commune / ～检察院 people's procuratorate / ～检察院检察长 chief procurator of the people's procuratorate / ～警察 the people's police / ～来信 letters from the masses / ～民主专政 people's democratic dictatorship / ～陪审员 [法] people's assessor / ～勤务员 servant of the people / ～群众 the masses / ～武装部 people's armed forces department / ～英雄纪念碑 the Monument to the People's Heroes / ～战争 people's war / ～政府 the People's Government / ～资本主义 people's capitalism

【人民币】Renminbi (RMB) ○外汇兑换券 foreign exchange certificate

【人民团体】mass organization; people's organization ○中华全国总工会(全总) All-China Federation of Trade Unions / 中国共产主义青年团(共青团) Communist Youth League of China / 中华全国青年联合会(全国青联) All-China Federation of Youth / 中华全国学生联合会(全国学联) All-China Students' Federation / 中国少年先锋队(少先队) China Young Pioneers / 中华全国妇女联合会(全国妇联) All-China Women's Federation / 中国科学技术协会(中国科协) China Association for Science and Technology / 中国文学艺术界联合会(文联) China Federation of Literary and Art Circles / 中华全国归国华侨联合会(全国侨联) All-China Federation of Returned Overseas Chinese / 中国福利会 China Welfare Institute / 中国红十字会 Red Cross Society of China

【人民性】affinity to the people

【人命】human life: ～案子 a case of homicide or manslaughter / ～关天。A case involving human life is to be treated with the utmost care.

【人莫予毒】no one dare harm me; an arrogant boast

【人怕出名猪怕壮】fame portends trouble for men just as fattening does for pigs

【人品】①(人的品质) moral standing; moral quality; character: ～很好 be a person of excellent character ②(仪表)[口] looks; bearing

【人情】①(人之常情) human feelings; human sympathy; sensibilities: 不近～ not amenable to reason; unreasonable ②(情面) human relationship: ～练达 experienced in the ways of the world / ～之常 natural and normal ③(恩惠) favor: 空头～ lip service / 做个～ do sb. a favor ④(礼物) gift; present: 送～ send gifts; make a gift of sth. ◇ ～味 human touch; human interest

【人情世故】worldly wisdom: 不懂～ not know

the ways of the world
【人穷志不穷】though one is poor, he has lofty aspirations; poor but with lofty ideas
【人穷志短】poverty chills ambition
【人权】human rights; rights of man ◇ ~法 Human Right Act / ~委员会 Human Right Commission
【人权宣言】①(法国的) Declaration of the Rights of Man and of the Citizen (1789) ②(联合国的) Declaration of Human Rights
【人群】crowd; throng; multitude
【人人】everybody; everyone: ~平等。Everyone is equal.(或:All men are equal.) / ~自危。Everyone finds himself in danger.
【人人为我,我为人人】All for one and one for all.
【人山人海】huge crowds of people; a sea of people: 观众~ a bumper audience / 广场上~。The square was a sea of people.
【人身】living body of a human being; person ◇ ~安全 personal safety / ~保护 physical protection / ~保险 personal insurance / ~不可侵犯 inviolability of the person; personal abuse / ~攻击 personal attack / ~事故 personal injury caused by an accident / ~侮辱 personal insult / ~虐待 body persecution / ~自由 freedom of person; personal freedom
【人参】ginseng: ~酒 ginseng liquor
【人生】life ◇ ~观 outlook on life / ~哲学 philosophy of life
【人声】voice: ~鼎沸 a hubbub of voices
【人士】personage; public figure: 爱国~ patriotic personage / 党外~ people outside the Party / 各界~ people of all walks of life / 官方~ official quarters / 体育界~ figures in the sports world / 文艺界~ people of literary and art circles / 消息灵通~ informed sources / 友好~ friendly personality / 知名~ well-known figures; celebrities
【人世】this world; the world: ~沧桑 tremendous changes in the world / 不在~ be no longer living; be no longer in the land of the living
【人事】①(人的处境) human affairs; occurrences in human life ②(人员的调动、安排等工作) personnel matters: ~调动 transfer of personnel / ~更迭 change of personnel ③(事理人情) ways of the world: 不懂~ not know the ways of the world ④(人的意识的对象) consciousness of the outside world: 不省~ lose consciousness ⑤(人力能做到的事) what is humanly possible: 尽~ do what is humanly possible; do one's best ◇ ~处 personnel division / ~档案 personal file / ~关系 organizational affiliation / ~管理 personal administration / ~制度 personnel system
【人手】manpower; hand: ~太少 short of hands; shorthanded
【人手一册】every one has a copy
【人寿保险】life insurance
【人寿年丰】the land yields good harvests and the people enjoy good health
【人体】human body ◇ ~残骸 human remains / ~模型 manikin ○头 head / 头顶 the top 〈crown〉 of the head / 发 hair / 额 forehead / 太阳穴 temple / 面部 face / 颧骨 cheekbone / 颊 cheek / 眼 eye / 眉毛 eyebrow / 睫毛 eyelash / 眼睑 eyelid / 眼球 eyeball / 瞳孔 pupil / 虹膜 iris / 耳 ear / 耳膜 ear drum / 听道 auditory canal / 鼻 nose / 鼻腔 nasal cavity / 鼻孔 nostril / 鼻梁 bridge of the nose / 人中 philtrum / 口 mouth / 唇 lip / 舌 tongue / 牙 tooth / 齿冠 crown / 齿龈 gum / 齿根 root; fang / 珐琅质 enamel / 硬腭 hard palate / 软腭 soft palate / 悬雍垂 uvula / 扁桃体 tonsil / 声带 vocal cords / 咽头 pharynx / 颏 chin / 颌 jaw / 颚骨 jawbone / 颈 neck / 项 nape 〈scruff〉 of the neck / 咽喉 larynx / 喉 throat / 躯干 trunk / 背 back / 肩 shoulder / 腋窝 armpit / 胸 chest; thorax / 胸部 chest / 乳房 breast / 乳头 nipple / 腰 waist; loins / 肚脐 navel; belly button / 腹 abdomen; belly / 上腹 the upper abdomen / 下腹 the lower abdomen / 腹股沟 groin / 臀部 buttocks; hips; bottom; backside / 生殖器 genitals; genital organ / 阴茎 penis / 阴囊 scrotum / 睾丸 testicle / 阴道 vagina / 肘 elbow / 腕 wrist / 手 hand / 手背 back of the hand / 手掌 palm of the hand / 手指 finger / 指甲 finger nail / 拇指 thumb / 小指 little finger / 中指 middle finger / 食指 forefinger; index finger / 无名指 ring finger / 腿 leg / 大腿 thigh / 小腿 lower leg; shank / 足 foot / 膝 knee / 膝关节 knee joint / 膝盖骨 kneecap / 膝弯 hollow of the knee; back of the knee / 腿肚 calf / 踝 ankle / 脚背 instep / 四肢 limbs / 臂 arm / 上臂 upper arm / 前臂 fore arm / 脚掌 sole of the foot / 后跟 hell / 趾 toe / 趾甲 toenail / 内脏 viscera; internal organs / 呼吸道 respiratory tract / 消化道 alimentary canal / 气管

trachea; windpipe／甲状腺 thyroid gland／淋巴节 lymph node／食管 gullet; esophagus／肺 lung／心脏 heart／动脉 artery／静脉 vein／毛细血管 blood capillary／横膈 diaphragm／胃肠道 gastrointestinal tract／胃 stomach／十二指肠 duodenum／肝 liver／胆囊 gall bladder／阑尾 appendix／肠 intestine／大肠 large intestine／结肠 colon／小肠 small intestine／直肠 rectum／肛门 anus／脾 spleen／胰 pancreas／生殖泌尿道 genitourinary tract／肾 kidney／输尿管 ureter／膀胱 urinary bladder／尿道 urethra／输精管 spermatic duct; seminal duct／精囊 seminal vesicle／输卵管 oviduct; fallopian tube／卵巢 ovary／子宫 womb; uterus／胚胎 embryo／胎儿 fetus／骨胳 bones and skeleton／头颅 skull／额骨 frontal bone／肩胛骨 shoulder blade／锁骨 clavide／肋骨 rib／肋软骨 costal cartilage／胸骨 breast bone: sternum／腕骨 caspal／掌骨 metacarpal／肘关节 elbow joint／桡骨 radius／尺骨 ulna／骨盆 pelvis／髋骨 hip-bone／髋关节 hip-joint／股骨 femur; thigh bone／坐骨 ischium／耻骨 pubis; pubic bone／指趾骨 phalange／腓骨 fibula／胫骨 tibia; shin bone／肌肉 muscle／面肌 facial muscle／二头肌 biceps／三头肌 triceps／胸肌 pectoral muscle／腹肌 abdominal muscle／横韧带 transverse ligament／三角肌 deltoid muscle／背伸肌 extensor of the back／臀肌 gluteus; gluteal muscle／腱 tendon／神经 nerve／中枢神经系统 central nervous system／周围神经系统 peripheral nervous system／脑 brain／大脑 cerebrum／小脑 cerebellum／脑神经 cranial nerve／脊髓神经 spinal nerve／交感神经 sympathetic nerve／感觉器官 sense organ／嗅神经 olfactory nerve／视神经 optic nerve／动眼神经 oculomotor nerve／滑车神经 trochlear nerve／三叉神经 trigeminal nerve／外展神经 abducent nerve／面神经 facial nerve／听神经 acoustic nerve／舌咽神经 glossopharyngeal nerve／迷走神经 vagus／副神经 accessory nerve／舌下神经 hypoglossal nerve／脊髓 spinal cord／颈神经 cervical nerve／胸神经 thoracic nerve／坐骨神经 sciatic nerve／股神经 femoral nerve

【人同此心,心同此理】everybody feels the same about this

【人头】①(人数) the number of people: 按～分 distribute according to the number of people ②(和人的关系) relations with people: ～熟 know a lot of people ③[方](品质) moral quality; character: ～儿次 be not much of a person ◇～税[旧] poll tax; capitation

【人微言轻】the words of the lowly carry little weight

【人为】artificial; man-made: ～的障碍 an artificially imposed obstacle ◇～地貌[军] culture features／～嬗变[物] artificial transmutation／～误差[气] personal equation／～噪声 man-made noise

【人为财死,鸟为食亡】Human beings die in pursuit of wealth, and birds die in pursuit of food. (或: The wages of avarice is death.)

【人为刀俎,我为鱼肉】be meat on sb.'s chopping block; be at sb.'s mercy

【人文科学】the humanities; humane studies

【人文主义】humanism

【人物】①(有代表性或突出特点的人) figure; personage: 大～ a big shot／杰出的～ an outstanding personage／历史～ a historical figure／小～ a nobody; a small potato／英雄～ a heroic figure; a hero or heroine ②(文学作品中描写的人) person in literature; character: ～塑造 characterization／典型～ typical character ◇～表 characters (in a play or novel)／～画 figure painting

【人像】portrait; image; figure ◇～靶 silhouette target

【人心】popular feeling; public feeling; the will of the people: ～丧尽 lose all popular sympathy／不得～ go against the will of the people／大快～ most gratifying to the people; to the great satisfaction of the people／得～ have the support of the people; enjoy popular support／深入～ strike root in the hearts of the people／收买～ curry favor with the public／振奋～ boost popular morale

【人心不古】Human heart are not what they were in the old days.(或: People are not so honest as their ancestors were.)

【人心不足蛇吞象】A man whose heart is not content is like a snake which tries to swallow an elephant.

【人心各异,犹如其面】Several men, several minds. (或: Men's hearts differ just as their faces do.)

【人心叵测】Man's heart is incomprehensible. (或: One's heart is past finding out.)

【人心所向】the popular sentiment; the feelings of the people

【人心向背】the feelings of the people: whether the people are for or against
【人行道】pavement; sidewalk
【人行横道】pedestrian crosswalk; pedestrian crossing
【人性】①(人的本性) human nature; humanity: 具体的～ human nature in the concrete／灭绝～ most barbarous; utterly inhuman ②(正常的感情和理性) normal human feelings; reason: 不通～ unfeeling and unreasonable
【人性论】the theory of human nature
【人选】person selected; choice of persons: 物色适当～ try to find a suitable person (for a job)
【人烟】signs of human habitation: ～稠密 be densely populated／～稀少 be sparsely populated／没有～ uninhabited; without a trace of human habitation
【人言可畏】gossip is a fearful thing
【人仰马翻】men and horses thrown off their feet; utterly routed
【人影儿】①(人的影子) the shadow of a human figure ②(人的形象或踪影) the trace of a person's presence; figure: 大街上看不到一个～。Not a soul was seen in the street.
【人员】personnel; staff: ～不足 understaffed; undermanned／技术～ technical personnel／教学～ the teaching staff; the faculty／全体～ the entire personnel; the whole staff
【人缘儿】relations with people; popularity: ～好 be very popular; enjoy great popularity
【人云亦云】echo the views of others; parrot
【人造】man-made; artificial; imitation ◇ ～宝石 imitation jewel／～冰 artificial ice／～革 imitation leather; leatherette／～红宝石 synthetic ruby／～黄油 margarine／～蓝宝石 synthetic sapphire／～棉 staple rayon／～石 [医] artificial stone／～石油 artificial petroleum／～丝 artificial silk; rayon／～树脂 artificial resin／～卫星 man-made satellite／～纤维 man-made fiber／～橡胶 artificial rubber; synthetic rubber／～羊毛 artificial wool
【人证】[法] testimony of a witness ◇ ～物证 human testimony and material evidence
【人之常情】the way of the world; what is natural and normal (in human relationships)
【人之将死,其言也善】A man's words are good when death is near. (或: When a man is near death he speaks from his heart.)
【人质】hostage: 被扣作～ be held as a hostage
【人中】philtrum
【人种】ethnic group; race ◇ ～学 ethnology
【人字呢】[纺] herringbone

壬 the ninth of the ten Heavenly Stems
【壬烷】nonane; nonyl hydride

仁 ①(仁爱) benevolence; kindheartedness; humanity: ～政 policy of benevolence; benevolent government ②(敏感) sensitive: 麻木不～ insensitive; apathetic ③(果仁) kernel: 核桃～ walnut kernel; walnut meat／花生～ shelled peanuts／虾～ shelled shrimps; shrimp meat
【仁爱】kindheartedness
【仁慈】benevolent; merciful; kind
【仁人君子】benevolent gentlemen; public-spirited people
【仁人志士】people with lofty ideals
【仁兄】my dear friend
【仁义道德】humanity, justice and virtue; virtue and morality
【仁者见仁,智者见智】the benevolent see benevolence and the wise see wisdom; different people look at a thing in diferent way
【仁政】policy of benevolence; benevolent government
【仁至义尽】do everything called for by humanity and duty; do what is humanly possible to help; show extreme forbearance

rěn

荏 ①(软弱)[书] weak; weak-kneed: 色厉内～ fierce of mien but faint of heart; threatening in manner but cowardly at heart ②[植] common perilla
【荏苒】[书] (of time) elapse quickly; pass imperceptibly; slip by: 光阴～,转瞬又是一年。Time zipped by and the year was soon over.

稔 [书] ①(庄稼成熟) harvest: 丰～ bumper harvest／一年两～ two crops a year ②(熟悉) be familiar with sb.: ～知 know sb. quite well／素～ have long been familiar with sb.

忍 ①(忍耐;忍受) bear; endure; tolerate; put up with: ～饥挨饿 endure the torments of hunger／～着眼泪 hold back one's tears／是可～,孰不可～? If this can be tolerated, what cannot? ②(忍心) be hardhearted enough to; have the heart to: 残～ cruel; ruthless／于心不～ not have the heart to
【忍不住】unable to bear; cannot help (doing sth.): ～哭 cannot help bursting into tears／～笑 cannot refrain from laughing

【忍冬】[植] honeysuckle
【忍饥挨饿】stay one's stomach; endure the torments of hunger
【忍俊不禁】cannot help laughing
【忍耐】exercise patience; exercise restraint; restrain oneself
【忍气吞声】swallow an insult; submit to humiliation
【忍让】exercise forbearance; be forbearing and conciliatory
【忍辱负重】endure humiliation in order to carry out an important mission
【忍辱偷生】live on, bearing one's shame
【忍受】bear; endure; stand: ～艰难困苦 endure hardships／～侮辱 pocket an insult／热得难以～ unbearably hot
【忍痛】very reluctantly: ～割爱 part reluctantly with what one treasures
【忍无可忍】be driven beyond (the limits of) forbearance; come to the end of one's patience
【忍心】have the heart to; be hardhearted enough to: 不～ cannot bear to; cannot stand the sight of

rèn

认 ①(认识) recognize; know; make out; identify: ～出某人 identify a person; recognize a person／他变化太大,几乎让人～不出了。He had changed so much that one could hardly recognize him. ②(建立关系) enter into a certain relationship with; adopt: ～师傅 apprentice oneself to sb. ③(承认;同意) admit; recognize; own: 承～ admit; recognize／否～ deny／公～ be generally acknowledged ④(应承) undertake to do sth. ～捐一笔款 undertake to contribute a sum ⑤(认吃亏) accept as unavoidable; resign oneself to: 这房子一定要建,多花钱我也～了。 I simply must have the house built, even if I have to spend a lot of money.
【认错】acknowledge a mistake; admit a fault; make an apology
【认得】know; recognize: 我～那家商店。I recognize the shop.／我～她。I know her.
【认敌为友】regard enemies as friends; take the enemy for one's own people; take a foe for a friend
【认定】firmly believe; maintain; hold: 我～这是错的。 I maintained that it was wrong.
【认定同一】establish sb.'s identity
【认购】offer to buy; subscribe: ～公债 subscribe for bonds／～股票 subscribed shares
【认股特权】subscription right
【认股证书】warrant; stock warrants
【认可】approve; accept; confirm
【认领】claim: ～失物 claim a lost article
【认清】see clearly; recognize; get a clear understanding of: ～形势 get a clear understanding of the situation
【认生】(of a child) be shy with strangers
【认识】①(确定某场) know; understand; recognize: ～某人 know somebody ②(头脑对客观世界的反映) understanding; knowledge; cognition: 感性～ perceptual knowledge／理性～ rational knowledge／～来源于实践。Knowledge originates in practice. ◇ ～过程 process of cognition／～功能 recognizing ability／～论 theory of knowledge; epistemology／～能力 cognitive ability／～水平 level of understanding／～系统 recognition system
【认输】admit defeat; throw in the sponge; give up
【认同作用】[心理] identification
【认为】think; consider; hold; deem: 你～怎样? What do you think of it?／我不～这是真的。I don't believe it to be true.／我们～所有国家都应一律平等。 We hold that all nations should be equal.
【认贼作父】take the foe for one's father; regard the enemy as kith and kin
【认帐】acknowledge a debt admit what one has said or done
【认真】①(不马虎) conscientious; earnest; serious: ～的自我批评 an earnest self-criticism／～执行政策 carry out the policy conscientiously ②(当真) take seriously; take to heart: 我是说着玩儿的,他就～了。 I was only joking, but he took it to heart.
【认证】[法] attestation; authentication ◇ ～费 certification fee／～遗嘱 prove a will
【认知】[心理] cognition
【认字】know or learn how to read
【认罪】admit one's guilt; plead guilty ◇ ～书 statement of confession／～态度 attitude toward admission of guilt

任 ①(任用) appoint: 新～的市长 the newly appointed mayor of the city ②(担任) assume a post; take up a job: ～教 take up teaching; work as a teacher ③(职务) official post; office: ～满 expiration of one's term of office／～内 during one's term of office／返～ resume charge of the office; return to one's post / 赴～ proceed to take up one's post / 就～ assume office / 离～ leave office / 离～期间 during one's absence / 上～ take up an official post; assume office ④[量]做过两～大使 have twice been ambassador ⑤(听凭) let; allow; give free rein to: ～你挑选 choose any as you like / ～其自流 let things run their course ⑥(无论) no matter (how, what, etc.): ～他怎么说,也别信他。Don't trust him, no matter what he says.

【任便】as you like; as you see fit: 你去不去～。You may go or not as you see fit.

【任何】any; whichever; whatever: 没有～理由拒绝这个建议。There's no reason whatsoever to turn down this suggestion. / 你可以从两个中间挑～一个。You may take either of the two.

【任劳任怨】work hard and not be upset by criticism; bear responsibility without grudge

【任免】appoint and remove: ～事项 appointments and removals / ～行政人员 appoint and remove administrative personal ◇ ～权 power of appointment and removal

【任命】commission; designate; appoint: ～他为校长 appoint him president (of the college) / 被～为驻…大使 be appointed ambassador to...

【任凭】①(听凭) at one's convenience; at one's discretion: 这事不能～他一人决定。This shouldn't be left entirely to his discretion. ②(不管) no matter (how, what, etc.) ～问题多复杂,我们也能弄清楚。We can solve the problem no matter how complicated it is.

【任凭风浪起,稳坐钓鱼船】sit tight in the fishing boat despite the rising wind and waves; hold one's ground despite pressure or opposition

【任期】term of office; tenure of office

【任人摆布】allow oneself to be ordered about

【任人唯亲】appoint people by favoritism

【任人唯贤】appoint people on their merits; appoint people according to their political integrity and ability

【任人宰割】allow oneself to be trampled upon; be partitioned by others at will

【任务】assignment; mission; task; job: 光荣而艰巨的～ a glorious but arduous task / 接受～ receive an assignment ◇ ～观点 get-it-over-and-done-with attitude; perfunctory attitude

【任性】capricious; self-willed; wayward; headstrong: ～恣意 indulge in emotions

【任意】wantonly; arbitrarily; willfully: ～捏造事实 indulge in pure fabrication / ～诬蔑 wantonly vilify ◇ ～常数 [数] arbitrary constant / ～条款 optional clause; permissive provision / ～卸货港交货 optional delivery

【任意球】①[足球] free kick ②[手球] free throw

【任用】appoint; assign sb. to a post

【任职】hold a post; be in office: ～期满 expiration of one's term of office / 在～期间 during one's tenure of office / 在…～ work in; hold an office

【任重道远】the burden is heavy and the road is long; shoulder heavy responsibilities

妊 be pregnant

【妊妇】pregnant woman

【妊娠】gestation; pregnancy: 输卵管～ tubal pregnancy ◇ ～反应 pregnancy reaction / ～期 gestational period / ～试验 pregnancy tests / ～中毒症 toxemias of pregnancy

刃 ①(刀口) the edge of a knife, sword, etc.; blade: 刀～ knife blade ②(刀) sword; knife: 利～ sharp sword / 白～战 bayonet fighting ③(用刀杀) kill with a sword or knife: 手～ stab sb. to death; kill with one's own hand

【刃具】[机] cutting tool

韧 pliable but strong; tenacious; tough

【韧带】[生理] ligament

【韧皮部】[植] bast; phloem

【韧皮纤维】[植] bast fiber

【韧性】toughness; tenacity

纫 ①(缝) sew; stitch ②(引线穿过针鼻儿) thread (a needle)

rēng

扔 ①(抛) throw; toss; cast: ～球给某人 toss a ball to sb. / 把那本书～给我。Throw me that book. ②(抛弃) throw away; cast aside: 把它～了吧。Throw it away.

【扔下】abandon; put aside; leave behind: 车坏得太利害了,他们把它～了。The car was badly damaged so they abandoned it. / 那个男人～了老婆和孩子。The man abandoned his wife and child.

réng

仍 ①(依照) remain: 一～其旧 remain the same; follow the beaten track ②(仍然) still; yet: ～未痊愈 have not yet recovered / ～须努

力 must continue to make efforts／～有效力 be still effective; be still in force／那事～未解决。The matter is still unsettled.
【仍旧】①(照旧) remain the same ②[副](仍然) still; yet: ～乐观 be still optimistic
【仍然】[副] still; yet: 他～很忙。He is still busy.

rì

日 ①(太阳) sun: ～出 sunrise／～落 sunset ②(白天) daytime; day: ～～夜夜 day and night; night and day／夜以继～ day and night ③(一昼夜; 天) day: 今～ today／多～不见了,你好吗? Haven't seen you for a long time. How are you? ④(每天) daily; every day; with each passing day: ～渐好转 get better every day ⑤(泛指一段时间) time: 春～ springtime; spring／来～ the days to come; the future／往～ former days
【日班】day shift: 上～ be on the day shift
【日报】daily paper; daily: 中国～ China Daily
【日本】Japan ◇ ～人 Japanese
【日本脑炎】encephalitis B; Japanese encephalitis
【日薄西山】the sun is setting beyond the western hills; declining rapidly; nearing one's end
【日不暇给】be fully occupied every day
【日常】day-to-day; everyday; daily: ～费用 current expense／～工作 day-to-day work; routine duties／～生活 everyday life; daily life／～用语 words and expressions for everyday use
【日场】day show; daytime performance
【日程】program; schedule: 访问～ itinerary of a visit／工作～ work schedule; program of work ◇ ～表 schedule
【日出星没】The sun blinds the stars.
【日戳】①(戳子) date stamp; dater ②(印下的戳记) datemark
【日珥】[天] prominence
【日工】①(白天的活儿) daywork ②(临时工作) day labor ③(临时工) day laborer
【日光】sunlight; sunbeam ◇ ～疗法 heliotherapy／～浴 sun bath／～浴室 solarium
【日光灯】fluorescent lamp; daylight lamp ◇ ～起动器 fluorescent lamp starter／～镇流器 fluorescent lamp ballast
【日晷】sundial
【日后】in the future; in days to come
【日积月累】accumulate over a long period
【日记】diary: 工作～ work diary; daily account of one's work／记～ keep a diary ◇ ～本 diary
【日记帐】journal; daybook
【日间】in the daytime; during the day
【日见】with each passing day; day by day: ～好转 get better every day／～衰败 decline day by day
【日渐】with each passing day; day by day: ～强壮 get stronger and stronger／～壮大 be growing steadily
【日界线】[天] date line
【日久】with the passing of time; in (the) course of time: ～天长 in (the) course of time; as the years go by; after a considerable period of time
【日久见人心】time reveals a person's heart; it takes time to know a person
【日理万机】attend to numerous affairs of state every day; be occupied with a myriad of state affairs
【日历】calendar ◇ ～变动 calendar variations／～年度 calendar year／～手表 calendar watch
【日冕】[天] (solar) corona ◇ ～仪 coronagraph
【日暮途穷】the day is waning and the road is ending; approaching the end of one's days
【日内】in a few days; in a day or two; in a couple of days
【日期】date: 出生～ date of birth／开会的～ the date for the meeting／没有～ without date／载明～ bear date／别忘了信上写明～。Don't forget to date your letters.／起程的～定了吗? Has the departure date been fixed?
【日前】a few days ago; the other day
【日趋】with each passing day; gradually; day by day: ～衰落 decline day by day／市场～繁荣。The market is becoming brisker day by day.
【日上三竿】the sun is three poles high; it's late in the morning
【日射】[气] insolation ◇ ～表 actinometer／～病[医] sunstroke; insolation
【日食】[天] solar eclipse: 日环食 annular eclipse／日偏食 partial solar eclipse／日全食 total solar eclipse
【日托】day care: ～托儿所 day nursery
【日心轨道】heliocentric orbit
【日心视差】[天] heliocentric parallax
【日心说】[天] heliocentric theory
【日新月异】change with each passing day: 祖

国面貌～。 The face of our motherland has been changing with each passing day.
【日夜】day and night; night and day; round the clock
【日夜安全弹簧锁】night latch
【日夜不宁】be restless by day and sleepless by night
【日夜操劳】work indefatigably day and night
【日夜商店】a shop open night and day; a round-the-clock shop; a day-and-night-service shop
【日以继夜】night and day; round the clock
【日益】increasingly; day by day: ～壮大 getting stronger day by day / 矛盾～尖锐。The contradictions are becoming increasingly acute.
【日用】①(日常生活费用) daily expenses ②(日常生活应用的) of everyday use ◇ ～必需品 daily necessities; household necessities / ～工业品 manufactured goods for daily use / ～开支 general expenses / ～品 articles of everyday use
【日语】Japanese (language)
【日预报】[气] daily forecast
【日元】Yen
【日月】life; livelihood: ～同辉 shine forever like the sun and the moon
【日月合璧】High in the sky hang sun and moon.
【日月经天】be as everlasting as the sun and the moon
【日月如梭】the sun and the moon move back and forth like a shuttle; time flies
【日月星辰】the sun, the moon and the stars; the heavenly bodies
【日晕】[气] solar halo
【日晕主雨】A solar halo means rain.
【日照】sunshine: 长～植物 long-day plant / 短～植物 short-day plant ◇ ～计 sunshine recorder / ～时间 sunshine time
【日臻完善】becoming better and approaching perfection day by day
【日志】daily record; journal: 工作～ daily record of work / 航海～ logbook; log
【日中】[书] noon; midday
【日中则昃】as soon as the sun reaches the meridian it declines; decline after reaching the zenith
【日子】①(日期) day; date: 定个～ fix a date ②(天数) days; time: 他走了有些～了。He's been away for some time. ③(生活) life; livelihood: ～好过 enjoy a happy life / 勤俭过～ lead an industrious and frugal life

róng

容 ①(容纳) hold; contain: 这个阅览室能～一百人。The reading room can hold a hundred people. ②(宽容) tolerate: ～人之过 be tolerant of other's faults / 宽～ be tolerant ③(允许) permit; allow: 不～耽搁 allow of no delay / 不～怀疑 admit of no doubt / 不～歪曲 brook no distortion / 详情～后再告。Permit me to give the details later. ④(脸上神情、气色) facial expression: 怒～ an angry look / 笑～ a smiling face ⑤(景象; 状态) appearance; looks: 市～ the appearance of a city / 阵～ lineup; battle array
【容差】allowance; tolerance
【容光焕发】one's face glowing with health
【容华绝代】be endowed with a rare and radiant beauty
【容积】volume ◇ ～吨 measurement ton / ～流量 volume flow / ～效率 volumetric flow rate
【容量】capacity
【容貌】appearance; looks
【容貌流盼】have alluring eyes and exquisite features
【容貌秀丽】be of charming appearance; pretty
【容纳】hold; have a capacity of; accommodate: ～不同意见 tolerate opinions different from one's own
【容器】container; vessel
【容情】show mercy: 我们对坏人是决不～的。We never show mercy to bad people.
【容人】tolerant towards others; magnanimous; broad-minded: ～之过 be tolerant of other's faults
【容忍】tolerate; put up with; condone: 我们不能～这种浪费现象。 We cannot tolerate such waste.
【容身】shelter oneself: ～之地 a place to stay / 无～之地 without a place to shelter oneself
【容限】[物] tolerance; allowance: 光学～ optical tolerance
【容许】①(许可) tolerate; permit; allow: 如条件～ if conditions permit / 情况不～我们再等待了。In such circumstances we can't afford to wait any longer. ②(或许) possibly; perhaps ◇ ～负载[电] allowable load / ～收缩量 shrinkage allowance / ～误差 admissible error / ～压力 allowable pressure / ～载荷 allowable load

【容颜】appearance; looks: ～憔悴 a sorrowful look
【容易】①(不难) easy: ～处理的局面 a situation easy to handle / 到那个地方是很～的。It's easy to get to that place. ②(可能性大) easily; likely; liable; apt: ～发生事故 be prone to accidents / ～晕船 be liable to seasickness
【容重】[水] unit weight

溶 dissolve: 盐～于水。Salt dissolves in water.
【溶化】dissolve: 把糖～在水中 dissolve sugar in water
【溶剂】[化] solvent; dissolvent; menstruum: ～精制油 solvent-refined oil
【溶胶】[化] sol
【溶解】dissolve: ～的物理显影 solution physical development ◇ ～度 solubility / ～热 heat of solution / ～物 dissolved matter / ～型 lysotype
【溶菌剂】bacteriolysant
【溶菌素】[医] bacteriolysin
【溶溶】[书] broad: 江水～。The river is broad and gentle.
【溶蚀】[地] corrosion
【溶性油】[化] soluble oil
【溶血】[医] hemolysis ◇ ～现象 hemolysis
【溶液】[化] solution: 当量～normal solution / 实在～ real solution ◇ ～处理 solution treating / ～分析 liquor analysis / ～电解 electrolysis of solutions / ～压力 solution pressure
【溶质】[化] solute

熔 melt; fuse; smelt
【熔池】[冶] (molten) bath
【熔滴】molten drop
【熔点】[物] melting point
【熔度】meltability
【熔断】[电] fusing ◇ ～器 fuse (box)
【熔化】melt: 热将使铁～。Heat will melt iron. ◇ ～炉 melting furnace / ～期 [冶] melting stage / ～速率 [冶] melting rate
【熔剂】[冶] flux
【熔解】[物] fuse; fusion ◇ ～热 heat of fusion
【熔炼】smelt: 闪速～ flash smelting ◇ ～炉 smelting furnace
【熔炉】①(熔炼金属的炉子) smelting furnace ②(锻炼思想的环境) crucible; furnace: 在革命的～中经受锻炼 be tempered in the furnace of revolution
【熔融】[化] melt ◇ ～纺丝 melting spinning / ～挤压法 [化纤] extrusion by melting
【熔丝管】cartridge fuse
【熔性】fusibility
【熔岩】[地] lava
【熔铸】founding; casting ◇ ～工 smelter

榕 [植] small-fruited fig tree; banyan

戎 [书] army; military affairs: 投笔从～ cast aside the pen to join the army; give up intellectual pursuits for a military career
【戎马】[书] army horse: ～生涯 army life; military life
【戎马倥偬】soldier and horses are in great haste; busily engaged in warfare
【戎装】[书] martial attire

绒 ①(绒毛) fine hair; down: 鸭～ eiderdown ②(绒布) cloth with pile: 长毛～ plush / 灯芯～ corduroy / 法兰～ flannel / 丝～ velvet ③(刺绣用丝) fine floss for embroidery
【绒布】flannelette; cotton flannel
【绒花】[工美] velvet flowers, birds, etc.
【绒裤】sweat pants
【绒毛】①(人、动物身体表面的软毛) fine hair; down ②[纺] nap; pile
【绒毛瘤】villioma; villoma
【绒毛膜】chorion; chorionic membrane ◇ ～癌 choriocarcinoma;
【绒膜癌】choriocarcinoma; chorioepithelioma
【绒面革】suede (leather)
【绒头绳】①(扎头发的绳) wool (for tying pigtails) ② [方](毛线) knitting wool
【绒线】①(粗丝线) floss for embroidery ②[方](毛线) knitting wool ◇ ～刺绣 crewelwork / ～衫 woolen sweater
【绒绣】[工美] woolen needlepoint tapestry; woolen embroidery ◇ ～地毯 finished needlepoint carpet
【绒衣】sweat shirt

荣 ①(茂盛) grow luxuriantly; flourish: 春～冬枯 grow in spring and wither in winter / 欣欣向～ flourishing; thriving; growing luxuriantly ②(光荣) honor; glory: ～获金质奖章 be awarded a gold medal / ～立一等功 be cited for meritorious service, first class / 引以为～ take it as an honor
【荣归】return in glory: ～故里 return to one's native place with honor
【荣华富贵】glory, splendor, wealth and rank; high position and great wealth

【荣获】have the honor to get or win: ～冠军 win the championship / ～英雄称号 be awarded the honorable title of hero
【荣辱】honor or disgrace
【荣幸】be honored: 如蒙光临,不胜～。We shall be greatly honored by your gracious presence. / 我～地通知你… I have the honor to inform you that....
【荣耀】honor; glory
【荣誉】honor; credit; glory: 为祖国赢得～ win honor for one's country ◇ ～感 sense of honor / ～学位 honorary degrees
【荣誉称号】honorary title: 获得～ win the honorary title
【荣誉军人】disabled soldier (wounded in revolutionary war)

蝾

【蝾螈】[动] salamander; newt

茸

茸 ①(草初生的细柔状) fine and soft; downy ②(鹿茸) young pilose antler
【茸茸】(of grass, hair, etc.) fine, soft and thick; downy: 绿草～ a carpet of green grass

融

融 ①(融化) melt; thaw: 春雪易～。Spring snow easily melts. ②(融合) blend; fuse; be in harmony: 水乳交～ blend as well as milk and water; be in perfect harmony
【融合】mix together; fuse; merge: 铜与锡的～ the fusion of copper and tin
【融化】melt; thaw: 雪已开始～。The snow is beginning to thaw. / 有太阳照射时,冰就～。The ice will melt when the sun shines on it. ◇ ～带[气] melting band
【融会贯通】achieve mastery through a comprehensive study of the subject
【融解】melt; thaw
【融洽】harmonious; on friendly terms: ～感情 blend with mutual feeling(s) and understanding
【融融】[书]①(和睦快乐的样子) happy and harmonious ②(暖和) warm: 春光～。Spring fills the air with warmth.
【融蚀形态】ablation form
【融通资金】circulate necessary funds
【融雪水】[水文] snowmelt

rǒng

冗

冗 ①(多余的) superfluous; redundant: ～词 superfluous words ②(烦琐) full of trivial details ③(繁忙的事) busyness: 拨～ find time in the midst of one's work
【冗长】tediously long; lengthy; long-winded; prolix: ～的讲演 a long and tedious speech
【冗词赘句】superfluous words and redundant sentences; verbose
【冗余】redundancy ◇ ～校验 redundancy check / ～率 redundancy rate / ～位 redundant digit
【冗员】redundant personnel
【冗杂】many and diverse; miscellaneous
【冗赘】verbose; diffuse: ～的句子 a redundant sentence

róu

柔

柔 ①(软) soft; supple; flexible: ～枝嫩叶 supple twigs and tender leaves ②(使变软) soften: ～麻 soften jute, hemp, etc. ③(柔和) gentle; yielding; mild: ～中有刚 firm but gentle / 温～ gentle and soft
【柔板】[乐] *adagio*
【柔肠寸断】the heart breaks thinking of one's love; brokenhearted
【柔道】*judo*
【柔和】soft; gentle; mild: ～的光线 soft light / 颜色～ a soft color
【柔毛】pubescence
【柔媚】gentle and lovely
【柔嫩】tender; delicate: ～的皮肤 delicate skin; soft skin / ～的幼芽 tender sprouts
【柔能克刚】softness can overcome the hardest; the soft can conquer the hard
【柔情】tender feelings; tenderness: ～似水 tender and soft as water / ～万种 infinitely affectionate
【柔韧】pliable and tough
【柔软】soft; lithe: ～的垫子 a soft cushion / ～的动作 lithe movements ◇ ～体操 callisthenics
【柔弱】weak; delicate: 身体～ in delicate health; weak; frail
【柔术】*jujitsu*
【柔顺】gentle and agreeable; meek
【柔性电路】flexible circuit
【柔性翼】flex-wing
【柔荑花序】[植] catkin; ament

糅

糅 mix; mingle
【糅合】mix; form a mixture

鞣

鞣 tan: ～皮子 tan hides
【鞣料】tanning material ◇ ～浸膏 tannin extract

【鞣酸】[化] tannic acid

揉 rub; knead: ～眼睛 rub one's eyes / 把纸～成团 crumple up a sheet of paper into a ball

【揉此万邦】subdue all other states; gain domination by war or force

【揉搓】rub; knead

【揉面】knead dough

蹂

【蹂躏】trample on; ravage; make havoc of; devastate: 遭到～ suffer devastation, oppression, outrages, etc.

ròu

肉 ①(人、动物的肉) meat; flesh: ～制品 meat products / 出～率 dressing percentage / 肥～ fat meat; fat / 净～ dressed carcass; dressed meat / 牛～ beef / 瘦～ lean meat / 鲜～ fresh meat / 熏～ smoked meat / 羊～ mutton / 猪～ pork ②(果肉) pulp; flesh (of fruit): 果～ pulp of fruit / 桂圆～ longan pulp / 椰～ coconut meat

【肉饼】meat pie

【肉搏】fight hand-to-hand ◇ ～战 hand-to-hand fight

【肉垂】[动] wattle

【肉店】butcher's (shop)

【肉丁】diced meat: 辣子～ diced pork with hot pepper

【肉冻】meat jelly; aspic: 鸡～ chicken in aspic

【肉冠】[动] comb

【肉桂】[植] Chinese cassia tree; cinnamon

【肉类加工厂】meat-packing plant

【肉林酒池】woods of flesh and ponds of wine; live in the world of wine and women; steeped in wine and surrounded by women

【肉瘤】[医] sarcoma

【肉麻】nauseating; sickening; disgusting: ～的吹捧 fulsome praise

【肉末】minced meat; ground meat

【肉排】steak

【肉皮】pork skin

【肉片】sliced meat

【肉色】yellowish pink

【肉食】①(肉类食物) meat ②(以肉类为食物) carnivorous ◇ ～动物 carnivorous animal; carnivore

【肉丝】shredded meat ◇ ～面 noodles with shredded meat

【肉松】dried meat floss

【肉穗花序】[植] spadix

【肉汤】broth

【肉体】the human body; flesh

【肉丸子】meatball

【肉馅】meat stuffing; chopped meat; ground meat

【肉刑】corporal punishment

【肉芽】[医] granulation

【肉眼】naked eye: ～看不到 be invisible to the naked eye

【肉眼凡胎】a short sighted and good-for-nothing person

【肉欲】carnal desire

【肉汁】gravy; (meat) juice

【肉中刺】a thorn in one's flesh

【肉赘】wart

rú

濡 [书] ①(沾湿) immerse; moisten: ～笔 dip a writing brush in ink ②(迟滞) linger

【濡染】immerse; imbue

【濡湿】soak; make wet

蠕 wriggle; squirm

【蠕变】creep; creep deformation ◇ ～断裂 creep rupture / ～极限 creep limit / ～强度 creep strength

【蠕虫】worm; helminth ◇ ～学 helminthology

【蠕动】①(爬动) wriggle; squirm ②[生理] peristalsis: ～波 peristaltic wave / ～消失 loss of peristalsis / ～音 peristaltic sound

【蠕蠕】wriggling; squirming: ～而动 crawl out

【蠕形动物】Vermes

儒 ①(儒家) Confucianism; Confucianist ②[旧](读书人) scholar; learned man: 腐～ pedantic scholar

【儒艮】[动] dugong

【儒家】the Confucian school; the Confucianists

【儒教】confusianism

【儒生】[旧] Confucian scholar

孺 child: 妇～ women and children

【孺子】[书] child

【孺子可教】The young man is worthy to be taught.

如 ①(依照) in compliance with; according to: ～命 in compliance with your instructions / ～数付清 pay in full ②(如同) like; as; as if: ～你所说 as you've said ③(比得上) can compare with; be as good as: 我不～他。 I'm not as good as he is. ④ (比如) for instance; such as; as ⑤ [连] if: ～遇火警,即按警铃。 In case of fire, ring the alarm bell.

【如常】as usual: 一切～。 Things are as usual.

【如出一辙】be exactly the same as; be no different from; be cut from the same cloth

【如此】so; such; in this way; like that: ～等等 and so on and so forth / 理当～。 Rightly so. / 情况就是～。 That's how things stand. / 似乎是～。 So it appears. / 我想是～。 I think so.

【如此而已】that's what it all adds up to; that's all there is to it

【如次】as follows: 他所写～。 He wrote as follows. / 原文～。 The text is as follows.

【如堕烟海】as if lost on a misty sea; all at sea; completely at a loss

【如法炮制】prepare herbal medicine by the prescribed method; follow a set pattern; follow suit

【如风过耳】turn a deaf ear to

【如鲠在喉】like a fish bone getting stuck in the throat; necessary to give vent to one's pent-up feelings

【如故】① (跟原来一样) as before: 依然～ remain the same as before; remain one's same old self ② (跟老朋友一样) like old friends: 一见～ feel like old friends at the first meeting; hit it off well right from the start

【如果】[连] if; in case; in the event of: ～不是他的话,我们就迷路了。 If it weren't for him, we would have gone astray. / ～她来的话,告诉我一声。 In case she comes, let me know.

【如何】how; what: 此事～办理? How are we to handle this matter? / 你近来情况～? How are things with you?

【如虎添翼】like a tiger that has grown wings; with might redoubled

【如花似玉】as pretty as a flower; as pretty as flower and jade

【如火如荼】like a raging fire

【如获至宝】as if one had found a treasure

【如饥似渴】as if thirsting or hungering for sth.; eagerly: ～地学习 study with great eagerness

【如胶似漆】stick to each other like glue or lacquer; remain glued to each other; be deeply attached to each other

【如今】nowadays; now

【如来佛】[佛教] Tathagata; Buddha

【如狼似虎】as ferocious as wolves and tigers; like cruel beasts of prey

【如雷贯耳】reverberate like thunder: 久闻大名,～。 Your name has long resounded in my ears.

【如临大敌】as if faced with a formidable enemy

【如履薄冰】as though treading on thin ice; be very careful; acting with extreme caution

【如梦初醒】as if awakening from a dream; beginning to see the light

【如鸟兽散】flee helter-skelter; be utterly routed

【如牛负重】like an ox carrying a heavy load

【如期】as scheduled; by the scheduled time; on schedule: 货物已～运到。 The goods arrived on schedule. / 任务已～完成。 The task has been accomplished according to schedule.

【如其】[连] if

【如切如磋】like "cutting and grinding" stones; to learn from each other

【如日中天】like the sun at high noon; at the apex 〈zenith〉 of one's power, career, etc.

【如入无人之境】like entering an unpeopled land; breaking all resistance

【如若】[连] if: ～不然 if not; otherwise

【如丧家之犬】like a dog with the tail between the legs; like a dog without its master

【如丧考妣】look as if one had lost one's parents; look utterly wretched

【如上】as above: ～所述 as stated above; as mentioned above

【如实】strictly according to the facts; as things really are: ～地反映情况 report the situation accurately; reflect things as they really are

【如释重负】as if relieved of a heavy load

【如数家珍】as if enumerating one's family valuables; very familiar with one's subject

【如数】exactly the number or amount: ～偿还 pay back in full / ～到齐 all present and correct

【如汤沃雪】like melting snow with hot water; easily done

【如同】like; as: ～草芥 be no more than the weeds by the roadside / 待我们～亲人 treat us like their kith and kin

【如下】as follows: 发表～声明 make the following statement / 计划的要点～。 The main points of the plan are as the follows. / 全文～。 The full text follows.

【如蚁附膻】like ants seeking sth. rank- smelling; a swarm of people running after unwholesome things or leaning on influential

people for support
【如意】as one wishes: 称心～ after one's own heart / 很难万事～。You can't expect everything to turn out as you wish.
【如意算盘】wishful thinking: 打～ indulge in wishful thinking
【如蝇逐臭】run after filth, as flies swarm around garbage; like flies taking to rottenness; like flies swarming about a bit of filth
【如影随形】like the shadow following the person; very closely associated with each other
【如鱼得水】feel just like fish in water; be in one's element
【如愿以偿】have one's wish fulfilled; achieve what one wishes
【如之奈何】What can be done about it?
【如坐针毡】feel as if sitting on a bed of nails; be on pins and needles; be on tenterhooks

茹 [书] eat: ～素 be a vegetarian
【茹毛饮血】eat birds and animals raw
【茹素吃斋】take vegetarian food

铷 [化] rubidium (Rb)
【铷锶测年】[地质] rubidium-strontium dating

rǔ

汝 [书] you: ～辈 you people; you

辱 ①(耻辱) disgrace; dishonor: 奇耻大～ galling shame and humiliation; terrible disgrace ②(使受辱) bring disgrace to; insult: 丧权～国 humiliate the nation and forfeit its sovereignty
【辱骂】abuse; call sb. names; hurl insults: ～和恐吓决不是战斗。 Hurling insults and threats is no way to fight.
【辱命】fail to accomplish a mission
【辱没】bring disgrace to; be unworthy of: 决不～先进集体的称号 will never be unworthy of the title of advanced collective

乳 ①(乳房) breast ②(奶汁) milk: 炼～ condensed milk ③(象奶汁的东西) any milk-like liquid: 豆～ bean milk ④(初生的) newborn (animal); sucking: ～猪 sucking pig; suckling pig
【乳白】milky white; cream color ◇ ～玻璃 opal glass; opalescent glass / ～灯泡 opal bulb
【乳钵】mortar
【乳齿】milk tooth; deciduous tooth
【乳臭未干】still smell of one's mother's milk; be young and inexperienced; be wet behind the ears
【乳儿】nursing infant; suckling
【乳房】①(指人的) breast; mamma ②(指动物的) udder ◇ ～切除 mastectomy
【乳化】[化] emulsification ◇ ～剂 emulsifying agent; emulsifier / ～原油 emulsified crude oil
【乳化液】[化] emulsion: 水包油～ oil-in-water emulsion / 油包水～ water-in-oil emulsion
【乳剂】[化] emulsion: 全色～[摄] panchromatic emulsion
【乳剂号】[摄] emulsion numbers
【乳胶】[化] emulsion ◇ ～漆 emulsion paint; latex paint
【乳酪】cheese
【乳酶】galactenzyme
【乳酶生】biofermin; lactasin
【乳糜】[生理] chyle ◇ ～尿 chyluria
【乳名】infant name; child's pet name
【乳母】wet nurse
【乳牛】dairy cattle; milch cow ◇ ～场 dairy farm
【乳酸】lactic acid ◇ ～钙 calcium lactate
【乳糖】milk sugar; lactosc ◇ ～酶 lactase
【乳头】①(奶头) nipple; teat; mammilla ②(奶头状物) papilla: 视神经～ optic papilla ◇ ～状瘤 papilloma
【乳腺】[生理] mammary gland ◇ ～癌 breast cancer / ～瘤 mastadenoma / ～炎 mastitis
【乳香】frankincense
【乳牙】deciduous teeth; primary teeth; milk teeth; temporary teeth
【乳罩】brassiere; bra
【乳汁】milk
【乳脂】butterfat ◇ ～糖 toffee; taffy
【乳制品】dairy products ◇ ～工业 dairy industry
【乳浊液】[化] emulsion

rù

溽 [书] humid; damp
【溽暑】sweltering summer weather

褥 cotton-padded mattress: 被～ bedding; bedclothes
【褥疮】[医] bedsore; decubitus ulcer; pressure sore
【褥单】bed sheet
【褥套】①(布套) bedding sack ②(褥子的棉花胎) cotton padding for a bedtick ③(褥子的布罩) mattress cover
【褥子】cotton-padded mattress

蓐 [书] straw mat or mattress

缛 elaborate; cumbersome: 繁文～节 unnecessary and overelaborate formalities; red tape

入 ①(进来;进去) enter: ～境 enter a country / ～冬以来 since winter has set in / ～夜 at nightfall / 火车进～隧道。The train entered a tunnel. ②(参加) join; be admitted into; become a member of: ～伍 enlist in the armed forces ③(收入) income; means: 岁～ annual income ④(合乎) conform to; agree with: ～时 fashionable

【入不敷出】run behind one's expenses; income falling short of expenditure; one's income is not adequate to one's needs

【入仓】be stored in a barn; be put in storage ◇ ～证书 warehousing entry

【入场】entrance; admission: 凭票～。Admission by ticket only. ◇ ～费 price of admission / ～券 (admission) ticket

【入超】unfavorable balance of trade; import surplus

【入定】[佛教]sit quietly and meditate

【入耳】pleasant to the ear: 不堪～ offensive to the ear; obscene; vulgar

【入伏】beginning of the hottest part of the summer: ～以来 since the dog days began

【入港】①[交](进港) enter a port ②(交谈投机) in full agreement; in perfect harmony: 他们谈得～。They are deep in conversation and in perfect agreement with one another. ◇ ～呈报表 bill of entry / ～费 inward charges / ～税 port duty

【入股】buy a share; become a shareholder

【入骨】to the marrow: 恨之～ bitterly hate; bear a bitter hatred for sb. or sth.

【入国问禁】ask about taboos and bans upon arrival in a foreign country

【入国问禁,入乡随俗】When in Rome do as the Romans do.(或: When you enter a country, inquire as to what is forbidden.)

【入画】suitable for a painting; picturesque: 黄山山景,处处可以～。Every bit of Huangshan Mountain scenery is worth painting.

【入伙】①(加入某集团) join a gang; join in partnership ②(入集体伙食) join a mess: 在我们食堂～ eat at our mess

【入籍】naturalization ◇ ～者 naturalized person

【入境】enter a country ◇ ～登记 entrance registration / ～护照 entry passport / ～签证 entry visa / ～申报单 customs declaration made at the time of entry / ～手续 entry formalities; entry procedures / ～许可证 entry permit / ～证书 entry certificate

【入境问俗】on entering a country, inquire about its customs

【入口】①(进入嘴中) enter the mouth: 难于～ have a nasty taste / 不可～! Not to be taken orally! ②(进入的门) entrance: 车站～处 entrance to the station ◇ ～程序 entry program / ～界灯 threshold lights / ～数据 entry data / ～指令 entry instruction

【入寇】invade

【入库】be put in storage; be laid up

【入款】income; receipts

【入殓】put a corpse in a coffin; encoffin

【入列】[军] take one's place in the ranks; fall in

【入门】①(初步学会) cross the threshold; learn the rudiments of a subject: 他是我的～师傅。He is the master who initiated me into the craft. ②(指初级读物;基本知识) elementary course; ABC:《摄影～》*The ABC of Photography*

【入梦】①(入睡) fall asleep ②(出现在梦中) appear in one's dream

【入迷】be fascinated; be enchanted: 看书看～了 be engrossed in a book

【入灭】[佛教] Nirvana; death

【入魔】be infatuated; be spellbound

【入木三分】①(书法有力) written in a forceful hand ②(议论深刻) penetrating; profound; keen

【入情入理】be fair and reasonable; perfectly logical and reasonable

【入侵】invade; intrude; make an incursion; make inroads: ～飞机 the intruding aircraft / 敌人～ the invasion of an enemy / 军事～ military incursion / 再次～ make another intrusion / 有军队～该国。Soldiers invaded the country.

【入射波】[物] incident wave

【入射角】[物] angle of incidence; incident angle

【入射线】[物] incident ray

【入神】①(兴趣浓厚) be entranced; be enthralled: 那紧张的故事使他看得～了。He was enthralled by the exciting story. ②(精妙) superb; marvellous: 这幅画画得真是～。This picture is really superb

【入手】start with; begin with; proceed from;

take as the point of departure: 从基本训练～ start with the basic training
【入睡】go to sleep; fall asleep
【入土】be buried; be interred: 快～了 have one foot in the grave
【入土为安】be laid to rest; have one's bones buried; burial brings peace to the deceased
【入托】start going to a nursery: 办理小孩～手续 enroll a child in a nursery
【入微】in every possible way; in a subtle way: 体贴～ show every possible consideration; be extremely thoughtful
【入味】①(有滋味) tasty: 菜做得很～。 The dish is very tasty. ②(有趣味) interesting
【入伍】enlist in the armed forces; join up
【入乡随俗】While in Rome, do as the Romans do.
【入席】take one's seat at a banquet, ceremony, etc.
【入选】be selected; be chosen
【入学】①(开始进小学学习) start school ②(开始进某学校学习) enter a school: 从～到毕业 from entrance to graduation ◇ ～考试 entrance examination / ～年龄 school age
【入眼】pleasing to the eye: 看不～ not to one's liking
【入药】[中药] be used as medicine
【入夜】at nightfall
【入狱】be put in prison; be sent to jail
【入院】be admitted to hospital; be hospitalized
【入帐】enter an item in an account; enter into the account book
【入赘】marry into and live with one's bride's family

ruǎn

朊 protein

阮

【阮囊羞涩】embarrassingly short of money; lacking sufficient funds to meet necessary expenses
【阮咸】[乐] a plucked stringed instrument

软 ①(质地不硬) soft; flexible; supple; pliable: ～床 soft bed / ～椅 soft chair / 柳条很～。 Willow twigs are pliable. ②(柔和) soft; mild; gentle: ～语 soft words ③(软弱) weak; feeble: 欺～怕硬 bully the weak and fear the strong / 两腿发～。 One's legs feel like jelly. ④(能力弱;质量差) poor in quality, ability, etc.: 工夫～ inadequate skill / 货色～ poor-quality goods ⑤(易被感动) easily moved or influenced: 耳朵～ credulous and pliable / 心～ tenderhearted
【软磁盘】disk; floppy disk ◇ ～系统 floppy disk system
【软刀子】soft knife; a way of harming people imperceptibly
【软底鞋】soft sole
【软调】[摄] soft
【软缎】soft silk fabric in satin weave
【软腭】[生理] soft palate
【软风】[气] light air
【软钢】mild steel; soft steel
【软膏】ointment; paste
【软骨头】a weak-kneed person; a spineless person; a coward
【软骨】[生理] cartilage ◇ ～病 osteomalacia / ～发育不全 achondroplasia / ～鱼 cartilaginous fish
【软管】flexible pipe or tube; hose: 铠装～ armored hose
【软焊】soft soldering; soldering
【软化】①(由硬变软) soften: 使硬水～ soften hard water / 态度～ become compliant ②(由坚定变为动摇) win over by soft tactics ③[皮革] bating
【软和】[口] ①(柔软) soft: ～的褥子 a soft mattress ②(柔和) gentle; kind; soft: 说几句～话 say some kind words
【软件】[计算机] software ◇ ～跟踪方式 software trace mode / ～库 software library / ～模型化 software modularity
【软禁】put sb. under house arrest; house arrest; house confinement
【软麻工艺】[纺] ①(亚麻的) bruising ②(黄麻的) batching
【软锰矿】pyrolusite
【软绵绵】①(柔软) soft: ～的枕头 a soft pillow / 这支歌～的。 This song is too sentimental. ②(软弱无力) weak: 她病好了,但身体仍然～的。 She is well now, but she still feels weak.
【软磨】use soft tactics
【软木】cork ◇ ～塞 cork (as a stopper)
【软泥】[地] ooze
【软片】(a roll of) film
【软弱】weak; feeble; flabby: ～可欺 be weak and easy to bully / ～无能 weak and incompetent
【软式飞艇】blimp; non-rigid airship

【软式推销法】soft sell
【软食】soft diet; soft food; pap
【软水】soft water
【软糖】soft sweets; jelly drops
【软梯】[口] rope ladder
【软体动物】 mollusk; mollusc ◇ ～学 malacology
【软席】[交] soft seat or berth; cushioned seat or berth ◇ ～车厢 railway carriage with soft seats or berths
【软线】[电] flexible cord
【软饮料】soft drinks
【软硬兼施】 use both hard and soft tactics; couple threats with promises
【软语温存】 have many caressing and affectionate words to say to one another
【软玉】[矿] nephrite
【软玉温香】as fair as jade; her body soft and warm, really enticing; the flesh and fragrance of a beauty
【软脂】palmitin ◇ ～酸 palmitic acid; palmic acid
【软着陆】soft landing
【软组织】soft tissue

ruǐ

蕊 stamen; pistil: 雌～ pistil / 雄～ stamen

ruì

瑞 auspicious; lucky
【瑞典】Sweden ◇ ～人 Swede; the Swedish / ～语 Swedish (language)
【瑞林】Rayleigh ◇ ～定律 Rayleigh's law / ～数 Rayleigh number / ～天平 Rayleigh balance
【瑞士】Switzerland ◇ ～人 Swiss
【瑞香】[植] winter daphne
【瑞雪】timely snow; auspicious snow
【瑞雪纷飞】A good snow is falling.(或: Snow falls thick and fast.)
【瑞雪兆丰年】A fall of seasonable snow gives promise of a fruitful year.(或: A timely snow promises a good harvest.)

蚋 buffalo gnat; blackfly

锐 ①(锐利) sharp; keen; acute: 尖～ pointed and sharp; acute ②(锐气) vigor; fighting spirit: 养精蓄～ conserve strength and store up energy; build up one's strength
【锐不可当】can't be held back; be irresistible: 以～之势 with irresistible force
【锐角】[数] acute angle
【锐利】sharp; keen: ～的笔锋 a sharp pen; a vigorous style / ～的匕首 a sharp dagger / ～的武器 a sharp weapon / ～的眼光 sharp-eyed; sharp-sighted
【锐敏】sensitive; keen: ～的嗅觉 a keen sense of smell
【锐器伤】[法] injury from sharp utensil; sharp instrument injury
【锐气】dash; drive: 挫敌～ take the edge off the enemy's spirit
【锐意改革】reform with keen determination

rùn

闰 [天] intercalary
【闰年】leap year; intercalary year
【闰日】leap day; intercalary day
【闰月】intercalary month in the lunar calendar; leap month

润 ①(细腻光滑) moist; smooth; sleek: 墨色很～ in dark full-bodied ink / 湿～ moist ②(加油或水) moisten; lubricate: ～一～嗓子 moisten one's throat ③(使有光彩) embellish; touch up ④(好处) profit; benefit: 分～ share in the benefit
【润笔】remuneration for a writer, painter or calligrapher
【润滑】lubricate: 飞溅～法 splash lubrication ◇ ～系统 lubricating system; lubrication system / ～油 lubricating oil; lubrication oil / ～脂 (lubricating) grease
【润色】polish; touch up: 这篇文章需要～一下。 This article needs polishing.
【润泽】①(滋润) moist; smooth; sleek: ②(使滋润) moisten; lubricate: 用油～轮轴 oil the axle

ruò

若 ①(好象) like; seem; as if: ～有所失 feel as if sth. were missing; look distracted ②[书][连](如果) if: 人不犯我,我不犯人;人～犯我,我必犯人。 We will not attack unless we are attacked; if we are attacked, we will certainly counterattack. ③[书](你) you: ～辈 people like you
【若虫】[动] nymph
【若非】if not; were it not for: ～我那时亲眼看到,我是不会相信的。 Were it not for the fact that I saw it with my own eyes, I wouldn't believe it.

【若干】①(一些) a certain number or amount: ～次 several times / ～地区 certain areas / ～年 several years ②(多少) how many; how much: 共得～? How many in all?
【若即若离】be neither friendly nor aloof; maintain a lukewarm relationship; keep sb. at arm's length
【若明若暗】have an indistinct picture of; have a hazy notion about
【若是】[连] if: 我～你,我就去。If I were you, I would go.
【若无其事】as if nothing had happened; calmly; casually
【若要人不知,除非己莫为】if you don't want others to know about it, don't do it
【若隐若现】partly hidden and partly visible; appear indistinctly
【若有所思】as if thinking of something; as if deep in thought; seem lost in thought

偌

such; so
【偌大】of such a size; so big: ～年纪 so old; so advanced in years

箬

【箬帽】a broad-rimmed bamboo hat
【箬竹】[植] indocalamus

弱

①(软弱) weak; feeble: ～国 a weak nation / 由～变强 go from weakness to strength ②(年幼) young: 老～ old and young ③(差) inferior: 他的能力并不比别人～。He's no less capable than the others. ④[书](丧失) lose (through death) ⑤(略少) a little less than: 五分之一～ a little less than one-fifth
【弱不禁风】too weak to stand a gust of wind; extremely delicate; fragile
【弱不胜衣】too frail to bear the weight of one's clothes
【弱点】weakness; weak point; failing
【弱定位】[航海] weak fix
【弱化】avianize
【弱碱】[化] weak base
【弱脉】[中医] weak pulse
【弱拍】[乐] weak beat; unaccented beat
【弱肉强食】the weak are the prey of the strong; the law of the jungle
【弱视】amblyopia; weak sight
【弱酸】[化] weak acid
【弱小】small and weak: ～民族 small and weak nations
【弱音器】[乐] mute; sordine

S

sā

撒 ①(放开；发出) cast; let go; let out: ～腿就跑 make off at once; take off at once／往水里～网 cast a net into the water ②(尽量施展出来) throw off all restraint; let oneself go: ～酒疯 in drunken brawls／～泼 be shrewish and make a scene
【撒旦】[宗] Satan
【撒欢儿】[方] gambol; frisk
【撒谎】[口] tell a lie; lie: 当面～ tell a barefaced lie; lie in one's teeth／你在～。 You are telling lies.
【撒娇】act like a spoiled child
【撒娇装嗔】act as a spoiled child; act in a pettishly charming manner
【撒赖】make a scene; act shamelessly; raise hell
【撒尿】[口] piss; pee
【撒泼】be unreasonable and make a scene
【撒气】①(轮胎等漏气) leak; go soft; get a flat: 前带～了。The front tire has got a puncture. ②(拿他人发泄怒气) vent one's anger or ill temper: 你别拿他～。 Don't take it out on him.
【撒手】let go one's hold; let go: ～不管 take no further interest in; refuse to take any further part in／抓住绳子，别～。 Hold the rope tight. Don't let it go.
【撒手归西】go to the Western Heaven; to die; pass away; pay the debt of nature
【撒手锏】an unexpected thrust with the mace; one's trump card
【撒腿】start (running): ～就跑 make off at once; scamper
【撒野】act wildly; behave atrociously: 对人～ be rude to sb.

仨 [口] three: ～苹果 three apples／我们哥儿～ we three brothers

sǎ

洒 sprinkle; spray; spill; shed: ～水扫地 sprinkle water and sweep the floor／别把酒～了。 Don't spill the wine.
【洒柏油柜车】tar-spaying tank
【洒泪】shed tears: ～告别 take a tearful leave
【洒扫】sprinkle water and sweep the floor; sweep: 黎明即起，～庭除。 Rise at dawn and sweep the courtyard.
【洒水车】watering car; sprinklers spraying car; water-barrow
【洒水管】sprinkler pipe
【洒脱】free and easy: 举止～ have an easy and unaffected manner

撒 ①(散布；扔出) scatter; sprinkle; spread: ～化肥 spread chemical fertilizer／～种子 scatter seed ②(洒落) spill; drop: 墨水～在书桌上了。 The ink has spilt on the desk.
【撒播】broadcast sowing ◇ ～机 broadcast seeder; broadcaster
【撒肥机】[农] fertilizer distributor; manure spreader
【撒粉】[农] dusting ◇ ～器 duster
【撒施】[农] spread fertilizer over the fields; broad cast (fertilizer)

sà

飒
【飒然】[书] soughing
【飒飒】[象] sough; rustle: 秋风～。 The autumn wind is soughing in the trees.
【飒爽】[书] of martial bearing; valiant: ～英姿 of valiant and heroic bearing; bright and brave

萨
【萨尔瓦多】El Salvador ◇ ～人 Salvadoran
【萨克管】[乐] saxophone
【萨克号】[乐] saxhorn
【萨摩亚】Samoa ◇ ～人 Samoan／～语 Samoan (language)
【萨其马】*saqima*, a kind of candied fritter

脎 [化] osazone

sāi

塞 ①(堵；填入) fill in; squeeze in; stuff: 把东西～进箱子 stuff 〈squeeze〉 things into a trunk／管子被脏东西～住了。 The pipe was clogged with dirt. ②(塞子) stopper: 软木～ cork
【塞规】[机] plug gauge
【塞环】ring of plug
【塞套】plug sleeve
【塞条】[医] tent
【塞子】stopper; cork; plug; spigot

噻
【噻草隆】benzthiazuron; gatinon
【噻吩】[化] thiophene ◇ ～甲基 thenyl

【噻唑】[化] thiazole ◇ ～染料 thiazole dye; thiazole dyestuff

腮 cheek

【腮帮子】[口] cheek

【腮垫】[乐] chin rest

【腮托】[乐] chin rest (of a violin or viola)

【腮腺】[生理] parotid gland

【腮腺炎】parotitis: 流行性～ mumps ◇ ～疫苗 mumps vaccine

【腮腺硬癌】parotidoscirrhus

鳃 gill; branchia: ～盖 gill cover

sài

塞 a place of strategic importance: 边～ frontier fortress

【塞拉利昂】Sierra Leone ◇ ～人 Sierra Leonian

【塞内加尔】Senegal ◇ ～人 Senegalese

【塞浦路斯】Cyprus ◇ ～人 Cypriot

【塞舌尔】Seychelles ◇ ～人 Seychellois

【塞外】beyond the Great Wall: ～风光 a northern-frontier scene / ～江南 lush southern-type fields north of the Great Wall

【塞翁失马,安知非福】when the old man on the frontier lost his mare, who could have guessed it was a blessing in disguise; a loss may turn out to be a gain

赛 ①(比赛) match: game; competition; contest: 安慰～ consolation event / 半复～ quarter-finals / 半决～ semi-finals / 表演～ exhibition match / 对抗～ dual meet / 复～ semi-finals / 锦标～ championship contest / 决～ finals / 联～ league / 淘汰～ elimination / 田～ field events / 团体～ team competition / 选拔～ tryouts / 小组循环～ group round robin / 循环～ round match / 邀请～ invitation tournament; invitational tournament / 预～ preliminary trials / 元老～ veterans' event / 足球～ football match; football game ② (胜过) be comparable to; surpass: 我这萝卜～梨。 These radishes of mine taste as good as pears.

【赛车】[体] ①(自行车、汽车等比赛) cycle racing; motorcycle race; auto mobile race ② (比赛用车) racing bicycle

【赛狗】dog race

【赛过】overtake; be better than; surpass; exceed: 上次比赛他～我了。 He was better than 〈surpassed〉 me in the last contest.

【赛力散】[农] phenylmercuric acetate

【赛璐珞】celluloid

【赛马】horse race; horse racing; racing: ～场 race course; race ground; turf / ～跑道 race track / 平道～ flat racing

【赛跑】race: 百米～ 100-meter dash / 长距离～ long-distance race; distance race / 短距离～ sprint: dash / 环城～ round-the-city race / 接力～ relay race; relay / 跨栏～ hurdles; hurdle race / 马拉松～ Marathon (race) / 越野～ cross-country race / 障碍～ steeplechase / 中距离～ middle-distance running / 与时间～ run a race against time ◇ ～运动员 runner

【赛艇】[体] ①(赛艇比赛) rowing ②(比赛用艇) racing boat; shell: 双桨～ scull

sān

三 ①(数) three: ～倍 three times; threefold; triple / ～幕喜剧 comedy in three acts / ～岁小孩 a child of three ② (多数;多次) more than two; several; many: ～思 think again and again; think twice (about doing sth.)

【三八妇女节】International Working Women's Day (March 8)

【三班制】three-shift system

【三宝】[佛教] Triratna ○佛 the Buddha / 法 the dharma / 僧 the sangha

【三倍体】[生] triploid

【三部曲】trilogy

【三彩】[考古] three-color glazed pottery: 唐～ three-color glazed pottery of the Tang Dynasty

【三叉戟】trident

【三叉神经】[生理] trigeminal nerve ◇ ～痛 trigeminal neuralgia

【三长两短】unexpected misfortune; sth. unfortunate, esp. death: 万一她有个～ if anything untoward should happen to her; in case she should die

【三朝元老】minister to three emperors; an official who stays in power under different regimes

【三重唱】[乐] (vocal) trio

【三重奏】[乐] (instrumental) trio

【三次方程】[数] cubic equation

【三寸不烂之舌】have s silver tongue

【三从四德】the three obediences and the four virtues ○未嫁从父 be obedient to father before marriage / 既嫁从夫 be obedient to husband after marriage / 夫死从子 be obedient to son after the death of husband / 妇德 morality / 妇言 proper speech / 妇容 modest manner / 妇功 diligent work

【三大纪律,八项注意】the Three Main Rules of Discipline and the Eight Points for Attention of the Chinese People's Liberation Army
【三等秘书】[外] third secretary
【三点式游泳衣】bikini (suit)
【三叠纪】[地] the Triassic Period
【三度空间】[哲] three-dimensional space
【三段论法】[逻] syllogism
【三番五次】again and again; time and again; over and over again; repeatedly
【三废】the three wastes: 从～中回收和提取大量有用物质 salvage large quantities of useful materials from the three wastes ◯废气 waste gas / 废水 waste water / 废渣 industrial residue
【三伏】①(初、中、末伏的总称) the three ten-day periods of the hot season: ～天 dog days ②(末伏) the last of the three periods of the hot season
【三副】[航海] third mate; third officer
【三纲五常】the three cardinal guides and the five constant virtues as specified in the feudal ethical code ◯君为臣纲 ruler guides subject / 父为子纲 father guides son / 夫为妻纲 husband guides wife / 仁 benevolence / 义 righteousness / 礼 propriety / 智 wisdom / 信 fidelity
【三个臭皮匠,合成一个诸葛亮】three cobblers with their wits combined equal Zhuge Liang the master mind; the wisdom of the masses exceeds that of the wisest individual
【三顾茅庐】make three calls at the thatched cottage; repeatedly request sb. to take up a responsible post
【三光政策】the policy of "burn all, kill all, loot all" ◯杀光 kill all the civilians / 烧光 burn down all the houses / 抢光 loot all that is movable
【三合板】three-ply board; plywood
【三合星】[天] triple star
【三合一】three-in-one
【三核苷酸】[生化] trinucleotide
【三化螟】[农] yellow rice borer
【三级风】[气] force 3 wind; gentle breeze
【三级跳远】[体] hop, step and jump; triple jump ◇～运动员 triple jumper
【三极管】[无] triode: 充气～ gas-filled triode / 晶体～ transistor
【三季稻】triple cropping of rice
【三尖瓣】[生理] tricuspid valve ◇～狭窄[医] tricuspid stenosis
【三件套】three-piece
【三角】①(象三角形的东西) triangle ②[数](三角学) trigonometry
【三角板】set square
【三角测量】triangulation; trigonometrical survey
【三角鲂】[动] triangular bream
【三角枫】[植] trident maple
【三角函数】trigonometric function
【三角肌】[生理] deltoid muscle
【三角裤】panties; briefs
【三角恋爱】love triangle
【三角贸易】triangular trade
【三角皮带】V belt; cone belt; triangle belt
【三角旗】pennant; pennon
【三角铁】① [乐] triangle ②(角钢) angle iron; L-iron
【三角形】[数] triangle: 不等边～ scalene triangle / 等边～ equilateral triangle / 等腰～ isosceles triangle / 直角～ right-angled triangle ◯边 side / 高 altitude / 中线 median / 底边 base / 顶点 vertex
【三角学】[数] trigonometry
【三角洲】[地] delta
【三脚架】tripod; tripod mounting; trivet
【三教九流】①(指儒、佛、道三教及其它各家) the three religions and the nine schools of thought ◯儒教 Confucianism / 佛教 Buddhism / 道教 Taoism / 儒家 the Confucians / 道家 the Taoists / 阴阳家 the *Yin-Yang* / 法家 the Legalists / 名家 the Logicians / 墨家 the Mohists / 纵横家 the Political Strategists / 杂家 the Eclectics / 农家 the Agriculturists ②(指宗教、学术中各流派) various religious sects and academic schools ③[贬](江湖上各色人等) people in various trades; people of all sorts
【三节棍】a cudgel of three linked sections; three section cudgel
【三九天】the third nine-day period after the winter solstice; coldest days of winter
【三句话不离本行】can hardly open one's mouth without talking shop; talk shop all the time
【三军】① [旧](军队) the army ②(陆、海、空三军) the three armed services
【三K党】Ku-Klux-Klan
【三棱尺】three-square rule; triangular scale
【三棱镜】[物] (triangular) prism
【三连音符】[乐] triplet
【三联单】triplicate form

【三联症】triad; trilogy
【三令五申】repeated injunctions
【三六九等】various grades and rank: 把工作分成～是错误的。It is wrong to regard different kinds of work as indications of rank or grade.
【三轮车】tricycle; pedicab
【三轮摩托车】motor tricycle
【三轮汽车】three-wheeled automobile
【三氯杀螨砜】[农] tetradiphon; tedion
【三昧】①[佛教] samadhi ②(诀要) secret; knack: 深得其中～ master the secrets of an art
【三民主义】the Three People's Principles ◯民族 Nationalism / 民权 Democracy / 民生 the People's Livelihood
【三七】[中药] pseudo-ginseng
【三秋】[农] the three autumn jobs ◯秋收 autumn harvesting / 秋耕 autumn ploughing / 秋播 autumn sowing
【三人成虎】Three people spreading reports of a tiger makes you believe there is one around. (或: A repeated slander makes others believe.)
【三人行, 必有我师】If three of us are walking together, at least one of the other two is good enough to be my teacher.
【三人知, 天下晓】When three know it, all know it.
【三三两两】in twos and threes; by twos and threes
【三色版】[音] three-color halftone; three-color block
【三色版印刷】[印] three-color printing
【三色堇】[植] pansy
【三生有幸】consider oneself most fortunate (to make sb.'s acquaintance, etc.)
【三十二分音符】[乐] demisemiquaver; thirty-second note
【三十二开】thirty-twomo; 32mo
【三十六计, 走为上计】of the thirty-six strategems, the best is running away; the best thing to do now is to quit
【三熟制】[农] triple-cropping system
【三思而行】think thrice before you act; look before you leap
【三天打鱼, 两天晒网】go fishing for three days and dry the nets for two; work by fits and starts; lack perseverance
【三通】①[机] tee; tee joint ②(指通邮、通商、通航) exchanges of mails, trade, air and shipping services ◇ ～管 three-way pipe
【三头六臂】(with) three heads and six arms; superhuman powers
【三维空间】three-dimensional space
【三位一体】①[基督教] the Trinity ②(三者联为一体) trinity; three forming an organic whole; three in one
【三五成群】in threes and fours; in knots
【三下五除二】neat and quick
【三夏】[农] the three summer jobs ◯夏收 summer harvesting / 夏种 summer planting / 夏管 summer field management
【三弦】[乐] *sanxian*, a three-stringed plucked instrument
【三项全能运动】[体] triathlon
【三相】[电] three-phase ◇ ～变压器 three-phase transformer / ～插座 three-phase socket
【三硝基甲苯】[化] trinitrotoluene (TNT)
【三心二意】①(犹豫不决) be of two minds; shilly-shally: 别～了, 就这样办吧。Don't shilly-shally. Go right ahead. ②(非全心全意) half-hearted
【三言两语】in a few words; in one or two words
【三氧化物】[化] trioxide: 三氧化二砷 arsenic trioxide
【三月】①(阳历) March ②(阴历) the third month of the lunar year; the third moon
【三灾八难】numerous adversities and calamities
【三战两胜】[体] the best of three games
【三只手】[方] pickpocket
【三趾鹑】[动] button quail

叁 three

sǎn

散 ①(松开; 无约束) come loose; fall apart; not hold together: 行李～了。The luggage has got loosened. ②(分散) scattered: ～～落落 scattered about ③[中药] medicine in powder form; medicinal powder: 健胃～ digestive powder
【散兵】[军] skirmisher ◇ ～壕 fire trench / ～坑 foxhole; pit / ～线 skirmish line
【散兵游勇】stragglers and disbanded soldiers
【散光】astigmatism ◇ ～眼镜 astigmatic glasses
【散记】random notes; sidelights
【散剂】[药] powder; pulvis
【散架】①(散脱) fall apart; fall to pieces ②(累极) (feel as if) all one's limbs are out of joint
【散居】live scattered
【散漫】①(随便) undisciplined; careless and

sloppy ②(分散) unorganized; scattered: ～无组织的状态 a disorganized state of affairs
【散射】[物] scattering ◇ ～波 scattered wave / ～层 scattering layer / ～光 scattered light / ～粒子 scattering particles / ～通信 scatter communication / ～线 scattered rays
【散文】prose
【散文诗】prose poem
【散运】loan in bulk
【散装】bulk; in bulk ◇ ～饼干 loose cookies / ～货船 bulk carrier / ～货物 bulk cargo; bulk freight / ～汽油 petrol in bulk / ～水泥 bulk cement / ～运输[石油] bulk transportation

伞 ①(雨伞;旱伞) umbrella: 缩折～ extension umbrella / 阳～ parasol / 打开～ put the umbrella up / 收～ put down the umbrella ②(伞状物) sth. shaped like an umbrella: 降落～ parachute
【伞兵】paratrooper; parachuter ◇ ～部队 parachute troops; paratroops
【伞齿轮】[机] bevel gear
【伞房花序】[植] corymb
【伞菌】agaric
【伞投】drop by parachute; parachute ◇ ～炸弹 parachute bomb; parabomb / ～照明弹 parachute flare
【伞形花序】[植] umbel
【伞形科】[植] carrot family
【伞形目】[植] Umbellales
【伞形天线】umbrella antenna
【伞衣】canopy

sàn

散 ①(由聚集而分离) break up; disperse: ～会。 A meeting breaks up. / 会还没有～。 The meeting is not over yet. / 乌云～了。 Dark clouds dispersed. ②(散布) distribute; tribute; give out: ～传单 give out handbills; distribute leaflets ③(排除) dispel; let out: 请打开窗户～～烟。 Please open the windows to let the smoke out.
【散播】disseminate spread: ～谣言 spread rumors
【散布】spread; disseminate; scatter; diffuse: ～流言蜚语 spread slanderous rumors ◇ ～存储 scatter storage / ～误差 dispersion error
【散步】take a walk; go for walk; go for a stroll
【散场】(of a theatre, cinema, etc.) empty after the show
【散发】①(发出) send out; send forth; diffuse; emit: 花儿～着清香。 The flowers sent forth a delicate fragrance. ②(分发) distribute; issue; give out: ～传单 distribute leaflets
【散会】(of a meeting) be over; break up: 宣布～ declare the meeting over
【散伙】(of a group, body or organization) dissolve; disband
【散开】spread out or apart; disperse; scatter: 人群纷纷～。 The crowd dispersed.
【散热器】radiator: 管式～ tubular radiator
【散热片】carbon fin; cooling fin; radiating fin; radiating rib
【散失】①(分散遗失) scatter and disappear; be lost; be missing: 防止图书～ prevent any loss of library books ②(消散失去) (of moisture, etc.) be lost; vaporize; dissipate; evaporate: 田里的水分～了。 The soil moisture in the fields was evaporated.
【散水】[建] apron
【散瞳药】mydriatic
【散心】drive away one's cares; relieve boredom
【散逸】[物] dissipation: 热～ heat dissipation

sāng

丧 funeral; mourning
【丧服】mourning apparel
【丧礼】obsequies; funeral
【丧事】funeral arrangements
【丧葬】burial; funeral ◇ ～费 funeral expenses
【丧钟】funeral bell; death knell; knell: 敲响～ sound the death knell

桑 white mulberry; mulberry
【桑蚕】silkworm ◇ ～丝 mulberry silk
【桑皮纸】mulberry (bark) paper
【桑葚】mulberry
【桑色素】morin
【桑树】white mulberry; mulberry
【桑榆暮景】the evening of one's life
【桑园】mulberry field
【桑梓】[书] one's native place: ～故里 one's native village / ～之情 the friendship of fellow countrymen

sǎng

搡 [方] push violently: 推推～～ pushing and shoving

嗓 ①(嗓子) throat; larynx ②(嗓音) voice
【嗓门儿】voice: 提高～ raise one's voice
【嗓音】voice: 他～洪亮。 He has resonant voice.

【嗓子】①(喉咙) throat; larynx: ～疼 have a sore throat ②(嗓音) voice: ～好 have a good voice / ～哑了 lose one's voice / 清一清～ clear one's throat

sàng

丧 lose

【丧胆】be terror-stricken; be smitten with fear: 闻风～ be panic-stricken at the news (of)

【丧魂落魄】be driven to distraction: 吓得～ be scared out of one's wits; be frightened out of one's life

【丧家之犬】stray dog; homeless dog: 惶惶如～ as panic-stricken as a stray dog

【丧尽天良】utterly devoid of conscience; conscienceless; heartless

【丧命】meet one's death; get killed

【丧偶】[书] bereft of one's spouse

【丧气】①(情绪低落) feel disheartened; lose heart; become crest fallen: ～话 demoralizing words ②[口](倒霉;不吉利) be unlucky; be out of luck; have bad luck

【丧权辱国】humiliate the country and forfeit its sovereignty; surrender a country's sovereign rights under humiliating terms: ～的条约 a treaty of national betrayal and humiliation

【丧失】lose; forfeit: ～警惕 lower one's guard; be off one's guard / ～劳动力 disability / ～理智 lose one's nerve / ～时机 miss the opportunity / ～信心 lose confidence

【丧心病狂】frenzied; unscrupulous; perverse; lose one's senses

sāo

臊 the smell of urine; foul smell

【臊气】foul smell; the smell of urine

搔 scratch: ～痒 scratch where it itches

【搔到痒处】scratch where it itches; to hit the nail on the head; scratch at the place that itches

【搔首】scratch one's head: ～踟蹰 scratch one's head in hesitation; hesitate

【搔头弄姿】[书] stroke one's hair in coquetry; be coquettish

骚 ①(扰乱) disturb; upset ②(屈原的《离骚》) short for *Li Sao*: ～体 poetry in the style of *Li Sao* ③(诗文) literary writings: ～人 poet ④(举止轻佻) coquettish

【骚动】①(扰乱) disturbance; commotion; ferment ②(动乱) be in a tumult; become restless: 全城～。The whole city was in a tumult.

【骚客】[书] poet

【骚乱】disturbance; riot; chaos: 引起～ cause a disturbance

【骚扰】harass; molest: ～破坏活动 harassing and wrecking activities

【骚人墨客】[书] men of letters; *literati*; a poet

缫 reel silk from cocoons; reel

【缫丝】silk reeling; filature ◇ ～厂 reeling mill; filature / ～工人 reeler / ～机 reeling machine; filature

sǎo

扫 ①(扫除;打扫) sweep: ～雪 sweep away the snow / 没有扫帚我无法去～。I can't sweep without a broom. ②(很快地移动) pass quickly along or over; sweep: 眼光～过人群 sweep one's eyes over the crowd ③(归拢在一起) put all together: ～数归还 the whole amount returned

【扫除】①(清除) cleaning; cleanup: 大～ general cleaning; through cleaning ②(消除) clear away; remove; wipe out: ～文盲 eliminate illiteracy; wipe out illiteracy / ～障碍 remove the obstacles

【扫荡】mop up: 粉碎敌人的～ smash the enemy's mopping-up operations

【扫地】①(打扫) sweep the floor ②(丧失名誉等) reach rock bottom; reach an all-time low; be dragged in the dust: 名誉～ be thoroughly discredited / 威信～ be shorn of one's prestige

【扫地出门】drive sb. out of his house and deprive him of everything; drive out sb. in dire poverty

【扫雷】mine sweeping ◇ ～机 mine-dredger / ～舰 minesweeper / ～器 mine-sweeping apparatus; minesweeper

【扫路机】road sweeper; street sweeper

【扫盲】eliminate illiteracy; wipe out illiteracy ◇ ～班 literacy class / ～运动 campaign to eliminate illiteracy; anti-illiteracy campaign

【扫眉才子】a gifted maiden; a girl poet; a female scholar

【扫描】[电] scanning: 飞点～ flying-spot scanning / 行～ line scanning ◇ ～器 scanner

【扫墓】sweep a grave; pay respects to a dead person at his tomb

【扫平】put down; crush; suppress: ～叛乱 put down a rebellion

【扫清】clear away: ～道路 clear the path; pave the way / ～障碍 clear away obstacles

【扫射】strafe; rake; swinging traverse
【扫尾】 wind up; round off: ～工作 rounding-off work
【扫兴】have one's spirits dampened; feel disappointed: 真叫人～! How disappointing!

嫂 ①(哥哥的妻子) elder brother's wife; sister-in-law ②(称呼年纪不大的已婚女子) sister
【嫂嫂】[方] elder brother's wife; sister-in-law
【嫂子】[口] elder brother's wife; sister-in-law

sào

扫
【扫帚】broom
【扫帚星】[天] comet

臊 shy; bashful: ～得满脸通红 blush to the ears / 害～ be bashful

sè

涩 ①(麻木味) puckery; astringent: 这柿子～。The persimmons tastes puckery. ②(不滑润) unsmooth; hard-going: 轮轴发～。The axle doesn't work smoothly. ③(文句难读、难懂) obscure; difficult: 文句艰～ make difficult reading
【涩脉】[中医] a weak, thready, uneven pulse

塞
【塞擦音】[语] affricate
【塞音】[语] plosive
【塞责】not do one's job conscientiously: 敷衍～ perform one's duty in a perfunctory manner

瑟 [乐] *se*, 25-stringed plucked instrument
【瑟瑟】[象] rustle: 秋风～。The autumn wind is rustling 〈soughing〉.
【瑟缩】curl up with cold; cower

啬 stingy; miserly

色 ①(颜色) color: 红～ red / 黄～ yellow / 蓝～ blue / 绿～ green / 原～ primary color ②(脸上的神气) look; countenance; expression: 满面喜～ beaming with joy / 面无惧～ not look at all afraid / 面有难～ show reluctance③(种类) kind; description: 各～人等 people of every description; all kinds of people ④(景色) scene; scenery: 湖光山～ a landscape of lakes and mountains ⑤(质量) quality (of precious metals, goods, etc.): 成～好 of good quality ⑥(妇女美貌) woman's looks: 姿～ good looks ⑦ [佛教] material appearance of things
【色标】color code
【色彩】color; hue; tint; shade: ～鲜明 in bright gay color / 地方～ local color / 感情～ emotional coloring / 文学～ literary flavor
【色层分析】[物] chromatographic analysis ◇ ～法 chromatography
【色层谱】[物] chromatogram
【色差】① [物] chromatism ② [纺] off color; off shade
【色淀】[纺] (color) lake: 绯红～ crimson lake
【色调】tone; hue: 暖～ warm tones
【色度】chromaticity; chrominance; colority
【色度计】[物] colorimeter: 光电～ photoelectric colorimeter
【色基】[化] color base
【色厉内荏】fierce of mien but faint of heart
【色盲】[医] achromatopsia; color blindness
【色品】[物] chroma; chromaticity
【色情】pornographic; sexy ◇ ～电影 pornofilm; wet dream / ～文学 pornography
【色球】[天] chromosphere
【色散】[物] chromatic dispersion
【色素】[生] pigment ◇ ～沉着 [医] pigmentation / ～细胞 chromatophere
【色泽】color and lustre: ～和谐 the colors match well / ～鲜明 bright and lustrous
【色授神与】communication between minds without use of words
【色温计】[摄] color temperature meter
【色香味美】good in color, smell and taste
【色织厂】[纺] yarn-dyed fabric mill
【色纸】colored paper
【色痣】mole

铯 [化] cesium (Cs)
【铯光灯】cesium vapor lamp
【铯原子钟】cesium-beam atomic clock; cesium-beam atomic oscillator
【铯源装置】cesium unit

sēn

森 ①(树多) full of trees ②[书](繁密; 众多) multitudinous; in multitudes: ～罗万象 myriads of things; everything under the sun ③(阴暗) dark; gloomy: 阴～ gloomy; grim
【森林】forest: ～地带 land covered with forests ◇ ～保护区 forestry reserve / ～保护员 forest ranger / ～调查 forest survey / ～法 forestry act; law of forestry / ～风 forest wind / ～抚

育 tending of woods / ～覆被率 percentage of forest cover / ～工业 timber industry / ～火灾 forest fire / ～学 forestry / ～资源 forest reserves

【森然】①(繁茂直立)(of tall trees) dense; thick: 林木～ thickly wooded with tall trees ②(森严可畏) awe-inspiring

【森森】dense; thick; luxuriant: 松柏～ dense pine and cypress trees

【森严】stern; strict; forbidding: 等级～ be rigidly stratified; form a strict hierarchy / 戒备～ heavily guarded

【森严壁垒】closely guarded; strongly fortified

sēng

僧 Buddhist monk; monk: 托钵～ fakir / 行脚～ palmer / 云水～ wandering monk

【僧多粥少】There are too many (Buddhist) monks and too little gruel; cannot meet the needs of the people.(或: The monks are many and the supply of gruel is meager; not enough to go around.)

【僧侣】monks and priests; clergy ◇ ～集团 theocracy / ～统治 hierarchy; hierocracy / ～政治 theocracy; sacerdotalism / ～主义 [哲] fideism

【僧帽瓣】mitral valve

【僧帽水母】Portuguese man-of-war

【僧尼】Buddhist monks and nuns

【僧尼层】Senni beds

【僧伽罗语】Singhalese

【僧俗】monks and laymen

【僧徒】Buddhist monks

【僧院】Buddhist temple; Buddhist monastery

shā

杉 [植] China fir

【杉篙】[建] fir pole

【杉木】China fir

沙 ①(沙子) sand: 在路上铺～ sand the road ②(象沙子的东西) granulated; powdered: 豆～ bean paste ③(声音不清脆) hoarse; husky

【沙癌】psammocarcinoma

【沙暴】sandstorm

【沙蚕】clam worm

【沙场】battlefield; battleground

【沙袋】sandbag

【沙丁鱼】sardine

【沙发】sofa; settee: 单人～ upholstered armchair

【沙岗】sand hill

【沙锅】earthenware pot; casserole

【沙果】[植] Chinese pear-leaved crabapple

【沙荒】sandy wasteland; sandy waste

【沙棘】sea buckthorn; sallow thorn

【沙鸡】sandgrouse

【沙金】alluvial gold; placer gold

【沙坑】[体] jumping pit

【沙拉】salad

【沙门】[佛教] Sramana; Buddhist monk

【沙弥】[佛教] Sramanera; acolyte

【沙梨】[植] sand pear

【沙里淘金】wash grains of gold out of the sands; extract the essential from a large mass of material; get small returns for great effort

【沙粒】grains of sand; sand

【沙砾】grit

【沙漏】hourglass: ～计时 use a sand filter for marking the hours

【沙罗周期】[天] saros

【沙漠】desert

【沙盘】[军] sand table ◇ ～作业 sand table exercise

【沙丘】(sand) dune: 流动～ moving dunes

【沙瓤】mushy watermelon pulp

【沙沙】[象] rustle

【沙参】[中药] the root of straight ladybell

【沙滩】sandy beach; sand beach; sea beach

【沙特阿拉伯】Saudi Arabia ◇ ～人 Saudi Arabian

【沙土】sandy soil

【沙文主义】chauvinism: 大国～ great-nation chauvinism

【沙哑】hoarse; husky; raucous: 声音～ have a husky voice

【沙眼】trachoma

【沙枣】[植] narrow-leaved oleaster

【沙蚤】[动] sand hopper

【沙洲】shoal; sandbar; sandbank

【沙柱】dust devil; sand column

【沙锥】[动] snipe

【沙子】①(细小的石粒) sand; grit ②(象沙的东西) small grains; pellets: 铁～ iron pellets; shot

【沙嘴】[地] sandspit

痧 [中医] ①(霍乱) cholera ②(中暑) heatstroke; sunstroke

【痧子】[方] measles: 出～have the measles

鲨 shark

【鲨鱼】shark

砂 sand; grit: 往米里掺～ sand the rice

【砂泵】sand pump

【砂布】emery cloth; abrasive cloth: 刚玉～ corundum cloth

【砂斗】sand hopper

【砂浆】[建] mortar: 石灰～ lime mortar／水泥～ cement mortar

【砂晶】sand crystal

【砂矿】placer deposit; placer ◇ ～开采 placer mining; alluvial mining; placering

【砂砾】gravel; grit

【砂轮】[机] emery wheel; grinding wheel; abrasive wheel ◇ ～机 grinder

【砂囊】[动] gizzard

【砂壤土】sandy loam

【砂糖】granulated sugar

【砂田】sandy land

【砂土】sandy soil; sand

【砂箱】[冶] sandbox; molding box

【砂型】[冶] sand mold ◇ ～心 sand core

【砂岩】[矿] sandstone

【砂眼】[冶] sand holes; blowholes; slag pin hole

【砂样】[石油] drilling mud cuttings

【砂纸】abrasive paper; sand paper: 玻璃～ glass paper／金刚～ emery paper

【砂质岩】[地] arenaceous rock

纱 ①(线) yarn: 粗支～ low count yarn／细支～ fine count yarn／中支～ medium count yarn／棉～ cotton yarn ②(经纬线很稀的织物) gauze; sheer: 铁～ wire gauze

【纱包线】cotton-covered wire

【纱布】gauze: ～绷带 gauze bandage

【纱厂】cotton mill

【纱橱】screen cupboard

【纱窗】screen window

【纱灯】gauze lantern

【纱锭】[纺] spindle

【纱巾】gauze kerchief

【纱笼】sarong

【纱罗】[纺] gauze

【纱染】yarn-dyed

【纱线】yarn: ～支数 count of yarn; yarn number

【纱罩】①(食物防蝇罩) gauze; screen covering (over food) ②(煤气灯白炽罩) mantle

杀 ①(杀死) kill; slaughter: ～敌 kill the enemy／他～ homicide／误～ manslaughter／侵略者把全镇居民都～了。 The invaders killed off all the inhabitants of the town. ②(战斗) fight; go into battle: ～出重围 fight one's way out of a heavy encirclement ③(削弱) weaken; reduce; abate: ～威风 deflate the arrogance／风势稍～。 The wind abated. ④(表示程度深) extremely; to death: 笑～人 laugh oneself to death

【杀草丹】[农] benthiocarb

【杀虫剂】[农] insecticide; pesticide

【杀敌】fight the enemy; engage in battle: 英勇～ be brave in battle; fight heroically

【杀风景】spoil the fun; be a wet blanket

【杀害】murder; kill ◇ ～妇女 femicide／～狂 phonomania／～人质 hostage murder／～婴儿罪 child-murder／～子女罪 filicide

【杀回马枪】give sb. a back thrust; wheel around and hit back

【杀鸡取卵】kill the hen to get the eggs; kill the goose that lays the golden eggs

【杀鸡吓猴】kill the chicken to frighten the monkey; punish someone as a warning to others

【杀鸡用牛刀】break a fly upon a wheel; use a steam-hammer to crack nuts

【杀价】offer to buy sth. cheap, knowing the seller needs cash

【杀菌】disinfect; sterilize ◇ ～剂 germicide; bactericide

【杀戮】massacre; slaughter: 惨遭～ be massacred in cold blood

【杀卵剂】[农] ovicide

【杀螨剂】[农] acaricide; miticide

【杀气】①(凶恶的气势) murderous look: ～腾腾 with a murderous look on one's face; be out to kill ②(出气) vent one's ill feeling

【杀人】kill a person; murder ◇ ～案 homicide case／～犯 murderer; manslayer; homicide; mankiller／～放火犯 homicide arsonist／～集团 homicide squad／～狂 blood thirstiness／～行为 homicidal act／～罪 bloodguilt

【杀人不见血】kill without spilling blood; kill by subtle means

【杀人不眨眼】kill without batting an eyelid; kill without blinking an eye

【杀人放火】murder and arson: ～, 无恶不作 commit murder, arson and every crime imaginable

【杀人可恕,天理难容】If murder might be pardoned, there would be no reason in the human

society.
【杀人灭迹】obliterate traces of murder
【杀人灭口】murder sb. to prevent divulgence of one's secrets; do away with a winess
【杀人如麻】kill people like flies
【杀人越货】kill a person and seize his goods
【杀婴】[法] infanticide
【杀伤】kill and wound; inflict casualties on: ～大批敌军 inflict heavy casualties on the enemy ◇ ～弹 fragmentation bomb; antipersonnel bomb
【杀身成仁】die to achieve virtue; die for a just cause
【杀身之祸】a fatal disaster
【杀鼠剂】rat poison; raticide
【杀鼠灵】warfarin
【杀头】behead; decapitate
【杀一儆百】execute one as a warning to a hundred; execute one man to warn a hundred
【杀真菌剂】fungicide

刹

put on the brakes; stop; check: ～车 brake a car; stop a car／～住歪风 check an unhealthy tendency／司机突然把车～住。The driver put on the brake suddenly.
【刹把】[机] brake crank
【刹车】①(用闸止住车辆行进) stop a vehicle by applying the brakes; put on the brakes ②(使机器停止运转) stop a machine by cutting off the power; turn off a machine ③(制动机件) brake

煞

①(停; 结束) stop; halt; check; bring to a close: ～住脚 stop short ②(勒紧) tighten: ～一～腰带 tighten one's belt
【煞笔】①(停笔) concluding lines of an article; ending of a piece of writing ②(结束语) write the final line
【煞车】firmly fasten a load secure with rope; lash down the load
【煞尾】①(收尾) finish off; round off; wind up ②(最后一段) final stage; end; ending

shǎ

傻

①(头脑糊涂) stupid; muddleheaded: 装～ act dumb; pretend not to know／你真～! How stupid you are.／他吓～了。 He was struck dumb with terror. ②(死心眼) think or act mechanically; tactless: 他为人太～, 不知怎样与人交往。 He has no tact in dealing with people.
【傻瓜】fool; blockhead; simpleton
【傻呵呵】simpleminded; not very clever
【傻劲儿】①(傻气) foolishness; stupidity ②(单凭力气干) with sheer enthusiasm; be foolhardy
【傻头傻脑】clumsy and stupid; foolish; silly
【傻笑】laugh foolishly; giggle; smirk
【傻眼】be dumbfounded; be stunned: 她一看考卷就～了。When she saw the examination paper she got a nasty shock.
【傻子】fool; blockhead; simpleton

shà

霎

a very short time; moment; instant: 一～ in a moment
【霎时间】in a twinkling; in a split second; in a jiffy

厦

a tall building; mansion: 高楼大～ tall buildings and great mansions

歃

[书] suck
【歃血】smear the blood of a sacrifice on the mouth; an ancient form of swearing an oath
【歃血为盟】smear the blood as a sign of the oath; smear the mouth with blood and swear fidelity

煞

①(凶神) evil spirit; goblin ②(很) very
【煞白】ghastly pale; deathly pale; pallid
【煞费苦心】rack one's brains; take great pains: ～地寻找借口 cudgel one's brains to find an excuse
【煞费周章】take much pain; this matter was full of difficulties
【煞有介事】make a great show of being in earnest; pretend to be serious (about doing sth.)

shāi

筛

①(筛子) sieve; sifter; screen ②(筛东西) sift; sieve; screen; riddle: ～煤 screen coal／～面 sieve flour; sift flour
【筛布】bolting cloth
【筛法】[数] sieve method
【筛分】screening; sieving ◇ ～机 screening machine
【筛管】[植] sieve tube
【筛号】screen size; screen mesh; mesh number
【筛绢】bolting-silk
【筛网】screen mesh
【筛选】screen; preparation
【筛子】sieve; sifter; screen: 粗～ riddle／煤～ coal screen

shǎi

色 [口] color: 这布掉～。This cloth fade.

【色子】dice: 掷～ play dice

shài

晒 ①(日照) shine upon: 日～雨淋 be exposed to the sun and rain ②(在阳光下吸收光和热) dry in the sun; bask: ～被子 air a quilt / ～粮食 dry grain in the sun / 我们在阳光充足的海滩上～太阳。We basked ourselves on the sunny beach.

【晒斑】[医] sunburn

【晒版】[印] plate burning; printing down

【晒垡】[农] sun the earth which has been ploughed up; sun the upturned soil

【晒坪】sunning ground

【晒台】flat roof (for drying clothes, etc.)

【晒图】make a blueprint; blueprint ◇ ～员 blueprinter / ～纸 blueprint paper

【晒像灯】printing lamp

【晒烟】sun-cured tobacco

【晒盐】evaporate brine in the sun to make salt

shān

潸 [书] in tears; tearfully

【潸然泪下】tears trickling down one's cheeks; shed silent tears; drop a few silent tears

扇 ①(摇动扇子等以加速空气流动) fan: ～火 fan a fire / ～扇子 fan oneself; use a fan ②(鼓动) incite; instigate; fan up; stir up: ～阴风 fan up an evil wind; secretly stir up trouble

【扇动】①(摇动) fan; flap: ～翅膀 flap the wings ②(鼓动别人干坏事) instigate; incite; stir up; whip up: ～叛乱 incite a rebellion / ～无政府主义 incite anarchism / ～种族主义情绪 whip up racist sentiments

【扇风点火】fan the flames; inflame and agitate people; stir up trouble

【扇风机】ventilating fan

【扇阴风，点鬼火】fan the winds of evil and spread the fires of turmoil; foment trouble

芟 ①(割草) mow (grass) ②(除去) weed out; eliminate

【芟除】①(除去) mow; cut down: ～杂草 weeding ②(删除) delete

苫 straw mat

【苫布】tarpaulin

山 ①(地面形成的高耸部分) hill; mountain: 爬上～ mount a hill / 珠穆朗玛峰是世界上最高的～。Mt. Qomolangma is the highest mountain in the world. ②(象山的东西) anything resembling a mountain: 冰～ iceberg ③(蚕蔟) bushes in which silkworms spin cocoons: 蚕上～了。The silkworms have gone into bushes to spin their cocoons.

【山坳】col

【山背】[气] yamase

【山崩】landslide; landslip

【山茶】[植] camellia

【山城】mountain city

【山川】mountains and rivers; land

【山慈姑】[植] edible tulip

【山村】mountain village

【山带】mountain belts

【山丹】[植] morningstar lily

【山道年】[药] santonin

【山地】①(多山地带) mountainous region; hilly area; hilly country ②(山上的农田) fields on a hill

【山顶】the summit of a mountain; hill top

【山洞】cave; cavern

【山豆根】[植] subprostrate sophora

【山风】[气] mountain breeze

【山峰】mountain peak

【山冈】low hill; hillock

【山高水长】(of nobility of character) as high as the hills and as long as the rivers; of lasting influence

【山高水低】death; unexpected misfortune; sth. unfortunate

【山歌】folk song

【山沟】gully; ravine; (mountain) valley

【山谷】mountain valley

【山河】mountains and rivers; the land of a country: 锦绣～ beautiful land

【山核桃】①[植] hickory ②(干果) hickory nut

【山洪】mountain torrents: ～暴发。Torrents of water rushed down the mountain.

【山货】①(山区的一般土产) mountain products ②(用竹、木、麻、陶瓷等制成的日用品) household utensils made of wood, bamboo, clay, etc.

【山鸡】[方] pheasant

【山鸡椒】[植] cubeb litsea tree

【山脊】ridge (of a mountain or hill)

【山涧】mountain stream

【山椒鸟】[动] minivet
【山脚】the foot of a hill
【山口】mountain pass; pass
【山梁】ridge (of a mountain or hill)
【山林】mountain forest; wooded mountain: ～地区 mountain and forest region; wooded and hilly lands ◇ ～所有权 ownership of mountains
【山岭】mountain ridge
【山麓】the foot of a mountain ◇ ～冲积平原 bajada / ～丘陵 foothills
【山峦】chain of mountains: ～起伏 undulating hills
【山脉】mountain range; mountain chain
【山猫】[动] leopard cat
【山毛榉】[植] beech
【山盟海誓】(make) a solemn pledge of love
【山明水秀】green hills and clear waters; picturesque scenery
【山南海北】south of the mountains and north of the seas; far and wide; all over the land
【山炮】mountain gun; mountain artillery
【山坡】hillside; mountain slope
【山墙】[建] gable
【山穷水尽】where the mountains and the rivers end; at the end of one's rope
【山区】mountain area
【山泉】mountain spring
【山雀】[动] tit
【山水】①(山上流下的水) water from mountain ②(山和水) mountains and rivers; scenery with hills and waters: ～相连 be linked by common mountains and rivers ③ [美术] traditional Chinese painting of mountains and waters; landscape ◇ ～画 mountains-and-waters painting; landscape painting
【山桃】[植] mountain peach
【山桐子】[植] idesia
【山头】①(山的上部) hilltop; the top of a mountain ②(宗派) mountain stronghold; faction: 拉～ form a faction
【山系】[地] mountain system
【山乡】mountain area
【山响】deafening; thunderous: 把鼓敲得～ drums beating thunderously
【山魈】[动] mandrill
【山鸦】[动] chough
【山崖】cliff
【山羊】①(羊的一种) goat: 公～ he-goat / 母～ she-goat / 小～ kid ② [体] buck ◇ ～胡子 goatee / ～绒 cashmere
【山腰】half way up the mountain
【山药】[植] Chinese yam
【山药蛋】[方] potato
【山阴】[气] ubac
【山雨欲来风满楼】the wind sweeping through the tower heralds a rising storm in the mountains; the rising wind forebodes the coming storm
【山鹬】[动] woodcock
【山岳】lofty mountains ◇ ～冰川 mountain glacier; alpine glacier / ～地区 mountainous region
【山楂】① [植] (Chinese) hawthorn ②(山楂的果实) haw ◇ ～糕 haw jelly
【山寨】mountain fastness; fortified mountain village
【山珍海味】delicacies from land and sea; dainties of every kind
【山中无老虎,猴子称大王】The monkey reigns in the mountains when the tiger is not there. (或: Among the blind the one-eyed man is king.)
【山庄】mountain villa
【山嘴】[地] spur

舢

【舢板】sampan

衫

衫 unlined upper garment: 衬～ shirt / 汗～ undershirt / 体恤～ T-shirt

杉

杉 [植] China fir

钐

钐 [化] samarium (Sm)
【钐中毒】samarium poisoning

膻

膻 the smell of mutton: 这羊肉味太～。This mutton has got a strong smell.

珊

【珊瑚】coral ◇ ～虫 coral polyp; coral insect / ～岛 coral island / ～礁 coral reef

栅

【栅极】[电] grid: 抑制～ suppressor grid
【栅距】pitch
【栅漏】grid leak

删

删 delete; leave out: 把长故事～短 cut a long story short / 有几个字被～掉了。Several words had been deleted. / 主编～去了最后一段。The editor in chief cut out the last paragraph.

【删除】delete; strike out; cut off; cut out; leave out
【删繁就简】simplify sth. by cutting out the superfluous
【删改】delete and change; revise finalized.
【删节】abridge; abbreviate: 本报略有～ slightly abridged by our editorial staff／这是由原本～的。It was abridged from the original work. ◇～本 abridged edition; abbreviated version／～号 ellipsis; suspension points; ellipsis dots (……) (...)

姗

【姗姗来迟】be slow in coming; be late

shǎn

闪 ①(闪避) dodge; get out of the way: ～到树后 dodge behind a tree／东躲西～ dodge about／往旁边一～ dodge swiftly to one side; jump out of the way ②(扭伤) twist; sprain: ～了腰 sprain one's back ③(闪电) lightning: 打～ flashes of lightning ④(突然出现) flash; sparkle; shine: 一～而过 flash past ⑤(丢下) leave behind: 你走时叫我一声，别把我～下。Please call for me when you go; don't leave me behind.
【闪避】dodge; sidestep
【闪挫】[中医] sudden strain or contusion of a muscle; sprain
【闪点】[化] flash point
【闪电】lightning: ～进攻 make a lightning attack ◇～战 lightning war; blitzkrieg; blitz
【闪光】①(突然一现的光亮) flash of light: 流星象一道～划过夜空。Like a flash of lightning, the meteor shot across the sky. ②(闪闪发光) gleam; glisten; glitter: 星星的～ twinkling of stars／露珠在～光。Dewdrops are glistening. ◇～对头焊 flash-butt welding
【闪光灯】[摄] flash lamp; photoflash: 万次～ electronic 〈multitime〉 flash unit ◇～灯泡 [摄] flash bulb; flash lamp; flash lamp bulb
【闪击战】lightning war; blitzkrieg; blitz
【闪开】get out of the way; jump aside; dodge: 车来了，快～! Look out! There's a bus coming.
【闪闪】sparkle; glisten; glitter: ～发光 sparkle; glitter
【闪身】①(闪避) dodge ②(侧身) sideways: ～进门 walk sideways through the door
【闪石】[矿] amphibole
【闪失】mishap; accident: 要是有个～，怎么办呢? What if anything should go wrong?
【闪烁】①(光亮动摇不定) twinkle; glimmer; glisten: 远处～着灯光。Lights glimmered in the distance. ②(不直说) evasive; vague; noncommittal: ～其词 speak evasively; hedge; dodge about; evade issues ③ [电] scintillation ◇～开关 flasher／～计数器 scintillation counter／～扫描 scintiscan／～系数 flicker factor
【闪现】flash before one: ～在眼前 flash before one's eyes
【闪锌矿】[矿] (zinc) blende; sphalerite
【闪岩】amphibolite
【闪耀】glitter; shine; radiate: 繁星～ glittering stars

shàn

擅 ①(擅自) arrogate to oneself; do sth. on one's own authority: ～作主张 make a decision without authorization ②(善于) be good at; be expert in: 不～辞令 lack facility in polite or tactful speech
【擅长】be good at; be expert in; be skilled in: 他～画山水画。He is good at landscape painting.
【擅离职守】be absent from one's post without leave; leave one's post without permission; unauthorized departure from official duty ◇～者 deserter
【擅自】do sth. without authorization: ～逮捕 make unwarranted arrest／～决定 make arbitrary decisions／～行动 act presumptuously／～作主 act arbitrarily; take an unauthorized action

嬗

【嬗变】①(演变) evolution ② [物] transmutation: 自然～ natural transmutation

善 ①(善良) good: 改恶从～ give up evil and return to good; mend one's ways／心怀不～ harbor ill intent ②(良好) satisfactory; good: ～策 a wise policy; the best policy ③(办好; 弄好) make a success of; perfect: 工欲～其事，必先利其器。A workman must sharpen his tools if he is to do his work well. ④(友好) kind; friendly: 友～ be friendly; be kind and helpful ⑤(擅长) be good at; be expert in: ～经营 be good at management ⑥(好好地) properly: ～为说辞 put in a good word for sb.／～自保重 take good care of yourself ⑦(容易; 易于) be apt to: ～变 be apt to change／～忘 be forgetful; have a short memory

【善罢甘休】(多用于否定) leave the matter at that; let it go at that; willing to let it go
【善本】reliable text; good edition: ～书 rare book
【善处】[书] deal discreetly with; conduct oneself well
【善恶】good and evil: ～分明 distinguish the good from the bad
【善后】cope with the aftermath of a disaster: ～事宜 matters concerning reconstruction; rehabilitation works
【善举】[书] philanthropic act or project
【善良】good and honest; kindhearted: ～的愿望 the best of intentions; good intentions / ～守法 good and lawful
【善男信女】devotees to Buddha; devout men and women
【善人】philanthropist; charitable person
【善始善终】start well and end well; do well from start to finish; see sth. through
【善心】mercy; benevolence: 发～ show kindness; become benevolent
【善意】goodwill; good intentions; bona fides; good faith: 出于～ out of goodwill; with the best intentions ◇ ～的错误陈述 innocent misrepresentation / ～的批评 well-intentioned criticism / ～第三人 third party acting in good faith / ～购得 innocent purchase / ～占有人 holder in good faith
【善有善报, 恶有恶报】good will be rewarded with good, and evil with evil
【善于】be good at; be adept in: ～辞令 one is skillful in making statement; eloquent in speech / ～歌舞 be good at singing and dancing
【善自保重】take good care of yourself
【善自为谋】be apt at devising a good plan for oneself; give a problem careful consideration
【善自珍摄】please take good care of yourself; coddle oneself
【善终】die a natural death; die in one's bed

膳 meals; board
【膳费】board expenses
【膳食】meals; food: 流质～ liquid diet
【膳宿】board and lodging: 安排～ arrage board and lodging

鳝 eel; finless eel

缮 ①(修补) repair; mend: 房屋修～ house repairing ②(缮写) copy; write out: ～清 make a fair copy
【缮写】write out; copy

禅
【禅让】abdicate and hand over the crown to another person

扇 ①(扇子) fan: 芭蕉～ palm-leaf fan / 电～ electric fan / 宫～ mandarin fan / 绢～ silk fan / 葵～ palm fan / 檀香～ sandalwood fan / 羽毛～ feather fan / 折～ folding fan / 通风～ draft fan ②(板状物) leaf: 门～ door leaf ③[量] 一～窗 a window / 一～门 a door
【扇贝】[动] scallop; fan shell
【扇车】winnowing machine; winnower
【扇骨子】the ribs of a fan
【扇面儿】the covering of a fan
【扇形】②(扇子的形状) fan-shaped: ～窗子 fan-shaped window / 他把牌在手上展成～。He fanned out the card in his hand. ②[数] sector ◇ ～齿轮 [机] sector gear / ～风[气] sector wind / ～构造 fan structure / ～激光束 fan-shaped laser beam / ～图 fan chart
【扇叶】[动] flabellum
【扇子】fan

骟 ①(割掉睾丸) castrate: ～马 castrate a horse ②(切除卵巢) spay: ～母马 spay a mare

苫 cover (with a straw mat, tarpaulin, etc.): 把麦子～上 cover up the wheat
【苫布】tarpaulin

赡 ①(赡养) support; provide for: ～家养口 support a family ②(丰富; 充足) [书] sufficient; abundant
【赡养】support; provide for: ～父母 support one's parents ◇ ～费 alimony; cost of support; maintenance / ～义务 duty of maintenance; duty to support

疝 hernia: 腹股沟～ inguinal hernia / 脐～ umbilical hernia ◇ ～带 truss
【疝气】[医] hernia ◇ ～带 hernia band; truss
【疝修合术】[医] herniorrhaphy

讪 ①(讥笑) mock; ridicule ②(难为情的样子) embarrassed; awkward; shamefaced: ～～地走开 walk away looking embarrassed / ～～地坐下 sit down feeling abashed
【讪笑】ridicule; mock; deride

shāng

商 ①(商量) discuss; consult: 面～ discuss with sb personally / 有要事相～。I have important matters to discuss with you. ②(商业) trade; commerce; business: 经～ be in business / 通～ have trade relations ③(商人) merchant; trader; businessman; dealer: 奸～ profiteer / 零售～ retailer / 批发～ wholesaler / 私～ businessman ④ [数] quotient

【商标】trade mark; brand ◇ ～法 trademark law / ～名称 brand name / ～权 trade mark privileges; trademark right / ～所有权 ownership of trademark / ～纸 label paper / ～注册 trade mark registration / ～注册人 trademark registrant / ～专用权 exclusive right to use trademark

【商飙徐起】a chilly wind began to blow

【商埠】[旧] commercial port

【商场】market; bazaar

【商船】merchant ship; merchantman: ～船长 master mariner ◇ ～航线 shipping route

【商店】shop; store: 百货～ department store / 儿童用品～ children's goods store / 日夜～ day-and-night shop / 友谊～ Friendship Store ◇ ～橱窗 shopwindow / ～售货员 shopman; shopwoman / ～营业时间 shop hours

【商定】decide through consultation; agree: 经～ it has been decided through consultation that / 双方～了谈判的日期。The two sides have agreed on the date of negotiations.

【商队】a company of travelling merchants; trade caravan

【商法】[法] commercial law; commercial act

【商贩】small retailer; pedlar

【商港】commercial port

【商行】trading company; commercial firm

【商号】shop; store; business establishment; corporate name ◇ ～注册法 registration of business names act

【商会】chamber of commerce

【商检证明书】commodity inspection certificate

【商界】business circles; commercial circles

【商量】consult; discuss; talk over: 与某人～某事 consult with sb. on sth.; discuss sth. with sb. / 这事好～。That can be settled through discussion.

【商品】commodity; goods; merchandise: 呆滞～ slow-selling goods / 对路～ marketable commodities / 冷背～ unsalable goods / 内销～ commodities for the home market ◇ ～拜物教 commodity fetishism / ～二重性 dual character of commodity / ～价格 commodity price / ～交换 exchange of commodities / ～检验 commodity inspection / ～检验法规 commodity inspection law / ～检验局 commodity inspection and testing bureau / ～交易所 commodity exchange / ～经济 commodity economy / ～库存 commodity stocks / ～粮基地 commodity grain base / ～流通 circulation of commodities; commodity circulation / ～流通费用 cost in commodity circulation / ～列名 commodity entry / ～目录 (descriptive) catalogue / ～清单 inventory / ～生产 commodity production / ～生产基地 bases for the production of commodities / ～输出 export of commodities / ～税 commodity tax / ～销售市场 outlet for goods / ～运输 transportation of goods / ～展览 trade shoe / ～制度 commodity system / ～住宅 commercial residential buildings

【商洽】arrange with sb.; take up (a matter) with sb.

【商情】market conditions ◇ ～预测 bushiness forecasting

【商榷】discuss; deliberate: 这一点值得～。This point is open to question.

【商人】businessman; merchant; trader

【商事】commercial affairs ◇ ～法庭 commercial act / ～纠纷 commercial dispute / ～诉讼 commercial action; commercial causes / ～行为 commercial act / ～仲裁 commercial arbitration

【商数】[数] quotient

【商谈】exchange views; confer; discuss; negotiate: ～对策 discuss a countermeasure / 与某人～某事 negotiate with sb. about <over> sth.

【商讨】discuss; deliberate over: 就发展两国关系进行有益的～ hold useful discussions on developing relations between the two countries

【商务】commercial affairs; business affairs ◇ ～参赞 commercial counsellor / ～处 commercial counsellor's office / ～代表 commercial representative; trade representative / ～代表处 trade representative's office; office of a trade delegation / ～秘书 commercial secretary / ～责任 commercial liability / ～仲裁 commercial arbitration / ～专员 commercial attaché

【商业】commerce; trade; business: 合作～ cooperative trade ◇ ～部门 commercial departments / ～贷款 commercial loans / ～道

德 business ethics / ～法 business law; commercial law / ～风险 commercial risks / ～惯例 business practice / ～汇票 commercial draft / ～机构 business organization; commercial undertaking / ～禁运 commercial embargo / ～票据 commercial bill; trade bill; business papers / ～区 business quarter; commercial district; business district; shoppy part of town / ～事务 business affair / ～网 commercial network; network of trading establishments / ～危机 commercial crisis / ～信贷 commercial credit / ～信用 commercial standing / ～信用卡 commercial letter of credit / ～信誉 commercial goodwill; commercial reputation / ～银行 commercial bank / ～预测 business forecasting / ～中心 commercial center; trading center; shopping center / ～折扣 commercial discount; trade discount / ～注册 business registration / ～资本 commercial capital; merchant capital

【商议】confer; discuss

【商约】commercial treaty

【商酌】discuss and consider; deliberate over

熵 [物] entropy; thermal charge

【熵产生】entropy production

墒 [农] moisture in the soil: 保～ preserve the moisture of the soil / 抢～ lose no time in sowing while there is sufficient moisture in the soil

【墒情】soil moisture content

伤 ①(受伤) wound; injury: 刀～ a knife wound / 轻～ a slight injury / 致命～ mortal wound; fatal wound / 十人死亡,三十人受～。Ten people were killed and thirty injured 〈wounded〉 ②(伤害) injure; hurt: ～了感情 hurt sb.'s feelings ③(悲伤) be distressed: 哀～ sad; sorrowful ④(厌烦) get sick of sth.; develop an aversion to sth.: 这孩子吃糖吃～了。The child has got sick of eating sweets. ⑤(妨碍) be harmful to; hinder: 无～大雅 involving no major principle; not matter much / 有～国体 bring discredit upon one's country

【伤疤】scar

【伤兵】wounded soldier

【伤病员】the sick and wounded; noneffectives

【伤残】permanent disability

【伤风】catch cold; have a cold ◇ ～病毒 common cold virus

【伤风败俗】offend public decency; corrupt public morals; breach of morality

【伤感】sick at heart; sentimental

【伤害】injure; harm; hurt: ～他人身体 [法] do an injury to sb.; injury; maiming / 不要～益鸟。Don't harm beneficial birds. ◇ ～保险 injury insurance

【伤寒】① [医] typhoid fever; typhoid ② [中医] diseases caused by harmful cold factors; febrile diseases; fevers ◇ ～杆菌 typhoid bacillus

【伤耗】damage

【伤痕】scar; bruise; hack: ～累累 scars of wounds strung together like bead

【伤筋动骨】be injured in the sinews or bones; have a fracture

【伤科】[中医] (department of) traumatology

【伤口】wound; cut: 洗～ bathe a wound

【伤脑筋】knotty; troublesome; bothersome: ～的问题 a knotty problem; headache / 这事真～。It's rather troublesome.

【伤神】overtax one's nerves; be nerve-racking

【伤食】[中医] dyspepsia caused by excessive eating or improper diet

【伤势】the condition of an injury: ～很重 be seriously wounded

【伤天害理】flagrantly wrongful; grossly unjust; do things offensive to God and reason

【伤亡】injuries and deaths; casualties: ～惨重 suffer heavy casualties ◇ ～报告 [军] return of losses / ～事故 [法] casualty accident

【伤心】sad; grieved; broken-hearted: ～落泪 shed sad tears; weep in grief

【伤心惨目】too ghastly to look at; tragic (scene)

【伤员】wounded personnel; the wounded

觞 [考古] wine cup; drinking vessel

【觞政娴习】be quite familiar with all sorts of wine-games

shǎng

赏 ①(奖赏) grant a reward; award: 领～ receive a reward ②(奖赏之物) reward; award: 有～有罚 mete out rewards or punishments as the case demands; duly mete out rewards and punishments ③(欣赏) admire; enjoy; appreciate: ～花 admire the beauty of flowers

【赏赐】grant a reward; award: 得到很多～ be given a handsome reward

【赏罚】rewards and punishments; deserving award and punishment: ～严明 be strict and fair in meting out rewards and punishments

【赏功罚罪】give rewards for good service and punishments for faults
【赏光】[套] 务请～ request the pleasure of your company
【赏鉴】appreciate: ～名画 appreciate the masterpieces in painting / ～一件艺术品 appreciate a work of art
【赏金】money reward; pecuniary reward
【赏景怡神】enjoy the landscape to one's satisfaction
【赏识】recognize the worth of; appreciate: ～某人 think highly of sb. / 受到某人的～ be in sb.'s good graces
【赏玩】admire the beauty of sth.; delight in; enjoy: ～古董 delight in antiques / ～山景 enjoy mountain scenery
【赏心悦目】find the scenery pleasing to both the eye and the mind
【赏月】enjoy the glorious full moon: ～观花 appreciate the moon and flowers

晌 ①(一天内的一段时间) part of the day: 前半～儿 morning / 晚半～儿 dusk ②[方](晌午) noon: 歇～ take a midday nap or rest
【晌饭】[方] ①(午饭) midday meal; lunch ②(农忙时的加餐) extra meal in the daytime during the busy farming season
【晌觉】[方] afternoon nap
【晌午】[口] midday; noon: ～饭 midday meal; lunch

shàng

尚 ①[副](还) still; yet: 一息～存 as long as one lives; so long as there is still breath left in one / 时机～未成熟。 The time is not yet ripe. ②(尊崇) esteem; value; set great store by: ～武 set great store by military affairs or martial arts
【尚付阙如】still wanting; not yet done
【尚且】[连] even: 大人～举不起来，何况小孩子。 Even grown-ups can't lift it, to say nothing of children.
【尚未字人】(she) is not yet betrothed to anyone.
【尚武精神】encourage a military spirit; militarism

绱 stitch the sole to the upper

上 ①(位置在高处) upper; up; upward: ～身 the upper part of the body / ～嘴唇 upper lip / 往～看 look up; look upward ②(等级或品质高) higher; superior; better: 中～水平 above the average; better than the average ③(次序或时间在前的) first (part); preceding; previous: ～半年 the first half of the year / ～半月 the first half of a month / ～册 the first volume; Volume One; Book One / ～集 the first part; Part One; Volume One ④(旧指皇帝) the emperor: ～谕 imperial decree ⑤(由低处到高处) go up; mount; board; get on: ～车 get on a car / ～船 go on board; go aboard a ship / ～飞机 get on a plane; board a plane / ～楼 mount stairs; go upstairs / ～坡 go up a slope / ～山 go up a hill; go uphill ⑥(到；去) go to; leave for: ～北京 leave for Beijing / 你～哪儿去? Where are you going? ⑦(呈递) submit; send in; present: 随函附～二十分的邮票一张。 Enclosed herewith is an tewnty-*fen* stamp. ⑧(向前进) forge ahead; go ahead: 快～! Go ahead. Quick! (指打球等); Hurry up and get on! (指上车等) ⑨ [体](上场) enter the court or field: 这后半场你～。 You play this second half. ⑩(加添) fill; supply; serve: ～货 replenish the stock of goods / 给水箱～水 fill the tank with water ⑪(安上) place sth. in position; set; fix: ～刀具 fix a cutting tool / ～螺丝 drive a screw (into sth.) ⑫(涂；搽) apply; paint; smear: ～肥 spread manure / ～药膏 apply ointment ⑬(登载) be put on record; be carried (in a publication): 她的名字～了报。 Her name has appeared on the newspaper. / 他的先进事迹都～了电视了。 His model deeds have been publicized on TV. ⑭(拧紧) wind; screw; tighten: 表该～了。 The watch needs winding. ⑮(开始工作，学习) be engaged (in work, study, etc.) at a fixed time: 我今天～夜班。 I'm on the night shift today. ⑯(达到) up to; as many as: ～百人 up to a hundred people ⑰(用在动词后，表示由低处向高处)爬～河堤 climb up to the top of the dike ⑱(用在动词后，表示达到目的)穿～外衣 put on a coat / 考～大学 be admitted to a college / 锁～门 lock the door; lock up ⑲(用在动词后，表示开始并继续)她爱～了草原。 She's fallen in love with the grasslands. ⑳(用在名词后，表示位置、范围或方面)会～ at the meeting / 理论～ in theory; theoretically / 事实～ in fact
【上岸】ashore; land; go ashore: 每到一个港口他都～。 He went ashore at every port ◇～证明书 landing certificate
【上班】go to work; start work; be on duty: 上午八点钟～ start work at eight in the

morning / 他～去了。He's gone to work. ◇～时间 work hours; office hours
【上半场】first half (of a game): ～比分多少? What was the score at half time? / ～比分是二比一。The score was 2 to 1 in the first half.
【上半身】the upper part of the body; above the waist
【上半夜】before midnight; the first half of the night
【上报】①(登载) appear in the newspapers ②(向上级报告) report to a higher body; report to the leadership
【上辈】①(祖先) ancestors ②(家族中的上一代) the elder generation of one's family; one's elders
【上臂】the upper arm
【上膘】(of animals) become fat; fatten
【上宾】distinguished guest; guest of honor
【上苍】Heaven; God
【上操】go out to drill; be drilling
【上策】the best plan; the best way out; the best thing to do
【上层】upper strata; upper levels ◇～分子 members of the upper strata; upper-class elements / ～人士 upper circles / ～社会 upper classes of society; upper-class society
【上层建筑】superstructure: ～领域 the realm of the superstructure
【上谄下骄】fawn upon one's superiors and look down upon those below; flatter those in high position and despise those of lower ranks
【上场】①(演员出场) appear on the stage; enter: 她在第二幕时才～。She doesn't go on until the second act. ②(运动员出场) enter the court or field; join in a contest; take part in the competition ◇～门 entrance (of a stage)
【上床】go to bed
【上窜下跳】run around on sinister errands
【上达】reach the higher authorities: 下情～ make the situation at the lower level known to the higher authorities
【上代】the previous generation; former generations
【上当】be taken in; be fooled; be duped: ～受骗 swallow a gudgeon; be tricked / 容易～ be easy to take in; be taken in easily
【上灯】light the lamp; light up ◇～时分 lighting-up time
【上等】first-class; first-rate; superior: ～货 first-class goods / ～料子 high-quality material / ～品 superior quality
【上低音号】[乐] baritone
【上帝】God
【上吊】hang oneself
【上冻】freeze
【上颚】maxilla (of a mammal); the upper jaw
【上方宝剑】the imperial sword
【上房】main rooms
【上访】apply for an audience with the higher authorities to appeal for help ◇～人员 visitors from the localities appealing to the higher authorities for help
【上坟】visit sb.'s grave; visit a grave to honor the memory of the dead
【上风】①(风刮来的那一方) windward: 站在～打农药 spray insecticide on the windward side ②(优势) advantage; superior position; upper hand: 占～ get the upper hand; win an advantage; prevail
【上腹部】epigastrium ◇～痛 epigastralgia
【上告】complain to the higher authorities or appeal to a higher court
【上工】go to work; start work
【上钩】rise to the bait; swallow the bait; get hooked
【上古】ancient times; remote ages ◇～史 ancient history
【上光】glazing; polishing ◇～机 glazing machine; glazer / ～蜡 wax polish
【上轨道】get on the right track; begin to work smoothly: 一切都已～了。Everything is in good order now.
【上好】first-class; best-quality; tip-top
【上颌】[生理] the upper jaw; maxilla
【上呼吸道】[生理] the upper respiratory tract ◇～感染 infection of the upper respiratory tract
【上火】①[方] get angry ②[中医] suffer from excessive internal heat
【上级】higher level; higher authorities: 报告～ report to the higher authorities; report to one's superior ◇～法院 court above; higher court; superior court / ～机关 higher bodies; higher authorities
【上浆】[纺] sizing: 棉布～ starching
【上将】general (陆军, 美空军); air chief marshal (英空军); admiral (海军)
【上交】hand in; turn over to the higher authorities
【上胶机】[纸] gluing machine
【上缴】turn over (revenues, etc.) to the higher authorities ◇～利润 that part of the profits

turned over to the state
【上街】①(到街上去) go into the street: ～示威 go on to the streets and demonstrate; going up in the streets ②(购物) go shopping
【上届】previous term or session; last: ～毕业生 last year's graduates
【上进】go forward; make progress: 不求～ not strive to make progress / 力求～ do one's best to advance ◇ ～心 the desire to do better; the urge for improvement
【上劲】energetically; with gusto; with great vigor: 越干越～儿 work with increasing vigor
【上敬下和】be respectful towards his superiors and kindly towards those under him
【上课】①(学生听课) attend class; go to class: 他昨天没来～。He didn't come to class yesterday. ②(教师讲课) conduct a class; give a lesson: 学校八点开始～。Classes begin at 8.
【上空】in the sky; overhead
【上口】fluent; smooth: 他把那首诗念得琅琅～。He can recite the poem fluently.
【上跨交叉】[交] overpass
【上款】the name of the recipient
【上蜡】[纺] waxing ◇ ～机 waxing machine
【上来】①(用在动词后，表示由低处到高处或由远处到近处来)乘电梯～ come up by lift / 影迷围～要他签名留念。The film fans gathered around him and asked for his autographs. ②(用在动词后，表示成功)这个问题你一定答得～。I'm sure you can answer this question. ③[方](用在形容词后，表示程度的增加)天气凉～了。The weather is getting cool.
【上联】the first line of a couplet on a scroll
【上梁】①(自行车的) cross bar; top tube ②(建筑物的) upper beam
【上梁不正下梁歪】If the upper beam is not straight, the lower ones will go aslant. (或: When those above behave unworthily, those below will do the same.)
【上列】the above-listed; the above: ～各项 the items listed above; the above-listed items
【上流】①(上游) upper reaches (of a river) ②(旧指社会地位高的) belonging to the upper circles; upper-class: ～社会 high society; polite society
【上路】set out on a journey; start off
【上马】①(骑上马背) mount a horse ②(口令) To horse! ③(着手办事) start: 这项工程明年～。The project will start next year.
【上门】①(登门) come or go to see sb.; call; drop in; visit: ～服务 come knocking at sb.'s door to offer one's service; make house calls / ～讨债 remain at someone's to demand payment of debt / 送货～ deliver goods to the doorstep / ～迎亲。The bridegroom goes to the bride's home to escort her to the wedding. ②(关门) shut the door for the night; bolt the door
【上面】①(位置较高的地方) above; over; on top of; on the surface of: 大型客机在云层～飞行。The air-liner flew above the clouds. / 河～有座桥。There is a bridge over 〈across〉 the river. ②(次序靠前的部分) above-mentioned; aforesaid; foregoing: ～所举的例子 the above-mentioned example ③(上级) the higher authorities; the higher-ups: 这是～的指示。These are the instructions from above. ④(方面) aspect; respect; regard: 他在外文～下了很多功夫。He has put a lot of effort into his study of foreign languages.
【上年纪】be getting on in years
【上皮癌】[医] epithelioma
【上皮组织】[生理] epithelial tissue
【上品】highest grade; top grade: 茅台是酒中～。*Maotai* is a top-grade spirit.
【上坡路】①(由低处通向高处的路) uphill road; upward slope ②(向好的方向发展的路) upward trend; steady progress
【上铺】upper berth
【上气不接下气】gasp for breath; be out of breath
【上腔静脉】[生理] superior vena cava; vena cava superior
【上去】①(由低处到高处) go up: 登着梯子～ go up 〈on〉 a ladder ②(用在动词后，表示由低到高，或由近及远，或由主体向对象)把大车推～ push the cart up / 爬～ climb up
【上染率】[纺] dye-uptake
【上任】take up an official post; assume office
【上色】color (a picture, map, etc.)
【上上】①(最好) the very best: ～策 the best plan ②(比前一时期再往前的) before last: ～星期 the week before last
【上身】①(身体的上半部) the upper part of the body: 光着～ be stripped to the waist ②(上衣) upper outer garment; shirt; blouse; jacket ③(初次穿在身上) start wearing
【上升】rise; go up; ascend: 气温～。The temperature is going up. / 我们的生产稳步～。Our production goes steadily up. ◇ ～角[航空] angle of climb / ～气流 ascending air; up current / ～失速[航空] advance stall / ～转弯[航

空] pull-up turn
【上声】[语] falling-rising tone
【上乘】① [佛教] Mahayana; Great Vehicle ② (文学艺术的上品) a work of a high order
【上士】sergeant first class (美陆军); staff sergeant (英陆军); petty officer first class (美海军); chief petty officer (英海军); technical sergeant (美空军); flight sergeant (英空军)
【上市】go 〈appear〉 on the market: 这是刚刚～的西红柿。These tomatoes have just appeared on the market.
【上视图】[机] top view
【上手】①(上首; 位置较尊的一侧) left-hand seat; seat of honor ② (开始) start; begin: 今天的球一～就打得很顺利。Today we were doing fine right after the game started.
【上书】submit a written statement to a higher authority; send in a memorial
【上述】above-mentioned; aforementioned; aforesaid: 达到～目标 achieve the aforementioned objectives
【上水】① (向上游航行) sail upstream: ～船 upriver boat ② (加水) feed water to a steam engine, radiator (of an automobile), etc.
【上水道】[建] water-supply line
【上税】pay taxes
【上司】superior; boss: 顶头～ one's immediate superior
【上诉】[法] appeal (to a higher court): 不服判决而～ appeal to a higher court; appeal against a decision / 提出～ lodge an appeal ◇ ～程序 procedure for appeal / ～法院 appellate court / ～费用 costs of appeal / ～理由 grounds of appeal / ～权 right of appeal / ～人 petitioner; appellant / ～审 trial on appeal; retrial on appeal / ～书 instrument of appeal / ～文件 instrument of appeal / ～委员会 board of review / ～要件 important conditions of appeal / ～中的附带诉讼 incidental to an appeal / ～种类 class of appeals / ～状 petition for appeal
【上算】paying; worthwhile: 烧煤气比烧煤～。It's more economical to use gas than coal.
【上台】①(登台) go up onto the platform; appear on the stage: ～演奏 go up onto the platform and play / ～讲话 mount the platform and address the audience ② (任职; 掌权) assume power; come to power
【上膛】①[军] (of a gun) be loaded: 子弹上了膛。The gun is loaded. ② (腭) palate
【上天】① (上苍) Heaven; Providence; God ② (升到天空) go up to the sky; fly sky-high: 人造卫星～。A man-made satellite has been launched.
【上天无路,入地无门】there is no road to heaven and no door into the earth; no way of escape; in desperate straits
【上吐下泻】vomit and have watery stools; vomiting and being purged
【上尉】captain (陆军,美空军); lieutenant (海军); flight lieutenant (英空军)
【上文】foregoing paragraphs or chapters; preceding part of the text: 见～ see above
【上无片瓦,下无寸土】have neither a tile over one's head nor a speck of land under one's feet; not to own a single brick or an inch of land
【上午】forenoon; morning
【上下】①(职务、辈分上的高低) high and low; old and young: 举国～ the whole nation / ～通气 full communication between the higher and lower levels ② (从上到下) from top to bottom; up and down: ～打量 look sb. up and down; scrutinize sb. from head to foot ③ (从高到低或从低到高) go up and down ④ (程度高低) relative superiority or inferiority: 不相～ equally matched; about the same / 难分～ hard to tell which is better ⑤ (用在数量词后面) about; or so; or thereabouts: 三十岁～ about thirty years old; thirty or so
【上下其手】practice fraud; league together for some evil end
【上下文】context
【上弦】① [天] first quarter (of the moon) ② (上钟表发条) wind up a clock or watch ◇ ～月 the moon at the first quarter
【上限】upper limit
【上校】colonel (陆军,美空军); captain (海军); group captain (英空军)
【上鞋】sole a shoe; stitch the sole to the upper
【上刑】put sb. to torture; torture
【上行】① [铁道] up; upgoing: ～列车 up train ② [航运] upriver; upstream: ～船 upriver boat
【上行下效】those in subordinate positions will follow the example set by their superiors
【上旋】[乒乓球] top spin
【上学】go to school; attend school; be at school: 去～ go to school / 这孩子～了吗? Is the child at school?
【上旬】the first ten-day period of a month
【上压力】[物] upward pressure
【上演】put on the stage; perform: 不日～ will be performed very soon / 那个剧场在～什么?

What's on at that theatre?
【上衣】upper outer garment; jacket
【上议院】upper house; the House of Lords (of Britain)
【上瘾】be addicted (to sth.); get into the habit (of doing sth.): 他抽烟～了。He's got into the habit of smoking. / 这东西吃多了会～吗? Is this habit-forming?
【上映】show (a film); screen; be on; run: 这部影片已～五天。The film has run for five days.
【上游】①(河流接近发源地的部分) upper reaches ②(先进) advanced position: ～无止境。One can always aim higher.
【上谕】imperial edict
【上涨】rise; go up: 河水～。The river has risen. / 物价～。The prices are going up.
【上帐】make an entry in an account book; enter sth. in an account
【上阵】go into battle; pitch into the work
【上肢】upper limbs
【上装】①(演员化装) make up (for a theatrical performance) ②[方](上衣) upper outer garment; jacket
【上座】seat of honor

shāo

烧 ①(使东西着火) burn: 干柴好～。Dry wood burns easily. / 他把熨衣板～了一个洞。He burned a hole on the ironing board. ②(加热) cook; bake; heat: ～点水 heat up some water / ～饭 cook food; prepare a meal ③(烹调方法) stew after frying or fry after stewing: ～茄子 stewed eggplant ④(烤) roast: ～鸡 roast chicken ⑤(发烧) run a fever; have a temperature: 病人～得厉害。The patient's running a high fever. ⑥(体温高) fever: ～退了。The fever is down.
【烧杯】[化] beaker
【烧饼】sesame seed cake
【烧焊】burn-in; freeze
【烧化】①(烧掉尸体) cremate ②(烧掉) burn (paper, etc. as an offering to the dead)
【烧荒】burn the grass on waste land
【烧毁】burn out; burn down; burn-through; overburning: 他家房子上星期～了。His house burnt down last week.
【烧火】make a fire; light a fire; tend the kitchen fire
【烧碱】[化] caustic soda
【烧结】sintering; agglomeration; agglutination ◇～厂 sintering plant / ～法 [冶] sintering process / ～剂 [化] agglutinant / ～品 sinter / ～箱 sinter box
【烧酒】spirit usu. distilled from sorghum or maize; white
【烧蓝】[工美] enameling; blueing
【烧卖】[食品] a steamed dumpling with the dough gathered at the top
【烧瓶】[化] flask
【烧伤】[医] burn: 三度～ third-degree burns / 医治～ healing of burns / 他在火灾中～致死。He died of the burns that he received in the fire.
【烧香】burn joss sticks (before an idol): ～拜佛 burn incense and pray / ～许愿 burn incense and make a vow to the god
【烧心】①[医] hearburn ②[方](of cabbages) turn yellow at the heart
【烧夷弹】incendiary bomb
【烧砖】bake bricks; fire bricks
【烧灼】burn; scorch; singe

鞘 whiplash

梢 tip; the thin end of a twig, etc.: 辫子～ the end of a plait / 树～ the top of a tree
【梢头】①(树枝的顶端) the tip of a branch ②[林] top log

捎 take along sth. to or for sb.; bring to sb.: 请把这张报～给他。Take this paper to him, please.
【捎带】incidentally; in passing: ～把这点提一下 mention this point in passing
【捎脚】pick up passengers or goods on the way; give sb. a lift

稍 [副] a little; a bit; slightly; a trifle: ～待片刻 stay a minute / ～～一怔 give a slight start / ～胜一筹 just a little better / ～事休息 have a short interval
【稍微】[副] a little; a bit; slightly; a trifle: ～搁点盐 put in a little salt / 觉得～有点冷 feel a bit cold
【稍息】[军] stand at ease: ～! (口令) At ease!
【稍纵即逝】transient; fleeting: ～的机会 a fleeting opportunity

艄 ①(船尾) stern ②(舵) rudder; helm: 掌～ be at the helm
【艄公】①(船尾掌舵人) helmsman ②(撑船人) boatman

sháo

勺 spoon; ladle: 长柄～ ladle; dipper

【勺子】ladle; scoop

芍

【芍药】[植] Chinese herbaceous peony; common peony

杓

【杓鹬】[动] curlew

韶 [书] splendid; beautiful

【韶光】①[书](美丽的春光) beautiful springtime ②(美好的青年时代) glorious youth

【韶光易逝】Time passes quickly.

【韶华不再】Spring time waits for no man. (或: Time flies.)

【韶华虚度】pouring one's life in the sand; wasting one's youth; idling one's youth away

shǎo

少 ①(数量小) few; little; less: ～出差错 made fewer mistakes / ～说空话 renounce empty talk / 以～胜多 defeat the many with the few / 你应该～喝酒。 You should drink less. ②(缺少) be short; lack: 人手～ be short of hands / 投票结果～三票不能成为多数。 The vote lacks three of being a majority. ③(丢失) lose; be missing: 看看～不～人。 See if anyone is missing. ④(暂时) a little while; a moment: 请～候。 Wait a moment, please. ⑤(不要) stop; quit: ～管闲事。 Mind your own business. / ～来这一套。 Cut it out.

【少安毋躁】don't be impatient, wait for a while

【少不得】cannot do without; cannot dispense with: 学英语, 词典是～的。 Dictionaries are indispensable in English study.

【少而精】smaller quantity, better quality; fewer but better: 教学内容要～。 Teaching content should be concise.

【少汗】hypohidrosis; olighidria

【少见多怪】consider sth. remarkable simply because one has not seen it before; comment excitedly on a commonplace thing: 用不着～。 There's nothing to be surprised at.

【少刻】after a little while; a moment later

【少量】a small amount; a little; a few

【少陪】[套] if you'll excuse me; I'm afraid I must be going now

【少顷】[书] after a short while; after a few moments; presently

【少时】after a little while; a moment later

【少食多餐】have more meals a day but less food at each

【少数】small number; few; minority: ～服从多数。 The minority is subordinate to the majority. / 他们是～。 They are in the minority.

【少数民族】minority nationality; national minority ◇ ～地区 areas inhabited by the minority nationalities; minority nationality regions / ～干部 minority nationality cadres / ～区域自治 regional autonomy of minority nationalities

【少数人】a small number of people; a few people; the few; the minority: 这种音乐只有～能欣赏。 The music appeals to the few.

【少许】[书] a little; a few; a modicum

【少有】rare; exceptional; seldom

【少云】[气] partly cloudy

shào

捎 drive (a cart) backwards; back (a cart)

【捎色】fade (in color)

哨 ①(哨所) sentry post; post: 岗～ sentry post / 观察～ observation post ②(鸟叫) warble; chirp ③(哨子) whistle: 吹～ blow a whistle

【哨兵】sentry; guard

【哨风傲月】whistling and swaggering in breeze and moonlight

【哨所】sentry post; post: 前沿～ forward post; outpost

【哨子】whistle

潲 ①(雨斜着落下来) slant in: 西边～雨。 The rain is driving in from the west. ②[方](洒水) sprinkle: 往菜上～水 sprinkle the vegetables with water ③[方](泔水) hogwash; swill: 猪～ hogwash; swill

少 ①(年纪轻) young: 男女老～ men and women, old and young ②(少爷) son of a rich family; young master: 恶～ young ruffian / 阔～ a profligate son of the rich

【少白头】①(头发过早花白) be prematurely grey ②(年纪轻而头发白的人) a young person with greying hair

【少不更事】young and inexperienced; green: ～者 a greenhorn

【少妇】young married woman

【少将】major general (陆军, 美空军); rear admiral (海军); air vice marshal (英空军)

【少奶奶】[旧]①(少妇) young mistress of the house ②(尊称别人的儿媳) your daughter-in-law
【少年】①(人从10岁至16岁的阶段) early youth ②(10岁至16岁的人) juvenile; young person ◇ ～单打 [体] boys' and girls' singles / ～读物 juvenile books; books for young people / ～法庭 children's court; juvenile court / ～犯罪 juvenile delinquency; juvenile crime; crime of children / ～犯罪案件 juvenile case / ～业余体校 children's sparetime sports school / ～运动员 juvenile athlete
【少年犯管教所】juvenile offender correctional institution
【少年宫】Children's Palace
【少年老成】①(年轻老练) an old head on young shoulders; young but prudent and capable ②(年轻而缺乏朝气) a young person lacking in vigor and drive
【少年之家】Children's Center; Children's Club
【少女】young girl
【少尉】second lieutenant (陆军,美空军); ensign (美海军); pilot officer; squadron leader (英空军); acting sublieutenant (英海军)
【少先队】[简] Young Pioneers ◇ ～辅导员 Young Pioneer counsellor / ～员 Young Pioneer
【少校】major (陆军,美空军); lieutenant commander (海军); squadron leader (英空军)
【少爷】[旧]①(主人的儿子) young master of the house: ～脾气 behavior of a spoilt boy ②(别人的儿子) your son
【少壮】young and vigorous: ～派 young guard / ～不努力,老大徒伤悲。If one does not exert oneself in youth, one will regret it in old age.

劭 [书]①(劝勉) encourage; urge; exhort ②(美好) excellent; admirable: 年高德～ of venerable age and eminent virtue; venerable

shē

奢 ①(奢侈) luxurious; extravagant: 穷～极欲 (indulge in) luxury and extravagance ②(过分的) excessive; inordinate; extravagant: ～望 extravagant hopes
【奢侈】luxurious; extravagant; wasteful: ～浪费 luxury and waste / 生活～ live in luxury
【奢侈品】luxury goods; luxuries ◇ ～税 luxury tax
【奢华】luxurious; sumptuous; extravagant: ～的旅馆 de luxe hotel / ～的宴会 sumptuous feast / 陈设～ be luxuriously furnished
【奢靡】extravagant; wasteful
【奢望】extravagant hopes; wild wishes

赊 buy or sell on credit
【赊购】buy on credit
【赊欠】buy or sell on credit; give or get credit
【赊售】credit sale
【赊销】sell on credit; charge sales
【赊帐】on credit; on account: ～交易 transaction on credit

猞
【猞猁】[动] lynx

shé

折 ①(断) break; snap: 腿骨～了 have broken one's leg / 小提琴的弦～了。The violin sting snapped. ②(亏损) lose money in business: 把本钱都～光了 have lost the capital altogether
【折本】lose money in business: ～生意 a losing business; a bad bargain
【折耗】damage (to goods during transit, storage, etc.); loss

蛇 snake; serpent: ～药 antidote for snake-bites / 多～的地方 a snaky place / 有些～是有毒的。Some snakes are poisonous.
【蛇豆】[植] snake gourd
【蛇毒】snake venom; venin
【蛇麻】[植] hop
【蛇莓】[植] mock-strawberry
【蛇皮管】[电] flexible metal conduit
【蛇丘】[地] esker
【蛇蜕】[中药] snake slough
【蛇纹石】[矿] serpentine
【蛇蝎】snakes and scorpions; vicious people: 毒如～ as vicious as a viper
【蛇行】[书] move with the body on the ground; crawl
【蛇形】snakelike; S-shaped ◇ ～管 coiled pipe coiler
【蛇足】feet added to a snake by an ignorant artist; sth. superfluous

舌 ①(舌头) tongue ②(舌形物) sth. shaped like a tongue: 火～ tongues of flame / 鞋～ the tongue of a shoe
【舌癌】tongue cancer
【舌敝唇焦】talk till one's tongue and lips are

parched; wear oneself out in pleading, expostulating, etc.
【舌根音】[语] velar
【舌尖】the tip of the tongue ◇ ～后音[语] blade-palatal / ～前音[语] dental / ～音[语] apical / ～中音[语] blade-alveolar
【舌剑唇枪】cross verbal swords with somebody; battle of wits
【舌面后音】[语] velar
【舌面前音】[语] dorsal
【舌神经】lingual nerve
【舌鳎】[动] tonguefish; tongue sole
【舌苔】[中医] coating on the tongue; fur: ～很厚 (tongue) heavily coated / ～厚腻 one's tongue is coated
【舌头】①(舌) tongue ②(活捉来的敌人) an enemy soldier captured for the purpose of extracting information: 侦察兵抓了个～。The scouts took a prisoner to get information.
【舌下神经】[生理] hypoglossal nerve
【舌炎】[医] glossitis
【舌战】have a verbal battle with; argue heatedly: 一场～ a heated dispute; a battle royal
【舌状花】[植] ligulate flower

shě

舍 ①(舍弃) give up; abandon: ～下妻儿 abandon one's wife and children / ～此别无他法。There is no other way than this. ②(施舍) give alms; dispense charity
【舍本逐末】attend to trifles to the neglect of essentials
【舍不得】hate to part with or use; grudge: ～花钱 hate to part with one's money
【舍得】be willing to part with; not grudge: ～下功夫 not grudge time spent on practice
【舍己为人】sacrifice one's own interests for the sake of others
【舍近求远】seek far and wide for what lies close at hand
【舍车保帅】give up a rook to save the king (in chess); make minor sacrifices to safeguard major interests
【舍命】risk one's life; sacrifice oneself
【舍弃】give up; abandon: ～个人利益 give up one's own interests
【舍去】[数] rounding; truncation
【舍入】[数] rounding off ◇ ～常数 round-off constant / ～误差 rounding error; round-off error / ～指令 round-off order / ～字符 separating character
【舍身】give one's life; sacrifice oneself: ～报国 sacrifice oneself for the country / ～成仁 die for the sake of the cause / ～救人 give one's life to rescue sb.; sacrifice oneself to save others
【舍生取义】lay down one's life for a just cause
【舍死忘生】disregard one's own safety; risk one's life
【舍我其谁】If I can't do it, who can? (或: Who but myself can do it?)
【舍项法】method of truncation
【舍项误差】[数] truncation error

shè

涉 ①(徒步过水) wake; ford: ～水过河 wade across a river; ford a stream / 远～重洋 travel all the way from across the oceans ②(经历) go through; experience: ～世不深 have scanty experience of life; have seen little of the world / ～险 go through dangers ③(涉及) involve
【涉笔成趣】write freely and well
【涉及】involve; relate to; touch upon: ～其他问题 touch upon other problems / ～重大原则性问题 involve major matters of principle
【涉猎】do desultory reading; read cursorily: ～书籍 dip into books for casual reading; wide and superficial reading
【涉禽】[动] wading bird; wader
【涉讼】be involved in a lawsuit: ～人 person involved in lawsuit
【涉外】concerning foreign affairs or foreign nationals: ～案件 foreign case
【涉嫌】be suspected of being involved; be a suspect; inviting suspicion
【涉足】[书] set foot in: ～其间 set foot there / ～仕途 start an official career; join the civil service

社 ①(集体组织) organized body; agency; society: 报～ newspaper office / 出版～ publishing house; press / 广告～ advertising agency / 通讯～ news agency ②(古代祭神的地方、日子、祭礼) the god of the land, sacrifices to him or altars for such sacrifices
【社会】society: 工业～ industrial society / 农业～ agricultural society / 人类～ human society ◇ ～保险 social insurance / ～保障 social security / ～必要劳动 socially necessary labor / ～不平等 social inequality / ～财富

wealth of society; public wealth / ～地位 social position; social status / ～调查 social investigation; social survey / ～发展 social development / ～发展史 history of social development; history of development of society / ～分工 division of labor in society / ～风气 social conduct; the standards of social conduct / ～福利 social welfare; public welfare / ～福利事业 social and welfare services / ～革命 social revolution / ～工作 work, in addition to one's regular job, done for the collective / ～公德 social morals / ～公证 notarization / ～环境 social environment / ～活动 social activities; public activities / ～基础 social base; social basis / ～交往 social interaction / ～阶级 social class / ～结构 social structure / ～进化 social evolution / ～科学 social sciences / ～力量 societal forces / ～名流 noted public figures / ～群体 social groups / ～认同 social identity / ～实践 social practice / ～体系 social system / ～问题 social problem / ～贤达 community leaders / ～效益 social benefits / ～心理学 social psychology / ～形态学 social morphology / ～一体化 social integration / ～舆论 public opinion / ～责任 community responsibility / ～震动 social repercussions / ～制度 social system / ～秩序 social order; the public order / ～组织 social organization

【社会党】Socialist Party

【社会关系】①(相互的) human relations in society; social relations ②(个人的) one's social connections

【社会化】socialization

【社会学】sociology: 比较～ comparative sociology / 家庭～ sociology of family / 政治～ political sociology◇ ～家 sociologist

【社会治安】security of society: ～条例 regulation on maintenance of social order

【社会主义】socialism ◇ ～道路 socialist road / ～国家 socialist state / ～觉悟 socialist consciousness / ～现代化建设 socialist modernization / ～制度 socialist system

【社稷】the god of the land and the god of grain; the state; the country

【社交】social intercourse; social contact ◇ ～秘书 social secretary

【社论】editorial; leading article; leader

【社团】mass organizations; body of persons; corporation ◇ ～法人 commonalty; juridical association

【社戏】village theatrical performance given on religious festivals in old times

设 ①(设立; 布置) set up; establish; found: ～圈套 set a trap / ～下天罗地网 spread a dragnet / ～一所学校 establish a school / 总公司～在上海 set up the head office in Shanghai ②(筹划) work out: ～计陷害 plot a frame-up; frame ③[数](假设) given; suppose; if: ～长方形的宽是七米。Suppose the width of a rectangle is seven meters. ④[书](倘若; 假如) if; in case: ～有困难,当助一臂之力。You can count on me to help in case of difficulty.

【设备】equipment; installation;l facilities: 电气～ electrical installations / 交通运输～ facilities for transport and communication / 冶金～ metallurgical equipment / 饭店～齐全。The hotel is well appointed. ◇ ～更新 updating equipment / ～类别 device class / ～利用率 utilization rate of equipment and installations / ～性能 equipment characteristic

【设法】think of a way; try; do what one can: ～帮助 find a way to help

【设防】set up defenses; fortify; garrison: 层层～ set up defenses in depth ◇ ～地带 fortified zone

【设计】design; plan: 建筑～ architectural design / 舞台～ stage design ◇ ～功率 design power / ～洪水 [水] design flood / ～能力 designed capacity / ～师 designer / ～图 design drawing / ～性能 design performance / ～院 designing institute

【设立】establish; set up; found: ～新的机构 set up a new organization

【设色】fill in colors on a sketch; lay paint on (canvas); color: ～柔和 painted in quiet colors

【设身处地】put oneself in sb. else's position; be considerate

【设施】installation; facilities: 防洪～ flood control installations / 集体福利～ collective welfare institutions / 军事～ military installations / 文化教育～ cultural and educational institutions / 医疗～ medical facilities

【设想】①(想象) imagine; envisage; conceive; assume: 不堪～ too ghastly to contemplate ②(想法) tentative plan; tentative idea ③(着想) have consideration for: 多为青少年～ give much thought to the needs of the younger generation

【设宴】give a banquet; fête: ～欢送 give sb. a send-off dinner / ～欢迎 give a welcome din-

ner to sb./ ～饯行 give a farewell party in honor of sb./ ～洗尘 give a dinner to someone on his return from travel/ ～招待贵宾 give a banquet in honor of the distinguished guests
【设营】[军] quartering; encampment ◇ ～地 camp site/ ～队 quartering party
【设置】set up; put up; install: ～专门机构 set up a special organization/ 课程～ courses offered in a college or school

赦 remit (a punishment); pardon: 大～ general pardon; amnesty/ 特～ special pardon
【赦令】decree for pardon
【赦免】remit (a punishment); pardon; absolve; excuse ◇ ～权 power of absolution
【赦罪】absolve sb. from guilt; pardon sb.

慑 [书] fear; be awed: ～于淫威 be awed by authority
【慑服】①(因恐惧而顺从) submit because of fear; succumb ②(使恐惧而服从) cow sb. into submission

摄 ①(吸收) absorb; assimilate ②(摄影) take a photograph of; shoot: ～下几个珍贵的镜头 take some superb shots ③(保养) conserve (one's health) ④(代理) act for: ～理 hold (an office) in an acting capacity
【摄动】[天] perturbation
【摄谱仪】[物] spectrograph
【摄取】①(吸收) absorb; assimilate; take in: ～营养 absorb nourishment ②(拍摄) take a photograph of; shoot: ～镜头 shoot a scene
【摄魂夺魄】hold spellbound; be under the spell
【摄生】[书] conserve one's health; keep fit: ～之道 path of sanitation
【摄氏度】Celsius; centigrade: 水在零～结冰。Water freezes at zero degrees centigrade.
【摄氏温标】Celsius temperature scale; centigrade temperature scale
【摄氏温度计】centigrade thermometer; Celsius thermometer
【摄像管】camera tube; pickup tube; television camera tube
【摄像机】pickup camera: 电视～ television camera
【摄影】①(照相) take a photograph: ～留念 take a photo as a moment; have a souvenir photograph taken/ 航空～ aerial photography/ 红外～ infrared photography ②(拍电影) shoot a film; film: 内景～ interior shooting/ 全景～ panoramic shooting/ 外景～ exterior shooting ◇ ～场 studio/ ～记者 press photographer; cameraman/ ～棚 film studio/ ～师 photographer; cameraman/ ～室 photographic studio; photo studio/ ～术 photography/ ～展览 photographic exhibition; photo exhibition
【摄影机】camera: 电影～ cinecamera; cinematograph/ 立体～ stereoscopic camera
【摄政】act as regent ◇ ～王 prince regent
【摄制】[电影] produce: ～组 production unit/ …电影制片厂～ produced by the ... Film Studio

舍 ①(房屋) house; shed; hut: 茅～ thatched hut/ 牛～ cowshed/ 校～ school buildings ②[谦] ～弟 my younger brother/ ～侄 my nephew
【舍间】[谦] my humble abode; my house
【舍利】[佛教] Buddhist relics ◇ ～塔 stupa; pagoda for Buddhist relics; Buddhist shrine
【舍亲】[谦] my relative
【舍下】[谦] my humble abode; my house

射 ①(用推力或弹送出) shoot; fire: ～靶 shoot at the target/ ～箭 shoot an arrow/ 慢～ slow fire/ 慢加速～ center fire/ 速～ rapid fire/ 扫～ strafe ②(喷射液体) discharge in a jet: 喷～ spout; spurt; jet/ 注～ inject ③(放出) send out (light, heat, etc.): 反～ reflect/ 光芒四～ radiate brilliant light ④(有所指) allude to sth. or sb.; insinuate: 影～ insinuate
【射程】range (of fire): 有效～ effective range
【射电天文学】radio astronomy
【射电望远镜】radio telescope
【射电星】radio star
【射电星体】radio star
【射电源】radio source
【射干】[植] blackberry lily
【射击】①(开枪、开炮) shoot; fire: 向敌人～ fire at the enemy/ 向目标～ shoot at a target ②[体] shooting 飞碟～ clay-pigeon shooting/ 多向飞碟～ transhooting/ 双向飞碟～ skeet shooting ◇ ～场 shooting range/ ～地境 sector of fire/ ～技术 marksmanship/ ～孔 embrasure
【射箭】①(用弓把箭射出) shoot an arrow ②[体] archery ◇ ～场 archery range/ ～手 archer; bow-man
【射角】angle of fire; elevation
【射界】area of fire; firing area

【射精】[生理] ejaculation ◇ ~管 ejaculatory ducts
【射孔】[石油] perforation
【射猎】hunting with bow and arrow or firearms
【射流】[物] jet; filament band; shooting flow; efflux ◇ ~技术 fluidics / ~喷口 efflux nozzle
【射门】[体] shoot (at the goal): 反弹~ bounce shot / 跳起~ jump shot / 鱼跃~ dive shot / 转身~ pivot shot; turn-around shot ◇ ~手 goal getter
【射频】[电] radio frequency ◇ ~放大器 radio frequency amplifier
【射手】shooter; marksman: 机枪~ machine gunner / 优秀~ sharpshooter
【射速】firing rate
【射线】[物] ray: α~ alpha ray / β~ beta ray / γ~ gamma ray / X~ X-ray / 阴极~ cathode-ray ◇ ~病 radiation sickness / ~疗法 radiotherapy
【射幸契约】aleatory contract
【射影几何】projective geometry

麝 ①(动物,似鹿而小) musk deer ②(麝香) musk
【麝馥兰香】breath of musk and the scent of orchids
【麝兰散馥】Sweet perfumes are constant diffused around.
【麝牛】musk-ox
【麝鼠】muskrat
【麝香】musk ◇ ~草 thyme / ~酮 musk ketone

shēn

深 ①(上下或里外距离大) deep: 雪~过膝 knee-deep snow / 进~十五米的大厅 a hall fifteen meters deep / 使运河加~ deepen a canal / 林~苔滑。The forest is thick and the mossy path is slippery. / 这个池塘~约五米。The pond is about five meters deep. ②(深奥) difficult; profound: 由浅入~ from the easy to the difficult ③(深刻; 深入) thoroughgoing; penetrating; profound: 功夫~ have put in a great deal of effort / 印象很~ have a deep impression of / 影响很~ of profound influence ④(深厚) close; intimate: 交情~ be on intimate terms ⑤(浓) dark; deep: ~红 deep red; crimson / ~蓝 dark blue / ~色 deep color ⑥(离开始的时间很久) late: ~秋 late autumn ⑦(很; 十分) very; greatly; deeply: ~表同情 show deep sympathy / ~得人心 enjoy immense popular support / ~恐 be very much afraid / ~受感动 be deeply moved; be greatly touched / ~知 know very well
【深奥】abstruse; profound; recondite: ~的哲理 abstruse philosophy; a profound truth / 这是一本~的书。This is a deep book.
【深不可测】have no bottom; too deep to be fathomable
【深藏若虚】be modest about one's talent or learning; not be given to boasting or showing off
【深长】profound; 意味~ pregnant with meaning; significant
【深沉】①(程度深) dark; deep: 在~的暮色中 in the deepening dusk / ②(低沉) deep; heavy; dull: ~的声音 dull sound ③(思想感情不外露) concealing one's real feelings: ~的人 man of great depth / 这人很~。He's a deep one.
【深成岩】[地] plutonic rock; plutonite
【深仇大恨】bitter and deep-seated hatred; profound hatred
【深处】depths; recesses: 在密林~ in the depths of the forest / 在内心~ in the depth of one's heart
【深度】①(深浅程度) degree of depth; depth: 河水的~ depth of the river ②(触及事物本质的程度) profundity; depth: 一篇表现出思想~的文章 an article that shows depth of thought ◇ ~计 depth gauge
【深更半夜】at dead of night; in the depth of night; in the middle of the night
【深耕】[农] deep ploughing ◇ ~细作 deep ploughing and intensive cultivation / ~易耨 ploughing deeply and weeding thoroughly
【深沟高垒】deep trenches and high ramparts; strong defense
【深闺】boudoir: ~幽阁 deep, hidden boudoir
【深海】deep sea ◇ ~测量 bathymetry / ~平原 abyssal plain / ~区 abyssal region / ~鱼 deep-sea fish / ~资源 deep-sea resources
【深厚】①(浓厚) deep; profound: ~的感情 deep feelings / 结成~的友谊 establish a profound friendship ②(坚实) solid; deep seated: ~的基础 a solid foundation
【深呼吸】deep breathing
【深化】deepen: 认识的~ deepening of cognition / ~改革 deepen reform
【深究】go into (a matter) seriously; get to the bottom of (a matter): ~某事 go deep into a problem / ~原因何在 get to the bottom of it

【深居简出】live in the seclusion of one's own home; live a secluded life
【深刻】deep; profound; deepgoing: ～的批评 incisive criticism / ～地阐明 expound profoundly / 进行～的分析 make penetrating analysis
【深厉浅揭】act according to circumstances; do the right thing in the right place
【深明大义】understand the important principle thoroughly; know clearly the right thing to do and the principles to follow
【深谋远虑】think deeply and plan carefully; be circumspect and farsighted
【深浅】①(深浅的程度) depth: 你知道这河水的～吗? Do you know how deep the river is? ②(分寸) proper limits (for speech or action); sense of propriety: 说话没～ speak without thought and often inappropriately ③(颜色的浓淡) shade: 颜色～不同 of different shades
【深切】heartfelt; deep; profound: ～关怀 bedeeply concerned about; show profound concern for / ～怀念 dearly cherish the memory of / ～同情 deep sympathy
【深情】deep feeling; deep love: ～厚谊 profound sentiments of friendship
【深入】①(达到事物的内部) go deep into; penetrate into: ～敌后 penetrate far behind enemy lines / ～基层 go down to the grass-roots units / ～群众 go into the midst of the common people / ～人心 strike root in the hearts of the people / ～生活 plunge into the thick of life / ～实际 go deep into the realities of life ②(深刻) thorough; deepgoing: ～进行调查研究 make a thorough investigation and study
【深入浅出】explain profound theories in simple language
【深山】remote mountains: ～老林 remote, thickly forested mountains / ～峡谷 deep mountain valleys
【深深】profoundly; deeply; keenly: ～地感谢 be deeply grateful (to sb for sth.) / ～感到 feel keenly
【深水】deepwater ◇ ～泊位 deep-water berths / ～港 deepwater port / ～码头 deepwater wharf / ～炸弹 depth charge; depth bomb
【深水流静】Deep waters run smooth.(或: Smooth waters run deep.)
【深思】think deeply about; ponder deeply over: ～练达 thoughtful and clearheaded / 值得～ be worth thinking deeply about; deserve careful pondering
【深思熟虑】careful consideration
【深邃】①(上下或里外距离大) deep: ～的山谷 a deep valley ②(深奥) profound; abstruse; recondite: 寓意～ have a profound message
【深文周纳】use every means to have an innocent person pronounced guilty; convict sb. by deliberately misinterpreting the law; frame up and punish severely
【深恶痛绝】have a deep-seated hatred; hate bitterly; abhor; detest
【深夜】late at night; in the small hours of the morning: 工作到～ work late into the night / 他一直学习到～。He went on studing deep into the night.
【深渊】abyss: 苦难的～ the abyss of suffering / 无底～ bottomless abyss
【深远】profound and lasting; far-reaching: ～的影响 far-rcaching influence / 具有～的历史意义 have profound historic significance
【深造】take up advanced studies; take a more advanced course of study or training; pursue advanced studies: 送到…～ be sent to... for further training
【深宅大院】a compound of connecting courtyards, each surrounded by dwelling quarters; imposing dwellings and spacious courtyards
【深湛】profound and thorough: 功夫～ consummate skill
【深重】very grave; extremely serious: 危机～ be in the grip of a crisis

莘

【莘莘】[书] numerous: ～学子 a great number of disciples; large numbers of students; numerous students

申

①(说明) state; express; explain: 重～前令 reiterate the previous order ②(地支的第九位) the ninth of the twelve Earthly Branches
【申报】①(向上报告) report to a higher body ②(向海关申报纳税等) declare sth. (to the customs) ◇ ～出生 declaration of birth / ～单 declaration form
【申辩】defend oneself; explain oneself; argue one's case: ～无罪 exculpate / 允许～ allow sb. to argue his case / 被告有权～。The accused has the right to defend himself.
【申斥】rebuke; reprimand
【申明】declare; state; avow: ～立场 declare

one's stand
【申请】apply for: ～补助 apply for an allowance / ～出境签证 apply for an exit visa / ～调动工作 apply for a transfer / ～入境签证 apply for an entry visa ◇ ～国 applicant country / ～回避 withdrawal by petition / ～离婚 petition for a divorce / ～人 applicant / ～书 (written) application; petition / ～外汇 application for exchange / ～中止诉讼者 caveator / ～作出判决 motion for decree
【申述】state; explain in detail: ～来意 explain the purpose of one's visit / ～意见 state 〈express〉 one's opinion
【申诉】appeal: 不服判决，提出～ appeal against a legal decision ◇ ～的权利 right of petition / ～人 declarant
【申讨】openly condemn; denounce
【申谢】acknowledge one's indebtedness; express one's gratitude
【申冤】①(洗雪冤屈) redress an injustice; right a wrong ②(申诉所受的冤屈) appeal for redress of a wrong

砷
[化] arsenic (As)
【砷化物】arsenide
【砷剂】arsenical
【砷疗法】arsenization; arsenotherapy
【砷中毒】arsenic poisoning

呻
【呻吟】groan; moan: 无病～ moan and groan without being ill; make a fuss about an imaginary illness

伸
stretch; extend: ～出手 extend one's hand / ～出友谊之手 stretch out a hand of friendship to sb. / ～大拇指 hold up one's thumb / ～胳臂 stretch one's arms
【伸长】elongation
【伸懒腰】stretch oneself
【伸手】①(伸出手) stretch out one's hand: ～不见五指 so dark that you can't see your hand in front of you; pitch dark / ～拿书 reach out for the book ②(向别人要东西) ask for help, etc.: ～派 a person who is in the habit of asking the higher level for help
【伸缩】①(引长和缩短) stretch out and draw back; expand and contract; lengthen and shorten: ～自如 be capable of expansion and contraction / 因温度关系而～ expand or contract according to the change of temperature ②(规模、数量的变动) flexible; elastic; adjustable: 没有～余地 leave one no latitude ◇ ～缝 [建] expansion joint / ～三角架 extension tripod / ～性 flexibility; elasticity
【伸腿】①(伸出腿) stretch one's legs ②(插足) step in (to gain an advantage) ③ [口](死亡) kick the bucket; turn up one's toes
【伸腰】straighten one's back; straighten oneself up
【伸展】spread; extend; stretch: 花园一直～到山脚下。 The garden extends as far as the foot of the hill.
【伸张】uphold; promote: ～正义 let justice prevail

绅
gentry
【绅士】gentleman; gentry: ～淑女 ladies and gentlemen

身
①(身体) body: ～高一米七五 1.75m. in height / 转过～去 turn round ②(生命) life: 以～殉职 die a martyr at one's post ③(自己;本身) oneself; personally: 以～作则 set a good example with one's own conduct ④(品格和修养) one's moral character and conduct: 立～处事 conduct oneself in society / 修～ cultivate one's mind ⑤(物的主要部分) the main part of a structure; body: 船～ the body of a ship; hull / 机～ fuselage / 汽车车～ the body of a motor car / 树～ trunk ⑥ [量] suit: 一～新衣服 a new suit
【身败名裂】lose all standing and reputation; bring disgrace and ruin upon oneself; be utterly discredited
【身板】[方] body; bodily health: ～儿挺结实 have a strong physique
【身边】①(身体近旁) at one's side ②(随身;身上) (have sth.) on one; with one: ～没带钱 have no money on one
【身不由己】involuntarily; in spite of oneself
【身材】stature; figure: 中等～ of medium stature / ～矮小 short and slight of stature / ～高大的人 a man of great stature / ～苗条 have a slender figure
【身残志坚】broken in body but firm in spirit
【身长】①(人体的高度) height ②(衣服的长度) length
【身段】①(女子身姿) figure: ～优美 have a graceful carriage / ～匀称 well-proportioned figure ②(演员的舞蹈化动作) posture
【身份】【身分】①(人的社会、法律地位) status; capacity; identity: 不合～ incompatible with

one's status / 以个人～发言 speak in a personal capacity / 以官方～发言 speak in an official capacity / 以私人～发言 speak in a private capacity ②(受人尊敬的地位) dignity: 有失～ be beneath one's dignity ◇ ～证 identity card; identification card; certificate of identification / ～证件 identity document / ～证据 evidence of identity / ～证明 identification paper

【身高】height (of a person): ～两米 two meters in height

【身故】die: 因病～ die of an illness

【身后】after one's death: ～无出 die without issue; without progeny after one's death

【身价】①(社会地位) social status: 突然～百倍 have a sudden rise in social statue ②(人身卖价) the selling price of a slave

【身教】teach others by one's example: ～胜于言教。Example is better than precept.

【身经百战】have fought a hundred battles: ～的老战士 a veteran who has fought countless battles; a battle-tested veteran; a seasoned fighter

【身量】[口] height (of a person); stature

【身临其境】be personally on the scene

【身强力壮】strong; tough sturdy

【身躯】body; stature: ～高大 tall of stature / 健壮的～ a sound body

【身上】①(身体上) on one's body: ～不舒服 do not feel with / ～穿一件白衬衫 wear a white shirt ②(随身) (have sth.) on one; with one: 我～没带名片。I haven't got calling cards with me.

【身世】one's life experience; one's lot: ～凄凉 have had a sad life

【身手】skill; talent: 大显～ fully display one's talents; exhibit one's skill

【身受】experience (personally): 感同～。I shall count it as a personal favor.

【身体】①(人体) body: ～上的伤害 infliction of body / 保持～平衡 keep one's balance ②(健康) health: ～好,学习好,工作好 keep fit, study well and work hard / 注意～ look after one's health ◇ ～缺陷 impediment

【身体力行】earnestly practice what one advocates

【身外之物】external things; mere worldly possessions

【身无寸功】have no merit; In my body there is not even an inch of merit.

【身无分文】be stone-broke; without a cash in one's pocket

【身无长物】have nothing

【身无已出】have no children of one's own

【身先士卒】lead one's men in a charge; charge at the head of one's men

【身陷囹圄】be thrown into jail

【身心】body and mind: ～健康 sound in body and mind; physically and mentally healthy / ～受到摧残 be physically injured and mentally affected / ～愉快 be in a good mood

【身心交瘁】be mentally and physically exhausted; in a state of complete bodily and mental prostration

【身心舒泰】body and mind at ease: there was peace in one's body and brain

【身影】a person's silhouette; form; figure

【身孕】pregnancy

【身在福中不知福】growing up in happiness, one often fails to appreciate what happiness really means; not appreciate the happy life one enjoys

【身子】[口] ①(身体) body: 光着～ be naked ②(身孕) pregnancy

参 ginseng: 人～ ginseng; Asiatic ginseng / 西洋～ American ginseng

shén

神 ①(神灵) god; deity; divinity: 尼普顿是海～。Neptune was the god of the sea. ②(高超出奇) supernatural; magical: ～效 magical effect; miraculous effect / 用兵如～ direct military operations with miraculous skill ③(精神) spirit; mind: 闭目养～ close one's eyes and rest one's mind / 耗～ take up one's energy / 走～ be absentminded ④(神气) expression; look: 眼～ expression in the eyes ⑤[方](机灵;聪明) smart; clever: 这家伙～了! This fellow is incredible!

【神不守舍】have ants in one's pants; out of one's mind

【神不知,鬼不觉】unknown to god or ghost; (do sth.) without anybody knowing it; in great secrecy

【神采】expression; look: ～奕奕 with a radiant look; glowing with health and radiating vigor

【神出鬼没】come and go like a ghost; appear and disappear mysteriously

【神父】【神甫】Catholic father; priest

【神怪】gods and spirits

【神乎其神】fantastic; wonderful; miraculous;

吹得～ laud sth. or sb. to the skies
【神化】deify
【神话】mythology; myth; fairy tale: 希腊～ Greek mythology / 这不过是个～。 It's only a myth.
【神魂】state of mind; mind: ～不定 be deeply perturbed / ～颠倒 be infatuated
【神机妙算】wonderful foresight (in military operations, etc.)
【神经】nerve: 感觉～ sensory nerve / 交感～ sympathetic nerve / 迷走～ vagus / 面～ facial nerve / 脑～ cranial nerve / 三叉～ trigeminal nerve / 视～ optic nerve / 坐骨～ sciatic nerve / ～错乱 be mentally deranged / ～紧张 be nervous ◇ ～冲动 nerve impulse / ～毒药 neuropoison / ～官能症 neurosis / ～胶质病 glioma / ～节 ganglion / ～末梢 nerve ending / ～衰弱 neurasthenia / ～损害 nervous lesion / ～痛 neuralgia / ～外科 neurosurgery / ～性皮炎 neurodermatitis / ～炎 neuritis / ～原 neuron / ～战 war of nerves / ～质 nervousness / ～中枢 nerve center
【神经病】①(神经系统中的疾病) neuropathy ②(精神病的俗称) mental disorder: 你简直是～。You are simply not in your right mind.
【神经过敏】① [医] neuroticism ②(多疑; 好大惊小怪) neurotic; oversensitive
【神经系统】nervous system: 中枢～ central nervous system / 周围～ peripheral nervous system
【神龛】a shrine for idols or ancestral tablets
【神来之笔】an inspired passage; a stroke of genius
【神力】superhuman strength; extraordinary power
【神灵】gods; deities; divinities
【神秘】mysterious; mystical: ～人物 a mysterious person; a person shrouded in mystery ◇ ～观念 mystery; solemn wonder / ～化 make a mystery of / ～主义 mysticism
【神妙】wonderful; marvellous; ingenious: ～的笔法 wonderful style of writing; ingenious brushwork
【神明】gods; deities; divinities: 奉若～ worship sb. or sth.; make a fetish of sth.
【神女】①(女神) goddess ② [旧](妓女) prostitute
【神炮手】crack gunner
【神奇】magical; mystical; miraculous: ～的效果 miraculous effect; magical effect
【神气】①(神情) expression; air; manner: 他脸上显出得意的～。 He had an air of complacency. ②(精神饱满) spirited; vigorous ③(骄傲得意) putting on airs; cocky; overweening: ～活现 very cocky / ～十足 putting on grand airs; very arrogant / 你别～! Don't give yourself airs.
【神气活现】very cocky; as proud as a peacock
【神枪手】sharpshooter; crack shot; expert marksman
【神情】expression; look: ～自若 with an easy grace / 露出愉快的～ look happy; wear a happy expression
【神权】①(神灵的权力) religious authority; theocracy ②(神赋予的权力) rule by divine right ◇ ～统治 thearchy / ～政治 theocracy
【神人共鉴】May it be taken as evidence both by gods and men.
【神色】expression; look: ～不对 look queer / ～慌张 look flustered / ～紧张 look hurried / ～自若 be unperturbed; show composure and presence of mind
【神圣】sacred; holy: ～权利 sacred right / ～职责 sacred duty
【神圣不可侵犯】sacred and inviolable; holy and inviolable
【神思】state of mind; mental state: ～不安 one's heart and mind were disturbed / ～不定 be in a fidget; be distracted / ～恍惚 be lost in a reverie / ～紊乱 one's mind was in a turmoil
【神似】be alike in spirit; be an excellent likeness: 不仅形似,而且～ be alike not only in appearance but also in spirit
【神速】marvellously quick; with amazing speed: 收效～ obtain instant results / 兵贵～。Speed is precious in war.
【神态】expression; manner; bearing; mien: ～悠闲 look perfectly relaxed / ～自若 look perfectly calm
【神通】remarkable ability; magical power: ～广大 be infinitely resourceful
【神童】child prodigy
【神往】be carried away; be rapt; be charmed: 为之～ be enchanted with
【神物】[书] ①(神奇的东西) wonder; prodigy; phenomenon ②(神仙) supernatural being; deity
【神仙】supernatural being; celestial being; immortal: ～下凡 a fairy becomes incarnate
【神仙葫芦】[机] chain block
【神像】the picture or statue of a god or

Buddha

【神学】［宗］theology ◇ ～生 theolog; theologue／～院 seminary; divinity

【神医】highly skilled doctor; miracle-working doctor

【神异】①(神怪) gods and spirits ②(神奇) magical; mystical; miraculous

【神游症】fugue

【神职人员】clergy

【神志】consciousness; senses; mind: ～不清 obnubilation／～昏迷 lose consciousness; be in a state of delirium／～清醒 be in one's right mind; remain fully conscious／～正常 imputability

【神州】the Divine Land; Cathay

什

【什么】①(表示疑问: 单用, 问事物)你找～? What are you looking for?／他说～? What did he say? ②(用在名词前面, 问人或事物)那是～颜色? What color is it?／你～时候见到她的? When did you see her?／他是～人? Who is he? ③(表示不肯定的事物)好象出了～事儿。 It seems something is amiss.／她～也没说。 She didn't say anything. ④(用在"也"或"都"前面, 表示所说的范围之内没有例外)～用处也没有 be quite worthless ⑤(两个"什么"前后照应, 表示由前者决定后者)有～就说～。 Just say what's on your mind. ⑥(表示惊讶或不满)～! 九点了, 车还没来! What's happened! 9 o'clock and the bus hasn't come yet! ⑦(表示责难)你说呀! 装～哑巴? Speak! Stop playing dumb. ⑧(表示不同意对方刚说的话)～不懂! 装糊涂就是了。 What do you mean — not understand? You're just pretending. ⑨(表示列举不尽)～网球啊, 羽毛球啊, 高尔夫球啊, 足球啊, 他都会。 He can play tennis, badminton, golf, soccer, anything.

【什么的】and so on; and so forth; and the like; and others; and what not: 她在菜市场买了些猪肉、土豆、青菜～。 She bought at the market pork, potatoes, green vegetables and the like

shěn

审

①(详细; 周密) careful: ～察 careful observation／～视 look closely at; gaze at; examine ②(审查) examine; go over: ～稿 go over a manuscript or draft ③(审讯) interrogate; try: ～案 try a case／公～ put sb. on public trial／受～ be tried ④[书](知道) know: 未～其详 not know the details ⑤[书](果然; 的确) indeed; really: ～如其言。 What he says is indeed true.

【审察】(with) careful observation

【审查】examine; investigate: ～经费 examine funds／提交…～ submit to ... for examination／～属实。 The fact was established after investigation. ◇ ～证书[外] examination of credentials

【审处】①(审判处理) try and punish: 交由法院～ hand over to the court for trial ②(审查处理) deliberate and decide

【审订】examine and revise: ～教材 revise teaching materials

【审定】examine and approve: ～文件 approve a document／由委员会～ be examined and approved by the committee

【审核】examine and verify: ～预算 examine and approve a budge

【审计】audit ◇ ～报告 audit report／～员 auditor／～制度 auditing system

【审理】[法] try; hear: ～案件 try a case; hear a case／～诉案 hold pleas

【审美】appreciation of the beauty ◇ ～观念 aesthetic standards／～能力 aesthetic judgment

【审判】bring to trial; try: 他因盗窃罪受到～。 He was on trial for theft. ◇ ～程序 judicial procedure／～独立 independence of trial and decision／～对象 object of adjudication／～工作 administration of justice／～规则 adjudication rule／～机关 judicial organ／～监督 adjudication supervision／～权 judicial authority; jurisdiction／～员 judge; judicial officer／～长 presiding judge; chief judge

【审批】examine and approve: 请你～。 This is subject to your approval.／这份报告已经领导～。 This report has been examined and approved by the leadership.

【审慎】cautious; careful; circumspect: ～从事 steer a cautious course／～考虑 consider cautiously; think over carefully.

【审时度势】judge the hour and size up the situation

【审问】interrogate; question; examine; hear; try: ～案件 hear a case／～被告 interrogate the accused ◇ ～者 examiner／～罪犯 pump a prisoner

【审讯】[法] interrogate; try; inquest: ～俘虏 interrogate prisoners of war ◇ ～笔录 hearing record / ～方法 method of interrogation / ～记录 record of trial / ～阶段 stage of trial / ～日期 hearing time / ～形式 mode of trial
【审议】consideration; deliberation; discussion; consider: 委员会～了这个问题。The committee has considered this matter.
【审阅】check and approve: ～稿件 go over a manuscript

婶 ①(婶母) aunt; wife of father's younger brother ②(对母辈妇女的亲热称呼) aunt; auntie
【婶母】wife of father's younger brother; aunt

shèn

慎 careful; cautious: 谨小～微 overcautious
【慎重】cautious; careful; prudent; discreet: ～处理 be prudent in dealing with sth. / 采取～的态度 adopt a prudent policy / 经过～考虑 after careful consideration

甚 ①(很;极) very; extremely: ～为痛快 find it most satisfying ②(超过;胜过) more than: 日～一日 get intensified with each passing day
【甚而】[连] even; (go) so far as to
【甚高频】very-high-frequency (VHF)
【甚或】[书][连] even; (go) so far as to; so much so that
【甚嚣尘上】cause a temporary clamour; make a great noise
【甚至】[连] even; (go) so far as to; so much so that: 她～连"再见"都没说就走了。She left without so much as saying "good-bye"

肾 [生理] kidney
【肾功能试验】[医] kidney function tests
【肾结石】[医] kidney stone; renal calculus
【肾结石切除术】[医] pyelolithotomy
【肾衰竭】[医] kidney failure; renal failure
【肾上腺】[生理] adrenal gland; suprarenal gland; adrenal
【肾上腺功能亢进】[医] hyperadrenalism; hypercorticism
【肾上腺素】[生化] adrenaline; epinephrine
【肾下垂】[医] nephroptosis
【肾炎】nephritis
【肾移植】[医] kidney transplant; renal transplant
【肾盂】[生理] renal pelvis ◇ ～炎 pyelitis
【肾盂肾炎】[医] pyelonephritis
【肾脏】kidney

胂 [化] arsine: ～酸 arsonic acid

蜃 [动] clam: 海市～楼 mirage
【蜃景】[气] mirage

渗 ooze; seep: ～水了。The water has oozed. / 她的伤口还在～血。Blood still oozed from his cut.
【渗沟】sewer
【渗坑】seepage pit
【渗漏】seepage; leakage ◇ ～损失 [水] seepage loss
【渗滤】[化] percolation filtration; percolation ◇ ～器 percolator
【渗入】①(慢慢地流入) permeate; seep into: ～地下 permeate the ground; seep into the ground ②(侵入) (of influence, etc.) penetrate; infiltrate
【渗色】[纺] bleeding
【渗碳】[冶] carburization; cementation ◇ ～钢 carburizing steel / ～体 cementite
【渗透】① [物] osmosis ②(透过) permeate; seep: 鲜血～了绷带。Blood soaked through the bandage. / 雨水～了泥土。The rain permeated the soil. ③(逐渐侵入) penetrate; infiltrate: 经济～ economic infiltration ◇ ～性 permeability / ～压力 osmotic pressure / ～战术 infiltration tactics

shēng

声 ①(声音) sound; voice: 传～ propagation of sound / 发～ production of sound / 脚步～ the sound of footsteps / 雨～ the sound of raindrops / 原～ primary sound / 低～地 in a hushed voice / 异口同～ with one voice / 这座古堡是黑暗和寂静无～的。The old castle was dark and voiceless. ②(发出声音) make a sound: 不～不响 not utter a word; keep quiet; making no sound ③(声母) initial consonant (of a Chinese syllable): 双～ alliteration ④(声调) tone: 四～ the four tones in classical and modern Chinese ⑤(声名) reputation: ～誉 reputation; fame; prestige ⑥ [量] 我喊了他两～。I called him twice.
【声辩】argue; justify; explain away
【声波】[物] sound wave; acoustic wave
【声部】[乐] part (in concerted music)
【声称】profess; claim; assert: ～实有其事 assert 〈declare〉 that it is true / ～已打破僵局 claim

to have broken the deadlock / 有人～… It is claimed that... ◇ ～被恶意中伤 allegation of malice / ～被欺诈 allegation of fraud / ～被强奸 allegation of rape / ～无效 allegation of invalidity
【声带】[生理] vocal cords ② [电影] sound track
【声调】①(音调) tone; note: ～低沉 in a low, sad voice / ～激昂 in an impassioned tone ② [语] the tone of a Chinese character
【声东击西】make a feint to the east and attack in the west; make a noise in the east while attacking in the west
【声光表演】son et lumiere
【声价】reputation: ～甚高 (of a person) be held in high repute; be held in high esteem
【声浪】voice; clamor: 抗议的～ a wave of protest
【声泪俱下】shedding tears while speaking; in a tearful voice
【声门】[生理] glottis
【声名】reputation: ～远扬 one's fame spread far and wide
【声名狼籍】disrepute; have a bad name; be notorious ◇ ～的人 gutter-bird
【声明】①(公开表明态度,说明真相) state; declare; announce: 发表～ make a statement / 口头～ a verbal statement / 庄严～ solemnly state ②(声明的文告) statement; declaration: 联合～ joint statement ◇ ～书 declaration; statement
【声母】[语] initial consonant (of a Chinese syllable)
【声纳】[物] sonar (sound navigation and ranging)
【声囊】[动] vocal sac
【声能学】[物] sonics
【声频】[物] acoustic frequency
【声谱】[物] sound spectrum ◇ ～仪 sound spectrograph
【声气】①(消息) information: 互通～ exchange information; keep in contact with each other ② [方](语气、声音) voice; tone: 小声小气地 in a low voice; in undertones
【声强】[物] sound intensity
【声容笑貌】a person's voice and expression
【声色】①(声音和脸色) voice and countenance: 不动～ maintain one's composure; stay calm and collected ② [书](歌舞和女色) woman and song
【声色俱厉】stern in voice and countenance
【声色犬马】drown oneself in sex and pleasure
【声色自娱】have one's fling; indulge oneself in song and with women
【声势】impetus; momentum: ～浩大 great in strength and impetus; powerful and dynamic / 虚张～ make a show of strength; bluff and bluster
【声势凌人】overwhelm the weaker with excessively strong language
【声势煊赫】one's influence is majestic and powerful.
【声嘶力竭】shout oneself hoarse; shout oneself blue in the face
【声速】[物] velocity of sound
【声讨】denounce; condemn: 愤怒～…的罪行 indignantly denounce the crimes of...
【声望】popularity; prestige: 有很高的～ enjoy great prestige
【声望日隆】one's reputation becomes more and more impressive.
【声威】renown; prestige: ～大震 gain great fame and high prestige; gain resounding fame
【声吸收】sound absorption
【声息】①(多用于否定) sound; noise: 没有一点～。Not a sound is heard. ②(消息) information: ～相闻 keep in touch with each other
【声响】sound; noise: 发出巨大～ make too much noise
【声学】acoustics: 超～ ultrasonics / 几何～ ray acoustics; geometrical acoustics / 建筑～ architectural acoustics
【声音】sound; voice: 沙哑的～ hoarse voice; husky voice / 没有一点～。Not a sound is heard. / 那是什么～? What was that sound? / 我们听到远处有各种奇怪的～。We heard strange sounds in the distance. ◇ ～干涉 interference of sound
【声誉】reputation; fame; prestige: ～扫地 fall into discredit / 维护国家的～ defend the honor of one's country
【声援】express support for; support: ～…的正义斗争 support sb. in their just struggles
【声乐】[乐] vocal music
【声韵学】[语] phonology
【声张】make public; disclose: 不要～。Hush up. (或: Don't breathe a word about it.)
【声障】[航空] sonic barrier; sound barrier
【声阻】acoustic resistance

生 ①(生育) give birth to; bear: ～孩子 give birth to a child / 新～儿 newborn baby / 她～了一个女孩。She has borne a girl. ②(出生)

be born: 他～于 1970 年。He was born in 1970. ③(生长) grow: ～根 take root / ～芽 sprout; put out buds ④(生存;活) existence; life: 一～ all one's life; one's lifetime / 栩栩如～ to the life / 有～之年 the remaining years of one's life ⑤(生计) livelihood: 谋～ earn one's livelihood; make a living ⑥(活的) living: ～物 living things ⑦(产生;发生) get; have: ～冻疮 get chilblains ⑧(使燃烧) light (a fire): ～炉子 light a stove ⑨ (未熟) unripe; green: ～瓜 unripe watermelon / ～苹果 a green apple ⑩(未煮熟的) raw; uncooked: ～肉 raw meat / 吃～虾 eat shrimps raw ⑪(没有加工或锻炼过的) unprocessed; unrefined; crude: ～皮 rawhide; (untanned) hide / ～石灰 quicklime ⑫(生疏) unfamiliar; unacquainted; strange: ～词 new word / 人～地不熟 a stranger in a strange place ⑬(生硬) stiff; mechanical: ～凑 mechanically put together (disconnected words and phrases); arbitrarily dish up (unrelated facts) ⑭(很) very: ～恐 for fear that / ～疼 very painful ⑮(学生) pupil; student: 师～关系 teacher-student relations ⑯(戏剧中的角色) the male character type in Beijing opera, etc.

【生搬硬套】copy mechanically in disregard of specific conditions; apply of copy mechanically

【生病】fall ill; get ill

【生不逢辰】it is unlucky to be born at such a time; be born under an unlucky star

【生财有道】expertly in making money

【生菜】[植] romaine lettuce; cos lettuce

【生产】①(创造产品) produce; manufacture: ～拖拉机 produce tractors ②(生孩子) give birth to a child: 他妻子快～了。His wife will be having her baby soon. ◇ ～成本 cost of production / ～定额 production quota / ～方式 mode of production / ～费用 expense of production / ～工具 tool of production / ～关系 relations of production; production relations / ～管理 production management / ～过程 process of production / ～过剩 overproduction / ～合作社 producers' cooperative / ～劳动 productive labor / ～力 productive forces / ～率 productivity / ～能力 production capacity / ～潜力 productive potentialities; latent productive capacity / ～设备 production facilities / ～手段 means of production / ～效率 production efficiency / ～要素 factor of production / ～指标 production quota / ～指数 index of production / ～资料 means of production / ～责任制 system of production responsibility / ～总值 total output value

【生辰】birthday: ～八字 the date of birth and the eight characters of a horoscope / ～忌日 the date of one's birth and the date of one's death

【生成物】[化] product; resultant

【生词】new word

【生存】subsist; exist; live ◇ ～竞争 struggle for existence / ～空间 vivosphere / ～农业 subsistence agriculture / ～曲线 survivorship curve / ～训练 survival training

【生地】①[农](从未耕种过的土地) virgin soil; uncultivated land ② [中药] the dried rhizome of rehmannia

【生动】lively; vivid: ～的描写 lively description

【生动活泼】lively; vivid and vigorous: ～的语言 vivid language

【生儿育女】bear sons and daughters; bear children

【生发油】hair oil

【生根】take root; strike root

【生花妙笔】(straight from) a gifted pen; (written with) a graphic pen

【生荒】[农] virgin soil; uncultivated land

【生活】①(为生存而进行的活动) life: 日常～ daily life ②(生存) live: 靠工资～ live on one's wages / 一个人脱离了社会就不能～下去。One cannot live cut off from society. ③(衣、食、住的情况) livelihood: ～俭朴 lead a thrifty and simple life / ～困难 be badly off ◇ ～必需品 necessaries of life; daily necessities / ～标准 standard of living / ～补助 extra allowance for living expenses / ～方式 way of life; life style / ～费用 living expenses; cost of living / ～福利 welfare; welfare benefits / ～环境 surroundings; environment / ～津贴 living allowance / ～经验 experience of life / ～来源 source of income / ～能力[生] viability / ～水平 living standard / ～条件 living conditions / ～习惯 habits and customs / ～细节 trifling matters of everyday life; domestic trivia / ～用品 articles for daily use / ～指数 index of living / ～周期 [生] life cycle / ～资料 means of subsistence; means of livelihood / ～作风 behavior; conduct

【生活费】①(生活费用) cost of living; living expenses ②[法] alimony ◇ ～指数 cost of living index

【生火】make a fire; light a fire

【生机】①(生存的机会) lease of life: 一线～ a

slim chance of survival; a gleam of hope ②(生命力) life; vitality: 充满～ full of vigor and vitality
【生计】means of livelihood; livelihood: 另谋～ try to find some other means of livelihood
【生姜】[口] ginger
【生境】[生] habitat
【生就】be born with; be gifted with: ～一张利嘴 have the gift of the gab
【生拉硬拽】①(用力拉扯) drag sb. along kicking and screaming ②(牵强附会) stretch the meaning
【生老病死】birth, age, illness and death
【生冷】raw or cold food: 忌食～。Avoid eating anything raw or cold.
【生离死别】part never to meet again; part for ever
【生理】physiology ◇ ～反应 physiological reaction / ～缺陷 physiological defect; physiological deficiency / ～心理学 physiological psychology / ～学 physiology / ～学家 physiologist / ～盐水[药] physiological saline; normal saline / ～作用 physiological action
【生力军】①[军] fresh troops ②(新参加某一工作的人) fresh activists; new force: 文艺战线上的一支～ a vital new force on the art and literary front
【生灵】[书] the people
【生灵涂炭】the people are plunged into an abyss of misery
【生龙活虎】doughty as a dragon and lively as a tiger; full of vim and vigor
【生路】means of livelihood; way out: 另谋～ try to find another job; look for a new means of livelihood
【生米煮成熟饭】the rice is cooked; what's done can't be undone
【生命】life: 冒着～危险 take one's life in one's hands / 丧失～ lose one's life / 为…献出～ lay down one's life for sth. / ～是怎样开始的? How did life begin?
【生命保障系统】life-support systems
【生命力】life-force; vitality: 具有强大的～ have great vitality
【生命线】lifeline; lifeblood: 贸易是大多数现代国家的～。 Trade is the lifeblood of most modern states.
【生命现象】biological phenomena
【生怕】for fear that; so as not to; lest
【生僻】uncommon; rare: ～的字眼 rarely used words
【生平】all one's life: ～事迹 one's life story / 作者～简介 a brief account of the author's life; a biographical note on the author
【生漆】raw lacquer
【生气】①(不愉快) take offense; get angry ②(生命力) life; vitality: ～勃勃 dynamic; vigorous; full of vitality
【生前】before one's death; during one's lifetime: ～友好 friends of the deceased / ～愿望 unrealized wish (of a person who has passed away)
【生前行为】[法] act of life-time; act *inter vives*
【生擒】capture (alive)
【生趣】joy of life: ～盎然 overflowing with joy of life
【生人】stranger
【生日】birthday
【生色】add color to; add luster to; give added significance to
【生涩】jerky; choppy; not smooth: 文字～jerky style of writing
【生杀予夺】hold power over sb.'s life and property; have sb. completely in one's power
【生身父母】one's own parents
【生事】make trouble; create a disturbance: 造谣～ spread rumors and make trouble
【生手】sb. new to a job
【生疏】①(不熟悉) not familiar: 人地～ be unfamiliar with the place and the people / 业务还～ be still not familiar with one's profession ②(不熟练) out of practice; rusty: 他的英语有点～了。 His English is getting rusty. ③(疏远) be not as close as before
【生水】unboiled water
【生丝】raw silk
【生死】life and death: ～搏斗 (be engaged in a) life-and-death struggle / ～不忘 in life and death I shall eternally be your debtor / ～存亡 (grave crisis between) life and death / ～关头 s moment when one's fate hangs in the balance / ～未卜 One's life is uncertain. / ～攸关 be a life-and-death matter / ～由命 every bullet has its billet / ～与共 share a common destiny; go through thick and thin together / ～之交 friends that are ready to die for each other
【生态】organisms' habits, modes of life and relation to their environment; ecology ◇ ～变异 ecocline / ～平衡 ecological balance / ～圈 ecosphere / 失调 ecological disturbance / ～系统 ecological system / ～型 ecotype / ～学

ecology / ～演替 ecological classification
【生铁】pig iron
【生土】[农] immature soil
【生吞活剥】swallow sth. raw and whole; accept sth. uncritically: 决不可～地搬用外国的经验。It's no good taking over the experience of foreign countries uncritically.
【生物】living things; living beings; organisms: 超显微镜～ ultramicroscopic organisms / 单细胞～ unicellular organism / 多细胞～ multicelluiar organism / 浮游～ plankton / 寄生～ parasites ◇ ～电流[生理] bioelectric current / ～发生律 biogenetic law; recapitulation theory / ～防治[农] biological control / ～分布 distribution of organisms / ～工程 genetic engineering / ～固氮 biological nitrogen fixation / ～合成 biosynthesis / ～化学 biochemistry / ～交互作用 biotic interations / ～节律 biological rhythm / ～科学 biological sciences / ～潜能 biotic potential / ～群落 biological community / ～武器[军] biological weapon / ～制品 biological product
【生物分类】biological classification ○界 kingdom / 门 phylum / 纲 class / 目 order / 科 family / 属 genus / 种 species
【生物分类学】taxonomy
【生物地理学】biogeography
【生物碱】[化] alkaloid
【生物膜】biomembrane
【生物气候学】bioclimatology
【生物圈】biosphere
【生物生态学】bioecology
【生物体】organism
【生物学】biology: 海洋～ marine biology / 实验～ experimental biology / 微～ micro- biology ◇ ～家 biologist
【生物岩】biogenic rock; biolith
【生物遥测器】biopack
【生物战】[军] biological warfare
【生物钟】biological clock; biochronometer; living clock
【生息】① (取得利息) bear interest ② [书] (生活; 生存) live; grow; propagate: 休养～ recuperate and multiply; rest and build up one's strength ◇ ～资本 interest-bearing capital
【生橡胶】raw rubber; caoutchouc
【生肖】any of the twelve animals, representing the twelve Earthly Branches, used to symbolize the year in which a person is born
【生效】go into effect; become effective: 签字后立即～ become effective immediately upon signature / 自签字之日起～ go into effect from the date of signature ◇ ～日期 effective date
【生性】natural disposition: ～固执 be stubborn by nature
【生锈】get rusty: ～的针 rusty needles / 经常擦油,以免～。 Oil it regularly to prevent rust.
【生涯】career; profession: 操笔墨～ write for a living / 舞台～ a stage career
【生药】crude drug; dried medicinal herbs
【生药学】pharmacognosy
【生疑】be suspicious
【生意】① (生机) tendency to grow; life and vitality: ～盎然 full of life ② (买卖) business; trade: 做～ do business / ～兴隆。Trade is brisk. ◇ ～经 the knack of doing business; shrewd business sense
【生硬】stiff; rigid; harsh: 态度～ be stiff in manner
【生油】① (没有熬过的油) unboiled oil ② (花生油) peanut oil
【生油层】[石油] source bed
【生于忧患,死于安乐】thrive in calamity and perish in soft living; life springs from sorrow and calamity, death comes from ease and pleasure
【生育】give birth to; bear: ～子女 bear children / 不能～ be unable to have children; be sterile / 过了～年龄 be past one's child-bearing age / 计划～ family planning; planned parenthood / 节制～ birth control ◇ ～酚 tocopherol / ～力 fertility / ～能力 fecundity / ～药 fertility drug
【生育率】fertility-rate; fertility: 净～ net fertility rate / 一般～ general fertility rate
【生源说】[生] biogenesis
【生造】coin (words and expressions) ◇ ～词 coinage
【生长】① (长大) grow: 水稻～良好。 The rice is growing well. ② (出生和成长) grow up; be brought up: 她～在城市。 She was born and brought up in the city. ◇ ～点[植] growing point / ～率 growth rate / ～期 growth period; growing period / ～曲线 growth curve
【生殖】reproduction: 单性～ parthenogenesis / 无性～ asexual reproduction / 有性～ sexual reproduction / 营养体～ vegetative reproduction ◇ ～回游[动] breeding migration / ～孔 gonopore / ～率 reproduction rate / ～系统 reproductive system / ～腺 gonad

【生殖器】reproductive organs; genitals ◇ ～崇拜 phallicism
【生猪】[商] live pig; pig; hog; pork on the hoof
【生字】new word ◇ ～表 (a list of) new words

甥 sister's son; nephew
【甥女】sister's daughter; niece

笙 sheng, a reed pipe wind instrument: ～箫管笛 flutes and pipes
【笙歌】[书] playing and singing ◇ ～不夜。Fluting and singing are heard all night. / ～聒耳。The sound of songs and every sort of music filled the ears without ceasing.

牲 ①(家畜) domestic animal ②(祭神用的牲畜) animal sacrifice
【牲畜】livestock; domestic animals ◇ ～车[铁道] livestock wagon; stock wagon; stock car
【牲粉】[化] animal starch; glycogen
【牲口】draught animals; beasts of burden ◇ ～贩子 cattle dealer / ～棚 stock barn; livestock shed

升 ①(由低往高移动) rise; hoist; go up; ascend: ～帆 hoist a sail / 太阳～起。The sun rises. ②(提高等级) promote: ～为教授 be promoted professor ③(容量单位) liter (l.): 一升啤酒 a liter of beer
【升班】go up (one grade in school)
【升船机】ship lift
【升调】[语] rising tune or tone
【升斗小民】poor people; peck and hamper people
【升高】①(向上升) go up; ascend: 我们看着飞机逐渐～。We watched the airplane ascend higher and higher. ②(等级提高) go up; promote; raise: 他的职位～了。He has been promoted. / 温度～了。The temperature has gone up.
【升格】promote; upgrade: 把外交关系～为大使级 upgrade diplomatic relations to ambassadorial level
【升汞】[化] mercuric chloride
【升轨】rail lift
【升官发财】win promotion and get rich; (be out for) power and money
【升号】[乐] sharp
【升华】①[物] sublimation ②(事物的提高与精炼) raising of things to a higher level; distillation; sublimation: 艺术是现实生活的～。Art is the distillation of life. ◇ ～干燥 lyophilization
【升级】①(晋升) upgrade ②(学生升级) go up (one grade, etc.) ③(战争的规模扩大、事态紧张程度加深) escalate: 使战争～ escalate the war
【升降舵】[航空] elevator
【升降机】elevator; lift
【升空】[航空] lift-off
【升力】[航空] lift ◇ ～特性 lift efficiency
【升幂】[数] ascending power ◇ ～级数 ascending power series
【升平】peace: ～世界 peaceful world; peaceful life
【升旗】hoist a flag; raise a flag ◇ ～典礼 flag-raising ceremony
【升水】[经] premium
【升堂入室】pass through the hall into the inner chamber; have profound scholarship; become highly proficient in one's profession
【升腾】leap up; rise: 火焰～。The flames leapt up.
【升限】[航空] ceiling ◇ ～高度 ceiling height
【升学】go to a school of a higher grade; enter a higher school ◇ ～率 proportion of students entering schools of a higher grade
【升压】[电] step up; boost ◇ ～变压器 step-up transformer / ～器 booster
【升值】①[经] revalue ②(提高价值;抬高价钱) appreciate
【升擢高任】be promoted; be appointed to a higher post

shéng

绳 ①(绳子) rope; cord; string: 粗～ thick rope / 钢丝～ steel cable; wire rope / 麻～ hemp rope / 细～ string ②(约束;制裁) restrict; restrain: ～以纪律 enforce discipline upon sb.
【绳鞭技】[杂技] (doing) tricks with a whip; (performing) feats with a whip
【绳锯木断】Little strokes fell great oaks.
【绳墨】①(木工用具) carpenter's line marker ②[书](规矩;法度) rules and regulations: 拘守～ stick to the rules
【绳索】rope; cord
【绳梯】rope ladder
【绳之以法】restrain by law; punish someone according to law
【绳子】cord; rope; string: 用～捆箱子 cord up a box

shěng

省 ①(节约) economize; save: ～力 save labor / ～钱 save money; economize money / ～时间 save time ②(免掉;减去) omit; leave out: 这个句子可以～去。This sentence can be omitted. ③(省份) province: 河北～ province of Hepai; Hepei Province / ～长 governor of a province

【省城】provincial capital

【省吃俭用】skimp and save; live frugally; save money on food and expenses

【省得】so as to save; so as to avoid: 我再说一遍,～你忘了。Let me remind you once again so that you won't forget.

【省份】province

【省会】provincial capital

【省界】provincial boundaries

【省力】save effort; save labor; economize labor

【省略】①(除去) leave out; omit: 第二段可以～。The second paragraph can be omitted. ②[语] ellipsis ◇～符号 ellipsis / ～规则 default rule / ～号 ellipsis; suspension points; ellipsis dots / ～句[语] elliptical sentence / ～值 default value

【省钱】save money; be economical

【省事】①(减少办事手续) save trouble; simplify matters: 这样可以省很多事。We can make it much simpler this way. ②(方便) it's more convenient to: 在快餐店就餐～。It's more convenient to eat in the fast food restaurant.

【省委】provincial Party committee

【省心】save worry

shèng

盛 ①(繁盛) flourishing; prosperous: 兴～ flourish / 菊花～开。The chrysanthemums bloom luxuriantly. ②(旺盛) vigorous, energetic: 年轻气～ young and aggressive ③(盛大) magnificent; grand: ～举 a grand occasion ④(深厚) abundant; plentiful: ～意 great kindness ⑤(盛行) popular; common; wide spread: ～传 be widely known; be widely rumored ⑥(程度深) greatly; deeply: ～夸 praise highly

【盛产】abound in; teem with: ～煤炭 abound in coal / 这条河～鱼虾。Fish and shrimp teem in this river. (或: This river is teeming with fish and shrimps.)

【盛大】grand; magnificent: ～的场面 a spectacular scenes / ～的欢迎 a rousing welcome / ～的欢迎会 a sgrand welcome party / ～的集会 a solemn assembly / ～的庆祝会 grand celebration rally

【盛典】grand ceremony

【盛会】distinguished gathering; grand meeting: 体育～ a magnificent sports meet

【盛极而衰】fell from the pinnacle of one's power

【盛极一时】be in fashion for a time; be all the rage at the moment

【盛况】grand occasion; spectacular event: ～空前 an exceptionally grand occasion

【盛名】great reputation: 负有～的人 man of great reputation / ～之下,其实难副。It is hard to live up to a great reputation .

【盛怒】rage; fury

【盛气凌人】domineering; arrogant; overbearing: ～的样子 imperious bearing

【盛情】great kindness; boundless hospitality: ～难却。It would be ungracious not to accept your invitation

【盛世】flourishing age; heyday: 太平～ times of peace and prosperity; piping times of peace

【盛暑】sweltering summer heat; very hot weather; the dog days: ～严冬 in sultry summer and in freezing winter

【盛衰】prosperity and decline; rise and fall; ups and downs

【盛衰荣辱】prosperity and decline, glory and humiliation;. rise and fall; ups and downs

【盛夏】the height of summer; midsummer

【盛行】be current; be in vogue: ～一时 be in vogue for a time; prevail for a time

【盛行风向】[气] prevailing wind direction

【盛行能见度】[气] prevailing visibility

【盛宴】grand banquet; sumptuous dinner: ～难再 difficult to have such a grand feast again / 节日～ a great festival feast / ～必散。Even the best party must have an end.

【盛意】great kindness; generosity

【盛誉】great fame; high reputation: 素有～ have long enjoyed high reputation

【盛赞】highly praise; speak of sb. in glowing terms

【盛装】splendid attire; rich dress: 穿着节日的～ be dressed in one's holiday best

乘 [史] a war chariot drawn by four horses: 千～之国 a state with a thousand chariots

剩 surplus; remnant; leave (over): ～货 surplus goods / 所～无几。There is very little

remained. (或: There is not much left.)
【剩磁】[物] residual magnetism; remanence; remanent magnetism ◇ ～测定 residual magnetism measurement / ～法 residual method; residual field method / ～感应 remanence
【剩下】be left (over); remain: ～多少? How much is left (over)? / 只～一个问题。There is only one problem left unsolved.
【剩余】surplus; remainder: 收支相抵,略有～。The reckoning up of revenue and expenditure shows a small surplus. ◇ ～产品 surplus products / ～磁场 remanent field / ～电离 residual ionization / ～电路 residual circuit / ～价值 surplus value / ～劳动 surplus labor / ～农产品 surplus agricultural commodities / ～物资 surplus materials

胜

胜 ①(胜利) victory; success: 获～ win; emerge the victor / 连～三局 win three sets in a row / 三局两～ best of three (games) / 险～ cliff-hanging win; nose out / 战～敌人 gain a victory over the enemy / 我们队～了他们队。Our team triumphed over their. ②(优越) surpass; be superior to; get the better of: 聊～于无 better than nothing ③(优美的) superb; wonderful; lovely: ～景 wonderful scenery ④(能承担或承受) be equal to; can bear: 力不能～ beyond one's ability / 数不～数 too numerous to count; countless
【胜败】victory or defeat; success or failure ◇ ～乃兵家常事。For a military commander, winning or losing a battle is a common occurrence. / ～未卜。It is uncertain whether (he) will succeed or fail.
【胜不骄,败不馁】not to get conceited because of victory or disheartened in case of defeat
【胜地】famous scenic spot: 避暑～ summer resort / 游览～ a scenic spot for tourists
【胜负】victory or defeat; success or failure: ～未定。Victor hangs in the balance.
【胜迹】famous historical site
【胜利】①(获胜) victory; triumph: 外交上的～ a diplomatic triumph / ～归来 return in triumph / ～果实 fruits of victory / 充满了～的信心 fully confident of victory / ～在望。Victory is in sight. ②(达到预定目的) successfully; triumphantly: ～会师 triumphantly join forces / ～完成任务 successfully carry out one's task ◇ ～者 victor; winner
【胜券】confidence in victory: 操～ be sure to win
【胜任】competent; qualified; equal to: ～工作 be competent at a job; prove equal to the task / ～愉快 be fully competent; prove more than equal to the work / 不能～ be unequal to one's task / 能～ be equal to one's task
【胜似】be better than; surpass
【胜诉】win a lawsuit; carry the cause; recover; win over ◇ ～当事人 successful party / ～的一方 prevailing party / ～人 winner of a lawsuit / ～债权人 judgement creditor
【胜算】[书] a stratagem which ensures success: 操～ be sure of success
【胜友如云】a cloud of good friends; a great many good friends
【胜仗】victorious battle; victory: 打～ win a battle; score a victory

圣

圣 ①(最崇高的) sage; saint ②(神圣的) holy; sacred: 神～领土 sacred territory ③(帝王) emperor: ～上 His Majesty; Her Majesty
【圣餐】[宗] Holy Communion
【圣城】the Holy City
【圣诞】the birthday of Jesus Christ
【圣诞节】Christmas Day ◇ ～贺片 Christmas card / ～礼物 Christmas present; Christmas box / ～前夜 Christmas Eve
【圣诞老人】Santa Claus
【圣诞树】Christmas tree
【圣地】① [宗] the Holy Land ②(具有重大历史意义和作用的地方) sacred place; shrine: 革命～ a sacred place of the revolution
【圣殿】[犹太教] Temple of God
【圣父】the Father
【圣公宗】[基督教] Anglicanism; the Anglican Church
【圣洁】holy and pure
【圣经】the Holy Bible; the Bible; Holy Writ
【圣灵】the Holy Ghost; the Holy Spirit
【圣灵节】Whitsunday
【圣灵降临节】Whitsunday; Pentecost
【圣马力诺】San Marino ◇ ～人 San Marinese
【圣母】①(女神) a female deity; goddess ②(圣母马利亚) the (Blessed) Virgin Mary; Madonna ◇ ～堂 the Chapel of Our Lady
【圣幕】[犹太教] Tabernacle
【圣人】sage; wise man: ～也有一分错。Sages, however wise, are human and can make mistakes. (或: Even a wise man makes some mistakes.)
【圣诗班】choir ◇ ～指挥 choirmaster

【圣所】[犹太教] Holy Place
【圣坛】chancel
【圣贤】sages and men of virtue: 人非~,孰能无过? Men are not saints, how can they by free from faults?
【圣旨】imperial edict
【圣子】the Son of God

shī

湿 wet; damp; humid: ~空气 humid air; moist air / ~衣服 wet clothes / 穿~衣服,你会着凉的。If you put on damp clothes, you will probably catch cold. / 你淋~了吗? Did you get wet in the rain? / 这里太~。There is too much damp.
【湿病】[中医] diseases caused by dampness
【湿度】humidity: 空气~ air humidity / 土壤~ soil moisture ◇ ~表 humidometer / ~调节器 humidistat
【湿度计】hygrometer; succulometer: 毛发~ hair hygrograph
【湿法冶金】hydrometallurgy
【湿纺】[纺] wet spinning
【湿寒】[气象] raw
【湿淋淋】dripping wet; drenched: ~的衣服 sopping wet clothes / 浑身~的 get dripping wet
【湿漉漉】wet; damp
【湿气】①(水蒸气) moisture; dampness ②[中医] eczema; fungus infection of hand or foot
【湿润】moist: ~的土壤 damp soil / 空气~ humid air
【湿透】wet through; drenched: 衣服~ wet to the skin / 她被雨淋得浑身~。She was drenched through with rain.
【湿选】[矿] wet separation
【湿疹】[医] eczema

诗 poetry; verse; poem: 写~ write a poem / 自由~ free verse
【诗歌】poems and songs; poetry: ~朗诵 recitation of poems; poetry readings
【诗话】notes on poets and poetry; notes on classical poetry
【诗集】collection of poems; poetry anthology
【诗经】*The Book of Songs*
【诗句】verse; line
【诗剧】drama in verse; poetic drama
【诗礼传家】a family of scholars
【诗篇】①(诗的总称) poem ②(好文章、动人的故事) inspiring story: 壮丽~ a magnificent epic (of)
【诗情画意】a quality suggestive of poetry or painting
【诗人】poet: 女~ poetess
【诗书门第】a scholarly family
【诗兴】urge for poetic creation; poetic inspiration; poetic mood: ~大发 feel a strong urge to write poetry; be in an exalted, poetic mood
【诗意】poetic quality or flavor: 饶有~ rich in poetic flavor; very poetic
【诗韵】①(做诗所押的韵) rhyme ②(韵书) rhyming dictionary

蓍 [植] alpine yarrow

师 ①(传授知识的人) teacher; master: 提倡尊~爱生 advocate students respecting teachers and teachers cherishing students ②(学习的榜样) model; example: 前事不忘,后事之~。Lessons learned from the past can guide one in the future. ③(掌握专门学术、技艺的人) a person skilled in a certain profession: 工程~ engineer / 技~ technician ④(指由师徒关系产生的) of one's master or teacher: ~母 the wife of one's teacher or master ⑤[军] division: 步兵~ infantry division ⑥(军队) troops; army: 正义之~ an army fighting for a just cause
【师表】[书] a person of exemplary virtue: 为人~ be worthy of the name of teacher; be a paragon of virtue and learning
【师部】[军] division headquarters
【师出无名】dispatch troops without just cause
【师道尊严】the teachers' dignity
【师弟】①(同一师傅而拜师在后的人) junior fellow apprentice ②(师傅的儿子中年龄小的人) the son of one's master (younger than oneself) ③(父亲的徒弟中年龄小的人) father's apprentice (younger than oneself)
【师法】①(效法) model oneself after (a great master); imitate ②(师徒相传的学问、技术) knowledge or technique handed down by one's master
【师范】①(培养师资的) teacher-training; pedagogical: ~大学 teachers university / ~学院 teachers college; teachers training college ②[简](师范学校) normal school
【师父】①(师傅) master; master worker ②(对和尚、尼姑等的尊称) a polite form of address to a monk or nun
【师傅】master worker
【师娘】[口] the wife of one's teacher or master

【师生】teachers and students: ～关系 teacher-student relationship
【师徒】master and apprentice ◇ ～关系 master- apprentice relation / ～合同 indentures
【师团】[军] division
【师兄】①(同一师傅而拜师在前的人) senior fellow apprentice ② (师傅的儿子中年龄大的人) the son of one's master (older than oneself) ③(父亲的徒弟中年龄大的人) father's apprentice (older than oneself)
【师爷】a private assistant attending to legal, fiscal or secretarial duties in a local *yamen*; private adviser
【师长】①[尊] teacher ②[军] division commander; divisional commander
【师专】[简](师范专科学校) teachers training school; normal school
【师直为壮】an army fighting for a just cause has high morale
【师资】persons qualified to teach; teachers: ～不足 shortage of teachers / 培训～ train teachers ◇ ～训练班 teachers training class

狮 lion
【狮身人面像】sphinx
【狮子】lion
【狮子鼻】pug nose
【狮子搏兔】not stint the strength of a lion in wrestling with a rabbit; go all out even when fighting a small enemy or tackling a minor problem
【狮子狗】pug-dog
【狮子头】[食品] large meatball
【狮子舞】lion dance
【狮子座】[天] Leo

嘘[叹](表示制止、驱逐等): ～声四起 hiss and boo everywhere / 被～下台 be hissed off the stage / ～,别出声! Sh! Keep quiet! / ～,别说话! Hush! Stop talking!

失①(丢掉) lose: 得～ gain and loss / ～而复得 lost and found again / 迷～方向 lose one's bearings ②(错过) miss; let slip: 坐～良机 let slip a good opportunity; lose a good chance ③(没有达到目的) fail to achieve one's end: 大～所望 be greatly disappointed ④(过失;错误) mishap· defect; mistake: ～之于烦琐 have the defect of being too detailed / 唯恐有～ fear that there may be some mishap ⑤(改变) deviate from the normal: ～色 turn pale ⑥(背弃) break (a promise); go back on (one's word): ～信 break one's promise
【失败】①(被打败) be defeated; lose (a war, etc.): 遭到了毁灭性的～ suffer a crushing defeat / 甲队第一局～了。Team A lost the first game. ②(没有达到预定目的) fail: ～是成功之母。 Failure is the mother of success. ◇ ～情绪 defeatist sentiments / ～主义 defeatism
【失策】unwise; inexpedient: 这样做非常～。It was a very unwise move.
【失察】neglect one's supervisory duties: 一时～ momentary oversight
【失常】not normal; odd: 精神～ be distraught; not be in one's right mind / 举动～ act oddly; behave strangely
【失宠】fall into disfavor; be out of favor; be in disgrace
【失传】not be handed down from past generations; be lost: 一种～的艺术 a lost art
【失措】lose one's presence of mind; lose one's head: 惊慌～ be panic-stricken
【失单】a list of lost articles
【失当】improper; inappropriate: 举止～ behave inappropriately / 这个问题处理～。This problem was not properly handled.
【失道寡助】an unjust cause finds scant support
【失地】lost territory: 收复～ recover lost territory
【失掉】①(原有的不再具有) lose: ～联系 lose contact with / ～民心 lose popular support / ～权力 be stripped of power ②(没有取得;没有把握住) miss: ～机会 miss a chance / ～战机 fail to grasp a good opportunity to engage the enemy
【失魂落魄】driven to distraction: 吓得～ be scared out of one's wits; be frightened out of one's life
【失火】catch fire; be on fire
【失脚】lose one's footing; slip: ～跌倒 lose one's footing and fall
【失节】①(失去气节) forfeit one's integrity; be disloyal ②(妇女失去贞操) lose one's chastity
【失禁】[医] incontinence: 大小便～ incontinence of feces and urine
【失敬】[套] sorry I didn't recognize you; sorry
【失控】out of control; runaway
【失口】a slip of the tongue
【失礼】breach of etiquette; impoliteness; discourtesy: 请原谅我的～。Please forgive me for being impolite.

【失利】suffer a setback：进攻～ suffer setback in attack／军事上的～ military reverses／战斗～ take it on the chin
【失恋】be disappointed in a love affair
【失灵】not work；not work properly；be out of order：开关～了。The switch is out of order.／刹车～。The brake is ineffective.（或：The brake doesn't work.）
【失落】lose
【失密】give away official secrets due to carelessness
【失眠】（suffer from）insomnia：他昨夜～。He had a sleepless night last night.
【失明】lose one's sight；go blind：双目～ lose the sight of both eyes
【失能性毒剂】［军］incapacitating agent
【失陪】［套］excuse me，but I must be leaving now
【失窃】have things stolen；suffer loss by theft
【失去】lose：～控制 out of hand／～时效 be no longer effective；cease to be in force／～信心 lose confidence／～知觉 lose consciousness／～作用 be ineffective
【失散】be separated from and lose touch with each other；be scattered
【失色】①（因受惊而脸色苍白）turn pale：大惊～ turn pale with fright ②（失去本来色彩）be eclipsed；be outshone：黯然～ be cast into the shade；be eclipsed；pale into insignificance
【失神】①（疏忽）inattentive；absent-minded ②（精神不振）out of sorts；in low spirits
【失慎】①（疏忽）not cautious；careless ②（失火）cause a fire through carelessness
【失声】①（不自主地发出声来）cry out involuntarily ②（哭不出声）lose one's voice：痛哭～ be choked with tears
【失时】miss the season；let slip the opportunity
【失实】inconsistent with the facts：报导～。The report gave a false picture of the situation.／传闻～。The rumor was unfounded.
【失势】lose power and influence；fall into disgrace
【失事】（have an）accident；wreck：飞机～ aviation accident；airplane crash
【失手】accidentally drop：他～打碎了一个玻璃杯。He lost hold of the glass and broke it.
【失守】fall：城市～ the fall of a city
【失速】［航空］stall ◇ ～滑翔 stalled glide／～状态 stall conditions
【失算】miscalculate；misjudge；be injudicious
【失态】forget oneself：酒后～ forget oneself in one's cups
【失调】①（失去平衡）imbalance；dislocation：供求～ imbalance of supply and demand／经济～ economic dislocation／雨水～ abnormal rainfall ②（失去调养）lack of proper care（after an illness，etc.）：产后～ lack of proper care after childbirth ③［无］maladjustment；detuning
【失望】①（感到没有希望）lose hope ②（因希望落空而不愉快）disappointed：令人～ disappointing
【失物】lost article；lost property：寻找～ look for lost articles ◇ ～招领处 Lost and Found Office；Lost Property Office
【失误】①（指打球、下棋）fault；muff：发球～ faulty service／接球～ muff a ball ②（因疏忽或水平不高而造成差错）slip up：他计算～。He slipped up in his calculations.
【失陷】（of cities，territory，etc.）fall；fall into enemy hands
【失效】①（失去效力）lose efficacy；cease to be effective：这药已～了。The medicine no longer has any effect. ②（无效）（of a treaty，an agreement，etc.）be no longer in force；become invalid：自动～ automatically cease to be in force／这张证明已～。This certificate is invalid. ◇ ～提单 stale bill of lading／～支票 stale check
【失笑】laugh in spite of oneself；cannot help laughing
【失谐】［无］detuning；mismatching
【失信】break one's promise；go back on one's word
【失修】be in bad repair；fall into disrepair：年久～ have long been out of repair；have been neglected for years
【失学】be deprived of education；be unable to go to school；be obliged to discontinue one's studies
【失血】lose blood：～过多 excessive loss of blood
【失言】make an indiscreet remark：酒后～ make an indiscreet remark under the influence of alcohol
【失业】lose one's job；be out of work；be unemployed ◇ ～补偿 unemployment compensation／～津贴 unemployment benefit／～率 rate of unemployment／～险 unemployment insurance／～者 the unemployed；the jobless

【失宜】[书] inappropriate: 处置～ handle improperly

【失意】be dejected; be frustrated; be disappointed

【失音】[医] aphonia

【失迎】[套] fail to meet (a guest): 昨天～了,很抱歉。 Sorry I was out when you called yesterday.

【失语症】[医] aphasia

【失约】fail to keep an appointment

【失真】①(跟原来的有出入) lack fidelity; not be true to the original ②[无] distortion: 频率～ frequency distortion

【失之东隅,收之桑榆】lose at sunrise and gain at sunset; make up on the roundabouts what you lose on the swings

【失之毫厘,谬以千里】A small discrepancy leads to a great error. (或: An error the breadth of a single hair can lead you a thousand *li* astray.)

【失之交臂】just miss the person or opportunity: 机会难得,幸勿～。 Don't let slip such a golden opportunity.

【失职】neglect one's duty; dereliction of duty ◇～罪 offense of misconduct in office

【失重】[物] weightlessness; zero gravity; weight loss: ～状态 the state of weightlessness

【失主】owner of lost property

【失踪】disappear; absence; be missing: 除了伤亡之外,还有许多人～。 In addition to the killed and wounded, many were missing. / (英国)陆军部发布～军人名单。 The War Office issued a list of the missing.

【失足】①(跌倒) lose one's footing; slip: ～落水 slip and fall into the water ②(堕落或犯罪) take a wrong step in life: 一～成千古恨。 One false step brings everlasting grief.

施 ①(实行) execute; carry out: 无所～其技 no chance (for sb.) to play his tricks ②(给予) bestow; grant; hand out: ～恩 bestow favor ③(强加) exert; impose: ～压力 exert pressure ④(在物体上加某物) use; apply: ～底肥 apply fertilizer to the subsoil

【施放】discharge; fire: ～催泪弹 fire tear-gas shells

【施肥】spread manure; apply fertilizer: 给稻秧～ apply fertilizer to rice seedlings

【施工】construction: ～重地,闲人免进。 Construction Site. No Admittance. / 水库正在～。 The reservoir is under construction. ◇～单位 unit in charge of construction / ～缝[建] construction joint / ～人员 builder; constructor / ～图 working drawing

【施惠不记心,受德莫忘恩】Do not remember favors you bestowed upon others, do not forget favors others bestowed upon you.

【施加】exert; bring to bear on: ～压力 bring pressure to bear on sb.; put pressure on sb.

【施礼】salute

【施力】[物] application of force ◇～点 point of application

【施舍】give alms; give in charity: 靠人～度日 live on charity; live on the alms given by others

【施事】[语] the doer of the action in a sentence; agent

【施威】exhibit one's power; show severity

【施行】①(执行) put in force; execute; apply: 自公布之日起～ come into force upon promulgation ②(做;实行) perform: ～急救 administer first aid / ～手术 perform a surgical operation ◇～细则 rules for implementation

【施用】use; employ; apply: ～化肥 apply fertilizer ◇～私刑者 lyncher

【施展】put to good use; give free play to: ～本领 put one's ability to good use; give full play to one's talent / ～才能 display one's ability / ～阴谋诡计 carry out schemes and intrigues

【施政】administration ◇～方针 administrative politics / ～纲领 administrative program

【施朱傅粉】paint and powder oneself

【施主】①(施舍财物的人) alms giver; benefactor ③[物] donor

尸 corpse; dead body; remains: 死～ dead body; corpse

【尸斑】cadaveric ecchymoses; livor mortis

【尸骨】skeleton

【尸横遍野】a field littered with corpses

【尸僵】cadaveric rigidity; rigor mortis

【尸蜡】[医] adipocere

【尸身】【尸首】corpse; dead body; remains

【尸体】corpse; dead body; remains ◇～检验 necropsy; postmortem examination / ～鉴定 identification of the dead / ～解剖 autopsy; postmortem (examination) / ～痉挛 cadaveric spasm / ～内部检查 internal postmortem examination / ～剖检 ptomatopsia; ptomatopsy; necropsy / ～现场 scene of death

【尸位素餐】hold down a job without doing a stroke of work

虱 louse
【虱病】pediculosis; phthiriasis
【虱传染】pediculation
【虱子】louse
鲺[动] carp louse; fish louse

shí

实 ①(实心) solid: 铁条是～心的,管子是空心的。 A iron bar is solid, a pipe is hollow./ 这球里面是～的。 This ball is solid. ②(真实) true; real; honest: 情况属～。 It's true. ③(实际) reality; fact: 名～相副 in reality as well as in name/ 名不副～。 The name falls short of the reality. ④(果实) fruit; seed: 开花结～ blossom and bear fruit
【实报实销】be reimbursed for one's actual expenses
【实词】[语] notional word
【实弹】[军] live shell; live ammunition ◇ ～射击 firing practice; range practice/ ～投掷 live grenade throw/ ～演习 practice with live ammunition
【实地】on the spot: ～勘测 field exploring/ ～考察 on-the-spot investigation
【实干】get right on the job; do solid work: 有～精神 with a spirit of working in earnest ◇ ～家 man of action
【实话】truth: ～实说 not mince words; not beat about the bush/ 说～ to tell the truth
【实惠】①(实际好处) material benefit: 从中得到～ really benefit from it ②(有实际好处) substantial; solid: 经济～的饭菜 inexpensive but substantial meals
【实际】①(客观存在的事物、情况) reality; practice: ～上 in fact; in reality; actually/ 符合～ correspond to reality/ 客观～ objective reality ②(实有的) practical; realistic: ～经验 practical experience ③(合乎事实的) real; actual; concrete: ～的例子 a concrete instance/ ～情况 the actual situation; reality/ 我不能举出～数字。 I can not give the actual figures. ◇ ～成本 actual cost; real cost/ ～工资 real wages/ ～汇价[经] effective rate/ ～价值 actual value/ ～控制线[军] line of actual control/ ～利率 true 〈actual〉 rate of interest/ ～利润 real profit/ ～伤害 actual harm; real injury/ ～收入 real income/ ～债务 actual debts/ ～支付 actual delivery/ ～资本 actual capital
【实价】actual price
【实践】①(有意识的活动) practice: ～出真知。 Genuine knowledge comes from practice./ ～是检验真理的唯一标准。 Practice is the sole criterion for testing truth. ②(实行) put into practice; carry out; live up to: ～诺言 keep one's word; make good one's promise ◇ ～性 practicality; practicalness
【实据】substantial evidence; substantial proof: 查无～ investigations show no evidence/ 查有～ investigation reveals valid evidence/ 提供～ produce factual evidence/ 真凭～ ironclad evidence
【实况】what is actually happening ◇ ～录音 on-the-spot recording; live recording/ ～转播 live broadcast; live telecast
【实力】actual strength; strength: ～相当 match each other in strength; be well matched in strength/ 军事～ military strength ◇ ～地位 position of strength
【实例】living example; example
【实模铸造法】[冶] cavityless casting
【实情】the true state of affairs; the actual situation; truth
【实权】real power
【实生苗】[农] seedling
【实施】put into effect; implement; carry out: 监督宪法的～ supervise the enforcement of the constitution ◇ ～法规 enforcement regulations/ ～机构 enforcement body/ ～条例 enforcement regulations/ ～者 executor
【实事求是】seek truth from facts; be practical and realistic: ～的批评 criticism based on facts
【实收款项】proceeds of sale
【实数】①(实在数字) the actual amount or number ②[数] real number
【实体】①[哲] substance ②[法] entity
【实物】①(真实具体之物) material object ②(代款之物) in kind ◇ ～地租 rent in kind/ ～工资 wages in kind/ ～幻灯机 epidiascope/ ～交易 barter/ ～赔偿 reparations in kind/ ～税 tax paid in kind/ ～证据 tangible evidence
【实习】practice; fieldwork; field trip: 进行教学～ do practice teaching ◇ ～法庭 moot court/ ～工厂 factory attached to a school/ ～审判 moot/ ～生 trainee/ ～医生 intern
【实线】hard-wire; actual line; real line
【实现】realize; achieve; bring about: ～改革 bring about a reform/ ～四个现代化 accomplish the four modernizations

【实像】[物] real image
【实效】actual effect; substantial results: 注重～ emphasize practical results
【实心】①(心地诚实) sincere: ～实意 honest and sincere ②(内部是实的) solid: 垒球是～的。The softball is solid. ◇～球[体] medicine ball / ～轴 solid shafting
【实行】put into practice; carry out; practise; implement: ～八小时工作制 institute an eight-hour (working) day / ～大赦 amnesty / ～民主集中制 put democratic centralism into practice
【实学】real learning; sound scholarship: 真才～ real ability and learning
【实验】experiment; test: 科学～ scientific experimentation / 做～ do an experiment; make a test ◇～报告 laboratory report / ～动物 animal used as a subject of experiment / ～室 laboratory / ～心理学 experimental psychology / ～员 laboratory technician / ～证据 experimental evidence
【实业】industry and commerce; industry ◇～家 industrialist
【实音】flatness
【实用】practical; pragmatic; functional: 既美观,又～ not only beautiful, but also practical
【实用主义】[哲] pragmatism ◇～法学 pragmatic jurisprudence / ～者 pragmatist
【实在】①(真实;不虚假) true; real; honest; dependable: ～的本事 real ability ②(的确) indeed; really; honestly: ～太好了 very good indeed ③(其实) in fact; as a matter of fact: 他装懂,～并没懂。He pretends to understand, but as a matter of fact he doesn't. ④[方](扎实) (of work) well-done; done carefully: 工作做得很～。The work is well-done.
【实在主义法学派】school of judicial realism
【实则】actually; in fact; in reality
【实战】actual combat: ～演习 combat exercise with live ammunition
【实至名归】fame follows merit
【实症】[中医] a case of a physically strong patient running a high fever or suffering from such disorders as stasis of blood, constipation, etc.
【实证主义】[哲] positivism ◇～法学 positivist jurisprudence / ～者 positivist
【实质】substance; essence: ～上 in essence; in substance; essentially / 问题的～ the central point at issue; the crux of the matter ◇～性变更 material alteration ～性陈述 material representation / ～性条款 substantive provision / ～性违反 material breach / ～性证据 material evidence
【实足】full; solid: ～年龄 exact age

识 ①(认识) know: 一字不～ not know a single character; absolutely illiterate ②(知识) knowledge: 学～ learning; knowledge
【识别】discriminate; distinguish; discern; spot: ～能力 capacity of discernment ◇～标记 identification marking
【识大体,顾大局】have the cardinal principles in mind and take the overall situation into account
【识货】know all about the goods; be able to tell good from bad; know what's what: 不怕不～,就怕货比货。Don't worry about not knowing much about the goods; just compare and you will see which is better.
【识荆】[书][敬] have the honor of making your acquaintance: ～恨晚。I regret to have made your acquaintance so late.
【识破】see through; penetrate: ～花招 see through a trick / ～骗局 see through a fraud
【识趣】know how to behave in a delicate situation: 不～ be insensible
【识善辨恶】know the difference between right and wrong; discern the difference between the good and evil
【识时务者为俊杰】Those who suit their actions to the time are wise. (或: A wise man submits to fate.)
【识途老马】an old horse which knows the way; a person of rich experience; a wise old bird
【识文断字】able to read; literate
【识相】[方] be sensible; be tactful
【识字】learn to read; become literate ◇～班 literacy class / ～课本 reading primer; elementary reader

十 ①(数目) ten: ～倍 ten times; tenfold / ～分之一 one tenth / ～周年纪念 the tenth anniversary ②(达到顶点) topmost: ～成 100 per cent
【十八般武艺】skill in wielding the 18 kinds of weapons; skill in various types of combat: ～,样样精通 be skillful in using each and every one of the 18 weapons; be versatile
【十大功劳】[中药] Chinese mahonia
【十滴水】[药] "10 drops", a popular medicine for summer ailments

【十冬腊月】the tenth, eleventh and twelfth months of the lunar year; the cold months of the year
【十恶不赦】unpardonable evil; guilty beyond forgiveness
【十二级风】[气] force 12 wind; hurricane
【十二进制】duodecimal notation
【十二平均律】[乐] twelve-tone equal temperament
【十二月】①(阳历) December ②(阴历) the twelfth month of the lunar year; the twelfth moon
【十二指肠】[生理] duodenum ◇ ～梗阻 duodenal ileus / ～溃疡 duodenal ulcer / ～炎 duodenitis
【十分】[副] very; fully; utterly; extremely: ～爱惜人力物力 use manpower and material resources most sparingly / ～宝贵 most valuable / ～猖狂 be on a rampage / ～仇视 harbor intense hatred for / ～高兴 be very pleased; be elated / ～难过 feel very sorry; feel very bad / ～有害 extremely harmful / ～注意 pay close attention to
【十级风】[气] force 10 wind; whole gale
【十进制】[数] the decimal system; decimalism
【十六分音符】[乐] semiquaver; sixteenth note
【十六进制】hexadecimal; sexadecimal
【十六开】[印] sixteenmo; 16mo
【十目所视,十手所指】with many eyes watching and many fingers pointing; one cannot do wrong without being seen
【十拿九稳】90 per cent sure; be very sure of; practically certain; in the bag: 这冠军,我们是～了。We're sure of winning the championship.
【十年寒窗】persevere ten years in one's studies in spite of hardships
【十年九不遇】not occur once in ten years; be very rare: 这样大的洪水真是～。A flood of this sort is really unprecedented.
【十年树木,百年树人】[谚] it takes ten years to grow trees, but a hundred to rear people
【十全十美】be perfect in every way; be the acme of perfection; leave nothing to be desired: 世上没有～的东西。Nothing in the world is flawless and perfect.
【十室九空】nine houses out of ten are deserted; a scene of desolation after a plague or war when the population is decimated
【十四行诗】sonnet
【十万火急】most urgent; extra-urgent
【十五个吊桶打水,七上八下】One's heart is like a well in which seven buckets are drawn up and eight dropped down.
【十项全能运动】[体] decathlon
【十一级风】[气] force 11 wind; storm
【十一月】①(阳历) November ②(阴历) the eleventh month of the lunar year; the eleventh moon
【十月】①(阳历) October: ～一日 October 1 ②(阴历) the tenth month of the lunar year; the tenth moon
【十之八九】in eight or nine cases out of ten; most probably; very likely: ～他是误会了。Most likely there is some misunderstanding on his part.
【十指连心】The nerves of the fingertips are linked with the heart. (或: What happens to children is of vital interest to parents.)
【十字镐】pick; pickaxe; mattock
【十字花科】[植] the mustard family; Cruciferae
【十字架】cross
【十字接头】cruciform joint; X-conn
【十字街头】crisscross streets; busy city streets
【十字军】①[史] the Crusades ②(指英美某种社会改革运动) crusade ③(指参加者) crusader
【十字路口】crossroads: 徘徊在～ hesitate at the crossroads
【十字轴】cross axle; centerpiece
【十足】①(纯粹的) 100 per cent; out-and-out; sheer; downright: ～的骗局 a sheer fraud / ～强盗逻辑 downright gangster logic / ～的强权统治 100% power politics ②(十分充足) full of: 干劲～ full of energy / 神气～ put on grand airs / 有～的理由 have ample reason

什 ①(多种的) assorted; varied; miscellaneous ②[书](多用于分数或倍数) ten: ～百 tenfold or hundredfold / ～一 one tenth
【什件儿】giblets: 炒～ fried giblets
【什锦】[食品] assorted; mixed: ～饼干 assorted biscuits / ～奶糖 assorted toffees / ～巧克力 assorted chocolates
【什物】articles for daily use; odds and ends; sundries

石 ①(岩石) stone; rock ②(石刻) stone inscription: 金～ inscriptions on ancient bronzes and stone tablets
【石斑鱼】grouper
【石板】[建] slabstone; flagstone; flag
【石版】[印] stone plate
【石碑】stone tablet; stele

【石笔】slate pencil
【石菖蒲】[植] grass-leaved sweetflag
【石沉大海】like a stone dropped into the sea; disappear for ever
【石担】[体] stone barbell
【石刁柏】[植] asparagus
【石雕】①(在石头上雕刻) stone carving ②(石雕艺术品) carved stone
【石墩】a block of stone used as a seat
【石方】①(一立方米石料) cubic meter of stone ②(石方工程) stonework: 两万～ twenty hundred cubic meters of stonework
【石膏】gypsum; plaster stone: 生～ plaster stone / 熟～ plaster; plaster of Paris ◇～绷带 plaster bandage / ～床 plaster bed / ～夹板 plaster splint / ～像 plaster statue; plaster figure
【石工】①(石料活) masonry ②(干石料活的人) stonemason; mason
【石拱桥】stone arch bridge
【石花菜】[植] agar
【石化作用】[地] petrifaction
【石灰】lime: 生～ quick lime / 熟～ slaked lime ◇ ～浆 lime white / ～砂浆 lime mortar / ～石 limestone / ～水 limewash / ～水泥砂浆 lime-and-cement mortar / ～窑 limekiln / ～质砂岩 calcareous sand stone
【石鸡】[动] chukar
【石匠】stonemason; mason
【石刻】①(刻着文字、图画的石制品) carved stone ②(刻的文字、图画) stone inscription
【石窟】rock cave; grotto
【石窟寺】[考古] the Cave Temple
【石块】stone; rock
【石蜡】paraffin wax ◇ ～油 paraffin oil
【石栗】[植] candlenut tree
【石硫合剂】[农] lime sulfur
【石榴】[植] pomegranate ◇ ～红 garnet (color) / ～石[矿] garnet
【石龙子】[动] skink
【石煤】bone coal
【石棉】asbestos ◇ ～板[建] asbestos board / ～布 asbestos cloth / ～衬里[机] asbestos lining / ～瓦[建] asbestos shingle; asbestos tile
【石末沉着病】silicosis
【石墨】graphite ◇ ～电极 graphite electrode / ～坩埚 black-lead crucible / ～润滑脂 graphite grease / ～铀堆 graphite-uranium pile
【石磨】stone mill
【石楠】[植] Chinese photinia
【石脑油】naphtha
【石破天惊】earth-shattering and heaven-battering; remarkably original and forceful (music, writing, etc.); great vibration or shock
【石器】①(石制工具、器具) stone implement; stone artifact ②(石制品;粗陶瓷) stone vessel; stoneware ◇ ～时代 the Stone Age
【石青】[矿] azurite
【石蕊】①[植] reindeer moss ②[化] litmus ◇ ～试纸 litmus paper
【石蒜】[植] short-tube lycoris
【石笋】[地] the stalagmite
【石锁】[体] a stone dumbbell in the form of an old-fashioned padlock
【石炭纪】[地] the Carboniferous Period
【石炭酸】[化] carbolic acid; phenol
【石头】stone; rock: 心里好象一块～落了地 feel as though a load has been taken off one's mind
【石头子儿】[口] small stone; cobble; pebble
【石印】lithographic printing; lithography ◇ ～机 lithographic press / ～石 lithographic stone / ～油画 oleograph / ～纸 lithographic paper
【石英】quartz ◇ ～玻璃 quartz glass / ～坩埚 silica crucible ～卤钨灯[摄] quartz tungsten halogen lamp / ～谐振器 quartz resonator / ～岩 quartzite / ～钟 quartz clock
【石油】petroleum; oil: 沥青基～ asphalt-base petroleum / 石蜡基～ paraffin-base petroleum ◇ ～产品 petroleum products / ～地质学 petroleum geology / ～工业 oil industry; petroleum industry / ～管路 petroleum pipeline / ～化工厂 petrochemical works / ～化学 petrochemistry / ～化学产品 petroleum chemicals / ～勘探 petroleum prospecting / ～沥青 petroleum pitch / ～输出国组织 the Organization of Petroleum Exporting Countries (OPEC) / ～运移 oil migration ○勘探 exploration; prospecting / 储油构造 oil-bearing structure / 含油层 oil-bearing formation / 矿苗露头 outcrop / 海相沉积 marine deposit; marine sediment / 陆相沉积 continental deposit; continental sediment / 大陆架 continental shelf / 海上油田 offshore oilfield / 海上平台 offshore platform / 油层 oil layer; oil horizon / 油藏 oil pool; oil deposit / 储油量 oil reserve / 可采储量 recoverable reserve / 含硫量 sulphur content / 油井 oil well / 油页岩 oil shale / 气田 gas field / 天然气 natural gas / 柴油 diesel oil / 重油 heavy oil / 煤油 kerosene / 机油 engine oil / 润滑油 lubrication oil; lubricant / 蜡 wax / 石蜡 paraffin / 凡士林

Vaseline／沥青 pitch／煤焦油 coal tar oil／挥发油 volatile oil／溶解油 soluble oil／裂化气 cracked gas／钻井 drilling／井架 oil derrick／钻台 derrick floor／钻具 drilling tools／钻头 drill bit／钻杆 drill rod; drill pipe／套管 casing／岩心 core／砂样 core sample／钻机 drilling rig／绞车 draw works／钻进速度 drilling rate／钻井记录 drilling record／进尺 drilling footage／自喷 blowing／井喷 blowout／重晶石 barite／泥浆 drilling mud／泥浆池 mud pit／泥浆泵 slush pump

【石油气】petroleum gas: 液化～ liquefied petroleum gas (LPG)

【石陨石】meteoric stone; meteorolite; stony meteorite

【石钟乳】stalactite

【石竹】[植] China pink

【石子】cobblestone; cobble; pebble: ～路 cobblestone street; cobbled road

拾 ①(捡起) pick up (from the ground); collect: ～柴 collect firewood／～麦穗 glean (stray ears of) wheat／～一块石头 pick up a stone／他～起烟灰缸,放到桌上。He picked up the ash tray and put it on the table. ②(十的大写) ten

【拾波】pick-up

【拾荒】glean and collect scraps (to eke out an existence)

【拾级而上】ascend the stairs

【拾金不昧】not pocket the money one picks up

【拾零】(多用于标题) news in brief; titbits; sidelights

【拾取】pick up; collect

【拾人牙慧】pick up phrases from sb. and pass them off as one's own

【拾遗】①(拾取失物) appropriate lost property: 路不～。No one pockets anything found on the road. ②(补充遗漏) make good omissions: ～补阙 make good omissions and deficiencies

【拾音】pickup: ～头 pick-up head

【拾音器】sound pick-up; pickup; adapter ◇ ～放大器 pick-up amplifier

时 ①(比较长的一段时间) time; times; days: 当～ at that time; in those days／古～ ancient times ②(规定的时候) fixed time: 准～到站 arrive at the station on time ③(计时单位) hour: 报～ announce the hour; give the time signal／从十～到十二～ from 10 to 12／上午九～ 9 a.m.／下午四～ 4 p.m.／钟已报～。The clock struck the hour. ④(季节) season: ～菜 delicacies of the season／应～鲜果 fruits in season ⑤(当前;现在) current; present: ～下 at present ⑥(时机) opportunity; chance: 失～ lose the opportunity; miss the chance ⑦(时常) now and then; occasionally; from time to time: ～有出现 occur now and then ⑧(叠用) now... now...; sometimes... sometimes...: ～断～续 on and off／～快～慢 sometimes fast, sometimes slow ⑨[语] tense: 过去～ the past tense

【时不我待】Time will not wait for me. (或: Time waits for no man.)

【时不宜迟】There's no time to be lost.

【时差】①(两地的时间差) time difference ②[天] equation of time

【时常】often; frequently

【时辰】one of the 12 two-hour periods into which the day was traditionally divided, each being given the name of one of the 12 Earthly Branches

【时代】①(时期) times; age; era; epoch: ～潮流 the tendency of the day; the trend of the times／划～的大事 the epoch-making event ②(生命中的某个时期) a period in one's life: 青年～ youth

【时而】①(事情重复发生) from time to time; sometimes: 蔚蓝色的天空中,～飘过几朵白云。Sometimes several clusters of white clouds drift across the blue sky. ②(叠用) now... now...; sometimes... sometimes...: 天气～热～冷。It is sometimes warm and sometimes cold.

【时分】time: 黄昏～ at dusk; at twilight

【时光】①(时间;光阴) time: ～不早了。It's getting late. ②(时期;日子) times; years; days

【时乖运蹇】be very hard up, and (in fact) be at a loose end; run into bad luck

【时过境迁】Circumstances change with the passage of time. (或: Once on shore, we pray no more.)

【时候】①(有起点和终点的一段时间) (the duration of) time: 农忙的～ a busy farming season ②(时间里的某一点) (a point in) time; moment: 现在什么～? What time is it?

【时机】opportunity; an opportune moment: ～一到 when the opportunity arises; at the opportune moment／错过～ miss an opportunity／等待～ wait for an opportunity; bide one's time

【时价】current price

【时间】time; hour: ～与空间 time and space／

上课～school hours／～到了。Time's up.／现在的～是十四点五分。The time now is five minutes past fourteen. ◇～表 timetable; schedule／～方位 time azimuth／～间隔 time separation／～效率 time efficiency／～压缩 time-lapse／～知觉 [心] time perception

【时间性】timeliness: 新闻报导的～强。News reports must be timely.

【时角】[天] hour angle

【时节】①(节令;季节) season: 春耕～ the season for spring ploughing ②(时候) time: 那～我们每年夏天都去游泳。At that time we used to go swimming every summer.

【时局】the current political situation

【时刻】①(时间里的某一点) time; hour; moment: 关键～ a critical moment／幸福的～ a happy moment／一生中最幸福的～ the happiest hour of one's life ②(每时每刻) constantly; always: ～保持清醒头脑 constantly keep a cool head／～准备着 be ready at all times ◇～表 timetable; schedule

【时空】space-time; space-time continuum

【时来运转】Fortune is smiling. (或: Time has moved in one's favor.)

【时令】season: ～不正 unseasonable weather／～正当 be in season ◇～病 seasonal disease

【时髦】fashionable; stylish; in vogue: ～的服装 fashionable clothes／衣着～ be in fashionable dress

【时期】period: 和平～ peacetime／社会主义建设～ the period of socialist construction／战争～ wartime／进入了一个新～ enter a new stage

【时区】time zone

【时时】often; constantly: ～想到 often recall or think about

【时势】the current situation; the trend of the times; the way things are going: ～造英雄。The times produce their heroes

【时事】current events; current affairs ◇～报告 report on current events／～述评 current events survey

【时速】speed per hour

【时态】[语] tense

【时务】current affairs; the trend of the times: 不识～ show no understanding of the times

【时鲜】(of vegetables,fruits, etc.) in season: ～果品 fresh fruits

【时限】time limit

【时效】①(在一定时间内能起的作用) effectiveness for a given period of time ②[法] prescription ◇～期限 length of limitation period; limitation period／～硬化 [冶] age-hardening／～中断 interruption; interruption of prescription／～终止 lapse of time

【时新】stylish; trendy; ～的式样 up-to-date style

【时兴】fashionable; in vogue; popular

【时宜】what is appropriate to the occasion: 不合～ be not appropriate to the occasion; be inappropriate; be out of keeping with the times

【时疫】epidemic

【时运】luck; fortune: ～不济 have bad luck; down on one's luck

【时针】①(钟、表面上的针形零件) hands of a clock or watch ②(钟、表上的短针) hour hand

【时至今日】at this late hour

【时钟】clock

【时装】fashionable dress; the latest fashion

【时作时辍】do something by fits and starts

鲥

【鲥鱼】hilsa herring; reeves shad

食

①(吃;吃饭) eat: 废寝忘～ (be so absorbed as to) forget to sleep and eat ②(食物) meal; food: 流～ liquid diet／面～ food made of fluor／肉～ meat ③(饲料) feed: 鸡～ chicken feed／猪～ pig feed ④(可食用的) edible: ～油 edible oil; cooking oil ⑤(天体现象) eclipse: 环～ annular eclipse／偏～ partial eclipse／全～ total eclipse／日～ solar eclipse／月～ lunar eclipse

【食变星】[天] eclipsing variable

【食不甘味】have no appetite for food (in deep sorrow); eat food but without knowing its taste

【食不果腹】have little food to eat; have not sufficient food to eat

【食不厌精】One does not object to the finest food.

【食草动物】herbivorous animal; herbivore

【食虫动物】insectivorous animal; insectivore

【食道】[生理] esophagus

【食而不化】eat without digesting; read without understanding

【食分】[天] totality

【食粪动物】coprophagous animal

【食腐动物】saprophagous animal; scavenger; saprozoic

【食古不化】swallow ancient learning without digesting it; be pedantic

【食管】[生理] esophagus ◇～癌 cancer of the

esophagus／～炎 esophagitis
【食火鸡】[动] cassowary
【食积】[中医] dyspepsia; indigestion
【食既】[天] second contact of an eclipse
【食具】tableware; dinner service
【食客】a person sponging on an aristocrat; a hanger on of an aristocrat
【食粮】grain; food: 精神～ spiritual food
【食量】capacity for eating; appetite
【食品】foodstuff; food; provisions: 罐头～ tinned food ◇ ～部 food department／～厂 bakery and confectionery; food products factory／～工业 food industry／～公司 food company／～加工 food processing／～商店 provisions shop／～添加剂 food additive
【食谱】recipes; cookbook
【食肉动物】carnivorous animal; carnivore
【食肉寝皮】want to eat sb.'s flesh and sleep on his skin; deep hatred for the enemy; tear sb. limb from limb
【食甚】[天] middle of an eclipse
【食宿】board and lodging: 安排～ make arrangements for board and lodging
【食堂】dining room; mess hall; canteen
【食糖】sugar
【食物】food; eatables; edibles ◇ ～储藏 food preservation／～环 food cycle／～金字塔 food pyramid／～链 food chain; food link／～摄入[动] food intake／～污染 food pollution／～中毒 bromatoxism; food-poisoning; ptomain poisoning
【食相】[天] phase of an eclipse
【食性】[动] feeding habits; eating patterns
【食血动物】sanguivorous animal
【食言】go back on one's word; break one's promise: ～而肥 fail to make good one's promise; break faith with sb.
【食盐】table salt; salt
【食蚁兽】anteater
【食用】edible: ～植物油 edible vegetable oil ◇ ～色素 food coloring
【食油】edible oil; cooking oil
【食欲】appetite: ～不振 have a jaded appetite; have a poor appetite
【食指】index finger; forefinger
【食之无味，弃之可惜】hardly worth eating but not bad enough to throw away
【食茱萸】[植] ailanthus prickly ash

蚀 ①(损失;损伤) lose: 亏～ lose (money) in business ②(腐蚀) erode; corrode: 风雨侵～ erosion by wind and rain ③(天体现象) eclipse
【蚀变】alteration
【蚀本】lose one's capital: ～出售 sell at a loss／～生意 a business running at a loss
【蚀沟】etched groove
【蚀防护罩】ablation shields
【蚀刻】etching ◇ ～机 etching machine／～剂 etchant／～印刷 etch printing

shǐ

史 history: 编年～ annals／断代～ dynastic history／古代～ ancient history／国际关系～ history of international relations／近代～ modern history／现代～ contemporary history／英国～ history of England／战～ annals of war／有～以来 since the beginning of recorded history
【史册】history; annals: 载入～ go down in history
【史抄】extracts from history
【史官】official historian; historiographer
【史迹】historical site or relics
【史籍】history; historical records
【史料】historical data; historical materials
【史前】prehistoric: ～时代 prehistoric age ◇ ～学[考古] prehistory
【史诗】epic
【史实】historical facts
【史书】history; historical records: 据～记载 according to historical records
【史无前例】without precedent in history; unprecedented
【史学】the science of history; historical science; historiography ◇ ～家 historian; historiographer

使 ①(派遣) send; tell sb. to do sth.: ～人去请医生 send sb. for a doctor ②(用) use; employ; apply: ～尽一切办法 employ all available means／这支笔很好～。This pen writes well. ③(让;致使) make; cause; enable: 虚心～人进步，骄傲～人落后。Modesty helps one to go forward, conceit makes one lag behind ④(使者) envoy; messenger: 特～ special envoy／信～ courier; messenger／出～国外 be accredited to a certain country; be sent abroad as an envoy ⑤(假如) if; supposing: 纵～ even if; even though
【使不得】①(不能用) can't be use ②(不可以) won't do; must not
【使出】use; exert: ～浑身解数 use all one's skill

【使得】①(可用) can be used; usable ②(能行; 可以) will do; workable; feasible: 这个计划可～? Is the plan workable? ③(致使) make; cause; render: ～家喻户晓 make known to everyone
【使乖弄巧】use strategy
【使馆】diplomatic mission; embassy: 美国驻华～ American Embassy in China / 在华盛顿的中国～ Chinese Embassy in Washington ◇ ～工作人员 the staff of a diplomatic mission; embassy personnel / ～馆长 head of a diplomatic mission
【使坏】[口] be up to mischief; play a dirty trick
【使唤】①(叫人替自己做事) order about: 爱～人 be in the habit of ordering people about; be bossy ②(使用)[口] use; handle
【使节】diplomatic envoy; envoy: 各国外交～ diplomatic envoys of various countries / 友好～ a goodwill mission
【使劲】exert all one's strength: ～干活 work hard / 再使把劲 put in more effort; put on another spurt
【使领馆】diplomatic and consular missions; embassies and consulates
【使命】mission: 完成历史～ accomplish (its) historical mission
【使女】maidservant; housemaid; chambermaid; maid
【使徒】[基督教] disciple
【使团】diplomatic corps ◇ ～团长 doyen <dean> of the diplomatic corps
【使性子】get angry; lose one's temper
【使眼色】tip sb. the wink; wink
【使用】make use of; use; employ; apply: ～方便 be easy to operate / ～自己的语言文字的自由 freedom to use one's own spoken and written languages / 学会～微机 learn to use a personal computer ◇ ～费 use fee / ～价值 [经] use value / ～假钞票者 smasher / ～率 rate of utilization; occupating coefficient / ～面积[建] usable floor area / ～年限 tenure of use / ～权[法] right of use; right to use a thing / ～税 royalties / ～寿命 service life (of machines) / ～说明书 operation instructions / ～支票骗取银行金钱罪 false checks
【使者】emissary; envoy; messenger: 友好～ a good will messenger
【使作废】invalidate

驶 ①(开动) sail; drive: ～出港口 sail out the harbor ②(飞快地跑) (of a vehicle, etc.) speed: 疾～而过 speed by; fly past

矢 ①(箭) arrow: 飞～ flying arrow ②(发誓) vow; swear
【矢车菊】[植] cornflower
【矢口否认】flatly deny; deny by oath; deny stoutly
【矢量】[数][物] vector: 风～ wind vector / 切变～ shear vector ◇ ～分析 vector analysis / ～空间 vector space / ～图 vectogram
【矢如雨下】The arrows come down like a shower.
【矢志不移】one's resolve is unshaken; vow to adhere to one's chosen course

屎 ①(粪) excrement; feces; dung; droppings: 鸡～ chicken droppings / 牛～ cow dung / 拉～ empty the bowels; shit ②(眼、耳等的分泌物) secretion: 耳～ earwax / 眼～ eye discharge
【屎壳郎】[方] dung beetle
【屎粒化石】[地质] casting; fecal pellets

始 ①(开始) beginning; start: 自～至终 from beginning to end; from start to finish ②[书][副] only then; not... until
【始爆器】primer
【始发站】starting station
【始沸点】bubble point
【始末】beginning and end; the whole story: 事情的～ the whole story
【始业】the beginning of the school year: 秋季～。 The school year begins in autumn.
【始终】from beginning to end; from start to finish; all along; throughout: ～不懈 unremitting; untiring / 宴会～充满亲切友好的气氛。 The banquet was full of cordial and friendly atmosphere from beginning to end. ◇ ～标记 sentinel
【始终不渝】unswerving; steadfast; consistently: ～地坚持原则 consistently adhere to the principles
【始终如一】constant; consistent; persistent; (forever) true to oneself
【始祖】first ancestor; earliest ancestor
【始祖鸟】[古生物] archeopteryx

shì

室 room: 办公～ office / 会客～ reception room / 卧～ bedroom / 住 2105～ live in Room 2105
【室间隔缺损】[医] ventricular septal defect

【室内】indoor; interior: ～布线 house wiring / ～空调设备 room conditioning / ～溜冰场 indoor skating rink / ～游泳池 indoor swimming pool / ～运动 indoor sport / ～照明 interior illumination / ～植物 houseplant / ～装饰 interior decoration
【室内乐】[乐] chamber music
【室女座】[天] Virgo
【室如悬磬】One's house is like an empty jar hanging up. (或: One's house is quite bare. / be very poor / living in poverty)
【室外】outdoor; outside: ～活动 outdoor activities ◇ ～天线 open aerial

市 ①(市场) market: 菜～ vegetable market / 上～ be on the market; be in season ②(城市) city; municipality: ～中心 the heart of the city; city center; downtown
【市场】①(商品交易场所) market house; marketplace ②(商品行销区域) market; bazaar: 股票～ effect market / 国内～ domestic markets / 国内外～ domestic and foreign markets / 国外～ foreign markets / ～繁荣。The market is brisk. ◇ ～分析 market analysis / ～调查 market research / ～活跃 brisk market; active market / ～价格 market price / ～潜力 market potential / ～索赔 market claim / ～调节 market regulation
【市集】①(集市) fair ②(市镇) small town
【市价】market price
【市郊】suburb; outskirts: 在北京～ in the suburbs of Beijing
【市井】[书] marketplace; town: ～小人 philistine
【市侩】sordid merchant: ～习气 sordid merchants' ways; philistinism
【市面】market conditions; business: ～呆滞 sick market / ～坚定 firm market / ～繁荣。Trade is flourishing.
【市民】residents of a city; townspeople
【市内电话中心局】local junction circuit
【市内中继线路】local central office
【市区】city proper; urban district
【市容】the appearance of a city: 保持～整洁 keep the city clean and tidy / 参观～ go sight-seeing in the city; have a look around the city
【市委】municipal Party committee
【市长】mayor
【市镇】small towns; towns; market town
【市政】municipal administration: ～府 municipal government ◇ ～工程 [建] municipal works; municipal engineering

柿 persimmon
【柿饼】dried persimmon
【柿霜】[中药] powder on the surface of a dried persimmon
【柿子】persimmon
【柿子椒】sweetbell redpepper

铈 [化] cerium (Ce)

视 ①(看) look at: 注～ look at closely ②(看待) regard; look upon: ～如仇敌 look upon sb. as one's enemy / ～为莫大光荣 regard as a great honor / ～为知己 look upon sb. as one's bosom friend / ～为至宝 look upon sth. as a priceless treasure ③(考察) inspect; watch: 巡～ go on an inspection tour; go around and inspect
【视差】[物] parallax; optical parallax ◇ ～角 angle of parallax
【视察】inspect: ～边防部队 inspect a frontier guard unit
【视唱】sightsinging ◇ ～练耳 solfeggio
【视程】visual range
【视地平】[天] apparent horizon
【视而不见】look but see not; turn a blind eye to: ～,听而不闻 look but see not, listen but hear not
【视轨道】[天] apparent orbit
【视角】angle of view; visual angle
【视觉】[生理] visual sense; vision; sense of sight ◇ ～辨认 visuognosis / ～疲劳 visual fatigue / ～缺陷 defects of vision / ～误差 collimation error / ～像[心] visual image / ～印象[心] eye impressions / ～暂留 persistence of vision / ～障碍 dysopia; dysopsia
【视力】vision; sight: ～测验 eyesight test / ～差 have poor eyesight / ～好 have good eyesight ◇ ～表 visual chart
【视亮度】[天] apparent brightness
【视频】[物] video frequency ◇ ～鉴别 video discrimination / ～时差 video time-base / ～增益 video gain
【视如敝屣】regard as worn-out shoes; cast aside as worthless
【视如粪土】look upon as filth and dirt; consider as beneath contempt
【视若草芥】regard as worthless
【视若无睹】take no notice of what one sees; shut one's eyes to; turn a blind eye to; ignore

【视神经】[生理] optic nerve: ～萎缩 optic atrophy

【视事】(of officials) attend to business after assuming office; assume office

【视死如归】look upon death as going home; look death calmly in the face; face death unflinchingly

【视听】seeing and hearing; what is seen and heard: 混淆～ throw dust in people's eyes; confuse the public / 以正～ so that the public may know the facts; so as to clarify matters to the public

【视同儿戏】treat (a serious matter) as a trifle; trifle with

【视同路人】regard as a stranger

【视图】[机] view: 侧～ side view / 前～ front view / 上～ top view

【视网膜】[生理] retina ◇ ～镜 retinoscope; skiascope / ～脱离 detachment of retina / ～炎 retinitis

【视线】line of vision; line of sight: 挡住～ obstruct the view

【视星等】[天] apparent magnitude

【视野】field of vision: 广阔的～ a wide field of vision

【视阈】[生理] visual threshold

【视紫质】[生理] visual purple

式 ① (样式) type; style; model: 新～ new type; new style / 新～汽车 a car of new model / 中～服装 Chinese-fashion garments ② (格式) pattern; form: 程～ pattern; form to be copied ③ (仪式) ceremony; ritual: 开幕～ opening ceremony ④ (自然科学中的一组符号) formula: 分子～ molecular formula ⑤ [语] mood; mode: 叙述～ indicative mood

【式样】style; type; model: ～美观 beautiful in style; stylish / 各种～的服装 clothes in different styles / 汽车的最新～ the latest model of the car

【式子】① (姿势) posture ② (自然科学中的一组符号) formula

试 ① (试验;尝试) try; test; attempt: ～穿 try on (a garment, shoes, etc.) / 尝～ try one's hand at / 让我～一下。 Let me have a try. / 她～穿新衣。 She tried on her new dress. ② (考试) examination; test: 笔～ written examination / 口～ oral examination

【试办】run an enterprise, etc. as an experiment; run a pilot scheme

【试表】[口] take sb.'s temperature

【试唱】audition

【试车】[机] test run; trial run; gree test: 开机～ put the machine to a trial run

【试点】① (先做小型试验) make experiments; conduct tests at selected points; launch a pilot project ② (做小型试验的地方) a place where an experiment is made; experimental unit

【试电笔】[电] test pencil; screw-driver with voltage tester

【试飞】test flight; trial flight ◇ ～驾驶员 test pilot

【试管】[化] test tube ◇ ～架 test-tube stand / ～婴儿 test-tube baby

【试航】① (用作名词) trial trip: 船舶～ trial voyage; shakedown cruise / 飞机～ trial flight; shakedown flight ② (用作动词) shake down: 船舶～ shake down a ship / 飞机～ shake down a plain

【试剂】[化] reagent

【试金石】touchstone: 检验友谊的～ the touchstone to test friendship

【试卷】examination paper; test paper: 批阅～ go over examination papers

【试射】[军] fire for adjustment; trial fire

【试探】sound out; feel out; probe; explore: ～他们的意图 sound out their intention

【试探性】trial; exploratory; probing: ～攻击 probing attack / ～气球 trial balloon / ～谈判 exploratory talks

【试题】examination questions; test questions

【试跳】① (田径) trial jump ② (跳水) trial dive

【试图】attempt; try

【试问】we should like to ask; it may well be asked; may we ask

【试想】(用于委婉的质问) just think: ～他竟然对那件事一无所知! To think of his not knowing anything about it!

【试销】trial marketing; trial sale: ～商品 commodities for trial marketing ◇ ～专柜 trial sale counter

【试行】try out: ～生产 trial production / 先～, 再推广 first try out, then popularize ◇ ～条例 proposed regulations

【试选样品】[机] pilot model

【试验】trial; experiment; test: 水力～ hydraulic test ◇ ～板 breadboard / ～场 proving ground; testing ground / ～车 instruction carriage / ～成本 experimentation cost / ～费用 testing expenses / ～机 testing machine / ～片 [摄] test film; test piece / ～田[农] experimental

plot／～系统 pilot system
【试样】(test) sample
【试映】[电影] preview
【试用】on trial; try out; on probation: ～人员 person on probation ◇～本 edition put out to solicit comments; trial edition／～品 trial products／～期 probation period
【试运行】pilot run; test run
【试运转】[机] test run; running-in
【试纸】[化] test paper: 姜黄～ turmeric test paper／石蕊～ litmus test paper／万用～ universal test paper
【试制】trial-produce; trial-manufacture: ～工作 development work／～小组 trial-production group
【试种】plant experimentally: ～水稻 growing rice on a trial basis

拭 wipe away; wipe: ～泪 wipe away tears
【拭擦效应】wiping effect
【拭接】[冶] wiped joint
【拭目以待】wait and see; rub one's eyes and wait

弑 [书] murder: ～父 murder one's father／～君 murder one's sovereign

示 show; notify; express; instruct: 暗～ hint; drop a hint／出～证件 produce one's papers／告～ notice／请～ ask for instructions／以～关怀 as an expression of solicitude
【示波管】[电] oscilloscope tube
【示波器】[电] oscillograph; osciloscope
【示范】set an example; demonstrate: 进行～教学 teach by demonstration／起～作用 play an exemplary role ◇～飞行 demonstration flight
【示功器】[机] indicator
【示功图】[机] indicator card; indicator diagram
【示警】give a warning; warn: 鸣枪～ fire a warning shot; give a warning by firing a shot
【示例】give typical examples; give a demonstration
【示弱】give the impression of weakness; take sth. lying down: 不甘～ not to be outdone
【示威】①(表示抗议或要求) demonstrate; hold a demonstration ②(显示力量) put on a show of force; display one's strength ◇～游行 demonstration; parade; march
【示悉】Your letter has been received.(或: Yours to hand.)
【示意】signal; hint; motion: ～某人做某事 give sb. the tip to do sth.／以目～ give a hint with the eyes
【示意图】diagrammatic sketch; abridged general view; map of the exhibition; sketch map; schematic diagram: 发动机～ diagram of the engine／水利工程～ sketch map of the water conservancy project
【示众】publicly expose; put before the public: 游街～ parade sb. through the streets
【示踪物】[物] tracer
【示踪元素】[物] tracer element
【示踪原子】labelled atom; tagged atom; tracer

士 ①(士人;读书人) scholar ②(军人) noncommissioned officer (NCO): 上～ staff sergeant (英); sergeant first class (美)／下～ corporal／中～ sergeant ③(某些种技术人员) a person trained in a certain field: 护～ nurse ④(美称) (commendable) person: 烈～ martyr／勇～ brave fighter; warrior ⑤(棋子) bodyguard, one of the pieces in Chinese chess
【士别三日,刮目相看】A scholar who has been away three days must be looked at with new eyes.(或: After a scholar's absence of three days, one will see in him a man changed for the better.)
【士兵】rank-and-file soldiers; privates
【士大夫】*literati* and officialdom (in feudal China)
【士女】young men and women: ～如云。Men and women gathered like clouds.
【士气】morale: 鼓舞～ boost morale／～大振。The martial spirit has been roused greatly.／～低落。The morale of the troops is sinking lower.
【士可杀,不可辱】A scholar prefers death to humiliation.
【士绅】gentry
【士为知己者死】The scholar dies for his bosom friend.
【士卒】soldiers; privates: 身先～ (of an officer) fight at the head of his men; lead the charge

仕 ①(做官) be an official: fiil an office: 学而优则～。A good scholar will make an official. ②(棋子) bodyguard, one of the pieces in Chinese chess
【仕女】[美术] traditional Chinese painting of beautiful women
【仕途】[书] official career: ～沉浮 the ups and downs of an official career

恃 rely on; depend on: ～势 rely on one's position / 有～无恐 secure in the knowledge that one has strong backing

【恃才傲物】be inordinately proud of one's ability; be conceited and contemptuous

【恃德者昌】Those who rely on virtue will thrive.

【恃强凌弱】use one's strength to bully the weak

【恃势凌人】trust to one's power and insult people; use one's power to bully others

侍 wait upon; attend upon; serve: ～立一旁 stand at sb.'s side in attendance

【侍从】[旧] attendants; retinue ◇ ～副官 aide-de-camp (A.D.C.); aide

【侍奉】wait upon; attend upon; serve: ～父母 look after one's parents

【侍奉箕帚】perform one's wifely duties

【侍候】wait upon; look after; attend: 病人有护士～。 The patient has a nurse attending (on) him.

【侍女】maidservant; maid

【侍卫】imperial bodyguard

【侍者】[书] attendant; servant; waiter

螫 sting

【螫针】[动] sting; stinger

世 ①(人的一辈子) lifetime; life: 今生今～ this present life / 一生一～ one's whole life; all one's life ②(一代又一代) generation: ～谊 friendship spanning many generations ③(有世交关系) ～叔 younger friend of one's father / ～兄 son of one's friend ④(时代) age; era: 当今之～ at present; nowadays ⑤(世界) world: 举～闻名 well known all over the world; world-famous ⑥[地] epoch: 古新～ the Palaeocene Epoch

【世仇】①(世代冤仇) family feud ②(世代仇人) bitter enemy (in a family feud)

【世传】be handed down through generations: ～秘方 secret recipe handed down from generation to generation

【世代】①(好几辈子) for generations; from generation to generation; generation after generation: ～相传 pass on from generation to generation ②[生] generation ◇ ～交替 [生] alternation of generations; degenesis

【世道】the manners and morals of the time

【世风日下】The moral degeneration of the world is getting worse day by day.

【世故】①(处世经验) the ways of the world: 老于～ worldly-wise; versed in the ways of the world / 人情～ worldly wisdom ②(处事、待人圆滑) worldly-wise: 这人很～。This chap is quite a smooth character.

【世纪】century: 二十～九十年代 the nineties of the 20th century; 1990's / 公元前十～ the 10th century B.C. ◇ ～末 end of the century

【世家】aristocratic family; old and well-known family

【世交】①(两代以上的交谊) friendship spanning two or more generations ②(上代就有交情的人家) old family friends

【世界】world: ～大事 world events / 闻名～的 world-famous ◇ ～霸权 world domination; world supremacy / ～博览会 World's Fair / ～冠军 world champion / ～贸易 world commerce / ～市场 world market / ～舆论 world opinion

【世界大战】world war; war of global proportions: 第二次～ World War II; the Second World War

【世界观】world outlook

【世界纪录】world record: 创造～ set a world record / 打破～ break a world record

【世界时】[天] universal time

【世界水平】world standard; world caliber; international level: 赶超～ catch up with and surpass world levels

【世界语】Esperanto

【世界主义】cosmopolitanism

【世面】various aspects of society; society; world; life: 见过～ have seen the world; have experienced life

【世人】common people

【世上】in the world; on earth: ～无难事,只怕有心人。 Nothing in the world is difficult for one who sets his heart on it.

【世事】affairs of human life

【世俗】①(流俗) common customs: ～之见 common views ②(非宗教的) secular; worldly

【世态】the ways of the world: ～炎凉 inconstancy of human relationships

【世外桃源】the Land of Peach Blossoms; a fictitious land of peace, away from the turmoil of the world; a haven of peace

【世袭】hereditary: ～财产 hereditary property; patrimony / ～制度 the hereditary system

【世系】pedigree; genealogy

莳 ① [方](移植) transplant: ～秧 transplant rice seedlings ②[书](栽种) plant; cultivate: ～花 grow flowers

事 ①(事情) matter; affair; thing; business: 公～ public business / 私～ private business; private affair / 无关重要的～ an affair of no consequence / 重要的～ an affairs of great moment / 还有一件～我要问你。 There's another thing I want to ask you about. / 我不愿再管这～了。 I am tired of the whole business. / 这件～我不太知道。This is a matter I know little about. ②(事故) trouble; accident: 出～ have an accident / 惹～ make trouble; stir up trouble / 平安无～。All is well. ③(职业;工作) job; work: 找～ look for a job ④(关系或责任) responsibility; involvement: 这件案子里还有他的～呢。 He was involved in the case too. ⑤(侍奉) wait upon; serve: ～父母 wait upon one's parents ⑥(从事) be engaged in: 不～生产 lead an idle life / 无所～～ doing nothing; loafing

【事半功倍】get twice the result with half the effort

【事倍功半】get half the result with twice the effort

【事必躬亲】see to everything oneself; take care of every single thing personally

【事变】①(重大事件) incident ②(重大变化) emergency; exigency: 若发生～ in case of emergency ③(泛指事物的变化) the course of events; events: 研究周围～的联系 look into the relations of events occurring around one

【事不关己,高高挂起】To let things drift if they do not affect one personally.

【事不宜迟】one must lose no time in doing it; we must attend to the matter immediately; the matter brooks no delay

【事出有因】there is good reason for it; it is by no means accidental

【事到临头】when things come to a head; when the situation becomes critical; at the last moment

【事到如今】under the circumstances

【事端】disturbance; incident: 挑起～ provoke incidents / 制造～ create disturbances

【事非经过不知难】You never know how hard a task is until you have done it yourself.

【事故】 accident; mishap: 无～ without accident / 医疗～ unskillful and faulty medical or surgical treatment; malpractice / 责任～ accident arising from sb.'s negligence ◇ ～报告书 accident report / ～死亡 death by accident / ～损伤 casualty loss / ～损失赔偿 accident compensation

【事过境迁】the affair is over and the situation has changed; the incident is over and the circumstances are different

【事后】after the event; afterwards ◇ ～承诺 after consent / ～从犯 accessory after the fact / ～审计 past audit / ～追认 subsequent ratification

【事后诸葛亮】be wise after the event

【事迹】deed; achievement: 英雄～ heroic deeds

【事假】leave of absence (to attend to private affairs); compassionate leave: 请两小时～ ask for two hours leave of absence

【事件】 incident; event: 边境～ frontier incident / 流血～ bloody incident / 制造～ create incident / 准备应付一切意外～ be prepared for all contingencies

【事界】event borizon

【事理】reason; logic: 明白～ be reasonable; be sensible

【事例】example; instance: 典型～ a typical case

【事略】biographical sketch; short biographical account

【事前】 before the event; in advance; beforehand: ～毫无准备 with no preparation at all / ～准备好 be prepared beforehand ◇ ～从犯 accessory before the fact / ～故意 antecedent intent; intent before the fact / ～审计 pre-audit

【事情】affair; matter; thing; business: ～的真相 the truth of the matter / 亟待解决的～ affairs to be settled right away / ～的经过是这样的。This is how it happened. / 那只有使～更加复杂。 That will only make things more complicated.

【事实】fact: 与～不符 not tally with the facts / ～俱在。The facts are all there. / ～胜于雄辩。 Facts speak louder than words ◇ ～材料 factual materials / ～错误 error of fact; mistake of fact / ～的陈述 affirmation; allegation / ～含糊 ambiguity upon the factum / ～问题 question of fact / ～证据 factual evidence / ～证明 factual proof / ～真象 sober truth

【事实上】in fact; in reality; as a matter of fact; actually: ～的承认 *de facto* recognition ◇ ～的错误 factual mistake / ～的法人团体 corporation *de facto* / ～的默示合同 implied con-

tract in fact / ～的所有人 practical owner
【事事】everything
【事态】state of affairs; situation: ～严重。The situation is serious. / ～有所缓和。Things are smoothing down a bit.
【事无巨细】All matters, big and small, (were handled by him).
【事务】①(事情) work; routine: ～繁忙 have a lot (of work) to do ②(总务) general affairs ◇～性工作 routine work; daily routine / ～员 office clerk / ～主义 routinism / ～主义者 a person bogged down in routine matters
【事物】thing; object: ～的矛盾法则 the law of contradiction in things
【事先】in advance; beforehand; prior: ～策划的暗杀 a premeditated murder / ～做好准备 get everything ready beforehand
【事项】item; matter: 章程中规定的～ items stipulated in the regulations / 注意～ matters needing attention; points for attention
【事业】①(从事的活动) cause; undertaking: 革命～ revolutionary cause / 文化教育～ cultural and educational undertakings ②(非企业) enterprise; facilities: 公用～ public utilities / 集体福利～ collective welfare facilities or services ◇～单位 institution / ～费 operating expenses / ～心 devotion to one's work; dedication
【事宜】(多用于公文、法令) matters concerned; arrangements: 商谈技术合作～ discuss matters relating to technical cooperation
【事由】①(事情的原委) the origin of an incident; particulars of a matter ②(公文用语) main content
【事与愿违】things go contrary to one's wishes; thing do not always turn out the way one wishes
【事在人为】it all depends on human effort; it is the man who disposes
【事主】the victim of a crime

誓

①(决心照说过的话实行) swear; vow; pledge: ～夺丰收 pledge to strive for bumper harvest ②(表示决心的话) oath; vow: 发～ take an oath; swear by
【誓不罢休】swear not to stop; swear not to rest: 不达目的,～。We'll never give up until we reach our goal.
【誓不两立】swear not to coexist with one's enemy; resolve to destroy the enemy or die in the attempt; be irreconcilable
【誓词】oath; pledge
【誓师】①(军队出征前的动员) a rally to pledge resolution before going to war ②(群众集会表示决心) take a mass pledge ◇～大会 a meeting to pledge mass effort; an oath-taking rally
【誓死】pledge one's life; dare to die: ～保卫祖国 pledge to fight to the death in defending one's country
【誓言】oath; pledge: 立下～ take a vow; make a pledge / 履行～ fulfill a pledge
【誓约】vow; pledge; solemn promise

逝

①(过去) pass: 时光易～。Time passes quickly. ②(死) die; pass away: 病～ die of illness
【逝世】pass away; die: 纪念…～十周年 commemorate the 10th aniversary of the death of
【逝水年华】Time passes like flowing water.(或: The light and moonlight followed each other like flowing water.)
【逝者如斯】it passes like this; to sigh over what has passed and warn against the repetition

势

①(势力) power; force; influence: 权～ (a person's) power and influence / 以～压人 overwhelm others with one's power ②(事物表现出来的趋向) momentum; tendency: 来～甚猛 come with tremendous force ③(自然界的现象或形势) the outward appearance of a natural object: 地～ physical features of the land; terrain / 山～ the lie of a mountain ④(形势) situation; state of affairs; circumstances ⑤(姿态) sign; gesture: 摆姿～ pose / 作手～ make a sign with the hand ⑥(雄性生殖器) male genitals: 去～ castration
【势必】certainly will; be bound to: ～如此。It is bound to come like this. / 饮酒过度,～影响健康。Excessive drinking will undoubtedly affect one's health.
【势不可当】irresistible
【势不两立】mutually exclusive; extremely antagonistic; irreconcilable
【势成骑虎】Circumstances make it difficult for one to back down.
【势均力敌】match each other in strength: 一场～的比赛 a close contest
【势力】force; power; influence: 社会～ the social forces ◇～范围 sphere of influence
【势利】snobbish: ～小人 snob
【势利眼】①(作风势利) snobbish attitude;

snobbishness ②(势利之人) snob
【势流】potential barrier; potential motion of a fluid
【势难从命】Circumstances make it difficult for me to comply with your request.
【势能】[物] potential energy
【势如累卵】be in a very perilous position; hazardous like a pile of eggs
【势如破竹】like splitting a bamboo; like a hot knife cutting through butter; with irresistible force: ~,所向披靡 smash all enemy resistance and advance victoriously everywhere
【势头】①(情势) impetus; momentum ②[口](形势) tendency; the look of things: 见~不对 find that the situation is unfavorable
【势在必行】be imperative (under the circumstances)

是 ①(对;正确) correct; right: 实事求~ seek truth from facts/ 似~而非 apparently right but actually wrong ②(答应的词) yes; right: ~,我就来。 Yes, I'm coming right now. ③[书](这个) this; that: ~日天气晴朗。 It was fine and clear that day. ④(表示两种事物同一,或后者说明前者) be: 我~一个学生。 I am a student. ⑤(表示陈述的对象属于"是"后面所说的情况) 院子里~冬天,屋子里~春天。 It was winter outdoors, but spring indoors. ⑥(与"的"字相应,有分类的作用) 我~来看老李的。 I came to see Lao Li. ⑦(表示存在) be; exist: 宿舍前面~花园。 There is a garden in front of the dormitory. ⑧(表示让步) be..., but: 文章~好文章,就是长了点。 It is a good article all right, but it's a bit too long. ⑨(表示所说的几桩事物互不相干) be: 对~对,错~错,不能一团和气。 Right is right, wrong is wrong, one must not keep on the right side of everyone. ⑩(表示适合) the right: ~时候 at just the right time/ 不~地方 not in the right place ⑪(表示"凡是""任何") all; every; any: ~人总要犯错误。 Every man is liable to error. (或: No human being can be free from mistakes.) ⑫(重读,表示坚决肯定) certainly; really: 天气~冷。 It's really cold. ⑬(用于问句) 你~累了不~? You're tired, aren't you? ⑭(用在句首,加重语气) ~谁告诉你的? Who told you? ⑮(认为正确) praise; justify: ~古非今 praise the past to condemn the present
【是的】yes; right; that's it
【是非】①(事理的正确与错误) right and wrong: ~问题 a matter of right and wrong/ 明辨~ distinguish clearly between right and wrong/ ~自有公论。 The public will judge the rights and wrongs of the case. ②(口舌) quarrel; dispute: 搬弄~ tell tales; sow discord/ 惹起~ stir up trouble; provoke a dispute
【是非颠倒】confound right and wrong
【是非曲直】rights and wrongs; truth and falsehood; merits and demerits: 不问~ not bother to look into the rights and wrongs of a case
【是非之地】a place where one is apt to get into trouble
【是否】whether or not; whether; if: ~符合实际 whether or not it corresponds to reality
【是可忍,孰不可忍】If this can be tolerated, what cannot?

嗜 have a liking for; be addicted to: ~酒 be addicted to drink
【嗜好】hobby addiction; habit: 她的~是打网球。 Her hobby is playing tennis./ 他没有吸烟的~。 He has no habit of smoking.
【嗜痂成癖】have an addiction for drugs etc.; an uncommonly low teste
【嗜眠】somnolence
【嗜血】bloodthirsty; bloodsucking ◇ ~杆菌 hemophilus
【嗜血成性】kill without batting an eyelid; be fond of killing

释 ①(解释) explain; elucidate: ~义 explain the meaning (of a word, etc.) ②(消除) clear up; dispel: ~疑 clear up doubts ③(放开;放下) let go; be relieved of: ~手 loosen one's grip; let go/ 如~重负 (feel) as if relieved of a heavy load ④(释放) release; set free: ~俘 set prisoners free; release prisoners ⑤[简](释迦牟尼) Sakyamuni ⑥(佛教) Buddhism: ~典 Buddhist sutra
【释放】①(恢复被拘押者的人身自由) release; set free: 取保~ be released on bail; be bailed out/ 刑满~ be released upon completion of a sentence ②[物] release: ~出能量 release energy ◇ ~裁定书 rulings of acquittal/ ~程序 release procedure/ ~出狱令 order of release/ ~令 order for discharge; order of acquittal/ ~囚犯 discharge of prisoner; release a prisoner/ ~证书 release certificate
【释古行今】justify one's own wishes and desires by abandoning ancient
【释迦牟尼】Sakyamuni, the founder of Bud-

dhism
【释能过程】exoergic process
【释然】[书] feel relieved; feel at ease

噬 bite: 反～ make a false countercharge; hurl back an accusation／吞～ swallow up
【噬菌体】[生] bacteriophage; phage

适 ①(适合) fit; suitable; proper: ～于儿童观看的影片 a film suitable for children ②(恰好) right; opportune: ～量 just the right amount／～逢休假。It happened to be a holiday. ③(舒服) comfortable; well: 感到不～ not feel well／舒～ comfortable ④(去;往) go; follow; pursue: 无所～从 not know what course to pursue; be at a loss what to do
【适才】[方] just now
【适当】suitable; proper; appropriate: ～的安排 proper arrangement／～的场合 suitable occasion／在～的时机 at an opportune moment ◇～处罚 proper punishment／～措施 adequate measures／～的权限 due authority／～的约因 adequate consideration
【适得其反】run counter to one's desire; be just the opposite to what one wished
【适度】appropriate measure; moderate degree: 长短～ be the right length
【适逢其会】happen to be present at the right moment
【适航性】①(飞机的) airworthiness ②(船舶的) seaworthiness
【适合】suit; fit: ～他的口味 suit his taste; be to his taste／～做这项工作 be fit for the job
【适可而止】stop before going too far; know when or where to stop; not overdo it
【适口】agreeable to the taste; palatable
【适龄】of the right age: (入学)～儿童 children of school age
【适时】at the right moment; in good time; timely: ～的劝告 timely advice／～播种 begin sowing in good time
【适宜】suitable; fit; appropriate: 这水不～饮用。The water is not fit to drink.
【适意】agreeable; enjoyable; comfortable
【适应】suit; adapt; fit: ～人民购买力的提高 meet the people's rising purchasing power／～社会需要 fit in with the needs of the society ◇～不良 maladjustment／～范围 accommodation／～亮度 adaptation brightness／～性[生] adaptability／～性免疫 adaptive immunity／～症[医] indication
【适用】suit; apply to: ～范围 scope of application; sphere of application／勤俭节约的原则～于一切事业。The principle of diligence and frugality applies to all undertakings.
【适者生存】[生] survival of the fittest
【适值】just when: 昨日来访,～外出,憾甚。I called on you yesterday, but unfortunately you were out.
【适中】①(适度) moderate: 雨量～ moderate rainfall／大小～的房间 room of moderate size ②(位置不偏于哪一面) well situated: 地点～ be conveniently situated; be nicely located

似
【似的】[助](表示跟某种事物或情况相似)像雪～那么白 as white as snow

饰 ①(装饰) decorations; ornaments: 窗～ window decorations／服～ clothes and ornaments ②(修饰) adorn; dress up; polish; cover up: 把文章修～一下 polish a piece of writing／文过～非 cover up one's mistakes ③(扮演) play the role of; act the part of; impersonate: 他～老头儿很出色。He played the part of old people very well.
【饰词】excuse; pretext
【饰带】braid
【饰面墙】faced wall
【饰物】②(首饰) articles for personal adornment; jewelry ②(装饰品) ornaments; decorations

氏 ①(姓) family name; surname: 陈～定理 Chen's theorem／摄～温度计 Celsius thermometer／王～兄弟 the Wang brothers ②(旧称已婚妇女) *nѐe*: 夫人李～ one's wife, *nѐe* Li
【氏族】clan ◇～公社 clan commune／～社会 clan society／～制度 clan system

shōu

收 ①(收到;接受) receive; accept: ～到信 receive a letter／～发电报 transmitting and receiving telegrams ②(收拢) put away; take in: ～工具 put the tools away／他～到了礼物,但不肯～下。He received the gift, but he declened to accept it. ③(收取) collect: ～税 collect taxes ④(收入) money received; receipts; income: 税～ tax revenue ⑤(收获) harvest; gather in: 秋～ autumn harvest ⑥(收缩) close: 伤～口了。The wound has healed. ⑦(结束;停止) bring to an end; stop: ～操 bring drill to an end / 时间不早了,今天就～了吧。It's getting late. Let's call it a day. ⑧(约束;控制) restrain; control: ～心 concentrate on more serious things

【收报机】telegraphic receiver; radiotelegraphic receiver

【收编】incorporate into one's own forces

【收兵】withdraw troops; call off a battle

【收藏】collect; store up: ～古画 collect old paintings / ～艺术珍品 collect art curiosities ◇ ～家 collector (of books, antiques, etc.)

【收场】①(结束) wind up; end up; stop: 草草～ wind up a matter hastily or perfunctorily ②(结局;下场) end; ending; denouement: 圆满的～ a happy ending / 戏快～了。 The play is coming to an end.

【收成】harvest; crop: ～不好 poor harvests; crop failures / 获得好～ reap a good harvest

【收存】receive and keep

【收到】receive; get; achieve; obtain: ～良好效果 achieve good results / ～一封信 receive a letter

【收发】①(收进和发出公文) receive and dispatch ②(担任收发工作的人) dispatcher ◇ ～室 office for incoming and outgoing mail

【收发报机】transmitter-receiver; transceiver

【收方】[簿记] debit; debit side

【收费】collect fees; charge: 不～ free of charge

【收费电视】pay TV

【收复】recover; recapture: ～失地 recover lost territory

【收割】reap; harvest; gather in: ～小麦 gather in the wheat ◇ ～机 harvester; reaper

【收工】stop work for the day; knock off; pack up: 该～了。It's time to knock off.

【收购】purchase; buy: ～农副产品 purchase farm produce and sideline products ◇ ～价格 purchasing price / ～站 purchasing station

【收归国有】be taken over by the state; be nationalized

【收回】①(取回) take back; call in; regain; recall: ～贷款 recall loans / ～投资 recoup capital outlay / ～主权 regain sovereignty ②(撤销) withdraw; countermand: ～成命 countermand an order; revoke a command / ～建议 withdraw a proposal

【收货凭单】consignment sheet

【收货人】consignee: ～地址 consignee's address

【收获】①(取得成熟的农作物) gather in the crops; harvest: ～量 harvest yield / 春天播种,秋天～ sow in spring and reap in autumn ②(成果) results; gains; fruits: 一次很有～的访问 a most rewarding visit / 不下苦功,便无～。No pains, no gains.

【收集】collect; gather: ～民歌 collect folk songs / ～资料 gather data

【收监】take into custody; put in prison; committed to prison

【收件人】addressee; consignee

【收缴】take over; capture: ～敌人的武器 take over the enemy's arms

【收紧】tighten up

【收据】receipt ◇ ～簿 receipt book

【收口】①(伤口愈合) close up; heal ②(编织物收口) binding off

【收款人】payee

【收敛】①(减弱或消失) weaken or disappear: 他～了笑容。 Her smile vanished. ②(约束言行) restrain oneself: 碰了钉子以后,他～些了。He has pulled in his horns since that setback. ③ [数] convergence ④ [医] astringent ◇ ～剂 astringent

【收殓】lay a body in a coffin

【收留】take sb. in; have sb. in one's care: ～孤儿 take in an orphan

【收拢】draw sth. in: 把网～ draw the net in

【收录】①[旧](吸收任用) employ; recruit; take on: ～几个职员 recruit some office workers ②(编集子时采用) include: 这本小说集～了他的作品。 His works are included in this anthology of stories. ③(录音) listen in and take down; take down; record: ～北京电台的英语广播 take down the English broadcasting from Radio Beijing ◇ ～两用机 radio cassette player

【收罗】collect; gather; enlist: ～人才 recruit qualified personnel / ～资料 collect data

【收买】①(收购) purchase; buy in: ～旧书 buy used books ②(笼络人心) buy over; bribe: ～人心 buy popular support ◇ ～赃物者 smasher / ～证人 tampering with witness

【收盘】[经] closing quotation (on the exchange, etc.) ◇ ～汇率 closing rate / ～价格 closing price

【收票员】ticket collector

【收起】pack up; cut out; stop: 你这些空话还是～为好。 You'd better stop this empty talk.

【收讫】①(收清) payment received; paid ②(如数收到) (on a bill of lading, an invoice, etc.) all the above goods received; received in full

【收清】received in full

【收容】take in; accept; house: ～难民 house

refugees ◇ ～所 collecting post; detention center

【收入】①(收进来的钱) income; revenue; receipts; earnings; proceeds: ～和支出 receipts and expenditures / 财政～ state revenue / 副业～ income from sideline occupations ②(收进) take in; include: 修订版～许多新词语。Many new words and phrases have been included in the revised edition. ◇ ～分类帐 receipts ledger / ～退还书 income refund notice / ～总额 gross income

【收拾】①(整理) put in order; tidy; clear away: ～残局 clear up a messy situation / ～屋子 tidy up the room ②(准备) get things ready; pack: ～行李 pack one's luggage; pack up one's things ③(修理) repair; mend: ～鞋子 mend shoes ④[口](整治) settle with; punish: 不出这个星期我们要～他。We'll settle with him before the week's out.

【收束】①(约束) bring together; collect: 把心思～一下 get into the frame of mind for work ②(结束) bring to a close ③(收拾) pack (for a journey)

【收缩】①(由大变小,由长变短) contract; shrink: 大多数金属遇冷就～。Most metals contract when they get cooler. ②(紧缩) concentrate one's forces; draw back ③[生理] systole ◇ ～压 systolic pressure

【收条】receipt

【收听】listen in: ～天气预报 listen in to the weather forecast / ～新闻广播 listen to the news broadcast

【收尾】①(结束) wind up: ～工作 winding up ②(文章末尾) ending (of an article, etc.)

【收文】incoming dispatches ◇ ～簿 register of incoming dispatches

【收效】yield results; produce effects; bear fruit: ～甚微 produce very little effect / ～显著 bear rich fruit; produce notable effect

【收心】①(把散漫的心收起来) get into the frame of mind for work; concentrate on more serious things ②(改变做坏事的念头) have a change of heart

【收信人】the recipient of a letter; addressee

【收押】take into custody; detain

【收养】take in and bring up; adopt: ～孤儿 adopt an orphan ◇ ～法与遗嘱法 law of adoption and of wills / ～关系 adoptive relationship / ～继承人 heir by adoption / ～人 consignee

【收益】income; profit; earnings; gains; avails ◇ ～债券 income bonds / ～帐户 income account

【收音】①(接收无线电广播的) reception: ～情况良好。Reception is good. ②(集中音波) (of an auditorium, etc.) have good acoustics ◇ ～电唱两用机 radiogramophone

【收音机】radio (set); wireless (set): 落地式～ console set

【收支】revenue and expenditure; income and expenses: ～平衡。Revenue and expenditure are balanced ◇ ～逆差 balance of payments deficit / ～预算 budget for revenues and expenditures / ～帐目 income and expenditure account

【收执】①(收下并保存) (of a certificate, etc.) be issued to the person concerned for safekeeping ②(凭证) receipt (issued by a government agency)

shǒu

守 ①(防守) guard; defend: ～城 defend a city / ～球门 keep goal / ～土有责 be duty-bound to defend the territory of one's country ②(守候;看护) keep watch: ～了一夜 keep watch for the whole night ③(遵守) observe; serve; abide by: ～规矩 behave well / ～纪律 observe discipline / ～信用 keep one's promise; be as good as one's word ④(靠近) close to; near: ～着这样好的老师,你为什么不学英语? Why don't you learn English when you have such a good teacher at your side?

【守备】perform garrison duty; be on garrison duty; garrison ◇ ～部队 garrison force; (holding) garrison

【守财奴】miser

【守场员】[棒、垒球] fielder

【守车】[铁道] guard's van; caboose

【守法】abide by the law; be law-abiding; keep the law

【守寡】remain a widow; live in widowhood

【守恒】conservation

【守恒定律】[物] conservation law: 能量～ the law of conservation of energy

【守候】①(等待) wait for; expect: ～机会 wait for one's opportunity ②(看护) keep watch: ～在病人身旁 keep watch by the patient's bedside

【守护】guard; defend ◇ ～神 [宗] patron saint

【守节】preserve chastity after the death of her husband; not remarry

【守旧】adhere to past practices; stick to old ways; be conservative ◇ ～派 old liners

【守军】defending troops; defenders
【守口如瓶】keep one's mouth shut; breathe not a single word; be tight-mouthed
【守垒员】[棒、垒球] baseman
【守灵】stand as guards at the bier; keep vigil beside the coffin
【守门】①(看门) be on duty at the door or gate ② [体] keep goal ◇ ～员 goalkeeper
【守势】defensive: 采取～ be on the defensive
【守岁】stay up late or all night on New Year's Eve
【守土】[书] defend the territory of one's country
【守望】keep watch ◇ ～台 watchtower
【守望相助】(of neighboring villages) keep watch and help defend each other; give mutual help and protection
【守卫】guard; defend
【守信】keep one's word; abide by one's word
【守业】maintain what has been achieved by one's forefathers or predecessors; safeguard one's heritage
【守夜】keep watch at night; spend the night on watch
【守则】rules; regulations: 工作～ work regulations / 学生～ students regulations
【守株待兔】stand by a stump waiting for more hares to come and dash themselves against it; trust to chance and windfalls

首 ①(头) head: 昂～ held one's head high / 搔～ scratch one's head ②(第一) first: ～批 the first batch ③(首领) leader; head; chief: 祸～ chief culprit ④(出头告发) bring charges against sb.: 出～ inform against sb. ⑤ [量] 诗一～ a poem / 一～歌 a song
【首倡】initiate; start: ～精神 initiative / ～者 initiator
【首车】first bus
【首创】initiate; originate;l pioneer: ～精神 creative initiative; pioneering spirit / ～一种学说 initiate a doctrine
【首次】for the first time; first: ～航行 maiden voyage / ～上映的影片 first run film; first showing of a film
【首当其冲】be the first to be affected (by a disaster, etc.); bear the brunt
【首都】capital
【首恶】chief criminal; principal culprit: ～必办，胁从不问，立功受奖。 The chief criminals shall be punished without fail, those who are accomplices under duress shall go unpunished and those who perform deeds of merit shall be rewarded.
【首犯】arch-criminal; chief criminal; main culprit
【首府】① [旧](省会所在地) the prefecture where the provincial capital is located ②(自治区、自治州政府所在地) an elderly the capital of an autonomous region or prefecture ③(附属国、殖民地最高当局所在地) the capital of a dependency or colony
【首肯】nod approval; nod assent; approve; consent
【首领】chieftain; leader; head
【首脑】head: 政府～ head of government ◇ ～会议 conference of heads of state or government; summit conference / ～人物 leading figure
【首屈一指】come first on the list; be second to none
【首任】the first to be appointed to an office: ～驻华大使 the first ambassador accredited to China
【首饰】(woman's personal) ornaments; jewelry ◇ ～盒 jewel case
【首鼠两端】be in two minds; shilly-shally
【首尾】①(起头部分和末尾部分) the head and the tail; the beginning and the end: ～不能相顾。 The vanguard is cut off from the rear. ②(始终) from beginning to end
【首位】the first place: 放在～ put in the first place; place before everything else; give first priority to
【首席】①(最高的席位) seat of honor: 坐～ be seated at the head of the table; be in the seat of honor ②(职位最高的) chief ◇ ～代表 chief representative / ～法官 chief judge; chief justiciar; chief justice / ～检察官 chief inspector; chief procurator / ～监事 chief supervisor / ～仲裁员 chief arbitrator
【首先】①(最先) first: ～发言 speak first ②(第一) in the first place; first of all; above all
【首相】prime minister
【首要】of the first importance; first; chief: ～任务 the most important task / ～原因 first cause / ～战犯 chief war criminal
【首战告捷】①(第一仗获胜) win the first battle ② [体] score a victory in the first game
【首长】leading cadre; senior officer: 团～ senior officers of the regiment

手 ①(人的手) hand: ～拉～ hand in hand / 伸出友谊之～ stretch out the hand of friendship / 握某人的手 shake sb.'s hand / 与某人～ shake hands with sb. ②(拿着) have in one's hand; hold: 人～一册。Everyone has a copy. ③(方便的) handy; convenient: ～册 handbook ④(亲手) personally: ～植 personally plant (a tree, etc.) ⑤(擅长某种技能的人) a person doing or good at a certain job: 能～ a skilled hand ⑥ [量](用于技能、本领)他真有两～。He really knows his stuff. ◇一～交钱,一～交货 cash on delivery

【手背】the back of the hand

【手臂】arm

【手笔】①(写的字) sb.'s own handwriting or painting ②(文字技巧的造诣) literary skill: 大～ a well-known writer; master

【手边】on hand; at hand

【手表】wrist watch

【手不释卷】always have a book in one's hand; be very studious

【手册】handbook; manual: 教师～ teacher's manual

【手抄本】hand-written copy

【手车】handcart; wheelbarrow

【手戳】[口] private seal; signet

【手到病除】bring back life to a patient; illness departs at a touch of the hand

【手倒立】[体] handstand

【手电筒】electric torch; flashlight

【手段】①(方法) means; medium; measure; method: 不择～ by fair means or foul; by hook or by crook; unscrupulously / 高压～ high-handed measures / 强制～ coercive method; coercion ②(不正当的方法) trick; artifice: 采用种种～ resort to all sorts of tricks;use every artifice / 耍～ play tricks

【手法】①(技巧) results; produce; technique: 独特的～ unique technique / 艺术的表现～ means of artistic expression ②(不正当的方法) trick; gimmick: 惯用～ habitual practice

【手风琴】accordion: 六角～ concertina

【手扶拖拉机】walking tractor

【手感】[纺] feel; handle

【手稿】original manuscript; manuscript

【手工】①(手做的工作) handwork: ～费 payment for a piece of handwork / 做～ do handwork ②(用手操作) by hand; manual: ～编织的 hand-knitted / ～操作 done by hand; manual operations ③[口](手工劳动的报酬) charge for a piece of handwork: 你这件上衣～多少? How much did you pay for the tailoring of this coat?

【手工业】handicraft industry; handicraft ◇～者 handicraftsman

【手工艺】handicraft art; handicraft ◇～工人 craftsman; artisan / ～品 articles of handicraft art; handicrafts

【手鼓】[乐] a small drum similar to the tambourine

【手迹】sb.'s original handwriting or painting

【手疾眼快】quick of eye and deft of hand

【手脚】①(动作) movement of hands or feet; motion: ～不干净 sticky-fingered; questionable in money matters / ～利落 nimble; agile ②[方](采取的行动) underhand method; trick: 一定是有人从中弄～。Someone must have juggled things.

【手紧】closefisted; tightfisted

【手劲儿】muscular strength of the hand

【手巾】towel ◇～架 towel rack

【手锯】handsaw

【手卷】[美术] hand scroll

【手绢】handkerchief

【手铐】handcuffs: 戴上～ be handcuffed

【手快】deft of hand: 眼明～ quick of eye and deft of hand

【手拉葫芦】[机] chain block

【手雷】[军] antitank grenade

【手力千斤顶】[机] hand jack

【手榴弹】hand grenade; grenade: 催泪～ tear gas grenade / 铝热剂～ thermite grenade / 燃烧～ incendiary grenade / 杀伤～ fragmentation grenade

【手炉】handwarmer

【手轮】[机] handwheel

【手忙脚乱】running around in circles; in a frantic rush; in a muddle

【手民】[书] typesetter: ～之误 misprint; typographical error

【手帕】[方] handkerchief

【手胼足胝】callosities found both on one's hands and feet; to have been working hard

【手旗】[军] handflag; semaphore flag

【手气】luck at gambling, card playing, etc.

【手钳】hand vice; pliers

【手枪】pistol: 标准～ standard pistol / 气～ air pistol / 自动～ automatic pistol / 自选～ free pistol / 左轮～ revolver ◇～弹创 pistol wound / ～慢加速比赛 [体] center-fire pistol / ～速射 [体] rapid-fire pistol / ～套 holster

【手巧】skillful with one's hands; deft; dexterous: 心灵～ clever and deft
【手勤】diligent; industrious; hardworking: 这徒弟～脚快。This apprentice is keen and quick in his work.
【手轻】not use too much force; handle gently
【手球】handball ◇～场 handball field; rink / ～运动员 handball player
【手软】be irresolute when firmness is needed; be softhearted
【手刹车】hand brake
【手势】gesture; sign; signal: 打～ make a gesture; gesticulate ◇～语 sign language
【手书】①(亲笔书写) write in one's own hand ②(手札) personal letter: 顷接～。I have just received your letter.
【手术】surgical operation; operation: 大～ major operation / 高频电刀～ high frequency electrotomy / 小～ minor operation / (病人)动～ undergo an operation / (医生)动～ perform an operation / 接受～ undergo operation / 由李医生作～ be operated under Dr. Li / 医生们决定立刻动～。The doctors decided to operate on the patient once.
【手术刀】scalpel: 激光～ laser scalpel ◇～刀包 surgical kit
【手术灯】operating lamp
【手术室】operating room; operating theatre
【手术台】operating table: 万能～ universal operation table
【手松】freehanded; openhanded
【手套】gloves; mittens: 皮里～ fur-lined gloves / 无指～ mitten
【手提】portable ◇～包 handbag; bag / ～打字机 portable typewriter / ～箱 suitcase / ～行李 hand baggage
【手头】①(伸手可拿到的地方) right beside one; on hand; at hand: ～工作挺多 have a lot of work on hand ②(经济状况) one's financial condition at the moment: ～紧 be short of money; be hard up / ～宽裕 be in easy circumstances; be quite well off at the moment
【手推车】handcart; wheelbarrow
【手腕】artifice; finesse; stratagem: 耍～ play tricks; use artifices
【手腕子】wrist
【手纹】lines of the hand ◇～鉴定 verification of hand-print
【手无寸铁】bare-handed; unarmed; defenseless
【手无缚鸡之力】lack the strength to truss up a chicken
【手舞足蹈】dance for joy
【手下】①(领导之下) under the leadership of; under: 在他～当技术员 be a technician under him ②(手头) at hand: 东西不在～。I haven't got the thing with me. ③(下手的时候) at the hands of sb.: ～败将 one's vanquished foe; one's defeated opponent
【手下留情】show mercy; be lenient
【手写体】handwritten form; script
【手心】①(手掌心) the palm of the hand ②(控制范围) control: 敌人逃不出我们的～。The enemy cannot escape from our control.
【手续】procedures; formalities: ～不完备 have not completed the formalities / 办～ go through formalities / 法律～ legal formalities ◇～费 service charge; commission / ～完备的保释 perfecting bail
【手癣】[医] tinea manuum; fungal infection of the hand
【手眼通天】exceptionally adept in trickery
【手痒】one's fingers itch; have an itch to do sth.
【手摇泵】hand pump
【手摇发电机】[电] hand generator
【手摇曲柄钻】brace and bit
【手艺】①(指技术) craftsmanship; workmanship: ～高 be highly skilled ②(手工艺) handicraft; trade: 学一门～ learn a trade ◇～人 craftsman
【手淫】masturbation
【手印】①(手留下的痕迹) an impression of the hand ②(印在契约上的指纹) thumb print; fingerprint
【手语】sign language; dactylology
【手札】[书] personal letter
【手掌】palm
【手杖】walking stick; stick
【手指甲】finger nail
【手纸】toilet paper
【手指】finger
【手重】use too much force; heavy-handed
【手镯】bracelet
【手足】brothers: ～之情 brotherly affection
【手足无措】all in a fluster; at a loss what to do; be bewildered
【手钻】hand drill

shòu

瘦 ①(肉少) thin; emaciated: ～得皮包骨头 be only skin and bones / 脸～ be thin in the face / 面黄肌～ be sallow and emaciated ②

(脂肪少) lean: ～肉 lean meat ③ (窄小) tight: 这件上衣他穿～了一点。 This jacket is a little too tight for him. ④ (不肥沃) not fertile; poor: ～土薄田 poor soil and barren land
【瘦长】long and thin; tall and thin; lanky: 他是～个儿。 He's a tall, lean chap.
【瘦果】[植] achene
【瘦骨嶙峋】become as emaciated as a fowl; a bony appearance
【瘦煤】lean coal; meager coal
【瘦弱】thin and weak; emaciated
【瘦小】thin and small: 身材～ slight of figure
【瘦型体质】ectomorph; ectomorphy
【瘦削】very thin; gaunt: ～的面孔 a haggard face
【瘦子】a lean person

兽 ① (哺乳动物的通称) beast; animal: 野～ wild animal ② (野蛮) beastly; bestial: 人面～心 a beast in human shape
【兽炭】[化] animal charcoal
【兽环】door-knocker
【兽类】beasts; animals
【兽力车】animal-drawn vehicle
【兽王】the king of beasts
【兽行】brutal act; brutality
【兽性】brutish nature; barbarity
【兽医】veterinary surgeon; veterinarian; vet ◇ ～学 veterinary medicine; veterinary science / ～站 veterinary station
【兽欲】animal desire

寿 ① (长命) longevity ② (生命;年岁) life; age: 长～ long life; longevity ③ (寿辰) birthday: 祝～ congratulate sb. on his birthday ④[婉](生前预备的) for burial: ～木 coffin (prepared before one's death)
【寿比南山,福如东海】May your age be as Mountain Tai and your happiness as the Eastern Sea
【寿材】a coffin prepared before one's death; coffin
【寿辰】birthday (of an elderly person): 八十～ eightieth birthday
【寿礼】birthday present (for an elderly person)
【寿面】noodles eaten on one's birthday; birthday noodles
【寿命】life-span; life: 机器的～ service life of a machine / 平均～ average life-span ◇ ～试验 length of life test / ～性能 life performance / ～周期 life cycle
【寿桃】peaches offered as a birthday present; (peach-shaped) birthday cake
【寿星】① (老人星) the god of longevity ② (被祝寿的人) an elderly person whose birthday is being celebrated
【寿衣】graveclothes; shroud; cerements
【寿终正寝】die in bed of old age; die a natural death

受 ① (接受) receive; accept: ～教育 receive an education / ～礼 accept gifts ② (遭受) suffer; be subjected to: ～法律制裁 be dealt with according to law / ～损失 suffer losses ③ (忍受) stand; endure; bear: ～不了 cannot bear; be unable to endure / 你～得住这疼痛吗? Can you stand the pain? / 真够～的。 It's really unbearable. ④ [方](适合) be pleasant: ～看 be pleasant to look at / ～听 be pleasant to hear
【受病】catch a disease; fall ill
【受潮】be affected with damp; become damp: 切勿～! Keep from moisture! / 营地的补给品在漫长的雨季里～了。 The camp supplies dampened during the long rainy season.
【受宠若惊】be overwhelmed by an unexpected favor; feel extremely flattered
【受挫】be foiled; be baffled; be thwarted; suffer a setback
【受罚】be punished
【受粉】[植] be pollinated
【受害】suffer injury; fall victim; be affected: ～不浅 suffer not a little; suffer a lot ◇ ～国 the country that's been wronged; victimized country / ～人 aggrieved party / ～外籍人 aggrieved alien / ～者 victim; suffer
【受寒】catch a chill; catch cold
【受话器】(telephone) receiver
【受欢迎】be well received; be well liked
【受欢迎的人】[外] persona grata: 不～ persona non grata
【受贿】accept bribes ◇ ～行为 act of accepting bribe / ～者 bribee; grafter / ～罪 acceptance of bribes
【受惠国】favored nation
【受奖】be rewarded: 立功者～。 Those who perform deeds of merit shall be rewarded.
【受戒】[佛教] be initiated into monkhood or nunhood
【受尽】suffer enough from; suffer all kinds of; have one's fill of: ～千辛万苦 undergo all kinds of hardships
【受惊】be frightened; be startled

【受精】be fertilized: 体内～ internal fertilization／ 体外～ external fertilization／异体～ cross-fertilization／自体～ self-fertilization ◇ ～卵 zygote
【受窘】be embarrassed; be in an awkward position
【受苦】suffer (hardships); have a rough time: ～受难 live in misery; have one's fill of sufferings
【受累】①(受到牵累) get involved on account of sb. else ②(受劳累) be put to much trouble; be inconvenienced: 让您～了。 Sorry to have given you so much trouble.
【受理】[法] accept and hear a case ◇ ～案件 accept and hear a case／～控诉 accept a complaint／～申诉 hear a claim／～诉状 hear a complaint; receive a petition
【受凉】catch cold; get a cold
【受命】receive instructions
【受难】suffer calamities or disasters; be in distress: 战争～者 war victim
【受骗】be deceived; be taken in
【受气】be bullied; suffer wrong ◇ ～包儿 a person whom anyone can vent his spite upon; one who always gets blamed
【受权】be authorized: ～发表声明如下 be authorized to issue the following statement／～宣布 announce upon authorization
【受热】①(受高温影响) be heated: 物体～则膨胀。 When matter is heated, it expands. ②(受暑) be affected by the heat; have heatstroke; suffer from sunstroke
【受辱】be insulted; be disgraced; be humiliated
【受伤】be injured; be wounded; sustain an injury
【受审】stand trial; be tried; be on trial ◇ ～人 person on trial
【受事】[语] the object of the action in a sentence; object
【受暑】suffer from heatstroke
【受损程度】extent of damage
【受胎】become pregnant; be impregnated; conceive ◇ ～率 [牧] conception rate
【受托】be commissioned; be entrusted (with a task): ～办理某事 be entrusted with sth.／～照看房子 be entrusted with the care of a house ◇ ～人 bailee; holder on trust; fiduciary
【受委曲】be upset by some unkindness
【受洗】[基督教] be baptized; receive baptism
【受降】accept a surrender
【受血者】receptor; donee; recipient
【受刑】be tortured; be put to torture
【受训】receive training
【受讯问】subject to cross-examination
【受益】profit by; benefit from; be benefited ◇ ～人 [法] beneficiary
【受遗赠能力】[法] capacity of legacy
【受遗赠人】[法] devisee; legatee
【受用】①(享受) benefit from; profit by; enjoy: ～不尽 benefit form sth. all one's life ②[方](身心舒服) feel comfortable: 今天身体有点不～。 I feel a bit under the weather today.
【受援】receive aid ◇ ～国 recipient country
【受孕】become pregnant; be impregnated; conceive
【受灾】be hit by a natural adversity: ～地区 disaster area; stricken area／严重～地区 badly stricken area
【受之无愧】deserve it; be worthy of it
【受之有愧】receive it with shame; I don't deserve it.
【受主】[电子] acceptor
【受罪】endure hardships, tortures, rough conditions, etc.; have a hard time.

授 ①(交付;给予) award; vest; confer; give: ～以全权 vest sb. with full authority ②(传授) teach; instruct: 函～ teach by correspondence; give a correspondence course
【授粉】[植] pollination: 人工～ artificial pollination／异花～ cross-pollination／自花～ self-pollination
【授计】confide a stratagem to sb.; tell sb. the plan of action
【授奖】award a prize ◇ ～仪式 prize-giving ceremony
【授精】insemination: 人工～ artificial insemination
【授课】give lessons; give instruction
【授命】①(下命令) give orders: ～组阁 authorize sb. to form a cabinet ②[书](献出生命) give one's life
【授旗】present (sb. with) a flag
【授权】empower; authorize; warrant: 经～的官员 authorized officer／谁～你做这件事? Who gave you authority to do this?／我～他在我不在的时候代理我的职务。 I have authorized him to act for me during my absence. ◇ ～范围 scope of authority／～立法 delegated legislation／～令 warrant／～行为 act of authorization／～证书 certificate of authority; vesting instrument

【授时】①[天](报告最精确的时间) time service ②(颁行历书) (in former times) issue the official calendar ◇ ～信号 time signal
【授受】grant and receive; give and accept: ～不亲 no physical contact between a man and a woman except between man and wife / 私相～ give and accept in private; illegally pass things between individuals
【授勋】confer orders or medals; award a decoration
【授意】incite sb. to do sth.; inspire; suggest: 某人～写的文章 inspired articles / 他这样干,是谁～的? Who got him to do that?
【授予】confer; award; grant; endow: 被～…称号 be given the title of... ◇ ～权 gift / ～人 grantor / ～物 grant

绶

【绶带】ribbon (attached to an official seal or a medal)
【绶带鸟】paradise flycatcher

售 ①(卖) sell: ～完 be sold out / 出～ put on sale / 按打出～ sell by the dozen / 亏本出～ sell at a loss / 廉价出～ sell at a bargain ②[书](施展) make (one's plan, trick, etc.) work; carry out (intrigues): 以～其奸 achieve one's treacherous purpose
【售货】sell goods ◇ ～机 vending machine / ～亭 kiosk
【售货员】shop assistant; salesclerk: 女～ saleswoman; salesgirl; shopgirl
【售价】selling price; price
【售票处】①(火车站的) ticket office; booking office ②(剧院等的) box office
【售票口】wicket
【售票员】①(公共汽车的) ticket seller; conductor ②(火车站的) booking office clerk ③(剧院等的) box office clerk

狩[书] hunting (esp. in winter)
【狩猎】hunting ◇ ～专业队 professional hunting team

shū

梳 ①(梳子) comb: 电热～ electric heat comb / 木～ wooden comb ②(梳理) comb: ～头发 comb one's hair
【梳理】[纺] carding: ～机 carding machine
【梳棉机】[纺] carding machine
【梳洗】wash and dress ◇ ～用具 toilet articles
【梳妆】dress and make up: ～打扮 deck oneself out; dress smartly; be dressed up ◇ ～台 dressing table
【梳子】comb

疏 ①(疏通) dredge (a river, etc.) ②(疏落) thin; sparse; scattered: ～林 sparse woods ③(关系远) (of family or social relations) distant: 不分亲～ regardless of relationship (family or social) ④(不熟悉) not familiar with: 人地生～ be unfamiliar with the place and the people; be a complete stranger ⑤(疏忽) neglect: ～于职守 negligent of one's duties ⑥(空虚) scanty: 才～学浅 have little talent and less learning / 志大才～ have great ambition but little talent ⑥(分散) disperse; scatter: 仗义～财 be generous in aiding needy people
【疏不间亲】A merest acquaintance should not say anything to estrange somebody from one who is dear to him.
【疏导】dredge
【疏果】[农] fruit thinning
【疏忽】carelessness; negligence; oversight: 我一时～,搞错了。 I made the mistake through an oversight.
【疏忽大意】be neglectful and careless; inattentive and heedless: ～的过失 [法] careless and inadvertent negligence
【疏花】[农] flower thinning
【疏浚】dredge: ～港口 dredge a harbor / 航道 dredge a channel / ～水道 dredge the waterways
【疏开】[军] extend; disperse; deploy: ～队形 dispersed formation; extended order; open order
【疏漏】careless omission; slip; oversight
【疏落】sparse; scattered: ～的村庄 scattered villages
【疏密】density; spacing: ～不匀 of uneven density
【疏散】①(疏落) sparse; scattered; dispersed: ～的村落 scattered villages ②(分散) evacuate: ～人口 evacuate inhabitants
【疏失】careless mistake; remissness: 这是我的～。 That's my fault.
【疏松】①(松散) loose: 土质～。 The soil is porous. ②(使松散) loosen: ～土壤 loosen the soil
【疏通】①(疏浚) dredge: ～河流 dredge a river / ～田间的排水沟 dredge the irrigation ditches in the fields ②(沟通;调解) mediate between two parties

【疏相】lean phase
【疏远】drift apart; become estranged: 他和朋友们～了。 He became estranged from his friends.

蔬 vegetables: 布衣～食 coarse clothes and simple fare
【蔬菜】vegetables; greens; greenstuff ◇ ～栽培 vegetable growing; vegetable farming ○园艺学 horticulture / 菜园 kitchen garden / 白菜 Chinese cabbage / 白薯(甘薯) sweet potato / 扁豆 lentil; hyacinth bean / 菠菜 spinach / 菜豆(豆角) French bean / 菜花 cauliflower / 菜苔 peduncle; flowering stalk / 长圆白萝卜 winter white radish / 慈菇 arrowhead / 大葱 green Chinese onion / 大蒜 garlic (clove) / 刀豆 sword bean / 冬瓜 Chinese wax gourd; white gourd / 冬笋 winter bamboo shoots / 番茄 tomato / 红圆萝卜 crimson round radish / 胡萝卜 carrot / 葫芦 calabash / 黄瓜 cucumber / 黄花菜 day-lily buds (dried) / 芥菜 leaf mustard / 芥兰 cabbage mustard / 韭菜 Chinese chive; fragrant-flowered-garlic / 韭黄 winter fragrant-flowered garlic / 空心菜 water spinach; swamp cabbage / 苦瓜 balsam pear / 辣椒 red pepper; chili; chilli; chile / 莲藕 lotus root / 莲籽 lotus seed / 菱角 water caltrop / 龙须菜 asparagus / 萝卜 radish / 马铃薯(土豆) potato / 毛豆 soybean / 蘑菇 mushroom / 木耳 edible fungus / 木薯 carrot / 南瓜 cushaw; pumpkin / 牛皮菜 leaf beet / 茄子 eggplant / 芹菜 celery / 青菜 greens / 青瓜 green cucumber / 山药 Chinese yam / 蛇瓜 serpent melon; snake melon / 生姜 ginger / 水萝卜 summer radish / 水田芹 watercress / 丝瓜 vegetable sponge; sponge gourd / 蒜苗 garlic sprouts;garlic stem / 塌棵菜(瓢儿菜) savoy / 豌豆 pea / 西葫芦 field pumpkin / 鲜榨菜 tuber mustard; mustard tuber preserved with chili / 香菜 coriander / 香椿 tender leaves of Chinese toon / 心里美 sweet pink-fleshed radish / 雪里蕻 potherb mustard / 洋扁豆 lima bean / 洋葱 onion / 银耳 tremella / 竹笋 bamboo shoots

枢 pivot; hub; center: 神经中～ nerve center
【枢杆】hinged arm
【枢接】pin joint
【枢纽】pivot; hub; axis; key position: ～工程 multi-purpose project; pivotal project / ～作用 a pivotal role / 交通～ a hub of communications
【枢轴】pivot; king journal; weigh bar shaft

叔 ①(叔父) father's younger brother; uncle ②(与父辈分相同而年轻的男人) uncle ③(丈夫之弟) husband's younger brother
【叔胺】tertiary amine
【叔伯】relationship between cousins of the same grandfather or great-grandfather: ～兄弟 first or second cousins on the paternal side; cousins
【叔父】father's younger brother; uncle
【叔母】wife of father's younger brother; aunt
【叔叔】[口] ①(叔父) father's younger brother; uncle ②(称呼与父辈辈分相同而年轻的男人) uncle
【叔祖】granduncle
【叔祖母】grandaunt

淑 [书] kind and gentle; fair: ～女 a fair maiden

菽 beans
【菽水承欢】be a dutiful son to one's parents even in poverty; poor but filial

输 ①(运输) transport; convey: 用卡车把货物～往山区 transport goods to the mountain regions by truck ②[书](捐献) contribute money; donate: 慷慨～将 make liberal contributions ③(失败) lose; be beaten; be defeated: ～了一局 lose one game in the set
【输出】①(从内部送到外部) export: 资本～ export of capital ② [电] output ◇ ～端数 fan-out / ～港 delivery port; outport / ～功率 output power / ～管制 export control / ～口岸 loading port / ～数据 dataout / ～限额 [经] export quota / ～许可证 export licence
【输电】transmit electricity ◇ ～网 power transmission network; grid system / ～线路 transmission line
【输精管】[生理] spermatic duct; deferent duct ◇ ～结扎术 [医] vasoligation / ～炎 deferentitis
【输理】be in the wrong
【输卵管】[生理] oviduct; Fallopian tube ◇ ～壶腹 ampulla tubae uterinae / ～结扎术 [医] tubal ligation / ～炎 salpingitis
【输尿管】[生理] ureter ◇ ～炎 ureteritis
【输入】①(从外部送到内部) import: ～新思想 the influx of new ideas ②[电] input ◇ ～程序 input routine; loading routine / ～端数

fan-in／ ～额 amount of imports／ ～功率 input power／ ～输出接口 input-output interface／ ～数据 data-in／ ～数字 input digit／ ～文件 input file／ ～限额[经] import quota／ ～许可证 import licence／ ～字段 input field／ ～字块 input block

【输送】carry; transport; convey: ～新鲜血液 infuse new blood ◇ ～带 conveyer belt／ ～机 conveyer

【输血】① [医] blood transfusion ② (给以支援) give aid and support; bolster up; give sb. a shot in the arm

【输氧】[医] oxygen therapy

【输液】[医] infusion; perfusion

【输油管】petroleum pipeline: 石油通过～输送到炼油厂。The oil is carried to the oil refinery by pipelines.

殊 ① (不同) different: 悬～ differ widely ② (特殊) outstanding; special; remarkable: 待以～礼 receive sb. with unusual ceremony／ 特～ remarkable efficacy ③ (很;极) very much; extremely; really: ～觉歉然 feel most regretful／ ～难相信 very difficult to believe; hardly credible

【殊不知】little imagine; hardly realize: 我以为她还在上海,～她已经走了。I thought she was still in Shanghai. I never dreamt that she had already left.

【殊深轸念】express deep solicitude; feel deeply concerned

【殊死】desperate; life-and-death: ～的搏斗 a life-and-death struggle

【殊途同归】reach the same goal by different routes

【殊勋】[书] outstanding merit; distinguished service

倏 swiftly

【倏忽】swiftly; in the twinkling of an eye: ～不见 quickly disappear

抒 express; give expression to; convey

【抒发】express; voice; give expression to: ～感情 express one's feelings

【抒情】express one's emotion ◇ ～散文 lyric prose／ ～诗 lyric poetry; lyrics

【抒写】express; describe

舒 ① (伸展; 宽解) stretch; unfold: ～了一口气 heave a sigh of relief ② (缓慢) easy; leisurely: ～徐 leisurely; in no hurry

【舒畅】happy; entirely free from worry: 心情～ have ease of mind; feel happy

【舒服】① (轻松愉快) comfortable: 这椅子坐着真～。One feels very comfortable in this chair. ② (无病) be well: 感到不～ feel very bad; not to feel well／ 我觉得不大～。I am not feeling very well.

【舒筋活络】[中医] stimulate the circulation of the blood and cause the muscles and joints to relax

【舒卷】[书] roll back and forth: 白云～。The white clouds mass and scatter.

【舒适】comfortable; cosy; snug: ～的生活 a comfortable life

【舒坦】comfortable; at ease

【舒展】① (不卷着) unfold; extend; smooth out: 舒眉展目 lift one's eyebrows and open one's eyes／ 荷叶～着。The lotus leaves are unfolding. ② (使舒适) limber up; stretch: ～一下筋骨 limber up one's muscles and joints

【舒张】[生理] diastole ◇ ～压 [医] diastolic pressure

书 ① (书写) write: 大～特～ record in letters of gold／ 振笔直～ take up the pen and write vigorously ② (字体) style of calligraphy; script: 楷～ regular script ③ (书籍) book: 写～ write a book／ 这～已绝版。The book is out of print. ④ (信) letter: 家～ a letter to or from home ⑤ (文件) document: 国～ letter of credence; credentials／ 批准～ instrument of atification／ 议定～ protocol／ 证～ certificate

【书包】satchel; schoolbag

【书报】books and newspapers

【书本】book: ～知识 book learning; book knowledge

【书橱】bookcase

【书呆子】pedant; bookworm

【书单】book list

【书挡】bookend; book end

【书店】bookshop; bookstore; bookseller's

【书法】handwriting; penmanship; calligraphy ◇ ～家 calligrapher／ ～练习 drills to improve one's handwriting

【书房】study

【书号】book number: 国际标准～ ISBN; International Standard Book Number

【书后】postscript

【书画】painting and calligraphy

【书籍】books; works; literature: 军事～ military literature

【书脊】spine (of a book)
【书记】①(党团的主要负责人) secretary: 总～ general secretary ②(文牍员) clerk ◇～处 secretariat / ～员 [法] clerk (of a court)
【书架】bookshelf; book case
【书简】letters; correspondence
【书经】*The Book of History*
【书局】publishing house; press
【书刊】books and periodicals ◇～登记 books registration / ～登记证 certification; books registration / ～审查(制度) censorship of book
【书靠】book end
【书库】stack room
【书立】book end
【书眉】the top of a page; top margin
【书面】written; in written form; in writing: ～报告 reading report / ～裁决 written decision / ～陈述 recitation / ～答复 written reply; answer in writing / ～控告 written accusation / ～判决 written judgement / ～契约 written agreement; written contract / ～声明 written statement / ～通知 written notice / ～语 written language; literary language / ～证据 documentary evidence
【书名】the title of a book; title ◇～号 punctuation marks used to enclose the title of a book or an article(《 》) / ～页 title page
【书目】booklist; title catalogue: 参考～ a list of reference books; bibliography
【书皮】book cover; book jacket; dust cover; cover: 塑料～ plastic cover ◇～纸 paper for covering books
【书评】book review
【书签】①(贴在线装书封面上的) a title label pasted on the cover of a Chinese-style thread-bound book ②(夹在书中的) bookmark
【书商】bookman
【书生】intellectual; scholar: ～之见 a pedantic view
【书生气】bookishness: ～十足 be bookish and naive
【书摊】bookstall; bookstand
【书套】slipcase
【书亭】book-kiosk; bookstall
【书香门第】a scholarly family; scholar-gentry families; family of scholars
【书写】write: ～标语 write slogans; letter posters ◇～错误 clerical error / ～规则 rules for writing / ～纸 writing paper
【书信】letter; written message: ～格式 form of a letter / 常有～往来 keep up a regular correspondence ◇～电 letter cable
【书信体】epistolary style ◇～小说 epistolary novel
【书页】page
【书院】academy of classical learning
【书札】[书] letters; correspondence
【书斋】study
【书桌】desk; writing desk

shú

孰 [书] ①(谁) who; which: ～胜～负? Who wins and who loses? / ～是～非? Which is right and which is wrong? ②(什么) what: 是可忍,～不可忍? If this can be tolerated, what cannot?

熟 ①(成熟) ripe: 一年两～ two crops a year ②(煮熟) cooked; done: ～肉 cooked meat / 半生不～ half-done; half-cooked ③(加工或锻炼过的) processed: ～铜 wrought copper / ～皮子 tanned leather ④(知道得清楚) familiar: ～朋友 familiar friend / 我和他不太～。I'm not on familiar terms with him. / 这口音听起来很～。The voice sounds familiar. ⑤(精通而有经验) skilled; experienced; practiced: ～手 practiced hand; old hand ⑥(程度深) deeply: ～睡 be in a deep sleep; be fast asleep
【熟谙】[书] be familiar with; be good at: ～水性 be an expert swimmer
【熟菜】cooked food; prepared food
【熟成机】[化纤] ripening machine
【熟地】①(经多年耕种过的土地) cultivated land ②[中药] prepared rhizome of rehmannia
【熟荒地】[农] once cultivated land; abandoned land
【熟记】learn by heart; memorize; commit to memory
【熟客】frequent visitor
【熟练】skilled; practiced; proficient: ～工人 skilled worker
【熟料】①[冶] grog; chamotte ②[建] clinker
【熟路】familiar route; beaten track: 熟门～ a familiar road and a familiar door; things that one knows well
【熟能生巧】skill comes from practice; practice makes perfect
【熟年】a year of good harvests; bumper year
【熟人】acquaintance; friend
【熟石膏】plaster of Paris

【熟石灰】slaked lime
【熟食】prepared food; cooked food
【熟视无睹】pay no attention to a familiar sight; turn a blind eye to; ignore
【熟识】be well acquainted with; know well: 彼此很～ know each other very well
【熟睡】sleep soundly; be fast asleep
【熟丝】[纺] boiled-off silk
【熟思】ponder deeply; consider carefully; deliberate
【熟铁】wrought iron
【熟土】[农] mellow soil
【熟悉】know sth. or sb. well; be familiar with; have an intimate knowledge of: ～内情 know the ins and outs of the matter; know the inside story of; be in the know
【熟习】be skillful at; have the knack of; be practiced in: ～业务 be practiced in one's field of work
【熟语】[语] idiom; idiomatic phrase
【熟知】know very well; know intimately
【熟字】words already learned; familiar words

塾 private school: ～师 tutor of a private school

秫 *kaoliang*; sorghum
【秫秸】*kaoliang* stalk
【秫米】husked sorghum

赎 ①(换回抵押品) redeem; ransom: ～回被绑架的人 ransom a kidnapped person / ～回抵押物 redeem a mortgage / 把东西～回来 redeem a pledge ③(抵销) atone for (a crime)
【赎当】redeem sth. pawned
【赎价】ransom price; ransom
【赎金】ransom money; ransom
【赎买】redeem; buy out ◇ ～政策 policy of redemption; buying-out policy
【赎身】(of slaves, prostitutes) redeem oneself; buy back one's freedom
【赎罪】atone for one's crime: 立功～ perform meritorious services to atone for one's crime ◇ ～日 [犹太教] Yom Kippur; Day of Atonement

shǔ

数 ①(查点) count: 倒着～ count down / 学～数 learn how to count ②(比较起来突出) be reckoned as exceptionally (good, bad, etc.): ～一～二的 one of the best; outstanding ③(列举) enumerate; list: 历～其罪 enumerate the crimes sb. has committed
【数不胜数】beyond count
【数典忘祖】give all the historical facts except those about one's own ancestors; forget one's own origins; be ignorant of the history of one's own country
【数九】the nine periods (of nine days each) following the winter solstice: ～寒天 the coldest days of the year
【数来宝】[曲艺] rhythmic storytelling to clapper accompaniment
【数落】[口] ①(列举过失而指责) scold sb. by enumerating his wrong-doings; reprove: 把他～一顿 give him a good scolding ②(不住嘴地列举着说) enumerate; cite one example after another

暑 heat; hot weather: 盛～ at the height of the summer; very hot weather / 中～ get sunstroke; get heatstroke; suffer heat exhaustion
【暑假】summer vacation
【暑期】summer vacation time ◇ ～训练班 summer course
【暑气】summer heat; heat
【暑热】hot summer weather
【暑天】hot summer days; dog days

署 ①(办公处所) a government office; office: 专员公～ prefectural commissioner's office ②(布置) make arrangements for; arrange ③(署理) handle by proxy; act as deputy: ～理部务 handle the ministry's affairs during the minister's absence ④(签署) sign; put one's signature to: 签～协定 sign an agreement
【署名】sign; put one's signature to: ～画押 sign one's name and affix one's seal to a document ◇ ～人 the undersigned / ～文章 a signed article

薯 potato; yam: 白～ sweet potato / 木～ cassava
【薯莨】[植] dye yam
【薯瘟锡】[化] fentinacetate
【薯蓣】[植] Chinese yam

曙 [书] daybreak; dawn
【曙光】first light of morning; dawn: 胜利的～ the dawn of victory ◇ ～恐怖 [心] auroraphobia
【曙红】[化] eosin; eosine
【曙色】light of early dawn: 灰白的～ the pale light of early dawn
【曙神星】[天] Aurora

蜀
【蜀葵】[植] hollyhock
【蜀犬吠日】in Sichuan dogs bark at the sun (because it's a rare sight in that misty region); an ignorant person makes a fuss about something which he alone finds strange
【蜀绣】Sichuan embroidery

黍 broomcorn millet

鼠 mouse; rat
【鼠辈】mean creatures; scoundrels
【鼠标器】[计算机] Genius mouse
【鼠疮】[中医] scrofula
【鼠窜】scamper off like a rat; scurry away like frightened rats: ～狼奔 run higher and thither like rats and wolves
【鼠道式排水沟】mole drain
【鼠肚鸡肠】suffer affronts without resentment; lacking courage or endurance
【鼠海豚】porpoise
【鼠笼式】[电] squirrel-cage ◇ ～电动机 squirrel-cage motor
【鼠目寸光】a mouse can see only an inch; see only what is under one's nose; be shortsighted
【鼠窃狗偷】filch like rats and snatch like dogs; play petty tricks on the sly
【鼠曲草】[植] affine cudweed
【鼠伤寒】mouse typhus
【鼠蹊】[生理] groin
【鼠咬热】[医] rat-bite fever
【鼠疫】the plague

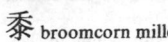

属 ①(类别) category: 金～ metals ② [生] genus: 亚～ subgenus ③(隶属) under; subordinate to: 所～单位和部门 subordinate units and departments ④(归属) belong to: 我们两国同～发展中国家。 Both our countries belong to developing countries. ⑤(亲属) family members; dependents ⑥(是) be: 查明～实 prove to be true after investigation / 实～无理 be really unreasonable ⑦(用十二生肖记生年) be born in the year of (one of the twelve animals): 我姐姐～马。 My sister was born in the year of the horse.
【属地】possession; dependency
【属国】vassal state; dependent state
【属性】[逻] attribute; property
【属性抽样】attribute sampling
【属于】belong to; be part of: ～同一范畴 fall into the same category / 民主～上层建筑。 Democracy is part of the superstructure.

shù

树 ①(木本植物) tree: 常绿～ evergreen / 苹果～ apple tree / 针叶～ conifer ○桉树 eucalyptus / 白腊树 ash / 白桦 birch / 白杨 popular; aspen / 柏树 cypress / 垂柳 weeping willow / 刺槐 locust / 枞树 fir / 冬青树 holly; evergreen; ilex / 番木瓜树 papaya / 凤凰树 flame-tree / 橄榄树 Chinese olive tree / 合欢 silk tree / 核桃树 walnut / 黑檀 ebony / 红木 mahogany / 红木树 redwood / 红松 Korean pine / 槐树 Chinese scholartree / 黄杨 boxwood / 加拿大云杉 cat spruce; white spruce / 栲树 evergreen chinquapin / 可可树 cocoa / 柳杉 cryptomeria / 柳树 willow / 落叶松 larch / 椴树 linden / 桧树 juniper / 没药树 myrrh / 木棉树 silk cotton tree / 泡桐树 paulowmia / 菩提树 pipal tree / 漆树 varnish tree / 榕树 banyan; banian / 桑树 mulberry / 水杉 metasequoia / 松树 pine / 檀香木 sandal wood / 铁树 sago cycas / 无花果树 fig-tree / 梧桐 Chinese parasol tree / 橡胶树 rubber tree / 橡树 oak / 雪松 white pine; cedar / 洋松 Oregon pine / 椰子树 coconut palm / 银白杨 white popular / 油茶树 tea-oil tree / 油松 Chinese pine / 油桐树 tung-oil tree / 油棕 oil palm / 榆树 elm / 云杉 spruce / 樟树 camphor tree / 梓树 catalpa / 棕榈 palm ②(栽培) plant; cultivate: 十年～木,百年～人。 It takes ten years to grow trees, but a hundred years to rear people. ③(树立;建立) set up; establish; up-hold: ～雄心 have lofty ambitions; aim high / ～正气 uphold healthy tendencies / 建～ achievement
【树碑立传】[贬] glorify sb. by erecting a monument to him and writing his biography; build up sb.'s public image
【树杈】crotch (of a tree)
【树丛】grove; thicket
【树大招风】[谚] a tall tree catches the wind; a person in a high position is liable to be attacked
【树倒猢狲散】Once the tree falls, the monkeys scatter. (或: When an influential person falls from power, his hangers-on disperse.)
【树敌】make an enemy of sb.; set others against oneself; antagonize: ～太多 make too many enemies; antagonize too many people
【树墩】tree stump; stump

【树蜂】[动] wood wasp
【树干】tree trunk; trunk
【树高千丈, 叶落归根】a tree may grow a thousand *zhang* high, but its leaves fall back to the roots; a person residing away from home eventually returns to his native soil
【树冠】crown (of a tree)
【树胶】gum (of a tree)
【树懒】[动] sloth
【树立】set up; establish: ～榜样 set an example / ～为人民服务的思想 adopt the idea of serving the people
【树林】woods; grove
【树苗】sapling
【树木】trees
【树皮】bark ◇ ～画 [美术] bark picture
【树梢】the tip of a tree; treetop
【树蛙】[动] tree frog
【树阴】shade (of a tree)
【树欲静而风不止】the tree may prefer calm, but the wind will not subside: class struggle is inevitable in class society
【树枝】branch; twig
【树脂】resin: 醇酸～ alkyd resin / 环氧～ epoxy resin / 合成～ synthetic resin / 离子交换～ ion exchange resin / 中性～ resinene ◇ ～胶合板 compo board; resin bounded plywood / ～漆 lacquer type organic coating / ～酸 resinic acid / ～整理 [纺] resin finishing

漱 gargle; rinse
【漱口】rinse the mouth; gargle: 用盐水～ gargle with salt water ◇ ～杯 a glass or mug for mouth-rinsing or teeth-cleaning; tooth glass / ～剂 gargle / ～药 mouth-wash / ～液 gargle

庶 ①(众多) multitudinous; numerous: ～物 every kind of creature; all things / 富～ rich and populous ②[旧](家庭的旁支) of or by the concubine: ～出 be born of a concubine ③[书](才可) so that; so as to: ～免误会 so as to avoid misunderstanding
【庶民】[书] the common people; the multitude
【庶母】concubine of one's father
【庶务】[旧] ①(杂项事务) general affairs; business matters ②(担任庶务的人) a person in charge of business matters

数 ①(数目) number; figure: ～以万计 number tens of thousands / 心中有～ have a good idea of how things stand; know what's what ②[数] number: 变～ variable / 常～ constant / 倒～ reciprocal / 分～ fraction / 负～ negative number / 奇～ odd numbers / 基～ cardinal numbers / 偶～ even numbers / 实～ real number / 未知～ unknown number / 无理～ irrational number / 小～ decimal / 虚～ imaginary number / 序～ ordinal numbers / 已知～ known number / 有理～ rational number / 整～ integer; whole number / 正～ positive number / 最大公约～ highest common divisor / 最小公倍～ lowest common multiple ③[语] number: 单～ singular number / 复～ plural number ④(几个;几) several; a few: ～百人 several hundred people / 连续～年 for several years running ⑤(天数) fate; destiny; God's will
【数词】[语] numeral: 基～ cardinal number / 序～ ordinal number
【数额】number; amount: 超出～ exceed the number fixed
【数据】data: 科学～ scientific data ◇ ～操作语言 data manipulation language / ～处理 data processing / ～存储系统 data-storage system / ～分析 data analysis / ～管理系统 data management system / ～检查 data checks / ～库 data base / ～输入 data entry; data input / ～纸 data sheet / ～终端 data terminal
【数控】[机] numerical control (Nc): 总体～ total numerical control
【数理逻辑】mathematical logic
【数理统计学】mathematical statistics
【数量】quantity; amount: ～上的差别 quantitative difference ◇ ～词 [语] numeral-classifier compound / ～管制 quantity control / ～界限 quantitative limits
【数论】number theory
【数码】①(数字) numeral: 阿拉伯～ Arabic numerals / 罗马～ Roman numerals ②(数目) number; amount
【数目】number; amount
【数系】[数] number system
【数学】mathematics ◇ ～归纳法 mathematical induction / ～家 mathematician / ～近似 mathematical approach / ～投影 mathematical projection
【数域】[数] number field
【数值】[数] numerical value ◇ ～天气预报 numerical weather forecast
【数轴】[数] number axis

【数珠】[佛教] beads
【数字】①(表示数目的文字) numeral; figure; digit: 阿拉伯～ Arabic numeral / 罗马～ Roman numeral / 天文～ astronomical figures ②(数量) quantity; amount: 不要单纯追求～。Don't just go after quantity. ◇～编码 digital coding; numeric coding / ～控制系统 numerical control system / ～计算机 digital computer / ～显示 digital display; digital presentation
【数罪并罚】concurrent punishment for several crimes; cumulative punishment
【数罪并合】joinder of offenses; merger of offenses

术 ①(技艺) art; skill; technique: 美～ the fine arts / 医～ the art of healing; doctor's skill / 不学无～ have neither learning nor skill ②(方法) method; tactics: 权～ political trickery / 战～ tactics
【术语】technical terms; terminology: 军事～ military terms / 医学～ medical terminology

述 state; relate; narrate: 陈～意见 state one's views / 如上所～ as stated above; as mentioned above
【述而不作】pass on the ancient culture without adding anything new to it
【述评】review; commentary: 每周时事～ weekly review of current affairs
【述说】state; recount; narrate: ～自己的观点 state one's own views / ～自己的经历 recount one's experiences
【述职】report on one's work; report: 回国～ go back for consultations

束 ①(捆;系) bind; tie: ～紧皮带 tighten one's belt ②[量] bundle; bunch; sheaf: 一～鲜花 a bunch of flowers ③(约束) control; restrain: ～手～脚 timid and hesitant / 无拘无～ without any restraint
【束缚】tie; bind up; fetter: ～手脚 bind sb. hand and foot; tie sb.'s hands; hamper the initiative of / 受传统的～ be fettered by tradition
【束射管】[电子] beam tube
【束身藏拙】hide one's inadequacy by keeping quiet
【束手】have one's hands tied; be helpless: ～就擒 allow oneself to be seized without putting up a fight
【束手待毙】fold one's hands and await destruction; helplessly wait for death; resign oneself to extinction
【束手无策】be at a loss what to do; feel quite helpless; be at one's wit's end
【束修之敬】in payment of a teacher's salary
【束之高阁】bundle sth. up and place it on the top shelf; lay aside and neglect; shelve; pigeonhole

戍 defend; garrison: ～边 garrison the frontiers

竖 ①(与地面垂直的) vertical; upright; perpendicular: ～着从上往下写 write vertically downward from the top / ～着放 place sth. vertically ②(使物体跟地面垂直) set upright; erect; stand: ～旗杆 erect a flagstaff ③(汉字的笔画) vertical stroke
【竖沟】flute
【竖井】[矿] (vertical) shaft; aven; cenote; pothole ◇～矿 shaft mine
【竖立】erect; set upright; stand: ～一座纪念碑 erect a monument
【竖起】hold up; erect: ～大拇指 hold up one's thumb in approval; thumbs up / ～衣领 turn up the collar
【竖琴】[乐] harp: ～手 harper; harpist
【竖蜻蜓】[方] handstand
【竖子】[书] ①(童仆) boy; lad ②(小子) mean fellow; fellow

恕 ①(不计较别人的过错) forgive; pardon; excuse: 宽～ forgive ②[套](请对方不要计较) excuse me; beg your pardon: ～我打扰。Excuse my troubling you. ③(忠恕) forbearance (as advocated by Confucius)
【恕不奉陪】I'm sorry but I cannot keep you company. (或: Excuse me for not keeping you company.)
【恕不拘礼】Pardon me for not standing on ceremony.
【恕难从命】We regret that we cannot comply with your wishes.
【恕罪】pardon an offense; forgive a sin

shuā

刷 ①(刷子) brush: 牙～ toothbrush ②(用刷子刷) brush; scrub: ～鞋 brush shoes ③(涂抹) daub; paste up: ～标语 paste up posters / 用石灰浆～墙 whitewash a wall ④[口](淘汰) eliminate; remove: 直到半决赛他才给～下来。He was not eliminated until the semifinals. ⑤[象]

swish; rustle: 杨树叶子被风吹得～～响。The poplar leaves rustled in the wind.
【刷布】brushing; napping
【刷弧】brush arc
【刷痕】brush mark
【刷洗】scrub: ～厨房用具 wash 〈clean〉 the kitchen utensils / ～地板 scrub the floor
【刷新】①(刷洗一新) renovate; refurbish: ～门面 repaint the front (of a shop, etc.); put up a new shopfront ②(创出新的) break: ～纪录 break a record
【刷牙】brush one's teeth
【刷子】brush; scrub

shuǎ

耍 ①[方](玩) play: 小男孩在院子里～。The boy is playing in the courtyard. / 这可不是～的! It's no joke! ②(玩弄) play with; flourish: ～刀 flourish a sword; give a performance of swordplay / ～猴 put on a monkey show ③(施展) play (tricks): ～鬼把戏 play dirty tricks
【耍把戏】①(玩杂耍) juggle (with) ②(耍花招) play tricks; be up to one's tricks
【耍笔杆】wield a pen; be skilled in literary tricks
【耍花招】①(卖弄武术技巧) display showy movements in *wushu*, etc. ②(施展欺诈手段) play tricks: 别～了! None of your tricks!
【耍滑】try to shirk work or responsibility; act in a slick way
【耍赖】act shamelessly; be perverse
【耍流氓】behave like a hoodlum; take liberties with women; act indecently
【耍弄】make fun of; make a fool of; deceive
【耍盘子】[杂技] plate-spinning; disc-spinning
【耍脾气】get into a huff; put on a show of bad temper
【耍贫嘴】[方] be garrulous
【耍手腕】play tricks; maneuver
【耍坛子】[杂技] juggling with jars; jar balancing act
【耍威风】make a show of authority; throw one's weight about; be overbearing
【耍无赖】act shamelessly; be perverse
【耍笑】①(随意说笑) joke; have fun ②(戏弄人以取笑) make fun of; play a joke on sb.
【耍心眼儿】exercise one's wits for personal gain; be calculating; pull a smart trick
【耍嘴皮子】①(卖弄口才) show off one's eloquence; talk glibly; be a slick talker ②(只说不做) merely chatter idly; mere empty talk; lip service

shuà

刷
【刷白】white; pale: 月亮把大地照得～。The earth turned white under the moon.

shuāi

衰 decline; wane: ～运 declining fortune / 年老力～ decrepit / 兴～ rise and decline / 风势渐～。The wind is falling. / 懒则～。Laziness leads to debility.
【衰败】decline; wane; be at a low ebb: ～没落 be on the decline
【衰变】[物] decay: 核～ nuclear decay ◇ ～毫居 millicurie-destroyed / ～类型 decay mode
【衰耗】pad control
【衰减】[电] attenuation ◇ ～器 attenuator / ～失真 attenuation distortion / ～信号 deamplification / ～因数 attenuation factor
【衰竭】[医] exhaustion; prostration: 心力～ heart failure
【衰老】old and feeble; decrepit; senile; slow death; senescence: 她显得十分～。She appears quite doddery.
【衰落】decline; be on the wane; go downhill: 日趋～ fall into decay with each passing day; be on the decline
【衰弱】①(失去强盛的精力) weak; feeble: 神经～ suffer from neurasthenia / 病人很～。The patient is very weak. ②(事物由强转弱) slack off; weaken: 攻势已经～。The offensive has slacked off.
【衰世之秋】in the time of the age of decadence
【衰颓】weak and degenerate
【衰退】fail; decline: 记忆力～ be losing one's memory / 经济～ economic recession / 视力～ failing eyesight
【衰亡】become feeble and die; decline and fall; wither away
【衰朽】[书] feeble and decaying; decrepit

摔 ①(跌倒) fall; tumble; lose one's balance: ～倒在地 fall down / 仰面～倒 fall on one's back ②(很快地往下落) hurtle down; plunge: 敌机～下来了。The enemy plane plunged to the ground. ③(打破) cause to fall and break; break: ～成两半 break in two / ～得粉碎 break into pieces ④(扔) cast; throw; fling: 把外衣往床上一～ throw one's coat onto the bed
【摔打】①(磕打) beat; knock: ～衣服上的灰尘

beat the dust off one's clothes ②(磨炼) rough it; temper oneself: 他经历了多年的～,如今更老练了。 Tempered through years' hardships, he has grown much more experienced than ever.
【摔跟头】①(跌倒) tumble; trip and fall ②(犯错误) trip up; come a cropper; make a blunder.
【摔交】①(跌倒) tumble; trip and fall ②(犯错误) trip up; come a cropper; blunder ③[体] wrestling ◇～台 platform / ～运动员 wrestler ○古典式 Greco-Roman Style / 自由式 freestyle / 48 公斤级 light flyweight / 52 公斤级 flyweight / 57 公斤级 bantamweight / 62 公斤级 featherweight / 74 公斤级 welterweight / 82 公斤级 middleweight / 90 公斤级 light heavyweight / 100 公斤级 heavyweight / 100 以上公斤 super heavyweight

shuǎi

甩 ①(挥动;抡) move backward and forward; swing: ～胳膊 swing one's arms / 把辫子往后一～ throw back one's plait ②(扔) throw; fling; toss: ～手榴弹 throw hand grenades / 把某物～出窗外 throw sth. out of the window ③(抛开) leave sb. behind; throw off: ～掉包袱 cast off a burden; get a load off one's back
【甩车】[铁道] uncouple a railway coach from the locomotive; uncouple
【甩卖】off a clearance sale; sale at a reduced price
【甩手】①(摆动手) swing one's arms ②(扔下不管) refuse to do; wash one's hand of: 你可不能～不管。You can't wash your hands of this

shuài

率 ①(带领) lead; command: ～代表团前往… head a delegation to leave for... / ～师 command troops / ～众前往 go (to a place) at the head of many people ②(不加思索) rash; hasty: 草～ careless; cursory / 轻～ hasty ③(直爽坦白) frank; straightforward: 坦～ frank / 直～ straightforward ④(大概) generally; usually: 大～如此。 This is usually the case.
【率尔】[书] rashly; hastily: ～而对 give a hasty reply; reply without thinking
【率领】lead; head; command:e case.
【率尔】[书] rashly; hastily: ～而对 give a hasty reply; reply without thinking
【率领】lead; head; command: ～代表团 lead a delegation / ～军队进攻敌人 head the troops to attack the enemy
【率土之滨】all this land; all within the boundaries; within the territory of a state
【率先】take the lead in doing sth.; be the first to do sth.
【率由旧章】follow the beaten track; act in accordance with established rules
【率真】forthright and sincere
【率直】straightforward; unreserved; blunt

帅 ①(军队最高指挥员) commander in chief: 统～ supreme commander / 挂～ take command ②(漂亮;英俊) beautiful; graceful; smart: 字写得～ write a beautiful hand ③(棋子) commander in chief, the chief piece in Chinese chess

shuān

闩 ①(门闩) bolt; latch: 门～ door bolt / 上～ fasten with a bolt ②(用闩插上) fasten with a bolt or latch: 把门～好 bolt the door
【闩柄】bolt handle; bolt lever
【闩锁】breech lock; latch
【闩体】breechblock
【闩托】latch bracket

栓 ①(可开关的机件) bolt; plug: 枪～ rifle bolt / 消火～ fire hydrant; fireplug ②(塞子) stopper; cork
【栓剂】[药] suppository
【栓皮】[林] cork
【栓皮栎】[植] oriental oak
【栓塞】[医] embolism: 肺～ pulmonary embolism / 静脉～ venous embolism / 脑～ cerebral embolism
【栓体】key
【栓子】[医] embolus

拴 tie; fasten: ～绳子晒衣服 put up a clothes line / 把马～在树上 tie a horse to a tree
【拴系点】tie-down point

shuàn

涮 ①(冲洗;漂洗) swill; rinse: ～瓶子 rinse a bottle / ～衣服 rinse the clothes ②(烫一下就吃) boil thin slices of meat instantly; scald thin slices of meat in boiling water; instant-boil: ～羊肉 instant-boiled mutton
【涮锅子】instant-boil slices of meat and vegetables in a chafing dish

shuāng

霜 ①(白色冰晶) frost: 地上有～。 There is

frost on the ground. ②（霜状物）frostlike powder: 柿～ powder on the surface of a dried persimmon / 糖～ frosting; icing ③（白色）white; hoar: ～鬓 grey temples
【霜冻】frost
【霜度】[气] degrees of frost
【霜害】frostbite; frost injury
【霜花】frostwork
【霜降】Frost's Descent (18th solar term)
【霜霉病】[农] downy mildew
【霜期】[气] frost season
【霜叶】red leaves; autumn maple leaves

孀 widow
【孀妇】widow
【孀居】be a widow; live in widowhood

双 ①（两个）two; twin; both; dual: ～发动机飞机 twin-engined plane / ～向交通 two-way traffic ②[量] pair: 一～鞋 a pair of shoes ③（偶数的）even: ～号座位 even-numbered seats / ～数 even numbers ④（加倍的）double; twofold: ～份 double the amount; twice as much
【双胞胎】twins
【双边】bilateral: ～会谈 bilateral talks / ～贸易 bilateral trade; two-way trade / ～条约 bilateral treaty / ～协定 bilateral agreement / ～支付 bilateral payments
【双层】double-deck; having two layers; of two thicknesses: ～玻璃窗 double window / ～床 double-decker bed 〈bunk〉/ ～公共汽车 double-deck bus / ～火车 double-decker / ～桥 double-decker bridge
【双重】double; dual; twofold: ～任务 double task; twofold task ◇ ～标准 double standard / ～代表权[外] dual representation / ～国籍 dual nationality / ～间谍 double agent / ～领导 dual leadership / ～人格 dual personality / ～性 dual nature / ～征税 double taxation
【双唇音】[语] bilabial (sound)
【双打】[体] doubles: 男子～ men's doubles / 男女混合～ mixed doubles / 女子～ women's doubles
【双方】both sides; the two parties: 缔约国～ both signatory states; the contracting parties / ～同意 by mutual consent / ～各执一词。Each side persisted in its own views.
【双峰驼】[动] two-humped camel; Bactrian camel
【双幅】double width: 这布是单幅还是～的? Is this cloth single or double width?
【双杠】[体] parallel bars
【双宫丝】[纺] doupion silk
【双关】having a double meaning: 一语～ a phrase with a double meaning ◇ ～语 pun
【双管】double-barrelled: ～猎枪 double-barrelled shotgun
【双管齐下】paint a picture with two brushes at the same time; work along both lines
【双轨】[交] double track ◇ ～铁路 double-track railway
【双号】even numbers (of tickets, seats, etc.)
【双簧】[曲艺] a two-man comic show: 唱～ give a two-man comic show; collaborate
【双簧管】[乐] oboe
【双季稻】double cropping of rice; double harvest rice
【双交】[农] double cross
【双联合同】indented deed; indenture
【双料】of reinforced material; extra quality: ～脸盆 special quality basin; extra good quality basin
【双轮】[体] double round: 五十米～射箭 50-meter double round archery event
【双面】two-sided; double-edged; double-faced; reversible: ～刀片 a double-edged razor blade ◇ ～绣 double-faced embroidery / ～摇纱机 double reeling frame / ～印刷机 perfecting press; perfector / ～织物 reversible cloth; reversibles
【双名法】binomial nomenclature
【双目失明】lose one's eyesight; lose the sight of both eyes
【双目显微镜】binocular microscope
【双亲】(both) parents; father and mother
【双球菌】[微] diplococcus
【双曲面】[数] hyperboloid: 单叶～ hyperboloid of one sheet / 双叶～ hyperboloid of two sheets
【双曲线】[数] hyperbola
【双全】complete in both respects; possessing both: 文武～ be adept with both the pen and the sword / 智勇～ possessing both wisdom and courage
【双人床】double bed
【双人舞】dance for two people; *pas de deux*
【双日】even-numbered days (of the month)
【双身子】[口] pregnant woman
【双生】twin: ～姐妹 twin sisters / ～兄弟 twin brothers / ～子 twins

【双声】[语] a phrase consisting of two or more characters with the same initial consonant; alliteration
【双式帐簿】double-entry bookkeeping
【双手】both hands: ～捧上 hand over in obeisance / 举～赞成 be all for it
【双数】even numbers
【双双】in pairs
【双糖】[化] disaccharide
【双体船】catamaran
【双筒枪】double-barrelled gun
【双筒望远镜】binoculars; field glasses
【双喜】double happiness: ～临门。A double blessing has descended upon the house.
【双线索道】bicable tramway
【双响】a firecracker which goes off twice; double-bang firecracker
【双向】two-way: ～电视 two-way television / ～飞碟射击 skeet shooting / ～交通 two-way traffic / ～开关 two-way switch
【双星】[天] double star
【双眼皮】double-fold eyelid
【双氧水】[药] hydrogen peroxide solution
【双翼机】biplane
【双音节词】[语] disyllabic word; disyllable
【双鱼座】[天] Pisces
【双元音】[语] diphthong
【双月刊】bimonthly
【双职工】man and wife both at work; working couple
【双周刊】biweekly; fortnightly
【双绉】[纺] *crêpe de Chine*
【双子叶植物】dicotyledon
【双子座】[天] Gemini
【双座】two-seater; double-seater: ～飞机 two-seater aircraft

shuǎng

爽 ①(明朗) bright; clear; crisp: 秋高气～。The autumn sky is clear and the air is crisp. ②(率直) frank; straightforward; openhearted: 豪～ straightforward; forthright ③(舒服) feel well: 身体不～ not feel well ④(差失) deviate: 毫厘不～ not deviating a hair's breadth; without the slightest error
【爽口】tasty and refreshing: 这道菜很～。This dish tastes very refreshing.
【爽快】①(舒服痛快) refreshed; comfortable: 心里很～ be quite at ease ②(直爽) frank; straightforward; outright: 为人～ be frank and straightforward ③(痛快) with alacrity; readily: 办事～ work readily and briskly
【爽朗】①(明朗而畅快) bright and clear: 天气～。It is serene. ②(开朗) hearty; candid; frank and open; straightforward: ～的笑声 hearty laughter / 她很～。She is open-minded.
【爽利】brisk and neat; efficient and able: 办事～ be brisk and neat in one's work
【爽然若失】[书] not know what to do; be at a loss
【爽身粉】talcum powder
【爽性】may just as well: 既然你开了头，～把它做完吧。Since you have started the job, you might as well finish it.
【爽约】[书] fail to keep an appointment; break an appointment
【爽直】frank; straightforward; candid

shuí

谁 ①(什么人) who: 他是～? Who is he? / 这是～的主意? Whose idea is it? ②(任何人) someone; anyone: 你有困难时～都会帮助你。Anyone will help you when you meet with difficulties. / 有～能帮助我就好了! If only someone could help me!

shuǐ

水 ①(无色、无臭、无味的液体) water: 淡～ fresh water / 硬～ hard water / 给街上洒～ water a street / 往牛奶里掺～ water down the milk ②(江、河、湖、海、洋) water; rivers, lakes, seas, etc.: ～陆运输 land and water transportation / ～上人家 boat dwellers / ～平如镜。The surface of the water is as smooth as a mirror. ③(稀的汁) a liquid: 桔子～ orangeade / 墨～ ink
【水坝】dam
【水半球】[地] water hemisphere
【水泵】water pump
【水笔】①(写小字的毛笔) a stiff-haired writing brush ②(画水彩画的笔) water-color paintbrush ③[方](钢笔) (fountain) pen
【水表】water meter
【水鳖】[植] frogbit
【水兵】seaman; sailor; bluejacket
【水玻璃】[化] water glass; sodium silicate
【水彩】watercolor ◇ ～画 watercolor (painting) / ～颜料 watercolors
【水草】①(有水源和草的地方) water and grass: ～丰美 (a place) with plenty of water and lush grass / 逐～而居 (of nomads) live where

there is water and grass; rove about seeking water and grass ②(水生植物) waterweeds; water plants
【水虿】[动] the nymph of the dragonfly, etc.
【水产】aquatic product ◇～品 aquatic product／～养殖 aquaculture／～业 aquatic products industry／～资源 aquatic resources
【水车】①(灌溉工具) waterwheel ②(运水工具) watercart; water wagon
【水成论】neptunian theory; neptunianism; neptunism
【水成岩】[地] aqueous rock
【水程】journey by boat; voyage
【水池】pond; pool; cistern
【水处理】[化] water treatment
【水到渠成】where water flows, a channel is formed; when conditions are ripe, success will come
【水道】①(河道) water course ②(水路) waterway; water route ③[体] lane; course: 自己的～ own water ◇～标志 marking of course／～裁判 course umpire
【水稻】paddy (rice); rice ◇～拔秧机 nursery planter puller; rice seedling puller／～插秧机 rice transplanter／～土 rice soil／～直播机 paddy planter
【水滴石穿】dripping water wears through rock; constant effort brings success
【水底电缆】submarine cable; subaqueous cable
【水电】water and electricity: ～供应 water and electricity supply ◇～费 charges for water and electricity
【水电站】[简](水力发电站) hydroelectric (power) station; hydropower station
【水貂】[动] mink
【水痘】[医] varicella; chicken pox
【水碓】water-powered trip-hammer (for husking rice)
【水飞蓟】[植] milk thistle
【水粉】soaked bean-noodles
【水粉画】[美术] *gouache*
【水分】①(物体所含的水) moisture content: 吸收～ absorb moisture ②(夸大的成分) exaggeration: 这个数字有～。This figure is inflated.
【水浮莲】[植] water lettuce; water cabbage
【水沟】ditch; drain; gutter
【水垢】scale; incrustation: 除去锅炉里的～ scour out a boiler
【水臌】[中医] ascites
【水管】waterpipe
【水果】fruit ◇～罐头 tinned fruit／～软糖 fruit jelly／～糖 fruit drops ○果园 orchard／果树栽培 pomiculture; fruit-growing／白果 gingko／槟榔 betelnut／菠萝 pineapple／草莓 strawberry／大樱桃 large cherry／番石榴 guava／瓜籽 melon seed／桂元(龙眼) longan／桂元肉 longan pulp／哈密瓜 Hamimelon／海棠果 flowering orange／核桃 walnut／红果 hill haw／京白梨 Beijing white pear／广柑(橙) orange／梨 pear／李子 plum／栗子 chestnut／荔枝 lychee／榴莲 jackfruit／芒果 mango／梅子 Japanese apricot／蜜桔 mandarin orange／蜜桃 honey peach; juicy peach／柠檬 lemon／枇杷 loquat／苹果 apple／蟠桃 flat peach／葡萄 grape／山楂果 hawthorn; haw／石榴 pomegranate／柿子 persimmon／桃 peach／无花果 fig／西瓜 watermelon／香瓜(甜瓜) muskmelon／香蕉 banana／香蕉苹果 banana- apple／杏子 apricot／杨梅 red bayberry／阳桃 carambola／椰子 coconut／樱桃 cherry／柚子 pomelo／枣 Chinese date; jujube [以上包括个别干果]
【水合】[化] hydration ◇～水 hydrate water／～物 hydrate
【水红】bright pink; cerise
【水壶】①(烧水用的) kettle ②(军用式水壶) canteen ③(浇水用的) watering can
【水葫芦】[植] water hyacinth
【水花】spray
【水患】flood; inundation
【水火】①(水和火) fire and water; two things diametrically opposed to each other: ～不相容 be incompatible as fire and water ②(灾难) extreme misery
【水火无情】floods and fires have no mercy for anybody
【水碱】scale; incrustation
【水浇地】[农] irrigated land
【水饺】boiled dumplings
【水解】[化] hydrolysis ◇～产物 hydrolysate／～蛋白[药] protein hydrolysate／～酶 hydrolase; hydrolytic enzyme／～质 hydrolyte
【水晶】crystal; rock crystal ◇～包[食品] a steamed dumpling stuffed with lard and sugar／～玻璃 crystal (glass)／～宫 the Crystal Palace (of the Dragon King)／～棺 crystal sarcophagus／～体[生理] crystalline lens
【水井】well
【水酒】watery wine (said by a host of his own wine)

【水坑】puddle; pool; water hole: 臭~ cesspool; cesspit
【水库】reservoir
【水牢】water dungeon
【水老鸦】[动] cormorant
【水涝】waterlogging ◇ ~地 waterlogged land
【水雷】[军] (submarine) mine: 磁性~ magnetic mine / 磁音混合起爆~ combination magnetic-acoustic mine / 敷设~ lay mines (in water) / 音响~ sonic mine; sound mine
【水冷】water-cooling: ~式发动机 water-cooled engine / ~式反应堆 watercooled reactor / ~式汽缸 water cooled cylinder / ~系统 water-cooling system
【水力】waterpower; hydraulic power ◇ ~采煤 hydraulic coal mining; hydro coal mining / ~发电 hydraulic electrogenerating / ~发电机 hydroelectric generator / ~发电设备 hydro-electric power generating equipment / ~发电站 hydroelectric (power) station; hydropower station / ~开采 [矿] hydraulic mining; hydraulicking / ~学 hydraulics / ~资源 hydroelectric resources; waterpower resources
【水利】①(利用水力资源) water conservancy: ~设施 water conservancy facilities ②(水利工程) irrigation works; water conservancy project: 兴修~ build irrigation works ◇ ~工程 irrigation works; water conservancy project / ~工程学 hydraulic engineering / ~灌溉网 irrigation network / ~化 bring all farmland under irrigation / ~纠纷 irrigation dispute / ~枢纽 key water control project / ~资源 water resources; hydro- power resources; water resources
【水疗】[医] hydrotherapy
【水灵】[方] ①(鲜美多汁而爽口) fresh and juicy ②(漂亮而有精神) bright and beautiful; radiant and vivacious: 两只~的大眼睛 a pair of bright, beautiful eyes
【水流】①(江河) rivers; streams; waters ②(水的流速) current; flow: ~迟缓 sluggish flow / ~湍急 rapid flow; rushing current ◇ ~量 discharge
【水流星】[杂技] spinning bowls of water; water meteors
【水龙】fire hose; hose
【水龙骨】[植] wall fern; golden locks
【水龙卷】[气] waterspout
【水龙头】(water) tap; faucet; bibcock: 关~ turn off the tap / 开~ turn on the tap / 随手关紧~。Don't leave the tap running after use.
【水陆】land and water: ~并进 proceed by both land and water; conduct a combined operation by army and navy ◇ ~交通线 land and water communication lines / ~联运 water-land transshipment / ~联运码头 a dock for joint land and water transport service / ~运输 transportation by land and water
【水陆两用】amphibious: ~车 amphibious vehicle / ~汽车 amphibian automobile; amphibious car / ~坦克 amphibious tank
【水路】waterway; water route: 由上海到武汉可以走~。One can travel from Shanghai to Wuhan by water.
【水铝矿】[矿] gibbsite
【水绿】light green
【水轮泵】(water) turbine pump ◇ ~站 (water) turbine-pump station
【水轮发电机】water turbogenerator
【水轮机】hydraulic turbine; water turbine
【水落管】[建] downspout; downpipe
【水落石出】when the water subsides the rocks emerge;the whole thing comes to light: 把事情辩个~ argue a matter out
【水煤气】[化] water gas
【水门】①(阀门) water valve ②(水闸) water gate
【水锰矿】manganite
【水蜜桃】honey peach
【水磨功夫】patient and precise work; painstaking work
【水磨石】[建] terrazzo ◇ ~地面 terrazzo floor
【水墨画】[美术] ink and wash; wash painting
【水磨】①(水力带动的磨) water mill ②(加水细磨) grind grain, etc. fine while adding water: ~年糕 New Year cake made from finely ground rice flour
【水母】[动] jellyfish; medusa
【水泥】cement: 矾土~ alumina cement / 高标号~ high-mark cement / 矿渣~ slag cement ◇ ~标号 strength of cement; cement grade / ~厂 cement plant / ~船 concrete boat; plastered boat / ~浆 cement mortar / ~瓦 cement tile
【水碾】water-powered roller (for grinding grain)
【水鸟】aquatic bird; water bird
【水牛】(water) buffalo; water ox
【水暖工】plumber
【水泡】①(水泡泡) bubble ②(水疱) blister: 脚上打了~ get blisters on one's feet

【水疱】blister
【水平】①(与水平面平行) horizontal; level: ～梯田 level terraced field; level terrace ②(政治、业务等方面达到的高度) standard; level: 认识～ level of one's understanding / 生活～ living standard / 文化～ standard of education; cultural level ◇ ～飞行 horizontal flight / ～轰炸 [军] horizontal bombing; level bombing / ～贸易[经] horizontal trade / ～面 horizontal plane; level (surface) / ～同步 horizontal synchronization / ～线 horizontal line / ～仪 level
【水泼不进,针插不进】water-tight and impenetrable
【水汽】vapor; steam; moisture ◇ ～浓度 [气] vapor concentration
【水枪】[矿] giant; (hydraulic) monitor: 水采～ hydraulic giant
【水禽】[动] waterfowl; water bird
【水清无鱼】Fish do not come when water is too clear. (或: When the water is too clear, there are no fish.) / One who is too clever has no friends.
【水情】[水] regimen
【水球】[体] ①(指运动项目) water polo ②(指球) water polo ball ◇ ～场 playing pool; bath / ～球门 cage
【水曲柳】[植] northeast China ash
【水渠】ditch; canal
【水溶性】[生化] water-soluble: ～维生素 B water-soluble (vitamin) B
【水溶液】[化] aqueous solution
【水乳交融】as well blended as milk and water; in complete harmony: 好的翻译可以使宾主谈得～。A fully qualified interpreter can help bring about a meeting of minds like milk mingling with water.
【水砂纸】waterproof abrasive paper
【水杉】[植] metasequoia
【水上芭蕾】water ballet
【水上飞机】seaplane; hydroplane; water plane
【水上飞行】overwater flight
【水上警察】harbor police; waterside police
【水上居民】boat dwellers
【水上派出所】water police station
【水上运动】[体] aquatic sports; water sports ◇ ～会 aquatic sports meet
【水蛇】[动] water snake
【水深】depth of water
【水深火热】deep water and scorching fire; an abyss of suffering; extreme misery
【水生动物】aquatic animal
【水生植物】water plant; hydrophyte
【水声学】[物] marine acoustics
【水虱】[动] beach louse
【水势】flow of water; rise and fall of floodwater: 密切注意～ keep a close eye on the flow of the water / ～减退。The flood subsided. / ～汹涌。The current is turbulent.
【水手】seaman; sailor ◇ ～长 boatswain
【水刷石】[建] granitic plaster
【水松】[植] China cypress
【水塔】water tower
【水獭】[动] otter
【水潭】puddle; pool
【水塘】pool; pond
【水天线】sky-line
【水天一色】The waters and skies are of one color. (或: The water and the sky merge in one color.)
【水田】paddy field; irrigated field ◇ ～耙 paddy field harrow / ～犁 paddy field plough
【水桶】pail; bucket
【水头】①(落差) head of water ②(洪峰) flood peak; peak of flow
【水土】①(水和土) water and soil: ～保持 water and soil conservation / ～流失 soil erosion / 保持～ conserve water and soil ②(环境和气候) natural environment and climate: ～不服 unaccustomed to the climate of a new place; not acclimatized
【水汪汪】(of eyes) bright and intelligent
【水网】a network of rivers
【水位】water level: 低～ low water level / 地下～ water table; groundwater level / 高～ high water level / 警戒～ warning line / 最高～ peak level ◇ ～计 fluviograph
【水文】hydrology ◇ ～测验 hydrologic survey / ～地理学 hydrography / ～地质 hydrogeology / ～地质学 hydrogeology / ～队 hydrological team / ～工作者 hydrologist / ～年鉴 Water Year Book / ～气象学 hydrometeorology / ～设计 hydrologic design / ～学 hydrology / ～预报 hydrologic forecast / ～站 hydrometric station; hydrologic station / ～资料 hydrological data
【水污染】water pollution; pollution of water
【水螅】[动] hydra
【水系】river system; water system; hydrographic net
【水仙】[植] narcissus
【水险】marine insurance; marine risk

【水线】waterline
【水乡】a region of rivers and lakes
【水箱】water tank
【水泻】[医] watery diarrhea
【水泄不通】not even a drop of water could trickle through; be watertight: 挤得～ be packed with people / 围得～ be so closely besieged that not a drop of water could trickle through
【水榭】waterside pavilion
【水星】[天] Mercury
【水性】①(游水的技能) ability in swimming: 这姑娘的～很好 This girl is a good swimmer. ②(水的深浅、特点) the depth, currents and other characteristics of a river, lake, etc.
【水性杨花】vamp; vampiness
【水锈】①(水垢) scale; incrustation ②(浸水痕迹) watermark (in water vessels)
【水压】hydraulic pressure ◇ ～机 hydraulic press
【水烟】shredded tobacco for water pipes: 抽～ smoke a water pipe ◇ ～袋 water pipe
【水杨】[植] bigcatkin willow
【水杨酸】[化] salicylic acid ◇ ～钠 sodium salicylate
【水翼船】hydrofoil
【水银】[化] mercury; quicksilver ◇ ～灯 mercury-vapor lamp / ～气压表 [气] mercury barometer / ～温度计 [气] mercury thermometer / ～蒸气灯 mercury-vapor lamp / ～整流器 mercury rectifier; vapor converter / ～中毒 mercurial poisoning / ～柱 mercury column
【水印】① [美术] watercolor block printing ② (纸上有明显纹理的图案) watermark
【水域】waters; water area; body of water: 国际～ an international body of water; international waters / 内陆～ inland waters
【水源】①(河流发源地) the source of a river; headwaters; water head ②(用水的来源) source of water: 寻找～ seek new sources of water
【水运】water transport ◇ ～码头 a port handling river cargo
【水灾】flood; inundation: 遭受～ suffer from floods; be affected by floods
【水葬】water burial
【水蚤】water flea
【水藻】algae
【水闸】sluice; water gate; waterlocks: 分～ distribution structure; diversion gate / 进～ head gate / 排～ drainage gate / 泄～ water release gate
【水涨船高】when the river rises the boat goes up
【水针疗法】[中医] acupuncture therapy with medicinal injection
【水蒸汽】steam; water vapor
【水至清则无鱼】When the water is too clear there are no fish. (或: One should not demand absolute purity.)
【水质】water quality ◇ ～保护 [环保] water quality protection / ～标准 water standard / ～分析 water analysis / ～污染 water pollution
【水蛭】[动] leech
【水中捞月】fish for the moon in the water; make impractical or vain efforts
【水肿】[医] oedema; dropsy
【水柱】water column
【水准】level; standard: 低于一般～ below average / 高于一般～ above average ◇ ～点 bench mark / ～面 level surface; level plane / ～器 spirit level / ～仪 surveyor's level; levelling instrument
【水渍险】[经] with particular average (W.P.A.)
【水族】aquatic animals ◇ ～馆 aquarium

shuì

说 try to persuade: 游～ go around urging rulers to adopt one's political views; peddle an idea; drum up support for a scheme or plan

税 tax; duty: 超额累进～ progressive tax of the income from wages and salaries in excess of specific amounts / 出口～ export duty / 房地产～ housing and land tax / 附加～ additional tax / 港口～ harbor dues / 间接～ indirect tax / 进口～ import duty / 累进～ progressive tax / 所得～ income tax / 营业～ business tax / 直接～ direct tax / 减～ reduce a tax; reduction of a tax / 免～ tax-free; duty-free; free of tax; free of duty; exemption from a tax / 纳～ pay taxes / 偷～ evade taxes ○扣缴义务人 withholding agent / 纳税人 taxpayer
【税额】the amount of tax to be paid
【税法】law of tax; tax law
【税款】tax payment; taxation
【税率】tax rate; rate of taxation; tariff rate
【税目】tax items; taxable items
【税收】tax revenue ◇ ～政策 tax policy
【税务机关】tax authority
【税务稽查员】inspector of taxes

【税务局】tax bureau
【税务员】tax collector
【税则】tax regulations
【税制】tax system; taxation: 累进～ progressive taxation
【税种】categories of taxes

睡 sleep: ～得好 sleep well / ～得晚 stay up late / 早～早起 early to bed and early to rise
【睡觉】sleep: 睡午觉 take a nap after lunch / 该～了。It's time to go to bed
【睡莲】[植] water lily
【睡帽】nightcap
【睡梦】sleep; slumber: ～状态 dream state / 在～中 in one's sleep
【睡眠】sleep: ～不足 not have enough sleep ◇ ～病 sleeping sickness / ～倒错 perversion of sleep / ～疗法 physiological sleep therapy / ～性麻痹 sleep paralysis / ～障碍 sleep-disorder; somnipathy
【睡醒】wake up: 她～了么? Has she waked up yet?
【睡衣】night clothes; pajamas
【睡衣裤】pajamas
【睡意】sleepiness; drowsiness: 有几分～ feel some what sleepy; be drowsy

shǔn

吮 suck: ～乳 suck the breast
【吮吸】suck

shùn

瞬 wink; twinkling: 转～之间 in a twinkling
【瞬时】[物] instantaneous ◇ ～速度 instantaneous velocity / ～性 instantaneity / ～值 instantaneous value
【瞬息】twinkling: ～间 in the twinkling of an eye / ～万变 undergoing a myriad changes in the twinkling of an eye; fast changing

顺 ①(向着同一方向) in the same direction as; with: ～流而下 go downstream / ～时针方向 clockwise ②(沿着) along: ～着大街走 follow the main street ③(使有条理) arrange; put in order: 这篇文章还得～一～。 This article needs polishing. ④(顺从) obey; yield to; act in submission to: ～从他的意愿 yield to his wishes ⑤(适合) suitable; agreeable: 不～他的意 not fall in with his wishes ⑥(趁便) take the opportunity to: ～致最崇高的敬意。 I avail myself of this opportunity to renew to you the assurances of my highest consideration. ⑦(依次) in sequence: 这些号码是～～的。 These are serial numbers.
【顺便】conveniently; in passing: ～说一句 by the way; incidentally
【顺差】favorable balance; surplus: 国际收支～ favorable balance of payments; balance of payments surplus / 贸易～ favorable balance of trade
【顺产】[医] natural labor
【顺畅】smooth; unhindered: 呼吸渐渐～ begin to breathe the more easily
【顺磁】[物] paramagnetic: ～共振 paramagnetic resonance
【顺次】in order; in succession; in proper sequence
【顺从】be obedient to; submit to; yield to
【顺当】smoothly; without a hitch
【顺丁橡胶】[化] butadiene rubber
【顺耳】pleasing to the ear
【顺风】①(与风同方向) have a favorable wind; have a tail wind: ～行船 sail with the wind / 一路～ a pleasant journey; bon voyage ②(顺利) favorable wind; tail wind
【顺风耳】①(旧小说中能听很远的人) a person who can hear voices a long way off ②(消息灵通的人) a well-informed person
【顺风转舵】trim one's sails; take one's cue from changing conditions
【顺竿儿爬】follow sb.'s cue and do everything to please him; readily fall in with other people's wishes
【顺口】①(念着流畅) read smoothly ②(未经考虑) speak casually; say offhandedly: ～答应 promise thoughtlessly ③[方](可口) suit one's taste: 这菜我吃着很～。 I like the taste of this dish.
【顺口溜】doggerel; jingle
【顺理成章】to write well, you must follow a logical train of thought; to do some work well, you must follow a rational line: 这显然是～的。 This is undoubtedly logical.
【顺利】smoothly; successfully; without a hitch: ～完成任务 have successfully done one's job
【顺路】①(顺着所走的路线到另一处去) on the way: ～到朋友家看看 call on one's friend on the way ②(道路没有曲折) direct route; regular route
【顺势】take advantage of an opportunity (as provided by an opponent's reckless move)
【顺手】①(顺利) smoothly; without difficulty:

事情办得相当～。It was done without a hitch. ②(随手) conveniently; without extra trouble: 请出去时～关门。 Would you close the door when you go out? ③(顺便) do sth. as a natural sequence or simultaneously ④(合用) handy; convenient and easy to use: 这把铁锹使起来挺～。This spade is very handy.

【顺手牵羊】lead away a goat in passing; pick up sth. on the sly; walk off with sth.

【顺水】downstream; with the stream

【顺水人情】a favor done at little cost to oneself

【顺水推舟】push the boat along with the current; make use of an opportunity to gain one's end

【顺藤摸瓜】follow the vine to get the melon; track down sb. or sth. by following clues

【顺天应人】follow the mandate of heaven and comply with the popular wishes of the people

【顺我者昌,逆我者亡】those who submit will prosper, those who resist shall perish

【顺心】satisfactory: 诸事～。All is well.

【顺行】[天] direct motion

【顺序】①(次序) sequence; order: 按年代～ in chronological sequence / 按字母～排列 in alphabetical order ②(依次) in proper order; in turn

【顺延】postpone: 遇雨～ subject to postponement in case of rain

【顺眼】pleasing to the eye: 看着不～ be offensive to the eye; be an eyesore

【顺应】comply with; conform to: ～民心 comply with the aspirations of the people / ～时势 go with the stream

shuō

说 ①(用话表达) speak; talk; say: ～得多做得少 talk much but do little / ～某人好话 speak well of sb. / ～某人坏话 speak ill of sb. / 用英语再～一遍。 Say it again in English. ②(解释) explain: 怎么也～不明白。 You can't make it clear, no matter how hard you try to explain. ③(言论;主张) theory; teachings; doctrine: 著书立～ write books to expound a theory ④(责备;批评) scold: 他爸爸～了他几句。Daddy gave him a talking-to.

【说白】spoken parts in an opera

【说不得】①(不能说) unspeakable; unmentionable ②(极其不堪,无从说起) scandalous

【说不定】perhaps; maybe: ～他已经走了。Maybe he's already left.

【说不过去】be hardly justified; cannot be explained away

【说不来】cannot get along (with sb.): 我跟他～。 I don't see eye to eye with him.

【说曹操,曹操就到】talk of the devil and he will appear

【说长道短】indulge in idle gossip; make captious comments

【说唱】a genre of popular entertainment consisting mainly of talking and singing

【说穿】tell what sth. really is; reveal; disclose: ～了,无非是想推卸责任。 To put it bluntly, this is shifting responsibility.

【说大话】brag; boast; talk big

【说到底】in the final analysis; at bottom

【说到痛处】sting to the quick

【说到做到】do what one says; match one's deeds to one's words; live up to one's word

【说得过去】justifiable; passable: 她的英语发音还～。 Her English pronunciation is passable.

【说得来】can get along; be on good terms: 找一个跟他～的人去动员他。 Get someone who is on good terms with him to try and persuade him.

【说定】settle; agree on: 他们已经～了时间和地点。 They have agreed on the time and place.

【说东道西】chatter away on a variety of things

【说法】①(措词) way of saying a thing; wording; formulation: ～不妥 inappropriate wording / 换一个～ say it in another way ②(意见) statement; version; argument: ～不一 have different versions / 这种～不对头。 Such statement isn't correct.

【说服】persuade; convince; prevail on; talk sb. over: ～教育的方法 the method of persuasion and education / 要耐心～他。 Talk to him patiently to bring him round ◇ ～疗法 pithiatism; pithiatry

【说合】①(从中介绍) bring two (or more) parties together: ～亲事 make a match ②(商议) talk over; discuss

【说和】mediate a settlement; compose a quarrel: 你去给他们～～。 Try to patch things up between them, will you?

【说话】①(用话表达) speak; talk; say: 有～的权利 have the right to speak / ～不算话 go back on one's word ②(闲谈) chat; talk: 找人～ have a chat with sb. ③(指责) gossip; talk: 你这样干,人家要～。 Considering what you've done, people will talk. ④[口](时间短) in a minute; in no time; right away: ～就得。 It'll be ready in a jiffy.

【说谎】tell a lie; lie
【说教】deliver a sermon; preach
【说客】a person often sent to win sb. over or enlist his support through persuasion; a persuasive talker
【说来话长】it's a long story
【说理】argue; reason things out: 与某人～ argue with sb.; reason things out with sb.
【说媒】act as matchmaker
【说明】①(解释明白) explain; illustrate; show: ～理由 give reasons / ～真相 give the facts ②(解释意义的话) explanation; directions; caption: 事实本身就是很好的～。 Facts speak for themselves. ◇ ～文 expository writing; exposition / ～语句 declarative statement / ～资料 detail file
【说明书】①(物品的文字说明) (a booklet of) directions ②(技术说明) manual; technical manual ③(影剧说明书) synopsis
【说亲】act as matchmaker
【说情】plead for mercy for sb.; intercede for sb.
【说三道四】make carding comments; make irresponsible remarks
【说书】storytelling
【说头儿】①(可谈之处) sth. to talk about; things to discuss ②(辩解的理由) excuse
【说妥】come to an agreement
【说笑】chatting and laughing
【说一不二】mean what one says; stand by one's word
【说嘴】①(自夸) brag; boast: 咱们谁也别～。 Let's not have any boasting. ②[方](争辩) argue; quarrel: 好和人～ like to quarrel with people

shuò

数 frequently; repeatedly
【数见不鲜】encountered with many times; common occurrence; nothing new
【数脉】[中医] rapid pulse

朔 ①(新月) new moon ②(朔日) the first day of the lunar month ②(北) north: ～风 north wind
【朔日】the first day of the lunar month
【朔望】the first and the fifteenth day of the lunar month; syzygy ◇ ～月 [天] lunar month; lunation; synodic month
【朔月】new moon

蒴
【蒴果】[植] capsule

硕 large
【硕大无朋】of unparalleled size; gigantic; exceptionally large
【硕果】rich fruits; great achievements: ～累累 have achieved great success
【硕果仅存】rare survival
【硕士】Master: 理学～ Master of Science (M.S.) / 文学～ Master of Arts (M.A.) ◇ ～学位 Master's degree
【硕学鸿儒】a great learned literate

烁 bright; shining: 闪～ twinkle; glimmer
【烁烁】glitter; sparkle

铄 [书] ①(熔化) melt (metal, etc.): ～石流金 sweltering ②(耗损) waste away; weaken

sī

斯 ①[书](这;此) this: ～人 this person / ～时 at this moment / 生于～,长于～ be born and brought up here ②[书](于是;就) then; thus
【斯里兰卡】Sri Lanka ◇ ～人 Sri Lankan
【斯瓦希里语】Swahili (language)
【斯威士兰】Swaziland ◇ ～人 Swazi
【斯文】refined; gentle: 举止～ refined in manner
【斯文扫地】disgrace one's scholarly dignity; a scholar's misbehavior; have the educated humbled to the dust

澌
【澌灭】totally disappear

厮 ①(男仆) male servant: 小～ page boy; page ②(对人轻蔑的称呼) fellow; guy: 那～ that guy ③(互相) with each other; together: ～混 fool around together
【厮打】come to blows; exchange blows; tussle
【厮杀】fight at close quarters (with weapons)

撕 tear; rip: ～成两半 tear in two; tear in half / ～得粉碎 tear to shreds / ～下布告 tear down the notice / ～下假面具 tear off the mask; unmask / ～下一张日历 tear off a leaf from a calendar / 把信～开 rip open a letter / 把信～得粉碎 tear the letter into tiny pieces / 从笔记本上～下一页 tear a leaf out of one's notebook
【撕毁】tear up; tear to shreds: ～协定 tear up an agreement; tear an agreement to shreds
【撕裂】laceration; tearing

【撕破脸皮】no longer spare sb.'s sensibilities
【撕脱】avulsion
嘶[书]①(叫) neigh: 人喊马～ men shouting and horses neighing ②(象) hiss ③(嘶哑) hoarse: 声～力竭 hoarse and exhausted
【嘶哑】hoarse: ～的嗓音 a hoarse voice
思①(思考;想) think; consider; deliberate: ～前想后 think over again and again / 多～ think more ②(思念) think of; long for: ～亲 think of one's parents with affection ③(思路) thought; thinking: 哀～ mourning / 文～ train of thought in writing
【思潮】①(思想潮流) trend of thought; ideological trend: 文艺～ trend of thought in literature and art ②(接二连三的思想活动) thoughts: ～起伏 disquieting thoughts surging in one's mind
【思忖】[书] ponder; consider
【思考】think deeply; ponder over; reflect on: ～问题 ponder a problem / 独立～ think things out for oneself; think independently ◇ ～型 thoughtful type
【思量】consider; turn sth. over in one's mind
【思路】train of thought; thinking: ～断了 lose the train of thought / 打断～ interrupt one's train of thought
【思虑】consider carefully; contemplate; deliberate
【思念】think of; long for; miss: ～战友 long for one's comrades-in-arms / 我多么～你呀! I miss you terribly!
【思索】think deeply; ponder: 用心～ do some hard thinking / 周密地～ consider carefully
【思维】[哲] thought; thinking ◇ ～奔逸 flight of ideas / ～方式 made of thinking / ～过程 thought process / ～破裂 split of thought / ～停顿 thought-stopping / ～障碍 disturbance of thought
【思贤若渴】desire greatly to win the support of the wise; One's love for able men is equal to one's thirst for water.
【思想】thought; thinking; idea; ideology: 军事～ military thinking / 正确～ correct ideas / 解除～顾虑 free one's mind of misgivings ◇ ～包袱 sth. weighing on one's mind / ～迟钝 bradyphrenia / ～动向 ideological trend / ～斗争 ideological struggle; mental struggle / ～方法 method of thinking / ～感情 thoughts and feelings; minds and sentiments / ～工作 ideological work / ～家 thinker / ～僵化 mental stagnation / ～解放 ideological emancipation / ～境界 ideological level / ～体系 ideological system; ideology / ～性 ideological content / ～意识 ideology
【思绪】①(思路) train of thought; thinking: ～纷乱 a confused state of mind; a confused train of thought ②(情绪) feeling: ～不宁 feel perturbed
锶[化] strontium (Sr): ～单位[核子] strontium unit; sunshine unit
【锶龄】strontium age
私 ①(个人的) personal; private: ～信 personal letter ②(私心) selfish: 无～ unselfish; selfless ③(暗地里) secret; private: ～话 confidential talk ④(非法的) illicit; illegal: ～卖 illicit sale / ～设公堂 set up an illegal court; set up a kangaroo court
【私奔】elopement
【私弊】corrupt practices
【私藏毒品】[法] possession of drug; unlawful possession of drug
【私藏武器】[法] unlawful possession of weapons
【私产】private property
【私娼】unlicensed prostitute
【私仇】personal enmity; personal grudge
【私邸】[旧] private residence
【私法】[法] private law
【私贩鸦片】opium trafficking
【私房】①(个人积蓄) private savings: ～钱 private savings of a family member ②(不想让外人知道的) confidential: 谈～话 exchange confidences
【私愤】personal spite: 泄～ vent personal spite
【私股】private share (in a joint state-private enterprise)
【私货】smuggled goods; contraband goods
【私交】personal friendship
【私立】[旧] privately run; private: ～学校 private school
【私利】private interests; personal gain: 不谋～ seek no personal gain / 图～ pursue private ends
【私了】compounding
【私囊】private purse: 饱～ line one's pockets; feather one's nest
【私情】personal relationships: 不徇～ not swayed by personal considerations
【私人】①(个人的) private; personal: ～访问

private visit / ～关系 personal relations / ～开业 private practice / ～汽车 private car ②(自己的人) one's own man: 任用～ fill a post with one's own man; practice nepotism ◇～财产 private goods and chattels / ～代表 personal representative / ～动产 chattels personal / ～借款 private loan / ～经济 private sector of the economy / ～劳动[经] individual labor / ～秘书 private secretary / ～企业 private enterprise / ～资本 private capital / ～侦探 private detective / ～支票 individual check

【私商】businessman; merchant; trader

【私生活】private life

【私生子】illegitimate child; bastard

【私事】privacy; private affairs; personal affairs: 我们不该打听他人的～。We must respect other's privacies.

【私塾】old-style private school

【私逃】abscond

【私通】①(私下勾结) have secret communication with: ～敌人 have secret communication with the enemy ②(通奸) illicit intercourse; adultery; fornication ◇～案件 case of fornication

【私下】in private; in secret; privately: ～商议 discuss a matter in private ◇～和解 compounding

【私相授受】privately give and privately accept; make an illicit transfer

【私心】selfish motives; selfishness: ～杂念 selfish ideas and personal considerations

【私刑】lynch; illegal punishment (meted out by a kangaroo court)

【私蓄】private savings

【私营】privately owned; privately operated; private: ～工商业 privately owned industrial and commercial enterprises / ～企业 private enterprise

【私有】privately owned; private ◇～财产 private property / ～财富 private wealth / ～观念 private ownership mentality / ～制 private ownership

【私语】①(小声说) whisper: 窃窃～ talk in whispers ②(私下说的话) confidence

【私欲】selfish desire

【私运进口】smuggle goods into a port

【私运军火】gun-running

【私运鸦片】opium smuggling

【私章】personal seal; signet

【私自】privately; secretly; without permission: ～逃跑 escape secretly

司 ①(主持;操作) take charge of; attend to; manage: 各～其事。Each attends to his own duties. ②(部级机关内之一部门) department: 外交部礼宾～ the Protocol Department of the Ministry of Foreign Affairs / 新闻～ Information Department

【司泵员】pump man; pumper

【司法】administration of justice; judicature ◇～部门 judicial departments; judiciary / ～程序 judicial process; judicial procedure / ～独立 judicial independence / ～鉴定 expert testimony / ～机关 judicial office / ～警察 judicial police / ～权 judicial powers / ～行为 judicial act / ～仲裁 judicial arbitration

【司号员】bugler; trumpeter

【司机】driver: 火车～ engine driver; locomotive engineer / 汽车～ driver ◇～室 cab

【司空见惯】a common sight; a common occurrence; a matter of repeated occurrence

【司令】commander;l commanding officer: 总～ commander in chief ◇～部 headquarters; command / ～员 commander; commanding officer

【司炉】stoker; fireman; chief stoker

【司务长】company quartermaster

【司线员】[体] linesman

【司药】pharmacist; druggist; chemist

【司仪】master of ceremonies

【司钻】(head) driller: 副～ assistant driller

丝 ①(丝绸;蚕丝) silk: ～绸之路 the Silk Road ②(丝状物) a threadlike thing: 肉～ meat cut into slivers; shredded meat / 铜～ copper wire / 钨～ tungsten filament / 一～亮光 a thread of light ③(极少、极小的量) a tiny bit; trace: 没有一～笑容 not a trace of a smile / 一～不差 not a bit of difference

【丝虫】[动] filaria ◇～病 filariasis

【丝绸】silk cloth; silk

【丝带】silk ribbon; silk braid; silk sash

【丝杠】[机] guide screw; leading screw ◇～车床 leading screw lathe

【丝糕】steamed corn cake

【丝瓜】towel gourd; dishcloth gourd ◇～络 loofah; vegetable sponge

【丝光】the silky luster of mercerized cotton fabrics ◇～机 mercerizing range / ～纱线 mercerized yarn

【丝毫】the slightest amount or degree; a bit; a

particle; a shred; an iota: ～不差 fit it to a hair; not err by a hair's breadth; tally in every detail; be just right / 拿不出～证据 cannot provide a shred of evidence
【丝极】[电] filament
【丝裂霉素】mitomycin
【丝绵】silk floss; silk wadding
【丝幕】[生] silkscreen; tent
【丝绒】velvet; velour
【丝丝入扣】all threads neatly tied up; (done) with meticulous care and flawless artistry
【丝网】[印] silk screen ◇ ～印刷 silk-screen printing; screen printing / ～印刷机 screen process press
【丝弦】silk string (for a musical instrument)
【丝线】silk thread (for sewing); silk yarn
【丝织品】①(丝纺织品) silk fabrics ②(用丝编织的衣物) silk knit goods
【丝竹】①(乐器总称) traditional stringed and woodwind instruments: ～乐 *ensemble* of such instruments ②(音乐) music
【丝锥】[机] tap: 粗制～ taper tap / 精～ bottoming tap / 中～ second tap

sǐ

死 ①(失去生命) die; pass away: 病～ die of illness / 打～ beat to death / 受伤而～ die from wound ②(拼死) to the death: ～战 fight to the death ③(达到极点) extremely; to death: ～要面子 save face at all costs / 高兴～了 be extremely happy / 渴得要～ be parched with thirst; be dying for a drink / 累～了 be tired to death / 讨厌～了 be dying of boredom ④(不可调和的) implacable; deadly: ～对头 sworn enemy ⑤(死板; 固定) fixed; rigid; inflexible: ～规矩 a rigid rule / ～教条 lifeless dogma ⑥(不能通过) impassable; closed: 把漏洞堵～ plug the holes; stop up loopholes
【死板】rigid; inflexible; stiff: 办事～ work in a mechanical way
【死不】would rather die than; stubbornly refuse to: ～认错 stubbornly refuse to admit one's mistake / ～要脸 be dead to all feelings of shame; be utterly shameless
【死不瞑目】not close one's eyes when one dies; die with a grievance or everlasting regret
【死产】[医] stillbirth; dead-birth
【死党】sworn followers; diehard followers
【死得其所】die a worthy death
【死敌】deadly enemy; mortal enemy; implacable foe
【死地】a fatal position; deathtrap: 置之～而后快 satisfied with nothing less than sb.'s death / 置之～而后生 confront a person with the danger of death and he will fight to live
【死而后已】until one's dying day; to the end of one's days: 鞠躬尽瘁, ～ have dedicated one's life to a cause; bend one's back to the task until one's dying day
【死光】death ray
【死鬼】devil: 你这个～ You devil!
【死胡同】blind alley; dead end
【死缓】[法] short for death sentence with a two-year reprieve and forced labor; stay of execution
【死灰复燃】dying embers glowing again; resurgence; revival
【死活】①(死或活; 活得下去活不下去) life or death; fate: 不顾别人～ leave other people to sink or swim / 不知～ have no idea of death or danger ②[口](无论如何) anyway; simply: 他～不承认。He simply won't admit it / 他～不让我走。I wanted to go, but he simply wouldn't hear of it.
【死火山】extinct volcano
【死记硬背】mechanical memorizing
【死寂】[书] deathly stillness: 夜深人静一片～。Night was deep and dead silence reigned everywhere.
【死角】①[军] dead angle; blind angle; dead space ②(影响尚未达到的地方) a spot as yet untouched by sth.
【死结】fast knot
【死扣儿】[口] fast knot
【死里逃生】escape from the jaws of death; have a narrow escape; barely escape with one's life
【死力】①(最大的力量) all one's strength: 出～ exert one's utmost effort ②(使出最大的力量) with all one's strength: ～抵抗 resist with might and main; fight tooth and nail
【死路】①(不通的路) blind alley ②(毁灭的途径) the road to ruin: 自寻～ court self-destruction; bring ruin upon oneself
【死马当作活马医】doctor a dead horse as if it were still alive; not give up for lost; make every possible effort
【死面】unleavened dough
【死命】①(必死的命运) doom; death: 制敌于～ send the enemy to his doom ②(拼命) desperately: ～挣扎 struggle desperately
【死难】die in an accident or a political incident

(esp. for a revolutionary cause): ～烈士 martyr
【死脑筋】one-track mind
【死皮赖脸】thick-skinned and hard to shake off; brazen-faced and unreasonable
【死棋】a dead piece in a game of chess; a hopeless case; a stupid move
【死气沉沉】lifeless; spiritless; stagnant
【死契】irrevocable title deed
【死囚】a convict sentenced to death; a convict awaiting execution; condemned prisoner
【死球】[体] dead ball
【死去活来】half dead; half alive; hovering between life and death: 哭得～ weep one's heart out
【死人】a dead person; the dead; (the) defunct
【死生由命】Life and death are fore ordained. (或: A man's life is governed by fate.)
【死尸】corpse; dead body
【死守】①(拼死守住) defend to the death; defend to the last; make a last-ditch defense: ～阵地 defend the position to the last ②(墨守) obstinately cling to; rigidly adhere to
【死水】stagnant water
【死胎】[医] stillborn foetus; stillbirth
【死亡】death; doom: 挣扎在～线上 struggle for existence on the verge of death; struggle to stave off starvation ◇ ～保险 insurance against death; mortality insurance / ～抚恤金 death benefits / ～鉴定 verification of death / ～率 death rate; mortality rate / ～诊断 medical diagnosis on death / ～证明书 death certificate; certificate of death
【死无对证】the dead cannot bear witness
【死无葬身之地】die without a burial place; come to a bad end
【死心】drop the idea forever; have no more illusions about the matter: 还不～ refuse to give up hope
【死心塌地】be dead set; be hell-bent
【死心眼儿】①(固执) stubborn; as obstinate as a mule ②(死心眼儿的人) a person with a one-track mind
【死信】①(无法投递的信) dead letter ②(死讯) news of sb.'s death
【死刑】[法] death penalty; death sentence; capital punishment; nameless death: 被处～ be put to death ◇ ～案件 capital case; death penalty case / ～缓刑判决 death sentence with reprieve / ～判决 death sentence; judgement of execution / ～囚犯 capital prisoner; condemned criminal / ～执行令 death warrant / ～执行人 executioner
【死讯】news of sb.'s death
【死因】cause of death ◇ ～调查 coroner's inquest; death inquiry / ～赠与 gift in prospect of death / ～证明书 certificate of cause of death
【死硬】①(呆板) stiff; inflexible ②(顽固) very obstinate; die-hard ◇ ～派 diehards
【死有余辜】even death would be too good for him; even death would not expiate all his crimes
【死于非命】die an unnatural death
【死者】the dead; the deceased; the departed
【死罪】capital crime; capital offense

sì

肆 ①(不顾一切) wanton; unbridled 大～攻击 wantonly vilify; launch an unbridled attack against ②("四"的大写) four ③[书](铺子) shop: 茶楼酒～ teahouses and wineshops
【肆口漫骂】rail and swear at wildly
【肆虐】indulge in wanton massacre or persecution; wreak havoc
【肆无忌惮】unbridled; brazen; unscrupulous: ～地攻击 make unbridled attacks
【肆意】wantonly; recklessly; willfully: ～挥霍 freely squander / ～歪曲事实 wantonly distort the facts / ～妄为 do what one wishes without restraint / ～诬蔑 wantonly slander

寺 temple: 清真～ mosque
【寺院】temple; monastery

四 (数目) four: ～分之一 a quarter; one-fourth / ～号 Number Four / 第～卷 Volume Four; the Fourth Volume
【四倍体】[生] tetraploid
【四边】(on) four sides; all sides: ～围着篱笆 be enclosed with a fence; have a fence on all sides
【四边形】quadrilateral
【四不象】① [动] David's deer; mi-lu ②(不伦不类的东西) nondescript; neither fish nor fowl
【四重唱】[乐] (vocal) quartet
【四重奏】[乐] (instrumental) quartet: 弦乐～ string quartet
【四处】all around; in all directions; everywhere: ～奔走 go hither and thither / ～碰壁 get into trouble on all sides / ～逃窜 flee in all directions
【四叠体】[生理] corpora quadrigemina
【四方】①(各处) the four directions (north,

south, east, west); all sides; all quarters: ～响应。 Response came from every quarter. ②(正方形) square ③(立方体的) cubic: 一块～木头 a wooden cubic

【四分五裂】fall apart; be rent by disunity; be all split up; disintegrate: 敌人营垒～。 The enemy camp is disintegrating.

【四分音符】[乐] crotchet, quarter note

【四个现代化】the four modernizations: 实现～ achieve the comprehensive modernization of agriculture, industry, national defense, and science and technology

【四海】the four seas; the whole country; the whole world: ～为家 make one's home wherever one is / ～之内皆兄弟。 All within the Four Seas are brothers.(或: Within the four seas all men are brothers.)

【四害】the four pests ◯老鼠 rat / 臭虫 bedbugs / 苍蝇 fly / 蚊子 mosquitoes

【四合院】*siheyuan*, a compound with houses around a courtyard; quadrangle

【四环素】[药] tetra vcline

【四级风】[气] forc 4 wind; moderate breeze

【四极管】[无] tetr de

【四季】the four seasons: ～不凋 bloom throughout the year / ～如春。 It's like spring all the year round. / ～宜人。 The view was delightful in all seasons. ◯春 spring / 夏 summer / 秋 autumn / 冬 winter

【四季豆】kidney bean

【四郊】suburbs; outskirts

【四脚蛇】lizard

【四开】[印] quarto ◇ ～本 quarto

【四邻】one's near neighbors

【四六风】[医] umbilical tetanus of newborn babies

【四氯化碳】carbon tetrachloride; tetrachloromethane

【四面】(on) four sides; (on) all sides: ～出击 hit out in all directions / ～受敌 be exposed to enemy attacks on all sides

【四面八方】all directions; all quarters; all around; far and near

【四面楚歌】be besieged on all sides; be utterly isolated; land oneself in a tight spot

【四旁】①(前后左右很近的地方) back and front, left and right; all around ②(四周) the "four sides" ◯屋旁 house side / 村旁 village side / 路旁 roadside / 水旁 waterside ◇ ～绿化 turning the "four sides" green

【四平八稳】①(稳当) very steady; well organized: 办事～ be dependable in work ②(缺乏创新精神) lacking in initiative and overcautious

【四起】rise from all directions: 歌声～。 Sounds of singing were heard from all around.

【四散】scatter in all directions

【四色印刷机】four-color press

【四舍五入】[数] rounding (off); to the nearest whole number

【四声】[语] ①(古汉语的字调) the four tones of classical Chinese phonetics ②(现代汉语普通话的字调) the four tones of modern standard Chinese pronunciation

【四时】the four seasons

【四书】The Four Books ◯《大学》*The Great Learning* / 《中庸》*The Doctrine of the Mean* / 《论语》*The Analects of Confucius* / 《孟子》*Mencius*

【四体不勤,五谷不分】can neither use one's four limbs nor tell the five grains apart

【四通八达】extend in all directions: 公路～。 Highways radiate in all directions.

【四项基本原则】the four cardinal principles; Four Fundamental Principles ◯社会主义道路 the socialist road / 人民民主专政 the people's democratic dictatorship / 马克思列宁主义毛泽东思想 Marxism-Leninism and Mao Zedong Thought / 中国共产党的领导 the leadership of the Communist Party of China

【四言诗】Chinese classical poem with lines of four characters each

【四野】the surrounding country; a vast expanse of open ground: ～茫茫,寂静无声。 All is quiet on the vast expanse of open ground.

【四月】①(阳历) April ②(阴历) the fourth month of the lunar year; the fourth moon

【四则】[数] the four fundamental operations of arithmetic ◇ ～运算 arithmetic; four fundamental rules ◯加 addition / 减 subtraction / 乘 multiplication / 除 division

【四诊】[中医] the four methods of diagnosis ◯望诊 observation / 闻诊 ausculatation and olfaction / 问诊 interrogation / 切诊 pulse feeling and palpation

【四肢】the four limbs; arms and legs

【四周】all around

【四足动物】quadruped; tetrapod

驷

【驷马】[书] a team of four horses: 一言既出,～难追。 Even four horses cannot take back what one has said.(或: What has been said cannot be unsaid.)

似 ①(像) similar; like: 白～雪 as white as snow／ 骄阳～火。 The sun was scorching hot. ②(似乎) seem; appear: ～曾相识 seem to have met before ③(表示超过)日子一年胜～一年。 Life has been getting better year by year.
【似懂非懂】have only vague idea
【似乎】it seems; as if; seemingly: ～明天要起风。 It looks as if it'll be windy tomorrow.／～要下雨了。 It looks like rain
【似是而非】apparently right but actually wrong; specious: ～的说法 a specious argument
【似水流年】As a fleeting wave, youth passes. (或: Years pass by quickly.)
【似笑非笑】a faint smile on one's face; with a spurious smile

嗣 ①(继承) succeed; inherit: ～位 succeed to the throne ②(子孙) heir; descendant: 后～ descendants
【嗣后】[书] hereafter; subsequently; afterwards; later on

伺 watch; await
【伺服】[电] servo ◇ ～传动 servo drive／～放大器 servo amplifier／～控制机构 servo-control mechanism／～制动器 servo brakes
【伺机】watch for one's chance: ～而动 wait for the opportune moment to go into action

饲 raise, rear
【饲槽】feeding trough
【饲草】forage grass
【饲料】forage; fodder; feed: 精～ (fodder) concentrates; fine fodder／ 青贮～ silage; ensilage／ 猪～ pig feed ◇ ～粉碎机 feed grinder／～加工厂 feed-processing plant／～作物 forage crop
【饲养】raise; rear: ～家禽 raise poultry／～牲畜 raise livestock
【饲养场】feed lot; dry lot; farm
【饲养员】①(饲养牲畜的) stockman; breeder ②(饲养家禽的) poultry raiser ③(动物园的) animal keeper (in a zoo)

巳 the sixth of the twelve Earthly Branches
【巳时】the period of the day from 9 a.m to 11 a.m.

祀 [书] offer sacrifices to the gods or the spirits of the dead

sōng

松 ①(松树) pine ②(松散) loose; slack: 把绳子再放～点儿。 Give the rope more play.／你的鞋带～了。 Your shoelace has come loose. ③(使松) loosen; relax; slacken: ～一～螺丝 loosen the screw a little／～一口气 relax a little ④(经济宽裕) not hard up: 现在手头～些 be better off ⑤(不坚实) light and flaky; soft: ～脆可口。 It's light and crisp.／这里的土质很～。 The soil here is very loose. ⑥(碎末食品) dried meat floss; dried minced meat: 肉～ dried minced pork
【松柏长青】remain evergreen as the pine and cypress; the pine and cypress stay evergreen
【松绑】untie a person
【松弛】①(不坚实) limp; flabby; slack: 肌肉～ flaccid muscles ②(不紧张) lax: 纪律～ lax discipline
【松貂】pine marten
【松动】①(不拥挤) become less crowded ②(经济宽裕) not hard up ③(活动) become flexible: 她的口气有点～。 She has become a bit more flexible.／我的一个牙齿～了。 I have a loose tooth.
【松果体】[生理] pineal body
【松花】【松花蛋】preserved egg
【松鸡】capercaillie; grouse
【松焦油】[化] pine tar
【松节油】[化] turpentine (oil); oil of turpentine
【松紧】①(松或紧的程度) degree of tightness ②(伸缩) elasticity ◇ ～带 elastic cord; elastic
【松劲】relax one's efforts; slacken (off): ～情绪 slack mood
【松口】①(张口放开) relax one's bite and release what is held ②(不坚持) become less intransigent; soften; relent
【松快】①(不拥挤) less crowded ②(轻松爽快) feel relieved: 吃药后感到～ feel relieved after taking the medicine ③(宽畅) relax
【松毛虫】pine moth
【松明】pine torches
【松球】pinecone
【松软】soft; spongy; loose: ～的表土 spongy topsoil
【松散】①(结构不紧密) loose: 文章结构～。 The article is loosely organized. ②(精神不集中) inattentive ③(使轻松舒畅) relax; take one's ease: 咱们出去～～吧。 Let's go out for a breath of air.

【松手】loosen one's grip; let go
【松鼠】squirrel
【松树】pine tree; pine
【松松垮垮】behave in a lax, undisciplined way; be slack and perfunctory
【松塔】① [方](松球) pinecone ② [中药] the cone of lacebark pine
【松涛】the soughing of the wind in the pines
【松土】[农] loosen the soil; scarify the soil ◇ ～机 loosener; scarifier
【松香】rosin; colophony ◇ ～油 retinol; rosin oil
【松懈】relax; slacken; slack: ～斗志 relax one's will to fight / 工作～ be slack in one's work / 学习～ slack off in one's studies
【松蕈】pine mushroom
【松烟墨】Chinese ink made from pine soot; pine-soot ink
【松针】pine needle
【松脂】rosin; pine resin
【松子】pine nut

嵩 (of mountains) high; lofty

sǒng

悚
【悚然】terrified; horrified: ～而立 stand up terrified / 毛骨～ with one's hair standing on end

怂
【怂恿】instigate; incite; egg sb. on; abet: ～支持 with the support and connivance of somebody

耸 ①(耸立) towering; lofty ②(使人吃惊) alarm; shock: 危言～听 exaggerate things just to frighten people
【耸动】①(向上动) shrug (one's shoulders) ②(使人震动) create a sensation: ～视听 create a sensation
【耸肩】shrug one's shoulders
【耸立】tower aloft: 纪念碑～在广场上。The monument towers aloft on the square. / 群山～。Mountains rise straight up.
【耸人听闻】deliberately exaggerate so as to create a sensation: ～的谣言 a sensational rumor
【耸入云霄】tower to the skies: ～的高山 a high mountain towering to the skies

sòng

宋
【宋体字】Song typeface; standard typeface of Chinese

送 ①(拿东西给人) deliver; carry: ～报 deliver newspaper / ～信 deliver a letter ②(赠送) give as a present; give: ～本书留作纪念 give sb. a book as a souvenir ③(伴送) see sb. off or out; accompany; escort: ～孩子上托儿所 take a child to the nursery / ～她回家 see her home / ～外宾去机场 accompany a foreign visitor to the airport
【送别】see sb. off
【送殡】attend a funeral; take part in a funeral procession
【送达】service ◇ ～传票 service of summons / ～地址 address for service / ～日期 date of service / ～文件 document for service / ～证书 certificate of delivery
【送弹手】[军] ammunition carrier
【送电】power transmission
【送风机】[机] forced draught blower; blower
【送话器】[电] microphone
【送还】give back; return
【送货】deliver goods
【送交】deliver; hand over: 把犯罪分子～法院审判 hand the criminal over to the court for trial
【送旧迎新】see off the old and welcome the new; ring out the Old Year and ring in the New
【送客】see a visitor out
【送礼】give sb. a present; present a gift to sb.: 请客～ give dinners or send gifts
【送命】lose one's life; get killed; go to one's doom
【送气】[语] aspirated ◇ ～音 aspirated sound
【送人情】①(给人好处) do favors at no great cost to oneself ②[方](送礼) make a gift of sth.
【送丧】attend a funeral; take part in a funeral procession
【送死】[口] court death
【送往迎来】see off those who depart and welcome those who arrive; speed the parting guests and welcome the new arrivals: 负责～事宜 be in charge of arrangements for receiving and seeing off guests
【送信儿】[口] send word; go and tell

【送行】①(送别) see sb. off; wish sb. bon voyage: 到机场给人～ go to the airport to see sb. off ②(饯行) give a send-off party
【送葬】take part in a funeral procession
【送终】attend upon a dying parent or other senior member of one's family; bury a parent

诵 ①(读出声来) read aloud; chant ②(背诵) recite
【诵读】read aloud; chant: ～诗篇 recite poems

讼 ①(打官司) bring a case to court ②(争辩是非) dispute; argue
【讼棍】legal pettifogger; shyster
【讼事】lawsuit; litigation

颂 ①(颂扬) praise; extol; eulogize; laud: 歌～ sing the praises of ②(颂歌) song; ode; paean; eulogy:《祖国～》*Ode to Our Motherland* ③(《诗经》中的祭祀歌词) a section in *The Book of Songs* consisting of sacrificial songs
【颂词】①(称赞功德) complimentary address; panegyric; eulogy ②[外] a speech delivered by an ambassador on presentation of his credentials
【颂歌】song; ode
【颂古非今】eulogize the past and condemn the present; extol the past to negate the present
【颂扬】sing sb.'s praises; laud; extol; eulogize

sōu

溲 [书] urinate
【溲血】[中医] haematuria

搜 search
【搜捕】track down and arrest
【搜查】search; ransack; rummage; seek ◇ ～合法[法] legality of search / ～赃物 search for stolen goods / ～证 document of search; search warrant
【搜刮】claw; extort; plunder; expropriate; fleece: ～人民的钱财 extort money from the people
【搜集】collect; gather: ～标本 collect specimens / ～民歌 collect folk songs / ～情报 gather information / ～意见 solicit opinions
【搜罗】collect; gather; recruit: ～人才 recruit qualified persons; scout for talent
【搜身】frisk; search the person; make a body search: ～检查 subject a person to a search; search a person
【搜索】hunting; search for; hunt for; scout around: ～前进 advance and reconnoiter / 到处～ search everywhere for ◇ ～范围 hunting zone / ～飞行 scouting flight / ～周期 search cycle
【搜索枯肠】rack one's brains (for fresh ideas or apt expressions)
【搜寻】search for; look for; seek; frisk
【搜腰包】search sb.'s pockets; search sb. for money and valuables
【搜章摘句】search for chapters and pick sentences

嗖 [象] whiz: 炮弹～的一声飞过去了。A shell whizzed past.

艘 [量] 两～巡洋舰 two cruisers / 一～油船 a tanker

馊 sour; spoiled: ～主意 rotten idea; lousy idea / 饭菜～。The food has spoiled. (或: The food smells a bit off.)

sǒu

薮 [书] ①(生长着很多草的湖) a shallow lake overgrown with wild plants ②(人、物聚集地) a gathering place of fish or beasts; den; haunt

嗾
【嗾使】instigate; abet: ～某人犯罪 abet sb. in a crime / ～某人干坏事 instigate sb. to do evil

sòu

嗽 cough
【嗽必妥】salbutamol

sū

苏 ①(苏醒) revive; come to: 死而复～ come back to life ②(苏维埃) Soviet
【苏打】soda ◇ ～饼干 soda biscuit; soda cracker / ～粉 soda ash
【苏丹】①(某些伊斯兰国家最高统治者的称号) sultan ②(国名) the Sudan ◇ ～人 Sudanese
【苏里南】Surinam ◇ ～人 Surinamese
【苏联】the Soviet Union ◇ ～人 Soviet citizen
【苏门羚】[动] serow
【苏维埃】Soviet
【苏醒】revive; regain consciousness; come to; come round: 使一个晕厥的人～过来 revive a person who has fainted
【苏伊士运河】the Suez Canal

【苏子】perillaseed

酥 ①(松而易碎) crisp; short: ~糖 crunchy candy ②(点心) short pastry; shortbread: 杏仁~ almond shortbread ③(肢体酥软) limp; weak; soft

【酥脆】crisp: ~的饼干 crisp biscuit

【酥麻】limp and numb: 两腿~ one's legs feel weak and numb

【酥软】limp; weak; soft

【酥油】butter ◇ ~茶 buttered tea

sú

俗 ①(风俗) custom; convention: 移风易~ break with old customs; bring about a change in morals and mores ②(普遍流行的) popular; common: 通~ popular (language, style, etc.) / 未能免~ have to do what others are doing ③(庸俗) vulgar: ~物 vulgarian ④(没出家的人) secular; lay: 僧~ monks and laymen; clergy and laity

【俗不可耐】unbearably vulgar

【俗话】common saying; proverb: ~说 as the saying goes

【俗名】popular name; local name

【俗气】vulgar; in poor taste

【俗套】conventional pattern; convention: 不落~ conform to no conventional pattern

【俗务羁身】Some business has detained me.

【俗语】common saying; folk adage

sù

宿 ①(住宿) lodge for the night; stay overnight: 借~ ask for a night's lodging ②[书](旧有的) long-standing; old: ~志 long cherished ambition ③[书](久于其事的; 年老的) veteran; old: ~将 veteran general

【宿根】[植] ①(多年生植物的根) perennial root ②(二年生植物的根) biennial root

【宿疾】chronic complaint; old trouble

【宿命论】[哲] fatalism ◇ ~者 fatalist

【宿舍】hostel; living quarters; dormitory: 学生~ students' hostel / 职工~ living quarters for staff and workers

【宿营】(of troops) take up quarters ◇ ~地 camping site

【宿怨】old grudge; old scores: 捐弃~ sink a feud

【宿愿】long-cherished wish

【宿主】[生] host: 中间~ intermediate host / 终~ final host

溯 ①(逆水行进) go against the stream ②(追溯) trace back; recall: 回~往事 recall past events

【溯流而上】going a up river; go upstream

【溯源】trace to the source: 追本~ track down the origin; trace to the source

塑 model; mold: ~像 mold a statue / 泥~ clay sculpture

【塑胶】plastic cement

【塑炼】plasticate ◇ ~机 plasticator

【塑料】plastics: 氟~ fluoroplastics / 工程~ engineering plastics / 泡沫~ foam plastics / 通用~ general-purpose plastics ◇ ~薄膜 plastic film / ~工业 plastics industry / ~胶布带 [电] plastic adhesive tape / ~凉鞋 plastic sandals / ~热合机 plastic welder / ~贴面板 plastic veneer / ~印版 [印] plastic (printing) plate / ~炸弹 plastic bomb

【塑像】①(塑造人像) mold a statue ②(塑成的人像) statue

【塑造】①(造型) model; mold: ~石膏像 mold a plaster figure ②(描写) portray: ~典型形象 create typical characters

诉 ①(说给人) tell; relate; inform: 告~ tell ②(倾吐) complain; accuse: 控~ accuse / 倾~ pour out (one's feelings, troubles, etc.); unbosom oneself of; unburden oneself of ③(控告) appeal to; resort to: 上~ appeal to a higher court

【诉苦】vent one's grievances; pour out one's woes: ~诉冤 voice one's grievances and state the wrong

【诉说】tell; relate; recount: ~苦衷 recount one's worries and difficulties; tell one's troubles

【诉讼】[法] lawsuit; litigation: 撤消~ withdraw an accusation; drop a lawsuit / 对某人提出~ take legal proceedings against sb. / 民刑~ civil and criminal lawsuits / 提出离婚~ take divorce proceedings ◇ ~案件 contentious case; court case / ~保险 insurance against litigation / ~程序 contentious procedure / ~代理人 law-agent; process attorney / ~费 costs; court costs / ~合并 joinder of causes of action / ~条例 rules of procedure / ~文书 charging document / ~自由 freedom of action

【诉讼法】procedural law: 刑事~ criminal procedure law

【诉因】cause; cause of action
【诉冤】inform
【诉诸法律】go to law; have recourse to law
【诉诸武力】resort to force; appeal to arms; have recourse to force
【诉状】[法] plaint; indictment: 向法院提出～ file a plaint at court

素 ①(白色;本色) white: ～绢 white silk ②(颜色单纯) plain; simple; quiet: ～色 plain color ③(蔬菜瓜果等食物) vegetable: 吃～ be a vegetarian / 两荤一～ two meat dishes and one vegetable dish ④(本来的) native: ～性 one's disposition; one's temperament ⑤(带有根本性质的物质) basic element; element: 毒～ poison / 色～ pigment / 维生～ vitamin ⑥(向来) usually; habitually; always: 我与她～不相识。I don't know her at all.
【素不相识】be strangers to each other; have never met before
【素材】source material (of literature and art); material: 搜集～ gather material
【素菜】vegetable dish
【素餐】①(素的饮食) vegetarian meal ②(吃素) be a vegetarian ③(不做事白吃饭) [书] not work for one's living: 尸位～ hold down a job without doing a stroke of work
【素常】usually; ordinarily: 和～一样 as usual / 起床比～晚 get up later than usual
【素淡】quiet (color)
【素服】white clothing (as a sign of mourning)
【素混凝土】plain concrete
【素净】plain and neat; quiet (color): 花色～ a pattern in quiet colors
【素酒】①(就着素菜而喝的酒) wine served at a vegetarian feast ②[方](素席) vegetarian feast
【素来】always; usually; all along: 他～不吸烟。He never smokes. / 他～守法。He always abides by the law.
【素昧平生】have never met before: 一个～的人 a complete stranger
【素描】①(不加色彩的画) sketch ②(不加渲染的描写) literary sketch
【素日】generally; usually: 他～不爱说话。He is usually very quiet.
【素食】①(素的饮食、点心) vegetarian diet ②(吃素) be a vegetarian ◇ ～者 vegetarian
【素数】[数] prime number
【素席】vegetarian feast
【素馨】[植] jasmine
【素雅】simple but elegant; unadorned and in good taste: 衣着～ be tastefully dressed in a simple style
【素养】accomplishment; attainment: 文学～ literary attainments / 艺术～ artistic accomplishment
【素因子】[数] prime factor
【素有大志】have always cherished a yearning for high enterprise
【素油】vegetable oil
【素质】①(事物本来的性质) quality; performance; level of competence ② [心] diathesis

嗉
【嗉子】crop (of a bird)

速 ①(迅速) fast; rapid; quick; speedy: 收效甚～ produce quick results; have a speedy effect / ～去～回。Go and return quickly. ②(速度) speed; velocity: 风～ wind speed / 声～ velocity of sound / 司机因超～而被罚款。The driver was fined for speeding. ③ [书](邀请) invite: 不～之客 uninvited guest; gate-crasher
【速成】speeded-up educational program ◇ ～班 accelerated course; crash course / ～教学法 quick method of teaching / ～识字法 quick method of achieving literacy
【速冻】quick-freeze: ～蔬菜 quick-frozen fresh vegetables / ～水果 quick-frozen fresh fruits
【速度】① [物] speed; velocity: 初～ initial velocity / 轨道～ orbital velocity / 逃逸～ [天] escape velocity / 巡航～ cruising speed / 匀～ uniform velocity ② [乐] *tempo* ③(快慢的程度) speed; rate; pace; tempo: 工业化的～ the pace of industrialization / 加快～ increase speed ◇ ～表 autometer / ～滑冰 [体] speed skating / ～极限 speed limit / ～计 speed indicator; speedometer
【速记】shorthand; stenography ◇ ～员 stenographer
【速决】quick decision: 速战～ fight a quick battle to force a quick decision ◇ ～战 war of quick decision
【速可眠】[药] secobarbital sodium; seconal
【速率】speed; rate: 冷却～ rate of cooling ◇ ～计 speedometer
【速射】[军] rapid fire ◇ ～炮 quick-firing gun; quick-firer
【速调管】[无] klystron
【速效】quick results ◇ ～肥料 quick-acting

fertilizer
【速写】①（绘画方法之一）sketch ②（一种文体）literary sketch

簌

【簌簌】①[象] rustle：风吹树叶～响。The leaves are rustling in the wind. ②（眼泪纷纷落下的样子）(tears) streaming down

夙[书] ①（早）early in the morning：～夜 morning and night ②（素有的）old; long-standing
【夙世冤家】a bitter enemy; an enemy of former life
【夙兴夜寐】rise early and retire late; hard at work night and day
【夙愿】long-cherished wish：～难偿。The long-cherished hope is hard to realize.
【夙怨】old grudge
【夙志】long-cherished ambition

肃 ①（恭敬）reverent; respectful ②（严肃）solemn：严～ solemn; serious; grave ③（肃清）eliminate; mop up：有反必～ counter-revolutionaries must be suppressed whenever they are found
【肃静】solemn silence：全场～无声。A solemn silence reigned.
【肃立】stand as a mark of respect：～致哀 stand silently mourning
【肃穆】solemn and respectful
【肃清】eliminate; clean up; mop up：～残敌 wipe out the remnants of the enemy／～封建势力 eliminate feudal forces
【肃然起敬】be filled with deep veneration：使我～ call forth in me a feeling of profound respect
【肃然生畏】be struck with awe

suān

酸 ①［化］acid：醋～ acetic acid／硫～ sulphuric acid／硝～ nitric acid／盐～ hydrochloric acid ②（酸味）sour; tart：～果 tart fruit ③（悲痛;伤心）sick at heart; grieved; distressed：令人辛～的往事 sad memories ④（迂腐）pedantic; impractical：～秀才 impractical old scholar; priggish pedant ⑤（酸痛）tingle; ache：腰～背痛 have a pain in the back; have a backache
【酸菜】Chinese sauerkraut; pickled Chinese cabbage
【酸处理】[石油] acid treatment; acidation
【酸楚】grieved; distressed
【酸度】[化] acidity
【酸酐】[化] acid anhydride
【酸解】[化] acidolysis
【酸辣汤】vinegar-pepper soup
【酸溜溜】①（酸味）tart; sour ②（酸痛）pain; tingle; ache：周身～ have a vague feeling of pain all over ③（心里难过的感觉）sad; mournful ④（说话尖刻）sharp-tongued; acrimonious ⑤（迂腐）pedantic
【酸马奶】koumiss
【酸梅】smoked plum; dark plum ◇～汤 sweet-sour plum juice
【酸木】sourwood; sorrel tree
【酸牛奶】yoghurt; sour milk
【酸软】aching and limp
【酸式盐】[化] acid salt
【酸甜苦辣】sour, sweet, bitter, hot; joys and sorrows of life
【酸痛】ache：浑身～ ache all over／四肢～ have aches in the limbs
【酸味】tart flavor; acidity
【酸洗】[冶] pickling; acid pickling ◇～试验[化] acid washing test／～液 pickle
【酸血】acidemia
【酸性】[化] acidity ◇～反应 acid reaction／～降水 acid precipitation／～染料 acid dyes／～试验 acid test
【酸雨】acid rain
【酸枣】wild jujube
【酸值】[化] acid value

suàn

蒜 garlic：青～ garlic sprout／一辫～ a braid of garlic
【蒜瓣儿】garlic clove
【蒜黄】blanched garlic leaves
【蒜苗】garlic bolt
【蒜泥】mashed garlic
【蒜头】the head of garlic

算 ①（计算数目）calculate; reckon; compute; figure：～一下生产成本 calculate the cost of production／能写会～ good at writing and reckoning ②（计算进去）include; count：我也～一个。Count me in. ③（谋划）plan; calculate：失～ miscalculate; make an unwise decision／暗～ plot against sb. ④（推测）think; suppose：我～他今天会来。I think he will come today. ⑤（当作）consider; regard as; count as：我～运气,赶上了最后一班汽车。I was lucky enough to catch the last bus. ⑥(算数) carry weight; count: 不应该由一两个人说了～。 One or two persons should not have the final say. / 他说了不～。 His words do not count. ⑦(作罢;总算) at long last; in the end; finally: 问题～解决了。 The problem is finally solved. ⑧(后面跟"了") let it be; let it pass: ～了,别说了。 That's enough! Let it go at that.

【算法】[数] algorithm ◇ ～语言 algorithmic language; ALGOL

【算计】①(计算数目) calculate; reckon ②(考虑) consider; plan: 我们～着改革生产流程。 We are considering a reorganization of the production process. ③(估计) expect; figure; guess: 我～他很快回来。 I figure he'll be back soon. ④(暗中谋划) scheme; plot: 暗中～别人 secretly scheme against others

【算命】fortune-telling ◇ ～先生 fortune-teller

【算盘】abacus: 打～ use an abacus / 他老打个人小～。 He is always calculating.

【算是】at last: 这一下你～猜着了。 At last you've guessed right.

【算术】arithmetic: 做～ do sums ◇ ～级数 arithmetic progression; arithmetic series / ～平均值 arithmetic mean

【算数】count; hold; stand: 个别情况不～。 Isolated instances do not count. / 我说话是～的。 I mean what I say.

【算学】①(数学) mathematics ②(算术) arithmetic

【算帐】①(计算帐目) do accounts; balance the books; make out bills: ～算得快 be quick at accounts ②(清算) square accounts with sb.; get even with sb.: 跟某人～ have an account to settle with sb. / 我回头找你～。 I will get even with you.

【算子】[数] operator: 微分～ differential operator

SUĪ

虽[连] though; although; even if: 他～很努力,但没成功。 For all his efforts, he didn't succeed. / 文章～短,但很有说服力。 The article is very convincing though it is short.

【虽然】[连] though; although

【虽说】[口][连] though; although

【虽死犹荣】honored though dead; have died a glorious death

【虽死犹生】live on in spirit

【虽则】[连] though; although

SUÍ

随①(跟) follow: ～我来。 Follow me. ②(顺从) comply with; adapt to: ～顺 yield and comply ③(任凭) let (sb. do as he likes): ～你的便。 Do as you please. ④(顺便) along with: 请你～手带上门。 Please close the door as you go out. ⑤[方](象) look like; resemble: 她长得～她母亲。 She looks like her mother.

【随笔】informal essay; jottings

【随便】①(不加限制) casual; random; informal: ～闲谈 chat; chitchat ②(不拘礼) do as one pleases: ～吃吧。 Help yourselves. ③(不加考虑) careless; slipshod: 说话～ not be careful about the way one talks ④(任意;任性) wanton; willful; arbitrary: ～歪曲事实 make willful distortion of the facts ⑤(无论) anyhow; any: 你～什么时候来都行。 You may come any time you like.

【随波逐流】drift with the tide

【随处】everywhere; anywhere

【随从】①(跟随) accompany (one's superior); attend ②(随员) retinue; suite; *entourage*

【随大溜】drift with the stream; follow the general trend

【随带】①(随同带去) going along with: 信外～书籍一包。 Accompanying the letter is a parcel of books. ②(随身携带) have sth. taken along with one: ～行李两件 two pieces of luggage which a passenger takes along with him

【随地】anywhere; everywhere: 不要～扔东西。 Don't litter. / 请勿～吐痰。 No spitting.

【随动件】[机] follower: 凸轮～ cam follower

【随风倒】bend with the wind; be easily swayed

【随风转舵】trim one's sails; take one's cue from changing conditions

【随函备忘录】covering memorandum

【随和】amiable; obliging: 脾气～ have an amiable disposition / 她人很～。 She is very amiable.

【随后】[副] soon afterwards: 你先走,我～就去。 You go first. I'll follow.

【随机】① [统计] random ② [数] stochastic ◇ ～抽样 random sampling / ～存取 random access / ～方案 randomizing scheme / ～过程 stochastic process / ～文件 random file

【随机应变】adapt oneself to changing conditions; act according to circumstances

【随即】immediately; presently

【随口】speak thoughtlessly or casually; blurt out whatever comes into one's head: ～答应 say "yes" absentmindedly; agree without thinking
【随人俯仰】be at sb.'s beck and call; follow sb. servilely
【随身】(carry) on one's person; (take) with one: ～武器 side arms / ～行李 personal luggage
【随声附和】echo what others say; chime in with others
【随时】①(任何时候) at any time; at all times: ～随地 at all times and places ②(有需要时) whenever necessary; as the occasion demands: ～调节温度 regulate the temperature whenever necessary
【随手】conveniently; without extra trouble: ～关门。Shut the door after you. / 勿忘～关灯。Don't forget turn the light off.
【随俗】comply with convention; do as everybody else does
【随…随…】～叫～到 be on call at any hour / 雪～下～化。The snow melted as it fell.
【随体】[细胞] satellite
【随同】be in company with; be accompanying
【随乡入乡】when in Rome do as the Romans do
【随想曲】[乐] caprice; *capriccio*
【随心所欲】follow one's inclinations; have one's own way; do as one pleases
【随行人员】*entourtage*; suite; party: 总统及其～ the President and his *entourage*
【随意】at will; as one pleases: 请～! Help yourselves! Please! ◇ ～肌 [生理] voluntary muscle
【随遇而安】feel at home wherever one is; be able to adapt oneself to different circumstances
【随遇平衡】[物] indifferent equilibrium
【随员】①(随行人员) suite; retinue; *entourage* ②[外] attaché
【随葬物】funerary objects; burial articles
【随着】along with; in the wake of; in pace with: ～生产的稳步上升 in pace with steady growth of production / ～时间的推移 as time goes on; with the lapse of time

绥[书] ①(安好) peaceful ②(安抚) pacify
【绥靖】pacify; appease ◇ ～政策 policy of appeasement

suǐ

髓 ①[生理] marrow: 脊～ spinal marrow ②[植] pith
【髓瘤】myeloma

suì

碎 ①(破成零片) break to pieces; smash: 玻璃杯打～了。The glass is smashed to pieces. ②(不完整的) broken; fragmentary: ～玻璃 bits of broken glass ③(唠叨) garrulous; gabby: 嘴太～ talk too much; be a regular chatterbox
【碎波】[海] breaker
【碎步儿】quick short steps
【碎颅术】cranioclasis
【碎尸万段】tear sb. to shreds; cut sb. to pieces
【碎石】[建] crushed stones; broken stones ◇ ～混凝土 crushed stone concrete / ～机 stone crusher / ～路 broken stone road; macadam road
【碎石术】[医] lithotrity
【碎屑岩】clastic rock
【碎音】[乐] *acciaccatura*

遂 ①(顺;如意) satisfy; fulfill: ～愿 have one's wish fulfilled ②(成功) succeed: 功成名～ have one's work done and one's name recognized / 所谋不～ fail in an attempt ③[书](于是;就) then; thereupon
【遂心】after one's own heart; to one's liking
【遂心如意】perfectly satisfied
【遂心所欲】one's liking; satisfy one's desire
【遂意】to one's liking

邃[书] ①(深远) remote (in time or space): ～古 remote antiquity ②(精深) deep; profound: 精～ profound

燧 ①(古代取火工具) flint ②(古代告警的烽火) beacon fire
【燧石】flint ◇ ～玻璃 flint glass

隧
【隧道】tunnel: 铁路～ a railway tunnel / 在河底下挖掘～ tunnel a river / 在山里开～ tunnel a hill
【隧道管】[电子] tunneltron
【隧道效应】[电子] tunnel effect

岁 ①(年) year: ～出 annual expenditure / ～末 the end of the year / ～人 annual income ②(年龄) year (of age): 七～的女孩儿 a seven-year-old girl; a little girl seven years old ③[书](年成) year (for crops): 歉～ lean year
【岁差】[天] precession of the equinoxes
【岁寒三友】three durable plants of winter ○松

pine / 竹 bamboo / 梅 plum blossom
【岁寒知松柏】only when the year grows cold do we see the qualities of the pine and the cypress; adversity reveals virtue
【岁暮】[书] the close of the year: ～天寒 Cold weather sets in as the year draws to its close.
【岁首】[书] the beginning of the year
【岁数】[口] age; years: 您多大～了? How old are you? / 她上～了。 She is getting on in years.
【岁星】an ancient name for the planet Jupiter
【岁月】years: ～不居。 Time and tide wait for no man. / ～不饶人。 Time and tide wait for no one. / ～如流。 Time is fleeting.

穗 ①(稻麦的穗) the ear of grain; spike: 麦～儿 the ear of wheat ②(下垂的装饰品) tassel; fringe
【穗选】[农] ear selection
【穗状花序】[植] spike
【穗子】tassel; fringe: 有～的旗 a banner fringed with tassels

祟 evil spirit; ghost: 作～ act like an evil spirit; haunt and plague

sūn

孙 ①(孙子) grandson: 祖～ grandfather and grandson ②(孙子以后的各代) generations below that of the grandchild: 玄～ great-great-grandson / 曾～ great-grandson ③(植物再生的) second growth of plants: ～竹 new shoots of bamboo from the old stump
【孙囊】grand-daughter cyst
【孙女】granddaughter ◇ ～婿 granddaughter's husband; grandson-in-law
【孙媳妇】grandson's wife; granddaughter-in-law
【孙子】grandson

sǔn

损 ①(减少) decrease; lose: 亏～ loss / 增～ increase and decrease ②(损害) harm ; damage: 有益无～ can only do good, not harm ③[方](挖苦人) sarcastic; caustic; cutting: 他爱～人。 He delights in making caustic remarks. ④[方](恶毒;刻薄) mean; shabby: 这法子真～。 That's a mean trick.
【损兵折将】suffer heavy casualties
【损公肥私】injure the public interest to benefit one's private interests; seek private gain at public expense; feather one's nest at public expense
【损害】harm; damage; injure; damnify: ～健康 impair one's health / ～视力 harm one's eyes ◇ ～保险 insurance of damage / ～程度 extent of damage / ～赔偿 compensation for damages; damages / ～赔偿金 damages / ～赔偿诉讼 damage suit / ～信用 injury to credit
【损耗】①(损失消耗) loss; wear and tear: 摩擦～ friction loss ②[商] wastage; spoilage ◇ ～费 cost of wear and tear / ～概率 loss probability / ～函数 loss function / ～率 [商] proportion of goods damaged
【损坏】breakdown; damage; injure; spoil: 如有～，照价赔偿 pay the full price for anything damaged ◇ ～程度 damaged condition / ～工程 spoiled work
【损人利己】harm others to benefit oneself; benefit oneself at the expense of others
【损伤】①(伤害) damnification; damnify; harm; damage; injure ②(损失) loss
【损失】①(失去) lose: ～金钱 lose one's money ②(失去的东西) loss: 我的～很大。 My losses have been very great. ◇ ～补偿保险 indemnity insurance / ～额 amount of loss / ～赔偿 compensation for damages / ～赔偿金 indemnification for loss
【损益】①(减少和增加) increase and decrease: 斟酌～ consider making necessary adjustments ②(赔和赚) profit and loss; gains and losses: ～相抵。 The gains offset the losses. ◇ ～比率 profit and loss ratio / ～分配 distribution of profit and loss / ～计算书 profit and loss statement
【损益周复】superabundance and scarcity follow each other in succession.

笋 bamboo shoot
【笋干】dried bamboo shoots
【笋瓜】[植] winter squash
【笋鸡】young chicken; broiler
【笋尖】tender tips of bamboo shoots

隼 [动] falcon

榫 tenon
【榫接】joggle; mortise joint
【榫头】tenon
【榫眼】mortise

SUŌ

莎

【莎草】[植] nutgrass flatsedge

桫

【桫椤】[植] spinulose tree fern

娑

【娑罗双树】[植] sal tree; meranti

蓑

【蓑衣】straw rain cape; palm-bark rain cape

羧[化] carboxyl

【羧基】[化] carboxyl; carboxyl group

【羧酸】[化] carboxylic acid

梭 shuttle: 无～织机 shuttleless loom

【梭标】spear

【梭梭】[植] sacsaoul

【梭巡】[书] move around to watch and guard; patrol

【梭鱼】(redeye) mullet

【梭子】① [纺] shuttle ② (子弹夹) cartridge clip ③ [量] clip: 打了一～子弹 fire a whole clip of ammunition

【梭子蟹】swimming crab

【梭子鱼】barracuda

唆 instigate; abet: 教～ instigate; abet

【唆使】instigate; abet ◇ ～者 instigator; abettor

缩 ① (收缩) contract; shrink: 热胀冷～ expand with heat and contract with cold ② (退缩) draw back; withdraw; recoil: 把手～回 withdraw one's hand back / 退～ flinch; shrink

【缩尺】reduced scale; scale ◇ ～图 scale drawing

【缩短】shorten; curtail; cut down: ～假期 curtail one's holidays / ～距离 reduce the distance; narrow the gap

【缩放仪】pantogragh

【缩合】[化] condensation ◇ ～反应 condensation reaction / ～物 condensation compound

【缩减】reduce; cut: ～军费 cut back military expenditure / ～开支 reduce spending / ～行政人员 retrench administrative staff

【缩聚】[化] condensation polymerization ◇ ～物 condensation polymer

【缩手】① (手缩回来) draw back one's hand ② (不敢再做下去) shrink (from doing sth.)

【缩手缩脚】① (因冷而四肢不舒展) shrink with cold ② (做事不大胆) be overcautious: 不要～。Don't be overcautious.

【缩水】shrink: 这布不～。This cloth will not shrink in the wash. ◇ ～率 shrinkage

【缩瞳】miosis; myosis ◇ ～剂 miotic / ～药 myotic; miotic

【缩头虫】[动] bamboo worm

【缩头缩脑】① (畏缩) recoil in fear; be timid; be fainthearted ② (怕负责任) shrink from responsibility

【缩微胶卷】microfilm; microfilm strip

【缩微胶片】microfiche

【缩微照片】microphotograph

【缩小】reduce; lessen; narrow; shrink: ～差距 narrow the gap / ～范围 reduce the scope; narrow the range

【缩写】① [语] abbreviation ② (缩短篇幅) abridge ◇ ～本 abridged edition / ～签字 initials

【缩印】reprint books in a reduced format

【缩影】epitome; miniature

SUǑ

索 ① (大绳;大链) large rope: 船～ ship's rigging / 绞～ (the hangman's) noose / 绳～ rope / 铁～桥 chain bridge ② (搜寻) search: 遍～不得 search high and low for sth. in vain ③ (要;取) demand; ask; exact: ～价 ask a price; charge / ～赔 claim damages / ～债 demand payment of a debt ④ [书] (孤单) all alone; all by oneself: 离群～居 live all alone ⑤ [书] (没有意味) dull; insipid

【索道】cableway; ropeway: 高架～ telpher

【索夹】cord clip; rope clip

【索具】rigging

【索马里】Somalia ◇ ～人 Somali / ～语 Somali (language)

【索密痛】[药] somidon

【索赔】claim indemnity; claimant; claimer: 要求延长～的期限 request to extend time-limit of claim ◇ ～期限 deadline for demanding compensation / ～时限 time limit for filing claims / ～证件 claims document; document for claims

【索取】ask for; demand; exact; extort: ～赔款 claim indemnity / ～样品 ask for a sample

【索然】dull; dry; insipid: ～寡味 flat and insipid / 兴致～ have no interest at all

【索性】[副] 既然已经做了,～就把它做完。

Since you have started the job, you might as well finish it.
【索引】index：标题～ subject index／卡片～ card index／书名～ title index／作者～ author index

琐 trivial; petty
【琐事】trifles; trivial matters：家庭～ household affairs
【琐碎】trifling; trivial
【琐谈症】circumstantiality
【琐闻】bits of news; scraps of information
【琐细】trifling; trivial

唢
【唢呐】*suona* horn, a woodwind instrument

锁 ①(一种安在开合处的器具) lock：挂～ padlock／弹簧～ spring lock ②(上锁) lock up：～门 lock a door／～得好好地 under lock and key／～在屋内 lock in／～在屋外 lock out／大门没～上。The gate wasn't locked.／这门～不住。This door won't lock. ③(皱眉) knit：双眉紧～ with knitted brows ④(缝纫法) lockstitch：～边 lockstitch a border／～眼 do a lickstitch on a buttonhold
【锁骨】[生理] clavicle; collarbone
【锁簧】[机] locking spring
【锁匠】locksmith
【锁紧】[机] locking：自～ self-locking ◇～螺母 locknut
【锁孔】lockhole; keyhole
【锁链】chain; shackles; fetters; chains
【锁钥】①(关键) key：解决问题的～ a key to the problem ②(军事要地) strategic gateway (to an important center or a major city)

所 ①(处所) place：住～ dwelling place／各得其～ each in his proper place／居无定～ without dcfinite residence ②(机构名称)研究～ research institute／诊疗～ clinic ③[量]两～学校 two schools／一～房子 a house ④(跟"为"或"被"合用,表示被动)为人～笑 be laughed at ⑤(跟动词连用,代表接受动作的事物)各尽～能 from each according to his ability／闻～未闻 unheard-of／无～不为 stop at nothing (in doing evil) ⑥(跟动词连用,动词后再用接受动作的事物的词)我～认识的人 the people I know ⑦(跟动词连用,动词后再用"者"或"的"代表接受动作的事物)～见者广 have wide experience ⑧(用在动词前)人～共知。It's known to all.
【所部】troops under one's command
【所长】what one is good at; one's strong point; one's forte
【所得】income; earnings; gains
【所得税】income tax：～抵免 income tax credit／～法 income tax law／～申报表 income tax return
【所见所闻】what one sees and hears; all that one saw and heard
【所属】①(统属之下的) what is subordinate to one or under one's command：通令～一体遵照。All the units are to be informed that the instructions should be carried out. ②(自己隶属的) what one belongs to or is affiliated with：向～派出所申报户口 apply with the local police station for residence
【所谓】①(所说的) what is called ②(某些人所说的) so-called：～"自由世界" the so-called "free world"
【所向披靡】carry all before one; sweep away all obstacles
【所向无前】be invincible; be irresistible; break all enemy resistance
【所以】①[连](表示因果关系) so; therefore; as a result：他有事,～没来。He hasn't come because he's got something else to do. ②[口](表示"原因就在这里")～呀,要不然我怎么会这么说呢？That's just the point, otherwise I wouldn't have said it. ③(表示实在的情由或适宜的举动)忘其～ forget oneself
【所以然】the reason why; the whys and wherefores：知其然而不知其～ know that sth. is so but not why it is so; know what is done but not why it is done
【所有】①(拥有) own; possess：这老人死后家产归谁～？Who did the property go to when the old man died? ②(领有的东西) possessions：尽其～ give everything one has; give one's all ③(一切;全部) all：～的学生 all the students ◇～格 [语] possessive case
【所有权】proprietary rights; ownership; title ◇～客体 object of the right of ownership／～证明书 certificate of title／～转移 passing of title
【所有制】ownership; system of ownership：封建土地～ feudal ownership of land／奴隶主～ slave-owning ownership ○公有制 public ownership／私有制 private ownership
【所在】①(处所) place; location：风景优美的～ a picturesque place; a scenic spot ②(存在的地方) where：这就是问题的～。That is where the question arises. ◇～地 location; seat; site

【所致】be caused by; be the result of: 这次事故是由于疏忽～。The accident was the result of negligence.

T

tā

它 it: ～的 its
【它们】they: ～的 their
铊 [化] thallium (Tl)
溻 [方] become soaked with sweat
塌 ①(倒下) collapse; Fall down; cave in: 房屋因地震而倒～。 The house collapsed on account of an earthquake. / 桥～了。 The bridge collapsed. ②(凹下) sink; droop ③(安定) settle down; calm down: ～下心去 set one's mind at ease
【塌鼻子】a flat nose; a snubby nose
【塌方】①(地层结构上的) landslide; landslip ②(指修筑上的) cave in; collapse
【塌棵菜】broadbeaked mustard
【塌实】【踏实】①(切实) steady and sure; dependable ②(安定) free from anxiety; having peace of mind
【塌台】collapse; fall from power
【塌下来】fall down; cave in; collapse: 天不会～。 The sky won't fall down.
【塌陷】subside; sink; cave in: 房基～。 The foundations have sunk.
踏
【踏实】①(切实; 不浮躁) dependable; steady and sure: 工作～ be steadfast in one's work; be a steady worker ②(安定) free from anxiety: 睡得～ have a sound sleep / 心里～了 feel relieved
趿
【趿拉】wear cloth shoes with the backs turned in; shuffle about with the backs of one's shoes trodden down
【趿拉板儿】[方] wooden slippers; clogs
他 ①(自己和对方以外的个人) he: ～的 his / ～父亲 his father / ～俩 the two of them / ～自己 himself ②(另外的; 其他的) other; another; some other: 调往～处 be transferred to another place / 留作～用 reserve for other use
【他方】①(另一方) the other party ②(别的地方) other places: ～求食 seek a livelihood in another region
【他妈的】[骂] damn it; blast it; to hell with it
【他们】they: ～俩 the two of them / ～学校 their school / ～自己 themselves
【他年】another year; sometime in the future
【他人】another person; other people; others: ～瓦上霜 none of one's business
【他日】[书] some other time; some (other) day; later on
【他杀】[法] homicide
【他山之力】trust in the counsel of another
【他山之石可以攻错】there are other hills whose stones are good for working jade; other people's good quality or suggestion whereby one can remedy one's own defects
【他乡】a place far away from home; an alien land: ～遇故知 run into an old friend in a distant land
她 she: ～的 her; hers / ～自己 herself
【她们】they: ～的 their; theirs / ～自己 themselves

tǎ

塔 ①(佛教的塔) pagoda ②(塔形建筑; 塔形物) tower: ～尖 spire / 纪念～ memorial tower / 了望～ watch tower / 水～ water tower / 象牙～ tower of ivory ③[化] column; tower: 蒸馏～ distillation column; distillation tower
【塔吊】tower crane
【塔夫绸】taffeta
【塔吉克人】(苏联) tadzhik
【塔吉克语】(塔吉克族语言) tajik
【塔吉克族】(中国民族) tajik (nationality)
【塔轮】[机] cone pulley; stepped pulley
【塔式起重机】tower crane
【塔塔尔族】tatar (nationality)
【塔台】[航空] control tower
鳎 [动] sole
獭 otter: 海～ sea otter / 旱～ marmot / 水～ otter

tà

挞 [书] flog; whip: 鞭～ flog; lash
【挞罚】flogging; corporal punishment
【挞伐】send troops to punish

拓 make rubbings from inscriptions, pictures, etc. on stone tablets or bronze vessels
【拓本】a book of rubbings
【拓片】rubbing

榻 a long, narrow and low bed; couch: 藤～ rattan couch／竹～ bamboo couch／同～ sleep in the same bed; share a bed／下～ stay at

沓 crowded; repeated: 纷至～来 come thick and fast; keeping pouring in
【沓沓多言】talking very much
【沓杂事物】complex things

踏 ①(踩) step on; tread; stamp: ～出一条通往…的路 tread a path to...／～上故土 set foot on one's native land ②(去现场) go to the spot
【踏板】①(机器等的) treadle; footboard; footrest: 楼梯～ stair tread ②(乐器的) pedal: 强音～ damper pedal／弱音～ soft pedal ③(脚蹬) footstool
【踏步】mark time: ～不前 make no headway
【踏歌】beat time to a song with the feet
【踏脚石】stepping-stone
【踏勘】make an on-the-spot; survey; investigate carefully
【踏看】go to the spot to make an investigation
【踏破铁鞋无觅处，得来全不费工夫】find sth. by chance after travelling far and wide in search of it
【踏青】go for a walk in the country in spring
【踏月吟诗】chant poems walking in the moonlight

tāi

胎 ①(幼体) foetus; embryo ②(生育次数) birth: 头～ firstborn; first baby ③(衬在衣被里的东西) padding; stuffing; wadding ④(坯子) roughcast ⑤(轮胎) tire: 内～ inner tube of a tire／外～ tire
【胎动】the quickening of the womb; movement of the foetus
【胎儿】foetus; embryo
【胎发】fetal hair; lanugo
【胎记】[医] birthmark
【胎教】prenatal education; prenatal influences
【胎具】mold; matrix
【胎毛】fetal hair; lanugo
【胎膜】[生理] fetal membrane
【胎盘】[生理] placenta ◇ ～球蛋白 placental globulin
【胎气】nausea; vomiting and edema of legs during pregnancy
【胎生】[动] viviparity ◇ ～动物 viviparous animal; vivipara／～学 embryology／～鱼 viviparous fish
【胎位】[医] position of a foetus ◇ ～不正 foetus in wrong position
【胎衣】afterbirth
【胎座】[植] placenta

tái

薹 ①(草本植物) sedge ②(蒜、油菜等中央的茎) the bolt of garlic, rape, etc.: 蒜～ garlic bolt
【薹草】[植] sedge

台 ①(平而高的建筑或设备) platform; stage; terrace: 讲～ platform／了望～ watch tower／舞～ stage／主席～ rostrum; platform ②(座子) stand; support: 蜡～ candlestick ③(象台子的东西) anything shaped like a platform: 窗～ windowsill; window sill ④(桌子及类似桌子的东西) table; desk: 梳妆～dressing table／写字～ writting desk ⑤[量] 一～机器 a machine
【台本】a playscript with stage directions
【台布】tablecloth
【台步】the gait of an actor or actress in Beijing opera, etc.
【台秤】platform scale; platform balance
【台词】actor's lines
【台灯】desk lamp; table lamp; reading lamp
【台地】[地] tableland
【台风】[气] typhoon: 强～ violent typhoon ◇ ～动向 typhoon movement／～警戒线 typhoon detective line／～路径 typhoon track／～眼 typhoon eye
【台虎钳】[机] bench vice
【台阶】①(踏级) flight of steps ②(避免因僵持而受窘的途径或机会) chance to extricate oneself from an awkward position: 给他们找个～下。Give them an out.
【台历】desk calendar
【台面】[电子] mesa ◇ ～型晶体管 mesa transistor
【台钳】[机] bench clamp
【台球】①(指台球游戏) billiards ②(指球) billiard ball ③(乒乓球) table tennis; ping-pong
【台柱子】pillar; mainstay
【台钻】[机] bench drill

苔 [植] liver mosses
【苔藓植物】bryophyte
【苔原】[地] tundra

抬 ①(举起) lift up; raise: ～头 raise one's head / 这箱子太重,我～不起来。This box is too heavy for me to lift. ②(用手或肩搬运) carry: ～担架 carry a stretcher
【抬秤】huge steelyard
【抬杠】①(争辩) argue for the sake of arguing; bicker; wrangle ②(抬运灵柩) carry a coffin on stout poles
【抬高】raise: ～价格 raise prices / ～物价 raise commodity prices; force up commodity prices / ～租金 raise the rent
【抬肩】half the circumference of the sleeve where it joins the shoulder
【抬举】praise or promote sb. to show favor; favor sb.: 不识～ not know how to appreciate favors
【抬头】①(昂起头) raise one's head: ～挺胸 chin up and chest out ②(不受压制) rise; look up; gain ground ③[商](指单据或发票的抬头) name of the buyer or payee
【抬头纹】wrinkles on one's forehead

鲐 chub mackerel

tài

泰 ①(平安) safe; peaceful ②(极;最) extreme; most
【泰斗】a leading authority
【泰而不骄】poised but not arrogant
【泰国】Thailand ◇ ～人 Thailander; Thai
【泰然】calm; composed; self-possessed: ～处之 take sth. calmly; bear sth. with equanimity / ～自若 behave with perfect composure; be self-possessed
【泰日之时】in the days of peace
【泰山北斗】Mount Taishan and the North Star; a respectful epithet for a person of distinction
【泰山不让土壤】a learned man never stops his pursuit of knowledge
【泰山压顶】bear down on one with the weight of Mount Taishan

太 ①[副](过分) too; over: ～忙 overbusy / ～晚 too late / 水～热,烫手。The water is too hot to touch. ②(极;最) highest; greatest; remotest ③(身分或辈份高) more or most senior ④[副](程度极高) too; extremely: ～感谢你了。Thanks a lot. ⑤(很) very; quite: 不～好 not very good
【太白星】[天] Venus; Vesper
【太公】[方] great-grandfather
【太公钓鱼,愿者上钩】like the fish rising to Jiang Tai Gong's hookless and baitless line; a willing victim letting himself be caught
【太古】remote antiquity ◇ ～代[地] the Archean Era / ～界[地] the Archean Group
【太后】mother of an emperor; empress dowager
【太监】eunuch
【太空】the firmament; outer space ◇ ～船 a space craft; space ship / ～人 astronauts; cosmonauts / ～时代 the space age
【太平】peace and tranquility: ～盛世 piping times of peace / ～无事 all is well ◇ ～龙头 fire hydrant; fire plug / ～门 exit / ～水缸 a vat filled with water for use in case of fire / ～梯 fire escape
【太平花】mockorange
【太平间】mortuary; reposing room
【太平鸟】waxwing
【太平天国】the Taiping Heavenly Kingdom (1851–1864)
【太平洋】the Pacific (Ocean)
【太婆】[方] great-grandmother
【太上皇】①(皇帝的父亲的称号) a title assumed by an emperor's father who abdicated in favor of his son ②(在幕后操纵的掌握实权者) overlord; supreme ruler
【太岁头上动土】provoke sb. far superior in power or strength
【太太】①(对已婚妇女的称呼) Mrs.; madame: 吴～ Mrs. Wu; Madame Wu ②(旧社会仆人尊称女主人) madame; lady
【太虚幻境】in a state of visionary and emptiness
【太阳】①[天] the sun; solar: ～在东方升起。The sun rises in the east. ②(太阳光) sunshine; sunlight: 晒～ bask in the sun; sun oneself ◇ ～灯[医] sunlamp; sunlight lamp / ～电池 solar cell / ～辐射[气] solar radiation / ～光 the sun's rays; sunlight; sunbeam; sunshine / ～光谱[物] solar spectrum / ～黑子[天] sunspot / ～历 solar calendar / ～炉 solar furnace / ～帽 sun helmet; topee / ～目视镜[天] helioscope / ～能 solar energy / ～年[天] solar year / ～系 the solar system / ～灶 solar energy stove; solar cooker

【太阳镜】sunglasses
【太阳鸟】sunbird
【太阳穴】the temples
【太医】imperial physician
【太阴】[天] lunar: ～历 lunar calendar / ～年 lunar year / ～月 lunar month; lunation
【太子】crown prince

态 ①(状态) form; appearance; condition: 形～ shape; morphology / 姿～ posture; attitude ②[语] voice: 主动语～ the active voice ③[物] state: 固～ solid state / 气～ gaseous state / 液～ liquid state
【态度】①(举止;神情) manner; bearing; how one conducts oneself: ～暧昧 be on the hedge / ～和蔼 amiable; kindly / ～激烈 violent attitude / ～冷淡 an indifferent attitude / ～娴雅 have refined manners / 耍～ lose one's temper; get into a huff ②(对事物的看法和采取的行动) attitude; approach: ～坚决 maintain a firm attitude / 劳动～ attitude towards labor
【态势】state; situation; posture: 军事～ military posture / 战略～ strategic situation

酞 [化] phthalein

钛 [化] titanium (Ti)
【钛白】[化] titanium white; titanium dioxide
【钛铁矿】ilmenite

肽 [化] peptide

tān

坍 collapse; fall; tumble
【坍塌】cave in; collapse

贪 ①(贪污) corrupt; venal: ～官 corrupt official ②(求多) greedy; have an insatiable desire for: ～财 be greedy for money ③(贪图) covet; hanker for: ～小便宜 keen on gaining petty advantages
【贪杯】be too fond of drink
【贪财害命】commit murder for money
【贪得无厌】be insatiably avaricious
【贪多嚼不烂】bite off more than one can chew
【贪婪】[书] avaricious; greedy; rapacious: ～的目光 greedy eyes
【贪恋】be reluctant to part with; hate to leave; cling to
【贪便宜】anxious to get things on the cheap; keen on gaining petty advantages
【贪生怕死】cravenly cling to life instead of braving death; care for nothing but saving one's skin; be mortally afraid of death
【贪天之功】arrogate to oneself the merits of others; claim credit for other people's achievements
【贪图】seek; hanker after; covet: ～安逸 seek ease and comfort / ～小利 covet small advantages; hanker after petty gains
【贪污】corruption; graft: ～盗窃 graft and embezzlement / ～渎职 corrupt and negligent about one's duties / ～腐化 corruption and degeneration; corruption ◇ ～分子 a person guilty of corruption; grafter; embezzler
【贪羡繁华】long for gaiety
【贪心】①(贪得的欲望) greed; avarice; rapacity ②(不知足) greedy; avaricious; insatiable; voracious: ～不足 insatiably greedy
【贪小失大】covet a little and lose a lot; seek small gains but incur big losses
【贪赃】take bribes; practice graft: ～枉法 take bribes and bend the law; pervert justice for a bribe
【贪嘴】greedy; gluttonous

滩 ①(沙滩) beach; sands: 海～ seabeach ②(险滩) shoal: ～多水急 with many shoals and rapids
【滩头堡】[军] beachhead
【滩头小调】a kind of boatman's song

摊 ①(摆开;铺平) spread out; unfold: ～开地图 unfold a map; spread out a map ②(分担) take a share in: 分～费用 share the cost ③(摊子) vendor's stand; booth; stall: 报～ newsstand / 水果～ fruit stall / 售货～ stall; stand ④[量] 一～稀泥 a mud puddle ⑤(一种烹饪方法) fry batter in a thin layer: ～鸡蛋 make an omelet ⑥(碰到;落到) befall: 他～上什么事了? What has befallen him?
【摊贩】street pedlar
【摊开】spread out; unfold
【摊款】share the burden of a payment
【摊牌】lay one's cards on the table; show one's hand; have a showdown: 迫使对方～ force one's opponent to show his hand; force a showdown
【摊派】apportion; quotas: ～费用 apportioned charges / ～捐税 assessment
【摊晒机】[农] tedder
【摊子】①(售货的摊子) vendor's stand; booth; stall ②(规模) the structure of an organization; setup: ～铺得太大 do sth. on too large a scale

瘫 paralysis: 吓～了 be paralysed with fright
【瘫痪】①(神经机能障碍引起的病变) paralysis; palsy: ～病人 paralytic ②(机构涣散) be paralysed; break down; be at a standstill: 机构～ the organ is paralysed
【瘫软】weak and limp
【瘫子】a person suffering from paralysis; paralytic

tán

痰 phlegm; sputum
【痰迷心窍】blinded judgment
【痰桶】[口] spittoon
【痰盂】spittoon; cuspidor

谈 ①(说话;讨论) talk; chat; discuss: ～家常 talk about everyday matters / ～生意经 discuss business ②(所说的话) what is said or talked about: 奇～ strange talk / 无稽之～ fantastic talk; sheer nonsense
【谈不上】out of the question; far from being: 他～是个优等生。 He is far from being an excellent student.
【谈到】speak of; talk about; refer to: 她从未～她母亲。 She never speaks of her mother.
【谈得来】get along well (with sb.)
【谈锋】volubility; eloquence: ～甚健 talk volubly; be a good talker; have the gift of the gab
【谈何容易】easier said than done; by no means easy
【谈虎色变】turn pale at the mention of a tiger; turn pale at the mere mention of something terrible
【谈话】①(多人在一起说话) conversation; talk; chat ②(用谈话的形式发表意见) statement; talk: 发表书面～ make a written statement
【谈论】discuss; talk about
【谈判】negotiations; talks: 和平～ negotiation for peace / 开始与某人进行～ open negotiations with sb. ◇ ～桌 conference table
【谈起】mention; speak of
【谈天】chat; make conversation: ～说地 talk of everything under the sun
【谈吐】style of conversation: ～高雅 have a refined style of conversation
【谈笑风生】talk cheerfully and humorously
【谈笑自若】go on talking and laughing as if nothing had happened: 沉着镇静，～ go on talking and laughing without turning a hair
【谈心】heart-to-heart talk
【谈言微中】speak tactfully but to the point; make one's point through hints
【谈助】topic of conversation: 足资～ serve as a good topic of conversation

坛 ①(古代大典用的台) altar: 日～ the Altar to the Sun ②(用土堆成的台) a raised plot of land: 花～ flower bed ③(某些组织) circles; world: 棋～ chess circles / 文～ the literary world ④(坛子) earthen jar; jug: 一～醋 a jar of vinegar
【坛坛罐罐】pots and pans; personal possessions
【坛子】earthen jar

昙 covered with clouds
【昙花】[植] broad-leaved epiphyllum
【昙花一现】flower briefly as the broad-leaved epiphyllum; last briefly; be a flash in the pan: ～的人物 a transient figure

檀 wingceltis
【檀板】[乐] hardwood clappers
【檀香】[植] white sandalwood; sandalwood ◇ ～木 sandalwood / ～扇 sandalwood fan / ～油 sandalwood oil / ～皂 sandal soap

潭 ①(深水池) deep pool; pond: 一～死水 a pond of stagnant water / 龙～虎穴 dragon's pool and tiger's den; a danger spot ②[方](坑) pit
【潭府】your residence

弹 ①(弹射) shoot; send forth ②(使变松) fluff; tease: ～棉花 fluff cotton; tease cotton ③(用手指弹出) flick; flip ④(拨弄) play; pluck: ～钢琴 play the piano ⑤(弹回) spring; leap ⑥(有弹性) elastic ⑦(抨击) accuse; impeach
【弹唱自如】play and sing as one pleases
【弹词】[曲艺] storytelling to the accompaniment of stringed instruments
【弹冠相庆】congratulate each other and dust off their old official's hats; congratulate each other on the prospect of getting good appointments
【弹劾】impeach: ～权 power to impeach
【弹花机】cotton fluffer
【弹簧】spring: 保险～ relief spring / 回动～ return spring ◇ ～秤 spring balance / ～床 spring bed / ～钢 spring steel / ～铰链 spring hinge / ～门 swing door / ～圈 spring coil / ～锁 spring lock

【弹力】elastic force; elasticity; resilience; spring ◇～尼龙 stretch nylon; elastic nylon / ～纱 stretch yarn / ～袜 stretch socks
【弹球】marbles
【弹射】[军] launch; catapult; shoot off; eject ◇～器 catapult; ejector / ～座舱 ejection capsule / ～座椅 ejection seat
【弹跳】bounce; spring: ～力好 have a lot of spring ◇～板[体] springboard
【弹涂鱼】[动] mudskipper
【弹性】①(弹力) elasticity; resilience; spring ②(事物的伸缩性) elastic; flexible: 这些规定是有～的。These regulations are elastic. ◇～垫圈 elastic washer / ～极限 elastic limit / ～计 elastometer / ～抗[物] elastic reactance / ～塑料 elastoplast / ～体 elastomer / ～系数 coefficient of elasticity
【弹压】suppress; quell
【弹指】a snap of the fingers: ～须臾 during the snapping of fingers / ～之间 in a flash; in an instant
【弹奏】play; pluck

tǎn

毯 blanket; rug; carpet: 地～ rug; carpet / 挂～ tapestry / 毛～ woolen blanket / 绒～ flannelette blanket
【毯子】blanket

忐
【忐忑】perturbed; mentally disturbed: ～不安 uneasy; fidgety / ～不决 vacillating and undecided

袒 ①(脱去或敞开上衣) leave uncovered; be stripped to the waist or have one's shirt unbuttoned: ～胸露臂 exposing one's neck and shoulders ②(袒护) give unprincipled protection to; shield; shelter: 偏～ give unpricipled support to; be partial to
【袒护】be partial to; shield: ～某人的过失 screen sb.'s faults / ～一方 be partial to one side
【袒露】expose; bare

坦 ①(平坦) level; smooth ②(坦白) open; candid ③(心里安定) calm; composed
【坦白】①(直率) honest; frank; candid: 襟怀～ openhearted; honest and aboveboard ②(如实说出) confess; make a confession; own up ◇～自首 surrender and confess one's crimes
【坦荡】①(宽广平坦) broad and level ②(心地纯洁) bighearted; magnanimous
【坦克】tank: 超轻型～ ultra light tank / 空降～ airborne tank / 轻型～ light tank / 水陆～ alligator tank; amphibious tank / 中型～ medium tank / 重型～ heavy tank ◇～兵 tank forces / ～乘员 tank crew / ～手 tankman
【坦然】calm; unperturbed; having no misgivings: ～自若 calm and confident; completely at ease / ～无疑 lay bare without suspicion
【坦桑尼亚】Tazania ◇～人 Tanzanian
【坦率】candid; frank; straightforward; outspoken: ～的批评 frank criticisms / ～地回答 give a straightforward reply / 为人～ be frank and open
【坦途】level road; highway

钽 [化] tantalum (Ta)

tàn

探 ①(试图发现) try to find out; explore; sound: ～路 explore the way / ～明情况 try to find out the situation / ～某人的口气 ascertain sb.'s opinions or feelings; sound sb. out ②(看望) visit; pay a call on: ～监 visit a prisoner ③(向前伸出) stretch out; stretch forward: 行车时不要～身窗外。Don't lean out of the window while the bus is in motion. ④(做侦察工作的人) scout; spy
【探病】visit the sick
【探测】survey; sound; probe: ～海底情况 survey the seabed / ～水深 take soundings / ～悬崖高度 gauge the height of a bluff ◇～器 sounder; probe; detector
【探端知绪】investigate the beginning and know the end
【探访】①(搜寻) seek by inquiry or search: ～珍禽异兽 seek rare birds and animals ②(探望) call on; visit; pay a visit
【探戈舞】tango
【探监】visit a prisoner
【探井】①[矿] prospect pit; exploring shaft ②[石油] test well; exploratory well
【探究】make a thorough inquiry; probe into: ～原因 look into the causes
【探空】[气] sounding ◇～气球 sounding balloon
【探矿】go prospecting; prospect
【探雷】detect a mine ◇～器 mine detector
【探明】ascertain; verify
【探囊取物】like taking something out of one's pocket; as easy as winking; as easy as falling off

a log
【探亲】go home to visit one's family; go to visit one's relatives: 到家乡～访友 return to one's homeland to visit one's relatives and friends ◇ ～假 home leave
【探求】seek; pursue; search after: ～真理 seek truth
【探伤】[冶] flaw detection; crack detection ◇ ～仪 flaw detector
【探视】visit: ～病人 visit a patient ◇ ～时间 visiting hours
【探索】explore; probe: ～究竟 try to find out the why and wherefore / ～真理 seek truth
【探讨】inquire into; probe into
【探听】try to find out, make inquiries: ～下落 inquire about the whereabouts of sb. or sth. / ～消息 make inquiries about sb. or sth.; fish for information
【探头探脑】pop one's head in and look about
【探望】①(看望) visit; pay a visit to; call on; look in: ～病人 visit a patient ②(伸出头去看) look about: 她不时向窗外～。She looked out of the window every now and then.
【探问】①(试探着询问) make cautious inquires about: ～情况 inquire about the situation ②(问候) inquire after
【探悉】ascertain; learn; find out: ～事实真相 learn the truth
【探险】explore; make explorations; venture into the unknown: 到南极洲去～ explore the Antarctic Continent ◇ ～队 exploring party; expedition / ～家 explorer
【探幽寻胜】visit scenic spots
【探友】visit friends
【探鱼仪】[渔] fish detector; fish-finder
【探照灯】searchlight: ～的灯光 searchlight beam
【探针】[医] probe
【探子】①(侦察人员) scout ②(管状用具) a thin tube used to extract samples of food grains, etc.

叹 ①(叹气) sigh ②(赞叹) exclaim in admiration: 赞～ highly praise
【叹词】[语] interjection; exclamation
【叹服】gasp in admiration: ～其智 praise and admire his intelligence / 令人～ compel admiration; command admiration
【叹气】sigh; heave a sigh: 唉声～ sigh in despair
【叹赏】admire; express admiration for
【叹为观止】acclaim sth. as the acme of perfection
【叹为奇迹】acclaim sth. as a marvel
【叹惜】sigh with regret; lament

炭 charcoal: 骨～ animal charcoal / 木～ charcoal / 烧～ make charcoal
【炭笔】charcoal pencil
【炭画】[美术] charcoal drawing; charcoal
【炭火】charcoal fire
【炭精棒】carbon stick
【炭疽】[医] anthrax ◇ ～病[农] anthracnose
【炭盆】charcoal brazier
【炭窑】charcoal kiln; charcoal burner

碳 [化] carbon (C)
【碳酐】[化] carbonic anhydride
【碳黑】[化] carbon black
【碳弧光灯】carbon lamp
【碳化】[化] carbonization ◇ ～钙 calcium carbide / ～硅 carborundum; silicon carbide / ～物 carbide
【碳精】[电] carbon ◇ ～棒 carbon rod; carbon / ～电极 carbon electrode
【碳氢化合物】[化] hydrocarbon
【碳刷】[电] carbon brush
【碳水化合物】[化] carbohydrate
【碳丝】[电] carbon filament ◇ ～灯 carbon lamp
【碳素钢】[化] carbon steel
【碳酸】[化] carbonic acid ◇ ～钙 calcium carbonate / ～钠 sodium carbonate; soda / ～气 carbon dioxide; chokedamp / ～氢钠 sodium bicarbonate; baking soda / ～盐 carbonate

tāng

汤 ①(热水; 开水) hot water; boiling water ②(温泉) hot springs ③(食物煮后的汁) soup; broth: 番茄～ tomato soup / 鸡～ chicken soup / 姜～ ginger tea
【汤包】steamed dumplings filled with minced meat and gravy
【汤池之固】impenetrable defense work
【汤匙】tablespoon; soupspoon
【汤壶】metal or earthenware hot-water bottle
【汤加】Tonga ◇ ～人 Tongan
【汤面】noodles in soup
【汤泉沐浴】bathe in a hot spring
【汤勺】soup ladle
【汤碗】soup bowl

【汤药】[中药] a decoction of medicinal ingredients
【汤圆】 stuffed dumplings made of glutinous rice flour served in soup

铴
【铴锣】 a small brass gong

羰 [化] carbonyl
【羰基】[化] carbonyl: ～键 carbonyl bond

趟 ①(从浅水走过) wade; ford: ～水过河 wade across a river ②[农](用犁把土翻开; 除去杂草) turn the soil and dig up weeds

耥 weed and loosen the soil
【耥耙】[农] paddy-field harrow

táng

唐
【唐花】 hothouse flower
【唐人街】 Chinatown
【唐三彩】[考古] tri-colored glazed pottery of the Tang Dynasty
【唐诗】 Tang poetry
【唐唐大国】 a great powerful nation
【唐突】 brusque; rude; offensive: ～的行动 a presumptuous act / 出言～ make a blunt remark

溏 half congealed; viscous
【溏心】 with a soft yolk: ～儿蛋 soft-boiled or soft-fried egg / ～儿松花 preserved egg with a jelly-like yolk

糖 ①(食糖) sugar: 白～ powdered sugar / 冰～ crystal sugar; rock candy / 方～ cube sugar / 红～ brown sugar / 原～ raw sugar / 蔗～ cane sugar / 你喜欢在茶里放～吗? Do you like sugar in your tea? ②(糖果) candy; sweets: 口香～ chewing-gum / 奶油薄荷夹心～ cream mint sweets / 奶油软～ cream candy / 柠檬硬～ lemon drops / 什锦水果软～ mixed fruit jelly / 水果硬～ fruit drops / 太妃～ toffee; taffy / 椰子～ coconut candy ③[化](糖类) carbohydrate: 麦芽～ maltose
【糖厂】 sugar refinery
【糖醋】 sugar and vinegar; sweet and sour ◇ ～排骨 sweet and sour spareribs / ～鱼 fish in sweet and sour sauce
【糖甙】[化] glucoside
【糖膏】 massecuite; fillmass
【糖果】 sweets; candy; sweetmeats ◇ ～店 sweet shop; candy store; confectionery
【糖葫芦】 sugarcoated haws on a stick
【糖化】[化] saccharification ◇ ～饲料 saccharified pig feed; fermented feed
【糖浆】 syrup
【糖姜】 sugared ginger; ginger in syrup
【糖精】 saccharin; gluside
【糖量计】[化] saccharometer; saccharimeter
【糖料作物】[农] sugar crop
【糖罗卜】①(甜菜) beet ②(蜜饯的胡萝卜) preserved carrot
【糖酶】[化] carbohydrase
【糖蜜】 molasses
【糖尿病】 diabetes ◇ ～患者 diabetic
【糖商】 a sugar merchant
【糖食】 sweetmeats; sugar
【糖水】 syrup: ～梨 pears in syrup
【糖蒜】 sweetened garlic; garlic in syrup
【糖业】 a sugar industry
【糖衣】 sugar-coat; sugarcoating: 这些药丸都裹有～。These pills are sugar-coated. ◇ ～炮弹 sugarcoated bullet
【糖原】[化] glycogen
【糖汁】 syrup

塘 ①(堤岸) dike; embankment ②(水池) pool; pond: 鱼～ fish pond ③(浴池) hot-water bathing pool: 澡～ bathhouse; public baths
【塘肥】 pond sludge used as manure
【塘鳢】[动] sleeper
【塘泥】 pond sludge; pond silt
【塘堰】 small reservoir

搪 ①(抵挡) ward off; keep out: ～风 keep out the wind ②(搪塞) evade; do sth. perfunctorily: ～帐 put off a creditor ③(把泥或涂料抹在炉灶等上) spread over; daub: ～炉子 line a stove with clay
【搪不过去】 unable to parry
【搪瓷】 enamel; porcelain enamellig ◇ ～茶缸 enamel mug / ～钢板[建] enamelled pressed steel / ～器皿 enamel ware; enamelware
【搪塞】 stall sb. off; do sth. perfunctorily: ～之词 a statement of excuse
【搪突陈说】 make a statement unceremoniously

堂 ①(大厅) a hall for a specific purpose: 人民大会～ the Great Hall of the People / 食～ dining hall ②(正房) the main room of a house ③(堂兄弟、姐妹) relations between cousins:

～兄弟 cousins ④(公堂) court of law: 过～ be tried; have a hearing ⑤[量](用于成套家具)一～家具 a set of furniture
【堂而皇之】a dignified bearing
【堂鼓】a kind of drum used in Chinese operas
【堂皇】grand; stately; magnificent: ～大方 dignified and liberal
【堂堂】①(形容容貌庄严大方) dignified; impressive: 仪表～ dignified in appearance ②(有志气或有气魄) having high aspirations and boldness of vision ③(阵容或力量大) imposing; awe inspiring; formidable
【堂堂大国】a great, powerful nation
【堂堂正正】①(光明正大) open and aboveboard ②(身材威武) impressive or dignified in personal appearance
【堂屋】central room

棠
【棠棣】[植] Chinese bush cherry
【棠梨】[植] birchleaf pear

樘 [建] door frame; window frame

螳 mantis
【螳臂当车】a mantis trying to stop a chariot; overrate oneself and attempt sth. impossible
【螳螂】[动] mantis
【螳螂捕蝉,黄雀在后】the mantis stalks the cicada, unaware of the oriole behind; covet gains ahead without being aware of danger behind

镗 [机] boring
【镗床】[机] boring machine; boring lathe; borer: 坐标～ jig boring machine
【镗刀】[机] boring cutter; boring tool
【镗孔】[机] bore hole; boring

膛 ①(胸膛) thorax; chest ②(器物的中空部分) an enclosed space inside sth.; chamber: 枪～ bore of a gun
【膛线】[军] rifling

tǎng

淌 drip; shed; trickle: ～口水 slaver; slobber / ～眼泪 shed tears

倘 [连] if; supposing; in case: ～我遗忘 in case I forget / ～有困难 if there should be any difficulty
【倘来之物】an unexpected or undeserved gain; windfall
【倘能如此】if this can be done
【倘能如愿】if one can satisfy his wishes
【倘若】[连] if; supposing; in case
【倘有不测】in the event of an accident; if anything untoward should happen

躺 lie; recline: 我想去～一会儿。 I think I'll go and lie down.
【躺倒】lie down: ～不干 stay in bed; refuse to shoulder responsibilities any longer
【躺柜】a long low box with a lid on top; chest
【躺椅】deck chair; sling chair

tàng

烫 ①(烫痛) scald; burn: ～了个泡 get a blister through being scalded ②(加热) warm; heat ③(温度高) boiling hot: 这水真～。 The water is boiling hot. ④(熨) iron; press: ～衣服 iron clothes; press clothes ⑤(烫发) perm; have one's hair permed: 冷～ cold wave
【烫发】give or have a permanent wave; perm: 去理发馆～ go to the hairdresser's for a perm
【烫金】[印] gilding bronzing: 布面～ cloth gilt ◇～机 gilding press; bronzing machine
【烫蜡】polish with melted wax; wax
【烫面】dough made with boiling water ◇～饺 steamed dumplings
【烫伤】[医] scald

趟 [量] ①(走动的次数)到重庆去了一～ have been to Chongqing once; have made a trip to Chongqing ②(用于成行的东西)两～桌子 two rows of tables

tāo

涛 great waves; billows: 惊～骇浪 terrifying crashing waves / 狂～ very high sea / 怒～ mountainous sea / 松～ the soughing of the wind in the pines

掏 ①(把东西弄出来) draw out; pull out; fish out: 从口袋里～出一枚硬币 fish out a coin from one's pocket / 河岸被流水～出洞来。 The river banks were hollowed out by rushing water. ②(挖) dig; scoop out; hollow out ③(从别人口袋里拿东西) steal from sb.'s pocket
【掏槽】[矿] cutting
【掏井】dredge a well
【掏腰包】①[口](出钱) pay out of one's own pocket; foot a bill ②(偷窃) pick sb.'s pocket

饕
【饕餮】①(一种传说的动物) a mythical fero-

cious animal ②(凶恶贪婪的人) a fierce and cruel person ③(贪吃的人) voracious eater; glutton; gourmand
【饕餮之徒】greedy persons; gluttons
叨 be favored with; get the benefit of
【叨光】[套] much obliged to you: ~不少 have received many favors
【叨教】[套] many thanks for your advice: ~多载 have been in receipt of your instruction for many years
【叨扰】[套] thank you for your hospitality
滔 inundate; flood
【滔滔】①(大水滚滚) surging; torrential; billowy: 白浪~ whitecaps surging ②(话多) keeping up a constant flow of words: ~不绝 pouring out words in a steady flow; incessant talk / ~雄辩 a torrent of eloquence
【滔天】①(形容波浪极大) dash to the skies; billowy ②(罪恶、灾祸极大) heinous; monstrous: ~大祸 a disaster like foaming billows dashing to the skies; a terrible disaster / ~大罪 a monstrous crime; a heinous crime
韬
【韬笔】let the pen idle; write no more: ~钳口 put away the pen and keep the mouth tight; used figuratively of a wise man
【韬光养晦】hide one's capacities and bide one's time
【韬晦】conceal one's true features or intentions; lie low: ~隐居 hide one's light and live in seclusion
【韬略】military strategy
绦 silk ribbon; silk braid
【绦虫】[动] tapeworm; cestode ◇ ~病[医] taeniasis; cestodiasis
【绦子】silk ribbon; silk braid

táo

逃 ①(逃跑) run away; escape; flee; take flight: 从敌人手中~出 escape from the enemy / 携细软潜~ run away with all one's valuables / 敌人溃败而~. The enemy were defeated and fled in disorder ②(逃避) evade; dodge; shirk; escape: ~不出人民的法网 be unable to escape the net of justice spread by the people / 罪责难~ cannot shirk responsibility for the crime
【逃奔】run away to: ~他乡 flee from one's native place
【逃避】escape; evade; shirk: ~法律制裁 evasion of legal sanction / ~现实 try to escape reality / ~责任 shirk responsibility / ~债务 dodge a creditor / ~职责 evade doing a duty
【逃兵】army deserter; deserter
【逃出重围】break out from a heavy siege
【逃出虎口】escape from a tiger's mouth; escape from a dangerous situation
【逃窜】run away; flee in disorder
【逃遁】flee; escape; evade: 仓皇~ flee in panic / 无处~ can find no hiding-place
【逃犯】escaped criminal; escape convict
【逃荒】flee from famine; get away from a famine-stricken area
【逃名避誉】avoid fame and praise
【逃命】run for one's life
【逃难】flee from a calamity; be a refugee
【逃匿】escape and hide; go into hiding: ~无踪 skulk; leaving no trace
【逃跑】run away; flee; take flight; take to one's heels ◇ ~主义 flightism
【逃散】become separated in flight
【逃生】flee for one's life; escape with one's life: 死里~ barely escape with one's life; have a narrow escape; have a close shave
【逃世深居】retire from the world and dwell in deep seclusion
【逃税】evade a tax; tax evasion ◇ ~人 tax dodger
【逃脱】succeed in escaping; make good one's escape; get clear of: ~责任 succeed in evading responsibility
【逃亡】become a fugitive; flee from home; go into exile ◇ ~者 a fugitive
【逃学】play truant; cut class
【逃逸】escape; run away; abscond ◇ ~临界高度 critical level of escape
【逃之夭夭】decamp; make one's getaway; show a clean pair of heels
【逃走】run away; flee; take flight; take to one's heels; make one's escape
桃 ①(桃树;桃子) peach ②(形状象桃的东西) a peach shaped thing: 棉~ cotton boll
【桃符】①(古代画门神的桃木板) peach wood charms against evil ②(春联) Spring Festival couplets
【桃脯】preserved peach
【桃红】pink
【桃花】peach blossom

【桃花心木】mahogany
【桃花汛】spring flood
【桃花鱼】minnow
【桃李】peaches and plums; one's pupils or disciples: ～满门 have many pupils / ～满天下 have pupils everywhere
【桃李不言,下自成蹊】the peach and the plum do not speak, yet a path is worn beneath them; a man of true worth attracts admiration
【桃色新闻】news of illicit love
【桃树】peach (tree)
【桃子】peach

淘 ①(洗去杂质) wash: ～米 wash rice ②(舀出污水等) clean out; dredge: ～阴沟 clean out a drain ③(耗费) tax ④[方](顽皮) naughty
【淘河】[动] pelican
【淘金】[矿] panning ◇ ～者 gold-digger
【淘箩】a basket for washing rice in
【淘气】naughty; mischievous: 爱～ be fond of pranks / 象猴子一样～ as mischievous as a monkey ◇ ～包 mischievous imp; a regular little mischief
【淘沙拣金】wash the sand for gold; choose or search for the very best
【淘神】[口] trying; bothersome: ～费力 worrying the mind and wasting energy
【淘汰】eliminate through selection or competition; be sifted out: ～出去 clean it out ◇ ～法 concentration / ～赛 elimination series

陶 ①(用黏土烧制的) pottery; earthenware: 彩～ painted pottery / 彩色釉～ color-glazed pottery ②(制造陶器) make pottery ③(比喻教育、培养) cultivate; mold; educate: 熏～ exert a gradual; uplifting influence on; nurture ④(快乐) contented; happy: 乐～～ feel happy and contented
【陶瓷】pottery and porcelain; ceramics ◇ ～工 potter / ～业 ceramics; ceramic industry
【陶管】[建] earthen-ware pipe
【陶粒】[建] ceramsite ◇ ～混凝土 ceramsite concrete
【陶器】pottery; earthenware
【陶情山水】relax one's mind in nature
【陶然】happy and carefree: ～忘机 contented and happy without any schemes in mind
【陶土】[矿] potter's clay; pottery clay; kaolin
【陶文】[考古] inscription on pottery
【陶冶】①(烧制陶器和冶炼金属) make pottery and smelt metal ②(给人思想、性格上的影响) exert a favorable influence; mold: ～性情 mold a person's temperament
【陶俑】pottery figurine; terracotta figurine
【陶铸】mold and educate persons
【陶醉】be intoxicated; revel in: 自我～ be intoxicated with self-satisfaction

tǎo

讨 ①(讨伐) send armed forces to suppress; send a punitive expedition against ②(索取) demand; ask for: ～帐 demand the payment of debt; dun ③(娶) marry: ～老婆 get married; take a wife ④(招惹) incur; invite: 自～苦吃 look for trouble / 自～没趣 court a rebuff; ask for a snub ⑤(讨论) discuss; study: 商～ discuss / 研～ deliberate
【讨伐】send armed forces to suppress; send a punitive expedition against
【讨饭】beg for food; be a beggar
【讨好】①(迎合别人) ingratiate oneself with; fawn on; curry favor with ②(得到好效果) be rewarded with a fruitful result; get good result: 费力不～ put in much hard work, but get very little result; undertake a thankless task
【讨还】get sth. back
【讨价】ask a price; name a price
【讨价还价】bargain; haggle
【讨教】ask for advice
【讨论】discuss; talk over: ～一个问题 discuss a question / 参加～ join in the discussion
【讨论会】symposium; discussion: 科学～ science symposium
【讨便宜】seek undue advantage; try to gain sth. at the expense of others; look for a bargain
【讨巧】act artfully to get what one wants; get the best for oneself at the least expense; choose the easy way out
【讨情】[方] plead for sb.; beg sb. off: ～告饶 plead for leniency; beg for pardon
【讨求宽恕】entreat one to forgive
【讨饶】beg for mercy; ask for forgiveness
【讨嫌】disagreeable; annoying
【讨厌】①(惹人厌烦) disagreeable; disgusting; repugnant: 苍蝇真～! The flies are a nuisance. ②(事情难办) hard to handle; troublesome; nasty: 这是一种很～的病。It is a nasty illness. ③(厌恶) dislike; loathe; be disgusted with: 她～蛇。She dislikes the snakes.
【讨债】demand repayment of a loan

tào

套 ①(套子) sheath; case; cover; sleeve: 封～ big envelop / 书～ slipcase / 椅～ slipcover for a chair ②(罩上) cover with; slip on: ～上一件外衣 slip a coat on ③(罩在外面的) that which covers: ～鞋 overshoes / ～袖 oversleeve ④(互相衔接或重叠) interlink; overlap ⑤(河流或山势弯曲的地方) the bend of a river or curve in a mountain range: 河～ the Great Bend of the Huanghe River ⑥(拴牲口的绳) traces; harness: 牲口～ harness for a draught animal ⑦(用套拴系) harness; hitch up ⑧(绳结成的环状物) knot; loop; noose ⑨(模仿) model on; copy: 生搬硬～ apply mechanically; copy indiscriminately ⑩(俗套子) convention; formula: 老一～ the same old stuff ⑪(引出) coax a secret out of sb.; pump sb. about sth.: 想法～出她的话 try to coax the secret out of her ⑫(拉拢) try to win ⑬[量] set; suit; suite: 一～房间 a suite of rooms / 一～家具 a set of furniture / 一～衣服 a suit of clothes

【套版】[印] registering

【套包】collar for a horse

【套裁】cut out a piece of cloth in a way with the minimum material to make two or more articles of clothing

【套车】harness an animal to a cart

【套购】fraudulently purchase: ～物资 illegally by up goods

【套管】[石油] casing pipe; casing ◇ ～程序 casing program

【套话】polite, conventional verbal exchanges: ～连篇 a link of conventional phrases

【套汇交易】arbitrage transaction

【套间】①(里边的屋子) a small room opening off another; inner room ②(一套相连的房间) apartment; flat

【套裤】trouser legs worn over one's trousers; leggings

【套曲】[乐] divertimento ◇ ～形式 cyclical form

【套色】[印] chromatography; color process ◇ ～版 process plate; colorplate / ～木刻 colored woodcut

【套衫】pullover

【套数】①(连贯成套的曲子) a cycle of songs in a traditional opera ②(系统的技巧和手法) a series of skills and tricks

【套索】lasso; noose

【套筒】[机] sleeve; muff: 气缸～ cylinder sleeve ◇ ～扳手 box spanner; socket wrench / ～联轴节 muff coupling / ～式减震器 direct-acting shock absorber

【套头裹脑】blind fold person

【套问】find out by asking seemingly casual questions; tactfully sound sb. out

【套印】[印] chromatography: 彩色～ process printing ◇ ～本 chromatograph edition

【套用】apply mechanically; use indiscriminately

【套语】polite formula

【套种】[农] interplanting

【套子】①(罩在外面的) sheath; case; cover: 沙发～ sofa cover ②(应酬的话) conventional remark; conventionality ③[方](被子里的棉絮) cotton padding; batting: 棉花～ cotton padding of a quilt

tè

铽 [化] terbium (Tb)

特 ①(特殊) special; particular; unusual; exceptional: 能力～强 unusually capable ②(特地) for a special purpose; specially ③(特务) secret agent; spy

【特别】①(与众不同) special; particular; out of the ordinary: ～的式样 peculiar style / ～会议 special meeting / 脾气～ be peculiar ②(格外) especially; particularly: ～注意 pay special attention to / 质量～好 be of extra fine quality ③(特地) specially ◇ ～案件 extraordinary case / ～保险 specific insurance / ～法 special law / ～法庭 special tribunal / ～海损 extraordinary average / ～开支 special expenses / ～快车 express train; express / ～留置权 particular lien / ～诉讼程序 special proceedings / ～提款权 special drawing rights (SDR) / ～条款 special clause / ～许可证 special license / ～预算 special budget / ～帐户 special account

【特产】special local product; specialty; speciality

【特长】what one is skilled in; strong point; speciality: 游泳不是她的～。 Swimming is not her strong point.

【特出】outstanding; prominent; extraordinary: ～的成绩 outstanding achievements / ～的作用 a prominent role

【特此】here by: ～通知 it is hereby announced that / ～作证 in testimony whereof; in witness whereof

【特大】especially big; the most: ～号服装 outsize garments / ～洪水 a catastrophic flood / ～喜讯 excellent news; most welcome news / ～自然灾害 extraordinarily serious natural calamities

【特等】special grade; top grade ◇ ～舱[交] stateroom; *de luxe* cabin / ～劳模 special-class model worker / ～射手 crack shot; expert marksman

【特地】[副] for a special purpose; specially: ～来请教 come specially to ask for advice / 我～来看你。I came here specially to see you.

【特点】characteristic; distinguishing feature; peculiarity; trait: 生理～ physiological characteristics

【特定】①(特别指定的) specially appointed; specially designated ②(某一个) given; specified; specific: ～的人选 a person specially designated for a post ◇ ～管辖权 limited jurisdiction; special jurisdiction / ～继承人 particular successor / ～期限 definite term / ～信用状 special credit / ～证据 particular evidence / ～罪行 particular crime

【特氟隆】[化] teflon

【特工】secret service ◇ ～人员 special agent; secret service personnel

【特惠】indulgence: ～关税 preferential tariff / ～条款 preferential clause / ～制度 preferential system.

【特混舰队】task force

【特级】special grade; superfine: ～茉莉花茶 superfine jasmine tea ◇ ～教师 teacher of a special classification / ～战斗英雄 special-class combat hero; combat hero special grade

【特急】extra urgent ◇ ～电 extra urgent telegram; flash message

【特辑】①(为特定目地编辑的文字资料) special number of a periodical ②(电影专辑) a special collection of short films

【特技】①(特殊技能) stunt; trick ②[电影] special effects ◇ ～飞行 stunt flying; aerobatics / ～镜头 trick shot / ～摄影 trick photography / ～跳伞 trick parachuting

【特价】special offer; bargai price: ～出售 sell at a bargain price

【特刊】special issue; special

【特快】[交] express

【特立尼达和多巴哥】Trinidad and Tobago

【特例】special case

【特留分继承权】forced heirship

【特留分继承人】forced heirs

【特命全权大使】ambassador extraordinary and plenipotentiary

【特命全权公使】envoy extraordinary and minister plenipotentiary

【特派】specially appointed ◇ ～记者 special correspondent; accredited journalist

【特遣部队】task force

【特屈儿】[化] tetryl

【特权】privilege; prerogative: ～地位 privileged position / 外交～ diplomatic privileges ◇ ～阶层 privileged stratum / ～思想 the idea that prerogatives and privileges go with position; the "special privilege" mentality / ～专利 special privilege monopoly

【特色】characteristic; distinguishing feature: 民族～ distinctive national features / 艺术～ artistic characteristics

【特设】ad hoc: ～委员会 ad hoc committee

【特赦】special pardon; special amnesty ◇ ～令 decree of special pardon / ～权 prerogative of mercy / ～证明书 charter of pardon

【特使】special envoy

【特殊】special; particular; peculiar; exceptional: ～情况 special circumstances; special case / 反对～化 oppose privileges / 受到～照顾 receive special care ◇ ～性 particularity; specific characteristics / ～折扣 channel discount

【特为】for a special purpose; specially; going out of one's way to: ～来看你 come specially to see you

【特务】①[军](特殊任务) special task ②(从事刺探情报等活动的人) special agent; spy ◇ ～活动 espionage / ～机关 secret service; espionage agency / ～组织 secret service; spy organization

【特效】specially good effect; special efficacy ◇ ～药 specific drug; specific; effective cure

【特写】①[文学] feature article or story; feature ②[电影] close-up ◇ ～镜头 close-up shot

【特性】specific property

【特许】special permission ◇ ～证书 special permit; letters patent / ～法人 chartered corporation / ～公司 chartered company

【特许权】chartered right ◇ ～使用费 royalty (for the right of special permission)

【特压】[化] extreme pressure ◇ ～添加剂 extreme pressure additive

【特邀】specially invite ◇ ～代表 specially invited representative

【特异】①(特别优异) exceptionally good; ex-

cellent; superfine ②(特殊) peculiar; distinctive ◇～体质[医] idiosyncrasy
【特有】peculiar; characteristic: ～财产 peculiar property
【特约】engage by special arrangement ◇～分摊 special condition of average / ～稿 special contribution / ～记者 special correspondent / ～经售处 special sales agency / ～评议员 special commentator / ～维修店 special repair shop / ～演员 guest actor / ～撰稿人 special contributor
【特征】characteristic; feature; trait: 地理～ geographical features / 中国人的～ characteristics of the Chinese
【特指】refer in particular to
【特制】specially made; made to order
【特种】special type; particular kind ◇～兵 the special arms / ～工艺 special arts and crafts; special handicraft products / ～技术部队 special technical units / ～诉讼案件 special conditions of an action / ～刑事犯 special kind of criminal offender / ～刑事案件 special criminal case / ～销售税 special sale tax / ～战争 special warfare

téng

疼 ①(痛) ache; pain; sore: 感到～ feel pain / 浑身都～ be aching all over / 嗓子～ have a sore throat / 头～ have a headache / 腿～ have a pain in the leg / 胃～ have a stomachache ②(疼爱) love dearly; be fond of; dote on
【疼爱】be very fond of; love dearly: 这孩子谁都～他。The child is dearly loved by all.
【疼痛】pain; ache; soreness: 全身～ ache all over; general aching / 伤口～。The wound ached.

誊 transcribe; copy out: 把稿子～一遍 make a clean-copy of the craft
【誊录】transcribe; copy out: ～文稿 copy out a manuscript
【誊清】make a fair copy of ◇～稿 fair copy
【誊写】transcribe; copy out ◇～版 stencil / ～钢版 steel plate for cutting stencils / ～蜡纸 stencil paper / ～油墨 stencil ink
【誊印社】mimeograph service

藤 ①(指用藤制的) cane; rattan: ～盔 rattan helmet / ～椅 cane chair; rattan chair / ～制品 rattan work ②(象藤一样的细茎) vine: 西瓜～ watermelon vine
【藤本植物】[植] liana; vine
【藤壶】[动] acorn barnacle; barnacle
【藤黄】①[植] garcinia ②(藤黄的树脂) gamboge
【藤萝】[植] Chinese wistaria
【藤牌】cane shield; shield
【藤条】rattan
【藤子】[口] vine

腾 ①(跑或跳) gallop; jump; prance: ～身而起 jump to one's feet ②(升) rise; soar: 飞～ soar ③(使空出) make room; clear out; vacate: ～出房间 vacate one's room ④(用在某些动词后面,表示反复)翻～ toss about; turn over and over / 扑～ throb
【腾贵】shoot up; soar; skyrocket
【腾空】soar; rise high into the air; rise to the sky: 五颜六色的气球～而起。The colorful balloons rose high into the air.
【腾挪】①(挪动款项) transfer to other use ②(挪动地方) move sth. to make room
【腾腾】steaming; seething: 烈焰～ raging flames / 热气～ steaming hot; seething with activity / 杀气～ full of bellicosity; murderous-looking / 烟雾～ hazy with smoke; smoke-laden
【腾笑】arouse laughter: ～海内 be laughed at in the whole country
【腾越】jump over: ～障碍 jump over obstacles
【腾云驾雾】①(乘云雾飞行) mount the clouds and ride the mist; speed across the sky ②(奔驰迅速) speed on as if flying in the air ③(头脑发胀) feel dizzy; feel giddy

tī

梯 ①(梯子) ladder; steps; stairs: 电～ elevator; left / 电动扶～ escalator ②(跟楼梯相似的设备) shaped like a staircase; terraced
【梯次队形】[军] echelon formation
【梯度】[数] gradient
【梯队】[军] echelon formation; echelon: 成～飞行 flying in echelon
【梯队结构】an orderly system of succession
【梯恩梯】[化] trinitrotoluene (TNT) ◇～当量 TNT equivalent
【梯级】stair; step ◇～船闸 chain of locks
【梯田】[农] terraced fields; terrace
【梯形】①(形状象梯子的) ladder-shaped ②[数] trapezoid; trapezium ◇～翼[航空] trapezoidal wing; tapered airfoil
【梯子】ladder; stepladder

锑
[化] antimony; stibium (Sb)

剔
①(刮) clean with a pointed instrument; pick: ～骨头 pick a bone / ～牙 pick one's teeth ②(剔除) pick out and throw away; reject

【剔除】reject; get rid of: 吸取精华，～糟粕 absorb the essence and reject the dross

【剔红】carved lacquerware

踢
kick: ～毽子 kick the shuttlecock / ～翻椅子 kick over a chair / ～进一个球 kick a goal / ～开大门 kick the gate open / ～足球 play football

【踢打交加】mingled kicking and beating

【踢蹬】kick at random

【踢脚板】[建] skirting board; skirtboard

【踢皮球】①(儿童游戏) play children's football ②(推卸责任) pass the buck: 互相～ pass the buck to each other

【踢踏舞】step dance; tap dance

体

【体已】①(亲近的) intimate; confidential: ～话 things one says only to one's intimates ②(家庭成员个人积蓄) private savings: ～钱 private savings of a family member

tí

啼
①(啼哭) cry; weep aloud ②(叫) crow; caw: 鸡～cocks crow

【啼饥号寒】cry from hunger and cold; cry out in hunger and cold

【啼哭】cry; wail

【啼笑皆非】not know whether to laugh or cry

蹄
hoof: 马～ horse's hoofs

【蹄冠炎】[兽] villitis

【蹄筋】tendons of beef, mutton or pork: 红烧～ tendons stewed in soy sauce

【蹄膀】[方] the upper part of a leg of pork

【蹄式制动器】[机] shoe brake

【蹄形磁体】horseshoe magnet

【蹄子】hoof: 猪～ leg of pork

鹈

【鹈鹕】pelican

题
①(题目) topic; subject; title; problem: 考～ examination questions / 话～ subject of conversation / 讨论～ topic for discussion / 文不对～ wide of the mark; irrelevant ②(写上) inscribe: ～诗 inscribe a poem / 某人～ inscription by sb.

【题跋】①(写在书籍、字画等前后的文字) preface and postscript ②(短评) short comments; annotations

【题材】subject matter; theme: ～范围 range of subjects

【题词】①(指写) write an inscription ②(指题的词) inscription ③(序文) foreword

【题花】title design

【题解】①(详细解答) key to exercises or problems:《代数～》*Key to Exercises in Algebra* ②(解释性文字) explanatory notes on the title or background of a book

【题名】①(写姓名) autograph: 请～留念。Please give your autograph as a moment. ②(题目的名称) title; subject

【题目】①(标题) title; subject; topic ②(习题) exercise problems; examination questions

【题签】①(书皮上的标签) a label with the title of a book on it ②(写标签) write the title of a book on a label to be stuck on the cover

【题字】①(为留纪念而写的字) inscription; autograph ②(题写) inscribe

提
①(垂手拿着) carry in one's hand: 手里～着个包 carry a bag in one's hand ②(使事物由下往上移) lift; raise; promote: ～价 raise the price ③(提出) put forward; bring up; raise: ～方案 suggest plans / ～抗议 lodge a protest / ～条件 put forward conditions / ～要求 make demands / ～意见 make a criticism; make comments or suggestions ④(提取) draw out; extract: ～款 draw money ⑤(谈起) mention; refer to; bring up: 上面～到的 the above-mentioned / 旧事重～ bring up a past event ⑥(一种器皿) dipper: 酒～ wine dipper ⑦(汉字的一种笔画) rising stroke

【提案】motion; proposal; draft resolution: ～被否决了。The proposal was rejected. / ～通过了。The motion was adopted. ◇ ～国 sponsor country; sponsor / ～审查委员会 motions examination committee

【提拔】promote: ～某人为… promote sb. to be...

【提包】handbag; shopping bag; bag; valise

【提倡】advocate; promote; encourage; recommend: ～计划生育 advocate family planning / ～晚婚 advocate late marriage

【提成】deduct a percentage (from a sum of money)

【提出】put forward; advance; pose; raise: ～程序问题 raise a point of order / ～控诉 enter a complaint / ～理由 produce reasons / ～上诉 entry of appeal; lodge an appeal / ～问题 raise a question / ～要求 raise a claim / ～一项合理化建议 put forward a rationalization proposal / ～证据 adduce evidence

【提纯】purify; refine: ～金属 purify metals / ～石油 purify oil ◇ ～复壮[农] purification and rejuvenation / ～器 purifier

【提词】[剧] prompt: 给演员～ prompt an actor ◇ ～人 prompter / ～箱 prompters' box

【提单】bill of lading (B / L): 联运～ through bill of lading / 直达～ direct bill of lading

【提到】mention; refer to

【提法】the way sth. is put; formulation; wording

【提防】be cautious or watchful; be on the alert; guard against

【提纲】outline: 拟定～ draw up an outline / 写发言～ make an outline for a speech

【提纲挈领】take a net by the headrope or a coat by the collar; concentrate on the main points; bring out the essentials

【提高】raise; heighten; enhance; increase; improve: ～产品质量 improve the quality of products / ～价格 raise prices / ～劳动生产率 raise labor productivity / ～文化水平 raise the cultural level / 不断～服务质量 steadily improve (their) service; improve the quality of service continually

【提供】provide; supply; furnish; offer: ～保证 give security / ～相反证据 produce proof to the contrary / ～证据 furnish evidence / ～证据的责任 liability to give evidence / 为市场～新产品 supply the market with new products

【提行】[印] begin a new line

【提花】[纺] jacquard weave ◇ ～枕巾 jacquard pillow cover / ～织机 jacquard loom / ～织物 jacquard

【提货】pick up goods; take delivery of goods ◇ ～单 bill of lading (B / L)

【提交】submit to; refer to: 把草案～委员会讨论 submit the draft to the committee for discussion / ～仲裁 dispute to arbitration

【提炼】extract and purify; abstract; refine: ～石油 refine oil / 从矿石中～金属 extract metal from ore

【提梁】handle; straps; hoop handle

【提名】nominate: ～某人为总统候选人 nominate sb. for Presidency

【提起】①(提到) mention; speak of ②(激起) raise; arouse; brace up: ～精神 raise one's spirit; brace oneself up

【提前】①(往前移) shift to an earlier date; move up; advance ②(事先) in advance; ahead of time; beforehand: ～释放 release before the sentence expires / ～一个月完成计划 fulfill the plan a month ahead of time

【提亲】a match-making; talk about a marriage

【提琴】[乐] the violin family: 大～ violoncello; cello / 低音～ double bass; contrabass / 小～ violin / 中～ viola

【提请】submit sth. to ◇ ～辩论 motion for debate / ～撤销案件 motion for dismissal / ～大会批准 submit to the congress for approval / ～复审 motion in error

【提取】①(取出) draw; pick up; collect: ～存款 draw deposits / ～行李 pick up one's luggage ②(提炼) extract; abstract; recover: 从废水中～有用物质 salvage useful materials from the waste water ◇ ～器 extractor / ～塔 extraction column

【提神】refresh oneself; give oneself a lift: 喝杯茶提提神 refresh oneself with a cup of tea

【提审】①(提讯) bring before the court; bring to trial ②(由上一级法院自行审判) take over by a higher court; review

【提升】①(提高职级) promote; advance: ～他当处长 promote him to be division director ②(向高处运) hoist; elevate ◇ ～设备 hoist; hoister / ～机 hoist; elevator

【提示】point out; prompt: 给某人～ prompt sb.

【提问】put questions to; quiz

【提线木偶】marionette

【提箱】suitcase

【提携】①(领着走) lead by the hand ②(扶植) guide and support; give guidance and help to

【提心吊胆】have one's heart in one's mouth; be on tenterhooks; be breathless with anxiety

【提醒】remind; warn; call attention to

【提选】select; choose: ～耐旱品种 select drought-resistant varieties

【提要】summary; abstract; epitome; synopsis: 本书内容～ capsule summary of the book / 本章～ the summary of this chapter

【提议】①(提出主张) propose; suggest; move: 我～现在休会。 I move the meeting be adjourned. ②(所提出的主张) proposal; motion: 根据某人～ on the motion of sb; in accordance with sb.'s proposal / 同意这个～ agree to the

proposal
【提早】shift to an earlier time; be earlier than planned or expected: ～半小时上班 go to work half an hour earlier than usual / ～出发 set out earlier than planned
【提制】obtain through refining; distil; extract

tǐ

体 ①(身体) body; part of the body: 人～构造 the structure of the human body ②(物体) substance; state of a substance: 固～ solid / 气～ gas / 液～ liquid ③(书写形式; 作品体裁) style; form; type of literature: 旧～诗 old-style poem; classical poetry ④ [语] aspect (of a verb) ⑤(亲身经验) personally do or experience sth.; put oneself in another's position ⑥(系统) system: 政～ system of government
【体壁】[动] body wall
【体裁】types or forms of literature: ～格律 style and meter of poetry and literary composition
【体操】gymnastics: 器械～ gymnastics on 〈with〉 apparatus / 柔软～ callisthenics / 徒手～ freestanding exercise / 现代韵律～ modern rhythmic gymnastics / 自由～ floor exercise ◇ ～表演 gymnastic exhibition / ～服 gym outfit / ～器械 gymnastic apparatus
【体察】experience and observe: 虚心～情况 be ready to look into matters with an open mind; not be prejudiced in sizing up situations
【体罚】corporal punishment; physical punishment
【体格】physique; build: ～魁梧的人 a man of great physique ◇ ～检查 health checkup; physical examination
【体会】know from experience; realize: ～到…的重要性 realize the significance of... / 你参加这次培训班有什么～? What have you learned from the training course?
【体积】volume; bulk: ～大 bulky / 容器的～ the volume of a container ◇ ～膨胀 [物] volume expansion
【体力】physical strength; physical power: 增强～ build up one's strength / 消耗～ be a drain on one's strength; consume one's strength / ～不支 feel run down; feel limp; feel weak / ～活动 physical exertion; physical activity; physical exercise
【体力劳动】physical labor: 参加～ take part in physical labor
【体例】stylistic rules and layout; style: 印刷～ style sheet; stylebook / ～繁多 regulations are numerous
【体谅】show understanding and sympathy for; make allowances for: ～别人 be considerate of others / ～别人的困难 make allowances for others' difficulties
【体面】①(体统) dignity; face: 有失～ be a loss of face ②(光荣) honorable; creditable; respectable: ～的外表 respectable appearance / 不～的行为 disgraceful conduct ③(好看) handsome; good-looking: 长得～ be handsome
【体魄】physique: 强壮的～ strong physique; vigorous health / 锻炼～ go in for physical training
【体腔】[生理] body cavity
【体虱】[动] body louse
【体式】①(文字的式样) form of characters or letters ②(体裁) style
【体视】[物] stereo-: ～望远镜 stereotelescope / ～显微镜 stereomicroscope
【体态】posture; carriage: ～轻盈 a graceful carriage
【体贴】show consideration for; give every care to: ～病人 show a patient every consideration / ～入微 look after with meticulous care; care for with great solicitude
【体统】decorum; propriety; decency: 不成～ most improper; downright outrageous / 有失～ be disgraceful; be scandalous
【体外受精】[动] external fertilization
【体温】temperature: 量～ take one's temperature / 试～ take one's temperature ◇ ～过低 hypothermia / ～计 thermometer
【体无完肤】①(浑身受伤) have cuts and bruises all over the body; be a mass of bruises ②(被驳倒) be thoroughly refuted
【体悉其艰】excuse and see through his difficulty
【体系】system; setup
【体现】embody; incarnate; reflect; give expression to
【体形】bodily form; build
【体型】type of build or figure
【体恤】understand and sympathize with; show solicitude for
【体恤衫】T-shirt
【体癣】[医] ringworm of the body
【体循环】[生理] systematic circulation; greater circulation
【体验】learn through practice; learn through one's personal experience: ～生活 observe and

learn from real life
【体液】body fluid
【体育】physical culture; physical training; sports: 军事～ military sports ◇～场 stadium／～道德 sportsmanship／～锻炼 physical training／～锻炼标准 standards for physical training／～馆 gymnasium; gym／～活动 sports activities／～界 the sporting world／～课 physical education (PE)／～疗法 physical exercise therapy／～用品 sports goods; sports requisites／～系 department of physical education／～中心 sports center; indoor stadium
【体育运动】sports; physical culture; physical education: 群众性～ mass sports activities
【体制】system of organization; system: ～改革 structural reforms
【体质】physique; constitution: ～差 have a poor constitution／～好 have a good constitution／增强人民～ build up the people's health
【体重】weight: ～减轻 lose weight／～增加 put on weight; gain weight

tì

涕 ①(眼泪) tears: 感激～零 be moved to tears of gratitude／痛哭流～ shed bitter tears; cry one's heart out ②(鼻涕) mucus of the nose; snivel
【涕泪俱下】tears and mucus flowing down together
【涕零如雨】tears streaming down like rain drops
【涕泣】weep: ～沾襟 wet the front part of one's garment with tears
【涕泗横流】tears and mucus flowing down rapidly

剃 shave: ～胡子 have a shave; shave oneself
【剃刀】razor
【剃头】①(剃光头) have one's head shaved ②(理发) have one's hair cut; have a haircut
【剃枝虫】[动] armyworm

绨 a silk and cotton fibric

替 ①(代替) take the place of; replace; substitute for ②(为) for; on behalf of: ～别人买几本书 buy some books for someone
【替代】substitute for; replace; supersede: 以电动机～蒸汽机 replace a steam engine with an electromotor ◇～法 method of substitution／～区 alternate area／～品 succedaneum／～债务人 expromissor
【替工】①(代替别人做工) work as a substitute ②(代替别人做工的人) temporary substitute (worker)
【替换】replace; substitute for; displace; take the place of: ～的衣服 a change of clothes
【替角儿】understudy
【替身】①(替身演员) substitute; replacement; stand-in ②(替罪羊) scapegoat
【替死鬼】[口] scapegoat; fall guy
【替罪羊】scapegoat

嚏
【嚏喷】sneeze

屉 ①(笼屉) steamer tray: 一～馒头 a trayful of steamed buns ②(抽屉) drawer: 三～桌 three-drawer desk
【屉子】①(成套的屉) a set of removable trays ②[方](抽屉) drawer

tiān

添 ①(增加) add; increase; have more: ～菜 have additional dishes／～饭 have another bowl of rice／～福～寿 add to your happiness and longevity／～一点水 add a little water ②[方](生小孩) have a baby
【添补】replenish; get more: ～不足 add in order to make up a deficiency
【添加】add to; increase ◇～剂 [化] additive
【添油加醋】add color and emphasis to (a narration); add inflammatory details to (a story)
【添枝加叶】embellish a story; falsify a story by coloring it up
【添置】add to one's possessions; acquire: ～仪器 buy more instruments
【添砖加瓦】do what little one can to help: 为社会主义建设～ do one's bit to help build socialism

天 ①(天空) sky; heaven: 蓝～ the blue sky／满～星 a star-studded sky ②(一昼夜; 白天) day: 后～ day after tomorrow／每～ every day／前～ day before yesterday／前几～ the other day／三～三夜 three days and three nights／一连好多～ day after day ③(位置在顶部的) overhead: ～棚 ceiling ④(时间) time; a period of time in a day: ～不早了。It's getting late. ⑤(季节) season: 春～ spring ⑥(天

气）weather：晴～ fine day ⑦（天然的）nature；elements：～灾 natural calamity／战～斗地 battle the elements ⑧（神）God；Heaven
【天安门】Tian An Men：～城楼 the rostrum of Tian An Men／～广场 Tian'anmen Square
【天崩地裂】heaven falling and the earth cracking
【天边】horizon；the ends of the earth；remotest places：远在～，近在眼前 seemingly far away，actually close at hand
【天兵】troops from heaven；an invincible army
【天不怕，地不怕】fear neither Heaven nor Earth；fear nothing at all；nothing daunted
【天才】①（才能）genius；talent；gift：有艺术～ have a genius for arts ②（有天才的人）man of genius
【天蚕】[动] giant silkworm；wild silkworm ◇～蛾 giant silkworm moth
【天长地久】enduring as the universe；everlasting and unchanging
【天长日久】after a considerable period of time
【天车】[机] shop traveller；crown block
【天成佳偶】a good match as if made in heaven
【天秤座】[天] Libra
【天窗】[建] skylight
【天赐良机】a godsend chance
【天从人愿】Heaven grants man's wish；by the grace of God
【天大】as large as the heavens；extremely big：～的好事 an excellent thing
【天敌】[生] natural enemy
【天底】[天] nadir
【天底下】[口] in the world；on earth
【天地】①（天和地）heaven and earth；universe；world：震天动地 shake the earth ②（活动范围）field of activity；scope of operation：～间 in this world／科学研究的新～ the new field for scientific research ◇～良心 from the bottom of my heart／～万物 works of God
【天地头】[印] top and bottom margins of a page；upper and lower margins of a page
【天电】[电] atmospherics；static ◇～干扰 statics；static disturbances
【天顶】[天] zenith
【天鹅】swan：小～ cygnet ◇～绒毛 swansdown
【天鹅绒】velvet
【天鹅座】[天] Cygnus
【天蛾】[动] hawkmoth；sphinx
【天翻地覆】heaven and earth turning upside down：～的变化 earthshaking changes
【天分】special endowments；natural gift；talent：～高 gifted；talented
【天府之国】the land of abundance；the land of plenty
【天赋】①（自然赋予）inborn；innate；endowed by nature ②（天资）natural gift；talent；endowments ◇～人权论 the theory of natural rights
【天干】the ten Heavenly Stems
【天罡星】[天] the Big Dipper
【天高地厚】①（恩情深）profound；deep ②（事物复杂性）immensity of the universe；complexity of things：不知～ not see much of the world；not understand things
【天各一方】live far apart from each other
【天工】work of nature；formed by nature：巧夺～ art beats nuture；one's works of art may overcome nature
【天公】the ruler of heaven；God：～不作美 Heaven is not cooperative；the weather isn't cooperating
【天公地道】absolutely fair
【天沟】[建] gutter
【天国】①[基督教]（上帝的所在）the Kingdom of Heaven ②（理想世界）paradise
【天寒地冻】the weather is cold and the ground is frozen
【天河】[天] the Milky Way；the Galaxy
【天候】weather：全～飞行 all-weather flight／全～公路 all-weather road
【天花】[医] smallpox
【天花板】ceiling
【天花乱坠】as if it were raining flowers；give an extravagantly colorful description
【天皇】[日本] Mikado
【天昏地暗】①（指景象）a murky sky over a dark earth；dark all around：～日月无光 a mass of murkiness which neither sunlight nor moonlight can penetrate ②（政治腐败）in a state of chaos and darkness
【天机】①（秘密）nature's mystery；something inexplicable ②（天意）God's design；secret：泄露～ give away a secret
【天极】[天] celestial pole
【天经地义】unalterable principle；right and proper；perfectly justified
【天井】①（院中空地）small yard；courtyard ②（天窗）skylight ③[矿] raise：通风～ air raise
【天空】the sky；the heavens
【天蓝】sky blue；azure
【天狼星】[天] Sirius

【天老儿】albino
【天理】justice: ~良心 the course of nature and one's conscience / ~难容 intolerable by the course of nature
【天良】conscience: 丧尽~ conscienceless
【天亮】daybreak; dawn: ~以前 before daybreak
【天灵盖】top of the skull; crown
【天龙座】[天] Draco
【天伦】[书] the natural bonds and ethical relationships between members of a family: ~之乐 family happiness
【天罗地网】an inescapable dragnet; nets above and snares below
【天马行空】a heavenly steed soaring across the skies; a powerful and unconstrained style
【天明】daybreak; dawn
【天命】God's will; the mandate of heaven; destiny; fate
【天幕】①(天空) the canopy of the heavens ②(舞台背景表示天空) backdrop
【天南地北】①(距离远) far apart; poles apart ②(地区不同) from different places or areas
【天南海北】①(到处) all over the country ②(东拉西扯) discursive; rambling
【天年】natural span of life; one's allotted span
【天牛】[动] longicorn; long-horned beetle
【天怒人怨】the wrath of God and the resentment of men; widespread indignation and discontent
【天疱疮】[医] pemphigus
【天棚】①[建](顶棚) ceiling ②(凉棚) awning or canopy; mat awning
【天平】balance; scales: 分析~[化] analytical balance
【天平动】[天] libration
【天气】weather: 好~ nice weather / ~转晴。It's clearing up. ◇ ~图 weather map; synoptic chart / ~形势预报 weather prognostics / ~学 synoptic meteorology / ~预报 weather forecast
【天堑】natural moat: 长江~ the natural moat of the Changjiang River
【天桥】overline bridge; platform bridge
【天琴座】[天] Lyra
【天青】reddish black
【天青石】[矿] celestine; celestite
【天清日晏】a peaceful, sunny day
【天穹】the vault of heaven
【天球】[天] celestial sphere ◇ ~赤道 celestial equator / ~仪 celestial globe / ~子午圈 celestial meridian / ~坐标 celestial coordinates
【天然】natural: ~产物 be made by nature ◇ ~磁铁 natural magnet / ~堤[地] natural levee / ~更新[林] natural regeneration / ~景色 natural scenery / ~牧地 natural pasture / ~淘汰 natural selection / ~资源 natural resources
【天然气】natural gas: 干~ dry gas; poor gas / 湿~ wet gas; rich gas
【天壤之别】a world of difference; poles apart
【天日】the sky and the sun; light: 重见~ once more see the light of day; be delivered from oppression or persecution
【天色】color of the sky; time of the day; weather: ~不早。It's getting late. / ~已晚。It's getting dark.
【天神】god; deity
【天生】born; inborn; inherent; innate: ~才子 born great artist; a genius / 这男孩~聋哑。The boy was born a deaf-mute.
【天生桥】[地] natural bridge
【天时】①(气候) weather climate ②(时机) timeliness; opportunity
【天使】[宗] angel
【天书】a book from heaven; abstruse writing; illegible writing
【天水相际】where the sky and water meet
【天堂】paradise; heaven
【天体】[天] celestial body ◇ ~光谱学 astrospectroscopy / ~力学 celestial mechanics / ~物理学 astrophysics / ~演化学 cosmogony / ~照相仪 astrograph
【天天】every day; daily; day in, day out: 好好学习，~向上。Study well and make progress every day.
【天庭】the middle of the forehead: ~饱满 a full forehead
【天头】the top margin of a page
【天王星】[天] Uranus
【天网恢恢】the net of Heaven has large meshes, but it lets nothing through; the mills of God grind slowly but surely; justice has a long arm
【天文】astronomy ◇ ~常数 astronomical constants / ~单位 astronomical unit / ~导航 astronavigation; celestial navigation / ~观测 astronomical observation / ~馆 planetarium / ~年历 astronomical yearbook; astronomical almanac / ~时 astronomical time / ~台 (astronomical) observatory / ~望远镜 astronomical telescope / ~仪器 astro-

nomical instruments / ～照相机 astronomical camera / ～制导 celestial guidance / ～钟 astronomical clock

【天文数字】astronomical figure; enormous figure

【天文学】astronomy: 航海～ nautical astronomy / 恒星～ stellar astronomy / 空间～ space astronomy / 球面～ spherical astronomy / 射电～ radio astronomy ◇ ～家 astronomer

【天无绝人之路】Heaven never seals off all the exits; there is always a way out

【天下】①(指中国或世界) China or the world; land under heaven: ～奇闻 a most fantastic tale; the most absurd thing in the world / ～太平 the world is at peace / ～无敌 be invincible throughout the world / ～无双 none such under heaven; unparalleled in the world ②(统治权) rule; domination: 打～ conquer the country; seize state power

【天下乌鸦一般黑】all crows are black; evil people are bad all over the world

【天仙】①(仙女) goddess ②(美女) a beauty

【天险】natural barrier

【天线】[无] aerial; antenna: 磁性～ ferrite rod aerial 〈antenna〉/ 定向～ beam antenna / 机内超高频～ built-in very high frequency aerial 〈antenna〉/ 可转定向～ rotatory beam 〈directional〉 antenna / 拉杆～ telescopic antenna / 偶极～ dipole antenna / 架设～ put up an aerial

【天香国色】heaven fragrance and national beauty

【天象】astronomical phenomena; celestial phenomena ◇ ～仪[天] planetarium

【天蝎座】[天] Scorpio; Scorpius

【天性】natural instincts; nature: ～乖戾 be sullen by nature

【天幸】a providential escape; a close shave

【天旋地转】the sky and earth were spinning round; very dizzy

【天涯】the end of the world; the remotest corner of the earth: ～海角 the remotest regions of the earth; the ends of the earth / 浪迹～ rove all over the world

【天衣无缝】a seamless heavenly robe; flawless

【天意】God's will; the will of Heaven

【天鹰座】[天] Aquila

【天有不测风云】a storm may arise from a clear sky; sth. unexpected may happen any time

【天渊】[书] high heaven and deep sea; poles apart: ～之别 a world of difference / 相去～ as far apart as the sky and the sea

【天灾】natural disaster: ～人祸 natural and man-made calamities

【天葬】celestial burial

【天造地设】created by nature; heavenly; ideal

【天真】innocent; artless; naive: ～活泼 innocent and lively / ～烂漫 innocent and artless; simple and unaffected

【天之宠儿】one specially blessed by Heaven

【天之骄子】God's favored one; an unusually lucky person

【天知地知】between you and me

【天职】bounden duty; vocation

【天轴】①[机] line shaft ②[天] celestial axis

【天诛地灭】stand condemned by God; be destroyed by heaven and earth

【天助我也】helped from heaven

【天竺鲷】[动] cardinal fish

【天竺葵】[植] fish pelargonium

【天竺鼠】[动] guinea pig; cavy

【天主教】Catholicism ◇ ～会 the Roman Catholic Church / ～徒 Catholic

【天资】natural gift; talent; natural endowments: ～聪颖 intelligent by natural endowments

【天子】the Son of Heaven; the emperor

【天字第一号】the greatest in the world; par excellence

【天作之合】a heaven-made match; a union made by heaven

tián

恬

【恬不为怪】not surprised at all

【恬不知耻】not feel ashamed; have no sense of shame; be shameless

【恬不知悔】quiet and devoid of any sense of repentance

【恬淡】indifferent to fame or gain

【恬和】quiet and gentle

【恬静】quiet; peaceful; tranquil

【恬漠无事】nonchalant and uneventful

【恬然】[书] unperturbed; calm; nonchalant

【恬逸自满】self-satisfied with quietness and indolence

填 ①(垫平;塞满) fill up; stuff: ～枕芯 stuff a pillow ②(填写) fill in; write: ～好一张表 fill a form out / 把文件～好 fill a paper up

【填报】fill in a form and submit it to the lead-

ership: 每周～教学进度 make a weekly progress report on teaching work
【填表】fill in a form
【填补】fill: ～空白 fill in the gaps / 牙科医生～龋齿。 A dentist fills decayed teeth.
【填充】①(一种测验方法) fill in the blanks ②(填补) fill up; stuff ◇ ～塔[石油] packed column / ～物 filler
【填词】compose a *ci* poem: ～做诗 making verses on a given rhyme and writing poems
【填凑一隅】crowding together in a corner
【填方】[建] fill
【填空】①(填补) fill a vacant position; fill a vacancy ②(填空测验) fill in the blanks
【填料】[机] packing; stuffing; filling; filler ◇ ～函 gland box; stuffing box
【填密】[机] packing: 液压～ hydraulic packing ◇ ～函 packing box
【填平】fill and level up: 把沟～ fill up a gully
【填塞物】tampon; stemming; wadding
【填写】fill in; write: ～表格 fill in a form
【填鸭】① (一种养鸭方法) force-feed a duck ②(用此方法饲养的鸭子) force-fed duck ◇ ～式教学方法 cramming method of teaching
【填字游戏】cross-word puzzle

田 field; farmland; cropland: 稻～ rice field / 煤～ coalfield / 水～ paddy field / 油～ oil field
【田鳖】[动] giant water bug; fish killer
【田产】real estate
【田地】①(种植用的土地) field; farmland ②(地步) plight; wretched situation: 落到这步～ get into such a plight
【田凫】[动] lapwing
【田埂】a low bank of earth between fields; ridge
【田鸡】frog
【田间】field; farm ◇ ～持水量[农] field capacity / ～管理 field management / ～劳动 field labor; farm work
【田径】[体] track and field ◇ ～队 track and field team / ～赛 track and field meet / ～赛项目 track and field events / ～运动 track and field sports; athletics / ～运动员 athlete
【田菁】[植] sesbania
【田鹨】[动] paddy-field pipit
【田螺】[动] river snail
【田赛】[体] field events
【田鼠】vole
【田野】field; open country
【田园】fields and gardens; countryside: ～荒芜 fields and gardens turn to a jungle / ～生活 idyllic life ◇ ～诗 idyll; pastoral poetry / ～诗人 pastoral poet / ～文学 pastoral literature
【田庄】country estate

钿 [方] coin; money: 铜～ copper cash; copper money

甜 ①(指味道) sweet; honeyed: ～味 sweet taste ②(睡得塌实) sound: 睡得真～ have a sound 〈sweet〉 sleep
【甜菜】①(一种草本植物) beet ②(这种植物的根) beetroot ◇ ～糖 beet sugar
【甜瓜】muskmelon
【甜酒酿】fermented rice
【甜美】①(甜) sweet; luscious ②(愉快; 舒服) pleasant; refreshing: ～的生活 a sweet and pleasant life; a comfortable life
【甜蜜】sweet; happy: ～的微笑 a sweet smile
【甜面酱】a sweet sauce made of fermented flour
【甜食】sweet food; sweetmeats
【甜水】①(不苦的水) fresh water ②(幸福) happiness; comfort: 这孩子是在～里长大的。The child's grown up in happy times.
【甜丝丝】①(有甜味) pleasant sweet ②(感到愉快) quite pleased; gratified; happy: 心里～的 feel quite happy
【甜酸苦辣】all sorts of joys and sorrows
【甜头】①(微甜的味道) sweet taste; pleasant flavor ②(好处) good; benefit: 尝到～ draw benefit from it
【甜味】sweet taste: 有点～ taste sweet; have a sweet taste
【甜言蜜语】sweet words and honeyed phrases; fine-sounding words

tiǎn

舔 lick; lap: 把调羹～干净 lick the spoon clean
【舔犊情深】very affectionate toward one's children
【舔犊之爱】the love of licking shown by cow for its calf; parental love

腆 ①(丰盛) sumptuous; rich ②[方](凸出) protrude; thrust out: ～着胸 stick out one's chest / 他的肚子～着。 His belly protrudes.

tiāo

挑 ①(挑选) choose; select; pick: 把那筐苹果

～一～ pick over that basket of apples ②(挑剔) pick; find: ～毛病 pick faults ③(挑担) carry on the shoulder with a pole; shoulder: ～水 carry water ④ [量] 一～白菜 two baskets of cabbage
【挑不动】too heavy to carry
【挑刺儿】[方] find fault; pick holes; be captious
【挑错】find faults; pick flaws
【挑大梁】play the leading role; shoulder the main responsibility
【挑担】carry a load
【挑肥拣瘦】pick the fat or choose the lean; choose whichever is to one's personal advantage
【挑拣】pick; pick and choose: 挑挑拣拣 be choosy
【挑三拣四】pick and choose; be choosy
【挑剔】nitpick; be hypercritical; be fastidious
【挑选】choose; select; pick out: ～和培养接班人 choose and train successors
【挑眼】[方] be fastidious
【挑字眼儿】find fault with the choice of words
【挑子】carrying pole with its load; load carried on a shoulder pole

tiáo

条 ①(细长的树枝) twig ②(狭长的东西) a long narrow piece; strip; slip: 布～ a strip of cloth / 纸～ a strip of paper ③(项目) item; article ④(层次) order: 有～不紊 in perfect order; orderly ⑤[量] 一～裤子 a pair of trousers / 一～香烟 a carton of cigarettes
【条案】a long narrow table
【条播】[农] drilling ◇ ～机 seed drill; drill
【条分缕析】make a careful and detailed analysis
【条幅】a vertically-hung scroll; scroll
【条钢】bar iron
【条痕】[矿] streak
【条件】①(客观的因素) condition; term; factor: 贸易～ terms of trade / 自然～ natural conditions ②(提出的要求) requirement; prerequisite; qualification: 提出～ put forward the requirement / 有～地同意 agree with qualifications ◇ ～刺激[生理] conditioned stimulus / ～反射[生理] conditioned reflex / ～句 conditional clause
【条款】clause; article; provision: 法律～ legal provision / 附加～ additional clause / 协定的～ articles of an agreement
【条理】proper arrangement or presentation; orderliness; method: ～分明 clear presentation / ～贯通 go through in regular sequence / 有～的人 an orderly person; a man with an orderly mind
【条例】regulations; rules; ordinances: 安全～ safety regulation / 组织～ organic rules
【条令】[军] regulations: 内务～ routine service regulations
【条目】①(规章中的) clauses and subclauses ②(词典中的) entry: 这词典有十四万多～。This dictionary has over 140000 entries.
【条绒】corduroy
【条施】[农] row replacement
【条鳎】[动] striped sole
【条条框框】[贬] rules and regulations; conventions: 打破～ break down conventions
【条文】article; clause: 这个文件包括许多～。This document includes many articles. ◇ ～范例 standard clause
【条纹】stripe; streak ◇ ～布 striped cloth; stripe
【条锈病】[农] stripe rust; yellow rust
【条约】treaty; pact: 边界～ boundary treaty / 多边～ multilateral pact / 双边～ bilateral treaty / 通商航海～ treaty of commerce and navigation / 友好合作～ treaty of friendship and cooperation / 友好和互不侵犯～ treaty of friendship and mutual non-aggression / 订立～ conclude a treaty
【条子】①(狭长的东西) strip ②(便条) a brief informal note

调 ①(配合适当) suit well; fit in perfectly: 风～雨顺 good weather for crops ②(使配合均匀) mix; adjust: ～匀 mix well ③(调解) mediate ④(挑逗) tease; provoke
【调处】mediate; arbitrate: ～争端 arbitrate a dispute
【调挡】[机] gear shift
【调幅】[无] amplitude modulation: ～广播 amplitude modulation broadcasting
【调羹】spoon
【调和】①(配合适当) be in harmonious proportion: 色彩～。Colors are in harmony. ②(调解) mediate; reconcile: 从中～ act as mediator ③(妥协) compromise; make concessions: 不～的斗争 uncompromising struggle
【调剂】①(配制药物) make up a prescription ②(调整) adjust; regulate: ～身心 provide physical and mental relaxation / ～生活 enliven one's life

【调浆】[纺] size mixing
【调焦】[摄] focusing ◇ ～镜头 focusing lens／～毛玻璃 focusing screen
【调节】 regulate; adjust: 空气～ air conditioning／～室温 regulate the room temperature／～水流 regulate the flow of water／这些桌椅可以随儿童的身高～。These desks and seats can be adjusted to the height of any child. ◇ ～税 regulating taxes
【调节器】 regulator; conditioner: 恒流～[电] constant current regulator／空气～ air conditioner
【调解】 mediate; make peace ◇ ～程序 conciliation proceedings／～法庭 conciliation court／～纠纷 mediation of disputes／～人 bridge builder; intermediator／～协议 reconciliation agreement
【调理】①(调养) nurse one's health; recuperate ②(照料) look after; take care of
【调料】 condiment; seasoning; flavoring
【调弄】①(戏弄) make fun of; tease: ～取笑 make fun of and excite with laughter ②(整理) arrange; adjust ③(调唆) instigate; stir up
【调配】 mix; blend: ～颜色 mix colors
【调皮】①(顽皮) naughty; mischievous: ～的孩子 a naughty child ②(不驯服) unruly; tricky: ～之人 a tricky fellow ③(耍小聪明) play tricks
【调频】[无] frequency modulation: ～广播 frequency modulation broadcasting
【调情】 flirt
【调色板】[美术] palette
【调色刀】[美术] palette knife; painting knife
【调色碟】[美术] color mixing tray
【调色】[美术] mix colors
【调试】[计算机] debugging
【调速器】[机] governor
【调唆】 incite; instigate: ～是非 stir up trouble
【调停】 mediate; intervene; act as an intermediary: ～纠纷 adjust disorder ◇ ～者 intervener
【调味】 flavor; season ◇ ～品 flavoring; seasoning; condiment
【调戏】 take liberties with; assail with obscenities: ～妇女 take liberties with a woman
【调笑】 make fun of; poke fun at; tease: ～为乐 take pleasure in jeering
【调谐】①(和谐) harmonious ②[电] turning ◇ ～范围 tuning range／～旋钮 tuning knob
【调压器】[电] voltage regulator
【调养】 take good care of oneself; build up one's health by rest and by taking nourishing food; be nursed back to health: ～身体 nurse one's body
【调音】[乐] tuning
【调匀】 mix evenly
【调整】 adjust; regulate; revise: 工资～ adjustment of wages／～供求关系 regulate supply and demand／～价格 readjust prices
【调整器】[机] adjuster: 自动松紧～ automatic slack adjuster
【调治】 recuperate under medical treatment: ～宿疾 attend and cure an old illness
【调制】[电] modulation: 音频～ voice modulation ◇ ～间隙 modulation gap

迢 far; remote

【迢迢】 far away; remote: 千里～ from a thousand *li* away; from afar
【迢迢旅途】 a remote journey
【迢迢远行】 take a long journey

苕

【苕帚】 whisk broom

tiǎo

挑 ①(支起) push sth. up with a pole or stick; raise ②(拨东西) poke; pick: ～火 poke a fire ③(挑拨; 挑动) stir up; instigate: ～起争端 stir up strife 〈trouble〉

【挑拨】 instigate; incite; sow discord: ～离间 set people by the ears; sow dissension／～是非 foment discord
【挑灯苦读】 raise the wick and study hard
【挑动】 provoke; stir up; incite: ～某人做某事 provoke sb. to do sth.／～骚乱 provoke a riot
【挑逗】 provoke; tease; tantalize
【挑花】 cross-stitch work
【挑弄是非】 arouse ill-will between two parties; stir up one side against the other
【挑起】 provoke; stir up; instigate: ～冲突 provoke a conflict 〈clash〉
【挑唆】 incite; abet; instigate: ～他人 stir up others
【挑衅】 provoke ◇ ～行为 provocative act／～者 provocateur
【挑战】①(挑衅) throw down the gauntlet; challenge to battle: 接受～ accept the challenge; take up the gauntlet ②(提出竞赛) challenge to a contest ◇ ～书 letter of challenge; challenge

tiào

眺 look into the distance from a high place: 远～ look far into the distance
【眺望】look into the distance from a high place: ～远景 gaze at the distant view

跳 ①(跳跃) jump; leap; spring; bounce; hop: ～起身来 jump to one's feet / 高兴得～了起来 jump for joy; jump with joy ②(跳动) move up and down; beat: 她的心～正常。Her heartbeat is normal. ③(越过) skip; make omissions: ～过两段 skip over two paragraphs / ～一针 drop a stitch
【跳板】①(供人走的长板) gangplank ②(跳水运动用的长板) springboard; diving board
【跳槽】abandon one occupation in favor of another
【跳虫】springtail; snowflea
【跳出】jump out; leap out
【跳弹】[军] ricochet ◇ ～轰炸 ricochet bombing; skip bombing
【跳动】move up and down; beat; pulsate: 她的心脏还在～。Her heart was still beating.
【跳房子】hopscotch
【跳高】[体] high jump: 背越式～ back style; Fosbury flop; flop / 撑竿～ pole vault; pole jump / 俯卧式～ straddle jump / 滚式～ Western layout / 剪式～ Eastern layout ◇ ～架 jumping standard / ～运动员 high jumper
【跳行】①(漏去一行) skip a line ②(改行) change to a new occupation
【跳级】skip a grade
【跳脚】stamp one's foot: 气得～ stamp with rage
【跳栏】[体] hurdle race; the hurdles
【跳梁小丑】a buffoon who performs antics; contemptible scoundrel
【跳马】[体] ①(指器械) vaulting horse ②(指项目) horse-vaulting
【跳蝻】[动] the nymph of a locust
【跳皮筋】rubber band skipping; skipping and dancing over a chain of rubber bands
【跳棋】Chinese checkers; Chinese draughts
【跳球】[篮球] jump ball
【跳伞】①(用降落伞往下跳) parachute; bale out ② [体] (指项目) parachute jumping: 定点～ accuracy jump; precision landing / 特技～ style jump / 造型～ relative work ◇ ～区 parachute drop zone / ～塔 parachute tower / ～运动员 parachutist; parachuter
【跳绳】rope skipping
【跳鼠】[动] jerboa
【跳水】[体] dive: 抱膝～ cannon ball; crouched jump / 臂立～ armstand dive; handstand dive / 低难度跳台～ plain high diving / 高难度跳台～ variety high diving / 立定～ standing dive / 面对板向内～ inward dive / 面对池反身～ reverse dive / 跳板～ springboard diving / 跳台～ platform diving / 向后～ back dive / 向前～ front dive / 燕式～ swallow dive; swan dive ◇ ～表演 diving exhibition / ～池 diving pool / ～动作 dive / ～运动员 diver
【跳台】diving tower; diving platform
【跳台滑雪】ski jumping
【跳舞】dance
【跳箱】[体] ①(指器械) box horse; vaulting box ②(指动作) jump over the box horse
【跳远】[体] long jump; broad jump: 三级～ hop, step and jump
【跳跃】jump; leap; bound
【跳跃着陆】[军] rebound landing
【跳蚤】flea
【跳闸】tripping operation ◇ ～开关[电] trip switch

tiē

帖 ①(服从) submissive; obedient: 服～ docile and obedient / ～服称臣 being submissive ②(妥当) well-settled; well-placed: 办事妥～ manage things fittingly

萜 [化] terpene

贴 ①(粘) paste; stick; glue: ～邮票 stick on a stamp / 把信封口～上 paste down an envelope②(紧挨) keep close to; nestle closely to: ～身衣服 underclothes ③(津贴) subsidies; allowance: 房～ housing allowance ④[量] 一～膏药 a piece of medicated plaster
【贴边】hem
【贴饼子】①(烤熟饼子) bake corn or millet cakes on a pan ②(饼子) baked corn; millet cakes
【贴补】subsidize; help (out) financially: ～家用 help out with the family expenses
【贴花】[纺] appliqué: ～织物 appliqué
【贴换】trade sth. in; trade-in
【贴金】①(贴金箔) cover with gold leaf; gild ②(美化) touch up; prettify: 别尽往自己脸上～。Don't put feathers in your cap.
【贴金漆】gold size

【贴近】press close to; nestle up against: 把耳朵～门边 press one's ear close to the door
【贴面砖】[材] furring brick
【贴切】apt; suitable; appropriate; proper: 措词～ aptly worded; well-put
【贴钱】pay out of one's own pocket (especially when one is not supposed to)
【贴身】next to the skin: ～衣服 underclothes; underclothing
【贴水】[商] agio
【贴题】relevant; pertinent; to the point: 不～ irrelevant; beside the point
【贴息】① [商] pay interest (in the form of a deduction when selling a bill of exchange, etc.) ② (付出的利息) interest so deducted; discount
【贴现】discount ◇ ～率 discount rate
【贴心】intimate; close: ～话 confidential talk; words spoken in confidence / ～朋友 close friend; bosom friend

tiě

帖 ①(帖子) invitation: 请～ invitation ②(字条) note; card: 谢～ card of thanks ③ [量] 一～药 a dose of herbal medicine

铁 ①(金属元素) iron (Fe): 废～ scrap iron / 马口～ galvanized iron; tin plate / 球墨铸～ nodular cast iron; nodular iron / 生～ pig iron; cast iron / 熟～ wrought iron / 铸～ cast iron ②(指刀枪等) arms; weapon: 手无寸～ completely unarmed ③(坚硬的) hard or strong as iron: ～拳 iron fist ④(坚定不移的) indisputable; unalterable: ～的纪律 iron discipline / ～的事实 hard fact; ironclad evidence ⑤(下决心) resolve; determine: ～了心 be unshakable in one's determination
【铁案如山】borne out by ironclad evidence
【铁板】iron plate; sheet iron
【铁板一块】a monolithic bloc
【铁笔】①(刻图章的刀) cutting tool ②(刻蜡纸用的笔) stencil pen
【铁箅子】①(炉箅) grate ②(烤肉架) gridiron; grill
【铁壁铜墙】iron walls and brass partitions; a stronghold
【铁饼】[体] ①(指项目) discus throw ②(指器械) discus
【铁蚕豆】roasted broad bean
【铁杵磨成针】an iron pestle can be ground down to a needle; little strokes fell great oaks
【铁窗】①(安铁栅的窗户) a window with iron grating ②(监狱) prison bars; prison: ～风味 prison life; life behind bars
【铁磁共振】[物] ferromagnetic resonance
【铁磁性】ferromagnetism
【铁道】railway; railroad: 地下～ underground; tube; subway ◇ ～炮兵[军] railway artillery
【铁道兵】[军] railway corps
【铁电现象】[物] ferroelectricity
【铁饭碗】iron rice bowl; a secure job
【铁工】①(制做、修理铁器的工人) ironworker; blacksmith ②(制做、修理铁器的工作) ironwork
【铁工厂】ironworks
【铁公鸡】iron cock; a stingy person; miser
【铁箍】iron hoop
【铁观音】*tie guanyin*, a variety of oolong tea
【铁管】iron pipe; iron tube
【铁轨】rail
【铁合金】[冶] ferroalloy
【铁黑】[化] ①(氧化铁) iron oxide black ②(一种颜色) iron black
【铁红】iron oxide red
【铁花】[工美] ornamental work of iron; iron openwork
【铁画】[工美] iron picture: ～银勾 vigorous touches and fine strokes in calligraphy
【铁环】iron hoop: 滚～ trundle a hoop; play with a hoop
【铁蒺藜】[军] caltrop
【铁甲】①(古代战衣) armor; mail ②(装甲的) armored ◇ ～车 armored vehicle; armored car / ～船 an ironclad
【铁匠】blacksmith; ironsmith ◇ ～铺 smithy; blacksmith's shop
【铁脚板】iron soles; toughened feet
【铁军】iron army; invincible army
【铁矿】iron mine ◇ ～石 iron ore
【铁力木】[植] ferrous mesua
【铁链】iron chain; shackles
【铁路】railroad; railway: ～运输 railway transportation; railway transport; shipping by rail / 国际～联运 international railway through transport ◇ ～电气化 railway electrification / ～干线 trunk railway / ～公路两用桥 combined bridge / ～路基 railway bed / ～网 railway network / ～线 railway line
【铁马】cavalry: 金戈～ shining spears and armored horses
【铁门】①(铁制的门) iron gate ②(铁花格) grille

【铁面无私】impartial and incorruptible
【铁木】[植] hophornbeam
【铁皮】iron sheet: 白～ tinplate; galvanized iron sheet / 黑～ black sheet (iron)
【铁器】ironware ◇ ～时代 the Iron Age
【铁锹】spade; shovel
【铁青】ashen; livid; ghastly pale: 气得脸色～ turn livid with rage
【铁纱】wire gauze; wire cloth
【铁砂】①(铁矿砂) iron sand ②(猎枪子弹) shot; pellets
【铁杉】[植] Chinese hemlock
【铁石心肠】be ironhearted; have a heart of stone; be hardhearted
【铁树】[树] sago cycas
【铁树开花】the iron tree in blossom sth. seldom seen or hardly possible
【铁水】molten iron
【铁丝】iron wire: ～网 wire netting; wire meshes; wire entanglement
【铁素体】[冶] ferrite
【铁索】cable; iron chain ◇ ～吊车 cable car / ～桥 chain bridge
【铁塔】①(钢铁造的塔) iron tower; iron pagoda ② [电](高压线的架子) pylon; transmission tower
【铁蹄】iron heel; cruel oppression of the people
【铁桶】metal pail; metal bucket; drum
【铁腕】iron hand ◇ ～人物 an ironhanded person; strong man
【铁锨】shovel; spade
【铁线订书机】wire stitcher; wire stitching machine
【铁线莲】[植] cream clematis
【铁屑】iron filings; iron chippings and shavings
【铁心】[电] core
【铁锈】rust
【铁血主义】blood-and-iron policy
【铁盐】molysite
【铁砚磨穿】long years of persistence
【铁陨石】iron meteorite
【铁砧】anvil
【铁证】ironclad proof; irrefutable evidence: ～如山 irrefutable, conclusive evidence

tiè

帖 a book of models: 碑～ a book of stone rubbings / 画～ painting models / 习字～ a book of models of calligraphy for copying

tīng

烃 [化] hydrocarbon: 闭链～ closed chain hydrocarbon / 开链～ open chain hydrocarbon
【烃气】[化] hydrocarbon gas

厅 ①(大房间) hall: 会议～ conference hall / 舞～ dance hall / 音乐～ concert hall ②(机关办事部门) office: 办公～ general office ③(省属机关名称) department: 河南省公安～ the Public Security Department of Henan Province

听 ①(用耳朵听) listen; hear: ～广播 listen in on the radio; listen to the radio / 他正在～音乐。He was listening to the music. / 我们在～总理的广播讲话。We were listening in to the Prime Minister's speech. ②(听从) heed; obey: ～指挥 obey orders / 不要～他的。Don't listen to him. ③(听凭) allow; let: ～其自然 let things take their own course ④[量] tin; can: 一～桃子罐头 a can of peaches
【听便】as one pleases; please yourself
【听从】obey; heed; comply with: ～劝告 accept sb.'s advice
【听得入神】completely absorbed
【听而不闻】hear but pay no attention; turn a deaf ear to
【听骨】[生理] ear bones
【听候】wait for; pending: ～裁判 waiting for a judgement / ～分配 wait for one's assignment / ～佳音 waiting for good news / ～拘押 impending apprehension
【听话】heed what an elder or superior says; be obedient
【听话儿】wait for a reply
【听见】hear: 我～他们在唱歌。I heard them singing.
【听讲】listen to a talk; attend a lecture: 一面～，一面记笔记 take notes while listening to a lecture
【听觉】[生理] sense of hearing
【听课】attend a lecture; sit in on a class
【听力】①(听的能力) hearing: 恢复～ regain one's hearing ②(听力课) aural comprehension
【听命】take orders from; be at sb.'s command: ～于人 be at one's beck and call
【听腻了】bored with listening to
【听凭】allow; let: ～别人摆布 be at the mercy of others

【听其言观其行】listen to what a person says and watch what he does; judge people by their deeds, not just by their words
【听其自便】let one take his own convenience
【听其自然】let things take their own course; let matters slide
【听起来】sound; ring: 她的话～是真诚的。Her words rang true.
【听取】listen to: ～汇报 hear reports; debrief / ～双方当事人的律师陈述 hear counsels on both side / ～证词 hear witnesses
【听任】allow; let: 不能～错误思想自由泛滥。No erroneous ideas should be allowed to spread unchecked.
【听神经】[生理] auditory nerve
【听说】be told; hear of: 我～他到哈尔滨去了。I hear he has gone to Harbin.
【听天由命】submit to the will of Heaven; resign oneself to one's fate; trust to luck
【听筒】①(电话听筒) receiver ②[电](耳机) headphone; earphone ③[医](听诊器) stethoscope
【听戏】go to the opera
【听写】[数] dictation
【听信】①(等候消息) wait for information ②(听到而相信) believe; believe what one hears: ～谗言 hear and believe slanders / ～谣言 believe rumors
【听诊】[医] auscultation ◇ ～器 stethoscope
【听之任之】let sth. go unchecked; take a laissez-faire attitude; let matters drift
【听众】audience; listeners
【听装】tinned; canned ◇ .～饼干 tinned biscuits

tíng

亭 ①(亭子) pavilion; kiosk ②(象亭子的小房子) kiosk: 书～ bookstall / 邮～ postal kiosk
【亭亭玉立】①(形容妇女) slim and graceful ②(形容花木) tall and erect
【亭午】midday; noon
【亭子间】[方] a small, dark back room over a kitchen; garret

停 ①(停止) stop; cease; halt; pause: 雨～了。The rain has stopped. ②(停留) stop over; stay ③(停放) be parked; lie at anchor: 汽车不能～在这里。You cannot park your car here.
【停摆】come to a standstill; stop
【停办】close down
【停泊】anchor; come to an anchor; berth: 意外～ unscheduled call / 在二号码头～ be berthed at No. 2 wharf ◇ ～处 anchorage; berth; dock / ～港 port of call
【停产】stop production ◇ ～期 idling period
【停车】①(车辆停止行驶) stop; pull up: 他在入口处停了车。He pulled his car up at the entrance. ②(停放车辆) park: 此处不准～! No Parking! ③(机器停止转动) stall; stop working: ～修理 stop working to undergo repairs ◇ ～场 car park; parking lot; parking area
【停当】ready; settled: 一切准备～。Everything's ready.
【停电】power cut; power failure
【停顿】①(停止) stop; halt; pause: 处于～状态 be at a standstill②(说话时的间歇) pause
【停放】park; place: 此处不准～自行车。Don't park 〈place〉 your bike here.
【停飞】[军] grounding of aircraft
【停工】stop work; shut down: ～待料 work being held up for lack of material
【停航】suspend air or shipping service
【停火】cease fire: ～协议 cease-fire agreement / ～休战 cease fire and make a truce
【停机坪】aircraft parking area; parking apron
【停刊】stop publication
【停靠】stop; berth ◇ ～港 port of call
【停课】suspend classes
【停灵】keep a coffin in a temporary shelter before burial
【停留】stay for a time; stop; remain: 在北京～一周 stay in Beijing for a week / 在南京作短暂～ have a brief stopover in Nanjing
【停留时间】[环保] retention perio
【停水】cut off the water supply; cut off the water
【停妥】be well arranged; be in order
【停息】stop; cease: 暴风雨～了。The storm has subsided.
【停歇】①(歇业) close down; stop doing business ②(停止) stop; cease ③(停下来休息) stop for a rest; rest
【停薪留职】stop payment of salary but retain office
【停学】stop going to school; drop out of school; suspend sb. from school
【停业】stop doing business; close down: 修理内部,～五天。Closed for five days for repairs.
【停战】cease-fire; armistice; truce; cessation of hostilities ◇ ～令 order to cease-fire / ～谈判

armistice talks／ ～协定 armistice; truce agreement
【停职】suspend sb. from his duties: ～检查 be temporarily relieved of one's post for self-examination
【停止】stop; cease; halt; suspend; call off: ～兑现 suspension of specie payment／～广播 go off the air; stop broadcasting／ ～ 支付 non-payment／～讲话! Stop talking! ◇～信号 break alarm
【停滞】stagnate; be at a standstill; bog down: ～不前 be at a standstill

庭 ①(庭院) front courtyard; front yard ②(法院) law court: 民～ a civil court／刑～ a criminal court
【庭外和解】settled out of court
【庭园】flower garden; grounds
【庭院】courtyard

tǐng

挺 ①(硬而直) straight; erect; stiff: 直～～地躺着 lie stiff ②(伸直) straighten up; stick out: ～起腰杆 straighten one's back; straighten up ③(勉强支持) endure; stand; hold out: 痛得～不住 cannot stand the pain ④(很) very; rather; quite: ～高兴 rather pleased／～好 very good ⑤[量] 一～机枪 a machine gun
【挺拔】①(直立而高耸) tall and straight ②(坚强有力) forceful
【挺拔不群】stiff upright and uncommon to all
【挺不住】cannot stand it; cannot take it any more
【挺杆】[机] tappet: 阀门～ valve tappet ◇～间隙 tappet clearance
【挺进】boldly drive on; press onward; push forward: ～敌后 boldly drive into the enemy areas／ 向前～ press onward
【挺举】[举重] clean and jerk
【挺立】stand upright; stand firm
【挺起胸膛】stick 〈throw〉 out one's chest
【挺然不群】be distinguished from fellow men
【挺身】straighten one's back: ～而出 step forward bravely; come out boldly
【挺秀】tall and graceful: ～超群 be eminent above the masses
【挺直】straight and upright

铤 (run) quickly
【铤而走险】risk danger in desperation; make a reckless move

艇 a light boat: 登陆～ landing craft／ 炮～ gunboat／汽～ steamboat

tōng

通 ①(可以穿过) open; through: 那条胡同是～的。 That is a through alley.／水管是～的。The waterpipe is not blocked.／这条胡同不～。 This is a blind alley. ②(使不堵塞) open up or clear out by poking or jabbing: ～一下炉子 give the fire a poke ③(有路达到) lead to; go to: 四～八达 extend in all directions; be linked by rail and road to various parts of the country／这趟列车直～沈阳。This train goes straitht to Shenyang.／这条路～哪儿? Where does this lead? ④(相往来) connect; communicate: 互～情报 exchange information／两个房间是～着的。 The two rooms are connected. ⑤(使知道) notify; tell: ～个电话 give sb. a ring; call sb. up ⑥(懂得; 了解) understand; know: 她～四种语言。 She knows four languages. ⑦(专家) authority; expert: 中国～ an expert of China; an old China hand ⑧(通顺) logical; coherent: 文理不～ ungrammatical and incoherent ⑨(普通) general; common: ～称 a general term ⑩(整个) all; whole: ～共 altogether; in all
【通报】①(书面通知) circulate a notice: ～表扬 give out a circular a notice of commendation／～批评 circulate a notice of criticism ②(文件) circular ③(刊物) bulletin; journal
【通便剂】[医] laxative; cathartic
【通病】common failing
【通才】an all-round person; a universal genius
【通常】general; usual; normal: 晚上我～在家。 I am usually at home in the evening.
【通畅】①(运行无阻) unobstructed; clear: 大便～ free movement of the bowels／道路～。The road is clear. ②(流畅) easy and smooth: 他文字～。 He writes easily and smoothly.
【通车】①(开始行车) be open to traffic ②(有车来往) have transport service
【通称】①(通常叫做) be generally called; be generally known as ②(通常的名字) a general term; common name; popular name
【通达】understand: ～人情 be understanding and considerate／～事理 be reasonable; be sensible／见解～ hold sensible views; show good sense
【通道】thoroughfare; passageway; passage

【通道式公共汽车】articulated bus
【通敌】collude with the enemy; have illicit relations with the enemy
【通电】①(通电流) set up an electric circuit; electrify; energize; be charged with electricity ②(公开发表电报) circular telegram: 大会~ the circular telegram of the conference / ~全国 publish an open telegram to the nation
【通牒】diplomatic note: 最后~ ultimatum
【通都大邑】large city; metropolis
【通读】read over
【通分】[数] reduction of fractions to a common denominator
【通风】①(使空气流通) ventilate; air: 让室内通~ ventilate a room ②(透气) be well ventilated ③(透露消息) divulge information; tip-off: 向某人~报信 give secret information to sb.; tip sb. off ◇ ~管道 ventilating duct / ~机 ventilator; fanner / ~井 ventilation shaft; air shaft / ~口 vent / ~设备 ventilation system / ~装置 ventilation installation
【通告】①(普遍通知) give public notice; anounce ②(文告) public notice; circular; announcement: 出~ give out a notice; make an announcement / 张贴~ put up a notice
【通共】in all; altogether; all told: 我们~十个人。There are ten of us altogether.
【通过】①(从一端到另一端) pass through; get past; traverse: ~边境 pass the frontier / ~城镇 pass through a city / ~障碍物 break the barriers ②(同意议案) adopt; pass; carry: ~决议 adopt a resolution; pass a resolution / 提案以压倒多数~。The motion was passed by an overwhelming majority. ③(以人、事物等为媒介) by means of; by way of; by; through: ~外交途径 through diplomatic channels / ~协商 through consultation / ~正当手段 by fair means ④(征得有关人员同意) ask the consent or approval of: ~群众 consult the masses
【通航】be open to navigation or air traffic: 开始~ open up navigation ◇ ~水域 navigable waters
【通红】very red; red through and through: 她的脸冻得~。Her face is red with cold.
【通话】①(通电话) communicate by telephone ②(彼此交谈) converse; talk with sb; hold conversation ◇ ~计时器 peg count meter
【通婚】be related by marriage; intermarry
【通货】[经] currency; current money ◇ ~膨胀 inflation / ~收缩 deflation / ~稳定 currency stabilization
【通缉】order the arrest of a criminal at large; list as wanted: ~犯 a criminal wanted by the law / ~在案 list as wanted; at large under a wanted-notice / 下~令 issue a wanted circular ◇ ~令 order for arrest / ~文告 hue and cry
【通奸】commit adultery ◇ ~者 fornicator / ~罪 crime of adultery
【通栏标题】banner headline; banner
【通力】concerted effort: ~合作 make a concerted effort; give full cooperation to
【通例】general rule; usual practice
【通亮】well-illuminated; brightly lit
【通量】[物] flux: 磁~ magnetic flux
【通令】①(发出同一个命令) issue a general order: ~全国 issue a general order to the whole nation ②(发出的同一个命令) a general order
【通路】①(往来的大路) thoroughfare; highway; highroad; route ②(通过的途径) passage; access
【通论】①(通达的议论) a sensible argument; a well-rounded argument ②(全面的论述) a general survey; an introduction: 《化学~》General Chemistry / 《史学~》An Introduction to History
【通明】well-illuminated; brightly lit: 灯火~ be ablaze with lights; be brightly it
【通年】throughout the year; all the year round
【通盘】overall; all-round; comprehensive: ~计划 overall planning; comprehensive planning / ~考虑 give an overall consideration; consider every possible angle
【通票】through ticket
【通气】①(使通空气) ventilate; aerate ②(互通声气) be in touch with each other; keep each other informed; have communication with ◇ ~孔 air vent; orifice; airflow
【通窍】understand things; be sensible or reasonable
【通情达理】showing good sense; reasonable
【通权达变】act as the occasion requires; adapt oneself to circumstances
【通融】①(给人方便) make an exception in sb.'s favor; stretch rules, get around regulations, etc., to accommodate sb.: 毫不~ make no exceptions ②(短期借钱) accommodate sb. with a short-term loan: 能不能~三千元钱? Could you lend me 3000 *yuan*? ◇ ~票据 accommodation notes
【通商】have trade relations: 订定~条约 conclude a trade treaty; sign a treaty of commerce ◇ ~口岸 trading port / ~贸易关系 trade re-

lations / ～许可 licence to trade
【通身】the whole body: ～是汗 sweat all over
【通史】comprehensive history; general history
【通式】[化] general formula
【通顺】clear and coherent; smooth: 文理～ coherent writing
【通俗】popular; common: ～易懂 easy to understand ◇ ～读物 books for popular consumption; popular literature / ～歌曲 popular song / ～化 popularization / ～小说 popular fiction
【通天】①(形容极高) exceedingly high; exceedingly great: ～的本事 exceptional ability; superhuman skill ②(指与上层有联系) direct access to the highest authorities
【通条】①(通炉子的) poker ②(通枪膛等的) clean rod
【通通】all; entirely; completely
【通途】[书] thoroughfare: 天堑变～ a deep chasm turned into a thoroughfare
【通脱木】[植] rice-paper plant
【通宵】all night; the whole night; throughout the night: ～不寐 couldn't sleep all night / ～达旦 all night long ◇ ～服务部 a shop that is open all night; an all-night shop
【通晓】thoroughly understand; be well versed in; be proficient in: ～本地情况 thoroughly understand local conditions / ～世故 be perfectly familiar with the ways of the world / ～西班牙语 have a good knowledge of Spanish / ～英国文学 be well versed in English literature
【通心粉】macaroni
【通信】communicate by letter; correspond ◇ ～保密[军] communication security / ～兵 signal corps; signalman / ～处 mailing address / ～鸽 homing pigeon; carrier pigeon / ～连 signal company / ～联络[军] signal communication; communications and liaison / ～频带 communication band / ～犬 messenger dog / ～枢纽 signal center / ～卫星 communication satellite; radio relay satellite / ～员 messenger; orderly
【通行】①(在交通线上通过) pass through: ～无阻 open thoroughfare; accessible to public / ～自由 can pass freely; have free passage / 禁止～! No thoroughfare! ②(通用) current; general: ～的办法 the current practice ◇ ～能力[交] traffic capacity / ～权 right of way / 税 transit duty
【通行证】pass; permit; safe-conduct; laissez-passer: 边境～ border pass / 军事～ military pass / 临时～ provisional pass
【通讯】①(传递消息) communication: 红外线～ infrared ray communication / 激光～ laser communication 微波～ microwave communication / 无线电～ radio communication ②(报道消息的文章) news report; news dispatch; correspondence; newsletter: 新华社～ Xinhua dispatches / ～报导 news report; news dispatch; news story / ～文学 reportage ◇ ～方法 means of communication / ～录 address book / ～社 news agency / ～设备 communication apparatus / ～卫星 communications satellite / ～线路 communication line / ～员 reporter; correspondent / ～自由 freedom of correspondence
【通用】①(普通使用) in common use; current general: 全国～教材 national textbooks ②(某些汉字彼此可换用) interchangeable ◇ ～插件 universal card / ～车辆 general-purpose vehicle / ～货币 current money / ～机 universal machine / ～机械厂 universal machine works / ～决算表 general purpose statement / ～软件 common software / ～语种 commonly used languages / ～月票 a monthly ticket for all urban and suburban lines
【通邮】accessible by postal communication
【通则】general rule; general provisions
【通知】①(把事情告诉人知道) notify; inform; give notice; let know: ～大家 notify everyone / 预先～ give advance notice ②(通知事项的文书) notice; circular; notification: 接到～ get notice; receive notice / 书面～ written notice ◇ ～单 advice note; letter of notice / ～人 notifier / ～书 advertisement; information; advice notice

tóng

童 ①(儿童) child: 牧～ cowherd; shepherd boy ②(未婚的) unmarried; virgin: ～男 virgin boy / ～女 maiden; virgin ③(未成年的仆人) page; page boy ④(秃) bare; bald; barren
【童工】child laborer
【童话】children's stories; fairy tales
【童年】childhood
【童山】bare hills: ～多石 bare hill with numerous stones / ～秃岭 bare hills and mountains
【童声】child's voice ◇ ～合唱 children's chorus
【童叟无欺】We are equally honest with aged and child customers.
【童心】childlike innocence; childishness; play-

fulness: ～未泯 still retain childlike innocence
【童颜皓首】ruddy complexion and hoary head
【童言无忌】knock on woods
【童谣】children's folk rhymes
【童贞】virginity; chastity
【童装】children's garments
【童子】boy; lad
【童子军】boy scouts: 女～ girl scouts
【童子鸡】[方] young chicken; broiler

瞳 pupil
【瞳孔】[生理] pupil: 放大～ have one's pupils dilated ◇ ～间距离 interocular distance / ～开大[医] mydriasis / ～缩小[医] myosis
【瞳人】【瞳仁】pupil

同 ①(相同) same; alike; similar: 异～ similarities and dissimilarities ②(跟…相同) be the same as: "辞典"～"词典"。"辞典" is the same as "词典" ③(共同) together; in common: ～呼吸,共命运 share a common fate; go through thick and thin together ④[介](引进动作的对象或比较的事物,同"跟") with: ～群众打成一片 become one with the masses ⑤[连](表示联合关系,同"和") and; with: 我～他是朋友。 He and I are friends.
【同案犯】a accomplice ◇ ～证言 testimony of accomplice
【同班】①(在一个班级里) in the same class ②(同班同学) classmate
【同伴】companion
【同胞】①(同一父母所生) born of the same parents ②(同一国或同一民族的人) fellow countryman; compatriot: 港澳～ compatriots in Hongkong and Macao / 台湾～ compatriots in Taiwan
【同辈】of the same generation
【同病相怜】those who have the same illness sympathize with each other; fellow sufferers commiserate with each other
【同步】[物] synchronism: 载波～ carrier synchronization ◇ ～齿轮 synchromesh gear / ～电动机[电] synchronous motor / ～回旋加速器[物] synchrocyclotron / ～加速器[物] synchrotron / ～卫星 syn- chronous satellite
【同仇敌忾】share a bitter hatred of the enemy
【同窗】①(同在一个学校里学习) study in the same school ②(在同一个学校学习的人) schoolmate
【同床异梦】share the same bed but dream different dreams; be strange bedfellows
【同道】people engaged in the same pursuit; people with the same ideals: ～中人 men of the same line
【同等】of the same class, rank, or status; on an equal basis: ～效力 equal effect; equal validity / ～学力 of the same educational level / ～约束力 equally binding / ～重要 of equal importance
【同恶相济】the wicked help the wicked; sharing the evil and assisting the cause
【同房】①(家族中同一支的) of the same branch of a family ②[婉](夫妻过性生活) sleep together; have sexual intercourse
【同分异构体】[化] isomer
【同父异母兄弟】consanguineous brothers
【同甘共苦】share weal and woe; share comforts and hardships
【同感】the same feeling
【同工同酬】equal pay for equal work: 男女～。 Men and women enjoy equal pay for equal work.
【同工异曲】different tunes rendered with equal skill; different in approach but equally satisfactory results
【同功酶】[生化] isoenzyme
【同归于尽】perish together; end in common ruin
【同行】①(行业相同) of the same trade or occupation ②(行业相同的人) a person of the same trade or occupation
【同呼吸,共命运】share a common fate; throw in one's lot with sb.
【同化】①(使不相同的事物逐渐相近或相同) assimilate ②[语] assimilation ◇ ～政策 the policy of national assimilation / ～作用[生] assimilation
【同伙】①(共同参加某种组织或从事某种活动) be in partnership; collude with ②(伙伴) partner; associate; confederate
【同居】①(住在一起) live together ②(男女未婚而在一起生活) cohabit
【同类】of the same kind; similar: ～相残 kill one's own kind ◇ ～项[数] similar terms / ～罪行 kindred offense
【同流合污】wallow in the mire with sb.; associate with an evil person
【同路】go the same way ◇ ～人 fellow traveller
【同盟】alliance; league ◇ ～罢工 joint strike / ～国 ally; allied nations; the Central Powers (第一次世界大战); the Allies (第二次世界大

战) / ～军 allied forces; allies / ～条约 treaty of alliance
【同名】of the same name: 他和我～。He is my namesake.
【同谋】①(参与谋划) conspire; be of complicity ②(参与谋划的人) accomplice; confederate; conspirator ◇ ～犯 accomplice / ～杀人罪 offense of homicide by conspiracy
【同母所生子女】children of the same venter
【同年】①(同一年) the same year ②(同岁) of the same age
【同期】①(同一个时期) the corresponding period: 产量比去年～增加百分之二十。There is an increase in production of 20% above that of the corresponding period of the last year. ②(同一届) the same term: 我们是～毕业的。We are graduates of the same year.
【同情】sympathize with; show sympathy for: ～他的不幸 sympathize with him in his misfortune / 博得～ win sympathy ◇ ～心 sympathy; fellow felling
【同上】ditto; *idem*
【同声传译】simultaneous interpretation: ～设备 facilities for simultaneous interpretation
【同声相应,同气相求】like attracts like
【同生死,共患难】share weal and woe
【同时】①(在同一时候) at the same time; simultaneously; meanwhile; in the meantime: ～并举 develop simultaneously / ～发生 happen at the same time; coincide; concur ②(并且) moreover; besides; furthermore
【同事】①(一同工作) work in the same place; work together: 我们过去～多年。We worked together for years. ②(在一起工作的人) colleague; fellow worker: 老～ an old colleague
【同室操戈】family members drawing swords on each other; internal strife; internecine feud
【同素异形】[化] allotrope; ◇ ～体 allotrope; allotropic substance
【同岁】of the same age: 我们都～。We are all the same age.
【同位角】[数] corresponding angles
【同位素】[化] isotope: 放射性～ radioisotope ◇ ～分离 isotope separation / ～扫描器 radioisotope scanner / ～探伤仪 isoscope
【同位语】[语] appositive
【同温层】[气] stratosphere
【同文照会】identic notes
【同系物】[化] homologue
【同乡】a person from the same village, town or province; a fellow villager; a townsman
【同心】①(齐心) with one heart: ～同德 with one heart and one mind; with one idea and will ②(具同一中心的) concentric ◇ ～度[机] concentricity / ～圆[数] concentric circles
【同心协力】work in full cooperation and with unity of purpose; work together with one heart; make concerted efforts
【同行】travel together
【同性】①(性别相同) of the same sex: ～子女 children of the same sex ②(性质相同) of the same nature or character: ～的电互相排斥。Two like electric charges repel each other.
【同性恋】homosexuality ◇ ～者 homosexual
【同姓】of the same surname
【同学】①(在同一学校学习) be in the same school; be a schoolmate of sb. ②(在一起学习的人) fellow student; schoolmate ③(称呼学生) comrade, a form of address used in speaking to a student
【同样】same; equal; similar: ～尺码的鞋子 shoes of the same size / 在～条件下 under the same conditions
【同一】same; identical
【同一律】[逻] the law of identity
【同一性】[哲] identity
【同义词】[语] synonym
【同意】agree; consent; approve: ～你的要求 consent to your request / ～他的条件 agree to his terms ◇ ～书 letter of consent / ～照会 note of approval
【同音词】[语] homonym; homophone
【同音异义词】homonym
【同余】[数] congruence ◇ ～数 congruent numbers
【同源多倍体】[生] autopolyploid
【同志】comrade
【同舟共济】cross a river in the same boat; people in the same boat help each other
【同轴】[电] coaxial ◇ ～电缆 coaxial cable
【同宗】of the same clan; have common ancestry

茼

【茼蒿】[植] crowndaisy chrysanthemum

桐

a general term for paulownia, phoenix tree and tung tree
【桐油】tung oil ◇ ～树 tung tree

酮

[化] ketone
【酮化】[化] ketonize

铜 copper (Cu)：～扣子 brass button／～丝 copper wire／这是～制的。This is made of copper.
【铜氨液】cuprammonia
【铜铵人造丝】[纺] cuprammonium rayon
【铜板】[方] copper coin; copper
【铜版】[印] copperplate ◇～画[美术] copperplate etching; copperplate／～印刷 copperplate printing／～印刷机 copperplate press; etching press／～纸 art paper; printing paper
【铜币】copper coin
【铜臭】the stink of money; profits-before-everything mentality：～之人 a wealthy but mean person
【铜鼓】[考古] bronze drum
【铜管乐队】brass band
【铜管乐器】brass-wind instrument; brass wind
【铜壶滴漏】[考古] copper clepsydra
【铜活】①(铜制器件) brass or copper fittings, accessories, etc. ②(制造铜器的工作) work in copper; coppersmithing
【铜匠】coppersmith
【铜镜】[考古] bronze mirror
【铜矿】copper mine
【铜绿】[化] verdigris
【铜模】[印] matrix; mold ◇～雕刻机 matrix cutting machine
【铜牌】bronze medal
【铜器】bronze, brass or copper ware ◇～时代 the Bronze Age
【铜钱】copper cash
【铜墙铁壁】bastion of iron; impregnable fortress
【铜像】bronze statue
【铜元】copper coin; copper
【铜子儿】[口] copper coin; copper

tǒng

筒 ①(粗大的竹管) a section of thick bamboo ②(较粗的管状器物) a thick tube-shaped: 笔～ brush pot／邮～ mailbox; pillar-box ③(衣服等的管状部分) the tube-shaped part of an article of clothing: 袜～儿 the leg of a stocking／袖～儿 sleeve ④[量] tin; can: 一～可乐 a can of coke／一～啤酒 a can of beer
【筒管】[纺] bobbin
【筒状花】[植] tubular flower
【筒子】tube or tube-shaped object: 枪～ barrel of a gun／竹～ bamboo tube

桶 tub; pail; bucket; keg; barrel: 一～牛奶 a pail of milk／～装啤酒 barrelled beer; draught beer／水～ water bucket／油～ oil barrel

捅 ①(戳；扎) poke; stab: ～了一个洞 poke a hole ②(碰；触动) poke; stir up: 她用胳膊～了我一下。She gave me a nudge. ③(揭露) disclose; give away; let out: 别把秘密～出去。Don't give away the secret.
【捅娄子】make a mess of sth.; make a blunder; get into trouble
【捅马蜂窝】stir up a hornets' nest; bring a hornets' nest about one's ears

统 ①(事物间连续的关系) interconnected system: 系～ system／血～ blood relationship ②(全部) all; together ③(总括) gather into one; unite ④(衣服等的筒状部分) any tube-shaped part of an article of 长～靴 high boots／短～袜 socks
【统舱】steerage: 搭乘～ go steerage; travel steerage ◇～旅客 steerage passenger
【统称】①(总起来叫) be called by a joint name ②(总的名称) a general designation
【统筹】plan as a whole: ～安排 overall arrangement／～规划 overall planning／～兼顾 unified planning with due consideration for all concerned; making overall plans and taking all factors into consideration／～全局 take the whole situation into account and plan accordingly
【统共】altogether; in all: 我们～十个人。We're altogether ten people.
【统购包销】unified purchase and sale
【统购统销】state monopoly for purchase and marketing
【统计】①(对有关数据的搜集、整理、计算和分析) statistics: 人口～ census; vital statistics／据官方～ according to official statistics ②(总括地计算) add up; count ～选票 count the votes ◇～抽样 statistical sampling／～地图 statistical map／～力学 statistical mechanics／～数列 statistical series／～数字 statistical figures; statistics／～推断[数] statistical inference／～图表 statistical graph／～学 statistics／～学家 statistician／～员 statistician／～资料 statistical data
【统铺】a wide bed for a number of people
【统属】subordination
【统帅】①(武装力量最高领导人) commander

in chief; commander ②(统辖率领) command ◇～部 supreme command
【统率】command: ～三军 command the country's armed forces
【统统】[副] all; completely; entirely
【统辖】have under one's command; exercise control over; govern
【统一】①(联成整体) unify; unite; integrate: ～战线 united front / 对立的～ the unity of opposites ②(一致的;整体的) unified; unitary; centralized: ～的多民族国家 unitary multi-national state / ～的意见 consensus of opinion / ～计划 unified planning ◇～币制 unification of the currency / ～发票 uniform invoice
【统一体】[哲] entity; unity
【统一性】[哲] unity
【统治】rule; dominate ◇～阶级 ruling class / ～权 dominion; sovereignty / ～者 ruler
【统制】control: 经济～ economic control / 军事～ military control

tòng

痛 ①(疼痛) ache; pain: 背～ have a pain in the back / 肚子～ have a stomachache; have a pain in the abdomen / 耳朵～ have a pain in the ear / 头～ have a headache / 胸～ have a pain in the chest / 腰～ have a pain in the loins / ～不～? Does it hurt? / ～得利害。It hurts badly. ②(悲伤) sorrow; sadness: 悲～ deep sorrow ③(尽情地;深切地) extremely; deeply; bitterly: ～饮 drink one's fill
【痛不可忍】unbearably painful
【痛不欲生】grieve to the extent of wishing to die; be overwhelmed with sorrow
【痛斥】bitterly attack; scathingly denounce: ～谬论 sharply denounce a fallacy
【痛楚】pain; anguish; suffering
【痛处】sore spot; tender spot
【痛打】beat soundly; belabor: 挨了一顿～ get a sound beating
【痛定思痛】recall a painful experience; draw a lesson from a bitter experience
【痛风】[医] gout
【痛改前非】sincerely mend one's ways; thoroughly rectify one's errors
【痛感】keenly feel: ～自己知识不足 keenly feel one's lack of knowledge
【痛恨】hate bitterly; utterly detest: ～在心 cherish bitter hatred in the mind
【痛经】[医] dysmenorrhoea
【痛觉】[生理] sense of pain
【痛哭】cry bitterly; wail: ～流涕 weep bitterly; cry one's heart out / ～一场 have a good cry / 放声～ be choked with tears
【痛苦】pain; suffering; agony: 减轻病人的～ alleviate the sufferings of the patient / 精神上的～ mental pain; mental agony / 肉体上的～ physical suffering
【痛快】①(高兴) very happy; delighted; joyful: 感到～ be filled with joy ②(尽兴) to one's heart's content; to one's great satisfaction: 吃个～ eat one's fill / 玩个～ have a wonderful time ③(爽快) simple and direct; forthright; straightforward: 说话很～ speak simply and directly / 做事很～ be a zealous and neat worker
【痛骂】scold severely; curse roundly; give a good scolding
【痛切】with intense sorrow; most sorrowfully: ～陈词 make a deeply felt plea
【痛恶】bitterly detest; abhor
【痛惜】deeply regret; deplore: ～这一巨大损失 lament this great loss
【痛心】pained; distressed; grieved: 对过去的错误感到～ deeply regret one's past mistakes; feel deep regret for one's past mistakes
【痛心疾首】with bitter hatred
【痛痒】①(疾苦) sufferings; difficulties: ～相关 share a common lot ②(紧要的事) importance; consequence: 无关～ a matter of no consequence
【痛阈】[医] threshold of pain
【痛自悔改】show deep repentance

通 [量](用于动作) 擂了三～鼓 three rolls of the drums / 说了他一～ give him a talking-to

tōu

偷 ①(偷窃) steal; pilfer; make off with ②(瞒着人作) stealthily; secretly; on the sly: ～看 steal a glance; peek; peep / ～越国境 cross the border illegally ③(抽出) find (time): ～空 take time off
【偷安】seek temporary ease: ～旦夕 take ease at dawn and evening / 苟且～ seek only temporary ease and comfort
【偷盗】steal; pilfer: ～公款 embezzle public money
【偷工减料】do shoddy work and use inferior material; scamp work and stint material; jerry-build

【偷鸡不着蚀把米】try to steal a chicken only to end up losing the rice; go for wool and come back shorn
【偷空】take time off: ～休息一下 snatch a rest
【偷懒】loaf on the job; be lazy
【偷垒】[棒、垒球] steal a base; steal
【偷梁换柱】steal the beams and pillars and replace them with rotten timber; perpetrate a fraud
【偷窃】steal; pilfer: ～情报 steal information ◇ ～汽车者 car napper / ～行为 stealing
【偷情】carry on a clandestine love affair
【偷生】drag out an ignoble existence: 忍辱～ pocket insults to remain alive
【偷税】evade taxes
【偷天换日】steal the sky and put up a sham sun; perpetrate a gigantic fraud
【偷听】eavesdrop; bug; tap
【偷偷】stealthily; secretly; covertly; on the sly: ～地告诉 tell sb. on the quiet / ～地溜走 sneak away
【偷偷摸摸】furtively; surreptitiously; covertly
【偷袭】sneak attack; sneak raid; surprise attack
【偷闲】①(挤出空闲时间) snatch a moment of leisure: ～度日 snatch leisure and pass the day / 忙里～ snatch a little leisure from a busy life; allow oneself a bit of time ②(偷懒;闲着) loaf on the job; be idle
【偷眼】steal a glance; take a furtive glance: ～相看 steal a glance at
【偷营】make a surprise attack on an enemy camp; raid an enemy camp
【偷越国境】illegally cross the national border
【偷嘴】take food on the sly

tóu

头 ①(脑袋) head: 低～ bow one's head / 摇～ shake one's head ②(头发式样) hair or hair style: 分～ parted hair / 平～ crew cut ③(物体顶端或末梢) top; end: 火柴～ head of a match / 山～ the top of a hill ④(起点或终点) beginning or end: 从～到尾 from beginning to end ⑤(物品的残留部分) stub; remnant; stump; end: 烟～ cigarette end ⑥(头目) head; chief; boss ⑦(方面) side; aspect: 两～兼顾 give consideration to both sides ⑧(第一;前面的) first: ～胎 firstborn ⑨(领头的) leading: ～马 lead horse ⑩(用在数量词前面) first: ～一遍 the first time ⑪[方] (用在"年"或"天"前面) previous; last: ～年 last year; the previous year / ～天 the day before; the previous day ⑫[量] 五十～牛 fifty head of cattle / 一～蒜 a bulb of garlic ⑬[语](词的后缀)案～ on the desk / 手～ at hand / 心～ at heart
【头版】front page
【头部】head
【头彩】the first prize
【头寸】①(银行等拥有的款项) cash: 缺～ be out of cash ②(银根) money market; money supply: ～紧。Money is tight.
【头灯】[矿] head lamp
【头等】first-class; first-rate: ～大事 a matter of prime importance; a major event / ～重要任务 a task of primary importance ◇ ～舱 first-class cabin / ～品 first-rate goods
【头顶】the top of the head; the crown of the head
【头发】hair: 他的～已灰白了。His hair has turned grey. ◇ ～夹子 hairpin
【头盖骨】【头骨】[生理] skull; cranium
【头号】①(第一号) number one; size one ②(最好的) first-rate; top quality: ～面粉 the best flour ◇ ～新闻 headline news; lead story in a paper
【头花】[工美] headdress flower
【头昏】dizzy; giddy: ～脑胀 overwhelmed with work; make one's head swim / ～眼花 dizzy; giddy
【头角】brilliance; talent: ～峥嵘 brilliant; very promising; outstanding / 初露～ begin to show ability or talent
【头巾】scarf; kerchief
【头额】[方] neck
【头盔】a helmet
【头里】①(前面) in front; ahead ②(事先) in advance; beforehand
【头颅】head: 抛～,洒热血 lay down one's life
【头鲈鱼】silver-spotted grunt
【头面人物】prominent figure; bigwig; big shot
【头目】head of a gang; ringleader; chieftain: 小～ head of a small group in a gang
【头脑】①(思想能力) brains; mind: ～不清 mixed-up; muddle-headed / ～简单,四肢发达 well-developed limbs but head of a moron / ～冷静 have a cool head / 很有～ have plenty of brains / 有数学～ have a head for mathematics ②(头绪) main threads; clue: 摸不着～ cannot make head or tail of sth.
【头皮】①(头上的皮肤) scalp: 硬着～走了进去 take courage and get in ②(头皮屑) dandruff; scurf ◇ ～癣 ringworm of scalp / ～屑

dandruff
【头破血流】head broken and bleeding; hurt one's head badly
【头虱】head louse
【头胎】the first-born
【头套】actor's headgear
【头疼】headache
【头疼脑热】headache and slight fever; slight illness
【头痛】headache: 患～ suffer from headache／～得利害 have a bad headache
【头痛医头，脚痛医脚】treat the head when the head aches, treat the foot when the foot hurts; treat symptoms but not the disease
【头头是道】clear and logical; closely reasoned and well argued: 他说的真是～。 Every word he said hit the nail on the head.
【头衔】title
【头像】①(塑像) head sculpture ②(画像) head portrait
【头绪】main threads: ～纷繁 have too many things to attend to／ 理出～ gather up the threads; get things into shape
【头癣】[医] favus of the scalp
【头羊】bellwether
【头油】hair oil; pomade
【头晕】dizzy; giddy: ～眼花 dizzy of head and dim of sight; feel dizzy
【头重脚轻】top-heavy
【头状花序】[植] capitulum; head
【头子】chieftain; chief; boss: 强盗～ bandit chief
【头足动物】cephalopod

投 ①(投扔) throw; fling; hurl: ～石 throw a stone ②(投入) put in; drop: 把卡片～到字纸篓里 fling a card into the waste basket ③(跳进去) throw oneself into: ～河 drown oneself in a river ④(投射) project; cast ⑤(寄给人) send; deliver: ～书 deliver a letter ⑥(迎合) fit in with; agree with; cater to: 意气相～ find each other congenial
【投案】give oneself up to the police: ～自首 voluntary surrender to court
【投保】insure ◇ ～单 insurance application; insurance cover note／～价值 insured value
【投奔】go to for shelter: ～亲戚 go to one's relatives for help
【投笔从戎】throw aside the writing brush and join the army; renounce the pen for the sword
【投币式自动售货机】slot machine
【投标】submit a tender; enter a bid: ～最高价 highest bid／ 我们～建一座大楼。 We bid for the erection of a multistory building. ◇ ～人 bidder; tenderer
【投产】go into operation; put into production
【投诚】surrender; cross over: 向我方～ cross over to our side
【投弹】①(空投炸弹) drop a bomb ②(投掷手榴弹) throw a hand grenades ◇ ～高度 release altitude／ ～角 dropping angle／ ～器 bomb rack control; bomb release mechanism／ ～手 bombardier; grenadier
【投敌】go over to the enemy; defect to the enemy
【投递】deliver: ～信件 deliver letters ◇ ～员 postman; letter carrier; mail carrier; mailman
【投放】①(放进) throw in; put in ②(提供资金或供应商品) put (money) into circulation; put (goods) on the market
【投稿】submit a piece of writing for publication; contribute: 欢迎～。 Contributes are welcome. ◇ ～人 contributor
【投合】①(合得来) agree; get along: 两人性格很～。 The two of them suited each other perfectly. ②(迎合) cater to: ～一般人的趣味 cater to popular tastes
【投机】①(见解相同) congenial; agreeable: 话不～ be disagreeable in conversation; do not see eye to eye with each other／ 谈得很～ talk very congenially ②(谋取私利) speculate: ～取巧 seize every chance to gain advantage by trickery; be opportunistic ◇ ～倒把分子 profiteer; speculator／ ～分子 opportunist; political speculator／ ～商 speculator; profiteer／ ～事业 speculative enterprise
【投考】sign up for an examination: ～大学 sign up for a college entrance examination
【投靠】go and seek refuge with sb.
【投劾退隐】give up an official post and retire in seclusion
【投篮】[体] shoot; shoot at the basket; shot: ～不中 miss a shot; miss the basket／ ～得分 shoot a basket／ 近距离～ close-in shot／ 单手～ one-hand shot／ 单手头上～ one-hand overhead shot／ 定位～ set shot; stop shot／ 勾手～ hook shot／ 远距离～ long shot／ 转身～ pivot shot／ 转身勾手～ back-up shot
【投票】vote; cast a vote: 无记名～ secret ballot／ ～表决 decide by ballot／ ～反对 vote against／ ～赞成 vote for; vote in favor of／ 去投票处～ go to the polls ◇ ～日 polling day／

～箱 ballot box / ～站 polling booth; the polls
【投其所好】cater to another's pleasure
【投枪】javelin; (throwing) spear
【投亲】go and live with relatives; seek refuge with relatives
【投入】put in; throw in: ～大量资金 invest much capital in / 积极～工作 throw oneself into work actively / 在这部作品上～了五年的劳动 put in five years' labor to the work ◇ ～资本 invested capital; vested capital
【投射】①(扔;掷) throw; cast ②(光线等射) project; cast ◇ ～点 incident point / ～角 angle of incidence / ～器 a projector / ～物 a projectile
【投身】throw oneself into: ～革命 join the revolutionary ranks; join in the revolution
【投生】be reborn; be reincarnated in a new body
【投师】seek instruction from a master: ～访友 learn from a master and call on friends to exchange knowledge or skills
【投石问路】throw a stone to clear the road
【投手】[棒、垒球] pitcher ◇ ～犯规 balk
【投鼠忌器】hesitate to pelt a rat for fear of smashing the dishes beside it; spare the rat to save the dishes; Burn not your horse to rid it of the mouse.
【投宿】seek temporary lodging; put up for the night
【投胎】reincarnation
【投桃报李】give a plum in return for a peach; return present for present; exchange gifts
【投纬】[纺] picking: 每分钟～数 picks per minute
【投降】surrender; capitulate: 无条件～ unconditional surrender / 堡垒中的守军～了。The men in the fort capitulated. / 上尉不得不向我军～。The captain had to surrender to our army. ◇ ～派 capitulationist clique; capitulators / ～主义 capitulationism
【投药法】medication
【投影】projection: 极～ polar projection ◇ ～几何学 projective geometry / ～图 projection drawing / ～仪 projecting apparatus
【投掷】throw; hurl: ～标枪 throw a javelin
【投资】①(投入资金) invest: ～于工业 invest in industrial enterprises ②(投入的资金) investment: 基本建设～ investment in capital construction / 削减～ cut back investment ◇ ～边际效率 marginal efficiency of investment / ～场所 outlet for investment / ～基金 investment funds / ～市场 investment market / ～收益 income from investment; income on investment / ～信用 investment credit / ～银行 investment bank

tòu

透 ①(透过) penetrate; pass through; seep through: ～过现象看本质 see through the appearance to get at the essence / 不～水 waterproof; impermeable to water / 风雨不～ impervious to wind and rain ②(暗地里告诉) tell secretly: ～个信儿 tip sb. off ③(透彻;达到充分的程度) fully; thoroughly; in a penetrating way: 恨～了他 hate him deeply / 湿～了 be wet through / 桃子熟～了。The peaches are quite ripe. ④(显露) appear; show: 她脸上～出幸福的微笑。A happy smile appeared on her face.
【透彻】penetrating; thorough; incisive: 分析得～ make an incisive 〈penetrating〉 analysis / 讲得～ give a thorough exposition
【透翅蛾】[动] clearwing
【透顶】[贬] thoroughly; downright; in the extreme; through and through: 荒谬～ highly absurd
【透风】①(让风通过) let in air; ventilate ②(透露风声) divulge a secret; leak
【透光】pervious to light
【透汗】perspire all over
【透镜】[物] lens: 凹～ concave lens / 凹凸～ convex-concave lens / 分光～ beam-splitting lens / 复合～ compound lens / 凸～ convex lens
【透亮】①(明亮) bright; transparent: 这间房子又向阳,又～。This room is sunny and bright. ②(明白) perfectly clear: 你这么一说,我心里就～了。Thanks to your explanation, it's clear to me now.
【透亮儿】allow light to pass through
【透漏】divulge; leak; reveal: 消息～出去了。The news has leaked out.
【透露】divulge; leak; disclose; reveal: ～机密 disclose a secret / 据报纸～ according to the information of the press; according to the press reports
【透明】transparent: ～的纱巾 diaphanous veil / 半～ translucent / 不～ opaque ◇ ～计 diaphanometer / ～胶带 scotch tape / ～漆 celluloid paint; clear lacquer / ～水印 watermark / ～体[物] transparent body / ～纸 cellophane paper; cellophane

【透明度】①(透过光线的程度) transparency; diaphaneity ②(政治方面的公开程度) open politics; openness of the politics
【透辟】penetrating; incisive; thorough: ～的分析 penetrating analysis
【透平】[机] turbine
【透气】①(空气通过) ventilate ②(呼吸空气) breathe freely: 跑得透不过气来 lose one's breath in running / 透一口气 catch a breath
【透熟性】[物] diathermancy; diathermaneity
【透射】[物] transmission: 定向～ regular transmission ◇ ～比 transmittance / ～率 transmissivity
【透视】①(在平面上表现立体的方法) perspective ②[医] fluoroscopy; roentgenoscopy: ～肺部 have one's chest X-rayed ◇ ～图 perspective drawing
【透水层】[地] pervious bed; permeable stratum
【透心凉】①(非常冷) penetrating coolness ②(非常失望) utterly disappointing
【透雨】saturating rain; soaker
【透支】①[经] overdraw; make an overdraft ②(预支工资) draw one's salary in advance: ～薪金 anticipate one's pay ◇ ～帐户 overdraft account

tū

突 ①(猛冲) dash forward; charge: 骑兵～入敌人阵地。 The cavalry charged into enemy positions ②(突然) sudden; abrupt: 气温～降。 The temperature suddenly dropped. ③(突起;突出) projecting; sticking out
【突变】①(突然急剧的变化) sudden change: 天气～变。 There was a sudden change in the weather. ②[生] mutation: 自发～ spontaneous mutation ◇ ～体[生] mutant
【突出】①(鼓出来) protruding; projecting; sticking out ②(明显;出众) outstanding; prominent: ～的成就 outstanding achievements / ～的例子 striking example ③(强调) stress; highlight; give prominence to: ～重点 lay stress on the key points / 他从不～自己。 He never pushes himself forward. ④(冲出) break out: ～重围 break through a tight encirclement ◇ ～部[军] salient
【突飞猛进】advance by leaps and bounds; advance with seven- league strides; make giant strides
【突击】①(猛烈急速地攻击) make a sudden and violent attack; assault ②(加快完成) make a concentrated effort to finish a job quickly; do a crash job: ～麦收 do a rush job of harvesting the wheat ◇ ～点 point of assault / ～队 shock brigade / ～任务 rush job; shock work / ～手 shock worker / ～战术 shock tactics
【突尼斯】Tunisia ◇ ～人 Tunisian
【突破】①(打开缺口) break through; make a breakthrough: ～一点 make a breakthrough at some single point / 癌症研究上的一项重大～ an important breakthrough in cancer ②(打破) surmount; break; top: ～定额 overfulfill a quota / ～难关 break the back of a tough job ◇ ～地区[军] area of penetration; area breakthrough / ～点[军] breakthrough point; point of penetration / ～口[军] breach; gap
【突起】①(突然发生;兴起) break out; suddenly appear ②(高耸) rise high; tower
【突然】suddenly; abruptly; unexpectedly: ～停止 suddenly stop / ～袭击 surprise attack / 他的死是～的。 His death was very sudden. ◇ ～事件 emergency / ～搜查 raid
【突如其来】arise suddenly; come all of a sudden
【突突】[象] ①(机器的声音) chug: 汽船～地前进。 The boat chugged along. ②(心跳等) pit-a-pat: 她的心～地直跳。 Her heart went pit-a-pat.
【突围】break out of an encirclement
【突兀】①(高耸) lofty; towering: ～的山石 towering crags ②(出乎意料) sudden; abrupt; unexpected
【突袭】surprise attack

凸 protruding; raised
【凸岸】[地] convex bank
【凸凹不平】rough and uneven in surface
【凸版】[印] relief printing plate ◇ ～轮转机 rotary letterpress machine / ～印刷 letterpress; relief printing
【凸出】bulge; protrude; project: ～的眼睛 protruding eyes
【凸镜】convex mirror; convex lens
【凸轮】[机] cam: 急升～ quick lift cam / 推动～ actuating cam ◇ ～轴 cam shaft
【凸面】convex surface
【凸面镜】[物] convex mirror
【凸透镜】[物] convex lens
【凸缘】[机] flange: 管～ pipe flange / 环状～ collar flange / 接头～ joint flange
【凸嘴凹鼻】projecting lips and a sunken nose

秃 ①(无毛发) bald; bare: ～头 bald head / ～尾巴的 dock-tailed ②(无树木;无枝叶) bare: ～山 bare hills / ～树 bare trees ③(失去尖端的) blunt; without a point ④(结构不完整) incomplete; unsatisfactory
【秃笔】①(无笔尖的毛笔) worn-out writing brush; bald writing brush ②(不高明的写作能力) poor writing ability
【秃疮】[口] favus of the scalp
【秃顶】bald
【秃发病】alopecia
【秃鹫】cinereous vulture
【秃子】baldhead

tú

屠 ①(宰杀) slaughter ②(屠杀) massacre; slaughter: ～城 massacre the inhabitants of a captured city
【屠刀】butcher's knife
【屠夫】①(屠户) butcher ②(屠杀人民的人) a ruthless ruler
【屠户】butcher
【屠戮】[书] slaughter; massacre: ～无辜 slaughter the innocents
【屠杀】massacre; butcher; slaughter: 侵略军在守军投降后仍将多人～。 The invading army had massacred many of the garrison after capitulation.
【屠烧】kill and burn on a conquered land: ～城池 massacre the inhabitants and burn the city
【屠宰】butcher; slaughter: ～牲畜 slaughter animals ◇ ～场 slaughterhouse / ～率 dressing percentage / ～税 tax on slaughtering animals
图 ①(图画) picture; drawing; chart; map: 草～ (rough) sketch; draft / 插～ illustration; plate / 地形～ topographic map / 航行～ air navigation chart / 蓝～ blueprint / 示意～ sketch map / 天气～ weather map; synoptic chart / 制～ make a drawing; make a chart ②(谋划) scheme; plan; attempt: 另有所～ have another plan ③(贪图) pursue; seek: 只～享乐 pursue pleasure solely / ～一时痛快 seek momentary satisfaction ④(意图) intention; intent
【图案】pattern; design: 几何～ geometrical pattern / 装饰～ decorative pattern; ornamental design ◇ ～操 callisthenic performance forming patterns
【图板】drawing board
【图版】plate
【图饱私囊】planning to fill one's own pocket
【图表】chart; diagram; graph: 统计～ statistical chart
【图财害命】murder for money
【图钉】drawing pin; thumbtack
【图画】drawing; picture; painting ◇ ～文字 picture writing / ～纸 drawing paper
【图鉴】illustrated handbook
【图解】①(用图形分析演算) diagram; graph; figure ②[数] graphic solution ③[印] pictorial; illustrated: ～词典 pictorial dictionary ◇ ～法 graphic method
【图景】view; prospect: 展现出一幅壮丽的～ open up a magnificent prospect
【图利】desire to make money or profit; plan to make money
【图例】legend; key
【图名】seek fame; for the sake of prestige: ～谋利 scheme for fame and seek for wealth
【图谋】plot; scheme; conspire: ～不轨 hatch a sinister plot / ～私利 seek personal interests
【图片】picture; photograph: ～说明 caption / ～展览 photo exhibition
【图谱】a collection of illustrative plates; atlas
【图穷匕首见】when the map was unrolled, the dagger was revealed; the real intention is revealed in the end
【图示曲线】diagrammatic curve
【图书】books: ～资料 books and reference materials ◇ ～目录 catalogue of books; library catalogue / ～室 a reading room ◇国际标准～编号 ISBN (International Standard Book Number)
【图书馆】library: 北京～ Beijing Library / 大学～ college library; university library; academic library / 国家～ national library / 区～ district library / 省～ provincial library / 市～ municipal library / 县～ county library / 学校～ school library ◇ ～学 library science ○馆长 chief librarian; librarian / 副馆长 associate librarian; deputy librarian; sub-librarian / 研究馆员 research librarian / 副研究馆员 associate research librarian / 馆员 librarian / 助理馆员 library assistant / 管理员 clerk
【图腾】totem ◇ ～崇拜 totemism
【图像】picture; image: 立体～[电子] stereopicture ◇ ～识别[电子] pattern recognition
【图形】graph; figure: 几何～ geometric figure
【图样】pattern; design; draft; drawing
【图章】seal; stamp

【图纸】blueprint; drawing: 施工～ working drawing

涂 ①(涂抹)spread on; apply; smear: ～上软膏 apply some ointment ②(抹去) blot out; cross out: 把名子从单子上～掉 blot out one's name from the list / ～掉几个字 cross out a few words ③(乱写) scribble; scrawl: 别在墙上乱～。 Don't scribble on the wall.

【涂层】coat; coating: 反雷达～ antiradar coating / 减磨～ friction coat

【涂毒人民】poison the people

【涂改】alter: ～文件 tamper with a document / ～无效 invalid if altered / ～帐目 obliteration of accounts

【涂料】coating; paint: 防腐～ anticorrosive paint / 耐火～ refractory coating

【涂抹】①(涂在东西上) daub; smear; paint ②(乱写;乱画) scribble; scrawl: 信笔～ doodle; scribble aimlessly

【涂片】[医] smear: 血～ blood smear

【涂饰】①(涂上油漆) cover with paint, lacquer, color wash, etc ②(抹灰泥;粉刷) daub on a wall; whitewash

【涂炭】[书] utter misery; great affliction; misery and suffering: ～生灵 the people are plunged into an abyss of misery

【涂消作废】invalidate by crossing or blotting out

【涂鸦】poor handwriting; scrawl

【涂脂抹粉】apply powder and paint; prettify; whitewash: 为自己～ try to whitewash oneself

途 way; road; route: ～中 on the way; en route / 半～而废 give up halfway / 沿～ along the way

【途程】road; way; course: 人类进化的～ the course of evolution of mankind

【途经】by way of; via

【途径】way; channel: 外交～ diplomatic channels / 通过非法～ through illegal ways / 通过合法～ through legal ways / 由另一条～去 go by another way

【途穷】at the end of one's resources

荼

【荼毒】[书] afflict with great suffering; torment: ～生灵 plunge the people into the depths of suffering

【荼蘼】[植] roseleaf raspberry

徒 ①(步行) on foot: ～涉 wade through; ford ②(空的) empty; bare: ～手 bare-handed; unarmed ③(仅仅) merely; only: ～具形式 be a mere formality / ～托空言 give nothing but lip service; make empty promises ④(徒然) in vain; to no avail: ～费唇舌 waste one's breath / ～费功夫 simply to cost toil and work / ～费光阴 fritter away one's time / ～费心机 go on a fool's errand ⑤(徒弟) apprentice; pupil: 学～ apprentice ⑥(同一派系的人) believer; follower: 教～ follower of a religion ⑦[贬](人) person; fellow: 好事之～ busybody / 无耻之～ a shameless person

【徒步】on foot: ～旅行 travel on foot

【徒弟】apprentice; disciple

【徒工】apprentice

【徒唤奈何】utter unavailing cries of despair

【徒劳】futile effort; fruitless labor: ～往返 make a futile trip; hurry back and forth for nothing

【徒劳无功】make a futile effort; work in vain; work to no avail

【徒劳无益】vain labor with no profit

【徒然】in vain; for nothing; to no avail: ～耗费精力 waste one's energy / ～耗费时间 waste of time

【徒手】bare-handed; unarmed: ～致富 become rich bare-handed / 激烈的～混战 a fierce, barefisted melee ◇ ～操 free-standing exercises

【徒孙】disciple's disciple

【徒刑】[法] imprisonment; sentence: 无期～ life imprisonment; life sentence / 有期～ specified sentence; fixed-term imprisonment / 判十年有期～ sentence sb. to ten years' imprisonment / 应判～的犯罪 an imprisonable offense ◇ ～制度 convict system

【徒行往返】go and return on foot

【徒有其表】save up appearance

【徒有其名】nominal; in name only: ～并无实学的人 those who enjoy a reputation unwarranted by any real learning; one who does not live up to his reputation

【徒长】[农] excessive growth; spindling

【徒子徒孙】[贬] disciples and followers; adherents; hangers-on and their spawn

tǔ

土 ①(土壤) soil; earth: 表～ topsoil / 冲积～ alluvial land / 腐殖～ humus / 黑～ black soil; black earth / 红～ red soil / 黄～ loess / 碱性～ alkaline soil / 粘～ clay / 壤～ loam / 心～ subsoil / 盐渍～ saline soil / 中性～

neutral soil / ～块 a lump of earth; clod / ～路 dirt road ②(土地) land; ground; 领～ territory; domain ③(地方性的) local; native: ～产 local product / ～风 local custom; localism / ～音 local accent ④(民间沿用的) homemade; indigenous: ～办法 indigenous methods / ～设备 indigenous equipment ⑤(不合潮流) unrefined; unenlightened: ～里～气 rustic; uncouth; countrified ⑥(未熬制的鸦片) opium: 烟～ opium

【土坝】[水] earth-filled dam; earth dam

【土包子】clodhopper; bumpkin

【土豹】[动] buzzard

【土崩瓦解】disintegrate; crumble; fall apart; collapse like a house of cards

【土鳖】[动] ground beetle

【土拨鼠】[动] marmot

【土布】handwoven cloth; homespun cloth

【土产】local product: ～公司 local produce corporation

【土地】①(田地) land; soil; ground: ～的附带权利义务 incident of tenure / 耕种～ till the land / 丈量～ measure land ②(疆域) territory ③(传说中管辖一个小地区的神) local god of the land; village god: ～庙 a tiny temple housing the village god ◇ ～报酬递减律 the law of diminishing returns / ～法 land law; agrarian law / ～分红 dividend on land shares / ～集中 concentration of landholdings / ～纠纷 land disputes / ～税 land tax / ～所有人 land holder / ～证 land certificate; land deed / ～制度 land system

【土地改革】land reform; agrarian reform

【土地革命战争】The Agrarian Revolutionary War (1927–1937)

【土豆】[口] potato: ～片 potato chips

【土耳其】Turkey ◇ ～人 Turk / ～语 Turkish (language)

【土法】indigenous method; local method: ～生产 produce by indigenous methods

【土方】①(民间药方) traditional cure; folk recipe ②(土方工程) earthwork ③(计量单位) cubic meter of earth

【土匪】bandit; brigand

【土豪劣绅】local tyrants and evil gentry

【土话】local, colloquial expressions; local dialect

【土皇帝】local despot; local tyrant

【土黄】color of loess; yellowish brown

【土鸡瓦狗】shape without soul; completely useless persons

【土炕】heatable adobe sleeping platform; adobe *kang*

【土霉素】[药] terramycin; oxytetracycline

【土木】building; construction: 大兴～ go in for large-scale building 〈construction〉 ◇ ～工程 civil engineering / ～工程师 civil engineer

【土坯】sun-dried mud brick; adobe

【土气】rustic; uncouth; countrified

【土壤】soil ◇ ～保持 soil conservation / ～肥力 soil fertility / ～改良 soil improvement; soil amelioration / ～结构 soil structure / ～普查 general survey of soil / ～温度 soil moisture / ～通气性 soil aeration / ～渗透性 soil permeability / ～学 soil science; pedology / ～学家 pedologist / ～质地 soil texture

【土色】ashen; pale: 面如～ turn deadly pale

【土生土长】locally born and bred; born and brought up on one's native soil

【土石方】cubic meter of earth and stone

【土司】chieftain

【土特产】local specialty; local speciality

【土头土脑】rustic; hillbilly; unsophisticated

【土豚】[动] earth pig

【土卫】[天] satellite of Saturn; Saturnian satellite

【土温】[农] soil temperature

【土星】[天] Saturn ◇ ～光环 Saturn's rings

【土洋并举】use both indigenous and foreign methods; use both traditional and modern methods

【土洋结合】combine indigenous and foreign methods; combine traditional and modern methods

【土音】local accent

【土语】local, colloquial expressions; local dialect

【土葬】burial in the ground

【土著人】original inhabitants; natives; aborigines

【土专家】self-taught expert; local expert

吐

①(从嘴里涌出来) spit ②(从口或缝里露出来) put: ～出新芽 put forth buds / ～穗 sprout ears; be earing ③(说出) say; tell; pour out: 不～不快 have to get sth. off one's chest

【吐根】[植] ipecace ◇ ～硷[药] emetine

【吐故纳新】get rid of the stale and take in the fresh

【吐露】reveal; tell: ～心腹 speak one's heart and bowels / ～真情 unbosom oneself; tell the truth

【吐气】①（松口气）feel elated after unburdening oneself of resentment; feel elated and exultant: 扬眉～ blow off steam in rejoicing ②[语]（送气）aspirated ◇～音 aspirated sound
【吐弃】spurn; cast aside; reject
【吐绶鸡】turkey
【吐穗】[农] earing; heading
【吐絮】[农] opening of bolls; boll opening

钍 thorium (Th)

tù

吐 ①（从嘴里吐出）vomit; throw up; spit: ～核 spit pits / 恶心想～ feel sick; feel like vomiting / 上～下泻 suffer from vomiting and diarrhea ②（被迫退还）give up unwillingly; disgorge: ～赃 disgorge ill-gotten gains
【吐剂】vomitory
【吐酒石】[药] tartar emetic
【吐沫】saliva; spittle; spit
【吐血】spit out blood; spitting blood; hematemesis
【吐泻】vomiting and diarrhea

兔 hare; rabbit: 家～ rabbit / 野～ hare
【兔唇】[医] harelip
【兔死狗烹】the hounds are killed for food once all the hares are bagged; trusted aides are eliminated when they have outlived their usefulness
【兔死狐悲】the fox mourns the death of the hare; like grieves for like
【兔崽子】[骂] brat; bastard
【兔子】hare; rabbit
【兔子不吃窝边草】a hare doesn't eat the grass near its own hole; a villain doesn't harm his nextdoor neighbors
【兔子尾巴长不了】[贬] the tail of a rabbit can't be long; won't last long; Their days are numbered.

tuān

湍
【湍急】rapid; torrential: 水流～。The current is swift.
【湍流】①（急流）swift current; rushing waters; torrent; rapids: ～滚下 torrents of water rolling down ②[物]（汹涌的流动）turbulent flow; turbulence

tuán

团 ①（圆形的）round; circular; roundish: ～脸 round face ②（团子）sth. shaped like a ball: 汤～ boiled rice dumpling ③（揉成球形）roll sth. into a ball; roll: 把绒线绕成～ roll the wool into a ball ④（工作或活动的集体）group; society; organization ⑤（会合在一起）unite; conglomerate ⑥[军] regiment ⑦[简]（中国共产主义青年团）the Communist Youth League of China; the League: ～费 League membership dues / ～支部 League branch / 入～ join the League ⑧[量] ball: 一～绒线 a ball of wool
【团拜】mass greetings; mass congratulations
【团粉】cooking starch
【团结】unite; rally: ～一致 unite as one / ～友爱 unity and fraternity / ～就是力量。Unity is strength.
【团聚】reunite 全家～ family reunion
【团矿】[冶] nodulizing; briquetting
【团粒】[农] granule ◇～结构 granular structure
【团脐】female crab
【团扇】a circular fan; moon-shaped fan; round silk fan
【团身】[体] tucked
【团体】organization; group; team: 群众～ mass organization / 社会～ a social organization ◇～操 group callisthenics / ～冠军 team title / ～票 group ticket / ～赛 team competition
【团团】round and round; all round: ～围住 surround completely; encircle / ～转 pace about in an agitated state of mind
【团鱼】soft-shelled turtle
【团员】①（共青团员）a member of the Communist Youth League of China; League member ②（代表团成员）member of a delegation
【团圆】reunion: 全家～ family reunion ◇～饭 family reunion dinner
【团圆节】the Mid-Autumn Festival
【团藻】[植] volvox
【团长】①[军] regimental commander ②（代表团的领导）head of a delegation
【团子】dumpling: 菜～ cornmeal dumpling with vegetable stuffing / 饭～ rice ball / 糯米～ dumpling made of glutinous rice

tuī

推 ①（使向前移动）push; shove: ～车 push a cart / ～铅球 shot put; put the shot / 把他～出门外去。Push him out door. / 请～门，不要拉。Push the door; don't pull. ②（磨、碾粮食）turn a mill or grindstone; grind: ～磨 turn a millstone ③（剪；削）cut; pare: ～头 have a haircut; cut sb.'s hair ④(使事情开展) push forward; promote; advance: ～动工作 push the work forward ⑤(推断) infer; deduce ⑥(推卸) push away; shirk; shift: 把责任～给某人 shift the resposibility on to sb. ⑦(推迟) put off; postpone ⑧(推选) elect; choose ⑨(推崇) hold in esteem; praise highly

【推本溯源】trace the origin; ascertain the cause

【推波助澜】make a stormy sea stormier; add fuel to the flames

【推不知情】pretend to be ignorant of the situation

【推测】infer; conjecture; guess: ～后果 calculate the effect／从效果～动机 infer a motive from an effect／根据～ by inference

【推陈出新】weed through the old to bring forth the new

【推诚相见】deal with sb. in good faith; treat sb. with sincerity

【推迟】put off; postpone; defer: ～六个月 be postponed for six months／把会议～到星期五开 put off the meeting till Friday／球赛因雨～了。 The ball game was postponed because of rain.

【推斥】[物] repulsion ◇ ～力 repulsive force

【推崇】hold in esteem; praise highly: ～备至 have the greatest esteem for

【推辞】decline

【推挡】[乒乓球] half volley with push

【推到】①(使倒下) push over; overturn ②(推翻) repudiate; cancel; reverse

【推动】push forward; promote; give impetus to ◇ ～力 motive force; driving force

【推断】infer; deduce: 从前提～出结论 deduce a conclusion from premises

【推翻】①(用武力打垮) overthrow; overturn; topple: ～殖民主义的统治 overthrow the rule of colonialism ②(根本否定) repudiate; cancel; reverse: ～协议 repudiate an agreement／～原来的想法 cancel the original idea

【推杆】[机] push rod

【推广】popularize; spread; extend: ～科学管理方法 spread the scientific method of management／～普通话 popularize the common spoken Chinese／～应用 application and dissemination

【推己及人】put oneself in the place of another; treat other people as you would yourself; be considerate

【推荐】recommend: ～信 letter of recommendation

【推进】①(推动前进) push on; carry forward; advance; give impetus to: 把工作向前～一步 carry the work a step forward ②[军] move forward; push; drive ◇ ～剂 propellant／～力 propulsive force; driving power／～器 propeller

【推究】examine; study: ～事理 study the whys and wherefores of things

【推举】①(推选) elect; choose ②[举重] clean and press; press

【推来推去】make all sorts of excuses

【推理】[逻] inference; reasoning: 用～方法 by inference／归纳～ inductive reasoning／间接～ indirect inference／类比～ reasoning from analogy／演绎～ deductive reasoning／直接～ direct inference

【推力】thrust: 螺旋桨～ propeller thrust／喷气发动机～ jet thrust

【推论】inference; deduction; corollary

【推铅球】[体] shot put

【推情度理】consider the circumstances and measure the reasons

【推敲】weigh; deliberate: ～词句 weigh one's words; seek the right word／经过反复～ after repeated deliberation

【推求】inquire into; ascertain: ～地面沉降的原因 inquire into the causes of surface subsidence

【推却】refuse; decline

【推让】decline

【推人犯规】[体] pushing

【推三阻四】decline with all sorts of excuses; give the runaround

【推算】calculate; reckon: ～价 computed price

【推土机】bulldozer

【推推搡搡】push and shove

【推托】offer as an excuse; plead: ～之词 words that make excuses

【推脱】evade; shirk: ～责任 shirk responsibility／～职责 evade doing a duty

【推挽】[电] push-pull ◇ ～电路 push-pull circuit

【推委】shift responsibility onto others

【推贤让能】cede to the worthy and yield to the able

【推想】imagine; guess; reckon

【推销】promote sales; market; peddle: ～陈货 peddle one's old wares／～商品 promote the sale of goods ◇ ～计划 sales program／～员 salesman／～政策 promotion policy

【推卸】shirk: ～责任，委过于人 shirk responsibility and shift the blame onto others

【推心置腹】repose full confidence in sb.; confide in sb.
【推行】carry out; pursue; practice: ～新的教育政策 carry out new education policies
【推选】elect; choose: ～他为委员会主席 elect him (to be) chairman of the committee
【推移】①(时间发展) elapse; pass: 随着时间的～ with the lapse of time; as time goes on ②(形势等的变化) develop; evolve ◇ ～质[水] bed load
【推子】hair-clippers; clippers

tuí

颓 ①(坍塌) ruined; dilapidated: ～垣断壁 crumbling walls and dilapidated houses ②(衰败) declining; decadent: ～风败俗 decadent customs ③(委靡) dejected; dispirited
【颓惰自甘】lazy and self-indulgent
【颓废】dispirited; decadent: ～情绪 decadent sentiments ◇ ～派 the decadent school; the decadents
【颓龄】closing years of one's life; the declining years
【颓靡】downcast; dejected; crestfallen
【颓然】[书] dejected; disappointed
【颓丧】dejected; dispirited; listless: ～不振 being defeated one fails to rouse
【颓势】declining tendency
【颓唐】dejected; dispirited: ～失意 failing and disappointing
【颓运】declining fortune

tuǐ

腿 ①(人或动物肢体的一部分) leg: 大～ thigh／ 后～ hind leg／ 前～ foreleg／ 小～ shank ②(似腿的东西) a leglike support: 椅子～ legs of a chair ③(火腿) ham
【腿肚子】[口] calf
【腿脚】legs and feet; ability to walk: ～不灵便 walking with difficulty／ ～利落 walking briskly
【腿腕子】ankle

tuì

蜕 ①(脱皮) slough off; exuviate; molt ②(脱下的皮) exuviae
【蜕变】①(发生质变) change qualitatively; transtorm; transmute ②[物] decay: 感生～ induced decay／ 自发～ spontaneous decay
【蜕化】①(脱皮) slough off; exuviate ②(腐化堕落) degenerate ◇ ～变质分子 degenerate element; degenerate
【蜕皮】[动] cast off a skin; exuviate

退 ①(向后移动) move back; retreat: 往后～ step back ②(使向后移动) cause to move back; withdraw; remove: ～兵 withdraw the troops ③(退出) withdraw from; quit: ～党 withdraw from a political party ④(减退) decline; recede; ebb: 潮水～了。 The tide has receded. ⑤(退还) return; give back; refund: ～稿 return the manuscript／ ～货 return merchandise ⑥(撤消) cancel; break off: ～订货 cancel an order
【退避】withdraw and keep off; keep out of the way
【退避三舍】give way to sb. to avoid a conflict
【退兵】①(撤退军队) retreat; withdrawal ②(使敌军撤退) force the enemy to retreat: ～之计 a plan for repulsing the enemy
【退步】①(落后) lag behind; retrogress: 他思想～了。 He has slipped back ideologically. ②(后步) room for maneuver; leeway: 留个～ leave some room for maneuver; leave some leeway
【退场式】march out
【退潮】ebb tide; ebb
【退磁】[电磁] demagnetization
【退出】withdraw from; secede; quit: ～比赛 withdraw from a competition／ ～会场 walk out of a meeting
【退党】withdraw from a political party
【退而求其次】seek what is less attractive than one's original objective
【退化】①[生] degeneration ②(由好变坏) degenerate; deteriorate; retrograde
【退还】return: ～空瓶 return empties／ ～原主 return to owner ◇ ～基金 retirement fund
【退换】exchange a purchase
【退回】①(退还) return; send back: 地址不符,原件～ original mail returned on grounds of incorrect address／ 收信人不在,原件～ original mail returned in the absence of the addressee ②(返回原地) go back ◇ ～股利 rescission of dividends／ ～支票 returned check
【退婚】break off an engagement
【退火】[冶] annealing
【退伙】cancel an arrangement to eat at a mess; withdraw from a mess
【退货】return of goods; returned purchase ◇ ～报告 return sales report／ ～单据 returned purchase invoice
【退款】reimburse

【退路】①(退回去的道路) route of retreat ②(回旋的余地) room for maneuver; leeway: 留个～ leave some leeway
【退赔】return what one has unlawfully taken or pay compensation for it
【退票】return a ticket; get a refund for a ticket
【退却】①[军] retreat; withdraw ②(畏难后退) hang back; shrink back; flinch
【退让】make a concession; yield; give in: 在原则问题上不能～ make no concessiones on matters of principle
【退色】fade
【退烧】bring down a fever: 她已经～了。Her fever is gone. ◇～药 antipyretic
【退税】[经] drawback
【退缩】shrink back; flinch; cower: ～不前 withdraw from advancing
【退团】withdraw from a youth league; give up league membership
【退位】give up the throne; abdicate
【退伍】retire or be discharged from active military service; be demobilized; leave the army ◇～军人 demobilized soldier; ex-serviceman; veteran
【退席】①(退出宴席) leave a banquet ②(退出会场) walk out: ～以示抗议 walk out in protest
【退省吾身】retire and consider oneself
【退休】retire: 达到～年龄 reach retiring age ◇～工人 retired worker / ～金 retirement pay; pension / ～年龄 retirement age / ～制度 retirement system
【退学】leave school; discontinue one's schooling
【退押】return a deposit
【退一步想】on second thought
【退役】retire or be released from military service ◇～军官 retired officer / ～军人 ex-serviceman
【退赃】give up ill-gotten gains
【退职】resign from office; be discharged from office; quit working

褪 ①(脱衣服) take off ②(脱颜色) fade ③(脱毛) shed (feathers)
【褪色】color fading

tūn

吞 ①(整个咽下) swallow; gulp down: ～金自尽 commit suicide by swallowing gold / ～下药丸 swallow the pills / ②(吞没) take possession of; annex: 独～ take exclusive possession of
【吞并】annex; gobble up; swallow up
【吞服】swallow; take
【吞没】①(据为已有) embezzle; misappropriate; take possession ②(淹没) swallow up; engulf: 被海浪～ be engulfed in the waves
【吞声】[书] gulp down one's sobs; dare not cry out: ～屏息 with breathless anxiety
【吞食】swallow; devour
【吞噬】swallow; gobble up; engulf ◇～细胞[生理] phagocyte / ～作用[生理] phagocytosis
【吞吐】①(大量进出) swallow and spit; take in and send out in large quantities ②(含混不清) hem and haw; mince one's words: ～其辞 hum and han one's words ◇～量 handling capacity; the volume of freight handled
【吞吞吐吐】hesitate in speech;; hem and haw
【吞云吐雾】blow a cloud

tún

屯 ①(储存) collect; store up ②(驻扎) station; quarter ③(村庄) village
【屯兵】station troops
【屯积货物】store goods
【屯聚】assemble; gather together: ～兵力 assemble the military strength
【屯垦】station troops to open up wasteland
【屯粮】hoard up grains; store up grain: ～救荒 hoard up grains for time of drought
【屯扎】station; quarter

囤 store up; hoard: ～货 store goods
【囤积】hoard for speculation; corner: ～居奇 hoarding and cornering; hoarding and speculation

豚 suckling pig
【豚鼠】guinea pig; cavy

臀 buttocks
【臀部】buttocks
【臀大肌】musculus glutaeus maximus
【臀鳍】[动] anal fin
【臀疣】[动] monkey's ischial callosities; monkey's seat pads

tùn

褪 slip out of sth.: ～下一只袖子 slip one's arm out of one's sleeve
【褪套儿】[方] ①(使身体摆脱绳索) break loose; free oneself; get oneself free ②(摆脱责任)

shake off responsibility
【褪头褪脑】slink away

tuō

脱 ①(脱落) shed; come off; peel: 头发～光了 lose all one's hair; become bald / 我的脸～皮了。My face peeled. ②(取下;除去) take off; cast off: ～鞋 take off one's shoes / ～衣 take off one's clothes ③(脱离) escape from; get out of: ～榫 be out of joint ④(漏掉) miss out: ～误 omissions and errors
【脱靶】miss the target in shooting practice
【脱班】①(上班迟到) be late for work ②(车、船、飞机误点) be behind schedule
【脱产】be released from production or one's regular work to take on other duties: ～学习 be released from work for study
【脱党】quit a political party; give up party membership
【脱发】[医] trichomadesis
【脱肛】[医] prolapse of the anus
【脱稿】be completed: 已～付印 have completed a work and sent it for printing
【脱钩】unhook
【脱轨】derail: 电车～了。The streetcar was derailed.
【脱焊】sealing off
【脱缰之马】a runaway horse; uncontrollable; running wild
【脱胶】①(用胶粘在一起的物体脱落) come unglued; come unstuck ②[化](失去胶质) degum
【脱节】come apart; be disjointed; be out of line with: 理论与实践不能～。Theory must not be divorced from practice.
【脱臼】[医] dislocation: ～复位 replace dislocated joints
【脱壳机】[农] huller; sheller
【脱扣】[机] trip
【脱口而出】say sth. unwittingly; blurt out; let slip
【脱蜡】[石油] dewaxing ◇ ～法 lost-wax process
【脱离】separate oneself from; break away from; be divorced from: ～本题 digress from the subject / ～关系 break off relations; cut ties / ～羁绊 kick over the traces / ～困境 off the hook / ～实际 lose contact with reality; be divorced from reality / ～危险 be out of danger
【脱粒】[农] threshing; shelling ◇ ～机 thresher; sheller
【脱磷】[化] dephosphorization
【脱硫】[化] desulphurization; sweetening
【脱漏】be left out; be omitted; be missing: 这里～了一句。A sentence is left out here. (或: A sentence is missing here.)
【脱落】drop; fall off; come off: 油漆～了。The paint has come off.
【脱毛】lose hair or feathers; molt; shed
【脱帽】take off one's hat: ～默哀 bare one's head and mourn in silence / ～致敬 take off one's hat in salutation
【脱敏】desensitization; hyposensitization: ～疗法 desensitization therapy
【脱模】[冶] drawing of patterns
【脱泡】[纺] deaeration ◇ ～桶 deaerator
【脱坯】mold adobe blocks
【脱皮】decortication
【脱氢】[化] dehydrogenation
【脱然无累】without a worry or care in the world
【脱色】①(除去原来的色素) decolor; decolorize ②(退色) fade ◇ ～剂 decolorant; decolorizer
【脱身】get away; get free; extricate oneself: ～而去 go away quickly and quietly / ～之计 a plan that helps one to slip away
【脱手】①(脱开手) slip out of the hand ②(卖出) get off one's hands; dispose of; sell
【脱水】①[医](人体液体减少) deprivation of body fluids; dehydration ②[化](物质失水分) dehydration; dewatering ◇ ～机 whizzer; hydroextractor / ～剂 dehydrating agent / ～蔬菜 dehydrated vegetables
【脱俗】free from vulgarity; refined: ～行为 an act that is free from mere conventionality
【脱胎】①[工美] a process of making bodiless lacquerware ②(由另一事物孕育而产生的) emerge from the womb of; be born out off ◇ ～漆器 bodiless lacquerware
【脱胎换骨】be reborn; cast off one's old self; thoroughly remold oneself
【脱逃】run away; escape; flee: 临阵～ flee from battle
【脱位】[医] dislocation
【脱险】be out of danger; escape danger: 经过抢救,病人终于～。The patient was finally out of danger after the emergency treatment.
【脱销】out of stock; sold out
【脱氧】[化] deoxidation; deoxidization ◇ ～剂 deoxidizer; deoxidant

【脱氧核糖核酸】deoxyribonucleic acid (DNA)
【脱颖而出】the point of an awl sticking out through a bag; talent showing itself
【脱羽】molt
【脱脂】de-fat; degrease ◇ ～剂[皮革] degreasing agent / ～棉 absorbent cotton / ～奶粉 de-fatted milk powder; nonfat dried milk / ～乳 skimmed milk / ～纱布 absorbent gauze

拖 ①(牵引) pull; drag; haul: ～后腿 hold sb. back; be a drag on sb. / ～着脚步走 drag oneself along / 把小船～到滩上 haul the boat up the beach ②(拖延) delay; drag on; procrastinate: 时间已晚，别～了。 It is getting late; don't delay.
【拖把】mop
【拖长】①(拉长) lengthen ②(拖延) drag on
【拖车】trailer
【拖船】tugboat; tug; towboat
【拖带】traction; pulling; towing
【拖家带眷】have family burden
【拖拉】dilatory; slow; sluggish: 做事～ be dilatory in doing things / ～作风 dilatory style of work
【拖拉机】tractor: 手扶～ walking tractor ◇ ～工厂 tractor plant / ～手 tractor driver / ～站 tractor station
【拖来拖去】pull and haul a man
【拖累】①(成为负担) encumber; be a burden on: ～家庭 be a burden to one's family ②(牵连) implicate; involve: ～他人 involve others into trouble
【拖轮】tugboat; tug; towboat
【拖泥带水】messy; sloppy; slovenly: 办事要利落，不要～。 Do things neatly, not sloppily.
【拖欠】be behind in payment; be in arrears; default: ～房租 be in arrears with the rent
【拖人下水】involve sb. in evil-doing; get sb. into trouble
【拖时间】stall for time; delay
【拖沓】dilatory; sluggish; laggard
【拖网】trawlnet; trawl; dragnet ◇ ～渔船 trawler
【拖鞋】slippers: 草～ straw slippers / 绣花～ embroidered slippers
【拖延】delay; put off; procrastinate: 不能再～ can't delay any more ◇ ～战术 dilatory tactics
【拖曳】drag; pull; tow

托 ①(向上承受着) hold in the palm; support with the hand or palm: 手～着下巴 cup one's chin in one's hand ②(托子) sth. serving as a support: 枪～ the stock of a rifle ③(陪衬) serve as a foil; set off: 衬～ make sth. stand out by contrast; set off ④(委托) ask; entrust: ～人买鞋 ask sb. to buy shoes for one ⑤(推托) plead; give as a pretext ⑥(依赖) rely upon; owe to: ～庇 rely upon one's elder or an influential person for protection ⑦(托儿所) nursery: 全～ boarding nursery / 日～ day nursery
【托病】on pretext of illness; plead illness
【托钵求食】begging for food
【托词】①(找借口) find a pretext; make an excuse ②(借口) pretext; excuse; subterfuge
【托儿所】nursery; child-care center; creche: 民办～ nursery run by the local people
【托福】[套] thanks to you: 托您的福，我们全家都好。 We are all very well, thank you.
【托付】entrust: commit sth. to sb.'s care: 把孩子～给保姆照料 leave the child to the nurse
【托故】give a pretext; make an excuse: ～不来 fail to show up on some pretext
【托管】trusteeship ◇ ～国 trustee / ～理事会 Trusteeship Council / ～领土 trust territory / ～制度 trusteeship
【托灰板】[建] hawk
【托架】[机] bracket: 发动机～ engine bracket ◇ ～臂 bracket arm
【托拉斯】[经] trust
【托盘】tray
【托人情】ask an influential person to help arrange sth.; gain one's end through pull; seek the good offices of sb.
【托收】collection: ～汇票 bill for collection / ～款项 items sent for collection / ～票据 collection bill; short bill
【托叶】[植] stipule
【托运】consign for shipment; check: 货物已交铁路～。 The goods have been consigned by rail. ◇ ～单 booking note / ～人 consignor / ～物 consignment
【托子】base; support: 花瓶～ vase support

tuó

沱 [方] a small bay in a river
【沱茶】a bowl-shaped compressed mass of tea leaves

坨
【坨子】①(成块) lump: 泥～ a lump of mud; clod ②(成堆) heap: 盐～ salt mound

柁 [建] girder

砣 ①(秤砣) the sliding weight of a steelyard ②(用砣子打磨玉器) cut or polish jade with an emery wheel ③(石头碾砣) stone roller

【砣子】an emery wheel for cutting or polishing jade

鸵 ostrich

【鸵鸟】ostrich ◇ ～政策 ostrichism; ostrich policy

陀

【陀螺】top: 抽～ whip a top

【陀螺仪】[航空] gyroscope; gyro

驼 ①(骆驼) camel: 单峰～ dromedary; Arabian camel / 双峰～ camel; Bactrian camel ②(后背弯曲) hunchbacked; humpbacked

【驼背】①(背弯) hunchback; humpback ②(背弯的人) hunchbacked; humpbacked

【驼峰】①(骆驼背)hump ②[铁路] (调车用的土坡) hump ◇ ～调车场 hump yard / ～调车法 hump switching

【驼鹿】[动] elk; moose

【驼绒】①(骆驼绒毛) camel's hair ②(骆驼绒布) camel hair cloth

【驼色】the color of camel's hair; light tan

鼍 [动] Chinese alligator

驮 carry on the back

【驮畜】pack animal

【驮筐】pannier

【驮马】pack horse

tuǒ

椭

【椭率】[数] ellipticity

【椭面】[数] ellipsoid

【椭球】[数] ellipsoid

【椭圆】[数] ellipse ◇ ～规 ellipsograph; elliptic trammel / ～截面 oval cross section / ～星云 elliptical nebula / ～柱面 elliptic cylinder / ～锥面 elliptic cone

妥 ①(妥当) appropriate; proper: 不～ not proper ②(齐备; 停当) ready; settled; finished: 事已办～。The matter has been settled.

【妥当】appropriate; proper: 把工作安排～ make proper arrangements for the work

【妥善】appropriate; proper; well arranged: ～安排 make appropriate arrangements / ～办法 a well-arranged undertaking

【妥贴】appropriate; fitting; proper: ～稳当 everything satisfactorily arranged on a sound basis

【妥为照料】take good care; make proper arrangement

【妥协】come to terms; compromise: ～绥靖 compromise and appease / 达成～ reach a compromise ◇ ～分子 an appeaser / ～性 a tendency towards compromise

tuò

拓 open up; develop: 开～边远地区 open up the border regions

【拓荒】open up virgin soil; reclaim wasteland ◇ ～者 pioneer; pathbreaker; trailblazer

【拓扑学】[数] topology

唾 ①(唾液) saliva; spittle ②(用力吐唾沫) spit

【唾骂】spit on and curse; revile

【唾沫】saliva; spittle: ～星子 sprays from coughs

【唾弃】cast aside; spurn: ～虚名 in contempt of and cast away a false reputation / 被人民所～ be cast aside by the people

【唾手可得】extremely easy to obtain

【唾液】saliva ◇ ～腺 salivary gland

W

wā

挖 dig; excavate; scoop (out): ～井 dig a well; sink a well / ～隧道 excavate a tunnel / ～通 dig through / 在沙中～一个洞 scoop out a hole in the sand / 一尊古希腊雕像被～出来了。An old Greek statue was dug up.
【挖补】mend by replacing a damaged part
【挖槽机】[机] groover
【挖掉】dig up: ～根子 uproot; unroot
【挖方】[建] ①(开挖的土石方) excavation (of earth or stone) ②(指工程量) cubage of excavation
【挖沟机】ditcher; trencher; trench digger
【挖掘】excavate; unearth: ～被掩埋的城市 excavate a buried city / ～地下宝藏 unearth buried treasure / ～古物 excavate ancient relics / ～潜力 tap the latent power
【挖掘机】excavator; navvy: 履带式～ caterpillar excavator / 迈步式～ walking excavator
【挖空心思】rack one's brains; cudgel one's brains
【挖苦】speak sarcastically or ironically ◇ ～话 ironical remarks; verbal thrusts
【挖泥船】dredger; dredge
【挖墙脚】undermine the foundation; cut the ground from under sb.'s feet
【挖肉补疮】rob one's belly to cover one's back; sacrifice sth. as a make-shift to tide over an imminent predicament
【挖树机】tree mover
【挖土机】excavator; navvy

洼 ①(凹陷) hollow; low-lying ②(凹陷的地方) low-lying area; depression: 水～儿 a waterlogged depression
【洼地】depression; low-lying land
【洼陷】(of ground) be sunken; be low-lying

哇
【哇啦】[象] hullabaloo; uproar; din

蛙 frog
【蛙人】frogman
【蛙式打夯机】frog rammer
【蛙泳】breaststroke ◇ ～蹬腿 frog kick

wá

娃 ①(小孩) baby; child ②(小动物) newborn animal: 猪～ pigling; piglet
【娃娃】baby; child: 胖～ a chubby child ◇ ～床 crib; cot
【娃娃鱼】giant salamander

wǎ

瓦 ①(屋瓦) tile: 无棱～ plain tile / 给屋顶铺～ tile a roof ②(用泥土烧成的) made of baked clay: ～盆 earthen basin / ～器 earthenware ③[电] watt
【瓦房】tile-roofed house
【瓦工】①(砌砖,盖瓦工作) bricklaying, tiling or plastering ②(建筑工人) bricklayer; tiler; plasterer
【瓦匠】bricklayer; tiler; plasterer
【瓦解】disintegrate; collapse; crumble: ～敌军 disintegrate the enemy forces
【瓦楞】①[建] rows of tiles on a roof ②(瓦楞状的) corrugated ◇ ～铁皮 corrugated sheet iron / ～纸 corrugated paper
【瓦砾】rubble; debris: ～堆 a heap of rubble / 成了一片～ be reduced to rubble
【瓦圈】rim (of a bicycle wheel, cart wheel, etc.)
【瓦时】[电] watt-hour
【瓦斯】gas ◇ ～爆炸 gas explosion / ～筒 gas cylinder
【瓦特】[电] watt ◇ ～计 wattmeter
【瓦屋顶】tile roof

wà

袜 socks; stockings; hose: 短～ socks / 长～ stockings / 连裤～ panty hose; tights / 女长丝～ silk stockings
【袜带】suspenders; garters
【袜底】sole of a stocking
【袜套】socks; ankle socks
【袜筒】the leg of a stocking
【袜子】socks; stockings; hose: 一双～ a pair of socks; a pair of stockings

瓦 cover (a roof) with tiles; tile
【瓦刀】(bricklayer's) cleaver

wāi

歪 ①(不正;斜) askew; crooked; inclined; slanting: ～鼻子 a wry nose / ～戴着帽子 wear one's hat askew; have one's hat on crooked / ～嘴 a wry mouth ②(不正当的) devious; underhand; crooked: ～道理 false reasoning / ～

主意 evil ideas; devil's advice
【歪打正着】hit the mark by a fluke; score a lucky hit; do something unintentionally, but harvest exactly what one wishes
【歪风】evil wind; unhealthy trend: ～邪气 evil winds and noxious influences; unhealthy trends and evil practices / 打击～，发扬正气 combat evil trends and foster a spirit of uprightness
【歪七扭八】twist around; jiggle body
【歪曲】distort; misrepresent; twist: ～别人的话 twist people's words / ～事实 distort the facts / ～真相 perversion of truth ◇ ～事实的人 twister / ～作者原意 misrepresent the author's meaning
【歪诗】inelegant verses; doggerel
【歪歪扭扭】crooked; askew; shapeless and twisted: 字写得～ write a poor hand; scrawl
【歪斜】crooked; askew; aslant

wǎi

崴 ①(山路不平) rugged (mountain path) ②(扭伤) sprain; twist: 脚～了 got a sprain in one's ankle; sprain one's ankle

wài

外 ①(外边;外边的) outer; outward; outside: ～屋 outer room / 窗～ outside the window / 室～ outside the room / ～出 go out / 出～散步 go out for a walk ②(指自己所在地以外的) other: ～省 other provinces ③(外国) foreign; external: ～商 foreign merchants / 对～贸易 foreign trade; external trade ④(母亲、姐妹、女儿方面的亲戚) (relatives) of one's mother, sisters or daughters: ～孙 daughter's son; grandson ⑤(关系疏远的) not of the same organization, class, etc.; not closely related: ～客 a guest who is not a relative / 见～ regard sb. as an outsider / 电话不～借。 This telephone is not for public use. ⑥(另外) besides; in addition; beyond: 此～ besides; into the bargain / 邮费和包装费在～ postage and packing extra. ⑦(非正式的) unofficial: ～传 unofficial biography
【外币】foreign currency ◇ ～兑换 foreign currency exchange / ～汇票 foreign currency bill
【外边】①(超出某一范围的地方) outside; out: 到～散步 go out for a walk ②(外地) another place; a place other than where one lives or works ③(表面) exterior; outside
【外表】outward appearance; exterior; surface: ～美观 have a fine exterior; look nice / 从～看人 judge people by appearances
【外宾】foreign guest; foreign visitor
【外部】①(某一范围以外) outside; external: ～世界 the external world / 事物的～联系 external relations of things ②(表面) exterior; outside; surface: ～装修 exterior decoration
【外埠】nonlocal; other ports; towns or cities other than where one is
【外层空间】outer space ◇ ～导弹 outer-space missile
【外差】[电] heterodyne: 超～ superheterodyne
【外场】[棒球、垒球] outfield ◇ ～手 outfielder
【外出血】[医] external hemorrhage
【外带】①(外胎) tire (cover) ②(又加上) as well; besides; into the bargain
【外敌】foreign enemy; enemy from outside
【外地】other places; nonlocal; parts of the country other than where one is
【外电】dispatches from foreign news agencies: 据～报导 according to reports from foreign news agencies
【外调】①(调出) transfer (materials or personnel) to other localities: ～物资 materials allocated for transfer to other places ②(外出调查) investigation mission outside the city or town
【外毒素】[医] exotoxin
【外耳】[生理] external ear ◇ ～炎 otitis externa
【外耳道】[生理] external auditory meatus
【外耳门】porus acusticus externus
【外分泌】[生理] exocrine; external secretion ◇ ～腺 exocrine gland
【外敷】[医] apply (ointment, etc.) ◇ ～药 medicine for external application
【外感】[中医] diseases caused by external factors
【外港】outport
【外功】exercises to benefit the muscles and bones
【外观】outward appearance; exterior: 这座大楼～很美。 This is a fine-looking building.
【外国】foreign country: ～留学生 foreign students / ～朋友 foreign friends / 从～回来 return from abroad / 到～学习 go abroad to study ◇ ～法人 foreign juristic person; foreign juridical person / ～公司 foreign company; foreign corporation / ～货 foreign goods; foreign product / ～汇票 foreign bill of exchange / ～侨民 foreign immigrant / ～投资 foreign investment / ～投资者 foreign

investor / ～语 foreign language / ～支票 foreign check / ～资本 foreign capital
【外国人】foreigner ◇～居留证 residence permit for foreigners
【外果皮】[植] exocarp
【外行】① (外行的人) layman; nonprofessional ② (不懂或没经验) lay; unprofessional ◇～话 lay language; a mere dabbler's opinion
【外号】nickname
【外患】foreign aggression: 内忧～ internal disturbance and foreign aggression
【外汇】foreign exchange ◇～偿付能力 ability to repay in foreign exchange / ～储备 foreign exchange reserve / ～兑换率 rate of exchange / ～兑换券 foreign exchange certificate / ～官价 official exchange rate / ～管理 (foreign) exchange control / ～行情 exchange quotations / ～交易 foreign exchange transaction / ～牌价 foreign exchange rate / ～牌价表 list of exchange rate quotations / ～平价 par of exchange; exchange parity / ～市场 foreign exchange market / ～收入 foreign exchange earnings 〈income〉 / ～限制 exchange restriction
【外货】foreign goods; imported goods ◇～进口报单 application for import of foreign goods
【外籍】foreign nationality: ～工作人员 foreign personnel
【外寄生物】ectoparasite
【外加】more; additional; extra: 这是～的东西。 This is sth. additional. ◇～电压 applied voltage
【外间】outer room
【外交】diplomacy; foreign affairs: 通过～途径解决 be settled through diplomatic channels ◇～庇护 diplomatic asylum / ～部 the Ministry of Foreign Affairs; the Foreign Ministry / ～部长 Minister of 〈for〉 Foreign Affairs; Foreign Minister / ～辞令 diplomatic language; diplomatic parlance / ～代表 diplomatic representative / ～代表机构 diplomatic mission / ～官 diplomat; diplomatist / ～官衔 diplomatic rank / ～惯例 diplomatic practice / ～护照 diplomatic passport / ～豁免权 diplomatic immunities / ～机关 diplomatic establishments / ～家 diplomat / ～礼节 diplomatic protocol / ～签证 diplomatic visa / ～人员 diplomatic personnel / ～使节 diplomatic envoy / ～使团 diplomatic corps / ～使团团长 dean 〈doyen〉 of the diplomatic corps / ～特权 diplomatic prerogatives; diplomatic privileges / ～途径 diplomatic channels / ～文书 diplomatic correspondence / ～衔 diplomatic rank / ～信袋 diplomatic pouch; diplomatic bag / ～信使 diplomatic courier / ～邮袋 diplomatic bag; diplomatic pouch / ～邮件 diplomatic mail / ～政策 foreign policy ○新闻司 Information Department / 礼宾司 Protocol Department / 大使馆 embassy / 公使馆 legation / 总领事馆 consulate-general / 领事馆 consulate / 代办处 office of the charge d'affaires / 武官处 military attache's office / 商务处 commercial counsellor's office / 新闻处 press section; information service / 特命全权大使 ambassador extraordinary and plenipotentiary / 大使 ambassador / 公使 minister; envoy / 教廷大使 nuncio / 教廷公使 internuncio / 公使衔参赞 counsellor with the rank of minister; minister-counsellor / 代办 charge d'affaires / 临时代办 charge d'affaires ad interim / 参赞 counsellor / 商务参赞 commercial counsellor / 文化参赞 cultural counsellor / 商务专员 commercial attache / 文化专员 cultural attache / 陆军武官 military attache / 海军武官 naval attache / 空军武官 air attache / 一等秘书 first secretary / 总领事 consul-general / 领事 consul / 巡回大使 roving ambassador / 无任所大使 ambassador at large / 互相承认 mutual recognition / 国书 letter of credence; credentials / 递交国书 present one's credentials / 招回公文 letter of recall / 不受欢迎的人 persona nongrata / 照会 note; present a note to
【外交关系】diplomatic relations: ～升格 upgrade diplomatic relations / 断绝～ sever diplomatic relations / 恢复～ resume diplomatic relations / 建立～ establishment of diplomatic relations / 建立大使级的～关系 establish diplomatic relations at ambassadorial level / 中断～ suspend diplomatic relations
【外角】[数] exterior angle
【外接圆】[数] circumscribed circle; circumcircle
【外界】① (某个物体以外的空间) the external world; the outside world: 对～的认识 knowledge of the external world ② (某个集体以外的社会) outside: ～影响 outside influence / 顶住～的种种压力 withstand all kinds of outside pressure / 向～征求意见 solicit comments and suggestions from people outside one's organization
【外景】outdoor scene; a scene shot on location; exterior: 拍摄～ film the exterior; shoot a scene

on location
【外径】[机] external diameter; outside diameter; outer diameter
【外科】surgical department ◇ ～病房 surgical ward／～手术 surgical operation; surgery／～学 surgery／～医生 surgeon
【外壳】outer covering; outer casing; shell; case: 电池～ battery case／热水瓶～ the outer casing of a thermos flask
【外快】extra income
【外来】outside; external; foreign: ～干涉 outside interference; foreign intervention; external intervention ◇ ～户 setters from other places／～人 a person from another place; nonnative／～语 word of foreign origin; foreign word; loanword
【外力】①(外部力量) outside force ②[物] external force
【外流】outflow; drain: 黄金～ gold bullion outflow／科技人员～outflow of scientific and technical personal／美元～ dollar outflow
【外贸】[简] (对外贸易) foreign trade; external trade ◇ ～经营权 the power to engage in foreign trade／～仲裁 foreign trade arbitration ○到岸价格 cost. insurance, and freight (c.i.f.)／离岸价格 free on board (f.o.b.)／付讫 paid／超收,多收 overcharge／空舱书 dead freight／佣金 commission／运费 freight (frt.)／重量吨和尺码吨 weight or measurement／重量 weight (wt.)／实际重量 actual weight／毛重 gross weight／净重 net weight／皮重 tare／启运重量 shipping weight／个;件 piece (单数)／个;件 pieces (复数)／直径 diameter／一打 dozen／最大 maximum／中 medium (med.)／最小 minimum／标志 mark／标准 standard／利息 interest／不包括利息 excluding interest／平均 average／约计 approximately (approx.)／信汇 mail transfer／电汇 telegraphic transfer／承兑交单 documents against acceptance／即期汇票 demand draft／发票 invoice／形式(形式发票) pro forma (invoice)／承兑 acceptance／本月 instant (the present month)／上月 ultimo; last month／下月 proximo／月底 end of month／季度 quarter／季末 end of season／电报 telegram／地址 address (add.)／转交 care of／经由 by way of (via.)／目的地 destination／船名前冠 steamship／保险 insurance／水渍险 with particular average／差错待查 errors and omissions excepted (E. & O.E.)／无商业价值 no commercial value
【外貌】appearance; exterior; looks
【外面】①(外表) outward appearance; exterior; surface: ～儿光 deceptively smooth appearance; outward show ②(外边) outside; out: 把花盆搬到～去。Take the flowerpots out.／今晚他们要在～吃饭。They'll eat out this evening.
【外强中干】outwardly strong but inwardly weak; strong in appearance but weak in reality; a weakling putting on a bold front
【外侨】foreign national; alien ◇ ～身份 alienism
【外切形】[数] circumscribed figure
【外勤】①(室外工作) work done outside the office or in the field; field-work ②(外勤工作人员) field personnel; field-worker
【外倾】[心] extroversion
【外圈】[体] outer lane; outside lane
【外人】①(局外人;无亲友关系的人) stranger; outsider: 别客气,我又不是～。Don't stand on ceremony. I'm no stranger. ②(外国人) foreigner; alien
【外商】foreign tradesman
【外伤】an injury or wound; trauma ◇ ～学 traumatology
【外生殖器】external genital organs
【外甥】sister's son; nephew
【外甥女】sister's daughter; niece
【外事】foreign affairs; external affairs: ～往来 dealings with foreign nationals or organizations ◇ ～部门 foreign affairs section
【外孙】daughter's son; grandson
【外孙女】daughter's daughter; granddaughter
【外胎】tire (cover)
【外逃】①(潜逃别处) flee to some other place ②(逃到国外) flee the country
【外套】①(大衣) overcoat ②(外衣) loose coat; outer garment
【外听道】[生理] external auditory meatus
【外头】outside; out; outdoors: 汽车在～。The car is outside.
【外围】periphery; outer- ring: 首都～ the periphery of the capital ◇ ～防线 outer defense line／～设备 peripheral equipment／～组织 peripheral organization
【外文】foreign language
【外屋】outer room
【外侮】foreign aggression; external aggression: 抵御～resist foreign aggression
【外务】①(本身职务以外的事) matters outside

one's job ②(外交) foreign affairs; external affairs
【外线】①[军] exterior lines: ～作战 fight on exterior lines; exterior-line operations ②(指电话) outside (telephone) connections
【外乡】another part of the country; some other place: ～口音 a nonlocal accent / ～人 nonnative
【外向型经济】export-oriented economy
【外销】for sale abroad or in another part of the country: ～产品 products for export; articles for sale in other areas / ～业务 foreign sales
【外心】①(因爱上别人而不忠于自己配偶的念头) unfaithful intentions (of husband or wife) ②[数] circumcenter
【外形】appearance; external form; contour
【外姓】(people) not of the same surname
【外延】[逻] extension
【外焰】[化] outer flame
【外衣】①(外套) coat; jacket; outer clothing; outer garment ②(遮盖物) semblance; appearance; garb: 披着…～ in the garb of; under the cloak of
【外阴】[生理] vulva ◇ ～搔痒 pruritus vulvae / ～炎 vulvitis
【外因】external cause
【外引内联】introduce investment from abroad and establish lateral ties at home
【外用】[药] external use; external application: ～药水 lotion / 只能～ for external use only
【外语】foreign language ◇ ～教学 foreign language teaching / ～学院 institute of foreign languages / ～专业 faculty of foreign languages
【外源河】exotic stream
【外援】foreign aid; outside help; external assistance
【外圆磨床】cylindrical grinder
【外圆内方】round outside but square inside; outwardly gentle but inwardly stern
【外在】external; extrinsic ◇ ～性 externalism / ～因素 external factor
【外债】external debt; foreign debt
【外展肌】[生理] abductor
【外长】[简](外交部长) Minister of 〈for〉 Foreign affairs; Foreign Minister
【外罩】outer garment; dustcoat; overall
【外痔】external piles; external hemorrhoids
【外置马达】outboard motor
【外资】foreign capital
【外族】①(本家族以外的人) people not of the same clan ②(外国人) foreigner; alien ③(本民族以外的民族) other nationalities
【外祖父】(maternal) grandfather
【外祖母】(maternal) grandmother

wān

豌

【豌豆】pea: 去皮干～ split peas
【豌豆黄】[食品] pea flour cake
【豌豆象】[动] pea weevil

剜 cut out; gouge out; scoop out

【剜肉补疮】cut out a piece of one's flesh to cure a boil; resort to a remedy worse than the ailment; resort to a stopgap measure detrimental to long-term interests

蜿

【蜿蜒】①(蛇类爬行的样子) wriggle: ～前进 wriggle one's way / 一条蛇～爬过道路。A snake wriggled across the road. ②(弯弯曲曲地延伸) wind; zigzag; meander: 那河～流入大海。The river winds its way to the sea. / 小溪～流过山谷。The stream winds through the valley.

弯 ①(弯曲) curved; tortuous; crooked; bent: ～～的月牙儿 a crescent moon ②(使弯曲) bend; flex: ～弓 bend a bow / ～着腰摘棉桃 bend over to pick cotton boll ③(弯子) turn; curve; bend: 拐～儿 go round curves; turn a corner / 这河突然向西转～。The river takes an abrupt bend to the west.

【弯路】①(不直的路) winding course; crooked road; tortuous path ②(多费的冤枉功夫) round-about way; detour: 少走～ avoid detours
【弯曲】winding; meandering; zigzag; crooked; curved: ～的木棍 a crooked stick / ～的山间小道 a winding mountain path
【弯头】[机] elbow; bend: 回转～ return bend / 接合～ joint elbow
【弯弯曲曲】having many bends or curves: ～的小溪 meandering stream
【弯子】bend; turn; curve

湾 ①(水流弯曲的地方) a bend in a stream: 河～ river bend ②(海湾) gulf; bay: 墨西哥～ the Gulf of Mexico ③(使船停住) cast anchor; moor: 在码头～住船 moor a ship at the pier

wán

完 ①（全；完整）intact; whole; unharmed: ～好 in good condition; intact ②（没有剩的；消耗尽）exhaust; finish; use up; run out: 书卖～了。All copies of the book were sold out.（或：The book is out of stock.）/ 纸用～了。The writing paper is used up. ③（完结）finish; complete; be over; be through: 下期续～ to be concluded in the next issue / 会议开～了。The meeting is over. / 我们把所有的肉都吃～了。We have finished all the meat. / 舞会还没～。The dance hasn't finished yet. ④（交纳）pay: ～税 pay taxes

【完备】complete; perfect: 一套～的工具 a complete set of tools / 指出不～之处 point out the imperfections

【完毕】fish; complete; end; be done: 他们工作～后游泳去了。After finishing their work, they went swimming.

【完璧归赵】return a thing intact to its owner; return sth. to its owner in good condition

【完成】accomplish; complete; fulfill; bring to success 〈fruition〉: ～工作 complete one's work / ～国家计划 fulfill the state plan / ～任务 complete one's mission; accomplish a task; discharge one's duty / ～生产指标 hit the production target; fulfill the production quota / ～使命 accomplish one's mission

【完蛋】[口] be finished; be doomed; be done for

【完稿】finish a piece of writing; complete the manuscript

【完工】complete a project, etc.; finish doing sth.; get through: 这大楼提前三个月～。The building was completed three months ahead of schedule.

【完好】intact; whole; in good condition: ～如初 as excellent as before / ～如新 as good as new / ～无缺 intact; undamaged / ～无损 excellent without damage

【完婚】(of a man) get married; marry; conclusion of marriage

【完结】end; be over; finish: 这案子终于～了。The case came to an end at last.

【完竣】(of a project, etc.) be completed

【完了】come to an end; be over

【完满】satisfactory; successful: 会议～结束。The meeting came to a satisfactory solution. / 问题终于～解决。The problem has been solved satisfactorily at last.

【完美】perfect; consummate: ～的计划 a perfect plan / ～无缺 perfect; flawless

【完清帐务】square all debts

【完全】①（不缺；齐全）complete; whole: 她话没说～。She didn't give a full picture. ②（全部）completely fully; wholly; entirely; absolutely: ～不可能 completely out of the question / ～不同 be totally different; have nothing in common / ～彻底为人民服务 serve the people heart and soul / ～错了 be completely wrong / ～同意 full accord with; complete agreement with; agree fully with / ～脱离现实 lose all touch with reality / ～相反 be the exact opposite / ～一致 completely in line; entirely at one; in full agreement / ～正确 perfectly right; absolutely correct ◇ ～变态 [生] complete metamorphosis / ～破产 dead broke; strong broke / ～燃烧 complete combustion / ～叶[植] complete leaf

【完人】perfect man: 金要足赤，人要～ gold must be pure and man must be perfect; perfectionism

【完善】perfect; consummate: 日趋～ be being perfected; be improving day by day / 设备～ very well equipped

【完事】finish; get through; come to an end: 她总算～了。She has eventually finished the job.

【完税货价】price duty paid

【完税凭证】duty paid proof

【完整】complete; integrated; intact: ～的工业体系 an integrated industrial system / ～的过程 the whole process / 维护领土～ safeguard territorial integrity ◇ ～文本 full copy

烷 alkane

【烷化】alkanisation; alkylation ◇ ～汽油 alkylation gasoline

【烷基】alkyl ◇ ～胺 alkylamine

玩 ①（玩耍）play; have fun; amuse oneself: ～牌 play cards / ～桥牌 play bridge / 到某地去～ make a trip to / 真好～! That's great fun! ②（玩弄）employ; resort to: ～儿邪的 employ underhand means; not play fair / ～手段 resort to crafty maneuvers; play tricks ③（玩忽）trifle with; treat lightly; toy with: ～法 trifle with the law ④（玩赏）enjoy; find pleasure in; appreciate: ～儿邮票 make a hobby of collecting stamps ⑤（供观赏的东西）object for appreciation: 古～ curio; antique

【玩忽】neglect; trifle with; ignore: ～职守 neg-

lect of duty; ignore one's duty
【玩花招】play tricks
【玩火】play with fire
【玩火自焚】He who plays with fire will get burnt. (或: Whoever plays with fire will perish by fire. / Sow the wind and reap the whirlwind.)
【玩剑者必死于剑下】All they take the sword shall perish with the sword
【玩具】toy; plaything: ～汽车 toy car / ～火车 toy train ◇ ～店 toyshop
【玩弄】①(调戏) dally with; flirt with: ～女性 philander; dally with women ②(搬弄) play with; juggle with: ～词句 juggle with words; go in for rhetoric ③(施展) resort to; employ: ～两面派手法 engage in double-dealing / ～新花招 employ some new tricks / ～于掌股之上 twist around one's finger / ～种种阴谋诡计 resort to all sorts of schemes and intrigues
【玩偶】doll; toy figurine
【玩儿命】gamble 〈play〉 with one's life; risk one's life needlessly
【玩儿完】the jig is up
【玩赏】enjoy; take pleasure 〈delight〉 in: ～风景 enjoy the scenery; admire the scenery
【玩世不恭】be cynical; disdain worldly affairs
【玩耍】play; have fun; amuse oneself
【玩味】ponder; ruminate; contemplate: 他的话值得～。 It is worth while to ruminate over his remarks.
【玩物】plaything; toy
【玩物丧志】riding a hobby saps one's will to make progress; excessive attention to trivia saps the will; be a playboy without ambitions
【玩笑】joke; jest: 开～ play a joke 〈prank〉 on; make jests / 我讲这话不是开～的。 I'm not saying it in fun.
【玩意儿】①(玩具) toy; plaything ②(指东西) thing: 新鲜～ newfangled gadget

顽

①(迟钝;愚笨) stupid; dense; insensate: ～石 hard rock; insensate stone ②(顽固) stubborn; obstinate: ～敌 stubborn enemy; inveterate foe ③(顽皮) naughty; mischievous
【顽磁】[物] magnetic retentivity
【顽钝】dull and obtuse; stupid; thickheaded
【顽固】①(思想保守) obstinate; stubborn; headstrong: ～不化 incorrigibly obstinate / ～地坚持错误立场 stubbornly cling to one's wrong position ②(立场反动) bitterly opposed to change; die-hard ◇ ～分子 diehard; die-hard element / ～派 the diehards
【顽抗】stubbornly resist
【顽皮】naughty; mischievous
【顽强】indomitable; staunch; tenacious: ～的革命精神 indomitable revolutionary spirit / 同干旱进行～的斗争 put up a tenacious fight against the draught
【顽石点头】(be so persuasive as to make) the insensate stones nod in agreement
【顽童】naughty child; urchin
【顽症】chronic and stubborn disease; persistent ailment

丸

①(球形的小东西) ball; pellet: 泥～ mud ball ②(丸药) pill; bolus: 每服两～ take two pills each time
【丸剂】pill
【丸药】pill 〈bolus〉 of Chinese medicine
【丸子】①(食品) a round mass of food; ball: 肉～ meatball / 鱼～ fish ball ②(药丸) pill; bolus

纨

fine silk fabrics
【纨绔子弟】profligate son of the rich; fop; dandy; playboy

wǎn

莞

【莞尔】smile: ～而笑 give a soft smile / 不觉～ cannot help smiling

宛

①(曲折) winding; tortuous ②(仿佛) as if: 音容～在 as if the person were still alive
【宛然】as if: 这里山清水秀, ～桂林风光。 The scenery here has great charm, reminding one of the land of Guilin.
【宛如】just like; as if

惋

sigh
【惋惜】feel sorry for sb. or about sth.; sympathize with; regret

碗

bowl: 摆～筷 put out bowls and chopsticks for a meal; lay the table / 铁饭～ iron rice bowl; a secure job
【碗橱】cupboard

婉

①(柔顺;婉转) gentle; gracious; tactful: ～商 consult with sb. tactfully 〈politely〉 / ～顺 complaisant; obliging ②(美好) beautiful; graceful; elegant: ～丽 beautiful; lovely
【婉辞】①(婉言) gentle words; euphemism ②(婉言谢绝) graciously decline; politely refuse

【婉言】gentle words; tactful expressions: ～相劝 gently persuade; plead tactfully / ～谢绝 graciously decline; politely refuse
【婉约】graceful and restrained
【婉转】①(温和而曲折) mild and indirect; tactful: 措词～ put it tactfully ②(抑扬动听) sweet and agreeable: 歌喉～ a sweet voice; sweet singing

挽 ①(拉) draw; pull: ～弓 draw a bow / 手～着手 arm in arm ②(向上卷) roll up: ～起袖子 roll up one's sleeves ③(哀悼) lament sb.'s death ④(盘绕) coil up
【挽歌】dirge; elegy
【挽回】retrieve; redeem: ～败局 retrieve a defeat / ～面子 save face / ～名声 redeem one's reputation / ～损失 retrieve a loss / ～影响 redeem one's reputation; retrieve one's reputation / 无可～ irredeemable; irretrievable
【挽救】save; remedy; rescue: ～病人的生命 save the patient's life
【挽联】elegiac couplet
【挽留】urge sb. to stay; persuade sb. to stay: 再三～ repeatedly urge sb. to stay; press sb. to stay
【挽马】[牧] draught horse
【挽诗】elegy

晚 ①(晚上) evening; night: 从早到～ from morning till night / 今～ this evening; tonight / 昨～ last night ②(时间靠后的) late; far on in time: 睡得～ go to bed late / 现在去还不～。 It's still not too late to go. / ～做总比不做好。 Better late than never. ③(后来的) younger; junior
【晚安】good night
【晚班】night shift: 上～ be on the night shift
【晚报】evening paper
【晚辈】the younger generation; one's juniors
【晚餐】supper; dinner
【晚场】evening show; evening performance
【晚车】night train
【晚稻】late rice
【晚点】late; behind schedule: 班机～了。 The airliner is late.
【晚饭】supper; dinner
【晚会】an evening of entertainment; soiree; social evening; evening party: 国庆～ National Day's evening party / 英语～ English evening
【晚婚】marry at a mature age; late marriage
【晚间】(in the) evening; (at) night
【晚节】integrity in one's later years: 保持革命～ maintain one's revolutionary integrity in one's later years
【晚景】①(傍晚的景色) evening scene ②(晚年的景况) one's circumstances in old age: ～不佳 the condition in one's closing years is no good
【晚年】old age; one's later 〈remaining〉 years
【晚期】later period: ～作品 sb.'s later works; the works of sb.'s later period / 十九世纪～ the late 19th century; the latter part of the 19th century / 她的病已到～。 Her illness has reached an advanced stage.
【晚秋】late autumn; late in the autumn ◇ ～作物 late-autumn crops
【晚上】(in the) evening; (at) night
【晚熟】late-maturing ◇ ～品种 late variety
【晚霜】late frost
【晚霞】sunset glow; sunset clouds
【晚香玉】[植] tuberose
【晚些时候】late; by and by; afterwards
【晚育】late childbirth

绾 coil up: ～个扣儿 tie a knot / 把头发～起来 coil one's hair

wàn

蔓 a tendrilled vine: 这棵黄瓜爬～了。 This cucumber plant is climbing.

万 ①(数目) ten thousand ②(很多) a very great number; myriad: ～里长空 vast clear skies / ～事～物 myriads of things; all nature ③(很;极;绝对) absolutely; by all means: ～不得已 out of absoiute necessity; as a last resort
【万般】①(各种各样) all the different kinds ②(非常;极其) utterly; extremely: ～无奈 have no alternative (but to); absolutely bored
【万般皆下品,唯有读书高】To be a scholar is to be the top of society.
【万变不离其宗】change ten thousand times without departing from the original aim or stand; remain essentially the same despite all apparent changes
【万不得已】as a last resort; out of absolute necessity; be forced to do it
【万次闪光灯】multitime flash lamp
【万端】multifarious: 变化～ multifarious changes; kaleidoscopic changes / 感慨～ all sorts of feelings welling up in one's mind
【万恶】extremely evil; absolutely vicious: ～的旧社会 the vicious old society / ～滔天 a myriad evils mount to heaven / ～之源 the source

of all evil; the root of all evil
【万方】①(各地) all places ②(多种多样) extremely; incomparably: 仪态～ incomparably graceful
【万分】very much; extremely: ～抱歉 be extremely sorry／～悲痛 be overtaken by heart-rending grief／～感动 be deeply touched／～感谢 thank you very much indeed／～高兴 be very happy; be highly pleased
【万夫不当】even 10000 men are not his match
【万夫之勇】be a host in himself
【万古】through the ages; eternally; forever: ～长存 last forever; be everlasting／～流芳 will be remembered throughout the ages; leave a good name that will live forever
【万古长青】remain fresh forever; be ever-lasting: 祝两国人民的友谊～! May the friendship between our two peoples last forever!
【万花筒】kaleidoscope
【万家灯火】a myriad twinkling lights (of a city); lamps and candles of myriad families
【万劫不复】beyond redemption
【万金油】①(药) a balm for treating headaches, scalds and other minor ailments ②(什么都能做,什么都不擅长的人) Jack of all trades and master of none
【万籁俱寂】all is quiet; silence reigns supreme
【万里长城】the Great Wall
【万里长征】a long march of ten thousand *li*
【万里无云】cloudless
【万绿丛中一点红】a single red in the midst of thick foliage
【万马奔腾】ten thousand horses galloping ahead; going full steam ahead
【万马齐喑】ten thousand horses stand mute
【万难】①(非常难于) extremely difficult; utterly impossible: ～照办 impossible to do as requested／～同意 can by no means agree ②(各种困难) all difficulties: 排除～ surmount all difficulties
【万能】①(无所不能) omnipotent; all-powerful ②(多用途的) universal; all-purpose ◇ ～材料试验机 universal testing machine／～工具机 all-purpose machine／～工作台 universal table／～胶 all-purpose adhesive／～磨床 universal grinder／～润滑脂 multipurpose grease／～拖拉机 multipurpose tractor／～铣床 universal milling machine／～钥匙 master-key／～圆规 universal compass
【万年】ten thousand years; all ages; eternity: 遗臭～ leave a bad name for generations to come
【万年历】perpetual calendar
【万年青】①[植] Japanese rohdea ②(泛指常绿植物) evergreen
【万念俱灰】all hopes dashed to pieces, tired of earthly life
【万千】multifarious; myriad: 变化～ eternally changing; changing all the time／思绪～ myriads of thoughts welling up in one's mind
【万全】perfectly sound; surefire: ～之策 prudential policy／～之计 a completely safe plan; a surefire plan
【万人坑】a large pit used as a common grave; mass grave; a pit of ten thousand corpses
【万人空巷】the whole town turns out
【万世】all ages; generation after generation: ～流芳 leave a good name to posterity／～师表 an exemplary teacher for all ages
【万事】all things; everything: ～不顺。All went awry.／～大吉。Everything is just fine. (或: All's well with the world.)／～亨通。Everything goes well.／～起头难。Everything's hard in the beginning.／～如意。May all your heart's wishes be fulfilled.
【万事俱备,只欠东风】everything is ready, and all that we need is an east wind; all is ready except what is crucial
【万事通】know-all
【万寿无疆】(wish sb.) a long life; many happy returns of the day to eternity
【万水千山】ten thousand crags and torrents; the trials of a long journey
【万死】die ten thousand deaths: ～不辞 willing to risk any danger to do one's duty／罪该～ deserve to die ten thousand deaths
【万岁】①(祝愿千秋万世) long live: 全世界人民大团结～! Long live the great unity of the people of the world! ②(指皇帝) the emperor; Your Majesty (对皇帝的直接称呼); His Majesty (对皇帝的间接称呼)
【万万】①(绝对,用于否定) absolutely; wholly: ～不可粗心大意 must never be negligent／我～没有想到。This idea never occurred to me. ②(数字;亿) hundred million
【万无一失】no danger of anything going wrong; no risk at all; perfectly safe; surefire
【万勿推辞】please do not refuse
【万物】all things on earth
【万向】[机] universal ◇ ～阀 universal valve／～节 universal joint／～联轴节 universal coupling／～轴 cardan shaft

【万象】every phenomenon on earth; all manifestations of nature: ～更新 everything takes on a completely new look; all things appear fresh and gay
【万幸】very lucky; very fortunate; by sheer luck
【万一】①(可能性极小的假设) just in case; if by any chance: ～我忘掉，请提醒我。 In case I forget, please remind me. ②(可能性很小的意外) contingency; eventuality: 防备～ be ready for all eventualities; be prepared for the worst ③(极小的一部分) one ten thousandth; a very small percentage
【万应灵丹】cure-all; panacea
【万用电表】[电] avometer; multimeter
【万有引力】[物] (universal) gravitation ◇～定律 the law of universal gravitation
【万丈】lofty or bottomless: ～高楼 lofty tower / ～光芒 shine with boundless radiance; shine in all its splendor / ～深渊 a bottomless chasm; abyss / 怒火～ a towering rage; a fit of violent anger
【万众】millions of people; the multitude: ～一心 millions of people all of one mind; all of one heart and one mind / 消息传来，～欢腾。 Millions of people rejoiced at the news.
【万状】in the extreme; extremely: 惊恐～ be frightened out of one's senses / 危险～ extremely dangerous
【万紫千红】a riot of color; a blaze of color; innumerable flowers of purple and red: ～才是春。 It is spring only when all the flowers are blooming.

腕

腕 wrist: ～力 wrist strength
【腕厚】wrist thickness
【腕宽】wrist breadth
【腕子】wrist: 手～ wrist
【腕足动物】brachiopod

wāng

汪 ①(液体聚集) collect; accumulate: 雨后路上～了好些水。 Water gathered in the street after the rain. ②[量] puddle; pool: 一～雨水 a puddle of rainwater ③[象] bark; bowwow
【汪汪】①(充满眼泪) tears welling up; tearful: 泪～的 with tearful eyes ②[象] bark; yap; bowwow: 狗～地叫。 A dog is barking.
【汪洋】(of a body of water) vast; boundless: ～大海 boundless ocean / 一片～ a vast expanse of water

wáng

亡 ①(逃跑) flee; run away: 出～ flee; live in exile / 逃～ run away from home ②(丢失;失去) lose; be gone: 唇～齿寒 If the lips are gone, the teeth will feel cold ③(死) die; perish: 阵～ die in battle; fall in battle ④(死去的) deceased: ～妻 deceased wife ⑤(灭亡) conquer; subjugate
【亡故】die; pass away; decease
【亡国】①(使国家灭亡) subjugate a nation; let a state perish: ～灭种 national subjugation and genocide ②(灭亡了的国家) a conquered nation: ～之民 the people of a conquered nation ◇～论者 subjugationist / ～奴 a slave of a foreign power; a slave without a country; a conquered people
【亡灵】the soul of a deceased person; ghost; specter
【亡命】①(逃亡;流亡) flee; seek refuge; go into exile ②(不顾性命) desperate: ～之徒 desperado
【亡羊补牢】mend the fold after a sheep is lost: ～，犹未为晚。 It is not too late to mend the fold even after some of the sheep have been lost.

王 ①(君主) king; monarch: 国～ king ②(大) grand; great: ～父 grandfather
【王八】①(乌龟) tortoise ②[骂] cuckold
【王八蛋】[骂] bastard; son of a bitch
【王朝】①(朝廷) imperial court; royal court ②(朝代) dynasty: 封建～ feudal dynasties
【王储】crown prince
【王道】kingly way; benevolent government
【王法】the law of the land; the law
【王公】princes and dukes; the nobility: ～大臣 princes, dukes and ministers / ～贵族 the nobility
【王宫】(imperial) palace
【王冠】imperial crown; royal crown
【王国】①(国家) kingdom ②(领域) realm; domain: 从必然～到自由～ from the realm of necessity to the realm of freedom / 独立～ independent kingdom; private preserve
【王侯】princes and marquises; the nobility
【王后】queen consort; queen
【王浆】royal jelly
【王牌】trump card ◇～军 *elite* troops; crack units
【王室】①(王族) royal family ②(朝廷) impe-

rial court; royal court
【王水】[化] *aqua regia*
【王孙】prince's descendants; offspring of the nobility
【王位】throne: 继承～ succeed to the throne
【王子】king's son; prince
【王族】persons of royal lineage; imperial kinsmen

wǎng

枉 ①(不合正道的事) crooked: 矫～ straighten sth. crooked; right a wrong ②(加以歪曲) twist; pervert: ～法 pervert the law ③(冤屈) treat unjustly; wrong: ～死 be wronged and driven to death / 冤～ wrong sb. (with false charges, etc.) ④(白白地;徒然) in vain; to no avail; vainly: ～花了许多时间 have spent a lot of time in vain; have wasted a lot of time
【枉法】pervert the law: 贪脏～ take bribes and pervert the law
【枉费】waste; try in vain; be of no avail: ～唇舌 be a mere waste of breath / ～心机 rack one's brains in vain; scheme without avail
【枉己正人】be crooked yet try to set others straight
【枉然】futile; in vain; to no purpose
【枉杀无辜】kill an innocent person unjustly

罔 ①(蒙蔽) deceive: 欺～ deceive; cheat ②(没有;无) no; not: 置若～闻 take no heed of; turn a deaf ear to

惘
【惘然】frustrated; disappointed: ～若失 feel lost

魍
【魍魉】demons and monsters

网 ①(网子或象网子的东西) net: 电～ electrified barbed wire / 发～ hair net / 排球～ volleyball net / 网球～ tennis net / 鱼～ fish net; fishing net / 蜘蛛～ cobweb / 编～ mesh a net / 用～捕鱼 net fish / 在河上张～ net a river ②(系统) network: 公路～ network of highways / 广播～ a network of broadcasting stations / 铁路～ railway network ③(用网捕捉) catch with a net; net: ～着一条鱼 net a fish ④(象网似地笼罩) cover or enclose as with a net: 眼里～着红丝 have bloodshot eyes
【网兜】string bag
【网获量】[渔] haul
【网开一面】leave one side of the net open; give the wrongdoer a way out
【网篮】a basket with netting on top
【网罗】①(网) a net for catching fish or birds; trap ②(搜寻招致) enlist the services of: ～人材 enlist able men
【网络】[电] network; electric network
【网球】①(指项目) tennis: 草地～ lawn tennis / 打～ play tennis ②(指用的球) tennis ball ◇ ～场 tennis court / ～拍 tennis racket ○发球线 service line / 前场 fore court / 后场 back court / 中点 center mark / 发球员 server / 接球员 receiver / 草地网球场 grass court
【网眼】mesh
【网状脉】[植] netted veins; reticulated veins: ～叶 net-veined leaf; reticulate leaf
【网子】①(网) net ②(发网) hairnet

往 ①(到;去) go: 来来～～ coming and going ②(向某处去) in the direction of; toward: ～西走去 go in an westward direction ③(过去的) former; past; previous: ～事 the past
【往常】habitually in the past; as one used to do formerly: 她比～来得晚。 She came later than usual.
【往返】①(来回) go there and back: ～要四个小时。 It takes four hours to go there and back. ②(反复) journey to and fro: ～于京津之间 travel to and fro between Beijing and Tianjin. ◇ ～运费 freight out and home
【往复】move back and forth; reciprocate: 循环～,以至无穷 repeat itself in endless cycles ◇ ～泵 reciprocating pump / ～式发动机 reciprocating engine / ～运动[机] reciprocating motion; alternating motion; advance and return movement
【往还】contact; dealings; intercourse: 经常有书信～ write to each other frequently; keep in contact by correspondence
【往来】①(去和来) come and go ②(互相访问;交际) contact; dealings; intercourse: 贸易～ trade contacts; commercial intercourse / 友好～ exchange of friendly visits; friendly intercourse ◇ ～帐户 current account; running account; open account / ～资产 quick assets
【往年】(in) former years
【往日】(in) former days; (in) bygone days
【往事】past events; the past: 回忆～ recollections of the past
【往往】often; frequently; more often than not:

～如此 it happens frequently that / ～有之 as often happens
【往昔】in the past; in former times

wàng

忘 ①(忘记) forget; escape one's memory: 我已～了这件事。I've forgotten about it. ②(忽略) overlook; neglect: 别～了你的责任。Don't neglect your duty.
【忘本】forget one's class origin; forget one's past suffering
【忘掉】forget; let slip from one's mind: 咱们～那些烦恼的事吧。Let's forget about those worries.
【忘恩负义】devoid of gratitude; ungrateful; be ungrateful and act contrary to justice
【忘乎所以】forget oneself: 心血来潮,～ be carried away by a sudden impulse; lose one's head in a moment of excitement
【忘怀】forget; dismiss from one's mind
【忘记】①(不记得) forget; slip from one's memory: 我～她的名字了。Her name has slipped from my memory. ②(忽略) forget; overlook; neglect; dismiss from one's mind: 不要～浇花儿。Don't neglect to water the plants.
【忘年交】①(指友情) friendship between generations ②(指人) good friends despite great difference in age
【忘情】①(无动于衷) be unruffled by emotion; be unmoved; be indifferent: 不能～ be still emotionally attached ②(不能节制自己的感情) let oneself go: ～地歌唱 let oneself go and sing lustily
【忘却】forget
【忘我】oblivious of oneself; selfless: ～的精神 spirit of selflessness / ～地工作 work selflessly; work untiringly
【忘形】be beside oneself (with glee, etc.); have one's head turned: 得意～ get dizzy with success; have one's head turned by success
【忘性】forgetfulness: ～大 be forgetful; have a poor memory

妄 ①(荒谬) absurd; preposterous: 狂～ wildly arrogant ②(胡乱) presumptuous; rash: ～加评论 make improper comments / ～作主张 make a presumptuous decision
【妄动】rash action; reckless action; ill-considered action: 轻举～ take rash action; take reckless action
【妄加猜测】make wild guesses
【妄加指责】make rash criticism
【妄念】wild fancy; improper thought
【妄求】inappropriate request; presumptuous demand
【妄图】try in vain; vainly attempt: ～掩盖事实真相 vainly attempt to conceal the facts; try in vain to cover up the truth
【妄想】vain hope; wishful thinking ◇ ～狂 nympholepsy; paranoia
【妄语】①(胡说;说假话) tell lies; talk nonsense ②(虚妄的话) wild talk; rant
【妄自菲薄】improperly belittle oneself; unduly humble oneself; underestimate one's own capabilities
【妄自尊大】have too high an opinion of oneself; be overweening; be self-important

望 ①(向远处看) gaze into the distance; look over; look far ahead: 登山远～ ascend a hill and look far ahead ②(探望) call on; visit: 拜～ call to pay one's respects / 看～ call on; visit ③(盼望;希望) hope; expect; look forward to; anticipate: ～回信。Awaiting your reply. / ～早日归来。Expect you to be back soon. ④(名望) reputation; prestige: 德高～重 be of noble character and high prestige ⑤(月相) full moon
【望板】[建] roof boarding
【望尘莫及】so far behind that one can only see the dust of the rider ahead; too far behind to catch up; too inferior to bear comparison
【望穿秋水】gaze with eager expectation; look forward with impatient expectancy
【望而却步】shrink back at the sight of (sth. dangerous or difficult); flinch
【望而生畏】be terrified <awed> by the sight of sb. or sth.; inspire awe even from distance: 令人～ awe-inspiring; forbidding
【望风】be on the lookout; keep watch
【望风而逃】flee at the mere sight of the oncoming force
【望风披靡】flee pell-mell <helter-skelter> at the mere sight of the oncoming force; flee at sight
【望江南】[植] coffee senna
【望楼】watchtower; lookout tower
【望梅止渴】quench one's thirst by thinking of plums; console oneself with false hopes; feed on fancies
【望日】the 15th day of a lunar month
【望文生义】take the words too literally; interpret without real understanding

【望眼欲穿】anxiously gaze till one's eyes are strained; have long been looking forward with eager expectancy; aspire earnestly
【望洋兴叹】lament one's littleness before the vast ocean; bemoan one's inadequacy in the face of a great task; feel powerless and frustrated
【望远镜】telescope: 反射～ reflecting telescope; reflector / 射电～ radio telescope / 双筒～ binoculars; field glasses / 天文～ astronomical telescope / (剧场用)小～ opera glasses / 折反射～ catodioptric telescope / 折射～ refracting telescope; refractor ◇～瞄准器 telescopic sight
【望月】full moon
【望子成龙】hope one's children will have a bright future
【望族】distinguished family; prominent family

旺 prosperous; flourishing; vigorous: 购销两～。Both purchasing and marketing are brisk.
【旺季】peak period; busy season: 西瓜的～ watermelon season
【旺盛】vigorous; exuberant: ～的生命力 exuberant vitality / 精力～的青年 a vigorous youth / 士气～ have high morale
【旺月】busy month (in business)

往 [介] to; toward: ～好处想 think of the better possibilities of a situation / ～坏处想 think of the unfavorable possibilities of a situation / ～南走 go southwards / ～前看 look forward / ～左转弯 turn to the left / 水～低处流。Water always flows downhill.
【往后】from now on; later on; in the future: ～的天气越来越冷了。It's getting colder and colder from now on.

wēi

威 impressive strength; might; power: ～震四方 known far and wide for one's military prowess / 军～ the might of an army; military prowess / 示～ demonstrate one's strength
【威逼】threaten by force; coerce; intimidate: ～利诱 alternate intimidation and bribery; combine threats with inducements
【威风】①(使人敬畏的声势或气派) power and prestige; high prestige backed up with power: 灭敌人的～ puncture 〈deflate〉 the enemy's arrogance ②(有威风) imposing; impressive; awe-inspiring; majestic looking
【威风凛凛】majestic-looking; awe-inspiring; one's manner is in a threatening aspect
【威风扫地】with every shred of one's prestige swept away; completely discredited; with one's dignity in the dust
【威吓】intimidate; threaten; bully: 不顾他人～ in defiance of sb.'s intimidation
【威力】power; formidable force; might: 舆论的～ the tremendous force of public opinion
【威名】fame based on great strength or military exploits; prestige; renown: ～天下扬 one's fame has spread far and wide
【威权】authority; power
【威慑】terrorize with military force; deter ◇～力量 deterrent force; deterrent / ～政策 deterrence policy
【威士忌】whisky
【威势】power and influence
【威望】prestige: 崇高的～ high prestige
【威武】①(权势;武力) might; force; power: ～不能屈 not to be subdued by force ②(力量强大) powerful; mighty: ～雄壮 full of power and grandeur
【威胁】threaten; menace; imperil: ～本地区的安全 threaten the security of this region ◇～利诱 intimidation and bribery; coercion and bribery / ～者 menace
【威信】prestige; popular trust: ～扫地 cause a loss of prestige; completely discredited / 使某人～扫地 make a clean sweep of sb.'s prestige; utterly discredit sb. / 他在群众中～很高。He has high prestige among the masses.
【威严】①(有威力而又严肃) dignified; stately; majestic; awe-inspiring ②(威风) awe; prestige; dignity: 保持～ keep up one's prestige
【威仪】impressive and dignified manner

煨 ①(用微火煮) cook over a slow fire; stew; simmer: ～牛肉 stewed beef ②(放在带火的灰里烤) roast in fresh cinders: ～白薯 roast sweet potatoes in fresh cinders

偎 snuggle up to; lean close to
【偎抱】hug; cuddle
【偎依】snuggle up to; lean close to

微 ①(细小;轻微) minute; tiny: ～火 slow fire / ～云 thin clouds / ～雨 drizzle / 细～ minute; tiny / 相差甚～。The difference is slight. ②(精深) profound; abstruse: 精～ subtle ③(衰落) decline: 衰～ on the decline ④(主单位的百万分之一) one millionth part of; micro-: ～米 micron (μ)

【微安】[电] microampere ◇ ～计 microammeter
【微波】[电子] microwave ◇ ～管 microwave tube / ～理疗机 microwave therapeutic apparatus / ～炉 microwave oven / ～区 microwave region / ～通信 microwave communication / ～遥感 microwave remote sensing / ～遥感器 microwave remote sensor
【微薄】meager; scanty: ～的贡献 little contribution / 收入～ have a meager income / 尽我们～的力量。 We'll exert what little strength we have. (或: We'll do what little we can.)
【微不足道】not worth mentioning; insignificant; inconsiderable; negligible
【微电子】microelectronics
【微法拉】[电] microfarad
【微分】[数] differential: 二项式～ binomial differential ◇ ～法 differentiation / ～分析 differential analysis / ～学 differential calculus
【微风】[气] gentle breeze
【微服出访】make a tour in disguise
【微观】microcosmic ◇ ～世界 microcosmos; microcosm / ～物理学 microphysics / ～现象 [物] microphenomenon
【微观搞活】micro-flexibility
【微乎其微】very little; next to nothing; as trifling as it is
【微积分学】[数] infinitesimal calculus; calculus
【微贱】humble; lowly
【微粒】① (微小的颗粒) particle ② [物] corpuscle ◇ ～回降 [物] fallout / (光的) ～说 corpuscular theory
【微量】trace; micro- ◇ ～分析 microanalysis / ～化学 microchemistry / ～天平 microbalance / ～元素 trace element
【微米】micron (μ)
【微妙】delicate; subtle: ～的处境 delicate situation / 关系～ subtle relations
【微末】trifling; insignificant: ～的成就 an achievement of minor importance / ～的贡献 an insignificant contribution
【微热】[医] low-grade fever
【微弱】faint; feeble; weak: ～的多数 a slender majority / ～的声音 a thin voice / 光线～ faint light; glimmer / 呼吸～ faint breath / 脉搏～ feeble pulse
【微生物】microorganism; microbe ◇ ～农药 microbial pesticide / ～学 microbiology
【微调】[电] fine tuning; trimming ◇ ～电容器 trimmer (condenser); padder
【微微】① (略微; 稍微) slight; faint: ～一笑 smile faintly ② (一万亿分之一) micromicro-; pico-: ～法拉 [物] micromicrofarad; picofarad / ～秒 picosecond
【微细】very small; tiny: ～的区别 fine distinction / ～的血管 very small blood vessels
【微小】small; little: ～的进步 meager progress / ～的希望 slender hopes / 极其～ infinitely small; infinitesimal
【微笑】smile: 甜蜜的～ a sweet smile
【微行】travel incognito
【微型】miniature; mini- ◇ ～公共汽车 minibus / ～化 microminiaturization / ～计算机 personal computer / ～汽车 minicar; mini / ～照相机 miniature camera; minicam / ～组件 [电] micromodule
【微血管】(blood) capillary
【微言大义】sublime words with deep meaning
【微恙】slight illness; indisposition
【微震】② (微小的震动) slight shock ② [地] [物] microseism

逶

【逶迤】winding; meandering: ～的山路 a winding mountain path / ～在群山之中 wind its way through the mountains

巍

towering; lofty
【巍峨】towering; lofty: ～的群山 lofty mountains
【巍然】towering; lofty; majestic; imposing: ～屹立 stand lofty and firm; stand rock-firm
【巍巍】towering; lofty: ～泰山 the towering Mount Taishan

危

① (危险) danger; peril: 居安思～ think of danger in times of peace ② (使处于危险境地; 损害) endanger; imperil: ～及生命 endanger one's life / ～及治安 with public security jeopardized ③ (病人快要死) dying: 病～ be critically ill; be dying ④ (高) high; precipitous: ～楼 a high tower / ～崖 a precipitous cliff ⑤ (端正) proper: 正襟～坐 sit up properly
【危殆】in great danger; in jeopardy; in a critical condition: 病势～ be dangerously ill; be critically ill
【危地马拉】Guatemala ◇ ～人 Guatemalan
【危害】harm; endanger; jeopardize: ～革命 endanger the revolution / ～公共利益 harm the public interest / ～公共卫生 impair public sanitation / ～农作物 harm the crops / ～治安 jeopardize public security ◇ ～程度 extent of

injury / ～公共安全罪 offense against public security / ～社会秩序罪 offense against social order / ～性 harmfulness; perniciousness

【危机】crisis: 经济～ economic crisis / 能源～ energy crisis / ～四伏 beset with crisis; crisis-ridden; danger lurks on every side / ～重重 bogged down in crisis; crisis-ridden

【危急】critical; in imminent danger; in a desperate situation: ～存亡之秋 at the moment of life and death; at the moment of crisis / ～关头 critical juncture; critical time; critical moment / ～之秋 critical time

【危局】a dangerous situation; a critical situation; a desperate situation

【危惧】worry and fear; be apprehensive

【危难】danger and disaster; calamity: 处于～之中 be in dire peril

【危如累卵】as precarious as a pile of eggs; in an extremely precarious situation

【危亡】in peril; at stake: 民族～的时刻 when the nation's existence is in peril; when the fate of the nation hangs in the balance

【危险】dangerous; perilous: 发出～信号 send out a danger signal / 冒生命～ at the risk of one's life / 脱离～ out of danger / 在～中 in danger of / 有电，～! Danger! Electricity! ◇～地带 danger zone / ～品 dangerous articles; dangerous goods / ～人物 a dangerous person; a danger / ～信号 danger signal

【危言耸听】say frightening things just to raise an alarm; exaggerate things just to scare people; use lofty words to excite one to listen

【危在旦夕】on the verge of death or destruction; in imminent danger

wéi

为 ①(做) do; act: 敢作敢～ decisive and bold in action; act with daring / 尽力而～ do one's best; exert one's utmost / 事在人～。Human effort is the decisive factor. ②(充当) act as; serve as: 选他～厂长 elect him director of factory / 以此～凭。This will serve as a proof. / 有诗～证。A poem testifies to that. ③(变成;成) become: 变沙漠～良田 turn the desert into arable land ④(是;合) be; mean; make: 1000公斤～一吨。One thousand kilograms make a ton.

【为非作歹】do evil; commit crimes; perpetrate outrages

【为富不仁】be rich and cruel; be one of the heartless rich

【为难】①(感到难以对付) feel embarrassed; feel awkward: ～的事 an awkward matter / 使人～ embarrass sb.; put sb. in an awkward situation ②(刁难;作对) make things difficult for: 故意～ deliberately make things difficult for sb.

【为期】(to be completed) by a definite date: ～不远。The day is not far off. (或: The day is in sight.) / 会议～三天。The meeting is scheduled to last three days. / 以三个月～ be done within the limit of three months

【为人】behave; conduct oneself: ～正直 be upright / 他～正派。He is a man of decency.

【为生】make a living: 以捕鱼～ make a living as a fisherman

【为时过早】premature; too early; too soon: 现在收割小麦～。It's still too early to harvest the wheat now.

【为首】with sb. as the leader; headed by; led by

【为数】amount to; number: ～不多 have only a small number / ～不少 come up to a large number; amount to quite a lot / ～甚微 amount to very little

【为所欲为】do as one pleases; do whatever one likes; have one's own way

【为伍】associate with: 羞与～ think it beneath one to associate with sb.

【为限】be within the limit of; not exceed; be good for ... only: 有效期以一个月～ be effective within the limit of a month; be good for one month only / 载重以五吨～。The load should be no more than five tons.

【为止】up to; till: 到去年年底～ up to the end of last year / 迄今～ up to now; so far / 今天的工作就到此～。That's all for today. (或: Let's call it a day.)

【为重】attach most importance to: 以大局～ put the general interest first / 以人民的利益～ value the interests of the people above everything else / 以友谊～ value the friendship highly

【为主】give first place to; give priority to; take sth. as the principal thing

韦

【韦伯】[物] weber (Wb)

违 ①(不依从;不遵照) disobey; violate: ～令 disobey orders ②(离别) be separated; part with: 久～了。I haven't seen you for ages.

【违碍】taboo; prohibition
【违拗】disobey; defy
【违背】violate; go against; run counter to: ～历史事实 be contrary to the historical facts / ～良心的合同 unconscionable bargain / ～人民的意志 go against the will of the people / ～条约 violate a treaty / ～原则 violate a principle / ～自己的诺言 go back on one's word
【违法】break the law; be illegal: ～乱纪 malfeasance; violate the law and discipline / ～行为 illegal activities; unlawful practice / 你～了。 You transgressed against the law. / 这是～的。 This is against the law. ◇ ～分子 law-breakers / ～搜查 extralegal search
【违反】violate; run counter to; transgress; infringe: ～党的政策 run counter to the policy of the Party / ～法律 breach of law; contrary to law; disregard of law / ～合同 breach of contract / ～交通规则 violate traffic regulations / ～决议的精神 be contrary to the spirit of the resolution / ～劳动纪律 violate labor discipline / ～诺言 break a promise / ～社会发展规律 go against the laws of social development / ～事实 fly in the face of facts / ～条约 transgress a treaty / ～信托义务 breach of trust / ～刑法 commit a criminal offense
【违犯】violate; infringe; act contrary to: ～纪律 violation of discipline; breach of discipline / ～宪法 violation of the constitution
【违禁】violate a ban ◇ ～品 contraband (goods)
【违抗】disobey; defy: ～命令 disobey orders; act in defiance of orders / ～上级 defy the higher leading body; defy one's superiors; be insubordinate
【违例】[体] breach of rules
【违误】(公文用语) disobey orders and cause delay: 迅速办理,不得～。 This is to be acted upon without delay.
【违宪行为】unconstitutional act
【违心】against one's will; contrary to one's convictions: ～的话 words uttered against one's conscience; obviously insincere talk; statement contrary to one's inner belief
【违约】①(违反条约、契约) break a contract; violate a treaty ②(失约) break one's promise; break off an engagement
【违章】break rules and regulations: ～操作 operate a machine contrary to its instructions / ～行驶 drive against traffic regulations

围 ①(环绕;四周拦挡起来) enclose; surround: ～着他问长问短 gather round him, asking all sorts of questions / ～着桌子坐 sit around a table / 包～ surround; encircle; close in on / 突～ break through an encirclement / 团团～住 completely surround; encircle; besiege ②(四周) all round; around: 四～都是花。 There are flowers all round.
【围城】①(包围城市) encircle a city; besiege a city ②(被围城市) besieged city
【围城打援】besiege a city to annihilate the enemy relief force
【围攻】①(攻击) besiege; lay siege to: ～要塞 lay siege to a fortress / 停止～ abandon a siege ②(用言语围攻) jointly speak or write against sb.; jointly attack sb.: 遭到～ come under attack from all sides; be caught in a cross fire
【围海造田】reclaim land from the sea
【围歼】surround and annihilate
【围剿】encircle and suppress
【围巾】muffler; scarf: 围上～ put a scarf round the neck
【围垦】enclose tideland for cultivation; (build dikes to) reclaim land from marshes
【围困】besiege; hem in; pin down
【围拢】crowd around
【围墙】enclosure; enclosing wall
【围裙】apron
【围绕】①(围着转动) round; around: 地球～太阳转。 The earth revolves round the sun. ②(以…为中心) center on; revolve round: ～中心任务安排其它工作 arrange other work around the central task
【围网】purse seine; purse net ◇ ～渔船 purse seiner; purse boat
【围魏救赵】relieve the besieged by besieging the base of the besiegers
【围岩】[矿] country rock; surrounding rock
【围堰】cofferdam; coffer
【围子】①(围绕村庄的障碍物) defensive wall; stockade surrounding a village: 土～ fortified village ②(帷子) curtain
【围嘴儿】bib

圩 dike; embankment: 筑～ build dikes
【圩田】low-lying paddy fields surrounded with dikes
【圩垸】protective embankments in lakeside areas

【圩子】①(防水护田堤岸) protective embankments surrounding low-lying fields ②(围绕村庄的障碍物) defensive wall; stockade surrounding a village

惟 ①(单单) only; alone: ～你是问。You'll be held personally responsible. ②(只是) but; only that ③(思想) thinking; thought: 思～ thinking

【惟独】only; alone: 她心里装着同志们,～没有她自己。She always bears her comrades in mind, scarcely does she ever think of herself.

【惟恐】for fear that; lest: ～落后 for fear that one should lag behind / ～天下不乱 desire to see the world plunged into chaos; be anxious to stir up trouble; desire to stir up trouble

【惟利是图】be bent solely on profit; put profit-making first; interested only in material gain

【惟命是从】always do as one is told; be absolutely obedient; receive absolutely one's instruction

【惟命是听】always do as one is told; be absolutely obedient; receive absolutely one's instruction

【惟我独尊】overweening; extremely conceited; no one is noble but me

【惟一】only; sole; unique: ～出路 the only way out / ～的理由 the sole reason / ～合法的政府 the sole legitimate government / ～可行的办法 the only feasible way / 实践是检验真理的～标准。Practice is the sole criterion of truth.

【惟有】only; alone: ～这样做,试验才能成功。Only in this way can the experiment succeed.

唯 only; alone

【唯成分论】the theory of the unique importance of class origin

【唯理论】[哲] rationalism

【唯美主义】aestheticism

【唯名论】[哲] nominalism

【唯能说】[物] energetics

【唯我主义】[哲] solipsism

【唯武器论】the theory that weapons alone decide the outcome of war

【唯物辩证法】[哲] materialist dialectics

【唯物论】[哲] materialism: ～的反映论 materialist theory of reflection

【唯物史观】[哲] materialist conception of history; historical materialism

【唯物主义】[哲] materialism: 辩证～ dialectical materialism / 机械～ mechanical materialism / 历史～ historical materialism / 庸俗～ vulgar materialism ◇ ～者 materialist

【唯心论】[哲] idealism

【唯心史观】[哲] idealist conception of history; historical idealism

【唯心主义】[哲] idealism: ～先验论 idealist apriorism

帷 curtain

【帷幕】【帷幔】heavy curtain

【帷幄】army tent: 运筹～ devise strategies within a command tent

【帷子】curtain: 床～ bed-curtain

维 ①(连接) tie up; hold together: ～系 hold together; maintain ②(保持;保全) maintain; safeguard; preserve: ～护 safeguard; defend; uphold ③(思想) thinking; thought: 逻辑思～ logical thinking ④[数] dimension: 三～空间 three-dimensional space

【维持】keep; maintain; preserve: ～秩序 keep order; maintain order / ～生活 support oneself or one's family / ～现状 maintain the *status quo*; let things go on as they are / ～原判 affirm the original judgement / ～治安 keep the peace; maintenance of peace ◇ ～费 maintenance / ～关税 preserving duties

【维护】safeguard; defend; uphold: ～法律面前人人平等的原则 uphold the principle that all men are equal before the law / ～国家主权 defend state sovereignty / ～民族尊严 vindicate national honor; defend national honor / ～人民的利益 safeguard the people's interests / ～团结 uphold unity

【维纶】[纺] polyvinyl alcohol fiber

【维棉】[纺] vinylon and cotton blend

【维妙维肖】remarkably true to life; absolutely lifelike

【维尼纶】vinylon

【维生素】vitamin: 复合～B vitamin B complex ◇ ～缺乏症 vitamin-deficiency; avitaminosis

【维数】[数] dimension; dimensionality ◇ ～论 dimension theory

【维系】hold together; maintain: ～人心 maintain popular morale

【维新】reform; modernization: 日本明治～ the Meiji Reformation of Japan (1868) / 戊戌～ the Constitutional Reform and Modernization of 1898

【维修】keep in (good) repair; service; maintain:

～得很好 be in good repair / ～房屋 maintain houses and buildings / ～汽车 service a car / 设备～ maintenance of equipment; upkeep of equipment ◇ ～费 upkeep; maintenance cost / ～工 maintenance worker / ～手册 servicing manual

桅 mast: ～顶 masthead / 船～ mast

【桅灯】①[航海] mast head light; range light ②(马灯) barn lantern

【桅杆】mast

【桅樯】mast

wěi

伪 ①(虚假的) false; fake; bogus: ～钞 counterfeit bank note; forged bank note / ～科学 pseudoscience / ～证 false witness / 去～存真 eliminate the false and retain the true ②(不合法的) puppet; collaborationist: ～政府 illegitimate government / ～政权 puppet regime

【伪币】①(伪造的) counterfeit money; counterfeit bank note; forged bank note; spurious coin ②(伪政府发行的) money issued by a puppet government

【伪军】①(指军队) puppet army; puppet troops ②(指士兵) puppet soldier

【伪君子】hypocrite

【伪善】hypocritical: ～的言词 hypocritical words ◇ ～者 hypocrite

【伪书】ancient books found to have been incorrectly dated, forged, or attributed to a wrong author; ancient books of dubious authenticity

【伪托】forge ancient literary or art works, or pass off modern works as ancient ones

【伪造】forge; falsify; fabricate; counterfeit: ～的文件 spurious document; fake document; forged document; pseudograph / ～公章 falsification of public seal / ～历史 fabricate history; falsify history / ～签名 forge a signature / ～现场 simulated scene / ～证件 forge a certificate / ～帐目 falsify accounts ◇ ～钞票罪 offense of counterfeiting bank notes / ～品 counterfeit; forgery / ～罪 forgery

【伪证】[法] perjury; false witness: 犯有～罪 be guilty of perjury / 作～ bear false witness ◇ ～人 perjured witness / ～罪 offense of false evidence

【伪装】①(假装) pretend; feign: ～进步 pretend to be progressive / ～中立 feign neutrality ②(假的装扮) disguise; guise; mask: 剥去～。Strip off the mask. ③[军] camouflage ◇ ～工事 camouflage works / ～网 camouflage net; garnished net

【伪足】[动] pseudopodium

苇 reed: 芦～ reed

【苇箔】reed matting

【苇塘】reed pond

【苇席】reed mat

【苇子】reed

伟 big; great: ～力 mighty force / 身体魁～ tall and broad-shouldered; gigantic in stature; stalwart / 雄～ magnificent

【伟大】great; mighty: / ～的胜利 a signal victory / ～的事业 a great undertaking / ～的作家 a great writer

【伟绩】great feats; great exploits; brilliant achievements: 丰功～ great achievements and abundant merits

【伟人】a great man; a great personage: 当代的～ a great man of our time

【伟业】great cause; exploit

纬 ①(横纱或横线) weft; woof ②[地] latitude: 北～四十度 forty degrees north latitude; forty degrees of latitude north (of the equator)

【纬编】[纺] weft knitting ◇ ～针织物 weft-knitted fabric

【纬度】[地] latitude: 低～ low latitudes / 高～ high latitudes

【纬纱】[纺] woof

【纬线】①[地] parallel ②[纺] weft

鲔 yaito tuna

唯 yea

【唯唯诺诺】be a yes-man; be obsequious

委 ①(把事情交给别人去办) entrust; appoint: ～以重任 entrust sb. with an important task; give one the bag to hold ②(抛弃) throw away; cast aside: ～弃 discard / ～之于地 cast sth. upon the ground ③(推萎) shift: ～过于人 put the blame on sb. else; lay the blame on others ④(曲折) indirect; roundabout: ～婉 mild and roundabout; tactful ⑤(末尾) end: 原～ the beginning and the end ⑥(不振作;无精打采) listless; dejected: ～靡 listless; dispirited ⑦(的确;确实) actually; certainly: ～系实情。This is really the case. ⑧(委员) committee member: 常～ member of a standing committee

⑨(委员会) committee; commission; council: 党～ Party committee／纪～ commission for inspecting discipline
【委顿】tired; exhausted; weary; fatigued
【委靡】listless; dispirited; dejected: ～不振 dispirited; in low spirits; dejected and apathetic／精神～ listless; dispirited and inert
【委内瑞拉】Venezuela ◇ ～人 Venezuelan
【委派】appoint; delegate; designate
【委曲】(of roads, rivers, etc.) winding; tortuous ◇ ～之词 an indirect statement
【委曲求全】compromise out of consideration for the general interest; stoop to compromise
【委屈】①(受冤屈) feel wronged; nurse a grievance: 诉～ pour out one's grievances 〈troubles〉 ②(使人受到冤屈) put sb. to great inconvenience: 对不起,～你了。Sorry to have made you go through all this.(或: Sorry to have trouble you with all such inconvenience.)
【委任】appoint: ～某人担任某职 appoint sb. to an office ◇ ～代理 agency by mandate; deputation／～书 certificate of appointment／～统治 mandate／～统治地 mandated territory／～行为 act of commission／～证书 certificate of appointment／～状 certificate of appointment
【委实】really; indeed: ～使我吃惊! That really astonishes me!
【委托】entrust; trust: ～某人担负这项工作 entrust sb. with the work／他把财产～给他的朋友看管。He entrusted his friend with the property.／我把事情～给一个有经验的律师。I trust my affairs to an experienced lawyer. (或: I trust an experienced lawyer with my affairs.)／这事就～你了。I leave this matter in your hands. ◇ ～保险 trust insurance／～代理 agency by agreement／～机构 associated agency／～人 client; commissaries; consignor／～商店 commission shop; commission house／～书 certificate of entrustment; commission; letter of attorney
【委婉】mild and roundabout; tactful: ～的语气 a mild tone
【委员】committee member; member of a committee: 中央～ member of the Central Committee ◇ ～长 chairman of a committee
【委员会】committee; commission; council: 常务～ permanent committee／筹备～ preparatory committee／调查～ committee of enquiry／国家计划～ State Planning Commission／纪律检查～ commission for inspecting discipline／接待～ reception committee／联络～ liaison committee／临时～ interim committee／起草～ drafting committee／特别～ ad hoc committee／执行～ executive board／中央～ Central Committee／专门～ special committee／组织～ organizing committee
【委罪】put the blame on sb. else

萎 wither; wilt; fade
【萎黄病】chlorosis
【萎蔫】[植] wilting
【萎缩】①(干枯) wither; shrivel; shrink ②(经济衰退) shrink; sag; contraction ③[医] atrophy: 肝～ hepatatrophy; atrophy of the liver／肌肉～ amyotrophy; muscular atrophy
【萎陷疗法】collapse therapy
【萎谢】wither; fade: ～的草木 withered plants

猥 ①(多;杂) numerous; multifarious: ～杂 miscellaneous ②(卑鄙;下流) base; obscene; salacious; indecent
【猥词】obscene language; salacious words
【猥琐】of wretched appearance; of dreadful appearance
【猥亵】①(下流言行;淫乱) obscene; salacious ②(做下流动作) act indecently towards (a woman) ◇ ～刊物 obscene publication／～行为 act of indecency; indecent conduct／～幼童 carnal abuse／～罪 crime of indecent act; indecent assault

尾 ①(尾巴) tail: 牛～ ox-tail／摇～ wag tail ②(末端) end: 从头到～ from beginning to end／排～ a person standing a the end of a line／首～相连 join end to end ③(未了结的事物) remaining part; remnant: 扫～工程 the final phase of a project
【尾巴】①(直义为动物的尾巴;转义为象征性的尾巴) tail: ～翘上了天 be very cocky／夹起～逃跑 run away with one's tail between one's legs／夹着～做人 behave oneself tuck one's tail between one's legs; pull one's head in ②(尾部) tail-like part: 飞机～ the tail of a plane／彗星～ the tail of a comet ③(无主见、完全随声附和的人) servile adherent; appendage ④(尾随者) a person shadowing sb.: 甩掉～ throw off one's tail
【尾巴主义】tailism
【尾大不掉】①(部下强大,不听指挥) leadership rendered ineffectual by recalcitrant subordinates ②(机构庞大,指挥不灵) too cumber-

some to be effective
【尾灯】tail light; tail lamp
【尾骨】[生理] coccyx
【尾鳍】tail fin; caudal fin
【尾气】[化] tail gas
【尾欠】①(有一小部分未偿还或未交纳) owe a small balance ②(未偿还或未交纳的小部分) balance due
【尾声】①[乐] coda ②(与序幕相对) epilogue: 序幕和～ prologue and epilogue ③(临近结尾) end: 接近～ be drawing to an end
【尾数】odd amount in addition to the round number
【尾随】①(跟在后面) tail behind; tag along after; follow at sb.'s heels; at the tail of: 有些孩子～在游行队伍之后。 Some boys tailed after the parade. ②(跟踪监视) tail; shadow: ～形迹可疑者 tail a suspect; shadow a suspect
【尾须】[动] cercus
【尾翼】[航空] tail surface; empennage
【尾蚴】[动] cercaria
【尾追】in hot pursuit; hot on the trail of

娓

【娓娓】(talk) tirelessly: ～不倦 talk tirelessly / ～动听 speak with absorbing interest; sound pleasing and attractive / ～而谈 talk volubly; talk effusively

wèi

为 ①(给;替) for; for the benefit of; in the interests of: ～顾客着想 think about the interests of the customers / ～他买车票 buy him a train ticket ②(为了) for; for the sake of: ～方便起见 for the sake of convenience / ～革命工作 work for the revolution / ～国增光 bring credit to one's country / 不～名,不～利 seek no personal fame or gain / ～她举行一次晚会。 A party was given for her. ③(因为) because; for; on account of: ～胜利而欢呼 hail a victory; cheer over the victory / ～救那女孩的生命他受到奖励。 He was rewarded for saving the girl's life.
【为此】to this end; for this reason; for this purpose; in this connection
【为国捐躯】lay down one's life for one's country
【为何】why; for what reason
【为虎傅翼】give wings to a tiger; assist an evildoer; add to the influence of a villain
【为虎作伥】play the jackal to the tiger; help a villain do evil
【为了】for; for the sake of; in order to: ～实现现代化 for the sake of modernization; in order to realize modernization
【为民除害】rid the people of an evil
【为民请命】plead in the name of the people; plead for the people
【为人作嫁】sew sb. else's trousseau; doing work for others with no benefit to oneself; toil just for the benefit of others
【为什么】why; why is it that; how is it that: ～不试一试? Why not have a try? / ～犹豫不决? Why hesitate?
【为我之物】[哲] thing-for-us
【为渊驱鱼】drive the fish into deep waters; drive one's friends to the side of the enemy: ～,为丛驱雀 drive the fish into deep waters and the sparrows into the thickets

未 ①(没) have not; did not; not yet: 意犹～尽 have not given full expression to one's views / 日期～定。 The date is not yet fixed. ②(不) not; no: ～见好转。 It isn't any better. / ～知可否 not know whether sth. can be done
【未爆弹】[军] dud
【未必】may not; not necessarily; not sure: 她～来。 She may not come.
【未便】not be in a position to; find it hard to: ～立即答复 find it difficult to give an immediate reply / ～擅自处理 cannot do it without authorization
【未卜先知】foresee; have foresight; be able to foresee the future
【未曾】have not; did not: ～听说过 never heard of it / 工业上～有过的发展 an unprecedented development in industry
【未尝】①(未曾) have not; did not ②[委婉语] ～没有可取之处 not without its merits / 那也～不可。 That should be all right.
【未成年】not yet of age; under age; immaturity ◇ ～子女 minor children / ～罪犯 minor offender
【未定】uncertain; undecided; undefined: 价格～。 The price is not yet fixed. ◇ ～稿 draft / ～界 undefined boundary; undemarcated boundary
【未分利益】undivided profit; undistributed profit
【未付】outstanding; unpaid: ～费用 outstanding expenses; unpaid expenses / ～股利 unpaid

dividend

【未敢苟同】beg to differ; cannot agree

【未婚】unmarried; single ◇ ～夫 fiancé / ～妻 fiancée

【未奸入】[法] non-penetration

【未决】unsettled; outstanding: 悬而～的问题 an outstanding issue; an open question; a pending question ◇ ～犯 prisoner awaiting trial; culprit; unconvicted prisoner

【未可】cannot: ～乐观 give no cause for optimism; nothing to be optimistic about / ～预卜 cannot foretell / 前途～限量 have a brilliant future

【未可厚非】be not altogether inexcusable; give no cause for much criticism

【未来】①(将要到来的) coming; approaching; next; future: ～的时代 future ages / ～的一年 the coming year; next year ②(将来) future; tomorrow: 美好的～ a glorious future; a bright future

【未来派】futurism

【未来学】futurology

【未老先衰】prematurely senile; old before one's time

【未了】unfinished; outstanding: ～的手续 formalities still to be complied with / ～的心愿 an unfulfilled wish / ～的债务 outstanding debts / ～事宜 unfinished business

【未免】rather; a bit too; truly: ～生硬了些 be rather harsh / ～太过分了 go a bit too far

【未能】fail to; cannot; haven't been able to: ～按时交货 fail to deliver as scheduled / ～实现 fail to materialize

【未能免俗】be unable to rise above the conventions; cannot but follow conventional practice

【未实现利润】unrealized profits

【未遂】not accomplished; abortive: 自杀～ an attempted suicide / 政变～。The coup d'état aborted. ◇ ～罪[法] attempted crime; attempt

【未完】unfinished: ～待续 to be continued

【未详】unknown: 本文作者～。The author of this article is unknown. / 病因～。What brought on the illness is not clear.

【未雨绸缪】repair the house before it rains; provide for a rainy day; take precautions

【未知量】[数] unknown quantity

【未知数】①[数] unknown number ②(未知事物) unknown; uncertain

味 ①(滋味) taste; flavor: 酒～儿 flavor of wine / 苦～儿 a bitter taste / 甜～儿 a sweet taste / 我伤风了，尝不出～儿来。I have a bad cold and can't taste. ②(气味) smell; odor: 臭～儿 an offensive smell; a foul smell; stench; stink / 焦～儿 a burnt smell / 香～儿 a sweet smell; fragrance; aroma ③(趣味) interest: 枯燥无～ dry and insipid / 语言无～ insipid language; colorless language ④(辨别味道) distinguish the flavor of: 玩～ ponder; ruminate / 细～其言 ponder his words ⑤[量] ingredient: 六～药 six medicinal herbs

【味道】taste; flavor: ～很好 delicious; taste nice / 没有～ tasteless / 心里有一股说不出的～ have an indescribable feeling / 让他尝尝鞭子的～。Give him a taste of the whip.

【味精】monosodium glutamate; gourmet powder

【味觉】sense of taste: 她的～非常灵敏。Her tastes is unusually keen.

【味蕾】[生理] taste bud

【味同嚼蜡】it is like chewing wax; insipid

畏 ①(畏惧) fear: 不～艰险 fear no hardship or danger / 不～强敌 stand in no fear of a formidable enemy / 大无～ fearless; dauntless ②(佩服) respect: 后生可～。Youth are to be regarded with respect.

【畏避】avoid sth. out of fear; recoil from; flinch from

【畏光】[医] photophobia

【畏忌】have scruples; fear; dread

【畏惧】fear; dread: 无所～ be fearless

【畏难】be afraid of difficulty: ～情绪 fear of difficulty

【畏怯】cowardly; timid; chickenhearted

【畏首畏尾】be full of misgivings; be overcautious

【畏缩】recoil; shrink; flinch: ～不前 shrink with fear; hesitate to press forward; hang back / 在困难面前从不～ never flinch from difficulty

【畏途】a dangerous road; a perilous undertaking: 视为～ regard it as a dangerous road to take; be afraid to undertake it

【畏罪】dread punishment for one's crime: ～潜逃 abscond to avoid punishment / ～自杀 commit suicide to escape punishment

喂 ①[叹](招呼的声音) hello; hey: ～，你上哪儿去? Hey, where are you going? ②(给东西吃) feed: ～猪 feed pigs / 给病人～饭 feed a patient / 给婴儿～奶 feed the baby on milk

【喂奶】breast-feed; suckle; nurse
【喂养】feed; raise; keep: ～家禽 keep fowls

胃 stomach
【胃癌】cancer of the stomach; gastric carcinoma
【胃病】stomach trouble; gastric disease
【胃肠炎】gastroenteritis
【胃穿孔】gastric perforation
【胃出血】gastrorrhagia; gastric hemorrhage
【胃蛋白酶】[生理] pepsin
【胃镜】gastroscope ◇ ～检查 gastrocopy
【胃口】①(食欲) appetite: ～不好 have poor appetite / ～好 have a good appetite / 没有～ have no appetite ②(兴趣) liking: 对～ to one's liking
【胃溃疡】gastric ulcer
【胃扩张】dilatation of the stomach; gastrectasis
【胃切除术】gastrectomy
【胃十二指肠溃疡】gastroduodenal ulcer
【胃舒平】gastropine
【胃酸】hydrochloric acid in gastric juice: ～过多 hyperchlorhydria; hyperacidity / ～过少 hypochlorhydria; hypoacidity
【胃痛】stomachache; gastralgia
【胃下垂】ptosis of the stomach; gastroptosis
【胃腺】gastric gland
【胃炎】gastritis
【胃液】gastric juice

谓 ①(说) say: 或～ someone says / 可～神速 may well be termed lightning speed ②(称呼; 叫做) call; name: 何～人造卫星? What is meant by man-made satellite? / 所～ so-called ③(意思;意义) meaning; sense: 无～的话 senseless talk; twaddle
【谓语】[语] predicate

位 ①(地点) place; location; site: 座～ seat ②(地位;职位) position: 名～ fame and position ③(王位) throne: 篡～ usurp the throne / 即～ come to the throne ④[数] place; figure; digit: 个～ unit's place / 十～ ten's place / 五～数 five-figure number; five-digit number / 小数～ decimal place / 计算到小数点后三～ calculate to three decimal places ⑤ [量] 几～朋友 some friends / 各～代表! Fellow Delegates!
【位卑言高】in humble station with high talk
【位次】precedence; seating arrangement ◇ ～卡 place card
【位能】[物] potential energy
【位势米】[气] geopotential meter
【位移】[物] displacement
【位于】be located; be situated; lie: ～太平洋沿岸 be situated on the Pacific coast / ～中国东部 be situated in the eastern part of China
【位置】①(所在地方) seat; place; site; location ②(地位) place; position: 在历史上占有重要～ occupy an important place in history
【位子】seat; place

尉
【尉官】junior officer ○上尉 Captain; Lieutenant(海军); Flight Lieutenant(英国空军) / 中尉 Lieutenant; Sub-Lieutenant(英国海军); Lieutenant Junior Grade(美国海军); Flying Officer(英国空军); First Lieutenant(美国空军, 美国海军陆战队) / 少尉 Second Lieutenant; Acting Sub-Lieutenant(英国海军); Ensign(美国海军); Pilot Officer(英国空军) / 准尉 Warrant Officer; Chief Warrant Officer(美国陆军, 美国空军)

慰 ①(安慰) console; comfort: ～勉 comfort and encourage ②(宽慰) be relieved; feel
【慰劳】bring gifts to, or send one's best wishes to, in recognitions of services rendered: ～解放军 bring gifts and greetings to the Chinese People's Liberation Army
【慰问】express sympathy and solicitude for; extend one's regards to; convey greetings to: ～灾区人民 express sympathy and solicitude for the people of disaster areas / 表示亲切～ express one's sincere solicitude ◇ ～袋 gift bag / ～团 a group sent to convey greetings and appreciation / ～信 a letter expressing one's appreciation or sympathy / ～演出 a special performance as an expression of gratitude or appreciation
【慰唁】condole with sb.

蔚 ①(茂盛;盛大) luxuriant; grand ②(有文采) colorful: 云蒸霞～。The rosy clouds are slowly rising.
【蔚蓝】azure; sky blue: ～的大海 the blue sea / ～的天空 a bright blue sky
【蔚然成风】become common practice; become the order of the day: 学习外语～。Studying foreign languages is the order of the day.
【蔚为大观】present a splendid sight; afford a magnificent view

卫 defend; guard; protect: 保家～国 protect our homes and defend our country / 自～ self-defense
【卫兵】guard; bodyguard
【卫道】defend traditional moral principles ◇ ～士 apologist
【卫队】squad of bodyguards; armed escort ◇ ～长 captain of the guard
【卫护】protect; guard
【卫矛】[植] winged euonymus
【卫生】hygiene; health; sanitation: 个人～ personal hygiene / 工业～ industrial hygiene / 公共～ public health / 环境～ environmental sanitation / 讲～ pay attention to hygiene / 劳动～ labor hygiene ◇ ～带 sanitary towel; sanitary napkin / ～队 medical unit; medical team / ～防疫站 sanitation and antiepidemic station / ～环境 hygienic conditions / ～局 public health bureau / ～科 health section / ～设备 sanitary equipment / ～室 clinic / ～学 hygiene; hygienics / ～员 health worker; medical orderly; medic / ～知识 hygienic knowledge
【卫生间】toilet
【卫生裤】sweat pants
【卫生球】camphor ball; mothball
【卫生衣】sweat shirt
【卫生纸】toilet paper
【卫士】bodyguard
【卫戍】garrison: 北京～区 the Beijing Garrison Command ◇ ～部队 garrison force
【卫星】①(天体) satellite; moon ②(人造卫星) artificial satellite; man-made satellite: 气象～ weather satellite; meteorological satellite / 通讯～ communications satellite / 发射～ launch an artificial satellite ◇ ～城 satellite town / ～国 satellite state; satellite country

wēn

温 ①(不冷不热) warm; lukewarm; tepid: ～水 lukewarm water ②(温度) temperature: 气～ atmospheric temperature / 体～ temperature (of the body) ③(稍微加温) warm up; heat up: 把酒～一下 warm up the wine ④(温习) review; revise: ～课 review one's lessons; revise one's lessons
【温饱】dress warmly and eat one's fill ◇ ～型 simply having adequate food and clothing
【温标】[物] thermometric scale: 华氏～ Fahrenheit's thermometric scale / 开氏～ Kelvin's thermometric scale / 摄氏～ Celsius' thermometric scale
【温差】difference in temperature; range of temperature
【温差电】[物] thermoelectricity ◇ ～堆 thermopile / ～检波器 thermodetector / ～偶 thermoelectric couple; thermocouple
【温床】①[农] hotbed ②(利于坏事发生、发展的环境) breeding ground; hotbed
【温存】①(殷勤抚慰) be attentive to; give tender attentions to ②(温柔体贴) affectionate; kind; gentle; tender
【温带】temperate zone: 北～ the north temperate zone / 南～ the south temperate zone
【温度】temperature: 室内～ indoor temperature / 室外～ outdoor temperature / 最低～ minimum temperature / 最高～ maximum temperature
【温度计】[气] thermograph
【温度表】thermometer: 华氏～ Fahrenheit thermometer / 摄氏～ centigrade thermometer; Celsius thermometer
【温故知新】gain new insights through restudying old material; reviewing the past helps one to understand the present; learn new things by reviewing old things
【温和】①(气候不冷不热) temperate; mild; moderate: 气候～ a temperate climate ②(平和) gentle; mild: 态度～ a kindly manner / 性情～ a gentle disposition / 语气～ a mild tone ③(不冷不热) lukewarm; warm: 粥还～呢。The porridge is still warm. ◇ ～派 moderates
【温厚】gentle and kind; good-natured
【温觉】[生理]sense of heat
【温暖】①(暖和) warm: ～的气候 warm climate / 天气～ warm weather ②(使感到温暖) warmth; kindness: 感到～ feel the kindness; sense the warmth
【温情】①(温柔的感情) tender feeling: ～脉脉 full of tender feeling ②(温和的态度) too softhearted ◇ ～主义 undue leniency; excessive tenderheartedness
【温泉】hot spring
【温柔】gentle and soft: ～敦厚 tender and gentle; gentle and kind
【温湿计】hygrothermograph; thermohygrograph
【温室】hothouse; greenhouse; glasshouse; conservatory: ～育苗 nurse young plants in hothouses ◇ ～效应 greenhouse effect
【温顺】docile; meek

【温文尔雅】gentle and cultivated; refined and cultivated
【温习】review; revise: ～功课 review one's lessons
【温血动物】warm-blooded animal
【温驯】(of animals) docile; meek; tame

瘟 acute communicable diseases
【瘟病】seasonal febrile diseases
【瘟神】god of plague
【瘟疫】pestilence

wén

文 ①(字) character; script; writing: 甲骨～ inscriptions on bones or tortoise shells / 中～ the Chinese language ②(文字) language: 英～ the English language ③(文章) literary composition; writing ④(文言) literary language: 半～半白 half literary and half vernacular ⑤(非军事的) civilian; civil: ～职 civilian post ⑥(自然界某些现象) certain natural phenomena: 水～ hydrology / 天～ astronomy ⑦(掩饰) cover up; paint over: ～过饰非 conceal faults and gloss over wrongs ⑧[量]: 一～不名 penniless / 一～不值 not worth a farthing
【文本】text; version: 《圣经》的权威～ the authorized version of the Bible / 本合同两种～同等有效。 Both texts of the contract are equally valid.
【文笔】style of writing: ～简洁 concise style; laconic style / ～流畅 write in an easy and fluent style; wield a facile pen
【文不对题】irrelevant to the subject; beside the point; wide of the mark
【文不加点】never blot a line in writing; have a facile pen
【文才】literary talent; aptitude for writing
【文采】①(华丽的色彩) rich and bright colors ②(文艺才华) literary grace; literary talent
【文昌鱼】lancelet
【文抄公】plagiarist
【文辞】diction; language: ～优美 exquisite diction; elegant language
【文从字顺】readable and fluent
【文牍】official documents and correspondence ◇～主义 red tape
【文法】grammar
【文房四宝】the four treasures of the study ○笔 writing brush / 墨 ink stick / 砚 ink slab / 纸 paper
【文风】style of writing: 整顿～ rectify the style of writing
【文风不动】absolutely still
【文稿】manuscript; draft
【文告】proclamation; statement; message
【文蛤】[动] clam
【文工团】song and dance ensemble; art troupe; cultural troupe
【文官】①(文职官员) civil official ②(总称) civil service
【文冠果】[植]shiny-leaved yellowhorn
【文过饰非】conceal faults and gloss over wrongs; gloss over one's faults; cover up one's errors; cover up and whitewash mistakes
【文豪】literary giant; great writer; eminent writer
【文化】①(精神财富) civilization; culture: 大众～ mass culture / 反～ counterculture / 非物质～ nonmaterial culture / 物质～ material culture / 希腊～ Greek culture / 亚～ subculture ②(知识) education; culture; schooling; literacy: 学～ acquire an elementary education; acquire literacy; learn to read and write ◇～参赞 cultural counsellor; cultural attaché / ～程度 educational level / ～迟滞 cultural lag / ～大革命 the Great Cultural Revolution / ～多元论 cultural pluralism / ～发源地 cultural hearth / ～宫 palace of culture; cultural palace / ～馆(站) cultural center / ～机关 cultural institution / ～交流 cultural exchange / ～教育事业 cultural and educational work / ～界 cultural circles / ～进化论 cultural evolutionism / ～决定论 cultural determinism / ～课 literacy class; general knowledge course / ～类型 culture type / ～落后 cultural lag / ～模式 culture pattern / ～飘移 cultural drift / ～平行 cultural parallelism / ～侵略 cultural aggression; cultural penetration / ～区 culture area / ～趋同 cultural convergence / ～人 cultural worker; intellectual / ～渗透 cultural infiltration / ～事业 cultural establishments; cultural undertakings / ～水平 cultural level; educational level / ～素养 artistic appreciation / ～停滞 cultural lag / ～相对论 cultural relativism / ～学 culturology / ～熏染 enculturation / ～遗产 cultural heritage; cultural legacy / ～遗址 a site of ancient cultural remains; remains of an ancient culture / ～用品 stationery / ～站 cultural center / ～中心 cultural center; culture center / ～专制主义 cultural tyranny

【文火】slow fire; gentle heat: ～焖四十分钟 simmer gently for forty minutes
【文集】collected works
【文件】①(公文、信件等) documents; papers; instruments: 机密～ classified documents／正式～ official documents ② [自] file ◇ ～备份 file backup／～编号 the reference or serial number of a document／～编辑 file edit／～查阅 file reference／～处理 file processing／～处理机 file processor／～存储 filestore／～袋 documents pouch; dispatch case／～段 file section／～格式 file layout／～柜 filing cabinet／～夹 file／～名称 file name／～识别 file identification／～转换 file conversion
【文教】[简](文化教育) culture and education ◇～界 cultural and educational circles／～事业 cultural and educational work; culture and education
【文静】gentle and quiet
【文具】writing materials; stationery ◇～店 stationer's; stationery shop
【文科】liberal arts ◇ ～学校 liberal arts school／～院校 colleges of arts
【文库】a series of books issued in a single format by a publisher; library
【文理】unity and coherence in writing: ～不通 be illogical and ungrammatical／～通顺 have unity and coherence; make smooth reading
【文盲】an illiterate person: 扫除～ wipe out illiteracy
【文明】①(文化) civilization; culture: ～古国 a country with an ancient civilization／精神～ spiritual civilization／物质～ material civilization ②(具有较高文化的) civilized: ～社会 civilized society
【文明礼貌月】Civic Virtues Month
【文墨】writing: 粗通～ barely know the rudiments of writing
【文鸟】[动] mannikin
【文痞】literary prostitute
【文凭】diploma
【文人】man of letters; scholar; *literati*: ～无行 a literary man of no conduct／～学士 scholars; men of letters／～相轻。 Scholars tend to scorn each other. (或: Writers like to disparage one another.)
【文如其人】the writing mirrors the writer
【文弱】gentle and frail- looking: ～书生 a frail scholar
【文身】tattoo
【文史】literature and history ◇～馆 Research Institute of Culture and History／～资料 historical accounts of past events
【文书】①(公文) document; official dispatch ②(文牍人员) copy clerk; clerical staff
【文思】the thread of ideas in writing; the train of thought in writing: ～枯竭 literary thought is dried up／～敏捷 have a ready pen; ability to think and write fast
【文坛】the literary world; the literary arena; the literary circles; the world of letters
【文体】①(文章的体裁) type of writing; literary form; style ②[简](文娱体育) recreation and sports ◇ ～活动 recreational and sports activities
【文武】civil and military: ～官员 civil and military officials／～双全 be well versed in both polite letters and martial arts
【文物】cultural relic; historical relic: ～保护 preservation of cultural relics; protection of historical relics／出土～ unearthed cultural relics
【文献】document; literature: 科技～ literature of science and technology／历史～ historical documents／马列主义～ Marxist-Leninist literature ◇ ～记录片 documentary (film)
【文选】selected works; literary selections: 活叶～ loose-leaf literary selections
【文学】literature ◇ ～家 writer; man of letters; *literati*／～流派 schools of literature／～批评 literary criticism／～作品 literary works
【文学语言】①[语] standard speech ②(文艺语言) literary language
【文雅】elegant; refined; cultured; polished: 举止～ refined in manner
【文言】classical Chinese ◇ ～文 writings in classical Chinese; classical style of writing
【文艺】literature and art ◇ ～创作 literary and artistic creation／～队伍 ranks of writers and artists／～复兴 the Renaissance／～工作 work in the literary and artistic fields／～工作者 literary and art workers; writers and artists／～会演 theatrical festival／～节目 program of entertainment; theatrical items; theatrical performance／～界 literary and art circles; the world of literature and art／～理论 theory of literature and art／～批评 literary or art criticism／～批评家 literary or art critic／～思潮 trend of thought in literature and art／～团体 literature and art organization; theatre company; theatre troupe／～作品 literary and artistic works／～座谈会 forum on literature

and art

【文娱】cultural recreation; entertainment ◇ ～活动 recreational activities

【文责】the responsibility an author should assume for his own writings; author's responsibility: ～自负。The author takes sole responsibility for his views.

【文摘】abstract; digest

【文章】①(短篇论著) essay; article ②(著作) literary works; writings: 一篇批评～ a critical essay ③(暗含的意思) hidden meaning; implied meaning: 其中大有～。There is a lot behind all this. (或: Thereby hangs a tale.)

【文职】civilian post; civil service: 解放军～干部 PLA civilian staff ◇ ～人员 nonmilitary personnel

【文质彬彬】gentle; suave

【文绉绉】genteel: 说话～的 speak in an elegant manner

【文竹】[植] asparagus fern

【文字】①(书写符号) characters; script; writing: 表意～ ideography / 拼音～ alphabetic writing / 楔形～ cuneiform characters / 象形～ pictograph; hieroglyph ②(语言的书写形式) written language: ～宣传 written propaganda / ～游戏 play with words; juggle with terms / 有～可考的历史 recorded history ③(文章形式) writing (as regards form or style): ～清通 lucid writing ◇ ～方程[数] literal equation / ～改革 reform of a writing system / ～学 philology

【文字狱】[史] imprisonment or execution of an author for writing sth. considered offensive by the imperial court; literary inquisition

蚊 mosquito: 驱～ drive off mosquitoes ○按蚊 anopheles / 库蚊 culex / 伊蚊 aedes

【蚊虫】mosquito

【蚊香】mosquito-repellent incense

【蚊帐】mosquito net; mosquito bar; mosquito-curtain ◇ ～布 mosquito netting

【蚊子】mosquito: ～会叮人。Mosquitoes bite. / ～嗡嗡叫。Mosquitoes buzz

纹 lines; veins; grain: 脸上的皱～ lines on one's face; furrows / 细～木 fine-grained wood

【纹理】veins; grain: ～细的木材 woods of fine grain / 有～的大理石 veined marble

【纹路儿】lines; grain

【纹丝不动】absolutely still; not a wrinkle was touched

【纹银】fine silver

闻 ①(听见) hear: ～讯 hear the news / 听而不～ listen but not hear; turn a deaf ear to ②(听到的事情;消息) news; story: 奇～ fantastic story / 要～ important news ③(有名望的) well-known; famous: ～人 well-known figure ④(名声) repute; reputation: 秽～ ill repute ⑤(嗅) smell: 这东西～起来真香。It smells fragrant.

【闻风而动】immediately respond to a call; go into action without delay

【闻风丧胆】become terror-stricken at the news

【闻过则喜】feel happy when told of one's errors; be glad to have one's errors pointed out; thankful for being told of one's errors

【闻名】①(有名) well-known; famous; renowned: ～全国 well-known throughout the country / ～全球 be well-known throughout the world / ～世界 world famous; world-renowned ②(听到名声) be familiar with sb.'s name; know sb. by repute

【闻人】well-known figure; famous man; celebrity

【闻所未闻】unheard-of; never heard of

wěn

紊 disorderly; confused

【紊乱】disorder; chaos; confusion: 新陈代谢功能～ metabolic disorder / 秩序～ in a state of chaos

稳 ①(稳当;稳定) steady; firm: 把桌子放～ make the table steady / 站～ stand steadily; stand firm / 站不～ be not firm on one's feet / 坐～ sit tight ②(稳妥) sure; certain: ～输 sure to lose / ～赢 sure to win

【稳步】with steady steps; steadily: ～前进 advance steadily; make steady progress / ～上升 go up steadily

【稳操胜券】have full assurance of success; be certain to win

【稳操胜算】have the ball at one's feet

【稳产高产】high and stable yields: ～田 land with high, stable yields

【稳当】reliable; secure; safe: ～的办法 a reliable method

【稳定】①(没有变动;稳固安定) stable; steady: ～的多数 a stable majority / ～收入 steady income / 不～的国际金融市场 a shaky international monetary market / 情绪～ be in a calm, unruffled mood ②(使稳定) stabilize; steady: ～情绪 set sb.'s mind at rest; reassure sb. / ～通货 stabilize currency / ～物价 stabilize commodity prices ◇ ～剂[化] stabilizer / ～平衡[物] stable equilibrium / ～伞 drogue / ～装置[化] stabilization plant

【稳固】firm; stable: ～的基础 a firm foundation; a solid foundation / ～的政权 a stable government / 地位～ hold a stable position

【稳健】firm; steady: 办事～ go about things steadily / 迈着～的步伐 walk with firm steps ◇ ～派 moderates

【稳如磐石】be as firm as a rock

【稳如泰山】as stable as Mount Taishan

【稳妥】safe; reliable: ～的办法 safe method / ～的计划 a safe plan / ～的消息来源 reliable sources of information

【稳压电源】regulated power supply

【稳扎稳打】go ahead steadily and strike sure blows; go about things steadily and surely

【稳重】steady; staid; sedate: ～寡言 earnest and taciturn / 他是个～的人。 He is a steady fellow.

刎 cut one's throat: 自～ cut one's own throat

吻 ①(嘴唇) lips ②(接吻) kiss: 互相亲～ kiss each other / 他们热烈地～着。 They kissed passionately. ③(动物的嘴) an animal's mouth

【吻合】be identical; coincide; tally: 意见～ have identical views / 证词与报告～。 The testimony tallies with the report.

【吻合术】[医] anastomosis

wèn

问 ①(询问) ask; enquire; inquire: ～路 ask the way / ～一个问题 ask a question ②(问到; 问候) ask after; enquire after: 询～ make enquires about / 你～过价钱吗? Did you ask the price? / 他～到我了吗? Did he enquire about me? ③(审讯) interrogate; examine: 审～ interrogate ④(追究) hold responsible: 出了差错唯你是～。 You'll be held responsible if anything goes wrong.

【问安】pay one's respects; wish sb. good health

【问案】try a case; hear a case

【问长问短】take the trouble to make detailed inquiries; ask many questions about other people's affairs

【问答】questions and answers ◇ ～练习 question-and-answer drills

【问道于盲】ask the way from a blind person; seek advice from one who can offer none

【问东问西】ask all sorts of questions

【问寒问暖】ask after sb.'s health with deep concern; be solicitous for sb.'s welfare

【问好】send one's regards to; say hello to; extend greetings to: 请代向你们老师～。 Please give my regards to your teacher. / 她向您～。 She wished to be remembered to you.

【问号】①(标点符号) question mark; interrogation mark (?) ②(疑问) unknown factor; unsolved problem: 这个案子还是个～。 This case is still unknown.

【问候】send one's respects 〈regards〉 to; extend greetings to: 致以亲切的～ extend cordial greetings

【问津】make inquiries (as about prices or the situation): 不敢～ not dare to make inquiries / 无人～ nobody cares to ask about sth.

【问荆】[植]meadow pine

【问明原委】find out origin of affair

【问世】be published; come out

【问俗问禁】enquire about the custom and prohibition

【问题】①(需回答的题目) question; problem: 解答～ solve a problem / 提～ ask a question ②(需研究解决的矛盾等) problem; matter: ～的关键 the heart of the matter / 共同关心的～ questions of common interest / 关键～ a key problem / 思想～ an ideological problem / 悬而未决的～ an outstanding issue / 原则～ a question of principle; a matter of principle / 枝节～ minor issue ③(事故或意外) trouble; mishap; something wrong: 没出什么～ without any mishap / 那机器有点～。 Something is wrong with the machine. ④(重要之点) the point; the thing: ～不在这里。That is not the point.

【问心无愧】have a clear conscience; feel no qualms upon self-examination

【问心有愧】feel a twinge of conscience; have a guilty conscience

【问讯】enquire; ask ◇ ～处 enquiry office; information desk

【问诊】interrogation; enquiry

【问罪】denounce; condemn: 兴师～ send a punitive force against; denounce sb. publicly for his crimes or serious errors

wēng

翁 ①(老头儿) old man: 渔～ an old fisherman ②(父亲) father ③(丈夫或妻子的父亲) father-in-law: ～姑 a woman's parents-in-law / ～婿 father-in-law and son-in-law

嗡 [象] drone; buzz; hum: 蚊子～～叫。 Mosquitoes buzz.

wèng

蕹

【蕹菜】water spinach

瓮 urn; earthen jar: 菜～ a jar for pickling vegetables / 水～ water jar

【瓮声瓮气】in a low, muffled voice

【瓮中之鳖】a turtle in a jar; trapped; bottled up

【瓮中捉鳖】catch a turtle in a jar; go after an easy prey

wō

涡 whirlpool; eddy: 水～ eddies of water

【涡虫】[动] turbellarian worm; turbellarian

【涡流】①(旋涡) the circular movement of a fluid; whirling fluid; eddy ②[物] eddy current; vortex flow

【涡轮】turbine ◇ ～泵 turbopump / ～泵组 gas-turbine-pump combination / ～发电机 turbogenerator / ～风扇发动机 turbofan / ～机 turbine / ～螺旋桨发动机 turboprop / ～喷气发动机 turbojet

【涡旋】[气] vortex: 大气～ atmospheric vortex

窝 ①(鸟兽昆虫住的地方) nest: 蜂～ beehive / 鸡～ hencoop; roost / 鸟～ bird's nest / 燕子在房檐下筑～。 The swallows nested under eaves. ②(坏人聚居的地方) lair; den: 土匪～ bandits' lair; bandits' nest / 贼～ thieves' den ③(人体凹进去的地方) a hollow part of the human body; pit: 夹肢～ armpit / 心～ the pit of the stomach / 眼～ eye-socket ④(窝藏) harbor; shelter: ～赃 harbor stolen goods ⑤(郁积不得发作或发挥) hold in; check: ～着一肚子火 be simmering with rage; be forced to bottle up one's anger ⑥(使弯或曲折) bend ⑦[量] litter; brood: 一～小鸡 a brood of chicken / 一～小猪 a litter of piglets

【窝藏】harbor; shelter: ～罪犯 give shelter to a criminal; harbor a criminal

【窝工】enforced idleness due to poor organization of work; holdup in the work through poor organization

【窝囊】①(因受委屈而烦闷) feel vexed; be annoyed: 受～气 be subjected to petty annoyances ②(怯懦;无能) stupid, cowardly and timid; good-for-nothing; hopelessly stupid

【窝囊废】good-for-nothing; worthless wretch

【窝棚】shack; shed; shanty

【窝头】steamed bread of corn

【窝主】a person who harbors criminals, loot or contraband goods

莴

【莴苣】lettuce

【莴笋】asparagus lettuce

蜗 snail

【蜗杆】[机] worm ◇ ～轴 worm shaft

【蜗居】humble abode

【蜗轮】[机] worm gear; worm wheel

【蜗牛】snail

喔 [象] cock's crow: ～～～! Cock-a-doodle-doo!

wǒ

我 ①(称自己) I: ～来了。 I'm coming. / ～在这儿。 I am here. ②(我们) we: ～方 our side; we ③(自己) self: 忘～的献身精神 selfless devotion / 自～牺牲 self sacrifice ④(我的) my: ～家 my family / ～母亲 my mother ⑤(我们的) our: ～国 our country / ～军 our army

【我们】①(包括自己的若干人) we: ～常到那里去。 We often go there. ②(我们的) our: ～祖国 our matherland / ～的是一间大房子。 Ours is a large house. / 这是～的教室。 This is our classroom.

【我行我素】persist in one's old ways; stick to one's old way of doing things

wò

沃 ①(肥) fertile; rich: ～土 fertile soil; rich soil / ～野千里 a vast expanse of fertile land ②(灌溉) irrigate: ～田 irrigate farmland

斡

【斡旋】①(调解) mediate: ～于两国之间 mediate between two nations ②[外] good offices

卧 ①(躺下) lie: 侧～ lie on one's side / 仰～ lie on one's back ②(动物趴) crouch; sit ③(睡

觉用的）for sleeping in：～铺 sleeping berth／～室 bedroom

【卧病】be confined to bed；be laid up

【卧车】①（设有卧铺的车厢）sleeping car；sleeping carriage；sleeper ②（小轿车）automobile；car；limousine；sedan

【卧床】lie in bed：绝对～ strict bed rest／因病～ keep to one's bed；be confined to bed；be laid up

【卧倒】drop to the ground；take a prone position：～！（口令）Lie down！

【卧具】bedding (provided on a train or ship)

【卧铺】sleeping berth；sleeper

【卧铺票】berth：订～ book a berth

【卧式】[机] horizontal：～发动机 horizontal engine／～镗床 horizontal boring machine

【卧室】bedroom

【卧薪尝胆】sleep on brushwood and taste gall；undergo self-imposed hardships so as to strengthen one's resolve to wipe out a national humiliation

【卧姿】[体] prone position

硪 a flat stone or iron rammer with ropes attached at the sides：打～ operate a rammer

肟 [化] oxime

握 hold；grasp

【握别】shake hands at parting；part

【握力】the power of gripping；grip ◇ ～器[体] spring-grip dumb-bells

【握拳】make a fist；clench one's fist

【握手】shake hands；clasp hands：～言欢 hold hands and chat cheerfully

龌

【龌龊】dirty；filthy：卑鄙～ sordid；foul

wū

污 ①（沾染的污垢）dirt；filth：血～ blood stains ②（脏）dirty；filthy；foul：～泥 mud；mire ③（不廉洁）corrupt：贪官～吏 corrupt officials ④（弄脏）defile；smear：玷～ stain；sully；tarnish

【污点】stain；spot；blemish；smirch：历史上的～ blemish in the past／洗去～ wash off stains

【污垢】dirt；filth

【污秽】filthy；foul：～不堪 intolerably dirty；intolerably filthy／～的语言 filthy language

【污迹】stain；smear；smudge

【污蔑】①（诬蔑）slander；calumniate ②（玷污）defile；sully；tarnish

【污泥】mud；mire；sludge

【污泥浊水】filth and mire：荡涤一切～ wash away all the sludge and filth

【污七八糟】in a filthy mess

【污染】pollute；contaminate：大气层～ atmosphere pollution／放射性～ radioactive contamination／工业粉尘～ industrial dust pollution／环境～ environmental pollution／空气～ air pollution／飘尘～ floating dust pollution；airborne dust pollution／水～ water contamination／噪音～ noise pollution ◇ ～等级 class of pollution／～物质 pollutant

【污辱】①（侮辱）humiliate；insult ②（玷污）defile；sully；tarnish

【污水】foul water；polluted water；waste water；sewage；slops：生活～ domestic sewage ◇ ～处理 sewage disposal；sewage treatment／～处理厂 sewage treatment plant／～管 sewage pipe；sewer (pipe)／～管道 sewage conduit；sewer line／～灌溉 sewage irrigation／～净化 sewage purification／～生物分解 sewage biolysis

【污浊】dirty；muddy；foul；filthy

巫 shaman；witch；wizard

【巫婆】witch；sorceress

【巫师】wizard；sorcerer

【巫术】witchcraft；sorcery

【巫医】witch doctor

诬 accuse falsely

【诬告】lodge a false accusation against；bring a false charge against：～和诽谤 malicious prosecution and defamation ◇ ～案件 frame-up；trumped-up case／～人 maliciously false accuser／～陷害罪 crime of false charge／～行为 action of malicious prosecution／～者 calumniator

【诬害】injure by spreading false reports about；calumniate；malign

【诬赖】falsely incriminate：～好人 incriminate innocent people

【诬蔑】slander；vilify；calumniate；smear：～不实之词 slander and libel／造谣～ rumormongering and mudslinging；calumny and slander

【诬陷】frame a case against；frame sb.：受～ be framed

乌 ①（乌鸦）crow ②（黑色）black；dark：～云 black clouds；dark clouds

【乌鲳】[动] black pomfret

【乌尔都语】Urdu
【乌饭树】oriental blueberry
【乌干达】Uganda ◇～人 Ugandan
【乌龟】① [动] tortoise ②(妻子有外遇的人) cuckold ◇～壳 tortoiseshell
【乌合之众】a disorderly band; a motley crowd; rabble; mob
【乌黑】pitch-black; jet-black
【乌桕】[植] Chinese tallow tree
【乌拉圭】Uruguay ◇～人 Uruguayan
【乌鳢】[动] snakehead; snakeheaded fish
【乌亮】glossy black; jet-black: ～的头发 dark, glossy hair; raven locks
【乌溜溜】dark and liquid: 一双～的眼睛 sparkling, black eyes
【乌龙茶】oolong (tea)
【乌梅】smoked plum; dark plum
【乌木】[植] ebony
【乌七八糟】①(十分杂乱) in a horrible mess; in great disorder ②(淫秽) obscene; dirty; filthy
【乌纱帽】①(古代的官帽) black gauze cap ②(比喻官位) official post: 丢～ be dismissed from office
【乌托邦】Utopia
【乌鸦】crow
【乌烟瘴气】foul atmosphere; pestilential atmosphere: 搞得～ foul up
【乌油油】shining black: ～的头发 shining black hair
【乌有】nothing; naught: 化为～ come to nothing
【乌鱼】[动] snakehead; snakeheaded fish
【乌云】black clouds; dark clouds
【乌枣】smoked jujube; black jujube
【乌贼】[动] cuttlefish; inkfish

呜 [象] toot; hoot; zoom: 远处传来～～声。The hoots came from the distance.
【呜呼】①(表示叹息) alas; alack ②(死亡) die: 一命～ give up the ghost
【呜呼哀哉】①(表示叹息) alas ②(死了) dead and gone ③(完蛋) all is lost
【呜咽】sob; whimper

钨 tungsten; wolfram (W)
【钨钢】wolfram steel; tungsten steel
【钨合金】tungsten alloy
【钨锰矿】huebnerite
【钨砂】tungsten ore
【钨丝】tungsten filament ◇～灯 tungsten lamp
【钨铁】ferrotungsten
【钨铁矿】ferberite

屋 ①(房子) house ②(屋子) room: 里～ inner room
【屋顶】roof; housetop ◇～花园 roof garden
【屋脊】ridge (of a roof): 世界～ the roof of the world
【屋架】roof truss
【屋面】[建] roofing: 瓦～ tile roofing ◇～板 roof boarding
【屋檐】eaves
【屋子】room: 三间～ three rooms

wú

无 ①(没有) nothing; nil; not have; without: ～一定计划 without a definite plan / 从～到有 grow out of nothing; start from scratch / 一～所知 know nothing about it ②(不) not: ～碍大局 not affect the situation as a whole / ～须乎着急。There's no need to get excited. ③(不论) regardless of; no matter whether, what, etc.: 事～大小,都有人负责。Everything, big and small, is properly taken care of.
【无伴奏合唱】*a cappella*
【无被选权】ineligible
【无比】incomparable; unparalleled; matchless: ～的毅力 tremendous determination / ～的优越性 incomparable superiority; unparalleled superiority / ～愤怒 furiously indignant / ～英勇 unrivaled in bravery
【无边无际】boundless; limitless; vast: ～的大海 a boundless ocean / ～的沙漠 a vast expanse of desert
【无柄叶】[植] sessile leaf
【无病呻吟】moan and groan without being ill; make a fuss about an imaginary illness
【无补】of no help; of no avail: 这样恐怕～于事。That would be of no avail.
【无不】all without exception; invariably: ～称快。All without exception were glad. / ～为之感动。None were unmoved.
【无产阶级】the proletariat: ～专政 dictatorship of the proletariat; proletarian dictatorship
【无产者】proletarian
【无常】①(变化无常) variable; changeable: 变化～的天气 changeable weather / 反复～ capricious; uncertain ②[佛教] impermanence
【无偿】free; gratis; gratuitous: ～援助 aid given gratis; aid given gratuitously / 提供～经济援助 render economic assistance gratis; give free

economic aid ◇～契约 naked contract／～取得 acquisition without consideration／～让与 voluntary conveyance
【无偿付能力】bankruptcy
【无偿还义务】without recourse
【无偿债能力】insolvency
【无承诺】without engagement ◇～能力 incapacity of consent
【无耻】shameless; brazen; impudent: ～谰言 shameless slander／～之徒 a person who has lost all sense of shame; a shameless person／～之尤 brazen in the extreme; the height of shamelessness
【无酬劳动】[经] unpaid labor
【无出其右】second to none; unequalled
【无从】have no way (of doing sth.); not be in a position (to do sth.): ～下手 have no way of doing it; not know how to start／～知道 be unable to find out
【无党派人士】a public figure without party affiliation; nonparty personage; patriots without party affiliation
【无敌】unmatched; invincible; unconquerable: ～于天下 unmatched anywhere in the world; invincible
【无底洞】a bottomless pit (that can never be filled)
【无地自容】can find no place to hide oneself for shame; feel too ashamed to show one's face; look for a hole to crawl into
【无的放矢】shoot an arrow without a target; shoot at random
【无冬无夏】be it winter or summer; throughout the year
【无动于衷】aloof and indifferent; unmoved; untouched; unconcerned
【无独有偶】it is not unique, but has its counterpart; not come singly but in pairs
【无毒不丈夫】ruthlessness is the mark of a truly great man
【无毒蛇】nonpoisonous snake
【无度】immoderate; excessive: 挥霍～ squander wantonly／饮食～ excessive eating and drinking
【无端】for no reason: ～侮辱 a gratuitous insult／～争吵 quarrelled for no reason
【无恶不作】stop at nothing in doing evil; stop at no evil; commit all manner of crimes
【无法】unable; incapable: ～解决的问题 an insolvable problem／～理解 unable to understand／～形容 beyond description／～应付 unable to cope with; at the end of one's resources ◇～履行 impossibility of performance
【无法无天】defy laws human and divine; become absolutely lawless; run wild
【无方】not in the proper way; in the wrong way; not knowing how: 经营～ mismanagement
【无妨】there's no harm; may as well; might as well: 你～试一试。There's no harm in having a try.／这样做并～。There is no harm in doing that.
【无纺织物】[纺] adhesive-bonded fabric
【无非】nothing but; no more than; simply; only: ～是老生常谈。It's no more than a platitude.
【无风】[气] calm
【无风不起浪】there are no waves without wind; there's no smoke without fire
【无风带】[气] calm belt; calm zone
【无缝钢管】seamless steel tube; seamless steel pipe ◇～厂 seamless (steel) tubing mill
【无干】have nothing to do with: 这事与你～。It's none of your business.(或: It has nothing to do with you.)
【无功受禄】get a reward without deserving it
【无辜】①(没有罪) innocent ②(没有罪的人) an innocent person ◇～杀人 innocent homicide
【无故】without cause or reason: ～迟到 be late without reason／～旷工 absent from work without reason／不得～缺席。Nobody may be absent without reason.
【无怪】no wonder; not to be wondered at
【无关】have nothing to do with; be unconcerned: ～大局 not affecting the general situation; insignificant／～紧要 of no importance; immaterial／～痛痒的话 comment without any bite; irrelevant or pointless remarks／～痛痒的自我批评 irrelevant or superficial self-criticism／这事跟我们～。It has nothing to do with us.
【无规】[物] random ◇～介质 random media／～取向 random orientation
【无轨电车】trackless trolley; trolleybus
【无国籍】[外] stateless ◇～者 a stateless person
【无过失】unerring; unimpeachable ◇～责任 liability without fault
【无官一身轻】One feels carefree when he is relieved of official duties.

【无害】harmless; do no harm to: ～于健康 be harmless to the health
【无害通过】[法] innocent passage
【无核化】denuclearize
【无核区】nuclear-free zone
【无核武器国家】nonnuclear country
【无花果】[植] fig
【无话不谈】keep no secrets from each other; be in each other's confidence
【无疾而终】die without known cause
【无机】[化] inorganic ◇ ～肥料 inorganic fertilizer; mineral fertilizer / ～化合物 inorganic compound / ～化学 inorganic chemistry / ～界 the inorganic world / ～物 inorganic substance; inorganic matter / ～盐 inorganic salts
【无稽】unfounded; fantastic; absurd: ～之谈 fantastic talk; sheer nonsense
【无级】[机] stepless ◇ ～变速 infinitely variable speeds / ～调速 stepless speed regulation
【无几】very few; very little; hardly any: 相差～ very little difference; almost the same / 所剩～。There's very little left.
【无脊椎动物】invertebrate
【无计可施】at one's wits' end; at the end of one's tether
【无记名股票】bearer stock certificate
【无记名投票】secret ballot
【无记名债券】bearer bonds
【无济于事】of no avail; to no effect
【无家可归】wander about without a home to go to; be homeless
【无价之宝】priceless treasure; invaluable asset
【无坚不摧】overrun all fortifications; carry all before one; be all-conquering
【无间】①(没有间隙) not keeping anything from each other; very close to each other: 亲密～的朋友 close friends; bosom friends ②(不间断) continuously; without interruption
【无精打采】listless; lackadaisical; in low spirits; out of sorts
【无拘束】unrestrained; unconstrained: ～地发表意见 express one's views freely / 在～的气氛中 in an unconstrained atmosphere
【无菌操作法】aseptic manipulation
【无菌隔离室】germfree isolator
【无菌生物学】gnotobiology
【无可比拟】incomparable; unparalleled
【无可非议】beyond reproach; above criticism; blameless: 他的行为～。His conduct is beyond reproach.
【无可奉告】no comment
【无可厚非】give no cause for much criticism
【无可讳言】there is no hiding the fact
【无可救药】incorrigible; incurable
【无可奈何】have no way out; have no alternative
【无可挽回】irretrievable; irredeemable; irrevocable
【无可无不可】not care one way or another: 我～。It makes no difference to me.
【无可争辩】indisputable; irrefutable
【无可争议】beyond controversy
【无可置疑】indubitable; unquestionable
【无孔不入】get in by every opening; be all-pervasive; seize every opportunity (to do evil)
【无愧】feel no qualms; have a clear conscience: 当之～ be worthy of the name / 问心～ feel no qualms upon self-examination
【无赖】①(不讲道理) rascally; scoundrelly; blackguardly: 耍～ act shamelessly ②(品行不端的人) rascal
【无礼】rudeness
【无理】unreasonable; unjustifiable: ～取闹 willfully make trouble; be deliberately provocative / ～要求 unreasonable demands / ～指责 unwarranted accusations; groundless charges / ～阻挠 unjustifiable obstruction
【无理数】[数] irrational number
【无力】①(没有气力) lack strength; feel weak: 四肢～ feel weak in one's limbs ②(没有力量) unable; incapable; powerless: ～抵抗 powerless to resist ◇ ～偿付 insolvency / ～支付 financial insolvency
【无利可图】profitless
【无量】measureless; immeasurable; boundless: 前途～ a boundless future
【无聊】①(闲得烦闷) bored ②(没有意义) senseless; silly; stupid: 过着～的生活 lead a dull life
【无论】no matter what, how, etc.; regardless of: ～什么 no matter what / ～谁 no matter who
【无论如何】in any case; at any rate; whatever happens; at all events: ～要完成任务。We must fulfill the plan, whatever happens.
【无米之炊】cook a meal without rice; make bricks without straw
【无名】①(没有名称的) nameless ②(名字不为人所知的) unknown ③(说不出所以然来的) indefinable; indescribable: ～的恐惧 an indefinable feeling of terror ◇ ～高地[军] an unnamed hill / ～氏 an anonymous person /

～死者 nameless dead / ～小卒 a nobody / ～英雄 an unknown hero / ～指 the third finger; ring finger

【无奈】①(无可奈何) cannot help but; have no alternative; have no choice: 出于～她才去。She couldn't help but go. ②(可惜) but; however

【无能】incompetent; incapable: 软弱～ weak and incompetent

【无能为力】powerless; helpless; incapable of action

【无期徒刑】life imprisonment: 判处～ be sentenced to imprisonment for life; be given a life sentence

【无牵无挂】have no cares; be free from care; carefree

【无巧不成书】without coincidences there would be no stories

【无情】merciless; ruthless; heartless: ～的打击 a merciless blow / ～的事实 harsh reality; hard facts

【无穷】infinite; endless; boundless; inexhaustible: ～的烦恼 endless troubles / ～的力量 inexhaustible power / ～的潜力 boundless potential / ～无尽的智慧和力量 inexhaustible wisdom and power / ～的忧虑 endless worries

【无穷大】[数] infinitely great; infinity

【无穷小】[数] infinitely small; infinitesimal

【无权】have no right: ～办理 have no right to handle / ～干预 have no right to interfere / ～追索[法] without recourse

【无人】①(无人驾驶的) unmanned: ～驾驶飞机 unmanned plane; pilotless plane; robot plane ②(无人居住的) depopulated: ～区 a depopulated zone; no man's land ③(自动售货的) self-service: ～售书处 self-service bookstall

【无人问津】no one shows any interest in

【无任】extremely; immensely: ～感激 be deeply grateful

【无任所大使】ambassador-at-large

【无日期票据】undated bill

【无商业价值】no commercial value

【无伤大雅】not affect the whole; not matter much

【无商业价值】no commercial value

【无上】supreme; paramount; highest: ～光荣 the highest honor / ～权力 supreme power

【无神论】atheism ◇ ～者 atheist

【无生代】[地] the Azoic Era

【无生物】inanimate object; nonliving matter

【无生殖力】agenesis; agensia

【无声】noiseless; silent ◇ ～打字机 noiseless typewriter / ～片 silent film / ～手枪 pistol with a silencer

【无声无臭】unknown; obscure: 不甘心～地过一辈子 hate to live out one's life unknown

【无师自通】self-taught; learned without teacher

【无时无刻】all the time; incessantly

【无事不登三宝殿】never go to the temple for nothing; would not go to sb.'s place except on business, for help, etc.; I wouldn't come to you if I hadn't sth. to ask of you.

【无事生非】make trouble out of nothing; be deliberately provocative; make the fur fly

【无视】ignore; disregard; defy: ～别国主权 disregard the sovereignty of other countries / ～人民的意志 defy the will of the people

【无熟料水泥】[建] clinker-free cement

【无数】①(难以计算) innumerable; countless: ～的事实 innumerable facts ②(不知道底细) not know for certain; be uncertain

【无双】unparalleled; unrivaled; matchless: 举世～ absolutely unrivaled

【无霜期】frost-free period

【无水】[化] anhydrous: ～酒精 absolute alcohol; anhydrous alcohol / ～溶剂 anhydrous solvent / ～酸 anhydrous acid

【无私】selfless; disinterested; unselfish: 给予～的援助 give disinterested assistance / ～才能无畏。Only the selfless can be fearless.

【无梭织机】[纺] shuttleless loom

【无所不包】all-embracing; all-encompassing

【无所不能】omnipotent

【无所不为】stop at nothing; do all manner of evil

【无所不用其极】resort to every conceivable means; stop at nothing; go to any length

【无所不在】omnipresent; ubiquitous

【无所不知】omniscient

【无所不至】①(没有达不到的地方) penetrate everywhere ②(所有的都做到了) spare no pains (to do evil); be capable of anything; stop at nothing: 威胁利诱,～ use intimidation, bribery and every other means

【无所措手足】be at a loss as to what to do

【无所事事】be occupied with nothing; have nothing to do; idle away one's time; do nothing all day

【无所适从】not know what course to take; be at a loss as to what to do; don't know whose suggestion to follow: 领导意见分歧,群众～。

The leaders are divided, and the rank and file don't know whom to turn to.
【无所畏忌】stop at nothing
【无所畏惧】fearless; dauntless; undaunted
【无所谓】①(说不上) cannot be designated as; not deserve the name of: 这种事～好坏。It cannot be said to be good or bad. ②(不在乎) be indifferent; not matter: 采取～的态度 adopt an indifferent attitude／这对我～。It doesn't matter to me.
【无所依仗】have nothing to rely upon
【无所用心】not give serious thought to anything: 饱食终日,～ be sated with food and remain idle
【无所作为】attempt nothing and accomplish nothing; be in a state of inertia
【无炭钢】carbon- free steel
【无题】no title ◇～诗 titleless poem
【无条件】unconditional; without preconditions ◇～承兑 absolute acceptance／～担保 absolute guaranty／～反射[生理] unconditioned reflex／～接受 absolute acceptance／～投降 unconditional surrender／～转让 absolute conveyance; absolute assignment
【无痛分娩】painless labor
【无头案】a case without any clues; unsolved mystery
【无土栽培】soilless culture
【无土农业】soilless agriculture
【无往不利】go smoothly everywhere; be ever successful
【无往不胜】ever-victorious; invincible
【无往不在】present everywhere; omnipresent
【无妄之灾】unexpected calamity; undeserved ill turn
【无微不至】meticulously; in every possible way: ～的关怀 perfect and minute care
【无为而治】govern by doing nothing that goes against nature
【无味】①(没有滋味) tasteless; unpalatable: 食之～,弃之可惜 unappetizing and yet not bad enough to throw away ②(没有趣味) dull; insipid; uninteresting: 枯燥～ dry as dust
【无畏】fearless; dauntless
【无谓】meaningless; pointless; senseless: ～的牺牲 a meaningless sacrifice; a senseless sacrifice／～的争吵 a pointless quarrel
【无…无…】～牵～挂 have no cares／～穷～尽 inexhaustible; endless／～依～靠 have no one to depend on; helpless／～影～踪 disappear completely; vanish without a trace／～忧～虑 free from care; carefree
【无息贷款】interest-free loan
【无隙可乘】no crack to get in by; no loophole to exploit; no weakness to take advantage of; no chink in sb.'s armor
【无暇】have no time; be too busy: ～考虑 have no time to think it over／～他顾 have no time to attend to other things
【无先例可援】have no precedent to go by
【无限】infinite; limitless; boundless; immeasurable: ～热爱 have boundless love for／～忠诚 absolute devotion to ◇～公司 unlimited company／～花序 indefinite inflorescence／～责任 unlimited liability
【无限大】[数] infinitely great; infinity
【无限期】indefinite duration: ～罢工 a strike of indefinite duration／～搁置动议 shelve a motion *sine die*／～休会 adjourn indefinitely; adjourn *sine die*
【无限小】[数] infinitely small; infinitesimal
【无限制】unrestricted; unbridled; unlimited
【无线】wireless: ～电报 wireless telegram; radiotelegram／～电话 radiotelephone; radiophone／～话筒 radio microphone／～装订 unsewn binding; thermoplastic binding
【无线电】radio ◇～波 radio wave／～测向器 radio direction finder; radio goniometer／～传真 radiofacsimile／～导航台 radio navigation station／～发射机 radio transmitter／～干扰 radio jamming／～跟踪 radio tracking／～收发两用机 transceiver／～收音机 radio receiver／～探空仪 radiosonde／～天文学 radio astronomy／～通信 radio communication; wireless communication／～遥控 wireless remotecontrol
【无效】of no avail; to no avail; invalid; null and void: 宣布合同～ declare a contract invalid; declare a contract null and void; invalidate a contract; nullify a contract／宣布选举～ nullify an election／医治～ fail to respond to medical treatment ◇～法律行为 nullity of juristic act／～分蘖[农] ineffective tillering／～诉讼 annulment suit／～条款 irritancy／～约因 infirmative consideration
【无懈可击】with no chink in one's armor; unassailable; invulnerable
【无心】①(没有心思) not be in the mood for: ～去看歌剧 be in no mood to go to the opera ②(不是故意) not intentionally; unwittingly; inadvertently: 我是～说的。I didn't say it in-

tentionally.
【无形】invisible：～的枷锁 invisible shackles／～的战线 invisible fronts ◇ ～财产 incorporeal property；non-visible property／～出口 invisible export／～价值 intangible value／～进口 invisible import／～利益 invisible gain／～贸易 invisible trade／～权利 intangible right／～商品 intangible goods／～损耗 invisible waste／～损失 invisible loss；non-physical loss／～遗产 incorporeal hereditaments／～资本 incorporeal capital
【无形中】imperceptibly；virtually：她～成了我们的顾问。 She has virtually become our adviser.
【无行为能力】incompetence；legal to go by
【无性】[生] asexual：～繁殖 vegetative propagation／～生殖 asexual reproduction／～世代 asexual generation／～杂交 asexual hybridization
【无休止】ceaseless；endless：～地争论 argue on and on
【无须】need not；not have to：～顾虑 need not worry／～过目 not have to read it／～细说。 It's unnecessary to go into details.
【无烟火药】smokeless powder；ballistite
【无烟煤】anthracite
【无颜见江东父老】no face or too ashamed to go back home to see one's elders
【无恙】in good health；well；safe：安然～ safe and sound／别来～？ I trust you've been in good health since we last met?
【无业游民】vagrant
【无疑】beyond doubt；undoubtedly：这～是错误的。 This is undoubtedly wrong.
【无遗嘱继承】intestate succession
【无遗嘱死亡】die intestate
【无以复加】in the extreme
【无依无靠】have no one to depend on；be helpless：～的孤儿 a helpless orphan
【无以自容】be ashamed of oneself
【无异】not different from；the same as；as good as：看上去和新的～。 It looks as good as new.
【无益】unprofitable；useless；no good
【无意】①(无意愿) have no intention (of doing sth.)；not be inclined to：～参加 have no intention of joining／～干涉 have no intention of interfering ②(不是故意的) inadvertently；unwittingly；accidentally：～中发现 discover accidentally ◇ ～遗漏 innocent omission
【无意识】unconscious：～的动作 an unconscious act；an unconscious movement／～的行为 unconscious conduct；automatism
【无翼鸟】kiwi
【无垠】boundless；vast：一望～的大草原 a boundless prairie
【无影灯】[医] shadowless lamp
【无影无踪】without a trace
【无庸讳言】not need for reticence
【无庸赘述】there is no need to go into details
【无用】useless；of no use
【无用能】[机] unavailable energy
【无由】not be in a position (to do sth.)；have no way (of doing sth.)
【无忧无虑】free from care；free from all anxieties
【无与伦比】unique；without equal；incomparable；unparalleled
【无原则】unprincipled：～纠纷 an unprincipled dispute／～妥协 unprincipled compromise
【无缘无故】without cause or reason；for no reason at all：～地发脾气 get into a temper for no reason at all
【无源】[无] passive ◇ ～天线 passive antenna
【无源之水】water without a source：～，无本之木 water without a source, a tree without roots
【无韵诗】blank verse
【无政府主义】anarchism ◇ ～者 anarchist
【无知】ignorant：～妄说 ignorant nonsense／出于～ out of ignorance
【无执照营业】interlope ◇ ～者 interloper
【无止境】have no limits；know no end
【无中生有】purely fictitious；fabricated
【无重力】agravic ◇ ～状态 null-gravity state
【无资格】disqualification ◇ ～的证人 incompetent witness
【无足轻重】of little importance；of little consequence；insignificant：～的人物 a nobody；a nonentity
【无罪】innocent；not guilty：～释放 set a person free with a verdict of "not guilty"；acquitted of a charge／宣判～ acquit sb. of a crime ◇ ～证据 evidence of innocence
【无座力炮】recoilless gun

芜 ①(草长得多而乱) overgrown with weeds：荒～ lie waste ②(乱草丛生的地方) grassland：平～ open grassland ③(杂乱) mixed and disorderly；miscellaneous：～词 superfluous words
【芜菁】[植] turnip

【芜杂】mixed and disorderly; miscellaneous

梧

【梧桐】Chinese parasol (tree)

鼯

【鼯鼠】flying squirrel

蜈

【蜈蚣】centipede
【蜈蚣草】ciliate desert-grass

毋

（表示禁止或劝阻）no; not: ～令逃逸。You must not let him escape.
【毋宁】rather...(than); (not so much...) as
【毋庸】need not: ～讳言 no need for reticence／～赘言 need not go into the details

wǔ

武

①（军事的）military: ～官 military officer ②（关于技击的）connected with boxing skill, swordplay, etc.: ～术 *wushu*, martial or physical arts such as shadowboxing, swordplay, etc. ③（猛烈;勇猛）valiant; fierce: 威～ martial-looking ④（武力）force; fight: 动～ use force; start a fight; come to blows
【武备】defense preparations; armaments and military provisions
【武昌鱼】blunt-snout bream
【武打】[戏] acrobatic fighting in Chinese opera or dance
【武断】arbitrary decision; subjective assertion: ～地说 arbitrarily assert／～行事 execute business dogmatically
【武工队】[简]（武装工作队）armed working team
【武官】①（军官）military officer ②[外] military attaché: 海军～ naval attaché／空军～ air attaché ◇～处 military attaché's office
【武库】armory; arsenal
【武力】①（强暴的力量）force ②（军事力量）military force; armed might; armed strength; force of arms: ～从事 resort to arms／～干涉 interfere through use of force／～镇压 armed suppression／诉诸～ resort to force
【武器】weapon; arms: 常规～ conventional weapons／放下～ lay down one's arms／核～ nuclear weapons／轻～ small arms／拿起～ take up arms ◇～装备 weaponry
【武士】①（宫廷卫士）palace guards in ancient times ②（有勇力的人）man of prowess; warrior; knight ◇～俑 warrior figure
【武士道】*bushido*
【武术】*wushu*, martial or physical arts such as shadowboxing, swordplay, etc. ○拳术 quanshu; barehanded exercise; Chinese boxing; pugilistic art／打太极拳 do taijiquan／练拳 practice shadow boxing／螳螂拳 mantis boxing／猴拳 monkey boxing／气功 qigong; breathing exercises／功夫 kung fu／拳 fist／拳法 fist position／指法 finger position／掌法 hand position／腿法 leg position／步法 footwork; steps／扫堂腿 ground sweeping／直刺 straight lunge／斜刺 diagonal stabbing／跳跃 jumping; leaping／扭身 twisting／踢腿 kicking／屈膝 bending of knee／弓腿 bending of leg／弓步 "bow step"／收腹 draw the abdomen in／收臀 keep buttocks pulled in／划弧 make an arc／勾手 hooked hand／实步 "solid step"／虚步 "empty step"／长兵器 long weapon／短兵器 short weapon／枪 spear／大刀 broadsword／单刀 short-hilled broadsword／剑 sword／双剑 double sword／棍 cudgel／三节棍 three-section cudgel／九节鞭 nine-section whip／钩 hook／双钩 double hook／戟 halberd; halbert／流星锤 meteor hammer
【武艺超群】one's military arts excel all
【武装】①（军事装备）arms; military equipment; battle outfit: 解除～ disarm／全副～ (in) full battle gear ②（武装力量）armed forces: ～夺取政权 seizure of power by armed force／人民～ the armed forces of the people ③（用武器装备）equip with arms; supply with arms; arm: ～到牙齿 be armed to the teeth／用马克思列宁主义～头脑 arm one's mind with Marxism-Leninism ◇～部队 armed forces／～冲突 armed conflict; armed clash／～带 Sam Browne belt／～斗争 armed struggle／～干涉 armed intervention／～力量 armed power; armed forces／～起义 armed uprising; armed insurrection／～泅渡 swim with one's weapons; swim in battle gear／～人员 armed personnel

妩

【妩媚】lovely; charming

五

five: ～倍 fivefold; quintuple／～分之一 one fifth／～十 fifty／～十年代 the fifties／～十周年 fiftieth anniversary
【五倍子】[中药] Chinese gall; gallnut ◇～虫 gall makers

【五边形】pentagon
【五步蛇】[动] long-noded pit viper
【五彩】①(五色) the five colors ○青 blue / 黄 yellow / 赤 red / 白 white / 黑 black ②(颜色多) multicolored: ～缤纷 colorful; blazing with color
【五重唱】[乐] (vocal) quintet
【五重奏】[乐] (instrumental) quintet
【五斗柜】chest of drawers
【五毒】the five poisonous creatures ○蝎 scorpion / 蛇 viper / 蜈蚣 centipede / 壁虎 house lizard / 蟾蜍 toad
【五分制】[教] the five-grade marking system
【五更】①(一夜分为五更) the five watches 〈periods〉 of the night ②(第五更) the fifth watch of the night; just before dawn: 起～,睡半夜 retire at midnight and rise before dawn
【五谷】①(指五种粮食) the five cereals: ～不分 cannot tell one cereal from the other ○稻 rice / 黍、稷 millet / 麦 wheat / 豆 beans ②(粮食作物) food crops: ～丰登 an abundant harvest of all food crops
【五官】①[中医] the five sense organs ○耳 ears / 目 eyes / 口 lips / 鼻 nose / 身 tongue ②(脸上的器官) facial features: ～端正 have regular features
【五光十色】①(色彩鲜丽) multicolored; bright with many colors ②(式样繁多) of great variety; of all kinds; multifarious
【五湖四海】all corners of the land
【五花八门】multifarious; of a wide variety
【五花肉】streaky pork
【五级风】[气] force 5 wind; fresh breeze
【五极管】[电子] pentode
【五加】[植] slender acanthopanax
【五讲四美】Five Stresses and Four Points of Beauty ○讲文明 stress on decorum / 讲礼貌 stress on manners / 讲卫生 stress on hygiene / 讲纪律 stress on discipline / 讲道德 stress on moral / 心灵美 beautification of the mind / 语言美 beautification of language / 行为美 beautification of behavior / 环境美 beautification of the environment
【五角星】five-pointed star
【五金】①(五种金属) the five metals ○金 gold / 银 silver / 铜 copper / 铁 iron / 锡 tin ②(金属) metals; hardware ◇ ～厂 hardware factory / ～店 hardware store / ～商 dealer in hardware; ironmonger
【五经】the Five Classics ○《诗经》The Book of Songs / 《书经》The Book of History / 《易经》The Book of Changes / 《礼记》the Book of Rites / 《春秋》The Spring and Autumn Annals
【五里雾】thick fog: 如堕～中 as if lost in a thick fog; utterly mystified
【五敛子】[中药] the fruit of carambola
【五氯硝基苯】[农] pentachloronitrobenzene; PCNB
【五内】[书] viscera: ～俱焚 be rent with grief
【五年计划】Five-Year Plan
【五日京兆】office held for a short time only
【五卅运动】the May 30th Movement (1925)
【五十步笑百步】one who retreats fifty paces mocks one who retreats a hundred; the pot calls the kettle black
【五四青年节】Youth Day (May 4)
【五四运动】the May 4th Movement (1919)
【五体投地】prostrate oneself before sb. in admiration: 佩服得～ admire sb. from the bottom of one's heart; worship sb.
【五味】①(五种味道) the five flavors ○甜 sweet / 酸 sour / 苦 bitter / 辣 pungent / 咸 salty ②(各种味道) all sorts of flavors
【五味子】[中药] the fruit of Chinese magnoliavine
【五线谱】[乐] staff; stave
【五香】①(五种调料) the five spices ○花椒 prickly ash / 八角 star aniseed / 桂皮 cinnamon / 丁香花蕾 clove / 茴香子 fennel ②(调料) spices ◇ ～豆 spiced beans / ～杨梅 spiced plum
【五项全能运动】[体] pentathlon
【五星红旗】the Five-Starred Red Flag
【五星上将】five-star general
【五刑】the five chief forms of punishment in ancient China
【五行】the five elements ○金 metal / 木 wood / 水 water / 火 fire / 土 earth
【五颜六色】of various colors; multicolored; colorful
【五业兴旺】prosperity in framing, forestry, animal husbandry, side-occupations and fishery
【五一国际劳动节】May 1, International Labor Day; May Day
【五音】[乐] the five notes of traditional Chinese music
【五月】①(阳历) May ②(阴历) the fifth month of the lunar year; the fifth moon ◇ ～节 the Dragon Boat Festival
【五脏】[中医] the five internal organs ○心 heart / 肝 liver / 脾 spleen / 肺 lungs / 肾

kidneys ◇ ～六腑 the vital organs of the human body
【五指】the five fingers ○拇指 thumb / 食指 index finger / 中指 middle finger / 无名指 third finger / 小指 little finger
【五子棋】gobang

伍 ①(五) five ②(队伍;军队) army: 革命队～ battalions 〈ranks〉 of the revolution / 入～ join the army / 退～ leave the army ③(伙伴) company: 羞与为～ be ashamed of sb.'s company

捂 seal; cover; muffle: ～鼻子 cover one's nose with one's hand / ～盖子 keep the lid on; cover up the truth / ～着耳朵 stop one's ears

午 noon; midday; meridiem
【午餐肉】pork luncheon meat: 火腿～ ham luncheon meat
【午饭】midday meal; lunch
【午后】afternoon; post meridiem (P.M.; p.m.): ～四点 at 4 p.m.
【午前】forenoon; before noon; morning; ante meridiem (A.M.; a.m.): ～八点 at 8 a.m.
【午睡】①(午觉) afternoon nap; noontime snooze ②(睡午觉) take 〈have〉 a nap after lunch
【午休】noon break; midday rest; noontime rest; lunch hour
【午夜】midnight

忤 ①(不顺从) disobedient: ～逆 disobedient (to parents) ②(不和睦) uncongenial: 与人无～ bear no ill will against anybody

舞 ①(舞蹈) dance: 红绸～ red silk dance / 集体～ group dance / 交谊～ social dancing; ballroom dancing / 秧歌～ *yangko* dance / 腰鼓～ drum dance / 她很会跳～。She dance well. ②(做出舞蹈动作) move about as in a dance: 手～足蹈 dance for joy / 雪花飞～ snowflakes dancing in the air ③(舞动) dance with sth. in one's hands: ～剑 perform a sword-dance / ～龙灯 perform a dragon lantern dance ④(挥舞) flourish; wield; brandish: 挥～大棒 brandish the big stick / 挥～指挥棒 wield the baton
【舞伴】dancing partner
【舞弊】malpractices; irregularities; fraudulent practices; embezzlement
【舞步】dance: 华尔兹和狐步是她跳得最好的～。The waltz and fox trot were the dances she knew.
【舞场】dance hall; ballroom
【舞蹈】dance ◇ ～动作 dance movement / ～家 dancer / ～设计 choreography
【舞蹈病】[医] chorea
【舞动】wave; brandish
【舞会】dance; ball; dance party: 举行～ hold a dance / 开～ give a dance / 他们请我参加～。They asked me to a dance.
【舞剧】dance drama; ballet
【舞弄】wave; wield; brandish: ～刀枪 brandish swords and spears
【舞女】taxi dancer; dancing girl; dance-hostess
【舞曲】dance music; dance
【舞台】stage; arena: 搬上～ present on the stage / 旋转～ revolving stage / 政治～ political arena; political scene; political stage / 在国际～上 in the international arena ◇ ～布景 (stage) scenery; *dècor* / ～工作人员 stagehand / ～幻灯机 cloudmachine / ～记录片 stage documentary / ～监督 stage direction / ～设计 stage design / ～效果 stage effect / ～艺术 stagecraft / ～照明 stage elimination; lighting
【舞厅】ballroom; dance hall
【舞文弄墨】①(歪曲法律条文作弊) pervert the law by playing with legal phraseology ②(玩弄文字技巧) engage in phrase-mongering
【舞榭歌台】entertainment setups

侮 insult; bully: 不可～ not to be bullied / 外～ foreign aggression
【侮慢】slight; treat disrespectfully
【侮辱】insult; humiliate; subject sb. to indignities: 莫大的～ gross insult / 忍受～ swallow an insult; bear an insult

wù

误 ①(错) mistake; error: 笔～ a slip of the pen / 口～ a slip of tongue / 失～ error ②(耽误) miss; fail to seize the right moment: ～了火车 miss a train / ～了农时 miss the farming season; let the farming season slip by ③(使受害) harm: ～人子弟 harm the younger generation; lead young people astray ④(不是故意) by mistake; by accident: ～伤 accidentally injure
【误差】error: 概然～ probable error / 平均～ mean error; average error / 仪器～ instrumental error / ～不超过千分之三 with a tolerance

of less than three-thousandths ◇ ～函数 error function / ～率 error rate
【误点】late; overdue; behind schedule: 班机～三小时。 The airliner was three hours late. / 火车～了。 The train is overdue. (或: The train falls behind its schedule.)
【误工】① (耽误了工作) delay one's work ② (耽误时间) loss of working time
【误会】① (误解对方的意思) misunderstand; mistake; misconstrue ② (对对方意思的误解) misunderstanding: 产生～ produce misunderstandings / 引起～ give rise to misunderstanding / 消除～ dispel 〈remove〉 misunderstanding
【误解】① (理解得不正确) misread; misunderstand: 你～了她的话。 You misunderstood what she said. ② (不正确的理解) misunderstanding: 消除双方的～ dissipate misunderstandings existing between both sides
【误判】[法] erroneous judgement
【误入歧途】[法] go astray; be misled
【误杀】[法] manslaughter
【误伤】① (无意中伤害人体) accidentally injure ② (无意中使人受的伤) accidental injury
【误事】① (耽误事情) cause delay in work or business; hold things up ② (把事情搞糟了) bungle matters

恶 loathe; dislike; hate: 好～ likes and dislikes / 可～ loathsome; hateful

痦
【痦子】naevus; mole

悟 realize; awaken: ～出其中的道理 realize why it should be so
【悟性】power of understanding; comprehension

焐 warm up: 把被褥～热 warm up the bedding / 用热水袋～一～手 warm one's hands with a hot-water bottle

晤 meet; interview; see
【晤面】meet; see
【晤谈】meet and talk; have a talk; interview

兀 ① (高高地突起) rising to a height; towering ② (秃) bald
【兀鹫】[动] griffon vulture
【兀立】stand upright
【兀自】still

务 ① (事务) affair; business: 公～ official business / 任～ task; job ② (从事) be engaged in; devote one's efforts to: ～农 be engaged in agriculture; be a farmer / 不～正业 not engage in honest work; not attend to one's proper duties ③ (务必) must; be sure to
【务必】must; be sure to: 你今晚～来一下。 Be sure to come here this evening.
【务实】deal with concrete matters relating to work
【务虚】discuss principles; discuss ideological guidelines

雾 ① [气] fog: 薄～ mist / 这个城市冬天有大～。 This city has bad fogs in winter. ② (象雾的许多小水点) fine spray: 喷～器 sprayer
【雾标】[交] fog buoy
【雾滴】[环保] droplet
【雾号】[交] fog signal
【雾化器】atomizer
【雾气】fog; mist; vapor
【雾凇】[气] (soft) rime ◇ ～雾 rime fog

勿 [副] (表示禁止或劝阻) 请～吸烟! No Smoking!

芴 [化] fluorene

物 ① (东西) thing; matter: 读～ reading matter / 废～ waste matter / 公～ public property / 矿～ minerals / 地大～博 vast territory and rich resources / 以～易～ barter ② (自己以外的人或跟自己相对的环境) the outside world as distinct from oneself; other people: 待人接～ the way one gets along with people ③ (内容;实质) content; substance: 言之无～ talk or writing devoid of substance
【物产】products; produce
【物故】pass away; die
【物归原主】return sth. to its rightful owner; things return to their proper owners
【物候学】phenology
【物换星移】things change with the passing of years; change of the seasons
【物极必反】things will develop in the opposite direction when they become extreme
【物价】(commodity) prices: ～波动 price fluctuation / 提高～ raise prices / 调整～ regulate prices; adjust prices / ～飞涨。 Prices skyrocketed. / ～稳定。 Prices remain stable.

◇～政策 pricing policy／～指数 price index
【物件】thing; article
【物尽其用】make the best use of everything; let all things serve their proper purpose
【物镜】[物] objective (lens)
【物理】①(事物的内在规律) innate laws of things ②(物理学) physics ◇～变化 physical change／～化学 physical chemistry／～疗法 physical therapy; physiotherapy
【物理学】physics: 低温～ low temperature physics／地球～ geophysics／高能～ high energy physics／固体～ solid state physics／理论～ theoretical physics／天体～ astrophysics／应用～ applied physics／原子～ atomic physics／原子核～ nuclear physics ◇～家 physicist ○力学 mechanics／热学 heat／热力学 thermodynamics／光学 optics; photology／声学 acoustics／电学 electricity／磁学 magnetism
【物力】material resources: 节约人力～ use manpower and material resources sparingly
【物美价廉】cheap and fine; excellent quality and reasonable price
【物品】article; goods: 贵重～ valuables／零星～ sundries; odds and ends／免税～ duty-free articles／违禁～ contraband／应上税～ dutiable articles／自用～ personal effects
【物色】look for; seek out; choose: ～人才 look for talented persons
【物伤其类】like feels for like
【物体】body; substance; object: 气态～ gaseous substance／透明～ a transparent substance; a transparent object／运动～ a body in motion
【物物交换】barter one thing for another
【物以类聚】things of one kind come together; like attracts like; birds of a feather flock together
【物以稀为贵】When a thing is scarce, it is precious. (或: A thing is valued if it is rare.)
【物议】criticism from the people: 免遭～ so as to avoid public censure; so as not to incur criticism by the masses
【物有本末,事有始终】there is a proper sequence of foundation and end-results
【物证】material evidence
【物质】matter; substance; material ◇～不灭定律 the law of conservation of matter／～财富 material wealth／～产品 physical product／～储备 reserve supply; stockpile／～刺激 material incentive／～鼓励 material reward／～基础 material base／～奖励 material award／～力量 material strength; material force／～利益 material benefits; material gains／～生活 material life／～世界 the material world; the physical world／～条件 material conditions or prerequisites／～文明 material civilization／～性[哲] materiality／～损失 material damage／～运动 the motion of matter／～资料 material goods
【物种】[生] species: ～起源 the origin of species
【物资】goods and materials ◇～调度 distribution of materials／～管理 handling of goods and materials／～交流 interflow of commodities／～消耗 consumption of materials

坞 a depressed place: 船～ dock／花～ sunken flower-bed

鹜 duck: 趋之若～ scramble for sth.

骛 go after; seek for: 好高～远 reach for what is beyond one's grasp; aim too high

X

xī

曦 sunlight: 晨～ early morning sunlight

熹 dawn; brightness

【熹微】 dim; pale: 晨光～ the dim light of dawn; first faint rays of dawn

嘻

【嘻嘻哈哈】 laughing and joking; laughing merrily; mirthful

嬉 play; sport

【嬉皮士】hippy; hippie

【嬉皮笑脸】 grinning cheekily; smiling and grimacing

【嬉戏】 play; sport

【嬉笑】 be laughing and playing: 孩子们的～声 the happy laughter of children at play

【嬉笑怒骂】 laughing merrily or cursing angrily

昔 former times; the past: 今～对比 contrast the past with the present

【昔日】 in former days; in the past; in the old days

惜 ①(爱惜) cherish; value highly; care for tenderly: 爱～ cherish; treasure ②(吝惜) spare; grudge; stint: 不～工本 spare neither labor nor money; spare no expense ③(惋惜) have pity on sb.; feel sorry for sb.: 惋～ feel sorry for

【惜别】 be reluctant to part; hate to see sb. go: 怀着～的心情 with reluctance to part

【惜力】 be sparing of one's energy; not do one's best: 他干活从不～。 He never spares himself in his work.

【惜指失掌】 stint a finger only to lose the whole hand; try to save a little only to lose a lot

熙

【熙熙攘攘】 bustling with activity; with people bustling about

析 ①(分开;散开) divide; separate: 分崩离～ fall to pieces; come apart ②(分析) analyse; dissect; resolve: ～义 analyse the meaning (of a word, etc.)

【析出】[化] separate out

【析象管】[电子] image dissector

【析疑】 resolve a doubt; clear up a doubtful point

晰 clear; distinct: 明～ clear; lucid / 清～ distinct

蜥

【蜥蜴】lizard

西 ①(方向) west: ～屋 west room / 华～ west China / 上海以～ to the west of Shanghai / 往～去 head west / 向正～航行 sail due west ②(西洋;西方) Occidental; Western: ～乐 Western music

【西班牙】 Spain ◇ ～人 Spaniard / ～语 Spanish (language)

【西半球】the Western Hemisphere

【西北】①(指方向) northwest ②(中国西北地区) northwest China; the Northwest ◇ ～风 northwest wind; northwesterly wind

【西部片】[电影] Western

【西餐】 Western-style food

【西点】 Western-style pastry

【西番莲】①(藤本观赏植物) passionflower ②(大丽花) dahlia

【西方】①(方向) the west; westward ②(欧美各国;欧洲资本主义国家和美国) the West; the Occident: ～国家 the Western countries

【西风】 west wind; westerly wind ◇ ～带[气] westerlies

【西服】 Western-style clothes ◇ ～料 suiting

【西府海棠】 midget crabapple

【西瓜】 watermelon ◇ ～子 watermelon seed

【西红柿】 tomato

【西葫芦】 pumpkin; summer squash

【西化】 occidentalize; occidentalizing

【西经】[地] west longitude: ～一百六十五度 longitude 165° W.

【西力生】[农] ceresan

【西南】①(指方向) southwest ②(中国西南地区) southwest China; the Southwest ◇ ～风 southwest wind; southwesterly wind

【西欧】 Western Europe

【西晒】 (of a room) with a western exposure; facing west, and hot on summer afternoons

【西式】 Western style: ～点心 Western-style pastry

【西天】[佛教] Western Paradise

【西维因】[农] sevin; carbaryl

【西文粗体字】 antiqua

【西文罗马体】roman
【西洋】the West; the Western world ◇ ～人 Westerner／～文学 Western literature
【西洋景】①(画片) peep show ②(故弄玄虚的事物或手法) hanky-panky; trickery: 拆穿～ expose sb.'s tricks; strip off the camouflage
【西洋参】American ginseng
【西药】Western medicine
【西医】①(指医学) Western medicine ②(指医生) a doctor trained in Western medicine
【西乐】Western music
【西藏野驴】kiang; kyang
【西装】Western-style clothes: ～革履 in Western dress and leather shoes

硒 [化] selenium (Se)
【硒光电池】selenium cell

牺

【牺牲】①(祭祀用的牲畜) a beast slaughtered for sacrifice; sacrifice ②(舍弃生命) sacrifice oneself; die a martyr's death; lay down one's life: 英勇～ die a heroic death ③(放弃或损害一方的利益) sacrifice; give up; do sth. at the expense of: ～个人利益 sacrifice one's personal interests／～休息时间 give up one's spare time (to do sth.) ◇ ～品 victim; prey

锡 [化] tin (Sn)
【锡箔】tinfoil paper
【锡匠】tinsmith
【锡金】sikkim: ～人 Sikkimese
【锡矿】tin ore
【锡石】[矿] cassiterite; tinstone
【锡杖】[宗] Nuddhist monk's staff
【锡纸】silver paper; tinfoil

吸 ①(引入体内) inhale; breathe in; draw: ～进新鲜空气 inhale fresh air／深深～一口气 draw a deep breath ②(吸收) absorb; suck up: 折信前,你先把信上的墨迹～干。Blot your letter before folding it.／这种纸不～水。This kind of paper does not absorb ink.
【吸杯】suction-cup
【吸尘器】dust catcher; dust collector: 真空～ vacuum cleaner
【吸虫】fluke: 肺～ lung fluke／肝～ liver fluke／血～ blood fluke ◇ ～病 fluke disease; trematodiasis
【吸毒】drug taking ◇ ～者 drug addict; drug-dependent
【吸附】[化] adsorption ◇ ～剂 adsorbent／～器 adsorber／～水 adsorbed water／～作用 adsorption
【吸力】suction; attraction: 地心～ force of gravity／相互～ mutual attraction ◇ ～计 suction gauge
【吸墨纸】blotting paper
【吸奶器】breast pump
【吸泥泵】dredge pump; scum pump
【吸盘】[动] sucking disc; sucker
【吸取】absorb; draw; assimilate: ～精华 absorb the quintessence／～教训 draw a lesson／～水分 absorb water
【吸热】absorption of heat ◇ ～反应 [化] endothermic reaction
【吸声】[建] sound absorption ◇ ～材料 sound-absorbing material; acoustic absorbent
【吸湿】moisture absorption ◇ ～剂 hygroscopic agent／～性 hygroscopicity
【吸食】suck; take in
【吸收】①(吸到内部) absorb; suck up; assimilate; imbibe; draw: ～水分 suck up moisture／～外资 absorb foreign capital／～养分 assimilate nutriment／～知识 absorb knowledge; imbibe knowledge ②(接受) recruit; enrol; admit: ～入党 admit into the Party／～新成员 recruit new members ◇ ～光谱 [物] absorption spectrum／～剂 absorbent／～率 absorptivity／～塔 [化] absorption tower／～作用 [物] absorption
【吸吮】suck; absorb
【吸铁石】magnet; lodestone
【吸血鬼】bloodsucker; vampire
【吸烟】smoke: 禁止～。No smoking. ◇ ～室 smoking room
【吸引】attract; draw; fascinate: ～注意力 attract attention／对观众很有～力 have a strong appeal to the audience ◇ ～器 [医] aspirator

奚

【奚落】scoff at; taunt; gibe

溪 small stream; brook; rivulet
【溪涧】mountain stream
【溪流】brook; rivulet

蹊 footpath
【蹊径】path; way

翕 ①(和顺) amiable and compliant ②(收敛) furl; fold; shut: ～张 furl and unfurl; close and open

希 ①(希望) hope: 敬～指教。 Your advice is hereby earnestly solicited. ②(稀少) rare; scarce; uncommon

【希伯来语】Hebrew (language)

【希罕】①(希奇) rare; scarce; uncommon ②(认为希奇而喜爱) value as a rarity; cherish: 你不～,我还～呢。You may not cherish it, but I do. ③(希罕的事物) rare thing; rarity: 看～儿 enjoy the rare sight of sth.

【希冀】hope for; wish for; aspire after

【希腊】Greece ◇ ～人 Greek / ～神话 Greek mythology / ～语 Greek (language) / ～正教 the Greek Orthodox Church / ～字母 the Greek alphabet

【希奇】rare; strange; curious: 这不是～的事情。This is nothing strange.

【希世之珍】a rare treasure

【希图】harbor the intention of; try to; attempt to: ～蒙混过关 try to wangle; try to get by under false pretenses / ～牟取暴利 go after quick profits

【希望】hope; wish; expect: ～落了空 fail to attain one's hope / 把～变成现实 turn hopes into reality / 有成功的～ have a hope of success / 有一线～。There is a ray of hope.

烯

【烯烃】[化] alkene

稀 ①(稀少) rare; scarce; uncommon: 物以～为贵。When a thing is scarce, it is precious. ②(稀疏) sparse; scattered: 地广人～。It is a vast area but sparsely populated. / 月明星～。The moon is bright and the stars are few. ③(稀薄;含水多) watery; thin: 粥太～了。The gruel is too thin.

【稀薄】thin; rare: ～的空气 thin air

【稀饭】rice or millet gruel; porridge

【稀客】rare visitor

【稀烂】①(极烂) completely mashed; pulpy: 肉煮得～。The meat was cooked to a pulp. ②(极破碎) smashed to pieces; broken to bits

【稀里糊涂】not knowing what one is about; muddleheaded

【稀少】few; rare; scarce: 人口～ a sparse population

【稀释】[化] dilute ◇ ～测定 dilution metering / ～剂 diluent; thinner

【稀疏】few and scattered; few and far between; thin; sparse: ～的晨星 a few scattered morning stars / ～的枪声 scattered shots; sporadic firing / ～的头发 thin hair; sparse hair / 树木～。The trees are sparse.

【稀树草原】savanna

【稀松】①(差劲) poor; sloppy: 你这活干得太～了。It's a sloppy job you're doing. ②(无关紧要) unimportant; trivial: 别把这些～的事放在心里。Don't take such trivial matters to heart.

【稀土金属】rare-earth metal

【稀土元素】rare-earth element

【稀稀拉拉】sparse; thinly scattered

【稀有】rare; unusual

【稀有金属】rare metal

【稀有元素】rare element

悉 ①(全;尽) all; entirely: ～力 go all out; spare no effort ②(知道) know; learn; be informed of: 惊～ be shocked to learn / 熟～ know very well

【悉数】①(数清;列举) enumerate in full detail: 不可～ too many to enumerate ②(全部) all; every single one: ～奉还 return all that has been borrowed or taken away

【悉听尊便】You are free to try anything you like.

【悉心】devote all one's attention; take the utmost care: ～研究 devote oneself to the study of sth. / ～照料病人 take the utmost care of the patient

蟋

【蟋蟀】cricket: 斗～ cricketfight

【蟋蟀草】yard grass

息 ①(呼吸进出的气) breath: 屏～ hold one's breath / 战斗到最后一～ fight to one's last breath ②(消息) news: 信～ news; message ③(停止) cease; stop: ～怒 cease to be angry / 经久不～的掌声 prolonged applause ④(休息) rest ⑤(滋生;繁殖) grow; multiply: 蕃～ multiply greatly ⑥(利息) interest: 存～ interest on deposit / 年～ annual interest / 无～贷款 interest-free loan

【息怒】cease to be angry; calm one's anger

【息票】interest coupon

【息肉】[医] polyp; polypus

【息事宁人】①(从中调解) patch up a quarrel and reconcile the parties concerned ②(在纠纷中自行让步) make concessions to avoid trouble; gloss things over to stay on good terms

【息息相关】be closely linked; be closely bound up; related as closely as each breath is to the

next
【息摺】interest pass-book

熄 extinguish; put out: ～灯 put out the light
【熄灯号】lights-out; taps
【熄灭】go out; die out: 火～了。The fire has gone out.

夕 ①(傍晚) sunset: 朝发～至 start at daybreak and arrive at sunset ②(晚上) evening; night: ～烟 evening mist / 除～ New Year's Eve / 旦～ this morning or evening; in a short time
【夕阳】the setting sun: ～西下 the sun is setting; the sun is going down; the evening sun sinks in the west
【夕照】the glow of the setting sun; evening glow

汐 tide during the night; nighttide

矽 [化] silicon (Si)
【矽尘】silicious dust
【矽肺】[医] silicosis
【矽钢】[冶] silicon steel

膝 knee: 双～跪下 kneel down (on both knees)
【膝盖】knee ◇ ～骨 kneecap; patella
【膝关节】knee joint ◇ ～炎 gonitis; gonarthritis
【膝腱反射】[医] knee jerk
【膝内翻】out knee; bowleg; bow legs
【膝外翻】in knee; knock knee

犀 rhinoceros: 黑～ black rhinoceros
【犀角】rhinoceros horn
【犀利】sharp; incisive; trenchant: ～的目光 sharp eyes / ～的批评 incisive criticism / 谈锋～ incisive in conversation / 文笔～ a trenchant pen
【犀鸟】hornbill
【犀牛】rhinoceros

xí

席 ①(编织的席子) mat: 编～ weave a mat / 草～ straw mat ②(席位) seat; place: 记者～ seats for the press / 来宾～ seats for visitors / 入～ take one's seat / 在议会选举中获得十～ win ten seats in the parliamentary election / 在议会选举中失去二十～ lose twenty seats in the parliamentary election ③(酒席) feast; banquet; dinner: ～间宾主频频举杯。Host and guests frequently raised their glasses during the feast.
【席不暇暖】not sit long enough to warm the seat; be in a tearing hurry; be constantly on the go
【席次】the order of seats; seating arrangement; one's place among the seats arranged: 按指定～入座 take one's assigned seat; sit down in one's place
【席地】on the ground: ～而坐 sit on the ground; be seated on the floor
【席卷】①(象卷席子一样把东西卷进去) roll up like a mat; carry everything with one; take away everything: ～而去 make off with everything that one can lay hands on; roll the mat with everything and run away ②(全部卷进去; 横卷) sweep across; engulf: 经济危机～了整个资本主义世界。An economic crisis engulfed the entire capitalist world.
【席棚】mat shed
【席位】seat

檄
【檄文】①(征召的文书) an official call to arms ②(声讨的文书) an official denunciation of the enemy

袭 ①(袭击;侵袭) make a surprise attack on; raid: 偷～ surprise attack; sneak raid / 夜～ night raid / 花香～人 the fragrance of flowers assails one's nose ②(照样做) follow the pattern of; carry on as before: 抄～ plagiarize / 因～ carry on (an old tradition, etc.)
【袭击】make a surprise attack on; surprise; raid: ～敌军阵地 make a surprise attack on the enemy positions / 受到台风的～ be hit by a typhoon
【袭取】①(出其不意地夺取) take by surprise ②(袭用) take over
【袭扰】[军] harassing attack
【袭用】take over (sth. that has long been used in the past): ～古方 take over an age-old recipe / ～老谱 follow old practice

媳 daughter-in-law
【媳妇】①(儿媳) son's wife; daughter-in-law ②(晚辈亲属之妻) the wife of a relative of the younger generation: 孙～ grandson's wife / 侄～ nephew's wife
【媳妇儿】①(妻子) wife ②(年轻已婚妇女) a young married woman

习 ①(温习;练习) practice; exercise; review: 复～ review (one's lessons)／自～ study by oneself ②(熟习) get accustomed to; be used to; become familiar with: ～闻 often hear／不～水性 be not good at swimming ③ (习惯) habit; custom; usual practice: 积～ old habit; longstanding practice／陋～ bad custom

【习非成是】accept what is wrong as right as one grows accustomed to it

【习惯】①(适应) be accustomed to; be used to; be inured to: 他～于做艰苦的工作。He is accustomed to hard work. ②(长期养成的行为、倾向) habit; custom; usual practice: 从小培养劳动～ cultivate the habit of doing manual labor from childhood／～成自然。Habit becomes second nature. (或: Once you form a habit, it comes natural to you.) ◇ ～法 common law; customary law／～继承人 heir by custom; heir custom／～势力 force of habit

【习惯性流产】habitual abortion

【习见】(of things) commonly seen: ～的现象 a common sight

【习气】bad habit; bad practice: 官僚～ habitual practice of bureaucracy; bad bureaucratic habits

【习俗】custom; convention

【习题】exercises (in school work) ◇ ～解答 key to the exercises

【习习】blow gently: 微风～。A gentle breeze is blowing.

【习性】habits and characteristics

【习焉不察】too accustomed to sth. to call it in question

【习以为常】be used to sth.; be accustomed to sth.

【习用】habitually use

【习与性成】habits become one's second nature

【习字】practice penmanship; do exercises in calligraphy ◇ ～帖 copybook; calligraphy model

【习作】①(练习写作) do exercises in composition ②(练习的作业) an exercise in composition, drawing, etc.

xǐ

喜 ①(快乐;高兴) happy; delighted; pleased: ～获丰收 reap a bumper harvest／～降瑞雪 (there is) a welcome fall of seasonable snow／心中暗～ secretly feel pleased ②(可庆贺的事) happy event (esp. wedding); occasion for celebration: 报～ report good news／大～的日子 a day of great happiness; a joyful occasion; an occasion for celebration ③(怀孕) pregnancy: 有～ be expecting; be in the family way ④(爱好) be fond of; like; have an inclination for: ～读书 be fond of reading／～交游 like to make friends

【喜爱】like; love; be fond of; be keen on: ～户外活动 be keen on outdoor activities

【喜报】a bulletin of glad tidings: 立功～ a bulletin announcing meritorious service

【喜不自胜】be in heaven; be delighted beyond measure

【喜冲冲】look exhilarated; be in a joyful mood

【喜出望外】be overjoyed; be pleasantly surprised

【喜从天降】an unexpected piece of good fortune; heaven－sent fortune

【喜好】like; love; be fond of; be keen on

【喜欢】①(喜爱) like; love; be fond of; be keen on: 他～在大河里游泳。He likes to swim in big rivers. ②(高兴) happy; elated; filled with joy: 让大家～ make everybody happy

【喜酒】①(指酒) wine drunk at a wedding feast ②(指酒席) wedding feast

【喜剧】comedy: 情节～ comedy of situation; situation comedy／性格～ comedy of character ◇ ～演员 comedian; comedienne

【喜马拉雅山】the Himalayas

【喜怒无常】subject to changing moods

【喜气洋洋】full of joy; jubilant

【喜庆】①(值得喜欢和庆贺的) joyous; jubilant: ～的日子 day of jubilation; festival day; happy occasion ②(值得喜欢和庆贺的事) happy event ③(喜欢地庆贺) celebrate: ～丰收 celebrate the bumper harvest

【喜鹊】magpie

【喜人】gratifying; satisfactory: ～的景象 scenes of joy／取得～的成果 achieve satisfactory results

【喜色】happy expression; joyful look: 面有～ wear a happy expression

【喜上眉梢】be radiant with joy

【喜事】①(高兴的事) happy event; joyous occasion ②(结婚) wedding

【喜闻乐见】love to see and hear; love: 人民～的艺术形式 the forms of art loved by the people

【喜笑颜开】light up with pleasure; be wreathed in smiles

【喜新厌旧】love the new and loathe the old; be

fickle in affection
【喜形于色】be visibly pleased; light up with pleasure
【喜讯】happy news; good news; glad tidings
【喜洋洋】beaming with joy; radiant
【喜雨】seasonable rain; a welcome fall of rain: 普降~ a widespread fall of seasonable rain; a seasonable fall of rain over a wide area
【喜悦】happy; joyous: 怀着万分~的心情 with a feeling of immeasurable joy
【喜滋滋】feeling pleased; filled with joy

洗 ①(用水等洗) wash; bathe: ~掉脸上的灰尘 wash the dust off one's face / ~干净 wash sth. clean / ~净患处 wash the affected part / ~伤口 bathe a wound / ~衣服 wash clothes / 碧空如~ a cloudless blue sky / 用水冲~汽车 wash down a car ②[宗] baptize: 受~ receive baptism; be baptized ③(除去) redress; right: ~冤 right a wrong; redress a grievance ④(杀光或抢光) kill and loot; sack: ~城 massacre the inhabitants of a captured city / 血~ plunge (the inhabitants) in a bloodbath; massacre ⑤(冲洗;显影定影) develop: ~胶卷 develop a film ⑥(洗牌) shuffle: ~骨牌 shuffle the dominoes
【洗必太】chlorhexidine; hibitane
【洗肠】intestinal lavage
【洗尘】give a dinner of welcome (to a visitor from afar)
【洗涤】wash; cleanse ◇ ~槽[化] washing tank / ~剂 detergent / ~器[化] scrubber / ~塔[化] washing tower
【洗耳恭听】listen with respectful attention; I'm all ears.
【洗耳器】aurilave
【洗发剂】shampoo
【洗剂】[药] lotion
【洗劫】loot; sack
【洗礼】①[宗] baptism ②(重大斗争的锻练和考验) severe test: 炮火的~ the baptism of fire / 受过战斗的~ have gone through (the test of) battle
【洗礼式】[宗] christening
【洗脸面巾】washcloth; washrag
【洗脸盆】washbasin; washbowl
【洗煤】coal washing ◇ ~厂 coal washery; coal cleaning plant / ~机 coal washer
【洗片】[摄] develop (a film) ◇ ~机 developing machine / ~夹 film clip / ~架 film hanger
【洗染店】cleaners and dyers; laundering and dyeing shop
【洗手】①(改邪归正) stop doing evil and reform oneself ②(不再干某项职业) wash one's hands of sth.
【洗刷】①(用水洗刷) wash and brush; scrub: ~地板 scrub the floor ②(除去耻辱、污点等) wash off; clear oneself of: ~自己的罪恶 wash away one's sin
【洗头】wash one's hair; shampoo
【洗碗碟机】washed-up; washing-up machine
【洗胃】gastric lavage
【洗心革面】turn over a new leaf; thoroughly reform oneself
【洗选】[矿] washing
【洗雪】wipe out (a disgrace); redress (a wrong)
【洗眼杯】eyecup
【洗衣】wash clothes; do one's washing ◇ ~板 washboard / ~店 laundry / ~粉 washing powder / ~刷 wash brush
【洗衣机】washing machine; washer: 双缸~ washer-dryer
【洗印】[摄] developing and printing; processing ◇ ~机 (film) processor
【洗澡】have a bath; take a bath; bathe: 她总是用冷水~。 She always washes in cold water.

铣 mill
【铣床】[机] milling machine; miller
【铣刀】[机] milling cutter
【铣工】①(指工作) milling (work) ②(指工人) miller; milling machine operator

徙 move (from one place to another)
【徙居】move house: ~内地 move up-country

玺 imperial or royal seal

xì

阋 quarrel; strife: 兄弟~于墙。 Brothers quarrel at home.

隙 ①(缝隙;裂缝) crack; chink; crevice: 墙~ a crack in the wall / 云~ a rift in the clouds ②(空闲) gap; interval: 农~ interval between busy seasons in farming ③(机会) loophole; opportunity: 乘~突围 seize an opportunity to break through the encirclement / 无~可乘 no loophole to take advantage of ④(感情上的裂痕) discord; rift: 并无嫌~ bear no ill will; bear no grudge
【隙地】unoccupied place; open space

系 ①(系统) system; series: 派～ faction / 太阳～ the solar system / 语～ (language) family ②(高校院系) department (in a college); faculty: 化学～ the department of chemistry / ～主任 dean ③(关联) relate to; bear on: 成败～于此举 stand or fall by this / 名誉所～ have a direct bearing on one's reputation ④(牵挂) feel anxious; be concerned: ～念 feel concerned about ⑤(栓;绑) tie; fasten: ～马 tether a horse ⑥(是) be: 纯～ be purely / 她讲的确～实情。What she says is the actual situation.

【系词】①[逻] copula ②[语] copulative verb; linking verb

【系列】series; set: 一～的问题 a series of problems / 一～政策 a whole set of policies / 运载～ vehicle series ◇ ～化 seriation

【系念】be anxious about; worry about; feel concerned about

【系谱】pedigree ◇ ～学 genealogy

【系数】[数] coefficient: 光学～ optical coefficient

【系统】①(按一定关系组成的同类事物) system: 财贸～ departments of trade and finance and affiliated organizations / 灌溉～ irrigation system / 通过组织～ through organizational channels ②(有条理的;有系统的) systematic: 作～的研究 make a systematic study / ～地说明 explain in a systematic way ◇ ～工程 systems engineering / ～化 systematize / ～性 systematicness

戏 ①(玩耍;游戏) play; sport: 嬉～ sport; have fun ②(开玩笑;嘲弄) make fun of; joke: ～言 say something for fun; joke ③(戏剧;杂技) drama; play; show: 京～ Beijing opera / 马～ circus show; circus performance / 去看～ go to the theatre

【戏班】theatrical troupe; theatrical company

【戏词】actor's part; actor's lines

【戏法】conjuring; juggling; tricks; magic: 变～ juggle; conjure; perform tricks / ～人人会变，各有巧妙不同。Every juggler has his own tricks. (或: Many are the magicians, but each has his own tricks.)

【戏剧】drama; play; theatre: 现代～ modern drama; the modern theatre / 一个富有～性的事件 a dramatic event ◇ ～家 dramatist / ～界 theatrical circles / ～评论 dramatic criticism

【戏迷】theatre fan

【戏目】theatrical program

【戏弄】make fun of; play tricks on; tease; kid: 你在～我。You're kidding me!

【戏曲】traditional opera: 地方～ local operas

【戏台】stage

【戏谑】banter; crack jokes

【戏院】theatre

【戏照】a photo of a person in stage costume

【戏装】theatrical costume; stage costume

细 ①(条状物横剖面小) thin; slender: ～铁丝 thin wire / ～线 fine thread / ～腰 slender waist ②(颗粒小) in small particles; fine: ～沙 fine sand ③(音量小) thin and soft: ～嗓子 a thready voice ④(精细) fine; exquisite; delicate: ～瓷 fine porcelain / 粗粮～作 make delicacies out of coarse food grain ⑤(仔细;详细;周密) careful; meticulous; detailed: ～看 examine carefully; scrutinize / ～问 make detailed inquiries; ask about details / 工作做得～ be meticulous in one's work ⑥(细微;细小) minute; trifling: ～节 minute detail / 分工很～ have an elaborate division of labor / 事无巨～ a matters, big and small

【细胞】cell: 植物的～ plant cells ◇ ～壁 cell wall / ～分裂 cell division / ～构造 cell structure / ～核 cell nucleus / ～呼吸 cellular respiration / ～膜 cell membrane / ～学 cytology / ～质 cytoplasm

【细布】fine cloth

【细部】detail (of a drawing)

【细长】long and thin; tall and slender: ～的身材 a tall and slender figure

【细大不捐】reject nothing, big or small

【细纺】finespun

【细高挑儿】①(细长身材) a tall and slender figure ②(身材细长的人) a tall, slender person

【细工】fine workmanship

【细活】a job requiring fine workmanship or meticulous care; skilled work

【细货】small wares

【细嚼慢咽】take one's time in eating

【细节】details; particulars

【细菌】germ; bacterium ◇ ～肥料 bacterial fertilizer / ～农药 bacterial pesticide / ～武器 bacteriological weapon; germ weapon / ～学 bacteriology / ～学家 bacteriologist / ～战 bacteriological warfare; germ warfare

【细粮】flour and rice

【细毛】fine, soft fur

【细毛羊】fine-wool sheep

【细密】①(质地仔密) fine and closely woven; close: ～的纹理 a close grain / 质地～ of close texture / 针脚～ in fine close stitches / ②(不疏忽大意;仔细) meticulous; detailed: ～的分析 a detailed analysis
【细木工】①(手艺) joinery ②(指人) joiner; cabinetmaker
【细目】detailed catalogue; specific item; detail
【细嫩】delicate; tender: ～的皮肤 delicate skin
【细腻】①(精细光滑) fine and smooth ②(细致入微) exquisite; minute: ～的表演 an exquisite performance / ～的描写 a minute description
【细巧】exquisite; dainty; delicate: ～的图案 an exquisite design
【细绒线】fingering yarn
【细软】jewelry, expensive clothing and other valuables
【细弱】thin and delicate; slim and fragile: ～的身子 of slim and delicate build / 声音～ a feeble voice
【细纱】[纺] spun yarn ◇ ～机 spinning frame
【细声细气】in a soft voice; soft-spoken
【细石器】[考古] microlith ◇ ～文化 microlithic culture
【细水长流】①(节约人力物力使经常不缺) economize to avoid running short ②(点滴地不间断地做某事) go about sth. little by little without a letup
【细说本末】recount the development from the beginning
【细碎】in small, broken bits: ～的脚步声 the sound of light and hurried footsteps
【细微】slight; fine; subtle: ～的变化 slight changes; subtle changes / ～差别 a fine distinction; a subtle difference
【细小】very small; tiny; fine; trivial: ～的零件 small parts (of a machine) / ～的事情 trivial matters / ～的雨点 tiny raindrops
【细心】careful; attentive: ～观察 carefully observe / ～护理伤员 nurse the wounded with care / ～听讲 listen to the lecture attentively
【细雨】drizzle; fine rain
【细则】detailed rules and regulations; by-laws
【细帐】itemized account
【细针密缕】in fine, close stitches; (work) in a meticulous way
【细枝末节】minor details; nonessentials
【细支纱】[纺] fine yarn
【细致】careful; meticulous; painstaking: 很～的活 a careful piece of work / 做～的思想工作 do painstaking ideological work

xiā

瞎 ①(丧失视觉) blind: ～了一只眼 blind in one eye ②(胡乱) aimlessly; groundlessly; foolishly; to no purpose: ～猜 make a wild guess / ～费劲儿 make a vain effort / ～干 go it blind / ～花钱 speak money foolishly / ～讲 speak groundlessly
【瞎扯】①(没有根据地乱说) talk irresponsibly; talk rubbish ②(没有中心地乱说) talk at random about anything under the sun; waffle; natter
【瞎话】untruth; lie: 说～ tell a lie; lie
【瞎猫逮死耗子】a blind cat caught a dead rat
【瞎闹】①(胡闹) act senselessly; mess about ②(没有事由或没有效果地做) fool around; be mischievous
【瞎炮】misfire ◇ ～孔眼 bootleg
【瞎说】talk irresponsibly; talk rubbish: ～八道 talk nonsense
【瞎指挥】issue confused orders; give arbitrary and impracticable directions; mess things up by giving wrong orders
【瞎抓】do things without a plan; go about sth. in a haphazard way
【瞎子】a blind person: ～点灯白费蜡 lighting a candle for a blind person; a sheer waste / ～摸象 the blindman feels an elephant; take a part for the whole / ～摸鱼 a blind person groping for fish; act blindly

虾 shrimp: ～群 a shoal of shrimps / 对～ prawn / 磷～ krill / 龙～ lobster / 油焖大～ braised prawns / 炸～球 fried prawn balls
【虾兵蟹将】shrimp soldiers and crab generals; ineffective troops
【虾酱】shrimp paste
【虾米】①(晒干的去头去壳的虾) dried, shelled shrimps ②(小虾) small shrimps
【虾皮】dried small shrimps
【虾仁】shelled fresh shrimps; shrimp meat: 芙蓉～ shrimps with eggwhite / 清炒～ saute of shrimps
【虾油】shrimp sauce
【虾子】shrimp roe; shrimp eggs ◇ ～酱油 shrimp-roe soy sauce

呷 sip: ～一口茶 take a sip of tea

xiá

峡 gorge: 长江三～ the Changjiang Gorges / 海～ strait
【峡谷】gorge; canyon
【峡湾】[地] fiord

侠
【侠客】chivalrous expert swordsman
【侠义】chivalrous

狭 narrow
【狭隘】①(宽度小) narrow: ～的山道 a narrow mountain path ②(不宽广;不宏大) narrow and limited; parochial: ～的看法 a narrow view / 心胸～ be narrow-minded ◇ ～民族主义 narrow nationalism; parochial nationalism / ～性 narrow-mindedness; parochialism
【狭长】long and narrow
【狭路相逢】(of adversaries) meet face to face on a narrow path; come into unavoidable confrontation
【狭小】narrow and small; narrow: ～的阁楼 a poky attic / 气量～ be narrow-minded; be intolerant / 走出～的圈子 step out of one's narrow circle
【狭义】narrow sense
【狭义相对论】special relativity
【狭窄】①(宽度小) narrow; cramped: ～的胡同 a narrow lane; a narrow alley ②(不宏大宽广) narrow and limited; narrow: 见识～ be limited in knowledge and narrow in experience / 心地～ be narrow-minded ③[医] stricture

辖 ①(车轴上的铁棍) linchpin ②(管辖) have jurisdiction over; administer; govern: 省～市 a municipality 〈city〉 under the jurisdiction of the provincial government / 下～两个装甲师 have two armored divisions under its command
【辖区】area under one's jurisdiction

黠 [书] crafty; cunning: 狡～ sly; crafty; cunning

匣 a small box; a small case; casket
【匣子】a small box; a small case; casket

狎 be improperly familiar with
【狎昵】be improperly familiar with

遐 ①(远) far; distant ②(长久) lasting; long: ～龄 advanced age
【遐迩】far and near: ～闻名 be well-known far and near; enjoy widespread renown
【遐想】reverie; daydream

霞 rosy clouds; morning or evening glow: 彩～ (the many hues of) rosy clouds / 晚～ the evening glow; sunset clouds / 朝～ morning glow
【霞光】rays of morning or evening sunlight: ～万道 a myriad of sun rays
【霞石】[矿] nepheline

瑕 ①(玉上面的斑点) flaw in a piece of jade ②(缺点) flaw; defect; shortcoming
【瑕不掩瑜】one flaw cannot obscure the splendor of the jade; the defects cannot obscure the virtues
【瑕疵】flaw; blemish
【瑕瑜互见】have defects as well as merits; have both strong and weak points

暇 free time; leisure: 无～兼顾 be too busy to attend to other things / 自顾不～ be unable even to fend for oneself (much less look after others); be busy enough with one's own affairs
【暇日】days of leisure
【暇时】in one's leisure time

xià

下 ①(位置在低处的) below; down; under; underneath: 零～二十度 twenty degrees below zero / 普天之～ every place under the sun / 山～ at the foot of the hill / 树～ under the tree ②(等级低的) lower; inferior: 分为上、中、～三等 divided into three grades: the upper, the middle and the lower ③(次序或时间在后的) next; latter; second: ～半辈子 the latter half of one's life; the rest of one's life / ～半夜 latter half of the night / ～半月 the latter 〈second〉 half of the month / ～册 the last of two or three volumes / ～星期 next week / ～星期一 next Monday / ～一班车 the next bus / ～一个航班 nest flight ④(向下面) downward; down: 物价～跌 prices dropped ⑤(表示属于一定范围、情况、条件等) under: 在…的帮助～ with the help of / 在党的领导～ under the leadership of the Party / 在这种情况～ under such circumstances ⑥(表示当某个时间或时节)年～ during the lunar New Year / 时～ in these days / 眼～ at the moment; at present ⑦(用在数字后,表示方面或方位)往四～一看 look all around; look about / 两～里都同意。

Both sides have agreed. ⑧(由高处到低处) descend; alight; get off: ～车 get off a car or bus / ～床 get out of bed / ～飞机 alight from a plane / ～楼 descend the stairs; go or come downstairs / ～山 descend the mountain / 顺流而～ sail downstream ⑨(雨、雪降落) fall: ～雹子了。 It's hailing. / 雪～得很利害。 The snow falls fast. / 雨不停地～。 The rain was falling steadily. ⑩(颁发;投递) issue; deliver; send: ～命令 issue orders; give orders / ～请帖 send an invitation ⑪(去;到) go to: ～馆子 go and eat in a restaurant; eat out ⑫(退场) exit; leave: 从右边门儿～ exit from the right door / 换人,七号～,五号上。 Substitution, No.5 for No.7. ⑬(放入) put in; cast: ～网打鱼 cast a net to catch fish / ～作料 put in the condiments ⑭(卸除;取下) take away; take off; dismantle; unload: 把纱窗～下来 take the screen window off ⑮(做出言论、判断等) form (an opinion, idea, etc.): ～定义 give a definition; define / ～结论 draw a conclusion / ～决心 make a resolution; be determined ⑯(使用) apply; use: ～力气 put forth strength; make an effort; exert oneself / 对症～药 prescribe the right remedy for an illness ⑰(动物生产) give birth to; lay: ～蛋 lay eggs / ～了一窝小猪 give birth to a litter of piglets ⑱(攻陷) capture; take: 连～数城 have captured several cities in succession ⑲(退让) give in; yield: 相持不～ neither side was ready to give in ⑳(到规定时间结束日常工作或学习等) finish; leave off: ～课 get out of class / ～夜班 come off night duty ㉑(少于) be less than: 不～三千人 no less than three thousand people

【下巴】①(下颌) the lower jaw ②(颏) chin

【下摆】①(长袍、上衣等的最下部分) the lower hem of a gown, jacket or skirt ②(指下摆的宽度) width of such a hem

【下班】come or go off work; knock off: 你什么时候～? When do you come off duty? / 她已经～了。 She is off duty now.

【下半场】second half (of a game)

【下半旗】fly a flag at half-mast

【下半夜】the time after midnight; the latter half of the night

【下辈】①(子孙) future generations; offspring ②(下一代) the younger generation of a family

【下笔】put pen to paper; begin to write or paint: 不知如何～ be at a loss as to how to begin writing or painting / ～千言,离题万里 write quickly but stray from the theme; A thousand words from the pen in a stream, but ten thousand *li* away from the theme

【下不来】①(降不下来) refuse to come down: 她的体温降～。 Her temperature won't come down. ②(不够) cannot be accomplished ③(在人面前受窘) feel embarrassed

【下不为例】not to be taken as a precedent; not to be repeated: 就此一回,～。 Just this once.

【下操】①(出操) have drills ②(收操) finish drilling

【下策】a bad plan; an unwise decision; the worst thing to do; a stupid move

【下层】①(指机构、组织) lower levels: 深入～ go to lower-level units; go down to the grassroots level ②(指阶层) lower strata

【下场】①[剧] go off stage; exit ②[体] leave the playing field ③(不好的结局) end; fate: 不会有好～ will certainly come to no good end / 遭到可耻～ come to a disgraceful end; meet with an ignominious fate ◇～门 exit (of a stage)

【下车伊始】the moment one alights from the official carriage; the moment one takes up one's official post

【下沉】①(沉没) sink; submerge: 船～了。 The boat has sunk. ②(陷下去) subside; sink; cave in: 地基～。 The foundations have subsided.

【下处】temporary lodging during a trip

【下穿交叉】[交] underpass; undercrossing

【下船】go ashore; disembark

【下垂】①(向下垂) hang down; droop ②[医] prolapse: 胃～ gastroptosis / 子宫～ prolapse of the uterus; metroptosis

【下次】next time; next

【下达】make known to lower levels; transmit to lower levels: ～任务 assign a mission / ～行动命令 issue orders of operation / ～指示 give instructions

【下地】①(到田地里去) go to the fields: ～劳动 go to work in the fields ②(下病床) leave a sickbed

【下碇】cast anchor: 停船～ come to an anchor

【下毒】empoison

【下毒手】strike a vicious blow; lay murderous hands on sb.: 在背后～ stab sb. in the back

【下颚】the lower jaw; mandible

【下凡】(of gods or immortals) descend to the world; come down to earth

【下饭】①(就着菜把主食吃下去) go with rice: 没有菜～。 There is no dish to go with rice. ②(适宜和饭一起吃) go well with rice: 这个菜～。 This dish goes well with rice.

【下放】①(权力下放) transfer (power) to a lower level: 权力～ transfer power to a lower level / 企业～ put an enterprise under a lower administrative level ②(干部下放) transfer (cadres, etc.) to work at the grass-roots level or to do manual labor in the countryside or in a factory ◇～干部 a cadre transferred to a lower level or to do manual labor in the countryside or in a factory
【下风】①(风向的下方) leeward: 在～的方向 on the leeward ②(不利的地位) disadvantageous position: 占～ be at a disadvantage
【下疳】chancre; primary lesion
【下岗】come or go off sentry duty
【下工】come or go off work; stop work; knock off
【下工夫】put in time and energy; concentrate one's efforts: 在学英语上～ devote a lot of time and energy to English study
【下跪】kneel down; go down on one's knees
【下海】go to sea; put out to sea: ～捕鱼 go fishing on the sea
【下颌】[生理] the lower jaw; mandible ◇～骨 lower jawbone; mandible
【下滑】gliding; letting down ◇～角 gliding angle
【下怀】one's heart's desire: 正中～ be exactly what one wants
【下货】unload goods; ship goods ◇～费 unloading hire / ～港 unloading port
【下级】①(指组织) lower level: ～服从上级。The lower level is subordinate to the higher level. ②(指人员) subordinate ◇～法院 court below / ～干部 junior cadre / ～机关 government office at a lower level / ～军官 low-ranking officer; junior officer / ～组织 subordinate organization
【下贱】low; mean; degrading
【下江】lower reaches of the Changjiang River ◇～人 a native of one of the provinces on the lower reaches of the Changjiang River
【下降】descend; go or come down; drop; fall; decline: 出生率～ a decline in the birth rate / 飞机正在～。The plane was descending. / 气温已经～。The temperature has dropped.
【下脚】①(插脚) get a foothold: plant one's foot: 没有～的地方 be unable to gain a foothold; have nowhere to plant one's foot ②(下脚料) leftover bits and pieces ◇～货 inferior goods / ～料 leftover bits and pieces (of industrial material, etc.) / ～棉 cotton waste
【下酒】①(就着菜喝酒) go with wine: 买个冷盘～ buy a cold dish to go with the wine ②(适宜和酒一起吃) go well with wine ◇～菜 a dish that goes with wine
【下课】get out of class; finish class
【下款】①(书画下款) name of the donor ②(信件下款) signature at the end of a letter
【下来】①(由高到低) come down: 我们看到他们从楼上走～。We saw them coming down stairs. ②(用在动词后) down; off: 从自行车上摔～ tumble off a bicycle / 古代流传～的寓言 fables handed down from ancient times / 坐～ sit down ③(用在形容词后) get; grow; become: 天渐渐地黑～了。It was getting dark.
【下里巴人】popular literature or art
【下联】the second line of a couplet
【下列】listed below; following: 应注意～几个问题。Attention should be paid to the following problems.
【下令】give orders; order
【下流】①(下游) lower reaches (of a river) ②(卑鄙龌龊) low-down; mean; obscene; dirty: ～的勾当 base acts / ～的谩骂 scurrilous attacks; coarse invectives / ～的玩笑 dirty jests; obscene jests; coarse jokes ◇～话 obscene language; dirty language; obscenities
【下落】①(寻找中的人或物的所在) whereabouts: 打听某人的～ enquire about sb.'s whereabouts / 他目前～不明。His present whereabouts is unknown. ②(下降) drop; fall 降落伞～的地点 the place where the parachute has fallen
【下马】①(从马上下来) get down from a horse; dismount from a horse ②(停止某项工程) discontinue: 这项工程～了。The project was abandoned.
【下马威】severity shown by an official on assuming office: 给他个～ deal him a head-on blow at the first encounter
【下面】①(位置较低的地方) below; under; underneath: 大桥～ under the bridge / 图片～的说明 the caption below the picture / 站在高墙～ stand under a high wall / 坐在树～ sit underneath a tree ②(次序靠后的部分) next; following: ～该谁了? Who is next? ③(下级) lower level; subordinate: 了解～的情况 find out about how things are at the lower levels
【下品】low-grade; inferior
【下坡路】downhill path; downhill journey; decline: 走～ go downhill; be on the decline
【下铺】lower berth

【下棋】play chess; have a game of chess
【下欠】①(尚欠) still owing: ～二十五元 with twenty five *yuan* still owing. ②(下欠的款项) a sum still owing
【下情】conditions at the lower levels; feelings or wishes of the masses: ～上达 make the situation at the lower levels known to the higher levels / 不了解～ not know what is going on at the lower levels / 了解～ find out what the masses are thinking
【下去】①(由高往低) go down; descend ②(用在动词后,表示由高到低) down: 太阳落～了。The sun went down. ③(用在动词后,表示继续) on: 请讲～。Please go on. ④(用在形容词后,表示程度继续增加) get; grow; become: 病人的情况在一天天坏～。 The patient is getting worse and worse.
【下身】①(身体下部) the lower part of the body ②(阴部) private parts ③(裤子) trousers
【下士】corporal(陆军或英空军); petty officer second class(英海军); petty officer third class (美海军)
【下手】①(动手;着手) put one's hand to; start; set about; set to: 不知从何～ not know where to start; not know how to set about a job ②(下首) right-hand seat: 坐在主宾的～ sit on the right hand of the chief guest ③(助手) assistant; helper: 打～儿 act as assistant
【下首】right-hand seat
【下属】subordinate
【下水】①(进入水中) enter the water; be launched: (使)船～ launch a ship ②(做坏事) take to evildoing; fall into evil ways: 拖人～ involve sb. in evildoing; entice 〈inveigle〉 sb. into evildoing ③(向下游航行) downriver; downstream: ～船 downriver boat ④(食用的牲畜内脏) offal: 猪～ pig's offal ◇ ～典礼 launching ceremony
【下水道】sewer
【下榻】stay (at a place during a trip)
【下台】①(下舞台或讲台) step down from the stage or platform ②(卸去公职) fall out of power; leave office: 被赶～ be driven out of office; be thrown out ③(多用否定式,比喻摆脱困难窘境) get out of a predicament or an embarrassing situation: 叫他下不了台 put him on the spot / 没法～ be unable to back down with good grace
【下体】①(身体的下部) the lower part of the body ②(阴部) private parts
【下同】(多用于附注) similarly hereinafter; the same below
【下文】①(文章某一段或一句以后的部分) what follows in the passage, paragraph, article, etc.: ～再作阐述 be explained in the ensuing chapters or paragraphs ②(事情的发展和结果) what follows; outcome; later development; sequel: 不见～ sequel unknown
【下午】afternoon
【下弦】[天] last quarter; third quarter ◇ ～月 the moon at the last 〈third〉 quarter
【下限】lower limit; prescribed minimum; floor level; floor: 不能低于～ should be kept above the prescribed minimum
【下乡】go to the countryside: ～知识青年 educated urban youth working in the countryside
【下行】①[铁道] down: ～列车 down train ②[航运] downriver; downstream ③(公文发往下级) to be issued to the lower levels
【下旋】[乒乓] underspin; backspin
【下旬】the last ten-day period of a month
【下药】①(医生用药) prescribe medicine: 对症～ prescribe the right remedy for an illness ②(下毒药) put in poison
【下野】(of a ruler) retire from the political arena; be forced to relinquish power
【下议院】①(众议院) lower house; lower chamber ②(英国下院) the House of Commons
【下意识】subconsciousness
【下游】①(河流下游) lower reaches ②(落后的地位) backward position: 甘居～ be resigned to being backward
【下狱】throw into prison; imprison
【下葬】bury; inter
【下肢】lower limbs; legs
【下中农】lower-middle peasant
【下种】sow (seeds)
【下逐客令】show a person the door
【下装】remove theatrical makeup and costume
【下坠】[医] straining (at stool); tenesmus
【下钻】[石油] run the drilling tool into the well ◇ ～速度 running speed
【下作】low-down; mean; obscene; dirty

吓 frighten; scare; intimidate: ～坏了 be terribly frightened; be overcome with fear / ～破了胆 be scared out of one's wits / 把我～一跳 give me a start 〈scare〉
【吓唬】frighten; scare; intimidate

夏 summer

【夏布】grass cloth; grass linen
【夏候鸟】summer resident
【夏季】summer
【夏枯草】selfheal
【夏历】the traditional Chinese calendar; the lunar calendar
【夏粮】summer grain crops
【夏令】①(夏季) summer; summertime ②(夏季气候) summer weather: 春行～ summer weather in spring; exceptionally warm days in spring ◇ ～商品 commodities for summer use / ～时 daylight saving time; summer time (英) / ～营 summer camp
【夏眠】[动] aestivation
【夏收】summer harvest ◇ ～作物 summer crops
【夏天】summer
【夏衣】summer clothing; summer wear
【夏至】the Summer Solstice

xiān

籼
【籼稻】long-grained nonglutinous rice; indica rice
【籼米】polished long-grained nonglutinous rice

仙 celestial being; immortal
【仙波】angels
【仙丹】elixir of life
【仙姑】①(仙女) female immortal; female celestial ②(女巫) sorceress
【仙鹤】red-crowned crane
【仙鹤草】hairyvein agrimony
【仙后座】[天] Cassiopeia
【仙境】fairyland; wonderland; paradise
【仙客来】[植] cyclamen
【仙女】female celestial; fairy maiden
【仙女座】[天] Andromeda
【仙人】celestial being; immortal
【仙人掌】cactus
【仙山琼阁】a jewelled palace in elfland's hills
【仙逝】pass away
【仙王座】[天] Cepheus

氙 [化] xenon (Xe)
【氙气灯】xenon lamp

先 ①(时间或次序在前的) earlier; before; first; in advance: ～付款 pay in advance / 她比我～到。 She arrived earlier than I did. ②(祖先) elder generation; ancestor: 祖～ ancestor; forefather ③(尊称死去的人) deceased; late: ～父 my late father ④(先前) earlier on; before: 你～怎么不告诉我? Why didn't you tell me before?
【先辈】elder generation; ancestors: ～遗训 teachings of our ancestors / 革命～ the older generation of revolutionaries
【先导】guide; forerunner; precursor
【先睹为快】consider it a pleasure to be among the first to read 〈see〉 sth.
【先发制人】gain the initiative by striking first; forestall the enemy: 采取～的手段 take pre-emptive measures
【先锋】vanguard; van: 打～ fight in the van; be a pioneer / 起～作用 play a vanguard role ◇ ～队 vanguard
【先后】①(先和后) early or late; priority; order: 办事应有个～次序。 Things should be taken up in order of priority. ②(前后相继) successively; one after another: 校队～胜了五场球。 The school team won five games successively.
【先己后人】put self before others
【先见之明】prophetic vision; foresight; ability to discern what is coming
【先进】advanced ◇ ～单位 advanced unit / ～分子 advanced element / ～个人 advanced individual / ～工作者 advanced worker / ～集体 advanced group / ～经验 advanced experience / ～事迹 meritorious deeds; exemplary deeds
【先决】prerequisite: ～条件 prerequisite; precondition
【先来后到】in the order of arrival; first come, first served
【先礼后兵】take strong measures only after courteous ones fail; try peaceful means before resorting to force
【先例】precedent: 开～ set a precedent; create a precedent / 有～可援 have a precedent to go by
【先烈】martyr: 革命～ revolutionary martyr
【先令】①(英国、肯尼亚、索马里、坦桑尼亚、乌干达等国货币) shilling ②(奥地利货币) schilling
【先期】earlier on; in advance: ～到达 have arrived at an earlier date / ～公布 have published in advance
【先前】before; previously: 她的胃病比～好多了。 Her stomach trouble is much better than before.

【先遣】sent in advance ◇ ～部队 advance troops; advance force／～队 advance party
【先驱】pioneer; forerunner; harbinger
【先人】①(祖先) ancestor; forefather ②(已故的父亲) my late father
【先人后己】put others before oneself; put other people's interest ahead of one's own
【先入为主】first impressions are strongest; preconceived ideas keep a strong hold; be prejudiced
【先入之见】preconception; preconceived idea; prejudice
【先声】first signs; herald; harbinger
【先声夺人】forestall one's opponent by a show of strength; overawe others by displaying one's strength
【先生】①(老师) teacher ②(对知识分子的称呼) mister(Mr.); gentleman; sir: 总统～ Mr. President／女士们,～们 ladies and gentlemen ③(医生) doctor ④[旧] 帐房～ bookkeeper／算命～ fortune-teller
【先手】on the offensive (in chess): ～棋 an offensive move
【先天】①(生来就有的) congenital; inborn: ～不足 be congenitally deficient; suffer from an inherent shortage／～的缺陷 congenital defect ②[哲] a priori; innate ◇ ～畸形 congenital malformation／～免疫 congenital immunity／～性心脏病 congenital heart disease／～愚型 mongolism
【先天下之忧而忧,后天下之乐而乐】A leader should plan and worry ahead of the people, and enjoy the fruits after the people.
【先头】①(位置在前) ahead; in front; in advance: ～部队 an advance party of soldiers; vanguard／走在最～ walk ahead of all other people ②(以前) before; formerly; in the past: 她～没来过。 She has never been here before.
【先下手为强】he who strikes first gains the advantage; to take the initiative is to gain the upper hand
【先下手为强,后下手遭殃】He who strikes first prevails, he who strikes late fails.
【先小人后君子】specify terms clearly at first and use a good deal of courtesy later
【先行】①(走在前面) go ahead of the rest; start off before the others ②(预先准备) beforehand; in advance: ～通知 notify in advance／～准备 make preparations beforehand
【先行官】commander of an advance unit or vanguard
【先行者】forerunner
【先验】[哲] a priori: ～知识 a priori knowledge ◇ ～论 apriorism
【先斩后奏】execute the criminal first and report to the emperor afterwards; act first and report afterwards
【先兆】omen; portent; sign; indication: 不祥的～ ill omen／成功的～ an omen of success／地震的～ indications of an impending earthquake／战争的～ portents of war ◇ ～流产 threatened abortion／～子痫 preeclampsia; toxemia of pregnancy
【先哲】a great thinker of the past; sage
【先知】①(对大事了解得较早的人) a person of foresight ②[宗] prophet
【先知先觉】①(对大事了解得较早的人) a person of foresight ②(对大事了解得较早) having foresight

酰 [化] acyl

鲜 ①(新鲜) fresh: ～蛋 fresh eggs／～花 fresh flowers／～蘑 fresh mushrooms／～奶 fresh milk／～肉 fresh meat ②(鲜明) bright-colored; bright: ～红色 bright red; scarlet／～绿色 vivid green ③(鲜美) delicious; tasty: ～汤 delicious soup／～味 delicate flavor ④(鲜美的食物) delicacy: 时～ delicacies of the season ⑤(鱼虾等水产) aquatic foods: 海～ seafood
【鲜花】fresh flowers; flowers
【鲜货】①(新鲜水果,蔬菜) fresh fruit or vegetables ②(新鲜鱼虾) fresh aquatic foods
【鲜美】delicious; tasty: ～的菜肴 delicious dish
【鲜明】①(颜色明亮) bright: 色彩～ in bright colors; bright-colored ②(显著) clear-cut; distinct; distinctive: ～的对照 a striking contrast; a sharp contrast／～的节奏 strongly accented rhythms／富有～的地方特色 be characterized by a distinctive local style or flavor／主题～ have a distinct theme
【鲜嫩】fresh and tender
【鲜血】blood
【鲜血淋漓】drenched with blood
【鲜艳】bright-colored; gaily-colored: ～夺目 dazzlingly beautiful; resplendent; attractively bright-colored／颜色～ in gay colors

掀 lift (a cover, etc.): ～掉盖子 take the lid off／～门帘 lift the door curtain／～一页过去 turn a page

【掀动】lift; start; set in motion
【掀翻】throw: 把对手～在地 throw the opponent off his balance / 马把骑马的人～下来。The horse threw its rider.
【掀起】①(揭起) lift; raise ②(翻腾) surge; cause to surge: 大海～了巨浪。Big wave surged on the sea. ③(大规模兴起) set off; start: ～建设的新高潮 set off a new upsurge of construction
锨 shovel
纤 fine; minute: ～尘 fine dust
【纤度】[纺] fiber number; size
【纤毛】[生] cilium ◇ ～运动 ciliary movement
【纤毛虫】[动] infusorian
【纤巧】dainty; delicate
【纤弱】slim and fragile; delicate
【纤维】fiber; staple: 合成～synthetic fiber / 人造～ man-made fiber; artificial fiber / 天然～ natural fiber ◇ ～板 fiberboard / ～长度 fiber length; staple / ～蛋白[生化] fibrin / ～植物 fiber plant
【纤维瘤】[医] fibroma
【纤维肉瘤】[医] fibrosarcoma
【纤维素】[化] cellulose
【纤细】very thin; slender; fine; tenuous: ～的头发 fine hair / ～的游丝 tenuous gossamer
【纤小】fine; tenuous

xián

涎 saliva
【涎皮赖脸】brazenfaced; shameless and loathsome; cheeky
【涎水】[方] saliva
舷 the side of a ship; board: 右～ starboard / 左～ port / 我看到右～有一艘轮船。I sighted a steamer to starboard.
【舷边】gunwale; gunnel
【舷窗】porthole
【舷梯】①(船边小梯) gangway ladder; accommodation ladder: 登上～ mount the gangway / 放下～ lower the gangway / 收起～ raise the gangway ②(上下飞机用的) ramp
弦 ①(弓弦) bowstring; string ②(乐器的弦) the string of a musical instrument: 一根～断了。One of the strings broke. ③(钟表发条) spring ④(连接圆周上两点的直线) chord ⑤(直角三角形的斜边) hypotenuse
【弦外之音】overtones; implication
【弦乐队】string orchestra; string band
【弦乐器】stringed instrument
闲 ①(没有事情;有空;没有活动) not busy; idle; unoccupied: ～不住 refuse to stay idle; always keep oneself busy / 不吃～饭 won't be an idler / 吃～饭 eat the bread of idleness ②(不在使用中) not in use; unoccupied; lying idle: ～房 unoccupied room; vacant room; unoccupied house; vacant house / 别让计算机～着! Don't let the personal computer stand idle. ③(闲空儿) spare time; free time; leisure
【闲扯】chat; engage in chitchat
【闲工夫】spare time; leisure
【闲逛】saunter; stroll: 一路～ saunter on the way
【闲话】①(与正事无关的话) digression: ～少说 enough of the digression; save your breath to cool your porridge ②(不满意的话) complaint ③(搬弄是非的话) gossip: 喜欢说人～ be fond of gossip
【闲话当年】chat about bygone days
【闲居】stay at home idle
【闲空】free time; spare time; leisure
【闲聊】chat: 我和他～。I had a chat with him.
【闲气】anger about trifles: 生～ get angry about trifles; lose one's temper over trifles
【闲钱】[口] spare cash; spare money
【闲情逸致】leisurely and carefree mood; leisure and mood for enjoyments
【闲人】①(无事的人) an unoccupied person; idler ②(无关的人) persons not concerned: ～免进。No admittance except on business. (或: Admittance to staff only.)
【闲散】①(空闲而无拘束) free and at leisure; at a loose end ②(闲着不使用的;没事干的) unused; idle: ～人员 idle personnel / ～土地 scattered plots of unutilized land / ～资金 idle capital
【闲事】①(与自己无关的事) a matter that does not concern one; other people's business: 爱管～ like to poke one's nose into other people's business / 别管～! Mind your own business. (或: None of your business.) ②(无关紧要的事) unimportant matter
【闲是闲非】irrelevant disputes about affairs
【闲书】light reading
【闲谈】chat; engage in chitchat
【闲暇】leisure

【闲心】leisurely mood: 没有～管这种事 not be in the mood to bother about such matters / 我没～开玩笑。I am in no mood for joking.
【闲杂】without fixed duties: ～人员 people without fixed duties; miscellaneous personnel
【闲置】leave unused; let sth. lie idle; set aside: ～的机器 idle machines
【闲置时间】standby time; idle time

娴 ①(文雅) refined ②(熟练) adept; skilled: ～于辞令 be gifted with a silver tongue
【娴静】gentle and refined
【娴熟】adept; skilled: ～的技巧 consummate skill / 弓马～ adept in archery and horsemanship
【娴雅】(of a woman) refined; elegant

咸 ①(咸味) salted; salty: ～蛋 salted egg / ～花生 salted peanuts / ～鱼 salt fish ②(全;都) all: 老少～宜 good for the old and the young / ～受其益。All benefited from it.
【咸菜】salted vegetables; pickles
【咸肉】salt meat; bacon
【咸水】salt water ◇ ～湖 saltwater lake / ～鱼 saltwater fish

贤 ①(有德;有才) virtuous; worthy; able: 任人唯～ appoint people on their merits ②(贤能;有德的人;有才的人) a worthy person; an able and virtuous person: 让～ relinquish one's post in favor of sb. better qualified ③[敬](旧时用于平辈或晚辈) ～弟 my worthy brother; your good self
【贤达】prominent personage; worthy: 社会～ social élite; élite of society
【贤惠】(of a woman) virtuous
【贤良】(of a man) able and virtuous
【贤明】wise and able; sagacious
【贤妻良母】a good wife and loving mother
【贤人】a person of virtue; worthy

衔 ①(用嘴含) hold in the mouth: ～着烟头 have a pipe between one's teeth ②(存在心里) harbor; bear: ～恨 harbor resentment; bear a grudge ③(级别) rank; title: 公使～参赞 counsellor with rank of minister / 有上校～ hold the rank of colonel
【衔接】link up; join: 两条公路在这里～起来。The two highways link up here.
【衔铁】[电] armature
【衔铁振动】armature chatter
【衔冤】nurse a bitter sense of wrong; have a simmering sense of injustice

嫌 ①(嫌疑) suspicion: 避～ avoid suspicion ②(嫌怨) ill will; resentment; enmity; grudge: 前～尽释。All previous ill will has been removed.(或: We have agreed to bury the hatchet.) ③(厌恶) dislike; mind; complain of: ～麻烦 not want to take the trouble; think it troublesome / 讨人～ get oneself disliked
【嫌气细菌】anaerobic bacteria; anaerobes
【嫌弃】dislike and avoid; cold-shoulder: 遭～ be cold-shouldered
【嫌恶】detest; loathe; disgust: 他的行为使每个人都～。His behavior disgusted everybody.
【嫌隙】feeling of animosity; enmity; ill will; grudge
【嫌疑】suspicion: 有间谍～ be suspected of being a spy ◇ ～犯 suspect / ～分子 suspected person; marked man
【嫌怨】grudge; resentment; enmity: ～未消。A grudge has not yet vanished.

xiǎn

显 ①(明显) apparent; obvious; noticeable ②(表现) show; display; manifest: ～身扬名 show one's mettle and make a name ③(有名声、有权势的) illustrious and influential
【显得】look; seem; appear: 她～十分快乐。She seems (to be) quite happy. / 他理发后～年轻些。He looks younger after the haircut.
【显而易见】obviously; evidently; clearly
【显赫】illustrious; celebrated: ～的名声 great renown / ～一时 celebrated for a while / ～的战功 illustrious war exploits / 声势～ have a powerful influence
【显花植物】phanerogam
【显见】obvious; self-evident; apparent: ～的理由 an obvious reason; an apparent reason
【显灵】(of a ghost or spirit) make its presence or power felt
【显露】become visible; appear; manifest itself: ～头角 show one's promises
【显明】obvious; manifest; distinct; marked: ～的对比 a sharp contrast / ～的道理 an obvious truth / ～的特点 a distinct characteristic; a marked characteristic
【显然】obvious; evident; clear: ～她是错了。It is obvious that she is wrong. / 他站着睡着了,～是太累了。His exhaustion was obvious when he fell asleep standing up. / 这是很～

的。It is quite obious.
【显身手】display one's talent or skill
【显圣】(of the ghost of a saintly person) make its presence or power felt
【显示】①(明显地表示) show; display; demonstrate; manifest: ～出大无畏的革命英雄主义 demonstrate dauntless revolutionary heroism / ～力量 make a show of force; display one's strength ②[石油] show; indication: 地面～ surface indications / 石油～ oil shows / 天然气～ gas shows
【显微胶片】①[摄] microfilm; microfiche ②(书页摄影用) bibliofilm
【显微镜】microscope: 红外线～ infrared microscope / …倍电子～ electron microscope with a magnification of... times ◇ ～盖片 cover glass / ～载物台 stage / ～载片 (glass) slide
【显微术】[物] microscopy
【显微阅读机】microfilm viewer; microfilm reader
【显微照片】micrograph
【显微照相术】microphotography
【显现】manifest oneself; reveal oneself; appear; show
【显像管】[电子] kinescope; picture tube: 彩色～ tricolor tube
【显形】show one's (true) colors; betray oneself
【显性】[生] dominance ◇ ～性状 dominant character
【显眼】conspicuous; showy: 穿得太～ be loudly dressed; be showily dressed
【显要】①(官职高、权力大) powerful and influential: ～人物 an influential figure ②(官职高、权力大的人) influential figure; important personage; VIP
【显影】[摄] develop ◇ ～机 developing machine / ～剂 developer / ～盘 developing dish / ～纸 developing-out paper
【显著】notable; marked; striking; remarkable; outstanding: ～的区别 marked difference / ～的特征 outstanding characteristics / 取得～的成就 achieve remarkable success / 收效～ yield notable results / 有～的进步 make marked progress

险 ①(险要) a place difficult of access; narrow pass; defile: 天～ natural barrier / 无～可守 have no tenable defense position; be strategically indefensible ②(危险) danger; peril; risk: 冒～ run a risk / 脱～ be out of danger / 遇～ meet with danger ③(阴险) sinister; vicious; venomous: 阴～ sinister ④(险些) by a hair's breadth; by inches; nearly: ～遭不测 barely escaped accident; have a narrow escape / ～遭不幸 come within an ace of death / ～遭毒手 was nearly killed
【险隘】strategic pass; defile
【险恶】①(凶险可怕) dangerous; perilous; ominous: 病情～ be dangerously ill / 处境～ be in a perilous position ②(邪恶的;恶毒的) sinister; vicious; malicious; treacherous: ～的用心 sinister intention; vicious intentions; evil motives
【险峰】perilous peak
【险境】dangerous situation: 脱离～ be out of danger
【险峻】dangerously steep; precipitous: ～难行 too dangerous and high to walk over
【险区】danger zone
【险胜】win by a narrow margin
【险滩】dangerous shoal; rapids
【险象环生】signs of danger appearing everywhere
【险象频生】dangerous images appear incessantly
【险些】narrowly; nearly: 我～误了那班火车。I nearly missed the train.
【险要】strategically located and difficult of access
【险诈】sinister and crafty
【险症】dangerous illness
【险阻】(of roads) dangerous and difficult: 崎岖～的山路 a dangerous and difficult mountain path / 不畏艰难～ not be afraid of dangers and difficulties

铣
【铣铁】cast iron

鲜 little; rare: ～见 rarely seen; seldom met with

藓 [植] moss

xiàn

宪 ①(法令) statute ②(宪法) constitution: 制～ draw up a constitution
【宪兵】military police; military policeman; gendarme ◇ ～队 gendarmerie; military police corps
【宪法】constitution; charter: ～草案 draft constitution / ～程序 constitutional process
【宪章】charter: 联合国～ the United Nations Charter

【宪政】constitutional government; constitutionalism

羡 admire; envy: 人人称～ be the admiration of everyone

【羡慕】admire; envy: 你真走运,我～你。You're lucky. I envy you. / 我们～他的工作能力。 We admire his capacity.

霰 [气] graupel

【霰弹】[军] case shot; canister (shot)

献 ①(奉献) offer; present; dedicate; donate: ～花圈 lay a wreath / ～血 donate blood ②(表现) show; put on; display: ～殷勤 show sb. excessive attentions; pay one's addresses

【献宝】①(献宝物) present a treasure ②(提供宝贵经验或意见) offer a valuable piece of advice or one's valuable experience ③(表现给人看) show off what one treasures

【献策】offer advice; make suggestions

【献丑】[谦](用于表演或写作时) show oneself up; show one's incompetence

【献词】congratulatory message: 新年～ New Year message

【献计】offer advice; make suggestions

【献技】show one's skill

【献礼】present a gift: 以优异的成绩向国庆～ greet the National Day with outstanding successes

【献媚】try to ingratiate oneself with; make up to

【献旗】present a banner

【献身】devote oneself to; dedicate oneself to; give one's life for: ～于四化 devote oneself to the four modernizations

县 county

【县城】county seat; county town

【县委】county Party committee

【县长】the head of a county; county magistrate

【县志】general records of a county; county annals

现 ①(现在;此刻) present; current; existing: ～阶段 the present stage / ～况 the existing situation; the present situation / ～政府 present government ②(临时;当时) (do sth.) in time of need; extempore: 她在晚会上～编了一首诗。 She improvised a poem at the evening party. ③(当时可以拿出来的) on hand: ～金 ready money; cash ④(现款) cash; ready money: 付～ pay cash ⑤(表露在外面,使人可以看见) show; appear: ～形 reveal one's true colors

【现场】①(出事地点) scene: 保护～ keep the scene intact / 作案～ the scene of a crime ②(工作地点) site; spot: 工作～ worksite / 试验～ testing ground ◇ ～表演 on-the-spot demonstration; live demonstration / ～采访 spot coverage / ～查验 view of the scene / ～会议 on-the-spot meeting / ～交货 ex point of origin / ～指导 on-the-spot guidance

【现成】ready-made: 吃～的 eat whatever is ready or prepared by others / 买～衣服 buy ready-made clothes; buy clothes off the peg

【现成饭】food ready for the table; unearned gain

【现成话】an onlooker's unsolicited comments; a kibizer's a comments

【现出】reveal; display: ～原形 come out in one's true colors

【现存】extant; in stock: ～的手稿 extant manuscripts / ～物资 goods and materials in stock

【现代】①(现代) modern times; the contemporary age ②(现代的) modern; contemporary: ～交通工具 modern means of communication / ～科学成就 modern scientific achievements / ～题材 contemporary theme / ～作家 modern writer; contemporary writer ◇ ～派 modernist school / ～史 contemporary history

【现代化】modernize: 实现四个～ achieve the four modernizations (of agriculture, industry, national defense, and science and technology) ◇ ～设备 sophisticated equipment

【现货】[商] merchandise on hand; spots ◇ ～供应 off the shelf / ～价格 spot price / ～交易 spot transaction; over-the-counter trading / ～市场 spot market

【现价】present price

【现浇】[建] cast-in-place; cast-in-situ ◇ ～混凝土 cast-in-place concrete; cast-in-situ concrete

【现今】nowadays; these days

【现金】①(现款) ready money; cash ②(银行库存的货币) cash reserve in a bank ◇ ～付款 cash payment; payment in cash / ～交易 cash transaction / ～收入 cash receipts / ～预算 cash budget / ～余额 cash balance / ～帐 cash account; cash book / ～支出 out-of-pocket expenses

【现款】ready money; cash

【现买现卖】sell something right after it is bought

【现钱】ready money; cash
【现任】①(现在担任) at present hold the office of: 她～公司经理。At present she holds the position of company manager. ②(现在任职的) currently in office; incumbent: ～校长是王教授。The present president is Professor Wang.
【现身说法】advise sb. or explain sth. by using one's own experience as an example; act as an example to others
【现时】now; at present
【现实】①(客观存在的事物) reality; actuality: 客观～ objective reality / 面对～ face the facts / 脱离～ be divorced from reality; be unrealistic ②(合乎客观情况的) real; actual: ～生活 real life; actual life / ～意义 practical or immediate significance / 采取～的态度 adopt a realistic attitude
【现实主义】realism ◇ ～文学 realistic literature / ～者 realist
【现世】①(今生) this life ②(出丑;丢脸) lose face; be disgraced; bring shame on oneself
【现下】now; at present
【现象】appearance (of things); phenomenon: 社会～ social phenomenon / 向不良～作斗争 combat unhealthy phenomena / 暂时～ transient phenomenon
【现行】①(现在施行的;现在有效的) currently in effect; in force; in operation: ～标准 current standard / ～法令 decrees in effect / ～规章制度 rules and regulations in force / ～政策 present policies ②(正在进行犯罪活动的) active ◇ ～反革命 active counterrevolutionary / ～犯 criminal caught in the act / ～罪 flagrant crime
【现形】reveal one's true features; betray oneself
【现眼】make a spectacle ⟨fool⟩ of oneself; lose face: 丢人～ make a fool of oneself; be a disgrace
【现洋】silver dollar
【现役】①(所服兵役) active service; active duty: 服～ be on active service ②(正在服兵役的) on active service; on active duty; active ◇ ～兵员 personnel on active service / ～军队 active military unit / ～军官 officer on the active list / ～军人 serviceman / ～年限 term of active service
【现有】now available; existing: ～材料 materials now available; materials now on hand; available information / ～人力 available manpower / ～设备 existing equipment
【现在】now; at present; today: 从～开始 from now on / 到～为止 up to now / 他～到外地度假去了。He is at present away on his holidays. / 我～不需要这本字典。I don't need the dictionary at present.
【现职】present job; current post; present employment
【现状】present situation; current situation; status quo; existing state of affairs: 安于～ be content with things as they are / 改变～ change the status quo / 维持～ maintain the status quo

苋 amaranth
【苋菜】three-colored amaranth

腺 gland: 汗～ sweat gland / 泪～ lachrymal gland / 唾液～ salivary gland
【腺瘤】[医] adenoma
【腺嘌呤】adenine; amidopurine

馅 filling; stuffing: 饺子～儿 stuffing for dumplings / ～肉儿 meat filling
【馅儿饼】meat pie

陷 ①(陷阱) pitfall; trap ②(掉进;陷入) get stuck; get bogged down: ～进泥里 get stuck in the mud / ～于被动 lose the initiative / ～于孤立 find oneself isolated / ～在日常事务堆里 get bogged down in everyday routine / 卡车～进泥里。The truck was bogged. (或:The truck bogged.) ③(凹进) sink; cave in: 地基下～。The foundations have sunk. / 她的眼睛～进去了。Her eyes have caved in. ④(陷害) frame (up): ～人于罪 frame sb. (up); incriminate sb. ⑤(被攻占;被占领) be captured; fall: 城～之日 the day the city fell ⑥(缺点) defect; deficiency: 缺～ defect; flaw
【陷害】frame (up); make a false charge against: ～好人 frame up an innocent person / 政治～ political frame-up
【陷阱】pitfall; pit; trap: 布设～ lay a trap
【陷坑】pitfall; pit
【陷落】①(下陷沉降) subside; sink in; cave in: 地壳的～ subsidence of the earth's crust ②(沦陷) fall into enemy's hands
【陷入】①(落在不利的境地) sink into; fall into; land oneself in; be caught in; get bogged down in: ～被动地位 fall into a passive position / ～重围 find oneself tightly encircled / ～困境 land in a predicament; be put in a tight spot; be cornered / ～僵局 come to a

deadlock; reach an impasse / ～罗网 fall into a snare / ～无休止的争论 be bogged down in endless debates ②(深深地陷入) be lost in; be immersed in; be deep in: ～沉思 be lost in thought; be deep in meditation

限 ①(范围;限度) limit; bounds: 期～ time limit / 以年底为～ set the end of the year as the deadline ②(指定范围) set a limit; limit; restrict: 年龄性别不～ put no restrictions on age or sex / 人数不～。There is no limit on the number of people.

【限定】prescribe a limit to; set a limit to; limit; restrict: ～参观人数为五十人 limit the number of visitors to fifty / ～时间完成 prescribe a time limit for fulfillment

【限度】limit; limitation: 超过～ go beyond the limit; exceed the limit / 减少到最低～ reduce sth. to a minimum / 最大～地调动人的积极性 bring people's initiative into full play

【限额】norm; limit; quota ◇ ～抵押 closed mortgage / ～交易 rationed exchange / ～输出 ration export / ～制 quota system

【限量】limit the quantity of; set bounds to: 前途不可～ have boundless prospects

【限令】order sb. to do sth. within a certain time

【限期】①(限定日期) within a definite time; set a time limit: ～报到 report for duty by the prescribed time / ～撤退 withdraw within a stated time ②(限定的日期) time limit; deadline: ～已满。The time limit has been reached.

【限于】be confined to; be limited to: ～个人水平 due to one's limited level / ～篇幅 as space is limited / ～时间 owing to the limitation of time

【限制】place restrictions on; impose restrictions on; restrict; limit; confine: ～数量 limit to a number or amount / 年龄～ age limit / 时间～ time limit / 对…实行～ impose restrictions on ◇ ～分配 restricted distribution / ～认付 qualified acceptance

【限制性】restricted; restrictive: ～会议 restricted meeting ◇ ～定语[语] restrictive attribute

线 ①(各种线) thread; string; wire: ～团 a ball of string; a reel of thread / 丝～ silk thread / 铜～ copper wire / 穿针引～ thread a needle ②[数] line: 直～ straight line ③(用棉线做的) made of cotton thread: ～手套 cotton gloves / ～毯 cotton blanket / ～衣～裤 cotton knitwear ④(交通路线) route; line: 供应～ supply route; supply line / 航～ airline or shipping line / 铁道～ railway line ⑤(边缘交界的地方) demarcation line; boundary: 边界～ boundary line / 海岸～ coastline / 军事分界～ military demarcation line ⑥(所接近的某种边际) brink; verge: 在饥饿～上 on the brink of starvation / 在死亡～上 on the verge of death

【线虫】nematode ◇ ～病 nematodiasis

【线电压】[电] line voltage

【线段】[数] line segment

【线规】[机] wire gauge

【线间】[乐] space

【线路】①[电] circuit; line: 电话～ telephone line / 供电～ supply line / 配电～ distribution line / 输电～ transmission lines ②[交] line; route: 公共汽车～ bus line / 航空～ airline ◇ ～图 circuit diagram

【线呢】cotton suitings

【线膨胀】[物] linear expansion

【线圈】[电] coil: 初级～ primary coil / 次级～ secondary coil

【线绳】cotton rope

【线速度】[物] linear velocity

【线索】clue; thread: 案子的～断了。The clue could not be followed up. / 故事的～ threads of a story / 破案的～ clues for solving a case

【线毯】cotton blanket; thread blanket

【线条】①[美术] line: 粗犷、雄浑的～ bold and vigorous lines ②(人体等的) lines: ～优美 of fine lines

【线头】①(线的一端) the end of a thread ②(短线) an odd piece of thread

【线形动物】round worm

【线形叶】[植] linear leaf

【线性】[数] linear: ～代数 linear algebra / ～方程 linear equation / ～规划 linear programming / ～函数 linear function

【线轴儿】①(缠线用的轴儿) a reel for thread; bobbin ②(轴线) a reel of thread; a spool of thread

【线装】traditional thread binding ◇ ～本 thread-bound edition / ～书 thread-bound Chinese book

xiāng

襄 assist; help: 共～义举 let everybody help to promote this worthy undertaking

【襄理】assistant manager

【襄助】assist

镶 ①(嵌入) inlay; set; mount: ～宝石 set gems; mount precious stones / 给窗子～玻璃 glaze a window / 金～玉嵌 inlaid with gold and jade ②(围在边缘) rim; edge; border: 给裙子～花边 edge a skirt with lace

【镶板】[建] panel

【镶嵌】inlay; set; mount: ～细工 inlaid work; marquetry; mosaic / ～银丝漆器 silverinlaid lacquerware

【镶牙】put in a false tooth; insert an artificial tooth

相 ①(互相) each other; one another; mutually: ～距太远 too far apart / 素不～识 not know each other ②(表示一方对另一方的动作)好言～劝 advise sb. with kind words / 另眼～看 look upon sb. with special respect or concern; view sb. in a new, more favorable light / 实不～瞒 to tell you the truth ③(亲自观看是否合意) see for oneself: ～女婿 take a look at one's prospective son-in-law

【相安无事】live in peace with each other

【相比】compare with: 与…～ in comparison with; compared with / 二者不能～。 There is no comparison between the two.

【相差】differ; difference between: 两者～甚远。 There is a great deal of difference between the two. / 两者～无几。 There is little difference between the two.

【相称】match; suit; be commensurate to; be worthy of: 与…的称号～ be worthy of the title of / 衬衫的颜色与上衣的不～。 The color of the shirt does not match that of the coat.

【相持】be locked in a stalemate: ～不下 be locked in a stalemate; be at a deadlock

【相处】get along with; live together: ～得很好 get on well with each other / 不好～ difficult to get along with

【相传】①(传说) tradition has it that...; according to legend ②(传递;传授) hand down; pass on: 世代～ hand down from generation to generation

【相当】①(两方面差不多;配得上或能够相抵) match; balance; correspond to; be equivalent to; be equal to; be commensurate with: 年龄～ of about the same age / 得失～。 The gains balance the losses. / 两个足球队实力～。 The two football teams are well-matched. ②(适宜;合适) suitable; fit; appropriate: ～的人选 suitable person; fit person / ～的字眼 suitable words; appropriate words ③(程度高) quite; fairly; considerably: ～成功 quite a success / ～好 fairly good

【相得益彰】each shinning more brilliantly in the other's company; bring out the best in each other; complement each other

【相等】be equal: 大小～ be equal in size / 价值～ be equal in value / 数量～ be equal in amount 〈quantity; number〉; be numerically equal

【相抵】offset; balance; neutralize each other; counterbalance

【相对】①(面对面) opposite; face to face: ～而坐 sit opposite 〈facing〉 each other; sit face to face / 丑与美是～的。 Beauty is the opposite of ugliness. ②(非绝对的) relative ③(比较的) relatively; comparatively: ～地说 comparatively speaking / ～稳定 relatively stable ◇ ～高度[测] relative altitude; relative height / ～价格 relative price / ～孔径 [摄] relative aperture / ～剩余 relative surplus / ～湿度[气] relative humidity / ～速度[物] relative velocity / ～误差[数] relative error / ～性 relativity / ～运动[物] relative motion / ～真理[哲] relative truth / ～值 relative value / ～主义[哲] relativism

【相对论】[物] the theory of relativity; relativity: 广义～ the general theory of relativity / 狭义～ the special theory of relativity

【相对论性】[物] the relativistic ◇ ～量子理论 relativistic quantum theory / ～物理学 relativistic physics

【相反】opposite; contrary; adverse; reverse: 朝～的方向驶去 drive off in the reverse direction / 持～的意见 hold opposite opinions / 正～ on the contrary

【相反相成】(of two things) be both opposite and complementary to each other; oppose each other and yet also complement each other

【相仿】similar; much alike; more or less the same: 内容～ be similar in content / 能力～ be more or less equal in ability / 年纪～ be about the same age / 颜色～ be similar in color

【相逢】meet (by chance); come across: 萍水～ have a casual, temporary meeting

【相符】conform to; tally with; agree with; correspond to: 她的单子与我的不～。 Her list does not tally with mine.

【相辅而行】coordinate; go together; be complementary to each other

【相辅相成】supplement each other; comple-

ment each other
【相干】①(多用于否定句或疑问句) have to do with; be concerned with: 那与她不～。 That does not concern her./ 这件事与他有什么～? What has this to do with him? ②[物] coherent ◇ ～散射 coherent scattering / ～性 coherence; coherency
【相隔】be separated by; be apart; be at a distance of: ～不久 soon afterwards; before long / ～多年 after a lapse of many years / ～十公里 be ten kilometers apart
【相关】be interrelated; be related to; be bound up with
【相好】①(关系密切) be on familiar terms; be on intimate terms ②(亲密的朋友) intimate friend ③(不正当的恋爱) have an affair with ④(不正当的恋爱的一方) lover (男); mistress (女)
【相互】mutual; reciprocal; each other: ～关系 mutual relation; interrelation / ～了解 have mutual understanding / ～配套 mutually reinforcing / ～牵连 be implicative of each other / ～需求 reciprocal demand / ～依赖 depend on each other; be interdependent / ～影响 influence each other; interact / ～作用 interaction; interplay
【相机行事】act when the time is opportune
【相继】in succession; one after another: ～而起 one event treads close on another / ～而亡 die one after another / ～发言 speak in succession
【相间】alternate with: 红白～ red alternating with white; in red and white check
【相见恨晚】regret we didn't meet sooner
【相交】①(相交叉) intersect: AB 与 CD 两直线～于 E 点。 The lines AB and CD intersect at E. ②(做朋友) be friends; make friends with
【相近】①(距离接近) close; near; in the neighborhood of ②(相似) be similar to
【相敬如宾】treat each other with respect
【相距】apart; at a distance of; away from: ～不远 not far from one another / ～无几 be near at hand / 两个工厂～三公里。 The two factories are three kilometers apart.
【相连】be linked together; be joined: 两个城镇有运河～。 The two towns are linked by a canal.
【相碰】collide with
【相亲相爱】be deeply attached to each other; love each other devotedly
【相劝】persuade; offer advice: 好意～ offer well-meaning advice
【相去无几】There is not much difference.
【相濡以沫】help each other when both are in humble circumstances
【相商】consult with; talk over with: 有要事～ have something important to consult with sb.
【相识】①(彼此认识) be acquainted with each other: 素不～ have never met; not be acquainted with each other ②(相识的人) acquaintance: 老～ an old acquaintance
【相视而笑】smile into each other's eyes
【相思】yearning between lovers; lovesickness: 单～ one-sided love; unrequited love ◇ ～病 lovesickness
【相思鸟】red-billed leiothrix
【相思子】①[植] jequirity; jequirity bean; love pea ②(指种子) ormosia seed
【相似】resemble; be similar; be alike: ～的情况 similar cases / 面貌～ look alike / 何其～乃尔! What a striking similarity! ◇ ～形 [数] similar figures
【相提并论】mention in the same breath; place on a par: 两者不能～。 The two cannot be mentioned in the same breath.
【相通】communicate with each other; be interlinked: 我们的感情是～的。 Our feelings are interlinked./ 这两个大厅有门～。 The two halls open onto each other.
【相同】identical; the same; alike
【相投】be congenial; agree with each other: 兴趣～ have similar tastes and interests; find each other congenial
【相象】resemble; be similar; be alike: 与…很～ resemble sth. closely; be very similar to sth./ 与…有～的地方 bear some likeness to sth.
【相销价值】opposing values
【相信】believe in; be convinced of; have faith in: ～党,～群众 have faith in the Party and the masses / ～真理 believe in truth
【相形见绌】prove definitely inferior; pale by comparison; be outshone
【相形之下】by contrast; by comparison
【相沿成习】become a custom through long usage
【相依】depend on each other; be interdependent: ～为命 depend on each other for survival / 唇齿～ be as close as lips and teeth; be closely related and mutually dependent
【相宜】suitable; fitting; appropriate
【相应】corresponding; relevant: 采取～的措施 take appropriate measures / 通过～的决议 pass relevant resolutions / 作～的改变 make

corresponding changes
【相映】set each other off; contrast with; form a contrast: ～成趣 form a delightful contrast; contrast finely with each other
【相与】①(相处) get along with sb.; deal with sb. ②(相互) with each other; together
【相约】agree (on meeting place, date, etc.); reach agreement; make an appointment
【相知】①(相互了解,感情深厚) be well acquainted with each other; know each other well: ～有素 have known each other long ②(相互了解,感情深厚的朋友) bosom friend; great friend
【相左】①(不相遇) fail to meet each other ②(不一致) conflict with each other; fail to agree; be at odds with: 意见～ cannot see eye to eye
厢 ①(厢房) wing; wingroom ②(类似房子隔间的地方) railway carriage or compartment; (theatre) box: 包～ (theatre) box / 车～ carriage ③(靠近城区的地方) the vicinity outside of a city gate: 城～ the city proper and areas just outside its gates / 关～ a neighborhood outside of a city gate
【厢房】wing; wing-room
【厢式车身】station wagon
箱 chest; box; case; trunk: 弹药～ ammunition chest / 货～ packing box / 垃圾～ dustbin; garbage can; ash can / 木～ wooden trunk; chest / 皮～ leather suitcase / 医药～ medicine chest / 他买了一～啤酒。he bought a case of beer. / 我给他两～蜜柑。I gave him two boxes of oranges.
【箱底】①(箱子底层) the bottom of a chest ②(不常动用的财物) valuables stowed away at the bottom of the chest; one's store of valuables
【箱子】chest; box; case; trunk
香 ①(气味好闻) fragrant; sweet-smelling; aromatic; scented: ～花 fragrant flowers / 不辨～臭 be unable to distinguish ②(食物味道好) savory; appetizing; delicious: 这道菜真～。 The dish is very appetizing. ③(胃口好) with relish; with good appetite: 吃得很～ eat with relish; enjoy the food / 吃饭不～ have no appetite ④(睡得塌实) soundly: 她睡得正～。She is sleeping soundly. ⑤(受欢迎) popular; welcome: 小面包车在城镇很吃～。Minibus is most popular in towns. ⑥(香料) perfume or spice: 麝～ musk / 檀～ sandalwood ⑦(烧的香) incense; joss stick: 盘～ incense coil / 蚊～ mosquito- repellent incense / ～灰 incense ashes
【香槟酒】champagne
【香菜】coriander
【香草】①[植] sweetgrass ②(香草香精) vanilla: ～冰激凌 vanilla ice-cream
【香草醛】vanillic aldehyde; vanillin
【香肠】sausage
【香橙】fragrant citrus
【香椿】[植] Chinese toon
【香榧】[植] Chinese torreya ◇ ～子 Chinese torreya nut
【香粉】face powder
【香馥馥】strongly scented; richly fragrant
【香干】smoked bean curd
【香菇】mushroom
【香瓜】muskmelon
【香蕉】banana
【香蕉插孔】[电] banana jack
【香蕉插头】[电] banana plug; split plug
【香蕉水】[化] banana oil
【香界】[宗] Buddhist temples collectively
【香精】essence: 合成～ compound essence / 食用～ flavoring essence ◇ ～油 essential oil
【香客】pilgrim
【香料】perfume; spice; condiment ◇ ～厂 perfumery
【香炉】incense burner
【香茅】[植] lemongrass ◇ ～醛[化] citronellal / ～油 citronella oil
【香喷喷】①(好闻) sweet-smelling ②(好吃) savory; appetizing
【香片】scented tea
【香蒲】[植] cattail
【香气】sweet smell; fragrance; aroma: ～扑鼻 fragrance striking the nose
【香水】perfume; scent
【香甜】①(又香又甜) fragrant and sweet: ～的瓜果 sweet melons and fruits / 味道～ delicious ②(睡得塌实) soundly: 睡得～ sleep soundly
【香味】sweet smell; fragrance; scent; perfume
【香烟】①(卷烟) cigarette: 过滤嘴～ filter-tipped cigarette ②(烧香的烟) incense smoke: ～缭绕 coiling incense smoke ◇ ～盒 cigarette case
【香油】sesame oil
【香橼】[植] citron
【香云纱】[纺] gambiered Guangdong gauze
【香皂】perfumed soap; scented soap; toilet

soap
【香獐】[动] musk deer
【香脂】①(冷霜) face cream ②(止痛用) balm; balsam

乡 ①(乡下) country; countryside; village; rural area: 城～ town and country; urban and rural areas ②(家乡) native place; home village; home town: 回～ return to one's native place; return to one's home village ③(行政单位) township
【乡巴佬】bumpkin
【乡村】village; countryside; rural area
【乡间】village; country: ～别墅 country villa
【乡里】①(家乡) home village or town ②(乡亲) fellow villager or townsman
【乡亲】①(同乡) a person from the same village or town; fellow villager or townsman ②(当地群众) local people; villagers; folks
【乡绅】country gentleman; squire
【乡思】homesickness; nostalgia
【乡土】native soil; home village; of one's native land; local: ～风味 local flavor ◇ ～观念 provincialism / ～志 local records; local annals
【乡下】village; country; countryside
【乡下人】country folk; country cousin
【乡音】accent of one's native place; local accent
【乡邮】rural postal service ◇ ～员 rural postman
【乡镇】①(乡和镇) villages and towns ②(小市镇) small towns ◇ ～企业 town and township enterprises

xiáng

详 ①(详细) detailed; minute: ～谈 speak in detail; go into details; have a detailed discussion ②(详细情况) details; particulars: ～见附录。 For details, see the appendix. ③(清楚) know clearly: 其生卒年月不～ sb.'s dates are unknown
【详尽】detailed; exhaustive; thorough: ～的调查 a thorough investigation / ～的记载 a detailed record / 进行～的研究 make an exhaustive study
【详密】elaborate; meticulous: ～的计划 a meticulous plan
【详明】full and clear; complete and explicit: ～的注解 full and clear annotations
【详情】detailed information; details; particulars
【详实】full and accurate: ～的材料 full and accurate data 〈material〉
【详图】detail (drawing)
【详细】detailed; minute: ～的报告 a detailed report / ～的记述 detailled account / ～情节 detailed circumstances; ins and outs / ～审计 detailed audit / ～条款 detailed provisions / ～地描述 give a minute description / ～了解情况 acquire detailed knowledge of the situation / ～说明 explain at some length / ～占有材料 collect all the available material; have all the relevant data at one's fingertips
【详详细细】in every detail and particular

翔 circle in the air; fly: 翱～ soar; hover
【翔实】full and accurate; detailed and accurate

降 ①(投降) surrender; capitulate ②(降伏) subdue; vanquish; tame
【降伏】subdue; vanquish; tame: ～劣马 break in a wild horse
【降服】yield; surrender
【降龙伏虎】subdue the dragon and tame the tiger; overcome powerful adversaries
【降顺】yield and pledge allegiance to

xiǎng

享 enjoy
【享福】enjoy a happy life; live in ease and comfort: 享清福 enjoy the happiness of leisure
【享乐】lead a life of pleasure; indulge in creature comforts ◇ ～思想 preoccupation with pleasure-seeking / ～主义 hedonism; pleasure-seeking
【享年】die at the age of: ～八十岁 died at the age of eighty
【享受】enjoyment; treat; enjoy: ～公费医疗 enjoy public health services / 贪图～ seek ease and comfort
【享用】enjoy the use of; enjoy: ～自己的劳动果实 enjoy the fruits of one's own labor
【享有】enjoy (rights, prestige, etc): ～崇高的威望 enjoy high prestige; be held in esteem / ～公民权 enjoy civil rights / ～盛名 enjoy high reputation / ～外交豁免权 enjoy diplomatic immunities

鲞 dried fish: 鳗～ dried eel

想 ①(思考) think: ～办法 think of a way; try to find a solution / ～得真周到 have really thought of everything / ～问题 think over a problem / 你在～什么? What are you thinking of? / 我在～下一步怎么办。I am thinking what to do next. ②(推测;认为) suppose; reckon; think; consider: 你～会下雨吗? Do you think it will rain? / 我～她会来的。I think she'll be coming. ③(希望;打算) want to; would like to; feel like (doing sth.): 你下课以后～干什么? What do you want to do after class? ④(怀念;想念) miss; remember with longing

【想必】presumably; most probably

【想不到】unexpected: 这真是～的事! This is sth. quite unexpected! / 真～在这里会见到你。Fancy seeing you here!

【想不开】take things too hard; take a matter to heart

【想不起来】unable to call to mind

【想当然】assume sth. as a matter of course; take for granted

【想到】think of; call to mind; have at heart

【想得到】(多用于反问) think; imagine; expect: 谁～会有这样的事! Who would have thought that such a thing could happen!

【想得开】try to look on the bright side of things; not take to heart; take philosophically

【想得要命】want to do or get something very badly; miss sb. very much

【想法】①(想办法) think of a way; do what one can; try ②(意见) idea; opinion; what one has in mind: 按我的～ in my opinion; to my mind / 你有什么～? What do you have in mind?

【想方设法】do everything possible; try every means; try by hook or by crook

【想见】infer; gather: 从这些小事上,你可以～他的为人。From these trifles you can gather what kind of person he is.

【想来】it may be assumed that; presumably: ～他是不会骗你的。I assume that he won't cheat you.

【想念】long to see again; miss; remember with longing

【想起】remember; recall; think of; call to mind: 那情景使我～了童年。The sight recalls the days of childhood to me.

【想入非非】indulge in fantasy; allow one's fancy to run wild

【想通】straighten out one's thinking; become convinced; come round: 你还没有～? Are you still not convinced? / 我已经～了。I've come round to the idea now.

【想头】①(想法;念头) idea ②(希望) hope: 没～了。There's no hope now.

【想望】desire; long for: 他一直～当一名汽车司机。He has longed to be a driver.

【想象】①(设想) imagine; fancy; visualize: ～不到的困难 unimaginable difficulties / 可以～的 conceivable; thinkable / 难以～ hard to imagine ②[心] imagination ◇ ～力 imaginative power; imaginative faculty; imagination

响 ①(响声) sound; noise: 听不到～声了。No more sound was heard. ②(回声) echo: ～应 respond; answer ③(发出声音) make a sound; sound; ring: 铃在～。The bell is ringing. ④(响亮) loud; noisy: 收音机太～了。The radio is too noisy.

【响板】[乐] castanets

【响鼻】(of a horse, mule, etc.) snort

【响彻】resound through; reverberate through: ～云霄 resound to the skies; resound across the heavens

【响动】sound of sth. astir

【响度】[物] loudness; volume

【响亮】loud and clear; resounding; resonant; sonorous: ～的回答 a loud and clear reply

【响器】[乐] Chinese percussion instruments

【响声】sound; noise: 沙沙的～ rustling sound

【响尾蛇】rattlesnake

【响音】[语] resonant

【响应】respond; answer: ～…的号召 respond to the call of

饷 provide dinner for; entertain: ～客 entertain a guest / 以～读者 offer to the readers

xiàng

项 ①(颈后部) nape (of the neck) ②[量] item: 各～政策 various policies / 四～基本原则 four basic principles; four fundamentals / 逐～讨论 discuss item by item ③(款项) sum: 进～ income / 欠～ liabilities ④[数] term: 外～ extreme term / ～值 term value

【项背】a person's back: ～相望 (walk) one after another in close succession / 不可望其～ cannot hold a candle to sb.

【项链】necklace

【项目】item: 出口～ goods for export; export items / 基本建设～ capital construction project / 男子～ men's event / 女子～ women's event / 讨论～ item for discussion / 田径～ track and field events / 团体～ team event /

训练～ training courses／营业～ items of business／援助～ aid project／正式～ title team／支出～ item of expenditure
【项圈】necklace; necklet
【项庄舞剑,意在沛公】act with a hidden motive
巷 lane; alley
【巷战】street fighting
相 ①(相貌) looks; appearance: 丑～ ugly looks／可怜～ pitiful appearance; sorry figure／凶～ fierce look／长～儿 one's appearance ②(坐立姿态) bearing; posture: 站没站～,坐没坐～ not know how to stand or sit properly／站有站～,坐有坐～ have a graceful carriage ③(观察;判断) look at and appraise: ～马 look at a horse to judge its worth／人不可以貌～ Never judge a person by his appearance ④(相位) phase: ～位 phase position／三～电动机 three-phase motor／调～ phase modulation ⑤(照片) photograph: 照个～ take a photo; have a photo taken ⑥(某些国家的部长) minister
【相册】photo album
【相机】①(察看机会) watch for an opportunity: ～而动 wait for an opportunity to act; bide one's time／～行事 act as the occasion demands; do as one sees fit ②(照相机) camera: 单镜头反光～ single lens reflex camera／一步成相～ instant camera
【相角】photo corner
【相貌】facial features; looks; appearance: ～端正 have regular features
【相面】practice physiognomy; tell sb.'s fortune by reading his face ◇～先生 physiognomist
【相片】photograph; photo
【相声】[曲艺] comic dialogue; cross talk
【相纸】(photographic) printing paper; photographic paper
向 ①(方向) direction: 风～ wind direction ②(对着) face; turn towards: ～东 face east／～阳 turn towards the sun ③(偏袒) favor; side with; take the part of; be partial to: ～着某人 take sb.'s part ④(表示动作的方向)从胜利走～胜利 march from victory to victory／～…告发某人 denounce sb. to／～群众学习 learn from the masses／～上级汇报 report to one's superior ⑤(向来) always; all along: ～无此例。There's no precedent for this.
【向背】support or oppose: 人心～ whether the people are for or against; the will of the people
【向壁虚构】make up out of one's head; fabricate
【向导】guide
【向光性】[生] phototropism
【向后】towards the back; backward: ～撤 withdraw／～看 look back
【向后转】(口令) About face! ◇～走! To the rear, march!
【向来】always; all along: ～不吸烟 have never smoked／～如此。It has always been so.／～守时 have always been punctual
【向量】[数] vector ◇～分析 vector analysis
【向前】forward; onward; ahead: 采取～看的态度 adopt a forward-looking attitude／奋勇～ forge ahead
【向前看】①(向前方看) look ahead ②(口令) Eyes front!
【向日葵】sunflower
【向日性】[植] heliotropism
【向上】upward; up
【向上爬】be intent on personal advancement; eager to climb up the social ladder: ～的思想 mentality of a careerist
【向水性】[生] hydrotropism
【向外】outward
【向往】yearn for; look forward to; be attracted toward: ～幸福生活 look forward to a happy life
【向下】downward; down: ～的压力 down pressure／～作调查 investigate conditions at a lower level
【向斜】[地] syncline ◇～谷 synclinal valley
【向心力】[物] centripetal force
【向性】[生] tropism
【向阳】①(对着太阳) exposed to the sun; sunny ②(朝南) with a sunny exposure; with a southern exposure: 这房子～。The house has a southern exposure.
【向右】towards the right
【向右转】(口令) Right face! ◇～走! By the right flank, march!
【向隅而泣】weep all alone in a corner; be left to grieve in the cold
【向着】①(朝着;对着) turn towards; face ②(偏袒) take the part of; side with; be partial to
【向左】towards the left
【向左转】(口令) Left face! ◇～走! By the left flank, march!
象 ①(大象) elephant: 亚洲～ Asiatic elephant ②(形状;样子) appearance; shape;

image：万～更新。All things take on a new aspect. ③（仿效；摹拟）imitate：～声 onomatopoeia ④（在形象上相同或有共同点）be like；resemble；take after：这兄弟俩长得很～。The two brothers are very much alike. ⑤（好象）seem；look as if；appear：～是要下雨了。It looks like rain. ⑥（比如；如）such as；like：～董存瑞这样的英雄 heroes such as Dong Cunrui

【象鼻】trunk；proboscis

【象鼻虫】weevil；snout beetle

【象话】reasonable；proper；right：你天天迟到，～么？Aren't you ashamed to come late every day?

【象皮病】[医] elephantiasis

【象棋】(Chinese) chess：下～ play at chess ◇国际～ chess

【象声】onomatopoeia ◇～词 onomatope

【象限】[数] quadrant ◇～仪[天] quadrant

【象形文字】pictograph；hieroglyph

【象牙】elephant's tusk；ivory：～雕刻 ivory carving／～制品 ivories／～之塔 ivory tower

【象样】up to the mark；presentable；decent；sound：她的针线活挺～的。Her needlework is quite presentable.

【象征】①（表现；意味）symbolize；signify；stand for ②（象征物）symbol；emblem；token：友谊的～ emblem of friendship；symbol of friendship／鸽子是和平的～。The dove is the symbol of peace. ◇～性 symbolic；emblematic；token

橡

①（橡树）oak ②（橡胶树）rubber tree

【橡胶】rubber：合成～ synthetic rubber／生～ raw rubber；caoutchouc／天然～ natural rubber ◇～厂 rubber plant／～促进剂 rubber accelerator／～防老剂 rubber antiox- idant／～海绵 cellular rubber；rubber sponge／～轮胎 rubber tire／～树 rubber tree／～种植园 rubber plantation

【橡胶草】Russian dandelion；kok-saghyz

【橡皮】①（硫化橡胶）rubber ②（文具）eraser；rubber ◇～船 rubber boat／～膏[医] adhesive plaster／～筋 rubber band／～泥 plasticine／～手套 rubber (operating) gloves／～艇 pneumatic boat；rubber dinghy／～图章 rubber-stamp／～外包线[电] rubber-sheathed wire

【橡实】acorn ◇～管[电子] acorn tube

像

①（图像；人像）likeness (of sb.)；portrait；picture：佛～ image of Buddha／画～ portrait／铜～ bronze statue ②[物] image：实～ real image／虚～ virtual image

【像差】[物] aberration

【像散】[物] astigmatism ◇～镜 astigmatoscope／～透镜 astigmatic lens

【像章】badge <button> with sb.'s likeness on it

xiāo

哓

【哓哓不休】argue endlessly

骁

valiant；brave

【骁勇】brave；valiant：～善战 brave and skillful in warfare

消

①（消失）disappear；vanish：肿已～了。The swelling has gone down. ②（消除）eliminate；remove；dispel：～愁解闷 divert oneself from boredom；dispel depression or melancholy／～去一个未知数 eliminate an unknown quantity／～痰 reduce phlegm／～烟除尘 eliminate smoke and dust ③（消遣）pass the time in a leisurely way；while away (the time)：～夏 pass the summer in a leisurely way ④（需要）need；take：不～说 needless to say；it goes without saying／只～几天功夫 it takes no more than a few days

【消沉】downhearted；low-spirited；dejected；depressed：意志～ demoralized；despondent

【消除】eliminate；dispel；remove；clear up：～分歧 eliminate differences／～顾虑 dispel misgivings／～误会 clear up a misunderstanding／～嫌疑 clear of suspicion／～一切疑虑 remove all doubt／～隐患 remove a hidden danger／船长愉快的笑声～了我们的恐惧。The captain's cheerful laugh dispelled our fears.

【消磁头】eraser；erasing head

【消毒】①（杀死致命微生物）disinfect；sterilize：用酒精～ sterilize in alcohol／sterilize with alcohol／用漂白粉～ disinfect with bleaching powder ②（清除流毒）eliminate the pernicious influence ◇～剂 disinfectant／～牛奶 sterilized milk；pasteurized milk

【消防】fire control；fire fighting；fire protection ◇～车 fire engine；fire truck／～队 fire brigade／～技术 fire prevention／～人员 fire fighter／～设备 fire-fighting equipment／～水龙 fire hose／～艇 fireboat／～演习 fire drill／～站 fire station

【消费】consume ◇～城市 consumer-city／～

合作社 consumer's cooperative / ～价格指数 consumer price index / ～品 consumer goods / ～水平 consumption level / ～税 consumption tax / ～者 consumer / ～资料 means of subsistence
【消耗】consume; use up; deplete; expend: ～精力 consume one's energy / 低～ low consumption (of raw material) ◇ ～热[医] hectic fever / ～战 war of attrition
【消化】digest: ～所学的东西 digest what sb. has learnt / 好～ digestible; easy to digest ◇ ～不良 indigestion; dyspepsia / ～道 alimentary canal; digestive tract / ～功能紊乱 disorders of digestion / ～酶 digestive ferment; digestive enzyme / ～系统 digestive system / ～药 digestant / ～液 digestive juice
【消火栓】fire hydrant; fire cock
【消极】①(否定的;反面的) negative: ～因素 negative factor / ～影响 negative influence ②(不求进取的;消沉的) passive; inactive: 情绪～ be dispirited / 态度～ take a passive attitude; remain inactive / ～怠工 be slack in work / ～抵抗 passive resistance / ～防御 passive defense / ～行为 act of omission
【消解】clear up; dispel
【消弭】put an end to; avert; prevent: ～水患 prevent floods / ～战祸 avert a war
【消灭】①(消失;灭亡) perish; die out; pass away: 自行～ perish of itself; die out of itself ②(使消灭;除掉) annihilate; eliminate; abolish; exterminate; wipe out: ～病虫害 wipe out insect pests and plant diseases / ～一切入侵者 wipe out all invaders / 我军战士～了三百名敌军。 Our soldiers annihilated a force of three hundred enemy troops.
【消灭时效】negative prescription
【消磨】①(逐渐消耗) wear down; fritter away: ～精力 fritter away one's energy / ～志气 sap one's will ②(度过) while away; idle away: ～时间 kill time; pass the time / ～岁月 while away the time
【消气】cool down; be mollified: 她～了。 She has cooled down.
【消遣】divert oneself; while away the time; pastime; diversion
【消融】melt: 冰雪～ melting of ice and snow
【消散】scatter and disappear; dissipate: 云消雾散。The clouds dispersed and the fog lifted.
【消色差】[物] achromatism ◇ ～透镜 achromatic lens; achromat
【消声】[建] noise elimination ◇ ～器 silencer; muffler; dissipative muffler / ～室 anechoic chamber
【消失】disappear; vanish; dissolve; die away; fade away: 船慢慢地在雾中～了。 The ship faded into the fog. / 他～在夜色中。 He disappeared into the night. / 许多动物已从地球上～了。 Many kinds of animals have vanished from the earth.
【消逝】die away; fade away; vanish; elapse
【消释】clear up; dispel: ～前嫌 remove a grudge; bury the hatchet / ～疑虑 dispel misgivings
【消受】①(多用于否定,享受) enjoy: 无福～ not have the luck to enjoy; be unable to enjoy ②(忍受) endure; bear
【消瘦】become thin; become emaciated: 显得有点～ look a bit emaciated
【消损】①(消磨而损耗) wear and tear ②(逐渐减少) wear down
【消亡】wither away; die out
【消雾】fog dispersal
【消息】①(情况报道) news; information: ～灵通人士 a well-informed source / 本地～ local news / 据新华社～ according to a Xinhua dispatch / 头版～ a front-page story ②(音信) tidings; news: 杳无～ have not heard from sb. since; have had no news of sb. or sth.
【消炎】[医] diminish inflammation; reduce inflammation; allay inflammation ◇ ～剂 antiphlogistic
【消焰器】flame damper
【消音器】muffler
【消长】growth and decline
【消帐】cross off account; balance account
【消肿】[医] detumescence; subsidence of a swelling

宵 night: 通～ all night; throughout the night
【宵禁】curfew: 解除～ lift a curfew / 实行～ impose a curfew

逍

【逍遥】free and unfettered: ～自在 be leisurely and carefree
【逍遥法外】go scot-free; be at large; outlawry

霄 ①(云) clouds: 高入云～ towering into the clouds ②(天空) sky; heaven
【霄汉】[书] the sky; the firmament; heavens
【霄壤】heaven and earth: 有～之别 be as far apart as heaven and earth; be as different as

heaven and earth

硝 ①[化] nitre; saltpetre ②[皮革] tawing

【硝化】[化] nitrify ◇ ～甘油 nitroglycerine / ～棉 nitrocotton / ～纤维素 nitrocellulose

【硝石】nitre; saltpetre: 智利～ Chile nitre; Chile saltpetre; sodium nitre

【硝酸】[化] nitric acid ◇ ～铵 ammonium nitrate / ～钙 calcium nitrate / ～甘油 monobel; nitroglycerin / ～钾 potassium nitrate / ～钠 sodium nitrate / ～盐 nitrate / ～银 silver nitrate

【硝烟】smoke of gunpowder

削 ①(切割) pare with a knife; peel with a knife: ～木头 whittle a piece of wood / ～苹果 pare an apple; peel an apple / ～铅笔 sharpen a pencil ②[乒乓球] cut; chop: ～球 cut; chop

销 ①(熔化金属) melt ②(除去;解除) cancel; annul: 注～ write off; cancel ③(销售) sell; market: 畅～ sell well / 代～ sales on a commission basis; sell goods on a commission basis; be commissioned to sell sth.; act as a commission agent / 试～ put on trial sale; trial sale / 展～ sales exhibition / 滞～ sell poorly / 自～ sales through one's own channel ④(消费) expend; spend: 开～ expenditure ⑤(销子) pin

【销案】close a case

【销毁】destroy by melting or burning: ～核武器 the destruction of nuclear weapons / ～罪证 destroy incriminating evidence

【销魂】be overwhelmed with sorrow or joy; feel transported

【销货】sales; merchandise sales ◇ ～单 sales slip / ～定单 sales order / ～发票 invoice for sale / ～记录 record of goods sold / ～利润 selling profits / ～凭证 sales voucher / ～收益 sales income / ～限额 sales quotas

【销假】report back after leave of absence

【销路】sale; market: ～很好 have a good sale; find a good market / 没有～ find no sale; find no market

【销声匿迹】keep silent and lie low; disappear from the scene

【销售】sell; market ◇ ～费用 marketing expenses / ～价格 selling price / ～量 sales volume / ～税 sales tax / ～折扣 discount on sales / ～总额 total sales; aggregate sales

【销行】sell; be on sale: ～各地 be on sale everywhere / ～二百万册 have sold two million copies

【销赃】disposal of stolen goods

【销帐】write off; cancel or remove from an account

【销子】pin; peg; dowel

哮 ①(急促喘气) heavy breathing; wheeze ②(吼叫) roar; howl: 咆～ roar; thunder

【哮喘】asthma

嚣 clamor; hubbub; din: 叫～ clamor

【嚣张】rampant; arrogant; aggressive: ～一时 run rampant for a time / 气焰～ swollen with arrogance

潇

【潇洒】natural and unrestrained

【潇潇】①(形容刮风下雨) whistling and pattering: 风雨～ the whistling of wind and pattering of rain ②(形容小雨) drizzly

萧

【萧墙】screen wall: ～之祸 trouble arising at home; trouble from within; internal strife / 祸起～ trouble arose from within

【萧瑟】①(指声音) rustle in the air: 秋风～。The autumn wind is soughing. ②(指景色) bleak; desolate

【萧索】bleak and chilly; desolate

【萧条】①(寂寞冷落;毫无生气)desolate; bleak: 一片～的景象 a desolate scene on all sides ②[经] depression: 经济～ economic depression; slump / 生意～ Business is bad

【萧萧】①(马鸣声) neigh; whinny ②(风声) whistle; sough

箫 *xiao*, a vertical bamboo flute

xiáo

淆 confuse; mix: 混～ mix up; confuse; obscure

【淆惑】confuse; bewilder

【淆乱】confuse; befuddle: ～视听 befuddle the minds of the public

xiǎo

晓 ①(天刚亮时) dawn; daybreak: 拂～ foredawn ②(知道) know: 家喻户～ known to all ③(使人知道) tell; let sb. know: ～以利害 warn sb. of the consequences

【晓得】know: 天～! God knows!

【晓示】tell explicitly; notify

【晓行夜宿】start at dawn and stop at dusk

小 ①(与"大"相对)small; little; petty; minor: ～姑娘 a little girl / ～狗 puppy / ～国 a small country / ～鸡 chick; chicken / ～猫 kitten / ～牛 calf / ～声说话 speak in a low voice / ～问题 a minor question / ～镇 small town ②(短时间地) for a while; for a short time: ～住 stay for a few days / ～坐 sit for a while ③(年纪小的) young: ～弟弟 younger brother / ～儿子 the youngest son / 一家老～ the whole family; old and young

【小白菜】pakchoi; a variety of Chinese cabbage

【小百货】small articles of daily use

【小班】the bottom class in a kindergarten

【小半】less than half; lesser part; smaller part

【小包邮件】parcel post

【小报】small-sized newspaper; tabloid

【小辈】younger member of a family; junior

【小本经营】business with a small capital; do business in a small way

【小便】①(排尿) urinate; pass water; make water; empty one's bladder ②(尿) urine ③(男性生殖器) penis ◇ ～池 urinal

【小辫儿】short braid; pigtail

【小辫子】a mistake or shortcoming that may be exploited by others; vulnerable point; handle: 有～给人抓 have vulnerable points that others may capitalize on / 抓某人的～ take it as a handle against sb.

【小标题】subheading; subhead

【小不忍则乱大谋】A little impatience spoils great plans.

【小步舞曲】[乐] minuet

【小才大用】a man of little ability in high capacity

【小菜】pickled vegetables; pickles

【小册子】booklet; pamphlet

【小产】miscarriage; abortion

【小肠】[生理] small intestine

【小潮】[地] neap (tide)

【小车】①(手推车) wheelbarrow; handbarrow; handcart; pushcart ②(小轿车) sedan; sedan car

【小乘】[佛教] Hinayana; Little Vehicle

【小吃】①(非正餐) snack; refreshments ②(西餐中的冷盘) cold dish; made dish ◇ ～部 snack counter; refreshment room / ～店 snack bar; lunchroom

【小丑】clown; buffoon: 扮演～角色 play the buffoon; act the clown ◇ ～跳梁 a contemptible wretch making trouble

【小葱】shallot; spring onion

【小聪明】cleverness in trivial matters; petty trick: 耍～ play petty tricks

【小刀】pocket knife

【小道】path; trail: 羊肠～ winding trail

【小道理】minor principle

【小道消息】hearsay; grapevine

【小调】①(民间小曲) popular tune; ditty ②[乐] minor: A～协奏曲 concerto in A minor

【小洞不补大洞叫苦】A little neglect may breed great mischief. (或: A small leak will sink a great ship.)

【小动作】petty action; little trick; little maneuver: 搞～ get up to little tricks

【小豆】red bean

【小队】team; squad

【小额活期存款】petty cash deposit

【小额钱币】small money; small change

【小额支票】small check

【小恩小惠】petty favors; small favors; economic sops; paltry charity: 对某人施～ bestow paltry charity to sb.; give petty favors to sb.

【小儿】①(儿童) children ②[谦] my son

【小儿科】[医] (department of) pediatrics ◇ ～医生 pediatrician

【小儿麻痹症】infantile paralysis; poliomyelitis; polio

【小贩】pedlar; vendor; hawker

【小费】tip; gratuity: 给～ give gratuities to sb.; give sb. a tip; tip sb.

【小腹】underbelly; lower abdomen

【小个子】little chap; small fellow

【小工】unskilled laborer

【小功率电动机】fractional electric motor

【小姑】husband's younger sister; sister-in-law

【小鼓】[乐] side drum; snare drum

【小褂】(Chinese-style) shirt

【小广播】spreading of hearsay information; grapevine

【小规模】small-scale: ～的农场 a samll-scale farm / ～的试验工厂 a small-scale pilot plant ◇ ～集成电路 small-scale integrated circuit

【小鬼】①(鬼神的差役) imp; goblin ②(对小孩的亲昵称呼) little devil

【小孩儿】child

【小号】[乐] trumpet

【小户】①(人口少的家庭) small family ②(无钱无势的家庭) family of limited means and without powerful connections

【小黄鱼】little yellow croaker
【小伙子】youngster; lad; young fellow; young chap
【小集团】clique; faction
【小家碧玉】daughter of a humble family; a pretty girl of humble birth
【小脚】bound feet: ～女人 woman with bound feet
【小轿车】sedan (car): 豪华～ limousine
【小节】①(琐碎事情) small matter; trifle: 不拘～ not bother about small matters; not be punctilious／生活～ matters concerning personal life ②[乐] bar; measure: ～线 bar line; bar
【小结】①(整个过程中一个段落的总结) brief summary; preliminary summary; brief sum-up; interim summary ②(做小结) summarize briefly
【小姐】①(尊称) Miss ②(少女) young lady
【小舅子】wife's younger brother; brother-in-law
【小看】look down upon; belittle
【小康】well-to-do; comfortably off: ～之家 a comfortable family; a well-to-do family ◇ ～型 being fairly well-off
【小考】mid-term examination; quiz
【小口径】small-bore ◇ ～步枪 small-bore rifle
【小老婆】concubine
【小两口】young couple
【小萝卜】radish
【小麦】wheat: 冬～ winter wheat ◇ ～秆锈病 stem rust of wheat／～黑穗病 wheat smut／～粒 wheat berry／～片 wheat groats／～田 wheat paddock／～线虫病 wheat nematode／～锈病 wheat rust
【小卖部】①(小店)canteen; small shop ②(食品小卖部) buffet; snack counter
【小熊猫】lesser panda
【小米】millet: ～粥 millet gruel
【小拇指】little finger
【小脑】[生理] cerebellum
【小年】[农] off year
【小农】small farmer ◇ ～经济 small-scale peasant economy
【小跑】trot; jog
【小朋友】①(孩子们) children ②(小孩) little boy or girl; child
【小批生产】small serial production
【小便宜】small gain; petty advantage: 贪～ go after petty advantages
【小品】a short, simple literary or artistic creation; essay; sketch: 广播～ short piece for broadcasting
【小品文】familiar essay; essay
【小评论】(short) comment
【小气】stingy; niggardly; mean
【小气候】[气] microclimate
【小前提】[逻] minor premise
【小钱不去大钱不来】If a little money does not go out, great money will not come in.
【小巧玲珑】small and exquisite
【小球藻】[植] chlorella
【小曲儿】ditty; popular tune
【小圈子】small circle 〈set〉 of people; small coterie: 搞～ form a small coterie ◇ ～主义 "small circle" mentality
【小犬座】[天] Canis Minor
【小群落】society
【小人】a base person; villain; vile character: ～得志 villains holding sway
【小人物】an unimportant person; a nobody; cipher; nonentity
【小日子】easy life of a small family
【小商品】small commodities ◇ ～经济 small commodity economy／～生产者 small commodity producer
【小舌】[生理] uvula
【小生产】small production; small-scale production ◇ ～者 small producer
【小狮座】[天] Leo Minor
【小时】hour
【小时候】in one's childhood; when one was young
【小市民】urban petty bourgeois
【小事】trifle; petty thing; minor matter: 计较～ fuss over trifles
【小事聪明,大事糊涂】penny wise and pound foolish
【小试锋芒】display only a small part of one's talent
【小手工业者】small handicraftsman
【小手小脚】①(不大方) ungenerous; stingy; mean ②(缩手缩脚) lacking boldness; timid; niggling
【小叔子】husband's younger brother; brother-in-law
【小数】[数] decimal: 循环～ recurring decimal; repeating decimal ◇ ～点 decimal point
【小说】novel; fiction: 长篇～ novel／短篇～ short story／科学幻想～ science fiction／中篇～ medium-length novel; novelette ◇ ～家 novelist; writer of fiction

【小苏打】[化] sodium bicarbonate
【小算盘】selfish calculations: 爱打～ be calculating
【小提琴】violin
【小提琴手】violinist: 首席～ concertmaster
【小题大作】make a fuss over a trifling matter; make a mountain out of a molehill
【小天地】one's own little world
【小艇】small boat; skiff
【小偷】petty thief; sneak thief; pilferer: ～小摸 pilfering
【小土地出租者】lessor of small plots
【小团体主义】cliquism; small-group mentality
【小腿】shank
【小巫见大巫】pale into insignificance in comparison with; feel dwarfed
【小五金】hardware; metal goods for domestic use; metal fittings
【小小不言】too trivial to talk about: ～的事,不必计较。The matter is too small to be worth niggling over.
【小写】small letter ◇ ～体 lower case
【小心】take care; be careful; be cautious: ～谨慎 be careful; be discreet / ～轻放! Handle with care! / ～,危险! DANGER! / ～油漆! Mind the wet paint!
【小心眼儿】narrow-minded; petty
【小心翼翼】with great care; cautiously
【小行星】[天] asteroid; minor planet; planetoid
【小型】small-sized; small-scale; miniature: ～公共汽车 minibus / ～摩托车 motor scooter / ～企业 small enterprise / ～摄影机 minican / ～拖拉机 baby tractor; small tractor / ～运动会 a small-scale athletic meet / ～照相机 miniature camera
【小熊座】[天] Ursa Minor
【小学】primary school; elementary school
【小学生】(primary school) pupil; schoolchild; schoolboy (男); schoolgirl (女)
【小循环】[生理] pulmonary circulation
【小样】[印] galley proof
【小业主】small proprietor; petty proprietor
【小夜曲】[乐] serenade
【小姨子】wife's younger sister; sister-in-law
【小意思】small token of kindly feelings; mere trifle
【小音阶】[乐] minor scale
【小阴唇】[生理] labium minus pudendi
【小引】introductory note; foreword
【小雨】light rain
【小照】small-sized photograph
【小指】little finger or toe
【小传】brief biography; biographical sketch; profile
【小资产阶级】petty bourgeoisie
【小字】small character
【小子】①(儿子) boy ②(用于男性,含轻蔑意) bloke; fellow; chap
【小组】group: ～会 group meeting / ～讨论 group discussion / ～长 group leader / 党～ group under a Party branch; Party group

xiào

校 ①(学校) school; college; university: 夜～ night school ②[军](校官) field officer
【校办工厂】school-run workshop; campus workshop
【校车】school bus
【校风】school spirit
【校官】field officer; field grade officer ◯ 上校 Colonel; Captain (海军); Group Captain (英国空军) / 中校 Lieutenant Colonel; Commander (海军); Wing Commander (英国空军) / 少校 Major; Lieutenant (海军); Squadron Leader (英国空军)
【校规】school regulations: 遵守～ abide by school regulations
【校徽】school badge: 佩带～ wear a school badge
【校刊】school magazine; college journal
【校历】school calendar
【校庆】anniversary of the founding of a school or college
【校舍】schoolhouse; school building
【校外】outside school; after school ◇ ～辅导员 after-school activities counsellor / ～活动站 after-school activities club
【校务】administrative affairs of a school or college
【校医】school doctor
【校友】alumnus (男); alumna (女)
【校园】campus; school yard
【校长】①(中小学) headmaster; principal ②(大专院校) president; chancellor
【校址】the location of a school or college

效 ①(效果) effect: 见～ produce an effect; prove effective / 生～ take effect / 无～ ineffective; of no effect / 有～ effective / 那法律仍然有～。The law is still in effect. ②(仿效) imitate; follow the example of: 上行下～。The subordinates will follow the example set by

their superiors. ③(献出) devote to; render a service to: ～死 ready to give one's life for a cause
【效法】follow the example of; model oneself on; learn from
【效果】①(产生的结果) effect; result: ～不大 not be very effective; produce little effect / ～显著 bring about a striking effect; be very effective / 取得良好的～ achieve good results / 医生开的那药没有～。 The prescribed medicine failed to take effect. ②[剧] sound effects
【效劳】work in the service of; work for: 乐于～ be glad to offer one's services / 为国～ serve one's country
【效力】①(效劳) render a service to; serve ②(效果) effect: 这药很有～。 The medicine is efficacious. ③(约束力) force; effect; avail: 两种文本具有同等～。 Both texts are equally authentic. ◇～范围 scope of validity
【效率】efficiency: ～低 inefficient / ～高 efficient / 提高～ promote efficiency
【效命】go all out to serve sb. regardless of the consequences: ～疆场 ready to lay down one's life on the battlefield
【效能】efficacy; usefulness: 充分发挥灌溉的～ make the best possible use of irrigation; make full use of irrigation
【效验】intended effect; desired result: 没有～ prove ineffective; fall flat
【效益】beneficial result; benefit: 经济～ economic benefit
【效应】[物] effect: 陀螺～ gyroscopic effect / 微观～ microeffect
【效用】effectiveness; usefulness ◇～递减律 law of diminishing utility
【效尤】knowingly follow the example of a wrongdoer: 以儆～ to warn others against following a bad example
【效忠】pledge loyalty to; devote oneself heart and soul to: ～于 be loyal to; be faithful to

孝 ①(孝顺) filial piety ②(孝服) mourning: 带～ in mourning
【孝服】mourning (dress)
【孝敬】give presents (to one's elders or superiors)
【孝顺】show filial obedience
【孝子贤孙】worthy progeny; true son

肖 resemble; be like: 维妙维～ absolutely lifelike
【肖像】portrait; portraiture ◇～画 portrait painting / ～画家 portraitist
【肖像名片】carte-de-visite

啸 ①(呼啸) whistle ②(野兽的叫声) howl; roar: 虎～ the roar of a tiger
【啸聚】band together; gang up: ～山林 go to the greenwood

笑 ①(欢笑) smile; laugh: ～出眼泪 laugh until tears come / ～得前仰后合 laugh oneself into convultion / ～着 with a laugh / 发～ give a laugh / 放声大～ laugh heartily / 哄堂大～ roar with laughter / 微～ smile / 一～了之 laugh out of court / 引人发～ raise a laugh / 她突然～起来了。 She broke into a laugh. ②(讥笑) ridicule; laugh at: 叫人～掉大牙 ridiculous enough to make people laugh their heads off
【笑柄】laughingstock; butt; joke: 这已成为全城的～。 It has become the laughingstock of the town.
【笑哈哈】laughingly; with a laugh
【笑话】①(引人发笑的故事;笑料) joke; jest: 闹～ make a fool of oneself; make a funny mistake / 说～ crack a joke ②(耻笑;讥笑) laugh at; ridicule
【笑话百出】make many ridiculous mistakes
【笑口常开】grinning all the time
【笑里藏刀】hide a dagger in a smile; with murderous intent behind one's smiles
【笑脸】smiling face: ～相迎 greet sb. with a smile; salute one with a smile / 陪～ meet rudeness with a flattering smile
【笑料】laughingstock; joke
【笑骂】deride and taunt
【笑眯眯】smilingly; with a smile on one's face
【笑面虎】smiling tiger; an outwardly kind but inwardly cruel person
【笑纳】kindly accept (this small gift of mine)
【笑气】[化] laughing gas; nitrous oxide
【笑容】smiling expression; smile: 慈祥的～ a kindly smile / 满面～的姑娘 a laughing girl / ～可掬 show pleasant smiles; be radiant with smiles / ～满面 be all smiles; have a broad smile on one's face
【笑声】laugh; laughter: 我听到隔壁屋里的～。 I heard sounds of laughter in the next room.
【笑谈】laughingstock; object of ridicule: 传为～ become a standing joke
【笑嘻嘻】grinning; smiling broadly
【笑逐颜开】beam with smiles; be wreathed in

smiles

xiē

楔

【楔形文字】cuneiform (characters); sphenogram

【楔子】①(榫头) wedge ②(木钉;竹钉) peg; wooden peg; bamboo peg ③(戏曲、小说的引子) prologue in some modern novels

些 some: 那么～ that much; that many / 这～ these / 好～人 a lot of people / 好～书 many books / 加～水 add a little water / 买～东西 do some shopping / 前～日子 a few days ago; sometime ago

【些微】slightly; a little; a bit

蝎 scorpion

【蝎虎】gecko; house lizard

【蝎子】scorpion

歇 ①(休息) have a rest: ～一会儿 have a short rest ②(停止) stop; knock off: ～脚 stop for a rest

【歇班】be off duty; have time off

【歇顶】get a bit thin on top; be balding

【歇伏】stop work during the dog days

【歇工】stop work; knock off

【歇晌】take a midday nap or rest

【歇斯底里】hysteria: ～大发作 go into hysterics; become hysterical

【歇息】①(休息) have a rest ②(住宿;睡觉) go to bed; put up for the night

【歇业】close a business; go out of business

xié

鞋 shoes: 布～ cloth shoes / 凉～ sandals / 棉～ cotton-padded shoes / 皮～ leather shoes / 皮便～ loafers / 球～ gym shoes; sneakers / 套～ overshoes / 拖～ slippers / 一双～ a pair of shoes / 雨～ rain shoes; rain boots; galoshes / 毡～ felt shoes

【鞋拔子】shoehorn; shoe lifter

【鞋帮】upper (of a shoe)

【鞋带】shoelace; shoestring

【鞋底】sole (of a shoe)

【鞋垫】shoe-pad; insole

【鞋跟】heel (of a shoe)

【鞋匠】shoe make; cobbler

【鞋刷】shoe brush

【鞋楦】last for shaping a shoe; shoe tree

【鞋油】shoe polish; shoe cream: 黑～ shoe-blacking

挟 ①(用胳膊夹住) hold sth. under the arm ②(挟制) coerce; force sb. to submit to one's will: 要～ coerce ③(心里怀着) harbor

【挟持】①(从两旁抓住) seize sb. on both sides by the arms ②(强迫对方服从) hold sb. under duress; coerce sb. into submission

【挟天子以令诸侯】have the emperor in one's power and order the dukes about in his name

【挟嫌】harbor resentment; bear a grudge: ～报复 bear resentment against sb. and retaliate

【挟制】take advantage of sb.'s weakness to enforce obedience; force sb. to do one's bidding

携 ①(携带) carry; take along: ～眷 bring one's wife and children along / ～款潜逃 abscond with funds / ～械投诚 come over from the enemy's side bringing weapons ②(拉着手) take sb. by the hand; hold sb. by the hand

【携带】carry; take along: ～方便 be easy to carry about / 随身～的物品 things carried on one's person

【携手】hand in hand: ～并进 go forward hand in hand

谐 ①(和谐) in harmony; in accord; in tune: 和～ harmonious ②(办好;商量好) settle; come to an agreement ③(诙谐) humorous: 诙～ humorous; jocular

【谐和】harmonious; concordant

【谐谑】banter: 语带～ speak somewhat jokingly

【谐谑曲】[乐] scherzo

【谐音】①(字词的音相同或相近) homophonic; homonymic ②[乐] partials

【谐振】[物] resonance: 空腔～ cavity resonance ◇～腔 resonant cavity

偕 together with; in the company of: ～行 travel together

【偕老】husband and wife grow old together: 白头～ stick to each other till the hair turns gray

【偕同】in the company of; accompanied by; along with

邪 evil; heretical; irregular: 改～归正 give up one's evil ways and return to the right path; turn over a new leaf / 我们决不搞歪的～的。 We never engage in irregularities.

【邪道】evil ways; depraved life; vice: 把某人引入～ lead sb. astray / 走～ lead a depraved

life; abandon oneself to evil ways
【邪恶】evil; wicked; vicious: ～的念头 wicked thoughts
【邪门儿】strange; odd; abnormal; irregular
【邪门歪道】crooked ways; crooked means; dishonest practices; dishonest methods
【邪魔】evil spirit; demon
【邪念】evil thought; wicked idea; evil intention: 某人起了～ a wicked idea came into one's head
【邪气】perverse trend; evil influence: 使正气上升,～下降 encourage healthy trends and check unhealthy ones
【邪说】heresy; heretical ideas; fallacy

斜 oblique; slanting; inclined; tilted: 她向旁边～看了一眼。 She gave an oblique look to one side. / 这根线向左～了。The line is slanting to the left. / 柱子有点～。 The pillar is a little tilted.
【斜边】①[数] hypotenuse ②[机] bevel edge
【斜长石】[矿] plagioclase
【斜方肌】[生理] trapezius muscle
【斜风细雨】light wind and drizzling rain
【斜高】[数] slant height
【斜角】①[数] oblique angle ②[机] bevel angle ◇～规[机] bevel square
【斜井】①[矿] inclined shaft; slope ②[石油] inclined well; slant hole
【斜路】wrong path: 走上～ take the wrong path; go astray
【斜率】[数] slope
【斜面】①[数] inclined plane ②[机] oblique plane; bevel (face)
【斜坡】slope
【斜射】[军] oblique fire
【斜视】①[医] strabismus ②(斜着眼看) look sideways; cast a sidelong glance: 目不～ not look sideways; refuse to be distracted
【斜视图】[机] oblique drawing
【斜体字】[印] italics
【斜纹】[纺] twill (weave) ◇～布 twill; drill
【斜线】oblique line ◇～号 slant (/)
【斜眼】①[医] strabismus ②(患斜视的眼睛) wall-eye or cross-eye ③(患有斜视的人) a wall-eyed or cross-eyed person
【斜阳】setting sun
【斜圆锥】[数] oblique cone
【斜轴线】[数] oblique axis

协 ①(共同) joint; common: ～办 do sth. jointly ②(协助) assist
【协定】①(协商而订的条款) agreement; accord: 贷款～ loan agreement / 军事停战～ armistice agreement / 君子～ gentlemen's agreement / 科学技术合作～ agreement on scientific and technical cooperation / 临时～ provisional agreement / 贸易～ trade agreement / 停火～ cease-fire agreement / 停战～ cease-fire agreement / 文化合作～ agreement on cultural cooperation / 一般～ general agreement / 支付～ payment agreement ②(协商订立) reach an agreement on sth. ◇～关税 conventional duty / ～配额 bilateral quota / ～税率 conventional tariff
【协会】association; society
【协理】①(协助办理) assist in the management (of an enterprise, etc.) ②(银行、大企业中协助经理主持业务的人) assistant manager ◇～员[军] political assistant
【协力】unite efforts; join in a common effort: ～进攻 launch a joint assault
【协商】consult; talk things over: ～会议 consultative conference / ～解决 negotiated settlement / ～一致的原则 the principle of reaching unanimity through consultation / 民主～ democratic consultation
【协调】coordinate; concert; harmonize; bring into line: ～发展 coordinated growth / ～我们的行动 coordinate our actions / ～优美的动作 harmonious and graceful movements ◇～委员会 coordination committee
【协同】work in coordination with; cooperate with: ～作战 fight in coordination
【协议】①(协商) agree on: 一致～的文件 a document unanimously agreed upon ②(协商取得的一致意见) agreement: 口头～ verbal agreement / 达成～ reach an agreement; come to an agreement / 撕毁～ tear up an agreement ◇～离婚 divorce by agreement
【协助】assist; help; give assistance; provide help
【协奏曲】[乐] concerto: 钢琴～ piano concerto / 小提琴～ violin concerto
【协作】cooperation; coordination; combined efforts; joint efforts: ～得很好 cooperate harmoniously / 经济～区 economically coordinated region

胁 ①(腋下到腰上的部分) the upper part of the side of the human body ②(胁迫) coerce; force: 裹～ force to take part; coerce / 威～ threaten

【胁持】abduct; hold under duress
【胁从】be an accomplice under duress ◇ ～分子 reluctant follower; unwilling follower; accomplice under duress
【胁肩谄笑】cringe and smile obsequiously
【胁迫】coerce; force

xiě

写 ①(用笔写) write: ～信 write a letter ②(写作) compose; write: ～剧本 write a play / ～论文 write a thesis / ～日记 make an entry in one's diary; keep a diary / ～诗 compose a poem ③(描写) describe; depict: ～景 describe the scenery ④(画) paint; draw: ～生 paint from life
【写稿】write for a magazine; contribute to a magazine: 为黑板报～ write for the blackboard newspaper
【写入】[计算机] read-in; write-in
【写生】[美术] paint from life; draw, paint or sketch from nature: 静物～ still life painting / 人物～ portrait from life / 外出去～ go out sketching
【写实】write or paint realistically
【写信】write: 她每星期都要给母亲～。 She writes every week to her mother.
【写意】[美术] freehand brushwork in traditional Chinese painting
【写照】portrayal; portraiture: 生活的真实～ be a true portrayal of the life
【写字间】office room
【写字台】writing desk; desk
【写作】writing: 从事～ take up writing as one's career ◇ ～班子 writing group / ～技巧 writing technique

血 blood: ～的教训 a lesson paid for in blood; a lesson written in blood
【血淋淋】dripping with blood; bloody
【血晕】bruise

xiè

亵 ①(轻慢) treat with irreverence; be disrespectful ②(淫秽) obscene; indecent
【亵渎】blaspheme; profane; pollute: ～神明 blaspheme the gods

泻 ①(很快地流) flow swiftly; rush down; pour out: 一～千里 rush down a thousand *li*; flow powerfully ②(腹泻) have loose bowels; have diarrhea: 上吐下～ suffer from vomiting and diarrhea
【泻肚】have loose bowels; have diarrhea
【泻湖】[地] lagoon
【泻盐】Epsom salts; salts
【泻药】laxative; cathartic; purgative

械 ①(器械) tool; instrument: 机～ machine; mechanism ②(武器) weapon: 军～ weapons; arms; ordnance / 缴～ lay down one's arms
【械斗】fight with weapons between groups of people

泄 ①(排出;排泄) let out; discharge; release: 象个～了气的皮球 like a deflated rubber ball; dejected ②(泄漏) let out; leak ③(发泄) give vent to; vent: ～恨 give vent to one's hatred / ～私愤 give vent to personal spite
【泄底】reveal or expose what is at the bottom of sth.
【泄愤】give vent to one's anger; vent one's resentment
【泄洪】[水] flood discharge ◇ ～道 flood-relief channel; floodway / ～隧洞 flood-discharge tunnel
【泄劲】①(失去信心) lose heart; feel discouraged; be disheartened ②(失去干劲) slacken one's efforts; slack off
【泄漏】leak; let out; divulge; give away: ～秘密 let out a secrets; divulge a secret; give away a secret
【泄露】let out; reveal; betray: ～秘密 disclose a secret ◇ ～国家机密罪 betrayal of state secrets; offense of betraying state secrets / ～军事机密罪 offense of disclosing military secrets
【泄密】divulge a secret; betray confidential matters: ～事件 case of leakage of a secret
【泄气】①(泄劲) lose heart; feel discouraged; be disheartened: 困难面前不～ keep one's end up in the face of difficulties / 别～! Don't lose heart. ②(讽刺低劣或无本领) disappointing; frustrating; pathetic: 真让人～! How pathetic!
【泄水】[水] sluicing ◇ ～道 sluiceway / ～工程 outlet work / ～孔 outlet / ～闸 sluice gate; sluice
【泄殖腔】[动] cloacal chamber; cloaca

卸 ①(从运输工具上搬下) unload; discharge; lay down: ～车 unload a vehicle ②(拆卸) remove; strip: ～零件 remove parts from a machine; strip a machine / 把门～下来 life a door off its hinges ③(推卸) get rid of; shirk: ～责

shirk the responsibility
【卸车】unload (goods, etc.) from a vehicle; unload
【卸担子】lay down the burden; put down a loan
【卸货】unload cargo; discharge cargo; unload: 从船上～ unload a ship; land goods from a ship ◇ ～承揽人 landing agent／～费 discharging expenses;landing charges／～港 port of delivery; port of discharge
【卸料装置】tripper
【卸磨杀驴】get rid of sb. as soon as he has done his job; kill the donkey the moment it leaves the millstone
【卸任】be relieved of one's office
【卸载传送机】unloading conveyor
【卸装】remove stage makeup and costume

谢 ①(感谢) thank: ～～你。Thanks. ②(认错;道歉) apologize; make an apology: ～过 apologize for one's fault ③(谢绝) decline ④(花、叶脱落) wither
【谢忱】gratitude; thankfulness: 谨致～ allow us to express our thanks for
【谢词】thank-you speech
【谢绝】refuse; decline: ～参观。Not open to visitors.／婉言～ politely decline; politely refuse
【谢幕】answer a curtain call; respond to a curtain call: 多次～ respond to repeated curtain calls
【谢世】pass away; die
【谢天谢地】thank goodness; thank heaven
【谢谢】thanks; thank you
【谢意】gratitude; thankfulness: 聊表～ as a token of my gratitude／预致～ thank you in anticipation
【谢罪】apologize for an offense; offer an apology

榭 a pavilion or house on a terrace: 歌台舞～ halls for the performance of songs and dances／水～ waterside pavilion

蟹 crab: 捕～ go crabbing; fish for crab／寄居～ hermit crab／沙～ ghost crab／梭子～ swimming crab
【蟹黄】the ovary and digestive glands of a crab
【蟹青】greenish-grey (color)
【蟹爪】crab claw
【蟹爪兰】[植] schlumbergera
【蟹爪式装载机】gathering arm loader

懈 slack; lax: 坚持不～ hold on unremittingly／松～ slacken; relax; let up／作不～的努力 make unremitting efforts
【懈怠】slack; sluggish

邂
【邂逅】meet unexpectedly; run into sb.; meet by chance: ～一位同窗 meet unexpectedly a school-mate

屑 ①(碎末) bits; scraps; crumbs: 金属～ metal fillings／煤～ (coal) slack／面包～ crumbs (of bread)／纸～ scraps of paper ②(琐碎) trifling ③(认为值得)不～ disdain to do sth.

xīn

辛 ①(辣) hot (in taste, flavor, etc.); pungent ②(辛苦) hard; laborious: 艰～ hardships ③(痛苦) suffering: ～酸 sad; bitter
【辛迪加】[经] syndicate
【辛苦】①(身心劳苦) hard; toilsome; laborious: ～的工作 hard work; laborious work／～地工作 toil at one's task ②(套语,用于求人做事) work hard; go to great trouble; go through hardships: 这事您还得～一趟。You'll have to take the trouble of going there to see about it.
【辛辣】pungent; hot; bitter: ～的讽刺 bitter irony; biting sarcasm／～的味道 a sharp flavor; a pungent flavor
【辛劳】pains; toil: 不辞～ spare no pains／日夜～ toil day and night
【辛勤】industrious; hardworking: ～劳动 work hard; labor assiduously
【辛酸】sad; bitter; miserable: ～的往事 sad memories; poignant memories／～泪 hot and bitter tears
【辛烷值】[化] octane number; octane value
【辛辛苦苦】take a lot of trouble; take great pains; work laboriously

锌 [化] zinc (Zn)
【锌白】[化] zinc white
【锌版】[印] zinc plate; zincograph ◇ ～印刷术 zincography
【锌钡白】[化] lithopone
【锌粉】[化] zinc powder

新 ①(跟"老"或"旧"相对) new; fresh; up-to-date: ～发明 a new invention／～风尚

new custom; new practice / ～技术 new technique; up-to-date technique / 最～消息 the latest news ②(新近) newly; freshly; recently: ～建的工厂 a newly built factory / ～上任的主任 the newly appointed director ③(新近结婚的) recently married: ～人 newlywed
【新版】new edition
【新兵】new recruit; recruit
【新陈代谢】①[生] metabolism ②(新事物代替旧事物) the new superseding the old
【新仇旧恨】new hatred piled on old; old scores and new
【新村】new residential quarter; new housing development
【新大陆】the New World; the Americas
【新房】bridal chamber
【新分析法学】new analytical jurisprudence
【新官上任三把火】a new official applies strict measures; a new broom sweeps clean
【新华社】the Xinhua News Agency
【新婚】newly-married: ～夫妇 newly-married; newlyweds / ～燕尔 happy wedding couple; newlyweds
【新纪元】new era; new epoch: 开创～ usher in a new epoch; open a new era
【新加坡】Singapore ◇～人 Singaporean
【新交】new acquaintance; new friend: 我和他是～。 I got acquainted with him only recently.
【新教】[宗] Protestantism ◇～徒 Protestant
【新近】recently; lately; in recent times
【新旧交替】the transition from the old to the new
【新居】new home; new residence
【新开发实用品】notions
【新开户头】new account
【新来乍到】newly arrived; a newcomer; have just arrived
【新郎】bridegroom
【新霉素】[药] neomycin
【新民主主义】new democracy ◇～革命 new-democratic revolution
【新名词】new term; new expression; vogue word; newfangled phrase
【新年】New Year: ～好! Happy New Year! / ～献词 New Year message
【新娘】bride
【新篇章】new page; new chapter
【新瓶装旧酒】old wine in a new bottle; the same old stuff with a new label
【新奇】strange; novel; new: ～的想法 a novel idea
【新人】①(具有新的道德品质的人) people of a new type: ～新事 new people and new things ②(某方面新出现的人物) new personality; new talent ③(新娘) bride ④(新郎新娘) newlywed
【新生】①(刚产生的) newborn; newly born: ～力量 newly emerging force; new force / ～事物 newly emerging things; new things / ～婴儿 newborn (baby) ②(新生命) new life; rebirth; regeneration: 获得～ be given a new life; be reborn ③(新入学的学生) new student
【新生代】[地] the Cenozoic Era
【新石器时代】the New Stone Age; the Neolithic Age
【新式】new type; latest type; new-style: ～服装 clothing of the latest fashion; fashionable dress / ～农具 new types of farm implements; improved farm implements / ～武器 modern weapons
【新收成】new crop
【新手】new hand; raw recruit
【新四军】the New Fourth Army
【新闻】news; 国际～ world news / 国内～ home news / 简明～ news in brief / 头版～ front-page news; front-page story / 有什么～? What is the news? ◇～报导 news report; news story; news coverage / ～处 office of information; information service / ～稿 press release; news release / ～工作 newspapering / ～工作者 journalist / ～公报 press communique / ～广播 newscast / ～记者 newsman; newspaperman; reporter; journalist / ～检查 censorship of press / ～检查官 press censor / ～简报 news summary / ～界 press circles; the press; newspaperdom / ～片 newsreel; news film / ～评论员 news analyst / ～社 news agency / ～司 department of information / ～图片橱窗 newsphoto display case / ～纸 newsprint / ～自由 freedom of information; freedom of the press
【新西兰】New Zealand ◇～人 New Zealander
【新鲜】① fresh: ～空气 fresh air / ～牛奶 fresh milk ②(新生的) new; novel; strange: ～经验 new experience; fresh experience
【新兴】new and developing; burgeoning: ～工业 the new and expanding industry / ～工业城市 a developing industrial city / ～社会力量 new and rising forces in society
【新星】[天] nova: 超～ supernova
【新型】new type; new pattern: ～材料 new ma-

terials
【新颖】new and original; novel: 式样～ in a novel style / 题材～ original in choice of subject 〈theme〉
【新约】[基督教] the New Testament
【新月】①(农历月初形状如钩的月亮) crescent ②[天] new moon
【新月形沙丘】[地] crescent dune; barchan
【新殖民主义】neocolonialism; new colonialism ◇～者 neocolonialist
【新装】new clothes

薪 ①(柴火) firewood; faggot; fuel ②(薪水) salary: 发～ pay out the salary
【薪金】salary; pay
【薪尽火传】as one piece of fuel is consumed, the flame passes to another; the torch of learning is passed on from teacher to student and from generation to generation
【薪水】salary; pay; wages
【薪炭林】[林] fuel forest

心 ①(心脏) the heart ②(思想、感情) heart; mind; feeling; intention: 爱国～ patriotic feeling; patriotism / 伤人的～ wound sb.'s feelings; hurt sb.'s feelings / 羞耻之～ sense of shame ③(中心) center; core: 白菜～ the heart of a Chinese cabbage / 核～ core; nucleus / 手～ the hollow of the palm / 圆～ the center of a circle
【心爱】love; treasure: ～的东西 treasured possession; prized possession / ～的人 one's beloved; loved one
【心安理得】feel at ease and justified; have an easy conscience; heart at rest and virtue in completion
【心包】[生理] pericardium
【心包炎】[医] pericarditis: 急性～ acute pericarditis
【心病】①(郁闷的心情) worry; anxiety ②(隐情或隐痛) sore point; secret trouble
【心搏】[生理] heartbeat ◇～过速 tachycardia
【心不在焉】absent-minded; inattentive; preoccupied (with sth. else)
【心裁】idea; conception;; mental plan: 独出～ show originality; be original
【心肠】①(用心;存心) heart; intention: ～软 have a soft heart; be softhearted / 好～ kindhearted / 坏～ evilminded / 她的～很好。 She has a very kind heart. ②(兴致) state of mind; mood
【心潮】a tidal surge of emotion; surging thoughts and emotions: ～翻滚 one's mind being in a tumult / ～澎湃 feel an upsurge of emotion
【心驰神往】one's thoughts fly to (a place or a person); have a deep longing for
【心慈手软】softhearted
【心胆俱裂】be frightened out of one's wits; be terror-stricken
【心得】what one has learned from work, study, etc.
【心地】a person's mind, character, moral nature, etc.: ～纯洁 pure in mind / ～单纯 simpleminded / ～光明 purehearted / ～糊涂 muddle- headed / ～善良 good- natured; kindhearted / ～坦白 candid; open / ～狭窄 narrow-minded
【心电描记器】[医] electrocardiograph
【心电图】[医] electrocardiogram: 做～ have an electrocardiogram ◇～机 electrocardiograph; ECG machine
【心动过速】[医] tachycardia: 阵发性～ paroxysmal tachycardia
【心动徐缓】[医] bradycardia
【心动周期】[生理] cardiac cycle
【心毒手辣】callous and cruel
【心耳】[生理] auricle
【心烦】be vexed; be perturbed: ～意乱 be terribly upset
【心房】[生理] atrium (of the heart) ◇～中隔缺损 auricular septal defect
【心肺杂音】cardiopulmonary murmur
【心浮】flighty and impatient; unstable
【心服】be genuinely convinced; acknowledge (one's defeat, mistake, etc.) sincerely: ～口服 be sincerely convinced; admit somebody's superiority
【心腹】①(亲信) trusted subordinate; henchman; reliable agent ②(藏在心里轻易不对人说的) confidential: ～事 a secret in the depth of one's heart / 说～话 tell sb. sth. in strict confidence; confide in sb.; exchange confidences
【心腹之患】disease in one's vital organs; serious hidden trouble or danger
【心甘情愿】be most willing to; be perfectly happy to
【心肝】①(良心;正义感) conscience: 没～ heartless ②(最亲热最心爱的人) darling; deary
【心广体胖】carefree and contented; fit and happy
【心寒】[方] be bitterly disappointed: 令人～

chill the heart; be bitterly disappointing

【心黑手辣】black-hearted and cruel

【心狠】cruel; merciless: ～手辣 cruel and evil; wicked and merciless

【心花怒放】burst with joy; be wild with joy; be elated

【心怀】①（心中存有）harbor; entertain; cherish: ～不满 feel discontented; nurse a grievance／～鬼胎 entertain dark schemes; have ulterior motives／～叵测 harbor dark designs; have evil intentions ②（心意）intention; purpose ③（心情）state of mind; mood

【心慌】be flustered; be nervous; get alarmed: ～意乱 be alarmed and nervous

【心灰意懒】the heart dispirited and the mind indolent; be disheartened; be downhearted

【心肌】[生理] cardiac muscle; myocardium ◇～梗塞 myocardial infarction／～炎 myocarditis

【心机】thinking; scheming: 费尽～ leave no stone unturned; try all ingenious ways; take great pains; cudgel one's brains／枉费～ rack one's brains in vain; make futile efforts

【心迹】the true state of one's mind; true motives or feelings: 表明～ lay bare one's true feelings

【心急】impatient; short-tempered: ～火燎 burning with impatience

【心计】calculation; scheming; planning: 工于～ adept at scheming; very calculating／做事有～ do things intelligently

【心悸】[医] palpitation

【心焦】anxious; worried

【心绞痛】[医] angina pectoris

【心惊胆战】tremble with fear; shake with fright

【心惊肉跳】palpitate with anxiety and fear; be filled with apprehension

【心境】state of mind; frame of mind; mental state; mood: ～不好 be in a bad mood／～愉快 be in a happy mood

【心静】calm

【心坎】the bottom of one's heart

【心口】the pit of the stomach

【心口如一】say what one thinks; be frank and unreserved

【心旷神怡】relaxed and happy; carefree and joyous

【心劳日拙】fare worse and worse for all one's scheming

【心理】psychology; mentality: 儿童的～ the children's psychology／这是一般人的～。This is how ordinary people feel about it. ◇～测验学 psychometry／～发育 psychological development／～分析 psychoanalysis／～健康 mental health／～疗法 psychotherapy／～卫生 mental hygiene／～因素 psychological factor／～战 psychological warfare

【心理学】psychology: 变态～ abnormal psychology／儿童～ child psychology／普通～ general psychology／社会～ social psychology／消费者～ consumer psychology ◇～法学 psychological theory of law／～家 psychologist

【心里】in the heart; at heart; in (the) mind: ～不安 not feel at ease／～不痛快 feel bad about sth.／～发闷 feel constriction in the area of the heart／～有事 have sth. on one's mind／记在～ keep in mind; bear in mind

【心里话】one's innermost thoughts and feelings: 说～ speak one's mind

【心力】mental and physical efforts: ～交瘁 be mentally and physically exhausted／费尽～ make strenuous efforts ◇～衰竭 [医] heart failure

【心连心】heart linked to heart; be of one mind with

【心灵】①（心思灵敏）clever; intelligent; quick-witted: ～手巧 clever and deft ②（内心）heart; soul; spirit: ～深处 deep in one's heart

【心领神会】the heart receives and spirit knows it; understand tacitly; readily take a hint

【心律】[医] rhythm of the heart ◇～不齐 arrhythmia／～失常 arrhythmia cordis

【心乱如麻】have one's mind as confused as a tangled skein; be utterly confused and disconcerted; be terribly upset

【心满意足】be perfectly content; be fully satisfied

【心明眼亮】see and think clearly; be sharp-eyed and clearheaded

【心目】mind; mental view; mind's eye: 在某人的～中 in sb.'s mind's eye; in sb.'s eye

【心皮】[植] carpel

【心平气和】even-tempered and good-humored; quiet in mind and peace in disposition: ～地交换意见 exchange views calmly

【心窍】capacity for clear thinking: 被…迷住了～ be obsessed by; be under an obsession of／权迷～ be obsessed by a lust for power

【心情】frame of mind; state of mind; mood: ～沉重 with a heavy heart／～激动 be excited; be thrilled／～舒畅 have ease of mind／～愉

快 be in a cheerful frame 〈state〉 of mind; be in a good 〈happy〉 mood; have a light heart

【心如刀割】feel as if a knife were piercing one's heart

【心软】be softhearted; be tenderhearted

【心神】mind; state of mind: ~不定 have no peace of mind; be distracted; feel ill at ease

【心声】heartfelt wishes; aspirations; thinking: 表达青年的~ voice the aspirations of the young people

【心室】[生理] ventricle ◇ ~肥大 ventricular hypertrophy / ~间隔缺损 interventricular septal defect

【心事】sth. weighing on one's mind; a load on one's mind; worry: 好象有什么~似的 seem to have sth. on one's mind / 了结一桩~ take a load off one's mind / ~重重 be laden with anxiety; be weighed down with care

【心输出量】[生理] cardiac output

【心术】intention; design: ~不正 harbor evil intentions; harbor evil designs

【心思】①(念头) thought; idea: 坏~ a wicked idea / 看透某人的~ gain an insight into sb.'s mind / 想~ ponder; contemplate ②(脑筋) thinking; thoughts: 白费~ rack one's brains in vain; make futile efforts / 用~ do a lot of thinking; think hard ③(心情) state of mind; mood: 没~去做某事 not be in the mood to do sth.

【心酸】be grieved; feel sad

【心算】mental arithmetic; doing sums in one's head

【心疼】①(疼爱) love dearly: 在你们姐妹中,我最~你。 Of all your sisters, I love you best. ②(惋惜) feel sorry; make one's heart ache: 这样浪费,真叫人看了~。 It makes one's heart ache to see such waste.

【心田】①(内心) heart ②(用心) intention

【心跳】palpitation

【心痛】cardiac pain

【心头】mind; heart: ~恨 rankling hatred / 记在~ bear in mind; keep in mind

【心窝儿】[口] the pit of the stomach

【心无二用】one cannot keep one's mind on two things at the same time; one should concentrate on one's work

【心细】careful; scrupulous: 胆大~ bold but cautious

【心弦】heartstrings: 动人~ tug at one's heartstrings

【心向往之】yearning for sb. or sth.

【心心相印】hearts and feelings find a perfect response; have mutual affinity; be kindred spirits

【心胸】breadth of mind: ~开阔 broad-minded; unprejudiced / ~狭窄 narrow-minded; intolerant

【心虚】①(怕人知道) afraid of being found out; with a guilty conscience: 做贼~ have a guilty conscience ②(缺乏自信心) lacking in self-confidence; diffident

【心绪】state of mind: ~不宁 in a disturbed state of mind; in a flutter / ~烦乱 emotionally upset; in an emotional turmoil

【心血】painstaking care; painstaking effort: 毕生~的结晶 the fruit of painstaking labor of one's whole lifetime / 费尽~ expend all one's energies

【心血管系统】cardiovascular system ◇ ~疾病 cardiovascular diseases

【心血管药】cardiovascular drug

【心血来潮】be prompted by a sudden impulse; be seized by a whim: ~,忘乎所以 be carried away by one's whims and act recklessly

【心眼儿】①(内心) heart; mind: ~小 oversensitive; petty / 打~里热爱祖国 love the motherland with all one's heart ②(心地;存心) intention: ~好 good-natured; kindhearted / 没安好~ have bad intentions; be up to no good ③(聪明机智) intelligence; cleverness: 他好像是缺~。 He seems slow-witted. / 她很有~。 She is alert and thoughtful. ④(不必要的顾虑) unfounded doubts; unnecessary misgivings: ~多 full of unnecessary misgivings; oversensitive

【心意】①(情意) regard; kindly feelings: 这点礼物是我们大家的一点~。 This little gift is a token of our regard. ②(愿望) intention; purpose: 我们明白了她的~。 We understood her intention.

【心音】[生理] heart sounds; cardiac sounds: 第二~ the second heart sound / 第一~ the first heart sound

【心硬】hardhearted; stonyhearted; callous; unfeeling

【心有灵犀一点通】heart which beat in unison are linked

【心有余而力不足】the spirit is willing, but the flesh is weak; unable to do what one wants very much to do

【心有余悸】one's heart still fluttering with fear; have a lingering fear

【心猿意马】restless and whimsical; fanciful and fickle; capricious; when one meant gibbon, he thinks of a horse
【心愿】cherished desire; aspiration; wish: 遂了他的～ have his wish fulfilled
【心悦诚服】feel a heartfelt admiration; be completely convinced
【心杂音】[医] heart murmur
【心脏】the heart: ～有杂音 have a heart murmur / 他的～停止了跳动。His heart stopped beating. ◇ ～搏动 heartbeat / 导管 cardiac catheter / ～地带 heartland / ～激活器 cardioactivator / ～畸形 heart malformations / ～起搏器 (cardiac) pacemaker; pacemaker electrotherapeutic apparatus / ～移植 heart transplant
【心脏病】heart disease; cardiopathy: ～发作 have a heart attack / 风湿性～ rheumatic heart disease / 先天性～ congenital heart disease
【心照】understand without being told; have an understanding: ～不宣 have a tacit understanding; understood but not expressed
【心之官则思】the office of the mind is to think
【心直口快】frank and outspoken; saying what one thinks without much deliberation
【心中无数】not know for certain
【心中有数】have a pretty good idea of; know fairly well; know what's what
【心轴】[机] mandrel: 花键～ splined mandrel
【心子】(of things) center; heart; core
【心醉】be charmed; be enchanted; be fascinated

馨 [书] strong and pervasive fragrance
【馨香】[书]①(芳香) fragrance ②(烧香的香味) smell of burning incense

欣 glad; happy; joyful: ～逢佳节 on the happy occasion of the festival / 欢～ happy; joyful
【欣然】[书] joyfully; with pleasure: ～接受 accept with pleasure / ～同意 gladly consent; readily agree
【欣赏】appreciate; enjoy; admire: ～风景 enjoy the scenery; admire the scenery / 音乐～ music appreciation
【欣慰】be gratified: 对结果感到～ be gratified at the results / 我们听到消息后感到～。We are relieved to hear the news.
【欣悉】be glad to learn; be happy to learn
【欣喜】glad; joyful; happy: ～若狂 be wild with joy; go into raptures
【欣欣向荣】thriving; flourishing; prosperous: ～的景象 scenes of flourishing life; picture of prosperity / 各行各业都～。Every trade is thriving.

xín

寻
【寻短见】commit suicide; take one's own life
【寻死】①(企图自杀) try to commit suicide; attempt suicide ②(自杀) commit suicide
【寻死觅活】repeatedly attempt suicide (in order to threaten)
【寻思】think sth. over; consider: 让我～～再说。Let me think it over first.

xìn

芯 core: 岩～ core
【芯子】① fuse; wick: 蜡烛～ candle wick ②(蛇的舌头) the forked tongue of a snake

信 ①(确实) true: ～史 trustworthy historical record ②(信用) confidence; trust; faith: 取～于民 win the people's confidence 〈trust〉/ 失～ break faith; break one's promise ③(相信) believe: ～不～由你 believe it or not / ～以为真 accept sth. as true; take sth. at face value ④(信奉) profess faith in; believe in: ～佛 profess Buddhism ⑤(听凭;随意;放任) at will; at random; without plan: ～笔写来 write freely without hesitation / ～步闲游 walk aimlessly ⑥(凭据) sign; evidence: ～号 signal / 印～ official seal ⑦(书信) letter; mail: 保价～ insured letter / 公开～ an open letter / 挂号～ registered letter / 航空～ air mail / 平～ ordinary letter / 双挂号～ double-registered letter / 证明～ certificate; certification / 寄～ post a letter / 送～ deliver a letter ⑧(信息) message; word; information: 口～ a verbal message; an oral message ⑨(引信) fuse
【信标灯】beacon light
【信步】take a leisurely walk; stroll; walk aimlessly: ～街市 stroll about the streets / ～所至 take to a place in aimless wandering
【信贷】credit: 长期～ long-term credit ◇ ～额度 line of credit / ～收支 credit receipts and payments / ～资金 credit funds; funds for extending credit
【信而有证】borne out by evidence
【信风】[气] trade (wind): 反～ antitrade wind; antitrades ◇ ～带 trade-wind zone
【信封】envelope

【信奉】believe in: ～基督教 be a Christian／～上帝 believe in God
【信服】completely accept; be convinced: 令人～的论据 convincing argument／使某人～ convince sb. of sth.
【信鸽】carrier pigeon; homing pigeon; homer ◇ ～比赛 pigeon racing
【信管】fuse: 触发～ contact fuse／近炸～ proximity fuse／延期～ delay-action fuse
【信号】signal: 臂板～[铁道] semaphore／灯光～ light signal／识别～ identification signal／遇难～[航海]distress signal; SOS／杂乱～[无] hash ◇ ～兵 signalman／～刺激[心理] signal stimulus／～弹 signal flare／～灯 signal lamp／～发生器 signal generator／～机 semaphore; teleseme／～枪[军] flare pistol; signal pistol; pyrotechnic pistol
【信汇】mail transfer (M／T)
【信笺】letter paper; writing paper
【信件】letters; mail
【信教】profess a religion; be religious
【信口雌黄】make irresponsible remarks; wag one's tongue too freely
【信口开河】wag one's tongue too freely; talk nonsense; talk irresponsibly; shoot one's mouth off
【信赖】trust; count on; have faith in: 他不是一个可～的人。He's not the sort of man to be trusted.／她是群众～的好干部。She's a good cadre trusted by the masses.
【信马由缰】ride with lax reins
【信念】faith; belief; conviction
【信任】trust; have confidence in: ～某人 have trust in sb.／不～某人 put no trust of sb.／得到…的～ enjoy the trust in sb. ◇ ～投票 vote of confidence
【信赏必罚】due rewards and punishments will be meted out without fail
【信使】courier; messenger: 外交～ diplomatic messenger ◇ ～证明书 courier's credentials
【信誓旦旦】pledge in all sincerity and seriousness; vow solemnly
【信手拈来】have words, material, etc. at one's fingertips and write with facility; take it off with the hand
【信守】abide by; stand by: ～不渝 be unswervingly faithful (to one's promise, etc.)／～诺言 keep a promise; be as good as one's word／～协议 abide by an agreement; stand by an agreement
【信天翁】[动] albatross
【信条】article of creed; article of faith; creed; precept; tenet
【信筒】pillar-box; mailbox
【信徒】believer; disciple; follower; adherent; devotee: 佛教～ Buddhist
【信托】trust; entrust ◇ ～存款 trust deposits／～公司 trust company／～基金 trust fund／～契约 deed of trust／～商店 commission shop; commission house; commission agent／～证书 trust certification
【信物】authenticating object; token; keepsake
【信息】①(音信,消息) information; news; message ②[数] information ◇ ～处理 information processing／～传递 information transmission／～技术 information technology／～检索 information retrieval／～容量 information capacity／～系统 information system／～转储 memory dump
【信息编码】[计算机] information encoding
【信息存储器】[计算机] information-storing device
【信息论】[数] information theory
【信息学】informatics
【信箱】①(信筒) letter box; mailbox ②(邮局内供人租用的) post-office box (P.O.B.)
【信心】confidence; faith: 满怀～ full of confidence
【信仰】faith; belief; conviction: 宗教～ religious belief／政治～ political conviction
【信义】good faith; faith: 无～ be perfidious／有～ act in good faith
【信用】①(信任) trustworthiness; credit: 讲～ keep one's word／失去～ lose one's credit／他的～好。His credit is good. ②[经] credit ◇ ～额度 line of credit／～风险 credit risks／～放款 unsecured loan; loan on credit／～合作社 credit cooperative／～汇票 credit bill／～卡 credit card／～投资 fiduciary contribution／～债券 debenture bonds
【信用证】letter of credit (L／C): 旅行～ traveller's letter of credit
【信誉】prestige; credit; reputation: ～卓著 enjoy high reputation
【信札】letters
【信纸】letter paper; writing paper

衅 quarrel; dispute: 挑～ provoke／寻～ pick a quarrel with sb.

xīng

兴 ①(兴盛;流行) prosper; rise; prevail; be-

come popular: ～衰 rise and decline; ups and downs ②(使盛行) encourage; promote: 大～调查之风 energetically encourage the practice of investigation and study ③(开始;发动) start; begin: ～兵 send an army / ～工 start construction
【兴办】 initiate; set up: ～工厂 set up factories / ～集体福利事业 initiate collective welfare work / ～社会主义新型企业 initiate the new socialist enterprises
【兴奋】①(振奋;激动) be excited: 令人～的时刻 an exciting moment / 令人～的消息 an exciting news / 我们都为这消息感到～。 We were all excited by the news. / 消息传来,人人为之～。 The news excited everybody. ②[生理] excitation ◇ ～剂 excitant; stimulant / ～性 excitability
【兴风作浪】 stir up trouble; make trouble; fan the flames of disorder
【兴建】 build; construct: ～化肥厂 build a fertilizer plant
【兴利除弊】 promote what is beneficial and abolish what is harmful
【兴隆】 prosperous; thriving; flourishing; brisk: 生意～。 Business is brisk.
【兴起】 rise; spring up; be on the upgrade: 一个建设的新高潮正在～。 A new upsurge in construction is in the making.
【兴盛】 prosperous; flourishing; thriving; in the ascendant: 国家～ prosperity of the nation
【兴师动众】 move troops about and stir up the people; drag in many people (to do sth.)
【兴师问罪】 send a punitive expedition against
【兴亡】 rise and fall (of a nation)
【兴旺】 prosperous; flourishing; thriving: ～的事业 a prosperous business / 到处呈现一片～景象。 Everywhere is a scene of prosperity. / 我们的党～发达,后继有人。 Our party is flourishing and has no lack of successors.
【兴修】 start construction (on a large project); build: ～水利 build water conservancy projects
【兴妖作怪】 conjure up a host of demons to make trouble; stir up trouble

星 ①(天空中的星) star: ～空 starlit sky; starry sky ②(天体) heavenly body: 白矮～ white dwarf / 变～ variable star / 超新～ supernova / 海王～ Neptune / 恒～ fixed star; star / 红巨～ red giant / 彗～ comet / 火～ Mars / 金～ Venus / 聚～ multiple star / 流～ meteor / 脉动～ pulsating star / 冥王～ Pluto / 木～ Jupiter / 食变～ eclipsing binary / 双～ binary / 水～ Mercury / 天王～ Uranus / 土～ Saturn / 卫～ satellite / 小行～ asteroid / 新～ nova / 行～ planet / 中子～ neutron star ③(细碎东西) bit; particle: 火～儿 spark / 一～半点 a tiny bit ④(明星) star: 球～ a star player / 影～ a movie star
【星表】[天] star catalogue
【星辰】 stars
【星虫】[动] siphon-worm
【星等】[天] (stellar) magnitude
【星斗】 stars: 满天～ a star-studded sky
【星号】 asterisk (*)
【星火】①(微小的火) spark: ～燎原 A single spark can start a prairie fire ②(流星的光) shooting star; meteor: 急如～ most urgent
【星际】 interplanetary; interstellar: ～飞行 interplanetary flight; space flight / ～分子 interstellar molecules / ～空间 interstellar space
【星罗棋布】 scattered all over like stars in the sky or men on a chessboard; spread all over the place
【星期】①(一周) week: 本～ this week / 上～ last week / 下～ next week / 今天～几? What day (of the week) is it today? ②(星期日) Sunday ◇ ～日(天) Sunday (Sun.) / ～一 Monday (Mon.) / ～二 Tuesday (Tues.) / ～三 Wednesday (Wed.) / ～四 Thursday (Thur.) / ～五 Friday (Fir.) / ～六 Saturday (Sat.)
【星球】 celestial body; heavenly body
【星球大战】(美) strategic defense initiative (SDI)
【星鲨】[动] gummy shark
【星体】[天] celestial body; heavenly body
【星图】[天] star chart; star map; star atlas
【星团】[天] (star) cluster
【星系】[天] galaxy: 河外～ extragalactic system / 总～ metagalaxy ◇ ～团 cluster of galaxies
【星星】①(细小的点) tiny spot ②(星) star
【星星点点】 tiny spots; bits and pieces
【星星之火,可以燎原】 a single spark can start a prairie fire
【星形航空发动机】 radial aeroengine
【星形空冷式发动机】 radial air-cooled engine
【星夜】 on a starlit night; on a starry night; by night: ～启程 set out by starlight; set out in great haste / ～行军 march by night; march in great haste
【星移斗转】 change in the positions of the stars;

change of the seasons; passage of time
【星云】[天] nebula: 网状～ network nebula / 蟹状～ Crab Nebula / 旋涡～ spiral nebula / 银河～ galactic nebula / 原始～ primordial nebula ◇～团 nebulous cluster
【星占】divine by astrology; cast a horoscope ◇～术 astrology
【星震】[天] starquake
【星座】[天] constellation

惺

【惺松】(of eyes) not yet fully open on waking up: 睡眼～ eyes still heavy with sleep; sleepy eyes
【惺惺】①(清醒) clearheaded; awake ②(聪明; 聪明的人) wise; intelligent: ～惜～。The wise appreciate one another.
【惺惺作态】be affected; simulate (friendship; innocence, etc.); have affected manners

腥

①(生肉;生鱼) raw meat or fish: 荤～ dishes of meat or fish ②(有腥气) having the smell of fish, seafood, etc.
【腥臭】stinking smell as of rotten fish; stench
【腥风血雨】sanguinary slaughter; great bloodshed
【腥黑穗病】[农] bunt
【腥气】①(腥味) the smell of fish, seafood, etc. ②(有腥味的) stinking; fishy; stenchy
【腥臊】smelling of urine; the offensive smell of a fox
【腥膻】smelling of fish or mutton
【腥味儿】smelling of fish; fishy

猩

orangutan
【猩红】scarlet; bloodred
【猩红热】[医] scarlet fever
【猩猩】[动] orangutan ○大猩猩 gorilla / 黑腥腥 chimpanzee
【猩猩草】[植] painted euphorbia

xíng

形

①(形状) form; shape: 不成～ shapeless; formless / 方～ square / 扇～ fan-shaped / 外～ outer form ②(形体) body; entity: 无～ intangible / 有～ tangible ③(显露;表示) appear; look: 喜～于色 look very pleased; beam with undisguised happiness ④(对照) compare; contrast: 相～之下 by comparison; by contrast
【形变】[物] deformation: 弹性～ elastic deformation ◇～作用 deformation effect
【形成】take shape; form; take form: ～风气 become a common practice / ～僵局 come to a deadlock / ～鲜明的对比 form a sharp contrast / 云在山顶上逐渐～。Clouds are forming on the top of the hill. ◇～层[植] cambium
【形单影只】a solitary form, a single shadow; extremely lonely; solitary
【形而上学】[哲] metaphysics
【形骸】the human skeleton; the human body: ～之外 out of the formality
【形迹】①(举动和神色) a person's movements and expression: ～可疑 of suspicious appearance; suspicious-looking / 不露～ betray nothing in one's expression and movements ②(礼貌) formality: 不拘～ without formality; not standing on ceremony
【形容】①[书](形体和容貌) appearance; countenance: ～憔悴 looking wan; thin and pallid ②(描述) describe: 难以～ difficult to describe; beyond description / 风景之美非笔墨所能～。The beauty of the scenery beggars description.
【形容词】[语] adjective
【形色仓皇】appear in a big hurry; as a fugitive
【形色自若】one's countenance remains as before
【形式】form; shape: ～和内容 form and content / 流于～ become a mere formality
【形式主义】formalism
【形式逻辑】[逻] formal logic
【形式上】in form; formal: ～的独立 nominal independence / ～的让步 nominal concession / ～的一致 formal unity
【形势】①(地势) terrain; topographical features: ～险要 strategically important terrain ②(事物的发展状况) situation; circumstances: 国际～ the international situation / 国内～ the domestic situation / ～逼人。The situation is pressing. (或: The situation demands immediate action.)
【形似】be similar in form or appearance
【形态】①(事物的形状或表现) form; shape; pattern: 社会经济～ social-economic formation; economic formation of society / 意识～ ideological form ②[语] morphology ◇～学 morphology
【形体】①(身体) shape (of a person's body); physique; body ②(形状和结构) form and structure: 文字的～ form of the written character
【形同虚设】exist in name only

【形象】image; form; figure: ～地表现革命精神 vividly depict the revolutionary spirit / 英雄～ heroic images
【形象化】in images; figuratively
【形象思维】thinking in (terms of) images
【形形色色】of every hue; of all shades; of all forms; of every description: ～的错误思想 erroneous ideas of every descriptions / ～的机会主义 all brands of opportunism
【形影不离】inseparable as body and shadow; always together
【形影相吊】body and shadow comforting each other; extremely lonely; sad and solitary
【形影相随】follow someone like his shadow
【形于辞色】show in one's words and expression
【形状】form; appearance; shape: 汽车的～已有很多改进。 There have been many improvements in form of cars.

刑 ①(刑罚) punishment: 服～ serve one's term in prison / 缓～ on probation / 减～ commute a sentence; reduce punishment; reduction of penalty / 量～ penal discretion / 免～ exempt from punishment / 判～ sentence; pass sentence / 死～ death penalty; death sentence; capital punishment / 死～缓期执行 stay of execution; short for death penalty with a two-year reprieve and forced labor / 徒～ imprisonment; prison sentence / 已被判处死～ be under sentence of death ②(对犯人的体罚) torture; corporal punishment: 受～ suffer corporal punishment / 用～ put sb. to torture; torture
【刑场】execution ground
【刑罚】[法] penalty; punishment; criminal penalty
【刑罚学】penology
【刑罚种类】kind of penalty
【刑法】① [法] penal code; criminal law ② (对犯人的体罚) corporal punishment; torture: 动了～ administer corporal punishment / 受了～ suffer corporal punishment
【刑法学】criminal jurisprudence; science of criminal law
【刑警】criminal police
【刑具】instruments of torture; implements of punishment
【刑律】[法] criminal law: 触犯～ violate the criminal law
【刑满】expiration of the imprisonment ◇ ～释放 be released after serving the full term of a sentence; be released after serving a sentence
【刑期】[法] term of imprisonment; prison term; term of penalty ◇ ～届满 expiration of the term sentence
【刑事】[法] criminal; penal ◇ ～案件 criminal case / ～处分 criminal sanction / ～法庭 criminal court / ～犯 criminal offender; criminal / ～犯罪 criminal offense; crime / ～管辖权 criminal jurisdiction / ～审判庭 criminal adjudication tribunal / ～诉讼 criminal suit / ～诉讼法 code of criminal procedure / ～侦察 criminal investigation / ～责任 responsibility for a crime; criminal responsibility / ～罪 criminal charge; penal offense / ～罪犯 prisoner at the bar
【刑讯】[法] inquisition by torture: ～逼供 extort a confession by torture; subject sb to the third degree

型 ①(模子) mold: 砂～ sand mold / 铸～ casting mold ②(样式) model; type; pattern: 句～ sentence pattern / 新～ new model / 血～ blood group / 流线～的 streamlined
【型板】[机] template; templet
【型钢】[冶] section (steel); shape ◇ ～轧机 shape (rolling) mill
【型号】model; type
【型砂】[机] molding sand; casting sand
【型心】[冶] core: 干砂～ baked core / 粘土～ loam core

行 ①(走) go: 步～ go on foot; walk / 寸步难～ unable to move an inch; cannot go forward at all ②(旅行) travel: ～程 route or distance of travel / 不虚此～ do not make this trip in vain ③(流动性的,临时性的) temporary; makeshift: ～灶 makeshift cooking stove ④(流通;推行) be current; prevail; circulate: 风～一时 be popular for a time; be in fashion for a time; be all the rage ⑤(做;办) do; perform; carry out; engage in: 简便易～ simple and easy to do / 可～ can be done; be practicable; can be carried out / 实～ carry out; put into effect ⑥(表示进行某项活动)即～答复 will give a prompt reply / 另～安排 make other arrangements / 另～通知 will notify separately ⑦(行为) behavior; conduct: 品～ character; conduct / 言～ words and deeds ⑧(可以) all right; O.K. ⑨(能干) capable; competent: 老王,你真～! Lao Wang, you're really terrific! / 她

教英语～吗? Is she competent to teach English?
【行百里者半九十】ninety li is only half of a hundred- li journey; the going is toughest towards the end of a journey; one must sustain one's effort when a task is nearing completion
【行板】[乐] andante
【行不通】won't do; won't work; get nowhere: 这个计划～。This plan won't work. / 这事永远～。This will never do.
【行车】drive a vehicle ◇ ～里程 distance travelled by a vehicle; mileage / ～速率 driving speed / ～执照 driver's license; driving license
【行成于思】a deed is accomplished through taking thought; success depends on forethought
【行程】①(路程) route or distance of travel: 艰难的～ hard journey / 漫长的～ long route ②[机] stroke; throw; travel: 活塞～ piston travel / 滑枕～ ram stroke
【行船】sail (a boat); navigate: ～到最近的港口 navigate the ship to the nearest port / 逆风～ sail against the wind / 顺风～ sail before the wind / 顺流～ navigate down the stream
【行刺】assassinate ◇ ～者 assassin
【行道树】shade tree
【行得通】will do; workable; practicable: 这样的方案是～的。Such a scheme is workable.
【行动】①(行走;走动) move about; get about: ～不便 have difficulty getting about / ～缓慢 move slowly; be slow-moving ②(为实现某种意图而活动) act; take action: ～起来 go into action / 采取～ take action; begin to act / 正义～ act of justice ③(行为;动作) action; operation: 军事～ military operations / 实际～ actual deeds ◇ ～纲领 program of action; guideline / ～准则 operative norm
【行方便】make things convenient for sb.; be accommodating
【行宫】imperial palace for short stays away from the capital; temporary dwelling place of an emperor when away from the capital
【行好】act charitably; be merciful; be charitable
【行贿】bribe; offer a bribe; resort to bribery: 这商人向他～。The merchant offers a bribe to him. ◇ ～受贿罪 offense of bribery
【行将】about to; on the verge of: ～崩溃 be about to collapse / ～灭亡 on the verge of extinction / ～完成 approach completion
【行将就木】be getting nearer and nearer the coffin; be fast approaching death; have one foot in the grave
【行劫】commit robbery; rob
【行进】march forward; advance
【行经】①(行程中经过) go by; pass by: 当汽车～村庄时,我看到了一群羊。I saw a flock of sheep when the car passed through a village. ②[生理] menstruate
【行径】act; action; move: 侵略～ act of aggression / 野蛮～ barbarous act
【行军】(of troops) march: 急～ rapid march / 夜～ night march; march by night ◇ ～床 camp bed; camp cot / ～锅 field caldron / ～壶 canteen / ～灶 field kitchen
【行乐】indulge in pleasures; seek amusement; make merry
【行礼】salute
【行李】luggage; baggage: 超重～ excess luggage / 超重～费 charge for over weight / 手提～ hand-luggage / 随身～ personal luggage / 托运的～ registered luggage / 先交送～ send the luggage in advance ◇ ～舱 baggage cabin; baggage compartment / ～车[铁道] luggage van; baggage car / ～袋 duffel bag / ～过磅处 baggage check-in counter / ～寄存处 check-room / ～架 luggage rack; baggage rack / ～牌 baggage claim tag / ～票 luggage check; baggage check / ～标签 label; tag
【行李卷儿】bedroll; bedding roll; bedding pack
【行骗】practice fraud; swindle; cheat; practice deception
【行期】date of departure
【行乞】beg one's bread; beg alms; beg
【行窃】steal; commit theft ◇ ～工具 implement of robbery
【行人】pedestrian
【行人横道】street crossing
【行若无事】behave as if nothing had happened
【行色】circumstances or style of departure: ～匆匆 in a hurry to go on a trip / 以壮～ to enable sb. to depart in style
【行善】do good works
【行商】itinerant trader; pedlar
【行尸走肉】a walking corpse; one who vegetates; an utterly worthless person; dead- alive person
【行时】①(流行) be in vogue; be all the rage ②(得势) be in the ascendent
【行使】exercise; perform: ～检查权 exercise procuratorial authority / ～权力 exercise power / ～权利 exercise of rights / ～职权 exercise authority

【行驶】(of a vehicle, ship, etc.) go; ply; travel: ～速度 running speed / 超速～ drive above the speed limit / 卡车向北～。The truck is going north. / 小船在河边～。The boat is sailing along the bank.
【行事】①(办事;做事) act; handle matters: 按计划～ act according to plan / 见机～ act according to circumstances ②(行为) behavior; conduct
【行署】[简](行政公署) administrative office (within a province)
【行头】①(戏装) actor's costumes and paraphernalia ②(泛指衣服) a person's wardrobe
【行为】action; behavior; conduct: ～卤莽 one's actions are rude / 不法～ illegal act / 正义的～ righteous action ◇ ～规范 code of conduct / ～疗法 behavior therapy / ～能力 capacity / ～遗传学 behavioral genetics / ～主义[心] behaviorism / ～准则 code of conduct; standard of conduct
【行文】①(组织文字) style or manner of writing: ～流畅 read smoothly ②(发公文) send an official communication to other organizations
【行销】be on sale; sell: ～全国 be on sale throughout the country / ～全球 be on sale all over the world
【行星】[天] planet ◇ ～际飞行 interplanetary flight / ～际物质 interplanetary matter
【行星齿轮】epicyclic gear; planet gear ◇ ～箱 epicyclic gearbox
【行刑】carry out a death sentence; execute
【行凶】commit physical assault or murder; do violence: ～杀人 commit murder
【行医】practice medicine (usu. on one's own)
【行营】field headquarters
【行辕】field headquarters
【行云流水】(of style of writing) like floating clouds and flowing water; natural and smooth
【行政】administration ◇ ～部门 administrative department; administrative unit; executive branch; administration / ～处分 disciplinary sanction / ～单位 administrative unite / ～当局 executive authorities / ～法 administrative law / ～管理 administration / ～管理费 expenditure on administration / ～机关 administrative organ / ～监督 administrative supervision / ～拘留 administrative attachment / ～命令 administrative decree; administrative order / ～区 administrative area / ～区域 administrative division / ～人员 administrative personnel; administrative staff / ～预算 administrative budget / ～长官 chief executive
【行之有效】effective (in practice); effectual: ～的办法 effective measures / ～的疗法 effectual remedy
【行止】①(行踪) whereabouts: ～不明 whereabouts unknown / ～无定 there's no telling where sb. is ②(行为) behavior; conduct
【行装】outfit for a journey; luggage: 整理～ pack (for a journey)
【行踪】whereabouts; track: ～不定 be of uncertain whereabouts / ～诡秘 mysterious movements / ～飘忽 no definite date of coming and going
【行走】walk; go on foot

xǐng

醒 ①(神志恢复正常) regain consciousness; sober up; come to: 她～过来了。She has come to. ②(睡醒) wake up; be awake: 半睡半～ between sleep and wake / 如梦初～ like awakening from a dream / 你平常什么时候～来? What time do you usually wake up? / 请在六点钟叫～我。Please wake me up at six. / 我～得很早。I woke up early. ③(清楚;明显) be clear in mind: 头脑清～ keep a cool head
【醒酒】dispel the effects of alcohol; sober up
【醒目】(of written words or pictures) catch the eye; attract attention; be striking: ～的标题 bold headlines / ～的标语 eye-catching slogans
【醒世之言】good advice that cautions the age
【醒悟】come to realize the truth, one's error, etc.; wake up to reality: 使某人～过来 awaken sb. to the reality

擤 blow (one's nose): ～鼻涕 blow one's nose

省 ①(检查自己的思想行为) examine oneself critically: 反～ make a self-examination; introspect / 内～ be critical of oneself ②(探望;问候) visit ③(醒悟;明白) become conscious; be aware: 不～人事 lose consciousness
【省察】examine one's thoughts and conduct; examine oneself critically
【省亲】pay a visit to one's parents or elders (living at another place)
【省视】①(探望) call upon; pay a visit to ②(察看) examine carefully; inspect

xìng

兴 mood or desire to do sth.; interest;

excitement: 酒～ excitement due to drinking / 诗～ an exalted, poetic mood / 游～ the mood for sight-seeing
【兴冲冲】(do sth.) with joy and expedition; excitedly
【兴高采烈】in high spirits; in great delight; jubilant
【兴会】a sudden flash of inspiration; brain wave
【兴趣】interest; taste: 不感～ be not interested in / 感～ be interested in / 各有各的～。Tastes differ.
【兴头】enthusiasm; keen interest: 在～上 at the height of one's enthusiasm / 她对体育活动～很大。 She is very keen on sport.
【兴味】interest: ～索然 uninterested; bored stiff
【兴致】interest; mood to enjoy: ～勃勃 full of zest

幸

①(运气好) good fortune; good luck: ～甚 very fortunate indeed / 有～ be lucky; have good fortune②(高兴) rejoice: 庆～ congratulate oneself; rejoice ③(侥幸) fortunately; luckily: ～未成灾。 Fortunately it didn't cause a disaster. ④(希望) pray; I hope; I trust: ～勿见怪。 I hope that you will not take offense at it. / ～勿推却。 I trust you will not refuse.
【幸存】survive: ～者为数不多。 There were not many survivors.
【幸而】luckily; fortunately
【幸福】①(好的境遇和生活) happiness; well-being: 为人民谋～ work for the well-being of the people / 祝你～。 I wish you happiness. ②(幸福的) happy: 过着～的生活 live a happy life
【幸好】fortunately; luckily: 他来看我时, 我～好在家。 Luckily I was at home when he called.
【幸亏】[副] fortunately; luckily: ～在我们动身前雨停了。 Fortunately the rain stopped before we started.
【幸免】escape by sheer luck; have a narrow escape: ～于难 escape death by sheer luck; escape death by a hair's breadth
【幸事】good fortune; blessing
【幸运】①(好运气) good fortune; good luck: 有些人似乎从来是～的。 Some people seem to be always lucky. ②(称心如意) fortunate; lucky
【幸运儿】fortune's favorite; lucky fellow; lucky dog: 你真是个～。 You are a lucky dog.
【幸灾乐祸】take pleasure in other's misfortune; gloat over other's misfortune

悻

【悻悻】angry; resentful: ～而去 go away angry; leave in a huff

杏

apricot: ～树 apricot (tree)
【杏脯】preserved apricot
【杏红】apricot pink
【杏黄】apricot yellow; apricot (color)
【杏仁】apricot kernel; almond ◇ ～茶 almond tea / ～豆腐 almond curd in syrup

性

①(性格) nature; character; disposition: 本～ inherent character; nature ②(性能) property; quality: 酒～ alcoholic strength / 药～ medicinal properties ③(性别) sex: 男～ the male sex / 女～ the female sex / 无～的 sexless ④(后缀,表示性质、状态、范围、方式等) 阶级～ class character / 可能～ possibility / 社会～ social nature / 正确～ correctness ⑤[语] gender: 阳～ the masculine gender / 阴～ the feminine gender / 中～ the neuter gender
【性本能】sex instinct
【性比率】sex rate
【性变态者】pervert
【性别】sexual distinction; sex ◇ ～差异 gender differences
【性病】venereal disease (V.D.) ◇ ～学 venereology
【性动机】sexual motivation; sex drive
【性反应周期】sexual response cycle
【性感】sexuality; sexy
【性高潮】orgasm; climax
【性格】nature; disposition; temperament: 内倾～ introversion / 外倾～ extroversion / ～不合 in compatibility of temperament / ～开朗 have a bright and cheerful disposition / ◇ ～培养 character building / ～特性 character trait / ～形成 character formation / ～学 characterology / ～训练 character training / ～障碍 character disorder
【性功能障碍】sexual dysfunction
【性关系】sexual intercourse; carnal knowledge: 与…发生～ have carnal knowledge of
【性激素】[生理] sex hormone
【性急】impatient; short-tempered
【性交】sexual intercourse
【性命】life

【性命交关】(a matter) of life and death; of vital importance
【性能】function (of a machine, etc.); performance; property: 反应堆～ reactor behavior / 金属的～ metal properties ◇ ～试验 performance test
【性偏离】sexual deviation
【性器官】[生理] sexual organs; genitals
【性情】disposition; temperament; temper: ～暴躁 have an irascible temperament; be short-tempered / ～平和 have a calm temper / ～温柔 have a gentle disposition
【性染色体】sex chromosome
【性染色体遗传】allosomal inheritance
【性生活】sexual life
【性腺】[生理] sexual gland; sex gland
【性行为】sex instinct act; sexual behavior
【性欲】sexual desire; sexual urge; libido ◇ ～缺乏 a sexuality
【性早熟】sexual precocity
【性征】sex character
【性质】quality; nature; character: 弄清问题的～ ascertain the nature of the problem
【性周期】sexual cycle
【性状】shape and properties; properties; character: 土壤的理化～ the physicochemical properties of soil
【性子】①(性情) temper: 急～ have a hot temper / 使～ get into a temper ②(酒、药等的刺激性) strength; potency: 这药～平和。This is a mild drug.

姓 surname; family name; clan name: 你贵～? What is your surname? / 我～李。My surname is Li.
【姓名】surname and personal name; full name
【姓氏】surname

xiōng

兄 ①(哥哥) elder brother: 胞～ elder brother of the same parents ②(对男性朋友的尊称) a courteous form of address between men: 老～ my dear friend
【兄弟】①(哥哥和弟弟) brothers ②(兄弟般的) fraternal; brotherly: ～党 fraternal parties / ～学校 brother schools ③(弟弟) younger brother ④(谦称,自己) your humble servant; I
【兄弟阋墙】quarrel between brothers; internal dispute: 兄弟阋于墙,外御其侮。Brothers quarrelling at home join forces against attacks from without. (或: Internal disunity dissolves at the threat of external invasion.)

凶 ①(不幸的) inauspicious; ominous: ～兆 ill omen / 吉～ good or ill luck / 吉～未卜 no one knows how things will turn out ②(年成很坏) crop failure: ～年 a year of crop failure or famine; a bad year ③(凶恶) fierce; ferocious: 他样子真～。He looks very fierce. ④(利害) terrible; fearful: 病势很～ terribly ill / 闹得太～了! What a terrific row! ⑤(杀害、伤人的行动) act of violence; murder: 行～ commit physical assault or murder
【凶暴】fierce and brutal
【凶残】fierce and cruel; savage and cruel
【凶多吉少】be fraught with grim possibilities; bode ill rather than well
【凶恶】fierce; ferocious; fiendish
【凶犯】one who has committed homicide or mayhem; murderer
【凶狠】fierce and malicious; malevolent
【凶猛】violent; ferocious; terrible
【凶器】tool or weapon for criminal purposes; lethal weapon
【凶杀】homicide; murder: ～案 a case of murder; murder case
【凶神恶煞】devils; fiends
【凶手】murderer; assassin; assailant (who has caused injury to sb.); cut-throat
【凶险】in a very dangerous state; critical: 病情～ dangerously ill; in a critical condition
【凶相】ferocious features; fierce look: ～毕露 look thoroughly ferocious; unleash all one's ferocity
【凶焰】ferocity; aggressive arrogance: ～万丈 extremely ferocious
【凶宅】haunted house; unlucky abode
【凶兆】ill omen; boding of evil

汹
【汹汹】①(波涛的声音) the sound of roaring waves ②(声势盛大) violent; truculent: 来势～ come foaming with violent temper / 气势～ blustering and truculent ③(争论的声音或纷扰的样子) tumultuous; agitated: 议论～ tumultuous debate; heated discussion
【汹涌】tempestuous; turbulent: 波涛～ turbulent waves; with waters raging ◇ ～澎湃 surging; turbulent; tempestuous

匈
【匈奴】Xiongnu; Hun

【匈牙利】Hungary ◇～人 Hungarian / ～语 Hungarian (language)

胸 ①(胸部) chest; bosom; thorax: 齐～高 breast-high / 齐～深 breast-deep / 挺～ throw out one's chest ②(心胸) mind; heart: ～怀祖国 have the whole country in mind; have one's own country at heart

【胸靶】[军] chest silhouette

【胸部】chest; thorax ◇～手术 thoracic operation

【胸大肌】[生理] pectoralis major

【胸骨】[生理] breastbone; sternum

【胸怀】mind; heart: 革命者的伟大～ a revolutionary's breadth of vision / ～开朗 the bosom is clear / ～全局 have the general interest at heart / ～坦白 openhearted; frank / ～狭窄 narrow-minded; small-minded

【胸肌】[生理] chest muscle

【胸襟】mind; breadth of mind: ～开阔 broad-minded; large-minded / ～狭窄 small-minded; narrow-minded

【胸口】the pit of the stomach

【胸膜】[生理] pleura ◇～炎 pleurisy

【胸脯】chest: 拍～保证 strike one's chest as a gesture of guarantee or reassurance / 挺起～ throw out one's chest

【胸鳍】[动] pectoral fin

【胸腔】thoracic cavity ◇～外科 thoracic surgery

【胸墙】[军] breastwork; parapet

【胸膛】chest: 枪口对准～ point the gun at sb.'s chest / 挺起～ throw out one's chest

【胸围】chest measurement; bust; chest circumference

【胸无城府】having nothing hidden in the mind

【胸无大志】with no ambition at all; want of lofty aspirations

【胸无点墨】unlearned; unlettered

【胸像】(sculptured) bust

【胸有成竹】have a well-thought-out plan, strategem, etc.

【胸中无数】have no idea of it at all; be ignorant of how things stand

【胸中有数】have a good idea of how things stand

【胸椎】[生理] thoracic vertebra

xióng

雄 ①(公的) male: ～鸡 cock / ～猫 male cat; tomcat ②(有气魄的) grand; imposing: ～伟 imposing; magnificent ③(强有力的) powerful; mighty: ～兵 a powerful army ④(强有力的人或国家) a person or state having great power and influence: 英～ hero

【雄辩】convincing argument; eloquence: ～地证明 prove incontrovertibly; be eloquent proof of / 事实胜于～ Facts speak louder than words

【雄才大略】(a man of) great talent and bold vision; (a statesman or general of) rare gifts and bold strategy

【雄蜂】[动] drone

【雄关】impregnable pass

【雄厚】rich; solid; abundant: ～的人力物力 rich human and material resources / 实力～ tremendous strength; enormous potentiality / 资金～ abundant funds

【雄花】[植] male flower; staminate flower

【雄黄】[矿] realgar; red orpiment

【雄浑】vigorous and firm; forceful: ～的诗篇 powerful poetry / ～高亢的乐曲 resounding music / 笔力～ vigor of strokes in calligraphy or drawing

【雄鸡】cock; rooster

【雄激素】androgen

【雄健】robust; vigorous; powerful: ～的步伐 vigorous strides

【雄赳赳】valiantly; gallantly: ～,气昂昂 valiantly and spiritedly

【雄蕊】[植] stamen

【雄师】powerful army: 百万～ a million bold warriors; a mighty army one million strong

【雄视一方】cut a conspicuous figure in a place

【雄图】great ambition; grandiose plan: ～大略 a great design and a big plan / ～大业 a grandiose and noble enterprise; a great cause

【雄伟】grand; imposing; magnificent; majestic: ～壮丽 grand; sublime; magnificent

【雄文】profound and powerful writing; great works

【雄心】great ambition; lofty aspiration: ～勃勃 very ambitious / ～壮志 high hopes and great ambition; lofty aspirations and great ideals / 树立～ set up high aims and lofty aspirations; set one's sights high

【雄性不育】[生] male sterility

【雄性先熟】[生] protandry

【雄蚁】[动] aner

【雄甾酮】androsterone

【雄壮】full of power and grandeur; magnificent; majestic: ～的军乐 majestic martial music

【雄资】majestic appearance; heroic posture

熊 ①[动] bear: 白～ polar bear / 狗～ Asiatic black bear / 黑～ black bear / 棕～ brow bear ②(斥责) rebuke; upbraid; scold

【熊蜂】bumblebee

【熊猫】[动] panda: 大～ giant panda / 小～ (lesser) panda

【熊熊】flaming; ablaze; raging: ～烈火 blazing fire

【熊掌】bear's paw (as a rare delicacy)

xiū

羞 ①(难为情;不好意思) shy; bashful: ～红了脸 blush / 怕～ feel bashful / 她有点害～。She is rather bashful. ②(羞耻) shame; disgrace: 遮～ conceal one's shame; hide one's shame ③(感到耻辱) feel ashamed: ～与为伍 consider it beneath one to associate with sb.

【羞惭】be ashamed: 满面～ be shamefaced

【羞耻】sense of shame; shame: ～之心 sense of shame / 真不知天下有～事 lose all sense of shame

【羞答答】coy; shy; bashful

【羞愤】ashamed and resentful

【羞愧】ashamed; abashed: ～地低着头 hang one's head for shame / ～难言 be ashamed beyond words

【羞明】[医] photophobia

【羞怯】shy; timid; sheepish: ～得说不出话来 be too shy to utter a word

【羞人】feel embarrassed or ashamed: 羞死人 simply die of shame; feel terribly embarrassed

【羞辱】①(耻辱) shame; dishonor; humiliation ②(使受耻辱) humiliate; put sb. to shame

【羞涩】shy; bashful; embarrassed

馐 delicacy; dainty: 珍～ a rare delicacy

修 ①(修饰) embellish; decorate: 装～商店门面 paint and decorate the front of a shop ②(修理;整治) repair; mend; overhaul: ～桥补路 repair bridge and mend roads / ～收音机 repair a radio / ～鞋 mend shoes ③(编写) write; compile: ～史 write history ④(学习和锻炼) study; cultivate: 进～ pursue one's studies / 自～ study by oneself ⑤(兴建;建筑) build; construct: ～渠 dig irrigation ditches / ～水库 construct a reservoir / ～铁路 build a railway ⑥(剪;削) trim; prune: ～指甲 trim one's fingernails; manicure one's fingernails ⑦(长) long; tall and slender: 茂林～竹 dense forests and tall bamboos ⑧[简](修正主义) revisionism

【修补】① mend; patch up; repair; revamp: ～篱笆 mend a fence / ～衣服 patch clothes / ～渔网 mend fishing nets ②[医] repair

【修布】[纺] mending; burling

【修长】tall and thin; slender: ～的身材 a slender figure

【修船厂】shipyard; dockyard

【修船坞】repair dock

【修辞】rhetoric

【修辞学】[语] rhetoric

【修道】cultivate oneself according to a religious doctrine

【修道院】[宗] ①(男修道院) monastery ②(女修道院) convent ◇ ～院长 abbot; the Father superior / 女～院长 abbess; the Mother superior

【修订】revise: ～教学计划 revise a teaching plan / ～条约 revise a treaty ◇ ～本 revised edition / ～者 reviser

【修复】①(修理使恢复完整) repair; restore; renovate ②[医] repair

【修改】revise; modify; amend; alter: ～合同 modification of a contract / ～计划 revise a plan / ～设计 alter a design / ～宪法 amend a constitution; revise a constitution / ～预算 revise a budget

【修改带】[计算机] change tape; transaction tape

【修好】①[书](国与国之间亲善友好) foster cordial relations between states ②(行好;行善) do good works

【修剪】prune; trim; clip: ～果枝 prune fruit trees / ～指甲 trim one's fingernails

【修建】build; construct; erect: ～机场 build an airport / ～纪念碑 erect a monument / ～桥梁 construct a bridge

【修脚】pedicure ◇ ～师 pedicurist

【修旧利废】repair and utilize old or discarded things

【修浚】dredge: ～河道 dredge a river

【修理】repair; mend; overhaul; fix: ～机器 repair a machine; fix a machine / 正在～ be under repair ◇ ～店 fix-it shop; repair shop / ～费 charge for repairs; repair cost / ～行业 repairing trades

【修女】[宗] nun; sister: 当～ become a nun; enter a convent

【修配】make repairs and supply replacements

◇ ～车间 repair and spare parts workshop
【修配厂】repair plant: 机车车辆～ rolling stock repair plant / 农具制造～ farm-tool manufacturing and repair plant / 汽车～ motor car repair and assembly plant
【修葺】repair; renovate: ～一新 take on a new look after renovation; be completely renovated
【修缮】repair; renovate: ～房屋 repair houses
【修身】cultivate one's moral character
【修身齐家】cultivate oneself and put family in order
【修身自省】look after one's conduct by self-examination
【修士】[宗] brother; friar
【修饰】①(整理;装饰) decorate; adorn; embellish ②(梳装打扮) make up and dress up: ～边幅 beautify one's outward appearance ③(修润文章) polish (a piece of writing) ④[语] qualify; modify ◇ ～剂[皮革] dressing agent / ～语 modifier
【修心养性】cultivate the heart and nature culture
【修行】practice Buddhism or Taoism: 出家～ become a Buddhist or Taoist monk or nun; strive for virtue and leave the home
【修修补补】patch up; tinker
【修养】①(理论、知识、艺术等的一定水平) accomplishment; training; mastery: 有文字～ well cultured in literature / 有艺术～ be artistically accomplished ②(待人处事的态度) self-cultivation: ～精神 form one's mind / 他很有～。He is quite cultivated.
【修业】study at school: ～年限 length of schooling / ～期满 finish one's school training / ～证书 certificate showing courses attended
【修造】build as well as repair: ～轮船 build or repair ships
【修整】①(修理) repair and maintain: ～农具 repair and maintain farm implements ②(修剪) prune; trim: ～果树 prune fruit trees
【修正】revise; amend; correct: ～草案 revised draft / ～错误 correct mistakes / 提出～意见 put forward amendments / 有待～ subject to correction ◇ ～案 amendment / ～量 correction
【修正液】correction fluid
【修正主义】revisionism ◇ ～思潮 revisionist trend / ～者 revisionist
【修枝】[农] pruning ◇ ～剪 pruning scissors; pruning shears
【修筑】build; construct; put up: ～堤坝 put up dikes / ～公路 build highways / ～工事 construct defenses; build fortifications; build defense works

休 ①(停止) stop; cease; end: 争论不～ argue ceaselessly ②(休息) rest ③(表示禁止或劝阻) don't: ～要自夸。Don't talk big.
【休耕】fallow; lie fallow: ～地 fallow ground
【休会】adjourn: ～期间 between sessions; when the meeting stands adjourned / 无限期～ adjourn indefinitely
【休假】①(用于工人学生) have a holiday; take a vacation; go on a vacation: ～一周 have a week's holiday ②(用于士兵、在国外工作的人) be on leave; be on furlough: 回国～ go home on furlough
【休克】[医] shock: 电～ electric shock / 病人～了。The patient is suffering from shock.
【休眠】[生] dormancy ◇ ～火山 dormant volcano / ～期[生] rest period / ～芽[植] resting bud; dormant bud
【休戚】weal and woe; joys and sorrows: ～相关 be bound by a common cause; be linked together in common joys and sorrows / ～与共 share weal and woe; stand together through thick and thin
【休憩】have a rest; take a rest; rest; repose
【休息】have a rest; take a rest; rest: ～一会儿 rest for a while; have a rest / ～一天 have a day off; take a day off / 课间～ break (between classes) / 幕间～ intermission; interval ◇ ～平台 [建] half space / ～室 lounge; lobby; vestibule; foyer
【休闲】[农] lie fallow ◇ ～地 fallow (land)
【休想】don't imagine that it's possible: 你～抵赖。Don't imagine you can deny that.
【休学】suspend one's schooling without losing one's status as a student
【休养】recuperate; convalesce: 到海边去～ go to the seaside to recuperate
【休养生息】(of a nation) recuperate and multiply; rest and build up strength; rehabilitate
【休养所】sanatorium; rest home
【休业】①(停止营业) suspend business; be closed down ②(学习暂告结束) come to an end of a short-term course
【休战】truce; cease-fire; armistice: ～状态 (state of) cease-fire
【休整】(of troops) rest and reorganization
【休止】stop; cease: 无～地重复 endless repeti-

tions / 无～地争论 argue ceaselessly
【休止符】[乐] rest

鸺

【鸺鹠】[动] owlet

xiǔ

宿 [量](用于计算夜)谈了半～ chat till midnight / 住一～ stay for one night

朽 ①(腐烂) rotten; decayed: 枯木～株 withered trees and rotten stumps ②(衰老) aged; senile: 老～ old and useless
【朽木】①(烂木头)rotten wood or tree ②(不可造就的人) a hopeless case; a good-for-nothing
【朽木粪土】rotten wood and dirt; a worthless person(指人); useless stuff(指物)

xiù

袖 ①(衣袖)sleeve: 长～ long sleeves / 短～ short sleeves / 套～ sleevelet ②(藏在袖子里) tuck inside the sleeve: ～着手 with hands in sleeves
【袖标】armband
【袖口】cuff (of a sleeve): 衬衫～ wristband ◇ ～纽扣 sleeve button
【袖扣】cuff-link
【袖手旁观】look on with folder arms; stand by with folder arms; look on unconcerned; not to stir a finger
【袖章】armband
【袖珍】pocket-size; pocket: ～半导体收音机 pocket-size transistor radio / ～计算器 pocket calculator / ～手枪 pocket pistol / ～照相机 vest-pocket camera / ～字典 pocket dictionary ◇ ～本 pocket edition
【袖子】sleeve: 卷起～ roll up one's sleeves / 拉某人的～ pull sb. by the sleeve / 他用～擦干脸上的汗水。 He sleeved the sweat off his face.

秀 ①(抽穗开花) put forth flowers or ears: ～穗 put forth ears ②(清秀) elegant; beautiful: 眉清目～ having well-chiselled features; handsome / 山清水～ beautiful hills and waters; lovely scenery ③(特别优异) excellent: 优～ excellent; first-rate
【秀才】scholar; skillful writer
【秀丽】beautiful; handsome; pretty: 这姑娘生得十分～。 This girl is very pretty.
【秀美】graceful; elegant: 书法～ beautiful handwriting
【秀气】①(清秀) delicate; elegant; fine: 眉眼生得～ have beautiful eyes ②(文雅) refined; urbane ③(小巧灵便) delicate and well-made

锈 ①(铜锈、铁锈等) rust: 防～的 rustproof / 铁～ iron rust ②(生锈) become rusty
【锈斑】rusty spot
【锈病】[农] rust: 小麦秆～ wheat stem rust
【锈蚀】rust-eaten

绣 ①(刺绣) embroider: 在桌布上～花 embroider flowers on a tablecloth ②(绣成的物品) embroidery: 湘～ Hunan embroidery
【绣花】embroider; do embroidery ◇ ～被面 embroidered quilt cover / ～丝线 floss silk / ～鞋 embroidered shoes / ～针 embroidery needle
【绣花枕头】①(枕套上有刺绣的枕头) a pillow with an embroidered case ②(徒有外表而无学识的人) an outwardly attractive but worthless person
【绣球】①(绸子结成的球形物) a ball made of strips of silk ②[植] bigleaf hydrangea
【绣像】tapestry portrait; embroidered portrait ◇ ～小说 novel with illnstrated fine-lined portraits of main characters

臭 odor; smell
【臭腺】[动] scent gland
【臭味相投】share the same rotten tastes, habits, etc.; be two of a kind

溴 [化] bromine (Br)
【溴仿】[化] bromoform; tribromomethane
【溴化物】[化] bromide
【溴化银】[化] silver bromide
【溴水】[化] bromine water
【溴酸】[化] bromic acid ◇ ～盐 bromate

嗅 smell; scent; sniff
【嗅觉】(sense of) smell; scent: ～很灵 have a keen sense of smell / 政治～灵敏 be politically sharp ◇ ～倒错 parosmia; parospheresis / ～器官 olfactory organ
【嗅神经】olfactory nerve

xū

需 ①(需要) need; want; require: 急～ need badly ②(需用物品) necessaries; needs: 必～品 necessaries / 军～ military supplies; military requirements
【需求】requirement; demand ◇ ～价格 demand price

【需水量】water requirement
【需氧量】oxygen demand; oxygen requirement
【需要】①(应该有或必须有) need; want; require; demand: ～立即采取行动 require immediate action／她～休息。She is in need of a rest.／这所房子～修理。This house needs repairing. ②(对事物的欲望或要求) needs: 经济上的～ economic needs／从群众的～出发 make the needs of the masses our starting point／满足市场～ meet the needs of the market ◇ ～功率[电] required power

吁 ① sigh: 长～短叹 sighs and groans; moan and groan ②[叹](表示惊异) why; oh
【吁吁】[象]气喘～～ pant; puff hard

须 ①(须要) must; have to: 务～注意下列事项。Attention must be paid to the following. ②(胡须) beard; mustache: 留～ grow a beard ③(须子) palpus; feeler(动物); tassel(植物)
【须发】beard and hair: ～皆白 white hair and beard
【须根】[植] fibrous root
【须要】must; have to: 教育儿童～耐心。It takes patience to educate children.
【须臾】moment; instant: ～不可离 cannot do without even for a moment／～之间 in an instant
【须知】①(一定要知道) one should know that; it must be understood that ②(必须知道的事项) points for attention; notice: 旅客～ notice to travellers; passengers, etc.／游览～ tourist guide; information for tourists
【须子】①(动物的) palpus: 虾～ feelers of a shrimp ②(植物的) tassel: 玉米～ tassels of maize

虚 ①(空虚) void; emptiness: 乘～而入 infiltrate by taking advantage of the other side's unpreparedness; exploit a weak point; act when one's opponent is off guard ②(空着) empty; void; unoccupied: 座无～席。There was no empty seat.(或: All seats were occupied.) ③(心虚; 勇气不足) diffident timid: 胆～ timid; milk-livered／心里有点～ feel rather diffident; feel somewhat diffident ④(徒然) in vain: 不～此行 have not made the trip in vain ⑤(虚假) false; nominal: ～名 false reputation／耳闻为～,眼见为实。What you hear about may be false; what you see is true. ⑥(虚心) humble; modest: 谦～ modest ⑦(虚弱) weak; in poor health: 气～ lacking in vital energy; sapless／身体很～ be very weak physically ⑧(政治、思想等方面的道理) guiding principles; theory: 务～ discuss principles or ideological guidelines ⑨[物] virtual: ～接地 virtual earth／～阴极 virtual cathode／～阻抗 virtual impedance
【虚报】make a false report; false declaration; false return: ～冒领 make a fraudulent application and claim／～年龄 lie about one's age／～税额 false declaration／～账目 cook accounts
【虚词】[语] function word; form word
【虚度】spend time in vain; waste: ～光阴 fritter away one's time; dream away one's time／～年华 pass the years in vain／～一生 dream away one's life; idle away one's life
【虚浮】impractical; superficial: ～的计划 an impractical plan／作风～ have a superficial style of work
【虚根】[数] imaginary root
【虚构】fabricate; make up: ～的情节 a made-up story／～的人物 a fictitious character／纯属～ an out-and-out fabrication; a sheer fabrication ◇ ～情节者 fabler
【虚汗】abnormal sweating due to general debility
【虚怀若谷】have a mind as open as a valley; be very modest; be extremely open-minded
【虚幻】unreal; illusory: ～的情景 a mere illusion／～境界 visionary world
【虚假】false; sham: ～的安全感 a false sense of security／～的可能性 spurious possibility／～的现象 fictitious phenomenon／～的友谊 hypocritical friendship
【虚价】[经] nominal price
【虚焦点】[物] virtual focus
【虚惊】false alarm: 受了一场～ be the victim of a false alarm
【虚夸】exaggerative; bombastic; boastful: ～的语言 exaggerative language; boastful words
【虚名】undeserved reputation
【虚拟】①(虚构的) invented; fictitious: 这个故事是～的。This is a fictitious story. ②(假设的) suppositional ◇ ～语气[语] the subjunctive mood
【虚胖】[医] puffiness
【虚情假意】false display of affection; hypocritical show of friendship
【虚荣】vanity: 爱～的人 vain person／不慕～ not affected by vanity; not vain
【虚荣心】vanity: 极度的～ vainglory

【虚弱】①(不结实) in poor health; weak; debilitated: 病后身体很～ suffer from general debility after an illness; be very weak after an illness ②(软弱;薄弱) weak; feeble: ～的本质 inherent weakness; intrinsic weakness／兵力～ weak in military strength
【虚设】nominal; existing in name only: 这个机构形同～。 This organization is but an empty shell.
【虚实】false or true; the actual situation (as of the opposing side): 探听～ try to find out about an opponent, etc.; try to ascertain the strength of
【虚数】①(虚假的数字) unreliable figure ②[数] imaginary number
【虚岁】nominal age
【虚脱】[医] collapse; prostration
【虚妄】unfounded; fabricated; invented: ～的故事 a fabricated story
【虚伪】sham; false; hypocritical: ～的情谊 hypocritical affection ◇～资本 fictitious capital／～资产 fictitious assets
【虚位以待】leave a seat vacant for sb.; save a seat for sb.; reserve a seat for sb.
【虚文】①(具文) dead letter: 规章制度一经订立,不应成为一纸～。 Rules and regulations, once set up, should not be turned into a dead letter. ②(没有意义的礼节) mere formalities; empty forms: ～浮礼 mere formalities; conventionalities
【虚无】nihility; nothingness ◇～主义 nihilism／～主义者 nihilist
【虚无缥缈】purely imaginary; entirely unreal; visionary; illusory
【虚线】①(由点构成的非实线) dotted line; line of dashes ②[数] imaginary line
【虚像】[物] virtual image
【虚心】open-minded; modest: ～使人进步,骄傲使人落后 Modesty helps one to go forward, whereas conceit makes one lag behind／～听取别人的意见 listen to people's criticisms with an open mind
【虚虚实实】sometimes false, sometimes true; seemingly false and real at the same time
【虚掩】with the door left unlocked or unlatched
【虚应故事】do sth. perfunctorily as a mere matter of form or as a routine practice
【虚有其表】look impressive but lack real worth; appear better than it is
【虚与委蛇】deal with sb. courteously but without sincerity; pretend politeness and compliance
【虚张声势】make an empty show of strength; bluff and bluster; be swashbuckling; make a deceptive show of power

墟 ruins: 废～ ruins

嘘 ①(嘘气) breathe out slowly ②(叹气) utter a sigh ③(火或蒸气的热力接触到) come into contact with sth.; scald; burn: 小心热气～着手。 Don't scald your hands. ④[象] sh; hush
【嘘寒问暖】inquire after sb.'s well-being; be solicitous about sb.'s health
【嘘气】breathe out slowly

xú

徐 slowly; gently: 清风～来。 A refreshing breeze is blowing gently.
【徐变】[电子] creep
【徐步】walk slowly; walk leisurely; stroll
【徐徐】slowly; gently: 一面红旗～升起。 A red flag slowly went up the pole.

xǔ

许 ①(称赞;承认优点) praise: 赞～ praise; commend ②(答应) promise: ～下诺言 have made a promise ③(允许) allow; permit: 此处不～吸烟。 Smoking is not permitted here.／只～前进,不～后退。 No retreat is allowed, only advance. ④(也许) maybe; perhaps; possibly
【许多】many; much; a great deal of; a lot of: ～人 many people／积累了～经验 have accumulated much experience
【许久】for a long time; for ages; for long
【许可】permit; allow: 凡是条件～的地方 wherever conditions permit／如时间～ if time permits
【许可信号】enabling signal
【许可证】licence; permit: 出口～ an export licence／进口～ an import licence／特别～ a special permit
【许诺】make a promise; promise
【许配】betroth one's daughter to; be betrothed to
【许愿】①(对神佛) make a vow (to a god) ②(对人) promise sb. a reward

诩 brag; boast: 自～为… style oneself...; boast that one is...

栩
【栩栩】vivid; lively: ～如生 lifelike; to the life
醑 [药] spirit: 樟脑～ camphor spirit
【醑剂】[药] spirit

xù

畜 raise (domestic animals)
【畜产】livestock products; animal products ◇ ～业 animal industry
【畜牧】raise livestock; rear livestock; rear poultry: 从事～ go in for animal husbandry ◇ ～场 animal farm; livestock farm; stock farm / ～业 animal husbandry; livestock husbandry; livestock farming
【畜养】raise; rear; keep
蓄 ①(储存;积蓄) store up; save up ②(留着而不剃掉) grow: ～发 wear one's hair long / ～须 grow a beard ③(心里藏着) entertain (ideas); harbor: ～念已久 have long entertained such ideas
【蓄电池】storage battery; accumulator ◇ ～车 battery car / ～充电器 battery charger; charger
【蓄洪】[水] store floodwater: ～防旱 shore floodwater for use against a drought ◇ ～坝 flood dam / ～工程 flood storage project / ～量 storage capacity
【蓄积】store up; save up: ～粮食 store up grain
【蓄谋】premeditate: ～迫害 harbor a design of persecuting sb. / ～已久 long premeditated
【蓄热】accumulation of heat
【蓄水】[水] retain water; store water ◇ ～池 cistern; reservoir / ～工程 (water) storage project
【蓄意】premeditated; deliberate: ～犯罪 intentional crime / ～进行破坏 deliberately sabotage / ～谋杀 willful murder; murder in the first degree / ～伤害 wounding with intent / ～挑衅 premeditated provocation / ～行骗 deliberate deception / ～遗漏 malicious omission
【蓄志】have long held the ambition
叙 ①(说;谈) talk; chat: ～家常 chitchat ②(记叙) narrate; recount; relate ③(评议等级次第) assess; appraise: ～功 assess service and give credit for it
【叙别】have a farewell talk
【叙旧】talk about the old days
【叙利亚】Syria ◇ ～人 Syrian
【叙事】narrate; recount ◇ ～曲 ballade / ～诗 narrative poem / ～文 narrative; narrative prose
【叙述】narrate; recount; relate
【叙说】tell; narrate
【叙谈】chat; chitchat; talk together
酗
【酗酒】excessive drinking: ～滋事 get drunk and create a disturbance
煦 warm; balmy: 春风和～ a balmy spring breeze
恤 ①(怜悯) pity; sympathize: 体～ understand and sympathize with ②(救济) give relief; compensate: 抚～ comfort and compensate a disabled person or a bereaved family
【恤金】pension for a disabled person or the family of the deceased
旭 brilliance of the rising sun
【旭日】the rising sun: ～东升 the sun rising in the eastern sky
序 ①(次序) order; sequence: 按年月为～ in chronological order / 按字母为～ in alphabetical sequence / 程～ procedure / 井然有～ in perfect order / 顺～ sequence ②(排次序) arrange in order: ～齿 arrange (seats, etc.) in order of age ③(开头的;在正式内容之前的) introductory; initial ④(序文) preface: 原～ the original preface
【序跋】preface and postscript
【序列】①(按次序排好的行列) alignment; array: 不成～ out of alignment / 战斗～ battle array; battle order ② [数] sequence
【序幕】prologue; prelude
【序曲】overture
【序数】ordinal number; ordinal
【序文】preface; foreword
【序言】preface; foreword
絮 ①(棉絮) (cotton) wadding ②(象棉絮样的东西) sth. resembling cotton: 柳～ (willow) catkin ③(往衣、被里铺棉花等) wad with cotton: ～被子 wad a quilt with cotton / ～棉衣 line 〈wad〉 one's clothes with cotton ④(絮叨) long-winded; garrulous
【絮叨】long-winded; garrulous; wordy
【絮棉】cotton for wadding
【絮凝剂】flocculant; flocculating agent
【絮片】flocculus

婿 ①(女婿) son-in-law ②(丈夫) husband: 夫～ husband / 妹～ younger sister's husband

绪 ①(丝的头) thread ②(事情的开端) order in sequence or arrangement: 头～ main threads (of a complicated affair); main lines / 准备就～ be all set ③(心情;思想) mental or emotional state: 心～不宁 be in a state of agitation ④(事业;功业) task; cause; undertaking: 续未竟之～ carry on an unfinished task; take up where another has left off

【绪论】introduction

【绪言】introduction

续 ①(接连不断) continuous; successive ②(接在后头) continue; extend; join: 待～ to be continued / ～会 extended session; follow-up meeting ③(添;加) add; supply more: 该～煤了。It's time to add some coal to the fire.

【续编】continuation (of a book); sequel: 小说～ a sequel of a novel

【续订】renew one's subscription

【续航力】①(飞机的) endurance ②(轮船的) cruising radius

【续集】continuation (of a book); sequel

【续假】extend one's leave of absence; extend leave: ～一个月 have one's leave extended for another month

【续借】renew: 书～两星期 renew a book for another two weeks

【续弦】remarry after the death of one's wife

xuān

宣 ①(宣布;宣告;传播) declare; proclaim; announce: ～示 declare; make known publicly / ～赦 proclaim a general amnesty / ～旨 announce an imperial decree ②(疏导) lead off (liquids); drain: ～泄洪水 drain off floodwater

【宣布】declare; proclaim; announce: ～表决结束 declare the vote closed / ～独立 declare independence; proclaim independence / ～会议开始 declare a meeting open; call a meeting to order / ～戒严 declare martial law; proclaim martial law / ～某人有罪 declare sb. guilty / ～投票结束 declare the ballot closed / ～无效 declare sth. invalid; declare sth. null and void / ～赦免 announce amnesty / ～为合法 legitimate / ～为非法 illegitimate / ～一件事 make an announcement

【宣称】assert; declare; profess

【宣传】conduct propaganda; propagate; disseminate; give publicity to: ～党的方针政策 publicize the Party's general and specific policies / ～群众 spread propaganda among the masses ◇ ～车 propaganda car / ～费 publicity expenses / ～工具 instrument of propaganda; means of publicity; mass media / ～工作者 propagandist / ～画 picture poster / ～机构 propaganda organ / ～机器 propaganda machine / ～品 propaganda material; publicity material / ～员 propagandist

【宣传队】propaganda team: 文艺～ performing arts propaganda team

【宣读】read out (in public): ～文件 read out a document

【宣告】declare; proclaim: ～成立 proclaim the founding of (a state, organization, etc.) / ～破产 declare bankruptcy; go bankrupt / ～失踪 declaration of disappearance / ～无效 declare sth. null and void / ～无罪 acquit; pronounce not guilty / ～为不合法 outlaw / ～有罪 condemn; pronounce guilty

【宣讲】explain and publicize

【宣判】[法] pronounce judgment: ～某人五年徒刑 pronounce a sentence of five years on sb. / ～无罪 pronounce sb. not guilty / ～有罪 pronounce sb. guilty / ～原告胜诉 pronounce a judgment for the plaintiff

【宣誓】take an oath; swear an oath; make a vow; make a pledge: ～就职 take an oath of office; be sworn in / 入党～ take the oath on being admitted to the Party / 庄严～ make a solemn vow

【宣泄】①(使积水流走) lead off (liquids); drain: ～洪水 drain off floodwater ②(舒散;吐露) get sth. off one's chest; unbosom oneself

【宣叙调】recitative

【宣言】declaration; manifesto: 人权～ Declaration of the Rights of Man / 中立～ declaration of neutrality

【宣扬】publicize; propagate; advocate; advertise

【宣战】declare war; proclaim war

萱

【萱草】tawny daylily

喧 noisy: 锣鼓～天 a deafening sound of gongs and drums

【喧宾夺主】a presumptuous guest usurps the host's role; the secondary supersedes the pri-

mary; minor taking precedence over a major issue
【喧哗】confused noise; hubbub; uproar: 请勿~! Quiet, please! / 笑语~ uproarious talk and laughter
【喧闹】noise and excitement; bustle; racket
【喧嚷】clamor; hubbub; din; racket: 人声~ a hubbub of voices; loud confused voices
【喧扰】noise and disturbance; tumult
【喧腾】noise and excitement; hubbub; uproar
【喧嚣】①(声音杂乱;不清静) noisy: ~的车马声 the noise of dense traffic ②(叫嚣;喧嚷) clamor; hullabaloo; din: 大肆~ raise a hullabaloo / ~鼓噪 make a clamor; stir up a commotion / ~战争 clamor for war /

轩

【轩昂】dignified; imposing: 气宇~ have an imposing appearance; have an impressive presence
【轩敞】spacious and bright
【轩然大波】a great disturbance; a mighty uproar
【轩轾】high or low; good or bad: 不分~ be equal; be on a par

xuán

旋 ①(旋转) revolve; circle; spin: 盘~ circle round / 觉得天~地转 feel as if heaven and earth were spinning round and round; feel one's head swim ②(返回;归来) return; come back: 凯~ return in triumph / ~里 return home ③(不久) soon: 他~即死去。He soon died.
【旋臂式起重机】swing lever
【旋耕】rotary tillage ◇ ~机 rotary cultivator; rotocultivator
【旋光性】[物] optical rotation
【旋律】melody
【旋毛虫】trichina
【旋钮】knob: 卷片~ film winder / 双套筒~ dual knobs / 调谐~ tuning knob / 音量控制~ volume control knob
【旋桥】[建] swing bridge
【旋绕】curl up; wind around
【旋塞】cock: 放水~ drain cock / 三通~ three-way cock
【旋梯】winding stair
【旋涡】whirlpool; vortex; eddy
【旋涡星云】[天] spiral nebula
【旋翼机】[航空] rotary-wing aircraft; rotorcraft
【旋踵】in an instant; immediately: ~即逝 vanish before one has time to turn round; disappear in the twinkling of an eye
【旋转】revolve; gyrate; rotate; spin: 逆时针方向~ counterclockwise rotation / 顺时针方向~ clockwise rotation / 轮子在~。The wheels are turning. ◇ ~门 revolving door / ~球[体] spinning ball / ~轴 revolving spindle
【旋转乾坤】effect a drastic change in nature or the established order of a country; be earthshaking

玄 ①(黑色) black; dark: ~狐 a black fox ②(深奥) profound; abstruse: ~理 a profound theory ③(玄虚;靠不住) unreliable; incredible: 这话真~。That's a pretty tall story
【玄妙】mysterious; abstruse
【玄青】deep black
【玄学】①[哲] metaphysics ②(魏晋时代的哲学) a philosophical sect in the Wei and Jin dynasties
【玄孙】great-great-grandson
【玄武玻璃】basalt obsidian; tachylite basalt glass
【玄武岩】[地] basalt
【玄虚】deceitful trick; mystery: 故弄~ purposely turn simple things into mysteries; be deliberately mystifying
【玄之又玄】mystery of mysteries; extremely mysterious and abstruse; the most mysterious of the mysterious

悬 ①(悬挂) hang; suspend ②(无着落;没结果) outstanding; unresolved: ~而未决的问题 an outstanding question ③(挂念) feel anxious; be solicitous: ~念 be anxious about (sb. who is elsewhere) ④(凭空设想) imagine: ~拟 imagine; conjecture ⑤(距离远;差别大) far apart: ~隔 be separated by a great distance ⑥(危险) dangerous
【悬案】①(没有解决的案件) unsettled law case ②(没有解决的问题) outstanding issue; unsettled question
【悬臂】[机] cantilever ◇ ~梁[建] cantilever beam / ~起重机 cantilever crane / ~桥 cantilever bridge
【悬灯结彩】adorn with lanterns and festoons; hang up lanterns and festoons
【悬而未决】hang in doubt; remain in suspense; unresolved
【悬浮】[化] suspension ◇ ~固体 [环保] sus-

pended solid / ～染色 suspension dyeing / ～体 suspended substance; suspension
【悬隔】far apart: 两地～ two places being far apart
【悬挂】①(挂着) hang; fly: ～在墙上 hang on the wall / 那船～着英国旗。 The ship was flying the British flag. ②[汽车] suspension: 前～ front suspension ◇ ～犁[农] mounted plough
【悬胶】[化] suspensoid ◇ ～态 suspensoid state
【悬空】①(悬在空中) hang in the air; suspend in midair ②(脱离实际) be divorced from reality ◇ ～索道 aerial conveyer
【悬梁】hang oneself from a beam: ～自尽 commit suicide by hanging oneself from a beam; hang oneself
【悬梁锥刺】tie to a beam and to prick the thing with an awl
【悬铃木】[植] plane tree; planetree
【悬念】①(挂念) be concerned about (sb. who is elsewhere) ② (电影、戏剧的) audience involvement ③ (文学作品的) reader involvement
【悬赏】offer a reward; post a reward: ～缉拿逃犯 offer a reward for the capture of a runaway criminal; set a price on a runaway criminal's head
【悬殊】great disparity; wide gap: 力量～ a great disparity in strength / 贫富～ wide gap between the rich and the poor
【悬索结构】[建] suspended-cable structure
【悬索桥】suspension bridge
【悬索铁路】suspension cable railway
【悬梯】hanging ladder
【悬腕】with the wrist raised
【悬想】imagine; fancy
【悬心吊胆】on tenterhooks; filled with anxiety or fear
【悬崖】overhanging cliff; steep cliff; precipice: ～绝壁 sheer precipice and overhanging rocks
【悬崖勒马】rein in at the brink of the precipice; wake up to and escape disaster at the last moment; pull back before it is too late
【悬雍垂】[生理] uvula

xuǎn

癣 tinea; ringworm

选 ①(挑选) select; choose; pick: 挑～ pick and choose ②(选举) elect: 当～ be elected / 普～ general election / 人～ be chosen; be selected; be elected ③(选出来的作品) selections; anthology: 民歌～ selections of folk songs / 诗～ selected poems / 文～ an anthology of prose
【选拔】select; choose: ～干部 select cadres / ～运动员 select athletes / ～最优秀的 choose the best ◇ ～赛 (selective) trials
【选本】anthology; selected works
【选材】select (suitable) material
【选读】selected readings: 文学～ selected readings in literature
【选购】pick out and buy; choose
【选集】selected works; selected writings; selections; anthology
【选举】elect: 间接～ indirect election / 无记名投票～ elect by secret ballot / 直接～ direct election ◇ ～程序 electoral procedure; electoral proceedings / ～单位 electoral unit / ～法 electoral law / ～结果 election results; election returns / ～权 the right to vote; franchise / ～人 voter; elector / 被～权 the right to stand for election
【选矿】[矿] ore dressing; mineral separation; beneficiation ◇ ～厂 ore dressing plant; concentration plant
【选民】①(指个人) voter; elector ②(指全体) constituency; electorate ◇ ～登记 registration of voters / ～名册 voting register / ～证 elector's certificate; voter's card
【选派】select; detail: ～代表参加会议 select sb. as representative to conference; depute sb. to attend a conference
【选票】vote; ballot; ballot ticket
【选区】electoral district; election district; electoral ward; constituency
【选曲】selected songs; selected tunes
【选取】select; choose
【选手】an athlete selected for a sports meet; (selected) contestant; player: 世界名～ world-famous athlete / 优秀～ topnotch athlete; ace athlete; top-ranking athlete / 种子～ seeded player; seed
【选修】take as an elective course: ～英语 take English as an elective course ◇ ～课 elective course; selective course
【选样】sampling; sample
【选育】[农] seed selection; breeding: ～良种小麦 wheat variety development by selection
【选择】select; choose; opt: ～日期 choose a date / 自然～ natural selection / 没有～的余地 have no choice at all ◇ ～场地 [体] choice

of ends
【选种】[农] seed selection

xuàn

渲
【渲染】play up; exaggerate; pile it on

楦 ①(鞋楦) shoe last ②(帽楦) hat block ③(用楦子填紧) shape with a last or block: ～鞋 last a shoe
【楦子】①(鞋楦) shoe last; shoe tree ②(帽楦) hat block

旋 ①(旋转的) whirl: ～风 whirlwind ②(切; 转圈地削) turn sth. on a lathe; lathe; pare: 把苹果皮～掉 pare an apple ③(临时做) at the time; at the last moment: ～用～买 buy sth. when you need it; buy for immediate use
【旋床】[机] (turning) lathe
【旋风】whirlwind
【旋工】lathe turner

炫 [书]①(晃眼) dazzle: 光彩～目 blindingly bright; dazzling splendor ②(夸耀) show off; display: 自～其能 show off one's ability
【炫示】show off; display
【炫耀】make a display of; show off; flaunt: ～财富 flaunt one's riches / ～力量 flaunt one's strength / ～武力 make a show of force / ～学问 parade one's learning

眩 ①(眼睛昏花) dizzy; giddy: 头晕目～ feel dizzy ②(迷惑;执迷) dazzled; bewildered: ～于名利 dazzled by the prospect of fame and wealth; obsessed with a desire for fame and wealth
【眩晕】dizziness: 一阵～ a fit of dizziness
【眩晕症】[医] vertigo: 内耳～ auditory vertigo; aural vertigo

绚 gorgeous
【绚烂】splendid; gorgeous: ～的朝霞 gorgeous morning clouds
【绚丽】gorgeous; magnificent: ～多彩 bright and colorful; gorgeous / ～的景色 magnificent scenery

xuē

削 (专用于合成词) pare; whittle; cut: 剥～ exploit
【削壁】precipice; cliff
【削价】cut prices; lower the price
【削减】cut (down); reduce; slash; whittle down: ～非生产性开支 cut down nonproductive expenditures; cut back on nonproductive spending / ～工资 reduce wages; cut wages / ～军费 cut down military expenditures / ～预算 slash the budget
【削弱】weaken; cripple: ～敌人的力量 cripple the enemy; weaken the enemy ～某人的地位 weaken the position of sb.
【削足适履】cut the feet to fit the shoes; act in a Procrustean manner

靴 boots: 马～ riding boots; high boots / 雨～ rubber boots
【靴襻】bootstrap
【靴楦】boot last; boot tree
【靴子】boots

xué

学 ①(学习) study; learn: ～当医生 study for the medical profession / ～开机器 learn to operate a machine / ～文化 acquire an elementary education; learn to read and write / ～外语 learn a foreign language; study a foreign language / ～游泳 learn to swim / 活到老,～到老。 Keep on learning as long as you live. ②(模仿) imitate; mimic: ～某人的走路样子 imitate sb.'s way of walking / ～猫叫 mimic the mewing of a cat ③(学问) learning; knowledge: 博～者 man of learning / 才疏～浅 have little talent and less learning ④(学科) subject of study; branch of learning: 数～ mathematics / 文～ literature / 政治经济～ political economy ⑤(学校) school; college: 小～ primary school / 中～ middle school / 大～ college; university / 复～ resume one's interrupted studies / 上～ go to school / 升～ go to a school of a higher grade; enter a higher school / 退～ leave school; discontinue one's schooling / 休～ suspend one's schooling without losing one's status as a student / 转～ transfer to another school
【学报】learned journal; journal
【学潮】student strike; campus upheaval
【学而不厌】have an insatiable desire to learn
【学而时习之】learn and review it from time to time
【学而知之】wisdom obtained by studies
【学阀】scholar-tyrant
【学非所用】one does not do that which one has learned

【学费】tuition fee; tuition
【学分】credit ◇ ～制 the credit system
【学风】style of study
【学府】seat of learning; institution of higher learning
【学好】learn from good examples; emulate good
【学会】①(学后能掌握) learn; master: ～新技术 master a new skill / 她～了弹钢琴。 She's learned to play the piano. ②(学术团体) learned society; institute
【学籍】one's status as a student; one's name on the school roll: 保留～ retain one's status as a student / 开除～ be dismissed from school / 取消～ be struck off the school roll
【学究】pedant ◇ ～气 pedantry
【学科】branch of learning; course; subject; discipline
【学力】knowledge; educational level; academic attainments: 具有同等～ have the same educational level
【学历】record of formal schooling; educational background
【学龄】school age: ～儿童 children of school age; school-age children / ～前儿童 preschool children; preschoolers
【学名】①(科学专名) scientific name ②(入学时的正式名字) formal name used at school
【学年】school year; academic year ◇ ～考试 year-end examination
【学派】school of thought; school
【学期】school term; term; semester
【学前教育】preschool education; infant school education
【学然后知不足】one discovers his ignorance only through learning
【学如逆水行舟,不进则退】Learning is like rowing upstream, not to advance is to drop back.
【学舌】①(模仿别人说话) mechanically repeat other people's words; parrot: 鹦鹉～ imitate mechanically; parrot ②(嘴不严) gossipy; loosetongued
【学生】①(在学校学习的人) student; pupil: 医科～ a medical student / ～时代 school days / 一派～腔 all in student jargon / 小学二年级～ second grade primary school pupil; second grader / 中～ middle school student ②(向老师或前辈学习的人) disciple; follower ◇ ～会 student union; student association / ～证 student's identity card
【学时】class hour; period
【学识】learning; knowledge; scholarly attainments: ～渊博 have great learning; be learned / ～浅薄 have little learning
【学士】①(文人) scholar: 文人～ scholars; men of letters ②(学位) bachelor: 理～ Bachelor of Science(B.S) / 文～ Bachelor of Arts (B.A.)
【学术】learning; science: ～成就 scholastic attainment / ～领域 sphere of learning / 国际～交流活动 international academic exchanges ◇ ～报告 learned report; academic report / ～界 academic circles / ～论文 research paper; scientific paper; thesis / ～讨论会 academic discussion; scientific conference; symposium / ～权威 academic authority / ～团体 learned society / ～研究 academic research / ～自由 academic freedom
【学说】theory; doctrine: 达尔文的进化论～ Darwin's theory of evolution / 马克思的国家～ Marx's teaching on the state
【学徒】apprentice; trainee: 在机床厂～ be an apprentice in a machine tool plant / ～期满 have served one's apprenticeship ◇ ～工 apprentice / ～合同 contract of apprenticeship
【学位】academic degree; degree: 博士～ doctor's degree; doctorate / 硕士～ Master's degree / 名誉～ honorary degree ◇ ～评审委员会 evaluation committee of academic degree / ～证书 diploma
【学问】learning; knowledge; scholarship: ～高深的人 a man of great learning; an erudite scholar / 大有～ take a lot of learning / 做～ engage in scholarship; do research
【学问无捷径】there is no royal road to learning
【学无止境】knowledge is infinite; there is no limit to knowledge
【学习】study; learn; emulate: 集中～ massed learning / ～某人的榜样 learn from sb.'s example; follow sb.'s example / ～文化 learn to read and write ◇ ～成绩 academic record; school record / ～动机 learning motivation / ～方法 learning method / ～能力 learning ability; learning capacity / ～年限 period of schooling / ～效率 learning efficiency / ～障碍 learning disorder
【学衔】academic rank; academic title ◇ ～评审委员会 evaluation committee of academic ranks
【学校】school; educational institution: 半日制～ half-day school / 工业～ school of technology / 高等～ institution of higher learning /

函授～ correspondence school／护士～ nurses' training school; nursing school／聋哑～ school for deaf-mutes／盲人～ school for the blind／全日制正规～ full-time regular school／少年业余体育～ youth spare-time sports school; youth amateur athletic school／师范～ teachers' school; normal school／职业～ vocational school／业余～ spare-time school／幼儿师范～ school for kindergarten teachers／中等专业～ secondary specialized school／中等技术～ secondary technical school／重点～ key school／专业～ specialized school

【学业】one's studies; school work

【学以致用】study for the purpose of application; study sth. in order to apply it

【学有专长】have acquired a specialty from study

【学员】student; trainee

【学院】college; academy; institute: 法律～ law school／工业～ engineering institute／建筑工程～ institute of civil engineering／教师进修～ teachers' college for vocational studies／军事～ military institute／美术～ school of art／师范～ teachers' college; teachers college／外语～ foreign languages institute／医～ medical college／音乐～ conservatory of music; academy of music

【学者】scholar; learned man; man of learning

【学制】①(教育制度) educational system; school system: ～改革 reform in the school system ②(学习年限) length of schooling: 缩短～ shorten the period of schooling

穴 ①(洞穴) cave; den; hole: 洞～ cave／虎～ tiger's lair／匪～ bandits' den／蚁～ ant hole ②(墓穴) grave ③(穴位) acupuncture point; acupoint

【穴播】bunch planting ◇～机 hill-drop drill

【穴居】live in caves ◇～人 cave dweller; troglodyte

【穴位】[中医] acupuncture point; acupoint

噱

【噱头】words to amuse; act meant to excite laughter

xuě

雪 ①[气] snow: 洁白如～ as white as snow／一场大～ a heavy fall of snow／路上积～很深。The roads are deep in snow.／下了一夜～。It snowed all right.／正在大～纷飞。It is snowing in great flakes. ②(洗掉耻辱、仇恨、冤枉等) wipe out (a humiliation); avenge (a wrong): ～耻 avenge an insult; wipe out a humiliation／昭～ right a wrong; clear sb. of an unjust or unfounded charge; rehabilitate

【雪白】snow-white; snowy white

【雪板】[体] ski

【雪豹】[动] snow leopard

【雪暴】snowstorm; blizzard

【雪崩】snowslide; avalanche

【雪堆】snow drift

【雪纺绸】[纺] chiffon

【雪恨】wreak vengeance; avenge: 报仇～ take revenge; avenge oneself

【雪花】snowflake

【雪花干扰】[电子] snow

【雪花膏】vanishing cream

【雪花石膏】alabaster

【雪鸡】[动] snow cock

【雪茄】cigar

【雪里红】[植] potherb mustard

【雪亮】bright as snow; shiny: 把汽车擦得～ put a good shine on the car／群众的眼睛是～的。The masses have sharp eyes.

【雪柳】[植] fontanesia

【雪盲】[医] snow blindness

【雪片】snowflake: 抗议书象～似的飞来。Written protests poured in.

【雪橇】sled; sledge; sleigh

【雪青】lilac (color)

【雪球】snowball: 那些小男孩在玩扔～。The little boys were snowballing.／要求增设新学校而签名的人象滚～似的越来越多。The number of signers of the petition for a new school snowballed.

【雪人】snowman ◇喜马拉雅～ yeti

【雪山】snow capped mountain

【雪上加霜】snow plus frost; one disaster after another

【雪蚀】nivation; snow patch erosion

【雪松】cedar

【雪兔】snow hare

【雪线】[地] snow line

【雪冤】clear sb. of a false charge; redress a wrong

【雪原】snowfield

【雪杖】[体] (ski) pole; (ski) stick

【雪中送炭】send charcoal in snowy weather; provide timely help

鳕 cod

xuè

谑 crack a joke; banter; tease: 戏～ banter; tease / ～而不虐 tease without embarrassing

血 ①(血液) blood: ～的教训 a lesson paid for with blood; a lesson written in blood / 抽～ draw blood / 出～ bleed / 换～ exchange blood transfusion / 流～ shed blood / 献～ blood donation / 验～ have one's blood tested / 止～ hemostasis; stop bleeding; stanch bleeding / 以～还～ demand blood for blood ②(有血统关系的) related by blood: ～亲 blood relation

【血案】murder case

【血本】principal; original capital

【血崩证】metrorrhagia

【血常规检查】blood routine examination

【血沉】[医] ESR; erythrocyte sedimentation rate ◇ ～试验 blood sedimentation test

【血钙】blood calcium

【血管】blood vessel: 人造～ artificial blood vessel ◇ ～瘤 hemangioma; angioma / ～造影 angiography

【血海】a sea of blood: ～深仇 a huge debt of blood; intense and deep-seated hatred; a blood feud

【血汗】blood and sweat; sweat and toil: ～钱 money earned by hard toil

【血红】blood red

【血红蛋白】[生化] hemoglobin; hemoglobin: ～10 克 have a hemoglobin of 10 grams

【血红蛋白尿】hemoglobinuria

【血迹】bloodstain: ～斑斑 bloodstained

【血浆】[生理] (blood) plasma

【血浆白蛋白】plasma albumin

【血口喷人】make unfounded and malicious attacks upon sb.; venomously slander

【血库】[医] blood bank

【血亏】[中医] anemia

【血淋淋】dripping with blood; bloody

【血流成河】blood flowed and has become a river

【血流如注】blood cascading down without stop; blood streaming down

【血尿】[医] blood in the urine; hematuria

【血泊】pool of blood

【血气】①(精力) animal spirits; sap; vigor: ～方刚 full of sap; full of vigor and vitality ②(血性) courage and uprightness: 有～的青年 a courageous and upright youth

【血亲】consanguinity ◇ ～关系 blood ties; relationship by consanguinity

【血清】[生理] (blood) serum ◇ ～病 serum sickness; serum disease

【血球】[生理] blood cell; blood corpuscle

【血肉】flesh and blood: ～之躯 the human body; flesh and blood

【血肉横飞】flesh and blood flying in all directions

【血肉模糊】badly mutilated

【血肉相连】as close as flesh and blood

【血色】redness of the skin; color: ～很好 have a high complexion / 脸上没有一点～ have no color in cheeks

【血色素】[生理] hemochrome

【血书】a letter written in one's own blood

【血栓】[医] thrombus ◇ ～形成 thrombosis

【血糖】[医] blood sugar

【血统】blood relationship; blood lineage; extraction; descent: ～工人 (industrial) worker of working-class parentage / 中国～的美国人 Americans of Chinese descent

【血吸虫】blood fluke ◇ ～病 snail fever; schistosomiasis

【血象】[医] blood picture; hemogram

【血小板】[生理] (blood) platelet

【血腥】reeking of blood; bloody; sanguinary: ～统治 sanguinary rule; bloodstained rule / ～味 smell of blood

【血型】[生理] blood group; blood type ◇ ～分类 typing of blood

【血胸】[医] hemothorax

【血循环】[生理] blood circulation

【血压】[生理] blood pressure: 低～ low blood pressure; hypotension / 高～ high blood pressure; hypertension / 量～ take one's blood pressure / 量不出～ cannot get a blood pressure reading / ～120～80。 The blood pressure is 120 over 80. ◇ ～计 sphygmomanometer

【血液】blood: 新鲜～ fresh blood ◇ ～鉴定 identification of blood

【血液病】blood disease

【血衣】bloodstained garment; clothes covered with gore

【血友病】hemophilia

【血雨腥风】a foul wind and a rain of blood; reactionary reign of terror

【血缘】ties of blood; consanguinity; blood relationship

【血债】a debt of blood: ～累累 have heavy blood debts／～要用血来还。Debts of blood must be paid in blood. (或: Blood must atone for blood.)

【血战】bloody battle; sanguinary battle: ～到底 fight to the last drop of one's blood; fight to the bitter end

【血肿】[医] hematoma

【血中毒】[医] blood poisoning

xūn

勋 merit; meritorious service; achievement: 功～ meritorious service; contribution

【勋绩】meritorious service; outstanding contribution

【勋爵】①(授予功臣的爵位) a feudal title of nobility conferred for meritorious service ②(英国贵族的名誉头衔) Lord

【勋劳】meritorious service: 卓著～ noted for meritorious service

【勋章】medal; decoration

熏 ①(用烟、气熏) smoke; fumigate: ～房间 fumigate a room／～蚊子 smoke out mosquitoes ②(熏制) treat with smoke; smoke: ～火腿 smoke ham／～肉 treat meat with smoke／～鱼 smoke fish

【熏染】exert a gradual, corrupting influence on: 受环境～ gradually influenced by environment

【熏肉】smoked meat

【熏陶】exert a gradual, uplifting influence on; nurture; edify: 起～作用 exert an edifying influence on／在…的～下成长 grow up under the nature of

【熏鱼】smoked fish

【熏蒸】①(闷热难受) stifling; suffocating: 暑气～ stifling summer heat ② [农] fumigate ◇ ～剂 [农] fumigant／～消毒 fumigation／～消毒器 fumigator

【熏制】smoke; fumigate sth. with jasmine; cure by smoke

醺 drunk: 微～ tipsy／醉～～的 dead drunk; tight

xún

驯 ①(顺服的;善良) tame and docile: ～象 a tame elephant ②(使顺服) tame; domesticate: ～马 break in a horse／～猛狮 tame a fierce lion

【驯服】①(顺从的) docile; tame; tractable: ～的马 a docile horse ②(使顺从) tame; break; domesticate: ～洪水 bring a flood under control／～野兽 domesticate a wild animal／马容易～。Horses are easy to tame.

【驯化】domestication; taming: ～动物 tame animals

【驯良】tractable; docile: tame and gentle

【驯鹿】[动] reindeer

【驯顺】tame and docile

【驯养】raise and train (animals); domesticate

循 follow; abide by: ～此前进 proceed along this line／～例 follow the usual practice; follow a precedent

【循规蹈矩】follow rules, orders, etc. docilely; conform to convention; toe the line

【循环】circulate; cycle: ～不息 move in endless cycles／恶性～ vicious circle／四季的～ the cycle of the seasons／血液～ blood circulation／体外～ extracorporal circulation ◇ ～论证[逻] argue in a circle／～赛[体] round robin／～系统[生理] the circulatory system／～小数[数] recurring decimal／～冷却 circulating cooling／～水 circulating water／～水泵 water circulating pump

【循环往复】move in circles: ～以至无穷 repeat itself in endless cycles

【循名责实】expect the reality to correspond to the name

【循序】in proper order or sequence: ～渐进 follow in order and advance step by step; proceed in an orderly way and step by step

【循循善诱】be good at giving systematic guidance; teach with skill and patience

旬 ①(十日) a period of ten days: 上～ the first ten days of a month／下～ the last ten days of a month／中～ the second ten days of a month ②(十岁) a period of ten years in a person's age

询 ask; enquire; inquire: 查～ make enquiries (about)

【询问】①(征求意见;打听) ask about; enquire: ～病状 enquire about sb.'s illness／～机器的效能 ask about the efficiency of the machine ②[法] examination ◇ ～表 questionnaire／～处 inquiry office

巡 ①(巡查;巡视) patrol; make one's rounds:

～夜 go on night patrol ② [量](遍) round of drink／酒过三～ The wine has gone round three time
【巡边员】[足球] linesman
【巡查】go on a tour of inspection; make one's rounds
【巡航】cruise ◇ ～半径 [军] cruising radius／～导弹 cruise missile／～速度 cruising speed
【巡回】go the rounds; tour; make a circuit of: ～演出 making a performance tour; be on tour ◇ ～报告 report tour／～大使 roving ambassador／～法庭 circuit court／～放映队 mobile film projection unit／～剧团 itinerant theatrical troupe／～医疗队 mobile medical team
【巡礼】①(朝拜圣地) visit a sacred land; go on a pilgrimage; make a pilgrimage to a holy place ②(观光;游览) tour; sight-seeing
【巡逻】go on patrol; patrol: 执行～任务 be on patrol duty; be on one's beat ◇ ～队 patrol party; patrol／～护卫舰 patrol escort／～艇 guard boat; patrol boat／～线 patrol route／～车 cruiser
【巡视】make 〈be on〉 an inspection tour; tour: ～各地 make an inspection tour of various places
【巡洋舰】cruiser
【巡弋】(of warships) cruise

寻

look of; search; seek: ～欢作乐 seek pleasure／～物 look for sth. lost
【寻常】ordinary; usual; common: 不～ out of the ordinary; unusual／异乎～的冷 extraordinarily cold
【寻的】[军] target-seeking; homing ◇ ～导弹 homing missile
【寻访】look for; try to locate; make inquiries about
【寻根究底】get to the bottom of things; inquire deeply into
【寻花问柳】go round singsong houses
【寻觅】seek; look for
【寻求】seek; explore; go in quest of: ～打开僵局的途径 explore possible paths for ending the stalemate／～解决问题的办法 explore ways of solving the problem／～真理 seek truth
【寻事生非】seek a quarrel; make trouble
【寻味】chew sth. over; ruminate; think over: 她这话耐人～ what she has said affords much food for thought
【寻衅】pick a quarrel; provoke: ～滋事 pick a quarrel and make trouble
【寻章摘句】seek; look for
【寻找】seek; look for: ～失物 look for sth. lost／～新的水源 search for new water sources

鲟

[动] sturgeon

xùn

训

①(教导;训诫) lecture; teach; train: ～他一顿 give him a lecture; give him a dressing down／受～ undergo training ②(准则) standard; model; example: 不足为～ not fit to serve as a model
【训斥】reprimand; rebuke; dress down: ～某人 read sb. a lecture
【训词】admonition; instructions
【训诂】explanations of words in ancient books: gloss ◇ ～学 critical interpretation of ancient texts
【训诫】①(教导和告诫) admonish; advise ②(对犯罪者进行公开的批评教育) rebuke; reprimand; be reprimanded
【训练】train; drill: ～部队 train troops／～有素的运动员 a well-trained athlete／实战～ exercises under battle conditions／适应性～ acclimatization training／专业～ professional training
【训练班】training class; training course: 短期～ short course／微机～ training course in personal computer

蕈

[植] gill fungus

汛

flood; high water: 伏～ summer flood; fresher flood prevention／秋～ autumn floods／防～ flood control
【汛期】flood season; high-water season

讯

①(讯问) interrogate; question: 审～犯人 interrogate a prisoner ②(消息;信息) message; dispatch: 电～ a telegraphic report; dispatch／据新华社～ according to a Xinhua dispatch／闻～ on hearing the news
【讯问】①(审问) interrogate; question: ～被告人 interrogate the defendant ②(问) ask about; enquire: ～病情 inquire about sb.'s illness／～道路 ask the way

迅

fast; swift
【迅即】immediately; at once: ～出发 march off on the instant／～处理 take immediate action on／～启程 start immediately
【迅疾】swift; rapid

【迅捷】fast; agile; quick
【迅雷不及掩耳】a sudden peal of thunder leaves no time for covering the ears; as sudden as lightning
【迅猛】swift and violent
【迅速】rapid; swift; speedy; prompt: ～取得成效 produce speedy results / ～作出决定 come to a prompt decision / 动作～ swift in action; quick-moving / 发展～ be expanding by leaps and bounds / 反应～ have a swift response

殉 ①(殉葬) be buried alive with the deal ②(牺牲) sacrifice one's life for
【殉国】die for one's motherland; give one's life for one's country
【殉难】die for (a just cause); die a martyr
【殉葬】be buried alive with the dead: ～的奴隶 slaves buried alive with their deceased masters ◇ ～品 funerary object; sacrificial object / ～制度 institution of burying the living with the dead
【殉职】die at one's post; die in the course of performing one's duty; die in line of duty

徇
【徇情】act wrongly out of personal considerations; practice favoritism; ～枉法 bend the law for the benefit of relatives or friends

逊 ①(让位) abdicate ②(谦虚;谦恭) modest: 出言不～ speak insolently / 谦～ modest ③(比不上) inferior: 稍～一筹 be slightly inferior
【逊色】be inferior: 毫无～ be by no means inferior
【逊位】abdicate

Y

yā

丫 bifurcation; fork

【丫杈】①(树枝分开的地方) fork; crotch ②(形容树枝歧出) ramified; crotched; forked

【丫头】①(女孩) girl ②(丫鬟) slave girl

桠 fork

【桠杈】①(树枝分出的地方) fork; crotch ②(树枝歧出) crotched; forked

压 ①(施压力) press; push down; hold down; weigh down: 用石块～住纸 press the paper with a stone／这箱子怕～。This chest won't stand much weight. ②(使稳定) keep under control; control; keep under; quell: 强～怒火 try hard to control one's anger ③(压制) bring pressure to bear on; suppress; daunt; intimidate: 以势～人 overwhelm people with one's power ④(逼近) approach; be getting near ⑤(积压) pigeonhole; shelve: 这个方案被～下来了。The scheme has been pigeonholed. ⑥(下赌注) risk on sth.: stake

【压扁】press flat; flatten

【压仓物】[航海] ballast

【压秤】be relatively heavy per unit volume

【压倒】overwhelm; overpower; prevail over: ～多数 an overwhelming majority／～优势 overwhelming superiority

【压低】bring down; lower; reduce; abate: ～价格 bring down the price／～声音 lower one's voice

【压电】[物] piezoelectricity ◇～晶体 piezocrystal; piezoelectric crystal／～器件 piezoelectric devices／～拾音器 piezoelectric pickup／～陶瓷 piezoelectric ceramic／～陶瓷拾音器 ceramic pick-up

【压锻】[冶] press forging

【压队】bring up the rear

【压服】force sb. to submit; bring sb. to his knees

【压盖】[机] gland ◇～填料 gland packing

【压花玻璃】pattern glass

【压坏】crush; squash

【压挤】[机] extrusion ◇～成形 extrusion molding

【压价】force prices down; demand a lower price

【压紧】compress tightly

【压惊】help sb. get over a shock

【压境】press on to the border: 大军～。A large enemy force is bearing down upon the border.

【压垮】collapse under pressure

【压力】①[物] pressure: 大气～ atmospheric pressure／外界～ ambient pressure ②(制伏人的力量) overwhelming force; pressure: 对某人施加～ bring pressure to bear on sb.／在舆论的～下 under the pressure of public opinion ◇～锅 pressure cooker／～机 press／～计 pressure gauge; manometer

【压路机】road roller; roller

【压平】flatten; even: 把路～ even a road; roll a road surface／把弄皱的纸～ flatten crumpled paper

【压迫】①(强制别人) oppress; repress: 哪里有～哪里就有反抗。Where there is oppression there is resistance. ②(对机体某部施加压力) constrict: 她胸部有～感。She feels a constriction in the chest. ◇～阶级 oppressor class／～者 oppressor

【压气】calm sb.'s anger

【压气机】compressor: compressor machine

【压强】[物] intensity of pressure; pressure ◇～计 pressure gauge

【压青】[农] green manuring

【压舌板】[医] depressor

【压水反应堆】pressurized water reactor

【压岁钱】money given to children as a lunar New Year gift

【压缩】compress; condense; reduce; cut down: ～非生产性开支 cut down non-productive expenses ◇～比 compression ratio／～饼干 ship biscuit; pilot bread; hardtack／～冲程 compression stroke／～点火 compression ignition／～空气 compressed air／～空气瓶 compressed air bottle

【压缩机】compressor: 空气～ air compressor

【压条】[农] layering: ～区 layering plot

【压痛】[医] tenderness

【压弯】bend: 重担～了扁担。The shoulder pole bent with heavy load.

【压线】[体] line ball

【压延】roll: ～能力 rolling capacity ◇～机 calender; mangler

【压抑】①(情绪、感情低落) constrain; inhibit; depress; hold back: 精神～ feel much depressed／内～ internal inhibition／外～ external inhibition ②(憋闷) oppressive; stifling:

胸口感到～ feel tight in the chest
【压韵】rhyme: ～诗 rhymed verse
【压载舱】[航海] ballast tank
【压榨】①(压取汁液) press; squeeze: ～葡萄汁 press the juice from grapes ②(剥削或搜刮) oppress and exploit; squeeze; bleed ◇ ～机 squeezer; mangle
【压纸型机】[印] stereotype press
【压制】①(强力限制) suppress; stifle; inhibit: ～合理化的建议 smother reasonable suggestions / ～群众的意见 stifle the opinions of the masses / ～新生力量 hold back new rising forces ②[机] (用压的方法制造) pressing ◇ ～板 pressboard / ～射击[军] neutralizing fire
【压轴戏】the last item but one on a theatrical program
【压铸】[冶] die-casting

呀 ①[叹](表示惊异) oh; ah ②[象] creak: 门～的一声开了。The door opened with a creak. (或: The door creaked open.)

鸦 crow
【鸦飞雀乱】utter disorder
【鸦鸣鹊噪】full of confused voices
【鸦片】opium: 戒～ give up opium-smoking / 吸～ smoke opium ◇ ～战争 the Opium War (1840—1842)
【鸦雀】[动] crow tit
【鸦雀无声】not even a crow or sparrow can be heard; silence reigns

押 ①(抵押) give as security; mortgage; pawn; pledge: 以戒指为～ leave a ring as security ②(扣留) detain; take into custody: 在～犯 criminal in custody ③(跟随照料) escort: ～送 escort; send under escort ④(签字) signature; mark in lieu of signature: 画～ sign; mark (a document) in lieu of signature
【押当】pawn sth.
【押解】send away under escort
【押金】cash pledge; deposit; security
【押款】[商]①(抵押借款) borrow money on security ②(用抵押方式借的款子) a loan on security
【押运】escort (goods) in transportation: ～货物 transport goods under the escort of sb.
【押租】rent deposit

鸭 duck: 公～ drake / 母～ duck / 绒～ eider duck / 填～ force-fed duck / 小～ duckling
【鸭蛋】duck's egg ◇ ～青 pale blue / ～形 oval / ～圆 oval
【鸭梨】a kind of pear grown in Hebei Province
【鸭绒】duck's down; eider down: ～被 eider down quilt; duck's down quilt / ～背心 duck's down waistcoat
【鸭舌帽】peaked cap
【鸭肫】duck's gizzard
【鸭子】[口] duck: ～的叫声 quack
【鸭嘴笔】drawing pen; ruling pen
【鸭嘴兽】platypus; duckbill; duckmole

yá

涯 margin; limit: 天～海角 the remotest corners of the earth

崖 precipice; cliff
【崖壁】precipice
【崖谷】valley between precipices
【崖盐】rock salt

牙 ①(牙齿) tooth: 假～ an artificial tooth / 乳～ temporary tooth / 拔～ extract a tooth; have a tooth pulled out / 补～ have a tooth filled; tooth-filling / 镶～ fix a false tooth; have a denture made / 小孩长～ cut a tooth ②(象牙) ivory: ～筷 ivory chopsticks ③(形状象牙齿的东西) tooth-like thing: 轮～ cog
【牙槽炎】alveolitis; dentoalveolitis
【牙碜】①(食物中夹砂子,嚼时不舒服) gritty ②(言语粗鄙不堪入耳) jarring; coarse
【牙齿】tooth
【牙床】①(齿龈) gum ②(有牙雕装饰的床) ivory-inlaid bed
【牙雕】ivory carving
【牙粉】tooth powder
【牙缝】slit between the teeth; crevice between teeth
【牙疳】[医] noma; cancrum oris
【牙膏】toothpaste
【牙垢】tartar; dental calculus
【牙关】mandibular joint: 咬紧～ clench one's teeth ◇ ～紧闭[医] lockjaw
【牙科】[医] dentistry ◇ ～医生 dentist; dental surgeon / ～诊疗所 dental clinic / ～综合治疗台 dental unit; universal dental engine
【牙口】①(指牲畜的年龄) the age of a draught animal ②(老年人的咀嚼能力) the condition of an old person's teeth
【牙轮】gear wheel; gear
【牙买加】Jamaica ◇ ～人 Jamaican

【牙签】toothpick ◇～筒 toothpick-holder
【牙石】dental calculus; dental deposit
【牙刷】toothbrush
【牙髓】[生理] dental pulp ◇～炎 pulpitis
【牙痛】toothache
【牙牙学语】learn to speak; babble out one's first speech sounds
【牙医】dentist
【牙龈】[生理] gum ◇～炎 gingivitis
【牙釉质】enamel; substantia adamantina
【牙周病】[医] periodontosis
【牙周炎】[医] periodontitis

芽 bud; sprout; shoot: 韭菜～ leek sprout
【芽孢】[生] gemma
【芽豆】sprouted broad bean
【芽接】[植] bud grafting; budding: 把蔷薇～在荆棘上 bud a rose on a brier
【芽眼】[植] eye

蚜
【蚜虫】aphid; aphis: 棉～ cotton aphid

yǎ

哑 ①(不能说话) mute; dumb ②(嘶哑) hoarse; husky: 喊得嗓子都～了 shout oneself hoarse
【哑巴】a dumb person; mute: 吃～亏 swallow a bitter pill in silence; have to keep one's grievances to oneself / ～吃黄连，有苦说不出 be unable to express one's discomfort, like a dumb person tasting bitter herbs; be compelled to suffer in silence
【哑剧】dumb show; pantomime ◇～演员 mummer
【哑口无言】be left without an argument; be rendered speechless; be dumb as a fish
【哑铃】[体] dumbbell
【哑谜】puzzling remark; enigma; riddle: 解～ solve a riddle
【哑然失笑】unable to stifle a laugh; can't help laughing
【哑然无声】silence reigns
【哑嗓子】hoarse voice; husky voice
【哑吒嘈杂】the noisy sound of a crowd

雅 ①(高雅) refined; elegant; graceful ②[敬] ～教 your esteemed opinion / ～意 your esteemed favor; your kindness ③(合乎规范的) standard; correct; proper
【雅观】refined; in good taste: 很不～ most unseemly; rather unsightly
【雅号】your name
【雅量】①(宽宏的气度) magnanimity; generosity ②(大的酒量) great capacity for liquor
【雅趣】refined tastes
【雅士】a refined scholar; a person of refined tastes
【雅俗共赏】appeal to both the more and the less cultured; suit both refined and popular tastes
【雅兴】aesthetic mood: ～不浅 be really in an aesthetic mood; have a really keen interest in sth.
【雅正】would you kindly point out my inadequacies
【雅致】refined; tasteful: 陈设～ tastefully furnished
【雅座】private room

yà

亚 ①(较差) inferior; second: 不～于人 not inferior to anyone ②(次于) sub-: ～科 subfamily / ～目 suborder / ～属 subgenus / ～正常温度 subnormal temperature / ～种 subspecies
【亚砜】[化] sulphoxide
【亚急性】[医] subacute ◇～病 subacute disease
【亚军】second place; runner-up: 获得～ get a second; win second place
【亚硫酸】[化] sulphurous acid
【亚麻】[植] flax ◇～布 linen (cloth) / ～精纺机 [纺] flax spinning frame / ～子 linseed / ～子油 linseed oil
【亚热带】subtropical zone; subtropics; semitropics ◇～园艺 subtropical gardening
【亚音速】[物] subsonic speed ◇～飞机 subsonic aircraft
【亚运会】Asian Games
【亚洲】Asia ◇～人 Asian

氩 [化] argon (Ar)

压
【压根儿】[口] from the start; in the first place; altogether: 她～就没来。She hasn't been here from the start.

揠 [书] pull up; tug upward
【揠苗助长】try to help the shoots grow by pulling them upward; spoil things by excessive enthusiasm

轧 ①(滚压) roll; run over: ～路 roll a road surface / ～碎 roll asunder; crush to pieces ②[象](机器声)铰链～～作响。The hinges squeaked.
【轧板机】[冶] mangle
【轧光】[纺] calendering ◇ ～机 calender
【轧花】[纺] cotton ginning ◇ ～厂 cotton ginning mill / ～机 cotton gin
【轧涩难言】difficult to speak with halting articulation; stammer
【轧伤】run over and injure
【轧死】run over and kill

砑 press and smooth; calender
【砑光】calendering
【砑光机】①[印] calender ②[纺] mangle

yān

湮
【湮灭】bury in oblivion; annihilate
【湮没】①(埋没) fall into oblivion; be neglected; be forgotten: ～无闻 sink into oblivion; drift into obscurity ②[物] annihilation: ～光子 annihilation photon

嫣
【嫣红】bright red: 姹紫～ beautiful and luxuriant
【嫣然】[书] beautiful; sweet: ～一笑 give a winsome smile

淹 ①(淹没) flood; submerge; inundate ②(皮肤被汗等浸的难受) be tingling from sweat
【淹博学问】deep and wide in learning; have a wide knowledge
【淹灌】[农] basin irrigation
【淹久异邦】be long delayed in a foreign country
【淹没】submerge; flood; inundate; drown: 被～的地区 inundated area / 洪水～了整个地区。The flood inundated the whole district. / 雨水～了庄稼。Heavy rainfalls flooded the crops.
【淹死】be drowned: 他跳进河里，去救将要～的人。He jumped into the river to save the drowning man.
【淹雅华美】cultured and beautiful
【淹雅之士】a deeply cultured and refined scholar

阉 castrate or spay ◇ ～牛 bullock / ～羊 wether / ～猪 hog
【阉割】①(割去睾丸或卵巢) castrate; spay ②(抽去主要内容) emasculate; deprive a theory

腌 preserve in salt; salt; pickle; cure: ～菜 pickled vegetables; pickles / ～肉 salted meat; bacon / ～鱼 salted fish

烟 ①(物质燃烧时产生的气体) smoke: 一团～ a cloud of smoke ②(象烟的东西) mist; vapor: ～霞 mist and clouds in the twilight ③(烟刺激眼睛) be irritated by smoke ④(烟草；烟卷) tobacco; cigarette: 鼻～ snuff / 雪茄～ cigar / 一包～ a pack of cigarette / 一听～ a tin of cigarette / 点～ light up a cigarette / 戒～ give up smoking; swear off tobacco / 吸～ smoke / 吸一支～ have a smoke / 你吸～吗? Do you smoke? ⑤(鸦片) opium: ～枪 opium pipe
【烟波】mist-covered waters: ～浩渺 mists and ripples of a river 〈lake〉
【烟草】tobacco
【烟囱】chimney; funnel; stovepipe
【烟袋】small-bowled long-stemmed pipe ◇ ～锅 the bowl of a pipe
【烟道】[建] flue ◇ ～尘 flue dust / ～气 flue gas
【烟斗】pipe: 吸～ smoke a pipe ◇ ～架 pipe rack / ～丝 pipe tobacco
【烟鬼】①(烟瘾大的人) heavy smoker ②(有鸦片烟瘾的人) opium addict
【烟海】a vast sea of fog; huge and voluminous: 浩如～ a tremendous amount; voluminous / 如堕～ be lost in a fog
【烟盒】cigarette case
【烟花柳巷】brothels
【烟灰】tobacco ash; cigarette ash ◇ ～缸 ashtray
【烟火】①(烟和火) smoke and fire: ～极盛 densely populated / 严禁～! Smoking or lighting fires strictly forbidden. ②(熟食) cooked food ③(焰火) fireworks ◇ ～探测器 smoke detector
【烟碱】[化] nicotine ◇ ～中毒 nicotinism
【烟具】smoking paraphernalia; smoking set
【烟煤】bituminous coal; soft coal
【烟幕】smoke screen: 放～ put up a smoke screen ◇ ～弹 smoke shell; smoke bomb
【烟丝】cut tobacco; pipe tobacco
【烟酸】[化] nicotinic acid; niacin ◇ ～缺乏症 [医] pellagra
【烟筒】chimney; funnel; stovepipe
【烟头】cigarette end

【烟土】crude opium
【烟雾】smoke; mist; vapor; smog: ～弥漫 full of smoke
【烟消云散】vanish like mist and smoke; completely vanish
【烟熏】smoke: ～蚊子 smoke mosquitoes
【烟叶】tobacco leaf; leaf tobacco: ～加工 stemming and cutting
【烟瘾】a craving for tobacco: ～大 smoke like a chimney; be a heavy smoker
【烟油】tobacco tar
【烟雨】misty rain
【烟柱】column of smoke
【烟嘴儿】cigarette holder

咽 [生理] pharynx: ～镜 pharyngoscope
【咽喉】①[生理] pharynx and larynx; throat: ～痛 have a sore throat ②(交通要道) key point; strategic passage: ～之地 a location of the throat / 扼守～要地 hold a key position
【咽头】[生理] pharynx
【咽峡炎】angina
【咽炎】pharyngitis

胭
【胭脂】①(一种红色的化妆品) rouge: 在脸上擦～ rouge one's cheeks ②(一种无脊椎动物) kermes ◇ ～红 carmine

殷
【殷红】blackish red; dark red

yán

颜 ①(脸) face; countenance ②(体面) face; prestige: 无～见人 too ashamed to face anyone ③(颜色) color
【颜料】pigment; color; dyestuff
【颜面】①(脸部) face: ～神经 facial nerve ②(体面) prestige; face: 顾全～ save face
【颜色】①(色彩) color: ～鲜艳 bright in color / 五颜六色 of various colors; multicolored ②(脸色) countenance; facial expression: 给他一点～看看 teach him a lesson / 看某人的～行事 take hint from sb.'s facial ③(颜料或染料) pigment; dyestuff ◇ ～失真[电子] cross-color

言 ①(话) speech; word: 无～以对 have nothing to say in reply / 一～不发 not utter a word ②(说) say; talk; speak: 换～之 in another word / 简而～之 in a word / 自～自语 talk to oneself ③(一个汉字) character; word: 全书近十万～。It is a book of nearly 100000 words.
【言必信,行必果】promises must be kept and action must be resolute; always be true in word and resolute in deed
【言必有中】whenever one speaks, one speaks to the point
【言不及义】never talk about anything serious; talk frivolously
【言不由衷】speak insincerely; speak with one's tongue in one's cheek
【言差语错】erroneous utterances
【言出必行】suit the action to the word
【言传】explain in words: 只可意会,不可～ only to be sensed, not explained
【言传身教】teach by personal example as well as verbal instruction
【言辞】one's words; what one says: ～不雅 use slang or low-class words / ～锋利 speak daggers; use sharp words / ～恳切 be sincere in what one says
【言而无信】fail to keep faith; go back on one's word
【言而有信】be true to one's word
【言梗喉间】words stuck in the throat
【言归于好】make it up with sb.; become reconciled
【言归正传】to come back to our story; to return to the subject
【言贵简洁】brevity is the soul of wit
【言过其实】exaggerate; overstate
【言和】make peace; become reconciled; bury the hatchet: 握手～ shake hands and make it up
【言简意赅】concise and comprehensive; compendious
【言教】teach by word of mouth; give verbal directions: ～不如身教。Example is better than precept.
【言近旨远】simple words but deep meaning
【言路】channels through which criticisms and suggestions may be communicated to thc leadership: 堵塞～ stifle criticisms and suggestions / 广开～ provide wide opportunities for airing views
【言论】opinion on public affairs; expression of one's political views; speech ◇ ～自由 freedom of speech
【言人人殊】different people give different views; each person offers a different version
【言谈】the way one speaks or what he says: ～举止 speech and deportment

【言听计从】always follow sb.'s advice; act upon whatever sb. says; have implicit faith in sb.
【言外之意】implication; what is actually meant
【言为心声】words are the voice of the mind; what the heart thinks the tongue speaks
【言行】words and deeds; statements and actions: ～不一 the deeds do not match the words / ～失检 let oneself loose / ～相悖 practice against what one preaches / ～一致 be as good as one's word
【言犹在耳】the words are still ringing in one's ears
【言有尽而意无穷】there's an end to the words, but not to their message
【言语】①(说的话) spoken language; speech: ～粗俗 be coarse and vulgar in speech / ～尖利 bitterness of speech / ～之争 a verbal warfare / ～支吾 prevaricated in speech ②[方](开口; 回答) speak; talk; answer: 他这个人不爱～。He is a man of few words.
【言者无罪,闻者足戒】blame not the speaker but be warned by his words
【言之成理】sound reasonable; speak in a rational and convincing way
【言之过早】premature to say
【言之无物】be devoid of substance; be just empty verbiage
【言之有据】speak on good grounds
【言之有物】having substance in a speech

阎

【阎罗】[宗] Yama
【阎王】①(阎罗) Yama; king of Hell: 见～ die ②(极凶恶的人) an extremely cruel and violent person ◇ ～殿 the Palace of Hell
【阎王帐】usurious loan; shark's loan

炎

①(极热) scorching; burning hot; blazing: 赤日～～ the blazing sun ②(炎症) inflammation: 嗓子发～ suffer from an inflammation of the throat
【炎黄子孙】all the children of the Yellow Emperor; the Chinese people
【炎凉】①(热和冷) hot and cold ②(势利) snobbishness: ～世态 the aspect of worldly affairs; now hot and now cold
【炎热】scorching; blazing; burning hot: 冒着～ braving the sweltering heat
【炎暑】hot summer; sweltering summer days; dog days
【炎威】oppressively imposing; oppressiveness
【炎症】inflammation
【炎症细胞】inflammatory cell

研

①(细磨) grind; pestle: ～成粉末 grind into fine powder ②(研究) study
【研钵】mortar
【研杵】pestle; pounder
【研究】①(探求事物的规律) study; research: ～生物学 study biology / ～物种起源 research into the origin of species / 详细～当地的风俗习惯 make a detailed study of local customs ②(商讨) consider; discuss; deliberate ◇ ～费 research fund / ～工作者 research worker / ～生 postgraduate (student); graduate student / ～所 research institute / ～员 research fellow / ～院 research institute; graduate school
【研磨】①(研成粉末) grind; pestle ②(磨光) abrade; polish ◇ ～粉 abrasive powder / ～料 abrasive
【研讨】deliberate; discuss ◇ ～会 a workshop; a seminar; a symposium
【研习】research and study
【研讯死因】make an enquiry of the cause of death
【研制】prepare; manufacture; develop: ～新产品 research and produce new varieties of products / ～新式武器 develop new weapons

盐

salt: 精～ refined salt / 食～ table salt
【盐巴】[方] salt; common salt
【盐层】salt deposit; salt bed
【盐场】saltern; saltworks
【盐池】salt pond
【盐肤木】[植] Chinese sumac
【盐罐】saltcellar
【盐湖】salt lake
【盐碱化】salinization
【盐碱土】saline-alkali soil
【盐井】salt well; brine pit: ～区域 salt-well regions
【盐矿】salt mine
【盐卤】bittern
【盐汽水】salt soda water
【盐泉】brine spring; salt spring
【盐霜】salt efflorescence
【盐水】salt solution; brine ◇ ～灌肠 saline enema / ～选种 seed sorting by salt water / ～针 saline infusion needle
【盐酸】[化] hydrochloric acid ◇ ～普鲁卡因 procaine hydrochloride
【盐田】salt pan; salina

【盐土】[农] solonchak; saline soil
【盐渍土】[农] salinized soil
【盐沼】salt marsh

严 ①(紧密) tight; close: 嘴～ tight-lipped; tight-mouthed / 门关得～～的。The door was shut tight. ②(严格) strict; severe; stern; rigorous: ～以律己，宽以待人 be strict with oneself and lenient towards others ③(指父亲) father: 家～ my father
【严办】deal with severely; punish with severity
【严惩】punish severely: ～不贷 punish severely without mercy; punish mercilessly
【严词】in strong terms; in stern words: ～拒绝 give a stern rebuff; sternly refuse / ～谴责 denounce in strong terms; sternly condemn
【严冬】severe winter
【严而不苟】strict but not harsh
【严防】be strictly on guard against; take strict precautions against
【严格】strict; rigorous; rigid; stringent: ～的规则 strict rules / ～规章制度 rigorously enforce rules and regulations / ～说来 strictly speaking; in the strict sense / ～要求自己 make strict demands on oneself; be strict with oneself ◇～限制 close confinement
【严寒】severe cold; bitter cold
【严加管束】control rigorously
【严加责备】haul a person over the coals
【严谨】①(严密谨慎) strict; rigorous: 说话～ be cautious of speaking; be exact in one's words ②(紧密) compact; well-knit: 文章结构～。The essay is well-knit.
【严禁】strictly forbid: ～吸烟！Smoking is strictly prohibited.
【严紧】tight; close: 防守～ guard carefully
【严峻】stern; severe; rigorous; grim: ～的斗争 a grim struggle / ～的考验 a severe test; a rigorous test / ～的事实 harsh facts
【严酷】①(严厉) harsh; bitter; grim: ～的战争时代 grim years of war ②(冷酷) cruel; ruthless: ～的剥削 cruel exploitation
【严厉】stern; severe: ～的惩罚 severe punishment / ～的处罚 stiff penalty / ～的批评 severe criticism / ～的谴责 stern rebuke / 采取～措施 take drastic measures / 他对孩子太～。He was too severe upon his child.
【严密】tight; close: ～防范 take strict precautions against / ～计划 thorough planning / ～监视 put under close surveillance; keep close watch over
【严明】strict and impartial: 纪律～ observe strict discipline; be highly disciplined
【严判】severe judgment
【严实】①(紧密) tight; close ②(藏得好) hide safely
【严师畏友】a severe teacher and a fearful friend
【严守】strictly observe; maintain strictly: ～秘密 keep secret / ～中立 strictly observe neutrality
【严霜】heavy frost
【严丝合缝】fit together perfectly; join tightly
【严肃】①(令人敬畏) serious; solemn; earnest: ～的纪律 stern discipline / ～的声调 severe tone of voice ②(认真) serious; earnest; grave: ～对待 take a grave view of the matter
【严刑】cruel torture: ～拷打 cruelly torture; cruelly beat up
【严刑峻法】severe law; draconian law
【严以律己,宽以待人】be strict with oneself and lenient towards others
【严阵以待】be ready in full battle array; stand in combat readiness
【严正】solemn and just; serious and principled; stern: ～立场 solemn and just stand / ～声明 solemn statement / ～指出… point out in all seriousness that
【严重】serious; grave; critical: ～程度 order of severity / ～错误 gross error / ～关头 critical juncture / ～过失 aggravated negligence / ～后果 serious consequences / ～警告 serious warning / ～人体伤害 grievous bodily harm / ～骚乱罪 felonious rioting / ～伤害 grievous injury / ～伤亡事故 casualty / ～失职 gross neglect of duty / ～损害 grievous injury / ～违法行为 outrage / ～猥亵罪 gross indecency / ～罪行 grave crime / 病情～ be seriously ill / 他的病很～。His illness was a severe one.

芫
【芫荽】[植] coriander

檐 ①(屋檐) eaves: 廊～ the eaves of a corridor ②(形状象屋檐的部分) ledge; brim: 帽～儿 the visor of a cap; the brim of a hat
【檐板】eaves board
【檐沟】[建] eaves gutter
【檐瓦】verge tile
【檐子】eaves

岩 ①(岩石) rock ②(岩峰) cliff; crag

【岩层】rock stratum; rock formation
【岩洞】grotto
【岩浆】[地] magma ◇ ~分异作用 magmatic differentiation / ~岩 magmatic rock / ~作用 magmatism
【岩居穴处】dwell in mountain caves
【岩羚】[动] chamois
【岩溶】[地] karst ◇ ~地形 karst topography
【岩石】rock ◇ ~力学 rock mechanics / ~圈 lithosphere / ~突出[矿] rock burst / ~学 petrology
【岩心】[地] (drill) core ◇ ~回收率 core recovery / ~筒 core barrel / ~样品 core sample
【岩穴】cave; cavern
【岩盐】rock salt; halite
【岩羊】blue sheep; bharal
【岩样】①[地](岩石标本) rock specimen ②[矿](矿石样品) core sample

延 ①(延长) prolong; extend; protract: 蔓~ spread ②(向后推迟) postpone; delay ③(聘请) engage; send for: ~师 engage a teacher
【延长】lengthen; prolong; extend: ~辩论时间 protract an argument / ~铁路线 extend a railway / 会议~了两天。The conference was prolonged for another two days. ◇ ~号[乐] pause / ~线 extension line
【延宕】procrastinate; delay; keep putting off
【延发】[军] delayed action ◇ ~引信 delay fuse
【延搁】procrastinate; delay
【延缓】delay; postpone; put off: ~工作进度 retard the progress of work / ~行期 delay the journey / ~作出决定 defer making a decision
【延及他人】have effect on others
【延年益寿】prolong life; promise longevity
【延期】postpone; defer; put off: ~付款 defer payment / ~释放 release after the sentence expires ◇ ~偿付 moratorium / ~交货 back order / ~装船 delay shipment
【延伸】extend; stretch; elongate ◇ ~火力[军] creeping fire / ~率[冶] percentage elongation
【延绳钓】[渔] longline fishing; long-lining
【延时器】delayed action; selftimer
【延时摄影】time-lapse photography
【延髓】[生理] medulla oblongata
【延误】incur loss through delay: ~时机 miss an opportunity because of a delay / ~时日 lose time
【延祥纳福】induce good luck
【延性】[物] ductility
【延续】continue; go on; last: 地震~了三天。The earthquake lasted for three days. ◇ ~性 continuity
【延音】[乐] tenuto

筵 feast; banquet: 喜~ a wedding feast
【筵席】①(指酒宴时的座位) seats arranged at a banquet ②(酒席) feast; banquet

沿 ①(顺着) along: ~门乞讨 go begging from door to door / ~着河边走 walk along the river ②(依照) follow: 世代相~ be handed down from generation to generation ③(镶边) trim ④(东西的边) edge; border: 床~ the edge of a bed
【沿岸】along the bank or coast; littoral or riparian: ~而行 go along the shore
【沿边儿】trim; braid: 用花边给衣服~ trim a dress with lace
【沿革】the course of change and development; evolution: 社会风俗的~ the evolution of social customs
【沿海】along the coast; coastal; littoral: ~一带 the region following the line of the sea ◇ ~城市 coastal city / ~岛屿 offshore islands / ~地区 coastal areas; coastland / ~国家 coastal state; littoral state / ~航船 coaster / ~航行 coastal navigation; cabotage / ~航运权 cabotage right / ~贸易 coasting trade; cabotage / ~渔业 inshore fishing / ~自然资源 the natural resources of coastal waters
【沿河】along the river: ~而上 follow the river up
【沿阶草】[植] dwarf lilyturf
【沿街叫卖】peddle something in the streets
【沿路】①(顺着路边上) along the road; on the roadside ②(一路上) on the way
【沿途】on the way; throughout a journey: ~见到许多趣事 see a lot of interesting things on the way / ~平靖 peace along the road ◇ ~贸易 way-port trade
【沿袭】carry on as before; follow: ~陈规 follow convention / ~旧习 follow the old routine
【沿线】along the line: 铁路~的村镇 villages and towns along the railway line
【沿用】continue to use: ~旧的规章制度 adopt old rules and regulations

yǎn

演 ①(演化) develop; evolve ②(发挥) deduce; elaborate: 推~ deduce ③(依照程式练习

或计算) drill; practice ④(表演;上演) perform; play; act: ~电影 show a film / ~木偶戏 put on a puppet show / ~杂技 perform acrobatics
【演变】develop; evolve: 社会的~ social evolution / 追溯一个词在词义上的~ trace the sense development of a word
【演唱】sing (in a performance)
【演出】perform; show; put on a show: 告别~ farewell performance / 首次~ first performance; first show ◇~单位 producer / ~节目 items on the program; program
【演出队】troupe: 巡回~ mobile troupe
【演化】evolution
【演技】acting
【演讲】give a lecture; make a speech; lecture
【演练】drill: 地面~[航空] ground drill
【演示】demonstrate
【演说】①(说明事理) deliver a speech; make an address ②(发表的见解) speech
【演算】perform mathematical calculations
【演习】maneuver; exercise; drill; practice: 海上~ exercises at sea / 军事~ military maneuver; war exercise / 实弹~ live ammunition maneuvers / 消防~ fire-fighting exercises / 野外~ field exercises
【演戏】①(演出戏剧) put on a play; act in a play ②(装假) playact; pretend; act: 别~了。Stop that playacting.
【演义】historical novel; historical romance
【演绎】[逻] deduction ◇~法 the deductive method
【演员】actor or actress; performer: 芭蕾舞~ ballet-dancer / 舞蹈~ dancer / 杂技~ acrobat ◇~表 cast
【演奏】give an instrumental performance; play a musical instrument: ~琵琶 play the pipa ◇~能手 virtuoso

偃

【偃旗息鼓】lower the banners and muffle the drums; cease all activities
【偃武修文】desist from armament and promote culture and education

奄

【奄忽】suddenly; quickly
【奄奄】feeble breathing: ~待毙 hang by a thread / 气息~ breathe feebly; be sinking fast; be dying

掩

①(遮盖) cover; hide: ~着脸 hide one's face in one's hands ②(关;合) shut; close: ~门 shut the door ③[方](被卡住) get squeezed while shutting a door, lid, etc.: 小心门~了手。Don't get your fingers caught in the door. ④(乘人不备进行袭击) launch a surprise attack
【掩鼻而过】pass by holding one's nose
【掩蔽】screen; shelter; cover; masking: ~目标 cover over the object ◇~部[军] shelter / ~物[军] screen / ~阵地 covered position
【掩藏】hide; conceal
【掩耳不闻】turn a deaf ear
【掩耳盗铃】plug one's ears while stealing a bell; deceive oneself; bury one's head in the sand
【掩耳却步】close the ears and step back
【掩盖】cover; conceal: ~不住内心的喜悦 can hardly conceal the joy in one's heart / ~事实真相 cover up the fact
【掩护】screen; shield; cover: ~撤退 cover the retreat / 空中~ air cover / 为某人的错误打~ shield sb. from blame ◇~部队 covering force / ~火力 covering fire
【掩口而笑】laugh in secret; hide one's smile
【掩埋】bury
【掩面而泣】cover one's face and weep
【掩旗息鼓】stop clamoring
【掩人耳目】deceive the public; hoodwink people
【掩饰】cover up; gloss over; conceal: ~过失 gloss over mistakes / ~外观 save appearances / 毫不~自己的感情 make no secret of one's feelings
【掩体】[军] blindage; bunker: 炮兵~ emplacement

眼

①(眼睛) eye: 假~ artificial eye / 左~ left eye / 两~无神 with dull eyes / 一夜没合~ have not slept a wink the whole night ②(小洞) small hole; aperture: 扎一个~ punch a hole ③(事物的关键所在) key point: 节骨~儿 critical juncture ④[围棋] trap ⑤(戏曲中的拍子) an unaccented beat ⑥(中心) eye: 台风~ typhoon eye ⑦[量] 一~井 a well
【眼巴巴】①(急切盼望) eagerly; anxiously: ~地等着回信 anxiously expecting a reply ②(无可奈何地) helplessly
【眼白】[方] the white of the eye
【眼不见为净】out of sight; out of mind
【眼不见心不烦】what the eye doesn't see, the heart doesn't grieve over
【眼馋】[方] covet; be envious
【眼眵】gum

【眼到手到】take down notes while reading
【眼底】[生理] eye ground; fundus oculi
【眼底镜】[医] ophthalmoscope
【眼底下】①(眼睛前) right before one's eyes ②(目前) at the moment: 先处理~的事。 Let's settle the business on hand first.
【眼点】[动] eyespot; stigma
【眼福】the good fortune of seeing sth. rare or beautiful: ~不浅 be lucky enough to see sth. / 一饱~ feast one's eyes on sth.
【眼干症】[医] xerophthalmia
【眼高手低】have grandiose aims but puny abilities; be fastidious but imcompetent
【眼高心傲】having a haughty look and a proud heart
【眼观六路,耳听八方】have sharp eyes and keen ears; be observant and alert
【眼光】①(视线) eye: 避开众人的~ escape public gaze / 投以惊奇的~ throw a startled look ②(观察事物的能力) sight; foresight; insight; vision: ~短浅 shortsighted / ~远大 farsighted ③(观点) view; way of looking at things: 用老~看某人 judge sb. by what he used to be / 用新~看事 view sth. in a new light
【眼红】①(羡慕;嫉妒) covet; be envious; be jealous ②(激怒的样子) furious
【眼花】have dim eyesight; have blurred vision: ~耳热 eyes blurred and ears hot / ~缭乱 be dazzled; dazzle the eyes / 使人~的灯光 dazzling lights
【眼尖】be sharp-eyed; have sharp eyes
【眼睑】[生理] eyelid
【眼见目睹】see with one's eyes
【眼见为信】seeing is believing
【眼角】the corner of the eye; canthus
【眼睫毛】[口] eyelash
【眼界】field of vision; outlook: 扩大~ widen one's field of vision; broaden one's horizon
【眼镜】glasses; spectacles: 戴~ wear glasses / 一副新~ a new pair of glasses ◇ ~盒 glasses case / ~框 spectacle-frame
【眼镜蛇】cobra ◇ ~毒 cobra venom
【眼睛】eye: 擦亮~ keep one's eyes open / 闭上你的~。 Shut your eyes.
【眼看】①(马上) soon; in a moment: ~天就要亮了。 It'll be daylight soon. ②(听凭) watch helplessly; look on passively
【眼科】[医] ophthalmology ◇ ~学 ophthalmology / ~医生 ophthalmologist; oculist; eye doctor
【眼库】eye bank
【眼眶】①(眼皮边缘) eye socket; orbit: ~里含着泪水 with tears in one's eyes ②(眼睛周围) rim of the eye: ~发黑 have livid rings round one's eye / 一拳把他的~打青了。 The blow gave him a black eye.
【眼泪】tears; eyedrop: ~汪汪 eyes brimming with tears / 擦干~ dry one's eyes
【眼力】①(视力) eyesight; vision ②(辨别力) judgment; discrimination: 他~过人。 He's a man of excellent judgement. / 他看人很有~。 He's good at sizing people up.
【眼里】within one's vision; in one's eyes: ~有活 see where there's work to be done / 在她~,你只是个孩子。 You're only a child in her eyes.
【眼帘】eye: 映入~ come into view; greet the eye
【眼明手快】quick of eye and deft of hand; sharp-eyed and quick-moving
【眼内压】[生理] intraocular pressure
【眼泡】upper eyelid
【眼皮】eyelid: ~浅 shortsighted / 上~ the upper eyelid / 下~ the lower eyelid
【眼前】①(跟前) before one's eyes ②(目前) at the moment; at present; now: ~利益 immediate interests / ~之欢 the pleasure of moment / 胜利就在~。 Victory is at hand.
【眼球】[生理] eyeball; bulbus oculi ◇ ~突出 exophthalmos; ocular proptosis; ophthalmoptosis
【眼热】covet; be envious
【眼如秋水】bright-eyed
【眼色】hint given with the eyes; meaningful glance; wink: 使~ tip sb. the wink; wink at sb.
【眼神】①(眼睛的神态) expression in one's eyes ②[方](眼力) eyesight
【眼生】look unfamiliar: 这位客人很~,我不认识他。 The visitor looks a stranger to me. I don't think I've met him before.
【眼屎】[方] gum
【眼熟】look familiar: 这人看着~。 That person looks familiar.
【眼竖目横】stare in anger or contempt
【眼跳】twitching of the eyelid: ~心惊 nervous apprehension
【眼窝】eye socket; eyehole
【眼下】at the moment; at present; now
【眼压】intraocular pressure
【眼压计】electronic tonometer; tonometer

【眼药】medicament for the eyes; eye ointment; eyedrops
【眼晕】feel dizzy
【眼罩】①(人用的)eyeshade ②(给牲畜用的)blinkers
【眼睁睁】helplessly; unfeelingly
【眼中钉】thorn in one's flesh
【眼中无人】haughty
【眼珠子】[口] eyeball

俨

【俨然】①(整齐) neatly arranged: 屋舍～ houses set out in neat order ②(很象) just like: 说起话来～是个圣人 speak just like a sage
【俨如】just like: ～白昼 as bright as day

鼹 mole

【鼹鼠】mole

衍

【衍变】develop; evolve
【衍射】[物] diffraction ◇ ～角 diffraction angle／～线 diffracted ray
【衍生物】[化] derivative: 纤维素～[纺] cellulose derivatives
【衍文】redundancy due to misprinting or miscopying
【衍沃之区】rich and fertile region
【衍溢四海】overflowing to the four seas

yàn

宴 ①(请人吃酒饭) entertain at a banquet: 国～ state banquet／～客 entertain guests at a banquet ②(酒席) feast; banquet: 赴～ attend a banquet／举行～ give a banquet／设～招待某人 entertain sb. at a banquet

【宴安鸩毒】seeking pleasure is like drinking poisoned wine
【宴会】banquet; feast; dinner party: 答谢～ reciprocal banquet／欢迎～ welcoming banquet ◇ ～厅 banquet hall
【宴请】entertain fete: ～贵宾 entertain the distinguished guests
【宴席】banquet; feast

谚 proverb; saying; adage; saw: 古～ old saw／农～ peasants' proverb; farmers' saying

【谚语】proverb; saying; adage; saw

艳 ①(色彩鲜明好看) gorgeous; colorful; gaudy ②(关于爱情的) amorous: ～史 love story

【艳丽】bright-colored and beautiful; gorgeous: ～夺目 of dazzling beauty／词藻～ flowery diction／服装～ be gorgeously dressed
【艳诗】love poem in a flowery style
【艳阳天】bright spring day; bright sunny skies

堰 weir

【堰塞湖】[地] barrier lake

燕 swallow: 家～ house swallow

【燕而新婚】marital happiness
【燕麦】oats ◇ ～片 oat meal
【燕雀】[动] brambling; bramble finch
【燕雀处堂】swallows and sparrows nesting in the hall, unmindful of the spreading blaze; oblivious of imminent danger
【燕式跳水】swallo dive
【燕婉之求】demand for an ideal husband
【燕尾服】swallowtail; swallow-tailed coat; tailcoat
【燕窝】edible bird's nest
【燕鱼】[动] Spanish mackerel
【燕子】swallow

酽 thick; strong: 茶太～。 The tea's too strong.

厌 ①(憎恶) be disgusted with; detest ②(因过多而不喜欢) be fed up with; be bored with; be tired of: 看～了 be sick of seeing sth. ③(满足) be satisfied: 贪得无～ be insatiably greedy

【厌烦】be sick of; be fed up with: ～琐碎 feel great hatred for trifles
【厌故喜新】dislike the old and take a delight in the new
【厌倦】be weary of; be tired of
【厌弃】detest and reject
【厌世】be world-weary; be pessimistic
【厌闻旧事】tired of hearing past things
【厌恶】detest; abhor; abominate; be disgusted with
【厌战】be weary of war; be war-weary ◇ ～情绪 war-weariness

雁 wild goose

【雁来红】[植] tricolor amaranth
【雁杳鱼沉】without news or letters

赝

【赝本】spurious edition or copy: ～书籍 a spurious copy of a book
【赝币】[书] counterfeit coin
【赝品】counterfeit; fake; sham

【赝真难辨】hard to distinguish the false from the true

唁 extend condolences

【唁电】telegram of condolence; message of condolence

【唁函】letter of condolence

咽 swallow: 细嚼慢～ chew carefully and swallow slowly

【咽气】breathe one's last; die

砚 inkstone; inkslab

【砚池】inkstone; inkslab

【砚台】inkstone; inkslab

【砚兄砚弟】school-mates

焰 flame; blaze: 烈～ blazing flames

【焰火】fireworks

验 ①(察看) examine; check; test: ～证件 examine the certificates ②(产生预期的效果) prove effective; produce the expected result: 屡试屡～ prove successful in every test／应～ come true

【验电器】electroscope

【验光】optometry

【验货】examine goods ◇ ～单 particular paper

【验明正身】make a positive identification of a criminal before execution; identify

【验枪】[军] inspect arms

【验湿器】hygroscope

【验尸】[法] postmortem; autopsy ◇ ～所 mortuary

【验尸官】coroner ◇ ～证书 certificate of coroner

【验收】check and accept; check before acceptance; check upon delivery: 逐项～ check item by item before acceptance ◇ ～单 receipt／～可靠水平 acceptable reliability level／～试验 acceptance test

【验算】[数] checking computations ◇ ～公式 check formula

【验血】blood test

【验证】test and verify

【验证载荷】[航空] proof load

yāng

央 ①(中心) center ②(恳求) entreat: ～请帮忙 entreat for help

【央告】beg; ask earnestly

【央求】beg; plead; implore: ～宽恕 beseech for forgiveness; beg for mercy

【央托】request; entrust

泱

【泱泱】①(水面广阔) vast ②(气魄弘大) ～大风 the impressive manner of a great country／～大国 a great country

殃 ①(祸害) calamity; disaster; misfortune: 遭～ meet with disaster; suffer disaster ②(使受祸害) bring disaster to: ～及池鱼 brings disaster to the fish in the moat／～及无辜 trouble involves the innocent people／祸国～民 bring calamity to the country and the people

秧 ①(幼苗) seedling; sprout: 黄瓜～ cucumber sprout／萝卜～ radish seedling ②(稻秧) rice seedling: 插～ transplant rice seedlings／育～ raise rice seedlings ③(茎) vine; stem: 瓜～ melon vines ④(小动物) young; fry: 鱼～ young fish

【秧歌】*yangko*, a popular rural folk dance: 扭～ do the *yangko* ◇ ～剧 *yangko* opera

【秧鸡】[动] water rail

【秧苗】rice shoot; rice seedling

【秧田】rice seedling bed

yáng

羊 sheep: 公～ ram／绵～ sheep／母～ ewe／肉用～ mutton sheep／山～ goat／小～ lamb／种公～ stud ram／护～狗 sheep dog／一群～ a flock of sheep

【羊肠线】[医] catgut

【羊肠小道】narrow winding trail; meandering footpath

【羊齿】[植] fern; bracken

【羊痘】[牧] sheep pox

【羊肚蕈】[植] morel

【羊羔】lamb

【羊倌】shepherd

【羊毫】writing brush made of goat's hair

【羊角锤】claw hammer

【羊角风】epilepsy

【羊疥】sheep mange

【羊圈】sheepfold; sheep pen

【羊痢疾】sheep dysentery

【羊毛】sheep's wool; fleece: 纯～的 of pure wool／剪～ shear the sheep ◇ ～衫 woolen sweater; cardigan／～商 wool stapler／～袜 woolen socks or stocking／～脂 wool oil

【羊毛出在羊身上】after all, the wool still comes from the sheep's back; in the long run,

whatever you're given, you pay for
【羊膜】[生理] amnion
【羊皮】sheepskin: ～袄 sheepskin jacket / 披着～的狼 a wolf in sheep's clothing
【羊皮纸】parchment
【羊群里头出骆驼】stand out like a camel in a flock of sheep
【羊绒衫】cashmere sweater
【羊肉】mutton: 烤～串 mutton cubes roasted on a skewer; kebab
【羊入狼群】in a perilous position
【羊水】[生理] amniotic fluid
【羊桃】[植] carambola
【羊痫风】epilepsy
【羊脂】suet ◇ ～玉 white jade
【羊质虎皮】a sheep in a tiger's skin; outwardly strong, inwardly weak

洋 ①(盛大) vast; multitudinous ②(海洋) ocean: 太平～ Pacific Ocean / 远涉重～ travel all the way from across the oceans ③(外国的) foreign: ～房 western-style house / ～货 imported goods ④(现代化的) modern: ～办法 modern methods
【洋八股】foreign stereotyped writing; foreign stereotypes
【洋白菜】cabbage
【洋财】a big fortune; a windfall: 他到了澳大利亚,不久便发了～。He went to Australia and soon made a big fortune.
【洋菜】agar
【洋车】[口] rickshaw
【洋葱】onion
【洋服】Western clothes; Occidental dress
【洋镐】pickax
【洋狗】a dog of foreign breed
【洋鬼子】foreign devil
【洋行】foreign firm
【洋槐】[植] locust
【洋里洋气】in an ostentatiously foreign style
【洋流】[地] ocean current
【洋楼】Western style building of two stories or more
【洋奴】slave of a foreign master; flunkey of imperialism; worshipper of everything foreign ◇ ～思想 slavish mentality towards all things foreign / ～哲学 slavish comprador philosophy: blind worship of everything foreign
【洋气】①(指西洋的样式) foreign flavor; Western style ②(洋里洋气) outlandish ways
【洋琴】dulcimer
【洋人】foreigner
【洋嗓子】a voice trained in Western style of singing
【洋娃娃】doll
【洋为中用】make foreign things serve China
【洋相】silly sight; awkward behavior: ～百出 make a spectacle of oneself
【洋烟】imported cigarette
【洋洋万言】run to ten thousand words; be very lengthy
【洋洋大观】spectacular; grandiose; imposing
【洋洋得意】proud and happy
【洋洋洒洒】voluminous; at great length
【洋洋自得】self-satisfied
【洋溢】be permeated with; brim with: ～着友好的感情 be permeated with friendly feelings / 热情～的讲话 a speech brimming with warm feeling
【洋油】kerosene; imported oil

佯 pretend; feign; sham: ～病缺席 pretend to be ill and be absent / ～死 feign death; play dead
【佯攻】[军] feign attack; make a feint
【佯为不见】shut one's eyes to another's faults
【佯言】tell a lie

阳 ①(太阳) the sun ②(山的南面,水的北面) south of a hill or north of a river ③(凸出的) in relief: ～文 characters cut in relief ④(外露的) open; overt: 阴一套,～一套 act one way in public and another in private; be engaged in double-dealing ⑤[物] positive: ～离子 positive ion; cation ⑥(与"阴"相对的) the masculine or positive principle in nature
【阳春】spring
【阳春白雪】①(古曲名) the Spring Snow ②(不通俗的文学艺术) highbrow art and literature
【阳地植物】sun plant
【阳电】positive electricity
【阳奉阴违】overtly agree but covertly oppose; comply in public but oppose in private; feign compliance
【阳沟】open drain; ditch
【阳关道】broad road; thoroughfare
【阳光】sunlight; sunshine: ～充足 full of sunlight; with plenty of sunshine; sunny
【阳极】[物] positive pole; positive electrode; anode ◇ ～板[电] positive plate / ～栅 anode grid / ～射线 positive ray
【阳间】this world

【阳离子电泳】cataphoresis
【阳历】solar calendar
【阳模】force plug; piston; plunger
【阳平】[语] rising tone
【阳畦】[农] seed bed with windbreaks; cold bed
【阳伞】parasol; sunshade
【阳台】balcony
【阳痿】[医] impotence
【阳文】characters cut in relief
【阳性】①[医](肯定的) positive: ～反应 positive reaction ②[语](与阴性相对的) masculine gender

杨 poplar
【杨柳】①(杨树和柳树) poplar and willow ②(柳树) willow
【杨梅】[植] red bayberry
【杨树】poplar
【杨桃】[植] carambola
【杨枝鱼】pipefish

扬 ①(高举) raise: ～鞭 flip the whip / ～起一片尘土 raise a cloud of dust ②(往上撒) throw up and scatter; winnow: 晒干～净 dry and winnow thoroughly ③(传播) spread; make known
【扬波】the swelling of waves
【扬长避短】make best use of the advantages and bypass the disadvantages; develop the strong points and avoid the weak points; foster strengths and circumvent weaknesses
【扬长而去】stalk off; swagger off
【扬场】[农] winnowing ◇ ～机 winnowing machine; winnower
【扬程】[水] lift: 高～抽水站 high-lift pumping station / 高～水泵 high-lift pump
【扬花】[农] flowering
【扬眉吐气】feel proud and elated
【扬名】make a name for oneself; become famous: ～于世 raise one's name to the world
【扬旗】[铁道] semaphore
【扬弃】①[哲] sublate ②(取其精华,去其糟粕) develop what is useful or healthy and discard what is not
【扬琴】dulcimer
【扬声器】loudspeaker: 低频～ woofer / 高频～ tweeter
【扬水】pump up water ◇ ～泵 lift pump / ～机 water raiser / ～站 pumping station
【扬汤止沸】try to stop water from boiling by scooping it up and pouring it back; an ineffectual remedy
【扬言】threaten: ～要进行报复 threaten to retaliate / ～于众 spread words to public
【扬扬】triumphantly; complacently: ～自得 be very pleased with oneself; be complacent / 得意～ be immensely proud; look triumphant
【扬子鳄】[动] Chinese alligator

yǎng

痒 itch; tickle: ～得难熬 too itchy to endure / ～而不痛 itching but not painful / 浑身发～ itch all over / 怕～ ticklish / 搔到～处 scratch where it itches; hit the nail on the head
【痒痒】[口] itch; tickle

氧 [化] oxygen (O): 含～酸 oxygen acid; oxyacid / 液态～ liquid oxygen
【氧割】[机] oxyacetylene metal-cutting
【氧合作用】[生理] oxygenation
【氧化】[化] oxidize; oxidate: ～过程 oxidizing process ◇ ～剂 oxidizer; oxidant / ～铁 ferric oxide / ～物 oxide / ～锌 zinc oxide / ～焰 oxidizing flame / ～抑制剂 oxidation retarder / ～作用 oxidation
【氧气】oxygen ◇ ～厂 oxygen installation / ～顶吹转炉 oxygen top-blown convertor / ～炼钢 oxygen steelmaking / ～面罩 oxygen mask / ～瓶 oxygen cylinder / ～枪[冶] oxygen lance / ～乙炔吹管[机] oxyacetylene blow pipe / ～帐篷[医] oxygan tent

养 ①(供养) support; provide for: ～家 support a family; keep a family ②(饲养或培植) raise; keep; grow; rear: ～蜂 keep bees / ～狗 keep a dog / ～花 grow flowers / ～猪 raise pigs ③(生育) give birth to: ～了一女 give birth to a daughter ④(领养的) foster; adoptive: ～父 foster father / ～母 foster mother / ～女 adopted daughter / ～子 adopted son ⑤(培养) form; acquire; cultivate: ～成优良作风 cultivate good style of work ⑥(调养) rest; convalesce; recuperate one's health; heal ⑦(养护) maintain; keep in good repair: ～路 maintain a road or railway
【养兵千日,用兵一时】maintain an army for a thousand days to use it for an hour
【养病】take rest and nourishment to regain one's health; recuperate
【养蚕】sericulture; silkworm breeding ◇ ～业 sericulture
【养成习惯】form a habit

【养而不教】bear children without educating them
【养分】nutrient: 土壤～ soil nutrient
【养蜂】bee-keeping; apiculture ◇ ～场 apiary; bee yard / ～人 bee keeper; apiarist / ～业 apiculture; beekeeping
【养虎遗患】to rear a tiger is to court calamity; appeasement brings disaster
【养护】maintain; conserve: ～公路 maintain public highway / ～设备 maintenance of equipment / 生物资源～ conservation of living resources
【养晦待时】dwell in retirement awaiting for opportunity
【养活】①(供养)support; feed: ～全家 keep the whole family alive / 他有妻室儿女要～。 He has a wife and family to keep. ②(饲养) raise ③(生育) give birth to: ～孩子 have a baby
【养鸡场】chicken run; chicken farm
【养精蓄锐】conserve strength and store up energy
【养老】①(奉养老人)provide for the aged ②(年老闲居休养) live out one's life in retirement ◇ ～保险 endowment insurance / ～金 old-age pension / ～院 a home for destitute old people
【养料】nutriment; nourishment: 从…吸取～ draw nourishment from
【养路】maintain a road or railway ◇ ～道班 road maintenance crew / ～费 road toll
【养马场】ranch
【养神】rest to attain mental tranquility; repose 闭目～ sit in repose with one's eyes closed
【养生】preserve one's health; keep in good health: ～要素 the essential elements of life / ～之道 the way to keep in good health
【养兔场】rabbit warren
【养痈成患】a carbuncle neglected becomes the bane of your life; leaving evil unchecked spells ruin
【养鱼】fish culture ◇ ～场 fish farm / ～池 fishpond
【养育】bring up; rear: ～之恩 the love and care from childhood / ～子女 bring up children; rear children
【养殖】breed: ～海带 cultivate kelp / 海水～ sea-farming; marine culture ◇ ～业 aquaculture
【养殖场】farm: 水产～ aquatic farm
【养猪场】pig farm; piggery
【养尊处优】enjoy high position and live in ease and comfort; live in clover

仰 ①(脸向上) face upward: ～起头 raise one's head / ～天一跤 fall flat on one's back / ～着睡 sleep on one's back ②(敬慕) admire; respect; look up to: 瞻～ look at with reverence; pay one's respects to ③(依靠) rely on; depend on: ～给于人 depend on others for living / ～食父兄 sponge on father and brother
【仰俯之间】between looking up and stooping down
【仰角】[数] angle of elevation
【仰面朝天】fall flat on one's back
【仰面求人】spy one's face and implore one
【仰慕】admire; look up to: ～大作 look up to your writing with admiration
【仰人鼻息】be dependent on the pleasure of others; be slavishly dependent
【仰视图】upward view
【仰天长叹】sign deeply; look up at heaven and lament
【仰天长啸】cry into the air
【仰望】①(抬头看) look up at ②(敬仰而有所期望) respectfully seek guidance or help from; look up to: ～终身 look up to one for the whole life
【仰卧】lie on one's back; lie supine ◇ ～起坐 [体] sit-ups
【仰泳】[体] backstroke
【仰仗】rely on; look to sb. for backing; ～权威 rely on powerful relatives

yàng

漾 ①(水面微微动荡) ripple ②(溢出) brim over
【漾奶】throw up milk; vomit milk from repletion

恙 ailment; illness: 安然无～ safe and sound; unscathed
【恙虫】[动] tsutsugamushi mite ◇ ～热 scrub typhus; tsutsugamushi disease

样 ①(形状) shape; form ②(样品) sample; model; pattern: 校～ proof sheet / 取～检验 take a sample for examination and test ③[量] kind; type: 三～菜 three dishes / 向市场供应多种多～的新商品 supply the market with various kinds of new commodities
【样板】①(板状样品)sample plate ②[工](板状

工具) templet ③(榜样) model; example
【样本】①(商品图样印本) sample book ②[印](样书) sample; specimen: 字体～ type specimen book
【样机】[航空] prototype
【样品】sample; specimen: 这和～相符。It comes up to sample. ◇～卡片 sample card
【样式】pattern; type; style; form: 各种～的羊毛衫 woolen sweaters in all styles / 最新～的外衣 a coat of the latest style / 照巴黎流行的～裁衣服 pattern a dress on a Parisian model
【样片】[摄] print; rush sample
【样图】master drawing
【样样】every kind; each and every; all: ～俱全 have all kinds of things
【样张】[印] specimen page
【样子】①(形状) appearance; shape ②(神情) manner; air: 我不喜欢她那～。I don't like her manner. ③(作为标准供模仿) sample; model; pattern: 衣服～ clothes pattern ④(趋势) tendency; likelihood: 天要下雨的～。It looks like rain.

怏

【怏怏】disgruntled; sullen: ～不乐 unhappy about sth.; morose

yāo

要 ①(求) demand; ask ②(强迫) force; coerce
【要买人心】win people's hearts by statecraft
【要求】ask; demand; require; claim: ～被告出庭的通知 notice to appear / ～赔偿损失 claim compensation for the losses / ～赔款的诉讼 action for indemnity / ～要点 points of claim / ～引渡 demand possession / ～再审的诉讼 action for re-trial / 领土～ territorial claims / 严格～自己 be strict with oneself
【要挟】coerce; put pressure on; threaten: ～对方 put pressure on the other party / ～他答应一件事 coerce him into consent / 对小国进行～ use coercion against small nations

腰 ①(腰部) waist; small of the back: ～酸背痛 have pains in the loins and back / 扭了～ sprain one's back muscles / 弯～ bent down; stoop / 细～ slender waist / 在齐～深的水中 be waist-deep in water ②(裤腰) waist: 裤～ waist of trousers ③(衣兜) pocket: 她～里还有些钱。She's got some money in her pockets. ④(事物的中间部分) middle: 山～ halfway up a mountain; on a hillside
【腰板脖硬】stiff or rigid in movements
【腰板儿】①(人的腰和背) waist and back ②(体格) body: 这老人～还挺硬朗。The old man is still strong-bodied.
【腰包】purse; pocket; wallet 把钱装进自己的～ pocket the money
【腰部】waist; small of the back
【腰缠万贯】very rich
【腰带】waistband; belt; girdle
【腰杆子】①(腰部) back: 挺起～ straighten one's back; be confident and unafraid ②(靠山) backing; support: ～不硬 have no strong backing; be unbolstered
【腰鼓】waist drum
【腰果】[植] cashew: ～仁 cashew nut
【腰花】scalloped pork or lamb kidneys: 炒～ stir-fried kidneys
【腰肌劳损】strain of lumbar muscles; psoatic strain
【腰扭伤】lumbar sprain
【腰身】waistline; waist; waist measurement; girth: 她胖得看不出～来。She has no waist.
【腰痛】lumbago
【腰围】waistline: 注意你的～。Watch you waistline
【腰眼】either side of the small of the back
【腰斩】cut sth. in half
【腰椎】[生理] lumbar vertebra
【腰子】[口] kidney

夭 die young
【夭亡】die young
【夭折】①(未成年而死) die young ②(事情中途失败) come to a premature end

妖 ①(妖怪) goblin; demon; evil spirit ②(邪恶而迷惑人的) evil and fraudulent ③(装束奇特) bewitching; coquettish
【妖风】evil wind; noxious trend
【妖怪】monster; bogy; goblin; demon
【妖精】①(妖怪) evil spirit ②(以姿色迷人的女人) alluring woman
【妖里妖气】seductive and bewitching
【妖媚】seductively charming; bewitching
【妖魔】evil spirit; demon
【妖魔鬼怪】demons and ghosts; monsters of every description
【妖术】black art; witchcraft; sorcery
【妖物】evil spirit; monster
【妖形怪状】grotesque
【妖言】heresy; fallacy: ～惑众 spread fallacies

to deceive people
【妖艳】pretty and coquettish
【妖冶】pretty and coquettish

邀 ①(邀请) invite; request: ～客 invite guests / ～游 invite to an excursion / 应～出席会议 attend a meeting on invitation ②(求得) solicit; seek
【邀功】[书] take credit for someone else's achievements: ～请赏 take credit and seek rewards for someone else's achievements
【邀集】invite to meet together; call together
【邀买人心】buy popularity
【邀请】invite: 发出～ send an invitation / 应某人～ at the invitation of sb. ◇ ～国 host country / ～赛[体] invitational tournament / ～书 invitation

幺 ①(一) one ②[方](排行最小的) youngest: ～妹 the youngest sister

吆
【吆喝】①(呼唤) cry out; call ②(叫卖) cry one's wares ③(赶牲畜) loudly urge on an animal

约 [口] weight: ～～多重。 See how much it weights.

yáo

尧 Yao, a legendary monarch in ancient China
【尧舜】Yao and Shun, ancient sages

肴 meat and fish dishes: 佳～ delicious dishes / 酒～ wine and meat at dinner
【肴馔】sumptuous courses at a meal

窑 ①(烧制砖瓦等的建筑物) kiln: 砖～ brickkiln ②(煤矿) pit: 煤～ coal pit ③(住人的窑洞) cave dwelling
【窑洞】cave dwelling
【窑灰钾肥】flue ash potash
【窑外分解法】the rotary kiln with a precalcinator
【窑子】[方] brothel

谣 ①(歌谣) ballad; rhyme: 民～ popular verse; ballad / 童～ children's rhyme ②(谣言) rumor: 辟～ refute a rumor / 造～ cook up a story and spread it around; start a rumors
【谣传】①(传播的谣言) rumor; hearsay ②(谣言传播) it is rumored: 据～ it is rumored that
【谣言】rumor; groundless allegation: ～惑众 delude the people with rumor / ～流传 gossip spreads / 戳穿～ give the lie to a rumor / 散布～ spread rumors

遥
【遥测】telemetering: 空间～ space telemetry ◇ ～计 telemeter / ～术 telemetry / ～温度计 telethermometer / ～心电图 telemetric radio cardiogram
【遥感】[电子] remote sensing: 红外～ infrared remote sensing ◇ ～技术 remote sensing technique / ～仪器 remote sensing instrument
【遥见】see at a distance
【遥控】remote control; telecontrol: 无线电～ radio telecontral ◇ ～飞机 remote control aircraft; telecontrolled airplane / ～开关 teleswitch / ～力学 telemechanics / ～无人驾驶飞机 drone
【遥望】look into the distance
【遥相呼应】echo each other at a distance; coordinate with each other from afar
【遥遥】far away; a long way off: ～领先 be far ahead; get a good lead / ～无期 not within the foreseeable future / ～相对 stand far apart facing each other
【遥远】distant; remote; faraway: ～的边疆 remote frontiers / ～的将来 the distant future / 路途～ a long journey; a long way to go

摇 shake; wave; rock; turn: ～铃 ring a bell / ～旗 wave a flag / 我的椅子能～。 My chair rocks.
【摇把】cranking bar
【摇摆】sway; swing; rock; vacillate: ～不定的人 a waverer / 左右～ vacillate now to the left, now to the right ◇ ～舞 rock and roll
【摇笔即来】be brimming with ideas
【摇臂】[机] rocker arm ◇ ～轴 rocker shaft / ～钻床 radial drilling machine
【摇床】[矿] table: 粗选～ roughing table / 选矿～ cleaning table
【摇唇鼓舌】flap one's lips and beat one's tongue; wag one's tongue; engage in loose talk
【摇荡】rock; sway: 树枝在风中微微～。 Branches sway gently in the wind.
【摇动】①(摆动) wave; shake: 地震时整个房子都～了。 The whole house shook during the earthquake. ②(摇东西使它动) sway; rock: 大树在狂风中～。 The big tree rocked in the strong wind.

【摇滚乐】big beat
【摇撼】give a violent shake to; shake to the root or foundation; rock
【摇晃】rock; sway; shake: ～的桌子 a shaky table／ 服药前～药瓶 shake the bottle well before use／ 波涛使船～。 The waves rocked the boat.
【摇来摇去】swing to and fro
【摇篮】cradle: 音乐的～ the cradle of the music／ 摇动～中的孩子 rock a baby in its cradle ◇ ～曲 [乐] lullaby; cradlesong
【摇耧】rock a drill barrow in planting; plant with a drill barrow
【摇蜜】[农] extract honey ◇ ～机 honey extractor
【摇旗呐喊】wave flags and shout battle cries; bang the drum for sb.
【摇钱树】a legendary tree that sheds coins when shaken; a ready source of money
【摇身一变】give oneself a shake and change into another form; suddenly change one's identity
【摇式卸车机】rocker unloader
【摇头】shake one's head
【摇头摆尾】shake the head and wag the tail; assume an air of complacency or levity
【摇头晃脑】wag one's head; look pleased with oneself; assume an air of self-approbation or self-conceit
【摇尾乞怜】wag the tail ingratiatingly; fawn
【摇摇摆摆】swagger along
【摇摇晃晃】vacillating and staggering
【摇摇欲坠】tottering; crumbling; on the verge of collapse
【摇曳】flicker; sway: ～的灯光 flickering light／ ～多姿 shaking slightly in many carriages／ 一开始蜡烛～不定，然后就灭了。 The candle flickered and then went out.
【摇椅】rocking chair
【摇钻】brace; bit-stock

鳐 [动] ray; skate

yǎo

窈
【窈窕】(of a woman) gentle and graceful

杳 distant and out of sight
【杳如黄鹤】disappear like the yellow crane; nowhere to be found
【杳无音信】there has been no news whatsoever about sb.; have never been heard of since
【杳无踪迹】disappear without a trace; vanish

咬 ①(用牙咬) bite; snap at: ～不动 too tough to bite／ 狗～他。 The dog bit him.／ 让我～一口面包。 Let me have a bite of the bread.／ 他把苹果～下一大口。 He bit off a large piece of the apple.／ 他的狗～人吗? Does his dog bite? ②(狗叫) bark ③(受责难或审讯时牵扯别人) incriminate another person when blamed or interrogated; make accusation: 反～一口 make false countercharges against the accuser ④(读字发音) pronounce; articulate: ～字清楚 clear articulation ⑤(过分计较字句) be nitpicking: ～字眼儿 be nitpicking on words ⑥(螺丝等互相卡住) grip; bite: 这个旧螺母～不住扣。 This old nut won't bite.
【咬定】insist: 一口～ assert emphatically; insist
【咬耳朵】[口] whisper in sb.'s ear; whisper
【咬钩】bite: 他钓了一上午的鱼，但没有一条鱼～。 He had been fishing all morning but hadn't had a single bite.／ 有鱼～吗? Do you have a bite?／ 有鱼～。 I have a bite.
【咬紧牙关】grit one's teeth
【咬破】break by the teeth; bite through
【咬伤】bite: 疯狗的～会致死的。 The bite of a mad dog is fatal.
【咬舌儿】①(发音不清楚) lisp ②(发音不清楚的人) lisper
【咬文嚼字】pay excessive attention to wording
【咬牙】①(咬紧牙关) grit one's teeth: 恨得直～ gnash one's teeth in hatred ②(熟睡时磨牙) grind one's teeth
【咬牙切齿】gnash one's teeth: ～地咒骂 curse between one's teeth

舀 ladle out; spoon up; scoop up: ～汤 ladle out soup
【舀子】dipper; ladle; scoop

yào

疟
【疟子】[口] malaria

药 ①(药物) medicine; drug; remedy: 安眠～ sleeping pill／ 避孕～ contraceptive drugs／ 补～ tonic／ 感冒～ a medicine for colds／ 化痰～ expectorant／ 抗癌～ anti-cancer drugs; cancer-fighting drugs／ 抗结核～ anti-tuberculous drug／ 口服避孕～ oral contraceptive; pill／ 良～ good medicine; a good remedy／ 轻泻～ laxative／ 退热～ antipyretic／ 内服～

for oral administration / 特效～ specific medicine; specific / 外用～ for external use / 预防～ preventive medicine; prophylactic / 镇静～ sedative / 服～ take medicine / 换～ change dressings / 煎～ decoct herbal medicine / (药剂师)配～ fill a prescription / (患者)抓～ have a prescription made up (filled) ②(某些有化学作用的物质) certain chemicals: 耗子～ rats poison; ratsbane / 杀虫～ insecticide ③(用药毒死) kill with poison: ～老鼠 poison rats

【药材】medicinal materials; crude drugs

【药草】medicinal herbs

【药厂】pharmaceutical factory

【药典】pharmacopeia

【药店】drugstore; chemist's shop; pharmacy

【药方】prescription: 开～ write out a prescription

【药房】①(医药商店) drugstore; chemist's shop; pharmacy ②(医院或诊所里的药房) hospital pharmacy; dispensary

【药费】expenses for medicine; charges for medicine

【药粉】(medicinal) powder

【药膏】ointment; salve: 上～ apply a plaster

【药罐子】①(煎中药用的罐子) a pot for decocting herbal medicine ②(经常生病的人) chronic invalid

【药衡】apothecaries' measure or weight

【药剂】medicament; drug ◇ ～师 pharmacist; druggist / ～士 assistant pharmacy / ～学 pharmaceutics; pharmacy

【药酒】medicinal liquor

【药理学】pharmacology

【药力】efficacy of a drug

【药棉】absorbent cotton

【药农】a peasant who cultivates or collects medicinal herbs; medicinal herb grower; medicinal herb collector; herbalist

【药片】tablet

【药品】medicines and chemical reagents

【药瓶】medicine bottle

【药铺】herbal medicine shop

【药石】remedies; medicines and stone needles for acupuncture: ～之言 exhortations; unpleasant but needed advice

【药水】①(液态的药) liquid medicine; medicinal liquid ②(洗液) lotion

【药丸】pill: 大～ bolus

【药味】①(中药方中的药) herbal medicines in a prescription ②(药的味道或气味) flavor of a drug

【药物】medicines; pharmaceuticals; medicaments: ～医治 heal with drugs ◇ ～过敏 drug allergy / ～学 materia medica / ～中毒 drug poisoning

【药箱】medical kit; medicine chest: 急救～ first-aid kit

【药性】property of a medicine

【药学院】pharmaceutical college

【药言可取】good advice is acceptable

【药浴】[牧] dipping: 羊～ sheep dipping ◇ ～池 dipping vat

【药皂】medicated soap

【药疹】[医] drug rash; drug eruption

要 ①(重要) important; essential: ～事 an important matter ②(希望得到) want; ask for; wish; desire: 各国人民都～和平。 All nations want peace. ③(请求;要求) ask sb. to do sth.: 那位老人～我替他写封信。 The old man asked me to write a letter for him. ④(想做) want to; wish to: 他～同市长谈一谈。 He wishes to talk to the mayor. ⑤(须要) must; should; it is necessary: 我们～向先进工作者学习。 We should learn from advanced workers. ⑥(将要) shall; will; be going to: 天～下雨了。 It is going to rain. ⑦(需要) need; take: 到那里～两个小时。 It takes two hours to get there. ⑧(表示估计;用于比较) might; must ⑨[连](如果) if; suppose; in case

【要隘】strategic pass

【要不】[连] otherwise; or else; or: 快走,～你赶不上火车了。 Be quick, or you'll miss the train.

【要不得】no good; intolerable: 这样做～。 It is no good doing that.

【要不是】if it were not for; but for

【要道】thoroughfare: 交通～ important line of communications; vital communications line

【要得】[方] good; fine; desirable

【要地】important place; strategic point

【要点】①(主要内容) main points; essentials; gist: 抓住～ grasp the main points ②(重要的据点) key strongpoint: 战略～ strategic point

【要犯】important criminal: ～落网 a principal criminal fell into the net

【要饭】beg ◇ ～的 beggar

【要害】①(身体上致命的部分) vital part; crucial point: ～部位 vital part / 击中～ hit home ②(重要地点或部门) strategic point: ～之地 a place of importance

【要好】①(感情融洽) be on good terms; be close friends ②(要求上进) eager to improve oneself; try hard to make progress: 这孩子很~。The kid is eager to make progress.
【要价】ask a price; charge: ~过高 demand an exorbitant price; ask too much
【要件】①(重要文件) important document ②(重要条件) important condition
【要紧】①(重要)important; essential: 我有点~的事跟他商量。I have sth. urgent to discuss with him. ②(严重) be critical; be serious; matter: 不过是轻伤,没什么~。It's only a slight wound, nothing serious.
【要领】①(要点)main points; essentials; gist: 不得~ fail to grasp the main points; not see what sb. is driving at; miss the point ②(基本要求) essentials: 掌握~ grasp the essentials
【要略】outline; summary
【要么】[连] or; either... or...
【要面子】be keen on face-saving; be anxious to keep up appearances
【要命】①(使丧生) drive sb. to his death; kill ②(程度达到极点) confoundedly; extremely; awfully; terribly: 冷得~ terribly cold / 热得~ awfully hot ③(令人讨厌) a nuisance: 真~,雨象是不会停了。What a nuisance! It seems the rain will never stop.
【要强】be eager to excel; be anxious to outdo others
【要求赔偿书】claim letter
【要人】very important person (V.I.P.); important personage
【要塞】fort; fortress; fortification
【要是】if; suppose; in case
【要素】essential factor; key element
【要闻】important news; front-page story: ~一览 a glance of the important news
【要务】important or urgent business
【要言不烦】terse; succinct
【要因合同】causative contract
【要因行为】causative action
【要账】demand payment of a debt; press for repayment of a loan: dun
【要职】important post: 身居~ hold an important post
【要旨】main idea; gist

鹞

【鹞鹰】sparrow hawk
【鹞子】①(雀鹰) sparrow hawk ②[方](风筝) kite: ~翻身 do a somersault / 放~ fly kites; kite-flying

钥

【钥匙】key: 开门的~ key to a door / 万能~ universal key / 一把~开一把锁 open different locks with different keys; use different methods to deal with different people or problems
【钥匙包】key case
【钥匙环】key ring
【钥匙孔】keyhole

耀

耀 ①(强光照射) shine; illuminate; dazzle: 照~ shine upon; illuminate ②(夸耀) boast of; laud ③(光荣) honor; credit
【耀斑】[天] solar flare
【耀武扬威】make a show of one's strength; swagger around
【耀眼】dazzling: 阳光~。The sunshine is dazzling.

yē

耶

【耶稣】[宗] Jesus ◇ ~会 the Society of Jesus; the Jesuits / ~基督 Jesus Christ
【耶稣教】Protestantism

椰

椰 [植] coconut palm; coconut tree; coco
【椰雕】[工美] coconut carving
【椰干】copra
【椰蓉】[食品] shredded coconut stuffing
【椰油】coconut oil; coconut butter
【椰枣】①(椰枣树) date palm ②(椰枣果实) date
【椰子】①(椰子树)coconut palm; coconut tree; coco ②(椰子果实) coconut ◇ ~肉 coconut meat / ~糖 coconut candy / ~汁 coconut milk

掖

掖 tuck in; thrust in between: ~~藏藏 clandestinely / 把信从门缝里~进去 slip a letter under the door / 给小孩把被子~好 tuck a child in bed / 他把衬衫下摆~进裤子里。He tucked his shirt in.

噎

噎 ①(食物堵住食管) choke: ~住 choke up / 被食物~住了 choke over one's food; be choked by one's food ②[方](说话顶撞人) render sb. speechless; choke off

yé

爷

爷 ①(祖父) grandfather ②(尊称长一辈或年长的男子) uncle: 大~ uncle ③(对官僚、财主等的称呼)老～ sir; master; lord / 少～ young master ④(对神的称呼) god: 老天～ God; Heaven
【爷儿俩】father and son; uncle and nephew; grandfather and grandson
【爷儿们】men of two or more generations
【爷们】①[方](男人)man; menfolk: 老～ man ②(丈夫) husband
【爷们儿】[口] men of two or more generation
【爷爷】[口]①(祖父) grandfather ②(与祖父同辈的男子) grandpa

yě

冶 smelt (metal)
【冶金】metallurgy ◇ ～粉尘 metallurgical dust / ～工业 metallurgical industry / ～焦炭 smelter coke / ～学 metallurgy
【冶炼】smelt ◇ ～操作 smelting operation / ～厂 smeltery / ～工 smelter / ～炉 smelting furnace / ～时间 duration of heat
野 ①(野外) open country; the open: 旷～ wilderness ②(界限) limit; boundary: 分～ line of demarcation / 视～ field of vision ③(处于不当政的地位) not in power; out of office: 下～ be forced to relinquish power / 在～党 a party not in power ④(野生的;非驯养的) wild; uncultivated; undomesticated; untamed: ～花 wild flower / ～牛 wild ox ⑤(蛮横;粗鲁) savage; rough; rude: 动作太～ rough play / 说话太～ speak rudely ⑥(不受约束) unrestrained; abandoned; unruly
【野菜】edible wild herbs
【野餐】picnic
【野草】weeds: ～丛生 be overgrown with weeds
【野传】[棒、垒球] wild throw ◇ ～球 passed ball
【野地】wild country; wilderness
【野果】wild fruit
【野火】prairie fire; bush fire: ～烧不尽,春风吹又生。 Not even a prairie fire can destroy the grass; it grows again when the spring breeze blows.
【野鸡】①[动] pheasant ②(妓女) street walker
【野菊花】mother chrysanthemum
【野驴】Asiatic wild ass; kiang
【野马】wild horse; mustang
【野蛮】①(不文明) uncivilized; savage ②(蛮横残暴) barbarous; cruel; brutal: ～的屠杀 brutal massacre / ～的种族主义 barbarous racism
【野猫】①(野生的猫) wildcat ②(无主的猫) stray cat
【野炮】field gun; field artillery
【野葡萄】bryony
【野蔷薇】[植] multiflora rose
【野生】wild; uncultivated; feral ◇ ～动物 wild animal; wildlife / ～蜂 wild bee / ～生物资源保护 wildlife conservation / ～植物 wild plant / ～种 wild species / ～状态 wild state
【野史】unofficial history
【野兽】wild beast; wild animal
【野头野脑】wild head and unruly brain
【野兔】hare
【野外】open country; field: ～旅行 take an excursion in an open country ◇ ～生活 outdoor life / ～演习[军] field exercise / ～作业 fieldwork; field operation
【野豌豆】[植] vetch
【野味】game (as food)
【野心】wild ambition; careerism: ～勃勃 be overweeningly ambitious; be obsessed with ambition / ～不死 cling to one's ambitious designs ◇ ～家 careerist / ～狼 a vicious wolf; a person of wicked ambition
【野性】wild nature; unruliness
【野鸭】wild duck
【野营】camp; bivouac: 出外～ go camping ◇ ～活动 camping / ～训练 camp and field training
【野战】[军] field operations ◇ ～仓库 field depot / ～工事 fieldwork / ～军 field army / ～炮 fieldpiece; field gun / ～医院 field hospital
【野猪】wild boar
也 [副]①(同样) also; too; as well; either: 她～懂英语。 She knows English as well. / 我～是教员。 I'm a teacher too. ②(叠用,强调两事并列或对等) as well as: 她～会打篮球,～会打网球。 She can play tennis as well as basketball. ③(表示转折或让步)即使失败十次,我～不灰心。 I'll never lose heart even if I should fail ten times. ④(表示委婉)我～只好如此。 I could not but do so. ⑤(表示强调) even: 连爷爷～到工地来了。 Even grandfather has come to the worksite.
【也罢】①(表示容忍或只得如此)～,既然他不愿做,就不要勉强他。 All right, don't force him to do it since he won't. ②[助](叠用,表示不以某种情况为条件) whether... or...; no matter whether: 刮风～,下雪,他都坚持长跑。 He

keeps up his long-distance running whether it's windy or snowy.
【也好】①(表示容忍) it may not be a bad idea; may as well: 说明一下～。 Better give an explanation. ②(叠用,表示不以某种情况为条件) whether ... or ...; no matter whether: 你去～,不去～,情况不会改变。 Whether you go or not, the situation will remain unchanged.
【也门】Yemen: ～民主人民共和国 People's Democratic Republic of Yemen(简称为: 民主～ Democratic Yemen)/ 阿拉伯～共和国 Yemen Arab Republic ◇ ～人 Yemeni; Yemenite
【也许】[副] perhaps; probably; maybe

yè

夜 night; evening: 日以继～ night and day / 一个寒冷的冬～ one cold winter evening / 昨～ last night
【夜班】night shift: 上～ go on the night shift / 值～ on night duty ◇ ～编辑 night editor / ～护士 night nurse
【夜半】midnight: ～更深 late at night
【夜不安枕】toss about in bed
【夜餐】night snack
【夜叉】a hideous, ferocious person
【夜长梦多】a long delay means many hitches; a long night is fraught with dreams
【夜场】evening show
【夜车】①(夜里开出、到达或经过的火车) night train ②(夜间工作或学习)开～ work or study deep into the night
【夜出动物】nocturnal animal
【夜大学】evening university
【夜盗】burglar
【夜蛾】[动] noctuid
【夜光杯】luminous wine glass
【夜光表】luminous watch
【夜光虫】noctiluca
【夜光云】noctilucent cloud
【夜航】night flight; night navigation
【夜壶】chamber pot
【夜间】at night: ～施工 carry on construction work at night ◇ ～飞行 night flying / ～磨牙 bruxism / ～入屋盗窃罪 burglary / ～演习 night exercise / ～遗精 nocturnal emission / ～战斗机 night fighter
【夜景】night scene
【夜来香】[植] cordate telosma
【夜阑人静】in the dead of night; in the still of the night
【夜郎自大】ludicrous conceit of the king of Yelang; parochial arrogance
【夜凉如水】the chilling autumn night
【夜盲】[医] nyctalopia; night blindness ◇ ～症 moon blindness; nyctalopia
【夜猫子】①[方](猫头鹰) owl ②(喜欢晚睡的人) a person who goes to bed late; night owl
【夜明珠】a legendary luminous pearl
【夜幕】curtain of night; gathering darkness: ～低垂。 The night screen has hung down.
【夜尿症】enuresis; bed-wetting
【夜勤】night duty
【夜色】the dim light of night: ～苍茫 twilight at dusk / 趁着～ by starlight or moonlight
【夜深】in the dead of night; late at night: ～人静 in the dead of night
【夜市】night fair; night market
【夜视瞄准器】[军] sniperscope
【夜视望远镜】[光] night-vision telescope
【夜晚】night
【夜望镜】[军] snooperscope
【夜袭】night attack taken late at night; midnight snack; night raid
【夜宵】food
【夜校】night school; evening school
【夜行军】night march
【夜遗尿】[医] nocturnal enuresis
【夜以继日】day and night; round the clock
【夜莺】nightingale
【夜鹰】goatsucker; nightjar
【夜游神】the legendary god on patrol at night; a person who is up and about at night; night owl
【夜游症】sleep-walking; somnambulism
【夜战】①[军](夜间作战) night fighting ②[口](夜间工作) work at night
【夜总会】nightclub; night club

液 liquid; fluid; juice: 体～ body fluid / 胃～ gastric juice
【液化】[化] liquefaction ◇ ～气[物] liquid gas / ～器 liquefier / ～石油气 liquefied petroleum gas (LPG) / ～天然气 liquefied natural gas (LNG)
【液晶】[物] liquid crystal
【液冷】[机] liquid cooling ◇ ～式内燃机 liquid cooled engine
【液力】[机] hydraulic ◇ ～变速箱 hydraulic transmission box / ～传动 hydraulic power / ～传动柴油机车 diesel-hydraulic locomotive / ～制动器 hydraulic brake

【液膜】liquid film
【液泡】[生] vacuole
【液态】[物] liquid state ◇ ～空气 liquid air / ～气体 liquid gas
【液体】liquid ◇ ～比重计 hydrometer / ～燃料 liquid fuel / ～燃料火箭发动机 liquid-rocket engine
【液压】hydraulic pressure ◇ ～泵 hydraulic pump / ～表 hydraulic pressure gauge / ～成形 hydroform / ～传动 hydraulic transmission / ～机车 hydraulic locomotive / ～联轴节 hydraulic coupling / ～升降机 hydraulic elevator; hydraulic lift / ～支柱 hydraulic prop / ～制动器 hydraulic brake

掖 ①(用手搀扶别人胳膊) support sb. by the arm ②(扶助; 提拔) help; assist; promote: 扶～ support; help: 奖～ encourage and help; reward and promote

腋 ① [生理](夹肢窝) axilla; armpit ②[植] axil
【腋臭】underarm odor
【腋毛】armpit hair
【腋窝】armpit
【腋芽】[植] axillary bud

页 ①(张) leaf; sheet: 活～簿 loose leaf notebook / 增～ supplementary sheet ②(面) page: 翻过一～ turn the page over / 一本有二十～的小册子 a 20-page pamphlet / 在十五～上的照片 the photographs on page 15 / 这本书差两～。 There are two leaves missing in this book.
【页边】margin
【页码】page number
【页心】[印] type page
【页岩】[地] shale ◇ ～油 shale oil

叶 ①(叶子) leaf; foliage: 长～ come into leaf / 长新～ put forth new leaves / 树在春天长～。 The trees leaf out in the spring. ②(象叶子的) leaf-like thing: 百～窗 shutter; blind ③(较长时期的分段) part of a historical period: 十九世纪末～ in the late 19th century; in the latter part of the 19th century
【叶斑病】[农] leaf spot
【叶柄】[植] petiole; leafstalk
【叶蜂】[动] sawfly
【叶公好龙】Lord Ye's love of dragons; professed love of what one really fears
【叶红素】[生化] phylloerythrin
【叶黄素】[生化] xanthophyll
【叶尖】[植] apex
【叶枯病】[农] leaf blight
【叶蜡石】[矿] pyrophyllite
【叶绿素】[生化] chlorophyll
【叶绿体】[生化] chloroplast
【叶轮】[机] impeller; vane wheel ◇ ～泵 vane pump
【叶落归根】falling leaves settle on their roots; a person residing elsewhere finally returns to his ancestral home
【叶脉】[植] vein
【叶片】① [植] blade ②[机] vane; 涡轮～ turbine blade ◇ ～式压缩机 vane compressor
【叶鞘】[植] leaf sheath
【叶肉】mesophyll
【叶酸】[药] folic acid
【叶锈病】[农] leaf rust
【叶序】[植] phyllotaxy; leaf arrangement
【叶芽】leaf bud
【叶针】leaf thorn
【叶子】leaf

业 ①(行业) line of business; trade; industry: 各行各～ all trades and professions; different trades and callings; all walks of life ②(职业) occupation; profession; employment; job: 就～ get a job; take up an occupation / 失～ be out of a job; be unemployed ③(学业) course of study: 结～ complete a course of study ④(事业) cause; enterprise: 创～ start an enterprise ⑤(产业) estate; property ⑥(从事) engage in ⑦(已经) already: ～已核实 have already been verified ⑧[佛教] karma; deed; action
【业绩】outstanding achievement: 光辉～ glorious achievements
【业经】already: ～批准 have been approved
【业务】vocational work; professional work; business: 钻研～ study diligently one's profession ◇ ～范围 scope of business / ～方针 business policy / ～过失 professional negligence / ～能力 professional ability / ～水平 professional skill; vocational level / ～协定 business agreement / ～学习 vocational study / ～知识 professional knowledge / ～专长 specialized skill
【业余】①(工作时间以外的) sparetime: 办～进修班 run sparetime class ②(非专业的) amateur ◇ ～爱好 hobby / ～爱好者 amateur / ～教育 sparetime education / ～理论学习 theoretical study after workhours / ～体育 ama-

teur sports / ～文艺工作者 amateur literary and art workers / ～无线电通讯 amateur radio / ～学校 sparetime school / ～作者 amateur writer
【业主】owner; proprietor

曳 drag; haul; tug; tow
【曳光穿甲弹】[军] armor-piercing tracer
【曳光弹】[军] tracer bullet or shell; tracer ◇ ～弹头 tracer bullet
【曳力】[物] drag force
【曳绳钓】[渔] trolling

yī

衣 ①(衣服) clothing; clothes; garment: ～单食薄 thinly clad, badly fed / 单～ unlined dress / 夹～ lined dress / 毛～ woolen sweater / 丰～足食 have ample food and clothing ②(包在物体外面的) coating; covering: 花生～ membrane of peanuts / 糖～药片 sugar-coated pills
【衣胞】afterbirth
【衣钵】a Buddhist monk's mantle and alms bowl which he hands down to his favorite disciple; legacy: 继承某人～ take over the mantle of sb.
【衣不遮体】wear rags; wear shabby clothes
【衣橱】wardrobe
【衣蛾】casemaking clothes moth
【衣服】clothing; clothes: 穿～ put clothes on / 脱～ take clothes / 一套～ a suit of clothes / 她在～上花钱太多。She spent too much money on clothes.
【衣钩】clothes hook
【衣冠】hat and clothes; dress: ～不整 be sloppily dressed / ～楚楚 decently dressed; be immaculately dressed / ～禽兽 a beast in human clothing / ～文物 civilization and culture
【衣冠冢】a tomb containing personal effects of the deceased
【衣柜】wardrobe
【衣夹】clothes-peg; clothespin
【衣架】①(衣帽架) clothes stand; clothes tree ②(挂衣架) hanger; clothes-rack
【衣锦还乡】return to one's hometown in silken robes; return home after making good
【衣来伸手饭来张口】live on the labor of others
【衣料】material for clothing; dress material
【衣领】collar
【衣帽架】clothes tree
【衣帽间】cloakroom; checkroom
【衣衫褴褛】be dressed in rags: 他～。He was clad in rags.
【衣裳】[口] clothing; clothes
【衣食住行】food, clothing, shelter and transportation; basic necessities of life
【衣刷】clothes brush
【衣物】clothing and other articles of daily use
【衣箱】suitcase; trunk
【衣着】clothing, headgear and footwear: ～讲究 turn oneself out smartly / ～整洁 be neatly dressed

铱 [化] iridium (Ir)
【铱金笔】iridium-point pen

依 ①(依靠) depend on; rely on: ～人篱下 be dependent on sb. for a living / 无～无靠 friendless and helpless ②(依从) comply with; listen to; yield to: 她要什么,她母亲都～她。Her mother complies with every wish of hers. ③(按照) according to; in the light of; judging by: ～法定程序 in a manner prescribed by law / ～我看 in my view; as I see it
【依傍】①(依靠) rely on; depend on: 互相～ rely on each other ②(模仿) imitate; emulate: ～前人 emulate the predecessors
【依出生地取得的国籍】nationality by birth
【依此类推】the rest may be deduced by analogy; and so on and so forth
【依次】in proper order; successively: ～发言 speak in turn / ～进来 come in one by one / ～上下车 get on and off in order
【依从】comply with; yield to
【依存】depend on sb. or sth. of existence: 相互～ be interdependent
【依法】according to law; by operation of law: ～逮捕 arrest in the name of the law / ～惩办 punish according to law / ～控诉 implead / ～判决 adjudge ◇ ～惩处 legal punishment / ～扣押 legal attachment / ～扣留债务人财产 legal custody of property of a debtor / ～宣告无罪 acquittal in law
【依父母国籍而取得的国籍】nationality by parenthood
【依附】depend on; attach oneself to; become an appendage to: ～权贵 attach oneself to bigwigs / ～于帝国主义 become an appendage to imperialism
【依合同】by contract
【依旧】as before; still: 她～是那个老样。She still looks her old self.

【依居住地取得的国籍】nationality by domicile
【依据】①(根据) according to; in the light of; on the basis of; judging by: ~上述意见 in accordance with the above views ②(基础) basis; foundation: 没有~ baseless; groundless / 提供科学~ provide scientific basis for sth.
【依靠】①(凭借) rely on; depend on: ~工资生活 live on one's wages / ~群众 rely on the masses ②(可以依靠的人或东西) something to fall back on; support; backing: 寻找~ seek support
【依赖】rely on; be dependent on: ~别人 be dependent on others ◇ ~思想 the dependent mentality / ~性 dependence
【依恋】be eluctant to leave; feel regret at parting from: ~故乡的一草一木 be attached to every tree and bush in hometown
【依凭】rely on; depend on
【依然】still; as before: ~如故 remain as before; remain unchanged / ~有效 still hold good; remain valid
【依然故我】①(情况没变) one's circumstances haven't changed much ②(人仍和从前一样) one is still one's same old self
【依人门下】take shelter under other's door
【依事实宣告无罪】acquittal in fact
【依顺】be obedient: 百依百顺 be all obedience
【依托】①(依靠) rely on; depend on ②(假借某种名义) support; prop; backing
【依违两可】shilly-shally
【依稀】vaguely; dimly: ~可见 faintly visible / ~如旧 similarly as old
【依样画葫芦】copy mechanically
【依依】reluctant to part: ~不舍 be reluctant to part; cannot bear to part
【依仗】count on; rely on: ~权势 rely on one's power and position; count on one's powerful connections / ~小策 resort to petty devices
【依照】according to; in the light of: ~法律规定的条件 under conditions prescribed by law / ~情况而定 decide as circumstances require

一 ①(数目) one: ~百公斤 one hundred kilogram / ~本书 a book ②(同一) same; one: 咱们是~家人。 We are of the same family. ③(另一) also; otherwise ④(全;满) whole; all; throughout: ~年到头 all the year round ⑤(专一) concentrated; wholehearted: ~心~意 heart and soul ⑥(表示动作是一次或短暂的) 笑~笑 give a smile / 歇~歇 have a rest ⑦(每一) every; each; per: ~日三餐 three meals a day / ~小时八十公里 at 80 kilometers per hour ⑧(与"就""即"等字连用) once; as soon as
【一把手】①(参加活动的一员) partner; participant; a member; a hand ②(能干的人) a good hand; able man: 她干活真是~。 She is a good hand. ③(主要负责人) head; chief: 她曾是我们厂的~。 She has been the head of our factory.
【一把钥匙开一把锁】open different locks with different keys; use different methods to solve different problems
【一把抓】①(样样不放手) take everything into one's own hands ②(不分轻重缓急) try to tackle all problems at once regardless of their relative importance; fail to put first things first
【一败涂地】suffer a crushing defeat
【一般】①(一样) same as; just like: 他们俩~高。 The two of them are the same height. ②(一种) sort; kind: 别有~滋味 have a distinctive taste ③(普通;通常) general; ordinary; common: ~常识 general knowledge / ~情况 ordinary circumstances / ~人 common people ◇ ~代理人 general attorney / ~等价物 universal equivalent / ~规律 universal law; general rule / ~过失 general negligence / ~继承人 heir general / ~留置权 general lien / ~民事纠纷 ordinary civil disputes / ~敲诈罪 crime of blackmail in general / ~损害赔偿费 general damages / ~刑事案件 ordinary criminal case / ~刑事犯 ordinary criminal offender
【一般化】vague generalization
【一般见识】lower oneself to the same level as sb.
【一般性】generality ◇ ~辩论 general debate
【一板一眼】following a prescribed pattern in speech or action; scrupulous and methodical
【一半】one half; half; in part: ~以上 more than half / 六的~是三。 Half of six is three. / 这些书有~是德语的。 Half of the books are in German.
【一半天】in a day or two
【一帮】a gang; a band: ~匪徒 a band of gangsters
【一报还一报】return like for like; pay sb. back in his own coin; measure for measure
【一辈子】all one's life; throughout one's life; as long as lives; a lifetime
【一本万利】a small investment brings a ten thousand-fold profit; make big profits with a small capital

【一本正经】in all seriousness; in dead earnest: 看来～ look as if butter won't melt in one's mouth / 装得～的样子 be sanctimonious
【一鼻孔出气】breathe through the same nostrils; sing the same tune
【一鼻子灰】meet rejection
【一笔勾销】write off at one stroke; cancel once for all: 前帐～ cancel all debts
【一笔抹杀】blot out at one stroke; condemn out of hand; totally negate
【一碧万顷】watery blue reaching far beyond the horizon
【一臂之力】a helping hand: 助我～ lend me a hand
【一边】①(一面;一方) one side: 三角形的～ one side of a triangle ②(旁边) by the side; beside; aside: 站在～ stand aside ③(一个动作跟另一个动作同时发生) while; as; at the same time; simultaneously: 他们～学习～工作。They learn and work at the same time.
【一边倒】①(倾向一方) lean to one side; side with sb. without reservation ②(一方占绝对优势) predominate; enjoy overwhelming superiority
【一表人才】with the appearance of a talent; a man of striking appearance
【一并】[副] along with all the others; in the lump: ～处理 be handled together / 书和有关资料～寄上。I'm sending you the book along with the material concerned.
【一病不起】take to one's bed and never leave it again; fall ill and die
【一波三折】full of twists and turns or ups and downs
【一波未平,一波又起】hardly has one wave subsided when another rises; one trouble follows another
【一不做,二不休】carry the thing through, whatever the consequences; in for a penny, in for a pound; stick it out
【一步不让】not yield an inch
【一步登天】reach the sky in a single bound; attain the highest level in one step; have a meteoric rise
【一步一个脚印】every step leaves its print; work steadily and make solid progress
【一草一木】every tree and bush
【一差二错】possible mistake or mishap
【一刹那】in a moment
【一场春梦】a spring dream; a fleeting illusion
【一场空】come to naught; all in vain; futile: 竹篮打水～ as futile as drawing water with a bamboo basket; a fruitless effort
【一唱百和】when one starts signing, all the others join in; meet with general approval
【一唱一和】sing a duet with sb.; sing the same tune; echo each other
【一朝天子一朝臣】every new sovereign brings his own courtiers; a new chief brings in new aids
【一尘不染】not soiled by a speck of dust; spotless; remain uncontaminated
【一成不变】immutable and frozen; invariable; unalterable: 世界上没有～的东西。Nothing in the world is immutable.
【一筹莫展】can find no way out; be at one's wits' end; be at the end of one's tether
【一触即发】may be triggered at any moment; be on the verge of breaking out: 形势～。It's an explosive situation.
【一触即溃】collapse at the first encounter
【一传十,十传百】spread from mouth to mouth; spread far and wide
【一串】a string: ～珍珠 a string of pearls / 一长串公共汽车 a string of buses
【一锤定音】set the tune with one beat of the gong; give the final word
【一次】once: 我只见过她～。I've met her only once.
【一次方程】[数] linear equation
【一次函数】[数] linear function
【一次性付款】lump-sum payment
【一蹴而就】accomplish in one move
【一寸光阴一寸金】time is gold; time is precious
【一搭一挡】hand in glove with each other; in collusion with each other
【一打】a dozen: ～铅笔 a dozen of pencils
【一大堆】a big heap of; a great pile of; a large amount of: ～谎言 a tissue of lies / ～垃圾 a great pile of dirty / ～问题 a crop of question
【一代名儒】a scholar of no equal of his time
【一旦】①(一天之间) in a single day; in a very short time: 毁于～ be destroyed in one day / 溃于～ collapse overnight ②(不确定的时间) once; in case; now that: ～时机成熟 once the time is ripe
【一刀两断】sever at one blow; make a clean break
【一道】together; side by side; alongside: 咱俩～走吧。Let's go together.
【一得之功】just an occasional, minor success
【一等】first-class; first-rate; top-grade ◇ ～功

Merit Citation, First Class / ～秘书 First Secretary / ～品 first-rate product
【一点儿】a bit; a little: ～都不知道 have not the faintest idea / ～也不累 not feel the least bit tired / ～用处也没有 utterly useless / 嫌小了～ a bit too small
【一点一滴】bit by bit; little by little
【一定】①(规定的,确定的) fixed; established; regular: ～的规章制度 established rules and regulations / ～的生产指标 fixed production quota ②(固定不变的) definite; constant: 每一种弹簧都有～的弹性。Every sort of spring has a definite elasticity. ③(必定) surely; certainly; be bound to: 他们～会成功。They will surely succeed. ④(特定的) certain; specific; given: 在～程度上 to a certain degree / 在～条件下 under given conditions / 在～意义上 in a certain sense
【一度】①(一次) once; at one time; for a time: 一年～ once a year; yearly; annually ②(有过一次) on one occasion: ～失业 be out of job on one occasion
【一端】①(一头) one end ②(一个方面) one respect
【一段】a section; a paragraph
【一堆】a pile; a heap: ～木头 a pile of wood / ～书 a heap of books
【一对】a pair; a couple: ～新婚夫妻 a new-married couple
【一而再,再而三】again and again; time and again; repeatedly
【一二】one or two; just a few; just a little: 略知～ know a little about; have some idea about
【一发千钧】a hundredweight hanging by a hair; in imminent peril; in a very precarious situation
【一帆风顺】plain sailing
【一反常态】depart from one's normal behavior; act out of character
【一方】①(一方面) one side; a party ②(一个地区) an area or region ◇ ～当事人 one party
【一方面】one side; for one thing; on the one hand
【一分钱一分货】The higher is the price the better is the quality of the merchandise.
【一分为二】[哲] one divides into two
【一风吹】scatter to the winds; dismiss altogether; cancel the whole thing
【一夫多妻制】polygamy; polygyny
【一夫一妻制】monogamy; monogyny
【一副】a pair; a set: ～眼镜 a pair of spectacles
【一概】[副] one and all; without exception; totally; categorically: ～不准 all forbidden / ～否定 totally negate / ～排斥 totally exclude
【一概而论】lump together all and sundry; lump under one head: 不可～ not to be lumped together
【一干二净】thoroughly; completely
【一干人犯】a bunch of criminals
【一竿子插到底】carry a task right down to the grass-roots level
【一个劲儿】continuously; persistently: ～搞试验 carry on experiments persistently
【一个心眼儿】①(专心一意) have one's heart set on sth.; devotedly: ～建设社会主义 devote oneself in socialist construction ②(固执) stubbornly; inflexibly
【一共】altogether; in all; all told: 我们班～三十人。There are altogether thirty comrades in our class.
【一股脑儿】[方] completely; lock, stock and barrel; root and branch
【一鼓作气】press on to finish without letup; get sth. done in one vigorous effort
【一官半职】an unimportant official post
【一贯】consistent; persistent; all along: ～思想 consistent idea / ～政策 consistent policy
【一棍子打死】knock sb. down at one stroke; finish off with one blow; completely negate
【一锅粥】a pot of porridge; a complete mess; all in a muddle: 乱成～ be all muddled up
【一国两制】one nation, two systems
【一哄而起】be aroused to precipitate action; rush headlong into mass action
【一哄而散】break up in a hubbub; disperse in a hubbub
【一呼百应】hundreds respond to a single call
【一环扣一环】closely linked
【一晃】①(一闪) flash ②(形容时间过得快) pass in a flash: ～就是五年。Five years have elapsed in a flash.
【一挥而就】a flourish of the pen and it's done; finish a piece of writing or a painting at one go
【一会儿】①(很短的时间) a little while: 过了～ after a while / 歇～ rest for a while / 请等～。Please wait for a moment. ②(在很短的时间内) in a while; in an instant; presently; soon: 我～就来。I'll be coming in a moment.
【一伙】a gang of; a band of: ～匪徒 a gang of gangsters
【一级风】[气] force 1 wind; light air
【一技之长】proficiency in a particular line;

professional skill; speciality
【一家人】all of the same family; one family: 我们是～。We are of the same family.
【一肩行李两袖清风】possess nothing but one's personal belongings
【一见如故】feel like old friends at the first meeting; become intimate at the first meeting
【一见钟情】fall inlove at first sight
【一箭双雕】shoot two hawks with one arrow; kill two birds with one stone
【一箭之地】a stone's throw
【一将功成万骨枯】One general achieves renown over the dead bodies of 10000 soldiers.
【一经】as soon as; once; immediately after
【一举】with one action; at one stroke; at one fell swoop: ～成名 become famous overnight; achieve instant fame / ～成名天下知 gain worldwide fame and success overnight
【一举两得】kill two birds with one stone
【一举一动】any move; any action
【一句话】in a word; in short
【一蹶不振】collapse after one setback; never be able to recover after a setback
【一决雌雄】compete for championship
【一绝】a special skill; a unique talent
【一刻千金】every minute is precious
【一孔之见】a peephole view; a narrow view; a limited view
【一口】①(表示口气坚决) with certainty; readily; flatly: ～答应 readily agree; readily promise / ～否认 flatly deny / ～回绝 flatly refuse / ～咬定 assert categorically; accuse definitely ②(纯正) pure: 讲～地道的英语 speak idiomatic English ③[量] a mouthful; a bite: ～蛋糕 a bite of cake / 今天我们～饭都没吃。We have not had a bite today. / 他要～把它吃掉。He wants to eat it up at a mouthful.
【一口吃不成胖子】You can't build up your constitution on one mouthful
【一口气】①(一口气息) one breath: 深吸～ draw a deep breath ②(不间断地) in one breath; without a break; at one go; at a stretch: ～跑到家 rush to one's home at one go
【一块】a piece; a block: ～地 a piece of land / ～面包 a piece of bread
【一块儿】①(同一个处所) at the same place: 他们在～上学。They study at the same school. ②(一同) together: 咱们～走吧。Let's go together.
【一览】general survey; bird's-eye view
【一览表】table; schedule: 火车行车时间～ railway timetable
【一览无余】take in everything at a glance
【一揽子】wholesale; package ◇ ～计划 package plan / ～交易 package deal
【一劳永逸】get sth. done once and for all: ～的解决办法 a solution that holds good for all time
【一类】of the same class, category or species
【一力】do one's best; do all one can: ～成全 do one's best to help sb. (to achieve his aim)
【一连】[副] in a row; in succession; running: ～下了几天雨。It rained for several days running.
【一连串】a succession of; a series of; a string of; a chain of: ～的不幸 a series of misfortunes / ～的谎言 a string of lies / ～的灾祸 a succession of disasters
【一了百了】①(主要问题解决了其他问题也跟着解决了) all troubles end when the main trouble ends ②(死结束了一个人的一切烦恼) death ends all one's troubles
【一鳞半爪】odd bits; fragments
【一溜儿】[方] ①(一排) row: ～房屋 a row of houses ②(附近一带) neighborhood; vicinity
【一溜烟】swiftly: 小男孩～跑了。The boy ran away swiftly.
【一路】①(在整个的行程中) all the way; throughout the journey: ～领先 lead all the way ②(同一类) of the same kind: ～货色 goods of a kind; be of the same stock ③(一起) go the same way; take the same route: 咱们是～吗? Are we going the same way?
【一路多保重】Take care of yourself on the journey.
【一路平安】Have a pleasant journey. (或: Have a good trip.)
【一律】①(一个样子) same; alike; uniform: ～对待 treat in the same way ②(无例外) all; without exception
【一落千丈】suffer a disastrous decline
【一马当先】take the lead; be in the forefront
【一马平川】a wide expanse of flat land; flat country
【一脉相承】come down in one continuous line; can be traced to the same origin
【一毛不拔】unwilling to give up even a hair; very stingy
【一面】①(物体几个面之一) one side ②(一个方面) one section; one aspect: 独当～ be solely responsible for one section ③(表示一个动作跟

另一个动作同时进行) at the same time; simultaneously: ~教,~学 learn while teaching ④[量] ~鼓 a drum / ~旗 a flag
【一面之词】the statement of only one of the parties
【一面之交】have met only once; be casually acquainted
【一鸣惊人】amaze the world with a single brilliant feat; set the world on fire
【一命呜呼】die; kick the bucket
【一模一样】exactly alike; as like as two peas
【一木难支】A single post cannot support a mansion.
【一目了然】be clear at a glance
【一目十行】take in ten lines at a glance; read rapidly
【一年半载】a year or so; in about a year
【一年到头】throughout the year; all the year round
【一年生】[植] annual ◇ ~植物 annual plant / ~作物 annual crop
【一年四季】throughout the year; all the year round
【一年一度】once a year
【一年之计在于春】the whole year's work depends on a good start in spring
【一念之差】a wrong decision made in a moment of weakness; momentary slip
【一诺千金】a promise that will be kept; that's a solemn promise
【一拍即合】fit in readily; chime in easily
【一派胡言】complete nonsense
【一盘散沙】a sheet of loose sand; in a state of disunity
【一批】a batch; a shipment: ~货 a batch of goods
【一偏之见】one-sided view
【一片丹心】a leaf of red heart
【一片焦土】with everything burned down and lying in ruins
【一瞥】①(用眼一看) a quick glance: 仇恨的~ shoot a look of hatred ②(一眼看到的情况) a glimpse; a brief survey
【一贫如洗】penniless; in utter destitution
【一品红】[植] poinsettia
【一暴十寒】work hard for one day and do nothing for ten; work by fits and starts
【一妻多夫制】polyandry
【一齐】[副] at the same time; simultaneously; in unison: ~出动 go into action simultaneously / ~鼓掌 clap hands in unison / ~努力 make a concerted effort
【一起】①(同一处所) in the same place: 坐在~ sit together ②[方](一同) altogether; in all: 这几件东西~多少钱? How much is that altogether?
【一气】①(不间断的) at one go; without a break; at a stretch: ~呵成 accomplish sth. at a stretch ②(声气相通) of the same gang; hand in glove: 串通~ work hand in glove ③(一阵) a spell; a fit: 胡吹~ tell tall stories
【一钱不值】not worth a penny; utterly worthless; mere trash
【一窍不通】know nothing about; lack the slightest knowledge of; be utterly ignorant of
【一切】all; every; everything: ~就绪 all in order / ~落空 a complete failure / 调动~积极因素 bring all positive factors
【一切险】all risks: ~条款 all risks clause
【一清二楚】perfectly clear; as clear as daylight
【一清早】early in the morning
【一穷二白】poor and blank
【一丘之貉】jackals from the same lair; birds of a feather
【一去不复返】gone for ever; gone never to return
【一群】a group; a crowd; a herd; a pack; a flock
【一人得道,鸡犬升天】when a man attains the Tao, even his pets ascend to heaven; when a man gets to the top, all his friends and relations get there with him
【一人做事一人当】a man must bear the consequences of his own acts; one should answer for what he does himself
【一仍旧贯】stick to the old practice; follow the old routine
【一日千里】a thousand *li* a day; at a tremendous pace; with giant strides
【一日三秋】one day apart seems like three years
【一如既往】just as in the past; as before; as always
【一扫而光】make a clean sweep of; clear off; finish off; get rid of sth. lock, stock and barrel
【一色】①(一样的颜色) of the same color: 水天~。The water and the sky are of one hue. ②(全部一样的) of the same type; uniform: ~的六层楼房 six-storeyed buildings of a uniform style
【一身】①(全身) the whole body; all over the body: ~冷汗 break out in cold sweat / ~泥巴 be covered all over with mud / ~是病 be af-

flicted by several ailments or diseases ②(一套) a suit: 穿着～工作服 wear a boiler suit ③(一个人) all alone
【一身是胆】know no fear; be absolutely fearless
【一审】first instance: ～辩论的终结 close of argument at the first trial
【一神教】monotheism
【一生】all one's life; throughout one's life
【一声不响】not say a word; not utter a sound
【一失足成千古恨】a single slip may cause lasting sorrow; the error of a moment becomes the regret of a lifetime
【一时】①(短时间) for a short while; temporary; momentary: ～冲动 momentary impulse / ～的热情 a moment's enthusiasm / ～糊涂 get confused temporarily ②(一个时期) a period of time: 此～,彼～。Times have changed. ③(时而) now... now...: 天气～冷,～热。The weather is now cold, now hot.
【一时半刻】a short time; a little while
【一时之计】temporizing measures of the time
【一事无成】accomplish nothing; get nowhere
【一视同仁】treat all men alike; treat equally without discrimination
【一手】①(技能或本领) proficiency; skill: 露～ show off one's skill ②(指要手段) trick; move: 他这～真毒辣! What a vicious trick he played! ③(一个人) single-handed; all by oneself; all alone: ～包办 take everything on oneself / ～扶植 foster single-handedly / ～交钱,～交货 cash on delivery; transaction in cash
【一手遮天】shut out the heavens with one hand; hide the truth from the masses; hoodwink the public
【一瞬】an instant; a flash; the twinkling of an eye: ～即逝 vanish in a flash
【一丝不苟】not be the least bit negligent; be scrupulous about every detail; be conscientious and meticulous
【一丝不挂】not have a stitch on; be stark-naked; nude
【一丝一毫】a tiny bit; an iota; a trace
【一塌糊涂】in a complete mess; in an awful state; in utter disorder: 屋子乱得～。The room was in a mess.
【一潭死水】a pool of stagnant water; a stagnant condition; a lifeless condition
【一套】①(固定方式) one's set way: 他干事有他的～。He has his own set way of doing things. ②(固定内容) convention; the same stuff: 老～ old convention; the same old stuff ③[量](衣物房间等) a set; a flat; a suite; a suit: ～房间 a suite of rooms / ～衣服 a suit of clothes
【一体】①(关系密切) an organic whole: 融成～ merge into an organic whole ②(全体) all people concerned; to a man
【一天】①(一昼夜) a day: ～又～ day after day ②(一个白天) one day ③[方](一天到晚) the whole day; all day long; from morning till night
【一天到晚】from morning till night; from dawn to dusk; all day long
【一条龙】①(一个行列) one continuous line ②(生产或工作程序的紧密联系和配合) a connected sequence; a coordinated process
【一条心】be of one mind; be at one: 军民～。Our fighters and people are of one mind.
【一通万通】master one and you'll master a hundred; grasp this and you'll grasp everything
【一同】[副] together; at the same time and place: ～工作 work together
【一统】unify: ～天下 unify the whole country
【一团和气】keep on good terms with everyone at the expense of principle; keep on the right side of everyone
【一团漆黑】pitch-dark; utterly hopeless
【一团糟】a complete mess; a chaotic state
【一网打尽】catch the whole lot in a dragnet; round up the whole gang at one fell swoop
【一往情深】be passionately devoted; be head over heels in love
【一往无前】press forward with indomitable will: ～的精神 indomitable spirit
【一望无际】stretch as far as the eye can see; stretch to the horizon
【一望无垠】look into a distance of no boundary
【一味】[副] blindly: ～地固执己见 stubbornly stick to one's own view / ～孤行 go one's own way / ～蛮干 persist in acting blindly / ～迁就 make endless concessions; make one concession after another / ～退让 retreat constantly
【一文不名】be penniless; be cleaned out
【一文不值】not worth a cent; of no use whatsoever
【一问三不知】not know a thing; be entirely ignorant
【一窝蜂】like a swarm of bees: 学生们～似地奔向操场。The students swarmed towards the playground.
【一无可取】be worthless; have nothing to rec-

ommend one
【一无是处】without a single redeeming feature; devoid of any merit
【一无所长】have no special skill; be Jack of all trades
【一无所获】have gained nothing
【一无所有】not own a thing in the world; not have a thing to one's name
【一无所知】know nothing about; not have the least inkling of; be absolutely ignorant of
【一五一十】systematically and in full detail
【一物降一物】there is always one thing to conquer another; everything has its vanquisher
【一误再误】make things worse by repeated delays
【一息尚存】so long as one still has a breath left; till one's last gasp
【一息奄奄】in the last gasp
【一席话】what one says during a conversation
【一系列】a series of: ～措施 a series of measures / ～的事件 a whole train of events / ～工作 a great deal of work / ～观点 a whole string of views / ～事实 a host of facts / ～问题 a whole series of questions
【一下】①(表示做一次动作或试着做) one time; once: 看～ have a look / 我来试～。Let me have a try. ②(短暂的时间) in a short while; all at once; all of a sudden: 需要睡～ need a little sleep / 天～阴了下来。It became overcast all of a sudden.
【一线】a ray of; a gleam of: ～生机 a slim chance of survival / ～希望 a gleam of hope
【一相情愿】one's own wishful thinking
【一向】①(过去的某一段时间) earlier on; lately: 前～你到哪里去? Where have you been lately? ②[副](一直) consistently; all along: 他～忠诚老实。He is always honest.
【一小撮】a handful: ～敌人 a handful of enemies
【一笑置之】laugh out of court; laugh off: 大家对他们的那些议论～。Their arguments were laughed out of court.
【一些】[量] a number of; certain; some; a few; a little: 有～国家 some countries / 作～适当的调整 make certain appropriate readjustments
【一泻千里】①(指水流) rush down a thousand *li*; flow down vigorously ②(文笔奔放) bold and flowing
【一蟹不如一蟹】each crab is smaller than the one before; each one is worse than the last
【一心】①(专心) wholeheartedly; heart and soul: ～为人民谋福利 work for the people's interests heart and soul ②(齐心) of one mind; at one: 万众～。Millions of people are all of one mind.
【一心一德】be of one heart and one mind
【一心一意】heart and soul; wholeheartedly
【一星半点】a tiny bit; a very small amount
【一行】a group travelling together; party: 他们～五人昨天到达北京。The five of them arrived in Beijing yesterday.
【一言不发】not say a word; keep one's mouth shut
【一言既出，驷马难追】a word once spoken cannot be overtaken even by a team of four horses; what is said cannot be unsaid
【一言难尽】it is hard to explain in a few words; it's a long story
【一言丧邦】A single wrong statement may bring disaster to the nation.
【一言堂】what I say goes; one person alone has the say; one person lays down the law
【一言为定】that's settled then
【一言兴邦】A timely warning may avert a national crisis.
【一言以蔽之】to sum up in a word
【一眼】a glance: ～看见某物 see sth. at a glance / 我看了她～。I gave her a glance.
【一氧化碳】[化] carbon monoxide; white damp ◇～中毒 carbon monoxide poisoning
【一氧化物】[化] monoxide
【一样】the same; equally; alike; as... as: ～大小的房屋 houses all of a size / 完全～ just the same; exactly alike / 他跑得跟他弟弟～快。He runs as fast as his brother.
【一叶障目，不见泰山】a leaf before the eye shuts out Mount Taishan; have one's view of the important overshadowed by the trivial
【一叶知秋】the falling of one leaf heralds the autumn; it is a straw in the wind; a small sign can indicate a great trend
【一一】one by one; one after another: ～告别 say good-bye to everyone / ～核对事实 check facts one by one / ～就座 seat themselves one after another / ～申辩 justify oneself point by point
【一衣带水】a narrow strip of water: ～的邻邦 close neighbors separated only by a strip of water
【一意孤行】cling obstinately to one's course; act wilfully; be bent on having one's own way

【一应俱全】complete in every line; everything needed is there
【一拥而上】rush up in a crowd
【一隅】a corner: ～之见 a glance from a corner
【一语道破】lay bare the truth with one penetrating remark; hit the nail on the head
【一语破的】hit the mark with a single comment
【一语双关】pun on a word
【一元方程】equation with one unknown quantity
【一元化】centralized; unified: ～的领导 centralized leadership
【一元论】①[哲] monism ②[免疫] monophyletic theory
【一月】January
【一再】time and again; again and again; repeatedly: ～高呼 shout again and again / ～警告某人 caution sb. repeatedly / ～宣称 declare time and again
【一早】[口] early in the morning
【一张一弛】stretched and relaxed; alternate tension with relaxation
【一长制】system of one-man leadership
【一着不慎,满盘皆输】one careless move and the whole game is lost
【一朝被蛇咬,十年怕井绳】Once bitten by a snake, one shies at a coiled rope for the next ten years. (或: Once bitten, twice shy.)
【一朝覆亡】collapse in one short day.
【一朝一夕】in one morning or evening; overnight; in one day: 非～之功 not the work of a single-day
【一朝之忿】a sudden outburst of anger
【一针见血】pierce to the truth with a single pertinent remark; hit the nail on the head
【一枕黄粱】a brief dream of grandeur
【一阵】in burst; a fit; a peal: ～狂风 a violent gust of wind / ～猛烈的炮声 a burst of furious shelling; a hail of shells / ～响亮的鼓掌声 a peal of applause
【一阵子】a period of time; a spell: 发了～烧 have a spell of fever
【一枝独秀】outshine others; One branch of the tree is particularly thriving.
【一枝之栖】get a branch to roost
【一知半解】have a smattering of knowledge; have scanty knowledge
【一直】①(不拐弯的) straight: ～往东 go straight eastward / ～走 go straight ahead; keep straight on ②[副](始终) continuously; always; all along; all the way: 我们～是同班同学。We have been classmates all along.
【一纸空文】a mere scrap of paper
【一致】showing no difference; identical; unanimous; consistent: ～同意 agree unanimously / 对某事意见～ be in agreement on sth. / 观点～ hold identical views; be of the same view / 体例前后不～。The style is inconsistent.
【一掷千金】throw away money like dirt; spend gold on one throw
【一专多能】expert in one thing and good at many
【一准】[副] sure; surely; certainly: 她～来。She is sure to come.
【一字长蛇阵】single-line battle array: 摆开～ string out in a long line
【一字尺】[图] parallel rule
【一字之差】the change of one word
【一字一板】unhurriedly and clearly
【一总】①(合并) altogether; all told; in all: 她～花了五十元。She spent fifty *yuan* in all. ②(全部) all: 这些任务～交给你干。Leave all the tasks to you.
【一走了之】evade the solution of a problem by walking away from where is exists

壹 one

医 ①(医生) doctor; physician: 军～ medical officer; surgeon / 缺～少药 shortage of medical services and medicines ②(医学) medical science; medical service; medicine: ③(医治) cure; treat: ～好某人的疾病 cure sb. of his illness / 给某人～病 give sb. medical treatment / 头痛～头,脚痛～脚 take stopgap measures
【医道】skill 〈knowledge〉 of a doctor
【医德】medical ethics
【医科】medical courses in general; medicine ◇ ～大学 medical university
【医理】principles of medical science; medical knowledge
【医疗】medical treatment: 公费～ public health services ◇ ～队 medical team / ～机构 medical establishment ～器械 medical apparatus and instruments / ～事故 unskillful and faulty medical or surgical treatment; malpractice / ～事故保险 malpractice insurance / ～体育 medico- athletics / ～卫生工作 medical and

health work
【医疗站】medical station; health center
【医生】doctor; medical man: 产科～ obstetrician／ 儿科～ pediatrician／耳鼻喉科～ ENT (ear- nose- throat) doctor／ 放射科～ radiologist／ 妇科～ gynecologist／ 骨科～ orthopedist／ 泌尿科～ urologist／ 内科～ physician／ 皮肤科～ dermatologist／ 实习～ intern／ 外科～ surgeon／ 牙科～ dentist; dental surgeon／ 眼科～ oculist; eye-doctor／ 整形外科～ plastic surgeon／ 肿瘤科～ oncologist／ 主治～ doctor in charge／住院～ resident doctor ◇ ～检查报告书 medical report／ ～提供的证据 medical evidence
【医师】doctor: 公共卫生～ public healthe doctor
【医士】practitioner with secondary medical school education
【医书】medical book
【医术】medical skill; art of healing
【医务】medical matters ◇ ～工作者 medical worker／ ～人员 medical personnel; public health worker
【医务所】clinic
【医学】medical science; medicine: 法～ forensic medicine; legal medicine／ 基础～ preclinical medicine／ 预防～ preventive medicine ◇ ～博士 Doctor of Medicine／ ～科学院 academy of medical sciences／ ～文献 medical literature／ ～学士 Bachelor of Medicine／ ～遗产 medical heritage／ ～院 college of medicine／ ～证人 medical witness
【医药】medicine ◇ ～常识 general medical knowledge／ ～费 medical expenses
【医院】hospital: 传染病～ hospital for infectious diseases／ 儿童～ children's hospital／ 妇产～ obstetrics and gynecology hospital／结核病～ tuberculosis hospital／ 口腔～ stomatological hospital／ 陆军～ army hospital／野战～ field hospital／中医～ hospital of Chinese medicine／ 肿瘤～ tumor hospital／ 综合性～ general hospital ○精神病院 mental hospital／ 麻风病院 hospital for lepers; leprosarium／ 疗养院 sanatorium／ 诊疗所 clinic／ 急救站 first-aid station／ 防疫站 quarantine station／ 内科 medical department; department of internal medicine／ 外科 surgical department; department of surgery／ 小儿科 pediatrics department／ 妇产科 obstetrics and gynecology department／ 眼科 ophtalmology department／ 牙科 dental department／ 耳鼻喉科 ENT (era-nose-throat) department／ 泌尿科 urology department／ 皮肤科 dermatology department; skin department／ 矫形外科 orthopedic surgery department／ 创伤外科 traumatology department／ 整形外科 plastic surgery／ 麻醉科 anesthesiology department／ 病理科 pathology department／ 心脏病科 cardiology department／ 精神病科 psychiatry department／ 骨科 orthopedics department／ 心脏外科 department of cardiac surgery／ 胸外科 department of cerebral surgery／ 神经科 neurology department／ 神经外科 neurosurgery department／ 脑外科 thoracic surgery department／ 中医科 department of traditional Chinese medicine／ 挂号处 registration office／ 门诊部 out-patient department／ 住院部 in-patient department／ 护理部 nursing department／ 诊室 consulting room／ 候诊室 waiting room／ 急诊室 emergency room／ 住院处 admitting office／ 手术室 operation room／ 放射科 X- ray department／ 血库 blood bank／ 药房 dispensary; pharmacy／ 病房 ward／ 化验室 laboratory／ 化验员 laboratory technician／ 护士 nurse／ 护士长 head nurse／ 麻醉师 anesthetist／ 药剂师 pharmacist; druggist／门诊病人 out-patient／ 急诊病人 emergency case
【医治】cure; treat; heal: ～烧伤 healing of burns／ ～无效 fail to respond to any medical treatment／ ～战争创伤 heal war wounds
【医嘱】doctor's advice: 遵照～服药 take medicine according to doctor's orders
【医助】assistant doctor

伊

she; he
【伊甸园】[基督教] the Garden of Eden: paradise
【伊拉克】Iraq ◇ ～人 Iraqi
【伊朗】Iran ◇ ～人 Iranian
【伊始】beginning: 就职～ upon assuming office／ 下车～ as soon as one alights from the official carriage; on arrival at a new post
【伊斯兰教】Islam; Islamism ◇ ～国家 Islamic country／ ～历 the Moslem Calendar／ ～徒 Moslem ○ 逊尼派 Sunni／ 逊尼派教徒 Sunnite／ 什叶派 Shiahs／ 什叶派教徒 Shiah／ 哈乃斐派 Hanafite school／ 古兰经 Koran／ 肉孜节 Festival of Fast-breaking／ 古尔邦节 Corban／ 拜兰节 bairam
【伊蚊】[动] yellow-fever mosquito; aedes

咿 ①[象] squeak; creak: ～的桨声 the squeak of oars in oarlocks ②(小孩学话的声音) prattle; babble

yí

宜 ①(合适) suitable; appropriate; fitting: 适～ fitting and proper ②(应当) should; ought to: 不～操之过急。 You should not act in haste. / 不～强加于人。You should not force your view on others.

【宜人】pleasant; delightful: 景物～ attractive scenery / 气候～ pleasant weather

颐 ①(保养) keep fit ②(腮;颊) cheek: 支～ cheek in palm

【颐和园】the Summer Palace

【颐养】take care of oneself

【颐指气使】order people about by gesture; be insufferably arrogant

咦 [叹] well; why

胰 [生理] pancreas

【胰蛋白酶】[生化] trypsin

【胰岛素】[药] insulin: 完全人工合成结晶牛～ total synthetic crystalline bovine insulin

【胰淀粉酶】[生化] amylopsin

【胰腺】[生理] pancreas ◇ ～炎 pancreatitis

【胰液】[生理] pancreatic juice

【胰脂酶】[生化] pancreatic lipase; steapsin

姨 ①(姨母) one's mother's sister; aunt ②(妻子的姐妹) one's wife's sister; sister-in-law

【姨表】maternal cousin: ～姐妹 female maternal cousins / ～兄弟 male maternal cousins

【姨父】uncle; the husband of one's maternal aunt

【姨母】maternal aunt; aunt

【姨太太】[口] concubine

遗 ①(遗失) lose; miss ②(遗失的东西) something lost: 路不拾～。 No one pockets anything found on the road. ③(遗漏) omit; slip over: 巨细无～ slip over nothing whether big or small ④(留下) leave behind; keep back; not give: 不～余力 spare no efforts ⑤(专指死人留下的) leave behind at one's death; bequeath; hand down: ～骨 remains / ～作 posthumous work ⑥(遗精) emission: 梦～ noctural emission

【遗笔】writings of a deceased person

【遗产】legacy; inheritance; heritage: 继承～ inherit a legacy / 历史～ a legacy of history / 文化～ cultural heritage ◇ ～管理人 administrator; manager of the deceased property / ～继承人 heir to property; inheritor / ～清册 account of executors / ～税 inheritance tax; succession duty / ～赠与者 devisor

【遗臭万年】leave a stink for ten thousand years; go down in history as a byword of infamy

【遗传】[生] heredity; inheritance: 交叉～ criss-cross inheritance / 这种病～吗? Is this disease hereditary? ◇ ～变异 genetic variation / ～病 hereditary disease / ～工程 genetic engineering / ～密码 genetic code / ～免疫 inherent immunity / ～生态学 genecology / ～特征 hereditary feature; heredity / ～信息 hereditary information / ～性精神病 hereditary insanity / ～学 genetics / ～学家 geneticist / ～因素 inherent cause / ～因子 genetic factor

【遗风】a custom left by a preceding generation

【遗腹子】posthumous child ◇ ～的领养 adoption posthumous

【遗稿】a manuscript left unpublished by the author at his death; posthumous manuscript

【遗孤】orphan

【遗骸】remains

【遗憾】regret; pity: 对此表示～ express regret over the matter / 一点不感到～ have no regrets / 终身～ a lifelong regret

【遗恨】eternal regret: ～绵绵 a lasting remorse

【遗迹】historical remains; vestige; traces: 古代文明的～ traces of an ancient civilization

【遗迹江湖】go away from home and roam about the world; retire to lake district

【遗教】instructions left by the dead

【遗精】[医] emission

【遗老】old fogy; old diehard; old adherent of the past dynasty

【遗留】leave over; hand down: 历史上～下来的边界问题 boundary questions left over by history

【遗漏】omit; leave out: 重要～ an important omisiom / 名单上把他名字～了。 His name was missing on the list. / 一个标点都没～。Not a comma was left out.

【遗墨】letters; manuscripts, scrolls of painting or calligraphy, etc. left behind by the deceased

【遗尿】[医] enuresis; bed-wetting

【遗弃】abandon; forsake; cast off ◇ ～案 case of abandonment／～物 abandon／～子女 abandonment of children／～罪 crime of abandonment
【遗容】①(人死后的容貌) remains: 瞻仰～ pay one's respects to the remains of sb. ②(遗像) a portrait of the deceased
【遗少】young man with the mentality of an old fogy; young diehard
【遗失】lose: ～招领 lost and found／我的字典～了。I've lost my dictionary.
【遗事】①(前代留下的事迹) incidents of the past ages ②(前人的事迹) deeds of those now dead
【遗书】①(后人刊印的前人书信) posthumous papers; writings of an author now dead ②(临终留下的书信) a letter or note left by one immediately before death
【遗孀】widow; relict
【遗俗】an old custom; traditional practices: ～之累 the burden of inherited customs
【遗体】remains: 向～告别 pay one's last respects to the remains
【遗忘】forget: 不要～了。Don't forget.
【遗忘症】amnesia
【遗物】things left behind by the deceased
【遗像】a portrait of a dead person
【遗笑后人】leave a laughing-stock for descendants; a butt for ridicule
【遗训】teachings of the deceased
【遗言】words of the deceased; last words
【遗业】business or career left behind by one's ancestors
【遗愿】unfulfilled wish of the deceased; last wish; behest
【遗赠】bequeath ◇ ～动产和不动产 devise and bequeath／～房产 demise premises／～人 devisor; legator／～者 donor; legator
【遗址】ruins; relics: 古城～ the ruins of an ancient city
【遗志】unfulfilled wish; behest; work bequeathed by the deceased: 继承先烈～ carry out the behest of the martyrs; continue the work left by the martyrs
【遗嘱】testament; will; dying words: 立～ make a will ◇ ～的公证 insinuation of a will／～的执行 execution of testament／～附录 codicillary／～认证 probate／～指定的监护人 guardian by testament／～捐赠 donation by will／～赠与 gift by will／～赠与继承人 heir by devise
【遗著】posthumous work

仪 ①(人的外表) appearance; bearing ②(礼节;仪式) ceremony; rite: 失～ commit a breach of etiquette／司～ master of ceremonies ③(礼物) present; gift: 贺～ present for birthday, wedding, etc ④(仪器) apparatus; instrument
【仪表】①(人的外表) appearance; bearing: ～不凡 handsome looks／～大方 poised and graceful／～堂堂 noble and dignified ②(各种测定仪) meter; instrument ◇ ～板 fascia board; instrument panel／～舱 guidance section／～厂 instrument and meter plant／～灯 gage lamp／～飞行 instrument flight／～进场着陆 instrument approach／～盘 gage board／～着陆 blind landing; instrument landing
【仪器】instrument; apparatus: 精密～ precision instrument ◇ ～厂 instrument plant／～制造工业 instrument-making industry
【仪容】looks; appearance: ～端庄 one's demeanor is upright
【仪式】ceremony; rite; function: 发奖～ prize-giving ceremony／火把点燃～ flame-lighting ceremony／举行隆重～ hold a grand ceremony／协定签字～ a ceremony for signing an agreement／宗教～ religious rites
【仪态】[书] bearing; deportment: ～万方 appear in all her glory／～万千 distinguished air of elegance and coquetry
【仪仗】flags, weapons, etc. carried by a guard of honor
【仪仗队】guard of honor; honor guard: 检阅～ review the guard of honor／陆海空三军～ a guard of honor of the three services

移 ①(移动) move; remove; shift ②(改变) change; alter
【移步】move one's steps; walk
【移调】[乐] transposition ◇ ～乐器 transposing instrument
【移东就西】make up deficiency by funds elsewhere
【移动】move; shift: 请你帮我把家具～一下。Will you help me to shift the furniture about please? ◇ ～摄影 tracking
【移防】be shifted elsewhere for garrison duty
【移风易俗】change prevailing habits and customs; transform social traditions
【移花接木】①(把带花的枝条嫁接到别的树木上) graft one twig on another; graft ②(暗中更

换人或事）stealthily substitute one thing for another
【移画印刷法】decalcomania
【移交】turn over; transfer; deliver into sb.'s custody; hand over one's job to a successor
【移居】move one's residence; migrate ◇ ～人境 immigration
【移苗】transplant seedlings
【移民】①(迁移至外地或外国) migrate; emigrate(移出); immigrate(移入) ②(迁移的人) migrant; immigrant(移入); emigrant(移出) ◇ ～点 settlement / ～法 immigration laws / ～局 immigration office / ～签证 immigrant visa / ～政策 immigration policy
【移情别恋】shift one's love to another person; have a new sweetheart
【移山倒海】remove mountains and drain seas; transform nature
【移项】transplant
【移植】①[农](移苗) transplant ②[医](机体移植) transplanting; grafting
【移樽就教】take one's wine cup to another person's table to seek his advice; go to sb. for advice

疑 ①(怀疑) doubt; disbelieve; suspect: 坚信不～ firmly believe; not have the slightest doubt / 无可置～ beyond doubt; undoubtedly ②(未能确定的) doubtful; uncertain: 存～ leave the question open
【疑案】①(疑难案件) disputed case ②(真相不明之事件) mystery
【疑兵】troops deployed to mislead the enemy; deceptive deployment
【疑病】[医] hypochondriasis
【疑点】doubtful point
【疑惑】feel uncertain; not be convinced: ～不解 feel puzzled; have doubts
【疑惧】apprehensions; misgivings
【疑虑】misgivings; doubt: ～莫决 suspicious and unable to decide / 消除心中的～ clear one's mind of doubt; free sb. from doubts and misgivings
【疑难】difficult; knotty: ～病症 difficult and complicated cases / ～问题 a knotty problem
【疑神疑鬼】be terribly suspicious; be even afraid of one's own shadow
【疑似】could be; suspected to be
【疑团】doubts and suspicions: 满腹～ be full of doubts and suspicions / ～尽释。 All the suspicions are cleared up.
【疑问】query; question; doubt: 毫无～ doubtless; without a doubt; without question ◇ ～句 interrogative sentence
【疑心】suspicious 起～ become suspicious / ～自误 make one's own mistakes for being suspicious / 我～他在说谎。 I suspect that he was lying.
【疑心病】a suspicious frame of mind
【疑信参半】half in belief and half in doubt
【疑义】doubt; doubtful point: 毫无～ no doubt
【疑云】misgivings or suspicion clouding one's mind
【疑阵】deceptive battle array to mislead the enemy; stratagem

怡 happy; cheerful; joyful
【怡和】delightful harmony; on very pleasant terms
【怡情养性】contribute to one's peace of mind or inner tranquility
【怡然】happy; contented: ～自得 happy and pleased with oneself
【怡颜悦色】a cheerful countenance and contented appearance

贻
【贻贝】[动] mussel
【贻害】leave a legacy of trouble: ～后世 entail woe on the after ages / ～无穷 entail untold troubles
【贻人口实】give occasion for scandal or gossip; be a source of ridicule
【贻误】affect adversely; bungle: ～大局 disrupt the general plan / ～工作 affect the work adversely / ～战机 bungle the chance of winning a battle; forfeit a chance for combat / ～终身 bring evil upon one's whole life
【贻笑大方】make a laughingstock of oneself before experts; incur the ridicule of experts
【贻赠】present; leave something to posterity

饴 maltose: 甘之如～ enjoy sth. bitter as if it were sweet as malt sugar; gladly endure hardships
【饴糖】maltose; malt sugar

yǐ

旖
【旖旎】charming and gentle: ～风光 a romantic or charming sight

椅 chair: 安乐～ easy chair / 单扶手～ tablet chair / 扶手～ armchair / 藤～ wicker chair / 摇～ rocking chair / 折～ folding chair / 转～ swivel chair / 坐在～上 sit on a chair
【椅背】the back of a chair
【椅垫子】a chair cushion
【椅套】chair cover

倚 ①(靠) lean on or against; rest on or against: ～门而立 stand leaning by the door / ～墙 lean against the wall ②(仗恃) rely on; count on: ～势欺人 take advantage of one's position to bully people ③(歪;偏) biased; partial: 不偏不～ unbiased; impartial
【倚靠】lean on or against; rest on or against
【倚老卖老】take advantage of one's seniority or old age; flaunt one's seniority
【倚势凌人】throw one's weight around
【倚仗】rely on; count on: ～权势 rely on one's power and position; count on one's powerful connections
【倚重】rely heavily on sb.'s service

蚁 ant: 兵～ soldier ant; dinergate / 雌～ gyne / 工～ ergate; worker ant / 雄～ aner
【蚁蚕】[动] newly-hatched silkworm
【蚁巢】ant nest
【蚁酸】[化] formic acid
【蚁丘】anthill
【蚁走感】formication

乙 ①(天干第二位) the second of the ten Heavenly Stems ②(二) second
【乙醇】ethanol; alcohol
【乙等】the second grade; Grade B
【乙二醇】[化] (ethylene) glycol
【乙醚】ether
【乙醛】acetic aldehyde; acetaldehyde
【乙炔】acetylene; ethyne ◇ ～发生器 acetylene generator / ～焊 acetylene welding / ～汽油 acetyl gasoline
【乙烷】ethane
【乙烯】ethylene: 聚～ polyethylene; polythene ◇ ～基 vinyl / ～树脂 ethylene resin; polyethylene / ～塑料 vinyl plastics
【乙酰】acetyl ◇ ～胆硷 acetylcholine / ～唑胺[药] acetazolamide; diamox
【乙型脑炎】encephalitis B; Japanese encephalitis
【乙种粒子】[物] beta particle
【乙种射线】[物] beta ray

钇 [化] yttrium (Y)

已 ①(停止) stop; cease; end: 争执不～ dispute endlessly ②(已经) already: ～成定局 be a foregone conclusion / 他～来过了。 He has already been.
【已承兑汇票】accepted bill
【已定罪者】convicted person
【已故】deceased; late: ～主席 the late chairman / 他的～的祖父 his deceased grandfather
【已婚】be married
【已结案件】chose jugée
【已经】already: ～到夏天了。 It's summer already. / 节目～完了。 The program is over.
【已决犯】[法] convicted prisoner; convict
【已决事件】[法] res judicata
【已死受害人】deceased victim
【已往】before; previously; in the past
【已消偿的债务】discharged obligation
【已证实的损害】proved damage
【已知数】[数] known number
【已知损失】known loss

尾 ①(马尾上的毛) hairs on a horse's tail ②(蟋蟀等尾部的针状物) spikelets on a cricket's tail

酏
【酏剂】[药] elixir

迤 go towards: …～东 to the east of
【迤逦】winding; tortuous; meandering

以 ①(用;拿) use; take: ～理服人 convince people by reasoning / ～其人之道,还治其人之身。 Pay a man back in his own coin. (或: Deal with a man as he deals with you.) ②(依) according to: ～成绩录取新生 enroll students according to their merits / ～姓氏笔画为序 be listed in the order of the number of strokes in their surnames ③(因) because of: 不～失败自馁,不～成功自满 not lose heart because of failure nor feel conceited because of success ④(表示目的) in order to; so as to: ～应急需 in order to answer an urgent need ⑤(表明时间、地位、方向或数量的界限) 二十年～前 twenty years ago or earlier / 五十～内 less than fifty
【以饱私囊】fill one's pocket
【以暴易暴】replace one tyranny by another
【以备不虞】be prepared for any contingency
【以便】so that; in order to; so as to; with the

aim of; for the purpose of
【以诚相与】be honest with
【以此】for this reason; on this account: ～为戒 take this as a lesson
【以次】①(依次序) in proper order ②(某处以下) the following: ～各章 the following chapters
【以德报怨】return good for evil; requite ingratitude with kindness
【以点带面】fan out from point to area; use the experience of selected units to promote work in the entire area
【以动产为担保的借贷】chattel loan
【以毒攻毒】combat poison with poison; use poison as an antidote for poison
【以杜后患】forestall future trouble
【以讹传讹】incorrectly relay an erroneous message
【以耳代目】rely upon hearsay instead of seeing for oneself
【以法律为准绳】take law as the criterion
【以防万一】be ready for any eventuality; be prepared for all contingencies
【以丰补歉】store up in fat years to make up for lean ones
【以攻为守】use attack as a means of defense; attack in order to defend
【以寡敌众】pit the few against the many; fight against heavy odds
【以观后效】see how the offender behaves
【以后】after; afterwards; later; hereafter: 从今～ from now on / 很久～ long afterwards
【以及】as well as; along with; and
【以己度人】judge others by oneself; measure other's corn by one's own bushel
【以假乱真】mix the spurious with the genuine
【以近】[交] up to: 武汉～的火车票 train tickets up to Wuhan
【以儆效尤】to warn others against following a bad example; as a warning to others
【以救眉急】answer urgent needs
【以来】since: 长期～ for a long time past / 三年～ in the past three years
【以礼相待】treat sb. with due respect
【以理服人】convince people by reasoning
【以邻为壑】use one's neighbor's field as a drain; shift one's troubles onto others
【以卵投石】throw an egg against a rock; court defeat by fighting against overwhelming odds
【以貌取人】judge people solely by their appearance
【以免】in order to avoid; so as not to; lest: 仔细检查～出错 check carefully to avoid mistakes
【以目示意】give a hint with the eyes; give a meaning look
【以内】within; less than; inside of: 长城～ within the Great Wall / 三年～ within three years
【以前】before; formerly; previously: ～各版 all the preceding editions / 很久～ a long time ago
【以勤补拙】amend stupidity by diligence
【以求】in order to; in an attempt to: ～一逞 in the hope of realizing one's ambition; in a bid for success / 梦寐～ crave for; long-dreamed of
【以色列】Israel ◇ ～人 Israeli; Israelite
【以上】①(某一个点以上) more than; over; above: 十二岁～的女孩 girls of twelve and upwards / 十人～ more than ten people ②(前面的) the above; the foregoing; the above-mentioned: ～各章 the above chapters
【以身试法】defy the law; challenge the law personally
【以身许国】dedicate oneself to country's cause
【以身殉职】die at one's post
【以身作则】set an example
【以事实为根据】base on fact
【以售其奸】in order to carry out an evil plot
【以太】[物] ether
【以退为进】retreat in order to advance; make concessions in order to gain advantages
【以外】beyond· outside; other than; except: 长城～ beyond the Great Wall / 离城三里～ three *li* beyond the town
【以往】before; formerly; in the past
【以为】think; believe; consider: 我～她是对的。 I believe she is right. / 我～这不可能。 I don't think it possible.
【以文会友】make friends by one's writing
【以下】①(在某一点之下) below; under: 零度～ below zero; sub-zero / 七岁～的儿童 children under seven ②(指下面的) the following: ～各章 the following chapters
【以小人之心,度君子之腹】gauge the heart of a gentleman with one's own mean measure
【以小易大】throw a sprat to catch a whale
【以信用担保的贷款】loan secured by credit
【以眼还眼,以牙还牙】an eye for an eye and a tooth for a tooth
【以一当十】pit one against ten
【以一警百】punish one to caution the others

【以逸待劳】wait at one's ease for an exhausted enemy or opponent
【以有易无】exchange what one has for what one has not
【以远】[交] beyond ◇ ～权 the right to extend a flying route
【以怨报德】return evil for good; requite kindness with ingratitude
【以正视听】in order to ensure a correct understanding of the facts
【以直报怨】justice in return for injustice; meet resentment with upright dealing
【以指纹认明罪犯】dactyloscopy
【以至】①(表示延伸) down to; up to: 循环往复,～无穷 repeat itself in endless cycles into infinity ②(表示结果) to such an extent as to...; so... that...
【以致】so that; as a result; with the result that; consequently
【以珠弹雀】make big investment for small returns
【以壮观瞻】strengthen the attraction
【以资】as a means of: ～鼓励 as an encouragement / ～弥补 to make up the deficit; to make up a shortage
【以子之矛,攻子之盾】set your own spear against your own shield; refute sb. with his own argument
【以罪定刑】punishment fits the crime

yì

意 ①(意思) meaning; idea: 文章的大～ the general idea of an article / 词不达～。The words fail to express the meaning. ②(心愿) wish; desire; intention: 恶～ bad intention / 好～ good will / 任～ at will; arbitrary / 善～ good intention / 中～ desirable ③(料想) expect; anticipate: 出其不～ catch sb. unawares ④(暗示) suggestion; hint; trace: 颇有秋～ make one feel that autumn has set in
【意大利】Italy ◇ ～人 Italian / ～语 Italian
【意到笔随】the pen follows where the mind reaches; write with ease
【意会】sense: 只可～,不可言传 can be sensed, but not explained in words
【意简言赅】a few simple ideas succinctly expressed
【意见】①(看法或相法) idea; view; opinion; suggestion: ～冲突 a clash of opinions / ～分歧 have a difference of opinion; disagree / ～一致 have identical ideas ②(反对或不满意) objection; differing opinion; complaint: 我对她很有～。I have a lot of complaints about her. ◇ ～簿 visitor's book; customers' book / ～箱 suggestion box
【意境】artistic conception
【意懒心灰】greatly discouraged
【意料】anticipate; expect: ～之外 unexpected happening / 这是～之中的事。That's to be expected.
【意念】idea; thought
【意气】①(意志和气概) will and spirit: ～昂扬 in high spirits / ～沮丧 in low spirits / ～自得 easy and dignified ②(志趣和性格) disposition; temperament: ～相投 be congenial with each other ③(不正常的情绪) personal feelings: ～之争 a dispute caused by personal feelings / 常闹～的人 one who is liable to be swayed by personal feelings
【意气风发】high-spirited and vigorous; daring and energetic
【意群】[语] sense-group
【意识】①[哲] consciousness: 潜～ subconsciousness / 同类～ consciousness of kind ②(觉察) be conscious of; awake to; realize: ～到自己的错误 be conscious of one's mistake ◇ ～流 stream of consciousness / ～能力 capacity of will
【意识形态】[哲] ideology: 社会～ social ideology / 在～领域 in the ideological sphere
【意思】①(语言文字的意义) meaning; idea: 文章的中心～ the central idea of an article ②(意见,愿望) opinion; wish; desire: 我的～还是不去为好。In my opinion it's better not to go. ③(礼物代表的心意) a token of affection, appreciation, gratitude, etc.: 一点儿小～ little gift as a token of one's appreciation ④(趋势) look like; seem: 天有点要下雨的～。It looks like rain. ⑤(情趣) interest; fun: 游泳很有～。Swimming is a lot of fun.
【意图】intention; intent: 领会～ understand the intention
【意外】①(意料之外) unexpected; unforeseen: ～的消息 unexpected news; be taken by surprise / 感到～ be taken by surprise ②(意外的不幸事件) accident; mishap: ～事件 an accident / 以免发生～ so as to avoid accidents ◇ ～保险 accident insurance / ～风波 unforeseen trouble / ～碰撞 accident collision / ～杀人 accidental homicide / ～伤亡 fortuitous casualty / ～事故死亡 accident death / ～事故损害 accident damage / ～收获 windfall / ～损

失 fortuitous casualty
【意味】①(含蓄的意思) meaning; significance; implication: ～深长 be full of meaning ②(情趣) interest; overtone; flavor: 带有文字～的信件 letters with a literary flavor
【意味着】signify; mean; imply: 这个数字～产量提高了一倍。This numeral means a onefold increase in production.
【意想】imagine; expect: ～不到的事情 unimaginable things
【意向】intention; purpose
【意兴】interest; enthusiasm: ～勃勃 be highly enthusiastic / ～阑栅 feel dispirited / ～索然 have not the least interest
【意义】①(意思) meaning; sense: 在本来的～上说 in the proper sense / 在某种～上 in a sense ②(价值；作用) significance; importance: 具有极其重要的现实～和深远的历史～ be of momentous current significance and far-reaching historical significance
【意译】free translation
【意欲为之】purpose to do something
【意愿】wish; desire; aspiration: 表达了人民的～ express the wishes of the people
【意在言外】the meaning is implied
【意旨】intention; wish; will: 秉承某人的～ in compliance with sb.'s wish
【意志】will: ～薄弱者 person of weak will / ～坚强 strong-willed / ～消沉 demoralized; despondent / 锻炼～ temper one's willpower
【意中人】the person one is in love with; person of one's heart
【意中之事】sth. that is expected

癔

【癔病】[医] hysteria ◇ ～患者 hysteriac / ～先兆 aura hysterica

薏

【薏苡】[植] Job's tears

镱 [化] ytterbium (Yb)

臆 ①(胸) chest ②(主观的) subjectively
【臆测】conjecture; surmise; guess
【臆断】assume; suppose
【臆说】assumption; supposition
【臆造】fabricate
【臆撰之文】an original essay

亦 also; too
【亦步亦趋】ape sb. at every step; imitate sb.'s every move; blindly follow suit
【亦非全坏】not altogether bad
【亦即】that is; i.e.; namely; viz.

奕

【奕奕】radiating power and vitality: 神采～ glowing with health and radiating vitality

益 ①(好处) benefit; profit; advantage: 得～于经验 profit by experience / 受～良多 derive much benefit ②(增加) increase: 延年～寿 prolong life ③(更加) all the more; increasingly: 多多～善 the more the better ④(有益的) beneficial
【益虫】beneficial insect
【益处】benefit; profit; good
【益发】increasingly; all the more: ～困难 increasingly difficult
【益加】increasingly
【益鸟】beneficial bird
【益寿】lengthen one's life
【益友】friend and mentor: 良师～ good teacher and helpful friend

溢 ①(充满而流出) overflow; spill: 河水～过两岸。The river overflowed its banks. / 水从桶中～出。Water spilled from the pail. ②(过分) excessive: ～美 undeserved praise; compliment
【溢出】spill over; overflow
【溢洪坝】[水] overflow dam
【溢洪道】[水] spillway
【溢洪闸】[水] spillway gate
【溢美之言】words of fulsome praise
【溢水管】warning pipe
【溢于言表】shows between the lines; in overtones

谊 friendship: 深情厚～ profound friendship

抑 ①(压制) restrain; repress; curb: ～强扶弱 curb the violent and assist the weak ②(可是) but ③(或是) or
【抑菌作用】[医] bacteriostasis
【抑塞】give no chance to: ～之才 a talent being repressed and hindered
【抑扬】rise and fall; modulate: ～顿挫 cadence; modulation in tone
【抑郁】depressed; despondent; gloomy: ～不平 feel disgruntled / ～成疾 so disheartened that one falls sick / 感到～ feel depressed ◇ ～症 [医] depression

【抑止】restrain; check
【抑制】①(控制) restrain; control; check: ~感情 suppress one's feelings / ~价格 check price / ~怒火 choke back one's anger; curb one's anger ②[生理] inhibition: 中枢~ central inhibition ◇ ~剂 inhibitor / ~神经 inhibitory nerve / ~作用 inhibition

易 ①(容易) easy: ~得罪人 be liable to give offense / ~学 easy to learn / ~做 can be done easily ②(平和) amiable: 平~近人 amiable and easy of access ③(改变) change: 移风~俗 change customs and habits; transform social traditions ④(交换) exchange: 以货~货 barter
【易变性】mutability
【易腐】perishable ◇ ~货 perishable goods / ~物 perishables
【易感者】[医] susceptible person; susceptible
【易货】barter: ~贸易 barter (trade) ◇ ~方式 barter system / ~协定 barter arrangement
【易燃】inflammable; combustible ◇ ~货物 inflammable cargo / ~物 combustibles; inflammables
【易熔】fusible: ~合金 fusible alloy
【易熔点】[物] eutectic point
【易如反掌】as easy as turning one's hand over; as easy as falling off a log
【易碎】breakable; fragile: ~物品 breakables ◇ ~性 fragility
【易损性】vulnerability
【易弯性】mutability
【易销货物】marketable goods

邑 ①(城市) city: 通都大~ big city; metropolis ②(县) county

义 ①(正义) justice; righteousness: 仁~ benevolence and righteousness ②(合乎正义或公益的) righteous; equitable; just: ~战 just war ③(情谊) human ties; relationship ④(抚养或拜认的) adopted; adoptive: ~父 adoptive father ⑤(意义) meaning; significance: 词~ the meaning of a word ⑥(人工制造的) artificial; false: ~齿 false tooth / ~肢 artificial limb
【义不容辞】have an unshirkable duty; be duty-bound
【义愤】righteous indignation; moral indignation: ~填膺 be filled with righteous indignation
【义举】a magnanimous act undertaken for the public good: ~济贫 relieve the poor for a good cause
【义卖】a sale of goods for charity or other worthy causes; charity bazaar
【义母】adoptive mother
【义女】adopted daughter; foster daughter
【义旗】the banner of an army fighting a just war; banner of righteousness: 举~ raise the banner of righteousness; rise against injustice
【义气】code of brotherhood; personal loyalty: 江湖~ code of the brotherhood / 讲~ be loyal to one's friends
【义士】a high-minded or chivalrous person; a person who upholds justice; righteous man
【义无反顾】honor permits no turning back; be duty-bound not to turn back
【义务】①(责任) duty; obligation ②(不要报酬的) volunteer; voluntary ◇ ~教育 compulsory education / ~劳动 voluntary labor
【义务兵】compulsory serviceman
【义务兵役制】compulsory military service; conscription
【义形于色】with indignation written on one's face
【义演】benefit performance
【义勇军】army of volunteers; volunteers ◇ ~进行曲 March of the Volunteers
【义正词严】speak sternly out of a sense of justice; speak with the force of justice
【义子】foster son; adopted child

议 ①(意见) opinion; view: 倡~ initiate; advocate / 动~ mover / 建~ propose; suggest; recommend / 提~ propose ②(商议) discuss; exchange views on; talk over: ~而不决 discuss sth. without reaching a decision
【议案】proposal; motion: 通过~ adopt the motion
【议程】agenda: 列入~ place on the agenda; include in the agenda / 临时~ provisional agenda
【议定的租金】agreed rent
【议定书】protocol: 附加~ additional protocol / 贸易~ trade protocol
【议订】negotiate: ~和约 negotiate a peace treaty
【议和】negotiate peace
【议会】parliament; legislative assembly: 解散~ dissolve parliament / 召开~ convene parliament ◇ ~党团 parliamentary groups / ~斗争 parliamentary struggle / ~迷 parliamentary cretinism / ~特权 parliamentary

privilege / ～制度 parliamentarism
【议价】①(议定价格)negotiate a price ②(议定的价格) negotiated price
【议决】resolve after deliberation; pass a resolution
【议论】comment; talk; discuss: ～不休 carry on endless discussions / 大发～ speak at great length
【议论文】argumentative writing; argumentation
【议事】discuss official business ◇ ～规则 rules of procedure; rules of debate / ～日程 agenda; order of the day
【议题】subject under discussion; topic for discussion
【议席】seat in legislative assembly
【议员】member of a legislative assembly; Member of Parliament (英); Congressman (美); Congresswoman (美) ◇ ～豁免权 parliamentary immunity / ～特权 parliamentary privilege
【议院】legislative assembly; parliament; congress: 参～ Senate / 上～ the House of Lords; the upper chamber / 下～ the House of Commons; the lower chamber / 众～ the House of Representatives
【议长】speaker; president

刈 cut; mow

【刈草机】mowing machine; mower

轶

【轶事】anecdote

屹

【屹立】stand towering like a giant; stand erect
【屹然】towering; majestic: ～不动 stand firm and erect

逸 ①(安乐) ease; leisure: 劳～结合 strike a proper balance between work and leisure; alternate work with relaxation ②(逃跑) escape; flee ③(散失) be lost ④(超过一般) excel all others: ～群 excel all others

【逸出常轨】run off the track
【逸出正道】err from the path of duty
【逸乐】comfort and pleasure
【逸趣横生】replete with humor; replete with refined interest
【逸失良机】let slip an opportunity
【逸事】anecdote
【逸文】ancient writings no longer extant

肄 study

【肄业】study in school; study at college

毅 firm; resolute: 刚～ fortitude

【毅力】willpower; will; stamina: 百折不回的～ indomitable will / 坚忍不拔的～ an unflinching will / 惊人的～ amazing willpower
【毅然】resolutely; firmly; determinedly: ～决然 resolutely and determinedly / ～振作 brace oneself up

疫 epidemic disease; pestilence: 防～ epidemic prevention / 鼠～ the plague

【疫病】epidemic disease
【疫苗】[医] vaccine
【疫情】information about and appraisal of an epidemic; epidemic situation ◇ ～报告站 station for reporting epidemic diseases

役 ①(需要出劳力的事) labor; service: 劳～ forced labor ②(兵役) military service: 现～军人 soldiers on service ③(役使) use as a servant ④(战争) battle

【役畜】draught animal; beast of burden
【役龄】enlistment age
【役使】work (an animal); enslave

忆 recall; recollect

【忆苦】recall past suffering
【忆及】call to mind; remember
【忆旧】recall the bygone days with nostalgia; recollect the past
【忆起】call to mind; remember; recall

亿 a hundred million

【亿万】hundreds of millions; millions upon millions: ～人民 hundreds of millions of people; the people in their hundreds of millions ◇ ～富翁 billionaire
【亿万斯年】time without end; eternity

艺 ①(技能) skill: 技～ technical skill ②(艺术) art: 文～ literature and art

【艺妓】[日本] geisha; geisha girl
【艺林】art circles
【艺龄】length of sb.'s artistic career
【艺名】stage name
【艺能】artistic skill; art; skill
【艺人】①(演员) actor or artist ②(手工艺工人) artisan; handicraftsman

【艺术】①(文艺) art: ～至上主义 art for art's sake ②(富有创造性的方法) skill; art; craft: 领导～ art of leadership ③(形状独特而美观的) conforming to good taste: 这座建筑装饰得很～。 The building is tastefully adorned. ◇～标准 artistic criterion / ～风格 artistic style / ～技巧 artistry; craftsmanship / ～家 artist / ～教育 art education / ～界 art circles / ～品 work of art / ～天才 gift for art; artistic talent / ～团 art ensemble; troupe of musicians and artists / ～形式 artistic form; forms of art / ～性 artistic quality; artistry / ～指导 art director / ～字 characters in a fancy style

【艺苑】the realm of art and literature; art and literary circles: ～奇葩 exquisite works of art

呓

【呓语】①(梦话) talk in one's sleep ②(胡言乱语) crazy talk; ravings: 狂人～ ravings of a madman

译

译 translate; interpret: ～成电码 coding / ～成密码 enciphering / ～成英语 translate into English / 笔～ written translation / 口～ oral interpretation / 意～ free translate / 音～ transliteration / 直～ literal translation / 把一部书从英文～成中文 translate a book from English into Chinese / 把一部英语书～成中文 translate an English book into Chinese / 这书～得好。 The book translates well.

【译本】translation

【译笔】the quality or style of a translation: ～流畅。 The translation reads smoothly.

【译电】①(把文字译成电码) encode; encipher ②(把电码译成文字) decode; decipher ◇～费 coding fee / ～员 decoder; code clerk; cryptographer / ～组 code and cipher section

【译解】decipher

【译码】decode; decipher ◇～器 decoder; decipherer

【译名】translated term or name

【译文】translated text; translation

【译音】transliteration

【译员】interpreter

【译者】translator

【译制】dub ◇～片 dubbed film

驿

【驿道】post road

【驿站】post

翼

翼 ①(翅膀) wing: 机～ wings of an airplane ②(侧) flank: 部队的左右两～ both flanks of an army / 左～ left wing

【翼庇】protect; patronize

【翼侧】[军] flank ◇～包围 outflank / ～攻击 flank attack / ～推进 lateral action / ～迂回 outflank

【翼尖】[航空] wing tip

【翼手动物】[动] bat; chiropter

【翼形螺母】[机] wing nut

【翼形螺栓】[机] wing bolt

【翼型】[航空] wing section; aerofoil

【翼翼】cautiously: 小心～ with exceptional caution; very carefully

【翼展】[航空] span; wing span

异

异 ①(有分别;不同) different: 求同存～ seek common ground while reserving differences ②(奇异;特别) strange; unusual; extraordinary: ～事 strange happening ③(惊奇) surprise ④(另外的) other; another: ～地 a strange land ⑤(分离) separate: 离～ divorce

【异邦】a foreign country

【异步】[物] asynchronous ◇～发电机 asynchronous generator / ～计算机 asynchronous computer

【异彩】extraordinary splendor

【异常】①(不同于寻常) unusual; abnormal: ～现象 abnormal phenomena / 神色～ not be one's usual self ②(非常) extremely; exceedingly; particularly: ～丰富 exceedingly rich

【异读】variant pronunciation

【异端】heterodoxy; heresy: ～邪说 heretical beliefs; unorthodox opinions

【异方殊俗】different customs in alien countries

【异父兄弟】half brothers

【异国】foreign country: ～情调 an exotic atmosphere

【异乎常人】different from the ordinary men

【异乎寻常】unusual; extraordinary: ～地热心 unusually enthusiastic

【异花传粉】[植] cross pollination

【异花受精】[植] allogamy; cross fertilization

【异化】①[哲](指向对方的转化) alienation ②[语](几个相同音中不同的发音) dissimilation ◇～作用[生] dissimilation

【异己】dissident; alien: 排除～ discriminate against those who hold different views; get rid of dissidents

【异教】paganism; heathenism ◇～徒 pagan; heathen

【异军突起】a new force suddenly coming to the fore
【异客】a stranger
【异口同声】with one voice; in unison
【异名】a different name
【异母所生子女】children by another venter
【异曲同工】different tunes rendered with equal skill; different in approach but equally satisfactory in result
【异体受精】[动] allogamy; cross-fertilization
【异体字】a variant form of a Chinese character: 废除～ eliminate complicated variants
【异同】similarities and differences
【异途同归】all roads lead to same goal
【异味】①(不寻常的美味) rare delicacy ②(不寻常的气味) peculiar smell
【异位妊娠】ectopic pregnancy
【异位素】heterotope
【异物】①[医] foreign matter; foreign body ②(奇异的物品) rarity
【异香异气】an unusual kind of pleasant smell; extraordinary fragrance
【异乡】foreign land; strange land: ～孤客 a stranger in a strange land
【异想天开】indulge in the wildest fantasy; have a very fantastic idea
【异形钢材】special-shaped steel
【异形管】[机] special pipe
【异性】①(性别不同的) the opposite sex ②(性质不同) different in nature
【异烟肼】isoniazid
【异言】[书] dissenting words: 并无～ raise no objection
【异样】①(不同) difference ②(不寻常) unusual; peculiar
【异议】objection; dissent: 独持～ be the only one to dissent / 提出～ raise an objection; take exception to; challenge
【异源多倍体】[生] allopolyploid
【异重流】[水] density current
【异族】different race or nation: ～通婚 mixed marriages

yīn

音 ①(声音) sound: 噪～ noise / 乐～ musical sound / 隔～ soundproof / 调～ tune up ②(消息) news; tidings ③[物](音质) tone: 纯～ pure tone; simple tone / 复～ complex tone
【音标】[语] phonetic symbol; phonetic transcription: 国际～ the International Phonetic Symbols
【音波】[物] sound wave
【音叉】tuning fork
【音程】[乐] interval
【音带】the vocal cords
【音调】①[乐] tone ②[理] pitch
【音符】[乐] note
【音高】[乐] pitch
【音阶】[乐] scale
【音节】[语] syllable ◇ ～文字 syllabic language
【音量】volume ◇ ～控制 volume control
【音律】[乐] temperament
【音名】[乐] musical alphabet
【音频】[物] audio frequency ◇ ～电路 voice-frequency circuit / ～振荡器 audio-frequency oscillator
【音强】①[乐] the loudness of a sound ②[理] intensity
【音儿】[方]①(声音) voice ②(话里的含意) implication: 听话听～ listen for the meaning behind sb.'s words
【音容】[书] the likeness of the deceased: ～宛在 as if the person were in the flesh / ～笑貌 one's lovely voice and happy countenance
【音色】tone color; timbre
【音书断绝】news and letters are broken off
【音素】[语] phoneme ◇ ～文字 phonemic language
【音速】[物] velocity of sound: 超～ supersonic / 高超～ hypersonic / 高超～风洞 hypersonic wind tunnel
【音位】[语] phoneme
【音位学】[语] phonology
【音响】sound; acoustics ◇ ～测距 sound ranging / ～水雷[军] sonic mine / ～效果 sound effects; acoustics
【音像同步装置】[工] moviola; movieola
【音信】mail; message; news: ～不通 have no communication / ～全无 have not heard from sb. ever since / 互通～ communicate with each other; be in correspondence with each other
【音译】transliteration
【音域】[乐] range; compass; register
【音乐】music: 古典～ classical music / 民间～ folk music / 轻～ light music / 舞蹈～ dance music ◇ ～会 concert / ～家 musician / ～节 Music Day / ～片 musical / ～厅 concert hall; music hall / ～形象 musical image / ～学院 conservatory of music / ～系 the department of music / ～演奏 musical performance
【音韵学】[语] phonology ◇ ～家 phonologist
【音障】[物] sound barrier; sonic barrier

【音值】[语] value
【音质】①(声音的属性) tone quality ②(声音的逼真程度) acoustic fidelity

因 ①(原因) cause; reason: 近～ immediate cause／内～ internal cause／外～ external cause ②(因为) because of; as a result of; for: ～诽谤提起的诉讼 action for libel／～侵犯专利权提起的诉讼 action for infringement of a patent／～违反合同而要求补偿 remedy for breach of contract
【因病】due to illness; because of illness: ～请假 ask for sick leave
【因材施教】teach students in accordance with their aptitude
【因此】therefore; for this reason; consequently; hence
【因地因时】in accordance with the time and the place
【因地制宜】suit measures to local conditions: ～地积极发展农业机械工业 energetically develop farm machinery industry in the light of local conditions
【因而】thus; as a result; with the result that
【因公】on duty; on business: ～出差 take an official trip／～出差到北京 go to Beijing on business／～殉职 death in line of duty
【因果】①(原因和结果) cause and effect ②[宗] karma; preordained fate ◇ ～倒置 inverting cause and effect／～关系 causality／～律 law of causation; law of causality／～论者 causationist／～性 causality
【因果报应】[佛教] karma; retribution for sin; punitive justice
【因祸得福】profit from a misfortune
【因陋就简】make do with whatever is available; do things simply and thriftily
【因人成事】rely on others for success in work
【因人借镜】learn from other persons
【因人设事】create jobs to accommodate some people
【因伤致死】die from a wound
【因式】[数] factor: ～分解 factorization
【因事制宜】circumstances alter cases
【因势利导】adroitly guide action according to circumstances
【因素】factor; element: 调动一切积极～ bring every positive factors into play／决定～ decisive factor／人的～ the human factor ◇ ～论 [哲] theory of factors
【因往推来】judge the future from the past
【因为】because; for; on account of
【因袭】follow; copy: ～陈规 follow outmoded rules／～前人 follow in the footsteps of one's predecessors
【因小失大】try to save a little only to lose a lot
【因循】①(沿袭) follow; continue in the same old rut: ～故辙 heedlessly following the old ruts／～守旧 stick to old ways; follow the beaten path ②(迟延拖拉) procrastinate: ～坐误 sit back and allow the situation to deteriorate; procrastinate until it is too late
【因噎废食】give up eating for fear of choking; refrain from doing sth. necessary for fear of a slight risk
【因由】reason; cause; origin
【因缘】①[佛教] cause; principal and subsidiary causes ②(缘分) predestined relationship
【因罪杀人】capital murder

洇 spread and sink in: 这种纸容易～。Ink blots on this paper.
【洇色】diffusion or running of coloring matter; bleeding

铟 [化] indium (In)

姻 ①(婚姻) marriage: 联～ connect by marriage ②(姻亲关系) relation by marriage: ～兄弟 brother-in-law
【姻亲】relation by marriage: ～故旧 marriage connections and old friends／～关系 relationship by marriage; affinity
【姻缘】the happy fate which brings lovers together: 美满～ a happy marriage; conjugal felicity

殷
【殷富人家】well-off family
【殷富之人】a man of ample means
【殷钢】[冶] invar
【殷切】ardent; eager: ～的期望 ardent expectations
【殷勤】eagerly attentive; solicitous: ～侍奉 wait on sb. hand and foot／受到～接待 be accorded solicitous hospitality／献～ do everything to please; pay one's addresses to
【殷实】well-off; substantial: ～的人家 well-off families／～的商号 a substantial firm
【殷殷垂问】inquire about anxiously

阴 ①(指天气) overcast; cloudy ②(无阳光处) shade: 在树～下 under the shade of trees

③(隐藏不露) hidden; secret; sinister: ～一套，阳一套 act one way in public and another in private; be engaged in double-dealing ④[物](带负电的) negative: ～离子 negative ion; anion ⑤(山的北面水的南面) north of a hill or south of a river ⑥(指月亮) the moon: ～历 lunar calendar
【阴暗】dark; gloomy: ～的角落 a dark corner / ～面 the dark side of things / ～无光 dark without light; obscure
【阴部】[生理] private parts; pudenda
【阴沉】cloudy; overcast; gloomy; sombre: ～的天气 dismal weather
【阴错阳差】a strange combination of circumstances
【阴丹士林】[化] indanthrene
【阴道】[生理] vagina ◇ ～隔膜 contraceptive diaphragm / ～炎 vaginitis
【阴蒂】[解] clitoris
【阴电】negative electricity ◇ ～子 negatron
【阴毒】insidious; sinister: ～的敌人 an insidious enemy
【阴风】①(寒风) chilly wind ②(从阴暗处来的风) sinister wind
【阴阜】[解] mons pubis; mons veneris
【阴干】be placed in the shade to dry; dry in the shade
【阴沟】sewer; drain: ～里翻船 fail miserably in a very easy task
【阴晦】shady; dark; dismal
【阴魂】soul; spirit
【阴极】[物] negative pole; negative electrode; cathode: 冷～ cold cathode ◇ ～激励 cathode drive / ～加热器 cathode heater
【阴极射线】[物] cathode ray ◇ ～管 cathode ray tube
【阴间】the nether world
【阴茎】[生理] penis; penes ◇ ～包皮 prepuce of penis / ～海绵体 corpus spongiosum / ～套 condom
【阴冷】①(阴沉而寒冷) gloomy and cold; raw ②(阴沉而冷酷) sombre; glum
【阴历】lunar calendar: ～正月 the first month of the lunar year
【阴凉】①(太阳照不到而凉爽) shady and cool: 把药放在～处 store the medicine in a cool, dark place ②(阴凉的地方) cool place; shade
【阴霾】haze
【阴谋】plot; scheme; conspiracy: ～诡计 schemes and intrigues / ～手段 conspiratorial means / ～篡权 scheme to usurp power / ～复辟 plot to restore the old order / ～破坏 plot sabotage / ～推翻政府 conspire to overthrow the government ◇ ～集团 conspiratorial clique / ～家 schemer; intriguer; conspirator / ～推翻政府罪 offense of conspiring to overthrow the government / ～武装叛乱罪 offense of plotting an armed rebellion
【阴囊】[生理] scrotum ◇ ～炎 scrotitis
【阴平】high and level tone
【阴森】gloomy; gruesome; ghastly: ～可怕 ghastly and bloodcurdling
【阴私】shameful secret
【阴天】overcast sky; cloudy day
【阴文】characters cut in intaglio
【阴险】sinister; insidious; treacherous: ～行为 an insidious act
【阴性】①[医](否定) negative: ～反应 negative reaction ②[语] feminine gender
【阴阳怪气】①(说话的方式奇特) mystifying; enigmatic; deliberately ambiguous ②(人难以琢磨) eccentric; queer; cynical
【阴影】①(阴暗的影子) shadow: 图画的～部分 shaded parts on a picture ②[医](斑痕) spot; shadow: 肺部发现有～ have a spot found on one's lungs
【阴雨】overcast and rainy: ～连绵 cloudy and drizzly for days on end; an unbroken spell of wet weather
【阴郁】gloomy; dismal; depressed: 感到～ feel gloomy
【阴云】dark clouds

荫 shade
【荫蔽】①(遮蔽) be shaded or hidden by foliage ②(隐蔽) cover; conceal
【荫庇】protect; patronize

yín

淫 ①(过多) excessive: ～雨 excessive rains ②(放纵) loose; wanton: 骄奢～逸 lordly, luxurious, loose and idle; wallowing in luxury and pleasure ③(关于色情的) obscene; pornographic: ～画 obscene picture / ～书 pornographic book ④(淫荡) licentious; lewd; lascivious: ～心 sexual lust
【淫荡】loose in morals; lascivious; licentious; lewd: ～无耻 licentious and shameless
【淫秽】obscene; salacious; bawdy ◇ ～诽谤罪 obscene libels
【淫乱】promiscuous; licentious
【淫靡】extravagant; extravagance

【淫威】abuse of power; despotic power
【淫猥】obscene ◇ ～之语 indecent words
【淫逸度日】pass days with luxurious ease

寅

【寅吃卯粮】eat next year's food; eat one's corn in the blade; anticipate one's income
【寅谊深厚】relationship among colleagues is deep and thick

吟

①（吟咏）chant; recite: 行～泽畔 sing while strolling on the lake front ②（一种古诗名）song, a type of classical poetry ③（一些动物的叫声）the cry of certain animals
【吟风弄月】sing of the moon and the wind; write sentimental verse
【吟缶之乐】the joy of singing folk songs
【吟诗】recite poems; hum verse
【吟诵】chant; recite
【吟味】recite with relish; recite with appreciation: 反复～ recite again and again in appreciation
【吟啸自娱】amuse oneself with loud singing
【吟咏】recite with a cadence; chant: ～作诗 chant and write poetry

龈

gum
【龈炎】[医] gingivitis

银

①（金属元素）silver: 整套～餐具 table silver ②（跟货币有关的）relating to currency; relating to money ③（银色的）silver-colored: ～发 silver hair / ～灰色 silver gray / ～色 silery
【银白】silvery white
【银白杨】white poplar
【银杯】silver cup
【银本位】[经] silver standard
【银币】silver coin
【银锭】silver ingot
【银耳】tremella
【银发】silver hair
【银根】[经] money market; money: ～紧 tight / ～松 easy
【银汉鱼】[动] silverside
【银行】bank: 储蓄～ savings bank / 发行～ bank of issue / 国家～ national bank / 汇兑～ exchange bank / 商业～ commercial bank / 贴现～ discount bank / 外汇指定～ authorized bank / 中国人民～ People's Bank of China / 中国～ Bank of China ◇ ～承兑 banker's acceptance / ～存款 bank deposit / ～存款凭证 bank deposit certificate / ～存折 bankbook; passbook / ～法 banking law / ～汇票 bank draft / ～家 banker / ～利率 bank rate / ～贴现 bank discount / ～透支 bank overdraft / ～信贷 bank credit / ～业 banking / ～准备金 bank reserves ○背书 endorsement; endorse / 本金 principal / 本票 cashier's order / 本息共计 sum total of the principal plus interest / 差额 balance / 出票人 drawer of a bill / 储蓄额 total savings deposit / 存款 deposit / 存入 deposit / 大写金额 amount in words / 贷款 loan / 单利 simple interest / 当年还本付息 repay principal and interest within the current year / 到期 maturity / 到期的 overdue / 到期本金 matured principle / 到期票据 mature / 法定利息 legal interest / 复利 compound interest / 付现 cash payment / 还本付息 repay capital with interest / 结帐 square accounts / 结转 carry-over / 金库 treasury / 开户头 open an account / 空白背书 blank endorsement / 利率 interest rate / 利息 interest / 期票 promissory note / 取出 draw / 贴现率 discount rate / 透支 make an overdraft; take out an overdraft / 小写金额 amount in figures / 预付 payment in advance / 支票 check; cheque / 止付 stop payment
【银河】[天] the Milky Way ◇ ～系 the Milky Way system; the Galaxy
【银狐】silver fox
【银花纷飞】silver flowers falling disorderly; falling of snow
【银婚】silver wedding: ～纪念 the 25th anniversary of one's wedding
【银匠】silversmith
【银两】tael
【银幕】screen; projection screen: 宽～ wide screen / 搬上～ make a screen version of
【银牌】silver medal
【银器】silverware
【银钱】money: ～交易 deal in cash
【银鼠】[动] snow weasel
【银团贷款】syndicated loan
【银杏】[植] ginkgo
【银心】[天] galactic center
【银鱼】[动] whitebait; salangid
【银圆】silver dollar ◇ ～外交 dollars diplomacy
【银质奖章】silver medal
【银朱】vermilion ◇ ～涂料 vermilion paint
【银子】silver

yin

饮 ①(喝) drink: ～水 drink water / 痛～ drink hard ②(可喝的东西) drinks: 冷～ cold drinks ③(心里存着) keep in the heart; nurse: ～恨 nurse a grievance

【饮茶】drink tea

【饮恨】[书] nurse a grievance: ～而终 die with a grievance in one's heart / ～吞声 harbor hatred in the heart

【饮酒】drink: ～不多 be a little drinker / ～消愁 drink sorrow down / 他太好～。He is too fond of drink.

【饮料】drink; beverage: 软～ soft drink

【饮泣】[书] weep in silence: ～吞声 swallow one's tears; weep silent tears

【饮刃而死】take one's own life by means of a sword; suicide

【饮食】food and drink; diet: 流质～ liquid diet / ～节制 be abstemious in eating and drinking / ～无味 distaste for food / 给病人规定～ put a patient on a diet ◇ ～店 eating house / ～疗法 dietotherapy; dietetic treatment / ～卫生 dietetic hygiene / ～业 catering trade

【饮水疗法】[医] dietotherapy; dietetic treatment

【饮水器】drinking bowl; drinker: 自动～ drinking fountain

【饮水思源】when you drink water, think of its source; never forget where one's happiness comes from

【饮用水】drinking water; potable water

【饮鸩止渴】drink poison to quench thirst; seek temporary relief regardless of the consequences

引 ①(牵引) draw; stretch ②(引导) lead; guide: ～路 lead the way / ～某人入座 usher sb. to his seat ③(离开) leave: ～避 make way for; keep clear of ④(伸着) stretch: ～项 crane one's neck ⑤(引起) cause; make ⑥(用来作证) quote; cite: ～某人的话 quote sb. ⑦(勾引) lure; attract: ～入圈套 lure into a trap; ensnare

【引爆】ignite; detonate ◇ ～剂 flashing composition / ～装置 igniter

【引柴】kindling

【引产】induced labor

【引出】draw forth; lead to: ～好结果 lead to good results / ～结论 draw a conclusion

【引导】guide; lead

【引逗】①(挑逗) tantalize; tease ②(引诱) lure; entice

【引渡】[法] extradite ◇ ～人 deliverer / ～方式 form of extradition / ～请求书 demandant of extradition / ～逃犯 surrender of fugitive / ～条约 extradition treaty / ～罪犯 delivery of offender

【引而不发】draw the bow but not discharge the arrow

【引发】[化] initiation ◇ ～剂 initiator

【引吭高歌】sing joyfully in a loud voice; sing heartily

【引航】[航海] pilotage

【引号】quotation marks: 单～ single quotation marks(‘ ’) / 双～ double quotation marks (“ ”)

【引河】①(为引水灌溉而开挖的河道) irrigation channel ②(减河) diversion canal

【引火】ignite; strike a fire: ～柴 kindling

【引火烧身】draw fire against oneself; make self-criticism to encourage criticism from others

【引疾求退】implore resignation under the pretext of illness

【引见】introduce; present

【引荐】recommend

【引进】①(进口) import; introduce from elsewhere: ～外国技术和设备 import foreign technology and equipment ②(引荐) recommend

【引经据典】quote the classics; copiously quote authoritative works

【引咎】[书] hold oneself responsible for a serious mistake; take the blame: ～自责 bear the blame and reproach oneself

【引狼入室】invite a wolf into the house; open the door to a dangerous foe

【引力】[物] gravitation; gravitational force; attraction: 核～ nuclear attraction / 万有～ universal gravitation ◇ ～场 gravitational field

【引流】[医] drainage: 切开～ incision and drainage / 十二指肠～ duodenal drainage ◇ ～管 drainage tube / ～术 drainage

【引起】give rise to; lead to; set off; touch off; cause; arouse: ～共鸣 evoke sympathy / ～哄堂大笑 set off roars of laughing / ～怀疑 arouse suspicion / ～连锁反应 set off a chain reaction / ～麻烦 cause trouble / ～骚乱 raise a disturbance / ～兴趣 provoke interest / ～严重后果 lead to grave consequences / ～争论 touch off a debate

【引桥】[交] bridge approach

【引擎】[机] engine ◇ ～盖 bonnet; hood

【引人发笑】make someone laugh

【引人入胜】absorbing; fascinating; enchanting;

bewitching: ～的故事 absorbing stories／这本书的风格本身就会～。The very style of the book fascinates.
【引人注目】noticeable; conspicuous; spectacular: 她是～的人物。She is a conspicuous figure.／这个班已有～的进步。The class has made noticeable improvement.
【引人注意】catch the attention
【引入】lead into; draw into: ～歧途 lead sb. onto a wrong path; lead sb. astray ◇～品种 [农] introduced variety
【引申】extend ◇～义 extended meaning
【引水】①(领港) pilot a ship into harbor ②(引出水) draw or channel water: ～上山 draw water uphill ◇～工程 diversion works／～员 pilot
【引退】retire from office; resign
【引为已任】take as one's own responsibility
【引文】quoted passage; quotation
【引线】①(线状的引信) threadlike fuse; lead ②(做媒介的人或事) go-between
【引言】foreword; introduction
【引以为戒】learn a lesson; take warning
【引以为荣】feel honored; count it an honor
【引以为慰】take comfort in something
【引用】①(援引) quote; cite: ～证言 adduce evidence ②(任用) recommend; appoint
【引诱】lure; seduce
【引证】quote or cite as proof or evidence ◇～者 citer
【引种】[农] introduce a fine variety
【引子】①[剧] an actor's opening words ②[乐] introductory music ③(引起正文的话) introductory remarks; introduction ④(中药的) an added ingredient

蚓

【蚓突】appendix
【蚓螈】caecilian

隐

隐 ①(隐藏不露) hidden from view; concealed: 时～时现 now visible; now hidden ③(潜伏的) latent; dormant; lurking: ～患 hidden danger
【隐蔽】conceal; take cover: ～不明的动机 ulterior motives／～罪恶 keep down the lid／他～在树后。He concealed himself behind the trees. ◇～色 cryptic coloration／～运动 concealed movement／～阵地 covered position
【隐藏】hide; conceal; remain under cover: ～的缺陷 concealed defect ◇～者 concealer
【隐恶扬善】cover up sb.'s faults and publicize his merits; hide sb.'s wrongdoing and praise his good deeds
【隐伏】lie concealed; lie low
【隐睾】[医] cryptorchidism
【隐花植物】[植] cryptogam
【隐患】hidden trouble; hidden danger; snake in the grass: 消除～ remove a hidden peril
【隐讳】avoid mentioning; cover up
【隐晦】obscure; veiled: 一段～的文章 an obscure passage
【隐疾】unmentionable disease
【隐居】live in seclusion; withdraw from society and live in solitude; be a hermit: ～田园 dwell in one's native place in seclusion
【隐瞒】conceal; hide; hold back: ～不报 concealed and unreported／～错误 conceal one's mistakes／～事实 withhold the truth; hide the facts／～真相 suppress the truth
【隐私】①(隐蔽不外露) conceal; hide ②(秘密事) secret
【隐匿】hide; go into hiding; lie low ◇～行为 act of concealment／～证据 suppress the evidence／～罪 misprision
【隐情】facts one wishes to hide
【隐忍】bear patiently; forbear: ～不言 forbear from speaking
【隐射】insinuate; hint; throw out innuendoes
【隐士】recluse; hermit
【隐私】one's secrets; private matters one wants to hide
【隐痛】secret anguish
【隐退】go and live in seclusion; retire from political life
【隐显墨水】invisible ink
【隐性】recessiveness ◇～基因 allogene／～性状 recessive character
【隐姓埋名】conceal one's identity; keep one's identity hidden; live incognito
【隐隐】indistinct; faint: ～可见 faintly visible／～约约 indistinct／感到～作痛 feel a dull pain
【隐语】enigmatic language; insinuating language
【隐喻】[语] metaphor
【隐约】indistinct; faint: ～其词 use ambiguous language; speak in equivocal terms
【隐衷】feelings or troubles one wishes to keep to oneself

瘾

瘾 ①(习惯性) addiction; habitual craving: 过～ satisfy a craving／酒～ drinking habit／

吸毒上～的人 drug addict ②(浓厚的兴趣) strong interest: 他看京剧上～了。 He is crazy about Beijing opera.

【瘾头】addiction; strong interest: 你们下棋的～可真不小。 You are really keen on playing chess.

yìn

窨 basement

【窨井】[建] inspection shaft; inspection well

荫 [口] shady; damp and chilly

【荫凉】shady and cool

【荫下之福】the happiness owned by a man under the care of his parents

印 ①(图章) seal; stamp; chop ②(痕迹) print; mark: 脚～ footprint / 手～ fingerprint ③(印上痕迹) print; engrave: ～一张宣传画 print a poster / 付～ put sth. into print / 已～好 be off the press / 那景象深深～在我的记忆中。 The scene is printed in my memory. / 这本书～得很好。 The book is well printed. ④(符合) tally; conform: ～证 verify

【印版】[印] plate

【印本】printed books

【印次】[印] impression

【印地语】Hindi

【印第安人】American Indian; Red Indian; Indian

【印度】India ◇ ～教[宗] Hinduism / ～人 Indian

【印度洋】the Indian Ocean

【印度尼西亚】Indonesia ◇ ～人 Indonesian / ～语 Indonesian

【印度支那】Indochina; Indo-China

【印发】print and distribute

【印盒】seal box

【印痕】impression

【印花】①(税票) stamp ② [纺] printing ◇ ～布 prints; printed calico / ～厂 printworks / ～绸 printed silk / ～机[纺] printing machine / ～棉布 cotton print / ～税 stamp duty; stamp tax / ～装饰布 cretonne

【印鉴】a specimen seal impression for checking when making payments: 核对～ identify the signature

【印泥】red ink paste used for seals

【印染】printing and dyeing ◇ ～厂 printing and dyeing mill / ～法 method of printing and dyeing

【印数】[印] printing; impression: ～四万册 an impression of 40000 copies

【印刷】printing: 彩色～ colored printing / 立体～ stereoscopic printing; three-dimensional printing / 三色版～ three-color halftone ◇ ～厂 printing house; press / ～错误 misprint; typographic error / ～电路 printed circuit / ～工人 printing worker; printer / ～品 printed matter / ～术 art of printing; printing / ～线路板 printed substrate / ～油墨 printing-ink / ～纸 printing paper

【印刷机】[印] printing machine; press: 滚筒～ cylinder press / 轮转～ rotary press / 双面～ perfecting press; perfector

【印刷体】block letter; print hand: 请用～写您的名字。 Print your name.

【印台】ink pad; stamp pad

【印相】[摄] print ◇ ～纸 photographic paper; print paper

【印象】impression: ～甚佳 give a favorable impression ◇ ～派 impressionist school; impressionist / ～主义 impressionism

【印油】stamp-pad ink

【印章】seal; signet; stamp ◇ ～学 sigillography

【印证】confirm; corroborate; verify: 有待～ yet to be confirmed

【印制电路】printed circuit

【印子】mark; trace; print

茚 [化] indene

饮 give water to drink; water: ～牲口 water the cattle

yīng

应 ①(答应) answer; respond: 喊不～ doesn't answer when called ②(答应做) promise; agree; accept ③(应该) should; ought to: 他做了事, ～给他钱。 Money is due to him for his work. / 我们的火车～于早八时到达北京。 Our train is due at Beijing at 8 a.m.

【应当】should; ought to

【应得】deserved; due: ～的惩罚 well-deserved punishment; just punishment / ～份 reserved portions

【应分】part of one's job: 帮助群众是干部～的事。 It's part of the duty as cadres to help the masses.

【应该】should; ought to; must

【应届毕业生】graduating students or pupils; this year's graduates

【应名儿】only in name; nominally

【应受惩处的罪行】punishable offense
【应有】due; proper; deserved: 做出～的贡献 make a due contribution
【应有尽有】have everything that one expects to find
【应允】assent; consent: 点头～ nod assent
鹰 hawk; eagle
【鹰鼻鹞眼】hawk-nosed and vulture-eyed; sinister and fierce-looking
【鹰钩鼻子】aquiline nose
【鹰犬】falcons and hounds; lackeys; hired thugs: ～爪牙 running dogs and attendants
【鹰视】fierce look: ～狼步 a wicked and fierce person
【鹰扬之胆】having great courage
莺 warbler; oriole
【莺歌燕舞】orioles sing and swallows dart; the joy of spring; a scene of prosperity
英 ①(花) flower: 落～缤纷 petals falling in riotous profusion ②(才智过人的人) hero; outstanding person: 群～会 a gathering of heroes; a conference of outstanding workers ③(英国) Britain: ～帝国 the British Empire ④(英语) English: ～译本 English translation / 汉译～ translate from Chinese into English
【英镑】pound sterling: ～结存 sterling balance ◇～区 the sterling area
【英才】person of outstanding ability
【英尺】foot
【英寸】inch
【英吨】long ton; gross ton
【英国】Britain; England: ～护照 British passport / ～史 English history ◇～管[乐] English horn / ～人 the British; Englishman or Englishwoman / ～英语 British English
【英汉词典】an English-Chinese dictionary
【英豪】heroes; outstanding figures
【英俊】①(才能出众) eminently talented; brilliant ②(容貌俊秀,又有精神) handsome and spirited; smart
【英里】mile
【英联邦】the British Commonwealth (of Nations)
【英烈】heroic; brave; valiant
【英灵】spirit of the brave departed; spirit of a martyr
【英名】illustrious name
【英明】wise; brilliant: ～的决策 wise decision / ～的论断 brilliant thesis
【英亩】acre
【英气】heroic spirit: ～勃勃 full of animated courage
【英式槌球】association croquet
【英式橄榄球】rugby
【英式台球】English billiards
【英属】British: ～殖民地 British colony
【英文】English
【英武】of soldierly bearing
【英仙座】[天] Perseus
【英雄】hero: 女～ heroine / ～本色 the true quality of a hero / ～肝胆 heroic spirit / ～气概 heroic spirit; mettle / ～所见略同 great minds think alike / ～主义 heroism
【英雄无用武之地】a hero with no place to display his prowess; no scope to exercise one's abilities; a fish out of water
【英勇】heroic; valiant; brave; gallant: ～不屈 show unyielding heroism / ～奋斗 fight heroically / ～就义 die a heroic death / ～善战 brave and skillful in battle / ～行为 heroic conduct
【英语】English: ～口语 spoken English / ～晚会 English Evening / 美国～ American English / 你会说～吗? Do you speak English?
【英制常衡】avoirdupois weight ○磅 pound / 盎司 ounce / 打兰 dram / 格令 grain
【英姿】heroic bearing: 飒爽～ valiant and heroic in bearing; bright and brave
罂 [书] small-mouthed jar
【罂粟】[植] opium poppy ◇～花 poppy flower
婴 baby; infant
【婴儿】baby; infant ◇～车 perambulator; baby carriage / ～猝死综合症 sudden infant death syndrome; crib death; cot death / ～湿疹 infantile eczema
樱 ①(樱桃) cherry ②(樱花) oriental cherry
【樱花】oriental cherry: 日本～ Japanese flowering cherry
【樱桃】cherry
嘤 [象] chirp
鹦
【鹦哥】parrot
【鹦哥绿】parrot green
【鹦鹉】parrot ○长尾～ parakeet / 虎皮～ budgerigar
【鹦鹉螺】nautilus

【鹦鹉热】[医] psittacosis; parrot fever
【鹦鹉学舌】repeat the words of others like a parrot; parrot
【鹦嘴鱼】parrot fish

缨 ①(穗状饰物) tassel: 红～枪 red-tasselled spear ②(泛指带状物) ribbon

yíng

赢 ①(胜) win; beat: ～局 win a game; score a game ②(获利) gain
【赢得】gain; win; obtain: ～时间 gain time / ～胜利 win victory / ～一分 score a point
【赢利】profit; gain
【赢余】surplus; profit: 略有～ with a small favorable balance

荧
【荧光】[物] fluorescence; fluorescent light ◇ ～灯 fluorescent lamp / ～点 phosphor dot / ～粉 phosphor powder / ～镜 fluoroscope / ～屏 fluorescent screen

萤 firefly; glowworm
【萤火虫】firefly; glowworm
【萤石】[矿] fluorite; fluorspar

营 ①(谋求) seek: ～利 seek profits ②(经营) operate; run; manage: 私～ privately owned; privately operated ③(军队编制单位) battalion ④(军队驻地) camp; barracks: 兵～ military camp
【营部】battalion headquarters
【营地】campsite; camping ground
【营房】barracks
【营火】campfire ◇ ～会 campfire party
【营建】construct; build
【营救】succor; rescue: ～某人脱险 rescue sb. from danger / 去～某人 go to sb.'s rescue ◇ ～者 rescuer
【营垒】①(军营和四周的围墙) barracks and the enclosing walls ②(阵营) camp: 革命～ revolutionary camp
【营利】seek profits; make money: ～目的 the object of profit-making
【营生】①(谋生活) earn a living; make a living: ～办法 ways and means for seeking a livelihood / 靠木工手艺～ earn a living as a carpenter ②[方](职业;工作) job: 找个～ look for a job
【营私】seek private gain; feather one's nest: ～舞弊 engage in malpractices for selfish ends; practice graft / ～自肥 plan private interest and enrich oneself / 结党～ form a clique for selfish purposes
【营养】nutrition; nourishment: 关心儿童～ take care of the nutrition of the children / 水果富有～。 Fruit is very nourishing. ◇ ～不良 malnutrition; undernourishment / ～价值 nutritive value / ～品 nutriment / ～缺乏病 deficiency disease / ～素 nutrient / ～性疾病 nutritional diseases / ～学 nutriology / ～增补剂 nutritional supplement
【营业】do business: 暂停～ business temporarily suspended / 照常～ business as usual ◇ ～报告 business report / ～部 sales department / ～额 turnover; volume of business / ～法 business law / ～范围 scope and field of business / ～利润 operating profit / ～年度 the business year / ～时间 business hours / ～收入 business income / ～税 business tax; sales taxes; transactions tax; turnover tax / ～外收入 non-business income / ～外支出 non-business expenditure / ～项目 items of business / ～许可证 business permit / ～员 shop employees / ～支出 operating expenses / ～执照 business licence
【营造】construct; build: ～房屋 construct houses / ～防风林 plant windbreak forests ◇ ～厂 construction firm / ～商 contractor; building contractor
【营长】battalion commander
【营帐】tent: 搭～ pitch a tent

萦
【萦怀】occupy one's mind
【萦回】hover; linger: ～脑际 linger in one's mind
【萦绕】hover; linger

潆
【潆洄】swirl

蝇 fly: 灭～ kill flies
【蝇拍】flyswatter; flyflap
【蝇头】small as the head of a fly; tiny: ～微利 petty profits / ～小楷 very small characters
【蝇营狗苟】shamelessly seek personal gain

盈 ①(充满) be full of; be filled with: 丰～ plentiful; abundant ②(多余) have a surplus of: ～千累万 in hundreds and thousands
【盈亏】①(月亮的圆和缺) the waxing and waning of the moon ②(赚钱和赔本) profit

and loss: 自负～ assume sole responsibility for its own profits or losses
【盈利】profit; gain: ～三百万元 net a profit of three million *yuan* ◇～性企业 profit-making enterprise / 这行业～不多。The profits in this business are not large.
【盈满】filled; full
【盈溢】brim over
【盈余】surplus; profit: ～七千元 have a surplus of 7000 *yuan* ◇～价值 surplus value / ～资金 surplus fund
【盈月】full moon

楹

principal columns of a hall
【楹联】couplet written on scrolls and hung on the pillars of a hall

迎

①(迎接) go to meet; greet; welcome; receive: ～宾曲 strains of welcome march / ～来了新的一年 welcome in a new year / 赴机场～客 greet a guest at the airport ②(对着;冲着) move towards; meet face to face: ～击 make a head-on attack / ～上前去握手 go forward and shake hand with sb.
【迎春】[植] winter jasmine
【迎阿权贵】please and toady those who have influence and standing
【迎风】①(对着风) facing the wind: ～航行 sail close to the wind / ～前进 haul upon the wind ②(随风) down the wind; with the wind: ～摇曳 bending before wind / ～招展 flutter in the wind ◇～面 windward side
【迎合】cater to; pander to: ～低级趣味 pander to low tastes; pander to the vulgar taste / ～对方心理 go along with the other side / ～顾客的需要 cater to the needs of customers / ～某人 cater to sb. / ～人意 fall in with the wishes of other persons / ～听众 suit one's speech to the audience
【迎候】await the arrival of; greet: 在机场～贵客 greet the distinguished guests at the airport
【迎虎于门】meet a tiger at the door; meet trouble half-way
【迎击】meet head-on: ～歪风 repulse the evil wind
【迎接】meet; welcome; greet: 他到门口去～客人。 He went to the door to receive his visitors. / 我要到火车站去～你。 I'll meet your train.
【迎来送往】welcome and speed the parting of guests
【迎面】head-on; in one's face: 微风～吹来 a breeze blowing in one's face
【迎刃而解】split all the way down once it's been chopped open; be readily solved
【迎头】head-on; directly: ～赶上 catch up forthwith / ～痛击 repulse a frontal attack; deal head-on blows / ～相撞 head-on clash / ～一棒 give a front blow with stick
【迎新】①(迎新年) see the new year in: 送旧～新 usher in the new and send off the old ②(迎接新来的人) welcome new arrivals ◇～会 welcome party; party to welcome the newcomers
【迎战】meet head-on
【迎着】in face of; towards: ～风雨前进 brave wind and rain / ～困难上 press forward in face of difficulties

yǐng

瘿

【瘿虫】[动] gall insect

影

①(影子) shadow: 树～ shadow of a tree ②(影像) reflection; image: 湖中的云～ clouds imaged in the lake / 桥在水中的倒～ reflection of a bridge in the water ③(模糊的形象) trace; vague impression; sign ④(照片) photograph; picture: 合～ group photo / 留～ have a photograph taken ⑤(电影) motion picture; film; movie: ～剧本 a film play
【影壁】①(做屏蔽的墙壁) screen wall ②(塑有各种形象的墙壁) wall with carved murals
【影集】photograph album
【影剧界】the entertainment world; movie and drama circles
【影迷】movie fan
【影片】①(放映的电影) film; movie; picture: 彩色～ color film ②(电影胶片) film; copy
【影评】film review
【影射】allude to; hint obliquely at; insinuate: ～攻击 attack by innuendo / 有意～某人 make pointed allusions to sb.
【影坛】the movie circle; moviedom: ～故事 stories of the film arena
【影条】[纺] shadow stripes: ～的确良 shadow stripe Dacron
【影响】①(对别人的思想或行为起作用) influence; affect: ～健康 affect one's health / ～交通 hold up traffic / ～学习 interfere with one's study / 受到气候～ be influenced by the weather / 天气～农作物。The weather influ-

ences crops. ②(对人或事物所起的作用) effect; influence: 社会～ social influence ◇～范围 coverage
【影像】an image; a portrait: ～艺术 the art of photo-taking
【影印】[印] photomechanical printing; photo-offset process: ～珍本书籍 photolithograph rare books ◇～版 process plate／～本 photo-offset copy; facsimile／～件 photocopy／～照相机 process camera／～制版 photomechanical process
【影影绰绰】vaguely; dimly; indistinctly
【影院】cinema; movie theatre
【影子】①(影) shadow; reflection ②(模糊的形象) trace; sign; vague impression ◇～内阁 shadow cabinet

颖 ①(麦粒等带芒的外壳) grain husk; glume ②(尖端) tip; point ③(聪明) clever; intelligent
【颖果】[植] caryopsis
【颖慧】clever; bright; intelligent

yìng

应 ①(回答) answer; respond; echo: ～声开门 answer the door ②(满足要求) comply with; grant: 以～急需 in order to fill an urgent need ③(顺应;适应) suit; respond to ④(应付) deal with; cope with
【应变】①(应付突然发生的事情) meet an emergency: 随机～ adjust to changing circumstances; act according to circumstances ②[物] strain: ～硬化 strain hardening ◇～措施 emergency measure／～计 strainometer
【应承】agree; promise; consent: ～履约 promising to carry out an agreement／把事情～下来 promise to do sth.
【应酬】engage in social activities; have social intercourse: 不善于～ be unskilled in social intercourse／说几句～话 exchange a few polite words ◇～信件 courtesy letter
【应答】reply; answer: ～如流 reply readily and fluently ◇～器 responder
【应敌】deal with the enemy; cope with the enemy
【应对】reply; answer: 善于～ good at repartee
【应付】①(对人对事采取方法措施) meet; cope with; deal with; handle: ～突然发生的危险 meet a sudden danger／难于～ be difficult to deal with ②(敷衍了事) do sth. perfunctorily; do sth. after a fashion: ～事儿 go through the motions ③(将就) make do: 我这个提包还可以～一阵子。I can make do with this handbag for some time.
【应急】meet an urgent need; meet an emergency: ～之用 meet the need of emergency ◇～措施 emergency measure／～电源 emergency supply／～计划 contingency plan; crash program
【应接不暇】have more visitors or business that one can attend to
【应景】do sth. for the occasion ◇～诗 occasional verses／～文章 routine article
【应考】take an examination
【应力】[物] stress: 内～ internal stress／预～ prestress／正～ direct stress
【应募】respond to a call for recruits; enlist; join up
【应聘】accept an offer of employment
【应允和解】give assent to a reconciliation
【应声】at the sound of: 手枪一响,逃犯～而倒。The escaped criminal fell at the sound of a pistol.
【应声虫】yesman; echo
【应时】①(适合时令的) seasonable; in season: ～对景 be fashionable and adapted to environment／～货品 seasonable goods／～水果 fruits of the season ②(立刻) at once; immediately: ～而动 act according to circumstance／～自鸣 struck at the proper time
【应验】come true; be confirmed; be fulfilled: ～～良方 an effective prescription／他的预言果然～了。His prophecy has now come true.
【应邀】at sb.'s invitation; on invitation: ～出席会议 attend a meeting by invitation
【应用】apply; use: 把理论～于实践 apply theory to practice／新技术在工业上的～ application of new technique to industry ◇～化学 applied chemistry／～科学 applied science／～物品 the various articles required
【应用文】practical writing
【应运而生】arise at the historic moment; emerge as the times require
【应战】①(跟进攻的敌人作战) meet an enemy attack ②(接受对方提出的挑战) accept a challenge ◇～书 letter accepting a challenge
【应诊】see patients
【应征】①(响应征兵号召) be recruited; enlist: ～当志愿兵 enlist as a volunteer ②(泛指响应某种征求) answer to calls; answer to requests: ～的稿件 contributions to a periodical, etc. at the editor's public invitation

硬 ①(坚硬) hard; stiff; tough: ～领 stiff collar / ～煤 hard coal ②(刚强) strong; firm; tough; obstinate: 态度～ be stiff in manner / 心肠～ hard-hearted ③(勉强) forcibly; arbitrarily; manage to do sth. with difficulty: ～撑着干 force oneself to work hard ④(能力强; 质量好) good; able: 货色～ goods of high quality / ～手 a good hand
【硬邦邦】very hard; very stiff
【硬币】coin; specie: 一个五分的～ a five-fen piece / 用～支付 payment in specie
【硬衬】[纺] buckram
【硬撑】hold on firmly despite extreme adversity, pain, etc.
【硬磁盘】[计算机] hard discs
【硬度】[物] hardness: 维氏～ Vickers hardness ◇ ～计 sclerometer
【硬干】do something in disregard of obstacles
【硬腭】[生理] hard palate
【硬功夫】great proficiency; masterly skill
【硬骨头】hard bone; a dauntless, unyielding person: ～精神 spirit of unyielding integrity
【硬汉】a dauntless, unyielding man; a man of iron
【硬化】①(变硬) harden; stiffen: ～为石 hardened into a rock / 经久～[机] age hardening ②[医](组织变硬) sclerosis; hardening: 动脉～ hardening of the arteries; arteriosclerosis / 血管～ vascular sclerosis
【硬件】[计算机] hardware
【硬结】①(结成硬块) indurate; harden ②[医](组织结成的硬块) scleroma
【硬朗】[口] hale and hearty
【硬煤】hard coal; anthracite
【硬锰矿】psilomelane
【硬模】[机] die ◇ ～铸造 diecasting
【硬木】hardwood ◇ ～家具 hardwood furniture
【硬碰硬】①(用强硬的态度对付强硬的态度) confront the tough with toughness ②(要求真本领的工作) demanding solid, painstaking work or real skill
【硬皮病】[医] chorionitis; dermato sclerosis
【硬皮书】[印] hardcover
【硬拼】fight recklessly
【硬如钢铁】as hard as steel
【硬石膏】anhydrite
【硬是】[方]①(实在是) actually; really ②(就是) just; simply: 他～不认错。 He just would not admit his fault.
【硬水】hard water
【硬说】stubbornly insist; assert; allege: 他～他不困。 He insisted that he wasn't sleepy.
【硬套】apply arbitrarily; apply mechanically
【硬挺】endure with all one's will; hold out with all one's might: 这场病他～过来了。 He succeeded in pulling through the illness.
【硬通货】hard currency
【硬席】[交] hard seats (on a train) ◇ ～卧铺 hard sleeper; sleeping carriage with hard berths
【硬橡胶】[化] hard rubber; ebonite; vulcanite
【硬性】rigid; stiff; inflexible: ～规定 rigid and inflexible rules; hard and fast rules
【硬玉】jadeite
【硬仗】tough battle; formidable task: 打～ fight a hard battle
【硬着头皮】toughen one's scalp; brace oneself; force oneself to do sth. against one's will: ～把这本书读完 brace oneself and read the book through
【硬脂】[化] tristearin; stearin ◇ ～酸[化] stearic acid / ～酸盐 stearate / ～油 stearine oil
【硬纸板】hardboard; cardboard
【硬质合金】[冶] hard alloy; hard metal ◇ ～刀具[机] hard alloy cutter; hard metal tool
【硬质塑料】[化] rigid plastics: 半～ semirigid plastics

映 ①(反映) reflect; mirror: ～入眼帘 greet the eye ②(放电影) project a movie: 首～ first showing
【映衬】set off
【映山红】[植] azalea
【映射】shine upon; cast light upon: 月光～着山谷。 The moonlight shines the valley.
【映照】shine upon; cast light upon

yōng

庸 ①(平庸) commonplace; mediocre: ～言～行 commonplace words and deeds ②(不高明) inferior; second-rate: ～～碌碌 mediocre and unambitious ③(用) need: 无～细述。 There is no need to go into details.
【庸才】[书] mediocre person; mediocrity
【庸夫俗子】ordinary persons and laymen
【庸腐之见】a simple and stale point of view
【庸碌】mediocre and unambitious: ～无能 mediocre and incompetent / ～之辈 an ordinary class of person
【庸人】mediocre person
【庸人自扰】worry about troubles of one's own

imagining
【庸俗】vulgar; philistine; low: ～的人 vulgarian / ～的作风 vulgar style ◇ ～化 vulgarize; debase
【庸俗进化论】[哲] vulgar evolutionism
【庸俗唯物主义】[哲] vulgar materialism
【庸言】a trite remark
【庸医】quack; charlatan: ～杀人 homicide by quackery
【庸庸之辈】the common run of men

鳙
【鳙鱼】variegated carp; bighead

雍 harmony
【雍容】natural, graceful and poised: ～华贵 elegant and poised; distingué / 态度～ have a dignified bearing

壅 ①(堵塞) stop up; obstruct; block ②(把土或肥料培在植物根上) heap soil or fertilizer over and around the roots: ～肥 heap fertilizer around the roots / ～土 hilling
【壅闭不通】obstructed and constipated
【壅塞】clogged up; jammed; congested: ～流 choked flow / ～窒息 blocked and suffocated

臃
【臃肿】①(过度肥胖,动作不灵) too fat to move: 穿得太～ be cumbersomely dressed; be encumbered by too much clothing ②(指机构庞大,调度不灵) overstaffed: ～的机构 overstaffed organizations

痈 [医] carbuncle: 养～遗患 neglecting a carbuncle will cause trouble; failure to deal sternly with evil-doers will lead to serious trouble
【痈疽】ulcer

拥 ①(拥抱) hold in one's arms; embrace; hug ②(围着) gather around ③(挤着走) crowd; throng; swarm: ～进大厅 throng into the hall ④(拥护) support ⑤(拥有) have; possess
【拥抱】embrace; hug; hold in one's arms: 朋友们热烈地～。The friends warmly embraced.
【拥被而卧】lie down and cover oneself with a quilt
【拥戴】support
【拥护】support; uphold; endorse: ～决议 uphold the decision / ～我国政府的声明 support our government statement ◇ ～者 supporter
【拥挤】crowd; push and squeeze: ～不堪 be crowded to capacity; be packed like sardines / ～的城市 crowded city / 不要～! Don't push!
【拥军爱民】support the army and cherish the people
【拥军优属】support the army and give preferential treatment to the families of the armyman and martyrs
【拥入】crowd into: ～公共汽车 crowd into a bus
【拥塞】jam; congest: 街道～ streets jammed with traffic
【拥上心头】well up: 欢乐拥上他的心头。Joy welled up in his heart.
【拥有】possess; have; own: ～丰富的矿藏 have mineral resources / ～核武器 possess nuclear weapons / ～核优势 retain nuclear superiority

佣 ①(雇用) hire ②(仆人) servant: 女～ woman servant; maid

yǒng

永 perpetually; forever; always: ～志不忘 will always bear in mind
【永别】part never to meet again; part forever; be parted by death
【永不】never: ～掉队 never drop behind / ～后悔 never look back
【永垂不朽】be immortal: 人民英雄～! The heroes of the people are immortal.
【永磁】[物] permanent magnetism ◇ ～发电机 magneto; permanent-magnet generator / ～体 permanent magnet
【永存】eternal; lasting forever; remain forever
【永恒】eternal; perpetual: ～的友谊 eternal friendship / ～的真理 eternal truth ◇ ～运动 [物] perpetual motion
【永久】permanent; perpetual; everlasting; forever: ～的和平 permanent peace; everlasting peace ◇ ～剥夺 permanently depriving / ～地址 permanent address / ～冻土 permafrost / ～积雪 firn / ～雪线 firn line / ～主权 permanent sovereignty / ～中立 permanent neutrality / ～中立国 neutralized state
【永久性】perpetual: ～法规 perpetual statute / ～条款 perpetual provision
【永生】①(哀悼死者用语) immortal: ～的战士 immortal fighter ②[宗](人死后灵魂永久不灭) eternal life
【永世】forever: ～不朽 last forever / ～长存 live for ever and ever / ～难忘 will never forget it for the rest of one's life / ～无尽 for end-

less ages
【永无宁日】never will there be days of peace
【永远】[副] always; forever; ever; in perpetuity: ～离开 go away forever
【永志不忘】will always bear in mind; will always cherish the memory of sb. or sth.

泳 swim: 侧～ side stroke／蝶 butterfly (stroke)／器～ swimming with breathing apparatus／潜～ underwater swimming／蛙～ breaststroke／仰～ backstroke／自由～ freestyle
【泳道】lane

咏 ①(诵读) chant; intone ②(用诗歌来叙述) sing of; express in poetic form
【咏叹】intone; chant; sing
【咏叹调】[乐] aria
【咏赞】sing the praises of; praise

甬
【甬道】①(甬路) paved path leading to a main hall or a tomb ②(走廊) corridor

涌 ①(水或云气冒出) gush; well; pour; surge: 泪如泉～ tears well up in one's eyes ②(从水或云气中冒出) rise; surge; emerge: 一排排巨浪～向石滩。 Huge waves surged over the rocks.
【涌泉潺潺】the bubbling of the spring
【涌现】emerge in large numbers; spring up; come to the fore: 许多新事物正在～。 Many new things are springing up.

蛹 pupa: 蚕～ silkworm chrysalis／蝶～ chrysalis

踊 leap up; jump up
【踊跃】①(跳跃) leap; jump: ～欢呼 leap and cheer ②(形容情绪热烈) vying with one another; eagerly; enthusiastically: ～参军 vie with one another to join the army／～响应号召 respond to a call enthusiastically

俑 wooden or earthen human figure buried with the dead in ancient times; tomb figure; figurine: 木～ wooden figurine／陶～ pottery figurine／武士～ warrior figure; terracotta warriors

勇 brave; valiant; courageous: ～攀高峰 be bold in scaling heights／智～双全 be both brave and wise
【勇不可挡】too courageous to be met with
【勇而无谋】be brave but have no plans
【勇敢】brave; courageous: ～的战斗精神 valiant fighting spirit／～善战 courageous and skillful in battle／～无匹 brave without equal／机智～ be brave and resourceful／勤劳～的人民 a brave and industrious people
【勇冠三军】peerless or matchless in bravery or valor
【勇猛】bold and powerful; full of valor and vigor
【勇气】courage; nerve: 鼓起～ pluck up one's courage
【勇士】a brave and strong man; warrior
【勇挑重担】be brave to take on heavy responsibilities
【勇往直前】march forward courageously; advance bravely
【勇于】be brave in; be bold in; have the courage to: ～创新 be bold in making innovations／～负责 be brave in shouldering responsibilities／～改过 be bold enough to reform／～自我批评 have the courage to practice self-criticism

yòng

用 ①(使用) use; employ; apply: ～法律来强制执行 enforcement at law／～强制手段 by strong arm／～石头铺路 pave a road with stone／～英语交谈 talk in English／古为今～ make the past serve the present／你会～筷子吗? Do you know how to use chopsticks?／我写字总～钢笔。 When I write, I always use a pen. ②(费用) expenses; outlay: 零～钱 pocket money／日～ daily expenses ③(用处) usefulness; use: 物尽其～ make the best use of everything／有关～处 be of some use ④(需要) need: 不～说 needless to say ⑤(吃; 喝) eat; drink; take
【用兵】①(使用武力) resort to arms; use military forces: 不得已而～ have no alternative but to resort to arms ②(指挥作战) command troops: ～如神 direct military operations with miraculous skill; work miracles in maneuvering troops／善于～ well versed in the art of war
【用不惯】unaccustomed to the use of sth.
【用不完】too many or too much for use; cannot be used up
【用不着】①(用不上) no need; have no use for ②(不需要) there is no need to; it is not worth while to
【用材林】commercial forest; timber forest

【用场】use; application: 派大～ be turned to good account / 派新～ be put to new uses / 有～ be useful
【用处】use; good: 没有～ be of no use / 抱怨有什么～? What's the use of complaining?
【用度】expenditure; expense; outlay: 他家人口多,～大。 He has a big family and many expenses.
【用法】use; usage: ～说明 direction / 学习工具的～ learn the use of tools
【用非所学】be engaged in an occupation not related to one's training
【用费】expense; cost
【用功】hardworking; diligent; studious: ～读书 be studious; study diligently; be diligent in one's studies
【用户】consumer; user: 电话～ telephone subscriber / 征求～意见 ask for consumers' opinions ◇ ～电报 telex / ～手册 user's manual
【用计陷害】use tricks for getting another in trouble
【用劲】exert oneself; put forth one's strength: ～推 push hard
【用尽】exhaust; use up: ～方法 use every possible means / ～心机 exhaust one's abilities / ～一切气力 exert every ounce of energy
【用具】utensil; apparatus; appliance: 办公～ office appliance / 炊事～ kitchen utensils / 救生～ lifesaving equipment / 消防～ fire-fighting apparatus
【用力】exert oneself; put forth one's strength: ～过度 make too strenuous efforts
【用品】appliance; articles: 盥洗～ toilet articles / 生活～ articles for daily use; daily necessities / 文化～ stationery / 体育～ sports goods; athletic equipment
【用钱】spend money: ～如水 spend money like water
【用人】①(选择与使用人员) choose a person for a job; make use of personnel: ～不当 not choose the right person for the job / 善于～ know how to choose the right person for the right job; know how to make proper use of personnel ②(需要人手) need hands ③(仆人) servant
【用事】act: 感情～ act impetuously / 意气～ be swayed by one's feelings and act rashly
【用途】use: ～广 be widely used / 具有多种～的电子仪器 an electronic device with many uses
【用武】use force; display one's abilities or talents: 大有～之地。 There's ample scope for one's abilities.
【用贤任能】use the capable and employ the skilled
【用项】items of expenditure; expenditures
【用心】①(集中注意力) diligently; attentively; with concentrated attention: ～去想一想 do some hard thinking / ～做习题 work hard at exercises ②(居心) motive; intention: ～良苦 have really given much thought to the matter / 别有～ have ulterior motives
【用眼示意】hint by the eye
【用意】intention; purpose: ～很好 with good intention
【用语】①(措词) choice of words; wording: ～不当 inappropriate choice of words; incorrect wording / ～简洁 be expressed in neat phrase ②(某方面的专用词语) phraseology; term: 外交～ diplomatic terms / 文学～ literary terms

佣

【佣金】commission; brokerage; middleman's fee

yōu

忧 ①(忧愁) worry; be worried: ～人民所～,乐人民所乐 share the people's joys and sorrows ②(使人忧愁的事) sorrow; anxiety; concern; care: 无～无虑 carefree; free from all anxieties
【忧愁】sad; worried; depressed: ～满面 mourning is all over the face
【忧愤】worried and indignant: ～累日 be sad and angry for days
【忧国忧民】be concerned about one's country and one's people
【忧患】suffering; misery; hardship: ～余生 survive countless distresses and worries / 饱经～ have gone through a good deal
【忧疾成病】be ill with anxiety and melancholy
【忧惧】worried and apprehensive
【忧苦以终】be distressed to death
【忧劳成疾】fall sick with grievance and toil
【忧虑】worried; anxious; concerned: ～过度 be worried to excess / 对前途的～ worries concerning the future / 深感～ feel extremely anxious; be very worried
【忧闷】depressed; feeling low; weighed down with cares
【忧伤】distressed; weighed down with sorrow; laden with grief
【忧心】[书] worry; anxiety: ～忡忡 heavyheart-

ed; care-laden; laden with anxieties / ～如焚 burning with anxiety; extremely worried

【忧形于色】sadness manifested on the countenance; look dismal and unhappy

【忧郁】melancholy; heavyhearted; dejected: ～不欢 anxious and unrest without joy / ～一生 depressed and melancholy for life ◇ ～症 melancholia

优 ①(优良) excellent ②(演戏的人) actor or actress

【优待】give preferential treatment ◇ ～券 complimentary ticket

【优等】high-class; first-rate; excellent ◇ ～品 high-class product / ～生 top student

【优点】merit; strong point; advantage; virtue

【优厚】munificent; liberal; favorable: 待遇～ excellent pay and conditions; liberal wages and benefits

【优惠】[经] preferential; favorable: 按～条件提供的经济援助 economic assistance given on liberal terms ◇～办法 preferential measure / ～贷款 loan on favorable terms / ～待遇 preferential treatment / ～关税 preferential duties / ～关税协定 preferential tariff agreements / ～汇率 preferential rate / ～价格 preferential prices / ～权 preferential rights / ～国 favored nation / ～利率 preferential interest rate / ～证 accommodation paper

【优良】fine; good; excellent ◇ ～传统 fine tradition / ～品质 fine quality / ～品种 good strains of seeds / ～作风 fine style of work

【优劣不分】no discrimination between good and bad

【优美】graceful; fine; exquisite: ～的音乐 beautiful music / ～绝伦 excellent beyond comparison / 风景～ fine scenery

【优柔寡断】irresolute and hesitant; indecisive

【优生学】[生] eugenics

【优胜】winning; superior: ～劣败 survival of the fittest ◇ ～红旗 championship red banner / ～奖 winning prize / ～者 winner; champion

【优势】superiority; preponderance; dominant position: 军事～ military superiority / 空中～ air supremacy / 在某方面占～ have an advantage in sth.

【优先】have priority; take precedence: ～考虑 give priority to ◇ ～抵押权 first mortgage / ～股本 capital stock-preferred / ～留置权 first lien; paramount lien / ～权 priority; preference

【优闲自得】with abundant leisure one is contented

【优秀】outstanding; excellent; splendid; fine: ～选手 topnotch player / ～作品 works of excellence / 成绩～ get good marks

【优选法】optimum seeking method; optimization

【优异】excellent; outstanding: ～的成绩 excellent results

【优裕】affluent; abundant: 生活～ be well-off; be well-to-do; live in affluence

【优越】superior; advantageous: ～条件 favorable conditions / 处于～的地位 be in an advantageous position ◇ ～感 sense of superiority; superiority complex

【优越性】superiority; advantage

【优哉游哉】[书] living a life of ease and leisure; leisurely and carefree; leisurely and unhurried

【优质】high quality; high grade: ～产品 products of quality / ～钢 high-quality steel

幽 ①(深远) deep and remote; secluded; dim: ～谷 a deep and secluded valley ②(隐蔽的) secret; hidden: ～居 live in seclusion ③(沉静) quiet; tranquil; serene ④(囚禁) imprison ⑤(阴间) of the nether world: ～魂 ghost

【幽暗】dim; gloomy

【幽闭】①(幽禁) put under house arrest ②(深居家中不能外出或不愿外出) confine oneself indoors

【幽愤】hidden resentment

【幽会】a secret meeting of lovers; a lovers' rendezvous; tryst

【幽寂】secluded and lonely

【幽禁】①(软禁) put under house arrest ②(囚禁) imprison; keep under detention

【幽静】quiet and secluded; peaceful

【幽灵】ghost; specter; spirit

【幽门】[生理] pylorus ◇ ～梗阻[医] pyloric stenosis

【幽默】humorous ◇～感 sense of humor / ～角色 humorous character / ～杂志 humor magazine / ～作品 a humorous production

【幽情】exquisite feelings: 发思古之～ muse over things of the remote past

【幽栖】live away from society: ～山林 dwell on a wooden-hill in seclusion

【幽囚终身】be imprisoned for life

【幽趣】the delightful serenity of seclusion

【幽深】deep and serene; deep and quiet

【幽思】①(沉静的深思) ponder; meditate ②(隐藏在内心的思想感情) thoughts on things remote
【幽闲】①(安详文雅) gentle and serene ②(闲适自得) leisurely and carefree
【幽香】a delicate fragrance
【幽雅】quiet and tastefully laid out: ～清静 retired and quiet
【幽幽】①(声音、光线等微弱) faint: ～的路灯 dim street lamps / ～啜泣 sob quietly ②(深远) faraway; distant; looming in the distance
【幽怨】hidden bitterness

悠 ①(久;远) long-drawn-out; remote in time or space ②(闲适) leisurely ③[口](悠荡) swing
【悠长】long; long-drawn-out
【悠荡】swing; sway
【悠久】long; long-standing; age-old: ～的传统 an age-old tradition / ～的文化 a civilization of long standing / 历史～ have a long history
【悠然】①(悠闲的样子) carefree and leisurely: ～自得 be carefree and content ②(遥远) long; distant; far away: ～神往 one's thoughts turn to things distant
【悠闲】leisurely and carefree: ～的生活 life of leisure / ～自在 leisurely and carefree
【悠扬】rising and falling; melodious
【悠悠】①(长久;遥远) long; long-drawn-out; remote: ～无期 at some distant time ②(从容不迫) leisurely; unhurried: ～自得 carefree and content
【悠悠荡荡】floating about
【悠悠忽忽】①(悠闲懒散) loiter ②(神态恍惚) be in a trance
【悠远】①(离现在时间长) a long time ago; long ago ②(距离远) far off; remote; distant: 山川～ mountains and rivers far, far away
【悠哉悠哉】free from restraint
【悠着】[方] take things easy: ～点劲,别太猛了。 Take it easy! Don't go at it so hard!

yóu

游 ①(游泳) swim ②(旅游;闲逛) rove around; wander; travel; tour: 周～世界 travel round the world ③(经常移动的) roving; itinerant ④(河段) part of a river; reach: 上～ the upper reaches / 下～ the lower reaches / 中～ the middle reaches
【游伴】travel companion
【游标】[机] vernier: ～卡尺 vernier caliper / ～千分尺 vernier micrometer
【游标卡尺】sliding gage
【游船】pleasure-boat
【游荡】loaf about; loiter; wander: ～无度 be dissipated without a limit
【游动】move about; go from place to place ◇ ～哨 a roving sentry; a patrol
【游逛】go sight-seeing; stroll about
【游击】guerrilla warfare ◇ ～队 guerrilla forces; a guerrilla detachment / ～队员 guerrilla; partisan / ～区 guerrilla area / ～战 guerrilla war / ～战术 guerrilla tactics
【游记】travel notes; travels
【游街】parade sb. through the streets
【游客】visitor; tourist; excursionist; sightseer
【游览】go sight-seeing; tour; visit ◇ ～车 tourist coach / ～地 place for sight-seeing; excursion center / ～图 tourist map / ～团体 tourist party
【游廊】covered corridor
【游乐园】amusement park
【游离】①(离开集体或附属的事物而存在) dissociate; drift away ②[化] free: ～基 free radical / ～酸 free acid / ～状态 free state
【游历】travel for pleasure; travel; tour
【游轮】belt tightening pulley; loose pulley
【游民】vagrant; vagabond: 无业～ vagrant
【游牧】move about in search of pasture; rove around as a nomad ◇ ～部落 nomadic tribe / ～民族 nomadic people / ～生活 nomadic life; nomadism
【游禽】natatorial bird
【游人】visitor; sightseer; tourist
【游刃有余】handle a butcher's cleaver skillfully; do a job with skill and ease; be more than equal to a task
【游山玩水】make a sightseeing tour; tour the scenic spots
【游手好闲】idle about; loaf: ～的人 idler; loafer / ～,不务正业 idle about and do no decent work
【游说】go canvassing; go about selling an idea; go about drumming up support for an idea
【游丝】①(飘在空中的) gossamer ②[机](钟表的) hairspring
【游艇】yacht; pleasure-boat
【游玩】①(游戏) amuse oneself; play ②(游逛) go sight-seeing; stroll about
【游戏】①(娱乐活动) recreation; game: 做～ play games ②(玩耍) play: 孩子们在花园里～。 The kids are playing in the garden.
【游行】parade; march; demonstration: 节日～

gala parade／抗议～ protest march／参加示威～ take part in a demonstration／举行～示威 hold a demonstration ◇～队伍 contingents of paraders or marchers; procession
【游兴】interest in going on an excursion
【游移】waver; vacillate; wobble: ～不定 keep on vacillating／～其辞 be hesitated in words
【游弋】cruise: 在海上～ cruise on the sea
【游艺】entertainment; recreation ◇～场 amusement park／～会 amusement gathering／～节目 program／～室 recreation room
【游泳】swim ◇～比赛 swimming contest／～馆 natatorium／～裤 swimming trunks／～帽 swimming cap／～运动员 swimmer
【游泳池】swimming pool: 室内～ indoor pool; swimming bath／室外～ outdoor pool; open air pool
【游泳衣】swimsuit; swimming suit: 三点式～ bikini
【游资】idle fund; idle money; floating capital
【游子】[书] man travelling or residing in a place far away from home

莜

【莜麦】[植] naked oats

尤

①(突出的) outstanding: 无耻之～ have absolutely no sense of shame; brazen in the extreme ②(尤其) particularly; especially ③(过失) fault; wrongdoing: 效～ knowing follow the example of a wrongdoer ④(怨恨) have a grudge against; blame: 怨天～人 blame god and man; blame everyone and everything but oneself
【尤其】[副] especially; particularly: 这一点～重要。 This is even more important

疣

[医] wart
【疣病】verrucosis

鱿

【鱿鱼】squid

犹

①(如同) just; as; like: ～如白昼 as bright as day; as if in the daylight ②(还; 尚且) still: 话～未了 have scarcely finished speaking／记忆～新 be still fresh in one's memory
【犹可】still all right
【犹如】just as; like; as if: ～饿虎 be like a hungry tiger
【犹太】Jew ◇～复国主义 Zionism／～复国主义者 Zionist
【犹太教】Judaism ◇～教士 rabbi／～教堂 synagogue／～教徒 Judaist ○逾越节 Passover／五旬节 Pentecost／结茅节 Feast of Tabernacles／赎罪日 Yom Kippur／安息日 Sabbath／先知 prophet; seer／割礼 circumcision
【犹太人】Jew (男); Jewess (女)
【犹言在耳】ring in one's ears
【犹疑】hesitate
【犹有可为】still retrievable
【犹有童心】an aged person with a youthful heart
【犹有余悸】even now one is scared
【犹豫】hesitate; be irresolute: ～不决 hesitate; remain undecided／毫不～ without the least hesitation／犹犹豫豫 shilly-shally

由

①(原由) cause; reason ②(由于) because of; due to ③(经过) by; through: ～公路来 come by the highway ④(顺从) follow; obey: 事不～已。 Things are beyond one's control. ⑤[介](归谁做) ～他负责这项工作。 Leave this work to him. ⑥[介](表示起点) from: ～表及里 proceed from the surface to the inner essence
【由不得】①(不得依从) not be up to sb. to decide; be beyond the control of: 这件事～他们。 It's not up to them. ②(不由自主) cannot help: 她～笑了起来。 She couldn't help laughing.
【由此】from this; therefrom; thus: ～看来 judging from this; in view of this／～前进 go forward from here
【由此可见】thus it can seen; this shows; that proves
【由简而繁】go from the simple to the complex
【由近及远】from the near to the distant
【由来】origin: 分歧的～ the origin of differences
【由来已久】long-standing; time-honored
【由浅入深】go from the easy to the difficult and complicated
【由浅入深, 由易到难】from the shallower to the deeper, from the easier to the more advanced
【由天而降】come from nowhere
【由头】pretext: 找～ find a pretext
【由一斑窥全豹】from one learn all
【由于】[介] owing to; thanks to; as a result of; due to; in virtue of
【由衷】from the bottom of one's heart; sincere; heartfelt: ～之言 words which come from the

bottom of one's heart; sincere words / 表示～的感激 extend one's heartfelt thanks

油 ①(脂肪;油脂) oil; fat; grease: 豆～ soybean oil / 菜子～ rape seed oil / 橄榄～ olive oil / 蚝～ oyster oil / 花生～ peanut oil / 棉子～ cotton seed oil / 沙拉～ salad oil / 松节～ oil of turpen tine / 香～ sesame oil / 椰子～ coconut oil / 植物～ vegetable oil / 猪～ pork fat; lard / 给锁上～ oil a lock ②(用油涂抹) apply tung oil or paint ③(被油弄脏) be stained with oil or grease ④(油滑) oily; glib

【油泵】[机] oil pump

【油饼】①(一种油炸的食品) deep-fried dough cake ②[农](油料作物炸油后的渣滓) oil cake

【油布】oilcloth; oilskin; tarpaulin: ～口袋 oilskin pouch

【油彩】greasepaint; paint

【油菜】rape ◇ ～籽 rapeseed

【油层】[石油] oil reservoir; oil layer; oil horizon ◇ ～动态 reservoir behavior / ～压力 reservoir pressure

【油茶】①[植] tea-oil tree; oil-tea camellia ②(一种食品) a gruel of sweetened, fried flour ◇ ～面儿 flour fried in beef fat with sugar and sesame

【油船】tanker; oil carrier

【油灯】oil lamp

【油地毡】linoleum

【油豆腐】fried bean curd

【油坊】oil mill

【油橄榄】[植] olive

【油膏】[药] ointment

【油瓜】[植] large-fruited hodgsonia

【油管】[石油] ①(输油管道) oil pipe: 铺设～ lay oil pipes ②(井架用的油管) oil tube: ～深度 tubing depth

【油罐】oil tank; storage tank ◇ ～汽车 oil car

【油光】glossy; shiny; varnished

【油壶】oilcan; oiler

【油滑】slippery; foxy

【油画】oil painting: 画～ paint in oils ◇ ～家 oil painter / ～颜料 oil paint; oil color

【油晃晃】very oily

【油灰】[建] putty

【油迹】oil stains; grease spots: ～斑斑 covered with grease spots

【油井】oil well: 钻一口～ drill a well

【油锯】[林] chain saw ◇ ～手 chain-saw operator

【油库】oil depot; tank farm: 转运～ oil terminal

【油矿】①(石油矿床) oil deposit ②(油田) oil field

【油亮】glossy; shiny

【油料作物】oil-bearing crops; oil crops; oil plant

【油绿】glossy dark green

【油轮】tanker; oil tanker

【油码头】oil jetty; oil wharf; tanker terminal

【油毛毡】[建] asphalt felt

【油门】[机] throttle

【油墨】printing ink: 快干～ quicksetting ink

【油母页岩】oil shale

【油泥】greasy filth; grease

【油腻】greasy; oily: ～的食品 greasy food; oily food

【油漆】①(涂料) paint: ～未干! Wet paint! ②(涂抹油漆) cover with paint; paint ◇ ～工人 painter

【油枪】[机] oil gun

【油腔滑调】glib; unctuous; have a smooth, oily tongue

【油浸法】immersion method; immersion technique

【油然】①(自然而然产生) spontaneously; involuntarily ②(形容云气上升) densely; profusely

【油然而生】arise spontaneously; be produced of itself

【油溶性染料】oil-soluble dyes

【油石】[机] oilstone

【油水】①(饭菜里所含的脂肪质) grease ②(比喻利己的好处) profit: ～不大 not very profitable

【油松】[植] Chinese pine

【油酥】short; crisp; flaky: ～点心 short pastry

【油酸】[化] oleic acid

【油田】oil field: ～开发 oil field development / 海上～ offshore oil field

【油条】deep-fried twisted dough sticks

【油桐】[植] tung oil tree; tung tree

【油桶】oil drum

【油头粉面】dressy or foppish; sleek-haired and creamy-faced

【油头滑脑】slick; flippant; oily: 一个～的家伙 a shifty-looking fellow

【油汪汪】①(油多) dripping with oil; full of grease ②(油光) glossy; shiny

【油位】[机] oil level ◇ ～表 oil gauge

【油污】greasy dirt

【油箱】oil box; fuel tank: 副～ auxiliary fuel

tank; auxiliary tank
【油压】[机] oil pressure ◇ ～泵 oil pressure pump / ～表 oil pressure gauge / ～传动 hydraulic transmission / ～机 hydraulic press / ～千斤顶 hydraulic jack; oil jack
【油烟】lampblack
【油盐酱醋】daily necessaries
【油页岩】oil shale
【油印】mimeograph: ～五百份 mimeograph five hundred copies ◇ ～机 mimeograph / ～讲义 mimeographed lecture notes / ～腊纸 stencil; stencil paper
【油渣】①(猪油的) dregs of fat ②[石油] oil residue
【油毡】[建] asphalt felt: 屋面～ roofing slate ◇ ～纸 lincrusta
【油脂】oil; fat: 动物～ animal fat; animal oil; tallow; grease / 植物～ vegetable fat; vegetable oil
【油纸】oilpaper
【油棕】[植] oil palm
【油嘴】①(说话油滑) glib ②(油嘴的人) glib talker
【油嘴滑舌】with a glib tongue; glib-tongued

柚
【柚木】[植] teak

蚰
【蚰蜒】[动] common house centipede
【蚰蜒草】alpine yarrow

铀
[化] uranium (U): 氟化～ uranium fluoride / 浓缩～ enriched uranium
【铀棒】uranium bar
【铀后元素】[化] transuranium element; transuranium
【铀矿类】uranite

邮
①(邮寄) post; mail: 我刚出来去～几封信。I've just come out to past some letter. ②(有关邮务的) postal; mail: ～局 post office
【邮包】postal parcel; parcel ◇ ～保险 parcel post insurance
【邮车】postal car
【邮船】ocean liner; liner; packet ship
【邮戳】postmark: ～为凭 post-mark serves as a proof / 纪念～ cachet / 一封盖有北京～的信 a letter postmarked Beijing
【邮袋】mailbag; postbag; pouch
【邮递】postal delivery ◇ ～区 postal delivery zone / ～员 postman; mailman
【邮电】post and telecommunications ◇ ～部 Ministry of Posts and Telecommunications / ～局 post and telecommunications office / ～业务 postal and telecommunications service
【邮费】postage: ～不足 insufficient postage / ～免收 post-free / ～未付 postage-unpaid / ～已付 postpaid
【邮购】mail-order ◇ ～部 mail-order department / ～服务 mail order service / ～商店 mail-order house
【邮汇】remit by post
【邮寄】send by post; post ◇ ～名单 mailing list
【邮件】postal matter; post; mail: 挂号～ registered post / 航空～ air mail / 留局待领～ post restate; general delivery / 小包 a postal packet ◇ ～爆炸物 letter bomb / ～编码号 postal number
【邮简】stamped envelope: 航空～ aerogram; air letter aerogram
【邮局】post office: 机场～ airport post office
【邮路】postal route
【邮票】postage stamp; stamp: 航空～ air mail stamp / 欠资～ postage-due stamp / 特种～ special stamp / 无齿孔～ imperforate / 一套纪念～ a set of commemorative stamps / 贴～ stick on the stamps ○新票 unused stamp / 旧票 used stamp / 齿孔 perforation / 面值 face value; denomination / 加盖 overprint / 双联 pair / 四联 block of four / 整张 sheet / 小本票 booklet pane / 小型张 souvenir card / 盖销 cancellation / 首日封 first day cover / 实寄封 used cover; used envelope / 集邮册 stamp album
【邮亭】postal kiosk
【邮筒】pillar-box; postbox; mailbox
【邮箱】postbox; mailbox
【邮运】deliver by mail; mail transportation
【邮政】postal service ◇ ～编码 postcode; zip code / ～储蓄 postal savings / ～储蓄凭证 postal savings certificate / ～代办所 postal agency / ～法 law of post / ～汇票 postal money order; postal order / ～局 post office / ～局长 postmaster / ～网 postal network / ～信箱 post-office box / ～支局 branch post office / ～总局 general post office
【邮资】postage: ～免付 postage free / ～已付 postage paid; postpaid / ～总付 postage paid by the lump / 国内～ postage paid for inland mail / 国外～ postage paid for overseas mail

yǒu

莠 ①(狗尾草) green bristlegrass ②(品质坏的人) bad people

酉 the tenth of the twelve Earthly Branches

【酉时】the period of the day from 5 p.m. to 7 p.m.

有 ①(表示领有) have; possess: ～空 be free／今天下午你～事吗? Have you anything on this afternoon.／他～勇气。He possessed courage.／我在乡下～一所房子。I have a house in the country. ②(存在) there is; exist: 教室里～三十名学生。There are thirty students in the classroom. ③(估量或比较) as... as: 他～他哥哥那么高。He is as tall as his brother. ④(表示发生或出现)孩子们～了许多进步。The children have made much progress. ⑤(表示多,大) ～经验 be experienced ⑥(用于泛指) one; some: ～人告诉我 someone told me

【有碍】be a hindrance to; get in way of; obstruct: ～交通 hinder traffic／～情面 impede one's feeling and face

【有案可查】be a matter of record; be on record; be documented: ～的事项 matters of record

【有把握】confident of success

【有板有眼】rhythmical; measured; orderly

【有备无患】where there is precaution, there is no danger; preparedness averts peril

【有背景】have powerful connections

【有本领】capable; talented; resourceful

【有鼻子有眼儿】with every detail vividly described

【有产阶级】propertied class

【有产者】a man of property

【有偿借贷】lending upon consideration

【有偿行为】non-gratuitous

【有吃有穿】have ample food and clothing; well provided with food and clothing

【有酬劳动】[经] paid labor

【有待】remain; await: ～改进 have much room for improvement／～解决 remain to be solved

【有的】some: ～赞成,～反对。Some are in favor of it, some are against it.

【有的是】have plenty of; there's no lack of

【有底】know how things stand and feel confident of handling them; be fully prepared for what is coming

【有的放矢】shoot the arrow at the target; have a definite object in view

【有点儿】①(不多) some; a little ②[副](略微) somewhat; rather; a bit: ～不好意思 be somewhat embarrassed／～疲倦 be a bit tired

【有毒】poisonous; venomous

【有犯罪嫌疑】suspected of being guilty

【有方】with the proper method; in the right way: 教导～ have the right way to educate and guide

【有分寸】having a sense of propriety; knowing how far to go and when to stop

【有夫之妇】married woman

【有福同享,有祸同当】share joys and sorrows; share weal and woe; stick together through thick and thin

【有负众望】disappoint those who had hopes or confidence in a person

【有妇之夫】married man

【有感而发】explode not without reason

【有功】have rendered great service; have performed meritorious service: ～部队 troops with a fine record service; meritorious army unit

【有功功率】active power; real power

【有骨气】adhering to moral principles; have integrity

【有关】have sth. to do with; have a bearing on; relate to; concern: ～部门 the department concerned／～当局 the authorities concerned; the proper authorities／～当事人 party concerned／～方面 the parties concerned／～事实 relevant facts／～委托机构 associated agency／～文件 the relevant documents／～证据 relevant testimony／～组织 related organization／这与你～。This has sth. to do with you.

【有光】①(上光的) glazed ②[纺](闪光的) bright ◇ ～人造丝 bright rayon／～纸 cap paper; glazed paper

【有轨电车】tramcar; streetcar

【有鬼】there's something fishy: 心里～ have a guilty conscience

【有过之无不及】go even farther than; out do

【有害】harmful; pernicious; detrimental: ～无益 having injury but no advantage／对健康～ harmful to one's health ◇ ～气体 pernicious gases

【有恒】persevering

【有机】①[化] organic ②(事物各部分的统一性) organic: ～的组成部分 an organic part ◇ ～玻璃 polymethyl methacrylate; plexiglass; perspex／～肥料 organic fertilizer or

manure／～耕作 organic farming／～硅 organosilicon／～合成 organic synthesis／～化合物 organic compound／～化学 organic chemistry／～磷化合物 organic phosphorus compounds／～溶剂 organic solvent／～酸 organic acid／～体 organism／～土 organic soil／～物质 organic matter／～盐 organic salt
【有机可乘】there's an opportunity to take advantage of; there's a loophole that can be used
【有计划】in a planned way; according to plan: ～地发展经济 develop the economy in a planned way
【有价证券】negotiable securities; securities
【有进无退】burn one's bridge
【有救】can be saved; can be cured
【有口皆碑】win universal praise; be universally acclaimed
【有口难辩】find it hard to vindicate oneself
【有口难言】cannot bring oneself to mention sth.; find it hard or embarrassing to bring up a matter
【有口无心】be sharp-tongued but not malicious
【有来有往】reciprocal
【有赖】depend on; rest on
【有理】①(有道理) reasonable; justified; in the right: ～、有利、有节 on just grounds, to our advantage and with restraint ②[数] rational: ～分式 rational fraction／～函数 rational function／～数 rational number
【有力】strong; powerful; forceful; energetic; vigorous: ～的打击 a forceful blow／～的证明 convincingly demonstrate
【有利】advantageous; beneficial; favorable: ～于集体 advantageous to the collective／～于健康 beneficial to health ◇～地形[军] favorable terrain／～时机 opportune time／～条件 favorable condition
【有利可图】have good prospects of gain; stand to gain; be profitable
【有两下子】[口] have real skill; know one's stuff
【有路子】have friends in high places
【有眉目】begin to take shape; about to materialize
【有门儿】①(得到决窍) get the hang ②(有希望) find the beginning of a solution; be hopeful
【有名】well-known; famous; celebrated: ～的戏剧评论家 a well-known drama critic／～的作家 a famous writer
【有名无实】in name but not in reality; merely nominal; titular
【有名有姓】a person who really exists; of established identity
【有目共睹】be there for all to see; be perfectly obvious
【有目共赏】have a universal appeal
【有奶便是娘】whoever suckles me is my mother; ready to serve anyone for material benefit
【有凭有据】having proof and evidence
【有期徒刑】[法] set term of imprisonment
【有其父，必有其子】like father, like son
【有起色】show signs of a rise
【有气无力】weakly; feeble
【有前科】previous conviction
【有钱】rich; wealthy: ～人 a rich man; the rich; the wealthy
【有求必应】respond to every plea; grant whatever is requested
【有趣】interesting; fascinating; amusing: 一本～的故事书 an interesting storybook
【有去无回】cross the Rubicon
【有权代理】authorized agency
【有日子】①(有好些天) for quite a few days; for days ②(有确定日期) have fixed a date
【有辱门楣】a disgrace to the house
【有色】colored: 戴着～眼镜看事情 look at things through colored spectacles; take a distorted view ◇～人种 colored race
【有色金属】nonferrous metal
【有伤体面】lower one's dignity
【有神论】theism ◇～者 theist
【有生理缺陷的人】handicapped person
【有生力量】effective strength; effectives
【有生以来】ever since one's birth: ～第一次 the first time in one's life
【有生之年】for the rest of one's life
【有声片】sound film; talkie; phonofilm
【有声有色】full of sound and color; vivid and dramatic; very impressive
【有失身份】beneath one's rank or station
【有识之士】a person with breadth of vision; a man of insight
【有时】sometimes; at times; now and then
【有史以来】since the beginning of history; throughout history
【有始无终】start sth. but fail to carry it through
【有始有终】carry sth. through to the end
【有事】①(有事情要做) be engaged; occupied; busy: 我今天下午～。I'm engaged this afternoon. ②(有问题) when problems crop up; if

sth. happens: ～去找老师商量 consult your teacher when problems crop up
【有恃无恐】secure in the knowledge that one has strong backing
【有数】①(不多) not many; only a few: 筐里只剩下～几个苹果了。There are just a few apples in the basket. ②(了解情况) know exactly how things stand; have a definite idea of what one's doing: 心中～ know fairly well; know what's what
【有损威严】impair one's dignity
【有所】to some extent; somewhat: ～借鉴 have certain experience to draw on / ～偏好 loves it specially / ～企图 have an axe to grind / ～追求 be after something / ～遵循 have something to go by
【有蹄类】[动] ungulate
【有条不紊】in an orderly way; methodically; systematically
【有条有理】systematically; in perfect order
【有头有尾】complete
【有为】promising: ～的青年 a promising young person / 年轻～ young and promising
【有味儿】①(好吃) tasty; delicious ②(美好) delightful ③(腐烂) stink
【有喜】[口] be pregnant; be expecting
【有隙可乘】there is a crack to squeeze through; there is a loophole to exploit
【有先例的】precedented
【有限】limited; finite: 为数～ limited in numbers not many / 文化水平～ have had little schooling ◇ ～公司 limited company; limited-liability company / ～花序[植] definite inflorescence / ～级数[数] finite progression; finite series / ～责任 limited liability
【有线】wired ◇ ～传真 wirephoto / ～电话 telephone / ～电报学 wire telegraphy / ～电视 cable television / ～广播 wire broadcasting; wired broadcasting; rediffusion on wire / ～广播站 wired broadcast station; rediffusion station / ～制导导弹 wire-guided missile
【有效】efficacious; effective; valid: 采取～措施 take effective measures / ～期为六个月 valid for six months ◇ ～保险 insurance in force / ～爆破 effective fragmentation / ～成分 active principle / ～范围 effective range / ～功率[电] effective power; useful power / ～荷载 useful load / ～能 available energy / ～票 valid ballot paper / ～期 term of validity / ～期满 exhaustion of effect / ～数字 significant digits / ～证件 perfect instrument / ～值 virtual value
【有些】①(有的) some: ～人 some people ②(有一些) somewhat; rather: ～不好意思 be somewhat embarrassed
【有心】(有某种想法) have a mind to; set one's mind on: ～人 one who sets his mind on doing sth. useful
【有心无力】①(有此心但无此力) having a heart but no strength ②(故意) intentionally; purposely
【有形】tangible; visible ◇ ～贸易 visible trade / ～损耗 material loss / ～资产 tangible assets; tangibles
【有性】[生] sexual ◇ ～生殖 zoogamy; sexual reproduction / ～杂交 sexual hybridization
【有血有肉】true to life; vivid: 这个故事～。The story is true to life.
【有言在先】make clear beforehand; forewarn
【有眼不识泰山】have eyes but fail to see Taishan Mountain; entertain and angel unawares
【有眼无珠】have eyes but not see
【有一手儿】have a remarkable skill
【有一说一】speak the whole truth; call a spade a spade
【有益】profitable; beneficial; useful: ～于健康 good for one's health / 上了～的一课 learn a useful lesson / 作出～的贡献 make valuable contributions
【有意】①(有心思) have a mind to; be inclined ②(故意) intentionally; deliberately; purposely: ～的侮辱 calculated insult / ～刁难 make things difficult for sb. on purpose / ～欺骗 being calculated to deceive / ～歪曲 deliberately distort
【有意识】consciously; knowingly; wittingly
【有意思】①(有意义) significant; meaningful ②(有趣味) interesting; enjoyable
【有瘾】be addicted to; have formed a habit
【有勇无谋】have valor but lack strategy; be brave but not resourceful; be foolhardy
【有余】①(有剩余) have a surplus; have enough and to spare: 粮食自给～ have grain enough and to spare / 绰绰～ more than enough; enough and to spare ②(有零) odd: 五十～ fifty odd
【有源】[电] active ◇ ～电路 active circuit / ～器件 active device; active parts
【有约在先】have a previous engagement
【有则改之，无则加勉】correct mistakes if you have made any and guard against them if you have not

【有朝一日】some day; one day
【有职有权】be entrusted with the responsibility and authority inherent in one's post
【有志者事竟成】where there's a will there's a way
【有志之士】a person of noble aspirations; a person with lofty ideals
【有助于】contribute to; be conducive to; conduce to
【有资格】capable; qualified: ～证明 capable of proof
【有组织的犯罪】organized crime
【有组织的抢劫】banditry
【有嘴无心】say what one does not mean
【有罪】culpability; guilt: ～行为 culpable / ～一方当事人 guilty party / ～证据 evidence of guilt

铕 [化] europium (Eu)

友 ①(朋友) friend: 男～ boy friend / 女～ girl friend ②(有友好关系的) friendly
【友爱】friendly affection; fraternal love: ～之谊 the friendliness of brothers / 团结～ fraternal unity
【友邦】friendly nation
【友好】①(好朋友) close friend; friend ②(亲近和睦) friendly; amicable: ～关系 friendly relations / ～往来 friendly intercourse ◇ ～代表团 goodwill mission / ～访问 friendly visit / ～人士 friendly personage; friend / ～使者 envoy of friendship / ～条约 treaty of friendship / ～协会 friendship association / ～邀请赛 friendship invitational tournament
【友情】friendly sentiments; friendship
【友人】friend: 外国～ foreign friend
【友谊】friendship: 深厚的～ profound friendship ◇ ～赛 friendly match

黝 black; dark
【黝黑】dark; swarthy: ～的脸 a swarthy face

yòu

又 [副]①(重复) again ②(表示几种情况同时存在)这些杂志～便宜～好。 These magazines are cheap but good. ③(再加上) and: 一～二分之一 one and a half ④(表示转折)我想去，～怕没时间。 I'd like to go, but I'm not sure if I can find the time
【又红又专】both red and expert
【又饥又渴】both hungry and thirsty
【又及】postscript
【又惊又喜】alarmed and happy at the same time
【又哭又笑】cry and laugh at the same time
【又要马儿好，又要马儿不吃草】expect a horse to run fast and yet won't feed it; eat one's cake and have it too

右 ①(跟"左"相对) the right side; the right: ～手 right hand ②(保守的) the Right ③(西) west
【右臂】the right arm
【右边锋】[足球] outside right; right wing
【右边】the right side; the right
【右侧通行】keep to the right
【右舵】right standard rudder; right rudder
【右锋】[篮球] right forward
【右后卫】[足球] right back
【右内锋】[足球] inside right
【右派】the Right; the right wing: ～分子 the Rightist
【右前轮】[汽车] off-front wheel
【右前卫】[足球] right halfback; right half
【右倾】Right deviation: ～保守 Right-deviationist conservatism / ～思想 Right-deviationist thinking ◇ ～机会主义 Right opportunism
【右首】the right-hand side; the right: 那天她坐在我～。 She was seated on my right that day.
【右心室】[生理] right ventricle
【右旋】[化] dextrorotation ◇ ～糖 corn sugar; grape sugar
【右翼】①[军](右侧部队) right wing; right flank ②(政治上，思想上保守的) the Right; the right wing ◇ ～分子 right-winger; member of the Right
【右座驾驶汽车】right-hand control car

囿 limited; hampered: ～于成见 blinded by prejudice / ～于习俗 constrained by custom / ～于一隅 be restricted to a narrow confine

柚 [植] shaddock; pomelo
【柚子】shaddock; pomelo

釉 glaze: 青～瓷器 blue glazed porcelain / 给瓷器上～ glaze porcelain
【釉瓷】vitreous enamel
【釉工】glazer
【釉料】frit
【釉面砖】[建] glazed tile
【釉陶】glazed pottery
【釉质】[生理] enamel

【釉子】glaze

鼬 [动] weasel: 白～ stoat

【鼬獾】ferret badger

幼 ①(小) young; under age: ～子 the youngest son / 年～无知 young and ignorant ②(小孩儿) children; the young: 扶老携～ bringing along the old and the young

【幼虫】[动] larva

【幼儿】child; infant ◇ ～教育 preschool education / ～期 infancy / ～湿疹 infantile eczema / ～园 kindergarten; nursery school; infant school

【幼龄林】[林] young growth

【幼苗】seedling

【幼年】childhood; infancy

【幼体】[生] the young; larva

【幼小】immature

【幼学壮行】learn while young and practice when strong

【幼芽】young shoot; bud

【幼稚】①(年龄小) young ②(头脑简单) childish; puerile; naive: ～可笑 childish and ridiculous

【幼稚病】①[心] infantilism ②(政治上的) infantile disorder

蚴 [动] the larva of a tapeworm or the cercaria of a schistosome: 毛～ miracidium / 尾～ cercaria

诱 ①(诱导) guide; lead; induce: 循循善～ be good at giving systematic guidance; teach with patience and skill ②(使用手段使人干事) lure; seduce; entice: ～人犯罪 induce others to break the law / ～人为恶 seduce others to evil

【诱捕】trap

【诱虫灯】moth-killing lamp

【诱导】①(劝诱教导) guide; lead; induce: ～他改过自新 guide him to mend his ways ②[医] revulsion: ～药 revulsive ◇ ～反应[化] induced reaction

【诱饵】bait

【诱发】bring out; induce; cause to happen

【诱供】trap a person into a confession; inducement leading to confession

【诱拐】abduct; carry off by fraud; kidnap ◇ ～者 abductor

【诱惑】①(使用手段引诱人做坏事) entice; tempt; seduce; lure: 受～ fall into temptation; be lured ②(吸引) attract; allure; fascinate: ～人的微笑 bewitching smiles / 一片～人的景色 an attractive scene

【诱奸】entice into unlawful sexual intercourse; seduce

【诱骗】inveigle; cajole; trap; trick

【诱杀】trap and kill; lure to destruction

【诱使犯罪】entrapment

【诱降】lure into surrender

yū

淤 ①(淤积) become silted up ②(淤积的泥沙) silt: 引～肥田 fertilize the soil with silt ③(血液不流通) stasis

【淤点】[医] petechiae

【淤灌】[农] warping

【淤积】silt up; deposit

【淤浆】slurry

【淤泥】silt; sludge; ooze

【淤塞】silt up; be choked with silt: 泥沙～了港口。The harbor has silted up.

【淤沙】silt

【淤血】extravasated blood

【淤滞】be retarded by silt; silt up: 疏通～的河道 dredge the sluggish river

【淤浊不清】muddy and unclear

迂 ①(曲折) circuitous; winding; roundabout ②(迂腐) clinging to outworn rules and ideas; pedantic: ～夫子 pedant / 这人～得很。He's very impractical.

【迂腐】stubborn adherence to outworn rules and ideas; pedantry: ～的见解 pedantic ideas / ～无用 doltish and useless

【迂缓】slow in movement; dilatory: ～误事 slowness causes trouble

【迂回】①(回旋) circuitous; tortuous; roundabout: ～其词 approach a subject in a round about way / ～曲折 full of twists and turns; circuitous; tortuous ②[军] outflank: ～战术 outflanking tactics

【迂见】absurd view; impractical opinion

【迂阔】high-sounding and impracticable: ～之论 impractical views

【迂论】impractical argument; unrealistic statement

【迂曲】tortuous; circuitous: ～的山路 a tortuous mountain path

【迂远而阔】impracticable and inapplicable

【迂直】impractical and artless

【迂拙】impractical and foolish: ～之道 an impractical and stupid doctrine

yú

于 ①[介](在) in; at; on: 出现～世界上 appear in the world／生～1950年 be born in 1950 ②(对于) for; to: 有益～人民 be beneficial to the people ③(自;从) from: 出～自愿 of one's own free will; of one's own accord ④(表示比较) than: 多～ more than

【于今】①(到现在) up to the present; since ②(如今) nowadays; today; now

【于是】[连] thereupon; hence; consequently; as a result

盂 a broad-mouthed receptacle for holding liquid; jar: 痰～ spittoon

舆 ①(地) area; territory ②(众人的) public; popular

【舆论】public opinion: 作～准备 prepare public opinion ◇～工具 mass media; the media／～界 the media; press circles

【舆情】public sentiment; popular feelings: 洞察～ know public sentiment well

虞 ①(猜测) supposition; prediction: 以备不～ be prepared for any contingency ②(忧虑) anxiety; worry ③(欺骗) deceive; cheat; fool: 尔～我诈 each trying to cheat the other

【虞美人】[植] corn poppy

娱 ①(使快乐) give pleasure to; amuse ②(快乐) joy; pleasure; amusement

【娱乐】amusement; entertainment; recreation ◇～场所 public place of entertainment／～活动 recreational activities; recreation

愚 ①(愚笨) foolish; stupid ②(愚弄) make a fool of; fool: ③[谦](用于自称的谦词) I: 据本人～见 in my humble opinion

【愚笨】foolish; stupid; clumsy: ～呆滞 stupid and silly

【愚不可及】a fool to the third degrees; be hopelessly stupid

【愚蠢】stupid; foolish; silly

【愚钝】slow-witted; stupid

【愚而好自用】stupid but willful

【愚见所及】as far as my humble opinion goes

【愚昧落后】ignorant and backward

【愚昧无知】benighted

【愚民政策】policy of keeping the people in ignorance; obscurantist policy; obscurantism

【愚弄】deceive; make a fool of; dupe: ～人民 dupe the people

【愚人】a fool; a simpleton

【愚妄】ignorant but self-important; stupid but conceited: ～之徒 an ignorant and senseless fellow

【愚者千虑,必有一得】even a fool occasionally hits on a good idea

【愚直】stupidly honest

【愚拙】stupid and clumsy: ～之极 foolish to the extreme

隅 ①(角落) corner; nook: 城～ the corner of a city wall ②(靠边沿的地方) outlying place; border: 海～ seaboard

渝 change: 始终不～ unswerving; consistent

愉 pleased; happy; joyful; cheerful

【愉快】happy; joyful; cheerful: ～的孩子们 joyous children／～的心情 merry heart／过得非常～ have a very good time／令人～的天气 pleasant weather

【愉悦】joyful; cheerful; delighted

逾 ①(超过) exceed; go beyond: ～额 exceed the allowed／饮酒～量 drink to excess ②(更加) even more

【逾假不归】absent over leave

【逾期】exceed the time limit; be overdue

【逾越】exceed; go beyond: ～常规 depart from the usual practice／～权限 overstep one's authority／不可～的鸿沟 an impassable gulf; an insurmountable barrier

榆

【榆树】elm

【榆叶梅】flowering plum

余 ①(剩下) surplus; spare; remaining ②(整数后的零头) more than; odd; over ③(以外或以后的时间) beyond; after

【余波】repercussions

【余存】balance; remainder

【余党】remnants of an overthrown clique; remaining confederates

【余地】leeway; margin; room; latitude: 留有～ leave some leeway／还有回旋～。There is enough space for maneuver.

【余毒】residual poison; pernicious vestige; pernicious influence: 封建～ pernicious vestiges of feudalism

【余额】①(名额中余下的空额) vacancies yet to be filled ②(帐目上剩余的金额) remaining

sum; balance ◇ ～表 balance sheet
【余忿未平】anger not yet appeased
【余辉】twilight at sunset; afterglow
【余悸】lingering fear: 心有～ have a lingering fear
【余烬】ashes; embers: 劫后～ a devastated waste of smoldering embers
【余可类推】the rest may be inferred by analogy
【余力】strength or energy to spare
【余粮】surplus grain: 无～户 household without grain reserves ◇ ～户 household with grain to spare; grain-surplus household
【余量】margin
【余年】one's remaining years: ～可数 one's failing years are numbered
【余孽】remaining evil element; leftover evil; surviving supporter of an evil cause: 封建～ dregs of feudalism
【余怒未消】feel the lingering anger
【余切】[数] cotangent
【余缺】①(指晚年) the remainder of one's life; one's remaining years: ～著书 write a book in the declining years of one's life ②(侥幸保全的性命) survival: 劫后～ be a survivor of a disaster
【余剩】surplus; remainder: ～所得 residual income
【余数】[数] remainder
【余威】remaining prestige or influence
【余味】agreeable aftertaste; pleasant impression: ～无穷 leave a lasting and pleasant impression or aftertaste
【余暇】spare time; leisure time; leisure
【余下】remaining: ～的时间 time left over
【余弦】[数] cosine ◇ ～定律 the cosine law
【余香】lingering fragrance
【余兴】①(未尽的兴致) lingering interest; a wish to prolong a pleasant diversion ②(附带的文娱活动) entertainment after a meeting or a dinner party
【余言未尽】unable to say all that one would like
【余音】lingering sound
【余勇可贾】still having plenty of fight left in one; with strength yet to spare
【余震】[地] aftershock

腴 ①(人胖) fat plump: 丰～ plump; well rounded out ②(肥沃) fertile: 膏～ fertile

鱼 fish: 白鲢～ silver carp / 比目～ turbot; flat fish / 草～ grass carp; Chinese ide / 鲳～ pomfret / 大马哈～ salmon / 带～ hairtail; cutlass fish / 鲽～ flounder / 凤尾～ long tailed anchovy / 桂～ mandarin fish / 河豚～ globe fish / 黄花～ yellow croaker / 鲫～ crucian carp / 金枪～ tuna / 鲤～ common carp / 鲈～ perch / 鳗～ sea eel / 墨～ cuttle fish / 鲇～ catfish / 胖头～(花鲢) big head; spotted silver carp / 鲆～ left-eyed flounder / 青～ black carp / 鲭～ mackerel / 沙丁～ sardine / 鲨～ shark / 鳝～ eel / 鲥～ shad / 梭子～ pike / 武昌～(团头鲂) blunt-snout / 鲟～ sturgeon / 银～ whitebait; salangid / 鱿～ squid / 元～(甲～) turtle / 章～ octopus / 鳟～ trout / 淡水养～ fresh-water fish-culture / 钓～ be fishing / 年捕～量 annual catch / 养～ fish culture; pisciculture / 养～池 fish pond / 养～场 fish farm / 一大网～ a bumper haul of fish; a good catch / 在河里钓～ fish a river
【鱼白】fish sperm; milt
【鱼鳔】air bladder
【鱼叉】fish spear; fishgig; fish fork
【鱼场】fish farm: 沿海～ in-shore fishing ground
【鱼翅】shark's fin
【鱼虫】water flea
【鱼刺】fishbone: 剔掉～ bone a fish
【鱼肚】maw
【鱼肚白】the whitish color of a fish's belly; grey down
【鱼饵】bait
【鱼粉】fish meal
【鱼肝油】cod-liver oil
【鱼竿】fishing rod
【鱼缸】fish jar; fish bowl
【鱼钩】fishhook
【鱼贯】one following the other; in single file: ～而入 enter in single file; file in
【鱼胶】fish glue; isinglass
【鱼雷】[军] torpedo: 空投～ aircraft torpedo / 人控～ hunter torpedo / 寻的～ target-seeking torpedo ◇ ～发射管 torpedo tube / ～快艇 torpedo boat / ～舰 mosquito boat
【鱼类学】ichthyology
【鱼类制品厂】fishworks
【鱼鳞】fish scale; scale: 刮去～ scrape the scales off a fish; scale a fish ◇ ～坑 fish-scale pits
【鱼龙混杂】dragons and fish jumbled together; good and bad people mixed up
【鱼露】[食品] fish sauce

【鱼卵】roe
【鱼米之乡】a land of fish and rice; land of abundance
【鱼苗】(fish) fry
【鱼目混珠】pass off fish eyes as pearls; pass off the sham as the genuine
【鱼片儿】slices of fish meat
【鱼漂】cork on a fishing line; float
【鱼群】shoal of fish
【鱼肉】①(鱼的肉) the flesh of fish ②(鱼和肉) fish and meat ③(压榨;宰割) oppress; victimize: ～百姓 cruelly oppress the people
【鱼石脂】[药] ichthammol; ichthyol
【鱼市】fish market
【鱼水和谐】marital harmony
【鱼水情深】be close as fish and water
【鱼松】[食品] dried fish floss
【鱼塘】fish pond
【鱼藤】[植] trifoliate jewelvine ◇ ～精[农] derris extract / ～酮[农] rotenone
【鱼网】fishnet; fishing net ○拖网 trawl; trawl net / 围网 purse seine
【鱼尾板】[铁路] fish plate
【鱼尾号】boldface square brackets (〖 〗; 【 】)
【鱼鲜】fish and shellfish as food; seafood
【鱼腥草】[植] cordate houttuynia
【鱼腥味】fishlike smell; fishy taste
【鱼雁往来】incoming and outgoing of epistolary correspondence
【鱼汛】fishing season
【鱼秧】fingerling
【鱼鹰】①(鹗的通称) osprey; fish hawk; sea eagle ②(鸬鹚的通称) cormorant
【鱼油】fish oil
【鱼游釜中】like fish swimming in a cooking pot; in imminent peril
【鱼跃】[排球] fish dive: ～救球 diving save
【鱼之失水】a fish is out of water; in extremities
【鱼子】[食品] roe ◇ ～酱 caviare

渔 ①(捕鱼) fishing ②(谋取) take sth one is not entitled to: ～利 reap unfair gains
【渔产】aquatic products
【渔场】fishing ground; fishery: 近海～ in-shore fishery / 远洋～ deep-sea fishery
【渔船】fishing vessel; fishing boat: 拖网～ trawler ○围网船 purse seiner; seine-boat / 机帆船 motorized junk
【渔村】fishing village
【渔港】fishing port; fishing harbor
【渔歌】fisherman's song
【渔鼓】a percussion instrument made of bamboo
【渔火】lights on fishing boats
【渔获量】catch
【渔具】fishing tackle
【渔利】①(谋取不正当的利益) reap unfair gains; profit at others' expense: 从中～ take advantage of a situation to benefit oneself; cash in on other people's efforts ②(谋取的不正当利益) easy gains; spoils: 坐收～ effortlessly reap the spoils of a contest fought by others; profit from others' conflict
【渔猎】fishing and hunting
【渔轮】fishing vessel
【渔民】fisherman; fisherfolk
【渔业】fishery ◇ ～区 fishing zone / ～协定 fisheries agreement / ～资源 fishery resources

yǔ

宇 ①(房檐) eaves ②(房屋) house ③(宇宙;世界) space; universe; world
【宇航员】astronaut; cosmonaut
【宇内】in the world: ～闻名 be reputed in the world
【宇宙】universe; cosmos
【宇宙尘】cosmic dust
【宇宙飞船】spaceship; space ship; spacecraft: 无人驾驶～ pilotless spaceship / 载人～ man-carrying spaceship
【宇宙飞行】space flight
【宇宙飞行员】astronaut; spaceman; cosmonaut
【宇宙观】world view; world outlook
【宇宙航行】astronavigation; space navigation ◇ ～学 astronautics / ～员 astronaut; spaceman; cosmonaut
【宇宙火箭】space rocket
【宇宙空间】cosmic space ◇ ～站 space station
【宇宙速度】cosmic speed

语 ①(话) language; tongue; words: 本族～ mother tongue / 汉～ Chinese (language) / 口～ colloquial speech / 鸟～ bird's twitter / 英～ English (language) ②(说) speak; say: 低～ speak in a low voice; whisper ③(谚语) set phrase; proverb; saying ④(代替语言表示意思的动作或方式) nonlinguistic means of communicating ideas; sign; signal: 旗～ flagsignal; semaphore / 手～ dactylology; sign language
【语病】faulty wording or formulation
【语词】words and phrases

【语调】[语] intonation
【语法】grammar: 传统～ traditional grammar／ 结构～ structural grammar／ 系统～ systemic grammar
【语汇】vocabulary
【语句】sentence: ～分析 analysis of sentences
【语录】recorded utterance; quotation
【语气】①(说话的口气) tone: manner of speaking: 用友好的～说 speak in a friendly tone ② [语] mood: 祈使～ imperative mood ◇ ～词 modal particle
【语塞】tongue-tied
【语态】[语] voice: 被动～ passive voice／主动～ active voice
【语体】[语] type of writing; style: 口语～ colloquialism
【语体文】prose written in the vernacular
【语文】①(中文) Chinese ②(语言文学) language ③(语言和文学) language and literature
【语无伦次】speak incoherently; talk in a confused manner
【语系】[语] family of languages; language family: 拉丁～ the Latin family of languages
【语序】[语] word order
【语焉不详】not speak in detail: not elaborate
【语言】language: ～无味 an expression of no taste／ ～与文字 spoken and written language／工作～ working language／正式～ official language ◇ ～地理 linguistic geography／ ～隔阂 language barrier／～规范化 standardization of speech／～教学法 language teaching method／ ～科学 linguistic science／～实验室 language laboratory
【语言心理学】linguistic psychology
【语言学】linguistics; philology: 工程～ engineering linguistics／ 共时～ synchronic linguistics／ 宏观～ macrolinguistics／计算～ computational linguistics／理论～ theoretical linguistics／ 历时～ diachronic linguistics／ 历史～ historical linguistics／ 普通～ general linguistics／ 社会～ sociolinguistics／ 微观～ microlinguistics／ 心理～ psycholinguistics／ 应用～ applied linguistics ◇ ～学家 linguist; philologist
【语义】semantic: ～结构学 semotactics／ ～扩展 semantic extension／ ～缩合 semantic condensation／ ～要素 semantic feature
【语义学】semantics: 结构～ structural semantics／ 解释～ interpretative semantics
【语意双关】having a double meaning; pun
【语音】①(说话的声音) speech sounds ②(发音) pronunciation ◇ ～课 phonetics class
【语音学】phonetics: 实用～ practical phonetics ◇ ～家 phonetician
【语源学】etymology
【语重心长】sincere words and earnest wishes
【语助词】[语] an auxiliary word that indicates

雨

rain: 暴风～ storm; tempest／大～ heavy rain／ 大暴～ torrential rain; downpour; cloudburst／ 地形～ orographic rain／ 黄梅～ plum rains／ 雷～ thunderstorm／ 毛毛细～ drizzle／ 倾盆大～ heavy downpour; torrential rain／ 细～ drizzle／ 阵～ shower／ 因～取消 rain out; be rained out／ ～很大。 The rain was very heavy.／ ～停了。 It has rained itself out.／ ～停了没有? Has the rain stopped?／ 在下～吗? Is it raining?
【雨布】waterproof cloth; waterproof
【雨层云】[气] nimbostratus
【雨带】[气] rain band
【雨点】raindrop
【雨过天晴】after gloom comes brightness; the sun shines again after the rain
【雨后春笋】bamboo shoots after a spring rain
【雨后送伞】give an untimely help
【雨季】rainy season
【雨夹雪】rain and snow mixed
【雨具】rain gear
【雨量】[气] rainfall: ～过多 excess of rain ◇ ～计 rain gauge／ ～强度 rainfall density／ ～站 precipitation station; rainfall station
【雨林】rainforest
【雨淋日晒】wet by the rain and burnt by the sun; the hardship of outdoor work; long travelling
【雨露】①(雨和露) rain and dew ②(恩惠) favor; grace; bounty
【雨帽】rain cap
【雨棚】[建] canopy
【雨伞】umbrella
【雨水】①(降雨而来的水) rainwater; rainfall; rain: 近几年来～最多的夏季 the wettest summer in the last few years／ 今年夏天～太多了。 We've had too much rain this summer. ② (节气) Rain Water ◇ ～管[建] downspout; downpipe; rain pipe
【雨丝风片】very light rain and wind
【雨天】rainy day; wet day
【雨鞋】rubber boots; galoshes; rubbers
【雨雪纷纷】confused falling of rain and snow
【雨燕】[动] swift

【雨衣】raincoat; waterproof
【雨意】signs of approaching rain

与 ①(给) give; offer; grant ②[介](跟) with; against: ～人民为敌 set oneself against the people ③[连](和) and; together with
【与虎谋皮】ask a tiger for its skin; request sb. to act against his own interests
【与理不合】not in accord with reason
【与其…不如】would rater... than...: ～苟且偷生，～一死 prefer to die rather than drag out an ignoble existence
【与人为善】aim at helping others; with good intentions toward others
【与日俱增】grow with each passing day; be steadily on the increase
【与日月争光】equal the sun and the moon in glory
【与世长辞】depart from the world for ever; pass away
【与世沉浮】follow the crowd without personal principles
【与世隔绝】seclude from the world
【与世无争】hold oneself aloof from the world; stand aloof from worldly success
【与之无关】have no truck with
【与众不同】out of the ordinary

屿 small island; islet: 岛～ islands and islets; islands

羽 feather; plume
【羽冠】crest
【羽化】①(成仙) ascend to heaven and become immortal ②[婉](道教徒称死) die ③[动](由蛹变为成虫) emergence; eclosion
【羽黄素】[脊椎] zoofulvin
【羽毛】feather; plume: ～丰满 become full-fledged / ～未丰 unfledged; still young and immature ◇ ～画 feather patchwork; feather picture
【羽毛球】①(球赛项目之一) badminton: ～拍 racket / ～赛 badminton match / ～网 badminton net ②(指此项运动用的球) shuttlecock ○前场 front court / 后场 back court / 发球 service / 高手击球 overhand stroke / 低手击球 underhand stroke / 封网 block the net / 扣球 smash; kill / 正手握拍法 forehand grip / 反手握拍法 backhand grip / 单打球场 singles court / 双打球场 doubles court
【羽毛扇】feather fan: 摇～的 the man with a feather fan; the mastermind behind an intrigue
【羽绒】eider down: ～衣 eider down outerwear
【羽纱】camlet
【羽扇纶巾】free from anxiety; calm; composed
【羽翼】①(鸟类的翅膀) wing ②(辅佐的人或力量) assistant
【羽族】the feathered tribe; birds

予 give; grant; bestow: 不～考虑 refuse to take into consideration / 免～处分 exempt sb. from punishment / 授～劳动模范称号 award sb. the title of model worker
【予人口实】give sb. a handle; give cause for gossip
【予以】give; grant: ～便利 offer convenience to sb. / ～表扬 commend sb.; give praise to / ～解释 throw some light on

yù

育 ①(生育) give birth to: 节～ birth control ②(养活) rear; raise; bring up; breed: ～婴 nurse a baby ③(教育) educate: 智～ intellectual education
【育雏】[牧] brood ◇ ～器 brooder
【育肥】[牧] fatten
【育苗】[农] grow seedlings ◇ ～区 nursery garden
【育秧】raise rice seedlings
【育种】[农] breeding: 单倍体～ haploid breeding / 多倍体～ polyploid breeding / 激光～ breeding by laser / 杂交～ crossbreeding

玉 ①(玉石) jade: 白～ white jade / 碧～ jasper / 黑～ jet / 黄～ topaz / 绿～ emerald / 青～ sapphire / 纹～ veined jade / 岫～ soapstone / 紫翠～ alexandrite ○蛋白石 opal / 电气石 tourmaline / 粉翠 Beijing jade / 翡翠 jadeite; green jade / 芙蓉石 rosy quartz / 缟玛瑙 onyx / 贵橄榄石 chrysolite / 海蓝宝石 aquamarine / 红宝石 ruby / 红玛瑙 carnelian / 红条纹玛瑙 sardonyx / 虎眼石 tigereye (stone) / 金绿宝石 chrysoberyl / 金星石 aventurine / 菊花石 chrysanthemum stone / 孔雀石 malachite / 绿晶石 green quartz / 绿松石 turquoise / 青金石 lapiz lazuli / 乳石英 milky quartz / 烟水晶 cairngorm / 蛇纹石 serpentine / 石榴石 garnet / 猫眼石 cat's eye (stone) / 月长石 moonstone / 血滴石 blood stone / 紫石英 violet quartz / 紫水晶 amethyst / 钻石 diamond ②(比喻洁白或美丽) pure; fair; hand-

some; beautiful: 亭亭～立 fair, slim and graceful ③[敬](指对方的身体或行动) your: ～照 your photograph
【玉帛】[书] jade objects and silk fabrics: 化干戈为～ turn hostility into friendship
【玉不琢,不成器】 one cannot become useful without being educated; spare the rod and spoil the child
【玉成其事】[敬] kindly help make a success of it; assistant another in accomplishing a task or attaining a goal
【玉带】 jade belt
【玉雕】 jade carving; jade sculpture ◇ ～工人 jade carver
【玉骨冰肌】 elegant demeanor and high personality
【玉洁冰清】 as pure as jade and as clean as ice; pure and noble
【玉兰】[植] yulan magnolia
【玉兰片】[食品] dried slices of tender bamboo shoots
【玉米】①(玉蜀黍) maize; Indian corn; corn ②(玉米穗) ear of maize ◇ ～花 popcorn / ～粒 kernel of corn; grain of corn / ～面 maize flour; cornmeal / ～螟 corn borer / ～脱粒机 maize sheller / ～芯 corncob; cob / ～粥 maize gruel / ～钻心虫 corn stalk borer
【玉器】 jade article; jade object; jadeware ◇ ～工厂 jade workshop
【玉色】 jade green; light bluish green
【玉石】 jade
【玉石俱焚】 jade and stone burned together; destruction of good and bad alike
【玉食锦衣】 sumptuous food and luxurious clothings
【玉玺】 imperial jade seal
【玉液琼浆】 top-quality wine
【玉宇】①(宇宙) the universe ②(华丽的宫殿) beautiful palace: ～琼楼 magnificent houses
【玉簪】①(用玉做成的簪子) jade hairpin ②[植] fragrant plantain lily

芋 ①[植](多年生草本植物) taro ②(泛指马铃薯等) tuber crops: 洋～ potato / 山～ sweet potato
【芋头】[植] taro

阈 ①[书](门坎儿) threshold; doorsill ②[生理](范围) threshold: 视～ visual threshold
【阈限】[心理] threshold
【阈限值】[医] threshold limit value (TLV)
【阈值】[自] threshold value

域 land within certain boundaries; territory; region: 领～ territory; domain; field; realm / 异～ foreign lands
【域中】 inside the country
【域外庇护】 extraterritorial asylum

郁 ①(香气浓厚) strongly fragrant ②(茂盛) luxuriant; lush ③(忧郁) gloomy; depressed
【郁闭】[林] closing: 林冠～ canopy closure ◇ ～度 canopy density
【郁积】①(积聚不得发泄) smoldering; pent-up: ～在心里的烦恼 pent-up vexation ②[医](郁结) stasis
【郁金香】[植] tulip
【郁闷】 gloomy; depressed: ～无聊 depressed and cheerless; have a thin time / ～之感 a feeling of oppression
【郁血】[医] stagnation of the blood; venous stasis
【郁郁不乐】 depressed; melancholy; joyless
【郁郁葱葱】 lush and green
【郁郁而亡】 die from sorrow

与 take part in; participate in
【与会】 participate in conference ◇ ～国 countries attending a conference; participating countries / ～者 conferee; participant
【与闻】 have a participant's knowledge of; be let ～其事 have a participant's knowledge of a matter; be in the know

誉 ①(名声) reputation; fame: 美～ good reputation / ～满全球 be famed all over the country ②(称赞) praise: ～不绝口 be praised all the while / 众口交～ win universal praise

寓 ①(居住) reside; live ②(住的地方) residence; abode ③(寄托) imply; contain
【寓居】 live; make one's home in: ～南京 live in Nanjing
【寓所】 residence; abode; dwelling place
【寓言】 fable; allegory; parable
【寓意】 implied meaning; moral; message; import: ～深刻 be pregnant with meaning

遇 ①(相逢) meet: 巧～ meet by chance ②(对待) treat; receive: 待～ treatment ③(机会) chance; opportunity: 机～ favorable circumstances; opportunity
【遇刺】 be attacked by an assassin: ～身死 be assassinated

【遇到】run into; encounter; come across: ～困难 come up against difficulties / ～埋伏 run into an ambush
【遇害】be murdered
【遇见】meet; come across: ～一位同学 meet a classmate
【遇救】be rescued; be saved
【遇难】①(意外死亡) die in an accident ②(被杀害) be murdered
【遇事】when anything crops: ～干涉 have an oar in every man's boat / ～生风 make trouble at every opportunity
【遇险】meet with a mishap; be in danger; be in distress: ～船只 ship in distress ◇ ～信号 distress signal; SOS

浴 bath; bathe: 芬兰蒸汽～ Sauna / 海水～ sea bathing / 冷水～ cold bath / 淋～ shower bath / 日光～ sunbath / 沙～ sand bath / 土耳其蒸汽～ Turkish bath
【浴场】outdoor bathing place: 海滨～ bathing beach
【浴池】①(洗澡的设备) common bathing pool ②(洗澡堂) public bathhouse; public baths
【浴巾】bath towel
【浴疗法】balneotherapy
【浴盆】bathtub
【浴室】bathroom; shower room: 公共～ bathhouse
【浴血】bathed in blood; bloody: ～奋战 fight a bloody battle
【浴衣】bathrobe

裕 abundant; plentiful: 富～ abundant; affluent; well-to-do; well-off
【裕国裕民】enrich the state and the people
【裕如】effortlessly; with ease: 应付～ handle wish ease

欲 ①(欲望) desire; longing; with: 名利～ desire for fame and wealth / 求知～ thirst for knowledge / 食～ a desire for food; appetite ②(希望) wish; want; desire: 畅所～言 pour out all that one wishes to say; speak one's mind freely / 为所～为 do whatever one likes; act willfully ③(将要) about to; just going to; on the point of: 摇摇～坠 on the verge of collapse; crumbling; tottering
【欲罢不能】try to stop but cannot; be unable to stop even thought one wants to
【欲盖弥彰】the more one tries to hide, the more one is exposed; try to cover up a misdeed, only to make it more conspicuous
【欲壑难填】greed is like a valley that can never be filled; avarice knows no bounds
【欲加之罪,何患无词】if you are out to condemn sb,; you can always trump up a charge
【欲擒故纵】leave sb. at large the better to apprehend him; allow sb. more latitude first to keep a tighter rein on him afterwards
【欲穷千里目,更上一层楼】ascend another story to see a thousand miles
【欲取姑与】give in order to take; make concessions for the sake of future gains
【欲速则不达】haste brings no success; more haste, less speed
【欲望】desire; wish; lust: 满足某人的～ satisfy sb.'s desire / 受～驱使 be stung with desire
【欲言又止】wish to speak but do not to on a second thought

愈 ①(痊愈) recover: 病～ recover from an illness ②(叠用表示越…越…) the more... the more; more and more: ～陷～深 sink more and more deeply into the mire / ～早～好。The sooner the better.
【愈合】[医] heal up
【愈加】increasingly; even more; all the more: 变得～麻烦 become increasingly troublesome

喻 ①(说明; 告知) explain; make clear; inform: ～以利害 explain the consequences of an action / 不可理～ won't listen to reason ②(明白; 了解) understand; know: 不言而～ it goes without saying that ③(比方) analogy: 比～ metaphor; analogy

御 ①(驾御) drive: ～者 driver ②(皇帝的) imperial: ～笔 emperor's hand writing ③(抵御) resist; keep out: ～寒 keep out the cold
【御夫座】[天] Auriga
【御花园】imperial garden
【御林军】palace guards
【御侮】resist invasion
【御用】①(帝王所用) employed by the emperor; for the use of an emperor ②(为反动势力服务) hired; in the pay of: ～报刊 hired press / ～文人 hired scribbler; hack writer
【御旨】imperial decree

狱 ①(监狱) prison; jail: 入～ be imprisoned; be put in prison / 越～ escape from prison; prison breach ②(官司) lawsuit; case: 冤～ an unjust charge; an unjust verdict

预 in advance; beforehand: ～约 make an appointment / ～祝成功 wish sb. success
【预报】forecast: 地震～ earthquake forecast; earthquake prediction / 天气～ weather forecast
【预备】prepare; get ready: ～饭 prepare a meal / ～考试 prepare for an exam ◇ ～党员 probationary Party member / ～队 reserve force; reserves / ～会议 preparatory meeting / ～金 reserve fund / ～期 probationary period / ～役[军] reserve duty
【预卜】augur; foretell: 战争胜负难以～。The outcome of the war is hard to foretell.
【预测】calculate; forecast: ～台风 detect a typhoon / 地震～ earthquake forecast
【预产期】[医] expected date of childbirth
【预处理】[自] pretreatment
【预订】subscribe; book; place an order: ～火车票 book a train ticket / ～一本书 place an order for a book / ～杂志 subscribe to a magazine
【预定】fix in advance; predetermine; schedule: ～的行动方针 fixed course of action / ～时间 scheduled time
【预防】prevent; take precautions against; guard against: ～疾病 prevent disease / ～药 preventive medicine / ～为主 put prevention first / ～与治疗相结合 combine treatment with prevention ◇ ～措施 precautionary measures / ～犯罪 anticrime / ～注射 inoculation; protective inoculation
【预付】payment in advance ◇ ～保证金 advance money on security / ～款 advance charge; advance payment / ～租金 advance deposit
【预感】① (事先感觉) have a premonition ② (事先感觉到的) premonition; presentiment: 不祥的～ an ominous presentiment
【预告】① (事先通知) announce in advance; herald: ～春天的到来 herald the coming of spring ② (事先通知的) advance notice
【预购】purchase in advance ◇ ～合同 forward purchasing contract
【预后】[医] prognosis: ～良好 favorable prognosis
【预计】calculate in advance; estimate: ～产量将提高 15%。It is estimated that the output will increase by 15%. ◇ ～成本 predetermined costs / ～产量 estimated output / ～到达时间[航海] estimated time of arrival (E.T.A.) / ～收入 anticipated revenue / ～数据 predicted data; scheduled data
【预加工】preparatory cut; prior operation
【预见】① (预先料到) foresee; predict: 在可以～到的将来 in the foreseeable future ② (能预先料到的见识) foresight; prevision: 科学的～ scientific prediction / 有～的人 a man of foresight ◇ ～性 foresight; farsightedness
【预缴所得税】advance payment of income tax
【预科】preparatory course
【预料】expect; predict; anticipate: ～不到地 unexpectedly / 如所～ as expected
【预谋】premeditate; plan beforehand: ～犯罪 calculated crime / ～杀人 premeditated murder; murder with malice prepense / 一次有～的行为 a deliberate action
【预期】expect; anticipate: ～利益 anticipated profits / 达到～的效果 achieve the desired results / 成绩比～的大。The success is greater than expected.
【预燃室】antechamber; energy cell
【预热】[机] preheat: ～炉 preheating furnace
【预赛】[体] preliminary contest; preliminary heats; preliminary; trial match
【预审】antecedent trial; preliminary hearing; inquiry ◇ ～程序 inquisitional procedure / ～法官 investigating magistrate / ～审判员 preliminary judge / ～庭 preliminary hearing court
【预示】betoken; indicate; presage; forebode: ～大风暴的来临 forebode a big storm / ～丰收 predict a good harvest
【预售】open to booking; advance booking: ～车票 tickets open to booking
【预算】budget: 1989 年度～ the 1989 budget ◇ ～程序 budget process / ～赤字 budget deficit / ～法 budget act / ～结余 budget surplus / ～年度 budget year / ～税额 estimated assessment / ～收入 budgetary receipts / ～收支平衡 budgeted revenue and expenditure in balance / ～项目 budget items
【预算内】budgetary: ～投资 budgetary investment / ～资金 budgetary funds
【预算外】extra-budgetary: ～投资 extra-budgetary investment / ～资金 extra-budgetary funds
【预习】prepare lessons before class

【预先】in advance; beforehand: ～感谢 thank sb. in anticipation／～警告 forewarn／～声明 state explicitly beforehand／～通知 notify in advance
【预想】anticipate; expect: ～的结果 anticipated results; objectives one has in mind
【预选】forward selection; preconcentration
【预言】①(预先说出) prophesy; predict; foretell ②(预先说出的) prophecy; prediction ◇～家 prophet
【预演】preview
【预应力】[物] prestressing force ◇～构件 prestressed component／～混凝土 prestressed concrete
【预约】①(预先约定时间) make an appointment ②(预订) order; subscribe
【预展】preview
【预兆】omen; presage; sign; harbinger: 心悸是心脏病的～。 Palpitation is a warming of heart trouble.
【预支】advance ◇～费用 prepaid expenses／～薪金 prepay salary
【预制】prefabricate ◇～房屋 prefabricated house／～构件 prefabricated components／～品 prefabrication／～装配式房屋 prefabricated house; prefab
【预祝】congratulate beforehand

鹬 sandpiper; snipe
【鹬蚌相争,渔人得利】when the snipe and the clam grapple, the fisherman profits; it's the third party that benefits from the tussle
【鹬鸵】[动] kiwi

驭 drive
【驭手】soldier in charge of pack animals; driver of a military pack train: 炮车～ gun-carriage driver

yuān

渊 ①(深水;潭) deep pool: 天～之别 as far apart as heaven from earth; worlds apart／为～驱鱼 drive the fish into deep waters ②(深) deep: ～泉 a deep spring
【渊博】broad and profound; erudite: 学识～ erudite; learned
【渊默寡言】profound and reticent
【渊深】profound; deep; erudite
【渊识博学】profound knowledge and extensive learning
【渊薮】a gathering place of fish or beasts; den; haunt: 盗贼的～ a den of bandits and thieves／罪恶的～ a hotbed of crime; a sink of iniquity
【渊雅高尚】elegant and upright
【渊源】origin; source: ～极深 be closely connected／历史～ historical origins／文明的～ origin of civilization

冤 ①(冤枉) wrong; injustice: 不白之～ unrighted wrong; unredressed injustice／喊～ call for redressing a grievance／伸～ redress a grievance; right a wrong ②(冤仇) feeling of bitterness; hatred; enmity ③[方](欺骗) kid; fool; pull sb.'s leg: 别～人。 Don't kid me. ④(吃亏) bad luck; loss; disadvantage
【冤案】case of injustice
【冤仇】rancor; enmity: 几代～ enmity for generations
【冤家】①(仇人) enemy; foe: ～对头 opponent and foe ②(似恨而实爱的人) one's destined love; sweetheart; lover
【冤家路窄】enemies are bound to meet on a narrow road; one can't avoid one's enemy
【冤假错案】unjust; framed-up and wrong cases
【冤情】grievance
【冤屈】①(冤枉) wrong; treat unjustly ②(不公平的待遇) wrongful treatment; injustice
【冤枉】①(受到不公平的待遇) wrong; treat unjustly: ～某人 wrong sb; do sb. wrong／别～好人。 Don't wrong an innocent person. ②(不值得) not worthwhile; not repaying the effort: 走～路 go the long way
【冤有头,债有主】every injustice has its perpetrator, every debt has its debtor
【冤狱】an unjust charge or verdict; a miscarriage of justice; frame-up: 平反～ reverse an unjust verdict
【冤冤相报】reprisal breeds reprisal

鸢 [动] kite
【鸢尾】[植] iris

鸳
【鸳鸯】①(水禽) mandarin duck ②(比喻夫妻) an affectionate couple

yuán

元 ①(开始的) first; primary: ～月 the first month of the year ②(为首的) chief; principal: ～首 head of state ③(主要) basic; fundamental: ～音 vowel ④(构成一个整体的) unit; component: 单～ unit ⑤(货币单位) yuan
【元宝】a gold or silver ingot used as money in

feudal China
【元宝车】[铁路] well car
【元旦】New Year's Day
【元古代】[地] the Proterozoic Era
【元件】element component; cell: 传输～ transfer element / 电光～ electrooptic cell / 电路～ circuit component / 加热器～ heater element / 敏感～ sensor ◇ ～盒 subassembly wrapper
【元老】senior statesman; founding member
【元年】the first year of an era or the reign of an emperor
【元配】a man's first wife
【元气】vitality; vigor: ～勃勃 one's constitution issues forth / ～消沉 be a cup too low / 大伤～ undermine one's constitution; sap one's vitality / 恢复～ regain one's strength
【元首】head of state
【元帅】①(军衔) marshal: Field Marshal(英陆军); Marshal of the Royal Air Force(英空军); Admiral of the Fleet(英海军) ②(统率全军的主帅) supreme commander
【元素】①(要素) element ②[数][化] element: 稀有～ rare element ◇ ～分析 ultimate analysis / ～周期表 periodic table of elements
【元宵】①(节目) the night of the 15th of the 1st lunar month: ～节 the Lantern Festival ②(食品) sweet dumplings made of glutinous rice flour
【元凶】prime culprit; arch-criminal
【元勋】a man of great merit; founding father: 开国～ founders of a state
【元音】[语] vowel
【元鱼】soft-shelled turtle
【元元本本】from beginning to end

芫
【芫花】[植] lilac daphne

园 ①(种蔬菜、花果、树木的地方) an area of land for growing plants: 菜～ vegetable garden / 种植～ plantation / 竹～ bamboo plantation ②(供人游览娱乐的地方) a place for public recreation: 动物～ zoological garden; zoo / 植物～ botanical garden
【园地】①(菜园、花园、果园等的统称) garden plot ②(活动范围) field; scope: 艺术的～ field of activities for arts
【园丁】gardener
【园林】gardens; park: 实现大地～化 turn the land green with parks and woods
【园圃】garden, ground used for growing vegetables, flowers or fruit
【园田】vegetable garden: 耕作～化 garden-style cultivation of farmland
【园艺】horticulture; gardening ◇ ～家 horticulturist / ～学 horticulture; gardening

鼋 [动] soft-shelled turtle
【鼋鱼】soft-shelled turtle

辕 shafts of a cart or carriage
【辕马】horse in the shafts; shaft-horse
【辕门】the outer gate of a government office

猿 ape: 从～到人 from ape to man / 类人～ anthropoid ape
【猿猴】apes and monkeys
【猿人】ape-man: 北京～ Peking man

原 ①(最初的) primary; original; former; at first: ～版 original edition / ～处 original place ②(没加工的) unprocessed; raw: ～粮 unprocessed food grains / ～煤 raw coal / ～油 crude oil ③(原谅) excuse; pardon 情有可～ excusable; pardonable ④(宽广平坦的地方) level; open country; plain: 雪～ snow-clad plain
【原案】original bill
【原班人马】the old cast; the former staff
【原版片】film in the original
【原本】①(原稿) original manuscript; master copy ②(原书) the original ③(本来) originally; formerly
【原璧归赵】return article to owner
【原病毒】provirus
【原材料】raw and processed materials ◇ ～工业 raw and semifinished materials industries
【原产地】country of origin ◇ ～证明书 certificate of origin
【原虫】[动] protozoon
【原抵押】original mortgage
【原定租金】original rent
【原订租约】original tenancy
【原动力】motive power; motivity
【原发价】original offer
【原封】with the seal unbroken; intact: ～不动 be left intact / ～退回 return to the sender a parcel or letter unopened / 钱～未动地被捡拾者送还。 The money was returned intact by its finder. ◇ ～烧酒 distilled wine
【原稿】original manuscript; master copy: ～架 copy holder
【原告】[法] ①(民事的) plaintiff ②(刑事的)

prosecutor ◇ ～的诉状 libel／～证人 evidence for the prosecution
【原国籍】nationality of origin
【原画】original painting
【原籍】ancestral home
【原价】cost price
【原件】script
【原来】①(起初) original; former ②(表示发现真实情况) so; turn out to be: ～如此! So that's how it is!／～是你! So it's you!
【原理】principle; tenet: 根本的～ a cardinal principle／数学的基本～ fundamentals of mathematics
【原谅】excuse; forgive; pardon: 请～,打扰您了。Excuse me for interrupting you.
【原料】raw material ◇ ～定单 material valuation／～生产 production of material／～消耗 material consumption
【原棉】[纺] raw cotton ◇ ～等级 grades of raw cotton
【原木】log
【原判】original judgement; original sentence
【原色】[物] primary colors
【原审】[法] first trial: 推翻～判决 reverse the judgement of the lower court／维持～判决 uphold the judgement of the lower court
【原生动物】protozoon ◇ ～学 protozoology
【原生矿物】primary mineral
【原生生物】protist
【原生植物】protophyte
【原生质】[生] protoplasm
【原始】①(最初的) original; firsthand: ～记录 original record／～凭证 original certificate／～所有人 original owner／～文件 original document／～协议 original agreement／～责任 original responsibility／～证据 original evidence／～资料 firsthand information; primary data ③(最古老的; 未开发的) primeval; primitive: ～部落 primitive tribe／～公社 primitive commune／～积累 primitive accumulation／～森林 primeval forest; virgin forest／～社会 primitive society／～文化 primitive culture／～状态 reset condition
【原诉】original suit
【原委】how a thing happened from beginning to end; the whole story; all the details
【原文】original text; the original: 读～ read in the original／和～核对一下 check it up with the original／忠实于～ be faithful to the original
【原物】original
【原先】former; original: ～的提纲 original outline／～是文盲 formerly an illiterate
【原形】original shape; the true shape under the disguise: ～毕露 be revealed for what one is; show one's true colors
【原型】[机] prototype
【原选区】home constituency
【原盐】crude salt
【原野】open country; champaign
【原意】meaning; original intention: 曲解～ distort
【原因】cause; reason: ～和结果 cause and effect／成功的～ reasons for the success／由于这个～ on this account
【原油】crude oil; crude: 低硫～ sweet crude／多蜡～ waxy crude／含硫～ sour crude／无蜡～ wax-free crude ◇ ～分馏塔 crude fractionating tower／～裂化设备 crude cracker
【原有】original: ～规定 already existing provisions／～让与人 original assigner／～文化水平 the original educational level
【原原本本】from beginning to end; the whole story
【原则】principle: ～分歧 differences in principle／～上同意 agree in principle／～问题 a matter of principle／～性的错误 mistake in matters of principle／～性立场 a principled stand
【原职】former post
【原址】former address
【原重】original weight
【原主】original owner: 物归～ return sth. to its owner
【原著】original work; original: 读莫泊桑～ read Maupassant in the original
【原状】original state; previous condition; status quo ante: 恢复～ restore the status quo ante; restore to the former state
【原子】atom: 标记～ labelled atom ◇ ～尘 fallout／～弹 atom bomb; atomic bomb; A bomb／～动力船 atomic-powered ship／～讹诈 atomic blackmail／～反应堆 atomic reactor; atomic pile／～核 atomic nucleus／～价 valence; atomicity／～量 atomic weight／～论 atomic theory; atomism／～炮 atomic gun／～数 atomicity／～武器 atomic weapon／～物理学 atomic physics／～序数 atomic number／～战争 atomic war
【原子能】atomic energy: ～发电站 atomic power station
【原罪】[天主教] original sin

【原作】original work; original

源 ①(水流起头的地方) source; fountainhead: 河～ source of a river / 饮水思～ when drinking water think of its source; remember where one's happiness comes from ②(来源) source; cause: 病～ cause of a disease / 财～ source of income / 货～ the source of supplies

【源编辑程序】[计算机] source editor

【源程序】[计算机] source program ◇ ～库 source library

【源代码】[计算机] source code

【源流】source and course; origin and development

【源泉】source; fountainhead

【源头】fountainhead; source

【源语句库】[计算机] source statement library

【源源】in a steady stream; continuously: ～不断的供应 an unfailing supply of sth / ～不绝 continue without end / ～而来 coming incessantly

【源远流长】a distant source and a long stream; of long standing and well established

员 ①(从事某种活动的人) a person engaged in some field of activity: 炊事～ cook / 教～ teacher / 人～ personnel / 售货～ shop assistant / 学～ student / 职～ staff member / 指挥～ commander ②(团体或组织中的成员) member: 党～ Party member / 团～ League member ③[量] 一～大将 an able general

【员额】specified number of personnel

【员工】staff; personnel: 师生～ teachers, students, administrative personnel and workers

圆 ①(圆形的) round; circular; spherical: ～锯 circular raw / ～孔 a round hole ②[数] circle: 半～ semicircle / 内切～ inscribed circle / 同心～ concentric circles / 椭～ ellipse / 外接～ circumscribed circle; circumcircle ③(周全) tactful; satisfactory ④(使圆满) make plausible; justify: 自～其说 make one's statement consistent; justify oneself ⑤(货币单位) yuan ⑥(圆形的货币) a coin of fixed value and weight: 铜～ copper coin / 银～ silver dollar

【圆材】[林] roundwood; log

【圆场】mediate; help to effect a compromise: 打～ mediate a dispute; smooth things over

【圆唇元音】[语] round vowel

【圆规】compasses: 一个～ a pair of compasses / 制图～ drawing compasses

【圆号】[乐] French horn; horn

【圆弧】an arc

【圆滑】smooth and evasive; slick and sly: ～手段 possessing tactful means

【圆寂】[佛教] Parinirvana; pass away

【圆括号】parentheses; curves (( ))

【圆脸】round face

【圆满】satisfactory: ～的答复 a satisfactory reply / ～解决 be settled satisfactorily / 两国谈判～结束。 The negotiation between the two countries came to a successful conclusion.

【圆盘】disc ◇ ～犁 disc plough / ～耙 disc harrow

【圆球】a ball; a globe

【圆圈】circle; ring

【圆润】mellow and full: ～的嗓音 a sweet, mellow voice

【圆熟】skillful; proficient; dexterous

【圆通】flexible; accommodating

【圆筒】a cylinder

【圆舞曲】[乐] waltz

【圆心】the center of a circle ◇ ～角[数] central angle

【圆形】circular; round ◇ ～建筑 round building; rotunda

【圆周】circumference ◇ ～角[数] circumferential angle / ～率 ratio of the circumference of a circle to its diameter (π) / ～运动 circular motion

【圆珠笔】ball-point pen; ball-pen

【圆柱】[数] cylinder

【圆锥】circular cone; taper ◇ ～根[植] conical root / ～花序[植] panicle / ～台[数] frustum of a cone / ～投影 conical projection

【圆桌】round table ◇ ～会议 round-table conference

【圆子】①(一种食品) dumpling ②[方](丸子) ball

【圆凿方枘】like a square tenon for a round mortise; at variance with each other

援 ①(以手牵引) pull by hand; hold: 攀～ climb up by holding on to sth. ②(引用) quote; cite ③(援助) help; aid; rescue: 孤立无～ isolated and cut off from outside help / 求～ ask for help; appeal for aid

【援笔疾书】take up a pen and write quickly

【援经引典】quote from classics and canons

【援救】rescue; save; deliver from danger

【援军】reinforcements; relief troops

【援例】cite a precedent

【援外】foreign aid: ～物资 materials in aid of a

foreign country
【援引】①(引用) quote; cite: ～法律规定 invoke a legal provision / ～条款 quote an article ②(推荐或任用跟自己有关系的人) recommend or appoint one's friends or favorites
【援用】quote; cite; invoke: ～成例 cite a precedent / ～条约 invoke a treaty
【援照】according to; adduce; cite
【援助】help; support; aid: 国际～ international support / 技术～ technical assistance / 经济～ economic aid / ～受难者 help the victims
缘 ①(缘故) reason: 无～无故 without rhyme or reason; for no reason at all ②(边缘) edge; fringe; brink: 外～ outer fringe ③(缘分) predestined relationship ④(沿着) along ⑤[佛教] conditioning cause: 结～ form cause; lay the basis for future relationship / 夙～ predestination from the past; fate inherited from previous incarnation
【缘分】①(发生联系的机会) lot or luck by which people are brought together ②[佛教] destiny 〈luck〉 as conditioned by one's past
【缘故】cause; reason
【缘木求鱼】climb a tree to catch fish; a fruitless approach
【缘由】reason; cause
【缘业】[佛教] a man's fate as conditioned by his past

yuǎn

远 far; distant; remote: ～处 faraway place / ～方来客 visitor from a distant land / ～～超过 far exceed / 差得～ be far behind / 敬而～之 stay at a respectful distance from sb.
【远程】long-range; long-distance: ～导航 long-distance navigation / ～航行 long voyage / ～轰炸机 long-range bomber / ～火箭 long-range rocket / ～运输 long-distance transport
【远大】long-range; broad; ambitious: ～的计划 a long-range plan; an ambitious plan / ～的理想 lofty ideals / 前途～ have a bright future / 眼光～ be farsighted; have a broad vision
【远道】a long way: ～传闻 a rumor originated from a distance / ～而来 come a long way; come from afar
【远地点】[天] apogee
【远方】distant place: ～的来客 a guest from afar
【远房】distantly related: ～亲戚 a distant relative; remote kinsfolk
【远隔重洋】be separated by vast oceans
【远古】remote antiquity: ～以来 from the time immemorial / 在～时代 in remote antiquity
【远光头灯】far-reaching headlamp
【远红外仪器】far infrared gear
【远见】foresight; vision: ～卓识 foresight and sagacity / 没有～的人 a man without vision / 有～的人 a man of foresight
【远郊】outer suburbs
【远近】①(近处和远处) far and near: ～闻名 be know far and wide ②(距离的长短) distance
【远景】①(距离的景物) distant view ②(将来的景象) long-range perspective; prospect: ～规划 a long-range plan ③[电影] long shot
【远距离操纵】remote control; telecontrol
【远距离操作】remote-controlled operation
【远距离射击】long-range shooting
【远距离投篮】[体] long shot
【远离家乡】far away from one's native village
【远虑】foresight; long view: 深谋～ think deeply and plan carefully; be circumspect and farsighted
【远期】at a specified future date; forward: ～汇价 forward rate / ～汇票 time bill / ～外汇 forward exchange / ～支票 credit check
【远亲】distant relative; remote kinsfolk
【远山近水】the faraway mountains and nearby water; beautiful scenery
【远涉重洋】travel across the high seas
【远射程炮】long-range gun
【远摄镜头】①[电影] long shot ②[摄影] telephoto lens
【远视】[医] long sight; farsightedness; hyperopia; hypermetropia ◇ ～眼镜 spectacles for long sight; spectacles for the farsighted
【远水解不了近渴】distant water cannot quench present thirst; the aid is too slow in coming to be of any help
【远水救不了近火】distant water won't put out a fire close at hand; a slow remedy can not meet an urgency
【远台】[乒乓球] far from the table ◇ ～防守 long defense; far-from-table defense / ～削球 off-table chop
【远行】go on a long journey
【远洋】①(大洋) ocean ②(距离大陆远的海洋) of the open sea beyond the littoral zone; oceanic ◇ ～班轮 ocean liner / ～船 ocean-going vessel / ～航行 oceangoing voyage / ～货轮 oceangoing freighter / ～渔业 deep-sea fishing

【远因】remote cause
【远游】travel faraway
【远缘杂交】[农] distant hybridization
【远征】expedition ◇～军 expeditionary army
【远走高飞】fly far and high; be off to distant parts
【远足】pleasure trip on foot; hike; walking tour
【远祖】remote ancestor

yuàn

院 ①(院子) courtyard; yard; compound: 场～ threshing ground / 居民大～ neighborhood compound ②(某些机关和处所的名称) a designation for certain government offices and public places: 电影～ cinema / 国务～ state council / 科学～ the academy of sciences ③(学院) college ④(医院) hospital: 出～ be out of hospital; be discharged from hospital / 收住～ admit to hospital / 送～ send to hospital / 住～ be in hospital; be hospitalized / 转～ transfer to another hospital / 医生建议让小孩住～。The doctor advised hospitalization for the child.
【院落】courtyard; yard; compound
【院士】academician
【院外调解】conciliation out of court
【院子】courtyard; yard: 用篱笆围～ enclose the courtyard with a fence

愿 ①(愿望) hope; wish; desire: 平生之～ a lifelong wish ②(愿意) be willing; be ready ③(愿心) vow: 还～ redeem a vow
【愿望】desire; wish; aspiration: 满足某人的～ satisfy sb.'s wish / 人民的共同～ the common aspiration of the people / 要求独立的～ the desire for independence / 主观～ subjective desire
【愿为牺牲】be willing to sacrifice for
【愿意】①(因符合心愿而同意) be willing; be ready: ～作出任何牺牲 be willing to make any sacrifice ②(希望) wish; like; want: 我不～打扰你。I don't want to bother you.

怨 ①(怨恨) resentment; enmity: 抱～ complain; grumble ②(责怪) blame; complain: 任劳任～ bear hardships and blame
【怨不得】①(不能抱怨) cannot blame ②(难怪) no wonder
【怨愤】discontent and indignation
【怨恨】①(对人或事强烈不满或仇恨) have a grudge against sb.; hate ②(强烈的不满或仇恨) resentment; grudge; enmity
【怨气】grievance; complaint; resentment: 出～ air one's grievances; vent one's resentment / 一肚子～ be full of complaints
【怨声载道】cries of discontent rise all round; complaints are heard everywhere
【怨天尤人】blame god and man; blame everyone and everything but oneself
【怨天怨地】murmur against heaven and earth
【怨言】complaint; grumble: 从未发过一句～ never utter a word of complaint / 毫无～ without a murmur; without a word of complaint

yuē

约 ①(提出或商量) make an appointment; arrange: ～某人做某事 make an appointment with sb. for sth ②(邀请) ask or invite in advance ③(约定的事) pact; agreement; appointment: 践～ keep an appointment / 立～ make a pact; come to an agreement ④(拘束) restrict; restrain ⑤(俭省) economical; frugal: 节～ economize ⑥[数](约分) reduction of a fraction ⑦(简要) simple; brief ⑧(大概) about; around: ～一个月 about a month
【约旦】Jordan ◇～人 Jordanian
【约定】agree on; appoint; arrange: ～的期限届满 expiration of the agreed term / ～会晤地点 agree on a meeting place / ～开会时间 fix a time for the meeting
【约定俗成】established by usage; accepted through common practice
【约法】provisional constitution: ～三章 agree on a three-point law; make a few simple rules to be observed by all concerned
【约分】[数] reduction of a fraction
【约会】appointment; engagement; date: 订～ make an appointment / 已有～ have previous engagement
【约集】meet by appointment; gather
【约略】rough; approximate: ～的估计 a rough estimate / ～地说来 roughly speaking / ～知道一些 know something about it
【约莫】about; roughly; or so: ～一小时光景 about an hour; an hour or so
【约期】fix a date; appoint a time: ～会晤 fix a date for the interview / 误了～ fail to keep the appointment
【约请】invite; ask
【约束】keep within bounds; restrain; bind: ～

严明 practice rigid discipline / ～自己 restrain oneself / ～自己的感情 restrain one's feelings / 自我～ selfdiscipline; self-control ◇ ～力 binding force

【约数】①(大约的数目) approximate number ②[数](能整除某个数的数) divisor: 公～ common divisor

【约言】promise; word; pledge: 实行～ redeem one's pledge / 违背～ break one's promise; go back on one's word

yuè

悦 ①(高兴) happy; pleased; delighted: 不～ displeased / 和颜～色 a kindly countenance; a genial look ②(使愉快) please delight: 取～于人 try to please sb.

【悦耳】music pleasing to the ear; sweet-sounding: ～的歌声 sweet song / ～的音乐 sweet music / ～之言 words that please the ear

【悦服】heartily admire

【悦目】pleasing to the eye; good-looking

阅 ①(看) read; go over: ～报 read newspapers / ～卷 go over examination papers ②(检阅) review; inspect ③(经历) experience; pass through

【阅兵】review troops ◇ ～场 parade ground / ～典礼 dress parade / ～式 military review; parade

【阅读】read: ～小说 read a novel ◇ ～材料 reading material / ～机 reader / ～器 reader

【阅览】read

【阅览室】reading room: 报刊～ newspaper reading room / 期刊～ periodical reading room

【阅历】①(亲身见过；听过或做过) see, hear or do for oneself ②(由经历得来的知识) experience: ～艰辛 have a rough time / ～浅 having little experience; inexperienced / ～颇深 have seen much of life

【阅世】see the world: ～甚深 be deeply experienced in worldly affairs

越 ①(跨过) get over; jump over: ～墙而入 enter by climbing over the wall ②(超出) exceed; overstep ③(昂扬) be at high pitch: 激～ intense; loud and strong

【越冬】live through the winter ◇ ～作物 winter crop; overwintering crop

【越发】all the more; even more: ～使人高兴 please sb. all the more

【越轨】exceed the bounds; transgress: ～行为 impermissible behavior; transgression

【越过】cross; surmount; negotiate: ～山顶 pass the summit of a mountain / ～障碍 surmount obstacles

【越级】①(越过直属一级到更高一级) bypass the immediate leadership: ～申诉 bypass the immediate leadership and present one's appeals and complaints to higher levels ②(不按照一般的次序) skip a grade or rank: ～提升 promote sb. more than one grade at a time

【越境】cross the boundary illegally; sneak in or out of a country

【越橘】[植] cowberry

【越来越…】more and more: ～放肆 become more and more intemperate / ～深入人心 penetrate deeper and deeper into the hearts and minds of the people

【越雷池一步】transgress the bounds

【越南】Viet Nam ◇ ～人 Vietnamese / ～语 Vietnamese

【越权】exceed one's power or authority: ～行为 act in excess of authority

【越位】[体] offside

【越野】cross-country ◇ ～汽车 cross-country vehicle / ～滑雪 cross-country skiing / ～赛车 offroad racing / ～识途比赛 orienteering / ～赛跑 cross-country race / ～行驶 cross-country run / ～载重车 cross-country cargo carrier; cross-country truck

【越狱】escape from prison; break prison ◇ ～犯 prison breaker / ～逃跑犯 prison escapee / ～罪 breach of prison; crime of escaping from custody

【越…越…】the more... the more: ～多～好 the more the better / ～快～好 the faster, the better / ～早～好 the earlier, the better

【越俎代庖】exceed one's functions and meddle in other's affairs; take sb. else's job into one's own hands

跃 leap; jump: ～居世界首位 leap to first place in the world / 飞～ by leaps and bounds

【跃进】make a leap; leap forward

【跃马飞驰】leap on a horse and ride swiftly

【跃然】appear vividly: ～纸上 full of life

【跃跃欲试】be eager to have a try; itch to have a go

岳 ①(高大的山) high mountain ②(称妻子的父母) wife's parents: ～家 family of one's wife's parents

【岳父】wife's father; father-in-law
【岳母】wife's mother; mother-in-law

月 ①(月球) the moon: 上弦～ first quarter of the moon / 朔～ new moon / 望～ full moon / 下弦～ last quarter of the moon ②(月份) month: 本～ this month / 逐～ month by month ③(每月的) monthly: ～报表 monthly report ④(圆的;形状象月亮的) round; full-moon shaped
【月白】bluish white; very pale blue
【月半】the 15the day of a month
【月报】①(月刊) monthly magazine; monthly ②(按月的汇报) monthly report
【月饼】moon cake
【月初】the beginning of a month
【月底】the end of a month
【月度】monthly: ～计划 a monthly plan
【月份】month: 上～ last month
【月份牌】calendar
【月宫】the palace of the moon; the moon
【月光】moonlight; moonbeam
【月桂树】[植] laurel; bay tree
【月华】①(日光) moonlight: ～如水 watery moonbeams; a flood of translucent moonlight ②[气] lunar corona
【月季】[植] Chinese rose
【月经】menses; menstruation; period ◇ ～不调 irregularity of menstruation / ～初现 menophania / ～带 sanitary belt; sanitary napkin / ～过多 menorrhagia; excessive menstruation / ～过少 hypomenorrhea / ～失调 menstrual disorder / ～周期 menstrual cycle
【月刊】monthly magazine; monthly
【月历】monthly calendar
【月亮】the moon ◇ ～门儿 moon gate
【月面行走】moonwalk
【月面学】selenography
【月偏食】[天] partial lunar eclipse
【月票】monthly ticket
【月琴】*yueqin*, a four-stringed plucked instrument with a full-moon-shaped sound box
【月球】[天] the moon ◇ ～火箭 moon rocket / ～探测 lunar probe
【月全食】[天] total lunar eclipse
【月色】moonlight
【月食】[天] lunar eclipse
【月台】railway platform ◇ ～票 platform ticket
【月息】monthly interest ◇ ～率 rate per mensem
【月下老人】the god who unites persons in marriage; matchmaker
【月薪】monthly pay
【月牙】[口] crescent moon
【月夜】moonlit night
【月影花痕】the shadow of the moon and flowers
【月晕】lunar halo
【月中】the middle of a month
【月子】①(妇女生育后的第一个月) month of confinement after giving birth to a child: 坐～ be in confinement ②(分娩的时期) time of childbirth; confinement ◇ ～病 puerperal fever

钥 key
【钥匙】key

乐 music: 交响～ symphony; symphonic music / 器～ instrumental music / 声～ vocal music / 室内～ chamber music / 奏～ play music
【乐池】orchestra pit; orchestra
【乐段】[乐] period
【乐队】orchestra; band: 管～ band / 管弦～ orchestra / 交响～ symphony orchestra / 军～ military band / 铜管～ brass band ◇ ～指挥 conductor; bandmaster
【乐理】[乐] music theory
【乐谱】music score; music ◇ ～架 music stand ○五线谱 staff / 简谱 numbered musical notation / 高音谱号 treble clef; G clef / 中音谱号 tenor clef; alto clef; C clef / 低音谱号 bass clef; F clef / 唱名 sol-fa syllables / 音名 musical alphabet / 休止符 rest / 附点 dot / 装饰音 grace note; grace; ornament / 二全音符 breve / 全音符 semi-breve / 二分音符 half note; minium / 四分音符 quarter note; crotchet / 八分音符 eighth note; quaver / 十六分音符 sixteenth note; semi-quaver / 三十二分音符 thirty-second note; demisemi-quaver / 三连音符 triplet / 切分音 syncopation / 小节 bar; measure / 反复记号 repeat / 拍 beat / 拍号 time signature / 二拍子 duple time / 三拍子 triple time / 四拍子 quadruple time / 调号 key signature / 转调 modulation / 移调 transposition / 升号 sharp / 降号 flat / 大调 major / 小调 minor / 曲调 tune / 旋律 melody / 节奏 rhythm / 和声 harmony / 和弦 chord / 音量 volume / 音程 interval / 音色 timbre; tone color / 音域 compass; range; regisfer / 音高 pitch / 音阶 scale / 速度 tem-

po／弱 p (piano)／强 f (forte)／稍弱 mp (mezzo piano)／稍强 mf (mezzo forte)／最弱 pp (pianissimo)／最强 ff (fortissimo)／渐弱 cresc. (crescendo)／渐强 dim. (diminuendo)

【乐器】musical instrument; instrument: 拨弦～ plucked instruments／打击～ percussion instrument／弓弦～ bowed stringed instruments／管～ wind instrument／击弦～ hammered string instruments／键盘～ keyboard instruments／木管～ wood-wind instruments; wood-winds／铜管～ brass-wind instruments; brass-winds; brasses／弦～ stringed instrument ◯小提琴 violin／中提琴 viola／大提琴 cello; violoncello／低音提琴 double bass; contrabass／竖琴 harp／曼陀林 mandolin／吉他 guitar／长笛 flute／短笛 piccolo／双簧管 oboe／英国管 English horn／大管 bassoon／单簧管 clarinet／圆号 French horn; horn／萨克管 saxophone／萨克号 saxhorn／小号 trumpet／短号 cornet／长号 trombone／大号 tuba／中音号 alto horn／军号 bugle／定音鼓 kettle drums; timpani／管风琴 pipe organ; organ／大钢琴 grand piano; concert piano／竖式钢琴 upright piano／簧风琴 harmonium; reed organ／手风琴 accordion／钢片琴 celesta／木琴 xylophone／排钟 chimes／三角铁 triangle／响板 castanets／大鼓 bass drum／小鼓 side drum; snare drum／铃鼓 tambourine／钹 cymbals／锣 gong

【乐曲】musical composition; composition; music

【乐师】musicians

【乐团】①(交响乐团) philharmonic society ②(交响乐团的管弦乐队) philharmonic orchestra: 交响～ symphony orchestra／民族～ national music emsemble

【乐舞】dance accompanied by music

【乐音】musical sound; tone

【乐章】[乐] movement

yūn

晕 ①(头脑发昏) dizzy; giddy ②(昏迷) swoon; faint: ～了过去 lose consciousness; faint; swoon

【晕车】car sickness

【晕船】sea sickness

【晕倒】fall in a faint; pass out

【晕厥】[医] syncope; faint

【晕头转向】confused and disoriented:

【晕晕忽忽】dizzy; giddy

yún

云 ①(说) say: 人～亦～ repeat what others say; parrot ②(云彩) cloud: 波状～ wave cloud／层～ stratus／飞～ scud／高层～ altostratus／高积～ altocumulus／积～ cumulus／积雨～ cumulonimbus／卷～ cirrus／卷层～ cirrostratus／卷积～ cirrocumulus／卷毛～ fleecy cloud／卷轴～ roll cloud／马尾～ mare's tails／蘑菇～ mushroom cloud／夜光～ luminous night cloud／雨～ nimbus／雨层～ nimbostratus／～端 high in the clouds／～块 cloud mass／～片 cloud sheet／多～ cloudy／少～ partly cloudy／无～ cloudless／一朵白～ a white cloud

【云层】cloud layer: 在～上面飞行 fly above the clouds

【云出无心】clouds arises without design; an inadvertent act; done without consideration

【云海】a sea of clouds

【云集】come together in crowds; gather; converge: 外商～广州。Many foreign merchants gathered in Guangzhou.

【云量】[气] cloudiness

【云母】[矿] mica ◇ ～电容器 mica condenser／～片 mica sheet; sheet mica／～铁矿 micaceous iron-ore

【云泥之别】clouds and mud; great difference in social standing

【云气】thin, floating clouds

【云雀】[动] skylark

【云山雾罩】extravagant and baseless talk

【云杉】[植] dragon spruce

【云梯】scaling ladder

【云图】[气] cloud atlas; cloud chart

【云雾】cloud and mist; mist ◇ ～天 soupy weather

【云霞】rosy clouds: ～之交 high-minded friendship

【云消雾散】the clouds melt and the mists disperse; vanish into thin air

【云霄】the skies: 响彻～ resound to the skies

【云烟】cloud and mist: ～过目 like cloud and smoke passing the eyes; be easily vanished

【云蒸霞蔚】radiant; colorful and flourishing

耘 weed: ～田 weed the fields

【耘锄】[农] hoe

【耘人之田】manage business on behalf of

others

芸

【芸草】a strong-scented herb; rue
【芸豆】kidney bean
【芸薹】[植] rape
【芸香】[植] rue
【芸芸】many; numerous
【芸芸众生】[佛教] all living things; all mortal beings

纭

【纭纭】numerous and disorderly; diverse and confused

匀 ①(均匀) even; smooth: 分得不～ be not evenly divided ②(使均匀) even up; divide evenly ③(分出一部分) spare: 你们能～一些化肥给我们吗? Would you spare us some chemical fertilizer?
【匀变速运动】uniformly retarded motion
【匀称】well-proportioned; well-balanced; symmetrical: 身材～ of proportional build
【匀出一份】portion out a share
【匀分不开】unable to divide equally
【匀加速度】[物] uniform acceleration
【匀减速运动】[物] uniformly variable motion
【匀净】uniform; even
【匀染】[纺] level dyeing ◇ ～剂 levelling agent
【匀实】[口] even; neat; uniform
【匀速运动】[物] uniform motion
【匀整】neat and well spaced; even and orderly

yǔn

殒 perish; die
【殒灭】[书] meet one's death; perish
【殒命】[书] meet one's death; perish

陨 fall from the sky or outer space
【陨落】fall from the sky of outer space
【陨石】[天] aerolite; stony meteorite ◇ ～雨 meteorite shower
【陨铁】[天] meteoric iron; iron meteorite; siderite
【陨星】[天] meteorite: 石～ aerolite; stony meteorite / 铁～ siderite; iron meteorite ◇ ～坑 meteorite crater / ～学 meteoritics

允 ①(允许) permit; allow; consent: 应～ consent ②(公平) fair; just: 公～ fair; equitable
【允诺】promise; consent; undertake: ～保释 assent to release on bail / 欣然～ readily consent / 请～我讲几句。 Permit me to make a few remarks. ◇ ～误差[机] allowable error; permissible error / ～载荷 allowable load
【允许】permit; allow: ～保释的 bailable

yùn

晕 ①(头脑发昏) dizzy; giddy; faint: 头～目眩 have a dizzy spell; be afflicted with vertigo ②[气] (日光或月光的光圈) halo: 日～ solar halo / 月～ lunar halo
【晕车】carsickness: 好～ be liable to carsickness
【晕船】seasickness
【晕机】airsickness

运 ①(运动) motion; movement ②(运输) carry; transport 货～ freight transport / 空～ air transport; airlift ③(运用) use; wield; utilize ④(运气) fortune; luck; fate: 不走～ be out of luck; have no luck / 好～ good luck
【运笔如飞】be quick in writing
【运筹帷幄】devise strategies within a command tent
【运筹学】operational research; operations research
【运动】①[物] motion; movement: ～中的物质 matter in motion / 均速直线～ uniform rectilinear motion / 直线～ rectilinear motion ②(体育活动) sports; athletics; exercise: 室外～ outdoor sports / 体育～ athletic sports ③(政治、文化、生产等运动) movement; campaign; drive: 爱国卫生～ patriotic public health movement / 国际共产主义～ international communist movement ◇ ～场 sports ground; playground / ～服装 sportswear / ～健将 master of sports / ～疗法 kinesiatrics; kinesitherapy fare / ～赛车 sport-racing car / ～学[物] kinematics / ～医学 sports medicine
【运动会】sports meet; athletic meeting; games: ～会旗 flag of the games / ～会歌 anthem of the games / ～会徽 emblem / 奥林匹克～ Olympic Games / 全国～ national games / 世界大学生～ World University Games; Universiade / 亚洲～ Asian Games
【运动员】sportsman; sportswoman; athlete; player: 业余～ amateur / 一级～ first grade sportsman / 用左手的～ left-hander
【运费】transportation expenses; freight; carriage: ～待收 freight to be collected ◇ ～表 freight list / ～单 freight note / ～率 freight rate / ～免付 carriage free / ～条款 freight

clause／ ～已付 freight paid／ ～预付 freight prepaid; advanced freight
【运河】canal: 大～ the Grand Canal ◇ ～税 canal dues
【运煤船】coal carrier; collier
【运气】①(把力气贯注到身体某部位) directing one's strength, through concentration, to a part of the body ②(命运;幸运) fortune; luck
【运球】[体] dribble
【运输】transport; carriage; conveyance: 陆上～ land transport; transport by land ◇ ～船 cargo ship; transport ship／ ～带 conveyer belt; travelling belt／ ～队 transport team／ ～费 traffic expense／ ～工具 means of transport; conveyance／ ～公司 transport company／ ～机 transport plane／ ～里程 transport mileage／ ～量 freight volume／ ～能力 transport capacity; carrying capacity／ ～网 transport network／ ～线 supply line／ ～业 transport service; carrying trade; transportation
【运送】transport; ship; convey: ～物资 ship goods and materials／ 用飞机～邮件 transport mail by airplane
【运算】[数] operation: 四则～ the four fundamental operations of arithmetic／ 每秒钟～一千万次 capable of 10,000,000 calculations per second ◇ ～分析 operational analysis／ ～微积分 operational calculus／ ～误差 arithmetic error
【运算器】[计算机] arithmetic unit
【运销】transportation and sale
【运行】move; be in motion: 在轨道上～ move in orbit
【运用】utilize; wield; apply; put to use: ～概念 use of concepts／ ～自如 handle very skillfully; have a perfect command of
【运载工具】means of delivery: 战略～ strategic vehicles
【运载火箭】carrier rocket
【运载技术】delivery technology
【运转】①(沿一定轨道运转) revolve; turn round ②(指机器转动) work; operate: 机器～正常。 The machine works smoothly.

酝

【酝酿】①(图谋) brew; ferment: ～起义 ferment an uprising ②(非正式讨论;仔细考虑) have a preliminary informal discussion; deliberate on

愠

【愠怒】be inwardly angry
【愠色】gloomy countenance

蕴

【蕴藏】hold in store; contain ◇ ～量 reserves; deposits
【蕴涵】[逻] implication
【蕴蓄】lie hidden and undeveloped; be latent

韵

①(好听的声音) musical sound; sweet tone ②(韵母) rhyme: 平～ rhymes in the even tone／ 押～ be in rhyme／ 仄～ oblique rhymes ③(情趣) charm: 风～ personal charm; graceful bearing
【韵脚】the rhyming word that ends a line of verse; rhyme
【韵律】①(诗歌中平仄格式) meter ②(押韵规则) rules of rhyming; rhyme scheme ◇ ～学 prosody
【韵母】[语] simple or compound vowel of a Chinese syllable
【韵事】①(风雅的事) literary or artistic pursuits; often with pretense to good taste and refinement ②(浪漫的事) romantic affair
【韵味】lingering charm; lasting appeal
【韵文】literary composition in rhyme; verse

孕

pregnant: 避～ contraception／ 她怀～了。 She is pregnant.／ 她有～。 She is big with child.
【孕妇】pregnant woman
【孕激素】progestational hormone
【孕期】[医] pregnancy; gestation
【孕穗】[农] booting ◇ ～期 boot stage
【孕酮】progesterone
【孕吐】[医] vomiting during pregnancy; morning sickness
【孕育】be pregnant with; breed: ～着新思想 be pregnant with new ideas

熨

iron; press: ～衣服 iron clothes／ 快干免～ wash and wear
【熨斗】flatiron; iron: 电～ electric iron

Z

zā

匝 [书] ①(周) circle; circumference: 绕树三～ circle a tree three times ②(茂密) dense; full: 密密～～ dense; thick

【匝地】[书] all over the ground; everywhere: 柳荫～。 The willow trees cast their shadows all around.

【匝月】[书] a full month

咂 ①(吸) sip; suck: ～一口酒 take a sip of wine ②(咂嘴) make clicks (of admiration, praise, etc.) ③(辨别滋味) savor carefully: taste carefully

【咂嘴】make clicks (of admiration, praise, etc.): ～声 smack

扎 tie; bind: ～小辫儿 tie up one's plaits; plait one's hair; wear one's hair in plaits / 腰上～一根皮带 tie a leather belt round the waist

zá

砸 ①(撞击;敲打) pound; tamp: ～了某人的脚 have one's foot squashed ②(打破) break; smash: 碗～了。 The bowl is broken ③[方](失败) fail; fall through; be bungled: 事儿办～了。 The job was bungled.

【砸饭碗】get the sack; be fired; lose one's work

【砸锅】[方] fail; fall through; be bungled

【砸碎】break into pieces; smash; shatter: 瓶子被～了。 The bottle was shattered to pieces.

杂 ①(多种多样) miscellaneous; sundry; mixed: ～而不乱 mixed but not confused / 复～的心情 mixed feelings ②(搀杂) mix; mingle: 夹～ be mixed up with; be mingled with

【杂拌儿】①(杂干果) assorted preserved fruits; mixed sweetmeats ②(杂凑而成的事物) mixture; miscellany; medley; hotchpotch

【杂草】weeds; rank grass: ～丛生 be overgrown with weeds / 除去园中的～ weed the garden

【杂凑】knock together; odds and ends

【杂费】①(零碎费用) incidental expenses; incidentals ②(额外费用) sundry fees; extras

【杂感】①(零星感想) random thoughts ②(杂记) a type of literature recording such thoughts

【杂烩】①(菜名) mixed stew; hotchpotch ②(杂凑的事物) mixture; miscellany; medley; hotchpotch

【杂货】sundry goods; groceries: 日用～ various household supplies ◇ ～店 grocery

【杂和面儿】maize flour mixed with a little soya bean flour

【杂环化合物】[化] heterocyclic compounds

【杂记】①(笔记) jottings; notes ②(一种文体) miscellanies

【杂技】acrobatics: ～表演 acrobatic show ◇ ～团 acrobatic troupe / ～演员 acrobat

【杂交】[生] hybridize; cross: 无性～ vegetative hybridization / 有性～ sexual hybridization ◇ ～繁育 crossbreeding / ～后代 filial generation / ～水稻 hybrid rice / ～物种 hybrid species / ～优势 hybrid vigor; hybrid heterosis / ～玉米 hybrid maize / ～育种 crossbreeding / ～种 hybrid strain

【杂居】(of two or more nationalities) live together: 几个民族～的地区 an area inhabited by several nationalities

【杂粮】coarse cereals; food grains other than wheat and rice

【杂乱】mixed and disorderly; in a jumble; in a muddle: 抽屉里的东西很～。 The things in the drawer were all in a jumble.

【杂乱无章】disorderly and unsystematic; disorganized

【杂念】distracting thoughts: 私心～ selfish ideas and personal considerations

【杂牌】a less known and inferior brand ◇ ～货 goods of an inferior brand

【杂品】sundry goods; groceries

【杂七杂八的东西】odds and ends; a medley

【杂散】[物] stray ◇ ～磁场 stray magnetic field / ～电流 stray current / ～幅射 stray radiation

【杂色】variegated; parti-colored; motley

【杂色砂岩】mottled sandstone

【杂食动物】omnivorous animal

【杂耍】variety show; vaudeville

【杂税】miscellaneous levies; sundry tax

【杂碎】chopped cooked entrails (of sheep or oxen)

【杂项】sundry ◇ ～费用 miscellaneous expenses / ～收益 miscellaneous income / ～支出 miscellaneous payments

【杂文】essay

【杂务】odd jobs; sundry duties

【杂音】①(不正常的声音) noise: 这收音机～很多。 There is much noise in this radio. ② [电] static ③[医] murmur: 心脏～ heart murmur ◇～表 psophometer
【杂志】magazine: 订阅～ subscribe to a magazine / 过期～ back number ◇ ～架 magazine rack
【杂质】①(不纯成分) impurity ②[化] foreign matter ◇～离子 foreign ion
【杂种】①[生] hybrid; crossbreed ②(骂人话) bastard; son of a bitch ◇ ～不育性 hybrid sterility / ～狗 mongrel dog / ～优势[生] hybrid vigor; heterosis / ～猪 cross-bred pig

zǎ

咋 [方] how; why: 你～不知道呢? Why don't you know? / 情况～样? How are things?

zāi

灾 ①(灾害) calamity; disaster: 虫～ plague of insects / 旱～ drought / 水～ flood / 天～ natural disaster / 受～ be stricken by disaster ②(不幸) personal misfortune; adversity: 没病没～ good health and good luck; with one's health all right and luck not bad
【灾害】calamity; disaster: 自然～ natural calamity
【灾患】[书] calamity; disaster: 屡经～ suffer calamity after calamity
【灾荒】famine due to crop failures: 闹～ suffer form famine
【灾祸】disaster; calamity; catastrophe: ～临头。A great disaster is befalling.
【灾民】victims of a natural calamity
【灾难】suffering; calamity; disaster; catastrophe: ～深重 disaster-ridden / 遭受～ suffer disaster
【灾情】the condition of a disaster: ～严重。The losses caused by the disaster were serious
【灾区】disaster area: 地震～ earthquake-stricken area / 重～ severely afflicted area

栽 ①(栽种) plant; grow: ～花 grow flowers / ～树 plant trees ②(插上) stick in; insert; plant: 在地上～根桩子 plant a stake in the ground ③(硬给安上) frame up; force sth. on sb.; impose: 把罪名～在某人身上 frame up a case against sb. ④(跌倒) tumble; fall: ～进河里 fall into a river / 从自行车上～下来 tumble off a bicycle
【栽跟头】①(跌倒) tumble; fall ②(失败或出丑) suffer a setback; come a cropper
【栽培】①(种植;培养) cultivate; grow: 棉花～ the culture of cotton ②(造就) foster; train; educate: ～人才 cultivate and nourish talent ③(提拔) help advance sb.'s career; patronize ◇ ～品种 cultivar / ～植物 cultivated plant
【栽赃】①(把赃物放在别人处,诬告他犯法) plant stolen or banned goods on sb. ②(硬给安上罪名) frame sb.; fabricate a charge against sb. ◇ ～陷害 place stolen goods on sb. to implicate him; incriminate sb. with "planted" evidence
【栽植】plant; transplant ◇ ～机 planting machine / ～器 planting apparatus
【栽种】plant; grow: ～果树 plant fruit trees / ～葡萄 grow grapes

哉 [书][助] ①(表示感叹)呜呼!哀～! Alas! ②(表示疑问或反诘)有何难～? What's so difficult about it?

甾 [化] steroid
【甾酮】ketosteroid

zǎi

宰 ①(杀) slaughter; butcher: ～鸡杀鹅 kill hens and geese / ～羊 butcher sheep ②(主管; 主持) govern; rule: 主～ dominate; dictate
【宰割】invade, oppress and exploit: 任人～ allow oneself to be trampled upon
【宰杀】slaughter; butcher
【宰相】prime minister (in feudal China); chancellor: ～之才 a talent with the ability to rule the country
【宰相肚里能撑船】broad-minded; The prime minister's heart is big enough to sail a boat in.

载 ①(年) year: 数～ several years / 一年半～ a year or so; in about a year ②(记载) put down in writing; record: ～入史册 go down in history

崽 [方] ①(儿子) son ②(幼小动物) young animal; whelp
【崽子】(骂人的话) whelp; bastard: 狗～ son of a bitch / 兔～ brat

仔 [方] ①(儿子) son ②(幼小动物) young animal; whelp: 下～ give birth to whelps

zài

载 ①(装载) carry; hold; be loaded with: ～客五十人 carry fifty passengers / 船上～满了

粮食。 The ship is fully loaded with grain. ②(充满道路) all over the road; everywhere along the way: 怨声～道 complaints are heard everywhere ③(又)and; as well as; at the same time: ～笑～言 talking and laughing at the same time

【载波】signal carrier; carrier wave; carrier: 三路～ three-channel carrier ◇ ～传输 carrier transmission / ～电报 carrier telegraphy / ～电话 carrier telephony / ～电流 carrier current / ～抑制 carrier suppression / ～再生 carrier regeneration.

【载歌载舞】festively singing and dancing

【载荷】load; lading ◇ ～变化 load fluctuation / ～装置 load device

【载货】carry cargo ◇ ～吨位 cargo tonnage / ～甲板 cargo deck / ～容积 cargo carrying capacity

【载频】[无] carrier frequency ◇ ～放大器 carrier amplifier / ～振荡器 carrier oscillator

【载人飞行器】manned vehicle

【载人航天飞行】manned space flight

【载体】[化] carrier: 催化剂～ catalyst carrier

【载誉归来】return home, enjoying tremendous popularity; come back winning high praise

【载运】convey by vehicles, ships, etc.; transport; carry

【载重】load; carrying capacity: 这艘船～多少? What's the carrying capacity of this ship? ◇ ～吨位 deadweight tonnage / ～量 loading capacity; deadweight capacity / ～轮胎 high-capacity tire / ～汽车 truck; lorry / ～线 load line; load waterline; Plimsoll line

【载舟之水足以覆舟】The water that bears the boat is the same that swallows it.

再 [副] ①(表示又一次) another time; again; once more: 一而～, ～而三 again and again; repeatedly / 把课文～读一遍。 Read the text once again. ②(表示更加)请～大声一点。 Still louder, please. ③(表示如果继续怎样)你～不走,就要迟到了。 You'll be late if you don't start now. ④(表示一个动作发生在另一个动作结束之后) then; before: 你仔细想想之后～回答。 Think carefully before you answer. ⑤(表示另外有所补充) ～不然 if not; or else; otherwise / ～则 moreover; furthermore; besides ⑥(再继续;再现) come back; return: 良机难～。 Opportunity knocks but once. / 青春不～。 One's youth never returns.

【再版】①(第二次出版) second edition ②(第二次印刷) reprint; second impression

【再保险】re-insurance ◇ ～人 reinsurer

【再不】[口] or else; or: 你可以领一套新工具,～就用我的吧。 You can get a new set of tools, or else you may use mine for the time being.

【再出口】[商] reexport: ～贸易 reexport trade

【再次】once more; a second time; once again: ～声明 restatement; state again

【再度】once more; a second time; once again: ～当选 be reelected / ～修订计划 revise the plan a second time

【再犯】repeat an offense

【再会】[套] good-bye; see you again

【再婚】remarry; marry again; deuterogamy; digamy

【再嫁】remarry

【再见】[套] good-bye; see you again

【再接再励】make persistent efforts; continue to exert oneself; work ceaselessly and unremittingly

【再进口】[商] reimport

【再拘捕】[法] rearrest

【再拘留】[法] rearrest

【再控告】[法] recharge

【再起】recurrence; resurgence; revival: 防止边境冲突～ prevent the recurrence of border clashes

【再三】over and over again; time and again; again and again; repeatedly: ～考虑 consider over and over again / ～嘱咐 bid or tell again and again

【再审】①(重新审查) examine again; review ②[法] retrial; new trial; reopening of a case; rehear

【再生】①(死而复生) revive; rise again ②[生] regeneration ③(回收再制) reprocess; regenerate ◇ ～稻 ratooning rice / ～检波器 [无] regenerative detector / ～纤维素 regenerated cellulose / ～橡胶 reclaimed rubber

【再生产】[经] reproduction: 简单～ simple reproduction / 扩大～ reproduction on an extended scale; extended reproduction / ～投资 reproductive investment of capital / ～周期 cycle of reproduction

【再生父母】one's great benefactor; one's savior

【再生障碍性贫血】aplastic anemia

【再生之德】the grace of rebirth; one's grateful acknowledgment

【再衰三竭】be nearing exhaustion; be weakened and demoralized

【再思有益】Second thoughts are best.

【再贴现】[经] rediscount; retrocession ◇～率 rediscount rate
【再现】reappear; be reproduced: ～在我脑海中 recur to my mind / ～在眼前 reappear before one's eyes
【再者】[书] moreover; furthermore; besides

在 ①(存在;生存) exist; be living: 他的父母都还健～。 His parents are still living and in good health. / 英灵永～。 The spirits of the brave departed will be immortal. ②(表示人或物的位置) ～山顶上 at the top of the mountain / 挂～墙上 hang on the wall ③(参加;属于) join or belong to an organization; be a member of an organization ④[介](表示时间、地点、情形、范围等)事情发生～上周。 It happened last week. ⑤[副](表示动作正在进行)他～工作。 He is at work. / 他们～打篮球。 They are playing basketball. ⑥(在于;决定于) rest with; depend on: 事～人为。 Human effort is the decisive factor.
【在案】be on record: 记录～ be put on record; be a matter of record
【在编人员】permanent staff; those on the regular payroll
【在场】be on the scene; be on the spot; be present; presence ◇～从犯 accessory at the plot; accessory during the fact
【在行】be expert at sth.; know a job, trade, etc. well: 这方面我不～。 I'm not expert at that.
【在乎】①(在意) care about; mind; take to heart: 满不～ not care a bit ②(在于) lie in; rest with
【在即】near at hand; shortly; soon: 完工～ will soon be completed; be nearing completion
【在家】①(在家里) be at home; be in: ～靠父母,出门靠朋友。 At home one relies on one's parents and outside on one's friends. ②[宗] remain a layman ◇～人[宗] layman
【在教】be a believer (in a religion, esp. Islam)
【在劫难逃】if you're doomed, you're doomed: there's no escape
【在理】reasonable; sensible; right: 他的要求～。 He is reasonable in his demands. / 这话说得～。 That's a perfectly reasonable statement.
【在世】be living: 他～的时候 in his lifetime
【在所不辞】will not refuse under any circumstances; will not hesitate to
【在所不惜】will not grudge; will never balk at
【在所难免】can hardly be avoided; be unavoidable
【在逃】[法] on the run; has escaped; be at large ◇～被告 absconding defendant / ～犯 criminal at large; escaped criminal
【在望】①(可以望见) be visible; be in sight; be in view: 山头隐隐～。 The mountain top was dimly visible. ②(即将到来) will soon materialize; be in sight; be in the offing: 丰收～。 A good harvest is in sight.
【在位】[旧] be on the throne; reign
【在握】be in one's hands; be within one's grasp; be under one's control: 大权～ with power in one's hands
【在先】formerly; in the past; before
【在线计算机】line computer
【在心】feel concerned; mind; be attentive
【在押】[法] behind bars; be under detention; be in custody; be in prison ◇～犯 criminal in custody; prisoner
【在野】not be in office; be out of office ◇～党 a party not in office
【在意】take notice of; care about; mind; take to heart: 这些小事他是不会～的。 He won't take such trifles to heart.
【在于】①(事物的本质所在) lie in; rest with: 事物发展的根本原因～其内部矛盾。 The fundamental cause of the development of a thing lies in its internal contradictoriness. ②(决定于) be determined by; depend on: 去不去,～你。 It rests with you to decide whether to go or not.
【在职】be on the job; be at one's post: ～期间 during one's tenure of office ◇～干部 cadres at their posts / ～培训 on-the-job training
【在座】be present (at a meeting, banquet, etc.): 所有～的同志 all comrades present here / 有客人～,她没好意思说。 She was too embarrassed to bring it up in front of the guest.

zān

糌
【糌粑】*zanba*, roasted *qingke* barley flour

簪 ①(簪子) hairpin: 碧玉～ emerald hairpin ②(插在头发上) wear in one's hair: ～花 wear flowers in one's hair
【簪子】hair clasp

zán

咱 ①(我们) we (including both the speaker and the person or persons spoken to) ②[方] I:

～是工人。 I am a worker.

【咱们】①(我们) we: ～商量一下。 Let's talk it over. / ～一块去。 Let's go together. ②[方] I: ～是矿工。 I am a coal miner.

zǎn

趱 ①(赶) hurry through: 紧～了一程 rush through one part of the journey ②(快走) urge; hasten: ～马向前 urge on a horse

攒 accumulate; hoard; save: ～钱 save up money / 他～钱要买辆摩托车。 He save money for a motorcycle.

zàn

暂 ①(时间短) of short duration: 短～ of short duration; brief ②(暂时) temporary; for the time being; for the moment: ～代 act for sb. / ～停营业 business suspended / ～住 stay temporarily (at a place)

【暂定】arranged for the time being; tentative; provisional: ～议程 tentative agenda / ～办法 provisional measures

【暂缓】postpone; put off; defer: ～答复 put off replying / ～作出决定 defer making a decision

【暂记帐】suspense account

【暂且】for the time being; for the moment: ～就这样吧。 That will do for the present. / 这件事我们～不考虑。 We wouldn't consider the matter.

【暂缺】(of a post) be left vacant for the time being; (of a commodity) be out of stock at the moment

【暂时】temporary; transient: ～的安排 provisional arrangement / ～的决定 tentative decision / ～的需要 temporary needs / ～现象 transient phenomenon

【暂停】①(短暂停止) suspend: ～付款 suspend payment / 谈判～。 Negotiations are suspended for the present. ②[体] time-out: 要求～ ask for time-out

【暂停起诉】suspend proceedings

【暂行】provisional; temporary: ～拘留 detain provisionally ◇ ～办法 interim procedures / ～法令 interim order / ～规定 temporary provisions / ～条例 provisional regulations / ～条文 transitional provisions

【暂牙】temporary teeth

錾 ①(凿;刻) engrave on gold or silver; carve; chisel: ～花 carve flowers or patterns / 在金属板上～字 engrave characters on a metal plate ②(凿子) engraving tool; chisel

【錾刀】(engraver's) burin; graver; engraving tool

【錾子】chisel

赞 ①(帮助) support; favor; assist: ～助 support; assistance ②(称赞) praise; commend: 盛～ highly praise ③(旧时文体之一) eulogy

【赞比亚】Zambia ◇ ～人 Zambian

【赞不绝口】be profuse in praise; be full of praise

【赞成】approve of; favor; agree with; endorse: ～意见 assenting views / ～政府的政策 approve the policies of the administration / 十票～，六票反对 ten votes for and six against / 投票～ cast an affirmative vote / 她父亲永远不会～她嫁给这样的懒人。 Her father will never approve of her marrying such a lazy fellow. ◇ ～票 affirmative vote

【赞歌】song of praise; paean

【赞美】praise; eulogize: ～四个现代化 sing the praises of the four modernizations ◇ ～诗[基督教] hymn; canticle; chant

【赞赏】appreciate; admire: 对这一友好行动表示～ express appreciation for this friendly act / 非常～她的才能 highly appreciate her talent

【赞颂】extol; eulogize; sing the praises of

【赞叹】gasp in admiration; highly praise: 令读者～不已 evoke muck admiration of the readers

【赞叹不已】praise somebody greatly; be full of praise

【赞同】approve of; agree with; endorse: 这一主张得到普遍的～。 This proposition met with general approval

【赞许】speak favorably of; praise; commend: 值得～ deserve commendation; be commendable; be worthy of praise / 他的作品受到读者的～。 His works won the enthusiastic approval of his readers.

【赞扬】speak highly of; praise; commend: ～声中不骄傲 be free from haughty amid a shower of praises / 公开～ sing sb.'s praise / 热情～技术革新的成果 warmly praise the success in technical innovation

【赞语】words of praise; praise

【赞助】support; assistance: ～某人 support sb.; give support to sb.

zāng

赃 ①(赃物) stolen goods; booty; spoils: 分～ share the booty; divide the spoils / 退～ return one's booty ②(贿赂) bribes: 贪～ take bribes; practice graft / 窝～ harbor stolen goods / 销～ disposal of stolen goods / 贪～枉法 take bribes and bend the law; pervert justice for a bribe

【赃官】corrupt official

【赃款】money stolen, embezzled or received in bribes; illicit money; filthy lucre

【赃物】①(盗窃来的东西) stolen goods; booty; spoils ②(受贿之物) bribes ◇～窝藏人 holder of stolen goods

脏 dirty; soiled; filthy: ～衣服 dirty clothes; dirty linen / ～水 filthy water; slops, sewage / 为了不弄～手，他带手套。He wore gloves in order not to dirty his hands. / 一些城镇里街道特别～。 In some cities and towns the streets are filthy.

【脏土】dust; garbage

【脏字】obscene word; swearword; dirty word: 话别带～。Don't swear.

臧 good; right

【臧否】[书] pass judgment (on people): ～人物 pass judgment on people

zàng

葬 bury; inter: 火～ cremation / 她～在公墓里。 She is buried in the cemetery.

【葬礼】funeral rites; funeral: 参加～ attend a funeral

【葬身】be buried: ～鱼腹 become fish food; be swept to a watery grave; be drowned / 死无～之地 die without a burial place; come to a bad end

【葬送】ruin; spell an end to: ～前途 ruin one's future

藏 ①(储存大量东西的地方) storing place; depository: 宝～ precious (mineral) deposits ②(佛教、道教经典的总称) Buddhist or Taoist scriptures: 道～ Taoist scriptures ③(西藏) Tibet

【藏红花】①[植] saffron crocus ②[中药] saffron

【藏蓝】purplish blue

【藏青】dark blue

脏 internal organs of the body: 肝～ liver / 脾～ spleen / 肾～ kidneys / 心～ heart

【脏壁层】splanchnic layer

【脏腑】[中医] internal organs; viscera ○心 heart / 肝 liver / 脾 spleen / 肺 lungs / 肾 kidneys / 胃 stomach / 胆 gall / 大肠 large intestine / 小肠 small intestine / 膀胱 bladder

【脏器】internal organs of the body; viscera

zāo

糟 ①(酒糟) distillers' grains; grains ②(用酒、糟腌制食物) be pickled with grains or in wine: ～鱼 fish pickled with grains or in wine; pickled fish ③(不结实) rotten; poor: 她身体很～。 She is in very poor health. / 这木头～了。The wood is rotten. ④(坏) in a wretched state; in a mess: ～得很! It's terrible! / 你把事情弄～了。You're made a mess of the job.

【糟糕】[口] how terrible; what bad luck; too bad: ～透顶 rotten to the core; rotten all the way through

【糟行】brewery; distillery; winery

【糟化石】[矿] saussurite

【糟化作用】saussuritization

【糟糠】poor men's foodstuffs; distillers' grains, husks, chaff, etc. ◇ ～之妻 wife who has shared her husband's hard lot

【糟粕】waste matter; dross; dregs: 剔除～，吸取精华 reject the dross and assimilate the essence

【糟蹋】①(浪费;损坏) waste; ruin; spoil: 不要～粮食。 Never waste grain.②(侮辱;蹂躏) insult; trample on; ravage; violate: ～妇女 violate a woman / 村子让敌人给～了。The villages were ravaged by the enemy.

【糟心】vexed; annoyed; dejected: 他因失败而～。He was vexed at his failure.

遭 ①(遇到) meet with (disaster, misfortune, etc.); suffer: ～难 meet with misfortune; suffer disaster / 惨～屠杀 be cruelly slaughtered ②[量](周) round: 走一～ make a trip ③[量](回;次) time; turn: 一～生 两～熟。 Strangers at first meeting become familiar at the next.

【遭到】suffer; meet with; encounter: ～反对 meet with opposition / ～拒绝 meet with refusal; be turned down / ～迫害 suffer persecution / ～失败 suffer defeat

【遭逢】meet with; come across; encounter: ～不幸 suffer misfortune / ～盛世 live in prosperous times

【遭劫】meet with catastrophe

【遭受】suffer; be subjected to; sustain: ～不白之冤 suffer an unrighted wrong / ～水灾 be hit by floods / ～损失 sustain losses

【遭殃】suffer disaster; suffer

【遭遇】①(碰上;遇到) meet with; encounter; run up against: ～不幸 meet with misfortune; have hard luck ②(遇到的事情) (bitter) experience; (hard) lot: 咱们有着共同的历史～。We have shared the same historical experiences. ◇～战 meeting engagement; encounter (action); contact battle

【遭灾】be hit by a natural calamity

záo

凿 ①(凿子) chisel ②(打孔;挖掘) cut a hole; chisel or dig: ～孔 bore a hole / ～山劈岭 tunnel through mountains and cut across ridges

【凿冰】make a hole in the ice

【凿缝】staking

【凿环】shearing bushing

【凿井】①(挖井;掘井) dig a well ②[矿] shaft sinking; pit sinking: 冻结法～ freeze sinking

【凿密】[机] caulking

【凿岩】[矿] (rock) drilling ◇～机 rock drill; quarrying machine

【凿子】chisel

zǎo

枣 jujube; (Chinese) date

【枣红】purplish red; claret

【枣泥】jujube paste

【枣树】jujube tree

【枣椰】date palm

早 ①(早晨) (early) morning: 从～到晚 from morning till night / 清～ early in the morning ②(很久以前) long ago; as early as; for a long time: 她～走了。She left long ago. ③(比一定的时间靠前) early; in advance; beforehand: ～作准备 get prepared in advance / 请尽～答复。Please reply at your earliest convenience. ④[套](问候) good morning: 先生～! Good morning, sir!

【早安】good morning

【早班】morning shift

【早餐】breakfast

【早操】morning (setting-up) exercises

【早产】[医] premature birth; premature labor; premature delivery

【早产儿】premature; premature infant

【早场】morning show; morning performance

【早车】morning train or coach

【早晨】(early) morning

【早春】early spring

【早稻】early (season) rice ◇～品种 varieties of early rice

【早点】(light) breakfast

【早发性痴呆】dementia praecox

【早饭】breakfast: 吃～ have a breakfast

【早寒武世】Lower Cambrian

【早花】[农] early blossoming

【早婚】marrying too early; early marriage

【早年】one's early years: 他～是农民。He was a peasant in his early years.

【早期】early stage; early phase: ～胃癌 early stage carcinoma ventriculi / ～症状 early symptom / ～作品 sb.'s early works; the works of sb.'s earlier period

【早期警报系统】early warning system

【早日】at an early date; early; soon: ～完工 complete the project as soon as possible / 请～答复。Your early reply is requested.

【早上】(early) morning

【早熟】①[生理] precocity: ～的孩子 a precocious child ②(生长期短、成熟快) early-maturing; early-ripe ◇～品种 early-maturing variety; early variety / ～作物 early-maturing crop; early crop

【早衰】[医] premature senility; early ageing

【早霜】early frost

【早退】leave earlier than one should; leave early

【早晚】①(早晨和晚上) morning and evening: ～各服一丸 take one pill in the morning and one in the evening ②(或早或晚) sooner or later: 他～得去。He'll have to go there sooner or later. ③[方](将来某个时候) some time in the future; some day ◇～服务部 before-and-after-hours shop

【早先】previously; in the past: ～这里是一片不毛之地。The place used to be a stretch of barren land.

【早泄】[医] premature ejaculation

【早已】long ago; for a long time: 我～告诉他们了。I told them that long ago.

【早知今日,悔不当初】If I'd known then what was going to happen, I wouldn't have acted as I did. (或: If you had known this day must come you would have repented long since of your evildoings.)

澡 bath: 洗～ take a bath; bathe

【澡盆】bathtub
【澡堂】public baths; bathhouse
【澡塘】common bathing pool (in a bathhouse); public baths

藻 ①(藻类植物) algae ②(水中绿色植物) aquatic plants ③(词藻) literary embellishment: 辞～ ornate diction
【藻红素】phycoerythrobilin
【藻井】[建] sunk panel; caisson ceiling
【藻类学家】algologist
【藻类植物】algae

蚤 flea: 沙～ beach flea

zào

灶 ①(生火、做饭的设备) kitchen range; cooking stove: 电～ electric stove / 煤气～ gas range ②(厨房) kitchen; mess; canteen: 学生～ students' dining room
【灶间】kitchen
【灶神】kitchen god
【灶台】the top of a kitchen range
【灶性感染】focal infection

燥 dry: ～热 hot and dry

噪 ①(鸟、虫叫) chirp: 蝉～ the chirping of cicadas ②(大声叫嚷) a confusion of voices: 鼓～ make an uproar; clamor / 名～一时 be a celebrity for a time
【噪鹃】Chinese koel
【噪声】noise; buzz; roar; rumble ◇ ～测试 noise testing / ～计 psophometer / ～污染 noise pollution
【噪音】noise; undesired sound: 低～马达 low-noise motor ◇ ～污染 sound pollution / ～抑制[无] noise suppression

躁 rash; impetuous; restless: 烦～ restless; fretful / 戒骄戒～ guard against arrogance and rashness
【躁动】move restlessly
【躁急】restless; uneasy
【躁狂】[医] mania ◇ ～者 maniac / ～症 mania / ～状态 manic state

造 ①(做;制作) make; build; create: ～表 draw up a form or list / ～房子 build a house / ～预算 make a budget ②(假编;捏造) invent; cook up; concoct: ～假帐 cook accounts / 捏～ fabricate; concoct ③(成就) train; educate: 深～ pursue advanced studies ④[书](前往;到) go to; arrive at: ～访 pay a visit ⑤(诉讼的两方) one of the two parties in a legal agreement or a lawsuit: 两～ both parties ⑥[方](收成或收成次数) crop: 晚～ late crops / 一年三～ three crops a year / 早～ early crops
【造币厂】a mint
【造成】create; cause; give rise to; bring about: ～混乱 cause confusion / ～假象 put up a facade; create a false impression / ～巨大损失 cause enormous losses
【造船】build a ship; shipbuilding ◇ ～厂 shipyard; dockyard / ～工业 shipbuilding industry
【造次】[书] ①(匆忙) hurried; hasty: ～之间 in one's hurry; in a moment of haste ②(鲁莽) rash; impetuous: ～行事 act rashly
【造反】rise in rebellion; rebel; revolt
【造访】[书] pay a visit; call on: 登门～ call at sb.'s house; pay sb. a visit
【造福】bring benefit to; benefit: 为后代～ benefit future generations / 为人民～ bring benefit to the people
【造化】①(运气;命运) good fortune; good luck: 有～ be born under a lucky star; be lucky ②(自然界的创造者) creator
【造价】cost (of building or manufacture)
【造就】①(培养使有成就) bring up; train: ～一代新人 bring up a new generation ②(造诣;成就) achievements; attainments
【造句】sentence-making
【造块】[冶] agglomeration
【造林】afforestation ◇ ～面积 afforestation area / ～学 silviculture
【造陆运动】[地] epeirogenic movement; epeirogeny; epeirogenesis
【造孽】[佛教] do evil; commit a sin
【造山运动】[地] orogenic movement; orogeny; orogenesis
【造物】the divine force that created the universe; Nature ◇ ～主[基督教] God; the Creator
【造像】[美术] statue
【造型】①(创造物体形象) modelling; moldmaking: ～艺术 plastic arts ②(创造出来的物体形象) model; mold ③[机] molding ◇ ～板[机] mold board / ～材料 molding material / ～压力 molding pressure
【造血】hematopoiesis; hemopoiesis ◇ ～器官 blood forming organ / ～系统 hemopoietic system / ～障碍性贫血 hypoproliferative

anemia
【造谣】cook up a story and spread it around; start a rumor: ～生事 start a rumor to create trouble; stir up trouble by rumormongering ◇～者 rumormonger / ～中伤 spread rumors to injure others' reputation
【造诣】(academic or artistic) attainments: ～很高 of great attainments
【造影】[医] radiography: 支气管～ bronchography
【造纸】papermaking ◇～厂 paper mill / ～机 paper machine
【造作】①(制造) make; manufacture ②(做作) affected; artificial

皂 ①(黑色) black ②(肥皂) soap: 香～ toilet soap / 药～ medicated soap
【皂白】black and white; right and wrong: ～不分 make no distinction between right and wrong
【皂化】[化] saponification ◇～剂 saponifier / ～作用 saponification
【皂荚】[植] Chinese honey locust
【皂石】bowlingite; soaprock
【皂素】saponin
【皂性】nature of soap
【皂洗机】[纺] soaper: 平幅～ open soaper
【皂质乳液】soap emulsion

zé

责 ①(责任) duty; responsibility: 负全～ assume full responsibility / 保卫祖国人人有～。Every one is duty-bound to defend his motherland. ②(要求达到标准) demand; require: 求全～备 demand perfection; nitpick / 严以～己,宽以待人 be strict with oneself and broad-minded towards others ③(责备) blame; reproach; reprove: 斥～ reprimand; rebuke; denounce ④(诘问) question closely; call sb. to account
【责备】reproach; blame; reprove; take sb. to task: 受到良心的～ feel a prick of conscience
【责成】instruct (sb. to fulfill a task); charge (sb. with a task); enjoin (sb. to do sth.): ～有关部门如期完成任务 instruct the departments concerned to fulfill their tasks according to schedule
【责罚】punish
【责怪】blame: ～他太粗心 blame him for his carelessness
【责令】order; instruct; charge
【责骂】scold; rebuke; dress down
【责难】censure; blame: 受到各方面的～ incur censure from various quarters
【责任】①(应做的事) duty; responsibility: ～重大 have a grave responsibility / 负完全～ assume full responsibility ②(应承担的过失) responsibility for a fault or wrong; blame: 追究～ ascertain where the responsibility lies ◇～保险承保人 liability insurer / ～保险费 liability insurance premium / ～范围 extent of liability / ～感 sense of responsibility / ～恐怖[心理] hypengyophobia / ～能力 capacity for duties / ～年龄 age of responsibility / ～事故 liability accident / ～条款 liability clause / ～心 sense of responsibility
【责任制】system of job responsibility: 联产承包～ output-related system of contracted responsibilities / 农业生产～ farm production responsibility system
【责问】call sb. to account
【责无旁贷】there is no shirking the responsibility; be duty-bound
【责有攸归】responsibility rests where it belongs

啧
【啧有烦言】there are a lot of complaints
【啧啧】①(咂嘴的声音) click of the tongue: ～称羡 click the tongue in admiration / ～叹赏 be profuse in one's praise ②(形容说话声)人言～。 There is a good deal of unfavorable comment.

则 ①(规范) standard; norm; criterion: 以身作～ set an example by one's own conduct / 准～ criterion; standard ②(规则) rule; regulation: 法～ law; rule / 总～ general rule ③[量] 新闻一～ an item of news / 寓言四～ four fables ④(表示因果条件等)有～改之,无～加勉。Correct mistakes if you have made any guard against them if you have not. ⑤(表示转折、对比等)今～不然。 However, things are quite different today ⑥[书](用于列举原因或理由)这篇故事不合适,一～太长,二～太乏味。 This story is unsuitable. For one thing it's too long, for another it's too dull.
【则声】make a sound; utter a word: 不～ keep silent / 没有人～。 None uttered a word.

泽 ①(聚水的地方) pool; pond: 湖～ lakes / 沼～ marsh; swamp ②(湿) damp; moist: 润～ moist; wet ③(光) luster; gloss: 光～ luster;

gloss; sheen / 色～ color and luster
【泽被万世】One's graces reached down to many generations.
【泽国】[书] ①(河流、湖泊多的地区) a land that abounds in rivers and lakes ②(受水淹的地区) inundated area: 尽成～ a whole area became submerged
【泽及亲仇】One's charity took in both friend and foe.

择 select; choose; pick: 不～手段 by hook or by crook; unscrupulously / 二者任～其一 choose either of the two
【择伐】[林] selective cutting
【择吉】[旧] pick an auspicious day (for a marriage, funeral, etc.)
【择交】choose friends: 慎重～ choose friends with care
【择日启程】fix a departure date
【择善而从】rate of increase; growth rate good
【择优定向】[地] preferred orientation
【择优录取】enroll only those who are outstanding; anyone who made a good enough score can be accepted
【择友】choose friends

zè

仄 ①(狭窄) narrow: 逼～ narrow; cramped ②[语] oblique tones
【仄声】[语] oblique tones (in classical Chinese pronunciation) ○上声 the falling-rising tone / 去声 the falling tone / 入声 the entering tone

zéi

贼 ①(偷东西的人) thief ②(做大坏事的人) traitor; enemy: 卖国～ traitor ③(邪的;不正派的) crooked; wicked; evil; furtive: ～眼 shifty eyes; furtive glance ④(狡猾) crafty; sly; cunning; deceitful: 老鼠真～。 Rats are really cunning. ⑤[方](很) extremely; disagreeably: ～冷 terribly cold / ～亮 disagreeably glossy or dazzling
【贼船】pirate ship: 上～ board the pirate ship; join a reactionary faction
【贼喊捉贼】a thief crying "Stop thief"
【贼眉鼠眼】shifty-eyed; thievish-looking
【贼去关门】lock the door after the thief has gone
【贼头贼脑】behaving stealthily like a thief; stealthy; furtive
【贼心】wicked heart; evil designs; evil intentions
【贼心不死】refuse to give up one's evil designs
【贼星】[口] meteor
【贼赃】stolen goods; booty; spoils

zěn

怎 [方] why; how: 你～不早点告诉他们呀? Why didn't you tell them earlier?
【怎么】①(询问性质、状况、方式、原因等)～办? What's to be done? / 这是～回事? What's all this about? ②(泛指性质或方式)你愿意～办就～。 Do as you please. ③(用于否定式,表示程度不够)这地方我不～熟悉。 I don't know much about the place.
【怎么得了】where will it all end; what a terrible thing it would be; this is one hell of a mess
【怎么样】①(询问性质、状况、方式等)你觉得这场演出～? How do you like the performance? ②(委婉说法,用于否定式)她绘画不～。 She's not much of a painter.
【怎样】①(询问性质、状况、方式等)这是～做的? How is it done? ②(泛指性质、状况或方式)想想从前～,再看看现在～ think of the past and look at the present

zèn

谮 [书] falsely charge; slander; calumniate: ～言 slander; calumny

zēng

曾
【曾孙】great-grandson
【曾孙女】great-granddaughter
【曾祖】(paternal) great-grandfather
【曾祖母】(paternal) great-grandmother

憎 hate; detest; abhor: 爱～分明 be clear about what to love and what to hate / 面目可～ repulsive in appearance
【憎称】derogatory name for sb. one hates or loathes
【憎恨】hate; detest
【憎恶】abhor; loathe; abominate

增 increase; gain; add: ～兵 throw in reinforcements / 与日俱～ grow with each passing day; be on the increase
【增白剂】[化] brightening agent; brightener
【增补】augment; supplement: ～人员 augment the staff ◇ ～本 enlarged edition / ～率 supplementary rate

【增产】increase production：～幅度 amount of increase in production／～节约 increase production and practice economy／挖掘～潜力 tap latent potentialities for increasing production
【增产率】rate of growth：几何级数～ geometric increase rate／年～ annual growth rate／平均年～ average annual rate of growth／算术级数～ arithmetic increase rate
【增订】revise and enlarge (a book)：～本 revised and enlarged edition
【增多】grow in number or quantity；increase：学生人数～。 The number of students increased.
【增光】add luster to；do credit to；add to the prestige of：为国～ do credit to one's country；为祖国～ win honors for the motherland
【增加】increase；raise；add：～抵抗力 build up one's resistance to disease／～复种面积 extend the area of double or triple cropping；enlarge the multiple-cropping area／～工资 get a raise in pay／～固定资产 additions／～积累 increase accumulation／～困难 add to the difficulties；multiply the difficulties／～生产 increase of production／～收入 increase income／～体重 put on weight
【增进】enhance；promote；further：～健康 improve one's health／～食欲 whet one's appetite／～友谊 promote friendship
【增刊】supplement (to a newspaper or periodical)；supplementary issue
【增量】[数] increment：～方式 incremental mode
【增强】strengthen；heighten；enhance：～斗志 raise one's morale／～防御力量 strengthen defense／～免疫力 develop immunity from disease／～人民体质 build up the people's health／～信心 heighten one's confidence ◇ ～塑料 reinforced plastics
【增删】additions and deletions；emendation
【增生】[医] hyperplasia；proliferation；multiplication
【增塑剂】[化] plasticizer；plastifier
【增添】add；increase：～麻烦 cause more trouble／～设备 get additional equipment
【增效】synergism；synergia；synergy ◇ ～剂 [化] synergist
【增压】pressure boost ◇ ～表 boost gauge／～法 supercharge method／～能量 pressurization energy
【增益】[电] gain：高～ high gain／分贝～ decibel gain ◇ ～放大 gain amplification／～控制 gain control／～衰减 gain reduction／～余量 gain margin
【增音机】[讯] repeater
【增音器】amplifier
【增援】[军] reinforce ◇ ～部队 reinforcements；reinforcing units
【增长】increase；rise；grow：～才干 enhance one's abilities／～三倍 increase by 300%／～知识 broaden one's knowledge／控制人口～ control population growth ◇ ～率 rate of increase；growth rate／～系数 growth factor
【增值】[经] rise in value；appreciation；increment：自然～ unearned increment ◇ ～保险 increased value insurance／～通货 appreciated currency
【增殖】①[医](增生) hyperplasia；proliferation；multiplication：细胞～ proliferation of cells ②(繁殖) breed；reproduce；multiply；propagate ◇ ～率 [牧] rate of increase／～体 vegetation／～腺 adenoids

zèng

甑 ①(古代炊具) an ancient earthen utensil for steaming rice ②(甑子) rice steamer ③(蒸馏或分解用的器皿) a utensil for distilling water, etc.：曲颈～ retort
【甑子】rice steamer

赠 give as a present；present as a gift：～书 present sb with a book
【赠答】present each other with gifts, poems, etc.
【赠股】donated stock
【赠款】grant ◇ ～人 donor
【赠品】(complimentary) gift：giveaway
【赠券】a complimentary ticket
【赠送】give as a present；present as a gift：向演员～花篮 present a basket of flowers to the performers ◇ ～仪式 presentation ceremony
【赠言】words of advice or encouragement given to a friend at parting：临别～ parting words of advice or encouragement
【赠与】favor；gift；grant ◇ ～人 donor／～证书 deed of gift
【赠阅】(of a book, periodical, etc.) given free by the publisher ◇ ～本 complimentary copy

锃 [方] (of utensils, etc.) polished：～亮 shiny

综 [纺] heddle；heald：～框 heald frame

zhā

渣 ①(渣滓) dregs; sediment; residue: 豆腐～ soya-bean residue (after making bean curd)/ 炉～ slag; cinder/ 煤～ coal cinder/ 蔗～ bagasse ②(碎屑) broken bits: 面包～儿 (bread) crumbs
【渣车】slag car; slag wagon
【渣瘤】slag nodule
【渣眼】slag blister
【渣油】[石油] residuum; residual oil ◇～路 residual-oil road
【渣滓】dregs; sediment; residue: 溶液的～ dregs of a solution/ 社会～ dregs of society

揸 [方] ①(用手指撮东西) pick up sth. with the fingers ②(把手指伸开) spread one's fingers

喳 [象] 喜鹊的～～声 the chatter of magpies

扎 ①(刺) prick; run or stick (a needle, etc.) into: ～一刀 stab with a knife/ 我手指头上～了一个刺。I've got a prick in my finger. ②[方](钻进去) plunge into; get into: ～到人群里 dash into the crowd
【扎除】removal by ligature
【扎耳朵】[口] grate on the ear; be earpiercing
【扎根】take root: ～于群众之中 take root among the masses
【扎花】[口] embroider
【扎接】butting
【扎猛子】[方] dive
【扎实】①(结实) sturdy; strong ②(实在) solid; sound; down-to-earth: 工作很～ do a solid job
【扎手】①(刺手) prick the hand: 玫瑰有刺,小心～。The roses are thorny. Take care not to prick your hands. ②(事情难办) difficult to handle; thorny: 这事真～。This is really a hard nut to crack.
【扎眼】①(刺眼) dazzling; offending to the eye; loud; garish: ～的颜色 loud color; color unpleasant to the eye ②(惹人注意) offensively conspicuous
【扎伊尔】Zaire ◇～人 Zairian
【扎针】[中医] give〈have〉an acupuncture treatment

咋
【咋呼】【咋唬】[方] ①(吆喝) bluster; shout blusteringly ②(炫耀) show off; make a display (of one's authority or ability)

zhá

闸 ①(水闸) floodgate; sluice gate ②(把水截住) dam up water ③(制动器) brake: 踩～ step on the brake/ 紧急～ emergency brake/ 捏～ apply the hand brake/ 气～ air brake/ 制动～ damper brake ④(电闸) switch: 扳～ operate a switch; switch on or off
【闸盒】fuse-box
【闸流管】[电子] thyratron; gastriode relay
【闸轮】brake wheel
【闸门】①(水闸) sluice gate; gate; water gate ②(船闸) lock gate ③[机] throttle valve
【闸瓦】[机] brake shoe

铡 ①(铡刀) hand hay cutter; fodder chopper ②(用铡刀切) cut up with a hay cutter: ～猪草 chop fodder for pigs
【铡草机】hay cutter; chaffcutter; chaff slicer; straw breaker
【铡刀】hand hay cutter; fodder chopper; straw chopper

炸 fry in deep fat or oil; deep-fry: ～糕 fried cake/ ～烹大虾 grilled prawns in sauce/ ～肉丸子 fried meat balls/ ～鱼 deep-fried fish
【炸鸡】fried chicken
【炸酱】fried bean sauce ◇～面 noodles served with such sauce

扎 pitch (a tent, etc.)
【扎营】pitch a tent or camp; encamp

札 ①(古时写字用的木片) thin pieces of wood used for writing on in ancient China ②[书](信件) letter: 适奉大～。I have just received your esteemed letter.
【札记】reading notes

轧 roll (steel)
【轧钢】steel rolling ◇～机 rolling mill
【轧钢厂】steel rolling mill: 大型～ heavy rolling mill
【轧辊】[冶] roll; roller ◇～调整装置 roll adjusting device
【轧机】[冶] rolling mill
【轧件】rolled piece
【轧屑】mill scale
【轧制】[冶] rolling ◇～钢 rolled steel/ ～钢轨 roll out steel rail/ ～公差 rolling tolerance

zhǎ

眨 blink; wink: 眼睛一～ blink (one's eyes)

【眨巴】[方] blink

【眨眼】very short time; wink; twinkle: 一～就不见了 disappear in a twinkling; disappear in the twinkling of an eye

砟 tiny fragments of stone, coal, etc : 炉灰～儿 cinder

zhà

栅 railings; paling; bars: 木～ paling; palisade / 铁～ iron railings; metal rails; iron bars

【栅栏】①(铁条或木条制成的篱笆状物) railings; paling; bars ②[军] boom: ～网 boom nets

【栅门】fence gate

乍 ①(起初;刚开始) first; for the first time: ～一听 at first hearing / 新来～到 be a newcomer; be a new arrival ②(忽然) suddenly; abruptly: 天气～冷～热。 The temperature changes abruptly. ③(张开) spread; extend: ～翅 spread wings

【乍得】Chad ◇ ～人 Chadian

【乍暖还寒】After suddenly getting warmer, the weather has turned cold again.

痄

【痄腮】[医] mumps

炸 ①(突然破裂) explode; burst: 那个有毛病的锅炉～了。 The defective boiler has exploded. / 一颗炸弹在离我们站的地方几米处～了。A bomb burst a few meters a way from where we stood. ②(用炸药、炸弹爆破、轰炸) blow up; blast; bomb: ～毁 blow up; blast to pieces; demolish / ～沉 bomb and sink ③[口] (因愤怒而激烈发作) fly into a rage; flare up: 肺都气～了 flare up; explode with rage

【炸弹】bomb: 爆破～ demolition bomb / 常规～ conventional bomb / 催泪～ tear bomb / 定时～ time bomb / 毒气～ gas bomb / 高爆～ high capacity bomb / 燃烧～ incendiary bomb / 热核～ thermonuclear bomb / 杀伤～ antipersonnel bomb / 深水～ depth bomb; diving torpedo; depth charge / 塑料～ plastic bomb / 未爆～ unexploded bomb / 细菌～ germ bomb / 液体～ liquid bomb ◇ ～坑 bomb-crater; crater

【炸胶】[化] blasting gelatine

【炸雷】[方] a clap of thunder

【炸药】explosive (charges); dynamite: 甘油～ dynamite / 可塑性～ plastic explosive / 烈性～ high explosive ◇ ～包 pack of dynamite; explosive package; satchel charges

诈 ①(欺骗) cheat; swindle: 欺～ swindle; cheat / ～人钱财 cheat a man out of his money; swindle money out of a person ②(假装) pretend; feign: ～败 feign defeat / ～死 feign death; play dead ③(用假话试探) bluff sb. into giving information: 他是拿话～我。 He was trying to draw me out.

【诈病】malinger; malingering ◇ ～者 malingerer

【诈称】jactitation ◇ ～者 jactitator

【诈唬】bluff; bluster

【诈骗】defraud; swindle: ～财物 defraud money and property ◇ ～犯 swindler / ～者 defrauder / ～罪 crime of fraud

【诈欺】deception ◇ ～取财罪 obtain money or goods by false pretense / ～罪 crime of fraud

【诈降】pretend to surrender; feign surrender

【诈语】lie; falsehood; fabrication

蚱

【蚱蜢】grasshopper

榨 ①(压出汁液) press; extract: ～甘蔗 press sugar cane / ～油 extract oil ②(器具) a press for extracting juice, oil, etc.

【榨菜】hot pickled mustard tuber

【榨棉机】cotton press

【榨取】squeeze; extort

【榨油机】oil press

【榨汁机】juicer

zhāi

斋 ①(斋戒) fast ②(素食) vegetarian diet adopted for religious reasons: 长～ permanent abstention from meat, fish, etc. / 吃～ practice abstinence from meat; be a vegetarian for religious reasons / 清～ fast; abstinence from meat ③(舍饭给僧人) give alms (to a monk) ④(屋子) room or building: 书～ study

【斋饭】vegetarian food

【斋戒】abstain from meat, wine, etc.; fast: 他们在某些天～。 They fast on certain days.

【斋戒节】Ramadan; the fast of Ramadan

【斋戒日】fast day

【斋戒沐浴】fast and ablution; fast and take a bath before a religious observance

【斋期】fast days; fast; Lenten
【斋月】[伊斯兰教] Ramadan; the month of fast

摘 ①(取) pick; pluck; take off: ～花 pluck flowers / ～苹果 pick apples ②(造取) select; make extracts from: ～译 translation of selected passages
【摘编】extract and compile
【摘抄】①(摘录) take passages; make extracts; extract; excerpt ②(摘录的文章) extracts; excerpts
【摘除】[医] excise; remove: ～腹部肿瘤 excise an abdominal tumor
【摘记】①(摘要记录) take notes: ～讲话要点 take down the gist of a speech ②(摘录) extracts; excerpts
【摘借】borrow money when in urgent need
【摘句成章】put sentences together in a composition
【摘录】①(选一部分记下来) take passages; make extracts; extract; excerpt ②(摘记的要点) extracts; excerpts: 文件～ extracts from a document
【摘要】①(摘录要点) make a summary: ～发表 publish excerpts of sth. ②(摘录下来的要点) summary; abstract: 谈话的～ the summary of a talk
【摘引】quote
【摘由】key extracts (of a document)

zhái

宅 residence; house
【宅邸】residence; abode; habitation
【宅门】①(深宅大院的大门) gate of an old-style big house ②(人家) family living in such a house
【宅院】a house with a courtyard; house

择 select; choose; pick: ～鸡毛 pick a chicken
【择不开】①(解不开) be unable to undo; past disentanglement: ～这团乱线 be unable to unravel the tangled threads ②(摆脱不开) cannot get away from: 一点儿工夫也～ be fully occupied; not have a moment to spare
【择菜】trim vegetables for cooking

zhǎi

窄 ①(不宽) narrow: ～胡同 a narrow lane ②(气量小) narrow-minded; petty; narrow: 心眼儿～ petty; oversensitive ③(生活不宽裕) hard up; badly off
【窄道】narrow path
【窄轨距】narrow gauge
【窄轨铁路】light railway; narrow- gauge railway
【窄片】[电影] narrow gauge; substandard film stock

zhài

寨 ①(栅栏) stockade ②(寨子) stockaded village ③(旧时驻兵的地方) camp: 营～ military camp / 安营扎～ pitch a camp; encamp ④(山寨) mountain stronghold
【寨子】stockaded village

债 debt: 还～ pay one's debt / 借～ borrow money / 内～ internal debt / 欠～ get into debt; be in debt / 外～ external debt / 负～容易还～难。It is easier to get into debt than to get out of debt.
【债户】debtor
【债款】loan ◇ ～收入 revenue from loans / ～支出 expenditure for loan payments
【债权】[法] creditor's rights; obligatory right ◇ ～国 creditor nation / ～人 creditor / ～人违约 default of creditor / ～诉讼 action in personam / ～证书 document of obligation
【债券】bond; debenture: 短期～ short-term liabilities / 偿还今年到期的～ redeem bonds for payment in this year / 他们热心购买政府～。They are eager to buy their government bonds. / 我把钱用于买～。My money is invested in debentures. ◇ ～持有者 bondholder; bond holder / ～股利 bond dividends / ～折价 bond discount
【债台高筑】be heavily in debt; be up to one's ears in debt; be debt-ridden
【债务】debt; liabilities; amount due ◇ ～额 amount of debt / ～国 debtor nation / ～纠纷 dispute over obligation / ～清结 performance of debt / ～人 debtor
【债项】amount due
【债主】creditor: ～权益 creditor's equity

zhān

占 practice divination: ～卦问卜 consult the oracle; inquire about by divination
【占卜】practice divination; divine
【占卦】divination
【占课】divine by tossing coins
【占梦】divine by interpreting dreams
【占星】divine by astrology; cast a horoscope ◇ ～术 astrology

沾 ①(浸湿) moisten; wet; soak: 泪～襟 tears wet the front of one's jacket ②(被附着上) be stained with: ～上了油 be stained with grease/～水 get wet ③(碰上;挨上) touch: 滴酒不～ not touch any alcoholic drinks ④(得到好处) get sth. out of association with sb. or sth.: 利益均～ share the benefit equally; have equal shares of the benefit

【沾边】①(略有接触) touch on only lightly: 这事他没～。He has nothing 〈sth.〉 to do with the matter. ②(接近应有的样子) be close to what it should be; be relevant: 你讲得一点也不～。What you said is wide of the mark.

【沾光】benefit from association with sb. or sth.

【沾花惹草】be promiscuous in sex relations

【沾亲带故】have ties of kinship or friendship

【沾染】be infected with; be contaminated by; be tainted with: ～恶习 slide into bad habits/伤口～了细菌。The wound was infected with germs. ◇～剂[军] contaminant/～监测器 contamination monitor/～区[军] contaminated area

【沾手】①(用于接触) touch with one's hand: 雪～就化。The snowflakes melt the moment they touch one's hand. ②(参与某事) have a hand in: 这事他也沾了手。He has a hand in the affair.

【沾沾自喜】feel complacent; be pleased with oneself: 不要～于一得之功。Don't be complacent over occasional success.

粘 glue; stick; paste: ～信封 stick down an envelope

【粘合剂】binder

【粘连】[医] adhesion ◇～现象 adhesion phenomenon

【粘贴】paste; stick: ～标语 paste up slogans

毡 felt: ～帽 felt hat

【毡衬】blanket; feltless packing

【毡垫】felt pan; feltless pad

【毡垫圈】felt; felt washer; feltless ring

【毡房】yurt

【毡子】felt; felt rug; felt blanket

谵 rave; be delirious

【谵妄】[医] delirium ◇～者 deliriant/～状态 delirious state

【谵语】[书] delirious speech; wild talk; ravings

瞻 look up or forward

【瞻顾】[书] look ahead and behind

【瞻念】look to; think of: ～前途 think of the future

【瞻前顾后】look ahead and behind; be overcautious and indecisive

【瞻望】look forward; look far ahead: ～未来 look to the future

【瞻仰】look at with reverence: ～遗容 pay respects to someone's remains

zhǎn

盏 ①(小杯子) small cup: 酒～ small wine cup ②[量] 一～电灯 an electric lamp

斩 ①(砍) chop; cut: 快刀～乱麻 cut the Gordian knot ②(杀头) behead; decapitate

【斩波】chopped wave ◇～放大器 chopper amplifier

【斩草除根】cut the weeds and dig up the roots; destroy root and branch; stamp out the source of trouble

【斩尽杀绝】kill all; wipe out the whole lot

【斩钉截铁】resolute and decisive; categorical: ～地回答 give a categorical answer

【斩假石】[建] artificial stone; imitation stone

【斩首】behead; decapitate: ～示众 cut off sb.'s head to display to the public

崭 ①[书](高峻;高出) towering (over) ②[方](优异) fine; swell

【崭露头角】cut a striking figure; stand out conspicuously; make oneself conspicuous

【崭新】brand-new; completely new: ～的机器 brand-new machine/～的阶段 a completely new stage

展 ①(张开) open up; spread out; unfold; unfurl: ～卷 open a book/愁眉不～ with a worried frown ②(施展) put to good use; give free play to: ～技 give full play to one's skill ③(展缓) postpone; extend; prolong: ～限 extend a time limit ④(展览) exhibition: 画～ painting exhibition/预～ preview

【展翅】spread the wings; get ready for flight: ～高飞 soar to great heights

【展出】put on display; be on show; exhibit

【展缓】postpone; extend; prolong: ～限期 extend the time limit

【展开】①(张开) spread out; unfold; open up: ～画卷 unfold a picture scroll ②(大规模地进行) launch; unfold; develop; carry out: ～攻势 unfold an offensive/～竞赛 launch an emulation drive ◇ ～法 method of development / ～角 angle of spread / ～面 developed surface

【展览】put on display; exhibit; show: 工业～ industrial exhibition / 花卉～ flower show / 美术～ art exhibition / 时装～ fashion display ◇ ～馆 exhibition center / ～会 exhibition / ～品 exhibit; item on display / ～室 exhibition room; showroom

【展品】exhibit; item on display: 请勿抚摸～。Please do not touch the exhibits.

【展平区】flattened region

【展期】①(往后推迟) extend a time limit; postpone: ～到月底截止 extend the deadline until the end of this month / 交易会～举行。The fair has been postponed. ②(展览的时期) duration of an exhibition; exhibition period

【展示】open up before one's eyes; reveal; show; lay bare: ～人物的内心世界 reveal a character's inner world

【展望】①(往远处看) look into the distance: 登上山顶向四周～ climb to the top of the mountain and get a view of the surrounding country ②(往将来看) look into the future; look ahead: ～未来 look forward to the future ③(对发展前途的预测) forecast; prospect: 二十一世纪～ prospects for the twenty-first century

【展现】unfold before one's eyes; emerge

【展性】[物] malleability

搌

搌 wipe or dab (with a soft dry object) to sop up liquid

【搌布】dishcloth; dish towel

辗

【辗转】①(身体翻来覆去) toss about (in bed): ～不能成眠 toss and turn (in bed); unable to go to sleep ②(经过许多人的手、许多地方) pass through many hands or places: ～流传 spread from place to place; pass through many places

【辗转不安】[医] jactitation

【辗转反侧】toss about (in bed); toss and turn restlessly

【辗转相告】pass from mouth to mouth

zhàn

湛

湛 ①(深) profound; deep: 精～ consummate; exquisite ②(清澈) crystal clear

【湛蓝】azure blue; azure

蘸

蘸 dip in: ～酱油 dip in sauce / ～墨水 dip in ink

【蘸火】[口] quenching

【蘸水钢笔】pen

【蘸液折射计】dipping refractometer

栈

栈 ①(栈房) warehouse: 货～ warehouse; storehouse ②(旅店) inn: 客～ inn ③(养牲畜的栅栏) shed; pen: 羊～ sheep pen

【栈道】a plank road built along the face of a cliff

【栈房】①(存放货物的地方) warehouse; storehouse ②[方](旅馆) inn

【栈桥】①(指港口的) landing stage ②(指火车站的) loading bridge

占

占 ①(占据) occupy; seize; take: 强～ forcibly occupy; seize / 这个建筑～了整整一个街区。The building occupies an entire block. ②(占有) constitute; hold; make up; account for: ～第一位 take the first place; rank first / ～多数 constitute the majority / ～少数 constitute the minority ③(占用) take (up): 这所房子～地不多。This building does not take up much land.

【占机信号】seizing signal

【占据】occupy; hold: ～有利地形 occupy a vantagepoint

【占领】capture; occupy; seize: ～城市 seize a city / ～要塞 capture a fort ◇ ～军 occupation army / ～区 occupied area / ～制度 occupation regime

【占便宜】①(取得额外利益) gain extra advantage by unfair means; profit at other people's expense: 占小便宜 gain petty advantages ②(有优越条件) advantageous; favorable: 你个子高，打排球～。In playing volleyball you have the advantage of being tall.

【占上风】gain the upper hand; have the advantage (over); come out top dog

【占先】take precedence; take the lead; get ahead of: 我们队在比赛中～。Our team led all the others in the contest.

【占线】[电话] the line's busy; busy line: ◇ ～通道[电] active channel / ～指示灯 busy lamp

【占用】occupancy; occupation ◇ ～公款 conversion of public money to one's own use

【占有】①(占据) own; possess; have: ～生产资料 have the means of production ②(处在) occupy; hold: 农业在国民经济中～重要地位。Agriculture occupies an important place in the national economy. ③(掌握) have; own: ～不动产 hold estate / ～财产 enter upon a property / ～第一手资料 have firsthand data ◇ ～期

间 occupancy / ～权 occupation; rights of possession / ～物 chose in possession / ～者 holder; possessor

【占着茅坑不拉屎】neither shit nor get off the pot; hold on to a post without doing any work and not let anyone else take over; be a dog in the manger

站 ①(站立) stand; be on one's feet; take a stand: 靠边～! Stand aside! / 请～起来。Stand up, please. / 他太虚弱了,～不起来。He was too weak to stand. ②(停留) stop; halt: 这车中途不～。This bus makes no stops along the way. ③(车站) station; stop: 公共汽车～ bus top; bus station / 火车～ railway station / 过～未停 go through the station without stopping / 每～都停 stop at every station ④(业务机构) station or center for rendering certain services: 保健～ health center / 服务～ service station

【站不住脚】unable to stand one's ground; cannot be justified; impossible to defend

【站得高,看得远】stand high and see far ahead; be farsighted

【站队】line up; fall in; stand in line

【站岗】stand guard; be on sentry duty; stand sentry: 站好最后一班岗 continue working hard till the last minute

【站柜台】serve as a shop assistant

【站立不能】[医] astasia

【站票】ticket for standing room ◇ ～观众 standee

【站台】platform (in a railway station) ◇ ～票 platform ticket

【站稳】①(停稳当) come to a stop: 等车～了再下。Don't get out until the car stops. ②(坚定) stand firm; take a firm stand: ～脚跟 get a firm foothold; stand firm

【站长】head of a station, center, etc.: 火车站～ stationmaster; station agent

【站住】①(停止行动) stop; halt: 谁? ～! Who's that? Don't move! ②(站稳) stand firmly on one's feet; keep one's feet: 他病刚好,虚弱得站不住。He has just been recovered from illness and is so weak that he can hardly keep his feet. ③(待下去) stand one's ground; consolidate one's position ④(理由成立) hold water; be tenable: 他的说法站不住。His opinion doesn't hold water.

【站住脚】①(停止行走) stop; halt ②(待在某地) stand one's ground; consolidate one's position ③(理由成立) hold water; be tenable: 这个论点站不住脚。This argument isn't tenable.

战 ①(战争) war; warfare; battle; fight: 模拟～ sham battle / 内～ civil war / 夜～ night fighting / 和…开～ make war on; wage war upon / 开～ go to war / 同…作～ be at war with; war with / 向某国宣～ declare war upon a country ②(进行战争、战斗) fight: ～而胜之 fight and defeat the enemy ③(发抖) shiver; tremble; shudder: 胆～心惊 tremble with fright / 冷得打～ shiver with cold

【战败】①(打败仗) be defeated; be vanquished; lose (a battle or war) ②(打胜) defeat; vanquish beat: ～敌人 defeat the enemy ◇ ～国 vanquished nation

【战报】war *communique*; battlefield report

【战备】war preparedness; combat readiness: 一级～ first- degree combat readiness ◇ ～等级 degree of combat readiness / ～工作 preparations against war / ～粮 grain stockpiled in case of war / ～状态 combat readiness

【战场】battlefield; battleground; battlefront: 打扫～ clean up battlefield / 开辟新～ open another front / 上～ go to the front

【战车】(war) chariot

【战刀】saber: 挥舞～ rattle the saber

【战地】battlefield; battleground; combat zone ◇ ～记者 war correspondent / ～指挥部 field headquarters

【战抖】tremble; shiver; shudder: 全身～ tremble all over

【战斗】①(武装冲突) fight; battle; combat; action: ～到底 fight to the finish / 英勇～ put up a heroic fight ②(斗争) militant; fighting: ～的友谊 militant friendship / ～的诗篇 militant poem ◇ ～编组 combat grouping / ～部队 combat forces / ～部署 tactical disposition / ～队形 battle formation / ～机 fighter plane; fighter / ～命令 combat orders / ～任务 combat mission; fighting task / ～性 militancy / ～意志 will to fight / ～英雄 combat hero / ～员 fighter

【战斗力】combat effectiveness; fighting capacity: ～强 high combat effectiveness / ～弱 low combat effectiveness

【战端】the beginning of a war: 重开～。War broke out again.

【战犯】war criminal

【战俘】prisoner of war (P.O.W.): 交换～ exchange prisoners of war / 遣返～ repatriate

prisoners of war ◇ ～收容所 prisoner-of-war collecting post / ～营 prisoner-of-war camp
【战歌】battle song; fighting song
【战功】meritorious military service; outstanding military exploit; battle achievement: 赫赫～ brilliant achievements in war
【战鼓】war drum; battle drum
【战果】results of battle; combat success; victory: 扩大～ exploit the victory
【战壕】trench; entrenchment
【战后】postwar: ～时期 postwar period / ～余殃 the aftermath of war
【战火】flames of war: 在那～纷飞的岁月里 in those war-ridden years
【战祸】disaster of war
【战机】opportunity for combat: 抓住～ seize the opportunity to win a battle
【战舰】warship
【战局】war situation
【战况】situation on the battlefield; progress of a battle
【战利品】spoils of war; captured equipment; war trophies
【战例】a specific example of a battle (in military science): 有名的～ a famous battle
【战列舰】battleship
【战列巡洋舰】battle cruiser
【战乱】chaos caused by war
【战略】strategy: 全球～ global strategy ◇ ～部署 strategic plan / ～储备 strategic reserves / ～反攻 strategic / counteroffensive / ～方针 strategic principle / ～和战术 strategy and tactics / ～核武器 strategic nuclear weapons / ～家 strategist / ～决战 decisive strategic engagement; strategically decisive battle / ～空军 strategic air command / ～思想 strategic thinking / ～物资 strategic materials / ～学 science of strategy / ～要地 strategic area; important strategic point
【战马】battle steed; war-horse
【战前】prewar: ～时期 prewar period / ～动员 mobilization before a battle
【战区】war zone; theatre of operations
【战伤外科】war-surgery
【战胜】defeat; triumph over; vanquish; overcome: ～对手 triumph over the opponent / ～顽敌 defeat a tough enemy / ～自然灾害 conquer natural disasters ◇ ～国 victorious nation
【战时】wartime ◇ ～编制 wartime establishment; war footing / ～活动 wartime activities / ～内阁 wartime cabinet / ～宣传 wartime propaganda
【战士】①(士兵) soldier; man ②(为正义而战的人) champion; warrior; fighter: 国际主义～ champion of internationalism
【战事】war; hostilities: ～结束 conclusion of the war; termination of hostilities
【战书】written challenge to war; letter of challenge
【战术】(military) tactics ◇ ～电子战 tactical electronic warfare / ～空军 tactical air command / ～空中支援 tactical air support / ～核武器 tactical nuclear weapons / ～学 science of tactics / ～训练 tactical training / ～演习 tactical maneuver / ～预备队 tactical reserve / ～指挥员 commander of a tactical operation
【战天斗地】fight against heaven and earth; combat nature; brave the elements
【战无不胜】invincible; ever-victorious; all-conquering
【战线】battle line; battlefront: 缩短～ shorten the battle line
【战役】campaign; battle: ～性的进攻 offensive campaign ◇ ～学 science of campaigns / ～指挥员 commander of a campaign
【战友】comrade-in-arms; battle companion: ～重逢 reunion of comrade-in-arms
【战云】war cloud: ～密布 gathering war clouds
【战战兢兢】①(怕得发抖) trembling with fear; with fear and trepidation ②(小心谨慎地) with caution; gingerly
【战争】war; warfare: ～的双方 the two sides in a war; both belligerents / 医治～创伤 heal the war wounds ◇ ～边缘政策 brink of war policy; brinkmanship / ～贩子 warmonger / ～机器 war machine; war apparatus / ～狂 war mania; war hysteria / ～冒险 war venture; war gamble / ～升级 war escalation / ～温床 hotbed of war / ～状态 state of war / ～罪行 war crimes

颤 tremble; shiver; shudder: 她打了一个寒～。A cold shiver ran down her spine.
【颤栗】tremble; shiver; shudder

绽 split; burst: 衣裳～线了。The seam has split

zhāng

章 ①(歌曲、诗文的段落) chapter; section: 第一～ the first chapter / 乐～ movement (of a symphony, etc.) ②(条理) order: 杂乱无～ disorderly and unsystematic ③(条目) point: 约法三～ agree on a three-point law ④(章程) rules; regulations; constitution: 规～制度 rules and regulations ⑤(图章) seal; stamp: 公～ official seal / 盖～ affix one's seal; seal; stamp ⑥(佩戴的标志) badge; medal: 臂～ arm band; armlet / 领～ collar badge

【章程】①(规程或条例) rules; regulations; constitution ②[方](办法) solution; way: 我心里还没个准～。 I'm not sure yet what's the best way.

【章动】[天] nutation: 黄经～ nutation in longitude / 倾角～ nutation in obliquity ◇ ～传动 nutation drive / ～器 nutator / ～轴 nutation shaft

【章法】①(文章结构) presentation of ideas in a piece of writing; art of composition: 文章严密,又很有～。 The article is compact and well organized. ②(办事的程序、规则) orderly ways; methodicalness: 他办事很有～。 He is quite methodical in his work.

【章回小说】a type of traditional Chinese novel with each chapter headed by a couplet giving the gist of its content

【章节】chapters and sections

【章句】①(古书的章节、句读) chapters, sections, sentences and phrases in ancient writings ②(对古书章句的解释) syntactic and semantic analysis of ancient writings

【章鱼】octopus

璋 a jade tablet

樟 camphor tree

【樟木】camphorwood

【樟脑】camphor ◇ ～丸 camphor ball; mothball / ～油 camphor oil

【樟树】camphor tree

蟑

【蟑螂】cockroach; roach

彰 clear; evident; conspicuous: 昭～ evident; clear

【彰明较著】become startlingly clear; very obvious; easily seen

【彰善瘅恶】exhibit virtue and expose vice; distinguish the good and punish the evil

【彰彰若是】as clear as that

【彰彰在人耳目】be clear for all to see

獐 river deer

【獐头鼠目】with the head of a buck and the eyes of a rat; repulsively ugly and sly-looking

【獐子】river deer

张 ①(分开;放开) open; spread; stretch: ～翅膀 spread the wings / ～网 spread a net ②(夸大) magnify; exaggerate: 夸～ exaggerate; overstate ③(看;望) look: 东～西望 gaze around ④(商店开业) opening of a new shop: 开～ open a business; begin doing business ⑤[量] 两～画 two pictures / 一～桌子 a table

【张本】①(预做的安排) an anticipatory action ②(预先说在前面的话) an anticipatory remark

【张驰】relaxation ◇ ～长度 relaxation length

【张大】magnify; exaggerate; publicize widely: ～其词 exaggerate / ～其事 publicize the matter widely

【张灯结彩】be decorated with lanterns and colored streamers

【张冠李戴】put Zhang's hat on Li's head; attribute sth. to the wrong person or confuse one thing with another

【张皇】alarmed; scared; flurried; flustered: ～失措 be in a flurry of alarm; lose one's head; get into a panic

【张角】field angle

【张口结舌】be agape and tongue-tied; be at a loss for words

【张狂】flippant and impudent; insolent

【张力】①[物] tension: 表面～ surface tension ②(拉力) pulling force ◇ ～计 tensiometer / ～摩擦 tenso-friction / ～系数 coefficient of tension / ～障碍 dystonia

【张量】[数] tensor ◇ ～分析 tensor analysis / ～计算 tensor calculus / ～空间 tensor space / ～微分 tensor differentiation

【张罗】①(料理) take care of; get busy about: 为婚事～ make arrangements for wedding ②(筹划) raise (funds); get together (money): ～一笔款 raise a sum of money ③(接待) greet and entertain; attend to: ～顾客 attend to customers / ～客人 entertain guests; look after the guests

【张目】①(睁大眼睛) open one's eyes wide ②(助长别人的声势) 为某人～ boost sb.'s arrogance; build up another

【张贴】put up: ～光荣榜 post an honor roll /

～海报 put up posters／～通告 post a notice／禁止～。 Post no bills.
【张望】①(从孔、缝里看) peep (through a crack, etc.) ②(向四周、远处看) look around: 四顾～ look around
【张牙舞爪】bare fangs and brandish claws; make threatening gestures; engage in saber rattling
【张扬】make widely known; make public; publicize: 四处～ publicize everywhere; spread (a story) all over the place／不要～! Don't blaze it around.
【张应力】[物] tensile stress
【张嘴】①(开口说话) open one's mouth (to say sth.) ②(向人借贷或请求) ask for a loan or a favor: 她倒好意思～! So she has the nerve to ask for a loan!

zhǎng

掌 ①(手掌) palm: 鼓～ clap hands; applaud ②(用手掌打) strike with the palm of the hand; slap: ～嘴 slap sb. on the face ③(掌握) hold in one's hand; be in charge of; control; wield: ～兵权 wield military power; have command of the armed forces ④(某些动物的脚掌) the bottom of certain animals' feet; pad; sole: 熊～ bear's paw／鸭～ duck's foot ⑤(鞋掌) shoe sole or heel: 后～ shoe heel／前～ shoe sole／给鞋打前～ have a shoe soled ⑥(马蹄铁) horseshoe: 这匹马该钉～了。 It's time for the horse to be shod.
【掌灯】①(举着灯) hold a lamp in one's hand ②(点油灯) light an oil lamp
【掌舵】be at the helm; operate the rudder; take the tiller; steer a boat ◇ ～人 helmsman; steersman
【掌骨】[生理] metacarpal bone; metacarpale
【掌故】anecdotes: 文坛～ literary anecdotes
【掌故之学】the study of historical records
【掌管】be in charge of; administer: ～财政 administer finances／～一项工程 take charge of a project／～钥匙 keep the keys
【掌柜】shopkeeper; manager (of a shop)
【掌节】propodite; propodus
【掌权】be in power; wield power; exercise control
【掌上明珠】a pearl in the palm; a beloved daughter
【掌声】clapping; applause: 热烈的～ warn applause／～雷动 applaud to the echo／博得经久不息的～ win prolonged applause
【掌握】①(支配;运用) grasp; master; know well: ～技术 master techniques／～新情况 keep abreast of new developments ②(控制;主持) have in hand; take into one's hands; control: ～会议 preside over a meeting／～主动权 have the initiative in one's hands
【掌纹】palmprint
【掌纹鉴定】identification by palm print
【掌心】the center of the palm: 跳不出某人的～ cannot escape from sb.'s control
【掌印】keep the seal; be in power
【掌子】[矿] face; work area ◇ ～面 face

长 ①(年纪较大) older; elder; senior: 你比他～三岁。 You are three years older than him. ②(排行最大) eldest; oldest: ～女 eldest daughter／～兄 eldest brother ③(领导人) chief; head: 代表团团～ head of a delegation／首～ leading cadre ④(生长) grow; develop: 树～得很快。 The trees grow fast. ⑤(生) come into being; begin to grow; form: ～疮 have a boil／～锈 get rusty ⑥(增进;增加) acquire; enhance; increase: ～才干 enhance one's abilities／此风不可～。 Such a tendency is not to be encouraged.
【长辈】elder member of a family; elder; senior
【长大】grow up; be brought up: ～成人 grow to manhood
【长官】senior officer or official; commanding officer
【长机】[军] lead aircraft; leader
【长见识】increase one's knowledge; gain experience
【长进】progress: 在学习上很有～ make good progress in one's studies
【长老】①[书](年纪大的人) elder ②(对和尚的尊称) elder of a Buddhist monastery
【长入】grow into
【长势】the way a crop is growing: 小麦～良好。 The wheat is doing well.
【长孙】son's eldest son; eldest grandson
【长相】[口] looks; features; appearance: ～好 be good-looking
【长者】①(年纪辈分高的人) elder; senior ②(年高有德的人) venerable elder
【长子】eldest son ◇ ～继承权 (right of) primogeniture; birthright

涨 (of water, prices, etc.) rise; go up: 河水～了两英尺。 The river has risen two feet.／物

价上～。The prices soared.
【涨潮】rising tide; flood tide
【涨风】upward trend of prices
【涨价】rise in price
【涨落】(of water, prices, etc.) rise and fall; fluctuate: ～无常 rise and fall irregularly / 价格的～ fluctuations of prices

zhàng

瘴 miasma
【瘴疠】communicable subtropical diseases, such as pernicious malaria, etc.
【瘴气】miasma

嶂 a screen-like mountain peak: 层峦叠～ peaks rising one higher than another

幛 a large, oblong sheet of silk with an appropriate message attached, presented at a wedding, birthday or funeral
【幛形天线】curtain antenna

障 ①(障碍) hinder; obstruct ②(遮挡的东西) barrier; block: 路～ roadblock; barricade
【障碍】①(阻碍) hinder; obstruct ②(阻挡前进的东西) obstacle; obstruction; barrier; impediment: 清除～ clear away obstacles / 制造～ erect barriers; create obstacles ◇ ～船[军] blockship / ～赛跑 steeplechase; obstacle race / ～物 obstacle; barrier; entanglement
【障蔽】block; obstruct; shut out: ～视线 obstruct one's view
【障眼法】cover-up; camouflage: 玩弄～ throw dust into people's eyes
【障子】a barrier made of reeds, sorghum stalks or closely planted shrubs; hedge

丈 ①(丈量) measure (land): 清～ make an exact measurement of the land ②(称呼) a form of address for certain male relatives by marriage: 姑～ the husband of one's father's sister; uncle / 姐～ elder sister's husband; brother-in-law / 岳～ wife's father; father-in-law ③(长度单位) *zhang* (= 3.33 meters)
【丈二和尚摸不着头脑】be completely at a loss; unable to make head or tail of it
【丈夫】①(妻子的配偶) husband ②(男子汉) man: ～气概 manliness
【丈量】measure (land): ～土地 measure land; take the dimensions of a field
【丈母娘】wife's mother; mother-in-law
【丈人】wife's father; father-in-law

杖 ①(手杖) cane; stick: 拐～ walking stick / 扶～而行 walk with a cane ②(棍棒) rod or staff used for a specific purpose: 擀面～ rolling pin

仗 ①[书](兵器总称) weaponry; weapons: 仪～ flags, weapons, etc. carried by a guard of honor / 明火执～ carry torches and weapons in a robbery; conduct evil activities openly ②(拿着) hold (a weapon): ～剑 hold a sword ③(凭借;倚仗) rely on; depend on: 狗～人势 like a dog threatening people on the strength of its master / 倚～权势 rely on one's power and position ④(战争) battle; war: 打～ go to war; be at war
【仗势欺人】take advantage of one's or sb. else's power to bully people; bully others on the strength of one's powerful connections or position
【仗恃】rely on (an advantage)
【仗义疏财】be generous in aiding needy people
【仗义执言】speak out from a sense of justice

涨 ①(涨大) swell after absorbing water, etc.: 豆子泡～了。The beans swelled up when soaked. ②(头部充血) (of the head) be swelled by a rush of blood: 头昏脑～ feel one's head swimming

帐 ①(帐子) curtain; canopy: 蚊～ mosquito net / ～钩 net hook ②(帐目) account: 记～ keep accounts ③(帐簿) account book ④(债) debt; credit: 还～ repay a debt / 赊～ buy or sell on credit
【帐簿】account book ◇ ～纸 account book paper; ledger paper
【帐单】bill; check
【帐房】[旧] ①(管理银钱物出入的处所) accountant's office ②(管钱物出入的人) accountant
【帐户】account: 非贸易～ noncommercial account / 在银行开立～ open an account with a bank
【帐面价值】book value
【帐面损失】book loss
【帐面盈余】book surplus
【帐面债务】book debts
【帐目】items of an account; accounts: ～不清 accounts not in order / 清理～ square accounts

【帐篷】tent: 拆～ strike a tent / 搭～ pitch a tent
【帐子】①(指张在床上的) bed-curtain ②(蚊帐) mosquito net
胀 ①(膨胀) expand; distend: 热～冷缩 expand when heated and contract when cooled ②(肿胀) swell; be bloated: 肿～ swollen
【胀管】expand tubes
【胀闸】expanding brake; expansion brake

zhāo

着 ①(下棋时走一步) a move in chess: 高～儿 a clever move / 一～不慎，满盘皆输。A single careless move and the game is lost. ②(计策;手段) trick; artifice; device; move: 这一～厉害。That's a shrewd move.
【着数】①(下棋时走一步或武术的动作) a move in chess or a movement in *wushu* ②(计策;手段) trick; device
朝 ①(早晨) early morning; morning: ～阳 morning sun ②(日;天) day: 今～ today; the present
【朝不保夕】not know in the morning what may happen in the evening; be in a precarious state
【朝不谋夕】be unable to plan out one's day; be preoccupied with the current crisis
【朝发夕至】start at dawn and arrive at dusk; a short journey
【朝晖】morning sunlight
【朝令夕改】issue an order in the morning and rescind it in the evening; make unpredictable changes in policy
【朝露】[书] morning dew; ephemeral; transitory
【朝气】youthful spirit; vigor; vitality: 有～ be full of vigor
【朝气蓬勃】full of youthful spirit; full of vigor and vitality; imbued with vitality: ～的青年 spirited young people
【朝秦暮楚】be fickle and inconstant; change one's loyalty frequently
【朝三暮四】blow hot and cold; play fast and loose; chop and change
【朝思暮想】yearn day and night
【朝夕】①(天天;时时) morning and evening; from morning to night; day and night; daily: ～相处 be together from morning to night; be closely associated ②(非常短的时间) a very short time: 只争～ seize the day, seize the hour; seize every minute
【朝霞】rosy clouds of dawn; rosy dawn
【朝阳】the rising sun; the morning sun
【朝朝暮暮】every morning and evening; day and night
招 ①(举手招人) beckon: 她～手让我进去。She beckoned me in. ②(招收) recruit; enlist; enroll: ～工 recruit workers ③(引来) attract; incur; court: ～人恨 incur the resentment of others / ～灾 court disaster; invite calamity ④(招惹) provoke; tease: 别～他。Don't tease him. ⑤(承认罪行) confess; own up: 不打自～ confess without being pressed ⑥(计策;一步) trick; device; move: 你这一～可真高。That was really a brilliant stroke of yours.
【招安】offer amnesty and enlistment to rebels
【招标】invite tenders; by tender; call for tenders; inviting bids: 采用国际竞争性～方式采购 purchase by way of international competitive bidding (ICB) / 这些汽车要用公开～的办法购买。These cars are to be procured through open tender. ◇ ～通告 invitation for bid / ～文件 bid document
【招兵】recruit soldiers; raise troops
【招兵买马】recruit men and buy horses; raise or enlarge an army; recruit followers
【招待】receive (guests); entertain; serve (customers): ～客人 entertain guests / 设宴～外宾 give a dinner for foreign guests ◇ ～费 entertainment allowance or expenses / ～券 complimentary ticket / ～所 guest house; hostel
【招待会】reception: 国庆～ National Day reception / 记者～ press conference / 举行～ give a reception
【招风】catch the wind; attract too much attention and invite trouble: ～惹草 get oneself into trouble
【招供】make a confession of one's crime; confess
【招股】raise capital by floating shares
【招呼】①(呼唤) call: 那边有人～你。Someone over there is calling you. ②(问候) hail; greet; say hello to: 热情地打～ greet warmly ③(吩咐) notify; tell: ～他赶快来。Tell him to come at once. ④(照料) care of: ～病人 take care of the patients
【招魂】call back the spirit of the dead
【招祸】court disaster
【招架】ward off blows; hold one's own: ～不住

unable to hold one's own / 只有～之功，并无还手之力 can only parry sb.'s blows without being able to hit back; only able to defend oneself but unable to hit back

【招考】give public notice of entrance examination; admit (students, applicants, etc.) by examination

【招徕】solicit (customers or business); canvass: ～顾客 solicit customers

【招揽】solicit (customers or business); canvass: ～生意 canvass business orders; drum up trade / ～主顾 solicit customers

【招领】announce the finding of lost property: 失物～ Found (布告标题) / 拾物～处 Lost and Found

【招募】recruit; enlist: ～入伍 enlist sb. for the army

【招女婿】take in a son-in-law to bear bride's family name

【招牌】shop sign; signboard: 打着裁军～扩充军备 build up armaments under the signboard of disarmament

【招聘】give public notice of a vacancy to be filled; invite applications for a job: ～工程师 advertise for engineers ◇ ～广告 want ad

【招惹】①(引起) provoke; incur; court: ～是非 bring trouble on oneself ②[方](触动；逗引) tease; provoke: 这人～不得。You'd better not provoke that fellow.

【招认】confess one's crime; plead guilty; shrift

【招生】enroll new students; recruit students ◇ ～考试 entrance exam / ～制度 enrollment system; admissions system

【招事】bring trouble on oneself; invite trouble

【招收】recruit; take in: ～打字员 recruit typists

【招手】beckon; wave: ～致意 wave one's greetings; wave back in acknowledgement

【招数】①(下棋走一步或武术动作) a move in chess or a movement in *wushu* ②(计策；手段) trick; device

【招贴】poster; placard; bill ◇ ～画 pictorial poster / ～纸 poster paper

【招贤纳士】invite to one's side men of wisdom and valor

【招降】summon sb. to surrender

【招降纳叛】recruit deserters and traitors

【招摇过市】swagger through the streets; blatantly seek publicity

【招摇撞骗】swindle and bluff; bluff and deceive ◇ ～者 moocher

【招引】attract; induce: 灯光～蛾子。Lights attract moths.

【招灾惹祸】court disaster; invite trouble: 这孩子老给父母～。This child is always making trouble for his parents.

【招展】flutter; wave: 彩旗迎风～。Colored flags are fluttering in the wind.

【招致】①(搜罗) recruit (followers); scout about for (talents, etc.) ②(引起) incur; bring about; lead to: ～失败 cause defeat / ～无穷后患 lead to endless trouble / ～重大损失 incur a heavy loss

【招赘】take in a son-in-law to bear bride's family name

【招租】let: 房屋～ House to Let

昭

昭 clear; obvious

【昭然若揭】abundantly clear; all too clear

【昭示】make clear to all; declare publicly: ～中外 declare to the whole world

【昭雪】exonerate; rehabilitate: 冤案得到了～。The wrong has been righted.

【昭信中外】show good faith to the nation and abroad

【昭彰】clear; manifest; evident: 罪恶～ have committed flagrant crimes

【昭著】clear; evident; obvious: 成绩～ have achieved signal successes / 臭名～ notorious; of ill repute

zháo

着 ①(接触) touch: 说话不～边际 not speak to the point ②(感受；受到) feel; be affected by (cold, etc.): ～风 become unwell through being in a draught ③(点亮；燃烧) burn; light: 灯都～了。The lights are all on. ④(表示动作已达到目的或有了结果)猜～了 have guessed right ⑤[方](入睡) fall asleep: 他躺下就～了。He fell asleep as soon as he lay down.

【着慌】become flustered; be thrown into a panic; get alarmed: 考试时不要～。Don't be nervous at examination.

【着火】inflammation; kindle; catch fire; be on fire: ～啦! Fire! ◇ ～点[化] fire point / ～温度 autogenous ignition temperature; ignition point / ～延迟 ignition lag

【着急】worry; feel anxious: 别～，安心养病。Just take care of yourself and don't worry.

【着凉】catch cold; catch a chill: 当心～。Take care not to catch cold.

【着陆】landing: 软～ soft landing / 软～装置 soft lander ◇ ～场 landing ground / ～导航设

备 approach aids／～灯 landing light／～航向信标 localizer beacon／～区 touch-down zone
【着忙】be in a hurry：不要～。Don't hurry.
【着迷】be fascinated；be captivated：观众越看越～。The audience watched the performance with growing fascination.
【着魔】be bewitched；be possessed：
【着色】coloration：～玻璃 stained glass；pigmented glass／～橡胶 tinted rubber

zhǎo

沼 natural pond
【沼气】biogas；sludge gas；marsh gas；firedamp；methane ◇～池 methane-generating pit
【沼煤】moor coal
【沼生植物】helophyte
【沼铁矿】bog iron ore；lake ore；marsh ore
【沼泽】marsh swamp ◇～地 swamp／～土 bog soil
找 ①(寻找) look for；try to find；seek：～出处 look for its source／～矿 look for mineral deposits／～到油田 discover an oil field ②(要求会见) want to see；call on；approach；ask for：有人～你。Someone wants to see you. ③(找钱) give change：这是～给你的钱。Here's your change.
【找补】make up a deficiency：他话没讲完，还得～几句。He hasn't finished what he had to say and is going to make it up.
【找碴儿】find fault；pick holes；pick a quarrel：～寻衅 fasten a quarrel upon somebody；pick a quarrel with somebody
【找对象】[方] look for a partner in marriage
【找麻烦】①(自找麻烦) look for trouble：自～ask for trouble；ask for it ②(给人添麻烦) cause sb. trouble：对不起，给你们～了。I'm sorry to have caused you so much trouble.
【找平】make level；level (out)
【找平层】[建] screed-coat
【找齐】①(使高低、长短相差不多) make uniform；even up ②(补足) make up a deficiency：先付一部分，差多少交货时～。We'll pay you part of the sum now. The balance will be paid on delivery of the goods.
【找钱】give change
【找事】①(寻找工作) look for a job ②(故意挑毛病，引起争吵) pick a quarrel
【找死】court death
【找寻】look for；seek
爪 claw；talon
【爪牙】talons and fangs；lackeys；underlings：帝国主义的～ a lackey of imperialism

zhào

肇 [书] ①(开始) start；commence；initiate ②(发生) cause (trouble, etc.)
【肇端】[书] beginning
【肇祸】cause trouble；cause an accident
【肇始】[书] start；commence；initiate
【肇事】cause trouble；create a disturbance：～后即逃遁的 hit-and-run／～后逃跑的司机 hit-and-runner ◇～者 a person who has created a disturbance；troublemaker
罩 ①(遮盖) cover；overspread；wrap：烟笼雾～ be shrouded in smog ②(罩子) cover；shade；hood；casing：保险～ protecting casing／玻璃～ glass cover／灯～ lampshade／发动机～(engine) hood／防尘～ dust guard ③(捕鱼用具) coop；a bamboo fish trap
【罩垫】hood pad
【罩盖】shroud
【罩面】mat coat
【罩棚】an awning over a gateway or a courtyard
【罩衫】overall；dustcoat
【罩袖】[方] oversleeve；sleevelet
【罩衣】dustcoat；overall
【罩子】cover；shade；hood；casing
兆 ①(预兆) sign；omen；portent：不祥之～ an ill omen ②(预示) portend；foretell：瑞雪～丰年。A timely snow promises a good harvest. ③(一百万) million；mega- ④(古代指一万亿) a million millions；billion
【兆安】megampere
【兆乏】megavar
【兆伏】crocodile；megavolt
【兆赫】[电子] megahertz；megacycle per second
【兆头】sign；omen；portent：好～ a good omen／坏～ a bad omen／有些人认为黑猫是不祥的～。Some people consider a black cat a creature of ill omen.
【兆周】[无] megacycle
笊
【笊篱】a bamboo, wicker or wire strainer
召 call together；convene；summon
【召唤】call；summon

【召回】recall: ～大使 recall an ambassador ◇ ～国书 letter of recall

【召集】call together; convene: ～会议 call a meeting / ～全体工人 call the workers together ◇ ～人 convener

【召见】①(上级约见下级) call in (a subordinate) ②[外] summon (an envoy) to an interview

【召开】convene; convoke: ～一次国际会议 convene an international conference / ～一次讨论会 hold a discussion

【召之即来】come as soon as called: ～,来之能战,战之能胜。 Be ready to assemble at the first call and be capable of fighting and winning.

诏 [书] ①(告诫) instruct ②(诏书) imperial edict

【诏书】imperial edict

照 ①(照射) shine; illuminate; light up: 拿手电～路 light the way a torch / 阳光普～ The sun illuminates everything. ②(照自己的影子) reflect; mirror: ～镜子 look in the mirror ③(照相) take a picture; photograph; film; shoot: 我们把它～下来。 Let's photograph it. (或: Let's take a photograph of it.) / 我想～一张相。 I want to have a picture taken. ④(相片) photograph; picture: 彩～ color photograph / 剧～ stage photo; still ⑤(执照) license; permit: 驾驶执～ a driving licence ⑥(照料) take care of; look after: ～应 look after; take care of ⑦(通知) notify: 知～ notify; inform ⑧(比照) contrast: 对～ contrast; check against ⑨(知晓) understand: 心～不宣 have a tacit understanding ⑩(向着) in the direction of; towards: ～这个方向走。 Go in this direction. ⑪(依照) according to; in accordance with: ～规章办事 act in accordance with the regulations

【照搬】indiscriminately imitate; copy: ～别人的经验 copy the experience of others

【照办】act accordingly; act in accordance with; act upon; comply with; follow: 她的要求你应～。 You must comply with her request.

【照本宣科】read item by item from the text; repeat what the book says

【照壁】a screen wall facing the gate of a house

【照常】as usual: ～营业 business as usual / 一切～进行 Everything proceeds as usual.

【照抄】copy word for word: 请你把这个材料～一份。 Please make a copy of this material.

【照出】break forth; emit rays of light; give light

【照度】[物] intensity of illumination; illuminance ◇ ～计 illumination meter; lightmeter

【照发】①(照例发给) issue as before: 女工产假期间工资～。 Women workers are entitled to maternity leave with full pay. ②(文件批语) approved for distribution

【照拂】[书] look after; care for; attend to: 请惠予～。 Please be so kind as to take good care of him.

【照顾】①(考虑到; 注意到) give consideration to; show consideration for; make allowance(s) for: ～多数 make allowance for the majority; think in terms of the majority / ～全局 take the whole into account; consider the situation as a whole / ～实际情况 in consideration of the actual conditions ②(关心;照料) look after; care for; attend to: ～伤员 look after the wounded

【照管】look after; tend; be in charge of: ～孩子 look after a child; mind a child / ～机器 tend a machine / 这件事由她～。 Trust her with this matter.

【照葫芦画瓢】draw a dipper with a gourd as a model — copy; imitate

【照护】look after

【照会】①(外交用语) present a note to (a government) ②(外交文件) note: 互换～ exchange notes / 普通～ note verbale; verbal note / 通知～ circular note / 正式～ personal note

【照价】according to the set price: ～付款 pay according to the arranged price

【照旧】as before; as usual; as of old: 一切～ Everything remains unchanged.

【照看】look after; attend to; keep an eye on: ～病人 attend to a patient

【照例】as a rule; as usual; usually

【照料】take care of; attend to; tend; mind: ～商店 tend a store

【照临】shine on; illuminate; light up: 曙光～大地。 The early sun bathes the land in light.

【照猫画虎】draw a tiger with a cat as a model; copy; imitate

【照面儿】①(露面) put in an appearance; show up; turn up: 互不～ avoid each other ②(不期而遇) encounter; come across: 打个～ come face to face with sb.; run into sb.

【照明】illumination; lighting: 舞台～ stage illumination ◇ ～弹 flare; star shell / ～灯 exciter lamp / ～电路 lighting circuit / ～装置 lighting installation

【照片】photograph; picture: 彩色～ color photograph
【照射】shine; illuminate; light up; irradiate: 用紫外线～ irradiate with ultraviolet rays
【照相】①(拍照) take a picture; take pictures; take photographs; take photos: 缩微～ microphotograph / 我给你照张相好吗? May I take your picture? / 我们在一起照了相。We were photographed together. ②(请别人给自己拍) have one's picture taken: 站(坐,蹲)好,准备～ pose for a picture ◇ ～版 process plate / ～簿 photo album / ～底版 photographic plates / ～复制 photocopy / ～馆 photo studio / ～用闪光灯 photoflash lamp / ～纸 photographic paper / ～制图 photomap
【照相凹版】photogravure; gravure
【照相凹版印刷】gravure printing
【照相机】camera: 反光～ reflex camera / 立体～ stereoscopic camera / 全景～ panoramic camera / 小型～ minicamera; minicam; miniature camera / 折叠式～ folding camera / 制版～[印] process camera
【照相胶版印刷】photo-offset process
【照相排字机】phototypesetter; photocomposer
【照相平版印刷】photolithography
【照相铜版】halftone
【照相凸版术】photoengraving
【照相制版】photomechanical process
【照样】①(依照样式) after a pattern or model: 照着样儿画 draw after a model ②(依然) in the same old way; all the same; as before: ～要感谢你。Thank you all the same.
【照样品定货】sample order
【照耀】shine; illuminate: ～我们前进的灯塔 a beacon lighting up out advance
【照应】①(呼应;配合) coordinate; correlate: 相互～ coordinate with each other / 文章要前后～。The ending of a composition should be correlated with its beginning. ②(照料) look after; take care of: 护士对病人～得很好。The nurses take good care of the patients.
【照章办事】act according to the rules; do everything by rule
【照准仪】[机] alidade; diopter; sight alidade

zhē

遮 ①(挡住) hide from view; cover; screen: 用帘子把窗户～起来 cover the window with curtains ②(拦住) block; obstruct; impede: 横～竖拦 impede in every way ③(掩盖) cover up: ～人耳目 try to hoodwink people
【遮蔽】①(挡住) hide from view; cover; screen ②(拦住) obstruct; block: ～视线 obstruct the view ③[军] defilade ◇ ～物[军] defilade / ～阵地 defiladed position
【遮藏】hide; conceal; cover up
【遮丑】gloss over one's blemishes; hide one's shame; cover up one's defect
【遮挡】shelter from; keep out: ～寒风 keep out the cold wind
【遮风挡雨】keep out wind and rain
【遮盖】①(遮住) cover; overspread: 雪～着大地。Snow covered the ground. ②(隐瞒) hide; conceal; cover up: 错误总是～不住的。Mistakes can never be hidden.
【遮光】shading ◇ ～板 gobo / ～玻璃 shade glass / ～框[摄] masking frame / ～片 anti-dazzling screen / ～罩[摄] lens hood; lens shade; sunshade
【遮拦】block; obstruct; impede
【遮羞】hush up a scandal; cover up one's embarrassment ◇ ～布 fig leaf
【遮掩】①(遮蔽) cover; overspread; envelop: 远山被云雾～着。The distant hills were enveloped in clouds and mist. ②(掩饰) cover up; hide; conceal: ～错误 cover up the mistakes
【遮眼法】cover-up; camouflage
【遮阳】sunshade ◇ ～板[建] sunshading board / ～甲板[船] awning deck

折 [口] ①(翻转) roll over; turn over: ～个跟斗 turn a somersault ②(倒来倒去) pour back and forth between two containers: 水太热,用两个杯子～一～就凉了。The water is too hot. Cool it down by pouring it one cup to another.
【折腾】[口] ①(翻来倒去) turn from side to side; toss about: 凑合着睡一会儿,别～了。Just try to get off to sleep for a while. Don't toss and turn restlessly. ②(反复做某事) do sth. over and over again ③(折磨) cause physical or mental suffering; get sb. down: 牙疼真～人。A toothache can get you down.

蜇 sting; irritate: 马蜂～人。Wasps sting. / 切洋葱～眼睛。Cutting onions irritates one's eyes.
【蜇针】[动] sting; stinger

zhé

谪 [书] ①(降职) relegate to a lower position; relegate a high official to a minor post in an outlying district; banish; exile ②(神仙受罚,降

到人间) (of fairies, etc.) be banished from Heaven ③(责备) censure; blame: 众口交～ be censured by everybody

【谪居】live in exile

折 ①(折断) break; snap: ～断胳膊 break one's arm / 把棍子～成两截 snap a stick in two ②(损失) lose; suffer the loss of: 损兵～将 suffer heavy casualties ③(弯曲;弯) bend; twist: 曲～ twists and turns ④(转变方向) turn back; change direction: ～向南方 turn southward ⑤(折服) be convinced; be filled with admiration: 心～ be deeply convinced; be filled with heartfelt admiration ⑥(折合) convert into; amount to: 这笔外币～成人民币是多少? How much does this sum of foreign money amount to in Renminbi? ⑦(折扣) discount; rebate: 打九～ give 10% discount; charge 90% of the original price ⑧(折叠) fold: ～被子 fold a guilt / 把信～好 fold the letter ⑨(折子) booklet in accordion form with a slipcase, used for keeping accounts, etc.; folder: 存～ bankbook; deposit book ⑩(汉字笔画) turning stroke (in Chinese character)

【折半】reduce (a price) by half; give 50% discount: 按原价～出售 sell at 50% discount; sell at half price

【折标】back-step; back-step marks

【折布机】[纺] folding machine

【折尺】folding rule

【折冲】[书] repulse or subdue the enemy: ～御侮 repel foreign aggression

【折冲樽俎】outmaneuver the enemy over glasses of wine; win by diplomacy; engage in diplomatic negotiations

【折叠】fold: 把地图～好 fold up the map ◇ ～床 folding bed / ～椅 folding chair / ～翼飞机 folding-wing aircraft

【折兑】exchange (gold or silver) for money; convert

【折服】①(说服) subdue; bring into submission: 强词夺理不能～人。 No one can be persuaded by sophistry. ②(信服) be convinced; be filled with admiration: 令人～ compel admiration

【折光】[物] refractive power ◇ ～度[物] dioptre / ～镜 enoscope

【折合】convert into; amount to: 把华氏温度～成摄氏温度 convert Fahrenheit degrees into Celsius / 二十市斤～十公斤。20 *jin* goes to 10 kilograms. ◇ ～率 conversion rate

【折回】turn back (halfway)

【折价】convert into money; evaluate in terms of money: ～退赔 pay compensation at the market price

【折旧】[经] depreciation ◇ ～费 depreciation charge / ～基金 depreciation fund / ～率 rate of depreciation / ～后净收益 net income after depreciation / ～总额 total depreciation

【折旧提成】amortization charge

【折扣】discount; rebate: 打～出售 sell at a discount / 这是打了～的价钱。 This is the discounted price. ◇ ～价格 prices at a discount / ～率 discount rate

【折门】[建] folding door; accordion door

【折磨】cause physical or mental suffering; torment; harassment: 受疾病的～ suffer severely from a lingering illness

【折扇】folding fan

【折射】[物] refraction; interception ◇ ～波 refracted wave / ～计 refractometer / ～角 angle of refraction; refraction angle / ～率 index of refraction; refracting power / ～望远镜 refracting telescope

【折实】①(打了折扣,合成实在数目) reckon the actual amount after a discount ②(折合成某种实物价格) adjust payment in accordance with the price index of certain commodities

【折算】convert ◇ ～率 conversion rate

【折线】[数] broken line ◇ ～图 broken-line graph

【折页】[印] folding ◇ ～机 paper folding machine

【折椅】folding chair

【折帐】pay a debt in kind

【折纸】paper folding

【折中】compromise: ～方案 a compromise proposal ◇ ～主义 eclecticism

【折子】booklet in accordion form with a slipcase, used for keeping accounts, etc.

【折子戏】highlights from operas

哲 ①(有智慧) wise; sagacious ②(有智慧的人) wise man; sage: 先～ the sages of old

【哲理】philosophic theory; philosophy

【哲人】[书]sage; philosopher

【哲学】philosophy ◇ ～家 philosopher / ～思想 philosophical thinking / ～系 philosophy department

蛰 [书] hibernate

【蛰伏】[动] dormancy; hibernation: ～过冬 lie dormant during the winter
【蛰居】live in seclusion: ～书斋 cloister oneself in one's study

辙 ①(车轮压出的痕迹) the track of a wheel; rut ②(韵) rhyme (of a song, poetic drama, etc.): 合～ in rhyme ③ [方](主意;办法) way; idea: 没～ can find no way out; be at the end of one's rope
【辙叉】[铁道] frog
【辙乱旗靡】 crisscross chariot tracks and drooping banners; signs of an army in headlong flight

辄 [书] ①(总是) always; often: 动～得咎 be frequently taken to task; be blamed for whatever one does／所言～听 always heed sb.'s advice ②(就) then: 饮少～醉 get drunk after a few sips

zhě

褶 pleat; crease: 百～裙 pleated skirt; accordion pleated skirt／这种料子容易起～ This material creases easily.
【褶点】plait point
【褶合】faltung
【褶积】[数] faltung
【褶皱】①(皮肤上的皱纹) wrinkle: 脸上的～ wrinkles on the face ②[地] fold: ～山 folded mountain／～作用 folding
【褶子】①(衣服经折叠而缝成的纹) pleat ②(折叠留下的痕迹) crease; fold; wrinkle ③(脸上的皱纹) wrinkle (on the face)

者 ①[助] (用于代替人或事物)爱国～ patriot／大～ the big one／革命～ revolutionist／老～ old man／胜利～ victor ②[助](表示从事某项工作或信仰某个主义的人)马克思主义～ Marxist／医务工作～ medical worker／作～ writer ③[助](指上文所说的几件事物)二～必居其一。It must be one or the other.／两～缺一不可。Neither is dispensable.

锗 [化] germanium (Ge)
【锗二极管】germanium diode; germanium rectifier
【锗晶体管】germanium transistor
【锗三极管】germanium triode
【锗石】germanite

赭 reddish brown; burnt ochre
【赭石】[矿] ochre

zhè

这 ①(指示代词) this: ～地方 this place／～可不行 This simply won't do.／～一回 this time／～是我的朋友。This is my friend. ②(这时候) now: 他～才知道自己的错误。Only now does he realize his mistakes.／她～就走。She is leaving right now.
【这般】such; so; like this: ～大小 this size; this big／如此～ thus and thus; thus and so
【这边】this side; here: 请～来。Come over here, please.／请～走 come this way please.
【这次】this time; present; current: ～会议 the present session／他～考试通过了。He succeeded in the examination this time.
【这个】this one; this: ～比那个好。This one is better than that one.
【这会儿】now; at the moment; at present: ～电话占线。The line's busy at the moment.／～你在干什么? What are you doing now?
【这里】here: 我们～没有商店。There is no shop here.
【这么】so; such; this way; like this: ～点儿水 so little water／应当～做。It should be done this way.
【这么着】like this; so: ～好。It's better this way.
【这儿】[口] ①(这里) here ②(只用在"打""从""由"后面) now; then: 从～以后,我们要好好学习。From now on we'll study hard.
【这山望着那山高】it's always the other mountain that looks higher; always think the grass is greener on the other side; never happy where one is
【这些】these: ～都是你的吗? Are these all yours?／～日子我们特别忙。We've been particularly busy these days.
【这样】so; such; like this; this way: ～的一个人 such a man／别走～快。Don't walk so fast.／照～做。Do it like this.

蔗 sugarcane
【蔗农】sugarcane grower
【蔗糖】①(有机化合物)sucrose ②(用甘蔗榨的糖) cane sugar ◇ ～酶 invertase; invertin; saccharase; sucrase／～蜜 cane molasses
【蔗渣】bagasse

鹧
【鹧鸪】[动] Chinese francolin; partridge
【鹧鸪病】partridge disease

柘 [植] three-bristle cudrania

zhe

着 [助] ①(表示动作或状态的持续)他们正谈~话呢。 They are having a talk. / 我今天闲~。 I am free today. ②(加强命令或嘱咐的语气)脚步轻~点儿。 Walk more quietly. / 快~点儿。 Be quick. / 你听~。 You just listen. ③(加在某些动词后面,使变成介词)挨~ next to / 朝~ towards / 沿~ along

zhēn

榛 [植] hazel

【榛鸡】hazel grouse

【榛莽】[书] luxuriant vegetation

【榛实】hazelnut

【榛子】①(榛树) hazel ②(指榛树的果实) hazelnut

臻 [书] attain (a high level): 方法日~完善。 The methods are being perfected.

斟 pour (tea or wine): 给她~一杯酒。 Pour her a glass of wine.

【斟酌】consider; deliberate: ~词句 weigh one's words / 很费~ call for much deliberation / 请~~。 Please think it over.

【斟酌尽善】consult about the most perfect way to do sth.; to consider the best plan

甄 [书] discriminate; distinguish; examine

【甄拔】select: ~人才 select people of talent

【甄别】①(审查辨别) examine and distinguish; screen; discriminate ②(考核鉴定) reexamine a case

【甄选】select

真 ①(真实) true; real; genuine: ~金 true gold / ~钻石 genuine diamond / 去伪存~ eliminate the false and retain the true / 这消息是~的吗? Is the news true? ②(的确) really; truly; indeed: ~感激 be truly grateful / 时间过得~快! How time flies! / 他~勇敢。 Truly, he is brave. ③(清楚确实) clearly; unmistakably: 字音咬得~ pronounce words distinctly / 你听得~么? Can you hear clearly?

【真北】[航空;航海] true north

【真不二价】have no second price

【真才实学】real ability and learning; genuine talent: 有~的科技人员 well-trained scientific and technical workers

【真诚】sincere; genuine; true: ~的友谊 true friendship / ~合作 sincerely cooperate / ~悔过 repent sincerely / ~无私 be sincere and selfless

【真刀真枪】real swords and spears; the real thing: ~地干起来 start a shooting war; start in real earnest

【真谛】true essence; true meaning: 人生的~ the true meaning of life

【真鲷】genuine porgy; red porgy

【真方位】[航空;航海] true bearing

【真分数】[数] proper fraction

【真果】[植]true fruit

【真迹】authentic work (of painting or calligraphy)

【真假】true and false; genuine and sham: ~莫辨 not to know the real from the false / 辨别~ tell the true from the false

【真金不怕火炼】true gold fears no fire; a person of integrity can stand severe tests

【真菌】[生] fungus ◇ ~病 fungous disease; mycosis

【真空】[物] vacuum: 未尽~ partial vacuum ◇ ~泵 vacuum pump / ~表 vacuum meter / ~处理[冶] vacuum treatment / ~弹道 vacuum trajectory; vacuum flight path / ~地带[军] no-man's-land / ~管[电子] vacuum valve / ~过滤 vacuum filtration / ~软管 vacuum hose / ~技术 vacuum technique / ~吸尘器 vacuum cleaner / ~压铸 vacuum die casting

【真核】eukaryon

【真理】truth: 绝对~ absolute truth / 客观~ objective truth / 相对~ relative truth / 坚持~,修正错误 uphold the truth and correct mistakes / 探求~ seek after truth

【真面目】true features; true colors: 露出~ show one's true features / 认清~ see sb. in his true colors

【真名实姓】real name

【真皮】[生理] derma; corium

【真凭实据】conclusive evidence; hard evidence

【真枪实弹】real guns and bullets; live ammunition: 进行~的演习 conduct exercises with live ammunition

【真切】vivid; clear; distinct: 看得~ see clearly

【真情】①(真实情况) the real situation; the facts: the actual state of affairs; truth ②(真诚的心情、感情) true feelings; real sentiments: ~的流露 a revelation of one's true feelings / ~实话 a round unvarnished tale

【真人真事】real people and real events; actual persons and events

【真善美】the true, the good and the beautiful
【真实】true; real; authentic: ～的材料 authentic data / ～的感情 true feelings; real sentiments ◇ ～感 sense of reality / ～供词[法] true statement / ～性 authenticity; factuality; truthfulness
【真率】sincere; unaffected; straightforward
【真丝】real silk
【真相】face; naked truth; truth; the actual state of affairs: 歪曲事实～ give a false account of the real situation / ～大白。 The whole truth has come out.
【真心】wholehearted; heartfelt; sincere: ～话 sincere words / ～悔改 sincerely repent and earnestly reform oneself / ～善意 sincerely and with good intentions; with sincerity and good intentions
【真心实意】genuinely and sincerely; truly and wholeheartedly; wholeheartedly; sincerely: ～地支持 wholeheartedly support
【真真假假】the true mingled with the false
【真正】genuine; true; real: ～的朋友 a true friend ◇ ～当事人[法] real party / ～嫌疑[法] genuine suspicion
【真知】genuine knowledge: 实践出～。 Real knowledge comes from practice.
【真知灼见】real knowledge and deep insight; penetrating judgment; high perspicacity
【真值表】[物] truth table
【真挚】sincere cordial: ～的同情 cordial sympathy / ～的友谊 sincere friendship
【真珠】pearl
【真主】[伊斯兰教] Allah

砧 hammering block; anvil: 台～ bench anvil / 锻～ smith anvil
【砧板】chopping block
【砧骨】[生理] incus; anvil
【砧木】[农] stock
【砧子】[口] hammering block; anvil

贞 ①(坚定不变) loyal; faithful: 坚～ staunch and faithful ②(女人的贞节) chastity; virginity
【贞操】①(贞节) chastity; virginity ②(忠贞) loyalty; moral integrity
【贞节】chastity; virginity
【贞洁】chaste and undefiled
【贞烈】ready to die to preserve one's chastity

桢 ①(硬木) hardwood ②(建墙时立的柱子) terminal posts used in building a wall in ancient times

侦 detect; scout; investigate
【侦查】[法] investigate (a crime): ～程序 procedure of investigation / ～员 investigator
【侦察】[军] reconnoiter; scout: ～敌情 gather intelligence about the enemy / 火力～ reconnaissance by fire ◇ ～兵 scout / ～部队 reconnaissance troops; scouting force / ～飞行 reconnaissance flight / ～机 reconnaissance plane; scout / ～卫星 reconnaissance satellite / ～员 scout
【侦缉】track down and arrest
【侦探】①(暗中侦察机密) do detective work ②(做侦探工作的人) detective; spy ◇ ～小说 detective story
【侦听】[军] intercept (enemy radio communications); monitor ◇ ～器 detectaphone / ～台 intercept station / ～员 monitor

箴 [书] ①(劝告;劝戒) admonish; exhort ②(古文体) a type of didactic literary composition
【箴言】admonition; exhortation; maxim

珍 ①(宝贵的东西) treasure: 奇～异宝 rare treasures ②(宝贵的) precious; valuable; rare: ～禽异兽 rare birds and animals ③(珍重) value highly; treasure
【珍爱】treasure; love dearly; be very fond of: 她～她的相册。 She treasures her photo album.
【珍宝】jewellery; treasure
【珍本】rare edition; rare book
【珍藏】collect (rare books, art treasures, etc.)
【珍贵】valuable; precious: ～的纪念品 precious souvenir / ～的历史文物 precious historical relics
【珍品】treasure: 艺术～ art treasure
【珍奇】rare: ～的动物 rare animals
【珍禽异兽】rare fowls and strange animals; rare animals and birds
【珍如拱璧】prize something like a piece of old jade; as precious as a piece of large jade
【珍赏】highly value and appreciate
【珍视】value; prize; cherish; treasure: ～我们之间的友谊 prize our friendship
【珍玩】rare curios
【珍闻】news titbits; fillers: 世界～ world briefs
【珍惜】treasure; value; cherish: ～劳动果实 treasure the fruits of labor / ～时间 value

one's time
【珍云母】ruby mica
【珍重】①(爱惜;珍爱) highly value; treasure; set great store ②(保重) take good care of yourself: 多多～。 Please take good care of yourself.
【珍珠】pearl ◇～贝 pearl shell; pearl oyster
【珍珠鸡】[动] guinea fowl
【珍珠梅】[植] false spiraea
【珍珠云母】margarite
胗 gizzard
【胗肝儿】gizzard and liver
针 ①(缝衣用的针) needle: 锈花～ embroidery needle ②(缝的一针) stitch: 伤口缝了六～ closed the wound with six stitches ③(针状物) anything like a needle: 分～ minute hand / 时～ hour hand / 松～ pine needle ④(针剂) injection; shot: 打～ give or have an injection ⑤(针刺) acupuncture: 耳～ ear acupuncture
【针鼻儿】the eye of a needle
【针布】[纺] card clothing
【针刺感】acanthesthesia; needling sensation
【针刺疗法】acupuncture treatment
【针刺麻醉】acupuncture anesthesia: ～心脏手术 heart surgery with acupuncture anesthesia
【针对】①(对准) be directed against; be aimed at; counter: 这个条约不～任何第三国。 The treaty is not directed against any third country. / 这些话不是～你说的。 These remarks are not aimed at you. ②(依照) in the light of; in accordance with; in connection with: ～目前情况 in view of existing situation
【针锋相对】give tit for tat; be diametrically opposed to: 进行～的斗争 wage a tit-for-tat struggle against
【针剂】[药] injection
【针尖】the point of a needle; pinpoint: ～儿对麦芒儿 a pin against an awn; diamond cut diamond
【针脚】stitch: ～很匀。 The stitches are neat.
【针灸】acupuncture and moxibustion ◇ ～医生 acupuncturist; doctor of acupuncture and moxibustion / ～针 acupuncture needle; acupuncture pin
【针头】[医] syringe needle; pip; needle head
【针线】needlework ◇ ～包 sewing kit / ～活 needlework; stitching; sewing
【针眼】①(针鼻儿) the eye of a needle ②(针刺的孔) pinprick ③[医] sty
【针鼹】[动] echidna; spiny anteater
【针叶树】coniferous tree; conifer
【针织】knitting: ～外衣 knitted coat ◇ ～长毛绒 knitted plush / ～厂 knitting mill; knit goods mill / ～机 knitting frame; knitting loom; knitting machine; needle loom / ～品 knit goods; knitwear; hosiery
【针状】acicular
【针状打字机】wire printer

zhěn

枕 ①(枕头) pillow ②(把头放在物体上) rest the head on: 他～着一块砖头睡着了。 He went to sleep with his head resting on a brick. ③[机] block: 靠～ backing block / 转～ swivel block
【枕部】occipitalia; occiput
【枕戈待旦】lie with one's head pillowed on a spear, waiting for day to break; be ready for battle; maintain combat readiness
【枕骨】[生理] occipital bone
【枕巾】a towel used to cover a pillow
【枕木】[铁道] tie; sleeper
【枕套】pillowcase; pillowslip
【枕头】pillow
【枕席】①(铺在枕头上的凉席) a mat used to cover a pillow; pillow mat ②(床榻) bed
【枕席不安】toss about in bed; cannot sleep peacefully
【枕席自荐】be willing to become a wife to someone
【枕心】pillow (without the pillowcase)
缜
【缜密】careful; meticulous; deliberate: ～的布置 meticulous arrangements / ～的计划 a deliberate plan / ～的准备 careful preparations
疹 rash: 荨麻～ nettle rash
【疹病】exanthema ◇ ～学 exanthematology
【疹子】[口] measles
诊 examine (a patient): 初～ first visit; first consultation / 出～ examine a patient at his home; visit; pay a call; be called out on a case / 触～ palpation / 复～ subsequent visit; subsequent consultation / 会～ consultation / 急～ emergency case; emergency treatment / 叩～ percussion / 确～ identify one's disease; make a definite diagnosis of; arrive at a correct diagnosis / 听～ auscultation / 误～ an erroneous diagnosis / 指～ touch

【诊病】diagnose a disease
【诊察】examine (a patient) examination ◇ ～室 consulting room
【诊断】diagnose: 物理～ physical diagnosis ◇ ～医师 diagnostician
【诊断床】diagnostic couch: 万能～ universal diagnostic table
【诊断书】medical certificate: 开～ make out a medical certificate
【诊疗】make a diagnosis and give treatment ◇ ～器械 medical instruments / ～室 consulting room / ～所 clinic; dispensary / ～学 diagnostics
【诊脉】feel the pulse
【诊室】consulting room
【诊所】clinic
【诊治】make a diagnosis and give treatment

袗 [书] ①(车后横木) the cross board at the rear of an ancient carriage ②(车) carriage ③(悲痛) sorrowful; distressed
【袗念】[书] sorrowfully cherish the memory of sb.; think anxiously about: 殊深～ express great solicitude for sb.

畛 [书] raised paths between fields
【畛域】[书] boundary: 不分～ make no distinctions

zhèn

鸩 ①(有毒的鸟) a legendary bird with poisonous feathers ②(毒酒) poisoned wine: 饮～止渴 quench a thirst with poison; seek quick relief regardless of the consequences ③[书](用毒酒害人) kill sb. with poisoned wine
【鸩毒】poisoned wine

震 ①(震动) shake; shock; vibrate; quake ②(情绪过分激动) greatly excited; deeply astonished; shocked: ～骇 shocked; stunned; astounded ③(地震) earthquake: 余～ aftershock of earthquake / 防～棚 shockproof shed
【震波】[地] seismic wave; earthquake wave ◇ ～图 seismogram
【震颤】tremble; quiver
【震颤性麻痹】paralysis agitans
【震荡】shake; shock; vibrate; quake: concussion
【震动】shake; shock; vibrate; quake: ～全国 reverberate through the whole country / 这消息使她大为～。She was very much shocked by the news. ◇ ～法 lash method / ～器 electromagnetic shaker; shaker / ～韧性 jolt toughness / ～引信 concussion fuse
【震耳欲聋】deafening: ～的雷声 deafening thunder
【震古烁今】[书] surpassing the ancients and amazing the contemporaries; earthshaking
【震撼】shake; shock; vibrate: ～世界 shake the whole world / ～天地 shake the skies and land
【震级】[地] (earthquake) magnitude: 里氏～ Reichter scale
【震惊】shock; amaze; astonish: ～中外 shock the country and the whole world
【震觉】seismesthesia
【震怒】be enraged; be furious
【震天憾地】shake the skies and land; rock the earth
【震源】[地] focus (of an earthquake)
【震中】[地] epicenter ◇ ～区 epicentral area
【震中对点】[地] anticenter; antiepicenter

振 ①(摇动) shake; flap: ～翅 flap the wings; flutter ②(奋起) rise with force and spirit; brace up: ～起精神 buoy up one's spirits / 精神不～ be in low spirits / 食欲不～ lose one's appetite; have a jaded appetite
【振臂】raise one's arm: ～高呼 raise one's arm and shout
【振笔直书】write with flying strokes; wield the pen furiously
【振翅高飞】flutter and soar high
【振荡】①[物] vibration ②[电] oscillation: 本机～ local oscillation / 寄生～ parasitic oscillation ◇ ～电路 oscillating circuit / ～管 oscillator valve / ～器 oscillator
【振捣器】[建] vibrator
【振动】[物] vibration: 等时～ isochronous vibration / 简谐～ simple harmonic vibration / 受迫～ forced vibration / 自由～ free vibration ◇ ～机 bobbing machine / ～计 vibration meter; vibrometer / ～频率 vibration frequency / ～载荷 oscillating load
【振奋】①(振作奋发) rouse oneself; rise with force and spirit; be inspired with enthusiasm: 我们要～起来。We should rouse ourselves up. ②(使振奋) inspire; stimulate
【振奋人心】inspire people; inspire the people with hope: ～的消息 heartening news
【振幅】[物] amplitude (of vibration): 脉冲～ pulse amplitude ◇ ～畸变 amplitude distortion
【振聋发聩】rouse the deaf and awaken the

unhearing; awaken the deaf; enlighten the benighted
【振兴】develop vigorously; promote: ～教育事业 vitalize education / ～中华 rejuvenating China; achieve China's rejuvenation
【振振有辞】speak plausibly and at length
【振作】bestir oneself; display vigor: ～精神 bestir oneself

赈 relieve: aid: 以工代～ provide work as a form of relief
【赈济】relieve; aid: ～难民 relieve the refugees / ～灾民 relieve the people in stricken areas; aid the victims of natural calamities
【赈款】relief fund
【赈灾】relieve the people in stricken areas

镇 ①(压;抑制) press down; keep down; ease: ～痛 ease pain ②(安定) calm; tranquil; at ease: ～静 calm ③(维持安定) guard; garrison: 坐～ assume personal command (of a garrison, etc.) ④(镇守的地方) garrison post: 军事重～ strategic post ⑤(城镇) town ⑥(使变凉) cool with cold water or ice: 冰～ cool with ice / 冰～啤酒 iced beer
【镇定】calm; cool; composed; unruffled: 保持～ keep cool; remain calm; keep one's head / 神色～ be calm and collected; show composure and presence of mind
【镇定自若】be perfectly calm and collected; be in possession of oneself
【镇静】calm; cool; composed; unruffled: 保持～ remain calm
【镇静剂】sedative; tranquilizer; palliative
【镇咳药】antitussive
【镇流管】[电] ballast tube
【镇流器】[电] ballast
【镇守】guard (a strategically important place); garrison: ～边疆 defend the frontier
【镇痛】①(抑制疼痛) ease pain ②[医] analgesia: 针刺～ acupuncture analgesia ◇ ～剂 analgesic / ～效果 analgesic effect
【镇压】①(用强力压制) suppress; repress; put down: ～叛乱 put down a rebellion ②[口](处决) execute (a counterrevolutionary) ③[农] rolling; compacting; tamping ◇ ～器[农] (land) roller
【镇纸】paperweight

朕 ①(皇帝自称) I, the sovereign ②[书](先兆) sign; omen
【朕兆】sign; omen; portent: 经济衰退的～ signs of economic depression

阵 ①(古代战术用语) battle array: 方～ square formation / 严～以待 stand ready in full battle array ②(阵地) front; position: 上～ go to the front ③(一段时间) a period of time: 病了一～儿 be ill for some time ④[量] 一～风 a gust of wind / 一～炮火 a burst of gunfire / 一～雨 a spatter of rain
【阵地】position; front: 前沿～ advance position / 进入～ get into position ◇ ～战 positional warfare
【阵点】lattice point
【阵发】paroxysm ◇ ～性痉挛 clonic spasm; clonospasm / ～性心搏过速 paroxysmal tachycardia
【阵风】[气] gust; puff of wind
【阵脚】①(阵的最前方) front line ②(用于比喻) position; situation; circumstances: 稳住～ secure one's position
【阵挛】clonus ◇ ～性 clonicity
【阵容】①(队伍外貌) battle array ②(人力的配备) lineup: ～强大 have a strong lineup
【阵势】①(作战的布置) battle array; a disposition of combat forces: 摆开～ deploy the ranks in battle array ②(情势;场面) situation; condition, circumstances
【阵痛】[医] labor pains; throes (of childbirth)
【阵亡】be killed in action; fall in battle
【阵线】front; ranks; alignment: 民族统一～ national united front
【阵雪】snow shower
【阵营】camp: 革命～ revolutionary camp
【阵雨】shower

zhēng

正
【正月】the first month of the lunar year; the first moon: ～初一 the lunar New Year's Day

症
【症结】crux; crucial reason: ～所在 the crux of the problem; where the trouble lies

怔 seized with terror; terrified; panic-stricken
【怔忡】[中医] palpitation

征 ①(走远路) go on a long journey ②(征讨) go on an expedition: 出～ go on an expedition / 南～北战 fight north and south ③(征

召) conscript; draft; recruit; levy (troops); call up: 新～的兵 new recruits ④(征收) levy; collect; impose: ～粮 impose grain levies; collect grain taxes / ～税 levy taxes ⑤(征求) ask for; solicit: ～稿 solicit contributions ⑥(证明) evidence; proof: 信而有～ borne out by evidence ⑦(迹象;现象) sign; portent: 象～ symbol; emblem

【征兵】conscription; draft; call-up ◇ ～法 conscription law / ～年龄 conscription age; age for enlistment / ～站 drafting center / ～制 universal military service; conscription system

【征程】journey: 踏上～ step on a long journey

【征调】requisition; call up: ～物资和人员 requisition supplies and draft personnel

【征伐】go on a punitive expedition

【征帆】a ship on a long journey

【征服】conquer; subjugate: ～自然 conquer nature / 用武力～ conquer by force of arms

【征购】requisition by purchase; requisition: ～粮食 purchase of grain by the state

【征候】sign; indication

【征集】①(收集) collect: ～历史文物 collect historical relics ②(征募) draft; call up; recruit: ～新兵 recruitment / 战时～ wartime draft

【征粮】requisitioning of grain

【征募】enlist; recruit

【征聘】give public notice of vacancies to be filled; invite applications for jobs; advertise for (a secretary, teacher, etc.)

【征求】solicit; seek; ask for: ～订户 solicit subscriptions / ～意见 solicit opinions; ask for criticisms

【征收】levy; collect; impose: ～赋税 levy taxes / ～进口税 impose import duties / ～烟酒税 impose taxes on tobacco and wine

【征税】levy taxes; taxation ◇ ～货物 dutiable goods

【征讨】go on a punitive expedition

【征途】journey: 艰险的～ a perilous journey

【征文】solicit articles or essays ◇ ～启事 a notice soliciting contributions for a special issue, etc.

【征象】sign; symptom

【征询】seek the opinion of; consult; ask advice of; probe

【征用】take over for use; commandeer; requisition: ～土地 make a requisition of land

【征战】go on an expedition

【征召】call up; enlist; draft; conscript: ～入伍 enlist in the army

【征兆】sign; omen; portent

丁

【丁丁】[书][象] 伐木～。 Clang, clang goes the woodman's axe.

争

①(力求得到、达到) contend; vie; strive: ～权 scramble for power / 不～一日之短长 not strive for only temporary superiority ②(争执;争论) argue; dispute: ～得面红耳赤 flush to the roots of one's hair in the hot dispute

【争霸】contend for hegemony; scramble for supremacy

【争辩】argue; debate; contend: 无可～ indisputable; incontestable

【争吵】quarrel; wrangle; squabble: 无谓的～ a pointless quarrel / ～不休 bicker endlessly

【争持】refuse to give in; stick to one's guns: 双方～不下 Both sides stuck to their own stand.

【争斗】fight; struggle; strife

【争端】controversial issue; dispute; conflict: 边界～ a border dispute / 国际～ an international dispute / 解决～ settle a dispute / 挑起～ provoke a controversy

【争夺】fight for; enter into rivalry with sb. over sth.; vie with sb. for sth.: ～市场 scramble for markets

【争风吃醋】fight for a man's 〈woman's〉 favors; quarrel from jealousy

【争光】win honor for: 为国～ win honor for one's country

【争衡】scramble for supremacy; strive for mastery; be in rivalry with

【争论】argument; controversy; dispute; debate; contention: ～的双方 the two contending sides / ～中的问题 question at issue / 没完没了的～ an endless debate

【争名夺利】strive for fame and wealth; spend one's energies in pursuit of fame and wealth

【争鸣】contend: 百家～。A hundred schools of thought contend.

【争气】try to make a good showing; try to win credit for; try to bring credit to: 为中国人～ win credit for the Chinese people

【争奇斗艳】compete with each other for beauty of looks; contend in beauty and fascination

【争球点】[体] face-off spot

【争球圈】[体] face-off circle

【争取】strive for; fight for; win over: ～时间 race against time / ～入团 strive to be admitted as a League member / ～主动 take the ini-

tiative
【争权夺利】scramble for power and profit
【争挑重担】rush to carry the heaviest load
【争先】try to be the first to do sth.: ～发言 try to have the floor before others
【争先恐后】strive to be the first and fear to lag behind; vie with each other in doing sth.
【争雄】contend for hegemony
【争议】dispute; controversy: 有～的地区 a disputed area / 有～的条款 a contentious clause
【争用】[讯] contention ◇ ～法 contention method / ～线路 contention
【争执】disagree; dispute; stick to one's position: ～的问题 question in dispute / ～不下。Each sticks to his own stand. ◇ ～点 point in dispute / ～事件 matter in dispute / ～事由 subject of dispute

挣

【挣扎】struggle: 进行垂死的～ put up a last-ditch struggle / 他～着要逃脱。He struggled to get free.

睁 open (the eyes): ～不开眼 cannot keep one's eyes open
【睁眼瞎子】be completely illiterate; illiterate person
【睁一眼,闭一眼】turn a blind eye to sth.; wink at sth.; pretend not to see
【睁着眼睛说瞎话】tell a bare-faced lie

峥

【峥嵘】①(山势高峻) lofty and steep; towering ②(才华超群) outstanding; extraordinary: ～岁月 eventful years

筝 *zheng*, a Chinese zither with 21 or 25 strings

铮

【铮铮】[象] clank; clang

狰

【狰狞】ferocious; savage; hideous: ～面目 ferocious features; a vile visage

蒸 ①(蒸发) evaporate ②(蒸熟) steam: ～饭 steam rice / 清～鱼 steamed fish
【蒸饼】steamed cake
【蒸发】evaporate ◇ ～计 evaporimeter / ～蒸腾 evapotranspiration
【蒸锅】a pot for steaming food; steamer
【蒸饺】steamed dumpling (with meat and vegetable stuffing)
【蒸馏】[物] distillation: 从废水中～出有用的溶剂 distill useful solvents from waste water / 拔顶～[化] topping distillation / 常压～[化] atmospheric distillation / 真空～[化] vacuum distillation ◇ ～釜 still / ～罐 retort / ～器 distiller; retort / ～水 distilled water / ～塔 distilling tower
【蒸笼】food steamer; steam box
【蒸气】vapor
【蒸汽】steam ◇ ～锤 steam hammer / ～处理 barffing / ～发生器 steam generator / ～鼓 dry drum / ～供暖 steam heating / ～锅炉 steam boiler / ～机 steam engine / ～机车 steam locomotive / ～消毒 moist-heat sterilization / ～浴 steam bath
【蒸腾】(of steam) rising: 热气～ steaming ◇ ～作用[植] transpiration
【蒸蒸日上】becoming more prosperous every day; flourishing; thriving

zhěng

整 ①(完整) whole; complete; full; entire: 十点～ ten o'clock sharp / ～夜不睡 lie awake all night; lie awake the whole night ②(整齐) orderly; in good order; neat; tidy: 衣冠不～ slovenly in one' dress; not properly dressed ③(整顿) put in order; rectify ④(修理) repair; mend; renovate: ～旧如新 repair sth. old and make it as good as new ⑤(使吃苦头) make sb. suffer; punish; fix: 挨～ be the target of criticism or attack / 我迟早要～他一下。I'll fix him sooner or later. ⑥[方](搞;弄) do; make; work: 铁丝给～断了。The iron wire was broken.
【整版】devote a full page to; full page
【整编】reorganize (troops)
【整饬】①(整顿) rectify; consolidate; straighten out: ～纪律 strengthened discipline ②(整齐) in good order; neat; tidy: 服装～ neatly dressed / 军容～ in fine battle array
【整除】[数] be divided with no remainder; divide exactly
【整党】consolidate the Party organization
【整地】[农] soil preparation
【整队】dress the ranks; line up: ～入场 file into the arena, auditorium, etc.
【整顿】rectify; consolidate; reorganize: ～币制 readjust the currency / ～纪律 strengthen discipline / ～党风 rectify the working style of the Party

【整改】rectify and reform
【整个】whole; entire: ～亚洲 the whole of Asia / ～说来 on the whole; as a whole
【整洁】clean and tidy; neat; trim: 衣着～ neatly dressed
【整旧如新】restore sth. to its original shape and appearance; repair sth. old and make it as good as new
【整理】put in order; straighten out; arrange; sort out: ～财政 regulate finances / ～参考资料 sort out the reference materials / ～行装 pack one's things for a journey
【整流】[电] rectification; commutation ◇ ～管 rectifier tube / ～器 rectifier / ～子 commutator
【整齐】①(有秩序;有条理) in good order; neat; tidy: 桌子摆得很～。The tables are in alignment. ②(大小,长短相差不多) even; regular: 步伐～ march in step / 出苗～。The seedlings come out evenly.
【整齐划一】uniform
【整人】give sb. a hard time
【整容】tidy up one's appearance ◇ ～术 cosmetic operation; face-lifting; cosmetology
【整数】①(不含分数、小数的数) integer; whole number ②(没有零头的数目) round number
【整肃纲纪】screw up discipline
【整套】a complete set of: ～设备 a complete set of equipment
【整体】whole; entirety: 从～上看形势 view the situation as a whole ◇ ～吊装[建] integral hoisting / ～汽缸发动机 monoblock engine
【整体法学】integrative jurisprudence
【整体观念】the concept of viewing the situation as a whole
【整体化】integration; integrate
【整天】the whole day; all day; all day long: 他干了五～。He worked for five whole days.
【整新】renewing
【整形】[医] plastic ◇ ～手术 plastic operation / ～外科 plastic surgery; plastics; orthopedics
【整修】rebuild; renovate; recondition: ～水利工程 rebuild water conservancy projects
【整训】train and consolidate (troops); training and consolidation
【整整】whole; full: ～三天 three whole days
【整枝】[农] training; pruning: 棉花～ pruning of cotton plants / 葡萄～ training of vines
【整治】①(修理;整理) renovate; repair; dredge (a river, etc.): ～航道 dredge waterways / ～河道 the realignment of a river ②(使吃苦头) punish; fix: 这坏蛋得～一下。That scoundrel needs to be punished.
【整装】get one's things ready (for a journey, etc.): ～待命 be ready for orders

拯 save; rescue; deliver
【拯救】save; rescue; deliver: ～人民于水深火热之中 save the people from untold miseries

zhèng

郑
【郑重】serious; solemn; earnest: ～表示 earnestly declare; solemnly state / ～声明 solemnly declare / ～其事 seriously; in earnest

正 ①(垂直或符合标准方向) straight; upright: ～北 due north / ～东 due east / ～南 due south / ～西 due west / 将地图挂～ put the map straight ②(位置居中) situated in the middle; main: ～门 main entrance / ～厅 main hall ③(时间正在那一点上) punctually; sharp: 十点～ at ten o'clock sharp ④(正面) obverse; right: 纸的～面 the right side of the paper ⑤(正直) honest; upright: 方～ upright; righteous / 公～ just; fair ⑥(正当) correct; right: ～理 correct principle; valid reason; the right thing to do ⑦(色、味纯正) pure; right: ～黄 pure yellow / 气味不～ not of the right smell ⑧(主要的) chief; principal: ～驾驶员 first pilot ⑨(图形各边、角都相等) regular: ～三角形 equilateral triangle ⑩[物] positive; plus: ～离子 positive ion; cation ⑪(端正;合乎法度) regular; be in conformity with the normal standard: ～楷 regular script / 五官不～ have irregular features ⑫[数](大于零的) positive: ～号 positive sign; plus sign ⑬(改正) correct; rectify; set right: ～误 correct mistakes / ～音 correct one's pronunciation ⑭(恰好) just; right; precisely; exactly: ～因为如此 precisely because of this / ～中奸计 fall right into the villain's trap ⑮(表示正在进行或持续)～下着雨呢。It's raining.
【正本】①(别于副本的) reserved copy (of a library book) ②(文件的正式的一份) original (of a document): 把～送存档案库 deposit the original in the archives
【正本清源】radically reform; thoroughly overhaul: 采取～的措施 take measures for thoroughgoing reform
【正比】[数] direct ratio: 成～ be in direct pro-

portion to
【正比例】[数] direct proportion
【正步】[军] parade step; goose step
【正步走】[口令] Parade step, march!
【正长石】orthoclase
【正常】normal; regular: 恢复～ return to normal／脉搏～ have a normal pulse／运转～ function normally ◇～价格 normal price／～亏损 normal loss／～损耗 normal loss／～需求 normal demand
【正常化】normalize; normalization: 两国关系的～ normalization of relations between the two countries／使两国关系～ normalize the relations between the two countries
【正大】upright; honest; aboveboard: ～光明 open and aboveboard; just and honorable
【正当】①(正处在) just when; just the time for: ～年 in the prime of life; in one's prime／～时 in the nick of time; the right season or time ②(合理合法的) proper; rightful: 通过～途径 through proper channels; in proper ways ◇～驳回 reasonable dismissal／～防卫 justifiable defense／～公告 formal announcement／～理由 good cause／～手段 proper means／～要求 just demand／～业务 legitimate business／～自卫 necessity of self-protection
【正当中】right in the middle
【正道】the right way; the correct path: 走～ follow the correct path
【正点】on schedule; on time; punctually: ～运行 running on schedule／飞机～到达。The plane arrived on time.
【正电】positive electricity
【正电荷】positive charge
【正电子】positive electron; positron
【正殿】main hall (in a palace or temple)
【正多边形】[数] regular polygon
【正法】execute (a criminal): 就地～ execute (a criminal) on the spot
【正犯】[法] principal offender
【正反】positive and negative: ～两方面的经验 both positive and negative experiences／～双方的看法 the pros and cons
【正方】square: ～盒子 a square box ◇～形 square
【正房】①(坐北朝南的正面的房子) principal rooms ②[旧](大老婆) legal wife
【正告】earnestly admonish; warn sternly
【正割】[数] secant
【正骨】[中医] bonesetting
【正规】regular; standard ◇～部队 regular troops; regulars／～化 regularize; standardize; be put on a regular basis／～教育 regular education; proper education／～军 regular army／～学校 regular school／～战争 regular warfare
【正轨】the right path: 纳入～ lead onto the correct path; put on the right track
【正好】①(恰好) just in time; just right; just enough: 你来得～。You've come just in time.／这帽子我戴着～。This hat fits me nicely. ②(碰巧) happen to; chance to; as it happens: ～我也在这里。It happened that I was also here.
【正号】positive sign; plus sign
【正极】[电] positive electrode; positive pole; anode ◇～板[电] positive plate
【正角】positive angle
【正教】the Orthodox Church
【正襟危坐】[书] straighten one's clothes and sit properly; be all seriousness
【正经】①(端庄正派) decent; respectable; honest: ～人 a decent person ②(正当的) serious: ～事儿 serious affairs ③(合乎标准的) standard: ～货 standard goods
【正经八百儿】[方] serious; earnest: 这是～的事。This is a serious matter.
【正楷】regular script
【正梁】[建] ridge purlin
【正路】the right way; the correct path: 走～ follow the correct path
【正门】front door; main entrance
【正面】①(区别于"侧面") front; frontage; facade: ～进攻 frontal attack ②(主要使用的一面) the obverse side; the right side: 皮革的～ the grain side of leather／硬币的～ the obverse side of a coin ③(好的、积极的一面) positive: ～教育 educate by positive measures or examples; positive education ④(直接) directly; openly: ～提意见 make a criticism directly ◇～人物 positive character／～图 front view／～战场 frontline battlefield
【正派】upright; honest; decent: ～人 a decent person／作风～ honest and upright in one's ways
【正片】①[摄] positive ②[电影](拷贝) copy ②[电影](区别于加映的短片) feature (film)
【正品】certified products; quality products
【正箕形纹】ulna
【正气】healthy atmosphere: 发扬～ encourage healthy trends; encourage standing up for what is right

【正巧】①(正好) happen to; chance to; as it happens ②(刚巧) just in time; in the nick of time; just at the right time
【正切】[数] tangent
【正确】correct; right; proper: ～的立场 a correct stand／～的判断 accurate judgement／～的政策 sound policy ◇ ～性 correctness; soundness; validity
【正人君子】a man of honor; gentleman: 打扮成～ masquerade as a gentleman
【正如】exactly as; just as
【正色】①[书](纯正的颜色) pure colors ②(严肃的神色) with a severe countenance: ～拒绝 refuse sternly
【正商标】original trademark
【正史】history books written in biographical style
【正式】formal; official; regular: ～列入记录 be officially placed on record／达成～协议 reach a formal agreement／代表团的～成员 a regular member of the delegation ◇ ～催告 peremptory call／～访问 official visit／～会谈 formal talks／～记录 official records／～警告 formal warning／～签署 official signature／～声明 official statement／～收据 official receipts／～文本 official text／～委托 official commission／～协议 formal agreement／～照会 official communication
【正视】face squarely; face up to; look squarely at: ～现实 look reality in the face／～自己的缺点 acknowledge one's shortcomings／～自己的责任 face up to one's responsibilities
【正视图】[机] front view; elevation
【正矢】[数] versed sine; versine
【正事】one's proper business: 现在咱们谈～。Now let's talk business.
【正手】[体] forehand: ～抽球 forehand drive
【正数】[数] positive number
【正题】①(主要题目) subject of a talk or essay: 转入～ come to the subject ②[哲] thesis
【正体】①(规范字形) standardized form of Chinese characters ②(楷书) regular script ③(印刷体) block letter
【正厅】①(正中大厅) main hall ②(剧场正对舞台的地方) stalls
【正统】①(封建王朝相继的系统) legitimism ②(党派、学派的嫡派) orthodox: ～观念 orthodox ideas ◇ ～派 orthodox party or school
【正文】main body; text: 词典～ the main body of a dictionary; the dictionary proper／书的～ the text of a book
【正午】high noon
【正误】correct (typographical) errors ◇ ～表 errata; corrigenda
【正弦】[数] sine ◇ ～波[电] sine wave／～曲线 double continuous curve
【正像】[光] erect image
【正凶】[法] principal murderer
【正颜厉色】look serious and severe; put on a stern countenance
【正业】regular occupation; proper duties: 不务～ not attend to one's proper duties; not engage in honest work
【正义】①(公正的道理) justice: ～之师 an army dedicated to a just cause／主持～ uphold justice ②(公正的) just; righteous: ～的立场 a just stand／～的事业 a just cause／～战争 just war
【正义感】sense of justice: 具有～ have a sense of justice
【正音】①(矫正语音) correct one's pronunciation ②(标准音) standard pronunciation ◇ ～法[语] orthoepy
【正在】[副] in process of; in course of: ～举行会议 now in session／～兴起 on the upsurge; on the rise
【正直】honest; upright; fair-minded: ～的人 an honest person; a person of integrity ◇ ～无私 upright; without partiality
【正中】middle; center: 把桌子放在屋子～。Put the table right in the middle of the room.
【正中下怀】be just what one hopes for; fit in exactly with one's wishes
【正中要害】hit the nail on the head; be to the point
【正宗】orthodox school

症 disease; illness: 不治之～ incurable disease／急～ acute disease／对～下药 suit the medicine to the illness
【症候】①(疾病) disease ②(症状) symptom
【症结所在】the crux of the problem; where the trouble lies
【症状】symptom: 前驱～ premonitory symptoms／早期～ early symptoms ◇ ～学 semiology

证 ①(证明) prove; demonstrate: 论～ demonstration／求～ seek to prove ②(证据) evidence; proof; testimony: 旁～ circumstantial evidence; collateral evidence／确～ proof positive; conclusive evidence／人～ testimony of a witness / 物～ material evidence; exhibit / 以此为～ take this as a proof / 作～ give evidence; bear testimony; testify; bear witness ③ (证件;证书) certificate; card: 出生～ birth certificate / 逮捕～ arrest warrant / 工作～ employee's I.D. card / 拘留～ order for provisional apprehension / 身份～ identification card / 许可～ permit

【证词】testimony

【证婚人】chief witness at a wedding ceremony

【证件】credentials; papers; certificate: 请出示～。 Please show your credentials

【证据】evidence; proof; testimony: 直接～[法] direct evidence / 搜集～ collect evidence / 提出～ offer testimony ◇ ～不足 lack of evidence; in sufficiency of evidence / ～充足 sufficient evidence / ～分类 classification of evidence / ～文件 instrument of evidence / ～作用 evidential function

【证明】①(表明、断定真实性) prove; testify; bear out: 充分～ fully prove / 雄辩地～ give eloquent proof of ②(证明书或信) certificate; identification; testimonial: 医生～ medical certificate ◇ ～人 authenticator / ～文件 certificate; testimonial; papers / ～无罪 proof of innocence / ～有罪 proof of guilt

【证明书】certificate; testimonial: 产地～ certificate of origin / 健康～ health certificate / 质量～ certificate of quality

【证券】bond; security; negotiable securities ◇ ～交易所 stock exchange; securities exchange

【证人】evidence; rapper; testation; witness: ～的作证资格 competence of witness ◇ ～对质[法] confrontation of both side / ～席 witness-box; witness stand

【证实】affirm; affirmance; demonstrate; confirm; verify: 有待～ remain to be confirmed / 消息得到了～。 The news has been confirmed. ◇ ～无可置疑[法] proof beyond a reasonable doubt / ～者 prover

【证书】certificate; credentials: 结婚～ marriage certificate; marriage lines / 毕业～ diploma ◇ ～审查委员会 credentials committee

【证物】[法] exhibit (produced in court as evidence)

【证言】deposition; oral testimony; testimony; verbal evidence; witness

【证章】badge

政 ①(政治) politics; political affairs: 议～ discuss political affairs ②(国家部门主管的业务) certain administrative aspects of government: 财～ (public) finance / 民～ civil administration / 邮～ postal service ③(事务) affairs of a family or an organization: 家～ household management

【政变】coup d'état; coup: 发动～ stage a coup d'état

【政策】policy: ～界线 demarcation line in policy / 执行～ implement a policy

【政党】political party

【政敌】political opponent

【政法】politics and law ◇ ～机关 political-legal organs / ～界 political-legal circles / ～人员 political-legal functionaries / ～学院 institute of political science and law

【政府】government ◇ ～部门 government departments / ～代表 government representative / ～公文 government documents / ～公债 government bonds / ～官员 government official / ～机构 government apparatus / ～机关 government bodies / ～人士 government circles / ～首脑 head of government

【政纲】political program; platform

【政躬违和】An official was indisposed while on duty.

【政绩】achievements in one's official career

【政见】political view

【政教分离】separation of religion from politics; separation of the church from the state

【政界】political circles; government circles: 退出～ withdraw from political life

【政局】political situation; political scene: ～不稳。 The political situation was unstable.

【政客】politician

【政令】government decree

【政论】political comment ◇ ～家 political commentator; political writer / ～文 political essay

【政权】state political power; political power; regime: ～机关 organs of state power / 国家～ state power / 武装夺取～ seize political power by armed forces

【政体】system of government; form of government

【政务】government affairs; government administration ◇ ～会 council / ～院 the Government Administration Council

【政协】[简](中国人民政治协商会议) the Chinese People's Political Consultative Conference (C.P.P.C.C.)

【政修年丰】The country enjoys a good gov-

ernment and bumper crops.
【政治】politics; political affairs ◇～庇护 political asylum／～避难 (political) asylum／～表现 political behavior or record／～部 political department／～待遇 political treatment／～犯 political offender; political prisoner／～纲领 political program; platform／～家 statesman／～掮客 political broker／～经济学 political economy／～空气 political atmosphere／～路线 political line／～面目 political affiliation or background／～派别 political grouping or faction／～骗子 political swindler／～权利 political rights／～态度 political attitude／～委员 political commissar／～舞台 political arena／～性 political nature／～嗅觉 political sense of smell; political acumen／～学 political science; government／～资本 political stock-in-trade／～罪行 political offense
【政治局】the Political Bureau ◇～常务委员 member of the Standing Committee of the Political Bureau／～委员 member of the Political Bureau
【政治信仰自由】freedom of political conviction

帧 [量](用于字画)一～油画 an oil painting
【帧频】[电视] frame frequency; picture frequency
【帧同步】[电视] frame synchronization

诤 [书] criticize sb.'s faults frankly; admonish; expostulate
【诤言】[书] forthright admonition
【诤友】[书] a friend who will give forthright admonition

挣 ①(用力摆脱束缚) struggle to get free; try to throw off: ～脱枷锁 throw off the shackles ②(用劳动换取) earn; make: ～钱养家 earn money to support one's family／你每月～多少钱? How much do you earn every month?
【挣命】struggle to save one's life
【挣钱】earn money

zhī

汁 juice: 胆～ bile; gall／果子～ fruit juice／橘子～ orange juice／墨～ prepared Chinese ink／牛肉～ beef extract／乳～ milk／椰子～ coconut milk

之 ①[代](代替人或事物)求～不得 all one could wish for／取而代～ replace someone ②(虚用,无所指)总之 sum up; in short; in a word ③[助](表示领属或修饰关系)一水～隔 be separated only by a river／原因～一 one of the reasons
【之后】later; after; afterwards: 从那～ since then／两周～ two weeks later／她毕业～ after her graduation
【之乎者也】pedantic terms; literary jargon; archaisms
【之间】between; among: 群山～的一个村庄 a village among the hills
【之类】and so on; and so forth; and the like; and what not
【之流】and his like: 希特勒～ Hitler and his ilk
【之内】in; within: 在法律许可的范围～ within the limits permitted by law／在五天～ in five days
【之前】before; prior to; ago: 五年～ five years ago／在她动身～ prior to her departure
【之上】over; above: 不要高居于群众～。Don't stand above the masses.
【之外】besides; except; beyond: 除了雨天～,她总骑自行车上班。She usually goes to work by bike except on rainy days.
【之下】under: 光天化日～ in broad daylight／在党的领导～ under the leadership of the Party
【之中】in; in the midst of; among: 生活在群众～ live in the midst of the masses

芝
【芝艾俱焚】the good perished with the bad
【芝麻】①(植物) sesame ②(芝麻种子) sesame seed: 拣了～,丢了西瓜 pick up the sesame seeds take no notice of the watermelons; penny wise and pound foolish
【芝麻酱】sesame paste
【芝麻开花节节高】sesame stalks putting forth flowers notch by notch, higher and higher; rising steadily
【芝麻油】sesame-seed oil; sesame oil

支 ①(撑) prop up; put up: ～帐篷 put up a tent／用柱子～住屋顶 prop a roof with poles ②(伸出;竖起) protrude; raise: ～着耳朵听 prick up one's ears ③(支持) support; sustain; bear: 疼得～不住了 cannot bear the pain any longer ④(指使) send away; put sb. off: 把孩子们～开 put the children off with excuses; send the children away upon some pretext ⑤(付出或领取) pay or draw money: 预～二十元 get an advance of twenty *yuan*／收～平衡. Income and expenses are balanced.⑥(分支) branch; offshoot: ～店 branch store/邮政～局 branch post office ⑦[量]三～队伍 three contingents of troops/一个四十～光的灯泡 a 40-walt bulb ⑧[纺] count: 六十～纱 60-count yarn

【支部】branch: 党～ Party branch/团～ League branch ◇～大会 general membership meeting of the branch/～书记 branch secretary/～委员 member of the branch committee

【支承】[机] supporting; bearing ◇～点 bearing point/～力 supporting force/～圈 backup ring; support ring

【支撑】①(抵抗住压力) prop up; sustain; support: ～危局 prop up a perilous situation ②[建] strut; brace ③(勉强维持) support; prop up: ～门面 maintain the front ◇～点[军] strong point; center of resistance

【支持】①(勉强维持) sustain; hold out; bear: ～不住 unable to keep up efforts ②(给予鼓励或赞助) support; back; stand by: 他们在工作中互相～。They support each other in their work.

【支出】①(付出去) pay; expend; disburse: ～十元钱 pay ten *yuan* ②(支付的款项) expenses; expenditure; outlay; disbursement: 国防～ expenditure on national defense/行政管理～ administrative expenditure/追加～ supplementary expenditure

【支绌】not enough; insufficient: 由于经费～ due to insufficient funds

【支点】[物] fulcrum

【支队】detachment: 游击～ a guerrilla detachment

【支付】pay (money); defray: ～水电费 pay for electricity and water/立即～ immediate payment/现金～ cash payment/全部货款用贷款～。All purchases shall be financed with the proceeds of loan. ◇～手段 means of payment/～协定 payments agreement/～日期 date of disbursement/～命令 payment order/～凭证 payment instrument/～损害赔偿金 payment of damages

【支行】subbranch

【支架】support; stand; trestle: 照相机～ camera stand/自行车～ prop stand of a bicycle

【支解】dismemberent: ～遗骸 dismember the remains

【支离破碎】torn to pieces; broken up; fragmented

【支流】①(河的支流) tributary; affluent ②(次要事物) minor aspects; nonessentials: 分清主流和～ distinguish between the essentials and the nonessential

【支脉】offshoot; branch range

【支派】①(分支) branch; sect; offshoot ②(支使) order; send; dispatch

【支配】①(安排) arrange; allocate; budget: 她善于～时间。She is good at budgeting her time. ②(控制) control; dominate; govern: ～某人 control sb./受人～ be controlled by others

【支票】check; cheque: 保付～ certified check/不记名～ bearer check/划线～ crossed check/记名～ order check/空白～ blank check/空头～ rubber check/来人～ bearer check/旅行～ traveller's check/开～ write a check; draw a check/开一张三千元的～ draw a check for three thousand *yuan* payable/用～兑现金 cash a check ◇～簿 checkbook/～票根 stub of a check; counterfoil

【支气管】bronchus ◇～癌 bronchiolar carcinoma/～出血 bronchorrhagia/～肺炎 bronchopneumonia/～扩张 bronchiectasis/～哮喘 bronchial asthma/～炎 bronchitis

【支渠】branch canal

【支取】draw: ～存款 draw one's deposit

【支使】①(命令人做事) order about: 不听他的～ refuse to be ordered about by him ②(让人到别处去) send away; put sb. off: 把他们～走 send them away

【支数】[纺] number; count: 纱线～ yarn number; yarn size

【支吾】prevaricate; equivocate; hum and haw: ～其词 speak evasively; hum and haw

【支线】branch line; feeder: 公路～ feeder highway/铁路～ feeder railway

【支应】①(应付) cope with; deal with ②(守候) wait on; attend to ③(搪塞) prevaricate; equivocate

【支援】support; assist; help: ～灾区 help the disaster area/互相～ help and support each other

【支柱】pillar; prop; mainstay: 矿用～ pit prop

【支柱根】[植] prop root

枝 ①(枝子) branch; twig: 树～ branches ②[量]一～玫瑰花 a spray of rose ②(杆状的东西)一～步枪 a rifle

【枝杈】branch; twig

【枝辞蔓语】a lengthy and confused talk

【枝根】ramose root

【枝接】[农] scion grafting
【枝节】①(次要的事) minor matters: ～问题 a minor problem; a side issue ②(麻烦事情) complication; unexpected difficulty: 横生～ raise unexpected difficulties; deliberately complicate an issue; create side issues
【枝蔓】branches and tendrils; complicated and confused
【枝条】branch; twig
【枝头】on the branch
【枝桠】branch; twig
【枝叶】①(枝子和叶子) branches and leaves ②(琐碎的事情) nonessentials; minor details
【枝状闪电】streak lightning
【枝子】branch; twig

肢 limb: 四～ the four limbs
【肢端肥大症】acromegaly; megalakria
【肢解】dismemberment
【肢势】[牧] standing posture
【肢体】①(四肢) limbs ②(四肢和躯干) limbs and trunk

只 ①(单独的) single; one only: ～字不提 not say a single word ②[量] 两～耳朵 two ears / 一～鸡 a chicken / 一～手提箱 a suitcase
【只身】alone; by oneself: ～在外 be away from home all by oneself
【只言片语】a word or two; a few isolated words and phrases: 未留下～ leave behind not even a single word / 只听见～ catch only a word or two
【只知其一,不知其二】one cannot see the wood for the trees

织 ①(纺织) weave: ～布 weaving cotton cloth / 纺～ spinning and weaving ②(编结) knit: ～毛衣 knit a sweater / ～毛袜子 knit wool into socks
【织补】darning; invisible mending
【织布】weaving cotton cloth; weaving ◇ ～工 weaver
【织布鸟】weaverbird
【织带机】inkle loom; webbing loom
【织机】loom: 多梭箱～ multiple box loom
【织锦】①(锦缎) brocade ②(织锦画) picture-weaving in silk: 风景～ landscape woven in silk ◇ ～厂 brocade mill
【织锦缎】tapestry satin
【织女】woman weaver
【织女星】[天] Vega
【织袜机】hosiery machine
【织物】fabric: 机织～ woven fabric
【织造】weaving ◇ ～厂 weaving mill
【织针】knitting needle

知 ①(知道) know; realize; be aware of: ～过必改 always correct an error when one becomes aware of it / 为…所熟～ be known to / 无所不～ know everything; omniscient ②(使知道) inform; notify; tell: 通～ inform; notify ③(知识) knowledge: 求～ seek knowledge / 对于历史他所～不多。 He has not much knowledge of history.
【知彼知己,百战不殆】know the enemy and know yourself, and you can fight a hundred battles with no danger of defeat
【知道】know; realize; be aware of: 不可能～ there is no knowing / 她～你的意思。 She knows what you mean. / 我怎么会～呢? How can I know?
【知底】know the inside story; be in the know
【知法犯法】knowingly violate the law; deliberately break the law
【知根知底】know through and through
【知更鸟】robin; redbreast
【知己】①(情谊深) intimate; understanding: ～的朋友 bosom friend ②(知己的人) bosom friend; intimate friend
【知交】bosom friend
【知觉】①(感觉) consciousness: 恢复～ recover consciousness; come to / 失去～ lose consciousness; pass out ②[心](感性认识) perception ◇ ～常性 perceptual constancies / ～异常 abnormal perception
【知了】[动] cicada
【知名】well-known; noted; celebrated; famous: ～人士 well-know personage; public figure; celebrity
【知难而进】press forward in the face of difficulties; advance despite difficulties
【知难而退】beat a retreat in the face of difficulties; shrink back from difficulties
【知其然,不知其所以然】know that sth. is so but not why is so; know the hows but not the whys
【知其一,不知其二】know only one aspect of a thing; have only a one-sided view
【知情】know the facts of a case or the details of an incident; be in the know: ～不举 conceal what one knows of a case
【知情人】person in the know; insider: ～的证据 state's evidence

【知情达理】reasonable; sensible
【知趣】know how to behave in a delicate situation; be sensible; be tactful
【知人善任】know one's subordinates well enough to assign them jobs commensurate with their abilities
【知人之长】be sensible of other's merits
【知人之明】ability to appreciate a person's character and capability; a keen insight into a person's character
【知人知面不知心】you may know a person's face but not his heart; one may know a person for a long time without understanding his true nature
【知识】①(认识和经验的总和) knowledge: ～就是力量。Knowledge is power. ②(有关学术文化的) pertaining to learning or culture; intellectual ◇ ～界 intellectual circles; the intelligentsia
【知识分子】intellectual; the intelligentsia
【知疼知热】feel for another person like oneself
【知无不言,言无不尽】say all you know and say it without reserve
【知悉】know; learn; be informed of: 业已～ have already learned of the matter
【知晓】know; be aware of; understand
【知心朋友】intimate friend
【知行】[哲] knowing and doing ◇ ～统一观 the theory of the unity of knowing and doing
【知音】a friend keenly appreciative of one's talents; bosom friend
【知足】be content with one's lot: ～常乐 happy is he who is content

蜘

【蜘蛛】spider ◇ ～丝 the thread of a spider web; cobweb / ～网 spider web; cobweb
【蜘蛛抱蛋】[植] aspidistra

栀

【栀子】[植] Cape jasmine

指

【指甲】nail: 手～ fingernail / 脚～ toenail / 剪～ clip the nails / 修～ have a manicure ◇ ～锉 nail clipper / ～刀 nail clippers / ～夹 nail clipper / ～剪 nail scissors / ～油 nail polish
【指甲花】[植] garden balsam

脂 ①(动植物所含的油质) fat; grease; tallow: 油～ fat; grease ②(胭脂) rouge
【脂蛋白】lipoprotein
【脂肪】fat: 动物～ animal fat / 植物～ vegetable fat ◇ ～肝 [医] fatty liver / ～酶 [生化] lipase / ～酸 fatty acid / ～心 fatty heart / ～组织 [生理] adipose tissue
【脂粉】rouge and powder; cosmetics
【脂膏】①(脂肪) fat; grease ②(比喻人的血汗和劳动果实) fruits of the people's labor; wealth of the people
【脂瘤】[医] lipoma
【脂血症】lipemia; lipidemia

zhí

职 ①(职务) duty; job: 做好本～工作 do one's job well ②(职位) post; office: 任～期间 tenure of office / 在～干部 cadres at their post
【职别】official rank
【职称】the title of a technical or professional post ◇ ～评审委员会 evaluation committee of professional titles
【职工】staff and workers; workers and staff members ◇ ～代表大会 workers congress / ～夜校 evening school for workers and staff members
【职能】function: 货币的～ the functions of money
【职权】powers of office; authority of office: ～的划分 division of functions and powers / 超越～ overstep one's authority; exceed one's powers / 行使～ exercise one's functions and powers
【职权范围】limits of one's functions and powers; terms of reference
【职守】post; duty: 擅离～ leave one's post without permission / 忠于～ be faithful in the discharge of one's duties
【职位】position; post: 他在银行谋得一个～。He got a position in the bank.
【职务】post; duties; job: 担任校长～ shoulder the duty of a headmaster
【职衔】post and rank
【职业】occupation; profession; vocation: ～上的特权 professional privilege ◇ ～安全法 job-safety law / ～病 occupational disease / ～道德 professional ethics / ～教育 vocational education / ～外交官 career diplomat / ～性过失 occupational negligence / ～学校 vocational school / ～运动员 professional athlete; professional / ～责任 professional responsibility / ～罪犯 career criminals
【职员】office worker; staff member; functiona-

ry：学校的教～ the teaching staff of a school
【职责】duty；obligation；responsibility：神圣的～sacred duty／应尽的～ bounden duty ◇～范围 scope of official duty

直 ①(成直线的) straight ②(垂直) vertical；perpendicular ③(挺直) straighten：～起腰来 straighten one's back；stand up straight ④(公正) just；upright：是非曲～ rights and wrongs；truth and falsehood／正～ upright；fair-minded ⑤(直爽) frank；straightforward：～抒已见 state one's views frankly／心～口快 plain-spoken and straightforward；frank and outspoken／我对他～说。I told him straight. ⑥(直接；一直) directly；straight：～射的太阳光 direct rays of the sun／这条路～通火车站。This road leads directly to the railway station. ⑦(不断地) continuously ⑧(汉字笔画，即"竖") vertical stroke ⑨(简直) just；simply
【直播】[农] direct seeding
【直肠】[生理] rectum ◇～癌 carcinoma of the rectum／～镜[医] proctoscope／～炎 proctitis；rectitis／～造口术 proctostomy；rectostomy
【直尺】straightedge
【直达】through；nonstop：～上海的火车 a through train to Shanghai ◇～车 through train or through bus／～车票 through ticket／～路线 through route
【直到】①(直到某时才做或才完成某事) until：她～晚上很晚才回来。She didn't come back until late in the evening. ②(直到某时还未完成) up to：～现在我还没收到她的信。I haven't received any letter from her up to the present.
【直根】[植] taproot
【直贡呢】[纺] venetian
【直观】directly perceived through the senses；audio-visual ◇～教具 aids to object teaching；audio-visual aids／～教学 object teaching／～教学课 object lesson
【直角】[数] right angle ◇～尺 square／～三角形 right triangle
【直接】direct；immediate：～会晤 meet sb. in person／～原因 immediate cause；direct cause／他～前往纽约。He went straight to New York without stopping anywhere. ◇～宾语[语] direct object／～成本 direct cost／～后果 immediate consequence／～反应 direct reaction／～教学法 direct method／～经验[哲] direct experience／～起飞 rolling takeoff；rolling start／～染料 direct dye／～税 direct tax／～损失 direct loss／～投资 direct investment／～推理 immediate reasoning／～选举 direct election／～责任 direct liability／～照明 direct lighting／～证据 direct evidence／～着陆 straight-in landing
【直截了当】straightforward；blunt；point-blank：～的回答 a point-blank answer／～地拒绝 flatly reject
【直径】diameter：大～ major diameter／小～ minor diameter
【直觉】[心] intuition ◇～主义 intuitionism
【直来直去】go and return without undue delay
【直立茎】[植] erect stem
【直流电】direct current (D.C.)
【直眉瞪眼】①(发脾气) stare in anger；fume ②(发呆) stare blankly；be in a daze；be stupefied
【直认不讳】plead guilty to a charge；admit frankly
【直上云霄】soar straight up into the sky
【直升机】helicopter；copter：乘～去试验场 helicopter to testing ground ◇～航空母舰 helicopter carrier／～机场 heliport／～驾驶员 helicopterist／～运载 helilift
【直视】look steadily at
【直抒己见】state one's views frankly
【直属】directly under；directly subordinate to：国务院～机关 departments directly under the State Council
【直率】frank；candid；straightforward
【直爽】frank；candid；straightforward；forthright：性格～ forthright in character
【直挺挺】straight；stiff；bolt upright：～地躺在地上 lie outstretched on the ground／～地站着 stand ramrod straight
【直系】direct line ◇～继承人 heir of line；heir of one's body／～亲属 lineal relative
【直体】[体] stretched
【直辖】directly under the jurisdiction of：文化部～机构 organizations directly under the Ministry of Culture ◇～市 municipality directly under the Central Government
【直线】①(不弯曲的线) straight line ②(急剧的) steep；sharp：生产～上升 sharp rise in production ◇～尺 straightedge rule／～飞行 rectilinear flight／～爬高 rectilinear climb／～形 rectilinear figure／～运动[物] rectilinear motion
【直性子】①(直爽) straightforward；downright；forthright ②(直爽的人) straightforward person

【直言不讳】speak without reservation; not mince words; call a spade a spade
【直言劝谏】using blunt words to remonstrate
【直译】literal translation; word-for-word translation

植 ①(栽种) plant; grow: ～树 plant trees / 移～ transplant ②(树立) set up; establish
【植保】[简] plant protection; crop protection ◇～机械 equipment for plant protection / ～员 plant protector
【植被】[植] vegetation
【植皮】[医] skin-grafting; make skin grafts ◇～刀 dermatome / ～术 skin grafting
【植入】embedding
【植树】[林] tree planting: ～机 tree-planting machine; tree planter / ～运动 afforestation campaign / ～造林 afforestation
【植物】plant; flora: 草本～ herbaceous plants / 单子叶～ monocotyledon / 蕨类～ ferns; pteridophyte / 双子叶～ dicotyledon / 苔藓～ mosses; bryophyte / 种子～ seed-bearing plants ◇～保护 plant protection / ～病毒 plant virus / ～病害 plant disease / ～病理学 plant pathology / ～分类学 plant taxonomy / ～化学 phytochemistry / ～检疫 plant quarantine / ～胶 vegetable gum; vegetable glue / ～界 plant kingdom; vegetable kingdom / ～净化[环保] plant purification / ～群落 plant community / ～群体 plant population / ～生态学 plant ecology / ～形态学 morphological botany / ～学 botany / ～学家 botanist / ～油 vegetable oil / ～育种 plant breeding / ～园 botanical garden
【植物性神经】[生理] autonomic nerve
【植物岩】phytogenic rocks
【植字】set types for printing ◇～板 a printing plate of set types

殖 breed; multiply: 繁～ multiply; propagate / 生～ breed; reproduce
【殖民】establish a colony; colonize: 非～化 decolonize; decolonization ◇～国家 colonialist power / ～战争 colonialist war
【殖民地】colony ◇～国家 colonial country / ～人民 people under colonial rule; colonial people
【殖民主义】colonialism: 新～ new colonialism; neocolonialism ◇～者 colonialist

值 ①(价值) value ②(货物和价钱相当) be worth: 不～一提 be not worth mentioning / 分文不～ be not worth a farthing; be worth nothing; not worth a cent ③(遇到) happen to: ～此危急之秋 at such a critical juncture / 我上次拜访,正～她外出。She happened to be out when I called. ④(值班) be on duty; take one's turn at sth.: ～夜 be on night duty; be on the night shift ⑤[数](用数字表示的量) value: 求 X 之～ determine the value of X
【值班】be on duty ◇～员 person on duty
【值得】be worth; merit; deserve: ～怀疑 be open to doubt / ～买 be worth buying / ～一提 to speak of; deserve to be mentioned / ～赞许 deserve commendation; be praiseworthy / ～仔细考虑 warrant careful consideration / 不～一驳 be not worth refuting / 我们费了很大力量,但是～。We worked hard but it was worth it.
【值钱】costly; valuable: 不～ worthless; valueless
【值勤】be on duty ◇～交通警 policeman on point duty / ～人员 personnel on duty
【值日】be on duty for the day; be one's turn to be on duty: 轮流～ take turns to be on duty ◇～表 rota; duty roster / ～生 student on duty
【值星】be on duty for the week

执 ①(拿着) hold; grasp ②(执掌) take charge of; direct; manage: ～教 be a teacher; teach ③(坚持) stick to; persist: ～意不肯 firmly refuse ④(执行) carry out; observe ⑤(凭单) written acknowledgement: 回～ receipt
【执笔】write; do the actual writing
【执法】enforce the law: ～如山 enforce the law strictly ◇～不公 denial of justice / ～机关 law enforcement agency / ～人员 law enforcement officials / ～者 law-executor
【执迷不悟】obstinately stick to a wrong course; be perverse; refuse to come to one's senses
【执拗】stubborn; pigheaded; willful
【执手同游】saunter holding each other's hand
【执行】carry out; execute; implement: ～法律 dispensing justice / ～决议 implement a decision / ～命令 execute an order / ～判决 enforcement of judgment / ～任务 carry out a task; perform a mission ◇～机构 executive body / ～机关 executive organ / ～秘书 executive secretary / ～委员会 executive committee / ～主席 executive chairman / ～者 executor
【执意】insist on; be determined to; be bent on:

她～要去游泳。She insists on going to swim.
【执掌】wield; be in control of: ～兵权 wield military power
【执照】license; permit: 申请驾驶～ apply for a driver's license ◇ ～税 fee of permit
【执政】be in power; be in office; be at the helm of the state ◇ ～党 the party in power; the ruling party

指

【指头】①(手指) finger ②(脚指) toe

踯

【踯躅】[书] walk to and fro; loiter around: ～街头 tramp the streets

侄 brother's son; nephew

【侄女】brother's daughter; niece
【侄孙】brother's grandson; grandnephew
【侄孙女】brother's granddaughter; grandniece
【侄媳妇】wife of brother's son; nephew's wife
【侄子】brother's son; nephew

zhǐ

止 ①(停止) stop; cease discontinue ②(到…为止) to; till: 到目前为～ to date; till now; up to the present / 到年底为～ by the end of the year ③(仅;只) only 不～一次 not just once; more than once

【止步】halt; stop; go no further: ～不前 halt; stand / 游人～ no visitors; out of bounds
【止喘药】antasthmatic neurosis; antiasthmatic
【止动杆】arresting lever; gag lever; stop arm
【止动机构】[机] stop motion
【止付】[经] stop payment ◇ ～通知书 stop-payment notice / ～支票 stopped check; stopping payment of check
【止回阀】check valve; non-return valve
【止境】end; limit: 进步无～。There is no end to progress.
【止咳】relieve a cough ◇ ～糖浆 cough syrup
【止渴】quench one's thirst
【止痛】relieve pain; stop pain ◇ ～药 anodyne; analgesic; pain-killer
【止息】cease; stop
【止泻药】antidiarrheal
【止血】stop bleeding; stanch bleeding ◇ ～带 tourniquet / ～棉 hemostatic cotton / ～器 hemostat / ～钳 hemostat / ～散 styptic powder / ～纱布 hemostatic gauze / ～药 hemostatic
【止痒】relieve itching ◇ ～药 antipruritic
【止住】stop; halt; desist; bring to a stop

趾 ①(脚指头) toe ②(脚) foot

【趾高气扬】strut about and give oneself airs; be swollen with arrogance
【趾骨】phalanx
【趾甲】toenail

只 [副] only; merely; just: 我～看到了他,但没时间和他说话。I just saw him, but had no time to speak to him. / 我～能告诉你我所知道的。I can only tell you what I know. / 我～问了他的姓名。I merely asked his name.

【只不过】only; just; merely; no more than: 那～是句玩笑话。That's only a joke. / 她～是个孩子。She is just a child.
【只得】have no alternative but to; be obliged to; have to
【只顾】①(专一不变) be absorbed in: 他～看书,连饭都忘了吃。He was completely absorbed in reading and forgot even to take his meal. ②(仅仅顾到) merely; simply; only cared for: 他～个人利益。He cares only for his own interests.
【只管】①(尽管) by all means; not hesitate to: 有什么问题～问。If you have any questions, don't hesitate to ask. ②(只顾) merely; simply
【只好】have to; be forced to: ～另想办法 cannot but seek other means / ～作罢 be forced to give up
【只怕】be afraid of only one thing
【只是】①(仅仅是) merely; only; just: 她来访～为了礼貌。She called on us merely for the sake of courtesy. / 这～托词而已。It is merely a pretense. ②(强调限于某种情况或范围) simply: 这～手续问题。This is simply a question of formalities. ③(但是) however; but then; only: 我很想去,～忙不过来。I would very much like to go there, however my hands are full.
【只许州官放火,不许百姓点灯】the magistrates are free to burn down houses, while the common people are forbidden even to light lamps; one may steal a horse while another may not look over the hedge
【只要】[连] so long as; provided: ～你做完工作就可以走了。You may go, provided your work is done.
【只要功夫深,铁杵磨成针】constant grinding can turn an iron rod into a needle; perseverance spells success

【只争朝夕】seize the day, seize the hour; seize every minute; race against time

枳 [植] trifoliate orange

咫

【咫尺】[书] very close: 近在～ close at hand

【咫尺山河】physically very near but separated as if by long distances

【咫尺天涯】a short distance away, and yet poles apart; see little of each other though living nearby

【咫尺之间】within a foot of

纸 ①(纸张) paper: 包装～ wrapper / 包装用牛皮～ kraft wrapping / 玻璃～ cellophane / 地图～ atlas paper / 复写～ carbon paper / 复写用薄～ manifold paper / 航空信～ air mail paper / 绘图～ drawing paper / 胶～ adhesive paper / 卡片～ card paper / 蜡光～ glazed paper / 美术～ art paper / 名片～ alabaster paper / 牛皮～ kraft paper / 砂～ abrasive paper / 晒图～ blueprint paper / 双面复写～ articulating paper / 水彩画～ art drawing paper / 吸水～ absorbent paper / 新闻～ newsprint / 帐簿～ account book paper / 证券～ bond paper / 皱纹～ crepe paper / 竹浆～ Indian paper / 一张白～ a blank sheet of paper ②[量]一～空文 a mere scrap of paper

【纸板】paperboard; cardboard: 波纹～ corrugated cardboard ◇ ～盒 cardboard case; cardboard box; carton

【纸版】[印] paper mold; paper matrix

【纸包不住火】you can't wrap fire in paper; there is no concealing the truth; truth will out

【纸币】paper money; paper currency; note: 不兑现～ fiat money / 发行～ issue notes

【纸带】[计算机] paper- tape ◇ ～穿孔机 paper- tape punch; tape punch / ～打印机 bound printer / ～校对机 paper-tape verifier / ～机 paper-tape unit; paper-tape equipment

【纸花】paper flower

【纸婚】the first wedding anniversary

【纸浆】paper pulp; pulp ◇ ～板 pulp board / ～厂 pulp mill / ～筛滤器 pulp strainer / ～原材 pulpwood / ～制造机 macerator

【纸捻】spill of rolled paper used to light a pipe, etc.; spill

【纸牌】playing cards

【纸片】scraps of paper

【纸上谈兵】fight only on paper; be an armchair strategist; engage in idle theorizing

【纸绳】paper string

【纸条】a slip of paper

【纸通货】paper currency

【纸型】[印] paper mold; paper matrix ◇ ～干燥机 scorcher

【纸烟】cigarette

【纸鱼】silverfish; fish moth

【纸张】paper

【纸醉金迷】luxury and dissipation

旨 ①(意义;用意;目的) purport; purpose; aim: 主～ purport; main purpose / ～在解放全人类 for the purpose of liberating mankind ②(滋味美) tasty; delicious: ～酒 excellent wine ③(意旨) decree: 圣～ imperial decree

【旨趣】[书] purport; objective: ～相同 essential thoughts are alike

【旨意】decree; order

酯 [化] ester: 聚～ polyester ◇ ～化 esterification / ～酶 esterase

指 ①(手指头) finger: 屈～可数 can be counted on one's fingers; very few ②(指向) point at; point to: 向南～ point to the south / 向前～ point forward ③(一个手指的宽度) fingerbreadth; digit: 二～宽 two digits wide ④(指点) indicate; point out; refer to ⑤(依靠) depend on; count on: 大家就～着你帮忙。We're counting on your help.

【指北针】compass

【指标】target; quota; norm; index: 生产～ production target / 质量～ quality index

【指不胜屈】too numerous to be counted on the fingers; a great many

【指出】point out; lay one's finger on: ～某人的弱点 lay one's fingers on sb.'s weak spot / ～正确方向 point out the correct way

【指导】guide; direct: 多谢你的～。I am very grateful to you for your advice. ◇ ～路线 guideline / ～思想 guiding ideology / ～作用 directive function

【指导员】[军] political instructor

【指点】give directions; show how: 多亏你的～ thanks to your directions

【指定】appoint; assign: ～某人任某职 designate sb. to an office / ～某人做某事 assign sb. to do sth. / ～一个日期 appoint a date ◇ ～代理人 authorized agent / ～管辖 designated jurisdiction / ～继承人 designated heir / ～监护

人 designated guardian / ～买主 nominee buyer
【指东说西】make concealed reference to sth.
【指法】[乐] fingering
【指骨】[生理] phalanx
【指猴】aye-aye
【指画】point at; point to
【指环】ring
【指挥】①(发令调度) command; direct; conduct: ～一个战役 direct a battle ②(发令调度的人) commander; director ③(乐队指挥) conductor ◇ ～棒[乐] baton / ～车 command car / ～刀 officer's sword / ～官 commanding officer; commander / ～所 command post / ～塔台[航空] control tower / ～系统 command system / ～员 commander
【指挥部】command post; headquarters: 防空～ air defense command / 前沿～ forward command post
【指尖】finger tip
【指教】[套] give advice or comments: 请您多加～。Kindly give us your advice.
【指靠】depend on; look to; count on
【指控】accuse; charge: ～某人犯罪 accuse sb. of a crime ◇ ～人 accuser
【指叩诊】finger percussion
【指令】①(指示;命令) instruct; order; direct ②[计算机](命令;指示) instruction: 遥控～ remote control command ◇ ～码 instruction code / ～系统 instruction repertoire; instruction set
【指令舱】[宇航] command module
【指令性计划】mandatory planning
【指鹿为马】call a stag a horse; deliberately misrepresent
【指路明灯】beacon light; beacon
【指路牌】signpost; fingerpost; guidepost
【指名】mention by name; name: ～道姓 name names; mention sb.'s name / ～漫骂 abuse sb. by mentioning his name
【指明】show clearly; demonstrate; point out: ～出路 point the way out / ～两者之间的差别 show clearly the difference between the two
【指南】guide; guidebook: 行动的～ guide to action
【指南针】compass
【指派】appoint; name; designate: ～某人任某职 designate sb. to an office
【指日可成】can finish in a few days
【指日可待】be just round the corner; can be expected soon
【指桑骂槐】point at the mulberry and abuse the locust; point at one but abuse another; make oblique accusations
【指使】instigate; incite; put sb. up to sth.: ～某人做坏事 instigate sb. to do evil / 受人～ under sb.'s instigation; act on sb.'s instigation
【指示】①(指给人看) indicate; point out ②(指示下级) instruct: ～某人做某事 instruct sb. to do sth. ③(指示的话或文字) directive; instructions: 发出～ give instructions / 接到～ receive instructions ◇ ～板[仪表] indicator board / ～代词[语] demonstrative pronoun / ～灯 pilot lamp; indicator lamp / ～功率[机] indicated power / ～管 indicator tube / ～剂[化] indicator / ～植物 indicator plant
【指示器】indicator: 刻度盘～ dial indicator / 液面～ level indicator
【指手划脚】①(用手势示意) make gestures; gesticulate: ～地高谈阔论 talk volubly with animated gestures ②(轻率地指点、批评) make indiscreet remarks or criticisms
【指数】①[经](比数) index number; index: 生活费～ cost of living index / 物价～ price index / 综合～ composite index ②[数] exponent: 负～ negative exponent / 正～ positive exponent ◇ ～变化 index movement / ～函数[数] exponential function / ～化 indexation / ～化制度 indexation system / ～律[数] index law
【指望】①(盼望) look to; count on; look forward to: 不～他帮忙 not look to him to help us; not count on him for help ②(盼头) prospect; hope: 他的病还有～。There is still hope of his recovery.
【指纹】①(手指上的纹理) loops and whorls on a finger ②(指纹留下的痕迹) fingerprint ◇ ～档案 fingerprint file / ～鉴定 fingerprint identification / ～学 dactylography
【指引】point (the way); guide; show: ～我们前进 guide us forward
【指印】fingerprint; finer mark: 按～ make a fingerprint
【指责】censure; criticize; find fault with: ～某人做某事 censure sb. for doing sth. / 横加～ make unwarranted charges
【指战员】officers and men
【指针】①(钟、表的针) indicator; pointer; needle ②(分辨正误的依据) guiding principle; guide: 今后行动的～ guiding principle for future work ◇ ～式温度计 dial thermometer
【指诊】[医] touch: 肛门～ rectal touch / 阴道～ vaginal touch

【指正】①(指出错误,使之改正) point out mistakes so that they can be corrected ②[套](请人批评自己的作品或意见) make a comment or criticism: 请大家给予～。 Please oblige me with your valuable comments.

【指指点点】gesticulating

zhì

滞 stagnant; sluggish

【滞洪】[水] flood detention ◇ ～区 detention basin; retarding basin

【滞留】be detained; be held up

【滞纳金】fine for delaying payment; fine for paying late; overdue fine

【滞期费】demurrage charges

【滞销】unsalable; unmarketable ◇ ～货 a drug on the market; unsalable goods

【滞滞泥泥】sticky in doing things

治 ①(治理) rule; govern; administer; manage: ～国 administer a country / ～家 manage a household / 分而～之 divide and rule / 自～ self-government; autonomy ②(安定或太平) order; peace ③(医治) treat; cure: 给某人～病 treat sb. for his illness; cure sb. of his disease ④(控制;消灭) control; harness; wipe out: ～蝗 eliminate locusts / ～碱 combat alkalinity / ～沙 sand-control / ～山～水 transform mountains and tame rivers / ～土改水 bring the water under control and improve the soil ⑤(惩办) punish

【治安】public order; public security: 维持～ maintain public order ◇ ～保卫人员 guardian of the peace; security personnel / ～保卫委员会 public security committee / ～机关 law enforcement office

【治本】effect a permanent cure; get at the root; take radical measures

【治标】merely alleviate the symptoms of an illness; bring about a temporary solution; take stopgap measures

【治病救人】cure the sickness to save the patient; help sb. mend his ways to save him

【治国安民】run the country well and give the people peace and security

【治军】direct military affairs; direct troops

【治理】①(统治;管理) administer; govern: ～国家 administer a country; run a state ②(处理;整修) harness; bring under control; put in order: ～黄河 bring the Huanghe River under control

【治疗】treat; cure: ～见效 respond well to treatment / 病人得到及时～ receive timely treatment / 对病人及时～ give timely treatment / 及时～ timely treatment / 精神～ psychotherapy / 住院～ be hospitalized / 新～法 new cure / 她正在进行～。 She is under treatment ◇ ～错误 malpractice / ～效果 therapeutic effect / ～学 therapeutics

【治丧】make funeral arrangements ◇ ～委员会 funeral committee

【治水】regulate rivers and watercourses; prevent floods by water control

【治外法权】extraterritoriality; exterritoriality; extrality

【治学】pursue one's studies; do scholarly research

【治愈率】[医] cure rate: ～达百分之九十七 a 97 percent cure rate / 提高～ improve the cure rate

【治罪】punish sb.

志 ①(志向) will; aspiration; ideal: 立～ be determined to / 人各有～。 Each has his own ideal. ②(记) keep in mind: 永～不忘 forever bear in mind ③(文字记录) records; annals: 县～ annals of a county ④(标志;记号) mark; sign

【志哀】indicate mourning: 下半旗～ fly a flag at half-mast as a sign of mourning

【志不可屈】indomitable will or ambition

【志大才疏】have great ambition but little talent; have high aspirations but little ability

【志得意满】fully satisfied or contented; complacent

【志坚如钢】have an iron will; have a will of steel

【志气】aspiration; ambition: 有～ have high aspirations

【志趣】aspiration and interest; inclination; bent

【志士】person of ideals and integrity: ～仁人 people with lofty ideals / 爱国～ noble-minded patriot

【志同道合】cherish the same ideals and follow the same path; have a common goal

【志向】aspiration; ideal; ambition: 远大的～ lofty ideals; lofty aspirations

【志愿】①(志向和愿望) aspiration; wish; idea ②(自愿) do sth. of one's own free will; volunteer ◇ ～兵 volunteer

【志愿军】people who volunteer to fight in an-

other country; volunteers
【志愿书】application form
【志在千里】one aims for the far future

痣 nevus; mole: 色～ pigmented mole / 胎～ birthmark ◇ ～癌 nevocarcinoma

痔 hemorrhoids; piles: 内～ internal piles / 外～ external piles
【痔疮】hemorrhoids; piles
【痔漏】[医] anal fistula
【痔切除术】[医] hemorrhodectomy

栉
【栉比】[书] placed closely side by side
【栉风沐雨】be combed by the wind and washed by the rain; travel or work despite wind and rain

至 ①(到) to; until; till: 由东～西 from east to west / 自春～冬 from spring till winter ②(极;最) extremely; most: 感人～深 deeply moving; extremely moving / 欢迎之～。 You are most welcome.
【至宝】most valuable treasure: 如获～ feel as if one had found a priceless treasure
【至诚】①(诚心诚意) complete sincerity: 出于～ in all sincerity; from the bottom of one's heart ②(诚恳) sincere; straightforward: ～的朋友 a sincere friend
【至迟】at the latest; no later than: 她～明天回来。 She will be back no later than tomorrow.
【至多】[副] at most; no more than: ～五个 at most five
【至高无上】most lofty; paramount; supreme: ～的权力 absolute power / ～的事业 most lofty undertaking
【至关紧要】of utmost importance
【至交】most intimate friend; best friend
【至今】up to now; to this day; so far
【至理名言】famous dictum; maxim; axiom; golden saying
【至亲】very close relative; close kin: ～好友 close relatives and good friends
【至上】supreme; the highest
【至少】at least: ～二百 at least two hundred
【至死】unto death; till death: ～不屈 not yield even unto death / ～不渝 will never change until death
【至友】the closest friend
【至于】①(表示达到某种程度) go so far as to: 她不～不来上学。 She wouldn't go so far as to refuse to attend school. ②(表示另提一件事) as for; as to

窒
【窒闷】close; stuffy
【窒塞】stop up; block
【窒息】stifle; suffocate ◇ ～弹 stifling bomb / ～死亡 death by suffocation / ～性毒气 asphyxiating gas / ～药 asphyxiant

桎
【桎梏】[书] fetters and handcuffs; shackles: 打碎精神上的～ smash spiritual shackles

蛭 leech
【蛭石】[矿] vermiculite

致 ①(给与) send; extend; deliver: ～电 send a telegram / ～欢迎词 deliver a welcoming speech ②(集中力量、意志等于某个方面) devote: 专心～志 be wholeheartedly devoted to; be wholly absorbed in ③(招致) incur; result in; cause: 招～重大损失 incur a heavy loss ④(精密;精细) fine; delicate: 精～ exquisite ⑤(情趣) manner or style that engages attention or arouses interest: 兴～ interest; mood for enjoyment
【致癌物质】[医] carcinogen; carcinogenic substance
【致病菌】[医] pathogenic bacteria
【致辞】make a speech: 新年～ New Year message / 请来宾向大会～ call upon a guest to address the conference
【致富】become rich; acquire wealth
【致幻剂】hallucinogen
【致敬】salute; pay one's respects to; pay tribute to: 向…～ salute to ◇ ～电 message of greeting
【致冷】refrigeration ◇ ～剂 refrigerant / ～循环 refrigeration cycle
【致力】devote oneself to; work for: ～力于科研工作 devote oneself to the scientific study
【致敏物质】sensitizer
【致命】causing death; fatal; mortal; deadly: ～的打击 a deadly blow / ～的弱点 fatal weakness / ～伤 a mortal wound / ～凶器 lethal weapon / ～之处 Achilles' heel / 给予～的打击 deal a deadly blow to
【致使】cause; result in
【致死】causing death; lethal; deadly: 一剂～的毒药 a lethal dose of poison ◇ ～剂量 lethal dose / ～性毒气 lethal gas / ～原因 cause of

death
【致谢】express one's thanks; extend thanks to
【致意】give one's regards; present one's compliments; send one's greetings: 点头～ nod a greeting / 挥手～ wave a greeting

掷 throw; cast: 弃～ cast aside; throw away / 投～ throw; fling
【掷币猜先】[体] toss
【掷标枪】[体] javelin throw
【掷弹筒】[军] granade launcher; grenade discharger
【掷界外球】[足球] throw-in
【掷铁饼】[体] discus throw

挚
【挚友】intimate friend; bosom friend: ～良朋 intimate friends and good companions

置 ①(搁;放) place; put: 安～ find a place for; help sb. to settle down / 搁～ put aside; shelve; pigeonhole ②(设立;布置) set up; establish; install: 设～障碍 set up obstacles ③(购置) buy: 购～ add to one's belongings
【置办】buy; purchase: ～家具 buy furniture / ～图书仪器 procure books and instruments
【置备】purchase
【置换】[化] displacement; replacement
【置若罔闻】turn a deaf ear to; pay no heed to
【置身】place oneself; stay: ～其间 be involved / ～事外 stay aloof from the affair; keep out of the business / ～险境 put one's head into the lion's mouth
【置信】believe: 难以～ hard to believe; unbelievable; incredible
【置疑】doubt: 不容～ allow of no doubt; not be open to doubt
【置于死地】expose someone to mortal danger; doom a person to death
【置之不顾】leave out of account; disregard; ignore
【置之不理】ignore; brush aside; pay no attention to
【置之度外】give no thought to; have no regard to
【置之脑后】banish from one's mind; ignore and forget
【置之死地而后快】will be content with nothing less than sb.'s destruction

制 ①(制造) make; manufacture: ～图 mapmaking; drafting / 炼～ refine / 中国～ made in China ②(拟订;制定) work out; formulate ③(用强力约束) restrict; control: 控～ control / 限～ restrict ④(制度) system: 公～ the metric system
【制版】[印] plate making: 平版～ lithographic plate making
【制备】[化] preparation: 氧的～ preparation of oxygen
【制币】standard national currency ◇ ～厂 mint
【制表】[统计] tabulation
【制裁】sanction; punish: 经济～ economic sanctions / 实行～ apply sanctions; impose sanctions / 受到法律～ be punished according to law
【制成品】finished products; manufactured goods; manufactures
【制导】control and guide (a missile, etc.) ◇ ～系统 guidance system / ～炸弹 guided bomb
【制订】work out; formulate: 方案 work out a scheme
【制定】lay down; draw up; formulate; draft: ～操作规程 lay down operating rules / ～法令 enact decrees / ～法律 make laws / ～计划 work out a plan / ～条例 institution of regulations / ～宪法 draw up a constitution / ～政策 formulate a policy
【制动】apply the brake; brake ◇ ～阀 brake valve / ～缸 checking cylinder / ～火箭 retro-rocket / ～距离[交] braking distance / ～器 brake / ～瓦 brake scotch / ～液 brake fluid / ～闸 damper brake
【制度】system; institution: 规章～ rules and regulation / 社会～ social system / 社会主义～ socialist system
【制伏】check; subdue; bring under control: ～风沙 check wind and sand / ～一匹马 bring a horse under control
【制服】①(有规定式样的服装) uniform: 穿～的 in uniform ②(用强力压服) subdue; bring under control
【制服呢】uniform cloth
【制高点】[军] commanding elevation
【制革】process hides; tan ◇ ～厂 tannery / ～工人 tanner
【制海权】[军] mastery of the seas; command of the sea
【制剂】[药] preparation: 标准～ standard preparation
【制空权】[军] control of the air; air domination
【制冷】refrigeration ◇ ～剂 cold-producing

medium
【制粒机】granulator
【制片人】[电影] producer
【制品】products; goods: 钢铁～ iron and steel products／ 奶～ dairy products
【制胜】get the upper hand of; subdue: 出奇～ defeat one's opponent by a surprise move
【制糖】refine sugar ◇ ～厂 sugar refinery
【制图】①(制地图) charting; map-making ②(制图样) drafting ◇ ～学 cartography／～仪器 drawing instrument; drafting instrument
【制图员】①[地] cartographer ②[机][建] draftsman
【制销】[机] cotter
【制药】pharmacy ◇ ～厂 pharmaceutical factory／～学 pharmaceutics
【制约】restrict; condition: 互相～ condition each other; interact／受历史条件的～ be restricted by historical conditions
【制造】①(对原材料进行加工) make; manufacture: ～机器 make machines／中国～ made in China ②(人为造成的气氛或局面等) engineer; create; fabricate: ～的过程 manufacturing operation／～重重障碍 raise one obstacle after another／～分裂 foment splits／～假象 put up a false front／～紧张局势 create tension／～纠纷 create trouble; sow dissension／～内乱 stir up internal strife／～事故 engineer an incident／～伪证 fabrication of false evidence／～谣言 fabricate rumors／～舆论 mold public opinion ◇ ～商 manufacturer／～伪币者 false-coiner／～业 manufacturing industry
【制止】check; curb; prevent; stop: ～流沙 curb shifting sand／～某人做某事 stop sb. from doing sth.／～侵略 put an end to aggression／～通货膨胀 check inflation ◇ ～令 inhibition
【制作】make; manufacture: ～家具 make furniture

秩

【秩序】order; sequence: ～大乱 in total disorder／～井然 in perfect or apple-pie order／工作～ sequence of work／维持社会～ maintain public order
【秩序册】[体] program

智

wisdom; resourcefulness; wit: ～德齐备 have both wisdom and virtue／～尽技穷 at the end of one's tether／斗～ a battle of wits／明～ sensible; wise／足～多谋 wise and resourceful
【智齿】[生理] wisdom tooth
【智多星】resourceful person; mastermind
【智慧】wisdom; intelligence: ～胜强权 wisdom is better than strength／吸取群众的～ draw on the wisdom of the masses
【智力】intelligence; intellect ◇ ～测量 intelligence measure／～测验 intelligence test／～开发 intellectual development／～落后 mental retardation／～年龄 intelligence age (IA)／～缺陷 mental deficiency
【智利】Chile ◇ ～人 Chilean
【智略】wisdom and resourcefulness
【智谋】resourcefulness: 靠勇敢也靠～ rely on both courage and resourcefulness
【智囊】brain truster ◇ ～团 brain trust
【智能】intellect; intelligence ◇ ～终端[计算机] intelligent terminal; smart terminal
【智取】take by strategy
【智商】intelligence quotient (IQ)
【智神星】[天] Pallas
【智勇双全】both intelligent and courageous; both brave and resourceful
【智育】intellectual education; intellectual development: 使学生在德育、～、体育几方面都得到发展 enable the students to develop morally, intellectually and physically
【智者千虑,必有一失】even the wise are not always free from error

雉

pheasant
【雉堞】crenelation
【雉鸠】turtledove

稚

young; childish: 幼～ childish
【稚虫】[动] naiad
【稚气】childishness: ～未脱 still possessing the innocence of childhood
【稚子】child

质

①(性质) nature; character: 本～ innate character; essence／实～ essence; substance ②(质量) quality: ～变 qualitative change ③(物质) matter; substance 流～食物 liquid food／木～纤维 wood fiber ④(朴素) simple; plain ⑤(询问) question: ～疑 call in question ⑥(抵押品) pledge: 人～ hostage ⑦(抵押) pawn; mortgage; pledge
【质变】[哲] qualitative change: 部分～ partial qualitative change

【质地】①(某种材料的结构性质) quality: ～好 high quality / ～坏 poor quality / ～细密 of close texture; fine-grained / 生产不同～的产品 manufacture goods in various qualities ②(人的品质) character; disposition

【质点】[物] particle

【质量】①[物] mass: 相对论～ relativistic mass ②(产品或物质的优劣程度) quality: ～差 of poor quality / ～高 of high quality ◇ ～标准 quality standard / ～检查制度 rules for testing quality / ～守恒[物] conservation of mass / ～要求 quality requirement / ～作用定律[物] law of mass action

【质朴】 simple and unadorned; unaffected; plain: ～无华 simple and unadorned / 为人忠厚～ be simple and honest; be unsophisticated

【质谱】[物] mass spectra ◇ ～分析 mass spectrographic analysis / ～仪 mass spectrometer; mass spectrograph

【质权】mortgage ◇ ～人 pawnee

【质数】[数] prime number

【质问】question; interrogate; call to account: 提出～ bring sb. to account

【质询】address inquiries to; ask for an explanation

【质疑】call in question; query: ～问难 raise doubts and difficult questions for discussion

【质子】proton ◇ ～轰击 proton bombardment / ～加速器 proton accelerator

炙

【炙手可热】if you stretch out your hand you feel the heat; the supreme arrogance of a person with great power

zhōng

中 ①(中心) center; middle: 华～ central China / 居～ in the center ②(中国) China ③(范围内;内部) in; among; amidst: 记在心～ keep in mind ④(中间) middle; mid-: 年～ in the middle of the year / 期～考试 mid-term exam ⑤(中等) medium; intermediate ⑥(不偏不倚) mean; halfway between two extremes: 适～ moderate ⑦(中人) intermediary ⑧(适于) fit for; good for: ～听 be pleasant to the ear / 不～用 good for nothing ⑨(在…过程中) in the process of; in the course of: 行军～ on the march / 在考虑～ under consideration

【中班】①(指幼儿园的班级) the middle class in a kindergarten ②(指工作的班次) middle shift: 上～ be on the middle shift; work the swing shift

【中饱私囊】batten on money entrusted to one's care; line one's pockets with public funds or other people's money; embezzle

【中保】middleman and guarantor

【中波】[无] medium wave

【中不溜儿】[口] fair to middling; middling

【中部】central section; middle part

【中餐】Chinese meal; Chinese food

【中草药】Chinese herbal medicine

【中策】the second best plan

【中层】middle-level ◇ ～干部 middle-level cadres

【中产阶级】middle class; middle bourgeoisie

【中常】middling; average: ～年景 average harvest

【中程】intermediate range; medium range ◇ ～弹道导弹 medium-range ballistic missile / ～轰炸机 medium bomber

【中垂线】[数] perpendicular bisector

【中词】[逻] middle term

【中道而废】stop halfway

【中道而弃】give up half way

【中道而行】follow the middle course

【中稻】semilate rice; middle-season rice

【中等】①(等级介于上下之间的) secondary: ～教育 secondary school education / ～专科学校 technical secondary school; polytechnic school / ～职业学校 secondary vocational schools ②(不高不低;不大不小的) medium; moderate; middling: ～城市 medium-sized city / ～身材 of medium height; of middle size

【中东】the Middle East

【中短波】 intermediate waves; medium high frequency wave

【中断】suspend; break off; discontinue: ～谈判 break off the negotiations / ～外交关系 suspend diplomatic relations

【中队】①(队伍编制) detachment: 交通～ a detachment of traffic police ②[军] squadron: 轰炸机～ bomber squadron

【中耳】[生理] auris media; middle ear ◇ ～炎 otitis media

【中饭】midday meal; lunch

【中锋】①[足球、水球] center forward ②[篮球] center

【中耕】[农] intertill ◇ ～机 cultivator / ～作物 intertilled crop

【中古】①(较晚的古代) the middle ancient times ②(指封建社会时代) medieval times; Middle Ages ◇ ～史 medieval history

【中国】China ◇ ～国籍 Chinese nationality／～话 the Chinese language; Chinese／～画 traditional Chinese painting／～人 Chinese／～字 Chinese characters; the Chinese written language
【中国共产党】the Communist Party of China ○全国代表大会 National Party Congress／一大 First National Congress／十一届三中全会 Third Plenary Session of the Eleventh Central Committee／十三大 Thirtieth National Congress／中央委员会 Central Committee／中央委员 member of the Central Committee／候补中央委员 alternate member of the Central Committee／政治局 Political Bureau／政治局委员 member of the Political Bureau／政治局候补委员 alternate member of the Political Bureau／政治局常务委员会 Standing Committee of the Political Bureau／政治局常委 member of the Standing Committee of the Political Bureau／中央书记处 Secretariat of the Central Committee／总书记 General Secretary／中央顾问委员会 Central Advisory Commission／中央纪律检查委员会 Central Commission for Discipline Inspection／中央军事委员会 Military Commission of the Central Committee／对外联络部 International Liaison Department／组织部 Organization Department／统战部 United Front Work Department／宣传部 Propaganda Department／代表大会 Congress／基层委员会 primary Party committee／总支 general Party branch／总支书记 secretary of the general Party branch／总支委员 member of the committee of the Party branch／支部 branch／支部书记 branch secretary／支部委员 member of the branch committee／入党志愿书 an application form for Party membership
【中国共产主义青年团】the Communist Youth League of China
【中国科学院】the Chinese Academy of Sciences
【中国人民解放军】the Chinese People's Liberation Army
【中国人民政治协商会议】the Chinese People's Political Consultative Conference
【中和】[化] neutralization ◇ ～剂 neutralizer
【中华】China: ～民族 the Chinese nation
【中华人民共和国】the People's Republic of China ◇ ～国务院 the State Council of the People's Republic of China
【中级】middle rank; intermediate ◇ ～人民法院 intermediate people's court
【中继】[电] relay ◇ ～器 repeater／～线 trunk line／～站 relay station
【中坚】nucleus; hard core; backbone: ～分子 backbone elements／～人物 key personnel
【中间】①(里面) among; between: 隐藏在树丛～ hide among the bushes ②(中心) center; middle ③(两者之间) intermediate; middle ◇ ～剥削 middleman's exploitation／～道路 middle road／～地带 intermediate zone／～阶层 intermediate strata／～力量 intermediate forces／～路线 middle road／～派 middle elements; intermediate sections／～人 the middle man; go-between／～线[机] medium line／～状态 intermediate state
【中间诉讼阶段】[法] mesne process
【中将】[军] lieutenant general; air marshal (英空军); vice admiral (英美海军)
【中介】intermediary; medium
【中景】[电影] medium shot
【中距离】[体] middle distance ◇ ～赛跑 middle-distance race
【中看】be pleasant to the eye: ～不中吃 look nice but not taste nice; be pleasant to the eye but not agreeable to the palate
【中肯之言】a statement that meets the exigency
【中栏】[体] intermediate hurdles
【中立】neutrality: 保持～ remain neutral／守～ observe neutrality／永久～ permanent neutrality ◇ ～地带 neutral zone／～法 law of neutrality; neutrality law／～港 neutral port／～公约 neutrality pact／～政策 policy of neutrality／～主义 neutralism
【中立国】neutral state ◇ ～船只 neutral ship
【中量级】[举重] middleweight
【中流】midstream: ～砥柱 firm rock in midstream; main stay
【中脑】[生理] mesencephalon; midbrain
【中年】middle age ◇ ～人 a middle-aged person／～知识分子 a middle-age intellectuals
【中农】middle peasant: 上～ upper-middle peasant／富裕～ well-to-do middle peasant／下～ low-middle peasant
【中篇小说】medium-length novel; novelette
【中频】[无] intermediate frequency ◇ ～道 mid band／～干扰 intermediate-frequency interference／～信号 intermediate-frequency signal／～增益 mid-frequency gain
【中期】mid-term; medium term: ～贷款 medium term loan／～计划 medium term plan／～目标 medium term objective／～审查

mid-term review / ～选举 mid-term election / ～预报 extended forecast; extended-range forecast
【中山狼】a person who repays good with evil
【中山装】Chinese tunic suit: 穿着一身灰色～ in gray tunic and trousers
【中生代】[地] Mesozoic Era; Mesozoic
【中士】[军] sergeant; staff sergent; petty officer second class (美海军); petty officer first class (英海军)
【中世纪】Middle Ages
【中式】Chinese style: ～服装 Chinese-style clothing
【中枢】center: 电讯～ telecommunications center / 领导～ leading center / 神经～ nerve center ◇ ～神经系统[生理] central nervous system
【中碳钢】[冶] medium carbon steel
【中堂】central scroll
【中提琴】viola
【中听】pleasant to the ear; agreeable to the hearer
【中途】halfway; midway: ～停留 stop halfway; stop over; break one's journey (at) / ～下汽车 get off the car midway
【中外】China and foreign countries; at home and abroad: ～驰名 well known both in China and abroad / 古今～ both ancient and modern, Chinese and foreign; at all times and in all countries ◇ ～合资经营企业 Chinese-foreign joint venture / ～合作经营企业 enterprise jointly managed by China and foreign countries
【中纬度】middle latitude
【中卫】[足球] center halfback
【中尉】[军] first lieutenant; lieutenant junior grade (美); lieutenant; sublieutenant (英陆军); flying officer (英空军)
【中文】the Chinese language; Chinese: ～书刊 books and magazines in Chinese
【中午】noon; midday
【中西】Chinese and Western: ～医结合 combine traditional Chinese and Western medicine
【中线】①[数] central line ②(篮、排球场的) center line ③(足球场的) halfway line
【中校】lieutenant colonel; commander (英美海军); wing commander (英空军)
【中心】center; heart; core; hub: 贸易～ a center of trade / 商业～ business center / 在城市～ at the center of the city / 抓住问题的～ get to the heart of the matter ◇ ～城市 key city / ～工作 central task / ～规[机] center gauge / ～环节 key link; central link / ～人物 central character / ～思想 central idea; gist / ～问题 central issue; crucial question / ～线[机] center line / ～项[哲] central term
【中兴】resurgence
【中型】medium-sized; middle-sized: ～词典 a medium-sized dictionary
【中性】①[化] neutral: ～反应 neutral reaction / ～树脂 neutral resin / ～土[农] neutral soil / ～盐 neutral salt ②[语] neuter: ～名词 neuter noun
【中性花】[植] neuter flower
【中学】middle school
【中旬】the middle ten days of a month
【中压轮胎】medium-pressure tire
【中央】①(中心) center; middle ②(最高领导机构) central authorities: 党～ the Central Committee of the Party; the Party Central Committee / ～各部门 departments under the Party's Central Committee and the State Council / ～领导同志 leading comrades of the central authorities ◇ ～处理机 central processing unit; main frame / ～工作会议 the Central Working Conference / ～顾问委员会 the Central Advisory Commission / ～纪律检查委员会 the Central Commission for Discipline Inspection / ～机构 central organs / ～全会 plenary session of the Central Committee / ～银行 central bank / ～直属机关 departments under the Party Central Committee
【中央集权】centralization (of authority): ～的国家 centralized state power / ～的君主制 centralized monarchy
【中药】traditional Chinese medicine ◇ ～铺 shop of traditional Chinese medicines; Chinese pharmacy / ～学 traditional Chinese pharmacology
【中叶】middle period: 二十世纪～ the middle of the 20th century; the mid-1900s
【中医】①(中国传统医学) traditional Chinese medical science ②(用中医理论和方法治病的医生) doctor of traditional Chinese medicine; practitioner of Chinese medicine ◇ ～学 traditional Chinese medicine / ～学院 college of traditional Chinese medicine / ～研究院 academy of traditional Chinese medicine
【中音号】[乐] althorn; alto horn
【中庸之道】the doctrine of the mean
【中用】of use; useful
【中游】①(河流中段) middle reaches ②(中间

状态) the state of being middling: 甘居～ be resigned to the middling state; be content to stay middling
【中雨】[气] moderate rain
【中原】Central Plains
【中原逐鹿】fight for hegemony
【中云】[气] medium cloud
【中止】discontinue; suspend; break off: 暂时～ break off for the present ◇ ～上诉 discontinuance of appeal / ～诉讼 abatement of action; stop a case / ～诉讼通知 notice of discontinuance
【中指】middle finger
【中转】[交] change trains ◇ ～站 transfer station
【中子】[物] neutron: 快～ fast neutron / 慢～ slow neutron / 热～ thermal neutron / 瞬发～ prompt neutron ◇ ～弹 neutron bomb / ～物理学 neutronics / ～星 neutron star / ～源 neutron source

衷

inner feelings heart: 由～之言 words spoken from the bottom of one's heart
【衷肠】[书] words right from one's heart: 畅叙～ pour out one's heart
【衷情】[书] heartfelt emotion; inner feelings
【衷心】heartfelt; wholehearted; cordial: ～欢迎 give a cordial welcome / ～拥护 give wholehearted support / 表示～的感激 express one's heartfelt gratitude

忠

loyal; devoted; honest
【忠臣】official loyal to his sovereign
【忠诚】loyal; faithful; staunch: ～可靠 both loyal and trustworthy
【忠肝义胆】having good faith, virtue and patriotism
【忠告】①(诚恳地劝告) sincerely advise; admonish ②(忠告的话) sincere advice; advice
【忠厚】honest and tolerant; sincere and kindly: ～待人 treat a man with honesty
【忠实】true; faithful: ～的朋友 devoted friend / ～的信徒 faithful disciple / ～的译文 translation true to the original / ～可靠 honest and reliable; trustworthy
【忠顺】loyal and obedient: ～的奴仆 willing servant
【忠心】loyalty; devotion: 赤胆～ ardent loyalty; wholehearted devotion; utter devotion
【忠心耿耿】loyal and devoted; most faithful and true
【忠言】sincere advice; earnest advice: ～逆耳 good advice jars on the ear
【忠于】true to; loyal to; faithful to; devoted to: ～职守 be devoted to one's duty / ～祖国,～人民 be loyal to one's country and people
【忠贞】loyal and steadfast: ～之士 a man of loyalty and chastity / ～不屈 staunch and indomitable / ～不渝 unswerving in one's loyalty

钟

①(响器) bell: 撞～ toll a bell; ...g a bell ②(计时的器具) clock: 挂～ wall clock / 闹～ alarm clock / 开～ wind a clock ③(指钟点) time as measured in hours and minutes: 六点～ six o'clock ④(感情集中) concentrate one's affections ⑤(盅) handleless cup
【钟爱】dote on (a child); cherish
【钟摆】pendulum
【钟表】clocks and watches; timepiece ◇ ～店 watchmaker's shop / ～油 watchmaker's oil
【钟点】[口] ①(指某个一定的时间) a time for sth. to be done or to happen: ～到了,我们上课吧。It's time. Let's begin our lesson. ②(小时;钟头) hour: 三个～ three hours
【钟楼】①(设置大钟的楼) bell tower; belfry ②(安装时钟的较高的建筑) clock tower
【钟情】be deeply in love: 一见～ fall in love at first sight
【钟乳石】[地] stalactite
【钟头】[口] hour: 连干几个～ work for hours at a stretch

螽

【螽斯】[动] katydid; long-horned grasshopper

终

①(最后;末了) end; finish: 期～ end of semester / ～成泡影 end up in a bubble; come to nothing / 善始善～ start well and end well; see sth. through / 自始至～ from beginning to end; from start to finish ②(指人死) death; end 临～ on one's deathbed; just before one dies ③(终归) eventually; after all; in the end ④(自始至终的整段时间) whole; entire; all: ～日 all day long / ～岁 the whole year; throughout the year
【终场】end of a performance or show
【终点】①(一段路程结束的地方) terminal point; destination: ～站 terminus②[体] finish ◇ ～裁判员 judge at the finish / ～带 finishing tape / ～线 finishing line; finishing tape
【终端】[电] terminal ◇ ～打字机 terminal writer / ～电缆 terminal cable / ～机 termi-

nating machine / ～局[邮] terminal station / ～用户 end user; terminal user

【终归】eventually; in the end; after all: 她～会明白的。She will understand eventually. / 他～是个小学生。After all he is still a schoolboy.

【终极】ultimate: ～目标 ultimate aim

【终结】end; final stage ◇ ～帐簿 book of final entry

【终究】eventually; in the end; after all: 她～会了解他的感情。In the end she will understand his feelings.

【终局】end; outcome: 战争的～ the outcome of a war

【终了】end: 学期～ the end of the term

【终南捷径】①(求官的最近门路) shortcut to high office; royal road to fame ②(达到目的的捷径) shortcut to success

【终年】①(全年) all the year round; throughout the year: ～积雪 be permanently covered with snow; be covered with snow all the year round ②(指人去世的年龄) the age at which one dies

【终曲】[乐] finale

【终日昏昏】drunk all day long

【终身】lifelong; all one's life: ～伴侣 lifelong companion / ～不忘 keep in memory throughout one's life span / ～大事 an important event in one's life / ～事业 one's lifework / ～徒刑的犯人 life termer ◇ ～残废 permanent disability / ～监护制 perpetual guardianship / ～监禁 life imprisonment / ～人寿险 whole life insurance / ～职务 freehold of office

【终审】[法] last instance; final judgment ◇ ～法院 court of last instance / ～判决 final adjudication; final judgement / ～权 power of final adjudication

【终生】all one's life; throughout one's life: ～难忘的教训 a lesson for life

【终霜】[气] latest frost

【终天之恨】lifelong regret

【终须】have to in the end; unavoidable in the long run

【终于】at last; in the end; finally: 冤案～得到了昭雪。At long last the wrong has been righted.

【终止】①(结束) stop; end: ～日期 closing date ②(停止) termination; annulment; abrogation: ～条约通知书 notice of termination of a treaty; notice of denunciation

【终止符】[乐] a full stop; a period

zhǒng

种 ①[生] (物种的简称) species: 本地～ endemic species / 外地～ exotic species / 亚～ subspecies ②(种类) kind; sort; type: 各～树 different kinds of trees ③(人种) race: 黄～人 the yellow race ④(种子) seed: 麦～ wheat seeds / 选～ strain selection ⑤(生物传代繁殖的物质) seed; strain; breed: 纯～ pure breed / 混血～ mixed blood / 良～牛 good breed of cattle / 同～繁殖 breed in and in / 异～繁殖 breed out and out / 杂交～ hybrid strain; halfbreed ⑥(指胆量或勇气) guts; grit: 这条汉子真有～。He is really a man with plenty of guts.

【种畜】[牧] breeding stock; stud stock

【种间杂交】[农] interspecific hybridization

【种类】kind; type; variety: ～繁多 a great variety

【种马】stud ◇ ～场 stud farm; stud

【种内杂交】[农] intraspecific hybridization

【种牛】bull kept for covering

【种皮】husk or shell of a seed; seed coat

【种种】all sorts of; a variety of: ～理由 all sorts of reasons / 用～手段 by hook or by crook; resort to every means / 由于～原因 for a variety of reasons

【种子】seed ◇ ～测定 seed testing / ～处理 seed treatment / ～队[体] seed team / ～清选机 seed cleaner / ～田 seed-breeding field / ～选手[体] seeded player; seed / ～植物 seed plant

【种族】race: ～平等 racial equality ◇ ～隔离 racial segregation; apartheid / ～关系 race relations / ～灭绝 genocide / ～偏见 racial prejudice / ～歧视 racial discrimination / ～顺应 accommodation / ～特征 racial traits / ～同化 assimilation / ～主义 racism; racialism / ～主义者 racist

肿 swelling; swollen

【肿大】enlargement; tumefaction; swelling: 肝～ enlargement of the liver; hepatomegaly / 淋巴结～ enlargement of lymph nodes

【肿块】phyma; tumor; mass: 腹内～ abdominal mass

【肿瘤】[医] tumor: 恶性～ malignant tumor / 良性～ benign tumor ◇ ～医院 tumor hospital

【肿胀】swelling

踵 heel: 接～而来 follow hard at heel
【踵事增华】[书] carry on a predecessor's task and make a greater success of it; take over and carry forward
【踵趾相接】follow the footsteps; one after another; in succession

zhòng

中 ①(正对上) fit exactly; hit: ～靶 hit the target / 击～要害 hit the most vulnerable point; hit home ②(受到;遭受) be hit by; fall into; be affected by; suffer: ～埋伏 fall into an ambush / ～煤气 be gassed / ～圈套 fall into a trap
【中弹】be struck by a bullet; get shot
【中毒】poisoning; toxicosis: 酒精～ alcoholism / 煤气～ gas poisoning / 食物～ food poisoning ◇ ～量 dosis toxica
【中风】apoplexy
【中计】be trapped; be victimized by a scheme
【中奖】draw a prizewinning ticket in a lottery; get the winning number in a bond
【中肯】①(抓住要点;正中要害) apropos; pertinent; to the point: ～的批评 pertinent comments / ～的评语 pertinent remarks ②[物](临界的) critical: ～质量 critical mass
【中签】be the lucky number
【中伤】slander; malign; vilify: 恶语～ malign sb. viciously / 造谣～ spread slanderous rumors
【中暑】①(日射病) heatstroke; sunstroke ②(得了日射病) suffer heatstroke; suffer sunstroke; be affected by the heat
【中选】be chosen; be selected
【中意】be to one's liking; catch the fancy of

种 grow; plant; cultivate: ～花 cultivate flowers / ～小麦 grow wheat / ～庄稼 plant crops
【种地】land; go in for farming
【种痘】vaccination: ～预防天花 be vaccinated against smallpox
【种瓜得瓜,种豆得豆】plant melons and you get melons, sow beans and you get beans; as you sow, so will you reap
【种植】plant; grow: 油菜～面积 rape-growing areas; areas sown to rape ◇ ～计划 planting scheme / ～季节 planting season / ～园 plantation / ～园主 plantation owner; planter

仲 ①(一季的第二个月) mid-; second: ～春 the middle of spring; second month of spring / ～冬 midwinter; second month of winter / ～秋 midautumn; second month of autumn / ～夏 midsummer; second month of summer ②(兄弟排行第二) the second: ～兄 the second eldest brother ③(地位居中的) middle; intermediate
【仲裁】arbitrate; arbitration: 将争端付诸～ submit a dispute to arbitration ◇ ～法庭 arbitration tribunal; court of arbitration / ～人 arbitrator / ～书 award / ～委员会 board of arbitration / ～协定 arbitration agreement
【仲继贸易】intermediary trade
【仲介贸易】merchanting trade

众 ①(许多) many; numerous: 寡不敌～ be hopelessly outnumbered; fight against hopeless odds / 人口～多的城市 multitudinous city / 人口～多的国家 a country with a large population ②(许多人) crowd; multitude: 大～ the multitude; the masses / 观～ spectators; audience / 听～ audience; listeners
【众多】multitudinous; numerous
【众寡悬殊】a great disparity in numerical strength
【众口难调】it is difficult to cater for all tastes
【众口铄金】public clamor can confound right and wrong
【众口一词】with one voice; unanimously
【众目睽睽】the eyes of the masses are fixed on sb. or sth.: ～之下 in the public eye; in the public gaze
【众目昭彰】the masses are sharp-eyed
【众怒难犯】you cannot afford to incur public wrath; it is dangerous to incur the anger of the masses
【众叛亲离】the masses are in rebellion and one's friends are deserting; be opposed by the masses and deserted by one's followers; be utterly isolated
【众擎易举】many hands make light work
【众人】everybody
【众生】[佛教] all living creatures
【众矢之的】target of public criticism
【众说纷纭】opinions vary: ～,莫衷一是。As opinions vary, no unanimous conclusion can be drawn.
【众所周知】as everyone knows; as is known to all; it is common knowledge that
【众望】people's expectations; popular confidence: ～所归 enjoy popular confidence / 不孚～ fall short of people's expectations

【众星拱月】a myriad of stars surrounding the moon; a host of lesser lights around the leading one

【众议院】①(美、澳、日等国的) House of Representatives ②(意、墨、智等国的) Chamber of Deputies

【众志成城】unity of will is an impregnable stronghold; unity is strength

【众醉独醒】all are besotted except one sober

重 ①(分量) weight: 净~ net weight / 毛~ gross weight / 过~ over weight / 你体~有多少? What is your weight? ②(重量大;比重大) heavy; weighty; important: 工作很~ have a heavy work load / 他比我~。He is heavier than me. ③(程度深) deep; heavy; serious: ~伤 severe wound / 患~病 be seriously ill / 情意~ deep affection ④(重要) important: 分别轻~ distinguish the trivial from the important ⑤(重视) lay stress on; attach importance to: ~调查研究 lay stress on investigation and study / ~友谊 set store by friendship; value friendship / ~预防 lay stress on prevention ⑥(不轻率) discreet: 慎~ careful; cautious / 自~ self-respect

【重办】severely punish

【重兵】[军] a large number of troops; massive forces: 派驻~ station massive forces / 有~把守 be heavily guarded

【重柴油】diesel fuel oil

【重创】inflict heavy losses on; maul heavily

【重大】great; weighty; major; significant: ~案件 important case / ~成就 significant achievement / ~过失 culpable negligence / ~事故 major accident / ~损失 heavy losses / ~嫌疑犯 major suspect / ~刑事犯 major criminal / ~责任 weighty responsibility

【重担】heavy burden; difficult task: 抢挑~ rush to shoulder heavy responsibilities

【重点】focal point; stress; emphasis: ~高校 key institutes of higher learning / ~工程 major project / 工作~ focal point of the work / 突出~ make the focal points stand out

【重读】[语] stress: ~音节 stressed syllable / 非~音节 unstressed syllable

【重负】heavy burden; heavy load: 如释~ feel as if relieved of a heavy load

【重工业】heavy industry

【重轨】[铁道] heavy rail

【重过磷酸钙】[农] double superphosphate

【重轰炸机】heavy bomber

【重金属】[化] heavy metal

【重晶石】[矿] barite; heavy spar

【重力】[物] gravity; gravitational force ◇ ~场 gravitational field / ~秤 gravity balance / ~加速度 acceleration of gravity / ~选矿 gravity separation / ~仪 gravity meter; gravimeter

【重利】①(很高的利息) high interest: ~盘剥 practice usury ②(很高的利润) huge profit: 牟取~ seek huge profit

【重量】weight: 启运~ shipping weight / 实际~ actual weight ◇ ~单 weight list / ~损失 loss in weight / ~证明书 weight certificate; surveyor's report on weight

【重量吨】weight: ~和尺码吨 weight or measurement

【重量级】[举重] heavyweight

【重任】important task; heavy responsibility: 身负~ be charged with important tasks

【重氢】[化] heavy hydrogen; deuterium

【重商主义】mercantilism

【重赏】handsome reward

【重视】attach importance to; pay attention to; think highly of; take sth. seriously; value: ~农业现代化 attach importance to the modernization of agriculture / ~舆论 take public opinion seriously

【重水】[化] heavy water

【重税】heavy taxation

【重托】great trust

【重武器】heavy weapons

【重心】①[物] center of gravity ②(事情的中心或主要部分) heart; core; focus ◇ ~位置 center-of-gravity position

【重刑】severe punishment; heavy penalty

【重型】heavy-duty; heavy ◇ ~机床 heavy-duty machine tool / ~卡车 heavy-duty truck; heavy truck

【重要】important; significant; major: ~关头 critical juncture / ~人物 important figure; prominent personage; very important person (VIP) / ~任务 vital task; important mission / ~遗漏 material omission / ~因素 key factor / ~原则 cardinal principle / ~政策 major policy

【重要性】importance; significance

【重义轻利】value justice above material gains

【重音】①[语] stress; accent: 次~ secondary accent / 句子~ sentence stress / 主~ primary accent ② [乐] accent ◇ ~符号 stress mark; accent

【重用】put sb. in an important position

【重油】heavy oil
【重元素】[化] heavy element
【重镇】place of strategic importance
【重子】[物] baryon
【重罪】capital felony; heavy offense ◇ ～犯 felon

zhōu

州 ①(指自治州) prefecture ②(美国的行政区划分范围) state

洲 ①(大陆) continent ②(沙洲) islet in a river; sand-bar
【洲际】intercontinental ◇ ～弹道导弹 intercontinental ballistic missile / ～导弹 intercontinental missile / ～火箭 ocean-spanning rocket

舟 [书] boat: 轻～ a light boat / 泛～ go boating
【舟车】①(船和车) vessel and vehicle ②(旅途) journey: ～劳顿 fatigued by a long journey; travel-worn / ～之苦 the hardship of a journey
【舟状窝】[解] fossa navicularis; fossa scaphoidea

周 ①(圈子) circumference; periphery; circuit ②(绕一圈) make a circuit; move in a circular course ③(普遍;全) all; whole; all over; all around: 众所～知 as is known to all ④(周到;完备) thoughtful; attentive: 计划不～ not well planned; not planned carefully enough / 招待不～ not be attentive enough to guests ⑤(星期) week ⑥[电] cycle: 千～ kilocycle / 兆～ megacycle
【周报】weekly publication; weekly
【周边】[机] periphery
【周波】[电] cycle
【周长】girth; circumference; perimeter: ～六米的烟筒 a funnel six meters in girth / 水库的～ the perimeter of a reservoir
【周到】attentive and satisfactory; thoughtful; considerate: 服务～ offer good service / 想得很～ be very thoughtful
【周而复始】go round and begin again; go round and round; move in cycles
【周济】help out (the needy); relieve
【周刊】weekly; weekly magazine; weekly publication
【周密】careful; thorough: ～的分析 a detailed analysis / ～的计划 a well-conceived plan / ～思考 think over carefully / 进行～的调查 carry out a thorough investigation
【周末】weekend
【周年】anniversary
【周期】period; cycle: 工作～[机] action cycle ◇ ～表[化] periodic table / ～律[化] periodic law
【周期性】periodicity; cyclicity: ～变动 cyclical movement / ～波动 cyclical fluctuations / ～发作 periodic attacks / ～经济危机 periodic economic crises / ～通货膨涨 cyclical inflation / ～循环 periodic return
【周全】①(周到; 全面) thorough; comprehensive: ～之计 a perfect plan ②(成全;帮助) help sb. attain his aim
【周身】the whole body; all over the body: ～疼痛 ache all over / ～湿透 be drenched all over
【周岁】one full year of life
【周围】around; round; about: 环顾～ look about; look around / ～环境 surroundings; environment ◇ ～神经系统[生理] peripheral nervous system / ～温度[机] environment temperature
【周详】comprehensive; complete; careful
【周效磺胺】sulfamethoxine
【周旋】①(打交道) mix with other people; socialize: ～其间 walk and talk to guests circuitously ②(较量) deal with; contend with: 与敌人～到底 fight the enemy to the finish
【周延】[逻] distribution
【周游】travel round; journey round: ～各地 travel far and wide / ～世界 travel round the world
【周折】twists and turns; setbacks: 大费～ cause a lot of trouble
【周转】①[经] turnover: 资金的～ capital turnover ②(开支调度情况) have enough to meet the need: ～不开 have not enough to go round / ～不灵 have not enough cash to meet the needs ◇ ～率 turnover rate / ～资金 working fund; revolving fund; circulating fund

粥 gruel; porridge; congee: 麦片～ oatmeal gruel / 米～ rice gruel / 小米～ millet gruel
【粥少僧多】the gruel is meager and the monks are many; not enough to go round

zhóu

轴 ①(圆柱形的零件) axle; shaft: 车～ car axle; axle / 曲～ crank shaft / 心～ spindle ②(圆柱形的用来往上绕东西的器物) spool; rod;

roller: 画～ roller for a scroll of Chinese painting ③[量] 一～山水画 a scroll painting of scenery / 一～线 a spool of thread
【轴衬】[机] axle bush
【轴承】[机] bearing: 滚珠～ ball bearing ◇～衬 bearing bush / ～钢 bearing steel / ～合金 bearing metal
【轴对称】[数] axial symmetry
【轴距】wheelbase
【轴流泵】axial-flow pump; axial pump
【轴流式喷气发动机】axial-flow jet engine
【轴套】[机] axle sleeve
【轴线】①[机] axis: 垂直～ normal axis ②(绕在轴子上的线) spool thread; spool cotton
【轴箱】axle box; axle housing
【轴向】[机] axial ◇～剖面 axial section / ～运动 axial motion
【轴心】①(轮轴) axle center ②(轴线) axis
【轴心国】Axis powers; the Axis
【轴重】gross rail load on axle

妯

【妯娌】wives of brothers; sisters-in-law

zhǒu

肘 elbow

【肘节】[机] toggle: 制动～ brake toggle / ～开关 toggle switch
【肘窝】fossa cubitalis
【肘腋之患】trouble coming from those closest
【肘子】①(作为食物的猪腿的最上部) upper part of a leg of pork ②(肘) elbow

zhòu

咒 ①(咒语) incantation: 念～ chant incantations ②(说希望人不顺利的话) curse; damn: 诅～ curse

【咒骂】curse; swear; abuse; revile
【咒语】incantation

皱 ①(皱纹) wrinkles; lines ②(起皱纹) wrinkle; crease; crumple

【皱眉头】knit one's brows; frown: 皱着眉头 with knitted brows
【皱胃】[动] abomasum
【皱纹】wrinkles; lines: 眼角的～ crow's-feet / 满脸～ have a wrinkly face ◇～法兰绒 crepe flannel / ～革 shrink leather / ～纸 crepe paper
【皱褶】fold
【皱皱巴巴】wrinkled; creased

绉 crape; crepe

【绉布】cotton crepe; crepe
【绉纱】crape
【绉纸】crepe paper

昼 daytime; daylight; day: ～伏夜出 hide by day and come out at night

【昼出动物】diurnal animal
【昼盲】hemeralopia
【昼夜】day and night; round the clock: ～不息 day and night without rest / ～相随 keep each other's company day and night

骤 ①(马快走) trot: 驰～ gallop ②(突然) sudden abrupt: 一阵～雨 a passing heavy shower

【骤然】suddenly; abruptly: ～离去 leave abruptly

zhū

朱

【朱陈之好】the union of two families
【朱顶雀】redpoll
【朱红】vermilion; bright red
【朱鹭】ibis
【朱门】vermilion gates; red-lacquered doors of wealthy homes: ～酒肉臭,路有冻死骨。Behind the vermilion gates meat and wine go to waste while out on the road lie the bones of those frozen to death.
【朱墨】①(黑和红) red and black ②(用朱砂制成的墨) ink made of cinnabar
【朱雀】rosefinch
【朱砂】cinnabar
【朱文】characters on a seal carved in relief
【朱颜】①(脸长得漂亮) peach blossom face of a beauty ②(年轻) youth
【朱颜鹤发】a hale aged man

诛 ①(杀) put to death ②(谴责处罚) punish; condemn

【诛求无厌】make incessant, exorbitant demands
【诛求无已】make endless exorbitant demands
【诛心之论】penetrating criticism; exposure of sb.'s ulterior motives

珠 ①(珠子) pearl: 养～ culture pearls ②(小的球形东西) bead: 泪～儿 teardrop / 露～ beads of dew; dewdrops / 算盘～ beads on an abacus

【珠蚌】the pearl oyster
【珠宝】pearls and jewels; jewelry ◇ ～店 a jeweller's / ～商 jeweller
【珠翠】pearls and jade; ornaments made with pearls and jade
【珠光宝气】be richly bejewelled
【珠光体】[冶] pearlite
【珠联璧合】strings of pearls and girdles of jade; a perfect pair; a happy combination
【珠母云】mother-of-pearl clouds; nacreous clouds
【珠算】reckoning by the abacus; calculation with an abacus
【珠圆玉润】round as pearls and smooth as jade; excellent singing or polished writing

株 ①(露在地面上的树木的根和茎) trunk of a tree; stem of a plant ②(植株) individual plant; plant: 幼～ young plant; sapling ③[量] 两～树 two trees
【株距】[农] spacing in the rows
【株连】involve (others) in a criminal case; implicate

蛛 spider
【蛛丝马迹】thread of a spider and trail of a horse; clues; traces
【蛛网】spider web; cobweb
【蛛蛛】spider
【蛛状痣】spider nevus; stellar nevus

铢
【铢积寸累】accumulate little by little; build up bit by bit
【铢两悉称】exactly equal in weight; have the same weight
【铢锱必较】stand on weight and measure

侏
【侏罗纪】[地] Jurassic Period
【侏儒】dwarf; midget; pygmy

诸 all; various
【诸多】[书] a good deal; a lot of: ～不便 a lot of trouble; great inconvenience
【诸侯】dukes or princes under an emperor
【诸如】such as
【诸如此类】things like that; such; and so on and so forth
【诸子百家】the various schools of thought and their exponents during the period from pre-Qin times to the early years of the Han Dynasty

猪 pig; hog; swine: 公～ boar / 怀孕母～ farrowed sow / 母～ sow / 小～ pigling; piglet / 养～ raise pigs
【猪草】pigweed
【猪场】pig farm; piggery
【猪丹毒】swine erysipelas; diamond-skin disease
【猪肝】pork liver: 炒～ stir-fried liver
【猪倌】swineherd
【猪獾】sand badger
【猪圈】pigsty; pigpen; hogpen
【猪笼草】[植] common nepenthes
【猪囊虫病】pork measles
【猪排】pork chop
【猪皮】pigskin; hogskin
【猪气喘病】swine enzootic pneumonia
【猪肉】pork: ～绦虫 pork tapeworm
【猪食】pig feed; pigwash; swill ◇ ～缸 (pig) trough
【猪瘟】swine fever; hog cholera
【猪油】lard
【猪鬃】bristles

潴 ①[书](积聚水) collect; accumulate; store ②(水积聚的地方) puddle; pool
【潴留】[医] retention: 尿～ retention of urine

zhú

烛 ①(蜡烛) candle: 蜡～ wax candle / ～心 candlewick ②(照亮) illuminate; light up
【烛光】[物] candlepower; candle
【烛花】snuff: 剪～ trim off the snuff
【烛台】candlestick

逐 ①(追赶) pursue; chase: ～水草而居 move from place to place in search of water and grass ②(驱逐) expel; drive out ③(挨着次序) one by one: ～项 item by item / ～月 month by month
【逐波而去】go over the waves
【逐步】step by step; progressively: ～加以解决 settle sth. step by step / ～降低生产成本 progressively reduce the production cost
【逐出门外】drive out of the door
【逐个】one by one
【逐行扫描】[电视] line by line scan; progressive scanning
【逐行倒相制】[电视] phase-alternation
【逐渐】gradually; by degrees

【逐客令】order for guests to leave: 下～ show sb. the door
【逐鹿】[书] chase the deer; fight for the throne; bid for state power: ～中原 fight among rivals for the throne
【逐年】year by year; year after year
【逐日】day by day; every day: ～追风 chase the sun and drive the wind
【逐字】word for word; verbatim: ～记录 verbatim record
【逐字逐句】word by word and sentence by sentence; word for word: ～的翻译 word-for-word translation; literal translation

竹 bamboo: ～林 bamboo forest; groves of bamboo / ～篓 bamboo crate
【竹板】bamboo clappers
【竹荚鱼】saurel; horse mackerel
【竹竿】bamboo pole; bamboo
【竹鸡】bamboo partridge
【竹简】bamboo slip
【竹节虫】stick insect; walkingstick
【竹篮打水一场空】draw water with a bamboo basket; all in vain
【竹篱茅舍】a bamboo fence around a thatched cottage; a simple dwelling
【竹帘画】[工美] painting on a bamboo curtain
【竹马】a bamboo stick used as a toy horse
【竹马青梅】companion in youth
【竹排】bamboo raft
【竹器】articles made of bamboo
【竹笋】bamboo shoots
【竹叶青】①(一种毒蛇) green bamboo snake ②(一种酒) bamboo-leaf-green liqueur
【竹芋】[植] arrowroot
【竹纸】paper made from young bamboo
【竹子】bamboo

zhǔ

主 ①(主人) host: ～队 home team; host team ②(权力或财物的所有者) owner; master: 房～ owner of a house / 企业～ proprietor of an enterprise / 物～ owner ③(当事人) person concerned; party concerned: 买～ buyer / 卖～ seller ④(基督教对上帝和真主的称呼) God; Lord ⑤(伊斯兰教对真主的称呼) Allah ⑥(重要的;主要的) main; primary: 预防为～ put prevention first ⑦(负主要责任;主持) manage; direct; be in charge of ⑧(主张) stand for; be in favor of; advocate: ～和 advocate peace; be for a peaceful settlement / ～战 stand for war ⑨(预示) indicate; signify
【主办】direct; sponsor: ～展览会 sponsor an exhibition
【主编】①(主要编辑) chief editor ②(负责编辑工作) edit
【主宾席】head table; seat for the guest of honor
【主持】①(负责处理) take charge of; manage; direct: ～日常事务 take care of routine matters ②(负责掌握) preside over; chair: ～会议 preside over a meeting ③(主张) uphold; stand for: ～正义 uphold justice
【主词】[逻] subject term; subject
【主次】primary and secondary: 分清～ differentiate what is primary from what is secondary
【主从】principal and subordinate: ～关系 the relationship between the principal and the subordinate
【主祷文】the Lord's Prayer
【主导】leading; dominant; guiding: 起～作用 play a leading role ◇ ～思想 dominant ideas; guiding ideology / ～主题[乐] *leitmotiv*
【主调音乐】homophony
【主动】①(能够造成有利局面) initiative: ～作某事 do sth. on one's own initiative; do sth. of one's own accord / 争取～ try to gain the initiative; contend for the initiative ②[机] driving ◇ ～齿轮 driving gear / ～轴 driving shaft; driving spindle
【主动脉】[生理] aorta ◇ ～弓 arch of aorta / ～炎 aortitis
【主发动机】[宇航] sustainer
【主犯】[法] prime culprit; principal criminal: ～和从犯 principal and accessory
【主峰】the highest peak in a mountain range
【主妇】housewife; hostess
【主干】①(植物的主要茎) trunk ②(主要的;起决定作用的) main force; mainstay
【主格】[语] subjective case; nominative case
【主根】[植] main root; taproot
【主攻】[军] main attack ◇ ～部队 main attack force / ～方面 main phase of attack / ～方向 main direction of attack
【主攻手】[体] spiker
【主顾】customer; client
【主观】subjective: ～努力 subjective efforts / ～愿望 desire; wishful thinking ◇ ～能动性[哲] subjective initiative; conscious activity / ～唯心主义[哲] subjective idealism / ～主义 subjectivism
【主管】①(负责管理) be responsible for; be in

charge of ②(主管人员) person in charge ◇ ～部门 department responsible for the work／～机关 competent authorities; responsible institution

【主航道】main channel ◇～中心线 center line of main channel

【主合同】master contract

【主机】①[军] lead plane; leader ②[机] main engine

【主见】ideas or thoughts of one's own; one's own judgment; definite view

【主讲】be the speaker; give a lecture

【主将】chief commander; commanding general

【主教】[宗] bishop: 大～ archbishop／红衣～ cardinal

【主井】[矿] main shaft

【主句】[语] main clause; principal clause

【主角】leading role; lead; protagonist: 女～ a leading lady

【主考】①(主持考试) be in charge of an examination ②(主持考试的人) chief examiner

【主课】main subject; major course

【主力】main force; main strength of an army ◇～兵团 main formations／～队员[体] top players of a team／～舰 capital ship／～军 main force

【主粮】staple food grain

【主流】①(干流) main stream; main current; mother current ②(事物发展的主要方面) essential or main aspect; main trend

【主谋】①(做主要谋划者) head a conspiracy; be the chief plotter ②[法] chief instigator; principle conspirator

【主脑】①(主要部分) control center; center of operation ②(首领) leader; chief

【主权】sovereign rights; sovereignty: 领土～ territorial sovereignty ◇～豁免 sovereignty immunity／～平等 sovereign equality／～行为 sovereign act

【主权国】sovereign state: 半～ half-sovereign state／部分～ part-sovereign state

【主人】①(财物所有人) owner ②(权力所有人) master ③(与客人相对的) host: 女～ hostess

【主人公】leading character in a novel, etc.; hero or heroine; protagonist

【主人翁】①(当家做主的人) master: 新中国的～ masters of New China ②(中心人物) hero; heroine

【主任】director; head; chairman: 车间～ director of a workshop／会计～ chief accountant／居民委员会～ head of the neighborhood committee

【主食】staple food; principal food

【主使】instigate; incite; abet

【主诉】action in chief; main suit; principle action

【主题】theme; subject; motif ◇～歌[电影] theme song

【主体】①(事物的主要部分) main body; main part; principal part ②[哲] subject ～和客体 subject and object; the perceiver and the world ◇～工程 principal part of a project

【主谓句】[语] subject-predicate sentence

【主文】[法] main body of a court verdict

【主席】①(主持会议的人) chairman ②(最高领导职位) chairman; president

【主席台】rostrum; platform: 在～前列就座的有… seated in the front row on the rostroum were

【主席团】presidium: 代表大会～ presidium of the congress

【主显节】[基督教] Epiphany

【主心骨】①(可依靠的人或事物) backbone; mainstay; pillar: 她是我们车间的～。She is the mainstay of our workshop. ②(主见) definite view; one's own judgment: 她没有～。She has no definite view of her own.

【主星】[天] primary (component)

【主刑】[法] principal penalty

【主修】specialize; major: ～科目 major subjects／她～化学。She majors in chemistry.

【主演】act the leading role

【主要】main; chief; principal; major: ～产品 major products; staple products／～动机 dominant motive／～环节 key-link／～目的 major objective／～设备 major installation／～条件 essential condition／～条款 main clause／～污染源 primary pollution source／～因素 primary factor／～责任 ultimate liability／～证据 main evidence／～指控 first charge／～罪行 major offense

【主义】doctrine; -ism: 达尔文～ Darwinism／个人～ individualism／社会～ socialism／唯物～ materialism／资本～ capitalism

【主意】①(办法) idea; plan: 好～ a good idea ②(主见) decision; definite view: 改变～ change one's mind／拿不定～ be in two minds

【主音】[乐] keynote; tonic

【主语】[语] subject

【主宰】dominate; dictate; decide: ～自己的命

运 decide one's own destiny
【主张】①(持有某种见解) advocate; stand for; maintain; hold ②(见解) view; position; stand; proposition
【主旨】purport; substance; gist
【主治医生】physician-in-charge; doctor in charge of a case
【主轴】[机] main shaft; spindle ◇ ～箱 spindle box
【主子】master; boss

煮 boil; cook: ～饭 cook rice / ～鸡蛋 boil eggs
【煮豆燃萁】burn beanstalks to cook beans fratricidal strife

嘱 enjoin; advise; urge
【嘱咐】enjoin; tell; exhort: ～某人做某事 exhort sb. to do sth. / 再三～ exhort again and again; din sth. into sb.
【嘱托】entrust: ～某人办件事 entrust sb. with a certain task

瞩 gaze; look steadily: 高瞻远～ stand high and see far; take a broad and long-term view; show great foresight
【瞩目】[书] fix one's eyes upon; focus one's attention upon: 举世～ be the focus of world attention
【瞩望】①(期待) look forward to ②(注视) gaze at; look long and steadily upon

zhù

苎
【苎麻】[植] ramie

贮 store; save; lay aside
【贮备】store up; have in reserve; lay aside
【贮藏】store up; lay in
【贮槽】storage tank
【贮存】store; keep in storage ◇ ～期 storage time
【贮木场】timber depot; timber yard; lumber yard
【贮蓄】save up

注 ①(灌入) pour: 把酒～到瓶里。Pour the liquor into the bottle. / 他汗流如～。The sweat was pouring off him. ②(集中) concentrate; fix: 全神贯～ concentrate on; be engrossed in; be preoccupied with ③(赌注) stakes: 孤～一掷 stake everything on a single throw ④(用文字来解释) annotate; explain with notes ⑤(注释文字) notes: 附～ annotations / 脚～ footnote ⑥(记载;登记) record; register: ～销 write off; cancel
【注册】register: ～新会员的名字 register the names of the new members ◇ ～处 registration office; registrar's office / ～标记 monomark / ～吨位 register ton / ～股票 registered certificate of shares / ～马力 registered horsepower / ～商标 registered trademark / ～条款 interpretation section / ～证书 registration certificate / ～资本 registered capital
【注定】be doomed; be destined: ～要失败 be doomed to failure / 命中～ decreed by fate; predestined
【注脚】footnote
【注解】①(用文字来解释) annotate; explain with notes ②(解释性文字) note; annotation
【注目】gaze at; fix one's eyes on: ～而视 look with fixed eyes / 行～礼 salute with eyes / 引人～ spectacular
【注入】pour into; empty into ◇ ～式教学法 the spoon-feeding way of teaching; cramming / ～式教育 spoon-fed education / ～井 [石油] injection well
【注射】[医] inject: 肌肉～ intramuscular injection / 静脉～ intravenous injection / 给病人～青霉素 give the patient an injection of penicillin / 往静脉里～葡萄糖 inject glucose into the veins ◇ ～器 injector; syringe / ～针头 syringe needle
【注视】look attentively at; gaze at: 密切～形势的发展 watch attentively the development of the situation
【注释】explanatory note; annotation ◇ ～读物 annotated readings / ～条款 interpretation section
【注水】[石油] water flooding: ～动态 flood performance / 边缘～ edgewater flooding
【注销】cancel; write off: 把借条～ cancel a written acknowledgment of a loan; cancel an I.O.U. ◇ ～户口 cancellation of household registration / ～登记 cancellation of registration
【注意】pay attention; take note to: ～后果 look to the end / 提请大家～ call everybody's attention to sth. ◇ ～标志 care mark; caution mark / ～广度[心] attention span; range of attention / ～力 attention / ～事项 matters needing attention; points for attention
【注音】[语] phonetic notation: ～字母 the na-

tional phonetic alphabet
【注油孔】oil filler point
【注油枪】grease gun; oil gun
【注重】lay stress on; pay attention to; attach importance to: ～调查研究 pay attention to investigation and study

柱 ①(柱子) post; upright; pillar; column ②(象柱子的东西) sth. shaped like a column: 水～ water column / 水银～ mercury column ③[数](柱面) cylinder
【柱面】[数] cylinder: 椭圆～ elliptic cylinder
【柱石】pillar; mainstay
【柱石岩】basalt
【柱头】①[植] stigma ②[建] column cap; column head
【柱子】post; pillar
【柱座】[建] column base

蛀 ①(蛀虫) moth ②(蛀虫咬坏) bore through; be moth-eaten
【蛀齿】decayed tooth; dental caries
【蛀虫】insect that eats books, clothes or wood; moth; borer
【蛀洞】cavity
【蛀心虫】borer

住 ①(居住) live; reside; stay; dwell: ～一夜 stay overnight / ～在乡下 dwell in the country / 他～在工厂附近。 He lives near the factory. / 她～在和平街十号。 She resides at 10 Peace Street. ②(停止) stop; cease: 雨～了。The rain has stopped. ③(做动词的补语) 记～ bear in mind / 忍不～ cannot help
【住持】[宗] abbot
【住处】residence; dwelling (place); lodging; quarters: 固定～ fixed residence / 永久 permanent residence / ～改变 change one's residence / 他们在村子里找到了～。 They found quarters in a village.
【住房】housing; lodgings: ～问题 the housing problem; accommodation
【住户】household; resident
【住口】shut up; stop talking
【住手】stay one's hand; stop
【住宿】stay; put up; get accommodation: 给学生安排～ arrange accommodation for the students / 他今晚可以在这里～。 He can put up here for the night.
【住所】swelling place; residence; domicile: ～的变更 change of residence / 永久～ permanent dwelling place
【住院】be in hospital; be hospitalized ◇ ～病人 inpatient / ～部 inpatient department / ～处 admission office / ～费 hospitalization expenses / ～医生 resident
【住宅】residence; dwelling; dwelling house: 建造～ build a dwelling ◇ ～区 residential quarters / ～转租 domestic subtenancy / ～租金 domestic rent / ～租赁 domestic tenancy
【住址】address: 永久～ permanent address / 请在信里给我写上～。Please address the letter for me.

驻 ①(停留) halt; stay: ～足 make a temporary stay ②(住在执行职务的地方) be stationed: ～华大使 ambassador to China / ～京记者 resident correspondent in Beijing / 中国～日本大使馆 Chinese Embassy in Japan
【驻波】standing wave; stationary wave
【驻地】①(部队或外勤人员住的地方) station; encampment ②(地方行政机关所在地) seat
【驻防】be on garrison duty; garrison: ～边疆 be garrisoned at the frontier / ～上海 garrison Shanghai ◇ ～部队 garrison
【驻守】garrison; defend
【驻外机构】institution functioning abroad
【驻颜有术】skilled in making oneself look youthful
【驻在国】[外] state to which a diplomatic envoy is accredited
【驻在期限】term of residence
【驻扎】be stationed; be quartered: ～重兵 station a huge force

祝 express good wishes; wish: ～你旅途愉快。 Have a pleasant journey. / ～你新年快乐。I wish you a happy New Year.
【祝词】congratulatory speech; congratulations
【祝福】blessing; benediction
【祝贺】congratulate: 致以衷心～ extend cordial greetings / ～你! Congratulations! / 请代我向她表示～。 Convey my congratulations to her. ◇ ～信 congratulatory letter
【祝捷】celebrate a victory ◇ ～大会 victory celebration
【祝酒】drink a toast; toast: 答谢～ respond to a toast / 向来宾们～ toast the guests / 致～辞 propose a toast
【祝寿】offer birthday congratulations
【祝愿】wish: ～某人快乐 wish sb. happiness / 谨致良好的 ～ with best wishes

著 ①(显著) marked; outstanding: 卓～ outstanding; distinguished ②(显出) show; prove: 颇～成效 prove rather effective ③(写作) write: ～书立说 write books to expound a doctrine ④(著作) book; work: 名～ a celebrated piece of writing; a famous work / 译～ a translation

【著称】[书] celebrated; famous: 以风景优美～ be celebrated for its scenic beauty

【著名】famous; celebrated; well-known: ～论断 a celebrated thesis / ～演员 famous actor

【著述】①(写作) write; compile ②(写作的成品) book

【著者】writer; author

【著作】①(作品) work; book; writings: ～等身 an author with many works to his credit ②(写作) write

【著作权】copyright ◇ ～诉讼 copyright action

助 help; assist; aid: ～成其事 help to finish a business / ～人为乐 find it a pleasure to help others / ～一臂之力 lend sb. a helping hand / ～纣为虐 help a tyrant to do evil

【助产士】[医] midwife

【助词】[语] auxiliary word

【助动词】[语] auxiliary verb

【助攻】[军] holding attack; secondary attack ◇ ～部队 holding element

【助教】assistant

【助理】assistant: 部长～ assistant minister ◇ ～法官 deputy judge / ～检察员 deputy procurator

【助跑】[体] run-up; approach

【助燃】[化] combustion-supporting ◇ ～气体 combustion-supporting gas

【助熔剂】flux

【助色团】[化] auxochrome

【助手】assistant; helper; aide

【助听器】audiphone; hearing aid; deaf-aid

【助推器】auxiliary boost; boost motor

【助威】boost the morale of; cheer: 给我们的球队～ cheer for our team

【助消化】aid digestion

【助兴】liven things up; add to the fun

【助学金】stipend; grant-in-aid: 领～的学生 a grant-aided student

【助战】①(协助作战) assist in fighting ②(助威) bolster sb.'s morale

【助长】[贬] encourage; abet; foster; foment: ～歪风邪气 encourage evil trends

筑 build; construct: ～堤 build a dike / ～路 construct a road

【筑埂机】[农] ridger

【筑室道谋】ask every passerby how to build one's house; have no idea or plan of one's own and accomplish nothing

铸 casting; founding: ～钟 cast a bell / ～钱 coin money

【铸币】coin; specie ◇ ～权 mintage

【铸成大错】make a gross error

【铸锭】[冶] ingot casting

【铸钢】[冶] cast steel

【铸工】[冶] ①(铸造工作) foundry work ②(铸造工人) foundry worker; founder ◇ ～车间 foundry (shop) / ～鼓风机 foundry fan

【铸件】[冶] cast; casting: ～和锻件 castings and forgings

【铸排机】[印] composing machines

【铸铁】①(炼铁) iron casting ②(生铁) cast iron: 球墨～ nodular cast iron

【铸型】casting mold

【铸造】casting; founding: 蜡模～ investment casting; lost wax casting / 无砂～ sandless casting ◇ ～厂 commercial foundry / ～车间 foundry; casting shop

【铸字】[印] typefounding; typecasting ◇ ～工场 typefoundry / ～工人 typefounder / ～机 typecasting machine

zhuā

抓 ①(拿在手中) grab; seize; clutch ②(挠) scratch: ～痒痒 scratch an itch ③(捕拿) arrest; catch; press-gang: 警察～住了小偷。The police arrested a thief. ④(特别着重) stress; pay special attention to: ～重点 stress the essentials ⑤(负责领导;主管) take charge of; be responsible for: 他是～农业的。He is in charge of agriculture. ⑥(吸引) attract; grip: ～住观众 grip the attention of the audience

【抓辫子】seize on sb.'s mistake or shortcoming; capitalize on sb.'s vulnerable point

【抓膘】fatten

【抓差】draft sb. for a particular task; press sb. into service

【抓点带面】draw experience from selected units to promote overall work

【抓耳挠腮】tweak one's ears and scratch one's cheeks; scratch one's head

【抓工夫】make good use of one's time; find time: 她～休息了二十分钟。She snatched twenty minutes' rest.

【抓紧】firmly grasp; pay close attention to: ～

时机 seize the opportunity／～时间 make the best use of one's time／～学习 attend to one's studies in earnest; study hard
【抓阄儿】draw lots
【抓举】[举重] snatch
【抓破脸】openly break relations
【抓瞎】[口] find oneself at a loss; be in a rush and muddle; be thrown off balance
【抓壮丁】press-gang able-bodied men

zhuǎ

爪 claw; talon
【爪尖儿】pig's trotters; pettitoes
【爪儿】[口]①(小动物的脚) paw of a small animal ②(某些器物的脚) foot of a utensil
【爪牙鹰犬】lackeves and hired ruffians
【爪子】claw; paw; talon: 猫～ a cat's paws／鹰～ an eagle's talons

zhuān

专 ①(集中在一件事上) concentrate: ～注 concentrate one's attention on; be absorbed in ②(独自掌管或占有) monopolize: ～卖 monopoly／～权 monopolize power ③(专门) special; specialized: ～差 special mission／～场 special performance／～程 special trip／～程拜访 pay a special visit to sb.／～号 special issue ③(专长) expert: 又红又～ be both red and expert; be both politically conscious and professionally competent
【专案】special case for investigation; exemplary case ◇～材料 material connected with a case; dossier／～法官 ad hoc judge／～组 special group for the examination of a case
【专长】specialty; special skill or knowledge: 没有～ have no special skill／学有～ have specialized knowledge of a subject; be expert in a special field of study／你的～是什么? What's your specialty?
【专车】special train; special car
【专电】special dispatch
【专断】make an arbitrary decision; act arbitrarily
【专攻】specialize in: 她～计算机科学。 She specializes in computer science.
【专横】imperious; peremptory; domineering: ～跋扈 imperious and despotic
【专机】①(某人专用机) private plane ②(班机之外专门飞行的飞机) special plane
【专家】expert; specialist: 水稻～ expert in rice-growing／眼科～ ophthalmologist; eye specialist ◇～顾问 expert advisor／～证人 expert witness
【专刊】①(报刊以某项内容编辑的一期) special issue or column ②(学术机关出版的单册著作) monograph
【专科学校】training school
【专款】special fund: ～专用 earmark a fund for its specified purpose only
【专栏】special column: 书评～ book review column ◇～作家 columnist
【专利】patent ◇～登记 patents registration／～调查 patent search／～法 patent law／～局 patent office／～品 patent; patented article／～许可证 patent licence
【专利权】patent right; patent ◇～立法 patent legislation／～所有人 patentee
【专卖】monopoly: ～货品 monopolized commodities／～权 right of monopoly
【专门】special; specialized ◇～机构 special agency; special organ／～人材 people with professional skill／～人民法院 special people's court／～人民检查院 special people's procuratorate／～术语 technical terms; nomenclature／～知识 specialized knowledge; expertise; technical know-how
【专名】[语] proper noun
【专区】prefecture; subprovincial administrative region
【专任】full-time; regular: ～教员 full-time teacher
【专属经济区】exclusive economic zone
【专属渔区】exclusive fishing zone
【专署】perfectural commissioner's office
【专题】special subject; special topic ◇～报告 report on a special topic／～调查 investigation of a special subject／～讨论 seminar／～研究 monographic study
【专线】①(专用铁路线) special railway line ②(专用电话线) special telephone line; line for special use
【专项资金】a special fund
【专心】be absorbed; concentrate one's attention: ～学习 study with undivided attention／～一意 close application／～致志 wholly absorbed; with single-hearted devotion
【专修】specialize in: ～化学 specialize in chemistry ◇～科 special (training) course
【专业】①(学校学科分类) special field of study; specialized subject; specialty; discipline ②(生产部门的各业务部门) specialized trade or profession; special line ◇～队伍 professional con-

tingent / ~化 specialization / ~课 specialized course / ~人员 personnel in a specific field / ~学校 vocational school / ~知识 professional knowledge
【专业户】specialized households
【专一】single-minded; concentrated: 爱情~ be constant in love / 心思~ with concentrated attention
【专用】for a special purpose: ~车床 special purpose lathe / ~电话 telephone for special use ◇ ~决算表 special purpose statements / ~商标 monopoly trademark / ~信箱 private letter box
【专有权】patent rights
【专员】assistant director; commissioner: 商务~ commercial attaché
【专政】dictatorship ◇ ~对象 object of dictatorship / ~工具 instrument of dictatorship / ~机关 organ of dictatorship / ~职能 functions of the dictatorship
【专职】①(由专人担任) sole duty; specific duty ②(专门从事) full-time: ~干部 full-time cadre
【专制】①(独自掌握政权) autocratic; despotic: 君主~ autocratic monarch ②(凭自己意志行事) autocracy ◇ ~政府 autocratic government / ~政体 autocracy / ~制度 dictatorial system
【专注】concentrate one's attention on; be absorbed in; devote one's mind to
【专著】monograph; treatise

砖 brick: 玻璃~ glass brick / 瓷~ ceramic tile / 方格~ checker brick / 空心~ hollow brick / 耐火~ refractory brick; firebrick / 砌~ lay bricks / 他们用~盖房子。 They build their houses with bricks. / 他用~砌起了一堵墙。 He bricked up a wall.
【砖茶】brick tea
【砖厂】brickfield; brickyard
【砖房】brick house
【砖坯】unfired brick
【砖墙】brick wall
【砖头】①(碎砖) fragment of a brick ②[方](砖) brick
【砖窑】brickkiln

zhuǎn

转 ①(改变) turn; shift; change: 好~ take a turn for the better / 晴~多云 change from fine to cloudy / 汽车~过街角。 The car turned round the corner. ②(转移) pass on; transfer: ~学 transfer to another school / 请把这个消息~告她。 Please pass on the news to her.
【转氨酶】transaminase
【转败为胜】turn defeat into victory
【转变】change; transform: ~立场 change one's stand; shift one's ground / ~态度 change one's attitude / 把一种能~为另一种能 transform one form of energy into another
【转播】relay ◇ ~台 relay station
【转产】change the line of production
【转车】change trains or buses; transfer to another train or bus
【转船】change to another ship; tranship ◇ ~提单 transshipment bill of lading / ~装运 transshipment
【转达】pass on; convey; communicate: 请向她~我的谢意。 Please convey my gratitude to her.
【转道】make a detour; go by way of
【转调】[乐] modulation
【转动】turn; move; turn round: ~手腕子 flex one's wrist / 在锁眼里~钥匙 turn a key in the lock
【转发】transmit: ~文件 transmit a document
【转告】pass on; communicate; transmit
【转化】[哲] change; transform: 向反面~ transform oneself into one's opposite; change into the reverse
【转换】change; transform: ~方向 change direction / ~话题 change the subject of conversation; switch the conversation to another subject ◇ ~开关 change-over switch
【转机】a favorable turn; a turn for the better
【转记帐簿】book of secondary entry
【转嫁】shift; transfer: ~罪责 shift the responsibility upon / 把责任~给他人 shift off one's responsibility; put the blame on someone else
【转交】pass on; transmit
【转角】street corner; corner
【转口】transit ◇ ~货物 transit goods / ~贸易 transit trade / ~税 transit duty / ~运输 transshipment
【转亏为盈】show a turn from loss to profit
【转脸】①(掉过脸) turn one's face ②(极短的时间) in no time; in the twinkling of an eye: ~不认帐 go back on one's word in the twinkling of an eye
【转卖】resell
【转念】reconsider and give up an idea; think better of

【转让】transfer the possession of; make over: 有偿～ pay for new technique to sb.; royalty ◇ ～合同 agreement of transfer / ～权利 assignment privilege / ～人 alienator; transferor / ～手续费 transfer charges / ～物 grant / ～证书 deed of release; letter of transfer
【转入】change over to; shift to; switch to: ～下页 carried forward / ～正常 return to normal
【转身】turn round; face about
【转手】①(转卖) sell what one has bought ②(不直接交) pass on ◇ ～交易 switch operation
【转述】report; relate sth. as told by another
【转瞬】in a twinkle; in a flash
【转送】①(转交) pass on; transmit ②(转赠) make a present of sth. given to one
【转体】[体] turn; twist ◇ ～跳 turning leap / ～跳水 twist dive
【转弯】turn a corner; make a turn: 来一个一百八十度的大～ make a 180-degree turn / 商店一～就到。The shop is just round the corner.
【转弯抹角】①(路弯弯曲曲) full of twists and turns ②(说话兜圈子) beat about the bush; speak in a round about way: 说话从来不～ never mince one's words
【转危为安】take a turn for the better and be out of danger; pull through
【转向】①(改变方向) change direction ②(改变政治立场) change one's political stand
【转向架】[铁道] bogie
【转向装置】steering gear
【转押】sub-mortgage
【转眼】in the twinkling of an eye; in an instant; in a flash: ～就睡着了 fall asleep in a trice / ～之间 in a trice
【转业】be transferred to civilian work ◇ ～费 military severance pay / ～军人 armyman transferred to civilian work
【转移】①(改换位置) shift; transfer; divert: ～视线 divert sb.'s attention ②(改变) change; transform: ～管辖 change of venue; change the venue ③[医](扩散) metastasis: 癌～ the metastasis of carcinoma
【转义】[语] transferred meaning; figurative sense
【转引】quote from a secondary source
【转运】①(运输) transport; transfer; transship ②(运气好转) have a change of luck; luck turns in one's favor ◇ ～港 port of transshipment / ～公司 transport company; forwarding agency / ～提单 transshipment bill of lading / ～站 transfer post
【转载】reprint sth. that has been published elsewhere; reprint
【转赠】make a present of sth. given to one
【转战】fight in one place after another: ～大江南北 fight successively in different parts north and south of the Changjiang River
【转帐】transfer accounts: ～凭单 transfer vouches
【转折】①(事物在发展过程中改变方向) a turn in the course of events ②(文意的转折) transition ◇ ～点 turning point
【转正】①(预备党员转正) become a full member ②(临时工转正) become a regular worker
【转注】[语] mutually explanatory or synonymous characters
【转租】sublet; sublease

zhuàn

转 ①(旋转) turn; revolve; rotate: 轮子～得飞快。The wheels were turning swiftly. / 月亮绕着地球～。The moon moves round the earth. ②[量](绕一圈) revolution: 每分钟一千六百五十～ 1650 revolutions per minute; 1650 r.p.m.
【转碟】[杂技] plate-spinning
【转动】turn; revolve; rotate: 没有动力,机器就不能～。You can't make a machine work without power supply.
【转来转去】walk back and forth
【转炉】[冶] converter ◇ ～钢 converter steel / ～炼钢法 converting process
【转轮手枪】revolver
【转门】[建] revolving door
【转盘】①[机] turntable ②[体] giant stride ③[杂技] disc-spinning ④[石油] rotary table: ～速度 rotary speed
【转石不生苔】a rolling stone gathers no moss
【转数】[机] revolution: 额定～ rated revolution / 每分钟～ revolutions per minute
【转速】rotational speed ◇ ～计 tachometer
【转台】①(指剧院的) revolving stage ②(指工具的) swivel table
【转椅】swivel chair; revolving chair
【转悠】[口] ①(转动) turn; move from side to side ②(漫步) stroll; saunter; take a leisurely walk
【转子】[机] rotor ◇ ～发动机[汽车] Wankel engine

传 ①(解释经文的著作) commentaries on classics ②(传记) biography: 列~ biographies in ancient Chinese history books／外~ unauthorized biography／自~ autobiography ③(叙述历史故事的作品) a novel or story written in historical style:《水浒》*Water Margin*

【传记】biography

【传略】brief biography; biographical sketch

赚 ①(获得利润) make a profit; gain: ~钱 make money; make a profit／在这笔买卖中他~了二百元。He gained two hundred *yuan* by this deal. ②[方](挣钱) earn: 他~很多钱。He earns a lot of money.

【赚头】[口] profit: 没多大~ not much profit

篆 ①(汉字形体的一种) seal character ②(指印章) seal

【篆刻】seal cutting

【篆书】seal character

撰 write; compose: 为报纸~稿 write articles for a newspaper

【撰著】write; compose

zhuāng

妆 ①(化妆) apply makeup; make up: 梳~ dress one's hair and apply makeup ②(女子或演员的装饰) woman's personal adornments; stage costume

【妆奁】trousseau

【妆饰】adorn; dress up; deck out

装 ①(修饰;打扮;化装) dress up; attire; deck; play the part of; act: ~点 deck; dress; decorate ②(服装) outfit; clothing: 冬~ winter clothing／军~ military uniform／童~ children's wear／身着盛~ be rigged out in one's best ③(化装) stage makeup and costume ④(假装) pretend; feign; make believe: ~蒜 pretend not to know／不懂~懂 pretend to know what one doesn't know ⑤(把东西放到容器里) load; pack; hold: ~船 load a ship／把衣服~进箱子 pack clothes into a trunk ⑥(安装) install; fit; assemble: 给屋子~上暖气设备 install heating system in the house

【装扮】①(打扮) dress up; attire; deck out ②(化装) disguise; masquerade: ~成外国商人 disguise oneself as a foreign merchant

【装备】①(配备) equip; fit out: ~精良 be well equipped／~新式设备 be equipped with up-to-date installation／军队~很好。The army is very goodly equipped. ②(配备的东西) equipment; outfit: 露营的~ a camp equipage／现代化~ modernized equipment

【装裱】mount

【装车机】carloader

【装船】shipment: 分批~ partial shipment ◇~标记 shipping mark／~单 shipping order／~费 shipping charges／~货样 shipment sample／~须知 shipping instructions

【装袋机】sack filling machine

【装订】binding; bookbinding: 布面~ cloth-bound／皮面~ bound in leather／把两本书~成一册 bind up two books into one volume ◇~车间 bookbindery; bindery／~工人 bookbinder／~机 bookbinding machine; binding machinery

【装疯卖傻】feign madness and act like an idiot; play the fool

【装湖涂】pretend not to know; feign ignorance

【装潢】①(装饰物品) mount; decorate; dress ②(物品的装饰) decoration; mounting; packaging

【装货】loading ◇~单 shipping order／~费用 loading charges／~港 port of shipment; port of loading／~容量 shipping capacity

【装机容量】[电] installed capacity

【装甲】①(装有防弹钢板的) armored ②(防弹钢板) plate armor ◇~兵 armored force／~列车 armored train／~汽车 armored automobile／~师 armored division

【装假】pretend; feign; make believe

【装具】harness

【装殓】dress and lay a corpse in a coffin

【装聋作哑】pretend to be deaf and dumb; pretend to be ignorant of sth.

【装门面】put up a front; maintain an outward show; keep up appearances

【装模作样】be affected; attitudinize; put on an act

【装配】assemble; fit together: ~机器 assemble a machine ◇~车间 assembly shop; fitting shop／~工 assembler; fitter／~件 assembly parts／~线 assembly line

【装瓶机】bottle filling machine

【装腔】behave affectedly; be artificial: ~作势 be affected or pretentious; strike a pose

【装饰】decorate; adorn; ornament; deck: 大厅~得富丽堂皇。The hall was sumptuously decorated. ◇~品 ornament／~图案 decora-

tive pattern / ～音[乐] grace note; grace; ornament
【装束】dress; attire: ～朴素 be simply dressed / ～入时 be dressed in the latest fashion; in fashionable dress
【装蒜】[口] pretend not to know; feign ignorance
【装卸】①(装到运载工具上或卸下) load and unload ②(装配和拆卸) assemble and disassemble ◇ ～工 loader; stevedore / ～机 charging crane / ～码头 shipping dock / ～时间[航运] lay days / ～作业 loading and unloading operation
【装修】①(在房屋工程上安装东西) fit up: ～门面 fit up the front of a shop ②(安装上的设备) fixtures
【装样子】put on an act; do sth. for appearance sake
【装运】shipment; loading: 分批～ partial shipment ◇ ～单据 shipping documents / ～口岸及目的地 loading port and destination / ～期 time of shipment
【装载】loading ◇ ～量 loading capacity
【装载机】fork-lift truck
【装帧】binding and layout
【装置】①(安装) install; fit ②(机器设备的配件) installation; unit; device; plant: 安全～ safety device / 暖气～ heating installation

庄 ①(村庄) village ②(规模较大的商店) a place of business: 布～ cloth store / 钱～ (old-fashioned) private bank
【庄户】peasant household ◇ ～人 peasant / ～人家 peasant family
【庄家】banker
【庄稼】crops: 种～ grow crops ◇ ～地 cropland; fields / ～汉 farmer; peasant / ～活儿 farm work / ～人 peasant; farmer
【庄严】solemn; dignified; stately: ～的誓言 solemn oath
【庄园】manor
【庄重】serious; grave; solemn

桩 ①(桩子) stake; pile ②[量] 一～大事 an important matter / 一～买卖 a business transaction
【桩子】stake; pile: 把马拴到～上 tie the horse to a stake

zhuàng

撞 ①(物体相碰) bump against; run into; strike; collide: ～车 collision of vehicles / ～墙 bump against a wall / 老人被自行车～倒了。The old man was knocked down by a bicycle. ②(碰见) meet by chance; bump into; run into ③(莽撞地行动) rush; dash; barge: 横冲直～ barge around; dash about madly
【撞大运】[方] try one's luck
【撞击】ram; dash against; strike
【撞见】meet or discover by chance; run across; catch sb. in the act
【撞骗】look about for a chance to swindle; swindle
【撞入】thrust into; burst into
【撞锁】spring lock
【撞针】[军] firing pin

壮 ①(强壮) strong; robust: 年轻力～ young and robust ②(加强) strengthen; make better: ～声势 to lend impetus and strength; to make it appear more vigorous and impressive ③(壮大) magnificent; grand: ～观 magnificent sight
【壮大】①(使强大) strengthen: ～社会主义经济 strengthen the socialist economy ②(变得强大) grow: 革命力量不断～。The revolutionary forces are growing steadily. ③(强壮粗大) big; strong
【壮胆】embolden; boost sb.'s courage
【壮工】unskilled laborer
【壮观】grand sight
【壮举】magnificent feat; heroic undertaking
【壮阔】vast; grand; magnificent; grandiose: 波澜～ surging forward with great momentum; unfolding on a magnificent scale
【壮丽】majestic; magnificent; glorious: ～的景色 magnificent scenery; majestic view
【壮烈】heroic; brave: ～牺牲 heroically give one's life; die a hero's death
【壮年】the more robust years of a person's life; prime of life
【壮士】heroic man; hero; warrior
【壮实】sturdy; robust
【壮志】great aspiration; lofty ideal: ～凌云 with soaring aspirations / ～未酬 with one's lofty aspirations unrealized

状 ①(形状) form; shape; appearance ②(状况) state; condition: 病～ condition of illness ③(陈述) describe: 不可名～ indescribable; nondescript; beyond description ④(指诉状) written complaint; plaint: 告～ lodge a complaint; file a suit ⑤(褒奖、委任等文件) certificate: 奖～ certificate of commendation

【状况】condition; state; state of affairs: 健康～ state of health; health / 经济～ financial situation; economic situation
【状态】state; condition; state of affairs: 混乱～ state of chaos / 心理～ psychology; state of mind / ～良好 in good condition / 处于紧急～ be in a critical situation ◇ ～图[冶] state diagram
【状语】[语] adverbial modifier; adverbial
【状元】①(本行业中成绩最好的) the very best: 行行出～。 Every profession produces its own topmost master. ②(科举时代的称号) Number One Scholar
【状子】[口] written complaint; plaint

zhuī

椎 vertebra: 颈～ cervical vertebra / 胸～ thoracic vertebra
【椎骨】[生理] vertebra
【椎间盘】intervertebral disc: ～突出症 protrusion of the intervertebral disc

锥 ①(锥子) awl ②(象锥子的东西) anything shaped like an awl ③(用锥子钻) bore; drill
【锥度】[机] taper ◇ ～规 taper gauge
【锥孔】make a hole with an awl
【锥栗】[植] chinquapin
【锥形】[机] taper; cone
【锥子】awl

追 ①(追赶) chase after; pursue; run after ②(追究) trace; look into; get to the bottom of ③(追求) seek; go after: ～名逐利 seek fame and wealth ④(回溯) recall; reminisce ⑤(事后补办) retroactively; posthumously: ～认 recognize retroactively
【追本溯源】trace to its source; get at the root of the matter
【追逼】①(追赶进逼) pursue closely ②(用强迫的方式追究或索取) press for; extort: ～欠款 press for repayment of debts
【追捕】pursue and capture
【追查】investigate; trace; find out: ～事故原因 investigate the causes of an accident / ～责任 find out where the responsibility lies
【追偿损失】recovery of loss
【追悼】mourn over a person's death ◇ ～会 memorial meeting
【追肥】[农] top application; topdressing
【追赶】quicken one's pace to catch up; run after; pursue
【追根】get to the bottom of sth.: ～问底 raise one question after another
【追怀故旧】bring old acquaintances to mind
【追回】recover: ～赃款 recover stolen money
【追悔】repent; regret: ～莫及 too late to repent / ～往事 repent of the past
【追击】pursue and attack; follow up: ～部队 pursuit troops / ～战 warfare of pursuit and attack
【追加】add to (the original amount): ～贷款 complementary financial facility / ～费用 additional expenses / ～判决 additional judgement / ～条款 rider / ～预算 supplement a budget; make a supplementary budget / ～支出 make an additional expenditure
【追缴税款】pursue the tax payment
【追究】look into; find out; investigate: ～祸源 investigate the origin of the misfortune / ～责任 find out who is to blame
【追求】①(争取) seek; pursue: ～名利 seek fame and fortune / ～幸福 pursue happiness ②(求爱) woo; court; chase; run after
【追认】①(事后认可) subsequently confirm or endorse; recognize retroactively ②(批准某人生前要求) admit or confer posthumously
【追溯】trace back to; date from ◇ ～既往 retroact / ～效力 retroactive effect
【追诉权】power to prosecute
【追诉时效】limitation of prosecution
【追随】follow: ～潮流 go with the tide ◇ ～者 follower; adherent
【追索办法】means of recourse
【追索财产者】recoverer
【追索权】recourse; right of recourse
【追索诉讼】recovery action
【追问】question closely; make a detailed inquiry; examine minutely: ～事实真相 make detailed inquiries about the facts
【追叙】①(追述) tell about the past; relate; recount ②(写作的一种手法) narration of earlier episodes; flashback
【追寻】pursue; search; track down
【追忆】recollect; recall; look back
【追赃】order the return of stolen money or goods; recover stolen money or goods; make sb. disgorge the spoils
【追赠】confer posthumously
【追逐】①(追赶) pursue; chase ②(追求) seek; quest
【追踪】follow the trail of; track; trace: ～嫌疑犯 trail a suspect

zhuì

赘 ①(多余的) superfluous; redundant ②(入赘) a son-in-law who lives in the household of the bride's parents ③[方](拖累) be burdensome; be cumbersome

【赘瘤】anything superfluous or useless

【赘述】give unnecessary details; say more than is needed

【赘婿】a son-in-law who lives in the home of his wife's parents

【赘言】verbosity

【赘疣】①(疣) wart ②(无用的东西) anything superfluous or useless

坠 ①(落) fall; drop: ～马 fall off a horse / 这门好象有点～了。The door seems to have dropped slightly. ②(下垂) weigh down ③(垂在下面的东西) weight; a hanging object: 铅～ plummet / 扇～儿 pendant of a fan

【坠毁】fall and break; crash

【坠落】fall; drop ◇ ～损害 sling loss

【坠入五里雾中】completely lost at sea

【坠子】①[方](坠) weight; plummet; pendant ②(耳坠) ear pendant

zhūn

谆

【谆谆】earnestly and tirelessly: ～告诫 repeatedly admonish / ～教导 earnestly instruct / ～教诲 inculcate

zhǔn

准 ①(准许) allow; grant; permit: ～假三天 grant sb. three days' leave / 获～ obtain permission ②(标准) standard; norm; criterion: 以此为～ take this as the standard ③(依据) in accordance with; follow ④(准确) accurate; exact: 我的表不～。My watch doesn't keep good time. ⑤(一定) definitely; certainly: 她～会来。Definitely she will come. ⑥(类似) quasi-; para-: ～军事组织 paramilitary organization / ～契约 quasi-contract

【准保】[副] certainly; for sure: 我们队～胜利。You can be sure that our team will win.

【准备】①(预先安排) prepare; get ready: 精神～ mental preparation / ～接待外国游客 prepare to receive foreign visitors ②(打算) intend; plan: 今年夏天你～到哪儿去? Where do you intend to go this summer? / 她～明年出国。She is planning on going abroad next year. ◇ ～阶段 preparatory stage

【准备活动】[体] warming-up exercise; limbering-up exercise

【准不动产】chattel real

【准将】brigadier (英陆军); air commodore (英空军); commodore (海军); brigadier general (美陆、空军)

【准确】accurate; exact; precise: 发音～ pronounce accurately ◇ ～度 degree of accuracy; accuracy / ～性 accuracy

【准绳】criterion; yardstick

【准时】punctual; on time; on schedule: 火车～到达。The train arrived on schedule. / 请～出席。Please be punctual.

【准尉】warrant officer

【准星】front sight (of a gun)

【准许】permit; allow

【准予】grant; approve; permit: ～保释 grant bail / ～入境 allow sb. to enter the country / ～休假 grant a leave

【准则】norm; standard; criterion: 党内政治生活的～ the guiding principles for inner-Party political life / 外交～ diplomatic norms / 行为～ code of conduct

【准直】[物] collimation ◇ ～透镜 collimating lens / ～仪 collimator

【准租金】quasi-rent

zhuō

捉 ①(握;抓) clutch; hold; grasp ②(捉拿) catch; capture: ～贼 arrest a thief in the act / 猫～老鼠。A cat catches mice.

【捉襟见肘】pull down one's jacket to conceal the raggedness, only to expose one's elbows; have too many difficulties to cope with; have too many problems to tackle

【捉迷藏】①(一种游戏) hide-and-seek; blindman's buff ②(说话兜圈子) be tricky and evasive; play hide-and-seek

【捉摸】fathom; ascertain: 不可～ unfathomable / 难以～ difficult to ascertain

【捉拿】arrest; catch: ～归案 bring sb. to justice / ～逃犯 arrest an escaped prisoner / 警察已～到凶手。The policeman has caught the murderer alive.

【捉弄】tease; make fun of; embarrass

【捉住要害】catch one on the hip

拙 ①(笨) clumsy; awkward; dull: ～于言词 be inarticulate; be clumsy in expressing oneself / 手～ be all thumbs ②[谦] my: ～笔

my poor writing／～稿 my poor manuscript／～工 a poor craftsman／～计 a foolish scheme／～见 my humble opinion
【拙笨】clumsy; dull; unskillful
【拙劣】clumsy; inferior: ～表演 a clumsy performance; a bad show／～手法 inferior tactics; clumsy trick
【拙嘴笨腮】slow of tongue and clumsy of utterance

卓 ①(高而直) tall and erect ②(高明) eminent; outstanding: ～见 brilliant idea; excellent opinion
【卓绝】unsurpassed; extreme; of the highest degree: 艰苦～ extreme hardships and difficulties／英勇～ extremely brave
【卓立一世】stand lofty in one's age
【卓然】outstanding; stately; eminent: ～而立 stand majestically／～自立 one is firmly established by oneself
【卓识】judicious judgment; sagacity: 远见～ foresight and sagacity
【卓有成效】fruitful; highly effective
【卓越】outstanding; brilliant; remarkable: ～的成就 remarkable achievements／～的人才 an eminent man of ability
【卓著】distinguished; outstanding; eminent: 成效～ distinguished results／声誉～ eminent fame

桌 ①(桌子) table; desk: 餐～ dining table／书～ writing desk／满～的客人 a tableful of guests ②[量]三～客人 three tables of guests／一～玩牌的人 a table of card-players
【桌布】tablecloth; table cover
【桌面】top of a table; tabletop
【桌面儿上】on the table; aboveboard; in public: ～的话 polite and unimpeachable remarks／把问题摆到～来 place problems on the table; bring problems out into the open
【桌子】table; desk: 把…放到～上 table sth.

zhuó

着 ①(穿) wear: 讲究穿～ be particular about one's clothing ②(接触) touch; come into contact with: 附～ adhere to ③(着落) whereabouts: 寻找无～ whereabouts unknown; nowhere to be found ④(派遣) send: 请～人前来领取。Please send someone here for it.
【着笔】put pen to paper; begin to write or paint
【着力】put forth effort; exert oneself: ～去干某事 take great pains to do sth.／～描写 concentrate one's efforts on depicting sb. or sth.
【着陆】land; touch down: 软～ soft landing／飞机在十分钟前～了。The plane landed ten minutes ago. ◇～舱 landing module／～场 landing field; landing ground／～接地 touchdown
【着落】①(下落) whereabouts: 遗失的孩子还没有～。The whereabouts of the missing child is still unknown. ②(可指望的来源) assured source: 经费无～ no funds available
【着色】put color on; color ◇～法 coloring／～剂 coloring agent; coloring material
【着实】①(确实) really; indeed: 这小孩～聪明。The boy is really clever. ②(分量重) severely: ～责备了她一顿 lecture her severely
【着手】put one's hand to; set about: ～进行一项任务 set about one's task／从技术革新～ start with technical innovation
【着手成春】cure every patients he treats
【着想】consider: 她是为你～才那样做的。She did that for your good.
【着眼】have sth. in mind; see from the angle of: ～于人民 have the people in mind／～于未来 have one's eyes on the future ◇～点 starting point; focus of attention; object in mind
【着意】act with care and effort; take pains: ～经营 manage with diligent care
【着重】stress; emphasize: ～指出 emphatically point out ◇～号 mark of emphasis

琢 chisel; carve: 精雕细～ exquisitely carved
【琢句雕词】polish sentences and to engrave phrases
【琢磨】①(雕刻打磨) carve and polish ②(加工使精美) improve; polish; refine

啄 peck
【啄木鸟】woodpecker

浊 ①(浑浊) turbid; muddy ②(声音低沉粗重) deep and thick: ～声～气 in a deep, raucous voice ③(混乱) chaotic; confused corrupted
【浊点】[化] cloud point
【浊酒】unstrained wine or liquor
【浊气】foul smell
【浊世】①[书] (黑暗混乱的时代) the corrupted world; chaotic times ②[佛] the mortal world
【浊音】[语] voiced sound: 浊辅音 voiced consonant

灼 ①(火烧) burn; scorch: ～伤 burn ②(明亮) bright; luminous
【灼见】profound view; penetrating view: 真知～ profound knowledge and penetrating insight
【灼热】scorching hot
【灼灼】[书] shining; brilliant: 目光～ with keen, sparkling eyes
酌 ①(斟;饮) pour out; drink: 独～ drink alone ②(酒饭) a meal with wine: 便～ informal dinner ③(斟酌) consider; think over; use one's discretion: ～办 act according to one's judgment; do as one thinks fit
【酌处权】discretion
【酌减】discretionary reduction
【酌量】consider; deliberate; use one's judgment
【酌情】take into consideration the circumstances; use one's discretion: ～处理 settle a matter as one sees fit; act at one's discretion / ～减刑 reduction at judicial discretion

zī

咨 consult; take counsel
【咨文】①[旧](平行机关的公文) official communication ②(国家元首提出的国情报告) report delivered by the head of a government on affairs of state: 国情～ State of the Union Message
【咨询】seek advice from; hold counsel with; consult ◇ ～服务 consulting service / ～机关 advisory body / ～小组 advisory group / ～意见 advisory opinion / ～委员会 consultative committee
资 ①(钱财;费用) money; expenses: 投～ put money in; invest ②(资助) subsidize; support ③(提供) provide; supply: 以～借鉴 that we can draw lesson from / 以～鼓励 by way of encouragement ④(资质) endowment; natural ability: 天～ natural endowments; aptitude ⑤(资格) qualifications; record of service ⑥(资方) capital: 劳～关系 labor-capital relations
【资本】①(经营工商业的本钱) capital: 固定～ fixed capital / 借贷～ loan capital / 金融～ financial capital / 流动～ current capital; floating capital / 生产～ production capital / 银行～ banking capital / 周转～ working capital ②(牟取利益的凭借) what is capitalized on; sth. used to one's own advantage ◇ ～过剩 surplus of capital / ～货物 capital goods / ～积累 accumulation of capital / ～亏损 capital deficit / ～市场 capital market / ～输出 export of capital / ～原始积累 primitive accumulation of capital / ～增值 increase in capital / ～周转 capital turnover; turnover of capital / ～转移 capital transfer / ～总额 capital investment
【资本家】capitalist
【资本主义】capitalism: 国家垄断～ state monopoly capitalism / 自由～ laissez-faire capitalism; non- monopoly capitalism / ～国家 capitalistic state; capitalist nation / ～经济 capitalistic economy / ～经济成份 capitalist sector of the economy / ～倾向 tendencies towards capitalism / ～社会 capitalist society / ～生产 capitalistic production / ～生产方式 capitalist mode of production / ～制度 capitalist system / ～总危机 general crisis of capitalism
【资财】capital and goods; assets: 清点～ make an inventory of the assets
【资产】①(财产) property ②(资金) capital fund; capital ③[经](资金的运用情况) assets: 流动～ liquid assets ◇ ～冻结 freezing of assets / ～负债 assets and liabilities / ～负债表 statement of assets and liabilities; balance sheet / ～决算 assets settlement / ～收益 assets income / ～总额 general assets
【资产阶级】the capitalist class; the bourgeoisie: 民族～ the national bourgeoisie / 官僚～ the bureaucrat- capitalist class ◇ ～分子 bourgeois element / ～革命 bourgeois revolution / ～个人主义 bourgeois individualism / ～化 become bourgeoisified / ～民主 bourgeois democracy / ～民主革命 bourgeois-democratic revolution / ～世界观 bourgeois world outlook / ～思想 bourgeois ideas; bourgeois ideology / ～专政 the dictatorship of the bourgeoisie / ～自由化 bourgeois liberalism; bourgeois liberalization
【资方】those representing capital; capital ◇ ～代理人 agent of the owner of a private enterprise; agent of a capitalist / ～人员 capitalists and their representatives
【资格】①(从事某种活动所必备的条件) qualification: 没有～参加比赛 be disqualified from taking part in the game / 有～参加投票 be qualified for the vote ②(从事某种活动长短所形成的身份) seniority: 摆老～ flaunt one's seniority ◇ ～审查委员会 credentials

committee／～证书 credentials
【资金】fund：建设～ funds for construction／流动～ current funds; circulating funds／积压国家～ let state funds lie idle／为国家积累～ accumulate funds for the state ◇～表 funds statement／～筹集活动 fund raising activities／～外流 capital flight／～运用 application of fund
【资力】financial strength：～雄厚 have a large capital; be financially powerful
【资历】qualifications and record of service
【资料】①(生产或生活的必需品) means：生产～ means of production; capital goods／生活～ means of livelihood; consumer goods ②(依据的材料) data; material ◇～处理 data processing／～库 data bank／～室 reference room
【资深望重】one's reputation is distinguished
【资源】natural resources; resources：人力～ manpower resources／开发天然～ exploit natural resources
【资质】natural endowments; intelligence：～平庸 natural disposition is ordinary
【资助】aid financially; subsidize

姿 ①(容貌) looks; appearance ②(姿势) gesture; carriage; posture; position：跪～ kneeling position／立～ standing position／卧～ prone position
【姿容】looks; appearance：～秀美 good-looking; pretty
【姿色】good looks：略有几分～ be rather good-looking
【姿势】posture; gesture：直立的～ an erect posture／～优美 have a graceful carriage
【姿态】①(姿势) posture; carriage; gesture ②(态度) attitude; pose：外交～ diplomatic gesture／友好～ friendly gesture／表现出高～ show magnanimity; be tolerant

滋 ①(滋生) grow; multiply：～事 create trouble ②(增添) more ③[方](喷射) spurt; burst：别～水。Don't spurt water.
【滋补】nourishing; nutritious：～食品 nourishing food; nourishment／～药品 tonic
【滋润】①(不干燥) moist：～的皮肤 soft skin ②(使不干燥) moisten：用唾液～喉咙 moisten one's throat with saliva
【滋生】①(繁殖) multiply; breed; propagate ②(引起) cause; create; provoke：～事端 cause trouble; create a disturbance
【滋事】cause trouble
【滋味】taste; flavor：这盘菜的～不错。This dish tastes good.
【滋养】①(供给养分) nourish ②(养分) nutriment; nourishment：丰富的～ rich nutriment ◇～品 nourishing food; nutriment; nourishment
【滋长】grow; develop

孳 multiply; propagate
【孳生】multiply; breed; propagate

吱 [象] ①(老鼠叫) squeak ②(小鸟叫) chirp; peep
【吱声】[方] utter sth.; make a sound

龇 [口] bare; show：～着牙 bare one's teeth
【龇牙咧嘴】①(凶狠的样子) show one's teeth; look fierce ②(疼痛难忍的样子) contort one's face in agony; grimace in pain

孜
【孜孜】diligent; industrious; hardworking：～以求 diligently strive after; assiduously seek
【孜孜不倦】diligently; assiduously; indefatigably

辎 an ancient covered wagon
【辎重】[军] impedimenta; supplies and gear of an army; baggage

锱 an ancient unit of weight
【锱铢必较】haggle over every penny; dispute over every detail

鲻 [动] mullet

zǐ

梓 ①[植](梓树) Chinese catalpa ②(刻板) cut blocks for printing：付～ send to the printers

紫 purple; violet
【紫菜】[植] laver
【紫草茸】[化] shellac; lac
【紫癜】[医] purpura
【紫貂】[动] sable
【紫丁香】lilac
【紫红】purplish red
【紫花地丁】[植] Chinese violet
【紫胶】shellac; lac ◇～虫 lac insect
【紫金牛】[植] Japanese ardisia
【紫禁城】the Forbidden City
【紫荆】[植] Chinese redbud
【紫罗兰】[植] violet; common stock

【紫茉莉】[植] four-o'clock
【紫杉】[植] (Japanese) yew
【紫石英】[矿] amethyst
【紫苏】[植] purple perilla
【紫穗槐】[植] false indigo
【紫檀】[植] red sandalwood; padauk
【紫藤】[植] Chinese wistaria
【紫铜】red copper
【紫外线】[物] ultraviolet ray ◇ ～灯 ultraviolet lamp
【紫薇】[植] crape myrtle
【紫阳花】[植] hydrangea
【紫药水】[药] gentian violet
【紫云英】[植] Chinese milk vetch
【紫竹】[植] black bamboo

姊 elder sister; sister
【姊妹】elder and younger sisters; sisters
【姊妹篇】companion volume

子 ①(儿女;儿子) son; child: 独生～ an only son; an only child／长～ eldest son ②(人) person: 男～ man; male person／女～ woman; female person ③(古代对有学问男子的美称) ancient title of respect for a learned or virtuous man ④(种子) seed: 花～儿 flower seed ⑤(卵) egg ⑥(幼小的) young; tender; small: ～鸡 chick ⑦(小而坚硬的东西) sth. small and hard: 枪～儿 bullet ⑧[量] 一～儿挂面 a bundle of fine dried noodles／一～儿毛线 a hank of knitting wool
【子程序】[计算机] subprogram; sub-routines ◇ ～库 sub-routine library
【子畜】young animal; newborn animal
【子代】[生] filial generation
【子弹】bullet; cartridge: 步枪～ rifle bullet／汽枪～ air-gun pellet／练习～ dummy cartridge ◇ ～带 cartridge belt; bandoleer／～箱 cartridge box
【子弟】sons and younger brothers; juniors; children: 纨绔～ profligate sons of the rich
【子房】[植] ovary
【子宫】[生理] uterus; womb ◇ ～出血 intermenstrual flow／～破裂 hysterorrhexis／～切除术 uterectomy／～脱垂 metroptosis; prolapse of uterus
【子宫颈】cervix ◇ ～炎 cervicitis
【子宫帽】cervical cap
【子宫外孕】ectopic pregnancy
【子宫托】pessary
【子公司】subsidiary
【子爵】viscount ◇ ～夫人 viscountess
【子粒】[农] seed; grain; kernel; bean: ～饱满 full grains
【子棉】unginned cotton
【子母弹】[军] shrapnel; canister shot
【子母扣儿】snap fastener
【子目】specific item; subtitle
【子目录】subdirectory
【子女抚养费】child-care expense
【子孙】children and grandchildren; descendants: ～后代 coming generations; descendants
【子通道】subchannel
【子午线】[地] meridian 本初～ the prime meridian
【子午仪】[天] meridian instrument ◇ ～卫星 Transit satellite
【子系统】subsystem
【子痫】[医] eclampsia
【子虚】[书] fictitious; unreal: ～乌有 no such man and nothing like that
【子叶】[植] cotyledon
【子夜】midnight
【子音】[语] consonant
【子语句】substatement
【子子孙孙】generation after generation of descendants

籽 seed: 菜～儿 vegetable seed／棉～儿 cotton seed
【籽棉】unginned cotton

仔 young: ～鸡 chick／～猪 piglet; pigling
【仔畜】newborn animal; young animal
【仔细】①(细心) careful; attentive: ～听 listen attentively; lend an attentive ear to ②(小心) be careful; look out ③(俭省) frugal; economical

zì

字 ①(文字) word; character: ～义 meaning of a word ②(字音) pronunciation ③(字体) form of a written or printed character; style of handwriting; printing type: 繁体～ the original complex form of a simplified Chinese character／黑体～ boldface; boldface type／简体～ simplified Chinese character／斜～ italic type／异体～ a variant form of a Chinese character／篆～ seal character ④(书法的作品) scripts; writings; calligraphy ⑤(字据) receipt; written pledge: 立～为凭 give a written pledge ⑥(许配) be betrothed: 尚未～人 be not betrothed yet

【字典】dictionary: 查～ consult a dictionary
【字调】[语] tones of Chinese characters
【字段长度】[计算机] field length
【字符】[计算机] character: ～移入 character shift in / ～组 character set
【字符串】[计算机] alphabetic string ◇ ～处理 string manipulation
【字号】the name of a shop
【字画】calligraphy and painting
【字汇】glossary; wordbook; lexicon
【字节】[计算机] byte ◇ ～处理 byte manipulation
【字迹】handwriting; writing: ～工整 neat writing / ～模糊 illegible handwriting
【字句】words and expressions; writing: 推敲～ weigh one's words and expressions
【字据】written pledge
【字块长度】[计算机] length of block
【字块输出】[计算机] block output
【字里行间】between the lines
【字轮】print wheel ◇ ～式打印机 wheel printer
【字谜】a riddle about a character or word
【字面】literal: ～上的意思 literal meaning / 照～翻译 translate literally
【字模】[印] matrix: 冲压～ punched matrix ◇ ～雕刻机 matrix cutting machine
【字母】letters of an alphabet; letter: 阿拉伯～ Arabic alphabet / 大写～ a capital letter / 梵文～ Sanskrit alphabet / 汉语拼音～ the Chinese phonetic alphabet / 拉丁～ Latin alphabet; Roman alphabet / 斯拉夫～ Slavonic alphabet / 希腊～ Greek alphabet / 英语～ the English alphabet / 按～顺序排列 be arranged in alphabetical order; be arranged alphabetically ◇ ～表 alphabet
【字幕】captions; subtitles
【字盘】[印] case: 大写～ upper case / 小写～ lower case
【字体】①(书写形式) form of a written or printed character; script; typeface ②(书法派别) style of calligraphy
【字条儿】brief note
【字帖】copybook
【字眼】wording; diction: 玩弄～ play with words / 挑～儿 cavil at sb.'s choice of words; quibble
【字样】①(字体规范) model of written characters ②(书写或排印的字句) printed or written words
【字斟句酌】choose one's words with great care; weigh every word
【字纸篓】wastepaper basket; wastebasket
【字字珠玑】every phrase a gem

恣 throw off restraint; do as one pleases
【恣情放纵】indulge in passions and run wild
【恣行无忌】act willfully and unscrupulously; behave recklessly
【恣意】unscrupulous; reckless; unbridled; willful: ～践踏 willfully trample on / ～任性 do as one feels like it / ～所欲 give rein to vehement desire / ～妄为 behave unscrupulously

自 ①(自己) self; oneself; one's own: ～成一家 have a style of one's own ②(自然) certainly; of course: ～当努力 will certainly do one's best ③(从;由) from; since: ～古以来 since ancient times / ～上而下 from above to below
【自爱】regard for oneself; self-respect
【自拔】free oneself; extricate oneself
【自白】make clear one's meaning or position; vindicate oneself ◇ ～书 a written confession
【自暴自弃】be backward and have no urge to make progress; be resigned to one's backwardness; give oneself up as hopeless
【自卑】feel oneself inferior; be self-abased: ～自贱 look down upon and despise oneself ◇ ～感 inferiority complex; sense of inferiority
【自备】provide for oneself: ～干粮 bring your own food with you / ～外汇 self-provided foreign exchange
【自变量】[数] independent variable
【自便】at one's convenience; as one pleases
【自不待言】it goes without saying; be self-evident
【自不量力】overestimate one's strength or oneself; not know one's own limitations
【自惭形秽】have a sense of inferiority or inadequacy; feel unworthy
【自差】[电] autodyne ◇ ～收音机 autodyne
【自称】①(自己称呼自己) call oneself ②(自己声称) declare oneself to be; claim to be: 被告～无罪。The defendant pleaded not guilty.
【自乘】[数] involution; squaring: 五～得二十五。The square of five is twenty-five. (或: 5 squared is 25.)
【自持】control oneself; restrain oneself; exercise self-restraint
【自筹资金】funds raised by oneself
【自吹自擂】blow one's own trumpet; crack

oneself up
【自从】since: ～去年冬天以来 since last winter
【自大】self-important; arrogant ◇ ～狂 megalomania
【自得】contented; self-satisfied: 安闲～ carefree and contented / ～其乐 be content with one's lot
【自动】①(自己主动) voluntarily; of one's own accord: ～参加 participate voluntarily ②(不凭借人为的力量) automatic ◇ ～步枪 automatic rifle / ～承保 automatic cover / ～打字机 automatic typewriter / ～扶梯 escalator / ～交换机 automatic exchange / ～控制 automatic control / ～铅笔 propelling pencil / ～摄影机 photomaton; strip-film camera / ～饲喂器[牧] self-feeder / ～险 voluntary insurance / ～线 transfer machine / ～音量控制 automatic volume control / ～饮水器[牧] automatic drinking bowl / ～装配线 automatic assembly line
【自动化】automation: 工艺过程～ process automation
【自动卸货卡车】dump truck; tip truck
【自发】spontaneous ◇ ～性 spontaneity
【自肥】fatten oneself; enrich oneself by misappropriating funds or material; feather one's nest
【自费】at one's own expense: ～旅行 travel at one's own expense
【自封】①(自命) proclaim oneself: 她～为专家。She proclaimed herself an expert. ②(限制自己) confine oneself; isolate oneself: 故步～ stand still and refuse to make progress; be complacent and conservative
【自负】①(自己负责) be responsible for one's own action ②(自以为对不起) think highly of oneself; be conceited
【自负盈亏】assume sole responsibility for its profits or losses ◇ ～制度 system of personal responsibility for profit and loss
【自甘堕落】wallow in degeneration; abandon oneself to vice
【自感应】[物] self-induction
【自高自大】self-important; conceited; arrogant
【自告奋勇】offer to undertake; volunteer
【自供】confess ◇ ～状 confession
【自顾不暇】be unable even to fend for oneself; be busy enough with one's own affairs
【自豪】have a proper sense of pride or dignity; be proud of sth.: 我们为伟大的祖国而～。We are proud of our great motherland. ◇ ～感 sense of pride
【自花传粉】[植] self-pollination
【自画像】self-portrait
【自己】oneself: ～动手 use one's own hands / 我～ my own self / 请在里面附一个写明你～地址的信封。Please enclose a self-addressed envelope. / 我～能办。I can do it myself.
【自己人】people on one's own side; one of us: 我们拿他当～。We took him as one of us.
【自给】self-sufficient; self-supporting: 粮食～有余 be more than self-sufficient in grain ◇ ～经济 self-supporting economy / ～率 degree of self-sufficiency
【自给自足】self-sufficiency; self-contained and self-sufficient; autarky
【自荐】recommend oneself; offer one's services
【自尽】commit suicide; take one's own life
【自咎】blame oneself; rebuke oneself
【自救】save oneself; provide for and help oneself: 生产～ provide for and help oneself by engaging in production
【自居】consider oneself to be; pose as: 以功臣～ give oneself the airs of a hero
【自决】self-determination: 民族～ national self-determination ◇ ～权 right to self-determination
【自觉】conscious; aware: ～地遵守纪律 conscientiously observe discipline ◇ ～性 consciousness
【自觉自愿】voluntarily; willingly; of one's own free will
【自绝】alienate oneself: ～于人民 alienate oneself from the people
【自夸】sing one's own praises; crack oneself up
【自宽自解】comfort and relieve oneself
【自来水】running water; tap water ◇ ～厂 waterworks
【自来水笔】fountain pen
【自理】take care of or provide for oneself: 伙食～ make one's own eating arrangements / 旅费～ pay one's own travelling expenses
【自立】stand on one's own feet; support oneself; earn one's own living
【自力更生】regeneration through one's own efforts; self-reliance
【自力霉素】[药] mitomycin C
【自量】estimate one's own ability or strength: 不知～ overrate one's abilities
【自流】①(自动地流) flow automatically; flow by itself ②(自由发展) take its natural course; do as one please ◇ ～灌溉[农] gravity irriga-

tion
【自流井】[地] artesian well
【自留畜】livestock for personal needs; privately owned livestock
【自留地】plot of land for personal needs; family plot
【自满】complacent; self-satisfied: ～情绪 complacency; self-satisfaction
【自鸣得意】show self-satisfaction; be very pleased with oneself; preen oneself
【自鸣钟】striking clock; chime clock
【自命】consider oneself; regard oneself as: ～不凡 consider oneself no ordinary being; have an unduly high opinion of oneself; think no end of oneself
【自馁】lose confidence; be discouraged
【自拍机】[摄] self-timer
【自喷井】[石油] flowing well; gusher well
【自欺欺人】deceive oneself as well as others
【自强不息】make unremitting efforts to improve oneself; constantly strive to become stronger
【自轻自贱】belittle oneself; lack self-confidence or self-respect
【自取灭亡】court destruction; take the road to one's doom
【自取其咎】bring blame on oneself; have only oneself to blame
【自然】①(自然界) natural world; nature: 改造～ transform nature ②(自由发展) naturally; in the ordinary course of events: 听其～ let things take their own course ③(理所当然) of course; naturally: 她认为这是～而然的事。 She took it as a matter of course. ④(不勉强) at ease; natural; free from affectation: 这小孩演得很～。 The acting of the little girl is natural and free from affectations. ◇ ～辩证法 dialectics of nature / ～保护区 nature preserve / ～博物馆 museum of natural history / ～规律 natural law / ～环境 natural environment / ～监护人 guardian by nature / ～界 natural world; nature / ～经济 natural economy / ～聚合 spontaneous aggregation / ～科学 natural science / ～力 natural forces / ～区域 natural regions / ～趋势 the natural trend / ～数 natural number / ～死亡 natural death / ～损耗 natural loss / ～淘汰 survival of the fittest / ～通风 natural draft / ～现象 natural phenomena / ～选择[生] natural selection / ～灾害 natural calamity / ～占有 natural possession / ～主义 naturalism / ～资源 natural resources
【自然而然】naturally; automatically; spontaneously; of oneself
【自然法】natural law ◇ ～学派 school of natural law
【自然人】[法] natural person: ～的权利能力 capacity for right of the natural person / ～的行为能力 disposing capacity of the natural person
【自然神论】[哲] deism ◇ ～者 deist
【自燃】[化] spontaneous combustion
【自认不如】consider oneself inferior to another
【自认不讳】confess without concealment
【自认晦气】accept bad luck without complaint; grin and bear it
【自如】[书] freely; smoothly; with facility: 运用～ handle and use with skill
【自若】[书] self-possessed; composed; calm and at ease: 神态～ appear calm and at ease; appear compose
【自杀】commit suicide; take one's own life
【自伤】self-wounding
【自上而下】from top to bottom; from above to below
【自身】self; oneself: ～难保 be unable even to fend for oneself
【自生自灭】emerge of itself and perish of itself; run its course
【自食其果】eat one's won bitter fruit; reap what one has sown
【自食其力】support oneself by one's own labor; earn one's own living
【自始至终】from start to finish; from beginning to end
【自恃】be self-assured for having sth. or sb. to rely on; count on; capitalize on: ～有靠山 count on sb.'s backing
【自首】①(犯法的人检举自己) voluntarily surrender oneself; confess one's crime; give oneself up: 投案～ surrender oneself to the police or judicial department ②(投敌) make a political recantation; surrender to the enemy: ～变节 recant and turn traitor ◇ ～书 confession
【自述】an account in one's own words
【自私】selfish; self-centered: 出于～的动机 from selfish motives
【自诉】[法] private prosecution; action initiated by an injured party without the participation of the public prosecutor ◇ ～人 private party who prosecutes a case by himself; party who initiates a private prosecution

【自讨苦吃】ask for trouble
【自讨没趣】invite ridicule rebuff; make oneself unwelcome
【自投罗网】hurl oneself willingly into the net; bite the hook
【自卫】defend oneself; self-defense ◇～反击战 counterattack in self-defense／～军 self-defense corps／～杀人 manslaughter in self-defense／～行动 an act in self-defense／～战争 war of self-defense
【自为阶级】class-for-itself
【自慰】console oneself: 聊以～ just to console oneself
【自刎】commit suicide by cutting one's throat; cut one's throat
【自问】①(自己问自己) ask oneself; examine oneself: 反躬～ examine one's conscience ②(自己衡量) reach a conclusion after weighing a matter: ～没有做错事。Upon self-examination, I can flatter myself that I have done no wrong.
【自我】self; oneself: ～介绍 introduce oneself; self-introduction／～陶醉 be intoxicated with self-satisfaction ◇～暴露 self-betrayal; self-exposure／～辩解 self-justification／～表现 self-expression／～吹嘘 self-glorification／～催眠 autohypnosis／～分析 autoanalysis; selfanalysis／～改造 self-remolding／～检查 self-examination; introspection／～教育 self-education／～批评 self-criticism／～牺牲 self-sacrifice／～欣赏 self-appreciation; self-admiration
【自我作古】be the first to do sth.; originate a method, etc.
【自习】study by oneself in scheduled time or free time ◇～时间 time for individual study
【自下而上】from bottom to top; from below
【自相残杀】kill each other; cause death to one another
【自相惊扰】alarm one's own group, etc.; create disturbance within one's ranks; raise false alarms
【自相矛盾】contradict oneself; be self-contradictory: ～的论据 inconsistent argument
【自卸吊车】clamshell car
【自卸卡车】dump truck; tip truck
【自新】turn over a new leaf; make a fresh start: 改过～ correct one's errors and make a fresh start; turn over a new leaf
【自信】self-confident: 有～心 have self-confidence; be sure of oneself
【自行】①(自己做) by oneself: 请～解决。Please settle it by yourself. ②(自动) of oneself; of one's own accord; voluntarily ③[天](天体自行运动) proper motion
【自行车】bicycle; bike: 骑～去看电影 go to the cinema by bike ◇～厂 bicycle plant／～架 bicycle stand／～棚 bicycle shed ○轻便车 light roadster; minibike／车把 handlebar／车灯 lamp／车铃 bell／车架 frame／前叉 front forks／前轮 front wheel／后轮 rear wheel／轮圈 rim／轮胎 tire; pneumatic tire／外胎 outer cover／内胎 inner tube／辐条 spoke／气门 (air) valve／气门盖 valve cap／气门芯 valve rubber tube／油门 lubrication hole; lubricator／油门卡 oiler clip／链轮 chain wheel; gear wheel／链条 chain; roller chain／(全)链套 gearcase; chaincase／(半)链罩 chain guard; chain cover／脚蹬 pedal／飞轮 freewheel; rear hub sprocket／行李架 luggage carrier; carrier／支架 prop stand／单支 kick stand／闸皮 brake block／线闸 cable brake; caliper brake／倒轮闸 back-pedal brake; coaster brake／手闸 hand brake／打气筒 pump; inflator／油壶 oil can／加快轴 three-speed gear／(轮胎)漏气 puncture／补车胎 mend a puncture
【自行车赛】[体] cycle racing: ～赛车 racing cycle; road racer／～赛车场 cycling track; cycle track; velodrome; saucer／～运动员 cyclist; rider／公路～ circuit race ○长距离赛 long-distance race; endurance race; staying race／长途分段赛 stage race／短距离赛 short-distance race; sprint／个人赛 individual race／计时赛 time race／赛车场赛 track cycling; road meeting／团体赛 team race／越野赛 cyclo-cross／追逐赛 pursuit race; chase race
【自行火炮】[军] self-propelled gun
【自行装卸搅拌车】loading concrete mixer
【自行其是】act as one thinks fit; go one's own way
【自修】①(自习) study by oneself; have selfstudy ②(自学) study on one's own; study independently
【自序】①(作者自己写的序言) author's preface; preface ②(叙述自己生平经历的文章) autobiographic note; brief account of oneself
【自选】[体] free; optional: ～动作 optional exercise
【自学】study on one's own; study independently; teach oneself ◇～课本 teach-yourself books

【自寻烦脑】bring vexation on oneself; worry oneself
【自寻死路】bring about one's own destruction
【自言自语】talk to oneself; think aloud; soliloquize
【自以为是】consider oneself in the right; regard oneself as infallible; be opinionated
【自用】①(自以为是) obstinately holding to one's own views; opinionated; self-willed: 刚愎～ obstinate; self-opinionated ②(私人使用) for private use; personal: ～物品 personal effects; personal belongings
【自由】①(自由权) freedom; liberty: 公民～权 civil liberties ②[哲] freedom: ～王国与必然王国 realm of freedom and realm of necessity ③(不受拘束) free; unrestrained: ～散漫 slack; lax in discipline ◇ ～兑换[经] convertibility / ～港 free port / ～过境 free transit / ～航行 free navigation / ～化 liberalization / ～汇率 free exchange rate / ～竞争 free competition / ～恋爱 freedom to choose one's spouse / ～贸易 free trade / ～民[史] freeman / ～诗 free verse; unorthodox verse / ～市场 free market / ～体操 free exercise; floor exercise; free calisthenics / ～通过权 passage / ～通行 free course / ～外汇 convertible foreign exchange / ～泳 freestyle (swimming); crawl / ～职业 profession / ～职业者 professional / ～资本主义 non-monopoly capitalism / ～资产阶级 non-monopoly bourgeoisie; liberal bourgeoisie
【自由主义】liberalism ◇ ～者 liberal
【自由自在】leisurely and carefree; free and unrestrained
【自由组合规律】[生] law of independent assortment
【自有资金】one's own reserves
【自圆其说】make one's statement consistent; justify oneself
【自怨自艾】①(悔恨自己的错误) repent and redress one's errors ②(后悔) be full of remorse
【自愿】voluntary; of one's own accord; of one's own free will: ～参加的原则 the principle of voluntary participation / ～做某事 volunteer to do sth. ◇ ～捐款 voluntary contribution; voluntary fund / ～遣返 voluntary repatriation / ～证人 independent witness
【自在】①(自由) free; unrestrained: 逍遥～ leisurely and carefree; free from trammels; at liberty to enjoy oneself ②(安闲舒适) comfortable; at ease: 觉得不～ feel ill at ease ③[佛教] freedom of worldly cares
【自找】suffer from one's own actions; ask for it: ～麻烦 look for trouble
【自知之明】self-knowledge: 无～ lack of self-knowledge
【自治】autonomy; self-government: 民族区域～ regional national autonomy ◇ ～国家 autonomous state / ～机关 organ of self-government / ～能力 capacity of self-government / ～区 autonomous region / ～权 autonomy / ～条例 regulations on the exercise of autonomy / ～县 autonomous county / ～州 autonomous prefecture
【自制】①(自己制造) made by oneself: 老师～的教具 teaching aid made by the teacher himself ②(克制自己) self-control; self-restraint: 失去～ lose self-control
【自重】①(注意自己的言行) conduct oneself with dignity; be self-possessed ②(机器等自身的重量) dead weight
【自主】act on one's own; decide for oneself; keep the initiative in one's own hands: 不由～ cannot help; involuntarily / 独立～ independently and with the initiative in one's own hands ◇ ～企业 autonomous enterprise / ～权 decision-making power
【自助餐】buffet
【自传】autobiography
【自转】[天] rotation: 地球的～ the rotation of the earth ◇ ～轴[天] axis of rotation
【自走式联合收割机】self-propelled combine harvester
【自尊】self-respect; self-esteem; proper pride: ～心 sense of self-respect / 民族～心 sense of national pride
【自作自受】suffer from one's own actions; as a man sows, so shall he reap; self do; self have
【自作聪明】think oneself clever: 许多人～乱服药,使病更难治。 Many compounded the mischief by harmful self-medication.
【自作多情】imagine oneself as the favorite of one of the opposite sex
【自作主张】act on one's own; decide for oneself; take matter into one's own hand

zōng

宗 ①(祖宗) ancestor: 列祖列～ successive generations of ancestors ②(家族) clan: 同～ of the same clan ③(宗派) sect; faction; school: 正～ orthodox school ④(宗旨) principal aim; purpose ⑤(效法) take as one's model; learn from ⑦ [量] 大～货物 a large quantity of goods / 大～款项 a large sum of money / 一～心事 a matter that worries one

【宗祠】ancestral hall; ancestral temple

【宗法】patriarchal clan system ◇ ～社会 patriarchal society / ～制度 patriarch system

【宗教】religion ◇ ～法庭 the Inquisition / ～改革 religious reform / ～戒律 religious taboo / ～派别 religious sect / ～信仰 religious belief / ～信仰自由 freedom of religious belief / ～仪式 religious rites; ritual / ～自由 religious liberty; freedom of religion ○基督教 Christianity / 天主教 Catholicism / 罗马公教 Roman Catholic Church / 东正教 Orthodox Eastern Church / 希腊正教 Greek Orthodox Church / 新教 (耶稣教) Protestantism; Reformed Church / 英国圣公会 Church of England / 中国基督教三自爱国运动委员会 Chinese National Christian Three-self Patriotic Movement Committee / 三自原则 the Principles of Self- Government, Self- Support and Self-Propagation / 中国天主教爱国委员会 Chinese Patriotic Catholic Association / 基督教徒 Christian / 天主教徒 Catholic / 基督教新教教徒 Protestant / 耶稣基督 Jesus Christ / 伊斯兰教 Islamism; Islam / 佛派 Buddhism / 喇嘛教 Lamaism / 道教 Taoism / 犹太教 Judaism / 犹太教徒 Judaist

【宗教界人士】people in religious conscience

【宗庙】ancestral temple of a ruling house

【宗派】faction; sect ◇ ～斗争 factional strife / ～活动 factional activities; sectarian activities / ～主义 sectarianism; factionalism / ～主义者 sectarian; factionalist

【宗师】[旧] master of great learning and integrity

【宗室】member of the royal clan: ～内幕 the inner stories of the Imperial house

【宗旨】aim; purpose: 办刊的～ aim of the publication of the magazine

【宗主国】suzerain; metropolitan state

【宗主权】suzerainty

【宗族】① (同一父系的家族) patriarchal clan ② (同一父系家族的成员) clansman ◇ ～观念 ancestral idea / ～纠纷 clan dispute

棕 ① (棕榈) palm ② (棕毛) palm fiber; coir: ～绳 coir rope

【棕绷】wooden bed frame strung with criss-cross coir ropes

【棕榈】palm ◇ ～酸 palmitic acid / ～油 palm oil; palm butter

【棕毛】palm fiber: ～蓑衣 palm rain cape

【棕壤】[农] brown earth

【棕色】brown ◇ ～森林土 [农] brown forest soil

【棕树】palm

【棕毯】coir matting

【棕熊】[动] brown bear

【棕竹】a variety of black bamboo

鬃 hair on the neck of a pig, horse, etc.: 马～ horse's mane / 猪～ pig's bristles

【鬃刷】bristle brush

踪 footprint; track; trace: 跟～ follow the tracks of; trail

【踪迹】trace; track: 不留～ not leave a trace

【踪影】trace; sign: 毫无～ leaving without a trace

综 put together; sum up: ～上所述 to sum up

【综观】make a comprehensive survey

【综合】① (综合起来) synthesize: ～有关资料 synthesize the relevant data ② (综合的) synthetical; comprehensive; multiple; composite: ～的研究 a synthetical study ◇ ～报导 comprehensive dispatch; news roundup / ～报告 comprehensive report; summing-up report / ～大学 university / ～规划 unified plan / ～经营 all- around development / ～开发 the comprehensive exploitation / ～考察 comprehensive survey / ～利用 comprehensive utilization; multipurpose use / ～平衡 overall balance / ～所得税 consolidated income tax / ～杂志 a catch- all magazine / ～指数 aggregative index number; composite index

【综合法】synthesis

【综合性工厂】multiple-producing factory

【综合语】[语] synthetic language

【综合治疗机】[牙科] dental unit

【综括】sum up: ～起来 to sum up; to state succinctly

【综览】a general or comprehensive survey; view generally

【综揽】be in overall charge

【综述】summarize; sum up: 来稿～ a summary of readers' contributions

zǒng

总 ① (总括) assemble; put together; sum up: ～起来说 to sum up ② (全部的;全面的) general; overall; total: ～罢工 general strike / ～产

量 total output / ～动员 general mobilization / ～根源 root cause / ～攻 general offensive / ～开关 master switch / ～趋势 general trend / ～危机 general crisis / 抓～ assume overall responsibility / ～的说来 generally speaking; by and large ③ (为首的) chief; head; general: ～编辑 editor in chief / ～参谋长 chief of the general staff / ～工程师 chief engineer / ～经理 general manager ④ (一直) always; invariably ⑤ (总归;毕竟) anyway; after all; inevitably; sooner or later

【总参谋部】 the Headquarters of the General Staff

【总得】 must; have to; be bound to: 无论如何, 任务～完成。 At any rate the task must be fulfilled.

【总代理】 general agency ◇ ～人 general agent

【总督】 governor-general; governor; viceroy

【总额】 total; the total amount: 存款～ total deposits / 工资～ total wages / 他的存款～达五万元。 His total deposits are 50000 *yuan*.

【总而言之】 in short; in a word; in brief; to make a long story short

【总方针】 general policy; general principle

【总纲】 general program; general principles: 宪法的～ the general principles of the constitution

【总工会】 federation of trade unions

【总共】 in all; altogether; in the aggregate: 这个班～有四十名学生。 There are altogether 40 students in this class.

【总管】 manager

【总归】 anyhow; eventually; after

【总行】 head office: 中国人民银行～ the Head office of the People's Bank of China

【总合同】 macrocontract

【总和】 sum; total; sum total: 人类知识的～ the sum total of man's knowledge

【总后勤部】 the General Logistics Department

【总汇】 ① (水流会合) come of flow together ② (汇合在一起的事物) confluence; concourse; aggregate

【总机】 switchboard; telephone exchange

【总计】 ① (总起来计算) amount to; add up to; total: 观众～达两万人。 The audience totalled 20000. ② [数] (总计数) grand total

【总监】 inspector general; chief inspector

【总结】 ① (分析研究经验做出结论) sum up; summarize: ～本课的要点 sum up the main points of the lesson ② (概括出来的结论) summary; summing up: 作～ make a summary ◇ ～会 summing-up meeting

【总括】 sum up: ～起来 to sum up; to state succinctly

【总揽】 assume overall responsibility; take on everything: ～大权 have overall authority; assume a dominant role

【总理】 premier; prime minister: 国务院～ the Premier of the State Council

【总领事】 consul general ◇ ～馆 consulate general

【总目】 comprehensive table of contents

【总平面图】 general layout

【总评】 general comment; overall appraisal

【总谱】 [乐] score

【总收入】 general income

【总数】 total; sum total: ～是一千二百。 The total is 1200.

【总司令】 commander in chief ◇ ～部 general headquarters

【总算】 [副] ① (终于) at long last; finally: 雨季～过去了。 The rainy season has ended at last. ② (大体上) considering everything; all things considered; on the whole: 天气～不坏。 On the whole the weather is not bad.

【总体】 overall; total ◇ ～规划 overall plan / ～预测 macroforecast / ～战 general war; total warfare

【总统】 president: 副～ vice-president ◇ ～当选人 president-elect / ～府 presidential palace; the residence and office of a president / ～候选人 presidential candidate / ～任期 presidential term / ～选举 presidential election / ～制 presidential government

【总务】 ① (机关、学校中的行政杂务) general affairs; general services: 他管～. He is in charge of the general affairs. ② (负责行政杂务的人) person in charge of general affairs ◇ ～科 general affairs section / ～人员 general affairs personnel / ～司 general service department

【总议定书】 general protocol

【总预算】 general budget

【总则】 general rules; general principles

【总章细则】 general rules and by-laws

【总政治部】 the General Political Department

【总之】 in a word; in short; in brief

【总支】 general branch: 党～书记 secretary of the general Party branch

【总值】 total value: 出口～ gross export value / 进口～ gross import value

【总指挥】 ① (指军队的) commander in chief ②

(其它方面的) general director ◇ ～部 general headquarters
【总状花序】[植] raceme

zòng

粽
【粽子】a pyramid-shaped dumpling made of glutinous rice wrapped in bamboo or reed leaves

纵 ①(地理上的南北方向) from north to south: ～贯铁路 north-south railway line ②(跟物体的长的一边平行的) vertical; longitudinal; lengthwise ③(释放) release; set free: ～囚 release a prisoner; set a prisoner free ④(放任) indulge; let loose; let oneself go: ～欲 indulge in sensual pleasures / 放～ let sb. have his way ⑤(纵身) jump up; jump into the air
【纵波】[物] longitudinal wave
【纵步】①(放开脚步) stride ②(向前跳跃的步子) jump; bound
【纵断面】vertical section
【纵队】column; file: ～队形 column formation / 第五～ the fifth column / 二路～ column of twos / 一路～ column of files; single file
【纵横】①(竖和横) in length and breadth; vertically and horizontally: ～交错的铁路网 a crisscross network of railways ②(奔放自如) with great ease; freely: ～驰骋 move about freely and quickly / ～四海 overrun the four seas / ～自如 capable of moving in any direction
【纵横捭阖】maneuver among various political groupings
【纵虎归山】let the tiger return to the mountain; cause calamity for the future
【纵火】set on fire; commit arson ◇ ～犯 arsonist / ～罪 crime of setting on fire; offense of arson
【纵酒】drink to cxccss
【纵览】look far and wide; scan; ～群书 read extensively
【纵目】look as far as one's eyes can see: ～远望 gaze far into the distance
【纵情】to one's heart's content; as much as one likes: ～歌唱 sing heartily / ～诗酒 give one's mind to poems and grant indulgence to wine / ～游乐 be on the spree
【纵然】even if; even though: ～任务艰巨,我们也要及时完成。We will manage to accomplish the task in time even though it is difficult.
【纵容】connive; wink at: ～行为 act of connivance / 受到～和庇护 be winked at and shielded / 在某人～下 with the connivance of sb.
【纵身】jump; leap: ～上马 leap onto a horse / ～一跳 jump into the air; jump up
【纵深】[军] depth: 向～发展 develop in depth ◇ ～防御 defense in depth
【纵使】even if; even though
【纵视图】longtitudinal view
【纵谈】talk freely: ～世事 talk worldly affairs freely～四化前景 talk freely about the prospects of the four modernizations
【纵欲】give way to one's carnal desires; indulge in sensual pleasures: ～无度 indulge in carnal pleasure without restraint
【纵坐标】[数] ordinate

zǒu

走 ①(走动) walk; go: ～近某人 walk up to a person / ～前人没有～过的路 break new paths / 向前～ walk straight ahead / 请这边～。Go this way, please. / 这小孩还不会～。The baby cannot walk yet. ②(跑) run; move: 奔～相告 run around spreading the news ③(离开) leave; go away ④(探望) visit; call on: ～亲戚 call on relatives ⑤(通过;由) through; from: 我们～这个门出去吧。Let's go out through this door. ⑥(改变原样) depart from the original; lose the original shape, flavor, etc.: ～样 go out of form; lose shape
【走步】[篮球] walk with the ball
【走道】①(供人行走的道路) pavement; sidewalk ②(走路) walk
【走调儿】out of tune
【走动】①(行走使身体活动) walk about; stretch one's legs: 在公园里～～ stretch one's legs in the park ②(拜访) visit each other: 这两家过去常～。The two families used to visit each other quite often.
【走读】attend a day school ◇ ～生 day student; nonresident student
【走访】①(访问) interview; have an interview with ②(拜访) pay a visit to; go and see
【走风】let out a secret; leak out
【走钢丝】[杂技] wirewalking
【走狗】running dog; lackey; flunkey; stooge; servile follower
【走过场】do sth. as a mere formality; go through the motions; do sth. perfunctorily or

superficially

【走后门】get in by the back door; secure advantages through pull or influence

【走火】①(枪走火) discharge ②(电线跑电) sparking ③(说话过头) go too far in what one says; put sth. too strongly; overstate

【走江湖】wander from place to place and earn a living by juggling, fortune-telling, etc.; become a vagrant

【走廊】corridor; passage; passageway

【走漏】①(泄漏) leak out; divulge: ～风声 leak information ②(走私漏税) smuggling and tax evasion

【走路】walk; go on foot: 我上学通常是～去。I go to school usually on foot.

【走马灯】trotting horse lamp

【走马疳】[医] noma; gangrenous stomatitis

【走马看花】look at flowers while riding on horseback; gain a superficial understanding through cursory observation

【走马上任】go to take up office; go to one's post

【走南闯北】roam all over the country; travel extensively

【走内线】go through private channels; take the inner line

【走禽】Cursores; cursorial birds

【走色】lose color; fade

【走绳】[杂技] ropedancing; ropewalking

【走失】wander away; be lost; be missing

【走兽】beast; quadruped; four-footed animal

【走私】smuggle: ～的货物 smuggled goods ◇ ～案件 smuggling case / ～犯 owler / ～集团 smuggling ring / ～漏税 booftegging / ～者 smuggler / ～罪 offense of smuggling

【走投无路】have no way out; be in an impasse; come to the end of one's tether: 逼得～ be driven to the wall

【走味儿】lose flavor

【走下坡路】go downhill; be on the decline

【走向】①(延伸的方向) run; trend; alignment: 海岸线～ the trend of the coastline / 矿脉的～ the run of the ore vein ②(向某方向发展) move towards; head for; be on the way to: ～光明 move towards the light / ～胜利 march to victory ③[地] (地质地层走向) strike: 矿层～ seam strike

【走样】lose shape; go out of form; be different from what is expected or intended

【走一步看一步】take one step and look around before taking another; proceed without a plan, or with caution

【走运】be in luck; have good luck: 她不～。She has bad luck. / 他真～。He is lucky indeed.

【走着瞧】wait and see

【走卒】pawn; cat's-paw; lackey; stooge

【走嘴】make a slip of the tongue; let slip an inadvertent remark

zòu

奏 ①(演奏) play; perform; strike up: 独～ solo / 二重～ duet / 三重～ trio / 四重～ quartet / 五重～ quintet / 弦乐钢琴四重～ string piano quartet / 弦乐四重～ string quartet / ～国歌 play the national anthem ②(取得) achieve; produce ③(向帝王陈述) present a memorial to an emperor

【奏捷】win a battle; score a success

【奏鸣曲】[乐] sonata: 小～ sonatina ◇ ～式 sonata form

【奏疏】memorial to the throne

【奏效】prove effective; be successful; get the desired result: 这药能立即～。This medicine will have immediate efficiency

【奏乐】play music; strike up a tune

zū

租 ①(租用) rent; hire; charter: ～房 rent a house / ～一辆大客车 hire a bus ②(出租) rent out; let out; lease: 本市出～的房屋不多。There are not many rentals available in this city. ③(租金) rent ④(田赋) land tax

【租船】chartering ◇ ～代理 chartering agent / ～合同 contract of affreightment / ～契约 charter party / ～人 charterer / ～市场 chartering market

【租佃】rent out land to tenants ◇ ～关系 tenancy relationship / ～制度 tenancy system

【租购】hire purchase: ～业务 hire purchase finance

【租户】①(租用房屋的人) tenant; lessee; leaseholder ②(租用物品的人) hirer

【租界】concession

【租借】rent; hire; lease ◇ ～地 leased territory; leasehold / ～期限 time limit of the lease / ～人 leaseholder; lessee; tenant; hirer / ～条件 lease terms

【租金】rent; rental: ～低的 low rented / ～高的 high rented / 她这房子年～是二千四百元。The yearly rental of her house is 2400 *yuan*.

【租赁】rent; lease; hire ◇ ～合同 contract of tenancy; lease contract / ～贸易 renting trade / ～业务 charter business
【租期】lease term
【租书处】book rental
【租税】[旧] land tax and other levies
【租用】rent; hire; take on lease: ～家具 rent furniture / ～礼堂 hire a hall ◇ ～人 leaseholder; tenant; lessee; hirer
【租与】grant
【租约】lease: ～的延续 continuation of tenancies / 取消～ cancel the deed of rental

zú

卒 ①(兵) soldier; private: 士～ the rank and file; soldiers ②(完毕) finish; end: ～业 finish school; graduate ③(终于) finally; at last ～偿夙愿 have one's wishes fulfilled at last ④(死) die: 暴～ die all of a sudden / 生～年月 dates of birth and death ⑤(一种象棋子) pawn
【卒然宣战】declare war suddenly
【卒业】[书] graduate; finish a course of study
【卒子】a soldier; a private

族 ①(家族) clan: ～人 clansman / ～长 clan elder ②(种族;民族) race; nationality: 汉～ Han nationality ③(事物的一大类) family: 猫～ the cat tribe / 水～ aquatic animals / 语～ a family of languages
【族居】live together as a clan: ～平原 dwell together on the plain
【族类】of the same clan or race: ～迥异 races are different / ～相聚 gather together all that belonging to the same class
【族谱】pedigree of a clan
【族权】clan authority; clan power

足 ①(脚;腿) foot; leg: 赤～ barefoot ②(足够) sufficient; enough; ample: 干劲十～ be full of vigor ③(足足) full; as much as ④(足以) enough; sufficient: 微不～道 not worth mentioning
【足不出户】never leave one's home
【足吃足喝】eat and drink to one's heart's content
【足赤】pure gold
【足够】enough; ample; sufficient: 有～的时间做这项工作 have enough time to do this work
【足迹】footmark; footprint; track
【足见】it serves to show; one can well perceive: ～这种意见是不对的。 It shows that such opinion is not correct.
【足金】pure gold; solid gold
【足龄】full age
【足谋寡断】resourceful but irresolute
【足球】①(球类运动) soccer; football; association football: 美国式～ American football / 英国式～ soccer; association football ②(指此项运动用的球) football ◇ ～队 football team; eleven / ～赛 football game / ～运动员 footballer; football player ◎守门员 goalkeeper; goaltender; goalie / 后卫 back / 左后卫 left back / 右后卫 right back / 中卫 center half back / 前卫 half back / 左前卫 left half back / 右前卫 right half back / 前锋 forward / 中锋 center forward; center / 右内锋 inside right forward / 右边锋 outside right forward; outside right / 左内锋 inside left forward / 左边锋 outside left forward; outside left / 中场 midfield / 中圈 kick-off circle / 中线 half-way line / 角球 corner ball; corner / 球门球 goal kick / 头球 header / 罚点球 spot kick / 罚任意球 free kick / 掷界外球 throw-in / 倒钩球 bicycle kick; overhead kick / 跳起顶球 flying header / 盘球 dribbling / 铲球 slide tackle / 射门 shoot / 越位 offside / 传球 pass the ball / 带球过人 break through; beat / 控制中场 control the midfield / 筑人墙 set a wall / 掌握进攻节奏 set the pace / 越位战术 off-side trap / 拖延战术 time wasting tactics / 红牌 red card / 黄牌 yellow card
【足色】of standard purity
【足衣足食】have no shortage of foot and clothing
【足以】enough; sufficient: ～糊口 enough to keep body and soul together / ～乱真 good enough to pass for genuine / 这些事实～证明她是对的。 These facts are enough to prove that she is right.
【足印鉴定】footprint identification
【足月】born after the normal period of gestation; mature
【足智多谋】wise and full of stratagems; resourceful

zǔ

诅
【诅咒】curse; swear; wish sb. evil; imprecate: 某人 swear at sb.; lay a curse upon sb.

祖 ①(祖父) grandfather ②(祖宗) ancestor: 远～ remote ancestors ③(事业或派别的首创

者) founder; originator: ～师爷 the founder
【祖辈】ancestors; forefathers; ancestry
【祖传】handed down from one's ancestors: ～古玩 curios handed down from forefather／～秘方 a secret prescription handed down in the family from generation to generation
【祖坟】ancestral grave
【祖父】grandfather
【祖国】one's country; homeland; native land; motherland; fatherland
【祖籍】original family home; ancestral home; the land of one's ancestors
【祖母】grandmother
【祖母绿】[矿] emerald
【祖师】①(学术或技术上创立派别的人) the founder of a school of learning, a craft, etc. ②(佛教、道教创立宗派的人) the founder of a sect of Buddhism or Taoism
【祖孙】grandparent and grandchild: ～三代 three generations
【祖先】ancestry; ancestors; forbears; forefathers
【祖宗】forefathers; ancestry; forbears
【祖祖辈辈】for generations; from generation to generation

阻 hinder; block; obstruct: 畅通无～ without let or hindrance; one can go through unimpeded
【阻碍】hinder; block; impede: ～交通 block the traffic／～生产力的发展 hinder the development of the productive forces
【阻挡】stop; stem; resist; obstruct: ～不住 unable to hinder ◇～犯规[篮球] blocking
【阻隔】separate; cut off: 山川～ be separated by mountains and rivers
【阻击】[军] block; check ◇～战 blocking action／～阵地 blocking position
【阻抗】[电] impedance: 反射～ reflected impedance ◇～匹配 impedance matching
【阻拦】stop; obstruct; bar the way: 让她走,别～她。Let her go. Don't stop her.
【阻力】①(阻碍事物发展的外力) obstruction; resistance: 冲破～ break through obstruction ②[物] resistance; drag: 空气～ air resistance／摩擦～ friction drag／迎面～ frontal resistance
【阻挠】obstruct; thwart; stand in the way; put a spoke in sb.'s wheel: 百般～ obstruct in every possible way／计划受到～ be thwarted in one's plan
【阻尼】[物] damping ◇～器 damper／～振荡 damped oscillation
【阻塞】block; obstruct; clog: 交通～。There is a traffic jam.／拥挤的车辆～了道路。Heavy traffic blocked the road. ◇～振荡器 blocking oscillator
【阻止】prevent; stop; hold back: ～事态的恶化 prevent the situation from deteriorating

组 ①(组织) organize; form: ～阁 form a cabinet ②(小组) group: 学习～ a study group ③[量] set; series; battery: 一～邮票 a set of stamps
【组胺】[化] histamine
【组成】form; make up; compose: 水的～ the composition of water／～统一战线 form a united front／他们～了一个英语初级班。They formed a group for beginners in English.
【组稿】commission authors to write on given topics; solicit contributions
【组歌】[乐] suite of songs
【组阁】form a cabinet
【组合】①(组织成为整体) make up; compose; constitute ②(组织起来的整体) association; combination ③[机][数] combination ◇～车床 combined lathe／～理论[数] combinatorial theory／～运价 combination joint rate／～钻床 combination drilling machine
【组合体】[机] assembly
【组曲】[乐] suite
【组织】①(组成) organize; form: ～一个歌咏队 organize a chorus ②(组织系统) organization; organized system: 党团～ Party and Youth League organizations／群众～ mass organizations ③[纺] weave: 平纹～ plain weave／斜纹～ twill weave ④[生理] tissue: 神经～ nerve tissue ◇～法 rules of organization; organic law; constituent act／～关系 credentials showing membership in an organization; membership credentials／～机构 institutional framework／～疗法 tissue therapy; histotherapy／～生活 regular activities of an organization／～条例 organic rules／～委员 committee member in charge of organizational work／～委员会 organizing committee
【组织学】[生理] histology
【组织液】[生理] tissue fluid
【组装车间】composing room
【组装船】fabricated ship
【组装生产】assembly line industries

【组装线】assembly line

zuān

钻 ①(钻眼) drill; bore: 在墙上～个洞 drill a hole in the wall ②(穿过;进入) get into; go through; make one's way into: 老鼠～进了洞。The rat got into the hole. ③(钻研) study intensively; dig into: ～业务 dig into one's job

【钻空子】avail oneself of loopholes; exploit an advantage: 不留空子给人钻 leave no loopholes

【钻牛角尖】①(费力研究不值得的问题) take unnecessary pains to study an insignificant or insoluble problem; split hairs ②(行不通) get into a dead end

【钻圈】[杂技] jumping through hoops

【钻探】drilling: 海底～ offshore drilling ◇ ～工 driller / ～机 drilling machine / ～设备 drilling equipment

【钻天入地】search for an opening for oneself by all possible means

【钻心虫】borer

【钻研】study intensively; dig into: ～业务 gain professional proficiency / 努力～技术 endeavor to perfect one's skill

【钻营】curry favor with sb. in authority for personal gain; secure personal gain

zuǎn

纂 [书] compile; edit: 编～词典 compile a dictionary

【纂辑】compile: ～史料 collect and edit historical materials

【纂修】compile; edit; prepare

zuàn

钻 ①(打眼用的工具) drill; auger: 风～ pneumatic drill / 环孔～ annular auger / 手摇～ hand drill; drill ②(钻石) diamond; jewel: ～戒 diamond ring / 十七～ 17 jewels ③(钻眼) bore; drill

【钻床】[机] drilling machine; driller: 龙门～ planer drilling machine

【钻杆】[石油] drill rod; drill pipe

【钻机】(drilling) rig; drilling machine: 套管～ casing drilling rig / 安装～ rig up / 拆卸～ rig down

【钻进速度】drilling rate

【钻井】well drilling: 海洋～ offshore drilling ◇ ～船 offshore drilling rig / ～队 drilling crew; drilling team / ～工人 driller / ～记录 drill log

【钻具】[石油] drilling tool; drilling rig

【钻孔机】drilling machine

【钻石】①(金刚石) diamond ②(宝石) jewel

【钻塔】[矿] boring tower; derrick

【钻头】bit: 三牙轮～ three cone bit / 卸下～ break out a bit / 装上～ brace a bit

zuǐ

嘴 ①(口) mouth: 张～ open one's mouth ②(形状象嘴的东西) anything shaped or functioning like a mouth: 茶壶～ the spout of a teapot / 瓶～儿 the mouth of a bottle / 烟～儿 cigarette holder

【嘴巴】①[口](嘴部附近) face: 打～ slap sb. in the face; box sb.'s ears ②[方](口) mouth

【嘴笨】inarticulate; clumsy of speech

【嘴不饶人】fond of making sarcastic remarks

【嘴馋】fond of good food

【嘴唇】lip: 上～ the upper lip / 下～ the lower lip

【嘴乖】[口] clever and pleasant when speaking to elders

【嘴尖】sharp- tongued; cutting in speech: ～舌巧 gifted with a quick and sharp tongue

【嘴角】corners of the mouth

【嘴紧】tight-lipped; close-mouthed; secretive

【嘴快】have a loose tongue: ～心直 jaws are quick and heart is straight

【嘴懒】not incline to talk much; taciturn

【嘴脸】[贬] look; features; countenance: 丑恶～ ugly features

【嘴皮子】[口] lips: 耍～ talk glibly

【嘴上无毛办事不牢】a man with downy lips is bound to make slips; young people cannot be trusted with important tasks

【嘴碎】loquacious; garrulous

【嘴甜】[口] ingratiating in speech; smooth-tongued; honeymouthed: ～心苦 sweet jaws and bitter heart

【嘴严】able to keep a secret; discreet in speech; cautious about speech

【嘴硬】stubborn and reluctant to admit mistakes or defeats: ～心软 firm in speech and soft in heart

zuì

醉 ①(饮酒过量,神志不清) drunk; intoxicated; tipsy: 喝～了酒 have had a glass too many / 烂～如泥 be dead drunk ②(用酒泡制) liquor-saturated; steeped in liquor

【醉鬼】drunkard; sot; inebriate

【醉汉】drunkard; drunken man
【醉生梦死】live as if intoxicated or dreaming; lead a befuddled life
【醉态】the state of being drunk; drunkenness
【醉翁之意不在酒】the drinker's heart is not in the cup; have ulterior motives
【醉心】be bent on; be wrapped up in: ～名利 be bent on seeking fame and gain
【醉醺醺】sottish; drunk; tipsy
【醉言不较】do not find fault with drunken talk
【醉意】signs or feeling of getting drunk: 他已有三分～了。He is a bit tipsy.

最 most: ～可靠的同盟者 the most reliable ally / 速度～快 the fastest
【最不中用】not of the least use
【最初】initial; first: ～的计划 the original plan; the first and earliest plan / ～的印象 first impressions / ～阶段 the initial stage
【最大】maximum: ～概率 maximum probability / ～高度 maximum height / ～功率 maximum power / ～宽度 maximum width / ～流量 maximum stream flow
【最大公约数】[数] greatest common divisor
【最低】lowest; minimum ◇ ～罚款额 minimum fine / ～纲领 minimum program / ～价格 the lowest price / ～年龄 minimum age / ～限度 the lowest limit; at least / ～援助拨款 minimum allocation of assistance
【最短刑期】minimum sentence
【最多】at most; maximum: 藏书～的图书馆 the most numerous library
【最高】highest; supreme; tallest: ～国家权力机关 the highest organ of state power / ～国家行政机关 the highest organ of state administration / ～监督机关 the highest supervisory office / ～司法机构 the highest judicial organs ◇ ～电压 ceiling voltage / ～法律效力 supreme legal authority / ～纲领 maximum program / ～国务会议 the Supreme State Conference / ～价格 ceiling price / ～年产量 peak annual output / ～权力 supreme power / ～速度 maximum speed / ～统帅 supreme commander / ～温度 maximum temperature / ～限额 ceiling / ～刑罚 maximum sentence
【最高级】①(等级最高的) highest; summit: ～会谈 top-level talks; summit talks / ～会议 summit conference ②[语] the superlative degree
【最好】①(顶好的) best; first-rate: ～的办法 the best way / ～的学生 the best student ②(最好还是) had better; it would be best: 我们今天～留在家里。We'd better stay at home today.
【最后】final; last; ultimate ◇ ～成本 final cost / ～结论 final submission / ～条款 final provisions; final articles / ～通牒 ultimatum; a final dispatch / ～协定 final agreement / ～议定书 final protocol
【最惠国】most-favored-nation: ～待遇条款 most-favored-nation treatment clause / ～条款 most-favored-nation clause / 给予～待遇 accord most-favored-nation treatment
【最佳】optimum: 育秧的～温度 optimum temperature for seedling nursing ◇ ～数 optimum number / ～谐振 optimum resonance
【最近】①(近来) recently; lately; of late: ～一期的《中国画报》the current issue of China Pictorial ②(距离最近的) nearest ③(最近的将来) in the near future; soon: 她～就要出国。She'll be going abroad soon.
【最先】the first; the earliest
【最小公倍数】[数] lowest common multiple
【最严厉的刑罚】harshest punishment
【最终】final; ultimate ◇ ～裁定 absolute decree / ～裁决 final ruling / ～目的地 final destination / ～判决 conclusive judgement / ～用户 final user; final customer

罪 ①(犯法的行为) crime; guilt: 轻～ misdemeanor / 重～ felony / 判～ condemn; pass sentence / ～上加～ be doubly guilty / 认～ admit one's guilt / 宣告无～ acquit sb. (of a crime); declare sb. innocent / 有～ guilty of a crime; guilty ②(过失) fault; blame: 归～于人 put the blame on others ③(苦难) suffering; pain; hardship: 受～ have a hard time; endure suffering
【罪案】details of a criminal case; case
【罪不容诛】even death cannot atone for the offense; be guilty of crimes for which even death is insufficient punishment
【罪大恶极】be guilty of the most heinous crimes
【罪恶】crime; evil: ～活动 criminal activities / ～累累 be guilty of innumerable crimes / ～昭彰 one's sins are known to all
【罪犯】criminal; offender; culprit: ～的供词 prisoner statement / ～的特征 characteristic of a criminal / 少年～ a juvenile offender
【罪该万死】be guilty of a crime for which one deserves to die ten thousand deaths; be guilty

of a crime for which even death cannot atone
【罪过】fault; offense; sin: 隐藏～ cloak a sin／这是我的～。That's my fault.
【罪魁】chief criminal; archcriminal: ～祸首 the chief offender and the ring leader
【罪名】charge; accusation: 偷窃的～ a charge of theft／罗织～ frame a case against sb.
【罪孽】wrongdoing that brings retribution; sin: ～深重 sinful
【罪人】guilty person; offender; sinner: 千古～ person condemned by history
【罪行】crime; guilt; offense: ～的种类 variety of crimes／～分类 classification of crimes／～细节 particulars of offense／～累累 cumulation of crimes
【罪有应得】culpable of punishment; deserve the punishment
【罪责】responsibility for an offense
【罪证】evidence of a crime; proof of one's guilt: ～确凿 proof of a crime beyond a shadow of a doubt
【罪状】facts about a crime; charges in an indictment

zūn

尊 ①(地位或辈分高) senior; of a senior generation: ～长 elders ②(敬重) respect; venerate; honor: 自～ self-respect ③[敬] ～夫人 your wife ④[量] 十～大炮 ten artillery pieces／一～佛像 a statue of a Buddha
【尊称】①(尊敬地称呼) address sb. respectfully ②(对人尊敬的称呼) a respectful form of address; honorific title
【尊崇】worship; revere; venerate
【尊贵】honorable; respectable; respected
【尊敬】respect; honor; esteem: 非常～某人 hold sb. in the greatest esteem; have the greatest esteem for sb.
【尊严】dignity; honor: 国家～ national dignity／人格～ dignity of human personality／维护法律的～ guard the sanctity of the law
【尊长】elders and betters
【尊重】respect; value; esteem: 互相～ respect each other／大家很～他。He was much esteemed.

遵 abide by; obey; observe; follow: ～命 obey your command／～纪守法 observing discipline and obeying the law
【遵从】defer to; comply with; follow: ～法令 act in accordance with the law
【遵法施行】obey the laws and to put into operation
【遵命】[敬] comply with your wish; obey your command
【遵守】observe; abide by; comply with: ～法律 abide by the law／～规则 comply with the rules／～合约 observe the agreement／～时间 be on time; be punctual
【遵循】follow; abide by; adhere to: ～愿望 comply with one's wishes
【遵医嘱】follow the doctor's advice
【遵照】obey; conform to; comply with; act in accordance with: ～党的政策办事 act in accordance with the policies of the Party

鳟 trout: 虹～ rainbow trout

zǔn

撙 save: ～下些钱 save some money
【撙节】retrench; practice economy: ～开支 retrench; cut down expenses

zuō

作 ①(从事某种活动) do ②(作坊) workshop
【作坊】workshop: 木工～ carpenters' workshop
【作弄】tease; make a fool of; play a trick on; poke fun at
【作死】seek death; take the road to ruin; look for trouble
【作揖】make a bow with hands folded in front

zuó

琢
【琢磨】turn sth. over in one's mind; ponder: ～出个办法 figure out a way／～问题 turn a problem over in one's mind

昨 yesterday: ～晚 yesterday evening; last night
【昨死今生】lead a new life from now on
【昨天】yesterday

作
【作践】①(糟蹋) spoil; waste ②(诽谤) disparage; run sb. down ③(侮辱) humiliate; insult
【作料】condiments; seasoning

zuǒ

左 ①(左边) the left side; the left: 向～转 turn to the left ②(邪;偏) queer; unorthodox;

heretical ③(错) wrong; incorrect: 你想~了。You are mistaken. ④(相反) different; contrary; opposite: 意见相~ hold different views; be at variance ⑤(激进的) the Left: ~派 the Left; the left wing

【左边锋】[足球] outside left; left wing

【左边】the left; the left side ◇ ~驾驶汽车 left-hand control car

【左不过】①(反正) anyway; anyhow; in any event ②(只不过) only; merely; just: 这本词典很有用,~贵了些。This dictionary is very useful, only it costs a lot of money.

【左道旁门】①(不正常的宗教派别) heretical sect; heterodox school ②(异端邪说) heresy; heterodoxy

【左舵】left standard rudder; left rudder

【左锋】[篮球] left forward

【左改右改】make changes over and over again

【左顾右盼】glance right and left

【左后卫】[足球] left back

【左计误事】an unsuitable plan spoils an affair

【左轮】revolver ◇ ~手枪 revolving pistol

【左面】the left side; the left-hand side

【左内锋】[足球] inside left

【左派】①(左派集团) the Left; the left wing ②(左派分子) Leftist

【左撇子】left-handed person; left-hander; lefty

【左前卫】[足球] left halfback; left half

【左倾】①(思想进步的) left-leaning; progressive; inclined towards the revolution ②(盲动的) "Left" deviation ◇ ~机会主义 "Left" opportunism / ~冒险主义 "Left" adventurism

【左劝右劝】try again and again to persuade sb.

【左首】the left-hand side

【左思右想】think from different angles; turn sth. over in one's mind

【左提右挈】give mutual help or assistance

【左舷】port

【左翼】①(作战时左侧的部队) left wing; left flank ②(倾向激进的一部分) the left wing; the Left ◇ ~分子 Leftist; Left-winger

【左右】①(左和右两方面) the left and right sides: ~摇摆 vacillate now to the left and now to the right ②(大约) about; or so; or thereabouts: 两个月~ two months or so / 这只手表价值八十元~。The watch is worth about eighty *yuan*. ③(支配;操纵) master; control; influence: 为某人所~ be controlled by sb. ④(随从) entourage; retinue; attendants

【左右奔走】beat up and down

【左右逢源】①(无论怎样进行都顺利) be able to achieve success one way or another ②(办事圆滑) gain advantage from both sides

【左右开弓】shoot first with one hand, then with the other; shoot first to one side, then to the other; use first one hand and then the other in quick succession

【左右手】right-hand man; valuable assistant

【左右为难】in a dilemma; in an awkward predicament

【左证】evidence; proof

【左支右绌】not have enough money to cover the expenses; be unable to cope with a situation; be in straitened circumstances

【左支右吾】speak in an evasive manner

佐 ①(辅佐) assist: ~理 assist sb. with a task ②(辅佐别人的人) assistant

【佐餐】be eaten together with rice or bread; go with rice or bread

【佐证】evidence; proof

zuò

凿 ①(明确) certain; authentic; irrefutable: 确~ authentic; conclusive ②(卯眼) mortise

【凿凿】[书] true; certain; verified: ~有据 with irrefutable evidence / 言之~ say sth. with certainty

坐 ①(坐下) sit; take a seat: 那车能舒舒服服地~五个人。The car will sit five people comfortably. / 她~在椅子上。She sat in a chair. ②(乘;搭) travel by: ~出租汽车去火车站 take a taxi to the railway station ③(背对着) have its back towards: 这房子~北朝南。The house faces south. ④(放) put: 把炒菜锅~在火上 put the frying pan on the fire ⑤(下沉) sink; subside ⑥(枪炮因反作用力而向后移动) recoil; kick back

【坐标】[数] coordinate ◇ ~轴 coordinate axis

【坐禅】[佛教] sit in meditation

【坐吃山空】sit idle and eat, and in time your whole fortune will be used up

【坐地分赃】take a share of the spoils without participating personally in the robbery

【坐垫】cushion

【坐骨】[生理] ischium ◇ ~神经 sciatic nerve / ~神经痛 sciatica

【坐观成败】wait to see what will come of another's venture; look on coldly; be a mere onlooker

【坐化】death (of Buddhist monks)
【坐禁闭】be placed in confinement
【坐井观天】look at the sky from the bottom of a well; have a very narrow view
【坐困】be confined; be walled in; be shut up: ～愁城 be walled in by one's own worries / ～一隅 be impoverished at one corner
【坐牢】be in jail; be imprisoned
【坐冷板凳】①(受冷遇) be cold-shouldered ②(久等) cool one's heels
【坐力】[物] recoil: 无～炮 recoilless gun
【坐立不安】feel uneasy whether sitting or standing; be fidgety; be on tenterhooks
【坐落】be situated; be located: ～在山上的一个小镇 a town situated on a hill / 学校～在小河附近。The school is located near the rivulet.
【坐山雕】[动] cinereous vulture
【坐山观虎斗】sit on top of the mountain to watch the tigers fight; watch in safety while others fight, then reap the spoils when both sides are exhausted
【坐失良机】let slip a golden opportunity
【坐视不救】sit back and watch without going to the rescue
【坐收渔利】reap the spoils of victory without lifting a finger; profit from others' conflict; reap third party profit
【坐探】agent provocateur
【坐位】①(供人坐的地方) a place to sit: 给老师留几个～。Reserve several seats for the teachers. ②(椅子;凳子等) a thing to sit on; seat: 请给她搬个～来。Please get her a seat.
【坐卧不宁】be unable to sit down or sleep at ease; feel restless; be on tenterhooks
【坐享其成】sit idle and enjoy the fruits of others' work; reap where one has not sown
【坐药】suppository
【坐以待毙】sit still waiting for death; await one's doom; resign oneself to death
【坐以待旦】sit up and wait for daybreak; remain awake till dawn
【坐月子】[口] confinement in childbirth; lying-in
【坐镇】personally attend to garrison duty; assume personal command
【坐庄】be the banker or dealer

座

①(坐位) seat; place: 就～ take one's seat / 入～ be seated ②(垫东西的座子) stand; pedestal; base: 花瓶～ vase stand ③[天](星座) constellation: 大熊～ the Great Bear / 狮子～ the Leo / 天琴～ the Lyra ④[量] 一～山 a mountain / 一～水库 a reservoir
【座舱】[航空] cockpit
【座舱盖】[航空] canopy
【座次】order of seats; seating arrangements
【座上客】guest of honor; honored guest
【座谈】have an informal discussion ◇ ～会 forum; symposium
【座位】seat; place
【座无虚席】all seats are occupied; there are no empty seats
【座右铭】motto; maxim
【座钟】desk clock
【座子】①(垫东西的座子) stand; pedestal; base ②(自行车等的车座) saddle

柞

oak
【柞蚕】tussah ◇ ～丝 tussah silk
【柞栎】[植] toothed oak
【柞丝绸】tussah silk; pongee

作

①(起) rise; get up: 日出而～ get up at sunrise ②(写作) write; compose: ～诗 write a poem ③(从事某种工作) do; make ④(作品) work; writings ⑤(装) pretend; affect: 故～镇静 pretend to be composed ⑥(当作) regard as; take sb. or sth. for: 把敌人看～朋友 regard an enemy as a friend; take an enemy for a friend ⑦(当) act as; be; become: ～向导 act as a guide ⑧(发作) feel; have: ～冷 feel a chill
【作案】commit a crime or an offense: ～时被捕 be caught in the act; be caught red-handed ◇ ～现场 scene of crime / ～凶器 weapon of offense
【作罢】drop; relinquish; give up: 这件事只好～了。The matter will have to be dropped.
【作伴】keep company
【作保】be sb.'s guarantor; go bail for sb.; sponsor sb. ◇ ～人 give security / ～证人 guarantee
【作弊】practice fraud; cheat; indulge in corrupt practices: 在考试中～ cheat in an examination
【作壁上观】stay behind the breastworks and watch others fight; be an onlooker; sit by and watch
【作操】do gymnastics; do calisthenics; do exercises
【作出裁决】give a ruling: 作出对某人有利的裁决 bring in a finding for sb. / 作出拒绝受理的裁决 bring in a finding of non liquet
【作出判决】enter a judgement; give judgement

【作词】write words
【作抵押】on mortgage
【作对】set oneself against; oppose: 她与我～。She opposes me.
【作恶】do evil: ～多端 do all kinds of evil; be steeped in iniquity
【作法】①(行法术) resort to magic arts; practice magic ②(做法) way of doing things ③(文章作法) art of composition: 散文～ art of writing essays
【作法自毙】make a law only to fall foul of it oneself; be hoist with or by one's own petard; get caught in one's own trap
【作废】become invalid: 宣布～ declare invalid / 过期～ become invalid after a specified date
【作风】style; style of work; way: 工作～ style of work / 实事求是的～ a practical and realistic way of doing things / ～正派 be honest and upright; have moral integrity
【作梗】obstruct; hinder; create difficulties
【作怪】do mischief; make trouble
【作家】writer ◇ ～协会 the Writers' Union
【作假】①(故作客套) behave affectedly ②(伪造) falsify; counterfeit ③(要花招) cheat; play tricks
【作假证】give false testimony
【作价】fix a price for sth.; evaluate
【作奸犯科】violate the law and commit crimes; commit offenses against law and discipline
【作茧自缚】in a cocoon around oneself; get enmeshed in a web of one's own spinning
【作乐】make merry; enjoy oneself; have a good time: 寻欢～ seek pleasure and make merry
【作乱】stage an armed rebellion; rise in revolt: ～造反 riot and rebel
【作难】①(为难) feel embarrassed; feel awkward ②(难为人) make things difficult for sb.
【作呕】①(要呕吐) feel like vomiting; feel sick: 令人～的食物 mauseating food ②(使人厌恶) disgusting; loathsome: 令人～地吹捧 nauseatingly extol / 我不得不从头到尾听那令人～的故事。 I had to listen to the whole nauseating story.
【作陪】help entertain the guest of honor; be invited along with the chief guest
【作品】works
【作曲】write music; compose ◇ ～家 composer
【作如是观】view the matter in this light
【作色】show signs of anger; get worked up: 愤然～ flush with indignation
【作祟】①(鬼神为难人) haunt ②(捣乱) make mischief; cause trouble; exercise evil influence
【作威作福】tyrannically abuse one's power; ride roughshod over others; act like a tyrant
【作为】①(行为) conduct; deed; action: 他的～是无可非议的。His conduct was blameless. ②(作出成绩) accomplish; do sth. worthwhile: 无所～ attempt nothing and accomplish nothing ③(当作) regard as; look on as; take for: ～知己 take sb. for a close friend ④(就人的身份或事物的性质来说) as: ～一个党员,应该全心全意为人民服务。As a party member one must serve the people wholeheartedly.
【作伪】fake; make an imitation; forge
【作伪证】forswear; give false testimony ◇ ～者 perjurer
【作文】①(写文章) write a composition ②(文章) composition
【作物】crop: 夏收～ summer crop / 油料～ oil-bearing crop
【作息】work and rest: 按时～ work and rest according to schedule ◇ ～时间表 daily schedule; timetable; work schedule
【作业】①(功课) school assignment: 家庭～ homework / 改～ correct students' papers / 做～ do one's assignment ②(军事活动或生产活动) work; task; operation; production: 水下～ underwater operation / 野外～ field work ◇ ～班 work team / ～灯 working light / ～段 job step / ～计划 production plan / ～区 operation area / ～线 production line
【作用】①(产生影响) act on; affect ②(产生影响的活动) action; function: 反～ reaction / 化学～ chemical action / 酸对金属所起的～ the action of acid on metal ③(影响;效果) effect: 副～ side effect / 积极～ positive role / 起～ take effect ④(用意) intention; motive ◇ ～范围 sphere of action / ～力 effort; applied force
【作贼心虚】uneasy lies the head of one with a guilty conscience
【作战】fight; conduct operations; do battle ◇ ～部队 combat troops / ～部署 operational preparations / ～地图 battle map; operation map / ～方案 battle plan; line of action / ～方式 mode of operations / ～方针 concept of operations; operational principles; operational policy / ～基地 operational base; base of operations / ～技术 fighting technique / ～命令 combat order / ～区域 theatre of war / ～效能 fighting efficiency / ～指挥部 operational headquarters

【作者】author; writer: ～不详 by an anonymous author; authorship unknown
【作证】testify; give evidence; bear witness: 为某人～ testify on behalf of sb. / 在法庭上～ bear witness in a lawcourt
【作主】①(作决定) decide; take the responsibility for a decision: 她作不了主。 She's not in position to decide. / 由你～。 It's up to you to. ②(支持) back up; support
做 ①(制造) make; produce; manufacture ②(写作) write; compose: ～诗 write a poem ③(从事某种工作或活动) do; act; engage in: ～生意 do business; carry on trade / ～手工 do handiwork ④(家庭庆祝) hold a family celebration: ～寿 hold a birthday party ⑤(用做) be used as ⑥(结成) form or contract a relationship: ～朋友 make friends with / ～盟友 become allies ⑦(担任) be; become: ～教师 be a teacher ⑧(烹调) cook; prepare: ～饭 do the cooking; prepare a meal
【做白日梦】build castles in the air
【做伴】keep sb. company
【做表面文章】work for appearance sake
【做到】accomplish; achieve: 说到～ be as good as one's word; abide by one's promise
【做东】play the host; host sb.; act as host to sb.
【做法】way of doing or making a thing; method of work; practice
【做工】①(从事劳动) do manual work; work ②(工作质量) workmanship: ～粗劣 of poor workmanship / ～精细 of exquisite workmanship ③(手工费) charge for the making of sth.
【做功】[剧] acting; business
【做官】 be an official; secure an official position: ～当老爷 act like high officials and overbearing bureaucrats; act as bureaucrats and overlords
【做广告】advertising: ～是值得的。 It pays to advertise.
【做鬼】play tricks; play an underhand game; get up to mischief
【做鬼脸】make faces; pull a face
【做假帐】manipulation of account
【做绝】leave no room for maneuver: 把事情～ get things into an impasse; leave oneself no avenue of retreat; pass the point of no return / 坏事～ perpetrate every kind of villainy
【做客】be a guest
【做礼拜】[基督教] go to church; be at church
【做媒】be a matchmaker
【做梦】①(睡觉做梦) have a dream; dream: 做恶梦 have a nightmare; dream a bad dream ②(比喻幻想) have a pipe dream; daydream: 做美梦 cherish fond hopes / 白日～ daydream
【做人】①(待人接物) conduct oneself; behave: 会～ get along with people quite well ②(当个正派人) be an upright person: 重新～ turn over a new leaf
【做人情】do sb. a good turn; give sb. a favor
【做声】make a sound: 别～! Keep quiet!
【做实事】perform real deeds
【做事】①(从事某种工作或处理某件事情) handle affairs; do a deed; act: ～得手 work smoothly / ～敏捷 quick and efficient / ～周到 do sth. up brown ②(工作) work; have a job
【做手脚】mess about with
【做寿】celebrate the birthday; hold a birthday party
【做文章】①(抓住一件事发议论) make an issue of: 在枝节问题上大～ kick up a rumpus over some side issue ②(写文章) write an essay
【做戏】①(演戏) act in a play ②(装假) put on a show; playact
【做一天和尚撞一天钟】go on tolling the bell as long as one is a monk; do the least that is expected of one; take a passive attitude towards one's work
【做贼心虚】have a guilty conscience
【做作】affected; artificial: 她演得太～。 She overdoes her acting

汉语拼音和威妥玛式拼音对照表

PINYIN	WADE		*PINYIN*	WADE		*PINYIN*	WADE		*PINYIN*	WADE	
a	a	1	chu	ch'u	119	er	erh	222	hu	hu	364
ai	ai	1	chua	ch'ua	—	fa	fa	225	hua	hua	369
an	an	3	chuai	ch'uai	127	fan	fan	230	huai	huai	375
ang	ang	7	chuan	ch'uan	127	fang	fang	236	huan	huan	376
ao	ao	7	chuang	ch'uang	130	fei	fei	241	huang	huang	380
ba	pa	9	chui	ch'ui	132	fen	fen	246	hui	hui	383
bai	pai	12	chun	ch'un	133	feng	feng	250	hun	hun	390
ban	pan	16	chuo	ch'o	135	fo	fo	256	huo	huo	392
bang	pang	21	ei	tz'u	135	fou	fou	256	ji	chi	399
bao	pao	22	cong	ts'ung	139	fu	fu	256	jia	chia	418
bei	pei	29	cou	ts'ou	140	ga	ka	269	jian	chien	425
ben	pen	33	cu	ts'u	141	gai	kai	269	jiang	chiang	436
beng	peng	35	cuan	ts'uan	142	gan	kan	271	jiao	chiao	440
bi	pi	36	cui	ts'ui	143	gang	kang	276	jie	chieh	448
bian	pien	43	cun	ts'un	144	gao	kao	279	jin	chin	459
biao	piao	48	cuo	ts'o	145	ge	ke(ko)	283	jing	ching	467
bie	pieh	51	da	ta	147	gei	kei	288	jiong	chiung	476
bin	pin	51	dai	tai	160	gen	ken	288	jiu	chiu	477
bing	ping	52	dan	tan	164	geng	keng	289	ju	chü	481
bo	po	56	dang	tang	169	gong	kung	291	juan	chüan	486
bu	pu	60	dao	tao	172	gou	kou	300	jue	chüeh	488
ca	ts'a	76	de	te	177	gu	ku	303	jun	chün	491
cai	ts'ai	76	dei	—	178	gua	kua	310	ka	k'a	495
can	ts'an	79	deng	teng	178	guai	kuai	312	kai	k'ai	495
cang	ts'ang	82	di	ti	181	guan	kuan	313	kan	k'an	499
cao	ts'ao	83	dian	tien	188	guang	kuang	318	kang	k'ang	501
ce	ts'e	84	diao	tiao	195	gui	kui	321	kao	k'ao	503
cen	ts'en	85	die	tieh	197	gun	kun	325	ke	k'e(k'o)	504
ceng	ts'eng	86	ding	ting	198	guo	kuo	326	ken	k'en	509
cha	ch'a	86	diu	tiu	201	ha	ha	334	keng	k'eng	509
chai	ch'ai	89	dong	tung	201	hai	hai	334	kong	k'ung	510
chan	ch'an	90	dou	tou	205	han	han	338	kou	k'ou	512
chang	ch'ang	92	du	tu	207	hang	hang	342	ku	k'u	515
chao	ch'ao	97	duan	tuan	210	hao	hao	344	kua	k'ua	517
che	ch'e	100	dui	tui	213	he	he(ho)	348	kuai	k'uai	518
chen	ch'en	102	dun	tun	215	hei	hei	355	kuan	k'uan	519
cheng	ce'eng	104	duo	to	216	hen	hen	356	kuang	k'uang	520
chi	ch'ih	110	e	e(o)	220	heng	heng	356	kui	ku'ei	522
chong	ch'ung	114	ei	ei	—	hong	hung	358	kun	k'un	524
chou	ch'ou	117	en	en	222	hou	hou	361	kuo	k'uo	524

la	la	526	nei	nei	636	ran	jan	731	tai	t’ai	858
lai	lai	528	nen	nen	637	rang	jang	731	tan	t’an	860
lan	lan	529	neng	neng	638	rao	jao	732	tang	t’ang	863
lang	lang	531	ni	ni	638	re	je	733	tao	t’ao	865
lao	lao	532	nian	nien	640	ren	jen	735	te	t’e	868
le	le	537	niang	niang	642	reng	jeng	742	teng	t’eng	870
lei	lei	537	niao	niao	642	ri	jih	743	ti	t’i	870
leng	leng	539	nie	nieh	643	rong	jung	744	tian	t’ien	874
li	li	541	nin	nin	644	rou	jou	746	tiao	t’iao	878
lia	lia	549	ning	ning	644	ru	ju	747	tie	t’ieh	881
lian	lien	549	niu	niu	645	ruan	juan	751	ting	t’ing	883
liang	liang	554	nong	nung	646	rui	jui	752	tong	t’ung	885
liao	laio	558	nu	nu	647	run	jun	752	tou	to’u	891
lie	lieh	560	nü	nü	648	ruo	jo	752	tu	t’u	895
lin	lin	561	nuan	nuan	648	sa	sa	754	tuan	t’uan	899
ling	ling	564	nüe	nüeh	649	sai	sai	754	tui	t’ui	899
liu	liu	568	nuo	no	649	san	san	755	tun	t’un	902
long	lung	573	ou	ou	650	sang	snag	758	tuo	t’o	903
lou	lou	574	pa	p’a	651	sao	sao	759	wa	wa	906
lu	lu	575	pai	p’ai	651	se	se	760	wai	wai	906
lü	lü	579	pan	p’an	654	sen	sen	760	wan	wan	910
luan	luan	581	pang	p’ang	656	seng	seng	761	wang	wang	915
lüe	lüeh	582	pao	p’ao	657	sha	sha	761	wei	wei	918
lun	lun	583	pei	p’ei	658	shai	shai	763	wen	wen	928
luo	lo	585	pen	p’en	660	shan	shan	764	weng	weng	933
ma	ma	588	peng	p’ang	661	shang	shang	768	wo	wo	933
mai	mai	590	pi	p’i	662	shao	shao	774	wu	wu	934
man	man	592	pian	pi’en	666	she	she	776	xi	hsi	946
mang	mang	595	piao	p’ao	668	shen	shen	780	xia	hsia	953
mao	mao	596	pie	p’ieh	669	sheng	sheng	786	xian	hsien	958
mei	mei	599	pin	p’in	670	shi	shih	794	xiang	hsiang	965
men	men	603	ping	p’ing	671	shou	shou	813	xiao	hsiao	972
meng	meng	604	po	p’o	675	shu	shu	821	xie	hsieh	979
mi	mi	606	pou	po’u	678	shua	shua	828	xin	hsin	982
mian	mien	609	pu	p’u	678	shuai	shuai	829	xing	hsing	988
miao	miao	612	qi	ch’i	681	shuan	shuan	830	xiong	hsiung	995
mie	mieh	614	qia	ch’ia	692	shuang	shuang	830	xiu	hsiu	997
min	min	614	qian	ch’ien	693	shui	shui	832	xu	hsü	999
ming	ming	614	qiang	ch’iang	700	shun	shun	837	xuan	hsüan	1003
miu	miu	621	qiao	ch’iao	703	shuo	shuo	838	xue	hsüeh	1006
mo	mo	621	qie	ch’ieh	705	si	szu	839	xun	hsün	1010
mou	mou	625	qin	ch’in	707	song	sung	845	ya	ya	1013
mu	mu	626	qing	ch’ing	709	sou	sou	847	yan	yen	1016
na	na	630	qiong	ch’iung	717	su	su	847	yang	yang	1024
nai	nai	631	qiu	ch’iu	718	suan	suan	850	yao	yao	1028
nan	nan	632	qu	ch’ü	720	sui	sui	851	ye	yeh	1032
nang	nang	634	quan	ch’üan	724	sun	sun	853	yi	i	1036
nao	nao	635	que	ch’üeh	728	suo	so	854	yin	yin	1056
ne	ne	636	qun	ch’ün	729	ta	t’a	857	ying	ying	1062

yo	yo	—	zao	tsao	1105	zhen	chen	1128	zhuo	cho	1173
yong	yung	1067	ze	tse	1108	zheng	cheng	1132	zi	tzu	1175
you	yu	1070	zei	tsei	1109	zhi	chih	1139	zong	tsung	1182
yu	yü	1080	zen	tsen	1109	zhong	chung	1152	zou	tsou	1185
yuan	yüan	1089	zeng	tseng	1109	zhou	chou	1159	zu	tsu	1186
yue	yüeh	1094	zha	cha	1111	zhu	chu	1160	zuan	tsuan	1189
yun	yün	1097	zhai	chai	1112	zhua	chua	1166	zui	tsui	1189
za	tsa	1100	zhan	chan	1113	zhuan	chuan	1167	zun	tsun	1191
zai	tsai	1101	zhang	chang	1117	zhuang	chuang	1170	zuo	tso	1191
zan	tsan	1103	zhao	chao	1121	zhui	chui	1172			
zang	tsang	1105	zhe	che	1125	zhun	chun	1173			

汉 语 词 汇 图 解

①马 horse
②水牛 water buffalo
③黄牛 cattle, ox (公牛), cow (母牛)
④猪 pig
⑤鹅 goose
⑥驴 donkey, ass
⑦鸡 chicken
⑧狗 dog
⑨猫 cat
⑩鸭 duck
⑪兔 rabbit
⑫山羊 goat
⑬绵羊 sheep

①虎 tiger
②猞猁 lynx
③豹 leopard
④狮 lion
⑤棕熊 brown bear
⑥狐 fox
⑦豺 jackal
⑧狼 wolf
⑨北极熊 polar bear

①鹿 deer
②象 elephant
③松鼠 squirrel
④骆驼 camel
⑤小熊猫 lesser panda
⑥黄鼠狼 weasel
⑦田鼠 field mouse
⑧麝 musk deer
⑨大熊猫 (giant) panda
⑩貂 marten
⑪獾 badger
⑫土拨鼠 marmot

①金丝猴 golden monkey
②猿 ape
③大猩猩 gorilla
④鲸 whale
⑤海豚 dolphin
⑥水獭 otter
⑦海豹 seal
⑧鳄鱼 crocodile
⑨青蛙 frog
⑩蜘蛛 spider
⑪蜻蜓 dragon-fly
⑫蚱蜢 grasshopper
⑬蝉 cicada
⑭螳螂 mantis
⑮蜜蜂 bee
⑯蚂蚁 ant
⑰蟋蟀 cricket
⑱蛇 snake

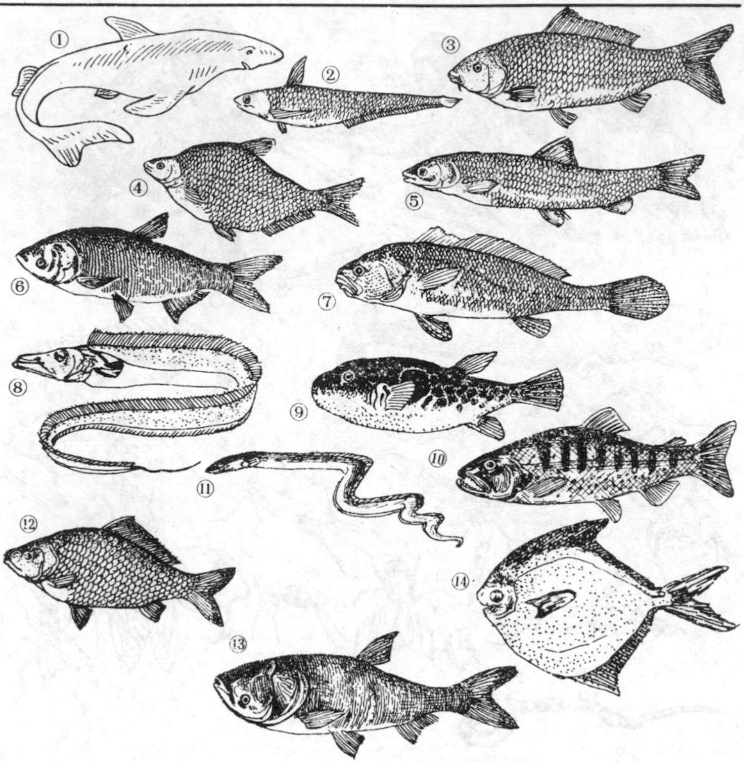

①鲨鱼 shark
②凤尾鱼 long tailed anchovy
③鲤鱼 common carp
④武昌鱼（团头鲂） blunt-snout
⑤青鱼 black carp
⑥白鲢鱼 silver carp
⑦黄花鱼 yellow croaker
⑧带鱼 hairtail; cutlass fish
⑨河豚鱼 globe fish
⑩大马哈鱼 salmon
⑪鳗鱼 sea eel
⑫鲫鱼 crucian carp
⑬胖头鱼（花鲢） big head; spotted silver carp
⑭鲳鱼 pomfret

①麻雀 sparrow
②画眉 thrush
③山雀 tit; titmouse
④百灵 (sky) lark
⑤杜鹃 cuckoo
⑥翠鸟 kingfisher; halcyon
⑦八哥 starling; mynah
⑧燕子 swallow
⑨啄木鸟 woodpecker
⑩鹦鹉 parrot
⑪丹顶鹤 (red-crested) crane
⑫猫头鹰 owl
⑬苍鹰 goshawk
⑭鹌鹑 quail
⑮大雁 wild goose
⑯雉 pheasant
⑰孔雀 peacock
⑱鸳鸯 mandarin duck

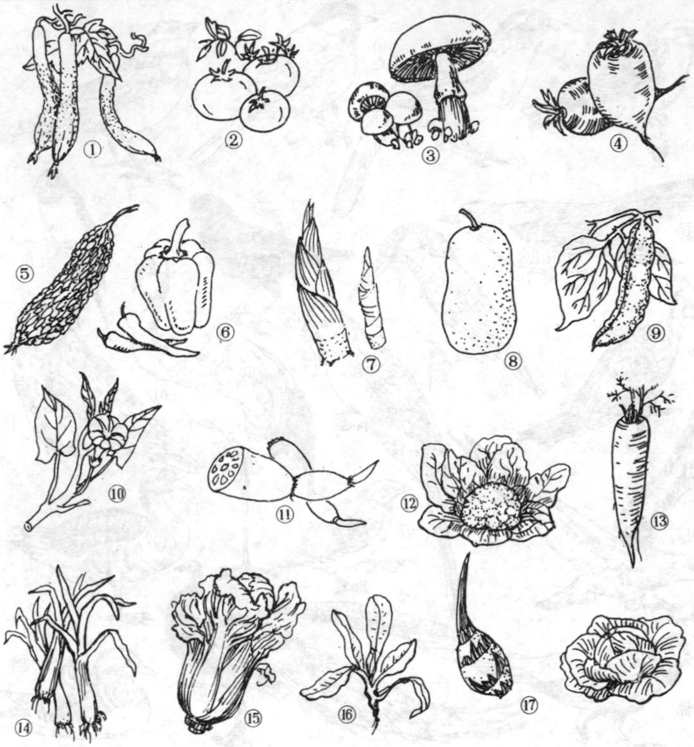

① 黄瓜 cucumber
② 番茄 tomato
③ 蘑菇 mushroom
④ 心里美 sweet pink-fleshed radish
⑤ 苦瓜 balsam pear
⑥ 辣椒 red pepper; chili; chilli; chile
⑦ 冬笋 winter bamboo shoots
⑧ 冬瓜 Chinese wax gourd; white gourd
⑨ 刀豆 sword bean
⑩ 空心菜 water spinach; swamp cabbage
⑪ 莲藕 lotus root
⑫ 菜花、cauliflower
⑬ 胡萝卜 carrot
⑭ 大葱 green Chinese onion
⑮ 白菜 Chinese cabbage
⑯ 菠菜 spinach
⑰ 慈姑 arrowhead

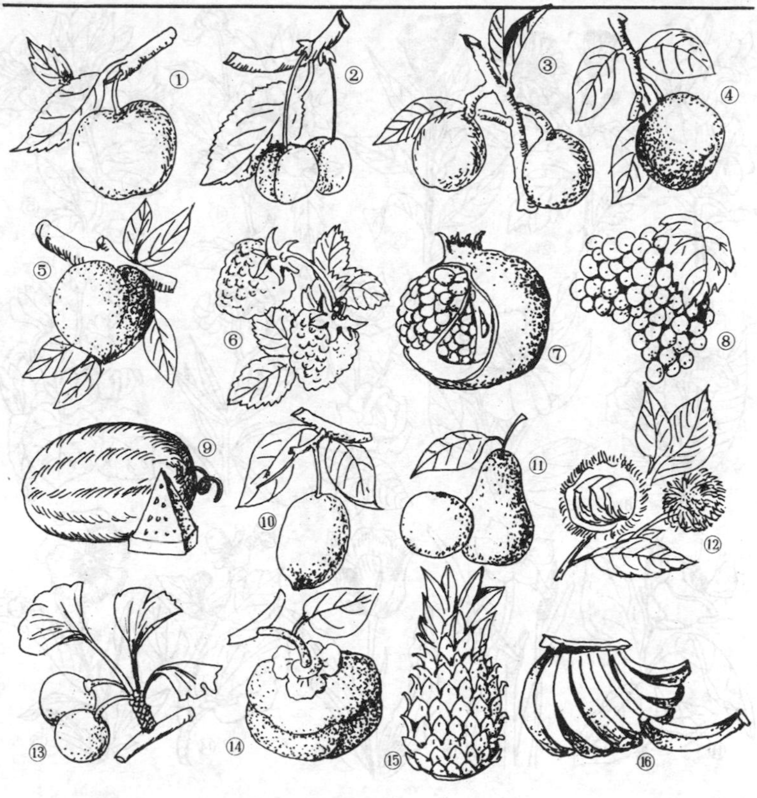

①苹果 apple
②樱桃 cherry
③枇杷 loquat
④蜜桔 mandarin orange
⑤桃 peach
⑥ 草 莓　strawberry
⑦ 石 榴　pomegranate
⑧ 葡 萄　grape
⑨ 西 瓜 watermelon
⑩柠檬 lemon
⑪梨 pear
⑫栗子 chestnut
⑬白果 gingko
⑭柿子 persimmon
⑮菠萝 pineapple
⑯香蕉 banana

①百合花 lily
②菊花 chrysanthemum; mum
③鸡冠花 cockscomb
④牡丹 peony
⑤杜鹃花 azalea
⑥美人蕉 canna; Indian shot
⑦芍药 herbaceous peony
⑧玫瑰 rose
⑨夹竹桃 oleander
⑩仙客来 cyclamen
⑪山茶花 camellia
⑫水仙 narcissus; daffodil
⑬郁金香 tulip
⑭秋海棠 begonia
⑮三色堇 pansy

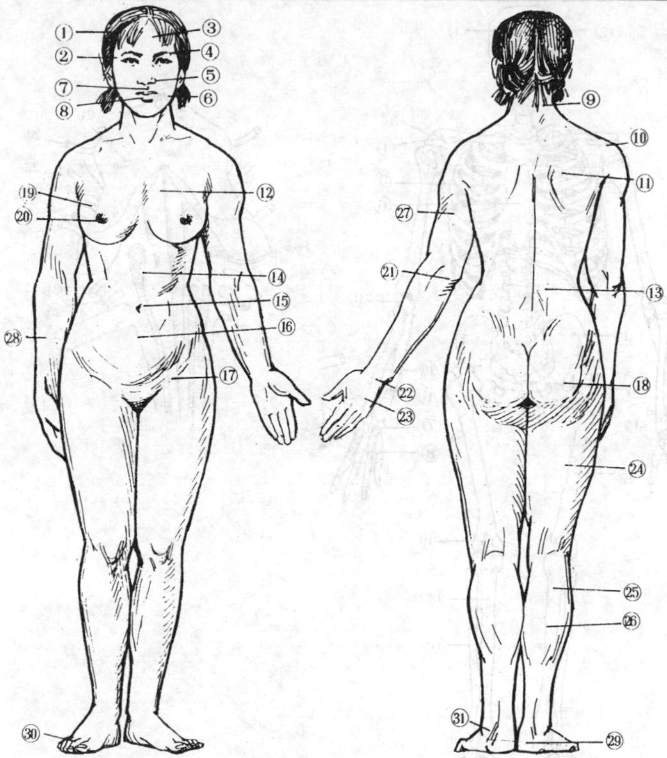

①头 head
②太阳穴 temple
③额 forehead
④眉毛 eyebrow
⑤鼻 nose
⑥颊 cheek
⑦人中 philtrum
⑧口 mouth
⑨颈 neck
⑩肩 shoulder
⑪背 back
⑫胸 chest; thorax
⑬腰 waist; loins
⑭上腹 the upper abdomen
⑮肚脐 navel; belly button
⑯下腹 the lower abdomen
⑰腹股沟 groin
⑱臀部 buttocks; hips; bottom; backside
⑲乳房 breast
⑳乳头 nipple
㉑肘 elbow
㉒腕 wrist
㉓手背 back of the hand
㉔大腿 thigh
㉕小腿 lower leg; shank
㉖腿肚 calf
㉗上臂 upper arm
㉘前臂 fore arm
㉙后跟 hell
㉚趾 toe
㉛踝 ankle

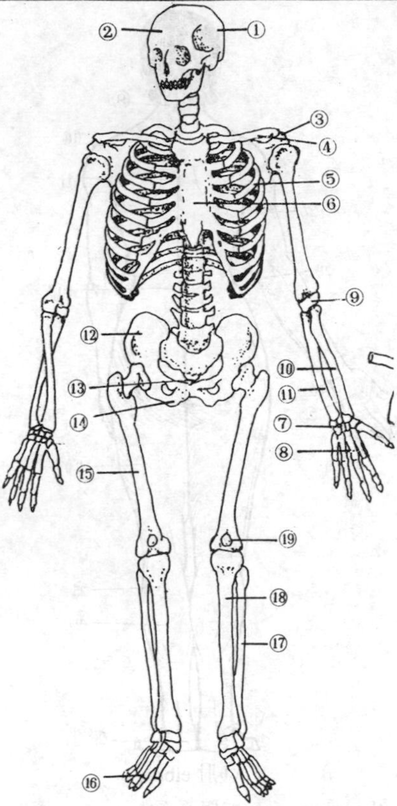

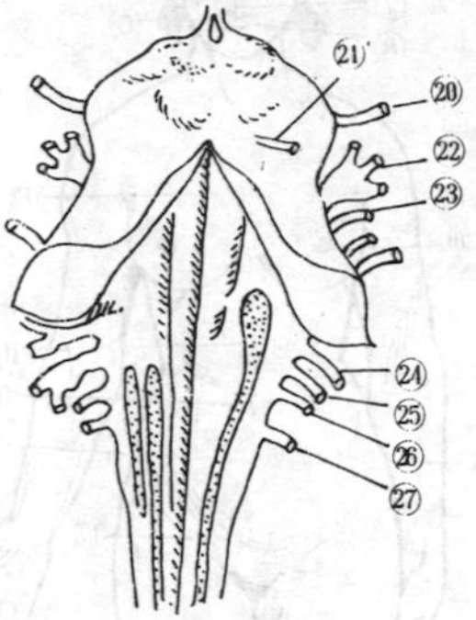

①头颅 skull
②额骨 frontal bone
③肩胛骨 shoulder blade
④锁骨 clavide
⑤肋骨 rib
⑥胸骨 breast bone; sternum
⑦腕骨 caspal
⑧掌骨 metacarpal
⑨肘关节 elbow joint
⑩桡骨 radius
⑪尺骨 ulna
⑫髋骨 hip-bone
⑬耻骨 pubis; pubic bone
⑭坐骨 ischium
⑮股骨 femur; thigh bone
⑯趾骨 phalange
⑰腓骨 fibula
⑱胫骨 tibia; shin bone
⑲膝盖骨 kneecap
⑳动眼神经 oculomotor nerve
㉑滑车神经 trochlear nerve
㉒三叉神经 trigeminal nerve
㉓外展神经 abducent nerve
㉔舌咽神经 glosso pharyngeal nerve
㉕迷走神经 vagus
㉖副神经 accesso ry nerve
㉗舌下神经 hypoglossal nerve

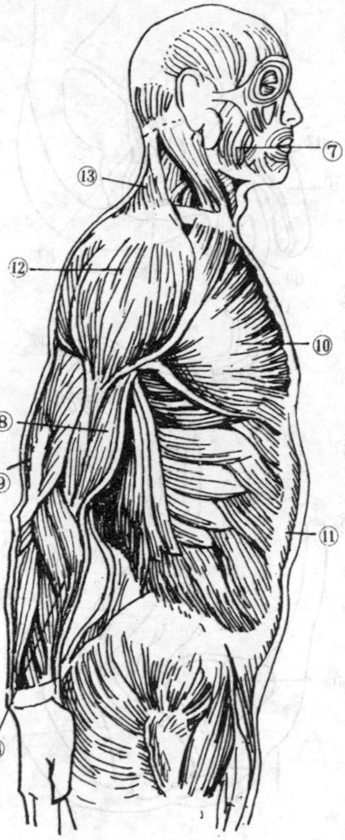

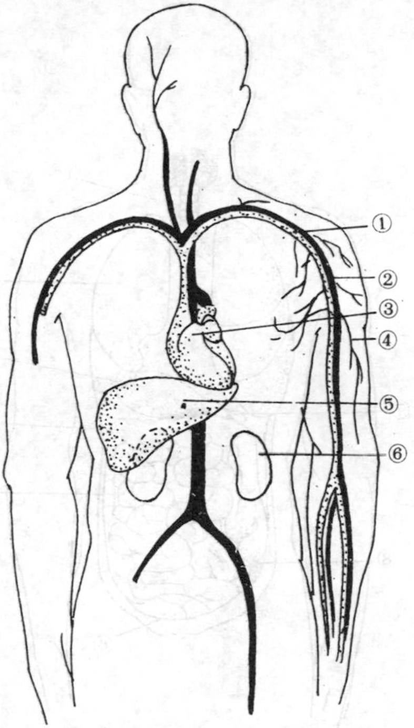

①动脉 artery
②静脉 vein
③心脏 heart
④毛细血管 blood capillary
⑤肝 liver
⑥肾 kidney
⑦面肌 facial muscle
⑧二头肌 biceps
⑨三头肌 triceps
⑩胸肌 pectoral muscle
⑪腹肌 abdominal muscle
⑫三角肌 deltoid muscle
⑬斜方肌 trapezius
⑭桡肌 brachioradialis

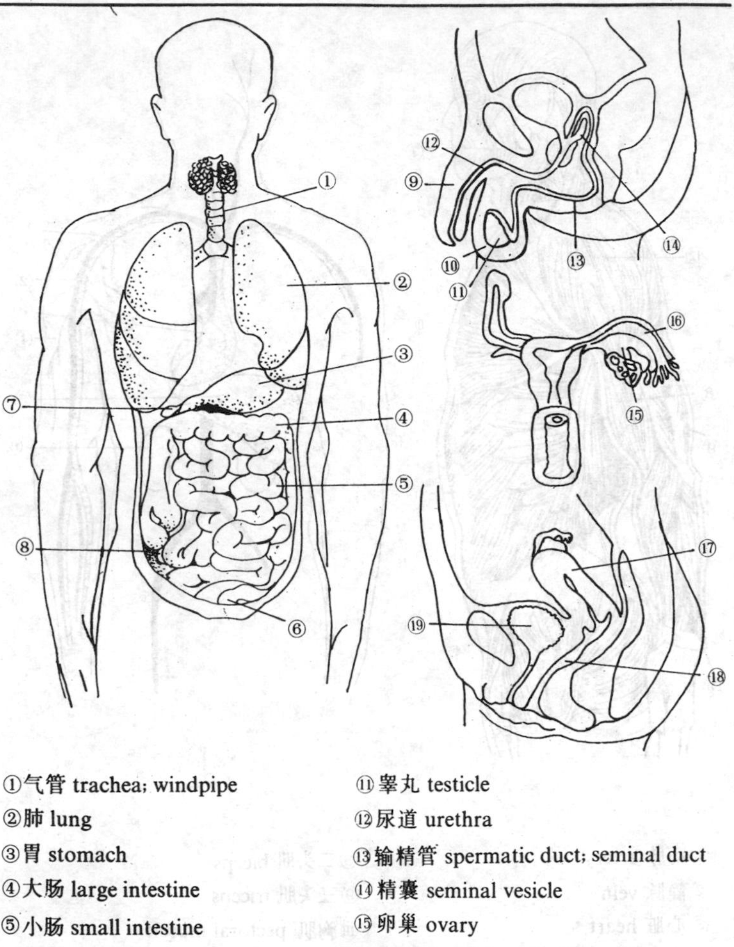

①气管 trachea; windpipe
②肺 lung
③胃 stomach
④大肠 large intestine
⑤小肠 small intestine
⑥直肠 rectum
⑦胆囊 gall bladder
⑧阑尾 appendix
⑨阴茎 penis
⑩阴囊 scrotum
⑪睾丸 testicle
⑫尿道 urethra
⑬输精管 spermatic duct; seminal duct
⑭精囊 seminal vesicle
⑮卵巢 ovary
⑯输卵管 oviduct; fallopian tube
⑰子宫 womb; uterus
⑱阴道 vagina
⑲膀胱 urinary bladder

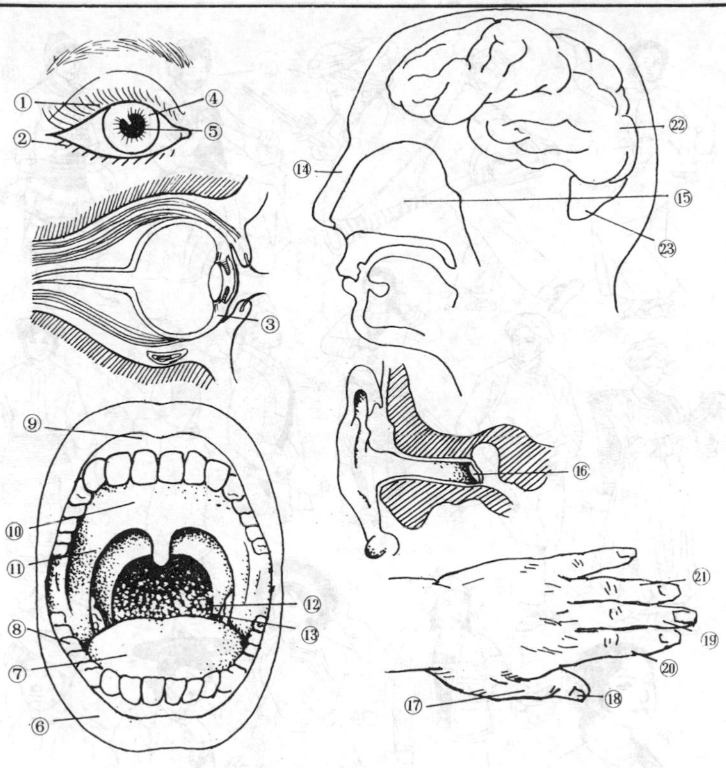

①眼睑 eyelid
②睫毛 eyelash
③眼球 eyeball
④虹膜 iris
⑤瞳孔 pupil
⑥唇 lip
⑦舌 tongue
⑧牙 tooth
⑨齿龈 gum
⑩硬腭 hard palate
⑪软腭 soft palate
⑫扁桃体 tonsil
⑬咽喉 larynx
⑭鼻梁 bridge of the nose
⑮鼻腔 na sal cavity
⑯耳膜 ear drum
⑰拇指 thumb
⑱指甲 finger nail
⑲中指 middle finger
⑳食指 forefinger；index finger
㉑无名指 ring finger
㉒大脑 cerebrum
㉓小脑 cerebellum

①傣族 Daizu
②傈僳族 Lisu
③佤族 Vazu
④拉祜族 Lahu
⑤景颇族 Jingpo
⑥布朗族 Blang
⑦崩龙族 Benglong
⑧仫佬族 Mulam
⑨基诺族 Jino
⑩毛难族 Maonan
⑪畲族 Shezu
⑫土家族 Tujia
(名称排列不分先后)

① 满族 Manzu
② 朝鲜族 Chaoxian
③ 鄂伦春族 Oroqen
④ 东乡族 Dongxiang
⑤ 保安族 Bonan
⑥ 蒙古族 Mongol
⑦ 达斡尔族 Daur
⑧ 赫哲族 Hezhen
⑨ 柯尔克孜族 Kirgiz
⑩ 鄂温克 Ewenki
(名称排列不分先后)

①回族 Huizu
②撒拉族 Salar
③土族 Tuzu
④裕固族 Yugur
⑤塔塔尔 Tatar
⑥锡伯族 Xibe
⑦塔吉克族 Tajik
⑧哈萨克族 Kazak
⑨维吾尔族 Uygur
⑩哈尼族 Hani
⑪京族 Ginzu
(名称排列不分先后)

①独龙族 Derung
②苗族 Miaozu
③高山族 Gaoshan
④仡佬族 Gelao
⑤布依 Bouyei
⑥水族 Suizu
⑦黎族 Lizu
⑧瑶族 Yaozu
⑨壮族 Zhuangzu
⑩侗族 Dongzu

(名称排列不分先后)

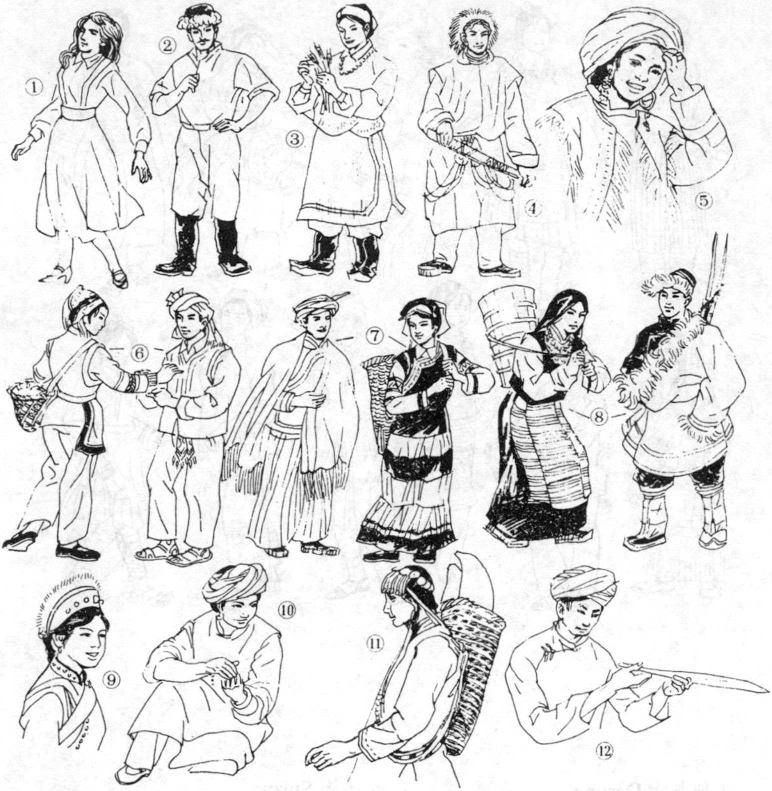

①俄罗斯族 Russ
②乌孜别克族 Uzbek
③门巴族 Monba
④珞巴族 Lhoba
⑤羌族 Qiangzu
⑥白族 Baizu
⑦彝族 Yizu
⑧藏族 Zangzu
⑨纳西族 Naxi
⑩普米族 Primi
⑪怒族 Nuzu
⑫阿昌族 Achang
(名称排列不分先后)

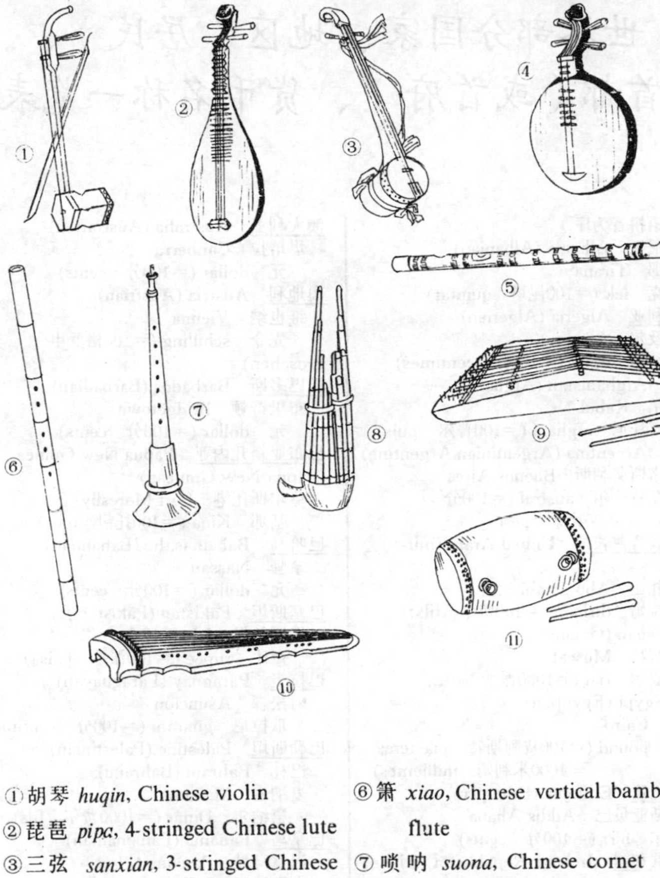

①胡琴 *huqin*, Chinese violin

②琵琶 *pipa*, 4-stringed Chinese lute

③三弦 *sanxian*, 3-stringed Chinese guitar

④月琴 *yueqin*, 4-stringed full-moon-shaped Chinese mandolin

⑤笛子 *dizi*, Chinese flute; 8-holed bamboo flute

⑥箫 *xiao*, Chinese vertical bamboo flute

⑦唢呐 *suona*, Chinese cornet

⑧笙 *sheng*, Chinese wind-pipe

⑨扬琴 *yangqin*, dulcimer

⑩古琴 *guqin*, Chinese zither

⑪腰鼓 waist drum

世界部分国家、地区（居民）、首都（或首府）、货币名称一览表

（按汉语拼音为序）

阿尔巴尼亚　Albania (Albanian)
　地拉那　Tirana
　　列克　lek (＝100昆塔　quintar)
阿尔及利亚　Algeria (Algerian)
　阿尔及尔　Algers
　　第纳尔　dinar (＝100分　centimes)
阿富汗　Afghanistan (Afghan)
　喀布尔　Kabul
　　阿富汗尼　afghani (＝100普尔　puls)
阿根廷　Argentina (Argentinian,Argentine)
　布宜诺斯艾利斯　Buenos Aires
　　奥斯特拉尔　austral (＝100分 centavos; ＝1000比索　pesos)
阿拉伯联合酋长国　United Arab Emirates,the
　阿布扎比　Abu Dhabi
　　迪拉姆　dirham (＝100费尔　fils)
阿曼　Oman (Omani)
　马斯喀特　Muscat
　　里亚尔　rial (＝1000派沙 baiza)
埃及　Egypt (Egyptian)
　开罗　Cairo
　　镑　pound (＝100皮阿斯特　piasters; ＝1000米利姆　milliemes)
埃塞俄比亚　Ethiopia (Ethiopian)
　亚的斯亚贝巴　Addis Ababa
　　比尔　birr (＝100分　cents)
爱尔兰共和国　Ireland, Republic of (Irishman,Irishwoman)
　都柏林　Dublin
　　镑　pound (＝100新便士　new pence)
安道尔　Andorra (Andorran)
　安道尔　Andorra la Vella
　　西班牙比塞塔　peseta,法国法郎　franc
安哥拉　Angola (Angolan)
　卢安达　Luanda
　　宽札　kwanza (＝100勒韦 lweis)
安提瓜和巴布达　Antigua and Barbuda
　圣约翰　St.John's
　　东加勒比元　EC dollar (＝100分 cents)
澳大利亚　Australia (Austrarian)
　堪培拉　Canberra
　　元　dollar (＝100分　cents)
奥地利　Austria (Austrian)
　维也纳　Vienna
　　先令　schilling (＝100格罗申 groschen)
巴巴多斯　Barbados (Barbadian)
　布里奇顿　Bridgetown
　　元　dollar (＝100分　cents)
巴布亚新几内亚　Papua New Guinea (Papua New Guinean)
　莫尔斯比港　Port Moresby
　　基那　Kina (＝100托伊　toea)
巴哈马　Bahamas,the (Bahamian)
　拿骚　Nassau
　　元　dollar (＝100分　cents)
巴基斯坦　Pakistan (Pakistani)
　伊斯兰堡　Islamabad
　　卢比　rupee (＝100派沙　paisa)
巴拉圭　Paraguay (Paraguayan)
　阿松森　Asuncion
　　瓜拉尼　guarani (＝100分　centimos)
巴勒斯坦　Palestine (Palestinian)
　巴林　Bahrain (Bahraini)
　麦纳麦　Manama
　　第纳尔　Dinar (＝1000费尔　fils)
巴拿马　Panama (Panamanian)
　巴拿马城　Panama City
　　巴波亚　balboa (＝100分　centesimos)
巴西　Brazil (Brazilian)
　巴西利亚　Brasilia
　　克鲁塞罗　cruzeiro (＝100分　centavos)
百慕达　Bermudas
　哈密尔顿　Hamilton
　　元　dollar (＝100分　cents)
保加利亚　Bulgaria (Bulgarian)
　索非亚　Sofia
　　列弗　lev,复数:列瓦　leva (＝斯托丁基 stotinki)
贝宁　Benin (Beninian)
　波多诺伏　Port Novo

非洲金融共同体法郎 CFAF (＝100分 centimes)
比利时 Belgium (Belgian)
布鲁塞尔 Brussels
法郎 franc (＝100分 centimes)
秘鲁 peru (Peruvian)
利马 Lima
索尔 sol (＝100分 centavos)
波多黎各 Puerto Rico
圣胡安 San Juan
美元 dollar
冰岛 Iceland (Icelander)
雷克雅未克 Reykjavik
克朗 krona,复数kronar (＝100奥拉 aurar)
波兰 Poland (Pole)
华沙 Warsaw
兹罗提 zloty (＝100格罗希 groszy)
玻利维亚 Bolivia (Bolivian)
拉巴斯 La Paz
比索 peso (＝100分 centavos)
伯利兹 Belize
贝尔莫潘 Belmopan
元 dollar (＝100分 cents)
博茨瓦纳 Botswana (单数: Botswana; 复数: Batswana)
哈博罗内 Gaborone
普拉 pula (＝100分 thebe)
不丹 Bhutan (Bhutanese)
廷布 Thimphu
努扎姆 Ngultrum (＝100切扎姆 chetrum)
布基纳法索 Burkina Faso
瓦加杜古 Ouagadougou
非洲金融共同体法郎 CFAF (＝100分 centimes)
布隆迪 Brundi (Brundian)
布琼布拉 Bujumbura
法郎 franc (＝100分 centimes)
朝鲜 Korea (Korean)
平壤 Pyongyang
元 won (＝100钱 chon)
南朝鲜 Korea,South (South Korean)
汉城 Seoul
元 won (＝100角 jeon)
赤道几内亚 Equatorial Guinea (Equatorial Guinean)
马拉博 Malabo
非洲金融共同体法郎 CFAF (＝100分 centimes)
丹麦 Denmark (Dane)
哥本哈根 Copenhagen
克朗 Krone (＝100欧尔 Ore)
德意志联邦共和国 Germany,Federal Republic of (West German)
波恩 Bonn
联邦德国马克 Deutsche mark (＝100芬尼 pfennig)
德意志民主共和国 German Democratic Republic (German)
柏林 Berlin
马克 mark (＝100芬尼 pfennig)
多哥 Togo (Togolese)
洛美 Lome
非洲金融共同体法郎 CFAF (＝100分 centimes)
多米尼加共和国 Dominican Republic
圣多明各 Santo Domingo
比索 peso (＝100分 centavos)
多米尼加联邦 Dominica,The Commonwealth of Dominican
罗索 Rosseau
东加勒比元 BC dollar (＝100分 cents)
厄瓜多尔 Ecuador (Ecuadorean)
基多 Quito
苏克雷 sucre (＝100分 centavos)
法国 France (Frenchman,Frenchwoman)
巴黎 Paris
法郎 franc (＝100分 centimes)
菲律宾 Philippines (Filipino,Filipina)
马尼拉 Manila
比索 peso (＝100分 centavos)
斐济 Fiji (Fijian)
苏瓦 Suva
元 dollar (＝100分 cents)
芬兰 Finland (Finn)
赫尔辛基 Helsinki
马克 markka (＝100盆尼 pennia)
佛得角 Cape Verde (Cape Verdean)
普拉亚 Praia
埃斯库多 escudo (＝100分 centavos)
冈比亚 Gambia,the (Gambian)
班珠尔 Banjul
达拉西 dalasi (＝100布图 bututs)
刚果 Congo (Congolese)
布拉柴维尔 Brazzaville
非洲金融共同体法郎 CFAF (＝100分 centimes)
哥伦比亚 Colombia (Colombian)
波哥大 Bogota
比索 peso (＝100分 centavos)

<u>哥斯达黎加</u> Costa Rica (Costa Rican)
圣何塞 San Jose
科朗 Colon (=100分 centimos)
<u>格林纳达</u> Grenada (Grenadian)
圣乔治 St.George's
东加勒比元 EC dollar (=100分 cents)
<u>古巴</u> Cuba (Cuban)
哈瓦那 Havana
比索 peso (=100分 centavos)
<u>圭亚那</u> Guyana (Guyanese)
乔治敦 Georgetown
元 dollar (=100分 cents)
<u>海地</u> Haiti (Haitian)
太子港 Portau Prince
古德 gourde (=100分 cents)
<u>荷兰</u> Netherlands (Dutchman,Dutchwoman)
阿姆斯特丹 Amsterdam
盾 gulden或 florin (=100分 cents)
<u>洪都拉斯</u> Honduras (Honduran)
特古西加尔巴 Tegucigalpa
伦皮拉 lempira (=100分 centavos)
<u>基里巴斯</u> Kiribati
塔拉瓦 Tarawa
澳大利亚元 Australian dollar; 英镑 pound sterling
<u>吉布提</u> Djibouti (Djiboutian)
吉布提 Djibouti
法郎 franc (=100分 centimes)
<u>几内亚</u> Guinea (Guinean)
科纳克里 Conakry
法郎 franc (=100分 centimes)
<u>几内亚比绍</u> Guinea—Bissau
比绍 Bissau
比索 peso (=100分 centavos)
<u>加拿大</u> Canada (Canadian)
渥太华 Ottawa
元 dollar (=100分 cents)
<u>加纳</u> Ghana (Ghananian)
阿克拉 Accra
塞迪 cedi (=100比塞瓦 pesewa)
<u>加蓬</u> Gabon (Gabonese)
利伯维尔 Libreville
非洲金融共同体法郎 CFAF (=100分 centimes)
<u>柬埔寨</u> Cambodia (Cambodian)
金边 Phnom Penh
瑞尔 riel
<u>捷克斯洛伐克</u> Czechoslovakia (Czechoslovak)
布拉格 Prague
克朗 koruna (=100赫勒 heller)
<u>津巴布韦</u> Zimbabwe (Zimbabwean)
哈拉雷 Harare
元 dollar (=100分 cents)
<u>喀麦隆</u> Cameroon (Cameroonian)
雅温得 Yaounde
非洲金融共同体法郎 CFAF (=100分 centimes)
<u>卡塔尔</u> Qatar (Qatari)
多哈 Doha
里亚尔 riyal (=100迪拉姆 dirhams)
<u>科摩罗</u> Comoros,the
莫罗尼 Moroni
法郎 franc (=100分 centimes)
<u>科特迪瓦</u> Ivory Coast,the
阿比让 Abidjan
非洲金融共同体法郎 CFAF (=100分 centimes)
<u>科威特</u> Kuwait (Kuwaiti)
科威特 Kuwait
第纳尔 dinar (=1000费尔 fils)
<u>肯尼亚</u> Kenya (Kenyan)
内罗毕 Nairobi
先令 shilling (=100分 cents)
<u>莱索托</u> Lesotho (单数: Mosotho,复数: Basotho)
马塞卢 Maseru
洛蒂 loti; 复数: 马洛蒂 maloti (=100利森特 lisente; 单数: 森特sente)
<u>老挝</u> Laos (Laotian)
万象 Vientiane
基普 kip (=100阿特 at)
<u>黎巴嫩</u> Lebanon (Lebanese)
贝鲁特 Beirut
镑 pound (=100皮阿斯特 piastre)
<u>利比里亚</u> Liberia (Liberian)
蒙罗维亚 Monrovia
元 dollar (=100分 cents)
<u>利比亚</u> Libya (Libyan)
的黎波里 Tripoli
第纳尔 dinar (=1000迪拉姆 dirham)
<u>列支敦士登</u> Liechtenstein (Liechtensteiner)
瓦杜兹 Vaduz
瑞士法郎 Swiss franc
<u>卢森堡</u> Luxemburg (Luxemburger)
卢森堡 Luxemburg
法郎 franc (=100分 centimes)
<u>卢旺达</u> Rwanda (Rwandese)
基加利 Kigali
法郎 franc (=100分 centimes)

罗马尼亚 Romania (Romanian)
布加勒斯特 Bucharest
列伊 leu,复数 lei (=100巴尼 bani)
马达加斯加 Madagascar (Madagascan)
塔那那利佛 Tananarive
法郎 franc (=100分 centimes)
马尔代夫 Maldives (Maldivian)
马累 Male
拉菲亚 rufiyaa (=100拉雷 laris)
马耳他 Malta (Maltese)
瓦莱塔 Valletta
镑 pound (=100分 cents; =1000米尔 mils)
马拉维 Malawi (Malawian)
利隆圭 Lilongwe
克瓦查 Kwacha (=100坦巴拉 tambala)
马来西亚 Malaysia (Malaysian)
吉隆坡 Kuala Lumpur
元 dollar (=100分 cents)
马里 Mali (Malian)
巴马科 Bamako
非洲金融共同体法郎 CFAF (=100分 centimes)
毛里求斯 Mauritius (Mauritian)
路易斯港 Port Louis
卢比 rupee (=100分 cents)
毛里塔尼亚 Mauritania (Mauritanian)
努瓦克肖特 Nouakchott
乌吉亚 Ouguiya (=5库姆斯 khoums)
美国 United States of America (American)
华盛顿 Washington
元 dollar (=100分 cents)
蒙古 Mongolia (Mongolian)
乌兰巴托 Ulan Bator
图格里克 tugrik (=100蒙戈 mongo)
孟加拉国 Bangladesh (Bangladeshi)
达卡 Dhaka (Dacca)
塔卡 taka (=100派士 paise)
缅甸 Burma (Burmese)
仰光 Rangoon
元 Kyat (=100分 pyas)
摩洛哥 Morocco (Moroccan)
拉巴特 Rabat
迪拉姆 dirham (=100分 centimes)
摩纳哥 Monaco (Monacan)
摩纳哥 Monaco—Ville
法国法郎 French franc
莫桑比克 Mozambique (Mozambican)
马普托 Maputo
梅蒂卡尔 metical (=100分 centavos)
墨西哥 Mexico (Mexican)
墨西哥城 Mexico City
比索 peso (=100分 centavos)
纳米比亚 Namibia
温得和克 Windhoek
南非兰特 South African rand
南非 South Africa (South African)
比勒陀利亚 Pretoria
开普顿 Cape Town
兰特 rand (=100分 cents)
南斯拉夫 Yugoslavia (Yugoslav)
贝尔格莱德 Belgrade
第纳尔 dinar (=100帕拉 paras)
瑙鲁 Nauru (Nauruan)
瑙鲁 Nauru
澳大利亚元 Australian dollar
尼泊尔 Nepal (Nepalese)
加德满都 Kathmandu
卢比 rupee (=100派沙 paisa)
尼加拉瓜 Nicaragua (Nicaraguan)
马那瓜 Managua
科多巴 cordoba (=100分 centavos)
尼日尔 Niger,the (Nigerois)
尼亚美 Niamey
非洲金融共同体法郎 CFAF (=100分 centimes)
尼日利亚 Nigeria (Nigerian)
拉各斯 Lagos
奈拉 naira (=100考包 kobo)
挪威 Norway (Norwegian)
奥斯陆 Oslo
克朗 krone,复数 kroner (=100欧尔 ore)
葡萄牙 Portugal (Portuguese)
里斯本 Lisbon
埃斯库多 escudo (=100分 centavos)
日本 Japan (Japanese)
东京 Tokyo
元 yen
瑞典 Sweden (Swede)
斯德哥尔摩 Stockholm
克朗 Krona (=100欧尔 ore)
瑞士 Switzerland (Swiss)
伯尔尼 Berne
法郎 franc (=100分 centimes)
萨尔瓦多 El Salvador (Salvadorean)
圣萨尔瓦多 San Salvador
科朗 colon (=100分 centavos)
塞拉利昂 Sierra Leone (Sierra Leonean)
弗里敦 Freetown
利昂 leone (=100分 cents)

塞内加尔 Senegal (Senegalese)
达喀尔 Dakar
非洲金融共同体法郎 CFAF (=100分 centimes)
塞浦路斯 Cyprus (Cypriot)
尼科西亚 Nicosia
镑 pound (=100分 cents)
塞舌尔 Seychelles (Seychellois,Seychelloise)
维多利亚 Victoria
卢比 rupee (=100分 cents)
沙特阿拉伯 Saudi Arabia (Saudi Arabian)
利雅得 Riyadh
里亚尔 riyal (=20库亚什 qursh; =100哈拉拉 halalas)
圣多美和普林西比 Sao Tome and Principe
圣多美 Sao Tome
多布拉 dobra (=100分 centimos)
圣克里斯托弗和尼维斯 St.Christopher and Nevis
巴斯特尔 Basseterre
东加勒比元 EC dollar (=100分 cents)
圣卢西亚 St.Lucia
卡斯特里 Castries
东加勒比元 EC dollar (=100分 cents)
圣文森特和格林纳丁斯 St.Vincent and St.Grenadines
金斯敦 Kingstown
东加勒比元 EC dollar (=100分 cents)
圣马力诺 San Marino (San Marinese)
圣马力诺 San Marino
意大利里拉 Italian lira
斯里兰卡Sri Lanka (Sri Lankan)
科伦坡 Colombo
卢比 rupee (=100分 cents)
斯威士兰 Swaziland (Swazi)
姆巴巴纳 Mbabane
里兰吉尼 lilangeni 复数: 埃马兰吉尼 emalangeni (=100分 cents)
苏丹 Sudan (Sudanese)
喀土穆 Khartoum
镑 pound (=100皮阿斯特 piastres; =1000米利姆 milliemes)
苏里南 Surinam (Surinamese)
帕拉马里博 Paramaribo
盾 guilder (=100分 cents)
苏联 Union of Soviet Socialist Republics (Russian, Soviet)
莫斯科 Moscow
卢布 rouble (=100戈比 kopecks)
所罗门群岛 Solomon Islands
霍尼亚拉 Honiara
元 dollar (=100分 cents)
索马里 Somalia (Somali)
摩加迪沙 Mogadishu
先令 shilling (=100分 cents)
泰国 Thailand (Thai)
曼谷 Bangkok
铢 baht (=100萨当 stangs)
坦桑尼亚 Tanzania (Tanzanian)
达累斯萨拉姆 Dar es Salaam
先令 shilling (=100分 cents)
汤加 Tonga (Tongan)
努库阿洛法 Nuku'alofa
潘加 pa'anga (=100分 seniti)
特利尼达和多巴哥 Trinidad and Tobago
西班牙港 Port of Spain
元 dollar (=100分 cents)
突尼斯 Tunisia (Tunisian)
突尼斯 Tunis
第纳尔 dinar (=1000米利姆 millimes)
图瓦鲁 Tuvalu
富纳富提 Funafuti
元 dollar (=100分 cents)
土耳其 Turkey (Turk)
安卡拉 Ankara
里拉 lira (=100库鲁 kurus)
瓦努阿图 Vanuatu
维拉港 Port Vila
瓦图 vatu (=100分 cents)
危地马拉 Guatemala (Guatemalan)
危地马拉 Guatemala City
格查尔 quetzal (=100分 centavos)
委内瑞拉 Venezuela (Venezuelan)
加拉加斯 Caracas
玻利瓦 bolivar (=100分 centimos)
文莱 Brunei
斯里巴加湾市 Bandar Seri Begawan
元 dollar (=100仙 sen)
乌干达 Uganda (Ugandan)
坎帕拉 Kampala
先令 shilling (=100分 cents)
乌拉圭 Uruguay (Uruguayan)
蒙得维的亚 Montevideo
比索 peso (=100分 centesimos)
西班牙 Spain (Spainard)
马德里 Madrid
比塞塔 peseta
西萨摩亚 Western Samoa

阿皮亚 Apia
塔拉 tala

<u>希腊</u> Greece (Greek)
雅典 Athens
德拉克马 drachma (＝100雷普塔 Lepta)

<u>新加坡</u> Singapore (Singaporean)
新加坡 Singapore
元 dollar (＝100分 cents)

<u>新西兰</u> New Zealand (New Zealander)
惠灵顿 Wellington
元 dollar (＝100分 cents)

<u>匈牙利</u> Hungary (Hungarian)
布达佩斯 Budapest
福林 forint (＝100菲勒 filler)

<u>叙利亚</u> Syria (Syrian)
大马士革 Damascus
镑 pound (＝皮阿斯特 piastres)

<u>牙买加</u> Jamaica (Jamaican)
金斯敦 Kingston
元 dollar (＝100分 cents)

<u>阿拉伯也门</u> Arab Yemem (Yemeni, Yemenite)
萨那 San'a
里亚尔 rial (＝100费尔 fils)

<u>民主也门</u> Democratic Yemen (Yemeni, Yemenite)
亚丁 Aden
第纳尔 dinar (＝1000费尔 fils)

<u>伊拉克</u> Iraq (Iraqi)
巴格达 Baghdad
第纳尔 dinar (＝1000费尔 fils)

<u>伊朗</u> Iran (Iranian)
德黑兰 Tehran
里亚尔 rial (＝100第纳尔 dinar)

<u>以色列</u> Israel (Israeli)
耶鲁撒冷 Jerusalem
谢克尔 shekel (＝100阿高洛 agora)

<u>意大利</u> Italy (Italian)
罗马 Rome
里拉 lira

<u>印度</u> India (Indian)
新德里 New Delhi
卢比 rupee (＝100派士 paise; 单数时为: 派沙 paisa)

<u>印度尼西亚</u> Indonesia (Indonesian)
雅加达 Jakarta
卢比,盾 rupiah (＝100仙 sen)

<u>英国</u>(联合王国) United Kingdom (Briton)
伦敦 London
镑 pound (＝100便士 pence)

<u>约旦</u> Jordan (Jordanian)
安曼 Amman
第纳尔 dinar (＝1000费尔 fils)

<u>越南</u> Vietnam (Vietnamese)
河内 Hanoi
盾 dong (＝10角 hao; ＝100分 xu)

<u>赞比亚</u> Zambia (Zambian)
鲁萨卡 Lusaka
克瓦查 kwacha (＝100恩韦 ngwee)

<u>扎伊尔</u> Zaire (Zairean)
金沙萨 Kinshasa
扎伊尔 zaire (＝100马库塔 malyta;
＝1000森吉 sengi)

<u>乍得</u> Chad (Chadian)
恩贾梅纳 Ndjamena
非洲金融共同体法郎 CFAF (＝100分 centimes)

<u>智利</u> Chile (Chilean)
圣地亚哥 Santiago
比索 peso (＝100分 cents)

<u>中非</u> Central Africa
班吉 Bangui
非洲金融共同体法郎 CFAF (＝100分 centimes)

<u>中国</u> China (Chinese)
北京 Beijing
元 yuan (＝10角 jiao; ＝100分 fen)

中国省、直辖市、自治区地名
汉英对照表

（按汉语拼音字母顺序排列）

省、直辖市、自治区名	简　称	省会、首府名
安徽　Anhui	皖　Wan	合肥　Hefei
北京　Beijing	京　Jing	
福建　Fujian	闽　Min	福州　Fuzhou
甘肃　Gansu	甘　Gan(Gān)	兰州　Lanzhou
广东　Guangdong	粤　Yue	广州　Guangzhou
广西　Guangxi	桂　Gui	南宁　Nanning
贵州　Guizhou	黔　Qian	贵阳　Guiyang
海南　Hainan	琼　Qiong	海口　Haikou
河北　Hebei	冀　Ji(Jì)	石家庄　Shijiazhuang
河南　Henan	豫　Yu	郑州　Zhengzhou
黑龙江　Heilongjiang	黑　Hei	哈尔滨　Harbin
湖北　Hubei	鄂　E	武汉　Wuhan
湖南　Hunan	湘　Xiang	长沙　Changsha
吉林　Jilin	吉　Ji(jí)	长春　Changchun
江西　Jiangxi	赣　Gan(Gàn)	南昌　Nanchang
江苏　Jiangsu	苏　Su	南京　Nanjing
辽宁　Liaoning	辽　Liao	沈阳　Shenyang
内蒙古　Nei Mongol	蒙　Meng	呼和浩特　Hohhot
宁夏　Ningxia	宁　Ning	银川　Yinchuan
青海　Qinghai	青　Qing	西宁　Xining
陕西　Shaanxi	陕　Shan	西安　Xi'an
山东　Shandong	鲁　Lu	济南　Jinan
山西　Shanxi	晋　Jin(Jìn)	太原　Taiyuan
上海　Shanghai	沪　Hu	
四川　Sichuan	川　Chuan	成都　Chengdu
台湾　Taiwan	台　Tai	台北　Taibei
天津　Tianjin	津　Jin(Jīn)	
西藏　Xizang	藏　Zang	拉萨　Lhasa
新疆　Xinjiang	新　Xin	乌鲁木齐　Urumqi
云南　Yunnan	滇　Dian	昆明　Kunming
浙江　Zhejiang	浙　Zhe	杭州　Hangzhou

中 国 历 史 年 表

五帝（公元前约26世纪初——公元前约22世纪末至公元前约21世纪初）
The Five Rulers
夏（公元前约22世纪末至公元前约21世纪初——公元前约17世纪初）
The Xia Dynasty
商（公元前约17世纪初至公元前约11世纪）
The Shang Dynasty
周（公元前约11世纪至公元前256年）
The Zhou Dynasty
西周（公元前约11世纪至公元前771年）
The Western Zhou Dynasty
东周（公元前770年至公元前256年）
The Eastern Zhou Dynasty
秦（公元前221年至公元前206年）
The Qin Dynasty
汉（公元前206年至公元220年）
The Han Dynasty
西汉（公元前206年至公元25年）
The Western Han Dynasty
东汉（25年至220年）
The Eastern Han Dynasty
三国（220年至280年）
The Three Kingdoms
魏（220年至265年）
The Wei Dynasty
蜀汉（221年至263年）
The Shu Han Dynasty
吴（222年至280年）
The Wu Dynasty
晋（265年至420年）
The Jin Dynasty
西晋（265年至317年）
The Western Jin Dynasty
东晋（317年至420年）
The Eastern Jin Dynasty
南北朝（420年至589年）
Northern and Southern Dynasties
南朝 Southern Dynasties
宋（420年至479年）
The Song Dynasty
齐（479年至502年）
The Qi Dynasty
梁（502年至557年）
The Liang Dynasty
陈（557年至589年）
The Chen Dynasty
北朝 Northern Dynasties
北魏（386年至534年）
The Northern Wei Dynasty
东魏（534年至550年）
The Eastern Wei Dynasty
北齐（550年至577年）
The Northern Qi Dynasty
西魏（535年至556年）
The Western Wei Dynasty
北周（557年至581年）
The Northern Zhou Dynasty
隋（581年至618年）
The Sui Dynasty
唐（618年至907年）
The Tang Dynasty
五代（907年至960年）
Five Dynasties
后梁（907年至923年）
The Posterior Liang Dynasty
后唐（923年至936年）
The Posterior Tang Dynasty
后晋（936年至947年）
The Posterior Jin Dynasty
后汉（947年至950年）
The Posterior Han Dynasty
后周（951年至960年）
The Posterior Zhou Dynasty
宋（960年至1279年）
The Song Dynasty
北宋（960年至1127年）
The Northern Song Dynasty
南宋（1127年至1279年）
The Southern Song Dynasty
辽（907年至1125年）
The Liao Dynasty
金（1115年至1234年）
The Jin Dynasty
元（1206年至1368年）
The Yuan Dynasty
明（1368年至1644年）
The Ming Dynasty
清（1616年至1911年）
The Qing Dynasty
中华民国（1911年至1949年）
The Republic of China
中华人民共和国（1949年10月1日成立）
The People's Republic of China

天干，地支

天干

甲 jia	乙 yi	丙 bing	丁 ding
戊 wu	己 ji	庚 geng	辛 xin
壬 ren	癸 gui		

地支	属相	黄道位置	时间
子 zi	鼠 rat	白羊宫 Aries	11-1a.m.
丑 chou	牛 ox	金牛宫 Taurus	1-3
寅 yin	虎 tiger	双子宫 Gemini	3-5
卯 mao	兔 hare	巨蟹宫 Cancer	5-7
辰 chen	龙 dragon	狮子宫 Leo	7-9
巳 si	蛇 serpent	室女宫 Virgo	9-11
午 wu	马 horse	天秤宫 Libra	11-1p.m.
未 wei	羊 sheep	天蝎宫 Scorpio	1-3
申 shen	猴 monkey	人马宫 Sagittarius	3-5
酉 you	鸡 cock	摩羯宫 Capricorn	5-7
戌 xu	狗 dog	宝瓶宫 Aquarius	7-9
亥 hai	猪 pig	双鱼宫 Pisces	9-11

天干地支纪年

甲子	(1984)	甲申	(2004)	甲辰	(2024)
乙丑	(1985)	乙酉	(2005)	乙巳	(2025)
丙寅	(1986)	丙戌	(2006)	丙午	(2026)
丁卯	(1987)	丁亥	(2007)	丁未	(2027)
戊辰	(1988)	戊子	(2008)	戊申	(2028)
己巳	(1989)	己丑	(2009)	己酉	(2029)
庚午	(1990)	庚寅	(2010)	庚戌	(2030)
辛未	(1991)	辛卯	(2011)	辛亥	(2031)
壬申	(1992)	壬辰	(2012)	壬子	(2032)
癸酉	(1993)	癸巳	(2013)	癸丑	(2033)
甲戌	(1994)	甲午	(2014)	甲寅	(2034)
乙亥	(1995)	乙未	(2015)	乙卯	(2035)
丙子	(1996)	丙申	(2016)	丙辰	(2036)
丁丑	(1997)	丁酉	(2017)	丁巳	(2037)
戊寅	(1998)	戊戌	(2018)	戊午	(2038)
己卯	(1999)	己亥	(2019)	己未	(2039)
庚辰	(2000)	庚子	(2020)	庚申	(2040)
辛巳	(2001)	辛丑	(2021)	辛酉	(2041)
壬午	(2002)	壬寅	(2022)	壬戌	(2042)
癸未	(2003)	癸卯	(2023)	癸亥	(2043)

二十四节气

节气名（日期）		相当于黄道十二宫的位置	
立春	the Beginning of Spring (Feb. 3.4.or 5)	宝瓶宫	Aquarius
雨水	Rain Water (Feb. 18,19 or 20)	双鱼宫	Pisces
惊蛰	the Waking of Insects (Mar. 5, 6 or 7)	双鱼宫	Pisces
春分	the Spring Equinox (Mar. 20, 21 or 22)	白羊宫	Aries
清明	Pure Brightness (Apr.4, 5 or 6)	白羊宫	Aries
谷雨	Grain Rain (Apr. 19, 20 or 21)	金牛宫	Taurus
立夏	the Beginning of Summer (May 5, 6 or 7)	金牛宫	Taurus
小满	Grain Full (May 20, 21 or 22)	双子宫	Gemini
芒种	Grain in Ear (June 5,6 or 7)	双子宫	Gemini
夏至	the Summer Solstice (June 21 or 22)	巨蟹宫	Cancer
小暑	Slight Heat (July 6,7 or 8)	巨蟹宫	Cancer
大暑	Great Heat (July 22, 23 or 24)	狮子宫	Leo
立秋	the Beginning of Autumn (Aug. 7, 8 or 9)	狮子宫	Leo
处暑	the Limit of Heat (Aug. 22, 23 or 24)	室女宫	Virgo
白露	White Dew (Sep. 7, 8 or 9)	室女宫	Virgo
秋分	the Autumnal Equinox (Sep. 22, 23 or 24)	天秤宫	Libra
寒露	Cold Dew (Oct. 8 or 9)	天秤宫	Libra
霜降	Frost’s Descent (Oct. 23 or 24)	天蝎宫	Scorpio
立冬	the Beginning of Winter (Nov.7 or 8)	天蝎宫	Scorpio
小雪	Slight Snow (Nov.22 or 23)	人马宫	Sagittarius
大雪	Great Snow (Dec.6, 7 or 8)	人马宫	Sagittarius
冬至	the Winter Solstice (Dec.21, 22 or 23)	摩羯宫	Capricorn
小寒	Slight Cold (Jan. 5, 6 or 7)	摩羯宫	Capricorn
大寒	Great Cold (Jan. 20 or 21)	宝瓶宫	Aquarius

中 国 百 家 姓

（按汉语拼音为序）

艾 ài
安 ān
敖 áo
白 bái
百里 bǎilǐ
班 bān
包 bāo
鲍 bào
贝 bèi
毕 bì
边 biān
卞 biàn
薄 bó
卜 bǔ
蔡 cài
曹 cáo
岑 cén
柴 chái
常 cháng
巢 cháo
晁 cháo
车 chē
陈 chén
成 chéng
程 chéng
储 chǔ
褚 chǔ
淳于 chúnyú
崔 cuī
戴 dài
邓 dèng
狄 dí
邸 dǐ
刁 diāo
丁 dīng

东方 dōngfāng
东郭 dōngguō
董 dǒng
窦 dòu
杜 dù
端木 duānmù
段 duàn
段干 duàngān
樊 fán
范 fàn
方 fāng
房 fáng
费 fèi
封 fēng
丰 féng
冯 féng
符 fú
伏 fú
富 fù
傅 fù
甘 gān
干 gān
高 gāo
戈 gē
盖 gě
葛 gě
耿 gěng
宫 gōng
龚 gōng
公 gōng
公孙 gōngsūn
公西 gōngxī
公羊 gōngyáng
公冶 gōngyě
巩 gǒng

贡 gòng
谷 gǔ
谷梁 gǔliáng
顾 gù
关 guān
管 guǎn
广 guǎng
桂 guì
郭 guō
韩 hán
杭 háng
郝 hǎo
何 hé
和 hé
贺 hè
洪 hóng
侯 hóu
后 hòu
呼延 hūyán
胡 hú
扈 hù
花 huā
华 huà
桓 huán
黄 huáng
皇甫 huángfǔ
惠 huì
霍 huò
嵇 jī
姬 jī
汲 jí
季 jì
纪 jì
计 jì
贾 jiǎ

简 jiǎn
江 jiāng
姜 jiāng
蒋 jiǎng
焦 jiāo
金 jīn
靳 jìn
荆 jīng
景 jǐng
鞠 jū
阚 kǎn
康 kāng
柯 kē
孔 kǒng
寇 kòu
蒯 kuǎi
匡 kuàng
隗 kuí
赖 lài
蓝 lán
郎 láng
劳 láo
雷 léi
冷 lěng
黎 lí
李 lǐ
郦 lì
历 lì
连 lián
廉 lián
梁 liáng
梁丘 liángqiū
廖 liào
林 lín
蔺 lìn
凌 líng
令狐 lìnghú
刘 liú
柳 liǔ
龙 lóng
娄 lóu
卢 lú
鲁 lǔ
陆 lù
路 lù
吕 lǘ
闾丘 lǘqiū
栾 luán
罗 luó
骆 luò
麻 má
马 mǎ
毛 máo
茅 máo
梅 méi
蒙 méng
孟 mèng
糜 mí
米 mǐ
宓 mì
苗 miáo
缪 miào
闵 mǐn
明 míng
莫 mò
万俟 mòqí
穆 mù
慕容 mùróng
倪 ní
聂 niè
宁 níng
牛 niú
钮 niǔ
欧 ōu
欧阳 ōuyáng
潘 pān
庞 páng
裴 péi
彭 péng
皮 pí
平 píng
濮 pú
濮阳 púyáng
蒲 pú
浦 pǔ
漆雕 qīdiāo
祁 qí
齐 qí
戚 qì
钱 qián
强 qiáng
乔 qiáo
秦 qín
丘 qiū
仇 qiú
裘 qiú
瞿 qū
屈 qū
麴 qū
全 quán
权 quán
冉 rǎn
饶 ráo
任 rèn
戎 róng
容 róng
荣 róng
茹 rú
阮 ruǎn
芮 ruì
桑 sāng
沙 shā
山 shān
单 shàn
尚 shàng
上官 shàngguān
邵 shào
申 shēn
申屠 shēntú
沈 shěn

慎 shèn
盛 shèng
施 shī
师 shī
石 shí
时 shí
史 shǐ
舒 shū
束 shù
水 shuǐ
司空 sīkōng
司马 sīmǎ
宋 sòng
苏 sū
孙 sūn
索 suǒ
谈 tán
谭 tán
澹台 tántái
汤 tāng
唐 táng
陶 táo
滕 téng
田 tián
童 tóng
屠 tú
拓跋 tuóbá
万 wàn
汪 wāng
王 wáng
危 wēi
微生 wēishēng
韦 wéi
魏 wèi
卫 wèi
温 wēn
文 wén
闻 wén
闻人 wénrén
翁 wēng

巫马 wūmǎ
乌 wū
邬 wū
毋 wú
吴 wú
伍 wǔ
武 wǔ
西门 xīmén
郗 xī
奚 xī
习 xí
席 xí
郤 xì
夏 xià
夏侯 xiàhóu
鲜于 xiānyú
向 xiàng
项 xiàng
萧 xiāo
谢 xiè
辛 xīn
邢 xíng
熊 xióng
须 xū
胥 xū
徐 xú
许 xǔ
宣 xuān
轩辕 xuānyuán
薛 xuē
荀 xún
燕 yān
阎 yán
颜 yán
严 yán
晏 yàn
杨 yáng
羊 yáng
羊舌 yángshē
养 yǎng

姚 yáo
耶律 yēlǜ
叶 yè
伊 yī
易 yì
阴 yīn
殷 yīn
尹 yǐn
印 yìn
应 yīng
雍 yōng
游 yóu
尤 yóu
鱼 yú
於 yú
虞 yú
俞 yú
余 yú
于 yú
庾 yǔ
宇文 yǔwén
禹 yǔ
郁 yù
喻 yù
尉迟 yùchí
元 yuán
袁 yuán
岳 yuè
乐 yuè
乐正 yuèzhèng
宰 zǎi
臧 zāng
曾 zēng
查 zhā
翟 zhái
詹 zhān
湛 zhàn
章 zhāng
张 zhāng
长孙 zhǎngsūn

赵 zhào
甄 zhēn
郑 zhèng
支 zhī
终 zhōng
钟 zhōng
钟离 zhonglí
仲 zhòng

仲孙 zhòngsūn
周 zhōu
诸 zhū
诸葛 zhūgé
朱 zhū
祝 zhù
颛孙 zhuānsūn
庄 zhuāng

卓 zhuō
子车 zǐché
宗 zōng
邹 zōu
祖 zǔ
左 zuǒ
左丘 zuǒqiū

数 字 的 用 法

(一) 基数词 cardinal numbers

1	one
2	two
3	three
4	four
5	five
6	six
7	seven
8	eight
9	nine
10	ten
11	eleven
12	twelve
13	thirteen
14	fourteen
15	fifteen
16	sixteen
17	seventeen
18	eighteen
19	nineteen
20	twenty
21	twenty-one
22	twenty-two
23	twenty-three
30	thirty
31	thirty-one
40	forty
50	fifty
60	sixty
70	seventy
80	eighty
90	ninety
100	one hundred
101	one hundred and one
200	two hundred
300	three hundred
400	four hundred
500	five hundred
600	six hundred
700	seven hundred
800	eight hundred
900	nine hundred
1000	one thousand
100 000	one hundred thousand
1 000 000	one million

(二) 序数词 ordinal numbers

第一	first
第二	second
第三	third
第四	fourth
第五	fifth
第六	sixth
第七	seventh
第八	eighth
第九	ninth
第十	tenth
第十一	eleventh
第十二	twelfth
第十三	thirteenth
第十四	fourteenth
第十五	fifteenth
第十六	sixteenth
第十七	seventeenth
第十八	eighteenth
第十九	nineteenth
第二十	twentieth
第二十一	twenty-first
第二十二	twenty-second
第二十三	twenty-third
第三十	thirtieth
第三十一	thirty-first
第四十	fortieth
第五十	fiftieth
第六十	sixtieth
第七十	seventieth
第八十	eightieth
第九十	ninetieth
第一百	(one) hundredth
第一百零一	(one) hundred and first
第二百	two-hundredth
第三百	three-hundredth
第四百	four-hundredth
第五百	five-hundredth
第六百	six-hundredth
第七百	seven-hundredth
第八百	eight-hundredth
第九百	nine-hundredth
第一千	(one) thousandth
第一万	ten thousandth
第一百万	(one) millionth

零 (0) nought; zero; O
一百二十三 (123) one hundred and twenty-three
一千三百 (1 300) thirteen hundred; one thousand three hundred
二千 (2 000) two thousand
四千五百零二 (4 502) four thousand five hundred and two
三万六千 (36 000) thirty-six thousand
四万五千七百八十九 (45 789) forty-five thousand seven hundred and eighty-nine
五十万 (500 000) five hundred thousand
三百一十二万四千五百 (3 124 500) three million, one hundred and twenty-four thousand five
八百万 (8 000 000) eight million
四千三百万 (43 000 000) forty-three million
九亿 (900 000 000) nine hundred million
十亿八千万 (1 080 000 000) one billion and eighty million
一百亿 (10 000 000 000) ten billion
三千亿 (300 000 000 000) three hundred billion
五万亿 (5 000 000 000 000) five million million

(三) 分数 fractions

二分之一 (1 / 2) one-half; a half
三分之一 (1 / 3) one-third
四分之三 (3 / 4) three-fourths
五分之一 (1 / 5) one-fifth
五分之二 (2 / 5) two-fifths
八分之七 (7 / 8) seven-eighths
十分之一 (1 / 10) one-tenth; a tenth
百分之一 (1 / 100) one-hundredth; one per cent
千分之一 (1 / 1 000) one-thousandth
万分之一 (1 / 10 000) one ten-thousandth
二又二分之一 two and one half; two and a half
四又三分之二 four and two-thirds

(四) 百分比 percentages

百分之百 (100%) one hundred per cent
百分之零点五 (0.5%) point five per cent
百分之零点三七 (0.37%) point three seven per cent

(五) 小数 decimals

4.03 four point nought three; four point O three
7.005 seven point nought nought five; seven point O O five
45.21 forty-five point two one

(六) 增加 go up by; shoot up by; rise by (to); be raised by; increase by

双倍; 加倍 (to) double
象…一样多 as many as; as much as
增加百分之一 to increase by one-hundredth
增加百分之十 to increase by one-tenth
增加百分之五十 to increase by one-half
增加百分之七十 an increase of seventy per cent; an seventy-percent increase
是去年的二倍(即增加一倍) to be double that of last year
增加百分之百(即增加一倍或增加到原数的二倍) to increase by one hundred per cent; to double; to increase to twice as many (much) as
比1987年增加一倍以上 to be more than double the 1987 figure
是…的四倍 four times as many (much) as
是去年的五倍 to be five times that of last year; to increase fivefold compared with last year; to increase by 400 per cent over last year
比1978年增加了十多倍 to be a dozen times that of 1978
粮食产量从1978年的30万吨上升到1983年的715 000 吨 grain output went up to 715 000 tons in 1983 from 300 000 tons in 1987
1983年粮食产量为38 728万吨,是1949年的3.42倍 in 1983 the output of grain was 387.28 million tons, 3.42 times that of 1949
1983年工农业总产值为 9 209 亿元,比1982年增加 10.2% the 1983 total combined output value of industry and agriculture was 920.9 billion *yuan*, 10.2 per cent higher than 1982
1979至1983年间工农业总产值平均每年的增长

率是 8.2% the average annual rate of increase of the total output value of industry and agriculture from 1979 through 1983 was 8.2 per cent

与1949年相比，大学生人数增加了63%，中学生增加了 1.5 倍，小学生增加了 1.1 倍 compared with 1949, the number of university students increased by 63 per cent, that of secondary school students by 1.5 times, and that of primary school pupils by 1.1 times

(七)减少(下降) go down by; fall by (to); reduce by

下降(减少)百分之六十 a reduction (fall) of sixty per cent

下降百分之一 to decrease by one-hundredth; to reduce by one-hundredth

下降百分之十(或十分之一) to decrease by ten per cent; to reduce by ten per cent; to reduce to nine-tenths; to be reduced to nine-tenths

下降百分之二十(或五分之一) to decrease by one-fifth; to reduce by one-fifth; to reduce to four-fifths; to be reduced to four-fifths

下降百分之二十五(或四分之一) to decrease by one quarter; to reduce by one quarter; to reduce to three-fourths; to be reduced to three-fourths

下降百分之五十(或一半) to decrease by one-half; to reduce by one-half; to reduce to one-half; to be reduced to one-half

(八)不确指的数字 indefinite numbers

几个 some; a few; several; a number of

十几个 more than ten; over a dozen; no more than twenty

几十个 dozens of

几十年 decades

四十好几 well over forty

好几百个 hundreds of

成千上万 thousands of

几十万 hundreds of thousands of

数以百万计的 millions of

亿万(人) hundreds of millions of

(九)时间 time

6:00	six o'clock; six a.m. (上午); six p.m. (下午)
7:10	ten past seven; seven ten
8:15	a quarter past eight; eight fifteen
9:25	twenty-five past nine; nine twenty-five
10:30	half past ten; ten thirty
11:35	eleven thirty-five; twenty-five to twelve
12:45	twelve forty-five
13:50	thirteen fifty
14:57	fourteen fifty-seven
15:00	fifteen

(十)年月日 dates

1643 年	sixteen forty-three
1755 年	seventeen fifty-five
1898 年	eighteen ninety-eight
1900 年	nineteen hundred
1989 年	nineteen eighty-nine
2000 年	twenty hundred

三月八日 the eighth of March

五月一日 the first of May

七月二十三日 the twenty-third of July

一九四九年十月一日 the first of October, nineteen forty-nine

(十一)电话号码 telephone numbers

117	one one seven
120	one two oh
699	six double-nine
2442	two double-four two
3271	three two seven one
893712	eight nine three seven one two
8123456	eight one two three four five six

(十二)罗马数字 Roman numeral

1	I
2	II
3	III
4	IV
5	V
6	VI

7	VII
8	VIII
9	IX
10	X
11	XI
12	XII
13	XIII
14	XIV
15	XV
16	XVI
17	XVII
18	XVIII
19	XIX
20	XX
21	XXI
22	XXII
23	XXIII
24	XXIV
30	XXX
31	XXXI
32	XXXII
40	XL
41	XLI
42	XLII
50	L
51	LI
52	LII
60	LX
61	LXI
62	LXII
70	LXX
80	LXXX
90	XC
100	C
200	CC
300	CCC
400	CCCC; CD
500	D
600	DC
700	DCC
800	DCCC
900	CM
1000	M
2000	MM
2500	MMD

常用计量单位和换算

(一)公制

长度

毫米 millimeter (mm) = 0.001 米
厘米 centimeter (cm) = 0.01 米
米 meter (m)
公里 kilometer (km) = 1000 米
  1 厘米 = 0.3937 英寸
  1 米 = 3.2808 英尺 = 1.0936 码
  1 公里 = 0.6214 英里

地积

公亩 are (a) = 100 平方米
公顷 hectare (ha) = 100 公亩
  1 公亩 = 0.0247 英亩
  1 公顷 = 2.4711 英亩

容积

毫升 milliliter (ml) = 0.001 升
升 liter (l)
  1 升 = 0.2600 加伦(美) = 0.2600 加伦(英)

重量

毫克 milligram (mg) = 0.001 克
克 gram (g)
千克(公斤) kilogram (kg) = 1000 克
公吨 ton (t) = 1000 千克
  1 克 = 0.0352 盎司(常衡)
  1 千克 = 2.2046 磅(常衡)
  1 公吨 = 1.1023 短吨(美吨)
  1 公吨=0.9842长吨(英吨)

(二)英美制

长度

英寸 inch (pl. inches) = 2.5400 厘米
英尺 foot (pl. feet) = 12 英寸 = 0.3048 米
码 yard = 3 英尺 = 0.9144 米
英里 mile = 1760 码 = 1.6093 公里

水程长度

链 cable length, cable's length = 185.2 米
海里 sea mile = 10 链 = 1.852 公里

地积

英亩 acre =40.4686 公亩

重量(常衡)

盎司 ounce = 28.3496 克
磅 pound = 16 盎司 = 0.4539 千克
长吨(英吨) long ton = 2240 磅
短吨(美吨) short ton = 2000 磅
  1 长吨(英吨) = 1016.0470 千克
  1 短吨(美吨) = 907.1849 千克

液量

加伦 gallon = 4.546 升(英) = 3.785 升(美)